Alphabetical List of the Elements

Element	Symbol	Atomic Number	Atomic Mass
Actinium	Ac	89	(227)
Aluminum	Al	13	26.981538
Americium	Am	95	(243)
Antimony	Sb	51	121.760
Argon	Ar	18	39.948
Arsenic	As	33	74.92160
Astatine	At	85	(210)
Barium	Ba	56	137.327
Berkelium	Bk	97	(247)
Beryllium	Be	4	9.012182
Bismuth	Bi	83	208.98038
Bohrium	Bh	107	(264)
Boron	B	5	10.811
Bromine	Br	35	79.904
Cadmium	Cd	48	112.411
Calcium	Ca	20	40.078
Californium	Cf	98	(251)
Carbon	C	6	12.0107
Cerium	Ce	58	140.116
Cesium	Cs	55	132.90545
Chlorine	Cl	17	35.453
Chromium	Cr	24	51.9961
Cobalt	Co	27	58.933200
Copper	Cu	29	63.546
Curium	Cm	96	(247)
Darmstadtium	Ds	110	(281)
Dubnium	Db	105	(262)
Dysprosium	Dy	66	162.500
Einsteinium	Es	99	(252)
Erbium	Er	68	167.269
Europium	Eu	63	151.964
Fermium	Fm	100	(257)
Fluorine	F	9	18.9984032
Francium	Fr	87	(223)
Gadolinium	Gd	64	157.25
Gallium	Ga	31	69.723
Germanium	Ge	32	72.64
Gold	Au	79	196.96655
Hafnium	Hf	72	178.49
Hassium	Hs	108	(269)
Helium	He	2	4.002602
Holmium	Ho	67	164.93032
Hydrogen	H	1	1.00794
Indium	In	49	114.818
Iodine	I	53	126.90447
Iridium	Ir	77	192.217
Iron	Fe	26	55.845
Krypton	Kr	36	83.798
Lanthanum	La	57	138.9055
Lawrencium	Lr	103	(262)
Lead	Pb	82	207.2
Lithium	Li	3	6.941
Lutetium	Lu	71	174.967
Magnesium	Mg	12	24.3050
Manganese	Mn	25	54.938049
Meitnerium	Mt	109	(268)
Mendelevium	Md	101	(258)
Mercury	Hg	80	200.59
Molybdenum	Mo	42	95.94
Neodymium	Nd	60	144.24
Neon	Ne	10	20.1797
Neptunium	Np	93	(237)
Nickel	Ni	28	58.6934
Niobium	Nb	41	92.90638
Nitrogen	N	7	14.0067
Nobelium	No	102	(259)
Osmium	Os	76	190.23
Oxygen	O	8	15.9994
Palladium	Pd	46	106.42
Phosphorus	P	15	30.973761
Platinum	Pt	78	195.078
Plutonium	Pu	94	(244)
Polonium	Po	84	(209)
Potassium	K	19	39.0983
Praseodymium	Pr	59	140.90765
Promethium	Pm	61	(145)
Protactinium	Pa	91	231.03588
Radium	Ra	88	(226)
Radon	Rn	86	(222)
Rhenium	Re	75	186.207
Rhodium	Rh	45	102.90550
Rubidium	Rb	37	85.4678
Ruthenium	Ru	44	101.07
Rutherfordium	Rf	104	(261)
Samarium	Sm	62	150.36
Scandium	Sc	21	44.955910
Seaborgium	Sg	106	(266)
Selenium	Se	34	78.96
Silicon	Si	14	28.0855
Silver	Ag	47	107.8682
Sodium	Na	11	22.989770
Strontium	Sr	38	87.62
Sulfur	S	16	32.065
Tantalum	Ta	73	180.9479
Technetium	Tc	43	(98)
Tellurium	Te	52	127.60
Terbium	Tb	65	158.92534
Thallium	Tl	81	204.3833
Thorium	Th	90	232.0381
Thulium	Tm	69	168.93421
Tin	Sn	50	118.71
Titanium	Ti	22	47.8
Tungsten	W	74	183.8
Uranium	U	92	238.0
Vanadium	V	23	50.9
Xenon	Xe	54	131.2
Ytterbium	Yb	70	173.0
Yttrium	Y	39	88.
Zinc	Zn	30	65.
Zirconium	Zr	40	91.

Numbers in parentheses indicate estimates of the atomic mass based on the longest-lived isotopes.

Applications and Chemical Encounters

Chapter	Page	Title	Chapter	Page	Title
1	6	Hydrogen and oxygen in the space shuttle fuel tank	7	277	CHEMICAL ENCOUNTERS: The Elements of Life
1	10	CHEMICAL ENCOUNTERS: Desalination	7	280	Treating heavy metal poisoning
1	30	The mass of orange juice	7	287	Electronegativity values and vitamins A and C
1	34	The frontiers of medicine	7	290	Iron and automobiles
2	51	CHEMICAL ENCOUNTERS: The Dover Boat and Radiocarbon Dating	7	291	Iron, hemoglobin, and oxygen
2	59	The mass spectrometer	7	292	CHEMICAL ENCOUNTERS: The Elements and the Environment
2	61	Grades and the weighted average	8	306	CHEMICAL ENCOUNTERS: The Uses and Behavior of Sodium Chloride
2	75	Chalk and the White Cliffs of Dover	8	309	The boiler scale and hot-water heater pipes
3	87	CHEMICAL ENCOUNTERS: Adrenaline in the Body	8	310	CHEMICAL ENCOUNTERS: Focus on Zeolites
3	95	CHEMICAL ENCOUNTERS: Cholesterol in the Blood	8	313	CHEMICAL ENCOUNTERS: Fluoridated Water and Tooth Decay
3	97	CHEMICAL ENCOUNTERS: Nitrogen in Fertilizers	8	317	CHEMICAL ENCOUNTERS: Tetraethyl Lead in Gasoline
3	99	CHEMICAL ENCOUNTERS: A Focus on Chocolate	8		
3	103	CHEMICAL ENCOUNTERS: Preparation of Aspirin	8	322	The creation of aspirin
3	104	Ethanoic anhydride in manufacturing processes	8	323	Industrial applications of hydrogen fluoride
3	113	CHEMICAL ENCOUNTERS: Cleaning the Air on Manned Spacecraft	8	329	The Breathalyzer
4	125	CHEMICAL ENCOUNTERS: Gold Mining and Cyanide Leaching	8	331	CHEMICAL ENCOUNTERS: Calcium Channel Blockers
4	129	CHEMICAL ENCOUNTERS: Sports Drinks and Electrolyte Balance	8	335	CHEMICAL ENCOUNTERS: Focus on Morphine
4	131	Water and long-distance runners	8	346	Crafting new drugs to be both polar and nonpolar solvents
4	135	CHEMICAL ENCOUNTERS: Maximum Levels of Chemicals in Drinking Water	9	356	CHEMICAL ENCOUNTERS: Molecular Structure and Eyesight
4	137	The fluoride ion in drinking water	9	359	Industrial applications of molecular hydrogen
4	140	Determining the amount of vitamin C in fruit juice	9	374	Retinal and light absorption in the eyes
4	145	CHEMICAL ENCOUNTERS: Focus on Lead Sulfide	9	384	CHEMICAL ENCOUNTERS: Benzene, Stability, and MO Theory
4	148	Gold mine drainage cleanup	9	385	Common solvents
4	156	Using ammonia to fertilize fields	10	401	CHEMICAL ENCOUNTERS: Dalton's Law and Food Packaging
4	157	CHEMICAL ENCOUNTERS: Revisiting the Maximum Levels of Chemicals in Drinking Water	10	403	Balloons and ozone analysis
			10	407	Boyle's law and food packaging
5	172	Chemical energy in space shuttle engines	10	419	CHEMICAL ENCOUNTERS: Automobile Air Bags
5	175	Ammonium salts in "cold packs"	10	420	CHEMICAL ENCOUNTERS: Acetylene
5	179	CHEMICAL ENCOUNTERS: Energy in Foods	10	426	Effusion and the preparation of uranium-based fuel rods
5	193	CHEMICAL ENCOUNTERS: Focus on Methane	10	427	The construction of condiment packages
5	197	CHEMICAL ENCOUNTERS: Energy Choices	10	427	CHEMICAL ENCOUNTERS: Ozone
6	219	CHEMICAL ENCOUNTERS: Simultaneous Determination of Elements in Water	10	431	CHEMICAL ENCOUNTERS: The Greenhouse Effect
6	224	CHEMICAL ENCOUNTERS: The Nature and Applications of Lasers	11	441	CHEMICAL ENCOUNTERS: Worldwide Water Use
6	229	Scanning electron microscopes	11	455	Canning at high pressure
6	237	CHEMICAL ENCOUNTERS: The Scanning Tunneling Microscope	11	461	CHEMICAL ENCOUNTERS: CO_2 as a Dry Cleaning Solvent
6	242	CHEMICAL ENCOUNTERS: Nuclear Spin and Magnetic Resonance Imaging	11	470	The EPA's maximum contaminant level in water
7	269	CHEMICAL ENCOUNTERS: Commercial Uses of Main-Group Elements	11	470	CHEMICAL ENCOUNTERS: Composition of Seawater
7	270	Some commercial uses of Group IA elements	11	471	CHEMICAL ENCOUNTERS: Impact of the Solubility of Oxygen in Fresh Water
7	271	Some commercial uses of the Group IIA elements	11	472	CO_2 and the preparation of soft drinks
7	271	Some commercial uses of the Group IIIA elements	11	473	Salt and the freezing point of water
7	273	Oxygen and human life	11	480	CHEMICAL ENCOUNTERS: Meeting Municipal Water Needs

(Continued at back of book)

AN ONLINE HOMEWORK SYSTEM YOU CAN RELY ON . . .

Developed by teachers, for teachers, **WebAssign®** is known for offering the *most flexible and stable* online homework system on the market, allowing instructors to focus on what really matters—*teaching*—rather than grading. Create assignments from our ready-to-use end-of-chapter questions, or write and customize your own exercises. WebAssign transforms the way your students learn!

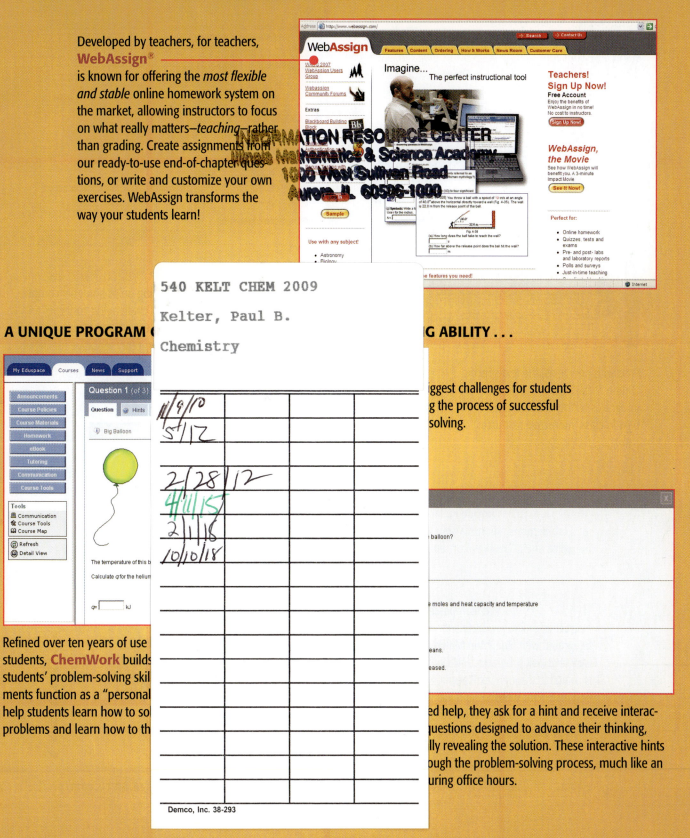

A UNIQUE PROGRAM O... ...G ABILITY . . .

...iggest challenges for students ...g the process of successful ...solving.

Refined over ten years of use ...
students, **ChemWork** builds ...
students' problem-solving skil ...
ments function as a "personal ...
help students learn how to sol... ...ed help, they ask for a hint and receive interac-
problems and learn how to th... ...questions designed to advance their thinking, ...lly revealing the solution. These interactive hints ...ough the problem-solving process, much like an ...uring office hours.

PREMIUM MEDIA RESOURCES REINFORCE KEY CONCEPTS . . .

Thinkwell® Video Lessons offer an engaging and dynamic way for students to understand core concepts. With over 45 hours of video, each mini-lecture combines video, audio, and whiteboard examples to address the various learning styles of today's student.

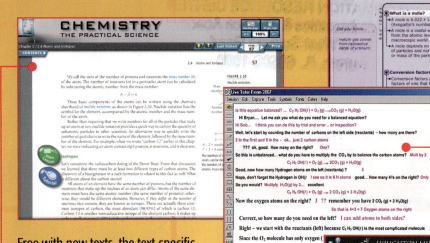

Free with new texts, the text-specific Online **Multimedia eBook** integrates textbook content with best in class interactive resources.

SMARTHINKING® live, online tutoring helps students comprehend challenging concepts and problems. Contact your Houghton Mifflin representative for details.

Interactive Tutorials and Visualizations provide molecular animations and lab demonstrations to help students visualize and review key concepts.

HM ChemSPACE encompasses the interactive online products and services integrated with Houghton Mifflin chemistry textbook programs. *HM ChemSPACE* is available through text-specific Student and Instructor websites and via **Eduspace®**, Houghton Mifflin's online course management system.

	HM ChemSPACE	HM ChemSPACE with Eduspace
Online Homework in WebAssign		✔
ChemWork		✔
SMARTHINKING		✔
Online Multimedia eBook	✔	✔
Thinkwell Video Lessons	✔	✔
Interactive Tutorials	✔	✔
Visualizations	✔	✔
ACE Practice Tests	✔	✔
Electronic Flashcards	✔	✔

To learn more about *HM ChemSPACE,* contact your Houghton Mifflin sales representative or visit **college.hmco.com/pic/kelterMEE**

Chemistry
The Practical Science

MEDIA ENHANCED EDITION

Paul Kelter
Northern Illinois University

Michael Mosher
University of Nebraska at Kearney

Andrew Scott
Perth College, UHI Millennium Institute

Contributing Writer
Charles William McLaughlin
University of Nebraska at Lincoln

Houghton Mifflin Company

Boston   New York

To Barb, Kristie, and Margaret,
who are with us through it all, and to Seth, Aaron, Jamie, David, and
Alison, for whom we do these things

To Jim Carr,
master craftsman of the finest teachers

Publisher: Charles Hartford

Senior Marketing Manager: Nicole Moore

Senior Development Editor: Rebecca Berardy Schwartz

Assistant Editor: Amy Galvin

Project Editor: Andrea Cava

Art and Design Manager: Jill Haber

Cover Design Manager: Anne S. Katzeff

Senior Photo Editor: Jennifer Meyer Dare

Senior Composition Buyer: Chuck Dutton

New Title Project Manager: James Lonergan

Editorial Associate: Henry Cheek

Marketing Associate: Kris Bishop

Cover photo © 2000 Richard Megna, Fundamental Photographs, NYC

Credits appear on pages A49–A52, which are considered an extension of the copyright page.

Printed in the U.S.A.

Library of Congress Control Number: 2007937002

ISBN-10: 0-547-05393-2
ISBN-13: 978-0-547-05393-6

123456789–CRK–11 10 09 08 07

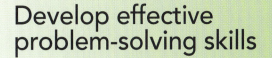

Develop effective problem-solving skills

This approach to problem solving emphasizes problem **strategies**, and then encourages students to look *beyond* the problem.

First Thoughts encourages students to think about a *strategy* before attempting the solution.

Solution demonstrates the detailed solution to the problem and effectively uses color to highlight key information.

Further Insights continues the discussion of the problem, illustrating applications of practical importance and linking multiple concepts throughout the book.

Practice includes additional practice problems for students to try on their own, with answers in the appendix.

EXERCISE 5.3 | The First Law of Thermodynamics

Calculate the change in the energy of a system if 51.8 J of work is done *by* the system with an associated heat loss of 12.3 J.

● **First Thoughts**

We must pay particular attention to the sign conventions for heat and work in this problem. In this case, work is done by the system. Is this work positive or negative? A useful system to keep in mind is you! That is, when you do work—by running, dancing, or even moving your textbooks from one class to the next—you are using energy. After the process of moving your body or your books, you have less energy than you had before, so the work has a "−" sign. Similarly, the 51.8 J of work done by the system means that its energy change, w_{system}, is −51.8 J. The energy loss as heat *by the system* (such as that accompanying your run as your body tries to stay cool) also has a "−" sign, so $q_{system} = -12.3$ J.

● **Solution**

From the standpoint of the system,

$$\Delta U = q + w$$

$$\Delta U = -12.3 \text{ J} + (-51.8 \text{ J}) = -64.1 \text{ J}$$

● **Further Insights**

We want to reinforce the point that there is no such thing as heat or work within a system. In other words, *a system does not contain heat or work*. Rather, heat and work exist only as *types of energy transfer* between the system and the surroundings. Work is done by or on a system to move it through a distance. A system can transfer 64.1 J of energy as heat and work. It does not contain 64.1 J of heat and work.

PRACTICE 5.3

Calculate the change in the energy of a system if 84.7 J of work is done on the system, with an associated loss of energy as heat of 39.9 J.

See Problems 11–14, 19, 20, and 89. ■

The problem-solving approach is more about reason than rote. Before jumping headfirst into an Exercise, students are asked to consider First Thoughts. *These remarks often ask students to apply their real-world experience to generate ballpark answers. They also call on prior course content to help students see the big picture and develop a plan of attack. A detailed solution is then followed by* Further Insights, *which compel students to not only ask themselves if their answer makes sense but to put the entire exercise in context and make connections to biology and numerous other disciplines. In short, students are asked to think like a scientist.*—Ed O'Sullivan, Parkland College

This way of doing things gives students real insights into how "real chemists or scientists" think.—Milton Johnston, University of South Florida

Engage students through integrated applications

A space shuttle launch is an awe-inspiring demonstration of the ability of energy to transport humans and material away from Earth's surface with an eye toward planetary exploration. The energy to launch the craft and its crew comes from the explosive violence of chemical reactions. When these reactions are used in a carefully controlled way, they can lift the massive shuttle (which weighs in at a robust 2,000,000 kg), the booster rockets, and the fuel tank and propel the ship into orbit hundreds of miles above Earth in a very brief 10-minute ride.

Two chemical reactions, indicated in Figure 5.1, power the launch of the space shuttle. The combustion of hydrogen gas (the combination of hydrogen and oxygen to form water) takes place in the main engines. At the same time, the solid rocket boosters a
the oxidat
primary e

Main engi

Boosters:

As the
ergy that
cle is quiet
booster ro

Application
CHEMICAL ENCOUNTERS:
Setting the Stage with
the Space Shuttle

Interwoven throughout the chapter, chemistry concepts are explained through the use of real-world examples. Application icons throughout the text show at a glance where chemical concepts are applied.

4. Diagram, using circles for atoms, a crystal of KCl versus the same crystal of KCl dissolved in water.

Chemical Applications and Practices

5. Earth's oceans contain tons of dissolved sodium chloride. Yet, when a ship develops an oil leak, almost none of the oil dissolves in the ocean. Explain this phenomenon.

6. When an ion dissolves, it is surrounded by a hydration sphere. If water molecules surrounded the ion so that the hydrogen portion of the water was closer to the ion, would the ion most likely be a cation or an anion?

7. Pure water does not conduct an electric current. However, aqueous solutions of some compounds do form solutions that conduct electricity. Explain why the presence of some solutes converts nonconducting water into a conducting solution.

8. Glycerin can be produced as a by-product in soap m
The compound dissolves so easily in water that it a
water from the air. This latter characteristic is why g
is often found in many skin lotions. As glycerin a
the water, the skin can be kept moist. Glycerin's stru
shown below. What aspects of glycerin's structure con
most to its ease of dissolving in water?

$$H-\overset{\overset{\displaystyle OH}{|}}{\underset{\underset{\displaystyle H}{|}}{C}}-\overset{\overset{\displaystyle OH}{|}}{\underset{\underset{\displaystyle H}{|}}{C}}-\overset{\overset{\displaystyle OH}{|}}{\underset{\underset{\displaystyle H}{|}}{C}}-H$$

Glycerin

9. A conductivity-testing apparatus, such as the one sh
this chapter, possesses a light bulb whose brightness is
to how much current is flowing through it (and also t
the solution.) A small, but measurable, amount of c
must be present before the bulb becomes visibly b
What effect would this characteristic of the appara

(a)   (b)

(c)   (d)

● Mg²⁺   ● Cl⁻   ● H₂O

12. Which of the following would best represent
water?

Applications are also found in How Do We Know? and NanoWorld / MacroWorld features, chapter openers, photos, and end-of-chapter exercises. Students will see how chemistry connects to their lives, their future careers, and their world.

NanoWorld / MacroWorld

Big effects of the very small: Color photography and the tricolor process

"Smile, you're on candid camera!" Everyone hopes never to be ambushed with those words. In the nineteenth century, when photography was in its infancy, candid photographs were not possible. In order to create a visual memory of an event, you had to visit the photographer and sit for a picture. "Sitting" meant holding a pose for up to 30 seconds. Any movement would show as a blur in the final picture. However, the idea that an inexpensive, yet realistic, picture could be created in such a short time brought people to the photographer's studio in droves.

Scientists in the early 1800s were intrigued by the idea of painting portraits without the artist's pen. In the summer of 1827, Joseph Nicéphore Niépce, using material that hardened on exposure to light exposed film for eight hours in order to get a picture. Because no one wanted to sit still that long, most of his pictures were landscapes. In 1839, however, two other photographic processes were developed, the daguerreotype (perfected by Louis Daguerre) and the calotype (Henry Fox Talbot). Although these techniques allowed pictures to be created with much shorter exposures, they still produced a black-and-white image (see Figure 9.30). This didn't matter much to the general public, and photographic portrait studios seemed to pop up on every corner across the globe. Despite some marginal success in producing faint color images during the 1850s and 1890s, the advent of the color photograph as an inexpensive, reproducible and stable process didn't occur until the 1920s.

Researchers at Kodak Research Laboratories in 1935 finally hit upon the process that would bring the color photograph to the world. Kodachrome®, their color

photography process, introduces color to a picture by separating the three primary colors into specific emulsion layers. How does it work? The emulsion layers contain a compound that reacts when light strikes it. A quick look at the chemistry behind the black-and-white photograph will give us some insight.

In black-and-white photography, a piece of transparent plastic is coated with silver bromide crystals (see Figure 9.31) and a sensitizer. The plastic sheet is then placed in a camera, and when the shutter is opened, the silver bromide absorbs photons of light. In the presence of the sensitizer, the silver cation is reduced by the sensitizer (gains an electron) and becomes silver metal:

$$Ag^+(s) + sensitizer \rightarrow Ag°(s) + sensitizer^+$$

Then the film is wound back into its container and sent to be developed. The technician rinses the plastic film to remove unreacted silver bromide. The silver metal remains in place on the plastic. The result is an inverted image called a negative. Next the technician shines light through the negative onto another silver bromide/sensitizer-coated support (typically paper this time) and rinses off the unreacted silver bromide. Violà! We have a positive image called a photograph.

Why is the film coated with silver bromide instead of silver chloride? We can answer this by examining the wavelength of light that is absorbed; see Figure 9.32. Comparison of the two shows that silver bromide absorbs a large amount of the visible light that enters our camera. To get the best picture, we want to absorb as much light as we can.

FIGURE 9.30
A daguerreotype of Michael Faraday (see Chapter 19) taken sometime between 1844 and 1860 at Mathew Brady's studio in New York City. Because the process required the subject to remain completely motionless for up to 10 seconds, most daguerreotypes were taken in well-lit studios. Even though the early daguerreotypes cost one month's wages for the average person, they were much cheaper than a painted portrait. This made them an instant success.

FIGURE 9.31
A magnified image of silver bromide crystals used in the photographic industry. Note that the crystals are mostly hexagonal in shape and flat so that more surface area is exposed to incoming photons.

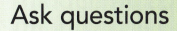

Ask questions

Inquiry-based learning is integral to this book. **Questions** are asked throughout the narrative to help students make the connection between what they are reading, concepts throughout the text, and their everyday lives.

Dilution

How do municipal water treatment plant workers prepare water so that it contains a fluoride ion concentration of about 1 part per million? They use a concentrated source of fluorine, hydrofluorosilicic acid, H_2SiF_6 ("HFSA"), which they then dilute in the drinking water. HFSA reacts with water in a fairly complex way that releases fluoride ion into the water. In Ireland, a company in the town of New Ross,

 Application

How do we know?

 ### The moons of Jupiter

The moons of the outer planets make most elegant chemical laboratories, because conditions on these worlds are so very different from those here at home. When we look at photos of Jupiter's moons, Europa and Ganymede (Figure 11.17), we see some vexing geological features. How do we know what caused these features? We can't visit the moons, at least not directly. However, our space probes can gather various types of electromagnetic radiation—including light, ultraviolet radiation, and X-rays—that give us data from which to draw conclusions.

The probes' data seem to show that deep within the surface of these moons, there are layers of ice, each in a different phase, mixed with rocks. The phase diagram of water that we showed as Figure 11.18 is inadequate to account for the low temperatures and massive pressures found within these moons. A more comprehensive phase diagram for water, emphasizing its solid phases, is given as Figure 11.20. The ice that is made in our kitchen freezer or found in an ice storm—what we might call "normal ice"—is technically called Ice Ih. Its structure is shown in Figure 11.21. There are 11 other forms of ice that have been made in the laboratory or simulated by computer.

The phase diagram shows that as the temperature and pressure inside the moons change, different types of ice are formed. Each ice has its own density. As layers form versions of ice of different densities, they form the geological features, such as cracking and

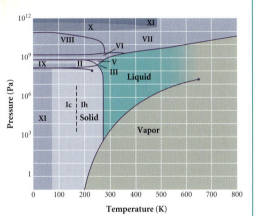

FIGURE 11.20

This is an expanded phase diagram for water, taking into account the phase changes among the various solid (ice) forms. These forms of ice are thought to occur in layers beneath the surface of some of the moons of the outer planets.

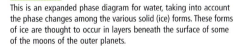

HERE'S WHAT WE KNOW SO FAR

- There are three natural forces: the gravitational, electroweak, and strong nuclear forces.
- Potential energy is the energy stored within a system.
- Kinetic energy is the energy associated with motion.
- The law of conservation of energy states that energy can be neither created nor destroyed. Instead, it just moves between the system and the surroundings and can be transformed from one form to another.
- Work and heat are two ways in which energy can move between the system and the surroundings.
- The flow of energy as heat from the system to the surroundings involves an exothermic process. An endothermic process involves the flow of energy as heat from the surroundings to the system.
- Changes in the energy of a reaction can be studied by examining a reaction profile diagram.

Highlighted questions appear throughout the chapter and prompt students to learn how to raise *meaningful* questions about topics.

How do we know? essays ask questions to probe the use of key chemical concepts in medicine, research, and other applications.

Students are also given many opportunities to reflect on what they have learned throughout the book. **Here's What We Know So Far** in-chapter summaries review complex topics.

Cold packs use endothermic reactions.

Dynamic, contemporary art program

This **contemporary art and photo program** appeals to tech-savvy students who expect exciting and visually appealing graphics regardless of medium.

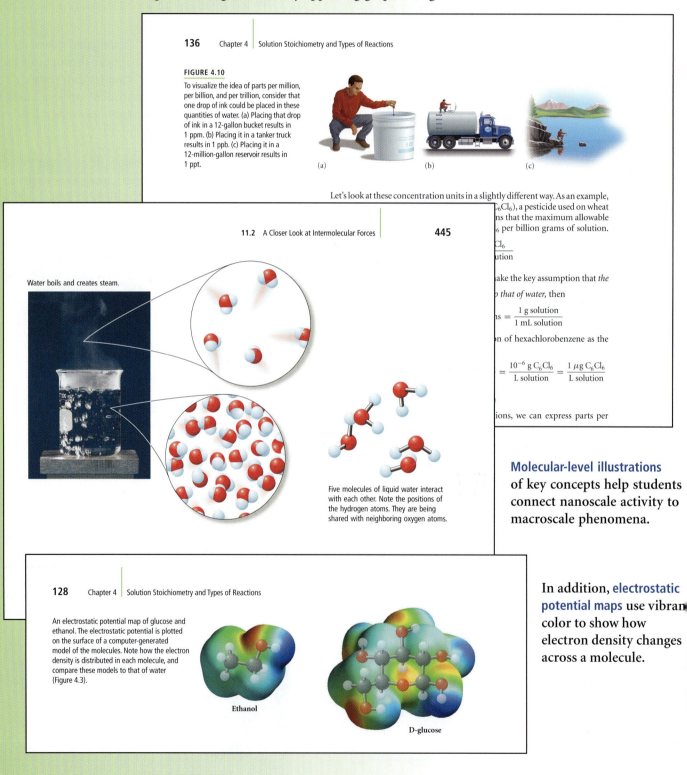

FIGURE 4.10
To visualize the idea of parts per million, per billion, and per trillion, consider that one drop of ink could be placed in these quantities of water. (a) Placing that drop of ink in a 12-gallon bucket results in 1 ppm. (b) Placing it in a tanker truck results in 1 ppb. (c) Placing it in a 12-million-gallon reservoir results in 1 ppt.

(a) (b) (c)

Let's look at these concentration units in a slightly different way. As an example, C_6Cl_6), a pesticide used on wheat ns that the maximum allowable $_6$ per billion grams of solution.

$$\frac{Cl_6}{\text{ution}}$$

ake the key assumption that *the o that of water*, then

$$\text{ns} = \frac{1 \text{ g solution}}{1 \text{ mL solution}}$$

n of hexachlorobenzene as the

$$= \frac{10^{-6} \text{ g } C_6Cl_6}{\text{L solution}} = \frac{1 \text{ } \mu\text{g } C_6Cl_6}{\text{L solution}}$$

ions, we can express parts per

Water boils and creates steam.

Five molecules of liquid water interact with each other. Note the positions of the hydrogen atoms. They are being shared with neighboring oxygen atoms.

Molecular-level illustrations of key concepts help students connect nanoscale activity to macroscale phenomena.

An electrostatic potential map of glucose and ethanol. The electrostatic potential is plotted on the surface of a computer-generated model of the molecules. Note how the electron density is distributed in each molecule, and compare these models to that of water (Figure 4.3).

Ethanol

D-glucose

In addition, **electrostatic potential maps** use vibran color to show how electron density changes across a molecule.

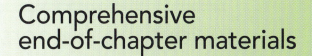

Comprehensive end-of-chapter materials

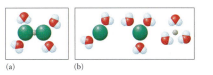

Focus Your Learning

The answers to the odd-numbered problems appear at the back of the book.

Section 4.1 Water—A Most Versatile Solvent

Skill Review

1. Explain how water molecules can dissolve both cations and anions.
2. Why doesn't pure water conduct an electric current?
3. Explain what is meant by the term *hydration sphere*?
4. Diagram, using circles for atoms, a crystal of KCl versus the same crystal of KCl dissolved in water.

Section 4.2 The Concentration of Solutions

Skill Review

11. Which of the following would best represent $MgCl_2$ dissolved in water?

(a)   (b)

A wealth of end-of-chapter problems test students' understanding of key concepts and problem-solving skills. They include **Skill Review** and **Chemical Applications and Practices**, organized by chapter sections and in matched pairs.

Comprehensive Problems

85. Individual atoms and molecules are so small that they have very low values of kinetic energy. However, given their mass and velocity, it is possible to calculate the value. What is the kinetic energy of an oxygen molecule (O_2) in air that you are breathing if its velocity is 460 m/s? Would you expect a nitrogen molecule (N_2) moving at the same speed to have more or less kinetic energy than the oxygen molecule? Explain.

86. Distinguish between the two terms in each of the following pairs:
    a. Heat and temperature
    b. System and surroundings
    c. Exothermic and endothermic
    d. q and ΔU

93. a. Suppose you are heating water (225 g) in a mug that you have placed in a microwave oven. As you wait to add the instant hot chocolate, please calculate, from the following data, the amount of energy as heat that the water has absorbed: The original water temperature was 15.0°C. When you remove the mug of hot water, you find that the temperature has risen to 98°C.
    b. What additional information would you need in order to determine the heat absorbed by the mug?

94. You have just removed a hot cheese pizza from the oven, and all of the ingredients are presumably at the same temperature. Without waiting for it to cool, you take a bite of the pizza. As you bite the pizza, the bread is hot on your tongue but does not burn. However, as you continue to bite, pizza sauce (mostly tomatoes and water) squeezes out and burns the roof of your mouth. Which has the higher specific heat,

Comprehensive Problems build on the students' mastery of concepts in the chapter and across *multiple* chapters in the textbook.

---

96. A student's coffee cup calorimeter, including the water it contains, has been calibrated in a manner similar to that described in Problem 46. The heat capacity was found to be 55.5 $\frac{J}{°C}$. If a 65.8-g sample of an unknown metal, at 100.0°C, was placed in the calorimeter initially at 25°C, and an equilibrium temperature of 29.1°C was reached, what is the specific heat of the metal?

97. The foods we eat provide fuel to keep us alive. Burning a 0.500-g sample of vegetable oil provides enough heat to raise the temperature of a calorimeter by 2.5 K. Assuming the heat capacity of the calorimeter to be 7.5 $\frac{kJ}{K}$, determine the heat of combustion for 1 g of the oil.

98. A student performs the experiment shown graphically here. What is the specific heat of the block of metal used in the experiment? (Assume that the heat capacity of the empty calorimeter is 7.5 $\frac{J}{°C}$.)

	—22.7°C	—24.1°C
	Water added	Metal block at 96.3°C
152.06 g = Mass of calorimeter	234.95 g = Mass of calorimeter and water	257.88 g = Mass of calorimeter, water, and metal block

Thinking Beyond the Calculation

99. Phosphoric acid is used in many soft drinks to add tartness. This acid can be prepared through the following reaction:

$$P_4O_{10}(s) + 6H_2O(l) \rightarrow 4H_3PO_4(aq)$$

Phosphoric acid

a. If the value of ΔH for the reaction is −453 kJ, what is the value of ΔH for the reverse of the reaction?
b. What is the value of ΔH for this reaction if 10.0 g of phosphoric acid is produced?
c. Is this reaction endothermic or exothermic?
d. If 1.50 g of $P_4O_{10}(s)$ and 2.50 mL of water were mixed, how many grams of phosphoric acid would result?
e. What is the enthalpy change for the process outlined in part d?
f. If 10.0 g of P_4O_{10} were mixed with 1.00 kg of water at 25.0°C, what would be the final temperature of the water?

Thinking Beyond the Calculation provides rigorous, cumulative, and conceptual problems based on multiple concepts.

About the Authors

Paul Kelter is Chair of the Department of Teaching and Learning at Northern Illinois University. Prior to that, he was most recently at the University of Illinois at Urbana-Champaign, where he was Professor of Chemistry and University Distinguished Teacher/Scholar. Dr. Kelter received his B.S. from the City College of the City University of New York in 1976 and was awarded a Ph.D. in Analytical Chemistry in 1980 from the University of Nebraska-Lincoln (UN-L). Before coming to Northern Illinois, he worked as an Educational Specialist for the National Aeronautics and Space Administration, was on the faculty of the University of Nebraska-Lincoln, and held the M. F. Rourk Chair in Chemistry and Chemical Education at the University of North Carolina at Greensboro, in addition to his time at Illinois. Professor Kelter has won many university and state teaching awards, including the University of Wisconsin-Oshkosh Distinguished Teaching Award, the Wisconsin Society of Science Teachers Regional Science Education Award, the Inaugural and Second University of Nebraska Student Body Outstanding Teacher of the Year Awards, and is a five-time recipient of the Certificate of Recognition to Students at UN-L. At Illinois, his student evaluations twice earned him recognition on the "Incomplete List of Teachers Ranked as Excellent by Their Students," which is given to teachers in the top ten percent of student rankings university-wide. He was the eighteenth faculty member inducted into the University of Nebraska-Lincoln Academy of Distinguished Teachers, and was the fifteenth faculty member honored with the University of Nebraska Outstanding Teaching and Instructional Creativity Award, the highest career-teaching award given by that university system.

Michael Mosher received his B.S. in Chemistry from the University of Idaho in 1988, his M.S. in Organic Chemistry from Dartmouth College in 1990, and his Ph.D. in Organic Chemistry from Texas Tech University in 1993. After teaching for two years at the University of Idaho as a Visiting Assistant Professor, he joined the faculty at the University of Nebraska at Kearney (UNK) in 1995. He currently holds the position of Professor and Chair of the Department of Chemistry at UNK. He is an active member of the American Chemical Society, the International Society for Heterocyclic Chemistry (ISHC), and serves on the executive board of directors for the International Center for First-Year Undergraduate Chemistry Education (ICUC). He has received numerous awards for teaching and research with undergraduate students, including the UNK Creative Teaching Award in 1999, 2001, and again in 2003. He is one of only three faculty members to have received two awards by the Pratt-Heins Foundation: the Faculty Award in Scholarship and Research in 2001 and the Faculty Award in Teaching in 2003. He is the author of numerous research publications and textbook ancillaries.

Andrew Scott is a science writer and lecturer in chemistry at Perth College of the UHI Millennium Institute in Scotland. He has a B.Sc. (Hons) in biochemistry from Edinburgh University (1977) and a Ph.D. in organic chemistry from Cambridge University (1981). He is the author of several textbooks and general science books, which have been translated into nine languages. Dr. Scott has also worked extensively as a science journalist and as a consultant on European education programs. He lives in the village of Dunning, in Perthshire, Scotland, with his wife Margaret. Their two children are currently at university in Edinburgh.

Contents

Why Is Chemistry of Practical Importance to Our Lives?   *xxii*
Features of the Textbook Program   *xxv*

Chapter 1

The World of Chemistry   1

Chemical Encounters: Chemists and Their Jobs   *2*
1.1   What Do Chemists Do?   *3*
1.2   The Chemist's Shorthand   *4*
Chemical Encounters: Desalination   *10*
1.3   The Scientific Method   *12*

How do we know?   How to control pain: An example of the scientific method in action   *14*

1.4   Units and Measurement   *16*
1.5   Conversions and Dimensional Analysis   *25*
1.6   Uncertainty, Precision, Accuracy, and Significant Figures   *28*
1.7   The Chemical Challenges of the Future   *34*

The Bottom Line   *37*
Key Words   *37*
Focus Your Learning   *39*

Chapter 2

Atoms: A Quest for Understanding   45

Chemical Encounters: How Old Is Life?   *46*
2.1   Early Attempts to Explain Matter   *46*
2.2   Dalton's Atomic Theory and Beyond   *49*
2.3   The Structure of the Atom   *51*
Chemical Encounters: The Dover Boat and Radiocarbon Dating   *51*
2.4   Atoms and Isotopes   *56*
2.5   Atomic Mass   *59*

NanoWorld / MacroWorld   Big effects of the very
small: Flavor analysis   *62*

2.6   The Periodic Table   *64*
2.7   Ionic Compounds   *67*
2.8   Molecules   *70*
2.9   Naming Compounds   *71*
2.10   Naming Acids   *76*

The Bottom Line   *77*
Key Words   *77*
Focus Your Learning   *79*

Chapter 3

Introducing Quantitative Chemistry   86

3.1   Formula Masses   *87*
Chemical Encounters: Adrenaline in the Body   *87*
3.2   Counting by Weighing   *89*

3.3   Working with Moles   *92*

Chemical Encounters: Cholesterol in the Blood   *95*

3.4   Percentages by Mass   *97*

Chemical Encounters: Nitrogen in Fertilizers   *97*

3.5   Finding the Formula   *99*

Chemical Encounters: A Focus on Chocolate   *99*

3.6   Chemical Equations   *103*

Chemical Encounters: Preparation of Aspirin   *103*

3.7   Working with Equations   *108*

Chemical Encounters: Cleaning the Air on Manned Spacecraft   *113*

Issues and Controversies   Everyday controversies of quantitative chemistry   *116*

The Bottom Line   *117*
Key Words   *117*
Focus Your Learning   *118*

Chapter 4

Solution Stoichiometry and Types of Reactions   124

Chemical Encounters: Gold Mining and Cyanide Leaching   *125*

4.1   Water—A Most Versatile Solvent   *126*

Chemical Encounters: Sports Drinks and Electrolyte Balance   *129*

4.2   The Concentration of Solutions   *131*

Chemical Encounters: Maximum Levels of Chemicals in Drinking Water   *135*

4.3   Stoichiometric Analysis of Solutions   *140*

How do we know?   How to test for small amounts of water   *142*

4.4   Types of Chemical Reactions   *144*

4.5   Precipitation Reactions   *145*

Chemical Encounters: Focus on Lead Sulfide   *145*

4.6   Acid–Base Reactions   *148*

4.7   Oxidation–Reduction Reactions   *152*

4.8   Fresh Water—Issues of Quantitative Chemistry   *157*

Chemical Encounters: Revisiting the Maximum Levels of Chemicals in Drinking Water   *157*

The Bottom Line   *159*
Key Words   *160*
Focus Your Learning   *161*

Chapter 5

Energy   167

Chemical Encounters: Setting the Stage with the Space Shuttle   *168*

5.1   The Concept of Energy   *169*

NanoWorld / MacroWorld   Big effects of the very small: Electrons capturing energy in photosynthesis   *173*

5.2   Keeping Track of Energy   *178*

Chemical Encounters: Energy in Foods   *179*

5.3   Specific Heat Capacity and Heat Capacity   *181*

5.4   Enthalpy   *186*
5.5   Hess's Law   *191*
Chemical Encounters: Focus on Methane   *193*
5.6   Energy Choices   *197*
Chemical Encounters: Energy Choices   *197*

    The Bottom Line   *200*
    Key Words   *200*
    Focus Your Learning   *202*

Chapter 6

Quantum Chemistry: The Strange World of Atoms   209

6.1   Introducing Quantum Chemistry   *210*
6.2   Electromagnetic Radiation   *211*

  Issues and Controversies   Putting the Squeeze on Radio Frequencies   *214*

6.3   Atomic Emission and Absorption Spectroscopy, Chemical Analysis and the Quantum Number   *216*
Chemical Encounters: Simultaneous Determination of Elements in Water   *219*
6.4   The Bohr Model of Atomic Structure   *220*
Chemical Encounters: The Nature and Applications of Lasers   *224*
6.5   Wave–Particle Duality   *225*
6.6   Why Treating Things as "Waves" Allows Us to Quantize Their Behavior   *227*
6.7   The Heisenberg Uncertainty Principle   *229*
6.8   More About the Photon—the de Broglie and Heisenberg Discussions   *230*
6.9   The Mathematical Language of Quantum Chemistry   *232*
6.10   Atomic Orbitals   *234*
Chemical Encounters: The Scanning Tunneling Microscope   *237*
6.11   Electron Spin and the Pauli Exclusion Principle   *242*
Chemical Encounters: Nuclear Spin and Magnetic Resonance Imaging   *242*
6.12   Orbitals and Energy Levels in Multielectron Atoms   *244*
6.13   Electron Configurations and the Aufbau Principle   *245*

    The Bottom Line   *251*
    Key Words   *251*
    Focus Your Learning   *253*

Chapter 7

Periodic Properties of the Elements   259

7.1   The Big Picture—Building the Periodic Table   *260*
7.2   The First Level of Structure—Metals, Nonmetals, and Metalloids   *265*

  NanoWorld / MacroWorld   Big effects of the very small: The diversity of steels   *267*

7.3   The Next Level of Structure—Groups in the Periodic Table   *269*
Chemical Encounters: Commercial Uses of the Main-Group Elements   *269*
Chemical Encounters: The Elements of Life   *277*
7.4   The Concept of Periodicity   *279*
7.5   Atomic Size   *280*

7.6 Ionization Energies *282*
7.7 Electron Affinity *286*
7.8 Electronegativity *287*
7.9 Reactivity *289*
7.10 The Elements and the Environment *292*
Chemical Encounters: The Elements and the Environment *292*

The Bottom Line *294*
Key Words *295*
Focus Your Learning *296*

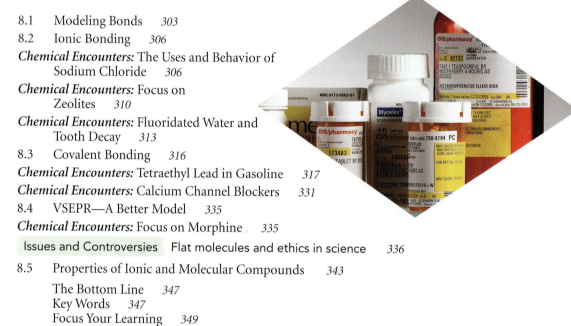

Chapter 8

Bonding Basics 302

8.1 Modeling Bonds *303*
8.2 Ionic Bonding *306*
Chemical Encounters: The Uses and Behavior of Sodium Chloride *306*
Chemical Encounters: Focus on Zeolites *310*
Chemical Encounters: Fluoridated Water and Tooth Decay *313*
8.3 Covalent Bonding *316*
Chemical Encounters: Tetraethyl Lead in Gasoline *317*
Chemical Encounters: Calcium Channel Blockers *331*
8.4 VSEPR—A Better Model *335*
Chemical Encounters: Focus on Morphine *335*
Issues and Controversies Flat molecules and ethics in science *336*
8.5 Properties of Ionic and Molecular Compounds *343*

The Bottom Line *347*
Key Words *347*
Focus Your Learning *349*

Chapter 9

Advanced Models of Bonding 355

Chemical Encounters: Molecular Structure and Eyesight *356*
9.1 Valence Bond Theory *358*
9.2 Hybridization *363*
9.3 Molecular Orbital Theory *374*
NanoWorld / MacroWorld Big effects of the very small: Color photography and the tricolor process *382*
9.4 Putting It All Together *384*
Chemical Encounters: Benzene, Stability, and MO Theory *384*
9.5 Molecular Models in the Chemist's Toolbox *386*

The Bottom Line *386*
Key Words *387*
Focus Your Learning *387*

Chapter 10

The Behavior and Applications of Gases   394

10.1   The Nature of Gases   *396*
10.2   Production of Hydrogen and the Meaning of Pressure   *397*
10.3   Mixtures of Gases—Dalton's Law and Food Packaging   *401*
Chemical Encounters: Dalton's Law and Food Packaging   *401*
10.4   The Gas Laws—Relating the Behavior of Gases to
Key Properties   *403*
Chemical Encounters: Balloons and Ozone
Analysis   *403*
10.5   The Ideal Gas Equation   *413*
10.6   Applications of the Ideal
Gas Equation   *418*
Chemical Encounters: Automobile Air Bags   *419*
Chemical Encounters: Acetylene   *420*
10.7   Kinetic-Molecular Theory   *422*
10.8   Effusion and Diffusion   *424*
10.9   Industrialization: A Wonderful, Yet Cautionary, Tale   *427*
Chemical Encounters: Ozone   *427*

NanoWorld / MacroWorld   The big effects of the very small: Accumulation of ozone
at the earth's surface   *430*

Chemical Encounters: The Greenhouse Effect   *431*

The Bottom Line   *433*
Key Words   *433*
Focus Your Learning   *434*

Chapter 11

The Chemistry of Water and the Nature of Liquids   440

Chemical Encounters: Worldwide Water Use   *441*
11.1   The Structure of Water: An Introduction to Intermolecular Forces   *443*
11.2   A Closer Look at Intermolecular Forces   *445*
11.3   Impact of Intermolecular Forces on the Physical Properties of Water, I   *451*
11.4   Phase Diagrams   *459*

How do we know?   The moons of Jupiter   *461*

Chemical Encounters: CO_2 as a Dry Cleaning Solvent   *461*
11.5   Impact of Intermolecular Forces on the Physical Properties of Water, II   *462*
11.6   Water: The Universal Solvent   *465*
11.7   Measures of Solution Concentration   *468*
Chemical Encounters: Composition of Seawater   *470*
11.8   The Effect of Temperature and Pressure on Solubility   *471*
Chemical Encounters: Impact of the Solubility of Oxygen in Fresh Water   *471*
11.9   Colligative Properties   *473*
Chemical Encounters: Meeting Municipal Water Needs   *480*

The Bottom Line   *482*
Key Words   *482*
Focus Your Learning   *484*

Chapter 12

Carbon 491

12.1 Elemental Carbon *493*
12.2 Crude Oil—the Basic Resource *494*
Chemical Encounters: Crude Oil—the Basic Resource *494*
12.3 Hydrocarbons *496*
12.4 Separating the Hydrocarbons by Fractional Distillation *506*
How do we know? Which hydrocarbons are in crude oil? *507*
12.5 Processing Hydrocarbons *508*
12.6 Typical Reactions of the Alkanes *510*
12.7 The Functional Group Concept *511*
12.8 Ethene, the C=C Bond, and Polymers *514*
12.9 Alcohols *516*
12.10 From Alcohols to Aldehydes, Ketones, and Carboxylic Acids *517*
12.11 From Alcohols and Carboxylic Acids to Esters *520*
12.12 Condensation Polymers *522*
12.13 Polyethers *524*
12.14 Handedness in Molecules *525*
12.15 Organic Chemistry and Modern Drug Discovery *527*

The Bottom Line *531*
Key Words *531*
Focus Your Learning *533*

Chapter 13

Modern Materials 539

Chemical Encounters: Materials in the Hospital *540*
13.1 The Structure of Crystals *540*
Chemical Encounters: X-ray Crystallography *546*
13.2 Metals *551*
Chemical Encounters: Photovoltaic Devices *555*
Chemical Encounters: Dental Amalgams *557*
13.3 Ceramics *557*

NanoWorld / MacroWorld Big effects of the very small: The chemistry of cement *558*

Chemical Encounters: Magnetic Resonance Imaging *561*
13.4 Plastics *563*
13.5 Thin Films and Surface Analysis *568*
Chemical Encounters: Heart Defibrillators *568*
13.6 On the Horizon—What Does the Future Hold? *571*
Chemical Encounters: "Green Chemistry" *571*

The Bottom Line *573*
Key Words *573*
Focus Your Learning *575*

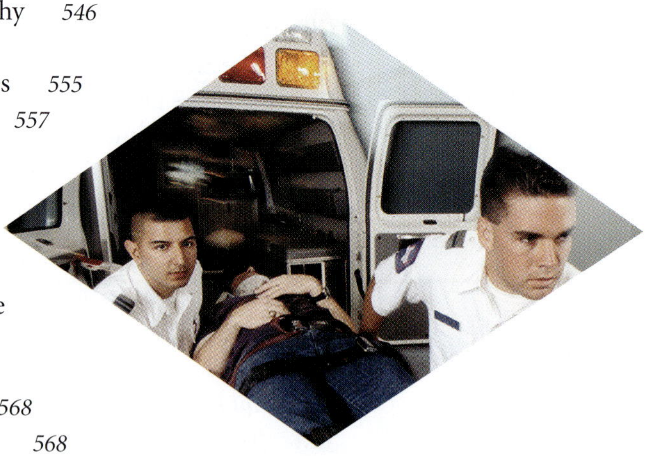

Chapter 14

Thermodynamics: A Look at Why Reactions Happen 579

Chemical Encounters: Bioenergetics   *580*
14.1   Probability as a Predictor of Chemical Behavior   *580*
14.2   Why Do Chemical Reactions Happen? Entropy and the Second Law of
          Thermodynamics   *585*
Chemical Encounters: Glycolysis   *585*
14.3   Temperature and Spontaneous Processes   *591*
Chemical Encounters: Psychrotrophs and Psychrophiles   *591*

NanoWorld / MacroWorld   Big effects of the very small:
          Industrial uses for the extremophiles   *592*

14.4   Calculating Entropy Changes in Chemical Reactions   *595*
Chemical Encounters: More Glycolysis   *595*
14.5   Free Energy   *600*
Chemical Encounters: Pyruvate and Lactate   *600*
14.6   When $\Delta G = 0$; A Taste of Equilibrium   *607*
Chemical Encounters: ATP Formation   *607*

          The Bottom Line   *612*
          Key Words   *613*
          Focus Your Learning   *613*

Chapter 15

Chemical Kinetics 619

Chemical Encounters: Atrazine and the Environment   *620*
15.1   Reaction Rates   *621*
15.2   An Introduction to Rate Laws   *627*
15.3   Changes in Time—The Integrated Rate Law   *631*
Chemical Encounters: Decomposition of DDT   *631*
Chemical Encounters: Persistent Pesticides   *636*
15.4   Methods of Determining Rate Laws   *641*
15.5   Looking Back at Rate Laws   *646*
15.6   Reaction Mechanisms   *647*
Chemical Encounters: Metabolism
          of Methoxychlor   *647*
15.7   Applications of Catalysts   *654*
Chemical Encounters: Destruction of Ozone   *655*

NanoWorld / MacroWorld   Big effects of the very small:
          Enzymes—nature's catalysts   *657*

          The Bottom Line   *658*
          Key Words   *658*
          Focus Your Learning   *660*

Chapter 16

Chemical Equilibrium 668

16.1   The Concept of Chemical Equilibrium   *670*
Chemical Encounters: Myoglobin in the Muscles   *670*

16.2   Why Is Chemical Equilibrium a Useful Concept?   *675*

Chemical Encounters: Important Processes That Involve Equilibria   *675*

16.3   The Meaning of the Equilibrium Constant   *676*

Chemical Encounters: The Manufacture of Sulfuric Acid   *676*

How do we know?   Equilibrium and chromatographic analysis   *678*

16.4   Working with Equilibrium Constants   *682*

16.5   Solving Equilibrium Problems—A Different Way of Thinking   *687*

16.6   Le Châtelier's Principle   *700*

Chemical Encounters: Catalysts in Industry   *704*

16.7   Free Energy and the Equilibrium Constant   *705*

The Bottom Line   *707*
Key Words   *708*
Focus Your Learning   *708*

Chapter 17

Acids and Bases   717

Chemical Encounters: Common Uses of Acids and Bases   *718*

17.1   What Are Acids and Bases?   *719*

17.2   Acid Strength   *723*

Chemical Encounters: Acids in Foods   *724*

17.3   The pH Scale   *730*

17.4   Determining the pH of Acidic Solutions   *736*

17.5   Determining the pH of Basic Solutions   *743*

Issues and Controversies   Nicotine and pH Control in Cigarettes   *746*

17.6   Polyprotic Acids   *747*

Chemical Encounters: Production and Uses of Phosphoric and Sulfuric Acids   *747*

17.7   Assessing the Acid–Base Behavior of Salts in Aqueous Solution   *752*

Chemical Encounters: Acid–Base Properties of Amino Acids   *755*

17.8   Anhydrides in Aqueous Solution   *757*

Chemical Encounters: Acid Deposition and Acid-Neutralizing Capacity   *757*

The Bottom Line   *759*
Key Words   *759*
Focus Your Learning   *760*

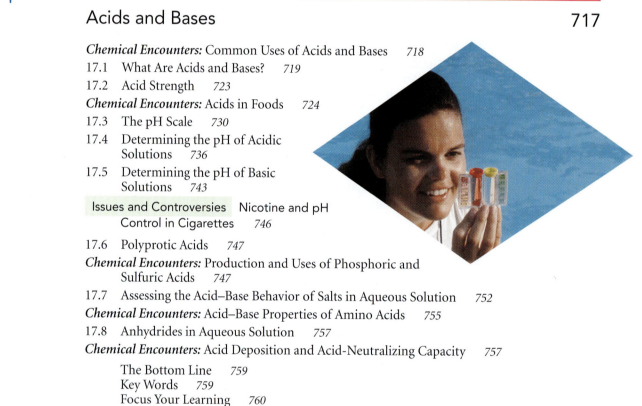

Chapter 18

Applications of Aqueous Equilibria   765

Chemical Encounters: Industrial and Environmental Applications of Titrations   *766*

18.1   Buffers and the Common-Ion Effect   *767*

Chemical Encounters: Scrubbing Sulfur Dioxide Emissions   *779*

NanoWorld / MacroWorld   Big effects of the very small: Buffers in biochemical studies and medicine   *785*

18.2   Acid–Base Titrations   *787*

Chemical Encounters: Anthocyanins and Universal Indicators   *801*

18.3   Solubility Equilibria   *802*

18.4   Complex-Ion Equilibria   *809*

Chemical Encounters: Commercial Uses of Aminopolycarboxylic Acid Chelating Agents   *810*

The Bottom Line   *815*
Key Words   *816*
Focus Your Learning   *817*

Chapter 19

Electrochemistry   823

19.1   What Is Electrochemistry?   *824*

How do we know?   DNA Profiling   *825*

19.2   Oxidation States—Electronic Bookkeeping   *826*

19.3   Redox Equations   *833*

19.4   Electrochemical Cells   *844*

Chemical Encounters: The Electric Eel   *844*

19.5   Chemical Reactivity Series   *849*

19.6   Not-So-Standard Conditions: The Nernst Equation   *851*

19.7   Electrolytic Reactions   *857*

Chemical Encounters: Electroplating   *859*

The Bottom Line   *860*
Key Words   *861*
Focus Your Learning   *862*

Chapter 20

Coordination Complexes   868

Chemical Encounters: Iron and Hemoglobin   *869*

20.1   Bonding in Coordination Complexes   *870*

20.2   Ligands   *873*

20.3   Coordination Number   *875*

How do we know?   What is the nature of the structure, bonding, and reactivity in cisplatin?   *876*

20.4   Structure   *876*

20.5   Isomers   *878*

20.6   Formulas and Names   *881*

20.7   Color and Coordination Compounds   *883*

Chemical Encounters: Transition Metals and Color   *884*

20.8   Chemical Reactions   *892*

Chemical Encounters: Cytochromes   *894*

The Bottom Line   *894*
Key Words   *895*
Focus Your Learning   *896*

Chapter 21

Nuclear Chemistry 900

21.1 Isotopes and More Isotopes *901*
21.2 Types of Radioactive Decay *904*
21.3 Interaction of Radiation with Matter *909*
Chemical Encounters: Radiation and Cancer *910*
21.4 The Kinetics of Radioactive Decay *912*
21.5 Mass and Binding Energy *915*
21.6 Nuclear Stability and Human-made Radioactive Nuclides *917*
21.7 Splitting the Atom: Nuclear Fission *920*
Chemical Encounters: Nuclear Weapons *920*
Chemical Encounters: Nuclear Reactors as a Vital Source of Electricity *921*
21.8 Medical Uses of Radioisotopes *924*
Chemical Encounters: Tracer Isotopes for Diagnosis *924*

How do we know? Imaging with Positron Emission Tomography (PET) *926*

The Bottom Line *928*
Key Words *928*
Focus Your Learning *929*

Chapter 22

The Chemistry of Life 934

22.1 DNA—The Basic Structure *935*
Chemical Encounters: The Human Genome Project *935*
22.2 Proteins *940*
22.3 How Genes Code for Proteins *941*
22.4 Enzymes *947*

NanoWorld / MacroWorld
Big effects of the very small: Vitamins and disease *950*

22.5 The Diversity of Protein Functions *953*
22.6 Carbohydrates *954*
22.7 Lipids *957*
22.8 The Maelstrom of Metabolism *958*
22.9 Biochemistry and Chirality *960*
22.10 A Look to the Future *961*

The Bottom Line *964*
Key Words *965*
Focus Your Learning *966*

Appendixes

Appendix 1   Mathematical Operations   *A1*

A1.1   Working with Exponents   *A1*

A1.2   Working with Logarithms   *A3*

A1.3   Solving the Quadratic Equation   *A5*

A1.4   Graphing   *A6*

Appendix 2   Calculating Uncertainties in Measurements   *A7*

Appendix 3   Thermodynamic Data for Selected Compounds at 298 K   *A9*

Appendix 4   Colligative Property Constants for Selected Compounds   *A13*

Appendix 5   Selected Equilibrium Constants at 298 K   *A13*

Appendix 6   Water Vapor Pressure Table   *A17*

Appendix 7   Standard Reduction Potentials at 298 K   *A17*

Appendix 8   Common Radioactive Nuclei   *A19*

Answers to Practice Exercises and Selected Exercises   A20

Credits   A49

Index/Glossary   A53

Why Is Chemistry of Practical Importance to Our Lives?

To the Student

Please allow us, for a moment, to share with you what happens on the first day of class in our own first-year chemistry classrooms. Often, we spend some time meeting as many of you as we can—beginning to know your names, learning your majors, and finding out why you have enrolled in our chemistry course. We then tell a few stories about ourselves, because part of studying together is seeing each other's very human side. During the entire year, we attempt to learn more about each other in order to make this new and exciting journey into chemistry a shared trip—where we ask each other questions, find answers, make connections, and discover why this wonderful subject is so useful for us to understand. By the same token, insofar as it is possible to share a bit of ourselves and our own sense of the beauty of this subject, we have tried to do so in this textbook.

In many ways, our presentation of material in this book is quite similar to the way we teach in class. We will share with you the big ideas of chemistry, and, because we all want to know "why we need to know this," we will discuss how the concepts of chemistry are applied to real-world issues, such as manufacturing processes, the blastoff of a rocket, the interaction of pharmaceuticals with the blood, and the age of life itself. We will then ask questions that lead to an explanation of how we know these things.

Our book has two overriding goals. The first is to help you appreciate the depth and breadth of chemistry. The second goal is to encourage you to make connections between concepts in chemistry and the world in which we live. "How do we know?" is one of the most vital questions, no matter what your field of study. In chemistry, we often say that asking good, focused questions is vital to "thinking like a chemist." *Knowing how to know* and *wanting to know* are two essential traits of successful, independent learning that not only are crucial to your study of chemistry but also will really pay off for a lifetime.

As you read this book (and we do believe that it can be read as a wonderful story of chemistry), look for the places where we raise questions. Ask yourself why we raise each question at the point where we do. What is the key idea? Why is this useful to know? What can I now figure out that I was not able to before? What connections can I make? What have I learned?

It is true that chemistry can be quite challenging and that persistence is necessary. Daily reading and study are the keys, and making a serious investment with your heart as well as your head will pay off in an understanding of chemistry as well as in the enjoyment of learning. If this book enhances your desire to learn more about the world around you, then we've been successful—and so have you.

These are the very things that we say to our own students. And then, together, we start doing chemistry. Let's go!

To the Instructor

We are excited to be presenting you with a different kind of textbook—one that is written from the standpoint of *how chemists really teach in the twenty-first century*. It is our aim to complement your teaching style by giving your students a teacher's viewpoint in print. This book is about the questions we ask when we teach chemistry. It focuses on the "big ideas" of chemistry that we, as chemists, found so appealing that we chose a career in this field. For example, we explore the periodic table, the energy exchanges that accompany all chemical changes, the ideas of quantum mechanics (science is about probability, not perfection), and how we can use Le Châtelier's principle to control the extent of chemical reactions. These are just some of the ideas we want students to grasp deeply when we say, "*Here's what it means to think like a chemist.*"

A Framework of Interwoven Applications

In addition to our commitment to sharing the big ideas that define science in general and chemistry in particular, we write from the belief that chemistry is vitally important to the world in which we live. In this text, we present chemistry in the context of how it is related to our everyday lives by interweaving the concepts of chemistry with their uses in the chemical industry, the human body, and the environment. For example, a discussion of the energy changes that occur in chemical reactions is entwined with how these concepts are used in the U.S. space shuttle program. The discussion of kinetics is applied to the fate of pesticides in the environment. We view chemistry's applications in society as fundamentally good, and we note that when the use of chemistry has led to unfortunate consequences (especially for the environment), chemistry has also been used to clean up the mess. This focus on the *vital role* that the ideas and practice of chemistry have in our day-to-day lives is written into the storyline of the text. We mean it when we say that this textbook is *applications-based*.

An Interrogative, Inquiry-Based Style

In our classrooms, we enjoy raising questions with our students, both because we like to hear their ideas and because raising questions is a key characteristic that defines the curious minds of scientists. We wrote our textbook to reflect an interrogative style, in which questions addressed to the student—often the same questions we ask in our own lectures—begin various topics. This approach involves students in the discussion, encourages them to pose questions about their world, and nurtures their curiosity about science and how it applies to society. This approach recurs throughout every chapter, as we continue to engage the student in what is most fundamental to practicing scientists: questioning the world around them.

Problem-Solving Approach

Our discussions are geared to science majors and unfold in the context of how they and their classmates will need to apply chemistry to their lives to solve problems. Our approach to problem solving helps students think critically by encouraging them to ask questions and frame them in such a way as to solve a problem effectively. Then, once the problem is solved, the text guides students in evaluating whether their answers make sense. The in-chapter *Exercises* include worked-out *Solutions* and are framed with a variety of pedagogical aids.

- *First Thoughts* engage the student.
- Worked-out *Solutions* guide the student step-by-step through the problem.
- *Further Insights* extend the concept.
- *Practice Problems* give students additional problems, to which answers are provided at the back of the book.
- The *list of corresponding end-of-chapter problems* at the end of each *Exercise* directs students to more practice.

These *Exercises* demonstrate how scientists think about problems and show how we work through a problem to arrive at an answer by asking questions.

Chapter Organization

Chapter 12 (Carbon) and Chapter 13 (Modern Materials) are strategically placed in the middle of this text for many reasons. The theme of the organic chapter (structure changes lead to changes in function and properties) logically follows the introduction of the topics on structural bonding models. The materials chapter logically follows the sequence of chapters on gases (Chapter 10) and liquids (Chapter 11). In addition, the placement of these chapters at the end of a typical semester of undergraduate chemistry allows the instructor the opportunity to end the semester on these applied topics of chemistry. The topical order is flexible, and these chapters can easily be taught in a different sequence than represented in the textbook.

The Bottom Line

Our applications-based and interrogative approach has resulted in a book that *students will actually read* and that reflects how *teachers really teach*. Here, then, as we say at the end of every chapter in the text, is *The Bottom Line*—in this case, a concise list of the main features of the textbook:

- **Applications** of chemistry are interwoven within the concepts of chemistry. Students often ask "Why do I need to know this?," so we have shown, at every opportunity, how chemistry is a part of our world.
- An **interrogative style** encourages students to be inquisitive about their world both locally and globally and involves them in discussing and learning concepts that are important to chemistry.
- Our **problem-solving approach** helps students to first think about a problem, to next approach the problem in a logical manner in order to arrive at a solution efficiently, and then to think beyond the calculation to uncover related information about the concepts that are being explored.
- A **practical, student-friendly pedagogy** includes writing that involves students and offers many opportunities for review (such as in the *Here's What We Know So Far* sections and *The Bottom Line* summaries). In addition, **visually engaging illustrations** clearly represent concepts for students and illustrate the vital connection of the world of the atom to the world in which we live.
- A **dynamic, contemporary art program** appeals to today's students, who expect exciting and visually appealing graphics. This program features molecular-level illustrations of key concepts to help students connect microscale activity to macroscale phenomena. In addition, electrostatic potential maps use vibrant color to show how electron density changes across a molecule.
- **Technology and print resources** accommodate a variety of student learning styles and help instructors more easily manage homework by creating assignable activities that can be graded automatically in an online environment.

We hope that you and your students will enjoy reading and working with this textbook as much as we enjoyed writing it.

Paul Kelter
Michael Mosher
Andrew Scott

Features of the Textbook Program

Feature	Purpose	Example	What Is It?
Integrated applications Application CHEMICAL ENCOUNTERS: Energy Choices	To demonstrate why students need to know each concept	See page 197.	■ An application opens each chapter. ■ *Chemical Encounters* present major applications that are emphasized as themes and listed in the Table of Contents and chapter outlines. ■ Application icons highlight major applications in the text narrative.
Key questions within the chapter about energy: **What is energy?** uttle store energy and release it hin molecules and compounds?	To model the interrogative approach in chemistry and to encourage students to think critically	See page 168.	■ Important questions are highlighted throughout the text.
Exercises within the chapter **EXERCISE 5.3** The First Law of Thermodynamics Calculate the change in the energy of a system if 51.	To help students think critically about questions and practice solving problems in chemistry	See page 178.	Features of in-chapter *Exercises*: ■ *First Thoughts* engage the student. ■ Worked-out *Solutions* are shown. ■ *Further Insights* extend the concept and connect the concept to additional applications. ■ Additional *Practice* problems are provided. ■ Corresponding end-of-chapter problems are listed.
Here's What We Know So Far **The Bottom Line**	To provide multiple opportunities for student review throughout each chapter	See pages 175 and 186. See page 200.	■ Key concepts are reviewed in bulleted lists throughout the chapter. ■ Important concepts are summarized at the end of each chapter.
Boxed features	To provide detailed applications of chemistry concepts outside of the flow of the text discussion	See page 14. See page 62. See page 116.	■ *How Do We Know?* features demonstrate the use of key chemical concepts in medicine, research, and other applications. ■ *NanoWorld / MacroWorld* essays focus on how the interactions at the molecular level translate into explanations of chemistry at the macro level. ■ *Issues and Controversies* essays explore current, controversial topics related to science.

(*continued*)

(*continued*)

Feature	Purpose	Example	What Is It?
Illustrations and photos **Methylhydrazine**	To engage visual learners, help clarify key concepts, offer examples of real-world applications, and demonstrate the connections between the macroworld and the microworld	See page 195.	■ Art and photos appear throughout each chapter and within end-of-chapter problems.
End-of-chapter problems	To promote student review and comprehension of material To provide instructors with numerous ways to meet student and course needs	See pages 118–123.	Features of end-of-chapter problems: ■ *Skill Review* helps students practice applying specific concepts via paired problems. ■ *Chemical Applications and Practices* provide paired problems within the context of the real world. ■ *Comprehensive Problems* test students' mastery of concepts in the chapter. ■ *Thinking Beyond the Calculation* provides rigorous, cumulative, and conceptual problems based on multiple concepts.
Appendixes	To provide students with a ready source of useful data in chemistry To provide instructors with data that can be used to construct problems that meet specific course needs	See page A1.	■ Eight appendixes appear at the back of the book.

3.3 Working with Moles

Skill Review

17. Which of these quantities of sodium chloride (NaCl) contains the greatest mass?

   0.100 mol 4.2×10^{23} formula units   1.60 g

18. Which of these quantities of acetaminophen ($C_8H_9NO_2$) contains the greatest mass?

   0.550

19. Conve

   a. 65.0

procedure produced 100 atoms of meitneriu
moles is this? What would be the mass of the s

Chemical Applications and Practices

29. Iron is essential for the transport of oxygen
    body and for energy production through sever
    cycles.
    a. What is the mass, in grams, of one atom of
    b. The recommended dietary allowance (RDA

tion. Student B obtains an 85.0% yield in the same reaction.
Can you now determine which student obtained the greater
mass of the product? Explain, or justify your answer.

Comprehensive Problems

87. Review the vitamin C controversy discussed in this chapter.
    How is it possible that a compound can be both good and
    bad for your health?

Thinking Beyond the Calculation

98. Xylene (ZIGH-leen) is an important organic molecule isolated from petroleum oil. It is often used as a thinner for oil-based paints.
    a. Elemental analysis of a sample of xylene shows that the mass percent of carbon is 90.51% and the mass percent of hydrogen is 9.49%. What is the empirical formula of xylene?

For the Student

An extensive print and media package has been designed to assist students in working problems, visualizing molecular-level interactions, and building study strategies to fully comprehend concepts.

Technology: For Students

■ **HM ChemSPACE™** is the portal to online student media resources (**college.hmco.com/pic/kelterMEE**) to help students prepare for class, study for quizzes and exams, and improve their grade. Students will have access to

- An **Online Multimedia eBook** integrates reading textbook content with embedded links to media activities and supports highlighting, note taking, zooming, printing, and easy navigation by chapter or page.
- **ACE practice tests**
- Electronic **flashcards**
- Over 50 hours of **video lessons** from Thinkwell®, segmented into 8- to 10-minute mini-lectures by a chemistry professor that combine video, audio, and whiteboard to demonstrate key concepts
- **Visualizations** (molecular animations and lab demonstration videos) give students the opportunity to review and test their knowledge of key concepts.
- Interactive **tutorials** allow students to dynamically review and interact with key concepts from the text.
- **General Chemistry resources**—interactive periodic table, molecule library of chemical structures, and careers in Chemistry

    HM ChemSPACE accompanies every new copy of the text. Students who have bought a used textbook can purchase access to *HM ChemSPACE* separately.

■ **HM ChemSPACE™ with Eduspace®, Houghton Mifflin's Complete Course Management System,** features all of the student resources available in *HM ChemSPACE* as well as randomized online homework, *ChemWork* assignments, and SMARTHINKING®—live, online tutoring. This dynamic suite of products gives students many options for practice and communication.

■ **Online Homework** Authored by experienced chemistry professors, *Chem-Work* exercises offer students opportunities to practice problem-solving skills that are different from the end-of-chapter homework provided in the text and online. These problems are designed to be used in one of two ways: The student can use the system to *learn* the problem-solving process (while doing actual homework problems) or the students can use the system as a *capstone* assignment to determine whether they understand how to solve problems (perhaps in final preparation for an exam). The *ChemWork* exercises test students' understanding of core concepts from each chapter. If a student needs help, assistance is available through a series of hints. If a student can solve a particular problem with no assistance, however, he or she can proceed directly to the answer and receive congratulations. The procedure for assisting the student is modeled after the way an instructor would help a student with a homework problem in his or her office. The hints are usually in the form of interactive questions that guide the student through the problem-solving process. Students cannot receive the right answer from the system, but it encourages them to continuing working on the problem through this system of multiple hints. (This default setting can be changed.) *ChemWork* problems also are automatically graded and recorded in the gradebook.

    Another important feature of *ChemWork* exercises is that each student in the course receives a unique set of problems. This is accomplished by using different methods to randomize the questions (such as using algorithms and pools of different versions). *ChemWork* problems also have the capability of checking for significant figures in calculations. Since it is a homework system, it is designed to tell the student if the significant figures are incorrect in their answer without marking the answer wrong. This feature encourages the student to pay attention to significant figures without causing them to lose their focus on the process of solving the problem.

In addition to *ChemWork,* Houghton Mifflin has partnered with WebAssign® to offer text-specific, end-of-chapter problems. In addition to the problems from this textbook, extra Comprehensive Problems are also offered online.

- **SMARTHINKING®—Live, Online Tutoring** *SMARTHINKING®* provides personalized, text-specific tutoring during typical study hours when students need it most. (Terms and conditions are subject to change; some limits apply.) With *SMARTHINKING,* students can submit a question to get a response from a qualified e-structor within 24 hours; use the whiteboard with full scientific notation and graphics; view past online sessions, questions, or essays in an archive on their personal academic homepage; and view their tutoring schedule. E-structors help students with the process of problem solving rather than supply answers. *SMARTHINKING* is available through *Eduspace* or, upon instructor request, with new copies of the student textbook.

Print Supplements: For Students

- **Study Guide,** by Gretchen M. Adams (University of Illinois at Urbana-Champaign) and Frank J. Torre (Springfield College): Expands on the problem-solving methods of the textbook with *First Thoughts, Solutions,* and *Further Insights* to help students further understand concepts that are particularly difficult. Each chapter of the guide includes 45–50 exercises for additional student practice. Tables and section descriptions also are included.

- **Student Solutions Manual,** by Scott A. Darveau (University of Nebraska, Kearney): Provides detailed solutions for half of the end-of-chapter exercises (designated by the blue question numbers) using the strategies emphasized in the text. This supplement has been thoroughly checked for precision and accuracy.

- **Lab Manual,** by James Almy (Cornell University): Offers a unique mix of both traditional and guided-inquiry experiments. This mix of experiments gives the instructor power in choosing the student's level of autonomy in the laboratory, and guidelines for use of these two types of experiments are discussed extensively in the "To the Instructor" section of the manual. Students will find each experiment to be clear, engaging and thought provoking. The front of the manual, with sections "To the Student," "Safety," and "How to Use Lab Equipment," and the appendix sections, give students a solid base of knowledge (and instructors a solid base of comfort) to perform well in the laboratory.

Complete Instructional Package for the Instructor

A complete suite of customizable teaching tools accompanies *Chemistry: The Practical Science.* Whether available in print, online, or via CD, these integrated resources are designed to save you time and help make class preparation, presentation, assessment, and course management more efficient and effective.

■ **HM Testing™ (powered by Diploma®)** combines a flexible test-editing program with comprehensive gradebook functions for easy administration and tracking. With *HM Testing,* instructors can administer tests via print, network server, or the web. Questions can be selected based on their chapter, section, level of difficulty, question format, algorithmic functionality, topic, learning objective, and five levels of key words. The *Complete Solutions Manual* files are also included on this CD.

With *HM Testing* you can

- Choose from the 2000 test items designed to measure the concepts and principles covered in the text.
- Ensure that each student gets a different version of the problem by selecting from over 777 algorithmic questions.
- Edit or author algorithmic or static questions that integrate into the existing bank, becoming part of the question database for future use.
- Choose problems designated as single-skill (easy), multi-skill (moderate), and challenging and multi-skill (difficult). Create questions, which then become part of the question database for future use.
- Customize tests to assess the specific content from the text.
- Create several forms of the same test where questions and answers are scrambled.

The *Complete Solutions Manual* files are included on this CD.

■ **HM ClassPresent™ General Chemistry** provides a library of molecular animations and lab demonstration videos, covering core chemistry concepts arranged by chapter and topic. The resources can be browsed by thumbnail and description or searched by chapter, title, or key word. Full transcripts accompany all audio commentary to reinforce visual presentations and to accommodate different learning styles.

■ **HM ChemSPACE™ (college.hmco.com/pic/kelterMEE)** is a website designed to be a "turnkey" solution that students can access and use quickly without any instructor set-up. The instructor version of the *HM ChemSPACE* website includes much of the content found in the *Eduspace* version but without any of the accompanying course management tools and services. It requires no instructor set-up and includes all the digital resources instructors need to develop lectures:

- *Lecture Outline PowerPoint™* presentations
- Virtually all of the text figures, tables, and photos (PPT and JPEG formats)
- *Instructor's Resource Manuals* for both the main text and the *Lab Manual* (PDF formats)
- Transparencies (PDF format)
- Animations and videos (also in PPT format)
- *Classroom Response System* (CRS) "clicker" content, offers a dynamic way to facilitate interactive classroom learning with students and is an excellent tool for teachers to gauge student success in understanding chapter material. This text-specific content is comprised of multiple-choice questions to test common misunderstandings, core objectives, and difficult concepts, all with an average time of one minute for feedback. Questions are conceptual and quantitative, and many also include applications. Students' responses display anonymously in a bar graph, pie chart, or other graphic and can be exported to a gradebook. (Additional hardware and software are required. Contact your sales representative for more information.)

■ **HM ChemSPACE™ with Eduspace®** Like all course management systems, *Eduspace* provides the powerful tools and premium services many instructors desire. *Eduspace* allows instructors to build and customize their own online courses, deliver homework and other assignments to students, and track student progress and results via a powerful gradebook. It includes:

- All instructor and student media included within *HM ChemSPACE*
- Online homework problems delivered through *WebAssign,* allowing you to choose the system that best meets your instructional needs
- *ChemWork* interactive assignments, which help students learn the process of problem solving through a series of interactive hints. These exercises are graded automatically.
- SMARTHINKING, live, online tutoring for students

■ **Online Course Content for Blackboard®, WebCT®, eCollege, and ANGEL** allows delivery of text-specific content online using your institution's local course management system. Through these course management systems, Houghton Mifflin offers access to all assets such as testbank content, tutorials, and video lessons. Additionally, qualified adoptions can use *PowerCartridges for Blackboard®* and *PowerPacks for WebCT®* to allow access to all *Eduspace* course content, including *ChemWork* and online homework, from your institution's local system.

■ **WebAssign®** is a Houghton Mifflin partner offering an online homework system with text-specific end-of-chapter problems. *WebAssign* was developed by teachers, for teachers. For information on this system, contact your HM representative. With *WebAssign,* you can

- Create assignments from a ready-to-use database of textbook questions or write and customize your own exercises
- Create, post, and review assignments 24 hours a day, 7 days a week
- Deliver, collect, grade, and record assignments instantly
- Offer more practice exercises, quizzes, and homework
- Assess student performance to keep abreast of individual progress
- Control tolerance and significant figures settings on a global and per-question basis

The *WebAssign* gradebook gives you complete control over every aspect of student grades. In addition, if you choose to enable it, your students will be able to see their own grades and homework, quiz, and test averages as the semester progresses, and even compare their scores with the class averages.

■ **Instructor's Resource Manual** (Mike Mosher, University of Nebraska, Kearney and Paul Kelter, University of Nebraska-Lincoln) Available online at *HM ChemSPACE,* this PDF manual offers suggestions for alternate teaching strategies, chapter and section summaries, important figures and tables from the text, and outside readings.

■ **Instructor's Resource Manual to the Lab Manual** (James Almy, Cornell University) Available online at *HM ChemSPACE,* this PDF manual provides instructors general information about the experiments to help them guide the lab, includes necessary equipment, and includes results and observations.

■ **Complete Solutions Manual** (Scott A. Darveau, University of Nebraska) Available online on the *HM Testing* CD, this complete version of the *Student Solutions Manual* contains detailed solutions to *all* of the end-of-chapter problems. This supplement is intended for the teacher's convenience.

Acknowledgments

We would like to thank the following friends and colleagues whose professional skills, understanding, and commitment have made this textbook possible.

Frank Torre and Gretchen Adams, dear friends both, thanks for your work on the study guide and for Frank's work on the PowerPoint slides. Frank, continue to work for peace and fly high. Gretchen, you throw like a guy.

Scott Darveau, how lucky we are to have you on our team. Thanks for putting together the solutions manual and offering so much valued counsel during the construction of this text.

Jeff Woodford and Jason Overby, many thanks for your excellent work on the test bank. Jeff, your work as a reviewer of the manuscript was marvelous. Mike Perona, we are grateful for your exceptional eye on the manuscript review as well.

Margaret Asirvatham, valued friend and colleague, thanks for your ability to peer into the future with your work on the clickers. Estelle Lebeau and Linda Bush, thank you for your many efforts in helping us to put our problems online into Eduspace. Laura Pence, teller of great stories, thank you for helping us to make our story come alive.

We also thank Laura McGinn, our marketing manager, for helping others to see what we have written; Andrea Cava, project editor, who was so good at keeping the process together over three continents; Connie Day, copyeditor, for her careful attention to detail; Cia Boynton, designer, for the attractive and flexible design; Jessyca Broekman, art development editor, to whom we owe much thanks for developing the lovely art program for the text; Naomi Kornhauser, for finding the photos that have added so much to the story; and Charles Hartford, vice president and publisher, for overseeing the project. Thanks also to Liz Hogan and Amy Galvin, assistant editors; Chip Cheek, editorial assistant; and Lynne Blaszak, senior media producer.

Special thanks to two who will be our friends forever: Richard Stratton and Katherine Greig, always with the right words to say and always there, no matter where our separate paths might take us.

Finally, thanks to Rebecca Berardy Schwartz, kindred spirit, confidante, soul-sister, and top-notch developmental editor. Thanks, Rebecca, for knowing so well that communicating science requires both the brain and the heart.

Reviewers

We sincerely appreciate our colleagues' help in reviewing the text and instructor and student resources.

Shawn B. Allin, *Spring Hill College*
Olujide T. Akinbo, *Butler University*
David Ball, *Cleveland State University*
Mufeed M. Basti, *North Carolina A&T State University*
Vladimir Benin, *University of Dayton*
Conrad Bergo, *East Stroudsburg University*
Fereshteh Billiot, *Texas A&M, Corpus Christi*
Richard E. Bleil, *Dakota State University*
Mary J. Bojans, *Pennsylvania State University*
Christine Bilicki, *Pasadena City College*
Sean Birke, *Jefferson College*
Iona Black, *Yale University*
Robert Blake, *Texas Tech University*
Jabe Breland, *St. Petersburg College*
James Carr, *University of Nebraska, Lincoln*
Annina Carter, *Adirondack Community College*
Tsun-Mei Chang, *University of Wisconsin–Parkside*
Paul J. Chirik, *Cornell University*
Kurt Christoffel, *Augustana College, Illinois*
Douglas S. Cody, *Nassau Community College*
Donald Cotter, *Mount Holyoke College*
Robert Cozzens, *George Mason University*
Janet DeGrazia, *University of Colorado*

Phillip Davis, *University of Tennessee, Martin*
Luther F. Elrod, *Piedmont College*
Dale D. Ensor, *Tennessee Tech University*
Cheryl Baldwin French, *University of Oklahoma*
Lois Hansen-Polcar, *Cuyahoa Community College*
Alton Hassell, *Baylor University*
HollyAnn Harris, *Creighton University*
Richard Hartmann, *Nazareth College*
Bert Holmes, *UNC, Asheville*
Martha Jackson-Carter, *Community College of Aurora*
Milton Johnston, *University of South Florida*
Don Jones, *Lincoln Land College*
Andy Jorgensen, *University of Toledo*
Phillip C. Keller, *University of Arizona*
Shahed Khan, *Duquesne University*
Bette A. Kreuz, *University of Michigan, Dearborn*
Shannon Lieb, *Butler University*
Art Landis, *Emporia State University*
Mark Masthay, *Murray State University*
Ursula Mazur, *Washington State University*
Craig McLauchlan, *Illinois State University*
Gary Mercer, *Boise State University*
Sally A. Meyer, *Colorado College*

The World of Chemistry

Contents and Selected Applications

Chemical Encounters: Chemists and Their Jobs

1.1　**What Do Chemists Do?**

1.2　**The Chemist's Shorthand**

Chemical Encounters: Desalination

1.3　**The Scientific Method**

1.4　**Units and Measurement**

1.5　**Conversions and Dimensional Analysis**

1.6　**Uncertainty, Precision, Accuracy, and Significant Figures**

1.7　**The Chemical Challenges of the Future**

Scientists are not especially different from anyone else. They are just inquisitive people who use a different set of tools to ask questions and formulate answers.

 Go to **college.hmco.com/pic/kelterMEE** for online learning resources.

More than 14,000 students graduate with chemistry degrees in the United States each year, according to the American Chemical Society, the world's largest professional organization for chemists. Although the majority of these students obtain bachelor's degrees (more than 10,000, of which about 50% are earned by women), many receive advanced degrees (nearly 4,000, of which 40% are earned by women). What do all of these new chemists do?

As shown in Figure 1.1, about 95,000 chemists are employed across the country in many sectors of the government, industry, and even academia. Chemists work with congressional leaders to develop new laws and provide information on topics of interest to the general public. Chemists design and make new drugs to treat maladies such as AIDS, attention deficit disorder, and sickle-cell anemia. Chemists design and manufacture new paints and coatings to keep our lawn chairs and cars rust-free. Chemists work in the energy industries developing new ways to harness the power of the Sun. Chemists prepare new laundry detergents that minimize harm to the environment. Chemists do analyses related to the law, an activity known as forensic chemistry. These are just some of the many areas where you'll find chemists at work. As you might also expect, chemists work in the world's oil fields as part of the vital international petroleum industry.

How much do chemists earn? Figure 1.2 shows average salary data from 2003 to 2004. The average starting salary for chemists entering the job market in 2003 with a bachelor's degree was $31,000. For those who had earned the Ph.D. degree, the average starting salary was $67,500.

More important, the rate of unemployment for the chemistry profession was only 3.1% nationwide in 2005. When compared to the unemployment rate of 5% in the United States at large, chemists do fairly well.

In this chapter, we'll take the first steps in learning what it means to think like a chemist. We will discuss what chemists do. We'll learn about the types of questions they ask, and we'll discover some of the most basic tools they use to answer those questions.

		2003	2004
Bachelor's	Industry	$63*	$65
	Government	58.7	62.4
	Academia	36.6	42.2
M.S.	Industry	76	80
	Government	74.5	74
	Academia	50.1	52
Ph.D.	Industry	100	103
	Government	95	98
	Academia	65.1	67.2

*In thousands of dollars.

FIGURE 1.2

An advanced degree in chemistry boosts average salary. Employment in industry is also more lucrative than working in academia.

FIGURE 1.1

Chemists are employed in many different sectors of the job market.

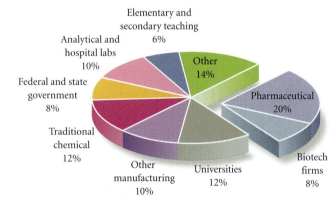

Chemistry Bachelor's Degrees

Elementary and secondary teaching 6%
Analytical and hospital labs 10%
Federal and state government 8%
Traditional chemical 12%
Other manufacturing 10%
Universities 12%
Other 14%
Pharmaceutical 20%
Biotech firms 8%

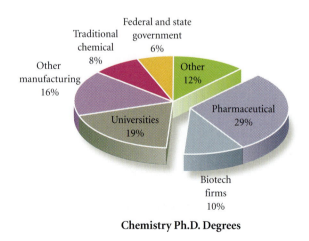

Chemistry Ph.D. Degrees

Federal and state government 6%
Traditional chemical 8%
Other manufacturing 16%
Universities 19%
Other 12%
Pharmaceutical 29%
Biotech firms 10%

1.1 What Do Chemists Do?

Judith Fairclough-Baity (Figure 1.3) is an industrial chemist who worked for 19 years at one of the largest chemical companies in the world, where she designed and made new tubing for heart–lung machines. This tubing carries blood from a patient undergoing open-heart surgery to a pump and then returns it to the patient. How did Fairclough-Baity approach her design of these tubes? Just like all chemists, she defined the key questions for which she needed answers. The experiments, data analyses, and conclusions that are part of this *scientific approach to knowing* led her and her coworkers to a fresh understanding—and to other questions. Eventually, after extensive experimentation and testing, she and her team arrived at a new design.

However, it is not just an inquisitive nature that makes a chemist. Chemists such as Fairclough-Baity also have to know which questions to ask. This means that chemists need to have a keen understanding of the basic principles behind what they do. By *knowing the core ideas of this science* and having a hunger to know, chemists can get to the bottom of a problem and come up with a possible solution. For example, questioning chemists discovered Teflon (the nonstick coating on your frying pan), Taxol (a very successful anticancer drug), and the plastic that makes up squeezable ketchup bottles (Figure 1.4). Their inquisitive approach has influenced every aspect of our lives. Today, chemists continue to work on the development of countless products, including environmentally friendly alternatives to gasoline in our cars, faster computer chips, and low-fat potato chips.

FIGURE 1.3

Judith Fairclough-Baity worked on developing new tubing for use in heart–lung machines.

FIGURE 1.4

Teflon, Taxol, and ketchup bottles.

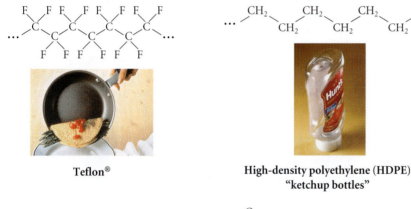

Teflon®

High-density polyethylene (HDPE) "ketchup bottles"

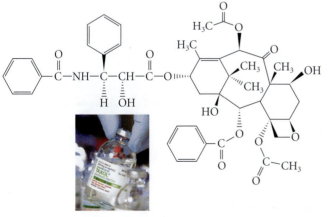

Taxol®

Environmentally friendly alternatives to gasoline in our cars, faster computer chips, and low-fat potato chips.

Not only do chemists solve problems, but they also share their questions, answers, and discoveries. They effectively communicate with each other via their common understanding of some fundamental ideas about substances and about how to make measurements on those substances. We will call this efficient way of communicating ideas the "chemist's shorthand."

1.2   The Chemist's Shorthand

Video Lesson: An Introduction to Chemistry

Chemistry *is concerned with the systematic study of the matter of our universe. This study involves the composition, structure, and properties of matter.* **Matter** *is the* "stuff" of which things are made. Chemists often say that **chemicals** make up matter. Unfortunately, in informal usage, the term *chemical* is often wrongly applied only to those materials produced by the chemical industry. In truth, the rainwater falling into a mountain lake is just as much a chemical as any industrially produced pesticide that is sprayed onto crops or any of the materials that are combined to manufacture a plastic soda bottle.

What are some examples of matter? This book is made up of matter. The desk, your pencil, the coffee in your cup, and the air in your lungs are all examples of matter. Matter is anything that occupies space and has mass. Then, what *isn't* made up of matter? Energy is not matter in the common sense. Heat and light,

Matter, the stuff of what things are made, is everywhere.

examples of energy, are not regarded as matter and do not occupy a specific volume of space in the conventional sense. On a more complex level, however, energy and matter are related, as we will learn later.

The chemicals that come together to make matter also occupy space. They, too, are matter. What makes up chemicals? They are formed from the combination of inconceivably tiny **atoms**, the chemically indivisible particles that chemists classify by their structure into the more than 90 naturally occurring **elements** that are known in the universe. Chemists and physicists have added to this list by preparing several additional, large elements that are not found in nature. You will find all the atoms, and therefore all the elements, listed in the periodic table of the elements on the inside front cover of this book. Many of the elements in the periodic table are familiar because we depend on them in our day-to-day lives. For example, the element copper, made from copper atoms, forms the copper wires that carry electricity around cities and into homes. The element helium, made from helium atoms, is the gas that enables weather balloons to rise into the stratosphere. The element gold, made from gold atoms, is highly prized, especially when it is encrusted by jewelers with precious gemstones. The element oxygen is an important part of the gas we must breathe into our lungs several times each minute in order to stay alive. However, oxygen gas is not composed of individual oxygen atoms. Instead, pairs of oxygen atoms are joined, or bonded, together to form oxygen molecules. A **molecule**, such as the oxygen gas shown in Figure 1.5, *is composed of two or more atoms bonded together*.

Some molecules, such as oxygen, contain identical atoms bonded together. Nitrogen, the most abundant molecule in the air that you are breathing as you read this, contains two atoms of the element nitrogen bonded together. However, many more kinds of molecules can be formed when atoms of *different* elements are bonded together. A **compound** is matter that contains different elements chemically bonded together. Antoine Lavoisier, a late-eighteenth-century French chemist, whom we will meet in the next chapter, called a compound any substance that can be decomposed into two or more other substances. For example, one type of powdered laundry bleach is a chemical compound composed of the elements chlorine, oxygen, and sodium. Ethanol, the "active" ingredient in beer, is a compound made up of carbon, hydrogen, and oxygen. Table sugar also has these three elements, but in different proportions than in ethanol. We call the relative proportions of the elements in a compound its **composition**, so we say that the composition of ethanol is different from that of table sugar.

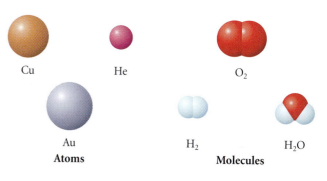

FIGURE 1.5

All of the chemical elements are listed in the periodic table. Each element is composed of one kind of atom. Atoms can become bonded together to form molecules of elements (such as oxygen) or molecules of compounds (such as water).

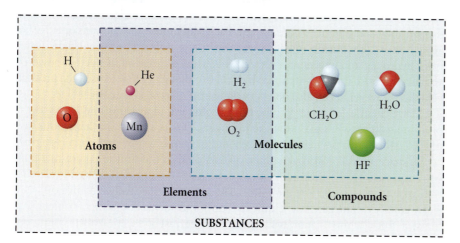

Video Lesson: Properties of Matter

Physical and Chemical Properties

Often, the chemist doesn't have access to the composition of a compound. As an example, assume we have two unlabeled jars. They both contain a **substance**, *a type of matter that exhibits* a fixed composition. The substance in one of the jars is called ethanol. The other jar contains the substance we commonly call table sugar. **How can we tell the difference between the two?** If we knew the composition of each compound, we could determine the answer. The chemist would rely on asking questions—some difficult, some easy, but all geared toward gathering the data necessary to solve this problem.

Can we tell the difference between the contents of two jars on the basis of our observations? Which one is a liquid? Does one have a characteristic odor? Answering these questions gives us information about the physical properties of the compounds. A **physical property** is a characteristic of a substance that can be determined without changing its chemical composition. In other words, physical properties are descriptions of a substance. In our example, the substance ethanol is a clear and colorless liquid. The table sugar is a solid substance. These are physical properties of the two compounds. On the basis of these observations, the chemist could determine which jar is more likely to contain the sugar and which would contain the ethanol. He or she could even go further by determining the temperature at which the compounds change from solid to liquid. The melting point, the temperature at which a substance melts, is an example of a **physical change** in which *the substance changes its physical properties without changing its chemical composition*. For instance, the sugar in the jar changes from a solid to a liquid at 185°C (365°F). Cellulose has a melting point of 260°C (500°F).

What if we compared a jar of powdered starch to a jar of powdered cellulose (plant fiber)? Figure 1.6 shows what we would see if we examined these compounds up close. Could we use their physical properties to distinguish clearly between them? At first glance they both appear to be powdery solids. Both of these solids are composed of the same elements. In addition, they appear to have generally similar physical properties. Their melting points are different, but we might ask another question to distinguish between these two substances: Do they behave in similar ways when combined with other substances? In other words, **do they have similar chemical properties?**

A **chemical property** is a characteristic of a substance that can be determined as it undergoes a change in its chemical composition. For instance, the ability of gasoline to burn is one its chemical properties. **Chemical changes,** the conversion of one substance into another due to a *change in composition or a reorganization of atoms,* often take place in what chemists call **chemical reactions**. For example, under the right conditions, the elements hydrogen and oxygen can react to form water in a chemical reaction that releases a great deal of energy (see Figure 1.7). This is a chemical change, because a new chemical, water, forms from the oxygen and hydrogen and is a different structural unit. Each water molecule contains one oxygen atom bonded (chemically joined) to two hydrogen atoms. The formation of water from hydrogen and oxygen is a very useful chemical reaction in electric-powered vehicles, because it is accompanied by a release of energy that is used to run such vehicles. However, it can also be a very dangerous chemical reaction.

Many chemical changes can be both useful and dangerous; the key is to exploit their utility while avoiding their dangers. One dramatic, but hazardous, use of the chemical properties of hydrogen and oxygen is to power the main engines of the space shuttle. Hydrogen and oxygen stored in the shuttle's main fuel tank combine explosively, after being ignited, to generate lots of thrust and release a trail of hot water vapor (Figure 1.8). The newly formed water comes out as a clear, colorless, hot gas, but it quickly cools and forms the water droplets that make a thick white "vapor trail" that you see just after the shuttle's launch. The dangers of this reaction were tragically demonstrated when the shuttle

FIGURE 1.6

A jar of cellulose and a jar of starch. Examining physical properties alone may not be sufficient to identify which of these jars contains cellulose and which contains starch.

Application

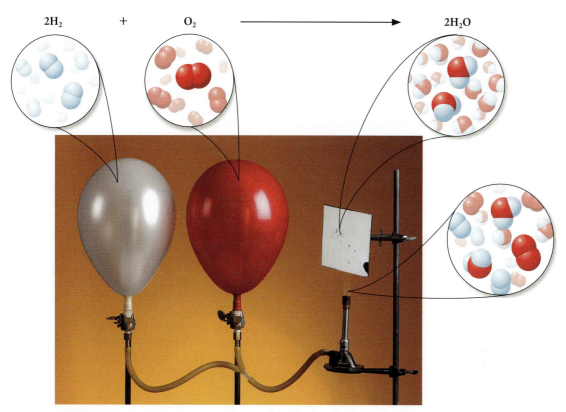

$2H_2$ + O_2 ⟶ $2H_2O$

FIGURE 1.7

Hydrogen gas and oxygen gas undergo a chemical change to make water. Note that the composition of water involves a combination of two hydrogen atoms and one oxygen atom. The chemical change is noted by the evolution of large amounts of heat, light, and sound.

Challenger exploded on January 28, 1986. The uncontrolled combination of hydrogen and oxygen, initiated by a fault in one of the solid fuel booster rockets, destroyed the spacecraft and killed the entire crew. This, combined with another deadly explosion upon reentry of the space shuttle *Columbia* on February 1, 2003, and funding concerns have placed the future of the shuttle program in doubt.

We now have enough background to answer our question about whether starch and cellulose have different chemical behavior by looking at our own biology. Starch undergoes a chemical change when humans eat it, via the processes known as digestion and metabolism. Humans, on the other hand, cannot digest cellulose. Using the physical and chemical properties of these substances, we can determine which is starch and which is cellulose.

EXERCISE 1.1   **Chemical and Physical Changes**

Define each of the following as a chemical or a physical change.

a. Burning automotive fuel in your car
b. Melting ice in a frying pan
c. Dissolving sugar in your tea
d. Digesting the sugar in a chocolate bar

First Thoughts

The key questions are these: What is a physical change? What is a chemical change? Which of these is represented by each process? Recall that a physical change is one in which the substance retains its chemical composition (the proportions in which

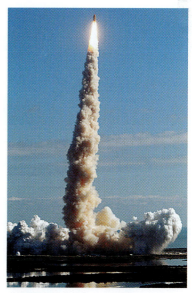

FIGURE 1.8

During a space shuttle launch, the reaction of the hydrogen and oxygen in the main fuel tank make a white vapor trail of water molecules and solid rocket fuel exhaust that can be seen for miles.

its elements appear in the substance). A chemical change results in a change in the chemical composition of a substance. Therefore, as we look at each process, we think about whether changes in chemical composition occur.

Solution

a. The fuel reacts with O_2 and is converted into carbon dioxide and water vapor (and some other gases as well). The reaction is accompanied by the release of energy. This is a chemical change.

b. Melting ice changes the physical condition of the water from solid to liquid. It is still water, so this is a physical change.

c. Dissolving sugar in your tea is a physical change. The sugar is still there; it is just mixed intimately with your tea.

d. Your body does a good job of utilizing the sugar in a chocolate bar. This is an example of a chemical change that releases energy as the sugar is converted into other substances.

Further Insights

As we learn more about how atoms are joined, or bonded, to other atoms, we will discover that many physical properties are based on the type and the ratio of the elements that make up a compound. In other words, physical properties such as color, density, and melting point are based on chemical composition. For example, we will learn why, at room temperature, water is a liquid, whereas methane, a compound of similar mass containing the elements carbon and hydrogen, is found as "natural gas."

PRACTICE 1.1

Indicate whether each process involves a chemical or a physical change.

a. The rusting of a copper statue

b. The yellowing of an old newspaper

c. Boiling water on the stove

d. Shaping a piece of wood into a post

e. Sharpening your lawn mower's blade

f. Developing a photograph

See Problems 19, 20, and 109–111.

Classification of Matter

We've discussed matter as anything that takes up space and possesses mass. However, chemists are often concerned with more specific classifications of matter. By classifying matter into groups with similar characteristics, we can reduce the number of questions that we need to ask in order to determine the chemical or physical properties of a substance.

Substances include elements and compounds. Chemists are often concerned with **mixtures** of substances, rather than pure substances. For example, the bottle of pure, vacuum-distilled water on the chemist's shelf contains only water. In contrast, tap water, well water, river water, and seawater contain mixtures of water and other substances (such as salt, dirt, small amounts of oxygen, and so on).

Mixtures themselves can be categorized as either homogeneous or heterogeneous (Figure 1.9). A **homogeneous mixture**, or **solution**, is uniform all the way through. Consider a bottle of tap water. Any part of the tap water sample is the same as any other part. A **heterogeneous mixture**, on the other hand, contains a different composition in different places. For example, if we throw some salt into a bucket of sand and stir the contents around for a short while, we will find that

FIGURE 1.9

Examples of homogeneous mixtures (tap water, gold ring, and grape juice) and examples of heterogeneous mixtures (concrete and salad dressing).

some parts of the mixture have more salt than sand, whereas some parts are mostly sand. The mixture is heterogeneous. Grape juice is a homogeneous mixture; it is the same throughout. The same is true of a 14-karat gold wedding ring—a homogeneous mixture of the elements gold, silver, and copper. On the other hand, cement is a heterogeneous mixture because some parts contain more sand and rock than other parts. Oil-and-vinegar salad dressing is also a heterogeneous mixture, no matter how hard you shake it. The relationships among the classifications of matter are shown in Figure 1.10.

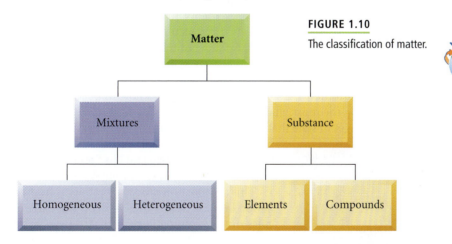

FIGURE 1.10

The classification of matter.

Visualization: Comparison of a Compound and a Mixture

Visualization: Comparison of a Solution and a Mixture

Visualization: Homogeneous Mixtures: Air and Brass

EXERCISE 1.2   **Describing the Chemistry of the Shuttle**

Figure 1.8 shows the propulsion system provided by the main engines of the space shuttle. Can you write a summary of this system that includes all these terms used in an appropriate way?

- physical change
- chemical change
- atom(s)
- element(s)
- compounds(s)
- mixture

Visualization: Structure of a Solid

Visualization: Structure of a Liquid

Visualization: Structure of a Gas

Video Lesson: States of Matter

Solution

There are many possible ways of phrasing a suitable answer. Here is one example: "The space shuttle's main fuel tank contains the elements oxygen and hydrogen. The oxygen is in the form of molecules, each composed of two oxygen atoms. The hydrogen is in the form of molecules, each composed of two hydrogen atoms. The oxygen and hydrogen are combined to form a mixture that, when ignited, reacts explosively to form molecules of water. This is a chemical change because the elements oxygen and hydrogen are being changed into the compound water. Each water molecule contains one oxygen atom and two hydrogen atoms. The water is formed as a gas, but as it leaves the back of the shuttle, it cools and undergoes a change into droplets of liquid water—an example of a physical change."

PRACTICE 1.2

Classify each as either a homogeneous or a heterogeneous mixture.
a. Hot tea
b. A milkshake
c. A root beer float
d. Gasoline
e. Bronze
f. A chef's salad

See Problems 15–18 and 21–22.

Physical Separations

Application

CHEMICAL ENCOUNTERS: Desalination

Pure water can be separated out of the homogeneous mixture called seawater by a process known as **desalination**, the removal of salts (Figure 1.11). Desalination employs physical, rather than chemical, changes. In Saudi Arabia, fresh water is so sparse that desalination of seawater has become essential to life. Just over 50% of Saudi Arabia's usable water supply is generated via desalination, and about 30% of the world's desalination capacity is in Saudi Arabia. The United States comes in second, with about 12%, and many other countries in the Middle East, North Africa, Asia, and the Caribbean account for the remainder. Desalination is a growing industry, and many companies are making a considerable investment in research to exploit significant improvements in the technology.

Figure 1.12 shows that the first crucial step in one important type of desalination process is to use heat to cause some of the water molecules in seawater to escape from the liquid into a gas. In fact, most liquids can be converted into gases if you supply enough heat. This process involves changing the **state** of the water sample. The common states of matter, which you know to be solid, liquid, and gas, are among the most vital physical characteristics of matter. Cooling can reverse the changes of state that can be driven forward by heating. This is what happens in the desalination plant. The gaseous water (water vapor) is cooled and changes back into liquid water for drinking water supplies.

This desalination process is known as heat desalination because heat is used to drive the water molecules out of seawater. Two other forms of desalination illustrate other ways of separating a component from a homogenous mixture (see Figure 1.13). In desalination by **electrodialysis**, the seawater is passed though a tube lined with a special porous membrane and surrounded by electrically charged plates. The salts in the seawater mixture are drawn through the membrane. The water molecules remain within the membrane and are not affected by the electric charge. In desalination by **reverse osmosis**, the seawater is pumped at high pressure through porous membranes. The high pressure forces water

FIGURE 1.11

A desalination plant 120 kilometers south of Jeddah, on Saudi Arabia's Red Sea coast.

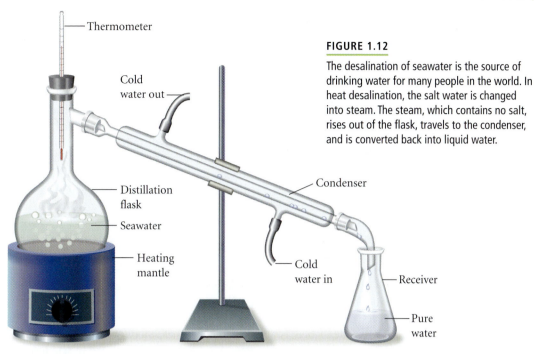

FIGURE 1.12

The desalination of seawater is the source of drinking water for many people in the world. In heat desalination, the salt water is changed into steam. The steam, which contains no salt, rises out of the flask, travels to the condenser, and is converted back into liquid water.

molecules through tiny pores in the membrane through which the salts are largely unable to pass. Both electrodialysis and reverse osmosis are likely to become increasingly viable methods of commercial desalination in the first few decades of the twenty-first century. We will have much more to say about these techniques in Chapter 11.

Desalination demonstrates that many important processes that include chemicals do not actually involve chemical changes. Chemistry can be about separating and changing the state of chemicals, as well as about using existing chemicals to make other chemicals. Chemists use their understanding of the way substances interact to interpret and change our world in predictable ways.

FIGURE 1.13

Separating pure drinking water from a homogeneous mixture of seawater. The separation of homogeneous mixtures is often more involved than the separation of heterogeneous mixtures.

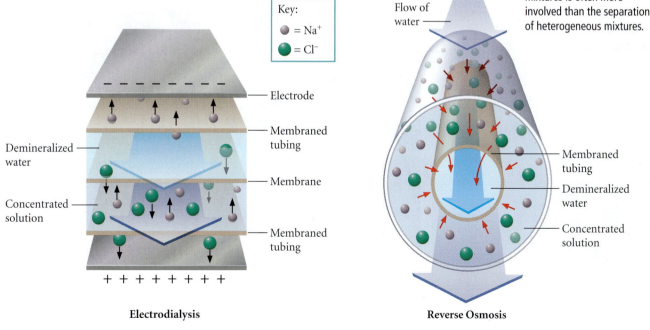

Electrodialysis   **Reverse Osmosis**

Video Lesson: The Scientific Method

1.3   The Scientific Method

As we have noted, chemists such as Judith Fairclough-Baity ask questions in order to discover and explore nature. This is how all scientists perform their craft. The *process of science* is all about asking questions and making careful observations. For example, the question "Is hydrogen gas explosive?" can be answered by preparing a little pure hydrogen, igniting it (from a safe distance!) in the presence of oxygen, and watching what happens. Although the question that Ms. Fairclough-Baity asks is a little more complex ("Can a silicone tube be created that doesn't fracture when it is being flexed?"), the process is the same as that which we use to answer all scientific questions (Figure 1.14).

We call this rational approach to knowing the **scientific method**. The scientific method is not a rigid set of rules, because the human mind is an intellectually flexible place. Rather, it is a way of asking questions and finding answers. Thinking about our world in this rational way ensures that we get meaningful results and steadily learn more about the topics we investigate. The method does not always provide the answer to a question immediately. Instead, many days, months, or even years may be needed to arrive at an adequate answer. The scientific method is often described as a series of steps; these steps are listed in Table 1.1. Figure 1.15, which illustrates the general procedure used to learn in science, conveys the same approach graphically.

The scientific method is a reasonable outcome of people's desire to learn about how nature works, what nature can and cannot do, and how we can chemically manipulate what nature has provided for us. There are two crucial points here. One is that the scientific method is the approach we use to test our proposed explanations and ideas about nature by looking for evidence, and by discarding or modifying explanations for which there is no evidence. The other point is that you can apply this way of understanding to *any* subject or topic, whether in the classroom or in life after college.

Step 5 in Table 1.1 shows us that the scientific method is a cyclical process. Each iteration of the steps allows the scientist to refine the hypothesis until it completely accounts for his or her observations. Many times, this refined and well-tested hypothesis provides the answer to the question of concern formulated in step 1.

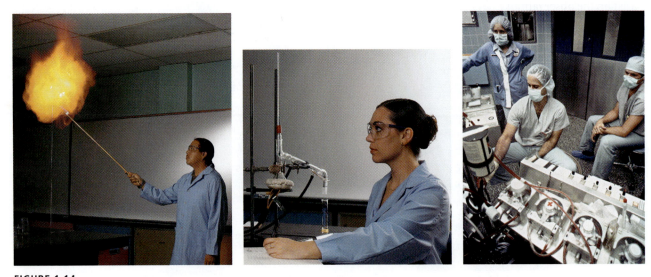

FIGURE 1.14

Asking questions and experimenting are the start of the scientific questioning process.

TABLE 1.1	The Scientific Method

Step 1: *Formulate a question that you would like answered.*

Step 2: *Find out what is already known about your question.* This is typically done by searching the scientific literature, which contains the work of other scientists.

Step 3: *Make observations—in other words, examine things.* The observations involve the collection of both qualitative (non-numeric) and quantitative (derived from measurement) data about the system under study. Recording the data in a systematic way is vital to this step.

Step 4: *Create a* **hypothesis***, or possible explanation that would account for your observations.* In other words, the hypothesis is a statement about what you think *might* be going on.

Step 5: *Design and perform experiments that are carefully and deliberately set up to test your hypothesis.* Specific testing of the hypothesis is needed to ensure that the possible explanation "holds up" to scrutiny. These experiments yield further observations and data for you to consider, and the method is joined again at step 3.

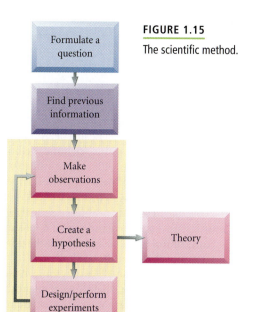

FIGURE 1.15

The scientific method.

Sometimes, patterns emerge in the data collected through experimentation. When these patterns are consistently noted across a series of different systems under study, a statement called a **scientific law** can be made. The scientific law *summarizes* the patterns that are observed in the data. They are succinct *descriptions* of the behavior of the natural world. Yet the scientific law does not attempt to explain these observations. For example, one scientific law that we will discuss later describes the behavior of a gas as the temperature, pressure, and volume of the gas are changed. This scientific law does not explain why the gas behaves the way it does; it simply tells us what to expect.

Once a law related to a particular phenomenon has been established using the scientific method, a theory can be constructed. A theory is based on the observations made from a variety of experiments, so it provides a *tested explanation*, developed from hypotheses that have been supported by the results of experiments. In short, a **theory** is *an attempt, based on scientifically acquired evidence, to explain observations that have been described as fact.* For example, the theory that an asteroid collided with Earth some 65 million years ago is an attempt to explain the demise of the dinosaurs.

A theory that explains the extinction of dinosaurs.

A theory is a powerful statement not to be taken lightly. It is not guesswork or supposition. It is based on thoughtful and structured questions, experiments, and data analysis. Some theories become trusted to such an extent that they almost acquire the status of certainties, because they are repeatedly confirmed by many different experiments. Other theories last for only a short time, until they are modified or even swept away by new evidence. For example, early theories on the origins of chemicals suggested that compounds inherent in living creatures, such as the molecule urea found in urine, contained a "life force" and that they could not be made from inanimate things such as dirt or rocks. The later discovery of evidence that this is not true refuted that early theory. All theories that stand the test of time do so because they continue to be supported by experimental evidence.

How do we know?

How to control pain:
An example of the scientific method in action

The intensive studying that goes on in college can occasionally give anyone a headache. A most beneficial result of the application of the scientific method is our ability to control the pain with readily available medicines. The systematic study of pain control began several thousand years ago, and now, centuries later, new words are still being added to our vocabulary as analgesic (pain-killing) drugs are discovered and tested. A look at the history of pain control helps us learn about the power and the limitations of the scientific method.

The question was "Can anything in the world around us be used to control pain?" Asian manuscripts from 2400 years ago offer some of the first examples of useful answers to that question, including the observation that infusions of willow tree bark can relieve pain and treat fevers.

This observation was the result of performing experiments, gathering data, and interpreting results in order to identify which naturally available plant materials might be useful in human health. The ancient manuscripts about willow bark are very early instances of the publication of results, allowing other people to know what experiments had been done, to check the results, and to make use of them. One example of such checking that results are replicable occurred in the 1830s, when a Scottish physician was able to confirm that extracts of willow bark relieved the pain of acute rheumatism.

As our understanding of chemistry and our ability to make better measurements developed, investigations on willow bark led scientists to create the hypothesis that one particular compound within the willow bark was responsible for the control of rheumatism pain. This led to the question "Which compound?" and to experiments designed to isolate the different substances in the willow bark. The answer came in the 1840s, when chemists and physicians determined that salicin (see Figure 1.16), isolated from both willow bark and a plant known as *spirea*, was the ingredient responsible for the pain control. In 1870, Professor von Necki in Basle, Switzerland, performed experiments and gathered data showing that salicin, when ingested, is converted into the salicylic acid shown in Figure 1.16. This led to the manufacture of salicylic acid. It was given directly to patients to test the hypothesis that it would relieve pain and fevers. One outcome of these experiments (some of the first examples of "controlled clinical trials") was that the salicylic acid, although effective, severely irritated the lining of the mouth, throat, and stomach. This led various chemists to form the hypothesis that modifying the salicylic acid in some way might yield compounds that were effective painkillers without having the nasty side effects of salicylic acid. In the 1890s, Felix Hoffman (see Figure 1.17) was working for the Bayer Company in Germany

The ancient Egyptians used reasoning to develop medicines for pain.

To ensure that the scientific method is working well, scientists perform their experiments several times. Repeating experiments time after time allows the scientist to prove that the data were collected properly, that the observations are correct, and that the new information he or she has discovered is real. By publishing the results of their experiments, scientists also expose their work to other people's scrutiny and criticism. Other scientists then have the opportunity to check the experiments, results, and hypotheses to make sure these are valid. *Communication of all types, whether via classroom teaching, writing books, or publishing journal articles, is as vital to science as performing experiments in the laboratory.* There are nearly 600,000 chemistry-related articles published each year worldwide; each article is intended to communicate ideas for others to consider.

The scientific method is powerful because it works. Over time, it has yielded hypotheses, theories, and laws that describe, explain, and predict much of the behavior of the natural world. It has enabled us to make use of fire and to extract

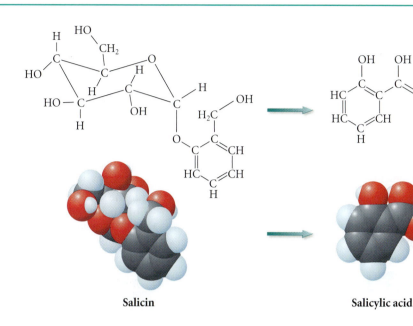

Salicin

Salicylic acid

FIGURE 1.17

Felix Hoffman and the wonder drug, aspirin.

FIGURE 1.16

Salicin can be converted into salicylic acid.

when he synthesized acetylsalicylic acid. This modification of salicylic acid became known as aspirin, today one of the most effective and widely used analgesic drugs.

In the twenty-first century we have many different analgesic drugs at our disposal. The development of each one involved the application of the scientific method to discover, test, and improve our ability to control pain. The powerful analgesic known as morphine, for example, is a natural product of the opium poppy. Heroin is a synthetic derivative of morphine, made by treating morphine with the chemical acetic anhydride. In experiments, heroin is observed to be up to eight times as ef-

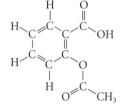

Acetylsalicylic acid

fective at relieving pain as morphine. Unfortunately, the side effects of these compounds include chemical dependence and the associated hazards of abuse. Therefore, clinically, they are used only under controlled conditions to alleviate acute pain.

pure minerals from the Earth. Through it we have learned about the interaction of individual molecules and have applied this understanding to form the massive three-dimensional molecules we call plastics. And it has allowed us both to understand the chemistry of life and to leave the Earth's surface to explore our moon, the other planets, and beyond.

EXERCISE 1.3   A Theory or a Law?

Earth's atmosphere is a mixture of gases—largely nitrogen (N_2), oxygen (O_2), and water vapor (H_2O), with much smaller amounts of argon (Ar) and a wide variety of other gases. Careful observation of these gases under controlled experimental conditions has shown that when the temperature is held constant, halving the pressure of the gas causes its volume to roughly double. As one goes down, the other goes up,

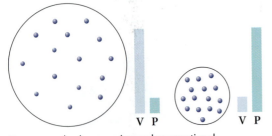

Pressure and volume are inversely proportional.

as though pressure and volume were at the two ends of a seesaw. Similarly, when the pressure is doubled ("see . . ."), the volume becomes half of the original value ("saw . . ."). We can describe this behavior by saying that *at constant temperature, the volume of a gas is inversely proportional to its pressure.* Most gases deviate slightly from this generalized statement, but they all adhere to it quite closely at the temperatures found in the atmosphere, and this behavior has been checked many times with many different gases. Is this an example of a hypothesis, a theory, or a scientific law?

Solution

The statement that "at constant temperature, the volume of a gas is inversely proportional to its pressure" is a *description* of the behavior of a gas. It is not an *explanation* for the behavior being described. Hence, this statement is an example of a scientific law rather than a theory. This particular law is actually known as Boyle's law, in honor of Robert Boyle (1627–1691), who discovered and publicized it.

PRACTICE 1.3

Describe each step a scientist would follow as he or she investigated the following question using the scientific method. Carefully explain each step, describing any specific experiments and observations the scientist would make. Will the result of this application of the scientific method be a hypothesis, a theory, or a scientific law?

There is a dark liquid in my cup. Is the liquid coffee and who put it there?

See Problems 25–28 and 33–34.

Video Lesson: The Measurement of Matter

1.4   Units and Measurement

Chemistry is a precise science that is based on our ability to properly measure the interactions of chemicals. When Judith Fairclough-Baity conducts performance experiments on a new piece of silicone tubing that she has constructed, she makes observations. She collects data that include the length of the tube, the diameter of the tube, the volume of blood it will hold, and the temperature range in which it is usable. The data she collects about her experiments are known as quantitative observations. In fact, making quantitative measurements of the results of an experiment is fundamental to all scientific inquiries.

Recently, Fairclough-Baity examined the usable life of three different types of silicone tubes. She set up an experiment that compressed and expanded the tube continuously, mimicking the use of the tube in a heart–lung pump. Her results were obtained on pieces of tubes with very well-defined masses. She reported those results to other chemists in an article she wrote. In order to communicate her data and results effectively, *she used units of measurement that are common among the sciences.* Before the eighteenth century, this was not done. Some scientists measured masses in karats; some in drams, grains, ounces, pennyweights, or pounds; and still others in scruples (an old pharmacist's measurement, in which 288 scruples = 12 ounces). The multitude of different units made comparing the results of experiments difficult. Today, only two major systems for measuring data are currently in use: the English system (used in several countries, including the United States, where it is more formally known as the United States Customary System, or USCS) and the metric system. In the 1960s, the metric system was modernized to what we use in scientific studies today.

TABLE 1.2	The Fundamental "Base Units" of the International System (SI)	
Physical Quantity	**Name of Unit**	**Symbol**
Mass	kilogram	kg
Length	meter*	m
Time	second	s
Temperature	kelvin	K
Amount of substance	mole	mol
Electric current	ampere	A
Luminous intensity	candela	cd

*Meter is spelled *metre* in most countries except the United States.

The *Système International* and Its Base Units

The modification of the metric system in the 1960s gave rise to the *Système International d'Unites*, also known as the **International System (SI)**. The SI defines seven **base units** used to measure seven fundamental physical quantities. These physical quantities, the corresponding base units, and their abbreviations are summarized in Table 1.2. From these seven base units, scientists can report measurements of every imaginable type and can effectively report those measurements to other scientists. During our exploration of chemistry, we will use the first five of these units extensively.

Suppose, for example, that we are in charge of a water desalination plant. Part of our daily routine at the plant would likely include checking the quality of the water that the plant produces. Some tests that measure the quality of the water require a fixed mass of the liquid. Using a balance, we can measure the mass of a water sample. To report our scientific measurement, we will combine the base unit for our measurement with a prefix. This prefix multiplies the number portion of the measurement by a power of 10. For example, the base unit of mass is the **kilogram (kg)**. The prefix *kilo-* in *kilogram*, one of the prefixes listed in Table 1.3, indicates that 1 kilogram is equal to 1000 grams (g). The kilogram is the only base unit that incorporates a prefix in its definition.

$$1\ \text{kg} = 1 \times 10^3\ \text{g} = 1000\ \text{g}$$

Unfortunately, the mass of an object in kilograms is often incorrectly referred to as a weight. In fact, there is a sharp distinction between the concepts of mass and weight. **Mass** is *a measure of the amount of matter in a body, determined by measuring its inertia, or resistance to changes in its state of motion.* **Weight** is *a measure of the downward force exerted on a body by gravity.* The mass of an object is a fundamental property of the object because it will not change, for instance, if we

TABLE 1.3	The SI Prefixes	
Multiple	**Prefix**	**Name**
10^{24}	Y	yotta
10^{21}	Z	zetta
10^{18}	E	exa
10^{15}	P	peta
10^{12}	T	tera
10^{9}	G	giga
10^{6}	M	mega
10^{3}	k	kilo
10^{1}	da	deka
10^{-1}	d	deci
10^{-2}	c	centi
10^{-3}	m	milli
10^{-6}	μ	micro
10^{-9}	n	nano
10^{-12}	p	pico
10^{-15}	f	femto
10^{-18}	a	atto
10^{-21}	z	zepto
10^{-24}	y	yocto

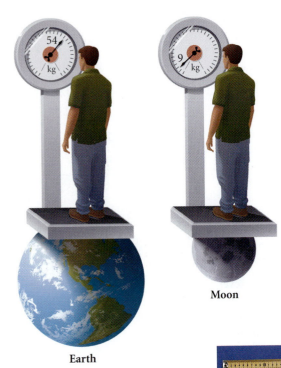

Earth

FIGURE 1.18

A person's weight on the Moon is different from his or her weight on Earth. The mass of the person hasn't changed. Only the force of gravity is different.

Moon

The mass of a raisin is about 1 g.

compare an object on Earth to the same object on the Moon. The weaker gravitational force on the Moon would cause the object to weigh less, but it would still have the same mass (Figure 1.18).

For many of our water quality measurements, we might only need a gram of water. How much is a gram? A raisin's mass is approximately 1 g. That is a pretty small amount, but 1000 raisins have about the same mass as a kilogram. A kilogram weighs about the same as 2.2 lb on Earth. However, masses with which you will become familiar in the laboratory are much smaller than the kilogram. In fact, most masses you will record in the laboratory will be grams or milligrams. In the chemical process industries, which produce large quantities of the chemicals used to manufacture consumer goods, masses are often measured in tons (2000 lb) or metric tons (1000 kg, 1 Mg, or 2200 lb).

Another important piece of information to include with our water quality measurements would be the distance to the farthest city that is served by the desalination plant. We would record this distance using the SI unit of *length*, the **meter (m)**. The meter is about 10% longer than a yard.

$$1 \text{ m} = 1.094 \text{ yd}$$

A meter and a yard.

To measure long distances, we often report them in kilometers (km).

$$1 \text{ km} = 1 \times 10^3 \text{ m} = 1000 \text{ m}$$

One kilometer is equal to 1000 m, or slightly more than 0.6 mi. A runner entered in a "5K" (5-km) road race will run 5000 m (a little more than 3 miles). In chemical analysis, though, scientists measure lengths much smaller than a meter. Because the particles of chemistry are extremely tiny compared to objects in the everyday world, their sizes are often measured in such tiny units as picometers, nanometers, and micrometers. One nanometer is equal to 0.000000001 m, so there are 1 billion nanometers (10^9 nm) in 1 m.

$$1 \text{ nm} = 1 \times 10^{-9} \text{ m} = 0.000000001 \text{ m}$$

How long it takes to desalinate enough water for the people served by the plant is also a key quantity. To report this value, we could use the SI unit of *time,* the **second (s)**. Luckily, the second has been used in virtually all measurement systems, so we are pretty familiar with it. The units of hour, minute, day, and year are often used in measurements of time. These are not SI units, but because they are so widely used, scientists don't often affix prefixes to the unit second to describe large time spans. For example, instead of reporting 3600 s as 3.6 ks (which is perfectly appropriate), the scientist would be more likely to indicate 1 hour (1 h),

$$3600 \text{ s} = 3.6 \times 10^3 \text{ s} = 3.6 \text{ ks}$$
$$3600 \text{ s} = 1 \text{ h}$$

Prefixes, however, are commonly used to indicate fractions of a second. One millisecond (1 ms) is one-thousandth of a second.

Temperature is a vital part of our day-to-day choices about what clothing to wear, what liquid to drink, and a host of other decisions. The SI base unit of temperature is the **kelvin (K)**. Notice that no "°" symbol is used with the abbreviation K. Two other temperature units are much more common, however. The English system uses the Fahrenheit scale, in which the unit of measure is the **degree**

Fahrenheit (°F). The metric system uses the Celsius scale, in which the unit of measure is the **degree Celsius (°C)**. All three systems can be used to accurately report temperatures, but they are based on different physical phenomena. The Fahrenheit scale was created by the German physicist Gabriel Daniel Fahrenheit in 1714. Fahrenheit used a mercury-based thermometer to define 0°F as the coldest temperature he could make from a mixture of water and ammonium chloride. He called his body temperature 100°F. (Linus Pauling, a chemist whose work we will discuss at many points later in this text, mused that Fahrenheit might have had a slight fever.) Using his two reference points of cold and warm, he found the boiling point of water to be 212°F.

The Swedish astronomer Anders Celsius developed the Celsius scale in 1742, based on dividing the difference in temperature from the freezing point, 0°C, to the boiling point, 100°C, of water into 100 individual units (degrees). In 1848, the British physicist Lord Kelvin (Figure 1.19) invented the temperature scale named for him in 1954 in which, fortunately, one kelvin is equal in magnitude to 1 degree Celsius, although the scales just start at different points. Although Kelvin did not recognize it at the time, the point we now call zero on the Kelvin scale equals −273 on the Celsius scale, so a temperature in Celsius can be converted into Kelvin by adding 273.

$$T_K = T_C + 273 \quad \text{and} \quad T_C = T_K - 273$$

More information and the exact definition of the Kelvin scale is presented later.

Converting between the Celsius and Fahrenheit scales is a little more complicated. The formula shows that the temperature of an object changes by 9 Fahrenheit degrees for every 5 Celsius degrees. This means that if you use a Fahrenheit thermometer to measure an 18-degree drop in temperature (9 × 2), a thermometer based on the Celsius scale will note a 10-degree drop (5 × 2). A Kelvin-based thermometer would *also* show a 10-degree drop in temperature, because 1 K equals 1°C (see Figure 1.20). Figure 1.20 also indicates the different

FIGURE 1.19

Lord Kelvin (1824–1907) was born William Thomson in Belfast. He was knighted in 1866 and given the title of Baron Kelvin of Largs in 1892.

FIGURE 1.20

The Fahrenheit, Celsius, and Kelvin temperature scales.

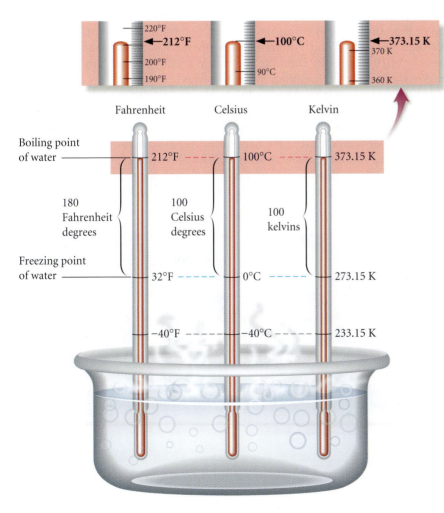

locations of the freezing point of water on the scales. Here are the formulas required for converting between temperatures on the Celsius and Fahrenheit scales:

$$T_C = (T_F - 32)/1.8 \qquad T_F = 1.8(T_C) + 32$$

Another property of our water sample, or of any other substance or mixture of substances, is the number of particles of each kind we have. These particles represent the amount of each substance in the sample. The amount of a substance is the number of entities, such as atoms, molecules, or ions (particles containing an electrical charge), present in a given sample of the substance. Determining the amount of each substance in a glass of tap water is just one part of a series of analyses that chemists do to make sure that a sample of water is wholesome to drink. For example, the analysis for the element arsenic in drinking water really deals with determining how many arsenic atoms are present in a known quantity of water. This idea is explored fully in Chapter 3.

The base unit for the amount of a substance is the **mole (mol)**. As a preview of Chapter 3, we will simply point out that 1 mole corresponds to a specific very large number of entities and that this number is approximately 6.02214×10^{23}. Thus when we talk of 1 mole of atoms, we mean 6.02214×10^{23} atoms; 1 mole of molecules is 6.02214×10^{23} molecules, and so on.

Other SI units may also be needed during our water quality measurements. For instance, the purity of desalinated water is inversely related to its ability to carry an electric current: The lower the electrical conductivity of the water, the purer the water (see Figure 1.21). The SI base unit of electric current is the **ampere (A)**, often called the amp. This unit is used a great deal in chemistry because some chemical reactions can be used to generate an electric current (within batteries, for example). Conversely, electricity can be used to make certain chemical reactions happen. We'll learn more about the ampere in the chapter on electrochemistry (Chapter 19).

The last SI base unit is the **candela (cd)**. This unit describes the **luminous intensity**, or brightness, of something under study. For example, the Sun, which radiates the light energy that powers photosynthesis, has an estimated luminous intensity of 10^{18} candelas. Light bulbs produce a vastly smaller luminous intensity. We won't deal with luminous intensity in this text.

We will refer to the world of individual atoms, molecules, and ions as the **nanoworld**, in contrast to the everyday, large-scale **macroworld** ("big world") with which we are familiar. There is a huge change in the units that we use to report quantities as we travel between the nanoworld and the macroworld. To make the trip smoothly, we must be able to work comfortably with exponential notation and related mathematical operations. Feel free to reinforce your understanding of exponential notation by consulting the appendix at the back of the book.

FIGURE 1.21

The conductivity of water solutions gives an indication of their purity.

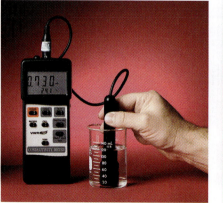 

One of the most important aspects of learning about chemistry is becoming familiar with the nanoworld of chemicals and being able to visualize what happens at that scale. The fundamental particles of chemistry are atoms (Figure 1.22). These range in size from a diameter of approximately 75 picometers (75×10^{-12} m = 7.5×10^{-11} m) for a hydrogen atom to about 500 picometers (5.00×10^{-10} m) for an atom of francium. **How can we get some idea of what these numbers mean?** Suppose we scale things up in our minds so that atoms become the size of pebbles that we can hold in our hand—some small, some larger, but averaging around 2 cm in diameter. A real pebble scaled up in size to the same extent would now be approximately 1000 kilometers (about 600 miles) in diameter. That's big enough to cover the entire Great Lakes (Figure 1.23). Even with these comparisons, this extraordinary difference in scale is difficult to visualize. Yet working with such comparisons should begin to give you some feeling for the scale of the nanoworld of chemistry. Atoms, molecules, and ions are very small indeed.

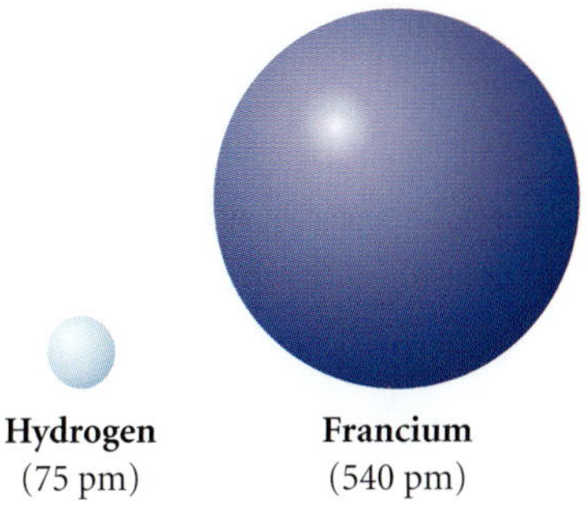

Hydrogen   **Francium**
(75 pm)   (540 pm)

FIGURE 1.22

The relative size of two atoms, hydrogen (75 pm) and francium (540 pm).

FIGURE 1.23

If an atom were scaled up to the size of a pebble on a beach, the beach pebble, scaled up to the same extent, would be 600 miles wide.

| EXERCISE 1.4 | Temperature Conversions |

The chemical reactions within the human body tend to proceed optimally at a temperature of around 37°C. Express this temperature using the Kelvin and Fahrenheit scales.

Solution

$$T_K = T_C + 273 \quad \text{so} \quad 37°C = 37 + 273 \text{ K} = 310 \text{ K}$$
$$T_F = 1.8 T_C + 32 \quad \text{so} \quad 37°C = 1.8(37) + 32°F = 99°F$$

Normal body temperature is commonly cited as equal to 98.6°F, which has more significant figures (see Section 1.6) than are justified in a comparison to 37°C.

| PRACTICE 1.4 |

Water is fed into the vaporization stage of a desalination plant at 220°F. Express this temperature using the Celsius and Kelvin scales.

See Problems 37c, 37f, 38c, 38f, 39c, 39f, 40c, 40f, and 53–54.

Derived Units

We often measure quantities other than the seven SI base quantities shown in Table 1.2. Our analysis of the sample of water we obtain during our daily check of water quality at the desalination plant or any municipal water treatment system

TABLE 1.4 | **Selected Derived Units**

Physical Quantity	Derived Unit	Name of Derived Unit
Volume	m^3	cubic meter
Density	$kg \cdot m^{-3}$	kilograms per cubic meter
Force	$kg \cdot m \cdot s^{-2}$	newton (N)
Pressure	$kg \cdot m^{-1} \cdot s^{-2}$ ($N \cdot m^{-2}$)	pascal (Pa)
Energy	$kg \cdot m^2 \cdot s^{-2}$ ($N \cdot m$)	joule (J)
Velocity	$m \cdot s^{-1}$	meters per second

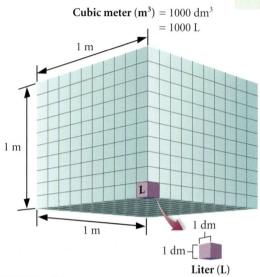

Cubic meter (m^3) = 1000 dm^3 = 1000 L

1 m
1 m
1 m

L

1 m
1 dm
1 dm
Liter (L)

FIGURE 1.24

The cubic meter, the SI unit for volume.

Video Lesson: CIA Demonstration: Differences in Density Due to Temperature

Volume: 1 dm^3
"1 liter" = 1 L = 1 dm^3
1000 cm^3 = 1000 mL

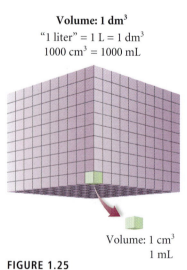

Volume: 1 cm^3
1 mL

FIGURE 1.25

One cubic centimeter is equal to a milliliter.

has several types of measurements associated with it. We've recorded the mass, the time spent to purify the water in the reverse osmosis column, the distance (using the SI unit of length) of the region served by the plant, and the temperature of the water, and now we also note the volume and density of the sample. These last two measures, along with many other common determinations, use **derived units**, *formed by the combination of SI base units*. Some of the more common derived units are shown in Table 1.4.

The volume of a water sample obtained from the desalination plant is a measure the space occupied by the sample. We could measure the volume of the sample in SI base units as **cubic meters (m^3)**. Specifically, a cubic meter is the volume described by a cube of matter measuring 1 m on each side (see Figure 1.24). The SI unit for volume, the cubic meter, is *derived* from the SI unit for length, the meter.

Although the cubic meter is the SI unit for volume, this derived unit is often much too large to reflect the sample sizes that we would obtain. Chemists tend to measure volumes using smaller derived units, such as the **cubic decimeter (dm^3)**, which is commonly used in most parts of the world, with the notable exception of the United States. As Figure 1.25 shows, a cubic decimeter is the volume of a cube that measures 10 cm on each side. This means that there are 1000 dm^3 in every cubic meter. Because we use the cubic decimeter in most of our measurements, a different name has been given to this unit. The cubic decimeter is also known as the **liter (L)**, which is the common volume measurement used by scientists in the United States.

Another common derived unit for volume is the cubic centimeter (cm^3). This unit describes a cube measuring 1 cm on each side (see Figure 1.25). In the health professions, one cubic centimeter is often abbreviated as 1 cc. There are 1000 cc in 1 L, so the cubic centimeter is also known as the **milliliter (mL)**: 1 cm^3 = 1 cc = 1 mL. Note that the unit for liters is a capital L, not a lowercase l.

$$1 \text{ L} = 1000 \text{ mL} = 1000 \text{ cc} = 1 \text{ } dm^3 = 0.001 \text{ } m^3$$

Judith Fairclough-Baity, in her report on the characterization of three types of silicone tubing, indicated the **density** of the tubes she made. Density reports *the mass of a substance that is present in a given volume of the substance*. During our analysis of the quality of water from a desalination plant, we might also report the density of the sample. The SI unit for measuring density is **kilograms per cubic meter (kg/m^3)**, although scientists often use smaller units, such as grams per cubic centimeter (g/cm^3). Because 1 cm^3 equals 1 mL, units of density can be reported for liquids and many solids in grams per milliliter. The tubing with which Fairclough-Baity worked had a density of between 1.1 and 1.2 g/mL.

$$\text{Density} = \frac{\text{mass}}{\text{volume}}$$

An interesting example of the use of density is in the comparison of three cubes of material. As shown in Figure 1.26, each block of material looks the same. However, one is silver-tinted wax, one is lead, and the third is aluminum. **How can we tell the wax, lead, and aluminum apart?** There are many physical properties that differ among the three blocks, such as color and hardness. Density is another. The wax has a density of 0.95 g/cm^3, the lead has a density of 11.4 g/cm^3, and the aluminum has a density of 2.70 g/cm^3. Therefore, although the volumes of all the cubes are identical, they differ in mass.

Extensive Versus Intensive Properties

Why does a log float on water? Mistakenly, we might say that the log is "lighter" than the water. Why does a pebble sink to the bottom of a lake? Again, we might mistakenly say that it is "heavier" than water. The heaviness or lightness of an object is a measure of the weight of the object. It is an **extensive property**, one that is dependent on how much sample you have. Length is another extensive property. If the sample is a precious metal, such as gold or platinum, then weight and mass are related to total cost, and all of these are extensive properties.

On the other hand, density, the mass of the metal per cubic centimeter (or milliliter) is an **intensive property**, because it remains the same irrespective of the amount of sample you have. Gold will still cost about $400 per ounce whether if you have 1 ounce (1 oz) or 1000 oz. The cost per ounce is an intensive property. The density of the gold, 19.3 g/mL, is the same no matter how much gold you have. However, the cost of a chunk of gold is an extensive property because it depends on the size of the chunk. Large chunks cost more than small chunks (Figure 1.27). Table 1.5 lists several intensive and extensive properties.

FIGURE 1.26

A block of silver-colored wax, a block of lead, and a block of aluminum. They look the same but are chemically quite different. We can tell which is which by measuring their densities.

Mass, length, and cost are extensive properties.

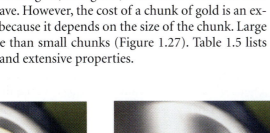

FIGURE 1.27

Extensive and intensive properties of gold. The weight and cost of a chunk of gold depend on the amount (extensive properties); the color and density of the gold remain independent of the amount (intensive properties).

TABLE 1.5	Intensive and Extensive Properties
Intensive Properties	**Extensive Properties**
Color	Mass
Density	Length
Temperature	Volume
Odor	Cost
Physical state	Energy

EXERCISE 1.5 Extensive Versus Intensive Properties

During an expedition to the top of a mountain, a camper wishes to cook some noodles for the entire group. In order to heat the noodles, the camper must boil an entire pot of water. Is the temperature at which water boils an intensive or an extensive property?

First Thoughts

Initially, this problem seems confusing because we know it will take longer to bring a pot of water to a boil than to heat a cup of water to its boiling point. Let's consider the question this way: Does the amount of water make a difference in how hot it must get before it boils?

Solution

The boiling point of water is an intensive property. The amount of water does not make a difference in how hot it must get in order to boil. In other words, the pot of boiling water and the cup of boiling water have the same temperature.

Further Insight

Why, then, does it take longer to boil a pot of water? Answering this question reveals another extensive property. The amount of water determines how much heat must be used to boil the water. The amount of heat used to boil the water is an extensive property of the water.

PRACTICE 1.5

Is the viscosity (the resistance of a liquid to flow, or the syrupiness of a liquid) of molasses an intensive or an extensive property?

See Problem 115.

We can apply this to our "floating log and sinking rock" discussion by noting that the extensive property of weight—that of the log, the rock, or the water—is not the reason why the log floats or the rock sinks. The explanation for these observations is related to the density—an intensive property—of the objects compared to the density of the water. Those objects with a density greater than the water will sink; those with a lesser density will float. Figure 1.28 illustrates this effect for six different objects.

Balsa wood
Hexane
Water
Oak wood

Chloroform

Lead

FIGURE 1.28

The densities of six different substances are represented in this density jar.

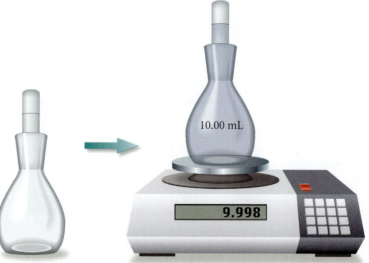

Experimental determination of density. A picnometer, which contains an exact volume, can be used to measure the density of a liquid.

EXERCISE 1.6 **Identifying a Metal via Density**

A chunk of metal has a mass of 81.76 g and is added to a graduated cylinder (a type of glassware that measures volume) containing 25.0 mL of water. The total volume of the metal and water rises to 34.2 mL. What is the identity of the metal?

Metal	Density in g/cm^3
Aluminum	2.7
Nickel	8.9
Tin	7.3
Zinc	7.1

First Thoughts

In this exercise, we are given only enough information to tell the metals apart by their densities. The real question, then, is "What is the density of the unknown metal?" The metal has a mass of 81.76 g. What volume does it occupy? The volume of water in the graduated cylinder increases from 25.0 mL to 34.2 mL when the metal chunk is added. The volume of the metal must then be $34.2 - 25.0 = 9.2$ mL. We have mass and volume. We can now solve for density.

Solution

$$\text{Density} = \frac{\text{mass}}{\text{volume}} = \frac{81.76 \text{ g}}{9.2 \text{ mL}} = 8.9 \text{ g/mL} = 8.9 \text{ g/cm}^3$$

The metal has the same density as nickel.

Further Insights

Knowing the density of metals is of great practical importance in manufacturing and commerce. For example, titanium is used as a component of expensive, light-weight bicycle frames because of its strength and low density of 4.50 g/mL. The acceleration that a rider can achieve on a bicycle is another derived unit, equal to force (itself a derived unit; see Table 1.4) per unit mass:

$$\text{Acceleration} = \frac{\text{force}}{\text{mass}} = \frac{\text{kg·m·s}^{-2}}{\text{kg}} = \text{m·s}^{-2}$$

The less mass (or weight) the bicycle has, the greater is the acceleration the rider can achieve using a constant force.

PRACTICE 1.6

How much water would be displaced if 81.76 g of aluminum, rather than of nickel, were added to the graduated cylinder discussed in this exercise?

See Problems 45–48 and 59–62.

1.5 Conversions and Dimensional Analysis

The SI units typically used by scientists are often quite different from some of the units in everyday use. For instance, visitors to Canada from the United States might wish to fill their cars with gasoline ("petrol" in Canada). Initially, they might be surprised that the gasoline is so inexpensive. Part of that has to do with the currency exchange rate: around $1.30 Canadian per $1.00 U.S. Another key is that the price of Canadian petrol is reported in liters, whereas in the United States, gasoline is sold by the gallon. Converting liters into gallons solves the difficulty caused by these differences. To do the conversion, we need an expression that relates one unit to another. Such a **conversion factor** is a mathematical expression of the ratio of one unit to another. For example, on Earth, a 2.2046-lb mass is equivalent to a 1-kg mass.

$$2.2046 \text{ lb} = 1 \text{ kg}$$

One kilogram weighs 2 lb, 3.30 oz (2.20 lb).

Because these two values are equal, we can use them as a conversion factor that can be multiplied by a number to convert its units. *The operation is identical to multiplying the original number by 1, because the numerator and denominator are equal to each other.* This equality can be written as a conversion factor in two ways:

$$\frac{1 \text{ kg}}{2.2046 \text{ lb}} = 1 \qquad \frac{2.2046 \text{ lb}}{1 \text{ kg}} = 1$$

For instance, suppose we are interested in learning the mass (in kilograms) of a book that weighs 3.5 lb. We can multiply the weight of the book by one of our conversion factors.

$$3.5 \, \cancel{\text{lb}} \times \frac{1 \text{ kg}}{2.2046 \, \cancel{\text{lb}}} = 1.6 \text{ kg}$$

Note that we chose to multiply the weight of the book by the conversion factor so that the unwanted units canceled. If we had multiplied by the other conversion factor, the answer would have been different, and the units would have been different, too! In this case, the units would not have canceled, and we would have been left with a meaningless answer:

$$3.5 \text{ lb} \times \frac{2.2046 \text{ lb}}{1 \text{ kg}} = 7.7 \, \frac{\text{lb}^2}{\text{kg}}$$

Video Lesson: Dimensional Analysis

This extremely important method of calculation can be used to solve a great variety of problems in chemistry. In fact, we will use this method for many of the calculations that we will do. It is known as **dimensional analysis** (or, sometimes, as the *unit-conversion method,* or the *factor-label method*) because canceling out the units (or "dimensions" or "factors") associated with each number enables us to arrive at the proper answer. Table 1.6 lists some of the more common conversion factors for mass, length, volume, and time. Others may be found on the inside back cover.

TABLE 1.6	Common Conversion Factors
Length	1 mi = 1.609 km
	1 in = 2.54 cm
	12 in = 1 ft
	1 mi = 5280 ft
Volume	1 qt = 0.9464 L
	1 gal = 3.785 L
	1 gal = 4 qt
	1 fl oz = 29.57 mL
Mass	2.2046 lb = 1.0 kg
	1 lb = 453.6 g
	1 oz = 28.35 g

Dimensional analysis can be applied to much more complex problems. For instance, suppose we would like to know the length of our textbook in millimeters, given that someone has measured its length to be 10.25 inches (in). We can perform a dimensional analysis and convert the units of inches into millimeters. However, examination of Table 1.6 doesn't show a direct conversion from inches to millimeters. But the table *does* show a conversion factor from inches to centimeters, and there is another conversion factor from centimeters to millimeters (which we know from Section 1.4). Here's our plan of attack:

$$\text{in} \xrightarrow{\text{inches to centimeters}} \text{cm} \xrightarrow{\text{centimeters to millimeters}} \text{mm}$$

Tutorial: Unit Conversions

Let's use this plan to construct a dimensional analysis of the units in one long problem. Although we could do it stepwise, it is often just as easy to do a single calculation. We'll just have to make sure that our unwanted units cancel at each step.

$$10.25 \text{ in} \times \frac{2.54 \text{ cm}}{1 \text{ in}} \times \frac{10 \text{ mm}}{1 \text{ cm}} = 260.4 \text{ mm}$$

These problems can even get longer, but they don't necessarily have to get harder. For example, suppose Judith Fairclough-Baity measured the density of her tubing as 1.154 g/mL. Assume she was interested in reporting this density in pounds per gallon. We can use conversion factors to solve this problem by converting one unit at a time into the units we desire. Here's our plan of attack:

$$\frac{\text{g}}{\text{mL}} \xrightarrow{\text{grams to kilograms}} \frac{\text{kg}}{\text{mL}} \xrightarrow{\text{kilograms to pounds}} \frac{\text{lb}}{\text{mL}} \xrightarrow{\text{milliliters to liters}} \frac{\text{lb}}{\text{L}} \xrightarrow{\text{liters to gallons}} \frac{\text{lb}}{\text{gal}}$$

Starting with the initial density in grams per milliliter, we set up the problem, taking care that each unwanted unit will cancel when we're done:

$$\frac{1.154\ g}{mL} \times \frac{1\ kg}{1000\ g} \times \frac{2.2046\ lb}{1\ kg} \times \frac{1000\ mL}{1\ L} \times \frac{3.785\ L}{1\ gal} = \frac{9.629\ lb}{gal}$$

The entire calculation converts the grams into pounds and the milliliters into gallons.

EXERCISE 1.7 Practice with Dimensional Analysis

Manufacturing on a global scale requires working with immense quantities of raw materials. As a close approximation of the actual amount, let's assume that 96 billion aluminum cans are produced in a given year. If each can requires 15 g of aluminum (Al), how many total kilograms of aluminum are used each day (assume 365 days in a year) to make aluminum cans?

First Thoughts

From the problem we can obtain these conversion factors:

$$\frac{365\ days}{1\ year} \qquad \frac{9.6 \times 10^{10}\ Al\ cans}{year} \qquad \frac{15\ g\ Al}{Al\ can} \qquad \frac{1\ kg\ Al}{1000\ g\ Al}$$

If we map out our steps, there are several routes we can follow. Here is just one:

$$\frac{cans}{year} \xrightarrow{\text{cans to grams}} \frac{g}{year} \xrightarrow{\text{grams to kilograms}} \frac{kg}{year} \xrightarrow{\text{years to day}} \frac{kg}{day}$$

The key with dimensional analysis is to show how each step allows you to cancel units so that you can end with the units you want. It doesn't matter what order we pick for the terms, as long as the units cancel.

Solution

$$\frac{9.6 \times 10^{10}\ cans}{year} \times \frac{15\ g\ Al}{1\ can} \times \frac{1\ kg\ Al}{1000\ g\ Al} \times \frac{1\ year}{365\ day} = \frac{3.9 \times 10^{6}\ kg\ Al}{day}$$

Further Insights

One of the most important questions we need to ask is **Does our answer make sense?** That is, did we calculate an answer that we "should be" getting? Our answer is quite large. Should it be so, or does it make more sense to use, perhaps, 0.001 kg of aluminum per day? If we prepare billions of cans per year, it seems reasonable that we should use quite a bit of aluminum each day, and our answer confirms that. This doesn't mean that the math was necessarily correct, but the answer is at least reasonable. Anticipating roughly what an answer should be, and then confirming that your answer is at least nearby, is certainly part of "thinking like a chemist."

PRACTICE 1.7

a. What is the mass in grams of 6.50 gallons of water that flow into a desalination plant every second? Assume the density of the water in this problem is 1.00 g/mL.

b. How many gallons of gasoline are in 58.6 L of gasoline?

c. How much would 11.2 gallons of gasoline cost in Canada if the price of gasoline in the United States is $1.34 per gallon and the exchange rate is $1.30 CN = $1.00 U.S.? (Assume the price of gasoline is the same in the United States and Canada.)

See Problems 65–69 and 71–80.

EXERCISE 1.8 **Powers in Dimensional Analysis**

A typical can of cola contains 355 mL of liquid. How many cubic meters would be occupied by the liquid from a 12-pack of cola?

First Thoughts

Let's map out how we will convert from milliliters into cubic meters. The conversion from cubic centimeters to cubic meters will be a little more involved than a straight conversion. We'll have to break the cm^3 term down into $cm \times cm \times cm$ in order to get those units to cancel.

$$\text{cans} \xrightarrow{\text{cans to milliliters}} \text{mL} \xrightarrow{\text{mL to } cm^3} cm^3 \xrightarrow{cm^3 \text{ to } m^3} m^3$$

Solution

$$12 \text{ cans} \times \frac{355 \text{ mL}}{\text{can}} \times \frac{1 \text{ cm}^3}{\text{mL}} \times \frac{\text{m}}{100 \text{ cm}} \times \frac{\text{m}}{100 \text{ cm}} \times \frac{\text{m}}{100 \text{ cm}} = 0.00426 \text{ m}^3$$

Further Insight

Note that in converting the unit cm^3 into m^3, we completed the conversion from cm to m three times. In the end, the units of $cm^3 = cm \times cm \times cm$, and the resulting unit m^3 was obtained. This practice for converting units of multiple powers can be used in any problem.

PRACTICE 1.8

A house requires 1200 yd^2 of carpet. How many square meters of carpet will it require? How many square inches of carpet will it require?

See Problems 70 and 81.

1.6 Uncertainty, Precision, Accuracy, and Significant Figures

The density of the silicone tubes used in heart–lung machines is a good indicator of the flexibility and strength of the tubing. Strong and flexible tubes typically have relatively low densities. Judith Fairclough-Baity knows this, so she measures densities to give an idea of the quality of a freshly made tube. Often she finds that the density of her new tubing is quite similar to that of the tubing already in use in hospitals. In some cases, the difference in density between good tubing and bad tubing is very small (1.2011 g/mL versus 1.2005 g/mL). What degree of confidence does she have in her measurements? How can she be sure that the numbers she obtains for her densities are correct? The confidence that Fairclough-Baity has in her measurements is based on the use of uncertainty and significant figures. This ensures that the numbers she calculates for the densities are meaningful.

Uncertainty

There are a few things in science, and in life, that we can know for certain, with no margin of error or room for doubt. For example, if you are asked how many U.S. presidents' faces are carved into Mt. Rushmore, you can be positively sure that the answer is 4. This is an **exact number**, a quantity that we know without any uncertainty. Indeed, all data that are determined by counting are exact.

Measurement is different from merely counting, however, because every measurement is associated with a certain degree of **uncertainty**. *There are no measurements made in which we are absolutely certain of the answer because we, and our*

measurement instruments, are imperfect. The uncertainty in a measurement, *how "sure" we are that our measurement is correct,* depends on the method used to obtain the measurement, on the quality of the measuring apparatus, and on the care taken by the person making the measurement. For example, if you are asked to measure the mass of an orange, you might obtain 151 g or 151.2 g or 151.20467 g, depending on the balance you use and on how well you know how to use the balance. Each measured value will be associated with a particular level of uncertainty. *No measurement is ever exact.*

How do scientists report the uncertainty in a measurement? One way is to use the symbol ±, which means "plus or minus," to indicate the limits within which we confidently expect the true value to lie. (This approach works only if we can experimentally determine those limits.) Here are two examples:

If we write a mass of 151 ± 1 g, we mean that we are certain that the mass lies between 150 g and 152 g, but we are not certain of exactly where it lies between these limits.

If we write a mass of 151.3 ± 0.2 g, we mean that we are certain the mass lies between 151.1 g and 151.5 g, but we are not certain of exactly where it lies between these limits.

Even though scientists do their best to minimize the uncertainty of every measurement, there is still a level of uncertainty in their answers. The two most common sources of uncertainty in measurements for properly designed experiments are

- human error and variability. None of us is perfect, and we can make errors from simple human clumsiness (spilling our reaction or improperly operating a piece of equipment). But it is also true that our eyes and hands and brains can manipulate labware, or read a balance, or interpret what we see *only so well*.

- instrument error and variability. Measurement devices are not perfect. Because of imperfections in their manufacture, faults, or limitations, instruments themselves can contribute to the error in a measurement.

Accuracy and Precision

As we noted in the previous section, the quality of the balance we use to measure mass is important in a measurement. More broadly, what is a high-quality measurement? We define this via the accuracy and precision of the measurements we make. To a chemist like Fairclough-Baity, these two terms mean completely different things. **Accuracy** relates how close a measurement is to its true value. **Precision** relates how close repeated measurements are to each other, whether they are accurate or not. A set of measurements can be accurate, can be precise, can exhibit both of these properties, or can offer neither (see Figure 1.29). Fine instruments tend to yield measurements that are both accurate and precise.

Video Lesson: Precision and Accuracy

Video Lesson: CIA Demonstration: Precision and Accuracy with Glassware

FIGURE 1.29

The application of precision and accuracy to a game of darts.

Neither precise nor accurate

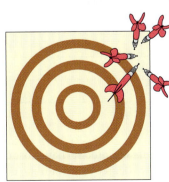

Precise, but not accurate

Accurate, but not precise

Precise and accurate

Application

Suppose four students need to measure the mass of a sample of orange juice prior to determining its vitamin C content. Let's assume that the mass of the orange juice is known to be 1.4683 g. The students all use the same sophisticated balance that provides four figures after the decimal point, but they forget that the balance must be set to zero before they use it. These students might get these results:

Student Group 1	
Amy	1.5674 g
Jorge	1.5673 g
Barb	1.5673 g
Mike	1.5675 g
Average mass	1.5674 g
True mass	1.4683 g

These are very precise answers because the masses are in close agreement and are reported to four decimal places. Unfortunately, however, they are not accurate answers, either individually or when averaged, because they all contain the same instrument error, due to the students' forgetfulness in not zeroing the balance before they used it.

Now another group of students measures the mass of the orange juice sample, and they remember to re-zero the balance before each use. Let's assume that they obtain these results:

Student Group 2	
Rebecca	1.4682 g
Maggie	1.4684 g
Winston	1.4685 g
Rob	1.4684 g
Average mass	1.4684 g
True mass	1.4683 g

The results obtained by Group 2 are both precise and accurate. Each value reported is in close agreement with the average, and all the values are accurate because they are very close to the true value for the mass.

A third group of students remembers to zero the balance before use, but they are less careful in making their measurements. This makes their measurements subject to human error. Let's assume this group of students obtains these results:

Student Group 3	
Zeb	1.4673 g
Jeff	1.4691 g
Elena	1.4675 g
Alice	1.4692 g
Average mass	1.4683 g
True mass	1.4683 g

The results they obtain are less precise than those of either Group 1 or Group 2, because they show a greater degree of variability. Each result is individually less accurate than the individual results obtained by Group 2, but they yield an average value that is accurate. Taking the average value of a set of results subject to random errors, such as many human errors, can yield a very accurate result. However, systematic instrument errors (caused by, for example, a balance that is tilted) cannot be overcome by averaging, because all the results suffer from the same instrument error.

What combination of precision and accuracy would the careless members of Group 3 have produced if they had also forgotten to zero the balance? They would have produced results that, compared to those of Group 2, were both imprecise and inaccurate.

Why is it important for us to know about uncertainties in our measurements and to worry about precision and accuracy? Lives and jobs and a lot of money can be at risk if our measurements don't do what we want them to do. The dosage of drugs must be measured carefully, within known limits, to avoid doing harm instead of good. The amount of fuel loaded onto a plane before takeoff is critical if the plane is to make it to its destination. Impure water puts lives at risk. And an accurate and precise measurement of the density of heart–lung machine tubing could mean the creation of a tube that doesn't crack or break apart during a long surgery.

Significant Figures

As we have just seen, it is useful to be able to assess the precision and accuracy of different values. Furthermore, we cannot claim that our answers are more precise than any of the measurements we used to obtain the answer. In practice, the results of calculations must contain at least as much uncertainty as is present in the measured values we used to obtain these results. This means that when we perform calculations, we cannot accept all of the numbers shown on the calculator display, because doing so would imply that we know the result to a degree of certainty that is not justified. A temperature of 325 K can be divided by 3 on a calculator, which might display an answer of 108.33333333333. However, we do not know the temperature to 14 digits, only to 3. Some of the numbers in the calculator's display are "not significant," in the sense that we simply cannot trust them; in other words, some of the numbers in the display have no meaning, so we should not use them. We can deal with this problem through the use of **significant figures**, *specific numbers in a measurement whose values we can trust.*

Let's follow Judith Fairclough-Baity as she measures the mass of a new piece of tubing she prepared. She uses a rough balance to obtain a mass of the little piece of tube. The balance's display says 5 g. What is the quality of that measurement? Specifically, the number means that the mass is closer to 5 g than it is to 4 g or to 6 g. This value has just one significant figure. Fairclough-Baity then uses a higher-quality balance to measure the mass. This balance says the mass of the tubing is 5.125 g. This number has four significant figures. The balance's display indicates that the mass of our sample is closer to 5.125 g than it is to either 5.124 g or 5.126 g. Some uncertainty about the value still exists, but there is less uncertainty than before. The more significant figures there are in any measurement, the less uncertainty there is about the measurement.

When Fairclough-Baity writes down any value in her notebook, all of the numbers other than zero are always implied to be significant. When a zero is encountered in a measurement, determining its significance is less straightforward. All zeros that come before the nonzero digits in a number are used to locate the position of the decimal point, so they do not count as significant figures. For example, 0.005 has only one significant figure, not four, because the zeros on the left just hold the position of the decimal point. She clears up any ambiguity by writing the value in scientific notation. For example, she writes 0.005 as 5×10^{-3} to show that there is only one significant figure.

Zeros that lie between nonzero digits, which are called "captive zeros," are always significant. Therefore, the number 505, for example, has three significant

Video Lesson: Significant Figures

When we divide 5.125 by 4.6 on a calculator, not all of the digits in its answer are significant.

Video Lesson: Scientific (Exponential) Notation

TABLE 1.7	Rules for Significant Figures	
Nonzero digits are always significant.	123 has three significant figures.
Zeros before the nonzero digits are never significant.	0.0123 has three significant figures.
Zeros between nonzero digits are always significant.	1.023 has four significant figures.
Zeros at the end of a number that does not contain a decimal point are ambiguous.	1230 has three significant figures.
When in doubt, write the number using exponential notation.	1230 written as 1.230×10^3 has four significant figures.
Counting numbers have infinite significant figures.	200 pencils is an exact number because it came from counting rather than measuring.

figures. The number 0.00505 also has only three significant figures. We can write the latter of these two numbers in exponential notation as 5.05×10^{-3}.

Zeros that come at the end of a number, which are sometimes called "trailing zeros," are truly ambiguous. For instance, if a technician reports a mass of 200 g, we don't know whether he or she means 2×10^2 or 2.0×10^2 or 2.00×10^2. We can take the most careful possible view and say the value is closer to 200 g than it is to either 300 g or 100 g. That's a fairly wide range of uncertainty, but we have no evidence to allow us to assume that the measurement is more precise. The only way to resolve the ambiguity is to write a measurement with ambiguous trailing zeros in scientific notation. If the technician records the value as 2.0×10^2, then the zero is after the decimal point, which makes it significant. This measurement unambiguously has two significant figures. If the value is written as 2.00×10^2, then the measurement has three significant figures. In this case, we would know that the actual value is between 199 g and 201 g. Table 1.7 summarizes the rules for determining the number of significant figures in a measured number.

Finally, note that counting numbers (such as the number of people in a room) and certain defined conversion factors and equalities (such as 4 qt = 1 gal) are considered to have an infinite number of significant figures. They are exact.

Calculations Involving Significant Figures

Judith Fairclough-Baity's mass for her piece of tubing can be used to determine the density of the tubing. She places the tubing into a graduated cylinder and notes the displacement of water. In the end, she determines that the volume displaced by the 5.125-g piece of tubing was 4.6 mL. She then divides the mass by the volume to obtain the density, which she reads from the calculator display as 1.1141304 g/mL.

$$\text{Density} = \frac{\text{mass}}{\text{volume}} = \frac{5.125 \text{ g}}{4.6 \text{ mL}} = 1.1141304 \text{ g/mL}$$

How many of the digits displayed on her calculator are significant? The rules we use to determine this (see Table 1.8), depend on whether the number is the result of addition, subtraction, multiplication, or division. These rules ensure that Fairclough-Baity's answer is no more precise than either of her two measurements.

Using these rules, Fairclough-Baity writes 1.1 g/mL in her notebook. This answer contains only two significant figures and is just as precise as the volume measurement that she obtained. If she needed to measure the density to four significant figures, she would have to measure more precisely the volume of water displaced by the piece of tubing, or she could use a different strategy for determining the density.

TABLE 1.8	Calculations Using Significant Figures	
Addition/ subtraction	The result has the same number of digits after the decimal point as the measurement with the least precision.	$21.3 + 1.045 = 22.3$
Multiplication/ division	The result contains the same number of significant figures as the measurement with the smallest number of significant figures.	$1.2 \times 4.2613 = 5.1$
Rounding	The final answer is rounded to the nearest significant digit. The digit is rounded down if the next digit is less than 5; it is rounded up if the next digit is greater than 5. If the last significant digit is odd and the next digit is exactly 5, the digit is rounded up; if the last significant digit is even and the next digit is exactly 5, the digit is not rounded.	$5.35 \div 2.68 = 1.996 = 2.00$ 8.26 rounds to 8.3 8.32 rounds to 8.3 8.25 rounds to 8.2 8.15 rounds to 8.2 8.251 rounds to 8.3 8.151 rounds to 8.2
Final answer	The final answer is the *only* number in a calculation that is rounded. Intermediate calculations are *never* rounded.	(3.9684167×1.8) $= 7.14315$ $\div 1.50 = 4.7621 = 4.8$

When to Round Off

If you are performing a series of calculations on a calculator, you should carry all the extra digits through until you obtain the final answer and then do the rounding off at the end. *You should not round off at intermediate stages,* because in doing so you might lose figures that affect the accuracy of the final result. This rule can sometimes cause problems—for example, if you perform in one step a calculation that a textbook or multimedia package has performed in two or more steps. If you get answers that differ very slightly from those in a textbook or multimedia package's solutions, don't automatically assume it is wrong. The problem may lie in the different approaches taken or in rounding off at different stages.

We'll emphasize this important rule about not rounding off during a calculation throughout this textbook. For instance, if we are working on a problem that involves many steps before the answer is obtained, we will not drop the next insignificant figure. Instead, the first insignificant figure will be highlighted in color, or otherwise noted at the end of the number. This will serve as a reminder that one shouldn't round off until the end of the calculation.

For example, let's consider the calculation of the density of a piece of metal. If we measured the volume of water initially as 25.25 mL and then added a 53.375-g chunk of metal, we note that the volume of water changes to 32.1 mL. To solve this problem, we determine the difference in the volumes and divide that into the mass of the metal. Mathematically, here are the steps we take:

$$\text{Density} = 53.375 \text{ g} \div (32.1 \text{ mL} - 25.25 \text{ mL})$$

$$= 53.375 \text{ g} \div (6.85 \text{ mL})$$

At this step, we've highlighted the insignificant digit in the subtraction so that we can easily count the number of significant figures for our next step. We haven't rounded anything yet because we are not finished with the calculation.

$$53.375 \text{ g} \div (6.85 \text{ mL}) = 7.79197 \text{ g/mL}$$

$$= 7.8 \text{ g/mL}$$

Here, approaching our final answer, we highlighted the insignificant figures, and then, because this was the final calculation, we rounded the final answer.

EXERCISE 1.9   **Calculations**

Provide the answer to this calculation. Be sure to indicate the appropriate number of significant figures.

$$\frac{288 \times 1.445}{7.9 \times 10^2} - 0.064 =$$

Solution

To start, let's complete the multiplication on the numerator. We retain at least one insignificant figure (the **2** in 416.**2**) in the intermediate calculation:

$$288 \times 1.445 = 416.2$$

Then we'll do the division. If this were our final computation, we would keep only two significant figures. But because we have more calculating to do, we retain a third, insignificant figure, highlighted in red:

$$\frac{416.2}{7.9 \times 10^2} = 0.527$$

Finally, we'll subtract the two numbers. We don't round at all until all calculations are finished. The presence of the highlighted digit represents all of the insignificant figures we retain.

$$0.527 - 0.064 = 0.463, \text{ which rounds to } \textbf{0.46}$$

Our answer only has two significant figures. If we had rounded at the previous step, 0.527 would have become 0.53. This would have led to an incorrect answer in the final step (0.53 − 0.064 = 0.466), because 0.466 rounds to 0.47. Remember, never round until you are completely done with a calculation.

PRACTICE 1.9

How many significant figures are there in each of these numbers?

   a. 1.3090   b. 3450   c. 0.0020   d. 2.000

How many significant figures will there be in the answer to this problem?

$$\frac{2.3 + 0.88}{79.4} =$$

See Problems 87–98.

 **1.7**   **The Chemical Challenges of the Future**

The chemist's ability to raise questions about everything around us, to design experiments, to make measurements, and to derive theories and laws via this process we call the scientific method has taught us a great deal about our world and the role of chemistry in all of its changes. The future is built by these changes, and they present many chemical challenges to us. Some of these challenges are problems caused by our use of chemistry in the past; others involve our continuing efforts to make life better, more enjoyable, and more fulfilling. Some information about a few of the future challenges for chemists follows.

 Application

■ *The frontiers of medicine.* Chemists will continue to search for new and more effective medicines to treat cancer, AIDS, heart disease, Alzheimer's disease, rheumatoid arthritis, and many other diseases. They will also be urgently trying to create new antibiotics that enable us to "stay ahead" of the pace at which disease-causing organisms become resistant to existing antibiotics.

- *New challenges in agriculture.* Chemists have already transformed agriculture once, through their development of artificial fertilizers and pesticides. They are now heavily involved in a new and controversial transformation involving the creation of genetically modified crops. This may enable us to create new "engineered" plants with larger and more reliable yields. Nevertheless, many people are concerned about the legal, ethical, and safety issues associated with genetically modified crops.

- *The challenge of pollution.* Oil spills, smog, and pollution in general are negative aspects of our use of chemical processes—aspects that often predominate in public perceptions of chemistry. Pollution of our environment has been the price we have paid for many of the comforts and conveniences of modern life. Chemists are working hard to reduce that price by developing alternative "green" technologies and by learning how to use chemistry more effectively to clean up the problems that still arise.

- *Global warming and ozone depletion.* Our desire for comforts and conveniences has created two major problems in the chemistry of the atmosphere. A rise in levels of carbon dioxide and other greenhouse gases has been implicated in global warming. And the destruction of the protective ozone layer by chemicals called chlorofluorocarbons, or CFCs, has contributed to increases in skin cancer. Chemists will continue to research both issues to determine the causes of, and solutions to, these environmental concerns. In addition, the development of new chemical technologies will help us learn more about the hazards that caused these problems, with the goal of reducing or eliminating them.

- *Better materials.* Materials are the chemicals we use to make things, such as clothes, cars, aircraft, homes, televisions, and computers. Every material has its own set of advantages and limitations for a given application. Chemists are continually trying to enhance the advantages, such as durability, flexibility, and efficiency as electrical conductors or insulators. They are also working to overcome the limitations of some materials, such as susceptibility to corrosion, high cost, inability to recycle and reuse, and so on. That work will continue indefinitely, hopefully yielding a steady supply of new and more versatile materials to make the things we need and want.

- *New ways to supply energy.* The modern world is sustained by huge supplies of materials—such as coal, oil, and gas—that can provide usable energy. Unfortunately, the chemical reserves of energy are limited. Moreover, their use contributes to environmental pollution and global warming. Chemists are learning how to get the energy we need in more sustainable and less polluting ways. Many alternative ways of supplying energy already exist, such as solar and wind power. But these ways need to be made more efficient, cost-effective, and powerful in order to find use in today's energy-hungry world.

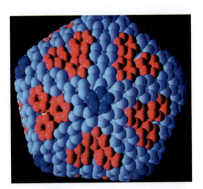

The frontiers of medicine involve finding the viruses that cause disease and identifying the compounds that make them.

Pollution in Mexico City.

FIGURE 1.30

Nanotubes, nanospheres, nanogears, and nanolevers. Such things might be the building blocks of a new industrial revolution in "nanotechnology."

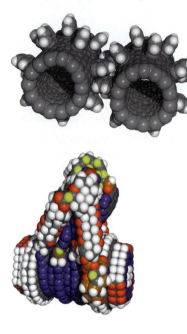

- *Nanotechnology.* In the 1970s people began to use the term **nanotechnology** to describe the manipulation and machining of matter at very small scales. Some developments in this field can be described as "molecular manufacturing," involving attempts to build tiny machines, materials, and medical devices by manipulating very small assemblies of matter, even molecule by molecule or atom by atom. Traditional chemistry works with huge numbers of particles. The twenty-first century might see a new industrial revolution, in which the precise manipulation of tiny numbers of atoms, molecules, and ions becomes a routine part of our technology. Many of the first steps toward that new revolution have already been taken (Figure 1.30).

- *Understanding life.* We understand a great deal about the chemistry of life, but many processes are not fully understood. Indeed, the chemistry of what we really are—the chemistry of consciousness—is still a complete mystery to us. We have uncovered the general chemical mechanisms at the heart of all life, but there are endless intricate details still to be learned. As we learn these details, and therefore learn more about the workings of life, we will be in a much better position to fix things when the chemistry of life goes wrong and makes us ill.

These are just a few of the "hot" chemical topics that will tap the creative energy of chemists tomorrow and into the distant future. Chemists will also be involved in the operation of oil refineries, plastics factories, pharmaceutical companies, food manufacturers, cosmetics companies, makers of paints, glues, and varnish, semiconductor plants, and much, much more (see Table 1.9). Their operation depends on the application of chemical knowledge, continuously, day and night, to keep the modern world running.

TABLE 1.9	Some Career Opportunities in Chemistry	
Agronomist	Food technologist	Pharmacist
Anesthesiologist	Geneticist	Pharmacologist
Biochemist	Geologist	Pharmacologist sales
Ceramics engineer	Industrial health	representative
Chemical engineer	engineer	Physician
Consumer protection	Internist	Professor
specialist	Laboratory analyst	Sanitarian
Dentist	Medicinal chemist	Science technician
Dietitian	Metallurgist	Technical writer
Educator	Nuclear scientist	Toxicologist
Food and drug analyst	Patent examiner	Wood scientist

Possible Employers of Chemistry Majors

Beverage companies	Local government
Centers for Disease Control	Medical laboratories
Chemical companies	Medical libraries
Chemical manufacturing firms	Medical supply companies
Colleges and universities	Mining companies
Cosmetics companies	National Institutes of Health
Department of Agriculture	Newspapers and magazines
Department of Defense	Petroleum refineries
Engineering firms	Pharmaceutical companies
Environmental Protection Agency	Research firms
Food companies	State and federal government
Food and Drug Administration	Technical libraries
Hospitals	Textile manufacturers
Journals	

The Bottom Line

- Our universe and everything in it are made of chemicals. Chemicals are involved in all of the changes that affect and sustain us. (Section 1.1)

- Atoms are the most fundamental particles of the chemical world. A substance containing only one kind of atom is called an element. The particles of chemistry—atoms and molecules—are incredibly tiny compared to the objects we see in the everyday world. (Section 1.2)

- Molecules are composed of two or more atoms chemically bonded together. A chemical compound is any substance that contains different elements chemically bonded together. (Section 1.2)

- Chemical changes occur when chemicals undergo reactions in which new chemical products are formed from the initial chemical reactants. Physical changes occur when chemicals undergo changes in their state. (Section 1.2)

- Scientists find out about nature, and learn how to change it, using the scientific method. (Section 1.3)

- The International System (SI) defines the fundamental base units, and a variety of derived units, that are used to measure physical quantities. (Section 1.4)

- The calculations in chemistry include the conversion of units by using factors that relate identical quantities. This can be done via a method known as dimensional analysis. (Section 1.5)

- Precision and accuracy are important to the discussion of chemistry. They relate the uncertainties in measurements. The number of significant digits a number contains is related to the precision of that number. (Section 1.6)

- Our understanding of chemistry can be used to solve problems in many fields, including medicine, agriculture, pollution control, and nanotechnology. In addition, we can broaden our understanding of many issues, including global warming, materials, meeting our energy needs, and life itself. (Section 1.7)

Key Words

accuracy The closeness of a measurement to the actual value. (*p. 29*)

ampere (A) The SI base unit of electric current. (*p. 20*)

atoms The smallest identifying unit of an element. (*p. 5*)

base units The set of seven fundamental units of the International System (SI). (*p. 17*)

candela (cd) The SI base unit of luminous intensity. (*p. 20*)

chemical *See* matter.

chemical changes Changes that involve chemical reactions in which existing chemicals, the reactants, are changed into different chemicals, the products. (*p. 6*)

chemical property A characteristic of a substance that can be determined as it undergoes a change in its chemical composition. (*p. 6*)

chemical reaction The combination or reorganization of chemicals to produce different chemicals. (*p. 6*)

chemistry The systematic study of the composition, structure, and properties of the matter of our universe. (*p. 4*)

composition The relative proportions of the elements in a compound. (*p. 5*)

compound A substance containing different elements chemically bonded together. (*p. 5*)

conversion factor A mathematical expression of the ratio of one unit to another, used to convert quantities from one system of units to another. (*p. 25*)

cubic decimeter (dm³) The derived unit commonly used to measure volumes in the laboratory. See liter. (*p. 22*)

cubic meter (m³) The derived SI unit used for volume. (*p. 22*)

degree Celsius (°C) The unit of temperature on the Celsius scale. (*p. 19*)

degree Fahrenheit (°F) The unit of temperature on the Fahrenheit scale. (*pp. 18–19*)

density The mass of a substance that is present in a given volume of the substance. The SI unit for measuring density is kilograms per cubic meter (kg/m^3). (*p. 22*)

derived units Units formed by the combination of SI base units. (*p. 22*)

desalination The process that removes dissolved salts from seawater to make potable water. (*p. 10*)

dimensional analysis An extremely useful method for performing calculations by using appropriate conversion factors and allowing units (dimensions) to cancel out, leaving only the desired answer in the desired units. (*p. 26*)

electrodialysis The process of removing unwanted salts from a solution through a series of semipermeable membranes by applying an electrical charge. (*p. 10*)

element A substance that contains only one kind of atom. All elements are listed in the periodic table of the elements. (*p. 5*)

exact number A number that can be known with absolute certainty. Exact numbers possess an infinite number of significant figures. (*p. 28*)

extensive property A characteristic of a substance that is dependent on the quantity of that substance. (*p. 23*)

heterogeneous mixture A mixture that is not uniformly mixed, so there are different proportions of the components in different parts of the mixture. (*p. 8*)

homogeneous mixture A mixture that is uniformly mixed so that it has the same composition throughout. Also known as a solution. (*p. 8*)

hypothesis A tentative explanation for an observation—that is, a statement about what we think might be an explanation for an observation. (*p. 13*)

intensive property A characteristic of a substance that is independent of the quantity of that substance. (*p. 23*)

International System *See* SI.

kelvin (K) The SI base unit of temperature. (*p. 18*)

kilogram (kg) The SI base unit of mass. (*p. 17*)

kilogram per cubic meter (kg/m³) The derived SI unit for density. (*p. 22*)

liter (L) A commonly used unit for volume; equal to 1 dm³. (*p. 22*)

luminous intensity Brightness, expressed in the SI unit candela (cd). (*p. 20*)

macroworld A term used to describe the "big world" of everyday experience, as opposed to the "nanoworld" of atoms, molecules, and ions, whose activities ultimately determine what happens in the macroworld. (*p. 20*)

mass A measure of the amount of matter in a body, determined by measuring its inertia (resistance to changes in its state of motion) and expressed in the SI base unit kilogram. (*p. 17*)

matter Anything that has a mass and occupies space. (*p. 4*)

meter (m) The SI base unit of length. (*p. 18*)

milliliter (mL) One thousandth of a liter. (*p. 22*)

mixture A sample containing two or more substances. (*p. 8*)

mole (mol) The SI base unit of amount of substance. One mole of entities, such as 1 mole of atoms, is approximately equal to 6.02214×10^{23} entities. (*p. 20*)

molecule A particle composed of two or more atoms bonded together. (*p. 5*)

nanotechnology Technology that depends on manipulating materials and their chemistry at the very small, "nanoscale" level of individual particles or assemblies of small numbers of particles. (*p. 36*)

nanoworld A term used to describe the "world of the very small" at the level of individual atoms, molecules, and ions, as opposed to the "macroworld" of everyday experience. (*p. 20*)

physical change Change in the physical state of a substance, such as a change between the solid, liquid, and gaseous states, that do not involve the formation of different chemicals. (*p. 6*)

physical property A characteristic of a substance that can be determined without changing its chemical composition. (*p. 6*)

precision The reproducibility of a measurement. (*p. 29*)

reverse osmosis The process of purifying water by passing it through a semipermeable membrane. Dissolved solutes are unable to pass through the membrane. (*p. 10*)

scientific law Concise description of the behavior of the natural world. (*p. 13*)

scientific method A reliable way to find out things about nature by making use of appropriate combinations of these key activities: making observations, gathering data, proposing hypotheses, performing experiments, interpreting the results of those observations and experiments, checking to ensure that the results are repeatable, publishing the results, establishing scientific laws, and formulating theories. (*p. 12*)

second (s) The SI base unit of time. (*p. 18*)

SI The International System (*Système International*) of agreed-upon base units, derived units, and prefixes used to measure physical quantities. (*p. 17*)

significant figures Those specific numbers in a measurement whose values we can trust. (*p. 31*)

solution A homogeneous mixture of solute and solvent. (*p. 8*)

state The physical appearance of a chemical, typically as a solid, liquid, or gas. (*p. 10*)

substance A type of matter that has a fixed composition. (*p. 6*)

theory A trusted explanation of an observation, based on a hypothesis that has been tested in experiments. (*p. 13*)

uncertainty A measure of the lack of confidence in a measured number. (*p. 28*)

weight A measure of the gravitational force exerted on a body. (*p. 17*)

Focus Your Learning

The answers to the odd-numbered problems and some selected problems appear at the back of the book, as represented by the blue numbering.

Section 1.1 What Do Chemists Do?

Skill Review

1. Define the term *chemistry*. Why is chemistry a science?

2. Explain why chemists need to be inquisitive.

3. In the section, we listed a number of things that chemists do. Look around the room in which you are sitting. List five additional things that you suspect are manufactured using chemistry.

4. Look on the label of ingredients for five products in your kitchen pantry. List some examples of substances that might be commercially prepared. Why did you choose these? List some substances that are added in more or less their natural form. Why did you choose these?

Chemical Applications and Practices

5. The directions for baking a particular cake include the following instruction: "Carefully add 1 cup of water to the powdered mix and stir." Use this statement to explain why measurements are important to the chemist.

6. A scientist has just identified a new drug for treating cancer. Use this statement to explain why a scientist communicating this discovery might need a chemical "shorthand."

Section 1.2 The Chemist's Shorthand

Skill Review

7. Define, differentiate between, and give three examples of an element and a compound.

8. What dispute might you have with the manufacturer of a cleanser that was labeled "chemical-free?"

9. How is it possible that oxygen, found in our atmosphere, can be called an element and a molecule but not a compound?

10. How is it possible that gold, found as flakes in a stream, can be called an element but not a molecule or a compound?

11. Name one characteristic that a mixture and a compound have in common.

12. Describe a critical difference between a mixture and a compound.

13. Which of these chemicals are elements and which are compounds? For the compounds, list the elements that are present. You may need to use the periodic table and a chemical encyclopedia to help you answer some of these.
    a. hydrogen gas
    b. sodium chloride (table salt)
    c. glucose
    d. neon
    e. copper sulfate
    f. titanium

14. Which of these chemicals are elements and which are compounds? For the compounds, list the elements that are present. You may need to use the periodic table and a chemical encyclopedia to help you answer some of these.
    a. nitrogen gas
    b. calcium chloride (sidewalk salt)
    c. aspirin
    d. helium gas
    e. silver metal
    f. water

15. Indicate whether each of these is a heterogeneous mixture or a homogeneous mixture.
    a. lake water
    b. yellow notebook paper
    c. marble
    d. soda
    e. milk
    f. dirt

16. Indicate whether each of these is a heterogeneous mixture or a homogeneous mixture.
    a. tap water
    b. apple juice
    c. beach sand
    d. paint thinner
    e. spaghetti sauce
    f. air

17. List at least two methods used to separate a homogeneous mixture.

18. Why are most heterogeneous mixtures easier to separate than most homogeneous mixtures?

19. For each of these processes, indicate whether a chemical change or a physical change is taking place.
    a. molding melted chocolate into a bar
    b. heating your home with a woodstove
    c. drying your clothes in the dryer
    d. snow melting in the heat of the sun

20. For each of these processes, indicate whether a chemical change or a physical change is taking place.
    a. the yellowing of an old newspaper
    b. making hard-boiled eggs
    c. magic ink appearing on a piece of paper
    d. making dirty clothes clean in the washing machine

Chemical Applications and Practices

21. Some ancient civilizations considered air an "element." Today we consider air a *mixture, compound* (select one). What is the basis of your choice?

22. Some ancient civilizations considered water an "element." Today we consider water a *mixture, compound* (select one). What is the basis of your choice?

23. A "health food" store has a poster in its window that says, "Eat natural food, not chemicals." Briefly explain what is wrong with this statement, and try to summarize, in a chemically accurate way, what the writer of the poster was really trying to say.

24. A box of noodles in a "health food" store indicates that the noodles are "free of chemicals." Briefly explain what is wrong with this statement, and try to summarize, in a chemically accurate way, what the writer of that phrase was really trying to say.

Section 1.3 The Scientific Method

Skill Review

25. Suppose you sit down at your computer to draft an e-mail to a friend. However, after you have written the e-mail, it appears that it cannot be sent from your computer. Explain how you might use parts of the scientific method to help you solve this problem.

26. Suppose you wanted to determine which grade, or type, of gasoline would provide the best miles-per-gallon ratio in your car. What would be the important variables that you

would control as you used the scientific method to draw your conclusion?

27. Which aspects of the scientific method are most likely to be subject to interpretation? Which, on the other hand, should be least ambiguous?

28. For what reasons, apart from the sharing of information, is it important to publish the results of scientific studies?

29. How do we distinguish a theory from a hypothesis?

30. We are familiar with how governments enact laws. How is a scientific law formed?

Chemical Applications and Practices

31. Several studies have been done to determine the effectiveness of various vitamins on specific health issues. Explain how it is possible for scientific studies on the same subject to produce conflicting results.

32. As a consumer, you encounter many claims about remedies for various human conditions from acne to balding. What aspects of the scientific method would you expect to see employed before such claims are made?

33. Assume that fish are dying in a river that runs through your town. The river begins in remote mountains, runs through many miles of farmland, and then passes through a large industrial area before entering town. What sorts of questions need to be should be asked in an attempt to learn why the fish are dying? What investigations could be carried out to answer the questions?

34. In an attempt to treat a wart on your hand, you buy a proprietary remedy, and two weeks later your wart has disappeared. Does this prove that the remedy is effective? What steps should be performed to examine the effectiveness of the remedy in a proper scientific manner?

Section 1. 4 Units and Measurement

Skill Review

35. Based on the table found in this chapter, arrange these distances in order from largest to smallest:
   1 millimeter   1 terameter   1 kilometer   1 nanometer

36. Based on the table found in this chapter, arrange these masses in order from largest to smallest:
   1 microgram   1 kilogram   1 milligram   1 picogram

37. Complete the conversions:
   a. 100 kg to grams   d. 3.20 mg to grams
   b. 25.9 m to kilometers   e. 9.11 nm to picometers
   c. 25°C to °F   f. 98.6°F to °C

38. Complete the conversions:
   a. 150 mL to liters   d. 2.33 L to milliliters
   b. 8.42 g to milligrams   e. 700 mg to grams
   c. 48.5°C to °F   f. −20°F to °C

39. Complete the conversions:
   a. 8.7 kg to milligrams   d. 3.20 dL to kiloliters
   b. 25.9 dm to meters   e. 9.11 s to nanoseconds
   c. 190°C to °F   f. 350°F to °C

40. Complete the conversions:
   a. 3.99 mL to deciliters   d. 14.5 L to megaliters
   b. 8.42 Mg to kilograms   e. 55.5 ks to gigaseconds
   c. −40°C to °F   f. 75°F to °C

41. A 2.00-L sample of nutrient agar would be able to be divided into how many equal 100-cm^3 media containers for a bacterial study?

42. Suppose the nutrient agar described in Problem 41 amounted to 32 containers, each holding 50 cm^3 of media. How many total liters of nutrient agar are in the containers?

43. Using scientific notation, express 327 kilometers in:
   a. meters   c. micrometers
   b. millimeters   d. nanometers

44. Using scientific notation, express 499 seconds in:
   a. milliseconds   c. deciseconds
   b. microseconds   d. kiloseconds

45. A 62.56-g sample of mercury was added to a graduated cylinder. It had a volume of 4.60 mL. What is the density of the mercury?

46. A 22.4-g sample of a substance was added to a graduated cylinder. It caused a 18.3-mL change in the volume of water in the cylinder. What is the density of the substance?

47. A 250.0-mL sugar solution had a density of 1.37 g/mL. An additional 30.0 g of sugar was added to the solution, raising the volume by 24.6 mL. What is the density of the resulting sugar solution?

48. A chemist added 17.8 g of salt to 150.0 mL of a salt solution of unknown density. The resulting solution had a final volume of 165.9 mL and a density of 1.22 g/mL. What was the density of the original salt solution?

49. List the fundamental units that you would combine to get these derived units (you may need to look up the meaning of some of the terms).
   a. velocity   c. volume
   b. acceleration   d. specific heat

50. List the fundamental units that you would combine to get these derived units (you may need to look up the meaning of some of the terms).
   a. density   b. pressure   c. energy   d. force

Chemical Applications and Practices

51. Which of the following rulers would provide the greatest number of significant figures in the measurement of the volume of a cardboard box?

52. Determine the volumes of liquid in each of these graduated cylinders, and report your answers with the correct number of significant figures.

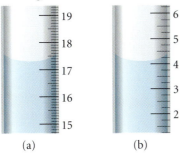

(a)　　　　　　　(b)

53. A fast-food restaurant wants to standardize the temperature of its coffee. It installs a coffee machine that is specified as delivering coffee at 75°C ± 3°C. What are the maximum and minimum temperatures of the coffee delivered by this machine in °F?

54. The temperature of the hot oil bath in a fast-food restaurant determines the quality of the french fries that are produced. The heater on the oil bath can regulate the temperature at 350°F ± 10°F. What are the maximum and minimum temperatures of the french fries delivered by this machine in °C?

55. Carbon atoms have an atomic radius of approximately 77 pm. If you drew a line that was 1 cm in length using a piece of charcoal, approximately how many atoms of carbon would be in this line? (For the purposes of this question, assume that the atoms are touching and that they form a straight line one atom wide.)

56. The same piece of charcoal (see Problem 55) was used to darken a box on a survey form. The box measures 5 mm on each side. How many atoms of carbon would be in this box? (For the purposes of this question, assume that the atoms are touching, that they line up into a square grid, and that they are only one layer thick on the paper.)

57. During the 1980s, several new elements were synthesized in Germany. These elements are made of unstable atoms. Half of the atoms in a sample of unstable atoms will decompose over a period of time known as the element's half-life. One of these newly discovered elements, Meitnerium (element 109) has a half-life of only 0.07 s.
a. What is its half-life in milliseconds?
b. What is the length of time of four of its half-lives in terms of microseconds?

58. Before filling, an empty, irregularly shaped container has a mass of 0.1956 kg. When this container is totally filled with water, the combined mass of the container and water is 305.6 g. What is the mass of water added to the container in kilograms?

59. A certain red tomato has a mass of 45.6 g. A green tomato also has a mass of 45.6 g. However, when placed in water, the green tomato floats and the red one sinks. (Try this. Depending on the degree of ripeness, it actually happens!) Which of the two is denser? Explain how two tomatoes with the same mass can have different densities.

60. Various plastics have identifiable density values. For example, you may have heard of high-density polyethylene and low-density polyethylene. If you had pieces of each that were equal in volume, which would have the greater mass?

61. Most samples of matter expand when heated. Assuming that this is the case with water at room temperature, would the density increase or decrease as the water warmed? Show the mathematical basis of your answer.

62. Ice floats on top of water. Using the information from Problem 61, explain how this could occur and show the mathematical basis of your answer.

63. In casual food preparation, one may be directed to add a pinch or a smidgeon or "just a bit" of a particular ingredient. If 10 pinches equaled 2 smidgeons and 10 "just a bits" equaled 1 smidgeon, how many "just a bits" of hot sauce should be added to a recipe that calls for 2 pinches?

64. Consider the tongue twister "Peter Piper picked a peck of pickled peppers." If two dozen pickled peppers are in a peck, and Peter Piper picked 8 pecks for his girlfriend Polly, how many peppers did Peter Piper pick?

Section 1.5 Conversions and Dimensional Analysis

Skill Review

65. Illustrate a plan of attack (in the same manner as indicated in the text) for converting units of miles per hour (mph) to meters per second.

66. Illustrate a plan of attack (in the same manner as indicated in the text) for converting units of kg·m/s² to lb·ft/s²

67. Osmium (element 76) is one of the most dense elements known. The standard density is 22.6 kg/L. What is the density in units of g/cm³? What would be the mass of a 0.50-L sample of osmium?

68. Mercury has a density of 13.6 g/cm³. What is the density in kg/L for mercury? Calculate the weight in pounds of 3.6 L of mercury.

69. Convert the following into the indicated units:
a. 35 mph to kph　　　c. 733 mi/gal to km/L
b. 22.4 L/s to cm³/s　　d. 4.184 g/°C to lb/°F

70. Convert the following into the indicated units:
a. 12 doz/lb to gross/kg　c. 14.4 lb/in² to kg/m²
b. 16 lb/gal to g/mL　　　d. 3.04 °C/min to °F/s

71. In the vacuum of space, light travels at a speed of 186,000 miles per second. Indicare how many miles light travels through space in:
a. 1 minute　　b. 1 day　　c. 1 year

72. The speed of sound varies greatly depending on the medium and conditions. If the speed of sound at sea level and zero degrees Celsius is approximately 1100 feet per second, indicate how far sound travels in:
a. 1 millisecond　　b. 1 dekasecond　　c. 1 microsecond

73. Perform these conversions:
a. 2.0 oz to pounds　　d. 96 in to meters
b. 4.0 qt to liters　　　e. 13 ft to centimeters
c. 160 lb to kilograms　f. 32 mi to kilometers

74. Perform these conversions:
a. 3.0 kg to pounds　　d. 134 oz to kilograms
b. 7400 s to days　　　e. 75.5 cm to feet
c. 50.34 mL to gallons　f. 2.88 L to quarts

75. A typical chemistry lecture lasts for approximately 50 minutes. What is that time in years? . . . in centuries? . . . in decades? Report each number using a proper metric prefix that enables you not to have any zeros in your final answer.

76. Some people live to be 100 years old. What is that time in seconds? . . . in minutes? . . . in hours? Report each number using a proper metric prefix that enables you not to have any zeros in your final answer.

Chemical Applications and Practices

77. The mass of 6.02×10^{23} atoms of gold is approximately 197 g.
a. What is the average mass, in grams, of just one atom of gold?
b. Select a prefix that could more appropriately be used to report the average mass, and express the mass in that unit.

78. The mass of 6.02×10^{23} atoms of carbon is approximately 12 g.
a. What is the average mass, in grams, of just one atom of carbon?
b. Select another prefix that could more appropriately be used to report the average mass, and express the mass in that unit.

79. Suppose a football "star" signs a contract for 200 megabucks over a five-year period. How many dollars is the star paid per second? . . . per three-hour game?

80. Which is longer; a 100-m soccer field or a 100-yd football field? Show the mathematical proof for your answer.

81. Air is a mixture that consists mostly of oxygen and nitrogen. The density of air varies, depending on temperature and pressure. However, for this problem let's assume a reasonable value for the density of air to be 1.20 mg/cm^3. Suppose the dimensions of your dorm room are approximately 12 ft × 14 ft × 9.0 ft. What would you calculate to be the approximate mass of the air in the room?

82. Modern pewter is a mixture of tin, antimony, and copper. Formerly, pewter also contained lead. If a 1.00-lb sample of a particular pewter alloy contained, by mass, 95.0% tin and 3.4 % antimony, how many grams of copper must be present in the sample?

Section 1.6 Uncertainty, Precision, Accuracy, and Significant Figures

Skill Review

83. Some chemists are playing horseshoes. In their version of the game, each gets four chances to throw a horseshoe to hit a stake. What can you state about the accuracy and precision of the chemist that threw the horseshoes to arrive at the outcome shown here?

84. Two more chemists playing horseshoes each take a turn. Which of the two outcomes shown here is more precise?

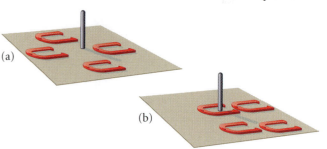

85. Three students weighed the same sample of copper shot three times. Their results were as follows:

Weighing	Student 1	Student 2	Student 3
1	17.516 g	15.414 g	13.893 g
2	17.888 g	16.413 g	13.726 g
3	19.107 g	14.408 g	13.994 g

a. Calculate the average mass of the sample, as determined by each student.
b. Which set of measurements is the most precise?
c. If the true mass of the copper shot is 15.384 g, which of the students was most accurate?
d. What are the main sources of error that could have caused the differences in the values?

86. The same three students measured the volume of the same sample of water four times. Their results were as follows:

Measurement	Student 1	Student 2	Student 3
1	25.55 mL	23.79 mL	25.02 mL
2	24.81 mL	24.01 mL	25.10 mL
3	23.03 mL	24.32 mL	25.07 mL
4	24.28 mL	24.19 mL	25.17 mL

a. Calculate the average volume of the sample, as determined by each student.
b. Which set of measurements is the most precise?
c. If the true volume of the water is 24.10 mL, which of the students was most accurate?
d. What are the main sources of error that could have caused the differences in the value?

87. Which of these values are quoted using exact numbers? How many significant figures are there in each of these numbers?
a. 12.000000 g   d. 12 L
b. 3125 students   e. 1 g
c. 12.2 L   f. 42 test tubes

88. Which of these values are quoted using exact numbers? How many significant figures are there in each of these numbers?
a. 15 apples   d. 44 mi
b. 500 people   e. 70°C
c. 3.2050 in   f. 10 g

89. In each of these numbers, underline any zeros that are considered significant.
    a. 0.700 cm
    b. 0.101 kg
    c. 100.0 cm
    d. 100 m
    e. 0.01010 g

90. In each of these numbers, underline any zeros that are considered significant.
    a. 1.000 cm
    b. 80.2 kg
    c. 2104 cm
    d. 0.56 m
    e. 3000 g

91. How many significant figures are there in each of these values?
    a. 6.07×10^{-15}
    b. 0.003840
    c. 17.00
    d. 8×10^8
    e. 463.8052
    f. 1406.20
    g. 0.0007
    h. 1600.0
    i. 0.0261140
    j. 1.250×10^{-3}

92. How many significant figures are there in each of these values?
    a. 6.022×10^{23}
    b. 1.79×10^{-19}
    c. 3.00×10^8
    d. 14
    e. 0.0035020
    f. 250
    g. 13.50
    h. 101.010
    i. 12.000
    j. 550.050

93. Determine the answer for each of these problems. Report your answer using the rules for significant figures.
    a. $3.44 + 6.2$
    b. $12.57 - 3.998$
    c. $2.534 + 1.23 + 2.0500$
    d. 12.54×5.0
    e. 84×100
    f. $45.6 \div 2.4$
    g. $(754 + 0.8) \div 1.3$
    h. $(49.53 \times 1.20) + 12$
    i. $(35.865 \div 84.2) + 2.3890$

94. Determine the answer for each of the problems below. Report your answer using significant figures rules.
    a. $120 + 6.77$
    b. $453 - 0.32$
    c. $51.8 + 7.225 + 2.01$
    d. 2.54×32
    e. 36.33×0.300
    f. $140 \div 2.9375$
    g. $(135 + 3.2) \div 1.332$
    h. $(2.78 \times 1.2) + 3.96$
    i. $(14.42 - 1.023) \div 2.3$

Chemical Applications and Practices

95. Suppose your chemistry textbook fell 10.0 cm from a backpack to a desk. Then it slipped off the desk and fell another 0.91 m from the desk to the top of someone's foot. (Ouch!) Finally, it fell the remaining 40 mm to the floor. What is the total distance, with the correct number of significant digits, that the textbook has fallen?

96. A piece of cheese has a mass of 250.67 g. Four people each remove a slice of cheese from the piece. The first removes 22.5 g of cheese, the second removes 10 g, the third removes 3.557 g, and the fourth takes 80.1 g. How many grams of cheese remain? Use appropriate significant figure rules to calculate your answer.

97. Molybdenum has a melting point of over 2600°C. It makes steel stronger, creates colors when added to molten glass, and is an integral component of some biological molecules known as enzymes. If you had a sample of the metal that had a mass of 14.56 g and a volume of 1.43 cm³, what would you calculate as the density of molybdenum? (Remember to follow the rules for significant figures.)

98. A student obtains the density of water. A 243-mL sample of water was noted to have a mass of 235.5 g. What does the student calculate as the density of the water sample? (Remember to follow the rules for significant figures.)

Section 1.7 The Chemical Challenges of the Future

Skill Review

99. One of the "future frontiers" mentioned in this chapter refers to the use of nanotechnology.
    a. To what does the prefix *nano-* refer?
    b. What would be the diameter, in centimeters, of a wheel 10 nm in diameter?

100. Ozone and other gases in the stratosphere are often measured in units known as parts per million (ppm). One ppm ozone is equivalent to 1 g of ozone per million grams of air.
    a. Determine the number of grams of ozone in 15 kg of air containing 2.00 ppm ozone.
    b. Which prefix would be best suited to identify this quantity of ozone?

Chemical Applications and Practices

101. Use current Internet resources, or current print media, to report one advantage and one disadvantage of genetic research on corn. How long have genetically selected plants been used in agriculture?

102. One method of cancer treatment involves chemotherapy. What is the chemical connection to this medical term?

103. Depending on their identity and how they are used, chemicals may be harmful or beneficial to our lives. Give two specific responses to the question "What has chemistry done for me?"

104. Ozone, like many other chemicals, may be helpful or harmful. The presence of ozone at high altitudes has advantages for us. However, at ground level it can cause several problems. Use other resources to explain how the same molecule can have such diverse impacts.

105. What are the chief energy problems facing the United States? How are these similar to or different from those facing other countries, both economically wealthy and poor?

106. Select one of your answers from Problem 105 and explain how your knowledge of chemistry might play a role in solving that problem.

Comprehensive Problems

107. What aspect of daily life do you predict will be most affected by chemistry in the generation to follow yours?

108. Name five areas of your life where chemistry plays a major role.

109. Describe two chemical changes and two physical changes that are important in growing food crops.

110. Someone remarks that the baking of a cake is a physical change. Another person says that the process is a chemical change. Who is right and why?

111. Write a sentence to describe what is happening when you dissolve some sugar in a cup of warm water.

112. The height of horses is often measured in "hands." A "hand" is considered to be 4 inches. Furthermore, up to three-tenths, each tenth of a hand is considered to be 1 inch. Any value over three-tenths of a hand is rounded up to the next hand.
     a. What is the height of a horse, in inches, that stands 14.2 hands?
     b. How many hands tall is a horse that stands, from the ground to the top of the withers, 1.6 m tall?

113. Numbers, and their meaning, are very important to the practicing chemist.
     a. How does the meaning of an exact number differ from that of a number taken from a measurement?
     b. Beverages are often sold in six-packs. Six six-packs would have how many individual containers?
     c. Those containers hold a total of 7200 mL of the beverage. Which of the value(s) represented in this problem are exact numbers?

114. What are the two most common sources of uncertainty in measurements? Of the two, which typically produces uncertainty in a consistent direction?

115. Describe two intrinsic and two extrinsic properties of seawater.

116. A chemist and her family go on vacation and ride the train to the mountains. They start their journey from Chicago and travel 1223 miles to Aspen, Colorado. The trip starts at 1:50 p.m. and the train arrives in Aspen at 1:53 p.m. the next day. How fast, in kilometers per hour, does the train average during the trip?

117. During a typical cross-country train trip, top speeds of 80.0 miles an hour are reached. What is that speed in kilometers per second?

118. A student notices that the contents of a 12 fl oz can of diet soda weigh 340 g.
     a. What is the density, in grams per milliliter, of the soda?
     b. Will this can of soda float on water or sink?

119. A dairy wishes to deliver its milk in large trucks to reduce the cost of transportation. Assuming that the delivery truck has a weight of 6500 pounds, how many gallons of milk ($d = 1.106$ g/mL) could be placed into the truck so that it could still make it over the small country bridge rated to hold a maximum of 10.0 tons?

Thinking Beyond the Calculation

120. In a recent study of an allegedly pre-Columbian (before Columbus) map of the North American continent, one researcher claimed that the map was a forgery because the black ink had a yellow tinge (older ink would sometimes do this) that could have been made by first laying down the yellow line and then copying a thinner black line over it. However, another researcher analyzed the difference between the boundaries of the black and yellow lines and claimed, after many measurements, that the differences were consistently so small that it could not have been done freehand.
     a. Was the second researcher using precision or accuracy to make his claim? Explain.
     b. Describe how the second researcher used the scientific method to study the map.
     c. If the black lines had an average width of 1.45 mm. What would that width be in inches? ... in meters? ... in yards?

121. Biological evolution is a topic that often sparks debate. Some say it is one of several theories, and others say it is a fact. The field of "chemical evolution," in which chemists study how the elements and simple compounds (such as hydrogen, methane, and ammonia) combined to form the molecules of life (including proteins and DNA) has been active for nearly a century. Look up the pioneering work of Miller and Urey using Internet or print resources. Do their experiments convince you that chemical evolution is possible? What sorts of experiments would you design to determine whether chemical evolution occurs? If so, would you call it a theory or a fact? Why?

122. There has long been a debate about the relationship between the mathematical preparation of students and their success in first-year chemistry. The debate goes beyond standardized test scores to social issues. If you were going to study the relationship between math preparation and success in first-year chemistry, how would you approach the problem? Although you would need to consider all aspects of the scientific method in your study design, we are especially interested in three parts. What are the main questions you would ask? What experiments would you design to answer your questions? And how would you factor other, non-quantitative data into your results?

Atoms: A Quest for Understanding

Contents and Selected Applications

Chemical Encounters: How Old Is Life?

2.1   **Early Attempts to Explain Matter**

2.2   **Dalton's Atomic Theory and Beyond**

2.3   **The Structure of the Atom**

Chemical Encounters: The Dover Boat and Radiocarbon Dating

2.4   **Atoms and Isotopes**

2.5   **Atomic Mass**

2.6   **The Periodic Table**

2.7   **Ionic Compounds**

2.8   **Molecules**

2.9   **Naming Compounds**

2.10 **Naming Acids**

Fossils can provide evidence that life existed millions of years ago. New technologies have been developed that can "see" evidence that life on Earth is more than 3.5 billion years old.

Go to **college.hmco.com/pic/kelterMEE** for online learning resources.

How old is life? Scientists have been debating that issue for quite some time. Xenophanes of Colophon, an ancient Greek philosopher who died about 490 B.C., examined fossils of marine life to generate a hypothesis—a possible explanation based on observations—about the history of Earth. Since then, scientists have been using fossils to date life on this planet. Until recently, however, virtually all fossils of creatures that lived before about 2.5 billion years ago had been nearly impossible to detect.

In the early 1990s, high-powered microscopes were used to detect and identify fossils in rocks that were 3.5 billion years old. Apparently, life had existed on Earth earlier than anyone had yet imagined. In 1996 a new technique was developed to examine rocks for the biosignatures of life. Biosignatures—traces of life left behind when a creature dies—were identified using this technique in rocks 3.9 billion years old. Faint traces of life have also been suggested in a Martian meteorite found in Antarctica (see Figure 2.1) and also in rocks found in Greenland and Australia. Is this evidence clear proof that life on Earth began over 3.5 billion years ago? Some scientists believe it is, but the debate rages on. Questions are raised. Experiments are designed and done. As we discussed in the last chapter, this is the nature of vibrant science.

FIGURE 2.1

An inside look at a meteorite, known simply as ALH84001, that is presumed to have come from the surface of Mars. In 1996, scientist discovered what appeared to be microfossils inside this meteorite. The meteorite is estimated to be about 4.5 billion years old.

What is the new technique that can be used to find the biosignatures of life? All organisms, whether they are multicellular or single-celled, leave traces where they have lived. In some cases, these traces can be seen as fossils. In other cases, particularly the more ancient ones, the biosignature is simply a slight excess of one kind of a type of matter known as the atom. All organisms are made up of characteristic concentrations of specific types of atoms. By measuring larger-than-normal concentrations of these atoms, scientists can claim to have identified a biosignature of life. The particular biosignature that led to the discovery of life in the 3.9-billion-year-old rock is a larger-than-normal concentration of carbon-12.

What is carbon-12? It is a specific set of particles that make up one type of carbon atom. Although several different sets of particles can give rise to a carbon atom, only one of these produces the carbon-12 isotope. In this chapter, we'll learn about the particles that make up the building blocks of matter called atoms. We'll uncover the basic structure of an atom, and we'll learn how atoms are arranged in the most important organizational chart used in chemistry: the periodic table of the elements.

2.1 Early Attempts to Explain Matter

As early as 600 B.C., philosophers developed a view that the world was made of a few basic "elements"—earth, air, water, and fire. The Greek philosopher Democritus (Figure 2.2) wondered how he could be sitting in one part of his house and detect that bread was baking elsewhere in the house. Could it be that small particles of the bread were breaking away from the loaf and traveling through the air and into his nose? Noticing that wet clothing gradually got drier and lighter led to the explanation that small, invisible pieces of water were gradually leaving the clothing. What was the smallest possible piece of this matter? In Democritus' culture, these pieces of matter were thought to be indivisible, unable to be broken down further, or uncuttable. In their language, the particles of the basic elements were *atomos*. This is the origin of the modern word *atom*.

The lack of certain instruments that we use routinely today made it impossible, in Democritus' time, to perform the experiments that have led to our current understanding of matter. For instance, the invention of the balance for measuring mass was one of the most significant experimental advances in the history of science. In fact, two important chemical laws were discovered with the use of the balance. These laws helped to shape our current understanding of the ideas put forth in Democritus' era.

The Law of Conservation of Mass

The French scientist Antoine Lavoisier (1743–1794) used the most precise balances of his time, shown in Figure 2.3, to examine a variety of chemical processes (chemical reactions). From his repeated measurements, Lavoisier concluded that the total mass of the **reactants** (what is consumed in a reaction) did not differ from the total mass of the **products** (what is produced in the reaction). In other words, *matter is neither created nor destroyed in a chemical reaction*. We call this conclusion the **law of conservation of mass**.

Strictly speaking, this law is not precisely true because of the tiny changes in mass resulting from something we'll discuss in Section 21.5: the mass-energy equivalence discovered by Albert Einstein. However, the law of conservation of mass is accurate enough to indicate a fundamental truth about chemical reactions: *The atoms with which you start are the same as the ones with which you end*. When atoms participate in chemical reactions, they are not destroyed or replaced by newly made atoms but, rather, are *rearranged*. DNA, the genetic code in our cells, is composed of atoms of carbon, hydrogen, oxygen, nitrogen, and phosphorus. Arranged in one way, these atoms are the genes that lie at the heart of the chemistry of life. Arranged in other ways, they are a mixture of water, carbon dioxide gas, and a fertilizer called ammonium phosphate. Arranged in still other ways, they form parts of biosignatures from billions of years ago.

The Law of Definite Composition

Another French scientist, Joseph Louis Proust (1754–1826; see Figure 2.4), studied the composition of various minerals. In 1799 he published his work about a compound that contained copper, carbon, oxygen, and hydrogen. His results indicated that every time he determined the composition of his mineral, *the components of the compound were always present in the same*

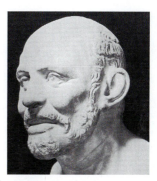

FIGURE 2.2

The Greek philosopher Democritus (460 B.C.–370 B.C.). His early deductions about the composition of the basic components of nature led him to believe in unseen and uncuttable particles, called *atomos*.

FIGURE 2.3

Using balances such as these, Lavoisier and others were able to reach conclusions about the nature of chemical reactions.

FIGURE 2.4

Joseph Proust (1754–1826) was born in Angers, France. He was a professor at different institutions in Spain. In 1797 he published his understanding of the Law of Definite Proportions. After returning to France in 1808, he spent considerable time investigating foodstuffs and discovered the molecule leucine, one of the building blocks of proteins.

FeS (iron pyrite).

ratio, by mass. This is called the **law of definite composition**. This law is true for all compounds, irrespective of the source of the compound. We can illustrate the law of definite composition using a compound containing the elements iron and sulfur, called iron(II) sulfide. If we analyze the composition of two different samples of iron(II) sulfide, we find that their mass ratios are identical.

Result from Analysis of Two Samples of Iron(II) Sulfide

Sample A: Compound mass = 42.0 g

Sample B: Compound mass = 100.0 g

Mass Ratios

Sample A: 26.7 g Fe / 15.3 g S = 1.74:1.00 = ratio of g Fe to g S

Sample B: 63.5 g Fe / 36.5 g S = 1.74:1.00 = ratio of g Fe to g S

We can also think of this law in the reverse fashion. Suppose we were interested in making an 80.0-g glass of water from its elements, hydrogen and oxygen. Given that the mass ratio of oxygen to hydrogen in water is 7.99:1.00, we would need to chemically combine 1.00 g of hydrogen for every 7.99 g of oxygen to make the water. In our specific example, then, we would need 8.90 g of hydrogen and 71.1 g of oxygen to make 80.0 g of water. Water is water; it is never made up of a different mass ratio of its components.

Preparation of Water (H_2O)

Hydrogen	8.90 g
Oxygen	71.1 g
Water	80.0 g

Mass Ratio

Water: 71.1 g oxygen / 8.90 g hydrogen = 7.99:1.00 g oxygen to g hydrogen

---

EXERCISE 2.1   **Investigating the Law of Definite Composition**

Suppose that you collected two samples of pure bottled water.

a. If you separated the water into hydrogen and oxygen and obtained the following results, what would you report as the ratio of the mass of oxygen to that of hydrogen in each sample?

>   Sample 1: 37.3 g of oxygen; 4.67 g of hydrogen
>
>   Sample 2: 69.3 g of oxygen; 8.67 g of hydrogen

b. Do these data support the law of definite composition?

c. If you obtained another sample of water and found it to contain 17.0 g of oxygen, how many grams of hydrogen would you expect to obtain?

Solution

a. The ratios of oxygen to hydrogen in these two samples are

$$\text{Sample 1:} \quad \frac{37.3 \text{ g oxygen}}{4.67 \text{ g hydrogen}} = \frac{7.99 \text{ g oxygen}}{1.00 \text{ g hydrogen}}$$

$$\text{Sample 2:} \quad \frac{69.3 \text{ g oxygen}}{8.67 \text{ g hydrogen}} = \frac{7.99 \text{ g oxygen}}{1.00 \text{ g hydrogen}}$$

b. Yes, the ratios are the same for different masses of the same compound, so these data support the law of definite composition.

c. Because the ratio of oxygen to hydrogen is constant in the compound called water, we would expect to have a ratio of 7.99 g of oxygen to 1.00 g of hydrogen. Using words, we say, "7.99 grams of oxygen is to 1.00 gram of hydrogen as 17.0 grams of oxygen is to how many grams of hydrogen?" Using equations, we can write

$$\frac{7.99 \text{ g oxygen}}{1.00 \text{ g hydrogen}} = \frac{17.0 \text{ g oxygen}}{? \text{ g hydrogen}}$$

Rearranging yields

$$\frac{17.0 \text{ g oxygen} \times 1.00 \text{ g hydrogen}}{7.99 \text{ g oxygen}} = ? \text{ g hydrogen}$$

$$= 2.13 \text{ g hydrogen}$$

PRACTICE 2.1

Given the information in Exercise 2.1, determine how many grams of oxygen will combine with 24.5 g of hydrogen to make water. How many grams of water will be made?

See Problems 9 and 10.

The law of conservation of mass and the law of definite composition provide two fundamental descriptions of the behavior of chemical processes. When they were discovered, there was no satisfactory explanation for them. After all, they are just chemical laws that describe observations of the natural world. Many scientists searched for explanations of these two laws. One of these was the English scientist John Dalton (1766–1844; see Figure 2.5).

FIGURE 2.5

John Dalton, a poor Quaker school-teacher and brilliant amateur meteorologist, developed the basic points of the atomic theory. He showed early signs of brilliance and even began teaching others in his small English hometown when he was only ten years old.

Video Lesson: Early Discoveries and the Atom

2.2   Dalton's Atomic Theory and Beyond

John Dalton performed experiments on compounds in which two elements made more than one type of compound. This is not unusual. There are many examples of compounds that contain the same elements but in different ratios. For example, we can combine oxygen and nitrogen to make many different compounds, three of which are shown in Table 2.1. Note how each has a different mass of oxygen that combines with a given mass of nitrogen.

The results demonstrate a general law formulated by Dalton: When the same elements can produce more than one compound, *the ratio of the masses of this element that combine with a fixed mass of another element is a small whole number.* This is known as the **law of multiple proportions**. Using the information in Table 2.1, we see that in nitric oxide, 14.0 g of nitrogen combines with 16.0 g of

TABLE 2.1	Compounds Containing Nitrogen and Oxygen: Comparing Oxygen to Oxygen		
Common Name	**Fixed Mass of Nitrogen**	**Mass of Oxygen**	**Ratio of Mass of Oxygen in Compound to Mass of Oxygen in Nitric Oxide**
Nitric oxide	14 g	16 g	1:1
Nitrogen dioxide	14 g	32 g	2:1
Nitric anhydride	14 g	40 g	5:2

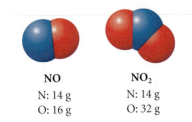

NO	**NO$_2$**
N: 14 g	N: 14 g
O: 16 g	O: 32 g

FIGURE 2.6

The ratio of the mass of oxygen in NO$_2$ to its mass in NO is a simple whole number.

TABLE 2.2	**Dalton's Atomic Theory**

- Every substance is made of atoms.
- Atoms are indestructible and indivisible.
- Atoms of any one element are identical.
- Atoms of different elements differ in their masses.
- Chemical changes involve rearranging the attachments between atoms.

oxygen. A different compound of these same two elements exists in which 14.0 g of nitrogen combines with 32.0 g of oxygen. The ratio of the masses of oxygen that combine with *the same mass of nitrogen* (14.0 g, in this case), is 32.0 g/16.0 g, or 2:1, a small whole number. The same ratio in nitric anhydride is 40.0 g/16.0 g, or 2.5:1. This is 5:2 in small whole numbers.

Why is the ratio of the components in such compounds based on small whole numbers? The implication of these results is illustrated in Figure 2.6. Dalton wondered whether these results could be explained by the existence, for each element, of some basic particle with a specific mass that would be the smallest part of every element. Moreover, his experiments were giving results that were entirely consistent with this idea. In 1803 he presented the results of his experiments before the Manchester Literary and Philosophical Society. His ideas, which came to be known as **Dalton's atomic theory**, are summarized in Table 2.2.

Dalton's atomic theory was a keystone in the foundation of chemistry, but as we shall see, not all of it is correct. However, his theory teaches us something very important about the scientific method: Ideas don't need to be completely correct to be very useful and influential. Dalton was limited by the measurements he was able to make at the turn of the nineteenth century. Our ability to know the mass of objects is orders of magnitude better now than in Dalton's time, and our understanding is therefore significantly better. Still, we view Dalton's atomic theory as a great step forward in chemistry, because it focused attention on the valid idea that compounds are made from little bits—atoms—combining in fixed proportions and that chemical reactions are rearrangements of these atoms.

Visualization: Oxygen, Hydrogen, Soap Bubbles, and Balloons

The Law of Combining Volumes

One example of the refinement of ideas in the light of new information is provided by the work of the French chemist Joseph Louis Gay-Lussac (1778–1850; see Figure 2.7). Gay-Lussac carefully measured the changes in volume when gases reacted to produce other gases. His results are summarized in what became known as the **law of combining volumes**, which states that *when gases combine, they do so in small whole-number ratios*, such as 1:1, 1:2, 1:3, and 3:2, provided that all the gases are at the same temperature and pressure. One example is illustrated in Figure 2.8.

FIGURE 2.7

Joseph Louis Gay-Lussac (1778–1850) was the eldest of five children. His colleagues considered him a careful, elegant experimentalist. He introduced the terms *pipette* and *burette* and was the first to isolate the element boron. He is most noted for publishing Charles's law (see Chapter 10) and formulating the law of combining volumes.

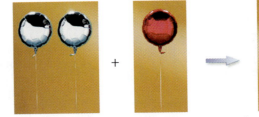

2 balloons of H$_2$ 1 balloon of O$_2$ 2 balloons of H$_2$O

FIGURE 2.8

Oxygen and hydrogen combine to form water vapor. According to the law of combining volumes, two volumes of hydrogen react with one volume of oxygen to make two volumes of water vapor.

Dalton could not accept the results in that example because it contradicted his understanding of what would have to happen at the level of individual particles. To fit with Gay-Lussac's law, it seemed to Dalton that one oxygen atom would have to split in two in order to react with two hydrogen atoms and produce two particles of water. Dalton's atomic theory said that atoms could not be split in two, so he presumed that either Gay-Lussac's results or his reasoning must be flawed. The apparent difficulty disappears, however, when we realize that oxygen gas consists of oxygen molecules (O_2), each composed of two oxygen atoms bonded together. We also now know that hydrogen gas is composed of H_2 molecules and that each water molecule contains two hydrogen atoms and one oxygen atom (H_2O). One molecule of oxygen can combine with two molecules of hydrogen to create two molecules of water as shown in Figure 2.9.

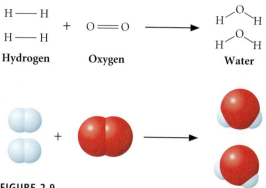

FIGURE 2.9

It is because the elemental forms of hydrogen and oxygen exist as diatomic substances that the law of combining volumes makes sense at the molecular level.

HERE'S WHAT WE KNOW SO FAR

The information gleaned by applying the scientific method to investigate the makeup of matter has provided us with a basic look at the nature of chemistry. Specifically, we know that

- Matter can be neither created nor destroyed in a chemical reaction.
- Matter is composed of small indestructible particles called atoms.
- Compounds are made by combining atoms in whole-number ratios.
- The components of a compound are always present in the same ratio, by mass.
- A chemical reaction rearranges the attachments that hold atoms together in a compound.

These points illustrate the scientific laws identified by Gay-Lussac, Lavoisier, Proust, and Dalton. In addition, Dalton's atomic theory helps to solidify our understanding of these laws and of the nature of the atom. In the next sections of this chapter, we will begin to examine the makeup of the atom.

2.3   The Structure of the Atom

Archaeology, the study of past human activities, often requires scientists to be able to assign ages to artifacts that have been unearthed. In some cases, such relics can be dated because they are buried underneath objects for which an age is known. In other cases, the archaeologists must perform laboratory experiments to estimate the age of the artifact. One exciting discovery that helped provide a wealth of information about ancient boat construction was found buried under 6 feet of earth in Dover, England. The "Dover Boat," as it has become known, was discovered underneath the footings of an ancient city wall (Figure 2.10).

Archaeologists knew the date that the wall was constructed from historical records and, judging from its location, surmised that the boat was probably older than the wall under which it was buried. To get an accurate age of the boat, though, they needed to perform "radiocarbon dating" on small pieces of wood from the boat. **What is the basis of this process, and what can it teach us about atomic structure?**

Some of the carbon atoms in living things emit certain particles and energy (radioactivity) that allows the atoms to become more energetically stable (see Chapter 21). It has been observed that the amount of radioactive carbon remains relatively constant while the organism is alive, as life's processes allow the

Application

CHEMICAL ENCOUNTERS:
The Dover Boat and
Radiocarbon Dating

FIGURE 2.10

The hull of the Dover Boat is partially excavated in this figure. Note the construction of the rails running down the center of the boat. These rails were used to fasten the two halves of the boat together.

exchange of atoms between, for example, a tree and its environment. Once death occurs, however, the radioactive carbon in the tree begins to diminish. If you can measure the amount of radioactive carbon remaining in a piece of wood, you can estimate the year in which the tree died. The results of radiocarbon dating on the Dover Boat indicated that it was constructed around 1550 B.C.

The previous discussion implies that some carbon atoms (those that are radioactive) are different from other carbon atoms (those that are not radioactive.) Dalton's claim that all atoms of any one element are identical is not correct. **What is it about the structure of the atom that gives rise to more than one type of the same element?** A brief look back in time will help us answer this question.

Electrons, Protons, and Neutrons

Video Lesson: Understanding Electrons

In the late 1800s and early 1900s, scientists were able to make measurements that changed our fundamental understanding of the structure of the atom. In the 1880s, Svante Arrhenius, a Swedish researcher working at the Academy of Sciences in Stockholm, found that some solutions contained large numbers of "electrically charged atoms." Arrhenius found, for example, that when copper chloride was placed in water and the positive and negative terminals of a power supply were immersed in the solution, copper collected at one electrode and chlorine collected at the other. He concluded that some atoms were themselves negatively charged and others positively charged, causing each to be attracted to the electrode carrying the opposite charge. This work provided some of the earliest evidence for the existence of what we now call ions.

In 1891, the English scientist G. Johnstone Stoney examined the physical properties of electricity. To represent a distinct unit of electricity, he proposed the name **electron**, from the Greek word *elecktron* (meaning "amber"), because rubbing a rod of amber on wool gives rise to what we now call static electricity. Were electrons small, charged pieces of atoms? If so, then atoms could no longer be thought of as small, *indestructible* or *indivisible* units of matter!

Many scientists noted that the discovery of the electron suggested that Dalton's atomic theory was inadequate. One such scientist, Joseph John ("J. J.") Thomson (1856–1940; see Figure 2.11), reasoned that if Dalton's ideas about the indestructibility of an atom were true, how could negatively charged parts of atoms (the electrons) be released from the atom? Furthermore, if negative particles were present, wouldn't positively charged pieces of atoms also exist? It makes sense that an atom with negatively charged particles inside must, in order to be

Static electricity.

FIGURE 2.11

Joseph John ("J. J.") Thomson (1856–1940) provided the first model of the structure of the atom, known as the plum pudding model. 1906, "in recognition of the great merits of his theoretical and experimental investigations on the conduction of electricity by gases," he was awarded the Nobel Prize in physics.

FIGURE 2.12

The plum pudding model of the atom.

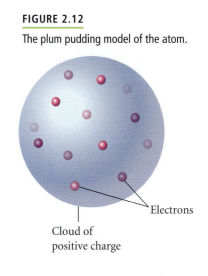

Electrons

Cloud of positive charge

electrically neutral overall, have a balancing positive charge associated with it. Thomson envisioned such an atom. In 1904, he wrote that "the atoms of the elements consist of a number of negatively electrified corpuscles (electrons) enclosed in a sphere of uniform positive electrification." His model of the atom (shown in Figure 2.12) was known as the "plum pudding model." This model helped scientists understand what an atom might look like.

Video Lesson: Understanding the Nucleus

Radioactivity and the Structure of the Atom

In 1909 Ernest Rutherford (1871–1937; see Figure 2.13), a New Zealand–born physicist and former student of J. J. Thomson, and his student Ernest Marsden (1889–1970), performed experiments in which positively charged radiation was directed toward a thin sheet of gold foil (Figure 2.14). If Thomson's plum pudding model of the atom were correct, most of the positively charged radiation would be expected to pass through the foil and undergo slight deflections as its positive charge interacted with the negatively charged electrons scattered throughout the atoms. Instead, most of the radiation went straight through the foil without deflection, some of the radiation was scattered at very wide angles, and—most amazingly—some was deflected nearly *straight back toward its source,* as shown in Figure 2.15. In 1911 Rutherford explained this startling result. His explanation led to a new model of the atom in which a small region of very concentrated positive charge existed within a large area of mostly empty space containing negatively charged electrons (see Figure 2.15).

Rutherford used the term **proton** (from the Greek *protos,* meaning "first") to describe the nucleus of the hydrogen atom. However, his calculations of the mass

FIGURE 2.13

Ernest Rutherford (1871–1937) was born in New Zealand after his family emigrated there from Scotland in the mid-1800s. As a student of Thomson at Cambridge, he developed an instrument that could detect electromagnetic radiation. Later, while he worked at McGill University (Montréal, Canada) and at the University of Manchester (England), he investigated the actions of alpha particles. For his discoveries, he was awarded the Nobel Prize in chemistry in 1908.

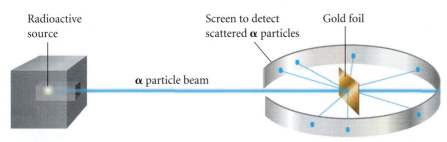

Radioactive source

Screen to detect scattered **α** particles

Gold foil

α particle beam

FIGURE 2.14

The Rutherford gold foil experiment. Positively charged radiation is directed toward a small piece of gold foil. Although most of the radiation passes through the foil, a noticeable percentage is scattered backward by the foil. This led Rutherford to develop a new model of the atom.

Visualization: Gold Foil Experiment

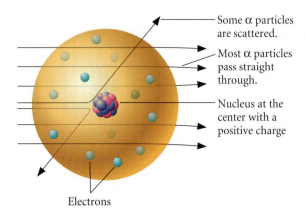

Some α particles are scattered.

Most α particles pass straight through.

Nucleus at the center with a positive charge

Electrons

FIGURE 2.15

Some of the positively charged radiation interacts with the positive nucleus of the atom.

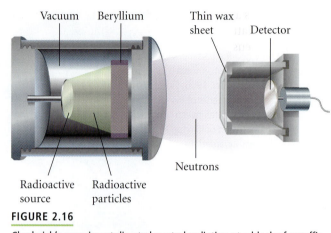

Vacuum   Beryllium   Thin wax sheet   Detector

Radioactive source   Radioactive particles

Neutrons

FIGURE 2.16

Chadwick's experiment directed neutral radiation at a block of paraffin wax. The energy of the particles that resulted from the collision of the radiation and the wax led to the discovery of the neutron. The radiation was generated by placing radioactive polonium near a sheet of beryllium metal. The wax, shaped into a thin sheet, was placed in front of the detector.

of an atom made from a nucleus of protons didn't agree with the experimentally determined mass of the atom. He theorized that some other kind of particle must be in the nucleus too and suspected that it was probably electrically neutral because the positive electric charge on the nucleus already balanced the negative charge of the electrons.

In 1932 James Chadwick (1891–1974) set up an experiment that allowed radiation from beryllium atoms to strike a block of paraffin wax. This resulted in the loss of protons from the wax. After calculating the force that would be needed to cause the ejection of the nuclei from the sample of paraffin, Chadwick realized that the radiation from the beryllium must be electrically neutral and have a mass approximately equal to that of the proton. Chadwick had discovered the **neutron**. The diagram in Figure 2.16, reproduced from the original article in which he published his results, illustrates the process that Chadwick used.

These discoveries also led to understanding of the different forms of radiation. In short, some atoms are unstable and can undergo **radioactive decay** by spontaneously emitting high-energy **radiation**, sometimes accompanied by fast-moving particles. Collectively, this phenomenon is known as **radioactivity**. The three most common types of radioactivity are **gamma rays** (γ rays), **alpha particles** (α particles) and **beta particles** (β particles) (Figure 2.17).

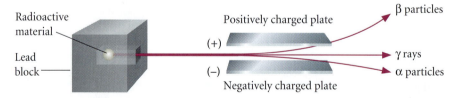

Radioactive material

Lead block

Positively charged plate

(+)

(−)

Negatively charged plate

β particles

γ rays

α particles

Visualization: Nuclear Particles

FIGURE 2.17

Radiation emitted from a radioactive element placed in a lead block. Alpha particles and beta particles are deflected by an electric field. Alpha particles are positively charged and therefore are pulled toward a negatively charged plate. Beta particles are negatively charged and hence are pulled toward a positively charged plate. Gamma rays are not charged and therefore are not deflected by the electric field.

Gamma rays are a very energetic form of electromagnetic radiation, with the same physical nature as visible light but a much higher energy. An alpha particle is just the nucleus of the helium atom. It is composed of two protons and two neutrons and carries a total charge of +2. Alpha particles were the type of radiation directed toward the foil in Rutherford's experiments. Beta particles are fast-moving electrons released from the nucleus of an atom, each one carrying its characteristic –1 charge.

The penetrating powers of radioactivity are put to many uses in modern life, including killing cancer cells, testing the integrity of welds in metal, and looking for hairline cracks in aircraft airframes. We will discuss the concepts and applications of radioactivity much more fully in Chapter 21.

HERE'S WHAT WE KNOW SO FAR

- Atoms can be broken down into smaller parts. Those parts include the proton, neutron, and electron.
- The proton and neutron occupy the center of the atom, called the nucleus.
- The electrons occupy the space around the nucleus.
- Protons and electrons have opposite charges; protons are positively charged and electrons are negatively charged.
- The total mass of the atom is the sum of all of the particles that make up the atom.
- Radiation is a result of the decay of atoms. Different decay products exist, such as the alpha particle, the beta particle, and the gamma ray.

Through the work we've just described, along with that done by other scientists of the time, we know that the nucleus of an atom contains the protons and neutrons. The proton has a tiny positive electrical charge (1.6×10^{-19} **coulombs**) and a similarly minute mass (1.6726×10^{-27} kg). Neutrons do not carry a charge and are a tiny bit more massive than the protons (1.6749×10^{-27} kg). The electrons travel around the remaining space in the atom. Although they are about 2000 times less massive (9.1094×10^{-31} kg) than the particles in the nucleus, they occupy the majority of the atomic volume, as shown in Figure 2.18. The electrons also carry a negative electrical charge, equal in magnitude to, but opposite in sign from, that on the proton. To simplify counting charges, we often report the charge on the proton as +1 and that on the electron as –1 (Table 2.3).

This model of the structure of the atom is sufficiently useful for us to understand the main differences among atoms and the most fundamental principles of chemical change. However, we must realize that it is just a model. Chemists use various models, which represent important aspects of reality in useful ways, but none of their models is likely to be entirely correct. Yet each model has a place in the toolbox of the chemist. We can use this model of the atom whenever it is most useful; and we can use more sophisticated models, about which we'll learn later, whenever those would be most useful. What model we use depends on what questions we are trying to answer.

FIGURE 2.18

The atom is mostly empty space. The nucleus constitutes about 1/10,000 the diameter of the atom.

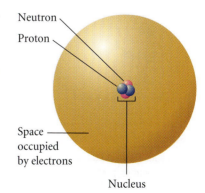

Neutron
Proton
Space occupied by electrons
Nucleus

TABLE 2.3	Subatomic Particles		
Particle	**Mass**	**amu***	**Charge**
Electron	9.1094×10^{-31} kg	5.486×10^{-4}	−1
Proton	1.6726×10^{-27} kg	1.0073	+1
Neutron	1.6749×10^{-27} kg	1.0087	0

*The atomic mass unit (amu) is an arbitrary unit used to set the masses reported in the periodic table. We will discuss it in the next section.

2.4 Atoms and Isotopes

The **periodic table of the elements** is arguably the single most important practical outcome in the history of chemistry. Because it is an essential tool for practicing chemists as well as chemistry students, it is located on the inside front cover of this book for your easy reference. The periodic table lists the atoms of just over 90 elements that are known to occur naturally, as well as a smaller number of heavy elements that have been synthesized.

One of the great beauties of the periodic table is that all of the atoms of the elements in the table are constructed of the same building blocks: protons and neutrons surrounded by a sea of electrons. Are the numbers and types of particles important in constructing an atom? The short answer is a definite "yes." The numbers of protons, neutrons, and electrons define each individual atom. The identity of the atom is determined by the quantity of protons in the nucleus, which is known as the **atomic number (Z)**. For instance, all hydrogen atoms have one proton ($Z = 1$), all helium atoms have two protons ($Z = 2$), and all iron atoms have twenty-six protons ($Z = 26$). Just as your university identification number uniquely identifies you at your school, the atomic number unequivocally identifies a specific element in chemistry. Potassium has an atomic number of 19, and any element with nineteen protons must be potassium.

The number of electrons in an *electrically neutral* atom is equal to the number of protons and, therefore, to the atomic number. However, when combined with atoms of other elements, an atom can gain or lose electrons to form a charged entity we call an **ion**. Although the number of electrons that an atom contains changes, the atomic number—the number of protons it contains—remains the same. For example, the sodium ion shown in Figure 2.19 is still sodium, even though its electron count is no longer the same as that of a neutral sodium atom. The key question is **How many protons does the atom contain?** *The number of protons always identifies the element.*

A positive charge indicates that there are more protons than electrons on the ion. We typically refer to this type of ion as a **cation** (pronounced CAT-ion). Conversely, a negative charge indicates that there are more electrons than protons. This type of ion is known as an **anion** (pronounced AN-ion). If we know the charge and the atomic number of an ion, we can determine the number of electrons that make up the ion.

2 electrons
2 protons
2 neutrons

He

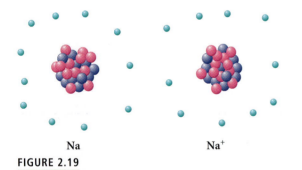

Na   **Na⁺**

FIGURE 2.19

The sodium atom and the sodium ion. Both of these combinations of electrons, protons, and neutrons are sodium, despite the difference in the number of electrons.

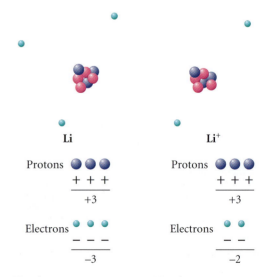

Li   **Li⁺**

Protons   ● ● ●   Protons   ● ● ●
        + + +   + + +
         +3   +3

Electrons   ● ● ●   Electrons   ● ●
        − − −   − −
         −3   −2

Net charge = +3 − 3 = 0   Net charge = +3 − 2 = +1

We call the sum of the number of protons and neutrons the **mass number (A)** of the atom. The number of neutrons (n) in a particular atom can be calculated by subtracting the atomic number from the mass number:

$$A - Z = n$$

These basic components of the atoms can be written using the chemist's shorthand of **nuclide notation**, as shown in Figure 2.20. Nuclide notation lists the symbol for the element, accompanied by the atomic number and the mass number of the atom.

Rather than requiring that we write numbers for all of the particles that make up an atom or ion, nuclide notation provides a quick way to convey the quantity of subatomic particles to other scientists. An alternative way to quickly write the number of particles is to write the name of the element, followed by the mass number of the element. For example, when we wrote "carbon-12" earlier in this chapter, we were indicating an atom containing 6 protons, 6 neutrons, and 6 electrons.

Isotopes

Let's reexamine the radiocarbon dating of the Dover Boat. From that discussion we learned that there must be at least two different types of carbon atoms. The discovery of a biosignature in a rock formation is related to this fact as well. What is different about the carbon atoms?

All atoms of an element have the same number of protons, but the number of neutrons that make up the nucleus of an atom can differ. Atoms of the same element must have the same atomic number (the same number of protons); otherwise, they would be different elements. However, if *they differ in the number of neutrons they contain*, they are known as **isotopes**. There are actually three common isotopes of carbon, the most abundant (98.93%) of which is carbon-12. Carbon-13 is another nonradioactive isotope of the element carbon; it makes up only a small amount (1.07%) of the carbon atoms in the universe. The carbon-14 isotope (2×10^{-10}% of all carbon atoms) is radioactive and is the element used to determine the age of ancient artifacts.

FIGURE 2.20

Nuclide notation.

Mass number

A **E** — Element symbol

Z

Atomic number

$$A \overline{} 7$$
$$Z \overline{} 3 \text{ Li}$$

$A - Z$ = number of neutrons
7 – 3 = 4 neutrons

The number of neutrons in a particular atom can be determined by subtracting the number of protons (Z) from the atomic mass number (A).

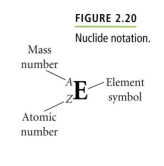

Video Lesson: Modern Atomic Structure

Tutorial: Isotopes

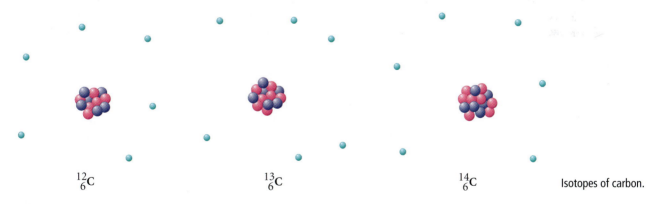

$^{12}_{6}$C $^{13}_{6}$C $^{14}_{6}$C   Isotopes of carbon.

Many examples of isotopes exist among the elements of the periodic table. In fact, it is rare for an element *not* to have isotopes. Some of the isotopes of hydrogen, carbon, and oxygen are shown in Table 2.4 in nuclide notation. Also included in the table is their relative natural abundance. For example, 99.985% of all hydrogen atoms are hydrogen-1, 0.015% are deuterium (the same single proton as hydrogen, but with one neutron), and only 10^{-18}% of them are tritium (one proton and *two* neutrons).

Many nuclei that have either too many or too few neutrons to be energetically stable tend to give off energy and particles to become stable. This is the reason for the radioactivity of carbon-14.

TABLE 2.4	Some Isotopes of Hydrogen, Carbon, and Oxygen				
Nuclide Notation	Relative Natural Abundance (%)	Name of Isotope	Protons	Neutrons	Electrons
$^{1}_{1}H$	99.985%	hydrogen	1	0	1
$^{2}_{1}H$	0.015	deuterium	1	1	1
$^{3}_{1}H$	10^{-18}	tritium	1	2	1
$^{12}_{6}C$	98.93	carbon-12	6	6	6
$^{13}_{6}C$	1.07	carbon-13	6	7	6
$^{14}_{6}C$	2×10^{-10}	carbon-14	6	8	6
$^{16}_{8}O$	99.762	oxygen-16	8	8	8
$^{17}_{8}O$	0.038	oxygen-17	8	9	8
$^{18}_{8}O$	0.200	oxygen-18	8	10	8

EXERCISE 2.2 **Atomic Bookkeeping**

Fill in the blanks for each neutral element in the following table, and write the elemental symbols using nuclide notation.

Element	Mass Number	Protons	Neutrons	Electrons
silicon	28	14		
molybdenum	96		54	
krypton			48	36

First Thoughts

To fill in our table, we will use the relationships that define the mass number and atomic number for an element. Specifically, we recall that

$$A - Z = n$$

A glance at the periodic table will be helpful, because the name of the element indicates the atomic number of the element. On the periodic table inside the front cover of this book, we can verify that the atomic number of silicon is in fact 14. We know that this is equal to the number of protons in atoms of the element. In the blanks for molybdenum, we are asked to find the number of protons and electrons present in an atom of the neutral element. We are given the number of neutrons (n) and the mass number (A), the sum of protons and neutrons. We can take the difference to solve for the number of protons (Z):

$$A - n = Z$$

This is also equal to the number of electrons in the electrically neutral atom. The key to filling in the blanks for krypton is to remember that the number of protons equals the number of electrons in a neutral atom.

Solution

Element	Mass Number	Protons	Neutrons	Electrons
silicon	28	14	**14**	**14**
molybdenum	96	**42**	54	**42**
krypton	**84**	**36**	48	36

Further Insights

As we go through the various numerical features of the elements, such as mass number and numbers of protons, neutrons, and electrons, please keep in mind that these numbers represent one of two ways in which we characterize the elements. The other way, which ultimately led to the formation of the periodic table of the elements, is via the physical and chemical characteristics of the elements.

PRACTICE 2.2

Indicate the number of protons, neutrons, and electrons in atoms of each of these elements:

$^{14}_{7}N$ $^{20}_{10}Ne$ $^{48}_{22}Ti$ Carbon-11   Lithium-7   Phosphorus-31

See Problems 25, 26, 35–40, 45, and 46.

2.5   Atomic Mass

Plants and other organisms have a slightly greater preference for the lighter isotopes of the carbon that they use as a raw material. When a plant absorbs carbon dioxide to make cell walls, it more efficiently uses the molecules that contain the carbon-12 isotope than those that contain carbon-13 or carbon-14. The preference isn't great at all, but the plant absorbs a slightly larger proportion of carbon-12 than of the other carbon isotopes. When the organism dies, it leaves behind some residue that is indicative of this preference. How does the scientist in search of the origins of life determine the amount of carbon-12 versus the amount of carbon-13 in a rock formation?

The distribution of isotopes in a sample of an element can be found using an instrument called a **mass spectrometer**. There are different kinds of mass spectrometers, but the simplest kind is illustrated in Figure 2.21. In the instrument, a sample is vaporized and then converted into positive ions when a beam of fast-moving electrons knocks some of the electrons out of the atoms of the sample. The ions that are formed are accelerated by their attraction to charged plates. Holes in these plates allow some of the ions to be projected through and into a magnetic field. The interaction between the charge on the ions and the magnetic field causes the ions to be deflected from their original path. *The extent of this deflection depends on the masses and charges of the ions.* The more massive an ion is, the less deflection it experiences, because of the greater momentum it possesses when heading out in its original straight-line path (much as a very heavy adult would not be blown so far off course by a fierce, gusty wind as a small child would).

Some atoms have two of their electrons knocked away by the beam that strikes them, converting them into cations with a +2 charge. These ions deflect

Application

Video Lesson: Mass Spectrometry: Determining Atomic Masses

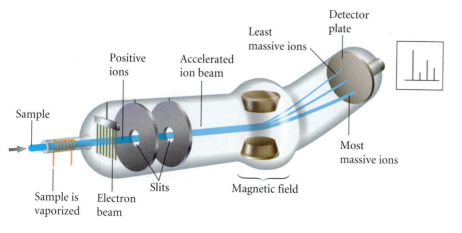

FIGURE 2.21

The mass spectrometer separates ions by exploiting their deflection to different extents in a magnetic field.

more than otherwise identical ions with a single positive charge because of the greater interaction between their charge and the magnetic field. Within a very short distance, the original coherent beam of ions spreads out into beams of distinct ions, each with a different mass-to-charge ratio.

At the end of the mass spectrometer, the beams of ions are detected. The results are displayed in a **mass spectrum**, as shown in Figure 2.22. Each line on the spectrum indicates the relative quantity of the different ion beams that result from the original sample. Using a standard that produces an ion beam with a known mass, we can reproduce the scale on the bottom of the mass spectrum.

In much the same way, we use a standard to determine the masses of the atoms in the periodic table. Which element is used as the standard? The most abundant isotope of carbon, ^{12}C, is defined as *exactly* 12 **atomic mass units (amu)**.

$$^{12}C = 12.000000000 \text{ amu}$$

An atomic mass unit, an arbitrary unit used to set the masses reported in the periodic table, is currently defined as follows:

$$1 \text{ amu} = 1.66053873 \times 10^{-27} \text{ kg} \qquad 1 \text{ kg} = 6.02214198 \times 10^{26} \text{ amu}$$
$$1 \text{ g} = 6.02214198 \times 10^{23} \text{ amu}$$

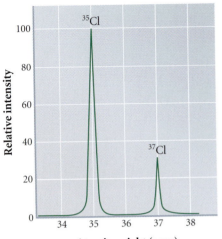

FIGURE 2.22

The mass spectrum of chlorine. From this spectrum we can determine the relative abundances of ^{35}Cl (75.77%) and ^{37}Cl (24.23%).

Strictly speaking, the amu is not an SI unit, but its use remains sufficiently common among scientists that we will use it in this textbook.

A detailed look at the mass of individual isotopes shows that they are typically *not* integer values. That is, the isotope ^{63}Cu does not have a mass of exactly 63. Only ^{12}C is defined with an exact mass. The isotope ^{63}Cu has a mass *number* of 63 but a mass of 62.9296011 amu. The isotope ^{96}Mo has a mass of 95.9046789 amu. Precise measurements have shown that neither a proton (roughly 1.0073 amu) nor a neutron (roughly 1.0087 amu) has a mass of 1 amu. Therefore, the isotope ^{238}U, which has an atomic mass of 238.0508 amu, has an atomic mass lower than the sum of the protons and neutrons, which have a total mass of 239.9418 amu, as shown.

$$\text{Mass of 92 individual protons of } ^{238}\text{U}$$
$$= 92 \times 1.0073 \text{ amu} = 92.6716 \text{ amu}$$

$$\text{Mass of 146 individual neutrons of } ^{238}\text{U}$$
$$= 146 \times 1.0087 \text{ amu} = 147.2702 \text{ amu}$$

$$\text{Total mass of individual protons and neutrons of } ^{238}\text{U} = 239.9418 \text{ amu}$$

Where did the rest of the mass go? When the nuclei of elements formed from the individual neutrons and protons, some of the mass—a different amount for each isotope of each element—was released as energy (see Section 5.7).

Unfortunately, the individual masses don't translate accurately into masses that we can measure for a macroscopic quantity of an element found on a laboratory shelf. The presence of isotopes complicates the mass even more. To illustrate this, assume we have a bottle of carbon powder. What mass should we use for the typical carbon atom found in this bottle, given that three different isotopes of carbon exist: carbon-12, carbon-13, and carbon-14? Fortunately, in nature, the relative abundance of each isotope remains largely constant. In addition, the relative natural abundances of each of the isotopes of the elements have been determined by scientists using mass spectrometers. Knowing the percentages of each of the isotopes (and their masses), scientists have calculated the weighted average mass of each element, and that is the atomic mass listed for it in the periodic table. For our bottle of carbon powder, we would use 12.011 amu as the average mass.

To better understand the concept of weighted average, we can draw an analogy to a student grade point average (GPA). Let's look at an "A student" who completed 10 credits of coursework in a semester. If 5 credits were A grades (4.0, on a scale of 0.0 to 4.0), 3 credits were B (3.0), and 2 credits were C (2.0), then her

Application

overall GPA would be the weighted average of the individual course grades. The A grade has the greatest weight because fully half the credits were in that course. The C grade would (thankfully) have the least weight, because that course represented only 20%, or 0.20, of the whole course load. The weighted average (here, the GPA) would be

$$(0.50 \times 4.0) + (0.30 \times 3.0) + (0.20 \times 2.0) = 3.3$$

contribution of: A grades   B grades   C grades

Does the answer make sense? Even though the student received one grade each of A, B, and C, *the A was present in much greater abundance*—5 credits—than the B or the C. It therefore makes sense that the GPA should be a little higher than 3.0, a straight B average. We can use the same strategy to calculate the atomic mass of an element. For example, mass spectrometry reveals what we have noted previously: that carbon-12 accounts for 98.93% of all carbon atoms. Carbon-13 makes up 1.07% of the carbon atoms, and carbon-14's contribution is negligible—about 2×10^{-10}%. The average mass of carbon-12 is then

$$(12.00000 \text{ amu} \times 0.9893) + (13.00335 \text{ amu} \times 0.0107) = 12.011 \text{ amu}$$

carbon-12   carbon-13

This agrees very nicely with the mass reported in the periodic table. The fact that any sample of carbon has a mass greater than 12 reflects the increased mass contribution made by the small proportion of carbon-13.

| **EXERCISE 2.3** | **Calculating the Atomic Mass Value** |

Naturally occurring chlorine is composed of two isotopes: chlorine-35 and chlorine-37. As the spectrum shown in Figure 2.22 reveals, the lighter isotope is substantially more abundant than the heavier one. Use the relative abundances given in the figure, along with the mass of each isotope, to calculate the atomic mass for chlorine. Compare the mass you calculated with what is reported in the periodic table of the elements.

First Thoughts

Analogies to common situations and experiences often help us visualize difficult concepts. Make use of the analogy to the grade point average that we discussed in the text. It is precisely the same calculation, except that the abundance of each isotope replaces the abundance of each grade.

Solution

Isotope	Mass	Abundance
chlorine-35	34.968852	75.77%
chlorine-37	36.965903	24.23%

$$\text{Atomic mass} = (34.968852 \times 0.7577) + (36.965903 \times 0.2423) = 35.45$$

There was more chlorine-35 than chlorine-37 present, so it makes sense that the mass should be closer to 35 than to 37. This value agrees with the value reported in the periodic table.

Further Insights

In addition to determining the masses of the isotopes of atoms, mass spectrometry can be used to reveal the masses of large and small molecules and to identify the presence of specific molecules. Some of the most important applications of mass spectrometry are found in biology and medicine, where it can be used to determine the masses of protein molecules or to analyze blood samples. One example—showing how mass spectrometry plays a role in testing for illegal drug use in athletics—is illustrated in Figure 2.23.

FIGURE 2.23

Drugs of abuse can be separated using a gas chromatograph and identified with a mass spectrometer.

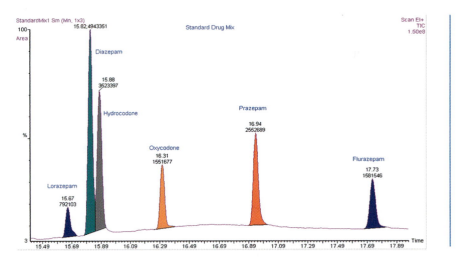

NanoWorld / MacroWorld

Big effects of the very small: Flavor analysis

"Organic" foods are big business. According to the U.S. Department of Agriculture, such foods are "(1) produced and handled without the use of synthetic chemicals and (2) *not* produced on land to which any prohibited substances, including synthetic chemicals, have been applied during the 3 years immediately preceding the harvest of agricultural products." Unlike organic foods, many of the other foods that we consume contain industrially produced chemical additives that increase their shelf life, lend them a particular texture, or enhance their flavor. Because of this, many consumers want products that not only are organically grown but also contain only natural additives. One useful way of characterizing additives as natural or synthetic is by looking at the atomic masses of their atoms with a mass spectrometer.

The key to the analysis, pioneered by Randy Culp and coworkers at the Center for Applied Isotope Studies at the University of Georgia, is that synthetic flavorings are typically made from organic chemicals derived from fossil fuels ("crude oil"). "Natural" flavorings are made from organic chemicals that have the same structure but are derived from plants. Vanilla flavoring, for example, can either come from vanilla beans or be made by combining individual compounds in an industrial process. Other than the method by which they are made, there is no difference between these two vanilla flavorings. Or is there?

Culp recognizes that plants were alive until they were harvested for the flavorings. Living things contain a constant and small amount of radioactive ^{14}C, which decreases slowly after they die. Fossil fuels, composed of formerly living material that has been chemically modified for *tens of millions of years*, contain essentially no ^{14}C.

Measurement of the radiation that is particular to ^{14}C indicates whether the chemicals in the flavoring are from fossil fuels or plants.

Once it is established that a flavoring was prepared from natural sources rather than fossil fuels, how can Culp's group tell whether the sample is made from a vanilla bean, say, or synthetically prepared from compounds derived from wood pulp? Each type of plant is unique in the way in which it processes carbon in photosynthesis, so the ratio of ^{13}C to ^{12}C distinguishes different sources of compounds. That is where a gas chromatograph (GC) and mass spectrometer (MS), shown in Figure 2.24, prove useful. The GC is a device that chemically separates the compounds in the vanilla flavor on the basis of their mass and structural properties. A separation of a vanilla sample is shown in Figure 2.25. After

FIGURE 2.24

A modern gas chromatograph/mass spectrometer, a PerkinElmer® Clarus® 500 GC/MS.

PRACTICE 2.3

Zinc has five isotopes, which have the masses (in amu) and the abundances given below. What is the average atomic mass of zinc?

Isotope	Mass	Abundance
zinc-64	63.9296011	48.63%
zinc-66	65.9260368	27.90%
zinc-67	66.9271309	4.10%
zinc-68	67.9248476	18.75%
zinc-70	69.925325	0.62%

See Problems 51–56 and 98.

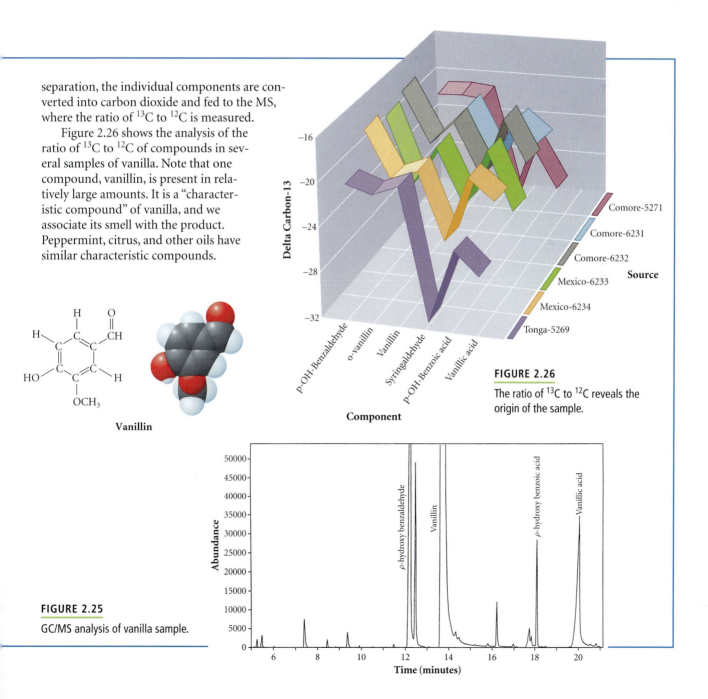

separation, the individual components are converted into carbon dioxide and fed to the MS, where the ratio of ^{13}C to ^{12}C is measured.

Figure 2.26 shows the analysis of the ratio of ^{13}C to ^{12}C of compounds in several samples of vanilla. Note that one compound, vanillin, is present in relatively large amounts. It is a "characteristic compound" of vanilla, and we associate its smell with the product. Peppermint, citrus, and other oils have similar characteristic compounds.

Vanillin

FIGURE 2.26

The ratio of ^{13}C to ^{12}C reveals the origin of the sample.

FIGURE 2.25

GC/MS analysis of vanilla sample.

2.6   The Periodic Table

The periodic table of the elements is the most significant document summarizing what we know about atoms. The table, shown in Figure 2.27 and on the inside front cover of this book, is the reference chart that shows the hierarchy of atomic structure. The layout of the table summarizes some of the most crucial ideas and discoveries of chemistry.

The periodic table is arranged in **periods** (rows) and **groups** (columns). If you read from left to right along the first period, then left to right along the second period, and so on, you will find the atomic numbers building up one proton at a time. The atoms generally get heavier as you move along the periods. The table seems to have a strange structure in places, with unexplained spaces we must jump over as we read. There are excellent reasons for the shape of the table, and we will discuss them in Chapter 6.

The arrangement of elements in the periodic table is linked to each element's chemical properties in ways that enable us to make many useful predictions. One of the specific arrangements is indicated by the heavy stair-step line on the right-hand side of the table. This line serves as the boundary between metal and non-metal elements. The **metals**, located to the left of the boundary, share common characteristics. They are lustrous (shiny), malleable (can be shaped or bent), and ductile (can be stretched into wires), and they act as conductors of heat and electricity. Many of these metals are commonplace in everyday life, such as iron (Fe), the main component of steel; copper (Cu), used to make electric wiring; and the silver (Ag) and gold (Au) used widely in jewelry.

FIGURE 2.27

The elements are arranged in the periodic table in columns that represent similar properties. Elements with similar properties are also color-coded.

Main-group elements / Transition metals / Main-group elements

Group																	
1 IA	2 IIA	3 IIIB	4 IVB	5 VB	6 VIB	7 VIIB	8	9 VIIIB	10	11 IB	12 IIB	13 IIIA	14 IVA	15 VA	16 VIA	17 VIIA	18 VIIIA
1 **H** 1.00794																	2 **He** 4.002602
3 **Li** 6.941	4 **Be** 9.012182											5 **B** 10.811	6 **C** 12.0107	7 **N** 14.0067	8 **O** 15.9994	9 **F** 18.9984032	10 **Ne** 20.1797
11 **Na** 22.989770	12 **Mg** 24.3050											13 **Al** 26.981538	14 **Si** 28.0855	15 **P** 30.973761	16 **S** 32.065	17 **Cl** 35.453	18 **Ar** 39.948
19 **K** 39.0983	20 **Ca** 40.078	21 **Sc** 44.955910	22 **Ti** 47.867	23 **V** 50.9415	24 **Cr** 51.9961	25 **Mn** 54.938049	26 **Fe** 55.845	27 **Co** 58.933200	28 **Ni** 58.6934	29 **Cu** 63.546	30 **Zn** 65.409	31 **Ga** 69.723	32 **Ge** 72.64	33 **As** 74.92160	34 **Se** 78.96	35 **Br** 79.904	36 **Kr** 83.798
37 **Rb** 85.4678	38 **Sr** 87.62	39 **Y** 88.90585	40 **Zr** 91.224	41 **Nb** 92.90638	42 **Mo** 95.94	43 **Tc** (98)	44 **Ru** 101.07	45 **Rh** 102.90550	46 **Pd** 106.42	47 **Ag** 107.8682	48 **Cd** 112.411	49 **In** 114.818	50 **Sn** 118.710	51 **Sb** 121.760	52 **Te** 127.60	53 **I** 126.90447	54 **Xe** 131.293
55 **Cs** 132.90545	56 **Ba** 137.327	57 **La*** 138.9055	72 **Hf** 178.49	73 **Ta** 180.9479	74 **W** 183.84	75 **Re** 186.207	76 **Os** 190.23	77 **Ir** 192.217	78 **Pt** 195.078	79 **Au** 196.96655	80 **Hg** 200.59	81 **Tl** 204.3833	82 **Pb** 207.2	83 **Bi** 208.98038	84 **Po** (209)	85 **At** (210)	86 **Rn** (222)
87 **Fr** (223)	88 **Ra** (226)	89 **Ac**** (227)	104 **Rf** (261)	105 **Db** (262)	106 **Sg** (266)	107 **Bh** (264)	108 **Hs** (277)	109 **Mt** (268)	110 **Ds** (281)	111 **Uuu** (272)	112 **Uub** (285)	114 **Uuq** (289)		116 **Uuh** (292)			

Key: 1 **H** 1.00794 — Atomic number / Symbol / Atomic weight

Inner transition metals

*Lanthanides	58 **Ce** 140.116	59 **Pr** 140.90765	60 **Nd** 144.24	61 **Pm** (145)	62 **Sm** 150.36	63 **Eu** 151.964	64 **Gd** 157.25	65 **Tb** 158.92534	66 **Dy** 162.500	67 **Ho** 164.93032	68 **Er** 167.259	69 **Tm** 168.93421	70 **Yb** 173.04	71 **Lu** 174.967
**Actinides	90 **Th** 232.0381	91 **Pa** 231.03588	92 **U** 238.02891	93 **Np** (237)	94 **Pu** (244)	95 **Am** (243)	96 **Cm** (247)	97 **Bk** (247)	98 **Cf** (251)	99 **Es** (252)	100 **Fm** (257)	101 **Md** (258)	102 **No** (259)	103 **Lr** (262)

Metals.

Nonmetals.

Metalloids. The metalloids silicon and boron have properties that are intermediate to the metals (see above) and the nonmetals (such as selenium).

The **nonmetals**, located to the right of the boundary, also share some common properties that are quite different from those of the metals. Typically, they are not lustrous, break easily, and act as insulators to heat and electricity. The nonmetals are a much smaller set of elements than the metals, but they include some of the most common elements found in living things. For example, the carbon (C), hydrogen (H), oxygen (O), nitrogen (N), and phosphorus (P) atoms that form DNA are all nonmetals.

The elements that form the stair-step boundary between metals and nonmetals are called **semimetals** or **metalloids**. This group of elements exhibits some properties associated with metals and other properties associated with nonmetals. Some of their properties are intermediate between the two. For example, they are lustrous but are often brittle. The semimetal silicon (Si) is called a semiconductor because its ability to conduct electricity is part way between that of a conductor and that of an insulator.

The elements are also located in another pattern that indicates similar properties (Figure 2.28). Each vertical column in the periodic table contains a group of elements with similar chemical and physical properties—hence the name *group*. For instance, the elements of a group may combine with other elements in a certain way, or they may have similar abilities to conduct or not conduct electricity. *All* of the elements in Group VIIIA, for example, are very stable gases that are unreactive except under extreme conditions. They are known as the **noble gases**—*noble* being used in the sense of having regal behavior. They are often referred to as the **inert gases**, although for those in the lower part of the group, this is not, strictly speaking, correct. The helium used to keep dirigibles (blimps) aloft and the neon used in neon lighting tubes are two common examples of noble gases. *All* of the elements in Group IA are highly reactive metals known as the **alkali metals** because they form compounds called alkalis when left in contact with water. Table 2.5 summarizes the titles most commonly used for some of the groups in the periodic table.

Elements in the A groups of the periodic table are historically known as the *main-group elements,* and those in the B groups are called *transition elements* or *transition metals* (Figure 2.28). There are two rows of elements separated from the main table known as the *inner transition elements.* We will have much more to say about the historical and chemical organization of the periodic table in future chapters. At this point, however, we have the background to begin learning about how the elements combine to form more complex structures, including compounds and molecules.

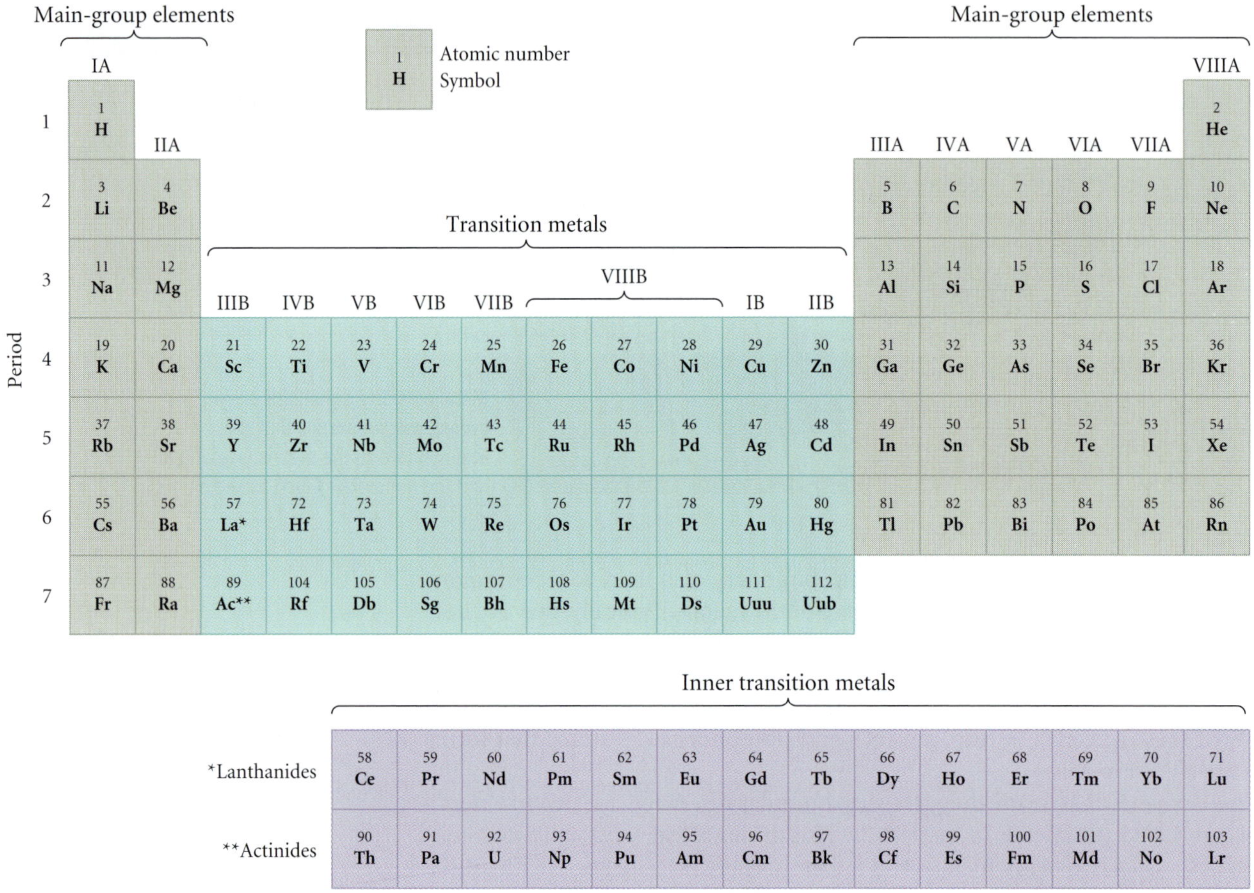

FIGURE 2.28

The main-group, transition, and inner transition elements.

TABLE 2.5	Some Groups in the Periodic Table with Characteristic Names	
Group	**Common Name**	**Members**
IA	Alkali metals	Li; Na; K; Rb; Cs; Fr
IIA	Alkaline earth metals	Be; Mg; Ca; Sr; Ba; Ra
IB	Coinage metals	Cu, Ag, Au
VIA	Chalcogens ("chalk formers")	O; S; Se; Te; Po
VIIA	Halogens ("salt formers")	F; Cl; Br; I; At
VIIIA	Noble gases (inert gases; "do not react easily")	He; Ne; Ar; Kr; Xe; Rn

2.7 Ionic Compounds

Chemical reactions consist of the rearrangement of atoms to make different substances. Although the nucleus of an atom remains unchanged in the course of a chemical reaction, the number of electrons surrounding that nucleus *can* change. When electrons are added to or subtracted from a neutral atom or group of atoms, a charged species known as an ion forms. For example, the neutral calcium-40 atom has 20 protons, 20 neutrons, and 20 electrons, and we denote this using the nuclide notation $^{40}_{20}\text{Ca}$. When calcium combines with oxygen, it loses two electrons, which the oxygen atom gains. We can write the nuclide notation for the resulting calcium *ion*, which now has 20 protons and only 18 electrons (two more positive than negative charges) as

$$^{40}_{20}\text{Ca}^{2+}$$

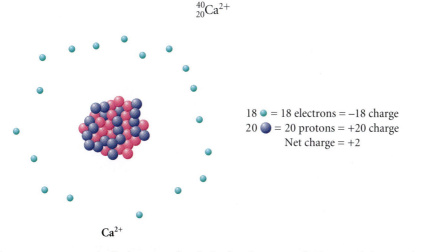

18 ● = 18 electrons = −18 charge
20 ● = 20 protons = +20 charge
Net charge = +2

Ca²⁺

Because we are typically interested only in the charge on the ion, and the number of protons is implicit in the symbol (that is, calcium *must* have 20 protons if it is to be called calcium), we simplify the notation even further and write Ca^{2+}. Using the same simplified notation, we denote the oxygen ion, which contains 8 protons and 10 electrons, as O^{2-}.

EXERCISE 2.4 | **Ions**

Fill in the numbers missing from the following table.

Symbol	Protons	Neutrons	Electrons	Charge
$^{32}_{16}\text{S}^{2-}$				
	56	81	54	
Cl^-		20		−1

Solution

Symbol	Protons	Neutrons	Electrons	Charge
$^{32}_{16}\text{S}^{2-}$	16	16	18	−2
$^{137}_{56}\text{Ba}^{2+}$	56	81	54	+2
$^{37}_{17}\text{Cl}^-$	17	20	18	−1

PRACTICE 2.4

Fill in the information missing from the following table.

Symbol	Protons	Neutrons	Electrons	Charge
$^{52}_{24}$ 6+				
	19	20	18	
	35	44		−1

See Problems 25, 26, 65, and 66.

Visualization: Determining Formulas for Ionic Compounds

Video Lesson: Describing Chemical Formulas

Tutorial: Determining Formulas for Ionic Compounds

FIGURE 2.29

Stibnite is a mineral with the chemical formula Sb_2S_3.

Some substances, known as **ionic compounds**, are composed entirely of ions. Calcium oxide (CaO) is one example. How do we know if the compound we are examining is an ionic compound? Although the distinction is clear with respect to the presence of ions in the structure, this is often not apparent at first glance because the formula for the substance doesn't include charges. However, the composition of the elements that make up the compound can provide a hint. Ionic compounds are often formed from the *combination of metals and nonmetals.*

Chemists use a **chemical formula**, containing element symbols and numbers, to represent the new combination of atoms or ions known as a chemical compound. A chemical formula is a kind of very precise chemist's shorthand. It shows the symbols for the elements found in the compound and the quantity of each of those atoms or ions. The ratios of the atoms in a chemical formula arise from the law of multiple proportions. For example, the formula NO tells us that each molecule of this compound contains one atom of nitrogen and one atom of oxygen. The formula NO_2 indicates one atom of nitrogen for every two atoms of oxygen in the molecule. The formula Sb_2S_3 tells us that two atoms of antimony (Sb) and three atoms of sulfur (S) make up the formula unit of the beautiful mineral stibnite (see Figure 2.29) used as a lubricant in antifriction alloys. Could we have determined this formula without knowing more than the identity of the ions in the compound?

When we are considering ionic compounds, the great clue to the ratio is *that the ions must combine in a ratio that makes the compound electrically neutral overall.* In sodium chloride, for example, each sodium cation has a +1 charge, and each chloride anion has a −1 charge, so these ions will combine in a 1:1 ratio to form the compound. Written as individual ions, they are Na^+ and Cl^-. When we write chemical formulas, we generally place any positive ion first, so the formula for sodium chloride is NaCl. Note that the charges are not shown. If no number follows the symbol of an element in a formula, it is assumed to contain just one atom of the element, as with CaO, discussed earlier. The formula CaF_2, for the chemical calcium fluoride, indicates that there is one calcium ion for every two fluoride ions.

Can we predict what the charge of an ion is likely to be in an ionic compound? As it so often does, the periodic table gives us insight. The main-group elements gain or lose electrons in predictable and consistent ways when they form ionic compounds. For example, the metals in Group IIA tend to *lose* two electrons when combining with nonmetals. This results in these atoms becoming cations with a charge of +2. The metals in Group IIIA tend to lose three electrons when forming ionic compounds, forming cations with a charge of +3. The nonmetals of Group VA tend to *gain* three electrons when forming ionic compounds, to form anions with a −3 charge. The nonmetals of Group VIA tend to gain two electrons to acquire a charge of −2. Keep in mind that these are tendencies—strong ones, in fact. However, as with many things in life and the chemistry that

FIGURE 2.30

Some common ions and their locations in the periodic table.

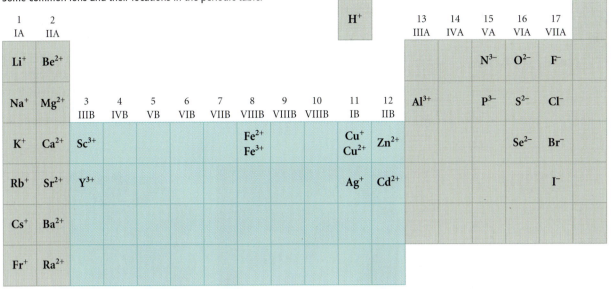

is a part of it, there are exceptions, especially with main-group metals in *Periods* 5 and 6. For example, lead commonly exists as Pb^{2+} or Pb^{4+}, depending on conditions. Still, it is a reasonable strategy to decide how many electrons a main-group element is likely to gain or lose on the basis of the group to which it belongs. On the other hand, many of the transition elements of the B groups can form ions with different charges. A list of some common ions is given in Figure 2.30. A summary of the charges found on the typical main-group ions is shown in Table 2.6.

TABLE 2.6	Charges on Typical Main-Group Ions
Group Number	**Most Likely Ionic Charge**
IA	+1
IIA	+2
IIIA	+3
IVA	+4/−4
VA	−3
VIA	−2
VIIA	−1
VIIIA	0

EXERCISE 2.5 Determining Formulas

1. The ionic compound used in some sidewalk de-icers is made from calcium and chlorine. What formula would represent that compound?
2. Magnesium combines with oxygen to give a bright flash of light used in some flares and fireworks. What formula would represent the resulting compound?

Solution

1. Because Ca is in Group IIA, it will form ions with a charge of +2. Chlorine is in Group VIIA and so will tend to form ions with a charge of −1. To form an electrically neutral ionic compound, there must be two chloride ions for every calcium ion. Therefore, the formula will be $CaCl_2$.

2. Magnesium is found in Group IIA, so it will form cations with a charge of +2. Oxygen is found in Group VIA and so will form anions with a charge of −2. A one-to-one combination produces an electrically neutral compound. Therefore, the formula will be MgO.

PRACTICE 2.5

What is likely to be the formula of a compound resulting from the atomic combination of cesium and chlorine?

See Problems 65–70.

Ionic compounds most often exist as part of large units such as the three-dimensional crystals of NaCl that you shake on your french fries. The crystal is composed of large numbers of sodium ions and chloride ions in equal amounts. These ionic compounds are typically brittle solids that are difficult to melt. For example, cesium fluoride (CsF) is an ionic compound with a melting point of 682°C. Ionic compounds vary in their solubility in water. Some, such as table salt (NaCl), are quite soluble. Others, such as strontium fluoride (SrF_2), are fairly insoluble. When soluble compounds such as NaCl are placed in water, their ions can separate and move around relatively independently. These mobile and electrically charged ions allow electricity to be conducted through the resulting solution.

Sodium chloride.

Molecular formula
H_2O

FIGURE 2.31

The molecular formula for water indicates that two hydrogen atoms are combined with one oxygen atom. Because water is a molecule, none of these atoms is an ion.

Side view of DNA.

2.8   Molecules

How is carbon dioxide (CO_2) different from sodium chloride (NaCl)—table salt? Sodium chloride has properties that we associate with ionic compounds, such as a high melting point and the ability to conduct electricity. Carbon dioxide is a gas that doesn't conduct electricity. These two substances are fundamentally different in their physical properties, and this is an indication of an essential difference in chemical structure. Sodium chloride is an ionic compound. Carbon dioxide is a molecule.

Molecules are distinct substances made up of two or more atoms linked together by sharing electrons between their nuclei, rather than by the transfer of electrons from one atom to another. We call bonds that are formed by sharing electrons between atoms **covalent bonds**. Many molecules, such as CO_2, water (H_2O), and methane (CH_4), are composed of only nonmetals that interact via covalent bonds. They differ from ionic compounds in that they are *not* made up of ions. When we write H_2O as the **molecular formula** for water, we mean that each molecule contains just two H atoms and one O atom (Figure 2.31). This is in contrast to the formulas for ionic compounds, which state *the ratio* in which ions are present without saying anything about the exact number of ions in any particular sample of the compound, because that number is an extensive property that varies depending on the size of the sample.

Molecules can range in size from diatomic molecules such as N_2 and O_2, the main components of air, to giant molecules containing many thousands, or even millions, of atoms. For example, the DNA that makes up the human genome and the protein molecules that control most of the chemistry of life are giant molecular compounds.

Iso-octane, a molecular compound, is important in the definition of the octane rating used to classify different grades of automobile gas. Iso-octane consists of molecules containing 8 carbon atoms and 18 hydrogen atoms and has the molecular formula C_8H_{18}. In some cases, we may be interested only in the smallest whole-number ratio that indicates the molecular formula. The simplest way to use whole numbers to express *the ratio* in which carbon and hydrogen atoms are present in iso-octane is C_4H_9. This is known as the empirical formula for octane. *Empirical* means "derived from experiment," and an **empirical formula** is the ratio in which elements are found by experiment to be present in compounds, regardless of the molecular structure of the compound. This distinction is shown in Figure 2.32.

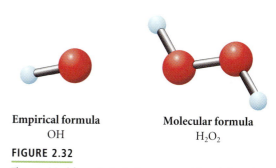

Empirical formula
OH

Molecular formula
H_2O_2

FIGURE 2.32

The empirical formula for hydrogen peroxide only lists the atoms in their simplest whole-number ratios. The molecular formula lists all of the atoms that make up the molecule.

Examination of the molecular and empirical formulas for iso-octane reveals something else about molecules. Even though molecules are electrically neutral like their ionic counterparts, they cannot be constructed using the rules for ionic compounds. Remember that these are two distinctly different types of compounds. The formula for many ionic compounds can be predicted from the typical charges found on the ions, especially in the A, or main-group, elements. For molecules, on the other hand, experimentation is necessary to determine the empirical and molecular formulas.

One of the themes of our study of chemistry is that *small differences in structure can lead to big differences in chemical behavior.* An important example of this is given by two forms of oxygen, one very common, the other less so. Oxygen is one of the **diatomic elements,** in which the normal state is in the form of molecules composed of two atoms bonded together. Other examples include H_2, N_2, F_2, Cl_2, Br_2 and I_2. The diatomic form of oxygen, O_2, accounts for approximately 20% of the volume of air and is the form of oxygen we need to stay alive. A different and much less common form of oxygen, O_3, contains three oxygen atoms bonded together. This form of oxygen is called ozone and accounts for just a few parts per billion of the composition of air. The extra atom of oxygen in ozone results in an **allotrope**, *another form of an element with different chemical and physical properties.* Ozone, when it occurs in the upper atmosphere, filters out hazardous ultraviolet radiation. Yet, when present at the earth's surface, O_3 is an irritant and poison that can cause serious damage to the tissues of the body, including the sensitive lining of our lungs, right where O_2 is the very molecule that is needed most!

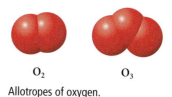

O_2 O_3

Allotropes of oxygen.

2.9 Naming Compounds

Communication between chemists is vital to effective application of the scientific method we discussed in Chapter 1. This communication relies on being able to convey information in writing, over the phone, and on the Internet. Complicating matters is that, according to the United Nations, over 6000 languages are spoken worldwide. How, then, does a chemist in Minden, Nebraska, tell a colleague in Mexico City about a particular compound? Luckily, some compounds have "common names" that are in widespread use across a country and around the world. For instance, benzene (C_6H_6, a compound found in petroleum distillates) and taxol (the anticancer compound from Chapter 1) are known by their common names in many different countries. But this isn't always true. As Table 2.7 illustrates for water, some compounds are so common that there is a different name for them in every language. For other compounds, there simply isn't a name that is commonly known. And moreover, without a system to follow, it would be difficult for those naming compounds to devise names that were meaningful to anyone else.

A simple solution to this problem was provided by the International Union of Pure and Applied Chemists (IUPAC), an organization formed in 1919 that continues to advise the scientific community on issues related to chemistry. IUPAC

TABLE 2.7	Water in Some Other Languages
Language	**Name of H_2O**
English	water
Spanish	agua
Hawaiian	wai
Urdu	pani
Swahili	maji
Cebuano	pagtubig
German	wasser
French	eau
Slovac	voda
Tagalog	tubig
Frisian	wetter
Japanese	mizu

TABLE 2.8 **Rules for Naming Binary Covalent Compounds**

The name of the compound mentions the elements in the order given in the formula. These rules apply to binary molecular compounds—those containing two elements that are both nonmetals.	Number	Prefix
	1	mono
	2	di
1. Name the first element using the exact element name.	3	tri
	4	tetra
2. Name the second element by writing the stem name of the element with the suffix *-ide*.	5	penta
	6	hexa
3. Add prefixes as shown in the list at the right, derived from Greek, to each element name to denote the subscript of the element in the formula. Generally, the prefix *mono-* is used only on the second element in the binary compound to distinguish it from other examples containing multiple atoms.	7	hepta
	8	octa
	9	nona
	10	deca

Video Lesson: Naming Chemical Compounds

has a set of rules that all scientists try to follow when naming compounds. Known as the rules of **chemical nomenclature**, they establish patterns to follow in naming existing compounds as well as compounds that have yet to be determined. The rules for naming **binary covalent compounds** (compounds composed of only two *nonmetal* elements, which have atoms that interact via covalent bonds) are listed in Table 2.8.

Consider the formula for NCl_3. This compound is a covalent compound (it has only covalent bonds) and is a binary compound (it has only two elements). Using the rules in Table 2.8, we write

nitrogen

as the first step of the rules for chemical nomenclature. In the second step, we write

nitrogen chloride

Then, to finish the name of the molecule, we add the prefixes. Note that we do not add the prefix *mono-* to the first element name. If it is not stated, *mono-* is assumed:

nitrogen trichloride

The following examples give you an opportunity to see these rules in practice:

N_2O_3 dinitrogen trioxide
HCl   hydrogen chloride, not monohydrogen monochloride
CO   carbon monoxide; *mono-* is needed here
CO_2 carbon dioxide
SF_4 sulfur tetrafluoride
SF_6 sulfur hexafluoride
ClO_2 chlorine dioxide
Cl_2O_7 dichlorine heptoxide
P_4O_6 tetraphosphorus hexoxide
S_2Cl_2 disulfur dichloride

EXERCISE 2.6 **Naming Binary Molecular Compounds**

Provide the IUPAC name for these binary molecular compounds:

PCl_5 NBr_3 $AsCl_3$ XeO_4 P_4S_3

Solution

PCl_5 phosphorus pentachloride XeO_4 xenon tetroxide
NBr_3 nitrogen tribromide P_4S_3 tetraphosphorus trisulfide
$AsCl_3$ arsenic trichloride

PRACTICE 2.6

Provide either the name or the formula for each of these binary molecular compounds:

sulfur tetrafluoride   carbon tetrachloride   diphosphorus pentoxide

PCl_3 N_2O OF_2

See Problems 77 and 78.

TABLE 2.9   Rules for Naming Binary Ionic Compounds

The name of the compound has the elements in the order given in the formula.

1. Name the first element using the exact element name.

2. Name the second element by writing the stem name of the element with the suffix *-ide.*

3. For metals that can have more than one stable charge, report the charge as part of the name. Do this by writing the charge, in roman numeral form, in parentheses immediately after the metal's name.

The rules for naming **binary ionic compounds** (compounds typically composed of a metal and a nonmetal element, which have ions that interact via electrostatic attractions) are similar to those for the binary molecular compounds. To summarize the rules in Table 2.9, a binary ionic compound can be named by stating the name of the atom that has become a cation (positive ion) and then adding the name of the anion (negative ion). The anion derives its name from the parent atom but ends in *-ide.* The name for a binary ionic compound does not tell us how many of each type of ion are present. How do we tell, from the name, how many of each type of ion are in the formula unit? Remember that ionic compounds contain ions whose charges must balance to result in an electrically neutral compound. Just use the charges to determine the number of each ion.

Sodium chloride (NaCl) offers a good example. We can tell that it is likely to be an ionic compound because it is composed of a main-group metal (Na) and a nonmetal (Cl). The procedure outlined in Table 2.9 indicates that we first give the name of the metal, the element listed first in the chemical formula. Therefore, we begin with

sodium

Then we add the name of the other element, which yields

sodium chloride

Because we know that sodium exists only as an ion with a +1 charge, we do not need to do step 3 of the rules. According to the IUPAC rules of chemical nomenclature, the name for NaCl, sodium chloride, is exactly as we expected.

EXERCISE 2.7   Naming Binary Ionic Compounds

Use the rules outlined in Table 2.9 to supply each name or formula missing from the list on the right.

Name	Formula
sodium fluoride	
calcium oxide	
	Al_2S_3
	$BaCl_2$

Solution

Name	*Formula*
sodium fluoride	NaF
calcium oxide	CaO
aluminum sulfide	Al_2S_3
barium chloride	$BaCl_2$

See Problems 79 and 80.

PRACTICE 2.7

Supply each name or formula missing from the following list.

Name	Formula
magnesium chloride	
lithium fluoride	
	NaBr
	Li$_2$O

Iron(III) chloride
FeCl$_3$

Iron(II) chloride
FeCl$_2$

Let's examine some cases where we need to use step 3 in the rules for naming ionic compounds. One such case involves the metals found between Groups IIA and IIIA in the periodic table. Most of these transition metals commonly exhibit more than one positively charged state. It is therefore possible for two bottles of iron chloride to contain two different ionic compounds. You might find FeCl$_2$ in one bottle and FeCl$_3$ in the other. If we considered only the first two steps in Table 2.9, we would conclude that they should both be named iron chloride. However, the first bottle contains an iron ion with a +2 charge. (Remember, it needs to be Fe^{2+} in order to balance the charges of the two Cl$^-$ atoms.) The second bottle contains Fe^{3+} for similar reasons. To name these two compounds in such a way as to distinguish between them, we follow step 3 in Table 2.9 and add the charge on the iron to the name of the compound. Thus FeCl$_2$ becomes iron(II) chloride, and the other bottle contains iron(III) chloride.

EXERCISE 2.8   **Naming Additional Compounds**

Use Table 2.9 to supply each name or formula missing from the following list.

Formula	Name
BaCl$_2$	
TiCl$_2$	
	copper(II) oxide
	manganese(IV) oxide

First Thoughts

To supply these names, we need to be familiar with the rules outlined in Table 2.9. More important, we will have to consult Figure 2.30 to determine the charges on each ion. For those with more than one charge, we must follow step 3 in Table 2.9.

Solution

Formula	Name
BaCl$_2$	barium chloride
TiCl$_2$	titanium(II) chloride
CuO	copper(II) oxide
MnO$_2$	manganese(IV) oxide

Further Insight

Writing the names of compounds requires, first of all, that we understand the nature of the atoms involved in the formulas. Often, it is easy to forget that ionic compounds and molecular compounds are named using different rules. We wouldn't want to accidentally name CuO as copper monoxide. This implies that the formula represents a molecule. And, as we saw earlier, molecules and ionic compounds have markedly different properties.

See Problems 83, 84, 89, and 90.

PRACTICE 2.8

Use Table 2.9 to supply each name or formula missing from the list.

Formula	Name
$CuCl_2$	
CrO_3	
	nickel(II) oxide
	palladium(IV) sulfide

Polyatomic Ions

Many compounds are made up of more than two elements. Some of these are still classified as ionic compounds, because they involve the association of ions. In some cases, the ion itself contains two or more atoms. The entire ion that results can carry a positive or negative charge and behave just like a monatomic ion. Ions such as this are known as **polyatomic ions** or **molecular ions**. Figure 2.33 shows the structure of some examples. Table 2.10 lists many others.

Polyatomic ions are everywhere. For example, the famous White Cliffs of Dover on the southern end of England are made of chalk, the same stuff we use to write on chalkboards. Chalk has the formula $CaCO_3$ and is made up of two ions, Ca^{2+} and CO_3^{2-}.

Using Table 2.10, we can arrive at the name *calcium carbonate* for this compound. The CO_3^{2-} ion in the compound is the polyatomic "carbonate" ion with a charge of -2. The calcium and carbonate ions combine to form the electrically neutral ionic compound calcium carbonate, but there are covalent bonds within the structure of the carbonate ions.

One common garden fertilizer has the formula $(NH_4)_3PO_4$. The ammonium ion, NH_4^+, has a $+1$ charge, and the phosphate ion, PO_4^{3-}, has a -3 charge. Three ammonium ions will be present for each phosphate ion. We must use parentheses in the formula to indicate that our subscript 3 applies to the entire ammonium ion, not just to the hydrogen atoms within it. What is the name of this compound? It is known as ammonium phosphate.

EXERCISE 2.9   Naming with Polyatomic Ions

Provide each name or formula missing from the table.

CaC_2O_4	
$Mg(NO_3)_2$	
	copper(II) sulfate
	aluminum hydroxide

Solution

Formula	Name
CaC_2O_4	calcium oxalate
$Mg(NO_3)_2$	magnesium nitrate
$CuSO_4$	copper(II) sulfate
$Al(OH)_3$	aluminum hydroxide

Application

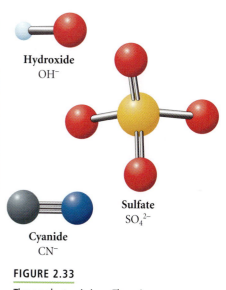

Hydroxide
OH^-

Sulfate
SO_4^{2-}

Cyanide
CN^-

FIGURE 2.33

Three polyatomic ions. These ions are made up of chemically bonded atoms. As a whole, they act as ions.

TABLE 2.10	Selected Polyatomic Ions		
	Formula	**Name**	**Charge for Ion**
Cation with a +1 Charge	NH_4^+	ammonium	+1
Anions with a −1 Charge	CH_3COO^-	acetate	−1
	ClO^-	hypochlorite	−1
	ClO_2^-	chlorite	−1
	ClO_3^-	chlorate	−1
	ClO_4^-	perchlorate	−1
	CN^-	cyanide	−1
	HCO_3^-	hydrogen carbonate (bicarbonate)	−1
	$H_2PO_4^-$	dihydrogen phosphate	−1
	HSO_4^-	hydrogen sulfate (bisulfate)	−1
	HSO_3^-	hydrogen sulfite (bisulfite)	−1
	MnO_4^-	permanganate	−1
	NO_2^-	nitrite	−1
	NO_3^-	nitrate	−1
	OH^-	hydroxide	−1
Anions with a −2 Charge	CO_3^{2-}	carbonate	−2
	$C_2O_4^{2-}$	oxalate	−2
	CrO_4^{2-}	chromate	−2
	$Cr_2O_7^{2-}$	dichromate	−2
	HPO_4^{2-}	monohydrogen phosphate	−2
	SO_3^{2-}	sulfite	−2
	SO_4^{2-}	sulfate	−2
	O_2^{2-}	peroxide	−2
	$S_2O_3^{2-}$	thiosulfate	−2
Anions with a −3 Charge	PO_4^{3-}	phosphate	−3

PRACTICE 2.9

What is the name of the compound with the formula $KMnO_4$? What is the formula for ammonium dichromate?

See Problems 81, 82, and 85–88.

 ## 2.10 Naming Acids

Some compounds, known as acids, are often written with a hydrogen at the start of their formula. These compounds have special names when they are in aqueous solutions. For those that contain only hydrogen and a halogen, the name is hydro_____ic acid, where the blank is replaced by the root name of the halogen. For instance, aqueous HCl is known as hydrochloric acid. The names of other acids come from the name of the polyatomic ion, by replacing the *-ate* ending of the ion with *-ic acid* or the *-ite* ending with *-ous acid*. For example, HNO_3 is nitric acid and HClO is known as hypochlorous acid.

The Bottom Line

- All matter is composed of atoms. (Section 2.1)

- The law of conservation of mass states that the mass of the chemicals at the start of a reaction is equal to the mass of the chemicals at the end of the reaction. (Section 2.1)

- The law of definite composition states that any particular chemical is always composed of its components in a fixed ratio, by mass. (Section 2.1)

- The law of multiple proportions states that when the same elements can produce more than one compound, the ratio of the masses of the element that combine with a fixed mass of another element corresponds to a small whole number. (Section 2.2)

- Dalton's atomic theory stated that every substance is made of atoms; atoms are indestructible; atoms of any one element are identical; atoms of different elements differ in their masses; and chemical changes involve rearranging the attachments between atoms. (Section 2.2)

- The law of combining volumes states that when gases combine, they do so in small whole-number ratios, provided that all the gases are at the same temperature and pressure. (Section 2.2)

- Atoms are composed of electrons, protons, and neutrons. The most common isotope of the hydrogen atom is the only exception. It contains only electrons and protons. (Section 2.3)

- The modern model of the atom indicates a tiny, but dense, positively charged nucleus surrounded by a diffuse electron cloud. (Section 2.3)

- Nuclei can contain both positively charged protons and neutral neutrons. Isotopes of the same element differ in their number of neutrons. All atoms of the same element have the same atomic number (the same number of protons). (Section 2.4)

- Atoms are electrically neutral and contain equal numbers of protons and electrons. Ions are charged particles and are formed from atoms or groups of atoms transferring electrons to other atoms or groups of atoms. Cations are positively charged ions, and anions are negatively charged ions. (Section 2.4)

- The atomic mass of an element is the weighted average of all the isotopes of that element. (Section 2.5)

- The periodic table of the elements lists all the known elements, arranged into periods and groups in a manner that reflects the chemical characteristics that particular elements share. (Section 2.6)

- A chemical formula makes use of the atomic symbols and subscripts, as needed, to represent the atoms in a chemical compound. (Section 2.7)

- Molecules are distinct substances made up of two or more atoms linked together by sharing electrons between their nuclei, rather than by the transfer of electrons from one atom to another. We call bonds that are formed by the sharing of electrons between atoms covalent bonds. (Section 2.8)

- There are systematic rules for naming compounds and listing their formulas. (Section 2.9)

Key Words

alkali metals The highly reactive metals found in Group IA of the periodic table, which form alkalis upon reaction with water. (*p. 66*)

alkaline earth metals The metals found in Group IIA of the periodic table. (*p. 66*)

allotropes Forms of an element that have very different chemical and physical properties, such as the allotropes O_2 and O_3. (*p. 71*)

alpha particle A fast-moving nucleus of helium (two protons and two neutrons) emitted during the decay of a radioactive element. (*p. 54*)

anion Ions with a negative charge. (*p. 56*)

atomic number (Z) The number of protons in the nucleus of an atom. (*p. 56*)

atomic mass unit The arbitrary unit of mass for elements based on the isotope carbon-12. One atom of carbon-12 is defined to have a mass of exactly 12.0000 amu. (*p. 60*)

beta particle A fast-moving electron emitted during the decay of a radioactive element. The beta particle originates from the nucleus of the decaying element. (*p. 54*)

binary covalent compounds Compounds composed of only two nonmetal elements. (*p. 72*)

binary ionic compounds Compounds typically composed of a metal and nonmetal element, which have ions that interact via electrostatic attractions. (*p. 73*)

cations Ions with a positive charge. (*p. 56*)

chalcogens The elements found in Group VIA of the periodic table of elements. (*p. 66*)

chemical formula A representation, in symbols, that conveys the relative proportions of atoms of the different elements in a substance. (*p. 68*)

chemical nomenclature A system of rules used to assign a name to a particular substance. (*p. 72*)

coulomb The unit of electrical charge. (*p. 55*)

covalent bonds Bonds that atoms form by sharing electrons. (*p. 70*)

Dalton's atomic theory The theory, developed by John Dalton, that all substances are composed of indivisible atoms. (*p. 50*)

diatomic elements Elements whose normal state is in the form of molecules composed of two atoms attached together (most notably H_2, N_2, O_2, F_2, Cl_2, Br_2, and I_2). (*p. 71*)

electron One of the subatomic particles of which atoms are composed. Electrons carry a charge of -1. (*p. 52*)

empirical formula A chemical formula indicating the ratio in which elements are found to be present in a particular compound, regardless of the molecular structure of the compound. (*p. 71*)

gamma ray High-energy electromagnetic radiation emitted from the decay of a radioactive element. (*p. 54*)

group One of the vertical columns in the periodic table of the elements. (*p. 64*)

halogens The elements of Group VIIA of the periodic table. (*p. 66*)

inert gases *See* noble gases. (*p. 65*)

ion An electrically charged entity that results when an atom has gained or lost electrons. (*p. 56*)

ionic compound A compound composed of ions. (*p. 68*)

isotopes Forms of the same element that differ in the number of neutrons within the nucleus. (*p. 57*)

law of combining volumes Scientific law stating that when gases combine, they do so in small whole-number ratios, provided that all the gases are at the same temperature and pressure. (*p. 50*)

law of conservation of mass Scientific law stating that the mass of chemicals present at the start of a chemical reaction must equal the mass of chemicals present at the end of the reaction. Although not strictly true, this law is correct at the level of accuracy of all laboratory balances. (*p. 47*)

law of definite composition Scientific law stating that any particular chemical is always composed of its components in a fixed ratio, by mass. (*p. 48*)

law of multiple proportions Scientific law stating that when the same elements can produce more than one compound, the ratio of the masses of the element that combine with a fixed mass of another element corresponds to a small whole number. (*p. 49*)

mass number (*A*) The total number of protons and neutrons in the nucleus of an atom. (*p. 57*)

mass spectrometer An instrument used to measure the masses and abundances of isotopes, molecules, and fragments of molecules. (*p. 59*)

mass spectrum The output of a mass spectrometer, in the form of a chart listing the abundance and masses of the isotopes, molecules, and fragments of molecules present. (*p. 60*)

metalloids A small number of elements, found at the boundary between metals and nonmetals in the periodic table, that have properties intermediate between those of metals and nonmetals. Also known as semimetals. (*p. 65*)

metals The largest category of elements in the periodic table. Metals exhibit such characteristic properties as ability to conduct electricity, shiny appearance, and malleability. (*p. 64*)

molecular formula A chemical formula indicating the actual number of each type of atom present in one molecule of a compound. (*p. 70*)

molecular ions Ions composed of two or more atoms attached together. Also known as polyatomic ions. (*p. 75*)

molecules Neutral compounds containing two or more atoms attached together. (*p. 70*)

neutron One of the subatomic particles of which atoms are composed, carrying no electrical charge and found in the nucleus. (*p. 54*)

noble gases The unreactive elements in the rightmost group of the periodic table (Group VIIIA). Also known as inert gases. (*p. 65*)

nonmetals The elements on the right-hand side of the periodic table. Nonmetals exhibit characteristic properties that are distinct from those of the metals. (*p. 65*)

nuclide notation Shorthand notation used to represent atoms, listing the symbol for the element, accompanied by the atomic number and the mass number of the atom. (*p. 57*)

period One of the horizontal rows in the periodic table of the elements. (*p. 64*)

periodic table of the elements A table presenting all the elements, in the form of horizontal periods and vertical groups; shown on the inside front cover of this book. (*p. 56*)

polyatomic ions Ions composed of two or more atoms covalently attached together. Also known as molecular ions. (*p. 75*)

products The materials that are produced in a chemical reaction. (*p. 47*)

proton One of the subatomic particles of which atoms are composed, carrying an electrical charge of $+1$ and found in the nucleus. (*p. 53*)

radiation The particles and/or energy emitted during radioactive decay. (*p. 54*)

radioactive decay The process by which an unstable nucleus becomes more stable via the emission or absorption of particles and accompanying energy. (*p. 54*)

radioactivity The emission of radioactive particles and/or energy. (*p. 54*)

reactants The materials that combine in a chemical reaction. (*p. 47*)

semimetals A small number of elements found at the boundary between metals and nonmetals in the periodic table. Also known as metalloids. (*p. 65*)

Focus Your Learning

The answers to the odd-numbered problems and some selected problems appear at the back of the book, as represented by the blue numbering.

Section 2.1 Early Attempts to Explain Matter

Skill Review

1. Exploring the unseen world often relies on indirect reasoning. Useful deductions led early philosophers and scientists to make conclusions about the possible existence of atoms. How would the formation of ice crystals on the branches of a bush support the ideas of Democritus?

2. Ancient practices, and some modern ones, involve burning incense. How would entering a room in which incense had recently been burned lead you to consider the world to be made of small, invisible particles?

3. The scientists who first used atoms to explain chemical changes assumed that the small, unseen pieces of matter could continually be rearranged into new combinations with new properties. Show how that same principle operates when the letters in the word *dormitory* are used to form several other words.

4. Taking the analogy in Problem 3 a bit further, explain what would have to happen in order for the exercise to illustrate the law of conservation of mass.

5. In your own words, explain the law of conservation of mass.

6. In your own words, explain the law of definite composition.

7. Modern science is based on asking questions. In what way does a researcher determining the age of Earth ask the same basic questions posed by Democritus?

8. What connection exists between the exploration and observations of the United States space program and Democritus' questions?

Chemical Applications and Practices

9. Some compounds absorb water from the atmosphere in such a way that the water is chemically combined with the compound. The absorbed water is called water of hydration, and the compounds with absorbed water are known as hydrates. If 14.7 g of calcium chloride hydrate, when heated, loses 3.6 g of water, how many grams of water are lost when a 23.4-g sample is heated? To which law(s) did you refer in obtaining these results?

10. The white appearance of several magnesium-containing compounds causes an apparent resemblance among the compounds. One such compound was shown to contain 60.0% magnesium. Another sample of a magnesium-containing compound was found. What quantity of magnesium, in grams, would have to be present in a 34.6-g sample of the second compound in order for us to claim that it is the same compound as the first sample?

Section 2.2 Dalton's Atomic Theory and Beyond

Skill Review

11. Lavoisier used carefully determined masses to make discoveries about changes that occurred in chemical reactions.
    a. Which fundamental law did his work lead him to discover?
    b. Suppose Lavoisier evaporated the water from 250.0 g of seawater. If the remaining salt had a mass of 2.2 g, what would you calculate as the mass of the water that had evaporated?

12. A 200.0-g sample of a pure substance was composed of 132.9 g of copper and 67.1 g of sulfur. Another sample of the same substance, this time with a mass of 150.0 g, was brought to Joseph Proust to analyze. How many grams of copper and sulfur would you expect to find in the sample? What basic law in chemistry are you employing in order to make your determination?

13. Fill in the missing information for a compound of copper and oxygen, assuming that it is composed of a fixed ratio of copper to oxygen of 3.97:1.00.

Copper	Oxygen
17.0 grams	grams
grams	11.5 grams

14. Fill in the missing information for a compound consisting of sodium and oxygen, assuming that is composed of a fixed ratio of sodium to oxygen of 1.44:1.00.

Sodium	Oxygen
6.0 grams	grams
grams	50.0 grams

Chemical Applications and Practices

15. Copper and oxygen combine in more than one ratio. In one such compound it is found that 65 grams of copper combine with 16 grams of oxygen. If you assumed, as John Dalton might have, that the combination represented one atom of copper combining with one atom of oxygen, what would you predict as the copper-to-oxygen ratio in another compound between the two?

16. Using the first mass ratio presented in Problem 15 and your answer for the other ratio, determine how many grams of oxygen would combine with 10.0 g of copper in both compounds.

17. When the calcium is properly extracted, you can expect to obtain 40 g of calcium from 100.0 g of calcium carbonate. A 10.0-g sample of pure calcium chloride yields 6.40 g of chlorine. Which of these two compounds is a better source of calcium?

18. Three magnesium compounds have the following percentages of magnesium: Compound A: 28.6% magnesium; Compound B: 60.0% magnesium; Compound C: 41.4% magnesium. Why, according to Dalton, can there be such a great difference in the percentages of magnesium?

19. Water and hydrogen peroxide are both clear liquids that are made up of only hydrogen and oxygen. Water contains 2 g of hydrogen for every 16 g of oxygen in water. In hydrogen peroxide there are 2 g of hydrogen for every 32 g of oxygen. Show how these data illustrate the law of multiple proportions.

20. Oxygen exists naturally in two combined forms. The diatomic oxygen that we utilize in breathing consists of two atoms combined into one unit. Ozone consists of three atoms of oxygen joined into one unit. The relative masses of the oxygen and ozone units are 32 and 48. Does this agree with the law of multiple proportions? Explain.

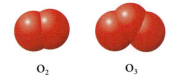

O_2 O_3

Section 2.3 The Structure of the Atom

Skill Review

21. If the discoverers of the particles that make up the atom had determined the charge on the electron to be -2, what would be the charge on the neutron and proton?

22. Describe what might have been the results of Rutherford's experiment if the charge on the electron had been reversed (that is, if it were $+1$ instead of -1).

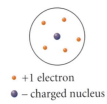

• +1 electron
● – charged nucleus

23. Draw diagrams of the models of the atom that J. J. Thomson and Rutherford, respectively, would have come up with if the electron had a $+1$ charge.

24. What are two objections to the explanation that positive protons compose a nucleus that is surrounded by negative electrons.

Section 2.4 Atoms and Isotopes

Skill Review

25. Fill in the information missing from the following table:

Isotope	Protons	Neutrons	Electrons	Charge
carbon-12		6	6	0
	13	14	10	
chlorine-35		18		-1

26. Fill in the information missing from the following table:

Isotope	Protons	Neutrons	Electrons	Charge
calcium-40	20		18	
	14	14		0
sulfur-32		16	14	

27. Indicate whether each of these phrases describes the proton, neutron, or electron. (You may need to assign more than one term to each phrase.)
    a. Determines the identity of an atom.
    b. Has about the same mass as a proton.
    c. The number of particles can be changed without changing the identity of the atom.

28. Indicate whether each of these phrases refers to protons, neutrons, and/or electrons. (You may need to assign more than one term to each phrase.)
    a. Is represented by the mass number.
    b. Is represented by the atomic number.
    c. Are always present in the same quantity in an atom with 0 charge.

29. What would you calculate for the mass, in grams, of an atom that contained 1 proton and 1 electron? . . . 1 proton, 1 neutron, and 1 electron?

30. What would you calculate for the mass, in grams, of an atom that contained 2 protons, 2 neutrons, and 2 electrons? . . . 1 proton, 1 neutron, and 2 electrons?

31. What would be the charge on each of the particles listed in Problem 29?

32. What would be the charge on each of the particles listed in Problem 30?

33. Assume that an atom contained 1 proton and 1 electron. By what percentage would the mass of this atom increase if 10 electrons were added to the atom?

34. Assume that an atom contained 1 proton and 1 electron. By what percentage would the mass of this atom increase if 10 neutrons were added to the atom?

35. Which of the most abundant isotope in each pair contains the greater number of neutrons? Assume the mass number of the most abundant isotope can be found by rounding the average atomic mass to the nearest whole number. (Use the periodic table of the elements inside the front cover of your textbook.)
    a. iron or chromium   c. cobalt or nickel
    b. tellurium or iodine   d. helium or neon

36. Which of the most abundant isotope in each pair contains the greater number of neutrons? Round the average atomic mass to the nearest whole number to obtain the mass number of the most abundant isotope. (Use the periodic table of the elements inside the front cover of your textbook.)
a. zinc or sodium c. phosphorus or sulfur
b. hydrogen or sodium d. bromine or selenium

37. Arrange the following atoms in order of their quantity of protons (least number to greatest number): iron, hydrogen, calcium, fluorine, aluminum, boron.

38. Arrange these atoms in order of their quantity of protons (least number to greatest number): sodium, magnesium, copper, iodine, tin, carbon, lithium.

39. Arrange these atoms in order of their quantity of electrons (least number to greatest number): iron, hydrogen, calcium, fluorine, aluminum, boron.

40. Arrange these atoms in order of their quantity of electrons (least number to greatest number): sodium, magnesium, copper, iodine, tin, carbon, lithium.

41. Using the periodic table, determine the first element that does not have the same number of protons as neutrons. Round the average atomic mass to the nearest whole number to obtain the mass number of the most abundant isotope.

42. Using the periodic table, indicate two atoms that have the same number of protons as neutrons. Round the average atomic mass to the nearest whole number to obtain the mass number of the most abundant isotope.

43. In a neutral atom, the number of electrons must equal the number of protons. However, if an atom had two more protons than electrons, what would you report as its charge? Would this ion be called an anion or a cation?

44. If an atom had two more electrons than protons, what would you report as its charge? Would this ion be called an anion or a cation?

Chemical Applications and Practices

45. Several radioactive isotopes have medical applications. Using the nuclide representation shown here, determine the number of protons, electrons, and neutrons in an atom of each of these elements.

Radioactive sodium is used in blood circulation studies: $^{24}_{11}\text{Na}$

Radioactive iodine can be used to examine the thyroid gland: $^{131}_{53}\text{I}$

Radioactive cobalt can be used to kill cancer cells: $^{60}_{27}\text{Co}$

Radioactive chromium can be used to measure total blood volume: $^{51}_{24}\text{Cr}$

Radioactive phosphorus is used in antileukemia therapy: $^{32}_{15}\text{P}$

46. Write nuclide notation for each of the following isotopes: carbon-12, carbon-13, phosphorus-32, chlorine-35, chlorine-37, iron-55.

Section 2.5 Atomic Mass

Skill Review

47. If the atomic mass of ^{12}C had been set equal to 6.000000 amu, what would be the mass of ^{16}O? . . . the mass of ^{1}H?

48. If the atomic mass of oxygen-16 were set equal to 4.000000 amu, what would be the mass of carbon-12? . . . the mass of copper-63?

49. Calculate the mass, in grams, of 1 atom of carbon-12. . . . of nitrogen-14. . . . of fluorine-19. (Assume that the mass number of these isotopes is equal to the atomic mass, rounded to the nearest whole number.)

50. Calculate the mass, in grams, of a gross (144) of atoms of carbon-12. . . . of nitrogen-14. . . . of fluorine-19. (Assume that the mass number of these isotopes is equal to the atomic mass, rounded to the nearest whole number.)

Chemical Applications and Practices

51. The metal indium is used to make heat-conducting alloys. Indium typically is found with two isotopes. Based on the following information, what would you calculate as the percent relative abundance of the two isotopes? Indium-113 has a mass of 112.9043, and indium-115 has a mass of 114.9041.

52. The element neon is widely used in brightly colored electronic advertising signs. There are three common isotopes of neon with masses of 21.99, 20.99, and 19.99. One of the three isotopes makes up approximately 90% of the atoms in a neon sample. Which isotope is most likely to be responsible for the majority of neon's atomic mass reported on the periodic table? Sketch a diagram of the probable appearance of the mass spectrum for neon that shows the masses and an approximation of the percentage abundance.

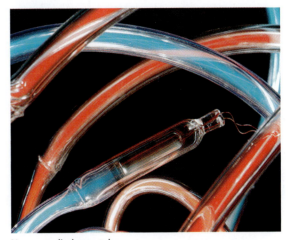

Neon gas discharge tube.

53. The sulfur-containing amino acids that help us make certain proteins are critically important to our diet. Sulfur samples typically consist of four isotopes. Use the following table to determine the average atomic mass of sulfur.

Isotope	Atomic Mass	Abundance
S-32	31.97	95.0%
S-33	32.97	0.76%
S-34	33.97	4.22%
S-36	35.97	0.014%

Sketch a mass spectrum that illustrates the masses observed for a natural sample of sulfur.

54. Silver has many uses in our world, only one of which is as a precious metal in coinage. Use the following table to determine the average atomic mass of silver.

Isotope	Atomic Mass	Abundance
Ag-107	106.9051	51.84%
Ag-109	108.9048	48.16%

Sketch a mass spectrum that illustrates the masses observed for a natural sample of silver.

55. Cerium has four naturally occurring isotopes. Use the following table to determine the average atomic mass of cerium.

Isotope	Atomic Mass	Abundance
Ce-136	135.907	0.19%
Ce-138	137.906	0.25%
Ce-140	139.905	88.48%
Ce-142	141.909	11.08%

Sketch a mass spectrum that illustrates the masses observed for a natural sample of cerium.

56. A hypothetical element was just discovered to have three isotopes. Use the following table to determine the average atomic mass of this hypothetical element.

Isotope	Atomic Mass	Abundance
A-100	99.754	35.25%
A-102	101.688	25.75%
A-103	102.599	39.00%

Sketch a mass spectrum that illustrates the masses observed for a natural sample of the element.

Section 2.6 The Periodic Table

Skill Review

57. Use the descriptions given in this chapter of the organization of the periodic table to identify each of these elements.
a. The element that is in the second column and in the second period.
b. The transition element that is in the seventh group and in the fourth row.
c. The third most massive noble gas.

58. Use the descriptions given in this chapter of the organization of the periodic table to identify each of these elements.
a. The lightest alkali metal.
b. The semimetal that is in the third group.
c. The halogen that is a liquid at room temperature.

59. How many elements are classified as halogens?

60. Which column contains the alkali metals? . . . the chalcogens?

61. Identify the name of the group for each of these elements.

sulfur   iodine   helium
beryllium   francium

62. Identify the name of the group for each of these elements.

lithium   barium   neon
oxygen   chlorine

63. Which of these elements would you predict to be shiny?

gold   silver   lead
silicon   carbon   iodine

64. Which of these elements would you predict to be brittle?

gold   silver   lead
silicon   carbon   iodine

Section 2.7 Ionic Compounds

Skill Review

65. One of the first considerations in writing ionic formulas is to determine which element is to be the cation and which is to be the anion. Decide which element in each of these pairs is more likely to become the positive ion.

Ca or Br   S or Al   Cl or Al
Sr or N   I or Be

66. Use the information from the previous question to determine what formula is most likely to result from the combination of each of these pairs of elements.

Ca and Br   S and Al   Cl and Al
Sr and N   I and Be

67. Write the formulas of the ionic compounds that form when chlorine combines individually with lithium, beryllium, sodium, calcium, and aluminum.

68. Write the formulas of the ionic compounds that form when oxygen combines individually with lithium, beryllium, sodium, calcium, and aluminum.

69. Ionic compounds typically dissociate into ions as they dissolve in water. For each of these compounds, predict the ions, including their charges, that will be produced when the ionic compound dissolves.

$MgBr_2$ $FeCl_3$
KI Na_2S

70. Ionic compounds typically dissociate into ions as they dissolve in water. For each of these compounds, predict the ions, including their charges, that will be produced when the ionic compound dissolves.

$CaCl_2$ TiO_2
ZnO $NiCl_2$

Section 2.8 Molecules

Skill Review

71. The two major classifications of compounds are ionic and molecular. Label the descriptions given below as characteristic of an ionic compounds or of a molecule
a. Results from a metal combining with a nonmetal.
b. Persists as a complete entity when dissolved in water.
c. Formula can often be determined by the identity of the atoms that combine.

72. The two major classifications of compounds are ionic and molecular. Label the descriptions given below as characteristic of an ionic compounds or of a molecule.
    a. Results as atoms share their electrons.
    b. Formula represents a ratio of ions involved rather than actual numbers of atoms combined.
    c. Results from the combination of two nonmetals.
    d. Separates into individual charged particles when dissolved in water.

73. Which of these compounds are *not* molecular compounds?
    a. CO
    b. BrO_3
    c. K_2O
    d. PCl_3
    e. WO_2

74. Which of these compounds are molecular compounds?
    a. SO_3
    b. NiO
    c. BCl_3
    d. Na_2S
    e. SiF_4

75. In these examples, the molecular and empirical formulas of several compounds are listed. In each case, decide which is the empirical formula.
    a. Hydrogen peroxide is a useful disinfectant. (H_2O_2 or HO)
    b. Butane is used as a heating fuel. (C_2H_5 or C_4H_{10})
    c. Teflon is made from tetrafluoroethylene. (C_2F_4 or CF_2)
    d. Glucose is a sweet-tasting molecule. (CH_2O or $C_6H_{12}O_6$)

76. In each of these examples, determine the correct empirical formula.
    a. Acetic acid is found in vinegar: $C_2H_4O_2$.
    b. Cane sugar is called sucrose: $C_{12}H_{22}O_{11}$.
    c. Paradichlorobenzene is used as a moth repellent: $C_6H_4Cl_2$.
    d. Benzene is a component of gasoline: C_6H_6.

Section 2.9 Naming Compounds

Skill Review

77. Provide the correct name for each of these binary molecular compounds.
    SO_2 N_2O_5 Cl_2O
    PCl_3 CCl_4

78. Provide the correct name for each of these binary molecular compounds.
    NO_2 PBr_5 OF_2
    N_2O P_2O_5

79. Provide the correct name for each of these binary ionic compounds.
    K_2O $CaBr_2$ Li_3N
    $AlCl_3$ BaS

80. Provide the correct name for each of these binary ionic compounds.
    CaO $MgCl_2$ Na_2S
    Al_2O_3 LiF

81. Provide the correct name for each of these ionic compounds.
    $CaBr_2$ $Fe(NO_3)_3$ $CaSO_4$
    NH_4Cl NaCl

82. Provide the correct name for each of these ionic compounds.
    $KHCO_3$ $Ca(CN)_2$ $Co(OH)_2$
    $Cu(NO_3)_2$ $MgSO_3$

83. Provide the correct formula for each of these compounds.
    copper(II) hydroxide
    chromium(III) oxide
    sulfur hexachloride
    carbon tetraiodide
    aluminum hydroxide
    magnesium sulfate
    sodium sulfite
    ammonium hydroxide
    boron tribromide
    sodium acetate

84. Provide the correct formula for each of these compounds.
    hydrogen chloride
    cobalt(VI) fluoride
    dinitrogen tetroxide
    calcium perchlorate
    barium nitrate
    lead(IV) chloride
    potassium dichromate
    sodium bicarbonate
    lithium hydroxide
    titanium(IV) carbonate

85. Using the metal ion Mn^{5+}, write the correct formula for a combination with each of the following: sulfate; chloride; nitrite; carbonate; and bisulfite.

86. Using the metal ion Cr^{6+}, write the correct formula for a combination with each of the following: sulfate; chloride; nitrite; carbonate; and bisulfite.

87. Write the correct formula for oxalate compounds of the following metals: Mn^{2+}; Cu^+; Fe^{3+}; Mn^{5+}; and Ti^{4+}.

88. Write the correct formula for nitrate compounds of the following metals: Mn^{2+}; Cu^+; Fe^{3+}; Mn^{5+}; and Ti^{4+}.

89. Name these compounds.
    $(NH_4)_2CO_3$ $NaHCO_3$
    $Cu(HSO_3)_2$ $Ca(OH)_2$
    $KMnO_4$ Na_3PO_4
    $Mg(CN)_2$ $LiClO_3$

90. Name these compounds.
    $Zn(CN)_2$ Na_2CO_3
    Na_2CrO_4 K_2HPO_4
    $Li_2Cr_2O_7$ $Ba(NO_2)_2$
    $SrSO_4$ $KClO$

91. Determine the charge on the metal cation in these compounds.
    $V(NO_3)_5$ $TiSO_4$ $W(C_2O_4)_3$
    AgOH $Ru(HCO_3)_3$

92. Determine the charge on the metal cation in these compounds.
    $Cr(NO_3)_3$ MnO_2 $Pd(C_2O_4)_2$
    $AuCl_3$ $Co(OH)_3$

Chemical Applications and Practices

93. A lab accident caused part of the label on a bottle of a chemical to be obscured. All that is visible is "$Fe(NO_3)$." Is this enough information to name the compound? What possible suggestions could you make to determine the identity of the compound?

94. The same lab accident caused part of the label on another bottle to be obscured as well. The visible part reads "NH₄C." Which of the following would you suggest as the most likely name of the compound in this bottle? Explain why you reject the others: ammonium carbonate, ammonium oxalate, ammonium chloride, ammonium copper(II).

Comprehensive Problems

95. What were the major objections to Dalton's atomic theory? Rewrite Dalton's atomic theory to take into account our current understanding of matter.

96. Match each chemical statement on the left-hand side of the following list with a correlating statement from Dalton's atomic theory on the right.

Chemistry

a. Graphite and diamond have different properties, but both are made of only atoms of carbon.

b. $2H_2 + O_2 \rightarrow 2H_2O$

c. It takes 6.02×10^{23} atoms of iron to make 55.85 g.

d. It takes 6.02×10^{23} atoms of gold to make 197 g.

Atomic Theory

(1) Elements are made up of tiny particles called atoms.

(2) Atoms of a given element are identical; atoms of different elements differ in some fundamental way.

(3) Compounds are formed when atoms combine; each compound always has the same numbers and types of atoms.

(4) Chemical reactions consist of the reorganization of the atoms involved; the atoms themselves are not changed.

97. When hydrogen gas combines with chlorine gas, a corrosive gas known as hydrogen chloride forms. According to Dalton, the volume ratio between the reactants and products should be 1:1:1. However, Gay-Lussac would claim a ratio of 1:1:2 (assuming that all gases are at the same temperature and pressure). Gay-Lussac's claim was proved to be valid. Which law is being used here? Explain why this explanation works better than Dalton's.

98. If the mass of a metal nut were 35.0 g and the mass of a matching threaded bolt were 7.0 g, a combined nut and bolt would have a mass of 42.0 g.
a. What would be the mass of the threaded bolts needed to attach to 175.0 g of nuts?
b. If someone decided to attach two bolts to a nut, what would be the resulting mass of the unit?
c. What would be the mass of the threaded bolts needed to attach to 175.0 g of nuts under the conditions stated in part b?
d. What is the mass ratio between bolts connected to nuts in the first unit (part a) and bolts connected to nuts in the second unit (part b)?

99. Robert Millikan determined the electrical charge for an electron, and its mass was easily calculated by using the charge-to-mass ratio. The following analogy applies roughly the same logic: Suppose you were buying a box of golf balls. What information would you need in order to know the price of each golf ball?

100. A tiny oil droplet is given an electrical charge by an electric current. The charge causes the oil droplet to be attracted to an oppositely charged metal plate above it. What two forces must be equal in order for an oil droplet to be suspended (neither rising nor falling) in the apparatus?

Problem 100

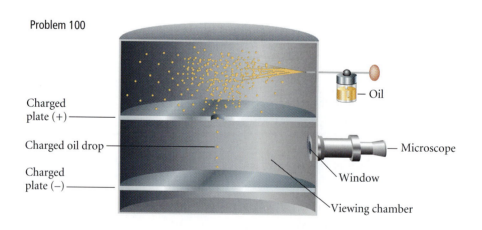

Charged plate (+)

Charged oil drop

Charged plate (−)

Oil

Microscope

Window

Viewing chamber

101. How much mass was lost from the combination of protons, neutrons, and electrons when ^{63}Cu (62.9296011 amu) was formed? What happened to this mass?

102. If the atoms in a 2.500 g sample of aluminum were lined up end to end, how long would the line of iron be in inches? (Assume the aluminum atoms are spheres with a radius of 143 pm.)

103. If a sample of carbon atoms contains 7.45×10^{15} atoms, how many gross of atoms are there in the sample (1 gross = 144 atoms)? If each of those atoms is 154 pm in diameter, would they stretch to 1 mile if they were lined up so that they were touching end to end?

104. During the analysis of an unknown sample of metal, a researcher places a 16.07 g chunk of the metal into a graduated cylinder containing 20.3 mL of water. The volume of the mixture was then recorded to be 22.3 mL. What is the density of the metal?

105. A new isotope of beryllium is created such that it contains 9 neutrons.
    a. How many protons and electrons would an atom of this isotope contain?
    b. What is the mass number for this new isotope?
    c. Use the information in Table 2.3 to determine the mass of the isotope in kilograms. (Assume that no mass is lost in the creation of the new isotope from the individual parts.)

Thinking Beyond the Calculation

106. In the produce section of a grocery store, bags of small tomatoes were being packaged. The weight of each bag was then measured, and the weights were found to be as follows (assume these are exact weights):

32 oz	28 oz
18 oz	22 oz
36 oz	14 oz

    a. What is the mass of each bag in grams?
    b. Judging on the basis of the masses recorded in ounces, and the simplifying assumption that the tomatoes are the same size, what would you estimate as the weight of just one tomato?
    c. Another bag had an exact weight of 20 oz; how does that influence your answer to part a? What is the chemical law that you are using to answer this question?
    d. Describe the process that you could follow to identify the volume of each of the tomatoes.

107. In this chapter, we discussed determining the age of objects using radioactive isotopes. Scientists are constantly refining these techniques to correct for the differences between conditions long ago and those at the present time. What conditions might have changed that could require correcting? In each case, what would be the nature of the correction (that is, whether the real age is greater or less) and why?

108. Scientists now think that much of the universe is composed of "dark matter." Investigate some scientific journals, such as *Scientific American*, to learn enough about dark matter to describe it. On the basis of your research, discuss the experiments that compelled scientists to draw this conclusion. Why is dark matter useful to understand?

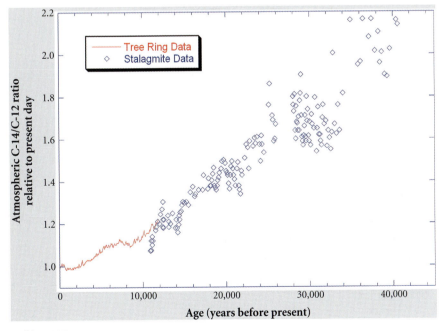

Problem 107

3

Introducing Quantitative Chemistry

The use of quantitative chemical equipment is important in the biomedical field.

Contents and Selected Applications

3.1 Formula Masses

Chemical Encounters: Adrenaline in the Body

3.2 Counting by Weighing

3.3 Working with Moles

Chemical Encounters: Cholesterol in the Blood

3.4 Percentages by Mass

Chemical Encounters: Nitrogen in Fertilizers

3.5 Finding the Formula

Chemical Encounters: A Focus on Chocolate

3.6 Chemical Equations

Chemical Encounters: Preparation of Aspirin

3.7 Working with Equations

Chemical Encounters: Cleaning the Air on Manned Spacecraft

Go to **college.hmco.com/pic/kelterMEE** for online learning resources.

A sudden, loud noise, the shock of some dramatic twist in a horror film, losing control of your car on an icy road—any one of these can cause the rapid physical responses we attribute to the release of adrenaline, including increasing heart rate, stronger heartbeat, sweating, and shifting of some blood flow from the skin to the internal organs. Adrenaline, also known as epinephrine (Figure 3.1), is normally present in your blood at approximately 2×10^{-8} g/L and is secreted from the adrenal glands near your kidneys. In times of extreme stress, your adrenaline level increases *a thousandfold*. The powerful effects of adrenaline demonstrate the vital significance of changing *quantities* of chemicals. The chemistry of the body provides many other examples. For instance, taking 600 mg of aspirin can effectively relieve pain. Taking 100 times that amount, 60 g, would kill you. Quantities in chemistry are crucial.

Up to this point, we have been dealing with individual atoms and molecules. However, in the lab we typically measure quantities in liters, grams, and other "macro-sized" units. What is the relationship between "nanoworld" and "macroworld" quantities? Specific questions arise: How many molecules of adrenaline will be in each liter of blood? How can we actually measure the level of adrenaline or any other chemical in blood? These are the kinds of quantitative issues we explore in this chapter.

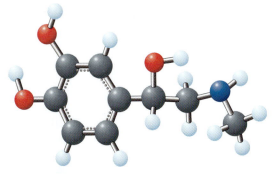

FIGURE 3.1

Adrenaline (epinephrine) has the chemical formula $C_9H_{13}NO_3$. In this "ball-and-stick" representation, carbon atoms are charcoal gray, hydrogen atoms are light blue, oxygen atoms are red, and nitrogen atoms are dark blue.

Aspirin

3.1 Formula Masses

In addition to being naturally present in the body, adrenaline is administered as a drug to stimulate the heart, to alleviate allergic reactions, and even to help break up fat cells during liposuction. As you might expect, control over the amounts administered is vital.

To make, use, or detect specific quantities of adrenaline, we need information about the mass of its molecules, so we may know how many molecules are in any given mass that is administered to a patient.

In Chapter 2 we looked at the method for determining the masses of atoms, and we saw that each element has a given mass (expressed in atomic mass units), which is the average mass of one atom of the element. We can work out the average **molecular mass** (or molecular weight) of a molecule, such as adrenaline, by adding together the average atomic masses of all the atoms present in the molecule. To simplify our calculations, we will use atomic masses rounded to two decimals, except for hydrogen, which we will round to three decimals.

Application

CHEMICAL ENCOUNTERS:
Adrenaline in the Body

$$\text{Formula of each adrenaline molecule} = C_9H_{13}NO_3$$

In the formula we have

9 carbon atoms @ 12.01 amu	= 108.09
13 hydrogen atoms @ 1.008 amu	= 13.104
1 nitrogen atom @ 14.01 amu	= 14.01
3 oxygen atoms @ 16.00 amu	= 48.00
Average molecular mass of adrenaline	= 183.204 = 183.2 amu

Tutorial: Formula Masses

This is also the average **formula mass** (or formula weight) of adrenaline. *Formula mass* is a more general term than *molecular mass*, because it also covers compounds that do not exist in the form of individual molecules, such as ionic compounds. For example, the ionic compound sodium chloride (NaCl) has an average formula mass of 58.44 amu, found by adding the atomic mass of sodium, 22.99 amu, to the atomic mass of chlorine, 35.45 amu.

NaCl

Na	22.99 amu
Cl	35.45 amu
Total	58.44 amu

The formula mass, or molecular mass, of adrenaline is the average mass of one molecule of adrenaline. Each molecule of the substance contains exactly the same number of each type of atom, but different individual molecules can have different masses, depending on which isotopes of carbon, oxygen, or hydrogen they contain.

EXERCISE 3.1   Calculating Formula Masses

1. The formula for aspirin is $C_9H_8O_4$. What is the formula mass of aspirin?

2. Many of us use the stimulant properties of the chemical caffeine in coffee to remain alert, but even just the *smell* of coffee can seem to perk us up in anticipation. One of the molecules that produce the aroma of coffee is C_5H_6OS (2-furylmethanethiol). Calculate the formula mass of 2-furylmethanethiol.

Solution

1. Aspirin = $C_9H_8O_4$

9 atoms of carbon @ 12.01 amu	= 108.09
8 atoms of hydrogen @ 1.008 amu	= 8.064
4 atoms of oxygen @ 16.00 amu	= 64.00
Formula mass	= 180.154 = 180.2 amu

2. 2-Furylmethanethiol = C_5H_6OS

5 atoms of carbon @ 12.01 amu	= 60.05
6 atoms of hydrogen @ 1.008 amu	= 6.048
1 atom of oxygen @ 16.00 amu	= 16.00
1 atom of sulfur @ 32.07 amu	= 32.07
Formula mass	= 114.168 = 114.17 amu

PRACTICE 3.1

Calculate the formula masses of these compounds:

a. Codeine, a painkiller with the formula $C_{18}H_{21}NO_3$

b. Magnesium sulfate, found in many medical ointments, with the formula $MgSO_4$

c. Trinitrotoluene (TNT), an explosive with the formula $C_7H_5N_3O_6$

See Problems 1–4.

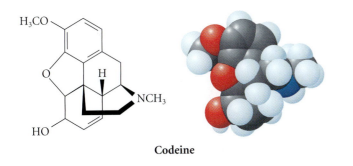

Codeine

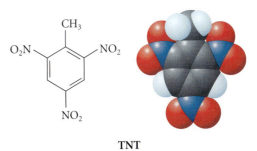

TNT

3.2 Counting by Weighing

When we do chemistry, we generally begin by measuring our chemicals using masses. However, we need to convert these masses into some measure of the *numbers* of atoms, molecules, or ions in order to relate the masses to what is actually happening in the chemical nanoworld. **Why can't we take a sample of adrenaline, which comes as a white powder, and use a pair of tweezers to separate out the number of individual molecules we need?** Adrenaline molecules, like all molecules, are stunningly small—on the order of 1 nm across. This means that you would need to line up *100 million adrenaline molecules* with your tweezers to get a sample as long as your index finger (about 10 cm)! This is why we need to establish the link between *weighing* things and *counting* them. Weighing a sample of adrenaline gives us insight into how many molecules of adrenaline we have.

We can see how this is done on a more familiar size scale by using one tablet of vitamin C that weighs 0.75 g. If all the tablets in a jar are placed onto a balance, and if they weigh a total of 45.0 g, how many tablets are in the jar? As shown in Figure 3.2, there are a total of 60 tablets in the jar.

$$45 \text{ g} \times \frac{1 \text{ vitamin C tablet}}{0.75 \text{ g}} = 60 \text{ vitamin C tablets}$$

We can do exactly the same thing with the atoms, molecules, and ionic compounds, the only difference being that the numbers are most often extraordinarily large or small.

Suppose you had just taken a pain reliever that contained aspirin, also known as acetylsalicylic acid (see Figure 3.3). The chemical formula for aspirin is $C_9H_8O_4$. If the tablet you used contained 500 mg (0.500 g) of aspirin, could you calculate how many *molecules* of pain reliever you had just ingested? You would

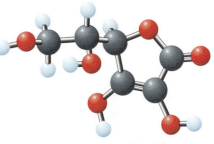

Vitamin C (ascorbic acid)

One vitamin C tablet weighs 0.75 g.

All the tablets in the jar weigh 45 g.

$$45 \text{ g} \times \frac{1 \text{ tablet}}{0.75 \text{ g}} = 60 \text{ tablets}$$

FIGURE 3.2

Counting by weighing—the basic logic.

FIGURE 3.3

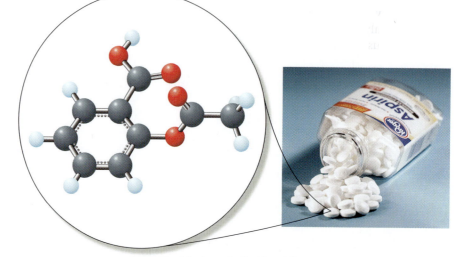

Aspirin (acetylsalicylic acid)
$C_9H_8O_4$

need to recall from Chapter 2 that 1 amu $= 1.6605 \times 10^{-24}$ g $(1.6605 \times 10^{-27}$ kg$)$. Here is the calculation:

The average molecular mass of aspirin, calculated in Exercise 3.1, is 180.2 amu.

$$\frac{180.2 \text{ amu}}{1 \text{ molecule}} \times \frac{1.66 \times 10^{-24} \text{ g}}{1 \text{ amu}} = 2.99 \times 10^{-22} \text{ g per molecule}$$

$$0.500 \text{ g} \times \frac{1 \text{ molecule}}{2.99 \times 10^{-22} \text{ g}} = 1.67 \times 10^{21} \text{ molecules}$$

This quantity is approximately 300 billion times more than the current human population of the earth! The comparison serves as a reminder of just how tiny molecules must be if that many are required to make up half a gram.

Knowing that the average molecular mass of aspirin is 180.2 amu, a chemist at a pharmaceutical plant could weigh out a 180.2-g sample. How many molecules would that sample contain? Let's do the calculations in order to prove that *the number of molecules of aspirin (molecular mass = 180.2 amu) in a 180.2-g sample must be the same as the number of amu's in 1 g* (see Figure 3.4). How many amu's are in 1 g? The mass of 1 amu $= 1.6605 \times 10^{-24}$ g, so we can set up this equation and solve it:

$$\frac{1 \text{ amu}}{1.6605 \times 10^{-24} \text{ g}} = \frac{x \text{ amu}}{1 \text{ g}}$$

$x = 6.022 \times 10^{23}$ amu in 1 g of any substance.

What a huge number that is!

FIGURE 3.4

The link between masses in amu and in grams.

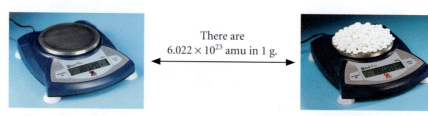

One aspirin molecule, too small to see!

There are 6.022×10^{23} amu in 1 g.

6.022×10^{23} aspirin molecules (known as 1 mol of aspirin)

amu per molecule = grams per mole

How many molecules of aspirin are in 180.2 g of aspirin? We'll use our dimensional analysis method to solve the problem. First we create a flowchart that will help us arrive at our answer.

$$\text{g aspirin} \xrightarrow{\frac{amu}{grams}} \text{amu aspirin} \xrightarrow{\frac{molecules}{amu}} \text{molecules aspirin}$$

Then we perform the calculation.

$$180.2 \text{ g aspirin} \times \frac{1 \text{ amu aspirin}}{1.6605 \times 10^{-24} \text{ g aspirin}} \times \frac{1 \text{ molecule aspirin}}{180.2 \text{ amu aspirin}}$$

$$= 6.022 \times 10^{23} \text{ molecules aspirin}$$

This is the same huge number as the number of amu's in 1 g of a substance!

By choosing to weigh out a *mass in grams that has the same numerical value as the formula mass in amu's* (180.2 grams and 180.2 amu, in the case of aspirin) we are able to ensure that the sample contains just as many molecules as there are amu's in 1 g. This gives us a very useful automatic link between masses and numbers of molecules (or atoms or ions) that works for any chemical.

This is a difficult construct, and it is entirely possible to read the preceding discussion several times and still feel uncertain about the mass-to-amu relationships. To help us reinforce the idea, let's do the same calculation with an element, rather than a compound. The element calcium is crucial to the manufacture of healthy bones and teeth. The average atomic mass of the element calcium is 40.08 amu, so we need to check how many calcium atoms are in 40.08 g of calcium:

$$40.08 \text{ g} \times \frac{1 \text{ amu}}{1.6605 \times 10^{-24} \text{ g}} \times \frac{1 \text{ atom}}{40.08 \text{ amu}} = 6.022 \times 10^{23} \text{ atoms}$$

That number 6.022×10^{23} appears again! This is a very important and useful number for us, because if we weigh out a mass *in grams* that is the same as an element's atomic mass or a molecule's formula mass *in amu,* we will always have 6.022×10^{23} atoms or formula units. The number 6.022×10^{23} is known as **Avogadro's number (N_A)** (or the Avogadro constant), after the Italian physicist Amadeo Avogadro (1776−1856; Figure 3.5) and is abbreviated as N_A.

FIGURE 3.5

Lorenzo Romano Amadeo Carlo Avogadro, conte di Quaregna e di Cerreto (1776–1856), was born into a family of well-established lawyers in Italy. He, too, prepared for a legal career, obtaining his law degree in 1792 when he was only 16 years old. However, in 1800 he began studying mathematics and physics privately, completing his first research project on electricity with his brother Felice in 1803. He was hired in 1809 at the College of Vercelli, where he began his most influential work in chemistry. He eventually obtained an appointment as a professor of mathematical physics at the University of Turin, where he worked until he retired in 1850.

Video Lesson: The Mole and Avogadro's Number

EXERCISE 3.2   How Many Molecules?

Codeine, with the formula $C_{18}H_{21}NO_3$, is often used as an analgesic (painkiller) for intense pain. How many molecules would be in a 0.10-g sample of codeine?

First Thoughts

Our first thoughts lead us to determining the average molecular mass of codeine. Once this mass is known, we should be able to determine the number of molecules in that mass, using the same strategy we employed to get the number of atoms of aspirin and the number of atoms of copper.

Solution

The molecular mass of codeine, $C_{18}H_{21}NO_3$, is 299.4 amu.

$$0.10 \text{ g codeine} \times \frac{1 \text{ amu codeine}}{1.6605 \times 10^{-24} \text{ g codeine}} \times \frac{1 \text{ molecule codeine}}{299.4 \text{ amu codeine}}$$

$$= 2.0 \times 10^{20} \text{ molecules codeine}$$

Further Insight

How many molecules are there in 299.37 g of codeine? Again, this would give Avogadro's number as the answer. This number must be rather important or we

Codeine

wouldn't keep running into it! In fact, in chemistry this number is vital to many of the calculations you will accomplish; accordingly, we give it a special name when we do calculations (see the next section).

PRACTICE 3.2

How many atoms are there in 100.0 g of gold-197?

See Problems 9–16.

The Quantity We Call the Mole

An amount of any substance that contains 6.022×10^{23} "elementary entities" such as molecules, atoms, or ions is given its own name in chemistry. We call it one **mole** (often abbreviated mol) of the substance. Perhaps it seems odd to give a special name to what is, in effect, just a number, but this is actually a very familiar practice:

$$1 \text{ mole corresponds to } 6.022 \times 10^{23}$$

just as

$$1 \text{ dozen corresponds to } 12$$

The mole is the basic counting unit used by chemists to indicate how many atoms, molecules, or ions they are dealing with. A chemist counts moles of particles just as a baker counts dozens of cupcakes.

To work out the mass of one mole of a chemical, you just look up its formula mass and convert the units of that number from amu into grams. The resulting quantity, expressed in grams per mole ($\frac{g}{mol}$), is known as the **molar mass**—*the mass of one mole of a substance.* Remember these relationships (illustrated by the example in Figure 3.4):

- The mass of 1 mol of any chemical is its formula mass expressed in units of grams. This is the molar mass of the substance. The molar mass of aspirin is $\frac{180.2 \text{ g}}{mol}$.

- 1 mol corresponds to 6.022×10^{23} entities, such as atoms, molecules, or ions.

- 6.022×10^{23} is known as Avogadro's number. It is often abbreviated as N_A.

Although you should learn to think of the mole as simply equal to 6.022×10^{23}, it is formally defined in a more complicated way by reference to the carbon-12 isotope. This formal definition states that *a mole of a substance is that amount of the substance that has the same number of atoms as exactly 12 g of the isotope carbon-12.*

Does this make sense in terms of what we have already said about the mole? We can check that the definition fits with the "1 mol = 6.022×10^{23}" rule by calculating the number of particles in exactly 12 g of the isotope carbon-12. The atomic mass of carbon-12 is *exactly* 12 amu.

$$12.00 \text{ g} \times \frac{1 \text{ amu}}{1.6605 \times 10^{-24} \text{ g}} \times \frac{1 \text{ atom}}{12 \text{ amu}} = 6.022 \times 10^{23} \text{ atoms}$$

The "dozen" is a counting unit. Its use in counting groups is similar to a pair (2), a ream (500), and a mole (6.022×10^{23}).

Video Lesson: Introducing Conversions of Masses, Moles, and Number of Particles

3.3 Working with Moles

Molecules to Moles and Back Again

Figure 3.6 illustrates the quantities of 1 mol of water, 1 mol of sodium chloride ("table salt"), and 1 mol of aspirin—three different chemicals of great significance to life and medicine. Each has a different mass and occupies a different volume, but each contains the same number of formula units of the compounds concerned. The containers in the figure contain 6.022×10^{23} molecules of water, 6.022×10^{23} molecules of aspirin and 6.022×10^{23} formula units of NaCl. (*Note:* Because each formula unit of NaCl includes one Na^+ ion and one Cl^- ion, our sample contains 6.022×10^{23} Na^+ ions and 6.022×10^{23} Cl^- ions!)

EXERCISE 3.3 Molecules and Moles

Really large numbers such as 4.33×10^{24} can be quite confusing. Let's show how using moles can simplify our work.

a. How many moles of water are there in 4.33×10^{24} molecules of water?

b. How many moles of codeine ($C_{18}H_{21}NO_3$) are there in 4.33×10^{24} molecules of codeine?

Solution

a. 4.33×10^{24} molecules $\times \dfrac{1 \text{ mol}}{6.022 \times 10^{23} \text{ molecules}} = 7.19 \text{ mol}$

This number is much easier to work with.

b. Note that the formula of the compound doesn't change the number of molecules that exist in 1 mol of a substance. Therefore, the answer is the same. 4.33×10^{24} molecules of any substance is equal to 7.19 mol, whether we are considering aspirin, adrenaline, copper, or codeine.

FIGURE 3.6

One mole of water (top), one mole of sodium chloride ("salt") (bottom), and one mole of aspirin (center). 1 mol = Avogadro's number = 6.022×10^{23}.

Sodium benzoate

PRACTICE 3.3

How many moles are there in 2.88×10^{20} formula units of sodium benzoate (C_6H_5COONa), a food preservative? How many formula units are there in 1.5 mol of sodium benzoate?

See Problems 19b, 19d, 20b, 20d, 23b, and 24b.

Grams to Moles and Back Again

Many of us sweeten our drinks by adding some "table sugar," the compound sucrose, with the formula $C_{12}H_{22}O_{11}$. If you added 10.0 g of sugar to a glass of tea, how many moles did you add? To answer this question, we must first determine the formula mass of sucrose, which tells us the mass of 1 mol and so gives us the mass-to-moles conversion factor that we need. Here is the calculation:

$$12 \text{ mol C} \times \frac{12.01 \text{ g}}{1 \text{ mol C}} = 144.12 \text{ g}$$

$$22 \text{ mol H} \times \frac{1.008 \text{ g}}{1 \text{ mol}} = 22.176 \text{ g}$$

$$11 \text{ mol O} \times \frac{16.00 \text{ g}}{1 \text{ mol}} = 176.0 \text{ g}$$

Molar mass of sucrose = 342.296 = 342.3 g

This means that the mass of 1 mol of sucrose is 342.3 g of sucrose. Mathematically, our moles-to-mass conversion factor is $\dfrac{1 \text{ mol sucrose}}{342.3 \text{ g sucrose}}$. Now, our sample of 10.0 g can be converted to moles:

$$10.0 \text{ g sucrose} \times \frac{1 \text{ mol sucrose}}{342.3 \text{ g sucrose}} = 0.02921 = 0.0292 \text{ mol sucrose}$$

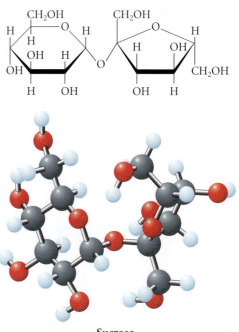

Sucrose

We can also determine how many *molecules* of sucrose have been added to our glass of tea. We do this using Avogadro's number, the number we call the mole, 6.022×10^{23}.

$$0.02921 \text{ mol sucrose} \times \frac{6.022 \times 10^{23} \text{ molecules sucrose}}{1 \text{ mol sucrose}}$$

$$= 1.759 \times 10^{22} \text{ molecules sucrose}$$

$$= 1.76 \times 10^{22} \text{ molecules sucrose}$$

We would have stirred this number of sugar molecules into our tea:

$$17,600,000,000,000,000,000,000$$

HCT

EXERCISE 3.4   Calculating Masses Corresponding to Moles

Hydrochlorothiazide (HCT) is a medicine used to lower blood pressure and deal with other illnesses related to fluid retention. Its molecular formula is $C_7H_8ClN_3O_4S_2$.

a. In a small-scale production process, a company needs 63.4 mol of the medicine. How many grams of hydrochlorothiazide is this?

b. A common individual dosage is 3.37×10^{-5} mol. How many milligrams are in this individual dose?

First Thoughts

We have two very different amounts to calculate. The first represents a small-scale industrial process, the second an individual dose. But the strategy is still the same. We can use the molar mass of the compound to go from moles to mass. The molar mass of the compound in grams per mole will be the same as the average molecular mass in amu.

Solution

The molar mass of hydrochlorothiazide is 297.72 g per mole. Again, we first write a flowchart to assist us in our calculations:

a. $\text{mol HCT} \xrightarrow{\frac{\text{g HCT}}{\text{mol HCT}}} \text{g HCT}$

$$63.4 \text{ mol HCT} \times \frac{297.72 \text{ g HCT}}{1 \text{ mol HCT}} = 1.89 \times 10^4 \text{ g HCT}$$

b. $\text{mol HCT} \xrightarrow{\frac{\text{g HCT}}{\text{mol HCT}}} \text{g HCT} \xrightarrow{\frac{\text{mg}}{\text{g}}} \text{mg HCT}$

$$3.37 \times 10^{-5} \text{ mol HCT} \times \frac{297.72 \text{ g HCT}}{1 \text{ mol HCT}} \times \frac{1 \text{ mg HCT}}{1 \times 10^{-3} \text{ g HCT}} = 10.0 \text{ mg HCT}$$

Further Insights

Do the answers make sense? We expect the industrial synthesis to result in much more of the product than is in the individual dose, so this answer makes sense. You may have noted from other medicines you have taken that a dose of perhaps 10 to 500 mg is standard. Therefore, our answers are reasonable.

PRACTICE 3.4

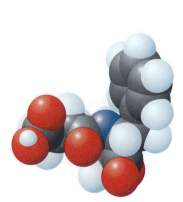

Aspartame

Calculate the mass in grams of 26 mol and of 0.0025 mol of each of these compounds: table salt (NaCl), water (H_2O), and the sweetener aspartame ($C_{14}H_{18}N_2O_5$).

See Problems 25 and 26.

EXERCISE 3.5 | Calculating Moles Corresponding to Masses

Methylphenidate hydrochloride, $C_{14}H_{19}NO_2 \cdot HCl$, is a medical drug used to stimulate the central nervous system. It is known as Ritalin, and its use in the treatment of attention deficit hyperactivity disorder (ADHD) has been a subject of controversy for years. A company sent a 2.00×10^4-g shipment to be processed into individual 20.0-mg tablets. How many moles of Ritalin are in the shipment and in each tablet? How many molecules of Ritalin are in each tablet? Note that the dot in the formula is used to indicate small molecules that are chemically part of the overall formula. Therefore, the formula methylphenidate hydrochloride could be written as $C_{14}H_{20}NO_2Cl$.

Solution

For the entire shipment, we calculate the chemical amount in moles as follows, after determining that the molar mass of Ritalin is 269.76 g/mol.

$$2.00 \times 10^4 \text{ g Ritalin} \times \frac{1 \text{ mol Ritalin}}{269.8 \text{ g Ritalin}} = 74.1 \text{ mol Ritalin}$$

We calculate the number of moles in each 20.0-mg tablet using the same general strategy, being careful to add the extra conversion factor that changes milligrams to grams before finding moles.

$$\frac{\text{mg Ritalin}}{\text{tablet}} \xrightarrow{\frac{g}{mg}} \frac{\text{g Ritalin}}{\text{tablet}} \xrightarrow{\frac{mole}{g \, Ritalin}} \frac{\text{mol Ritalin}}{\text{tablet}}$$

$$\frac{20.0 \text{ mg Ritalin}}{\text{tablet}} \times \frac{1 \text{ g Ritalin}}{1000 \text{ mg Ritalin}} \times \frac{1 \text{ mol Ritalin}}{269.8 \text{ g Ritalin}}$$

$$= 7.413 \times 10^{-5} = 7.41 \times 10^{-5} \text{ mol Ritalin per tablet}$$

We can convert from moles to molecules of Ritalin in a 20.0-mg tablet:

$$\frac{\text{mol Ritalin}}{\text{tablet}} \xrightarrow{\frac{molecules}{mole}} \frac{\text{molecules Ritalin}}{\text{tablet}}$$

$$\frac{7.413 \times 10^{-5} \text{ mol Ritalin}}{\text{tablet}} \times \frac{6.022 \times 10^{23} \text{ molecules Ritalin}}{1 \text{ mol Ritalin}}$$

$$= \frac{4.46 \times 10^{19} \text{ molecules Ritalin}}{\text{tablet}}$$

The individual dose has considerably fewer moles than the industrial shipment, so the answer makes sense. The number of molecules in the individual dose is exceedingly high, which is also reasonable because molecules are so small.

PRACTICE 3.5

Calculate the number of moles in 5.3 g and 0.0022 g these compounds: lactic acid, $C_3H_6O_3$, and sulfuric acid, used in car batteries (H_2SO_4).

See Problems 19a, 19c, 20a, and 20c.

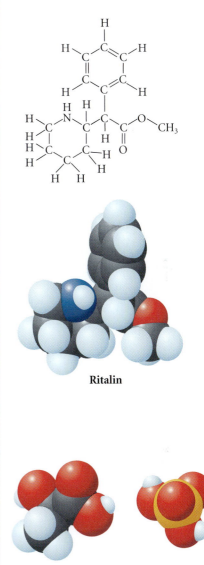

Ritalin

Lactic acid   **Sulfuric acid**

As researchers have investigated the causes of heart-related health problems, one particular molecule, cholesterol, has been found to be associated with certain types of circulatory problems. Giving a blood sample for the analysis of your cholesterol level is a routine part of a medical checkup. The situation is not as simple as cholesterol being "bad." Cholesterol is essential for life—without it, you would die. Cholesterol plays a crucial role in maintaining the thin membranes that enclose every cell of your body. It also aids in the production of hormones,

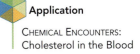

Application

CHEMICAL ENCOUNTERS:
Cholesterol in the Blood

FIGURE 3.7

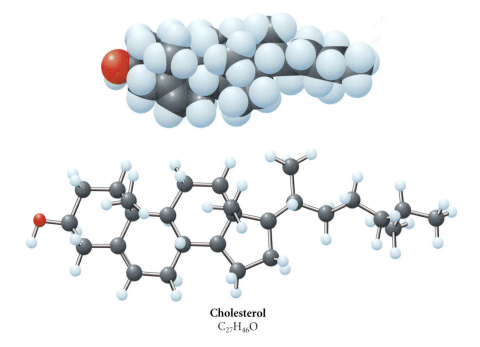

Cholesterol
$C_{27}H_{46}O$

including estrogen and testosterone. So again, we find that the *quantity* of a chemical is crucial. The chemical structure of cholesterol is shown in Figure 3.7. Its formula is $C_{27}H_{46}O$. If a blood analysis showed a cholesterol level of 195 mg per deciliter of blood, which is equal to 1950 mg per liter, how many molecules of cholesterol would be present per liter?

In this case you can see that the starting measurement is in milligrams, but the question asks about the number of molecules. *Remember that the "nano" (molecules, in this case) and "macro" (moles, here) worlds are connected via Avogadro's number.* The strategy is to express the mass as a quantity of moles of cholesterol and then convert moles to molecules. Our dimensional analysis flowchart for the task looks like this:

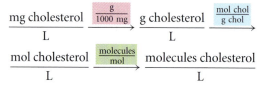

When we have finished our calculation, our answer should make sense. **What sort of answer should we expect?** We know that there are a great many molecules in even the smallest sample of cholesterol-containing blood. We would therefore expect a very large number of molecules. Answers such as 10^{-24} molecule/L or even 10^{2} molecules/L would not make sense.

Because cholesterol has a known molecular formula, $C_{27}H_{46}O$, we can determine its molar mass as 386.6 g per mole (check that you agree). We will follow our flowcharted strategy to get the answer.

$$\frac{1950 \text{ mg } C_{27}H_{46}O}{L} \times \frac{1 \text{ g } C_{27}H_{46}O}{1000 \text{ mg } C_{27}H_{46}O} \times \frac{1 \text{ mol } C_{27}H_{46}O}{386.6 \text{ g } C_{27}H_{46}O}$$

$$\times \frac{6.022 \times 10^{23} \text{ molecules } C_{27}H_{46}O}{1 \text{ mol } C_{27}H_{46}O} = \frac{3.04 \times 10^{21} \text{ molecules } C_{27}H_{46}O}{L}$$

Does the answer make sense? The quantity 195 mg is much less than a mole of cholesterol (386.6 g per mole), so we would have expected an answer far less than Avogadro's number, but because we are counting our sample in molecules, the number should still be quite large. Our value meets these conditions, so our answer makes sense.

EXERCISE 3.6 Sugar Is Sweet

Sucrose ($C_{12}H_{22}O_{11}$) is a wonderful molecule used to sweeten our cup of coffee. If a packet of sucrose contains 1.50 g of sucrose, how many molecules is this?

Solution

$$1.50 \text{ g sucrose} \times \frac{1 \text{ mol sucrose}}{342.3 \text{ g sucrose}} \times \frac{6.022 \times 10^{23} \text{ molecules of sucrose}}{1 \text{ mol sucrose}}$$

$$= 2.639 \times 10^{21}$$

$$= 2.64 \times 10^{21} \text{ molecules of sucrose}$$

PRACTICE 3.6

How many molecules are there in 1.50 g of water? Would you need a bathtub, a swimming pool, or a cup to hold 5.33×10^{29} molecules of water?

See Problems 23a and 24a.

3.4 Percentages by Mass

Just like humans, plants undergo chemical changes that are a part of life. They have genetic material that directs the synthesis of proteins, some of which are part of cell structure and some of which make possible the chemical reactions that sustain life. Nitrogen is a vital element in the genetic material and the proteins in plants, so in order to grow, plants need plenty of nitrogen. For farms throughout the world, fertilizer supplies that nitrogen in the form of manure or industrially produced products.

Some plants need more nitrogen than others. Environmental factors, such as temperature, rainfall, and the type of soil, also affect the ability of plants to use the nitrogen in fertilizer. This means that a farmer must apply the correct amount of nitrogen for each particular situation. Among the most important factors in deciding what fertilizer to use is the *percent by mass*, or **mass percent**, of nitrogen in the product (see Figure 3.8). Mass percent values are determined by dividing the mass of the component of interest by the total mass of the entire sample. The result, multiplied by 100%, is the mass percent of the component in the sample. The component can be a compound (such as ammonium nitrate, NH_4NO_3, which is found in many liquid and solid fertilizers) or it can be an ion, atom, or element.

$$\text{Mass percent} = \frac{\text{total mass component}}{\text{total mass whole substance}} \times 100\%$$

Which of the following two compounds would supply plants with the more concentrated source of nitrogen: ammonium nitrate, NH_4NO_3, or ammonium sulfate, $(NH_4)_2SO_4$?

$$NH_4NO_3 \quad \text{or} \quad (NH_4)_2SO_4$$

We can answer the question by calculating the mass percent of nitrogen present in each compound. We need to express the mass of nitrogen in each one as a percentage of the total mass.

Application

CHEMICAL ENCOUNTERS:
Nitrogen in Fertilizers

FIGURE 3.8

The mass percent of nitrogen is often reported on the ingredients label of a bag of fertilizer.

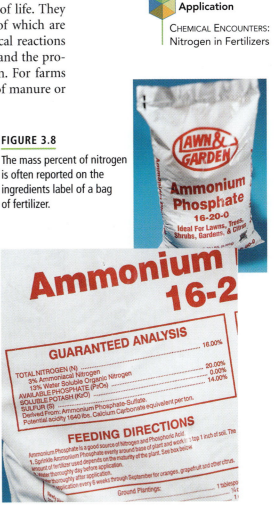

For NH₄NO₃

$$2 \text{ mol N} \times \frac{14.01 \text{ g N}}{1 \text{ mol N}} = 28.02 \text{ g N}$$

$$4 \text{ mol H} \times \frac{1.008 \text{ g H}}{1 \text{ mol H}} = 4.032 \text{ g H}$$

$$3 \text{ mol O} \times \frac{16.00 \text{ g O}}{1 \text{ mol O}} = 48.00 \text{ g O}$$

$$\text{Molar mass} = 80.05 \text{ g}$$

$$\text{Mass percent nitrogen} = \frac{\text{mass total N}}{\text{mass total compound}} \times 100\%$$

$$= \frac{28.02}{80.05} \times 100\% = 35.00\%$$

For (NH₄)₂SO₄

$$2 \text{ mol N} \times \frac{14.01 \text{ g N}}{1 \text{ mol N}} = 28.02 \text{ g N}$$

$$8 \text{ mol H} \times \frac{1.008 \text{ g H}}{1 \text{ mol H}} = 8.064 \text{ g H}$$

$$1 \text{ mol S} \times \frac{32.07 \text{ g S}}{1 \text{ mol S}} = 32.07 \text{ g S}$$

$$4 \text{ mole O} \times \frac{16.00 \text{ g O}}{1 \text{ mol O}} = 64.00 \text{ g O}$$

$$\text{Molar mass} = 132.15 \text{ g}$$

$$\text{Mass percent nitrogen} = \frac{\text{mass total N}}{\text{mass total compound}} \times 100\%$$

$$= \frac{28.02}{132.15} \times 100\% = 21.20\%$$

The results indicate that any chosen mass of NH_4NO_3 contains a significantly greater percentage of nitrogen (35.00%) than does the same mass of $(NH_4)_2SO_4$ (21.20%).

EXERCISE 3.7   Calculating Mass Percent

Phosphorus is another element that is vital for plant growth. Many fertilizers therefore also contain phosphorus and are sold on the basis of the percentage of phosphorus in the fertilizer. One example is calcium dihydrogen phosphate, $Ca(H_2PO_4)_2$. The fertilizer, commonly known in the agriculture industry as "triple superphosphate," is a granular substance that is made from phosphoric acid and phosphate rock and is used in fertilizing grain fields and sugar cane. What is the mass percent of phosphorus in triple superphosphate?

Solution

The compound's molar mass is 234.05 g/mol. The 2 phosphorus atoms contribute $30.97 \times 2 = 61.94$ g per formula unit. The mass percent of phosphorus is

$$\text{Mass percent} = \frac{61.94 \text{ g P}}{234.05 \text{ g compound}} \times 100\% = 26.46\% \text{ P}$$

PRACTICE 3.7

Some mineral supplement tablets contain potassium in the form of potassium chloride (KCl). What is the mass percent of potassium in this compound?

See Problems 37–48.

HERE'S WHAT WE KNOW SO FAR

- Formula masses are calculated by adding the average atomic masses of the individual atoms in a formula.
- Chemists count molecules in a sample by measuring its mass.
- Masses can be converted into moles by using the molar mass of a compound.
- One mole of any substance is 6.022×10^{23} entities.
- The mass percent of a component in a substance is the total mass of the component divided by the total mass of the substance, and then multiplied by 100%.

3.5   Finding the Formula

For many people, chocolate is close to being essential for life! Per-person consumption in 2000 was 22.4 lb in Switzerland and 11.6 lb in the United States. Chocolate is big business. For instance, in 2002, U.S. manufacturers shipped over 1.5 billion kg (3.3 billion lb) of chocolate, worth $8.5 billion. This includes not only candy bars but also powdered chocolate for drinks and baking. How is chocolate made? It is harvested from cacao beans from a tree aptly named *Theobroma cacao*. *Theobroma* means "food of the gods" (Figure 3.9).

A multitude of compounds make up the flavor of chocolate. One of these is named 2,5-dimethylpyrazine. To work with this compound in chemistry, it is necessary to have its empirical and molecular formulas. These are found by experimentally determining the mass percent of the elements present in a sample of the compound. 2,5-Dimethylpyrazine, for example, contains, by mass, 66.62% carbon, 7.47% hydrogen, and 25.91% nitrogen. How can we use these results to calculate the molecular formula of 2,5-dimethylpyrazine? We can begin

Application

CHEMICAL ENCOUNTERS:
A Focus on
Chocolate

Video Lesson: Finding Empirical
and Molecular Formulas

FIGURE 3.9

Theobroma cacao and the food of the gods.

by calculating the ratio in which the atoms of the elements are present in the compound. To do this, we convert the mass ratio obtained from the analysis experiment into a mole ratio. The process is made simpler because the mass percent data tell us how many grams of each element are present in 100 g of the compound. That is, 66.62% carbon means that in every 100 g of 2,5-dimethylpyrazine, 66.62 g are carbon. This is our starting point for the conversion into moles.

$$66.62 \text{ g C} \times \frac{1 \text{ mol C}}{12.01 \text{ g C}} = 5.547 \text{ mol C}$$

$$7.47 \text{ g H} \times \frac{1 \text{ mol H}}{1.008 \text{ g H}} = 7.41 \text{ mol H}$$

$$25.91 \text{ g N} \times \frac{1 \text{ mol N}}{14.01 \text{ g N}} = 1.849 \text{ mol N}$$

If we divide all these mole values by the smallest mole value (1.849), we will obtain the simplest whole-number ratio of the atoms that make up the substance.

$$\frac{5.547}{1.849} = 3 \qquad \frac{7.41}{1.849} = 4 \qquad \frac{1.849}{1.849} = 1$$

In this case, the division step yields the simplest whole-number ratio immediately. We know that it is the simplest whole-number ratio because the ratio 3:4:1 cannot be reduced further. We can now write the empirical formula for the compound as

$$C_3H_4N$$

The numbers will not always work out as neatly as that, sometimes because of experimental errors and sometimes because the formula is a little more complex. For example, if for a different compound we experimentally obtained the results

1.506 mol C   3.007 mol H   1.008 mol O

they could reasonably be simplified to this ratio:

1.5 mol C   3 mol H   1 mol O

This ratio is not the simplest *whole-number* ratio that we need in order to write an empirical formula. However, it can be converted into the simplest whole-number ratio by multiplying all the numbers by 2. When we do this, we get 3 mol C, 6 mol H, and 2 mol O, which gives the empirical formula $C_3H_6O_2$. This illustrates that when we're trying to find the simplest whole-number ratio, we can multiply or divide the set of values by any number we choose, because the ratio will remain the same provided that we perform the same multiplication or division on each value. However, *we always choose the number that provides the simplest whole-number ratio of atoms in the formula.*

The empirical formula of 2,5-dimethylpyrazine, the chocolate flavoring molecule, is C_3H_4N. This formula tells us only the ratio in which the elements are present. It does not tell us the molecular formula, which is the actual number of each type of atom in a molecule of the compound. To determine the molecular formula, we would need to know the average molecular mass of the compound, which can be determined experimentally via the technique of mass spectrometry, introduced in Section 2.5. This would reveal that the average mass of one molecule of the compound is 108.1 amu, corresponding to a molar mass of 108.1 g. What is the average molecular mass of the empirical formula?

If we add together the masses of the atoms in the empirical formula, we get a total of 54.07 amu (or g/mol).

$$3 \text{ mol C} \times \frac{12.01 \text{ g}}{1 \text{ mol C}} = 36.03 \text{ g}$$

$$4 \text{ mol H} \times \frac{1.008 \text{ g H}}{1 \text{ mol H}} = 4.032 \text{ g}$$

$$1 \text{ mol N} \times \frac{14.01 \text{ g N}}{1 \text{ mol N}} = 14.01 \text{ g}$$

$$= 54.07 \text{ g C}_3\text{H}_4\text{N per mole}$$

Empirical formula
C_3H_4N
3 ● 4 ○ 1 ●

Molecular formula
$C_6H_8N_2$
6 ● 8 ○ 2 ●

2,5-Dimethylpyrazine

The average molecular mass is greater than this value, so we have to work out *how many times that value must be multiplied to obtain the average molecular mass.*

Because $2 \times 54.07 = 108.1$, the molecular formula for 2,5-dimethylpyrazine must be $C_6H_8N_2$, derived from $(C_3H_4N) \times 2$. There are two empirical formula units per molecule of 2,5-dimethylpyrazine.

EXERCISE 3.8 | **Alkynes Rule the World: Empirical and Molecular Formulas**

An alkyne is a special class of molecules that have a particular arrangement of atoms that make them useful in the manufacture of pharmaceuticals and polymers. One alkyne with a molar mass of 54.09 g was determined to have 11.18% hydrogen and 88.82% carbon. What is the empirical formula? What is the molecular formula?

Solution

We first assume that we have 100 g of the alkyne. Then

$$11.18 \text{ g H} \times \frac{1 \text{ mol H}}{1.008 \text{ g}} = 11.09 \text{ mol H} \qquad 88.82 \text{ g C} \times \frac{1 \text{ mol C}}{12.01 \text{ g}} = 7.396 \text{ mol C}$$

We next determine the simplest whole-number ratio between these two values.

$$\frac{11.09 \text{ mol H}}{7.396 \text{ mol C}} = 1.499 \text{ H} \qquad \frac{7.396 \text{ mol C}}{7.396 \text{ mol C}} = 1.000 \text{ C}$$

Multiplying these by 2, we get the empirical formula for the alkyne:

$$C_2H_3$$

The molar mass of this empirical formula is 27.04 g. Given that we need 54.09 g to obtain the molecular formula, we can reason that the correct molecular formula must be C_4H_6.

Butyne

PRACTICE 3.8

What are the empirical formula and the molecular formula of a compound that contains 92.26% carbon and 7.74% hydrogen, if the molar mass of the substance is 78.11 g/mol?

See Problems 49–52.

EXERCISE 3.9 | **Determining the Formula of Heme**

Oxygen is carried around your body bound to iron ions that are part of a "heme" chemical group, which is part of the protein we call hemoglobin. The heme portion of hemoglobin has a molar mass of 616.5 g. The percent composition of heme, rounded to three significant figures, is

66.4% C   5.24% H   9.09% Fe   9.09% N   10.4% O

From these data, calculate the empirical and molecular formulas for this key biochemical molecule.

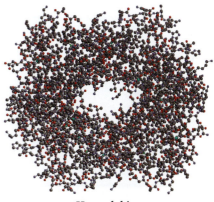

Hemoglobin

Solution

We first convert the percentages listed into grams by assuming that we have a 100.0-g sample of heme and then convert the grams into moles by dividing each quantity by the molar mass of the appropriate element. We are retaining an extra (insignificant) figure throughout the intermediate calculations. We drop all insignificant figures at the end.

$$66.4 \text{ g C} \times \frac{1 \text{ mol C}}{12.01 \text{ g C}} = 5.529 \text{ mol C}$$

$$5.24 \text{ g H} \times \frac{1 \text{ mol H}}{1.008 \text{ g H}} = 5.198 \text{ mol H}$$

$$9.09 \text{ g Fe} \times \frac{1 \text{ mol Fe}}{55.85 \text{ g Fe}} = 0.1628 \text{ mol Fe}$$

$$9.09 \text{ g N} \times \frac{1 \text{ mol N}}{14.01 \text{ g N}} = 0.6488 \text{ mol N}$$

$$10.4 \text{ g O} \times \frac{1 \text{ mol O}}{16.00 \text{ g O}} = 0.6500 \text{ mol O}$$

Divide each number by the smallest to determine the simplest whole-number ratio:

$$\frac{5.529 \text{ mol C}}{0.1628} = 33.96 \text{ mol C}$$

$$\frac{5.198 \text{ mol H}}{0.1628} = 31.93 \text{ mol H}$$

$$\frac{0.1628 \text{ mol Fe}}{0.1628} = 1.000 \text{ mol Fe}$$

$$\frac{0.6488 \text{ mol N}}{0.1628} = 3.985 \text{ mol N}$$

$$\frac{0.6500 \text{ mol O}}{0.1628} = 3.993 \text{ mol O}$$

These numbers round off to this empirical formula:

$$C_{34}H_{32}FeN_4O_4$$

The molecular formula for heme either will be the same as the empirical formula or will correspond to the empirical formula multiplied by some whole number. If we calculate the molar mass corresponding to the empirical formula, we get

$$(12.01 \times 34) + (1.008 \times 32) + (55.85 \times 1) + (14.01 \times 4) + (16.00 \times 4)$$

$$= 616.5 \text{ grams per mole.}$$

Because we were told that the molar mass of heme is 616.5, we can say that the molecular formula for heme is also $C_{34}H_{32}FeN_4O_4$.

PRACTICE 3.9

A common component of several wines that have gone bad—they taste like vinegar—is a compound with an average molecular mass of 60.06 amu and the composition 40.0% C, 6.70% H, 53.3% O. Calculate the molecular formula of this compound, which is called ethanoic acid (or, commonly, acetic acid).

See Problems 53–58 and 96.

3.6   Chemical Equations

We now have sufficient background to explore how the techniques introduced up to this point in the chapter are put to use in the study of chemical change. *The fundamental process of change in chemistry is the chemical reaction.* For example, the final step in the manufacture of aspirin is the reaction shown below.

Application

CHEMICAL ENCOUNTERS: Preparation of Aspirin

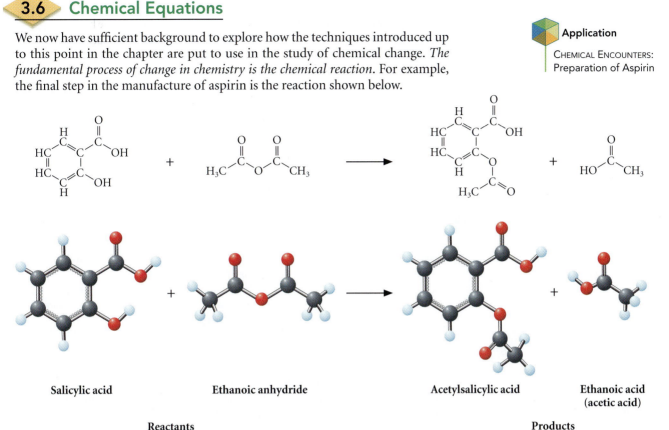

| Salicylic acid | Ethanoic anhydride | Acetylsalicylic acid | Ethanoic acid (acetic acid) |

Reactants   Products

Salicylic acid and ethanoic anhydride (acetic anhydride) are the **reactants** of this reaction, which means they are *the chemicals present at the start of the reaction.* Aspirin and ethanoic acid (acetic acid) are the **products**, *the chemicals produced as a result of the reaction.* **How do we know how much of our reactants we must combine to get the quantity of aspirin we want to produce?** For this or *any* other chemical reaction, the key is to understand the proportions in which the reactants react and in which the products are formed. That information, and all the other key aspects of a reaction, can be summarized in a **chemical equation**, another example of the "chemist's shorthand."

Video Lesson: Balancing Chemical Equations

Video Lesson: An Introduction to Chemical Reactions and Equations

A chemical equation uses chemical formulas to indicate the reactants and products of the reaction. The equation also indicates *the proportions* in which the chemicals involved react together and are formed. The first step in writing a chemical equation is to write the formulas or structures of all the reactants and products, with an arrow between them indicating the course of the reaction. For the synthesis of aspirin, this beginning is

$$C_7H_6O_3 + C_4H_6O_3 \rightarrow C_9H_8O_4 + C_2H_4O_2$$

We know that during a reaction, atoms cannot be created or destroyed but are simply rearranged. We describe this by saying that equations must be **balanced** to be accurate descriptions of a chemical change. This means that all of the atoms that appear on the left-hand side of the equation, within the reactants, must also appear on the right-hand side, within the products. We can check whether an equation is balanced by counting the atoms on each side of the equation (Figure 3.10).

Therefore in this case, the equation we found in our beginning step is balanced, because the number of each type of atom in the reactants is equal to the number of that type of atom in the products. The equation represents a rearrangement of the atoms, but no atom is created or destroyed; matter in the

FIGURE 3.10

A chemical equation has the same number and type of each atom on each side of the arrow. Note that the mass of the products equals the mass of the reactants. Ethanoic anhydride and ethanoic acid are commonly known as acetic anhydride and acetic acid, respectively.

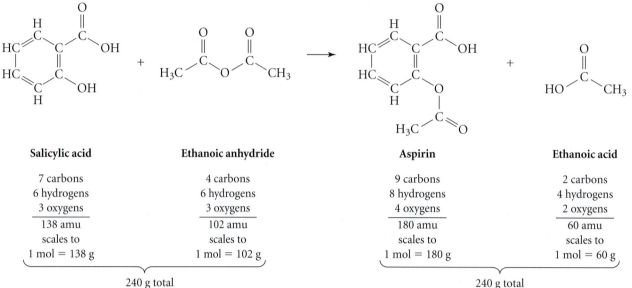

Salicylic acid	Ethanoic anhydride	Aspirin	Ethanoic acid
7 carbons	4 carbons	9 carbons	2 carbons
6 hydrogens	6 hydrogens	8 hydrogens	4 hydrogens
3 oxygens	3 oxygens	4 oxygens	2 oxygens
138 amu	102 amu	180 amu	60 amu
scales to	scales to	scales to	scales to
1 mol = 138 g	1 mol = 102 g	1 mol = 180 g	1 mol = 60 g

240 g total   240 g total

Visualization: Conservation of Mass and Balancing Equations

Tutorial: Balancing Chemical Equations

equation is conserved. Many equations, however, are *not* balanced when we simply write down the formulas for the reactants and products. In order to fix this problem and balance these reactions, we have to use some reasoning to discover the true proportions in which the chemicals react and are formed.

Balancing Equations

Application

The ethanoic anhydride used in the synthesis of aspirin is a very versatile compound that is used in many other chemical-manufacturing processes, such as the production of fibers, plastic materials, and pharmaceuticals. One way to make ethanoic anhydride on an industrial scale is by heating ethanoic acid in the presence of quartz or porcelain chips.

This causes a "dehydration" reaction, *in which water is removed from the reactants and becomes an additional product.* To analyze the quantitative relationships among the chemicals involved in the manufacture of ethanoic anhydride, we need to write the balanced chemical equation for the process. The structures of the reactants and products are shown below.

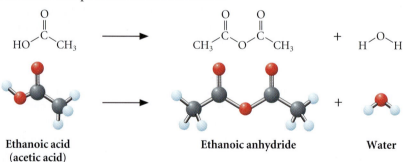

Ethanoic acid (acetic acid)   Ethanoic anhydride   Water

The first step in balancing the equation is to write down the formulas of the reactants and products, as follows:

$$C_2H_4O_2 \rightarrow C_4H_6O_3 + H_2O$$

But this equation is not balanced (Figure 3.11).

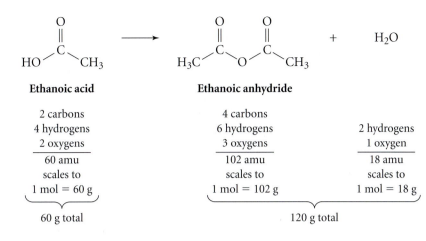

FIGURE 3.11

This equation is not balanced.

To make this equation mathematically valid, we must balance it. We'll do this by adjusting the numbers of the compounds until the numbers of atoms of each type are the same on both sides of the equation. We cannot simply change any of the formulas we have written; *doing so would change the substances in the reaction!* The only thing we can do to balance an equation is to use *multiples* of the correct formulas in order to find the proportions that would produce a balanced equation. We indicate the desired multiples by placing numbers *in front of* the formulas. We call these numbers the **coefficients** in an equation. For example, we could change the equation for the formation of ethanoic anhydride to read

$$2C_2H_4O_2 \rightarrow C_4H_6O_3 + H_2O$$

We chose to add the coefficient 2 because we noticed that there were 4 carbons on the right-hand side and only 2 on the left-hand side of the unbalanced equation. Placing the coefficient 2 before the formula for ethanoic acid changes the meaning of the equation. In this case, it indicates that 2 molecules of ethanoic acid react to produce 1 molecule of ethanoic anhydride and 1 molecule of water. Alternatively, we can say 2 mol of ethanoic acid react to produce 1 mol of ethanoic anhydride and 1 mol of water. Check that you agree that the equation is now balanced, as shown in Figure 3.12. You should find that each side of the equation now contains 4 carbon atoms, 8 hydrogen atoms, and 4 oxygen atoms. We have determined the proportions in which these reactants and products can be combined into a feasible chemical reaction.

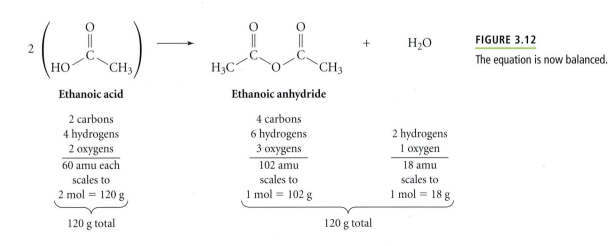

FIGURE 3.12

The equation is now balanced.

EXERCISE 3.10   **Balancing Equations**

1. Consider the production of hydrogen via the reaction of ethane (C_2H_6) with steam. This process is completed in industry to make a mixture of carbon monoxide (CO) and hydrogen gas (H_2) known as "synthesis gas." The hydrogen can be separated from the other gases by passing it through a filter that selectively adsorbs the CO. This process is known as pressure swing adsorption. What is the balanced equation for the reaction of ethane with steam?

2. Butane (C_4H_{10}) is a fuel that reacts with oxygen in the air to produce carbon dioxide and water. What is the balanced equation describing the reaction of butane and oxygen?

Solution

1. The chemical reaction is

$$C_2H_6 + H_2O \rightarrow CO + H_2$$

In this reaction, all of the hydrogen appears in two reactants, and all other elements are in just one compound on each side. Therefore, we avoid hydrogen and balance elements that appear only once. Let's begin with carbon.

- Balancing carbon: There are 2 C's on the left and one on the right, *so we multiply the CO by 2.*

$$C_2H_6 + H_2O \rightarrow 2CO + H_2$$

- Balancing oxygen: There is 1 O on the left and 2 O's on the right, so we must *multiply H$_2$O by 2.*

$$C_2H_6 + 2H_2O \rightarrow 2CO + H_2$$

- Balancing hydrogen: There are now 10 H's on the left ($6 + 2 \times 2$) and 2 on the right, *so we must multiply H$_2$ by 5* ($5 \times 2 = 10$) to get 10 H's on the right.

$$C_2H_6 + 2H_2O \rightarrow 2CO + 5H_2$$

- Checking our work: We have 2 C's, 10 H's, and 2 O's on the left. We have 2 C's, 10 H's, and 2 O's on the right. The reaction is, in fact, now an equation as written. Note that throughout this process, we multiplied only entire compounds. We did not change their structure or add compounds that were not already part of the reaction.

$$C_2H_6 + 2H_2O \rightarrow 2CO + 5H_2$$

2. The reaction described is

$$C_4H_{10} + O_2 \rightarrow CO_2 + H_2O$$

Following the same procedure, we note that oxygen appears in two compounds of the products, so we avoid oxygen and balance the other elements first.

- Balancing carbon: There are 4 C's on the left side and 1 C on the right. We multiply CO$_2$ by 4 to balance carbon.

$$C_4H_{10} + O_2 \rightarrow 4CO_2 + H_2O$$

- Balancing hydrogen: There are 10 H's on the left and 2 on the right, so we must *multiply H$_2$O by 5.*

$$C_4H_{10} + O_2 \rightarrow 4CO_2 + 5H_2O$$

- Balancing oxygen: There are now 2 O's on the left and 13 ($2 \times 4 + 5$) on the right, so we must multiply O$_2$ by $\frac{13}{2}$ ($\frac{13}{2} \times 2 = 13$) to get 13 O's on the left.

$$C_4H_{10} + \frac{13}{2}O_2 \rightarrow 4CO_2 + 5H_2O$$

Although the equation is now mathematically balanced (and perfectly acceptable), it is common practice to clear the fraction. In this case, we multiply all quantities by 2, giving the final equation:

$$2C_4H_{10} + 13O_2 \rightarrow 8CO_2 + 10H_2O$$

After balancing, we have 8 C's, 20 H's, and 26 O's on the left, and we have 8 C's, 20 H's, and 26 O's on the right. All is well.

PRACTICE 3.10

Balance this equation for the reaction that produces phosphoric acid (H_3PO_4), which can be used as a fertilizer, and calcium sulfate, forming gypsum used in housing wallboard.

$$Ca_3(PO_4)_2 + H_2SO_4 \rightarrow CaSO_4 + H_3PO_4$$

See Problems 59–64.

More Information from the Equation

As you will learn as you continue your chemistry studies, the chemical equation contains a great wealth of information about a reaction. The chemical equation can also provide information about the states, or phases, of reactants and products. In Chapter 1, we discussed the states of matter. **Phase** is another term for the state of matter, a part of matter that is chemically and physically homogeneous. This is often indicated by using italic letters in parentheses after each formula in the equation.

(*s*) = the solid phase
(*l*) = the liquid phase
(*g*) = the gaseous phase
(*aq*) = aqueous phase solutions (solutions in water)

For example, the equation we balanced in Exercise 3.10, butane gas reacting with oxygen gas to produce carbon dioxide gas and liquid water, could be written to include this information. This equation specifically says that 13 mol of oxygen gas reacts with 2 mol of butane gas to produce 8 mol of carbon dioxide gas and 10 mol of liquid water.

$$2C_4H_{10}(g) + 13O_2(g) \rightarrow 8CO_2(g) + 10H_2O(l)$$

The Meaning of a Chemical Equation

An equation is a precise quantitative summary of a reaction. It indicates the ratios in which the reactants react and the products form. This means that the chemical equation shows the exact number of molecules (or atoms or ions) that react and the exact number of product molecules (or atoms or ions) that are formed. For correctly balanced equations, the total masses on both sides of the equation (taking the coefficients into account!) should be equal. For example, we can summarize this quantitative information contained in the equation for the synthesis of aspirin as follows:

Equation	$C_7H_6O_3(s)$	+	$C_4H_6O_3(l)$	$\rightarrow$	$C_9H_8O_4(s)$	+	$C_2H_4O_2(l)$
Coefficients	1 mol	+	1 mol	$\rightarrow$	1 mol	+	1 mol
Molar mass of component	138.118 g	+	102.088 g	$\rightarrow$	180.154 g	+	60.052 g

Total reactant mass = 240.206 g   Total product mass = 240.206 g

Ideally, we would follow significant-figure rules in the above calculations. We have not done so here in order to make a point about the conservation of mass in an equation.

This is useful information for a chemist who wants to make aspirin. For example, the equation indicates that it would be appropriate to add 102.088 g of ethanoic anhydride for every 138.118 g of salicylic acid to prepare 180.154 g of aspirin. That corresponds to the simple mole ratio of 1 mol of ethanoic anhydride per 1 mol of salicylic acid, as indicated by the equation. In the context of a large chemical process plant, in which millions of aspirin tablets are made daily, the reactions may require megagram quantities.

3.7 Working with Equations

Video Lesson: Stoichiometry and Chemical Equations

Tutorial: Introduction to Stoichiometry

Video Lesson: CIA Demonstration: Magnesium and Dry Ice

The formation of ethanoic anhydride and water from ethanoic (acetic) acid can be used to explore further how we put equations to work, using them to calculate things we want to know. Remember that in addition to indicating what chemicals are involved in the reaction, the equation establishes the mole ratios among all the chemicals. We found that the balanced chemical equation indicates that 2 mol of ethanoic acid ($C_2H_4O_2$) produce 1 mol of ethanoic anhydride ($C_4H_6O_3$) and 1 mol of water (H_2O).

$$2C_2H_4O_2 \rightarrow C_4H_6O_3 + H_2O$$

$$\boxed{2 \text{ mol}} \rightarrow \boxed{1 \text{ mol}} + \boxed{1 \text{ mol}}$$

The mole ratios that indicate how compounds react and form products are known as **stoichiometric ratios** after the Greek *stoicheon* for "element" and *metron* for "measure." The stoichiometric ratios are vital when we want to answer questions about the quantity of reactant that reacts or the quantity of product that forms. This process of asking and answering mathematical questions based on balanced chemical equations is an aspect of **stoichiometry**, which is essentially the study and use of quantitative relationships in chemical processes.

For instance, the equation relating the formation of ethanoic anhydride describes not only the compounds involved in the reaction but also the stoichiometric ratios of those compounds. We can write all the mole ratios in the form of conversion factors that will allow us to convert among the numbers of moles of the chemicals concerned:

$$\frac{2 \text{ mol ethanoic acid}}{1 \text{ mol ethanoic anhydride}} \qquad \frac{1 \text{ mol ethanoic anhydride}}{2 \text{ mol ethanoic acid}}$$

$$\frac{2 \text{ mol ethanoic acid}}{1 \text{ mol water}} \qquad \frac{1 \text{ mol water}}{2 \text{ mol ethanoic acid}}$$

$$\frac{1 \text{ mol ethanoic anhydride}}{1 \text{ mol water}} \qquad \frac{1 \text{ mol water}}{1 \text{ mol ethanoic anhydride}}$$

Note that the ratios can be written either "right side up" or "upside down."

For example, let's assume we wish to know the maximum mass of ethanoic anhydride that could be formed from 300.0 kg of ethanoic acid. The molar mass of ethanoic acid is 60.05 g and that of ethanoic anhydride is 102.1 g. We can obtain the answer using the appropriate conversion factors. Note that this follows our flowchart diagram in converting mass to moles, then moles to moles, then moles to mass.

$$\text{kg } C_2H_4O_2 \xrightarrow{\frac{1000 \text{ g}}{\text{kg}}} \text{g } C_2H_4O_2 \xrightarrow{\frac{\text{mol } C_2H_4O_2}{\text{g } C_2H_4O_2}} \text{mol } C_2H_4O_2 \xrightarrow{\frac{\text{mol } C_4H_6O_3}{\text{mol } C_2H_4O_2}}$$

$$\text{mol } C_4H_6O_3 \xrightarrow{\frac{\text{g } C_4H_6O_3}{\text{mol } C_4H_6O_3}} \text{g } C_4H_6O_3$$

$$300.0 \text{ kg ethanoic acid} \times \frac{1000 \text{ g}}{1 \text{ kg}} \times \frac{1 \text{ mol ethanoic acid}}{60.05 \text{ g ethanoic acid}} \times \frac{1 \text{ mol ethanoic anhydride}}{2 \text{ mol ethanoic acid}}$$

$$\times \frac{102.1 \text{ g ethanoic anhydride}}{1 \text{ mol ethanoic anhydride}} = 255,000 \text{ g } (255.0 \text{ kg}) \text{ ethanoic anhydride}$$

Three hundred kilograms of ethanoic acid can produce 255,000 g (255.0 kg) of ethanoic anhydride by this reaction.

EXERCISE 3.11 **Determining Products from Reactants**

Hydrogen gas is used extensively in the food industry in a process called hydrogenation, in which the gas is added to compounds called unsaturated fatty acids. This hardens the fatty acids of vegetable oils, making them solids at room temperature, like the fats found in butter. It also converts the fatty acids into forms that are less prone to the "oxidation" processes that give them an unpleasant rancid taste and odor. An example of hydrogenation is the reaction of hydrogen gas with the fatty acid called oleic acid:

$$C_{18}H_{34}O_{2(s)} + H_{2(g)} \rightarrow C_{18}H_{36}O_{2(s)}$$
Oleic acid   Stearic acid

What is the maximum amount of stearic acid that can be produced from the reaction of 425 g of hydrogen gas with an excess of oleic acid?

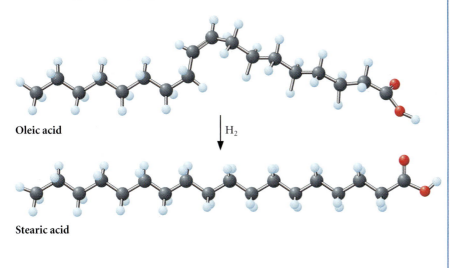

Oleic acid

H_2

Stearic acid

First Thoughts

The key strategy in solving problems such as this is to establish the conversion factor that describes how many moles of the product are formed from how many moles of reactant. In this case, the mole ratio is one-to-one, based on the balanced equation. That is,

$$\frac{1 \text{ mol stearic acid}}{1 \text{ mol hydrogen}}$$

Our strategy for the conversion is

$$g\ H_2 \xrightarrow{\frac{\text{mol } H_2}{g\ H_2}} \text{mol } H_2 \xrightarrow{\frac{\text{mol } C_{18}H_{36}O_2}{\text{mol } H_2}} \text{mol } C_{18}H_{36}O_2 \xrightarrow{\frac{g\ C_{18}H_{36}O_2}{\text{mol } C_{18}H_{36}O_2}} g\ C_{18}H_{36}O_2$$

Solution

$$425 \text{ g } H_2 \times \frac{1 \text{ mol } H_2}{2.016 \text{ g } H_2} \times \frac{1 \text{ mol } C_{18}H_{36}O_2}{1 \text{ mol } H_2} \times \frac{284.5 \text{ g } C_{18}H_{36}O_2}{1 \text{ mol } C_{18}H_{36}O_2}$$

$$= 6.00 \times 10^4 \text{ g } C_{18}H_{36}O_2$$

Further Insights

Demand for hydrogen gas is increasing rapidly. By the year 2007, almost 15 trillion ft^3 of H_2 gas is expected to be used in the chemical industry, with over 93% used in the production of ammonia-based agricultural fertilizers and "petrochemicals," chemicals derived from petroleum and natural gas. Although food and drink processing, including hydrogenation reactions, make up a relatively small percentage of the total demand, their percentage is rising as a consequence of increasing consumer demand for products containing hydrogenated fatty acids.

PRACTICE 3.11

How many grams of oleic acid are needed to produce 72.55 g of stearic acid?

See Problems 65–68.

EXERCISE 3.12 Determining Products from Reactants: Focus on Mole Ratios

Propane

The most common route used for the industrial synthesis of hydrogen is the reaction of water as steam with hydrocarbons—substances that contain only hydrogen and carbon, such as ethane (C_2H_6), as in Exercise 3.10, or propane (C_3H_8). The reactants are passed over a catalyst such as nickel, a substance that greatly speeds the process. The production of hydrogen from the reaction of steam with propane is as follows:

$$C_3H_8(g) + 3H_2O(g) \xrightarrow{\text{Ni(700-900°C)}} 3CO(g) + 7H_2(g)$$

How much hydrogen can be produced from the reaction of 30.0 g of propane with excess steam?

Solution

In this case, our goal is to find grams of hydrogen from propane, so the pertinent conversion factor is

$$\frac{7 \text{ mol } H_2}{1 \text{ mol } C_3H_8}$$

Note that because the conversion factor was determined by counting the number of moles of each compound, the resulting factor has an infinite number of significant digits. The strategy can be diagrammed as follows:

$$\text{g } C_3H_8 \xrightarrow{\frac{\text{mol } C_3H_8}{\text{g } C_3H_8}} \text{mol } C_3H_8 \xrightarrow{\frac{\text{mol } H_2}{\text{mol } C_3H_8}} \text{mol } H_2 \xrightarrow{\frac{\text{g } H_2}{\text{mol } H_2}} \text{g } H_2$$

$$30.0 \text{ g } C_3H_8 \times \frac{1 \text{ mol } C_3H_8}{44.09 \text{ g } C_3H_8} \times \frac{7 \text{ mol } H_2}{1 \text{ mol } C_3H_8} \times \frac{2.016 \text{ g } H_2}{1 \text{ mol } H_2} = 9.60 \text{ g } H_2$$

Even though hydrogen is a very light gas, the 7-to-1 mole ratio of hydrogen to propane means that a good deal of hydrogen will be produced from the reaction of propane with steam.

PRACTICE 3.12

Using the equation in Exercise 3.11, determine how many kilograms of stearic acid can be formed from the reaction of 673 kg of oleic acid with excess hydrogen gas.

See Problems 69a, 69b, 69c, 70a, and 71.

EXERCISE 3.13   Calculations Using Equations

Photosynthesis in plants generates sugars such as glucose ($C_6H_{12}O_6$) and oxygen from carbon dioxide and water. How many kilograms of oxygen are produced from the photosynthesis of 330 kg of carbon dioxide?

First Thoughts

We need to obtain a balanced equation before performing any stoichiometric calculations in chemistry. For photosynthesis, the information given in the question enables us to work out that the equation is unbalanced:

$$CO_2(g) + H_2O(l) \rightarrow C_6H_{12}O_6(s) + O_2(g)$$

After balancing the equation, we get

$$6CO_2(g) + 6H_2O(l) \rightarrow C_6H_{12}O_6(s) + 6O_2(g)$$

One aspect of good problem solving is to *know what you are given, what you want, and how to get there.* We are starting with kilograms of CO_2 and want to find kilograms of O_2. We can navigate from one to the other in this way:

$$kg\ CO_2 \xrightarrow{\frac{1000\ g}{1\ kg}} g\ CO_2 \xrightarrow{\frac{mol\ CO_2}{g\ CO_2}} mol\ CO_2 \xrightarrow{\frac{mol\ O_2}{mol\ CO_2}} mol\ O_2 \xrightarrow{\frac{g\ O_2}{mol\ O_2}} g\ O_2 \xrightarrow{\frac{1\ kg}{1000\ g}} kg\ O_2$$

Solution

In our equation we have a 1-to-1 mole ratio of CO_2 to O_2.

$$330\ kg\ CO_2 \times \frac{1000\ g\ CO_2}{1\ kg\ CO_2} \times \frac{1\ mol\ CO_2}{44.01\ g\ CO_2} \times \frac{6\ mol\ O_2}{6\ mol\ CO_2}$$

$$\times \frac{32.00\ g\ O_2}{1\ mol\ O_2} \times \frac{1\ kg\ O_2}{1000\ g\ O_2} = 240\ kg\ O_2$$

Further Insights

One mole of glucose ($C_6H_{12}O_6$) is produced for every 6 mol of carbon dioxide (CO_2) consumed in the reaction. If we wanted to calculate the mass of glucose produced to accompany the calculated amount of oxygen, the problem-solving strategy would be the same, except we would use glucose and its mole-to-mole ratio to carbon dioxide instead of that of oxygen to carbon dioxide.

$$330\ kg\ CO_2 \times \frac{1000\ g\ CO_2}{1\ kg\ CO_2} \times \frac{1\ mol\ CO_2}{44.01\ g\ CO_2} \times \frac{1\ mol\ C_6H_{12}O_6}{6\ mol\ CO_2}$$

$$\times \frac{180.16\ g\ C_6H_{12}O_6}{1\ mol\ C_6H_{12}O_6} \times \frac{1\ kg\ C_6H_{12}O_6}{1000\ g\ C_6H_{12}O_6} = 230\ kg\ C_6H_{12}O_6$$

Keep in mind the key parts of stoichiometry problem solving: the balanced equation, where you start, where you want to end up and, how you get there.

PRACTICE 3.13

What mass of CO_2 gas is released when 26 g of methane (CH_4) burns in oxygen (O_2) to generate CO_2 and water (H_2O)? To solve this question, you need to use several of the techniques presented in this chapter—specifically, working out molar masses, writing balanced equations, and performing a stoichiometric calculation.

See Problems 75a, 76a, and 85.

Limiting Reagent

We have seen that equations indicate how many moles of each reactant will react together, but chemists do not usually combine reactants in the exact proportions shown in the chemical equation. Instead, they generally use an *excess amount* of one or more of the reactants, perhaps the least expensive one, in order to encourage as much of the other reactants as possible to react. In this situation, the reactant that will be used up first is called the **limiting reagent**, or limiting reactant, because it is the quantity of this reagent that imposes a limit on how much product can be formed.

An analogy that is helpful in understanding what we mean by a limiting reagent is the preparation of hamburgers for a frozen food company. Each hamburger must be like every other because the company's customers expect a consistent product. Each hamburger, shown in Figure 3.13 both in parts and completed, contains *only*

FIGURE 3.13

A hamburger.

Video Lesson: Finding Limiting Reagents

Visualization: Limiting Reactant

Video Lesson: CIA Demonstration: Self-Inflating Hydrogen Balloons

1 patty
5 pickles
2 slices of cheese
1 bun

In this analogy, we can write the preparation of the product from the "reactants" as follows:

1 patty + 5 pickles + 2 cheese + 1 bun → 1 hamburger

Our quota for each hour of work is 10 completed hamburgers. We have on the shelf 10 patties, 50 pickles, 20 slices of cheese, but only 3 buns. How many hamburgers can we make? As Figure 3.14 shows, we can only prepare 3 complete hamburgers, because *the number of buns limits our production*. The buns are the limiting reagent in this case. How many slices of cheese are in excess—that is, how many slices of cheese are left over? In a manner analogous to expressing a mole ratio, we can say that according to the equation for sandwich preparation, each bun requires 2 slices of cheese, giving us a ratio of

$$\frac{1 \text{ bun}}{2 \text{ cheese}}$$

We can write also other valid reactant "mole ratios," such as

$$\frac{1 \text{ bun}}{5 \text{ pickles}} \qquad \frac{5 \text{ pickles}}{1 \text{ patty}} \qquad \frac{2 \text{ cheese}}{5 \text{ pickles}} \qquad \frac{1 \text{ patty}}{1 \text{ bun}}$$

We asked how many slices of cheese are left over from our original stack of 20 slices if we use 3 buns. First, we can ask, "How many cheese slices are actually used to combine with 3 buns?" The ratio of buns to cheese is

$$\frac{1 \text{ bun}}{2 \text{ cheese}} \quad \text{or} \quad \frac{2 \text{ cheese}}{1 \text{ bun}}$$

FIGURE 3.14

Analogy of limiting reagent to limiting ingredient in the construction of a sandwich.

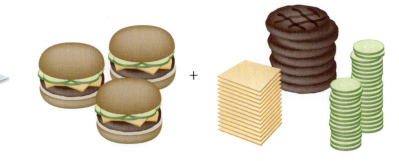

The number of cheese slices used, then, is

$$3 \text{ buns} \times \frac{2 \text{ cheese}}{1 \text{ bun}} = 6 \text{ cheese}$$

The number of cheese slices in excess is the number we had on the shelf (20 cheese slices) minus the number we used (6 cheese slices).

$$20 \text{ cheese} - 6 \text{ cheese} = 14 \text{ slices of cheese in excess}$$

This process of determining how much of a reagent is used and how much is left over is about the same as the one we use with chemical systems. The only difference is that we think in terms of moles rather than slices of cheese, patties, pickles, or buns. Let's carefully build on our understanding of the concept by using plastic models to assemble a water molecule, as shown in Figure 3.15.

FIGURE 3.15

Model atoms typically used in teaching laboratories can be used to illustrate the principle of the limiting reagent.

The formula for water is H_2O, so we need 2 model hydrogen atoms and 1 model oxygen atom per water molecule. If we have available a pile of 10 oxygen atoms and a pile of 60 hydrogen atoms, we will be able to assemble only 10 water molecules before we run out of oxygen atoms. In this situation oxygen is the limiting reagent because the number of oxygen atoms present limits to 10 the number of water molecules we can form. Hydrogen atoms, on the other hand, are "in excess," so some of them are left over once all the oxygen has been used up.

We can apply our understanding of limiting reagents to a chemical system in which ensuring that the correct reactant is the limiting one is literally a matter of life or death. In a "manned" spacecraft, it is vital that the carbon dioxide gas breathed out by the astronauts not be allowed to accumulate in the cabin. If it were to accumulate, it would soon rise to levels that would cause the occupants to become dizzy and confused and eventually to slip into unconsciousness. One process that has been used by the U.S. space program to remove carbon dioxide involves the reaction of the gas with lithium hydroxide (LiOH) as it is drawn through special filters:

Application

CHEMICAL ENCOUNTERS: Cleaning the Air on Manned Spacecraft

$$2LiOH(s) + CO_2(g) \rightarrow Li_2CO_3(s) + H_2O(l)$$

In this application, it is essential that there be more than enough lithium hydroxide available to react with all of the carbon dioxide that the astronauts will produce. Therefore, it is vital that carbon dioxide be the limiting reagent and that lithium hydroxide be present in excess. Will 5.0 kg of lithium hydroxide be sufficient to remove the carbon dioxide released by one astronaut per day (typically 1.0 kg of CO_2)?

We can answer the question by determining the limiting reagent in these circumstances. Here is the basic information we need:

$$2LiOH(s) \;+\; CO_2(g) \rightarrow Li_2CO_3(s) + H_2O(l)$$

2 moles + 1 mole

23.95 g/mol   44.01 g/mol

We can calculate the amount of lithium hydroxide needed to react with the average 1.0 kg (1.0×10^3 g) of carbon dioxide released per day as follows. Remember our flowchart diagram:

$$1.0 \times 10^3 \text{ g } CO_2 \times \frac{1 \text{ mol } CO_2}{44.01 \text{ g } CO_2} \times \frac{2 \text{ mol LiOH}}{1 \text{ mol } CO_2} \times \frac{23.95 \text{ g LiOH}}{1 \text{ mol LiOH}}$$

$$= 1100 \text{ g LiOH per } 1.0 \times 10^3 \text{ g } CO_2$$

If we have 5.0 kg (5.0×10^3 g) of LiOH available per day, the carbon dioxide is going to be the limiting reagent, as we wish, with a big margin of error to be on the safe side. The LiOH will be in excess, and there will be plenty of LiOH left over at the end of the day.

EXERCISE 3.14 **Calculating Yield and Limiting Reagent**

Sodium hydroxide reacts with phosphoric acid to give sodium phosphate and water. Assume that 17.80 g of NaOH is mixed with 15.40 g of H_3PO_4.

a. How many grams of Na_3PO_4 can be formed?

b. How many grams of the excess reactant remain unreacted?

First Thoughts

We must first write a balanced chemical equation for this reaction:

$$3NaOH(aq) + H_3PO_4(aq) \rightarrow Na_3PO_4(aq) + 3H_2O(l)$$

In part (a), as in all limiting-reagent problems, the basic question to be addressed is **Which compound limits the amount of product formed?** One useful strategy is to calculate the amount of product that would be formed by *each* reactant if it were fully consumed. *The reactant that forms the least product is limiting*, and that total amount of product will be formed, theoretically:

$$g \text{ of reactant} \xrightarrow{molar\ mass} \text{moles of reactant} \xrightarrow{mole\ ratio}$$

$$\text{moles of product} \xrightarrow{molar\ mass} g \text{ of product}$$

Given your understanding of mole-to-mole relationships, you might see several ways of dealing with this problem, including not going all the way to grams but, rather, making your decision on the basis of moles. We present here one of several ways of solving this problem. In part (b), to determine grams of excess reactant, you must calculate how many moles of excess reactant were actually used.

$$\# \text{ moles excess} = \# \text{ moles original} - \# \text{ moles used}$$

Then convert moles to grams.

Solution

a. Determination of limiting reagent

$$g\ Na_3PO_4 \text{ from NaOH} = 17.80\ g\ NaOH \times \frac{1\ mol\ NaOH}{40.00\ g\ NaOH} \times \frac{1\ mol\ Na_3PO_4}{3\ mol\ NaOH}$$

$$\times \frac{163.94\ g\ Na_3PO_4}{1\ mol\ Na_3PO_4} = 24.32\ g\ Na_3PO_4 \text{ from NaOH}$$

$$g\ Na_3PO_4 \text{ from } H_3PO_4 = 15.40\ g\ H_3PO_4 \times \frac{1\ mol\ H_3PO_4}{97.99\ g\ H_3PO_4} \times \frac{1\ mol\ Na_3PO_4}{1\ mol\ H_3PO_4}$$

$$\times \frac{163.94\ g\ Na_3PO_4}{1\ mol\ Na_3PO_4} = 25.76\ g\ Na_3PO_4 \text{ from } H_3PO_4$$

Therefore, NaOH is the limiting reagent, and 24.32 g of Na_3PO_4 are formed.

b. Determination of grams of excess reactant

H_3PO_4 is in excess. If 24.32 g of Na_3PO_4 is formed, the moles of H_3PO_4 used is given by

$$mol\ H_3PO_4 \text{ used} = 24.32\ g\ Na_3PO_4 \times \frac{1\ mol\ Na_3PO_4}{163.94\ g\ Na_3PO_4} \times \frac{1\ mol\ H_3PO_4}{1\ mol\ Na_3PO_4}$$

$$= 0.1483\ mol\ H_3PO_4 \text{ used}$$

The number of moles of H_3PO_4 originally present is

$$15.40\ g\ H_3PO_4 \times \frac{1\ mol\ H_3PO_4}{97.99\ g\ H_3PO_4} = 0.1572\ mol\ H_3PO_4 \text{ originally present}$$

$$\text{mol } H_3PO_4 \text{ excess} = \text{mol } H_3PO_4 \text{ originally present} - \text{mol } H_3PO_4 \text{ used}$$

$$= 0.1572 \text{ mol} - 0.1483 \text{ mol}$$

$$= 0.0089 \text{ mol } H_3PO_4 \text{ excess}$$

$$\text{g } H_3PO_4 \text{ excess} = 0.0089 \text{ mol } H_3PO_4 \times \frac{97.99 \text{ g } H_3PO_4}{1 \text{ mol } H_3PO_4}$$

$$= 0.87 \text{ g } H_3PO_4 \text{ excess}$$

Further Insights

We often choose to perform a chemical reaction with one reactant in excess. Yet there is an important class of reactions, "titrations," in which the goal is to add precisely enough moles of one reactant to just consume the other. Using titrations, we can determine the amount of many types of substances, including acids, bases, and metals. We will learn more about titrations in the next chapter, and we will take an in-depth look at the technique in Chapters 16–18.

PRACTICE 3.14

Calcium hydroxide reacts with hydrogen chloride to produce water and calcium chloride. If 12.33 g of calcium hydroxide is placed in a flask with 32.15 g of hydrogen chloride, how many grams of calcium chloride will be formed?

See Problems 70, 73, 74, 77, and 80.

Chemical reactions are rarely 100% efficient, because there are always losses due to unwanted side reactions, to some of the reactants remaining unreacted, and/or to some of the products being converted back into reactants once they are formed. Chemists designing a synthesis reaction can compare the actual masses obtained under various conditions with the masses predicted from the equation. Doing so enables them to calculate the percentage yield of a reaction and to adjust the conditions until the maximum percentage yield is obtained. The maximum amount of any chemical that could be produced in a chemical reaction, which can be calculated from the equation for the reaction, is called the **theoretical yield**. The amount of product obtained experimentally is called the **actual yield**. The **percentage yield** of a reaction equals the actual yield expressed as a percentage of the theoretical yield:

Video Lesson: Theoretical Yield and Percent Yield

$$\text{Percentage yield} = \frac{\text{actual yield}}{\text{theoretical yield}} \times 100\%$$

This first look at the uses of equations should help you realize that *equations provide useful information that chemists must have if they are to do their jobs of predicting and producing desired products,* whether on a relatively small scale in a laboratory or on an industrial process scale in a huge manufacturing facility.

EXERCISE 3.15   **Percentage Yield**

In the previous exercise, we determined that 24.32 g of sodium phosphate could theoretically be prepared under the conditions given. If only 20.07 g were obtained from the reaction, what would be the percentage yield of the reaction?

Solution

$$\text{Percentage yield} = \frac{\text{actual yield}}{\text{theoretical yield}} \times 100\% = \frac{20.07 \text{ g}}{24.32 \text{ g}} \times 100\% = 82.52\% \text{ yield}$$

Video Lesson: A Problem Using the Combined Concepts of Stoichiometry

PRACTICE 3.15

Assuming that the yield of sodium phosphate from sodium hydroxide and phosphoric acid can never be greater than 82.52%, how many grams of sodium hydroxide would be needed to produce 100 g of sodium phosphate?

See Problems 69d, 75b, 76b, 81, 82b, 84, and 86.

Issues and Controversies

Everyday controversies of quantitative chemistry

In everyday life, we are bombarded by conflicting messages from different sides of debates about quantitative chemistry. We may not realize that quantitative chemistry is at the heart of these debates, but it is. In considering our diet, we often pose questions such as: How much food should we eat each day? How many grams of fat should be in our diet? What levels of each vitamin are too low for good health (see Figure 3.16), and what levels are dangerously high? We know that too much alcohol in beverages is bad for us, but are small, regular amounts beneficial, or is total abstinence the most healthful choice?

These issues are complex because most chemicals have more than one effect on the body. In other words, a single chemical can participate in several different chemical reactions within the body. Some of these reactions may bring clear benefits, whereas others can pose

dangers. These are all issues of quantitative chemistry, revolving around what quantities of a chemical are required to produce a certain level of benefit and what quantities produce a given level of harm. The benefits and the dangers are all caused, directly or indirectly, by the stoichiometry of the chemical reactions in which the chemicals participate.

Vitamin C is an excellent example. We know that we must consume a certain amount of this vitamin, which is normally obtained from fresh fruits and vegetables, for good health. A deficiency of vitamin C causes a variety of problems known collectively as "scurvy," which is characterized by fatigue, sore joints and muscles, hemorrhaging, and anemia. The minimum level of vitamin C needed to prevent these problems can be determined by the quantitative analysis of people's diets and by the association between these diets and the appearance of symptoms of vitamin C deficiency. Even this issue, however, is linked to controversy, because different countries use different recommended levels, and nutritionists regularly debate whether the levels should be revised upward or downward. The current recommended dietary allowance (RDA) of vitamin C in the United States is 60 mg per day for adults. Go into a drugstore, however, and you will easily find tablets available that each deliver a massive 1000-mg (1-g) dose of vitamin C. These are sold because some scientists, most notably Linus Pauling, have promoted the view that huge "megadoses" of vitamin C can protect against colds and even cancer. Others caution, however, that excessive consumption of vitamin C may cause problems such as nausea, diarrhea, iron overload—and possibly even cancer.

The debate about the appropriate amount of vitamin C to consume is mirrored by similar debates about every other vitamin and about many other components of our diet. One of the most important activities in the attempts to resolve these debates is the careful quantitative analysis of the many effects each chemical can have on the chemistry of life.

FIGURE 3.16

The quantities of chemicals we consume are important—and what quantities are ideal is the subject of much debate.

Supplement Facts
Serving Size 1 tablet

Amount Per Tablet	% DV	Amount Per Tablet	% DV
Vitamin A 3,000 I.U.	60%	Iodine 150 mcg	100%
Vitamin C 120 mg	200%	Magnesium 100 mg	25%
Vitamin D 400 I.U.	100%	Zinc 15 mg	100%
Vitamin E 50 I.U.	167%	Selenium 25 mcg	36%
Vitamin K 25 mcg	31%	Copper 2 mg	100%
Thiamin 1.5 mg	100%	Manganese 2 mg	100%
Riboflavin 1.7 mg	100%	Chromium 120 mcg	100%
Niacin 20 mg	100%	Molybdenum 25 mcg	33%
Vitamin B6 2 mg	100%	Chloride 36 mg	1%
Folic Acid 400 mcg	100%	Potassium 40 mg	1%
Vitamin B12 6 mcg	100%	Boron 150 mcg	*
Biotin 30 mcg	10%	Nickel 5 mcg	*
Pantothenic Acid 10 mg	100%	Silicon 2 mg	*
Calcium 100 mg	10%	Tin 10 mcg	*
Iron 9 mg	50%	Vanadium 10 mcg	*
Phosphorus 77 mg	8%	Lutein 250 mcg	*

*Daily Value (DV) not established.

Oranges are a good source of Vitamin C.

The Bottom Line

- Quantities in chemistry are crucial. Different amounts of the same chemical can have very different effects on chemical systems such as living things, environmental systems, and industrial processes. (Chapter opening)

- The formula mass (formula weight) of a chemical is the total mass of all the atoms present in its formula, in atomic mass units (amu) or in grams per mole. (Section 3.1)

- The average molecular mass (molecular weight) of a molecule is the total mass of the molecule, in atomic mass units (amu) or in grams per mole. (Section 3.1)

- The mole is the basic counting unit of chemistry—the chemist's "dozen"—and 1 mol of any chemical contains Avogadro's number (6.022×10^{23}) of molecules, atoms, or formula units (which entity to use depends on the chemical concerned). (Section 3.2)

- We use the mole to convert between molecules and grams of a substance. (Section 3.3)

- The percent, by mass, of an element in a compound is called its mass percent. (Section 3.4)

- The empirical formula for a compound indicates the simplest whole-number ratio in which its component atoms are present. The molecular formula indicates the actual number of each type of atom in one molecule of the compound. (Section 3.5)

- A chemical equation uses chemical formulas to indicate the reactants and products of a reaction and uses numbers before the formulas to indicate the proportions in which the chemicals involved react together and are formed. (Section 3.6)

- Stoichiometry is the study and use of quantitative relationships in chemical processes. (Section 3.7)

- The limiting reagent in a reaction is the one that is consumed first, causing the reaction to cease despite the fact that the other reactants remain "in excess." (Section 3.7)

- The percentage yield of a reaction equals the actual yield expressed as a percentage of the theoretical yield:
$$\text{Percentage yield} = \frac{\text{actual yield}}{\text{theoretical yield}} \times 100\%$$
(Section 3.7)

- Chemistry is a quantitative science. The practice of chemistry in the real world demands mastery of the quantitative skills introduced in this chapter. (Issues and Controversies)

Key Words

actual yield The experimental quantity, in grams, of product obtained in a reaction. (*p. 115*)

Avogadro's number (N_A) The number of particles (6.022×10^{23}) of a substance in 1 mol of that substance. By definition, Avogadro's number is equal to the number of carbon-12 atoms in exactly 12.0000 g of carbon-12. (*p. 91*)

balanced Appropriate coefficients have been added such that the number of atoms of each element are the same in both reactants and products. (*p. 103*)

chemical equation A precise quantitative description of a reaction. (*p. 103*)

coefficients The numbers placed in front of the substances in a chemical equation that reflect the specific numbers of units of those substances required to balance the equation. (*p. 105*)

formula mass The total mass of all the atoms present in the formula of an ionic compound, in atomic mass units (amu), or the mass of one mole of formula units in grams per mole. (*p. 88*)

limiting reagent The reactant that is consumed first, causing the reaction to cease despite the fact that the other reactants remain "in excess." Also known as the limiting reactant. (*p. 112*)

mass percent The percent of a component by mass. (*p. 97*)
$$\text{Mass percent} = \frac{\text{total mass component}}{\text{total mass whole substance}} \times 100\%$$

molar mass The total mass of all the atoms present in the formula of a molecule, in atomic mass units (amu) or in grams per mole. (*p. 92*)

mole The quantity represented by 6.022×10^{23} particles. (*p. 92*)

molecular mass The mass of one molecule, expressed in atomic mass units (amu), or the mass of one mole of molecules in grams per mole. (*p. 87*)

percentage yield The actual yield of a reaction divided by the theoretical yield and then multiplied by 100%. (*p. 115*)

phase A part of matter that is chemically and physically homogeneous. (*p. 107*)

products The substances located on the right-hand side of a chemical equation. (*p. 103*)

reactants The substances located on the left-hand side of a chemical equation. (*p. 103*)

stoichiometric ratios The mole ratios relating how compounds react and form products. (*p. 108*)

stoichiometry The study and use of quantitative relationships in chemical processes. (*p. 108*)

theoretical yield The maximum amount of any chemical, in grams, that could be produced in a chemical reaction. This value can be calculated from the equation for the reaction. (*p. 115*)

Focus Your Learning

The answers to the odd-numbered problems and some selected problems appear at the back of the book, as represented by the blue numbering.

3.1 Formula Masses

Skill Review

1. Determine the mass, in amu, for each:
   a. Carbon monoxide, CO
   b. Silicon dioxide (the principal component in sand), SiO_2
   c. Ammonia, NH_3
   d. Sodium thiosulfate (photographer's "hypo" solution), $Na_2S_2O_3$
   e. Tristearin (a type of animal fat), $C_{57}H_{110}O_6$

2. Determine the mass, in amu, for each:
   a. Water, H_2O
   b. Sodium hydroxide, NaOH
   c. Fructose ("fruit sugar")
   d. Potassium dichromate, $K_2Cr_2O_7$
   e. Ammonium phosphate (a fertilizer ingredient), $(NH_4)_3PO_4$

Fructose

3. Arrange these in order from lightest to heaviest on the basis of formula masses.

C_6H_6	H_2O	C_2H_4OH
$CaCl_2$	CO	

4. Arrange these in order from heaviest to lightest on the basis of formula masses.

NaBr	KBr	$KMnO_4$
LiOH	Na_3PO_4	

Chemical Applications and Practices

5. Saccharin ($C_7H_5NO_3S$) and aspartame ($C_{14}H_{18}N_2O_5$) have both been used as sugar substitutes.
   a. What is the mass, in amu, of one molecule of each sugar substitute?
   b. What is the mass ratio of aspartame to saccharin?
   c. What mass, in grams, of saccharin has the same number of molecules as 42.0 g of aspartame?

6. As you are reading this, you are breathing in some molecular oxygen (O_2).
   a. Express the mass of an oxygen molecule in amu.
   b. Convert that mass into grams per molecule of O_2.
   c. What is the mass of 6.022×10^{23} molecules of O_2?

7. The price of gold fluctuates daily. However, if you paid $10,000 for 1 kg of gold, how much would you be paying per one atom of gold?

8. Sodium bicarbonate ($NaHCO_3$) is a useful ingredient employed in baking.
   a. What is the formula mass of $NaHCO_3$?
   b. If you paid $1.50 for 250 g of $NaHCO_3$, how much would you be paying for each formula unit?

3.2 Counting by Weighing

Skill Review

9. Convert these to grams:
   a. 6.02×10^{22} atoms of silver
   b. One trillion atoms of gold
   c. 2 dozen water molecules
   d. One molecule of propane (C_3H_8)

10. Convert these to grams:
    a. 3.54×10^{21} atoms of gold
    b. A gross (12 dozen) of atoms of mercury
    c. 1.2×10^{27} molecules of glucose ($C_6H_{12}O_6$)
    d. One molecule of carbon dioxide (CO_2)

11. Indicate how many molecules there are in each of the following:
    a. 12.01 g of water
    b. 68.3 g of sodium bicarbonate ($NaHCO_3$)
    c. 100 g of methane (CH_4)
    d. 2.3 mg of glucose ($C_6H_{12}O_6$)

12. Indicate how many atoms there are in each of the following:
    a. 100 g of silver
    b. 100 g of gold
    c. 12.01 g of carbon
    d. 5.3 kg of water

13. a. One of the main waste products of animal metabolism is urea. What is the mass, in amu, of one molecule of urea (CH_4ON_2)?
    b. What is the mass of 6.022×10^{23} molecules of urea?
    c. What would be the total mass of 6.022×10^{22} molecules of urea?

Urea

14. Ammonium nitrate is often used as a fertilizer. Express these amounts of NH_4NO_3 in grams.
    a. one million formula units
    b. ten mol
    c. 0.500 kg

Chemical Applications and Practices

15. Acetaminophen, a nonaspirin pain reliever, has the formula $C_8H_9NO_2$. A dose of 0.500 g of acetaminophen would contain how many molecules of the pain reliever?

16. a. Methane (CH_4) is used in many laboratory burners. If a student burned 10.0 g of methane, how many molecules were consumed?
    b. Would 10.0 g of butane contain more or fewer molecules?

Butane

3.3 Working with Moles

Skill Review

17. Which of these quantities of sodium chloride (NaCl) contains the greatest mass?

    0.100 mol 4.2×10^{23} formula units   1.60 g

18. Which of these quantities of acetaminophen ($C_8H_9NO_2$) contains the greatest mass?

    0.550 mol 9.1×10^{23} molecules   80.4 g

19. Convert these to moles:
    a. 65.0 g of CO_2
    b. 1.5×10^{22} atoms of neon
    c. 25 g of propane (C_3H_8)
    d. 4.5×10^{24} molecules of ammonia (NH_3)

20. Convert these to moles:
    a. 123 g of H_2O
    b. 9.72×10^{23} atoms of gold
    c. 25.6 mg of methane (CH_4)
    d. 3.22×10^{10} molecules of hemoglobin ($C_{2954}H_{4508}N_{780}O_{806}S_{12}Fe_4$)

21. Which of these would contain the greatest number of *atoms*?
    454 g of gold
    56.0 g of O_2
    245 g of graphite

22. Which of these would contain the greatest number of *atoms*?
    100 g of tin
    17.9 mg of glucose ($C_6H_{12}O_6$)
    2 g of water

23. Convert these to the correct number of formula units:
    a. 15.0 g of $CaCl_2$
    b. 15.0 mol of $CaCl_2$
    c. 15.0 mL of a $CaCl_2$ solution that contains 0.42 mol per liter

24. Convert these to the correct number of formula units:
    a. 1.0 g of K_2CO_3
    b. 15.0 mol of K_2CO_3
    c. 7.33 kg of K_2CO_3

25. Express these quantities in grams:
    a. 1500 mol of Ne
    b. 12.5 mol of Na
    c. 0.42 mol of N_2

26. Express these quantities in grams:
    a. 0.250 mol of NaCl
    b. 0.350 mol of $(NH_4)_3PO_4$
    c. 1.5×10^{21} molecules of CO_2

27. Radioactive samples of elements are used in both the treatment and the detection of certain diseases. One such element is element 43, technetium. Nuclear medicine applications of this element include bone scans, thyroid monitoring, and brain monitoring. If a pharmacist were preparing doses to be given to patients that involved using 5.00 g of radioactive technetium, how many atoms of technetium would actually be present?

28. Two elements have been named in honor of female scientists, curium (element 96) and meitnerium (element 109). Element 109 is a relatively recent discovery. Often when synthesizing newer elements, scientists are working with extremely small samples, even only a few atoms. Suppose a certain procedure produced 100 atoms of meitnerium. How many moles is this? What would be the mass of the sample?

Chemical Applications and Practices

29. Iron is essential for the transport of oxygen in the human body and for energy production through several biochemical cycles.
    a. What is the mass, in grams, of one atom of iron?
    b. The recommended dietary allowance (RDA) for iron is approximately 15 mg. How many atoms of iron is this?
    c. Finally, express the RDA in moles of iron.

30. How many caffeine molecules are found in each of these drinks?
    a. a cup of coffee with 95 mg of caffeine
    b. a cup of tea with 0.065 g of caffeine
    c. a soda with 0.038 mol of caffeine

Hydrogen · Carbon · Oxygen · Nitrogen

Caffeine

31. Cobalt is a metal required for human health. We normally ingest small amounts of cobalt in the food we eat. Typically, you may have about 1.5 mg present in your body.
    a. How many moles is this?
    b. How many atoms of cobalt are present?

32. Male silkworms are attracted to an organic molecule, secreted by the females, that acts as an attractant for mating purposes. If the mass per mole of the compound is 238 g, and a female releases 4.20×10^{-6} g, how many molecules are available to be detected by nearby males?

33. Some elements exist in more than one molecular form. This property is called allotropy. For example, phosphorus, which is very reactive in its elemental form and is used in some explosives and some types of matches, has three different

forms. The structure of white phosphorus has interlocked tetrahedrons. The simple formula of white phosphorus is P_4.

a. The white streams of smoke from military bombings may be due to the reaction of white phosphorus with oxygen. How many grams of phosphorus would be used if an explosive device contained 10.5 mol of P_4?

b. The most stable of the allotropic forms of sulfur is called orthorhombic sulfur (S_8.) The structure resembles a puckered crown. How many moles of S_8 are equivalent to 454 g of S_8?

34. The ability of nonmetal atoms to form attachments among themselves is called catenation. Carbon exhibits this property more than any other element. Catenation is responsible for the huge variety of carbon compounds.

a. Determine the number of moles in 100.0 g each of the following carbon compounds. (In each compound, the carbon atoms are joined to each other). C_3H_8 (propane), C_4H_{10} (butane), C_5H_{12} (pentane)

b. Silicon, which is found directly below carbon in the periodic table of the elements, also has the ability to catenate. Determine the number of grams in 0.100 mol of each of these silicon–hydrogen compounds (called silanes): Si_2H_6 and Si_6H_{14}.

35. A vitamin that helps activate many enzymes in humans is pyridoxin (also called vitamin B_6). Its chemical formula is $C_8H_{11}NO_3$. Vitamin B_6 can be found in good supply in wheat and legumes. If a nutritionist found 0.156 g of pyridoxine in a food sample, how many moles would be reported? How many molecules would be reported? How many atoms of carbon would be contained in the vitamin sample?

36. Carbon can be found in three elemental forms: graphite, diamond, and buckministerfullerene. Because all the carbon atoms in a diamond in a ring are connected in linked tetrahedrons, a single diamond can be considered to be one large molecule (often called a macromolecule). If a diamond has a mass of 0.50 carat, how many atoms of carbon make up the ring? (1 carat = approximately 0.20 g)

3.4 Percentages by Mass

Skill Review

37. Calculate the percent of carbon in each of these organic compounds:

a. C_3H_6 c. C_6H_6

b. CH_3OH d. $C_{12}H_{22}O_{11}$

38. Calculate the percent of oxygen in each of these organic compounds:

a. CO_2 c. C_3H_8O

b. $C_6H_{12}O_6$ d. H_2CO_3

39. Arrange these sulfur-containing compounds from the greatest mass percent sulfur to the least.

a. H_2S c. $Na_2S_2O_4$

b. SO_2 d. H_2SO_3

40. Arrange these oxygen-containing compounds from the greatest mass percent oxygen to the least.

a. H_2O c. $Na_2C_2O_4$

b. CO_2 d. $C_6H_{12}O_2$

41. Boron forms many compounds, often with some unusual bonding properties. In each of these, determine the mass percent of boron: $B_{13}C_2$, Ti_3B_4, CaB_6.

42. Determine the mass percent of sodium in each compound: $NaCl$, $NaHCO_3$, $NaOH$.

43. Which compound in each pair has the greater percent oxygen?

a. $FeSO_4$ or Na_2SO_4 c. $Fe(NO_3)_3$ or HNO_3

b. K_2FeO_4 or $KMnO_4$

44. Which compound in each pair has the greater percent carbon?

a. CO_2 or CO c. C_3H_8O or $C_5H_{10}O_2$

b. $NaCN$ or KCN

Chemical Applications and Practices

45. The main component in kidney stones is CaC_2O_4. What is the mass percent of each element found in CaC_2O_4?

46. Hydrates are compounds that contain water molecules as part of their structure. For example, sodium carbonate decahydrate ($Na_2CO_3 \cdot 10H_2O$), washing soda, has been used as a grease remover. What percent of the compound is water? (*Note:* The dot between Na_2CO_3 and H_2O indicates that water is loosely attached within the crystal—in this case in a 1:10 ratio.)

47. Unsaturated fatty acids are responsible for the liquid nature of many plant oils. In this case, the term *unsaturated* refers to the presence of double bonds between two or more carbon atoms in the molecule. Saturated hydrocarbons, which are associated with animal fats, contain only single bonds between carbon atoms. Unsaturated compounds can be made saturated by adding hydrogen to the double-bonded areas. After such a reaction, often called hydrogenation, would the percent by mass of carbon in the saturated compound be increased or decreased over its percent by mass in the unsaturated hydrocarbon? Explain the basis for your answer.

48. The sidewalk salt that we spread on our walkways is mostly $CaCl_2$. A stable hydrate of this salt is $CaCl_2 \cdot 2H_2O$ (see Problem 46). What is the mass percent of calcium in each of these salts?

3.5 Finding the Formula

Skill Review

49. Suppose an orchestra has this composition: 24 violinists, 18 brass instrumentalists, 6 cellos, and 3 percussionists. What is the "empirical formula" of this orchestra?

50. Suppose a lasagna is made from 2 boxes of noodles, 2 packages of ground beef, 2 tomatoes, and 4 packages of cheese. What is the "empirical formula" of this lasagna?

51. A compound contains only carbon and hydrogen. The mass percent of carbon in the compound is 85.7%. What is the empirical formula of the compound?

52. A compound contains only carbon, hydrogen, and oxygen. The mass percent of carbon is 52.1% and that of hydrogen is 13.1%. What is the empirical formula of the compound? If the molar mass of the compound is 92, what is the molecular formula of the compound?

Chemical Applications and Practices

53. a. Based on this mass percent composition, determine the empirical formula of the compound: 57.1% carbon, 4.76% hydrogen, 38.1% oxygen.

b. Which of these could possibly be the molar mass of the compound: 56, 84, or 116? (Show proof for your answer.)

54. The compound found in many small portable lighters is butane. An analysis of butane yields the following mass percent values. From these data, determine the empirical formula of this useful fuel: 82.76% C, 17.24% H.

55. Potassium argentocyanide is a toxic compound that is used in the important industrial process of silver plating. It has these mass percent values: 19.6% potassium, 54.3% silver, 12.1% carbon, and 14.7% nitrogen. What is the empirical formula of this compound?

56. In humans, ingested ethyl alcohol is enzymatically decomposed into acetaldehyde. The mass percent values of the components of acetaldehyde are 54.55% C, 9.09% H, and 36.36% O. From these data, determine the empirical formula of acetaldehyde.

57. Linoleic acid is a fatty acid that contains unsaturated carbon bonding. Generally, this trait is found in many compounds that compose vegetable oils. The molar mass of linoleic acid is 280. The mass percent values of the elements in linoleic acid are 77.1% carbon, 11.4% hydrogen, and 11.4% oxygen. What is the molecular formula of this helpful agricultural compound?

58. During extreme exercise, lactic acid is produced when aerobic metabolism is not available for glucose breakdown. The mass percent values of the components in lactic acid are 40.0% C, 6.72% H, and 53.3% O. The average molecular mass of lactic acid is 90.1. What are the empirical and molecular formulas for this important biological compound?

3.6 Chemical Equations

Skill Review

59. Provide the proper coefficients to balance each chemical equation.

a. __ N_2H_4 + __ N_2O_4 → __ N_2 + __H_2O

b. __ $Pb(C_2H_3O_2)_2$ + __ KI → __PbI_2 + __$KC_2H_3O_2$

c. __ PCl_5 + __ H_2O → __ H_3PO_4 + __ HCl

d. __Ba_3N_2 + __ H_2O → __ $Ba(OH)_2$ + __ NH_3

60. Provide the proper coefficients to balance each chemical equation.

a. __ N_2 + __ O_2 → __ NO

b. __ CH_4 + __ O_2 → __ CO_2 + __ H_2O

c. __ H_2SO_4 + __ KOH → __ K_2SO_4 + __ H_2O

d. __ CO_2 + __ H_2O → __ $C_6H_{12}O_6$ + __ O_2

61. Write the proper formulas and balance the equation: Calcium chloride reacts with sodium phosphate to produce calcium phosphate and sodium chloride.

62. a. Balance this decomposition reaction:

__ $(NH_4)_2Cr_2O_7$ → __ Cr_2O_3 + __ N_2 + __ H_2O

b. How many atoms of nitrogen appear in the reactant side of the balanced equation? How many nitrogen molecules appear on the product side of the balanced equation?

63. This equation shows the proper stoichiometric ratio for all the components but one. Fill in the correct coefficient and compound to complete the equation.

$TiCl_4$ + 2H_2O → TiO_2 + __ ____

64. One method that could be used to make a chlorofluorocarbon (CFC) is based on this reaction:

$CCl_4(g)$ + $SbF_3(g)$ → $CCl_2F_2(g)$ + $SbCl_3(s)$

Freon-12

Use coefficients to balance the reaction. How many total moles of gases are involved in the balanced equation?

Chemical Applications and Practices

65. The banned insecticide DDT could be produced using this reaction:

__ C_6H_5Cl + __ C_2HOCl_3 → __ $C_{14}H_9Cl_5$ + __ H_2O

a. Balance the equation.

b. How many grams of DDT ($C_{14}H_9Cl_5$) would be expected from the complete reaction of 45.0 g of C_6H_5Cl?

66. A key ingredient for many over-the-counter antacids is aluminum hydroxide, $Al(OH)_3$. Use the following chemical reaction to predict the maximum number of grams of aluminum hydroxide that could be produced from 5.20 g of $Al_2(SO_4)_3$ and unlimited NaOH.

$Al_2(SO_4)_3$ + NaOH → $Al(OH)_3$ + Na_2SO_4 (not balanced)

67. The "fizz" formed when some pain relief tablets are placed in water is caused by the following chemical reaction. The bubbles are produced from the release of gaseous carbon dioxide. How many grams of $NaHCO_3$ would have to be present in a tablet to react with 0.487 g of citric acid ($H_3C_6H_5O_7$)?

$NaHCO_3(aq)$ + $H_3C_6H_5O_7(aq)$ →
$CO_2(g)$ + $Na_3C_6H_5O_7$ + $H_2O(aq)$

68. If some sugar from grapes were to be fermented to make wine, one of the primary reactions would be

__ $C_6H_{12}O_6$ → __ C_2H_6O + __CO_2

a. Balance the equation.

b. How many grams of sugar would you need to make 100.0 g of ethanol (C_2H_6O)?

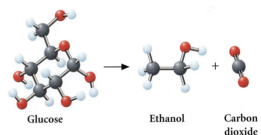

Glucose   Ethanol   Carbon
                                   dioxide

3.7 Working with Equations

Skill Review

69. a. Balance this equation:

__ NH_3 + __ O_2 → __ NO + __H_2O

b. Starting with 34.0 g of NH_3 and excess O_2, what mass of NO would be expected to be produced?

c. How much H_2O would form?

d. If the percent yield of NO was only 35.0%, how many grams of NO were formed?

70. a. Balance this equation:

__ Fe_2O_3 + __ C → __ Fe + __ CO

b. If the reaction were started with 1.00 kg of Fe_2O_3 and 0.250 kg of C, which reagent would you consider to be the limiting component?

c. Under the conditions described in part b, what would be the theoretical yield of Fe?

71. a. Balance this equation:

$$\_\_\ Ca(OH)_2 + \_\_\ H_3PO_4 \rightarrow \_\_\ Ca_3(PO_4)_2 + \_\_\ H_2O$$

 b. How many grams of $Ca(OH)_2$ are required to form 0.567 g of $Ca_3(PO_4)_2$?

 c. How many grams of H_3PO_4 would be required for the stoichiometry described in part b?

72. This reaction takes place when some types of matches are struck:

$$\_\_\ KClO_3 + \_\_\ P_4 \rightarrow \_\_\ P_4O_{10} + \_\_\ KCl$$

 a. Balance the equation.

 b. If 52.9 g of $KClO_3$, in the presence of excess P_4, produced 25.0 g of P_4O_{10}, what would be the calculated percentage yield of the process?

73. A fast-food restaurant serves this sandwich: 2 pieces of bread, 1 meat patty, 2 pieces of cheese, and 4 pickles. If the restaurant currently had on hand the following inventory, what would be the maximum number of specialty sandwiches that it could prepare? What ingredient would limit the sandwich production? Inventory: 200 pieces of bread, 200 meat patties, 200 cheese slices, 200 pickles.

74. A certain laboratory set-up for general chemistry requires four 250-mL beakers, six test tubes, and two graduated cylinders.

 a. Write an "equation" that shows the correct combination of lab equipment to form the arrangement.

 b. If a lab assistant set up 50 such arrangements, how many test tubes would be required?

 c. Would 100 beakers and 100 graduated cylinders provide enough equipment for these 50 arrangements? Explain.

Chemical Applications and Practices

75. One method used to produce soaps includes the reaction between NaOH and animal fat. The following equation represents that process:

$$C_3H_5(C_{17}H_{35}CO_2)_3 + 3NaOH \rightarrow$$
$$C_3H_5(OH)_3 + 3NaC_{17}H_{35}CO_2$$

 a. How many grams of soap, $NaC_{17}H_{35}CO_2$, would theoretically be produced by starting with 125 g of fat, $C_3H_5(C_{17}H_{35}CO_2)_3$?

 b. If the actual amount of soap produced were 31 g, what would be the percentage yield of the reaction?

76. The compound $Na_2S_2O_3$ is used in developing photographs. It can be synthesized by using this chemical reaction:

$$Na_2CO_3 + 2Na_2S + 4SO_2 \rightarrow 3Na_2S_2O_3 + CO_2$$

 a. Predict the mass of $Na_2S_2O_3$ (hypo) that could be produced from 48.5 g of Na_2S and unlimited quantities of the other reactants.

 b. If the procedure produced only a 78.9% reaction yield, how many grams of hypo would you expect to obtain?

77. When acid reacts with a base, the typical reaction yields water and a salt product formed from the anion of the acid and the cation of the base. For example, hydrochloric acid added carefully to sodium hydroxide forms water and NaCl. Use the balanced equation shown here to calculate the maximum yield of NaCl that can be formed when a solution containing 0.155 mol of HCl reacts with 4.55 g of NaOH.

$$HCl(aq) + NaOH(s) \rightarrow NaCl(aq) + H_2O(l)$$

78. A "magic show" demonstration often performed to illustrate volcanic action is to react vinegar with baking soda. The re-

lease of carbon dioxide gas that results produces a frothing action of bubbles. How many moles of carbon dioxide could be produced from the complete reaction of 10.0 g of baking soda ($NaHCO_3$) with excess acetic acid ($HC_2H_3O_2$) in vinegar?

$$HC_2H_3O_2(aq) + NaHCO_3(s) \rightarrow$$
$$H_2O(l) + CO_2(g) + NaC_2H_3O_2(aq)$$

79. Different baking powders are used in cooking to achieve various results in food textures as carbon dioxide gas is released. One such baking powder contains cream of tartar ($KHC_4H_4O_6$) and sodium hydrogen carbonate ($NaHCO_3$). Typically, starch is also added to keep moisture away from these active ingredients. When water from the recipe contacts these two compounds, this reaction takes place:

$$KHC_4H_4O_6 + NaHCO_3 \rightarrow NaKC_4H_4O_6 + H_2O + CO_2$$

 What is the maximum amount of $NaKC_4H_4O_6$ (also known as Rochelle salt) that could be formed if a baking powder sample contained 10.0 g each of the starting ingredients?

80. Often in dying cloth, a substance must be added that helps the dye to adhere to the cloth. Such a substance is known as a mordant. One such compound is aluminum sulfate ($Al_2(SO_4)_3 \cdot 18H_2O$). One method to obtain this useful compound is as follows:

$$\_\_\ Al(OH)_3 + \_\_\ H_2SO_4 + \_\_\ H_2O \rightarrow Al_2(SO_4)_3 \cdot 18H_2O$$

 Balance the equation and determine the maximum yield that could be obtained from reacting 25.0 g of $Al(OH)_3$ with 50.0 g of H_2SO_4 and unlimited water.

81. The following equation shows the conversion of ethanol in a wine sample into acetic acid through the action of oxygen from the air. This reaction drastically changes the taste of the wine. If 100.0 mL of wine initially contained 12.0 g of ethanol (C_2H_5OH), and after a period of time 4.00 g of acetic acid ($C_2H_4O_2$) were detected, what percent of the ethanol had been oxidized? $C_2H_5OH + O_2 \rightarrow C_2H_4O_2 + H_2O$

82. Sodium hypochlorite (NaOCl) is the active ingredient in most household bleach products. One method of preparing this important compound is as follows:

$$Cl_2(g) + 2NaOH(aq) \rightarrow NaOCl(aq) + NaCl(aq) + H_2O(l)$$

 a. Starting with 50.0 g of $Cl_2(g)$ and 50.0 g of NaOH, predict the theoretical yield of NaOCl.

 b. If the reaction, under a given set of conditions, formed only 38.9 g of NaOCl, what would you report as the percentage yield?

83. The following reaction can be used to represent the rusting of iron. (Annually, about 20% of the steel manufactured is used to replace steel corroded as a consequence of the oxidation of iron.) Determine how many grams of oxygen and iron reacted if 454 g of Fe_2O_3 were formed. (First balance the equation.)

$$Fe(s) + O_2(g) \rightarrow Fe_2O_3(s)$$

84. You may have noticed an athletic trainer spraying an injured player with a quickly evaporating liquid during a time out. The liquid is probably chloroethane, which quickly cools down an injured area with a slight numbing sensation. Chloroethane can be synthesized via the following reaction. If the reaction were begun using 85.0 g of ethene (C_2H_4) and 100.0 g of hydrogen chloride (HCl) but had only a 68.0% yield, how many grams of chloroethane (C_2H_5Cl) would be produced?

$$HCl(g) + C_2H_4(g) \rightarrow C_2H_5Cl(g)$$

85. Adding yeast cells to glucose can cause the glucose ($C_6H_{12}O_6$) to be converted to ethanol (C_2H_5OH) and CO_2. Bakers make use of this process when the CO_2 gas causes bread to rise during the baking process. Balance the following equation, and determine the number of moles of CO_2 that could be produced from 25.0 g of glucose.

$$\_\_ C_6H_{12}O_6(s) \rightarrow \_\_ C_2H_5OH(l) + \_\_ CO_2(g)$$

86. Two students perform a chemical synthesis in their general chemistry lab. Student A obtains a 90.0% yield in the reaction. Student B obtains an 85.0% yield in the same reaction. Can you now determine which student obtained the greater mass of the product? Explain, or justify your answer.

Comprehensive Problems

87. Review the vitamin C controversy discussed in this chapter. How is it possible that a compound can be both good and bad for your health?

88. What government agency is most concerned with setting recommended amounts of vitamins and minerals?

89. Explain why it is not totally correct to use the atomic masses given in the periodic table on the inside front cover of this book to express the mass of one molecule of any compound.

90. How big is Avogadro's number? If it were possible to place circles on notebook paper to represent atoms, how many pages would be required to complete the task? (Assume 25 lines on each side, 32 circles per line, and use of both sides of the paper.)

91. How many grams of argon would contain the same number of atoms as 10.0 g of neon?

92. Copper, silver, and gold are often referred to as the "coinage metals." If you had 454 g (approximately 1 lb) of each, which sample would contain the greater number of atoms?

93. The human hemoglobin molecule is quite large. It is known that each hemoglobin molecule contains four atoms of iron. If the iron makes up only 0.373% of the total mass of the molecule, what is the mass of 1 mol of hemoglobin?

94. In green plants, the compound chlorophyll a assists in the production of plant products and the oxygen essential for life on Earth. The compound contains one atom of magnesium. The mass percent of magnesium per molecule is 2.72%. What is the molar mass of this important compound?

95. When 1.00 g of one of the main components of gasoline was completely combusted, it produced 3.05 g of CO_2 and 1.50 g of H_2O. On the basis of this information, determine the mass percent values of carbon and hydrogen in the compound and the empirical formula of the compound.

96. The main compound responsible for the characteristic aroma of garlic is allicin. From the following mass percent values, determine the empirical formula of this familiar compound: 44.4% C, 6.21% H, 39.5% S, and 9.86% O.

97. When we examine balanced equations, we find that total of the coefficients on the left-hand side of the equation arrow does not always equal the total of the coefficients on the right-hand side. Explain why this does not violate the law of conservation of mass.

98. Examine the formula of salicylic acid in Section 3.6. How many molecules of salicylic acid would be required to react with 0.247 mL of ethanoic anhydride ($d = 1.080$ g/mL)?

99. A sample of ore is found to contain 14.5% aluminum oxide by mass. How many pounds of the sample would be required to make exactly 100 grams of pure aluminum? (Assume the reaction to make aluminum metal and oxygen gas from aluminum oxide occurs with only a 58% yield.)

100. Zinc oxide is often used in sunscreens to help protect your skin from harmful UV rays. If a particular sunbather wishes to coat 0.75 m^2 of his exposed skin with a layer of zinc oxide that is 1.0 mm thick, how many grams of zinc oxide would be needed? (Assume the sunscreen lotion has a density of 1.10 g/mL, and that the lotion contains 18% zinc oxide by mass.)

101. Palladium(II) nitrate is a reagent that can be used to link two smaller molecules together with a covalent bond. This reagent can be made by treating metallic palladium with nitric acid.
    a. What is the balanced reaction for the preparation of the reagent?
    b. How many grams of palladium(II) nitrate can be made from 10.0 grams of palladium metal and 5.6 mL of 15% (by mass) nitric acid whose density is 1.056 g/mL?

102. A researcher claims that a 1.00 gram sample of gold(II) chloride contains a greater mass of chloride than does a 1.00 gram sample of gold(IV) chloride.
    a. Is the researcher correct? Explain your answer.
    b. Which sample contains a greater mass of gold?
    c. If a 2.0 cm cube of gold(II) chloride was spread so that it was only 1/8 inches thick, what area would the compound occupy in in^2?

Thinking Beyond the Calculation

103. Xylene (ZIGH-leen) is an important organic molecule isolated from petroleum oil. It is often used as a thinner for oil-based paints.
    a. Elemental analysis of a sample of xylene shows that the mass percent of carbon is 90.51% and the mass percent of hydrogen is 9.49%. What is the empirical formula of xylene?
    b. The molar mass of xylene is 106.17 g/mol. What is the molecular formula of xylene?
    c. In the laboratory setting, xylene burns in limited oxygen environments to produce carbon monoxide and water vapor. Write the balanced reaction for this reaction.
    d. If 12.5 g of xylene were combusted using the equation from part c, how many grams of oxygen gas would be required to react completely with the xylene?
    e. A vessel containing 45.8 mL of xylene (density = 0.8787 g/mL) and 31.0 g of O_2 produced carbon monoxide and water. Which reagent is the limiting reagent in this reaction? How many grams carbon monoxide are produced in the reaction?
    f. If 1.40 g of CO were actually isolated from the reaction in part e, what would be the percentage yield of the reaction?

Solution Stoichiometry and Types of Reactions

View of a gold mine in the Black Hills of South Dakota. The gold ore is placed in "heap-leach pads," where cyanide solution is added. The solution percolates through the pads and collects in the ponds. Gold metal can be obtained from these ponds.

Contents and Selected Applications

Chemical Encounters: Gold Mining and Cyanide Leaching

4.1 Water—A Most Versatile Solvent

Chemical Encounters: Sports Drinks and Electrolyte Balance

4.2 The Concentration of Solutions

Chemical Encounters: Maximum Levels of Chemicals in Drinking Water

4.3 Stoichiometric Analysis of Solutions

4.4 Types of Chemical Reactions

4.5 Precipitation Reactions

Chemical Encounters: Focus on Lead Sulfide

4.6 Acid–Base Reactions

4.7 Oxidation–Reduction Reactions

4.8 Fresh Water—Issues of Quantitative Chemistry

Chemical Encounters: Revisiting the Maximum Levels of Chemicals in Drinking Water

Go to **college.hmco.com/pic/kelterMEE** for online learning resources.

"There's gold in them thar' hills!" That statement, and others like it, conjured up images of great wealth in a worldwide gold rush that began in the late 1840s, stretching from the hills of California to Australia. Whether our excitement comes from finding a nugget in a mountain stream, buying a necklace at the store, or just watching a show about the bullion bars stored at Fort Knox, we are fascinated with gold. The use of gold is not a recent phenomenon; it has been known for a large part of recorded history. In fact, gold was first used as coinage in 700 B.C., and it continued to flourish in coin form until the early 1900s.

In our modern world, gold is prized for its rarity, beauty, and high price. In recent years, its price per troy ounce (31.1 grams, or about 1.1 "avoirdupois ounces" in the units commonly used in the United States) has been as low as US$ 102 (1976) and as high as US$ 850 (1980). Currently it commands around US$ 500 per troy ounce. Gold is a fine electrical conductor, is not very reactive, and is quite malleable, making it ideal for use in electronic products, including computers. It is also highly reflective, so it is used as a shielding coating to protect spacecraft from solar radiation (Figure 4.1).

Gold has been used as currency for a very long time.

Gold used to be found in large nuggets as relatively pure metal, and miners would dig, sort, and harvest visible chunks of gold by almost every means possible. These days, in both large- and small-scale operations, gold ore is mined, pulverized into particles the size of grains of sand, and then mixed with a solution containing cyanide ion (CN^-).

$$4Au(s) + 8NaCN(aq) + O_2(g) + 2H_2O(l) \rightarrow 4NaAu(CN)_2(aq) + 4NaOH(aq)$$

The gold in the ore is dissolved by the solution, and the waste ore is discarded. Metallic gold is then obtained by precipitation and/or electroplating (see Chapter 19)

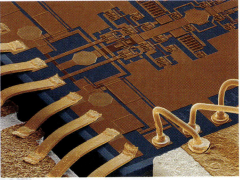

FIGURE 4.1

Gold has many uses in today's society. It blocks harmful UV rays in an astronaut's visor and serves as corrosion-resistant connections in a computer.

the solution. The precipitation reaction produces gold powder, although any silver and copper in the solution is also deposited with the gold.

$$Zn(s) + 2NaAu(CN)_2(aq) \rightarrow 2Au(s) + Na_2Zn(CN)_4(aq)$$

The extraction of gold using this method, known as the cyanide leaching method, takes place in a *water-based* or aqueous solution. In fact, much of the chemistry of life, and much of the chemistry of the environment, takes place in aqueous solution. Each cell of the body is a watery sac full of reacting chemicals. Rivers, lakes, and oceans contain complex solutions, each with its own special chemistry. At home we use solutions to cook and clean. And in large-scale industrial processes, such as gold mining, aqueous chemistry is important to the economic "bottom line."

Working with aqueous chemistry means understanding the stoichiometry (see Chapter 3) of the reactions that take place in the solution. For example, although too little cyanide (CN^-) in a given volume of water means an inefficient extraction of gold in the mining process, too much of it, or any compound in the blood, can be fatal. For example, long-term exposure to too much cyanide can lead to health problems in those who process the gold, including central nervous system, heart, and thyroid gland damage, and even death. Chemical reactions that require us to look at measures of concentration, such as how much cyanide is in a given volume of water, deal with "solution stoichiometry," the focus of the ideas, new terms, and techniques we will explore in this chapter.

4.1 Water—A Most Versatile Solvent

Thinking about the range of chemicals that must dissolve in our blood reveals water to be an extraordinary solvent—a compound that typically makes up the majority of a homogeneous mixture of molecules, ions, or atoms. We find ions such as potassium (K^+) and chloride (Cl^-); larger molecules including glucose ($C_6H_{12}O_6$) and, to a lesser extent, cholesterol ($C_{27}H_{46}O$); and giant protein molecules composed of thousands of atoms dissolved in our blood. Water, the solvent that dissolves these things in blood, has historically been called the universal solvent because many things dissolve in it.

A solution is a homogeneous mixture (see Figure 4.2). Those molecules, ions, or atoms that dissolve (or are *soluble*) in the solvent are known as solutes. The solvent is typically present in the larger amount in a solution, and the solute is typically the minor component. Exceptions arise when large amounts of solute can dissolve in small amounts of solvent, as with a solution of sugar in water. Considerably more than 1000 grams of sucrose ($C_{12}H_{22}O_{11}$, table sugar) can

Video Lesson: Properties of Solutions

FIGURE 4.2

In solution, the particles of a solute (such as the ions of copper sulfate, a sample of which is shown here) form a homogeneous mixture with the particles of the solvent (water molecules, in this case). The blue color of the solute is evenly distributed.

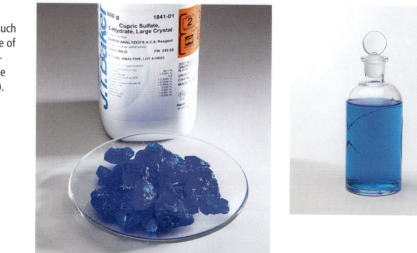

Sucrose is very soluble in water.

Water + Sucrose ⟶ Sugar water

dissolve in 1000 grams of water, but we still refer to the water as the solvent because it retains its phase. As we work with solutions here and throughout our study of chemistry, we tend to be keenly interested in the behavior and impact of the various *solutes,* such as cadmium or nitrate ions, dissolved in the water, the solvent.

Solutes dissolve in solvents because of the energetically favorable interactions between the two components of the solution. Water is a particularly versatile solvent because its structure permits it to interact with a wide variety of other chemicals. The resulting mixture is more energetically stable than if the compounds remained separate (such as layers of oil and water.) **Why is this so?** A closer look at the structure of water will help us answer this question.

Water is a molecule that contains an oxygen atom rich in electrons. This oxygen atom, therefore, possesses a partial negative charge. The hydrogen atoms attached to the oxygen atom have a lower concentration of electrons, which results in a partial positive charge. This is illustrated by the computer-drawn electrostatic potential map shown in Figure 4.3. When an ionic compound dissolves in water, it dissociates into its composite ions and interacts with the partial charges on the water molecule.

As an example, let's explore the dissolution of table salt (sodium chloride, NaCl) in water. The solid salt is composed of equal numbers of sodium ions (Na^+) and chloride ions (Cl^-). Figure 4.4 shows that as the solid salt dissolves, the opposite charges of the ions and the water molecules attract. The water molecules cluster around the ions in a way that allows the ions to separate and mix with the water. The salt dissolves. The cage of water molecules around the ions is known as the **hydration sphere** of the ions, and we say that the Na^+ and Cl^- ions become hydrated. The ions are not static, nor are they held in place by the water molecules. Instead, the ions separate and move about independently, though

Video Lesson: Factors Determining Solubility

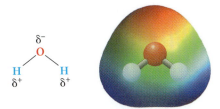

FIGURE 4.3

The electrons in the water molecule are distributed toward the oxygen atom. The oxygen end of water is electron-rich (δ^-) and the hydrogen end of water is electron-poor (δ^+). This can be represented with a colored map of electrostatic potential. Red areas indicate high electron density, and blue areas represent low electron density. The colors in between (ranging from yellow to green to sky blue) indicate varying degrees of electron density.

FIGURE 4.4

When an ionic compound (such as NaCl) dissolves, the uneven charge distribution in water molecules facilitates the process by interacting with the positive and negative charges on the ions.

Visualization: Dissolution of a Solid in a Liquid

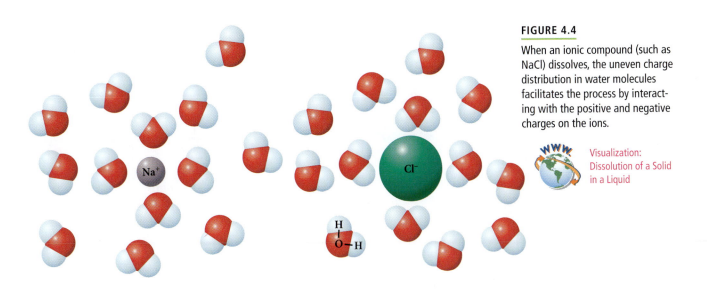

An electrostatic potential map of glucose and ethanol. The electrostatic potential is plotted on the surface of a computer-generated model of the molecules. Note how the electron density is distributed in each molecule, and compare these models to that of water (Figure 4.3).

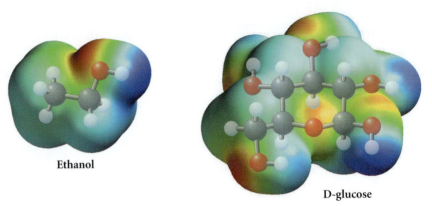

Ethanol

D-glucose

surrounded by their individual hydration spheres. We can represent the hydration sphere in a chemical equation by writing the symbol (*aq*) after the hydrated molecules, ions, or atoms to specify their aqueous phase:

$$NaCl(s) \rightarrow Na^+(aq) + Cl^-(aq)$$

There are many substances that are not made up of cations and anions. Some of them dissolve in water, such as the nutrient molecule glucose ($C_6H_{12}O_6$) and the molecule ethanol (C_2H_6O) found in alcoholic beverages. Why do they dissolve in water? These molecules possess a feature similar to water: The electrons are polarized toward specific regions of the molecule. This leaves a slightly positive charge and a slightly negative charge on different regions of the molecules. These partial charges can interact with the partial charges on water molecules. The strong interaction allows the molecules to mix easily with water and dissolve as shown in Figure 4.5. And, as is true with ions, a cage of water molecules surrounds the dissolved solute molecules.

$$C_6H_{12}O_6(s) \rightarrow C_6H_{12}O_6(aq)$$

Acids and bases are common substances, and many of them dissolve readily in water. For example, nitric acid, HNO_3, dissolves in water to produce the nitrate ion, $NO_3^-(aq)$, and the hydrogen ion, $H^+(aq)$. According to one definition of these common substances, acids produce a hydrogen ion when dissolved in water. On

FIGURE 4.5

The partial charges on polar covalent bonds, such as the O—H bonds in ethanol and glucose, can interact with the partial charges on water molecules and allow molecules to dissolve in water.

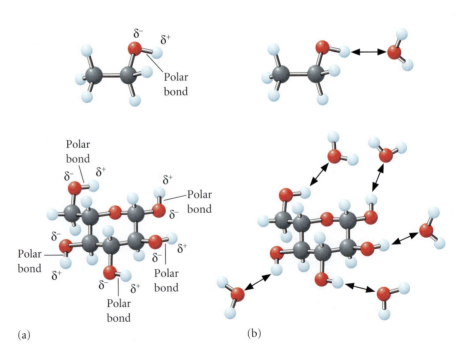

(a)

(b)

the other hand, sodium hydroxide (NaOH) is a common base that dissolves in water, resulting in $Na^+(aq)$ and $OH^-(aq)$. According to the same definition, bases produce hydroxide ion in solution.

Electrolytes

You may have encountered sports drinks sold with the claim that they help maintain a healthy electrolyte balance by supplying ions, sugar (to supply calories), and water in an appropriate combination. The key ingredients quoted on the label of a leading brand of sports drink can be found in Figure 4.6. The sodium and potassium in the drink are present as hydrated Na^+ and K^+ ions. They are accompanied by hydrated anions such as Cl^- (chloride ions) so that the entire solution has no net electrical charge. Ionic compounds, such as sodium chloride and potassium chloride, which dissociate in water to release free ions, are known as **electrolytes**. The name comes from the ability of solutions containing electrolytes to conduct electricity because of the presence of mobile ions that can carry the electric current.

Electrolytes are of great medical importance, because our blood and cells must contain an appropriate mixture of electrolytes to maintain good health. The presence of electrolytes in blood and cells allows the body to absorb the right amount of water from the gut and to excrete water via the kidneys. Maintaining the optimal balance of electrolytes is especially important after strenuous exercise, in which sweat containing water, sodium ions, and potassium ions leaves the body as part of a vital cooling mechanism. The ions of electrolytes also play a central role in creating the nerve impulses that allow our sense organs to work, our muscles to move, and our brains to think.

Substances such as sodium chloride, which *dissociate completely into ions* when they dissolve, are called **strong electrolytes**. In fact, most ionic compounds that dissolve in water fall into this category of electrolyte. A few molecules, such as hydrogen chloride (HCl), are also considered strong electrolytes. The equation below shows that gaseous HCl first dissolves, and then dissociates in the water, to form hydrated ions.

$$HCl(g) \rightarrow HCl(aq) \rightarrow H^+(aq) + Cl^-(aq)$$

Other electrolytes, such as acetic acid (vinegar, CH_3COOH), are called **weak electrolytes**, because even though they may dissolve completely in water, *the formation of ions occurs to a much lesser extent*, as shown in Figure 4.7. Note the different type of arrow, $\rightleftharpoons$, that defines a reaction that does not proceed completely to products.

$$CH_3COOH(l) \rightarrow CH_3COOH(aq) \rightleftharpoons H^+(aq) + CH_3COO^-(aq)$$

Those substances, such as glucose ($C_6H_{12}O_6$), that *dissolve in water without forming ions* are called **nonelectrolytes**. The nonelectrolytes dissolve and associate with water to form a hydration sphere, but they do not form ions. The equation that represents a nonelectrolyte's addition to water does not show the dissociation step found in the previous two equations.

$$C_6H_{12}O_6(s) \rightarrow C_6H_{12}O_6(aq)$$

FIGURE 4.6

The key ingredients of a sports drink include sodium and potassium cations.

Visualization: Electrified Pickle

Video Lesson: CIA Demonstration: The Electric Pickle

Application

CHEMICAL ENCOUNTERS: Sports Drinks and Electrolyte Balance

HCl

Electrostatic potential map of HCl. Note the very low electron density at the hydrogen end of the molecule.

FIGURE 4.7

Weak electrolytes (such as acetic acid) do not dissociate completely in water.

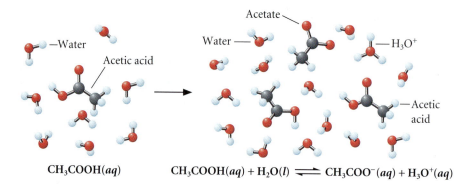

$CH_3COOH(aq)$

$CH_3COOH(aq) + H_2O(l) \rightleftharpoons CH_3COO^-(aq) + H_3O^+(aq)$

Defining Electrolytes

How do we know whether a compound is a strong electrolyte, a weak electrolyte, or a nonelectrolyte? Often the type of compound, assuming it dissolves in water, can tell us whether it will dissociate. For example, most ionic compounds that dissolve in water are strong electrolytes, as are the strong acids and bases. Weak acids and bases do not dissociate completely in aqueous solution and are therefore weak electrolytes. Nonelectrolytes include the water-soluble but nonionic compounds. Although there are many exceptions to these rules, the basic trends, summarized in Table 4.1, are valid.

Experimentation can provide evidence as to whether a particular solution contains a strong electrolyte, a weak electrolyte, or a nonelectrolyte. One such experiment is shown in Figure 4.8. Using a solution as the connector in a light bulb circuit, we can measure the ability of the solution to conduct electricity. The

FIGURE 4.8

The effect that each type of electrolyte has on the conductivity of a solution as measured by the brightness of a light bulb. From left to right are solutions of sodium chloride, acetic acid, and glucose.

Visualization: Electrolyte Behavior

Visualization: Electrolytes

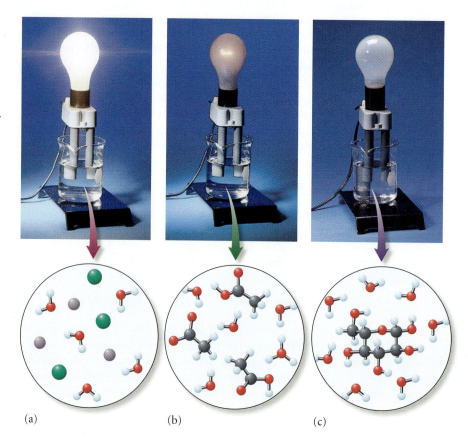

(a)   (b)   (c)

TABLE 4.1	Electrolytes and Types of Compounds			
	Ionic Compounds	**Strong Acids/Bases**	**Weak Acids/Bases**	**Molecular Compounds**
Strong Electrolyte	Yes	Yes	No	Sometimes
Weak Electrolyte	Sometimes	No	Yes	Sometimes
Nonelectrolyte	No	No	No	Sometimes

Strong and weak acids and bases are described in Section 4.6.

brightness of a light bulb demonstrates the type of electrolyte in solution. What's happening in Figure 4.8? Because ions can "carry" an electrical charge, the solution containing the greatest number of ions will have the brightest light bulb.

4.2   The Concentration of Solutions

Hyponatremia is a serious condition characterized by a low sodium level in the blood. The organization USA Track and Field recently released new guidelines calling for long-distance runners to avoid drinking excessive amounts of water during long runs, because doing so could lead to excessively diluted blood and a severely reduced sodium level. The result, hyponatremia, gives rise to a high fever, nausea, and, ultimately, heat stroke. Hyponatremia occurs when the sodium concentration is less than 130 mmol (mmol $= 10^{-3}$ mol) of sodium ions per liter of blood.

The **concentration**, the amount of solute per volume of solution, is often quoted in moles of solute per liter of solution, in grams of solute per liter of solution, or in various other ways. For example, long-distance runners should consume sports drinks instead of water. A typical sports drink has 110 mg Na^+ per 8-ounce serving, which won't reduce the sodium level in your blood. An 8-ounce serving of this drink contains 110 mg sodium ions. A larger quantity of the drink—say, 1 quart (32 ounces)—contains 440 mg Na^+, *but the concentration remains the same:* 110 mg Na^+ per 8-ounce serving. Concentration is an intensive property (see Section 1.4) because it doesn't change with the volume of solution.

A useful concentration unit is known as **molarity**. This unit specifically indicates the moles of solute per liter of solution, as illustrated in Figure 4.9.

$$\text{Molarity} = \frac{\text{moles of solute}}{\text{liter of solution}} = M$$

Application

Video Lesson: Concentrations of Solutions

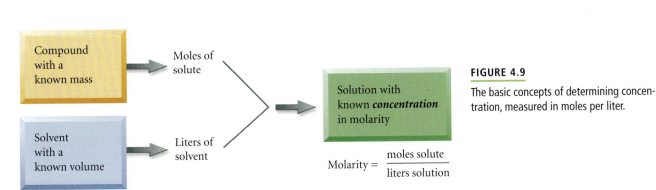

FIGURE 4.9

The basic concepts of determining concentration, measured in moles per liter.

Compound with a known mass → Moles of solute

Solvent with a known volume → Liters of solvent

Solution with known *concentration* in molarity

$$\text{Molarity} = \frac{\text{moles solute}}{\text{liters solution}}$$

For example, 600.0 mg of aspirin in a soluble aspirin tablet might be dissolved in orange juice to form 150.0 mL (150.0 cm^3) of solution. We can calculate the concentration of the aspirin solution once we have calculated the number of moles of aspirin and converted the volume of the juice into liters. We can use dimensional analysis to convert the units of milligrams of aspirin per milliliter of solution into moles of aspirin per liter of solution:

$$\frac{600.0 \text{ mg aspirin}}{150.0 \text{ mL solution}} \times \frac{1 \text{ g aspirin}}{1000 \text{ mg aspirin}} \times \frac{1 \text{ mol aspirin}}{180.2 \text{ g aspirin}} \times \frac{1000 \text{ mL}}{1 \text{ L}}$$

$$= \frac{0.02220 \text{ mol aspirin}}{\text{L solution}}$$

The molarity is 0.02220 mole of aspirin per liter, or 0.02220 **molar** (written 0.02220 M).

The relationship $\text{Molarity} = \dfrac{\text{moles of solute}}{\text{liter of solution}}$ can be used to calculate the number of moles of a solute if you know the molarity and the volume. In fact, this relationship can be used in many ways as a dimensional analysis factor. The exercises and practices that follow will give you an opportunity to use this new factor to answer questions.

EXERCISE 4.1   **Calculating Molarity**

Sodium hydroxide (NaOH) is one of the most important substances used in the chemical industry, with over 9.5 billion kg manufactured in 2004, ranking it among the top ten chemicals produced in the United States. It is used in chemical manufacturing processes, such as in making soaps and detergents. It is also used to break down the lignin that holds cellulose together in wood so that the cellulose can be made into paper. Dilute solutions of NaOH are often used in industry to verify the quality of medicines produced.

a. What is the molarity of NaOH in a solution that contains 3.48 g of sodium hydroxide in 500.0 mL of solution?

b. How many grams of NaOH are required to prepare 617 mL of 1.200 M NaOH solution?

Solution

a. Molarity is expressed in units of $\dfrac{\text{moles of solute}}{\text{liter of solution}}$. We can solve for the molarity by converting the units of $\dfrac{\text{g NaOH}}{\text{mL soln}}$ into $\dfrac{\text{moles of NaOH}}{\text{liter of solution}}$ using dimensional analysis. First, we construct a flowchart such as we introduced in Chapter 1:

$$\frac{\text{g NaOH}}{\text{mL soln}} \xrightarrow[\;\;1\text{ L}\;\;]{1000\text{ mL}} \frac{\text{g NaOH}}{\text{L soln}} \xrightarrow[\;\;\text{g NaOH}\;\;]{\text{mol NaOH}} \frac{\text{moles of NaOH}}{\text{liter of solution}}$$

Then we complete the calculation:

$$\frac{3.48 \text{ g NaOH}}{500.0 \text{ mL solution}} \times \frac{1000 \text{ mL}}{1 \text{ L}} \times \frac{1 \text{ mol NaOH}}{40.00 \text{ g NaOH}} = \frac{0.174 \text{ mol NaOH}}{\text{L}}$$

$$= 0.174 \; M \text{ NaOH}$$

b. Given the molarity and the volume, we can solve for moles of NaOH and then change to grams of NaOH using the molar mass of the compound. Again, we'll do it all in one step by converting units.

$$\text{mL soln} \xrightarrow{\frac{\text{L soln}}{\text{mL soln}}} \text{L soln} \xrightarrow{\frac{\text{mol NaOH}}{\text{L soln}}} \text{mol NaOH} \xrightarrow{\frac{\text{g NaOH}}{\text{mol NaOH}}} \text{g NaOH}$$

Then we complete the calculation:

$$617 \text{ mL solution} \times \frac{1 \text{ L}}{1000 \text{ mL}} \times \frac{1.200 \text{ mol NaOH}}{1 \text{ L solution}} \times \frac{40.00 \text{ g NaOH}}{\text{mol NaOH}} = 29.6 \text{ g NaOH}$$

PRACTICE 4.1

a. What is the molarity of K^+ in the typical muscle cell, which contains 6.5 g K^+ per liter of solution?

b. What is the molarity of Na^+ in a sports drink that contains 110 mg Na^+ per 8-ounce serving? (1 ounce = 29.6 mL)

c. How many grams of ethanol (C_2H_6O) do we need to prepare 600.0 mL of a 1.200 molar ethanol solution?

See Problems 15–17, 19–20, 29, 31, 32, and 34.

EXERCISE 4.2   Calculating Moles

a. How many moles of calcium hydroxide, $Ca(OH)_2$, are found in 250 mL of a 0.800 molar solution of this compound?

b. If we assume that $Ca(OH)_2$ is a strong electrolyte, how many moles of Ca^{2+} ions will be present in 250 mL of 0.800 M $Ca(OH)_2$? How many moles of OH^- ions will be present?

First Thoughts

This problem is different from the previous one because it asks for the number of moles in a solution that is already prepared, rather than asking for the concentration of a solution (part a of Exercise 4.1) or the mass of a reagent required to prepare a solution (part b). Still, the problem-solving approach is similar. We note what we start with, what we want, and how we get there. You can use a flowchart to help you stay on track.

Solution

a. Again, we use dimensional analysis to solve the problem. Note that we must convert the milliliters of solution into liters so that we can use the molarity term. Our flowchart for this calculation is constructed first:

$$\text{mL soln} \xrightarrow{\frac{1 \text{ L}}{1000 \text{ mL}}} \text{L soln} \xrightarrow{\frac{\text{mol Ca(OH)}_2}{\text{L soln}}} \text{mol Ca(OH)}_2$$

Then

$$250 \text{ mL solution} \times \frac{1 \text{ L}}{1000 \text{ mL}} \times \frac{0.800 \text{ mol Ca(OH)}_2}{1 \text{ L solution}} = 0.20 \text{ mol Ca(OH)}_2$$

b. We start by writing the balanced equation that describes what happens to $Ca(OH)_2$ when it is added to water.

$$Ca(OH)_2(s) \rightarrow Ca^{2+}(aq) + 2OH^-(aq)$$

Then we can calculate the answer by performing a dimensional analysis. Our calculation can be accomplished using the mole ratio relating the number of moles of $Ca(OH)_2$ to the number of moles of Ca^{2+}. Our flowchart is written first:

$$\text{mL soln} \xrightarrow{\frac{1 \text{ L}}{1000 \text{ mL}}} \text{L soln} \xrightarrow{\frac{\text{mol Ca(OH)}_2}{\text{L soln}}} \text{mol Ca(OH)}_2 \xrightarrow{\frac{\text{mol Ca}^{2+}}{\text{mol Ca(OH)}_2}} \text{mol Ca}^{2+}$$

Then

$$250 \text{ mL solution} \times \frac{1 \text{ L}}{1000 \text{ mL}} \times \frac{0.800 \text{ mol Ca(OH)}_2}{1 \text{ L solution}} \times \frac{1 \text{ mol Ca}^{2+}}{1 \text{ mol Ca(OH)}_2}$$

$$= 0.20 \text{ mol Ca}^{2+}$$

Again, we need to use the mole ratio to indicate the relationship between $Ca(OH)_2$ and OH^-. We'll use the same flowchart for the Ca^{2+} determination, but we'll modify the last step to show the mole ratio between $Ca(OH)_2$ and OH^-. The calculation is

$$250 \text{ mL solution} \times \frac{1 \text{ L}}{1000 \text{ mL}} \times \frac{0.800 \text{ mol Ca(OH)}_2}{1 \text{ L solution}} \times \frac{2 \text{ mol OH}^-}{1 \text{ mol Ca(OH)}_2}$$

$$= 0.40 \text{ mol OH}^-$$

Further Insights

Although it might seem like "busy work" to include the mole ratio in a calculation that can be determined by counting the number of ions in the formula, it is very important to include mole ratios along with the equation that shows the dissolution process. If we avoid this step now, we will find it harder to solve more complex problems in later chapters.

PRACTICE 4.2

How many moles of the strong electrolyte sodium phosphate (Na_3PO_4) will be in 5.00 L of a 0.77 M solution? How many moles of hydrated sodium ions will be in the solution?

See Problems 13, 14, 37, and 38.

EXERCISE 4.3 Calculating Volumes

What volume of a 0.150 M solution of ethanol (C_2H_6O) will contain 12.5 moles of ethanol?

First Thoughts

This is another unit conversion problem that we can solve using dimensional analysis. However, in this case, we're using the molarity in a different way: to determine the volume rather than moles, grams, or molarity, as in previous problems.

Solution

First, we write our flowchart:

$$\text{mol ethanol} \xrightarrow{\frac{\text{L soln}}{\text{mol ethanol}}} \text{L ethanol}$$

Then

$$12.5 \text{ mol ethanol} \times \frac{\text{L solution}}{0.150 \text{ mol ethanol}} = 83.3 \text{ L solution}$$

Further Insights

In questions like this, the identity of the solute is irrelevant *provided we are dealing with moles, rather than masses.* For example, the answer we obtained for the *volume* of 0.150 *M* ethanol containing 12.5 moles of ethanol would be equally valid for the volume of a 0.150 *M* solution of glucose containing 12.5 moles of glucose. Both require a volume of 83.3 L. In a similar way, if we were asked to calculate the number of moles of a compound in a given volume of solution, we would not need to know the identity of the compound if we were given the molarity of the solution. However, if we were asked about *specific numbers of moles of ions of a strong electrolyte*, we would need the formula of the compound to determine the number of times the ions or atoms appear in the formula.

PRACTICE 4.3

a. What volume of a 3.40 *M* solution of copper sulfate will contain 4.76 moles of copper sulfate?

b. What volume of a 2.25 *M* solution of $Ca(NO_3)_2$ will contain 5.5 moles of calcium nitrate? What volume will contain 5.5 moles of nitrate ions?

See Problems 18, 21, and 22–24.

Parts per Million, Parts per Billion, and so on

Drinking water standards, which indicate the maximum permissible level of harmful contaminants in the water that we consume, are dictated in the United States by the Environmental Protection Agency (EPA). Other countries have their own agencies responsible for water standards, such as La Secretaría de Medio Ambiente y Recursos Naturales (SEMERNAT, The Federal Agency of the Environment and Natural Resources) in Mexico. If water in the United States has more than these maximum levels, the EPA declares it unsafe to drink. For example, as of January 23, 2006, the maximum allowed level of arsenic in drinking water is 1.3×10^{-7} *M*. We can use molarity to discuss the concentrations of pollutants such as arsenic, but the resulting numbers are, in many cases, very small. When the concentrations get this small, we often find it easier to use an alternative measure of concentration. Specifically, we can talk about the concentration of arsenic in terms of **parts per million (ppm)**, or **parts per billion (ppb)**, or even **parts per trillion (ppt)**.

We are already familiar with the related, but larger, unit "percent." *What does percent mean?* In a compound that is "1 percent nitrogen by mass," the nitrogen contributes 1 gram out of every 100 grams to the total mass. In the same way, a level of one part per million (1 ppm), means the chemical contributes 1 gram out of every million grams of the total mass. Similarly, one part per billion (1 ppb) corresponds to a level of 1 gram out of every billion grams of the total. One part per trillion (ppt) corresponds to 1 gram out of every trillion grams.

These concentration units correspond to very small levels indeed. We can get an idea of exactly how small from the following approximate comparisons, as illustrated in Figure 4.10:

■ One part per million is roughly equivalent to a drop of ink in a 12-gallon bucket of water.

■ One part per billion is roughly equivalent to a drop of ink in a large tanker truck full of water.

■ One part per trillion is roughly equivalent to a drop of ink in a 12-million-gallon reservoir of water.

Application

CHEMICAL ENCOUNTERS: Maximum Levels of Chemicals in Drinking Water

An environmental chemist sampling the water in a river.

FIGURE 4.10

To visualize the idea of parts per million, per billion, and per trillion, consider that one drop of ink could be placed in these quantities of water. (a) Placing that drop of ink in a 12-gallon bucket results in 1 ppm. (b) Placing it in a tanker truck results in 1 ppb. (c) Placing it in a 12-million-gallon reservoir results in 1 ppt.

(a)　　　　　　　　　(b)　　　　　　　　　(c)

Let's look at these concentration units in a slightly different way. As an example, the EPA maximum level for hexachlorobenzene (C_6Cl_6), a pesticide used on wheat in the United States until 1965, is 1 ppb. This means that the maximum allowable level in drinking water should be 1 gram of C_6Cl_6 per billion grams of solution.

$$1 \text{ ppb } C_6Cl_6 = \frac{1 \text{ g } C_6Cl_6}{10^9 \text{ g solution}}$$

If we use $\dfrac{1 \text{ g water}}{1 \text{ mL water}}$ as the density of water and make the key assumption that *the solution is so dilute that its density is about equal to that of water,* then

$$\text{Density of very dilute aqueous solutions} = \frac{1 \text{ g solution}}{1 \text{ mL solution}}$$

This means that we can express the concentration of hexachlorobenzene as the number of grams of C_6Cl_6 per liter of solution:

Hexachlorobenzene

$$\frac{1 \text{ g } C_6Cl_6}{10^9 \text{ g solution}} \times \frac{1 \text{ g solution}}{1 \text{ mL solution}} \times \frac{10^3 \text{ mL solution}}{1 \text{ L solution}} = \frac{10^{-6} \text{ g } C_6Cl_6}{\text{L solution}} = \frac{1 \text{ } \mu g \text{ } C_6Cl_6}{\text{L solution}}$$

$\qquad\quad$ ↑ $\qquad\qquad\qquad$ ↑ $\qquad\qquad\qquad$ ↑
(one ppb C_6Cl_6)　(density of the solution)　(convert mL to L)

The bottom line is that in dilute aqueous solutions, we can express parts per billion as

$$\text{ppb solute} = \frac{\mu g \text{ solute}}{\text{L solution}}$$

This relationship, the number of grams of solute per liter of solution, is more conceptually understandable than the actual definition of parts per billion. For our hexachlorobenzene example, 1 liter of solution that contains 1.0 μg hexachlorobenzene is said to have a concentration of 1.0 ppb hexachlorobenzene. We can even use this definition to convert from other concentrations to ppb. What is the maximum level of arsenic, in ppb, if the molarity of a solution at this maximum level is 1.3×10^{-7} M arsenic?

Our flowchart looks like this:

$$\frac{\text{mol As}}{\text{L solution}} \xrightarrow{\frac{\text{g As}}{\text{mol As}}} \frac{\text{g As}}{\text{L solution}} \xrightarrow{\frac{1 \mu g}{10^{-6} g}} \frac{\mu g \text{ As}}{\text{L solution}} = \text{ppb As}$$

Then

$$\frac{1.3 \times 10^{-7} \text{ mol As}}{1 \text{ L solution}} \times \frac{74.92 \text{ g As}}{1 \text{ mol As}} \times \frac{1 \mu g \text{ As}}{10^{-6} \text{ g As}} = \frac{9.7 \mu g \text{ As}}{\text{L solution}} = 9.7 \text{ ppb}$$

Note the relationships among parts per million, parts per billion, and parts per trillion in Table 4.2.

TABLE 4.2	Common Units Used in Expressing "Parts per" Concentration of "X"	
Unit	Mass-to-Mass Relationship	Mass-to-Volume Relationship
parts per million (ppm)	$\dfrac{\text{g X}}{10^6 \text{ g solution}}$	$\dfrac{\text{mg X}}{\text{L solution}}$
parts per billion (ppb)	$\dfrac{\text{g X}}{10^9 \text{ g solution}}$	$\dfrac{\mu\text{g X}}{\text{L solution}}$
parts per trillion (ppt)	$\dfrac{\text{g X}}{10^{12} \text{ g solution}}$	$\dfrac{\text{ng X}}{\text{L solution}}$

Mass-to-volume relationship assumes a density of 1g/mL.

Although we more often hear about how small concentrations of compounds in water can be harmful, very small levels of some chemicals in drinking water can actually be helpful to human health. For example, the presence of tiny amounts of fluoride (F^-) in drinking water is recognized as beneficial in the prevention of tooth decay. The U.S. Public Health Service recommends a level of between 0.7 and 1.2 ppm F^-. The effects of such small levels in preventing tooth decay have been reported as "striking" by the American Dental Association.

EXERCISE 4.4 Converting ppm to Molarity

An acceptable midrange value for fluoride ion (F^-) in drinking water is 1.0 ppm. To what concentration of fluoride, in M, does this correspond?

First Thoughts

Using the mass-per-volume relationship we established in Table 4.2 enables us to solve the problem with $1.0 \text{ ppm} = \dfrac{1.0 \text{ mg F}^-}{\text{L solution}}$.

Solution

Remember to construct the flowchart before you start the calculation.

$$\frac{1.0 \text{ mg F}^-}{\text{L solution}} \times \frac{10^{-3} \text{ g F}^-}{1 \text{ mg F}^-} \times \frac{1 \text{ mol F}^-}{19.00 \text{ g F}^-} = 5.3 \times 10^{-5} M \text{ F}^-$$

Further Insights

High concentrations of fluoride ion can be very harmful to your health. In fact, some would argue that fluoride at any level is harmful. The debate surrounding fluoridation of drinking water continues.

PRACTICE 4.4

Cyanide ion (CN^-) concentrations of 0.200 ppm in drinking water are considered the upper limit for human consumption. What is this concentration in M?

See Problems 25–28, 32, 35, and 36.

Dilution

How do municipal water treatment plant workers prepare water so that it contains a fluoride ion concentration of about 1 part per million? They use a concentrated source of fluorine, hydrofluorosilicic acid, H_2SiF_6 ("HFSA"), which they then dilute in the drinking water. HFSA reacts with water in a fairly complex way that releases fluoride ion into the water. In Ireland, a company in the town of New Ross,

 Application

FIGURE 4.11

New Ross, Ireland, is about 100 miles (160 km) from Dublin.

about 160 km south-southwest of Dublin (Figure 4.11), receives periodic shipments of a concentrated aqueous HFSA solution from a chemical manufacturer in Spain. The concentration of this HFSA solution, shipped in 4000-gallon containers, is about 22 *M* HFSA. The solution is then diluted with water to 7.8 *M* and shipped to cities and towns all over Ireland for use in the nation's many water treatment plants. At these plants, the HFSA solution is further diluted in the water supply to a final fluoride ion concentration of 1 ppm.

How do the workers know how much to dilute? There are really two questions here. One is a mathematics question, and one is a chemical process question ("How do I prepare this solution?"). In terms of the chemical process, it has long been known that unless you are adding two of the exact same liquids to each other, the volumes are not additive. That is, adding 100 mL of ethanol to 100 mL of water does not give 200 mL of solution. Rather, it gives a volume that is a little bit *less than* 200 mL. The practical outcome of this is that for laboratory-scale preparation of a solution of known molarity, we must use volumetric glassware (shown in Figure 4.12), in which the etched mark on the flask indicates the listed volume to a very high accuracy, typically within $+/-0.05\%$ of the stated volume. We add the reagent to the flask and then dilute with solvent (water, in this case) to the mark.

We can now deal with the mathematics question. Let's assume we wish to prepare 500.0 mL of 7.8 *M* HFSA solution from a sample of 22 *M* HFSA. To do this, we need to calculate the volume of 22 *M* HFSA that can be added to the volumetric flask and diluted with water. How many moles of HFSA will be in the final solution? We are given the molarity (7.8 *M*) and volume (500.0 mL) of the final solution, so we can do the following calculation:

$$\frac{7.8 \text{ mol HFSA}}{\text{L HFSA solution}} \times 0.5000 \text{ L solution} = 3.9 \text{ mol HFSA}$$

What volume of the concentrated (22 *M*) HFSA solution must we dispense into the volumetric flask and dilute with water in order to give us 3.9 moles of HFSA?

$$3.9 \text{ mol HFSA} \times \frac{1 \text{ L HFSA}}{22 \text{ mol HFSA}} = 0.1773 \text{ L HFSA solution}$$

$$= 177.3 \text{ mL HFSA solution} = 180 \text{ mL HFSA solution (to 2 significant figures)}$$

This means that we can measure 180 mL of HFSA solution and dilute to 500.0 mL in our volumetric flask and, after mixing, have a 7.8 *M* HFSA solution.

Visualization: Dilution

Video Lesson: CIA Demonstration: Dilutions

Tutorial: Dilution

FIGURE 4.12

Using a volumetric flask. A solute is added to the volumetric flask and dissolved in water. The resulting solution is then diluted to the mark on the neck of the flask. The result is a precise and accurate volume of solution.

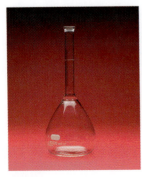

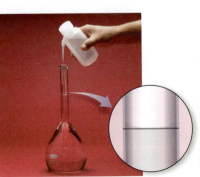

Note that in our calculations, we have compared the number of moles of solute in the concentrated solution that we are measuring to the number of moles of the diluted solution that we want to prepare. Mathematically, this can be written:

$$\text{number of moles } \textit{measured} \text{ from the concentrated solution} = \text{number of moles } \textit{needed} \text{ in the dilute solution}$$

Because we calculated the number of moles *needed* in the dilute ("final") solution by multiplying its concentration (C_{final}) by its total solution volume (V_{final}), and because this is equal to the number of moles of concentrated ("initial") solution *measured*, we can say that

$$C_{initial} \times V_{initial} = C_{final} \times V_{final}$$

Our goal is to find the volume of the initial solution of HFSA to dilute, so we can solve for $V_{initial}$.

$$V_{initial} = \frac{C_{final} \times V_{final}}{C_{initial}}$$

Using our data for HFSA, we find that

$$V_{initial} = \frac{7.8\ M \times 0.5000\ L}{22\ M} = 0.177\ L = 180\ mL$$

The total volume of 22 M HFSA needed to prepare large, intercity shipments of 7.8 M HFSA is far larger, but the problem-solving strategy is the same.

EXERCISE 4.5   **Practice with Dilution of Solutions**

Hydrochloric acid is typically shipped as a 12 M aqueous solution. If we are to do a laboratory analysis that requires 250.0 mL of a 1.0 M solution, how many milliliters of the concentrated solution should be diluted?

First Thoughts

What range of volumes would make sense as an answer? We want to dilute our solution by a factor of 12, to go from 12 M to 1.0 M. If the total volume of the solution is to be 250.0 mL, and we are diluting a solution with water by a factor of 12, this would suggest that we need about 20 mL of the concentrated solution (roughly 1/12 of 250.0 mL). The remainder of the solution is water, added to dilute the concentrated acid.

Solution

$$V_{initial} = \frac{C_{final} \times V_{final}}{C_{initial}} = \frac{1.0\ M \times 250.0\ mL}{12\ M} = 20.8\ mL = 21\ mL$$

Further Insights

It is quite common for industrially prepared concentrated acids to be labeled for shipment on the basis of grams of acid per grams of solution, $\dfrac{\text{g acid}}{\text{g solution}}$. This is known as weight-to-weight percent, or (w/w). Hydrochloric acid, for example, is shipped as a 37% (w/w) solution and has a density of 1.18 grams of acid per milliliter of solution. How do we relate w/w to molarity? Here are the flowchart and the calculation.

$$\frac{\text{g HCl}}{\text{g soln}} \xrightarrow{\frac{\text{g soln}}{\text{mL soln}}} \frac{\text{g HCl}}{\text{mL soln}} \xrightarrow{\frac{1000\ \text{mL}}{1\ \text{L}}} \frac{\text{g HCl}}{\text{L soln}} \xrightarrow{\frac{\text{mol HCl}}{\text{g HCl}}} \frac{\text{mol HCl}}{\text{L soln}}$$

$$\frac{37\ \text{g HCl}}{100\ \text{g solution}} \times \frac{1.18\ \text{g solution}}{1\ \text{mL solution}} \times \frac{1000\ \text{mL solution}}{1\ \text{L solution}} \times \frac{1\ \text{mole HCl}}{36.5\ \text{g HCl}} = 12\ M$$

4.3   Stoichiometric Analysis of Solutions

The concepts of concentration and solution stoichiometry are put to important use every day to examine all manner of solutions. In medicine, they are used to analyze bodily fluids for evidence of illness and to prepare pharmaceuticals in appropriate amounts. In industry, they are used to ensure that the correct quantities of dissolved reactants are mixed together. In forensic chemistry they can determine the levels of drugs and poisons in a victim or suspect of crime. And in the food industry they make it possible to measure and modify the nutritional content of foods.

Application

For example, the amount of vitamin C in a sample of fruit juice can be determined by adding iodine to the fruit juice via the following reaction:

$$C_6H_8O_6 \quad + \quad I_2 \quad \rightarrow \quad C_6H_6O_6 \quad + \quad 2HI$$

vitamin C   brownish-red   colorless   colorless
colorless

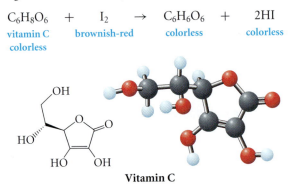

Vitamin C

The controlled addition of a solution of known concentration to react with the substance of interest to determine its concentration is called a **titration**. In the titration of fruit juice with iodine shown in Figure 4.13, we slowly add iodine solution from a **buret** (a laboratory device used to precisely and accurately add small known quantities of solution) to a sample of fruit juice. When all the vitamin C in the sample has reacted, excess iodine will begin to accumulate. The point at which the reactants have been *completely* consumed is known as the **equivalence point** (very close to the visual **end point**) of the titration. We can detect when the reaction has reached this stage by including a compound, in this case starch, that reacts with the excess iodine to produce a color change. Although the iodine itself is colored, it is often difficult to see this color, especially in fruit juice. However, the addition of starch results in a very obvious color change.

$$\text{starch} \quad + \quad I_2 \quad \rightarrow \quad I_2\text{—starch}$$

colorless   brownish-red   deep blue

Starch–iodine end point.

Here is how we can use this reaction to determine the vitamin C content of a fruit juice in moles per liter and percent by mass. We begin by preparing a solution of a known concentration of iodine that turns out to be, for example, 0.01052 M I_2. This is known as a **standard solution**, because we use it as a standard of known concentration against which other solutions can be tested. We then measure out 50.00 mL (0.05000 L) of fruit juice and add some starch to react with excess iodine. By adding our standard solution of iodine from a buret, we titrate the fruit juice until a color change is observed. Typically, this process is repeated several times to minimize error. In our hypothetical titrations, let's assume that

we averaged 16.97 mL of iodine solution before the color change was observed. How much vitamin C is in the fruit juice? We can quantify the vitamin C content of the fruit juice using the following strategy:

1. Write the balanced chemical equation relating the reaction of iodine with vitamin C, $C_6H_8O_6$.

2. Use dimensional analysis to convert the units of 16.97 mL of iodine solution into grams of vitamin C.

$$C_6H_8O_6 \quad + \quad I_2 \quad \rightarrow \quad C_6H_6O_6 \quad + \quad 2HI$$
$$\text{1 mole} \quad + \quad \text{1 mole} \quad \quad \text{1 mole} \quad + \quad \text{2 moles}$$

$$16.97 \text{ mL } I_2 \text{ solution} \times \frac{1 \text{ L } I_2 \text{ solution}}{1000 \text{ mL } I_2 \text{ solution}} \times \frac{0.01052 \text{ mol } I_2}{1 \text{ L } I_2 \text{ solution}} \times \frac{1 \text{ mol vitamin C}}{1 \text{ mol } I_2}$$

$$\times \frac{176.1 \text{ g vitamin C}}{1 \text{ mol vitamin C}} = 0.03144 \text{ g vitamin C}$$

That is the number of grams of vitamin C in our 50.00 mL (0.05000 L) sample of fruit juice. We can determine the amount of vitamin C in 1 L by doing the following calculation:

$$\frac{0.03144 \text{ g vitamin C}}{50.00 \text{ mL fruit juice solution}} \times \frac{1000 \text{ mL fruit juice solution}}{1 \text{ L fruit juice solution}} = \frac{0.6288 \text{ g vitamin C}}{\text{L fruit juice solution}}$$

Our calculations reveal that there are 0.03144 g of vitamin C in our 50.00 mL sample of fruit juice. Alternatively, this is equal to 0.6288 g of vitamin C per liter of fruit juice. If we assume that the density of the fruit juice is $\frac{1.0 \text{ g}}{\text{mL}}$, what is the concentration of vitamin C in ppm?

Recall from Table 4.2 that parts per million (ppm) of vitamin C can be expressed as $\frac{\text{mg vitamin C}}{\text{L fruit juice solution}}$. We know the vitamin C concentration in $\frac{\text{g}}{\text{L}}$, so all we need to do is convert from grams to milligrams of vitamin C.

$$\frac{0.6288 \text{ g vitamin C}}{\text{L fruit juice solution}} \times \frac{1000 \text{ mg vitamin C}}{1 \text{ g vitamin C}} = \frac{628.8 \text{ mg vitamin C}}{\text{L fruit juice solution}}$$

$$= 628.8 \text{ ppm vitamin C}$$

How do we know?

How to test for small amounts of water

Water is such a precious resource that countries have been known to threaten one another with war over water rights. In the United States, individual states quarrel and even sue each other over access to fresh water. For example, Nebraska and Kansas have fought for decades over who owns the water that flows from Colorado through Nebraska and into Kansas via the Republican River.

We are so used to thinking of water as essential and beneficial that it is easy to overlook the many chemical situations in which too much water can be very undesirable. What would be the effect of too much water in the Republican River? On a smaller scale, what would be the effect of too much water in a chocolate bar or potato chips? Canned cooking fats and oils certainly do not benefit from the presence of water; and there are limits to the amount of water that can be in pharmaceutical products that are ingested as tablets, gelcaps, and caplets. Therefore, it is important to be able to determine how much water is present in samples of foods, medicines, and other industrial products, even if the quantities of water involved are very small.

The Republican River supplies Harlan County Reservoir with water used for irrigation and recreation.

Hot oil spatters violently when water, such as that on the surface of this turkey, is present.

One chemical method used to quantify the water content of a wide range of samples, such as chewing gum, jelly beans, and peanuts, is the Karl Fischer titration, named after the scientist who devised the basic method in 1935. The Karl Fischer titration makes use of the reaction of iodine (I_2) with the water in the sample being analyzed. The reaction is performed in the presence of organic solvents such as pyridine (C_5H_5N) and methanol (CH_3OH). The precise details of the chemical reaction are quite complex, to the extent that even now, more than 70 years after the method was first developed, the exact processes involved in the reaction are still the subject of research. We can summarize it, however, as follows:

$$(\text{Reactants}) + I_2 + H_2O \rightarrow (\text{products})$$

The chemical details vary with the actual method used. In all cases, however, the crucial quantitative fact that allows the water content to be measured is that *the iodine and water always react in a known mole ratio*, which is 1:1 in the case shown above. The end point of the titration, when we can tell that all the water has been consumed, is marked by a distinctive color change, which enables us to determine the quantity of iodine required to achieve that color change. The amount of water that must have reacted with that quantity of iodine can then be calculated.

Modern laboratory instrumentation has allowed the titration process to be automated, as shown in Figure 4.14. Instead of direct observation of a color change, the automated instruments may rely on measurement of a flow of electrons to excess iodine as soon as all the water has reacted. This flow of electrons is due to the process $I_2 + 2e^- \rightarrow 2I^-$. Another option is to use the reversal of the process shown above—namely $2I^- \rightarrow I_2 + 2e^-$—to generate the iodine needed to react with the water.

The details vary depending on the machines used, but automated Karl Fischer titrations allow accurate determination of the water content in tiny samples containing hardly any water at all. What do we mean by "hardly any" and "tiny"? The method is useful down to the level of 50 ppm water in a sample that can have a volume as small as 10 microliters.

FIGURE 4.14

Automated Fischer titration apparatus.

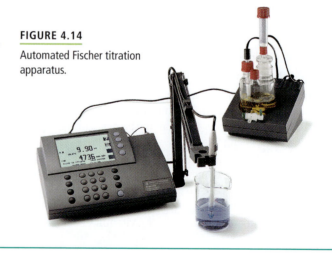

| EXERCISE 4.6 | Practice with Titration |

Video Lesson: Solving Titration Problems

A common type of titration is the reaction of a solution of sodium hydroxide (with a known concentration) with a solution of an acid (with an unknown concentration). The result can be used to determine the molarity of the acid solution. What is the molarity of a solution of 50.00 mL of HCl if 31.98 mL of 0.1253 M NaOH is required to react with it? The products of this reaction are sodium chloride and water.

First Thoughts

We should begin this problem as we begin all stoichiometry problems in chemistry, with a balanced chemical equation. From the problem we determine the equation for the titration of hydrochloric acid with sodium hydroxide:

$$HCl(aq) + NaOH(aq) \rightarrow H_2O(l) + NaCl(aq)$$

Our flowchart for the dimensional analysis can be expressed as follows:

$$mL\ NaOH \rightarrow mol\ NaOH \rightarrow mol\ HCl \rightarrow M\ HCl$$

Remember to change the milliliters of NaOH and HCl to liters in order to work with molarity in the proper units.

Solution

From the equation, 1 mole of HCl reacts with 1 mole of NaOH. Therefore, we can calculate the molarity of the HCl. (Remember to construct the flowchart first.)

$$31.98\ mL\ NaOH \times \frac{1\ L\ NaOH\ solution}{1000\ mL\ NaOH\ solution} \times \frac{0.1253\ mol\ NaOH}{1\ L\ NaOH\ solution} \times \frac{1\ mol\ HCl}{1\ mol\ NaOH}$$

$$= 0.004007\ mol\ HCl$$

Now that we know the quantity of HCl in moles, we can determine the molarity by dividing by the volume of the HCl.

$$\frac{0.004007\ mol\ HCl}{50.00\ mL\ HCl\ solution} \times \frac{1000\ mL\ HCl\ solution}{1\ L\ HCl\ solution} = \frac{0.08014\ mol\ HCl\ solution}{1\ L\ HCl\ solution}$$

$$= 0.08014\ M\ HCl$$

Since the titration used more of the HCl solution than of the NaOH solution, we would expect the HCl solution to be more dilute. Our answer makes sense.

Further Insights

Titration is just one (albeit one very important) technique for finding out how much of something you have. We will learn about many other "quantitative analysis" procedures in this textbook, based on physical properties such as color, electrical conductivity, and the formation of a precipitate.

| PRACTICE 4.6 |

The equation for the reaction between a sodium hydroxide solution and dilute sulfuric acid is

$$2NaOH(aq) + H_2SO_4(aq) \rightarrow Na_2SO_4(aq) + 2H_2O(l)$$

What is the molarity of a solution of 25.00 mL of H_2SO_4 if 22.25 mL of 0.100 M NaOH is required to reach equivalence?

See Problems 47–54.

4.4 Types of Chemical Reactions

The vast number of aqueous reactions that have been identified since the beginning of chemistry would stun the ancient alchemists. Fortunately, most of these reactions fall within certain general types or categories because of key similarities among them. Three of the more important types of chemical reactions are

- Precipitation reactions
- Acid–base reactions
- Oxidation–reduction reactions

In order to more fully understand the three types of processes, we will take this opportunity to build on our ability to write chemical equations in aqueous solutions. Our first step is determining what happens to the individual components when they are added to water. We'll use our knowledge of electrolytes to assist us in this process.

Molecular and Ionic Equations

The reaction that was the focus of Exercise 4.6 includes three strong electrolytes: hydrochloric acid (HCl), sodium hydroxide (NaOH), and sodium chloride (NaCl). The reaction equation, *containing the individual components written as compounds,* is known as the **molecular equation**. The molecular equation illustrates that the individual substances exist as hydrated compounds in the aqueous solution. The molecular equation representing the reaction of hydrochloric acid and sodium hydroxide can be written

$$HCl(aq) + NaOH(aq) \rightarrow H_2O(l) + NaCl(aq)$$
molecular equation

However, as we have already discussed, strong electrolytes dissolve in water, dissociate into their component ions, and become relatively independent within the solution. For this reason, all the individual ions can be shown separately, in what is known as a **complete ionic equation**.

$$H^+(aq) + Cl^-(aq) + Na^+(aq) + OH^-(aq) \rightarrow Na^+(aq) + Cl^-(aq) + H_2O(l)$$
complete ionic equation

Note that only the strong electrolytes have been written as individual ions. Unlike the electrolytes, molecules such as water do not dissociate into ions to any appreciable extent. They are not written as individual ions.

Complete ionic equations list all substances as they exist in the solution, whether or not they participate in the reaction. For example, in the complete ionic equation above, the sodium (Na^+) and chloride (Cl^-) ions are present at both the start and the end of the reaction, yet they do not participate at all during the reaction; they "sit on the sidelines," like spectators at a sporting event. For this reason, they are known as **spectator ions**. The real action during this reaction is the combination of hydrogen ions (H^+) and hydroxide ions (OH^-) to form water (H_2O). If we remove the spectator ions from the equation, we can simplify the equation to better show the combination of hydrogen ions and hydroxide ions:

$$H^+(aq) + \cancel{Cl^-(aq)} + \cancel{Na^+(aq)} + OH^-(aq) \rightarrow \cancel{Na^+(aq)} + \cancel{Cl^-(aq)} + H_2O(l)$$

The result represents *the overall, or "net," chemical change* that occurs in this reaction and is known as the **net ionic equation**:

$$H^+(aq) + OH^-(aq) \rightarrow H_2O(l)$$
net ionic equation

HERE'S WHAT WE KNOW SO FAR

- *Molecular equations* show the formulas of all reactants and products but do not indicate whether any of the compounds really exist as ions in solution.
- *Complete ionic equations* show all of the dissolved ions present in an equation individually, along with all other reactants and products.
- *Net ionic equations* show only those dissolved ions, and other reactants and products, that actually participate in or result from the reaction concerned.

EXERCISE 4.7 **The Three Equations**

Aqueous hydrochloric acid and aqueous sodium carbonate react to form aqueous sodium chloride, water, and gaseous carbon dioxide. Write the molecular, complete ionic, and net ionic equations for this reaction.

Solution

The molecular equation is

$$2HCl(aq) + Na_2CO_3(aq) \rightarrow 2NaCl(aq) + H_2O(l) + CO_2(g)$$

In the complete ionic equation, all ionic compounds are written as separated ions in solution:

$$2H^+(aq) + 2Cl^-(aq) + 2Na^+(aq) + CO_3^{2-}(aq)$$
$$\rightarrow 2Na^+(aq) + 2Cl^-(aq) + H_2O(l) + CO_2(g)$$

The net ionic equation is completed by removing the spectator ions:

$$2H^+(aq) + CO_3^{2-}(aq) \rightarrow H_2O(l) + CO_2(g)$$

PRACTICE 4.7

Identify the molecular, complete ionic, and net ionic equations that describe the reaction of potassium oxalate ($K_2C_2O_4$) and nitric acid (HNO_3) to form potassium nitrate (KNO_3) and oxalic acid ($H_2C_2O_4$). Assume that oxalic acid is the only nonelectrolyte in the reaction.

See Problems 43–46, 59–62, and 91.

Video Lesson: Precipitation Reactions

Tutorial: Precipitation Reactions

Visualization: Precipitation Reactions

4.5   Precipitation Reactions

An important challenge for chemists trying to clean up industrial wastewater is to remove the ions of "heavy metals" (often very dense metals, 5 g/mL or greater) such as lead, mercury, and cadmium, which are toxic to humans and other life. This is particularly important in mining operations, where aqueous waste streams from the mine can contain abnormally high levels of heavy metals. One way to clean up the wastewater is to treat it with sulfide ions (S^{2-}), because most nonalkali metal ions will combine with sulfide ions to form a solid "precipitate." The word **precipitate** is both a noun and a verb. It is used as a noun to refer to any solid material that forms within a solution, and as a verb to describe that process in action. We can say that metal sulfides form a precipitate (noun). Alternatively, we can say that metal sulfides will precipitate (verb) out of solution as soon as they form. We say that a precipitate is **insoluble** because not very much of it can dissolve into the solvent. On the other hand, a relatively large amount of a **soluble** substance can dissolve into the solvent. About *2 kg* of sucrose, table sugar, can dissolve in a liter of water at 25°C. That is highly soluble compared to silver chloride, which has a solubility of about *2 mg* per liter of water at 25°C.

Application

CHEMICAL ENCOUNTERS: Focus on Lead Sulfide

Water drainage from a mine is often tainted with high levels of sulfide, sulfate, and other ions.

FIGURE 4.15

Sulfate-reducing bacteria. Spring water in the floodplain of the Prairie Dog Town Fork of the Red River that flows through the panhandle of Texas has exposed black mud. The black sediment is formed from the interaction of dissolved metal ions and sulfide produced from sulfate-reducing bacteria.

FIGURE 4.16

Lead sulfide precipitates from the combination of its soluble ions.

Visualization: Solubility Rules

Industrial wastewater can be treated with sulfate-reducing bacteria. These bacteria convert sulfate ions into sulfide ions as part of their natural biological activity (Figure 4.15). In the presence of heavy metals, the sulfide generated by the bacteria can help to reduce the amount of soluble heavy metals by making insoluble precipitates. For example, the net ionic equation for the reaction of lead(II) ions with sulfide ions is

$$Pb^{2+}(aq) + S^{2-}(aq) \rightarrow PbS(s)$$

The ions are initially in the aqueous phase and so are dissolved in water. When they meet and combine, they form the insoluble solid precipitate lead(II) sulfide. This is an example of a **precipitation reaction**, *a reaction in which an insoluble precipitate is formed from soluble reactants.* We can precipitate lead(II) sulfide in the laboratory by mixing a solution of a soluble lead(II) compound with a solution of a soluble sulfide. This is illustrated in Figure 4.16, which shows a solution of ammonium sulfide being added to a lead(II) nitrate solution. The visible precipitate forms via the following molecular equation:

$$(NH_4)_2S(aq) + Pb(NO_3)_2(aq) \rightarrow 2NH_4NO_3(aq) + PbS(s)$$

The lead(II) sulfide is a solid that essentially doesn't dissolve or ionize, so we note this in all written forms of the equation. For example, the complete ionic equation is

$$2NH_4^+(aq) + S^{2-}(aq) + Pb^{2+}(aq) + 2NO_3^-(aq) \rightarrow 2NH_4^+(aq) + 2NO_3^-(aq) + PbS(s)$$

The net ionic equation more clearly shows this precipitation reaction. The ammonium and nitrate ions are spectator ions, so the net ionic equation reduces to

$$Pb^{2+}(aq) + S^{2-}(aq) \rightarrow PbS(s)$$

which is simply the precipitation of lead(II) sulfide.

How can we predict when a precipitation reaction will occur? For most common ions, the answer has been determined experimentally and can be summarized by a set of *solubility rules*. Table 4.3 illustrates these rules in detail. Compounds generally follow these rules with a few exceptions, some of which are also noted in the table. We can use this information to write and balance the molecular equation for a reaction, determine whether any of the compounds form insoluble precipitates, and write the complete ionic and net ionic equations.

TABLE 4.3	Solubility Rules for Ionic Compounds
Soluble	**Insoluble**
Group IA and ammonium compounds are soluble.	Most carbonates are insoluble except Group IA carbonates and $(NH_4)_2CO_3$.
Acetate, chlorate, perchlorate, and nitrate compounds are soluble.	Most phosphates are insoluble except Group IA phosphates and $(NH_4)_3PO_4$.
Most chlorides, bromides, and iodides are soluble except those of Ag^+, Hg_2^{2+}, and Pb^{2+}.	Most sulfides are insoluble except Group IA sulfides and $(NH_4)_2S$.
Most sulfates are soluble except those of Ca^{2+}, Sr^{2+}, Ba^{2+}, Ag^+, Hg_2^{2+}, and Pb^{2+}.	Most hydroxides are insoluble except Group IA hydroxides and $Ca(OH)_2$, $Sr(OH)_2$, and $Ba(OH)_2$.

EXERCISE 4.8 | **Identifying Precipitation Reactions**

Which combinations of the following aqueous solutions will produce precipitates: aluminum bromide, barium hydroxide, magnesium sulfate, and nickel(II) iodide? Use the solubility rules to guide you in your decision.

First Thoughts

You need to consider all the possible combinations of positive and negative ions in order to identify those that will produce an insoluble product. The four solutions in this question contain Al^{3+}, Br^-, Ba^{2+}, OH^-, Mg^{2+}, SO_4^{2-}, Ni^{2+}, and I^-.

Solution

Table 4.3 reveals the following solubilities for all the possible combinations of ions:

aluminum hydroxide—*insoluble*   aluminum sulfate—soluble
aluminum iodide—soluble   barium bromide—soluble
barium sulfate—*insoluble*   barium iodide—soluble
magnesium bromide—soluble   magnesium hydroxide—*insoluble*
magnesium iodide—soluble   nickel(II) bromide—soluble
nickel(II) hydroxide—*insoluble*   nickel(II) sulfate—soluble

The solubility rules indicate that we could make four insoluble precipitates, in the following precipitation reactions:

▎ Add aluminum bromide solution to barium hydroxide solution, to form a precipitate of aluminum hydroxide.

$$Al^{3+}(aq) + 3OH^-(aq) \rightarrow Al(OH)_3(s)$$

▎ Add barium hydroxide solution to magnesium sulfate solution, to form a precipitate of barium sulfate, mixed with a precipitate of magnesium hydroxide.

$$Mg^{2+}(aq) + 2OH^-(aq) \rightarrow Mg(OH)_2(s) \text{ and } Ba^{2+}(aq) + SO_4^{2-}(aq) \rightarrow BaSO_4(s)$$

▎ Add nickel(II) iodide solution to barium hydroxide solution, to form a precipitate of nickel(II) hydroxide.

$$Ni^{2+}(aq) + 2OH^-(aq) \rightarrow Ni(OH)_2(s)$$

Further Insights

The result when we add a barium hydroxide solution to a magnesium sulfate solution reveals a complication. Precipitation reactions may produce a mixture of precipitates if the cations of each dissolved compound form insoluble products with the anions of the other compound with which they are mixed. When we prepare a compound by precipitation, we must ensure that the only precipitate that forms is the desired one. We will discuss ways of doing this in Chapter 18.

PRACTICE 4.8

Which combination of the following solutions will produce a precipitate: $AgNO_3(aq)$, $NaCl(aq)$, $Na_2S(aq)$, $ZnSO_4(aq)$?

See Problems 57–66, 89, and 90. ▊

EXERCISE 4.9 | **The Stoichiometry of a Precipitation Reaction**

Calculate the mass of the precipitate that is formed when 1.30 L of 0.0200 M $AlBr_3$ solution is added to 3.00 L of 0.0350 M NaOH solution.

First Thoughts

First we map out our plan of attack. We will do the following steps:

1. Write the balanced molecular equation (we will not need to write the net ionic equation).

2. Determine the limiting reagent for the reaction.
3. Calculate the number of grams of product formed.

Solution

The balanced molecular equation is

$$AlBr_3(aq) + 3NaOH(aq) \rightarrow Al(OH)_3(s) + 3NaBr(aq)$$

Examination of the solubility rules indicates that the aluminum hydroxide is insoluble in water. How many grams of $Al(OH)_3(s)$ are produced if $AlBr_3$ is the limiting reagent?

$$1.30 \text{ L } AlBr_3 \text{ solution} \times \frac{0.0200 \text{ mol } AlBr_3}{1 \text{ L } AlBr_3 \text{ solution}} \times \frac{1 \text{ mol } Al(OH)_3}{1 \text{ mol } AlBr_3} \times \frac{78.00 \text{ g } Al(OH)_3}{1 \text{ mol } Al(OH)_3}$$

$$= 2.03 \text{ g } Al(OH)_3$$

How many grams of $Al(OH)_3(s)$ are produced if NaOH is the limiting reagent?

$$3.00 \text{ L NaOH solution} \times \frac{0.0350 \text{ mol NaOH}}{1 \text{ L NaOH solution}} \times \frac{1 \text{ mol } Al(OH)_3}{3 \text{ mol NaOH}} \times \frac{78.00 \text{ g } Al(OH)_3}{1 \text{ mol } Al(OH)_3}$$

$$= 2.73 \text{ g } Al(OH)_3$$

Our calculations indicate that the $AlBr_3$ is the limiting reagent because it, rather than the NaOH, limits the mass of $Al(OH)_3$ formed. Mixing the two reagents together will produce 2.03 grams of aluminum hydroxide as a precipitate.

PRACTICE 4.9

What mass of precipitate is produced when 0.35 L of 0.25 M sodium carbonate (Na_2CO_3) reacts with 0.55 L of 0.35 M barium chloride?

See Problems 67, 68, and 91.

4.6 Acid–Base Reactions

Drainage from a gold mine is often very acidic (see Figure 4.17). In fact, the water from a gold mine can be acidic enough to seriously burn anyone who touches it. To make this water suitable for the environment, cleanup crews often add a compound that decreases the amount of acid in the waste. What is an **acid**? A good working definition is that an acid is a substance that releases hydrogen ions (H^+) in a solution. The pain of "acid indigestion," familiar to most of us, is caused by too many hydrogen ions in the gastric fluid that fills the stomach. The chemical opposite of an acid is a base. A **base** is a substance that releases hydroxide ions (OH^-) in a solution. A base that is soluble in water is also called an alkali. Acids, bases, and their solutions are explored in detail in Chapters 17 and 18. In this chapter we introduce some principles that govern the reactions between them and examine the stoichiometry of their reactions.

An **acid–base reaction** is the reaction between an acid and a base. The result of this reaction is neutralization of the acid by the base, and vice versa, resulting in a solution that is neither acidic nor basic. For this reason, acid–base reactions are often referred to as *neutralization reactions*.

Strong Acids and Bases

One very common acid is hydrochloric acid (HCl). In the industrial world, the startling quantities of HCl produced (5.0 billion kg prepared in 2004, easily

Application

Tutorial: Neutralization Reactions

Video Lesson: Acid–Base Reactions

Video Lesson: Strong Acid–Strong Base and Weak Acid–Strong Base Reactions

Video Lesson: Strong Acid–Weak Base and Weak Acid–Weak Base Reactions

Visualization: Neutralization of a Strong Acid by a Strong Base

Visualization: Proton Transfer

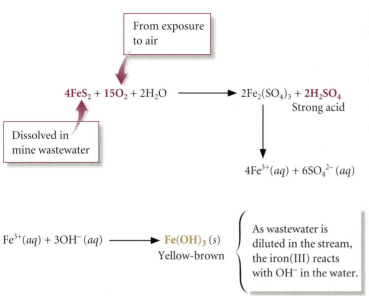

$$4FeS_2 + 15O_2 + 2H_2O \longrightarrow 2Fe_2(SO_4)_3 + 2H_2SO_4$$

From exposure to air

Dissolved in mine wastewater

Strong acid

$$4Fe^{3+}(aq) + 6SO_4^{2-}(aq)$$

$$Fe^{3+}(aq) + 3OH^-(aq) \longrightarrow Fe(OH)_3(s)$$
Yellow-brown

As wastewater is diluted in the stream, the iron(III) reacts with OH⁻ in the water.

FIGURE 4.17

Water that fills a gold mine dissolves small amounts of FeS$_2$ (iron pyrite, also known as fool's gold) that is often found in gold deposits. This compound undergoes oxidation when it reaches the air outside the mine. The result is a mixture of insoluble Fe(OH)$_3$ and sulfuric acid. Gold mine runoff typically has a rust color and extremely high acidity.

within the top 20 amounts of industrial chemicals produced) are used in the isolation and cleaning of metals and in the production of solvents and chlorinated organic compounds, as well as in acid–base reactions. As we have noted, HCl is a strong electrolyte. When HCl molecules dissolve in water, they completely "ionize" (dissociate completely into ions) to form hydrogen ions and chloride ions:

$$HCl(g) \rightarrow HCl(aq) \rightarrow H^+(aq) + Cl^-(aq)$$

Because HCl is acidic and *ionizes completely in aqueous solution* (i.e., because it is a strong electrolyte), we call it a **strong acid**. Examples of other strong acids are given in Table 4.4.

A **strong base** is one that *ionizes completely to produce OH⁻ in water*. Sodium hydroxide (NaOH), which we first introduced in Exercise 4.1, is one of the more common strong bases. It is often used in bathroom cleaners, in drain uncloggers, and in chemical reactions that are used to prepare soaps. Because it is a strong electrolyte, it dissociates completely when it is added to water:

$$NaOH(s) \rightarrow NaOH(aq) \rightarrow Na^+(aq) + OH^-(aq)$$

Examples of other strong bases are given in Table 4.4.

TABLE 4.4	Common Strong Acids and Bases
Strong Acids	
HCl, HBr, HI, HNO$_3$, HClO$_4$, H$_2$SO$_4$	
Strong Bases	
Hydroxides of Group IA metals, such as LiOH, NaOH, and KOH	

Drain decloggers contain strong bases such as sodium hydroxide.

Weak Acids and Bases

A **weak acid** differs from a strong acid in that it *does not dissociate completely in aqueous solution*. Most acids (except those in Table 4.4) are considered weak acids. For example, acetic acid is added to water to make vinegar. The reaction illustrating its acidity is shown below. Note that the reaction doesn't proceed completely to products, and because of this, we refer to weak acids as weak electrolytes.

$$CH_3COOH(l) \rightarrow CH_3COOH(aq) \rightleftharpoons CH_3COO^-(aq) + H^+(aq)$$

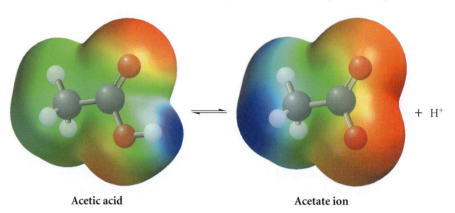

Acetic acid   **Acetate ion**

Weak bases don't react extensively to produce hydroxide ions. When ammonia is added to water, only a small proportion of the ammonia reacts to produce the ammonium ion (NH_4^+) and hydroxide ion (OH^-). *Most bases are weak.* In a fashion similar to the weak acids, weak bases are known as weak electrolytes.

$$NH_3(aq) + H_2O(l) \rightleftharpoons NH_4^+(aq) + OH^-(aq)$$

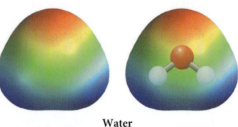

Water

When Acids and Bases Combine

In a reaction between a strong acid and a strong base, hydrogen ions (H^+) and hydroxide ions (OH^-) come together to form water. This is illustrated by writing the chemical equation for the reaction of HCl with NaOH. In the molecular equation, we can see that the products are an ionic compound (in this case, NaCl) and water.

$$HCl(aq) + NaOH(aq) \rightarrow NaCl(aq) + H_2O(l)$$

The complete ionic equation shows exactly which ions are involved in the reaction.

$$H^+(aq) + Cl^-(aq) + Na^+(aq) + OH^-(aq) \rightarrow Na^+(aq) + Cl^-(aq) + H_2O(l)$$

The sodium and chloride ions are spectator ions, so we can simplify to the net ionic equation:

$$H^+(aq) + OH^-(aq) \rightarrow H_2O(l)$$

This is the formation of water from its ions. This is the typical outcome for strong-acid–strong-base reactions; the formation of water as the net ionic equation for the reaction indicates a neutralization reaction. In other words, the acid (H^+) and the base (OH^-) react to make a compound that is neither acidic nor basic but, rather, the neutral liquid water (H_2O).

Hydrochloric acid is also known as a **monoprotic** acid, because it produces just one mole of hydrogen ions (H^+, which is a proton) from each mole of HCl when it dissociates. This results in one mole of water forming when the acid is completely neutralized in an acid–base reaction. Other common acids can be **diprotic** or **triprotic** and so generate two or three moles of water, respectively, for each mole of acid neutralized. Compare the molecular and net ionic equations of hydrochloric acid (HCl), sulfuric acid (H_2SO_4), and phosphoric acid (H_3PO_4), each reacting with sodium hydroxide:

$$HCl(aq) \quad + \quad NaOH(aq) \quad \rightarrow \quad NaCl(aq) \quad + \quad H_2O(l)$$

hydrochloric acid (monoprotic)   1 mole NaOH required   1 mole water formed per mole acid

$$H_2SO_4(aq) \quad + \quad 2NaOH(aq) \quad \rightarrow \quad Na_2SO_4(aq) \quad + \quad 2H_2O(l)$$

sulfuric acid (diprotic)   2 mole NaOH required   2 mole water formed per mole acid

$$H_3PO_4(aq) \quad + \quad 3NaOH(aq) \quad \rightarrow \quad Na_3PO_4(aq) \quad + \quad 3H_2O(l)$$

phosphoric acid (triprotic)   3 mole NaOH required   3 mole water formed per mole acid

---

EXERCISE 4.10	Stoichiometry of a Neutralization Reaction

What volume of a 0.200 M H_2SO_4 solution is needed to neutralize 25.0 mL of a 0.330 M NaOH solution?

First Thoughts

As usual, our first step in answering a question about a reaction is to write the balanced equation for the reaction. Then we can use dimensional analysis to determine the volume of sulfuric acid solution that is needed. Remember, it's a good idea to map out the process before you begin the calculation.

Solution

The balanced molecular equation is

$$H_2SO_4(aq) \quad + \quad 2NaOH(aq) \quad \rightarrow \quad Na_2SO_4(aq) \quad + \quad 2H_2O(l)$$

sulfuric acid (diprotic)   2 mole NaOH required   2 mole water formed per mole acid

Using unit conversion, we develop the entire flowchart to calculate the number of liters of sulfuric acid solution that will react with the base:

$$\text{mL NaOH} \xrightarrow{\frac{L}{mL}} \text{L NaOH} \xrightarrow{\frac{\text{mol NaOH}}{\text{L NaOH solution}}} \text{mol NaOH} \xrightarrow{\frac{\text{mol } H_2SO_4}{\text{mol NaOH}}} \text{mol } H_2SO_4$$

$$\xrightarrow{\frac{\text{L } H_2SO_4 \text{ solution}}{\text{mol } H_2SO_4}} \text{L } H_2SO_4 \text{ solution}$$

$$25.0 \text{ mL NaOH} \times \frac{1 \text{ L}}{1000 \text{ mL}} \times \frac{0.330 \text{ mol NaOH}}{\text{L NaOH solution}} \times \frac{1 \text{ mol } H_2SO_4}{2 \text{ mol NaOH}}$$

$$\times \frac{1 \text{ L } H_2SO_4 \text{ solution}}{0.200 \text{ mol } H_2SO_4} = 0.0206 \text{ L } H_2SO_4 \text{ solution}$$

A total of 0.0206 L (or 20.6 mL) of 0.200 M H_2SO_4 is required. Note that the mole ratio of 2 moles of NaOH to 1 mole of H_2SO_4 is used because sulfuric acid is diprotic, having two protons that can react with sodium hydroxide.

Further Insights

We didn't need to determine the net ionic equation to answer how many liters of an acid will react with a particular amount of base. However, we did need to determine the balanced molecular equation. In general, stoichiometry calculations require only that we know one form of the balanced chemical equation.

PRACTICE 4.10

What volume of a 0.550 M H_3PO_4 solution is needed to neutralize 50.0 mL of a 0.250 M NaOH solution?

See Problems 69–72 and 87.

Video Lesson: Oxidation–Reduction Reactions

Visualization: Sugar and Potassium Chlorate

Metal is protected from the reactants that cause rusting.

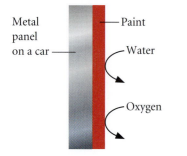

Metal panel on a car — Paint

Water

Oxygen

Painted surfaces of a car are protected from oxidation.

4.7   Oxidation–Reduction Reactions

Living in an atmosphere rich in oxygen makes life possible, but it also gives us the continual challenge of preventing unwanted reactions between oxygen and chemicals on which we rely. One kind of very fast reaction with oxygen, called combustion, yields fire, and we go to great lengths to protect ourselves against unwanted fires. A slow but almost equally troublesome reaction of metals with oxygen causes corrosion, such as that which occurs when iron rusts. Rust forms when iron reacts with oxygen (in the presence of water) to form various forms of iron oxide, such as Fe_2O_3. The presence of water is crucial for this reaction to occur at a significant rate, because the oxygen must be in solution. This explains why rusting of iron and steel can be prevented by protecting the metal from water and oxygen with paint, plastic coatings, and so on.

When chemists examine what happens in such **oxidation** reactions, they find that electrons are lost from the reactant that is being **oxidized**. Many oxidations do involve oxygen, but what is really happening is the transfer of electrons in the reaction. In forming rust, for example, elemental iron atoms lose electrons to become iron ions:

$$Fe \rightarrow Fe^{3+} + 3e^-$$

All processes in which chemicals lose electrons, *regardless of whether or not oxygen is involved,* are known as oxidations. An oxidation cannot happen on its own, because the electrons must have somewhere to go after they are lost from the reactants that are oxidized. That is, they must be gained by some other chemical species. When rust forms, for example, the electrons from the iron atoms are transferred to molecular oxygen to form oxide ions:

$$O_2 + 4e^- \rightarrow 2O^{2-}$$

Visualization: Oxidation of Zinc with Iodine

Video Lesson: CIA Demonstration: The Reaction between Al and Br_2

We say the chemical that accepts the electrons has been **reduced** or has undergone **reduction**. A reduction is any process in which *electrons are gained by a chemical.* One useful mnemonic to help you remember oxidation and reduction is **OIL RIG**:

Oxidation **I**nvolves **L**oss of electrons, **R**eduction **I**nvolves **G**ain of electrons

Some prefer the mnemonic **LEO says GER**:

Loss of **E**lectrons is **O**xidation, **G**ain of **E**lectrons is **R**eduction

Combustion and corrosion are redox processes.

No matter which mnemonic you prefer, the chemical point remains the same: Oxidation and reduction processes must occur together in **oxidation–reduction reactions** (also called **redox reactions**, for **red**uction–**ox**idation). The chemicals concerned must react in the proportions that allow the number of electrons lost via oxidation to equal the number of electrons gained via reduction. The processes that we discussed above do not exist alone. Instead, each is one-half of the complete electron-exchange process that produces iron(III) oxide (Fe_2O_3) from iron and oxygen. Still, we find it convenient to separate the overall reaction into individual oxidation and reduction **half-reactions**. Let's examine these half-reactions more closely.

The oxidation half-reaction is: $Fe \rightarrow Fe^{3+} + 3e^-$ (3 electrons are lost)

The reduction half-reaction is: $O_2 + 4e^- \rightarrow 2O^{2-}$ (4 electrons are gained)

To obtain the complete oxidation–reduction reaction that results from the iron and oxygen half-reactions, we need to make sure that the number of electrons involved is the same on both sides of the equation. *The number of electrons lost by the iron must be equal to the number gained by the oxygen.* Charge, like mass, is conserved.

Multiplying the entire oxidation half-reaction by 4 and the entire reduction half-reaction by 3 will allow 12 electrons to be lost from iron as the same 12 electrons are gained by oxygen.

$$4Fe \rightarrow 4Fe^{3+} + 12e^-   \text{(oxidation)}$$
$$3O_2 + 12e^- \rightarrow 6O^{2-}   \text{(reduction)}$$

The overall oxidation–reduction equation is the sum of the two half-reactions:

$$4Fe \rightarrow 4Fe^{3+} + \cancel{12e^-}$$
$$\underline{3O_2 + \cancel{12e^-} \rightarrow 6O^{2-}}$$
$$\mathbf{4Fe + 3O_2 \rightarrow 4Fe^{3+} + 6O^{2-}}$$

When the electrons are transferred in this reaction, the resulting iron ions and oxide ions combine to form Fe_2O_3.

Oxidation Numbers

In business, bookkeepers are hired to keep track of the money coming in to and that paid out by the company. They keep the financial books. In chemistry, our

Video Lesson: Oxidation Numbers

electronic bookkeeping tool is called an **oxidation number**, which we assign to individual atoms on the basis of where electrons in a bond are likely to be found. For example, in the ionic compound NaCl, in which sodium has given an electron to chlorine, we say that the sodium ion has an oxidation number of $+1$, because it has one less electron than the sodium atom. The chloride ion has an oxidation number of -1, because it has one more electron than the chlorine atom. We use oxidation numbers to keep track of electrons as they move among atoms, molecules, and ions in redox reactions and, in fact, in all types of reactions. The term *oxidation number* is often used interchangeably with the term **oxidation state**.

In the iron and oxygen reaction that produces iron(III) oxide,

$$4Fe \rightarrow 4Fe^{3+} + 12e^- \qquad \text{(oxidation)}$$

$$3O_2 + 12e^- \rightarrow 6O^{2-} \qquad \text{(reduction)}$$

each iron started as a neutral atom (sometimes noted with a superscript "0", Fe^0) and was oxidized, losing three electrons, to form Fe^{3+}. The oxidation number (or oxidation state) of the iron is now $+3$. The oxygen molecule includes two oxygen atoms that share electrons equally—that is, neither atom exerts a strong preference for the bonding electrons—so each oxygen atom is assigned (remember, we are bookkeepers here!) an oxidation state of 0. When oxygen reacts with iron, the electrons gained produce oxide ions (O^{2-}) that have an oxidation number of -2.

When atoms are combined to make a molecule, they neither lose nor gain electrons to form ions. Instead, the electrons in molecules are shared, to one degree or another, between the atoms. Overall, the molecule is electrically neutral, with a net charge of zero and, therefore, a net oxidation number of 0. However, in many molecules, there is a tendency for electrons to be closer to the nuclei of some atoms than of others. For example, in water (H_2O), there is a marked tendency for the electrons in each of the two hydrogen-to-oxygen bonds to be found closer to the oxygen nucleus than to the hydrogen nucleus (Figure 4.3). This behavior is indicated in our electron bookkeeping system by giving oxygen an oxidation state of -2. The electron from each hydrogen atom is drawn away from its nucleus more of the time, and we denote this by giving each hydrogen atom an oxidation number of $+1$. Our chemical understanding has led to a host of bookkeeping conventions to assign oxidation numbers to the individual atoms that make up a molecule. These rules are listed in order of importance in Table 4.5. In cases where the rules appear to contradict each other, follow the rule that comes first in the table.

What oxidation numbers would we assign to the carbon and hydrogen atoms of methane (CH_4)? By following the rules in Table 4.5, we determine that each hydrogen atom has a $+1$ oxidation number. Because the molecule is neutral overall, we assign the oxidation number -4 to this carbon atom. The carbon atom in carbon monoxide (CO), on the other hand, is assigned a different oxidation number. According to the rules, the oxygen atom has an oxidation number of -2. The carbon atom must, then, have an oxidation number of $+2$. Carbon monoxide is more oxidized than methane. The oxidation number of the carbon atom in each compound bears this out.

It is important that we understand the difference between the oxidation numbers of atoms in molecules and the oxidation numbers of ions in ionic compounds. In molecules, assigning oxidation states is simply an accounting procedure used to keep track of electrons; it does *not* imply that electrons have really been lost or gained by any atoms. Remember that we have just assigned the oxidation numbers on the basis of some simple rules that hypothetically assume the electrons are transferred when the molecule is made. In ionic compounds, on the other hand, the oxidation number of an ion is the real charge on the ion.

TABLE 4.5	Rules for Assigning Oxidation Numbers
1. The oxidation number of an atom in an element is 0.	Na, H_2, N_2, O_2, O_3, and He, all have atoms with oxidation number $= 0$
2. The oxidation number of a monatomic ion is the same as its charge.	The oxidation number of Li^+ is $+1$, of Ca^{2+} is $+2$, of Cl^- is -1, and of O^{2-} is -2.
3. The oxidation number of a Group IA metal in a compound is always $+1$; the oxidation number of a Group IIA metal in a compound is always $+2$.	The oxidation number of sodium in NaCl is $+1$. The oxidation number of calcium in $CaBr_2$ is $+2$.
4. The oxidation number of fluorine in its compounds is always -1.	In HF, the oxidation number of F is -1 and of H is $+1$.
5. The oxidation number of hydrogen is $+1$ in its nonmetal compounds. It is -1 when combining with many metals in compounds known as hydrides.	In HCl, the oxidation number of H is $+1$ and of Cl is -1. In H_2O, the oxidation number of H is $+1$ and of O is -2. However, in NaH, the oxidation number of H is -1 and of Na is $+1$.
6. The oxidation number of oxygen in its covalent compounds is usually -2. Exceptions include peroxides such as hydrogen peroxide (H_2O_2), in which oxygen is in the -1 oxidation state.	In CO_2, the oxidation number of oxygen is -2 and of carbon is $+4$. In CO, the oxidation number of oxygen is also -2, but that of carbon is $+2$.
7. In binary compounds containing metals, Group VIIA elements typically have an oxidation number of -1; Group VIA elements have an oxidation number of -2; and Group VA elements have an oxidation number of -3.	The oxidation number of sulfur in FeS is -2. The oxidation number of nitrogen in Mg_3N_2 is -3.
8. The sum of the oxidation numbers of all atoms in the molecule or ion must equal the total charge of the molecule (zero) or ion.	In the water molecule (H_2O), each H has an oxidation number of $+1$ and the O has an oxidation number of -2, so the sum of all the oxidation numbers $= (+1 \times 2) + (-2) = 0$. In the ionic compound calcium fluoride (CaF_2), each Ca^{2+} ion has an oxidation number of $+2$ and each F^- ion has an oxidation number of -1, so the sum of all the oxidation numbers $= (+2) + (2 \times -1) = 0$. In the NH_4^+ ion, each H has an oxidation number of $+1$ and the N has an oxidation number of -3, so the sum of all the oxidation states $= (-3) + (+1 \times 4) = +1$, which is the charge on the polyatomic ion.

EXERCISE 4.11	Assigning Oxidation States

Assign oxidation states to all the atoms in the following elements, compounds and ions:

a. He   c. Al_2O_3 e. NO_3^- g. F_2

b. C_6H_6 d. Zn   f. SF_6 h. CO_3^{2-}

Solution

a. This is an element, so the oxidation number is 0.

b. This is a neutral compound. Each hydrogen atom has an oxidation number $+1$ *per atom,* and because there are the same number of carbon atoms as of hydrogen atoms, *each carbon atom* must have an oxidation number of -1.

c. This is a neutral compound. Oxygen has an oxidation number of -2, so in this case the aluminum must have an oxidation number of $+3$, because there are three oxygen atoms (total $= -6$), and the oxidation numbers must sum to 0 overall.

d. This is an element, so the oxidation number $= 0$.

e. This is an ion with a -1 charge overall, so the oxidation number of the atoms present must sum to -1. Oxygen has an oxidation number of -2, so the three oxygen atoms contribute a total of -6. Nitrogen must therefore have the oxidation number of $+5$.

f. This is a neutral compound. Fluorine has an oxidation number of -1, and there are six fluorine atoms (total $= -6$), so the sulfur must have an oxidation number of $+6$.

g. This is an element, so each atom has an oxidation number of 0.

h. This is an ion with a -2 charge overall, so the oxidation number of the atoms present must sum to -2. Oxygen has an oxidation number of -2, so the three oxygen atoms contribute a total of -6. The carbon, therefore, must have an oxidation number of $+4$.

PRACTICE 4.11

Use Table 4.5 to assign the oxidation numbers of each of the species in the compounds below.

KCl Fe_2O_3 P_4 CH_2Cl_2 Al PBr_3 HCN

See Problems 81 and 82.

Identifying Oxidation–Reduction Reactions

 Application

Redox reactions are often disguised and can be difficult to identify without close examination. Let's examine a redox reaction to see how it is identified as such. The reaction of hydrogen and nitrogen to make ammonia, known as the Haber process, is a redox reaction that is vital to farming. The ammonia made by this process is used to fertilize fields for the production of most crops, including corn, wheat, and sorghum (a versatile crop used to make syrup, hay, flour, animal feed, and even brooms!). The reaction is

$$3H_2(g) + N_2(g) \rightarrow 2NH_3(g)$$

A first look at the reaction does not reveal that any electrons are being transferred in the process, so at first we might incorrectly assume that this is not a redox reaction. However, if we examine the oxidation numbers of the individual atoms involved, we can see that reduction and oxidation are taking place.

H_2 Hydrogen is assigned the oxidation number 0.
N_2 Nitrogen is assigned the oxidation number 0.
NH_3 Hydrogen is assigned the oxidation number $+1$, nitrogen -3.

Overall, the hydrogen changes its oxidation state from 0 (in H_2) to $+1$ (in NH_3), losing electrons in this reaction as it is oxidized. The nitrogen changes its oxidation state from 0 (in N_2) to -3 (in NH_3), gaining electrons as it is reduced. Because the Haber process includes reduction and oxidation, it is a redox reaction.

Redox reactions are characterized by the exchange of electrons, which makes balancing these equations difficult to do by trial and error. We will discuss a method for balancing these reactions in Chapter 19.

EXERCISE 4.12 | **Identifying Redox Reactions**

Is the combustion of ethane a redox reaction? Prove your answer.

$$2C_2H_6(g) + 7O_2(g) \rightarrow 4CO_2(g) + 6H_2O(g)$$

First Thoughts

What should we look for that is common to all redox reactions? The single common feature is the exchange of electrons. How do we know whether there is an electron exchange? The best way is to assign oxidation numbers to each atom and see whether there are changes going from reactants to products.

Solution

Judging on the basis of the oxidation number rules in Table 4.5, we can make the following assignments:

$$2C_2H_6(g) + 7O_2(g) \rightarrow 4CO_2(g) + 6H_2O(g)$$

$$\begin{array}{ccccc} -3 \ +1 & 0 & +4 \ -2 & +1 \ -2 \end{array}$$

Carbon goes from -3 to $+4$.   (oxidation)
Oxygen goes from 0 to -2.   (reduction)
Hydrogen does not change oxidation number.

Because electrons have been exchanged, this a redox reaction.

Further Insights

Combustion is the high-temperature reaction of oxygen with another compound. All combustion reactions are redox reactions, because electron exchange is occurring. However, not all redox reactions are combustion reactions. For example, batteries work because of redox reactions that include metals such as cadmium, lead, or lithium, along with acids such as sulfuric acid or bases such as potassium hydroxide.

PRACTICE 4.12

Is the reaction of aqueous solution of phosphoric acid and sodium hydroxide a redox reaction?

See Problems 55, 56, 83, and 84.

4.8   Fresh Water—Issues of Quantitative Chemistry

Nothing is more important to us than our supplies of fresh water, which we largely draw from rivers, lakes (Figure 4.18), and underground aquifers. Managing these precious freshwater resources requires our understanding of quantitative chemistry and aqueous solutions. If fresh water is to be of use to us as a supply of drinking water or for agricultural use, we must ensure that it is clean, meaning that certain dissolved solutes must occur only in quantities that are within the acceptable limits for good health. Table 4.6 lists, for selected chemicals, the EPA's maximum permitted level, below which there is minimal risk to human health (this is called the maximum contaminant level goal, or MCLG).

Chemists in water collection and treatment facilities check for compliance with safe water standards by quantitative analysis of the water. According to those safe-water standards, they also administer appropriate quantities of disinfecting chemicals. One such disinfectant is chlorine, which is toxic to freshwater life above about 19 parts per billion. The good news is that chlorination of drinking water has been instrumental in reducing the risk of microbial disease transmitted through a water supply.

For instance, cholera (a bacterial infection of the intestines that causes vomiting, diarrhea, and dehydration) claims thousands of lives each year and is particularly invasive in countries in which the population has been uprooted as a consequence of civil war and grinding poverty. In 2005, cholera outbreaks were

Application

CHEMICAL ENCOUNTERS:
Revisiting the Maximum Levels of Chemicals in Drinking Water

FIGURE 4.18

The quantitative chemistry of fresh water is a vital consideration in ensuring the safety of our water supplies and the health of the environment.

A water technician checks the chlorination process at a water treatment plant.

Application

The Danube sturgeon.

TABLE 4.6	Maximum Contaminant Level Goal for Selected Substances in Safe Water		
Substance	MCLG (mg/L = ppm)	Potential Human Organ Damage Due to Exceeding the MCLG	Source
As	0.010	Skin and circulatory system	Runoff from orchards and glass-manufacturing plants
Cd	0.005	Kidney	Corrosion of pipes, discharge from used batteries
CN^-	0.200	Thyroid and nerve	Discharge from gold mining, fertilizer, and plastics manufacturing
Hg	0.002	Kidney	Discharge from refineries
Dioxin ($C_{12}H_4Cl_4O_2$)	0.00000003	Reproductive system	By-products of smelting, bleaching, and pesticide manufacture

reported by the World Heath Organization in the African countries of Benin, Burkina-Faso, Guinea, Guinea-Bissau, Mali, Mauritania, Niger, and Senegal (Figure 4.19). In those countries that disinfect their water supply with chlorine, however, cholera has been all but eradicated. Unfortunately, wastewater treatment effluent (runoff from the cities back into lakes, streams, rivers, etc.) has resulted in chlorine concentrations between 1 and 5 parts per million. Any effect these levels of chlorine have on the environment has yet to be determined.

The gold miners we mentioned at the start of this chapter have to be careful with their cyanide leaching solution, because the MCLG for CN^- is 0.200 ppm (200 ppb). Larger quantities of this ion in water can be very harmful to the environment. In the winter of 2000, an accidental spill in Romania resulted in approximately 22 million gallons of cyanide waste from a gold mine flowing into the nearby Tisza River. This river flows into the Danube River and through Belgrade, Yugoslavia, on its way to the Black Sea. The initial cyanide spill killed most of the life in the Tisza River and harmed much of the life in the Danube River. To compound the disaster, the Tisza River is home to 17 of Hungary's 29 protected species of fish, including the last known species of Danube sturgeon. It will be decades before the environment recovers from this accident.

FIGURE 4.19

Cholera outbreaks in 2005 were reported in many western Africa nations.

What must rivers and lakes contain in order to be healthful habitats for freshwater life? They must contain sufficient oxygen, hold appropriate quantities of nutrients (too little or too much of these can adversely affect the water quality), and have an appropriate acid–base balance. If pollution problems are suspected in any river or lake, chemists and biologists must analyze the water and perhaps also the flesh and blood of fish, birds, and other organisms that live in or around the water. Chemical water analysis includes working with measures of concentration, including parts per million, parts per billion, and molarity, and investigating each specific type of reaction that occurs in the aqueous environment.

The Bottom Line

- Water is an extremely versatile solvent, partly thanks to its polarity, which is due to the uneven distribution of electrons in the molecule. (Section 4.1)

- When ionic compounds dissolve in water, the ions dissociate and become surrounded by water molecules—a process known as hydration. (Section 4.1)

- The concentration of a solution can be expressed in units known as molarity. This term indicates how many moles of the chemical concerned would be present if we had one liter of the solution. (Section 4.2)

$$\text{Molarity} = \frac{\text{moles of solute}}{\text{liter of solution}} = M$$

- Chemicals present at very low levels are often measured in terms of parts per million (ppm), parts per billion (ppb), or parts per trillion (ppt). (Section 4.2)

- The quantitative analysis of the chemicals present in solutions is of great importance in medicine, industry, and environmental science. (Sections 4.3, 4.8)

- Precipitation reactions involve an insoluble precipitate forming when soluble chemical species combine. (Section 4.5)

- Acid–base reactions involve acids and bases reacting in ways that can neutralize the acidic and basic character of each. (Section 4.6)

- Oxidation–reduction reactions are electron transfer processes in which some reactants lose electrons (are oxidized), while others gain electrons (are reduced). (Section 4.7)

Key Words

acid A compound that produces hydrogen ions (H^+) when dissolved in water. (*p. 148*)

acid–base reaction The reaction between an acid and a base. The products are water and an ionic compound. (*p. 148*)

aqueous Water-based; also implies that a dissolved substance has a sphere of hydration. (*p. 126*)

base A compound that produces hydroxide ions (OH^-) when dissolved in water. (*p. 148*)

buret A laboratory device used to precisely and accurately add small known quantities of solution. (*p. 140*)

complete ionic equation A chemical equation that indicates all of the ions present in a reaction as individual entities. (*p. 144*)

concentration An intensive property of a solution that describes the amount of solute dissolved per volume of solution or solvent. The typical concentration units include molarity, ppm, and w/w. (*p. 131*)

diprotic Can produce 2 mol of H^+ when it dissolves. (*p. 151*)

electrolyte A compound that produces ions when dissolved in water. (*p. 129*)

end point In a titration, the volume of the added reactant that causes a visual change in the color of the indicator. (*p. 140*)

equivalence point In a titration, the point at which all reactants have just been completely consumed. (*p. 140*)

half-reaction An incomplete equation that describes the oxidation or reduction portion of a redox reaction. (*p. 153*)

hydration sphere The shell of water molecules surrounding a dissolved molecule, ion, or other compound. This shell arises because of the force of attraction between the water molecules and the solute. (*p. 127*)

insoluble Not capable of dissolving in a solvent to an appreciable extent. (*p. 145*)

molar (M) The "shorthand" method of describing molarity, as in "that is a 3 molar HCl solution." (*p. 132*)

molarity (M) A specific concentration term that reflects the moles of solute dissolved per liter of total solution. (*p. 131*)

molecular equation A chemical equation that shows complete molecules and compounds. (*p. 144*)

monoprotic Can produce 1 mol of H^+ when it dissociates. (*p. 151*)

net ionic equation A complete ionic equation written without the spectator ions. (*p. 144*)

nonelectrolyte A compound that doesn't dissociate into ions when it dissolves. (*p. 129*)

oxidation The process of losing electrons. Such a substance is said to be oxidized. (*p. 152*)

oxidation number A "bookkeeping" number that reflects the charge on an ion. (*p. 154*)

oxidation–reduction reactions Reactions that involve the transfer of electrons from one species to another. Also known as redox reactions. (*p. 153*)

oxidation state See oxidation number. (*p. 154*)

oxidized The species that has lost electrons in a redox reaction. (*p. 152*)

parts per billion (ppb) One gram of solute per billion grams of solution. (*p. 135*)

parts per million (ppm) One gram of solute per million grams of solution. (*p. 135*)

parts per trillion (ppt) One gram of solute per trillion grams of solution. (*p. 135*)

precipitation reaction A reaction involving the formation of a solid that isn't soluble in the reaction solvent. (*p. 146*)

precipitate Any solid material that forms within a solution; the action describing the formation of a solid. (*p. 145*)

redox reactions See oxidation–reduction reactions. (*p. 153*)

reduced The species that has gained electrons in a redox reaction. (*p. 152*)

reduction The process of gaining electrons. Such a substance is said to be reduced. (*p. 152*)

soluble The ability of a substance to dissolve within a solution. (*p. 145*)

solutes Molecules, ions, or atoms that are dissolved in a solvent to form a solution. (*p. 126*)

solution (Defined in Chapter 2.) A homogeneous mixture of solute and solvent. (*p. 126*)

solvent A compound that typically makes up the majority of a homogeneous mixture of molecules, ions, or atoms; dissolves the solute. (*p. 126*)

spectator ions Ions that do not participate in a reaction. (*p. 144*)

strong acid An acid that completely dissociates in solution. (*p. 149*)

strong base A base that completely dissociates in solution. (*p. 149*)

strong electrolyte Any compound that completely dissociates in solution. (*p. 129*)

standard solution A solution with a well-defined and known concentration of solute. (*p. 140*)

titration The process of adding one reactant to an unknown amount of another until the reaction is complete; used to determine the concentration of an unknown solute. (*p. 140*)

triprotic Can produce 3 mol of H^+ when it dissolves. (*p. 151*)

weak acid An acid that partially dissociates in solution. (*p. 150*)

weak base A base that partially dissociates in solution. (*p. 150*)

weak electrolyte Any substance that only partially dissociates in solution. (*p. 129*)

Focus Your Learning

The answers to the odd-numbered problems and some selected problems appear at the back of the book, as represented by the blue numbering.

Section 4.1 Water—A Most Versatile Solvent

Skill Review

1. Explain how water molecules can dissolve both cations and anions.

2. Why doesn't pure water conduct an electric current?

3. Explain what is meant by the term *hydration sphere*?

4. Diagram, using circles for atoms, a crystal of KCl versus the same crystal of KCl dissolved in water.

Chemical Applications and Practices

5. Earth's oceans contain tons of dissolved sodium chloride. Yet, when a ship develops an oil leak, almost none of the oil dissolves in the ocean. Explain this phenomenon.

6. When an ion dissolves, it is surrounded by a hydration sphere. If water molecules surrounded the ion so that the hydrogen portion of the water was closer to the ion, would the ion most likely be a cation or an anion?

7. Pure water does not conduct an electric current. However, aqueous solutions of some compounds do form solutions that conduct electricity. Explain why the presence of some solutes converts nonconducting water into a conducting solution.

8. Glycerin can be produced as a by-product in soap making. The compound dissolves so easily in water that it absorbs water from the air. This latter characteristic is why glycerin is often found in many skin lotions. As glycerin absorbs the water, the skin can be kept moist. Glycerin's structure is shown below. What aspects of glycerin's structure contribute most to its ease of dissolving in water?

$$\begin{array}{c} \quad \text{OH} \ \text{OH} \ \text{OH} \\ \quad | \quad\; | \quad\; | \\ \text{H} - \text{C} - \text{C} - \text{C} - \text{H} \\ \quad | \quad\; | \quad\; | \\ \quad \text{H} \;\;\; \text{H} \;\;\; \text{H} \end{array}$$

Glycerin

9. A conductivity-testing apparatus, such as the one shown in this chapter, possesses a light bulb whose brightness is related to how much current is flowing through it (and also through the solution.) A small, but measurable, amount of current must be present before the bulb becomes visibly brighter. What effect would this characteristic of the apparatus have on the classification of solutions containing strong electrolytes, containing weak electrolytes, and containing non-electrolytes?

10. When dissolved in water, which of the following would you expect to cause a conductivity tester, shown in the chapter, to produce a very bright light? (Assume that 0.50 mole of each is placed in 1.0 L of solution.)
   a. C_2H_5OH (ethanol)   d. KCl
   b. NaOH   e. $BaSO_4$
   c. Na_2CO_3

Section 4.2 The Concentration of Solutions

Skill Review

11. Which of the following would best represent $MgCl_2$ dissolved in water?

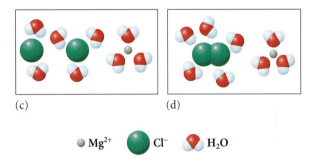

(a)   (b)

(c)   (d)

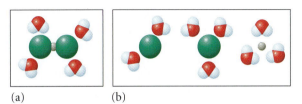

● Mg^{2+} ● Cl^- ◆ H_2O

12. Which of the following would best represent H_2 dissolved in water?

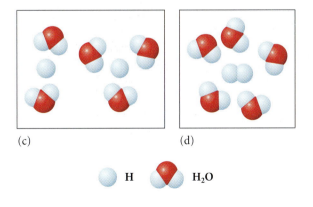

(a)   (b)

(c)   (d)

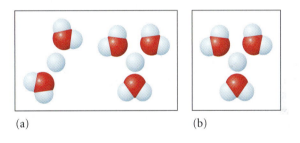

○ H   ◆ H_2O

13. If 0.100 mol $MgCl_2$ is dissolved in water, how many total moles of ions are there in the resulting solution?

14. If 0.100 mol H_2 is dissolved in water, how many total moles of ions are there in the resulting solution?

15. The formula of vitamin C is $C_6H_8O_6$. Calculate the molarity of each of the following vitamin C solutions:
    a. 0.150 g of vitamin C dissolved in enough water to produce a 1.50 L solution
    b. 0.250 g of vitamin C dissolved in enough water to produce a 0.500 L solution
    c. 3.50 g of vitamin C dissolved in enough water to produce a 2.0 L solution

16. The formula of ethanol is C_2H_6O. Calculate the molarity of each of the following ethanol solutions:
    a. 0.150 g of ethanol dissolved in enough water to produce a 1.50 L solution
    b. 25 g of ethanol dissolved in enough water to produce a 0.500 L solution
    c. 100 g of ethanol dissolved in enough water to produce a 2.0 L solution

17. Determine the mass of glycine $(C_2H_5NO_2)$ in each of the following solutions:
    a. 100.0 mL of 0.015 M glycine
    b. 125.0 mL of 0.0145 M glycine
    c. 74.6 mL of 1.44 M glycine

18. Determine the volume of solution needed to provide the following amounts:
    a. 4.50 g ethanol (C_2H_6O) from a 2.50 M ethanol solution
    b. 63.7 g HCl from a 6.0 M HCl solution
    c. 3.0 g glucose $(C_6H_{12}O_6)$ from a 0.150 M solution

19. Calculate the molarity of
    a. calcium ions in 1.00 L of solution containing 24.55 g of calcium chloride
    b. chloride ions in 1.00 L of solution containing 24.55 g of calcium chloride
    c. water in pure water

20. Calculate the molarity of
    a. HBr if 45.0 g is dissolved in 1.0 L.
    b. NaOH if 32.5 g is dissolved in 500 mL
    c. chloride ions in a 5.0 L solution containing 2.00 g NaCl.

21. One important source of bromine, in the form of $Br^-(aq)$, is seawater. If the molarity of Br^- in seawater is approximately 0.00081 M, how many liters of seawater would be required to obtain 1 mole of Br^-?

22. Chloride can be obtained from seawater. If the molarity of Cl^- in seawater is approximately 0.522 M, how many liters of seawater would be required to obtain 1 mole of Cl^-?

23. Determine the volume of a 0.125 M solution of $MgCl_2$ needed to provide
    a. 0.10 mol $MgCl_2$
    b. 0.10 mol Mg^{2+}
    c. 2.33 mol Cl^-

24. Determine the volume of a 0.0357 M solution of Na_2SO_4 needed to provide
    a. 0.10 mol Na_2SO_4
    b. 0.10 mol Na^+
    c. 4.30 mol SO_4^{2-}

25. Determine the concentration, in ppm, for each of the following solutions.
    a. 2.5 mg NaCl per 1000.0 L solution
    b. 5.25 mg Cu^{2+} ions per 500.0 mL solution
    c. 12.5 mg Cl^- ions per 300.0 mL solution

26. Determine the concentration, in ppm, for each of the following solutions.
    a. 1.20 mg KCl per 1000.0 mL solution
    b. 0.100 g Co^{2+} ions per 500.0 mL solution
    c. 1.00 micrograms CN^- ions per 450.0 mL solution

27. Perform the indicated conversions.
    a. 1.20×10^{-4} M NaCl to ppm
    b. 5.33 ppm CN^- to ppb
    c. 170 ppm NaOH to mass percent

28. Perform the indicated conversions.
    a. 0.00250 M $CuCl_2$ to mass percent
    b. 19 ppm CN^- to M
    c. 0.0011% NaOH by mass to M

Chemical Applications and Practices

29. When growing certain bacteria, biologists must often control the acidity level of the growth medium. To do this, a variety of compounds, referred to as buffer systems, may be added in specific amounts. Which of the following has the greater molarity?
    a. 42.0 g of KH_2PO_4 dissolved in 250.0 mL of solution
    b. 21.0 g of KH_2PO_4 dissolved in 125.0 mL of solution

30. If a student seeking to prepare a 1.0 M solution of KOH added 56 g of solid KOH to exactly 1.0 L of water, would the solution be greater or less than 1.0 M? Explain the basis of your conclusion.

31. Adding antifreeze to aqueous automobile cooling systems lowers the freezing point of the resulting solution. The main ingredient in most antifreeze products is ethylene glycol $(C_2H_6O_2)$. How many kilograms of ethylene glycol are dissolved in a 10.0-liter solution that is 16.0 M?

32. Oxygen dissolved in water is critical for aquatic life. The level of dissolved oxygen is often used to report the quality of water in the environment. If 0.0090 g of O_2 is dissolved per liter of solution, what would be reported as (a) the molarity, and (b) the concentration in ppm, of dissolved oxygen?

33. Occasionally students enjoy a cup of coffee or tea while studying chemistry. The amount of caffeine in those beverages varies greatly. At a temperature of 65°C, the maximum amount of caffeine $(C_8H_{10}N_4O_2)$ that can be dissolved in water is 455g/L. What would be the maximum molarity of caffeine in a 425-mL cup of 65°C coffee?

34. The compound potassium permanganate forms intensely purple solutions that are used to react with other solutions that contain iron. How many grams of potassium permanganate $(KMnO_4)$ would you need to dissolve into 0.500 L of solution if you wanted to prepare a 0.00100 M solution of $KMnO_4$?

35. The pesticide atrazine $(C_8H_{14}N_5Cl)$ is so slightly soluble in water that its concentration is not usually reported in molarity. Health advisory warnings are issued if the atrazine concentration is higher than 3.0 parts per billion. How many

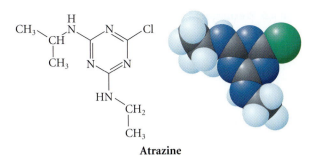

Atrazine

grams of atrazine would you expect to find in 1 L of water that is 3.0 ppb in atrazine? What would be the molarity of the solution?

36. A can of nondiet soda may contain 45 g of sucrose ($C_{12}H_{22}O_{11}$) per 450 g of soda. What is the sugar concentration reported as ppm? Approximately how many grams of water would you have to add to the soda to reduce this sugar concentration to only 1 part per million?

37. A certain sports drink has a NaCl concentration of 0.20 M. If the container held 250 mL of the liquid, how many moles of Na^+ would you consume when swallowing 50.0 mL of the drink?

38. One method used to detect the presence of the toxic heavy metal barium is to precipitate the barium +2 ion as barium sulfate. If a 10.0-mL sample produced 0.565 g of barium sulfate (and assuming that no more Ba^{2+} remained dissolved) what was the original molarity of Ba^{2+} in the solution?

Section 4.3 Stoichiometric Analysis of Solutions

Skill Review

39. Assume 1:1 reaction stoichiometry in each of the following:
   a. 25.0 mL of HCl required 35.0 mL of 0.155 M NaOH to neutralize. What is the molarity of the HCl solution?
   b. 50.0 mL of $Sr(OH)_2$ required 35.0 mL of 0.0200 M sulfuric acid to neutralize. What is the molarity of the $Sr(OH)_2$ solution?
   c. 40.5 mL of niric acid (HNO_3) required 25.0 mL of 0.35 M KOH to neutralize. What is the molarity of the HNO_3 solution?

40. Assume 1:1 reaction stoichiometry in each of the following:
   a. 33.0 mL of HCl required 17.6 mL of 2.50 M NaOH to neutralize. What is the molarity of the HCl solution?
   b. 10.25 mL of $Sr(OH)_2$ required 15.56 mL of 0.00150 M sulfuric acid to neutralize. What is the molarity of the $Sr(OH)_2$ solution?
   c. 100.0 mL of nitric acid (HNO_3) required 84.30 mL of 1.562×10^{-4} M KOH to neutralize. What is the molarity of the HNO_3 solution?

41. The label on a solution was partially obscured. It was either 0.10 M CuCl or 0.10 M $CuCl_2$. When the entire 25.0 mL of the solution was evaporated to dryness, 0.336 g of solid remained. What should the label read?

42. The label on a solution was partially obscured. It was either 0.10 M $FeCl_2$ or 0.10 M $FeCl_3$. If the solution was actually 0.10 M $FeCl_2$, how many grams of $FeCl_2$ would be obtained if 100.0 mL of the solution were evaporated to dryness?

43. First complete the following reaction using words rather than chemical symbols. Then write the molecular, complete ionic, and net ionic equations that describe this aqueous solution reaction.

Potassium hydroxide + hydrochloric acid →

44. List the spectator ions for the following reaction:

Sulfuric acid + potassium hydroxide →

45. Write the molecular, complete ionic, and net ionic equations for the following reaction:

Sodium chloride + calcium nitrate
→ sodium nitrate + calcium chloride

46. Write the molecular, complete ionic, and net ionic equations for the following reaction:

Magnesium chloride + sodium bromide
→ sodium chloride + magnesium bromide

Chemical Applications and Practices

47. Using the stoichiometry found in the chapter for the vitamin C and I_2 reaction, determine which of the following two drinks contains the greater concentration of vitamin C. Which sample contains the greater amount of vitamin C?
   a. Brand A: 250.0 mL solution required 10.5 mL of 0.0855 M I_2
   b. Brand B: 300.0 mL solution required 12.0 mL of 0.0855 M I_2

48. Using the stoichiometry found in the chapter for the vitamin C and I_2 reaction, determine which of the following two drinks contains the greater concentration of vitamin C. Which sample contains the greater amount of vitamin C?
   a. Brand A: 150.0 mL solution required 8.5 mL of 0.0650 M I_2
   b. Brand B: 100.0 mL solution required 12.5 mL of 0.0250 M I_2

49. One of the more common agents for standardizing solutions of sodium hydroxide (NaOH) is potassium acid phthalate ($KHC_8H_4O_4$), or, more accurately, potassium hydrogen phthalate. The compound is often abbreviated KHP, where the P designates the phthalate ion ($C_8H_4O_4^{2-}$), not the phosphorus atom. From the following data, determine the average molarity of a NaOH solution that was standardized with KHP.
   Trial A: 45.12 mL of NaOH solution neutralized 0.5467 g of KHP
   Trial B: 44.89 mL of NaOH solution neutralized 0.5475 g of KHP
   Trial C: 46.50 mL of NaOH solution neutralized 0.5501 g of KHP

50. One of the more common agents for standardizing solutions of potassium hydroxide (KOH) is potassium acid phthalate ($KHC_8H_4O_4$), or, more accurately, potassium hydrogen phthalate. From the following data, determine the average molarity of a KOH solution that was standardized with KHP.
   Trial A: 45.12 mL of KOH solution neutralized 0.2573 g of KHP
   Trial B: 48.89 mL of KOH solution neutralized 0.2250 g of KHP
   Trial C: 45.50 mL of KOH solution neutralized 0.2502 g of KHP

51. Hydrochloric acid has many industrial uses. The steel industry uses hydrochloric acid in a process known as "pickling steel." This is done to steel prior to galvanizing. To analyze the concentration of the "pickle liquor," a titration with sodium hydroxide may be used. Write the net ionic reaction between hydrochloric acid and sodium hydroxide, and then determine the concentration of the HCl in the pickling liquor if a 10.00 mL sample of the hydrochloric acid required 45.55 mL of a 0.9876 M solution of NaOH to completely react.

52. Nitrogen for increased yields on farms can come from a variety of sources. One common nitrogen-containing fertilizer is ammonium sulfate, $(NH_4)_2SO_4$. The reaction between ammonia (NH_3) and sulfuric acid (H_2SO_4) produces this important fertilizer.
    a. Balance the reaction.
    b. How many liters of a 1.55 M solution of sulfuric acid would be needed to completely react with 1.00 kg of ammonia?

53. One method that field geologists use to test for the presence of carbonates is to drip hydrochloric acid on a sample and note the formation of carbon dioxide gas bubbles.
    a. Balance the reaction between calcium carbonate ($CaCO_3$) and hydrochloric acid (HCl.)
    b. If 5.00 mL of 0.500 M HCl completely reacted with calcium carbonate in a rock sample, how many grams of calcium carbonate did the sample contain?

54. Hydrogen peroxide, in very dilute concentrations, can be used as a disinfectant. The concentration of hydrogen peroxide (H_2O_2) can be determined in a titration experiment using potassium permanganate ($KMnO_4$) as shown in the following balanced equation:

$$2MnO_4^-(aq) + 5H_2O_2(aq) + 6H^+(aq)$$
$$\rightarrow 5O_2(g) + 2Mn^{2+}(aq) + 8H_2O(l)$$

If a 25.0 mL sample of hydrogen peroxide required 25.2 mL of 0.353 M $KMnO_4$ solution to react with all the hydrogen peroxide, what was the molarity of the hydrogen peroxide solution?

Section 4.4 Types of Chemical Reactions

Skill Review

55. Predict the products of the following incomplete reactions. After balancing the reaction, label each as a precipitation, acid–base, or redox reaction. For those reactions where it is possible, write the net ionic equation.
    a. $C_4H_{10}(g) + O_2(g) \rightarrow$
    b. $Ca(OH)_2(aq) + HNO_3(aq) \rightarrow$
    c. $Pb(NO_3)_2(aq) + NaCl(aq) \rightarrow$

56. Predict the products of the following incomplete reactions. After balancing the reaction, label each as a precipitation, acid–base, or redox reaction. For those reactions where it is possible, write the net ionic equation:
    a. $HCl(g) + Ca(OH)_2(aq) \rightarrow$
    b. $CH_4(g) + O_2(aq) \rightarrow$
    c. $Ba(ClO_4)_2(aq) + Na_2S(aq) \rightarrow$

Section 4.5 Precipitation Reactions

Skill Review

57. Which of the following salts would qualify as soluble in water?
    a. $CuCO_3$ d. KOH
    b. NiS   e. lead acetate
    c. $(NH_4)_2CO_3$

58. Which of the following salts would qualify as soluble in water?
    a. $NaNO_3$ d. $PbBr_2$
    b. $Ba(OH)_2$ e. AgI
    c. $MgSO_4$

59. Predict the products and write the net ionic equations for each of the following:
    a. $BaCl_2(aq) + NaNO_3(aq) \rightarrow$
    b. $Fe(NO_3)_3(aq) + (NH_4)_2SO_4(aq) \rightarrow$
    c. $CaCl_2(aq) + K_2SO_4(aq) \rightarrow$

60. Predict the products and write the net ionic equations for each of the following:
    a. $MgCl_2(aq) + KNO_3(aq) \rightarrow$
    b. $AgNO_3(aq) + NH_4Cl(aq) \rightarrow$
    c. $CaCl_2(aq) + NaOH(aq) \rightarrow$

61. Write out the formulas of each of the following reactants. Then predict the result of mixing the aqueous solutions for each situation. Finally, report the net ionic equation for each.
    a. Copper(II) nitrate + potassium hydroxide →
    b. Sodium carbonate + aluminum chloride →
    c. Ammonium phosphate + zinc chloride →

62. Write out the formulas of each of the following reactants. Then predict the result of mixing the aqueous solutions for each situation. Finally, report the net ionic equation for each.
    a. Barium nitrate + sodium hydroxide →
    b. Ammonium carbonate + aluminum bromide →
    c. Silver nitrate + zinc bromide →

63. Name two soluble ionic compounds that could be used to produce each of the following insoluble salts:
    a. BaS   b. $Cu(OH)_2$ c. $PbSO_4$

64. Name two soluble ionic compounds that could be used to produce each of the following insoluble salts:
    a. AgBr   b. $Fe(OH)_2$ c. PbS

Chemical Applications and Practices

65. Suppose you have a water sample that is to be analyzed. You know the sample contains Cu^{2+}, Ba^{2+}, and Ag^+ ions. Suggest a sequence for adding other aqueous electrolyte solutions that could separate each of these by selective precipitation.

66. The presence of metal ions in aqueous systems can often cause environmental complications. Suppose it was necessary to remove Cu^{2+} ions from a sample of water. Suggest at least two reagents that you could add to the aqueous system to remove the copper ions.

67. A traditional method for the analysis of dissolved ions is called *gravimetric analysis*. This name is derived from the process in which the ion to be analyzed is precipitated and filtered as gravity separates the liquid from the precipitate.

Suppose that you found that all the silver ion (Ag^+) in a solution, present as $AgNO_3(aq)$, reacted with 25.0 mL of 0.242 M NaCl to form solid AgCl.
a. Write the molecular and net ionic equations for the precipitate formation.
b. Determine the grams of precipitate formed and the grams of silver present in the original solution.

68. One method of preparing the compound AgBr, used in photographic films, is to precipitate it from a solution of KBr. After first writing and balancing the molecular and net ionic equations, calculate many grams of AgBr precipitate when 100.0 mL of 2.00 M KBr is mixed with 100.0 mL of 1.00 M $AgNO_3$.

Section 4.6 Acid–Base Reactions

Skill Review

69. What is the net ionic reaction of the following acid–base reaction? You'll have to provide the balanced molecular equation in order to begin this problem.

Hydrogen bromide + magnesium hydroxide
→ water + magnesium bromide

70. What acid–base reaction would produce each of the following salts? Write the balanced chemical equation in each case.
a. NaCl   b. K_2CO_3 c. Na_2SO_4 d. $Al(NO_3)_3$

71. a. What would be the molarity of a KOH solution if, during a titration with 0.50 M HCl, 34.5 mL of the HCl neutralized 22.4 mL of the KOH solution?
b. How much of a 0.50 M solution of H_2SO_4 would be needed to neutralize the same amount of the KOH solution?

72. If a student mixes 25.0 mL of 0.255 M H_2SO_4 with 50.0 mL of 0.115 M KOH (assume the volumes are additive), which reagent will be in excess? What will be the concentration of the excess reagent?

Chemical Applications and Practices

73. Oxalic acid can be found in rhubarb plants. If a 0.255-g sample of purified oxalic acid required 25.7 mL of NaOH to neutralize, what would you report as the molarity of the NaOH solution? (Oxalic acid: $H_2C_2O_4$)

74. Phosphoric acid is a very versatile acid with widespread uses, from making fertilizer to soft drink ingredients.
a. Balance the reaction between phosphoric acid (H_3PO_4) and ammonium hydroxide (NH_4OH).
b. How many milliliters of 0.459 M ammonium hydroxide would be needed to neutralize 33.5 mL of 0.100 M phosphoric acid?

75. Carbonic acid is formed when gaseous carbon dioxide is pumped into soft drinks to establish their carbonation. Suppose you are employed at the new "ChemCola" beverage company. You must titrate the carbonic acid in the soft drink using 0.1445 M NaOH.

Carbonic acid

a. Write the molecular and net ionic equations.
b. What is the molarity of carbonic acid if a 50.00 mL sample required 38.98 mL of 0.1445 M NaOH?

76. Acetic acid is the ingredient in vinegar that gives it its vinegary taste and smell. If a particular brand claims to be 5.00%, by mass, acetic acid ($HC_2H_3O_2$), how many milliliters of 0.255 M NaOH would be required to neutralize the acetic acid in a 25.0 mL vinegar sample?

77. Stomach acid (HCl) is neutralized by a variety of commercial antacids. For each of the following, determine how many grams of active antacid ingredient would be necessary to neutralize the HCl in 50.00 mL of 0.0100 M HCl (an approximation of the concentration of HCl in the stomach).
a. $Al(OH)_3$ b. $Mg(OH)_2$ c. $CaCO_3$

78. Sulfuric acid is the acid ingredient in automobile battery acid solutions. The sulfuric acid content of such a solution can be determined through a lab analysis by reacting it in a titration with potassium hydroxide. The unbalanced reaction is

$$H_2SO_4(aq) + KOH(aq) \rightarrow K_2SO_4(aq) + H_2O(l)$$

Use the balanced reaction to fill in the missing data in the following table.

Molarity of H_2SO_4 (M)	Volume of H_2SO_4 (mL)	Molarity of KOH (M)	Volume KOH (mL)
0.25	75.0		28.6
	28.9	0.36	35.8
0.88		1.11	27.5
1.76	22.0		49.7

Section 4.7 Oxidation–Reduction Reactions

Skill Review

79. In a compound made up only of each of the following pairs of elements, select the atom that is more likely to carry a negative or slightly negative charge.
a. C, H   d. P, Fe
b. O, F   e. Ca, O
c. Na, O

80. In a compound made up only of each of the following pairs of elements, select the atom that is more likely to carry a negative or slightly negative charge.
a. K, H   d. N, Ca
b. Li, F   e. N, Mg
c. Na, S

81. Assign oxidation numbers to all the elements in each of the following molecules or ions:
a. N_2O_5 d. N_2
b. PO_4^{3-} e. H_2SO_3
c. $CuCO_3$

82. Assign oxidation numbers to all the elements in each of the following molecules or ions:
    a. CH_4 c. $KHCO_3$ e. $KMnO_4$
    b. SO_4^{2-} d. $Na_2Cr_2O_7$

Chemical Applications and Practices

83. The following balanced equation depicts a reaction that can be used for the determination of iron in a steel sample. Is this a redox reaction? Prove it by showing which elements change oxidation state.

$$6Fe^{2+}(aq) + 14H^+(aq) + Cr_2O_7^{2-}(aq)$$
$$\rightarrow 6Fe^{3+}(aq) + 2Cr^{3+}(aq) + 7H_2O(l)$$

84. The following reaction depicts one of the steps in obtaining the important steel alloying ingredient titanium.

$$TiCl_4 + 2Mg \rightarrow Ti + 2MgCl_2$$

    a. Identify the substance being oxidized.
    b. Identify the substance being reduced.

Comprehensive Problems

85. Why is water called the universal solvent?

86. An interesting demonstration often used by chemistry teachers utilizes several of the principles that you have been reading about in this chapter. First, a solution of barium hydroxide is shown to conduct electricity by having the electrodes from a conductivity tester immersed in the solution and observing that a light begins to shine. Then a solution of sulfuric acid is slowly added. A white precipitate begins to form, and the light bulb dims. Eventually, the addition of the sulfuric acid causes the bulb to go out. Finally, continued addition of the sulfuric acid solution brings the bulb back on to bright light.
    a. What is the identity of the white precipitate?
    b. What is the net ionic equation for the reaction between barium hydroxide and sulfuric acid?
    c. What is the significance of the point at which the bulb goes out completely?
    d. Explain why continued addition of the sulfuric acid causes the light to come back on.

87. Some chemical reactions are best done in solution. If you had 100.0 mL of 0.230 M NaOH, how many milliliters of 0.530 M HCl would you need to have the same number of moles as found in the NaOH solution?

88. Extreme ozone pollution is described as any concentration greater than 0.28 ppm ozone, C, O_3. If the density of a sample of air containing that concentration of ozone were 1.30 g/L, what would be the molarity of ozone in the sample? How many molecules of ozone would be in 1 L of the air?

89. Using Table 4.3, determine the identity and formula of any and all precipitates that are likely to form when a solution containing NaOH and $(NH_4)_2CO_3$ is mixed with a solution that contains $CuNO_3$ and $BaCl_2$.

90. a. An unlabeled solution may contain either Ag^+ ions or Al^{3+} ions. Using Table 4.3, determine a suitable anion solution that, through selective precipitation, could be added to identify the cation present in the solution.
    b. Another unlabeled solution contains either nitrate ions or sulfate ions. Using Table 4.3, determine a solution that

contains a cation that could be used in a selective precipitation to determine the identity of the anion in the unlabeled solution.

91. Barium-containing "milkshakes" are often used to obtain X-rays of patients suffering from intestinal problems. Barium compounds can also be toxic. Barium sulfate is insoluble in water, so it can be given to patients without concern that it would be absorbed. It is also opaque to X-rays. Write the molecular and ionic balanced equations for the formation of $BaSO_4(s)$ from $Ba(NO_3)_2(aq)$ and $Na_2SO_4(aq)$. How many grams could be obtained from mixing 125.0 mL of 0.567 M $Na_2SO_4(aq)$ and 75.0 mL of 0.786 M $Ba(NO_3)_2(aq)$?

92. The reaction of gaseous dinitrogen trioxide with water can provide aqueous nitrous acid as the product.
    a. Write the balanced chemical reaction.
    b. If 12.5 grams of dinitrogen trioxide treated with excess water produced 12.5 grams of nitrous acid, what is the percent yield of the reaction?
    c. To make the nitrous acid as described in part b, the dinitrogen trioxide was bubbled into 1.55 gallons of water. Assuming that the amount of water remains constant during the reaction, what is the concentration of nitrous acid after the reaction?

93. A chef wishes to make a very lightly sweetened tea by dissolving 1.00 g fructose ($C_6H_{12}O_6$) in 12 fluid ounces of tea.
    a. What is the concentration of fructose in the tea in molarity?
    b. What is the concentration of fructose in ppm?
    c. How many carbon atoms are there in 1.00 g of fructose?
    d. How many water molecules are in the serving of tea?

Thinking Beyond the Calculation

94. Oxalic acid can be found in a variety of plants. This compound is considered toxic. Therefore, it is often important to determine the quantity in a sample. This can be done through a redox titration.

$$5C_2O_4^{2-}(aq) + 16H^+(aq) + 2MnO_4^-(aq)$$
$$\rightarrow 2Mn^{2+}(aq) + 10CO_2(g) + 8H_2O(l)$$

    a. Judging on the basis of the structure of oxalic acid shown here, would you expect it to be soluble or insoluble in water?
    b. Write equations that illustrate the dissolution and dissociation of oxalic acid in water.
    c. Which of the species in the balanced redox reaction above is more oxidized, oxalate ($C_2O_4^{2-}$) or carbon dioxide?
    d. Which of the species in the balanced redox reaction is more oxidized, permanganate or manganese ions?
    e. How many electrons are being transferred among the reactants in the balanced redox reaction?
    f. Determine which species is oxidized, and which is reduced, in the balanced redox reaction.
    g. If a properly prepared plant sample required 33.5 mL of a 0.00976 M $KMnO_4$ solution to react, calculate the number of grams of oxalic acid ($H_2C_2O_4$) present.

Oxalic acid

Energy

Contents and Selected Applications

Chemical Encounters: Setting the Stage with the Space Shuttle

5.1   **The Concept of Energy**

5.2   **Keeping Track of Energy**

Chemical Encounters: Energy in Foods

5.3   **Specific Heat Capacity and Heat Capacity**

5.4   **Enthalpy**

5.5   **Hess's Law**

Chemical Encounters: Focus on Methane

5.6   **Energy Choices**

Chemical Encounters: Energy Choices

The space shuttle lifts off from its launch pad in Florida. The tremendous amount of energy released by the chemical reactions in the solid rocket boosters and main engine results in a force strong enough to oppose the force of gravity acting on the shuttle.

Go to **college.hmco.com/pic/kelterMEE** for online learning resources.

A space shuttle launch is an awe-inspiring demonstration of the ability of energy to transport humans and material away from Earth's surface with an eye toward planetary exploration. The energy to launch the craft and its crew comes from the explosive violence of chemical reactions. When these reactions are used in a carefully controlled way, they can lift the massive shuttle (which weighs in at a robust 2,000,000 kg), the booster rockets, and the fuel tank and propel the ship into orbit hundreds of miles above Earth in a very brief 10-minute ride.

Application

CHEMICAL ENCOUNTERS: Setting the Stage with the Space Shuttle

Two chemical reactions, indicated in Figure 5.1, power the launch of the space shuttle. The combustion of hydrogen gas (the combination of hydrogen and oxygen to form water) takes place in the main engines. At the same time, the solid rocket boosters attached to the sides of the shuttle release a host of products resulting from the oxidation of aluminum by ammonium perchlorate, though we show only the primary equation here.

Main engines: $2H_2(g) + O_2(g) \rightarrow 2H_2O(g)$

Boosters: $10Al(s) + 6NH_4ClO_4(s) \rightarrow 5Al_2O_3(s) + 6HCl(g) + 3N_2(g) + 9H_2O(g)$

As the shuttle sits on the launch pad, it is hard to believe that the energy that will launch it with such a spectacular display of chemical muscle is quietly present within the main fuel tank and the solid fuel of the booster rockets. Chemicals can store huge amounts of energy in this way—and then release it, with very dramatic effects, as soon as a chemical reaction begins.

These thoughts raise many questions about energy: **What is energy?** The rocket fuels used to lift the space shuttle store energy and release it during takeoff. How is energy stored within molecules and compounds? Can we calculate how much energy will be released from a molecule or compound during a shuttle launch so that enough will be harnessed to allow the shuttle to climb to exactly the intended orbit? These questions can be answered by focusing our attention on a branch of science known as **thermodynamics**. Thermodynamics is concerned with the interconversion of different forms of energy. The specific area of thermodynamics most relevant to chemical reactions is known as **thermochemistry**, the study of energy changes and exchanges in chemical processes.

In order to properly describe energy exchanges, we must clearly state where they occur. We define the **system** as that object in which we are interested. In our discussion, the main shuttle craft is the system. Everything else is defined as the **surroundings**. Energy that is lost by the shuttle's main engines in our system is gained by the atmosphere, the surroundings. The system and the surroundings combine to make up the **universe**. These terms are fundamental to our discussion of energy in chemistry.

$2H_2(g) + O_2(g) \rightarrow 2H_2O(g)$

$10Al(s) + 6NH_4ClO_4(s) \rightarrow$
$5Al_2O_3(s) + 6HCl(g) +$
$3N_2(g) + 9H_2O(g)$

FIGURE 5.1

The reactions that provide the energy to lift the space shuttle into orbit.

Universe = System + Surroundings

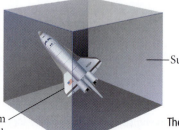

Surroundings

System
(shuttle,
payload, people)

The system and the surroundings make up the universe.

5.1   The Concept of Energy

The simplest commonly used definition of energy is that it is "the capacity to do work." Often it is defined as "the capacity to do work or produce heat." In order to fully understand the definition of energy, we need to define work.

Work is done whenever any force is used to move an object some distance. For example, suppose we use a chemical reaction to propel a heavy weight, such as the space shuttle, upward. We are using the reaction to provide the force that makes the shuttle move. The chemicals that are combined in the reaction are doing work on the shuttle, so they are releasing energy. *Why does it take so much energy to lift the shuttle?* You might say, "Because the shuttle is very heavy," but that is only the beginning of the answer. The shuttle is heavy because of the force of gravitational attraction between the particles in this system and the rest of planet Earth. To raise the shuttle, we must make it move *against* the force of gravity, one of the three known fundamental forces of nature listed in Table 5.1. Since the force of gravitational attraction is so large between the shuttle and Earth, we must supply a lot of energy to raise the shuttle into orbit. The result is that a tremendous amount of work is done on the shuttle.

We can therefore also describe energy as *the capacity to move something against a fundamental force.* This idea can help us develop a good understanding of what energy is all about. In terms of the system and the surroundings, we can make the following statement:

Energy is absorbed by the system from the surroundings in order to oppose a natural attraction.

This means, for example, that energy must be added to two magnets that are stuck together (a system) in order to pull them apart, because we are opposing their natural magnetic attraction. Similarly, when we pull an electron away from a nucleus in an atom (another system), energy is required (must be added to the system) because opposite charges attract and we are opposing that natural attraction. Figure 5.2 illustrates the energy exchange between the system and surroundings in this case.

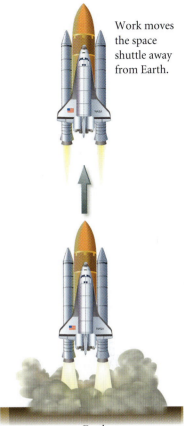

Work moves the space shuttle away from Earth.

Earth

Tutorial: Work, Heat, and Energy Flow

TABLE 5.1   The Three Known Fundamental Forces of Nature

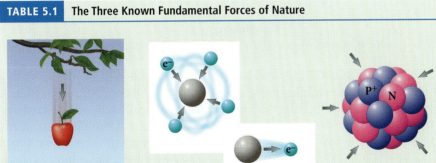

Gravitational force causes all objects with mass to be attracted to one another.

Electroweak force is responsible for the attraction between objects with opposite electric charges and the repulsion between objects with the same electric charge (when it is known as the *electric force*), the phenomena of magnetism and light, and some transformations within subatomic particles.

Strong nuclear force binds protons and neutrons together within atomic nuclei.

FIGURE 5.2

Energy exchange between the system and the surroundings during the ionization of an atom.

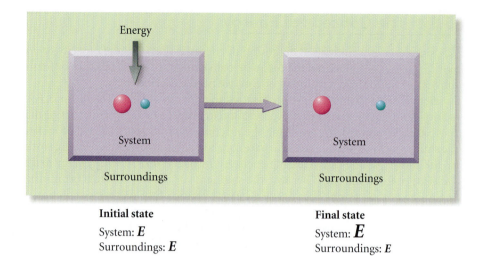

Initial state
System: *E*
Surroundings: *E*

Final state
System: *E*
Surroundings: *E*

Because energy is needed to oppose a natural attraction, the reverse process allows us to draw the following opposite, perhaps counterintuitive, conclusion:

> **Energy is released from the system to the surroundings when a natural attraction occurs.**

This happens when, for example, magnets are made to stick together, an electron is attracted to an atom, or a ball falls to Earth toward the planet's center of gravity.

Kinetic Energy Versus Potential Energy

 Video Lesson: The Nature of Energy

Energy can be relegated to two basic forms: kinetic energy and potential energy. In addition to these two forms, energy can also be propagated through space in the form of **electromagnetic radiation**, such as light, which we will discuss in detail in Chapter 6.

Kinetic energy is the energy of movement. Everything that is moving has an amount of kinetic energy that depends on its mass and velocity, as described by the equation

> **Kinetic energy $= \frac{1}{2}mv^2$**
> **where** $m =$ mass in kilograms, $v =$ velocity in meters per second.

For example, a 36,000-kg 18-wheeled truck moving at the *same speed* as an 800-kg subcompact automobile has more kinetic energy than the car because its mass is greater. A truck moving faster than another, identical truck has more kinetic energy than the slower truck because its velocity is greater. The relationship of kinetic energy to mass and velocity is true for a car, a truck, a space shuttle—or a molecule.

If movement is associated with energy, then everything that is moving must, in appropriate circumstances, be able to do work. You can convince yourself of this by thinking about the steel ball shown in Figure 5.3. If the ball is moving toward the slope, its energy of movement (kinetic energy) allows it to rise a certain distance up the slope. The height to which it rises depends on how fast it was originally moving. Kinetic energy is associated with the work of raising the ball, and the amount of kinetic energy that the rolling ball possesses will determine how high it can rise in defiance of the gravitational force.

The second fundamental form of energy is **potential energy**, which is the energy that objects possess because of their *positions*. We often think of potential energy as stored energy, which, if released, can be transformed into kinetic energy. For example, any object raised up against the force of gravity gains an amount of

Assuming they have equal mass, the faster truck has greater kinetic energy.

potential energy because of its position relative to the gravitational attraction to Earth's center. Recall our key idea that *energy is absorbed by the system from the surroundings in order to oppose a natural attraction.* The farther we pull our object away from Earth's center of gravity, the more energy is required, and the greater will be the object's potential energy.

As our steel ball in Figure 5.3 rises upward, it begins to slow down. In fact, if we ignore the loss of energy due to friction, the steel ball steadily gains potential energy in an amount that is equal to the kinetic energy it loses because it is slowing down. By the time the ball becomes stationary at the top of its climb, kinetic energy due to the rolling of the ball has been converted into potential energy. What happens next? The potential energy stored in the steel ball due to its relatively high altitude is released as the ball begins to run downhill. The amount of potential energy is reduced as the kinetic energy, and therefore the velocity of the ball, increases. At the bottom of the hill, this potential energy has been completely converted into kinetic energy, and the ball is traveling at its maximum velocity.

This example of the interconversion between kinetic and potential energy illustrates one of the most significant fundamental laws of nature. It is known as the **law of conservation of energy**:

> **Energy is neither created nor destroyed. It is only transferred from place to place and transformed from one form into another.**

The energy contained in molecules and compounds, such as those that power the space shuttle, is a constantly interconverting mixture of kinetic and potential energy. The particles in the system are moving, bouncing off one another, vibrating, and rotating as the bonds between atoms stretch in and out, atoms and larger groups rotate around bonds, and entire molecules cartwheel through space. All of this motion is associated with a corresponding amount of kinetic energy. The available potential energy in chemical compounds is stored in the positions and arrangements of the electrons and nuclei that make up those compounds. For

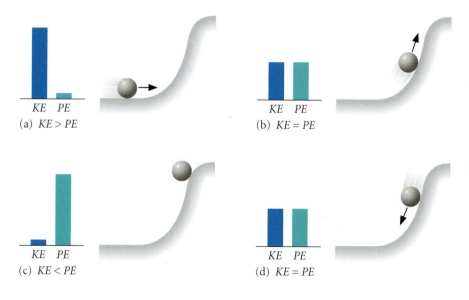

FIGURE 5.3

A steel ball moving toward a hill demonstrates that energy can do work and that kinetic energy and potential energy can be interconverted. As the ball moves up the slope, its kinetic energy is converted into potential energy. When it reaches its maximum height and begins to fall back, that potential energy is converted back into kinetic energy. Work is done when the mass of the ball rises up the slope.

(a) $KE > PE$

(b) $KE = PE$

(c) $KE < PE$

(d) $KE = PE$

FIGURE 5.4

The gravitational force. Energy is required to lift an apple from the ground. The opposing force is gravity.

FIGURE 5.5

Chemical energy. The positions of the nuclei and electrons, as well as the motions of the particles, are part of the energy stored in a molecule.

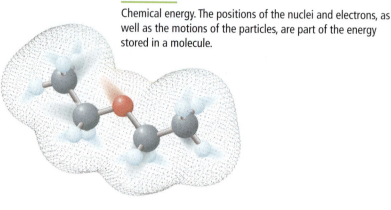

instance, as we discussed previously, it takes energy to move an electron away from an atom's nucleus, against the pull of the electric force, in just the same way as it takes energy to lift a weight up from the ground against the pull of the gravitational force illustrated in Figure 5.4.

It also requires an influx of energy to a system of atoms in order to force two or more (like-charged) electrons closer together against the repulsion due to the electric force, or to force two nuclei together against that force. The electrons and nuclei of chemicals have potential energy as a consequence of their positions in the electric force field, as shown in Figure 5.5, just as objects we lift and throw around have potential energy due to their positions in the gravitational force field.

> Any set of compounds, at a given temperature and pressure, contains **chemical energy** stored as a result of the motions and positions of their atomic nuclei and electrons.

 Application

When a chemical reaction occurs in a system, new chemicals are formed with different internal motions, different electron arrangements, and a different total energy content. Any excess energy must be released to the surroundings, usually as heat and possibly as light and/or sound. Any energy acquired by the new chemicals must come from the surroundings. For example, when hydrogen and oxygen react to form water in the main engines of the space shuttle, a tremendous amount of chemical energy is released. This release of energy occurs because the structure of water molecules embodies much less energy than that of the oxygen and hydrogen molecules used to form the water. A small fraction of the released energy is converted into light and sound. The majority of the released energy dramatically increases the speed of motion of the particles involved, causing them to undergo a massive and explosive expansion as all of the very fast-moving particles bounce off one another. This chemical cyclone of particles within the expanding gas pounds against the inner upper surface of the shuttle engines and literally pushes the shuttle upward while water vapor escapes from the opening at the bottom of the engines, as shown in Figure 5.6.

FIGURE 5.6

Forces acting on the shuttle's main engine.

Gravity

Net force

Force of expanding gases

EXERCISE 5.1   **Energy in Various Forms**

In our day-to-day chats and political or environmental discussions, we speak of many different kinds of energy, such as that from wind, waves, or fossil fuels. Can you explain how these forms of energy (wind, waves, and fossil fuels) are related to the fundamental forms of energy discussed above?

Solution

Wind energy is the energy associated with the movement of air, so it is a form of kinetic energy. Similarly, wave energy is the energy associated with the movement of

NanoWorld / MacroWorld

Big effects of the very small: Electrons capturing energy in photosynthesis

The trapping of the energy of sunlight by green plants is one of the most significant chemical processes on Earth. What makes it even more interesting is that the process is achieved by the smallest chemical change. That tiny change, powered by the absorption of energy from the Sun, is the transfer of an electron from one molecule to another. This change begins the process that provides energy to the plant and allows it to grow.

Why are plants green? They contain molecules of chlorophyll, which absorb light in the red and blue parts of the visible spectrum (the range of visible colors) more than in the green parts, leaving the green light to reach our eyes. There are actually several different types of molecules known as chlorophyll. Chlorophyll *a* is shown in Figure 5.7.

The absorption spectra of some chlorophylls, meaning the extent to which they absorb light of different wavelengths, is shown in Figure 5.8. Because of differences in structure, every atom and molecule has its own absorption fingerprint, which we will discuss more fully in Chapter 6.

When a molecule absorbs light energy, it usually raises one or more of the electrons into a higher energy configuration. When this happens to the chlorophyll in a plant, an excited higher-energy electron can be transferred to an adjacent receptor molecule. This electron transfer requires an input of energy as the negatively charged electron moves away from a positively charged atomic nucleus within chlorophyll, against the pull of the electric force. The transferred electron carries much of this energy to the receptor molecule.

This is the tiny nanoscale event that powers photosynthesis. The transferred electron then moves among a series of other molecules involved in photosynthesis. The complex series of steps that each of the molecules goes through allows some of the energy originally absorbed from the Sun to be stored within other compounds. Ultimately, the energy from the Sun converts carbon dioxide and water into carbohydrate and oxygen—the overall energy-requiring process of photosynthesis.

The path the energy takes in completing this task is long. The chemical processes are very complex, and the involvement of huge numbers of electrons is necessary to generate significant amounts of carbohydrate and oxygen. But the crucial event begins with a single electron.

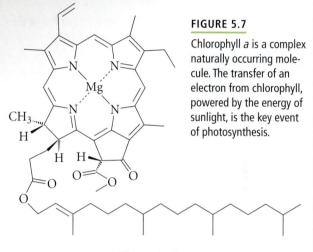

FIGURE 5.7

Chlorophyll *a* is a complex naturally occurring molecule. The transfer of an electron from chlorophyll, powered by the energy of sunlight, is the key event of photosynthesis.

Chlorophyll *a*

FIGURE 5.8

The absorption spectra of a series of chlorophylls.

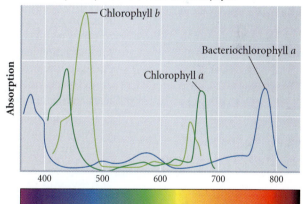

Wavelength (nm)

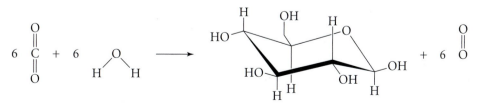

Carbon dioxide and water combine to make carbohydrates and oxygen. This reaction is driven by the transfer of an electron from chlorophyll.

water, so it is a form of kinetic energy. Fossil fuel energy is the energy that we release from the fossil fuels—coal, oil and natural gas—when we burn them. The energy is stored within the chemicals involved as a mixture of kinetic and potential energy, associated with the movement and positions of the particles of the chemicals.

PRACTICE 5.1

Explain how the "activities" of plants are related to the fundamental forms of energy. Include in your discussion the interaction of plants with the environment and their ultimate fate as food.

See Problems 1, 2, 5–8, 17, and 18.

Heat and Reaction Profile Diagrams

During a chemical change, energy is exchanged between the system and the surroundings. Much of this energy exchange is as heat. **Heat (q)** is the energy that is exchanged between a system and its surroundings because of a difference in temperature between the two. The temperature of a system is proportional to the kinetic energy of its particles, and the energy flow as heat between a system and its surroundings is essentially a transfer of kinetic energy due to the collisions between particles.

A forest fire results from many different exothermic reactions.

We can portray the change in the internal energy content of chemicals during a reaction using "reaction profile diagrams" like those shown in Figure 5.9. These diagrams illustrate the changes in the energy of the components of a reaction as it progresses. The horizontal axis in the diagram is known as the reaction coordinate, an arbitrary scale that denotes the progress of the reaction from start (left-hand side) to finish (right-hand side). The vertical axis represents the relative energy of the individual components of the system. By comparing the energy of the starting materials with the products, we can determine the direction of the flow of energy in the reaction (either from the system to the surroundings or from the surroundings to the system.)

From the diagram, we note that the curve of the reaction rises to a peak as the reaction proceeds. This indicates an input of energy needed to drive the reaction toward products. Virtually all reactions begin with an input of energy, known as the **activation energy**. This energy may be supplied when the overall kinetic energy of the moving particles is converted into internal potential and kinetic energy. The chemicals, after colliding, lose energy as they settle into the new chemical arrangement of the products.

A chemical reaction that releases energy as heat to the surroundings, like that shown in Figure 5.10, is known as an **exothermic reaction**. Energy, as heat, flows out of the system and into the surroundings. A forest fire includes all manner of

FIGURE 5.9

Reaction profile diagrams. The exothermic reaction on the left involves a loss of heat. The endothermic reaction on the right requires the absorption of heat.

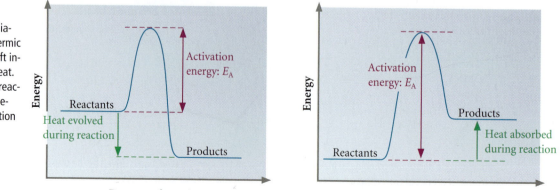

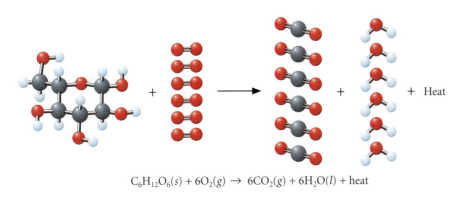

$$C_6H_{12}O_6(s) + 6O_2(g) \rightarrow 6CO_2(g) + 6H_2O(l) + \text{heat}$$

The overall equation describing cellular respiration.

FIGURE 5.10

An exothermic process. During chemical reactions, energy is transferred between the chemical system and its surroundings. Exothermic reactions release energy from the system into the surroundings.

exothermic reactions, releasing torrents of energy as heat to the surroundings. Cellular respiration, a complex process that we can summarize with the above equation, is also exothermic. Note that in an exothermic reaction, we can think of heat as a product of the reaction.

A chemical reaction that absorbs energy as heat is known as an **endothermic reaction**. In this process, energy as heat flows into the system from the surroundings, as illustrated in the melting of ice cream (Figure 5.11). In an endothermic reaction, such as the formation of oxygen and hydrogen from water, we can think of heat as a reactant.

Many ammonium salts, such as ammonium nitrate, endothermically dissolve in water, absorbing energy as heat from the surroundings. Chemical "cold packs" that are used to cool the injured knee or elbow of an athlete use this concept.

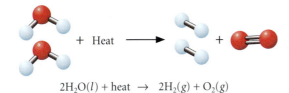

$$2H_2O(l) + \text{heat} \rightarrow 2H_2(g) + O_2(g)$$

An endothermic reaction requires the addition of heat. In addition to boiling water, the addition of heat can, under certain conditions, induce a chemical reaction that produces hydrogen and oxygen.

Application

FIGURE 5.11

An endothermic process. Some chemical reactions require the transfer of energy from the surroundings to the system. This is an endothermic process.

HERE'S WHAT WE KNOW SO FAR

- There are three natural forces: the gravitational, electroweak, and strong nuclear forces.

- Potential energy is the energy stored within a system.

- Kinetic energy is the energy associated with motion.

- The law of conservation of energy states that energy can be neither created nor destroyed. Instead, it just moves between the system and the surroundings and can be transformed from one form to another.

- Work and heat are two ways in which energy can move between the system and the surroundings.

- The flow of energy as heat from the system to the surroundings involves an exothermic process. An endothermic process involves the flow of energy as heat from the surroundings to the system.

- Changes in the energy of a reaction can be studied by examining a reaction profile diagram.

Cold packs use endothermic reactions.

A Closer Look at Work

A system doesn't contain work. Rather, we can calculate the amount of work that a system *does* in moving an object through a distance. Mathematically, we can define work as the product of the applied force and the distance the object moved.

$$\text{Work } (w) = \text{force } (F) \times \text{distance } (d)$$

To relate this equation to a chemical system, as is done by the rocket scientist, the force exerted by an expanding gas is directly related to the product of the pressure of a system and the area in which it is expanding. Pressure, as we'll discover in Chapter 11, is the force exerted by a gas per unit area on the walls of its container.

$$\text{Force } (F) = \text{pressure } (P) \times \text{area } (A)$$

Substituting this equation into that for the calculation of the amount of work, we get

$$\text{Work } (w) = P \times A \times d$$

This equation can be simplified by considering the product of the area (A) multiplied by the height (h), a measure of distance. As shown in Figure 5.12, the base of the cylinder in the initial figure has a known area (A). Moreover, the cylinder has a known height (h), the distance from the bottom to the top of the cylinder. Therefore, the gas inside the cylinder has a known volume (V), because $V = A \times h$. In the final state in Figure 5.12, the piston has been displaced by a known distance (Δh). *We use the Greek letter Δ (delta) to represent the change in some quantity.* Therefore, Δh means "the change in height." The resulting three-dimensional measurement is the change in the volume of the system (ΔV). Specifically, for this case, ΔV is the final volume of the system minus the initial volume of the system: $\Delta V = V_{\text{final}} - V_{\text{initial}}$. *In other words, $A \times \Delta h$ for a reaction is best represented by ΔV.*

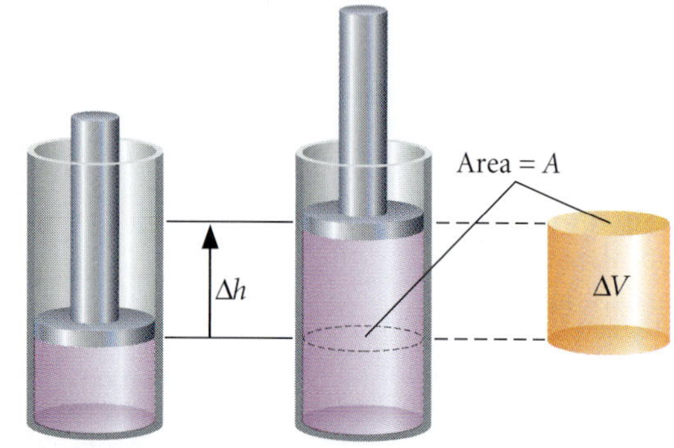

$$\text{Work } (w) = \text{pressure } (P) \times \text{change in volume } (\Delta V)$$

Our final modification to the equation that defines work is an adjustment to account for the loss of energy when a system does work and the gain of energy when a system has work done on it. We add a negative sign to the equation to illustrate this fact:

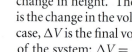

This equation indicates that *when the volume of a system expands (i.e., ΔV is positive), work is done by the system, and, as a result, energy is lost from the system* ($w_{\text{system}} = $ "$-$"). Conversely, *when the volume of a system contracts (i.e., ΔV is negative), the system gains energy, and work is done on the system* ($w_{\text{system}} = $ "$+$"). Because pressure is recorded in atmospheres and volume in liters, the units for work (energy) are in L·atm. This is mathematically related to the SI unit of energy known as the joule (J), in honor of the English physicist James Prescott Joule (1818–1889). We will derive this relationship later, but for now we can assume that 1 L·atm = 101.3 J.

FIGURE 5.12

Gas expansion does work. The initial volume of the gas increases by an amount ΔV. When the pressure remains constant, this becomes $P\Delta V$, or work.

(a) Initial state   (b) Final state

Area = A

Δh

ΔV

EXERCISE 5.2   **Work, Work, and More Work**

Calculate, in L·atm and joules, the work associated with the contraction of a gas from 75.0 L to 30.0 L at a constant external pressure of 6.20 atm. (1 L·atm = 101.3 J)

First Thoughts

The road signs concerning work can be confusing. When a system *expands,* work is done *by* the system ($w_{\text{system}} = $ "$-$") *on* the surroundings ($w_{\text{surroundings}} = $ "$+$").

Conversely, when the system *contracts,* the surroundings do work *on* the system (w_{system} = "+"). In this case, the work (from the standpoint of the system) is positive because the gas is being compressed. In other words, *work is being done on the system.*

Solution

$$\Delta V = \text{change in volume} = V_{final} - V_{initial} = 30.0\ L - 75.0\ L = -45.0\ L$$

$$w = -P\Delta V = -6.20\ atm \times (-45.0\ L) = +279\ L \cdot atm$$

converting this to joules, we get

$$w = +279\ L \cdot atm \times \frac{101.3\ J}{L \cdot atm} = +2.83 \times 10^4\ J = 28.3\ kJ$$

Further Insights

The sign on the value of the work term is important to us. Not only does it reflect how much the internal energy of the system changes, but it also indicates the direction in which it changes. Mathematically, the sign was determined by the direction of the change in the volume. In our example, the volume of the system is reduced, so the work done *on the system* is positive.

PRACTICE 5.2

Calculate, in L·atm and joules, the work associated with the expansion of a gas from 30 L to 300 L at a constant external pressure of 1.0 atm.

See Problems 9, 13d, and 14d.

Internal Energy

The reaction of hydrogen and oxygen in the shuttle's main engines causes energy transfer as heat and as work. The clearest result of the work done by this system is that the shuttle rises upward through Earth's atmosphere. The reaction also releases energy that heats the internal parts of the craft, along with many molecules in the atmosphere that come into contact with the hot exhaust vapors from the reaction.

The total energy of the space shuttle, just as in any other system, is known as the **internal energy (U)** of the system. The internal energy is made up of the sum of the system's kinetic and potential energies. Unfortunately, we cannot calculate the absolute value for the internal energy of a system, because it is extremely difficult to account for all the individual energies that give rise to the kinetic and potential energies. However, because the *change* in the internal energy (ΔU) in a chemical process is measurable—if we can calculate the exchange of energy as heat (q) within, and the work done (w) by the system—we can determine the energy change within the system. This is the **first law of thermodynamics:**

$$\boxed{\Delta U = q + w}$$

A word of caution: We must take care with the way in which we attribute positive or negative values to q and w. If we are looking at things from the point of view of the system, the loss of heat from the system to the surroundings will carry a negative sign (q_{system} = "−"). Similarly, if the system does work on the surroundings, w_{system} will have a negative sign (w_{system} = "−"). Conversely, the flow of energy as heat (q) *into* the system will be a positive value (q = "+"), and if work (w) is done *on* the system by the surroundings, w_{system} will be positive (w_{system} = "+").

Because energy can be neither created nor destroyed in a process, the total amount of energy transferred as heat and work to or from the system is exactly equal and opposite to that transferred from or to the surroundings.

$$\Delta U_{system} = -\Delta U_{surroundings}$$

The reaction of hydrogen and oxygen in the main shuttle engine produces heat and does work on the surroundings.

Video Lesson: The First Law of Thermodynamics

Video Lesson: Work

Video Lesson: Heat

Visualization: Work vs. Energy Flow

Reorganizing this equation by adding the change in internal energy of the surroundings to both sides, we get

$$\Delta U_{system} + \Delta U_{surroundings} = 0$$

Mathematically, this is a restatement of the first law of thermodynamics. If we know how much energy the system loses, we automatically know that the surroundings will gain that same amount. If we know how much energy the system gains, we know that the surroundings must have lost that same amount. The total energy change in the universe (system and surroundings) is zero. Energy is conserved. The bottom line is that *the law of conservation of energy is also the first law of thermodynamics.*

EXERCISE 5.3   **The First Law of Thermodynamics**

Calculate the change in the energy of a system if 51.8 J of work is done *by* the system with an associated heat loss of 12.3 J.

First Thoughts

We must pay particular attention to the sign conventions for heat and work in this problem. In this case, work is done by the system. Is this work positive or negative? A useful system to keep in mind is you! That is, when you do work—by running, dancing, or even moving your textbooks from one class to the next—you are using energy. After the process of moving your body or your books, you have less energy than you had before, so the work has a "−" sign. Similarly, the 51.8 J of work done by the system means that its energy change, w_{system}, is −51.8 J. The energy loss as heat *by the system* (such as that accompanying your run as your body tries to stay cool) also has a "−" sign, so $q_{system} = -12.3$ J.

Solution

From the standpoint of the system,

$$\Delta U = q + w$$
$$\Delta U = -12.3 \text{ J} + (-51.8 \text{ J}) = -64.1 \text{ J}$$

Further Insights

We want to reinforce the point that there is no such thing as heat or work within a system. In other words, *a system does not contain heat or work.* Rather, heat and work exist only as *types of energy transfer* between the system and the surroundings. Work is done by or on a system to move it through a distance. A system can transfer 64.1 J of energy as heat and work. It does not contain 64.1 J of heat and work.

PRACTICE 5.3

Calculate the change in the energy of a system if 84.7 J of work is done on the system, with an associated loss of energy as heat of 39.9 J.

See Problems 11–14, 19, 20, and 89.

A solid rocket motor is test fired. Studies such as this allow scientists to measure the temperature changes to the sensitive parts of the motor.

5.2 Keeping Track of Energy

In designing complex engineering equipment such as a space shuttle, an aircraft, or a chemical plant, the scientist or engineer needs to know how hot the parts that are exposed to exothermic chemical reactions will become. The scientist also needs to know that the shuttle's engine components, the parts within an aircraft engine, or the containers that are used to hold the hot products of chemical manufacturing processes won't melt. This raises a whole series of questions that need answering, such as "How can we quantify the amount of energy being produced?" "How much heat will a chemical reaction generate?" and "How quickly will the surrounding materials heat up, and to what temperature?" Similar questions are raised in all practical investigations of energy, such as studies into the amount of energy provided to us by food or the amount of energy needed to heat a building or to operate a train or an automobile.

The Units of Energy

We often talk about the chemist's shorthand and the importance of having a common set of units with which to communicate. We can start our examination of units by doing a calculation of kinetic energy. Let's suppose you are riding a bicycle along a level road at 4.40 meters per second (about 10 miles per hour) and that the total mass of both you and the bicycle is 85.0 kg. The kinetic energy of forward motion possessed by this system (you plus the bicycle) would be

$$\text{Kinetic energy} = \frac{1}{2}mv^2$$
$$= \frac{1}{2} \times 85.0 \text{ kg} \times \left(\frac{4.40 \text{ m}}{\text{s}}\right)^2$$
$$= 823 \text{ kg·m}^2\text{·s}^{-2}$$

This calculation yields the units of energy ($\text{kg·m}^2\text{·s}^{-2}$) in terms of SI base units. However, we typically refer to this collection of units as the **joule**, and this is the unit that we will use most often when we discuss energy exchanges in chemical reactions.

$$\boxed{1 \text{ joule (J)} = 1 \text{ kg·m}^2\text{·s}^{-2}}$$

The kinetic energy in our example is therefore 823 J.

From where did you and the bicycle get this energy? The immediate source was the series of chemical reactions in your muscles that allowed you to pedal the bike up to speed. These chemical reactions released energy that was extracted from your food. The original source of that energy is the energy of sunlight that fell on the plants that were used to make your food.

An alternative unit used in energy exchange is related to our diet; many of us use this unit as we "count calories" to assess the amount of energy available in the food we eat. We do this by looking at the nutritional information on the food package's label, such as the one shown in Figure 5.13. That label gives us a measure of the energy content of food in units of **Calories (Cal)**—note the capital C. In a slightly confusing habit of nomenclature, the Calorie (with a capital C) is equal

A bicycle and rider possess kinetic energy. The source of this kinetic energy is from the food the bicyclist consumed.

Video Lesson: Energy, Calories, and Nutrition

Application

CHEMICAL ENCOUNTERS: Energy in Foods

FIGURE 5.13

Food labels contain information about the chemical content of the food and the amount of energy it supplies to the body. This cereal supplies 180 Calories per 52-g serving.

to a **kilocalorie** (kcal); that is, it is 1000 times larger than the **calorie (cal)**—note the small c.

$$1 \text{ Cal} = 1000 \text{ cal} = 1 \text{ kcal}$$

The values listed on most foods, even though you will often see the spelling *calories* on the label, are given in terms of the "big" Calorie, rather than the "little" calorie. In any case, these alternative units for energy have entered everyday language, in talk of "calorie-conscious lifestyles," "high-calorie food," "low-calorie snacks," and so on. How much is a calorie? One calorie (with a small c) equals the amount of energy needed to raise the temperature of one gram of water by one degree Celsius (for example, from 14.5°C to 15.5°C). The calorie is not an SI unit, but it can readily be converted into joules:

$$1 \text{ cal} = 4.184 \text{ J}$$

The calorie is rather a small unit to use for measuring the energy available to our bodies from the food we eat. A typical slice of bread, for example, can provide our bodies with approximately 80,000 calories of usable energy. This equals 80 kilocalories, or 80 Calories.

$$1 \text{ calorie (cal)} = 4.184 \text{ joules}$$
$$1 \text{ kilocalorie} = 4184 \text{ joules} = 4.184 \text{ kilojoules} = 1 \text{ Calorie}$$

In many countries, food energy is stated in kilojoules, as in the box of Zucaritas (Frosted Flakes) from Mexico that is shown in Figure 5.14.

FIGURE 5.14

A food nutrition label from a box of Zucaritas (Frosted Flakes) from Mexico.

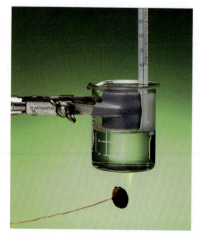

Peanuts contain a lot of calories.

EXERCISE 5.4 **The Power of Peanuts**

Peanuts are a compact source of energy coming from proteins, fats, and carbohydrates found within. When these chemicals from one brand of peanuts are combined with oxygen in the cells of the body, they release 625 Calories (note the capital C) of usable energy per 1.00×10^2 g of peanuts. The following two questions refer to the food energy in a 1.00-ounce (oz.) bag of peanuts. There are 28.4 g in 1.00 oz.

a. How many joules and how many kilojoules of energy are (ideally) available to the body from the bag of peanuts?

b. A 100-watt (W) light bulb (1 W = 1 J/s) requires 100 joules to run for 1 second. How many hours would the light bulb run if the energy from the peanuts were used to light the bulb?

Solution

a. There are 4.184 kJ in 1 Cal (remember that the capital C signals kilocalories) and 28.4 g of peanuts per bag. We can do our calculation as follows:

$$\text{Kilojoules} = \frac{625\text{ Cal}}{100\text{ g}} \times \frac{4.184\text{ kJ}}{1\text{ Cal}} \times \frac{28.4\text{ g}}{1\text{ bag}} = 743\text{ kJ/bag}$$

$$\text{Joules} = \frac{743\text{ kJ}}{\text{bag}} \times \frac{1000\text{ J}}{1\text{ kJ}} = 743{,}000\text{ J/bag}$$

b. We determined from part 1 that we have 743,000 J of energy available from the peanuts. We can solve for time in hours knowing that there are 100 J/s, or 360,000 J/h.

$$\text{Hours} = \frac{1\text{ s}}{100\text{ J}} \times \frac{1\text{ h}}{3600\text{ s}} \times 743{,}000\text{ J} = 2.06\text{ h} \approx 2\text{ h}$$

PRACTICE 5.4

Apples supply us with approximately 30 Cal per 100 g, and an average apple weighs about 200 g. How many apples would you need to eat to obtain the same amount of energy as is supplied by the bag of peanuts?

See Problems 21–26, 90, and 91.

Video Lesson: CIA
Demonstration: Cool Fire

5.3   Specific Heat Capacity and Heat Capacity

A scientist working with the space shuttle engines is often concerned with the heat generated during takeoff. The amount of heat given off by the burning rocket fuel will need to be absorbed by something if the rocket is to remain intact. The amount of insulation, and the type, will be of great importance. A chef faces the same concerns when deciding on a method by which to pick up a hot pan. Whether dealing with the space shuttle or the choice of metal cookware, the general question is **How great will be the change in temperature when a specific material absorbs or releases a certain amount of heat?**

Every substance has a particular **specific heat capacity (c)**, often shortened to **specific heat**, which is defined as the amount of energy as heat needed to raise the temperature of one gram of the substance by one degree Celsius (or one kelvin) when the pressure is constant (see Table 5.2 for some representative examples of specific heat). Substances with large values for their specific heat require more energy to raise their temperature than substances with small specific heat values. For example, 1 g of water requires more than four times the energy to raise its temperature 1.0°C than does aluminum.

The specific heat capacity of the water in your teapot at home is $\frac{4.184\text{ J}}{\text{g}\cdot{}^\circ\text{C}}$. This means that it will require 4.184 joules (or 1 calorie) to raise the temperature of 1 gram of the water by 1°C. Because the change in temperature in degrees Celsius is equal to the change in temperature in kelvins, the specific heat capacities can also be reported in units of $\frac{\text{J}}{\text{g}\cdot\text{K}}$. Given the units of the specific heat capacity, we can see how the heat required to raise the temperature of any compound can be determined using the following equation:

$$q = m \times c \times \Delta T$$

where, q = heat c = specific heat capacity
m = mass ΔT = change in temperature, in °C or K

The piece of pie stays hot, while the metal pie plate stays cool.

TABLE 5.2	Specific Heat Capacity of Selected Materials
Substance	**Specific Heat Capacity J / g·°C**
Water	4.184
Ethanol	2.460
Aluminum	0.902
Copper	0.385
Lead	0.128
Sulfur	0.706
Iron	0.449
Silver	0.235

This equation is useful in determining the amount of heat absorbed, or released, during a temperature change for a substance. The amount of heat needed to raise the temperature of any particular amount of a substance by 1°C is known as that substance's **heat capacity (C)**, in units of $\frac{J}{°C}$. The heat required to raise the temperature of this substance by a given amount is

$$q = C \times \Delta T$$
$$J = \frac{J}{°C} \times °C$$

Let's consider our teapot of water to help illustrate the use of these equations. Assume that we determine the heat capacity of the entire quantity of water in your teapot, when it is filled to the maximum, to be 6.27 kilojoules per °C = $(6.27 \times 10^3$ J/°C$)$. We can use dimensional analysis to work out the mass of water in the teapot, remembering that the specific heat capacity of water is 4.184 J/g·°C.

$$\text{g water} = \frac{\text{g·°C}}{4.184 \text{ J}} \times \frac{6.27 \times 10^3 \text{ J}}{°C} = 1.50 \times 10^3 \text{ g}$$

We can also use this equation to determine the amount of heat required to change the temperature of a particular substance. For instance, how much energy as heat is required to raise the temperature of that teapot of water from a cold 5.00°C to the boiling point? (The change in temperature for the water will be $\Delta T = T_{\text{final}} - T_{\text{initial}} = 100.0°C - 5.00°C = 95.0°C$.) Note that in finding our answer, we will not determine the energy needed to vaporize the water. Instead, we will calculate the heat required to arrive at a teapot full of 100.0°C water.

The required energy as heat is $q = C \times \Delta T$. Thus

$$\frac{6.27 \times 10^3 \text{ J}}{°C} \times 95.0°C = 5.96 \times 10^5 \text{ J} = 596 \text{ kJ}$$
$$1000 \text{ J} = 1 \text{ kJ}$$

Note how the dimensions work out to give you the proper units.

Another useful quantity that relates the heat capacity of a substance is the **molar heat capacity**, which is the heat capacity of one mole of the substance, in $\frac{J}{\text{mol·°C}}$. Knowing the specific heat of water, we can calculate the molar heat capacity of water using dimensional analysis:

$$\frac{4.184 \text{ J}}{\text{g·°C}} \times \frac{18.0 \text{ g H}_2\text{O}}{\text{mol H}_2\text{O}} = \frac{75.3 \text{ J}}{\text{mol·°C}}$$

Calorimetry

The thermochemist (a chemist who studies energy transfers in chemical reactions) uses the information about specific heat capacity to determine the amount of energy that is either gained or released by reactions. For example, the thermochemist could provide information about the reaction that takes place in the main engine of the space shuttle:

$$2H_2(g) + O_2(g) \rightarrow 2H_2O(g)$$

The scientist could provide the automobile designers with information about a reaction that takes place in a gasoline engine:

$$2C_8H_{18}(l) + 25O_2(g) \rightarrow 16CO_2(g) + 18H_2O(g)$$

How does he or she accomplish this task? By performing these reactions under carefully controlled conditions, and trapping and measuring the energy as heat given off or absorbed, the thermochemist can obtain information about energy transfers. This procedure is known as **calorimetry**, and the apparatus in which it is performed is called a **calorimeter**. To illustrate the key principles, we can construct a calorimeter from two Styrofoam cups as shown in Figure 5.15. The cups act as insulation to (ideally) prevent energy exchange between the contents of the inner cup and the rest of the universe.

To understand the process, we can begin with a hot piece of iron (the system) that we add to water (the surroundings), *both within Styrofoam cups*. Everything outside the cups consititutes the rest of the universe. While crude looking, this apparatus does a pretty good job of measuring the energy as heat transferred from the system (the piece of iron, in this case) to the surroundings (the water).

Within the *perfect* Styrofoam calorimeter, *the energy as heat lost by the system equals that gained by the surroundings.* Mathematically, this is represented by

$$q_{system} = -q_{surroundings}$$

$$m_{system} \times c_{system} \times \Delta T_{system} = -m_{surround} \times c_{surround} \times \Delta T_{surround}$$

After all of the energy as heat has been transferred, *the final temperatures of the system (iron) and that of the surroundings (water) within the calorimeter are identical.* For example, let's determine the change in temperature of a water bath when a 155-g piece of iron at 95.0°C is added to 1.000 kg of water at 25.0°C. We keep in mind that the energy as heat lost by the system and that gained by the surroundings are equal and opposite in magnitude; $q_{system} = -q_{surroundings}$. Here are the data.

The Block of Iron

$m_{system} = 155$ g of iron

$c_{system} = \dfrac{0.449 \text{ J}}{\text{g} \cdot °\text{C}}$ (from Table 5.2)

$\Delta T_{system} = ?$ We do not know the final temperature of the system, but we know that it will be less than 95.0°C, because the surrounding water is cooler than the piece of iron. Let's call the change in temperature, ΔT_{system}, "$T_f - 95.0°C$." The change in temperature is "$-$" because the final temperature is less than the initial temperature.

The Water

$m_{surround} = 1.000 \times 10^3$ g of water

$c_{surround} = \dfrac{4.184 \text{ J}}{\text{g} \cdot °\text{C}}$

$\Delta T_{surround} = ?$ We do not know the final temperature of the surroundings, but we know that it will be greater than 25.0°C, because the piece of iron (the system) is hotter than the water (surroundings,) and will therefore transfer energy as heat into the water. Let's call the change in temperature, $\Delta T_{surround}$, "$T_f - 25.0°C$." The change in temperature is "$+$" because the final temperature is greater than the initial temperature.

Thermometer

Stirrer

FIGURE 5.15

A Styrofoam calorimeter. The inner cup of the calorimeter contains both the system and the surroundings. The outer cup provides additional insulation to keep the surroundings within the set-up.

Video Lesson: Constant Pressure Calorimetry

We can now set the equations for the energy exchanges as heat between the system and the surroundings equal to each other (but opposite in sign!) and solve:

$$m_{system} \times c_{system} \times \Delta T_{system} = -m_{surround} \times c_{surround} \times \Delta T_{surround}$$

$$155 \text{ g} \times \frac{0.449 \text{ J}}{\text{g·°C}} \times (T_f - 95.0\text{°C}) = -1.000 \times 10^3 \text{ g} \times \frac{4.184 \text{ J}}{\text{g·°C}} \times (T_f - 25.0\text{°C})$$

$$\frac{69.595 \text{ J}}{\text{°C}} \times (T_f - 95.0\text{°C}) = \frac{-4184 \text{ J}}{\text{°C}} \times (T_f - 25.0\text{°C})$$

$$69.595 T_f - 6611.525 = -4184 T_f + 104600$$

$$4253.595 T_f = 111211.525$$

$$T_f = 26.145 = 26.1\text{°C}$$

The final temperature of the water (and the piece of iron) will be 26.1°C.

However, our "perfect" Styrofoam calorimeter isn't perfect. It still allows some energy as work to be transferred from the system to the surroundings. In addition, the calorimeter itself can participate in the transfer of energy as heat, distorting the calculations. We can address these problems and eliminate small errors in our measurements if we arrange things so that no work is done by the system or on the system ($w = 0$). In cases like this, the heat for any reaction (q) will be equal to the total change in the energy of the system, because the equation $\Delta U = q + w$ will simplify to $\Delta U = q$. We can achieve this using a technique called **constant-volume calorimetry**, in which the reacting system is sealed within a steel chamber of fixed volume, called a **bomb calorimeter**, shown in Figure 5.16. When a combustion reaction is ignited inside a bomb calorimeter, heat is transferred between the reaction and the surrounding water and calorimeter chamber. We can then calculate the heat released by the reaction.

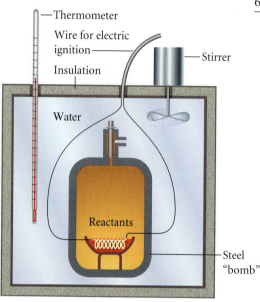

Thermometer
Wire for electric ignition
Insulation
Stirrer
Water
Reactants
Steel "bomb"

FIGURE 5.16

A bomb calorimeter. The combustion reaction under study is conducted within the steel bomb submerged in the water bath. The heat given off by the reaction is measured by the change in the temperature of the water bath.

Video Lesson: Bomb Calorimetry (Constant Volume)

EXERCISE 5.5 **Calculations with a Calorimeter**

Each bomb calorimeter is different, but its own heat capacity can be determined experimentally using a substance that releases a known amount of energy. Once calibrated in this way, the calorimeter can be used to determine the heat output of other chemicals.

a. Glucose ($C_6H_{12}O_6$), also known as "blood sugar," is the main sugar that serves to transport chemical energy through the blood and distribute it to the body's cells. Glucose is known to release $\dfrac{2.80 \times 10^3 \text{ kJ}}{\text{mol}}$ when combined with excess oxygen at 298 K (25°C). A sample of glucose weighing 5.00 g was burned with excess oxygen in a bomb calorimeter. The temperature of the calorimeter rose by 2.40°C. Calculate the heat capacity of the calorimeter in joules per degree Celsius.

b. Propane gas has the formula C_3H_8 and can be used as a source of heat for cooking and domestic heating. For storage purposes, it is liquefied and stored in canisters, often under the name of liquefied petroleum gas (LPG). A 4.409-g sample of propane was burned with excess oxygen in the bomb calorimeter calibrated in part 1. The temperature of the calorimeter increased by 6.85°C. Calculate how much energy as heat is released per mole of propane burned under these conditions.

c. If the energy released from 1 mole of propane determined in part 2 were used to heat 90.0 kg (9.00×10^4 g) of water originally at 30.0°C, what would be the final temperature of the water? The specific heat capacity of water is $\dfrac{4.184 \text{ J}}{\text{g} \cdot °\text{C}}$. (Remember to convert the energy released per mole of propane from kilojoules to joules!)

Solution

a. We first need to calculate the number of moles of glucose ($C_6H_{12}O_6$) in the 5.00 g used. The molar mass of glucose is 180.0 g/mol. We calculate the number of moles used, and then use this to calculate the energy released per degree Celsius, as follows:

$$\text{Moles } C_6H_{12}O_6 = 5.00 \text{ g } C_6H_{12}O_6 \times \frac{1 \text{ mol } C_6H_{12}O_6}{180.0 \text{ g } C_6H_{12}O_6} = 0.0278 \text{ mol}$$

$$\text{Heat capacity of calorimeter} = \frac{2.80 \times 10^3 \text{ kJ}}{1 \text{ mol}} \times 0.0278 \text{ mol} \times \frac{1}{2.40°C} = \frac{32.4 \text{ kJ}}{°C}$$

b. The molar mass of propane (C_3H_8) is 44.09 g/mol. Knowing this enables us to calculate the number of moles of propane burned and, therefore, the amount of energy released. We'll need to use the heat capacity of the calorimeter to correct for the effect of the calorimeter:

$$\text{Moles } C_3H_8 = 4.409 \text{ g } C_3H_8 \times \frac{1 \text{ mol } C_3H_8}{44.09 \text{ g } C_3H_8} = 0.1000 \text{ mol } C_3H_8$$

$$\text{Energy change per mole } C_3H_8 = \frac{32.4 \text{ kJ}}{°C} \times 6.85°C \times \frac{1}{0.1000 \text{ mol } C_3H_8}$$

$$= \frac{-2.22 \times 10^3 \text{ kJ } C_3H_8}{\text{mol } C_3H_8}$$

We add a negative sign to the value because energy is *released* from this reaction; that is, the reaction is exothermic.

c. Remember that $q = c \times m \times \Delta T$. This rearranges to

$$\Delta T = \frac{q}{c \times m}$$

$$\Delta T = \frac{2.22 \times 10^6 \text{ J } C_3H_8}{\dfrac{4.184 \text{ J}}{\text{g} \cdot °\text{C}} \times (9.00 \times 10^4 \text{ g})} = 5.90°C$$

The temperature of the water will rise to nearly 36°C in a total of 90.0 L (about 24 gal) of water!

PRACTICE 5.5

A sample of benzoic acid (C_6H_5COOH) weighing 2.442 g was reacted with excess O_2 in a bomb calorimeter. The temperature rose from 26.34°C to 39.20°C. The heat capacity of the calorimeter was $\dfrac{5.02 \text{ kJ}}{°C}$. Calculate the energy released in kilojoules per mole as heat for the reaction.

See Problems 27, 28, 31, 32, 37–46, 51, 52, 95, 96, and 98.

- Changes in the internal energy of a system are determined by the sum of the work and the energy change as heat for the process. The signs on these terms are vital, because they define whether energy is transferred to the system or to the surroundings.

- The heat capacity of a substance reflects its ability to absorb heat. A large heat capacity indicates that the addition of a large amount of energy is required in order to change the temperature of the substance.

- A calorimeter can be used to measure the amount of heat transfer between the system and the surroundings. We may use the equation $m_{system} \times c_{system} \times \Delta T_{system} = -m_{surround} \times c_{surround} \times \Delta T_{surround}$ to determine the specific situation of the heat transfer.

Video Lesson: Heats of Reaction: Enthalpy

5.4 Enthalpy

Many of the reactions we do in the laboratory do not occur under constant-volume conditions. Any gases that are generated are often free to expand outward into the atmosphere. These reactions occur under constant-pressure conditions, because the release of gas or any other expansion of volume will occur until the pressure of the products of the reaction becomes equal to the atmospheric pressure.

Under constant-pressure conditions, we cannot use the simple relationship $\Delta U = q$ but instead have to work with the slightly more complex $\Delta U = q + w$ to take into account any work done by the system or done on the system. When a reaction occurs under constant-pressure conditions, the only type of work the system will be able to do on the surroundings is called "pressure–volume work" (see Section 5.1), such as the work done by the system when a released gas is allowed to expand. Under these conditions of constant pressure, the work done by the system is equal to $-P \times \Delta V$ (pressure multiplied by the change in volume), so the equation $\Delta U = q + w$ can be written as

$$\Delta U = q_p - P\Delta V$$

where q_p is the **heat of reaction** at constant pressure. We can rearrange that equation to get

$$q_p = \Delta U + P\Delta V$$

We introduce a new term, **enthalpy**, which is symbolized by H. Enthalpy is measured in the units of energy (joules) and is defined as the sum of the internal energy and the pressure–volume product of a system:

$$H = U + PV$$

A change in enthalpy (ΔH) can be defined as $\Delta H = \Delta U + \Delta(PV)$ and is equal to q_p.

Moreover, if the pressure of the system is constant, PV can change only as a consequence of changes in volume, so $\Delta(PV)$ is equal to $P\Delta V$. Under these constant-pressure conditions, the definition of the change in enthalpy becomes exactly the same as the definition for the heat of reaction at constant pressure, namely q_p:

At constant pressure: $q_p = \Delta U + P\Delta V$

And $\Delta H = \Delta U + P\Delta V$

So $\boldsymbol{\Delta H = q_p}$

Heats of reaction measured under constant-pressure conditions are known as changes in enthalpy. It is important to remember that this unfamiliar term really just refers to the familiar idea of energy exchange as heat for a reaction that proceeds under constant-pressure conditions.

In situations where only very small amounts of work are done by a system or on a system, the value of w is very small compared to q_p. In these situations, the easily measured heat of reaction (q_p), which also equals the enthalpy change (ΔH), is approximately equal to the total energy change of the system ΔU_{system}. This is useful when studying the chemistry of living things. For instance, most biochemical reactions occur in body fluids in which there are negligible changes in volume and, therefore, negligible contribution to the energy change of the system from work. However, if a chemical reaction produces a gas, then the work component is not necessarily negligible because the change in volume becomes large.

The most interesting and useful value to us when we are studying energy and chemistry is the total energy change (ΔU) of the chemicals in the system as they react. This is not always easy to measure directly, so one of the most significant facts about enthalpy change (ΔH) values is that they provide a readily measured approximation to the ΔU values in which we are really interested.

In reactions that involve gases, ΔH does not equal ΔU because the reaction does work on the surroundings. Explosions are an extreme example of reactions that generate gases and do work on the surroundings.

Standard Enthalpies of Reaction

Comparing changes in enthalpies for reactions is a tricky business. Not only do the enthalpies need to be measured at the same temperature, but to be meaningful, they also require the same conditions. Comparisons are often made with heats of reaction obtained when all of the reactants and products are in their **standard states**, as illustrated in Figure 5.17. Then the enthalpy of the reaction becomes known as a **standard enthalpy of reaction ($\Delta_{rxn}H°$)**. What is the standard state of a reactant or product? The *most commonly used* standard states for thermodynamic work are as follows:

■ For a pure solid, liquid, or gas, the standard state is the state of the substance at a pressure of exactly 1 atmosphere (1 atm), which equals 101,325 Pa. (IUPAC has adopted 100,000 Pa, known as 1 bar, as the standard pressure, but 1 atm is still in widespread use.)

■ For any substance in solution, the standard state is at a concentration of exactly 1 molar.*

We indicate that a thermodynamic value has been determined under standard conditions by using the degree sign (°), so a standard enthalpy of reaction would be indicated as $\Delta_{rxn}H°$.

Any substance subjected to these standard conditions is said to be in its standard state. For example, water is present all around us in three main forms: as the liquid water that runs from our taps, as the water vapor in the air, and as the solid water such as the ice in our freezers. The *standard state of water*, however, is the liquid form in which pure water appears at 1 atm of pressure. Most of the water around us is not "pure water" because it has other chemicals dissolved in it. To be in its standard state, a substance must be pure. The **reference form** of an element is *the most stable form of the element at standard conditions*. For the element oxygen at 1 atm and 25°C, the reference form is O_2, rather than the less stable allotrope O_3.

Ethanol
$C_2H_6O(l)$

Chlorine
$Cl_2(g)$

Bromine
$Br_2(g, l)$

Sodium chloride
$NaCl(s)$

Glucose
$C_6H_{12}O_6(s)$

FIGURE 5.17

Examples of some compounds in their standard states.

*Standard states for thermodynamic properties are often tabulated at 25°C. However, the definition for a substance in its standard state does *not* require the temperature to be 25°C. For example, one could calculate the enthalpy of reaction at 350°C, although to do so, one would need access to thermodynamic values tabulated at this temperature.

Water as gas, liquid, and solid.

Many different standard enthalpies of reactions have been defined. Common ones include the standard enthalpy of formation and the standard enthalpy of combustion. The **standard enthalpy of formation, $\Delta_f H°$** (also known as the **standard heat of formation**) of a substance is the enthalpy change for the formation of 1 mole of the substance in its standard state from its elements in their *reference forms*. For example, the standard enthalpy of formation of carbon dioxide from carbon and oxygen in their reference forms, which for carbon is graphite, can be summarized as follows:

This is an exothermic reaction.

$$C(s) + O_2(g) \rightarrow CO_2(g) \qquad \Delta_f H° = -394 \text{ kJ/mol}$$

We have to show just 1 mole of the product, because that value is embodied in the definition. This can force us to use fractional amounts of moles of some of the reactants, as in the next example, which indicates the standard enthalpy of formation of water:

$$H_2(g) + \frac{1}{2}O_2(g) \rightarrow H_2O(l) \qquad \Delta_f H° = -286 \text{ kJ/mol}$$

We must show the oxygen in its standard state as $O_2(g)$. We cannot simply use O (because O is not the reference form of oxygen), so we must indicate half a mole of O_2. Keep in mind that when you are writing standard enthalpy equations, it is acceptable to leave equations with fractional stoichiometric values, rather than multiplying by some common factor to convert all the fractions to whole numbers.

EXERCISE 5.6   To the Moon

The combustion of gaseous hydrazine (N_2H_4) is a reaction that provides thrust for rockets. This reaction is also used on the space shuttle for minor attitude changes when the shuttle is in orbit. Write the equation that describes the standard enthalpy of formation for this compound. Given that for hydrazine $\Delta_f H° = +95.40$ kJ/mol, calculate the change in enthalpy accompanying the formation of 125 g of hydrazine from its elements in their reference forms.

First Thoughts

The equation that we prepare needs to describe the formation of just 1 mole of hydrazine. Because hydrazine contains nitrogen and hydrogen, we can show its formation from those elements in their reference forms. We can then convert the number of grams into moles and use that value to determine the enthalpy for the reaction.

Solution

The equation for the formation of gaseous hydrazine from its elements in their reference forms is

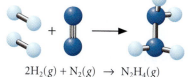

$$2H_2(g) + N_2(g) \rightarrow N_2H_4(g)$$

The number of moles of hydrazine can then be used to calculate the enthalpy.

$$125 \text{ g N}_2\text{H}_4 \times \frac{1 \text{ mol N}_2\text{H}_4}{32.05 \text{ g N}_2\text{H}_4} = 3.90 \text{ mol N}_2\text{H}_4$$

$$3.90 \text{ mol N}_2\text{H}_4 \times \frac{95.40 \text{ kJ}}{1 \text{ mol N}_2\text{H}_4} = 372 \text{ kJ}$$

Further Insights

The amounts of materials that we use in reactions determine the change in enthalpy of the reaction. Conversely, the enthalpy change in a reaction can be used to determine the amount of material consumed or produced in the reaction.

PRACTICE 5.6

Calculate the enthalpy change for the formation of 38.0 g of water. (Use the appendix to find the standard enthalpy change for water.)

See Problems 47–50.

Elements in their reference forms are the basic starting materials for all the reactions associated with standard enthalpies of formation. *The elements themselves are not formed from any simpler chemicals, so the standard enthalpies of formation of elements in their reference forms are all equal to zero.* This is to say that, using calcium as an example,

$$\text{Ca}(s) \rightarrow \text{Ca}(s) \qquad \Delta_f H° = 0 \text{ kJ/mol}$$

Note that $\Delta_f H° = 0$ kJ/mol does not mean that $U = 0$ kJ/mol. Also, recall that only one form of an element is its reference form. For example, $O_2(g)$ is the reference form for oxygen. When O_2 forms the allotrope ozone, O_3, in the atmosphere, $\Delta_f H°$ is *not* equal to 0 kJ/mol.

$$\tfrac{3}{2}O_2(g) \rightarrow O_3(g) \qquad \Delta_f H° = 143 \text{ kJ/mol}$$

The **standard enthalpy of combustion, $\Delta_c H°$** (also known as the **standard heat of combustion**) of a substance is the enthalpy change when 1 mole of the substance in its standard state is completely burned in oxygen gas. For example, consider the combustion of propane gas (C_3H_8), which we first discussed in Exercise 5.5.

$$C_3H_8(g) + 5O_2(g) \rightarrow 3CO_2(g) + 4H_2O(l) \qquad \Delta_c H° = -2202 \text{ kJ/mol}$$

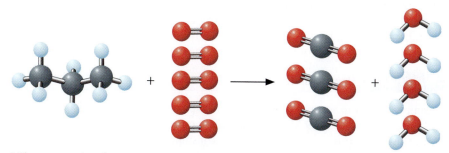

When we write the equation to illustrate the standard enthalpy of combustion, we must show exactly 1 mole of the reactant, propane, being burned. The coefficients must be adjusted to fit that requirement, even as the mole *ratios* of all reactants and products necessarily remain the same. Consider, as well, the equation for the standard enthalpy of combustion of carbon. Note that the equation is the same as the equation for the standard enthalpy of formation of carbon dioxide.

$$C(s) + O_2(g) \rightarrow CO_2(g) \qquad \Delta_cH° = -394 \text{ kJ/mol} = \Delta_fH° \text{ (CO}_2\text{)}$$

Manipulating Enthalpies

When we know the value for any standard enthalpy change of a reaction, we also automatically know the value for the enthalpy change of the reverse reaction. For example, converting carbon dioxide gas back into solid carbon and oxygen gas would be the reverse of the forward reaction, the combustion of carbon to form CO_2. It should therefore be accompanied by an enthalpy change that is equal to that of the forward reaction but opposite in sign, as shown in Figure 5.18. To help us understand what the signs on the enthalpy terms mean, we can consider this manipulation as though the enthalpy change were part of the equation. Let's examine the following equations and see how this applies to the end result. The enthalpy of combustion is negative, so it must be released from the reaction and, hence, appear on the same side as the products in this exothermic process.

$$C(s) + O_2(g) \rightarrow CO_2(g) + 394 \text{ kJ/mol}$$

The reverse of this reaction is

$$CO_2(g) + 394 \text{ kJ/mol} \rightarrow C(s) + O_2(g)$$

which shows that the enthalpy is gained by the reaction.

FIGURE 5.18

Negative enthalpy changes can be thought of as adding enthalpy to the product side of the equation. Reversing the direction of the equation changes the sign of the enthalpy change.

$$C(s) + O_2(g) \rightarrow CO_2(g) + \boxed{394 \text{ kJ/mol}} \qquad \Delta_cH° = \boxed{-394} \text{ kJ/mol}$$

$$\boxed{394 \text{ kJ/mol}} + CO_2(g) \rightarrow C(s) + O_2(g) \qquad \Delta H = \boxed{+394} \text{ kJ/mol}$$

In some cases, we will find it useful to multiply all of the equation coefficients by some number. Let's see what happens when we multiply the coefficients in the "combustion of carbon" equation by 2:

$$C(s) + O_2(g) \rightarrow CO_2(g) + 394 \text{ kJ/mol}$$

$$2C(s) + 2O_2(g) \rightarrow 2CO_2(g) + 2 \times (394 \text{ kJ/mol})$$

Note that although the mole *ratios* of the reactants and products remain the same, the enthalpy change is affected by this multiplication. Remembering that enthalpy as a product indicates a negative change in enthalpy, we note that ΔH of this reaction is now $(2 \times -394 \text{ kJ/mol}) = -788 \text{ kJ/mol}$.

This ability to reverse equations and automatically know the corresponding enthalpy change is a consequence of what we know as Hess's law, which is the subject of the next section. It enables us to determine the enthalpy changes of reactions that might be very difficult to actually perform.

EXERCISE 5.7   **How Much Heat?**

A camper wishes to warm a pot of water on a propane stove. She calculates that she'll need to produce 209 kJ to increase the temperature of her pot of water by 50°C. How much propane, in moles, will be consumed to produce this much heat?

$$C_3H_8(g) + 5O_2(g) \rightarrow 3CO_2(g) + 4H_2O(l) \qquad \Delta_cH° = -2202 \text{ kJ/mol}$$

First Thoughts

We must incorporate the enthalpy of combustion into the equation and determine the stoichiometric relationship between moles of propane ($C_3H_8(g)$) and the change in enthalpy. Accordingly, we can write

$$1 \text{ mol } C_3H_8(g) = -2202 \text{ kJ}$$

or

$$\frac{1 \text{ mol } C_3H_8}{2202 \text{ kJ}}$$

Solution

$$209 \text{ kJ} \times \frac{1 \text{ mol } C_3H_8}{2202 \text{ kJ}} = 0.0949 \text{ mol } C_3H_8(g)$$

Note that we drop the negative sign from the enthalpy change because the amount of heat released from the system is the amount of heat needed to warm the water.

Further Insights

The thermodynamic sign conventions can be a little misleading and confusing at first. In the present example, the quantity of enthalpy released (209 kJ) by the combustion reaction (the system) is equal to the quantity of enthalpy gained by the surroundings (209 kJ), assumed here to be the pot of water, so although the quantities are equal, the signs are opposite: $\Delta H_{system} = -209 \text{ kJ}$; $\Delta H_{surround} = +209 \text{ kJ}$.

PRACTICE 5.7

Calculate how many moles of the compound printed in boldface type will be needed to produce a change in enthalpy of 250 kJ for each of these reactions.

a. **$2NH_3(g)$** $+ 3N_2O(g) \rightarrow 4N_2(g) + 3H_2O(l)$ $\Delta H = -1012 \text{ kJ}$

b. $2N_2O(g) \rightarrow$ **$O_2(g)$** $+ 2N_2(g)$ $\Delta H = -164 \text{ kJ}$

See Problems 34, 35, 71b, 71c, 72b, 72c, 76, and 99b.

5.5 Hess's Law

The search for alternative fuel sources to supplement what we currently use is a high priority in all fields of transportation, including commercial aviation, personal automobile use, and even rocketry. The search for such fuels could involve the examination of lots of new reactions, some of which might never have been performed before. Fortunately, one of the features of enthalpy change values works to our benefit. We can calculate the energy or enthalpy changes of reactions without having to perform them. The reason for this ease of calculation lies in the fact that energy (U) and enthalpy (H) are state functions. *A* **state function** *is a property of a system that depends only on its present state, not on how it got there.* State functions are usually represented by an italic capital letter.

For an illustration of this idea, consider the different paths that the space shuttle might take to reach the international space station, orbiting at an altitude of 240 miles above sea level, as graphically shown in Figure 5.19. One shuttle mission may be launched directly up to a 240-mile-high orbit, ready to rendezvous with the space station. A later shuttle mission might first go into a higher orbit to release a satellite and then come back down to the 240-mile-high orbit of the space station. The eventual altitude of each shuttle is the same: 240 miles above sea level. In short, the final altitude is independent of the path taken. The distances traveled by the two shuttles in reaching their final altitude are very different, however, because one took the direct route and the other took a more circuitous route. Altitude is a state function; distance traveled is not.

Video Lesson: Hess's Law

Video Lesson: Enthalpies of Formation

Tutorial: Hess's Law

FIGURE 5.19

Two shuttles visiting the International Space Station will end up at the same altitude, regardless of the route taken. In this example, altitude is a state function; its value depends only on the initial and final states, not on the route taken.

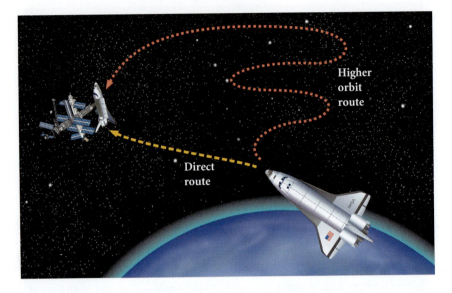

The change in the value of any state function depends *only* on the initial and final states between which it is changing. In the case of enthalpy,

$$\Delta H = H_{final} - H_{initial}$$

If a chemical reaction is carried out by two different chemical paths, the overall enthalpy *change* associated with each path will be the same. That basic principle is known as **Hess's law** and is summarized diagrammatically in Figure 5.20.

Enthalpy is a state function, which also justifies our earlier statement that the enthalpy *change* for any reaction will have the same value as the enthalpy change for the reverse reaction but will be opposite in sign. If the forward reaction has an enthalpy change of −100 joules, the reverse reaction will have an enthalpy change of +100 joules. This is summarized diagrammatically in Figure 5.21.

Visualization: Hess's Law

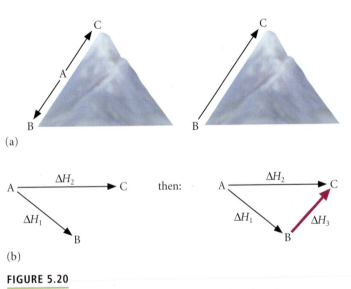

(a)

(b)

FIGURE 5.20

A diagrammatic summary of Hess's law. (a) If the distance down the mountain from A to B and the distance up to the top from A to C are known, the distance from B to C can be determined. (b) The enthalpy change for a series of reactions is analogous to this. If the enthalpy of reaction from A to B and from A to C is known, the enthalpy of reaction from B to C can be calculated.

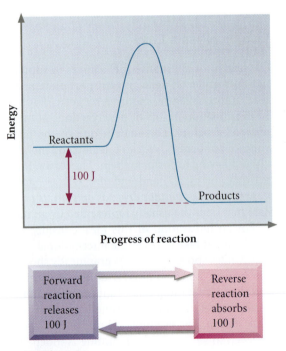

FIGURE 5.21

A forward reaction has an enthalpy that is equal, but opposite in sign, to that of its reverse reaction.

Using Hess's Law

On the Salt River-Pima Maricopa Indian Community in Arizona, methane (CH_4) collected from a 10-million-ton garbage landfill is used to fuel a "green" (environmentally benign) power plant that provides electricity for 2000 homes. Methane is a chemically very simple and convenient fuel, arising from the biological degradation of foodstuffs and other wastes in the landfill. It also makes up about 95% of the "natural gas" found above oil deposits, is burned to provide energy for domestic cooking and heating, and is as a source of heat in many industries.

Table 5.3 and Figure 5.22 list the important sources of methane emissions into the atmosphere. The data in the table suggest that it is not possible to accurately quantify the production of a gas that is emitted every day from so many sources. The numbers should be seen as rough projections of quantities that cannot be measured with great certainty. Nonetheless, harvesting some of this methane, as at the Salt River Project in Arizona, could represent an important addition to the world's energy supply.

Although methane found in natural gas is formed in a variety of different biological and chemical ways, it can be described by the following chemical equation:

$$C(s) + 2H_2(g) \rightarrow CH_4(g)$$

If we use Hess's law, we can calculate the standard enthalpy change for this reaction without actually performing the reaction. To complete the calculation, we need to know the standard enthalpy changes of other reactions that will allow us to consider forming methane *indirectly*. We use reference tables to obtain these reactions and their corresponding standard enthalpies. These values are

$$C(s) + O_2(g) \rightarrow CO_2(g) \qquad \Delta H° = -394 \text{ kJ/mol}$$

$$H_2(g) + \tfrac{1}{2}O_2(g) \rightarrow H_2O(l) \qquad \Delta H° = -286 \text{ kJ/mol}$$

$$CH_4(g) + 2O_2(g) \rightarrow CO_2(g) + 2H_2O(l) \qquad \Delta H° = -890 \text{ kJ/mol}$$

Application

CHEMICAL ENCOUNTERS: Focus on Methane

Methane gas can be harvested from landfills.

TABLE 5.3	Projected U.S. Methane Emissions in the Year 2020
Source	**Teragrams ($1 \text{ Tg} = 10^{12}$ g) of CH_4**
Landfills	7.0
Coal mining	5.5
Natural gas/oil drilling	7.0
Livestock manure	4.5
Livestock digestion releases ("enteric fermentation")	6.5
Other	0.5
TOTAL	31.0

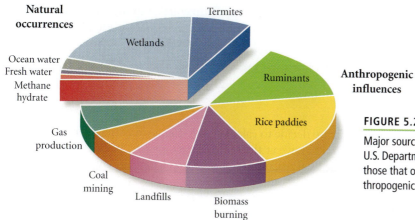

Natural occurrences

Termites

Wetlands

Ocean water

Fresh water

Methane hydrate

Ruminants

Anthropogenic influences

Rice paddies

Gas production

Coal mining

Landfills

Biomass burning

FIGURE 5.22

Major sources of atmospheric methane as reported by the U.S. Department of Energy. Sources can be broken down into those that occur naturally and those that occur through anthropogenic influences (that is, through human activity).

FIGURE 5.23

A diagrammatic representation of how Hess's law can be used to calculate the enthalpy of formation of methane.

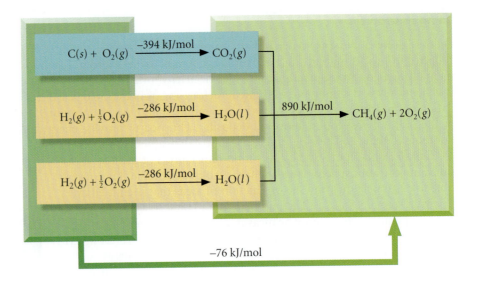

We arrange these reactions in such a way that they allow us to calculate the enthalpy of formation of methane. This is shown diagrammatically in Figure 5.23. Alternatively, we can rearrange the chemical reactions to mathematically arrive at the same answer. The underlying logic to Hess's law calculations is that we can *multiply, and/or reverse, and then combine* any chemical equations in whatever way will achieve the overall chemical change in which we are interested. For example, to calculate the enthalpy of formation of methane, we can manipulate the three combustion equations as follows:

$$C(s) + O_2(g) \rightarrow CO_2(g) \qquad \Delta H° = -394 \text{ kJ/mol}$$

$$2 \times [H_2(g) + \tfrac{1}{2}O_2(g) \rightarrow H_2O(l)] \qquad 2 \times [\Delta H° = -286 \text{ kJ/mol}]$$
$$= 2H_2(g) + O_2(g) \rightarrow 2H_2O(l) \qquad\qquad = -572 \text{ kJ}$$

$$-1 \times [CH_4(g) + 2O_2(g) \rightarrow CO_2(g) + 2H_2O(l)] \qquad -1 \times [\Delta H° = -890 \text{ kJ/mol}]$$
$$= CO_2(g) + 2H_2O(l) \rightarrow CH_4(g) + 2O_2(g) \qquad\qquad = +890 \text{ kJ}$$

Adding the three equations together, we calculate the overall enthalpy change:

$$C(s) + O_2(g) \rightarrow CO_2(g) \qquad \Delta H° = -394 \text{ kJ}$$
$$2H_2(g) + O_2(g) \rightarrow 2H_2O(l) \qquad \Delta H° = -572 \text{ kJ}$$
$$CO_2(g) + 2H_2O(l) \rightarrow CH_4(g) + 2O_2(g) \qquad \Delta H° = +890 \text{ kJ}$$

$$C + O_2 + 2H_2 + O_2 + CO_2 + 2H_2O \rightarrow CO_2 + 2H_2O + CH_4 + 2O_2$$
$$\Delta H° = -76 \text{ kJ}$$

When we cancel the chemicals that appear in equal amounts on both sides of the equation (O_2, CO_2, H_2O), this simplifies to the equation for the formation of methane:

$$C + \cancel{O_2} + 2H_2 + \cancel{O_2} + \cancel{CO_2} + \cancel{2H_2O} \rightarrow \cancel{CO_2} + \cancel{2H_2O} + CH_4 + \cancel{2O_2}$$
$$C(s) + 2H_2(g) \rightarrow CH_4(g)$$

$$\Delta H° = -76 \text{ kJ/mol}$$

Whichever way we decide to calculate the change in enthalpy for the reaction of interest, we should arrive at the same correct answer. In either case, we don't have to perform the final reaction to be able to calculate its enthalpy change.

EXERCISE 5.8   **Hess's Law**

Given the following data, calculate the value of ΔH for the reaction shown below.

$$P_4(s) + 10Cl_2(g) \rightarrow 4PCl_5(g) \qquad \Delta H = -2139 \text{ kJ}$$
$$PCl_3(g) + Cl_2(g) \rightarrow PCl_5(g) \qquad \Delta H = -155 \text{ kJ}$$
$$\overline{P_4(s) + 6Cl_2(g) \rightarrow 4PCl_3(g) \qquad \Delta H = \text{?? kJ}}$$

Solution

$$\boxed{1} \times [P_4(s) + 10\,Cl_2(g) \rightarrow 4PCl_5(g)] \qquad \boxed{1} \times \Delta H = -2139 \text{ kJ}$$
$$\boxed{-4} \times [PCl_3(g) + Cl_2(g) \rightarrow PCl_5(g)] \qquad \boxed{-4} \times \Delta H = +620 \text{ kJ}$$
$$\overline{P_4(s) + 6Cl_2(g) \rightarrow 4PCl_3(g) \qquad \Delta H = -2139 \text{ kJ} + 620 \text{ kJ} = \mathbf{-1519\,kJ}}$$

PRACTICE 5.8

Given the following data, calculate the value of ΔH for the reaction shown below.

$$6Fe(s) + 4O_2(g) \rightarrow 2Fe_3O_4(s) \qquad \Delta H = -1787 \text{ kJ}$$
$$2Fe_3O_4(s) + \tfrac{1}{2}O_2(g) \rightarrow 3Fe_2O_3(s) \qquad \Delta H = -186 \text{ kJ}$$
$$\overline{3Fe_2O_3(s) \rightarrow 6Fe(s) + \tfrac{9}{2}O_2(g) \qquad \Delta H = \text{?? kJ}}$$

See Problems 75–78, 81, and 82.

Reaction Enthalpies from Enthalpies of Formation

The energy needed to maneuver the space shuttle when it is in orbit is provided by the reaction between methylhydrazine (CH_3NHNH_2) and dinitrogen tetroxide (N_2O_4):

$$4CH_3NHNH_2(l) + 5N_2O_4(l) \rightarrow 4CO_2(g) + 9N_2(g) + 12H_2O(l)$$

One of the requirements for a chemical reaction that can be used as part of an effective propulsion system is that it have a large negative ΔH value, indicating that it releases a lot of energy. How can we determine the value of ΔH for this reaction?

One answer, based on our earlier discussion, is to perform the reaction in a calorimeter—something that we may not wish to do, especially if we are studying rocket fuel. Alternatively, we could use Hess's law to calculate the enthalpy of the reaction. However, it is often difficult to find all of the enthalpies that are needed to complete the circuitous route from reactants to products. Instead, there is another way to determine the enthalpy of the reaction. We can calculate ΔH of the reaction by means of a useful general rule about enthalpies of formation if other appropriate values of ΔH are known:

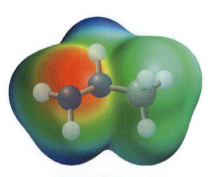

Methylhydrazine

> The standard enthalpy change for a reaction can be calculated by subtracting the sum of the enthalpies of formation of the reactants from the sum of the enthalpies of formation of the products.

This can be expressed mathematically as follows:

$$\Delta H° \text{ reaction} = \Sigma n_p \Delta_f H°(\text{products}) - \Sigma n_r \Delta_f H°(\text{reactants})$$

where the Greek symbol Σ (sigma) means "the sum of the following terms."

n_p = the number of moles of each product in the reaction equation

n_r = the number of moles of each reactant in the reaction equation

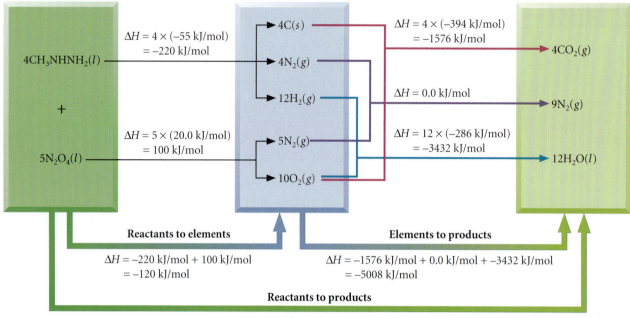

FIGURE 5.24

We can consider performing a reaction by taking the indirect route, via the elements in their reference forms. Then, by Hess's law, the reactions shown here can allow us to determine the enthalpy of the reaction.

Figure 5.24 clarifies why this procedure works.

$$\Delta_f H°(CO_2(g)) = -394 \text{ kJ/mol} \qquad \Delta_f H°(N_2(g)) = 0.0 \text{ kJ/mol}$$

$$\Delta_f H°(H_2O(l)) = -286 \text{ kJ/mol}$$

$$\Sigma n_p \Delta_f H°(\text{products}) = [4 \times -394] + [9 \times 0.0] + [12 \times -286] = -5008 \text{ kJ}$$

$$\Delta_f H°(CH_3NHNH_2(l)) = +55 \text{ kJ/mol} \qquad \Delta_f H°(N_2O_4(g)) = -20.0 \text{ kJ/mol}$$

$$\Sigma n_r \Delta_f H°(\text{reactants}) = [4 \times +55] + [5 \times -20.0] = +120 \text{ kJ}$$

$$\Delta H°\text{reaction} = \Sigma n_p \Delta_f H°(\text{products}) - \Sigma n_r \Delta_f H°(\text{reactants})$$

$$= (-5008 \text{ kJ}) - (120 \text{ kJ}) = -5128 \text{ kJ}$$

We find, as expected, that this reaction used to provide propulsion to maneuver the space shuttle in orbit has a relatively large negative ΔH value equal to -5128 kJ/mol.

EXERCISE 5.9 | **Thermite in Space**

One reaction that may prove useful for the welding and brazing that are necessary when building platforms in space is the "thermite reaction" between aluminum and iron oxide:

$$2Al(s) + Fe_2O_3(s) \rightarrow 2Fe(s) + Al_2O_3(s)$$

Calculate the standard enthalpy change of this reaction, given the following enthalpies of formation:

$$\Delta_f H°(Fe_2O_3) = -824 \text{ kJ/mol}; \qquad \Delta_f H°(Al_2O_3) = -1676 \text{ kJ/mol}$$

First Thoughts

Why don't reference tables provide heat of formation values for aluminum and iron? They are in their reference forms, and by definition, their heat of formation is zero. This means that the enthalpy change of the reaction (remember, it's a state

The thermite reaction in action. This photo shows Austrian railroad workers in 1910 using the thermite reaction. The molten iron produced by the reaction drips between the sections of rail and welds the track together.

function!) is equal to the heat of formation of the aluminum oxide (Al_2O_3) minus the heat of formation of the iron oxide (Fe_2O_3).

Solution

$$\Delta H° \text{reaction} = \Sigma n_p \Delta_f H°(\text{products}) - \Sigma n_r \Delta_f H°(\text{reactants})$$

$$\Delta H° \text{reaction} = (-1676 + 0) - (-824 + 0) = -852 \text{ kJ/mol}$$

Further Insights

In "manned" space flight, many other, less dramatic reactions occur some exothermic and some endothermic. In particular, the reaction of molecular hydrogen and oxygen gases on board the space shuttle to form water and usable electricity is exothermic, whereas many of the scientific experiments, such as growing plants, require the input of energy, rather than its release, to run.

Video Lesson: CIA Demonstration: The Thermite Reaction

Visualization: Thermite Reaction

PRACTICE 5.9

Calculate the enthalpy change for the reaction

$$4NH_3(g) + 3O_2(g) \rightarrow 2N_2(g) + 6H_2O(l)$$

given the following reactions and ΔH values:

a. $2NH_3(g) + 3N_2O(g) \rightarrow 4N_2(g) + 3H_2O(l)$ $\Delta H = -1012 \text{ kJ}$
b. $2N_2O(g) \rightarrow O_2(g) + 2N_2(g)$ $\Delta H = -164 \text{ kJ}$

See Problems 75–78, 81, and 82.

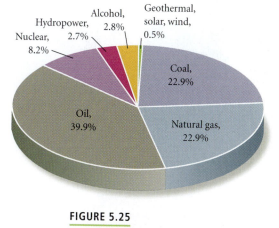

Smog is a serious urban pollutant in almost every major city in the world.

5.6 Energy Choices

Application

CHEMICAL ENCOUNTERS: Energy Choices

Humans have always had a variety of energy sources to choose from, and choices based on energy issues have always been important. In ancient times, people could keep warm by burning wood, moving south to sunnier climes, or minimizing energy losses from their bodies by wrapping themselves in animal skins and furs. Ancient people also learned to make use of wind energy to power great journeys by boat across seas and oceans. Nowadays, the energy choices facing us are much more complex, and the possible effects of making unwise choices are much greater.

The development of modern civilization was powered largely by burning things to release heat. This heat could keep people warm and could also be used to boil water, generating the steam to power steam engines and steam turbines. The engines and turbines were used to power machinery, trains, and ships and eventually to make electricity. The greatest leaps forward in technology came with the discovery of the fossil fuels—coal, oil, and natural gas—that could be harvested from the Earth in seemingly limitless quantities and burned with few obvious disadvantages, apart from the annoyance of a bit of smoke and the loss of lives in the process of extracting these natural resources. Today, the burning of fossil fuels provides about 70% of the energy needed to sustain a complex industrialized country such as the United States, as shown in Figure 5.25. However, we now know that the supplies of fossil fuels are not limitless and that burning them to provide energy also generates problems. The most significant problem might be the apparent warming of the planet, the "global warming" effect, caused in part by the carbon dioxide gas that is a by-product of burning these fuels.

Because fossil reserves are finite, the burning of these fuels causes environmental problems, and their processing and distribution have had a significant impact on the world's economic and political

FIGURE 5.25

U.S. Energy use in 2004.

The Hoover Dam harnesses water to generate electricity.

landscape. The alternative strategy of harnessing energy from nuclear processes was eagerly adopted in many countries in the latter half of the twentieth century. The source of this energy will be discussed fully in Chapter 21.

One of the most attractive options is to make increasing use of **renewable sources of energy**, which can be rapidly replaced by natural processes. This is in contrast to nonrenewable sources, particularly the fossil fuels: coal, oil, and natural gas. Staff at the U.S. National Renewable Energy Laboratory in Colorado have studied the feasibility of using renewable energy sources to meet all of America's energy needs. The United States is the largest single user of energy in the world, accounting for approximately 50% of global consumption. A feasible plan to meet all or most of U.S. energy needs with renewable sources of energy would encourage countries around the world to make a similar commitment.

According to the Renewable Energy Laboratory, the main renewable energy sources that could meet the needs of the United States are

- **Solar cells** (also known as **photovoltaic systems**, or PVs), which convert the energy of sunlight into electricity.

- **Solar thermal systems**, which covert the energy of sunlight into heat.

- **Biomass conversion**, which releases energy from the chemical conversion— often the burning—of plants and trees that can be grown as quickly as they are consumed.

Solar energy can be used to make electricity.

Windmills can convert wind into electricity.

Photovoltaic cells can be used to power almost anything.

- **Hydroelectric systems**, which use the power of falling water is used to turn turbines that generate electricity.

- **Wind power**, which uses turbines driven by the force of the wind to generate electricity.

- **Geothermal systems**, which involve drilling deep into the earth and exploiting the flow of heat from the interior of the earth out toward the surface.

Each of these systems could readily make a significant contribution to power generation in the United States. They already contribute 8%, as shown in Figure 5.25, but that could rise toward 50% or even 100% if social and political incentives were in place. One of the most interesting points made in the Renewable Energy Laboratory's report is summarized in Figure 5.26, which shows the total area that would have to be devoted to solar cells sufficient to generate all the energy required by the United States. It is a large area, equivalent to more than 10% of the area of the state of Nevada, but that is less than 25% of the area of the country that is currently paved over with roads and streets. So equipped, the United States could garner enough energy from the sun to meet all its needs.

Solar energy arrives free of charge, although providing the technology to capture, distribute, and use it entails significant costs. The energy is also very "clean" in the sense that its use does not release anywhere near the amount of pollutants released by the burning of fossil fuels. Some pollution would inevitably be associated with the manufacturing processes and other activities involved in making and maintaining the solar cells, but the overall effect of a switch to solar is likely to be a huge reduction in pollution. One of the main challenges of the future might be to develop the chemical systems needed to make full use of the free energy that floods down on the planet every day from the Sun.

When the possible contributions from biomass, hydroelectric, wind, and geothermal power are added to the equation, it all amounts to a persuasive argument for continuing to explore the use of "renewables" as an energy source for the future.

FIGURE 5.26

Total area required for a solar cell power plant to meet the total U.S. annual electrical power demand is represented by the square on this map.

The Bottom Line

- Chemical changes are accompanied by the gain or release of energy. (Section 5.1)

- Chemicals can store huge amounts of energy and then release it as soon as a chemical reaction begins. (Section 5.1)

- Thermodynamics is the study of energy changes and exchanges. (Section 5.1)

- Energy comes in two basic forms, which we call kinetic energy (the energy of motion) and potential energy (positional energy). (Section 5.1)

- Energy is never created or destroyed; it is only transferred from place to place and converted from one form into another. This is the law of conservation of energy. (Section 5.1)

- A chemical reaction that releases energy from the chemicals involved is known as an exothermic reaction, because energy is flowing out of the system and into the surroundings. A reaction that absorbs energy into the chemicals involved is known as an endothermic reaction. (Section 5.1)

- All chemical reactions begin with an input of energy, needed to "jolt" the chemicals into reacting. The energy required to make this happen is known as the activation energy of the reaction. (Section 5.1)

- The total change in the energy of a chemical system, as it undergoes a chemical reaction, is equal to the heat flow (q), known as the heat of reaction, and the work done (w): $\Delta U = q + w$. (Section 5.1)

- The SI unit of energy is the joule, which we can relate to the more familiar units of calories and (in the context of food) Calories. (Section 5.2)

- Each substance has a particular specific heat capacity (c), often simply called its specific heat, which is the amount of heat needed to raise the temperature of 1 gram of the substance by 1 degree Celsius (or 1 kelvin) when the pressure is constant. (Section 5.3)

- Hess's law states that the enthalpy change of a chemical reaction is independent of the chemical path or mechanism involved in the reaction. (Section 5.5)

- The standard enthalpy change for a reaction can be calculated by subtracting the sum of the enthalpies of formation of the reactants of the reaction from the sum of the enthalpies of formation of the products of the reaction. (Section 5.5)

- Making appropriate choices about our energy sources will be an important part of building a successful and sustainable future for humanity. (Section 5.6)

Key Words

activation energy The energy that is required to initiate a chemical reaction. (*p. 174*)

biomass conversion The release of energy from the chemical conversion—often simply the burning—of plants and trees. (*p. 198*)

bomb calorimeter Apparatus in which a chemical reaction occurs in a closed container, allowing the energy released or absorbed to be measured. (*p. 184*)

Calorie (Note capital C.) Unit of energy equal to 1000 calories (1 kilocalorie) and to 4184 joules. (*p. 179*)

calorie (Note lowercase c.) Unit of energy equal to 4.184 joules. (*p. 180*)

calorimeter An apparatus in which quantities of heat can be measured. (*p. 183*)

calorimetry The study of the transfer of heat in a process. (*p. 183*)

chemical energy Energy that is stored in a substance as a result of the motions and positions of its atomic nuclei and their electrons. (*p. 172*)

constant-volume calorimetry A form of calorimetry in which the reacting system is sealed within a chamber of fixed volume, and the only way the system can release or gain energy is by the exchange of heat with the surroundings. (*p. 184*)

electromagnetic radiation Energy that propagates through space. Examples include visible light, X-rays, and radiowaves. (*p. 170*)

electroweak force The fundamental force responsible for the attraction between objects carrying opposite electric charges, for the repulsion between objects carrying the same electric charge, and for some transformations within subatomic particles. It is also responsible for the phenomena of magnetism and light. (*p. 169*)

endothermic reaction A reaction that absorbs energy from the surroundings. (*p. 175*)

enthalpy A thermodynamic quantity symbolized by H and defined as $H = U + PV$. (*p. 186*)

exothermic reaction A reaction that releases energy into the surroundings. (*p. 174*)

first law of thermodynamics The total change in the closed system's energy in a chemical process is equal to the heat flow (q) into the system and the work done (w) on the system:

$$\Delta U = q + w$$

Energy is neither created nor destroyed; it is only transferred from place to place and converted from one form into another: $\Delta U_{system} + \Delta U_{surroundings} = 0$. (*p. 177*)

geothermal systems Power-generating systems that involve drilling deep into the earth and exploiting the flow of heat from the interior of the Earth out toward the surface. (*p. 199*)

gravitational force The fundamental force that causes all objects with mass to be attracted to one another. (*p. 169*)

heat (q) The energy that is exchanged between a system and its surroundings because of a difference in temperature between the two. (*p. 174*)

heat capacity The amount of heat needed to raise the temperature of any particular amount of a substance by 1°C. (*p. 182*)

heat of reaction The energy as heat released or absorbed during the course of a reaction. (*p. 186*)

Hess's law Thermodynamic law stating that the enthalpy change of a chemical reaction is independent of the chemical path or mechanism involved in the reaction. This enables us to determine the enthalpy changes of reactions that might be very difficult to actually perform. (*p. 192*)

hydroelectric systems Power-generating systems in which the power of falling water is used to generate electricity. (*p. 199*)

internal energy (U) The energy of a system defined as the sum of the kinetic and potential energies. Absolute internal energy is difficult to measure. (*p. 177*)

joule (J) The SI unit of energy. In terms of base units, 1 Joule = 1 kg·m^2·s^{-2}. (*p. 179*)

kilocalorie Unit of energy equal to 1000 calories and to 1 Calorie (note capital C). (*p. 180*)

kinetic energy The energy things possess as a result of their motion. (*p. 170*)

law of conservation of energy Law stating that energy is neither created nor destroyed but is only transferred from place to place or converted from one form into another. (*p. 171*)

molar heat capacity The heat capacity of one mole of a substance. (*p. 182*)

photovoltaic systems (PVs) Fabricated systems that convert the energy of sunlight into electricity. Also known as solar cells. (*p. 198*)

potential energy The energy things possess as a result of their positions, such as their position in a gravitational or electromagnetic field. (*p. 170*)

reference form The most stable form of the element at standard conditions. (*p. 187*)

renewable sources of energy Energy sources that can be rapidly replaced by natural processes. (*p. 198*)

solar cells Fabricated systems that convert the energy of sunlight into electricity. Also known as photovoltaic systems or PVs. (*p. 198*)

solar thermal systems Fabricated systems that convert the energy of sunlight into heat. (*p. 198*)

specific heat The amount of heat needed to raise the temperature of one gram of a substance by one degree Celsius (or one kelvin) when the pressure is constant. Also known as the substance's specific heat capacity. (*p. 181*)

specific heat capacity (c) The amount of heat needed to raise the temperature of one gram of a substance by one degree Celsius (or one kelvin) when the pressure is constant. Also known as the substance's specific heat. (*p. 181*)

standard enthalpy of combustion ($\Delta_c H°$) The enthalpy change when one mole of a substance in its standard state is completely burned in oxygen gas. Also known as the substance's standard heat of combustion. (*p. 189*)

standard enthalpy of formation ($\Delta_f H°$) The enthalpy change for the formation of one mole of a substance in its standard state from its elements in their reference form. Also known as the standard heat of formation. (*p. 188*)

standard enthalpy of reaction ($\Delta_{rxn} H°$) Enthalpy change of a reaction in which all of the reactants and products are in their standard states. (*p. 187*)

standard state The state of a chemical under a set of standard conditions, usually at 1 atmosphere of pressure and a concentration of exactly 1 molar for any substances in solution. Standard states are often reported at 25°C. (*p. 187*)

state function A property of a system that depends only on its present state, not on the path by which it reached that state. (*p. 191*)

strong nuclear force The force that holds protons and neutrons together in atomic nuclei. (*p. 169*)

surroundings Everything in contact with a chemical system. Energy may flow between the surroundings and system. Strictly speaking, the surroundings of a chemical system comprise the entire universe except the system. (*p. 168*)

system Any set of chemicals whose energy change we are interested in. (*p. 168*)

thermochemistry The study of energy exchange in chemical processes. (*p. 168*)

thermodynamics The study of energy exchange. (*p. 168*)

universe The space consisting of both the system and the surroundings. (*p. 168*)

wind power The use of the energy of the wind to generate electricity. (*p. 199*)

work (w) Force acting on an object over a distance. (*p. 169*)

The answers to the odd-numbered problems and some selected problems appear at the back of the book, as represented by the blue numbering.

Section 5.1 The Concept of Energy

Skill Review

1. Imagine a skateboarder at the top of a half-pipe. Describe the changes in potential energy and in kinetic energy as the skater goes from the top of one side of the half-pipe to the top of the other side.

2. Describe the changes in potential and kinetic energy of a snow-boarder going over a hill. If there were another uphill slope waiting at the bottom of the first hill, would the snowboarder's potential energy increase or decrease as she or he came down the first hill and immediately began to climb the second?

3. Calculate the kinetic energy of the following:
   a. A 185-lb chemistry professor jogging at 8.0 mi/h.
   b. A 42-g hackey sack being kicked at 0.78 m/s.
   c. A molecule of CO_2 moving at 560 m/s.

4. Calculate the kinetic energy of the following:
   a. A 2-ton truck lumbering along at 60.0 mi/h.
   b. A 200-g apple falling from a tree at 20 mi/h.
   c. A 0.25-kg ball moving at 1.75 m/s.

5. Describe the following atomic phenomena as primarily a function of the potential or kinetic energy of the system.
   a. The attractive force between an electron in a hydrogen atom and its nucleus.
   b. The vibration of two hydrogen atoms bonded in a molecule of H_2.
   c. The movement of a hydrogen molecule inside a balloon.

6. Describe the following atomic phenomena as primarily a function of the potential or kinetic energy of the system.
   a. The attraction of a hydrogen atom's electron to oxygen in water.
   b. The repulsion of a hydrogen atom's electron from the other hydrogen electron in H_2.
   c. The collision of two molecules as they react to generate a product.

7. Review the definition of heat. What is being transferred when heat moves between a system and its surroundings? How is this transfer accomplished?

8. What are the three main forms in which energy can be transferred between a system and its surroundings? Give an example of each.

9. When gasoline in an automobile engine undergoes combustion, new compounds and heat are produced. Identify the system and the surroundings in this process. Is the system gaining or losing energy in this process? What is the sign on w for this process?

10. When a cold pack is applied to a sports injury, the athlete may remark that the pack feels very cold. Identify the system and surroundings associated with the cold pack. Is the system gaining or losing energy in this process? What is the sign on q for this process?

11. In a chemical reaction, the value of q was determined to be +24 joules (J). The work function was determined to be +12 J. What is the total energy change for this system? Did the surroundings gain or lose energy?

12. In a chemical reaction, the value of q was determined to be −200 joules (J). The work function was determined to be +158 J. What is the total energy change for this system? Did the surroundings gain or lose energy?

13. Calculate the total energy change for each of the following systems. Do the surroundings gain or lose energy in each process?
   a. $q = +45$ J; $w = +45$ J
   b. $q = -266$ J; $w = 1.2$ kJ
   c. $q = 23.4$ kJ; $w = -14$ kJ
   d.

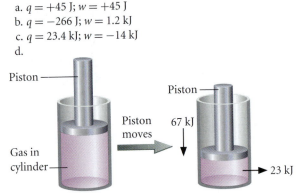

14. Calculate the total energy change for each of the following systems. Does the system gain or lose energy in each process?
   a. $q = -23$ J; $w = -37$ J
   b. $q = -88$ J; $w = +36$ J
   c. $q = +105$ J; $w = -133$ J
   d.

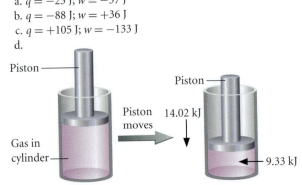

Chemical Applications and Practices

15. Burning hydrocarbon fuels and digesting carbohydrates both involve the formation of CO_2 molecules and the release of chemical energy. Which type of compounds—hydrocarbons, carbohydrates, or carbon dioxide—would release the least amount of energy? Explain how you came to this conclusion.

16. During photosynthesis, green plants take in CO_2 and combine it with water to make carbohydrates and other compounds. This process is driven by energy from the Sun. When these food molecules are digested, energy is released. (This energy can be used to maintain healthy body temperature and to run other reactions.) Is the energy released via digestion the same as the energy that absorbed from the Sun? Explain.

17. On a surface excursion during *Apollo 14,* Alan Shephard hit a few golf balls. While he originally stated that the balls flew for miles and miles, he later and more accurately estimated the distance as 200 to 400 yards. How would the amount of force needed to drive a ball 300 yards on the Moon differ from that needed to drive a ball the same distance on Earth? Base your explanation on the differences between the environments.

18. There are many forces at work in our everyday commute to work. Identify the typical forces that you'd expect to see evidence of as you drive an automobile down the street.

19. In the catabolic biochemical pathways in cells, adenosine triphosphate (ATP) is produced as part of an endothermic reaction. What is the sign for q in the reaction that represents the production of ATP? Later, the ATP can be used to help the cells do work on the surroundings. What is the sign for w in the reaction that represents that work?

20. In each of the following situations, supply the correct sign for q and w in the reaction that represents the process.
    a. In order to make soft drinks have fizz, or become carbonated, gaseous CO_2 at room temperature must be pumped, under pressure, into an aqueous solution at lowered temperatures. The system may be considered to be the resulting carbonated solution.
    b. Water placed in a microwave oven can be made to boil away into steam. Consider the water in its container to be the system.

Section 5.2 Keeping Track of Energy

Skill Review

21. Suppose a 275-g apple and a 175-g orange are moving with the same kinetic energy. If the apple is moving at 15 m/s, how fast is the orange moving? (Maybe we can compare apples and oranges after all!)

22. Some of the best tennis players may serve a 56.9-g tennis ball at a velocity of approximately 115 mi/h. What is the kinetic energy of the tennis ball in such a serve?

23. A molecule of N_2 moving through a room may have a velocity of approximately 420 m/s. What is the kinetic energy of a molecule of N_2?

24. An automobile is moving at a rate of 31 m/s. At this rate it has a kinetic energy of 436 kJ. What is the mass of this automobile?

25. A low-fat popcorn snack advertises that one serving provides 90 Cal. One serving is considered to be 34 g. What is the energy provided in Calories per gram? What is the energy provided in calories per gram? What is the energy provided in joules per gram?

26. A different food item advertises that one serving provides 320 Cal. One serving is considered to be 80 g. What is the energy provided in Calories per gram? What is the energy provided in calories per gram? What is the energy provided in joules per gram?

27. How much energy as heat is needed to raise 50.0 g of water from 23.0°C to 37.0°C?

28. How much energy as heat is required to raise 77.0 g of water from 18.0°C to 25.0°C?

29. The British thermal unit is used on some heating and cooling devices. The BTU is defined as the amount of energy as heat that will raise one pound of water from 58.5°F to 59.5°F. Express that amount of heat in joules.

30. A popular instant coffee-flavored drink provides 25 food calories per serving. Express that quantity in calories, joules, and kilojoules.

Chemical Applications and Practices

31. One of the reactions used in some hand-warmer packets is the oxidation of iron to form rust. The formation of 1 mole of rust produces approximately 410 kJ of heat.
    a. If this heat were absorbed by 2000.0 g of water at 22.0°C in a calorimeter, how hot could the water be made?
    b. If only 0.10 mole of rust were formed, how hot could 200.0 g of water at 22.0°C become?

32. Ammonium nitrate can sometimes be used in the chemical cold pack applied to some sports injuries. If 10.0 g of ammonium nitrate were placed in a coffee cup calorimeter that contained 100.0 g of water and the temperature changed from 25.0°C to 18.0°C, what was the heat transferred in kilojoules?

33. Ethanol (C_2H_5OH) is being blended with gasoline mixtures to produce a fuel for automobiles called gasohol. If the combustion of 10.0 g of ethanol produces 268 kJ of heat, how much heat will be produced in burning 1 mole of ethanol?

34. If the combustion of 48.8 g of methane (CH_4) produces 855 kJ of heat, how much heat will be produced in burning 1 mole of methane?

Section 5.3 Specific Heat Capacity and Heat Capacity

Skill Review

35. Most of the definitions that you have learned in chemistry have very specific terminology. Explain why the definition of specific heat allows you to report the value in either °C or K.

36. In order to clarify the differences among specific heat, heat capacity, and molar heat capacity, provide a brief definition of each.

37. Which of the following involves the greater amount of heat transfer?
a. 10.0 kg of water in an automobile cooling system changes from 45.0°C to 10.0°C as it cools overnight.
b. In preparation for the annual "chili cook-off," 8.0 L of water is heated from 22.0°C to 99.0°C.

38. If the heat capacity of a metal coffee pot filled with water were known to be 5.87 $\frac{kJ}{°C}$, what would you calculate to be the amount of heat transferred from a campfire when the temperature of the coffee pot changes from 22.0°C to 95.0°C?

39. Fill in the values missing from the following table.

Heat, q	Specific Heat $\left(\frac{J}{g \cdot °C}\right)$	Mass (g)	ΔT (°C)
10.0 joules	4.184		10.0
	0.115	10.0	5.0
15.5 joules		42.5	15.0

40. Fill in the values missing from the following table.

Heat, q	Specific Heat $\left(\frac{J}{g \cdot °C}\right)$	Mass (g)	ΔT (°C)
	4.184	100.0	40.0
450 joules	0.315	15.0	
48.5 joules		48.5	15.0

Chemical Applications and Practices

41. Some research is being done on the production of nickel compact discs. These disks could store tremendous amounts of information and last an unusually long time. Like most metals, nickel has a low specific heat. If 35.0 g of nickel absorbs 311 J, it increases in temperature by 20.0°C. What is the specific heat of nickel?

42. If 50.0 g of a metal alloy absorbs 471 J, it increases in temperature by 30.0°C. What is the specific heat of this metal alloy?

43. In order to calibrate a bomb calorimeter, a 2.000-g sample of benzoic acid (C_6H_5COOH; molar mass = 122.12) was combusted. The calorimeter temperature rose by 1.978°C. What is the heat capacity of this calorimeter? The molar energy of combustion for benzoic acid is known to be approximately −3227 kJ.

44. A different bomb calorimeter was calibrated by combusting a 2.250-g sample of methanol (CH_4O). The calorimeter temperature rose by 0.522°C. What is the heat capacity of this calorimeter? The molar energy of combustion for methanol is known to be approximately −890.3 kJ.

45. a. The heat capacity of a certain bomb calorimeter was calibrated at 28.9 $\frac{kJ}{°C}$. When 1.500 g of an unknown sugar was combusted in the calorimeter, the temperature rose by 2.56°C. What was the energy of combustion of the sugar?
b. What additional information would we need in order to report the molar heat of combustion for the sugar?

46. A student constructs a crude "coffee cup" calorimeter that contains 94.1 g of water, at 22.0°C, in a double cup set up with a thermometer and cork cover. When an 85.8-g piece of copper at a temperature of 100.0°C was placed in the calorimeter, the temperature was noted to equilibrate at 28.0°C. The specific heat of copper is approximately 0.386 $\frac{J}{g \cdot °C}$.
a. Calculate the heat gained by just the water.
b. Determine the heat capacity, in J/°C, for the empty calorimeter.

47. The molar heat of combustion of propane is −2.2 × 10³ kJ. How many grams of propane would have to be combusted to raise 1.0 kg of water (the amount you might boil to prepare a tasty macaroni and cheese dinner) from 22.0°C to 100.0°C? Assume that no heat is absorbed by the container or the air.

48. The specific heat of water is 4.184 J/g°C. How many grams of water at 85.0°C would have to be added to raise 1.00 kg of water from 25.0°C to 50.0°C? Assume that no heat is absorbed by the container or the air.

49. The heat of combustion for acetylene (C_2H_2) is −1300 kJ/mol. Methane has a heat of combustion of −890 kJ/mol. Decide, by calculation, which provides more energy as heat per gram.

50. The heat of combustion for glucose ($C_6H_{12}O_6$) is −2803 kJ/mol. Tristearin ($C_{57}H_{110}O_6$), a typical fat, has a heat of combustion of −37,760 kJ/mol. Decide, by calculation, which provides more energy as heat per gram.

51. One way to test a metallic sample to see whether it is made of gold is to heat it up and place it in a calorimeter to determine the specific heat of the sample. The specific heat of gold is approximately 0.13 $\frac{J}{g \cdot K}$. What temperature change would prove that the metal was gold, if 15.0 g of the sample at 99.0°C were placed in a calorimeter initially at 25.0°C? The heat capacity of the calorimeter is 25.0 $\frac{J}{K}$.

52. In the chapter, a bomb calorimeter was discussed that had a heat capacity of 32.4 $\frac{kJ}{°C}$. If 0.550 g of benzoic acid ($C_7H_6O_2$) were combusted in the calorimeter, what would be the expected change in temperature? (The molar energy of combustion for benzoic acid is −3227 kJ/mol.)

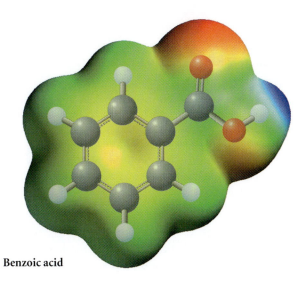

Benzoic acid

Section 5.4 Enthalpy

Skill Review

53. Explain how chemical reactions that generate gases can be thought of as constant-pressure situations but the volume of the products can be thought of expanding?

54. When the equation $\Delta U = q_p + w$ is recast as $\Delta U = q_p - P\Delta V$, the sign between the heat and work function changes. Explain why it is now proper to express the $P\Delta V$ with a negative sign.

55. Without using any symbols or numbers, define the term *enthalpy* as it is applied in chemistry.

56. What is the chief thermodynamic difference among the expressions known as "heat of reaction," "standard heat of reaction," "standard heat of formation," and "standard heat of combustion"?

57. In an exothermic reaction, does the system or the surroundings gain energy as heat?

58. What sign would characterize the value of ΔH for the reverse of an exothermic reaction?

59. What are the conditions being described when a substance is taken to be in its standard state?

60. What is the physical state of carbon dioxide when it is in standard state conditions?

61. Of the following, which would *not* be considered appropriate equations to represent "standard heat of formation" processes? Explain the reasons for your choices.
    a. $C(s) + \frac{1}{2}O_2(g) \rightarrow CO(g)$
    b. $N_2(g) + 3H_2(g) \rightarrow 2NH_3(g)$
    c. $CS_2(g) \rightarrow CS_2(l)$
    d. $C(g) + 4H(g) \rightarrow CH_4(g)$
    e. $4CO_2(g) + 5H_2O(l) \rightarrow C_4H_{10}(g) + \frac{13}{2}O_2(g)$

    f.
    (gas)   (gas)   (gas)   (gas)

62. Of the following, which would *not* be considered appropriate equations to represent "standard heat of formation" processes? Explain the reasons for your choices.
    a. $C(s) + O_2(g) \rightarrow CO_2(g)$
    b. $H_2(g) + O_2(g) \rightarrow H_2O_2(g)$
    c. $\frac{1}{2}H_2(g) + O_2(g) \rightarrow H_2O(g)$
    d. $CH_4(g) + 2O_2(g) \rightarrow CO_2(g) + 2H_2O(g)$
    e. $2H(g) + O(g) \rightarrow H_2O(l)$

    f.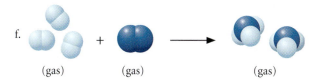
    (gas)   (gas)   (gas)

Chemical Applications and Practices

63. Explain why the values of ΔH and ΔU for reactions such as the explosion of nitroglycerin, represented here, could vary significantly.

$$4C_3H_5N_3O_9(l) \rightarrow 12CO_2(g) + 10H_2O(g) + 6N_2(g) + O_2(g)$$

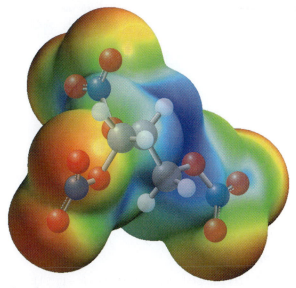

Nitroglycerin

64. Consider the reaction for nitroglycerin decomposition shown in Problem 63. Cite two reasons why the reverse of the reaction would not qualify as the standard heat of formation for nitroglycerin.

65. a. Write out the balanced chemical equation for the standard formation of carbon monoxide.
    b. If the value for the standard heat of formation for carbon monoxide were -110.5 kJ/mol, would you consider the reaction endothermic or exothermic?
    c. What would be the ΔH value for the reverse of the reaction?

66. a. Write out the balanced chemical equation for the standard formation of gaseous hydrogen peroxide (H_2O_2).
    b. If the value for the standard heat of formation for hydrogen peroxide were -136.1 kJ/mol, would you consider the reaction endothermic or exothermic?
    c. What would be the ΔH value for the reverse of the reaction?

67. The hydrocarbon fuel butane (C_4H_{10}) is used in small portable lighters. Write the reaction for the standard heat of combustion for this reaction. (Recall that the two products of hydrocarbon combustion are carbon dioxide and water.)

68. Pentane (C_5H_{12}) can be blended with other hydrocarbons and additives to form gasoline. Write the reaction that depicts the standard heat of formation of pentane.

69. The primary energy molecule in cells is ATP (adenosine triphosphate, $C_{10}H_{15}N_5O_{13}P_3$). Write out the standard heat of formation reaction for this important biomolecule.

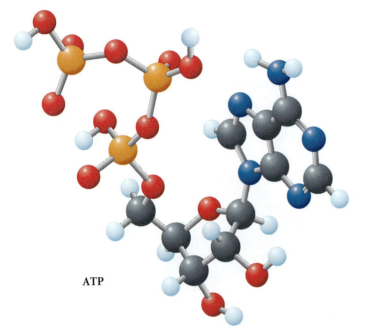

ATP

70. Glucose ($C_6H_{12}O_6$) is used by our bodies as an energy source. Write out the standard heat of formation reaction for this important molecule.

Section 5.5 Hess's Law

Skill Review

71. The ΔH value for the following reaction is -1012 kJ.
$$2NH_3(g) + 3N_2O(g) \rightarrow 4N_2(g) + 3H_2O(l)$$
    a. What is the value of ΔH for the reverse of the reaction?
    b. What is the value of ΔH for 1 mole of NH_3 reacting?
    c. What is the value of ΔH of 4 moles of NH_3 reacting?

72. The ΔH value for the following reaction is $+284.6$ kJ.
$$3O_2(g) \rightarrow 2O_3(g)$$
    a. What is the value of ΔH for the reverse of the reaction?
    b. What is the value of ΔH for 1 mole of O_2 reacting?
    c. What is the value of ΔH of 4 moles of O_2 reacting?

Chemical Applications and Practices

73. The fuel used in many rural settings is propane (C_3H_8). Write out the heat of formation reaction for this important fuel. Now use the combustion reaction sequence shown in the text to develop a Hess's law scheme that would allow you to calculate the ΔH of formation for propane.

74. The sugar arabinose ($C_5H_{10}O_5$) is obtained from plants with the polysaccharide gum arabic. In wheat plants, this sugar helps form important cell wall structures. The heat of formation reaction for arabinose is
$$5C(s) + 5H_2(g) + \tfrac{5}{2}O_2(g) \rightarrow C_5H_{10}O_5(s)$$
Without using actual kJ values, show the reactions that you could use, and the way you would use them, to obtain the arabinose formation reaction.

75. Given the following reactions and ΔH values:

$B_2O_3(s) + 3H_2O(g) \rightarrow B_2H_6(g) + 3O_2(g)$	$\Delta H = +2035$ kJ
$2H_2O(l) \rightarrow 2H_2O(g)$	$\Delta H = +88$ kJ
$H_2(g) + \tfrac{1}{2}O_2(g) \rightarrow H_2O(l)$	$\Delta H = -286$ kJ
$2B(s) + 3H_2(g) \rightarrow B_2H_6(g)$	$\Delta H = +36$ kJ

Calculate ΔH for
$$2B(s) + \tfrac{3}{2}O_2(g) \rightarrow B_2O_3(s) \qquad \Delta H = ?$$

76. Use the heat of formation reaction data for water and carbon dioxide and the following reaction for the combustion of acetylene, employed in high-temperature welding applications, to find the heat of formation of acetylene (C_2H_2).
$$C_2H_2(g) + \tfrac{5}{2}O_2(g) \rightarrow 2CO_2(g) + H_2O(l)$$
$$\Delta_cH = -1300 \text{ kJ/mol}$$

77. a. Ethanol (C_2H_5OH) is being added to gasoline to create the fuel "gasohol." Using heat of formation data and Hess's law, determine the heat of combustion, in kJ/mol, for ethanol.

    Combustion of ethanol:
    $$C_2H_5OH(l) + 3O_2(g) \rightarrow 2CO_2(g) + 3H_2O(l) \quad \Delta_cH = ?$$
    Formation of ethanol:
    $$2C \text{ (graphite)} + 3H_2(g) + \tfrac{1}{2}O_2(g) \rightarrow C_2H_5OH(l)$$
    $$\Delta H = -278 \text{ kJ}$$
    b. Using your answer from part a, determine the heat released when 100.0 g of pure ethanol is combusted.

78. The oxidation of sulfur has many important environmental connections. Notably, acid rain is formed from sulfur oxides reacting with moisture in the air. Use the following two reactions to determine the enthalpy change when sulfur dioxide reacts with oxygen to form sulfur trioxide.
    a. $S(s) + O_2(g) \rightarrow SO_2(g)$ $\qquad \Delta H° = -296.8$ kJ
    b. $2S(s) + 3O_2(g) \rightarrow 2SO_3(g)$ $\qquad \Delta H° = -791.4$ kJ
    $\quad 2SO_2(g) + O_2(g) \rightarrow 2SO_3(g)$ $\qquad \Delta H° = ?$

79. The octane rating on gasoline is a method of comparing fuel energy values. What is the value for the heat of combustion of octane? ($\Delta_fH°(C_8H_{18}) = -249.9$ kJ/mol)
$$C_8H_{18}(l) + \tfrac{25}{2}O_2(g) \rightarrow 8CO_2(g) + 9H_2O(l)$$

80. A typical component of gasoline is pentane (C_5H_{12}). The standard enthalpy of combustion of pentane is -3537 kJ/mol. Write the combustion reaction for pentane, and determine the standard enthalpy of formation for pentane.

81. Calcium metal will react in water to form calcium hydroxide, $Ca(OH)_2$. Use the thermochemical data that follow and Hess's law to calculate the value of $\Delta H°$ for the reaction

$$Ca(s) + 2H_2O(l) \rightarrow Ca(OH)_2(s) + H_2(g)$$

   a. $H_2(g) + \frac{1}{2}O_2(g) \rightarrow H_2O(l)$ $\Delta H° = -286$ kJ
   b. $CaO(s) + H_2O(l) \rightarrow Ca(OH)_2(s)$ $\Delta H° = -64$ kJ
   c. $Ca(s) + \frac{1}{2}O_2(g) \rightarrow CaO(s)$ $\Delta H° = -635$ kJ

82. Ozone is reduced by hydrogen to produce water. Use the thermochemical data that follow and Hess's law to calculate the value of $\Delta H°$ for the reaction

$$3H_2(g) + O_3(g) \rightarrow 3H_2O(g)$$

   a. $H_2(g) + \frac{1}{2}O_2(g) \rightarrow H_2O(l)$ $\Delta H° = -286$ kJ
   b. $2H_2(g) + O_2(g) \rightarrow 2H_2O(g)$ $\Delta H° = -483.6$ kJ
   c. $3O_2(g) \rightarrow 2O_3(g)$ $\Delta H° = +284.6$ kJ

Section 5.6 Energy Choices

Chemical Applications and Practices

83. Most of the world's energy consumption for power is based on fossil fuels. This source, however, is considered nonrenewable. Look back at the list of renewable energy sources and cite one advantage, in addition to this renewability, that each would have over the petroleum-based fuels used today.

84. Assuming that the fossil fuels are completely depleted at some point in the future, the world will need to depend on renewable sources of energy. Look back at the list of renewable energy sources and describe one disadvantage of using each source as the sole source of energy for a country.

Comprehensive Problems

85. Individual atoms and molecules are so small that they have very low values of kinetic energy. However, given their mass and velocity, it is possible to calculate the value. What is the kinetic energy of an oxygen molecule (O_2) in air that you are breathing if its velocity is 460 m/s? Would you expect a nitrogen molecule (N_2) moving at the same speed to have more or less kinetic energy than the oxygen molecule? Explain.

86. Distinguish between the two terms in each of the following pairs:
   a. Heat and temperature
   b. System and surroundings
   c. Exothermic and endothermic
   d. q and ΔU

87. What is the role of activation energy in a chemical process? Is it possible for the activation energy of a reaction to be greater than the overall change in energy for a chemical reaction? Explain.

88. The average temperature of a healthy human is approximately 37°C. The average room temperature may be about 25°C. Explain how we are able to keep our body temperature so much higher than that of our environment.

89. A 400-mL glass beaker contains 250 g of water at room temperature. As several NaOH pellets are placed in the water and begin to dissolve, you notice a warming sensation in the hand in which you are holding the beaker. Answer the following questions.
   a. If the NaOH and the water make up the system, would you consider the process endothermic or exothermic? Explain.
   b. Is the beaker part of the system or part of the surroundings?
   c. During this process, is energy flowing into or out of the system?
   d. During this process, has the kinetic energy of the water molecules been raised or lowered?
   e. What work is being done during this process?

90. A serving of Italian rice—risotto—provides 150 food Calories. What would this value be in kilojoules? Assume that this quantity of energy would be available to do the work of lifting 2.5-kg chemistry textbooks from the floor to a height of 1.5 m. How many such chemistry textbooks could, theoretically, be lifted through that height? (And later, of course, after that work, brought to class.)

91. Express the energy content of a 2.00-oz. candy bar that contains 247 Calories in Calories per gram, in joules per gram, and in kilojoules per gram. If this energy were efficiently used to provide the kinetic energy to move a 1.5-pound chemistry book, how fast, in meters per second, could the book move?

92. One of the main reasons for eating is to obtain a supply of energy. A low-fat apple muffin may provide 170 food Calories for each 50-g muffin. How many muffins (or fractions of muffins) are needed to provide 1 joule?

93. a. Suppose you are heating water (225 g) in a mug that you have placed in a microwave oven. As you wait to add the instant hot chocolate, please calculate, from the following data, the amount of energy as heat that the water has absorbed: The original water temperature was 15.0°C. When you remove the mug of hot water, you find that the temperature has risen to 98°C.
   b. What additional information would you need in order to determine the heat absorbed by the mug?

94. You have just removed a hot cheese pizza from the oven, and all of the ingredients are presumably at the same temperature. Without waiting for it to cool, you take a bite of the pizza. As you bite the pizza, the bread is hot on your tongue but does not burn. However, as you continue to bite, pizza sauce (mostly tomatoes and water) squeezes out and burns the roof of your mouth. Which has the higher specific heat, the bread or the sauce? Explain the basis of your answer using $q = m \times c \times \Delta T$.

95. One way to determine the heat capacity of a constant-volume calorimeter is to burn measured amounts of pure carbon in the presence of oxygen gas to form carbon dioxide. From the following experimental data, determine the heat capacity of the calorimeter. A 0.200-g sample of carbon, when completely combusted, raised the temperature of the water and the contents of the entire calorimeter from 24.0°C to 25.5°C. It is known that under these conditions, the heat released from the complete combustion of 1 mole of carbon is 392 kJ.

96. A student's coffee cup calorimeter, including the water it contains, has been calibrated in a manner similar to that described in Problem 46. The heat capacity was found to be $55.5 \frac{J}{°C}$. If a 65.8-g sample of an unknown metal, at 100.0°C, was placed in the calorimeter initially at 25°C, and an equilibrium temperature of 29.1°C was reached, what is the specific heat of the metal?

97. The foods we eat provide fuel to keep us alive. Burning a 0.500-g sample of vegetable oil provides enough heat to raise the temperature of a calorimeter by 2.5 K. Assuming the heat capacity of the calorimeter to be $7.5 \frac{kJ}{K}$, determine the heat of combustion for 1 g of the oil.

98. A student performs the experiment shown graphically here. What is the specific heat of the block of metal used in the experiment? (Assume that the heat capacity of the empty calorimeter is $7.5 \frac{J}{°C}$.)

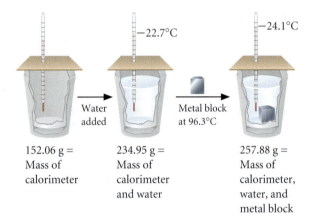

Water added — Metal block at 96.3°C

−22.7°C   −24.1°C

152.06 g =
Mass of
calorimeter

234.95 g =
Mass of
calorimeter
and water

257.88 g =
Mass of
calorimeter,
water, and
metal block

99. The reaction of the gases ethane (C_2H_6) and oxygen gives gaseous carbon dioxide and water vapor.
   a. Write the balanced chemical reaction for this combustion.
   b. What is the enthalpy change for this process?
   c. If 250.0 grams of ethane is consumed in the reaction, how many grams of carbon dioxide would be produced?
   d. Why should precautions be taken to avoid performing the reaction outlined in part c in an enclosed space?

100. Under certain conditions, the reaction of chlorine gas with metallic iron can produce iron(III) chloride.
   a. Write the balanced chemical reaction for this reaction.
   b. If 8.44 grams of iron(III) chloride are produced, how many grams of iron were required?
   c. What is the enthalpy change required to produce 8.44 grams of iron(III) chloride? ($\Delta_f H(FeCl_3) = -399$ kJ/mol; $S(FeCl_3) = 142$ J/kmol; $\Delta_f G = -344$ kJ/mol.)

101. Sodium hydroxide pellets can be used to unclog a drain.
   a. What is the enthalpy change when 15.0 g sodium hydroxide is added to 1.00 L water?
   b. What is the molarity of this solution? (Assume that the volume change is negligible.)

102. A student wishes to prove the existence of silver ions in a particular solution that was supposedly made by dissolving silver nitrate in water.
   a. Explain how this could be determined using a solution of sodium sulfide, and write a reaction that illustrate this method.
   b. What is the molar enthalpy change for the reaction of a solution of silver nitrate and sodium sulfide?

103. A 0.10 lb sample of sodium metal is added to 1.00 gallon water.
   a. Assuming no change occurs in the volume of the sample, what would be the concentration of the resulting solution of sodium hydroxide?
   b. What is the enthalpy change for this process?

Thinking Beyond the Calculation

104. Phosphoric acid is used in many soft drinks to add tartness. This acid can be prepared through the following reaction:

$$P_4O_{10}(s) + 6H_2O(l) \rightarrow 4H_3PO_4(aq)$$

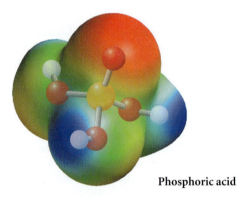

Phosphoric acid

   a. If the value of ΔH for the reaction is -453 kJ, what is the value of ΔH for the reverse of the reaction?
   b. What is the value of ΔH for this reaction if 10.0 g of phosphoric acid is produced?
   c. Is this reaction endothermic or exothermic?
   d. If 1.50 g of $P_4O_{10}(s)$ and 2.50 mL of water were mixed, how many grams of phosphoric acid would result?
   e. What is the enthalpy change for the process outlined in part d?
   f. If 10.0 g of P_4O_{10} were mixed with 1.00 kg of water at 25.0°C, what would be the final temperature of the water?

Quantum Chemistry: The Strange World of Atoms

Scanning tunneling microscope (STM) image showing iron atoms adsorbed on a copper surface forming a "quantum corral" at a very low temperature (4 K). The image shows the contour of electron density. The corral is about 14.3 nm in diameter.

Contents and Selected Applications

6.1   **Introducing Quantum Chemistry**

6.2   **Electromagnetic Radiation**

6.3   **Atomic Emission and Absorption Spectroscopy, Chemical Analysis and the Quantum Number**

Chemical Encounters: Simultaneous Determination of Elements in Water

6.4   **The Bohr Model of Atomic Structure**

Chemical Encounters: The Nature and Applications of Lasers

6.5   **Wave–Particle Duality**

6.6   **Why Treating Things as "Waves" Allows Us to Quantize Their Behavior**

6.7   **The Heisenberg Uncertainty Principle**

6.8   **More About the Photon—the de Broglie and Heisenberg Discussions**

6.9   **The Mathematical Language of Quantum Chemistry**

6.10   **Atomic Orbitals**

Chemical Encounters: The Scanning Tunneling Microscope

6.11   **Electron Spin and the Pauli Exclusion Principle**

Chemical Encounters: Nuclear Spin and Magnetic Resonance Imaging

6.12   **Orbitals and Energy Levels in Multielectron Atoms**

6.13   **Electron Configurations and the Aufbau Principle**

Go to **college.hmco.com/pic/kelterMEE** for online learning resources.

Funny things happen when you shrink the scale of observation down below that which you can see with your eye, even when it is aided by the most powerful optical microscope. Things that once appeared to have a single value, a single space on the lab bench, or a single speed at which they move through space now get smeared out over time and space so that you can discuss only *the probability of where they might be* at a given time or *how fast they might be traveling* when you finally catch up with them. Nothing at this level seems absolute anymore.

The field of chemistry that models the behavior of atoms and molecules at the atomic level is called quantum chemistry, and the physical description of this model is called quantum mechanics. Because our everyday experiences do not easily prepare us for what we observe in experiments done on the atomic level, the whole idea of quantum mechanics might at first appear strange or even incomprehensible. However, the rules, or postulates, of quantum theory are elegant and precise, of great utility, and readily understandable if we put aside our "macroscopic" expectations. **Why is it useful for us to know about quantum chemistry?**

Quantum chemistry is important because we use it to predict chemical reactivity and other kinds of chemical properties. For example, we need quantum chemistry to answer questions such as, How is the energy of sunlight captured by plants? and Why do leaves and flowers have distinctive colors? There is so much richness in our macroworld that is truly revealed only by looking deep within, at the nanoworld that underlies it.

The colors of these flowers, and the fact that sunlight helps them grow, can be explained in terms of the basic interactions of light with matter. These interactions are governed by quantum chemistry.

6.1   Introducing Quantum Chemistry

A liter of water can be divided in half to give two half-liters of water. Each of those half-liters can itself be divided into halves (two quarter-liters), and so on. But the process cannot go on indefinitely because we will eventually be left with a single water molecule. We could break this molecule down into two hydrogen atoms and an oxygen atom, but the pieces would no longer be water. The H_2O molecule is the smallest unit, or *quantum*, of water. Energy can be quantized, too. The quantum of light is called a photon, the smallest possible amount of light energy there can be. In general, a quantum (plural *quanta*) is a single indivisible unit of something. Many familiar things also come in quanta, including eggs, cats, and chemistry books. Yet it is the quantization of light energy into photons that will play a key part in our understanding of atomic and molecular structure—and in all the chemistry we can explain and predict as a result. Light energy is a type of electromagnetic radiation, so let us begin by looking at it from that point of view.

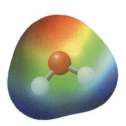

The smallest unit of water is H_2O.

6.2 Electromagnetic Radiation

Application

When you open your eyes in the morning, information carried by light, a type of electromagnetic radiation, floods into your brain. Light lets you see where you are, notice what sort of day it is, and perhaps locate the button to turn on the television news. The television signal arrives from the television station—or, in more recent times, from a communication satellite—in the form of electromagnetic radiation, as does most of the other information we need to live our lives. Studying electromagnetic radiation from stars enables scientists to figure out the structure of the universe. Studying the interaction of atoms with electromagnetic radiation led to the modern view of atomic structure, the subject of this chapter.

Electromagnetic radiation, such as visible light, can be described as electromagnetic waves that transmit energy through space. Each wave is just a part of the **electromagnetic spectrum**, the *entire* range of electromagnetic radiation that is possible. These waves have a wavelength, λ (the Greek letter lambda), frequency, ν (the Greek letter nu), and speed (see Figure 6.1). The **wavelength** is defined as *the distance from the top of one crest of the wave to the top of the next crest*. In the past, wavelengths were measured using a unit known as the **angstrom (Å)**, where $1\text{ Å} = 10^{-10}$ m. The use of this unit of length has been supplanted by units in the SI system, such as nanometers. The **frequency** of electromagnetic radiation is *defined as the number of waves that pass a given point per second*. So, the frequency is reported in units of $1/s$ (s^{-1}), known as a hertz (Hz).

Video Lesson: The Wave Nature of Light

The speed of light in a vacuum is given the symbol c and is a fundamental physical constant equal to 299,792,458 meters per second (this is usually rounded to 3.00×10^8 m·s^{-1}). How fast is the speed of light? We can get a handle on this by considering that the Sun is, on average, 93 million miles, or 1.5×10^{11} m, from the Earth. Sunlight takes a little more than 8 minutes to reach our planet.

$$1.5 \times 10^{11}\text{ m} \times \frac{1\text{ s}}{3.00 \times 10^8\text{ m}} = 500\text{ s } (= 8.3\text{ min})$$

We can also use the speed of electromagnetic radiation to tell us about distances on smaller scales. New global positioning system (GPS) devices receive signals from at least three of the more than two-dozen GPS satellites in orbit 20,200 km (12,500 mi) above the Earth's surface. At that distance, it takes about

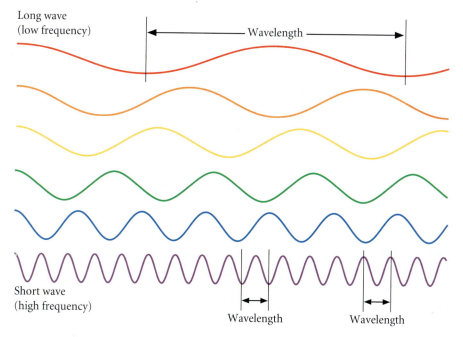

FIGURE 6.1

Sine waves of similar amplitude, with wavelength and frequency illustrated.

FIGURE 6.2

A GPS watch and how it works. The watch receives signals from at least three satellites. On the basis of the time it takes each signal to reach the watch, the distance from each satellite is calculated. The exact position of the watch is then triangulated from the three distances.

(latitude, longitude)

0.07 s for signals from each satellite to reach the receiver. Each satellite's signal will take a slightly different amount of time than the others to reach the running watch or car receiver, because each is a little farther from or closer to you, depending on your location. The differences in time of receipt of the signal—on the order of milliseconds—from just three of these satellites can be used to locate your position within a meter or two in three dimensions (Figure 6.2).

Wavelength, frequency, and speed are interrelated by the equation

$$\nu = \frac{c}{\lambda}$$

Note the inverse relationship between frequency, ν, and wavelength, λ. As one increases, the other decreases. The illustration of the waves in Figure 6.1 also indicates the **amplitude** of a wave, which is the distance between its highest point and zero.

EXERCISE 6.1 **Conversion Between Frequency and Wavelength**

What frequency of light has a wavelength of 585 nm, which appears orange to human eyes?

Solution

Because the speed of light is expressed in meters per second, it is useful to convert the wavelength, 585 nm, to meters:

$$585 \text{ nm} \times \frac{1 \times 10^{-9} \text{ m}}{1 \text{ nm}} = 5.85 \times 10^{-7} \text{ m}$$

We can use the relationship among the speed of light, wavelength, and frequency to solve for frequency:

$$\nu = \frac{c}{\lambda}$$

$$\frac{3.00 \times 10^{8} \text{ m·s}^{-1}}{5.85 \times 10^{-7} \text{ m}} = 5.13 \times 10^{14} \text{ s}^{-1}$$

We can also do the equivalent calculation by dimensional analysis:

$$\frac{3.00 \times 10^{8} \text{ m}}{\text{s}} \times \frac{1}{5.85 \times 10^{-7} \text{ m}} = 5.13 \times 10^{14} \text{ s}^{-1}$$

PRACTICE 6.1

What is the wavelength of light that has a frequency of 7.45×10^{14} Hz? Should the wavelength be shorter or longer than that in the exercise?

See Problems 5a, 6a, and 11–14.

Even though we mention that light is a wave, no *matter* is actually "waving" as the light travels through space. When electromagnetic radiation travels through matter, such as water, air, or a pane of glass, the electrons and nuclei of the matter tend to follow the oscillatory motion of the electromagnetic radiation, slowing the wave a little. The end result is that electromagnetic radiation travels a bit more slowly through matter than in a vacuum.

If we allow an electromagnetic wave to strike a more dense transparent material at an angle, the radiation will change direction as it enters. We use this effect in lenses made for eyeglasses, cameras, or telescopes. The angle through which a ray of light or other electromagnetic ray bends depends on its frequency; higher-frequency (shorter-wavelength) rays are bent more than lower-frequency ones. This is why a glass prism, shown in Figure 6.3, is used to separate the frequencies of light to generate a spectrum, familiar to us as the colors of a rainbow.

The visible light spectrum of Figure 6.3 represents only a small fraction of the entire electromagnetic spectral range, shown in Figure 6.4. Our eyes do not respond to anything outside the visible range, but our technology can detect and use the invisible frequencies. For example, very long wavelengths are used for radio transmission and are called radio waves. The shortest radio waves, at tenths of meters, are called microwaves and are used in cellular phone communications, radar systems, and microwave ovens.

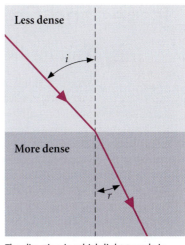

The direction in which light travels is bent as the electromagnetic radiation moves from a substance of given density to another.

FIGURE 6.3

The glass prism separates out the frequencies of light from the electromagnetic spectrum by wavelength.

Visualization: Refraction of White Light

FIGURE 6.4

Electromagnetic spectrum as a function of energy.

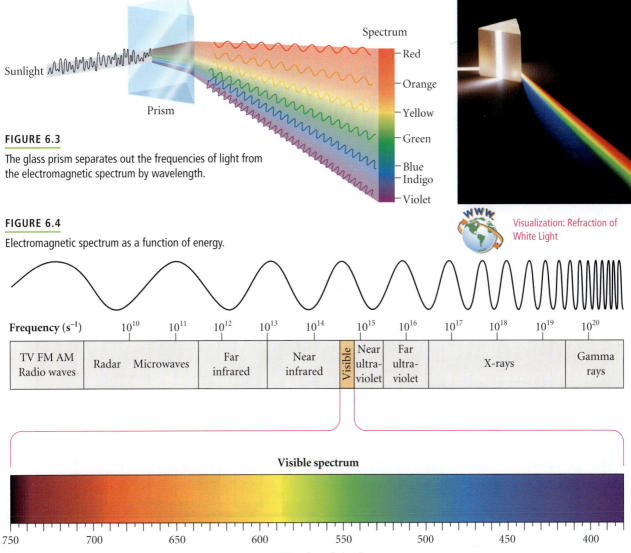

Issues and Controversies

Putting the Squeeze on Radio Frequencies

When we think of traffic jams, images of cars stuck in a sea of smog and blaring horns come to mind. Yet with the furious surge of technological innovation in consumer products such as cellular phones and other two-way mobile communications devices, a new electromagnetic traffic jam has been produced. The frequencies of the radio wave part of the electromagnetic spectrum are becoming uncomfortably cluttered, and the assignment of the frequencies in this range of about 3 kilohertz (3000 s^{-1}) to 300 gigahertz (3.00 × 10^{11} s^{-1}) is becoming what one newspaper reporter called "a high-wire act achieved through technological advances and the careful management of the radio spectrum."[1]

The International Telecommunications Union (ITU), headquartered in Geneva, Switzerland, is responsible for developing standards and procedures related to the assignment of radio frequencies "based on the principle of equal rights of all countries, large or small, to equitable access to these bands."[2]

In the United States, there is a dual system of administration. Federal use is administered by the National Telecommunications and Information Administration (NTIA). All other uses are managed by the Federal Communications Commission (FCC). About 93.1% of the total spectrum has shared use; only 5.5% is exclusively allocated for private use and 1.4% for government use.

Figure 6.5 shows the division of the radio spectrum into 450 bands, many of which have specific allocations. To keep up with the extraordinary demand for new radio frequency assignments, and to resolve the occasionally complex issues that accompany these assignments, the ITU began to meet every 2 years starting in 1993 instead of the prior meeting schedule of every 20 years. In the United States, one of the most complex issues is the allocation of frequencies in response to ever more requests. How can we make it technologically possible to stuff more users into a finite wavelength region by reducing the necessary "bandwidth" (how wide the frequency band needs to be for the transmission) for each user? Is it possible to create communications devices that use alternative frequencies? The personal cellular phone was just a dream 30 years ago. Digital audio was not even on the horizon. What advances might we anticipate in the next 20 years that will squeeze the frequency range even more?

Cell phones, video phones, and text messaging are just some of the modern conveniences that are squeezing the available bandwidth.

1. McClain, Dylan Loeb, "Directing Traffic in the Radio Spectrum's Crowded Neighborhood," *New York Times*, February 24, 2000 p. D7.

2. International Telecommunications Union, http://www.itu.int/ (accessed October 2005).

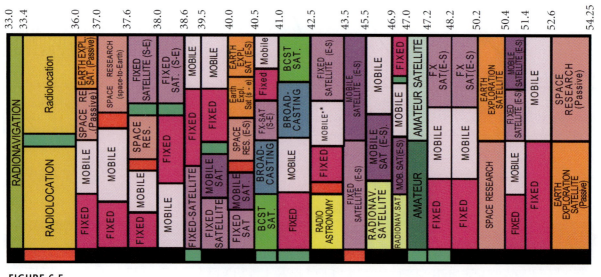

FIGURE 6.5

The radiofrequency band is packed with assigned frequencies. How will we add more as new technologies become available?

TABLE 6.1	Some Commercial Uses of Electromagnetic Radiation
Region	**Example of Use**
Radio wave	Communication
Microwave	Cooking and reheating foods
Infrared	Heating, drying, and cameras for finding thermal "hot spots" in products
Visible	Detection of low concentrations of elements in steel, milk, water, etc.
Ultraviolet	Germicide for killing bacteria in foods, in water, and on surfaces
X-ray	Skeletal imaging, inspecting luggage at airports
Gamma ray	Finding defects in pipe welds and castings, inspecting containers

An infrared picture of a home can show where most of the heat is lost. Note that most of the heat loss in this home, shown as bright yellow colors, occurs through its windows.

How do we use the different regions of the electromagnetic spectrum? Infrared radiation (IR), spans about 10^{-6} to 10^{-3} m in wavelength and can be detected with special "heat-sensing" scopes. IR rays are associated with the vibrations of molecules, so you can feel the warmth they generate in your skin.

The Sun's ultraviolet radiation (UV) at approximately 10^{-7} m can give you a tan, or a burn, depending on the wavelength and how long you sit in the sun or under a UV lamp. The short wavelength of X-ray radiation, at about 10^{-10} m, is approximately the same size as atomic distances. This allows X-rays to be used to detect the spatial arrangements of atoms. Gamma rays have the shortest waves in the electromagnetic spectrum and are also the most energetic. They are a by-product of many nuclear reactions and are very damaging to most materials, including those that make up living things. Table 6.1 gives selected commercial (that is, related to commerce) uses for radiation from various regions of the electromagnetic spectrum.

The shorter the wavelength, or the higher the frequency, of the electromagnetic radiation, the greater the energy associated with it. This relationship between wavelength, or frequency, and the energy of a single photon was determined in 1900 by Albert Einstein (1879–1955) by extending the work of Max Planck (1858–1947).

$$E = h\nu = \frac{hc}{\lambda}$$

In this equation, E is the energy carried by *each photon* of the electromagnetic ray, and h is **Planck's constant**, equal to $6.6260693 \times 10^{-34}$ J·s. This equation enables us to calculate the energy of a single quantum of electromagnetic energy. As we'd predict, the energy each photon carries is very small.

EXERCISE 6.2	The Energy of Electromagnetic Radiation

Compare the energy of a photon of visible radiation at 605 nm, the red light that we detect with our eyes, and that of the X-ray radiation at 2.00×10^{-10} m that we use in medical diagnosis.

First Thoughts

Which has greater energy, a photon of red light or an X-ray? Our own experience tells us that the calculation should result in a higher energy value for the X-ray.

Solution

We use Planck's constant and the speed of light to obtain the energies of each wavelength:

$$E = \frac{hc}{\lambda} = \frac{(6.626 \times 10^{-34} \text{ J·s}) \times (3.00 \times 10^8 \text{ m·s}^{-1})}{6.05 \times 10^{-7} \text{ m}}$$

$$= 3.29 \times 10^{-19} \text{ J for visible light}$$

$$E = \frac{hc}{\lambda} = \frac{(6.626 \times 10^{-34} \text{ J·s}) \times (3.00 \times 10^8 \text{ m·s}^{-1})}{2.00 \times 10^{-10} \text{ m}}$$

$$= 9.94 \times 10^{-16} \text{ J for X-rays}$$

Further Insights

The energy of the X-ray is 3000 times greater than that of the visible light. **Does our answer make sense?** Yes, it does, because our experience and our discussion up to this point agree that X-rays are much more powerful than visible radiation. This also explains why X-rays can do a lot more damage to living tissue than visible light. Used with caution, however, they can penetrate the body and help physicians diagnose broken bones and tumors. Their damaging powers can also be used to destroy cancerous tissue.

PRACTICE 6.2

What is the energy associated with microwaves of wavelength 8.00 mm?

See Problems 5b, 5c, 6b, 6c, and 15–24.

6.3 Atomic Emission and Absorption Spectroscopy, Chemical Analysis and the Quantum Number

Electromagnetic radiation can be used to transfer energy to and from atoms and molecules. Energy emitted from or absorbed by an atom or molecule is often found as a collection of discrete wavelengths that depend on the structure of the atom or molecule. Using the technique known as **spectroscopy**, chemists can identify and quantify elements in all types of samples—from food and water to brass and stainless steel. Spectroscopy, *the study of how substances interact with electromagnetic radiation as a function of wavelength or frequency,* helps us determine the molecular composition of both synthetic and natural drugs, as well as the kinds of organic molecules present in interstellar space or a sample of blood.

Figure 6.6 shows the solar spectrum, the range of visible light that is emitted from the Sun. Note in the figure the discrete dark lines. These are an all-important feature in atomic spectroscopy because, much like each person's unique fingerprint in a crime scene analysis, each element leaves its unique fingerprint that we can use to identify it unequivocally via **emission spectroscopy** (studying radiation *emitted* vs. wavelength) or **absorption spectroscopy** (studying radiation *absorbed* vs. wavelength).

Spectra have been obtained from the Sun since the beginning of the nineteenth century, and elemental analysis via emission spectroscopy has been done since the late 1850s. For example, Kirchhoff and Bunsen, who published their discovery of the elements cesium (in 1860) and rubidium (in 1861), did so by looking at the emission spectra of these elements in a flame. However, they did not understand the relationship of the wavelengths of light that they observed to the atomic structure of the elements. Unraveling this connection was left to

Video Lesson: Absorption and Emission

Johann J. Balmer, a mathematician who numerically analyzed in detail the hydrogen spectrum obtained by the physicist Anders Ångstrom and others, and who published his results in 1885. The next part of our discussion shows how we can get a hydrogen spectrum and how Balmer's conclusions advanced the understanding of atomic structure.

Obtaining the Hydrogen Spectrum

The hydrogen emission spectrum can be obtained using a quartz tube with a metal electrode at each end. If the tube is then filled with hydrogen gas (H_2) and a very high voltage is applied across the electrodes, the tube glows as shown in Figure 6.7. In this apparatus, electrons jump from the negative electrode to the positive electrode, transferring energy to the hydrogen molecules in the tube. Enough energy is transferred to make some of the molecules dissociate into energetically excited hydrogen atoms (H^*). The extra energy can be released if the atoms emit light, represented by the expression

$$H^* \rightarrow H + h\nu$$

where ν is the frequency of the light emitted by the excited hydrogen (H^*) and h is Planck's constant. If you wished, you could write the energy term as hc/λ:

$$H^* \rightarrow H + \frac{hc}{\lambda}$$

In this case, you would talk instead about the *wavelength* of light emitted.

After they lose energy by emitting radiation, the hydrogen atoms (H) eventually recombine to form H_2, the stable form of hydrogen under normal conditions.

$$H + H \rightarrow H_2$$

Each of the recombined molecules is then able to absorb more energy from the electrical discharge to dissociate into two more excited atoms, and the cycle begins again. This makes the tube glow continuously as electrical energy is converted into electromagnetic energy.

If we pass the light emitted by the excited hydrogen atoms through a prism, we can separate the light into individual wavelengths to get the hydrogen **emission spectrum**, as shown in Figure 6.7. Note that unlike the Sun's emission, only specific, discrete wavelengths of light are emitted in the hydrogen emission spectrum.

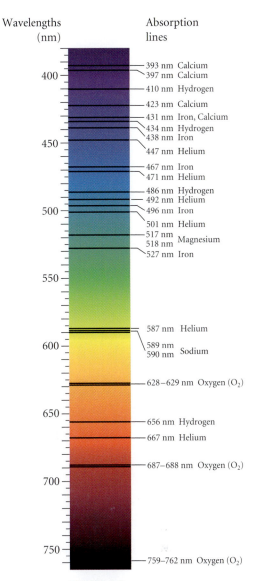

FIGURE 6.6

The spectrum of light from the Sun includes dark bands at certain frequencies, caused by particular elements in the Sun absorbing light at these frequencies. This phenomenon can be used to identify which elements are present in the sun. The existence of helium was postulated by this method.

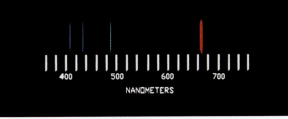

FIGURE 6.7

The hydrogen emission spectrum. An electric current is passed through a tube of hydrogen gas, which causes some of the molecules to dissociate to form excited hydrogen atoms (H*). The H* atoms lose the energy by emitting radiation, which is separated by wavelength with a glass prism. The intensity of the emitted radiation can then be observed directly with your eye or can be captured on film with electronic detectors such as a photomultiplier tube, or a change-coupled device.

Balmer's Analysis and Atomic Structure

Balmer found that all of the wavelengths emitted from the hydrogen atom in the visible portion of the spectrum could be fitted to an equation that relates the wavelength (in nanometers) to an integer (n):

$$\frac{1}{\lambda} = 1.0968 \times 10^{-2} \left[\frac{1}{2^2} - \frac{1}{n^2} \right] \text{nm}^{-1}$$

The most significant thing about this equation is that it introduces to us a **quantum number**, n, that can vary only in *whole, positive increments*. For example, the equation predicts an infinite number of wavelengths in the spectrum generated by $n = 3, 4, 5 \ldots$ all the way to infinity. If you don't quite understand where the quantum number comes from, you are in good company. Balmer did not understand it either, but he used the equation, complete with quantum numbers, because it fits the data from the hydrogen atom. He had no theory or model that predicted this fit and had no good reason to propose the quantum numbers—apart from the fact that they worked.

Although in concept an infinite number of wavelengths can be emitted in the Balmer spectrum, many wavelengths of light *cannot* be emitted. For example, the longest wavelength of visible light emitted from the hydrogen atom (656.5 nm) is found when $n = 3$. The next longest wavelength in the Balmer emission spectrum (486.1 nm) is at $n = 4$. No light is emitted with a wavelength between 486.1 nm and 656.5 nm.

Although Balmer's series is predominantly in the visible region of the electromagnetic spectrum, other hydrogen emission series exist that can all be fitted to a similar form of an equation. This generalized equation was developed by the Swedish physicist Johannes Rydberg in 1888 and is known as the Rydberg formula. The main feature of the Balmer series occurs when we set $n_f = 2$. The other quantum number is variable, and we'll call it n_i.

The Rydberg Formula: $\dfrac{1}{\lambda} = 1.0968 \times 10^{-2} \left[\dfrac{1}{n_f^2} - \dfrac{1}{n_i^2} \right] \text{nm}^{-1},$

The Balmer Equation: $\dfrac{1}{\lambda} = 1.0968 \times 10^{-2} \left[\dfrac{1}{2^2} - \dfrac{1}{n^2} \right] \text{nm}^{-1},$

Remember that Balmer didn't attach a physical meaning to the numbers. He was just looking at the data from the standpoint of a mathematician.

The physicist Theodore Lyman reported observing a series of wavelengths in the ultraviolet and X-ray region. Each of those wavelengths could be calculated if n_f were held constant at 1 and if n_i were varied from 2 to 3 to 4 to 5 to ∞. In addition, there is a series with $n_f = 3$, with quantum numbers $n_i = 4, 5, 6, \ldots ∞$, which begins in the infrared region; another with $n_f = 4$ and quantum numbers $n_i = 5, 6, 7, \ldots ∞$; another with $n_f = 5$ and $n_i = 6, 7, 8, \ldots ∞$; and so on, with n_f increasing in units of one up to infinity. Table 6.2 lists these first four series for the hydrogen atom.

TABLE 6.2 Spectral Series in the Hydrogen Atom			
Spectral Series	n_f	n_i	**Wavelength Range**
Lyman	1	2, 3, 4, . . .	X-ray and UV
Balmer	2	3, 4, 5, . . .	Visible
Ritz-Paschen	3	4, 5, 6, . . .	Short-wave infrared
Brackett	4	5, 6, 7, . . .	Long-wave infrared

EXERCISE 6.3 **Using the Balmer Equation**

The longest wavelength for the Balmer series is found when $n = 3$:

$$\frac{1}{\lambda} = 1.0968 \times 10^{-2} \left[\frac{1}{2^2} - \frac{1}{3^2} \right] nm^{-1} = 0.0015233 \; nm^{-1}$$

$$\lambda = \frac{1}{0.0015233 \; nm^{-1}} = 656.47 \; nm$$

What is the smallest energy of radiation for this series?

Solution

The smallest energy corresponds to the longest wavelength, because energy and wavelength are inversely proportional. We just calculated the longest wavelength to be 656.47 nm, and if we put this into SI units as 6.5647×10^{-7} m, we can convert it directly into the energy:

$$E_{smallest} = \frac{hc}{\lambda} = \frac{(6.626 \times 10^{-34} \; J \cdot s) \times (3.00 \times 10^8 \; m \cdot s^{-1})}{6.5647 \times 10^{-7} \; m} = 3.03 \times 10^{-19} \; J$$

PRACTICE 6.3

Calculate the wavelength of a photon corresponding to the collapse of an electron from $n = 4$ to $n = 3$; from $n = 6$ to $n = 4$; from $n = 6$ to $n = 5$.

See Problems 25, 26, 30–34, and 111.

The equation describing the wavelengths of light found in the hydrogen emission spectrum has been **empirically derived**, which means that it has been derived from experiments and observations rather than from theory. In other words, Balmer, Lyman, and other mathematicians and spectroscopists of the late nineteenth century performed comprehensive measurements on the hydrogen emission spectrum and *fitted a mathematical expression to the data*. Empirical fits to data generally raise more questions than they answer, because scientists want to know *why* the equation works. **Why, for example, is the hydrogen emission spectrum quantized to give only certain wavelengths?** Figure 6.8 shows the visible region of the emission spectra of several elements. Note that the wavelengths of light emitted by each metal are *specific to that element*. For instance, if we focus an instrument at one of those wavelengths and look specifically for that signal, we will observe it only if that particular element is present in the sample. In other words, if we tune our instrument to one of the wavelengths in the copper spectrum, our instrument will be able to detect if copper is present. This process is quite useful to the environmental chemist. By exciting atoms within a small sample of river water, the chemist can detect the presence of copper. In addition, the intensity of the light detected by the instrument can be related to the quantity of copper in the river water sample. It is currently possible to measure nearly two dozen elements in water *at the same time* at the parts per billion level (ppb, micrograms, μg, of element per liter of solution) or parts per trillion level (ppt, ng/L) by simultaneously focusing on one discrete wavelength given off by each element.

Table 6.3 gives several of the elements, the emission wavelengths typically used for detection, and the minimum concentration of the element that can be

Application

CHEMICAL ENCOUNTERS:
Simultaneous
Determination of
Elements in Water

TABLE 6.3	**Simultaneous Determination of Elements in Water**	
Element	**Wavelength (nm)**	**Detection Limit (ppb = μg/L)**
Al	396.152	1.5
As	188.980	3.5
Ba	455.403	0.04
Ca	315.887	1.5
Cd	214.439	0.3
Cr	267.716	0.5
Cu	324.754	0.3
Mo	202.032	0.8
Pb	220.353	3.0
Zn	213.857	0.3

Source: Fast Analysis of Water Samples Comparing Axially and Radially Viewed CCD Simultaneous ICP-AES, Varian, Inc., http://www.varianinc.com/image/vimage/docs/products/spectr/icpoes/atworks/ icpes028.pdf (accessed February 2006).

FIGURE 6.8

The emission spectra of different metals. Each wavelength emitted is unique to that element, which makes it possible to analyze the element's concentration at very low concentrations.

Video Lesson: CIA
Demonstration: Flame Colors

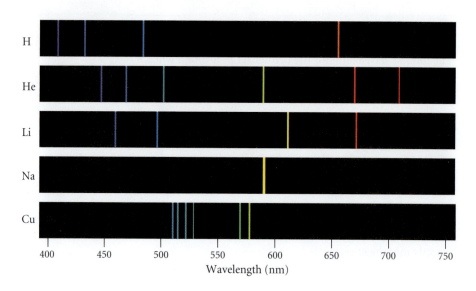

reliably detected ("detection limit"). Note that the majority of these wavelengths are in the ultraviolet region of the electromagnetic spectrum. Atomic emission, absorption, and related spectroscopy techniques are among the most important in chemistry for the analysis of elements in all manner of samples from brass to blood. It is through atomic spectra that we learned core ideas about the fundamental structure of the atom.

Video Lesson: The Bohr Model

6.4   The Bohr Model of Atomic Structure

In Chapter 2, we saw how the model of the atom evolved from the indivisible spheres proposed by Dalton all the way to the "nuclear atom" view that emerged from Rutherford's discovery of the nucleus. We also discussed the way in which scientific models develop and noted that useful information is gleaned even from models that eventually turn out to be incorrect or only partially correct. Like the earlier models, the Bohr model we are about to consider provides an incomplete and oversimplified view of how atoms are really constructed. But it is useful to examine this model and to appreciate how it solved certain puzzles of atomic emission and contributed to scientists' development of the quantum picture of the atom. The work is so important that it earned Niels Bohr the Nobel Prize in physics in 1922.

The simplest atom is that of hydrogen with a single electron and a nucleus of a single proton. According to the Bohr model of the hydrogen atom, the electron is found some distance from the nucleus, whirling around it in a circular orbit, as shown in Figure 6.9. The electron can be in any one of an infinite number of possible orbits around the nucleus, each orbit at a particular distance, or radius, from the nucleus. Not just any radius will do, however. The orbits are *spatially* quantized. Much like the layers of an onion, they are confined to specific locations, in spheres of ever-increasing radii, r, expressed in nanometers, according to the equation:

$$r = 0.052917\, n^2 \text{ nm}$$

where n, our quantum number for the Bohr model, can be 1, 2, 3, 4, ..., ∞.

Electrons in orbits close to the nucleus travel much more rapidly than electrons in the outside orbits. This means that the kinetic energy of an

FIGURE 6.9

Schematic of the Bohr model of the hydrogen atom.

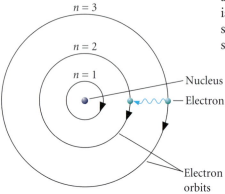

electron in the orbits close to the nucleus is greater than the kinetic energy of those in the outer orbits. This is true for satellites that orbit the earth just as it is for electrons. The space shuttle circles the planet Earth many times during a voyage in a low Earth orbit of between 185 and 650 km; it must travel at a very high speed, about 27,900 km/h, in order to maintain its orbit. On the other hand, geosynchronous satellites, such as weather and communications satellites, are 35,800 km above the Earth's surface and travel at a speed of only 11,300 km/h. They appear to remain stationary over a specific place because their movement just matches the rotation of our planet.

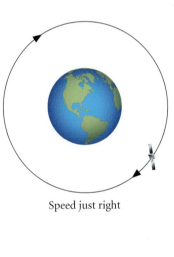

In 1922 Niels Bohr (1885–1962) was awarded the Nobel Prize in physics for his work on the structure of the atom and his explanation of how atoms emit light.

The increase in kinetic energy of the negatively charged electron closest to the nucleus is offset by a greater decrease in the potential energy due to its position relative to the positive nucleus. In other words, because opposite charges attract, the closer the electron and the nucleus are to each other, the lower the potential energy of the electron. *The key outcome is that electrons closer to the nucleus have lower energy than those farther away.*

We call the allowed orbits **energy levels**, because the electron in a given orbit will have a constant total energy according to this equation:

$$E_n = -\frac{2.1786 \times 10^{-18} \text{ J}}{n^2}$$

What is especially interesting about this equation is that the quantum number n is the same quantum number we discovered in the equation for the radius. Note that this equation calculates the energy of a given orbit as a negative number. This is because *a completely free electron (corresponding to $n = \infty$) is assigned an energy of zero.* When electrons become bound within atoms, their energy falls, and because it is falling from zero, it must become negative. This means that as an electron falls from higher orbits to lower ones, *its energy has increasingly larger negative values relative to zero, the value for the free electron.* The lower values of the quantum number have more negative energies than the higher values. The most negative value for energy occurs when $n = 1$. At this orbit ($n = 1$), the electron is as strongly bound to the atom as is possible, and it is in its lowest energy state in the atom. Unless energy is supplied to the hydrogen atom, its electron will tend to be in the most strongly bound level, closest to the nucleus. As we noted above, for the hydrogen atom, this is the smallest quantum number, $n = 1$. Any atom that contains all its electrons in their lowest possible energy levels is said to be in the **ground state**, like the ground floor of a building.

Speed just right

Larger quantum numbers, $n = 2, 3, 4, \ldots, \infty$, represent progressively higher atomic energy levels, just as the second, third, fourth, . . . floors of a building represent progressively higher gravitational energies. When electrons are moved from the ground state into these higher energy levels, we say that they occupy **excited states**. Atoms and molecules in excited states generally tend to relax back down to their ground states after a short period of time.

Energy must be supplied to the hydrogen atom to promote its electron from the ground state ($n = 1$) to any other state ($n > 1$). Conversely, the atom must lose energy when the electron falls from any state to a lower energy state. These **electronic transitions**, as they are called, can be explained very well by the quantum picture of the Bohr atom. The electron must gain or lose the exact amount of energy that separates two energy levels (two orbits) in order to jump between the levels. The change in energy is expressed mathematically as

Speed too slow

The speed of a satellite is related to its distance from the planet it circles. Slower-moving objects must have higher orbits or they will crash back to the surface.

$$\Delta E = E_f - E_i$$

where E_f is the energy of the final state of the electron, E_i is the energy of the initial state of the electron, and ΔE is the difference in energy between these two levels.

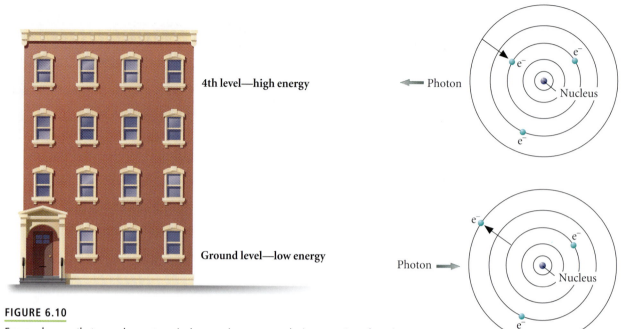

4th level—high energy

Ground level—low energy

FIGURE 6.10

Energy changes that occur in an atom. An increase in energy results in promotion of an electron to a higher energy level. A release of energy is observed when the electron returns to its original state.

The change in energy made during an electronic transition can be either positive or negative as long as the energy is conserved. If ΔE is positive, energy is supplied to the atom to make the electron jump from a lower to a higher state. If ΔE is negative, energy is released from the atom because the electron falls from a higher to a lower energy level, as shown in Figure 6.10.

One way to supply or release the required energy is for the atom to absorb or emit electromagnetic radiation. This is where the hydrogen emission spectrum proves useful. If we write the equation for hydrogen emission in terms of energy instead of wavelength, we can show that the emission lines are the result of electronic transitions from the higher energy levels (or states) of the excited hydrogen atom to lower energy states. We call the quantum number of the initial energy level, n_i and that of final energy level n_f:

$$\Delta E = -2.1786 \times 10^{-18} \text{ J} \left[\frac{1}{n_f^2} - \frac{1}{n_i^2} \right]$$

Note that the energy change is a discrete value. In other words, *the energy is quantized.* For the electron to move from one energy level to a lower energy level, it must lose this discrete amount of energy. Knowing that the energy is related to the frequency indicates that $h\nu$, the electromagnetic energy, is equal to the energy lost in the transition, ΔE:

$$h\nu = \frac{hc}{\lambda} = \Delta E$$

This relationship ($\Delta E = h\nu$), although somewhat similar to that proposed by Einstein, was proposed by Niels Bohr in an effort to correlate the energy emitted from a hydrogen atom during an electronic transition with the frequency of observed light. The relationship is referred to as the Bohr frequency condition.

If we combine the last two equations to calculate the wavelength of radiation predicted for the hydrogen emission spectrum and convert from meters to nanometers ($1 \text{ m} = 10^9$ nm), we get

$$\frac{1}{\lambda} = 1.097 \times 10^7 \left[\frac{1}{n_f^2} - \frac{1}{n_i^2} \right] \text{m}^{-1} = 1.097 \times 10^{-2} \left[\frac{1}{n_f^2} - \frac{1}{n_i^2} \right] \text{nm}^{-1}$$

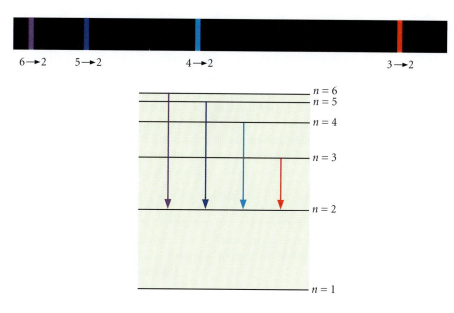

FIGURE 6.11

The visible electronic transitions of the hydrogen emission spectrum. The energy of the radiation observed must exactly match the difference between energy of the excited state and that of the ground state (most energetically stable state).

Visualization: The Line Spectrum of Hydrogen

This is the equation that Rydberg, Balmer, and others found by analyzing the hydrogen emission spectrum and empirically fitting an equation to the results. When $n_f = 1$, we get the Lyman series. When $n_f = 2$, we get the Balmer series. We can continue this process until we generate all of the energies of electromagnetic radiation observed to be emitted from the hydrogen atom. These electronic transitions in the visible region are depicted schematically in Figure 6.11.

EXERCISE 6.4 Energy-to-Frequency Conversion in the Hydrogen Atom

A hydrogen atom in one of its excited states has an energy of -1.5129×10^{-20} J. What frequency of radiation is emitted when the atom relaxes down to its ground state $(n = 1)$?

First Thoughts

An important idea here is that an atom moving from its excited state to its ground state *releases* energy (ΔE is $-$), typically as light, whereas an atom that goes from the ground state to an excited state *absorbs* energy (ΔE is $+$).

Solution

The energy of the photon *emitted* must exactly match the energy *lost* by the hydrogen atom as the electron moves from a higher energy level to its ground state:

$$E_{\text{photon}} = -\Delta E_{\text{electron}}$$
$$h\nu = -(E_f - E_i)$$
$$\nu = \frac{-(E_f - E_i)}{h}$$

The final state energy is given by

$$E_f = -\frac{2.1786 \times 10^{-18} \text{ J}}{n^2} = -\frac{2.1786 \times 10^{-18} \text{ J}}{1^2} = -2.1786 \times 10^{-18} \text{ J}$$

$$\nu = -\frac{(E_f - E_i)}{h} = -\frac{(-2.1786 \times 10^{-18} \text{ J}) - (-1.5129 \times 10^{-20} \text{ J})}{6.626 \times 10^{-34} \text{ J·s}}$$

$$\nu = -\frac{(-2.163 \times 10^{-18} \text{ J})}{6.626 \times 10^{-34} \text{ J·s}} = 3.265 \times 10^{15} \text{ s}^{-1}$$

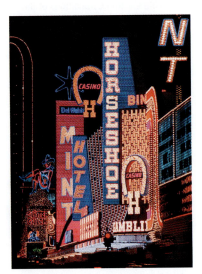

A neon sign emits light when an electric current is passed through the tube.

Further Insights

Our discussion is focusing on the hydrogen atom. Helium, with its two electrons, was discovered in 1868 by the British astronomer Norman Lockyer and, in separate observations of the emission spectrum from the Sun, Pierre Janssen of France. The solar spectrum showed lines never before seen, indicating a previously unknown element. Lockyer named the element helium after the Greek word for Sun, *Helios*. The discovery was confirmed in 1895 by the Scottish chemist William Ramsey, who saw the same spectrum when looking at gases coming from the heating of the uranium-containing mineral cleveite. Ramsey is known for discovering all of the noble gases except radon.

PRACTICE 6.4

What wavelength of radiation needs to be supplied to a hydrogen atom in its ground state to raise it to its excited state that has an energy of -1.5129×10^{-20} J?

See Problems 35–40 and 43–46.

We make use of the emission of light from excited atoms everyday. For example, when electric current is passed through a tube containing a small amount of neon gas, it causes the neon atoms to become excited. As they return to the ground state, they emit light. **What is this contraption?**

When the tube is bent to spell words, we refer to the contraption as a neon sign. Another example of light emission from excited atoms is found in **lasers**. Lasers (the acronym stands for "**l**ight **a**mplification by **s**timulated **e**mission of **r**adiation") provide a powerful light source because they involve the simultaneous collapse of a large population of excited atoms to the ground state.

The light emitted by each excited atom has the same phase (the peaks and troughs of the light waves line up), giving rise to a powerful emission of light. Table 6.4 sheds some light on common lasers and their uses.

Application

CHEMICAL ENCOUNTERS: The Nature and Applications of Lasers

TABLE 6.4	Lasers and Their Applications	
Lasing Medium	**Wavelength**	**Application**
Diode	635 nm (red) 670 nm (deep red) 780 nm, 800 nm, 900 nm, 1550 nm (Infrared)	CD players, DVDs, laser printers, laser pointers, high-speed fiber optics, dental surgery
Helium/neon (HeNe)	632.5 nm (red)	Alignment, barcode scanners, laser pointers, light shows, blood cell counting
Argon and krypton	457.9 nm (violet-blue) 488 nm (blue) 514 (green) 646 (red)	Forensic medicine, high-performance printing, general and ophthalmic surgery, holography, light shows
CO_2	1060 nm (mid-IR)	Industrial metal cutting, welding, surgery
HeCd	442 nm (violet blue) 325 nm (UV)	Nondestructive testing and spectroscopy
Solid state Nd:YAG Ruby	 1064 nm (IR) 694 nm (red)	Materials processing (drilling, cutting, welding), target designation, surgery, hair removal
ArK excimer	193 nm (UV)	Laser eye surgery

A laser pointer, one example of the many uses of lasers.

HERE'S WHAT WE KNOW SO FAR

- Electromagnetic radiation is characterized by wavelength, frequency, and energy. These values can be interconverted.
- Electromagnetic radiation is quantized into photons, the smallest unit of radiation.
- An atom can gain or lose discrete units of energy when an electron jumps between the orbits allowed by the Bohr model.
- Each line in the hydrogen emission spectrum is created by the light released when an electron falls between two of the allowed orbits in the Bohr model.
- Each time an electron falls between two levels, one photon of light is released, carrying exactly the amount of energy equal to the difference between the two energy levels.
- The frequencies (and therefore wavelengths) of light emitted are determined by the energy levels available for electrons to fall between. In other words, the light coming out of the atom is a consequence of the atom's electronic structure.

6.5 Wave–Particle Duality

We can dig deeper into the explanations that enable us to make sense of atoms if we look at some apparent contradictions between the way the world appears on the basis of everyday experience and the way it behaves at the level of individual atoms. These two views of the world are known as the classical mechanical and quantum mechanical views. In **classical mechanics** we have particles and we have waves, each with their own distinct characteristics. The particles do not act like waves, and the waves do not act like particles. This is the kind of behavior we see on a macroscopic level, as shown in Figure 6.12. In quantum mechanics, we can still talk about particle and wave behavior, but the distinction between particles and waves becomes fuzzier, although it is still of great importance, as we shall see.

FIGURE 6.12

A person is very different from a wave. In the quantum realm the nature of people (particles) and waves become intertwined in strange but very significant ways.

According to classical mechanics, particles have exact locations, and their positions can be defined precisely in space. Classically, a particle can move at any speed (below the speed of light), provided that it is supplied with the necessary energy. Also, if a particle is moving, it carries momentum (p), which is defined as mass multiplied by velocity (v):

$$p = m \times v$$

A 100-ton coal car moving at 30 mi/h has considerably more momentum than a 1-ton compact car moving at the same speed. A coal car traveling at 30 mi/h has more momentum than another coal car that is crawling along the tracks. Both mass and velocity contribute to momentum.

One of the main descriptors a particle has is mass. In addition, because all particles can carry momentum, they *must* have mass. This means that we can use momentum to test whether something is behaving like a classical particle or not.

The Bohr model of the hydrogen atom defies the classical logic of the behavior of particles. Although electrons have momentum and therefore carry mass, they are quantized in the Bohr model both in terms of where they can be and in terms of how much energy they can have. The quantized nature of space and energy is not the kind of behavior we expect from particles on a *macroscopic* level. For example,

The lighter of two cars has much less momentum, even if the vehicles are traveling at the same speed. Consequently, it usually suffers the greater amount of damage in a collision.

FIGURE 6.13

A tuned violin string plays a note when it vibrates at a specific frequency, when a standing wave is set up between the end points of the string. A different note can be played when the player's finger is used to vary the length of the string. But only certain lengths—certain positions of the player's finger—will allow a proper note to be played when a standing wave of a different frequency occurs. The notes are a set of quantized frequencies.

when we think of a moving race car, we envision it as capable of possessing any velocity or mass within a continuous range. This is not true for the quantum mechanical view of the atom.

However, quantized energy states and locations have long been known in everyday life. For example, musicians who play stringed instruments quantize the energy of the vibration of the strings by shortening or lengthening the string being played. This changes the frequency of the string's vibration by setting the wavelength of the vibration, as shown in Figure 6.13. A wave vibrating between two fixed endpoints is called a **standing wave**.

The crucial word here is *wave*. All of the cases that were known to create quantized states prior to the Bohr atom were systems that could be described by "waves," oscillating forms of energy that had a specific wavelength and frequency. To clarify the distinction between the ideas of particles and waves, we can first list the characteristic classical properties of particles as follows:

In the Classical Mechanical View of the World:

- Particles have a continuous range of available energies and positions (such as a person going out for a walk).

- Particles have a single value of energy and position measured at any one time (such as a person seated in a chair).

- Particles carry momentum (have mass, such as the mass of the person).

- Particles are quantized (exist as individual particles, such as the person himself or herself).

We can now build up a similar list of criteria for the classical mechanical view of waves. In the view of classical mechanics, waves are neither composed of single units nor located at exactly one place in space as are individual particles. As we can see in our picture of the violin string in Figure 6.13, the wave exists everywhere between the bridge and the player's finger. The amplitude changes between these two sites, but the wave exists along the whole vibrating part of the violin string. It is smeared out over a pretty large volume of space.

Waves can also be quantized by confining them to a specific region in space, as we do when we tie the two ends of the violin string down with the bridge and our finger. In order for the amplitude of the wave to go to zero at the ends, the number of wavelengths that must fit between the two ends must be some half-integral number of the wavelength:

$$\text{Length of string} = n\lambda/2$$

In this example, the length of the string, l, sets the possible wavelengths that the vibration can have to $\lambda = 2l, l, (2/3)l, (1/2)l, (2/n)l, \ldots, 0$. This is illustrated in Figure 6.14, where we can see the first few allowed standing waves. Note that whereas $2l$ ($n = 1$) and l ($n = 2$) are permissible values, no wavelengths between these two values are allowed.

We can now list the characteristic classical mechanical properties of waves:

In the Classical Mechanical View of the World:

■ Waves have specific energy values (such as the specific notes on a violin string).

■ Waves occupy a range of spatial positions *all at the same time* (such as the positions of the vibrating violin string).

■ Waves do not carry momentum (have no mass; the sound waves from the violin don't have a mass).

■ Waves have a continuous range of amplitudes (smallest, single wave units; such as the loudness of each note from the violin).

In our everyday (macroscopic) world, the characteristics of particles and waves are separate. But here is a truly strange thing: If we investigate matter and energy on a *quantum* level, where dimensions are on the order of atomic spacing, *things we once regarded as particles, such as electrons, can also behave like waves, and things that were traditionally thought of as waves, such as light, can also behave like particles.* The electrons will still behave like particles, and the electromagnetic radiation will still retain the properties of waves. But the distinction between particles and waves is no longer absolute. *All quantum "things"—electrons, protons, photons of electromagnetic radiation, whatever—have the properties of both waves and particles simultaneously.*

This schizophrenic behavior is called the **wave–particle duality**, which proposes that at very small dimensions, things behave like both waves and particles, regardless of how they behave on the macroscopic level. The earliest experiments performed on the electron investigated particle behavior by measuring the electron mass (the Millikan oil drop experiment, Chapter 2) and position (J. J. Thomson and the cathode ray tube, Chapter 2). However, we can show that the energy quantization observed in the hydrogen emission spectrum or predicted by the Bohr model results from the wave nature of the electron. The electron behaves as both a particle and a wave. It has mass, carries momentum, and is found as a discrete single unit, which are properties expected of a classical particle, but when contained in an atom or molecule it is spatially and energetically quantized, which are properties expected of a classical wave.

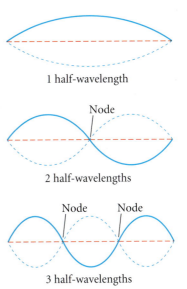

1 half-wavelength

2 half-wavelengths

3 half-wavelengths

FIGURE 6.14

Overtones of a violin string. In addition to the half-wavelength standing wave, the violin also produces integral multiples of the half-wavelength. These vibrations are more energetic than the half-wavelength vibration and are not deliberately played by a violinist.

6.6   Why Treating Things as "Waves" Allows Us to Quantize Their Behavior

To see why the wave nature of the electron allows its energy levels to be quantized, look at Figure 6.15, where we have drawn a "cross section" of several possible orbits for the Bohr model. The Bohr model is three-dimensional, with spherical orbits, but that would be hard to show schematically, so we will show a circular cross section of the orbit.

Much like the standing wave vibrating on the violin string, only certain wavelengths will fit on a circle. If the radius of the circle is r, we can quantize the allowed wavelengths, λ, to be integral multiples of the circumference, which you might recall from your high school geometry class has a length of $2\pi r$:

$$\text{Circumference} = 2\pi r = n\lambda$$

In other words, we can fit n waves of wavelength λ around the circumference of the circle, as shown in Figure 6.16.

FIGURE 6.15

Circular cross-sections of the atomic energy levels of the Bohr atom.

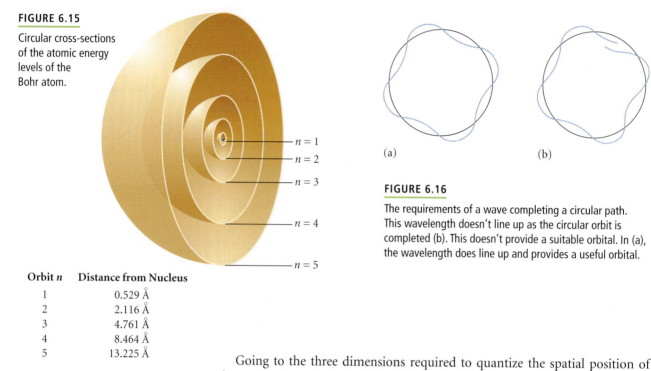

$n = 1$
$n = 2$
$n = 3$
$n = 4$
$n = 5$

Orbit n	Distance from Nucleus
1	0.529 Å
2	2.116 Å
3	4.761 Å
4	8.464 Å
5	13.225 Å

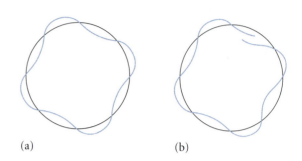

(a)   (b)

FIGURE 6.16

The requirements of a wave completing a circular path. This wavelength doesn't line up as the circular orbit is completed (b). This doesn't provide a suitable orbital. In (a), the wavelength does line up and provides a useful orbital.

www Video Lesson: The Wave Nature of Matter

Going to the three dimensions required to quantize the spatial position of electrons in the Bohr hydrogen atom is considerably more difficult. Bohr himself never actually assigned a wavelength to his electrons, which remained decidedly particle-like in his own model. Bohr merely postulated that the momentum and radius of permissible levels were quantized because they had to be if his model were to explain the hydrogen emission spectrum. But the *reason* why space and energy are quantized for the hydrogen atom arises from the wave-like nature of the electron.

If electrons behave like waves, how do we find their wavelengths? We can do this by using the de Broglie formula, proposed by Louis de Broglie in 1924:

$$\lambda = \frac{h}{p} = \frac{h}{mv}$$

The equation, which shows that the wavelength, λ, is inversely proportional to the momentum, p, through Planck's constant h, connects wave (λ) and particle (p = momentum) properties in a single formula. Again, note that the momentum term, p, uses the velocity (v). This symbol looks similar to that for the frequency (ν), but it is vastly different. The velocity has units of m/s, whereas frequency has units of 1/s.

What is the purpose of the de Broglie equation? It illustrates the basic principle of the wave–particle duality: *Moving particles exhibit definite wave-like characteristics.* The values of wavelengths calculated by this equation are meaningful for small, fast-moving particles such as the electron. The calculation is less meaningful but still demonstrative for massive, slow-moving objects.

Louis-Victor de Broglie (1892–1987), a French physicist and prince, published his wave–particle duality theory in 1924.

EXERCISE 6.5 **Working with the de Broglie Equation**

An electron in the ground state of the hydrogen atom has a velocity of about 2.1×10^6 m·s⁻¹. What is its de Broglie wavelength?

Solution

$$\lambda = \frac{h}{p} = \frac{h}{mv} = \frac{6.626 \times 10^{-34} \text{ J·s}}{(9.109 \times 10^{-31} \text{ kg}) \times (2.1 \times 10^6 \text{ m·s}^{-1})}$$

Because 1 J $= 1$ kg·m^2·s^{-2}, substituting yields

$$\frac{6.626 \times 10^{-34} \text{ kg·m}^2\text{·s}^{-2} \text{ (s)}}{(9.109 \times 10^{-31} \text{ kg}) \times (2.1 \times 10^{6} \text{ m·s}^{-1})} = 3.5 \times 10^{-10} \text{ m} = 3.5 \text{ Å, or } 0.35 \text{ nm}$$

PRACTICE 6.5

Calculate the de Broglie wavelength for a 1200-lb car traveling 75 mi/h.

See Problems 59 and 60.

One application of the wave nature of electrons lies in their very small wavelength. Microscopes magnify and allow us to see objects with visible light. The limit to their magnification is the wavelengths of electromagnetic radiation (in this case, visible light). Objects must be larger than the wavelengths of the radiation in order to be seen. Using an optical microscope, we are able to see objects as small as about 200 nm (the radius of a hydrogen atom is 25 picometers, or 0.025 nm). We are able to see objects much smaller than this if we use electrons to image the objects. This is the principle behind the scanning electron microscope (SEM). Scanning electron microscopes operate much like compound light microscopes, but they use a beam of electrons to generate an image of the sample rather than using light to image the sample. Because the wavelength of an electron is about 1000 times smaller than that of visible light, and given instrumental imperfections, a scanning electron microscope can image objects or features as small as about 3 nm. This means that the SEM can achieve a magnification of up to 300,000-fold. Unfortunately, this is still too large to "see" individual atoms. For that we need to use a scanning tunneling microscope (STM) (see Figure 6.17). The STM takes advantage of another quantum effect, tunneling, which we will discuss in Section 6.10.

 Application

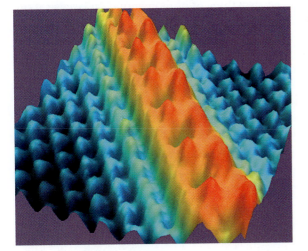

An atom is smaller than the visible wavelengths of light and cannot be seen with an optical microscope.

FIGURE 6.17

Scanning tunneling microscope image of cesium atoms lined up in zigzag fashion on a GaAs surface. An STM can allow us to "see" objects at a much higher magnification than an optical microscope.

6.7   The Heisenberg Uncertainty Principle

 Video Lesson: The Heisenberg Uncertainty Principle

How does wave–particle duality alter our ideas of how the electron is moving within the hydrogen atom? The Bohr model regards the electron as a particle allowed to orbit the nucleus at only certain precisely quantized distances. However, if it is also to behave as a wave, the electron must be spread out over all of its orbit in the hydrogen atom. We can reconcile these two very different ideas by considering scale. Assume that we measure the electron's position as having a specific value of x, y, and z in our macroscopic laboratory setting. If the electron is bound in a hydrogen atom but is somehow spread out, some uncertainty will be introduced into our measurement of x, y, and z. But the uncertainty is very small, on the order of 0.1 nm or so. Our macroscopic rulers are not this precise, so it looks to us as if the electron is at a specific and absolute position in space.

If we then look at the electron on the microscopic level, we find that the best we can do to establish its position is to say it is *somewhere* in the atom, within a certain range from of the atom's center. We see a particle's behavior of spatially discrete positions with continuous possible values when we are looking at things on a macroscopic scale, and we see a wave's behavior of being allowed only in specific regions of space but spread out to fill the "container" of the atom when we make microscopic measurements.

This leads us to another of the "Funny things happen" phenomena that we observe only when we are in the microscopic quantum world. On a macroscopic level, you can measure both the position and the momentum of a particle as precisely as your instruments will allow. Quantum mechanics, however, proposes that, regardless of the instrument used to measure them, there is an ultimate uncertainty in both the position and the momentum of a particle, and these uncertainties are connected. This is called the **Heisenberg uncertainty principle** after Werner Heisenberg, who proposed it in 1925. Mathematically, we can state the Heisenberg uncertainty principle as

$$\Delta x \Delta p_x \geq \frac{h}{4\pi}$$

where Δx is the uncertainty in position, Δp_x is the uncertainty in momentum along the x-direction, and h is Planck's constant.

There is no limitation on how precisely we measure either x or p_x. Instead, the limitation is on the product of their uncertainties $\Delta x \Delta p_x$. In other words, as the uncertainty in x becomes smaller, the uncertainty in p_x gets larger. Therefore, as Δx approaches zero and we know exactly where the particle is positioned, Δp_x becomes infinite and we do not know anything of its momentum. In essence, it can have any value of momentum possible, including infinity. If we know the mass of the particle, then infinite momentum implies infinite velocity. In short, limitations are set on real-world behavior by the Heisenberg uncertainty principle, which among other things limits the maximum achievable resolution of microscopes, the ultimate size of computer chips, and, in the nanoworld, the behavior of actual atoms and molecules.

The position and momentum of a racing car are known with enough precision for the macroworld.

6.8   More About the Photon—the de Broglie and Heisenberg Discussions

We have focused on the electron in our previous discussion, but the principles apply to all particles, although the effect will really be *evident* only on the microscopic level. Any small particle such as the electron, proton, or neutron would be a good candidate for testing the Heisenberg uncertainty principle. Bowling balls, elephants, and cars would not. But what about phenomena that are wave-like in our macroscopic world? What happens to waves on the quantum scale?

Electromagnetic radiation is wave-like on a macroscopic level. Light, in a manner similar to chemical substances, can be decreased only so far. Eventually, you will come to its smallest, indivisible unit. For light, the smallest unit is the photon, and, like particles, photons have been shown to carry momentum, another of our particle-like characteristics. In fact, we can calculate the momentum using the de Broglie equation. **What does such a calculation tell us?** Photons, although we think of them primarily as waves, act like particles too. Again, the calculation verifies the particle–wave duality.

EXERCISE 6.6   **Photons and Momentum**

How much momentum is carried by a single green photon and by a mole of green photons? Assume that green photons have wavelengths of 530 nm.

Solution

We can use the de Broglie equation to calculate the momentum of a single photon. Because $\lambda = \dfrac{h}{p}$,

$$p = \frac{h}{\lambda} = \frac{6.626 \times 10^{-34} \text{ J·s}}{5.30 \times 10^{-7} \text{ m}}$$

$$= \frac{6.626 \times 10^{-34} \text{ kg·m}^2\text{·s}^{-2} \times \text{s}}{5.30 \times 10^{-7} \text{ m}} = 1.25 \times 10^{-27} \text{ kg·m·s}^{-1}$$

This is for a single photon. If we want the momentum for a mole of photons, we must multiply this value by Avogadro's number:

$$p = 1.25 \times 10^{-27} \text{ kg·m·s}^{-1}\text{·photon}^{-1} \times (6.022 \times 10^{23} \text{ photons·mol}^{-1})$$

$$= 7.53 \times 10^{-4} \text{ kg·m·s}^{-1}\text{·mol}^{-1}$$

Does our answer make sense? In comparison, a 68-kg (150-lb) person ambling down the road at 2 mi/h has a momentum of 60 kg·m·s^{-1}, so you can see that even a mole of green photons has a much smaller momentum.

PRACTICE 6.6

Calculate the momentum of a mole of photons with a wavelength of 200 nm, of 700 nm, and of 1000 nm. Is there a difference? Explain.

See Problems 65–70.

A word of caution about this type of calculation: The wavelength will have the same value, as will the frequency, whether you have one photon or a mole of photons. These properties of a photon are *intrinsic properties*—they do not change with a change in amount. The *total* momentum and energy carried by the electromagnetic radiation, however, are *extrinsic properties* because both depend on the number of photons you have. When we use the formula $E = h\nu$, we always use the energy of a single photon. Similarly, with the de Broglie equation, we always use the momentum of a single photon.

We also need to be careful of the answer that the de Broglie equation gives us for the mass of a photon. The mass that appears in the de Broglie formula is known as the **rest mass**, the mass the particle would have if it were not moving at all. Photons, however, are moving at 2.998×10^8 m/s, and slowing them down to zero speed is just not feasible. The rest mass you calculate for photons from the de Broglie formula does not make any sense for a photon at the speed of light. Blame this discrepancy on Einstein's relativistic effect, if you like.

HERE'S WHAT WE KNOW SO FAR

- Small particles behave both as waves and as particles.
- Photons have both wave and particle characteristics.
- The de Broglie equation illustrates the relationship between wave and particle characteristics.
- The Heisenberg uncertainty principle further verifies that small particles have wave-like characteristics.

Erwin Schrödinger (1887–1961) was awarded the 1933 Nobel Prize in physics for his development of the theory we use to describe the structure of an atom.

FIGURE 6.18

William Hamilton (1805–1865), developer of the Hamiltonian operator.

FIGURE 6.19

The wave functions, energy levels, and probability functions for a particle confined to a one-dimensional box. The particle is defined by a wave function that has nodes at each end of the wave, indicating that the particle cannot be found at the wall or outside the box.

6.9   The Mathematical Language of Quantum Chemistry

The language of quantum chemistry is mathematics. In quantum chemistry, knowing a few of the more important relevant mathematical concepts will pave the way for us to explore the current understanding of atomic structure, which has developed far beyond the Bohr model. The heart of our current understanding of the atom lies within a simple-looking mathematical equation:

$$\hat{H}\Psi_n = E_n\Psi_n$$

This is the time-independent **Schrödinger equation**. This equation allows us to calculate the exact energies, E_n, *available to the electrons in an atom*, if we know the mathematical description, Ψ_n, called the **wave function**, *that describes the positions and paths of the electron in its given energy level (n)*. Although it appears harmless, the wave function, Ψ_n, is a complex equation containing variables, constants, and quantum numbers. This function tells us everything we can possibly know about our system from a quantum mechanical perspective. It is called a wave function because it is derived from equations developed to predict the properties of waves. The $\hat{H}$ in the Schrödinger equation is known as the **Hamiltonian operator** after William Hamilton (1805–1865), an Irish mathematician who early on did work that hinted at wave–particle duality (Figure 6.18). In simple terms, an operator is a mathematical term, such as $+$, $-$, $*$, $/$, or (in this case) $\hat{H}$, that tells us how to proceed—what operation to do—in a mathematical equation. The Hamiltonian is one of the more complex operators that we'll encounter; it is too complex for us to carry out, in this discussion, the operation it calls for. However, *the conclusions that result from the operation are fundamental to our proper view of atomic structure.*

To illustrate the implications and utility of the Schrödinger equation, let's examine the values of a wave function, shown in Figure 6.19, for a particle constrained to move in only one dimension.

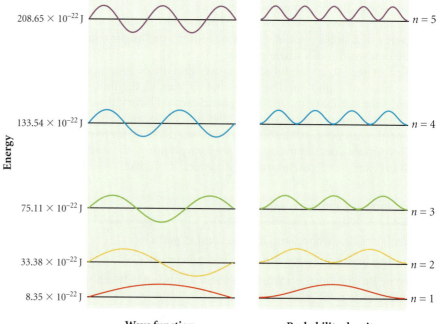

Wave function   Probability density

In this example, our particle can move between two points within that one dimension. Think of this as a one-dimensional container or box that will confine a particle. If we use the Schrödinger equation and solve for the energy of the particle, we find the first three wave functions shown starting from the bottom of the box in Figure 6.19. These are called standing waves, which oscillate in time just like the string on a violin, and we show them as a snapshot in time at their greatest amplitude. The first level is described by half the wavelength of a sine wave. The second level is described by a full sine wave and has a **node** in the center of the wave function—a point in the wave function where the amplitude is always zero. The next level, represented by one and a half sine wavelengths, has two nodes that divide the wave function into three segments. The wave function with $n = 4$ would have an additional node for a total of three, and the wave function $n = 5$ would have still another for a total of four.

Additionally, complex calculations of a particle in a one-dimensional box result in an equation illustrating the *permitted energy levels* for this model:

$$E_n = \frac{n^2 h^2}{8 \, mL^2}$$

where n is the quantum level ($n = 1, 2, 3, \ldots$), h is Planck's constant, m is the mass of the particle, and L is the length of the box. Although the one-dimensional model that gave rise to this equation is simple (it is based on a box containing a particle), it explains the intense colors observed in leaves and flowers as well as other materials. Certain molecules within a flower petal contain delocalized electrons (electrons that are able to move across a molecule; we'll have much more to say about this in Chapter 9). The molecule itself acts as a one-dimensional "box" because of its bonding structure. When a photon is absorbed by one of these electrons, it becomes excited and moves to a higher energy level described by our simple one-dimensional model. *The energy differences between the lower energy level and the excited energy level correspond to the color we observe. The length of the bond or bonding structure defines the length of the box, L in the above equation.*

If we calculate the energy of an electron confined to a box the size of a single covalent bond (1.5×10^{-10} m), the wavelength of the energy needed to excite the electron corresponds to ultraviolet light. If several atoms define the "box," as is seen in some types of molecules common within flower petals, then the energy is smaller (because L is bigger), and the electrons can absorb visible light. For example, cyanidin-based compounds, shown in Figure 6.20, are responsible for the red color of apples, autumn leaves, roses, strawberries, and cranberry juice. Changing the structure of the molecule changes the size of the box, which changes the color of the photons absorbed.

Introducing Orbitals

One of the more interesting outcomes of the Schrödinger equation is that we can use the wave function to reveal *where an electron is most likely to be found within an atom.* We say "most likely" because we can never pin the electron down exactly, as the Heisenberg uncertainty principle tells us. Instead, we can only determine the *probability* of finding it at each point in space. The space that the electron is allowed to occupy in a given energy level on an atom is called an **orbital**.

For example, if we plot the wave function of an electron around a nucleus, we get a figure similar to that shown in Figure 6.21. We can mathematically manipulate the wave function to create a new function depicted as $\Psi^*\Psi$, which looks a lot like the square of the wave function Ψ. The great usefulness of $\Psi^*\Psi$ is that *it is proportional to the probability of finding the particle at a particular point in space.* The nodes in Ψ remain at the same places in $\Psi^*\Psi$, but all values of $\Psi^*\Psi$ are positive. In short, *the probability function describes the spatial distribution of the electron around an atom;* $\Psi^*\Psi$ describes the shape of an orbital.

The colors we see in a flower are due to the molecules that are present in the petals. These molecules can be thought of as examples of the one-dimensional particle in a box.

Cyanidin

FIGURE 6.20

This molecule is responsible for many of the colors found in plants.

FIGURE 6.21

A plot of the wave function (blue line) represented as the distance from the nucleus. The probability function ($\Psi^*\Psi$; red line) represents the most likely shape of the orbital resulting from the wave function.

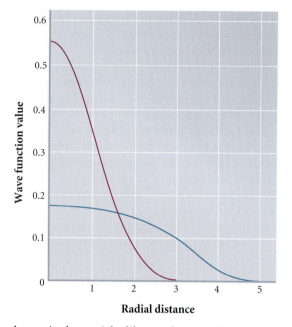

The idea that a single particle, like an electron, doesn't exist at a single point in space, but is merely more likely to be found at certain places and less likely to be found at others, is admittedly rather odd. But it fits the ideas presented here and years of measurements that confirm these ideas. Note that there are nodes where we can guarantee that the particle will *never* be found. At these nodes, the probability function, $\Psi^*\Psi$, is equal to zero. An electron can never be found at a node. The electron can be found on either side of the node, but never at the node. **How does it go from one side to the other, if it is never allowed to be on the node?** It does this by behaving like a wave, which has amplitude on both sides of the node: a clear demonstration of wave–particle duality.

Video Lesson: Atomic Orbital Size

6.10 Atomic Orbitals

Both the Hamiltonian operator and the wave functions for the hydrogen atom are known, and they allow us to develop both mathematical and visual representations of the orbitals that electrons can occupy in the three-dimensional space around the hydrogen nucleus. Specialists in quantum chemistry describe this three-dimensional system in terms of a radial part and an angular part. **Radial** means "along the radius" and **angular** means "as a function of the angles that describe the orientation of the radius." Because the calculations of these three-dimensional orbitals are quite complicated, we present only the results here. We will use these results to explain, over the next several chapters (and, from time to time, for the remainder of the textbook), how and why atoms and molecules interact the way they do.

Quantum Numbers and *s* Orbitals

There are *four different quantum numbers* in the wave function for the electron in the hydrogen atom, and they are given the symbols n, l, m_l, and m_s. The n is known as the **principal quantum number** and can be any whole number $n = 1, 2, 3, \ldots, \infty$. We are already familiar with the principal quantum number. This n is what Bohr referred to as the energy levels in his atom. Based on the possible values of n, there are an infinite number of energy levels for the hydrogen atom. The energy of each is given by

$$E_n = -\frac{2.1786 \times 10^{-18} \text{ J}}{n^2}$$

which is just what was predicted by the Bohr model. In atoms with more than one electron, any electrons with the same value for the principal quantum number (n) are said to occupy the same **principal shell**, but only the energy levels in the hydrogen atom are given by this equation.

Two of the other three quantum numbers, l and m_l, depend on the value of n. The **angular momentum quantum number**, l, can be any number from zero to $n - 1$ in whole-number steps:

$$l = 0, 1, 2, \ldots, n - 1$$

This quantum number is considered to represent the shape of the electron orbital.

In the first energy level, $n = 1$, the equation used to calculate the angular momentum quantum number reveals that $l = 0$. This implies that only one shape of orbital is possible in the first energy level. In the second energy level, $n = 2$, the value of l can be 0 or 1. This means that there are two different shapes of the orbitals in the second energy level. When $n = 3$, l can be 0, 1, or 2 (three different shapes). And so on, all the way to infinity. Any electrons that share the same values for n and l are said to occupy the same **electron subshell**.

The m_l quantum number is the **magnetic quantum number** (also known as the orbital angular momentum quantum number). It is based on the value of l and can have any value from l to $-l$ in whole numbers.

$$m_l = -l, \ldots, -2, -1, 0, 1, 2, \ldots, l$$

This quantum number can be thought of as defining the direction in which the individual electron orbitals are pointed. For example, a specific orbital could be oriented along the x axis or the y axis. The value of m_l indicates this direction.

The number of possible values for the magnetic quantum number indicates the number of specific orbitals of a specified shape, as illustrated in Figure 6.22. For example, in the second principal shell, when $n = 2$, we determined that there are two values for l, $l = 0$ and $l = 1$. We need to consider each of these values to generate all the possible sets of quantum numbers. When $n = 2$ and $l = 0$, m_l can only be 0 (one orbital). When $n = 2$ and $l = 1$, m_l can be -1, 0, or $+1$. In this case, we have three specific orbitals of the same shape oriented in three different directions. Calculating all of the possible orbitals within an energy level can be a

n	l	m_l	Subshell notation	Number of orbitals in the subshell	Number of electrons needed to fill subshell	Maximum possible number of electrons in shell
1	0	0	1s	1	2	2
2	0	0	2s	1	2	
2	1	1, 0 –1	2p	3	6	8
3	0	0	3s	1	2	
3	1	1, 0, –1	3p	3	6	
3	2	2, 1, 0, –1, –2	3d	5	10	18
4	0	0	4s	1	2	
4	1	1, 0, –1	4p	3	6	
4	2	2, 1, 0, –1, –2	4d	5	10	
4	3	3, 2, 1, 0, –1, –2, –3	4f	7	14	32

FIGURE 6.22

Allowed values for n, l, m_l, and m_s quantum numbers for hydrogen atomic orbitals.

Video Lesson: Atomic Orbital Shapes and Quantum Numbers

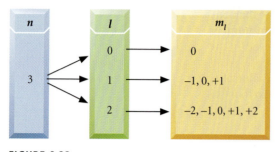

FIGURE 6.23

Outline of n, l, and m_l values.

daunting task because the number of orbitals gets large fairly quickly, as shown in Figure 6.23. But as long as you are *systematic* about what is essentially a bookkeeping task, confusion can be kept to a minimum.

The final quantum number, m_s, is known as the **electron spin quantum number** (it is also known as the spin angular momentum quantum number). Fortunately for our bookkeeping, the only possible values for m_s are $+\frac{1}{2}$ and $-\frac{1}{2}$. The value of m_s can be either $+\frac{1}{2}$ or $-\frac{1}{2}$ for every set of n, l, and m_l allowed.

We can use shorthand to indicate the four quantum numbers that describe the position of electrons in an atom using the notation (n, l, m_l, m_s). As shown in Figure 6.24, for $n = 1$ the sets of quantum numbers for the electrons are $(1, 0, 0, \frac{1}{2})$ and $(1, 0, 0, -\frac{1}{2})$. For $n = 2$ the sets are $(2, 1, 1, \frac{1}{2})$, $(2, 1, 1, -\frac{1}{2})$, $(2, 1, 0, \frac{1}{2})$, $(2, 1, 0, -\frac{1}{2})$, $(2, 1, -1, \frac{1}{2})$, $(2, 1, -1, -\frac{1}{2})$, $(2, 0, 0, \frac{1}{2})$, and $(2, 0, 0, -\frac{1}{2})$. If you try $n = 3$, you should get 18 sets of quantum numbers, and for $n = 4$ you should get 32. Note how this follows a general trend. The number of possible quantum numbers in each energy level is $2n^2$.

How do these values help us model the electronic structure of atoms? Although the specific set of four quantum numbers represents a convenient label for a specific hydrogen atomic wave function, quantum numbers are much more than that. The principal quantum number, n, gives the energy of the system and sets the values of the l and m_l quantum numbers. Together, the n, l, and m_l tell us about the spatial distribution of the electron and define the orbital shape and size. For example, as shown in Figure 6.25, the radial component of the orbitals with $n = 1$, 2, and 3 illustrates the difference in size of the atomic orbitals.

Orbitals that have equal energies are said to be **degenerate orbitals**, and the number of orbitals having the same energy is called the degeneracy. In the hydrogen atom, orbitals with the same value of n and l have the same specific amount of energy, and these orbitals are therefore degenerate. Consider the orbitals within the second energy level ($n = 2$). When $n = 2$, l can be either 0 or 1. When $l = 0$, the value of m_l can be only 0. There is only one orbital, so it can't be degenerate. However, when $l = 1$, m_l can be -1, 0, or $+1$. In this case, three orbitals with the same value for l are possible. These three orbitals have the same amount of energy and are degenerate. The third energy level contains two sets of degenerate orbitals. One set contains three degenerate orbitals ($l = 1$, $m_l = -1, 0, +1$); the other set contains five degenerate orbitals ($l = 2$, $m_l = -2, -1, 0, +1, +2$).

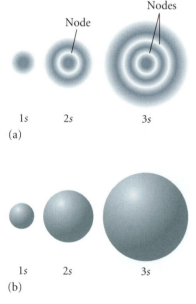

FIGURE 6.25

(a) The radial component of the probability for the hydrogen atomic orbitals with $n = 1$, 2, and 3. (b) The spatial component of those orbitals at 90% probability.

(n, l, m_l, m_s)	
Energy Level 1:	$(1, 0, 0, +\frac{1}{2})$; $(1, 0, 0, -\frac{1}{2})$
Energy Level 2:	$(2, 0, 0, +\frac{1}{2})$; $(2, 0, 0, -\frac{1}{2})$
	$(2, 1, 1, +\frac{1}{2})$; $(2, 1, 0, +\frac{1}{2})$; $(2, 1, -1, +\frac{1}{2})$
	$(2, 1, 1, -\frac{1}{2})$; $(2, 1, 0, -\frac{1}{2})$; $(2, 1, -1, -\frac{1}{2})$
Energy Level 3:	$(3, 0, 0, +\frac{1}{2})$; $(3, 0, 0, -\frac{1}{2})$
	$(3, 1, 1, +\frac{1}{2})$; $(3, 1, 0, +\frac{1}{2})$; $(3, 1, -1, +\frac{1}{2})$
	$(3, 1, 1, -\frac{1}{2})$; $(3, 1, 0, -\frac{1}{2})$; $(3, 1, -1, -\frac{1}{2})$
	$(3, 2, 2, +\frac{1}{2})$; $(3, 2, 1, +\frac{1}{2})$; $(3, 2, 0, +\frac{1}{2})$; $(3, 2, -1, +\frac{1}{2})$; $(3, 2, -2, +\frac{1}{2})$
	$(3, 2, 2, -\frac{1}{2})$; $(3, 2, 1, -\frac{1}{2})$; $(3, 2, 0, -\frac{1}{2})$; $(3, 2, -1, -\frac{1}{2})$; $(3, 2, -2, -\frac{1}{2})$

FIGURE 6.24

Systematic bookkeeping for quantum numbers.

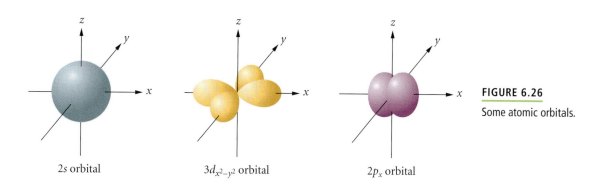

FIGURE 6.26
Some atomic orbitals.

2s orbital $3d_{x^2-y^2}$ orbital $2p_x$ orbital

Quantum chemists have been able to show us pictures that describe what the quantum numbers tell us. For example, Figure 6.26 also shows us shapes of many of the orbitals that we have described with quantum numbers. Remember that at the atomic level, we can no longer speak in terms of absolute locations but can only give the *probability* of the quantum particle being in a given location. The pictures therefore can show us only the most probable place to find the electron in the orbital. This means that if you were in Champaign, Illinois, studying a hydrogen atom in an atom trap, and you asked a friend to see whether he could find your atom's electron in Perth, Scotland, there would be a finite probability (albeit outrageously small) that he would be able to find it there. We do not have instruments that are sensitive enough to measure the tiny probability of an electron existing in Scotland when its atom's nucleus is in Illinois. To be honest, we would be hard pressed to measure the probability even *a few millionths of a millimeter* from the atom.

The Scanning Tunneling Microscope

The scanning tunneling microscope (STM) allows us to draw a picture that represents the surface of a material. This microscope is so powerful that it can even "see" single molecules and atoms. The STM basically uses an atomic-sized needle that runs across the surface of a material. As the tip moves, it runs up and over molecules or atoms, and changes in the position of the tip of the needle are recorded on a graph. It works because if we bring the tip of the needle close enough to the sample material, the wave function of the atoms in the tip will overlap with those of the sample. This allows electrons to move from the tip into the material by a process called quantum tunneling. The electrons that "tunnel"

Application

CHEMICAL ENCOUNTERS:
The Scanning
Tunneling
Microscope

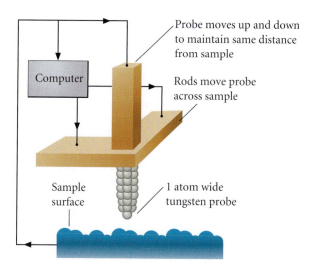

Probe moves up and down to maintain same distance from sample

Computer

Rods move probe across sample

Sample surface

1 atom wide tungsten probe

FIGURE 6.27

An STM tip scans over an object to produce an image.

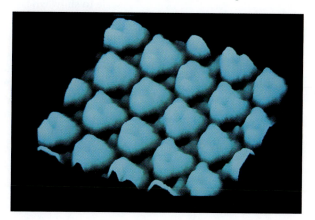

from the tip to the material set up an electric current (the "tunneling current") that increases as we bring the tip closer to the atom and decreases as we pull away. This allows us to image the atomic positions (though only to a level of precision limited by Heisenberg's uncertainty principle) by scanning the tip over the material and seeing just what kind of overlap we get. An example image is shown in Figure 6.27. The bigger the tunneling current is, the more electrons there are in proximity to the tip. As we pull the tip away from the surface, the tunneling current falls off exponentially with distance. Because we assume that the atom is in the center of the greatest electron density, we get an image of the atoms at the surface of the material. The image that results from these measurements conveys information about the positions of electrons in atoms. These are not photographs of atoms, but they are impressive testimony to our ability to detect where individual atoms are, within the constraints of the uncertainty principle. Even more impressive, the image that opens this chapter shows the genuine wavelike nature of matter at the quantum scale. Look at the ripples: direct evidence that wave–particle duality is real!

The STM is a practical application of the fact that we don't actually have a natural end to the radial distribution function. It goes on forever and ever, just with increasingly smaller and smaller amplitude. According to classical mechanics the STM should not work, because electrons should not be able to "tunnel" into the places classical mechanics says they cannot go. The STM is practical proof that quantum mechanics is a valid description of nature.

s Orbitals

What can we say about the shapes of the orbitals within an atom? The orbital shown in Figure 6.26 that has $l = 0$ is radially symmetric and has a spherical shape. That is, as we look out from the nucleus, the probability distribution function looks the same in every direction and varies only with r, the distance from the nucleus. All orbitals with $l = 0$ are called *s* **orbitals** and can be written as *s*. The principal quantum number is added to this as a prefix when we write the name of this specific orbital. For example, when $n = 1$ and $l = 0$, we write 1*s*. When $n = 2$ and $l = 0$, we write 2*s*, and so on. The 1*s* orbital, with $n = 1$, has no nodes and represents the lowest energy level, or ground state, for an electron in a hydrogen atom. The 2*s* orbital has one node that is concentric around the nucleus. The 3*s* orbital has two radial nodes, the 4*s* has three, and so on, with the number of concentric nodes increasing along with the principal quantum number.

The "solid spheres" of Figure 6.28 are useful in comparing relative sizes. As we expect, the 1*s* orbital is the smallest, and the size increases as n increases. This representation does not give us any help in visualizing the nodes present; the two-dimensional cross sections are more helpful in this respect. The last method of representing the *s* orbitals takes a more probabilistic approach and uses a gray scale to show where the probability of finding the electron is large and where it is small. The nodes are indicated by the lightest part of the picture. All *s* orbitals are spherical, with the number of radial nodes increasing from zero in the 1*s* orbital to $n - 1$ in the *ns* orbital. Remember that *s* orbitals all have orbital angular momentum quantum numbers of $l = 0$ and magnetic quantum numbers of $m_l = 0$.

We still have to consider the Heisenberg uncertainty principle as we construct our pictures. In order to draw sensible pictures of atomic sizes, we arbitrarily assign the boundary of an atom to lie at some distance r from the center of the atom, so that when we look for the electron within that value of r from the nucleus, we find the electron a large portion (90%) of the time. We show the 1*s*, 2*s*, and 3*s* orbitals in Figure 6.28 at the 90% probability distribution function for the electron and in several different formats, each of which is meant to emphasize some aspect of the orbital shape.

An *s* orbital.

Visualization: 1*s* Orbital

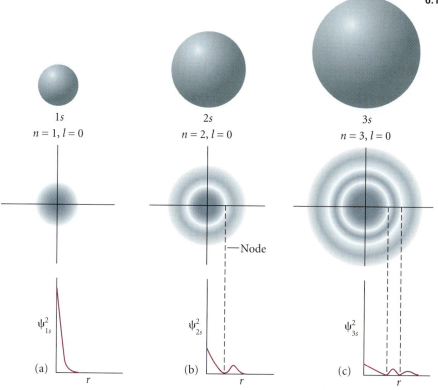

FIGURE 6.28

The 1s, 2s, and 3s spatial component of the hydrogen atomic orbitals at 90% probability. Note the presence of nodes in the cutaway plots.

Visualization: $2p_x$ Orbital

Visualization: $2p_y$ Orbital

Visualization: $2p_z$ Orbital

p Orbitals

For orbitals with $l = 1$, which are known as **p orbitals** are abbreviated p, we have a different spatial arrangement, as shown in Figure 6.29. Again we use the convention of writing the principal quantum number as a prefix to the letter abbreviation for the orbital. However, because $l = 1$ allows for three possible m_l values ($+1$, 0, and -1), there will be three degenerate orbitals that are symmetrically equivalent. We can choose to orient the three equivalent orbitals along the x, y, and z axes, and when we do so, we call them the $2p_x$, $2p_y$, and $2p_z$ orbitals. This is shown for the lowest-energy $2p$ orbitals in Figure 6.29. The $2p$ orbitals have no radial nodes, but each does have a planar node, as illustrated in Figure 6.30. For the $2p_x$ orbital, this node is the yz plane; for the $2p_y$ orbital, this node is the xz plane; and for the $2p_z$ orbital, this node is the xy plane. We have set the size of the orbital to include 90% of the probability for finding the electron within it, just as we did with the s orbitals.

When we increase the primary quantum number to $n = 3$, we have three $3p$ orbitals with the same kind of planar node structure (Figure 6.31), but now we also have a radial node. The three $4p$ orbitals have the planar node plus two radial nodes, the three $5p$ orbitals have the planar node plus three radial nodes, and so on.

FIGURE 6.29

The $2p_x$, $2p_y$, and $2p_z$ spatial component of the hydrogen atomic orbitals at 90% probability.

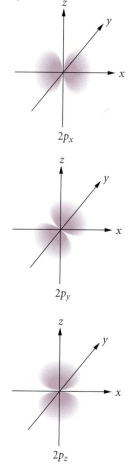

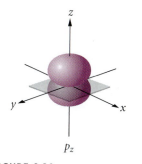

FIGURE 6.30

The location of the planar node found in the $2p_z$ orbital. Although $n = 2$ is used in the example, all p orbitals have a single planar node in an analogous position.

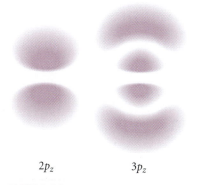

$2p_z$ $3p_z$

FIGURE 6.31

The $2p_z$ and $3p_z$ spatial component of the hydrogen atomic orbitals at 90% probability.

FIGURE 6.32

The $3d_{xy}$, $3d_{yz}$, $3d_{xz}$, $3d_{x^2-y^2}$, and $3d_{z^2}$ spatial component of the hydrogen atomic orbitals at 90% probability.

Visualization: $3d_{x^2-y^2}$ Orbital

Visualization: $3d_{xy}$ Orbital

Visualization: $3d_{x^2}$ Orbital

Visualization: $3d_{yz}$ Orbital

Visualization: $3d_{z^2}$ Orbital

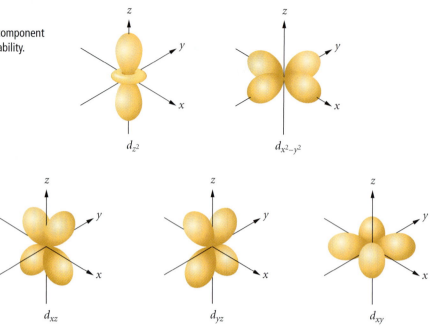

d Orbitals

Orbitals with $l = 2$ are called ***d* orbitals** (abbreviated as *d*) and have two nonradial nodes. The first set of quantum numbers for which l can equal 2 occurs when $n = 3$. Five *d* orbitals occur, because m_l takes the values of 2, 1, 0, −1, −2 when $l = 2$. For all but the $m_l = 0$ *d* orbital, the nodes are planar and are at 90° to each other. These orbitals have four lobes that are labeled d_{xy}, d_{yz}, d_{xz}, and $d_{x^2-y^2}$ and are oriented as shown in Figure 6.32. The remaining *d* orbital is labeled d_{z^2} and looks a bit different but is equivalent in energy (degenerate) to the other four *d* orbitals provided that it has the same principal quantum number. The d_{z^2} orbital has two hyperbolic nodes that give the orbital a dumbbell structure with a little toroidal "donut." When $n = 3$, no radial nodes occur, as shown in Figure 6.33, but $n = 4$ has one radial node, $n = 5$ has two radial nodes, and *d* orbitals with higher principal quantum numbers have $n - 3$ radial nodes.

FIGURE 6.33

The location of the two planar nodes found in the $3d_{xy}$, $3d_{yz}$, $3d_{xz}$, and $3d_{x^2-y^2}$ lobes and the two hyperbolic nodes orbitals of the $3d_{z^2}$ lobe. Although $n = 3$ is used in the example, all *d* orbitals have two planar nodes in analogous positions.

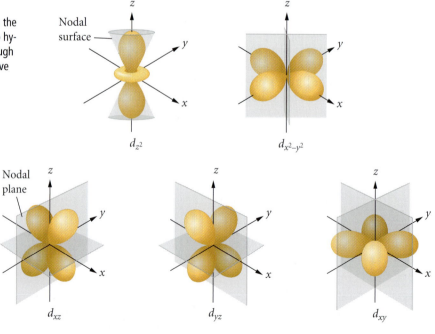

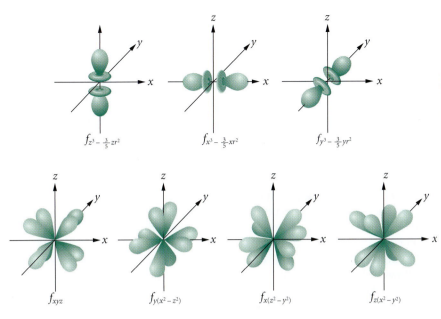

FIGURE 6.34

The 4*f* spatial component of the hydrogen atomic orbitals at 90% probability.

f Orbitals

Much of our use of orbitals focuses on the *s*, *p*, and *d* levels. The one exception is the **f orbitals** (abbreviated *f*) with $l = 3$ and $m_l = 3, 2, 1, 0, -1, -2, -3$. These seven orbitals will be used in our description of heavy atoms, such as those in the lanthanide and actinide series of the periodic table. They are shown in Figure 6.34. Note that they are multilobed and geometrically quite complicated.

At last we have arrived at our current "quantum mechanical" or "wave mechanical" model of atomic structure, although we can depict it only by showing the places around the nucleus where an electron in any orbital will *most probably* be found. In painting a more accurate picture of what electrons are doing, we have had to acknowledge the uncertainty and strangeness—the funny things that happen—down in the quantum world.

EXERCISE 6.7 Quantum Numbers

Indicate the possible values for the quantum numbers (n, l, m_l) that would correspond to the 3*d* orbital.

Solution

A single 3*d* orbital would correspond to one of these sets of quantum numbers:

$n = 3, \ l = 2, \ m_l = -2$
$n = 3, \ l = 2, \ m_l = -1$
$n = 3, \ l = 2, \ m_l = 0$
$n = 3, \ l = 2, \ m_l = +1$
$n = 3, \ l = 2, \ m_l = +2$

PRACTICE 6.7

Indicate the orbital described by the quantum numbers, $n = 2, \ l = 1, \ m_l = 0$.

See Problems 86 and 117.

Wolfgang Ernst Pauli (1900–1958) was awarded the 1945 Nobel Prize in physics for his discovery of what we now know as the Pauli exclusion principle.

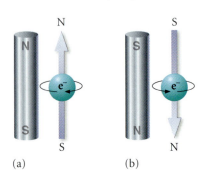

FIGURE 6.35

Electrons in the spin up and spin down states.

Application

CHEMICAL ENCOUNTERS: Nuclear Spin and Magnetic Resonance Imaging

Video Lesson: Understanding Electron Spin

FIGURE 6.36

MRI scanner.

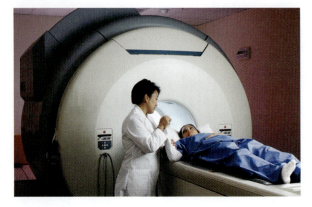

6.11 Electron Spin and the Pauli Exclusion Principle

An electron *behaves* as though it were spinning about an axis. A "spin" in one direction corresponds to an electron spin quantum number (m_s) of $+^1/_2$, and a "spin" in the opposite direction corresponds to an electron spin quantum number (m_s) of $-^1/_2$. These are the only two values of m_s available; in other words, spin is quantized into these two states, and an electron must be in one of them. The "spin" of an electron makes the electron behave as though it were a tiny magnet. For example, if we put hydrogen atoms into a magnetic field, electrons with $m_s = ^1/_2$, called the "up" state, will line up parallel to the field; and those with $m_s = -^1/_2$, called the "down" state, will line up antiparallel to it, as shown in Figure 6.35. We sometimes represent the up state by an arrow pointing up: ↑. Similarly, the down state is represented by an arrow pointing down: ↓.

One vital consequence of this idea of electron "spin" is that it allows two electrons to occupy the same atomic orbital. Wolfgang Pauli, in 1925, proposed a rule that allows only two electrons to occupy the same atomic orbital. According to the **Pauli exclusion principle**, no two electrons can have the same set of the four quantum numbers (n, l, m_l, and m_s) in a given atom. **What does this tell us about the electronic structure of the atom?** For two electrons to occupy the same orbital, they must have the same values for the quantum numbers n, l, and m_l. The only way that two electrons can have these quantum numbers is if their spin quantum numbers (m_s) are different. In essence, one of the electrons in the orbital is in the spin up state, and the other is in the spin down state. *Because it requires four quantum numbers to describe an electron on an atom, it is not possible for more than two electrons to occupy the same orbital.*

Nuclear Spin and Magnetic Resonance Imaging

An MRI scanner consists of a large magnet with a cylindrical tube bored through the center. An example is shown in Figure 6.36. When a person is scanned, the body part to be imaged needs to be in the center of the magnetic field. The technique focuses on the *nuclei* of all of the hydrogen atoms in the body, because they have a relatively large magnetic moment (a measure describing the magnitude and direction of the nuclei's magnetic field). Placing the body in the magnet causes the hydrogen nuclei to align with the magnetic field. Just as with electrons, there are nuclei with both up and down "spins," so most of the nuclear "spins" cancel each other out. But there are enough "unpaired" nuclei to result in a nice image.

The image is created by directing radio frequency pulses toward the body part to be examined. The radio pulses produce a miniature magnetic field aligned in a direction perpendicular to the field of the magnet around the body. This miniature magnetic field disrupts the alignment of the atoms in the body part. The atoms are then said to occupy an excited state. When the radio pulse is turned off, the atoms relax to their original state of alignment with the magnet and release energy. The energy is detected by the MRI and converted into an image of the body part.

How does the computer in the MRI distinguish the structures in the body? It does so on the basis of the amount of time it takes for the excited hydrogen nuclei to relax back to their original alignment with the external magnetic field. The realignment of the excited nuclei can last anywhere from a few hundred milliseconds to a few seconds. The time it takes

Low Energy   **High Energy**

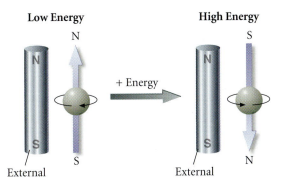

Adding energy to a nucleus in a magnetic field can cause it to change its orientation with the field. After time has passed, the excited nucleus returns to the low energy state and emits a photon of energy in the radio frequency range.

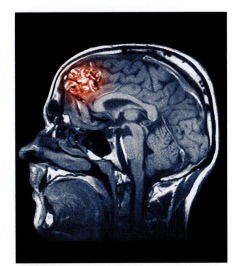

FIGURE 6.37

MRI brain image.

to realign is heavily dependent on the environment surrounding the hydrogen nuclei. For example, hydrogen nuclei in water (which is the primary component in blood and spinal fluid) relax much more rapidly than hydrogen nuclei that make up tissues, which in turn are faster than hydrogen atoms in fats. The difference is primarily due to the fact that hydrogen atoms are attached to different types of atoms in each of these three components of the body. It is the variation in relaxation times, combined with the differing concentrations of hydrogen atoms in different tissues and body fluids, that allows for the exceptional contrast observed using this technique. Figure 6.37 shows an MRI image of a brain, where color has been added to help distinguish the different tissues. This image can be used to locate the position of a brain tumor and the major blood vessels leading to it. MRI scanners are used to image many different areas of the body without the use of invasive surgeries.

Reviewing the Significance of Quantum Numbers

We have seen that the four quantum numbers in any electron's wave function specify everything we need to know about the electron's location in the atom. They are a bit like an electron's address in the atom. Table 6.5 summarizes the significance of all the quantum numbers.

TABLE 6.5	The Quantum Numbers
n = principal quantum number	Indicates which of the major energy levels or electron shells an orbital is in. Also, the larger the value of n, the greater the total volume of the orbital
l = angular momentum quantum number	Indicates the shape of the orbital and which quantum number type of subshell the orbital is in, which will be an s, p, d, or f subshell
m_l = magnetic quantum number	Quantum number indicates the orientation of the orbital in space
m_s = electron spin quantum number	Indicates the orientation of the spin of the electron in the orbital

6.12 ◆ Orbitals and Energy Levels in Multielectron Atoms

Unfortunately, the orbitals calculated in the manner that we have discussed are only *absolutely* correct for one-electron atoms and ions. You can use them for hydrogen and its isotopes deuterium and tritium. You can also use them for He^+, Li^{2+}, or even Hg^{79+} if you adapt the equation slightly.

Problems arise when you put a second electron into an atom or ion. Because the two electrons have the same charge, they repel each other. The interaction between the two electrons makes it impossible to solve for their energies and orbital wave functions exactly. Fortunately, ways exist to *approximate* the correct energies and electron orbitals. The approximations also work for atoms containing three or more electrons, and the results can be quite good. How do we describe the orbitals on an atom with more than one electron? We take the hydrogen atomic orbitals, which are approximately correct because they describe the effect the nucleus has on a single electron, and we change them a little to approximate the effect of other electrons. The results are useful both because they accurately predict the physical properties of atoms, such as ionization energies and emission spectra, and because we can use the multielectron orbitals to form **molecular orbitals**, occupied by electrons in molecules (discussed in Chapter 9).

The shapes we derive for orbitals in multielectron atoms are very similar to those in the hydrogen atom. Helium contains a 1s orbital, a 2s orbital, three 2p orbitals, and so on, just as we found previously for the hydrogen atom. Each of these can hold up to two electrons, one with spin up and one with spin down. The four quantum numbers n, l, m_l, and m_s are called the principal, angular momentum, orbital angular momentum, and spin angular momentum quantum numbers, just as before. The total number of nodes is $n - 1$, and the number of nonradial nodes is $n - l - 1$, just as for the hydrogen atom. The kinds of the nodes and their approximate positions are the same, and the shapes of the orbitals are similar enough that we can use the same pictures we used above.

The one thing that will be different is the energies of the electron orbitals. These are so complicated that we will not be able to calculate meaningful values for them without the aid of a computer, so we will talk about energy only qualitatively here. One big difference in orbital energy levels is that whereas the energy depends only on the principal quantum number, n, for the hydrogen atom, levels with different l values will now have different energies for the multielectron atom.

This is because of something called **electron shielding** (it is also known as electron screening). When a second electron is present in an atom, some of the time this electron is between the first electron and the nucleus. Because electrons are negatively charged, the positive nuclear charge experienced by the first electron is smaller that it otherwise would be, and the attractive force between that electron and the nucleus is decreased proportionately. The first electron also sometimes spends time closer to the nucleus than its partner, and when this happens, that second electron will be shielded too. For the two 1s electrons in helium, both are shielded the same amount, and they remain degenerate.

When we focus on orbitals with different n and/or different l values, all electrons in an atom shield all other electrons, but not necessarily to the same extent. The 2s electrons are, on average, closer to the nucleus than are 2p electrons, as shown in Figure 6.38, and the 2s electrons will screen the 2p electrons far better than the 2p will shield the 2s electrons. The nuclear charge will therefore seem to be smaller to the 2p electrons than to the 2s electrons as a result of the greater shielding. Because the apparent nuclear charge has been decreased, the attractive

www

Video Lesson: Electron Shielding

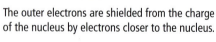

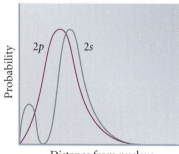

The outer electrons are shielded from the charge of the nucleus by electrons closer to the nucleus.

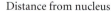

FIGURE 6.38

The relative distance from the nucleus of the electron density in the 2s and 2p orbitals.

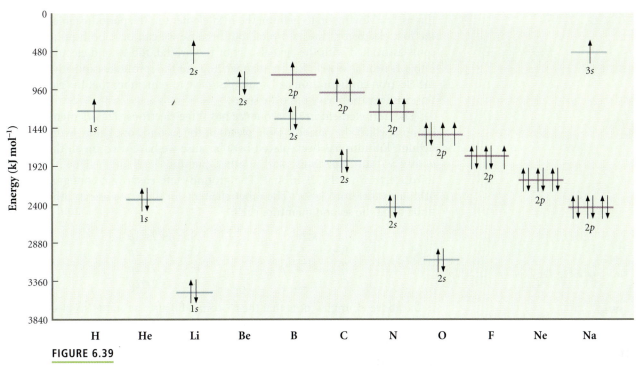

FIGURE 6.39

Relative energy levels for multielectron atoms. The relative placement of the energies of the orbitals is only approximate and varies with *Z*, the nuclear charge, which increases to the right across the periodic table.

forces between the nucleus and the electron will also decrease. The bottom line is that the 2*p* electrons will no longer be as low in energy as the 2*s* electrons. This is shown schematically in Figure 6.39. The $2p_x$, $2p_y$, and $2p_z$ orbitals will still be degenerate, however, because they are all at the same average distance from the nucleus and will all experience the same shielding from other electrons in the atom. For *n* = 3, the next principal quantum number up in energy, the 3*s*, will be lower in energy than the 3*p*, which will be lower than the 3*d*. For the next set of orbitals, in which *n* = 4, the 4*s* will be below the 4*p*, which will be below the 4*d*, which will be below the 4*f*.

Visualization: Orbital Energies

6.13   Electron Configurations and the Aufbau Principle

Now that we have a complete set of orbitals and a general scheme for ranking their energies, we can list all the orbitals occupied by the electrons of any atom in the periodic table. The complete list of filled orbitals is called the **electron configuration** of an atom. We are often interested in the lowest energy configuration, which represents the ground state of the atom. This can be generated by adding one electron at a time to an unoccupied atomic orbital, starting with the lowest energy orbital and working our way up in energies until all electrons in the atom have been assigned an orbital. This method of filling hydrogen atomic orbitals to get multielectron configurations operates in accordance with the **Aufbau principle**, from the German word for "building up." The Aufbau principle says that as protons are added to the nucleus, electrons are successively added to orbitals of increasing energy, beginning with the lowest-energy orbitals. In essence, we build up the electron configuration one electron at a time until all electrons are accommodated. To convey the maximum information in the minimum space, we use a chemical shorthand to indicate which orbitals are occupied. In its ground state, the hydrogen atom will have its lone electron in the 1*s* orbital,

Tutorial: Aufbau Principle

H	$1s^1$
He	$1s^2$

this as $1s^1$. Moving on to helium, the next element in the periodic table, we place the first electron in the $1s$ orbital, but because we can have both up and down spins associated with the spatial part of the orbital, the second electron will go into a $1s$ orbital as well. Rather than writing out the spin state explicitly, we write the electron configuration as $1s^2$ and just have to remember that the electrons in this configuration have opposite spins.

The next element in the sequence has three electrons. This element must be lithium, which the periodic table places in the second row, below hydrogen and helium. Lithium has been placed here because we have used up all the orbitals with $n = 1$ and now must start another quantum level, or shell, by placing the third electron in the $2s$ orbital. The ground-state electronic configuration for lithium is therefore $1s^2 2s^1$. The other elements in the same row as lithium continue to fill the second quantum shell:

Li:	$1s^2 2s^1$	N:	$1s^2 2s^2 2p^3$
Be:	$1s^2 2s^2$	O:	$1s^2 2s^2 2p^4$
B:	$1s^2 2s^2 2p^1$	F:	$1s^2 2s^2 2p^5$
C:	$1s^2 2s^2 2p^2$	Ne:	$1s^2 2s^2 2p^6$

H	$1s^1$
Ne	$1s^2 2s^2 2p_x^2 2p_y^2 2p_z^2$ = $1s^2 2s^2 2p^6$

Video Lesson: Electron Configurations through Neon

In the above examples, note that the p orbitals fill after a total of six electrons because $2p_x$, $2p_y$, and $2p_z$ can each accommodate two electrons, one with spin up and one with spin down. Remember that the p orbitals are referred to as occupying their own subshell.

When we get to the end of the row, we will have completely filled another shell. The next row must start filling the $3s$ orbitals because there is no more room for electrons in the $n = 2$ shell. Now you can begin to understand how *the structure of the periodic table shows a splendid consistency, whether we are classifying elements in terms of physical and chemical properties, as did Mendeleev and Meyer, or by the quantum mechanical solution to the Schrödinger equation in the order of increasing electron energy, as did the quantum chemists and physicists.*

In order to proceed to the next row, or period, of the periodic table, we still need to know a few more things about how electrons fill the orbitals. Because of their negative charge, electrons tend to remain as far apart as possible in a manner that minimizes repulsive interactions among them. When filling the $2p$ orbitals for boron through neon, we could keep the electrons farther apart on the average by placing them into different $2p$ orbitals for as long as possible. We could also minimize the repulsive electron–electron interactions by keeping the spins of the electrons parallel—that is, with their respective spin axes pointing along the same direction. To illustrate this, we will fill the electrons from the second period of the chart one more time. The first three, shown in Figure 6.40, are unambiguous.

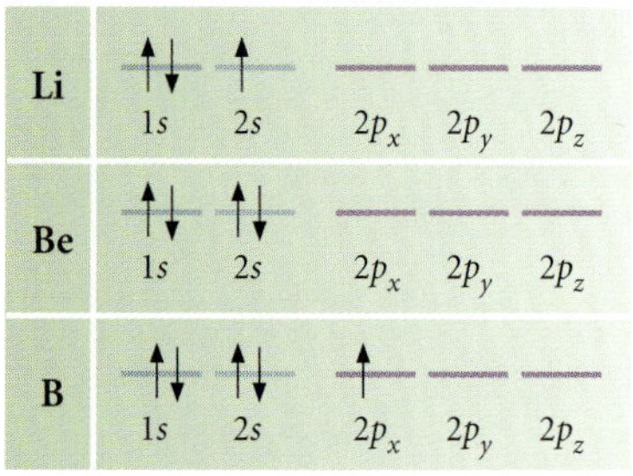

For boron, the $2p_x$ orbital choice is arbitrary. Either the $2p_y$ or the $2p_z$ orbital would have been equally valid. To extend the series to carbon, we now have to decide whether we will place the two $2p$ electrons in the same orbital or in different orbitals. Three unique choices are possible, as shown in Figure 6.41, all of which represent possible electronic states for the carbon atom.

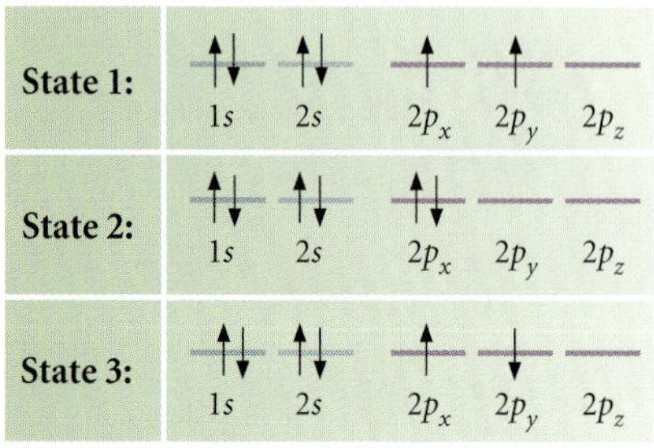

The ground state for the carbon atom is state 1, which has the two $2p$ electrons in different orbitals with the spins aligned in the same direction. It is the lowest energy state for carbon, and it is therefore the state in which you will find a carbon atom unless you supply it with energy—by irradiating it with electromagnetic radiation, for example.

To understand why we need to keep spins aligned the same for partially filled levels, remember that no two electrons in an atom can have the same four quantum numbers n, l, m_l, and m_s (Pauli exclusion principle). Giving the two $2p$ electrons the same spin gives them the same value of m_s and requires that they be placed in different $2p$ orbitals. This keeps the two electrons

FIGURE 6.40

The electron configuration of the first three elements of the second period.

	$1s$	$2s$	$2p_x$	$2p_y$	$2p_z$
Li	↑↓	↑	—	—	—
Be	↑↓	↑↓	—	—	—
B	↑↓	↑↓	↑	—	—

FIGURE 6.41

Three possible electron configurations for carbon.

	$1s$	$2s$	$2p_x$	$2p_y$	$2p_z$
State 1:	↑↓	↑↓	↑	↑	—
State 2:	↑↓	↑↓	↑↓	—	—
State 3:	↑↓	↑↓	↑	↓	—

spatially separated and therefore minimizes the repulsion between the two. Overall, this behavior is summarized by what is known as **Hund's rule**. Proposed by the German physicist Friedrich Hund in 1925, this rule states that when orbitals of equal energy are available, the lowest energy configuration for an atom has the maximum number of unpaired electrons with parallel spins.

If we continue adding the electrons to the hydrogen-like atomic orbitals, we will have the maximum number of unpaired $2p$ electrons for nitrogen, as shown in Figure 6.42.

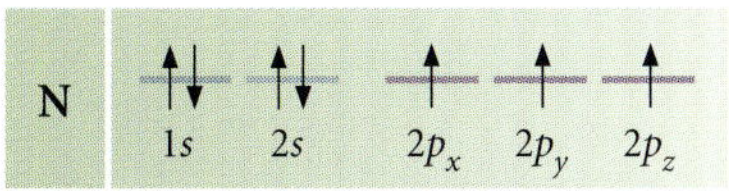

We must again begin pairing them to continue adding electrons to the orbitals, as shown in Figure 6.43. By the time we get to neon, the entire second shell is filled.

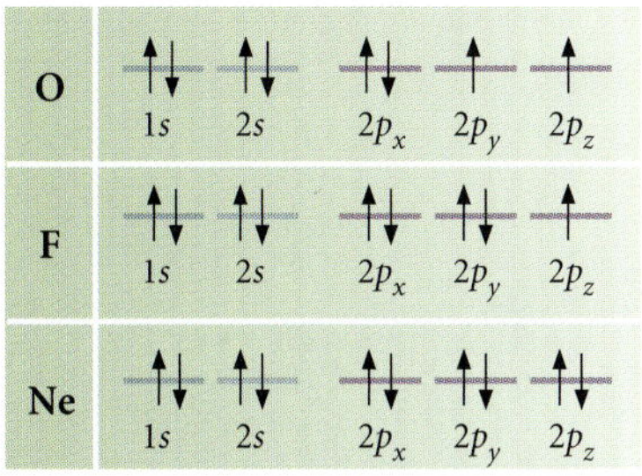

To proceed to sodium, we need to add the next electron to the $3s$ level. Even with our shorthand notation, the electron configurations are beginning to get unwieldy. However, we can say that all atoms in the third row of the periodic table must have filled the $n = 1$ and $n = 2$ shells before electrons can be added to the $n = 3$ level. This allows us to write the electron configuration in one of two ways, as given in the example for sodium as shown in Figure 6.44. The first notation shows all of the electrons explicitly, and the second uses [Ne] to take the place of $1s^2 2s^2 2p^6$. We can say that the configuration of sodium is $3s^1$ with a neon core. The electrons listed after the **core electrons** ([Ne]) are called **valence electrons** and will be very important in establishing the chemical reactivity of the element.

Similarly, the ground-state electron configuration for potassium could be written either explicitly as $1s^2 2s^2 2p^6 3s^2 3p^6 4s^1$ or as $[\text{Ar}]4s^1$. The latter case indicates only the electrons in the outermost shell of the configuration and places the less reactive, more strongly bound filled-shell electrons into an argon core. This does not imply that the nucleus of potassium has been replaced with argon. Argon has one fewer proton and, for the most abundant of their isotopes, two more neutrons than does potassium. The [Ar] core notation signifies that the atomic configuration has the same hydrogen-like orbitals occupied with electrons *plus* all the others noted after it.

You might have noticed that the argon ground-state electron configuration has been given as $1s^2 2s^2 2p^6 3s^2 3p^6$ but that when we added one more electron to the set to represent the occupied orbitals of potassium, the next element in the periodic table, we placed it into a $4s$ level. This makes sense from the position that potassium occupies in the periodic table: right under sodium, as shown in Figure 6.45. And rubidium, which is directly under potassium, has the electron configuration of $[\text{Kr}]5s^1$. In fact, all the elements in the first column of the

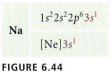

FIGURE 6.42

The electron configuration for nitrogen.

FIGURE 6.43

The electron configuration for the last three elements of the second period.

Na $\quad 1s^2 2s^2 2p^6 3s^1$

$\quad$ [Ne]$3s^1$

FIGURE 6.44

The electron configuration for sodium can be written two ways.

www

Video Lesson: Electron Configurations Beyond Neon

FIGURE 6.45

Position of potassium and sodium on the periodic table.

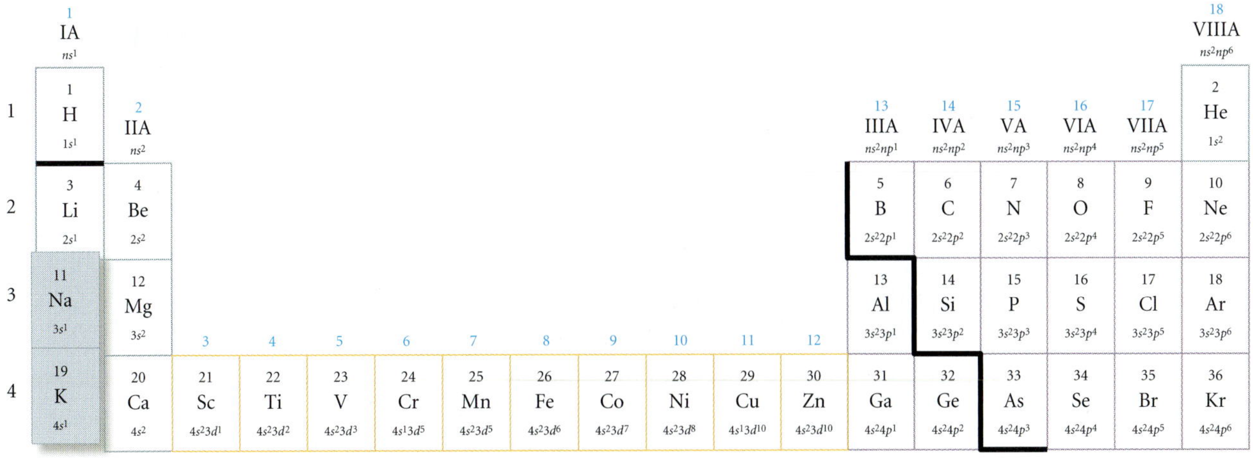

	1 IA ns^1																	18 VIIIA ns^2np^6
1	1 H $1s^1$	2 IIA ns^2											13 IIIA ns^2np^1	14 IVA ns^2np^2	15 VA ns^2np^3	16 VIA ns^2np^4	17 VIIA ns^2np^5	2 He $1s^2$
2	3 Li $2s^1$	4 Be $2s^2$											5 B $2s^2 2p^1$	6 C $2s^2 2p^2$	7 N $2s^2 2p^3$	8 O $2s^2 2p^4$	9 F $2s^2 2p^5$	10 Ne $2s^2 2p^6$
3	11 Na $3s^1$	12 Mg $3s^2$	3	4	5	6	7	8	9	10	11	12	13 Al $3s^2 3p^1$	14 Si $3s^2 3p^2$	15 P $3s^2 3p^3$	16 S $3s^2 3p^4$	17 Cl $3s^2 3p^5$	18 Ar $3s^2 3p^6$
4	19 K $4s^1$	20 Ca $4s^2$	21 Sc $4s^2 3d^1$	22 Ti $4s^2 3d^2$	23 V $4s^2 3d^3$	24 Cr $4s^1 3d^5$	25 Mn $4s^2 3d^5$	26 Fe $4s^2 3d^6$	27 Co $4s^2 3d^7$	28 Ni $4s^2 3d^8$	29 Cu $4s^1 3d^{10}$	30 Zn $4s^2 3d^{10}$	31 Ga $4s^2 4p^1$	32 Ge $4s^2 4p^2$	33 As $4s^2 4p^3$	34 Se $4s^2 4p^4$	35 Br $4s^2 4p^5$	36 Kr $4s^2 4p^6$

periodic table have the configuration [core]ns^1, where [core] is the electron configuration of the noble gas directly preceding the element in the periodic table. The n is the row number that the element occupies in the periodic table, and n is also the principal quantum number of the ns electron. It is this similarity in ground-state electron configuration that gives all of the elements in this column similar chemical properties, including metallic behavior, malleability, tendency to form ions with a charge of $+1$, and high reactivity with water.

What happened to the $3d$ electrons? When we were talking about the effect of adding more than one electron to the orbital energies of the hydrogen atom, we made the point that interactions among the electrons removed the degeneracy among the electrons in the n-shell orbitals, so that orbitals with increasing value of l had increasing energy. The $3s$ is lower in energy than the $3p$, which, in turn, is lower than the $3d$. The $3d$ orbital energy is so destabilized by the electron–electron interactions that it actually rises in energy to be just a little higher in energy than the $4s$ orbital for the atom in the gas phase. To form the lowest energy electron configuration, the $4s$ subshell fills before the $3d$ subshell. The same problem is noted when we begin filling the f orbital subshell. The ground-state configurations for the elements as gas-phase atoms are given in Figure 6.46.

FIGURE 6.46

Periodic table with the electron configurations of all atoms in the gas phase. The "partial" electron configurations are shown when the cores of the electron configuration can be deduced from the earlier periods.

	IA (ns^1)	IIA (ns^2)	3	4	5	6	7	8	9	10	11	12	IIIA (ns^2np^1)	IVA (ns^2np^2)	VA (ns^2np^3)	VIA (ns^2np^4)	VIIA (ns^2np^5)	VIIIA (ns^2np^6)
1	1 H $1s^1$																	2 He $1s^2$
2	3 Li $2s^1$	4 Be $2s^2$											5 B $2s^22p^1$	6 C $2s^22p^2$	7 N $2s^22p^3$	8 O $2s^22p^4$	9 F $2s^22p^5$	10 Ne $2s^22p^6$
3	11 Na $3s^1$	12 Mg $3s^2$											13 Al $3s^23p^1$	14 Si $3s^23p^2$	15 P $3s^23p^3$	16 S $3s^23p^4$	17 Cl $3s^23p^5$	18 Ar $3s^23p^6$
4	19 K $4s^1$	20 Ca $4s^2$	21 Sc $4s^23d^1$	22 Ti $4s^23d^2$	23 V $4s^23d^3$	24 Cr $4s^13d^5$	25 Mn $4s^23d^5$	26 Fe $4s^23d^6$	27 Co $4s^23d^7$	28 Ni $4s^23d^8$	29 Cu $4s^13d^{10}$	30 Zn $4s^23d^{10}$	31 Ga $4s^24p^1$	32 Ge $4s^24p^2$	33 As $4s^24p^3$	34 Se $4s^24p^4$	35 Br $4s^24p^5$	36 Kr $4s^24p^6$
5	37 Rb $5s^1$	38 Sr $5s^2$	39 Y $5s^24d^1$	40 Zr $5s^24d^2$	41 Nb $5s^14d^4$	42 Mo $5s^14d^5$	43 Tc $5s^14d^6$	44 Ru $5s^14d^7$	45 Rh $5s^14d^8$	46 Pd $4d^{10}$	47 Ag $5s^14d^{10}$	48 Cd $5s^24d^{10}$	49 In $5s^25p^1$	50 Sn $5s^25p^2$	51 Sb $5s^25p^3$	52 Te $5s^25p^4$	53 I $5s^25p^5$	54 Xe $5s^25p^6$
6	55 Cs $6s^1$	56 Ba $6s^2$	57 La* $6s^25d^1$	72 Hf $4f^{14}6s^25d^2$	73 Ta $6s^25d^3$	74 W $6s^25d^4$	75 Re $6s^25d^5$	76 Os $6s^25d^6$	77 Ir $6s^25d^7$	78 Pt $6s^15d^9$	79 Au $6s^15d^{10}$	80 Hg $6s^25d^{10}$	81 Tl $6s^26p^1$	82 Pb $6s^26p^2$	83 Bi $6s^26p^3$	84 Po $6s^26p^4$	85 At $6s^26p^5$	86 Rn $6s^26p^6$
7	87 Fr $7s^1$	88 Ra $7s^2$	89 Ac** $7s^26d^1$	104 Rf $7s^26d^2$	105 Db $7s^26d^3$	106 Sg $7s^26d^4$	107 Bh $7s^26d^5$	108 Hs $7s^26d^6$	109 Mt $7s^26d^7$	110 Ds $7s^26d^8$	111 Rg $7s^16d^{10}$	112 Uub $7s^26d^{10}$						

Representative elements — *d*-Transition elements — Representative elements — Noble gases

Period number, highest occupied electron level

f-Transition elements

*Lanthanides	58 Ce $6s^24f^15d^1$	59 Pr $6s^24f^35d^0$	60 Nd $6s^24f^45d^0$	61 Pm $6s^24f^35d^0$	62 Sm $6s^24f^65d^0$	63 Eu $6s^24f^75d^0$	64 Gd $6s^24f^75d^1$	65 Tb $6s^24f^95d^0$	66 Dy $6s^24f^{10}5d^0$	67 Ho $6s^24f^{11}5d^0$	68 Er $6s^24f^{12}5d^0$	69 Tm $6s^24f^{13}5d^0$	70 Yb $6s^24f^{14}5d^0$	71 Lu $6s^24f^{14}5d^1$
**Actinides	90 Th $7s^25f^06d^2$	91 Pa $7s^25f^26d^1$	92 U $7s^25f^36d^1$	93 Np $7s^25f^46d^1$	94 Pu $7s^25f^66d^0$	95 Am $7s^25f^76d^0$	96 Cm $7s^25f^76d^1$	97 Bk $7s^25f^96d^0$	98 Cf $7s^25f^{10}6d^0$	99 Es $7s^25f^{11}6d^0$	100 Fm $7s^25f^{12}6d^0$	101 Md $7s^25f^{13}6d^0$	102 No $7s^25f^{14}6d^0$	103 Lr $7s^25f^{14}6d^1$

The order of filling is systematic and is illustrated by the diagram shown in Figure 6.47. For example, we can follow the arrows in the figure and add the electrons to each subshell to obtain the ground-state electron configuration of sulfur (16 electrons):

$$S: \quad 1s^2 2s^2 2p^6 3s^2 3p^4 = [Ne]3s^2 3p^4$$

Using this information, we can obtain the electron configuration of iodine:

$$I: \quad 1s^2 2s^2 2p^6 3s^2 3p^6 4s^2 3d^{10} 4p^6 5s^2 4d^{10} 5p^5 = [Kr]5s^2 4d^{10} 5p^5$$

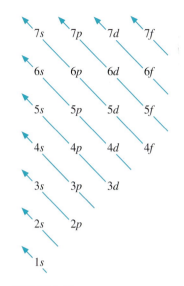

FIGURE 6.47

The order in which atomic orbitals are filled. The order is determined by writing the orbitals found in each energy level in rows and then drawing diagonal lines up and to the left. The order is 1*s*, 2*s*, 2*p*, 3*s*, 3*p*, 4*s*, 3*d*, 4*p*, 5*s*, 4*d*, 5*p*, 6*s*, 4*f*, 5*d*, 6*p*, ...

EXERCISE 6.8   Practice with Electron Configurations

Write the ground-state electron configuration for vanadium and for tellurium.

First Thoughts

The structure of the periodic table gives us guidance for writing the ground-state electron configuration of an element. The two key questions we ask are "In what period is our element?" and "Is its highest partially filled energy orbital an *s*, *p*, *d*, or *f* orbital?"

Solution

Vanadium, atomic number 23, is in Period 4 and is the third element in the 3*d* series that begins with scandium, as shown in Figure 6.46. It will therefore have an argon electron core with two 4*s* and three 3*d* electrons. Alternatively, you can use the triangle in Figure 6.47 and fill the orbitals to an electron count of 23.

$$V \ (23 \ electrons): \quad [Ar]4s^2 3d^3 = 1s^2 2s^2 2p^6 3s^2 3p^6 4s^2 3d^3$$

Tellurium, atomic number 52, is in Period 5 and Group VIA, in which *p* orbitals are being filled. It will have a krypton electron core, with (reading across Group V in Figure 6.46) two 5*s* electrons, ten 4*d* electrons, and six 5*p* electrons. Alternatively, you can use Figure 6.47, as with vanadium, although counting electrons in this way can get clerically messy.

$$Te \ (52 \ electrons): \quad [Kr]5s^2 4d^{10} 5p^4 = 1s^2 2s^2 2p^6 3s^2 3p^6 4s^2 3d^{10} 4p^6 5s^2 4d^{10} 5p^4$$

Further Insights

Write the ground-state electron configuration of chromium. You will get $[Ar]4s^2 3d^4$. It is often said that one of the *s* electrons will move to the *d* orbital to give an energetically more favorable (half-filled) $3d^5$ configuration (that is, $[Ar]4s^1 3d^5$). This reorganization is noted when we're talking about chromium atoms in the gas phase. Molybdenum atoms ($[Kr]5s^1 4d^5$) and copper atoms ($[Ar]4s^1 3d^{10}$) are other examples where the reorganized electron configuration is observed in the gas phase.

It should be noted that these cases are specific to the gas-phase electron configuration, where interactions with other atoms cannot stabilize the predicted electron configurations. Because these elements are not found in nature or generally used as individual gas-phase atoms, we can use the predicted electron configurations in our normal work.

PRACTICE 6.8

Write the ground-state electron configuration for gallium and strontium.

See Problems 99–108 and 118.

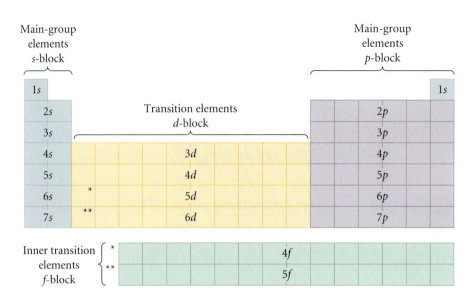

FIGURE 6.48

The periodic table can be divided into regions (known as blocks) that illustrate the type of atomic orbital that is being filled.

A Summary of the Ground-State Electron Configuration–Based Structure of the Periodic Table

The periodic table can be divided into four regions based on the filling of orbitals of the elements; see Figure 6.48.

- *Region 1:* This region includes the alkali and alkaline earth metals, in which the s orbitals are being filled. These Group IA and IIA elements, such as potassium and calcium, are part of the main-group elements.

- *Region 2:* This region includes the remainder of the main-group elements, Groups IIIA through VIIIA, such as carbon, nitrogen, and chlorine, in which the p orbitals are being filled.

- *Region 3:* This region includes the transition elements, in which the d orbitals are being filled. These include metals such as chromium, silver, and mercury.

- *Region 4:* This includes the inner-transition elements, such as uranium and plutonium, in which the f orbitals are being filled.

The structure of the periodic table is a very satisfying one for chemists because it gives rhyme and reason to the physical properties and chemical reactivity of the elements. Although the periodic table was an outgrowth of Mendeleev's periodic law formulated in 1869, this early grouping of elements with similar properties was done empirically, just by seeing how the material behaved. The odd shape of the periodic table, however, simply begs for an explanation, and quantum mechanics has given it! By solving the Schrödinger equation, we have generated wave functions with unique sets of quantum numbers. By generating the wave functions and quantum numbers, we have created a complete set of electron orbitals, and by filling these orbitals according to the Aufbau principle, we have generated the complete periodic table—its form, the properties of its elements, everything!

The Bottom Line

- We can interconvert among the wavelength, frequency, and energy of electromagnetic radiation. (Section 6.2)

- Atoms and molecules absorb and emit electromagnetic radiation to gain and lose energy, but only certain wavelengths of radiation can be absorbed and emitted. (Section 6.3)

- The wavelengths of electromagnetic radiation absorbed and emitted by an atom are characteristic of that atom and can be used to probe the atomic structure. (Section 6.3)

- One of the first models of atomic structure, the Bohr model, quantized the energies and spatial locations of the electrons to explain the hydrogen emission spectrum. Although the model was not correct in other details, quantum energy and orbit locations were breakthrough concepts and are key concepts in the modern picture of the atom. (Section 6.4)

- Electronic transitions between the quantized energy levels are responsible for atomic absorption and emission spectra. The energies are given by $\Delta E = E_f - E_i$, where the subscripts i and f stand for initial and final states. (Section 6.4)

- The energy of electromagnetic radiation absorbed or emitted in an electronic transition must exactly match the energy of the transition. (Section 6.4)

- At the atomic scale, all things show both wave and particle behavior. This concept is known as wave–particle duality. (Section 6.5)

- The wave and particle natures of quantum objects, such as electrons and photons, are linked in the de Broglie equation: $\lambda = h/p$. (Section 6.6)

- The Heisenberg uncertainty principle, $\Delta x \Delta p \geq h/4\pi$, places limits on how precisely we can simultaneously measure position and momentum. (Sections 6.7 and 6.8)

- All quantum systems, like the hydrogen atom, can be completely described by the Schrödinger equation: $\hat{H}\Psi_n = E_n\Psi_n$. (Section 6.9)

- The probability distribution function $\Psi_n^*\Psi_n$ tells us where to find electrons in the atom, and the shape described by this function is called the atomic orbital. (Section 6.9)

- The orbitals are associated with the probability of finding the electron within a certain region of space. (Section 6.9)

- Each electron wave function has a set of four quantum numbers associated with it. They are n, the principal quantum number; l, the angular momentum quantum number; m_l, the magnetic quantum number; and m_s, the electron spin quantum number. (Section 6.10)

- The principal quantum number n indicates the electron shell that an electron is in, and n, l, and m_l determine the shape and orientation of the orbital in three-dimensional space. (Section 6.10)

- Only two possible electron spin states exist: $m_s = +^1/_2$ and $m_s = -^1/_2$. (Section 6.10)

- The Pauli exclusion principle states that in a given atom, no two electrons can have the same set of the four quantum numbers — n, l, m_l, and m_s. (Section 6.11)

- We can envisage multielectron atoms being assembled by placing electrons into the orbitals one by one, starting with the lowest energy level, according to the Aufbau principle. The complete listing of occupied orbitals is called the ground-state electron configuration. (Section 6.13)

- Hund's rule for ground-state electron configurations states that when orbitals of equal energy are available, the lowest energy configuration for an atom is the one with the maximum number of unpaired electrons with parallel spins. (Section 6.13)

Key Words

absorption spectroscopy The measurement of how atoms and molecules absorb electromagnetic radiation as a function of the wavelength or frequency of the radiation. (p. 216)

amplitude The distance between the highest point and the midpoint (zero) of a wave. (p. 212)

angstrom A unit of distance equal to 10^{-10} m. It is symbolized by Å. (p. 211)

angular As a function of the angles that describe the orientation of the radius in spherical polar coordinates. (p. 234)

angular momentum quantum number The quantum number, l, that describes the shape of the orbital; l can be any whole-number value from zero to $n - 1$. (p. 235)

Aufbau principle As protons are added to the nucleus, increasing the atomic number of the atom, electrons are added successively to the next highest energy orbital. The method of filling hydrogen atomic orbitals to get the ground-state electron configuration of a multielectron atom by adding the electrons one at a time until all are accommodated; from the German phase for "building up." (*p. 245*)

classical mechanics The physical description of macroscopic behavior based on Newton's equations of motion. (*p. 225*)

core electrons Electrons in the configuration of an atom that are not in the highest principal energy shell. (*p. 247*)

***d* orbital** An orbital with quantum number $l = 2$. (*p. 240*)

degenerate orbitals Orbitals that have equal energies. (*p. 236*)

electromagnetic spectrum The entire range of radiation as a function of wavelength, frequency, or energy. (*p. 211*)

electron configuration The complete list of filled orbitals in an atom. (*p. 245*)

electron shielding The ability of electrons in lower energy orbitals to decrease the nuclear charge felt by electrons in higher energy orbitals. (*p. 244*)

electron spin quantum number The quantum number, m_s, that describes the spin of an electron. Also known as the spin angular momentum quantum number. (*p. 236*)

electron subshell The energy level occupied by electrons that share the same values for both n and l. (*p. 235*)

electronic transition A change in atomic or molecular energy level made by an electron bound in an atom or molecule. (*p. 221*)

emission spectroscopy The measurement of how atoms and molecules give off electromagnetic radiation as a function of the wavelength or frequency of the radiation. (*p. 216*)

emission spectrum A plot of the intensity of radiation as a function of wavelength or frequency in an emission experiment. (*p. 217*)

empirically derived Derived from experiments and observations rather than from theory. (*p. 219*)

energy levels The allowed orbits that electrons may occupy in an atom. (*p. 221*)

excited state Any higher energy state than the ground state characterized by the existence of an electron in an orbital that violates Hund's Rule and/or the Aufbau principle. (*p. 221*)

***f* orbital** An orbital with quantum number $l = 3$. (*p. 241*)

frequency A descriptor of electromagnetic radiation. Defined as the number of waves that pass a given point per second, in units of 1/s (s^{-1}). (*p. 211*)

ground state The lowest energy state of an atom. (*p. 221*)

Hamiltonian operator A mathematical function that is used in the Schrödinger equation. (*p. 232*)

Heisenberg uncertainty principle There is an ultimate uncertainty in the position and the momentum of a particle. Reducing the uncertainty in one increases the uncertainty in the other. (*p. 230*)

Hund's rule When orbitals of equal energy are available, the lowest energy configuration for an atom has the maximum number of unpaired electrons with parallel spins. (*p. 247*)

laser An acronym for "light amplification by stimulated emission of radiation." (*p. 224*)

magnetic quantum number The quantum number, m_l, that describes the orientation of the orbital; m_l can be any whole-number value from $-l$ to $+l$. Also known as the orbital angular momentum quantum number. (*p. 235*)

molecular orbitals Electron orbitals that are appropriate for describing bonding between atoms in a molecule. (*p. 244*)

node A point in the wave function where the amplitude is always zero. (*p. 233*)

orbital The volume of space to which the electron is restricted when it is in a bound atomic or molecular energy level. (*p. 233*)

***p* orbital** An orbital with quantum number $l = 1$. (*p. 239*)

Pauli exclusion principle In a given atom, no two electrons can have the same set of the four quantum numbers — n, l, m_l, and m_s. (*p. 242*)

photon A single unit of electromagnetic radiation. (*p. 210*)

Planck's constant A fundamental constant equal to 6.62608×10^{-34} J·s. (*p. 215*)

principal quantum number The quantum number, n, that describes the energy level of the orbital; n can be any whole-number value from 1 to infinity. (*p. 234*)

principal shell The energy level that is occupied by electrons with the same value for the principal quantum number (n). (*p. 234*)

quantum A single unit of matter or energy. (*p. 210*)

quantum chemistry The study of chemistry on the atomic and molecular scale. (*p. 210*)

quantum mechanics The physical laws governing energy and matter at the atomic scale. (*p. 210*)

quantum number A number used to arrive at the solution of an acceptable wave function that describes the properties of a specific orbital. (*p. 218*)

radial Along the radius in spherical polar coordinates. (*p. 234*)

rest mass The mass a particle has when it is stationary. (*p. 231*)

***s* orbital** An orbital with quantum number $l = 0$. (*p. 238*)

Schrödinger equation The mathematical expression that relates the wave function to the energy in a quantum system. (*p. 232*)

spectroscopy The measurement of how atoms and molecules interact with electromagnetic radiation as a function of the wavelength or frequency of the radiation. (*p. 216*)

standing wave The constructive interference of two or more waves that results in the presence of nodes in fixed locations. (*p. 226*)

valence electrons The electrons in an atom, ion, or molecule that are in the highest principal energy shell. (*p. 247*)

wave function The mathematical equation describing a system, such as an electron, atom, or molecule, that contains all physical information that can be obtained for the system by quantum mechanics. (*p. 232*)

wave–particle duality The quantum mechanical theorem that states that all things have both wave and particle natures simultaneously and that both of these natures can be observed on the atomic scale. (*p. 227*)

wavelength A descriptor of electromagnetic radiation. Defined as the distance from the top of one crest to the top of the next crest of the electromagnetic wave. (*p. 211*)

Focus Your Learning

The answers to the odd-numbered problems and some selected problems appear at the back of the book, as represented by the blue numbering.

Section 6.1 Introducing Quantum Chemistry

Skill Review

1. At the submicroscopic level we typically discuss the probability of events and objects rather than their location and individual speeds. Describe one event or object that is best explained using probability. Is this event considered "macro" or "micro"?

2. Use the analogy of flipping a coin to explain the term *probability*.

Section 6.2 Electromagnetic Radiation

Skill Review

3. As you walk through the produce section of a supermarket, you will see various colors of vegetables. Say the grocer decided to organize the vegetables in order of increasing frequencies of their reflected light—a novel marketing concept. Indicate in what sequence these vegetables would appear: a red tomato, a green zucchini squash, a purple eggplant, and a yellow spaghetti squash.

4. The three colors of a traffic light are red, yellow, and green. The green light is placed at the bottom, with yellow in the center and red on top. Are these colors in order, from bottom to top, by frequency or by wavelength? Justify your answer.

5. a. What is the frequency of an X ray with a wavelength of 1.5×10^{-2} nm?
   b. What is the energy, in joules, associated with a photon of this frequency?
   c. What would be the energy of a mole of such photons?

6. a. What is the frequency of visible light with a 400-nm wavelength?
   b. What is the energy, in joules, associated with a photon of this frequency?
   c. What would be the energy associated with a mole of these photons?

7. In what region on the electromagnetic radiation spectrum would a wavelength of 2.5×10^4 nm be placed? What would be the frequency and energy of this radiation?

8. In what region on the electromagnetic radiation spectrum would a wavelength of 5.8×10^2 nm be placed? What would be the frequency and energy of this radiation?

9. Calculate your height in nanometers, meters, and light-years. Which do you find the most convenient unit for this application?

10. Calculate the length of a 12-in ruler in nanometers, meters, and lightyears. Which do you find the most convenient unit for this application?

11. Cell phones operate at frequencies of 824 to 894 MHz. What range of wavelengths is this? To which part of the electromagnetic spectrum does this correspond?

12. Helium–neon (HeNe) lasers are both cheap and common. They are even sold as attachments to novelty key chains. If the helium–neon mixture lases at 0.632 μm, what is its frequency? To which part of the electromagnetic spectrum does this correspond?

13. Radio stations broadcast with frequencies given in megahertz, where 1 MHz = 10^6 s^{-1}. This means that when a station advertises itself as Radio WXYZ located at 98.3 on your dial, it is broadcasting at 98.3 MHz. What wavelength does Radio WXYZ use in its broadcast?

14. A wireless Internet connection broadcasts its signal in the 1200-MHz range. To what wavelength does this correspond?

15. Chlorophyll, the green pigment found in plants, absorbs visible radiation best in the red region (at about 675 nm) and in the blue-violet region (at about 440 nm). What are the energies of the photons collected by plants at these two wavelengths?

16. What are the energies of photons emitted from a 100-MHz magnetic resonance imager?

17. How much energy resides in 1 mol of photons whose wavelength is 440 nm? . . . 675 nm?

18. How much energy resides in 2 mol of photons with a frequency of 1.5×10^{14} Hz? . . . in 0.75 mol?

Chemical Applications and Practices

19. a. The process of photosynthesis is quite complex. However, one of the main considerations is that plant pigments absorb visible light to power reactions that convert carbon dioxide and water into food and produce oxygen. If a plant absorbs blue light that has a wavelength of 565 nm, what is the energy per photon that is being absorbed?

    b. A typical ratio between photons absorbed and oxygen molecules produced is 8:1. How much energy is required to produce one molecule of oxygen in this manner?

20. Ozone (O_3) is important in our upper atmosphere because it aids in filtering out harmful ultraviolet rays. Ultraviolet rays may be classified as either UV-A, UV-B, or UV-C. The UV-B rays cause the most problems for earth-based organisms. For example, higher incidences of "jumping genes" that cause mutations may be related to exposure to UV-B.

    a. What is the energy in 1 mol of UV-B photons that have a wavelength of 312 nm?

    b. What is the energy of 1 mol of photons with a wavelength of 600 nm? To what type of radiation does this correspond?

21. UV-B radiation is responsible for "sunburn" in humans. A helpful advancement in technology is the personal UV detector. This small device uses a photoelectric response. An example is a gallium-based device to convert absorbance into an electrical signal. If the device gave a maximum reading at 290 nm, what energy is being absorbed? To what frequency does this correspond?

22. a. Some snakes have the ability to detect infrared radiation (IR). Are they detecting energy that is higher or lower in energy than human eyes can see?

    b. What is the source of typical IR wavelengths?

    c. A television remote control may use IR with a frequency of 1×10^{13} cycles per second. To what energy does this correspond?

23. One way to gain information about the origin and functions of stars within the universe is to study the origin and distribution of the hundreds of gamma ray sources in the sky. If one such source were producing high-energy gamma rays of 1.6×10^{-8} J, what wavelength would astronomers have detected? What is the frequency of this radiation?

24. If an astronomer recorded energy from a distant star at 3.6×10^{-10} J, what wavelength would he have detected? What is the frequency of this radiation?

Section 6.3 Atomic Emission and Absorption Spectroscopy, Chemical Analysis and the Quantum Number

Skill Review

25. When bombarded with high-energy electrons, copper metal gives off radiation with a wavelength of 1.54 Å. What is the frequency of this radiation? To which range of the electromagnetic spectrum does this correspond?

26. Sodium arc lamps, which are used as automobile headlights and street lights, are colored by the sodium doublet: electromagnetic radiation produced by excited sodium atoms found at 5895 Å and 5904 Å. What color are these lights?

27. Much of the radiation striking the earth from the sun has a wavelength of approximately 500 nm. Express this wavelength in meters, angstroms, centimeters, and inches.

28. Express 280-nm ultraviolet radiation in meters, angstroms, centimeters, and inches.

29. According to the Balmer equation, the wavelength of emitted light from hydrogen can be calculated from the whole-number values of n. The difference between $n = 5$ and $n = 4$ is only 1. The difference between $n = 2$ and $n = 1$ is also only 1. Why don't the two conditions produce the same wavelength of emitted light in hydrogen?

30. The Balmer equation can be used to calculate the wavelength of light emitted from excited hydrogen. To what initial value of n in hydrogen would an emitted wavelength of 5547 Å correspond? Explain why this value of n is never noted for hydrogen.

31. The first two wavelengths of the Balmer series of the hydrogen emission spectrum are 6562.1 Å, 4860.8 Å. What are the next three values in this series? What are the frequencies and energies of the emission lines?

32. What is the highest frequency of the Balmer series of the hydrogen emission spectrum? What is the lowest frequency of this series?

Chemical Applications and Practices

33. The presence of cadmium in drinking water is undesirable because exposure to large amounts has been associated with weakening of bones and joints. The wavelength of electromagnetic radiation strongly absorbed by cadmium is 214.439 nm.

    a. What is the frequency of that light?

    b. In what range of the electromagnetic spectrum would you classify this frequency?

    c. Which other element has an absorbance wavelength closest to (and therefore possibly difficult to distinguish from) cadmium? Consult Table 6.3 for additional information.

34. An adaptation of the Balmer equation makes it possible to calculate other emitted wavelengths from excited hydrogen. For example, if the final value for n is 3, then emitted light is in the infrared area of the spectrum. These line spectra are known as the Paschen series. Calculate the wavelength emitted when an electron in hydrogen drops from the fourth Bohr level to the third.

Section 6.4 The Bohr Model of Atomic Structure

Skill Review

35. Calculate the energy of an electron in each of these Bohr energy levels:

    a. $n = 1$ b. $n = 3$ c. $n = 5$ d. $n = 7$

36. Calculate the energy of an electron in each of these Bohr energy levels:

    a. $n = 2$ b. $n = 4$ c. $n = 6$ d. $n = 8$

37. Calculate the energy a photon released from a hydrogen atom in each of these transitions:

    a. $n = 4$ to $n = 1$ c. $n = 5$ to $n = 4$

    b. $n = 3$ to $n = 1$ d. $n = 7$ to $n = 2$

38. Calculate the energy a photon released from a hydrogen atom in each of these transitions:

    a. $n = 2$ to $n = 1$ c. $n = 4$ to $n = 3$

    b. $n = 4$ to $n = 2$ d. $n = 8$ to $n = 2$

39. Calculate the wavelength of a photon that would cause these transitions:

a. $n = 1$ to $n = 2$
c. $n = 2$ to $n = 4$
b. $n = 3$ to $n = 5$
d. $n = 1$ to $n = 6$
e.
f.

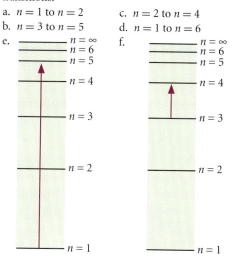

40. Calculate the frequency of a photon that would cause these transitions:

a. $n = 8$ to $n = 10$
c. $n = 4$ to $n = 8$
b. $n = 3$ to $n = 6$
d. $n = 4$ to $n = 5$
e.
f.

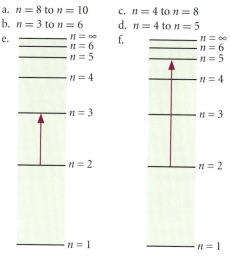

41. What is the shortest wavelength that can be emitted by the hydrogen atom in the Brackett spectral series? What is the longest wavelength in this series? Consult Table 6.2 for assistance with this problem.

42. According to the Bohr model of the hydrogen atom, what is the closest an electron can get to the nucleus? What is the farthest it can get from the nucleus? (If the latter answer seems absurd, it might amuse you to know that it is the same value predicted by present-day quantum chemistry.)

Chemical Applications and Practices

43. The emission spectrum being given out by a mixture of gases, possibly including hydrogen, includes an emission line with an energy of 4.84×10^{-19} J per photon. Is this emission possible for hydrogen? (To make this problem more manageable, consider transitions within the first eight energy levels only.)

44. The emission spectrum for a specific sample of gas includes an emission line with an energy of 7.38×10^{-19} J per photon. Is this emission possible for hydrogen? (To make this problem more manageable, consider transitions within the first eight energy levels only.)

45. To ionize hydrogen is to remove its single electron. We consider a free electron to have, with respect to an electron in hydrogen, zero energy. In other words, the initial state is $n = 1$ and the final state is $n = \infty$. Calculate the energy needed to ionize an electron from the ground state.

46. Calculate the energy released when an electron is added to a hydrogen nucleus. Assume the transition is $n = \infty$ to $n = 1$.

Section 6.5 Wave–Particle Duality

Skill Review

47. In a classical sense, compare matter and waves in terms of permissible energy values, spatial positions, and momentum.

48. How did the quantum view change the classical distinction between matter and waves? What is the quantum view called?

Chemical Applications and Practices

49. What is the momentum of an electron that is moving at 68% of the speed of light? (Use 9.1×10^{-31} kg as the mass of the electron.)

50. What is the speed of an electron that has a momentum of 9.7×10^{-23} kg $\cdot$ m $\cdot$ s^{-1}?

51. A freight train locomotive weighs about 415,000 lb and has a top speed of about 100.0 mi/h. What is its momentum?

52. The average speed of an electron in the ground state of the hydrogen atom is 2.19×10^6 m/s. What is the average momentum of an electron in this state?

53. What is the momentum of a photon with a wavelength of 540 nm?

54. What is the momentum of a photon with a frequency of 1.0×10^{14} Hz?

Section 6.6 Why Treating Things as "Waves" Allows Us to Quantize Their Behavior

Skill Review

55. List two particle-like properties of electrons.

56. List two wave-like properties of electrons.

Chemical Applications and Practices

57. The average radius in the second Bohr orbit ($n = 2$) is 2.116×10^{-10} m. Using the equation $2\pi r = n\lambda$, calculate the wavelength for the standing electron wave.

58. The approximate wavelength for an electron in hydrogen's first energy level has been calculated to be 3.3×10^{-10} m. What is the ratio of this wavelength to the diameter of ten hydrogen atoms (7.4×10^{-11} m)? What might be some reasons why the diameter isn't closer in size to the wavelength?

59. In badminton, the object being struck by the racket is called a shuttlecock. Although the shuttlecock may be made of various materials, it mass must be close to 5.00 g. What is the wavelength of a served shuttlecock that is moving at 78 mi/h? Without doing exact calculations, indicate whether the wavelength of a softball moving at the same speed would be larger or smaller.

A shuttlecock.

60. If a proton, mass = 1.67×10^{-27} kg, were moving as fast as the electron in the ground state of hydrogen (2.1×10^6 m/s), what would be the wavelength?

Section 6.7 The Heisenberg Uncertainty Principle

Skill Review

61. In the uncertainty relationship, what do the symbols p and x represent? Explain the importance of noting that Δp and Δx multiplied must be equal to or greater than a constant.

62. Taking a photograph of a moving object—for example, a person sprinting—will cause some blurring of the actual person's position when the photograph is developed. Therefore, the more blur, the better you represent movement. If the entire scene were re-shot using a faster shutter speed, what information would you gain and what would you lose in the photo?

Chemical Applications and Practices

63. Using the mass of an electron as 9.11×10^{-31} kg and velocity as 2.1×10^6 m/s, what would you calculate as the uncertainty in the position, Δx, for the electron if the uncertainty in the velocity were 5.0%? What would be the answer to this if the uncertainty in the velocity were 10.0%? How does this uncertainty compare to the radius of the hydrogen atom (3.7×10^{-11} m)?

64. Which of these diagrams of an atom most agrees with the Heisenberg uncertainty principle? Explain your answer, and indicate why the other two choices do not conform to the Heisenberg uncertainty principle.

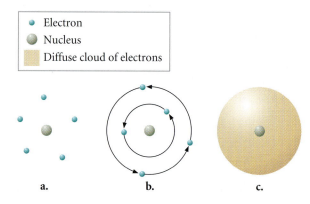

- Electron
- Nucleus
- Diffuse cloud of electrons

a.   **b.**   **c.**

Section 6.8 More About the Photon—the de Broglie and Heisenberg Discussions

Skill Review

65. What is the momentum of an X-ray that has a wavelength of 6.5×10^{-10} m?

66. How does the momentum of the X-ray in Problem 65 compare to the momentum of a photon of green light that has a wavelength of 560 nm?

67. What is the momentum of a mole of X-rays with wavelength 6.5×10^{-10} m?

68. What is the momentum of a mole of photons with wavelength 560 nm?

Chemical Applications and Practices

69. What is the momentum of a 190-lb chemistry professor walking through a chemistry lab at 3.0 mi/h?

70. In order for you to see the strolling professor of Problem 69, light photons would have to bounce off the professor and enter your eyes. If some of that light had a wavelength of 565 nm, what would be the momentum of the photon?

Section 6.9 The Mathematical Language of Quantum Chemistry

Skill Review

71. Acceptance of the Schrödinger equation literally knocked the electrons out of Bohr's orbit concept. However, the term *orbital* was maintained as a connection to the quantization of energy states proposed in Bohr's model. Explain the meaning of an electron orbital.

72. a. What is meant by the term *node* when it is applied to waves?
b. What is the relationship among nodes, wavelength, and energy in the context of standing waves?

Chemical Applications and Practices

73. In both the Bohr model of electronic structure and the Schrödinger model, electrons appear to change energy levels by a method that does not show the electron making a gradual transition. Using the wave model, explain why a stable standing wave is not possible between two adjacent energy levels. Does this explain why there is no gradual transition between states?

74. a. What is a photon?
b. In what way does a photon resemble particles, and in what way does it resemble light?

Section 6.10 Atomic Orbitals

Skill Review

75. The sequence in each line that follows represents values for the quantum numbers for an electron in a hydrogen atom. Select any sequence(s) that are not possible and explain the problem(s).

	n	l	m_l
a.	3	−1	0
b.	2	+2	+1
c.	3	+2	+3
d.	1	+1	+1
e.	4	+3	−2

76. The sequence in each line that follows represents values for the quantum numbers for an electron in a hydrogen atom. Select any sequence(s) that are not possible and explain the problem(s).

	n	l	m_l
a.	3	+1	0
b.	2	+2	+1
c.	3	0	−3
d.	5	+4	0
e.	2	+3	−2

77. When an electron is in the fifth energy level, how many sublevels are possible? How many orbitals are possible?

78. When an electron is in the fourth energy level, how many sublevels are possible? How many orbitals are possible?

79. When $l = 3$, how many degenerate orbitals are possible?

80. When $l = 6$, how many degenerate orbitals are possible?

81. How many sets of quantum numbers are possible when $n = 5$?

82. How many sets of quantum numbers are possible when $n = 3$?

83. We often use the phrase *the shape of an orbital*. Indicate what restrictions apply when the phrase is used, and explain why the phrase is a bit of an abstraction.

84. Name two distinguishing features that the three p orbitals in the same level have in common. What property allows them to be identified separately?

85. Define and give an example of both a radial node and a planar node.

86. In which energy level would the f orbitals make their first appearance? (Use a quantum mechanical proof for your answer.)

Chemical Applications and Practices

87. a. The explanation for scanning tunneling microscopic pictures relies on the concept presented in the quantum mechanical atomic model. When the tip of the STM probe nears an atom, what actually touches?
b. What is meant by quantum tunneling?

88. The "photograph" that is produced by a scanning tunneling microscope is not a photograph in the traditional sense. In what way does the image differ from a traditional photograph?

Section 6.11 Electron Spin and the Pauli Exclusion Principle

Skill Review

89. The four quantum numbers representing an electron may be shown as: $(3, 1, 0, \uparrow)$. What is the value of the fourth quantum number? What is the name given to the fourth quantum number?

90. If the four quantum numbers for an electron were $(3, 2, 1, -\frac{1}{2})$ what would be the four quantum numbers for an electron in the same orbital as the first electron?

Chemical Applications and Practices

91. The magnetic properties of elements are related to the number of unpaired electrons in the atoms. Of the following, which would have unpaired electrons?
C   Ca   O   Ne   Zn

92. The magnetic properties of elements are related to the number of unpaired electrons in the atoms. Of the following, which would have unpaired electrons?
N   S   Na   He   Sc

93. If the Pauli exclusion principle were not used, show how the configuration of the electrons in oxygen would appear to allow no unpaired electrons.

94. Among the elements from 21 to 30, which would have the highest number of unpaired electrons?

Section 6.12 Orbitals and Energy Levels in Multielectron Atoms

Skill Review

95. Name three things that will be the same in multielectron atoms' orbital descriptions, and name one critical difference.

96. How many radial and how many nonradial nodes, respectively, will be possible for each of these electron orbitals? What will be the letter designation for each of the represented orbitals?
a. $n = 3; l = 2$ b. $n = 3; l = 0$ c. $n = 4; l = 3$

Chemical Applications and Practices

97. Would you expect it to be easier to remove the outermost electron from He or He^+? Explain the basis of your answer.

98. Would an electron in the $3p$ sublevel experience more nuclear pull than an electron in the $3d$ sublevel? (Assume that the electron is in the same atom and that the atom's sublevels up to $3s$ are filled.)

**Section 6.13
Electron Configurations and the Aufbau Principle**

Skill Review

99. a. What is the written notation for the ground state of the nitrogen atom?
b. Nitrogen commonly forms a -3 ion. What is the electron configuration of the ion?

100. In the ground state of manganese, there are five electrons that would occupy the $3d$ sublevel. Show the way these electrons could be configured to follow Hund's rule and a way that would violate Hund's rule.

101. Report, for each of the following, which element is being represented.
a. $[Ne]3s^2 3p^2$ b. $[Ne]3s^2 3p^5$ c. $[Ar]4s^1$ d. $[Kr]5s^2$

102. Report, for each of the following, which element is being represented.
a. $[Ne]3s^2$ b. $[He]2s^2 2p^5$ c. $[He]2s^1$ d. $[Ar]4s^2$

103. Write the electron configuration for element 21. Explain why the correct configuration shows the $4s$ sublevel filling with electrons before the $3d$ sublevel.

104. Write out the ground-state electron configuration for element 19.

Chemical Applications and Practices

105. Which, if any, of the following contain unpaired electrons in their ground state?
K   Ca   Fe   Zn   Ne

106. Which, if any, of the following contain unpaired electrons?
K^+ Ca^{2+} Fe^{3+} Zn^{2+} Ne^+

107. Evidence indicates that copper has no unpaired electrons in the $3d$ sublevel. What would have to be the ground-state electron configuration of copper to make this possible?

108. Iridium (element 77) is one of the metals that can be found in the Earth's crust not combined with other elements. It is a brittle, lustrous metal and has a melting point over 2400°C.

    a. Judging on the basis of iridium's position in the periodic table, what other elements would it most resemble?

    b. Does iridium have any unpaired electrons?

    c. Using the Aufbau principle, determine to what sublevel the 25th electron was added to the configuration of iridium.

Comprehensive Problems

109. a. The light produced in the explosion of a dramatic fireworks display reaches you before the sound of the explosion. Use another reference to determine the speed of sound. Then determine the ratio of the speed of light to the speed of sound.

    b. What is the speed of light in miles per hour?

110. When you observe the striking colors of the fireworks at a special occasion, you are observing emission spectra. Describe, chemically, what has transpired in the atoms of the elements to cause the emission of light.

111. What is the longest wavelength of the Lyman series of the hydrogen emission spectrum?

112. Explain how both the volume and the energy of an electron associated with an atom are quantized.

113. Using the equation that allows calculation of the radii of energy levels for single-electron situations in hydrogen, compare the distance between the first and second energy levels to the distance between the third and fourth energy levels and to the distance between the fifth and sixth energy levels. What trend do you notice?

114. The equations used by Bohr are valid for the hydrogen atom, but when they are applied to helium, a significant error shows up. And when they are applied to lithium and elements with higher atomic numbers, the error becomes so large that the equations offer little. However, the equations can be applied to helium and lithium *ions* with a fair agreement with experimental values. What helium and lithium ions would be most like the hydrogen atom?

115. When applying wave properties to electrons in atoms, we use the expression $2\pi r = n\lambda$. Explain why n can have only integer values.

116. Define the terms *discrete* and *continuous*. List five everyday items that have some property that is discrete and five that have a property that is continuous on a macroscopic scale.

117. Assign possible quantum numbers for each of these orbital pictures. Assume that each orbital is in the lowest possible principal shell.

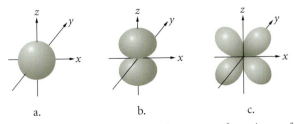

    a.   b.   c.

118. Compare the ground-state electron configurations of potassium and argon. Explain why it is easier to remove the outer electron of potassium than that of argon, even though potassium has more positive charge in its nucleus.

119. A 385 nm photon of light strikes a sheet of a particular metal and ejects an electron at a velocity of 6.1×10^5 m/s. What is the energy, in kJ/mol, associated with this photon?

120. A 0.30 L cup of water is placed in a microwave and irradiated with microwaves of 12.0 cm wavelength. The temperature of the water raises from 25°C to 80°C. How many photons are used to heat the water?

121. a. What is the frequency and energy associated with a photon of light from a ruby laser (see Table 6.4)?

    b. What mass, in ounces, would a particle exhibiting that wavelength have if it were traveling at half the speed of the photon?

Thinking Beyond the Calculation

122. A photon with wavelength $\lambda = 2165$ nm strikes an excited hydrogen atom.

    a. What energy, in joules, is associated with this photon? What energy, in joules, is associated with a mole of these photons?

    b. What is the frequency of the photon?

    c. Which region of the electromagnetic spectrum contains photons of this energy?

    d. If the photon is absorbed by the hydrogen atom, to what energy level is the electron promoted? To make the calculation easier, assume that the electron begins in the $n = 4$ energy level.

    e. If the hydrogen atom relaxes back to the $n = 2$ energy level from the $n = 7$ level, what would be the wavelength of the photon that was emitted? Which region of the spectrum contains this photon?

    f. What could you write as the electron configuration for the hydrogen atom described in part e, after emission of the photon?

Periodic Properties of the Elements

Contents and Selected Applications

7.1 The Big Picture—Building the Periodic Table

7.2 The First Level of Structure—Metals, Nonmetals, and Metalloids

7.3 The Next Level of Structure—Groups in the Periodic Table

Chemical Encounters: Commercial Uses of the Main-Group Elements

Chemical Encounters: The Elements of Life

7.4 The Concept of Periodicity

7.5 Atomic Size

7.6 Ionization Energies

7.7 Electron Affinity

7.8 Electronegativity

7.9 Reactivity

7.10 The Elements and the Environment

Chemical Encounters: The Elements and the Environment

Our planet is composed of a wide variety of elements.

Go to **college.hmco.com/pic/kelterMEE** for online learning resources.

We live on and within a big globe of chemicals that have interacted for well over 4 billion years to form the materially closed system that we call the planet Earth. Our atmosphere provides the oxygen molecules that interact with hemoglobin in our blood and support life on this planet. The foods we eat help sustain this life, and they include molecules with common names like carbohydrates, proteins, and vitamins, as well as salts such as sodium chloride. Our quality of life is enhanced by our clothing, which is often made by combining molecules processed from crude petroleum found deep beneath the seabed. The materials that we have produced for cooking, cleaning, killing, and healing through the ages, from the Stone Age to the Bronze Age to the Iron Age all the way through to the industrial revolution of the late eighteenth and nineteenth centuries and on to the Information Age that has defined the twenty-first century, have a common origin.

Gold.

Throughout the ages, the stuff of life and of our way of life have been based on the set of chemical elements listed in the periodic table.

Why is the oxygen (O_2) that we must breathe every minute a gas, whereas the gold we can dig up from the Earth to use as jewelry is a dense solid? Why does this gold last for thousands of years, unchanged, while the oxygen reacts so readily with many other elements and compounds? Why is the oxygen that is carried around by our blood bound to the iron ions that form part of the protein hemoglobin in our red blood cells? What makes iron so well suited to this task? We need answers to these types of questions if we are to understand the natural environment in which we live, its effects on us, our effects on it, and how it is that we interact with it to acquire and process materials that are so much a part of our day-to-day lives.

Our focus in this chapter is on the elements themselves and on how we use the understanding of their basic structure that we developed in the last chapter to gain insight into their chemical behavior. In the next two chapters, we will look at how and why the elements interact with each other to form the compounds that support our twenty-first-century life. As we continue our four-chapter atomic and molecular tour, we keep in mind that the greatest single statement of our understanding of the behavior of the elements is the periodic table into which they have been organized. It is there that we begin our discussion.

Oxygen.

Video Lesson: Periodic Relationships

7.1 The Big Picture—Building the Periodic Table

The structure of the modern periodic table was initially conceived in the mid-nineteenth century by scientists trying to make sense of the properties and reactivities of all the elements found in the natural environment. It developed through the recording of *experimental* results. That its structure is consistent with the quantum mechanical understanding of the electronic structure of the elements that scientists arrived at in the early twentieth century, as we discussed in Chapter 6, indicates the crucial role of electron arrangement in determining an atom's reactivity. It also confirms *the power of the periodic table as a classification system that becomes more meaningful and valid with each new chemical discovery.*

The structure of the periodic table includes "blocks" defined in terms of which type of orbital is being filled via the Aufbau principle. This gives us the *s*-block, *p*-block, *d*-block, and *f*-block, illustrated in Figure 7.1. Elements in the *s*-block, such as sodium, potassium, and calcium, are naturally found as positive ions, such as the Na^+ ion found in seawater and the Ca^{2+} ion that characterizes "hard" well water. Elements in the *p*-block often form negative ions; examples include the Cl^- ions in blood and the S^{2-} ion that is a part of the minerals shown in Figure 7.2, such as pyrite (FeS_2), called "fool's gold" for its goldlike appearance, and galena (PbS), mined as the main source of lead metal. Many elements in the *d*-block can form positive ions with different charges. For example, iron can form the fairly stable Fe^{2+} and Fe^{3+} ions and also, as part of more complex molecules and ions, the somewhat less stable Fe^{4+}, Fe^{5+}, and Fe^{6+} ions.

The careful recording of experimental results made it possible to determine physical properties of the elements that occur in the natural environment.

FIGURE 7.1

The main ways in which we divide the periodic table into different sections. The *s*-, *p*-, *d*-, and *f*-blocks contain elements with outer electrons in the same type of orbital. The horizontal rows are referred to as periods. The vertical columns are called groups. The color coding indicates the metals, metalloids, and nonmetals. Helium is misplaced in the typical periodic table; it should be located next to hydrogen in the *s*-block.

FIGURE 7.2

Iron pyrite, FeS_2 (left), and galena PbS (right). The sulfur in these two minerals has gained electrons and exists as S^{2-} in PbS and S_2^{2-} in FeS_2.

Day 1

Helium atoms are small enough to escape from a balloon.

Day 8

We see, then, that there is some relationship between the blocks in which elements reside and their chemical and physical properties.

The horizontal "period" of the periodic table to which an element belongs indicates *how many energy levels contain electrons in the ground state of atoms of the element*. Elements in Period 1 have electrons in only the first energy level (principal quantum number $n = 1$). Elements in Period 2 have electrons in the first two energy levels (principal quantum numbers $n = 1$ and $n = 2$), and so on. The more occupied energy levels an atom has, the bigger will be the atom. Helium atoms, from Period 1, are tiny compared to radon atoms, from Period 6. The tiny size of helium atoms explains why a party balloon filled with helium deflates in just a few days, as the atoms escape through little pores in the material of the balloon. Again we see that position in the periodical table is related to what elements do.

Another key link between electron arrangement and position in the periodic table is that *elements in any one main group (Groups IA through VIIIA) have the same number of electrons in their highest energy level*, so that the members of Group VIIA all contain seven electrons in their highest energy, or valence, level.

Periods and groups in the periodic table.

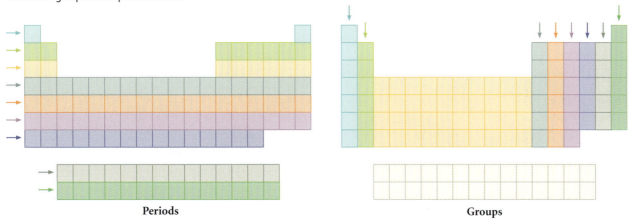

Periods

Groups

That is essentially why elements from each main group share some very signifi-cant chemical characteristics—why helium and neon are both unreactive gases, for example.

HERE'S WHAT WE KNOW SO FAR

- The periodic table is organized into blocks. The *s*-block, *p*-block, *d*-block, and *f*-block indicate the orbitals that are partially full in the highest energy level of the element.
- Rows in the periodic table are known as periods. The period number indicates the highest energy level in which electrons are found in the ground state on each element.
- Columns in the periodic table are known as groups. Elements in the same group have the same number of electrons in their highest energy level.
- Elements in the same group have similar properties.

EXERCISE 7.1	**Identifying Elements**

Give the block, the period number, and the group number for each of these elements.

K, Sc, Al, C, Br

Solution

Potassium is an *s*-block metal found in Period 4 and Group IA.

Scandium is a *d*-block metal found in Period 4 and Group IIIB.

Aluminum is a *p*-block metal found in Period 3 and Group IIIA.

Carbon is a *p*-block nonmetal found in Period 2 and Group IVA.

Bromine is a *p*-block nonmetal found in Period 4 and Group VIIA.

PRACTICE 7.1

Identify each of these elements.

a. A *p*-block element in Period 5 and Group IVA.

b. A *d*-block metal from Period 6 and Group VB.

c. An *s*-block element in Period 1 and Group IA.

See Problems 3–6.

The Historical Development of the Periodic Table

We have just summarized the structure of the periodic table from a modern "elec-tron arrangement" viewpoint. However, the first recorded steps in building the table were made from the quite different viewpoint of the experimental chemist observing what the elements actually do when they react with one another. These first steps were taken by the German scientist Johan Dobereiner (1780–1849) when he identified several groups (he called them triads) of elements whose properties were very similar. Dobereiner's "triads" included lithium, sodium, and potassium (now found in Group IA of the modern periodic table), calcium, strontium, and barium (now in Group IIA), and chlorine, bromine, and iodine (Group VIIA).

In 1864, Englishman John Newlands (1837–1898) arranged some elements in order of increasing atomic mass and found that similarities in properties

The three triads of elements discovered by Dobereiner.

John Newlands noticed that every eighth element had similar properties, a principle that he called the law of octaves. Eight consecutive white keys on a piano comprise an octave.

occurred in elements eight places apart in his scheme. Newlands's system of elements, arranged in what he called octaves (an octave is a musical interval of eight tones, as in the eight white keys on the piano from one C note to another) provided an early glimpse of the regular repetition, or periodicity, in properties that gives the periodic table its name. We know today that the elements have to be arranged in order of increasing *atomic number* for the periodic nature of the periodic table to be observed.

The modern form of the table first began to take shape through the work of German chemist Julius Lothar Meyer (1830–1895) and Russian chemist Dmitri Mendeleev (1834–1907). They added many more elements to the table, and Mendeleev took the crucial step of *predicting that certain elements must exist* to occupy the gaps that were left in the table. In 1869, Mendeleev published the periodic table shown in Figure 7.3. Within a few years of Mendeleev's predictions, three elements were discovered that matched these predictions and filled some of the gaps. For example, Mendeleev predicted the existence of the element "ekasilicon," which we now know as germanium; it was discovered in an ore called argyrodite by the German chemist Clement Winkler in 1886. Table 7.1 gives some of germanium's properties as predicted by Mendeleev and as later measured after the element was discovered. The other two elements discovered were ekaboron (scandium) and ekaaluminum (gallium).

Mendeleev's realization that the periodic table could be used as a *predictive* tool is probably the main reason why Mendeleev, rather than Meyer, is now

FIGURE 7.3

Mendeleev's periodic table. The atomic mass of each of the elements is recorded in the table. Dashes in the table indicate elements that Mendeleev proposed but had not yet been discovered. The symbols at the top of the table (R^2O, RO, R^2O^3, etc.) are chemical formulas written in the style of the 1870s.

Video Lesson: Creating the Periodic Table

T a b e l l e II.

Reihen	Gruppe I. — R^2O	Gruppe II. — RO	Gruppe III. — R^2O^3	Gruppe IV. RH^4 RO^2	Gruppe V. RH^3 R^2O^5	Gruppe VI. RH^2 RO^3	Gruppe VII. RH R^2O^7	Gruppe VIII. — RO^4
1	H=1							
2	Li=7	Be=9,4	B=11	C=12	N=14	O=16	F=19	
3	Na=23	Mg=24	Al=27,3	Si=28	P=31	S=32	Cl=35,5	
4	K=39	Ca=40	—=44	Ti=48	V=51	Cr=52	Mn=55	Fe=56, Co=59, Ni=59, Cu=63.
5	(Cu=63)	Zn=65	—=68	—=72	As=75	Se=78	Br=80	
6	Rb=85	Sr=87	?Yt=88	Zr=90	Nb=94	Mo=96	—=100	Ru=104, Rh=104, Pd=106, Ag=108.
7	(Ag=108)	Cd=112	In=113	Sn=118	Sb=122	Te=125	J=127	
8	Cs=133	Ba=137	?Di=138	?Ce=140	—	—	—	
9	(—)	—	—	—				
10	—	—	?Er=178	?La=180	Ta=182	W=184	—	Os=195, Ir=197, Pt=198, Au=199.
11	(Au=199)	Hg=200	Tl=204	Pb=207	Bi=208	—	—	
12	—	—	—	Th=231	—	U=240	—	—

Dmitri Mendeleev (1834–1907), the youngest of the 17 children in his family, was born in Serbia. After his father died, his mother moved the family to St. Petersburg, Russia, so that Mendeleev could get the best education possible. After 20 years of work, he developed the basic layout of the periodic table that we still use today.

TABLE 7.1	**Predicted (Ekasilicon) and Measured Properties of Germanium**
Ekasilicon (Ek)	**Germanium (Ge)**
Predicted in 1871	Discovered in 1886
Atomic mass = 72	Atomic Mass = 72.6
Density = 5.5 g/cm^3	Density = 5.47 g/cm^3
Density of EkO$_2$ = 4.7 g/cm^3	Density of GeO$_2$ = 4.70 g/cm^3
EkCl$_4$ will be liquid, density = 1.9 g/cm^3	GeCl$_4$ is a liquid, density = 1.89 g/cm^3

recognized as the "father of the periodic table." What we now appreciate as positively inspired reasoning was, at one time, dismissed as Mendeleev's madness—a telling illustration of how the scientific method requires time for important insights to be identified among other, less fruitful ideas.

7.2   The First Level of Structure—Metals, Nonmetals, and Metalloids

The periodic table hangs in a highly visible place on the wall of chemistry laboratories all over the world. It is not there for reasons of history or adornment. It is there because *it is the most important and useful work developed in the history of science for describing what we know about the chemistry of the elements.* There are many versions of the table, and they vary in precisely what information is given about each element. The basic structure of the table, however, in terms of the positioning of elements within vertical groups and horizontal periods, is generally the same. A chemistry student from Russia, South Africa, Mexico, or Scotland can look at the table and communicate its ideas with any other chemistry student in the world. The language of chemistry, as illustrated by the periodic table, is universal. What does this basic structure tell us about the elements?

The elements in the periodic table are arranged into three main sections—the metals, nonmetals, and metalloids (or semimetals)—as shown in Figure 7.1. The majority of elements are metals, which possess a variety of "metallic" characteristics, including those in this list.

Copper is not very reactive—hence its use in electrical wire and pipes.

Metals are generally

- shiny in appearance
- solids at room temperature and pressure (apart from mercury, the only metal that is a liquid at room temperature and pressure)
- good conductors of electricity
- malleable—that is, able to be hammered into various shapes
- likely to form positively charged ions when they react to form ionic compounds

Metals, including iron, aluminum, and copper, supply us with some of our most widely used and versatile materials for fabrication. For example, copper is used in electric wires and in plumbing throughout your home. Iron is the basic

TABLE 7.2	Selected Metals, Their Sources, and Their Uses	
Metal	**Major Sources in Nature**	**Uses**
Al	Bauxite, clay, mica, feldspar, alumina	Food wrap, kitchenware, beverage cans
Ca	Lime (CaO), limestone ($CaCO_3$), feldspar, apatite	Manufacture of vacuum tubes, alloys, preparation of other metals
Cu	Chalcopyrite (copper iron sulfide), pure metal	Coinage metal, electric wires, plumbing pipes
Fe	Hematite (Fe_2O_3), limonite ($Fe_2O_3 \cdot 3H_2O$)	Steel
Na	Salt (NaCl), Borax ($Na_2B_4O_7 \cdot 10H_2O$)	Street lamps, nuclear reactor coolant
Ni	Pentlandite and pyrrhotite (nickel–iron sulfides), garnierite (nickel–magnesium silicate)	Coinage metal, stainless steel, NiCad batteries, heating elements
Sn	Casserite (SnO_2)	Tin cans, many alloys, including bronze
Ti	Rutile (TiO_2), ilmenite	Aircraft parts, lightweight tank armor
W	Wolfram ochre (WO_3)	Filaments in light bulbs

constituent of a homogenous mixture of metals (known as an **alloy**) that make up steel. Steel is used throughout our modern world as a structural component of cars, trains, buildings, and bridges.

What are the industrial and commercial uses of steel? Chromium and nickel are used to create the familiar stainless steels. These steels resist corrosion and are commonly used in making tableware. Tellurium and selenium promote the machinability of steel, its ability to be easily turned and shaped into bolts and screws. Manganese makes steels that are very resistant to wear as well as to chemical reaction with water. Molybdenum is used to create hard steels for use in bearings. Careful addition of silicon creates electrical steels used in the generation and transmission of electricity. The exploration of ways in which subtle changes in the composition of steels brings about significant changes in their properties is a continuing process. That huge research and manufacturing effort all stems from the simple observation, hundreds of years ago, that letting hot charcoal mix in with molten iron improved its usefulness.

Stainless steel is commonly used as flatware in kitchens.

EXERCISE 7.2 | Heavy Metals, Both Necessary and Toxic

If you follow environmental issues, you may have heard of the term *heavy metals,* which some think of as those with atomic masses greater than or equal to 63.546 g/mol. Some of these metals are essential for life, but many of them are toxic. Some of the heavy metals that are vital for life in low concentrations can be quite toxic in high concentrations. Heavy metals in water and soil are a major focus of environmental concern.

1. According to the definition given above, what is the lightest of the heavy metals?
2. Using the periodic table, can you identify some other heavy metals that are commonly discussed as environmental contaminants?
3. Chromium is included in some lists of heavy metals. What does that indicate about the definition of the term *heavy metals?*
4. Can you suggest why heavy metals can be particularly persistent environmental contaminants, posing problems that are difficult to correct?

Solution

1. According to the definition supplied, copper is the lightest of the heavy metals.
2. The metallic elements heavier than copper in the periodic table include many that have received publicity as heavy metal contaminants of the environment. Examples which feature prominently in news reports on this issue are cadmium, mercury, and lead.
3. Chromium has a lower atomic mass than copper, so its inclusion in lists of heavy metals suggests that there is no universally accepted definition of the term *heavy metal.* This is in fact the case. It is a term used rather loosely to describe metallic elements of relatively high atomic mass that are also toxic.
4. Heavy metals are particularly persistent environmental contaminants because, being elements, they cannot be degraded into simpler, less toxic components. This is in contrast to toxic *compounds,* some of which are readily degraded into less harmful compounds or their component elements.

PRACTICE 7.2

Which of these elements are metals? Which of these are heavy metals?

Li, Si, Ni, Ce, Ge, Al, Po, Se, Rb, Cu

See Problems 11 and 12.

NanoWorld / MacroWorld

Big effects of the very small: The diversity of steels

Steel, shown in Figure 7.4, is one of our most versatile construction materials, with 1.05 billion metric tons (2.42 *trillion* pounds) manufactured in 2004. What exactly is steel?

The most basic definition tells us that steel is a mixture of iron and small amounts of carbon, suggesting that it is a metal with a little nonmetal blended into it. Hundreds of years ago, it was discovered that the metal element iron could be changed into a tougher and more resilient material by allowing carbon from wood fires to become mixed in with it. Since then, we have discovered that adding various other elements in differing proportions can be used to vary the properties of steel in a great many ways. In fact, nowadays, we can no longer talk of steel as though it were a single thing. There are actually *over 3500 different grades of steel*, each with a particular mixture of elements added to the iron that forms the basis of them all. Looking at the differences among these grades reveals how small changes in the atoms present in the steel can have very significant effects on its properties.

There are three main types of steel, known as carbon-steel, low-alloy steel, and high-alloy steel. Carbon-steels have as little as 0.1% and sometimes more than 2% carbon added to their iron, but only very small amounts of other elements. Over 90% of the world's steel is carbon-steel, which is further categorized as high-carbon, medium-carbon, low-carbon, ultra-low carbon, and so on, depending on exactly how much carbon it contains. Alloy steels have homogeneous mixtures of metals, called alloys, added to them. The metals most commonly added to alloy steels are manganese, aluminum, copper, nickel,

chromium, cobalt, molybdenum, vanadium, tungsten, titanium, niobium, zirconium, and tellurium. Steels also often contain some added nonmetals, such as silicon, selenium, nitrogen, and sulfur. The compositions of three standard steels are shown in Table 7.3.

TABLE 7.3	The Compositions of Selected Steels		

The elements alloyed with iron to make selected steels. The numbers are reported as the mass percent of the total composition.

	Tool Steel	Basic Electric Steel	Stainless Steel
C	0.864	0.215	0.225
Mn	0.341	0.393	0.544
P	0.012	0.016	0.030
Si	0.185	0.211	1.00
Cu	0.088	0.211	0.226
Ni	0.230	0.248	8.76
Cr	4.38	0.017	16.7
V	1.83	0.003	0.176
Mo	4.90	0.038	0.24
W	6.28	—	—
Sn	0.029	—	—
Al	—	0.002	—
Co	—	—	0.127

The enhanced strength of the iron–carbon alloy was used initially to create swords and armor.

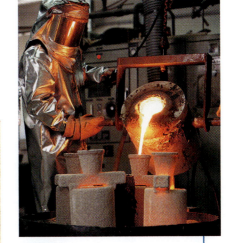

FIGURE 7.4

Steel, such as that being poured here, has a variety of uses in the manufacture of building materials, office equipment, and automobile parts.

TABLE 7.4	Selected Nonmetals, Their Sources in Nature, and Some of Their Industrial Uses	
Nonmetal	**Sources in Nature**	**Industrial Uses**
C	Coal, graphite, diamonds	Steel manufacture, pencil "lead"
Cl	Salt (NaCl), briny water	Water purification, manufacture of dyes and explosives
N	Air	Fertilizers, gunpowder, low-temperature reactions, inert atmospheres
O	Air	Medical field, steel manufacture, combustion
P	Apatite (calcium hydroxy phosphate)	Fertilizer, chemical warfare agent, rat poison, steel manufacture
S	Iron pyrite, galena, barite, pure element	Black powder, fertilizer, fireworks, rubber production

Video Lesson: General Properties of Nonmetals

Nonmetals possess characteristics that are generally quite different from those of metals. **Nonmetals** are generally

- gases or dull, brittle solids at room temperature and pressure
- poor conductors of electricity (with the exception of the form of carbon known as graphite)
- likely to form negatively charged ions when they react to form ionic compounds

Despite being small in number, the nonmetals include some of the most abundant elements found in living things, particularly the carbon, hydrogen, oxygen, nitrogen, phosphorus, and sulfur atoms from which the chemicals of life are largely made.

Between the metals and nonmetals in the periodic table we find elements known as **metalloids**, or **semimetals**. Specifically, boron, silicon, germanium, arsenic, antimony, tellurium, and astatine make up the list of elements we call metalloids. These elements share some of the characteristic properties of both the metals and nonmetals, making it difficult to place them in either of these two main categories. For example, silicon and germanium have properties of both metals and nonmetals. Unlike the metals, these elements are not good conductors of electricity, but unlike the nonmetals, they are not very poor conductors of electricity. Their ability to conduct electricity lies somewhere in between that of the

The metalloids.

The nonmetals, shown in green, have properties that are opposite those of the metals, shown in blue.

Period	1 IA	2 IIA	3 IIIB	4 IVB	5 VB	6 VIB	7 VIIB	8	9 VIIIB	10	11 IB	12 IIB	13 IIIA	14 IVA	15 VA	16 VIA	17 VIIA	18 VIIIA
1	1 H																	2 He
2	3 Li	4 Be											5 B	6 C	7 N	8 O	9 F	10 Ne
3	11 Na	12 Mg											13 Al	14 Si	15 P	16 S	17 Cl	18 Ar
4	19 K	20 Ca	21 Sc	22 Ti	23 V	24 Cr	25 Mn	26 Fe	27 Co	28 Ni	29 Cu	30 Zn	31 Ga	32 Ge	33 As	34 Se	35 Br	36 Kr
5	37 Rb	38 Sr	39 Y	40 Zr	41 Nb	42 Mo	43 Tc	44 Ru	45 Rh	46 Pd	47 Ag	48 Cd	49 In	50 Sn	51 Sb	52 Te	53 I	54 Xe
6	55 Cs	56 Ba	57 La*	72 Hf	73 Ta	74 W	75 Re	76 Os	77 Ir	78 Pt	79 Au	80 Hg	81 Tl	82 Pb	83 Bi	84 Po	85 At	86 Rn
7	87 Fr	88 Ra	89 Ac†	104 Rf	105 Db	106 Sg	107 Bh	108 Hs	109 Mt	110 Uun	111 Uuu	112 Uub						

*Lanthanide series	58 Ce	59 Pr	60 Nd	61 Pm	62 Sm	63 Eu	64 Gd	65 Tb	66 Dy	67 Ho	68 Er	69 Tm	70 Yb	71 Lu
†Actinide series	90 Th	91 Pa	92 U	93 Np	94 Pu	95 Am	96 Cm	97 Bk	98 Cf	99 Es	100 Fm	101 Md	102 No	103 Lr

metals and that of the nonmetals. We say that they are semiconductors, which makes their use as materials in the computer-manufacturing industry ideal.

The subdivision into metals, nonmetals, and metalloids is the first level of structure found within the periodic table. The next, and more fundamental, level is the subdivision into the vertical columns called groups and the horizontal rows called periods.

7.3   The Next Level of Structure— Groups in the Periodic Table

Elements within any of the vertical columns of the periodic table, which are known as groups, share some significant chemical properties. There is a clear explanation for the chemical similarities of elements within each group based on each atom's electron arrangements. For instance, elements in any of the **main groups**, traditionally labeled IA through VIIIA, all have the same number of electrons in their outer energy level, which corresponds to the highest principal quantum number. The outer electrons—the **valence electrons**—are very significant in determining how an atom interacts with and reacts with other atoms. For those elements known as transition elements, the definition of *valence electrons* actually includes electrons in two energy levels.

Application

CHEMICAL ENCOUNTERS: Commercial Uses of the Main-Group Elements

Recent modifications to the periodic table have been made by the International Union of Pure and Applied Chemistry (IUPAC). One such modification is the standardization of the numbers at the top of each of the columns of the periodic table. In this system, the columns have been numbered from 1 to 18 from left to right. Although this numbering system has eliminated confusion based on the number of the column, vital information about the elements in the particular groups of the periodic table has been lost. The authors prefer the use of the historically common system of numbering the groups (IA, IIA, IB, etc.). You can compare this time-honored system to the new IUPAC numbering system in the periodic table on the inside front cover of this text.

Group IA: Hydrogen and the Alkali Metals

Video Lesson: Hydrogen, Alkali Metals, and Alkaline Earth Metals

Video Lesson: Hydrogen

Video Lesson: The Alkali Metals

The elements in Group IA, apart from hydrogen, are known as the **alkali metals** because they are highly reactive elements that combine with water to form chemicals called alkalis. [*Al qili* is from the Arabic, describing the burning of the saltwort plant, which produces an ash that forms an alkaline (basic) solution. Alkali metals form alkaline solutions when they react with water. We will discuss acids and bases in Chapter 17.] The alkali metals all have one valence electron in an *s* orbital, which they generally lose when reacting, to form ions with a +1 charge. We don't find the alkali metals as free elements in the environment because they are so chemically active that their atoms have all reacted to form compounds or dissolved ions. Sodium is not found in its elemental state but, rather, in the form of sodium ions (Na^+). Sodium ions are the most abundant positive ions in seawater, and they are also plentiful in the internal environment of our blood and intercellular fluids.

Hydrogen shares some characteristics with the alkali metals, particularly its tendency to react by forming a +1 ion (H^+). In other respects, however, it is quite different from the alkali metals. It is not a metal; it can also form a negative ion, the "hydride" ion (H^-), when combining with other alkali metals; and two hydrogen atoms can combine by covalent bonding to form the hydrogen molecule (H_2).

| 3
Li
6.941 |
| 11
Na
22.990 |
| 19
K
39.098 |
| 37
Rb
85.468 |
| 55
Cs
132.905 |
| 87
Fr
(223) |

The alkali metals.

TABLE 7.5	Some Commercial Uses of Group IA Elements
Hydrogen	Fertilizers, plastics, pharmaceuticals, fuel, fuel cells
Lithium	Glass making, television tubes, battery electrolytes, lithium–aluminum alloys in the aerospace industry, greases
Sodium	Nuclear reactors, reagent, salt, washing soda
Potassium	Fertilizers, soaps and detergents, explosives, glass and water purification
Rubidium	Photoelectric cells
Cesium	Special glass, radiation-monitoring equipment, atomic clock
Francium	None (too rare)

Video Lesson: The Alkaline Earth Metals

Group IIA: The Alkaline Earth Metals

The elements in Group IIA are known as the **alkaline earth metals**. These are also reactive metals, although generally less reactive than the alkali metals. They have two valence electrons within an *s* orbital, which they generally lose when reacting, to form ions with a +2 charge. Compounds of these metals combined with oxygen are reasonably common in the environment. Because these metal oxides

TABLE 7.6	Some Commercial Uses of Group IIA Elements
Beryllium	Telecommunications equipment, automotive electronics, computers, undersea communications equipment, pipe products in the oil and gas industry
Magnesium	Automotive industry, bicycles, luggage
Calcium	Metallurgic applications, lead and aluminum industries, nuclear applications, cement, soil conditioner, water treatment
Strontium	Cathode-ray tubes, automotive industry, special glass, fireworks, flares
Barium	Given to patients with digestive disorders to highlight the bowels in an X-ray, drilling fluid, oil for gas wells
Radium	Cancer treatment, luminous paint for watches and clocks

Video Lesson: Aluminum

Video Lesson: CIA Demonstration: The Reaction between Al and Br_2

were so readily available in the Earth, and because they form alkalis when dissolved in water, the elements in Group IIA were originally given the name *alkaline earth metals*. For example, calcium oxide (CaO) is an alkaline compound used in the manufacture of cement and steel.

| 4 Be 9.012 |
| 12 Mg 24.305 |
| 20 Ca 40.078 |
| 38 Sr 87.62 |
| 56 Ba 137.327 |
| 88 Ra (226) |

The alkaline earth metals.

Group IIIA

The elements in Group IIIA are substantially less reactive metals than those in either Group IIA or Group IA. They all have three valence electrons, two in an *s* orbital and one in a *p* orbital. When they react to form ionic compounds, they generally do so by losing the three outer electrons to form ions with a +3 charge. Note that as the number of electrons that an atom loses to form an ion increases, the reactivity of the element *decreases*. This is our first example of a *trend* within the periodic table—a characteristic that varies in a logical and regular manner as we move through the table. The most significant Group IIIA element that we gather from the environment is aluminum (23 million metric tons were produced worldwide in 2005, according to the International Aluminum Institute), found in the ore known as bauxite, which is composed of aluminum oxide (Al_2O_3) combined with varying amounts of water.

| 5 B 10.811 |
| 13 Al 26.982 |
| 31 Ga 69.723 |
| 49 In 114.818 |
| 81 Tl 204.383 |

Group IIIA.

A conveyor belt at the Rio do Norte mining company in the Amazon forest in Brazil piles up bauxite ore, a source of aluminum oxide. Mining bauxite can be very devastating to the environment, but the Rio do Norte mining company has been successful in reforesting the land that they've mined, as evidenced by the green area in the middle of the photo. Their efforts challenge the notion that the Amazon forest must be destroyed to tap its riches and offer an opportunity to learn how to repair degraded forests.

TABLE 7.7	Some Commercial Uses of the Group IIIA Elements
Boron	Glass making, soaps and detergents, nuclear reactor control rods
Aluminum	Packaging, especially soft drink containers, transportation (lightweight components in motor vehicles), window frames, aircraft parts, engines, kegs, cooking oils, indigestion tablets
Gallium	Semiconductors, microwave equipment
Indium	Display devices, low-melting-point alloys and solders, semiconductors, fire sprinkler systems
Thallium	Rat poisons and hair removers (now banned)

TABLE 7.8	Some Commercial Uses of the Group IVA Elements
Carbon	Commercial and military aircraft, fibers, thermoplastic matrix materials, petrochemicals, clothing, dyes, fertilizers, fuels, pharmaceuticals
Silicon	Aluminum alloys, silicones, silicon chips used in computers, semiprecious stones
Germanium	Semiconductors, transistors, catalysts (substances that greatly increase the rate of a reaction without being consumed) in polymer production, glass for infrared devices
Tin	Coatings for other metals, bronze, soft solder, pewter, special paint used on boats to prevent barnacles
Lead	Storage batteries, cable covering, radiation shielding, pipes, pewter, pottery, additive in gasoline, lead crystal glass

Group IVA.

6	C	12.011
14	Si	28.086
32	Ge	72.64
50	Sn	118.710
82	Pb	207.2

Video Lesson: General Properties of Carbon

Video Lesson: Silicon

Group IVA

The elements in Group IVA, with four valence electrons, two in an *s* orbital and two in individual *p* orbitals, occupy the "center ground" of the main-group elements, and they include elements of central importance to living things and the fabricated materials with which modern society survives. Carbon, at the top of the group, can be regarded as the basic elemental "building block" of life, because the chemicals of life are largely based on chemical chains and rings of carbon atoms with various other atoms attached. Carbon does not generally form ions. Instead, it forms covalent bonds (we will discuss these in Chapter 8), in which electrons are more or less shared between two atoms. Group IVA also includes silicon, the basis of the "silicon chips" of the computer industry, and the semiconductor germanium, also used in the manufacture of computer chips.

Video Lesson: Nitrogen

Video Lesson: Phosphorus

Group VA

The elements in Group VA have five valence electrons, two sharing an *s* orbital and three occupying individual *p* orbitals. When they form ions, these elements generally gain three electrons to form ions with a −3 charge. The Group VA elements also readily participate in covalent bonding. Nitrogen, for example, makes up 80% of the volume of the Earth's atmosphere in the form of N_2 molecules. Nitrogen is also a crucial component of many covalently bonded compounds required for life, such as DNA, proteins, and many vitamins and hormones. Phosphorus is also crucial for life, being one of the atoms found in DNA, for example, and being part of the compounds found in fertilizers that are used to grow the food we consume.

TABLE 7.9	Some Commercial Uses of the Group VA Elements
Nitrogen	Fertilizers, plastics, explosives, dyes
Phosphorus	Fertilizers, matches, detergents, coating to prevent corrosion
Arsenic	Pesticides, wood preservatives, semiconductors, special glass
Antimony	Flame retardants, pigments, lubricants, ammunition, used to harden other metals
Bismuth	Industrial and laboratory chemicals, pharmaceuticals, cosmetics, replacement for lead in steel alloys and aluminum in ceramics, high-temperature superconductors, indigestion tablets

Group VA.

7	N	14.007
15	P	30.974
33	As	74.922
51	Sb	121.760
83	Bi	208.980

Group VIA: The Chalcogens

The elements in Group VIA are known as the **chalcogens**, a name derived from the Greek *khalkos* ("copper"), because the copper ores contain the elements in this group. All of the elements in Group VIA have six valence electrons, two in an *s* orbital and four distributed within three *p* orbitals. When they form ions, the Group VIA elements gain two electrons to form ions with a −2 charge.

In looking for Group VIA elements of significance to ourselves and the environment, we immediately turn to oxygen. Oxygen gas, in the form of O_2 molecules, makes up about 21% of the atmosphere and is the gas we must breathe in order to stay alive. We need it because it combines with hemoglobin, as we noted in the chapter opening, and reacts with food molecules to release the energy that powers all life. We have also met, in Chapter 2, the relatively rare form of oxygen known as ozone (O_3) that is a vital part of our environment high in the atmosphere, where the ozone layer absorbs harmful UV rays, but is a troublesome environmental pollutant at ground level.

Group VIA also contains sulfur, another element that is both vital for life and associated with harmful pollution. We need the sulfur that is a part of proteins and some vitamins and other important biochemicals, as well as fertilizers and industrial chemicals such as sulfuric acid (H_2SO_4), but oxides of sulfur such as sulfur dioxide (SO_2) are pollutants that can cause "acid rain."

Video Lesson: Oxygen
Video Lesson: Sulfur

 Application

TABLE 7.10	Some Commercial Uses of the Group VIA Elements
Oxygen	Steel making, metal cutting, chemical industry
Sulfur	Lead–acid storage batteries, fertilizers, water treatment, petroleum refining, drying agents
Selenium	Photoreceptors, glass colorant, pigments, metallurgic and biological applications, photoelectric cells, photocopiers, semiconductors
Tellurium	Additive in steel and other metal alloys, catalyst in synthetic rubber production
Polonium	Source of alpha radiation, heat source in space vehicles

The chalcogens.

| 8 O 15.999 |
| 16 S 32.065 |
| 34 Se 78.96 |
| 52 Te 127.60 |
| 84 Po (209) |

Sulfur can be found near volcanoes as yellow rocks.

TABLE 7.11	Some Commercial Uses of the Group VIIA Elements
Fluorine	Welding metals, frosting glass, used in the nuclear power industry, insulating gas for high-power electricity transformers, as an additive to municipal water supplies
Chlorine	Chemical warfare, bleach, PVC plastic, water purification
Bromine	Flame retardants, pharmaceutical intermediates, swimming pool disinfectants, industrial water treatment, air conditioning absorption systems, precious metal leaching, fuels, additives, insecticides, photography
Iodine	Conductive polymers, fuel cells, dyes, photography, pharmaceuticals, industrial catalyst
Astatine	Nuclear reactors

The halogens.

Video Lesson: Halogens

Video Lesson: Aqueous Halogen Compounds

Group VIIA: The Halogens

The elements in Group VIIA are known as the **halogens** (from the Greek *hals + gen*, meaning "salt makers") because they are reactive nonmetals that can combine with metal ions to form the chemicals we call salts. All of the elements in Group VIIA have seven valence electrons, two in an *s* orbital and five distributed among three *p* orbitals. When they form ions, as happens when they become part of salts, they gain one electron per atom to form ions with a -1 charge. The most common example of a salt is sodium chloride (NaCl), which we know as common table salt, and we will meet many other examples throughout this book. Chloride ions (Cl^-) are abundant in seawater and in the blood and intercellular fluids of the body. Tiny amounts of fluoride ions (F^-) are important in making our teeth resistant to decay.

Video Lesson: Properties of Noble Gases

Group VIIIA: The Noble Gases

The elements of Group VIIIA are all largely unreactive gases, and the group is collectively known as the **noble gases**, in the sense of aloof nobility—being distinguished from the other, much more reactive elements. The noble gases, which are also known as the **inert gases**, are present in very small amounts in the environment, but through their remarkable stability, they reveal to us one of the most significant secrets of chemistry. All the noble gases apart from helium have eight valence electrons distributed between a full *s* orbital and three full *p* orbitals. The

TABLE 7.12	Some Commercial Uses of the Group VIIIA Elements
Helium	Used by divers to dilute the oxygen they breathe, balloons, low-temperature research
Neon	Filling discharge tubes, ornamental lighting
Argon	Filling discharge tubes, provide an inert atmosphere for high-temperature metallurgic processes, light bulbs
Krypton	High-quality light bulbs
Xenon	Research purposes
Radon	Gives off alpha particle radiation

The noble gases.

fact that atoms of these elements do not naturally react either with themselves or with any other elements (although some compounds of xenon, krypton, and radon have been made under extreme conditions) indicates that they have an exceptionally stable valence electron arrangement. This arrangement is known as a stable **octet** because it contains eight valence electrons. Helium, with only two valence electrons, also has an exceptionally stable electron arrangement because it has a completely full energy level. It is the "odd one out" among the noble gases in that, with only two electrons, it does not have a stable octet but nevertheless shares the remarkable stability of the other inert gases. When we examine chemical bonding in more detail in Chapter 8, we will find that many chemical reactions can be understood in terms of the participating atoms acquiring electron arrangements that are similar to the stable valence electron arrangements of the noble gases.

Although they are very rare, we have found various good uses for the noble gases of the environment. Helium, for example, is well known as the "lighter than air" gas within party balloons, weather balloons, and airships that allows them to rise upward. Neon has found fame as the gas within "neon" lights. In light bulbs, argon, mixtures of argon and nitrogen, or, in newer "halogen" bulbs, krypton and xenon are used as inert gases to prevent the tungsten filament from oxidizing, and xenon has similar uses in flash photography.

Uses of the noble gases.

EXERCISE 7.3   **Group Fun**

How many electrons are found in the valence shell of each of these elements?

   a. Be   b. As   c. I   d. In

Solution

   a. Two, both in the $2s$ orbital

   b. Five, two in the $4s$ orbital and three in the $4p$ orbitals

   c. Seven, two in the $5s$ orbital and five in the $5p$ orbitals

   d. Three, two in the $5s$ orbital and one in the $5p$ orbitals

---

PRACTICE 7.3

Indicate which element is described by each of these phrases.

   a. Two valence electrons, both in the $6s$ orbital

   b. One valence electron, in the $3s$ orbital

   c. Six valence electrons, two of them in the $2s$ orbital

   d. Four valence electrons, and the element is a nonmetal

See Problems 23 and 24.

The Transition Elements

In the middle of the periodic table we find elements known as the **transition elements**, with the **inner transition elements** usually shown as a separate block, as in Figure 7.1. The transition elements are arranged in short groups, historically labeled IB through VIIIB in the case of the main transition elements, and currently unlabeled for the inner transition elements.

The elements in the first period of the inner transition elements are known as the **lanthanides**, and the second period forms the **actinides**. These inner transition elements are shown separately, primarily for the visual ease of having the periodic table fit into a convenient space. A more logical version of the periodic table is shown in Figure 7.5, with the inner transition elements fitted where they belong

Video Lesson: Transition Metals and Nonmetals

Video Lesson: Properties of Transition Metals

Video Lesson: CIA Demonstration: Copper One-Pot Reactions

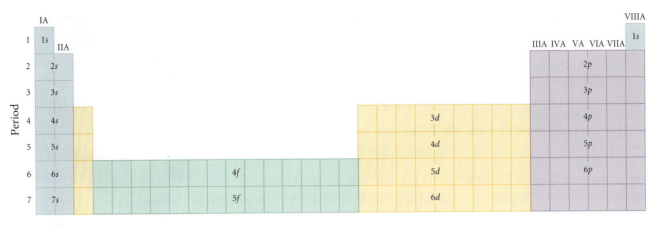

FIGURE 7.5

Periodic table with the lanthanides and actinides where they belong in terms of electron arrangement.

Transition metals, such as the copper atom (green), are an essential part of some enzymes. This three-dimensional model is spinach plastocyanin, an enzyme involved in energy production in some plants. The hydrogen atoms have been omitted from the model to make it easier to see the other atoms.

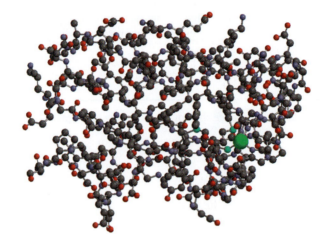

in terms of their atomic numbers. The inner transition elements make up the *f*-block, and they have some specific chemical characteristics due to electrons in *f* orbitals acting as valence electrons in addition to those in the outer *s* orbital. We most often think of the inner transition elements such as uranium and plutonium in terms of their nuclear rather than their chemical properties.

The transition elements of the *d*-block have electrons in *d* orbitals that can count as valence electrons, in addition to the electrons in the *s* orbital of their highest energy level. This situation of having valence electrons in two energy

The *f*-block elements.

58 Ce 140.116	59 Pr 140.908	60 Nd 144.24	61 Pm (145)	62 Sm 150.36	63 Eu 151.964	64 Gd 157.25	65 Tb 158.925	66 Dy 162.500	67 Ho 164.930	68 Er 167.259	69 Tm 168.934	70 Yb 173.04	71 Lu 174.967
90 Th 232.038	91 Pa 231.036	92 U 238.029	93 Np (237)	94 Pu (244)	95 Am (243)	96 Cm (247)	97 Bk (247)	98 Cf (251)	99 Es (252)	100 Fm (257)	101 Md (258)	102 No (259)	103 Lr (262)

levels arises because the energies of the electrons involved are very close. It results in some interesting and important chemical characteristics, such as the ability to readily form ions with different values of charge.

The special chemistry of the transition elements, such as iron, cobalt, copper, and zinc, makes many of them vital for life. Biochemists, for example, have become quite used to the idea that when a difficult feat of chemistry is achieved within a living thing, it is very often catalyzed by an enzyme (a biological substance that significantly increases the rate of a process; see Chapter 22) that includes a transition metal ion at the "active site," where the chemical reaction occurs. Transition elements are also widely used as catalysts in the chemical industry and, in the automotive industry, as rhodium- and platinum-based catalytic converters in cars.

The Elements of Life

Figure 7.6 shows the periodic table with all of the elements that make up the human body highlighted in color and their relative abundances indicated, and Table 7.13 lists the relative abundance of these elements within us. Note the location of these elements in the periodic table. Can you suggest any generalizations about their locations? Can you offer any explanations for your suggestions?

The two most clear-cut generalizations are that the human body is largely constituted of elements only from the upper right-hand portion of the periodic table and that the noble gases are not part of the chemistry of life.

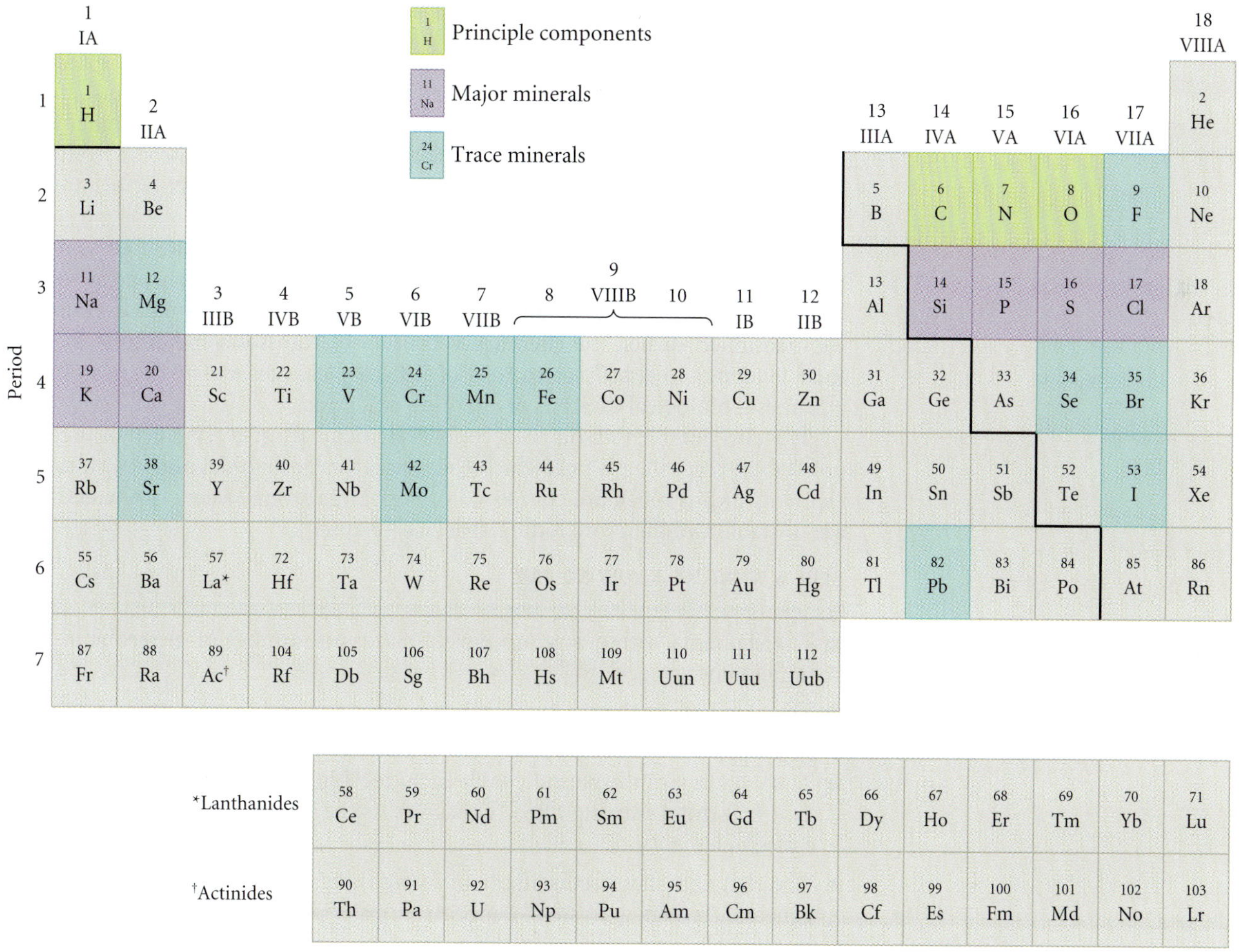

Application

CHEMICAL
ENCOUNTERS:
The Elements
of Life

FIGURE 7.6

The elements that make up *you*.

TABLE 7.13	Elements of the Human Body
Element	**Abundance (percent, by mass)**
Oxygen	61.0
Carbon	23.8
Hydrogen	10.0
Nitrogen	2.6
Calcium	1.4
Phosphorus	1.1
Sulfur	0.2
Silicon	0.02
Potassium	0.02
Sodium	0.014
Chlorine	0.012
Fluorine	0.004
Magnesium	0.003
Iron	0.0004
Strontium	0.00005
Bromine	0.00003
Lead	0.00002

Trace amounts of Mn, I, Cr, Mo, Se, and V

A seven-base-pair segment of double-stranded DNA. The phosphorus atoms are yellow.

We can explain the absence of noble gases by looking at their chemical reactivity—or lack of it. Life is a very complex process of chemical *change* involving regulated chemical reactions. The noble gases generally do not participate in chemical change because they are so unreactive, so we would not expect to find them playing any part in the chemistry of life.

The fact that life is made primarily of the smaller elements, from a small section of the periodic table, may be linked to the fact that these elements are generally more abundant on Earth, and in the universe at large, than elements with very large atoms. As we'll see in Chapter 21, larger atoms can be built from smaller ones. Although there are exceptions, the larger atoms are generally less abundant than smaller ones in nature. In any case, nearly all of the mass of a human is made up of atoms of hydrogen (atomic number = 1), carbon (atomic number = 6), nitrogen (atomic number = 7), and oxygen (atomic number = 8). We are largely made from some of the smallest atoms of the environment. Phosphorus (atomic number = 15) and sulfur (atomic number = 16) are also important as part of proteins, DNA, and RNA.

Another feature of the elemental composition of human life that you may have identified is that a relatively high number of the elements found within us consists of transition metals, although we contain these in very small amounts. As we mentioned earlier, the chemical versatility of transition metal ions, such as their tendency to readily form ions of differing charges, makes them useful in enzymes—biological catalysts of chemical reactions.

It is also notable that humans contain elements from every block of the periodic table except the *f*-block, which comprises the lanthanides and actinides. The elements in this block are composed of very large atoms. Many of the actinides are unstable and therefore radioactive (see Chapter 21).

HERE'S WHAT WE KNOW SO FAR

- The elements within a group have a common number of electrons in their highest energy level.
- The elements in the periodic table are arranged into three main sections: the metals, nonmetals, and metalloids (or semimetals).
- The structure of the periodic table includes "blocks" defined in terms of which type of orbital is being filled. This gives us the *s*-block, *p*-block, *d*-block, and *f*-block elements.
- The elements that predominate in living things are carbon, hydrogen, nitrogen, and oxygen, with small concentrations of phosphorus, sulfur, and transition metals.

7.4 The Concept of Periodicity

Why is the structure of the periodic table useful to know? Once we understand the trends as we traverse the groups and periods in the periodic table, we can make *reasonable predictions* about the chemical and physical behavior of any element in the periodic table. We will add to our understanding in subsequent chapters, even learning how to assess the likely *nuclear* behavior of an element. Once we are armed with this understanding, the periodic table serves as a most wonderful guide to the formation and interaction of substances.

The basis of the periodic table is **periodicity**, which will be revealed in many of the properties we explore in the remainder of this chapter. When chemists talk of "periodic properties" among the elements, they mean the way in which characteristic properties *recur in a periodic manner* as we move through the periodic table. For example, element number 3 in the table, lithium, is a very reactive metal that forms an alkali on reaction with water. Element 4 (beryllium), is less reactive, and as we move through elements 5 (boron), 6 (carbon), 7 (nitrogen), 8 (oxygen), 9 (fluorine), and 10 (neon), we find elements that become steadily less like lithium in chemical reactivity and physical properties. Then suddenly, with element 11 (sodium), we find another very reactive metal that forms an alkali on reaction with water. The similarities between the reactivities of lithium and sodium are so striking that it is clear that we are observing some significant *repeating* feature of reactivity as we move through the periodic table. If we then move on to elements 12 through 19, we find the same thing happening again. Elements 12 through 18 have properties less and less like those of sodium and lithium, and then suddenly, with element 19 (potassium), the property of being a very reactive metal that forms an alkali on reaction with water recurs. These chemical characteristics, which we call the characteristics of the alkali metals, recur in a periodic manner as we move through the periodic table.

That is the basic concept of chemical periodicity, and we could have chosen various other properties to make the same point. For example, the property of being a very unreactive gaseous element that exists as free individual atoms occurs in elements with atomic numbers 2, 10, 18, 36, 54, and 86. As with the alkali metals, we have a chemical property—the lack of reactivity, in this case— that recurs in a systematic, or periodic, manner as we move through the periodic table. The number of elements we have to pass by before a characteristic property recurs is not constant. It is 8, 8, 18, 18, 32 in the series above (see Exercise 7.4), but the crucial fact is that each characteristic property does recur *periodically* as we move through the periodic table.

Visualization: Periodic Table Trends

EXERCISE 7.4   Explaining the Periods Between Periodicities

The periodic property of being an "inert gas" recurs after we move on to 8, then 8, then 18, then 18, then 32 intervening elements in the periodic table. Can you suggest the underlying physical reason why the periodicity follows this particular pattern?

Solution

The pattern is a consequence of the way in which electrons fill up orbitals before a particular electron arrangement associated with a particular property recurs. The fundamental characteristic associated with being an inert gas is to have a stable octet of eight electrons in the atom's highest energy level, or two electrons in a completely full energy level in the case of helium. The Aufbau principle indicates that after helium, the stable octet will recur after the $2s$ and $2p$ orbitals have been filled, which means eight electrons must be added before we arrive at neon, then another eight before we arrive at argon. As we then move through Period 4, ten $3d$ electrons must be added before the $4p$ orbitals become filled, so this time we must move through 18

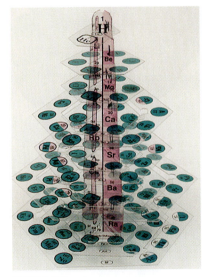

FIGURE 7.7

This three-dimensional periodic table, known as the ElemenTree, was developed by Canadian Fernando Dufour to emphasize the periodic relationships of the elements.

elements before the stable octet recurs. Similar reasoning applies to the gap between krypton and xenon. Then the gap jumps up to 32, between xenon and radon, as a consequence of the filling of f orbitals that must occur before the $6p$ orbitals of radon can be filled. This explanation is most clearly visualized by looking at the long version of the periodic table in Figure 7.5 or the pyramidal version in Figure 7.7.

PRACTICE 7.4

Judging on the basis of the periodicity of the periodic table, indicate what would be the hypothetical atomic number of a metal that would be more reactive than Fr. How many electrons would it take to fill a noble gas in the hypothetical Period 8 of the periodic table?

See Problems 29–34.

Video Lesson: Periods and Atomic Size

Application

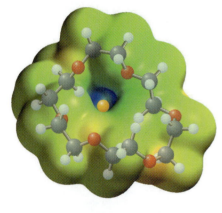

The chelating agent known as 18-crown-6 binds to sodium cations.

Atomic radii can be measured by measuring the distance between the nuclei of atoms in a metal. Measuring the distance between nuclei in a molecule gives the covalent radius in molecules.

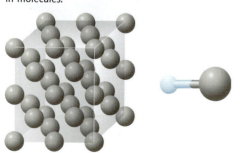

7.5 Atomic Size

Heavy metal poisoning is unfortunately a common occurrence, particularly among children living in homes painted with lead-based paints. In the hospital, once the poisoning is recognized, the patient is treated with a chelating agent. Chelating agents (see Chapter 18) specifically grab ions of a particular size and help to remove them from the body. The size of the ion is extremely important in how well it binds to the chelating agent. If it is too small or too large, it will not associate with the chelating agent and be excreted from the body in that manner.

Figure 7.8 reveals the key trends and periodicities found in the size of atoms. **What do you note about the size of the atoms in the periodic table?** Atomic size decreases as we move from left to right along any period, and it increases as we move down any group. Figure 7.8 uses atomic radii as a measure of atomic size. The **atomic radius** is defined as half the distance between the nuclei in a molecule consisting of identical atoms. It is also known as the **covalent radius** of an atom, because it indicates the size of the atom when that atom is involved in covalent bonding. Not all elements form such molecules. Some atomic radii values must be estimated via indirect methods, such as comparing the distances between atomic nuclei when the atoms are bound within chemical compounds instead of in a molecule made of identical atoms. Values for the radii of metal atoms can also be obtained by analyzing the distance between the nuclei of the atoms within the solid structure of the metal concerned. These values are called **metallic radii**.

The uncertainties involved in measuring atomic radii, and the various methods that can be used, mean that the values we obtain should be regarded as *approximate,* and you may find slightly different values quoted in different sources. The basic trends, however, are absolutely clear: *Atomic size decreases along periods and increases down groups.* How can we explain these data? As we move down the group, we encounter atoms with increasing numbers of occupied electron energy levels. Each new occupied energy level, corresponding to a higher principal quantum number, includes electron orbitals of greater average radius than those of the previous level, so the atoms grow larger as we move down a group.

It might seem more difficult to explain why we find atoms of decreasing size as we move along any period, because the atoms actually contain more matter as we travel from left to right, thanks to the steady addition of protons, electrons, and neutrons. However, the additional electrons are being added to energy levels already present in previous elements of the period, and they are held around nuclei whose positive charge is steadily increasing across the period. This increased positive charge draws the electrons of the occupied energy levels closer to the nucleus as we move along a period, so the atomic radius steadily decreases.

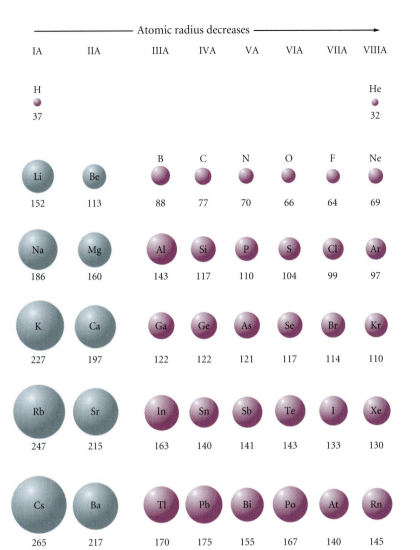

Atomic radius decreases →

Trend in atomic radius.

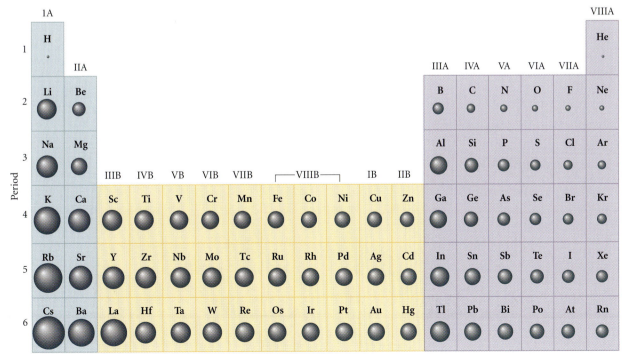

Video Lesson: Ionization Energy

7.6 Ionization Energies

Most of the elements in the periodic table are metals. When they react with other elements to form compounds, they generally do so by losing one or more electrons to form positive ions. An input of energy is required to remove one or more electrons from an atom, and the energy involved is known as **ionization energy**. The smaller the ionization energy, the more likely it is that ionization will occur. All elements, not just the metals, have ionization energy values. By examining the ionization energies of the elements, we can see *why* metals are particularly prone to form positive ions, and we can see how the differing reactivities of metals matches their differing ionization energies.

The **first ionization energy** (I_1) of an element is the energy required to remove the highest energy electron from an atom of the element in the gaseous ground state. The first ionization energy is typically quoted per mole of atoms being ionized and can be symbolized in general terms, using X to represent any element, as

$$X(g) \rightarrow X^+(g) + e^-$$

For example, the first ionization energies for sodium and chlorine are

$$Na(g) \rightarrow Na^+(g) + e^- \qquad I_1 = 495 \text{ kJ/mol}$$
$$Cl(g) \rightarrow Cl^+(g) + e^- \qquad I_1 = 1255 \text{ kJ/mol}$$

Why is the first ionization energy of sodium so much smaller than that of chlorine? Note the resulting ion's electron configuration. The sodium ion has a noble gas electron configuration. The chlorine ion does not. The first ionization energies of selected main-group elements are listed in Figure 7.9.

The **second ionization energy** (I_2), of an element is the energy required to remove one electron from each singly charged +1 ion of an element in the gaseous state:

$$X^+(g) \rightarrow X^{2+}(g) + e^-$$

Third, fourth, and all subsequent ionization energies can be similarly defined, as successive electrons are removed.

We can see the link between ionization energies and reactivities by examining some of the alkali metals. Lithium, sodium, and potassium are all reactive metals that form alkalis on reaction with water. These metals, when they react, form ions of Li^+, Na^+, and K^+, respectively. We can get an indication of how readily the formation of these ions occurs by considering these ionization energies:

$$Li(g) \rightarrow Li^+(g) + e^- \qquad I_1 = 520 \text{ kJ/mol}$$
$$Na(g) \rightarrow Na^+(g) + e^- \qquad I_1 = 495 \text{ kJ/mol}$$
$$K(g) \rightarrow K^+(g) + e^- \qquad I_1 = 419 \text{ kJ/mol}$$

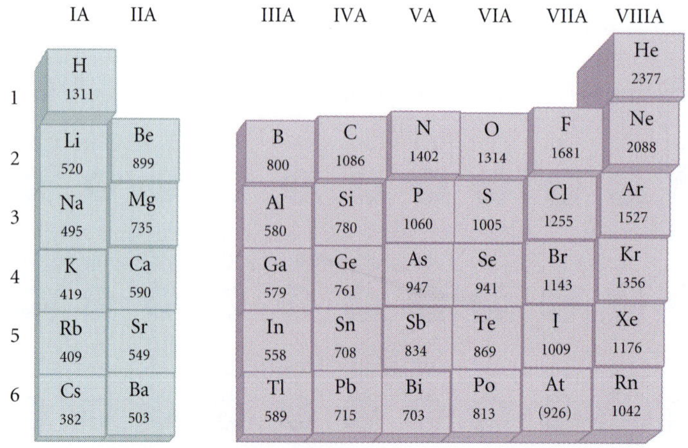

FIGURE 7.9

First ionization energies for selected elements.

	IA	IIA		IIIA	IVA	VA	VIA	VIIA	VIIIA
1	H 1311								He 2377
2	Li 520	Be 899		B 800	C 1086	N 1402	O 1314	F 1681	Ne 2088
3	Na 495	Mg 735		Al 580	Si 780	P 1060	S 1005	Cl 1255	Ar 1527
4	K 419	Ca 590		Ga 579	Ge 761	As 947	Se 941	Br 1143	Kr 1356
5	Rb 409	Sr 549		In 558	Sn 708	Sb 834	Te 869	I 1009	Xe 1176
6	Cs 382	Ba 503		Tl 589	Pb 715	Bi 703	Po 813	At (926)	Rn 1042

The metals lithium (left), sodium (center), and potassium (right).

EXERCISE 7.5   **Implications of Ionization Energy Trends**

The reactions of lithium, sodium, and potassium metals with water to form an alkaline solution are as follows:

$$2Li(s) + 2H_2O(l) \rightarrow 2Li^+(aq) + 2OH^-(aq) + H_2(g)$$

$$2Na(s) + 2H_2O(l) \rightarrow 2Na^+(aq) + 2OH^-(aq) + H_2(g)$$

$$2K(s) + 2H_2O(l) \rightarrow 2K^+(aq) + 2OH^-(aq) + H_2(g)$$

Given the values listed in the text for the first ionization energies of these metals, which metal will react most vigorously if added to water? If we were to put a chunk of cesium metal in water, how might it react?

First Thoughts

Ionization energy is a measure of the energy required to remove the highest energy electron from an atom. It is an endothermic process, so when comparing elements in an otherwise equivalent reaction, we find that the lower the ionization energy, the less energy will be required, and the more readily the reaction will occur.

Solution

The potassium metal has the lowest ionization energy of the three, so it will react most readily with water; sodium will react less readily and lithium the least impressively. Figure 7.10 shows the reactions of these metals with water. Note the flame that potassium makes upon reaction with water. What about cesium? Its ionization energy of 376 kJ/mol means that it will react even more vigorously than potassium. In fact, the reaction with water is explosive.

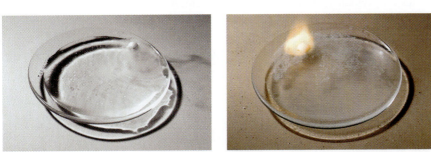

Sodium

Potassium

Calcium

Gold

FIGURE 7.10

The relative reactivities of metals can sometimes be quite clearly visualized.

Further Insights

Francium would be expected to react even more vigorously than cesium. However, francium is exceedingly rare and has no stable isotopes. One place that prepares samples for scientific study is the Nuclear Structure Laboratory at the State University of New York at Stony Brook. An isotope known as francium-210 (along with five neutrons) is prepared by slamming oxygen-18 nuclei atoms into gold-197 nuclei using a nuclear accelerator:

$$^{197}Au + {}^{18}O \rightarrow {}^{210}Fr + 5\ n$$

This, and other isotopes of francium, can be produced at the rate of about 1 million atoms per second. Keep in mind, though, that 1 million atoms are not very many compared to the 6.02×10^{23} atoms that make up a mole. The francium that is collected can be studied to determine its physical and chemical properties, but the nuclei decay so quickly that it is impractical to collect francium-210 for other uses.

PRACTICE 7.5

Which of the metals in Group IIA would you expect to react most vigorously with water? Explain your answer in terms of the ionization energies of the elements.

See Problems 45, 46, 73, and 75.

We have just identified one of the general trends linking ionization energy values with an element's position in the periodic table. *Ionization energies generally decrease moving down any group.* We can explain this trend by noting that as we move down a group, the valence electrons being removed are coming from energy levels of successively higher principal quantum numbers. Therefore, as we move down a group, the electrons being removed are screened from the nucleus by additional inner energy levels occupied with electrons.

The other main trend in ionization energies is that they generally increase from left to right along any period. We can rationalize this as due to the extra energy needed to release an electron from the pull of the increasing nuclear charge found as we move along a period.

These trends in ionization energy can be clearly seen in the laboratory if we choose metals with significantly different reactivities (see Figure 7.10). The trends can also be plotted graphically, as shown in Figure 7.11, in a way that makes the *periodicity* in this property apparent. As we move through Figure 7.11, we see the ionization energies rising and falling in a periodic manner.

FIGURE 7.11

The values of first ionization energies for the elements in the first six periods.

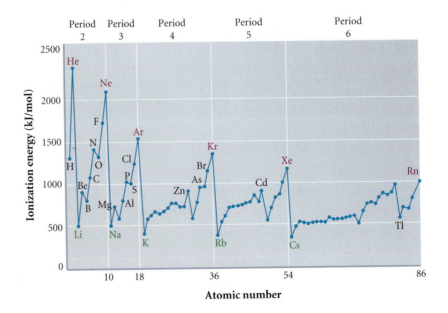

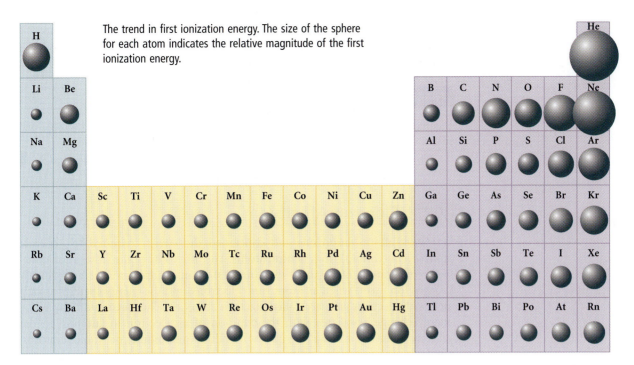

The trend in first ionization energy. The size of the sphere for each atom indicates the relative magnitude of the first ionization energy.

EXERCISE 7.6   Differences in Ionization Energies

Here are the first three ionization energies of sodium and magnesium, in kilojoules per mole:

	I_1	I_2	I_3
Sodium	495	4560	6920
Magnesium	735	1445	7730

There is a very large difference between the first and second ionization energies of sodium. With magnesium, however, the very large difference is seen between the second and third ionization energies. Why do the large differences in ionization energies occur where they do?

First Thoughts

What does the ionization energy tell us? It is the energy required to remove an electron from the atom. The key question, then, is "What is it about the electron configuration of each element that results in such a sudden increase in ionization energy?"

Solution

Sodium atoms have one outer-shell electron ($[Ne]3s^1$), whereas magnesium atoms have two ($[Ne]3s^2$). Once sodium has lost one electron, the next electron must come from a deeper electron shell, closer to the nucleus, so a great deal more energy is required to remove that second electron. In the case of magnesium, the transition to removing electrons from an inner shell does not occur until the two outer electrons have been removed, so the big jump in ionization energy is seen between the second and third ionization energies, rather than between the first and second.

Further Insights

Iron is an example of a transition element that has several possible oxidation states, including Fe^{2+} and Fe^{3+}. A most wonderful and trustworthy website to retrieve ionization energy data (and so much more!) is the WebElements site at http://www.webelements.com. Look up the ionization energy for the ions of iron at that site.

Where are the sudden increases in ionization energy, and how do they compare to those found in a nonmetal such as phosphorus? Can you explain the differences between these two elements?

PRACTICE 7.6

Predict, and draw a graph of, the ionization energy versus electron ionized for sulfur. Compare your graph to the measured values at the WebElements site.

See Problems 47, 48, 53, 54, 96, and 99.

Video Lesson: Electron Affinity

7.7 Electron Affinity

Our look at ionization energies concerned the energy required to generate *positive ions* (*cations*) from neutral atoms. The other side of the "ionization coin" is the generation of *negative ions* (*anions*) from neutral atoms. A negatively charged ion forms when an atom *gains* electrons, and the energy changes associated with electron gain are called electron affinities. The **electron affinity** is the energy change associated with the addition of an electron to an atom in the gaseous state, and as usual, we quote the values in kilojoules per mole of atoms. For example,

$$F(g) + e^- \rightarrow F^-(g) \qquad -328 \text{ kJ/mol}$$
$$Cl(g) + e^- \rightarrow Cl^-(g) \qquad -349 \text{ kJ/mol}$$
$$Br(g) + e^- \rightarrow Br^-(g) \qquad -325 \text{ kJ/mol}$$
$$I(g) + e^- \rightarrow I^-(g) \qquad -295 \text{ kJ/mol}$$
$$At(g) + e^- \rightarrow At^-(g) \qquad -270 \text{ kJ/mol}$$

The electron affinity values for Group VIIA reveal one of the general trends in electron affinity and also one of the complications. The values indicate to us that electron affinities generally become *less negative* (or more positive), corresponding to less energy being released, as we move down a group. **Does this trend make sense?** Because added electrons are farther from the nuclei of larger atoms, they interact less with the positive charge of the nucleus. The trend makes sense. There are many exceptions, however, such as more energy released when a chloride ion forms, in the above list, than upon formation of the fluoride anion. Why does fluorine not fit the trend? In searching for a possible explanation, we should note that fluorine is the smallest of this group of atoms, so the electrons in the outer *p* orbitals, to which the extra electron is being added, will be much closer together than in the other atoms of this group. The larger electron–electron repulsions between these closely spaced electrons, compared to those of other halogen atoms, will be associated with increased potential energy.

Figure 7.12 shows many more electron affinity values. It allows us to confirm the general trend just identified within groups, in addition to revealing exceptions. Figure 7.12 also confirms that elements that form stable negative ions have large negative electron affinity values, whereas those that form stable positive ions have much lower negative electron affinity values, or even positive values.

Note, as well, that the values in Figure 7.12 indicate that electron affinity values generally get more negative as we move from the left to the right along a period in the periodic table. **Does this trend make sense?** We know from our understanding of the trend associated with atomic radii that atoms get smaller as we move to the right along the period table. For the same reasons we discovered as we move down a group, the electron affinity generally tends to get more negative as we move to the right along a period.

1 IA	2 IIA		13 IIIA	14 IVA	15 VA	16 VIA	17 VIIA	18 VIIIA
Li −60	Be >0		B −27	C −154	N ~0	O −141	F −328	Ne >0
Na −53						S −200	Cl −349	
K −48						Se −195	Br −325	
Rb −47						Te −190	I −295	
Cs −46						Po −183	At −270	

FIGURE 7.12

Electron affinity values, in kilojoules per mole (kJ/mol), for selected elements. Electron affinity values generally decrease as we go up and to the right in the periodic table. More negative values indicate a greater affinity for electrons.

Many of the periodic properties identified in the periodic table are related to each other. Is there a relationship between electron affinity and the number of valence electrons?

First Thoughts

We should first construct a crude periodic table with arrows pointing to the most negative electron affinity. Superimposing the number of valence electrons on this drawing will reveal any relationship.

Solution

As the number of valence electrons gets larger, the atomic radius gets smaller across a period. As the radius gets smaller, the electron affinity becomes more negative. Therefore, as the number of valence electrons increases, the electron affinity becomes more negative. This trend, however, is valid only within a given period in the periodic table, because the electron affinity changes and the number of valence electrons remain constant as we move down a group.

Further Insights

Periodic trends are abundant in the periodic table. Size, electron affinity, reactivity, and ionization potential are all related to position on the periodic table. This table has trends in many other properties, such as the formation of acids or bases upon reaction of elemental oxides with water, the tendency to form chlorides, the tendency to behave like metals, and many more.

PRACTICE 7.7

Is there a relationship between the trends observed in ionization energy and electron affinity? Explain.

See Problems 57–62, 74, 76, 79, and 80.

7.8 Electronegativity

Video Lesson: An Introduction to Electronegativity

In our discussion about electron affinities, we are actually considering a rather contrived situation, because most elements rarely exist in the form of free gaseous atoms. A more meaningful characteristic of elements, one that is related to their interaction with additional electrons and therefore is more firmly rooted in real chemical behavior, is their electronegativity. The **electronegativity** of an atom is a measure of the ability of the atom in a molecule to attract shared electrons to itself. Atoms in molecules share electrons to create covalent bonds. However, if the atoms sharing electrons are different, they will contain nuclei with different charges and different electron arrangements. The result is that the electrons are not shared equally. Electronegativity values have been calculated for every atom in the periodic table. However, we generally do not quote electronegativity values for the noble gases (Group VIIIA) because, as a consequence of their exceptional stability, they do not readily form bonds.

Electronegativity values for the elements in the periodic table are one of the most powerful quantities chemists have that explain and predict the behavior of molecules and ions. Their importance cannot be overstated. In 1932, Linus Pauling defined the concept in terms of bond energies. The Pauling scale of electronegativities and the variations of it are the bases of much of our discussion about bonding in subsequent chapters. To illustrate the utility of electronegativity values, let's consider two essential vitamins that we must regularly consume: vitamin A and vitamin C. Deficiencies in these vitamins are responsible for serious maladies.

Application

Vitamin A and vitamin C differ in the colors displayed on their electrostatic potential maps. The areas of blue and red in vitamin C's map indicate differences in the electronegativity of the atoms in this molecule. These differences result in the water solubility of vitamin C.

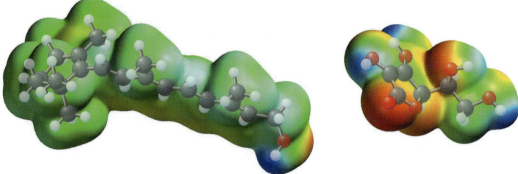

However, our bodies require different amounts of each vitamin as part of our daily diet. Why? Vitamin A is a fat-soluble compound that our bodies retain. Vitamin C, on the other hand, is a water-soluble compound that is readily excreted from our bodies. This means that we must consume vitamin C in larger quantities than vitamin A and on a more regular basis. As we will discuss in the next chapter, electronegativities are the basis for understanding the solubility of these vitamins.

Pauling derived the electronegativity values by comparing the energies required to break various bonds, as we will discuss more fully in Chapter 8. The values shown in Figure 7.13 reveal two general trends. The first of these trends is that *electronegativities generally increase as we move along periods* all the way to Group VIIA. The second trend indicates that *electronegativities generally decrease as we move down groups*. This means that the lowest electronegativity values are at the bottom of Group IA and that the element fluorine has the highest. **Do these trends make sense?** The increase in electronegativity along periods occurs because the atoms are progressively smaller. In addition, atoms on the right-hand

FIGURE 7.13

Electronegativity values for selected elements. Electronegativity increases as we go up and to the right in the periodic table.

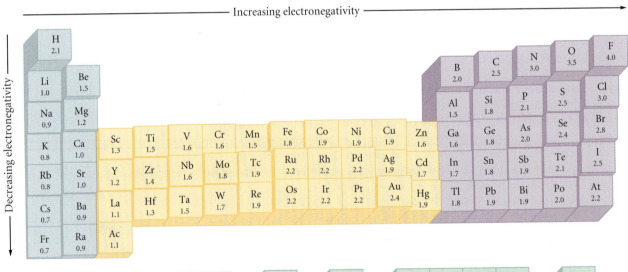

side of each period possess an increased nuclear charge. Both of these factors cause electrons to be attracted to the nucleus more strongly.

The decrease in electronegativity that we note as we move down groups makes sense because the atoms are growing larger. With an extra intervening shell of electrons at each step, the nuclear charge is being screened from the electrons shared in a covalent bond. The effect of the increased size and electron screening outweighs the increase in nuclear charge as we move down a group.

The trends in electronegativity, electron affinity, and ionization energy are not without exceptions, so they are general trends rather than absolute rules. They do, however, reveal one very clear message about which atoms are most likely to attract electrons and become negative ions, and which are most likely to lose electrons and become positive ions. *Atoms on the extreme left of the periodic table, especially the lower extreme left, will most readily form positive ions. Atoms on the extreme right of the periodic table, especially the upper extreme right (always remembering that Group VIIIA is excluded), will most readily form negative ions.*

EXERCISE 7.8   **Trends in Electronegativity**

Without examining the actual Pauling electronegativity values, predict which of these elements in each pair is more electronegative.

   Li or Be   N or P   S or I

Solution

Be is closer in the periodic table to fluorine. It is more electronegative.

N is only two elements away from fluorine, whereas P is three elements away. Nitrogen is more electronegative.

S is two elements away from fluorine. It is more electronegative.

PRACTICE 7.8

Which of the following in each pair is more electronegative?

   P or Na   Ne or Cl   N or C

See Problems 68–70.

7.9   Reactivity

The concept of **reactivity** is a descriptive but very useful notion in chemistry. When we describe an element as "highly reactive," we mean that it readily participates in chemical reactions to form compounds. When we describe an element as "unreactive," we mean it does not readily participate in chemical reactions to form compounds. As we have already discovered, most highly reactive elements are, in their natural state, combined with other elements in the form of compounds. You will never find a hunk of soft, shiny sodium in nature. Rather, you will find it in ionic form in seawater or in salts on land. Unreactive elements *can* be found in their pure form. This has great bearing on the uses we make of elements. The metals sodium and potassium, for example, would be of little use for making automobiles or jewelry, because they react so vigorously and readily with water. Gold, on the other hand, is an ideal metal for jewelry. Because it is so unreactive, it can be found in its free form within the earth. The scarcity of gold makes it unsuitable for the construction of automobiles, even though such automobiles might last for a very long time.

Gold in its natural state.

We compromise when making automobiles, and build their metal components mostly out of steel, which is largely composed of the element iron, as we noted in the NanoWorld/MacroWorld feature in Section 7.2. Iron is not totally unreactive; it corrodes into rust (iron oxide) when combined with the oxygen found in air or dissolved in water. Iron lasts long enough, however, to form the basic materials of automobiles that will last quite a few years.

 Application

Can the periodic table tell us which elements are the most reactive? The elements that react most readily generally do so by either losing or gaining electrons. The most reactive elements will be those with the greatest tendency to form positive ions (those with the lowest ionization energies), or with the greatest tendency to form negative ions (those with the largest negative electron affinities or the largest electronegativities). These highly reactive elements are found at the extreme left of the periodic table (Groups IA and IIA) and on the far right in Groups VIA and VIIA. The least reactive elements of all are found in Group VIIIA, the noble gases; and those with intermediate and low reactivities are found between the extremes just mentioned, in the central portion of the periodic table. In Group IB, we find gold and silver, unreactive elements we use in jewelry, and also copper, which is sufficiently unreactive to be used in copper electric wires and the pipes of plumbing systems.

Our discussion now enables us to find more general degrees of logic in the periodic table. We now have enough understanding to give at least rudimentary answers to the questions we raised in the chapter opening about why oxygen reacts with so many elements and why gold has a very low reactivity. These questions can be answered by considering the location of these elements in the periodic

Periodic table with the most reactive elements highlighted in green.

Period	1 IA	2 IIA	3 IIIB	4 IVB	5 VB	6 VIB	7 VIIB	8	9 VIIIB	10	11 IB	12 IIB	13 IIIA	14 IVA	15 VA	16 VIA	17 VIIA	18 VIIIA
1	1 H																	2 He
2	3 Li	4 Be											5 B	6 C	7 N	8 O	9 F	10 Ne
3	11 Na	12 Mg											13 Al	14 Si	15 P	16 S	17 Cl	18 Ar
4	19 K	20 Ca	21 Sc	22 Ti	23 V	24 Cr	25 Mn	26 Fe	27 Co	28 Ni	29 Cu	30 Zn	31 Ga	32 Ge	33 As	34 Se	35 Br	36 Kr
5	37 Rb	38 Sr	39 Y	40 Zr	41 Nb	42 Mo	43 Tc	44 Ru	45 Rh	46 Pd	47 Ag	48 Cd	49 In	50 Sn	51 Sb	52 Te	53 I	54 Xe
6	55 Cs	56 Ba	57 La*	72 Hf	73 Ta	74 W	75 Re	76 Os	77 Ir	78 Pt	79 Au	80 Hg	81 Tl	82 Pb	83 Bi	84 Po	85 At	86 Rn
7	87 Fr	88 Ra	89 Ac†	104 Rf	105 Db	106 Sg	107 Bh	108 Hs	109 Mt	110 Uun	111 Uuu	112 Uub						

*Lanthanides	58 Ce	59 Pr	60 Nd	61 Pm	62 Sm	63 Eu	64 Gd	65 Tb	66 Dy	67 Ho	68 Er	69 Tm	70 Yb	71 Lu
†Actinides	90 Th	91 Pa	92 U	93 Np	94 Pu	95 Am	96 Cm	97 Bk	98 Cf	99 Es	100 Fm	101 Md	102 No	103 Lr

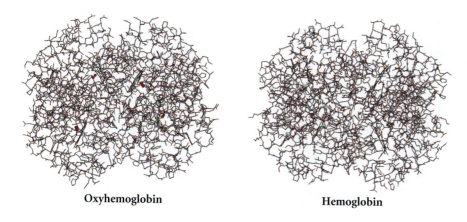

Oxyhemoglobin   **Hemoglobin**

The four oxygen molecules in oxyhemo-globin can be seen near the center of the model. Hydrogen atoms have been omitted for clarity.

table and all that this means. Another question posed in our introduction was what makes iron so well-suited to the task of carrying oxygen around the body, bound to the iron-containing protein we call hemoglobin. The iron in hemo-globin is in the form of Fe^{2+} ions, which bind *reversibly* (that is, they can continue to bind and then release, depending on conditions) to oxygen molecules (O_2). The binding has to be readily reversible, because hemoglobin must combine with oxygen where oxygen is abundant, in the blood vessels of the lungs; but must re-lease the oxygen to the tissues of the body in which oxygen levels are much lower. A positive ion can perform the binding task by interacting with the electrons of the oxygen molecule. In order to do that in a *readily reversible* way, however, the ion and the atom from which it is derived must have neither too strong nor too weak an attraction for electrons. A suitable ion will be from an element of inter-mediate electronegativity, likely to be found around the middle of the periodic table. Iron, in the middle of the periodic table, can do the job very nicely. This is only one of several reasons why iron ions serve within hemoglobin molecules as the oxygen carriers of life, but it is surely a significant one. Other reasons include the availability of iron in the environment in which life originated and the suit-ability of iron to combining with the other components of hemoglobin.

One idea that we did not discuss here but will consider in the next two chap-ters is that judgments about the reactivity of an element cannot be made in isola-tion. That is, atoms react with other types of atoms, and *the extent to which a reaction occurs depends on the properties of both, or even many, types of atoms.* More broadly speaking, the reaction environment must be considered in decid-ing the nature of chemical behavior.

Application

Video Lesson: The Activity Series of the Elements

The Reactivity Series of Metals

The relative reactivities of some of the most abundant and most useful elements in the environment are very significant to us. For example, in deciding which metal is suitable for a particular industrial use, we find that the ease with which it corrodes (oxidizes in the presence of substances in the environment, often lead-ing to deterioration in the properties of the metal) will be very significant. The most common form of corrosion is reaction with oxygen, and all forms of corro-sion are chemical reactions of one kind or another. The most reactive metals will generally corrode most quickly. Thinking about such issues led to the idea of list-ing metals in a **reactivity series** (or **activity series**), which ranks selected metals in order of reactivity (see Table 7.14).

Such a list can be made by exposing metals to a range of substances, such as hydrochloric acid, water, and air, and observing how readily they react. There are some problems with this approach. For example, aluminum reacts more quickly with oxygen than does iron. The aluminum soon becomes coated, however, with a cohesive layer of aluminum oxide that protects the aluminum beneath from further corrosion. That is why aluminum cookware is quite long-lived. Iron

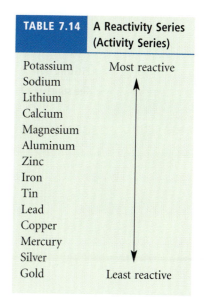

TABLE 7.14	A Reactivity Series (Activity Series)
Potassium	Most reactive
Sodium	
Lithium	
Calcium	
Magnesium	
Aluminum	
Zinc	
Iron	
Tin	
Lead	
Copper	
Mercury	
Silver	
Gold	Least reactive

Rusty aluminum (left) and rusty iron (right).

oxide, on the other hand, does not form any protective layer but instead produces the crumbling, weak structure we know as "rust," which breaks away to reveal fresh corrosion-prone metal underneath. Nevertheless, the idea of ranking metals by reactivity has proved very useful.

An alternative, and in some ways more satisfactory, way of ranking metals by reactivity is found in the "electrochemical series" that we will discuss in Chapter 19. This is related to the tendency of each metal to give up electrons. Remember that when metals react, they generally do so by losing outer electrons to form ions. The most reactive metals are the ones that lose electrons and form positive ions most readily.

7.10   The Elements and the Environment

Application

CHEMICAL
ENCOUNTERS:
The Elements and
the Environment

The elements of planet Earth are distributed throughout the environment in places and forms that are a consequence of their physical and chemical properties, and these physical and chemical properties can be related to the elements' positions in the periodic table. The **environment** in this sense comprises the *entire* world around us, including the Earth, its atmosphere, and ourselves.

Geologists describe the structure of the Earth in terms of five distinct regions: the core, mantle, crust, hydrosphere, and atmosphere. These are shown diagrammatically in Figure 7.14. Although nobody has traveled to the Earth's core or has even been able to obtain samples, we know its structure from the accumulation of

FIGURE 7.14

The structure of planet Earth.

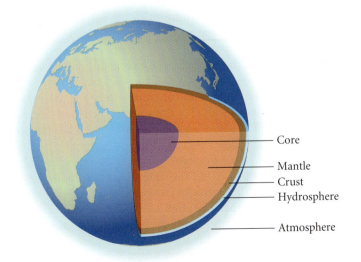

Core
Mantle
Crust
Hydrosphere

Atmosphere

years of indirect evidence. There is no doubt that the core is composed largely of iron, mixed with smaller amounts of the metals nickel and cobalt and lighter elements such as carbon and sulfur. The core is largely molten, so these elements form a molten alloy (a homogeneous mixture of metals) with some nonmetals mixed within the alloy. The predominance of iron at the center of the Earth is a consequence of its high density in relation to most of the other elements that make up our planet. When the Earth formed, it went through a molten phase, during which the densest materials were drawn toward the center. The less dense materials rose outward, in just the same way as a mixture of cooking oil and water will settle into layers, the dense water beneath the less dense oil.

The mantle is another region of the Earth that has never been directly sampled, although a few different types of volcanic rocks are believed to have been derived from this region. The evidence of these rocks, combined with other indirect evidence, suggests that the mantle is composed largely of oxides of the elements iron, magnesium, silicon, calcium, and aluminum. The low-density element oxygen, which is a gas at typical atmospheric temperatures and pressures, has been retained within the Earth's mantle because of its chemical reactivity. This has caused it to react with various metals to form the much more dense compounds found in the mantle.

The Earth's crust makes up less than 1 percent of the mass of the planet and, as shown in Figure 7.15, it has a ratio of elements that differs greatly from that of the Earth as a whole. However, it is the only solid part of the Earth that we can analyze directly. The crust is also the part of the environment from which we draw nearly all of our raw materials for use in industrial fabrication and as sources of energy. The most abundant elements of the crust are listed in Table 7.15. These elements are largely found bound within compounds, oxides being the most predominant. Oxides of silicon are among the most common components of the crust. For example, silicon dioxide (SiO_2) is the principal component of sand and of the many types of rock from which sand is derived.

The hydrosphere is the name given to the waters of the Earth, comprising oceans, seas, lakes, rivers, and underground aquifers. In addition to the hydrogen and oxygen that make up water, the hydrosphere contains a great variety of dissolved substances that influence its properties and suitability for various uses. Seawater is unsafe for humans to drink because of its high concentration of salts, such as sodium chloride and magnesium chloride. Earth's atmosphere contains the low-density, gaseous materials that nevertheless are sufficiently dense to be retained by the gravitational pull of the planet. As shown in Table 7.16, the atmosphere is largely composed of nitrogen and oxygen, with much smaller amounts of argon, carbon dioxide, and other rare gases. Some of the most significant components of the environment, however, are too rare even to show up in most tables of the atmosphere's composition. Gases such as sulfur dioxide and

— Oil

— Vinegar

Oil and vinegar, which is mostly water, are immiscible; they do not mix. The denser vinegar sinks to the bottom of this bottle.

FIGURE 7.15

Comparison of the elements in the Earth's crust with those in the entire Earth.

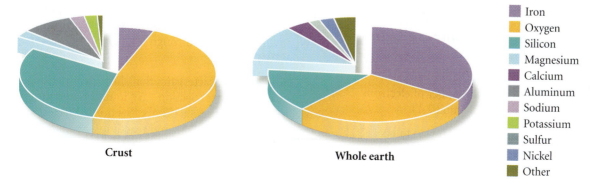

Crust

Whole earth

- Iron
- Oxygen
- Silicon
- Magnesium
- Calcium
- Aluminum
- Sodium
- Potassium
- Sulfur
- Nickel
- Other

TABLE 7.15	The Abundance of Elements in the Earth's Crust
Element	**Abundance (percent, by mass)**
Oxygen	46.60
Silicon	27.72
Aluminum	8.13
Iron	5.00
Calcium	3.63
Sodium	2.83
Potassium	2.59
Magnesium	2.09
Titanium	0.44
Hydrogen	0.14
Phosphorus	0.12
Manganese	0.10
Copper	0.0070
Gold	0.00000005

TABLE 7.16	Composition of Dry Air
Substance	**Volume Fraction**
N_2	0.7808
O_2	0.2095
Ar	0.00934
CO_2	0.00034
Ne	1.82×10^{-5}
He	5.82×10^{-6}
CH_4	2×10^{-6}
Kr	1.1×10^{-6}
H_2	5×10^{-7}
N_2O	5×10^{-7}
Xe	8.7×10^{-8}
SO_2	$< 1 \times 10^{-6}$
O_3	$< 1 \times 10^{-7}$
NH_3, CO, NO, I_2	$< 1 \times 10^{-8}$

nitrogen dioxide are extremely rare overall, but they cause problems over areas where they are present as environmental pollutants.

Finally, it is worth pointing out that the element hydrogen (H_2), the most abundant element in the universe, is of such low density that any hydrogen released into the atmosphere will eventually escape into space, unless it reacts with some other chemicals on the way. The only way for hydrogen to be retained on our planet is within chemical compounds, such as water (H_2O), methane (CH_4), or any one of millions of other compounds, including most of those found within living things. If chemical changes had not locked up hydrogen within such chemical compounds, we would never have evolved to ponder and study these chemical changes.

The Bottom Line

- The structure of the periodic table was initially developed by scientists trying to make sense of the differing reactivities of all the elements found in the natural environment. (Section 7.1)

- The structure of the periodic table includes "blocks" defined in terms of which type of orbital is being filled as we imagine filling up the available orbitals using the Aufbau principle. This gives us the s-block, p-block, d-block, and f-block. (Section 7.1)

- The horizontal "period" of the periodic table to which an element belongs indicates how many energy levels are either fully or partially occupied by electrons in atoms of that element. (Section 7.1)

- Elements in any one main group (Groups IA through VIIIA) have the same number of electrons in their highest energy level. (Section 7.1)

- The modern form of the table first began to take shape through the work of the German Julius Lothar Meyer and the Russian Dmitri Mendeleev. (Section 7.1)

- The elements in the periodic table are arranged into three main sections: the metals, nonmetals, and metalloids (or semimetals). (Section 7.2)

- Metals and nonmetals exhibit specific types of physical properties. (Section 7.2)

- The periodic table is divided into groups, and elements in each group are similar in electronic structure, physical properties, and chemical reactivity. (Section 7.3)

- Periodicity is apparent in the way important physical and chemical characteristics recur in a periodic manner as we move through the periodic table. (Section 7.4)

- Examples of characteristics that show clear periodicity are atomic size (Section 7.5), ionization energy (Section 7.6), electron affinity (Section 7.7), electronegativity (Section 7.8), and reactivity (Section 7.9).

- The distribution of the elements on planet Earth is a result of their chemical reactivities. (Section 7.10)

Key Words

actinides The second set of inner transition elements. The actinides include elements 90–103. (*p. 275*)

activity series A series of elements ranked in order of reactivity. Also known as a reactivity series. (*p. 291*)

alkali metals The Group IA elements, excluding hydrogen. (*p. 270*)

alkaline earth metals The Group IIA elements. (*p. 270*)

alloy A homogeneous mixture of metals. (*p. 266*)

atomic radius Half the distance between the nuclei in a molecule consisting of identical atoms. Also known as the covalent radius. (*p. 280*)

catalyst A substance that greatly speeds up the rate of a reaction without being consumed by that reaction. (*p. 272*)

chalcogens The Group VIA elements. (*p. 273*)

covalent radius Half the distance between the nuclei in a molecule consisting of identical atoms. Also known as the atomic radius. (*p. 280*)

electron affinity The energy change associated with the addition of an electron to an atom in the gaseous state. (*p. 286*)

electronegativity The relative ability of an atom participating in a chemical bond to attract electrons to itself. (*p. 287*)

environment The world around us, comprising the Earth and its atmosphere. We ourselves are part of the environment. (*p. 292*)

first ionization energy (I_1) The energy required to remove one electron from each atom of an element in the gaseous state. (*p. 282*)

halogens The Group VIIA elements. (*p. 274*)

inert gases The Group VIIIA elements. Also known as noble gases. (*p. 274*)

inner transition elements The elements in the *f*-block of the periodic table, consisting of the lanthanides and actinides. (*p. 275*)

ionization energy The energy needed to ionize an atom (or molecule or ion) by removing an electron from it. (*p. 282*)

lanthanides The first set of inner transition elements. The lanthanides include elements 58–71. (*p. 275*)

main groups Groups IA through VIIIA of the periodic table. Groups IB through VIIIB are not main groups. (*p. 269*)

metallic radius Half the distance between the nuclei of the atoms within the solid structure of a metal. (*p. 280*)

metalloids A small number of elements that have characteristics midway between those of the metals and those of the nonmetals. Also known as semimetals. (*p. 268*)

metals A large number of elements that share certain typical characteristics, including shiny appearance, malleability, and ability to conduct electricity. (*p. 265*)

noble gases The Group VIIIA elements. Also known as inert gases. (*p. 274*)

nonmetals A collection of elements that shares certain characteristics that are in contrast to those of the metals, including being gases or dull, brittle solids at room temperature. (*p. 268*)

octet Eight electrons in the valence shell of an atom. (*p. 275*)

periodicity The recurrence of characteristic properties as we move through a series, such as the elements of the periodic table. (*p. 279*)

reactivity The ability of an element or compound to participate in chemical reactions. *Reactivity* is an imprecise but useful term, particularly when we are discussing relative reactivities under defined conditions. (*p. 289*)

reactivity series A series of elements ranked in order of reactivity. Also known as an activity series. (*p. 291*)

second ionization energy (I_2) The energy required to remove one electron from each singly charged +1 ion of an element in the gaseous state. (*p. 282*)

semimetals A small number of elements that have characteristics midway between those of the metals and those of the nonmetals. Also known as metalloids. (*p. 268*)

transition elements Metal elements from Groups IB through VIIIB found in the middle of the periodic table. (*p. 275*)

valence electrons The electrons that occupy the outermost energy level of an atom, which interact with those of other atoms. (*p. 269*)

The answers to the odd-numbered problems and some selected problems appear at the back of the book, as represented by the blue numbering.

Section 7.1 The Big Picture—Building the Periodic Table

Skill Review

1. Cite two examples wherein two elements' locations on the current periodic table would be reversed if they were placed in accordance with Mendeleev's atomic mass system of organization.

2. Some may argue that providing a list of elements in alphabetical order along with their properties would be the most useful arrangement of the known elements. Explain what advantage there is to arranging them in the periodic manner displayed in our traditional periodic table format.

3. Identify the "block" location in the periodic table for each of these elements.
   a. Sr   c. S   e. Se   g. Sb
   b. Sc   d. Sn   f. Sm

4. Identify the "block" location in the periodic table for each of these elements.
   a. Rb   c. Ru   e. Re   g. Ra
   b. Rn   d. Rh   f. Rf

5. Identify the "block" location in the periodic table for each of these elements.
   a. C   c. Cu   e. Cs   g. Ce
   b. Ca   d. Cr   f. Cl

6. Identify the "block" location in the periodic table for each of these elements.
   a. He   c. Fe   e. Xe   g. Te
   b. Be   d. Ge   f. Ne

Chemical Applications and Practices

7. The following portion of a hypothetical periodic table shows the approximate melting points of elements that are neighbors to a missing element. On the basis of the data provided, predict the melting point of the missing element.

	2610°C	
3000°C	?	3180°C

8. The following portion of a hypothetical periodic table shows the densities of elements that are neighbors to a missing element. On the basis of the data provided, predict the density of the missing element.

	10.2 g/cm^3	
16.7 g/cm^3	?	21.0 g/cm^3

9. In each of these pairs, select the larger of the two elements.
   a. Mg or Ca   c. Cl or Br
   b. P or S   d. Cs or Ba

10. In each of these pairs, select the larger of the two elements.
    a. K or Ca   c. Se or Br
    b. O or S   d. Ra or Ba

Section 7.2 The First Level of Structure—Metals, Nonmetals, and Metalloids

Skill Review

11. Classify each of these as a metal, nonmetal, or metalloid.
    a. Zn   c. Ge   e. Si   g. Se
    b. Ga   d. Sn   f. P   h. As

12. Classify each of these as a metal, nonmetal, or metalloid.
    a. Li   c. Co   e. C   g. I
    b. Ne   d. Te   f. Bi   h. Sb

13. Most elements, at room temperature, are either solids or gases. However, two elements, at room temperature, are classified as liquids. What are these two elements?

14. Most of the key ingredients to alloy with iron to make various types of steel are metals. However, this is not always the case. What are two nonmetals that can be alloyed with steel?

Chemical Applications and Practices

15. a. In the production of stainless steel, small amounts of several elements are alloyed with iron. Two of the chief ingredients are nickel and chromium. Use the chart given in Table 7.3 to report the number of grams of each metal found in 100.0 g of stainless steel.
    b. The atoms of the added metals can become part of the iron structure, which changes the properties of iron. How many atoms of nickel and chromium, respectively, are present in the 100.0 g of stainless steel?

16. Assume that the other elements (besides chromium, nickel, and iron) found in stainless steel are insignificant. Based on the calculations in Problem 15, how many iron atoms are present in the sample for every atom of chromium?

17. Locate the element antimony in the periodic table. Would this element be classified as a metal, nonmetal, or metalloid? Antimony is used in alloys with lead or tin, is combined with other substances to make fireproof fabrics, is mixed with some types of glass and ceramics, and serves as the active component of some medicines. Antimony, however, does have some toxic properties. One way to obtain it is as follows:

$$Sb_2S_3 + Fe \rightarrow FeS + Sb$$

Balance this equation. The annual worldwide production of antimony is approximately 6.00×10^4 tons. How many moles of antimony is this?

18. Locate the element gold in the periodic table. Would this element be classified as a metal, nonmetal, or metalloid? Gold is used not only in jewelry but also as a wire in some computer parts and in other situations where a noncorrosive metal is needed. Gold can be solubilized in water by using a mixture of very strong acids, as follows:

$$Au(s) + HCl(aq) + NO_3^-(aq) \rightarrow$$
$$[AuCl_4]^-(aq) + H_2O(l) + NO(g)$$

Balance this equation. How many grams of gold must be solubilized to produce 1 mol of NO(g)?

Section 7.3 The Next Level of Structure—Groups in the Periodic Table

Skill Review

19. Using the most common oxidation number for each, write out the formulas of the compounds that would be most likely to form when Li, Be, B, C, N, O, and F, respectively, combine with oxygen.

20. Write out the chemical formulas for Li, Na, K, Rb, and Cs, respectively, combining with oxygen. What accounts for the similarities in the formulas of these compounds?

21. If an element from Group VIA were going to form an ionic compound, which type of element (metal or nonmetal) would be most likely to form the other part of the compound? Explain your answer.

22. If an element from Group IIA were going to form an ionic compound, which type of element (metal or nonmetal) would be most likely to form the other part of the compound? Explain your answer.

23. How many unpaired electrons are found in the elements that make up Group VIIA? How many unpaired electrons are found in the most common ions of those elements?

24. How many unpaired electrons are found in the elements that make up Group IIIA? How many unpaired electrons are found in the most common ions of those elements?

Chemical Applications and Practices

25. The listing in Table 7.13 of elements in the human body is based on mass. Considering only the first three (oxygen, carbon, and hydrogen), indicate whether the ranking would change if the table were based on number of atoms instead of mass. Explain the basis of your answer.

26. The listing in Table 7.13 of elements in the human body includes calcium. What use does our body make of this element? In addition, calculate the quantity of calcium in a typical human in units of milligrams per kilogram of body weight.

27. Many dietary supplements contain a list of metals present. This list may include zinc, iron, cobalt, and others. Name one important role that transition metals play in human biochemistry.

28. Iron plays several roles in humans. One of its important roles involves the electron transport chain. Iron can easily change oxidation states between the +2 and +3 charges. What characteristic feature of iron's electron configuration makes this possible?

Section 7.4 The Concept of Periodicity

Skill Review

29. The noble gases, except for helium, have a stable, filled octet. Explain why helium is still considered a noble gas even though it does not have a stable, filled octet.

30. The alkali metals have a complete shell of a noble gas plus an extra electron. Explain why hydrogen is often associated with this group of the periodic table.

Chemical Applications and Practices

31. Examine the full electron configuration of calcium and potassium. Explain why calcium is not as reactive as potassium.

32. The same reasoning used to answer Problem 31 applies to the relative reactivity of magnesium compared to sodium, but why is sodium not as reactive as potassium?

33. In some of the original periodic table research, tellurium was placed after iodine due to its slightly greater relative mass. However, this placed tellurium in Group VIIA directly below bromine, chlorine, and fluorine. After examining the electron configuration of tellurium and iodine explain why the present classification is logical.

34. Nickel is placed on the periodic table immediately after cobalt. However, its average mass is less than that of cobalt. Explain why the placement is logical.

Section 7.5 Atomic Size

Skill Review

35. These three spheres can be used to compare the relative sizes of sodium, magnesium, and sulfur. Which sphere would best represent each atom?

36. These three spheres can be used to compare the relative sizes of potassium, rubidium, and cesium. Which sphere would best represent each atom?

37. If the bond length for a C—C bond were 154 pm, what would you determine as the radius of a carbon atom?

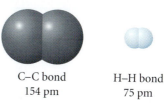

C–C bond
154 pm

H–H bond
75 pm

38. If the bond length for a H—H bond were 75 pm, what would you determine as the approximate radius for an atom of hydrogen?

39. If the bond length for a C—Cl bond were 171 pm, what would you determine as the approximate radius for an atom of chlorine? (Use the answer from Problem 37 to assist you in determining the radius of a chlorine atom.)

40. If the bond length for a H—F bond were 92 pm, what would you determine as the approximate radius for an atom of fluorine? (Use the answer from Problem 38 to assist you in determining the radius of a fluorine atom.)

Chemical Applications and Practices

41. Sodium and potassium are often found as ions in sports drinks. Which is the larger element? Explain.

42. Helium and hydrogen have both been used in dirigibles (also known as blimps or airships). Which of these two element's atoms is larger? Explain.

Section 7.6 Ionization Energies

Skill Review

43. For this calculation, use the first ionization values of sodium and potassium found in the chapter. How many grams of potassium could be ionized using the same energy that is required to ionize 1.00 g of sodium?

44. For this calculation, use the first ionization values of calcium and potassium found in the chapter. How many grams of calcium could be ionized using the same energy that is required to ionize 5.00 g of potassium?

45. Within each of these two lists, rank the elements listed from lowest to highest first ionization energy:
    a. Al, Si, P   b. Ne, Ar, Kr

    Review your rankings. Would you make any changes if asked to do the rankings by second ionization energy? If, so describe the reasons for any changes.

46. Within each of these two lists, rank the elements listed from lowest to highest first ionization energy:
    a. Li, Be, B   b. K, Rb, Cs

    Review your rankings. Would you make any changes if asked to do the rankings by second ionization energy? If, so describe the reasons for any changes.

47. Based on the following ionization energy (I) information, predict which element(s) are most likely metals and which are nonmetals. Also, suggest the most likely periodic table group for each of these "unknown" elements. (Note that these values, all of which are in units of kilojoules per mole, may not reflect actual ionization energies of known elements.)

	I_1 (kJ/mol)	I_2 (kJ/mol)	I_3 (kJ/mol)	I_4 (kJ/mol)
Element A	740	1440	7730	8670
Element B	2100	3230	4400	5500

48. On the basis of the following ionization energy (I) information, predict which element(s) are most likely to be metals and which to be nonmetals. Also, suggest the most likely periodic table group for each of these "unknown" elements. (Note that these values, all of which are in units of kilojoules per mole, may not reflect actual ionization energies of known elements.)

	I_1 (kJ/mol)	I_2 (kJ/mol)	I_3 (kJ/mol)	I_4 (kJ/mol)
Element A	500	4600	5800	7460
Element B	1060	1900	2900	5000

Chemical Applications and Practices

49. Some claims have been made regarding the production (discovery) of element 118. However, those claims have been disputed. If element 118 does exist, to which group in the periodic table does it belong? Would its ionization energy be expected to be higher or lower than that of its next neighbor above it in the periodic table? Explain your answer.

50. Element 119 could be discovered in the near future. To which group in the periodic table would it belong? Would its ionization energy be expected to be higher or lower than that of its neighbor directly above it in the periodic table? Explain your answer.

51. The flow of potassium and sodium ions into and out of nerve cells makes it possible for signals to be sent throughout our bodies. The fact that you are reading right now is dependent on this flow of ions.
    a. What are the charges on the potassium and sodium ions, respectively?
    b. Which is the larger of the two ions, sodium or potassium?
    c. Which of the two ions, sodium or potassium, is easier to produce from their neutral atoms?

52. Calcium ions also have great importance in cells and move into and out of other channels in cell membranes. What would be the charge on a calcium ion? Would this ion be larger or smaller than a potassium ion? Explain.

53. Explain why the first ionization energy of sodium is lower than magnesium's first ionization energy, but the second ionization energy of sodium is higher than that of magnesium.

54. Examining the first ionization energies of the elements in Period 4 of the periodic table reveals the general tendency for an increase in ionization energy from left to right. However, there is a slight drop in ionization energy from calcium to gallium. Study the electron configuration of both of these elements and offer an explanation for this apparent deviation from the trend.

55. Many elements may be found in nature in a variety of oxidation states. However, these statements express some limitations. Explain the basis for each of these limitations.
    a. Magnesium is never found naturally as the +3 ion.
    b. Fluorine is never found naturally as the +1 ion.
    c. Hydrogen is never found naturally in the +2 oxidation state.
    d. Aluminum tends to "prefer" the +3 ion naturally.

56. Explain the basis for each of these limitations on the oxidation states in which certain elements appear.
    a. Arsenic can often be found with a charge of +5 but never with a charge of +6.
    b. Titanium could be found with either a +2 or a +4 charge but not with a +5 charge.
    c. Potassium is never found naturally as a neutral element.
    d. Tin can be found with a +2 or a +4 charge but not with a +5 charge.

Section 7.7 Electron Affinity

Skill Review

57. Which of these would have the best chance of forming a stable anion: S or Xe? Justify your choice.

58. Which of these would have the best chance of forming a stable anion: Cl or Ar?

59. Which of these has the more exothermic value for electron affinity: Cl or Ar?

60. Which of these has the more exothermic value for electron affinity: O or F?

61. Arrange the following atoms on the basis of electron affinity, from least negative to most negative: Na, Mg, Al. Now arrange them on the basis of their ability to form anions. Is the order any different? If so, explain any changes you made in the ordering.

62. Arrange the following list in order from least negative electron affinity value to most negative electron affinity value: Br, Br$^-$, K.

63. Because they all have a negative charge, the electrons in atoms repel each other. Explain why the electron repulsion in a fluorine atom is such a large factor in determining its unusually small electron affinity value.

64. Explain why ionization energy is always a positive quantity, whereas electron affinity may be positive or negative.

Chemical Applications and Practices

65. The first ionization energy of sodium is +495 kJ/mol. The electron affinity of chlorine is −349 kJ/mol. When sodium metal and chlorine gas are placed near each other, a violent reaction takes place. After a sodium atom has lost an electron and a chlorine atom has gained an electron, both are oppositely charged and have the stable configuration of noble gases.
   a. What is the total energy change that occurs in these two processes? Is this exothermic or endothermic?
   b. Two additional energy changes occur in this reaction: the dissociation of a chlorine atom from Cl_2 (+158.8 kJ/mol) and the energy released when the ions are combined (−1030 kJ/mol). What can you determine about the spontaneous, violent reaction between sodium and chlorine when all of the energy changes in this process are considered?

66. The first ionization energy of lithium is 5.392 eV. (1 eV = 96.485 kJ/mol). The electron affinity of fluorine is −328 kJ/mol. When lithium metal and fluorine gas are placed near each other, a reaction takes place. After the ionization of lithium and the formation of anionic fluorine, both are oppositely charged and have the stable configuration of noble gases.
   a. What is the total energy change that occurs in these two processes? Is this exothermic or endothermic?
   b. Two additional energy changes occur in this reaction: the dissociation of a chlorine atom from F_2 (+158.8 kJ/mol), and the energy released when the ions are combined (−1030 kJ/mol). What can you determine about the spontaneous, violent reaction between lithium and fluorine when all of the energy changes in this process are considered?

Section 7.8 Electronegativity

Skill Review

67. Arrange this list of atoms in a correct ranking from least electronegativity to highest electronegativity: Na, F, As, Li, S.

68. Arrange this list of atoms in a correct ranking from least electronegativity to highest electronegativity: K, P, O, Br, N.

69. Each of these situations depicts two atoms bonded to each other. In each case, which atom is more likely to attract electrons toward itself within the bond?
   a. N—O   b. Se—S   c. Br—Ge   d. Cl—O

70. Each of these situations depicts two atoms bonded to each other. In each case, which atom is more likely to attract electrons toward itself within the bond?
   a. F—O   b. P—S   c. H—C   d. C—S

Chemical Applications and Practices

71. Using the information in the chapter, determine the change in electronegativity from Ga to Se. This change arises as the number of protons is increased by three. The change in proton number from Sc to Zn is nine. What is the change in electronegativity from Sc to Zn? Based on the change in number of protons, is this what you would expect? Why or why not?

72. In an early section we noted that the trend in electron affinity from fluorine to chlorine was a bit different than we might expect. The trend in electronegativity is, however, what we would expect; that is, it decreases from top to bottom in the group. What aspect of the definition of electronegativity helps explain the difference between the two trends?

Section 7.9 Reactivity

Skill Review

73. Arrange this list of metals in order from generally least reactive to most reactive: Na   Mg   Rb

74. Arrange this list of nonmetals in order from generally least reactive to most reactive: S   Cl   I

75. What is the relationship between first ionization energy and reactivity in metals and in nonmetals?

76. What is the relationship between electron affinity and reactivity in metals and in nonmetals?

Chemical Applications and Practices

77. The "coinage metals" are copper, silver, and gold.
   a. Describe the location of these elements in the periodic table.
   b. Where are they located in the activity series of metals used in the chapter?
   c. Is the location of Group IB consistent with periodic reactivity and the activity series?

Coins made from copper, silver, and gold. In the past, your pocket might contain coins made from these three metals. Due to the value of silver and gold, the U.S. mint currently uses other metals and mixtures of metals known as alloys.

78. Aluminum metal is considered a fairly active metal. It certainly reacts with oxygen more vigorously than does iron. The reaction of both metals with oxygen produces oxides that have different characteristics. Contrast the properties of the two oxides.

79. Nonmetals such as chlorine and oxygen are both considered very reactive. However, they may also react with each other. Using any available information, decide which of the two would probably become more negative in the reaction? Explain the basis for your answer.

80. Until the 1960s the noble gases were considered chemically inert, or totally unreactive. Eventually, however, some noble gases were made to react. Among the first compounds of noble gases, fluorine was typically a reactant. Explain why fluorine was such a good candidate for this role of reactivity.

Section 7.10 The Elements and the Environment

Skill Review

81. What explanation can be given for the fact that iron is found in much greater abundance in the Earth's crust than in the mantle?

82. What substance that is common in the Earth's crust accounts for most of the silicon and oxygen (the two highest-ranking elements in the crust?)

83. a. What is the third most abundant substance, by volume, in the Earth's atmosphere?
b. How many particles of that substance would be found in one mole of dry air?

84. a. What is the fourth most abundant substance, by volume, in the Earth's atmosphere?
b. How many molecules of that substance would be found in one mole of dry air?

Chemical Applications and Practices

85. Table 7.15 lists the ranked abundance of elements in the Earth's crust by mass. Which element(s) in that list are most likely to be found in an uncombined state? (*Uncombined*, in this case, means "not combined with other elements.")

86. In Table 7.15, titanium is listed as more abundant, by mass, than hydrogen. Convert the mass to moles and number of atoms. If the ranking were to be redone by number of atoms, would hydrogen still be below titanium? Prove your answer.

Comprehensive Problems

87. For each of these, provide a brief summary of his contribution to organizing the properties of elements into a useful arrangement.
a. Johan Dobereiner
b. John Newlands
c. Dmitri Mendeleev

88. Indium (element 49) is an important element, but it is not commonly used in textbook examples. Use the Internet or another text reference to determine a chemical and a physical property that indium has in common with gallium, element 31.

89. a. What is the most common form of steel used in the world?
b. Based on composition, what is the chief difference between this most common type of steel and other types of steel?

90. Based on the characteristics listed, which type of element (metal, nonmetal, or metalloid) is being described?
a. At room temperature the sample is a dull, brittle solid. The element is most likely to assume a negative charge in ionic compounds.
b. At room temperature the sample is a solid that is able to conduct electricity.

91. Given the information depicted in the illustration below, what would you predict for the bond length between a carbon and a hydrogen atom?

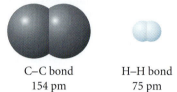

C–C bond   H–H bond
154 pm   75 pm

92. Contrast the meaning of the term *valence electrons* in the context of calcium and in the context of chromium.

93. a. Using the first 18 elements on the periodic table, prepare a graph with number of valence electrons on the *y* axis and group number on the *x* axis. Is this a periodic function? Explain why or why not.
b. Prepare another graph of the same elements, using the most common oxidation number on the *y* axis and the group number on the *x* axis. Is this a periodic function? Explain why or why not.

94. The definition of the atomic radius of an atom is basically straightforward; it is one-half the distance between the nuclei of a molecule made of two identical atoms.
a. What problems would arise if we were to define the radius as one-half the diameter of an atom?
b. How does metallic radius differ from covalent radius?

95. Explain why, when comparing the radius of two atoms, it would be important not to base the comparison solely on which atom has the greater number of protons.

96. a. This reaction depicts an atom being ionized. To which side of the reaction (left or right) should you show the energy term for the reaction?

$$X \rightarrow X^+ + e^-$$

b. Select the correct response for each of these situations: The smaller the value for ionization energy, the (*less, more*) easily ionization will occur. The smaller the value for ionization energy, the (*less, more*) chemically reactive a metal will be.

97. Seldom do we get to describe a chemistry situation as "always" happening without exception. However, this statement is always true: Successive ionization energies in an atom are always higher than previous values. What underlying factors make this statement true?

98. *Electron shielding* or *electron screening* provides a descriptive way to explain how the attraction for an electron by the nucleus can vary in different electron arrangements. However, electron screening can have a slightly different description when applied to elements in the same row on the periodic table than when applied to elements in the same group. Contrast the type of screening, and its effectiveness, when the term is applied to a period and when it is applied to a group.

99. Arsenic and selenium are next to each other in Period 4 of the periodic table. In general the trend, from left to right on the table, is in favor of an increase in ionization energy. However, when you compare the values for As and Se, you note that the ionization drops slightly instead of increasing. What is the basis for this situation?

100. Radioactive fallout from nuclear testing can contain significant amounts of unstable strontium. This can be particularly harmful to young children, who have rapidly growing bone structures. What common charge would strontium ions have? What element is an important component in bone development? Why is exposure to radioactive strontium such a grave danger to young children?

101. a. The electron affinity of chlorine is approximately -350 kJ/mol. Does this indicate an exothermic or an endothermic reaction?
    b. Write out the reaction, including placing the energy term in the equation, that depicts the "electron affinity reaction" for the process described in part a.

102. When summarizing the periodic trends in ionization energy, atomic radius, electronegativity, and (to some extent) electron affinity, we can typically indicate the trend with a one-directional arrow to show the trend increasing or decreasing from left to right in a row, or from top to bottom in a group. Explain why the trends for increasing reactivity start from the center of the periodic table and move outward in both directions.

103. In the chapter, we note that tool steel contains 4.38% by mass of chromium and 0.864% by mass of carbon.
    a. What is the electron configuration of chromium?
    b. If a steelmaker wishes to make 1.0 kg of tool steel, which element (carbon or chromium) would require a larger number of moles?
    c. How many atoms of chromium would be present in an 88.7 g sample of tool steel?

104. The halogens are so-named because they react with metals to make salts.
    a. How many valence electrons do each of the atoms in this group contain?
    b. Write the balanced reaction that occurs between sodium metal and iodine crystals.

105. According to Table 7.13, the human body contains the same mass percent of silicon and potassium. Which occurs in the body in a greater number of moles?

106. If the periodic table was arranged, in order, from smallest atom (based on atomic radius) to largest atom
    a. What atom would be listed first?
    b. What atom would be listed last?
    c. Would the noble gases still be aligned in a column?
    d. How many helium atoms would be needed to create a line of atoms 1.0 inches long (assuming that the atoms are just touching and placed in a straight line)?

107. In the chapter, we mention that francium-210 can be made from gold-197 by bombarding the gold with oxygen-18.
    a. How many protons, neutrons, and electrons are found in one atom of francium-210?
    b. How many grams of francium-210 could be made from 2.50 g gold-197?
    c. Which has a larger atomic radius—francium-210 or gold-197? (Assume specific isotopes do not differ in their atomic radius.)

Thinking Beyond the Calculation

108. An aqueous solution of sodium bicarbonate ($NaHCO_3$) reacts with aqueous hydrochloric acid to produce carbonic acid (H_2CO_3) and sodium chloride (NaCl).
    a. Write a balanced equation illustrating this reaction.
    b. Indicate the group and period for each of the elements involved in the reaction.
    c. Indicate those elements in the reaction that can be characterized as metals.
    d. Carbonic acid decomposes in solution to produce carbon dioxide and water. Write the balanced equation for this reaction.
    e. If 10.0 mg of sodium bicarbonate is added to 275 mL of water and reacted completely with a stoichiometric amount of HCl, what concentration (in ppm) of sodium chloride will result? (Assume no volume change.)
    f. Using the information from part e, determine what the concentration (in ppm) of sodium ions will be.
    g. A pastry chef may wish to perform a reaction similar to this using sodium bicarbonate and acid. Why would the chef wish to perform this reaction?

Reaction of sodium bicarbonate and HCl.

8

Bonding Basics

Bottles in the medicine chest. The active ingredients in these medications are a mix of molecules and ionic compounds, two classes of compounds that differ in the way their atoms are bound together.

Contents and Selected Applications

8.1   Modeling Bonds

8.2   Ionic Bonding

Chemical Encounters: The Uses and Behavior of Sodium Chloride

Chemical Encounters: Focus on Zeolites

Chemical Encounters: Fluoridated Water and Tooth Decay

8.3   Covalent Bonding

Chemical Encounters: Tetraethyl Lead in Gasoline

Chemical Encounters: Calcium Channel Blockers

8.4   VSEPR—A Better Model

Chemical Encounters: Focus on Morphine

8.5   Properties of Ionic and Molecular Compounds

Go to **college.hmco.com/pic/kelterMEE** for online learning resources.

Americans spend more than \$250 billion per year on prescription and over-the-counter (OTC) medicines, about one-half of the world's total. Our medicine cabinets are fairly well stocked. If you and your family are typical consumers, you may have headache pills, muscle relaxers, antacids, cough syrup, and a couple of old bottles of antibiotics. A quick check of the active ingredients reveals that these bottles contain such compounds as ibuprofen, magnesium salicylate, sodium bicarbonate, and pseudoephedrin. One of the more common ingredients in a pain-relieving OTC medicine is aspirin (a molecular compound). Sodium bicarbonate (an ionic compound) is typically used as an antacid. These two compounds work in different ways to relieve common maladies, and they are fundamentally different in their chemical makeup. They do not have the same numbers and types of atoms, although this alone isn't enough to explain why these compounds differ radically in many properties, including melting point, boiling point, solubility in water, and chemical reactivity. The main difference lies in their designation as molecular compounds or ionic compounds, and this designation is made on the basis of the way in which the atoms are bound together.

Biochemists, medicinal chemists, and **pharmacognocists** (who study the properties of drugs, especially focusing on those from natural sources) are familiar with molecular and ionic compounds. One of the fundamental questions asked by these scientists is **How do the bonds within a compound, the shape of the compound, and the physical properties of the compound contribute to its ability to treat a disease or common malady?** In this chapter, we will answer this question and others as we explore how atoms are held together, and how this bonding determines the shapes and properties of compounds.

8.1 Modeling Bonds

We will begin our discussion by looking at the ways in which researchers view the compounds with which they work. Although the formula of a potential drug can be useful for determining the number and types of atoms it contains, it doesn't say much about the molecular shape. Aspirin—acetylsalicylic acid—which we discussed in Chapter 3, has the formula $C_9H_8O_4$. The formula alone provides no information about how aspirin structurally interacts with the body to alleviate a headache. To find out more, we must look deeper into the atom—into the role that the electron plays in determining shapes. Our first key idea is that *the different arrangements of the electrons in a compound help determine the shape and the properties of that compound.*

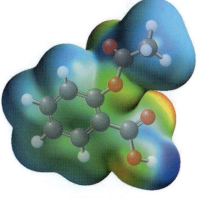

Acetylsalicylic acid
$C_9H_8O_4$

Three Kinds of Bonds

In their search for marketable pharmaceuticals, biochemists, medicinal chemists, and pharmacognocists use a wide range of **molecular models**, three-dimensional depictions of the structure of molecules, as tools to examine the shape and properties of compounds. To be ideal for the marketplace, a compound should

- effectively treat a particular disease, malady, or ailment
- have no serious side effects
- be inexpensive to mass-produce

The molecular models that researchers use in their search for marketable drugs serve as representations of the **chemical bonds**, the *forces that hold atoms together*, within a compound. Chemical bonds arise between atoms when some of the outermost electrons on the bonding atoms interact. In some cases, the electrons tend to congregate on one of the atoms of the bond. In other cases, the electrons are shared more or less equally between the atoms. A vital point to keep in mind is that *there are no absolutes*. There are many degrees of electron sharing, from more or less complete ownership by one atom to about equal sharing by both atoms, and every possibility on the scale of sharing can occur. Still, we classically think of three types of chemical bonds: the covalent bond, the ionic bond, and the metallic bond.

Sodium chloride, the stuff we sprinkle on our French-fries, is an example of a compound with an ionic bond. The **ionic bond** lies at one end of the spectrum of chemical bonds, where the atoms are held together by the force of attraction of opposite charges. We say that one or more electrons have been removed from one atom (remember, there are no absolutes!) and congregate on the other atom in the compound.

Visualization: Covalent Bonding

Tutorial: Covalent and Ionic Bonding

Aspirin is an example of a compound in which the atoms are held together by covalent bonds. The **covalent bond** lies at the other end of the spectrum of chemical bonds, where electrons are shared between the atoms. The positively charged nuclei on either end of the bond attract the negatively charged electrons. It is this force that holds the atoms together.

Purely   "Real" bonds   Purely
covalent   ionic

Covalent and ionic bonds lie at opposite ends of the bonding spectrum. Real bonds between different elements are neither purely ionic nor purely covalent.

The aluminum atoms in a soda can are held together with a metallic bond. The **metallic bond** is a special type of bond in which metal cations are spaced throughout a sea of mobile electrons. We'll learn more about this bonding pattern in Chapter 13.

Lewis Dot Symbols

The first step in the construction of the molecular model of a compound, such as aspirin, sodium chloride, or aluminum, is drawing an atom itself. We recognize that the "business part" of the atom is its set of valence electrons. In 1916, G. N. Lewis (Figure 8.1) developed a useful shorthand representation employing

FIGURE 8.1

The American chemist G. N. Lewis (1875–1946) developed a model of atomic bonding that is still used today. These notes, written by Lewis in 1902, illustrate his thinking on how electrons were arranged around an atom. The Lewis dot symbols that we use today are slightly modified from this original work.

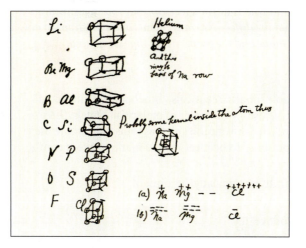

Lewis dot symbols. Typically, the first two dots, representing valence electrons in the *s* orbital, are placed one at a time, as a pair, to the side of the element symbol. The next three dots, representing valence electrons in the *p* orbitals, are placed individually on the other sides of the element symbol. The remaining electrons are placed alongside of the first three dots to indicate pairs of electrons in the *p* orbitals. The Lewis dot symbols for the elements of Period 2 are shown below.

Video Lesson: Valence Electrons and Chemical Bonding

$$•\text{Li} \quad :\text{Be} \quad :\overset{}{\text{B}}• \quad :\overset{}{\text{C}}• \quad :\overset{•}{\underset{•}{\text{N}}}• \quad :\overset{•}{\underset{•}{\text{O}}}: \quad :\overset{•}{\underset{••}{\text{F}}}: \quad :\overset{••}{\underset{••}{\text{Ne}}}:$$

Lewis dot symbols of elements 3 to 10.

Lewis dot symbols can be used to represent ions as well. Because an ion is just an atom (or a group of atoms) with a different number of electrons than the uncharged species, we can draw its Lewis dot symbol. For instance, the fluoride ion is a fluorine atom plus an electron (F^-). To distinguish ions from Lewis dot symbols of atoms, we place brackets around the drawing, and we place the charge on the ion outside the brackets. Note in the diagram below that the nitrogen anion has three extra electrons and the sulfur anion has two extra electrons. All three have eight valence electrons and the same electron configuration as the period's noble gas (Ar for S and Ne for F and N). This idea is important in **Lewis dot structures**, and we'll revisit it later in this chapter.

$$\left[:\overset{••}{\underset{••}{\text{S}}}:\right]^{2-} \quad \left[:\overset{••}{\underset{••}{\text{N}}}:\right]^{3-} \quad \left[:\overset{••}{\underset{••}{\text{F}}}:\right]^{-}$$

Lewis dot symbols of some monatomic anions.

Lewis dot symbols can also be drawn for cations. In a monatomic cation, electrons have been removed from the atom (or group of atoms,) resulting in a positive charge. When we write the Lewis dot symbols, we indicate the number of valence electrons and the resulting positive charge.

$$\left[\text{Na}\right]^{\oplus} \quad \left[\text{Al}\right]^{3+} \quad \left[:\overset{•}{\text{N}}•\right]^{\oplus}$$

Lewis dot symbols of cations.

Electron Configuration of Ions

We determined the electron configuration of the elements in Chapter 6 using the Aufbau principle. Electron configurations can also be used to describe which atomic orbitals in ions contain electrons. For example, the electron configuration of the sodium ion contains one electron less than that of the sodium atom. This ion, which is necessary for proper contraction of heart tissue and electrolyte balance inside and outside of the body's cells, has an electron configuration that lacks the valence electron from the sodium atom.

$$\text{Na: } 1s^2 2s^2 2p^6 3s^1 \qquad \text{Na}^+\text{: } 1s^2 2s^2 2p^6$$

In anions such as the fluoride ion (F^-, found in toothpaste and added to most U.S. municipal water supplies to help prevent tooth decay), the electron configuration shows the addition of another electron to the valence shell:

$$\text{F: } 1s^2 2s^2 2p^5 \qquad \text{F}^-\text{: } 1s^2 2s^2 2p^6$$

We also know from our discussion of ionization energy that valence electrons are the most accessible of the electrons in an atom. The addition of an electron to an atom to make an anion occurs in the valence shell. Similarly, cations can be made by removing valence electrons. **How do the electron configurations of Na$^+$ and F$^-$ compare with each other and with the noble gas nearest to them in atomic number on the periodic table?** They have the same electron configuration as neon, $1s^2 2s^2 2p^6$. We say that the electron configuration of Na$^+$ is **isoelectronic** with (has the *same* electron configuration as) F$^-$ and Ne.

The position of the electrons as either valence or core electrons on an atom can help us understand the structure of a molecule such as aspirin. Knowing where the electrons on the atom reside and how those electrons behave is of utmost importance in developing our three-dimensional model of a molecule. As we'll see shortly, the number of valence electrons on the atom is also important to building our best possible model.

Octet Rule

In any chemical reaction, one of the driving forces is the ability of each atom to reach a stable electron configuration. Electrons shuffle around until each atom has its most energetically stable arrangement of electrons. As in F^- and Na^+, the most stable electron configuration of the main-group elements is isoelectronic with a noble gas. We refer to this as the **octet rule** because the stable arrangement for all of the noble gases beyond helium has eight valence electrons. For the elements H and He and the ions Li^+ and Be^{2+}, the rule is also known as the **duet rule** because of the need for only two electrons to fill the valence shell of the first row elements. In other words, *main-group atoms typically react by changing their number of electrons in such a way as to acquire the more stable electron configuration of a noble gas.* A full valence shell around an atom is a good situation for an atom, because it then has the same electron configuration as a noble gas. We'll use this rule a lot as we put together atoms to make ionic compounds and molecules.

EXERCISE 8.1   **Writing Lewis Dot Symbols**

Write the electron configuration, and the Lewis dot symbol, for both P^{3-} and Al^{3+}. With which element are they isoelectronic?

Solution

The P^{3-} anion has three electrons more than the phosphorus atom. Therefore, the electron configuration is $1s^2 2s^2 2p^6 3s^2 3p^6$. This is the same electron configuration as argon. The Lewis dot symbol shows an octet of electrons.

The Al^{3+} cation is missing three electrons, compared to the neutral atom. Its electron configuration is $1s^2 2s^2 2p^6$, which is isoelectronic with neon. The Lewis dot symbol also illustrates an octet of electrons.

PRACTICE 8.1

Write the electron configuration, write the Lewis dot symbol, and determine which atoms are isoelectronic with S^{2-}, F^-, Mg^{2+}, and Br^-.

See Problems 1–4 and 9–14.

 ## 8.2 Ionic Bonding

Application

CHEMICAL ENCOUNTERS: The Uses and Behavior of Sodium Chloride

We noted before that sodium chloride is an ionic compound. In ancient times, it was a highly sought-after seasoning used in cooking and pickling. According to the United States Geological Survey (USGS), 210 million metric tons of sodium chloride were harvested from salt water or mined from deposits in the ground in 2005 (see Figure 8.2). In the United States, most of the salt that is harvested is used to manufacture chlorine gas and sodium hydroxide. About 37% of the total is used to de-ice highways. Only 3% is used in the food industry. Even so, the average American consumes more than 1.5 kg of salt each year!

The human body requires only about 500 mg of sodium per day, yet the average American ingests between 2300 and 6900 mg each day. This high level of sodium consumption can contribute to hypertension, sleep apnea, and other disorders.

Sodium chloride (NaCl) is a compound held together by ionic bonds. As we saw in Chapter 4, sodium chloride, like all other Group IA salts, dissociates essentially completely into its ions when added to water. It is a strong electrolyte and is a good example of a typical ionic compound. Other examples of important ionic compounds are shown in Table 8.1 and Figure 8.3. Note that some ionic compounds contain ionic and covalent bonds. For example, calcium carbonate contains an ionic bond between the calcium ion and the carbonate ion. The carbon and oxygen atoms in the carbonate ion are covalently bonded to each other. What makes sodium chloride a "typical" ionic compound? The answer lies in the nature of its bonding and properties, which we now explore.

FIGURE 8.2

Harvested sodium chloride is stored in piles near the purification plant.

TABLE 8.1	Important Ionic Compounds	
Compound	**Formula**	**Selected Use**
Calcium carbonate	$CaCO_3$	Limestone, chalk
Calcium chloride	$CaCl_2$	Sidewalk salt
Iron(III) oxide	Fe_2O_3	Pigment
Magnesium hydroxide	$Mg(OH)_2$	Milk of magnesia
Sodium carbonate	Na_2CO_3	Glass, soaps, and detergents
Sodium bicarbonate	$NaHCO_3$	Baking soda
Sodium chloride	$NaCl$	Production of chlorine and sodium hydroxide, seasoning, saline solutions
Sodium fluoride	NaF	Toothpaste

FIGURE 8.3

Important ionic compounds.

Video Lesson: Ionic Bonds

Visualization: Structure of an Ionic Solid (NaCl)

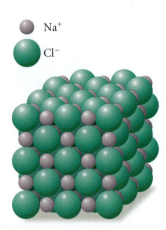

FIGURE 8.4

Crystal structure of sodium chloride. In the crystal structure of NaCl, we note that each sodium ion is surrounded by six chloride ions and that each chloride ion is surrounded by six sodium ions. The resulting crystal possesses a regular close-packed pattern of ions (see Chapter 14).

Description of Ionic Bonding

Sodium chloride is commercially mined or processed from brine (salt water, such as from the oceans or the Great Salt Lake). In the laboratory, we can combine sodium metal and chlorine gas in a violent reaction to make the salt. The reaction, illustrated below, involves the transfer of electrons from sodium atoms to

$$Na(s) + \tfrac{1}{2}Cl_2(g) \rightarrow NaCl(s) \qquad \Delta H = -410 \text{ kJ}$$

chlorine. If we look at this reaction more closely, we note that one of the atoms (the metal) loses electrons to become a cation. In our reaction, the sodium atom

$$Na\bullet \rightarrow [Na]^{\oplus} + e^-$$

loses an electron to become the sodium cation (Na^+). The other atom (the nonmetal) gains electrons to become an anion. Addition of an electron to a chlorine atom forms the chloride anion (Cl^-). The charges on the sodium and chloride

$$\bullet\ddot{\underset{\bullet\bullet}{Cl}}\colon + e^- \longrightarrow \left[\colon\ddot{\underset{\bullet\bullet}{Cl}}\colon\right]^{\ominus}$$

ions attract each other, causing the ions to associate with each other in an **ion pair** (Na^+Cl^-).

But it doesn't stop there. Other sodium cations are also attracted to the negative charge on the chloride. Other chloride anions are attracted to the positively charged sodium. The end result is a collection of alternating sodium and chloride ions arranged in a solid **crystalline lattice**, a highly ordered, three-dimensional arrangement of atoms, ions, or molecules (explained in detail in Chapter 13). Within a sodium chloride crystal, each of the sodium cations has six neighboring chloride anions, and each of the chloride anions has six neighboring sodium cations. The forces of attraction combine to provide a neatly packed crystal of alternating sodium and chloride ions, as shown in Figure 8.4.

Examples of Ionic Bonding

We can illustrate the formation of an ionic compound such as sodium chloride (NaCl) through the use of Lewis dot symbols. As reflected in their positions on the periodic table, sodium is a metal and chlorine is a nonmetal. Sodium has a relatively low ionization energy and loses an electron to achieve the electron configuration of a noble gas.

$$Na\bullet \rightarrow [Na]^{\oplus} + e^-$$
$$1s^2 2s^2 2p^6 3s^1 \rightarrow 1s^2 2s^2 2p^6$$

Chlorine, which has a very favorable electron affinity, gains an electron to achieve a noble gas electron configuration.

$$\bullet\ddot{\underset{\bullet\bullet}{Cl}}\colon + e^- \longrightarrow \left[\colon\ddot{\underset{\bullet\bullet}{Cl}}\colon\right]^{\ominus}$$
$$1s^2 2s^2 2p^6 3s^2 3p^5 \rightarrow 1s^2 2s^2 2p^6 3s^2 3p^6$$

In the formation of NaCl from Na metal and Cl_2 gas, the electron from the sodium atom is added to the valence shell of the chlorine atom. We can represent this process in equation form by showing the movement of the electron from the sodium atom to the chlorine atom with a fishhook-shaped arrow. The result is a sodium cation and a chloride anion that are attracted to each other. The ionic crystal that results has the formula NaCl.

Energy must be added to the system to remove the electron from the sodium atom. *Energy, however, is released when that electron is added to the chlorine atom, and even more energy is released when the ionic bond forms.* The net result from the formation of a crystalline lattice of sodium and chloride ions is a large release of energy.

$$Na\cdot \;+\; \cdot\ddot{\underset{\cdot\cdot}{Cl}}\colon \longrightarrow [Na]^{\oplus} \;+\; \left[\colon\!\ddot{\underset{\cdot\cdot}{Cl}}\colon\right]^{\ominus} \longrightarrow [Na]^{\oplus}\left[\colon\!\ddot{\underset{\cdot\cdot}{Cl}}\colon\right]^{\ominus}$$

EXERCISE 8.2   Ionic Compound Formation

Calcium chloride, $CaCl_2$, is spread on our sidewalks to melt ice. Because of its ability to absorb moisture, it is used to dry gases and prevent dust buildup in mining and highway maintenance. Using Lewis dot symbols, show the transfer of electrons to form $CaCl_2$ from calcium metal and chlorine atoms.

Solution

We start by drawing the Lewis dot symbols for the calcium and chlorine atoms. Then we show the transfer of an electron from the less electronegative calcium to the more electronegative chlorine atom (arrow). We show a similar transfer with the other calcium electron to a different chlorine atom. We end with three ions (Ca^{2+} and two Cl^-) that have noble gas electron configurations. Note that the ions end up with either a completely full valence electron configuration (and become isoelectronic with a noble gas) or a completely empty valence shell (and become isoelectronic with a noble gas). This is the octet rule in action.

$$\overset{\cdot}{\underset{\cdot}{Ca}} \;+\; \begin{matrix}\cdot\ddot{\underset{\cdot\cdot}{Cl}}\colon \\[8pt] \cdot\ddot{\underset{\cdot\cdot}{Cl}}\colon\end{matrix} \longrightarrow \left[Ca\right]^{2+} \;+\; \left[\colon\!\ddot{\underset{\cdot\cdot}{Cl}}\colon\right]^{\ominus} \;+\; \left[\colon\!\ddot{\underset{\cdot\cdot}{Cl}}\colon\right]^{\ominus}$$

PRACTICE 8.2

Use the Lewis dot structure model to show the formation of sodium oxide from atomic sodium and oxygen atoms.

See Problems 21, 22, and 27–30.

Medicinal chemists, biochemists, and pharmacognocists know about the driving force for the formation of ionic bonds. Ethidium bromide, shown in Figure 8.5, is a toxic compound that is used by researchers to stain DNA in order to make it easier to see. This compound exhibits its effects by inserting itself into DNA strands, causing structural changes to the DNA molecule. The negative charges on the DNA and the positive charge of the ethidium cation help hold the compound in place. These forces of attraction explain why this compound has such a strong association with DNA. Another example of the importance of ionic bonding accounts for the source of calcium in milk. Casein, one of the major proteins found in milk, contains phosphate anions that form ionic bonds with calcium cations.

Sizes of the Ions

Water from many wells typically contains relatively high concentrations of calcium, often along with some magnesium and iron. When the water is heated, dissolved calcium bicarbonate, $Ca(HCO_3)_2$, decomposes to form rock-like deposits of calcium carbonate, $CaCO_3$, also known as **boiler scale**. The boiler scale

$$Ca(HCO_3)_2(aq) \rightarrow CaCO_3(s) + H_2O(l) + CO_2(g)$$

coats the inside of the pipes that carry the water in the hot-water heater and throughout the house, so such water is called "hard." Hard water is not hazardous

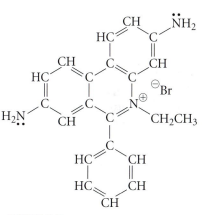

FIGURE 8.5

Ethidium bromide. The stabilizing force that holds the molecule in place in DNA is the attraction between the anionic charges on the DNA and the positive charge on ethidium bromide.

 Application

"Hard" water contains relatively high concentrations of calcium, typically along with magnesium and iron. The rock-like deposits of calcium carbonate that form on the inside of hot water pipes give hard water its name.

Application

CHEMICAL ENCOUNTERS: Focus on Zeolites

Visualization: Ionic Radii

to our health. It is just not the best thing that could happen to our household pipes. If the buildup is bad enough, the pipes can become clogged, much as heart arteries become clogged with fatty plaques, reducing blood flow in a human. Many homeowners and some cities "soften" water by passing it through a bed of **zeolites**, which exchange the calcium ions for sodium ions that (remember our solubility rules) do not form insoluble salts. How do these ion-exchanging zeolites work?

Zeolites like that shown in Figure 8.6 are composed of aluminum and silicon oxide subunits containing Group IA and IIA cations. The resulting honeycomb arrangement of subunits results in a structure full of atom-sized holes. Ions and molecules small enough to fit through these holes can enter the zeolite, and if they are just about the same size as the holes in the zeolite, they get stuck inside. Interactions of the ions or molecules with the zeolite help to hold them inside. Substances that are too small easily flow in and out of the zeolite, whereas ions and molecules that are too big can't enter the zeolite in the first place. Consequently, a zeolite retains only specific sizes of ions and molecules. In exchange for the trapped ions, the zeolite releases ions of the same total charge. For example, when a calcium ion is taken up by the zeolite, it typically releases two sodium ions.

By carefully constructing the zeolite, researchers have been able to develop an "ionic sponge" that grabs only the ions of interest to them. This is vital, in the chemical industry, to the formation of better reaction catalysts (compounds that significantly increase the rate of a chemical reaction without being consumed) the filtration of polluted air, and the cleanup of hazardous wastes, among other applications. Experiments with the zeolite known as clinoptilolite, with a pore size of 0.4 nm, indicated that it was capable of removing radioactive cesium (^{134}Cs and ^{137}Cs) from cows affected by the 1986 Chernobyl (Ukraine) nuclear accident.

By far the most common use of zeolites is as an ingredient called a "builder" in laundry detergents for the removal of calcium ions from hard water. Size is a critical factor in the behavior of ionic compounds. Not only does size suggest what type of zeolite can trap a particular ion, it also plays a major role in determining the structure of the ionic crystal and the strength of the ionic bond.

What contributes to the size of an ion? We can examine the electrons in an atom more closely to determine the answer. In our previous discussion of the shape of the orbitals (Chapter 6), we learned that the electrons penetrate deep into the atom. Each electron helps to balance the charge of the protons in the nucleus and shields the other electrons from the nuclear charge. When an electron

FIGURE 8.6

Ions can enter, and they can leave: The structure of natrolite and sodalite. These zeolites have pores that ions of the right size can enter.

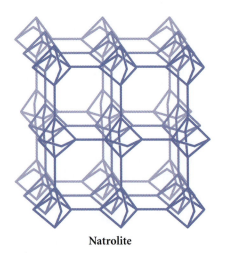

Natrolite

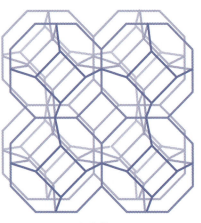

Sodalite

TABLE 8.2 The Radius of Some Common Ions and Atoms (in picometers)

Period	Group IA	Group IIA	Group IIIA	Group VA	Group VIA	Group VIIA
2	Li 152 Li^+ 60	Be 113 Be^{2+} 31	B 88 B^{3+} 20	N 70 N^{3-} 171	O 66 O^{2-} 140	F 64 F^- 133
3	Na 186 Na^+ 99	Mg 160 Mg^{2+} 65	Al 143 Al^{3+} 50	P 110 P^{3-} 212	S 104 S^{2-} 184	Cl 99 Cl^- 181
4	K 227 K^+ 133	Ca 197 Ca^{2+} 99			Se 117 Se^{2-} 198	Br 114 Br^- 195
5	Rb 247 Rb^+ 148	Sr 215 Sr^{2+} 113			Te 143 Te^{2-} 221	I 133 I^- 216
6	Cs 265 Cs^+ 169	Ba 217 Ba^{2+} 135				

is removed from an atom, the amount of shielding is reduced. We say that the effective nuclear charge experienced by each electron increases. The total number of electron–electron repulsions decreases, so the electrons are pulled closer to the nucleus. The net effect is a reduction in the size of the electron cloud around the cation. *The size of the cation is less than the size of the atom* (see Table 8.2).

The opposite happens when an electron is *added* to an atom. The extra electron increases the number of electron–electron repulsions, thereby reducing the effective nuclear charge felt by each electron. This causes the electron cloud to swell in size. *The anion, then, is larger than the atom from which it was made.* These effects continue as more electrons are removed from or added to an atom. All of the ions shown in Figure 8.7 have the same electron configuration, but they differ in the charge of the nucleus. We can see that as the nuclear charge increases, the size of the ion decreases.

We noted that the removal of the first electron greatly decreases the size of an ion. Similar reasoning holds for the removal of subsequent electrons. *In general,*

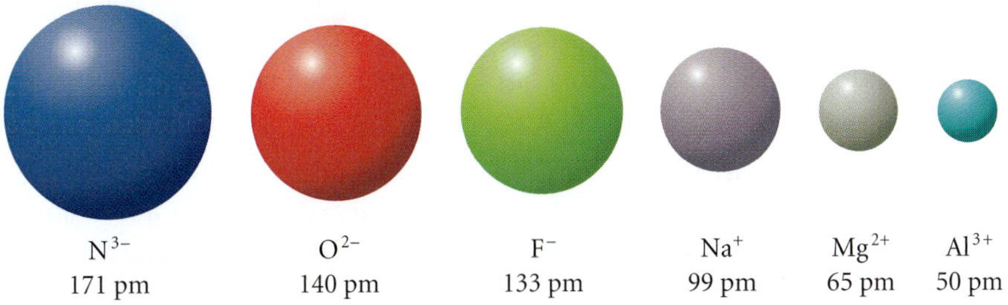

FIGURE 8.7
The radii of a series of isoelectronic ions.

the cation gets smaller as more electrons are removed. Removal of the first electron from an atom greatly decreases the radius of the resulting ion. Removal of the second and all subsequent electrons from the ion has the same effect, but in a less pronounced way. This is due to the fact that electrons in the same subshell do not shield each other very effectively from the nuclear charge.

As you descend within a group in the periodic table, the principal quantum number of the valence electrons increases. The distance from the nucleus increases with an increasing principal quantum number. The size of the ion increases. This trend is shown in Table 8.2.

EXERCISE 8.3   Ions and Their Sizes

Argon is a noble gas found in the atmosphere in relatively large concentration (0.934% by volume). It is used as an inert atmosphere in the chemical laboratory for reactions that would react with normal atmospheres. List four ions that are isoelectronic with argon, and arrange them in order of increasing size.

First Thoughts

The question is really asking us to complete two tasks. The first is to find four atoms that can be made into ions with the same electron configuration as argon. Therefore, we should limit our search to those atoms that have an atomic number close to that of argon. Second, the question asks us to organize the ions we've created by size. We know that cations are smaller than atoms and anions are larger.

Solution

$$Ca^{2+} < K^+ < Ar < Cl^- < S^{2-}$$

Further Insight

Knowing the size and electron configuration of an ion is helpful in constructing the bigger picture for a model of a compound. The five species we've arranged here have the same number of electrons around their nuclei, but they differ greatly in size. Moreover, their having the same number of electrons doesn't imply that these ions have any properties in common. They are completely different in their reactivity and in their physical properties.

PRACTICE 8.3

Arrange four ions that are isoelectronic with Ne in order of increasing size.

See Problems 23, 24, 31, and 32.

HERE'S WHAT WE KNOW SO FAR

- The electron configuration of an ion determines the number of electrons in its outer shell.
- We can use the electron configuration of an ion to draw the Lewis dot symbol for the ion.
- Anions, atoms with extra electrons, are larger than their corresponding atoms.
- Cations, atoms missing electrons, are smaller than their corresponding atoms.
- Ionic compounds are the combination of anions and cations to make an electrically neutral compound. The force of attraction between an anion and a cation is the ionic bond.

Energy of the Ionic Bond

According to the American Dental Association, the addition of fluoride to tooth-paste and public water systems has brought about a nationwide reduction in tooth decay. The Centers for Disease Control suggest a "safe, effective and inexpensive" municipal waterway fluoride concentration of between 0.7 and 1.2 parts per million. The mineral portion of teeth is hydroxyapatite, $Ca_5(PO_4)_3(OH)$. When you drink fluoridated water or brush your teeth with toothpaste containing fluoride, some of the hydroxide anions in your teeth are replaced with fluoride.

$$Ca_5(PO_4)_3(OH) \longrightarrow Ca_5(PO_4)_3(OH,F)$$

Hydroxyapatite   Fluorapatite

The new mineral, called fluorapatite, is much stronger than hydroxyapatite. What accounts for the added strength? Does the size of the fluoride and hydroxide ions have something to do with the strength of the mineral?

The strength of the ionic bond is usually referred to as the **lattice enthalpy** of the ionic solid. The lattice enthalpy of a molecule is the amount of energy required to separate 1 mol of a solid ionic crystalline compound into its *gaseous* ions (see the following equation). The lattice enthalpies of some common compounds can be found in Table 8.3.

$$MX(s) \rightarrow M(g)^+ + X(g)^-$$

Although qualitative statements can be made about the relative size of the lattice enthalpy on the basis of ionic radii, most chemists use lattice enthalpy to make some *quantitative* statements about the strength of an ionic compound. Unfortunately, it is quite difficult to measure lattice enthalpy accurately in an ionic crystalline solid. However, we can calculate the lattice enthalpy using Hess's law (see Chapter 5) in a process known as the **Born–Haber cycle**, named after two Nobel Prize–winning German scientists (Max Born, 1882–1970, and Fritz Haber, 1868–1934).

The Born–Haber cycle is a diagrammatic representation of the formation of an ionic crystalline solid. Figure 8.8 illustrates the Born–Haber cycle for the formation of sodium chloride from its elements in their standard states. Although chemistry is a process in which all kinds of things happen concurrently and continuously, for clarity we often break down the Born–Haber cycle into a series of steps.

TABLE 8.3	Lattice Enthalpies for Some Common Ionic Solids

Values in the table are in kilojoules per mole for the simple ionic compounds (for example, the lattice energy of K_2O is 2238 kJ/mol).

	F^-	Cl^-	Br^-	I^-	OH^-	O^{2-}
Li^+	1030	834	788	757	1039	2799
Na^+	923	787	747	704	887	2481
K^+	821	701	682	649	789	2238
Rb^+	785	689	660	630	766	2163
Cs^+	740	659	631	604	721	—
Mg^{2+}	2913	2326	2097	1944	2870	3795
Ca^{2+}	2609	2223	2132	1905	2506	3414
Ba^{2+}	2341	2033	1950	1831	2141	3029
Sc^{3+}	5096	4874	4711	4640	5063	13,557
Al^{3+}	5924	5376	5247	5070	5627	15,916

FIGURE 8.8

The Born–Haber cycle for the formation of sodium chloride.

Visualization: Born–Haber Cycle for NaCl(s)

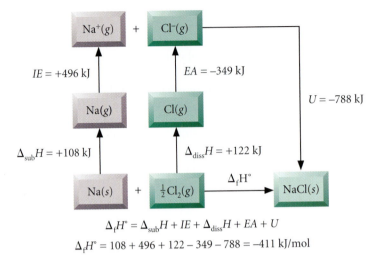

$$\Delta_f H° = \Delta_{sub}H + IE + \Delta_{diss}H + EA + U$$
$$\Delta_f H° = 108 + 496 + 122 - 349 - 788 = -411 \text{ kJ/mol}$$

- At the starting point to the cycle, we sublime solid sodium metal into gaseous sodium metal. The enthalpy change for this process is known and is recorded on the cycle.

- Next, we dissociate the chlorine molecule into individual chlorine atoms, using the equation that relates to the bond dissociation energy for chlorine. Because we need only one chlorine atom, we take only half of the enthalpy change for this process.

- In the next steps, the gaseous atoms are ionized to the gaseous ions. Sodium atoms are ionized to sodium cations with an enthalpy change that corresponds to the first ionization enthalpy. Chlorine atoms are converted into chloride anions with an enthalpy change that corresponds to the electron affinity for chlorine.

- In the final step, the sodium and chloride ions are brought together to make the ionic crystalline solid. This enthalpy change corresponds to the negative of the lattice enthalpy for the ionic solid.

When the heat of formation for the ionic solid is known by direct measurement, the direct route from the starting point to the end of the Born–Haber cycle (the lattice enthalpy) can be calculated by summing all of the individual enthalpy changes. Because the individual enthalpy changes are often known (heat of formation, first ionization, bond dissociation, and electron affinity), this provides a convenient method for determining the lattice enthalpy of the ionic crystalline solid.

Calculation of the **lattice energy** (not lattice enthalpy) of an ionic compound can also be accomplished using **Coulomb's law**. Coulomb's law states that the force between two particles is proportional to the product of the charges (Q) on each particle divided by the square of their distance of separation (d). The French physicist Charles Augustin Coulomb (1736–1806) proved this relationship by experiment. Because the force of attraction between two ions is related to the lattice energy, a modification of Coulomb's potential gives us the lattice energy:

$$\text{Lattice energy} = k \left(\frac{Q^+ \times Q^-}{d} \right)$$

In this equation, k is a proportionality constant, Q^+ and Q^- are the charges on the ions, and d is the distance between the ions. The lattice energy is related to the lattice enthalpy. The difference is that the lattice energy ($E_{lattice}$) is the amount of energy released as the solid ionic crystal is formed from its gaseous ions, whereas the lattice enthalpy ($\Delta H_{\text{lattice}}$) is the amount of heat at constant pressure necessary to separate the solid ionic crystal into its gaseous ions. The good news is that these two values are just about the same, though opposite in sign; lattice energy describes a release in energy ($E_{\text{lattice}} = $ "$-$"), whereas lattice enthalpy describes energy that is absorbed ($\Delta H_{\text{lattice}} = $ "$+$").

$$\Delta H_{\text{lattice}} \cong -E_{\text{lattice}}$$

The magnitude of the lattice energies or lattice enthalpies increases as the charges on the ions increase, as shown in Table 8.3. In the series $ScCl_3$, $CaCl_2$, and KCl, the charges decrease on the cation: ($+3$), ($+2$), and ($+1$), respectively, with a decrease in lattice enthalpies in the order $ScCl_3$ (4874 kJ/mol), $CaCl_2$ (2223 kJ/mol), KCl (701 kJ/mol). Coulomb's law also indicates that as the distance between the ions decreases, the lattice enthalpies increase. Evidence of this is found by comparing LiBr (788 kJ/mol), LiCl (834 kJ/mol), and LiF (1030 kJ/mol). The charges on the ions are the same in this series of compounds, but the size of the anion decreases from bromide to fluoride. Because the bromide ion (Br^-) is larger than Cl^- or F^-, LiBr has the smallest lattice enthalpy. In general, lattice enthalpies are greatest for ionic compounds that are made up of small, highly charged particles.

Let's revisit our section-opening question of why fluoride in our toothpaste is important. As we noted then, the replacement of a hydroxy group in the hydroxyapatite, $Ca_5(PO_4)_3(OH)$, mineral that makes up our teeth gives rise to a new mineral called fluorapatite, $Ca_5(PO_4)_3(OH,F)$. Because F^- is smaller than OH^-, Coulomb's law dictates that the force of attraction between the Ca^{2+} and the F^- should be greater than that between Ca^{2+} and OH^- in this mineral. This is the case with, for example, nearly all binary ionic salts of fluoride compared to the metal hydroxide, so that the lattice enthalpy of, for example, silver fluoride (AgF), which is 953 kJ/mol, is greater than the lattice enthalpy of silver hydroxide (AgOH), 918 kJ/mol. It is therefore reasonable to consider that the lattice enthalpy of fluorapatite is greater than that of hydroxyapatite. This is one of several reasons why fluorapatite is more stable than hydroxyapatite when bathed in our saliva—and more resistant to the formation of cavities. Fluorapatite formation is even more compelling when fluoride is present in our mouths from municipal water fluoridation or dental treatments; this is related to a concept called chemical equilibrium, which we will consider in Chapters 16–18.

EXERCISE 8.4   **Predicting Lattice Enthalpies**

Use the relationship of ionic sizes to predict whether calcium fluoride (found in toothpaste) or calcium chloride (sidewalk salt) has the greater lattice enthalpy. Also predict whether aluminum chloride or sodium chloride has the greater lattice enthalpy.

First Thoughts

Comparing the lattice enthalpies of two ionic compounds can be accomplished by realizing that the magnitude of the lattice enthalpy is inversely proportional to the distance between the individual ions and directly proportional to the size of the nuclear charge on the ions. The distance between the ions is directly related to the individual ionic radii.

Solution

Calcium fluoride (CaF_2) and calcium chloride ($CaCl_2$) differ in the size of the anion bound to the calcium cation. From the discussion of atomic size, we noted that fluorine is smaller than chlorine. Moreover, the radius for both anions also follows this trend; fluoride is a smaller anion than chloride. Coulomb's law says that CaF_2

has a larger, more positive lattice enthalpy. And indeed it does (2609 kJ/mol versus 2223 kJ/mol).

Aluminum chloride ($AlCl_3$) and sodium chloride (NaCl) differ in the charge of the cation. The larger charge on the aluminum cation (+3) indicates that aluminum chloride should have the larger lattice enthalpy. It does (5376 kJ/mol versus 787 kJ/mol).

Further Insights

Which has a greater influence on the lattice enthalpy, ionic radius or ionic charge? The examples we've examined suggest that the ionic charge has a much greater effect. We can reason, and rightly so, that the electrostatic force of attraction between a cation and an anion is a very powerful force. This powerful force assists some proteins as they fold into a biologically active molecule.

PRACTICE 8.4

Which has the greater lattice enthalpy, $FeCl_3$ or $FeCl_2$?

See Problems 25, 26, 33, 34, 100, 101, and 103.

8.3 Covalent Bonding

In addition to NaCl (table salt), sucrose ($C_{12}H_{22}O_{11}$, common sugar) is another compound we see at the dinner table (Figure 8.9). What is the difference between table salt and sugar? Placed side by side, they appear relatively similar to the naked eye. Both are colorless crystalline compounds. When added to water, they both dissolve. However, as we saw in Chapter 4, solutions of these two compounds act differently. Table salt, a strong electrolyte, dissociates into sodium cations and chloride anions when dissolved in water. Sucrose, a nonelectrolyte, doesn't dissociate when it dissolves. This is why electric current passed through the salt solution and lit the bulb, while the bulb over the sugar solution remained unlit. Sucrose is an example of a compound containing only covalent bonds. The atoms that make up sugar are firmly held together; bonds between these atoms don't dissociate upon addition to water. Because of this, an aqueous solution of sucrose doesn't conduct electricity. These bonds differ from the bonds in sodium chloride, an ionic compound that dissociates in water, because the electrons in sucrose are shared.

Knowing the type of the bonds that make up a molecule is fundamental to those who do chemistry for a living. For example, pharmacognocists understand that whereas many ionic compounds are readily soluble in water and can be

FIGURE 8.9

Common table sugar. Sugar can be crystallized in large chunks that look much like the crystals of table salt.

Cough syrups often contain ethanol (listed on the ingredient label as alcohol) to increase the solubility of the active compounds in the medicine that would otherwise not be sufficiently soluble in water.

FIGURE 8.10

Examples of compounds that contain covalent bonds include CO_2, H_2O, H_2, and CH_4.

administered to patients in aqueous solutions, many covalent compounds have limited solubility in water. Drugs that are insoluble in water must be administered in other solvents, which explains why cough syrups often contain ethanol. In this section, we will examine the covalent bond up close, draw pictures of molecules that utilize this bonding scheme, and calculate the forces that hold these bonds in place.

Description of Covalent Bonding

Opposite to the ionic end of the bonding spectrum is the covalent bond. In the ionic bond, valence electrons gather on one of the atoms in the bond, leaving the other atom with a deficiency of electrons. In the covalent bond, the atoms share valence electrons to differing degrees. This occurs because the electronegativity of the atoms on either end of the covalent bond are similar in magnitude, so the atoms do not form ions. However, in order to participate in a bond and at the same time obtain a noble gas electron configuration, the atoms in the covalent bond must share their electrons. Examples of compounds that contain covalent bonds include H_2, CH_4, CO_2, and H_2O (Figure 8.10).

Because of the similarities in electronegativity, the majority of covalent bonds exist between nonmetals. There are exceptions, however, such as the covalent bonds between lead and each of the four of the carbon atoms in tetraethyl lead, $Pb(C_2H_5)_4$, a compound that used to be added to gasoline to improve engine performance. Why is this an exception? The four lead–carbon covalent bonds in the compound exist as a consequence of the relatively small difference in the electronegativity between the lead and carbon atoms. Because the emissions-reducing catalytic converters required on all cars and light trucks in the United States since 1981 are destroyed by lead, and because the combustion of the lead-containing gasoline releases the metal into the environment, this type of automobile fuel was banned in the United States in 1986. All of the automotive gasoline sold today in the United States is unleaded gasoline. And as a result, the amount of lead pollution of the environment has decreased. Table 8.4 lists the 47 countries that have completely eliminated the use of leaded gasoline in motor vehicles. A World Bank Regional Conference on "The Phase-out of Leaded Gasoline in Sub-Saharan Africa," held in Senegal in 2001, set as a goal the elimination of leaded gasoline from that entire continent by 2005. Recent United Nations and World Bank data show that this goal has largely been achieved.

Application

CHEMICAL ENCOUNTERS: Tetraethyl Lead in Gasoline

TABLE 8.4	Countries That Have Eliminated Tetraethyl Lead Use in Motor Vehicles	
Albania	Ecuador	Netherlands
Antigua	Egypt	New Zealand
Argentina	El Salvador	Nicaragua
Austria	Finland	Norway
Bahrain	Germany	Philippines
Bangladesh	Guatemala	Saudi Arabia
Belgium	Haiti	Singapore
Belize	Honduras	Slovakia
Bolivia	Hungary	South Korea
Brazil	Iceland	Sweden
Canada	India	Switzerland
Colombia	Ireland	Taiwan
Costa Rica	Jamaica	Thailand
Denmark	Japan	United Kingdom
Dominican Republic	Luxembourg	United States
	Mexico	Vietnam

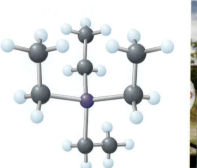

Tetraethyl lead, $Pb(C_2H_5)_4$, delivered from gas pumps similar to today's, has been banned as an additive to vehicular fuel in many countries because of the harmful effects of lead on people and the environment.

The simplest example of a compound containing a covalent bond is hydrogen gas (H_2), used in the food industry to make partially saturated oils, which contain mostly carbon–carbon and carbon–hydrogen single bonds, as well as in the manufacture of ammonia intended for agricultural use as a fertilizer. The bond between the hydrogen atoms in H_2 results from the sharing of one electron from each of the atoms. The atoms on either end of the bond have the same affinity for electrons, so they must be shared if each hydrogen is to have the electron configuration of a noble gas. In other words, sharing the electrons between the two atoms effectively gives each hydrogen a $1s^2$ electron configuration. The sharing of bonding electrons is the defining characteristic of a covalent bond. What holds the two atoms together in a covalent bond? Each nucleus on either end of the bond exerts a force of attraction on the pair of bonding electrons. This attractive force pulls the nuclei close to form a bond. If we examine the electron cloud around the two nuclei in molecular hydrogen, we note an interesting feature of the covalent bond, which is shown in Figure 8.11. The density of electrons is concentrated between the two atoms, but a significant amount of electron density surrounds *each* of the two nuclei. As shown in Figure 8.12, the picture of electron density in HF, a compound used to etch glass, is remarkably different because the

FIGURE 8.11

Electrons hold the nuclei together in this model of hydrogen gas. Note that the electron density, shown in red, encircles both nuclei.

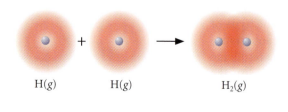

FIGURE 8.12

Compare the electron density distribution in HF with that of H_2.

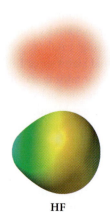

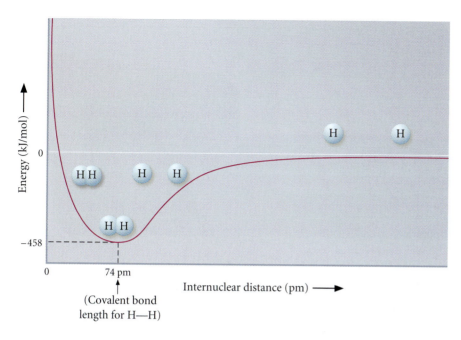

FIGURE 8.13

The energy profile of a covalent bond in H_2 as a function of distance. Note how the attractive forces and repulsive forces balance each other at a distance of 74 pm. At this distance the interaction has the lowest energy. To break the covalent bond in H_2 and separate the atoms would require the addition of 458 kJ/mol.

Visualization: Bonding in H_2

bonding electrons are not shared quite equally, although they are shared to an important extent, and the bond is still considered to be covalent.

What keeps the nuclei from bumping into each other? While the attractive force of the nuclei for the shared electrons pulls the atoms together, a repulsive force of the like-charged nuclei pushes them apart. This repulsion keeps the nuclei from getting too close. It is the combination of attractive and repulsive forces that holds the nuclei apart at a particular average distance that we refer to as the **bond length**. A plot of the potential energy of two hydrogen atoms as a function of their distance, shown in Figure 8.13, helps to illustrate this fact.

Electronegativity and the Covalent Bond

Why are electrons shared unequally in some covalent bonds such as H—F? Some atoms have a greater attraction for shared electrons than do others. The electron density in the covalent bond is pulled closer to the atom with the greater attraction for the bonding electrons. We say that some covalent bonds are **polarized** toward one of the atoms. In hydrogen fluoride, the attraction of the fluorine atom for electrons causes the bonding electrons to spend more time at the fluorine end of the molecule, resulting in a net **polarization** of the bond. The lack of electron density at the hydrogen end of the bond means that some of the total nuclear charge isn't balanced on the atoms, so each of the atoms in the polar covalent bond possesses a partial charge. We usually indicate the charge separation with a lowercase Greek letter delta (δ) and a plus or minus sign. We'll discuss how to arrive at the bonding model for HF later, but the charge separation can be represented like this:

$$\overset{\delta+}{H}\!-\!\overset{\delta-}{\ddot{\underset{\cdot\cdot}{F}}}\!:$$

Methods have been developed to attempt to identify the type of bond in a molecule on the basis of the electronegativity of the bonded atoms. **Electronegativity** was defined in Chapter 7 as *the ability of an atom in a molecule to attract shared electrons to itself* (Figure 8.14). The atom with the stronger attraction for the electrons pulls the bonding electrons closer to itself. By examining the difference in the electronegativity values of the two atoms involved in a bond, we can

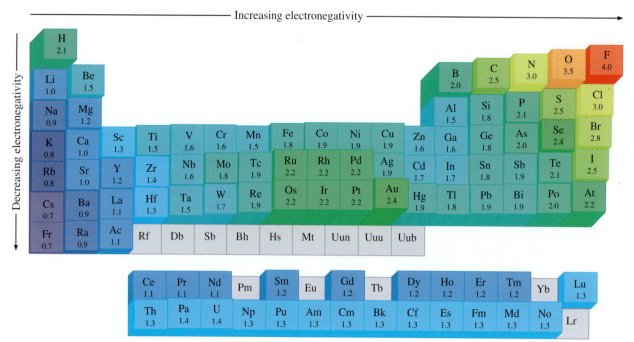

← Increasing electronegativity →

FIGURE 8.14

Pauling's electronegativity values for the elements of the periodic table.

assess the polarization of the bonding electrons (also known as the bonding electron pair). For example, fluorine is much more electronegative than hydrogen, so the electrons spend more time on the fluorine end of the bond in HF.

The Pauling electronegativity scale is just one of the many attempts scientists have made to develop a general trend in the polarization of electrons in bonds. Pauling's electronegativity scale, the most popular among several such scales, is based on bond energies of diatomic molecules. Other scientists have developed useful electronegativity trends based on ionization energies and electron affinities (the Mulliken scale), atomic energies and covalent radii (the Allred–Rochow scale), the "compactness of an atom's electron cloud" (the Sanderson scale), dielectric properties (the Phillips scale), and quantum defects (the St. John–Bloch scale). All of these are based on the experimentally determined properties of some standard compounds, and this makes the scales subject to some variation. Even though all of these scales have different values for the electronegativity of the elements, *the general trends are identical among the different systems*. Pauling's electronegativity scale, however, is the most important historically, it has proved quite simple to use, and the conclusions drawn from it are reasonable. We will be on safe ground using the Pauling scale.

Types of Covalent Bonding

How can we use electronegativity to determine the degree of polar character in a bond? In molecular hydrogen (H_2), the difference in the Pauling values for electronegativity (Δ) between the two atoms is zero ($\Delta = 2.1 - 2.1 = 0$). In a molecule of HCl, the electronegativity difference is less than 1 electronegativity unit ($\Delta = 0.9$). Between hydrogen and fluorine in HF, the difference is fairly large ($\Delta = 1.9$). And in table salt (NaCl), the difference in electronegativity is 2.1. Molecular hydrogen is an example of a molecule with a covalent bond that is not polarized. We say that it contains a **nonpolar covalent bond** ($\Delta < 0.5$). Hydrogen chloride (HCl) and hydrogen fluoride (HF), on the other hand, contain fairly polarized covalent bonds. We say that these are **polar covalent bonds** ($0.5 < \Delta < 2.0$). If the difference in electronegativity is very large (>2.0), the

Compare the electron density distributions of H_2, HF, and HCl with their electronegativity values.

more electronegative atom's ability to attract electrons in the bond overcomes the tendency to share electrons, and we call this an ionic bond. The electronegativity difference between sodium and chlorine in table salt ($\Delta = 2.1$) is characteristic of an ionic bond. The difference in electronegativity can be useful in determining the type of bond in a compound, but the values shown here are useful guidelines only for the determination of the bonding pattern. We can say this in another way: There is no fixed cutoff. For example, it is not chemically reasonable to say that two bonds with electronegativity differences that are within, say, 0.1 of each other are significantly different in bonding character. *Electronegativity differences describe a continuum*, not a discrete "on-or-off," as with a light switch.

Let's revisit our example of a covalent bond between a metal and a nonmetal in tetraethyl lead. Does the electronegativity difference between the lead and carbon atoms indicate that the bond should be covalent? Examination of the table of electronegativity values indicates that the difference is only 0.6 electronegativity unit. On the basis of this information, we can say that the Pb—C bonds in tetraethyl lead are polar covalent bonds and that these bonds are not as polarized as the bond in HF or HCl. We'll discuss the implications of this fact in the next section.

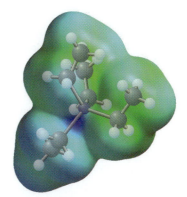

Here is the electron density distribution in tetraethyl lead. How does it compare to those of other covalent molecules?

EXERCISE 8.5   What Type of Bond Is It?

For each of these compounds, indicate whether the bond is an ionic bond, a polar covalent bond, or a nonpolar covalent bond.

a. KCl (a replacement for table salt in low-sodium diets)

b. H_2O (water)

c. Br_2 (used in the manufacture of brominated vegetable oils)

First Thoughts

To answer this question, we need to consider the electronegativity difference between the bonded atoms or ions.

Solution

a. We'd predict an ionic bond between potassium and chloride. The two atoms differ by 2.2 electronegativity units.

b. A polar covalent bonding pattern is predicted (the difference is 1.4 electronegativity units). Hydrogen is less electronegative than oxygen, so the majority of the electron density in the bonds lies closer to the oxygen.

c. A nonpolar covalent bond is indicated here (no difference in electronegativity).

Further Insights

Differences in electronegativity can be used to identify the type of bond in a compound. However, is it correct to consider that NaCl ($\Delta = 2.1$) contains a purely ionic bond and that NaBr ($\Delta = 1.9$) contains a polar covalent bond? What do we say about the bond in LiCl ($\Delta = 2.0$)? Questions such as these help illustrate that bonding is a spectrum, from the "purely" ionic to the "purely" nonpolar covalent. Almost all bonds have some ionic and some covalent character. We'd surmise that the bond in NaCl is more ionic than the bond in either LiCl or NaBr, but we'd realize that it, too, has some covalent character.

PRACTICE 8.5

Predict the type of bond (covalent, polar covalent, or ionic) present in each of these binary compounds: NO, F_2, MgO.

See Problems 51 and 52.

Application

Video Lesson: Lewis Dot Structures for Covalent Bonds

Video Lesson: Predicting Lewis Dot Structures

Modeling a Covalent Bond—Lewis Structures

Felix Hoffman, a chemist hired by the Friedrich Bayer & Co. fabric dye plant, was charged in the 1890s with finding some better products that could be made by the company. One of the ideas that came to Hoffman was related to his father's rheumatism. At the time, this ailment and other pains were treated with large doses of salicylic acid, found in the inner bark of several types of willow trees. Unfortunately, Hoffman's father was unable to take this pain medication because its acidity irritated his stomach and throat. In 1897, Hoffman set about trying to reduce the acidity of the medicine. To do this, he needed to know how the arrangement of the atoms made the compound acidic. After some experimentation, Hoffman was able to produce a compound that reduced this acidity. His product, acetylsalicylic acid ($C_9H_8O_4$) later named aspirin, reduced the harmful irritation by chemically modifying one of the acidic parts of salicylic acid to make it less acidic. Knowing how the atoms are attached, and what kind of bonds link the atoms, is important to understanding how compounds react and interact in the body. One of the ways to illustrate how atoms are attached in covalent bonds is to build a model of the molecule using Lewis dot structures. After we discuss the basic ideas, we will draw the Lewis dot structure of aspirin.

Rules describing how the atoms and electrons are placed in a Lewis dot structure enable us to draw compounds in a systematic way. These rules, found in Table 8.5, include drawing a skeletal picture of the molecule and then placing extra valence electrons in the skeleton until the model of the compound is complete. We will use these rules as we draw a Lewis dot structure of molecular hydrogen (H_2).

According to Table 8.5, the first step is to count all of the valence electrons on the atoms in the molecule. Because valence electrons are involved in bonding, knowing the total number of these electrons helps us determine the number of **bonding pairs** (electrons involved in bonding) and the number of **lone pairs** (electrons not involved in bonding). In hydrogen, there are two total valence electrons (one from each atom). In step 2 (Table 8.5), we draw a skeleton for the molecule. The guidelines (steps 2a–e) for drawing a skeletal picture of a molecule are based on preferences that atoms have for particular locations in a molecule. There are only two atoms in the molecule, so the skeleton of H_2 is

H   H

TABLE 8.5   A Set of Rules for Drawing Lewis Dot Structures

1. Determine the total number of valence electrons.

2. Determine the skeletal structure.
   a. Hydrogen atoms are on the edges of the molecule.
   b. The central atom has the lowest electronegativity. (There are many exceptions.)
   c. In oxoacids, hydrogens are usually on the oxygens.
   d. Think compact and symmetric.
   e. Use intuition.

3. Draw the Lewis dot structure.
   a. Draw the bonds connecting the atoms.
   b. Determine the number of electrons remaining.
   c. Place the remaining electrons as lone pairs, beginning on the most electronegative atoms, until each atom has an octet.
   d. All remaining lone pairs go on the central atom.
   e. Assign formal charges, and redraw bonding electrons if necessary.

In step 3a, we place pairs of electrons to indicate bonding pairs between each atom. For molecular hydrogen, the electrons are placed as follows;

$$H \cdot\cdot H$$

Most often, we use shorthand to show bonding pairs of electrons by replacing the bonding electrons with a line. We draw the line to show that the atoms are chemically bonded together in the molecule.

$$H—H$$

In steps 3b and 3c, we place all remaining electrons around the more electronegative atom first until the octet rule (or duet rule, for hydrogen) is satisfied. Because we do not have any remaining electrons, and because the duet rule is satisfied for both atoms, we are finished.

Hydrogen fluoride is either a fuming gas or a liquid, depending on the temperature of use (its boiling point is 19.5°C, about 67°F), and has a host of industrial applications. For example, it is used as a raw material in the production of chlorofluorohydrocarbons (CFCs), insecticides, and fertilizers; as a catalyst in the production of pharmaceuticals; in the manufacture of semiconductors; and in etching glass (see Figure 8.15). Let's use the rules in Table 8.5 to prepare a Lewis dot structure for HF. Step 1 requires us to count the valence electrons on all the atoms. Hydrogen has one valence electron, and fluorine has seven, so we have a total of eight valence electrons with which to work. Next, in step 2, we draw the best skeletal structure of the molecule:

$$H \qquad F$$

There are only two atoms in the molecule, so we'll draw them next to each other. Our bonding pair of electrons is represented as a line connecting the two atoms (step 3a).

$$H—F$$

We've used two electrons in our skeleton to represent the bond between the two atoms, so we have six electrons remaining (step 3b). These are placed in pairs around the more electronegative atom until the octet rule (or duet rule) is satisfied (step 3c). If any electrons remain (step 3d), the pairs are placed around the other atoms.

$$H—\overset{\cdot\cdot}{\underset{\cdot\cdot}{F}}\!:$$

Because we used all of the electrons to fulfill the octet rule around fluorine, and because both atoms now satisfy the octet rule (or duet rule), we have completed the structure. Both fluorine and hydrogen in HF have electron configurations of a noble gas.

Formal Charges

Methanol (also known as methyl alcohol and wood alcohol, CH_4O) is a compound being advanced as a potential substitute for gasoline, because it has a high-octane-rating equivalent. Methanol is also a renewable resource obtained from the fermentation of cellulose-containing materials (such as wood). The following three structures satisfy the octet rule (or duet rule) for every atom. Only one of them is methanol.

 Application

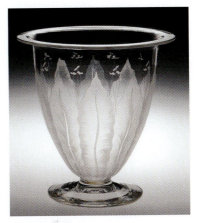

FIGURE 8.15

This punch bowl (American, 1918–1919) was made from blown glass. The pattern was applied by etching the glass using a dilute hydrogen fluoride solution.

Video Lesson: Formal Charge

Which structure is it? Chemists use formal charges on atoms as one important piece of evidence in determining the most reasonable structure for a compound in which more than one structure might be possible.

What is a formal charge (see step 3e in Table 8.5) and how does it help us to select the most reasonable structure? The **formal charge** on an atom is the difference between the number of valence electrons on the free atom and the number of electrons assigned to the atom when it is part of a molecule. The use of formal charges on atoms indicates that the atoms exhibit an imbalance between the number of electrons and the number of protons on the atom. Although this imbalance can be estimated by assigning an oxidation number to each of the atoms (see Chapter 4), the method of calculating formal charges lends us additional guidance. Oxidation numbers provide an indication of the charge an atom would have if it were completely ionic. Formal charges, on the other hand, provide the charge on an atom assuming there is no difference in electronegativity among the atoms in a structure. In other words, formal charges assume that electrons are shared equally, as is the case in the covalent bonding pattern.

Formal charge = valence electrons − # bonds − # nonbonded electrons

Mathematically, the formal charge equals the number of valence electrons on the free atom minus the number of bonds to that atom minus the number of nonbonded electrons. As a check of our math, the sum of the formal charges on all the atoms should equal the total charge of the molecule (0) or ion (+ or −). We can use this information to calculate the formal charge on each atom in a simple molecule, our Lewis dot structure of HF. The fluorine atom has seven valence electrons (in Group VIIA, noted from the periodic table). Subtracting the number of bonds in the structure (one) and the number of nonbonded electrons (six) from this number gives a formal charge of zero for the atom. Similar calculations can be done with the hydrogen atom, which also has a formal charge of zero.

When we draw Lewis dot structures, the best structure is one that satisfies the largest number of formal charge rules. The best structure:

■ has the smallest magnitude for all of the formal charges

■ places negative formal charges on the more electronegative atoms

■ has the smallest number of nonzero formal charges

We can now consider the second part of our section-opening question: **Why is it useful to know the formal charges on atoms within a particular Lewis dot structure?** Let's return to our discussion of methanol. Calculating the formal charges on the central atoms in each of the possible structures enables us to choose the one on the left as the correct structure. Every atom in both structures has an octet of electrons, but only the structure on the left shows a formal charge of zero on each atom. This is one piece of evidence that the structure on the left is likely to be the most energetically stable structure of the three.

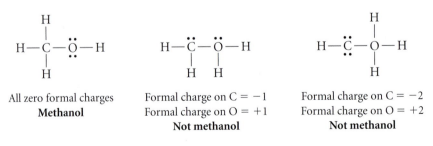

All zero formal charges
Methanol

Formal charge on C = −1
Formal charge on O = +1
Not methanol

Formal charge on C = −2
Formal charge on O = +2
Not methanol

EXERCISE 8.6 **Return to Basics**

Pioneers in the American West made most of their everyday items from natural sources. On the treeless plains of the Midwest they built homes of sod, burned dried buffalo dung in the stove for heat, and made lye soap. Lye (a mixture of sodium hydroxide, NaOH, and potassium hydroxide, KOH) obtained from fireplace ashes was used to make soap from animal fat. What is the Lewis dot structure model for the hydroxide ion (OH^-)? On which atom in lye does the nonzero formal charge reside?

Solution

The hydroxide ion can best be modeled by drawing the skeleton of the molecule with a single bond between the oxygen and the hydrogen. Placing the remaining electrons in pairs around the most electronegative element gives the structure shown below. The brackets indicate that the negative charge has not been assigned to a particular atom.

$$\left[:\ddot{O}-H \right]^{\ominus}$$

Formal charge (F.C.) calculations indicate that the charge must reside on the oxygen (F.C. $= 6 - 1 - 6 = -1$) and not on the hydrogen (F.C. $= 1 - 1 - 0 = 0$), which makes sense because the oxygen atom has a much higher electronegativity (3.5) than the hydrogen atom (2.1). Our structure can be drawn as follows:

$$^{\ominus}:\ddot{O}-H$$

PRACTICE 8.6

What is the formal charge on each of the atoms in OCl^-? in H_3C-NH_2?

See Problems 55 and 56.

Multiple Bonds

We often discuss carbon dioxide in this text because it is important to life in photosynthesis and cell respiration, and it is one product of combustion. The Pauling electronegativity values of the atoms indicate that CO_2 probably contains polar covalent bonds. Following our guidelines for constructing a Lewis dot structure of CO_2, we count the valence electrons in all atoms (4 in carbon and 6 in each of the two oxygen atoms = 16 valence electrons) and build the skeleton. The less electronegative carbon atom is the central atom in this molecule, and the oxygen atoms are symmetrically placed about the carbon. Next, the bonding pairs of electrons are placed between the atoms. The remaining 12 electrons are then placed on the more electronegative oxygen atoms to fill their octets. The result is shown below. Is this a satisfactory Lewis dot structure?

$$:\ddot{O}-C-\ddot{O}:$$

No. Both oxygen atoms have octets of electrons, but the carbon has only four electrons (all bonding electrons) around it. What are the formal charges on each atom in the structure?

$$:\ddot{O}-C-\ddot{O}:$$

Formal charges -1 $+2$ -1

Formal charge (O) $= 6 - 1 - 6 = -1$
Formal charge (C) $= 4 - 2 - 0 = +2$

Our calculations reveal a lot of nonzero formal charges. Although the formal charges add up to zero, the size and number of formal charges indicate that something may be amiss. Because energy is required to cause a separation of charges (as indicated by the presence of nonzero formal charges in a molecule), the better structure is generally that which minimizes formal charges. You also might have noted that the octet rule is disobeyed for carbon in the structure. To eliminate the charge on the oxygen atoms, let's move a lone pair from each oxygen and place them as a bond between the oxygen and the adjacent carbon. These are called double bonds.

$$\ddot{\text{O}} = \text{C} = \ddot{\text{O}}$$

Formal charges   0   0   0

The resulting model has no nonzero formal charges and allows each atom to satisfy the octet rule. Overall, the molecule is neutral. How is this model different from the structures we discussed previously? Does the existence of more than one pair of electrons between two atoms indicate something about the properties of this molecule? We'll discuss the answers to these questions in the next section. *The best sign that we've constructed a good model is its agreement with observed properties for the compound.* In the case of CO_2, the Lewis dot structure model does agree with the experimentally determined shape and polarity of carbon dioxide. Based on differences in electronegativity, we observe that the electrons are polarized toward each end of the molecule.

Multiple covalent bonds occur in many molecules. Carbon dioxide has two **double bonds**. Each oxygen atom shares four bonding electrons with a carbon atom. In a molecule of nitrogen, three pairs of electrons are used to satisfy the octet rule for each of the atoms. The resulting molecule contains two nitrogen atoms that share six electrons. A **triple bond** links the atoms. Nitrogen, which makes up nearly 80% of the air we breathe, is a very stable molecule that contains a particularly strong triple bond, with the Lewis structure shown in Figure 8.16.

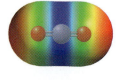

FIGURE 8.16

(Top) The double bonds in carbon dioxide (CO_2). (Bottom) The triple bond in nitrogen (N_2).

EXERCISE 8.7   **Chemical Warfare and Bonding**

Phosgene ($COCl_2$), a highly toxic gas, was used in World War I and several subsequent wars to kill soldiers who were hiding in places that bombs couldn't penetrate. The gas, now used in industry to make polymers, pharmaceuticals, herbicides, and other useful compounds, reacts with water and other electron-rich molecules (molecules with lone pairs of electrons). Draw the best Lewis dot structure for this molecule.

First Thoughts

All of the bonds in phosgene are covalent ($COCl_2$ contains only nonmetals), but a Lewis dot structure model of the compound may help explain why phosgene reacts with electron-rich molecules.

Solution

The skeletal picture of the molecule places the carbon in the center (least electronegative). The other atoms are placed symmetrically about the carbon. Of the 24 total electrons, 6 (three pairs) are used to connect the atoms. The remaining 18 electrons (nine pairs) are placed as lone pairs around the more electronegative atoms. The result is that all of the atoms, except the carbon, have a full octet. We need to share a lone pair of electrons with carbon to satisfy the octet rule on each atom. Calculation of the formal charges on each of the atoms indicates that the oxygen and the carbon atoms should share another pair of electrons.

$$\begin{array}{c}
:\overset{\displaystyle ..}{O}: \\
| \\
:\overset{..}{\underset{..}{Cl}}-C-\overset{..}{\underset{..}{Cl}}:
\end{array}$$

F.C. (O) $= 6 - 1 - 6 = -1$
F.C. (C) $= 4 - 3 - 0 = +1$
F.C. (Cl) $= 7 - 1 - 6 = 0$

The best model of this molecule therefore shows a double bond between the carbon and the oxygen atoms and single bonds between the carbon and the chlorine atoms.

$$\begin{array}{c}
:O: \\
\| \\
:\overset{..}{\underset{..}{Cl}}-C-\overset{..}{\underset{..}{Cl}}:
\end{array}$$

F.C. (O) $= 6 - 2 - 4 = 0$
F.C. (C) $= 4 - 4 - 0 = 0$
F.C. (Cl) $= 7 - 1 - 6 = 0$

We could have drawn a Lewis dot structure that placed a double bond between the carbon and one of the chlorine atoms. This would have changed the formal charge on carbon to zero, but the formal charge on the chlorine atom would become $+1$, and the formal charge on oxygen would remain -1. The resulting structure would not be the best Lewis structure because of the existence of nonzero formal charges.

Further Insights

Does this explain why phosgene is so reactive? By noting how the bonds in the molecule are polarized, we can identify why this molecule reacts with electron-rich compounds. According to the electronegativity values of the atoms, each of the bonds is polarized away from the carbon atom. Electron-rich molecules such as water (H_2O), ammonia (NH_3), and other compounds containing lone pairs can react with the electron-starved carbon atoms in phosgene.

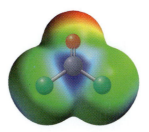

PRACTICE 8.7

Draw the Lewis dot structure for C_2H_4. for CH_3N.

See Problems 39–48, 61–66, and 99.

Resonance Structures

 Video Lesson: Resonance Structures

Carbonate ion ($CO_3{}^{2-}$) is a common polyatomic ion found in limestone, baking powder, and baking soda. Addition of acid to the carbonate ion causes the formation of carbonic acid (H_2CO_3), which decomposes rapidly into water (H_2O) and carbon dioxide (CO_2). In baking, the carbon dioxide that is released causes the bread to rise and makes its texture lighter.

Our first attempt at drawing the Lewis dot structure of the carbonate ion results in the structure shown below. Carbonate has 24 electrons, 2 of them responsible for the -2 charge, probably electrons from calcium ($CaCO_3$), sodium (Na_2CO_3), or whatever salt resulted in a cation that donated electrons to the carbonate anion. The carbon atom in our structure still needs to share electrons to satisfy the octet rule. Which atom is most likely involved in sharing electrons?

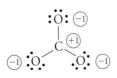

Using the formal charges on the atoms, we could reconfigure our electrons to participate in a double bond with the carbon. At this point, the positive charge on the carbon atom is gone, and all of the valences are filled (the octet rule is satisfied). The sum of the formal charges is equivalent to the charge on the carbonate ion. This is a good Lewis dot structure for carbonate.

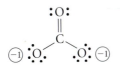

We could have shown the lone pair on one of the other atoms involved with satisfying the octet rule for carbon. In fact, there are two other structures that also seem to be satisfactory Lewis dot structures. Turn your textbook from "high noon" to 4 o'clock (or "120 degrees") to verify this fact. **Which one is correct?**

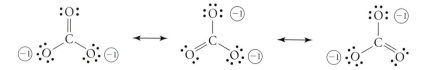

Each one of the three structures is an equally good representation of the carbonate ion. The only difference among these models is that the positions of the electrons have changed. The relative positions of the atoms did not change.

A **resonance structure** is a model of a molecule in which the positions of the electrons have changed, but the positions of the atoms have remained fixed. All of the resonance structures for a molecule are correct ways to draw the Lewis dot structure. We show the relationship among these resonance structures by drawing a double-headed arrow between them. *However, we shouldn't consider a single resonance structure to be a discrete entity.* Because the only difference is the location of the electrons, the resonance structures must be considered together as the model of the molecule. *A combination of all of the resonance structures is the best model.* The resulting model is called the **resonance hybrid**. The resonance hybrid is an equal or unequal (based on experimental evidence) combination of all of the resonance structures for a molecule. For the carbonate ion, the resonance hybrid looks like this:

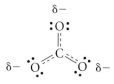

Instead of full charges on two of the oxygen atoms, a partial charge (about −0.67) exists on each of the oxygen atoms that adds up to the overall −2 charge for the ion. Partial double bonds (1.33 times as much of a bond as a single bond) are drawn to show how the three resonance structures combined to make one resonance hybrid.

We now have a sufficiently complete model to draw the Lewis dot structure for aspirin, or acetylsalicylic acid ($C_9H_8O_4$).

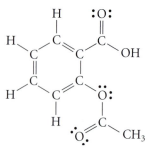

We note that all the octets are filled and that the formal charges of all of the atoms in the structure sum to 0.

HERE'S WHAT WE KNOW SO FAR

■ The energy of an ionic bond is related to the charges on the ions and the distance separating the ions.

- Lewis dot structures can be used to draw a model of a compound. The octet rule drives our prediction of the best model for a compound. The resulting model represents the locations of electrons and atoms in a compound.
- Electrons are shared in covalent bonds.
- The atoms in a covalent bond are held at a distance that reflects the attraction for the bonding electrons and the repulsion of the adjacent nuclei.
- The Pauling electronegativity scale is the most widely used among several that indicate the polarization of bonding electrons.
- We can classify covalent bonds as polar and nonpolar, depending on the electronegativity difference between the atoms in the bond.
- Formal charges help us to select the most likely structure among viable alternatives.
- Resonance structures exist in a molecule when electrons can shift positions among different atoms while the positions of the atoms remain fixed.

Exceptions to the Octet Rule

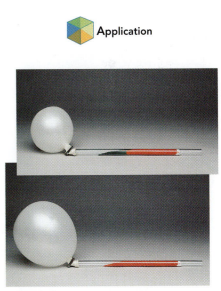

Application

In 1954, Robert Borkenstein, a captain in the Indiana State Police, developed a Breathalyzer like that shown in Figure 8.17. A device similar to this was first used by police to determine alcohol levels in drunk drivers. The Breathalyzer contains potassium dichromate and sulfuric acid that converts exhaled ethanol (CH_3CH_2OH) into acetic acid (CH_3COOH). The conversion causes the chemical reagent in the Breathalyzer to change color from yellow-orange to blue-green. The rest of the compounds in the reaction are colorless. After the test is performed, the presence of the blue-green color indicates that the suspect has been drinking. What is the Lewis dot structure for sulfuric acid?

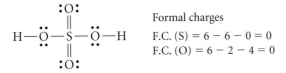

$$2K_2Cr_2O_7 + 8H_2SO_4 + 3CH_3CH_2OH \rightarrow 2Cr_2(SO_4)_3 + 2K_2SO_4 + 3CH_3COOH + 11H_2O$$

Yellow-orange   Blue-green

Drawing the structure, we see that the sulfur atom has the smallest formal charge when there are six bonds from the sulfur to the adjacent atoms. Can this be correct? There is still debate about the existence of two S=O bonds in sulfuric acid (rather than single bonds to all atoms, which gives decidedly nonzero formal charges—can you prove this?) but the structure does seem to fit our understanding of Lewis dot structures and formal charges. Some exceptions to the octet rule must be considered if we are to accept this Lewis dot structure model.

FIGURE 8.17

The Breathalyzer. This device can be used to estimate how much alcohol a person has consumed. Shown here is a demonstration of the process that occurs in the instrument.

Formal charges

F.C. (S) = 6 − 6 − 0 = 0
F.C. (O) = 6 − 2 − 4 = 0

In 1962, Neil Bartlett, then at the University of British Columbia, reported the first inert gas compound ($XePtF_6$). This was quickly followed by reports of other inert gas compounds. For example, researchers at the Argonne Laboratories prepared xenon tetrafluoride (XeF_4). In the Lewis dot structure of XeF_4, the xenon atom has twelve total electrons around it. What are the formal charges on the atoms in this model?

How can sulfur and xenon have more than eight electrons around them in the Lewis dot structure? Consider the orbitals that are available to valence electrons. In the valence shell of oxygen, for example, there are $2s$ and $2p$ orbitals. In the valence shell for sulfur, however, there are $3s$, $3p$, and $3d$ orbitals. The $3s$ and $3p$ orbitals contain electrons. A higher-energy unfilled set of orbitals (the $3d$ orbitals) can be used if needed to hold extra electrons. This means that the sulfur atom in sulfuric acid (H_2SO_4) can have eight electrons in the $3s$ and $3p$ orbitals and place the extra four electrons in the previously empty $3d$ orbitals. Xenon, in XeF_4, can also accommodate the extra electrons by placing them in the empty $5d$ orbitals. This is a general rule for atoms of the third row and higher of the periodic table. *Atoms can have **expanded octets**.* Often this occurs when the valence electron shell includes unfilled d orbitals.

www

Video Lesson: Electronegativity, Formal Charge, and Resonance

Boron trifluoride (BF_3) is an example of a compound in which the central atom has fewer than eight electrons around it. What is the formal charge on the boron atom in BF_3?

Even though the formal charges on each atom indicate that the Lewis dot structure is satisfactory, the boron atom does not satisfy the octet rule. We'd predict this molecule, then, to be quite reactive with molecules that are electron-rich. In fact, boron trifluoride reacts quite violently with molecules containing a lone pair of electrons (such as water and ammonia). The bond that forms is an example of a **coordinate covalent bond**. This bond is special because both electrons that make the bond were donated from just one of the two atoms involved. However, after the bond is made, it is indistinguishable from a normal covalent bond.

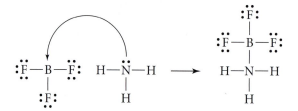

The coordinate covalent bond is fairly common, particularly in reactions involving acids and bases. For instance, when we write the equation describing the dissolution of gaseous HCl in water, we describe the formation of a coordinate covalent bond.

$$HCl(g) + H_2O(l) \rightarrow H^+(aq) + Cl^-(aq) + H_2O(l) \rightarrow H_3O^+(aq) + Cl^-(aq)$$

The hydronium ion (H_3O^+) contains a coordinate covalent bond between the oxygen and one of the hydrogen atoms. As the strong hydrochloric acid dissolves in the water, it ionizes into H^+ and Cl^-. The hydrogen cation combines with a pair of electrons from the oxygen of water, and a covalent bond results. In the end, that bond is indistinguishable from the other O—H bonds in the hydronium ion.

Superoxide (O_2^-) is an extremely reactive anion that can cause severe damage to living tissue. To protect itself, the body has evolved an enzyme, superoxide dismutase, that quickly hunts down and destroys superoxide. What is the Lewis dot structure for superoxide, which is more accurately called the dioxygen (−1) ion? In drawing the structure, we note the presence of an odd number of total valence electrons. Because bonding electrons and lone-pair electrons are pairs of electrons, an odd number of total valence electrons in a Lewis dot structure implies that there is a *lone unpaired electron* (called a **radical**). The existence of a radical electron in a molecule often makes a molecule quite reactive. By placing a radical electron in a Lewis dot structure, we generate an atom that doesn't complete an octet. Note the location of the formal charge in superoxide:

$$\cdot \ddot{O} - \ddot{O} : ^{\ominus}$$

Energy of the Covalent Bond

Angina, a sudden pain in the chest, is a symptom of a heart that doesn't have an adequate flow of oxygenated blood to work properly. Treatments for this type of heart disease require administration of drugs that dilate the coronary artery and increase the flow of blood. Calcium channel blockers are a class of medicines used to treat this disease. These compounds fit neatly into "pockets" within proteins that control the flow of calcium ions into muscle cells. As the rate at which calcium ion passes into the heart muscle is greatly reduced, the heart relaxes and its arteries dilate. The result is a heart that has enough oxygen to function adequately. One problem with calcium channel blockers is that the body breaks specific bonds in these drugs and makes them into biologically inactive compounds (i.e., the drugs are rapidly metabolized). For example, nifedipine, a common calcium channel blocker, persists in the body for only 4–8 hours (Figure 8.18). Patients must continuously take the drug to prevent heart damage due to lack of oxygenated blood.

Medicinal chemists are interested in designing drugs that are similar in structure to nifedipine but resistant to metabolism. To do this, the medicinal chemist must make a compound that fits into the same pocket in the protein but contains bonds that do not break as easily as nifedipine. The shape of the new drug, then, must be very similar to the shape of nifedipine. The shape can be determined by building a model of the compound. **How do scientists know which bonds are susceptible to reactions within the body?** Because reactions break

Application

CHEMICAL ENCOUNTERS:
Calcium Channel
Blockers

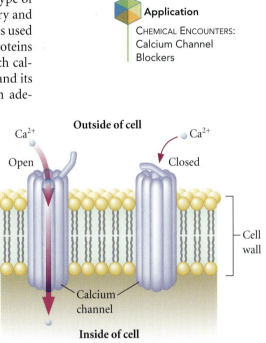

The heart muscle can relax when the flow of calcium ions to the heart is reduced. Calcium channel blockers fit into the pockets of proteins that control calcium intake, reducing the flow of calcium to the heart.

FIGURE 8.18

Nifedipine, a calcium channel blocker prescribed for some patients with heart trouble. Carbon atoms are black, oxygen atoms are red, nitrogen atoms are blue, and hydrogen atoms are white.

bonds, the answer to this question is related to the strength of the different bonds in a compound.

Video Lesson: Bond Properties

How strong are covalent bonds? Is there a relationship between the type of bond and the strength of the bond? Is a multiple bond stronger than a single bond? To answer these questions, we must rely on our ability to estimate the amount of energy that it takes to dissociate a bond. The **enthalpy of bond dissociation** ($\Delta_{diss}H$), or, simply, the **bond dissociation energy**, is the energy required to break 1 mol of bonds in a *gaseous species*. The equation that describes bond dissociation is shown below.

$$X—Y(g) \rightarrow X(g) + Y(g)$$

Bond breakage requires energy, so bond dissociation enthalpies are *always endothermic* and the values for bond energy will always have a positive sign. Table 8.6 lists the values for several important bonds. From the values in the table, we note that single bonds require less energy to break than their corresponding double or triple bonds. Compare the C—C single bond ($\Delta_{diss}H = 347$ kJ/mol), a C=C double bond ($\Delta_{diss}H = 614$ kJ/mol), and a C≡C triple bond ($\Delta_{diss}H = 839$ kJ/mol). Also note from the table that the length of the covalent bond shrinks as we go from a single to a double to a triple bond for a given element or pair of elements. For diatomic molecules, the bond dissociation energy can be measured directly in the laboratory. The process is a little different for atoms that do not form diatomic molecules. Other atoms close to the bond influence how electrons are distributed in a bond and affect the energy it takes to break the bond. The data in Table 8.6, then, are *average* bond dissociation energies.

The overall enthalpy change of a reaction ($\Delta_{rxn}H$) can be estimated using bond dissociation energies. Because the enthalpy change (a state function, independent of path) in a reaction is the sum of all of the energy as heat added to the system minus the sum of all of the energy as heat removed from the system at constant pressure, the enthalpy change of a reaction can be determined by measuring the enthalpies of bond breakage and bond formation. Breaking a bond requires an addition of energy to the system. Forming a bond results in a release of energy from a system. If we add up all of the enthalpies for the bonds that are

TABLE 8.6   **Average Bond Dissociation Energies and Bond Lengths for Some Common Covalent Bonds**

Bond	Energy (kJ/mol)	Length (pm)	Bond	Energy (kJ/mol)	Length (pm)	Bond	Energy (kJ/mol)	Length (pm)
H—H	432	75	N—H	391	101	F—F	154	142
H—F	565	92	N—F	272	136	F—Cl	253	165
H—Cl	427	127	N—Cl	200	175	F—Br	237	176
H—Br	363	141	N—Br	243	189	F—I	273	191
H—I	295	161	N—N	160	145	Cl—Cl	239	199
C—H	413	109	N=N	418	125	Cl—Br	218	214
C—F	485	135	N≡N	941	110	Cl—I	208	232
C—Cl	339	177	O—H	467	96	Br—Br	193	228
C—Br	276	194	O—F	190	142	Br—I	175	247
C—I	240	214	O—Cl	203	172	I—I	149	267
C—C	347	154	O—I	234	206	S—H	363	134
C=C	614	134	O—O	146	148	S—F	327	156
C≡C	839	120	O=O	495	121	S—Cl	253	207
C—N	305	143	C—O	358	143	S—Br	218	216
C=N	615	138	C=O	745*	120	S—S	226	205
C≡N	891	116	C≡O	1072	113			

*C=O bond energy in CO_2 is 799 kJ/mol.

broken and subtract all the enthalpies for the bonds that are formed, we can arrive at the enthalpy change of the reaction.

$$\Delta_{rxn}H = \Sigma(\Delta_{diss}H)_{\text{breaking bonds}} - \Sigma(\Delta_{diss}H)_{\text{making bonds}}$$

To use this equation, we have to know which bonds are broken and which are formed in a reaction. This is where Lewis dot structure models come in. By examining the Lewis dot structures of the products and reactants, we can determine which bonds are broken and which are formed. Then we can tally up the energy required to break all of the bonds in the reactants, and all of the energy released when the bonds in the products form. *The difference between the two* is the enthalpy change for the reaction. As we go through this process, we must keep in mind that using bond enthalpies to calculate the approximate enthalpy change in a reaction is *strictly a mathematical tool;* chemical reactions do not actually occur by breaking bonds into individual atoms and then having the atoms combine in different configurations. The actual chemistry is much more subtle and is continuous, consisting of a series of energy changes along the way as bonds are broken and formed. We will discuss the subtleties of reaction mechanisms in some detail in Chapter 15.

The reaction of water and phosgene illustrates how we use this handy tool in the chemist's toolbox to determine approximate bond enthalpy changes during reactions:

$$H_2O(l) + COCl_2(g) \rightarrow CO_2(g) + 2HCl(aq)$$

We must draw the reaction and the Lewis dot structure for each compound. In this reaction we break one C=O bond, two C—Cl bonds, and two O—H bonds, and form two C=O bonds and two H—Cl bonds (Figure 8.19).

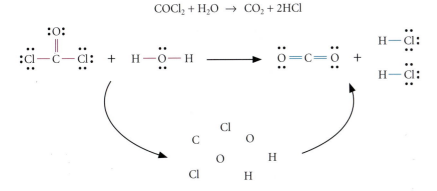

FIGURE 8.19

Reaction of phosgene with water. Energy is added to the system to break all of the bonds in the reactants. Energy is released when the bonds in the products are formed.

The enthalpy change of the reaction is calculated on the basis of this information. The enthalpy change for the reaction is negative. When the reaction occurs, heat is released.

Video Lesson: Using Bond Dissociation Energies

$$1 \text{ mol C=O bonds} \times 745 \text{ kJ/mol} = +745 \text{ kJ}$$
$$2 \text{ mol C—Cl bonds} \times 339 \text{ kJ/mol} = +678 \text{ kJ}$$
$$2 \text{ mol O—H bonds} \times 467 \text{ kJ/mol} = \underline{+934 \text{ kJ}}$$
$$\text{Total bonds broken} = +2357 \text{ kJ}$$

$$2 \text{ mol H—Cl bonds} \times (-427 \text{ kJ/mol}) = -854 \text{ kJ}$$
$$2 \text{ mol C=O bonds} \times (-799 \text{ kJ/mol}) = \underline{-1598 \text{ kJ}}$$
$$\text{Total bonds formed} = -2452 \text{ kJ}$$

$$\text{Total energy absorbed} = +2357 \text{ kJ}$$
$$\text{Total energy released} = \underline{-2452 \text{ kJ}}$$
$$\text{Net energy change } (\Delta H) = -95 \text{ kJ}$$

The difference between the energy absorbed and the energy released = net energy change (ΔH) = $-$ 95 kJ.

EXERCISE 8.8 **Calculating the Enthalpy of Combustion**

Calculation of the enthalpy of combustion can be useful in determining whether a compound could make a good fuel. Calculate the enthalpy of combustion for methane (found in natural gas) using bond dissociation enthalpies.

$$CH_4(g) + 2O_2(g) \rightarrow CO_2(g) + 2H_2O(g)$$

First Thoughts

This problem requires that we first draw the Lewis dot structures of all of the compounds in the reaction. After correctly drawing the structures, we can determine the type of bond that breaks or forms. **What sort of answer do we expect?** We know that the combustion of methane is highly exothermic, supplying heat to countless homes as natural gas is consumed in heaters.

Solution

$$
\begin{array}{r}
4 \text{ mol C—H bonds} \times 413 \text{ kJ/mol} = +1652 \text{ kJ} \\
2 \text{ mol O}_2 \text{ bonds} \times 495 \text{ kJ/mol} = +\ 990 \text{ kJ} \\
\hline
\text{Total bonds broken} = +2642 \text{ kJ}
\end{array}
$$

$$
\begin{array}{r}
2 \text{ mol C}{=}\text{O bonds} \times (-799 \text{ kJ/mol}) = -1598 \text{ kJ} \\
4 \text{ mol O—H bonds} \times (-467 \text{ kJ/mol}) = -1868 \text{ kJ} \\
\hline
\text{Total bonds formed} = -3466 \text{ kJ}
\end{array}
$$

$$
\begin{array}{r}
\text{Total energy absorbed} = +2642 \text{ kJ} \\
\text{Total energy released} = -3466 \text{ kJ} \\
\hline
\text{Net energy change } (\Delta H) = -\ 824 \text{ kJ}
\end{array}
$$

The reaction is exothermic. Heat is given off during the combustion of methane, so our answer makes sense.

Further Insight

Calculations such as this can be used to find a more efficient alternative fuel source. For example, toluene (C_7H_8, a major component of gasoline) releases 3933 kJ/mol, or 43 kJ/g, during its combustion. Based on this information, methane, releasing 824 kJ/mol, or about 52 kJ/g, gives off nearly 20% more energy as heat than does toluene on a gram-to-gram basis. The best alternative on a gram-to-gram basis may well be hydrogen, for which the energy released is about 280 kJ/mol, or 140 kJ/g! This is one of several reasons why hydrogen is being developed as an alternative to fossil fuels.

PRACTICE 8.8

Calculate the enthalpy change for the reaction of methanol and hydrogen bromide.

$$CH_3OH(aq) + HBr(aq) \rightarrow CH_3Br(aq) + H_2O(l)$$

See Problems 72–74.

HERE'S WHAT WE KNOW SO FAR

- Atoms can exceed the octet rule if they have available *d* orbitals in their valence shell.

- The enthalpy of a reaction can be determined from the energy absorbed during bond breakage or released during bond formation.

- We can combine the net changes in energy from bond-breaking and bond formation because bond enthalpy is a state function.

- Between two particular atoms, breaking multiple bonds requires more energy than breaking single bonds.

8.4   VSEPR—A Better Model

Knowing the strength of bonds in molecules such as nifedipine is useful for determining information about the biological processes that break down the molecule and render it useless as a pharmaceutical. However, the three-dimensional structure of molecules is even more important in determining the biological activity of the compound.

For instance, the opium poppy has long been used for its ability to alleviate pain. The ancient Sumerians in 3400 B.C. enjoyed its use so much that they referred to it as the "joy plant." It wasn't until 1827 that the active ingredient, morphine, was isolated from the poppy and produced commercially. The analgesic effects of morphine are one of the reasons why it continues to be used today to treat postoperative pain. However, morphine has serious side effects, which include respiratory depression, constipation, muscle rigidity, physical dependence, and a high potential for abuse. The medicinal chemist, the pharmacognosist, and the biochemist have been working together to try to prepare compounds that have beneficial properties similar to those of morphine but lack its harmful side effects. To develop a compound with a similar mode of action, they must determine not only the shape of the morphine molecule but also the shape of the active site on the receptor (the biological molecule to which morphine binds in the human body). Recent efforts have determined that the shape of the active site must accommodate a structure with particular dimensions, as shown in Figure 8.20. Because there are so many different probable locations for the active site, computers were used to complete most of the work mapping the shape of the active site on the receptor.

Application

CHEMICAL ENCOUNTERS: Focus on Morphine

WWW

Tutorial: VSEPR Theory

FIGURE 8.20

The structure of morphine and the opioid receptor binding requirements.

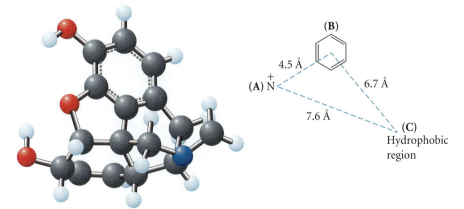

Issues and Controversies

Flat molecules and ethics in science

In the 1800s, Jacobus Henricus van't Hoff (Dutch physical chemist, 1852–1911) and Achille Le Bel (French chemist, 1847–1930) independently came to the conclusion that molecules must exist as three-dimensional structures. Despite the fact that established thought on the structure of molecules asserted that they were flat, these researchers advanced their theory for public and professional scrutiny. Using careful theoretical considerations of well-studied molecules (such as tartaric acid), van't Hoff concluded that molecules had to occupy a three-dimensional structure. It was a very compelling argument for some. For others, it was heresy.

Adolph Wilhelm Hermann Kolbe (1818–1884), one of the prominent German chemists of the time, vehemently discounted the theory of three-dimensional molecules. His flat models seemed to provide the best explanation for the properties that he observed. He took such an aggressive stance on this issue that he attempted to discredit van't Hoff and his colleagues. Without experimenting to determine whether the theory was correct, he published letters in prominent chemistry journals just to tarnish the new theory. In one of his published articles, Kolbe wrote as follows:

> I have recently published an article giving as one of the reasons for the contemporary decline of chemical research in Germany the lack of well-rounded as well as thorough chemical education. Many of our chemistry professors labor with this problem to the great disadvantage of our science. As a consequence of this, there is an overgrowth of the weed of the seemingly learned and ingenious but in reality trivial and stupefying natural philosophy. This nat-

ural philosophy, which had been put aside by exact science, is at present being dragged out by pseudoscientists from the den which harbors such failings of the human mind, and is dressed up in modern fashion like a freshly rouged prostitute whom one tries to smuggle into good society where she does not belong.

> A J. H. van't Hoff of the Veterinary School in Utrecht has, as it seems, no taste for exact chemical investigation. He has thought it more convenient to mount Pegasus (apparently on loan from the Veterinary School) and to proclaim in his "La chimie dans l'espace" how, during his bold flight to the top of the chemical Mount Parnassus, which he ascended in his daring flight, the atoms appeared to him to have grouped themselves throughout the Universe.

Journal für praktische Chemie, 1877

The responsibility of scientists to maintain objectivity in the methodical practice of science carries over into the public realm as well. To practice science with less than complete objectivity results in professional ridicule and increased public skepticism. Tens of thousands of chemistry-related articles are published worldwide each year. Highly regarded journals use the system of prepublication "peer review," in which knowledgeable scientists evaluate the validity of the research recounted by the authors of articles. While not perfect, it is the best system we know of to keep the scientific process as honest as possible. The best indication that chemistry is overwhelmingly done by ethical workers is that there are so many advances in our discipline every year, and these advances are based on communicating meaningful results of chemical research.

Lewis dot structures would not be very helpful in determining the shape of the morphine molecule or the active site in a case like this. **What are the limitations of the Lewis dot structure models?** Does a Lewis dot structure tell us anything about the three-dimensional shape of a molecule? The Lewis dot structure might lead us to believe that all molecules are flat, planar structures. But people, gerbils, roses, and guitars all occupy three dimensions, and all of these things are composed of molecules, so it makes sense that molecules occupy three-dimensional space. Even so, the idea of a "three-dimensional molecule" was fervently debated when it was first introduced. Many chemists could not believe that molecules would be anything other than flat.

The orientation of atoms in a molecule plays a major role in determining its properties. For instance, the Lewis dot structure of water (H_2O) can be drawn in two different ways. In one, it is a linear molecule, in the other, it is bent. Which is the correct way to draw water? If water were a linear molecule, it would have completely different properties from those we observe. The consequences of this slight change would have drastic effects in the real world. An ocean of linear water molecules would dissolve oxygen quite well and probably kill most of the life in the sea. Even the beading of water as it runs down a window would not be possible if water were a linear molecule. Water is a bent molecule. Can we prove this with a model—one that will pass the test of properly *predicting* the shapes of molecules with which we are unfamiliar? *As we continue this discussion, please keep in mind that models are, by their nature, simplifications that help us predict chemical behavior. Because they simplify often subtle chemistry, they do not always properly predict what happens.* Yet we use models because the best ones give us insight and some degree of predictive power. Such is the case with the VSEPR model, which we describe next.

Description of VSEPR

Ronald J. Gillespie (1924–) and Sir Ronald Nyholm (1917–1971) introduced the **valence shell electron-pair repulsion model** (VSEPR; pronounced "VES–per") in 1957 to facilitate the construction of three-dimensional molecular structures. This model is based on the repulsion between pairs of valence electrons (which have similar charges). The key assumption in the **VSEPR** model is that bonding pairs and lone pairs of electrons move away from each other and orient themselves in three-dimensional space to give minimum repulsions (lowest energy configurations.) To assist in visualizing this process, imagine pairs of electrons as balloons tied together (Figure 8.21). Two balloons tied together orient in such a way that they are opposite each other. Three balloons tied together orient themselves in a triangular shape, and four balloons push themselves to occupy the corners of a tetrahedron. What happens if you distort the shapes by pushing the balloons together and then letting go? The balloons push away again to return, ideally, to the original shape.

VSEPR models determine the shape of a molecule on the basis of the number of electron groups about the central atom. The resulting model of the molecule represents the positions of the electron groups in three dimensions (the **electron-group geometry**). Electron groups include lone pairs and bonding electrons; a single bond, a double bond, or a triple bond counts as a single electron group. However, when we determine the experimental shape of a molecule, we actually find the positions of the different nuclei (the **molecular geometry**). The actual shape of the molecule can be different from the electron-group geometry, but that doesn't mean that the molecular geometry is unattainable from the electron-group geometry. In fact, we can use the VSEPR model's electron-group geometry to determine the molecular geometry if we consider the geometry of the atoms to contain "invisible" lone pairs. The result is the molecular geometry. Let's consider some examples of the VSEPR model to illustrate this process. Table 8.7 shows the shapes of each of the electron-group geometries that we will discuss.

Examples of Three-Dimensional Structures Using the VSEPR Model

Beryllium chloride ($BeCl_2$) is a toxic white solid primarily used to make beryllium metal. The Lewis dot structure of this compound indicates to us that the beryllium atom is surrounded by two bonding pairs of electrons. By analogy, we can consider the central atom in $BeCl_2$ by tying two balloons together. The balloons push apart to occupy opposite orientations, so VSEPR predicts the electron-group geometry to be linear with a Cl—Be—Cl bond angle of 180°. The molecular geometry is also linear because there are no lone-pair electrons around the central beryllium atom.

This type of geometry is also predicted for carbon dioxide (CO_2). The Lewis dot structure of CO_2 indicates that there are only two electron groups in the form of double bonds that extend from the central carbon atom. Carbon dioxide is a linear molecule.

Visualization: VSEPR

Video Lesson: Valence Shell Electron-Pair Repulsion Theory

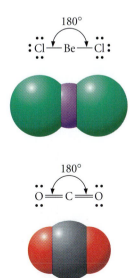

FIGURE 8.21

Balloon analogy of the VSEPR theory.

Visualization: VSEPR: Two Electron Pair

Visualization: VSEPR: Three Electron Pair

Visualization: VSEPR: Four Electron Pair

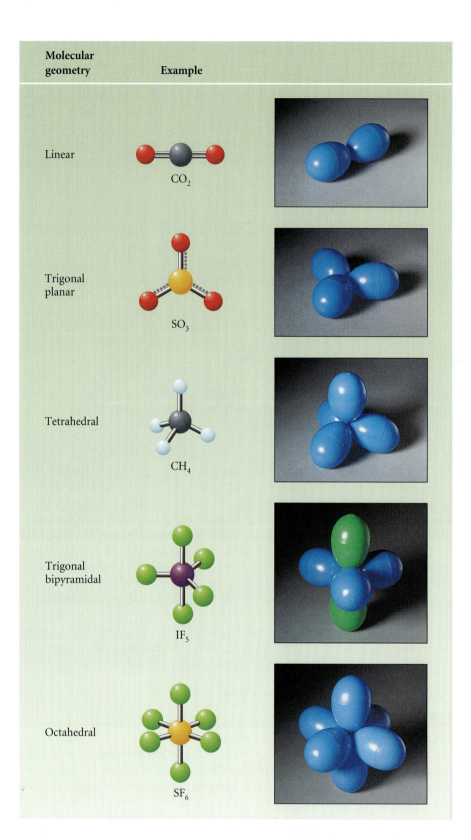

Molecular geometry	Example
Linear	CO_2
Trigonal planar	SO_3
Tetrahedral	CH_4
Trigonal bipyramidal	IF_5
Octahedral	SF_6

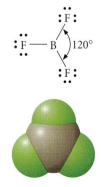

Boron trifluoride (BF_3), a compound used as a flux for soldering magnesium, is shown below as a Lewis dot structure. This model shows that three electron pairs (the three bonds) radiate from the central boron atom. The VSEPR model dictates that the three electron pairs should occupy the corners of a triangle. Therefore the electron-group geometry has trigonal planar geometry. The

TABLE 8.7   Shapes of the Electron-Group Geometries Predicted by VSEPR

Number of Electron Groups	Electron Group Geometry	Number of Lone Pairs	Molecular Geometry	Number of Electron Groups	Electron Group Geometry	Number of Lone Pairs	Molecular Geometry
2	Linear	0	G—A—G 180° Linear	5	Trigonal bipyramidal	1	See-saw
3	Trigonal planar	0	120° Trigonal planar	5	Trigonal bipyramidal	2	T-shaped
3	Trigonal planar	1	Bent/angular	5	Trigonal bipyramidal	3	Linear
4	Tetrahedral	0	109.5° Tetrahedral	6	Octahedral	0	90° Octahedral
4	Tetrahedral	1	Trigonal pyramidal	6	Octahedral	1	Square pyramidal
4	Tetrahedral	2	Bent/angular	6	Octahedral	2	Square planar
5	Trigonal bipyramidal	0	120° 90° Trigonal bipyramidal				

molecular geometry is also trigonal planar because there are no lone pairs on the boron. Bond angles for F—B—F are 120°.

Methane (CH_4) is a molecule containing four electron pairs around the central carbon atom. As a consequence, VSEPR predicts that the bonding pairs of electrons should point to the vertices of a tetrahedron. Similarly, because of the lack of lone-pair electrons on the carbon, the molecular geometry is tetrahedral with H—C—H bond angles of 109.5°. A dashed wedge is used to illustrate that the bond points behind the plane of the paper; the filled wedge protrudes in front of the paper.

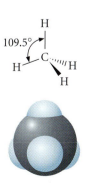

Video Lesson: Molecular Shapes for Steric Numbers 2–4

Video Lesson: Molecular Shapes for Steric Numbers 5 and 6

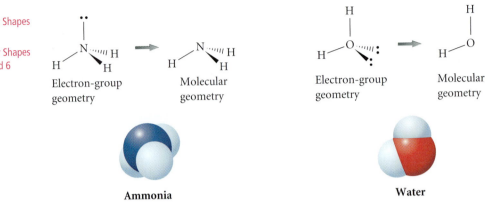

Electron-group geometry   Molecular geometry   Electron-group geometry   Molecular geometry

Ammonia   **Water**

Ammonia (NH_3), a compound used extensively as a fertilizer in farming, possesses three **bonding electron pairs** and one lone pair. VSEPR models predict a tetrahedral electron-group geometry. Because one of the electron groups is a lone pair, the *molecular geometry* of ammonia is predicted to be trigonal pyramidal. Similarly, the Lewis dot structure model of water indicates that there are two bonding pairs and two lone pairs on the central oxygen atom. The VSEPR model dictates that the electron groups should be arranged in a tetrahedral shape. However, the molecular geometry, determined by considering that the lone pairs are invisible, is bent, or angular. We'll discuss the subtleties of bond angles later in this section.

For molecules that contain an expanded octet, there may be five or six electron groups about the central atom. Consider the structure of PCl_5. This compound is used as a catalyst in the manufacture of acetylcellulose (the plastic film on which motion pictures are printed). The Lewis dot structure of phosphorus pentachloride shows five electron groups attached to the central atom. VSEPR indicates that the molecule has a trigonal bipyramidal electron-group geometry, and the molecular geometry is also trigonal bipyramidal. This model has some interesting features. In particular, there are two distinct positions for chlorine atoms within the trigonal bipyramid. Two locations, with bond angles of 90° from one chlorine to the next nearest neighbor, occupy the **axial positions** (up and down), and three locations occupy the **equatorial positions** (outward) with bond angles of 120° between them. Molecules that contain six electron groups occupy the octahedral electron configuration. Each of the positions in the octahedron is the same. All atoms have 90° bond angles to the next closest neighbor. An example of this shape is found in SF_6.

EXERCISE 8.9   **Ozone and VSEPR**

Ozone is an important molecule in our atmosphere because it protects us from the harsh ultraviolet rays of the sun. What is the shape of ozone (O_3)?

First Thoughts

Although a Lewis dot structure might provide the answer, we should apply the VSEPR model to assist us in developing the best model possible at this point in our discussion.

Solution

The Lewis dot structure model shows a single bond and a double bond with a lone pair on the central oxygen atom. Each of these is an electron group.

$$:\ddot{O}—\ddot{O}=\ddot{O}$$

With three electron groups, ozone has a trigonal planar electron-group geometry. The VSEPR model indicates that the molecular geometry is bent.

Further Insights

We should further develop our model of ozone to show resonance structures. Because it is known that chlorofluorocarbons (CFCs) react with ozone, it might be useful to consider the reactivity of ozone. Knowing the strengths of the polar covalent bonds could help us determine the reactivity.

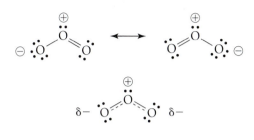

Ozone
O_3

Video Lesson: Predicting Molecular Characteristics Using VSEPR Theory

---

PRACTICE 8.9

Indicate the VSEPR model and bonding angles for HNO_3, CCl_4, and NH_3.

See Problems 75–80, 85, 86, and 88.

Advanced Thoughts on the VSEPR Model

VSEPR theory illustrates how electron pairs repel each other because they occupy space and have similar charge. Does the degree of repulsion depend on the type of electron pair that we're examining? To put it another way, is there a difference in the size and shape of different electron pairs? Lone pairs of electrons are big compared to bonding pairs of electrons. Gillespie illustrated this fact by noting the repulsions in terms of the size of the electron pair. The result is that the different types of electron pairs can be ordered in terms of the *three-dimensional space they require:*

Lone pairs > triple bonds > double bonds > single bonds

Note that the *space required* in the VSEPR model is not the same as the *bond lengths*, in which a triple bond between, for example, two carbon atoms is shorter than a double bond, which, in turn, is shorter than a single bond between the atoms.

Why do we need to discuss the space that bonds and lone pairs require? Let's consider the physical shape of a molecule of ammonia. According to the VSEPR rules, ammonia (NH_3) has a tetrahedral electron geometry. The lone pair on the nitrogen, however, is much more repulsive than the bonding pairs of electrons. In response to the repulsions and three-dimensional space requirements, the bonding pairs move closer together. The result is that the H—N—H bond angles in ammonia, at 107°, are smaller than those of the true tetrahedron. The same reasoning holds for the experimentally measured bond angle in water (104.5°). The angle is severely pushed by the presence of two lone pairs on the central oxygen atom.

Gillespie noted that multiple bonds require more space than single bonds. Let's examine the structure of formaldehyde (H_2CO) as an example. The Lewis dot structure shown in Figure 8.22 predicts three electron groups (one double bond and two single bonds) surrounding the central carbon atom.

VSEPR correctly predicts a trigonal planar electron-group geometry. Experimentally, though, the bond angles in formaldehyde are not 120°. Because the double bond is larger than the single bond, we predict the H—C—H bond angle to be less than 120°. Experimental evidence suggests that the actual angle is 116°. Similarly, the H—C—O bond angle should be larger than 120°. Experimentally, it has been measured at 122°.

Formaldehyde

FIGURE 8.22

Multiple bonds affect bond angles in formaldehyde (CH_2O).

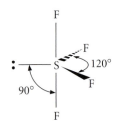

FIGURE 8.23

Sulfur tetrafluoride is a molecule with see-saw geometry.

The molecular geometries of CIF$_3$ (T-shaped), BrF$_5$ (square pyramidal), and XeF$_4$ (square planar) are affected by the presence of lone pairs around the central atom in each molecule.

Using this information, we can determine the molecular geometry for sulfur tetrafluoride (SF$_4$) with 34 valence electrons. The molecule is typically used in the laboratory to make other fluorine-containing compounds. The VSEPR model predicts the trigonal bipyramidal electron-group geometry shown in Figure 8.23. However, because one of the electron groups is a lone pair, the molecular geometry is not a trigonal bipyramid. Which one of the positions, axial or equatorial, should be occupied by the lone pair of electrons? Because the lone pair requires more space than the bonding pairs of electrons when we are using the VSEPR model, the lone pair will occupy the position with the fewest repulsions. In the trigonal bipyramid, the most unhindered site is one of the equatorial positions. In fact, lone pairs in trigonal bypyramids will always occupy the equatorial positions. The space-eating lone pairs then push against the other electron pairs and distort them from their ideal bond angles. The result for SF$_4$ is referred to as a see-saw structure. The same logic can be applied to the construction of ClF$_3$ (T-shaped molecular geometry), BrF$_5$ (square pyramidal molecular geometry), and XeF$_4$ (square planar molecular geometry).

EXERCISE 8.10   **A Closer Look at Ozone**

When we apply the VSEPR rule to ozone, our initial picture is that it is a bent molecule with an O—O—O bond angle of 120°. Given that different electron groups have different degrees of repulsions, what do you predict to be the actual bond angle in ozone?

Solution

The lone pair on the central oxygen occupies more space than the bonding pairs, so the bond angle becomes smaller between the adjacent oxygen atoms. The experimentally measured angle (O—O—O) is 116.8°.

PRACTICE 8.10

Which molecule has larger bond angles, SO$_2$ or H$_2$O?

See Problems 81–84 and 87.

FIGURE 8.24

The VSEPR model allows us to understand and interpret the three-dimensional structure of morphine, as described in the text.

Let's use what we now know to examine the three-dimensional structure of morphine. The formula reveals little: C$_{17}$H$_{21}$NO$_3$. As we noted at the start of this section, the Lewis dot structure, just like the formula, fails to illustrate the three-dimensional structure of morphine. The structure contains trigonal planar and tetrahedral carbon atoms arranged to make a scaffold that holds the OH groups pointed at one end of the molecule. The nitrogen resides at the opposite end of the molecule as shown in Figure 8.24. The scaffold holds these groups at specific distances, allowing morphine to fit into the opioid receptor quite well. We can get a handle on the structure of large molecules such as morphine by analyzing the three-dimensional structure at each carbon atom. *By examining each atom in the structure and determining the geometry about it, we can build the overall shape of a molecule.*

Medicinal chemists need to be able to determine the shapes of the molecules they construct in order to assess the fit with an appropriate receptor in the body. This application stems from the simple rule that "structure follows function." That is, *the three-dimensional arrangement of atoms within a molecule is just as important to the function (and chemical properties) of a molecule as are the identities of the atoms that make up the molecule.* If the geometry of one of the bonds in morphine were not aligned just right for the opioid receptor, morphine wouldn't interact well as an analgesic agent.

8.5   Properties of Ionic and Molecular Compounds

The cells within our body are held together by membranes that not only define the limits of the cell but also regulate the passage of materials into and out of the cell. Because most targets that interact with drugs are located inside the cell, any newly developed pharmaceutical agent's activity is related to the drug's ability to cross the cell membrane. A cross section of the cell membrane exposes three distinct regions with which a drug must be able to have favorable interactions in order to pass across the membrane. The first and last regions of the cell membrane will interact well only with compounds that have an overall polarization of electrons toward one end of the molecule. The middle portion of the membrane interacts well with molecules that lack this overall polarization, and this difference makes the cell membrane difficult to cross. Any new drug, then, must be carefully designed if its target lies inside the cell. How do medicinal chemists know whether their new drug will be able to enter cells? In this section, we will discuss how the polarization of electrons can be used to determine this ability and some of the other properties associated with molecules.

Bond Dipoles

The polarization of electrons in a bond is commonly referred to as a **bond dipole**. A difference in the electronegativity between the atoms on either end of the bond can be used to illustrate this polarization. Earlier in this chapter, we noted that polarization of the bonding electrons gave rise to a polar covalent bond. If the polarization were large enough, an ionic bond would result. In order to incorporate this information into the structure of a compound, we typically write delta plus $(\delta+)$ and delta minus $(\delta-)$ on the atoms in the bond. Alternatively, the bond

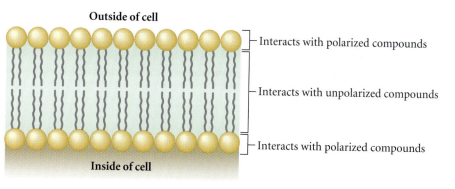

Cross section of a cell membrane.

The electron density distribution diagram of HF shows the positive (in blue) and negative (in red) ends of the bond dipole.

dipole can be represented by drawing an arrow over the bond pointing to the more electronegative atom. For example, H_2 lacks a bond dipole because of the lack of polarization of the electrons in the bond. Hydrogen fluoride (HF), however, contains a strongly polarized bond and can be drawn with the bond dipole to illustrate this.

$$H \overset{\longleftrightarrow}{-} \overset{\cdot\cdot}{\underset{\cdot\cdot}{F}}:$$

We can also draw an electron density distribution diagram of HF that shows the dipole in a more visually appealing and detailed way.

Dipole Moment

The *net polarization of electrons in a molecule* is known as the **dipole moment**. For a molecule, this is a function of the orientation and the magnitude of each of the individual bond dipoles. Mathematically, the dipole moment (μ) is defined as

$$\mu = Q \times d$$

in which the charge at either end of the dipole (Q) times the distance between the charges (d) yields the dipole moment in Coulomb·meters (C·m). In honor of Peter Debye (1884–1966), who won the 1936 Nobel Prize in chemistry for his work on molecular structures, we usually refer to dipole moments in terms of the number of debyes (D):

$$3.34 \times 10^{-30} \text{ C·m} = 1 \text{ D (debye)}$$

Bond dipoles can be observed experimentally in binary molecules as the dipole moment. When the molecule is placed in an electric field, the unequal distribution of charge in the molecule ($\delta+$ on one end of the bond and $\delta-$ on the other end) causes the molecule to orient itself in such a way as to maximize electrostatic attractions. In this way, the molecule behaves similarly to a small bar magnet. The resulting orientation can be measured and the degree of polarization calculated. The trends observed in dipole moments of some common binary molecules (Table 8.8) seem to follow our expectations for the polarization of the bonds. As we would predict, hydrogen fluoride has the greatest dipole moment of all the hydrogen halides (HF, HCl, HBr, HI) because the electronegativity difference in this series is greatest for HF.

For molecules with more than two atoms, however, an individual bond dipole is harder to measure. Instead, the dipole moment that is measured is related to the sum of all of the bond dipoles in the molecule. For example, hydrogen cyanide (HCN) contains two bond dipoles. Both bond dipoles point in the same

$$H \overset{\longleftrightarrow}{-} C \overset{\longleftrightarrow}{\equiv} \overset{}{N}:$$

TABLE 8.8	Dipole Moments of Some Common Binary Compounds		
Compound	Dipole Moment (D)*	Compound	Dipole Moment (D)*
H_2	0.00	ClF	0.88
HF	1.91	BrF	1.29
HCl	1.07	NO	0.16
HBr	0.79	CO	0.13
HI	0.38	O_2	0.00

*D stands for debye.

direction, giving rise to an overall polarization of the electrons in the molecule, and a molecule that aligns itself in an electric field and has a dipole moment of about 3.0 D. Carbon dioxide also has two bond dipoles, shown in Figure 8.25. However, they point opposite to each other. It is as though you and an equally strong friend are involved in a tug of war. Your pull cancels out your friend's pull, and the flag in the middle doesn't move. The molecule lacks the ability to align itself in an electric field, and the dipole moment of CO_2 is 0.00 D.

The dipole moment is a measure of the **polarity** of a molecule. In some cases, the bond dipoles cancel each other out, as in CO_2. The result is a **nonpolar molecule**. In other cases, the bond dipoles don't cancel out, as in HCN. The net result is a **polar molecule**. At the start of Section 8.4, we noted that the shape of a molecule is very important in determining its properties. In particular, we mentioned that some of water's important properties result from its shape. Is water a polar or nonpolar molecule? Water has a tetrahedral electron-group geometry and, because of the existence of two lone pairs on the oxygen atom, has a bent molecular geometry. The bond dipoles of water point toward the oxygen, but because the molecule is bent, the net dipole moment is not zero. In fact, water has a dipole moment (1.85 D) and is a polar molecule (Figure 8.26).

The dipole moment can be used by the medicinal chemist, who can measure the "fingerprint" of a molecule with an infrared (IR) spectrophotometer. How does this instrument work? The IR spectrophotometer records the energy associated with the vibrations of a molecule, as long as the vibration produces a change in the dipole moment. The resulting information can be used to classify compounds in terms of the types of bonds they contain, and it also reveals the identity of compounds through a comparison to known samples. Other professionals have recently applied IR spectroscopy to determine the quality of motor oil, to find the amount of carbon monoxide in automobile exhaust, and to identify different polyester fibers from crime scenes.

FIGURE 8.25

Carbon dioxide contains two bond dipoles, but overall the molecule lacks a dipole moment.

The electron density distribution diagram of HCN shows that this is a polar molecule.

FIGURE 8.26

The bond dipole in water. Note the large dipole moment (blue arrow) that arises from the sum of the two bond dipoles.

Visualization: Polar Molecules

EXERCISE 8.11   **Does CHCl₃ Have a Dipole Moment?**

Does $CHCl_3$ (chloroform, a compound formerly used as the major component in cough syrups) have a dipole moment? Give the Lewis dot structure for chloroform, and show bond dipoles and the dipole moment for the molecule, if any exists.

Solution

The Lewis dot structure model of $CHCl_3$ indicates that the central carbon atom contains four electron pairs. Because they are all bonding pairs of electrons, chloroform has a tetrahedral geometry. By placing individual bond dipoles on the molecule, we note that there is a net dipole moment (1.04 D) to the molecule. We would therefore predict that chloroform is a polar molecule.

PRACTICE 8.11

Predict the bond dipoles and overall dipole moment for N_2 and NH_3.

See Problems 95 and 96.

Polar Versus Nonpolar

As we mentioned before, we can use the dipole moment of a molecule to illustrate its overall polarity. In a polar molecule, *a dipole moment exists*. In a nonpolar molecule, *no dipole moment exists*. However, it isn't really that cut and dried. Molecules reside on a scale of polarities from the purely nonpolar to the highly polar. Water, with its two polar covalent bonds and resulting dipole moment, is a polar molecule. However, it isn't the most polar molecule. Why do we need to know about polarity? The polarity of a molecule can be a useful predictor of the solubility of the molecule. In other words, "like dissolves like." Polar molecules generally dissolve in polar solvents. Nonpolar molecules generally dissolve in nonpolar solvents. And polar molecules typically do *not* dissolve in nonpolar solvents. However, like all rules, this one has exceptions.

One exception is that not all molecules of the same polarity dissolve in each other. For example, water and chloroform ($HCCl_3$) do not mix together, in spite of the fact that they are both polar. **Why is this so?** In short, solubility is affected by more factors than relative molecular polarities. The structure of a molecule is very much a part of the equation, but the rules of solubility are best determined by considering the forces of interaction between dissolving molecules. See Chapter 12 for more information about polarity, structure, and solubility.

One of the major components in cell membranes is sphingomyelin (Figure 8.27). This molecule contains large bond dipoles at one end and very small bond dipoles at the other. Because the molecule is so long, one end of the molecule is polar and the other end is nonpolar. A cell membrane is made up of two layers of sphingomyelins with the nonpolar ends pointing toward each other. A medicinal chemist interested in developing a new drug must consider the implications of this fact. As we discussed earlier in this section, a new drug that must penetrate a cell to cause a biological response must cross the outer polar region of the cell membrane, pass through the large nonpolar region of the membrane, and then cross the inner polar region of the membrane. In other words, a drug that targets a location inside a cell must be able to dissolve in polar *and* nonpolar solvents. It must have some polar and some nonpolar character.

Just as the construction engineer must model a building in such a way as to take into account all expected mishaps, we must try to address all of the observed properties when we construct models of molecules. The models discussed in this chapter address the shape, dipole moment, and polarity of a compound, but they do not adequately explain bond lengths, bond strengths, and the reactivity of molecules. In Chapter 9, we will examine models that do a more accurate job of explaining some of the properties of molecules.

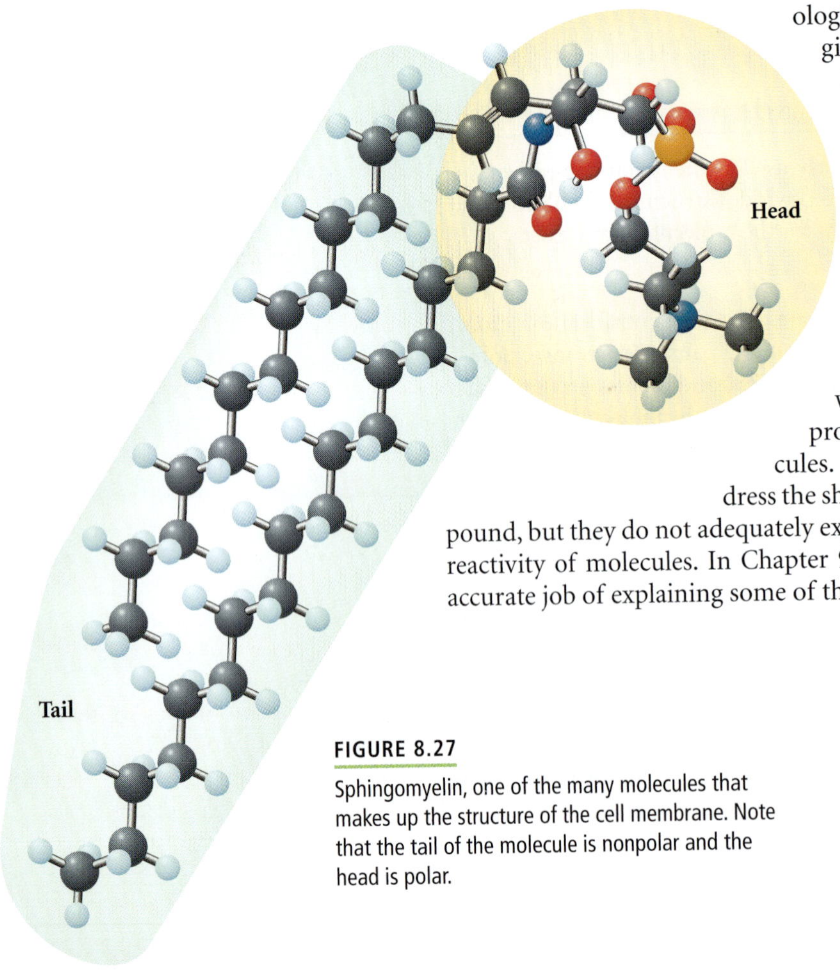

Application

Head

Tail

FIGURE 8.27

Sphingomyelin, one of the many molecules that makes up the structure of the cell membrane. Note that the tail of the molecule is nonpolar and the head is polar.

The Bottom Line

- Models are an important tool that chemists use to help them determine the properties of molecules. (Section 8.1)

- Estimation of the properties of a molecule on the basis of the structure of that molecule is only as good as the model of that molecule. (Section 8.1)

- Bonding can range across the full spectrum between equal sharing and complete transfer of electrons. (Section 8.1)

- The three main types of bonds are ionic, polar covalent, and metallic. (Section 8.1)

- An anion is always bigger than the atom from which it is derived. A cation is always smaller than the atom from which it is derived. (Section 8.2)

- The lattice enthalpy is an important expression of the energetic stability of a salt. (Section 8.2)

- We can use bond enthalpy calculations to determine the approximate energy change involved in a reaction. (Section 8.3)

- Lewis dot structures are useful in constructing a simple model showing the location of atoms within a molecule. (Section 8.3)

- Resonance hybrids offer an overall picture of a molecule. Individual resonance structures do not adequately describe a molecule. (Section 8.3)

- VSEPR theory describes the shapes of molecules better than Lewis dot structures. The model drawn using VSEPR provides a three-dimensional picture of the molecule. (Section 8.4)

- The polarity of a molecule is related to the overall forces of the individual bond dipoles in the molecule *and* to the molecule's three-dimensional shape. (Section 8.5)

Key Words

axial position The position of a group when it is aligned along the *z* axis of a molecule. (*p. 340*)

boiler scale The deposit of calcium carbonate (or calcium sulfate) on the inside of water pipes. (*p. 309*)

bond dipole The polarization of electrons in a bond that results in a separation of partial charges in the bond. (*p. 343*)

bond dissociation energy The energy required to break 1 mol of bonds in a gaseous species. Also known as the enthalpy of bond dissociation. (*p. 332*)

bonding electron pairs Pairs of electrons involved in a covalent bond. Also known as bonding pairs. (*p. 340*)

bonding pairs Pairs of electrons involved in a covalent bond. Also known as bonding electron pairs. (*p. 322*)

bond length The average distance between the nuclei of bonded atoms. (*p. 319*)

Born–Haber cycle A diagrammatic representation of the formation of an ionic crystalline solid using Hess's law. The cycle reveals the lattice enthalpy, which is difficult to obtain by direct measurement. (*p. 313*)

chemical bonds A sharing of electrons between two adjacent atoms. This sharing can be complete, partial, or ionic in nature. (*p. 304*)

coordinate covalent bond A covalent bond that results from the donation of two electrons from one of the two atoms involved in the bond. The resulting bond is indistinguishable from other covalent bonds. (*p. 330*)

Coulomb's law The force between two particles is proportional to the product of the charges (*Q*) on each particle divided by the square of their distance of separation (*d*). (*p. 314*)

covalent bond A sharing of electrons between two adjacent atoms. This sharing can be complete or partial. (*p. 304*)

crystalline lattice A highly ordered, three-dimensional arrangement of atoms, ions, or molecules into a solid. (*p. 308*)

dipole moment The polarization of electrons in a molecule that results in a net unequal distribution of charges throughout the molecule. (*p. 344*)

double bond A covalent bond consisting of two individual bonding pairs of electrons. (*p. 326*)

duet rule The exception to the octet rule involving the atoms H and He. A full valence shell for the atoms H and He. (*p. 306*)

electron-group geometry The positions of the groups of electrons (lone pairs and bonding pairs) around a central atom in three dimensions. (*p. 337*)

electronegativity The ability of an atom in a molecule to attract shared electrons to itself. (*p. 319*)

enthalpy of bond dissociation The enthalpy change related to breaking 1 mol of bonds in a gaseous species. Also known as the bond dissociation energy. (*p. 332*)

equatorial position The position of a portion of a molecule when it is arranged along the *x* axis or *y* axis of the molecule. (*p. 340*)

expanded octet The existence of more than eight electrons in the valence shell of an atom. Atoms of atomic number greater than 12 are capable of this feat. In the third row of the periodic table, it typically occurs in sulfur and phosphorus. (*p. 330*)

formal charge The difference between the number of valence electrons on the free atom and the number of electrons assigned to that atom in the molecule. (*p. 324*)

ion pair Ions of opposite charges that exist in solution as an ionically bonded pair. (*p. 308*)

ionic bond A strong electrostatic attraction between ions of opposite charges. (*p. 304*)

isoelectronic Having the same electron configuration. (*p. 305*)

lattice energy The amount of energy released as a solid ionic crystal is formed. (*p. 314*)

lattice enthalpy The amount of energy required to separate 1 mol of a solid ionic crystalline compound into its gaseous ions. (*p. 313*)

Lewis dot structure A drawing convention used to determine the arrangement of atoms in a molecule or ion. (*p. 305*)

Lewis dot symbol A drawing convention used to determine the number of valence electrons on an atom or monatomic ion. (*p. 305*)

lone pairs Pairs of electrons that are not involved in bonding. (*p. 322*)

metallic bond A bond in which metal cations are spaced throughout a sea of mobile electrons. (*p. 304*)

molecular geometry The geometry of the atoms in a molecule or ion. The molecular geometry comes from the electron-group geometry but considers the lone pairs "invisible." (*p. 337*)

molecular models Models used by chemists to explain the three-dimensional nature of molecules. Although many can be made from plastic or wood, these models can also be constructed using computers. (*p. 303*)

nonpolar covalent bond A covalent bond with a very small difference in electronegativity between the bonded atoms. The result is minimal or no charge separation in the bond. (*p. 320*)

nonpolar molecule A molecule in which there is no net charge separation and no net dipole moment. (*p. 345*)

octet rule The existence of eight electrons in the valence shell of an atom, creating a stable atom or ion. (*p. 306*)

pharmacognocist A person who studies natural drugs, including their biological and chemical components, botanical sources, and other characteristics. (*p. 303*)

polar covalent bond A covalent bond in which the atoms differ in electronegativity, resulting in an unequal sharing of electrons between two adjacent atoms. (*p. 320*)

polar molecule A molecule in which there exists a charge separation resulting in a net dipole moment. (*p. 345*)

polarity The location of a compound on a scale from polar to nonpolar. The dipole moment is a measure of the polarity of a compound. (*p. 345*)

polarization A molecule or bond containing electrons that are unequally distributed. (*p. 319*)

polarized A molecule or bond that contains an unequal distribution of electrons. (*p. 319*)

radical A molecule that contains an unpaired nonbonding electron. (*p. 331*)

resonance hybrid An equal or unequal (based on experimental evidence) combination of all of the resonance structures for a molecule. (*p. 328*)

resonance structure A model of a molecule in which the positions of the electrons have changed, but the positions of the atoms have remained fixed. (*p. 328*)

triple bond A covalent bond in which three electron pairs (six electrons) are shared between adjacent atoms. (*p. 326*)

valence shell electron-pair repulsion model A model that proposes the three-dimensional shapes of molecules on the basis of the number of electron groups attached to a central atom. Also known as VSEPR. (*p. 337*)

VSEPR The valence shell electron-pair repulsion model. (*p. 337*)

zeolite A porous solid with a well-defined structure, typically made of aluminum and silicon. Zeolites are capable of binding to ions of a specific radius based on the relative size of its pores. (*p. 310*)

Focus Your Learning

The answers to the odd-numbered problems and some selected problems appear at the back of the book, as represented by the blue numbering.

Section 8.1 Modeling Bonds

Skill Review

1. Use Lewis electron dot structures to answer these questions about aluminum and nitrogen.
   a. Which neutral atom, Al or N, has the greater number of unpaired valence electrons?
   b. Which neutral atom, Al or N, has the greater number of valence electrons?
   c. Which neutral atom, Al or N, is more likely to gain electrons to form an octet?

2. Use Lewis electron dot structures to answer these questions about carbon and sulfur.
   a. Which neutral atom, C or S, has the greater number of unpaired valence electrons?
   b. Which neutral atom, C or S, has the greater number of valence electrons?
   c. Which neutral atom, C or S, is more likely to gain electrons to form an octet?

3. Of the following, which, if any, would *not* have the same configuration as a hydrogen atom with two electrons?
   a. C^{2+} b. B^{3+} c. N^{3-} d. H^+

4. Of the following, which, if any, would *not* have the same configuration as a fluorine atom with eight valence electrons?
   a. Ne   b. Na^+ c. Br^- d. S^{2-}

5. To which group does each of these ions belong?
   a. Ion of an element with a -2 charge and a full octet
   b. Ion of an element with a $+2$ charge and a full octet
   c. Ion of an element with a -3 charge and a full octet

6. To which group does each of these ions belong?
   a. Ion of an element with a $+1$ charge and a full octet
   b. Ion of an element with a -1 charge and a full octet
   c. Atom of an element with no charge and a full octet

7. The charges have been omitted from these Lewis structure diagrams of ions. From your knowledge of the ground state for atoms, determine the number of extra electrons that each structure is showing, and provide the proper charge for each ion.
   a. $\left[:\overset{..}{\underset{..}{S}}:\right]$ b. $\left[:\overset{..}{\underset{.}{P}}:\right]$ c. $\left[:\overset{..}{\underset{..}{Cl}}:\right]$

8. The charges have been omitted from these Lewis structure diagrams of ions. From your knowledge of the ground state for atoms, determine the number of extra electrons that each structure is showing, and provide the proper charge for each ion.
   a. $\left[:\overset{..}{\underset{..}{O}}:\right]$ b. $\left[:\overset{..}{\underset{..}{Br}}:\right]$ c. $\left[:\overset{..}{\underset{..}{N}}:\right]$

9. Write Lewis electron dot diagrams for each of these ions.
   a. Se^{2-} b. I^- c. Sr^{2+} d. Sc^{3+} e. Si^{2+}

10. Write Lewis electron dot diagrams for each of these ions.
   a. S^{2-} b. Br^- c. Ti^{2+} d. Ti^{4+} e. B^{3+}

11. Which of these ion(s) would have the same electron configuration as the noble gas neon?
   Cl^- Na^+ F^- C^{2+} Al^{3+}

12. Which of these ion(s) would have the same electron configuration as fluoride (F^-)?
   Ne^- O^{2-} N C^{2+} H^+

13. Which of these, if any, are isoelectronic?
   Ca^{2+} Sc^+ S   Ar Cl^-

14. Which of these, if any, are isoelectronic?
   Mg^{2+} Na Be^{2+} Ar F^-

15. Which neutral atoms could be represented by the Lewis dot symbol shown below?

   $$:\overset{..}{X}\cdot$$

16. Which cations with a $+1$ charge could be represented by the Lewis dot symbol shown below?

   $$:X$$

Chemical Applications and Practices

17. Lithium, sodium, and potassium are all very reactive alkali metals.
   a. Diagram the Lewis electron dot structure for each neutral atom.
   b. Predict the formula of the oxide compound of each.

18. Calcium, magnesium, and barium are members of the alkaline earth metals.
   a. Diagram the Lewis electron dot structure for each neutral atom.
   b. Predict the formula of the oxide compound of each.

Section 8.2 Ionic Bonding

Skill Review

19. Consider each statement below and determine whether it is true or false for compounds with ionic bonding.
   a. They are typically composed of a nonmetal and a metal.
   b. Electrons are shared among atoms in the compound.
   c. The metal typically becomes a cation.
   d. Like charges repel each other.

20. Consider each statement below and determine whether it is true or false for compounds with ionic bonding.
   a. They require only one metal and one nonmetal.
   b. The metal is always written first in the formula.
   c. The nonmetal must have a -1 charge.
   d. The metal and nonmetal must be in the same period.

21. Using the octet rule, explain why aluminum loses three electrons in both aluminum oxide and aluminum chloride, yet the ratio of aluminum to the nonmetal is 1:3 in aluminum chloride and 2:3 in aluminum oxide.

22. Using the octet rule, explain why magnesium loses two electrons in both magnesium oxide and magnesium chloride, yet the ratio of magnesium to the nonmetal is 1:2 in magnesium chloride and 1:1 in magnesium oxide.

23. Within each of the series presented, rank the atoms or ions from smallest to largest radius.
    a. I^{5+} I I^- c. C^+ C C^-
    b. S^{6+} S^{4+} S^{2-} d. Fe Fe^{3+} Fe^{2+}

24. Within each of the series presented, rank the atoms or ions from smallest to largest radius.
    a. Cu^{2+} Cu^+ Cu c. N^- N^{3-} N^+
    b. Cr^{6+} Cr^{4+} Cr d. Pd Pd^{4+} Pd^{2+}

25. Using Coulomb's law, arrange these alkali halides from strongest to weakest attraction of anion and cation: KCl, NaCl, CsCl, LiCl, RbCl.

26. Using Coulomb's law, arrange these alkali halides from strongest to weakest attraction of anion and cation: LiCl, LiF, LiBr, LiI.

Chemical Applications and Practices

27. Potassium chloride (KCl) has been used as a sodium substitute in some commercial salt (NaCl) products. Show the formation of this important ionic compound using the Lewis dot symbol model.

28. When some metals oxidize, the product formed may be detrimental to the strength of the metal structure. This is the case when iron corrodes. However, aluminum forms an oxide that resists further reaction and adheres to the metal, forming a protective covering. Show the formation of this ionic compound using the Lewis dot symbol model.

29. When placed in water, the oxide compounds of many metals form alkaline, or basic, solutions. This takes place when the metal oxide reacts with water to produce hydroxide ions. Diagram the Lewis dot structure of sodium oxide, and diagram the structure of the compound that forms when sodium oxide reacts with water.

30. Calcium oxide is often called lime. Lime has a number of uses, including reacting with acidic gases from coal plants. Diagram the Lewis dot structure for this compound, and diagram the structure of the hydroxide ion that forms when lime is placed in water.

31. If Mn^{5+} and Mn^{4+} were in a solution together, which would fit into a smaller zeolite hole?

32. Which of these copper ions would fit into a smaller zeolite hole, Cu^+ or Cu^{2+}?

33. The melting point of KBr is 734°C. The melting point of CsBr is 640°C. Which of the two compounds do you predict to have the higher lattice enthalpy? Explain your answer in terms of Coulomb's law.

34. Of NaCl, KCl, and $MgCl_2$, which would you predict to have the highest lattice enthalpy? Explain your answer in terms of Coulomb's law.

Section 8.3 Covalent Bonding

Skill Review

35. The terms *electronegativity* and *electron affinity* are both used to explain some bonding concepts. Compare and contrast these two important terms.

36. The expression *covalent bond* is actually used in three different contexts within the chapter. Explain the differences among nonpolar covalent bonds, polar covalent bonds, and coordinate covalent bonds. Provide an example of a compound that illustrates each of these three types of covalent bonds.

37. Diagram the outline of a periodic table. Using arrows to indicate a direction of increase, show the general periodic trend for electronegativity within a group and within a period.

38. Diagram the outline of a periodic table. Using arrows to indicate a direction of increase, show the general periodic trend for atomic size within a group and within a period.

39. Diagram the Lewis structure for each of these neutral compounds. Identify any that contain an odd number of (radical) electrons
    H_2O NO CO NO_2
    HCl PCl_2 NBr_3

40. Diagram the Lewis structure for each of these neutral compounds. Identify any that contain an odd number of (radical) electrons.
    CS_2 N_2O PCl_3 SO_2
    H_2O_2 SeH_2 SF_4

41. Diagram the Lewis structure for each of these ions. Identify any that contain an odd number of (radical) electrons.
    OH^- NO_2^- Br^- PO_4^{3-}
    SO_3^{2-} CO_3^{2-} BrO_4^-

42. Diagram the Lewis structure for each of these ions. Identify any that contain an odd number of (radical) electrons.
    SH^- NO_3^- F^- IO_4^-
    BO_3^{3-} CN^- CrO_4^{2-}

43. Diagram the Lewis structure for each of these compounds. Identify any that contain an odd number of (radical) electrons.
    C_2H_6 C_3H_6 C_2H_4 C_3H_8
    C_4H_{10}

44. Diagram the Lewis structure for each of these compounds. Identify any that contain an odd number of (radical) electrons.
    C_2H_2 C_3H_4 C_3H_8O C_2H_6O
    CH_2Cl_2

45. It is possible to diagram three resonance structures for the nitrate (NO_3^-) ion. Show, by giving formal charges, that each contributes equally to the overall hybrid structure that depicts bonds of equal strength between the three oxygen atoms and one nitrogen atom.

46. It is possible to diagram many resonance structures for the phosphate (PO_4^{3-}) ion. Show, by giving formal charges, that each contributes equally to the overall hybrid structure that depicts bonds of equal strength between the four oxygen atoms and one phosphorus atom.

47. Under the proper conditions, carbon atoms can bond to form closed ring structures when the carbon atoms are bonded to each other. These are called cyclic compounds. Three four-carbon ring compounds (C_4H_4, C_4H_6, and C_4H_8) are known to exist. Diagram the Lewis dot structure of each and predict which, if any, may have a resonance structure. Diagram the resonance structures of any possible results.

48. Three straight-chain four-carbon compounds that are not cyclic (C_4H_{10}, C_4H_6, and C_4H_8) are known to exist. Diagram the Lewis dot structure of each and predict which, if any, may have a resonance structure. Diagram the resonance structures of any possible results.

49. The compounds N_2O, N_2, and N_2H_4 all contain nitrogen to nitrogen bonds. Diagram the Lewis electron structure of each. Then match these bond energies for the nitrogen-to-nitrogen bond in each. (946 kJ/mol, 160 kJ/mol, 418 kJ/mol)

50. In which structure would you expect to find the shorter carbon-to-oxygen bond, carbon monoxide or carbon dioxide? Explain the basis of your choice. In which would you expect to find the higher value for bond energy? Explain the basis of your choice.

51. Each of these compounds contains both ionic and covalent bonding. Using their Lewis dot structures, indicate where each type of bonding occurs.
a. Na_3PO_4 b. $CaCO_3$ c. $Fe(NO_3)_2$

52. Each of these compounds contains both ionic and covalent bonding. Using their Lewis dot structures, indicate where each type of bonding occurs.
a. NaOH   b. $NaHCO_3$ c. K_2CrO_4

53. This list gives bonds that could form in compounds. Arrange them in order from least polar bond to most polar bond.
Ca—N   H—C   Ca—H
C—C   N—O

54. This list gives bonds that could form in compounds. Arrange them in order from least polar bond to most polar bond.
H—F   O—P   Li—F
S—F   N—S

55. a. Calculate the formal charges on the carbon atoms in ethane (C_2H_6).
b. If one of the hydrogen atoms is replaced with an OH group, the compound becomes ethanol. What if any changes in the formal charge of the carbon atoms takes place when this conversion is made?
c. What is the formal charge on the oxygen atom in ethanol?
d. If the hydrogen atom were then removed from the oxygen atom, with no other atom taking its place, what would you calculate as the formal charge for oxygen?

56. Diagram the Lewis dot structure of the sulfate ion (SO_4^{2-}) without an expanded octet and with an expanded octet. According to the reasoning about formal charges suggested in the chapter, which of the two structures is more likely to represent the actual structure of this important ion?

Chemical Applications and Practices

57. Ammonium chloride (NH_4Cl) is used in the manufacture of some dry cell batteries. Show the structure of this important ionic battery component using the Lewis dot symbol model.

58. One of the major manufactured compounds containing iodine is hydrogen iodide. It is used to produce other compounds that contain the iodide anion. Diagram the Lewis dot structure of this compound.

59. Diagram the Lewis structure for the air pollutant nitrogen dioxide (NO_2). Determine the formal charges on each atom in the structure. Circle and label any electrons that would be considered bonding, lone-pair, or radical.

60. Many states are blending ethanol with gasoline to produce "gasohol" products to use in automobiles. This mixture allows us to extend our dwindling gasoline supplies and improve the octane rating of the gasoline. Diagram the Lewis structure for this renewable energy extender (C_2H_5OH). Determine the formal charges on each atom in the structure. Circle and label any electrons that would be considered bonding, lone-pair, or radical.

61. The cyanide ion (CN^-) plays a critical role in reacting metals from ores in the process of producing pure metals. Diagram the Lewis structure of this important ion, and determine the formal charges on both the carbon and nitrogen atoms. When the ion reacts with H^+, hydrocyanic acid is formed. On the basis of your calculations of formal charges, would it be more likely for the H^+ to attach to the carbon side or the nitrogen side of the ion? Explain your choice.

62. Hypochlorous acid is the acid that can be used to make the salt sodium hypochlorite that is found in most commercial bleach preparations. Hypochlorous acid consists of one atom each of hydrogen, chlorine, and oxygen. Diagram three different arrangements of the atoms in the molecule, and use formal charge considerations to predict which is most likely.

63. Sulfurous acid (H_2SO_3) is one of the molecules that contributes to acid rain. The molecule has three oxygen atoms attached to the central sulfur atom. Diagram the Lewis structure for the acid, and predict the formal charge on the sulfur atom. Is the formal charge the same as the oxidation number? Explain any differences.

64. Dimethyl sulfoxide is a solvent used in some veterinary applications. The molecule consists of a central sulfur atom bonded to an oxygen atom and two carbon atoms. Each carbon atom has three bonded hydrogen atoms attached. Each atom follows the Lewis electron dot rules. Diagram the Lewis structure and determine the formal charge on the sulfur atom.

65. The concentrations of both nitrite ion (NO_2^-) and nitrate ion (NO_3^-) must be monitored in well water. Both can be quite harmful for humans. Diagram the Lewis dot structures of both (the nitrogen atom is in the center of both, and there are no oxygen-to-oxygen bonds), and assign a formal charge to the nitrogen atom in each.

66. Phosphoric acid (H_3PO_4) is used in the production of phosphate fertilizers and is often found in soft drinks (check the labels). In the molecule, phosphorus is at the center, and the hydrogen atoms are attached to the oxygen atoms. Phosphorous acid (H_3PO_3) differs slightly in that one of the hydrogen atoms is attached to the central phosphorus atom. Diagram the Lewis electron dot structure of both, and compare the formal charges on the phosphorus atoms.

67. Of the approximately 20 naturally occurring amino acids, glycine, the principal component in silk, has the simplest structure. Show the Lewis dot structure of glycine. (It is composed of two atoms of carbon, one of which is bonded to two hydrogen atoms, and a nitrogen, which is also bonded to two hydrogen atoms. The other carbon atom is bonded to two oxygen atoms, one of which is bonded to an atom of hydrogen. Examine the structure. Is there a possible resonance structure to draw for the one you have shown? If so, diagram the other resonance form.

68. Oxalic acid ($C_2O_4H_2$) is a toxin found in rhubarb leaves. It can also be used in general chemistry labs, in its pure form, as a standard acid with which to react unknown base solutions. In the structure, the two carbons are connected by a single bond. Each carbon is connected to two oxygens (one with a double bond and one with a single bond). The hydrogens are on opposite sides of the structure. Diagram the structure and determine whether a possible resonance structure exists for the compound. If so, draw all of the possible resonance structures.

69. The three simplest two-carbon hydrocarbon compounds are the heating fuel ethane (C_2H_6), the plant hormone ethene (C_2H_4), and the welding gas ethyne (C_2H_2). Diagram the Lewis structure of each, and predict which would have the highest bond energy. Which would have the longest carbon-to-carbon bond?

70. Absorbed light can be sufficient to break chemical bonds. Compare the bonding of Cl_2 and O_2. In which would the light absorbed have to be greater in energy? Use Planck's constant to calculate the frequency of the energy needed to break apart the molecule you selected. What would the energy need to be in order to break 1 mol of those bonds?

71. Butane (C_4H_{10}) is the fuel typically used in small, hand-held lighters to provide heat for other reactions through combustion. Use bond energies to determine the heat of the reaction between butane and oxygen to produce water and carbon dioxide. When another determination for the heat of this reaction was made, using a calorimeter rather than tabular values, the answer was slightly different. Explain why the two values may not agree, even though they were obtained for the same combustion reaction.

72. One way to produce ethanol (CH_3CH_2OH) is by the reaction of ethene (C_2H_4) and water. If the heat of this reaction was −37 kJ/mol, what would you calculate as the bond energy for the carbon double bond in ethene? How does this compare with the tabular value for the $C=C$?

73. Using hydrogen gas and oxygen gas, calculate the heat of reaction both for water and for the bleaching agent hydrogen peroxide (H_2O_2). Judging on the basis of your calculation, which of the two should be more stable? Explain the basis of your choice.

74. Small portable tanks often contain propane (C_3H_8) as the fuel to provide heat for camping trips. Use bond energies and the balanced chemical equation for the combustion of propane to determine the heat of the reaction between oxygen and propane to produce carbon dioxide and water. On a per-gram basis, which fuel provides more heat, propane or butane (C_4H_{10})?

Section 8.4 VSEPR Theory—A Better Model

Skill Review

75. For those compounds listed in Problem 39 that contain more than two atoms, determine the electron group and molecular geometry of the central atoms.

76. For those compounds listed in Problem 40 that contain more than two atoms, determine the electron group and molecular geometry of the central atoms.

77. For those compounds listed in Problem 41 that contain more than two atoms, determine the geometry of the central atoms.

78. For those compounds listed in Problem 42 that contain more than two atoms, determine the geometry of the central atoms.

79. For those compounds listed in Problem 43, determine the geometry of each carbon atom.

80. For those compounds listed in Problem 44, determine the geometry of each carbon atom.

81. Use Lewis dot structures and the VSEPR theory to explain the bond angles in ClO_2.

82. Use Lewis dot structures and the VSEPR theory to explain the bond angles in SF_2.

83. Use VSEPR modeling to arrange this list in order from largest to smallest Cl-to-element-to-Cl bond angle:

$BeCl_2$ $AlCl_3$ CCl_4 $XeCl_4$ NCl_3

84. Use VSEPR modeling to arrange this list in order from largest to smallest F-to-element-to-F bond angle:

MgF_2 NF_3 CF_4 PF_6 IF_3

Chemical Applications and Practices

85. Carbon tetrachloride follows the "octet rule" and has tetrahedral geometry. However, another tetrachloride compound, $SeCl_4$, has a different geometry and does not follow the octet rule. Predict the shape of the second compound, and diagram its Lewis dot structure.

86. Noble gases already have an octet of valence electrons. Therefore, any combination with another atom is likely to produce an exception to the octet rule. Predict the shape of the compound XeF_2, and diagram its Lewis dot structure.

87. The compound N_2H_2 has a double bond between the two nitrogen atoms. Use the VSEPR model to predict the H—N—N bond angle in the molecule. If $N_2H_2^{2+}$ had a triple bond between the two nitrogen atoms, how would that affect the same bond angle? Explain the basis of your prediction.

88. The bicarbonate ion (HCO_3^-) is essential as a buffer (discussed in Chapter 18) in your blood.
a. Diagram the Lewis dot structure for this important ion.
b. Is there a possibility for resonance in the structure? If so, show the example(s).
c. In VSEPR modeling, what shape would this ion have?

Section 8.5 Properties of Ionic Compounds and Molecules

Skill Review

89. Examine each of the bonds depicted as dashes between atoms here. Rank them from least polar to most polar. In addition, rewrite the bond, showing an arrow in the direction of the more negative atom.

C—Cl	C—N	C—O
C—H	C—B	C—Mg

90. Examine each of the bonds depicted as dashes between atoms here. Rank the bonds from least polar to most polar. In addition, rewrite the bond, using the $\delta+$ and $\delta-$ symbols to indicate the polarity of the bond.

H—O	H—C	H—F
H—N	H—Na	H—H

91. Diagram the Lewis structures of both SF_5 and SF_6. Use the VSEPR model to predict the shape and the resulting polarity of each.

92. Diagram the Lewis structures of both PBr_5 and PBr_3. Use the VSEPR model to predict the shape and the resulting polarity of each.

Chemical Applications and Practices

93. This table illustrates some properties of three unknown substances. On the basis of the clues provided, decide whether each compound listed in the table is an ionic compound, a nonpolar molecule, or a polar molecule.

Compound	Melting Point	Solubility In		Electrical Conductivity As	
		Water	Hexane	Solid	Molten
A	Very high	Yes	No	No	Yes
B	Very low	No	Yes	No	No
C	Medium	Yes	No	No	No

94. Refer to the clues in the previous problem and match these substances as possible identities with A, B, and C.
C_3H_8
KCl
citric acid, $C_3H_5O(COOH)_3$

95. The disinfectant hydrogen peroxide (H_2O_2) has a dipole moment greater than zero. Explain why this information supports suggesting that the molecule is not linear. If the molecule were linear, what would you predict about its dipole moment?

96. The formula $C_2H_2Cl_2$ can be drawn many different ways. Provide the Lewis dot structure that would give a molecule with the largest dipole moment. Provide the Lewis dot structure that would give a molecule that did not have a dipole moment.

Comprehensive Problems

97. These diagrams represent a hydrogen atom with different numbers of electrons. Determine which diagram represents hydride (H^-), which hydrogen (H), and which a proton (H^+).

98. In various compounds, nitrogen can be shown to form single, double, or even triple bonds. Diagram the Lewis dot structure of a nitrogen atom. What arrangement do you see

that would allow this variety in bonding? Diagram the structure of the N_2 molecule.

99. When you inhale a breath of air, you are taking in oxygen and nitrogen molecules. Examine the Lewis dot structure of an N_2 molecule (see Problem 98). Suggest a reason why the N_2 that you inhale is unaffected and is soon exhaled as the same N_2 that you inhaled.

100. Prepare a diagram, using circles with charges inside, showing a relative size comparison between magnesium oxide (used as an abrasive) and calcium oxide (used in some plaster formulations). On the basis of your representations, explain how Coulomb's potential could be used to predict which compound contains the larger lattice energy.

101. These circles represent cations and anions. From the choices provided, select the cation and anion combination that would have the lowest lattice energy. Explain your choice.

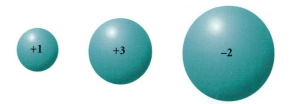

102. a. In your own words explain the lattice energy formula.
b. Lithium forms many useful salts. For example, lithium fluoride is used extensively in industry to assist in purifying aluminum. Lithium bromide has applications in the air–conditioning industry. Explain which of the two compounds has the stronger lattice energy.

103. On the basis of your arrangement of the ionic compounds in Problem 101, which compound would you predict to have the highest lattice enthalpy?

104. a. What is the qualitative relationship between lattice enthalpy and melting point?
b. On the basis of that relationship, which would you predict to have a higher melting point, MgF_2 (lattice enthalpy = 2913 kJ/mol) or NaCl (lattice enthalpy = 787 kJ/mol)? Use another reference such as the Merck Index or the Chemical Rubber Company's Handbook of Chemistry and Physics to obtain the melting points of the two compounds to check your assumptions.

105. Most ionic compounds have very high melting points.
a. Use Coulomb's law to explain this property.
b. Using Table 8.2, select the ionic combination that would be likely to have the highest melting point and the one that would be likely to have the lowest melting point. Explain the basis of both selections.

106. Magnesium metal will burn so vigorously in air that it will even combine with nitrogen.
a. What is the formula of magnesium nitride?
b. Would the compound be ionic or covalent?
c. Diagram the Lewis dot structure of the compound.

107. a. Explain why the bond energy listed in tables is only an approximation, based on averages, of the bond energy for a bond within a molecule.

b. In which case would the carbon-to-carbon bond be expected to deviate more from the average, H_3C—CH_3 or H_3C—CF_3? Explain the basis for your reasoning.

108. The environment of an electron in a bond varies considerably depending on the type of bonding arrangement in which the electron finds itself. Electrons are attracted to the positive nucleus of an atom or ion. Using the examples of covalent, polar covalent, ionic, and metallic bonds, describe the degree of attraction of an electron to its "original" atom once it has become part of a bond with another atom.

109. Iron metal can react with oxygen gas to make iron(III) oxide.
    a. Write the balanced equation for this reaction.
    b. How many unpaired electrons are in an isolated atom of iron(III)?
    c. What is the mass percent of iron in this compound?

110. People on low-sodium diets sometimes use a salt substitute for their meals. Suppose the salt substitute contains potassium chloride.
    a. What is the electron configuration of potassium ions?
    b. Would you expect potassium chloride to taste similar to sodium chloride? Explain.
    c. Which is larger, the potassium cation or the sodium cation?
    d. How many potassium cations are there in 1.00 g of potassium chloride?

111. Calcium carbonate is a very common mineral found in nature.
    a. Is this compound soluble in water?
    b. What is the mass percent carbon in this compound?
    c. When hydrochloric acid is added to this compound, calcium chloride and water are formed. What is the Lewis dot structure of the other product?

112. A fertilizer salesperson remarks that potassium nitrate provides more nitrogen per pound than calcium nitrate.
    a. Based on the mass percent nitrogen, is the salesperson correct?
    b. How many grams of nitrogen would be present in 1.00 lb of the potassium nitrate fertilizer?
    c. Write the Lewis dot structure for the nitrate ion.

113. The structure of nifedipine, Figure 8.18, contains four different types of atoms.
    a. What is the molar mass of this useful drug?
    b. Draw at least one other resonance structure for this molecule.
    c. When this molecule is treated with an acid, a proton (H^+) is added to one of the nitrogen atoms in the structure. To which nitrogen does that proton bond?

Thinking Beyond the Calculation

114. Someone proposes the use of dichloroethane ($C_2H_4Cl_2$) as a solvent to remove small wax (nonpolar hydrocarbon) buildups.
    a. Diagram a way to depict the structure that possesses an essentially zero dipole moment.
    b. Diagram another structure that would yield a dipole moment.
    c. Which of the compounds, that in part a or that in part b, would be best suited to dissolve wax?
    d. Dichloroethane can be prepared by adding chlorine (Cl_2) to ethene (C_2H_4). Draw a Lewis dot structure for ethene.
    e. Would you predict a nonzero dipole moment for ethene?
    f. If 2.0 g of dichloroethane was prepared using the method in part d, how many grams of ethene was consumed in the reaction?

Advanced Models of Bonding

Contents and Selected Applications

Chemical Encounters: Molecular Structure and Eyesight

9.1   **Valence Bond Theory**

9.2   **Hybridization**

9.3   **Molecular Orbital Theory**

9.4   **Putting It All Together**

Chemical Encounters: Benzene, Stability, and MO Theory

9.5   **Molecular Models in the Chemist's Toolbox**

At the back of the human eye are millions of photoreceptor cells known as rods and cones. These cells are home to rhodopsin, a molecule that uses 11-cis-retinal to catch light and initiate a nerve impulse. The double bonds in 11-cis-retinal enable it to interact with light in the visible region of the electromagnetic spectrum. This chapter will focus on the construction of models that help explain this property.

Go to **college.hmco.com/pic/kelterMEE** for online learning resources.

We begin our study of advanced models of bonding by peering down into a most remarkable organ, the human eye. Unlike our hands, our eyes enable us to interact with our environment without direct contact. Light reflected from other objects enters the eye through a small opening called the iris. Once there, the photons hit the retina, where they are collected, converted into electrical impulses, and sent to the brain for processing (Figure 9.1). The retina contains two types of photoreceptor cells (rods and cones) that collect the light. The number and type of rods and cones distinguish the human eye from that of other species.

Over 120 million rods occupy the human retina. These cells are extremely sensitive to light but do not distinguish color. It is these rods that allow humans some limited nighttime vision. The 7 million cones, which are much less sensitive to light than the rods, allow humans to perceive color.

How are rods and cones related to molecular structure and advanced bonding models? It has to do with the way they absorb light. Located in each rod and cone is a protein called opsin. The protein itself doesn't absorb visible light. However, when a molecule known as 11-*cis*-retinal (see Figure 9.2) is attached to the opsin, the resulting protein, rhodopsin, becomes able to absorb wavelengths near 500 nm. A photon that strikes the rhodopsin causes a reaction that converts 11-*cis*-retinal to the more stable all-*trans*-retinal. This results in a change in the position of the atoms in the retinal molecule, causing the rhodopsin to alter its shape. This change assists in the generation of a nerve impulse and the release of *trans*-retinal from the protein. After the impulse, the *trans*-retinal is converted back into 11-*cis*-retinal and attached to another opsin—ready for the next photon of light to begin the process again (see Figure 9.3). **What is it about rhodopsin that makes it capable of absorbing photons?** Why does 11-*cis*-retinal absorb light near 500 nm and not near, say, 850 nm?

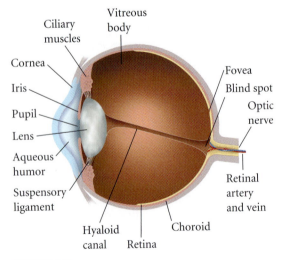

FIGURE 9.1

The human eye is a complex organ. Light enters the front of the eye through the cornea and is focused by the lens on the retina at the back of the eye.

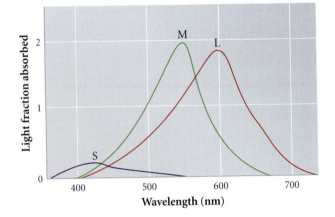

There are three kinds of cones in the human eye (they are known as S, M, and L). About 64% are sensitive to light in the red region of the visible spectrum, 32% are green-sensitive, and 2% respond best to blue light.

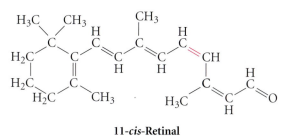

11-cis-Retinal

FIGURE 9.2

11-*cis*-Retinal absorbs light with wavelengths near 500 nm. The absorption of light causes an electron in the double bonds to become excited, allowing the molecule to rotate about the double bond in red.

The answer to these questions largely lies in the structure of the retinal molecule. However, this is best understood when we can examine more advanced models of bonding. Although the VSEPR model and Lewis dot structures are useful for determining the shape and connectivity of the atoms within a molecule, they do not adequately predict many of its properties. For instance, the VSEPR model of water does predict that the H—O—H bond angle should be less than 109.5°. However, Lewis dot structure and VSEPR models do not properly explain why the C—C bond in ethane (CH_3—CH_3) is longer than the C—C bond in butadiene (CH_2=CH—CH=CH_2). Moreover, *VSEPR doesn't explain how we can end up with equally spaced bonds.* For example, in methane (CH_4), the bonding electrons occupy *s* and *p* orbitals. VSEPR doesn't even hint at why electrons in *p* orbitals (90° apart) would, in reality, form 109.5° bond angles. In order to explain the reasons behind the observed bond angles, and to more accurately assess and predict the properties of molecules, *we need to make better models.* Our goal in this chapter is to introduce you to some of the better models.

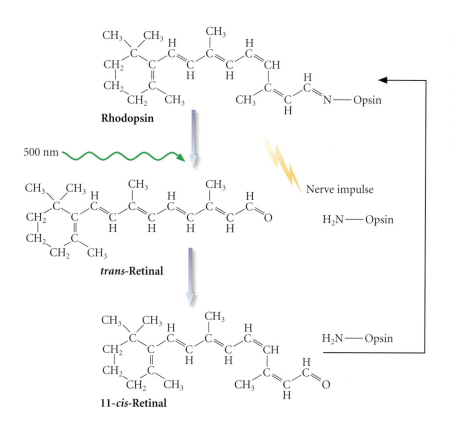

Rhodopsin

500 nm

trans-Retinal

11-cis-Retinal

Nerve impulse

H_2N——Opsin

H_2N——Opsin

FIGURE 9.3

Transformation of 11-*cis*-retinal to *trans*-retinal. After the absorption of light and the reorganization of the double bond, the rhodopsin molecule undergoes a conformational change that brings about a nerve impulse. The *trans*-retinal is recycled by conversion back to 11-*cis*-retinal and reattached to an opsin molecule.

Video Lesson: Valence Bond Theory

9.1 Valence Bond Theory

Just as light can be absorbed by the molecule known as retinal, it can also be absorbed by the diatomic halogens. Chlorine (Cl_2) and fluorine (F_2) both absorb light according to the following reactions:

$$Cl_2 + \text{light } (h\nu) \rightarrow 2\ Cl\cdot$$

$$F_2 + \text{light } (h\nu) \rightarrow 2\ F\cdot$$

The resulting very reactive atoms are called radicals (remember these from Chapter 8) because they contain one unpaired electron. It is this reactivity that accounts for chlorine's effects on the ozone layer and fluorine's ability to react with normally unreactive atoms such as xenon.

The Lewis dot structures of the halogens indicate that they contain nonpolar covalent bonds and similar bond lengths. However, on the basis of the different wavelengths of light (remember that energy is inversely proportional to wavelength!) required to break the bond between these atoms (Figure 9.4), we can reason that chlorine and fluorine must have different bond *strengths*. It appears that they have some differences in their bonds that the Lewis dot structure model does not identify.

Knowing the exact makeup of a bond can give us a good picture of its strength. For instance, we discussed in Chapter 8 how bond strengths can be used to arrive at a rough approximation of the enthalpy of a reaction. We also noted that bond energies were averaged to get the values we saw in the table. The fact that they were averaged implies that *not all bonds between the same atoms have the same energy*. We can explore this idea by studying the combustion reactions of both ethane (CH_3CH_3) and acetylene ($CH\equiv CH$). Experimentally, as shown in Table 9.1, the combustion of acetylene is accompanied by an enthalpy change of −1300 kJ/mol, compared to the −1559 kJ/mol associated with the combustion of ethane. Using the bond energy values from Chapter 8, we end up calculating enthalpy changes that are different than these values. Even though the C—C bond in ethane (376.1 kJ/mol) is much weaker than the C≡C bond of acetylene (962 kJ/mol), this difference doesn't account for the experimentally determined difference in the enthalpy of combustion between ethane and acetylene. **Why are our numbers so different?** Some of the missing energy is accounted for by the differences in C—H bond energy. Experimentally, researchers have determined that the energy of the C—H bond in ethane (410 kJ/mol) is much smaller than that of the C—H bond in acetylene (536 kJ/mol). However, according to Lewis dot structures and to the VSEPR models, there should be no difference in energy between the bonds because they are both C—H bonds. We need to construct a *better* model of these compounds that takes this difference into account.

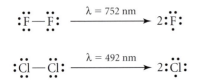

FIGURE 9.4

The interaction of light with chlorine and fluorine imply that these two molecules possess different bond strengths.

TABLE 9.1	Calculated Versus Experimental $\Delta_c H$

The experimental and calculated $\Delta_c H$ values for the fuels listed do not agree. The temperature of the combustion is listed only for reference.

Compound	Flame Temperature (K)	$\Delta_c H$ (kJ/mol) Experimental	$\Delta_c H$ (kJ/mol) Calculated
Hydrogen (H_2)	2490	−242	−242
Methane (CH_4)	2285	−890	−803
Ethane (CH_3CH_3)	2338	−1559	−1429
Ethylene ($CH_2\!=\!CH_2$)	2643	−1411	−1324
Acetylene ($CH\equiv CH$)	2859	−1300	−1257

One such "better model" of a covalent bond is the valence bond model. In the 1930s, Linus Pauling (whom we remember from Chapters 7 and 8), devised valence bond theory (VB theory) to address the inadequacy of the bonding models of G. N. Lewis.

What Is a Valence Bond?

Pauling's valence bond theory envisions a bond as the overlap of atomic orbitals on adjacent atoms. Because two electrons interact in a bond, overlapping two half-filled orbitals supplies these electrons, which are covalently shared. If all of the bonds were treated this way, an atom would be able to satisfy the octet (or duet, for hydrogen) rule.

Molecular hydrogen's use in industry to make such products as the hydrogenated vegetable oils used in processed foods and its possible use as an alternative to gasoline in motor vehicles makes it a meaningful compound to study. The electron configuration for a hydrogen atom is $1s^1$. When the $1s$ orbital on the hydrogen atoms overlap, a bond results. Because the electrons are shared by both orbitals, each of the hydrogen atoms possesses the $1s^2$ electron configuration, as shown in Figure 9.5, which is a full $1s$ orbital (the duet rule is satisfied for both hydrogen atoms).

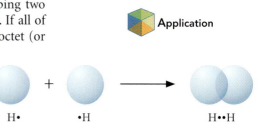

Application

H•   •H   H••H

FIGURE 9.5

Overlap of orbitals makes the bond in H_2. As two hydrogen atoms approach, their $1s$ orbitals overlap to form a bond. The overlap allows both hydrogen atoms to satisfy the duet rule.

| **EXERCISE 9.1** | **Modifications to the Structure of HF** |

We noted in Chapter 8 that hydrogen fluoride is often used by master glassworkers as they etch a design into a piece of art. Some of the world-famous Steuben art has been made via this etching technique. Using valence bond theory, describe which orbitals overlap to form a covalent bond between the hydrogen and fluorine atoms in HF.

First Thoughts

Our first thoughts bring to mind the Lewis dot structure of HF. We note that the structure includes three lone pairs around the fluorine atom and the one bonding pair of electrons between the two atoms. To construct the valence bond model for HF, we need to consider the electron configurations of these atoms. How many p orbitals are there in the second principal energy level of fluorine? Which type of orbital on fluorine might overlap with the hydrogen $1s$ to share a needed electron?

Solution

The configuration of the valence electrons in fluorine is $2s^2 2p^5$. Overlap of the hydrogen $1s$ orbital with one of the $2p$ orbitals allows the electrons to be shared between the two atoms. The valence bond model of HF in Figure 9.6 shows a $1s$–$2p$ orbital overlap that constitutes the covalent bond.

H   F   H-F

FIGURE 9.6

Orbital overlap in HF.

Further Insights

Does the valence bond model we constructed imply anything about the properties of the bond? What is the strength of the bond? Does the s–s orbital overlap in H_2 result in a stronger bond than the s–p orbital overlap in HF? Does the overlap of the s orbital in the first principal energy level and the p orbital in the second principal energy level indicate anything about the bond? We answer these questions next.

PRACTICE 9.1

Using valence bond theory, describe the orbital overlap in F_2, a very reactive molecule used to manufacture Teflon $((C_2F_4)_n)$ and other fluorinated compounds.

See Problems 2, 9, and 10.

Application of Valence Bond Theory

Because the degree of electron sharing is related to the strength of a bond, orbitals that exhibit more overlap result in stronger covalent bonds. What determines the amount of overlap? The keys are the relative energy, as shown in Figure 9.7, and the size of the atomic orbitals. More specifically:

■ Smaller orbitals overlap more than larger orbitals.

■ Orbitals with similar sizes overlap more than orbitals with mismatched sizes.

■ Orbitals with similar energies overlap more than orbitals with very different energies.

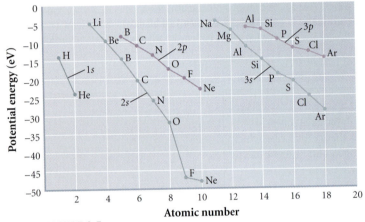

FIGURE 9.7

Relative energy of the atomic orbitals. The energy level of the atomic orbitals decreases with increasing nuclear charge, and the atomic orbital becomes more stable because it has a lower potential energy.

The hydrides LiH, NaH, and KH—used as bases in chemical reactions, in the removal of oxide coatings on metals, and in processes to make purified hydrogen gas—make excellent case studies. In each of these substances, the bonds between adjacent atoms result in covalent overlap of a $1s$ orbital of the hydrogen atom and the ns valence orbital of the metal atom. The valence bond in LiH results from a $2s–1s$ orbital overlap. The valence bond in NaH results from the overlap of a $3s$ (Na) and a $1s$ (H) orbital. Similarly, overlap in KH results from a $4s$ orbital and a $1s$ orbital. Because the energy and the size of the metal's s orbital are *dramatically greater* in potassium than in lithium, the overlap of the potassium $4s$ and hydrogen $1s$ orbitals isn't well matched (see Figure 9.8). We would therefore predict the bond in LiH to be much stronger (better overlap) than the bond in KH.

Table 9.2 lists the bond energies for LiH, NaH, and KH as 238 kJ/mol, 185.7 kJ/mol, and 174.6 kJ/mol, respectively. Note in the table the relatively low F—F bond energy. Although the $2p–2p$ orbital overlap is expected to be quite good, the electronegativity of each halogen atom competes with the orbital overlap. The electrons that participate in the bond between the two fluorine atoms are held more tightly to the atoms. This results in a decreased electron density between the atoms in F_2—and an unusually low bond energy.

Valence bond theory also addresses any misconceptions reflected in the Lewis dot structure and VSEPR models about the lengths of bonds. There doesn't appear to be any difference in the bond length for H_2 compared to F_2 *if we use only Lewis dot structures as our model*. Experimentally, however, we know that the bond lengths are different. Which is longer, the bond in the hydrogen molecule or that in fluorine? The difference in bond lengths results from a difference in the orbitals that overlap to form the covalent bond. Orbitals that extend farther from the nucleus result in bonds that are longer. The key question then, is which orbital

FIGURE 9.8

Overlap of atomic s orbitals.

TABLE 9.2	Bond Energies Resulting from Orbital Overlap

As the difference in size of the overlapping orbitals increases, the strength of the resulting bond decreases. The data shown here are for diatomic molecules.

Bond	$\Delta_{diss}H$ (kJ/mol)	Orbital Overlap	Representation of Orbital Overlap
H—H	435.8	$1s$–$1s$	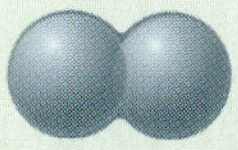
Li—H	238.0	$1s$–$2s$	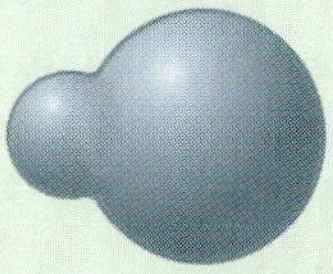
Na—H	185.7	$1s$–$3s$	
K—H	174.6	$1s$–$4s$	
Rb—H	167.0	$1s$–$5s$	
H—F	569.9	$1s$–$2p$	
H—Cl	431.9	$1s$–$3p$	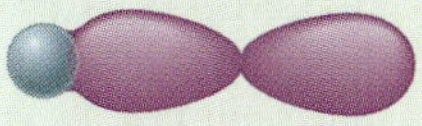
H—Br	366.3	$1s$–$4p$	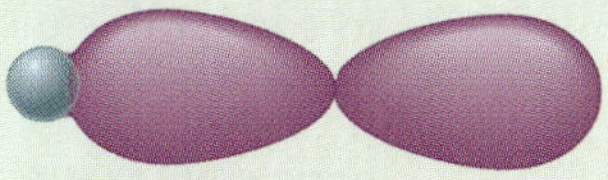
H—I	298.4	$1s$–$5p$	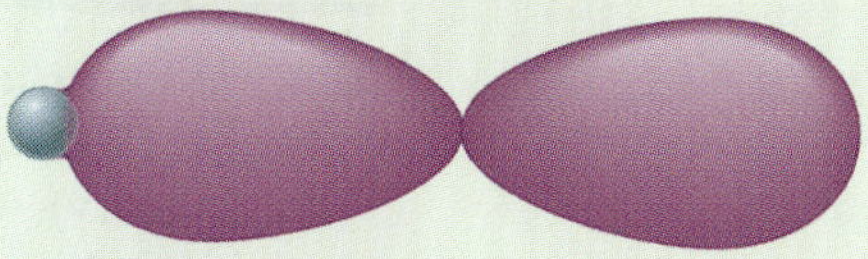
F—F	158.3	$2p$–$2p$	

reaches farther from the nucleus, an s or a p? The end-on overlap of two p orbitals makes a longer bond than the overlap of two s orbitals, because the average electron density lies farther from the nucleus of the atom. For this reason, the bond in F_2 is longer (141.7 pm) than the bond in H_2 (74.6 pm). Overlap of two $1s$ orbitals makes a shorter bond than the overlap of two $2s$ orbitals for the same reason. We'll discuss this in greater detail in Section 9.2.

EXERCISE 9.2   **Using Valence Bond Theory: Beam Me Up, Scotty**

According to *Star Trek* fans, dilithium crystals (Li_2) power much of the universe of the future. Judging on the basis of valence bond theory, does this compound actually exist? Does Na_2 exist? Which would have a longer bond?

Solution

Because each of the lithium atoms has a half-filled $2s$ orbital, we'd predict that their overlap should provide a stable dilithium molecule. Dilithium does exist, but because the formation of metallic lithium is so favorable, dilithium can be observed only as a gas at high temperatures. Two sodium atoms can have overlap of their $3s$ orbitals to make a bond. So, theoretically, disodium should also exist. Because the disodium molecule is made up of bigger orbitals, we'd also predict it to have a longer bond than dilithium.

PRACTICE 9.2

Which of these molecules has the longest bond: Cl_2, HCl, or F_2?

See Problems 13–15.

What's Wrong with This Model?

Methane (CH_4) is the major component of natural gas. Piped into our homes, it undergoes combustion to heat our water, cook our food, and warm our rooms. Let's build a valence bond model of methane. We start by writing the valence electron configurations of the atoms in methane. Immediately, we note that the valence electron configurations of carbon ($2s^2 2p^2$) and hydrogen ($1s^1$) indicate a problem. The valence shell of carbon contains a completely full $2s$ orbital, two partially filled p orbitals, and one completely unfilled p orbital. Valence bond theory indicates that we should be able to make only two bonds to the hydrogen atoms resulting from the overlap of the two partially filled p orbitals with the hydrogen $1s$ orbitals, as shown in Figure 9.9. Something is wrong here.

We know that the Lewis dot structure model of methane correctly shows four bonds. VSEPR models represent the molecule as a tetrahedral structure with four equal bonds. Therefore, each of the C—H bonds in methane must be made up of the same types of orbitals. But unless we modify the valence bond theory to account for our *experimental* evidence, we're sure to build a structure that will fail. In the next section, we will describe a modification that enables us to correct this discrepancy.

HERE'S WHAT WE KNOW SO FAR

- Lewis dot structure and the VSEPR model do not properly explain such properties of molecules as bond lengths and bond energies. Valence bond theory (VB theory) was introduced by Linus Pauling to better explain molecular properties.

- A valence bond is seen as the overlap of atomic orbitals on adjacent atoms.

Atomic orbitals on four hydrogen atoms

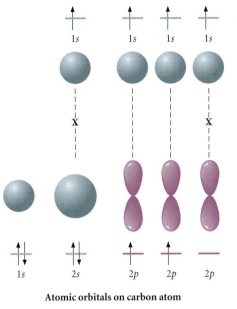

Atomic orbitals on carbon atom

FIGURE 9.9

Valence bond model of methane. Atomic orbitals that contain only one electron can overlap to form a bond. These are shown with a dotted line. The orbitals that cannot overlap to form a bond (either too many electrons or not enough electrons) have an "X" on the dotted line. The model needs some corrections because there aren't enough orbitals to make four bonds.

- In VB theory, smaller orbitals overlap more than larger orbitals.
- In VB theory, orbitals similar in size overlap more than orbitals with mismatched sizes.
- In VB theory, orbitals with similar energies overlap more than orbitals with very different energies.
- Bond length and bond energy in simple molecules can be explained by valence bond theory.
- VB theory does *not* properly explain bond angles in molecules, so *we must modify it to make a better model.*

9.2 Hybridization

Linus Pauling (we have already discussed his contributions in the areas of electronegativity and valence bond theory) advanced the theory of hybridization to address the problems associated with VB theory. The result fixes the problems with the valence bond approach to model construction. In 1954, Pauling was awarded the Nobel Prize in chemistry for his years of research into the nature of the chemical bond.

Tutorial: Hybridization

Hybridization Defined

A hybrid is a mixture of two species. Hybrid roses, tulips, marigolds, and lilies add beauty to the garden. Much as red roses and white roses can be interbred to give pink hybrids, we can mathematically combine two or more orbitals to provide new orbitals of equal energy in the process of **hybridization**.

Pinocchio rose

Crimson Glory rose

Fashion rose

Just as gardeners combine red and white roses to obtain pink hybrids, chemists mathematically combine two or more orbitals to represent new orbitals of equal energy in the process of hybridization. The Pinocchio rose and the Crimson Glory rose can be combined to give the Fashion rose.

To make pink paint, we can mix red paint and white paint. The resulting paint is a hybrid—a product that has characteristics of both paints.

Video Lesson: An Introduction to Hybrid Orbitals

Recall from Chapter 6 that orbitals with the same energy are known as degenerate. What types of orbitals can be mixed together? The degree to which an orbital mixes with another is directly related to the difference in energy of the two orbitals. Typically, we hybridize only orbitals of the same subshell, such as a hybrid made using 2s and 2p orbitals, resulting in a set of new orbitals that have the properties of all the orbitals from which they were mixed. The preparation of pink paint offers an analogy to hybridization. To make a good pink paint, we must mix red and white paint that have the same base (i.e., latex or oil). We don't get a good mixture by combining one can of latex paint and one can of oil paint.

How many orbitals do we make when we hybridize atomic orbitals? Just as if we were to mix one can of red paint and one can of white paint to get two cans of pink paint, we should expect to get two hybridized orbitals if we mix two atomic orbitals. The number of orbitals that are hybridized determines the number of new orbitals that are made. The orbitals that result from this mixing will be degenerate (have the same energy) and will have the same shape, but they will be oriented in different directions. **What is the energy of the resulting hybridized orbitals?** Consider our paint example again. We expect the color of the mixed paint to be the weighted average of the colors that we added. Similarly, the energy of the resulting hybridized orbitals should be the weighted average of the energies of the atomic orbitals that were mixed.

sp, sp^2, and sp^3 Orbitals

To determine the hybridized orbitals used by an atom, we follow a brief series of steps that convert atomic orbitals into hybridized orbitals. These steps are outlined in Table 9.3, and we will show in detail how the process works for methane (CH_4). The Lewis dot structure of methane indicates that we should have one bond from each hydrogen atom to the carbon atom.

Step 1: Write the electron configuration of the carbon atom. The electron configuration of carbon ($1s^2 2s^2 2p^2$) indicates that there are only two half-filled orbitals.

Step 2: In order to place four hydrogen atoms in degenerate bonds around the carbon atom, we need to make four orbitals on carbon that will be able to interact with the four hydrogen 1s orbitals. How do we show that the carbon makes four bonds?

Step 3: We hybridize the existing orbitals to make four orbitals. Knowing that we need to mix four orbitals to make four orbitals, we take the available 2s orbital and the three 2p orbitals and mix them to make four new degenerate orbitals, as shown in Figure 9.10.

TABLE 9.3	Algorithm for Hybridizing Atomic Orbitals
Step 1: *Electron Configuration*	Write the electron configuration for the atom that will be hybridized.
Step 2: *Bonds Needed*	Note the number of attached atoms and lone pairs, and select the orbitals to be hybridized.
Step 3: *Hybridize*	Mix the orbitals to make an identical number of new orbitals. Note the energy of the hybridized orbitals.
Step 4: *Rewrite Electron Configuration*	Redraw the electron configuration and include the hybridized orbitals. Place the electrons in the new orbitals according to the Aufbau principle, Hund's rule, and the Pauli exclusion principle.
Step 5: *Form Valence Bonds*	Identify which orbitals will overlap to form a bond.
Step 6: *Examine Structure*	Describe the geometry and bond lengths for the new bond.

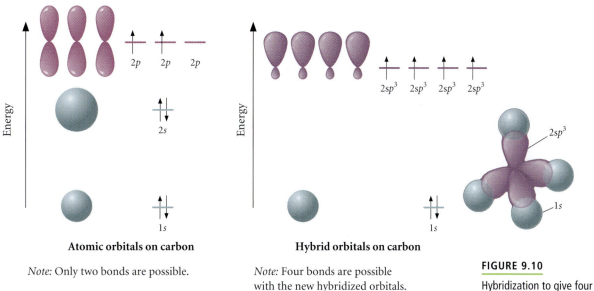

Atomic orbitals on carbon

Note: Only two bonds are possible.

Hybrid orbitals on carbon

Note: Four bonds are possible with the new hybridized orbitals.

FIGURE 9.10

Hybridization to give four orbitals.

The new orbitals are given a name in order to distinguish them from the $2s$ and $2p$ orbitals, but to show their relationship to these orbitals, we call them the $2sp^3$ orbitals. The name of the new orbitals illustrates that one $2s$ orbital and three $2p$ orbitals are mixed to make four new orbitals.

Step 4: The next step in Table 9.3 is to show the electron configuration of our hybridized carbon atom. We can write $1s^2(2sp^3)^4$ to show the hybrid orbitals. Carefully observe how this is written because the notation is tricky. Each of the four $2sp^3$ orbitals on our carbon atom contains only one electron and is able to participate in bonding.

Step 5: The new shape of the orbitals (large at one end and small at the other) is shown in Figure 9.10. Does the geometry of this hybridized carbon atom—that is, with what are now four equivalent pairs surrounding it—agree with that obtained from the VSEPR model? In other words, **does our answer make sense?**

Step 6: Let's look at the energy changes that occurred with this hybridization. Energy was added to two $2s$ electrons on carbon in order to promote them to the new $2sp^3$ orbitals. A small amount of energy was removed when two $2p$ electrons moved to the new orbitals. The net result is that the carbon atom did increase its energy in making the new hybrid orbitals. However, when the hybridized carbon combines with the four hydrogen atoms, a large amount of energy is released as the new C—H bonds are formed. Although the hybridization process itself is energetically *uphill*, the end result (four equal bonds) is still energetically *downhill* (and quite favorable). Why? What reduces the energy? In methane, the s–sp^3 overlap between each hydrogen and the central carbon atom forms the four bonds (Figure 9.10). The outcome of the resulting valence bond model now agrees with the Lewis dot structure and the VSEPR model. Moreover, we also have an idea of the relative bond lengths and strengths in methane.

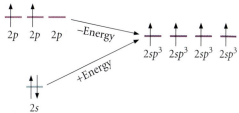

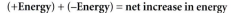

(**+Energy**) + (**−Energy**) = **net increase in energy**

Boron trifluoride (BF$_3$) was mentioned in Chapter 8. This compound is particularly useful in enhancing the reactivity of molecules during the synthesis of new pharmaceutical agents. Using the ideas of hybridization and valence bond theory, what is the structure of BF$_3$? The electron configuration of boron is $1s^2 2s^2 2p^1$. The Lewis dot structure of boron trifluoride indicates that we should have three single bonds, so we must hybridize the orbitals on boron to make three

FIGURE 9.11

Hybridization to give three orbitals in BF₃.

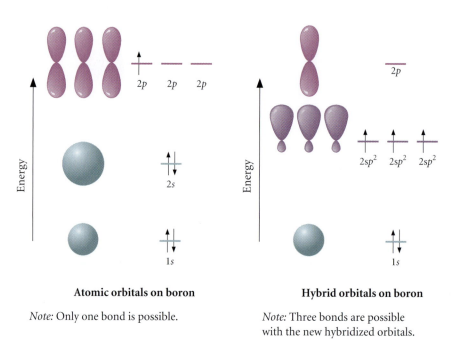

Atomic orbitals on boron

Note: Only one bond is possible.

Hybrid orbitals on boron

Note: Three bonds are possible with the new hybridized orbitals.

orbitals. Following the steps in Table 9.3, we mix the $2s$ orbital and two of the $2p$ orbitals (all from boron) to make the new orbitals. The three resulting orbitals have the designation $2sp^2$, as shown in Figure 9.11. We didn't hybridize one of the $2p$ orbitals because it wasn't needed to make a bond (i.e., the empty unhybridized $2p$ orbital is still available on the boron atom). We rewrite the electron configuration for boron trifluoride to end up with three orbitals that are available to bond with the half-filled $2p$ orbital on each unhybridized fluorine atom (p–sp^2 overlap).

We can build a model of beryllium hydride (BeH₂) showing covalent bonds between the beryllium ($1s^2 2s^2$) and hydrogen atoms. There are no electrons in the atomic $2p$ orbitals. Because we know that there are two bonds in BeH₂, we hybridize two orbitals to make two new $2sp$ orbitals, as shown in Figure 9.12. Rewriting the electron configuration shows that we can now make bonds to the hydrogen atoms, resulting in an overlap of the $1s$ and $2sp$ orbitals. In beryllium hydride, there are two empty, unchanged $2p$ orbitals.

FIGURE 9.12

Hybridization to give two orbitals in BeH₂.

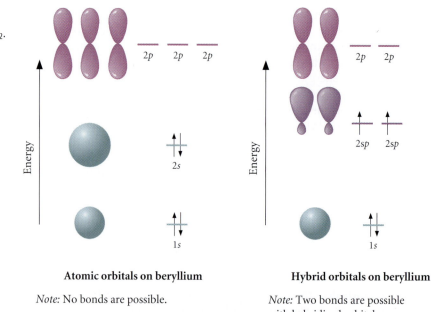

Atomic orbitals on beryllium

Note: No bonds are possible.

Hybrid orbitals on beryllium

Note: Two bonds are possible with hybridized orbitals.

EXERCISE 9.3   **Hybridization in Common Molecules**

Describe the hybridization of the central atom in each of these molecules:

a. H_2O b. NH_3 c. AlH_3

Solution

a. Lewis dot structure models for water indicate that the central oxygen has two bonds and two lone pairs. Mixing the four orbitals (one $2s$ and three $2p$) on the oxygen atom allows us to have two lone pairs of equal energy and two equal bonds to the adjacent hydrogen atoms. The oxygen atom possesses $2sp^3$ hybridized orbitals.

b. In a similar fashion, the nitrogen atom requires four things (three bonds and one lone pair) in ammonia. Mixing the four orbitals in its valence shell gives us three bonds to the hydrogen atoms and one lone pair (all degenerate in energy). The nitrogen atom is $2sp^3$ hybridized.

c. Aluminum has only three electrons in its valence shell. On the basis of its electron configuration, only one bond would be allowed. Because three points of attachment are needed (one to each hydrogen), we mix the $3s$ and two of the $3p$ orbitals to make an $3sp^2$ hybridized aluminum.

PRACTICE 9.3

What is the hybridization of the central atom in each of the following substances?

a. OF_2 b. H_2S c. $NH_4{}^+$

See Problems 23, 24, 33, 34, 43, and 44.

Shapes of the Hybrids

Visualization: Hybridization: *sp*
Visualization: Hybridization: *sp²*
Visualization: Hybridization: *sp³*

Hybrid orbitals have a shape that results from the mixing of the corresponding atomic orbitals. If we add an *s* orbital and a *p* orbital together, the result is one of the two hybrid *sp* orbitals. If we subtract the *s* orbital from the *p* orbital, we get the other hybrid *sp* orbital as shown in Figure 9.13. The angles between the resulting orbitals agree with what we expect from VSEPR rules. The *sp* hybridized atom has bonds with 180° angles, the *sp²* hybridized atom forms bonds with 120° angles, and the *sp³* hybridized atom contains bonding angles of 109.5° (see Figure 9.14; see also Table 9.4).

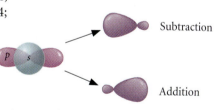

Subtraction

Addition

FIGURE 9.13

Mixing an *s* and a *p* orbital. Addition and subtraction of the *s* and *p* orbitals gives two *sp* orbitals.

TABLE 9.4   **Selected Hybrid Orbitals**

	Hybrid	Example	C—H Bond Length (pm)	Angle	C—C Bond Length (pm)
Orbital Energy ↑	sp^3	CH_3—CH_3	109	109.5°	154
	sp^2	CH_2=CH_2	108	120°	134
	sp	$CH{\equiv}CH$	106	180°	120

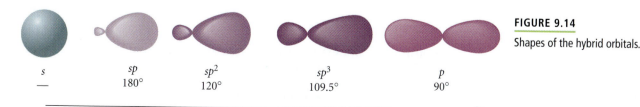

FIGURE 9.14

Shapes of the hybrid orbitals.

s   *sp* 180°   *sp²* 120°   *sp³* 109.5°   *p* 90°

Increasing length of orbital

EXERCISE 9.4 Shapes of the Molecules

For each of the molecules in Exercise 9.3, indicate the geometry around the central atom.

Solution

a. H_2O geometry: sp^3 hybridized atoms adopt a tetrahedral geometry. Because two of the sp^3 orbitals contain lone pairs, the VSEPR model indicates that the molecule has an overall bent geometry. The bond angle should be less than 109.5° because the lone pairs repel each other more than the bonding pairs. The angle H—O—H is actually 104.5°.

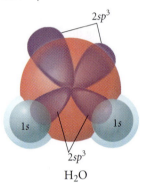

H_2O

b. NH_3 geometry: Again, we should have sp^3 hybridized orbitals with a tetrahedral geometry. Because one of the sp^3 orbitals contains a lone pair, ammonia has a trigonal pyramidal geometry. The bond angle (H—N—H) should be less than 109.5° because of the repulsions between the lone pair and the bonding pairs of electrons (the angle is actually 107°).

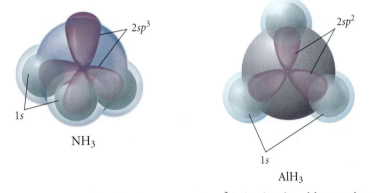

NH_3

AlH_3

c. AlH_3 geometry: The aluminum atom is sp^2 hybridized and has a trigonal planar geometry. The bond angle is 120°.

PRACTICE 9.4

For each of the molecules in Practice 9.3, indicate the geometry about the central atom.

See Problems 25–28, 35, and 36.

sp^3d and sp^3d^2 Orbitals

Hybrid orbitals can also be constructed to account for expanded octets. A good example is phosphorus pentachloride (PCl_5), a reactive molecule used to make compounds containing chlorine. The Lewis dot structure model of PCl_5 indicates that there should be five bonds to the phosphorus atom. Because the combination of all of the s and p orbitals will allow only four bonds, we must include a d orbital in the hybridization scheme. Five orbitals on the phosphorus can be made by

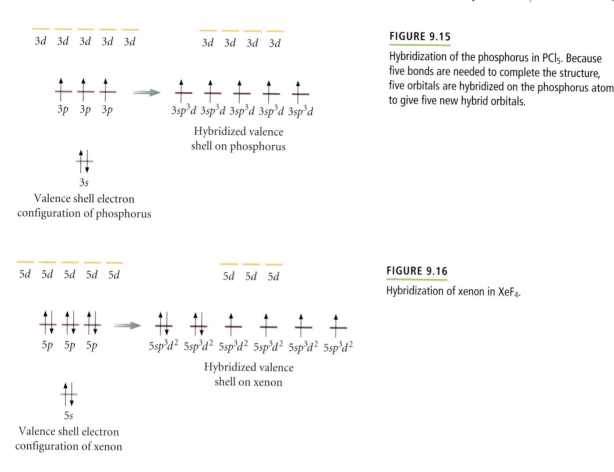

FIGURE 9.15

Hybridization of the phosphorus in PCl_5. Because five bonds are needed to complete the structure, five orbitals are hybridized on the phosphorus atom to give five new hybrid orbitals.

FIGURE 9.15

Hybridization of the phosphorus in PCl_5. Because five bonds are needed to complete the structure, five orbitals are hybridized on the phosphorus atom to give five new hybrid orbitals.

FIGURE 9.16

Hybridization of xenon in XeF_4.

hybridizing the $3s$, three $3p$, and one of the $3d$ orbitals. The result is five degenerate $3sp^3d$ orbitals. We can mix the d orbital because it is in the same principal energy level and is similar in energy to the other orbitals. The resulting valence bond model of PCl_5 shows overlap of a $3p$ orbital from the chlorine atom with one of the $3sp^3d$ orbitals from the phosphorus atom (Figure 9.15). The space-filling model of the molecule shows how the orbitals fit together to give the molecule its essential structure.

Other examples of expanded octets are found in compounds containing xenon. The first example of a compound containing xenon, a noble gas, was made in 1962. Shortly after the discovery that xenon could make compounds with other elements, XeF_2, XeF_4, and XeF_6 were prepared. The orbital overlap required to make xenon tetrafluoride (XeF_4) shows why these compounds are possible. The Lewis dot structure of XeF_4 requires six orbitals (four for bonds and two for lone pairs). We hybridize a $5s$, three $5p$, and two $5d$ orbitals to make six degenerate $5sp^3d^2$ orbitals. Each bond in XeF_4 results from overlap of a $2p$ orbital from the fluorine atom with a $5sp^3d^2$ orbital from the xenon atom, with the two lone pairs on the xenon atom occupying two $5sp^3d^2$ orbitals (Figure 9.16).

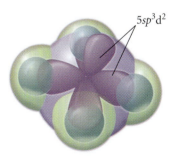

Sigma and Pi Bonds

Our discussion so far has centered on the use of valence bond theory to address the formation of single bonds in molecules. However, we know from Chapter 8 that atoms can participate in multiple bonds. **How is this possible if hybridization provides only orbitals that are oriented directly at the atom with which they are bonded?** The answer lies in the different ways in which orbitals can overlap in valence bond theory. **Sigma bonds (σ bonds)**, which make up the framework of a molecule, result from an *end-on overlap* of orbitals. Almost every "single bond" is a sigma bond. In fact, every bond we've discussed thus far in this chapter results from end-on overlaps of orbitals, such as the four σ bonds in CH_4, each of which

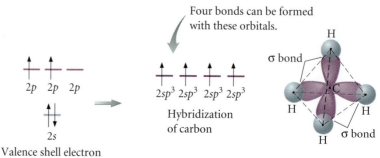

Four bonds can be formed
with these orbitals.

FIGURE 9.17

Methane hybridization
and sigma bonds.

is an end-on overlap of an sp^3 orbital and an s orbital. The resulting σ bond does not possess a nodal plane along the axis of the bond (Figure 9.17) and has a shape reminiscent of a hot dog. **Nodal planes** are flat, imaginary planes passing through the bonded atoms where the orbital does not exist. An example of a nodal plane is found in many molecules that contain double bonds.

Video Lesson: Pi Bonds

Pi bonds (π **bonds**), which react with other molecules as if they are electron-rich bonds, result from the *side-to-side overlap* of orbitals. This type of bond occurs primarily when p orbitals on adjacent atoms overlap, as is found between atoms containing a double or triple bond. The resulting π bond has one nodal plane along the axis of the bond and has the shape of the two halves of a hot dog bun separated by empty space. That is, there is a plane that exists between the atoms that is not occupied by π electrons. Acetylene (C_2H_2) is a molecule that contains two π bonds resulting from the overlap of p orbitals (Figure 9.18).

FIGURE 9.18

Acetylene pi bonds.
There are two π bonds
in acetylene that result
from the overlap of two
p orbitals from each
carbon atom.

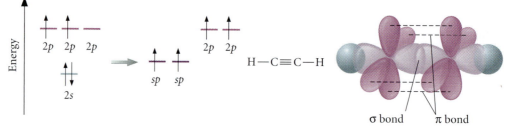

Why do we show the $2sp$ and $2p$ orbitals in the diagrams in Figure 9.18 with one electron each? Shouldn't we fill the $2sp$ orbitals before filling the slightly higher energy $2p$ orbitals? Hund's rule would seem to suggest this, but because the energy difference between the hybridized and unhybridized orbitals of the same shell is relatively small, the electrons can be temporarily placed in higher energy $2p$ orbitals prior to bond formation. The bonds that are created end up with lower energies than they would have had in the hybridized orbitals. Keep in mind that because each hybridized orbital will overlap to form a bond, no more than one electron can be placed in each of the hybridized orbitals. The neighboring atom will provide the second electron used in the bond. In addition, it is important to note that although we have discussed the hybridization process as a series of steps, no hybridization will take place without the formation of bonds. That is, in the absence of any other atoms, hybridization does not occur in an individual atom.

EXERCISE 9.5 **Polymers and Pi Bonds**

Ethylene (C_2H_4) is used as the starting material for the preparation of polyethylene, a common polymer used in the production of plastic bags, plastic sheeting, and the like (see Chapter 13 for more on plastics). Describe the bonding patterns in ethylene using the valence bond theory.

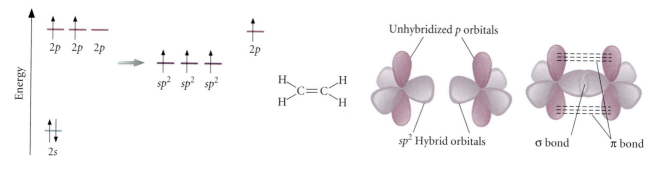

FIGURE 9.19

Ethylene and valence bond theory. The overlap of the orbitals between the carbons gives rise to a sigma bond and a pi bond. The sigma bond results from $2sp^2$–$2sp^2$ overlap. The pi bond results from the side-to-side overlap of the $2p$ orbitals.

Visualization: Formation of C=C Double Bond in Ethylene

Solution

After hybridization of the two carbon atoms in ethylene, we observe that the $2sp^2$ orbitals on the carbons can overlap to form a sigma bond. In addition, each carbon atom still contains a half-filled $2p$ orbital. Such orbitals can overlap to create a pi bond between the atoms. Ethylene, shown in Figure 9.19, has one pi bond and a total of five sigma bonds.

PRACTICE 9.5

Describe the valence bond theory for formaldehyde (CH_2O), a compound used in the preservation of biological tissues.

See Problems 11, 12, 37–40, 96, and 99.

Application of Hybridization Theory

The hybridization of orbitals allows us to address the same geometric features that are explained in the VSEPR approach to modeling. Why do we hybridize the orbitals in water? **Does it give a better model?** Valence bond theory indicates that if we didn't hybridize the orbitals of the central oxygen in water, there would still be two bonds. However, the bonds that would result from the lack of hybridization would have 90° angles between each other (overlap of a hydrogen s orbital with an oxygen p orbital). Hybridization is necessary for the model of water to agree with the *experimentally* measured bond angles.

The use of hybridized orbitals also provides information about the relative lengths of bonds in molecules. We've already discussed how s–p orbital overlap produces a bond that is longer than one containing s–s orbital overlap. This discussion can be carried further to illustrate the lengths of hybridized orbital bonds. Table 9.4 lists the relative sizes of these orbitals and some features of the resulting hybridization. From this table we see that the sp^3 hybrid orbital makes bonds that are longer than the sp^2. The sp^2 hybrid orbital is longer than the sp. Apparently, mixing more p orbitals into the hybrid produces longer hybrid orbitals. And we note that a bond made from sp^3–sp^3 orbital overlap is longer than one made from s–sp^3 overlap. The values in the table are *average* values (bond lengths and bond angles) based on actual measurements from a collection of similar molecules.

EXERCISE 9.6 **Bond lengths of C—C Single Bonds**

The Lewis dot structures of butane (often used as lighter fluid), 1–butene (a compound used to make plastic wraps), and 1,3-butadiene (the starting material used to make synthetic rubber for your car's tires) are shown in Figure 9.20. Which has the longest C—C bond? Which has the shortest C—C bond?

FIGURE 9.20

Butane, 1-butene, and 1,3-butadiene. (The bond length in question is indicated with an arrow.)

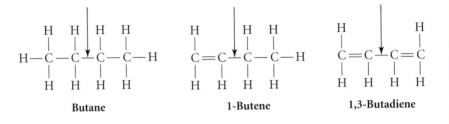

First Thoughts

Our first goal here is to determine the hybridization of the atoms in each of the molecules. Then, using this information, we should put together a geometric picture of each molecule that shows the types of orbital overlap that we expect for each bond.

Solution

The shortest C—C bond is found in 1,3-butadiene. It results from the overlap of an sp^2 hybridized orbital with another sp^2 hybridized orbital. The longest bond is found in butane (sp^3–sp^3 orbital overlap).

Further Insights

The C—C bond length in 1,3-butadiene indicates that the electrons in the double bonds can travel throughout the entire molecule. We can illustrate this by drawing resonance structures showing a double bond between the two central carbon atoms.

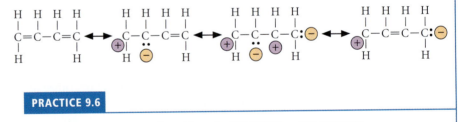

PRACTICE 9.6

Which has a longer N—O bond length, CH_2NOH or CH_3NHOH?

See Problems 13–15.

Advantages and Disadvantages of Hybridization

We must gain some advantage from spending time developing the hybridized model of a molecule. In fact, we do. With this model, we are able to explain bond angles, the lengths of bonds in molecules, and the existence of pi bonds. Unfortunately, valence bond theory and hybridization have some shortcomings. The relative size of many bond lengths can be accurately predicted, though some are "mystifyingly" unable to be accurately estimated. For example, why is the C=C bond longer in 1,3-butadiene than in 1-butene? An *even more refined* picture of the orbitals on the molecule needs to be considered.

Organic chemists do put hybridization to good use. Often they use it to describe the three-dimensional structure of a molecule to others. This approach enables organic chemists to talk to other organic chemists without drawing pictures on cocktail napkins. It also allows the organic chemist to describe specific structural features of a molecule that play an important role in the chemical and physical properties of that molecule.

EXERCISE 9.7   **Using Hybrid Orbital Theory: Will the Molecule Be Solid or Liquid?**

Animal fats, such as lard, tend to be solid. Vegetable oils, such as corn oil, tend to be liquid. Fats and oils are made up of compounds known as triacylglycerols. In turn, the triacylglycerols are constructed from molecules known as fatty acids. Look at the two fatty acids, stearic and oleic acids, depicted below. Describe the shapes of these molecules using valence bond theory, and use this information to determine which molecule is more likely to exist as a solid at room temperature.

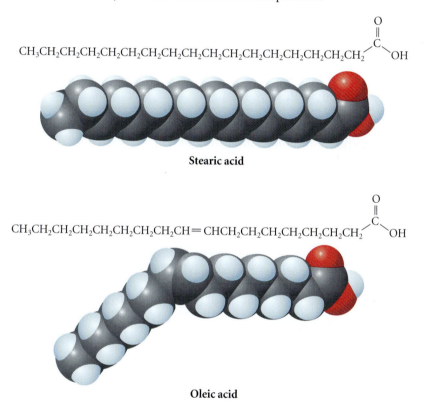

$CH_3CH_2CH_2CH_2CH_2CH_2CH_2CH_2CH_2CH_2CH_2CH_2CH_2CH_2CH_2CH_2CH_2$ —C(=O)—OH

Stearic acid

$CH_3CH_2CH_2CH_2CH_2CH_2CH_2CH_2CH$ = $CHCH_2CH_2CH_2CH_2CH_2CH_2CH_2$ —C(=O)—OH

Oleic acid

First Thoughts

What makes a compound likely to be solid at room temperature? If the molecules can pack easily and tightly together, they will be more likely to form a solid. If the molecules are more rigid and less able to get close together, they will be less likely to form a solid at room temperature.

Solution

Stearic acid is made up of 17 carbon atoms that are sp^3 hybridized and 1 carbon atom that is sp^2 hybridized. The sp^3 hybridized carbons involve single bonds to adjacent carbons and can rotate to form a zig-zag pattern of atoms. The sp^2 hybridized carbon is more rigid in its trigonal planar configuration. Oleic acid contains mostly sp^3 hybridized carbon atoms, but it also has three sp^2 hybridized carbons. The presence of two of the sp^2 hybridized atoms in the middle of the long carbon chain causes a kink in the zig-zag structure. Steric acid can pack tightly with other steric acid molecules, so it is a solid at room temperature. Oleic acid, because of its bent, rigid structure doesn't stack together well with other molecules. Oleic acid is a liquid at room temperature.

Further Insights

Knowing the three-dimensional structure of a molecule can provide information about the melting temperature of a molecule. But the three-dimensional structure of a molecule can also yield a wealth of other information. Physical properties, such as boiling point, melting point, color, taste, and smell, and chemical properties, such as reactivity with acids or bases, are all dependent on the shape of the molecule.

PRACTICE 9.7

Describe the three-dimensional shape of each of the following molecules. Solely on the basis of its shape, which do you predict will have the highest boiling point?

    a. $CH_3CH_2CH_2CH_2CH_3$ (pentane)

    b. $CH_3CH_2CH(CH_3)_2$ (2-methylbutane)

    c. $C(CH_3)_4$ (2,2-dimethylpropane)

See Problems 41, 42, 47, 48, 53, and 54.

HERE'S WHAT WE KNOW SO FAR

- We can mathematically combine two or more orbitals to provide new orbitals of equal energy in the process of hybridization.

- The models of many molecules agree with the experimentally measured bond angles when hybridization is taken into account.

- Although the hybridization process itself "runs uphill" energetically, the end result is still energetically downhill.

- Hybrid orbitals can be constructed to account for atoms that observe the octet rule as well as for those with expanded octets.

- The sigma (σ) bond does not possess a nodal plane along the axis of the bond and has a shape reminiscent of a hot dog. A pi (π) bond results from the side-to-side overlap of orbitals and is found between atoms containing a double or triple bond. It has one nodal plane along the axis of the bond.

9.3 Molecular Orbital Theory

Application

Let's revisit the molecule of retinal (the light-absorbing portion of the rhodopsin protein in our eyes) that we introduced in the chapter opener. 11-*cis*-Retinal is a molecule containing an extended series of alternating single and double bonds known as **conjugated π bonds** (or **conjugation**), as shown in Figure 9.21. Electrons in π bonds (whether conjugated or not) can absorb a photon of light. These electrons become excited (more energetic) when the photon is absorbed and the electrons are promoted to a higher energy state. The models we've already discussed say nothing about these higher energy states. Here are the key ideas we will explore in this section:

1. Higher energy states can be reached by the absorption of energy.

2. All electrons in a molecule can absorb photons of light.

FIGURE 9.21

11-*cis*-retinal is a conjugated molecule. Note that there are alternating single and double bonds through a large portion of the molecule.

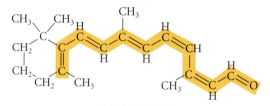

11-*cis*-Retinal

3. To become excited, core electrons require more energy (such as that of X-rays) than do valence electrons (for which the energy of UV or visible light is enough).

4. Sigma-bonding electrons require more energy (UV less than 200 nm) than pi-bonding electrons (visible between 200 and 700 nm) to become excited.

5. Electrons in conjugated π bonds require even less energy than unconjugated π bonds to promote an electron in a π bond to the higher energy state.

6. As the conjugated π system gets longer, the energy of the photon required to excite the electrons is reduced, moving from the blue toward the red region of the visible electromagnetic spectrum.

This final property is used by photography chemists in the design of molecules that can absorb a particular wavelength of light. Molecular orbital theory can also be used to tell us what happens to bonds within a molecule when they absorb light, why hydrogen exists as a diatomic molecule and helium does not, and why some molecules (such as O_2) are attracted to magnetic fields and why some molecules (such as N_2) are not.

Molecular Orbital Theory Defined

At about the same time that Linus Pauling worked out valence bond theory, Robert S. Mulliken (1896–1986) began thinking about how bonds could arise from delocalized valence orbitals. Erwin Schrödinger (1887–1961) further developed this approach by devising a mathematical equation that described the hydrogen atom. This theory, known as **molecular orbital (MO) theory**, is based on the principles of quantum mechanics that we discussed in Chapter 6. It treats electrons not as particles, but as waves that encompass the entire molecule. As in the quantum mechanical description of atomic orbitals, molecular orbitals encircle the atoms in a molecule. The Schrödinger equation defines the energy of each of the orbitals of an atom or molecule. Unfortunately, the Schrödinger equation cannot be solved exactly, except for a few very simple systems, so we usually make some approximations that allow us to arrive at a usable solution. Probably the most commonly used approximation is known as the **linear combination of atomic orbitals–molecular orbitals (LCAO–MO) theory**. In this theory, atomic orbitals are added together (both constructively and destructively) to make *molecular* orbitals.

We must raise several points about the formation of molecular orbitals from atomic orbitals (see Table 9.5). The most important point is that the combination of two orbitals gives two new orbitals, just as in the hybridization of atomic orbitals. One of the new molecular orbitals, called the **bonding orbital**, results from the *addition* of two overlapping atomic orbitals; the other, called the **antibonding orbital**, results from the *subtraction* of two overlapping atomic orbitals. The bonding orbital is lower in energy than either of the two atomic orbitals (the formation of a molecular orbital with lower energy than the atomic orbitals drives the formation of the bond). The bonding orbital indicates that there is some electron density between adjacent nuclei (a bond exists). The antibonding orbital is higher in energy than the two atomic orbitals from which it is formed and is typically represented with an asterisk (*) to distinguish it from the bonding orbital. The antibonding orbital indicates a lack of electron density between adjacent nuclei (no bond exists). Each new molecular orbital (both bonding and antibonding) can contain two electrons.

A similarly important point in mixing atomic orbitals to make molecular orbitals is that only atomic orbitals of similar symmetry (shape and orientation) and energy provide significant overlap. This rule means that only 1s orbitals overlap to a great degree with 1s orbitals. The $2p_z$ orbital overlaps best with a $2p_z$ orbital. Their symmetry and size (i.e., energy) are similar. The 1s orbital doesn't

Video Lesson: Molecular Orbital Theory

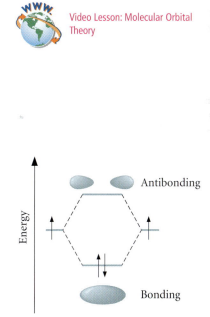

In LCAO–MO theory, atomic orbitals are added together, both constructively and destructively, to make molecular orbitals. The resulting bonding orbital is of lower energy, and the antibonding orbital is of higher energy, than the individual atomic orbitals.

TABLE 9.5	Key Features of MO Diagram Construction

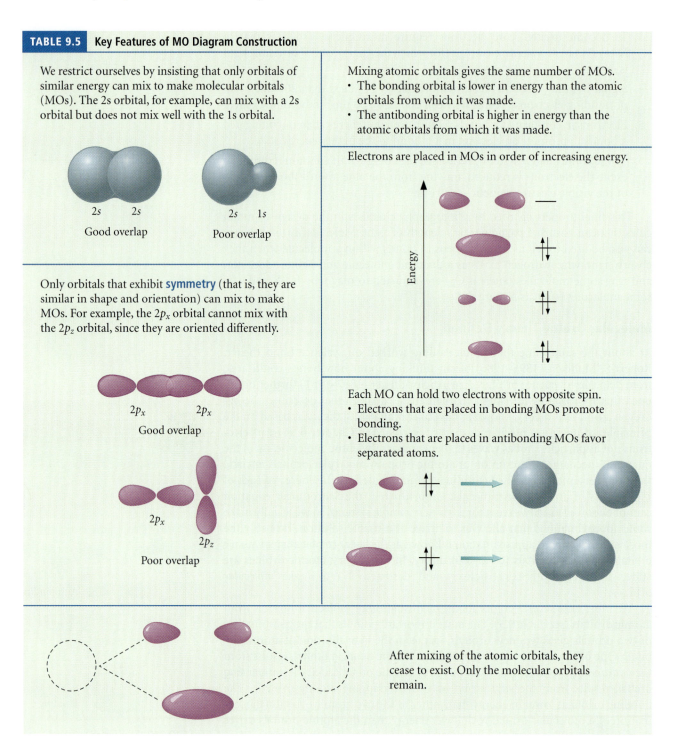

We restrict ourselves by insisting that only orbitals of similar energy can mix to make molecular orbitals (MOs). The 2s orbital, for example, can mix with a 2s orbital but does not mix well with the 1s orbital.

2s   2s
Good overlap

2s   1s
Poor overlap

Only orbitals that exhibit **symmetry** (that is, they are similar in shape and orientation) can mix to make MOs. For example, the $2p_x$ orbital cannot mix with the $2p_z$ orbital, since they are oriented differently.

$2p_x$ $2p_x$
Good overlap

$2p_x$
$2p_z$
Poor overlap

Mixing atomic orbitals gives the same number of MOs.
- The bonding orbital is lower in energy than the atomic orbitals from which it was made.
- The antibonding orbital is higher in energy than the atomic orbitals from which it was made.

Electrons are placed in MOs in order of increasing energy.

Energy

Each MO can hold two electrons with opposite spin.
- Electrons that are placed in bonding MOs promote bonding.
- Electrons that are placed in antibonding MOs favor separated atoms.

After mixing of the atomic orbitals, they cease to exist. Only the molecular orbitals remain.

overlap well with a 3s orbital (the size isn't a good match). The $2p_x$ orbital does not overlap well with the $2p_z$ orbital (the symmetry doesn't allow them to line up properly).

We can think of this by imagining that overlapping orbitals are a game of Tetris. In Figure 9.22, we can see that the best game piece to use to fill the 4 × 4 hole at the bottom of the screen is a 4 × 4 rectangle. A 3 × 3 rectangle wouldn't do a very good job of filling the hole, just as a 1s orbital and a 3s orbital don't overlap very well. They have the same symmetry (shape), but the size isn't right.

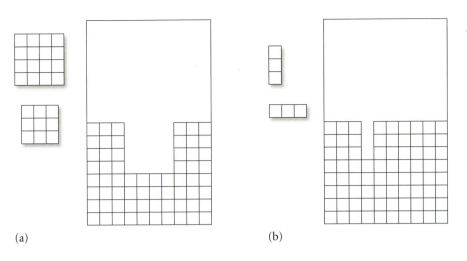

(a)   (b)

FIGURE 9.22

Tetris analogy to overlapping orbitals: (a) The 4 × 4 square piece has the same size and shape as the hole. The 3 × 3 square has the same shape but the wrong size. (b) The 1 × 3 rectangle has the same size and shape as the hole. The horizontal 1 × 3 rectangle has the same size but is the wrong orientation to fit into the hole.

In the second image of Figure 9.22, we need a vertical 1 × 3 rectangle to fill the vertical 1 × 3 hole. A horizontal rectangle of the same size wouldn't work, just as a $2p_z$ orbital doesn't overlap with a $2p_x$ orbital. The symmetry (shape) of the two orbitals doesn't match, even though their size does.

Shapes of Orbitals

What about the shapes of the molecular orbitals? Molecular orbitals encompass entire molecules, and the shapes of the molecular orbitals often look similar to portions of the entire molecule. We know about their existence by calculation using some approximations of the Schrödinger equation, but how do we really know what the shape of a molecule is? X-ray crystallography (see Section 13.1) is one of the methods used to take a "snapshot" of a molecule. But it relies on determining the average location of the atoms in a crystal. The scanning tunneling microscope that we discussed in Chapter 6 may actually be the closest we can get to taking a picture of a single molecule.

The strength of the bond between two atoms can be represented by the **bond order**, the number of electrons in bonding orbitals minus the number of electrons in antibonding orbitals, divided by 2.

$$\text{Bond order} = \frac{\text{bonding electrons} - \text{antibonding electrons}}{2}$$

A larger bond order indicates a greater bond strength. The bond order typically agrees with the Lewis dot structure's expected number of bonds between adjacent atoms. However, the bond order between two atoms doesn't have to be a whole number.

To summarize, molecular orbital theory holds that when the atoms of a molecule come together, atomic orbitals on each atom that are similar in symmetry and energy not only overlap but are *transformed* into a series of new orbitals that surround the entire molecule. These molecular orbitals have completely different shapes, sizes, and energies than the atomic orbitals from which they came. Although the atomic orbitals cease to exist after the creation of the molecular orbitals, the new molecular orbitals follow the same rules that guide us in placing electrons in the atomic orbitals.

Visualization: Pi Bonding and Antibonding Orbitals

MO Diagrams

Just as the civil engineer diagrams the layout of a building with a blueprint, we represent the placement of electrons in molecular orbitals with our own type of blueprints called **molecular orbital (MO) diagrams**. These diagrams graphically

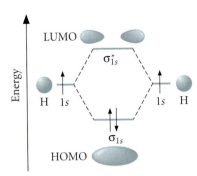

FIGURE 9.23

MO diagram for H_2. This diagram graphically illustrates the energies and the resulting orbital shapes of the different molecular orbitals in the molecule.

Visualization: Sigma Bonding and Antibonding Orbitals

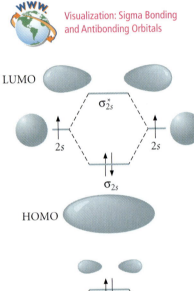

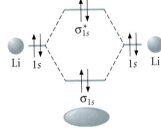

FIGURE 9.24

Dilithium (Li_2) MO diagram. The MO configuration can also be written in shorthand format as $\sigma_{1s}^2 \sigma_{1s}^{*2} \sigma_{2s}^2$.

Video Lesson: Applications of the Molecular Orbital Theory

illustrate the energy of each of the bonding and antibonding orbitals and assist in determination of the bond order for a molecule. Graphically, the shapes of each of the molecular orbitals can be added to the diagram to get the overall picture of the molecule.

We can draw the blueprint—the MO diagram—for molecular hydrogen (H_2) to illustrate how this is done. We show in Figure 9.23 that as the two hydrogen atoms approach each other, their atomic $1s$ orbitals mix to provide two new molecular orbitals. The overlap of these orbitals possesses symmetry, which means that they are equal in size and shape about the axis between the atoms (σ). One of the molecular orbitals resulting from this overlap is lower in energy (the bonding orbital, σ_{1s}, read as "sigma one s") and the other is higher in energy (the antibonding orbital, σ^*_{1s}, read as "sigma one s star"). Placement of the two electrons in H_2 follows the Pauli exclusion principle (two electrons with opposite spin per orbital) and Hund's rule (fill lowest-energy orbitals first) to fill the bonding orbital. The antibonding molecular orbital is left empty. The **highest-energy occupied molecular orbital (HOMO)** is the molecular orbital that contains electrons (in this case, the bonding orbital). The antibonding orbital ends up as the **lowest-energy unoccupied molecular orbital (LUMO)**. The bond order for hydrogen can be calculated ((2 bonding electrons − 0 antibonding electrons)/2) as 1. We therefore predict that molecular hydrogen has a single bond between the two nuclei. **What would happen to the bond order for this molecule if one of the electrons were promoted to the LUMO?** The bond order for such a molecule would be zero ((1 bonding electron − 1 antibonding electron)/2). With no net bond between the two atoms, the molecule would break into the individual hydrogen atoms.

EXERCISE 9.8 MO Theory: More Power, Scotty

We showed in Exercise 9.2 that dilithium (of *Star Trek* fame) can be explained by the valence bond model. What does MO theory say about dilthium (Li_2)?

Solution

The MO diagram for dilithium is shown in Figure 9.24. The HOMO is the σ_{2s} orbital and the LUMO is the σ^*_{2s} orbital. The bond order for this molecule is 1 ((4 − 2)/2), so MO theory suggests that Li_2 should exist as a molecule with one bond between the two lithium atoms. The molecule does exist, but it quickly reacts with other lithium atoms to make lithium metal. Because of this reactivity, it is not possible to have a bottle of Li_2.

PRACTICE 9.8

Use MO theory to determine whether He_2 is a stable molecule. What is the bond order in this molecule?

See Problems 67, 68, and 71–74.

More About MO Diagrams

Lewis dot structures and valence bond theory agree that there are six bonding electrons (that is, a triple bond) between the nitrogen atoms in gaseous nitrogen (N_2). The MO diagram for the molecule should verify that the theory also gives rise to a bond order of 3. In molecules containing more than just the s orbitals, the MO diagram becomes much more complicated. Figure 9.25 shows that the overlap of the $2p_y$ orbitals on each N and the $2p_z$ orbitals on each N, produces four orbitals: two π and two π^*. The overlap of the $2p_x$ orbitals gives a sigma molecular orbital. If we now place the electrons in the diagram, we see that they fill four bonding orbitals and one antibonding orbital. From the MO diagram, we

FIGURE 9.25

MO diagram for N_2.

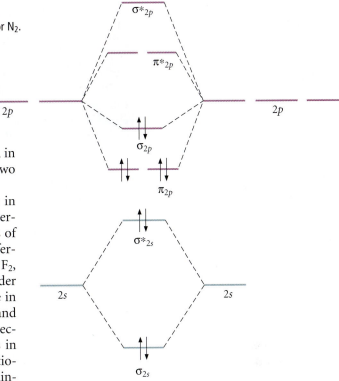

can calculate a bond order of 3 for molecular nitrogen, in which the HOMO is the σ_{2p} orbital and there are two LUMOs (the degenerate π_{2p}^* orbitals).

The energy of the molecular orbitals in N_2 results in the ordering shown in Figure 9.25. This is the same ordering that is observed in most of the diatomic molecules of the second period of the periodic table. However, a different ordering of the molecular orbitals is seen in O_2, F_2, and Ne_2, as shown in Figure 9.26. The change in the order of the molecular orbitals arises because the difference in energy between the $2s$ and $2p$ atomic orbitals on O, F, and Ne is relatively greater than the other elements in the second row (see Figure 9.7). Note that the two electrons in the HOMO are not paired in O_2. This allows us to rationalize a property of molecular oxygen that is unexplainable by other bonding theories. The other models of molecular oxygen correctly assign two bonds between the oxygen atoms, but oxygen possesses properties that we would not predict on the basis of those models. Oxygen, shown in Figure 9.27, is an example of a paramagnetic molecule. The term **paramagnetism** refers to the ability of a substance to be attracted into a magnetic field. This attraction arises because of the presence of *unpaired* electrons within the molecule. A special case of this is **ferromagnetism** (so named because the effect is especially strong in iron), in which the paramagnetic atoms are close enough together that they reinforce their attraction to the

FIGURE 9.26

MO diagram for O_2. Note that molecular oxygen has a different order for its molecular orbitals. There are also two unpaired electrons in the MO diagram. What is the bond order for O_2?

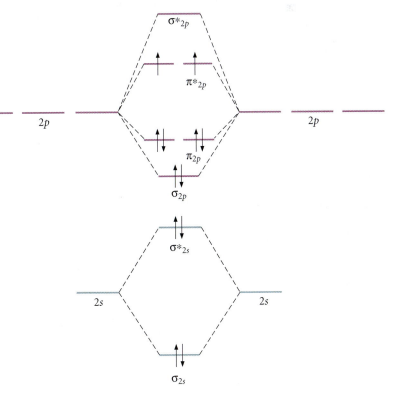

FIGURE 9.27

The results of paramagnetism and diamagnetism. Molecular oxygen is paramagnetic. Because of this, liquid oxygen is attracted to a magnetic field (a). Liquid nitrogen is diamagnetic and is repelled from the magnetic field (b).

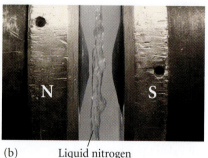

(a) Liquid oxygen (b) Liquid nitrogen

Visualization: Magnetic Properties of Liquid Nitrogen and Oxygen

Video Lesson: CIA Demonstration: The Paramagnetism of Oxygen

magnetic field, so the whole is, in effect, greater than the sum of its parts. The opposite of paramagnetism is **diamagnetism**. A diamagnetic molecule's electrons are *paired*, resulting in the molecule being repelled from a magnetic field. Nitrogen (N_2) is an example of a diamagnetic molecule because all of its electrons are paired (see Figure 9.25).

Constructing models of molecules made with more than two atoms or molecules made from two different atoms (heteronuclear diatomic molecules) complicates the molecular orbital diagram because similar atomic orbitals on each atom have different energy levels. A further complication is added because the orbitals encompass the entire molecule and therefore must be constructed from the linear combination of the atomic orbitals on every atom in the molecule. For a 10-atom molecule, that would mean that each molecular orbital would arise from the combination of 10 different atomic orbitals. Such a diagram would be very complex indeed!

EXERCISE 9.9 **MO Theory: The Power of a Photon**

The halogens (F_2, Cl_2, Br_2, and so on) are extremely reactive compounds. In fact, their reaction can be initiated by a single photon of light, as we note at the start of this chapter when we discussed 11-*cis*-retinal in the eye. Given that the photon will promote one electron from a bonding orbital to an antibonding orbital of similar symmetry, illustrate the reaction of F_2 with that photon using MO diagrams. The MO diagram for F_2 is similar to that for O_2.

Solution

The MO diagram for F_2 is shown in Figure 9.28. Promotion of an electron from the π_{2p} orbital to an MO of the same symmetry, the σ^*_{2p} orbital, would cause the bond order for F_2 to become zero. The photon would effectively break the bond between the fluorine atoms. The reaction of photons with chlorofluorocarbons in the ozone layer similarly breaks C—Cl bonds. The resulting chlorine atoms are extremely reactive toward ozone (O_3).

FIGURE 9.28

MO diagram of F_2. The promotion of an electron from a π MO to the σ^* MO reduces the bond order from 1 to 0. When this happens, there is no net bond remaining between the atoms. Addition of a photon of light to the molecule with a wavelength of 754 nm causes the molecule to break in two.

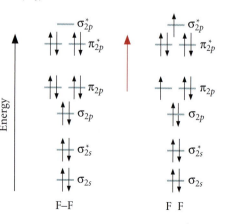

PRACTICE 9.9

Illustrate the interaction of H_2 and O_2 with a single photon of light. After the promotion of an electron, which molecule has the larger bond order?

See Problems 65 and 66.

Looking back to 11-*cis*-retinal, we can produce an MO diagram that illustrates the location of the electrons in the conjugated π system. Although the diagram is unusually complex, the energy levels of the conjugated π system are more closely spaced as the π system increases in length. Promotion of an electron requires less and less energy as the π system lengthens, as indicated by the increasing wavelength of absorbed light in the molecules of Figure 9.29. This occurs because it takes less energy to promote an electron to a closer molecular orbital. We pointed out earlier that this phenomenon is used by the photographic industry in the development of color print film. Here's how. By constructing a molecule that has a conjugated π system of just the right length, photo manufacturers can make molecules that absorb exactly the color needed to make your pictures really stand out. We develop this idea further in the accompanying NanoWorld/MacroWorld feature.

FIGURE 9.29

The MOs of conjugated systems. As the length of the conjugated π system increases, the energy required to promote an electron decreases. Evidence of this is observed in the wavelength of light absorbed by the molecules. Longer π systems require less energy and a larger wavelength (energy and wavelength are inversely proportional, $E = \frac{hc}{\lambda}$).

1,3-Butadiene
$\lambda = 217$ nm

1,3,5-Hexatriene
$\lambda = 268$ nm

1,3,5,7-Octatetraene
$\lambda = 304$ nm

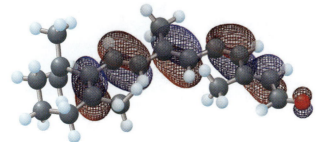

cis-Retinal
$\lambda = 325$ nm

NanoWorld / MacroWorld

Big effects of the very small: Color photography and the tricolor process

"Smile, you're on candid camera!" Everyone hopes never to be ambushed with those words. In the nineteenth century, when photography was in its infancy, candid photographs were not possible. In order to create a visual memory of an event, you had to visit the photographer and sit for a picture. "Sitting" meant holding a pose for up to 30 seconds. Any movement would show as a blur in the final picture. However, the idea that an inexpensive, yet realistic, picture could be created in such a short time brought people to the photographer's studio in droves.

Scientists in the early 1800s were intrigued by the idea of painting portraits without the artist's pen. In the summer of 1827, Joseph Nicéphore Niépce, using material that hardened on exposure to light exposed film for eight hours in order to get a picture. Because no one wanted to sit still that long, most of his pictures were landscapes. In 1839, however, two other photographic processes were developed, the daguerreotype (perfected by Louis Daguerre) and the calotype (Henry Fox Talbot). Although these techniques allowed pictures to be created with much shorter exposures, they still produced a black-and-white image (see Figure 9.30). This didn't matter much to the general public, and photographic portrait studios seemed to pop up on every corner across the globe. Despite some marginal success in producing faint color images during the 1850s and 1890s, the advent of the color photograph as an inexpensive, reproducible and stable process didn't occur until the 1920s.

Researchers at Kodak Research Laboratories in 1935 finally hit upon the process that would bring the color photograph to the world. Kodachrome®, their color photography process, introduces color to a picture by separating the three primary colors into specific emulsion layers. How does it work? The emulsion layers contain a compound that reacts when light strikes it. A quick look at the chemistry behind the black-and-white photograph will give us some insight.

In black-and-white photography, a piece of transparent plastic is coated with silver bromide crystals (see Figure 9.31) and a sensitizer. The plastic sheet is then placed in a camera, and when the shutter is opened, the silver bromide absorbs photons of light. In the presence of the sensitizer, the silver cation is reduced by the sensitizer (gains an electron) and becomes silver metal:

$$Ag^+(s) + sensitizer \rightarrow Ag^\circ(s) + sensitizer^+$$

Then the film is wound back into its container and sent to be developed. The technician rinses the plastic film to remove unreacted silver bromide. The silver metal remains in place on the plastic. The result is an inverted image called a negative. Next the technician shines light through the negative onto another silver bromide/sensitizer-coated support (typically paper this time) and rinses off the unreacted silver bromide. Violà! We have a positive image called a photograph.

Why is the film coated with silver bromide instead of silver chloride? We can answer this by examining the wavelength of light that is absorbed; see Figure 9.32. Comparison of the two shows that silver bromide absorbs a large amount of the visible light that enters our camera. To get the best picture, we want to absorb as much light as we can.

FIGURE 9.30

A daguerreotype of Michael Faraday (see Chapter 19) taken sometime between 1844 and 1860 at Mathew Brady's studio in New York City. Because the process required the subject to remain completely motionless for up to 10 seconds, most daguerreotypes were taken in well-lit studios. Even though the early daguerreotypes cost one month's wages for the average person, they were much cheaper than a painted portrait. This made them an instant success.

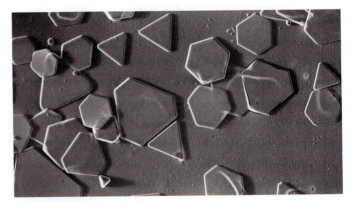

FIGURE 9.31

A magnified image of silver bromide crystals used in the photographic industry. Note that the crystals are mostly hexagonal in shape and flat so that more surface area is exposed to incoming photons.

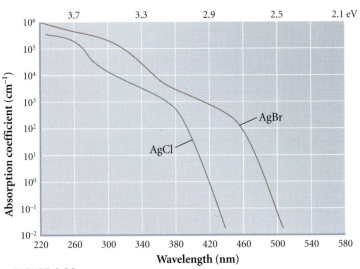
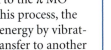

FIGURE 9.32

Spectrum of silver bromide and silver chloride. Silver bromide absorbs more of the visible spectrum than silver chloride. What color is silver bromide?

excited molecules, many fates exist for the promoted electron. First, the molecule could release a photon and return the promoted electron to the π MO, which would result in the original dye molecule. The released photon would have the same energy as the absorbed photon. Second, the electron could collapse back to the π MO through thermal excitation. In this process, the dye molecule would release the energy by vibrating. Third, the electron could transfer to another MO on another molecule. This is the fate that gives rise to the photographic reaction.

The excited electron transfers from the dye molecule to an antibonding orbital on the silver bromide. And as it does so, our silver cation is reduced to silver metal. Even though the silver bromide was unable to absorb the non-blue photon of light, it still ended up as silver metal. Therefore, all we have to do to make a color photograph is find two dye molecules that can transfer electrons to the silver bromide. One should absorb yellow light and the other should absorb red light. The exact structures of these molecules are a closely guarded secret in the photography industry, but the compounds shown in Figure 9.33 have been used to do this in the past. Each has an extended pi system that allows the molecules to absorb light in the visible region of the spectrum.

Current advances in the photography industry include the development of four-color processes. These processes make possible better definition of objects and a more brilliant picture. Development of better ways to coat the silver bromide evenly on plastic and the preparation of silver bromide crystals that give the best interaction with light are at the forefront of current research. One day these developments may be able to completely eliminate "red eye" and make it easier to tolerate being the subject of a candid shot.

Silver bromide, though, doesn't absorb the entire spectrum of visible light. What color of light does silver bromide absorb? Figure 9.32 shows that it absorbs blue light. In order for us to make a full-color photograph, we need to make some of the silver bromide on the plastic absorb yellow light and some absorb red light. By mixing these primary colors (blue, yellow, and red), we can recreate the entire spectrum of colors. This is what happens in the tricolor film process. **How, then, can we make silver bromide absorb a different color of light?**

The process requires coupling the silver bromide with a dye molecule. These molecules are designed to absorb different colors by varying the length of their conjugated pi bonds. When the photon is absorbed into this system, an electron in a π MO is promoted to a π^* MO, and the dye molecule becomes excited. As with all

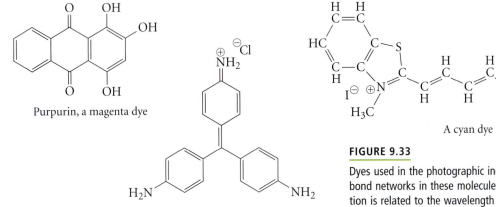

Purpurin, a magenta dye

Basic Red 9, a magenta dye

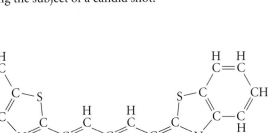

A cyan dye

FIGURE 9.33

Dyes used in the photographic industry. Note the extended pi bond networks in these molecules. The length of the conjugation is related to the wavelength of light that these molecules can absorb.

9.4 Putting It All Together

Visualization: Delocalized
Pi Bonding in the Nitrate Ion

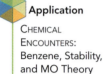

Application

CHEMICAL
ENCOUNTERS:
Benzene, Stability,
and MO Theory

FIGURE 9.34

Delocalized π bonding in benzene.
Lewis dot structures and valence bond
models do a good job of describing
the sigma bond network in benzene.

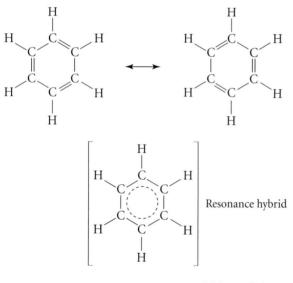

Resonance hybrid

FIGURE 9.35

Delocalized π bonding in benzene according to MO theory.

The models of each of the bonding theories provide increasingly better correlation with the observed properties of many covalent compounds. Valence
bond theory works well at showing how the electrons in sigma bonds come together to create the skeletal framework of a molecule. At the same time, the localized bonding picture described by valence bond theory conveniently determines,
among other things, the lengths of bonds in a molecule. The rules of VSEPR
modeling help identify the three-dimensional shape of the molecule, and Lewis
dot structures rapidly reveal the connectivity of atoms in a molecule. However,
molecular orbital theory works best at describing the behavior of electrons located in the bonds. An approach to drawing a molecule that uses all of these models is often followed. We can show this approach using benzene (C_6H_6).

Benzene is a carcinogenic compound that is widely used in industry as a nonpolar solvent and as a starting compound in the manufacture of other useful
products such as phenol (an antiseptic compound) and nylon. Experiments have
revealed that benzene is a hexagon of six carbon atoms, each bonded to one hydrogen atom. It has also been observed experimentally that the distances between
carbon atoms in the molecule are identical. In the laboratory, benzene reacts as
though it has C=C bonds at each position in the molecule.

No single Lewis dot structure of benzene can be drawn to illustrate the experimentally determined bond lengths and reactivity in the molecule. Considering
that benzene is best shown as a hybrid of two resonance structures, we can address the experimental observation that all C—C bonds in benzene are the same
length as shown in Figure 9.34. Valence bond theory can
also be used to indicate that each carbon is sp^2 hybridized. The sigma bond network that makes up the
molecule defines 120° bond angles at each flat sp^2 hybridized carbon. Bonds between carbon atoms result
from the overlap of a $2sp^2$ hybrid orbital with another
$2sp^2$ hybrid orbital. Each C—C bond is 140 pm in length.
Bonds between the carbon and hydrogen atoms are the
result of $2sp^2$–$1s$ orbital overlap. In addition, each carbon
atom contains a half-filled $2p$ orbital perpendicular to the
plane of the molecule.

However, the six half-filled $2p$ orbitals in benzene are
best handled by molecular orbital theory. The σ bond
network is separated from the π bond network in this approach. Doing so makes preparing a MO diagram a more
manageable task. The six $2p$ orbitals on the carbon atoms
in benzene have the same symmetry and energy, so they
can be mixed to provide six new π MOs. Three bonding
MOs and three antibonding MOs result (see Figure 9.35). We place the six electrons into the new MOs, filling the three bonding MOs. The electrons in the MO
diagram occupy a π bond at every carbon atom, implying that there isn't any difference in the carbons of benzene. Each C—C bond has a bond order of 1.5. The
approach we have taken to draw the best model of benzene has provided us with
the best (and quickest) way to represent the experimentally determined properties of benzene (such as the C—C bond lengths and the reactivity of benzene). We
no longer have to imagine different structures (arrived at from different models)
of benzene to account for the observed properties.

If we more closely examine the molecular orbital model of benzene, we find
that the electrons in the π orbitals are spread out over the six atoms in the structure. This conjugated π system is said to be **delocalized**. In fact, delocalized systems occur any time the electron density in a molecule can be distributed among

Molecular orbital diagram for benzene.

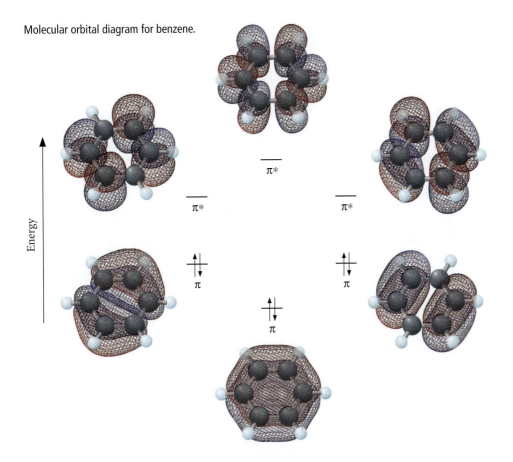

more than two atoms. Delocalization also can be seen in the carbonate ion (CO_3^{2-}) and in 11-*cis*-retinal. However, in the case of benzene (C_6H_6) and related compounds, the delocalization of the electron density imparts a particular stability to the molecule (more so than the delocalization of electrons in the carbonate ion and 11-*cis*-retinal). The stability is so important to the molecule that it is often written as shown in Figure 9.36, with a circle to illustrate the delocalization of three π bonds in the molecule. Benzene's stability makes it a good choice for use as a reaction solvent. Unfortunately, this stability also makes it difficult to metabolize if accidentally ingested. One of the major problems with using benzene is that it causes cancer in humans. Table 9.6 lists certain properties of benzene and of some alternative solvents that have been used in place of it. Their properties aren't identical to those of benzene, but their substitution for benzene has done a great deal to safeguard the health and lives of industrial chemists.

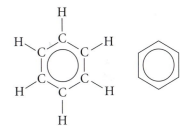

FIGURE 9.36

Shorthand notations for benzene.

TABLE 9.6	Common Solvents			
Solvent	**Structure**	**Boiling Point**	**Dipole Moment**	**Solubility in Water**
Benzene	C_6H_6	80.1°C	0	Very slight
Hexane	C_6H_{14}	68.7°C	0	Insoluble
Dichloromethane	CH_2Cl_2	40°C	1.60	Slight
Tetrahydrofuran	C_4H_8O	66°C	1.63	Miscible
Toluene	C_7H_8	111°C	0.36	Very slight

9.5 Molecular Models in the Chemist's Toolbox

Just as the best models in civil engineering give the best information on the outcome of a construction project, our happiness with a particular molecular model is based on our satisfaction with the information that the model provides. The more rigorous models provide better information, but at a cost in time and calculations that we may not be willing to accept. Valence bond theory quickly determines the skeletal framework of a molecule by considering separate parts of the molecule. It does not explain the molecule as a whole, nor does it consider how electrons can be delocalized between more than two atoms. Molecular orbital theory, on the other hand, considers the molecule as a whole. It explains the bonding patterns in detail, providing information about the energy of the electrons in a particular orbital and the properties of the molecule. However, detailed molecular orbital theory is difficult to solve mathematically. Using approximations of molecular orbital theory, such as the LCAO–MO model, considerably reduces the amount of time involved in determining the model, but at the expense of some of the detail provided by the theory. All of the models we have discussed—Lewis dot structures, VSEPR, valence bond theory, and molecular orbital theory—have a place in the toolbox of the chemist.

The Bottom Line

- The molecular models that chemists construct increase in complexity from Lewis dot structures, to the VSEPR model, to valence bond theory, to MO theory. This increase in complexity is accompanied by an increase in satisfactory agreement with observed properties for the molecule. (Section 9.1)

- Valence bond theory, originated by Linus Pauling, defines bonds as the overlap of atomic orbitals. Sigma bonds result from end-on overlap of orbitals; pi bonds result from side-to-side overlap of orbitals. (Section 9.1)

- Hybridization of atomic orbitals gives rise to new orbitals that help explain bonding. Hybridization also gives structures consistent with the VSEPR model. (Section 9.2)

- Mixing n atomic orbitals gives n hybridized orbitals. (Section 9.2)

- Hybridization can be used to explain the existence of sigma (σ) and pi (π) bonds in molecules. (Section 9.2)

- Molecular orbital theory defines bonding with orbitals that are not confined to a single atom. Bonds result from molecular orbitals in a molecule that encompass the atoms. (Section 9.3)

- Mixing n atomic orbitals gives n molecular orbitals. One-half n of these are bonding; the other half are antibonding. Bonding MOs are lower in energy than the atomic orbitals from which they are constructed. Antibonding MOs are higher in energy than the atomic orbitals from which they are constructed. (Section 9.3)

- The Pauli exclusion principle and Hund's rule must be obeyed when placing electrons in the new MOs. (Section 9.3)

- MO diagrams can be used to identify the number of bonds between atoms. (Section 9.3)

- Paramagnetism results from unpaired electrons in a molecule. Diamagnetism results from complete pairing of all electrons in a molecule. (Section 9.3)

- Electrons in conjugated π orbitals are said to be delocalized, because the electron density can be distributed among more than two atoms. (Section 9.4)

Key Words

antibonding orbital An orbital that indicates a lack of electron density between adjacent nuclei (no bond exists when the orbital is occupied). The antibonding orbital arises from the subtraction of two overlapping atomic orbitals. (*p. 375*)

bond order The number of electrons in bonding orbitals minus the number of electrons in antibonding orbitals, divided by 2. The bond order indicates the degree of constructive overlap between two atoms. (*p. 377*)

bonding orbital An orbital that indicates the presence of electron density between adjacent nuclei (a bond exists when the orbital is occupied). The bonding orbital arises from the addition of two overlapping atomic orbitals. (*p. 375*)

conjugated π bonds An extended series of alternating single and double bonds. (*p. 374*)

conjugation The presence of conjugated π bonds—that is, a series of at least two double bonds alternating with single bonds. (*p. 374*)

delocalized Term used to describe a π system wherein the electron density in a molecule can be distributed among more than two atoms. (*p. 384*)

diamagnetism The ability of a substance to be repelled from a magnetic field. This property arises because all of the electrons in the molecule are paired. (*p. 380*)

ferromagnetism A property of a compound that occurs when paramagnetic atoms are close enough to each other (such as in iron) that they reinforce their attraction to the magnetic field, such that the whole is, in effect, greater than the sum of its parts. (*p. 379*)

highest-energy occupied molecular orbital (HOMO) The most energetic molecular orbital that contains at least one electron. (*p. 378*)

hybridization The mathematical combination of two or more orbitals to provide new orbitals of equal energy. (*p. 363*)

linear combination of atomic orbitals–molecular orbitals (LCAO–MO) theory An approximation of molecular orbital theory wherein atomic orbitals are added together (both constructively and destructively) to make molecular orbitals. (*p. 375*)

lowest-energy unoccupied molecular orbital (LUMO) The least energetic molecular orbital that contains no electrons. (*p. 378*)

molecular orbital (MO) diagram A diagram used to illustrate the different molecular orbitals available on a molecule. (*p. 377*)

molecular orbital (MO) theory A theory that mathematically describes the orbitals on a molecule by treating each electron as a wave instead of as a particle. (*p. 375*)

nodal planes Flat, imaginary planes passing through bonded atoms where an orbital does not exist. (*p. 370*)

paramagnetism The ability of a substance to be attracted into a magnetic field. This attraction arises because of the presence of unpaired electrons within the molecule. (*p. 379*)

pi bond (π bond) A covalent bond resulting from side-to-side overlap of orbitals. The pi bond possesses a single nodal plane along the axis of the bond. (*p. 370*)

sigma bond (σ bond) A covalent bond resulting from end-on overlap of orbitals. The sigma bond does not possess a nodal plane along the axis of the bond. (*p. 369*)

symmetry The property associated with orbitals that have similar size and shape. (*p. 376*)

valence bond theory The theory that all covalent bonds in a molecule arise from overlap of individual valence atomic orbitals. A modification of this theory allows valence atomic orbitals to include hybridized orbitals. (*p. 359*)

Focus Your Learning

The answers to the odd-numbered problems and some selected problems appear at the back of the book, as represented by the blue numbering.

Section 9.1 Valence Bond Theory

Skill Building

1. a. Professional scientific journal articles, textbooks, and popular writing have all depicted chemical bonds as a typed dash between two atomic symbols. Although this does communicate that the atoms are associated with each other, cite one reason why it does not accurately describe a chemical bond.

   b. Why do chemists seek to provide better models to describe the makeup of a chemical bond?

2. The VSEPR model of chemical bonding has been very successful for some descriptions of chemical bonding, but what is the major weakness of this popular model?

3. Write out the ground-state configuration of silicon. According to the VSEPR model, how many bonds should one atom of Si be able to form? Is your answer consistent with the formula of $SiCl_4$ (a compound used in the formulation of some smoke screens)?

4. Write out the ground-state configuration of aluminum. According to the VSEPR model, how many bonds should one atom of Al be able to form? Is your answer consistent with the formula of $AlCl_3$?

5. Write the ground-state electron configuration for each of the following atoms. According to the VSEPR model, how many bonds should each atom be able to form?

    Cl   Se   B

6. Write the ground-state electron configuration for each of the following atoms. According to the VSEPR model, how many bonds should each atom be able to form?

    Ge   Ne   Mg

Chemical Applications and Practices

7. Carbon tetrachloride (CCl_4) is seldom used as a cleaning solvent. It formerly saw widespread use, but more recently its ill effects on human health have led to its restriction. What is the ground-state designation for carbon's valence electrons? Explain why this configuration does not lend itself to forming the four equal bonds found between carbon and chlorine in this compound.

8. Tin can form compounds such as $SnCl_2$ and $SnCl_4$. What is the ground-state designation for tin's valence electrons? Explain why this configuration does not lend itself to forming the equal bonds found between tin and chlorine in one of these compounds.

Section 9.2 Hybridization

Skill Building

9. a. Write out the electron configuration of hydrogen and bromine.
    b. What valence orbitals are involved in the overlap that forms the bond between H and Br?
    c. Compare the strength of this bond to the strength of the HF bond described earlier in the chapter.

10. a. Write out the electron configuration of hydrogen and chlorine.
    b. What valence orbitals are involved in the overlap that forms the bond between H and Cl?
    c. Compare the strength of this bond to the strength of the HF bond described earlier in the chapter.

11. Using valence bond theory, describe the orbital overlap in Cl_2.

12. Using valence bond theory, describe the orbital overlap in F_2.

13. Using the concept of orbital overlap, predict in each pair of compounds which bond will be longer:

    H_2 or Cl_2 Br_2 or Cl_2 HCl or HBr

14. Using the concept of orbital overlap, predict in each pair of compounds which bond will be longer:

    H_2 or Li_2 F_2 or Cl_2 LiF or HF

15. Hypochlorous acid has excellent bleaching properties and must be handled with care. Using valence bond theory, show what orbitals are likely to be overlapped in HOCl. Which bond would you predict to be longer, the bond between H and O bond or the bond between O and Cl? Explain the basis of your prediction.

16. Look at the ground-state electron configurations of the atoms in the HOCl compound mentioned in the previous problem. Why do you think that the structure presented as HOCl more logically represents the bonding than HClO?

17. Predict the formula of potassium hydride. Which hydride would you predict to have the stonger bonds, potassium hydride or cesium hydride? Explain the basis of your prediction.

18. What would you predict as the formula of calcium hydride? Which hydride would you predict to have the stronger bonds, calcium hydride or magnesium hydride? Explain the basis of your prediction.

19. When explaining and demonstrating orbital hybridization to chemistry students, a teacher prepares "orbital omelets." This analogy involves mixing one, two, or three eggs with 100 mL of milk. How could the three omelets be used to describe three ways in which s and p orbitals can hybridize?

20. Develop an analogy similar to that in Problem 19 using an "orbital milkshake." Mix one, two, or three scoops of ice cream with one glass of milk. How could the three resulting milkshakes be used to describe three ways in which s and p orbitals can hybridize?

21. In some situations d orbitals may become involved in hybridization.
    a. If an atom's hybridization were designated sp^3d^2, how many orbitals would be involved?
    b. What would be the same and what would be different about these orbitals compared to hybrid orbitals made from only s and p orbitals?

22. In many sciences, including chemistry, we are sometimes forced to use the same symbols to mean different things. In each of the following examples, distinguish the meaning of the two sets of similar symbols:
    a. $2s^1 2p^1$ and $(2sp)^1$ b. $(2sp^3)$ and $2p^3$

23. a. Show the ground state for the valence electrons of silicon. What type of hybridization would the orbitals of silicon undergo to produce SiH_4?
    b. What would be the resulting geometric shape of a SiH_4 molecule?

24. BCl_3 and NH_3 have the same basic formula (one central atom with three attached atoms), but they differ in shape. Using the hybridization of their valence orbitals, explain the basis for their different shapes.

25. The bonding in H_2O, NH_3, and CH_4 can be explained by using sp^3 hybridization of the valence electrons in oxygen, nitrogen, and carbon, respectively. However, all three molecules have different shapes. What are the bond angles for the three molecules and what is the basis for each?

26. $SiCl_4$ and SCl_4 both contain a central atom attached to four other atoms. After examining the ground state of Si and S and bonding four chlorine atoms to each, would you predict the two molecules to have the same shape? Explain the basis for your conclusion.

27. The carbon atom in both methane (CH_4) and chloroform ($CHCl_3$) has undergone sp^3 hybridization. However, the two molecules do not have exactly the same shape. Explain the cause and result of any differences.

28. The oxygen atom in both H_2O and HOF has undergone sp^3 hybridization. However, the two molecules do not have exactly the same shape. Explain the cause and result of any differences.

29. How many hybrid orbitals are formed from mixing an s orbital with a p orbital?

30. How many hybrid orbitals are formed from mixing an s orbital with two p orbitals?

31. For each hybridization listed below, indicate the number of bonds that could be formed by overlapping with the hybridized orbitals.

$$sp \qquad sp^2 \qquad sp^3$$

32. For each type of hybridization listed below, indicate the idealized geometric shape that would be produced around a central atom having that type of hybridization.

$$sp \qquad sp^2 \qquad sp^3$$

33. What type of hybridization would be found in the central atom of each of the following?

a. OF_2 b. CCl_4 c. BCl_3 d. $BeCl_2$

34. What type of hybridization would be found in the central atom of each of the following?

a. CS_2 b. H_2S c. $CSCl_2$ d. SO_2

35. Indicate what geometric shape is produced when the number of orbitals around a central atom is:

a. 2   b. 3   c. 4   d. 5   e. 6

36. Indicate what bonding angle is produced when the number of orbitals around a central atom is:

a. 2   b. 3   c. 4   d. 5   e. 6

37. Diagram the hybridization for the following three hydrocarbons:

$$C_2H_2 \qquad C_2H_4 \qquad C_2H_6$$

Energy is required to break chemical bonds. In which of the three molecules would the *least* amount of energy be required to break the carbon-to-carbon bond. Explain, or justify your answer.

38. Diagram the hybridization of the carbon in each of the following compounds:

$$CO_2 \qquad CO \qquad CO_3^{2-}$$

Energy is required to break chemical bonds. In which of the three molecules would the *least* amount of energy be required to break the carbon-to-oxygen bond. Explain, or justify your answer.

39. Sulfur is capable of many different oxidation states. In each of the following, determine the hybridization of sulfur and the resulting shape of the molecule or ion.

a. SO_3 b. SO_3^{2-} c. SF_6 d. S_8

40. Nitrogen is capable of many different oxidation states. In each of the following, determine the hybridization of nitrogen and the resulting shape of the molecule or ion.

a. NO_3^- b. NO   c. NO_2 d. N_2

41. Predict the hybridization of the underlined atom and the overall shape of the following molecules.

a. $\underline{N}H_4^+$ b. $\underline{Xe}F_4$ c. $\underline{S}F_4$ d. $\underline{N}O_2^-$

42. Predict the hybridization of the underlined atom and the overall shape of the following molecules.

a. $H\underline{C}N$ b. $\underline{Cl}F_3$ c. $\underline{Cl}F_5$ d. $\underline{I}Cl_3$

43. Predict the hybridization of each of the atoms in the following molecule.

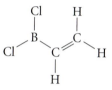

44. Predict the hybridization of each of the atoms in the following molecule.

Chemical Applications and Practices

45. In the realm of science fiction, as we noted earlier, dilithium was used to propel the matter–antimatter drive of the starship *Enterprise*. (Some of the details are still a bit elusive.) Suppose a Klingon engineer suggests that dipotassium might also be a possible fuel. Would you argue that K_2 would have a stronger or a weaker bond than that found in Li_2? What would be the basis of your argument? (It will be best if you can cite some sound basis in valence bond theory for your argument. Klingons can be defensive when challenged.)

46. Several noble gas compounds have been synthesized under controlled conditions. Explain, using ground-state electron configurations and valence bond theory, why no diatomic molecules such as He_2, Ne_2, and the like have been made.

47. One of the first synthesized "natural" organic compounds was the nitrogen-based solid found in mammal waste: urea. From examining the formula shown here, what would you predict for the hybridization of the nitrogen atoms? What would be the bond angle from hydrogen to nitrogen to carbon?

$$H_2N - \underset{\underset{O}{\|}}{C} - NH_2$$

48. Three compounds composed of carbon and hydrogen can be used to illustrate the importance of hybridization and molecular shape. Diagram the Lewis dot structure for each case shown here. Then predict the hybridization in the carbon atoms, and predict the bond angle from H to C to C in each structure.

a. C_2H_2 (acetylene, used in high-temperature welding)

b. C_2H_4 (ethylene, a plant hormone that hastens ripening)

c. C_2H_6 (ethane, a heating fuel and an ingredient in polymers)

49. The structure of the artificial sweetener aspartame is shown on the next page. How many pi bonds are shown in the structure? What would be the hybridization and bond angle around the carbon atom double-bonded to oxygen at the top of the molecule?

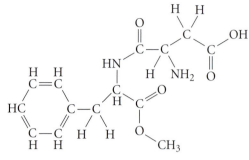

50. The male sex hormone testosterone is synthesized from cholesterol. Examine the structure shown here.

a. How many pi bonds are present in the molecule?

b. There are two carbon atoms that are bonded to oxygen. Describe the hybridization present in these two carbon atoms.

c. What are the bond angles around those two carbon atoms?

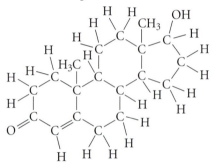

51. Antihistamines are often taken to counteract the effects of the amino acid histamine that is released in allergic reactions. Examine the structure of histamine shown below. Compare the two nitrogen atoms in the ring portion of the molecule on the basis of lone-pair electrons, bond angle, and hybridization.

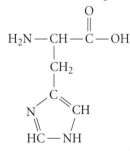

52. Vitamin B$_6$ (pyridoxine) is important in the metabolism of carbohydrates, proteins, and fats because it enhances the action of several enzymes. After examining the structure of this important molecule, answer the following questions.

a. How many pi bonds are present in the structure?

b. Would the bond between the carbon atoms within the ring be shorter or longer than the bond between a carbon atom in the ring and one outside the ring? Explain.

c. Are the bond angles around the CH$_3$ group the same as or different from the bond angles around the carbon in the CH$_2$OH group? Explain.

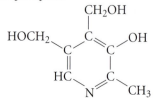

53. One form of the molecule POCl$_3$ is shown below. What hybridization would be found in the phosphorus atom? What are the bond angles around the phosphorus?

54. Biacetyl is one of the additives included in margarine to enhance the butter-like taste. The diagram of biacetyl shown here does not accurately depict the bond angles. What would be the correct bond angles around the two central carbon atoms? Redraw the structure to show those angles.

$$H_3C-C-C-CH_3$$

with the structure showing two C=O (O atoms double-bonded below the two central carbons).

55. Capsaicin is one of the key ingredients in chili peppers that produce the spicy taste sensation sought by chefs. Examine its structure and answer the following questions.

a. Which carbon atoms have sp^3 hybridization?

b. How many H atoms would have to be removed to produce a double bond between the two carbon atoms that are shown farthest to the right?

c. How many pi bonds are shown in the molecule?

d. How many lone pairs of electrons are needed around the oxygen atom that is double-bonded to the carbon atom?

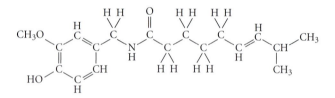

56. Estradiol is the active steroid used in estrogen replacement therapy.

a. Which carbon atoms have sp^2 hybridization?

b. How many H atoms would have to be removed to produce a double bond between the oxygen and carbon atoms that are shown farthest to the right?

c. How many pi bonds are shown in the molecule?

d. How many lone pairs of electrons are needed around each oxygen atom?

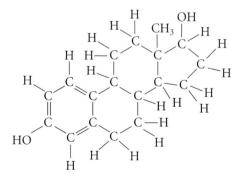

Section 9.3 Molecular Orbital Theory

Skill Building

57. Working from left to right, create a table in which the name of each model is correctly aligned with the scientist who devised it and a general summary of the model.

Model	Scientist	General summary
VB	G. Lewis	Subtract and add overlap to create π and σ bonds.
MO	L. Pauling	Distribute bonding and nonbonding electron pairs.
VSEPR	E. Schrödinger	Overlap and hybridization create new orbitals.

58. The development of molecular orbital theory gave chemists some advantages in explaining chemical bonding.
    a. What is the theoretical basis of MO theory?
    b. What main advantage over VSEPR does this treatment of bonding offer?
    c. What is a disadvantage of MO theory?

59. Oxygen and sulfur are both found in Group VIA on the periodic table. Consequently, they share many properties, including some bonding characteristics. However, the larger size of the p orbitals in sulfur prevents them from overlapping in an effective way to form pi bonds. Use orbital diagrams to show how this can be used to explain why O_2 is more stable than S_2.

60. Nitrogen and phosphorus are both found in Group VA on the periodic table. Consequently, they share many properties, including some bonding characteristics. However, the larger size of the p orbitals in phosphorus prevents them from overlapping in an effective way to form pi bonds. Use orbital diagrams to show how this can be used to explain why N_2 is more stable than P_2.

61. Cite two differences between bonding and antibonding orbitals.

62. Cite two similarities between bonding and antibonding orbitals.

63. Most sciences have their share of acronyms. In this chapter you have encountered several important ones related to chemical bonding. Supply the full name and meaning of each of the following:
    a. LCAO   b. HOMO   c. LUMO

64. In this chapter you have encountered several important symbols related to chemical bonding. Supply the full name and meaning of each of the following:
    a. π b. π^* c. σ

65. Removing or adding electrons to molecules can change several properties of the bonding within the molecule. Provide complete MO diagrams for N_2^-, N_2, and N_2^+.
    a. Calculate the bond order for each.
    b. Indicate which, if any, are paramagnetic.
    c. Rank the three nitrogen species in order from shortest to longest nitrogen-to-nitrogen distance.

66. Removing or adding electrons to molecules can change several properties of the bonding within the molecule. Provide complete MO diagrams for O_2^-, O_2, and O_2^+.
    a. Calculate the bond order for each.
    b. Indicate which, if any, are paramagnetic.
    c. Rank the three oxygen species in order from shortest to longest oxygen-to-oxygen distance.

67. A diatomic homonuclear +3 ion has the following molecular orbital configuration:

    $$(\sigma_{1s})^2\ (\sigma_{1s}^*)^2\ (\sigma_{2s})^2\ (\sigma_{2s}^*)^2\ (\sigma_{2p})^2\ (\pi_{2p})^4\ (\pi_{2p}^*)^3$$

    a. What is the identity of the element used to make this ion?
    b. Is the ion paramagnetic or diamagnetic?
    c. What is the bond order of the ion?

68. A diatomic homonuclear +2 ion has the following molecular orbital configuration:

    $$(\sigma_{1s})^2\ (\sigma_{1s}^*)^2\ (\sigma_{2s})^2\ (\sigma_{2s}^*)^2\ (\sigma_{2p})^2\ (\pi_{2p})^2$$

    a. What is the identity of the element used to make this ion?
    b. Is the ion paramagnetic or diamagnetic?
    c. What is the bond order of the ion?

69. Indicate the total number of sigma and pi bonds in the following molecule.

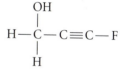

70. Indicate the total number of sigma and pi bonds in the following molecule.

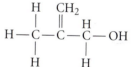

71. What is the bond order between the indicated atoms in the following molecule?

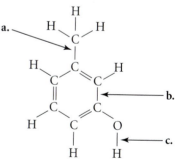

72. What is the bond order between the indicated atoms in the following molecule?

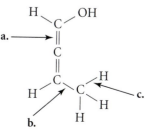

Chemical Applications and Practices

73. Typically phosphorus may be found in nature as P_4 (among other forms). However, P_2 is also known to exist at sufficiently lower temperatures. Using the outer valence electrons, produce the MO diagram for P_2. (You may assume the same sequence of orbitals as found in N_2.)
    a. What is the bond order of P_2?
    b. How many electrons would be found in the highest π^*?

74. Br_2 is a liquid at room temperature. Using the outer valence electrons, produce the MO diagram for Br_2. (You may assume the same sequence of orbitals as found in F_2.)
    a. What is the bond order of Br_2?
    b. How many electrons would be found in the highest π?

Section 9.4 Putting It All Together

Skill Building

75. a. The term *orbital overlap* was used throughout this chapter. Describe the meaning of this important term. What exactly is being "overlapped"?
    b. Diagram the orbital overlap typical of an *s–s*, *s–p*, and *p–p* overlap.

76. a. The term *hybrid* was used throughout this chapter. Describe the meaning of this important term.
    b. Draw the shapes of at least three different hybrid orbitals.

77. a. Why is it necessary to draw resonant hybrids when representing the structure of ions and molecules that have delocalized electrons?
    b. Using the diagram of acetylsalicylic acid shown in Problem 87 as one example, show another example of a resonance hybrid of the molecule.

78. Using the diagram of the steroid illustrated in Problem 56, show another example of a resonance hybrid of the molecule.

Use the compound shown below to answer Problems 79–82.

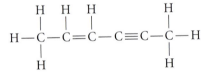

79. Numbering the carbon atoms from right to left, what would be the hybridization for the second carbon?

80. Using the same numbering system, between which three atoms would you predict the bond angle to be largest?

81. How many pi bonds are in the molecule? How many sigma bonds are present?

82. Bonds between atoms can rotate if they lack a nodal plane aligned along the bond's axis. Using the same right-to-left numbering system, which carbon atom(s) would be free to rotate without affecting the movement of any other carbon atoms?

Use the compound shown below to answer Problems 83–86.

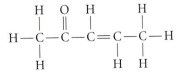

83. Numbering the carbon atoms from right to left, what would be the hybridization for the second carbon?

84. Using the same numbering system, between which three atoms would you predict the bond angle to be largest?

85. How many pi bonds are in the molecule? How many sigma bonds are present?

86. Bonds between atoms can rotate if they lack a nodal plane aligned along the bond's axis. Using the same right-to-left numbering system, which carbon atom(s) would be free to rotate without affecting the movement of any other carbon atoms?

Chemical Applications and Practices

87. The structure for the common pain reliever acetylsalicylic acid (found in aspirin products) is shown below.
    a. Indicate which carbon atoms have associated delocalized electrons.
    b. What type of hybridization do these carbon atoms have?
    c. What types of orbitals are directly involved with the π bonding?

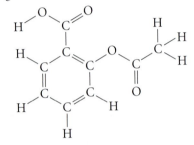

88. The versatile carbonate ion, found in seashells, in chalk, and as part of our blood buffering system, is made from three oxygen atoms bonded to a central carbon atom and two additional electrons. Diagram three resonance structures for the ion, and show the resonance hybrid. What is the approximate bond order for a carbon-to-oxygen bond in the ion?

89. Naphthalene ($C_{10}H_8$) is the active ingredient in moth balls, giving them their characteristic odor. Naphthalene contains two equal-sized rings of carbon atoms that share a common side. Every carbon atom is part of a ring. Diagram the structure of this molecule and indicate the hybridization of each of the atoms.

90. Draw two resonance structures for naphthalene; see Problem 89. Then indicate the resonance hybrid for the molecule. What is the approximate bond order for a C—C bond in naphthalene?

Comprehensive Problems

91. How many unpaired electrons would be found in one atom of carbon in each of the following conditions?
    a. ground state   c. sp^2 hybridization
    b. sp hybridization   d. sp^3 hybridization

92. Magnesium hydride is one of the few covalent metallic hydrides. Write out the ground-state electron configuration for magnesium.
    a. What type of hybridization would be necessary to form equal bonds to the two hydrogen atoms?
    b. What orbitals would overlap to form the bonds?
    c. What shape would this molecule have?

93. Using the designated lone pairs and bonded pairs to describe electrons around a molecule's central atom, predict the shape that would be produced in each of the following:
    a. Three bonded pairs and one lone pair
    b. Six bonded pairs and no lone pairs
    c. Two bonded pairs and two lone pairs
    d. Five bonded pairs and no lone pairs

94. Nitrogen and phosphorus are both found in Group VA on the periodic table. As you would expect, they have some reactions and properties in common. For example, both nitrogen and phosphorus form trihalides such as NCl_3 and PCl_3. However, phosphorus also forms PCl_5 but nitrogen does not. Explain the reason behind this difference.

95. The nitrate ion (NO_3^-) and the nitrite ion (NO_2^-) can both be drawn showing a double bond between one oxygen atom and the central nitrogen atom. However, the bond angle is not the same in the two compounds. Which would have the smaller angle? Explain the basis for your choice.

96. The air you inhale is mostly nitrogen gas (N_2). The air you exhale is mostly the unchanged N_2 that you just inhaled. Use ground-state electron configurations and valence bond theory to show why N_2 is so stable that your biochemical processes cannot change its structure.

97. Two molecules found in petroleum oil are the straight-chain octane and the highly branched compound 2,2,3,3-tetramethylbutane. The structures of both are shown here. From what you can deduce about the three-dimensional shape of these structures, decide which one is most likely be able to associate closely with more molecules of itself and to have a higher boiling point. Explain the basis of your answer.

$$CH_3—CH_2—CH_2—CH_2—CH_2—CH_2—CH_2—CH_3 \quad \text{(octane)}$$

$$(CH_3)_3C—C(CH_3)_3 \quad \text{(2,2,3,3-tetramethylbutane)}$$

98. Examine the structure of oleic acid shown below (H atoms are omitted from the structure). You may have read the label on a food product that described "hydrogenated" or "partially hydrogenated" oils as an ingredient. These terms refer to the addition of hydrogen atoms to the area of the double bond(s) in an oil. How many hydrogen atoms have been left off the oleic acid structure? How many more hydrogen atoms would have to be added in order to fully hydrogenate oleic acid and remove the double bond?

$$C—C—C—C—C—C—C—C—C{=}C—C—C—C—C—C—C—C—COOH$$

(oleic acid)

99. The compound formamide (CH_3NO) has one hydrogen atom, the nitrogen, and the oxygen attached to the carbon. In addition, the remaining two hydrogen atoms are attached to the nitrogen. Every atom satisfies the octet rule. Diagram the structure of this molecule, and explain the hybridization of both the carbon and nitrogen atoms. If you built a model of this molecule, would it lie flat on a table (planar) or have a "puckered" structure?

100. Dinitrogen monoxide, also known as nitrous oxide (N_2O), has had some use as an anesthetic and, because of some side effects, is occasionally called "laughing gas." Diagram two suitable Lewis dot structures for the compound, and state the hybridization of the nitrogen atoms in each.

101. In the opening discussion of the chapter, the molecule retinal is introduced. This molecule contains a series of *conjugated* double bonds—double bonds that are separated by single bonds.
    a. Which is a longer bond in the molecule, the C=C bond in the ring, or the C—C bond on the opposite side of the ring?
    b. This molecule absorbs light at about 500.0 nm. What is the energy of a single photon of light with this wavelength?
    c. What is the energy of a mole of photons of this wavelength?
    d. To what color does this wavelength correspond?

102. The text of the chapter indicates that the chlorine–chlorine bond can be cleaved with 492 nm light.
    a. What is the energy of a mole of photons of this wavelength?
    b. Using Lewis dot diagrams, show how the chlorine molecule breaks apart into two chlorine atoms.
    c. What is the difference between a chlorine atom and a chloride ion?
    d. What is the hybridization of the chlorine atom in a molecule of chlorine?

103. Oleic acid, Exercise 9.7, can be treated with hydrogen gas to make stearic acid.
    a. Write the balanced molecular equation for this process.
    b. How many grams of hydrogen are needed to prepare 1.0 lb stearic acid?
    c. Stearic acid can form a film on the surface of water. In doing so, the molecule stands straight up—one end points toward the water and one end points toward the sky. Which end of stearic acid would you expect to point toward the water? Explain.

Thinking Beyond the Calculation

104. Hydrazine (N_2H_4) has many uses. For example, it is employed as a rocket fuel and as a starting material in the production of fungicides.
    a. Draw the Lewis dot structure for hydrazine.
    b. What is the VSEPR shape of the nitrogens in this molecule?
    c. What is the hybridization of each nitrogen atom?
    d. Would you expect this molecule to have a color that we can see?
    e. Hydrazine can be used as rocket fuel. The products of combustion are nitrogen gas and water vapor. Write a balanced equation showing this reaction, and calculate the enthalpy of combustion for this reaction using the bond energies from Chapter 8.

10

The Behavior and Applications of Gases

Earth's atmosphere is a very thin layer of gases (only 560 km thick) that surrounds the planet. Although the atmosphere makes up only a small part of our planet, it is vital to our survival.

Contents and Selected Applications

10.1 **The Nature of Gases**

10.2 **Production of Hydrogen and the Meaning of Pressure**

10.3 **Mixtures of Gases—Dalton's Law and Food Packaging**

Chemical Encounters: Dalton's Law and Food Packaging

10.4 **The Gas Laws—Relating the Behavior of Gases to Key Properties**

Chemical Encounters: Balloons and Ozone Analysis

10.5 **The Ideal Gas Equation**

10.6 **Applications of the Ideal Gas Equation**

Chemical Encounters: Automobile Air Bags
Chemical Encounters: Acetylene

10.7 **Kinetic-Molecular Theory**

10.8 **Effusion and Diffusion**

10.9 **Industrialization: A Wonderful, Yet Cautionary, Tale**

Chemical Encounters: Ozone
Chemical Encounters: The Greenhouse Effect

Go to **college.hmco.com/pic/kelterMEE** for online learning resources.

Our planet is surrounded by a relatively thin layer known as an atmosphere. It supplies all living organisms on Earth with breathable air, serves as the vehicle to deliver rain to our crops, and protects us from harmful ultraviolet rays from the Sun.

Although it does contain solid particles, such as dust and smoke, and liquid droplets, such as sea spray and clouds, much of the atmosphere we live in is actually a mixture of the gases shown in Table 10.1. These gases, and the rest of the components in the atmosphere of planet Earth, are vital to our survival.

In addition to forming the atmosphere surrounding planet Earth, gases are vital to our society in many other ways. Gases that are used in manufacturing, medicine, or anything else related to the economy are called industrial gases, shown in Table 10.2. Some, such as nitrogen and oxygen, can simply be separated from the air. Others, including sulfur dioxide, hydrogen, and chlorine, are produced by chemical reactions. No matter what uses we make of them, nearly all of these common gases behave in about the same way when present at low density. The fact that almost all gases share this characteristic enables us to work with them predictably and successfully.

TABLE 10.1 Composition of Dry Air

Component	Percent by Volume	Parts per Million by Volume
N_2	78.08	780,800
O_2	20.95	209,500
Ar	0.934	9,340
CO_2	0.033	330
Ne	0.00182	18.2
He	0.000524	5.24
CH_4	0.0002	2
Kr	0.000114	1.14
H_2	0.00005	0.5
N_2O	0.00005	0.5
Xe	0.0000087	0.087
O_3	$<1 \times 10^{-5}$	<0.1
CO	$<1 \times 10^{-6}$	<0.01
NO	$<1 \times 10^{-6}$	<0.01

Source: Chemical Rubber Company. *Handbook of Chemistry and Physics,* 70th ed. CRC Press: Boca Raton, FL, 1990.

TABLE 10.2 Industrial Uses of Some Common Gases

Gas	Use	U.S. Production (2004) in Metric Tons (1 metric ton = 1000 kg)
Cl_2	Preparation of bleaches and cleansing agents, purification of water, preparation of pesticides	12.2 million
CO_2	Refrigeration, preparation of beverages, inert atmosphere in chemical reactions and food packaging, fire extinguishers	8.1 million (estimated)
H_2	Hydrogenation of oils in food and other unsaturated organic molecules, energy source in vehicles, petroleum-refining reactions, manufacturing of resins used in plastics	17.7 bcm (gas + liquid) (bcm = billion m^3)
NH_3	Fertilizer and preparation of other fertilizers	10.8 million
N_2	Inert atmosphere in chemical reactions and food packaging, electronics, and metalwork; manufacture of ammonia; refrigerant	30.3 million (not incl. NH_3)
O_2	Oxidizer in many chemical processes, such as production of steel and acetylene; oxidizer in reaction of fuels in rockets; in hospitals (breathing); pulp and paper industries for bleaching	25.5 million
SO_2	Production of sulfuric acid, bleaching agent in food and textile industries, controls fermentation in wines	300,000

Source: Chemical and Engineering News, July 11, 2005.

Video Lesson: Properties of Gases

10.1 The Nature of Gases

Many of the chemists in the eighteenth century became aware of a very interesting phenomenon as they worked with gases. They noted that dissimilar gases appeared to behave in relatively similar ways. **Why is this possible?** It all comes down to the common features of gases. Under ambient conditions, electrostatic interactions make water a liquid and sodium chloride a solid. Gases have few such interactions, and when they occur, they are often weak and fleeting. In fact, *the atoms or molecules of gases most often behave as individual units.* Gases that behave as though each particle has no interactions with any other particles are called **ideal gases**; their properties are listed in Table 10.3. Most gases behave nearly ideally under normal conditions, so unless otherwise noted, the quantitative relationships we will discuss assume ideal behavior. Notably, deviations from ideal behavior occur when gases are examined under conditions that encourage the interactions between molecules of a gas: high pressures and low temperatures.

Gases have an exceedingly low density—that is, the particles are relatively far apart. For example, at 25°C, liquid water has a density of 1.00 g/mL, whereas dry air at sea level has a density of 0.00118 g/mL. The low density of gases means that they can be significantly compressed in order to save space during shipping, as shown in Figure 10.1.

Solids and liquids are nearly incompressible and have volumes that change very little with temperature and pressure. This difference, too, is related to the nature of gases. Because gas molecules are far apart compared to molecules within a liquid or solid, the volume of a gas can be greatly affected by changes in temperature and pressure. As we will discuss later, we make use of the compressibility of gases, and the energy exchanges that accompany it, in applications from refrigeration to rocketry.

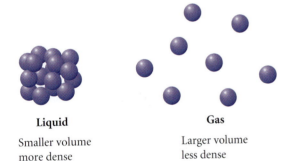

Liquid

Smaller volume
more dense

Gas

Larger volume
less dense

Gas molecules occupy a much larger volume than the same number of molecules of a liquid. This results in a much lower density for gases than for liquids.

FIGURE 10.1

Gases can be significantly compressed, which makes it possible to ship large quantities of industrial gases via truck in compressed gas cylinders.

TABLE 10.3   The Ideal Gas

The ideal gas is one that follows each of these rules:
- The individual gas particles do not interact with each other.
- The individual gas particles are assumed to have no volume.
- The gas strictly obeys all of the simple gas laws.
- The gas has a molar volume of 22.414 L at 1 atm of pressure and 273.15 K.

Gases are highly compressible compared to liquids and solids.

Solid

Tightly packed
least compressible

Liquid

Loosely packed
slightly compressible

Gas

Widely spaced
highly compressible

Visualization: Electrolysis of Water

10.2 Production of Hydrogen and the Meaning of Pressure

Nearly 18 billion cubic meters of hydrogen gas were produced in the United States in 2004. Because hydrogen gas is a very minor component of air (see Table 10.1), it must be generated via chemical processes. One method is the decomposition of water by **electrolysis**, in which an electric current passing through a solution causes a chemical reaction. The resulting hydrogen gas can be collected separately from the oxygen gas. Let's assume that the process takes place at constant temperature and that the collection vessel, which has a constant volume, is empty as hydrogen begins to enter, as shown in Figure 10.2.

What happens as H_2 flows in? The molecules begin to collide with the walls of the container and with each other, as shown in Figure 10.3. The **force** that each molecule applies during these collisions is equal to the mass of the particle times its acceleration:

$$\text{Force} = \text{mass} \times \text{acceleration} = kg \times \frac{m}{s^2} = kg \cdot m \cdot s^{-2}$$

The SI unit of force is the **newton (N)**, which is equal to 1 $kg \cdot m \cdot s^{-2}$. Because each of us has mass and is accelerated toward the center of the Earth by the pull of gravity $\left(\frac{9.81 \text{ m}}{s^2}\right)$, each of us exerts a force on

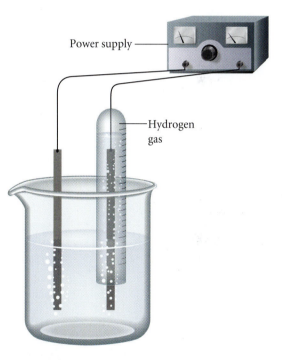

Power supply

Hydrogen gas

FIGURE 10.2

Our hypothetical reaction set-up for the electrolysis of water. The hydrogen gas can be isolated from the reaction and sent to the collection chamber.

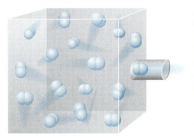

= H_2 molecule

FIGURE 10.3

Molecules entering a vacant container collide with each other and with the walls of the container. The molecules have mass, acceleration, and a resulting force.

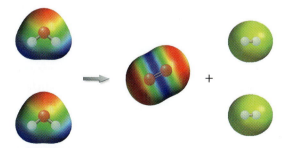

Hydrogen and oxygen can be produced by the electolysis of water.

9.8 m/s^2

The force of gravity on a person.

the floor upon which we are standing. In describing our own mass × acceleration, we most often speak of force by using the familiar unit pounds, where 1 lb = 4.47 N.

EXERCISE 10.1   Calculating Force

Suppose that your mass is 55 kg and you are on the Moon, where the acceleration due to gravity is about $\dfrac{1.6 \text{ m}}{\text{s}^2}$, roughly one-sixth that on Earth. What force would you exert on a scale ("How much would you weigh") in newtons and in pounds?

First Thoughts

Mass is a measure of the amount of substance, whereas force is the effect of acceleration, in this case as gravity, on the mass. We would therefore expect the mass to remain the same on the Moon (or on any other celestial body), whereas the force would change.

Solution

$$55 \text{ kg} \times \frac{1.6 \text{ m}}{\text{s}^2} = \frac{88 \text{ kg·m}}{\text{s}^2} = 88 \text{ N}$$

$$88 \text{ N} \times \frac{1 \text{ lb}}{4.47 \text{ N}} = 20 \text{ lb}$$

Further Insights

Jupiter and Mars reveal the effects of gravitational strength on planetary atmospheres. Jupiter has a much greater gravitational pull than Earth because it is more massive. This prevents light gases, such as hydrogen, from escaping, so Jupiter's atmosphere is largely hydrogen. Earth's much lower gravitational pull cannot retain hydrogen or helium, so our atmosphere is composed mostly of gases relatively higher in molar mass, such as oxygen and nitrogen. Mars, which exerts an even lower gravitational pull than Earth, has an atmosphere rich in carbon dioxide.

PRACTICE 10.1

A space traveler visiting the hypothetical planet X wishes to determine the gravitational pull of that planet. A 10.0 kg block of lead has been found to weigh 15 lb on planet X. What is the pull of gravity in m/s^2 on planet X?

See Problems 7, 8, 11, and 12. ▮

    The force that each molecule exerts on its container is very small, because the mass of each molecule in kilograms is very small. But when large numbers of molecules exert that force, as is common in the laboratory, the net result is readily measurable. Scientifically, we say that the **pressure** is the amount of force applied to a given area. If we substitute the SI base units for force and area, we see that the pressure of a gas can be expressed in this way:

$$\text{Pressure} = \frac{\text{force}}{\text{area}} = \frac{\text{N}}{\text{m}^2} = \frac{\text{kg·m·s}^{-2}}{\text{m}^2} = \text{kg·m}^{-1}\text{·s}^{-2}$$

The SI unit of pressure is the **pascal (Pa)** = 1 kg·m^{-1}·s^{-2}, named after the French mathematician Blaise Pascal (1623–1662). Measuring pressure in pascals is becoming ever more common, with kilopascals (kPa) used in maritime weather reporting as well as on tire pumps.

    What would happen to the force and pressure you exerted on the surface on which you were standing in our previous exercise if you stood on one foot instead of two? The force (88 N) would remain the same because your mass and the acceleration haven't changed. However, the pressure would double because the area on which you exerted your force would be cut in half (one foot rather than two). In short, you can increase pressure either by increasing force *or* by decreasing the area on which that force is exerted.

We can extend this idea to hydrogen molecules being produced by electrolysis. As we increase the number of molecules flowing into the collection vessel, more will collide in the given area. Therefore, the force per unit area—the pressure—will increase as shown in Figure 10.4. This suggests an important relationship that we will look at a bit later: At constant volume and temperature, the pressure exerted by a gas is proportional to the amount of gas present.

We have noted that the pascal is the SI unit of pressure, but other units of pressure are often used in scientific measurements and in day-to-day conversations. For instance, we have discussed previously the **atmosphere (atm)**, which is used extensively in scientific work. One atmosphere is roughly equal to the pressure exerted by the earth's atmosphere at sea level on a typical day. Because scientific work requires a more precise standard, we can define a **standard atmosphere**, exactly 1 atm, as the pressure that supports a 760-mm column of mercury (**mm Hg**) in a barometer—a device, like that shown in Figure 10.5, that measures the pressure of the atmosphere. Measurements have indicated that the standard atmosphere is equal to 1.01325×10^5 Pa.

The first barometer was made by Evangelista Torricelli in 1643, who used it to determine that atmospheric pressure changes with changes in the weather. In recognition of his work, we say that 1 atm is equal to 760 **torr**. Therefore, one standard atmosphere is equal to 760 mm Hg, to 760 torr, and to 1.01325×10^5 Pa.

Still other units are also used to describe pressure. For example, weather forecasters in some countries speak of atmospheric pressure in **bars**. A bar is equal to 1×10^5 Pa, a little less than 1 atm. In the United States, however, weather forecasters report atmospheric pressure in units of inches of mercury (which is abbreviated **in Hg**). Another unit with which you may be familiar is used when we inflate car or bicycle tires to their recommended pressure. That unit is known as **pounds per square inch gauge (psig)**, or, commonly, gauge pressure. This is the pressure of the gas in the tire *in excess* of the standard atmospheric pressure of 14.7 **pounds per square inch (psi)**. This means that a tire inflated to 32.0 psig really contains air exerting a pressure of 46.7 psi ($32.0 + 14.7$) on the inside walls of the tire. Table 10.4 gives the relationships among different units of pressure.

For many years, scientists defined a **standard pressure** as 1 atm so that a standard reference was available at a common pressure. In 1982, IUPAC

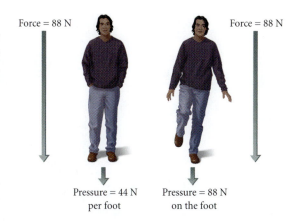

Force = 88 N   Force = 88 N

Pressure = 44 N   Pressure = 88 N
per foot   on the foot

Pressure and force are related.

FIGURE 10.4

The pressure of the hydrogen container increases as more molecules enter the vessel. The pressure is related to the force each molecule exerts on the walls of the container.

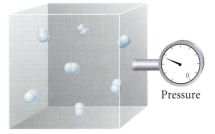

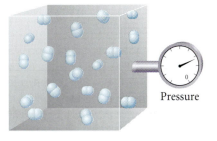

Pressure   Pressure

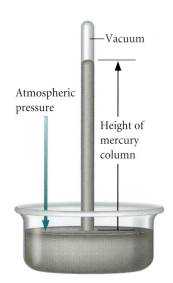

Vacuum

Atmospheric pressure

Height of mercury column

FIGURE 10.5

The pressure exerted by air at 1 atm supports a column of mercury 760 mm (29.9 inches) high. As the atmospheric pressure changes, so does the height of the mercury in the column. This is a simple barometer.

TABLE 10.4	**Pressure Unit Conversion Factors**

1 standard atmosphere is equal to . . .

- 760 mm Hg (millimeters of mercury)
- 760 torr
- 14.7 psi (pounds per square inch)
- 101,325 Pa (pascals)
- 1.01325 bar
- 0.0 psig (pounds per square inch gauge)
- 29.921 in Hg (inches of mercury)

recommended that standard pressure be defined as equal to exactly 1 bar. However, most chemists still commonly use 1 atm as the standard pressure, and we will do so in this textbook.

Scientists also define a **standard temperature** for gases as 0°C, or 273 K. Gases at these conditions are said to be at **standard temperature and pressure (STP)**. Increasingly, scientists are beginning to use **standard ambient temperature and pressure (SATP)**, or 1 bar and 298 K, to discuss gases, although in this book we will use STP as the standard. Why use either of these standards? Each is a convenient reference point at which to compare the properties of gases. Did you note that the standard temperature in STP conditions, 0°C, is different from that often used in thermodynamics, 25°C? Keep this in mind as we learn more about gases.

EXERCISE 10.2	**Conversion Among Measures of Pressure**

The gas in a volleyball is measured to have a *gauge pressure* of 8.0 psig. What is the *total pressure* exerted by the gas in units of atmospheres and torr?

First Thoughts

In this problem, we are given the gauge pressure, which is the pressure *in excess* of the atmospheric pressure, 14.7 pounds per square inch (psi). To determine the total pressure of the gas in the volleyball, we need to add 14.7 psi to the gauge pressure.

$$P_{gas} = P_{gauge\ pressure} + P_{volleyball} = (8.0 + 14.7)\ psi = 22.7\ psi$$

We can now do our unit conversions.

Solution

$$22.7\ psi \times \frac{1\ atm}{14.7\ psi} = 1.54\ atm$$

(We retain an extra figure because we will use this value in the next calculation.)

$$1.54\ atm \times \frac{760\ torr}{1\ atm} = 1.2 \times 10^3\ torr$$

Further Insights

Airlines caution their maintenance staff to always use a pressure regulator, a device that safely delivers gas at a desired pressure, when filling aircraft tires with nitrogen gas. The servicing tanks contain nitrogen gas at pressures as high as 3000 psi (200 atm). If the tank is hooked up *directly* to a tire that can tolerate only 200 psi (14 atm), the tire may explode. Several workers have been killed or severely injured in such accidents.

PRACTICE 10.2

A weather report provides the current barometric pressure as 29.30 in Hg. What is this pressure in atm, torr, bar, and Pa?

See Problems 9, 10, 15, and 16.

10.3 Mixtures of Gases— Dalton's Law and Food Packaging

Using mixtures of industrial gases to prevent spoilage in food products, especially poultry, is a rapidly growing segment of the food-processing industry. Microbes, especially *Pseudomonas* bacteria, spoil the flavor and color of food and make it unhealthful to eat. Carbon dioxide gas, which in small quantities is not particularly harmful to the food or to humans, has been found to inhibit the growth of many types of microbes. Food scientists discovered this fact and have used it to develop **modified atmosphere packaging (MAP)**—packaging the food item in an atmosphere that is different from air (recall Table 10.1). The results are impressive. For instance, meat and poultry stored in air lasts only a few days, but food stored under MAP can have a shelf life as long as a month, with proper refrigeration.

Nitrogen and argon gases are also used in meat-based MAP, and sulfur dioxide gas inhibits the growth of microbes in some beverages. Typical MAP includes mixtures containing a mole ratio of 30–35% CO_2 and 65–70% N_2 for meats. This mixture brings up an interesting and useful point. If the total pressure of the gas mixture is equal to 0.97 atm, how much does each gas contribute to the pressure?

John Dalton (remember his work from Chapter 2) conducted experiments that led to what we now know as **Dalton's law of partial pressures**. This law states that for a mixture of gases in a container, the total pressure is equal to the sum of the pressures that each gas would exert if it were alone. Each individual gas within a mixture contributes only a part of the total pressure, so we say that each gas exerts a **partial pressure**. Dalton's law can be summarized by the following equation:

$$P_{total} = P_1 + P_2 + \cdots + P_n$$

where P_{total} = the total pressure exerted by a mixture of gases

    P_1 = the partial pressure exerted by gas 1

    P_2 = the partial pressure exerted by gas 2

    P_n = the partial pressure exerted by gas n

For example, let's assume that the total pressure of the MAP gas that surrounds cuts of chicken in a package is 0.97 atm and that the gas is composed of 35% CO_2 and 65% N_2 by volume. Using Dalton's law of partial pressures, we see that 35% of the pressure, or 0.34 atm (0.35 × 0.97 atm) is exerted by CO_2 and 0.63 atm (0.65 × 0.97 atm) is exerted by the N_2 gas. The calculation of the partial

Application

CHEMICAL ENCOUNTERS: Dalton's Law and Food Packaging

Video Lesson: Partial Pressure and Dalton's Law

Food is easily spoiled by naturally occurring bacteria.

Ground turkey can be kept longer if stored using MAP.

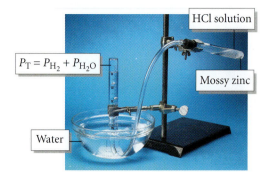

$P_T = P_{H_2} + P_{H_2O}$

HCl solution

Mossy zinc

Water

FIGURE 10.6

Hydrochloric acid solution can be mixed with pieces of zinc metal to produce hydrogen gas. The gas is collected over water, so the total pressure, P_T, of the collected gas is equal to the sum of the partial pressures of the H_2 and the water vapor.

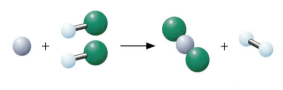

$$Zn(s) + 2HCl(aq) \rightarrow ZnCl_2(aq) + H_2(g)$$

pressures is straightforward, but Dalton's law is truly correct only for ideal gases. The actual pressures of the gases are a bit different from what we calculate because N_2 and CO_2 are not ideal gases; we will discuss corrections for nonideal behavior in Section 10.5. A key point for our current discussion is that *the closer gases come to ideal behavior, the more closely they follow Dalton's law of partial pressures, and deviations occur when gases are at high pressure, are at low temperature, or are otherwise concentrated enough to exhibit intermolecular interactions with each other.*

We can apply our understanding of partial pressures to the production of hydrogen gas, our focus at the opening of Section 10.2. A convenient way to produce small amounts of hydrogen gas is by the reaction of zinc with hydrochloric acid. The hydrogen gas produced by this reaction can be collected over water, as illustrated in Figure 10.6. As we'll discover in Chapter 11, a sample of water is accompanied by a certain amount of gaseous water—water vapor—above it. When we collect a gas by bubbling it through water or by leaving it in contact with moisture, the total pressure above the liquid is the sum of the partial pressure of the water vapor and the partial pressure of the gas. At 20°C, the vapor pressure of water is 17.5 torr. Therefore, for example, if the total pressure of the gases is 750 torr, then the partial pressure of the H_2 can be calculated as follows:

$$P_{total} = P_{H_2O} + P_{H_2}$$
$$750 \text{ torr} = 17.5 \text{ torr} + P_{H_2}$$
$$732 \text{ torr} = P_{H_2}$$

EXERCISE 10.3 | **Partial Pressure of Gases in the Atmosphere**

A jet is cruising at 11,500 ft (3500 m) above sea level, where the atmospheric pressure outside the plane is 493 torr (0.649 atm). The plane, normally pressurized to about 650 torr (0.85 atm), suddenly has a loss of pressure until the cabin pressure equals the pressure outside. What is the partial pressure of oxygen gas (see Table 10.1) when the pressure of the gas in the plane is lowered? Most people unaccustomed to low-oxygen environments will lapse into unconsciousness, and eventually die, if the partial pressure of oxygen falls below 30 torr. With that figure in mind, if the pressure in the plane isn't quickly restored, can the passengers survive this accident?

First Thoughts

We can use the law of partial pressures to determine the partial pressure of oxygen gas in air at 3500 m above sea level. Oxygen gas makes up about 21% of air. Therefore, we multiply the air pressure by 0.21 to obtain the partial pressure due to O_2.

Solution

$$P_{O_2} = P_{air} \times 0.21 = 493 \text{ torr} \times 0.21 = 1.0 \times 10^2 \text{ torr}$$

The partial pressure of O_2 at this altitude, about 100 torr, is less than the sea-level O_2 pressure of about 160 torr and the plane's normal O_2 pressure of about 140 torr, but it is still easily sufficient for survival.

Further Insights

The partial pressure of O_2 falls below 30 torr at an altitude of about 12,000 m, or 39,000 ft. This is higher than Mount Everest (8848 meters, or 29,028 ft), which explains why climbers without oxygen supplies can survive on the top of the world's highest peak ($P_{O_2} = 60$ torr). In fact, a stowaway is known to have survived an airplane flight in the wheel well of a jumbo jet that flew at 38,000 ft (11,600 m), where $P_{O_2} = 32$ torr.

PRACTICE 10.3

What is the partial pressure of nitrogen gas under the same conditions that the stowaway mentioned above endured?

See Problems 17–24.

HERE'S WHAT WE KNOW SO FAR

- Gases exert a force on the system that holds them. That force is known as pressure.

- Pressure is force distributed over an area. There are many units that describe the pressure of a system.

- A barometer is used to measure the pressure of the atmosphere.

- STP is the abbreviation for standard pressure and temperature. For chemists, this means exactly 1 atm pressure and 0°C.

- The total pressure of a system is the sum of all of the individual pressures of the component gases. This is Dalton's law of partial pressures.

- An ideal gas is a hypothetical gas with no intermolecular forces of attraction, where the particles of the gas have no volume. No gases are ideal, but they approach ideal behavior at low pressure and high temperature.

Video Lesson: Application of the Gas Laws

10.4 The Gas Laws—Relating the Behavior of Gases to Key Properties

Atmospheric scientists at the South Pole routinely measure the ozone (O_3) content in our atmosphere as a function of altitude. To complete this task, they employ a helium-filled balloon, such as that shown in Figure 10.7, to carry sophisticated instruments 35 km up into the atmosphere. The balloon rises because the density of the helium gas it contains is much less than that of the air outside. As the balloon climbs, the pressure and temperature of the surrounding atmosphere change, and the balloon expands. Eventually, the balloon expands so much that it bursts and falls back to Earth, allowing the scientists to recover the scientific instruments. However, in order to have a successful flight, the balloon must be filled with just the right amount of helium. Moreover, the scientists need to know how the helium gas will be affected by changes in temperature and pressure as the balloon rises through the atmosphere, so that the balloon can carry the instruments up to where they are needed. Too much gas and the balloon will burst before it reaches 35 km; too little gas and the balloon may never burst. To make their decisions, atmospheric scientists rely on some of the most fundamental properties of gases, which investigators have been aware of for hundreds of years. These

Application

CHEMICAL ENCOUNTERS: Balloons and Ozone Analysis

FIGURE 10.7

A helium-filled balloon is used to carry sophisticated instruments into the atmosphere to monitor the levels of atmospheric ozone (O_3). The balloon rises because the helium gas it contains is much less dense than the air outside.

long-established laws describing the behavior of ideal gases—the "gas laws"—help us to study modern concerns about gases, such as the level and fate of ozone in our atmosphere.

Avogadro's Law

A fundamental relationship that atmospheric scientists understand addresses the relationship between the amount of a gas and its volume. Let's explore this connection by taking an empty balloon and attaching it to a cylinder containing hydrogen gas like that shown in Figure 10.8. The gas, which is stored inside the cylinder at a pressure of 14 MPa, or about 140 atm, can be dispensed safely into the balloon with a pressure regulator. The temperature of the room is 25°C (298 K), and the pressure is 1.0 atm.

As we begin to fill the balloon, it expands because of the addition of the gas. The Mylar balloon has relatively little tension until it is nearly full, so the balloon skin doesn't significantly affect the pressure of the gas within. We find that at a volume of 1.2 L, the balloon holds 0.050 mol of H_2. If we double the number of moles of the gas to 0.10, the volume of the balloon also doubles, to 2.4 L, as shown in Figure 10.9.

Now let's take another balloon and hook it up to a cylinder containing oxygen gas, also at a pressure of 140 atm. If we add 0.050 mol of O_2 we find, as with H_2, that the volume of the balloon is 1.2 L, also shown in Figure 10.10. Double the number of moles, and the volume doubles to 2.4 L, just as in the hydrogen-filled balloon. This linear relationship holds irrespective of the nature of the gas, assuming ideal behavior. **Avogadro's law**, named after its discoverer Amadeo Avogadro (1776–1856), states that equal numbers of molecules are contained in equal volumes of all dilute gases under the same conditions. An implication of this law is that *volume is directly proportional to the amount of a gas expressed in moles, at constant temperature and pressure*. We represent this as follows:

$$V = kn$$

which can be rewritten as

$$\frac{V}{n} = k \qquad \text{(at constant } T, P)$$

where V = volume of a gas
n = amount of the gas expressed in moles
k = a constant

FIGURE 10.8

A Mylar balloon can be filled with gas from a cylinder.

WWW
Video Lesson: Avogadro's Law

FIGURE 10.9

At a volume of 1.2 L, the balloon on the left holds 0.050 mol of H_2. If we double the number of moles of the gas to 0.10, the volume of the balloon (shown on the right) also doubles, to 2.4 L.

FIGURE 10.10

Mylar balloons of O_2 and a balloon of H_2, containing 0.050 mol of gas, each has the same volume.

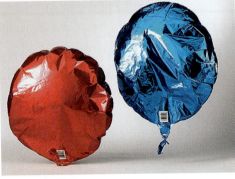

O₂   H₂

Because the constant k is the same value for a given temperature and pressure, we can write

$$\frac{V_{initial}}{n_{initial}} = k = \frac{V_{final}}{n_{final}}$$

or

$$\frac{V_{initial}}{n_{initial}} = \frac{V_{final}}{n_{final}} \qquad \text{(at constant } T, P\text{)}$$

The conclusion seems to make sense, because our experience tells is that the more gas we add to a balloon, the larger it gets, until its skin resists.

EXERCISE 10.4   Avogadro's Law

Much of the chlorine gas produced industrially is manufactured from the electrolysis of aqueous Group IA chlorides, such as sodium chloride. The reaction is

$$NaCl(aq) + H_2O(l) \rightarrow NaOH(aq) + \tfrac{1}{2}H_2(g) + \tfrac{1}{2}Cl_2(g)$$

Sodium hydroxide is formed along with hydrogen gas at one electrode, and chlorine gas is formed at the other electrode. If after such a reaction, 22.4 L each of hydrogen and chlorine gases, equaling 1.0 mol of each gas, is collected at STP, and then reacted to form hydrogen chloride gas, what would be the final volume of HCl gas at STP?

$$H_2(g) + Cl_2(g) \;\rightarrow\; 2HCl(g)$$

Solution

According to Avogadro's law, if 1.0 mol of each gas occupies 22.4 L at STP, then the 2.0 mol of reactants have a total volume of 44.8 L. Because 2.0 mol of gas product is produced from 2.0 mol of reactants, the product would occupy 44.8 L at STP. The numerical value would change only if the number of moles of gaseous products differed from the number of moles of reactants.

PRACTICE 10.4

If a 12.8-L sample of He gas contains 6.4 mol of He, how many moles would there be in a 1.5-L sample of He at the same temperature and pressure?

See Problems 25, 26, and 35.

Boyle's Law

Two long-time friends, one of whom teaches at a U.S. college located at sea level (air pressure = 1.00 atm) and the other at a university in Mexico City (7340 ft above sea level, air pressure = 0.764 atm), prepare to show their students the link between moles of hydrogen and oxygen gas and the volume they occupy. Each of them adds 0.050 mol of hydrogen at 25°C to a balloon. The U.S. professor notes that his balloon has a volume of 1.2 L. His colleague in Mexico City measures the volume of his balloon as 1.6 L. *Except for the pressure of the gas in each balloon, which equals the atmospheric pressure in each city, the conditions are identical.* The Mexico City balloon contains gas at lower pressure and, as a result, larger volume than the balloon at sea level in the United States. Take this to extremes and the increase in volume can cause a balloon to burst, as we described at the beginning of this section.

Visualization: Boyle's Law: A Graphical View

Visualization: Boyle's Law: A Molecular-Level View

Atmospheric pressure causes changes in the volume of these balloons.

This inverse relationship between pressure and volume of a given amount of gas at constant temperature has been understood since 1662, when Robert Boyle (1627–1691; Figure 10.11) demonstrated what is now known as **Boyle's law**. This law can be summarized as follows:

$$PV = k' \qquad \text{(at constant } n, T\text{)}$$

where P = pressure of the gas

V = the volume occupied by the gas

k' = a constant (different from the constant in Avogadro's law)

Using the same logic as we did for Avogadro's law, we can rewrite Boyle's law to relate the pressure and volume of the same gas under two different conditions. For a change in either volume or (as in the case of the two balloons) pressure, we can say,

$$P_{\text{initial}} V_{\text{initial}} = P_{\text{final}} V_{\text{final}} \qquad \text{(at constant } n, T\text{)}$$

Boyle investigated the relationship between pressure and volume using the apparatus shown in Figure 10.12. Mercury was added to the open end of a U-shaped tube so that air was trapped between the mercury and the closed end. The height of the trapped air was indicative of the volume, V, of the gas. During his studies, Boyle noted that the difference in the heights of the mercury on both sides of the U-tube, plus the height of mercury at atmospheric pressure, 29⅛ in.

FIGURE 10.11

Robert Boyle (1627–1691) was the youngest of fourteen children born to a wealthy Irish family. He was interested in advancing his understanding of the world at a very early age. He kept very accurate observations about his experiments and was one of the first to build and use a vacuum pump.

 Video Lesson: Boyle's Law

FIGURE 10.12

Robert Boyle investigated the relationship between pressure and volume using the apparatus shown. Mercury was added to the open end of a U-shaped tube so that air was trapped between the mercury and the closed end. The height (*h*) of the trapped air was indicative of the volume, *V*, of the gas. The difference between the heights of the mercury on the two sides of the U-tube, plus the height of mercury at atmospheric pressure, 29⅛ in. (740 mm), was indicative of the pressure, *P*.

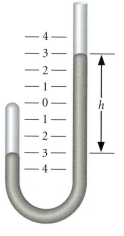

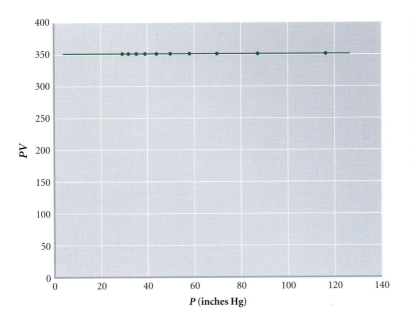

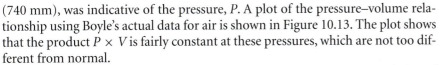

FIGURE 10.13

A plot of the pressure–volume relationship using Boyle's actual data for air. The fact that the product $P \times V$ is fairly constant at these pressures testifies to the care with which Boyle did these experiments. The pressure was measured as the height of a mercury column in inches, which is proportional to the pressure of the system. The volume was measured as the height of a column of air in inches, which is directly proportional to the volume of the system.

(740 mm), was indicative of the pressure, P. A plot of the pressure–volume relationship using Boyle's actual data for air is shown in Figure 10.13. The plot shows that the product $P \times V$ is fairly constant at these pressures, which are not too different from normal.

We began Section 10.3 by using food packaging to illustrate Dalton's law of partial pressures. Food packagers also acknowledge Boyle's law when they allow for the effect of higher altitudes (lower atmospheric pressures) on food packages. In fact, the food packagers adjust the pressure in their packages for the specific destination of the food items. For example, Figure 10.14 shows a bag of potato chips in Estes Park Colorado, at an altitude of 2440 m (8000 ft). The bag, bought in a store near sea level, is nearly ready to burst!

FIGURE 10.14

This bag of potato chips is a lot bigger in Estes Park, Colorado, at 2440 m, than at sea level. This is one reason why food manufacturers express the amount of product that their packages contain in terms of weight, rather than volume.

 Application

EXERCISE 10.5 Boyle's Law

A teacher wants to know what volume her balloon will need to be if it is to hold 0.050 mol of hydrogen at 25°C (as with the previous balloons) at a pressure of 1.3 atm. Using data for either sea level or Mexico City (provided in the main text), what is the minimum volume her balloon must hold?

First Thoughts

In solving the problem, the key questions to consider are "What do we expect to happen to the volume of the balloon as a result of increasing the pressure? Is it likely to increase or decrease?" Framing good answers to these questions *before we proceed* will give us an indication of whether our answer makes sense.

Solution

Using the Mexico City data, we get

$$P_{\text{initial}} = 0.764 \text{ atm} \qquad V_{\text{initial}} = 1.6 \text{ L}$$

$$P_{\text{final}} = 1.3 \text{ atm} \qquad V_{\text{final}} = ??? \text{ L}$$

$$P_{\text{initial}}V_{\text{initial}} = P_{\text{final}}V_{\text{final}}$$

$$0.764 \text{ atm}(1.6 \text{ L}) = 1.3 \text{ atm}(V_{\text{final}})$$

$$V_{\text{final}} = 0.94 \text{ L}$$

Visualization: Changes in Gas Volume, Pressure, and Concentration

Further Insights

Does the answer make sense? The pressure of the gas is greater than in either of the other balloons. At constant temperature and with the same amount of gas in each balloon, the only way to increase the pressure of the gas is to decrease the volume of the balloon, so that the force per unit area remains the same.

PRACTICE 10.5

How would the volume of the balloon in the previous exercise change if the pressure were decreased to 0.975 atm?

See Problems 27, 28, 34, and 36.

Visualization: Charles's Law: A Graphical View

Visualization: Charles's Law: A Molecular-Level View

Video Lesson: Charles's Law

Charles's Law

As we continue to look at the gas laws, we fill a balloon *at STP* with equal moles each of hydrogen and oxygen gases, so the total volume of the large Mylar balloon is, according to Avogadro's law, 2.0 L. That means we've added 1.0 L of each gas. We then take the balloon outside into a frosty winter's morning where the temperature is −10.0°F (−23.3°C). We notice that the balloon seems a bit smaller, as shown in Figure 10.15. Why?

There are several factors to consider. One of these factors involves the temperature of the gases in the balloon. Temperature is a measure of the kinetic energy of a system—the gas particles in our discussion. When the temperature drops, the gas particles travel at lower velocities. This means that there are fewer particle collisions with the walls of the balloon in a given time (also, each collision will occur with a lower kinetic energy and velocity, and so each collision has less impact). Because each gas particle collision with the balloon's inner wall adds to the force exerted by the gas on the balloon, *a smaller number of collisions per unit time indicates that the pressure of the gas is lowered*. As a result, the balloon shrinks until the pressure of the gas inside the balloon equals the pressure outside the balloon. In this process, the pressure was the same at the beginning as at the end. When the temperature is lowered, the volume of the balloon gets smaller. The general result that *the volume of a gas is directly proportional to the temperature (at constant pressure and number of moles)* was first reported by Jacques Alexander Charles (Figure 10.16) in 1787 and is known as **Charles's law**.

In the early 1800s, Joseph Louis Gay-Lussac (Figure 10.17) further studied the effect of temperature on the volume of a gas and reported that gases expand by

FIGURE 10.15

The relationship between temperature and volume of a gas (constant quantity and pressure) can be illustrated by comparing the volume of a balloon at 0°C, on the left, with that of the balloon on the right, at −23.3°C.

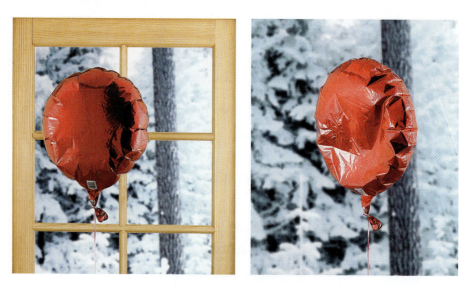

1/273 of their volume at 0°C for each increase in temperature of 1°C. This relationship is shown graphically at standard pressure and constant amount of gas in Figure 10.18. If we follow the line down to where the volume would become zero, we note that the temperature would be −273.15°C. In fact, such an extension would not be reasonable, because any gas would deviate substantially from ideal behavior and liquefy well before that point.

In any case, **does −273.15°C seem familiar?** It might. It is the temperature we know as **absolute zero**, the lowest possible temperature. This value serves as the basis for the Kelvin temperature scale, named after William Thomson (Lord Kelvin), who created this scale in 1848. On the Kelvin scale, a reading of zero kelvin (0 K) is equal to −273.15°C. At this temperature, there is no kinetic energy available, and therefore all atomic translations (movement in the x, y, or z directions) stop. Is such a temperature actually achievable? Even in the far reaches of deep space, residual heat from the Big Bang keeps even the coldest objects at 3 K. In the laboratory, however, scientists have been able to cool atoms down to within a few billionths of a degree of absolute zero.

Charles's law illustrates the relationship between volume and temperature. These two properties of a gas are directly proportional, so we can write

$$V = k''T \qquad \text{(at constant } n, P)$$

which can be rearranged to

$$\frac{V}{T} = k'' \qquad \text{(at constant } n, P)$$

where V = volume occupied by an ideal gas

T = temperature of the gas in kelvins

k'' = a constant relating the two quantities (different from the prior constants)

Just as we did with the other gas laws, we can set the two ratios of volume-to-temperature equal to each other if only the conditions of the gas have changed:

$$\frac{V_{\text{initial}}}{T_{\text{initial}}} = \frac{V_{\text{final}}}{T_{\text{final}}} \qquad \text{(at constant } n, P)$$

We need to keep in mind that the relationship between volume and temperature is an approximate one (because no gas is truly "ideal") and varies from substance to substance.

FIGURE 10.16

Jacques Alexander César Charles (1746–1823), a French inventor and scientist, was the first to take a voyage in a hydrogen balloon (to a height of 550 meters). He invented many scientific instruments and used them in his studies, including his confirmation of Benjamin Franklin's experiments with electricity.

FIGURE 10.17

Joseph Louis Gay-Lussac (1778–1850), a French chemist and physicist, is known for his work explaining the behavior of gases. He and Jean-Baptiste Biot were the first to ride in a hot-air balloon in 1804 to a height of 5 km. He was one of the discoverers of boron in 1808, which was later shown to be a new element.

Visualization: Liquid Nitrogen and Balloons

Visualization: Collapsing Can

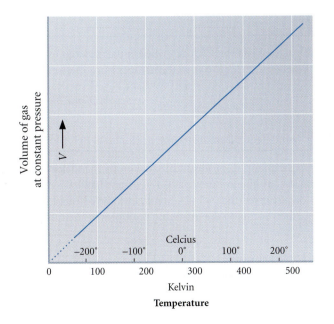

FIGURE 10.18

In the early 1800s, Gay-Lussac found that ideal gases expand by 1/273 of their volume for each increase of 1°C in temperature. If we follow the line down, we intersect the "zero volume" point at −273°C. In reality, such an extension would not be reasonable, because any gas would deviate substantially from ideal behavior and, in fact, liquefy well before that point. This does, however suggest that there is an absolute zero temperature.

EXERCISE 10.6   **Charles's Law**

As a demonstration of Charles's law, a Mylar balloon is filled with 1.00 L of helium gas at −12°C and held over a flame until the temperature of the gas within the balloon reaches 150.0°C. What is the new volume of the balloon, assuming constant pressure?

First Thoughts

What do we expect to happen to the volume of the balloon as a result of warming the gas inside? According to our discussion about Charles's law, it would make sense for the volume to increase. We also note that we must change the temperatures to the kelvin scale.

Solution

$$\frac{V_{\text{initial}}}{T_{\text{initial}}} = \frac{V_{\text{final}}}{T_{\text{final}}}$$

$$\frac{1.00\ \text{L}}{261\ \text{K}} = \frac{V_{\text{final}}}{423\ \text{K}}$$

$$\frac{1.00\ \text{L} \times 423\ \text{K}}{261\ \text{K}} = V_{\text{final}}$$

$$V_{\text{final}} = 1.62\ \text{L}$$

Further Insights

What would we have calculated our final volume to have been if we had kept our temperatures in Celsius degrees?

$$\frac{V_{\text{initial}}}{T_{\text{initial}}} = \frac{V_{\text{final}}}{T_{\text{final}}}; \qquad \frac{1.00\ \text{L}}{-12°\text{C}} = \frac{V_{\text{final}}}{150.0°\text{C}}$$

$$\frac{1.00\ \text{L} \times 150.0°\text{C}}{-12°\text{C}} = V_{\text{final}}$$

$$V_{\text{final}} = -13\ \text{L}$$

This outcome of a "negative volume" sharply illustrates why the Celsius scale cannot be used in gas law calculations.

PRACTICE 10.6

Assume the volume of the Mylar balloon described in the previous exercise is heated until its volume reaches 2.00 L. What is the temperature of the gas within the balloon?

See Problems 29, 30, 37, and 38. ▪

EXERCISE 10.7   **Mendeleev's Demonstration of Nonideal Behavior**

Based on his own experiments done around 1870, Dmitri Mendeleev noted that for real gases, Charles's law (originally known as Gay-Lussac's Law) was not entirely correct. Mendeleev's data showed that when several gases were heated (constant n, P) from 0°C to 100°C, they occupied the following multiples of their original volume:

$$\text{Air} = 1.368$$
$$\text{"Carbonic anhydride" (CO}_2\text{)} = 1.373$$
$$\text{Hydrogen} = 1.367$$
$$\text{Hydrogen bromide (HBr)} = 1.386$$

If the gases were truly ideal, what fraction of their original volume should they occupy when heated from 0°C to 100°C? Which of these gases most nearly exhibits ideal behavior?

First Thoughts

Which of these gases is likely to be behave closest to ideally? Why? We note that HBr occupies the greatest volume among the four gases, with CO_2 close behind. Why might this be? The best way to begin is perhaps to find the increase in volume of an ideal gas when heated from 0°C to 100°C. Let's keep our calculations straightforward by assuming an initial volume at 0°C (273 K) of 1.000 L. Mendeleev used one extra figure in his work, so we will do the same.

Solution

$$V_{initial} = 1.000 \text{ L} \qquad T_{initial} = 273 \text{ K} \quad \text{and} \quad V_{final} = ? \text{ L} \qquad T_{final} = 373 \text{ K}$$

$$\frac{V_{initial}}{T_{initial}} = \frac{V_{final}}{T_{final}} \qquad \text{(constant } n, P)$$

For an ideal gas,

$$\frac{1.000 \text{ L}}{273 \text{ K}} = \frac{V_{final}}{373 \text{ K}}$$

$$V_{final} = \frac{1.000 \text{ L} \times 373 \text{ K}}{273 \text{ K}} = 1.366 \text{ L}$$

Further Insights

Hydrogen, at 1.367 L, is the closest to ideal behavior. This makes sense because it is the smallest molecule and takes up the least space. Hydrogen bromide is a much larger molecule, requiring space for itself in order to exert the same pressure as hydrogen in, for example, a balloon at the higher temperature. All four gases deviate from ideality, the more so the greater their molar mass.

PRACTICE 10.7

In addition to its having volume, give another reason why HBr gas deviates from ideal gas behavior.

See Problem 104.

Combined Gas Equation

We can combine Avogadro's law,

$$\frac{V_{initial}}{n_{initial}} = \frac{V_{final}}{n_{final}}$$

and Boyle's law,

$$P_{initial} V_{initial} = P_{final} V_{final},$$

and Charles's law,

$$\frac{V_{initial}}{T_{initial}} = \frac{V_{final}}{T_{final}}$$

to come up with an equation called the **combined gas equation**. Why can we do this? The fact that all of these relationships use a proportionality constant makes this process easy. Our final equation takes volume, pressure, amount, and temperature into account:

$$\frac{P_{initial} V_{initial}}{n_{initial} T_{initial}} = \frac{P_{final} V_{final}}{n_{final} T_{final}}$$

Video Lesson: The Combined Gas Law

Video Lesson: CIA Demonstration: The Potato Cannon

We can use this equation to solve for the pressure, volume, amount, or temperature of a gas, as the gas changes conditions from one state to another. If something remains constant, that condition drops out of the equation and simplifies our calculations. Typical use of this equation implies that we use atmospheres for the pressure, liters for the volume, moles for the amount, and kelvins for the temperature.

EXERCISE 10.8 Combined Gas Equation

Recall our discussion of modified atmosphere packaging (MAP) in Section 10.3. If a chicken is packaged within a nitrogen–carbon dioxide gas atmosphere at STP and a volume of 300.0 mL of gas, what will be the volume of the gas (assuming no leakage through the packaging) when it is stored in a grocery freezer at a temperature of −8.0°C and a pressure of 1.04 atm?

First Thoughts

We are given initial pressure, temperature, and volume, and we change the pressure and temperature. We must find the final volume. What about the amount of the gas? Apparently, it doesn't change in the process. Therefore, we can substitute what we know into the combined gas equation. Before doing so, we should ask ourselves, "Would it make sense for the volume to increase or decrease as a result of these changes?" The temperature is decreasing from 273 K to 265 K. Charles's law suggests that the volume would decrease as a result. The pressure is increasing from 1.00 atm to 1.04 atm, which, via Boyle's law, would also lead to a slight reduction in volume. On balance, then, the volume should decrease.

Solution

$$P_{\text{initial}} = 1.00 \text{ atm} \qquad T_{\text{initial}} = 273 \text{ K} \qquad V_{\text{initial}} = 0.3000 \text{ L}$$
$$P_{\text{final}} = 1.04 \text{ atm} \qquad T_{\text{final}} = 265 \text{ K} \qquad V_{\text{final}} = ?$$
$$n_{\text{initial}} = n_{\text{final}}$$

$$\frac{P_{\text{initial}} V_{\text{initial}}}{n_{\text{initial}} T_{\text{initial}}} = \frac{P_{\text{final}} V_{\text{final}}}{n_{\text{final}} T_{\text{final}}}$$

We can rearrange variables to solve for V_{final}:

$$\frac{P_{\text{initial}} V_{\text{initial}} n_{\text{final}} T_{\text{final}}}{n_{\text{initial}} T_{\text{initial}} P_{\text{final}}} = V_{\text{final}}$$

Because $n_{\text{initial}} = n_{\text{final}}$, they cancel, and we have

$$\frac{P_{\text{initial}} V_{\text{initial}} T_{\text{final}}}{T_{\text{initial}} P_{\text{final}}} = V_{\text{final}}$$

$$\frac{1.00 \text{ atm} \times 0.3000 \text{ L} \times 265 \text{ K}}{273 \text{ K} \times 1.04 \text{ atm}} = 0.280 \text{ L} = V_{\text{final}} = 280 \text{ mL}$$

Further Insights

Does our answer make sense? We expected the volume of gas to decrease, and it did. When the physical properties change in ways that yield opposite effects—as, for example, when both temperature and pressure increase—we can judge what the outcome might be before calculating by seeing which change is greater, and that change will dominate.

PRACTICE 10.8

Assuming the same initial conditions as in Exercise 10.8, solve for the volume when the final pressure is 0.85 atm and the final temperature is −11.0°C.

See Problems 31 and 32.

TABLE 10.5	The Gas Laws	
Common Name	**Equation**	**Summary**
Avogadro's law	$\dfrac{V}{n} = k$ (constant P, T)	Equal numbers of molecules are contained in equal volumes of all dilute gases under the same conditions.
Boyle's law	$PV = k'$ (constant n, T)	There is an inverse relationship between pressure and volume of a given amount of gas at constant temperature.
Charles's law	$\dfrac{V}{T} = k''$ (constant n, P)	The volume of a gas is directly proportional to the temperature at constant pressure and moles.
Combined gas equation	$\dfrac{PV}{nT} = $ constant	The combination of Avogadro's, Boyle's and Charles's laws relates initial and final pressure, volume, amount, and temperature.

To this point, we have discussed relationships among the pressure, volume, temperature, and amount of a gas. We have seen that we can take into account changes in any one of these and find its affect on another variable. These relationships are summarized in Table 10.5.

10.5 The Ideal Gas Equation

So far we have looked at three laws describing the behavior of gases, the laws named in honor of Avogadro, Boyle, and Charles. The balloon that carries aloft the ozone-detecting equipment (as was shown in Figure 10.7) bursts as a consequence of the *combined* effects of temperature and pressure on the volume of the gas within the balloon. In order to understand why this bursting occurs, we need to examine how these three variables are related.

The three gas laws were summarized in the previous section as follows:

$$\frac{V}{n} = k \qquad \text{(constant } P, T\text{)}$$

$$PV = k' \qquad \text{(constant } n, T\text{)}$$

$$\frac{V}{T} = k'' \qquad \text{(constant } n, P\text{)}$$

Setting each equation equal to V yields

$$V = kn \qquad \text{(constant } P, T\text{)}$$

$$V = \frac{k'}{P} \qquad \text{(constant } n, T\text{)}$$

$$V = k''T \qquad \text{(constant } n, P\text{)}$$

We can combine all the variables on the right-hand side, because they are all proportional to V via the constants k, k' and k''. To clean it up nicely, we then combine all three of these constants into a single constant called R. This gives us this equation:

$$V = \frac{RnT}{P}$$

Visualization: The Ideal Gas Law, $PV = nRT$

Video Lesson: The Ideal Gas Law

Tutorial: Ideal Gas Law

The denominator can be cleared to give the common form of the **ideal gas equation**:

$$PV = nRT$$

The name of this equation includes the term *ideal* because the equation assumes ideal behavior of the gas. In fact, we know that gases will deviate from the ideal, yet the ideal gas equation will give us meaningful approximate results unless the gases are at very high pressures and/or low temperatures.

What is the value of R? It can be determined by measuring each of the properties (P, V, T, and n) of a gas and solving for the combined gas law shown in Table 10.5. For example, at STP ($T = 273.15$ K, $P = 1.0$ atm), 1 mol of an ideal gas occupies 22.414 L. The value of R, then, is 0.08206 L·atm·mol^{-1}·K^{-1}.

$$\frac{PV}{nT} = R = \frac{(1.000 \text{ atm})(22.414 \text{ L})}{(1.000 \text{ mol})(273.15 \text{ K})} = 0.08206 \frac{\text{L·atm}}{\text{mol·K}}$$

This is a very important number. It is known as the **ideal gas constant**, and it is used quite often in chemistry. It is helpful to commit not only the number, but also the units, to memory. Why? The units of the constant help us remember which units to use for the other variables in the equation, and they also help us confirm the units of our answers.

EXERCISE 10.9   Using the Ideal Gas Equation

A sample containing 0.631 mol of a gas at 14.0°C exerts a pressure of 1.10 atm. What volume does the gas occupy?

Solution

We can use the ideal gas equation to solve for volume. By placing what we know into the equation, we can solve for the quantity that we don't know.

$$n = 0.631 \text{ mol} \qquad T = 287.2 \text{ K} \qquad P = 1.10 \text{ atm}$$

$$PV = nRT$$

$$1.10 \text{ atm} \times V = 0.631 \text{ mol} \times \left(\frac{0.08206 \text{ L·atm}}{\text{mol·K}} \right) \times 287.2 \text{ K}$$

$$V = 13.5 \text{ L}$$

PRACTICE 10.9

A 2.50-L sample of an ideal gas at 25°C exerts a pressure of 0.35 atm. How many moles of this gas are there?

See Problems 45, 46, 51, 52, 58, and 72.

EXERCISE 10.10   The Ideal Gas Equation and Compressed Gas

How much pressure must a 48.0-L steel oxygen gas cylinder tolerate if it contains 9.22 kg of O_2 at a temperature of 25.0°C? Given the conditions, does the gas show approximately ideal behavior?

First Thoughts

There are two differences between this exercise and Exercise 10.9. Here, we are given mass, in kilograms, instead of the amount of moles. Second, the sample is much larger than laboratory-sized samples. The question asks about ideal gas behavior. The unknown here is the pressure. Gases approach ideality if the pressure is low. High pressure creates the conditions for a great deal of interaction among gas molecules and with the container walls.

Solution

We find it convenient first to convert mass into moles and then to find pressure using the ideal gas equation. We will follow this strategy.

$$n_{O_2} = 9.22 \text{ kg O}_2 \times \frac{1000 \text{ g O}_2}{1 \text{ kg O}_2} \times \frac{1 \text{ mol O}_2}{32.0 \text{ g O}_2} = 288.1 \text{ mol O}_2$$

$$PV = nRT$$

$$P \times 48.0 \text{ L} = 288.1 \text{ mol} \times \left(\frac{0.08206 \text{ L·atm}}{\text{mol·K}} \right) \times 298 \text{ K}$$

$$P = 146.8 \text{ atm}$$

$$P = 147 \text{ atm}$$

Further Insights

The pressure in the steel cylinder is quite high. We would therefore expect the oxygen gas within the cylinder to deviate significantly from ideal behavior.

PRACTICE 10.10

A 0.250-mol sample of gas occupies 8.44 L at 28.7°C. What is the pressure of the system?

See Problems 42, 63, and 64.

We assumed ideal behavior in the previous exercise, yet we suspect that this is not likely for oxygen gas at a pressure of 147 atm. One of the themes in our discussion of gases has been the recognition that deviations, even small ones, from ideality are expected. We have chosen to neglect them in our calculations, assuming that the approximate answers we get are meaningful. However, as shown in Figure 10.19, the ratio PV/nRT, is not constant, and the deviation becomes more severe at high pressures. Is there a way to correct for relatively high pressure or low temperatures, such as we had in Exercise 10.10 (compressed O_2)?

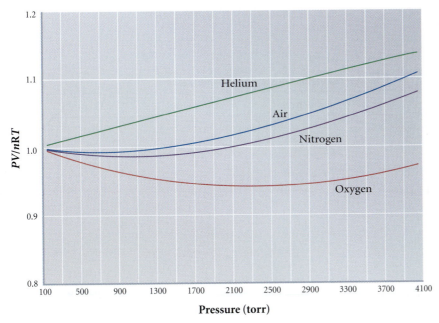

FIGURE 10.19

"Ideal" is just that—a standard of perfection that no gas can ever meet. If $PV = nRT$, then $\frac{PV}{nRT} = 1.00$. Plotting actual values shows that all gases deviate from this equation, especially as the pressure increases. Moreover, the deviation is different for each gas.

Video Lesson: Comparing Real and Ideal Gases

Accounting for the Behavior of Real Gases

Volume Correction

Real gas molecules occupy discrete volumes, and larger atoms and molecules take up more space than smaller ones. Therefore, the available volume of the container is not as large as if it were empty or if we assumed that the gas particles had no volume. The more moles of gas in a given space, the less free volume compared to ideal conditions, as shown in Figure 10.20. To take this into account, we can express the available volume as

$$V_{available} = V_{container} - nb$$

in which b is a constant roughly related to the size of the molecules.

Pressure Correction

The molecules in a real gas interact with each other. Larger atoms and molecules, and those with dipole moments, exhibit stronger interactions than smaller ones. When particles interact with each other, they do not collide as often with the walls of the container, and their force per unit area—their pressure—is lower (see Figure 10.20). The reduction in observed pressure can be given by

$$P_{actual} = P_{ideal} - a\left[\frac{n}{V}\right]^2$$

in which a is a constant with a magnitude that is loosely indicative of the intermolecular forces that are possible in molecules. Note that the values of a in Table 10.6 are large in the relatively large molecules that have more polarizable electrons than the smaller molecules and atoms. The constant, a, is multiplied by $\left[\frac{n}{V}\right]^2$. As the volume, V, gets larger, the correction goes down, because the molecules are farther apart. Similarly, when the number of moles, n, is low, the correction term is reduced.

The ideal gas equation assumes the use of P_{ideal}. Therefore, in our modified model, we can substitute for P_{ideal} as follows:

$$P_{ideal} = P_{actual} + a\frac{n^2}{V^2}$$

This model of the pressure and volume deviations was put into equation form in 1873 by J. D. van der Waals. The equation he produced is

$$\left(P + a\frac{n^2}{V^2}\right)(V - nb) = nRT$$

↑ Pressure correction   ↑ Volume correction

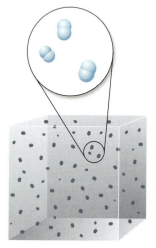

FIGURE 10.20

As more and more molecules of a gas are added to a given volume, the amount of free space within the volume decreases and the gas deviates greatly from "ideal." The van der Waals equation takes this into account, as we saw in the text.

TABLE 10.6	van der Waals Constants for Some Gases		
Gas	**Formula**	**a (atm L^2·mol^{-2})**	**b (L·mol^{-1})**
Acetylene	C_2H_2	4.39	0.0514
Ammonia	NH_3	4.17	0.0371
Argon	Ar	1.35	0.0322
Carbon dioxide	CO_2	3.59	0.0427
Chlorine	Cl_2	6.49	0.0562
Ethane	C_2H_6	5.49	0.0638
Helium	He	0.034	0.0237
Hydrogen	H_2	0.244	0.0266
Methane	CH_4	2.25	0.0428
Nitrogen	N_2	1.39	0.0391
Oxygen	O_2	1.36	0.0318
Sulfur dioxide	SO_2	6.71	0.0564

Source: Maron and Prutton, *Principles of Physical Chemistry.* New York: MacMillan, 1959, p. 33.

As we noted previously, the constant b is a correction indicating that real molecules do have a finite volume and are not as far apart as we claim in ideal behavior. The more moles of gas, n, the greater will be the correction to volume, $V - nb$. The values for b are only roughly proportional to molecular size.

We can rearrange the **van der Waals equation** to solve for pressure:

$$P = \frac{nRT}{V - nb} - a\frac{n^2}{V^2}$$

The van der Waals equation is the most often cited mathematical model for use with real gases because it works fairly well at normal pressures and temperatures. However, it doesn't fit the experimental data well at very high pressures and low temperatures. Therefore, as pressure and temperature change, different values of a and b must be chosen. Other mathematical models that fit the data even better (and are correspondingly more complex) have been derived.

EXERCISE 10.11   **The van der Waals Equation**

We now have the tools to better answer Exercise 10.10. Assuming the same volume, temperature, and number of moles of O_2 gas in the cylinder, what is a more accurate estimate of the pressure of the gas?

First Thoughts

We expect the actual pressure exerted by the oxygen gas to be less than the pressure exerted by an ideal gas because the gas molecules are interacting more with each other and colliding (and, therefore, interacting—the a term) less often with the walls of the container. On the other hand, the nb correction term would make the pressure greater by making the actual open volume less. The a-term correction is much more significant, so we expect the pressure to be lower than the ideal value.

Solution

$$P = \frac{nRT}{V - nb} - a\frac{n^2}{V^2}$$

$$P = \frac{288.1\ \text{mol}\ O_2 \times 0.08206\ \text{L·atm·mol}^{-1}\text{·K}^{-1} \times 298\ \text{K}}{48.0\ \text{L} - (288.1\ \text{mol}\ O_2 \times 0.0318\ \text{L·mol}^{-1})}$$

$$- 1.36\ \text{L}^2\text{·atm·mol}^{-2}\ \frac{(288.1\ \text{mol}\ O_2)^2}{(48.0\ \text{L})^2}$$

$$= 181.4\ \text{atm} - 49.0\ \text{atm} = 132\ \text{atm}$$

Further Insights

We expected a lower pressure, and the van der Waals equation confirmed this. The answer of 132 atm therefore makes sense. Keep in mind that *this equation is an empirical model and fits many, but not all, situations well.* It does, however, come a lot closer than the ideal gas equation to assessing the behavior of real gases correctly.

PRACTICE 10.11

What is the pressure of 1.00 mol of O_2 gas in a 1.00-L balloon at 25°C? Calculate the pressure using the ideal gas law and the van der Waals equation. Is there a difference? Then repeat the same calculations for ammonia.

See Problems 102 and 103.

10.6 Applications of the Ideal Gas Equation

Atmospheric chemists not only know the relationships that dictate how gases behave but also understand the applications of these relationships as ways to solve real-world problems. Many of their calculations—such as those that determine how high a balloon will travel—can be done using the mathematical adjustments of the van der Waals equation, but close approximations can be obtained using the ideal gas equation. Other applications of this equation are equally useful. They can be broken down into two main types: those that address physical properties and those that address chemical reactions. Examples of these applications follow.

Physical Change Applications

The key questions to ask when applying the ideal gas equation are the same as in any problem solving. "What do I want? How do I get there? What answer do I expect? Does my answer make sense?" The following two exercises illustrate how the ideal gas equation, density, and the molar mass of gases are related.

EXERCISE 10.12 **An Application of Gas Density**

The density of liquid nitrogen at $-196°C$ is 0.808 g/mL. What volume of nitrogen gas at STP must be liquefied to make 20.0 L of liquid nitrogen?

First Thoughts

We can work backwards, via the density, to find how many moles of liquid nitrogen there are in 20.0 L of liquid nitrogen. This is the amount of gas that must be liquefied. We can then use the ideal gas equation to convert from moles of N_2 to liters of the gas.

Solution

$$\text{g liquid N}_2 = 20.0 \text{ L liquid N}_2 \times \frac{1000 \text{ mL liquid N}_2}{1 \text{ L liquid N}_2} \times \frac{0.808 \text{ g liquid N}_2}{1 \text{ mL liquid N}_2}$$

$$= 16{,}160 \text{ g N}_2$$

$$n_{N_2} = 16{,}160 \text{ g N}_2 \times \frac{1 \text{ mol N}_2}{28.0 \text{ g N}_2} = 577 \text{ mol N}_2$$

$$PV = nRT$$

$$1.00 \text{ atm} \times V = 577 \text{ mol} \times \left(\frac{0.08206 \text{ L·atm}}{\text{mol·K}} \right) \times 273 \text{ K}$$

$$V = 1.29 \times 10^4 \text{ L N}_2$$

Further Insights

Many gases, such as oxygen, ammonia, and nitrogen, are converted to liquids for shipping and then used on-site as gases. As we see in this example, a lot of liquid nitrogen can be shipped in this way. Alternatively, as we learned previously, gases are shipped under pressures approaching 150 atm in compressed gas tanks.

PRACTICE 10.12

A gas at 16°C and 1.75 atm has a density of 3.40 g/L. Determine the molar mass of the gas. (*Hint:* Begin by assuming 1.00 L of the gas, and then determine the number of moles that would occupy this volume.)

See Problems 47, 50, 53, 55, and 67.

Dimensional analysis can lead us to a useful formula that we can use to determine the molar mass of a gas, as was requested in Practice 10.12. The symbol M is the variable used to represent molar mass, and it is also used to represent molarity. We will underline the variable for molar mass ($\underline{M}$) to help distinguish between the two. Translating the quantities back into the variables they represent, we find that the relationship between molar mass and density is

$$\underline{M} = \frac{\overset{\overset{d}{\downarrow}}{3.40 \text{ g L}^{-1}} \overset{\overset{R}{\downarrow}}{(0.08206 \text{ L·atm·mol}^{-1}\text{·K}^{-1})} \overset{\overset{T}{\downarrow}}{(289 \text{ K})}}{\underset{\underset{P}{\uparrow}}{1.75 \text{ atm}}}$$

$$\underline{M} = \frac{dRT}{P}$$

We see from this equation that *the molar mass of an ideal gas is directly proportional to its density*. This relationship approximately holds for real gases that have close to ideal behavior, but it does *not* hold for liquids or solids. As this equation indicates, you can determine the molar mass of a gas if you can measure its density.

Chemical Reactions—Automobile Air Bags

When a car crash occurs, the car stops suddenly. The occupant, however, continues to move with great velocity toward the steering wheel or dashboard. The deployed air bag distributes over a large area (that of the air bag) the force that the occupant would otherwise exert on the dashboard, minimizing the pressure on any particular part of the occupant's upper body. The result is an accident that causes less bodily injury than an accident in which no air bag is deployed.

Inside the most common style of air bag is a capsule of sodium azide (NaN_3), iron(III) oxide (Fe_2O_3), and a small detonator cap that is rigged to start the decomposition of NaN_3:

$$2NaN_3(s) \rightarrow 2Na(s) + 3N_2(g)$$

The nitrogen gas inflates the bag to its full volume (74 L for front air bags) within 50 ms of a crash. The sodium metal that is also generated in the reaction reacts quickly with the iron(III) oxide to form sodium oxide. This second reaction is necessary to remove dangerous sodium metal.

$$6Na(s) + Fe_2O_3(s) \rightarrow 3Na_2O(s) + 2Fe(s)$$

Application

CHEMICAL ENCOUNTERS:
Automobile
Air Bags

Air bags save lives by redistributing the force of a person's body over a larger area.

| **EXERCISE 10.13** | **Filling the Air Bag** |

A technician is designing an air bag for a new model car. She wants to use 72.0 g of sodium azide (NaN_3) to inflate the bag at 22.0°C and 1.00 atm. What is the maximum volume of the air bag under these conditions?

First Thoughts

The air bag reaction is notable not only because the bag is filled with nitrogen gas, not air, but also because the stoichiometry of the reaction indicates that 3 mol of N_2 are formed from the decomposition of 2 mol of NaN_3. This must be taken into account in solving the problem.

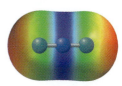

Azide ion

Solution

$$\text{mol NaN}_3 = 72.0 \text{ g NaN}_3 \times \frac{1 \text{ mol NaN}_3}{65.0 \text{ g NaN}_3} = 1.108 \text{ mol NaN}_3$$

$$\text{mol N}_2 = 1.108 \text{ mol NaN}_3 \times \frac{3 \text{ mol N}_2}{2 \text{ mol NaN}_3} = 1.662 \text{ mol N}_2$$

$$PV = nRT$$

$$1.00 \text{ atm} \times V = 1.662 \text{ mol} \times \left(\frac{0.08206 \text{ L·atm}}{\text{K·mol}} \right) \times 295 \text{ K}$$

$$V = 40.2 \text{ L}$$

Further Insights

The amount of chemistry that occurs in 50 ms is startling, and the effects are profound. Between 1987 and 2001, nearly 8400 lives in the United States alone were saved by air bag deployments in motor vehicle accidents.

PRACTICE 10.13

How many grams of sodium azide would be required if the volume of the air bag were to be 30.9 L at 31°C and 1.00 atm?

See Problems 59 and 60.

Chemical Reactions—Acetylene

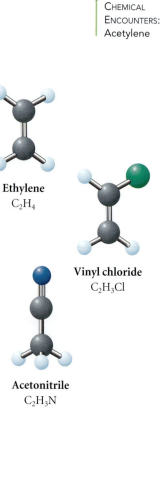

Ethylene
C₂H₄

Vinyl chloride
C₂H₃Cl

Acetonitrile
C₂H₃N

Application

CHEMICAL
ENCOUNTERS:
Acetylene

Acetylene (C_2H_2) is an industrial gas used in welding and in the manufacture of vinyl chloride (CH_2CHCl), acetonitrile (CH_3CN), and other industrial chemicals, including polymers such as "neoprene." The gas was first prepared inadvertently in the early 1890s by Thomas Willson, who owned a small aluminum-making company. In his quest to more effectively convert bauxite, composed mostly of aluminum oxide (Al_2O_3) to aluminum metal, he combined the ore with lime (CaO), and coal tar (mostly carbon compounds) in a furnace. A gray solid was formed that, when reacted with water, gave off a flammable gas. After receiving a sample of the solid from Willson, Francis Venable of the University of North Carolina–Chapel Hill determined that Willson had made calcium carbide (CaC_2)

$$CaO(s) + 3C(s) \rightarrow CaC_2(s) + CO(g)$$

The calcium carbide reacted with water to form acetylene (C_2H_2) which had been discovered by Edmund Davy in 1836.

$$CaC_2(s) + 2H_2O(l) \rightarrow Ca(OH)_2(s) + C_2H_2(g)$$

Acetylene has a great many industrial and commercial uses. Acetylene torches are commonly used to weld and cut metals. The demand for acetylene is currently met using a process different from that discovered by Willson and Davy. Acetylene can be made by pyrolysis (adding heat to convert organic solids into gases and liquids) of methane, the main component of natural gas:

$$2CH_4(g) \rightarrow C_2H_2(g) + 3H_2(g)$$

Acetylene burns with a much brighter flame than many other fuels, leading to the use of acetylene as a light source in many areas. Its use as a lamp by miners as they work underground is most notable. Acetylene also burns very hot, at about 3000°C, when reacted with oxygen, giving rise to its use in welding and cutting equipment.

Miners use calcium carbide lamps mounted on their helmets to see in the dark recesses of a mine. Within the lamp, water is slowly dripped onto a piece of calcium carbide. The reaction produces acetylene gas, which is ignited in the center of a mirrored reflector.

EXERCISE 10.14   Production of Acetylene

A small chemical company is setting up the facilities to produce acetylene. Planners want to know what volume of C_2H_2 will be produced from the reaction of 4.25×10^3 g of methane if the product is to be stored at 38°C and 1.00 atm. Assume a yield of 34%.

First Thoughts

A good problem-solving strategy is to determine the amount of acetylene that can be formed assuming a 100% theoretical yield, and then reduce it to account for the 34% actual yield. We can start by converting grams of methane to moles of methane and then to moles of acetylene, remembering the 2-to-1 mole ratio of methane to acetylene. We can use the ideal gas equation to find the volume of C_2H_2 assuming 100% yield and then, finally, adjust for the lower yield.

Solution

$$\text{mol CH}_4 = 4250 \text{ g CH}_4 \times \frac{1 \text{ mol CH}_4}{16.0 \text{ g CH}_4} = 265.6 \text{ mol CH}_4$$

$$\text{mol C}_2\text{H}_2 = 265.6 \text{ mol CH}_4 \times \frac{1 \text{ mol C}_2\text{H}_2}{2 \text{ mol CH}_4} = 132.8 \text{ mol C}_2\text{H}_2$$

$$PV = nRT$$

$$1.00 \text{ atm} \times V = 132.8 \text{ mol} \times \left(\frac{0.08206 \text{ L·atm}}{\text{mol·K}}\right) \times 311 \text{ K}$$

$$V = 3.39 \times 10^3 \text{ L}$$

The volume of 3.39×10^3 L assumes a 100% yield. Because the actual yield is 34%, we must multiply 3.39×10^3 L by 0.34, to get the final volume of 1.15×10^3 L of C_2H_2.

Further Insights

We have discussed the preparation of several industrial gases. Some are shipped as compressed gases, others as liquids. Using the Internet, you can gain additional food for thought by finding out which ones are shipped in what way and why.

PRACTICE 10.14

How many grams of methane would be needed to produce 5.00×10^3 L of C_2H_2? Again, assume that the actual yield of the reaction is only 34% under the same storage conditions.

See Problems 65 and 66.

HERE'S WHAT WE KNOW SO FAR

- Avogadro's, Boyle's, and Charles's laws can be combined into a equation that relates the initial and final states of a gas.

$$\frac{P_{\text{initial}} V_{\text{initial}}}{n_{\text{initial}} T_{\text{initial}}} = \frac{P_{\text{final}} V_{\text{final}}}{n_{\text{final}} T_{\text{final}}}$$

- The ideal gas law provides information on the current state of a gas.

$$PV = nRT$$

- The ideal gas constant (R) relates the value and the units needed to mathematically operate the ideal gas law. $R = 0.08206 \text{ L·atm·mol}^{-1}\text{·K}^{-1}$

- The molar mass of a gas can be determined using a modification of the ideal gas equation.

$$\underline{M} = \frac{dRT}{P}$$

Video Lesson: The Kinetic-Molecular Theory of Gases

Visualization: Kinetic-Molecular Theory/Heat Transfer

10.7 Kinetic-Molecular Theory

The ideal gas laws are elegant in their simplicity and profound in their meaning. Because of their simplicity, many scientists, including Bernoulli, Clausius, Maxwell, Boltzmann, and van der Waals, sought to prove that the laws governing the behavior of ideal gases were a result of the interactions among atoms and molecules (collectively called "molecules" in the theory). The beauty of the resulting **kinetic-molecular theory** is that *it enables us to arrive at the same conclusions by thinking about the nature of molecules as we do from performing well-chosen experiments*. The experimental work and the theoretical work are consistent. Both are based on the following statements that model gas behavior:

1. Gases are composed of particles. The particles are negligibly small compared to their container and to the distance between each other.

2. Therefore, intermolecular attractions, which are exhibited at small distances, are assumed to be nonexistent.

Visualization: Visualizing Molecular Motion: Single Molecule

Visualization: Visualizing Molecular Motion: Many Molecules

3. Gases are in constant, random motion, *colliding with the walls of the container and with each other*.

4. Pressure is the force per unit area caused by the molecules colliding with the walls of the container.

5. *Because pressure is constant in a container over time*, molecular collisions are assumed to be perfectly elastic. *That is, no energy of any kind is lost upon collision*.

6. The average kinetic energy of the molecules in a system is linearly proportional to the absolute (Kelvin) temperature.

Using the Assumptions About Gas Behavior to Rationalize the Gas Laws

Boyle's Law (*PV* = constant)

Collisions with the walls of the container give rise to pressure. What would happen if the volume of the container were increased at a constant temperature for a given number of molecules? Because temperature is a measure of the average kinetic energy of the molecules in the system, a constant temperature means that the velocity of the molecules would stay the same. The same number of molecules, traveling at the same speed, are bouncing off a larger total surface area, so the pressure on a unit of surface area would fall.

Charles's Law (*V/T* = constant)

In this case, both the pressure and the number of molecules are constant. If the volume is lowered, how can the number of collisions per unit time (a measure of pressure) stay the same? The only way this can happen is if the velocity of the molecules becomes slower, via lowering of the temperature. As a result, the more slowly traveling molecules, which also have less kinetic energy, hit the now-closer walls at the same rate as the more rapidly moving molecules did when the walls were farther out.

Avogadro's Law (*V/n* = constant)

How is it possible to keep the pressure and the average kinetic energy (temperature) of the molecules constant if the volume is increased? We need to add more molecules to the system—that is, increase *n*—so that the number of collisions per unit time in the larger system remains the same.

EXERCISE 10.15 **The Importance of Assumptions**

According to point 1 of the kinetic-molecular theory, "Gases are composed of molecules that are negligibly small compared to their container and to the distance between molecules." Strictly speaking, this statement is not true for real gases. How would the gas laws change if point 1 were not included?

Solution

One answer is that if the more realistic picture of gases were taken into account, then the volume would depend on the nature of the gas. Gases composed of larger atoms and molecules take up more volume and would therefore cause more deviation from Boyle's law. The *b* term in the van der Waals model for nonideal gas behavior is one attempt to quantitatively account for this deviation.

PRACTICE 10.15

Explain, using the equations for kinetic energy, why two different molecules must differ in speed if the molecules have identical kinetic energy. If their kinetic energies are the same, which would be moving faster, a low-molecular-mass molecule or a high-molecular-mass molecule?

See Problems 73–76.

A Closer Look at Molecular Speeds

Video Lesson: Molecular Speeds

The kinetic energy of a molecule is equal to $\frac{1}{2}mv^2$, in which *m* is the mass and *v* is the velocity of that molecule. For a system with one type of molecule (that is, *m* is constant), the temperature is proportional to the square of the velocity, v^2. Molecules move faster at higher temperatures, slower at lower ones.

For a gas containing Avogadro's number of molecules, there are several additional ideas we must introduce. First, the term *velocity* means the rate of travel—the speed—in a specified direction. If we have Avogadro's number of molecules, they will be traveling in all different directions, and the net velocity (including the direction!) may well be zero meters per second, even if the individual molecules are moving very rapidly. *When we talk about the rate of travel of large numbers of molecules, we are interested in their speed (how fast), not their velocity (how fast and in what direction).* Also, the collisions will not be perfectly elastic, and some energy will, in fact, be transferred. There will be a distribution of kinetic energies among the many molecules and, therefore, a range of speeds that can be determined. Just as the SAT or ACT scores for students throughout the United States have a fairly wide range, the speeds of individual molecules in a sample also have

FIGURE 10.21

Molecular speeds are related to the mass of the gas and to the temperature, as shown here for helium and methane. Why do methane molecules move more slowly, on average, than those of helium at the same temperature?

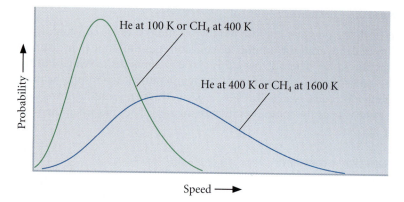

a range of values. Just as there is an average SAT or ACT score, we can define the **average speed** of i molecules as we would any average value; it is the sum of the speeds divided by the number of molecules:

$$\text{Speed}_{\text{average}} = (\text{speed}_1 + \text{speed}_2 + \cdots + \text{speed}_i)/i$$

A related value, often used in discussions about the statistics of molecular behavior, is the **root-mean-square (rms) speed**, u. This is the speed that a molecule with the average kinetic energy will have, and it is equal to the square root of the sum of the squares of the individual speeds divided by the number of molecules—that is,

$$u = \left[\frac{\left(\text{speed}_1{}^2 + \text{speed}_2{}^2 + \cdots + \text{speed}_i{}^2\right)}{i} \right]^{1/2}$$

The average score on the SAT or ACT is not necessarily the most probable; it is just the average. So it is with gases, for which there exists the **most probable speed**, α, of a gas. The root-mean-square speed, the average speed, and the most probable speed are related as 1.22 : 1.12 : 1.00.

The distribution of molecular speeds at different temperatures for helium and that for methane are shown in Figure 10.21. Why do methane molecules move more slowly, on average, than those of helium at the same temperature? Again we revisit the factors that account for kinetic energy ($\frac{1}{2}\,mv^2$). If two gases are at the same temperature, their average kinetic energies will also be equal. Methane (16.0 g/mol) is heavier than helium (4.00 g/mol), so its molecules must be moving more slowly.

10.8 Effusion and Diffusion

Video Lesson: Effusion and Diffusion

Tutorial: Effusion and Diffusion of Gases

At a given temperature (constant molecular kinetic energy = $\frac{1}{2}\,mv^2$), heavier gas molecules move more slowly than their lighter counterparts. Although the mathematical derivation is fairly involved, it is possible, using the ideas of the kinetic-molecular theory, to state that for any sample of gas, the rms speed is related to the molar mass of the gas as follows:

$$u_{\text{rms}} = \sqrt{\frac{3RT}{M}}$$

where T = the temperature in K

$\underline{M}$ = the molar mass of the gas in kg·mol^{-1}

R = the ideal gas constant, 8.314 J·K^{-1}·mol^{-1}

= 8.314 $\underline{\text{kg·m}^2\text{·s}^{-2}}$·mol^{-1}·K^{-1}

$\updownarrow$

J

because

$$P \times V = \left[\frac{\text{force}}{\text{area}} \times V \right]$$

$$= \left[\text{mass} \times \frac{\text{acceleration}}{\text{area}} \times \text{volume} \right]$$

$$= \frac{\text{kg·m·s}^{-2}}{\text{m}^2} \times \text{m}^3$$

$$= \text{kg·m}^2\text{·s}^{-2} = \text{J}$$

EXERCISE 10.16   The rms Speed of Gases

Compare the rms speeds of methane and helium at 25°C.

First Thoughts

What do we expect and why? If you wanted to use an analogy when teaching a class about the relationship between molar mass and speed, what might it be? A useful analogy might be to compare the running speed of a 125-lb world-class marathon runner to that of a 600-pound sumo wrestler, where the helium is the marathon runner and the methane is the sumo wrestler.

Solution

For methane,

$$u_{rms} = \sqrt{\frac{3RT}{M}} = \sqrt{\frac{3 \times 8.314 \text{ kg·m}^2\text{·s}^{-2}\text{·mol}^{-1}\text{·K}^{-1} \times 298 \text{ K}}{0.0160 \text{ kg·mol}^{-1}}}$$

$$= \sqrt{4.645 \times 10^5 \text{ m}^2\text{·s}^{-2}} = 682 \text{ m/s}$$

For helium,

$$u_{rms} = \sqrt{\frac{3RT}{M}} = \sqrt{\frac{3 \times 8.314 \text{ kg·m}^2\text{·s}^{-2}\text{·mol}^{-1}\text{·K}^{-1} \times 298 \text{ K}}{0.00400 \text{ kg·mol}^{-1}}}$$

$$= \sqrt{1.858 \times 10^6 \text{ m}^2\text{·s}^{-2}} = 1360 \text{ m/s}$$

Further Insights

Note that the rms speed of methane is half that of helium, and that methane has a molar mass four times the atomic mass of helium. This suggests that the speed of gas molecules is inversely proportional to the square root of the molar mass. Note as well the incredibly high speeds of these molecules!

PRACTICE 10.16

Which gas has a higher average speed at 20°C, F_2 or CO_2?

See Problems 75–85.

In 1829, Thomas Graham (1805–1869) made measurements of the relative rates at which gases pass through small openings into a very low-pressure region, a process called **effusion** (Figure 10.22). His results, like those that led to the other gas laws, predated the kinetic-molecular theory and confirm our conclusion regarding the relative speeds of gas particles. His measurements led to what

FIGURE 10.22

Gases effuse through a small opening in their container. The faster molecules escape through the opening more rapidly than the slower molecules.

Visualization: Effusion of a Gas

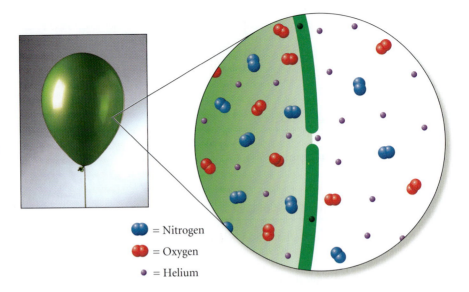

• = Nitrogen

• = Oxygen

• = Helium

we now know as **Graham's law of effusion**. We can state this law mathematically. For two gases of differing molar mass, $\underline{M}_1$ and $\underline{M}_2$, the ratio of their effusion rates (re_1/re_2) is equal to the square root of the inverse ratio of their molar masses ($\underline{M}_2/\underline{M}_1$).

$$\frac{re_1}{re_2} = \sqrt{\frac{\underline{M}_2}{\underline{M}_1}}$$

Application

Effusion can be used to separate mixtures of gases that would be difficult to separate by other methods. Effusion has been used to separate ^{235}U and ^{238}U from the natural mixture of uranium isotopes (0.72% ^{235}U and 99.28% ^{238}U). The separation of these two naturally occurring uranium isotopes from the uranium ore called pitchblende (UO_2) makes possible the construction of atomic fuel rods for use in nuclear power plants as well the production of atomic weapons. The fundamental process requires converting the ore into solid uranium hexafluoride (UF_6). At the processing facility, the solid is heated, and the UF_6 gas, containing the ^{235}U and ^{238}U isotopes, is fed into a series of vessels. At each stage, the lower-molar-mass UF_6 strikes the walls of the vessel more frequently than does the high molar mass UF_6. The vessel has semipermeable walls, so more of the lighter, ^{235}U-containing UF_6 can flow through. Hundreds of stages later, the resulting "enriched" mixture contains about 5% ^{235}U. The mixture is then solidified for use in nuclear power plants. Nuclear weapons require an enrichment to at least 90% ^{235}U.

Visualization: Diffusion of Gases

Visualization: Gaseous Ammonia and Hydrochloric Acid

A related phenomenon is called **diffusion**, which involves the mixing of one gas with another or with itself. On a molecular level diffusion is far more complex, because in mixing, molecules collide with each other after moving only a very short distance—typically 5×10^{-8} m, only hundreds of times the diameter of a gas molecule itself. Therefore, mixing is a chaotic, rather slow process of the dilution of two or more gases.

The food industry has a particular interest in the ability of gases to pass through small openings, such as the pores of plastic wrap that surround meats and other food items. The ability of oxygen to pass through these pores to the food lowers its shelf life unless the packaging can maintain an inert atmosphere, as discussed in Section 10.3. Small condiment packages of ketchup and mustard, such as those in Figure 10.23, are usually packaged in several layers that include plastic and foil. The foil helps prevent the exchange of air with the food, and the plastic keeps the metals in the foil away from the food, with which they might otherwise react.

A latex balloon filled with a mixture of N_2 and He is observed. Which gas would effuse through the pores in the latex balloon more rapidly? How much more rapidly would it effuse?

Solution

The gas with the lower molecular mass, He, would effuse more rapidly. The ratio can be calculated with Graham's law of effusion:

$$\frac{re_1}{re_2} = \sqrt{\frac{M_2}{M_1}}$$

$$\frac{re_{He}}{re_{N_2}} = \sqrt{\frac{28.0}{4.00}}$$

$$\frac{re_{He}}{re_{N_2}} = 2.65$$

Therefore, the helium gas would effuse from the balloon 2.65 times faster than the nitrogen gas.

FIGURE 10.23

Small condiment packages of ketchup and mustard are usually packaged in several layers that include plastic and foil. The foil helps prevent the exchange of air with the food, and the plastic keeps the metals in the foil away from the food, with which they might otherwise react.

 Application

Which effuses more rapidly under identical conditions, He or H_2? How much more rapidly?

See Problems 87, 88, 97, and 98.

10.9   Industrialization: A Wonderful, Yet Cautionary, Tale

The glory of science and technology is our ability to modify nature's bounty in ways that enable us to sustain a world of nearly 7 billion people. As consumers, we often ignore the impact that gases have on us. However, the chemical changes that inhaled oxygen makes possible in our bodies sustain life. The chemical change of combustion moves motorized vehicles of all shapes and sizes along our highways, on our waterways, and on a fluid bed of air in the sky. Gases can wreak havoc in the form of tornadoes, hurricanes, and natural gas explosions, or lend calm to a cool evening. They can kill when used as "nerve agents" in warfare or help heal when used as anesthetics in surgery. As we modify nature, we recognize the changes in the atmosphere that our 150-year-long focus on industrialization has produced. These changes bear watching, for in the relative blink of an eye on the global time scale, we risk unraveling the protective blanket that has taken nature 2 billion years to create. Two of the greatest concerns are changes in ozone concentration and the greenhouse effect.

Ozone

This pale blue gas, O_3, is an allotrope of oxygen and is found in two of the four layers of the atmosphere, the stratosphere and the surface layer called the troposphere. These layers and the important gases within them are shown in Figure 10.24. In the stratosphere, ozone absorbs much of the UV radiation that would otherwise be harmful to us. UV radiation is typically categorized into three regions based on the energy of the corresponding wavelengths. The UV-A

 Application

CHEMICAL ENCOUNTERS:
Ozone

FIGURE 10.24

Two of the four layers of the atmosphere contain ozone, O_3.

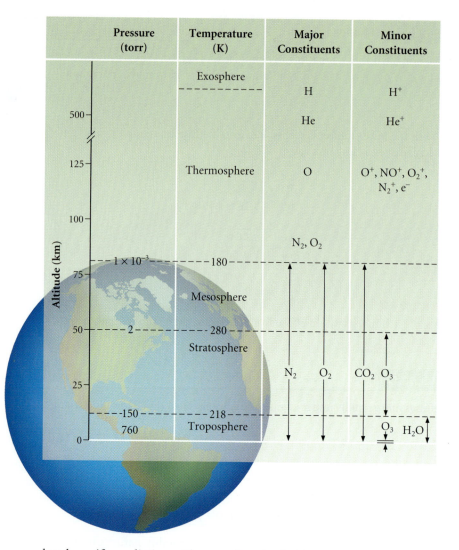

Pressure (torr)	Temperature (K)	Major Constituents	Minor Constituents
	Exosphere		
500	⸺⸺⸺	H	H^+
		He	He^+
125	Thermosphere	O	O^+, NO^+, O_2^+, N_2^+, e^-
100			
		N_2, O_2	
1×10^{-3} / 180			
	Mesosphere		
2 / 280			
	Stratosphere		
		N_2 O_2	CO_2 O_3
150 / 218			
760	Troposphere		O_3 H_2O

band specifies radiation with a wavelength of 400 nm to 320 nm; UV-B indicates wavelengths of 320 nm to 290 nm; and UV-C, the most energetic, is the classification of wavelengths from 290 nm to 100 nm. The destruction of ozone occurs by **photodissociation**—that is, UV light of sufficient energy (UV-B) causes the molecule to dissociate, forming oxygen gas and atomic oxygen:

$$O_3(g) \xrightarrow{\text{UV}} O_2(g) + O(g)$$

There is also a mechanism for the formation of ozone, which also contributes to the filtering of UV radiation from the sun. Shorter-wavelength UV radiation (UV-C) provides sufficient energy for the O_2 to separate into oxygen atoms:

$$O_2(g) \xrightarrow{\text{UV}} 2O(g)$$

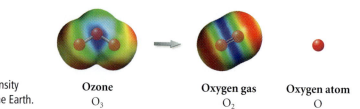

The reaction of ozone with UV-B light reduces the intensity of UV-B at the surface of the Earth.

Ozone
O_3

Oxygen gas
O_2

Oxygen atom
O

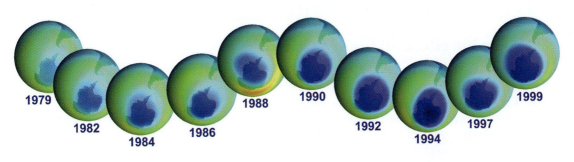

These highly reactive individual atoms then recombine with O_2 to form ozone along with the release of heat

$$O_2(g) + O(g) \rightarrow O_3(g)$$

The resulting combination of these three equations gives rise to the Chapman mechanism, a useful description of the chemistry involving oxygen and ozone in the middle stratosphere.

The natural balance between ozone photodissociation and ozone formation keeps most harmful UV radiation away from the Earth's surface. Within the last 30 years, however, activities related to industrialization have led to a significant reduction in the stratospheric ozone layer over many parts of the planet, as shown in Figure 10.25. Most notable is the discovery of a "hole" in the ozone layer over Antarctica.

Sherwood Rowland and Mario Molina suggested in 1974 that **chlorofluorocarbons (CFCs)**, used in refrigeration and in the formation of polymer foams, were the culprits in the destruction of the ozone layer. Table 10.7 lists common CFCs and related compounds, called **halons**, along with their uses.

CFCs were initially much sought after as refrigerants. This class of compounds is chemically inert, nonflammable, inexpensive, and quite stable at the Earth's surface. These properties make them very suitable for use in homes and automobiles. However, as they are released into the atmosphere, they travel upward to the stratosphere without reacting with other compounds. Then, in the stratosphere, they are bombarded with high-energy UV radiation and do react. This produces highly reactive chlorine atoms that react with ozone as well as with individual oxygen atoms that come from ozone.

$$Cl(g) + O_3(g) \rightarrow ClO(g) + O_2(g)$$

$$\underline{ClO(g) + O(g) \rightarrow Cl(g) + O_2(g)}$$

overall reaction: $O_3(g) + O(g) \rightarrow 2O_2(g)$

FIGURE 10.25

Total Ozone Mapping Spectrometer (TOMS) data from satellites orbiting the Earth give us a picture of the damage done by ozone depleting substances. These images from the period 1979 to 1999 show the development of a large area of decreased ozone in the stratosphere over Antarctica.

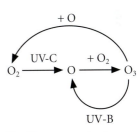

The Chapman mechanism.

TABLE 10.7	Common Chlorofluorocarbons and Halons	
Formula	**Common Symbol**	**Major Uses**
CCl_3F	CFC-11	Polymer foams, refrigeration, air conditioning
CCl_2F_2	CFC-12	Polymer foams, refrigeration, air conditioning, aerosols, food–freezing solvents
CCl_2FCClF_2	CFC-113	Solvent
$CBrClF_2$	halon-1211	Portable fire extinguishers
$C_2Br_2F_4$	halon-2402	Fire extinguishers

NanoWorld / MacroWorld

The big effects of the very small: Accumulation of ozone at the earth's surface

As we monitor the decrease in stratospheric ozone with growing concern, we also note the localized *increase* in surface ozone, because this gas is an irritant to the eyes, nose, and throat and damages cells in plants. Where does tropospheric ozone come from? We typically note its presence in the metropolitan areas of industrialized nations. More specifically, the reaction mechanism for the *accumulation* of ozone at the surface involves the release of nitric oxide from automobiles without catalytic converters. This is due to the unwanted combustion of nitrogen gas (present in the air) inside the combustion chambers of automobiles:

$$N_2(g) + O_2(g) \rightarrow 2NO(g)$$

Unfortunately, even automobiles with catalytic converters produce small amounts of NO. The released NO reacts with oxygen to make nitrogen dioxide:

$$2NO(g) + O_2(g) \rightarrow 2NO_2(g)$$

Then, when the Sun shines on a city filled with NO_2, the molecule decomposes to produce highly reactive oxygen atoms:

$$NO_2(g) \xrightarrow{\text{sunlight}} NO(g) + O(g) \quad \text{(reaction A)}$$

The atomic oxygen combines with oxygen gas in the same way that it does in the stratosphere to produce ozone:

$$O_2(g) + O(g) \rightarrow O_3(g) \quad \text{(reaction B)}$$

Then ozone is consumed in this reaction:

$$NO(g) + O_3(g) \rightarrow NO_2(g) + O_2(g) \quad \text{(reaction C)}$$

The resulting combination of reactions A, B, and C that produce and consume ozone at ground level is known as a null cycle. No net reaction occurs in a null cycle.

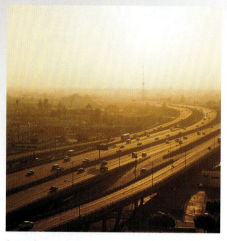

Smog is a common annoyance—and often a hazard—in industrialized cities. Its production is governed by a null cycle that produces ozone in the presence of sunlight.

Instead, the reactions are governed specifically by the relative concentrations of the individual reactants and products. Especially important is the presence of organic (carbon-based) pollutants and heat, such as are present on a hot summer day. These pollutants create highly reactive organic compounds that react *faster* with NO than does O_3. In the presence of these organic pollutants, the null cycle is modified and excess O_3 remains! What is the effect of this "nanoworld" process on the "macroworld"? This is the impact of heat and pollution in urban areas. Ozone accumulates at the Earth's surface, especially in large cities that have lots of motor vehicle traffic. Even smaller metropolitan areas, such as the "Triad" cities in North Carolina shown in Figure 10.26, have experienced an increase in ozone concentrations due to automobile exhaust.

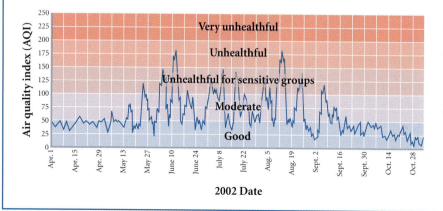

FIGURE 10.26

The ozone level in the "Triad" of central North Carolina from April to October 2002. Even in this midsized region (including the cities of Greensboro, High Point, and Winston-Salem), air quality is a significant concern.

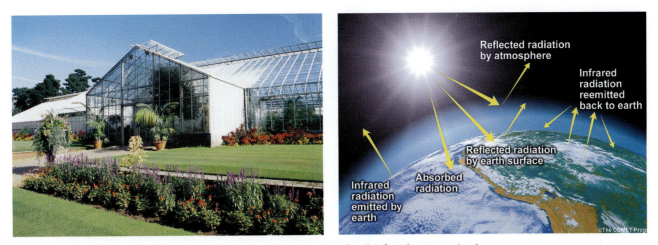

The greenhouse effect keeps the Earth warm. The gases in the atmosphere absorb infrared energy emitted from the Earth and reemit it to the Earth, much as the panes of glass keep heat in a greenhouse.

The chlorine atom is a catalyst in the destruction of ozone. Each chlorine atom can catalyze the breakdown of perhaps hundreds of thousands of ozone molecules.

Rowland and Molina shared the 1995 Nobel Prize in chemistry for their groundbreaking research and predictions about the fate of the ozone layer. Since that time, substitute compounds called HCFCs have been developed, including CHF_2Cl and CF_3CFH_2. These react in the lower atmosphere to release HCl and HF, never making it to the stratosphere. The loss of ozone in the stratosphere, combined with the generation of unhealthful ozone levels in large cities, is a continued cause for concern and a focus of research. The very good news is that as a result of worldwide action to reduce CFC emissions, we may have turned the corner. In 2005, scientists found a slight worldwide increase in the amount of ozone in the stratosphere. Many decades of continued diligence will be required before the damage to the ozone layer can be completely reversed, yet we have made an important start.

The Greenhouse Effect

The **greenhouse effect** is caused by the accumulation, in the atmosphere, of gases that permit light to enter but prevent some energy as heat from exiting, much like a plant greenhouse. Methane and CFCs are examples of greenhouse gases, although they are not present in sufficient concentration to be of real concern at this time. A potent new greenhouse gas named trifluoromethyl sulfur pentafluoride (SF_5CF_3) has recently been identified.

Carbon dioxide is entering the atmosphere in much greater amounts than are being used up, and this makes CO_2 the most important greenhouse gas.

The **carbon cycle**, discussed in Chapter 12, has historically kept the carbon in the atmosphere, seas, and land in balance over the long term. Human razing of forests has radically curtailed the number of plants that use up CO_2 in photosynthesis:

$$6CO_2(g) + 6H_2O(g) \rightarrow C_6H_{12}O_6(s) + 6O_2(g)$$

Burning carbon-based fuels, especially coal, oil, and natural gas, has added CO_2 to the environment, and the annual release of carbon into the atmosphere continues to rise (see Figure 10.27). There is general agreement among scientists

Application

CHEMICAL ENCOUNTERS:
The Greenhouse
Effect

FIGURE 10.27

The amount of carbon released as CO_2 into the atmosphere continues to rise along with our seemingly insatiable appetite for fossil fuels.

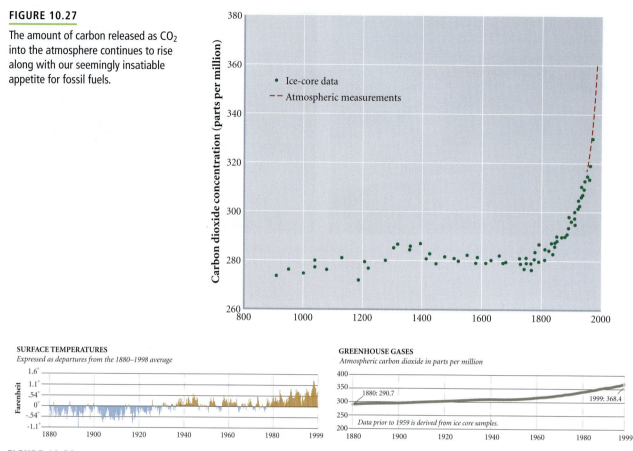

FIGURE 10.28

There is general, though not universal, agreement among scientists that the 100-year increase in the Earth's average surface temperature, shown here, is at least partially caused by the greenhouse effect.

that the 100-year increase in the Earth's average surface temperature, shown in Figure 10.28, is at least partially caused by the greenhouse effect.

It is often said that the problems caused by the use of chemistry can also be fixed by the use of chemistry. Some good signs are on the horizon. As shown in Figure 10.29, the fraction of global energy consumption derived from the burning of coal and wood is declining.

The interaction among gases in the atmosphere is so complex that we cannot confidently predict the long-term atmospheric effects of industrialization. The best models indicate that as we produce gases that meet the needs and the wants of the nearly 7 billion people on Earth, we must be mindful of the possible impact of our current consumption. We are, after all, the keepers of the global commons.

FIGURE 10.29

Advances in chemistry have changed the way in which energy is produced and consumed.

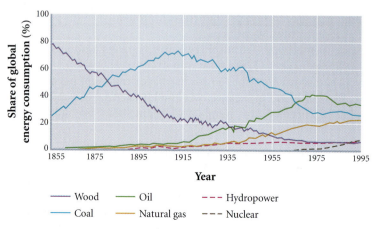

Source: International Institute for Applied Systems Analysis.

The Bottom Line

- Industrial gases are used in manufacturing, medicine, and other industries. (Section 10.1)

- Gases present at low density all behave in about the same way. This permits us to generalize about their behavior in a manner that is not possible with solids and liquids. (Section 10.1)

- Gases that behave as though each particle has no interactions with any other are called ideal gases. (Section 10.1)

- High pressure, low temperature, and intermolecular forces of attraction between gas molecules cause gases to deviate from ideal behavior. (Section 10.1)

- Real gases deviate from ideal behavior, especially at low temperatures and high pressures. (Section 10.1)

- Pressure is a measure of force per unit area. Pressure is the result of collisions of gas molecules with the walls of a container. (Section 10.2)

- Several units of pressure exist, including the pascal (the SI unit), atmosphere, mm Hg, in Hg, torr, bar, psi, and psig. (Section 10.2)

- In current usage, the standard atmosphere (1 atm) is different from standard pressure (1 bar, or about 0.987 atm). (Section 10.2)

- Dalton's law of partial pressures is concerned with the contribution of each gas to the pressure of the entire gas mixture. (Section 10.3)

- Avogadro's law deals with the relationship between the volume and number of moles of ideal gases, at constant temperature and pressure. (Section 10.4)

- Boyle's law expresses the relationship between volume and pressure of an ideal gas, at constant number of moles and temperature. (Section 10.4)

- Charles's law expresses the relationship between the volume and temperature of an ideal gas, at constant number of moles and pressure. (Section 10.4)

- The ideal gas equation combines the gas laws to interrelate the pressure, volume, amount, and temperature of an ideal gas. (Section 10.5)

- The ideal gas equation can be applied to find the molar mass and the density of an ideal gas. (Section 10.5)

- Several scientists have made mathematical models to account for nonideal behavior. J. D. van der Waals's model is the most commonly used because it is relatively simple and takes into account corrections for pressure and volume. (Section 10.5)

- The molar mass and the density of an ideal gas are directly proportional. (Section 10.6)

- The kinetic-molecular theory shows how it is possible to start from some elementary constructs about gas behavior and derive the gas laws, therefore showing the consistency between theory and experiment. (Section 10.7)

- Graham's law of effusion shows the inverse relationship between the speed of a gas and its molar mass. (Section 10.8)

- The effects of industrialization on the atmosphere are currently being debated. Ozone levels and the greenhouse effect are two issues of greatest social concern. (Section 10.9)

Key Words

absolute zero The temperature obtained by extrapolation of a plot of gas volume versus temperature to the "zero volume" point. The lowest temperature possible, which is 0 K, or −273.15°C. (*p. 409*)

atmosphere (atm) A unit used to measure pressure. 1 atm = 760 mm Hg. (*p. 399*)

average speed The sum of the molecular speeds of each molecule divided by the number of molecules. (*p. 424*)

Avogadro's law Equal amounts of gases occupy the same volume at constant temperature and pressure. (*p. 404*)

bar A unit used to measure pressure. 1.01325 bar = 1 atmosphere. (*p. 399*)

Boyle's law The volume of a fixed amount of gas is inversely proportional to its pressure at constant temperature. (*p. 406*)

carbon cycle The natural process that describes how carbon atoms are moved among the land, sea, and atmosphere. (*p. 431*)

Charles's law The volume of a fixed amount of gas is directly proportional to its temperature in kelvins at a constant pressure. (*p. 408*)

chlorofluorocarbons (CFCs) Compounds containing only carbon, chlorine, and fluorine. CFCs are typically used as solvents and refrigerants and are known to be harmful to the ozone layer. (*p. 429*)

combined gas equation The combination of Avogadro's law, Boyle's law, and Charles's law. (*p. 411*)

Dalton's law of partial pressures The total pressure of a mixture of gases is the simple sum of the individual pressures of all the gaseous components. (*p. 401*)

diffusion The process by which two or more substances mix. (*p. 426*)

effusion The process by which a gas escapes through a small hole. (*p. 425*)

electrolysis A process in which an electric current passing through a solution causes a chemical reaction. (*p. 397*)

force The mass of an object multiplied by its acceleration. (*p. 397*)

Graham's law of effusion The rates of the effusion of gases are inversely proportional to the square roots of their molar masses. (*p. 426*)

greenhouse effect The re-radiation of energy as heat from gases in the atmosphere back toward Earth. An increase in the greenhouse effect leads to an increase in global temperatures. (*p. 431*)

halons Compounds containing carbon and halogens. Halons are typically used in fire extinguishing applications but are considered harmful to the ozone layer. (*p. 429*)

ideal gases Gases that behave as though each particle has no interactions with any other particle and occupies no molar volume. (*p. 396*)

ideal gas constant R, the constant used in the ideal gas equation. $R = 0.08206$ L·atm/mol·K. (*p. 414*)

ideal gas equation The equation relating the pressure, volume, moles, and temperature of an ideal gas. (*p. 414*)

in Hg A unit used to measure pressure. 29.921 in Hg equals 1 atmosphere. (*p. 399*)

industrial gases Gases used in industrial or commercial applications. (*p. 395*)

kinetic-molecular theory A theory explaining the relationship between kinetic energy and gas behavior. This theory makes possible derivation of the ideal gas laws. (*p. 422*)

mm Hg A unit used to measure pressure. 760 mm Hg represents the height of a column of mercury that can be supported by 1 atm pressure. (*p. 399*)

modified atmosphere packaging (MAP) The replacement of the air within food packaging with gases in order to prolong the shelf life of the food. (*p. 401*)

most probable speed The most likely speed of a molecule from among a group of molecules. (*p. 424*)

newton (N) The basic SI unit of force. 1 N $= 1$ kg·m·s^{-2}. (*p. 397*)

partial pressure The pressure exerted by a single component of a gaseous mixture. (*p. 401*)

pascal (Pa) The basic SI unit of pressure. 1 Pa $= 1$ N·m^{-2}. (*p. 398*)

photodissociation A chemical reaction where light, of sufficient energy, causes the cleavage of a bond in a molecule. (*p. 428*)

pounds per square inch (psi) A unit of pressure. 14.7 psi $= 1$ atm. (*p. 399*)

pounds per square inch gauge (psig) A unit of pressure expressing the pressure of a gas in excess of the standard atmospheric pressure. Also known as gauge pressure. 14.7 psig $= 2$ atm. (*p. 399*)

pressure The force per unit area exerted by an object on another. (*p. 398*)

root-mean-square (rms) speed The square root of the sum of the squares of the individual speeds of particles divided by the number of those particles. (*p. 424*)

standard atmosphere Exactly 1 atm, the pressure that supports 760 mm of mercury. (*p. 399*)

standard pressure Defined as 1 bar, although chemists typically use 1 atm. (*p. 399*)

standard temperature For gases, defined to be 0°C, or 273 K. (*p. 400*)

standard temperature and pressure (STP) 1 atm and 273 K. (*p. 400*)

standard ambient temperature and pressure (SATP) 1 bar and 298 K. (*p. 400*)

torr A unit used to measure the pressure of gases. 1 torr $= 1$ mm Hg. (*p. 399*)

van der Waals equation The equation that corrects the gas laws for gases that deviate from ideal behavior. (*p. 417*)

Focus Your Learning

The answers to the odd-numbered problems and some selected problems appear at the back of the book, as represented by the blue numbering.

10.1 The Nature of Gases

Skill Review

1. Define the intermolecular forces typically experienced by molecules in the gaseous state.

2. Explain why we never see signs on delivery trucks that read "Danger—Compressed Liquid."

3. What conditions tend to favor ideal behavior in gases?

4. What are some characteristics common in nonideal gases?

10.2 Production of Hydrogen and the Meaning of Pressure

Skill Review

5. If your chemistry textbook had a mass of 0.89 kg, how much force, in newtons, would the textbook exert on a desk in your classroom?

6. If your textbook (mass = 0.89 kg) measured 27 cm × 23 cm, what pressure, in pascals, would the book exert on the desk?

7. How much force, in newtons, would an 86-kg man exert on the ground?

8. A 115-lb flight executive wears spike heels every day to work. If each heel has an area of 4.0 cm^2, what is the pressure exerted by each heel in pascals? (Assume all of the woman's weight is placed on the heels of her shoes.)

9. The pressure of a gas is measured to be 797 mm Hg. Convert this pressure into units of Pa, kPa, atm, torr, bar, and in Hg.

10. The pressure of a gas is measured to be 0.750 atm. Convert this pressure into units of Pa, kPa, torr, bar, and mm Hg.

11. An astronaut weighs 145 lb on the planet Earth, where the gravitational pull is 9.8 m/s^2. She is sent on a mission to a planet whose gravitational pull is only 0.33 m/s^2. What would a scale on this planet indicate that the astronaut weighs?

12. What is the mass in kg and weight in lb of a person who exerts a force of 375 N on a scale?

13. The air in a car tire exerts a pressure of 22.3 psig. What is the force exerted by the air if the total area inside the tire is 567 cm^2?

14. A full book bag that exerts a force of 12 N on a table takes up an area that measures 25 cm × 36 cm. Calculate the amount of pressure exerted by the bag.

Chemical Applications and Practice

15. Small inflatable sleeping mattresses make camping enjoyable for many outdoor enthusiasts. If the air pressure of a comfortable mat is measured as 17.5 psig, what is the actual pressure that the air is exerting on the inside walls of the mat? What is the pressure if reported in units of atmospheres?

16. As the altitude increases, the air pressure decreases. To prove this fact, a mountain climbing group measured the air pressure on Mount Everest at 0.67 atm. What would this pressure be if it were reported in units of kPa, bar, mm Hg, and psi?

10.3 Mixtures of Gases—Dalton's Law and Food Packaging

Skill Review

17. The partial pressure of O_2 in a sample of air on a mountaintop is 115 mm Hg. The oxygen makes up 21% of the gas in the atmosphere. What is the total atmospheric pressure on the mountaintop?

18. A gaseous mixture contains 54% N_2, 39% O_2, and 7% CO_2 by volume. If the total pressure is 813 mm Hg at STP, what is the partial pressure of each of the gases?

19. A total of 12.4 g of N_2 and 12.4 g of O_2 are combined in a mixture that exerts a pressure of 1.23 atm. Calculate the partial pressure of N_2 and that of O_2 in the mixture.

20. A mixture of gases contains CO_2 and O_2 in a 3-to-1 ratio. If the partial pressure of CO_2 in the mixture is 2.34 atm, what is the partial pressure of O_2? What is the total pressure of the mixture?

Chemical Applications and Practices

21. A common way to generate small amounts of oxygen in the lab is by heating potassium chlorate in the presence of a catalyst and collecting the oxygen by bubbling it through water. If the total pressure of the collected gas was 785 mm Hg at 27°C, what is the partial pressure of the collected oxygen? The pressure of gaseous water at 27°C is 27 torr.

22. Producing hydrogen gas is becoming a very important technical and economic concern for food processing, and potentially as a fuel for future transportation needs. If a chemical engineer were developing a model for this purpose and collected 4.25 g of H_2 over water at 25°C that had a total pressure of 1.15 atm, what would be the partial pressure of the H_2 collected? The pressure of gaseous water at 25°C is 23.8 torr.

23. In a helium–neon laser, a mixture of helium and neon gas is contained in a small tube used to generate coherent light. In some applications, this type of laser is used to scan price codes at the grocery store. If the internal gas mixture is 8.95% Ne and 91.15% He by volume and the total pressure is maintained at 3.42 mm Hg, what are the partial pressures of the gas components?

24. A balloon is filled with H_2 and O_2 for a demonstration. The gases are added such that they will react completely when a flame is brought in contact with the balloon. Write the equation describing this reaction. If the total pressure of the balloon is 3.4 atm, what would the partial pressure of each gas need to be in order for the reaction to go to completion?

10.4 The Gas Laws—Relating the Behavior of Gases to Key Properties

Skill Review

25. At STP, a 2.5-L sample of gas contains 4.5 mol of Ne. How many moles would the sample contain if the volume increased to 5.0 L? . . . if the volume decreased to 1.0 L?

26. A balloon containing 0.50 mol of He is found to have a volume of 1.75 L. What volume would the balloon have if an additional 0.15 mol of He were added? What volume would the balloon have if 0.23 mol of He were removed from the balloon? Assume that the temperature and pressure remain constant.

27. A container holds 4.70 L of air at a pressure of 861 torr. If the gas in the container were lowered to standard pressure, what would be its volume? If the pressure of the gas were reduced to 400 torr, what would be its volume? Assume the temperature remains constant.

28. A balloon holds 2.50 L of air at a pressure of 19.5 psi. If the balloon were squeezed to a volume of 1.00 L, what would be the new pressure of the gas? If the balloon were stretched to a volume of 5.00 L, what would be the new pressure of the gas? Assume the temperature remains constant.

29. A balloon containing 1.00 L of CO_2 at 72°F is placed in a freezer at −10°F. What is the new volume of the balloon? If the balloon is placed in the oven at 250°F, what is the new volume of the balloon? Assume the pressure remains constant.

30. The volume of a fixed quantity of gas at 25°C is changed from 0.750 L to 5.57 L. What is the resulting temperature of the gas if the pressure remains constant? If the volume of the original sample were reduced to 0.150 L, what would be the resulting temperature? Assume the pressure remains constant.

31. A 0.47-mol sample of gas at 37°C occupies 3.20 L at 2839 mm Hg. What volume would the same gas occupy at STP?

32. Consider this experiment, wherein a gas is heated in a sealed chamber of fixed volume. What is the new pressure of the gas?

273 K — 1 atm — 5.44 mol of gas → 500°C — ? atm — 5.44 mol of gas

33. A balloon containing 0.15 g of H_2 at STP is pressurized to 32.0 psi. What is the new temperature of the gas inside the balloon? Assume the volume remains constant.

34. A CO gas cylinder possesses a volume of 45.0 L of gas at 2.20×10^3 psi. If the valve breaks and the gas is immediately released from the cylinder, what is the new volume of the gas? Assume the temperature remains constant and standard pressure is attained.

Chemical Applications and Practice

35. Pressurized N_2O gas can be used to provide the "inflating" power for canned whipped cream. If such a can that is used to provide the topping for a sundae contains 0.58 g of N_2O and corresponds to 0.217 L, how many liters of N_2O will 0.33 g of N_2O fill at the same temperature and pressure?

Whipped cream in a bottle uses N_2O as the propellant. Does the electrostatic potential map for N_2O tell us anything about how to draw the Lewis dot structure?

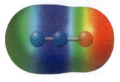

36. Inside the cylinder of an automobile engine, a mixture of gasoline vapor and oxygen is combusted by a spark. However, the gaseous mixture is first pressurized by the action of a moving piston that decreases the volume of the gas. If the initial volume of the cylinder is approximately 485 mL and the pressure is 0.988 atm, what pressure will be required to compress the gas mixture to 50.0 mL? Assume the temperature remains constant.

37. The molecule nitrogen monoxide (NO) plays several vital roles in animal physiology. Nitrogen monoxide helps regulate blood pressure, influences blood clotting, and also influences the immune system. A researcher studying this important gas isolates a small, 0.150-mL sample at 37°C. If this sample is cooled to room temperature (25°C), what volume will the sample have if the pressure remains constant?

38. Because ozone (O_3) has beneficial properties at high altitudes and harmful properties at low altitudes, atmospheric scientists often study it. If a 50.0-mL sample of ozone collected at high altitude has a temperature of −25°C is brought back to the lab to study (where the temperature is 25°C), what will the new volume of the gas be? Assume the pressure remains constant.

10.5 and 10.6 The Ideal Gas Equation and Its Applications

Skill Review

39. Determine the value of the ideal gas constant, R, to four significant figures if 1.00 mol of gas at 1.00 atm and 273 K occupies 22.4 L. Repeat the calculation, but use each of the following as the label for the equivalent amount of pressure.
a. Pa   b. mm Hg   c. psi

40. A researcher explores the ideal gas equation. He measures the pressure of a system at different temperatures (keeping the volume and moles of the gas constant) and creates a plot of P versus T. What is the slope of the line that results if the volume is 1.0 L when 1.0 mol of gas is used in the exploration?

41. When inflated with 6.0×10^6 g of helium, a weather balloon has a pressure of 0.955 atm at 22.6°C. What volume does the balloon occupy?

42. What would be the pressure of H_2 in a tank that has a volume of 25 L, if it contained 45 g of H_2 at 25°C?

43. What is the volume of 44.0 g of CO_2 at STP? . . . at 39°C and 0.500 atm?

44. What is the temperature of 85.0 g of N_2 if the volume is 7.49 L at 850 mm Hg?

45. How many moles of Ar would be found in a 4.33-L balloon at STP?

46. A gas is trapped in a 1.00-L flask at 27°C and has a pressure of 0.955 atm. The mass of the gas is measured at 1.95 g. Of these, which is most likely to be the identity of the gas: NO, NO_2, or N_2O_5?

47. An unknown gas has a density of 0.600 g/L at 743 mm Hg and 66°C. What is the molar mass of the unknown gas?

48. What is the density of CO_2 at STP? . . . of N_2 at STP?

49. At STP, the density of methane (CH_4) is 0.714 g/L. What is the density of methane at 25°C and 1.15 atm?

50. If a 500.0-mL sample of air at 26.5°C and 698 mm Hg has a mass of 0.543 g, what is the average molar mass of air?

51. A typical breath may have a volume of 450 mL. If the air you breathed were 21% oxygen gas, how many molecules of oxygen would you inhale at 37°C and 0.922 atm?

52. If you were able to inhale 2.50 L of air at STP, how many moles of N_2 would you be breathing? Assume that air is composed of 79% nitrogen and 21% oxygen.

53. A gas at 301 K and 0.97 atm has a density of 4.48 g/L. What is the molar mass of the gas?

54. At 27.0°C and 802 torr, a gas has a molar mass of 62.0 g/mol. What is the density of this gas? What would be the density of the gas if the pressure were changed to 1.00 atm?

55. A 5.00-g sample of a gas has a volume of 85.0 mL at 20.0°C and 1.00 atm. What is the molar mass of this gas? Would the molar mass change if the temperature were increased to 23°C?

56. What is the density of CO_2 at 50°C and 0.44 atm?

Chemical Applications and Practice

57. A student working in the lab forgets to record the temperature at which she obtained 0.675 g of O_2 (g). She did, however, report the pressure as 745 mm Hg and the volume as 478 mL. What would have been the temperature at those conditions?

58. Ozone (O_3) in the stratosphere helps reduce the amount of ultraviolet radiation reaching the surface of the Earth. Assuming a temperature of −25°C and a partial pressure due to ozone of 1.2×10^{-7} atm, how many ozone molecules would be present in 1.00 L of air in the stratosphere?

59. The metabolism of glucose is the main source of energy for humans. This equation shows the overall reaction for the process:

$$C_6H_{12}O_6(s) + 6O_2(g) \rightarrow 6CO_2(g) + 6H_2O(g)$$

Calculate the volume of CO_2 produced at 37°C and 1.00 atm when 10.0 g glucose is oxidized.

60. Small portable lighters use compressed butane (C_4H_{10}) as their fuel. When butane is combusted, this reaction takes place:

$$2C_4H_{10}(g) + 13O_2(g) \rightarrow 8CO_2(g) + 10H_2O(g)$$

a. At 27.0°C and 0.888 atm, how many liters of oxygen are required for every 1.00 g of butane to react completely?

b. If the system is compressed to 3.50 atm, keeping the temperature at 27.0°C, how many liters of oxygen will be required to react completely with 1.00 g of butane?

61. An industrial process can be used to change ethylene into ethanol. The process uses a catalyst, high pressures (6.8 MPa), and temperatures of 3.00×10^3 K. What is the density of ethanol vapor under these conditions?

62. The familiar aroma of garlic comes mainly from the compound diallyl disulfide. Based on these measurements, determine the molar mass of this pungent compound. At 298 K, a 105-mL container holds 0.618 g of diallyl disulfide with a pressure of 0.987 atm.

63. Argon gas is being used in some incandescent light bulbs to extend the life of the tungsten filament. If a light bulb has a volume of 200.0 mL and contains 0.100 g of Ar at 25°C, what is the pressure inside the bulb?

64. An aerosol can contains carbon dioxide as its propellant. If the can has an internal pressure of 1.3 atm at 25.0°C, what pressure would it have if the temperature were accidentally raised to 450°C? (Assume that the can has not yet burst, although this would probably be a very close call.)

65. The Haber process is used to produce ammonia (NH_3) for use in fertilizing corn fields in the Midwest. The reaction is

$$N_2(g) + 3H_2(g) \rightarrow 2NH_3(g)$$

a. Which, if either, is the limiting reagent when 2.00 L of N_2 at STP is prepared to react with 2.00 L of H_2 at 1120 mm Hg and 21.0°C?

b. If 2.00 L of the gases, both at STP, reacted via the Haber process, how many liters of ammonia would be produced?

66. Much of the electricity in the United States is produced from the combustion of coal. However, some of the coal deposits contain FeS_2 (iron pyrite) as a contaminant. During the combustion of coal, this impurity also burns and produces harmful sulfur dioxide gas (SO_2). How many liters of SO_2 are produced from the burning of 1.00 kg of FeS_2 at 758 torr and 275°C?

$$FeS_2(s) + 2O_2(g) \rightarrow Fe(s) + 2SO_2(g)$$

67. A gaseous sulfur compound causes the pungent aroma of rotten eggs. If 1.60 g of the gas were collected in a 1.00-L vessel at 1065 torr and 89.0°C, what would be the molar mass of the trapped gas?

68. One common way to obtain metal samples from their impure oxide ores is to react the oxide with carbon. Generically, the equation can be written this way:

$$2MO(s) + C(s) \rightarrow 2M(s) + CO_2(g)$$

If 5.00 g of an unknown metal oxide (MO) reacted with excess carbon and formed 0.738 L of CO_2 at 200.0°C and 0.978 atm, what is the identity of the metal?

69. The halogen gases are caustic and toxic. Therefore, they must be handled with extreme care. Which halogen gas has a density of approximately 1.70 g/L at STP?

70. Hydrogen has an isotope known as deuterium (D) that is made up of a proton, a neutron, and an electron. Deuterium is present in measurable amounts in all hydrogen-containing substances. At STP, what are the densities of the three compounds that could exist in a hydrogen gas sample: H_2, HD, and D_2?

71. Neon lights actually do contain neon gas. But from where do we get neon? Very small amounts of Ne are present in air, less than 2.0×10^{-3}% by volume. However, when air is liquefied, the sample can carefully be separated on the basis of boiling points. This must be controlled, but because N_2, O_2, and Ar boil away in the range from 77 to 97 K, neon (with a boiling point of 27 K) can be separated. If a manufacturer collected 1.00 lb of neon at 25°C and 789 torr, what volume would the Ne occupy?

72. Assume you are pumping air into the tires of your mountain bike before a ride. If the tire volume is 1.50 L at 28.5°C and 6.55 atm, how many moles of air have you put inside the tire? How many molecules of air are in the tire? (Refer to Problem 50 for the average molar mass of air.)

10.7 Kinetic-Molecular Theory

Skill Review

73. Using the tenets of the kinetic-molecular theory, explain why the absolute temperature is directly proportional to the pressure of a trapped volume of gas.

74. Does the kinetic-molecular theory assist in the explanation of Avogadro's law? If so, how?

75. Two gases known for their bleaching power are ClO_2 and Cl_2. Assume you have separate 1-L containers containing 5 g of each at 25°C.

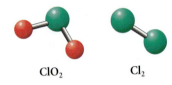

      ClO_2 Cl_2

    a. Compare the average kinetic energy of the molecules of each gas.
    b. Compare the average speed of the molecules of each gas.
    c. Which, if either, would be exerting more pressure?
    d. If the samples were placed in the same container, which, if either, would have the larger partial pressure?

76. Equal masses of two gases, B_2H_6 and C_2H_2, are placed in separate containers at the same temperature.

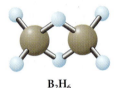

      B_2H_6

    a. Compare the average kinetic energy of the molecules of each gas.
    b. Compare the average speed of the molecules of each gas.
    c. Which, if either, would exert a greater pressure?
    d. If the samples were placed in the same container, which, if either, would have the larger partial pressure?

77. Assuming that a single molecule of each of these gases is moving at a speed of 1100 m/s, what would you calculate as its kinetic energy?
    a. CO_2 b. NH_3 c. Ne

78. Assuming that a single molecule of each of these gases is moving at a speed of 1100 m/s, what would you calculate as its kinetic energy?
    a. H_2 b. C_2H_6 c. Ar

79. Arrange these gases in order from lowest to highest average molecular speed (assume that there are 1.00 mol of each at the same temperature and pressure).
    a. H_2 b. C_2H_6 c. Ar

80. Arrange these gases in order from lowest to highest average molecular speed (assume that there are 1.00 mol of each at the same temperature and pressure).
    a. CO_2 b. NH_3 c. Ne

81. At 25.0°C, what would be the root-mean-square speed of propane (C_3H_8) used in bottle gas fuel systems?

82. At extremely low temperatures, molecular motion also becomes relatively slow. What would the temperature be when the root-mean-square speed of He was 100 m/s?

83. Calculate the rms speed of CO_2 and of H_2O at 250.0 K.

84. What is the rms speed of a gas with a molar mass of 16.0 g/mol if the temperature is 45°C?

Chemical Applications and Practice

85. A device sometimes used in teaching the gas laws consists of small marbles trapped inside a glass-walled container. The container is open at one end, where it is fitted with a moveable piston. Using each of the six points of the kinetic-molecular theory, discuss how this device compares to a gas sample in a container. In addition to the marbles obviously being larger than gas molecules, where does the analogy fit the kinetic-molecular theory and where does it not?

86. During exercise, our bodies heat up as a consequence of the increase in metabolism. Using the kinetic-molecular theory, explain how panting (breathing faster) can help reduce this increase in temperature.

10.8 Effusion and Diffusion

Skill Review

87. Arrange these gases in order from lowest to highest rate of effusion.
    a. N_2 b. SO_3 c. CO_2 d. Xe

88. Arrange these gases in order from lowest to highest rate of effusion.
    a. O_2 b. Ar   c. F_2 d. HF

89. A gas effuses at a rate of 0.300 m/s. A second gas has a molar mass of 2.02 and effuses 6.00 times faster than the first gas. What is the molar mass of the first gas?

90. Gas A diffuses 1.47 times as fast as gas B. If gas B has a molar mass of 54.6 g/mol, what is the molar mass of gas A?

91. If a noble gas diffuses 0.317 times as fast as does helium at the same pressure, which noble gas is it?

92. Which would take longer to effuse, 1.00 mol Xe gas or 2.00 mol CO_2 gas? Assume both are initially at the same temperature and pressure. How much faster would 1.00 mol Xe gas effuse if there were 10.0 mol CO_2 gas at the same temperature and pressure?

93. Would it be possible to separate a mixture of propane (C_3H_8) and carbon dioxide (CO_2) using Graham's law of effusion? Explain your answer.

94. Would it be possible to separate a mixture of diborane (B_2H_6) and acetylene (C_2H_2) using Graham's law of effusion? Explain your answer.

Chemical Applications and Practice

95. An interesting visual demonstration to show diffusion rates of gases is to introduce some NH_3 gas at one end of a hollow tube, while simultaneously placing some HCl at the other end. Within a few minutes, the two gases will diffuse through the air inside the tube and react to form a white NH_4Cl precipitate when they meet. If the tube was 39 cm in

NH$_3$ and HCl are released at opposite ends of a glass tube.

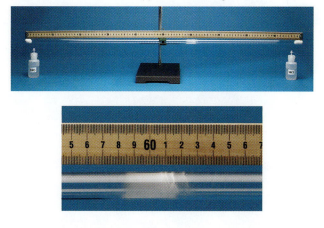

length, approximately where would the white precipitate form inside the tube?

96. The demonstration described in Problem 95 was repeated, this time using HF instead of HCl. Where would the precipitate form inside the tube?

97. During the Manhattan project, efforts were made to separate ^{238}U from the fissionable isotope ^{235}U. The uranium sample was converted to UF_6, which is a gas at low pressure. How much faster does $^{235}UF_6$ effuse than $^{238}UF_6$?

98. Atmospheric scientists have recently detected very low concentrations of a long-lived greenhouse gas containing sulfur and fluorine, SF_5CF_3. What is the ratio of the rates of effusion of this gas and of its suspected precursor SF_6, a material used in high-voltage insulators?

Comprehensive Problems

99. Examine Table 10.2 and answer these questions.
a. Assuming SATP conditions, how many liters of ammonia gas were produced in 2004?
b. How many liters of oxygen gas at SATP were produced?
c. How many metric tons of hydrogen gas were produced at SATP?
d. How many moles of N_2 gas were produced in the United States in 2004 at SATP?

100. When you exhale, you are releasing carbon dioxide, all of the nitrogen you inhaled, and unused oxygen gas. Under the same conditions of temperature and pressure, rank these three gases in order from
a. least dense to most dense.
b. fastest rate of diffusion to slowest rate of diffusion.

101. During exercise, we breathe faster and obtain more oxygen for metabolism. If a handball player absorbs oxygen at a rate of approximately 75 mL per kilogram of body mass per minute, how many molecules of oxygen would a player with a mass of 86.8 kg absorb during a 30.0-min match at SATP?

102. Although the ideal gas equation is sufficient for most gas law calculations, there are times when it needs to be modified. Explain the specific physical purpose behind the van der Waals constants a and b in the modified equation.

103. Hydrogen gas continues to gain attention as a possible fuel for modified cars of the near future. Compare the pressure of H_2 in a tank with a volume of 25 L that contains 45 g of H_2 at 25°C, using calculations from the ideal gas equation and the van der Waals modification to the gas equation.

104. In a quality test, one tennis ball is filled with N_2 gas. Another tennis ball of the same volume is filled to the same pressure as the first with air. If the two tennis balls were at the same temperature, what else would also be the same?

105. If you collected some oxygen in an experiment at 298 K, how much of a change in the temperature (in kelvins), at constant pressure, would be needed to quadruple the volume? If that same sample of oxygen had its temperature lowered by 25°C, what volume change, at constant pressure, would be expected?

106. If the CO_2 trapped above your favorite carbonated drink is exerting a pressure of 7.2 atm, yet the total pressure above the aqueous drink is 7.9 atm, what are the partial pressure and percent of water in the trapped space? (Assume that no other gas is present.)

107. The helium in a pressure tank at a carnival may be pressurized to 21.0 atm. If the temperature of a 27.0-L tank is 29.5°C, how many grams of helium are in the tank? If it was your job to inflate 1.50-L balloons with the helium, how many could you inflate so that each of them would have 1.11 atm of pressure at 27.5°C?

108. In 2004, 17.7 billion m^3 of hydrogen gas was produced in the United States, as shown in Table 10.2. Most of the gas is delivered via trucks, either as a compressed gas or as a liquid stored at very low temperature. If, for example, 50% of the volume of gas produced were liquefied and stored at 20 K, and each delivery truck held 6000 gal of liquid hydrogen, how many truckloads of hydrogen would be needed in one year? (Assume the density of liquid hydrogen is 0.070 g/cm^3.)

Thinking Beyond the Calculation

109. In Problem 108 we noted that hydrogen gas can be transported via tanker trucks either as a compressed gas (at about 400 atm) or as a liquid at 20 K. If you are charged with deciding whether to ship the hydrogen as a compressed gas or as a liquid, what chemical, physical, demographic, economic, and other considerations would factor into your decision? Use the Internet and examine how some large gas production companies make these decisions.

110. Under STP conditions, 100.0 mL of an unknown hydrocarbon gas was combusted in excess oxygen. The only products of the combustion were 300.0 mL of carbon dioxide and 400.0 mL of water vapor.
a. What are the intermolecular forces common to hydrocarbon gases?
b. Does the behavior of these gases tend to resemble ideal gas behavior? How might these gases be expected to deviate from this behavior?
c. How many moles of carbon dioxide are produced in the combustion?
d. What is the formula of the hydrocarbon?
e. If the combustion were performed at 2.50 atm and 500°C, how many liters of water vapor would you expect to produce?

11

The Chemistry of Water and the Nature of Liquids

Water is a vital part of our lives. In the developed world, most of us tend to take clean water for granted. But many people, like these girls in Nicaragua, are constantly preoccupied with how they'll obtain each day's supply of water.

Contents and Selected Applications

Chemical Encounters: Worldwide Water Use

11.1 The Structure of Water: An Introduction to Intermolecular Forces

11.2 A Closer Look at Intermolecular Forces

11.3 Impact of Intermolecular Forces on the Physical Properties of Water, I

11.4 Phase Diagrams

Chemical Encounters: CO_2 as a Dry Cleaning Solvent

11.5 Impact of Intermolecular Forces on the Physical Properties of Water, II

11.6 Water: The Universal Solvent

11.7 Measures of Solution Concentration

Chemical Encounters: Composition of Seawater

11.8 The Effect of Temperature and Pressure on Solubility

Chemical Encounters: Impact of the Solubility of Oxygen in Fresh Water

11.9 Colligative Properties

Chemical Encounters: Meeting Municipal Water Needs

Go to **college.hmco.com/pic/kelterMEE** for online learning resources.

The Tampa Bay area has a severe water short-age brought about by a fivefold increase in popula-tion since 1950. With over 2.5 million residents and population growth estimated at over 50,000 per year, even the abundant rainfall of Florida's coastal areas cannot keep up with the skyrocketing demand for water. Tampa Bay is not at all unique in the world, or even in the United States, in its thirst for this pre-cious natural resource. Figure 11.1 shows the expected trends in water use by continent through the year 2025. Most of the world's increase in water consump-tion will be driven by population growth and rapid industrialization in Asian coun-tries. Localized dramatic increases are also predicted in other areas with pockets of rapid population growth, such as southern Florida and the southwestern states of Nevada and Arizona. This will further stress the already limited water supplies.

What uses are there for water, and why is it so important in our lives? Domestic use for washing, cooking, and drinking generates part of the demand, but agricul-ture accounts for most of our consumption of water (Figure 11.2). Large amounts of water are also used for industrial production, including the generation of electric-ity. The patterns of water demand also reveal a great deal about the development of a country's economy. Examining these data across the countries of the world, as is done in Figure 11.3, illustrates that water is used differently in developed countries than in underdeveloped countries. For example, Figure 11.4 shows that the United States earmarks most of its water supply for power generation and agriculture.

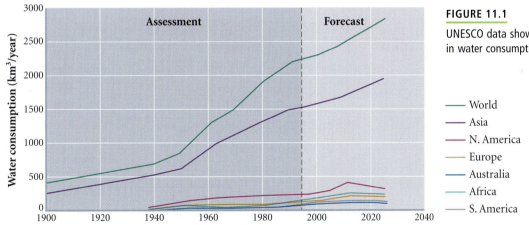

FIGURE 11.1

UNESCO data showing the trends in water consumption.

— World
— Asia
— N. America
— Europe
— Australia
— Africa
······ S. America

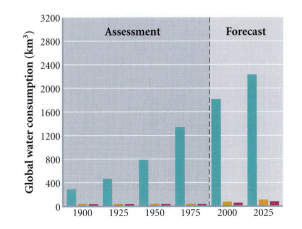

FIGURE 11.2

Agriculture is a major consumer of water around the world.

■ Agricultural
■ Domestic
■ Industrial

441

Water is so fundamental to life that the search for it has extended to the outer reaches of the solar system. Detecting water elsewhere in the universe could provide important information about the chemistry of the universe and, perhaps, the origin of life.

What is it about the structure and properties of water that makes it so pervasive and so vital to our world? How can these properties help us to understand water's many and

FIGURE 11.3

The use of water can be broken down into agricultural, industrial, and domestic (personal) uses.

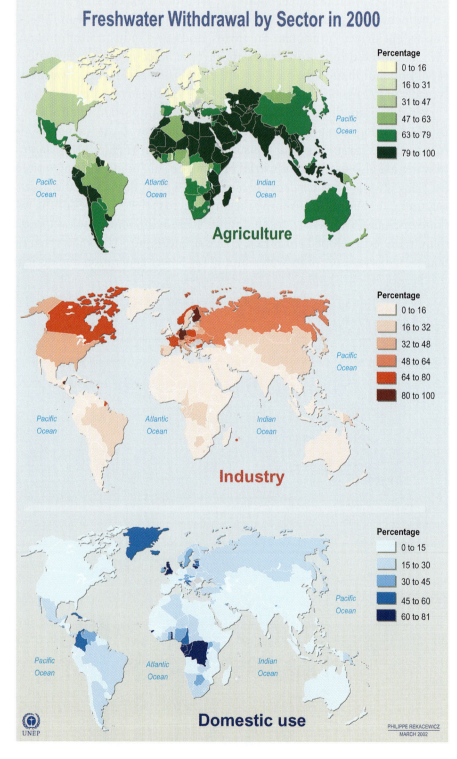

varied uses—and assist us in our efforts to provide clean water for places like Tampa Bay? This chapter focuses on the nature of water. We will compare it to other liquids in order to highlight its own special character. As you might sense, water is truly unique, vital in so many ways, and it's worth knowing why.

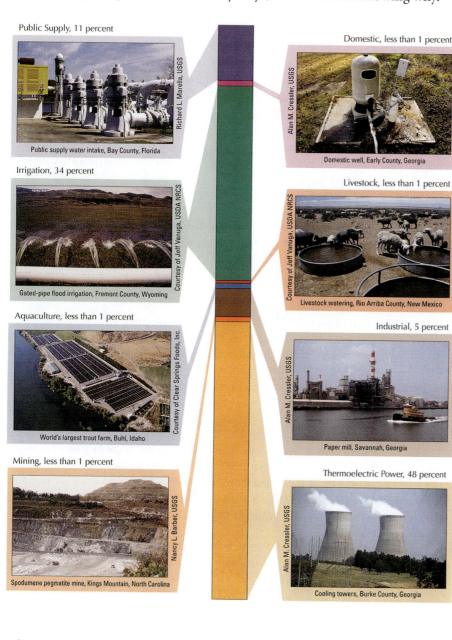

Public Supply, 11 percent

Public supply water intake, Bay County, Florida

Richard L. Marella, USGS

Irrigation, 34 percent

Gated-pipe flood irrigation, Fremont County, Wyoming

Courtesy of Jeff Vanuga, USDA NRCS

Aquaculture, less than 1 percent

World's largest trout farm, Buhl, Idaho

Courtesy of Clear Springs Foods, Inc.

Mining, less than 1 percent

Spodumene pegmatite mine, Kings Mountain, North Carolina

Nancy L. Barber, USGS

Domestic, less than 1 percent

Domestic well, Early County, Georgia

Alan M. Cressler, USGS

Livestock, less than 1 percent

Livestock watering, Rio Arriba County, New Mexico

Courtesy of Jeff Vanuga, USDA NRCS

Industrial, 5 percent

Paper mill, Savannah, Georgia

Alan M. Cressler, USGS

Thermoelectric Power, 48 percent

Cooling towers, Burke County, Georgia

Alan M. Cressler, USGS

FIGURE 11.4

According to the U.S. Geological Survey (USGS), more than half of the water used in the United States in the year 2000 was used for thermoelectric power generation and agriculture.

11.1 The Structure of Water: An Introduction to Intermolecular Forces

When we get thirsty, we can pour ourselves a glass of water. Huge numbers of water molecules stream out of the faucet in the **liquid** state. They fall into our glass and, like the molecules of all liquids, conform to the shape of their container. We can freeze water to make it a **solid**, which has its own shape, or boil it, forming a **gas**, which completely fills a container no matter how much is present. To make our water hot, we pipe the water into a water heater, which may be operated using another substance vital to our way of life, **natural gas** (mostly methane, CH_4).

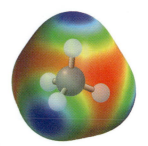

FIGURE 11.5

There are several ways to represent a water molecule: (a) ball-and-stick model, (b) Lewis dot structure, (c) space-filling model, and (d) electron dot surface model.

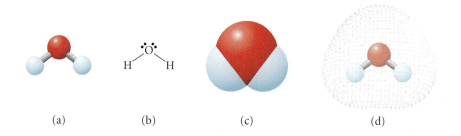

(a)   (b)   (c)   (d)

VSEPR structure of methane.

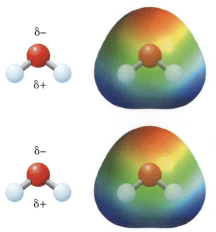

Dipoles attract each other.

Video Lesson: An Introduction to Intermolecular Forces and States of Matter

Both water and natural gas arrive at our homes via a system of underground pipes. Even though they possess similar molar masses (18 g/mol for H_2O and 16 g/mol for CH_4), water is a liquid and methane is a gas. **Why is this so?** The properties, chemical behavior, and day-to-day uses of these compounds are based on their structures. As we've noted in previous discussions, the structure of a molecule determines its properties.

Figure 11.5 provides several representations of a water molecule. Recall from Chapter 7 that oxygen is substantially more electronegative than hydrogen, leading to poles of charge on the individual O—H bonds in water. Using the VSEPR model from Chapter 8, we saw that water is a bent molecule, with an H—O—H bond angle of 104.5°. This led to our discovery of a net dipole moment for the polar molecule. Unlike water, methane is made up of atoms with similar electronegativity values. Additionally, the VSEPR structure of CH_4 is tetrahedral, and the molecule lacks a net dipole moment.

As we've seen in previous discussions, opposite charges attract. This is apparent in the structure of table salt (NaCl), in which the positive charge on the sodium ion is attracted to the negative charge on the chloride ion. But this attraction isn't limited to complete charges. Even slight distortions of electron distribution give rise to this attraction. If we recall the concept of energy (Chapter 5) as that which is needed to oppose a natural force, we can deduce that *energy is released when opposite poles attract*. This concept helps us to understand why water is liquid under normal conditions: The hydrogen atoms at positive poles of the dipole moment on water molecules can interact with the negative poles of oxygen atoms on other molecules. The resulting release of energy is the key to the stability of liquid water at room temperature. This interaction is an example of an **intermolecular force**. Unlike water, methane, does not have sufficiently strong intermolecular forces to exist as a liquid or solid at normal conditions, so it travels to our homes as a gas.

The result of the attraction between opposite poles in a sample of water is the interaction of 3 to 6 water molecules with each other at any one time, with an average of about 4.5. These molecules change partners constantly as they swirl about in the sample exchanging intermolecular forces of attraction. Although the *number* of interactions among the molecules of a substance does say something about its properties, the relative strength of each *individual* intermolecular interaction is particularly important.

How do *inter*molecular (between molecules) forces compare with the *intra*molecular (within molecules) forces that we call bonds? Consider what happens when we boil water. We produce water vapor ($H_2O(g)$), but we do *not* produce H_2 and O_2 gas. In boiling water, the individual intramolecular forces (bonds) *within* each water molecule have *not* been broken. Instead, the *inter*molecular interactions *between* molecules have been disrupted. **Why is this so?** About 44 kJ of energy is required to convert 1 mol of liquid water to the vapor, via the breaking of intermolecular attractions. It takes about 940 kJ to break the O—H bonds *within* a mole of water. This idea can be reinforced by examining similar forces in a sample of methane. About 9 kJ of energy is needed to vaporize

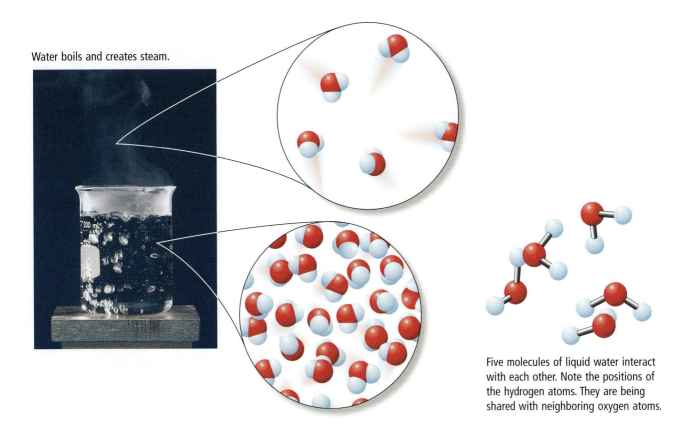

Water boils and creates steam.

Five molecules of liquid water interact with each other. Note the positions of the hydrogen atoms. They are being shared with neighboring oxygen atoms.

a mole of liquid methane, compared to 1650 kJ to break the four C—H bonds in a mole of methane! We can extend this information to make the more general statement that *intermolecular interactions are much weaker than intramolecular interactions*. Less energy is required to boil a liquid than to break it into its component elements.

11.2   A Closer Look at Intermolecular Forces

Video Lesson: Intermolecular Forces

Can we use our understanding of intermolecular forces to explain the nature of liquids other than water? In short, yes; the forces by which molecules come together as liquids are based on the intermolecular interaction of oppositely charged poles. They are collectively known as **van der Waals forces**, after the Dutch physicist Johannes Diderik van der Waals, who noted their existence in 1879 and whose correction for the behavior of real gases we discussed in Chapter 10. van der Waals forces can be quite weak, as reflected in the low boiling points of methane, −164°C (109 K), and N_2, −196°C (77 K), at 1 atm. van der Waals forces connecting molecules, such as water, are different in nature and much stronger. The stronger forces result in a much higher boiling point for water (boiling point = 100°C at 1 atm). What are these forces and how do they vary among types of molecules?

London Dispersion Forces—Induced Dipoles

In our comparison of water and methane, we noted that the polarity of water explains why it exists as a liquid at normal temperatures. Yet the ability of nonpolar methane to liquefy at very low temperatures (boiling point = −164°C) suggests that there is *something* even in nonpolar substances that can hold molecules together. The molecule octane (C_8H_{18}) is nonpolar, yet it is a liquid at room

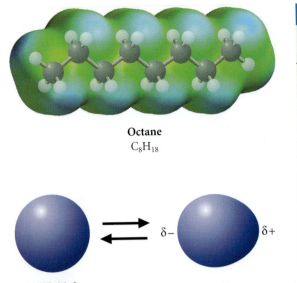

Octane
C_8H_{18}

TABLE 11.1	Boiling Points of Compounds at 1 atm		
Compound	Name	Molar Mass (g/mol)	Boiling Point (°C)
F_2	Fluorine	38	−188.1
Cl_2	Chlorine	71	−34.6
Br_2	Bromine	160	58.8
I_2	Iodine	254	184.4
CH_4	Methane	16	−164
C_2H_6	Ethane	30	−88.6
C_4H_{10}	Butane	58	−0.5
C_6H_{14}	Hexane	86	69
C_6H_{14}	2,2-Dimethylbutane	86	50
C_6H_{14}	2,3-Dimethylbutane	86	58
$C_{10}H_{22}$	Decane	142	174.1
$C_{24}H_{50}$	Tetracosane	339	391

FIGURE 11.6

In this representation, instantaneous induced dipoles are shown on atoms.

Visualization: Intermolecular Forces: London Dispersion Forces

temperature. It has a boiling point of 125.7°C, which is higher than that of our small polar water molecule. How can we reconcile these two observations?

The physicist Fritz London, who defined a concept we now call **London forces**, gave the explanation to this question in 1929. The key to his concept is a property called **polarizability**, the extent to which electrons can be shifted in their location by an electric field. Polarized electrons produce an **induced dipole**, an uneven distribution of charge caused (or "induced") by the electric field. In the same way, atoms or molecules can have *temporary dipoles* induced by instantaneous distortions in neighboring electron positions, as shown in Figure 11.6. The larger the number of electrons in the system, the larger the polarizability. The larger polarizability results in a larger induced dipole moment. In short, stronger attractive forces between molecules are observed when the induced dipole moment is larger.

Do London forces explain the differences among boiling points of nonpolar substances? Table 11.1 shows that the boiling points of substances are quite reliably related to their size, such that, for example, fluorine is a gas at room temperature, whereas bromine is a liquid and iodine is a solid. Similarly, ethane (C_2H_6) is a gas, whereas hexane (C_6H_{14}) is a liquid and tetracosane ($C_{24}H_{50}$) is a solid.

However, the structure of a molecule does have some effect on its polarizability. From Table 11.1 we see that the boiling point of hexane is 69°C. However, 2,2-dimethylbutane, an isomer of hexane (that is, it has the same chemical formula but a different structure) shown in Figure 11.7, boils at 50°C, and 2,3-dimethylbutane boils at 58°C. **Why is this so?** Polarizability is highly related to distance. Within a molecule of hexane, electrons are relatively close to those on

FIGURE 11.7

The "linear" molecule hexane has a higher boiling point than the two isomers in which some polarizable electrons are "hidden."

$$CH_3 - CH_2 - CH_2 - CH_2 - CH_2 - CH_3$$

Hexane
bp 68°C

$$CH_3 - CH_2 - \underset{\underset{CH_3}{|}}{\overset{\overset{CH_3}{|}}{C}} - CH_3$$

2,2-Dimethylbutane
bp 50°C

$$CH_3 - \underset{}{CH} - \underset{\underset{CH_3}{|}}{CH} - CH_3 \quad \overset{CH_3}{|}$$

2,3-Dimethylbutane
bp 58°C

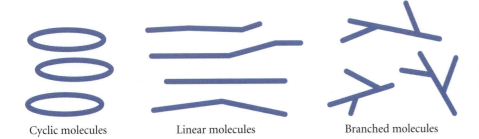

Cyclic molecules   Linear molecules   Branched molecules

Cyclic, linear, and branched molecules interact through London forces differently. Those molecules that can get closer to each other have greater intermolecular forces of attraction and higher boiling points.

other hexane molecules. On the other hand, some of the carbon atoms in the isomers of hexane are "hidden," farther away from other neighboring molecules, so the polarizability of compact molecules is not as great. Moreover, within a series of molecules of approximately the same molecular mass, those molecules that can interact more with their neighbors have a higher boiling point. Branched molecules lack some of this interaction; linear molecules have more such interaction but, because of their flexibility, do not interact perfectly; cyclic molecules have the most interaction with their neighbors and can stack together like dinner plates. This greater intermolecular interaction leads to higher boiling points.

The data in Table 11.1 also show us that the molar mass (related to the number of electrons in a molecule) of *nonpolar* substances is a major factor in their boiling points. The impact of molar mass on the boiling point is much greater than the effect observed in the isomers of hexanes. By comparing the boiling points of the straight-chain molecules in the table (such as ethane, −88.6°C; butane, −0.5°C; hexane, 69°C; decane, 174.1°C), we can see that as the molar mass gets larger, the boiling point increases. In fact, with a sufficiently large molar mass, it is possible for a nonpolar molecule to have a boiling point higher than that of water. Why is it that water itself has such a high boiling point? Water must have additional intermolecular forces of attraction that the nonpolar molecules do not possess.

EXERCISE 11.1   **Molar Mass, Structure, London Forces, and Boiling Point**

One of the following *nonpolar* substances is a gas at 200°C. All the others are liquids. Which one has the lowest boiling point at 1 atm, and why?

$$CH_3 - CH_2 - CH_2 - CH_2 - CH_2 - CH_2 - CH_2 - CH_2 - CH_2 - CH_2 - CH_2 - CH_2 - CH_2 - CH_2 - CH_3$$

Pentadecane
$C_{15}H_{32}$

$$CH_3 - \overset{\overset{\displaystyle CH_3}{|}}{\underset{\underset{\displaystyle CH_3}{|}}{C}} - CH_2 - \overset{\overset{\displaystyle CH_3}{|}}{CH} - CH_2 - \overset{\overset{\displaystyle CH_3}{|}}{\underset{\underset{\displaystyle CH_3}{|}}{C}} - CH_3$$

2,2,4,6,6-Pentamethylheptane
$C_{12}H_{26}$

$$CH_3 - CH_2 - CH_2 - CH_2 - CH_2 - CH_2 - CH_2 - CH_2 - CH_2 - CH_2 - CH_2 - CH_3$$

Dodecane
$C_{12}H_{26}$

First Thoughts

The two most important criteria in determining the relative boiling points of *nonpolar* substances are their molar masses and their structure. When we evaluate the structures to see which substance has the lowest boiling point, what we are looking for is an especially low molar mass, or perhaps a nonlinear structure.

Solution

Comparing the 12-carbon compounds, we note that dodecane has a straight-chain structure, whereas 2,2,4,6,6–pentamethylheptane is highly "branched," with many of the atoms hidden from other molecules, so dodecane will have the higher boiling point of the two. Pentadecane is a longer straight-chain molecule than dodecane, so its boiling point will be the highest among the group. The actual boiling points of the compounds are (from lowest to highest)

177.8°C	2,2,4,6,6-Pentamethylheptane
216.3°C	Dodecane
270.6°C	Pentadecane

Further Insights

Other factors affect the boiling (and melting) points of organic chemicals. One is the way in which molecules fit together. Molecules that only have single bonds between carbon atoms fit together nicely, leading to higher melting and boiling points than those that have double or triple bonds. The presence of other types of atoms can also substantially affect boiling points.

PRACTICE 11.1

Propane (C_3H_8) is a gas at room temperature, hexane (C_6H_{14}) is a liquid, and dodecane ($C_{12}H_{26}$) is a solid. Explain why these three molecules have their specific properties.

Propane C_3H_8 $CH_3 — CH_2 — CH_3$

Hexane C_6H_{14} $CH_3 — CH_2 — CH_2 — CH_2 — CH_2 — CH_3$

Dodecane $C_{12}H_{26}$

$CH_3 — CH_2 — CH_2 — CH_2 — CH_2 — CH_2 — CH_2 — CH_2 — CH_2 — CH_2 — CH_2 — CH_3$

See Problems 11, 25, 26, and 27a. ■

Visualization: Intermolecular Forces: Dipole–Dipole Forces

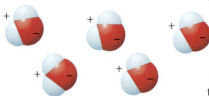

FIGURE 11.8

Interaction of the oppositely charged poles of water molecules makes water a liquid at room temperature.

Visualization: Intermolecular Forces: Hydrogen Bonding Forces

Permanent Dipole–Dipole Forces

The forces that give strength to the interactions among water molecules are a result of the permanent dipoles that exist in water. Water molecules organize to maximize the energetic stability gained by attractions and to minimize the repulsions, as shown in Figure 11.8. Dipole–dipole interactions are relatively weak compared to covalent bonds, but they become stronger in molecules with relatively large dipole moments. However, even permanent dipoles in a molecule may not lead to significant intermolecular bonding at normal conditions, as with the relatively small molecules HCl and H_2S, which have boiling points of only –85°C and –61°C.

Hydrogen Bonds

One very special type of interaction, the hydrogen bond, is of great importance. Knowing about **hydrogen bonds**, a term first used in 1912, furthers our understanding of why water is a liquid at room temperature. This knowledge also helps us explain some very elegant and important ideas about protein and DNA structure (Chapter 22) that add to our collective insight into human biology.

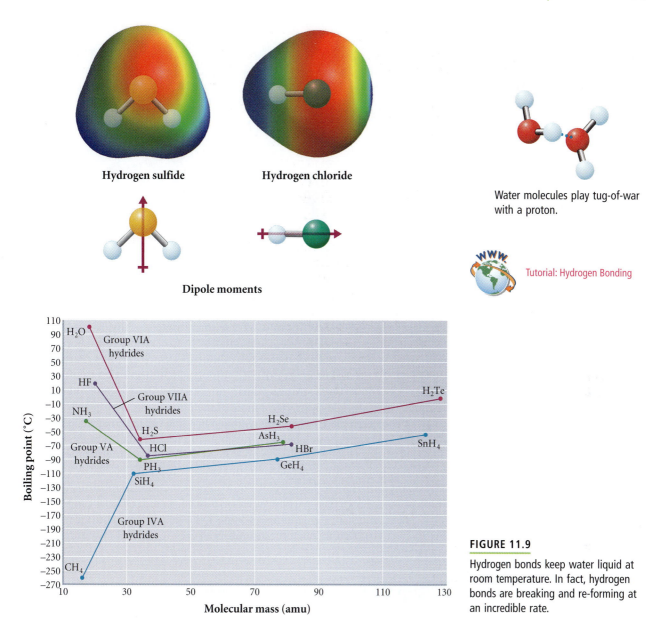

Hydrogen sulfide   Hydrogen chloride

Dipole moments

Water molecules play tug-of-war with a proton.

Tutorial: Hydrogen Bonding

FIGURE 11.9

Hydrogen bonds keep water liquid at room temperature. In fact, hydrogen bonds are breaking and re-forming at an incredible rate.

Judging on the basis of its low molar mass alone, we would never have assumed that water is a liquid at room temperature. However, water isn't unique in this regard. Other molecules, such as some of those shown in Figure 11.9, exhibit boiling points higher than their molar masses would lead us to expect. Why are their boiling points higher than expected? Each possesses hydrogen bonds.

A hydrogen bond is largely a dipole–dipole interaction of unusual strength compared to other intermolecular forces. It is formed when a hydrogen nucleus (a proton) is shared between two highly electronegative atoms of oxygen, nitrogen, or fluorine. These electronegative atoms interact with the proton through their available lone pair (or pairs) of electrons. In essence, the electronegative atoms play tug-of-war with the proton, creating a very strong intermolecular force of attraction. We can illustrate the hydrogen bond by drawing a series of dots from one of the electronegative atoms to the hydrogen.

Recent data have shown that hydrogen bonds are at least partly—as much as 10%—covalent in nature. Why is a hydrogen bond so strong? The partial positive charge on the tiny hydrogen atom gives it a relatively large charge-to-size ratio. That helps it pack a large attractive punch. This permits a strong interaction with

H — O — H • • • O — H
                   |
                   H

FIGURE 11.10

Compounds such as water, ammonia, and hydrogen fluoride have much higher boiling points than compounds of similar size, as a consequence of hydrogen bonding.

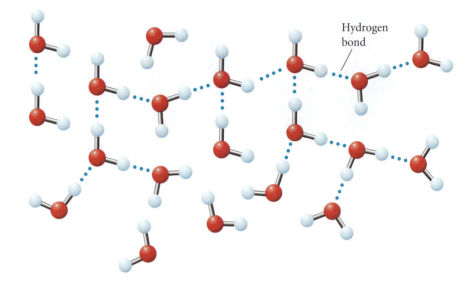

Hydrogen bond

the highly electronegative fluorine, oxygen, or nitrogen atoms. How strong are hydrogen bonds compared to covalent bonds? The $O \cdots H$ hydrogen bond in liquid water has a measured bond energy of about 23 kJ/mol, compared to the average O—H covalent bond energy of 470 kJ/mol, which means that this hydrogen bond is only about 5% as strong as the intramolecular O—H bond, but stronger than the 9 kJ required to vaporize a mole of methane.

What does our understanding of hydrogen bonding tell us about the properties of water? Hydrogen bonds between neighboring water molecules, as shown in Figure 11.10, keep water in the liquid state at STP. Hydrogen bonds break and re-form billions of times per second, but on average, enough hydrogen bonds exist at any one time to keep water liquid. This process of bonds breaking and re-forming means that a particular set of three atoms, HOH, will not stay together very long. Covalent bonds become hydrogen bonds, and vice versa. A flurry of activity is going on at the atomic level, even in a glass of water resting on a tabletop.

EXERCISE 11.2 **Comparing the Boiling Points**

Ethane (C_2H_6) is an important starting material for the industrial production of polyethylene plastics, used in items such as soft drink bottles. Ethanol (C_2H_6O) second only to water as an industrial solvent, is used in the synthesis of other compounds and in some blends of gasoline. Ethylene glycol ($C_2H_6O_2$) is the main component in conventional automobile antifreeze. Judging on the basis of their structures, arrange these compounds from lowest to highest boiling point.

First Thoughts

Hydrogen bonds have a substantial impact on boiling point, especially when the molecule itself is small, with many hydrogen-bonding sites. Molecules that have —OH groups are likely candidates for hydrogen-bonding interactions.

Ethane
CH_3—CH_3

Ethanol
CH_3—CH_2—OH

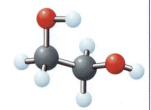

Ethylene glycol
HO—CH_2—CH_2—OH

Solution

Ethylene glycol has two polar sites at which hydrogen bonds can form, compared to one on ethanol and none on ethane. Ethane therefore has the lowest boiling point, −88.6°C, ethanol is next at 78.5°C, and ethylene glycol boils at 198°C.

Further Insights

An important idea in chemistry is that small changes in structure can lead to large changes in properties. Having an alcohol group (—OH, see Chapter 12) substitute for a hydrogen atom increases the boiling points of the compounds by over 100°C for each such substitution. The effect is less in larger compounds. Why is this so? Think about the types of intermolecular interactions that become possible.

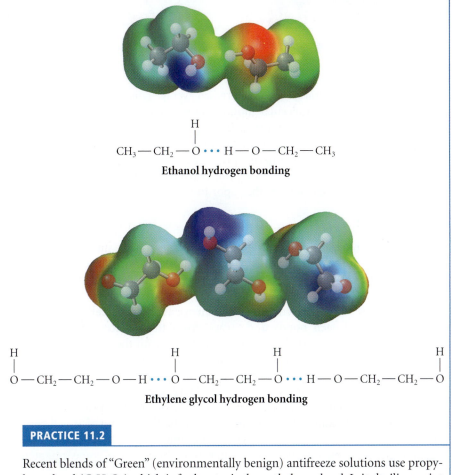

$$CH_3 — CH_2 — O \cdots H — O — CH_2 — CH_3$$

with H above the first O

Ethanol hydrogen bonding

$$O — CH_2 — CH_2 — O — H \cdots O — CH_2 — CH_2 — O \cdots H — O — CH_2 — CH_2 — O$$

with H above each terminal O

Ethylene glycol hydrogen bonding

PRACTICE 11.2

Recent blends of "Green" (environmentally benign) antifreeze solutions use propylene glycol ($C_3H_8O_2$) which is far less toxic than ethylene glycol. Is its boiling point higher or lower than that of ethylene glycol?

See Problems 12, 19–23, 27b, and 103.

Propylene glycol

$$HO — CH_2 — CH_2 — CH_2 — OH$$

11.3 Impact of Intermolecular Forces on the Physical Properties of Water, I

Video Lesson: Properties of Liquids

A First Look at Phase Changes

A glass of pure water resting on a tabletop seems to be just that: at rest. But on a molecular level, the system is deliriously active. The water molecules are moving randomly and at great speed. Intermolecular hydrogen bonds are being broken and re-formed at an incredible rate. Water molecules on the surface of the liquid have enough energy to break free of their hydrogen bonds and go into the vapor

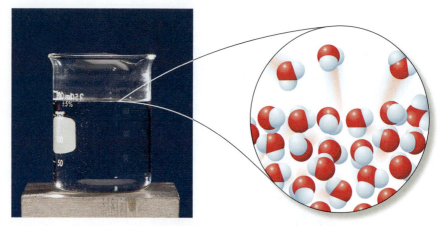

Evaporation of water from the surface of the liquid.

state. With higher temperature comes increased average kinetic energy of the system. That means greater molecular motion, more breaking of hydrogen bonds, and more molecules leaving the surface in the process we call **evaporation**. Some of the surrounding molecules return from the vapor to the liquid, attaching to the surface via hydrogen bonding, in the process of **condensation**. The corresponding escape of molecules from a solid (such as ice) to the vapor is called **sublimation**, and their return from the vapor to the solid is known as **deposition**.

Vapor Pressure

If we were to let our glass of pure water sit long enough, more molecules would evaporate than would condense, and we would be left with an empty glass. Let's change the set-up by enclosing the water in a sealed flask, as shown in Figure 11.11. How will the behavior of the system change?

Initially, those water molecules with sufficient energy will break free of their intermolecular hydrogen bonds and escape from the surface. The resulting vapor will collide with air molecules in the flask and with the walls of the container many times each second, exerting a force per unit area that we measure as **pressure**. Recall from Chapter 10 that a pressure of 760 torr equals 1 atm.

In our sealed flask, as illustrated in Figure 11.12, the more vapor that exists, the more collisions per second with the surroundings and the greater the pressure. As the concentration of vapor builds up in the flask, some of it will condense. Eventually, the air in the flask will hold as much vapor as it possibly can, and, as might occur on a sultry summer afternoon, the air will become **saturated**

Video Lesson: CIA Demonstration: Boiling Water at Reduced Pressure

FIGURE 11.11

How will the behavior of the liquid and vapor change if we use a stoppered (sealed) flask?

FIGURE 11.12

In a sealed flask, the system will come to equilibrium, with the rate of evaporation being equal to the rate of condensation.

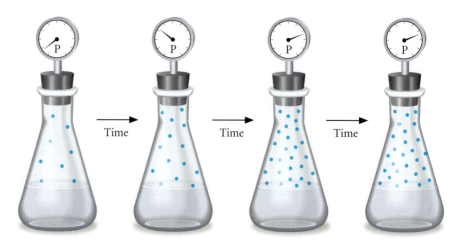

with water vapor. At this point, the rate of evaporation from the surface of the water will equal the rate of condensation back to the surface. When the rates of the forward (evaporation) and reverse (condensation) processes are equal, they are said to be in **equilibrium**. We indicate that both evaporation and condensation can occur by using double arrows.

As with every process we study, bond breaking and bond forming are always vigorously occurring, even though to our eyes all seems quiet. We often indicate that the molecular-level equilibrium is dynamic, rather than static, by calling it **dynamic equilibrium**. The pressure exerted by a vapor in equilibrium with its liquid is called the **equilibrium vapor pressure**, or, commonly, the **vapor pressure**, of the liquid at a given temperature. The vapor pressure of hot water at 80°C is 355.1 torr; that of cool water at 20°C is 17.5 torr; this means that more water molecules are in the gas phase at relatively higher temperatures than at lower temperatures. *The vapor pressure of a liquid increases with temperature.* Figure 11.13 and Table 11.2 show this dependence for water.

Does this make sense at the molecular level? A higher system temperature means higher average kinetic energy. Water molecules are moving faster than at lower temperature, so more molecules have enough energy to break their hydrogen bonds and evaporate.

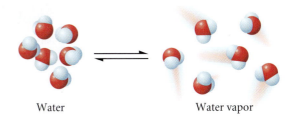

Water ⇌ Water vapor

TABLE 11.2	Vapor Pressure of Water at Selected Temperatures
Temperature (°C)	**Vapor Pressure (torr)**
0	4.58
10	9.21
15	12.8
20	17.5
21	18.7
22	19.8
23	21.1
24	22.4
25	23.7
26	25.2
30	31.8
35	41.2
40	55.3
50	92.5
60	149.4
70	233.7
80	355.1
90	525.8

Source: Chemical Rubber Company. *Handbook of Chemistry and Physics,* 66th ed.; CRC Press: Boca Raton, FL, 1986.

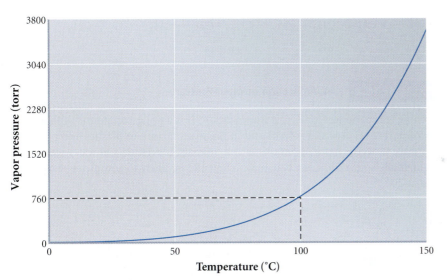

FIGURE 11.13

The vapor pressure of liquids is related to temperature, as shown for water. The line on the graph indicates the boiling point of water at 1 atm.

EXERCISE 11.3 | **Comparing Vapor Pressures**

Which substance, water or methanol (CH₃OH), would have a *lower* vapor pressure at 50°C? Explain your choice.

Solution

The molecule that has more extensive hydrogen bonding will have the lower vapor pressure at a given temperature. The entire structure of water encourages hydrogen bonding. Methanol can also hydrogen-bond, but it also has nearly nonpolar C—H bonds. The vapor pressure of water is 93 torr at 50°C, compared to about 400 torr for methanol.

PRACTICE 11.3

Name a substance that would have a lower vapor pressure at 50°C than either water or methanol. Explain your reasoning.

See Problems 33–35 and 36a.

We now know that the ability to form hydrogen bonds is one important factor leading to lower vapor pressure at a given temperature, but it is not the only factor. Molar mass and structure are also important because, as we have already learned, London forces can be significant in relatively large molecules. In order to develop a vapor pressure of 400 torr, octane (C_8H_{18}) must be heated to 104°C. The branched molecule 2,3,3–trimethylpentane, shown in Figure 11.14, has this vapor pressure at 92.7°C, and water reaches it at 83.0°C.

FIGURE 11.14

The London forces are greater in octane than in the highly branched 2,3,3-trimethylpentane. Which would require a higher temperature to reach a vapor pressure of 400 torr?

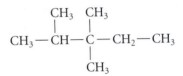

2,3,3-Trimethylpentane

$$CH_3 - CH_2 - CH_2 - CH_2 - CH_2 - CH_2 - CH_2 - CH_3$$

Octane

EXERCISE 11.4   **An Implication of Vapor Pressure**

Acetone (propanone, C_3H_6O), is shown below. It is a polar compound with a variety of industrial uses as a solvent and in the manufacture of plastics and pharmaceuticals. It occurs naturally in plants and animals. If acetone and water were both accidentally spilled on a lab bench, which would be likely to completely evaporate first at 20°C?

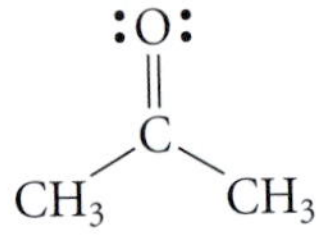

First Thoughts

To "completely evaporate" means to fully change from the liquid to the gaseous state. Even though evaporation is not an equilibrium process (that is, molecules are leaving the liquid's surface more rapidly than they are condensing back), we can use the vapor pressure as a guide. What contributes to a high vapor pressure? A low-molar-mass molecule, even a polar one, that cannot form hydrogen bonds to itself will probably have a higher vapor pressure than that of water.

Solution

Acetone at 20°C has a vapor pressure of about 185 torr, much greater than water's 17.5 torr. Acetone (boiling point = 57°C) will completely evaporate first.

Further Insights

Looking a bit more deeply, we note that there are hydrogen bond donors, such as water, that can use a hydrogen atom to engage in hydrogen bonding with, for example, another water molecule. Water is also a hydrogen bond acceptor, because its oxygen can accept a hydrogen bond from another water or, for example, from ethanol. On the other hand, acetone is only a hydrogen bond acceptor (as in the present example), so it cannot internally hydrogen-bond. However, it can form such bonds with water and is soluble in it. There are millions of organic (carbon-based) compounds that have a vast range of properties. Chapter 12 will introduce the compounds and chemistry that make organic chemistry well worth knowing.

PRACTICE 11.4

Compare acetone and hexane (C_6H_{12}) using the same lab bench scenario as in Exercise 11.4.

See Problems 29 and 30. ■

HERE'S WHAT WE KNOW SO FAR

- The vapor pressure of a liquid increases with temperature.

- More hydrogen bonding leads to lower vapor pressure.

- All other things being equal, heavier molecules have lower vapor pressures than lighter molecules.

- Straight-chain molecules have lower vapor pressures than their branched isomers.

- When several of these factors come into play, it is hard to predict which will dominate. We then run experiments or look up information in data tables.

Visualization: Boiling Water with Ice Water

Boiling Point

Let's return to our glass of pure water. As we heat it, the vapor pressure of the liquid increases along with the temperature. If we are at sea level on a day when the surrounding pressure is 1 atm, the liquid will start to bubble from within as the temperature approaches 100°C. As it does so, the vapor pressure of the liquid will edge ever closer to the atmospheric pressure. When the temperature reaches 100°C, bubbles burst forth throughout the water in a familiar phenomenon we call **boiling**. We discussed boiling earlier in the chapter, and you understand its general meaning from all the years you have been boiling water to make tea or cook vegetables. Now we are ready to look at it in more depth. Boiling is not just a surface process like evaporation, because it involves the entire liquid. We define the **boiling point** as the temperature at which the pressure of the liquid's vapor (rather than the vapor pressure, which is defined for an equilibrium process) is equal to the surrounding pressure. If that pressure is 1 atm, at or near which so many of life's normal activities take place, the temperature at which a liquid boils is called its **normal boiling point**.

A good portion of the world's population does not live at sea level, and for them, "normal" is anything but. In mile-high Denver, the atmosphere is less dense than at sea level, so the atmospheric pressure is correspondingly lower, about 0.82 atm (620 torr). If our glass of water were heated in Denver, it would boil at about 95°C. The difference in boiling point with pressure is even more dramatic in Mexico City, at 2240 m (7340 ft), where the atmospheric pressure is only 0.76 atm (580 torr). The boiling point at that altitude is only about 90°C (194°F). Figure 11.15 shows the decrease in the boiling point of water as the altitude increases and atmospheric pressure consequently decreases. Food manufacturers take advantage of the *increase* in boiling point with pressure when they process foods by canning them at *high* pressure, allowing the food item to be heated to a relatively high temperature, typically 107°C, for at least 3 minutes, killing any bacteria within. A quick way to cook soup is to use a pressure cooker, which increases the pressure within from 1 to 2 atm, raising the boiling point of the soup

Water boils at 100°C at 1.0 atm. At this temperature, the vapor pressure equals the atmospheric pressure.

Video Lesson: Vapor Pressure and Boiling Point

FIGURE 11.15

The boiling point of water goes down as the altitude rises. This is because air pressure is lower at higher altitude (remember the definition of boiling point).

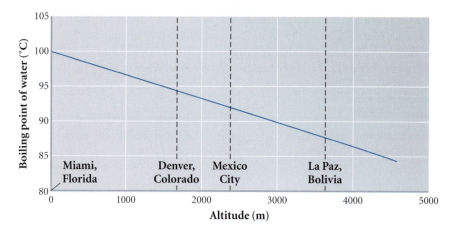

Directions for baking a cake at high elevations.

to about 120°C (250°F). Commercial cake mixes often have two sets of instructions: one for cooking the batter at sea level (normal boiling point) and one for higher altitudes.

Heating Curves

We opened this chapter by looking at drinking water in the Tampa Bay area. Now let's look north, perhaps above the Arctic Circle, to a group of hikers who want to get drinking and washing water by melting some ice and then purifying it by boiling. What happens as they heat a 10.0-kg block of the ice at −10.0°C? To fully understand the process that occurs, we must consider four changes: (1) warming the ice; (2) melting the ice; (3) heating the water, and (4) boiling the water. As we proceed, think about what happens to the energy we add, the molecular motion, and the resulting system temperature. We will assume that all of the heat goes into the ice (the system), not into the surroundings.

Change 1: Warming the Ice

The temperature of our ice is −10.0°C. As we add heat, the molecules that are fairly rigidly held in place begin to move a bit more. We still have ice, but the average energy has increased, so the temperature rises. Figure 11.16 displays a **heating curve**, a plot of the temperature of a compound versus time as it is heated. The plot indicates the specific temperature ranges for solid, liquid, and gas phases.

The heat needed to warm 1 g of a substance enough so that its temperature rises 1.0°C is its **specific heat** (see Section 5.4), which for ice is equal to 2.05 J/g·°C. Raising the temperature of our 10.0-kg block of ice from −10.0°C to its **melting point**, at which it changes from a solid to a liquid, requires that 205 kJ of heat, q, be added to the system.

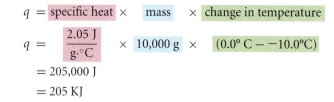

$$q = \text{specific heat} \times \text{mass} \times \text{change in temperature}$$

$$q = \frac{2.05 \text{ J}}{\text{g·°C}} \times 10{,}000 \text{ g} \times (0.0° \text{C} - -10.0°\text{C})$$

$$= 205{,}000 \text{ J}$$

$$= 205 \text{ KJ}$$

This much heat would be supplied by, for example, burning about 4 g of methane (natural gas is typically over 90% methane) in air.

Change 2: Melting the Ice

Our ice is now at its melting point of 0.0°C. As we add more heat, water molecules move more freely and randomly, beginning the transformation from the solid

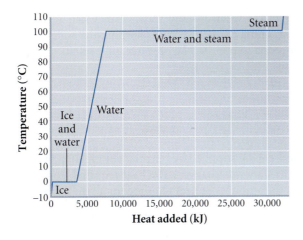

FIGURE 11.16

As water is heated, it goes through changes from solid to liquid to gas. Why are some regions sloped and others flat? Why are areas of constant temperature of different lengths?

Visualization: Changes of State

Video Lesson: CIA
Demonstration: Boiling Water in a Paper Cup

crystal to the liquid state. The temperature at this change of state, the melting point, *is constant* (Figure 11.16), because the added heat is going into breaking hydrogen bonds, rather than increasing the average kinetic energy of the water molecules. The amount of heat needed to convert the solid to liquid at the melting point at constant pressure is called the **heat of fusion ($\Delta_{fus}H$)**, which for water is 334 J/g, or 6.01 kJ/mol. Melting is an *endothermic* process, which means that freezing (the opposite process) is an *exothermic* process. Heat is released when a liquid freezes.

How much heat is needed to melt 10.0 kg of ice? The calculation shown below indicates that we need to add 3340 kJ to the system.

$$q = \quad \text{mass} \quad \times \quad \Delta_{fus}H$$

$$10{,}000 \text{ g ice} \times \frac{334 \text{ J}}{\text{g ice}} = 3{,}340{,}000 \text{ J} = 3340 \text{ kJ}$$

This is about the amount of heat supplied by burning about 60 g of methane in air.

Change 3: Heating the Liquid

The heat going into the (pure liquid) water increases molecular motion, raising the temperature of the liquid (Figure 11.16). The heat needed to raise the temperature of our 10.0-kg water sample from 0.0°C to 100.0°C can be calculated using the specific heat of water, 4.184 J/g·°C.

$$q = \text{specific heat} \times \quad \text{mass} \quad \times \quad \text{change in temperature}$$

$$\frac{4.184 \text{J}}{\text{g·°C}} \quad \times 10{,}000 \text{ g} \times \quad (100.0°C - 0.0°C)$$

$$= 4{,}180{,}000 \text{ J} = 4180 \text{ kJ}$$

The total of 4180 kJ can be supplied by about 75 g of methane burning in air.

Change 4: The Boiling Point

As we continue to heat the water, it boils as the liquid is converted to vapor at constant temperature (Figure 11.16). Analogous to the heat of fusion for melting is the **heat of vaporization ($\Delta_{vap}H$)** for converting the liquid water to vapor at the normal boiling point. The value of $\dfrac{2.44 \text{ kJ}}{\text{g water}}$ is *over 7 times that of the heat of fusion* $\left(\dfrac{0.334 \text{ kJ}}{\text{g ice}}\right)$. This indicates that a lot more energy is required to overcome the intermolecular forces when boiling water than when melting ice. Boiling causes all of the remaining hydrogen bonds keeping the water as a liquid to be broken. The heat needed to boil 10.0 kg of water is 24,400 kJ.

$$q = \text{mass} \times \Delta_{vap}H$$

$$10{,}000 \text{ g} \times \frac{2.44 \text{ kJ}}{\text{g water}} = 24{,}400 \text{ kJ}$$

This amount of heat can be supplied by burning about 440 g of methane, over 3 times the total of the previous changes! The heat required from start to finish is equal to the sum of the amounts of heat required for all the change along the way.

$$\text{Total heat} = q_{\text{ice warming}} + q_{\text{ice melting}} + q_{\text{water warming}} + q_{\text{water boiling}}$$
$$= 205 \text{ kJ} + 3340 \text{ kJ} + 4180 \text{ kJ} + 24{,}400 \text{ kJ}$$
$$= 32{,}100 \text{ kJ}$$

It is possible to heat the water vapor to quite a high temperature, and the amount of heat needed is related to the specific heat of water vapor, 1.84 J/g·°C. This is shown in Figure 11.16.

EXERCISE 11.5 **Energy Changes Upon Heating the Water**

Determine the heat required to raise the temperature of a 2.50×10^2 g block of ice (equivalent to about a cup of water) in a 425 g aluminum pot from 0.0°C to room temperature, 20.0°C. Roughly how many grams of methane must be burned in air to supply this much heat? The heat of combustion of methane is −890 kJ/mol. Solid aluminum has a specific heat of 0.880 J/g·°C. Assume that all heat goes to the ice and the aluminum pot.

First Thoughts

What processes occur? The input of heat will melt the ice at constant temperature (change of state) and then warm the ice from 0.0°C to 20.0°C. The aluminum pot will be warmed from 0.0°C to 20.0°C.

Solution

$$\text{Total heat} = q_{\text{melting ice}} + q_{\text{warming liquid}} + q_{\text{warming aluminum}}$$

$$= \left(\frac{334 \text{ J}}{\text{g ice}} \times (2.50 \times 10^2 \text{ g ice})\right) + \left(\frac{4.184 \text{ J}}{\text{g ice·°C}} \times (2.50 \times 10^2 \text{ g ice})\right.$$

$$\left. \times (20.0°\text{C} - 0°\text{C})\right) + \left(\frac{0.880 \text{ J}}{\text{g Al·°C}} \times 425 \text{ g Al} \times (20.0°\text{C} - 0°\text{C})\right)$$

$$= 83{,}500 \text{ J} + 20{,}920 \text{ J} + 7480 \text{ J} = 111{,}900 \text{ J} = 112 \text{ kJ}$$

$$\text{Grams of CH}_4 \text{ required} = 112 \text{ kJ} \times \frac{1 \text{ mol CH}_4}{890 \text{ kJ}} \times \frac{16.0 \text{ g CH}_4}{1 \text{ mol CH}_4} = 2.0 \text{ g CH}_4$$

Further Insights

Even though the specific heat of the aluminum pan is a lot less than that of water, it is a heavy pan, and a lot of heat is required to raise its temperature. A more realistic picture would need to take into account the heat loss, and lots of it, to the air.

PRACTICE 11.5

How much energy does it take to convert 130.0 g of ice at −40.0°C to steam at 160.0°C?

See Problems 31, 32, 37, and 38.

11.4   Phase Diagrams

We now understand the changes that occur in water as we raise its temperature from below freezing to the boiling point at normal pressure. However, much of the known universe is nowhere close to our own "normal" pressure or temperature, and that affects the form water takes beyond our world. Figure 11.17 shows recent photographs of the moons of Jupiter taken by NASA's Galileo space probe. Why do the moons look so different from one another, from their host planet, and from anything else we've seen in the solar system? One reason apparently has to do with ice—not the type of ice we find in the refrigerator, but ice that exists under pressures of tens of thousands of atmospheres. Under these immense pressures, the molecules of ice that form large parts of the interiors of the "Jovian" moons are highly distorted. Hydrogen bonds are squeezed and shifted. Atoms are brought closer together and bond lengths shortened so that the ice becomes unusually dense. Temperature changes deep inside these moons also lead to changes in the nature of the ice. The consequences of all these changes are that the moons contract and expand, causing huge cracks and grooves. We add to our understanding of the solar system by looking at the changes in ice as temperature and pressure change.

A graph showing the way the phase of a substance, or a mixture of substances, is related to temperature and pressure is called a **phase diagram**. The heating curves we discussed previously describe how water changes with temperature *at a constant pressure*. However, as with the changes in the moons of Jupiter, a great many processes do not happen at constant pressure, and a phase diagram gives us a more comprehensive picture of these changes. In the chemical industry, phase diagrams have many uses, including determining the conditions required for manufacturing ceramics from clays (see Chapter 13). The composition of homogeneous mixtures of metals, called alloys, can also be changed on the basis of the information in phase diagrams.

The phase diagram that includes the various types of ice in the core of planets is fairly complex, so let's work first with a small portion of it, that for water at more Earth-like temperatures and pressures, shown in Figure 11.18. Note in the

Tutorial: Phase Diagrams

FIGURE 11.17

The moons of Jupiter as taken by NASA's Galileo space probe.

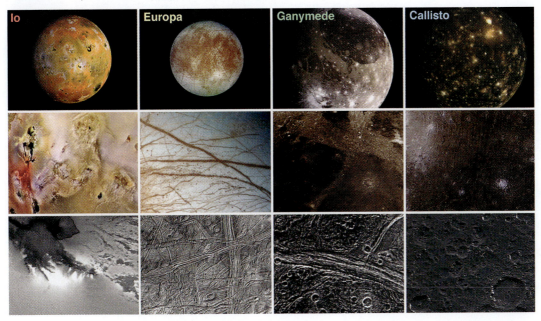

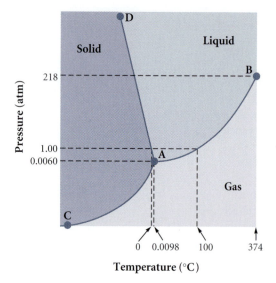

FIGURE 11.18

Phase diagram of water.

Video Lesson: Phase Diagrams

figure the three normal phases: solid, liquid, and gas (water vapor). There are also bold lines, called **phase boundaries**, at which two phases exist in equilibrium. How does water change state as we heat it from well below 0°C and at 1 atm (the dotted line on Figure 11.18)? Keep in mind that these are the common changes that we experience every day as we interact with water by boiling it to make tea or coffee or cause it to melt when we add it as ice to a soda.

Our water sample begins as ice. We raise the temperature to 0°C, where it arrives at a phase boundary at which ice and liquid water exist in equilibrium. This is the normal melting (and freezing) point of water. As we continue to raise the temperature, the water remains liquid until 100°C, the boiling point, which is at the liquid/gas phase boundary. Above 100°C, the water will always be a gas if the pressure is kept constant at 1 atm. This is the same information we gleaned from the heating curve of water, but a phase diagram gives us so much more. We can tell, for example, that there is a **triple point** of water (Figure 11.18) where all three phases exist in equilibrium. For water, this occurs at 0.01°C and 4.6 torr. The triple points of elements and compounds are useful because they serve as reference points with which temperature scales for thermometers are defined, and against which the highest-quality thermometers are calibrated. If we follow the liquid/gas phase boundary (the curved line segment AB, the boiling point) toward increased temperature and pressure, we get to a point above which the substance can no longer be liquefied and the gas and liquid have the same density—that is, we have a **supercritical fluid**. This point (B) is called the **critical point**, and it is associated with a **critical temperature**. The pressure at the critical temperature is the **critical pressure**.

EXERCISE 11.6   Reading Phase Diagrams

What does the line segment AD on the phase diagram of water represent? What does line AC indicate?

Solution

Line segment AD represents the point of equilibrium between solid and liquid—the melting (freezing) point at which the solid and liquid exist in equilibrium. Line segment AC is point of equilibrium between solid and gas for water; sublimation and deposition occur at equilibrium anywhere along this line.

FIGURE 11.19

Phase diagram of carbon dioxide.

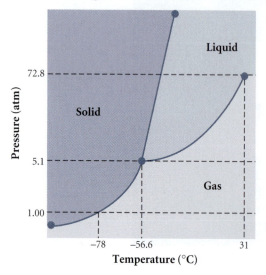

PRACTICE 11.6

What is the approximate boiling point of water when the external pressure is 0.70 atm? Does this agree with our previous discussion about boiling points at high altitudes?

See Problems 39–48.

The phase diagram of CO_2, shown in Figure 11.19, looks similar to that of water, with a triple point, critical temperature and pressure, and several phase boundaries. At 1 atm, CO_2 changes directly between the solid and gas phases (sublimation/deposition). You might have seen dry ice, which does not melt but, rather, sublimes directly from a solid into a gas. At pressures greater than about 5 atm, CO_2 will liquefy from the solid phase. This liquid is used for dry-cleaning clothing to replace

How do we know?

The moons of Jupiter

The moons of the outer planets make most elegant chemical laboratories, because conditions on these worlds are so very different from those here at home. When we look at photos of Jupiter's moons, Europa and Ganymede (Figure 11.17), we see some vexing geological features. How do we know what caused these features? We can't visit the moons, at least not directly. However, our space probes can gather various types of electromagnetic radiation—including light, ultraviolet radiation, and X-rays—that give us data from which to draw conclusions.

The probes' data seem to show that deep within the surface of these moons, there are layers of ice, each in a different phase, mixed with rocks. The phase diagram of water that we showed as Figure 11.18 is inadequate to account for the low temperatures and massive pressures found within these moons. A more comprehensive phase diagram for water, emphasizing its solid phases, is given as Figure 11.20. The ice that is made in our kitchen freezer or found in an ice storm—what we might call "normal ice"— is technically called Ice Ih. Its structure is shown in Figure 11.21. There are 11 other forms of ice that have been made in the laboratory or simulated by computer.

The phase diagram shows that as the temperature and pressure inside the moons change, different types of ice are formed. Each ice has its own density. As layers form versions of ice of different densities, they form the geological features, such as cracking and grooves, that we see on the surface of these incredible satellites. Part of the task of scientists is to develop phase diagrams that tell us about the conditions at which each type of ice exists.

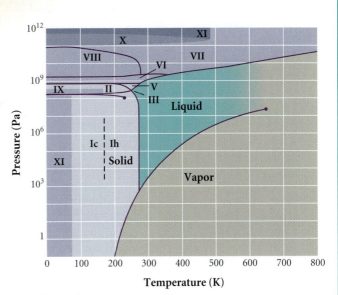

FIGURE 11.20

This is an expanded phase diagram for water, taking into account the phase changes among the various solid (ice) forms. These forms of ice are thought to occur in layers beneath the surface of some of the moons of the outer planets.

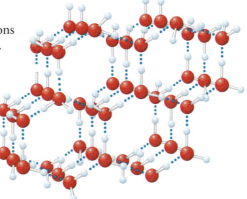

FIGURE 11.21

The arrangement of water molecules within a crystal of "normal" ice (Ice Ih).

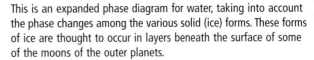

Application

CHEMICAL ENCOUNTERS:
CO_2 as a Dry Cleaning Solvent

nonpolar, but environmentally hazardous, solvents such as tetrachloroethylene (C_2Cl_4). These solvents rid clothing of nonpolar grease and dirt.

Liquid CO_2 is an excellent and relatively benign reusable alternative. Clothes are cleaned in CO_2 at 15°C and 50 atm pressure, which is along the liquid/gas boundary in the CO_2 phase diagram. The clothes are kept close to room temperature, and about 98% of the CO_2 is recycled. The critical temperature and pressure of CO_2 are 304.2 K and 70.8 atm. Supercritical carbon dioxide, typically in the range of 310 to 400 K and 75 to 100 atm, can be used to clean industrial machinery.

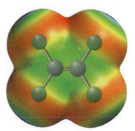

Tetrachloroethylene
C_2Cl_4

Motor oil viscosity is rated for use on the basis of the type of engine and the outside temperature.

11.5 Impact of Intermolecular Forces on the Physical Properties of Water, II

Viscosity

Making chocolate at home is difficult, because so much can go wrong, including unintended air bubbles, cracking, and a greasy look and feel to the final product. Having these things occur on an industrial manufacturing scale can be especially damaging. One of the most important properties that cause these and other inconsistencies in chocolate is called **viscosity**, which is a measure of resistance to flow. In the chocolate process industries, a viscometer is used to constantly monitor the viscosity of fully melted chocolate so that it flows into the molds and produces a final product that is consistent from batch to batch. Viscosity measurements are also important to the fuel industry, where the ability of fuels to flow is important to consider when storing, pumping and, within engines, injecting the fuel. Viscosity is reported in units of millipascal seconds (mPa·s). A highly viscous liquid, ethylene glycol ($C_2H_6O_2$), is combined with water to form commercial antifreeze and has a viscosity of 16.1 mPa·s at 25°C and 1 atm. The water with which it is combined has a viscosity of 0.890 mPa·s at the same conditions. Why does the viscosity differ among liquids?

As with so many things related to chemistry, we gain understanding by looking at behavior at the molecular level. The molecules that constitute viscous liquids do not move well relative to each other. **Why is this so?** Part of the answer has to do with intermolecular forces. More such interactions tend to increase viscosity. Size is also important. It is hard to move large molecules. And, as with boiling point, molecular shape is relevant because molecules must be able to move well among each other in order to flow easily. Octane (C_8H_{18}) has a viscosity of 0.508 mPa·s at 25°C and 1 atm, whereas the viscosity of pentane (C_5H_{12}) measures 0.224 mPa·s at these conditions. Temperature is also a key factor because, as we have seen, molecules with high average energy move relatively quickly and can overcome intermolecular forces, leading to low viscosity. For example, the viscosity of water changes from 0.890 mPa·s at 25°C to 0.378 mPa·s at 75°C. That's why chocolate, as well as oil in your car's engine, flows more smoothly at higher than at lower temperatures.

Surface Tension

The water strider glides from place to place along the water's surface, as shown in Figure 11.22. Why can the insect travel on the water's surface? The key is a property of liquids called **surface tension**. This property is also responsible for the

FIGURE 11.22

A water strider (*Gerris remigis*) on the surface of a pond. Note that the water indents as it supports the weight of the water strider.

beads of water that run down your window in the rain or across the surface of your car after it has been waxed. Surface tension is a measure of the energy per unit area on the surface of a liquid. To better understand what this measure means, we note that hydrogen bonding in water makes it more energetically stable. Water molecules on the interior of the solution can hydrogen-bond in all directions, whereas surface molecules can form these bonds only toward the solution itself. In order to maximize the number of hydrogen bonds (and, therefore, the energetic stability of the liquid), water forms a shape with the minimum surface area, a sphere. This is why small droplets of water form beads and water striders can glide on water.

The chemical industries pay close attention to surface tension in the design of many consumer products. For example, soaps and detergents are designed to have very low surface tension in order to interact well with clothing and dishes. Surface tension is also considered in the design of pharmaceuticals and plays a large role in how tablets and gelcaps dissolve after they are swallowed.

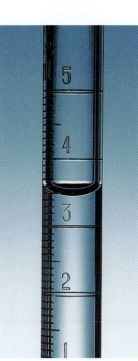

Water beads on a freshly waxed car.

Capillary Action

A **buret**, shown in Figure 11.23, is an essential piece of glassware for measuring volume when performing laboratory analyses with liquids. Water has been added to the buret in Figure 11.24. Why does the water seem to rise up the sides of the glass? And given this curvature, what do we say is the volume of water in the buret? The answer to the first question again lies in one of our chemical themes, the impact of hydrogen bonding on the structure of water. The forces that keep the large group of water molecules together are called **cohesive forces**. Then there is the glass in which the water sits. The glass is largely silicon dioxide (SiO_2) but with a surface containing many polar oxygen and hydrogen atoms. As is shown in Figure 11.25, the hydrogen atoms on water interact with the oxygen atoms on the surface of the glass. The net result is that some molecules of water adhere to the glass, a process called, **adhesion**. The effect in which water seems to crawl up the sides of a thin tube such as a buret is called **capillary action**. In this case, the adhesive forces between water and the glass are stronger than the cohesive forces within water itself. The weight of the water prevents it from rising up even higher

FIGURE 11.23

A buret is essential labware in the analytical laboratory.

FIGURE 11.24

The meniscus of water within a buret is readily observable.

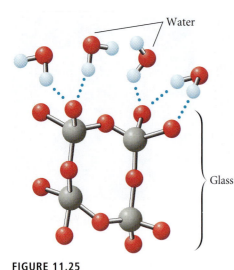

FIGURE 11.25

The interaction of water with the surface of glass results in adhesion. Hydrogen bonds between the water molecules and the glass surface are illustrated.

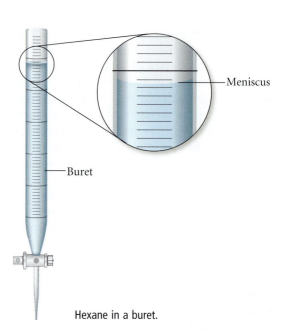

Hexane in a buret.

within the buret. Mercury in a glass tube (such as in a thermometer) shows the opposite behavior, bulging upward within the glass, in response to the dominance of cohesive over adhesive forces.

Adhesive forces lead to the formation of a **meniscus**, the concave shape at the surface of the water in the buret. By convention, the volume of liquid delivered from the buret is read at the bottom of the meniscus, as shown in Figure 11.24, so our reading of the buret is 40.03 mL. Because burets are graduated from the top (0 mL) down, there are 9.97 mL of water remaining in our 50.00-mL buret.

The balance between forces is not always easy to assess. For example, hexane (C_6H_{14}) shows a concave meniscus similar to that of water, though not quite so pronounced. **Would we have predicted this fact?** The surface of the glass can, presumably, induce a dipole in hexane, so we can explain the behavior. However, making predictions when several effects are important is sometimes quite challenging.

HERE'S WHAT WE KNOW SO FAR

- A liquid fills its container from the bottom up and flows easily.

- Intermolecular forces of attraction determine the properties of a substance. The number and strength of these forces contribute to the attractions between adjacent molecules.

- London forces are typically the weakest forces between molecules. Because every molecule has the ability to become polarized, every molecule possesses London forces. In very large molecules, these forces can be significant.

- Dipole–dipole interactions are much stronger than London forces. They exist in molecules that contain a nonzero dipole moment, and their strength is related to the strength of the dipole moment.

- Hydrogen bonds are the strongest of the van der Waals forces. They exist primarily between the interaction of a hydrogen atom and two atoms of F, O, and/or N.

- The strength of the intermolecular forces of attraction determines the shape of a meniscus, the surface tension of the liquid, its vapor pressure, and other physical properties associated with the liquid.

11.6 Water: The Universal Solvent

In the introduction to this chapter, we raised three questions related to the projected water shortage in Tampa Bay: What is it about water that makes it unique among molecules? How can we use our understanding of the nature of water to explain its properties? And how can we use these properties to help in the search for clean water in places like Tampa Bay? We have answered the first question, focusing especially on the polar nature of water and its ability to form the special interaction called a hydrogen bond. This enabled us to explain a variety of properties of water, including its unusually high boiling point, its vapor pressure, and its phase diagram, giving us the understanding to answer much of the second question.

Implicit in the third question about the search for clean water is recognition that the water found in nature is not pure. It is filled with all manner of **solutes**, substances that dissolve in it. Water is often called the **universal solvent** because of its ability to dissolve so many chemicals to form homogeneous mixtures called **solutions**. **Aqueous solutions** (water-based solutions) can have solutes that are gases, liquids, or solids. A special class of large-molar-mass molecules that finely disperse in solution are called colloids.

This water contains many dissolved salts, compounds, and molecules. It also contains particles too big to dissolve; these particles remain suspended and make the water cloudy.

Why Do So Many Substances Dissolve in Water?

We can answer the question that opens this section by looking at seawater, in which dissolved sodium chloride is one of the primary solutes. Chemically, we can understand what happens when salt dissolves in water by considering the process of dissolving as a series of discrete steps, as shown in Figure 11.26. Although the process does not, in fact, occur in steps, we may use this procedure to enhance our understanding of the dissolution process. At each of the steps along our imaginary solution-forming pathway, ask yourself, "Is this likely to be an endothermic or an exothermic process?"

■ *Solute separation:* In the first step, the salt must separate into sodium ions and chloride ions. The electrostatic forces that keep the ions together must be overcome, so energy is required. This part of the process is *endothermic* ($\Delta H = +$).

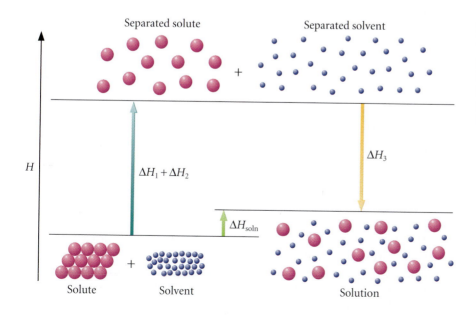

FIGURE 11.26

The processes necessary, along with their energy exchanges, for the dissolution of a salt, such as NaCl, in water. The overall change in energy can be $+$ or $-$, depending on the ion–dipole interaction.

FIGURE 11.27

The interaction of the sodium ion with the negative end of the dipole on water and that of the chloride ion with the positive end of the dipole on water contribute to the energetic stability that permits NaCl to dissolve in water. Note in the electrostatic potential maps how the water molecule's electron potential changes as it interacts with the ions.

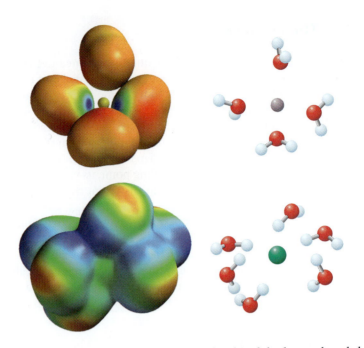

- *Solvent separation:* Energy is required to break hydrogen bonds holding water molecules together in order to accommodate the incoming ions from the solute. This part of the process is *endothermic* ($\Delta H = +$).

- *Electrostatic interaction between the solvent and solute ions:* Electrostatic attraction is the great stabilizer. As shown in Figure 11.27, sodium cations are attracted to the negatively charged dipole at oxygen on each water molecule. Conversely, chloride anions are attracted to the positively charged hydrogen poles. Although there is constant ion movement in the solution, at any one time it is typical to have up to six water molecules surrounding each ion. This **ion–dipole interaction** contributes to the solution process in two ways. First, for the process involving sodium chloride and water, the interaction is *highly exothermic* ($\Delta H =$ "−"). In order to have favorable formation of a solution, the value of the exothermic interaction of the solute and solvent is typically larger than the endothermic processes in the first two steps. Second, dissolving salt ions in water results in dispersal of the cations and anions throughout the solution. This ion–solvent interaction is called **solvation,** and when the solvent is water, it is known as **hydration**. The **heat of solution** ($\Delta_{sol}H$) can be viewed as the sum of the heats of solute separation, solvent separation, and ion–dipole interaction.

$$\Delta_{sol}H = \Delta_{solute}H + \Delta_{solvent}H + \Delta_{solvation}H$$

An *overall* exothermic heat of solution ($-\Delta_{sol}H$) favors solution formation but is *not* by itself an indicator of solubility. The additional dispersal of the ions in the solvent (see Chapter 14) is also important. For example, NaCl has a value of $\Delta_{sol}H = +3$ kJ/mol, yet it dissolves in water. So does potassium hydroxide, KOH ($\Delta_{sol}H = -57.6$ kJ/mol). In fact, the heat of solution of KOH is so large that adding 100 g to a liter of water will raise its temperature to nearly boiling, with $\Delta_{sol}H = -103$ kJ for the 100 g.

As a rule of thumb, *like dissolves like*. Polar molecules typically dissolve many salts via ion–dipole interactions. Similarly, polar liquids mix as a consequence of the stability gained from dipole–dipole and, when possible, hydrogen-bonding interactions. Nonpolar substances, such as the industrial solvents benzene (C_6H_6) and toluene (C_7H_8) are often **miscible** (mix in all proportions) with other nonpolar substances. Part of the reason lies with London forces and part with the stability gained by dispersion as the liquids mix.

EXERCISE 11.7 Predicting Solubility

Which of the following substances is (are) soluble in water: ethanol (C_2H_5OH), ethylene glycol ($C_2H_6O_2$), and cyclohexane (C_6H_{12})?

First Thoughts

The ability to form electrostatic attractions to water, including hydrogen bonds, dictates the solubility of a substance in water. Which of these compounds can have such attractions?

Solution

Ethanol and ethylene glycol are infinitely soluble in water. In fact, both can act as a solvent *or* solute with water. Cyclohexane is not appreciably soluble in water.

Further Insights

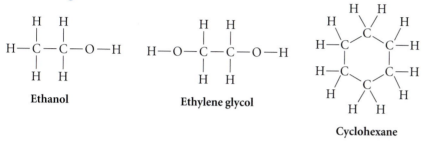

| Ethanol | Ethylene glycol |

Cyclohexane

Both benzene (C_6H_6) and toluene (C_7H_8) are excellent solvents for cyclohexane. Industrial manufacturing of polymer-based plastics often involves dissolving nonpolar solutes in nonpolar solvents.

PRACTICE 11.7

Which of the following substances is (are) soluble in hexane: water, carbon tetrachloride (CCl_4), and iodine crystals (I_2)?

See Problems 58–60 and 62.

The maximum amount of a solute that can dissolve in a solvent is its **solubility**, often given in grams of solute per 100 mL of solvent ($\frac{g\ solute}{100\ mL\ solvent}$). If we added any more solute, it would not dissolve. At that point, we say that the solution is **saturated** (compare this with the use of the term *saturated* for the vapor of the pure liquid). Sometimes the solubility of a substance in water depends on the relative size of the polar and nonpolar parts of a molecule of the substance. Table 11.3 shows that the series of alcohols gets progressively less soluble in water as the nonpolar end of each becomes larger. However, even substances that we

Visualization: Saturated Solution Equilibrium

Video Lesson: Types of Solutions

Methanol

Ethanol

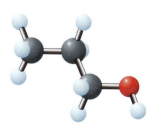

n-**Propanol**

TABLE 11.3	Solubility of Some Alcohol Compounds in Water	
Compound	**Name**	**Solubility (g/L)**
CH_3OH	Methyl alcohol (methanol)	Infinite
C_2H_5OH	Ethyl alcohol (ethanol)	Infinite
C_3H_7OH	Propyl alcohol (1-propanol)	Infinite
C_4H_9OH	Butyl alcohol (1-butanol)	79
$C_5H_{11}OH$	Pentyl alcohol (1-pentanol)	27
$C_6H_{13}OH$	Hexyl alcohol (1-hexanol)	5.9
$C_7H_{15}OH$	Heptyl alcohol (1-heptanol)	0.9
$C_{10}H_{21}OH$	Decyl alcohol (1-decanol)	<0.04

Source: Kelter, Carr, and Scott. *Chemistry: A World of Choices;* McGraw-Hill: New York, 2003.

classify as "insoluble" in water, our universal solvent, can dissolve at least a little. One such chemical is oxygen (O_2), and that is fortunate for fish and other sea life that depend on it. Another such chemical is carbon tetrachloride, a nonpolar industrial solvent; that is *not* so fortunate, because even in low concentrations in water, it is considered an environmental hazard.

11.7 Measures of Solution Concentration

We have said that seawater is largely an aqueous solution of dissolved sodium chloride. However, there are many other components of this solution, including a good deal of Mg^{2+}, SO_4^{2-}, and Ca^{2+}; smaller quantities of iron, phosphorus, and copper; and really small amounts of dissolved oxygen, cadmium, and even gold. What do we mean by "a good deal," "smaller quantities," and "really small amounts"? It depends on whom you ask. And "it depends" is too vague in a scientific community that requires clarity when communicating the results of measurements. We need descriptions of concentration that have consistent meaning to everyone reading the data.

Measures Based on Moles

Video Lesson: Molarity and the Mole Function

Molarity (*M*)

We first examined **molarity** in Chapter 4 as a measure of moles of solute per liter of solution.

$$M = \frac{\text{mol solute}}{\text{L solution}}$$

Molarity is a standard concentration unit in the chemical laboratory. We can speak of the *initial* molarity of a solute added to a solution, as in "What is the initial molarity of the sodium chloride?" We can also discuss the actual molarity of *each ion* after the solute has dissolved, as in "What is the molarity of the sodium ion in the solution?"

Video Lesson: Molality

Molality (*m*)

Molality is a measure of moles of solute per kilogram of *solvent*.

$$m = \frac{\text{mol solute}}{\text{kg solvent}}$$

The molality of a solute in solution is independent of temperature because it is based on *measuring the mass of the solvent, rather than the volume of the solution.* It is useful in exploring properties at a variety of temperatures, as we shall discuss in Section 11.8. And because we can measure mass accurately and precisely, molality can be determined to many significant figures, if necessary.

Mole Fraction (χ_i)

The **mole fraction** of a substance is the ratio of the number of moles of a substance present per total moles of all substances in the solution. If there are three solutes, i, j, and k, in aqueous solution, then the mole fraction of solute i is

$$\chi_i = \frac{\text{mol i}}{\text{mol i} + \text{mol j} + \text{mol k} + \text{mol water}}$$

It is important, when we calculate the mole fraction, to include the contribution of *all* components in the system. Note that the denominator in the equation indicates that we add the number of moles of i, j, k, and water to obtain the total number of moles in the system.

EXERCISE 11.8   **Practice with Mole-Based Units of Concentration**

Fructose is one of the three important "simple sugars," the other two being galactose and glucose. What are the values for molarity, molality, and mole fraction of 36.0 g of fructose ($C_6H_{12}O_6$) in 250.0 mL of a fruit-flavored drink? The density of the water-based drink is 1.05 g/mL, and you may neglect the presence of flavorings.

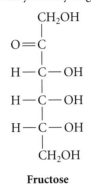

Fructose

Solution

The addition of fructose changes the density of the solution compared to pure water, so we would expect the molarity (which is based on the solution volume) and molality (based on the solvent mass) to be different. When we calculate the mole fraction, we must take into account both the moles of fructose and the moles of the solvent, water. All three measures of concentration require us to know the number of moles of fructose ($C_6H_{12}O_6$).

$$\text{mol fructose} = 36.0 \text{ g fructose} \times \frac{1 \text{ mol fructose}}{180.0 \text{ g fructose}} = 0.200 \text{ mol fructose}$$

$$\text{Molarity of fructose} = \frac{\text{mol fructose}}{\text{L solution}} = \frac{0.200 \text{ mol fructose}}{0.2500 \text{ L solution}} = 0.800 \text{ } M$$

The molality calculation requires that we know the mass of water. We can find this via the mass of the solution and its density.

$$\text{Mass of solution} = 250.0 \text{ mL solution} \times \frac{1.05 \text{ g solution}}{\text{mL solution}} = 262.5 \text{ g solution}$$

(Note that we keep the extra figure for this calculation and round only the number that will be our final answer.)

$$\text{Mass of water} = \text{mass of solution} - \text{mass of fructose} = 262.5 \text{ g} - 36.0 \text{ g}$$
$$= 226.5 \text{ g water}$$

$$\text{Molality} = \frac{\text{mol fructose}}{\text{kg solvent}} = \frac{0.200 \text{ mol fructose}}{0.2265 \text{ kg water}} = 0.883 \text{ } M$$

We can covert the mass of water to moles for our mole fraction calculation.

$$\text{mol water} = 226.5 \text{ g water} \times \frac{1 \text{ mol water}}{18.0 \text{ g water}} = 12.58 \text{ mol water}$$

$$\text{Mole fraction of fructose} = \chi_{\text{fructose}} = \frac{\text{mol fructose}}{\text{mol fructose} + \text{mol water}}$$

$$= \frac{0.200}{0.200 + 12.58} = 0.0156$$

PRACTICE 11.8

What mass of potassium hydroxide (KOH) is required to prepare 600.0 mL of a 1.40 M KOH solution? What is the mole fraction of water in this solution? (Assume that the density of the solution is 1.07 g/mL.)

See Problems 67, 70, 71, and 73.

Measures Based on Mass

We discussed the two useful sets of mass-based concentration measures, weight percent and **parts per million**, **parts per billion** and **parts per trillion**, in Section 4.2. We present them again here for purposes of review and for application in the context of our discussion of seawater.

Weight Percent (wt %)

Weight percent is a measure of the mass fraction of a substance in a solution, expressed as a percentage.

$$\text{wt \%} = \frac{\text{g substance}}{\text{g solution}} \times 100\%$$

If you were to prepare a reference solution that has the same weight percent of sodium chloride as seawater, it would contain 29.5 g of sodium chloride per 1.00×10^3 g of solution, a weight percent of 2.95% NaCl.

$$\frac{29.5 \text{ g NaCl}}{1.00 \times 10^3 \text{ g solution}} \times 100\% = 2.95\% \text{ NaCl}$$

Other related units are weight-to-volume and volume-to-volume.

Parts per Million, Billion, and Trillion (ppm, ppb, ppt)

 Application

In discussing the concentration of possible health hazards in water, the Environmental Protection Agency (EPA) cites its **maximum contaminant level** (MCL) the highest acceptable level in a solution, as 10 parts per million for nitrate and 5 parts per billion for cadmium. Very low solute concentrations (sometimes called trace concentrations) are often expressed this way. The density of very dilute aqueous solutions is close enough to that of pure water, 1.0 g/mL, that we may use these volume-based conversions that we derived in Section 4.2.

$$\text{ppm} = \frac{1 \text{ g solute}}{10^6 \text{ g solution}} \approx \frac{1 \text{ mg solute}}{\text{L solution}}$$

$$\text{ppb} = \frac{1 \text{ g solute}}{10^9 \text{ g solution}} \approx \frac{1 \text{ μg solute}}{\text{L solution}}$$

$$\text{ppt} = \frac{1 \text{ g solute}}{10^{12} \text{ g solution}} \approx \frac{1 \text{ ng solute}}{\text{L solution}}$$

The average concentration, in parts per million, of the major ions that are present in seawater are listed in Table 11.4.

TABLE 11.4	Average Composition of Major Ions in Seawater	
Element (main form in seawater)	**Parts per Million (mg/L)**
Chlorine (Cl^-)	19,000
Sodium (Na^+)	10,500
Magnesium (Mg^{2+})	1,250
Sulfur (SO_4^{2-})	900
Calcium (Ca^{2+})	400
Potassium (K^+)	380
Bromine (Br^-)	65
Bicarbonate (HCO_3^-)	30
Strontium (Sr^{2+})	12

Source: HPS Certified Reference Materials.

 Application

CHEMICAL ENCOUNTERS:
Composition of Seawater

EXERCISE 11.9 | **Conversion Between Mole and Mass Concentration Units**

The maximum concentration of O_2 in seawater is 2.2×10^{-4} M at 25°C. What is this concentration in parts per million of oxygen?

Solution

We can solve this via dimensional analysis, as follows.

$$\frac{2.2 \times 10^{-4} \text{ mol } O_2}{\text{L seawater}} \times \frac{32 \text{ g } O_2}{\text{mol } O_2} \times \frac{1000 \text{ mg } O_2}{1 \text{ g } O_2} = 7.0 \text{ ppm } O_2$$

Does the answer make sense? We expect the solubility of (nonpolar) oxygen to be quite low in water, and our answer confirms that.

PRACTICE 11.9

The MCL for arsenic in drinking water is 10 ppb, according to Environmental Protection Agency guidelines. Convert this value to molarity.

See Problems 74–76, 79, 80, and 111.

11.8 The Effect of Temperature and Pressure on Solubility

Video Lesson: Temperature Change and Solubility

Temperature Effects

We saw in Exercise 11.9 that oxygen is nearly insoluble in seawater, yet 7.0 ppm at 25°C is still enough to allow the seas to teem with life. The concentration of oxygen in the seas and in freshwater lakes, ponds, and rivers varies as natural processes such as photosynthesis and respiration cycle oxygen into and out of the water. The other important factor that determines oxygen's solubility in water (see Table 11.5) is *temperature*. Note the trend, followed by all gases, that *solubility of a gas decreases with temperature.* Figure 11.28 shows this behavior for several common gases.

Small changes in temperature do not dramatically affect the solubility of oxygen, but a large increase in temperature can significantly lower the oxygen concentration in a waterway, and the harm done can be felt throughout the aquatic food chain. The artificial raising of the ambient water temperature is called **thermal pollution** and is of concern in the design of nuclear power plants, in which river water is used to cool the nuclear core of the reactor (see Chapter 21). The heated river water is passed through a **cooling tower** (Figure 11.29, on page 472) before it flows back to the river.

Application

CHEMICAL ENCOUNTERS: Impact of the Solubility of Oxygen in Fresh Water

FIGURE 11.28

The solubility of gases decreases with temperature. This is especially important with O_2, where "thermal pollution" can have important consequences for the aquatic food chain.

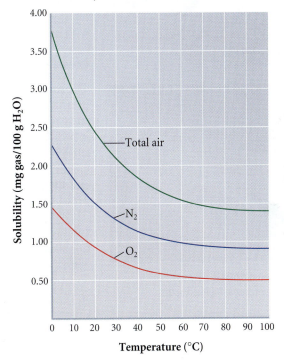

TABLE 11.5	Solubility of O_2 in Fresh Water at Several Temperatures (in air at 1 atm)
Temperature (°C)	**Parts per Million (mg/L)**
0	14.6
5	13.1
10	11.3
15	10.1
20	9.1
25	8.3
30	7.6
35	6.9

FIGURE 11.29

A cooling tower lowers the temperature of water that has been heated as part of the operation of power plants. This water is close to the temperature of the waterway from which it was taken and to which it will be returned, so the impact of thermal pollution is minimized.

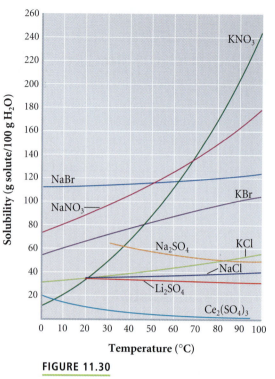

FIGURE 11.30

The solubility of some salts in water as a function of temperature. Note that the solubility of the majority of these salts increases as the temperature increases. Li_2SO_4 decreases solubility as the temperature increases.

In contrast to gases, the solubility of ionic solids in water generally *increases* with temperature, as shown in Figure 11.30. For example, the solubility of sodium chloride in water at 0°C is 35.7 g/100 mL, and at 100°C it is 39.1 g/100 mL. Some solids show a much more marked increase in solubility. Sodium carbonate decahydrate ($Na_2CO_3 \cdot 10H_2O$) has an aqueous solubility of 22 g/mL at 0°C to 420 g/mL at 100°C, a 22-fold increase. A few salts, such as manganese(II) sulfate hexahydrate ($MnSO_4 \cdot 6H_2O$) and sodium sulfate (Na_2SO_4), have lower solubility with increasing temperature.

 Application

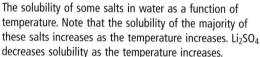
Video Lesson: Pressure Change and Solubility

Pressure Effects

The solubilities of solid solutes in water are not especially responsive to modest pressure changes. The solubility of gases in water, on the other hand, is quite sensitive to the external pressure. This is important in the preparation of soft drinks, in which CO_2 is combined with water, sweetener, and other flavorings at pressures between about 6 and 15 atm. The "fizz" in the soda is due to the dissolved CO_2. When the can is opened, the CO_2 above the soda escapes from the container, and the pressure of the can suddenly drops to atmospheric pressure. The lower pressure decreases the solubility of the CO_2 in the soda, and bubbles of CO_2 form.

William Henry (1775–1836), an English chemist, noted this behavior. **Henry's law** says that at constant temperature, the solubility of a gas is directly proportional to the pressure that the gas exerts above the solution.

$$P_{gas} = k_{gas}C_{gas}$$

Carbon dioxide is released from a glass of freshly opened soda.

where P_{gas} = the pressure of the gas above the solution
C_{gas} = the concentration of the gas in the solution
k_{gas} = a constant relating the gas pressure above the solution and its concentration

In pure water at 1 atm and 0°C, the solubility of CO_2 is 3.48 g/L. According to Henry's law, if we triple the CO_2 pressure, the solubility will roughly triple. Overall, this law holds best for gases such as N_2 and O_2 that do not significantly interact with the solvent. It does not hold well for HCl, which ionizes in water to form the hydrated ions $H^+(aq)$ and $Cl^-(aq)$. However, as we have noted, even molecules such as O_2 interact with water to some degree.

EXERCISE 11.10   Henry's Law and CO_2 in Soda

In air in which the pressure of CO_2 is 3.4×10^{-4} atm at 0°C, the solubility of CO_2 in water is 1.18×10^{-3} g/L water. If the pressure of the CO_2 above the water is increased to 6.00 atm, what will be the solubility of the CO_2 in the water? Do these data lead to a conclusion consistent with our prior assertion that sodas go "whoosh" when they are opened?

Solution

The CO_2 pressure is being increased by a huge factor, so we would expect the solubility to increase about proportionately. Assuming proportionality, we can eliminate the need to calculate the Henry's law constant and just solve the problem with ratios.

$$\frac{1.18 \times 10^{-3} \text{ g/L}}{3.4 \times 10^{-4} \text{ atm}} = \frac{x \text{ g/L}}{6.00 \text{ atm}}$$

$$x = 21 \text{ g/L}$$

The solubility of the gas increased sharply at high external CO_2 pressure, in accordance with Henry's law. As is consistent with much of our discussion on solutions, the formulas we cite work best for very dilute solutions. When the can of soda is opened, the pressure of CO_2 above the liquid sharply decreases, lowering the solubility of the gas and therefore contributing to the escape of gas that we hear when we open the soda.

PRACTICE 11.10

A researcher adds 22.7 g of NaCl to 55.0 mL of water. Examine Figure 11.30. Will all of the salt dissolve in the given amount of water at 25°C? If not, how much water will need to be added to just dissolve the salt completely?

See Problems 83 and 84.

Application

11.9   Colligative Properties

In the dead of winter at Lake Riley in Minnesota, it is common to see people brave the harsh winter weather (Figure 11.31) to sit around meter-deep fishing holes in the ice. The ice fishing season here typically runs from late January through mid-March. We take it for granted, but ice fishing is possible only because the water in the freshwater lake has a relatively low concentration of dissolved solutes and has a freezing point just below 0°C. If Lake Riley were as salty as the Red Sea or the Persian Gulf, with sodium and chloride ion concentrations equal to 40 parts per thousand, its freezing point would be about −2.5°C, and the ice fishing season would be shorter (if not nonexistent), as anglers waited for sufficient ice to form to make the lake safe for the fishing shacks and vehicles that transport them. Why does the presence of the salt lower the freezing point of water? Does the amount of salt affect the freezing point, and does the nature of the salt matter? Are there any other solution properties that are affected in this way?

FIGURE 11.31

Ice fishing on Lake Riley in Minnesota can be rewarding.

Let's answer the last question first. Properties of a solution that approximately depend only on the number of nonvolatile solute particles, irrespective of their nature, are called **colligative properties** (from the Latin *colligatus,* which means "collected together"). There are four useful colligative properties: vapor pressure lowering, freezing-point depression, boiling-point elevation, and osmotic pressure. Understanding vapor pressure lowering will help us answer our questions about dissolving salts in water, as well as give us insight into the other three colligative properties.

Vapor Pressure Lowering

Visualization: Vapor Pressure Lowering: Liquid/Vapor Equilibrium

Visualization: Vapor Pressure Lowering: Addition of a Solute

Visualization: Vapor Pressure Lowering: Solution/Vapor Equilibrium

Video Lesson: Vapor Pressure Lowering

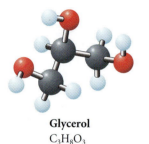

Glycerol
$C_3H_8O_3$

1,2,3-Propanetriol ($C_3H_8O_3$) is the systematic name for the nonvolatile substance we commonly call glycerol or glycerin. The colorless, viscous liquid is used as a lubricant and moistener, especially in cosmetics, and to reduce swelling in medical procedures, such as eye examinations. The presence of three OH groups on the molecule leads to significant hydrogen bonding, making glycerol completely soluble in water. We noted in Section 11.3 that water has a vapor pressure equal to 23.8 torr at 25°C. (Judging on the basis of our discussion in that section about intermolecular forces, structure, and vapor pressure, **does it make sense that water should be volatile, whereas glycerol is nonvolatile?**) Glycerol has essentially no vapor pressure at room temperature. When glycerol and water are mixed, the total vapor pressure of the resulting solution is *dependent only on the vapor pressure of pure water,* $P°_{H_2O}$, *multiplied by its mole fraction,* χ_{H_2O}, *in the solution.*

$$\text{Vapor pressure of the solution} = P_{solution} = \chi_{H_2O} P°_{H_2O}$$

For example, if we add enough glycerol to water so that the mole fraction of the water is reduced to 0.900, the resulting vapor pressure of the solution will be reduced. At 25°C, the vapor pressure of the solution would be

$$P_{solution} = \chi_{H_2O} P°_{H_2O}$$

$$P_{solution} = 0.900 \times 23.8 \text{ torr} = 21.4 \text{ torr}$$

The relationship of the vapor pressure of the solution, $P_{solution}$, to the mole fraction, $\chi_{solvent}$, and vapor pressure, $P°_{solvent}$, of the volatile solvent holds true for any *ideal* solution containing a nonvolatile solute. It is known as **Raoult's law**, named after the French chemist Francois-Marie Raoult (1830–1901).

$$P_{solution} = \chi_{solvent} P°_{solvent}$$

An **ideal solution** exists when the properties of the solute and solvent are not changed by dilution. This means that other than being diluted, combining solute and solvent in an ideal solution does not release or absorb heat, and the total volume in the solution is the sum of the volumes of the solute and solvent. Only very dilute solutions approach ideal behavior, so although Raoult's law is a good first approximation, actual measurements are required to properly describe vapor pressure changes in mixtures of solutions. Figure 11.32 shows the general trend: The vapor pressure is depressed with the addition of a nonvolatile solute.

FIGURE 11.32

The vapor pressure of water (red line) is lowered by the addition of a nonvolatile solute. This is described for an ideal solute by Raoult's law.

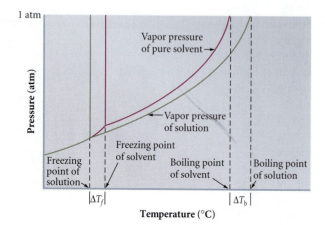

EXERCISE 11.11 **Vapor Pressure Lowering**

You stir 3 teaspoons (45.0 g) of sucrose, table sugar ($C_{12}H_{22}O_{11}$, molar mass = 342 g/mol), into a cup of tea containing 250.0 mL of water at 90.0°C (density = 0.965 g/mL). What is the new vapor pressure of the solution? $P°_{H_2O}$ = 526 torr at 90.0°C.

First Thoughts

The key problem-solving hurdle is calculating the mole fraction of water in the solution. To do this, we must calculate the number of moles of each component. We will retain an extra, nonsignificant figure until the end of the calculations.

Solution

$$\text{mol sucrose} = 45.0 \text{ g} \times \frac{1 \text{ mol}}{342 \text{ g}} = 0.1316 \text{ mol sucrose}$$

$$\text{g water} = 250.0 \text{ mL} \times \frac{0.965 \text{ g water}}{1 \text{ mL water}} \times \frac{1 \text{ mol water}}{18.02 \text{ g water}} = 13.39 \text{ mol water}$$

$$\chi_{\text{water}} = \frac{\text{mol water}}{\text{mol glucose} + \text{mol water}} = \frac{13.39}{0.1316 + 13.39} = 0.9903$$

$$P_{\text{solution}} = \chi_{\text{water}} P°_{\text{water}}$$

$$P_{\text{solution}} = 0.9903 \times 526 \text{ torr} = 520.9 \text{ torr} \approx 5.2 \times 10^2 \text{ torr}$$

Further Insights

As expected, the vapor pressure of the solvent decreases as a consequence of the addition of the sucrose. Note the use of the symbol $\approx$, which means "is approximately equal to." Raoult's law is strictly followed only for ideal solutions. This solution has enough sucrose in it that it does not approach ideal behavior.

PRACTICE 11.11

The vapor pressure of ethanol (C_2H_5OH) at 40°C is 135.3 torr. Calculate the vapor pressure of a solution containing 26.8 g of glycerin ($C_3H_8O_3$) a nonvolatile solute, in 127.9 g of ethanol.

See Problems 87, 88, 93, 95, and 96.

Boiling-Point Elevation

At 1 atm, the temperature of any solution must be elevated until its vapor pressure equals 760 torr in order to reach the boiling point. Concentrated solutions possess lowered vapor pressures (which we discussed earlier), because the solution must have a vapor pressure equal to the atmospheric pressure in order to boil. The more solute is dissolved in a solution, the more the vapor pressure of the solution is lowered, and the higher the boiling point. The temperature of the solution must be elevated for the vapor pressure to reach that of the surroundings. For fairly dilute solutions of nonelectrolytes, the following formula *approximately* describes the **boiling-point elevation** due to the addition of a nonvolatile solute.

$$\Delta T_b = K_b m$$

where ΔT_b = the change in boiling point in °C
 K_b = the boiling-point elevation constant, which depends on the solvent, in units of °C/m
 m = the molality of the solute in moles of solute per kilogram of solvent

Visualization: Boiling-Point Elevation: Liquid/Vapor Equilibrium

Visualization: Boiling-Point Elevation: Addition of a Solute

Visualization: Boiling-Point Elevation: Solution/Vapor Equilibrium

Video Lesson: Boiling-Point Elevation and Freezing-Point Depression

TABLE 11.6	Boiling-Point Elevation Constants for Several Liquids	
Solvent	K_b (°C/m)	T_b (°C)
Acetone	1.7	56.5
Benzene	2.6	80.1
Carbon tetrachloride	5.0	76.7
Ethanol (ethyl alcohol)	1.2	78.5
Methanol (methyl alcohol)	0.80	64.7
Water	0.51	100.0

The value of K_b for water is 0.512°C/m. This means that a 2.00 molal aqueous sugar solution would have a boiling-point elevation of approximately 1.02°C (2.00 m × 0.512°C/m) and a boiling point of about 101°C. We say "about" because the boiling-point elevation constant begins to deviate significantly from the ideal solution value at higher concentrations. Boiling-point elevation constants for several liquids are listed in Table 11.6.

Colligative properties depend on the *number*, not the *nature*, of the particles in the solution. We would expect solutions containing compounds at the same concentration to have the same elevated boiling point. However, we must consider the total number of particles that result from making the solution. For example, what would happen to the boiling point of our aqueous sucrose solution if our solute were 2.00 molal cobalt(II) chloride ($CoCl_2$) instead of sucrose? Because cobalt(II) chloride is a strong electrolyte that dissociates to form cobalt ion and chloride ion,

$$CoCl_2(s) \xrightarrow{H_2O} Co^{2+}(aq) + 2Cl^-(aq)$$

we would expect three moles of particles (ions, in this case) for every mole of $CoCl_2$ added to the solution. That is, the concentration of ions in the solution should be 6.00 molal, and the boiling point of the solution should be raised by (6.00 m × 0.512°C/m) = 3.07°C. This suggests that when dealing with strong electrolytes, we can modify our boiling-point elevation formula to take into account the dissociation of strong electrolytes into i particles.

$$\Delta T_b = iK_b m$$

For $CoCl_2$, $i = 3$ if the solution behaves ideally. The actual boiling-point elevation for this solution is 4.6°C, not 3.1°C, which tells us that at this relatively high concentration (2.00 m), the solution does *not* behave even close to ideally. The value i is known as the **van't Hoff factor**, after J. R. van't Hoff (1852–1911), a chemist from the Netherlands who suggested its use in the 1880s. Table 11.7 shows the van't Hoff factors for several electrolytes. Note that there is *significant deviation* from the expected values as the solute concentration increases, so results are only approximate even at relatively low concentrations.

Jacobus Henricus van't Hoff (1852–1911) was a Dutch chemist who initially shook the basic ideas of chemistry with his description of the three-dimensional nature of molecules. In 1901 he won the first Nobel Prize in chemistry for his work on solutions.

Video Lesson: Boiling-Point Elevation Problem

EXERCISE 11.12 Boiling-Point Elevation

Recipes for cooking spaghetti often call for putting a little table salt in the water before boiling it and adding the spaghetti. Does this help the spaghetti cook faster? Assume that we add 10.0 g of NaCl to 6.00 L of water and that the density of the solution is 1.00 g/mL. Also assume that this very dilute solution behaves ideally, so $i = 2$. The value of K_b for water is 0.512°C/m.

Solution

$$\text{mol NaCl} = 10.0 \text{ g NaCl} \times \frac{1 \text{ mol NaCl}}{58.5 \text{ g NaCl}} = 0.171 \text{ mol NaCl}$$

$$\text{kg water} = 6.00 \text{ L water} \times \frac{1.00 \text{ kg water}}{\text{L water}} = 6.00 \text{ kg water}$$

$$\text{Molality of NaCl} = \frac{\text{mol NaCl}}{\text{kg water}} = \frac{0.171 \text{ mol NaCl}}{6.00 \text{ kg water}} = 0.0285 \text{ } m \text{ NaCl}$$

$$\Delta T_b = iK_b m$$

$$= 2.0 \times \left(\frac{0.512°C}{m}\right) \times 0.0285 \text{ } m$$

$$\Delta T_b = 0.03°C$$

TABLE 11.7	van't Hoff Factors for Several Electrolytes						
Compound	Expected Value of i	$m = 0.005$	$m = 0.01$	$m = 0.05$	$m = 0.10$	$m = 0.20$	$m = 1.00$	$m = 2.00$
HCl	2	1.95	1.94	1.90	1.89	1.90	2.12	2.38
NH_4Cl	2	1.95	1.92	1.88	1.85	1.82	1.79	1.80
$CuSO_4$	2	1.54	1.45	1.22	1.12	1.03	0.93	—
$CoCl_2$	3	2.80	2.75	2.64	2.62	2.66	3.40	4.58
K_2SO_4	3	2.77	2.70	2.45	2.32	2.17	—	—

This shows that there is nearly no elevation in the boiling point of water with the addition of a little table salt. The salt is added for taste.

Video Lesson: Colligative Properties of Ionic Solutions

PRACTICE 11.12

What is the predicted boiling point of each of these solutions? (Assume that each follows ideal behavior.)

a. 1.00 m NaCl in water

b. 0.35 m $FeCl_3$ in water

c. 1.50 m KCl in methanol

See Problems 89, 90, and 97.

In contrast to the minimal impact of adding a dash of table salt to water when cooking spaghetti, making an aqueous solution that is 40% by volume ethylene glycol ($C_2H_6O_2$) elevates the boiling point to 105°C (221°F). This solution, called "antifreeze," helps protect your automobile engine in hot weather.

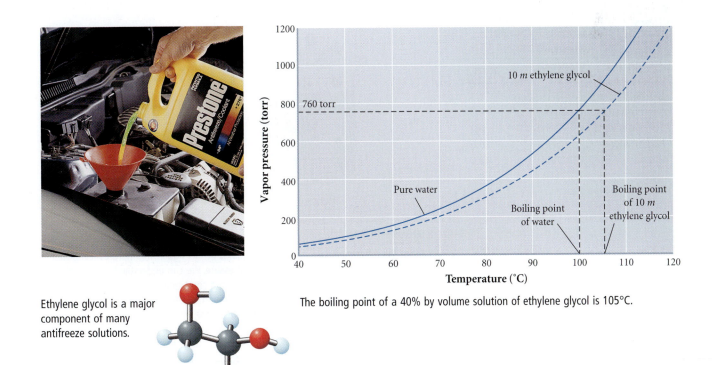

Ethylene glycol is a major component of many antifreeze solutions.

The boiling point of a 40% by volume solution of ethylene glycol is 105°C.

Visualization: Freezing-Point
Depression: Solid/Liquid
Equilibrium

Visualization: Freezing-Point
Depression: Addition of a Solute

Visualization: Freezing-Point
Depression: Solid/Solution
Equilibrium

Freezing-Point Depression

The same antifreeze solution that raises the boiling point does double duty, because it also lowers the freezing point by about 18°C. This **freezing-point depression** is another of the colligative properties and, like boiling point elevation, is approximately proportional to the molal concentration of the solute:

$$\Delta T_f = i K_f m$$

where ΔT_f = the change in freezing point in °C

K_f = the freezing-point depression constant, which depends on the solvent, in units of °C/m

m = the molality of the solute in moles of solute per kilogram of solvent

i = the van't Hoff factor

Raoult himself, in 1883, was the first to note that the lowering of the freezing point was the same for any nonelectrolyte solute in a given solvent. Table 11.8 lists freezing-point depression constants for several liquids. Water has a relatively low value of K_f; freezing-point depressions are much greater in other solvents. Historically, the freezing-point depression was used to determine the molar masses of substances.

Why are freezing points depressed whereas boiling points are elevated? As with all colligative properties, the key is the vapor pressure. In Figure 11.33, the solution has a lower vapor pressure than the pure solvent. When the solvent freezes, its vapor pressure must be lowered to equal that of the solution. In order to reach that vapor pressure, the solution must be cooled below the freezing point of the pure solvent. The result is a new freezing point that is lower than the freezing point of the solvent.

One important consequence of freezing-point depression is the need to keep food freezers well below 0°C, because the food within contains all manner of solutes, especially various salts and sugars, so the freezing points of products such as meats and ice cream are significantly less than 0°C.

TABLE 11.8	Freezing-Point Depression Constants for Several Compounds	
Solvent	K_f (°C/m)	T_f (°C)
Benzene	5.1	5.5
Cyclohexane	20.0	6.5
Formic acid	2.8	8.4
Naphthalene	6.9	80.0
Phenol	7.4	43.0
Water	1.86	0

Pure solvent   Solvent with
                      nonvolatile solute

FIGURE 11.33

Note the differences between the two liquids in this figure. The more concentrated solution on the right has a lower vapor pressure. We can understand this by spraying a little perfume at the front of a room. In a short time, everyone in the room will smell the perfume. Why is this so? There is a natural tendency for things to become more dilute. The perfume vapors spread out across the room as they become more dilute. The same situation occurs in vapor pressure lowering. If solvent molecules left the container on the right, the solution would become more concentrated, which would violate the natural tendency of the solution to become more dilute. Therefore, fewer molecules leave the container on the right.

Osmosis

Our final colligative property is osmosis (from the Greek *osmos*, meaning "to thrust"). Recalling the question with which we opened the chapter (How do we meet the long-term water needs of rapidly growing communities?), we now have the payoff of our discussion.

To answer the question, let's set up the following experiment in the same manner as experiments that were done as early as the year 1748. A solution of salt water is placed in a two-chambered container separated by a **semipermeable membrane** from a sample of pure water, as shown in Figure 11.34. The membrane has very small holes in it so that small water molecules can travel through, but not larger solute particles. Why won't small ions, such as the potassium cation, travel through the semipermeable membrane? Although the holes are bigger than a potassium cation (so that the water molecules can go through), they are smaller than the entire hydrated ion of potassium. Remember that solutes dissolved in water contain a sphere of hydration that surrounds the solute and helps keep it dissolved. The sphere is typically much larger than a hole the size of a water molecule.

What will happen in our experiment? According to Raoult's law, the vapor pressure of the saltwater solution will be lower than that of the pure water. In an attempt to equalize the vapor pressures, water will flow through the membrane to the solution side, diluting it and making the vapor pressures equal. In fact, water will keep flowing until the weight of the solution (the solution's "hydrostatic pressure") becomes so great that it stops the flow (see Figure 11.34). Water can travel through the membrane in both directions, but the drive toward the most energetically stable equilibrium position makes the overall direction of flow toward dilution. This process is known as **osmosis**.

If we run the experiment again, but this time apply pressure to the saltwater solution in an amount that just prevents the osmosis, there will be no net change, as shown in Figure 11.35. The pressure that must be applied on the solution side to prevent osmosis of the solvent into it is called the **osmotic pressure**, Π. In 1887, van't Hoff determined that osmotic pressure is proportional to the temperature and the molarity of the solution:

$$\Pi = iMRT$$

where Π = osmotic pressure, in atmospheres
  M = molarity
  T = temperature in kelvins
  i = the van't Hoff factor
  R = the universal gas constant, 0.08206 L·atm·mol^{-1}·K^{-1}

Note that this colligative property uses molarity, not molality, to express the concentration of solute in the solution. Additionally, the gas constant R is the same one we introduced in Chapter 10 when we discussed the behavior of gases.

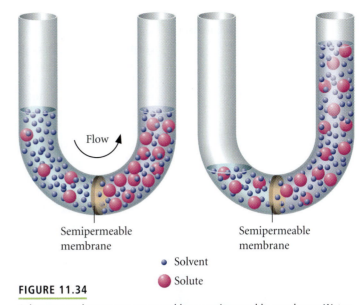

Semipermeable membrane   Semipermeable membrane

• Solvent
• Solute

FIGURE 11.34

Salt water and water are separated by a semipermeable membrane. Water molecules pass through, with more going from the purer water (on the left) to the more concentrated solution (on the right) than the other way, therefore raising the solution level. The pressure necessary to prevent this shift in solution level via osmosis is called osmotic pressure.

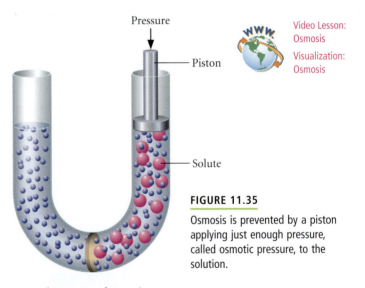

Pressure

Piston

Solute

Video Lesson:
Osmosis

Visualization:
Osmosis

FIGURE 11.35

Osmosis is prevented by a piston applying just enough pressure, called osmotic pressure, to the solution.

Application

The original measurements of osmotic pressure used animal membranes, with limited success. More reliable membranes were developed in the 1860s, incorporating a film of copper(II) ferrocyanide, $Cu_2Fe(CN)_6$, on a porous tube, and for decades, many quantitative measurements up to several hundred atmospheres were made in this way. As with our other colligative properties, we are mindful that the osmotic pressure equation holds well only for very dilute solutions. But with a dilute solution, one can measure the molar mass of very large polymers via the osmotic pressure experiment. Descriptive applications of this process are also very meaningful.

One example relates to red blood cells, which have a salt concentration equal to their surrounding fluid (which identifies the solutions as **isotonic**). If an intravenous solution given to a patient has a higher solute concentration than blood plasma (and is called a **hypertonic** solution), water will leave the blood cell membranes by osmosis. Hypertonic solutions are sometimes intentionally administered to patients when they have too much water in their blood plasma, such as in water intoxication, liver disease, or congestive heart failure. If the blood plasma is too concentrated, as happens in severe diarrhea or excess sweating, a **hypotonic** solution (a solution that has a lower solute concentration than blood plasma) is needed so that water will flow into the cells. Many intravenous solutions given to patients in hospitals are isotonic with blood serum. Because these isotonic solutions have the same solute concentration as blood serum, they will not cause cells in the patient's body to shrink or swell. A solution **isosmotic** with blood plasma (same osmotic pressure) is 0.154 M each in Na^+ and Cl^- ions.

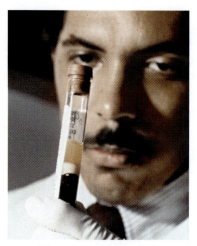

Many different ions are dissolved in blood serum, the light yellow solution that can be isolated by centrifuging whole blood.

EXERCISE 11.13   Osmosis and You

What is the osmotic pressure of a solution at 25°C that is isosmotic with blood plasma, 0.154 M NaCl?

Solution

Sodium chloride dissociates into two ions, Na^+ and Cl^-, when dissolved in water. That means that the van't Hoff factor for this solution should be equal to 2.0.

$$\Pi = iMRT$$
$$= 2.0 \times 0.154\ M \times 0.08206\ \text{L·atm·mol}^{-1}\text{·K}^{-1} \times 298\ \text{K}$$
$$= 7.532\ \text{atm} = 7.53\ \text{atm}$$

Osmosis is a fairly powerful property of solutions. If we placed this solution in an osmotic pressure apparatus with pure water, we would need 7.53 atm (110 psi) of pressure just to stop the osmosis.

PRACTICE 11.13

What is the osmotic pressure of each of the these solutions at 298 K?

  a. 0.100 M sucrose   b. 0.375 M CaCl$_2$ c. 1.250 M (NH$_4$)$_2$SO$_4$

See Problems 94, 101, and 102.

Reverse Osmosis

 Application

CHEMICAL
ENCOUNTERS:
Meeting
Municipal
Water Needs

What would happen if our osmosis experiment were run in reverse? Instead of using osmotic pressure merely to prevent the natural flow from the water to the saltwater side, let's apply *additional* pressure to actually push water from the saltwater solution through the semipermeable membrane, in opposition to the natural tendency toward dilution by osmosis. What will happen? The solution is losing solvent, so it will become more concentrated (saltier). Water is pushed through the membrane, so the amount of pure water increases. We have used **reverse osmosis** to desalinate water, producing more fresh water.

TABLE 11.9	Applications of Reverse Osmosis
Industry	**Application**
Cosmetics	Product preparation
Desalination	Potable water, beverage preparation
Drinking water	Mineral removal
Electronics	Water low in impurities for manufacturing processes
Food	Low sodium and organic chemicals in food preparation
Laboratories	Rinsing glassware
Pharmaceutical	Pure water for large-scale production
Restaurants	Spot-free rinses, drinking water

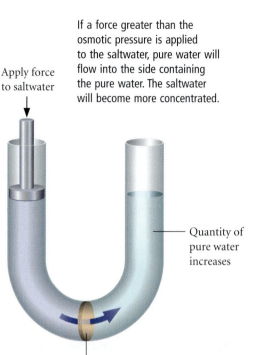

If a force greater than the osmotic pressure is applied to the saltwater, pure water will flow into the side containing the pure water. The saltwater will become more concentrated.

Apply force to saltwater

Quantity of pure water increases

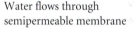

Water flows through semipermeable membrane

Back to the Future

After quite a long discussion on many aspects of water as a liquid and a solvent, we now answer the third and final question with which we started the chapter: "How can we use these properties to help in the search for clean water in places like Tampa Bay?" An important part of the answer is "by reverse osmosis." This will help meet Tampa Bay's burgeoning water demand by supplying nearly 25 million gallons per day (mgd) of desalinated water, according to the "Master Water Plan" approved by the Tampa Bay Water Authority. The desalinating plant opened in January 2003, although its operation has been sporadic and not without controversy, including the debated hazard of putting water with a relatively high salt concentration back into Tampa Bay. (Food for thought: **Will it harm the ecosystem?**) To date, other U.S. municipal desalination plants have been used only in water emergencies. They include a 2-mgd plant in Key West, Florida; a 1-mgd facility on Marathon Key, Florida; and a 6.7-mgd plant in Santa Barbara, California.

But the process doesn't have to be used on such a large scale in order to be useful. Reverse osmosis units find applications in manufacturing, food processing, printing, and even car washes, as shown in Table 11.9. The use of long-lasting nylon or polyamide semipermeable membranes gives the factories, such as the one shown in Figure 11.36, the ability to filter up to 120,000 gals per day. Usable water, this precious resource, is becoming ever more scarce. It is through our understanding of its structure and its properties that we can create safe, long-term solutions that will keep fresh water flowing.

FIGURE 11.36

Reverse osmosis units can filter up to 120,000 gal of salt water per day. Such units are likely to play an increasingly important role in supplying growing communities with fresh water. This plant in Saudi Arabia contains many reverse osmosis units and produces 210 million gal of fresh water each day for the kingdom's capital city of Riyadh.

A technician inspects the reverse osmosis equipment at a desalination plant in Catalina, California.

Drinking water can be purified by reverse osmosis.

The Bottom Line

- Water and other liquids have in common intermolecular forces called van der Waals forces that hold their molecules together. These forces are weaker than covalent or ionic bonds. (Section 11.2)

- Among these forces is dipole–dipole interactions. (Section 11.2)

- A hydrogen bond is an especially strong interaction that occurs when a hydrogen attached to an oxygen, nitrogen, or fluorine atom is close to a different atom of oxygen, nitrogen, or fluorine. (Section 11.2)

- Weaker, but collectively important intermolecular interactions in large molecules are known as London forces. (Section 11.2)

- Water is special among liquids because of intermolecular hydrogen bonds. (Section 11.3)

- Intermolecular interactions can be used to explain many physical properties of water, including its viscosity, surface tension, and capillary action. (Section 11.5)

- The pressure exerted by the evaporation of a liquid in equilibrium with its surroundings is called its vapor pressure. The boiling point is reached when the vapor pressure is equal to the surrounding pressure. (Section 11.3)

- A heating curve describes the changes in phase and temperature that occur as a substance is heated at constant pressure. (Section 11.3)

- A phase diagram describes the changes in phase as pressure and temperature are changed. (Section 11.4)

- Water is known as the universal solvent because of its ability to form solutions with many substances. (Section 11.6)

- Solution formation can be understood in terms of specific types of energy changes. (Section 11.6)

- Solution concentration can be expressed in a variety of units, including molarity, molality, and parts per million. (Section 11.7)

- Raoult's law describes the change in vapor pressure of a solvent as solute is added to a solution. (Section 11.9)

- Solubility is affected by pressure and temperature in, as a first approximation, predictable ways. (Section 11.8)

- Colligative properties are based on the number of particles in a solution and can be understood in terms of Raoult's law. (Section 11.9)

- Colligative properties include vapor pressure lowering, boiling-point elevation, freezing-point depression, and osmosis. (Section 11.9)

- It is possible that reverse osmosis will become a vital process for supplying clean water to large cities. (Section 11.9)

Key Words

adhesion The interaction of a compound with the surface of another compound, such as the interaction of water with the surface of a drinking glass. (*p. 463*)

aqueous solution A homogeneous mixture of a solute dissolved in the solvent water. (*p. 465*)

boiling The process that occurs when the vapor pressure of a liquid is equal to the external vapor pressure. (*p. 455*)

boiling point The temperature at which the pressure of a liquid's vapor is equal to the surrounding pressure. (*p. 455*)

boiling-point elevation The increase in the boiling point due to the presence of a dissolved solute. (*p. 475*)

buret A piece of laboratory glassware used to measure volume and to add discrete amounts of solution in a repetitive fashion. (*p. 463*)

capillary action The effect seen when a liquid rises within a narrow tube as a consequence of adhesive forces being stronger than cohesive forces. (*p. 463*)

cohesive forces The collection of intermolecular forces that hold a compound within its particular phase. (*p. 463*)

colligative properties Physical properties of a solution that depend only on the amount of the solute and not on its identity, including boiling-point elevation, freezing-point depression, vapor pressure lowering, and osmotic pressure. (*p. 474*)

condensation The process of a gas undergoing a phase transition to a liquid; the opposite of evaporation. (*p. 452*)

cooling tower A large industrial apparatus used to condense gaseous vapor into liquid or to cool water that has been heated before it is discharged. (*p. 471*)

critical point The temperature and pressure at which the physical properties of the liquid phase and vapor phase of a substance become indistinguishable. (*p. 460*)

critical pressure The highest pressure at which a liquid and gas coexist in equilibrium. (*p. 460*)

critical temperature The highest temperature at which a liquid and a gas coexist in equilibrium. (*p. 460*)

deposition The process of a gas undergoing a phase transition to a solid; the opposite of sublimation. (*p. 452*)

dynamic equilibrium A term sometimes used in chemistry as synonymous with *equilibrium* to emphasize that molecular-level equilibrium is not static. (*p. 453*)

equilibrium In a reaction or process, a condition wherein the rates of the forward and backward reactions are equal. The amounts of the reactants and products do not change, but the forward and backward reactions are still proceeding. (*p. 453*)

equilibrium vapor pressure The pressure exerted by the vapor of a liquid or solid under equilibrium conditions. (*p. 453*)

evaporation The process of a liquid undergoing a phase transition to a gas; the opposite of condensation. (*p. 452*)

freezing-point depression The lowering of the melting point of a compound due to the presence of a dissolved solute. (*p. 478*)

gas The phase of a substance characterized by widely spaced components exhibiting low density, ease of flow, and the ability to occupy an enclosed space in its entirety. (*p. 443*)

heat of fusion ($\Delta_{fus}H$) The enthalpy change associated with the phase transition from liquid to solid. (*p. 457*)

heat of solution ($\Delta_{sol}H$) The enthalpy change associated with the dissolution of a solute into a solvent. (*p. 466*)

heat of vaporization ($\Delta_{vap}H$) The enthalpy change associated with the phase transition from a liquid to a gas. (*p. 457*)

heating curve A plot of the temperature of a compound versus time or energy as heat is added at constant pressure. The plot indicates the specific temperature ranges for solid, liquid, and gas phases. (*p. 456*)

Henry's law The solubility of a gas is directly proportional to the pressure that the gas exerts above the solution. (*p. 472*)

hydration The interaction of water (as the solvent) with dissolved ions. (*p. 466*)

hydrogen bond A particularly strong intermolecular force of attraction between F, O, and/or N and a hydrogen atom. (*p. 448*)

hypertonic Containing a concentration of ions greater than that to which it is judged against. (*p. 480*)

hypotonic Containing a concentration of ions lower than that to which it is judged against. (*p. 480*)

ideal solution A solution in which the properties of the solute and solvent are not changed by dilution. This means that other than being diluted, combining solute and solvent in an ideal solution does not release or absorb heat, and the total volume in the solution is the sum of the volumes of the solute and solvent. (*p. 474*)

induced dipole A dipole produced in a compound as a consequence of its interaction with an adjacent dipole. (*p. 446*)

intermolecular force A force of attraction between two molecules. (*p. 444*)

ion–dipole interaction An attraction between an ion and a compound with a dipole. (*p. 466*)

isotonic Containing the same concentration of ions. (*p. 480*)

isosmotic Possessing the same osmotic pressure. (*p. 480*)

liquid A phase of a substance characterized by closely held components. The properties of a liquid include a medium density, the ability to flow, and the ability to take the shape of the container that holds it by filling from the bottom up. (*p. 443*)

London forces The weakest of the intermolecular forces of attraction, characterized by the interaction of induced dipoles. (*p. 446*)

maximum contaminant level (MCL) The highest acceptable level of a contaminant in a particular solution, according to the Environmental Protection Agency. (*p. 470*)

melting point The temperature of the phase transition as a compound changes from a solid to liquid. (*p. 456*)

meniscus The concave or convex shape assumed by the surface of a liquid as it interacts with its container. (*p. 464*)

miscible Mixable. Two solvents that are miscible dissolve in each other completely. (*p. 466*)

molality A concentration measure defined as moles of solute per kilogram of solvent. (*p. 468*)

molarity A concentration measure defined as moles of solute per liter of solution. (*p. 468*)

mole fraction A concentration measure defined as moles of solute per total moles of solution. (*p. 468*)

natural gas A complex mixture of gases extracted from the Earth. The major component in many cases is methane. (*p. 443*)

normal boiling point The temperature at which a liquid boils when the pressure is 1 atm. (*p. 455*)

osmosis The flow of solvent into a solution through a semipermeable membrane. (*p. 479*)

osmotic pressure The pressure required to prevent the flow of a solvent into a solution through a semipermeable membrane. (*p. 479*)

parts per billion A concentration defined as the mass of a solute per billion times that mass of solution. (*p. 470*)

parts per million A concentration defined as the mass of a solute per million times that mass of solution. (*p. 470*)

parts per trillion A concentration defined as the mass of a solute per trillion times that mass of solution. (*p. 470*)

phase boundary On a phase diagram, a set of specific temperatures and pressures at which a compound undergoes a phase transition. (*p. 460*)

phase diagram A plot of the phases of a compound as a function of temperature and pressure. (*p. 459*)

polarizability The extent to which electrons can be shifted in their location by an electric field. (*p. 446*)

pressure The force exerted per unit area by a gas within a closed system. (*p. 452*)

Raoult's law Describes the change in vapor pressure of a solvent as solute is added to a solution. (*p. 474*)

reverse osmosis A purification process wherein pure solvent is forced (with pressure) to flow through a semipermeable membrane away from a solution. (*p. 480*)

saturated When considering a pure substance—the surrounding atmosphere contains the maximum possible amount of vapor, expressed as the vapor pressure. When considering a solution—the solution contains the maximum concentration of dissolved solute. (*p. 452, 467*)

semipermeable membrane A thin membrane (typically made from a polymer) with pores big enough to allow solvent to pass through, but small enough to restrict the flow of dissolved solute particles. (*p. 479*)

solid A phase of a substance characterized by high density, rigid shape, and the inability to flow. (*p. 443*)

solubility A property of a solute; the mass of a solute that can dissolve in a given volume (typically 100 mL) of solvent. (*p. 467*)

solute The minor component in a solution. (*p. 465*)

solution A homogeneous mixture of two or more substances. (*p. 465*)

solvation The interaction of a solvent with its dissolved solute. (*p. 466*)

specific heat The heat needed to warm 1 g of a substance so that its temperature rises 1.0°C. (*p. 456*)

sublimation The process of a solid undergoing a phase transition to a gas; the opposite of deposition. (*p. 452*)

supercritical fluid This phase is seen only when the pressure and temperature of the substance are greater than the critical pressure and critical temperature. It is characterized with a density midway between that of a liquid and a gas, a viscosity similar to a gas, and the ability to completely fill the volume of a closed container. (*p. 460*)

surface tension A measure of the energy per unit area on the surface of a liquid. (*p. 494*)

thermal pollution An artificial increase in the temperature of water. (*p. 471*)

triple point The temperature and pressure at which the gas, liquid, and solid phases of a substance are in equilibrium. (*p. 460*)

universal solvent Water is often called the universal solvent because of its ability to dissolve so many chemicals to form homogeneous mixtures. (*p. 465*)

van der Waals forces The intermolecular forces of attraction that result in associations of adjacent substances. (*p. 445*)

van't Hoff factor A factor that modifies the colligative properties on the basis of their ability to dissociate in solution. (*p. 476*)

vapor pressure The pressure of vapor above a liquid in a closed system. (*p. 453*)

viscosity A measure of resistance to flow. (*p. 494*)

weight percent A concentration unit based on the mass of the solute per total mass of solution, reported in percent. (*p. 470*)

Focus Your Learning

The answers to the odd-numbered problems and some selected problems appear at the back of the book, as represented by the blue numbering.

Section 11.1 The Structure of Water: An Introduction to Intermolecular Forces

Skill Review

1. Describe the differences between intermolecular and intramolecular forces.

2. Describe the similarities between intermolecular and intramolecular forces.

3. Explain how intermolecular forces between molecules arise.

4. How is the strength of an intermolecular force related to the phase of matter?

5. Using electronegativity differences, rank these bonds from least to most polar:

   O—H   C—H   S—H   N—H

6. Using electronegativity differences, rank these bonds from least to most polar:

   O—C   Cl—C   F—H   P—O

7. In each of these statements, insert the term *intermolecular* or *intramolecular* to describe the process taking place.
   a. Water freezes when _____ attractions are forming.
   b. The production of carbon and sulfur from CS_2 indicates that _____ bonds have been broken.
   c. When dry ice sublimes, _____ bonds are broken.
   d. In general, _____ bonds are stronger than _____ bonds.

8. The terms *polar* and *polarizable* have both been used in discussions of intermolecular forces.
   a. Distinguish between these two important terms.
   b. Which of these contains the more polar bond, CO or NO?
   c. Which of these is the more polarizable atom, He or Ne?

Chemical Applications and Practices

9. In a sample of water there are two types of hydrogen-to-oxygen attractions taking place. In one attraction the average distance between the hydrogen and oxygen is 0.101 nm. In the other, the distance is 0.175 nm. Which distance represents an intermolecular attraction, and which represents an intramolecular attraction? Which of the two is the stronger attraction?

10. Carbon dioxide is generated as a result of metabolism and as a result of the combustion of carbon-based materials. This substance is a gas under "normal" conditions. Explain why this is so.

Section 11.2 A Closer Look at Intermolecular Forces

Skill Review

11. In each of these pairs of organic compounds, select the one that would have the higher boiling point. Then, using intermolecular forces, describe the basis for your selection.
    a. Pentane or 2,2-dimethylpropane
    b. Hexane or cyclohexane
    c. Pentane or hexane
    d. Pentane or water
    e. Pentane or methane

$$CH_3 - CH_2 - CH_2 - CH_2 - CH_3$$

Pentane

$$CH_3 - CH_2 - CH_2 - CH_2 - CH_2 - CH_3$$

Hexane

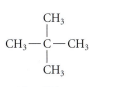

2,2-Dimethylpropane   **Cyclohexane**

12. In each of these pairs of organic compounds, select the one that would have the higher boiling point. Then, using intermolecular forces, describe the basis for your selection.
    a. Methanol (CH_3OH) or ethanol (CH_3CH_2OH)
    b. Water or hydrogen sulfide
    c. Pentane or ethanol
    d. Hydrogen sulfide or hydrogen fluoride
    e. Ammonia (NH_3) or methanol

13. In these pairs, compare the polarity of the bonds shown. Decide which bond in each pair is more polar, and show in which direction the electron density would be shifted.
    a. B—O and B—S
    b. P—S and P—Cl
    c. O—H and O—C

14. In these pairs, compare the polarity of the bonds shown. Decide which bond in each pair is more polar, and show in which direction the electron density would be shifted.
    a. C—O and C—S
    b. N—S and N—H
    c. C—H and C—Cl

15. Name the type of intermolecular forces that must be overcome in order to change each of these from the liquid state to the gas state.
    a. CH_3CH_2OH c. SF_6
    b. CO_2 d. HF

16. Name the type of intermolecular forces that must be overcome in order to change each of these from the liquid state to the gas state.
    a. CH_3OH c. CCl_4
    b. CS_2 d. COS

17. Select the molecule in each pair that is less polarizable.
    a. CF_4 or CCl_4
    b. H_2O or H_2Se

18. Select the molecule in each pair that is less polarizable.
    a. HI or HF
    b. CH_4 or COS

19. Which of these compounds are likely to participate in hydrogen bonding? For those that do not qualify, explain what is lacking.
    a. CH_3OH (methanol)
    b. NH_3 (ammonia)
    c. C_6H_6 (benzene)
    d. $C_2H_5OC_2H_5$ (diethyl ether)

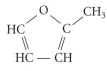

Diethyl ether

20. Which of these compounds are likely to participate in hydrogen bonding? For those that do not qualify, explain what is lacking.
    a. H_2S c. B_2H_6
    b. NCl_3 d. CH_3NH_2

Chemical Applications and Practices

21. "Wake up and smell the coffee!" is a phrase that you may have used to catch someone's attention. Perhaps less compelling, but more accurate, would be "Wake up and smell the 2-methylfuran." This compound is one of several that are typically responsible for the aroma of coffee. Its structure is shown here.
    a. Which bond in the molecule do you think is the most polar?
    b. What intermolecular forces exist between the molecules of 2-methylfuran?

22. Answer these questions on the basis of your answer to Problem 21.
    a. Would you predict that this compound would have a high or a low normal boiling point?
    b. Would you predict that this compound would be a solid, a liquid, or a gas at 1 atm and 25°C? Explain.

23. Suppose a small lab accident involved spilling the same amounts of hexane and hexanol. Which compound would you expect to evaporate first? Using intermolecular forces, explain the rationale behind your answer.

$$CH_3 — CH_2 — CH_2 — CH_2 — CH_2 — CH_3$$

Hexane

$$CH_3 — CH_2 — CH_2 — CH_2 — CH_2 — CH_2 — OH$$

Hexanol

24. Would you predict a cup of water or a cup of Everclear (pure ethanol) to evaporate more rapidly? Using intermolecular forces, explain the rationale behind your answer.

25. Under the proper conditions, London forces are sufficient to liquefy the noble gases. Of Ne, Ar, and Kr, which would possess the stronger London forces? Explain the basis of your selection.

26. Which of these compounds would you predict to have the greatest intermolecular forces, F_2, Cl_2, Br_2, or I_2?

27. The Group V elements form compounds with hydrogen as follows:

	NH_3	PH_3	AsH_3	SbH_3	BiH_3
Boiling point:	−33°C	−88°C	−63°C	−17°C	16°C

   a. Temporarily excluding NH_3, explain the general trend shown here.
   b. Explain why NH_3 deviates from the general trend.

28. HF and H_2O are both compounds that exhibit hydrogen bonding. The hydrogen bonding that takes place between HF molecules is stronger than those in H_2O, yet H_2O has the higher boiling point. Explain why.

Section 11.3 Impact of Intermolecular Forces on the Physical Properties of Water, I

Skill Review

29. Explain why the heat of vaporization of water is so much larger than the heat of vaporization of methane.

30. Would you expect the heat of vaporization of CO to be closer to that of water or of CH_4? Explain.

Chemical Applications and Practices

31. Water's relatively high specific heat (4.184 J/g°C) makes it an ideal coolant for automobile engines. How much heat from an engine would be absorbed if 12.0 kg of cooling system water were heated from 25°C to 75°C?

32. Ethanol (C_2H_5OH) is the alcohol found in intoxicating beverages. It has a heat of vaporization of approximately 39.0 kJ/mol. How many joules would be released when 42.0 g of ethanol vapor condensed into the liquid phase?

33. Propane ($CH_3CH_2CH_3$) is a fuel typically used to operate our outdoor barbecue grills. This compound is often found inside pressurized cylinders.
    a. What type of intermolecular forces of attraction would you predict to exist in propane?
    b. Given this information, would you predict propane to exist as a gas or as a liquid at room temperature and 1.0 atm pressure?
    c. Why is propane often found in liquid form (as it is inside the barbecue grill cylinder)?

34. Chlorine exists, at typical room conditions, as a diatomic gas (Cl_2). However, under higher pressures (greater than 250 kPa), the gas can be converted to a liquid. Chlorine has many uses, chief of which is to produce bleaching agents.
    a. What type of bond exists between two chlorine atoms in Cl_2?
    b. What type of intermolecular force(s) are responsible for the gas molecules entering the liquid phase?

35. Oxygen and tellurium both form dihydrogen compounds (H_2O and H_2Te.) Explain why water molecules participate in hydrogen bonding, whereas H_2Te does not.

36. a. Commercial rubbing alcohol is typically sold as a 70% solution of 2-propanol ($CH_3CHOHCH_3$). Explain why, when it is in contact with your skin, you soon feel a cooling sensation.
    b. The heat of vaporization of 2-propanol is approximately 42 kJ/mol. How much heat would be required to evaporate 5.0 g of 2-propanol?

37. Using the values given in the chapter, calculate the amount of heat needed to change a 15.0-g ice cube at −5.0°C to 15.0 g of steam at 100.0°C.

38. Using the values given in the chapter, calculate the amount of heat that is liberated when a 1.0-kg bucket of steam at 100.0°C is cooled to give a 1.0-kg bucket of water at 25.0°C.

Section 11.4 Phase Diagrams

Skill Review

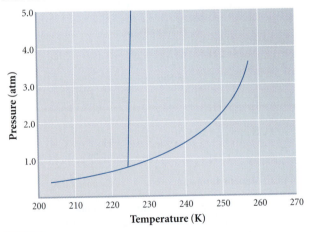

39. Using the phase diagram above, determine the following:
    a. The normal boiling point
    b. The triple point
    c. In what phase would you find the compound at −25°C and 1 atm?

40. Using the phase diagram shown on page 486, determine the following:
   a. The critical pressure
   b. The critical temperature
   c. In what phase would you find the compound at −63°C and 3 atm?

41. Using the phase diagram shown on page 486, indicate what phase transition(s) occur as the compound goes from 220 K and 0.5 atm to 220 K and 3 atm.

42. Using the phase diagram shown on page 486, indicate what phase transition(s) occur as the compound goes from 220 K and 1.0 atm to 240 K and 1.0 atm.

43. Using the phase diagram shown on page 486, indicate what phase change will occur when the compound is heated from −45°C to −53°C at 1 atm.

44. Using the phase diagram shown on page 486, indicate what phase change, if any, will occur when the pressure is adjusted from 2.0 atm to 1.0 atm at constant temperature of −33°C.

45. Using the phase diagram shown on page 486, is it possible to liquefy a gas sample of this compound at −33°C using increasing pressure? Explain why or why not.

46. Using the phase diagram shown on page 486, is it possible to form a solid from a liquid sample when the temperature is kept constant at −53°C? Explain your answer.

Chemical Applications and Practices

47. Ethylene (C_2H_4) is one of the most widely used compounds in the manufacture of synthetic materials. From the following data, construct a labeled phase diagram similar to the one shown for Problems 39–46. The normal boiling point is approximately −104°C. The critical point, at 50 atm, is approximately 9.8°C. The triple point is −169°C and 0.0012 atm. The normal melting point is −169°C.

48. Using the diagram you constructed in Problem 47, indicate the phase in which ethylene would be found at typical room conditions.

Section 11.5 Impact of Intermolecular Forces on the Physical Properties of Water, II

Skill Review

49. Explain in your own words the differences among surface tension, viscosity, and capillary action.

50. Would you predict there to be a relationship among surface tension, viscosity, and capillary action?

51. Draw a diagram of a meniscus that would form from a buret filled with mercury.

52. Draw a diagram of a meniscus that would form from a buret filled with honey.

53. Arrange these substances in order of their increasing viscosity. Explain your ordering.

   honey   water   gasoline

54. Which of these compounds do you predict will rise highest within a capillary tube: water, hexane, or ethanol? Explain your prediction.

Chemical Applications and Practices

55. Lubricating motor oils are rated on the basis of their viscosity. The oils in these lubricants are nonpolar molecules. Explain why these hydrocarbon-based oils have a relatively high viscosity.

56. Honey has a very high viscosity. It contains (in large part) many different carbohydrates similar in molecular structure. What specific atoms would you predict to exist in a carbohydrate? Explain your reasoning.

Section 11.6 Water: The Universal Solvent

Skill Review

57. The solubility of NH_4Cl at 20°C is approximately 39 g per 100 g of water. At 80°C, 68 g per 100 g of water dissolves.
   a. Is the value for the enthalpy of solution of NH_4Cl positive or negative?
   b. Assuming a near linear relationship over the temperature range, if you saturated a solution at 40°C, how many grams of NH_4Cl would be dissolved?

58. Organic alcohols have varying solubilities in water. Examine the following list, and explain the cause of the trend you observe.

1–Butanol	C_4H_9OH	79 g/1000 mL of water
1–Pentanol	$C_5H_{11}OH$	27 g/1000 mL of water
1–Hexanol	$C_6H_{13}OH$	5.9 g/1000 mL of water

59. Water is often considered the universal solvent. Explain this designation in your own words.

60. What properties would you associate with a solvent that was considered to be universal?

61. a. Is it possible for a compound to have a $\Delta_{soln}H = 0$?
   b. Why would a solute and solvent form a solution if the value for $\Delta_{soln}H$ were 0?

62. Cyclohexane (C_6H_{12}) is often used as a nonpolar solvent. Of the following solutes, which would best dissolve in cyclohexane, and which would best dissolve in water?

   NaCl C_6H_6 CH_3OH $C_6H_{12}O_6$ $CH_3(CH_2)_{16}COOH$
   Salt Benzene Methanol Dextrose   Stearic acid

Chemical Applications and Practices

63. Two burets are filled with different liquids. Which meniscus is more likely to be representative of heptane (C_7H_{16}) and which of water?

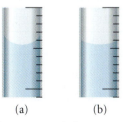

   (a)   (b)

64. If you were to extract caffeine from some cola beans, you would find that you could extract much more using hot water than using cold water. Caffeine is more soluble in hot water than in cold. Explain what this fact tells you about the heat of solution for caffeine.

65. Water and ethanol are said to be miscible.
    a. Distinguish between a solute that is miscible with water and a solute that is soluble in water.
    b. What do miscible substances have in common?

66. What would you predict would occur if a saturated solution of sodium chloride were warmed? What would happen if a saturated solution of lithium sulfate were warmed?

Section 11.7 Measures of Solution Concentration

Skill Review

67. Calculate the molarity (M) of these solutions:
    a. 42.0 g of NaOH dissolved in enough water to form 0.500 L of solution
    b. 10.0 g of $C_6H_{12}O_6$ (dextrose) dissolved in enough water to form 0.250 L of solution
    c. 25.0 g of urea (NH_2CONH_2) dissolved in enough water to form 100.0 mL of solution

68. Using the information in Problem 67, calculate the molality of each solution (assume the density of the solution is 1.0 g/mL).

69. Calculate the molality (m) of these solutions:
    a. 12.5 g of ethylene glycol antifreeze (CH_2OHCH_2OH) dissolved in 0.100 kg of water
    b. 53.0 g of sucrose ($C_{12}H_{22}O_{11}$) dissolved in 500.0 g of water
    c. 4.55 g of sodium bicarbonate ($NaHCO_3$) dissolved in 250.0 g of water

70. Using the information in Problem 69, calculate the molarity of each solution (assume the density of each solution is 1.0 g/mL).

71. Determine the mole fraction of solute in the following:
    a. 22.7 g of benzene (C_6H_6) dissolved in 67.5 g of cyclohexane (C_6H_{12})
    b. 15.0 g of formic acid (HCOOH, found in ants) dissolved in 100.0 g of water
    c. 0.195 g of acetaldehyde (C_2H_4O, found as the product of ethanol metabolism) dissolved in 25.0 g of water

72. Using the information in Problem 71, calculate the molality of each solutions (assume the density of water is 1.0 g/mL).

73. Fill in the missing information.

Compound	Grams of Compound	Grams of Water	Mole Fraction of Solute	Molality
NH_4Cl	12.5 g	95.0 g		
KNO_3		125 g		0.155
$C_6H_{12}O_6$		250.0 g	0.115	

74. Fill in the missing information.

Compound	Grams of Compound	Grams of Water	Mass %	ppm
NH_4Cl	1.0 g	99.0 g		
KNO_3		125 g		125
$C_6H_{12}O_6$		300.0 g	15.8	

Chemical Applications and Practices

75. A typical cup of coffee may contain 75 mg of caffeine per 200.0 mL of water. If the density of the solution were approximately 1.09 g/mL, what would you calculate as the mass percent caffeine, the molarity, and the molality? (*Hint:* You'll need the formula of caffeine to answer this problem.)

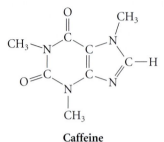

Caffeine

76. Perspiration has a slight acidity because of the presence of lactic acid ($C_3H_6O_3$). Suppose we analyze the perspiration of a chemistry student running late to class. If the density of the perspiration were 1.15 g/mL and the mass percent lactic acid were found to be 4.88%, what would you calculate as the mole fraction, molarity, and molality of the lactic acid?

77. The saline solution used in some medical procedures is a 5.00% NaCl solution. What would be the molarity of sodium ions (Na^+) in such a solution if the density were assumed to be approximately 1.02 g/mL?

78. Sodium alkylbenzene sulfonate ($C_{18}H_{29}SO_3Na$) is often used as a synthetic detergent. This ingredient helps prevent scale buildup in "hard" water applications. If a detergent solution were 0.100 M in sodium alkylbenzene sulfonate, how many grams would be dissolved per liter of solution?

79. *Hard water* is the name given to water containing relatively high concentrations of metal ions, such as calcium and magnesium. For instance, if a particular water sample were said to contain 130.0 ppm Ca^{2+}, it would be classified as hard water.
    a. What would be the mass percent of calcium ion containing 130.0 ppm Ca^{2+}?
    b. What would be the same concentration expressed as parts per billion?

80. The concentration of potassium ion in human blood cells varies over a healthy range. If the concentration of K^+ in a red blood cell were listed at 0.95 μmol/mL, what would you report as the molarity and parts per million?

Section 11.8 The Effect of Temperature and Pressure on Solubility

Skill Review

81. This represents a system at equilibrium: $O_2(aq) \rightarrow O_2(g)$
    a. In which direction will the equilibrium be shifted by an increase in pressure?
    b. In which direction will the equilibrium be shifted by an increase in temperature?

82. This a system at equilibrium: $CO_2(g) \rightarrow CO_2(aq)$
    a. In which direction will the equilibrium be shifted by an increase in pressure?
    b. In which direction will the equilibrium be shifted by an increase in temperature?

Chemical Applications and Practices

83. Underwater activities require assisted breathing techniques. When using pressurized air, divers must be aware of Henry's law. Assuming that air contains 78% nitrogen, what is the concentration of N_2 in blood at 1 atm? (The Henry's law constant for $N_2(g)$ at 25°C is 1540 atm/M.)

84. What would be the concentration of N_2 in blood if the N_2 pressure were increased to 3.0 atm? (The Henry's law constant for $N_2(g)$ at 25°C is 1540 atm/M.)

Section 11.9 Colligative Properties

Skill Review

85. In theory, what concentration of ions would you expect to find dissolved in 500.0 mL of a 0.00100 m solution of $CaCl_2$?

86. What value is expected for the van't Hoff factor in very dilute solutions of each of these?
    a. $AlCl_3$
    b. $(NH_4)_3PO_4$
    c. $Mg(OH)_2$
    d. $C_6H_{12}O_6$

87. Although several factors are necessary to explain the magnitude of the vapor pressure of a sample, some general trends can be noted. Arrange these substances in order of decreasing vapor pressure.

    Water (H_2O)
    Glycerol ($HOCH_2CHOHCH_2OH$)
    Pentane ($CH_3CH_2CH_2CH_2CH_3$)

88. a. Complete this sentence by inserting the appropriate term in the blank: As external air pressure _____ (increases, decreases) the boiling point of a liquid decreases.
    b. Explain why the term you selected is correct.

89. For each of these solutions, determine the boiling point of the solution. You may need to use Table 11.6 to help with this problem. (Assume these solutions behave ideally.)
    a. 0.75 m NaCl in water
    b. 0.040 m glucose ($C_6H_{12}O_6$) in water
    c. 0.250 M $CaCl_2$ in water (Assume the density of the solution is 1.00 g/mL.)
    d. 0.25 mol fraction naphthalene in benzene (Assume the density of the solution is 0.88 g/mL.)

90. For each of these solutions, determine the boiling point of the solution. You may need to use Table 11.6 to help with this problem. (Assume these solutions behave ideally.)
    a. 1.00 m sucrose ($C_{12}H_{22}O_{11}$) in ethanol
    b. 0.14 m Na_2SO_4 in water
    c. 0.250 M glucose ($C_6H_{12}O_6$) in water (Assume the density of water is 1.00 g/mL.)
    d. 500.0 ppm $FeCl_3$ in methanol

91. For each of these solutions, determine the melting point of the solution. You may need to use Table 11.8 to help with this problem. (Assume these solutions behave ideally.)
    a. 0.500 m glucose ($C_6H_{12}O_6$) in water
    b. 0.055 m LiOH in water
    c. 0.125 m methanol in phenol
    d. 1.20 m benzoic acid in water

92. For each of these solutions, determine the melting point of the solution. You may need to use Table 11.8 to help with this problem. (Assume these solutions behave ideally.)
    a. 0.200 mol fraction glucose ($C_6H_{12}O_6$) in water
    b. 0.055 m $CoCl_2$ in water

c. 0.125 m benzene in phenol
d. 1.20 m HCl in water

93. For each of these solutions, determine the vapor pressure of the solution at STP. (Assume these solutions behave ideally.)
    a. 0.200 mol fraction glucose ($C_6H_{12}O_6$) in water
    b. 0.055 mol fraction $CoCl_2$ in water
    c. 0.125 mol fraction benzoic acid in water
    d. 1.20 m HCl in water

94. For each of these solutions, determine the osmotic pressure of the solution at STP. (Assume these solutions behave ideally.)
    a. 0.200 M $FeCl_2$ in water
    b. 0.055 m $CoCl_2$ in water (Assume the density of the solution is 1.0 g/mL.)
    c. 0.125 mol fraction glucose ($C_6H_{12}O_6$) in water (Assume the density of the solution is 1.02 g/mL.)
    d. 0.0945 mol fraction NaCl in water (Assume the density of the solution is 1.10 g/mL.)

Chemical Applications and Practices

95. An amount of water in a closed container will have a certain vapor pressure when the temperature is held constant. Varying the shape of the container can change the surface area of the volume of water. However, this does not change the vapor pressure. Explain this observation, and reconcile it with the observation that the surface area of a sample of water *does* change the rate of evaporation in an open container.

96. The vapor pressure of water at 25°C is 23.76 mm Hg. What would you calculate as the new vapor pressure of a solution made by adding 50.0 g of ethylene glycol ($HOCH_2CH_2OH$, antifreeze) to 50.0 g of water? You may assume that the vapor pressure of ethylene glycol at this temperature is negligible.

97. Explain why 0.10 m solutions of nonelectrolytes have approximately the same boiling point regardless of the identity of the solute. Why doesn't this same statement apply to electrolytes?

98. Two solutions are placed in a −1.0°C storage cooler. One solution is labeled 5.0% $C_6H_{12}O_6$ and the other 15.0% $C_6H_{12}O_6$. When they were to be retrieved the next day, the labels of both had loosened and fallen off the containers. Would you be able to identify the solutions on the basis of their freezing points? Prove your answer.

99. How many grams of antifreeze ($HOCH_2CH_2OH$) would have to be added to 5.0 kg of water in an automobile cooling system to keep it from freezing during a cold Nebraska winter with temperatures of −32°C?

100. Freezing-point depression can be used to determine the molecular mass of a compound. Suppose that 1.00 g of an unknown molecule were added to 20.0 g of water and the freezing point of the solution determined. If the new freezing point of water were found to be −1.50°C, what would you predict to be the molecular mass of the compound?

101. a. The outer membrane of many fruits acts as a semipermeable membrane through which osmosis can take place. Diagram the cell of a cucumber before and after it has been set in a highly concentrated salt solution.

b. Why is it important to know the osmotic pressure of human fluids before administering any fluids to a patient?

102. One of the first stages in healing a small cut takes place when a blood clot begins to form. A key enzyme in this process is thrombin. This large enzyme has a molar mass of nearly 34,000 g/mol. What would be the osmotic pressure of a solution that contained 0.20 g of thrombin per milliliter, at 37.0°C?

Comprehensive Problems

103. The compounds guanine and cytosine are two bases that make up part of the structure of your DNA. Use the structures shown to indicate how three hydrogen bonds can form between these two molecules.

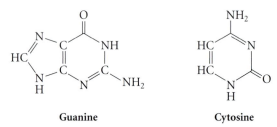

Guanine   **Cytosine**

104. Explain why you would expect the heat of vaporization for water to be so much larger than the heat of fusion.

105. If the combustion of methane were used as the source of heat for Problem 37, how many grams of CH_4 would be needed? $\Delta_c H$ for $CH_4 = 55.5$ kJ/g.

106. When preparing some solutions for analysis, chemists must be aware that some solution processes are very exothermic. Sodium hydroxide solutions are often used when analyzing acid solutions.
    a. The heat of solution for NaOH is −44.5 kJ/mole. Using three factors, describe the dissolving of solid NaOH pellets in water.
    b. What temperature change would you calculate for a solution, starting at 25.0°C, made when 100.0 g of water is mixed with 10.0 g of NaOH? Assume the heat capacity of the solution is the same as that for water.

107. Carbon tetrachloride (CCl_4) and bromoform ($CHBr_3$) both have tetrahedral structures. One has more than twice the surface tension of the other. Which has the higher surface tension, and what factor gives rise to this difference?

108. The viscosity of glycerol is over a thousand times larger than that of chloroform ($CHCl_3$). Explain what factor accounts for this large difference.

109. Suppose you are on a camping trip in the mountains of Colorado. While boiling water to cook some dried food, your friends notice, with their handy thermometer, that the water is boiling at only 90°C. How would you explain to your perplexed friends that turning up the gas on the stove will not increase the temperature of the water?

110. The final production of the writing paper you may be using to solve this problem involves several chemical treatments. One compound used in the process is aluminum sulfate. What would be the molarity of an industrial solution if 3250 g of $Al_2(SO_4)_3$ were dissolved in enough water to make 20.0 L of solution?

111. Sulfuric acid (H_2SO_4) solutions are used as the primary electrolyte in the lead storage batteries found in automobiles. One of the most common ways to check the charge level in such a battery is to determine the density of the solution. A sulfuric acid solution had a density of 1.58 g/mL and was known to contain 35.6% by mass H_2SO_4. What is the molarity of the solution?

112. According to Figure 11.1, the world's water consumption in 2020 will be 2700 km^3/year. Assuming a density of 1.00 g/mL, how many metric tons of water will be consumed each year?

113. Hexane is a common liquid used in the chemistry laboratory as a solvent.
    a. How many carbon atoms are there in hexane?
    b. Is there a relationship between the number of carbon atoms in a straight-chain (normal) alkane and the boiling point of the alkane? If so, what is it?
    c. What intermolecular forces are present in hexane?

114. In the electron density maps shown throughout the text, the red color indicates regions of the molecule that possess partial negative change (greater electron density). The blue color indicates regions of the molecule with partial positive charges (reduced electron density). Compare the electron density maps of water (page 127) and of hydrogen bonded water (page 449). Do you notice any striking differences between the two images? If so, what are those differences and what does that imply about the effects of hydrogen bonding?

115. As a beaker of water boils, bubbles develop at the bottom of the beaker and rise to the surface (see the figure on page 455).
    a. What is the origin of the bubbles?
    b. How many liters of water vapor could be produced from a beaker containing 250 mL water ($d = 1.00$ g/mL) if the beaker is boiled to dryness? (Assume the temperature of the water vapor is 100°C at 0.95 atm.)
    c. How much heat would be required to complete the conversion of 250 mL water at 25°C into water vapor at 100°C?

Thinking Beyond the Calculation

116. The fuel most commonly used in portable lighters is butane (C_4H_{10}). The normal boiling point of butane is −0.50°C.
    a. Draw the Lewis dot structure for butane where all of the carbons are in a row.
    b. What types of intermolecular forces of attraction would you predict for this molecule?
    c. Does the low boiling point of butane make sense?
    d. Is this compound soluble in water?
    e. If 3.50 g of butane were burned in oxygen, how many moles of water would be formed? (The other product is carbon dioxide.)
    f. Considering that −0.50°C is so much lower than typical room temperatures, why doesn't the butane in the lighter boil?
    g. Relative to butane, estimate the normal boiling point of pentane (C_5H_{12}) and propane (C_3H_8).

Carbon

Contents and Selected Applications

12.1 Elemental Carbon

12.2 Crude Oil—the Basic Resource
Chemical Encounters: Crude Oil—the Basic Resource

12.3 Hydrocarbons

12.4 Separating the Hydrocarbons by Fractional Distillation

12.5 Processing Hydrocarbons

12.6 Typical Reactions of the Alkanes

12.7 The Functional Group Concept

12.8 Ethene, the C══C Bond, and Polymers

12.9 Alcohols

12.10 From Alcohols to Aldehydes, Ketones, and Carboxylic Acids

12.11 From Alcohols and Carboxylic Acids to Esters

12.12 Condensation Polymers

12.13 Polyethers

12.14 Handedness in Molecules

12.15 Organic Chemistry and Modern Drug Discovery

Oil, like that stored aboard this tanker, is responsible for heating our homes, running our cars, and serving as the feedstock for production of an extraordinary number of organic chemical compounds. The processing of oil provides plastics that we use every day. From grocery bags to the shirts on our backs, oil is vital to our current way of life.

Go to **college.hmco.com/pic/kelterMEE** for online learning resources.

Deep beneath the surface of the Earth, a spinning drill bit from an oil exploration platform breaks through the hard rock and hits pay dirt—oil. The dark liquid escapes from its high-pressure vault to emerge at the surface as a gushing fountain of chemical possibilities. The famous "gushers" of the past (see Figure 12.1) may be rare now, but oil remains the chemical foundation of a huge industry sustaining our modern way of life. The gas and diesel fuels that power our vehicles, along with plastics, paints, modern textiles, and a vast range of medicines, are derived from the petrochemical industry. The raw material that flows out of the Earth and into that industry is crude oil, petroleum. The element at the heart of the chemistry of oil is carbon. This carbon is not pure but is bonded together with hydrogen and other elements to form a rich mixture of molecules that can be separated, modified, and exploited to make so many of the products that sustain our modern way of life.

Coal, another carbon-based treasure, is mined from natural outcroppings both above and below ground, such as the one shown in Figure 12.2. The use of this resource helped fuel the industrial revolution of eighteenth and nineteenth centuries. Since then, coal has been employed to make a wide range of useful carbon-based substances, including gasoline, diesel and jet fuels, and some chemicals. Coal also remains one of the most important fuels for generating electricity (see Table 12.1).

Our planet is made most interesting by the chemistry of carbon. In addition to its vital role as the primary component of fossil fuels, the chemistry of carbon is also important to perhaps the most precious treasure of all, life. All of the key molecules of life are built around a framework of bonded carbon atoms. These carbon-based molecules, were initially known as **organic compounds** because this class of compounds was found in living organisms. It was later determined that organic compounds do not appear exclusively in living organisms, but the name stuck. In fact, the oil and coal we use as a source of carbon-based compounds were themselves formed from the carbon-based chemicals in ancient animals and plants. For this reason, the study of carbon compounds, known as **organic chemistry**, focuses on the reactions and properties of a vast number of organic compounds.

FIGURE 12.1

Black gold. An oil derrick becomes engulfed in a spray of crude oil.

TABLE 12.1	2004 Fuel Sources of Energy in the United States
Source	**Percent of Total BTU***
Coal	32.2%
Natural gas	27.5
Petroleum	16.4
Hydroelectric	3.9
Nuclear	11.7
All others	8.3
Total Production = 70.369 quadrillion BTU (quadrillion = 10^{15})	

*1 BTU = 1055 joules
Source: U.S. Department of Energy, Energy Information Agency.

FIGURE 12.2

Coal is obtained from huge open-pit mines and from deep underground in large pockets. This natural resource is useful in heating homes, in fueling electric plants, and in making steel.

What is the nature of these organic chemicals? How do we make them? How do we chemically manipulate them to produce the stunning range of goods based on the crude oil that generates over a quarter of a trillion dollars in sales for oil and gasoline companies per year?

12.1 Elemental Carbon

Carbon is a very versatile element that can be used to prepare a nearly endless array of carbon-based compounds. It also exists in three different and distinct forms as the element. Different forms of the same element, known as **allotropes**, are not exclusive to carbon. Other elements, such as phosphorus, oxygen (as O_2 and O_3,) and sulfur, also can be found in a variety of allotropes. The allotropes of carbon are diamond, graphite, and fullerenes.

Diamond

Figure 12.3 illustrates the first of the three allotropes of carbon, diamond. The carbon within a diamond is bonded to four neighboring carbons in a giant covalent network, as shown in Figure 12.4. The bonds in a diamond are the result of the overlap of identical sp^3 hybridized orbitals that are arranged tetrahedrally around each atom. This makes diamond one of the hardest substances known, because the network of many strong covalent bonds must be disrupted to break or distort a piece of diamond. The strong bonding in diamond explains why it forms such an enduring gemstone and why it can be used to cut other materials.

FIGURE 12.3

Diamond, an allotrope of carbon that humans find particularly valuable when cut and polished.

Graphite

Compare the structure of diamond to that of graphite, another of the allotropes of carbon shown in Figure 12.5. The sp^2 hybridized carbon atoms in graphite share covalent bonds to three neighboring atoms, creating a repeating hexagonal network of carbon atoms in extended layers. Each atom in this structure contributes one unbonded electron to a system of delocalized π electrons above and below each hexagonal layer. Unlike the sp^3 hybridized carbons in diamond, the delocalized and mobile electrons in graphite allow it to conduct electricity. This means that graphite can be used as the electrodes in industrial processes such as the Hall–Heroult process for producing aluminum (discussed in Chapter 19).

The layered structure of graphite also makes it soft and slippery. Each layer is only weakly attracted to the layers above and below. This enables us to use graphite as pencil "lead"; portions of the layers will slide off the tip of the pencil and onto the paper. Graphite's slipperiness also makes it useful as a solid lubricant in specialized machinery. Diamond, by contrast, in which the entire structure is a network of covalent bonds, is a terrible lubricant but an excellent grinding agent.

FIGURE 12.4

The covalent network structure of diamond.

Fullerenes

A third allotrope of carbon, shown in Figure 12.6, was discovered in 1985. This allotrope consists of sp^2 hybridized carbon atoms folded into structures that resemble balls and tubes. The core member of this class of molecules is called Buckminsterfullerene (C_{60}) because of its similarity to the geodesic domes designed by the architect Buckminster Fuller. Buckminsterfullerene (also known as

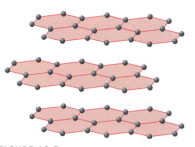

FIGURE 12.5

The structure of graphite.

Fullerenes, one of the allotropes of carbon, includes ball-shaped structures and hollow tube structures. Each of the carbon atoms in these structures is *sp*^{2} hybridized.

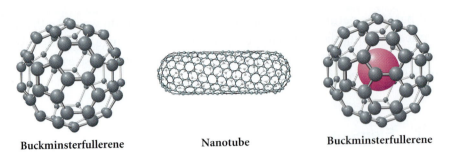

Buckminsterfullerene   **Nanotube**   **Buckminsterfullerene**

Geodesic domes have been used to construct spacious homes.

"buckyball") is prepared when an electric discharge arcs between two graphite rods surrounded by helium gas at high pressure. Although C_{60} and graphite are similar in the presence of *sp*^{2} hybridized carbon atoms, the carbon atoms in buckminsterfullerene are held in both six-membered and five-membered rings. The five-membered rings cause a curvature in the overall structure of the compound.

Buckminsterfullerene is a member of an increasingly varied group of related structures known as **fullerenes**. These include cage-like molecules both smaller and larger than the C_{60} cage, as well as tubes of bonded atoms that can occur either alone or in concentric nested patterns, as shown in Figure 12.6. Many research teams are busy trying to find medical and technological applications for these structures.

12.2   Crude Oil—the Basic Resource

Application

CHEMICAL ENCOUNTERS: Crude Oil—the Basic Resource

Crude oil, also known as petroleum, is not an allotrope of carbon. Instead, it is a very complex mixture of hundreds of compounds containing carbon and other atoms. In general terms, crude oil contains many different kinds of **hydrocarbon** molecules (molecules composed of only carbon and hydrogen atoms), together with smaller amounts of molecules derived from hydrocarbons, often including atoms of sulfur, nitrogen, oxygen, and various metals. Table 12.2 lists the typical classes of compounds that can be found in crude oil.

Crude oil is generally found deep beneath the surface of the Earth, trapped between layers of rock in pockets like that shown in Figure 12.7. It is found nearly everywhere in the world, under both land and sea. Unfortunately, removing the oil from these pockets is an involved and expensive process, but it is done because of the incredible usefulness of the compounds found in crude oil. Table 12.3 on page 496 lists the countries that export the most crude oil and indicates how much oil they ship *each day*. Note how many of these countries are global "hot spots" of regional conflict.

Why do we so passionately seek out oil? Why are nations willing to go to war over it? Why are national economies dependent on the prevailing price of oil? The

FIGURE 12.7

Geology of a crude oil reservoir.

▨ Sandstone

▨ Impervious rock such as shale

▨ Oil

(a) Anticlinal trap

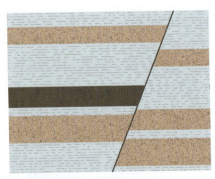

(b) Fault displacement trap

TABLE 12.2	Composition of Crude Oil	
Class of Compound	**Percent of Oil**	**Typical Example**
Hydrocarbons		
Aliphatics (hydrocarbons containing no double or triple bonds)	25%	
Aromatics (containing aromatic groups, first defined in Chapter 9)	17%	
Naphthenes (cyclic hydrocarbons with all single-bonded carbons)	47%	
Nonhydrocarbons		
Sulfurous	<8%	
Nitrogenous	<1%	
Oxygenated	<3%	
Metals	<<<1%	Fe, Mn, Zn, V

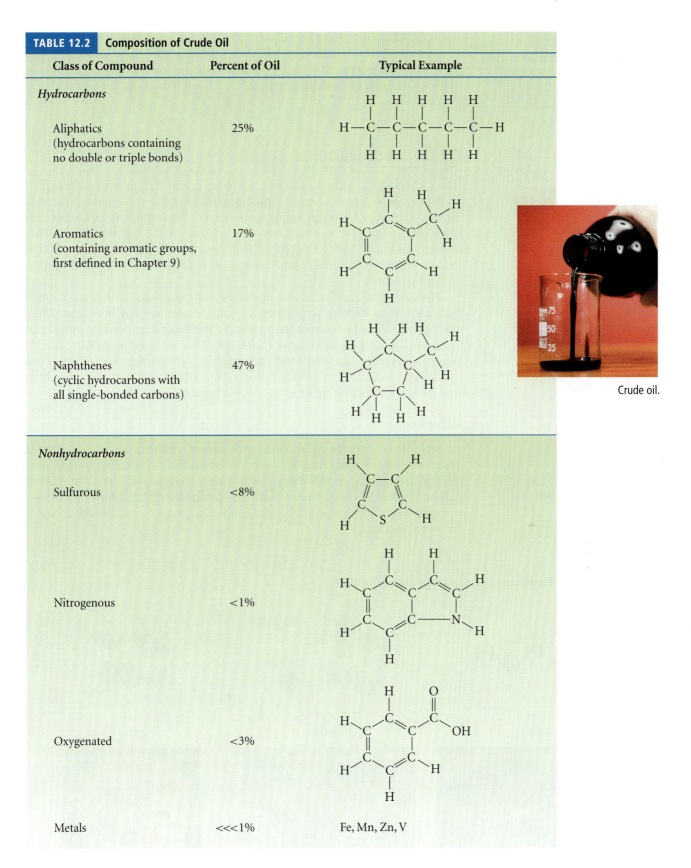

Crude oil.

TABLE 12.3	Top Daily Producers of Crude Oil in 2004	
Country	**Million Barrels** *per Day*	
Saudi Arabia	8.73	
Russia	6.67	
Norway	2.91	
Iran	2.55	
Venezuela	2.36	
United Arab Emirates	2.33	
Kuwait	2.20	
Nigeria	2.19	
Mexico	1.80	
Algeria	1.68	
Iraq	1.48	
Libya	1.34	
Kazakhstan	1.06	
Qatar	1.02	

Source: Energy Information Administration, Non-OPEC Fact Sheet.

Video Lesson: Alkanes

answers include our heavy use of oil to produce the gasoline and other fuels for cars, trucks, trains, boats, and jet airplanes. But the uses of oil go well beyond these things. We are surrounded by products made from oil, including plastics, clothing, furniture, carpeting, eyeglasses, and compact disks; the list is almost endless.

12.3 Hydrocarbons

The carbon atoms of hydrocarbons are bonded into straight chains, branched chains, rings, or more complex combinations of these three basic structures (Figure 12.8). The hydrogen atoms are bonded to the carbon atoms so that each carbon has a total of four covalent bonds. Some of the carbon atoms in hydrocarbons can be bonded together by carbon–carbon double bonds or carbon–carbon triple bonds, reducing the number of hydrogen atoms that can bind to the carbon atoms.

Hydrocarbons that contain no double or triple bonds are known as **saturated hydrocarbons**, or **aliphatic compounds**. They are "saturated" with hydrogen, which means that the carbon atoms are bonded to the maximum possible number of hydrogen atoms, because none of the electrons are involved in double or triple bonds. This is the origin of the term *saturated fats,* which might be familiar to you from food labels. All fats include hydrocarbon chains as part of their structure, and the saturated fats have no C=C double bonds.

Alkanes

Saturated hydrocarbons are also known as **alkanes**, and they make up the largest fraction of any class of crude oil compounds (see Table 12.2). The simplest possible alkanes are the **normal alkanes**, which have no branched chains or rings of carbon atoms. The simplest normal alkane—and also the simplest hydrocarbon—contains a single carbon atom bonded to four hydrogen atoms. This molecule is known as methane.

FIGURE 12.8

The three main classes of hydrocarbons: straight-chain, branched, and cyclic.

Straight-chain

Branched

Cyclic

The three classes of hydrocarbons have structures that we see elsewhere in nature. A needlefish, tree branches, and a sand dollar are representative of straight-chain, branched-chain, and cyclic hydrocarbons.

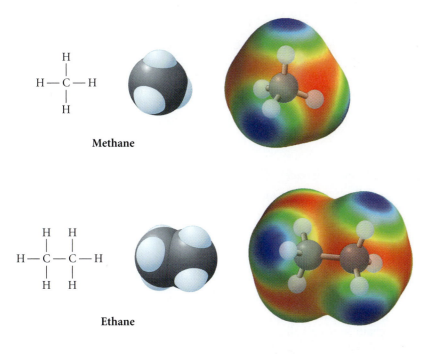

Methane

Ethane

Methane is the principal component of "natural gas," which is commonly found trapped above petroleum deposits and is piped into homes, offices, and factories as a fuel for heating and cooking.

The next-simplest hydrocarbon has two carbon atoms bonded together, with each carbon atom bonded to three hydrogen atoms. This is ethane. As you can see from Table 12.4, the normal alkanes form a regular series in which each member has one more —CH_2— group, a **methylene group**, than the preceding member of the series. These alkanes are members of a **homologous series** of compounds; this means they all share the same general formula, $C_nH_{(2n+2)}$ for the normal alkanes. They have similar chemical and physical properties, which change in a gradual manner as we move through the series.

Although we call the normal alkanes **straight-chain alkanes,** in reality the carbon chain zig-zags, as you can see in Figure 12.8. This is because the four bonds of each carbon atom are formed when four sp^3 hybridized orbitals of the atom overlap with the orbitals of neighboring carbon atoms and hydrogen atoms. The bonds around each carbon atom adopt a tetrahedral arrangement, as we discussed in Chapter 9.

The major direct use we make of the alkanes in petroleum is as fuels. Hydrocarbons burn in oxygen (or in air, which is one-fifth oxygen) to generate mostly carbon dioxide and water and release considerable amounts of heat. The combustion of hydrocarbons helps us heat our homes, cook our food, and power our cars. Like most of the other hydrocarbons in crude oil, the alkanes also serve as valuable and versatile "feedstock" molecules—molecules that can be modified by the chemical industry to generate many useful products.

Video Lesson: Organic Nomenclature

TABLE 12.4	The First Ten Normal Alkanes			
Name	Number of Carbons	Molecular Formula	Structural Formula	Boiling Point (°C)
Methane	1	CH_4	CH_4	−161.5
Ethane	2	C_2H_6	CH_3CH_3	−88.6
Propane	3	C_3H_8	$CH_3CH_2CH_3$	−42.1
Butane	4	C_4H_{10}	$CH_3CH_2CH_2CH_3$	−0.5
Pentane	5	C_5H_{12}	$CH_3(CH_2)_3CH_3$	36.0
Hexane	6	C_6H_{14}	$CH_3(CH_2)_4CH_3$	68.7
Heptane	7	C_7H_{16}	$CH_3(CH_2)_5CH_3$	98.5
Octane	8	C_8H_{18}	$CH_3(CH_2)_6CH_3$	125.6
Nonane	9	C_9H_{20}	$CH_3(CH_2)_7CH_3$	150.8
Decane	10	$C_{10}H_{22}$	$CH_3(CH_2)_8CH_3$	174.1

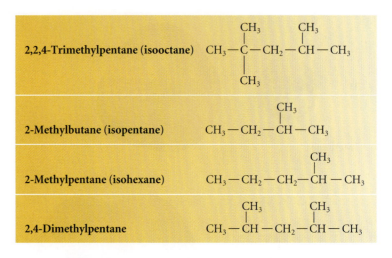

Branched-Chain Alkanes: Isomers of the Normal Alkanes

Crude oil also contains a significant proportion of alkanes with "branched" chains. Each branched-chain alkane has the same formula as a corresponding straight-chain (normal) alkane, so each one can be regarded as a **structural isomer** of a normal alkane. Structural isomers always have the same formula but differ in the way the atoms are attached. We say that they share the same molecular formula but have different molecular structures. Structural isomers do not have the same properties. For instance, the structural isomers pentane and 2,2-dimethylpropane have the same formula (C_5H_{12}). Their boiling points, however, are much different: 36°C for pentane and 10°C for 2,2-dimethylpropane.

EXERCISE 12.1 **Finding the Isomers**

How many structural isomers of pentane (C_5H_{12}) exist? How many structural isomers are there for hexane (C_6H_{14})?

First Thoughts

Isomers must have the same atoms arranged in different ways. Isomers of alkanes differ in their skeleton of bonded carbon atoms. We can find all the isomers by drawing the possible different carbon skeletons then adding the hydrogen atoms needed to ensure that each carbon atom has four bonds.

Solution

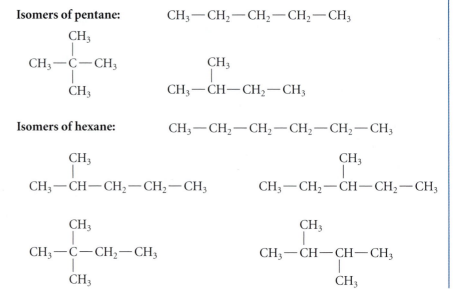

Further Insights

Why are the two structures below not counted as different isomers of hexane? Although at first glance they look different, they are actually the same molecule just flipped over.

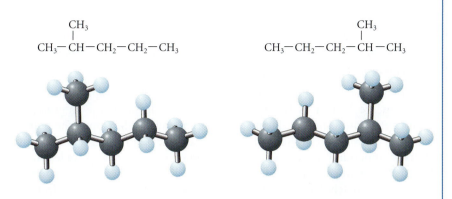

Here are two more examples of structures that initially appear different but are in fact identical. You can tell that they're identical by noting the attachments of the carbon atoms. In each case, there are four carbons in a row, with the branched carbons emanating from the middle two carbon atoms. You need to be careful to eliminate such identical structures when searching for isomers. Working with three-dimensional models, rather than two-dimensional drawings, can make this easier to follow.

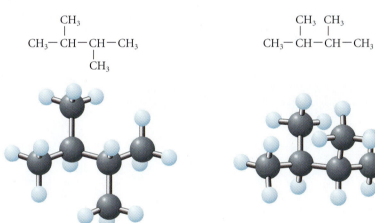

Draw all of the isomers of heptane (C_7H_{16}).

See Problems 13, 14, 21, and 22.

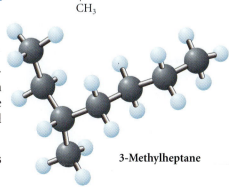

3-Methylheptane

Naming the Alkanes

With over 16 million known organic chemicals, it is useful to have an internationally agreed-upon set of nomenclature rules. We hinted at such a system with the first 10 normal alkanes in Table 12.4. These names form the basis of the more complex names given to branched-chain alkanes. This molecule is called 3-methylheptane. Why?

1. We identify and name the longest straight chain in the molecule, which in this case is derived from *heptane*.

2. We name the branch, using the name of the alkane with the same number of carbon atoms as are in the branch (methane, in this case). However, we replace the *-ane* ending with *-yl* to indicate that the group is a branch. The word *methyl* then is added to the front of our molecule's name, which becomes methylheptane.

3. We identify the location of the branch by associating it with the numbered carbon atom of the main chain to which it is attached. We number the main chain from the end that gives the branch point the lowest number. The final name is *3-methylheptane*.

Various other rules guide us in naming more complex alkanes. These rules, collectively, are known as IUPAC nomenclature rules after the International Union of Pure and Applied Chemistry, whose members developed the naming system. All compounds in the world can be named using IUPAC nomenclature, but many compounds have less structured names. For instance, as we learned in Chapter 2, dihydrogen monoxide is commonly known as water. The common names, unfortunately, have few rules—and many exceptions. It is best to learn the common names as you are exposed to the compound, though with all but the most common compounds, we will use the IUPAC nomenclature.

Cyclic Alkanes

Carbon atoms can be bonded into rings, to form **cyclic alkanes.** The four simplest cyclic alkanes are shown below. Cyclohexane is used in the preparation of nylon fiber and is a nonpolar solvent. Cyclopentane is increasingly employed as a foaming agent—a substance that is needed to create foam—in the preparation of insulation for use in refrigerators. It replaces some of the chlorofluorocarbons implicated in global warming.

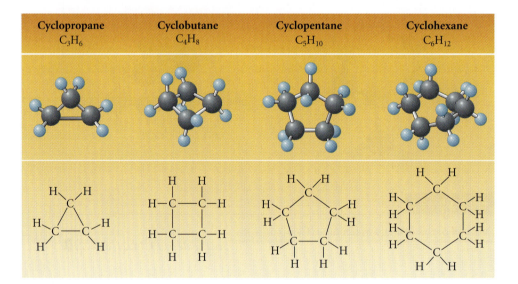

Cyclopropane C_3H_6	Cyclobutane C_4H_8	Cyclopentane C_5H_{10}	Cyclohexane C_6H_{12}

 Video Lesson: Alkenes and Alkynes

Alkenes

Alkenes are hydrocarbons that contain at least one carbon-to-carbon double bond (C=C). The members of this class of compounds are important to the chemical industry as starting materials in the manufacture of a host of organic compounds. Only tiny amounts of alkenes are found in crude oil, but certain alkenes are made from crude oil in huge quantities, in processes that we will consider shortly. The simplest three alkenes are shown at the top of the next page.

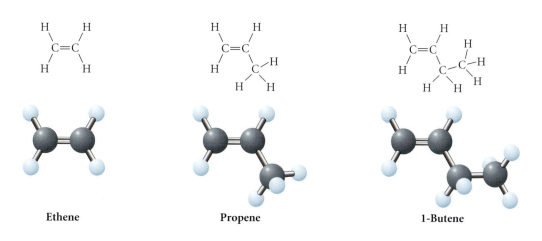

Ethene   Propene   1-Butene

The names of the alkenes are derived from the names of the corresponding alkanes, but with *-ene* at the end instead of *-ane*. The simple alkenes, those with only one double bond, form a homologous series with the general formula C_nH_{2n}. This general formula is also the same as that of the cyclic alkanes.

Alkenes are known as **unsaturated hydrocarbons** because, unlike the alkanes, they are not "saturated" with hydrogen. Each C=C bond in an alkene can react with one hydrogen molecule, under suitable conditions, to generate the corresponding alkane. For example, the **hydrogenation** (addition of a molecule of hydrogen) of ethene gives ethane. This is a general reaction of the alkenes. Platinum metal is a catalyst in this reaction and therefore is indicated over the arrow in the chemical equation that represents the reaction.

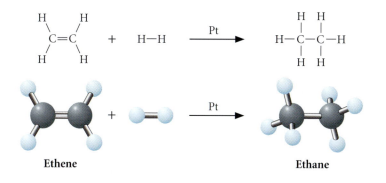

Ethene   Ethane

Alkynes

Hydrocarbons with a carbon-to-carbon triple bond are known as **alkynes**. Those with only one triple bond form a homologous series with the general formula $C_nH_{(2n-2)}$. The first three members of this series are shown below. Ethyne (acetylene) is burned with oxygen to release the energy that generates very high temperatures in oxy-acetylene welding, shown in Figure 12.9.

$H-C\equiv C-H$

Ethyne

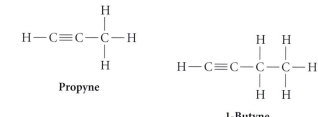

Propyne

1-Butyne

FIGURE 12.9

Photograph of oxy-acetylene welding. High temperatures are obtained during the combustion of acetylene. This melts the metals, allowing a strong weld to form.

EXERCISE 12.2   Care with General Formulas

We have seen that the general formula for the alkenes is C_nH_{2n}. Does this mean that every hydrocarbon with that general formula must be an alkene?

Solution

Can you draw any hydrocarbon structure that has twice as many hydrogen atoms as carbon atoms, but no double bonds? It cannot be done using straight chains or branched chains of carbon atoms, but it *can* be done if the carbon chain bonds back on itself, to form a cyclic hydrocarbon. As you can verify from the examples below, cyclic alkanes share the general formula C_nH_{2n} with alkenes.

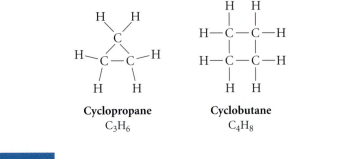

Cyclopropane   Cyclobutane
C_3H_6 C_4H_8

PRACTICE 12.2

Provide a structural drawing for a cyclic hydrocarbon that would be an isomer of hexene.

See Problems 23 and 24.

The general formula C_nH_{2n} is *not* unique to alkenes. We saw in Exercise 12.2 that each alkene with three or more carbon atoms will have a cyclic alkane as one of its isomers. For example, cyclobutane is an isomer of butene:

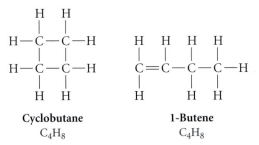

Cyclobutane   1-Butene
C_4H_8 C_4H_8

Isomers do not always have to belong to the same homologous series; they just have to have the same atoms bonded in different ways.

The naming of alkenes and alkynes is complicated by the need to distinguish exactly where the double or triple bond lies within the molecule. This is handled by attaching a number in front of the name of the molecule. The number corresponds to the location of the first carbon in the double or triple bond. Just as in naming branched alkanes, the number we use must be the lowest number that can be generated by numbering the longest carbon chain. For example, $CH_3CH=CHCH_2CH_3$ is 2-pentene, not 3-pentene. In this way, we see that there are only two alkenes with the formula C_4H_8: 1-butene ($CH_2=CHCH_2CH_3$) and 2-butene ($CH_3CH=CHCH_3$).

Geometric Isomers

Video Lesson: Isomers

Geometric isomers are common in the alkenes. What is a **geometric isomer**? Looking at the boiling points of two of the isomers of C_4H_8 shown at the top of the next page will help us arrive at an answer to this question.

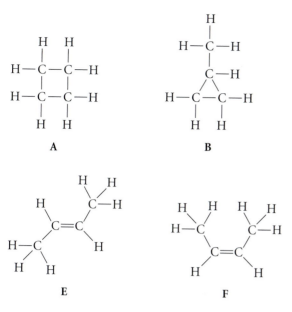

A   B   C   D

E   F

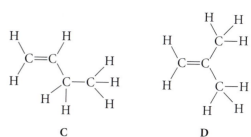

We see that two cyclic molecules, cyclobutane (**A**) and methylcyclopropane (**B**), are the only *alkanes* we can draw. The remainder of the molecules with the formula C_4H_8 are *alkenes*, including 1-butene (**C**), 2-methylpropene (**D**), and two different 2-butenes (**E** and **F**). Are the two 2-butenes really different? We could imagine that if we rotated the C=C bond around, we would have the same structures. However, *the C=C bond doesn't allow rotation to occur*, so these two molecules cannot be redrawn to show the same structure. In fact, their physical properties indicate to us that they are completely different. This is where the boiling points prove important. Molecule (**F**) has a boiling point of 3.7°C, whereas molecule (**E**) has a boiling point of 1°C, giving evidence for two different molecules.

To distinguish between these two different molecules, we use the designations *cis* and *trans* to describe their structures, as illustrated in Figure 12.10. These prefixes are always given in italics. In the *cis* form, the groups on either side of the C=C are on the same side of the molecule (Figure 12.10). (As a remembering device, think of *cis* as "sisters.") The molecule with the 3.7°C boiling point is *cis*-2-butene. In the *trans* form, the groups on either side of the C=C are on opposite sides of the molecule (Figure 12.10). (Think about the groups having to *transfer* from one side to the other.) This molecule is *trans*-2-butene.

In general, geometric isomers are isomers that differ in the location of the atoms only because of the geometry of the carbon to which they are attached. *Cis* and *trans* isomers are geometric isomers because the carbons of the alkene group have a fixed geometry.

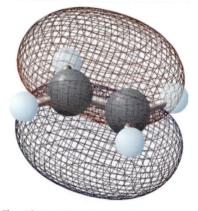

The side-to-side overlap of adjacent *p* orbitals within the C=C bond gives rise to the pi bond between the carbon atoms. As an atom on one end of the bond begins to rotate the *p* orbital overlap decreases. Thus, the pi bond restricts rotation of the atoms at either end.

FIGURE 12.10

Cis and *trans* isomers of the alkenes. The *cis* isomer contains groups on the same side of the alkene. The *trans* isomer contains groups on opposite sides of the alkene.

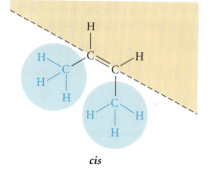

cis

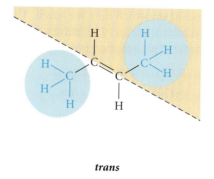

trans

Consuming *trans* fatty acids is considered bad for your health. They arise when vegetable oil is hydrogenated. The resulting partially hydrogenated vegetable oils contain a small amount of *trans* fats. Is the compound below a *cis* or a *trans* fatty acid?

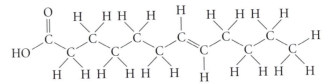

Solution

This compound is a *trans* fatty acid. The groups on either end of the C=C are on opposite sides of the molecule

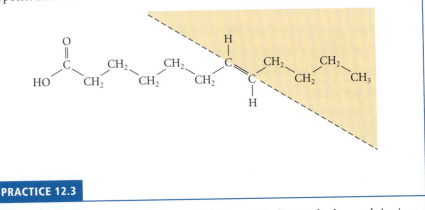

PRACTICE 12.3

Draw all of the alkenes with the formula C_5H_{10} and indicate whether each is *cis* or *trans*.

See Problems 25 and 26.

Video Lesson: Aromatic Hydrocarbons

Aromatic Hydrocarbons

Aromatic hydrocarbons contain one or more aromatic rings, which were discussed in Chapter 9. A common aromatic compound is benzene. This compound and its derivatives are used in the production of a vast array of consumer products such as polystyrene and other plastics, nylon, detergents, and pharmaceuticals. The molecule consists of three alternating carbon-carbon double bonds in a six-carbon ring. Benzene and several more complex aromatic compounds are shown on page 505.

Alkyl Groups

Many hydrocarbons can be regarded as being composed of a main chain of carbon atoms with various hydrocarbon branches attached. These branches are known as **alkyl groups**, and many have a name derived from the name of the alkane on which they are based. The names of other structures are derived from common names. The names of alkyl groups are placed in front of the parent name in alphabetical order.

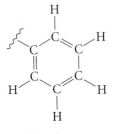

Phenyl group

—CH_3 is a methyl group.

—C_2H_5 is an ethyl group.

—C_3H_7 is a propyl group.

—C_4H_9 is a butyl group.

—C_5H_{11} is a pentyl group.

—C_6H_{13} is a hexyl group.

—C_7H_{15} is a heptyl group.

—C_8H_{17} is an octyl group.

—**Ph** (C_6H_5) is a phenyl group.

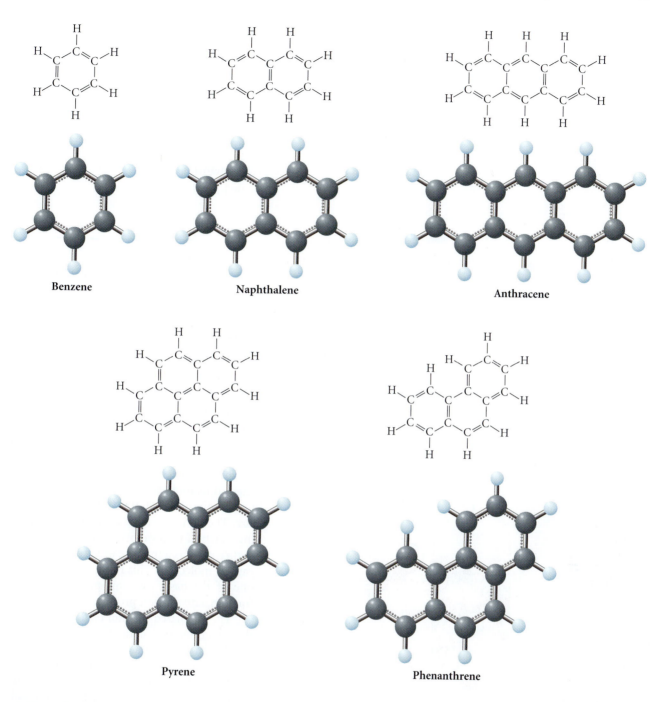

Benzene

Naphthalene

Anthracene

Pyrene

Phenanthrene

EXERCISE 12.4   Naming with Alkyl Groups

Fill in the blanks, corresponding to the type of alkyl group, to complete the names of the hydrocarbons shown.

a.

$$CH_3-CH-CH_2-CH_3$$ with CH_3 above CH 2-[   ] butane

b. $CH_3-CH_2-CH_2-CH-CH_2-CH_2-CH_2-CH_3$ with CH_3-CH_2 above CH 4-[   ] octane

c.

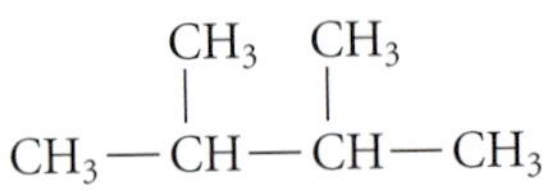

$$CH_3—CH—CH—CH_3$$

with CH_3 groups on carbons as drawn

2,3-[   ] butane

d.

$$CH_3—CH_2—CH—C—CH_3$$

with $CH_3—CH_2$ and CH_3 groups and CH_3 below

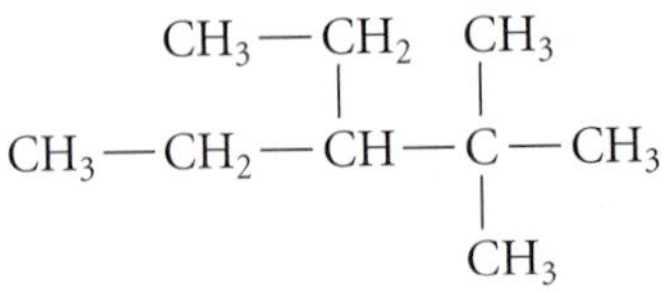

3-[   ]-2,2-[   ] pentane

Solution

a. 2-**methyl**butane

b. 4-**ethyl**octane

c. 2,3-**dimethyl**butane

d. 3-**ethyl**-2,2-**dimethyl**pentane

PRACTICE 12.4

Draw the structure of 2,2-dimethyl-4-propyloctane.

See Problems 17 and 18.

12.4 Separating the Hydrocarbons by Fractional Distillation

Crude oil contains a wide variety of carbon-based compounds. Separating those compounds so that they can be used to make fabrics, medicines, and other compounds we need everyday is an important procedure. The first step in this process is to separate the oil into several "fractions" using fractional distillation, as shown in Figure 12.11. Each of the fractions contains a selection of hydrocarbons with generally similar physical properties. Because a hydrocarbon's boiling point is broadly related to its molecular size, each fraction contains molecules that are similar in molecular size. The oil is heated to around 400°C, causing much of it to vaporize. The tarry residue that remains is a mixture of very large hydrocarbon molecules. This residue is collected as one of the basic fractions produced by the fractional distillation process.

The vaporized portion of the oil rises up the fractionating column (Figure 12.11), which can be around 60 m in height. The vapor cools as it rises, causing the various fractions of the oil to condense back into liquid form at different heights up the column, depending on their boiling points. At selected levels, the hydrocarbons that are in liquid form gather on collecting trays and are piped away. The smallest (and lowest-boiling) hydrocarbons do not condense into liquid at all but emerge at the top of the column as "refinery gas."

This primary fractional distillation process is typically used to separate the crude oil into six major fractions, summarized in Table 12.5. These fractions can be subjected to secondary fractional distillation or can be fed

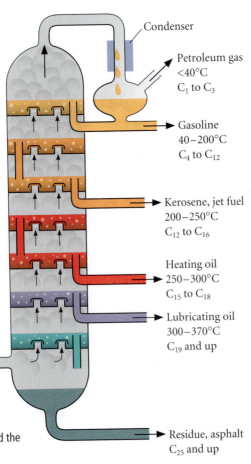

Condenser

Petroleum gas
<40°C
C_1 to C_3

Gasoline
40–200°C
C_4 to C_{12}

Kerosene, jet fuel
200–250°C
C_{12} to C_{16}

Heating oil
250–300°C
C_{15} to C_{18}

Lubricating oil
300–370°C
C_{19} and up

Hot petroleum
(crude oil)

Residue, asphalt
C_{25} and up

FIGURE 12.11

Fractional distillation—a diagram of the process and the main fractions.

How do we know?

Which hydrocarbons are in crude oil?

On March 24, 1989, the oil tanker *Exxon Valdez* ran aground in Prince William Sound, Alaska, spilling over 40 million liters of crude oil into the sea. Despite the massive clean-up effort, small amounts of crude oil are still found along the shores near the accident. Toward the end of 1990, samples of these oil residues were collected as part of research to monitor the fate of the oil as the environment slowly recovered. It turned out, however, that some of the samples could not have come from the *Exxon Valdez*, because they contained the wrong mixture of hydrocarbons. The debris from older oil spills was complicating the picture. How was this determined? It was done using a technique called **gas chromatography (GC)**, one of the three most common methods for analyzing the hydrocarbon content of crude oil.

In gas chromatography, a sample such as crude oil that is to be analyzed is converted into vapor by heating and then carried by a flow of inert gas (such as He) through a heated column packed with solid powder such as silica, as shown in Figure 12.12. In some cases, the particles of solid in the column may be coated with a nonvolatile liquid, a technique known as **gas–liquid chromatography (GLC)**. As the sample flows through the column, the components of the mixture travel at different speeds because of the differing extents to which they are adsorbed onto the solid phase or are soluble in the liquid phase (in GLC). The components emerge from the column separately and are detected, often by their effect on a detecting flame. The apparatus will have previously been calibrated using a range of known hydrocarbons, so specific hydrocarbons in the sample can be identified by comparison with the results of the calibration.

An example of the results of an analysis of one type of crude oil is shown in Figure 12.13. The technique is very powerful, particularly in providing a GC "fingerprint" of different samples from different origins, but it has significant limitations. In particular, it cannot readily distinguish between some of the different structural isomers of hydrocarbons. For this reason, more thorough analysis of the hydrocarbons in crude oil also makes use of the techniques of mass spectrometry and nuclear magnetic resonance spectroscopy.

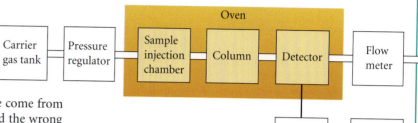

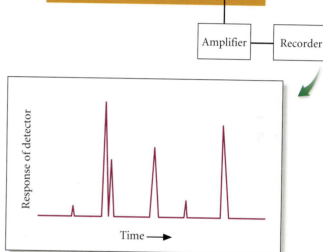

FIGURE 12.12

The operating principles of gas chromatography and a sample readout (a gas chromatogram).

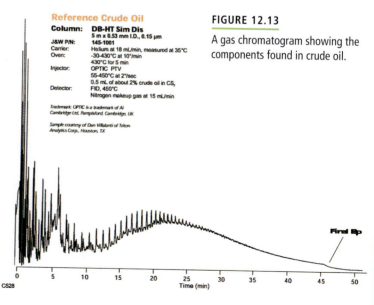

FIGURE 12.13

A gas chromatogram showing the components found in crude oil.

TABLE 12.5	Products of the Fractional Distillation of Crude Oil and Their Uses		
Fraction	**Formulas**	**Uses**	**Boiling Point (°C)**
Natural gas	CH_4 to C_4H_{10}	Fuel, cooking gas	<0
Petroleum ether	C_5H_{12} to C_6H_{14}	Solvent for organic compounds	30–70
Gasoline	C_6H_{14} to $C_{12}H_{26}$	Fuel, solvent	70–200
Kerosene	$C_{12}H_{26}$ to $C_{16}H_{34}$	Rocket and jet engine fuel, domestic heating	200–300
Heating oil	$C_{15}H_{32}$ to $C_{18}H_{38}$	Industrial heating, fuel for producing electricity	300–370
Lubricating oil	$C_{16}H_{34}$ to $C_{24}H_{50}$	Lubricants for automobiles and machines	>350
Residue	$C_{20}H_{42}$ and up	Asphalt, paraffin	Solid

into further processing steps to convert them into useful products. Table 12.5 also lists some of the major uses made of each fraction. We will explore some of these uses in more detail, and in doing so we will investigate the chemical properties and reactivities of the hydrocarbons in each fraction.

12.5 Processing Hydrocarbons

The hydrocarbons within the fractions obtained by distillation are normally subjected to further processing before being used as fuels or as feedstocks for the chemical industry. The two major types of processing are known as cracking and reforming.

Cracking

Many of the most useful hydrocarbons are the smaller ones, with fewer than 10 carbon atoms. At the refinery, long-chain hydrocarbons are isolated from the crude oil by fractional distillation. Then they are broken into a mixture of shorter-chain hydrocarbons, including some branched and unsaturated hydrocarbons, by means of a process called **cracking**. This process uses heat to break some of the carbon–carbon bonds in the long-chain hydrocarbons. The products of cracking often contain short saturated and unsaturated hydrocarbons. There are various types of cracking:

- **Thermal cracking** uses heat alone.
- **Catalytic cracking** uses heat plus a catalyst.
- **Hydrocracking** uses heat plus a catalyst in the presence of hydrogen gas, encouraging formation of saturated rather than unsaturated products.
- **Steam cracking** uses heat plus steam and a catalyst.

Reforming

Another crucial process at the refinery is called **reforming**. In this process hydrocarbons distilled from petroleum are passed through a catalyst bed and converted into more useful forms. When a mixture is reformed, the predominantly straight-chain hydrocarbons are made into a mixture containing more aromatic and branched-chain hydrocarbons.

Reforming is a key step in making the gasoline we use as automobile fuel. Gasoline must contain a suitable proportion of branched and aromatic hydrocarbons to prevent the problem known as knocking. This is the rapid explosion of the gas–air mixture as it is compressed in the automobile cylinder *before* the spark plug creates the spark that is intended to cause the mixture to explode. In some cases knocking damages a car's engine, but most often it results only in poor engine performance and reduced gas mileage. Addition of reformed distillates to gasoline reduces knocking.

Gasoline

The gasoline we use to power automobiles is a complex mixture of well over a hundred different hydrocarbons, which can include molecules such as pentane, benzene, ethylbenzene, toluene (methylbenzene), and xylene (dimethylbenzene). The precise composition of the mixture varies with the brand of gasoline, the location, and the time of year, but the hydrocarbon molecules in automobile gas generally contain between 5 and 17 carbon atoms, many being in the C_5-to-C_9 range. One factor influencing the optimum mixture of hydrocarbons is the need to produce a fuel with a vapor pressure that allows it to evaporate to an appropriate degree under the conditions in which it will be used.

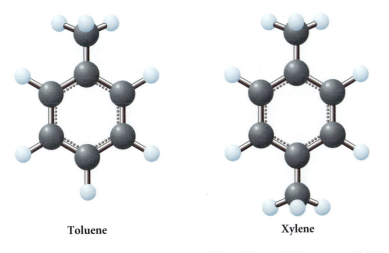

Toluene   **Xylene**

   The extent to which a particular type of gasoline will burn smoothly in an engine (without knocking) is indicated by using the octane rating. This is based on the early twentieth-century discovery that pure isooctane produced minimal knocking in an engine, whereas pure heptane produced a very high level of knocking. On the octane scale, 2,2,4-trimethylpentane (also known as isooctane) is given the octane rating of 100, and heptane is given the octane rating of 0. By comparing mixtures of isooctane and heptane to different formulations of gasoline, an octane rating for the gasoline can be assigned. For example, a high-quality gasoline, which produces a level of knocking equivalent to a mixture of 90% isooctane and 10% heptane, is assigned an octane rating of 90. Low-quality gasoline can be improved by adding other compounds with octane ratings that are higher than 100, such as ethanol (octane rating 112) or toluene (octane rating 118). Brands of automobile gasoline are typically available with octane ratings in the range of 83 to 98.

$$CH_3-\underset{\underset{\textstyle CH_3}{|}}{\overset{\overset{\textstyle CH_3}{|}}{C}}-CH_2-\underset{\overset{\textstyle CH_3}{|}}{CH}-CH_3 \qquad \textbf{2,2,4-Trimethylpentane (isooctane)}$$

$$CH_3-CH_2-CH_2-CH_2-CH_2-CH_2-CH_3 \quad \textbf{Heptane}$$

12.6 Typical Reactions of the Alkanes

Globally, the most significant reaction of the alkanes is their combustion in air. As we have already seen, the complete combustion of alkanes generates carbon dioxide and water vapor and releases considerable amounts of energy. The energy can be used to provide heat, both for warmth and for cooking, to generate electricity, or to power vehicles.

The chemical industry also makes substantial use of alkanes in **substitution reactions**, in which one or more hydrogen atoms are replaced by other types of atoms, most commonly halogens such as bromine and chlorine. These substitution reactions of the alkanes are promoted by ultraviolet light, which causes the halogen molecules to split into two atoms:

$$Cl_2 \xrightarrow{hv} Cl\cdot + Cl\cdot$$

The halogen atoms generated in this process are known as **free radicals**, each having an unpaired electron. The presence of the unpaired electron makes them highly reactive. In the presence of an alkane, the free radicals can readily attack the C—H bonds. The result is a halogenated alkane and a molecule of the hydrogen halide.

For example, the simplest hydrocarbon, methane, can be converted into four different organic products and HCl by successive substitution reactions with chlorine:

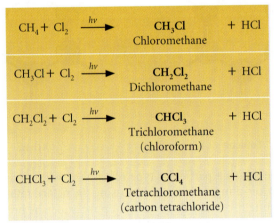

These four products can be put to a variety of uses. Chloromethane is used in the manufacture of silicone-based polymers and as a solvent in the production of certain types of rubber. Dichloromethane is widely used as a solvent in the chemical industry and within products such as paint strippers.

Trichloromethane, also known as chloroform, was one of the first anesthetics, pioneered as such in 1847 by the Scottish physician Sir James Simpson. Because of the sweetness of this molecule, it was also used as the solvent in early cough medicines. However, this compound was found to be both toxic and carcinogenic, and ethanol has taken its place in medicines. Chloroform is still used as an industrial solvent. Tetrachloromethane, also known as carbon tetrachloride, once was the solvent most commonly used for the dry cleaning of clothes and other fabrics. It was also used as a fire extinguisher agent because it is not flammable. Unfortunately, it too is toxic and carcinogenic. It has now largely been abandoned for dry cleaning and fire extinguishing in favor of less toxic compounds such as liquid carbon dioxide (see Chapter 10). However, it is still used as a solvent in various industrial processes.

Another important reaction of the alkanes is their **dehydrogenation** to produce the corresponding alkenes. In a dehydrogenation reaction, a molecule of

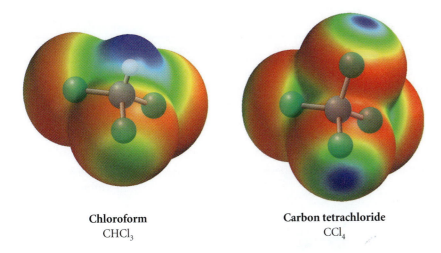

Chloroform
$CHCl_3$

Carbon tetrachloride
CCl_4

hydrogen is removed from the starting material. The result is an unsaturated hydrocarbon, an alkene. For example, dehydrogenation of ethane produces ethene. The dehydrogenation of propane produces propene, and so on:

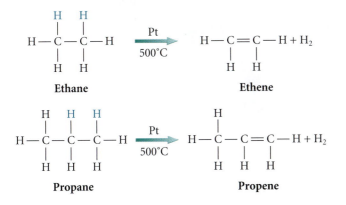

Ethane　　　　　　　　　　　　　　**Ethene**

Propane　　　　　　　　　　　　　　**Propene**

12.7 The Functional Group Concept

The processes of adding chlorine atoms or C=C double bonds to the basic structure of alkanes are examples of adding **functional groups** to a hydrocarbon framework. A functional group is an atom or group of atoms with a characteristic set of chemical properties. Chemists often make sense of the infinite variety of chemical reactions in organic chemistry by treating each molecule as a hydrocarbon framework with a particular set of functional groups incorporated within that framework.

A chlorine atom bonded to a hydrocarbon is an example of an **alkyl halide** functional group, a term that covers any of the *halogen* atoms bonded to a hydrocarbon framework. These and many other examples of functional groups are listed in Table 12.6. For example, the C=C double bond (the alkene functional group) is a functional group whose properties depend on the type of bonding between the atoms involved. The C≡C triple bond (the alkyne functional group) is another example.

Much of the chemistry done with hydrocarbons from oil involves the addition of specific functional groups. Chemists can modify the structure of the basic hydrocarbon chain, adjusting its length and degree of branching, and then add whatever selection of functional groups will create the molecules they are seeking.

TABLE 12.6 Selected Functional Groups

Structure	Group	Example	3-D Structure	Name	Selected Uses
$\diagdown C = C \diagup$	Alkene	$CH_2{=}CH_2$		Ethene (ethylene)	Refrigerant, production of polyethylene
$-C{\equiv}C-$	Alkyne	$HC{\equiv}CH$		Acetylene	Welding, cutting, brazing
$-\overset{|}{\underset{|}{C}}-O-H$	Alcohol	CH_3CH_2OH		Ethanol	Liquors, industrial solvent
$\diagdown C - O - C \diagup$	Ether	$CH_3CH_2OCH_2CH_3$		Diethyl ether	Industrial solvent
aldehyde structure	Aldehyde	$CH_2{=}O$		Formaldehyde	Bactericide, fungicide, chemicals production
ketone structure	Ketone	$CH_3-\overset{O}{\overset{\|}{C}}-CH_2-CH_3$		Methyl ethyl ketone	Solvent in synthetic rubber industry
carboxylic acid structure	Carboxylic acid	$H-\overset{O}{\overset{\|}{C}}-OH$		Formic acid	Manufacture of textiles, pesticides; electroplating
ester structure	Ester	$CH_3-\overset{O}{\overset{\|}{C}}-O-CH_3$		Methyl acetate	Paint remover, pharmaceutical manufacture
amine structure	Amine	$CH_3(CH_2)_3NH_2$		Butylamine	Manufacture of rubber, insecticides

(Continued)

TABLE 12.6   (Continued)

Structure	Group	Example	3-D Structure	Name	Selected Uses
—C≡N	Nitrile	CH_3—C≡N	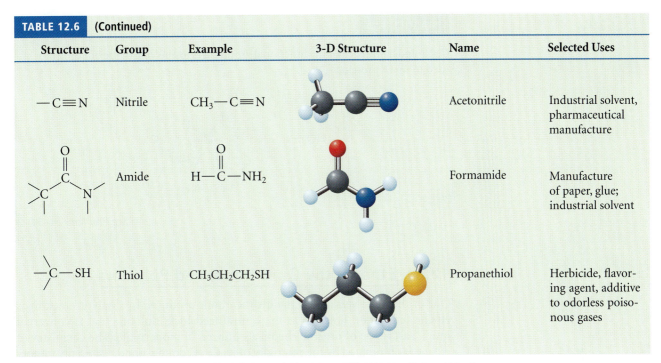	Acetonitrile	Industrial solvent, pharmaceutical manufacture
![amide structure]	Amide	![formamide example]		Formamide	Manufacture of paper, glue; industrial solvent
—C—SH	Thiol	$CH_3CH_2CH_2SH$		Propanethiol	Herbicide, flavoring agent, additive to odorless poisonous gases

Source: Kelter, Scott, and Carr. *Chemistry, A World of Choices,* 2nd Ed; McGraw-Hill, 2003: New York; and New Jersey Department of Health & Senior Services Right-to-Know Program Fact Sheets.

EXERCISE 12.5   **Functional Groups and You!**

Circle and name each of the functional groups in cyclexanone, a pharmaceutical agent used to prevent coughing (an antitussive agent).

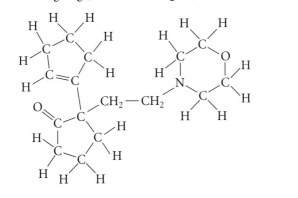

Solution

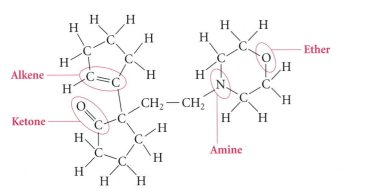

Circle and name each of the functional groups in tyrosine, one of the building blocks used to make proteins.

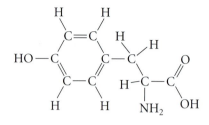

See Problems 51–54.

12.8 Ethene, the C=C Bond, and Polymers

One of the most versatile of the hydrocarbons produced by catalytic cracking is ethene (ethylene). It serves as the central feedstock molecule that is converted into a vast range of organic chemicals utilized in modern life. The chemistry of ethene is dominated by the reactivity of its C=C double bond. Especially important is the **addition reaction**, in which two parts of another chemical species become added to the atoms at either end of the double bond. The C=C of ethene is converted into a single bond in the process. The *pi* electrons of the double bond end up participating in the new bonds holding the groups that have become added "across" the double bond. We've already examined one of the addition reactions with ethene, hydrogenation.

As another example, ethene can be converted into ethanol (CH_3CH_2OH) by reaction with water in the presence of phosphoric acid (H_3PO_4) catalyst at 330°C. Industrially, ethanol can be made in large quantities using this process. However, because of the value of ethene as a feedstock for other organic molecules, much of the ethanol that is made in the United States today comes from the fermentation of sugars found in corn. Ethanol is an example of an **alcohol**, a compound carrying the **hydroxyl group** (—OH group).

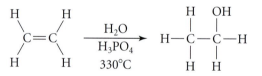

Another addition reaction converts ethene into dichloroethane (ethylene dichloride), an intermediate in the production of the seemingly ubiquitous material polyvinyl chloride (PVC).

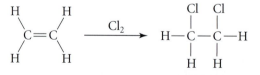

Ethene is the starting material for a great many of the plastics that are such a common feature of modern life (see Chapter 13). The simplest of these plastics is **polyethene**, which is made when huge numbers of ethene molecules participate in addition reactions *with themselves* to generate long chain-like polyethene molecules. In the reaction, one molecule of ethene adds to an adjacent molecule of ethene, which adds to another molecule of ethene, and so on. The result is a very long chain of carbon atoms.

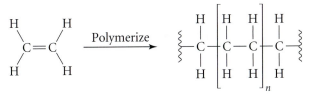

Polyethene (polyethylene)

Video Lesson: Organic Polymers

Polyethene, which is more commonly known as **polyethylene**, is an example of a **polymer** (literally, "many units"), a compound composed of large molecules made by the repeated bonding together of smaller **monomer** (literally, "one unit") molecules, ethene in this case. Because the reaction that bonds the ethene monomers together is an addition reaction, polyethylene is known as an **addition polymer**, formed by the process of **addition polymerization**.

Different types of polyethylene can be made by varying the temperatures and pressures of the reaction, as well as changing the type of catalysts and initiating substances used. These different forms are classified into two main types: **high-density polyethylene (HDPE)**, in which the molecules are long and linear with no branch points, and **low-density polyethylene (LDPE)**, with shorter, branched-chain molecules, as shown in Figure 12.14. The length of the polyethylene chain imparts different properties to the resulting molecule. For instance, because we know that shorter alkanes have lower melting points, you might expect a shorter polyethylene chain to have a lower melting point than a longer polymer.

The high-density part of HDPE is a result of the close packing of the linear molecules. This makes HDPE a strong plastic, used to make such things as blow-molded bottles and toys. LDPE has a more flexible, less crystalline structure; it is used to make such things as plastic trash and grocery bags, packaging films, insulation sleeves around electrical cables, and squeeze bottles. The properties of polyethylene can be modified by incorporating some other atoms, such as chlorine or sulfur, into its structure as it is formed. This can make the resulting modified polyethylene more resistant to oxidation, for example, or more flexible. Many of the polyethylene-based materials around us are actually composed of such modified polyethylenes.

We can prepare polymers that have properties different from those of polyethylene, and yet are closely related to it in chemical structure, by using monomers in which one or more of the hydrogen atoms of ethene are replaced with other atoms or larger chemical groups. Table 12.7 lists some of these.

FIGURE 12.14

HDPE and LDPE. The molecular structures of HDPE and LDPE are similar, but their processing gives them unique properties for use in different materials.

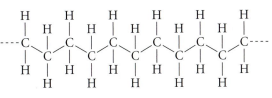

TABLE 12.7 **Polymerization of Some Common Monomers**

Monomer	3-D Structure	Polymer	Uses
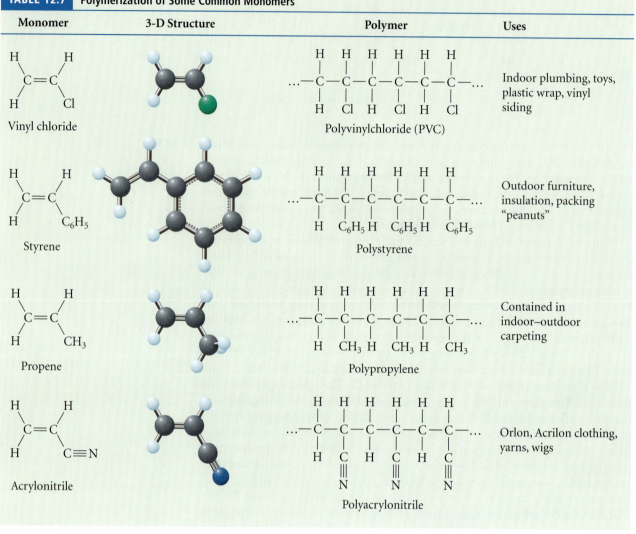			Indoor plumbing, toys, plastic wrap, vinyl siding

Vinyl chloride / Polyvinylchloride (PVC)

Styrene / Polystyrene — Outdoor furniture, insulation, packing "peanuts"

Propene / Polypropylene — Contained in indoor–outdoor carpeting

Acrylonitrile / Polyacrylonitrile — Orlon, Acrilon clothing, yarns, wigs

Video Lesson: Alcohols, Ethers, and Amines

12.9 Alcohols

As the alcohol found in alcoholic beverages, ethanol is the best known of the alcohols. Ethanol can be made by hydrating ethene. However, because of the value of ethene as a feedstock for other organic molecules, and because of the toxicity of some of the by-products of this process, much of the ethanol that is typically used for human consumption is made by the biological activity of yeast cells. In this reaction, the yeast consumes glucose in a process known as **fermentation** (see Chapter 14). The waste products of the process are ethanol and carbon dioxide. These waste products are toxic to the yeast and are excreted from the cell. Fortunately, the industrial chemist can easily harvest the ethanol by distilling it from the solution. The chemist also can capture the CO_2 gas given off in the reaction in order to make other products.

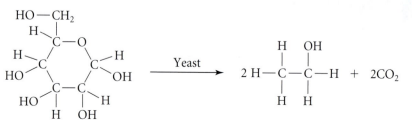

The ethanol that is distilled from the fermentation mixture unavoidably contains a small amount (roughly 5%, by volume) of water. For beverages, this water is not a problem. However, if the ethanol is to be used as an additive to fuel used in automobiles, the water must be removed. The process used to make 100% ethanol by fermentation often leaves behind traces of toxic compounds, which render the pure ethanol undrinkable.

Other alcohols of note include methanol, 2-propanol, the "dihydroxy alcohol" 1,2-ethanediol, and the "trihydroxy alcohol" 1,2,3-propanetriol, whose structures and uses can be found in Table 12.8. Note the way in which numbers are used to indicate which carbon atom carries the —OH (hydroxy) group or groups in alcohol structures. We can also use the structures themselves to explain the way in which alcohols are classified into categories known as primary, secondary, or tertiary alcohols, a type of classification that can be applied to other functional groups as well.

Primary alcohols have only one carbon atom bonded to the carbon atom that carries the hydroxyl group. This means that ethanol is a primary alcohol. **Secondary alcohols** have two carbon atoms bonded to the carbon atom that carries the hydroxyl group, so 2-propanol is a secondary alcohol. **Tertiary alcohols** have three carbon atoms bonded to the carbon atom that carries the hydroxyl group, so 2-methyl-2-propanol is a tertiary alcohol. This classification system is well correlated with the reactivity of the alcohols. For instance, primary alcohols undergo many reactions more rapidly than do secondary or tertiary alcohols. For example, primary alcohols react fastest, and tertiary alcohols slowest, in the formation of esters.

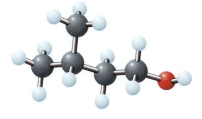
Primary alcohol

Secondary alcohol

Tertiary alcohol

12.10 From Alcohols to Aldehydes, Ketones, and Carboxylic Acids

Video Lesson: Carbonyl-Containing Functional Groups

In addition to determining the rate of reaction, the distinction between primary, secondary, and tertiary alcohols is important in determining what types of molecules can be made from the alcohol. For example, primary alcohols can undergo oxidation to form **aldehydes**, which carry a **carbonyl functional group** (C=O) with at least one H atom bonded to the carbonyl carbon atom. The oxidation is usually performed with a highly oxidized metal such as chromium (VI). In the lab, however, special reactions must be performed to prevent the overoxidation of the aldehyde. Methanol, for example, can be oxidized to the aldehyde methanal (H_2C=O), also known as formaldehyde. The solution formed when formaldehyde is dissolved in water is most commonly known as formalin, used to preserve biological specimens.

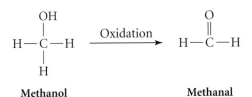

Methanol   **Methanal**

TABLE 12.8   Structures and Uses of Selected Alcohols

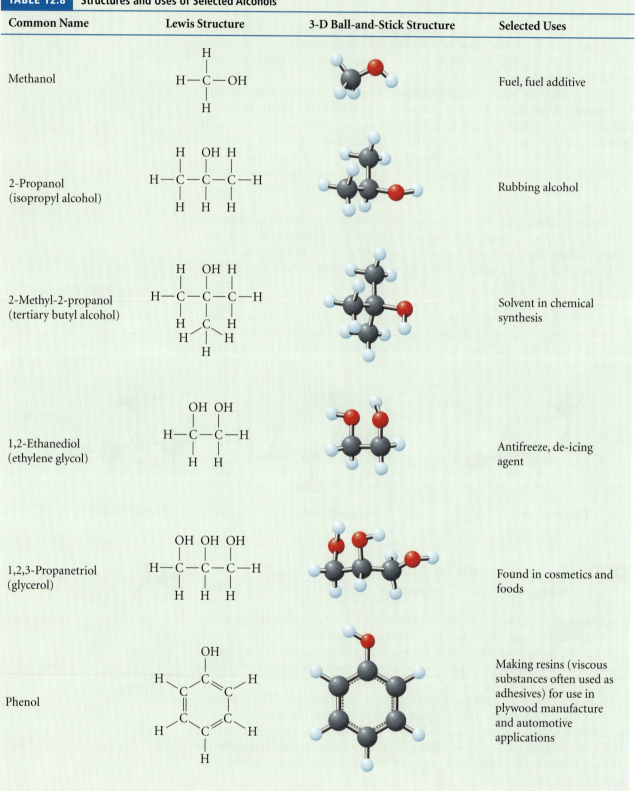

Common Name	Lewis Structure	3-D Ball-and-Stick Structure	Selected Uses
Methanol			Fuel, fuel additive
2-Propanol (isopropyl alcohol)			Rubbing alcohol
2-Methyl-2-propanol (tertiary butyl alcohol)			Solvent in chemical synthesis
1,2-Ethanediol (ethylene glycol)			Antifreeze, de-icing agent
1,2,3-Propanetriol (glycerol)			Found in cosmetics and foods
Phenol			Making resins (viscous substances often used as adhesives) for use in plywood manufacture and automotive applications

Note that this reaction is similar to the dehydrogenation reactions of the alkanes. As another example, ethanol can be oxidized to ethanal. Ethanal, also known as acetaldehyde, is used primarily in the production of polymers (Section 12.12) and medicines.

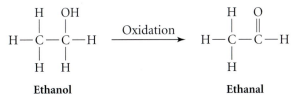

An aldehyde is named by taking the name of the alcohol from which it is derived and replacing the ending -*ol* with -*al*. No number is needed as part of the name to indicate where the aldehyde group occurs in the compound, because it always occurs at the end of a carbon chain.

The hydroxyl group of *secondary* alcohols can be oxidized to a carbonyl group. In this case, compounds called **ketones** are formed. Ketones differ from aldehydes in that they have two carbon atoms bonded to the carbonyl carbon atom. For example, 2-propanol can be oxidized to propanone (acetone). Propanone is an excellent solvent because of its ability to dissolve both polar and nonpolar molecules. It has a low boiling point and a low toxicity, finding use as a solvent for paints and as a fingernail polish remover. For these reasons, over 3 billion kilograms of propanone are used each year in industry.

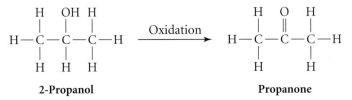

Tertiary alcohols do not react with oxidizers to make carbonyl compounds. For instance, there is no reaction when we attempt to oxidize 2-methyl-2-propanol under the conditions typical for primary and secondary alcohols. This lack of reactivity occurs because the carbon bearing the —OH group does not have a hydrogen that can be removed to make the carbonyl (C=O).

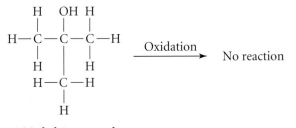

Aldehydes have a characteristic biting and often unpleasant aroma. Ketones, on the other hand, have an aroma that is not so unpleasant, but more medicinal. Chemically, though, the important difference between aldehydes and ketones is that the aldehydes can be oxidized further to form **carboxylic acids**. The carboxylic acids carry the **carboxyl functional group**, the combination of a carbonyl and a hydroxyl group. Ethanal, for example, can be easily oxidized to ethanoic acid.

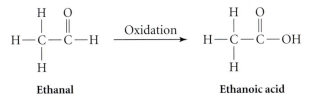

Ethoxide

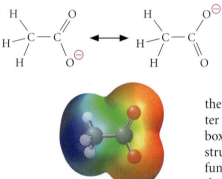

Ethanoate

Ethanoic acid (also known as acetic acid) is the organic acid that we use as vinegar when it is in the form of a weak solution (typically about 4 to 5% by mass) in water. Some bacteria can perform the conversion of ethanol to ethanoic acid, which can cause wines to spoil when they "turn to vinegar" due to bacterial contamination. However, this process can also be exploited purposefully to produce the wine vinegars that are widely used in cooking. Industrially, ethanoic acid is used in the manufacture of plastics, pharmaceuticals, dyes, and insecticides.

The carboxyl functional group is weakly acidic due to the dissociation of a hydrogen ion from the oxygen. The result is the formation of a **carboxylate anion** and a hydrogen ion (i.e., a proton). Although it would appear that any hydrogen on an oxygen could dissociate in a similar manner, the carboxylic acids have a special feature that alcohols do not have (hence the latter are not acidic). Let's explore this property further. **Why does the carboxyl functional group dissociate?**

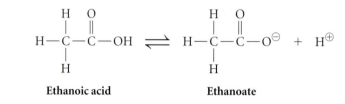

Ethanoic acid   **Ethanoate**

When we draw Lewis dot structures of compounds, we must always consider the possibility that the electrons could be placed in different locations (see Chapter 8). This would generate a resonance structure of the compound. The carboxylate anion possesses this ability. In fact, we can draw two good resonance structures for this ion. The resonance spreads the large negative charge across the functional group and helps to stabilize the carboxylate by lowering the energy of the anion.

Why doesn't the alcohol functional group also dissociate? If a hydrogen ion were to leave the molecule, a negative charge would reside on the oxygen. Unfortunately, there aren't any other resonance structures that we can draw for this **alkoxy anion**. Therefore, the electrons cannot be spread to other atoms by resonance, and the resulting alkoxyl anion is not energetically stabilized.

12.11 From Alcohols and Carboxylic Acids to Esters

An extremely important reaction, widespread throughout both natural and industrial chemistry, takes place between the alcohol and carboxylic acid functional groups. These two groups can react, often with an acid catalyst such as sulfuric acid, to *eliminate a molecule of water* in a reaction known as a **condensation reaction**. In the reaction, the carbonyl carbon from the carboxyl group becomes bonded to the oxygen atom from the alcohol group in what is known as an ester linkage or ester bond. Compounds containing this arrangement of atoms (carbonyl–oxygen–carbon) are known as **esters**. The simplest possible ester is methyl methanoate (also known as methyl formate), formed when methanol reacts with methanoic acid. Note that the first part of the name of an ester is derived from the alcohol that can be used in its formation, and the second part of the name is derived from the carboxylic acid component.

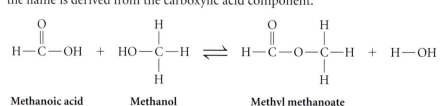

Methanoic acid   **Methanol**   **Methyl methanoate**

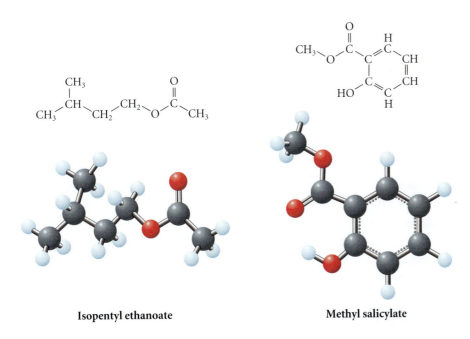

Isopentyl ethanoate

Methyl salicylate

Compounds that contain the ester functional group often have a very fruity aroma and in nature are found in plant oils. For example, isopentyl ethanoate has the odor of ripe bananas, and methyl salicylate smells like wintergreen. For this reason, esters are often used in industry to lend flavors and odors to products that we buy. Esters are also used in industry as solvents, due to their ability to dissolve both polar and nonpolar solutes. For example, ethyl ethanoate (commonly known as ethyl acetate) is found in cleaners and glues.

EXERCISE 12.6 | **Nomenclature of Esters**

Provide names for each of these esters.

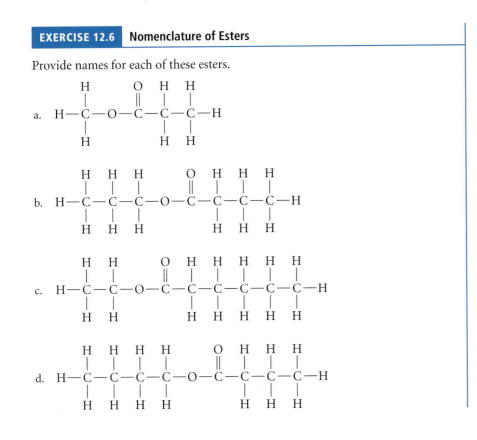

First Thoughts

All ester names are in the form "____yl ____anoate." The first part of the name is derived from the part of the molecule that was originally an alcohol, and the second part of the name is derived from the part of the molecule that was originally a carboxylic acid. Therefore you need to find the alcohol component, which is the alkyl group linked to the O atom, count its carbon atoms to get the first part of the name, and then count the carbon atoms in the rest of the molecule to get the second part of the name.

Solution

The names are (a) methyl propanoate, (b) propyl butanoate, (c) ethyl hexanoate, and (d) butyl butanoate.

Further Insights

Working in reverse, you can figure out the name of an ester that will form between any alcohol and any carboxylic acid, just by linking their names in the correct order and manner. Trace through that process for the esters considered above.

PRACTICE 12.6

Shown below is the structure of benzoic acid. Can you draw the structure of ethyl benzoate? . . . of benzyl ethanoate?

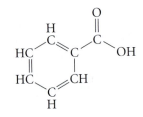

See Problems 81–84.

12.12 Condensation Polymers

Industrial preparation of polyethylene, polyvinylchloride, and polystyrene occurs by addition polymerization. But that is not the only way to make a polymer. A major class of polymers that are very useful in our lives are constructed by a condensation reaction. Condensation reactions entail the removal of a small molecule as two larger molecules are connected. If a polymer is the product of the reaction, we call the process **condensation polymerization**.

Esters can be made by condensation reactions. The process, **esterification**, can be used to link a variety of monomers into long polymer chains. The products of the reaction are **polyesters**, which may be familiar to you as parts of clothes, bedding, fabrics, upholstery, ropes, belts, and so on. They are called polyesters because of the many ester linkages that hold the monomer units within the structure of the polymer. One of the most common polyesters is polyethylene terephthalate (PETE), which as used for such things as soft drink and mouthwash bottles.

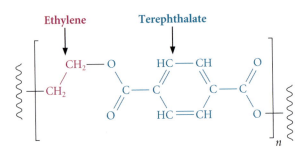

Another functional group that is a part in many important condensation polymerization reactions is the **amino functional group**, —NH$_2$. Many **amines**, unfortunately, have a particularly foul odor. For instance, the common names of some of the amines indicate the types of environments from which they were first isolated; putrescine (1,4-butanediamine) and cadaverine (1,5-pentanediamine) are examples. When an amine group participates in a condensation reaction with a carboxylic acid group, an **amide bond** is formed.

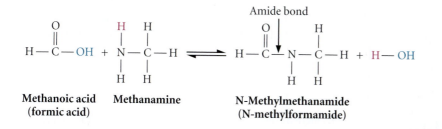

Methanoic acid   **Methanamine**   **N-Methylmethanamide**
(**formic acid**)   (**N-methylformamide**)

The amide bond holds together the monomer units of polymers known as the **polyamides**. The fabrics we know as nylons, for example, are polyamides. Nylons are named with numbers to help indicate the structure of the polymer. For example, Nylon [6,6] is made of monomers that each contain six carbon atoms. Nylon [6,6] can be made when an acid chloride functional group (COCl) on one molecule reacts with an amine functional group on another molecule. This reaction is accompanied by the release of HCl. Carbon chains of different lengths can be found within other nylon polymers.

Visualization: Synthesis of Nylon

Video Lesson: CIA

Demonstration: The Synthesis of Nylon

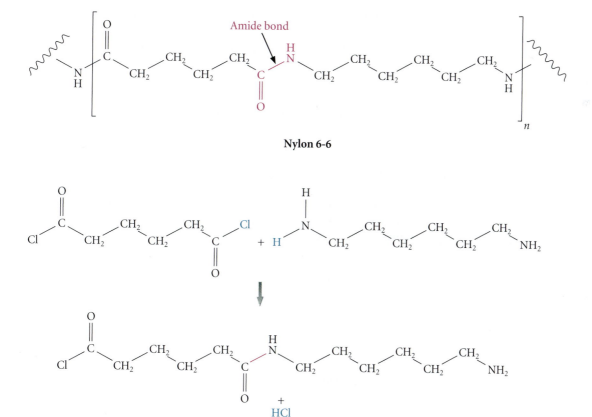

Nylon 6-6

— Acid chloride solution

— Diamine solution

12.13 Polyethers

Another important category of polymers contains the **ether** functional group at regular intervals along the polymer chain. The ether group is —C—O—C—, an oxygen atom linking two carbon atoms. Three simple examples of ethers are ethyl methyl ether, diethyl ether, and methyl t-butyl ether (MTBE). Diethyl ether is the "ether" that, like the chloroform mentioned earlier, was used as an anesthetic in the early days of medicine. It is often still used in biology classes to anesthetize and determine the characteristics of *Drosophila* flies. MTBE is the ether used as a fuel additive. Its octane rating is 116, and for this reason it can help boost the rating on otherwise poor-quality gasoline.

$$CH_3-CH_2-O-CH_3$$

Ethyl methyl ether

$$CH_3-CH_2-O-CH_2-CH_3$$

Diethyl ether

$$CH_3-\underset{\underset{CH_3}{|}}{\overset{\overset{CH_3}{|}}{C}}-O-CH_3$$

Methyl t-butyl ether
(MTBE)

The properties of ethers make them ideal as solvents. Because they lack the ability to form hydrogen bonds with themselves, they have boiling points similar to those of the alkanes. However, the presence of an oxygen atom within the structure indicates that they can participate in hydrogen bonding as acceptors with other molecules that can donate a hydrogen atom. This means that they make good solvents for chemical reactions.

Polyethers, which incorporate repeating ether linkages within very long chain-like molecules, are widely used in adhesive resins. An example of a polyether containing the ether functional group is found in the **epoxy resins**. These

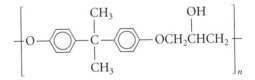

rather viscous polymers, used mostly as construction and household adhesives, contain an ether linkage formed between one carbon atom that is part of a phenyl ring, and another that is part of a methylene (CH_2) group. The monomers that make up an epoxy resin contain the phenol functional group and a triangular ether structure known as an **epoxide**.

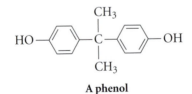

A phenol

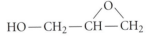

An epoxide

- The major components of crude oil include hydrocarbons.

- Crude oil has many uses, including serving as fuel and supplying feedstock molecules for industrial processes.

- Manipulation of the hydrocarbons can yield a variety of molecules containing different functional groups.

- Functional groups are common specific arrangements of atoms within a molecule that impart similar physical and chemical properties to the molecule.

- Organic molecules are named by applying a set of rules established by the International Union of Pure and Applied Chemistry (IUPAC).

12.14   Handedness in Molecules

Symmetry abounds in nature. Take, for instance, the symmetry of a person. The left-hand side of your friend appears to be a mirror image of the right-hand side. Proof of this can be found in your own hands. Hold your hand to a mirror such as that shown in Figure 12.15. What do you see? Your other hand is the mirror image. What's more interesting is that your left and right hands (despite being mirror images) are not superimposable. That is to say, your left hand and your right hand cannot be lined up perfectly. Try placing a left-handed glove on your right hand. It just doesn't fit.

Molecules can also exhibit this handedness. In chemistry, we say that "handed" molecules are **chiral**, a term derived from the Greek *cheir* ("hand"). A chiral object has a nonsuperimposable mirror image. Those molecules that lack **chirality** are said to be **achiral**, and they have a superimposable mirror image. Every chiral molecule possesses a twin that has the same chemical formula, the same structure, but a different arrangement of atoms in three-dimensional space.

What makes a molecule chiral? *Carbon-based molecules are chiral when they contain a carbon atom that is attached to four different groups.* These **chiral centers** exist in many molecules in nature. Let's examine a derivative of methane to explain this feature. Bromochlorofluoromethane (CHBrClF) contains a carbon atom attached to four different groups. A mirror image of this molecule (see Figure 12.16) shows this molecule to have handedness. These two molecules are **stereoisomers**. They are a special kind of structural isomers that differ in their 3-D arrangements of atoms, rather than in the order in which the atoms are bonded.

The right-handed molecule and the left-handed molecule have *almost* identical physical properties because they have the same chemical structure. They are different, however. For instance, the receptors in your nose perceive different aromas based on their interaction with molecules. Carvone, shown in Figure 12.17, a molecule your nose might encounter in the kitchen, possesses a chiral carbon and therefore exists as two stereoisomers. One of the isomers smells like caraway seeds; the other smells like spearmint.

Chiral molecules also exhibit the ability to rotate the plane of polarized light. Light, passed through a polarizing filter (such as the one in most sunglasses), is composed of electromagnetic waves that are aligned in one direction (such as the up-and-down direction). This polarized light interacts with the two stereoisomers of a chiral molecule to the same degree, but one of the isomers rotates the light to the left and the other to the right, as shown in Figure 12.18.

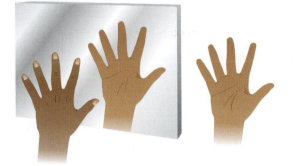

FIGURE 12.15

Chiral objects have a nonsuperimposable mirror image. The mirror image of an achiral object is superimposable.

FIGURE 12.16

Some organic molecules are also chiral. The image and the original are non-superimposable.

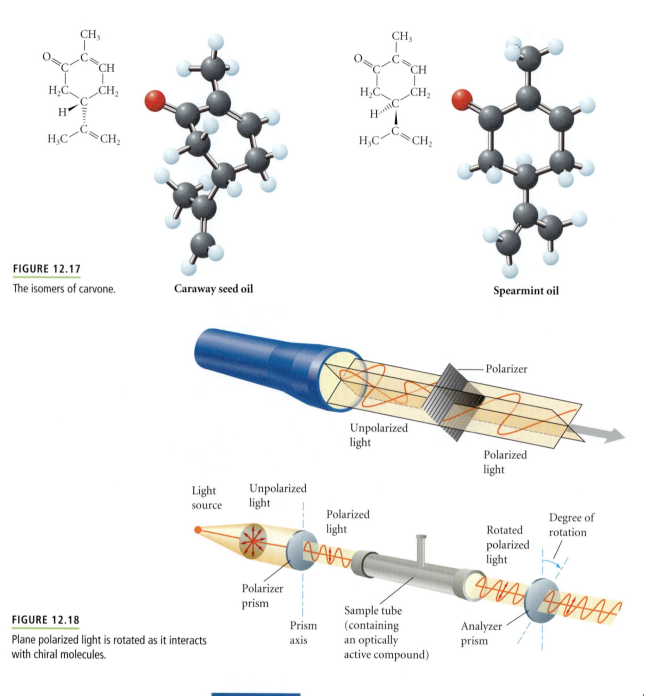

FIGURE 12.17

The isomers of carvone.

Caraway seed oil

Spearmint oil

FIGURE 12.18

Plane polarized light is rotated as it interacts with chiral molecules.

EXERCISE 12.7 **Chiral Drugs**

Thalidomide was used as a drug that relieved morning sickness in pregnant women in the 1950s and 1960s. Although the FDA was hesitant (because of some neurological side effects) to approve the drug for use in the United States, thalidomide found widespread use in Europe and Canada. On the basis of the structure given here, identify the chiral center in the molecule.

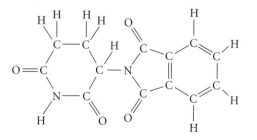

First Thoughts

We'll examine the molecule, looking for carbons that are bonded to four different groups. A helpful starting point is to ignore all carbons that contain double bonds or more than one hydrogen atom.

Solution

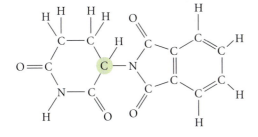

Further Insight

The existence of a chiral center in thalidomide implies that there are two molecules with the same structure and physical properties. It also implies that each of the two stereoisomers might have different interactions with receptors in the body. In fact, this is the case. Whereas one of the stereoisomers does exert a powerful sedative effect and reduce nausea, the other isomer is a powerful teratogen. Teratogens cause terrible birth defects. Women who used the drug during pregnancy gave birth to babies with a wide variety of defects, such as deformed or missing limbs. Thalidomide isn't used to treat morning sickness anymore, though it is used to treat some people with leprosy.

PRACTICE 12.7

Identify the chiral centers in each of the molecules below.

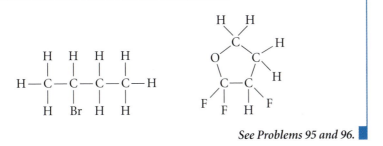

See Problems 95 and 96.

12.15   Organic Chemistry and Modern Drug Discovery

Perhaps one of the most profound impacts of organic chemistry on modern society has been its key role in pharmaceutical science. Historically, chemists and medical researchers have analyzed natural product extracts for effectiveness in curing diseases. Many times these plant extracts are the same ones identified in folk medicine as curatives for common maladies. Ultimately, there must be some component or collection of compounds present in the extract mixture that is responsible for their effectiveness. Organic chemists can purify the active constituents of complex mixtures originating from such natural sources, and very often, this active constituent turns out to be a single compound. These compounds are nearly always based on a carbon framework. Organic chemists can determine the structure of these compounds and can even synthesize them using simple starting materials. They can also make clever *improvements* on these natural products by rationally changing structural features (and, therefore, specific properties) of the molecule.

How do drugs work? Cellular processes, such as intercellular communication, metabolism (Chapter 14), and construction of the cell structure can be disrupted by the binding of drug molecules to specific proteins and other biomolecules essential to those functions. The organic chemical Taxol™ is arguably one of the most important anticancer compounds to have been found in a plant extract over the last 50 years. Let's briefly trace the discovery of Taxol to show how drug development occurs in industry.

Taxol

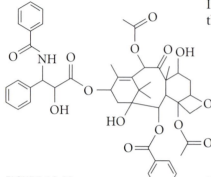

FIGURE 12.19

The chemical structure of Taxol. Can you identify each of the 11 chiral centers?

In 1962, National Cancer Institute scientists found that extracts from the bark of the Pacific yew tree (*Taxus brevifolia*) had the ability to kill some types of cancer.

Nine years later, the compound responsible for this promising activity, named paclitaxel or Taxol, was isolated from the yew extract, and its structure was determined to be the one shown in Figure 12.19. Taxol acts by tight association with cellular microtubules, which are protein assemblies essential to cell division. The idea behind using Taxol for anticancer chemotherapy is to stop the uncontrolled cell division responsible for tumor growth.

Given Taxol's importance, scientists needed large quantities of this compound for experimental treatment of cancer patients and also for basic research. When large amounts of a natural product are required, chemists can either purify it in bulk from its natural source or synthesize it from readily available materials. In the case of Taxol, neither of these two avenues alone was feasible.

At the time of its discovery, the only known source of Taxol was the bark of mature Pacific yew trees. Unfortunately, the yield of Taxol obtained from the Pacific yew is 100 mg per kilogram of bark—a very small amount. Even worse, bark removal results in the death of the tree. The Pacific yew tree is mainly found in the Pacific Northwest, and large-scale harvesting of these trees for Taxol would quickly result in extinction of the tree species. Another way to produce Taxol was desperately needed.

A potential solution to this problem might have been to synthesize Taxol from scratch (an approach known as **total synthesis**), therefore eliminating the threat to the Pacific yew tree. The problem is that the structure of Taxol is enormously complex (compare the structure in Figure 12.19 to the structures of the heptane isomers from Practice 12.1). What was needed was the development of an *efficient* total synthesis of Taxol, and synthetic organic chemists accepted the challenge. After several years of effort, total syntheses of Taxol were accomplished. Yet even though this was a great feat for modern organic chemistry, these total syntheses did little to address the supply problem. All of these total syntheses were long, technically complex, very costly, and consequently incapable of cheaply producing large quantities of Taxol. In summary, neither isolation from the Pacific yew tree nor total synthesis was an environmentally or economically viable method for the mass production of Taxol.

During the course of Taxol research, chemists found that a biosynthetic intermediate of Taxol (10-deacetylbaccatin III, shown in Figure 12.20) could be isolated from the needles of a relative of the Pacific yew tree, in yields of up to 1 g per kilogram of needles. The tree first synthesizes the biosynthetic intermediate 10-deacetylbaccatin III and then converts it to Taxol. And crucially, harvesting these needles for 10-deacetylbaccatin III is not fatal to the tree. Chemists could then take advantage of this *renewable* resource for 10-deacetylbaccatin III by finding a way to convert it into Taxol. This is one reason why the efforts to achieve total synthesis were so important to Taxol production. All of the previous total syntheses combined the "core" with the "side chain" to give the full structure of Taxol. Pharmaceutical scientists adopted this same strategy by converting 10-deacetylbaccatin III into something synthetically useful and then attaching

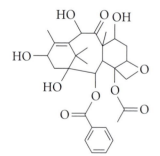

FIGURE 12.20

The structure of 10-deacetylbaccatin III.

FIGURE 12.21

The semisynthesis of Taxol.

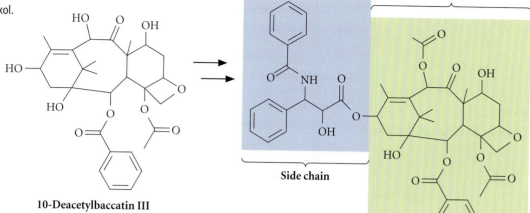

10-Deacetylbaccatin III

Side chain

Core

the "side chain," as shown in Figure 12.21. This method is called a semisynthetic method: A portion of the structure of the desired natural product is isolated from a natural source and then converted into the natural product via synthetic chemistry.

In 1992, Taxol was approved for use in the United States to treat ovarian cancer, breast cancer, and certain types of lung cancer. Along with the drug cisplatin (see Chapter 20), Taxol is often the first drug used in chemotherapy against cancer. Today, the pharmaceutical industry uses a version of the semisynthetic method to produce Taxol.

EXERCISE 12.8   **Taxol as a Drug**

One problem often associated with drugs is side effects. This problem can come from a lack of complete specificity for the target. In treating cancer, we would like to kill the cancer cells but leave the patients' normal cells alone. Cancer tissue grows very rapidly, but other than that, these cells don't have many other unusual features compared to normal cells. This is one of the reasons why cancer is such a difficult disease to treat. Why might Taxol have side effects?

First Thoughts

Cancer cells have acquired mutations in their genetic code that enable them to grow uncontrollably. Despite this, all of the normal cellular machinery is still present in cancer cells. First consider the mode of action of Taxol and then think about how this might affect normal cells.

Solution

Taxol acts by binding to microtubules, which are proteins directly involved in cell division. When Taxol binds to the microtubules, the cancer cells cannot divide to give two daughter cells. This can stop the growth of tumors, which are masses of cells that divide uncontrollably. Because it is nearly impossible to administer Taxol so that only cancer cells in the patient's body receive it, the drug is likely to interact with normal cells as well and bind to *their* microtubules. As a result, these normal cells cannot divide, which could result in the disruption of otherwise healthy tissue.

Further Insights

Taxol has revolutionized cancer chemotherapy, but it does have a few major problems. The first was mentioned before: It is in very short supply. Another problem is its poor formulation properties. Taxol is not very soluble in water, so it is difficult to administer effectively to patients. In addition, some cancers eventually develop drug

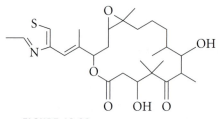

FIGURE 12.22

The chemical structure of epothilone B.

resistance to Taxol. These cancer cells have developed proteins that dispose of Taxol by shooting it back out from inside the cell, where it can do no good for the patient.

Fairly recently, another class of compounds called the epothilones (one is shown in Figure 12.22) has been discovered. The epothilones are also effective against cancer cells, and they have the *same* mode of action as Taxol. Unlike Taxol, however, epothilones are made by a bacterium that is easy to culture on a very large scale. Therefore, there is no serious supply problem for epothilone. The structure of the epothilones is simpler than that of Taxol, so organic chemists can more easily make changes to improve the properties of the drug. Finally, the epothilones are more water soluble than Taxol and also are effective against Taxol-resistant cancer cell lines. All of these characteristics have made the epothilones very important compounds in contemporary cancer research.

PRACTICE 12.8

m-Amsacrine is another anticancer drug that exhibits biological activity by binding to DNA inside cancer cells. Once bound, it causes proteins to break the DNA molecule into fragments. The result is the death of the cancer cell. Would you predict m-amsacrine to show side effects?

See Problems 97 and 98.

EXERCISE 12.9 **Functional Groups of Taxol**

Taxol is a very complicated molecule, and organic chemists typically cope with structural complexity by first looking at its *most* important features: the functional groups. Look at the structure of Taxol in Figure 12.19, and point out all of the functional groups you have learned about in this chapter.

Solution

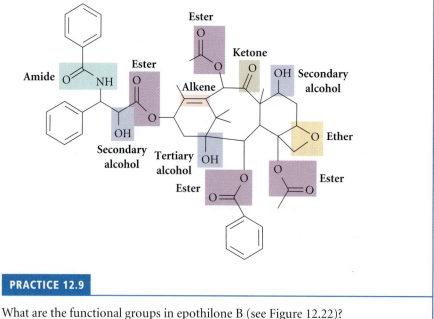

PRACTICE 12.9

What are the functional groups in epothilone B (see Figure 12.22)?

See Problems 51–54.

The Bottom Line

- Organic chemistry is the study of carbon-containing compounds, particularly those derived from living things or fossil fuels. (Section opener)

- The element carbon occurs in the form of three allotropes: the diamond, graphite, and fullerene forms. (Section 12.1)

- Crude oil is our major source of carbon-containing compounds, which are processed by the chemical industry into a wide range of products. (Section 12.2)

- Hydrocarbons are compounds composed of carbon and hydrogen only. They include the alkanes, alkenes, alkynes, and aromatic hydrocarbons. (Section 12.3)

- Organic compounds can exist as structural and geometric isomers. (Section 12.3)

- Hydrocarbons are separated into fractions by fractional distillation of crude oil. (Section 12.4)

- Cracking and reforming are two important methods of preparing specific organic molecules. (Section 12.5)

- Alkanes can undergo substitution and dehydrogenation reactions. (Section 12.6)

- All organic compounds can be regarded as carrying zero, one, or more distinctive functional groups on a hydrocarbon frame. (Section 12.7)

- Each functional group is associated with a distinctive set of chemical characteristics. (Section 12.7)

- Many modern materials are polymers composed of thousands of organic monomer units linked together. (Section 12.8)

- Molecules may exhibit chirality, which may lead to different chemical properties in two stereoisomers. (Section 12.14)

- Organic chemists can isolate and purify the active constituents of complex mixtures in order to make products such as pharmaceuticals. (Section 12.15)

Key Words

achiral A term used to describe molecules that have a superimposable mirror image. (*p. 525*)

addition reaction A type of reaction in which atoms are added to a double or triple bond. (*p. 514*)

addition polymer A polymer formed by the process of addition polymerization. (*p. 515*)

addition polymerization A type of polymerization involving addition reactions among the monomers involved. (*p. 515*)

alcohol A chemical containing the hydroxyl (—OH) group. (*p. 514*)

aldehydes Compounds that carry a carbonyl functional group (C=O) with at least one H atom bonded to the carbonyl carbon atom. (*p. 517*)

aliphatic compounds Compounds in which all of the carbon atoms are sp^3 hybridized and contain only single bonds. (*p. 496*)

alkanes Saturated hydrocarbons. (*p. 496*)

alkenes Hydrocarbons containing at least one C=C double bond. (*p. 500*)

alkyl group A hydrocarbon group, such as —CH_3 or —C_2H_5, within the structure of an organic compound. (*p. 504*)

alkyl halide A functional group in which a halogen atom is bonded to an alkyl group. (*p. 511*)

alkoxy anion An anion that contains an alkyl group attached to an oxygen anion, such as CH_3O^- or $CH_3CH_2O^-$. (*p. 520*)

alkynes Hydrocarbons containing a C≡C bond. (*p. 501*)

allotropes Different forms of the same element. (*p. 493*)

amide bond A bond formed by a condensation reaction between an amino group (—NH_2) and a carboxylic acid group (—COOH). (*p. 523*)

amine A compound containing the —NH_2 functional group. (*p. 523*)

amino functional group A functional group containing an —NH_2 attached to an sp^3 hybridized carbon atom. (*p. 523*)

aromatic hydrocarbons Hydrocarbons that contain one or more aromatic rings. (*p. 504*)

carbonyl functional group A functional group containing the group C=O attached to either hydrogen atoms (as in the aldehydes) or carbon atoms (as in the ketones). (*p. 517*)

carboxyl functional group The —COOH functional group. (*p. 519*)

carboxylate anion An anion resulting from the removal of a hydrogen ion from the oxygen in a carboxylic acid functional group. (*p. 520*)

carboxylic acids Organic acids that carry the carboxyl functional group (—COOH). (*p. 519*)

catalytic cracking A type of cracking wherein heat and a catalyst are used. (*p. 508*)

chiral A term used to denote molecules that have a nonsuperimposable mirror image. (*p. 525*)

chiral center A carbon attached to four different groups. (*p. 525*)

chirality The quality of being chiral—that is, of having a nonsuperimposable mirror image. (*p. 525*)

cis A molecule containing groups that are fixed on the same side of a bond that cannot rotate. (*p. 503*)

condensation polymerization Formation of a polymer by condensation reactions. (*p. 522*)

condensation reaction A reaction in which two molecules become joined in a process accompanied by the elimination of a small molecule such as water or HCl. (*p. 520*)

cracking The breaking up of hydrocarbons into smaller hydrocarbons, using heat and in some cases steam and/or catalysts. (*p. 508*)

cyclic alkanes Alkanes that possess a group of carbon atoms joined in a ring. (*p. 500*)

dehydrogenation The removal of hydrogen. (*p. 510*)

epoxide A compound containing a triangular ring of two carbon atoms linked by an oxygen atom, known as the epoxide group. (*p. 524*)

epoxy resins Polymer resins containing the epoxide group, in which two carbons atoms and an oxygen atom form a cyclic ether ring. (*p. 524*)

esterification The process by which an ester is formed when the —OH group of an alcohol participates in a condensation reaction with the —COOH group of a carboxylic acid. (*p. 522*)

esters Compounds formed when an alcohol and a carboxylic acid become bonded as a result of a condensation reaction. (*p. 520*)

ether Any compound containing the ether functional group (—C—O—C—). (*p. 524*)

fermentation The biological process that converts glucose into ethanol and carbon dioxide. (*p. 516*)

free radicals Chemical species that carry an unpaired electron and so are highly reactive. (*p. 510*)

fullerenes Forms of carbon based on the structure of buckminsterfullerene, C_{60}. (*p. 494*)

functional groups Groups of atoms, or arrangements of bonds, that bestow a specific set of chemical and physical properties on any compound that contains them. (*p. 511*)

gas chromatography (gc) Analytical technique in which components of a mixture are separated by their different flow rates through a chromatography column, driven by a carrier gas. (*p. 507*)

gas–liquid chromatography (glc) A form of gas chromatography in which the solid phase is covered with a thin coating of liquid. (*p. 507*)

geometric isomers Molecules that differ in their geometry but have identical chemical formula and points of attachment. See also *cis* and *trans*. (*p. 502*)

high-density polyethylene (HDPE) A form of polyethylene in which the molecules are linear, with no branch points. (*p. 515*)

homologous series A series of compounds sharing the same general formula. (*p. 497*)

hydrocarbons Compounds composed of only hydrogen atoms and carbon atoms. (*p. 494*)

hydrocracking A type of cracking that uses heat plus a catalyst in the presence of hydrogen gas. (*p. 508*)

hydrogenation The addition of a molecule of hydrogen to a compound. (*p. 501*)

hydroxyl group The —OH functional group. (*p. 514*)

ketones Compounds that have two carbon atoms bonded to a carbonyl carbon atom. (*p. 519*)

knocking The rapid explosion of a gas–air mixture as it is compressed in an automobile cylinder *before* the spark plug creates the spark that is intended to cause the mixture to explode. (*p. 509*)

low-density polyethylene (LDPE) A form of polyethylene with branched-chain molecules that are shorter than the unbranched chains found in HDPE. (*p. 515*)

methylene group The —CH_2— group. (*p. 497*)

monomer One of the small chemical units that combine with other monomers to form polymers. (*p. 515*)

normal alkanes Alkanes that have no branched chains or rings of carbon atoms. (*p. 496*)

octane rating A number assigned to motor fuel (gas) on the basis of its "anti-knock" properties. (*p. 509*)

organic chemistry The chemistry of carbon-containing compounds, especially those derived from living things or fossil fuels. (*p. 492*)

organic compound Carbon-based compounds. (*p. 492*)

polyamides Polymers whose monomers are bonded together by amide bonds. (*p. 523*)

polyesters Polymers in which the monomers are held together by repeated ester linkages. (*p. 522*)

polyethenes Organic polymers made when huge numbers of ethene molecules participate in addition reactions with themselves to generate long chain-like molecules. Also known as polyethylenes. (*p. 574*)

polyethylenes See polyethenes. (*p. 515*)

polymer A chemical composed of large molecules made by the repeated bonding together of smaller monomer molecules. (*p. 515*)

primary alcohols Alcohols that have no more than one carbon atom bonded to the carbon atom that carries the hydroxyl group. (*p. 517*)

reforming A process that changes a mixture of predominantly straight-chain hydrocarbons into a mixture containing more aromatic and branched-chain hydrocarbons. (*p. 508*)

saturated hydrocarbons Hydrocarbons that do not contain double or triple bonds. (*p. 496*)

secondary alcohols Alcohols that have two carbon atoms bonded to the carbon atom that carries the hydroxyl group. (*p. 517*)

steam cracking A form of cracking in which heat, steam, and a catalyst are used. (*p. 508*)

stereoisomers Structural isomers that differ in their three-dimensional arrangements of atoms, rather than in the order in which the atoms are bonded. (*p. 525*)

straight-chain alkanes Molecules that contain only hydrogen atoms and sp^3 hybridized carbon atoms arranged in linear fashion. (*p. 497*)

structural isomers Molecules with the same formula but different structures. (*p. 498*)

substitution reactions Reactions in which one or more atoms (often hydrogen atoms) are replaced by other types of atoms (often halogens or a hydroxyl group). (*p. 510*)

tertiary alcohols Alcohols that have three carbon atoms bonded to the carbon atom that carries the hydroxyl group. (*p. 517*)

thermal cracking A form of cracking that uses heat alone. (*p. 508*)

total synthesis the preparation of a molecule from readily available starting materials. Typically the total synthesis involves many steps. (*p. 528*)

trans A molecule containing groups that are fixed on opposite sides of a bond that cannot rotate. (*p. 503*)

unsaturated hydrocarbons Hydrocarbons that contain $C=C$ or $C\equiv C$ bonds. (*p. 501*)

Focus Your Learning

The answers to the odd-numbered problems and some selected problems appear at the back of the book, as represented by the blue numbering.

12.1 Elemental Carbon

Skill Review

1. List the three allotropes for carbon, and indicate the hybridization of the carbon atoms in each.

2. Explain the differences in the properties of amorphous (lacking a defined crystalline shape, disordered) carbon and graphite.

Chemical Applications and Practices

3. Carbon in the diamond form and carbon in the graphite form have widely differing uses. We pay hundreds, even thousands, of dollars for diamond jewelry, but only pennies for carbon in pencils. Explain, using the differences in bonding, why diamonds are held together more strongly than the carbon atoms in graphite.

4. Using hybridization, explain how carbon forms both sigma and pi bonds in graphite, but only sigma bonds in the diamond form.

5. Diamonds are typically measured in carats (1 carat is 200 mg). If an industrial diamond contained only carbon, how many atoms of carbon would be found in a 2.00-carat diamond?

6. Graphite can be converted to diamond with great pressure. The type of diamond used in polishing powder can be produced with pressures on the order of 1000 MPa.
   a. What is this pressure in terms of atmospheres?
   b. Why is pressure a factor in conversion of the graphite form of carbon to the diamond form?

12.2 Crude Oil—the Basic Resource

Skill Review

7. Judging on the basis of Table 12.2, what is the main difference between the aliphatic and the aromatic compounds found in crude oil?

8. Judging on the basis of Table 12.2, what is the difference between the structures and formulas for compounds in the aliphatic and naphthene classes?

Chemical Applications and Practices

9. Sulfurous compounds occur to a minor extent in crude oil, and their presence is indicative of the source of the crude. Explain how the decomposition of biological material could give rise to sulfur-containing compounds.

10. Why do you think that there is such a wide diversity of compounds in crude oil?

12.3 Hydrocarbons

Skill Review

11. Provide the correct name for each of these:
    a. $CH_3(CH_2)_5CH(CH_3)_2$ c. $CH_3CH_2CH(CH_2CH_3)_2$
    b. $CH_3CH_2CH_2CH(CH_3)_2$ d. $CH_3CH_2CH_2CH_2CH_3$

12. Provide the correct name for each of these:
    a. $CH_3CH(CH_3)_2$ c. $CH_3CH_2CH_2C(CH_3)_3$
    b. $CH_3CH_2CH_3$ d. $(CH_3)_2CHCH_2CH(CH_3)_2$

13. Diagram the structure of the compounds listed in Problem 11.

14. Diagram the structure of the compounds listed in Problem 12.

15. Complete this table:

	Alkane	Alkene	Alkyne
Type of C to C bond			
Type of hybridization			

16. Complete this table:

	Alkane	Alkene	Alkyne
Angles between bonds to other atoms			
If made of only two carbons, will have __ total hydrogen atoms			
Saturated or unsaturated?			

17. a. Diagram the structure of each of these:

    pentane

    2-methylbutane

    2,2-dimethylpropane

    Also give the empirical formula of each.

    b. Remembering that intermolecular forces affect boiling point, match each structure with its corresponding boiling point (all are in degrees Celsius): 9.5, 36, 28.

18. a. Diagram the structure of each of these:

    3-ethyl-2-pentene

    cyclooctene

    3-hexyne

    Also give the molecular formula of each.

    b. Remembering that intermolecular forces affect boiling point, match each structure with its corresponding boiling point (all are in degrees Celsius): 82, 94, 148.

19. One of the high-molecular-mass hydrocarbons that helps form the protective waxy skin of an apple contains 28 carbon atoms. What is the formula of this noncyclic alkane?

20. What is the formula of a saturated noncyclic hydrocarbon containing 8 carbons? Assume that a different hydrocarbon contains two C=C groups and has a total of 10 carbon atoms. What is its formula?

21. Diagram five isomers of octane that all have a six-member continuous carbon chain.

22. Diagram three isomers of octane that all have a five-member continuous carbon chain.

23. Which, if any, of these hydrocarbons could be cycloalkanes?

    a. C_2H_6
    b. C_6H_{12}
    c. $C_{12}H_{26}$
    d. C_7H_{14}

24. How many hydrogen atoms are found in one molecule of cyclodecane?

25. Identify each of these molecules as a *cis* or a *trans* isomer.

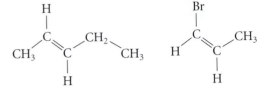

26. Identify each of these molecules as a *cis* or a *trans* isomer.

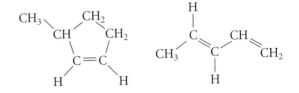

Chemical Applications and Practices

27. Ethyne (also known as acetylene) can be produced by the reaction between calcium carbide (CaC_2) and water. In addition to ethyne, calcium hydroxide is a product. Write and balance the equation for the production of this important alkyne.

28. The combustion of methane is an important reaction that heats our homes. Incomplete combustion, however, involves the reaction of methane with oxygen to produce carbon monoxide and water. Write and balance the equation for the incomplete combustion of methane.

29. Cyclohexane and benzene both have a ring structure containing six carbon atoms. Compare the two structures in terms of:

    a. The number of hydrogen atoms in each
    b. Carbon orbital hybridization
    c. The angles between carbon atoms
    d. The number of double bonds
    e. Resonance structures

30. Four alkene isomers that contain two C=C can be drawn for C_5H_8. Draw these four isomers and identify the one that is stabilized most by resonance structures.

12.4 Separating the Hydrocarbons by Fractional Distillation

Skill Review

31. Mixtures of hydrocarbons, and of other types of compounds, can be separated into their individual components by gas chromatography. Briefly explain the main steps by which this technique makes possible the identification of compounds in a mixture.

32. For specific uses, hydrocarbon fractions obtained from crude oil may be further modified. The two major industrial processes that provide useful hydrocarbon products are cracking and reforming. Compare and contrast these two important techniques.

33. What property of hydrocarbons serves as the basis for a separation, during fractional distillation, of the components in crude oil? What is the relationship between this property and the structure of the components?

34. Although gas chromatography is often used to resolve mixtures of hydrocarbons, it cannot be relied on in every case. What is a chief limitation of GC when applied to hydrocarbon separations?

12.5 Processing Hydrocarbons

Skill Review

35. The octane rating system has been used as a standard to rate the burning efficiency of other fuels. What is the formula of octane? Use the formula to write and balance the reaction for the combustion of octane.

36. The actual isomer of octane used in the rating system is "isooctane." The descriptive name for the compound is 2,2,4-trimethypentane. Diagram the structure of this compound and write a balanced reaction for the combustion of isooctane.

Chemical Applications and Practices

37. Another compound, besides isooctane, used in fuel studies was found to have the following percentage composition: 84.0% carbon and 16.0% hydrogen. The molar mass of the alkane is 100.0. From these data, what would you calculate as the molecular formula of the compound?

38. Calculate the mass percent carbon and the mass percent hydrogen of isooctane.

39. Currently, many automobiles run on fuel with a minimum octane rating of 87. To what mixture of isooctane and heptane does this fuel correspond?

40. Some additives to gasoline have a much higher octane rating. For instance, a particular alcohol has an octane rating of 116. What quantity of this alcohol would need to be added to crude gasoline (octane rating 55) to give a product with an octane rating of 92?

12.6 Typical Reactions of the Alkanes

Skill Review

41. Using methane and bromine, show the balanced equation to form dibromomethane.

42. Using butane as a reactant, show the balanced equation that produces 2-butene.

43. Provide the structural diagram and the name for all of the products that would result from the substitution reaction between Cl_2 and ethane.

44. Provide the structural diagram and the name for all of the products that would result from the dehydrogenation of butane.

45. What are the typical products that result from the combustion of an alkane?

46. A reaction produces a molecule with a C=C bond. If the product was made from an alkane, what would be the classification of this reaction?

Chemical Applications and Practices

47. The combustion of propane to form carbon dioxide and water can produce approximately 2200 kJ per mole. If a burning propane torch used 42.0 g of propane, how many kilojoules would be released?

48. Why do arctic hikers use propane as fuel for their cook stoves instead of pentane?

12.7 The Functional Group Concept

Skill Review

49. Provide a suitable chemical formula and an example structure of a molecule for each of these functional groups.
a. alcohol   b. aldehyde   c. alkene   d. ketone

50. Provide a suitable chemical formula and an example structure of a molecule for each of these functional groups.
a. carboxylic acid   b. alkyne   c. ether   d. cyclic alkane

Chemical Applications and Practices

51. The structure of the common analgesic acetaminophen follows. Identify any and all functional groups shown in the structure.

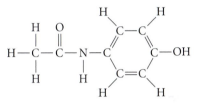

52. The structure for tetracycline follows. Identify any and all functional groups shown in the structure.

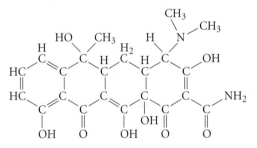

53. The partially complete structure (showing all atoms except hydrogen) for the pesticide malathion follows. Reproduce the structure and add the missing hydrogen atoms to the molecule.

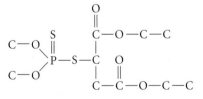

54. The partially complete structure (showing only carbons and oxygens) for carvone follows. Reproduce the structure and add the missing hydrogen atoms to the molecule.

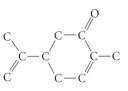

12.8 Ethene, the C=C Bond, and Polymers

Skill Review

55. How many carbon–carbon π bonds are present in each of these noncyclic molecules?
    a. C_3H_8 b. C_6H_8 c. $C_{10}H_8$

56. How many carbon–carbon π bonds are present in each of these noncyclic molecules?
    a. C_4H_8 b. C_5H_8 c. C_4H_6

57. Diagram the structure of each of these compounds:
    a. 3-Methyl-2-pentene
    b. 2-Methyl-4-propyl-3-heptene

58. Name these compounds:

    a.
    ```
        H   H   H   H
        |   |   |   |
    H — C — C = C — C — H
        |           |
        H           H
    ```

    b.
    ```
                CH2 — CH3
                 |
    CH2 = CH — CH — CH — CH3
                      |
                      CH3
    ```

Chemical Applications and Practices

59. One gram of an important hydrocarbon, when completely combusted, produced 2.93 g of carbon dioxide and 1.80 g of water. The molar mass of the compound is 30.0. Is this compound an alkane or an alkene? What is the formula of the compound?

60. Chemists often refer to the addition reaction of alkenes as "adding across the double bond." Explain what aspect of the double bond between carbon atoms favors this process.

61. a. Diagram the structure of propene.
    b. Show the balanced equation that represents the conversion of propene into the alcohol 1-propanol.

62. One way to distinguish an alkane from an alkene is to halogenate the double bond. If propene reacted with Br_2, what compound would be formed?

63. Polypropylene is an addition polymer with many uses, one of which is in indoor–outdoor carpeting.
    a. What is the structure of the monomer used in this polymer?
    b. Diagram a section of polypropylene using four joined monomers.

64. Clothing materials made from Orlon® contain the addition polymer polyacrylonitrile shown here.

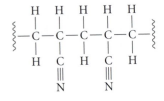

    After examining the structure, diagram the structure of the repeating monomer.

65. HDPE and LDPE are both plastics that can be recycled. HDPE is coded as 2, and LDPE is coded as 4. What contrasting structural factor in the molecules of these two polymers causes their differing properties?

66. Indicate which of these statements is(are) true? For those that are false, explain why.
    a. Addition reactions always involve compounds with multiple bonds.
    b. All addition reactions produce polymers, but not all polymers are addition polymers.
    c. All addition polymers are based on reactions between molecules with multiple bonds.
    d. Addition polymers must contain double bonds.

12.9 Alcohols

Skill Review

67. After looking at the structure of a compound, a student incorrectly named it 2,5-pentanediol. Provide the correct name for the compound.

68. Another compound was mislabeled as 1,4-propanediol. Explain the mistake in this name.

69. Diagram the structure of 1,3-pentanediol. Classify each —OH group as primary, secondary, or tertiary.

70. Diagram the structure of 2-butanol. Would this alcohol be classified as a primary, secondary, or tertiary alcohol?

Chemical Applications and Practices

71. Phenol is an example of an aromatic alcohol containing six carbons, six hydrogens, and one oxygen. It is used in producing disinfectants and some polymers. Diagram the structure of this important compound.

72. Alcohols can undergo a reaction known as dehydration. In this reaction, the alcohol loses a molecule of water to form an alkene. Provide the name of the alkene formed from the dehydration of each of these alcohols:
    a. Ethanol
    b. 1-Propanol
    c. 2-Propanol
    d. 1-Butanol

12.10 From Alcohols to Aldehydes, Ketones, and Carboxylic Acids

Skill Review

73. Draw the structure of 1-propanol, and identify the product that would result from oxidation of this compound.

74. Diagram the structure of butanal. What alcohol can be oxidized to produce this compound?

75. Propanoic acid is used to make a type of mold inhibitor. Diagram the structure of propanoic acid. What aldehyde could be oxidized to form this acid?

76. Isopropyl alcohol (2-propanol) is used as the ingredient in rubbing alcohol. This compound can be oxidized to a ketone, but not to a carboxylic acid. Draw the structure of the alcohol and the ketone, and explain why it can't be used to produce the corresponding three-carbon carboxylic acid.

Chemical Applications and Practices

77. Methyl ethyl ketone is often used as a solvent for organic compounds. Diagram the structure of this important organic compound. What secondary alcohol could be oxidized to form this compound?

78. An unknown compound was oxidized to form pentanoic acid. The unknown compound was either 2-pentanone or pentanal. Identify the starting compound.

79. Carboxylic acids are generally considered only weakly acidic. Show the acid dissociation reaction of propanoic acid in water.

80. An unknown compound was found to produce an acidic solution and contained four carbon atoms. The compound could be prepared from a straight-chain alcohol. Identify the unknown compound and the starting alcohol.

12.11 From Alcohols and Carboxylic Acids to Esters

Skill Review

81. The ester that produces a banana aroma is called 3-methylbutyl ethanoate. Diagram the structures of the starting materials that produce the ester, and show the balanced equation for the production of isopentyl acetate.

82. An apple-like aroma can be produced from the combination of methanol and butanoic acid. Diagram the structure and name this ester.

83. An apricot aroma can be produced from the ester pentyl butanoate. Using structures, show the balanced equation that produces this pleasant fruity aroma.

84. Tristearin is the animal fat associated with beef. The formula of stearic acid is $CH_3(CH_2)_{16}COOH$. Show the reaction between stearic acid and 1,2,3-propanetriol to form the triple ester known as tristearin.

Chemical Applications and Practices

85. Linoleic acid is used to form the major oil found in corn. This unsaturated acid has two double bonds per molecule. The double bonds are found between carbon atoms number 9 and 10 and between carbon atoms number 12 and 13. The formula is $C_{18}H_{32}O_2$. Draw the diagram of this linear unsaturated carboxylic acid.

86. Oleic acid (a major component of animal fats) contains a *cis* alkene and 18 carbon atoms and has the formula $C_{18}H_{34}O_2$. The alkene is found between carbons number 9 and 10. Draw the diagram of this linear carboxylic acid.

12.12 Condensation Polymers

Skill Review

87. The simplest amino acid is glycine. Use the structure shown below to diagram a condensation polymer made from three glycine monomers. What molecule has been split out as the polymer forms?

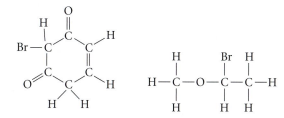

88. Why is the formation of a condensation polymer impossible in these cases?
    a. Ethanoic acid + $NH_2CH_2CH_2CH_2CH_2NH_2$
    b. Ethanoic acid + $CH_3CH_2NH_2$
    c. $HOOC-CH_2-COOH + CH_3CH_2NH_2$

Chemical Applications and Practices

89. Thiokol, dating back to the 1920s, was the first synthetic rubber commercially produced in the United States. It can be formed as a condensation polymer from the monomer $ClCH_2CH_2OCH_2CH_2Cl$. The reaction takes place when sodium polysulfide (Na_2S_2) reacts with the monomer. Assuming that Na_2S_2 is a source of —S—S—, draw the structure of the polymer. What smaller compound is also produced from the reaction?

90. In addition to water, a by-product from the condensation polymerization of 5-aminopentanoic acid is common. What is the structure of this by-product?

    $NH_2CH_2CH_2CH_2CH_2COOH$ (5-aminopentanoic acid)

12.13 Polyethers

Skill Review

91. Using the information and examples in the chapter, devise the name of these ethers.

    CH_3-O-CH_3 $CH_3-O-CH_2-CH_2-CH_2-CH_3$

92. Using the information and examples in the chapter, draw the structure for these ethers.
    a. ethyl methyl ether
    b. propyl methyl ether
    c. butyl ethyl ether

12.14 Handedness in Molecules

Skill Review

93. Indicate whether chirality exists in each of these objects.
    a. a pencil   c. your textbook
    b. your feet   d. a fork

94. Indicate whether chirality exists in each of these objects.
    a. a CD   c. a bottle
    b. a butterfly   d. a sportscar

95. Identify the chiral center in each of these molecules.

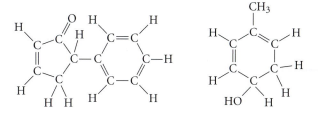

96. Identify the chiral center in each of these molecules.

12.15 Organic Chemistry and Modern Drug Discovery

Chemical Applications and Practices

97. The total synthesis of a potential drug requires nine steps, each with a 95% yield. What is the overall yield of the final drug made from this synthesis?

98. Acetylsalicylic acid, also known as aspirin, is a derivative of a natural compound found in willow trees. Explain how a researcher might have originally surmised that an active substance was to be found in the leaves and bark of the willow tree.

Comprehensive Problems

99. Why do gasoline manufacturers adjust the gasoline mixture for automobiles in accordance with the season of the year?

100. Using information from this chapter, describe how the acidity of salicylic acid, shown below, can be reduced by modification of the compound via a chemical reaction.

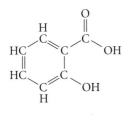

Salicylic acid

101. Gasoline is delivered to a farmer for use in the farm's combines. A cylindrical storage tank with a circumference of 6.0 ft is used to store the gasoline until it is used. The farmer uses a yardstick to measure the depth of the gasoline in the storage tank.
    a. How many gallons of gasoline are in the tank if the yardstick measures the gasoline to a depth of 22 inches?
    b. 2,3-dimethylhexane is one of the components in the gasoline. Draw the structure of this compound.

102. The compound responsible for the odor of wintergreen was discussed in the chapter.
    a. What two compounds (an alcohol and a carboxylic acid) could be condensed to prepare methyl salicylate?
    b. If a company that manufactures a muscle pain relief cream requires 352 kg of methyl salicylate, how many liters of the alcohol would be needed? (Assume the density of the alcohol is 0.789 g/mL.)

103. In Table 12.5, we mention that one of the fractions distilled from crude oil is known as petroleum ether. This fraction is used in the organic research lab to recrystallize reaction products.
    a. Draw and name at least two compounds that could be found in this distillation fraction.
    b. Why does the name of this fraction appear to be a misnomer?

104. The addition of a halogen to an alkene is a good way to produce an alkyl halide in the laboratory. But often, a researcher is interested in adding only one halogen atom to the compound. In such a case, a hydrogen halide is used for the addition reaction.
    a. If a research student fills a 2.5 L flask with 0.25 g gaseous hydrogen chloride and 0.48 g of gaseous *trans*-2-butene, what mass of the addition product can be made?
    b. What is the name of that product?
    c. What is the concentration, in molarity, of the product in the flask?

105. The organic chemistry industry of the late part of the 1800s was interested in creating new dyes for fabrics.
    a. What features of an organic compound would make it suitable for use as a dye?
    b. If a particular organic dye was yellow in color, what range of wavelengths would be reflected by the compound?
    c. What range of energies, in kJ/mol, is associated with these wavelengths?

Thinking Beyond the Calculation

106. An unknown organic compound is identified. Reaction of this unknown with two molecules of ethanol produces a diester. The unknown organic compound can be prepared by oxidizing 1,2-ethanediol.
    a. What is the structure of the unknown?
    b. What properties (chemical and physical) would you predict this unknown to possess?
    c. If 10 g of the unknown produce 10 g of the diester, what is the percent yield of the reaction?
    d. When the diester is placed in aqueous solution containing a small amount of catalyst, the solution becomes acidic. Provide a balanced reaction that explains this result.
    e. A polymer can be made when the unknown organic compound is condensed with ethylene glycol ($HOCH_2CH_2OH$). Draw the structure of this polymer.

13

Modern
Materials

Modern materials affect our lives every day. Their use in bulletproof vests, automobiles, and computers saves and protects us from harm. And, without their use in the health care industry, we would surely suffer. Let's examine a trip to the hospital to see how modern materials are used.

Contents and Selected Applications

Chemical Encounters: Materials in the Hospital

13.1 The Structure of Crystals
Chemical Encounters: X-ray Crystallography

13.2 Metals
Chemical Encounters: Photovoltaic Devices
Chemical Encounters: Dental Amalgams

13.3 Ceramics
Chemical Encounters: Magnetic Resonance Imaging

13.4 Plastics

13.5 Thin Films and Surface Analysis
Chemical Encounters: Heart Defibrillators

13.6 On the Horizon—What Does the Future Hold?
Chemical Encounters: "Green Chemistry"

Go to **college.hmco.com/pic/kelterMEE** for online learning resources.

A trip to the hospital offers interesting insight into the products of chemistry. Look around and you'll see chemistry that is central to the mission of the hospital and to the quality of our lives. The emergency room (ER) nurse who gives an injection, the radiologist who uses the magnetic resonance imager (MRI), and the doctor who replaces a hip joint all depend on the products of materials science to do their jobs. Materials science is concerned with *the chemistry and development of substances employed to make the items we use everyday*. What types of materials do you see in the hospital? The emergency room syringe is made of plastic, as are the bag and the tubing that holds the intravenous (IV) fluid. The hospital MRI requires a powerful magnetic field generated with a special type of ceramic material called a superconductor. Hip replacements are often made from metal alloys and plastic polymers that are stronger than the bone they replace. Each of these products (the plastic, the superconductor, and the metal alloy) is a recently developed material.

In this chapter we will discuss the chemical features of modern materials. Knowing about bonding in solids, the properties of metals, and substances called liquid crystals will lead us to a better understanding of materials science. Along the way, we'll discover some new materials that are of great practical use, and we'll find out that the *old* materials we know are also part of the field of materials science. Because many of the modern materials are solids, we begin our study of modern materials with the structure of crystals.

Application

CHEMICAL ENCOUNTERS: Materials in the Hospital

13.1 ▷ The Structure of Crystals

Application

Sodium chloride (table salt) plays an important role in our diet. Medical workers recognize the essential role that sodium plays in maintaining blood pressure, allowing the transmission of nerve signals and the transporting of nutrients into cells. When a nurse reaches for an intravenous saline solution, a bag of sterile 0.9% (w/v) sodium chloride is what she or he gets. Why choose a solution of sodium chloride? Blood contains 0.9% (w/v) sodium chloride, and this concentration keeps the blood in a patient at the same osmotic pressure (Chapter 11).

An IV saline solution.

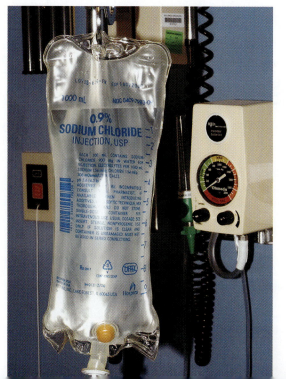

TABLE 13.1	Different Types of Solids and Their Properties			
Type of Solid	**Particles in Solid**	**Forces Between Particles**	**Physical Properties**	**Examples**
Ionic	Cations and anions	Electrostatic attraction	Hard and brittle, high melting point, poor thermal and electrical conductivity	NaCl, CaO, $MgBr_2$
Molecular	Atoms or molecules	London dispersion, dipole–dipole, and/or hydrogen bonding	Fairly soft, low to moderate melting point, poor thermal and electrical conductivity	CH_4, $C_6H_{12}O_6$ (glucose), H_2O, Kr
Network covalent	Atoms	Covalent bonds	Very hard, very high melting point, typically poor thermal and electrical conductivity	C (diamond), SiO_2 (quartz), SiC, BN
Metallic	Atoms	Metallic bond	Soft to hard, low to high melting point, good thermal and electrical conductivity	Na, Fe, Au, Ag, Al

In Chapter 8, we discussed the formation of sodium chloride from its elements. We noted the strength of the ionic bond by focusing on the lattice energy, 786 kJ/mol. To learn more about sodium chloride and other crystalline ionic compounds, however, we need to explore how the ions come together to make an ionic crystal. What can the shape of an ionic solid tell us? Knowing the locations of the ions within an ionic solid enables us to determine the density, the nature (crystalline or amorphous), and the properties of the solid (see Table 13.1).

Types of Solids

Why are some compounds, such as table salt or sugar, solid rather than liquid or gaseous under normal conditions? The atoms, ions, or molecules of a solid are not free to move significantly relative to each other. The materials are solids because the intermolecular forces of attraction are strong enough to hold the particles in a rigid structure. In a **crystalline solid** (Figure 13.1), the atoms, ions, or molecules are highly ordered in repeating units over long ranges. Although they have fairly rigid and fixed locations of the atoms, ions, or molecules, **amorphous solids** such as the glass in a window pane lack the high degree of long-range order (they still have short-range order) found in a crystalline solid. Sodium chloride is an example of a crystalline solid (Figure 13.2), because *it has an ordered structure with fixed locations for the sodium and chloride ions.*

Crystalline solid

Amorphous solid

FIGURE 13.1

Solids can be crystalline or amorphous.

Video Lesson: Types of Solids

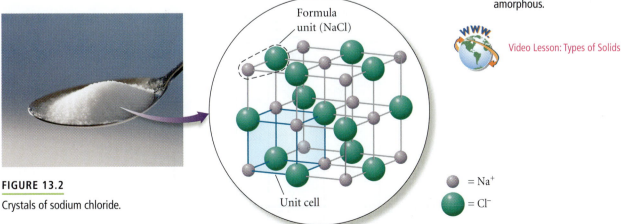

FIGURE 13.2

Crystals of sodium chloride.

Formula unit (NaCl)

Unit cell

$= Na^+$

$= Cl^-$

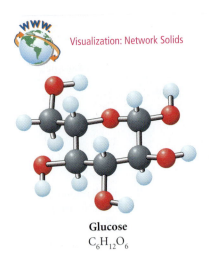

Glucose
$C_6H_{12}O_6$

What does a crystal look like? Try using a magnifying glass or microscope to look at crystals of table salt or sugar. These two crystalline solids appear as solid cubes, whereas other solids take the shape of needles, plates, and a host of other faceted designs. The shape of the crystal you see is a result of the way in which the atoms and ions are arranged at the atomic level, known as the crystalline lattice. Some solids, such as table salt (NaCl), are **ionic solids**, made up of ionic compounds held together by strong electrostatic forces of attraction between adjacent cations and anions. Because these forces are formidable, the ionic solids typically have high melting points. Solutions made from ionic solids conduct electricity, as we saw in Chapter 4. We referred to them as electrolytes.

On the other hand, some solids, such as glucose ($C_6H_{12}O_6$), are **molecular solids**, made up of *molecules* held in a rigid structure. Despite the structural similarity to an *ionic* solid, the intermolecular forces of attraction between adjacent molecules are not as strong as the electrostatic forces found in ionic solids. Therefore, melting points of the molecular solids are not very high. In addition, because most molecules do not dissociate into ions when they are dissolved, solutions made from molecular solids typically do not conduct electricity. As we'll see in this chapter, the properties of molecular and ionic solids are different in other important ways.

Crystal Lattices

A structure known as a crystal lattice defines the shapes of crystalline solids. This **crystal lattice** *marks the position each of the atoms, ions, or molecules within the crystal*. The 3-D grid comprises three pairs of parallel planes that intersect to make parallelepipeds (three-dimensional boxes), and in Figure 13.3 we see that repeating the parallelepiped in three dimensions gives us the lattice for a crystal. The simplest parallelepiped that can be repeated to produce the entire crystal lattice is the **unit cell** (*the smallest repeating unit of the lattice*). In other words, the unit cell is the smallest repeating unit of atoms, ions, or molecules that would describe the entire crystal. Whenever possible, we try to arrange the corners of the lattice so that they line up with the centers of the atoms, ions, or molecules in the solid.

If the parallel sets of planes intersect at right angles to make a cube (as they do in Figure 13.3), the lattice is known as a cubic lattice. The most basic of the cubic lattices is made up of the **simple cubic unit cell** (also known as the primitive unit cell), in which the centers of the atoms, ions, or molecules are located only on the corners of the unit cell. Polonium, a radioactive metal that shows promise as a thermoelectric power source in space satellites, crystallizes in the simple cubic structure. The presence of an additional atom, ion, or molecule at the center of

FIGURE 13.3

Parallelopiped in a cubic lattice.

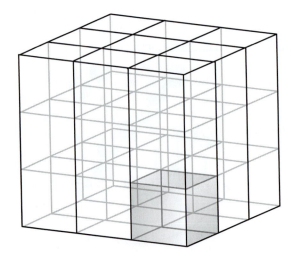

Unit cell	Lattice structure	Space-filling unit cell	Example

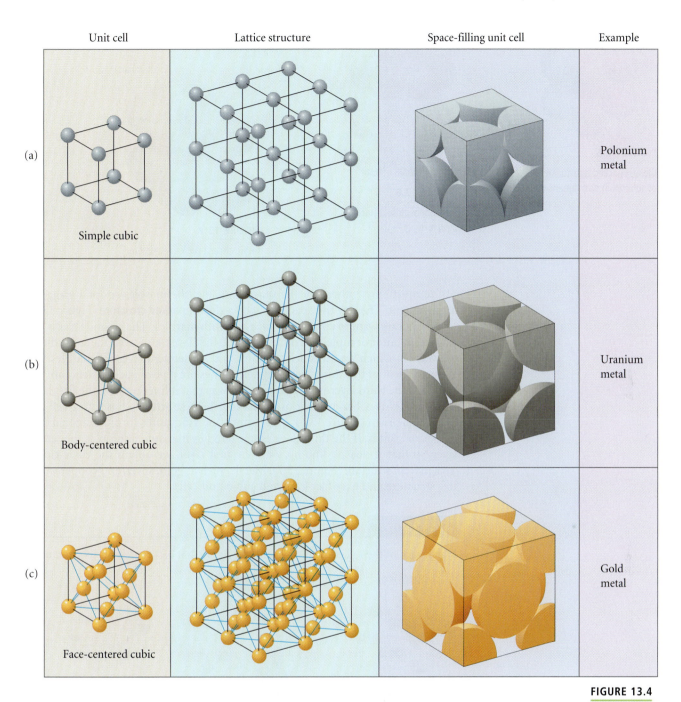

(a) Simple cubic — Polonium metal

(b) Body-centered cubic — Uranium metal

(c) Face-centered cubic — Gold metal

FIGURE 13.4

Cubic unit cells.

the simple cubic unit cell gives a **body-centered cubic (bcc) unit cell**. Iron metal crystallizes as a body-centered cubic structure. If an atom, ion, or molecule is located on each of the faces of the cubic lattice, the lattice is called a **face-centered cubic (fcc) unit cell**. When cooled below −182°C, methane (the gas that heats many homes) forms crystals that have a face-centered cubic structure. Figure 13.4 illustrates each of the unit cells in the cubic lattice group.

Video Lesson: Crystal Packing

Metallic Crystals

Solid metals are made up of atoms of an element arranged in a crystalline lattice. If we imagine that the atoms are hard spherical balls, we can get a picture of how they might be arranged into a crystal. We can visualize this by placing marbles in a box. Put in just enough marbles to cover the bottom and what happens. To

FIGURE 13.5

Closest packed arrangement of spheres in two dimensions.

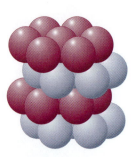

FIGURE 13.6

Hexagonal closest packed structure.

FIGURE 13.7

Cubic closest packed structure.

utilize most efficiently all of the space available, the marbles adopt a staggered arrangement. The way they pack is called a **closest packed structure** (Figure 13.5).

What if we want to completely fill a box with marbles? The second layer is also staggered, but it occupies the dimples in the first layer. However, there are different ways in which we can stack the third and subsequent layers of marbles. It is the placement of this third layer that determines the type of packed structure that results. If the third layer is placed in such a way that it lines up with the first layer, what results is a **hexagonal closest packed (hcp) structure** whose unit cell is a hexagonal prism (Figure 13.6). Magnesium and zinc atoms form hexagonal closest packed solids. If the third layer is staggered in the same fashion as the second layer, so that it doesn't line up with the first layer, we obtain the **cubic closest packed (ccp) structure** (Figure 13.7). Metals, such as gold, silver, and copper, adopt this structure and have face-centered cubic unit cells.

EXERCISE 13.1 | **Matching**

Match each of the following metals with the figure that best represents its structure.

   a. copper (fcc)   b. iron (bcc)   c. iron (hcp)

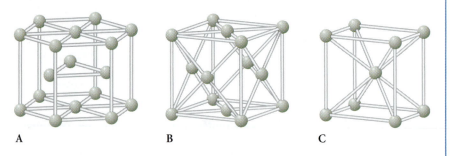

A   B   C

Solution

Copper (face-centered cubic) is **B**. Iron (body-centered cubic) is **C**, and iron (hexagonal closest packed) is **A**.

PRACTICE 13.1

Provide a diagram for each of the following elements in the given unit cell.

   a. manganese (simple cubic)   b. nickel (fcc)   c. tungsten (bcc)

See Problems 5 and 6.

Some metals do not form closest packed structures. For instance, the alkali metals stack together into body-centered cubic unit cells. Although we can know how a metal packs into a particular crystalline lattice, the reason why metals pack in the way they do is not well understood. Predicting the crystalline lattice structure of a metal is difficult. **Why is it useful to know the structure of the unit cell?** One answer has to do with the enzymes within us. Read on.

HERE'S WHAT WE KNOW SO FAR

- Solids can be either amorphous or crystalline.

- Crystalline solids result from highly structured packing of the material.

- The arrangement of particles within the crystal is as a repeating unit called the unit cell.

- The simple cubic, body-centered cubic, and face-centered cubic are examples of unit cells.

- In addition, substances can also pack into closest packed structures, which include the hexagonal closest packed and the cubic closest packed structures.

Crystallography and the Unit Cell

In 1997, the Nobel Prize in chemistry was awarded to Paul D. Boyer, John E. Walker, and Jens C. Skou "for their elucidation of the enzymatic mechanism underlying the synthesis of adenosine triphosphate (ATP)." Essential to their discovery was the identification of the crystalline structure of an enzyme found in cows, bovine F1 ATP synthase (Figure 13.8). By determining the structure of this enzyme, the researchers were able to understand how ATP (a source of energy in living things) is made. This is just one example of the utility of knowing the crystal structure of a solid. Medical researchers use crystal structures to get a clear picture of the location where reactions take place, which is called the "active site" of the enzyme. This information makes it possible to design drugs that will best influence the activity of the enzyme. For instance, the structure of the active site of HMG-CoA reductase (an enzyme that makes cholesterol) was used in the development of Rosuvastatin, a drug that binds to the enzyme and helps slow the formation of cholesterol. The binding of Rosuvastatin leads to lower serum cholesterol levels in patients.

The structures of these enzymes were determined by X-ray crystallography. How does this technique help us determine the structure of a crystal? We begin by examining the behavior of light. When light waves pass through a grating (a series of closely spaced narrow slits),

Application

FIGURE 13.8

Cartoon of the crystal structure of bovine F1 ATP synthase. Because there are so many atoms in this enzyme, each strand of the protein is represented as a ribbon of a different color. Specific atoms are not included in the structure.

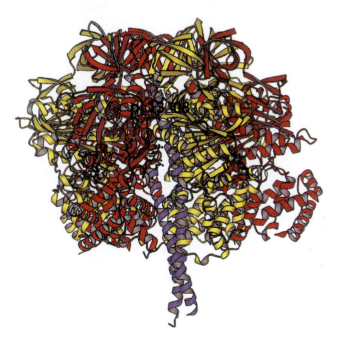

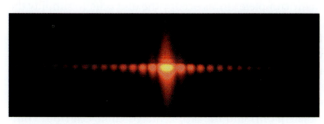

A single light beam from a laser, when passed through a narrow opening, makes a pattern of dots due to constructive and destructive interference of the individual photons.

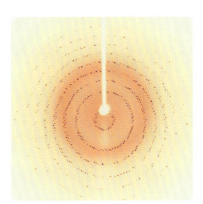

FIGURE 13.9

A pattern of reflected X-rays is captured after interaction with a crystal. The location of the dots in this diffraction pattern can be used to determine the structure of a compound.

Video Lesson: Crystal Structure

Application

CHEMICAL ENCOUNTERS: X-ray Crystallography

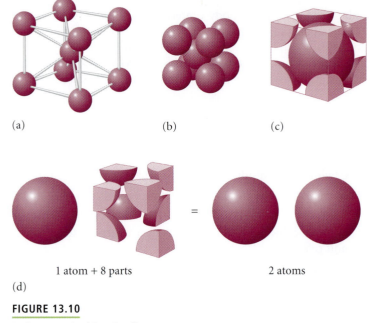

(a)   (b)   (c)

1 atom + 8 parts   =   2 atoms

(d)

FIGURE 13.10

Body-centered cubic unit cell.

constructive and destructive interference of the waves produces a pattern of light and dark regions. This **diffraction pattern** can be used to determine the position and widths of the slits. The most useful diffraction pattern occurs when the wavelength of the light is similar to the widths of the slits. In X-ray crystallography, a diffraction pattern is generated by using a crystal instead of a series of slits.

Look at the unit cells of the crystals mentioned in the previous section. Instead of slits, the unit cells contain a lot of empty space. Because the spaces in the crystal are on the order of 100 picometers (1 pm = 10^{-12} m), radiation of similar wavelengths is used. X-rays—electromagnetic radiation with wavelengths near 100 pm—are directed through the crystal and are diffracted as shown in Figure 13.9. The resulting pattern of high- and low-intensity X-rays is used to determine the position and size of the atoms in the crystal. Our ability to calculate the positions of the atoms from the crystal diffraction pattern is based on the work of William Henry Bragg (1862–1942) and William Lawrence Bragg (1890–1972), a father-and-son team who shared the Nobel Prize in physics in 1915 for their work in X-ray crystallography.

Electromagnetic radiation can pass through a crystal because of the existence of empty spaces in the unit cell. **How much empty space is there in a crystal?** We talk about how the spheres are packed as closely together as possible, but in our unit cells we seem to have a lot of gaps. To determine the amount of empty space, we need to know how much space the atoms in a unit cell occupy and how much space the entire unit cell occupies. The difference between these two values is the empty space. How many complete atoms occupy the unit cell? To simplify the process, consider the atoms as hard spheres. Figure 13.10 shows the bcc unit cell, which contains one complete atom in the center of the unit cell. At the corners of the cell are ⅛ "pieces" of atoms. Because there are eight corners in the unit cell, there must be the equivalent of one complete atom. The unit cell for a body-centered cubic structure contains two complete atoms: the central one and the eight ⅛'s that make up the corners. We can go through a similar process for the primitive cell (one complete atom) and the face-centered cubic cell (four complete atoms).

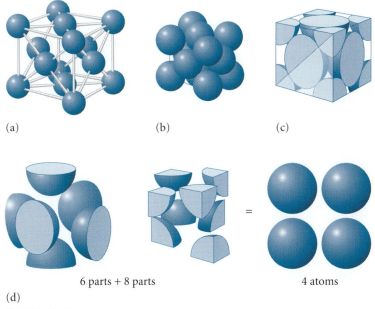

(a)

(b)

(c)

6 parts + 8 parts

4 atoms

(d)

FIGURE 13.11

Face-centered cubic unit cell.

Because we can determine how many atoms occupy a unit cell, knowing the volume of the unit cell enables us to find the amount of empty space in the cell. What is the volume of a unit cell? We can calculate this for each type of unit cell. In the face-centered cubic unit cell (Figure 13.11), we see that the diagonal across one face of the cube is 4 times the radius of the atom ($4r$) and that the height and width of the cell are the same distance (d). To find the length d, we use the Pythagorean Theorem ($a^2 + b^2 = c^2$) and solve for the distance:

$$d^2 + d^2 = (4r)^2$$

$$2d^2 = 16r^2$$

$$d^2 = 8r^2$$

$$d = r\sqrt{8} = 2.828r$$

Therefore, the distance along one of the faces of the fcc unit cell is 2.828 times the radius of the atoms. Note that the distance along the edge of the unit cell is greater than twice the radius of the atoms. This implies, correctly, that the atoms in the corners of the unit cell do not touch the atoms in the other corners.

The volume of any parallelepiped is determined by multiplying the length times the width times the height. For the fcc unit cell, the volume would then be $(r\sqrt{8})^3$, or $22.63r^3$. To illustrate this equation, gold metal crystallizes in the cubic closest packed structure (with fcc unit cells). The volume of the unit cell for gold metal whose radius is 144 pm is

$$r = 144 \text{ pm}$$

$$\text{Volume} = (144\sqrt{8})^3 = 6.76 \times 10^7 \text{ pm}^3$$

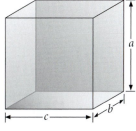

Volume $= a \times b \times c$

The volume of any parallelepiped is equal to the height times the width times the length.

EXERCISE 13.2	**There's a Lot of Room in Here**

Copper, silver, and gold are known as the coinage metals because of their use in making coins. Each of these metals forms a face-centered cubic solid. What percentage of a fcc unit cell is occupied by empty space?

First Thoughts

For us to answer this question, we must think about how much space a face-centered cubic unit cell occupies. Then, if we know how much space is actually taken up by atoms, we can determine the amount of free space in the unit cell.

Solution

Because we haven't specified any particular example, let's consider a generic fcc unit cell. The volume of a fcc unit cell is $(r\sqrt{8})^3$. We know that a face-centered cubic unit cell has four complete atoms (each with a volume of $\frac{4}{3}\pi r^3$), so the total volume of the spheres is $4 \times \frac{4}{3}\pi r^3$ or $\frac{16\pi r^3}{3}$. The ratio of the two volumes will tell us the percentage of the unit cell that is taken up by atoms:

$$\% \text{ occupied} = \frac{\frac{16\pi r^3}{3}}{(r\sqrt{8})^3} = \frac{\frac{16\pi}{3}r^3}{(\sqrt{8})^3\, r^3} = \frac{\frac{16\pi}{3}}{(\sqrt{8})^3} = \frac{16.755}{22.627}$$

$$= 0.7405 \times 100\% = 74.05\%$$

The percent of occupied space in the unit cell is 74.05%, which leaves 25.95% of the unit cell vacant. That's a lot of empty space. Note that the value of r cancels out of the equation, so our answer is valid for any ccp crystal.

Further Insights

Is there really this much empty space in a solid? Yes, according to our calculations, there are a lot of holes in a solid. Is the space really empty? As we'll see later in this chapter, it isn't completely empty in a metal. Electrons occupying large molecular orbitals occupy a lot of this empty space.

> **PRACTICE 13.2**
>
> Polonium forms the simple cubic unit cell. How much of a primitive unit cell is empty space?
>
> *See Problems 7, 8, 11, 12, and 17.* ■

The ease of calculating the volume of the unit cell enables us to determine the density of a crystal without measuring it. Mathematically, the density of a solid is the mass divided by the volume. Because we've already determined the volume of the unit cell, we need to determine its mass. How do we do this calculation? If we know how many atoms make up the unit cell and we know the mass of one atom, then we can calculate the mass of a unit cell. For example, gold crystallizes in the cubic closest packed structure (fcc unit cell). The atomic mass of gold is 197.0 g/mol, or 197 amu per atom. As we've done before, we'll keep a couple of extra figures in each calculation, rounding off only at the end.

$$\frac{197.0 \text{ g Au}}{\text{mol}} \times \frac{1 \text{ mol Au}}{6.022 \times 10^{23} \text{ atoms}} = \frac{3.2713 \times 10^{-22} \text{ g}}{\text{atom Au}}$$

Then

$$\frac{3.2713 \times 10^{-22} \text{ g}}{\text{atom Au}} \times \frac{4 \text{ atoms Au}}{\text{fcc unit cell}} = 1.3085 \times 10^{-21} \text{ g/unit cell}$$

What is the density of a block of gold? Because we can calculate the volume and the mass of any unit cell, we can determine the density of a crystalline solid if we know the unit cell. For instance, gold has a mass of 1.3085×10^{-21} g / unit cell and a unit cell volume of 6.756×10^7 pm^3. The density (mass/volume) can be calculated:

$$6.756 \times 10^7 \text{ pm}^3 \times \left(\frac{1.0 \times 10^{-12} \text{ m}}{\text{pm}}\right)^3 \times \left(\frac{100 \text{ cm}}{\text{m}}\right)^3 = 6.756 \times 10^{-23} \text{cm}^3$$

$$\frac{1.3085 \times 10^{-21} \text{ g}}{6.756 \times 10^{-23} \text{ cm}^3} = 19.4 \text{ g/cm}^3$$

The value we've calculated for gold (19.4 g/cm³, based solely on physical measurements of the unit cell) agrees quite well with the experimental value for the density of gold (19.3 g/cm³). Table 13.2 lists the important calculations for the face-centered cubic unit cell.

Other information about the crystalline solid can be determined by examining the unit cell. For example, the number of species that surround a particular ion can be determined from the packing arrangement.

TABLE 13.2	Face-Centered Cubic Unit Cell Calculations
Length of side	$(r\sqrt{8})$
Volume of unit cell	$(r\sqrt{8})^3$
Length of diagonal	$4r$
Mass of unit cell	$4 \times$ mass of one atom

EXERCISE 13.3 He Isn't Heavy, He's Dense

Some metals are much denser than others. What makes them so dense? The arrangement of the atoms in the solid, coupled with the mass of the element, accounts for the different densities we observe. Let's calculate the density of an iron bar. Iron, whose atomic radius is 124 pm, forms bcc unit cells.

Solution

We can determine the density of a block of iron by dividing the mass of the appropriate number of atoms by the volume of the unit cell. The body-centered cubic unit cell has the volume $\left(r\dfrac{4}{\sqrt{3}}\right)^3$. For an iron body-centered cubic unit cell, the volume is

$$\left(124\ \text{pm} \times \frac{4}{\sqrt{3}}\right)^3 = (286.4)^3 = 2.348 \times 10^7\ \text{pm}^3$$

or

$$2.348 \times 10^7\ \text{pm}^3 \times \left(\frac{1\ \text{m}}{10^{12}\ \text{pm}}\right)^3 \times \left(\frac{100\ \text{cm}}{1\ \text{m}}\right)^3 = 2.348 \times 10^{-23}\ \text{cm}^3$$

In a bcc unit cell containing two atoms of iron, the total mass is

$$2\ \text{atoms} \times \frac{55.847\ \text{g}}{\text{mol}} \times \frac{1\ \text{mol}}{6.022 \times 10^{23}\ \text{atoms}} = 1.855 \times 10^{-22}\ \text{g}$$

The density of iron is the mass divided by the volume:

$$\frac{1.855 \times 10^{-22}\ \text{g}}{2.348 \times 10^{-23}\text{cm}^3} = 7.90\ \text{g/cm}^3$$

The experimentally measured density of iron is 7.86 g/cm³.

PRACTICE 13.3

Lead is used as fishing weights because of its malleability and density. What is the density of lead (Pb)? Lead has an atomic radius of 175 pm and crystallizes in a fcc arrangement.

See Problems 16, 18, 21, and 22.

Calculating the positions of the atoms in a unit cell based on an X-ray diffraction pattern used to be a tedious task that could take months or years to complete. Today, scientists use the X-ray diffractometer to generate X-rays and direct them at a single crystal at a variety of different angles. Powerful computers collect the data over a period of hours and store them until the diffraction pattern is complete, at which time the computers go to work deciphering the pattern into a picture of the crystal. Depending on the quality of the crystal, the location of each of the atoms in the solid can be accurately determined overnight.

FIGURE 13.12

Liquid crystals. The liquid crystal has properties that resemble the liquid and properties that resemble the solid. It is a fluid, yet highly organized phase.

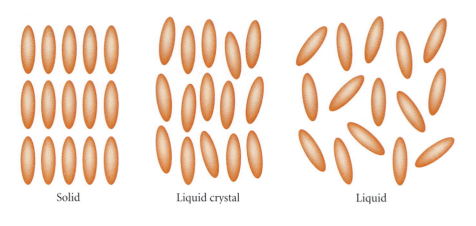

Solid   Liquid crystal   Liquid

Liquid Crystals

 Application

We use liquid crystals in the displays of watches, pocket calculators, computers, and privacy windows. What is a liquid crystal and how does it work? The liquid crystal is a transitional phase between a solid and a liquid. There are some exceptions, but the molecules of the liquid crystal generally have long, rigid structures with a strong dipole moment that point along a common axis (Figure 13.12) to align into highly ordered crystalline solids. An example of a solid that forms a liquid crystal is cholesteryl benzoate.

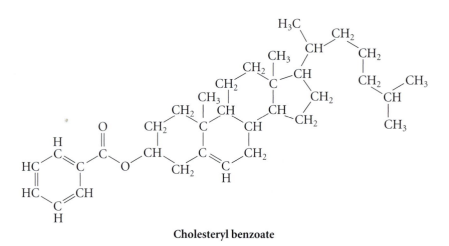

Cholesteryl benzoate

If the temperature of a solid with these characteristics is increased, the long, rigid structures of the molecules and the strong dipole moment continue to maintain a high degree of order within the structure. In fact, even well into the liquid state, order is maintained despite the increase in translational freedom (the ability to move in many directions). When the temperature gets hot enough, the degree of order in the structure is eliminated. In other words, in the transitional state between completely solid and completely liquid, the molecules are highly ordered and yet quite fluid (the **liquid crystal** state).

The strong dipole moment is the key to the behavior of liquid crystals. From Chapter 9, we know that dipole moments align themselves within an electric field. When such a field is applied to the liquid crystal, its molecules orient themselves with the electric field. This highly ordered structure is transparent to light directed along the axis of orientation. If the electric field is removed, the translational freedom of the molecules scrambles their alignment. Visually, a liquid crystal in this state is opaque because the slight randomness scatters light. The net effect is a material that can be made transparent or opaque with the flick of a switch.

- The dimensions and the density of a solid can be determined if the mass and the radii of the components (be they ions, molecules, or atoms) are known.

- X-ray crystallography is a useful technique that can also determine the arrangements of atoms in the unit cell.

- The liquid crystalline state is a transitional phase that exists between the solid and liquid phase for some molecules. Typically, these molecules have long, rigid structures with strong dipole moments.

Video Lesson: Metallurgical Processes

Application

13.2 Metals

To fill a cavity in a tooth, a dentist cleans out the hole and packs it with a composite material made of glass and plastic resin, or with a "silver" filling, to arrest the growth of the enamel-eating bacteria. The composite filling is popular, but because the silver filling is much harder, more durable, and relatively inexpensive, there is still a need for it. Why did we put quotation marks around the word *silver*? The filling is actually is a mixture of about 50% mercury and 50% other metals (including silver, tin, and copper). This is one of countless applications of metals in our lives. In this section, we will examine the properties of metals and learn why these elements are so useful.

Using a cast iron skillet to cook your eggs and bacon used to be the rule, not the exception. You had to have a mitten handy to pick the skillet up off the stove, or your hand would be badly burned. Metals, such as the iron in the skillet, have long been known to conduct heat and electricity (see Chapter 5). **Why do metals conduct heat?** One of the theories that explains these properties (and others) is called **band theory**. This theory addresses how adjacent metal atoms interact. For example, two lithium atoms can interact with each other by sharing their electrons. As we discussed in Chapter 9, this gives rise to our postulate that dilithium can exist. A closer look at the molecular orbitals of dilithium reveals that the valence electrons come from the 2s orbital. A bonding orbital lies below the energy of the 2s orbital, and an antibonding orbital lies above the energy of the 2s orbital (Figure 13.13). The valence electrons are placed in the molecular orbitals in such a way as to fill the bonding orbital and leave the antibonding orbital empty.

What happens if *three* lithium atoms participate in bonding? The result of mixing three atomic valence orbitals gives rise to three valence molecular orbitals (Figure 13.14). One of these orbitals is bonding, one is antibonding, and one is nonbonding. A **nonbonding molecular orbital** can be thought of as one that neither enhances nor diminishes the molecular bonding. The nonbonding molecular orbital arises when there is an odd number of atomic orbitals that mix in a given energy level. The result of this mixing is an odd number of MOs in the energy level (one of the orbitals is the nonbonding MO). Three lithium atoms contribute a total of three valence electrons to the valence molecular orbital diagram. Placing them in the diagram results in a full bonding MO and a half-filled nonbonding MO.

The overlap of *four* lithium atoms results in two bonding orbitals (completely full of electrons) and two antibonding orbitals (empty), as shown in Figure 13.15. An equal number of bonding molecular orbitals and antibonding molecular orbitals are formed. And the bonding orbitals reside at an energy level below that of the antibonding orbitals.

If we increase the number of atoms that participate in forming molecular orbitals to *eight*, we find that half of the orbitals (the bonding orbitals) fill with electrons. The other half of the molecular orbitals is comprised of the empty antibonding orbitals. In addition, the orbitals within the bonding and antibonding

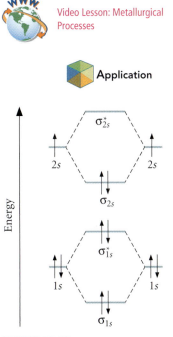

FIGURE 13.13

Molecular orbital diagram of dilithium. Two bonding and two antibonding MOs make up the diagram for dilithium. Placing electrons in this MO diagram provides a full bonding MO and an empty antibonding MO.

FIGURE 13.14

Molecular orbital diagram of trilithium. Note that the energy of the MOs derived from 2s orbitals is approaching the energy of the MOs derived from 1s orbitals.

Video Lesson: Band Theory of Conductivity

levels begin to have noticeably *different* energies. This difference arises due to the requirements of the symmetry for molecular orbitals; as the size of the molecular orbital increases, the number of nodes within the molecular orbitals increases and results in a decrease in the magnitude of the molecular orbital's energy. Overall, the energy difference between the antibonding and the bonding molecular orbitals becomes smaller (bonding MOs increase in energy and antibonding MOs decrease in energy).

As more and more atoms combine to make a metallic crystal, the difference in energy between bonding and antibonding orbitals becomes negligible. The result is a band of molecular orbitals constructed from the overlap of *s* orbitals. In metals, the range of energies in this band is often wide enough to overlap with similarly constructed bands of *p* and *d* orbitals. Many of these bands are empty, or only partially filled, with electrons. For any metal not at absolute zero, many of the electrons occupy the unfilled portions of the higher energy bands. Because the molecular orbitals encompass the entire metallic crystal, the electrons are free to roam from one end of the metal crystal to the other.

Crystals made from nonmetal elements, however, have a slightly different arrangement. As we see from Figure 13.15, the lower energy bands are accompanied by an empty higher energy band that doesn't overlap. For instance, carbon has a band of molecular orbitals that results from the overlap of the 2*s* and 2*p* atomic orbitals. Higher in energy lies a band of molecular orbitals that result from the overlap of the 3*s* and 3*p* atomic orbitals. The lower-energy, partially filled bonding molecular orbital band is known as the **valence band**. This band contains the valence electrons. The empty higher-energy molecular orbital band is the **conduction band**. Some nonmetal elements have conduction bands that are much greater in energy than their valence band. Still others have conduction bands that overlap or nearly overlap with the valence bands.

Band theory describes a metal as a lattice of metal cations spaced throughout a sea of delocalized electrons. How does this theory explain why metals are ductile, malleable, conductive, and shiny? Imagine, for example, deforming a metal bar with a hammer and displacing some of the metal ions. The sea of delocalized electrons can immediately adjust to the change, so the overall structure doesn't become significantly weaker as it is bent. Metals, then, are **ductile** (able to be pulled) and **malleable** (able to be pounded) to different degrees. Additionally, because electrons can easily move from one end of the molecular orbital to the other, metals can conduct electricity.

Why can metals conduct heat? Kinetic energy can be conveyed easily across a crystalline lattice that consists of positive ions within a sea of electrons; metals

FIGURE 13.15

Molecular orbital diagram of lithium metal. As the number of lithium atoms participating in bonding increases, the regions of anti-bonding and bonding molecular orbitals become continuous.

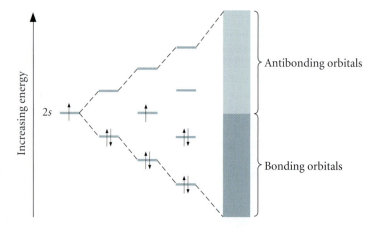

conduct heat easily. In other words, kinetic energy can be transferred from positive ion to positive ion with ease. This stands in contrast to the difficulty involved in transferring kinetic energy across an ionic or molecular solid, where the kinetic energy must be transferred from cation to anion to cation (or from molecule to molecule). This explains why metal ice cube trays feel colder than plastic ice cube trays. The metal trays, as thermal conductors, are better able to pull the heat from your hand. The plastic ice cube trays are made of a thermal insulator (there isn't enough thermal energy available to promote electrons into the conduction band in the plastic). Or, to say it differently, *the valence electrons in plastic are tied to covalent bonds specifically joined to a given pair of atoms and are not free to move around.*

Band theory also explains why metals are shiny. Because the molecular orbitals within the valence bond are so plentiful and close together, electrons can absorb and release almost any wavelength of light from the infrared to the ultraviolet. The absorption promotes the electron into an unfilled molecular orbital within the band; the release of a photon returns the electron to its original location in the band. The result is that metals reflect light, giving rise to the luster that we associate with them. White paper also reflects light but does so diffusely (in all directions). Some metals have a distinctive color. For example, gold has a yellow hue. This is due to the inefficiency of the metal when it reflects certain frequencies of absorbed light. These metals are still shiny; they just don't reflect all wavelengths of light equally.

Band theory is used in determining the color of paints, identifying the insulating value of materials in your home, and (remembering the opening scenario of the chapter) in the hospital as well, in the form of heart monitors, IV pumps, calculators, and many other instruments. **What do such instruments have in common?** They are all controlled by small computers, and at the heart of all computers is the element silicon. Why silicon? Its **band gap** is the key. A band gap is a small energy gap that exists between the valence band and the conduction band, as illustrated in Figure 13.16. The existence of a band gap in silicon suggests that a certain amount of energy is needed in order to promote an electron from the valence band into the conduction band. If the gap is relatively small, as it is in silicon, the metal is a **semiconductor**. In these cases, the amount of energy needed to promote an electron into the conduction band is relatively small. (See Table 13.3 for a list of some compounds and their band gaps.) When the temperature

TABLE 13.3	Band Gaps of Some Common Materials	
	Material	**Band Gap**
Insulator (>300 kJ/mol)	Carbon (diamond)	502 kJ/mol
	Zinc sulfide	350 kJ/mol
Semiconductor (50–300 kJ/mol)	Silicon	106 kJ/mol
	Selenium	215 kJ/mol
	Germanium	64 kJ/mol
Conductor (<50 kJ/mol)	Lithium	0 kJ/mol
	Tin (gray tin)	7.7 kJ/mol

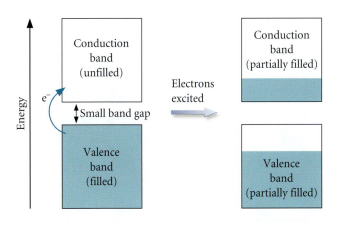

FIGURE 13.16

The band gap in silicon.

Video Lesson: Intrinsic Semiconductors

is low, there isn't enough kinetic energy to promote many electrons into the conduction band, and the metal doesn't conduct much electricity. Increase the temperature of the semiconductor, and the conductivity increases. Silicon (band gap = 106 kJ/mol) and germanium (band gap = 64 kJ/mol) are good examples of semiconductors.

If the band gap is sufficiently large that there are barely any electrons at room temperature in the conduction band, the material is an **insulator**. Only a tiny amount of electrical conductivity can be found in an insulator. However, as an insulator becomes hotter, its conductivity increases. Our plastic ice cube tray is a good example of an insulator. Diamond is another good insulator (the band gap in diamond is 502 kJ/mol). In short, when the band gap is large, the material acts as an insulator. Remove the gap, and the material is a **conductor**. For example, compare the band gap in diamond to the amount of energy required to promote an electron into an unfilled orbital in a conductor such as lithium, roughly 10^{-45} kJ/mol.

Adding small amounts of other elements to the metal can modify the behavior of a semiconductor. For example, if we **dope** (add an atom or other compound to) silicon (Group IVA) with an element that has more valence electrons than silicon (such as phosphorus, Group VA), the extra valence electrons must be placed into the conduction band (Figure 13.17). As a consequence, the semiconductor becomes capable of conducting current. The result is called an **n-type semiconductor** because the addition of the phosphorus adds extra *n*egative charges (electrons) to the system. However, if we dope silicon with an element containing fewer valence electrons (such as aluminum, Group IIIA), there will not be enough valence electrons to fill the valence band. The result: Electrons in the valence band can be easily excited to move to an unoccupied molecular orbital. This semiconductor is electrically conductive if we apply a small electric field. We call it a **p-type semiconductor** because the addition of gallium can be thought of as introducing *p*ositive holes in the valence band. The properties of the n- and p-type semiconductors make them quite useful for the production of small electronic devices.

Video Lesson: Doped Semiconductors

FIGURE 13.17

Doping the semiconductor. An n-type semiconductor contains extra electrons added to the conduction band. A p-type semiconductor is missing some electrons from the valence band.

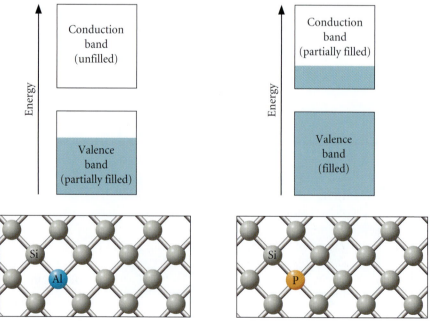

The mineral greenockite (CdS) is best known as cadmium yellow and is used as a yellow pigment in paints. Cadmium sulfide has a medium-sized band gap (4.2×10^{-19} J/photon). Judging on the basis of the information in Table 13.3, is greenockite an insulator, a semiconductor, or a conductor? What color light does cadmium sulfide absorb?

Solution

The band gaps in Table 13.3 are listed in units of kilojoules per mole, whereas this exercise gives the size of the band gap for greenockite in joules per photon. We can use Avogadro's number to make the units the same.

$$\frac{4.2 \times 10^{-19}\ \text{J}}{\text{photon greenockite}} \times \frac{6.022 \times 10^{23}\ \text{photons}}{\text{mol greenockite}} \times \frac{1\ \text{kJ}}{1000\ \text{J}} = \frac{250\ \text{kJ}}{\text{mol greenockite}}$$

This value places cadmium sulfide in the semiconductor region. The promotion of an electron from the valence band to the conduction band actually requires 4.165×10^{-19} J. Using the equations $E = h\nu$ and $c = \lambda\nu$, we find a frequency of 6.286×10^{14} Hz and a wavelength of 476.9 nm. This corresponds to the absorption of blue light. A solution of cadmium sulfide appears yellow to the eye.

What is the largest wavelength of light, in nanometers, that can be absorbed by a particular semiconductor if the band gap is determined to be 0.95 eV? (1 eV/atom = 96.485 kJ/mol).

See Problems 27, 28, 35, and 36.

Perhaps the greatest benefit that a semiconductor offers is its ability to act as a **photovoltaic device**, in which light energy can be used to generate an electric current. Photovoltaic cells capable of over 30% efficiency have been designed. How is the sunlight converted into electricity? When silicon is doped with two elements such as phosphorus (Group VA) and gallium (Group IIIA), the semiconductor contains both electron-rich and electron-deficient areas. Light energy throughout the visible and IR range (the particular range of wavelengths depends on the chemical composition of the semiconductor) provides enough energy for electrons to "jump" the band gap and enter the conduction band. This enables both negative charges (electrons) and positive holes to move so that the flow of electrons can be harvested as energy.

Photovoltaic cells have been used in many applications, from the solar-powered calculator to solar-powered water heaters, space stations, and satellites. The use of photovoltaic cells to make a solar-powered car, such as that shown in Figure 13.18, is the next big hurdle. Because our world demands fossil fuels as a source of energy for transportation, and because of our limited supply of this natural resource, alternative-energy automobiles will one day be commonplace. Currently, auto makers and researchers have been able to develop solar-powered cars that can travel an average of 50 miles an hour relying solely on the photovoltaic cell. However, despite the estimated 60% decline in costs to implement the use of photovoltaic cells over the last 20 years, sales of solar-powered vehicles are still dwarfed by sales of those powered by fossil fuels; solar-powered autos made up less than 1% of the total auto sales in 2003.

Application

Chemical Encounters: Photovoltaic Devices

FIGURE 13.18

The solar-powered car is currently being tested as a replacement for the gasoline auto that we use today. This solar-powered prototype is not too far from what we may be driving in the near future. Note the photovoltaic cells on the roof of the car.

Alloys

We return now to our opening theme of materials used in the hospital by noting that the limited properties of the metals don't lend themselves well to any more than routine medical tasks. For example, surgeons have a relatively limited number of choices in materials when replacing a hip joint. They need something as strong and durable as metal, but they don't use solid copper, silver, or gold. **Why don't they use a coinage metal for a hip joint?** Many of the pure metals are too malleable and ductile to withstand the forces in a human body. A gold hip joint would easily bend when a patient walked with it. Other factors, such as the cost and toxicity of some of the pure metals, are also important. In fact, it's rare that we see *pure* metals used in the "real world." Because there are only a limited number of pure metals, and because the properties of those pure metals are often not what is required, we look to *mixtures* of metals to give us the properties we need.

A homogeneous mixture of two or more metals is known as an **alloy**. The elements are typically mixed together in the molten state and then allowed to cool to form the alloy. Alloys are useful because their properties can be adjusted by changing the ratio in which the elements are combined. (The Bronze Age is named after the discovery of one of the first of these metal blends.) Typically, alloys are less ductile and less malleable than pure metals. Table 13.4 lists some of the common alloys, along with a typical composition for each and several uses to which each is put. Manufacturers vary the actual composition of each type of alloy to change its properties slightly.

Two different classes of alloys exist, the **substitutional alloys** and the **interstitial alloys**, illustrated in Figure 13.19. Substitutional alloys include metals such as brass, sterling silver, pewter, and solder. In a substitutional alloy, specific atoms within the lattice structure of a metal are replaced with other metal atoms. In the example of sterling silver, an average of 7 silver atoms out of every 100 are replaced with copper atoms. Interstitial alloys are made of metals with "impurities" trapped in the spaces of the lattice structure. By far the most important of these is steel. Steel, which we discussed in Section 7.2, contains carbon atoms located in the spaces between the atoms in an iron atom lattice.

Cu   Ag

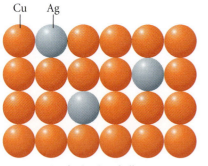

A substitutional alloy

Fe   C

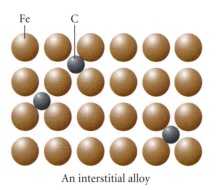

An interstitial alloy

FIGURE 13.19

Substitutional versus interstitial alloys.

TABLE 13.4	Common Alloys and Their Uses	
Name	**Common Composition (mass %)**	**Uses**
Substitutional		
Brass	90 Cu : 10 Zn	Decorative fittings
Pewter	85 Sn : 7 Cu : 6 Bi : 2 Sb	Plates, jewelry, figurines
Medal bronze	92–97 Cu : 1–8 Sn : 0–2 Zn	Medals, artwork
14-carat gold	58 Au : 30 Ag : 12 Cu	Jewelry
Sterling silver	92.5 Ag : 7.5 Cu	Jewelry, kitchenware
Coinage silver	90 Ag : 10 Cu	Silver coins
Plumber's solder	67 Pb : 33 Sn	Pipe solder
Interstitial		
Steel	98–99.9 Fe : 0–2 C	Building materials
Stainless steel	>10.5 Cr : <89.5 Fe	Eating utensils, structural materials

Amalgams

An amalgam is an alloy containing the metal mercury. Dentists use an amalgam for filling teeth by mixing a powder containing silver and tin and then adding mercury. The amalgam immediately begins to form at room temperature, making a complex and very hard mixture of Ag_2Hg_3 and Sn_7Hg.

$$14Ag_3Sn + 65Hg \rightarrow 21Ag_2Hg_3 + 2Sn_7Hg$$

Other amalgams do exist. For instance, the amalgamation of gold with mercury is one way in which gold dust was extracted from powdered gravel in the California gold rush era of the mid- to late nineteenth century. Because of the hazards of mercury poisoning, the use of mercury has been supplanted by other means of extracting gold (see Chapter 4). Unfortunately, it continues to be used by individuals and small companies in the poorer countries of the developing world. Its use has led to mercury poisoning of the workers and the local water supplies. Other amalgams include zinc and cadmium, used in batteries to increase their lifetime (see Chapter 19).

Application

CHEMICAL ENCOUNTERS: Dental Amalgams

13.3 Ceramics

According to the American Academy of Orthopedic Surgeons, most of us will have two bone fractures in our lifetime, and many of these accidents will require immediate medical attention. Some of the casts used to hold fractured bones in place as they heal are made of fiberglass or plastic, but the less expensive plaster cast is still widely used. The plaster, also known as plaster of Paris, is a ceramic material made from calcium sulfate hemihydrate ($CaSO_4 \cdot \frac{1}{2} H_2O$) and water. **Why does plaster harden?** This is the subject of the accompanying NanoWorld/MacroWorld feature.

Ceramics are a class of nonmetallic materials that are made of inorganic compounds. They can be either amorphous or crystalline in structure. Examples of ceramics include glass, clay pottery, cement, bricks, and china dinner plates. Because ionic bonds exist within a ceramic material, the band gap is typically quite large. These compounds therefore act as insulators with low electrical conductivity, high melting point, and a lack of malleability and ductility, but they do have good resistance to corrosion. With the exception of glass, melting does not reshape ceramics. Typically, a mixture of the finely ground ceramic and some binders is used to mold the material into the desired shape. After heating of the new shape (called firing), the ceramic becomes hard. Compare the properties of the ceramics with those of other materials (see Table 13.5 on page 559).

Research into the uses of ceramics as bone replacements or supplements is currently under way. One goal of the research is the ability to replace badly damaged bones with synthetic material that later becomes bone, rather than using bone from elsewhere in the patient's body. Ceramics made from hydroxyapatite (see Chapter 8) show some promise in this area because they contain the same sponge-like structure as real bone. This enables new bone to grow and integrate into or replace the ceramic material.

The large band gap and the relatively low coefficient of thermal expansion in a ceramic material allow design engineers to use ceramics in places where insulation against extremely high temperatures is needed. Of particular interest is the use of ceramics as insulation on the space shuttle. During a typical mission, the space shuttle's hull is subjected to temperatures of −156°C during orbit and up to 1650°C as the shuttle reenters the Earth's atmosphere.

Video Lesson: Ceramics and Glass

Application

Hydroxyapatite (above) can be formed into replacements for bones, as seen in the X-ray (left).

NanoWorld / MacroWorld

Big effects of the very small: The chemistry of cement

Whereas metals and plastics can bend and stretch, ceramic materials are less ductile. In fact, their ability to withstand compression makes them ideal candidates for supporting weight, so ceramics are commonly used as building materials such as plaster, mortar, and cement.

The formation of plaster and that of mortar are similar in that a natural material is converted into a different compound that can be molded before it returns to its natural state. In plaster, the reactions that take place are based on the dehydration and hydration chemistry of calcium sulfate ($CaSO_4$). To make plaster, the mineral gypsum ($CaSO_4 \cdot 2H_2O$) is mined and ground into a powder. This powder, when heated at 150°C to remove some of the water, is converted into calcium sulfate hemihydrate ($CaSO_4 \cdot \frac{1}{2}H_2O$). The hemihydrate, also known as plaster of Paris (after a gypsum deposit near Paris, France, that had been mined since the seventeenth century), is then mixed with water to make a paste. The hemihydrate reabsorbs the water to re-form the gypsum, which is now in a shape that the manufacturer has molded.

$$CaSO_4 \cdot 2H_2O + heat \rightarrow CaSO_4 \cdot \tfrac{1}{2}H_2O + \tfrac{3}{2}H_2O$$

$$CaSO_4 \cdot \tfrac{1}{2}H_2O + \tfrac{3}{2}H_2O \rightarrow CaSO_4 \cdot 2H_2O + heat$$

Construction workers use mortar to hold bricks together in the shape of a wall like that shown in Figure 13.20. Mortar is made from calcium carbonate ($CaCO_3$, limestone) by heating to a high temperature in a kiln. The calcium oxide (CaO, lime) that results is mixed with sand and added to water. This wet paste is sandwiched between the bricks in a wall. Slowly, the lime in the paste reacts to form calcium hydroxide ($Ca(OH)_2$, slaked lime). Over time, as the mortar is "curing," the reaction of carbon dioxide in the air returns the slaked lime to limestone.

$$CaCO_3 + heat \rightarrow CaO + CO_2$$

$$CaO + H_2O \rightarrow Ca(OH)_2$$

$$Ca(OH)_2 + CO_2 \rightarrow CaCO_3 + H_2O$$

The mortar also reacts with the SiO_2 that is the main chemical component of sand to form complex calcium silicates via an acid–base reaction between the lime (CaO; base) and the sand (SiO_2; acid).

Cement is a much more complex material than plaster or mortar, but the chemistry of its hardening is very similar. Made by heating shale and limestone at 1500°C, the product (called clinker) is ground up in large rotating chambers containing steel balls until it has the consistency of flour. Gypsum is often added to the mixture to improve its hardening properties. The result is dry cement. Specific amounts of water, sand, and pebbles are combined with the dry cement to make concrete. The reactions, shown below, that harden the cement then begin, trapping the sand and pebbles inside the mixture. Some of the grains of the mixture begin absorbing water and react with it (hydration). Although the rate of hydration is relatively slow for cement, a poured driveway is often hard enough to use within a few days. Most cements continue to harden for up to a year.

$$Ca_3SiO_5 + 3H_2O \rightarrow Ca_2SiO_4 \cdot 2H_2O + Ca(OH)_2$$

$$2Ca_3SiO_5 + 7H_2O \rightarrow Ca_3Si_2O_7 \cdot 4H_2O + 3Ca(OH)_2$$

$$Ca_3Al_2O_6 + 3CaSO_4 \cdot 2H_2O + 24H_2O \rightarrow$$
$$Ca_6Al_2(SO_4)_3(OH)_{12} \cdot 24H_2O$$

A ball mill in a cement plant grinds the material into clinker. This is just one of the steps taken to produce cement.

FIGURE 13.20

Mortar is used to hold these bricks together.

TABLE 13.5 **Comparison of Ceramics with Other Materials**

Hardness is a measure of the relative ability of a material to scratch another; it is expressed using the Mohs hardness scale. The higher the hardness value, the harder the material (talc = 1, gypsum = 2, diamond = 10). The coefficient of expansion indicates the size change upon heating or cooling. A larger value indicates a greater change in the size of the material as it is heated or cooled. Note that the ceramics (alumina and zirconia) have relatively high melting points, are hard, and don't expand much when heated.

Material	Melting Point (°C)	Density (g/cm³)	Hardness (Mohs)	Coefficient of Expansion (°C⁻¹ × 10⁻⁶)
Ceramics				
Alumina (Al_2O_3)	2050	3.8	9	8.1
Zirconia (ZrO_2)	2660	5.6	8	6.6
Alloys				
Steel	1370	7.9	5	15
Brass	930	8.6	4	20
Metals				
Zinc	420	7.1	2.5	35
Sodium	98	0.97	0.4	70

Coating the hull with materials capable of withstanding the heat and resisting expansion protects the astronauts and the shuttle itself. Even though ceramics alone can satisfactorily insulate the bottom of the shuttle, they are often glazed with silica glass. This improves the thermal insulation and the durability of the tiles.

Application

Glass

The **glass** used to make test tubes for the hospital laboratory is an amorphous solid ceramic material. Solids such as obsidian, a natural glass ejected from volcanoes, shown in Figure 13.21, lack order in the arrangement of the particles that constitute them. Human-made glass has been known throughout history. For example, the ancient Mesopotamians and Egyptians made a very successful business out of the preparation of glass objects. Today, we use glass in many different applications, as illustrated in Table 13.6. The windows on the doors into the hospital, the screens on the computers at the nurses' stations and the drinking glasses in the cafeteria are all made of glass.

The simplest glass is made from quartz (crystalline silica). Figure 13.22 illustrates the very ordered structure of SiO_2 units in quartz. When heated to the melting point (1600°C), some of the silicon–oxygen bonds break and the ordered structure is disturbed. Rapid cooling of this material results in a highly disordered structure. The resulting silica glass is composed of SiO_2 units connected in a very disordered array. Because of the disorder, glass does not have a definite melting point. Instead, silica glass has a transitional temperature where it begins to soften. When it is in this softened state, glass can be molded into many different shapes.

Of particular use to the endoscopist at the hospital is the fiber optic line, a hair-thin wire of plastic-coated glass that can be used to transport light along its length (Figure 13.23). On the end of the line are a small camera and light. The Internet that connects the emergency room computer to data sources across the world also uses fiber optic lines to transmit information. How does light bend along the path taken by the fiber optic wire? John Tyndall (1820–1893) noted that light could be bent inside a bent stream of water by a property known as total internal reflectance. Total internal reflectance of the light ensures that any light inside the glass pipe reaches the end of the fiber optic wire. Because a fiber optic

FIGURE 13.21

Obsidian, a natural glass. Ancient peoples used obsidian as a tool because of its sharpness and durability.

Application

TABLE 13.6 **The Six Main Types of Glass**

Type of Glass	Example of Use	Resistance to Temperature Shock	Resistance to Corrosive Chemicals	Sample Composition (in addition to silica, SiO_2), Mass %
Soda-lime glass	Most common (90% of all glass)	Not good	Fair	16% Na_2O, 7% CaO, 3% MgO, 1% Al_2O_3, 0.3% K_2O
Lead crystal	Good electrical insulator, brilliance when cut	Not good	Fair	24+% PbO
Borosilicate	Light bulbs, bakeware	Good	Good	12% B_2O_3, 5% Na_2O, 5% Al_2O_3
Aluminosilicate	Resistors	Good	Good	7.0% B_2O_3, 10.4% Al_2O_3, 21.0% CaO, 1.0% Na_2O
Silica	Chemical glassware	Excellent	Good	96% SiO_2
Fused silica	Cuvettes, space shuttle windows	Excellent	Excellent	100% SiO_2

FIGURE 13.22

Silica glass (SiO_2), also known as quartz glass, or fused silica. While this type of glass is far superior in many ways to other types of glass, it is much more expensive and much more difficult to mold into useful apparatii.

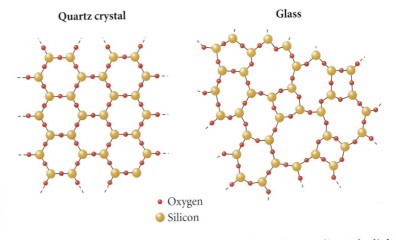

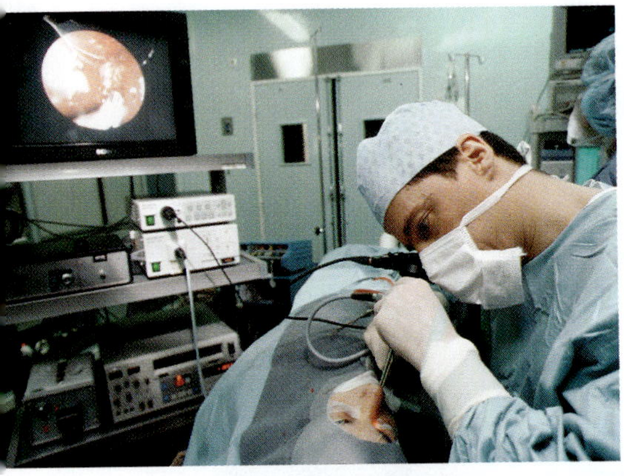

Quartz crystal   Glass

- ● Oxygen
- ● Silicon

wire can direct light around corners, the endoscopist can direct the light to any part of a patient's interior without the need for large incisions. The incredible flexibility of fiber optic wire has increased its use as telephone wire and for connections between computers and other electronic communication systems.

FIGURE 13.23

The endoscope in action. A fiber optic cable can be used to explore inside a body with no need for large incisions.

EXERCISE 13.5 **Classifying Common Materials**

Classify each of the following common household items as a metal, an alloy, a glass, or a ceramic: A sidewalk, a 1-oz silver dollar, a bicycle frame, and a perfume bottle.

Solution

The sidewalk is made from concrete. Concrete is the mixture of cement and small stones or sand. Cement is a ceramic material.

Some coins, such as the 1-oz U.S. silver dollar, are made of nearly pure silver (>99% pure silver metal). However, most coins (such as the U.S. Golden Dollar), are made of alloys to help reduce the expense of their manufacture.

The bicycle frame, in order to be light yet durable, is made out of a metal alloy. The top bicyclists use only aluminum, magnesium, or titanium alloys.

The perfume bottle is typically made from glass. The transparency, esthetic beauty, and low cost of glass make it an excellent material for bottles.

PRACTICE 13.5

Try some on your own. What class of materials makes up the rims on your car? . . . a casserole dish? . . . the toilet in your bathroom? . . . a wedding band?

See Problems 51 and 52.

Superconductors

Some patients in the hospital enter through the emergency room as cases of trauma. Often, the patient presents symptoms that are not easily identifiable by external examination of the body. At this point, physicians call upon devices that can "see" inside a patient, such as an X-ray machine, which can take images of bones. Another instrument that can investigate the structure of the organs in the body is the magnetic resonance imager (MRI) shown in Figure 13.24.

The MRI works by measuring the absorption of energy by a particular type of nucleus in the radio frequency range of the electromagnetic spectrum. However, in order for the absorption to be noticed, the nucleus must be aligned with a strong magnetic field. The size of the magnetic fields required is much greater that what can be obtained with a bar magnet. In fact, the bulk of the MRI is occupied by a superconducting magnet that can generate these large magnetic fields. What is a superconductor and how does it find use as a magnet?

A **superconductor** is a type of ceramic material that acts quite differently than a metal. The conduction of electrons in metals is met with resistance. We use this property to toast bread when the NiChrome (an alloy of nickel, chromium, and, often, iron and other metals) coils of a toaster get quite hot and exhibit their characteristic orange glow, as a consequence of their high electrical resistance. We might imagine that copper wires could be used to toast bread, too. However, copper melts when you put enough voltage through it to get it hot enough to toast bread. NiChrome wire, on the other hand, has a high melting point and won't melt inside your toaster.

Why do ceramics and metals such as NiChrome get hot when electrons flow through them? The resistance is due to a combination of impurities in the alloy and to the existence of induced quantized vibrations (called **phonons**) in the metal lattice. Resistance

Video Lesson: CIA Demonstration: Superconductivity

Application

CHEMICAL ENCOUNTERS: Magnetic Resonance Imaging

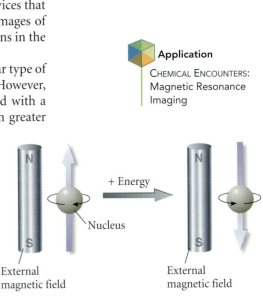

A nucleus acts as a tiny bar magnet and aligns with an external magnetic field. If the nucleus then absorbs energy in the radio frequency range, it can flip over and align against the external magnetic field. This higher energy state is unstable, and as the nucleus realigns with the external magnetic field, it gives off the energy it absorbed.

FIGURE 13.24

The magnetic resonance imager at work. The MRI generates powerful magnetic fields using a superconducting ceramic.

Visualization: Magnetic Levitation by a Superconductor

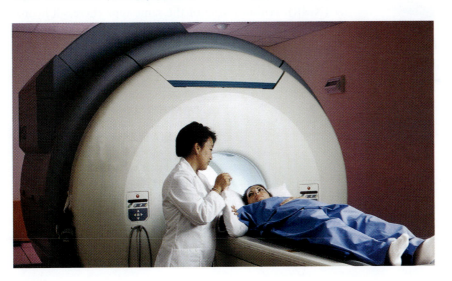

TABLE 13.7	Comparison of Superconducting Materials

The record for the highest superconducting temperature for a ceramic material is held by a complex mixture of atoms ($Hg_{0.8}Tl_{0.2}Ba_2Ca_2Cu_3O_{8.33}$). The noninteger subscripts are used in these formulas to show a simple ratio of the atoms. In order for us to observe the superconductive properties of the commonly used superconductors, they must be immersed in liquid helium (boiling point = −268.9°C, or 4.2 K).

Material	Highest Superconducting Temperature (K)
$Hg_{0.8}Tl_{0.2}Ba_2Ca_2Cu_3O_{8.33}$	138 (record)
$HgBa_2Ca_2Cu_3O_8$	133–135
YPd_2B_2C	23
$LuNi_2B_2C$	16.6
$TmNi_2B_2C$	11
$ErNi_2B_2C$	10.5
$Nb_{0.6}Ti_{0.4}$	9.8
Lead (fcc)	7.196
Tantalum (bcc)	4.47
Aluminum (fcc)	1.175
Platinum (fcc)	0.0019

increases as the vibrations of the metal lattice increase (the metal gets hotter), and eventually the metal can reach its melting point or combust. For instance, if too much current is placed into the NiChrome wires in your toaster, they will catch fire.

When metals are cooled, the resistance to the flow of electrons drops until it reaches a constant positive value. A superconductor acts differently. As the temperature of the superconducting ceramic is lowered, the resistance to the flow of electrons decreases, as in metals. However, at the **transition temperature**, the resistance to the flow of electrons becomes negligible. **What does "negligible" mean to us and how does that make a superconductor useful?** No resistance implies that no energy is lost in the movement of electrons along the superconductor's lattice. If the coil in your toaster were made from a superconducting wire, the electrons could pass through it without resistance, and it wouldn't get hot.

It wouldn't toast either. With a superconducting wire, electricity could be delivered to your home efficiently and very inexpensively (gone would be the need for the step-down transformer on the telephone pole near your home). This same lack of resistance also implies that a superconductor can carry extremely large electric currents. This is exactly what is needed in the MRI.

To generate the sizable magnetic field needed in the MRI, a large electric current is placed in a coil of superconducting ceramic wire. As the electrons travel around the coil, they generate the magnetic field perpendicular to the flow of the electrons. In order to be superconducting, the ceramic must be kept cold, typically at the temperature of boiling liquid helium (4 K). However, ongoing research on "high-temperature" superconductors may one day eliminate the need for liquid helium. This would be quite beneficial because helium is a nonrenewable resource. Table 13.7 lists some superconductors and the temperature at which each is superconducting.

The MRI is actually based on a key instrument in chemical research: the nuclear magnetic resonance spectrophotometer (NMR). Within nuclei of the same type (here, all hydrogen nuclei) a particular range of frequencies is commonly absorbed. Moreover, specific electromagnetic environments

(a)

(b)

A nuclear magnetic resonance spectrophotometer can be used to obtain information on the structure of a molecule. A hydrogen atom spectrum of ethylbenzene illustrates the information that is obtained. This information can be used to build the structure of a molecule.

within a molecule cause individual nuclei to absorb slightly different frequencies of radio waves. When a compound is placed in the NMR, these specific frequencies are recorded. Then they can be used to identify the environments within a molecule. With a very powerful magnetic field, chemists can use this information from the NMR to determine the structural formula for a molecule.

13.4   Plastics

Today's hospitals are completely different from those of 50 years ago. Advances in the field of medicine from open-heart surgery to the use of more powerful and useful medicines are among the most dramatic changes. Walk into a hospital and what do you see? Plastics. Everywhere you look, **plastics** have found a use. The nurse, the physician, and the staff work with plastics in the form of gloves, smocks, masks, and surgical booties. During the time in the hospital, the patient is exposed to plastic in the form of syringes, tubing, sterile packaging, bandages, and even the chairs in the waiting room.

What is plastic? All plastics are polymers (see Chapter 12), but not all polymers are plastics. The DNA and protein within us, for example, are naturally occurring polymers that are not plastics. Plastics are polymers that can be molded into a shape and then hardened in that form. In 1907, the U.S. chemist Leo Baekeland prepared the first completely synthetic plastic. Named after its inventor, *bakelite* could be formed easily into almost any shape. After it hardened, bakelite was tough, durable, and resistant to heat. Bakelite was used in the early half of the 1900s as a lightweight counterpart to steel. Like most plastics, it also works as a good insulator, which increases its utility as handles for frying pans, spatulas, electrical plate covers, and other household items.

The successes with Bakelite prompted the preparation of many more types of plastics, including polyethylene, saran, Teflon, nylon, neoprene, and a host of others. Today, our world is inundated with plastics, which are used wherever possible because they are light in weight and low in cost (see Table 13.8).

Why are there so many different types of plastics? The short answer is that different types of plastics have different properties. Differing properties suit different uses. A rigid plastic would not make a good climbing rope because it does not bend. A stretchable plastic wouldn't make a good ruler or calculator housing. Therefore, we make different plastics for different specific applications. The properties of a plastic are related to the way it is manufactured and to its composition. By controlling the polymerization reactions, we can make short polymer chains (with lower melting points) or long polymer chains. **Crosslinking**, or linking the chains of adjacent polymer strands together, bestows strength on the overall plastic. Orienting the chains in parallel makes stretchable fibers.

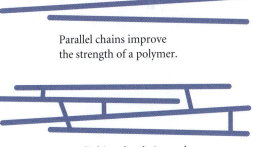

Parallel chains improve the strength of a polymer.

Crosslinking the chains makes the polymer even stronger.

Fibers

Some airplanes are made largely of plastic—not the same type of plastic that is in your soda bottle, but a plastic that is made in roughly the same manner. The type of plastic that makes up the wings of some planes is a polymeric fiber. A **fiber**, some examples of which are shown in Table 13.9 on page 566, is a polymer whose chains are aligned in one direction. Fibers have a high **tensile strength** (they stretch without breaking when you pull them), a property that is good for, among other products, airplane wings. Airplane wings need to be able to stretch when the plane rides through turbulence. If the plastic weren't stretchable, the airplane's wings could snap off.

TABLE 13.8	Selected Plastics and Their Chemical Composition	
Plastic (and representative uses)	**Structure**	**Monomer**

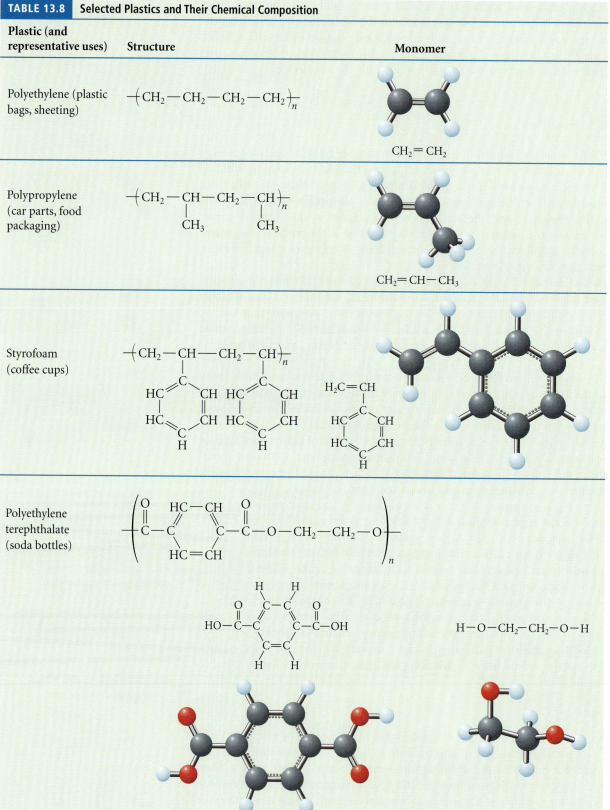

Polyethylene (plastic bags, sheeting)

$-(CH_2-CH_2-CH_2-CH_2)_n$

$CH_2=CH_2$

Polypropylene (car parts, food packaging)

$-(CH_2-CH-CH_2-CH)_n$
$\quad\quad\quad\; CH_3 \quad\quad\; CH_3$

$CH_2=CH-CH_3$

Styrofoam (coffee cups)

$H_2C=CH$

Polyethylene terephthalate (soda bottles)

$H-O-CH_2-CH_2-O-H$

TABLE 13.8	*(Continued)*

Plastic (and representative uses)	Structure and Monomer

Polyurethane
(foam, gaskets,
bearings)

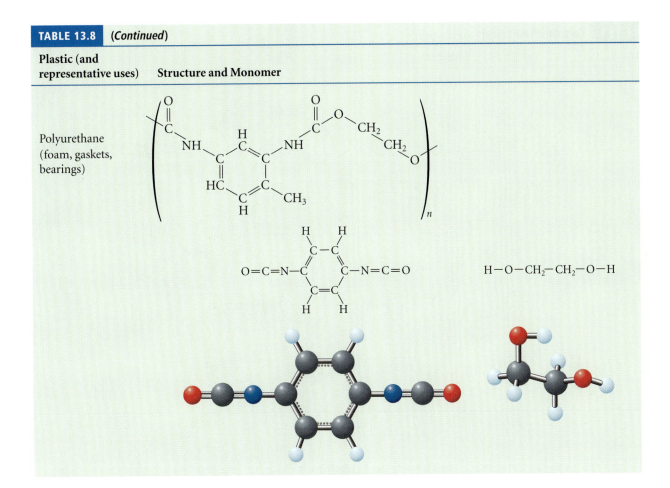

Other plastic fibers can be spun into thread and then woven into textiles. Put a price tag and a label on them, and you could sell them in the clothing store. Nylon and polyester, when made into clothing like that shown in Figure 13.25, are two very common plastic fibers. Fibers do have drawbacks, however. They stretch only in the direction in which they are aligned, so if you pull on a fiber in the wrong direction, it tends to come apart quite easily. To alleviate this problem, the fiber can be manufactured with the addition of a different material. The result is a **composite material** that possesses the strengths of both components. Composite materials don't need to be made from only plastics. In fact, a brick wall is an example of a composite material (cement and ceramic bricks). Another example is the 50/50 cotton/polyester blend used to make shirts.

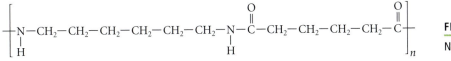

Nylon [6,6]

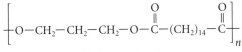

Polyester [3,16]

FIGURE 13.25

Nylon and polyester.

TABLE 13.9	Common Plastic Fibers
Fiber	**Structure**
Kevlar (note interactions between strands)	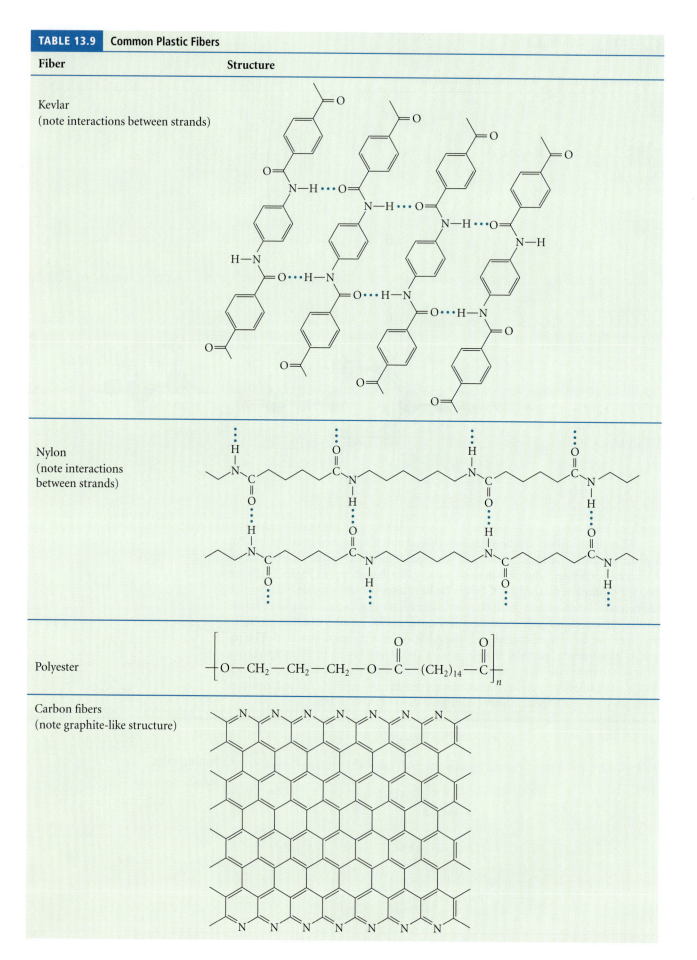
Nylon (note interactions between strands)	
Polyester	
Carbon fibers (note graphite-like structure)	

Aramid (short for aromatic amide) plastics such as Kevlar®, shown in Figure 13.26, are an example of a composite fiber that is particularly strong. What makes this material so strong? The adjacent polymer chains in the material are held together by hydrogen bonds. If the fiber is mixed with an epoxy resin when it is manufactured, it becomes even stronger. The fibers are then woven together to make fabric that is light, flexible, and so strong that it can even stop bullets, enabling police departments to use them as bulletproof vests. High tensile strength and low density make Aramid plastics better than steel.

FIGURE 13.26

Aramid polymers and the bulletproof vest.

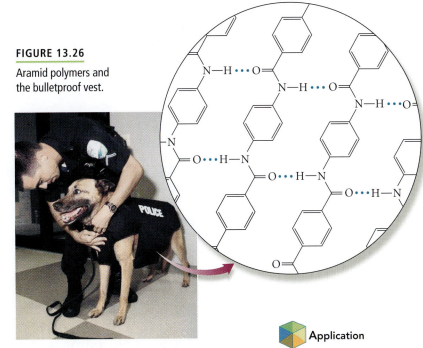

Application

Carbon fiber is another of the composite materials. A polymerization reaction of acrylonitrile gives the polymer polyacrylonitrile (PAN) via the reactions illustrated in Figure 13.27. Heating PAN in air to 300°C oxidizes the polymer. To finish the preparation of a carbon fiber, the oxidized PAN is heated to 3000°C, resulting in a structure that is very similar to graphite and is extremely strong. It is three times stronger than high-tensile steel, but only one-sixth as dense. This makes carbon fiber composites the perfect choice for airplanes, boats, and other structural components (light, yet extremely strong).

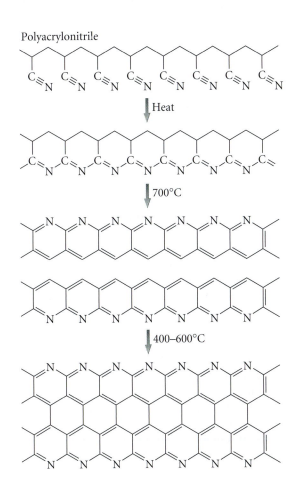

Polyacrylonitrile

FIGURE 13.27

Reactions that make carbon fiber.

Examine and identify the following items from everyday life. Indicate whether they are made of a composite material, a plastic, a metal, an alloy, or a ceramic: an iron bar–reinforced concrete wall, the light switch plate, and a brass doorknocker.

Solution

The reinforced concrete wall is made from iron bars (called re-bar) buried in a mixture of cement and pebbles. Therefore, it is a composite material. The light switch plate is made from Bakelite, a plastic. Its insulating ability makes it a good choice for an electrical plate. The doorknocker is made from brass, an alloy. The sheen and durability of polished brass makes this item luxurious and useful.

PRACTICE 13.6

Identify the materials that make up a floppy disk, the nonstick coating in your frying pan, and the windshield of your car.

See Problems 55 and 56.

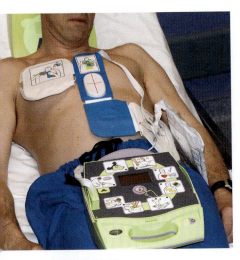

FIGURE 13.28

The portable cardiac defibrillator. The defibrillator stores electrical energy in capacitors made with thin films.

Application

CHEMICAL ENCOUNTERS: Heart Defibrillators

13.5 Thin Films and Surface Analysis

People who experience sudden cardiac arrest (disruption of the normal function of the heart) are often treated with a portable cardiac defibrillator like that shown in Figure 13.28. This instrument delivers to the cardiac victim's heart a powerful and sudden shock, which can resynchronize the beating cycle of the heart muscle. The defibrillator contains devices called capacitors that can store large amounts of electrical energy. At the push of a button a circuit is closed, and the capacitors release the stored energy as one strong electrical pulse.

One of the most durable capacitors on the market is made with a thin film of aluminum on either side of a plastic sheet. The two sheets of thin aluminum act as the positive and negative terminals of a battery. Because a plastic sheet separates them, the terminals do not allow electricity to pass until a large voltage is applied. The thin film capacitors are particularly useful in portable devices because they are self-healing. If a short circuit develops inside the capacitor, the heat generated causes the aluminum to vaporize, making that portion of the capacitor ineffective. This reduces the power of the capacitor, but it is still able to function. In a standard capacitor, a short circuit often destroys the capacitor, which would not be desirable if you were answering a cardiac arrest emergency.

A thin film is generally defined as a film of any material, typically only 0.1 μm to 300 μm thick, much thinner than a layer of paint. However, like paint, a thin film must adhere strongly to the surface on which it is applied. Thin films are carefully prepared so that the chemical composition is well defined (uniform thickness, homogeneity, and so on).

How do we make a film of a material so thin? A film so thin is very fragile unless it is made directly on a surface. Many techniques have been designed to do just that. The most common techniques are physical deposition, sputtering, and chemical-vapor deposition.

Physical deposition relies on sublimation and deposition to place a thin film on a substrate. A chamber is attached to a vacuum. The material to be made into a thin film is placed at one end of the chamber, and the substrate is placed at the other end. The vacuum is turned on, and heat is applied to the material to be vaporized. When the appropriate heat and pressure have been obtained, the material sublimes and moves through the chamber, where it deposits on the substrate.

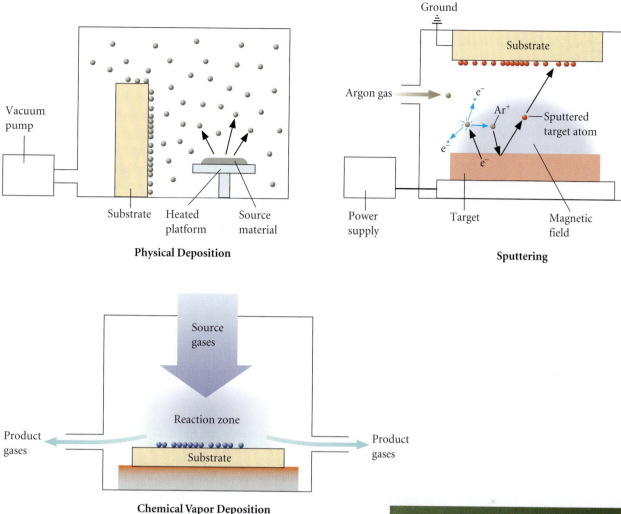

Physical Deposition

Sputtering

Chemical Vapor Deposition

Despite the rotation of the substrate, uniform thickness of the thin film is difficult to obtain. Typically, thin films of inorganic materials such as MgF_2 and SiO_2 are made by this method.

In **sputtering**, very high voltage is used to move a material from a negative terminal (the cathode) to a positive terminal (the anode, see Chapter 19), both surrounded by a low-pressure argon atmosphere. Inside the device, the high voltage forms argon cations that race to the cathode. Their impact dislodges metal atoms from the cathode with extremely high kinetic energy. These ejected atoms fly in all directions. Some strike the anode and form a thin film. Typically, thin films of metals (such as silicon, titanium, aluminum, gold, and silver) are made via sputtering techniques.

In **chemical-vapor deposition**, a thin film is prepared when a compound placed on a surface undergoes a reaction. For example, scratch-resistant sunglasses can be made by coating the lenses with a carbon-based thin film. In this process, a mixture of methane (CH_4) and hydrogen (H_2) is passed over the lenses. Application of intense microwave radiation to this mixture causes the molecules to break apart into their individual atoms. The carbon atoms re-form into a diamond-like thin film. The hydrogen atoms react with any carbon that starts to form graphite-like films. Other methods include the reaction of a metal(IV) halide with hydrogen gas to prepare metal films and the reaction of silane (SiH_4) and ammonia

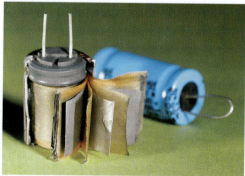

Capacitors can be made from thin films.

FIGURE 13.29

Soichi Noguchi, a Japanese astronaut, took a walk in space during the August, 2005, mission of the space shuttle *Discovery*. The face shield on his suit is lowered, showing the gold thin film.

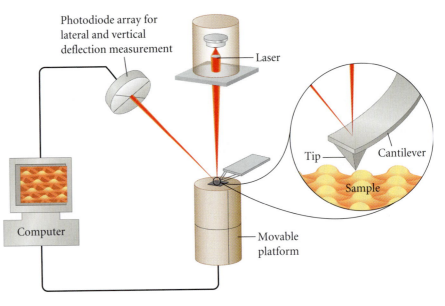

FIGURE 13.30

A block diagram illustrating atomic force microscopy (AFM). The atomic force microscope measures the deflection of an atomic-sized needle from the surface of a substrate using a laser. The data obtained can be used to draw a picture of the hills and valleys on a substrate.

(NH_3) to make silicon nitride ceramic thin films. This method makes thin films of metals, alloys, ceramics, and polymers.

Applications of thin films range from the fabrication of microelectronic devices to use as novel catalysts and as protective coatings. For instance, thin films of gold that absorb harmful UV radiation from the sun are used to create protective eyewear for astronauts in spaceflight (see Figure 13.29).

Thin films are analyzed using a variety of techniques, including scanning tunneling microscopy (STM) and atomic force microscopy (AFM). Atomic force microscopy is similar to STM (Chapter 9) in that an atomic-sized needle (with a diameter less than 10 nm) is used. However, in AFM, the needle rests on the surface to be explored. A laser is directed at the top of the needle and is used to calculate the distance of the needle from the surface. As the surface is scanned, the needle moves up and down through the surface's peaks and crevices. The laser beam records the changing distance to the surface, and a picture of the surface is produced (see Figure 13.30). This method is very useful for exploring surface defects. Modifications of AFM can provide specific information about the surface, such as the type of atom and the thickness of the film.

EXERCISE 13.7 **That's Pretty Thin**

How many atoms of tin (atomic radius = 141 pm) are needed to make a one-atom-thick thin film that covers an area equal to that of a sheet of paper measuring 8.5 by 11 in? Assume the atoms are aligned in rows and columns as they cover the piece of paper.

First Thoughts

To answer the question, we need to determine the area of the paper. Then we'll consider the area occupied by one atom (assuming it is a square that is 282 pm on each side—the diameter of the tin atom).

Solution

$$8.5 \text{ in} = 21.59 \text{ cm}; \ 11 \text{ in} = 27.94 \text{ cm}$$

$$\text{Area of paper} = 21.59 \text{ cm} \times 27.94 \text{ cm} = 603.22 \text{ cm}^2$$

$$\text{Area of atom} = 282 \text{ pm} \times 282 \text{ pm} = 79524 \text{ pm}^2 \times \left(\frac{1 \times 10^{-12} \text{ m}}{1 \text{ pm}} \times \frac{100 \text{ cm}}{1 \text{ m}} \right)^2$$

$$= 7.952 \times 10^{-16} \text{ cm}^2$$

$$\text{Number of atoms} = \frac{603.22 \text{ cm}^2}{7.952 \times 10^{-16} \text{ cm}^2} = 7.6 \times 10^{17} \text{ atoms}$$

Further Insight

That is quite a few atoms, though it is much less than a mole. In fact, this number of atoms amounts to 1.3×10^{-6} mol, or 1.5×10^{-4} g, of tin. The change in mass of the paper would be less than a milligram!

PRACTICE 13.7

How many moles of nickel (atomic radius = 125 pm) would be needed to coat a photograph measuring 4.0 in × 6.0 in? Assume that the nickel atoms are arranged neatly in rows and columns as they cover the photo. How many layers of nickel atoms would be needed to increase the mass of the photograph by 1.00 g? . . . by 20.0 g? Is it possible to place a thin film of gold on a photograph measuring 4.0 in × 6.0 in with only 0.050 mg of Au (atomic radius = 144 pm)?

See Problems 61–66 and 68.

13.6   On the Horizon— What Does the Future Hold?

Health science professions utilize cutting-edge technology wherever possible to provide the best care for their patients. Visit the doctor's office, have your annual dental exam, or stop by the emergency room, and you'll notice the use of modern materials that improve the quality of your care. These materials are so ingrained in our society that it really would be hard to live our current lifestyle without them. In fact, fifty years ago we would never have been able to believe that plastics would stop bullets, that a ceramic bone would be used to repair a fracture, or that a plastic heart could keep someone alive. With recent advances in the technology of modern materials, it appears that anything is possible. Just read your newspaper to see what's coming.

"Green Chemistry"

One of the main thrusts in modern materials science is the development of environmentally benign technologies, or **Green Chemistry**. Replacing toxic catalysts in manufacturing processes and using aqueous solutions instead of flammable organic solvents are just two of many things that can be done to make the chemical manufacturing industry more environmentally friendly. Replacing hazardous pesticides with the application of plant proteins to trigger the production of substances that defend against insects is another approach to helping clean up the environment. Twelve principles to help guide chemists in crafting a more environmentally friendly science were articulated by Paul Anastas of the White House Office of Security and Technology Policy and John Warner of the University of Massachusetts–Boston. Six of those principles are waste prevention, atom economy (using as much of each reagent as possible in a reaction), reducing the

Application

CHEMICAL
ENCOUNTERS:
"Green Chemistry"

Zeolites show promise as useful materials. They contain pockets that can be engineered to hold small molecules.

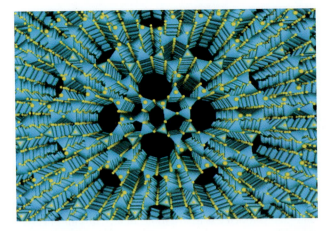

use of hazardous chemicals, energy efficiency, use of catalytic reactions, and pollution prevention.

The use of specially engineered zeolites (see Chapter 8) as catalyst components is promising. One area where a zeolite catalyst could have great impact is in the manufacture of phenol (primarily used as to make the resin in plywood sheets). Research on the use of a zeolite to directly oxidize benzene (C_6H_6) to phenol (C_6H_5OH) is currently under way. Typically, the manufacture of phenol from benzene is a two-step process. The preparation of phenol via this method generates a lot of by-products. Zeolites may be the perfect solution to this problem, reducing the large amount of waste generated.

Biopolymers

Research into the modification of *natural polymers* (**biopolymers**) has the potential to have tremendous impact in the health professions. Biopolymers include nucleic acids (DNA and RNA), proteins (such as enzymes), and some carbohydrates (such as cellulose, chitin, and starch). These compounds can be chemically modified in the laboratory to make biological polymers with interesting properties.

Biopolymers are environmentally friendly because they degrade (break down to chemically more benign compounds) in the environment and are often reabsorbed by living creatures. For example, dead trees (composed of cellulose and other natural polymers) decay and become food for other organisms. Cellulose is biodegradable. The recent use of cellulose instead of Styrofoam "peanuts" as packaging materials has been an excellent application of Green Chemistry. The two materials do the same job, acting as cushions for our breakables when shipped. However, the cellulose peanut can be biodegraded (which greatly reduces its volume when added to water), and large quantities of nonrenewable resources are not needed to manufacture it. Replacing synthetic polymers with biopolymers helps promote Green Chemistry.

Biopolymers can assist in the delivery of drugs to patients by helping to make the drugs more readily absorbable by the body. Another recently issued patent illustrates the use of a biopolymer to enhance the flavor of chewing gum. Because compounds made by biological systems are often chiral (see Section 12.14), the use of biopolymers as catalysts in a chemical reaction can provide an environmentally benign method to make medicines.

Aerogels

A special type of glass known as an **aerogel** has attracted attention lately. The aerogel shown in Figure 13.31 is, as its name implies, a silica glass whose structure is mostly air. We can think of the aerogel as a ceramic foam. It is produced by

FIGURE 13.31

Hot flame and cool flower. The aerogel is an excellent insulator for this flower.

crystallizing the porous silica structure in a solvent and then removing the solvent to give a nearly transparent silica structure that has the appearance of soapsuds.

What makes aerogels so interesting is that they have tremendous surface area. A 1-g sample of an aerogel can have as much as 1000 m^2 of surface area—equivalent to about 1.4 acre of farmland or one-seventh of the street area taken up by the Empire State Building in New York City. Combine the high surface area and extremely low density (0.003 to 0.35 g/mL) with the high resistance to temperature common among the ceramics (some aerogels are completely stable up to 3000°C), and you have a perfect insulator for a spacecraft. In fact, these materials have significantly greater insulating power than fiberglass. Research is currently being conducted on the manufacture of flexible aerogels that could be used to make insulation for homes. Just imagine a firefighter's coat as thin and light as a sheet but insulating enough to protect its wearer against the hottest fires.

The Bottom Line

- In a crystalline solid, the atoms, ions, or molecules are highly ordered in repeating units that make up the lattice of the crystal. Amorphous solids, although their atoms, ions, or molecules inhabit rigid and fixed locations, lack the long-range order in a crystal. (Section 13.1)

- The unit cell is the basic repeating unit that, by simple translation in three dimensions, can be used to represent the entire crystalline lattice. (Section 13.1)

- The simple cubic unit cell, the body-centered cubic unit cell, and the face-centered cubic unit cell make up most of the crystalline lattices of the metallic elements. (Section 13.1)

- Metals often adopt a closest packed structure. These structures include the hexagonal closest packed structure and the cubic closest packed structure. (Section 13.1)

- We can calculate the volume and density of a unit cell by means of simple geometry. In doing so, we assume the atoms are hard spheres. (Section 13.1)

- Band theory describes why metals are electrical and thermal conductors, shiny, malleable, and ductile. (Section 13.2)

- The valence band and the conduction band result from the nearly infinite number of atomic orbitals that overlap to form the molecular orbitals in a metal. The band gap can be used to assess whether a compound is a conductor, a semiconductor, or an insulator. (Section 13.2)

- Alloys are mixtures of two or more metals. They include the interstitial and substitutional alloys. Amalgams are special alloys of mercury. (Section 13.2)

- Ceramics, which include glasses and superconductors, are compounds that contain both ionic and covalent bonds. (Section 13.3)

- Plastics, named after their ability to be molded into shape, are polymeric materials. (Section 13.4)

- Composite materials are made of an intimate combination of two or more materials. (Section 13.5)

- "Green Chemistry" is the practice of chemistry via environmentally benign methods. (Section 13.6)

Key Words

aerogel A silica glass whose structure is mostly air; also known as a ceramic foam. (*p. 572*)

alloy A solution of two or more metals. (*p. 556*)

amalgam A solution made from a metal dissolved in mercury. (*p. 557*)

amorphous solid A solid whose atoms, ions, or molecules occupy fairly rigid and fixed locations but that lacks a high degree of order over the long term. (*p. 541*)

band gap A small energy gap that exists between the valence band and the conduction band (*p. 553*)

band theory A metal is a lattice of metal cations spaced throughout a sea of delocalized electrons. (*p. 551*)

biopolymer A polymer of biological materials, such as DNA or cellulose. (*p. 572*)

body-centered cubic (bcc) unit cell A unit cell built via the addition of an atom, ion, or molecule at the center of the simple cubic unit cell. (*p. 543*)

ceramic A nonmetallic material made from inorganic compounds. (*p. 557*)

chemical-vapor deposition A thin film prepared when a compound placed on a surface undergoes a reaction. (*p. 569*)

closest packed structure A structure wherein the atoms, molecules, or ions are arranged in the most efficient manner possible. (*p. 544*)

composite material A material made from two or more different substances. (*p. 565*)

conduction band The collection of empty molecular orbitals on a metal. (*p. 552*)

conductor A substance that contains a very small band gap. (*p. 554*)

crosslinking Reactions that covalently bond two or more individual strands of a polymer together. (*p. 563*)

crystal lattice A highly ordered framework of atoms, molecules, or ions. (*p. 542*)

crystalline solid A solid made from atoms, ions, or molecules in a highly ordered long-range repeating pattern. (*p. 541*)

cubic closest packed (ccp) structure A closest packed structure made by staggering a second and third row of particles so that none of the rows line up. (*p. 544*)

diffraction pattern A pattern of constructive and destructive interference after EMR passes through a solid material. (*p. 546*)

dope To add an impurity to a pure semiconductor to alter its conductive properties. (*p. 554*)

ductile Able to be pulled or drawn into a wire. (*p. 552*)

face-centered cubic unit cell A unit cell built via the addition of an atom, ion, or molecule at the center of each side of the simple cubic unit cell. (*p. 543*)

fiber A polymer whose chains are aligned in one direction. (*p. 563*)

glass An amorphous solid ceramic material. (*p. 559*)

Green Chemistry The practice of chemistry using environmentally benign methods. (*p. 571*)

hexagonal closest packed (hcp) structure A closest packed structure made by staggering a second and third row of particles in such a way that the first and third rows line up. (*p. 544*)

insulator A substance that contains a very large band gap. (*p. 554*)

interstitial alloy An alloy made by the addition of solute metals placed in the existing spaces within a solvent metal. (*p. 556*)

ionic solids Solids made up of ionic compounds held together by strong electrostatic forces of attraction between adjacent cations and anions. (*p. 542*)

liquid crystal A transitional phase that exists between the solid and liquid phases for some molecules. Typically, these molecules have long, rigid structures with strong dipole moments. (*p. 550*)

malleable Capable of being shaped. (*p. 552*)

molecular solids Solids made up of molecules that are held together by intermolecular forces. (*p. 542*)

nonbonding molecular orbital A molecular orbital that is not involved in bonding or antibonding. (*p. 551*)

n-type semiconductor A semiconductor that contains electrons (negative charges) in the conduction band. (*p. 554*)

phonons The induced vibrations produced in a metal lattice. (*p. 561*)

photovoltaic device A device in which light energy can be used to generate an electric current. (*p. 555*)

physical deposition The process of sublimation and deposition used to place a thin film on a substrate. (*p. 568*)

plastics A thermally moldable polymer. (*p. 563*)

p-type semiconductor A semiconductor that is electrically conductive if we apply a small electric field. The semiconductor contains positive holes in the valence band. (*p. 554*)

semiconductor A metal that contains a small band gap. (*p. 553*)

simple cubic unit cell An arrangement of particles within a crystalline solid comprised of six particles occupying the corners of a cubic box. (*p. 542*)

sputtering The process of producing a thin film on a substrate by using high voltage to move a material from a negative terminal (the cathode) to a positive terminal (the anode). (*p. 569*)

substitutional alloy An alloy made by directly replacing solvent atoms with solute atoms. (*p. 556*)

superconductor A ceramic whose resistance to the flow of electrons drops to zero as the temperature is reduced. (*p. 561*)

transition temperature The temperature of a superconductor at which the resistance to the flow of electrons becomes negligible. (*p. 562*)

tensile strength Resistance to breaking when a substance is stretched. (*p. 563*)

unit cell The repeating pattern that makes up a crystalline solid. (*p. 542*)

valence band The collection of filled molecular orbitals on a metal. (*p. 552*)

Focus Your Learning

The answers to the odd-numbered problems and some selected problems appear at the back of the book, as represented by the blue numbering.

13.1 The Structure of Crystals

Skill Review

1. What key feature distinguishes a crystalline solid from an amorphous solid?

2. Cite an example of a crystalline solid and of an amorphous solid.

3. Use the descriptive term *parallelepiped* to define both a crystal lattice and a unit cell.

4. Cite an example of an object made from parallelepipeds that are visible to the naked eye.

5. Which of these diagrams represents the body-centered cubic structure of iron?

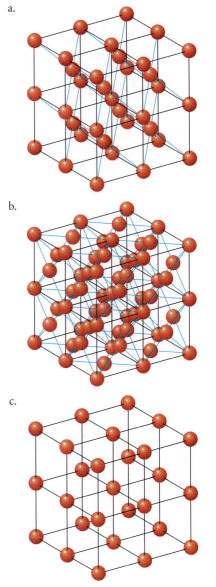

a.

b.

c.

6. Which of these diagrams represents the face-centered cubic structure of strontium?

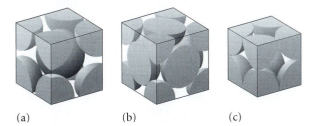

(a)   (b)   (c)

7. How many complete atoms would you expect to find in a unit cell made from the simple cubic unit cell, but with additional atoms placed along every edge of the unit cell?

8. How many complete atoms would you expect to find in a unit cell made from the simple cubic unit cell, but with additional atoms placed only at the face of two of the sides of the unit cell?

9. Without any further information, indicate whether each of the following three solids is most likely to be molecular, ionic, or metallic: Solid A is the only one of the three that conducts electricity in the solid state. Solid B melts when placed in boiling water. Solid C dissolves in water, and the resulting solution conducts electricity.

10. As part of a display of fruit, a grocer places two types of apples (red and yellow) in a pattern. The apples are to be placed in a slanted display with a wooden frame. The first layer is made of red apples, and the second layer is placed above the first in the "dimples" or gaps. Show how the third layer of red apples would be placed if the grocers, remembering their chemistry, wanted to make the display resemble hexagonal closest packing.

11. Calculate the volume of a fcc unit cell with each of the following atomic radii:
    a. 125 pm   b. 210 pm   c. 302 pm

12. Calculate the volume of a bcc unit cell if it is composed of atoms with each of the following radii:
    a. 200 pm   b. 183 pm   c. 164 pm

Chemical Applications and Practices

13. Suppose the label on a can of soup reports 220 mg of sodium per serving of soup. How many grams of NaCl will be present in one serving of the soup?

14. When passing through the small openings within crystal structures, X-rays diffract into a pattern. Say a researcher is deriving the structure of an interesting protein by using X-rays with a frequency of 1.94×10^{19} s^{-1}. What is the wavelength, in nanometers, of the X-rays?

15. The metal nickel has several remarkable uses. When alloyed with other metals, it can form a type of "memory metal" called nitinol, which is a nearly 1:1 mixture of nickel and titanium. Nickel crystallizes in a face-centered cubic arrangement. The density of nickel is 8.90 g/cm³. From this information, calculate the radius of an atom of this coinage metal? (Currently the percentage of nickel in a U.S. "nickel" coin is approximately 25%.)

16. The element iridium is currently an important part of an extinction theory concerning dinosaurs. Unusual deposits of iridium indicate that a large, iridium-containing meteor may have hit our planet, resulting in a dust cloud from the impact blocking the light from the sun. Iridium crystallizes in a face-centered cubic structure. What is the density of this corrosion-resistant metal whose radius is 136 pm?

17. Titanium metal and its oxide have a wide variety of uses. Its relatively light weight and strength make it ideal in some alloys. Titanium dioxide is found in paints and some lipsticks. Pure metallic titanium crystallizes in a body-centered cubic structure. Titanium has a density of 4.5 g/cm³. What is the radius of an atom of titanium? What percent of the unit cell is empty space?

18. Metallic chromium has an atomic radius of 125 pm. Judging on the basis of the density of chromium (7.2 g/cm³), which of three cubic unit cells (simple, face-centered, or body-centered) would you predict for chromium?

19. One way to verify the numerical value of Avogadro's number is to use X-ray data and unit cell calculations. Although the reasoning is the same for any selection, use the following specific information about copper to make the verification. Copper crystallizes with a face-centered unit cell with an edge length of 3.6×10^{-10} m. The density of copper is approximately 8.9 g/cm³.

20. A hypothetical metal was determined to have a fcc unit cell and a density of 0.185 mol/cm³. What is the calculated radius of the metal?

21. Assuming that the following metals crystallize in the fcc unit cell, which would you predict to have the greater density, Co (radius = 125 pm) or Rh (radius = 135 pm)?

22. Assuming that the following metals crystallize into a bcc unit cell, which would you predict to have the greater density, V (radius = 131 pm) or Cr (radius = 125 pm)?

13.2 Metals

Skill Review

23. Identify the metals in the following list of elements.

    B, Na, Al, Ca, Cr, Se, Sb, Bi

24. List two physical characteristics that would enable you to differentiate between calcium and carbon (graphite).

25. How many valence electrons are found in an atom of potassium? How many molecular orbitals are found when three atoms of potassium combine?

26. How many valence electrons are found in an atom of calcium? How many molecular orbitals are found when three calcium atoms combine?

27. The following three energies represent the band gap values for three samples. Which of the values in this list is most likely to be that of a metal? . . . that of a semiconductor? . . . that of an insulator? 85 kJ/mol, 450 kJ/mol, 2.5 kJ/mol.

28. Which of these energies would you expect copper metal to possess as its band gap: 0.06 kJ/mol, 73 kJ/mol, or 510 kJ/mol?

29. The particular types of alloys known as amalgams use mercury as a component. Explain, using the delocalized electron bonding model, how the metal components in any alloy allow the atoms of different metals to bond.

30. How does the amalgam picture you created in Problem 29 allow for varying ratios of components, unlike the set ratio of typical compounds?

Chemical Applications and Practices

31. A grocery stocker wishes to illustrate the two types of alloys to his friends by using red and green apples to repressent atoms. Explain how this would be accomplished.

32. A grocery stocker creates a stack of soup cans to display two different kinds of soups. Judging on the basis of the pattern in which they are stacked, which type of alloy do they appear to represent?

33. Sodium metal is very reactive, yet under controlled conditions its ability to absorb heat on a per-gram basis can be very useful. Some nuclear reactor designs make use of sodium instead of water as a heat transfer agent.
    a. How many valence electrons does sodium have?
    b. Using band theory, explain why thermal energy causes the electrons in sodium, like other metals, to populate antibonding orbitals.

34. Show the construction of a molecular orbital diagram from the combination of two sodium atoms, of three sodium atoms, and of four sodium atoms.

35. Semiconductors can also be used in lasers. If a laser were constructed and it produced a wavelength of 820 nm, what would be the band gap energy (in joules per photon)?

36. What would be the band gap energy (in kilojoules per mole) for a laser whose wavelength was 720 nm?

37. Doping silicon allows for the production of n-type and p-type semiconductors. Would doping silicon with indium make an n- or a p-type semiconductor? Explain the basis for your answer.

38. Silicon and selenium both have semiconductor properties. Explain why doping both with arsenic would have opposite results in terms of producing n-type or p-type semiconductors.

39. The blue color of the Hope diamond is due to small amounts of boron in the carbon lattice. Predict what electronic properties this diamond must have.

40. What properties would you expect to find in a diamond doped with small quantities of arsenic in the carbon lattice?

13.3 Ceramics

Skill Review

41. Explain why the bonding and resulting "band gap" in ceramic materials make them good insulators.

42. Describe the difference between the bonding in the glass in a window and the plaster in a wall.

43. Most compounds melt over a very small temperature range that can even be used to help identify the compound. However, glass materials do not have such a small temperature range for melting. Explain why the melting temperature for glasses is replaced with a transition temperature.

44. Color, brittleness, and other properties of glass can be influenced by the presence of impurities. However, the glass used in fiber optics must be based on very pure SiO_2. Explain how this material, in the form of glass fibers, can be used to send messages using light.

45. The frequency used for a typical MRI scan is in the range of 250 MHz. What are the wavelength and energy (in joules per photon) for this technique?

46. a. Metals are known for their ability to conduct electricity. However, the structure and bonding in metals also cause resistance to conductance. Explain how phonons contribute to the resistance.
    b. What causes resistance to change at the transition temperature in superconductors?

Chemical Applications and Practices

47. Silicates are formed from the connections between tetrahedral structures made of a silicon atom bonded to four oxygen atoms. Diagram this tetrahedral anion with a −4 charge.

48. Joining of the tetrahedral silicates typically occurs through sharing of one of the oxygen atoms from one tetrahedron to another. Draw a diagram of how the SiO_4^{-4} ion could be used to produce two different substances.

49. Glass containers are known to be relatively inert. In fact, corrosive acids are typically stored in glass containers. However, hydrofluoric acid (HF) will react with glass and must be stored in polyethylene containers. This property of HF has been used as a method of glass etching. (However, the fumes produced are extremely harmful.) In the reaction shown here, how many grams of glass would react with 10.0 g of HF?

$$4HF(aq) + SiO_2(s) \rightarrow 2H_2O(l) + SiF_4(g)$$

50. Safety glass is found in automobiles manufactured in the United States. It consists of a "polymer sandwich" of a layer of plastic between two glass layers. Using the known properties of plastics and glass, explain why this arrangement may help reduce some impact injuries in automobile accidents.

51. During a typical day, you are likely to come in contact with the materials discussed in this chapter. Classify each of the following as a metal, an alloy, a glass, a plastic, or a ceramic (more than one answer may be appropriate).
    a. the body of the pen you use;
    b. a contact lens;
    c. the frames on a pair of glasses.

52. Classify each of the following as a metal, an alloy, a glass, a plastic, or a ceramic:
    a. a milk jug;
    b. the body of the spark plug in an automobile engine;

c. the frame of an automobile;
d. an etched plaque.

Section 13.4 Plastics

Skill Review

53. What distinguishing property of plastics forms the basis of this quote? "All plastics are polymers, but not all polymers are plastics."

54. Define the term *plastic*. Provide an example of a plastic and an example of a polymer that isn't a plastic.

55. Determine whether each of the following everyday items contains a plastic, a metal, an alloy, a glass, or a ceramic (more than one answer may be appropriate).
    a. the frame of a compact mirror;
    b. a mirror;
    c. the body of the computer you use;
    d. a drinking cup used at the dentist's office.

56. Determine whether each of the following everyday items contains a plastic, a metal, an alloy, a glass, or a ceramic:
    a. a CD case;
    b. the outer body of a cell phone;
    c. a leisure suit;
    d. a burner on a stove.

Chemical Applications and Practices

57. Some manufacturers are using composites to obtain desirable properties for the exterior of new automobiles. Describe what advantages a composite material might provide for automobile exteriors. What disadvantages might exist?

58. Why would a manufacturer consider constructing an airplane wing using composite materials?

59. Most hydrocarbon polymers make poor electrical conductors, although there are some that do conduct electricity. Explain (noting the types of bonds in the polymers) why this property arises in polymers.

60. The construction of a polymer made from ethyne (C_2H_2) could conceivably be quite advantageous. Considering the fact that a resulting polymer could contain a nearly infinite string of alternating single and double bonds, what advantage could you foresee to such a polymer?

Section 13.5 Thin Films and Surface Analysis

Skill Review

61. Approximate the number of gold atoms that would make up the thickness of a gold film on an astronaut's visor. Assume that the film is 215 nm thick and that the radius of a gold atom is 144 pm.

62. How many moles of gold would be needed to prepare a thin film that is 576 pm thick covering an area similar in size to a postage stamp (1 in by 1 in)? The radius of a gold atom is 144 pm.

63. How many atoms of silver (atomic radius = 145 pm) must be deposited to completely cover an area that measures 2.54 cm by 2.54 cm? Assume the layer is only one atom thick.

64. A notecard that measures 3 in by 5 in is coated on one side with a thin film of copper (atomic radius = 128). How much will this coating add to the mass of the notecard?

65. Consider the calculation completed in question 63. How many grams of silver must be used to cover the area with a thin film that is one atom thick? . . . three atoms thick?

66. Consider the copper-coated notecard in question 64. How thick must the thin film be to increase the mass of the notecard by 5.00 g?

Chemical Applications and Practice

67. A recent advance in thin films has involved the use of titanium dioxide on window glass. The film helps keep dust from depositing on the windows. The exact technique for applying the thin film of TiO_2 is proprietary information (meaning that the information is not available to the public). But the process involves heating the glass and passing the film material as a gas stream over the glass. The resulting interaction forms TiO_2 as a film during cooling. Which of the three techniques described in this chapter most resembles this process?

68. The thin film in the application described in problem 67 is approximately 5.00×10^2 angstroms thick (1 angstrom = 1×10^{-10} m). Describe how this measurement is made. How much additional mass does the thin film add to a 2.500-ft^2 piece of glass? (Assume the density of the film is 4.26 g/cm^3.)

Section 13.6
On the Horizon—What Does the Future Hold?

Skill Review

69. Uses for zeolites, both naturally formed from clays and specially synthesized, vary greatly but typically depend on what property of these versatile substances?

70. Recycling is an important part of the process of conserving and reusing resources. List two common items that are typically recycled.

71. Instead being thrown away, your old TV can be recycled. What types of materials make up the TV? Which of these can be recycled?

72. Many used automobiles can be recycled. What are some of the materials that could be harvested from a modern automobile?

73. Name and briefly describe the three classes of biopolymers.

74. Describe the chemical and physical makeup of an aerogel. Provide at least one application for an aerogel.

Chemical Applications and Practices

75. Zeolite catalysts may be employed in the production of the hydrocarbons used in gasoline. Would this be an example of a heterogeneous or a homogeneous catalyst? How can a catalyst be used to reduce waste?

76. Chemical companies now advertise many products, such as pharmaceuticals, as containing one chiral isomer. What is the advantage of using a chiral catalyst rather than a nonchiral catalyst when synthesizing a product that must be only one type of isomer?

77. In terms of density, insulation, and resistance to cracking, what advantages does an aerogel offer over other ceramic materials?

78. One of the guidelines for practicing Green Chemistry is to reduce the amount of chemical waste that occurs in manufacturing compounds. From where does this waste arise? What is typically done with chemical waste in "nongreen" processes?

Comprehensive Problems

79. How do molecular and ionic solids, in both the molten and the dissolved state, differ in their ability to conduct an electric current?

80. When selecting molecules that make good candidates for liquid crystal displays, chemists search for (compact; long rigid) molecules with (strong; weak) dipole moments. Explain why each of your circled choices is appropriate for liquid crystals.

81. Metals such as silver, chromium, and platinum are known for their shiny appearance. Most metals, in their pure state, also have this appearance. Use the delocalized electron model to explain this property of metals.

82. Superconductor materials are able to conduct electricity, typically with pairs of electrons, as a consequence of the arrangement of sheets or layers within their structure. These ceramic materials are continually being studied to make it possible to have superconducting ceramics at higher temperatures. Use Internet resources to obtain an example of the chemical composition and structure of a ceramic superconductor.

83. Use the Internet or other sources to find out how Charles Goodyear used crosslinking with sulfur to change the properties of some rubber substances he was working with.

84. Use the Internet to find the twelve principles of Green Chemistry. Which of these principles are used in your chemistry laboratory?

85. Replacing hazardous pesticides with less persistent (but just as effective) compounds is a common goal. What are the hazards associated with pesticides if they persist in the environment?

Thinking Beyond the Calculation

86. A chemist prepares a polymer from methyl acrylate (also known as methyl propenoate). The resulting polymer was made by dissolving a small amount of the monomer in solution and then forming the polymer. The polymer was very gel-like in nature, with a spongy appearance.

Methyl acrylate   Acrylic acid   Methanol

a. Draw any potential resonance structures for the monomer.
b. Draw the polymer that would result from the polymerization of methyl acrylate.
c. What uses would such a polymer find?
d. If the monomer is treated with water and acid, it reacts to form acrylic acid and methanol. How many grams of methanol would be obtained from the reaction of 1.50 g of methyl acrylate with excess water?
e. What properties that are not found in a polymer of methyl acrylate would you expect a polymer of acrylic acid to possess?

Thermodynamics: A Look at Why Reactions Happen

Contents and Selected Applications

Chemical Encounters: Bioenergetics

14.1 Probability as a Predictor of Chemical Behavior

14.2 Why Do Chemical Reactions Happen? Entropy and the Second Law of Thermodynamics

Chemical Encounters: Glycolysis

14.3 Temperature and Spontaneous Processes

Chemical Encounters: Psychrotrophs and Psychrophiles

14.4 Calculating Entropy Changes in Chemical Reactions

Chemical Encounters: More Glycolysis

14.5 Free Energy

Chemical Encounters: Pyruvate and Lactate

14.6 When $\Delta G = 0$; A Taste of Equilibrium

Chemical Encounters: ATP Formation

Sugary between-meal snacks are a good source of energy. Each contains sucrose, which is broken down in the body into glucose and fructose. Both of these molecules undergo complete oxidation and provide energy for the body.

Go to **college.hmco.com/pic/kelterMEE** for online learning resources.

After a hearty breakfast, we leave for work full of energy and ready to conquer the day. However, the midmorning hours can be difficult to get through because our energy level drops a couple of hours after we eat. This is especially true if we had a big bowl of sugar-coated, sugar-injected cereal for breakfast. To make matters worse, the mid-afternoon hours are no easier—they seem to be the longest of the day. Why does this happen and what can we do to give ourselves that "burst of energy" we need when we feel so tired? Enter the snack. Whether it takes the form of a chocolate bar, a donut, or a bottle of juice, it has the effect of raising our energy level. Just like breakfast, lunch, and dinner, the snack provides our bodies with a source of glucose. How does the consumption of glucose give us energy?

Antoine Lavoisier (the scientist we met in Chapter 3, whose measurements led to the formulation of the law of conservation of mass before he was guillotined in the French Revolution) noticed that living things consume foods and transform them into the energy that maintains life. Lavoisier's views on the process seem rudimentary given our modern understanding, but they were quite revolutionary in his day. The addition of glucose to living cells is an example of this food-to-energy transformation process. To the biochemist, this is part of the broader field of **bioenergetics**, the study of the energy changes that occur within a living cell.

In this chapter we will discuss some of the chemical reactions in the field of bioenergetics. Although we'll soon introduce terms such as *entropy, spontaneity, free energy,* and *equilibrium,* the underlying concept of probability demands our immediate attention. Chemists benefit from understanding these topics. They use probability in many ways: to locate electrons, to determine the macroscopic properties of compounds and mixtures, and to predict the outcome of chemical reactions. *We will use probabilities to discover why chemical processes occur*—including the chemical transformations of the body and their relationship to our ability to live.

Application

CHEMICAL ENCOUNTERS: Bioenergetics

14.1 Probability as a Predictor of Chemical Behavior

FIGURE 14.1

An interesting arrangement of the students in a classroom.

If you entered a classroom or movie theater and noticed all the males seated on one side and all the females on the other, as shown in Figure 14.1, would you think that some announcement, rule, or social convention had dictated that arrangement? Perhaps. But this seating pattern could also have emerged from purely random choices. In fact, there are numerous ways in which a room full of people could be sitting, and "separate seating" is just one of many possibilities.

Let's shift our focus to the room in which you are now sitting as you read this. As we know, the air in the room is a mixture of primarily oxygen and nitrogen gases. Oxygen and nitrogen molecules do not chemically interact with each other, so they can occupy essentially any unfurnished position in the room. Is it *possible* that these gases could be arranged such that all of the oxygen molecules were in one corner of the room and all of the nitrogen molecules in another, as shown in Figure 14.2? Possible? Yes. Likely? No. Is there something that these two seemingly unrelated situations—the room of people and the room of molecules—share?

The short answer is yes, the two situations have a lot in common. To understand how this could be so, let's introduce some common terminology to help in our discussion. We refer to the **macrostate** of a system (whether seated people or

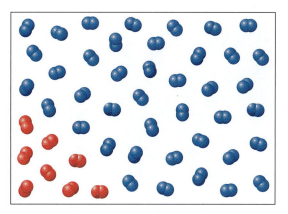

FIGURE 14.2

Is this arrangement of oxygen molecules within a room possible?

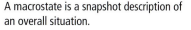

A macrostate is a snapshot description of an overall situation.

A microstate is a snapshot description of a particular representation of individual particles. Each of these three pictures includes a different microstate for a single person in a movie theater.

molecular positions in a room) when we want to take a *snapshot* of the overall situation. The macrostate of the theater seating is that it is sorted by gender. The dispersed oxygen and nitrogen molecules in the room exert a certain pressure and have a certain composition; this is all part of the macrostate of the gases in the room.

If we wanted to describe the *individuals* in the theater, we might discuss which seat they are in, which way they are facing, and whether their arms or ankles are crossed. Each of these individual descriptions is referred to as a **microstate**. For instance, each arrangement of people in the theater indicates a different microstate. How does this apply to molecules? *The total of the microstates in a system defines, or describes, the macrostate of the system.* For individual molecules, the microstate can include their translational motion (changing from one location to another), their vibrational–rotational state (this involves the atomic movements around a chemical bond), and their electron configuration (ground state versus the excited state, even the oxidized state).

How might you describe the macrostate of the gases in your room? You could measure their temperature and pressure. You could also assess the number of molecules of each component in the room. Given the huge number of molecules of gas in the room, we would predict an even larger number of microstates for the gas. We mentioned the hypothetical condition of all the oxygen molecules being in one corner. To bring about that unlikely macrostate of air, the oxygen molecules (being restricted to the corner) would have to assume a limited number of the multitude of possible microstates in the room.

Why have you never encountered the macrostate that has all the oxygen molecules stuck in one corner of the room? One answer can be found in looking at

FIGURE 14.3

George Mallory and Andrew Irvine on Mt. Everest pack heavy cylinders of oxygen on their backs. This photo, taken in 1924 during their ascent to the summit, is the last known image of the pair.

the *probability* that such an event could occur. Out of all the possible macrostates, **what are the chances that "all O_2 on one side, all N_2 on the other side" would occur?** In short, there isn't a very good probability that this situation would happen. (Read on for a more detailed answer!)

Think back to the seating choices for a person entering a room. Which situation has more choices, the strict division by gender or "open seating"? Which situation would you consider to have the greater number of microstates that satisfy the conditions for a particular macrostate? Ludwig Boltzmann (1844–1906), an Austrian mathematician/physicist, dealt with this idea, along with its powerful implications, mathematically. We can get a sense of what he did by looking inside the tanks of oxygen that scuba divers and mountain climbers carry with them as they set off on their quests. Look closely at Figure 14.3. This photo, taken of mountain climbers George Leigh Mallory and Andrew Irvine before their fateful June 1924 attempt to scale the summit of Mt. Everest, shows the climbers each with two tanks of oxygen strapped to their backs. Let's assume that these tanks are connected by a valve and that one of the tanks is empty and the other full. What will happen to the distribution of the gas if the valve is opened? The final volume of the two tanks, which we'll call V_f, is double the initial volume of just one tank, V_i. We can also express this by saying that the ratio $\frac{V_f}{V_i} = 2$. How many ways can the individual molecules be distributed within the two tanks? (That is, how many "microstates" are there?) To make things a little simpler, let's assume that the tanks contain only two molecules of a gas. If we call the molecules A and B, we can have a total of four microstates, as shown in Figure 14.4. We will call the final number of microstates W_f (for the German word *Wahrscheinlichkeit,* "probability"), and $W_f = 4$ in this case. The initial number of microstates, W_i, is equal to 1, representing A and B both present in one tank, before the valve was opened allowing the gases to spread apart. This means that for two molecules in two tanks, the ratio $\frac{W_f}{W_i} = \frac{4}{1}$.

How is the number of microstates, $\frac{W_f}{W_i}$, related to the volume, $\frac{V_f}{V_i}$? For N molecules,

$$\frac{W_f}{W_i} = \left(\frac{V_f}{V_i}\right)^N$$

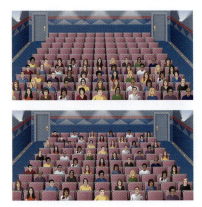

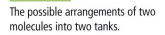

Two individual macrostates.

FIGURE 14.4

The possible arrangements of two molecules into two tanks.

Microstate	Tank 1	Tank 2
1	AB	
2		AB
3	A	B
4	B	A

That is the equation that Boltzmann derived. Using our numbers,

$$4 = 2^2$$

What if our tanks contained three molecules ($N = 3$) instead of two?

$$\frac{W_f}{W_i} = \left(\frac{V_f}{V_i}\right)^N \qquad \frac{W_f}{W_i} = 2^3 = 8$$

In this case, there would be eight possible microstates. *More molecules in the tanks make available more microstates—more ways in which the molecules can distribute themselves among the two tanks.* This means that the probability of all of the molecules being in only one tank—a single microstate among all other possibilities—becomes smaller. This is the key point: *As the number of molecules increases, the likelihood of their all being in one tank decreases sharply, and the probability of the gases tending toward an equal distribution increases.* Let's look further into this point and discuss its vital implications.

Let's put four oxygen molecules in the full tank. Here Boltzmann's equation indicates that there should be 16 ($2^N = 2^4 = 16$) different microstates of the oxygen molecules in the two-tank system. This is exactly what we predict by drawing each of the arrangements as in Figure 14.5. But if we look closely, we see that some of the microstates would give the same macrostate for the system. Figure 14.5 shows that only 5 unique *macro*states could arise from the 16 different *micro*states. Only 6 of these 16 microstates describe the equal distribution of the gases. The probability is therefore 6/16, or 37.5%, that the four molecules will arrange themselves into the two tanks equally. The probability of *exactly* equal distribution has diminished. However, there are many more microstates at or near equal distribution than microstates that are all on one side of either tank. Only 2 of the 16 microstates, or 12.5%, describe all of the molecules being on one side or the other.

As illustrated in Figure 14.6, greatly increasing the number of molecules in our experiment (from 8 to 32 to 128) also greatly increases the number of microstates that are available. As a result, the probability *of equal or nearly equal distribution of the molecules between the two tanks increases.* At the same time, the probability of *all* of the molecules being in one tank or the other is approaching zero. For even small real-world-sized samples of molecules (such as 0.01 mol or 0.1 mol), *equal or nearly equal distribution of the gases becomes the most likely outcome.* The bottom line is that probability—a mathematical construct—governs physical behavior, and in this case, probability says that the gases will spread from one tank to occupy both tanks evenly.

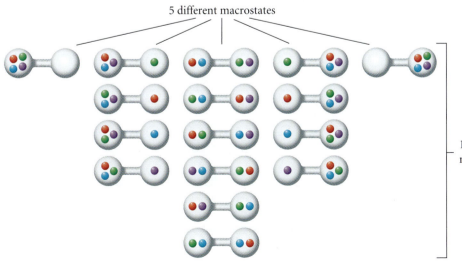

5 different macrostates

16 different microstates

FIGURE 14.5

Microstates describing the possible arrangements of four molecules between two tanks.

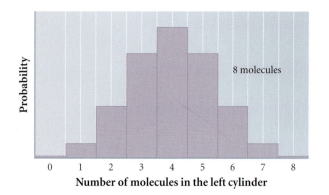

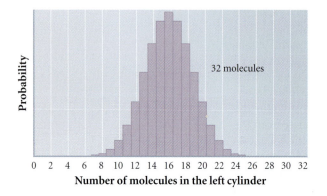

FIGURE 14.6

Probability of distribution of a gas between two cylinders. As the number of molecules increases in the system, the tanks' containing equal numbers of molecules becomes the more likely event. Note that the probability of equal distribution is a bell-shaped (Gaussian) curve.

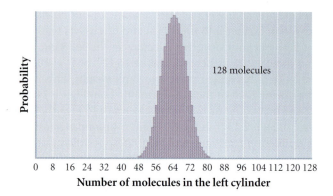

A valve connects two glass jars. One of the jars contains six atoms of gaseous helium, and the other is empty. Determine the probability of the macrostate in which each jar will fill equally (that is, each jar will contain three atoms of helium) after the valve is opened. For the same system, what is the probability of finding four molecules in the left-hand jar and two in the right? What about the probability of finding all six molecules in the left-hand jar?

Solution

Because there are six molecules in this example, there are $2^6 = 64$ possible microstates. Drawing each of them out and grouping them into similar macrostates leads us to the conclusion that there are only 7 macrostates (see Figure 14.7). There are 20 microstates that indicate equal distribution, so the probability of equal distribution of the six molecules of gas is 20/64, or 31%. The probability of finding two molecules of gas in the right-hand jar and four in the jar on the left is 15/64, or 23%. The probability of finding all of the molecules in the left-hand jar is 1/64, or 1.6%.

FIGURE 14.7

Probability distribution for six molecules of gas in two tanks.

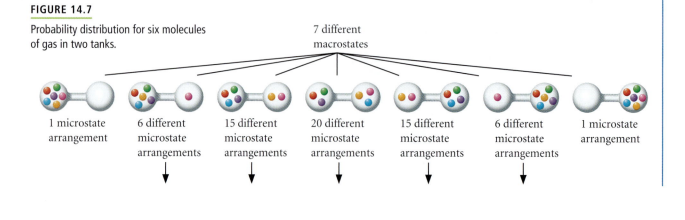

7 different macrostates

1 microstate arrangement | 6 different microstate arrangements | 15 different microstate arrangements | 20 different microstate arrangements | 15 different microstate arrangements | 6 different microstate arrangements | 1 microstate arrangement

PRACTICE 14.1

Let's use the same valve and glass jar set-up, with one jar containing eight atoms of gaseous helium. Then the valve is opened. Determine the probability of the macrostate that shows three atoms of He in one jar and five atoms of He in the other jar.

See Problems 1–4, 7, and 8.

Application

CHEMICAL ENCOUNTERS: Glycolysis

14.2 Why Do Chemical Reactions Happen? Entropy and the Second Law of Thermodynamics

When sucrose enters the body, it is broken down into two simpler molecules, glucose and fructose. Glucose and fructose are structural isomers, both with the formula $C_6H_{12}O_6$.

$$C_{12}H_{22}O_{11}(s) + H_2O(l) \rightarrow C_6H_{12}O_6(aq) + C_6H_{12}O_6(aq)$$

    Sucrose   Fructose   Glucose

Both molecules are used by the body to generate energy. For example, glucose in a cell undergoes **glycolysis**, the series of ten chemical transformations, shown in Figure 14.8, that produces two molecules of pyruvate. The pyruvate is further converted, releasing energy via a different set of transformations. In addition, during glycolysis two new energy storage molecules known as ATP (adenosine triphosphate, Figure 14.9) are produced. Glucose provides a major source of energy for living cells via the glycolysis pathway, which is the sole source of metabolic energy in some mammalian tissues and cell types. Many anaerobic (non-oxygen-consuming) microorganisms depend entirely on glycolysis for energy to carry out other biological reactions and survive.

Glycolysis is an example of a **spontaneous process**. The rusting of iron nails, the melting of ice in a glass, and the decay of wood buried in moist soil are also spontaneous processes. That is, these processes occur *without continuous outside intervention*. Say a brand new deck of cards falls off the kitchen table. The principles of probability say that there are many more ways in which the cards can land out of order than ways in which they can land sequentially. Furthermore, some of the cards will land face up and some face down. This disorder happens without our assistance—the process occurs spontaneously.

The reverse of a spontaneous process is known as a **nonspontaneous process**. Nonspontaneous processes *do not occur without continuous outside*

FIGURE 14.8

Glucose enters the glycolysis pathway on its way to complete oxidation. The reactions along the pathway produce two molecules of pyruvate and energy.

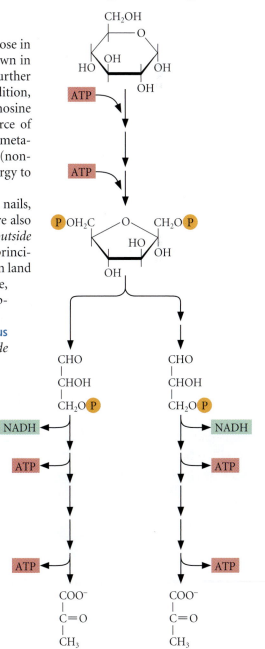

FIGURE 14.9

Adenosine triphosphate (ATP)

A disordered pile of playing cards.

Application

WWW
Video Lesson: Spontaneous Processes

Visualization: Spontaneous Reaction of Phosphorus (Barking Dogs)

intervention. A rusty nail does not revert to a polished iron nail without continuous help. Decayed wood buried in moist soil does not re-form freshly cut pieces of wood, and a deck of cards spread out on the ground does not leap into numerical and suit-based order. Just like these transformations, the reactions in chemistry can be considered either spontaneous or nonspontaneous.

What happens to a copper or bronze statue when it is exposed to the environment? As illustrated by the Newfoundland memorial erected to the memory of those lost in the worst plane crash in Canadian history (see Figure 14.10), a green patina forms on the surface. This patina is a complex mixture of colored compounds, such as antlerite ($Cu_3SO_4(OH)_4$, blue-green), brochantite ($Cu_4(SO_4)(OH)_6$, pale green), chalcanthite ($CuSO_4 \cdot 5H_2O$, blue-green), cuprite (Cu_2O, dark red), and tenorite (CuO, black). All of these compounds include either Cu^+ or Cu^{+2}, so they are oxidation products of the copper metal. Although the chemistry is much more complex, we can represent this patina by showing its formation as the simple oxidation of copper:

$$2Cu(s) + O_2(g) \rightarrow 2CuO(s)$$

A chemical reaction causes the green patina to form without our assistance, so this is a spontaneous process. The reverse process,

$$2CuO(s) \rightarrow 2Cu(s) + O_2(g)$$

is nonspontaneous; it does not occur without some outside assistance. *The reverse reaction of a spontaneous process is always nonspontaneous.*

The spontaneous oxidation of metals due to exposure to the environment is quite common. Table 14.1 lists some of the patinas that form on other metals.

Does the spontaneity of the oxidation of copper imply anything about the speed of the process? The short answer is no. It takes years for the green patina on a new copper roof to form completely. Under standard conditions, the conversion of diamond into graphite is also a spontaneous process, but luckily for people with diamond jewelry, the rate of this reaction is incredibly (almost immeasurably) slow. Rust forming on a nail and the decay of a buried log, even though they are spontaneous, are also slow processes. Biochemical reactions of glucose, on the other hand, are spontaneous and very rapid. The combustion of methane used to heat a pot of soup on the stovetop is even faster. **Thermodynamics**, the study of the changes in energy in a reaction (see also Chapter 5), determines *whether* a process is possible. In Chapter 15, we'll study the chemical kinetics of these processes to determine *how fast, and by what mechanism, they occur.*

FIGURE 14.10

A green patina on the bronze Silent Witness Memorial near Gander, Newfoundland, Canada.

TABLE 14.1	Patinas That Spontaneously Form on Metals		
Metal	**Element Symbol or Composition**	**Natural Color**	**Patina Color**
Aluminum	Al	Silvery white	Light gray
Brass	copper and zinc	Gold	Dark brown to black
Bronze	copper and tin	Yellow to olive brown	Dark brown to black
Copper	Cu	Light red brown	Green
Iron	Fe	Lustrous silvery white	Reddish brown
Silver	Ag	White to gray	Black

EXERCISE 14.2	Spontaneity in Common Processes

Which of these processes are spontaneous?

a. Ice melting in a hot oven

b. Carbon dioxide and water reacting at STP to form methane and oxygen

c. A basketball player jumping to dunk a basketball

d. NaOH(aq) and HCl(aq) reacting when combined in a beaker

Solution

The processes described in parts (a) and (d) are spontaneous, because they happen without continuous outside intervention. The combination of sodium hydroxide and hydrochloric acid is often used in chemical analysis precisely because the reaction is not only spontaneous but also rapid. We will have much more to say about reaction speed in the next chapter, in which we discuss chemical kinetics. The process described in part (b) is nonspontaneous, which suggests that the reverse process, the combustion of methane, is spontaneous. The basketball player in part (c) must intervene (by adding energy to oppose the force of gravity) in order to dunk the basketball. This process is nonspontaneous.

PRACTICE 14.2

Indicate whether each of these processes is spontaneous or nonspontaneous.

a. Potassium and water reacting
b. Leaves falling from a tree
c. A puddle evaporating from the sidewalk
d. Photosynthesis

See Problems 15–16.

Entropy

The overall **catabolism** (the biological degradation of molecules to provide smaller molecules and energy to an organism) of glucose via glycolysis is a spontaneous process similar to the combustion of glucose. When glucose is burned in air, six molecules of oxygen gas and one molecule of glucose combine to produce six molecules of water vapor and six molecules of carbon dioxide gas.

$$C_6H_{12}O_6(s) + 6O_2(g) \rightarrow 6CO_2(g) + 6H_2O(g)$$

When this reaction occurs, the number of gaseous molecules increases, which corresponds to a dramatic increase in the number of microstates that describe the system. Because of this increase, the probability that the reaction will produce products is greater than the probability that carbon dioxide and water will spontaneously form glucose and oxygen. In short, *an increase in the number of microstates favors spontaneous reactions.* We can describe this principle in more practical terms by introducing entropy.

Entropy (S) can be thought of as a measure of how the energy and matter of a system are distributed throughout the system. Investigations related to the concept of entropy began in the 1820s and 1830s with Nicolas Léonard Sadi Carnot (1796–1832) and Benoit Paul Emile Clapeyron (1799–1864). However, the concept wasn't mathematically developed until Rudolf Julius Emmanuel Clausius (1822–1888) worked on it in 1865. And although Clausius properly illustrated entropy, its relationship to the molecular level wasn't illuminated until Boltzmann did so several decades later.

Entropy isn't an easy concept to master, but we can gain important insight by considering the probabilities that have been the focus of our discussion. If the **multiplicity**—the number of microstates—increases, the number of ways in

Glucose

Visualization: Entropy

Video Lesson: Entropy and the Second Law of Thermodynamics

Tutorial: Positional Entropy

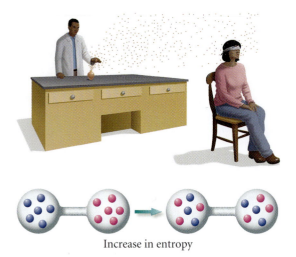

Increase in entropy

FIGURE 14.11

An experiment to explain entropy. Diffusion of gases is driven by entropy.

which energy and matter can be distributed also increases. Probability predicts that if the multiplicity of the system increases, there should be a corresponding increase in the number of ways in which energy and matter can be distributed in the system. This growth in the number of microstates increases the entropy of a system. *In other words, the more probable outcome of a spontaneous process is that an increase in entropy occurs.*

We can perform a simple experiment to help explain entropy. Say we have a friend place a bottle of perfume at one end of a room while we sit blindfolded in a chair on the opposite side of the room, as illustrated in Figure 14.11. Our friend opens the bottle. Still blindfolded, can we tell that the bottle has been opened? At first we would say it is still closed. Given a little time, though, we begin to notice the fragrance of the perfume and conclude that the bottle was opened. Why do we smell the perfume? The fragrance was released from the bottle on the opposite side of the room, so shouldn't it remain near the opened bottle? You know from experience that this isn't the case. What causes the perfume molecules to diffuse throughout the room and, eventually, into the noses of people at its most distant points? There is no pressure difference across the room, but still the perfume mixes spontaneously with the air. We can assume from the kinetic molecular theory that the attractive forces between the molecules of perfume and of those in the air are negligible, so there should be no significant change in the enthalpy of the process. Diffusion is neither exothermic nor endothermic for ideal gases. Instead, diffusion increases the distribution of the molecules throughout the room. Diffusion also increases the distribution of energy. The entropy of the system has increased.

We need to be extremely careful when we think of entropy. In processes represented by the perfume experiment, it appears that the level of disorder of the molecules has increased. Sometimes, we *incorrectly* think of entropy as a measure of the disorder of a system. But even though the increase in entropy often parallels the increase in disorder, entropy is not a measure of how disorganized a system has become. Disorder is a *macroscopic* description of a system, whereas entropy is related to the number of *microstates*.

The Second Law

Application

Inside the cells of your body lie the enzymes (polymers of amino acids; see Chapter 12) that release energy from glucose. As shown in Figure 14.12, these large polymers start out as long flexible strands but, shortly after being made, fold into a small globular shape that contains pockets for binding glucose, ATP, ADP, water, and other compounds. The structure and type of amino acids around the binding pockets determine what type of reaction the enzyme will catalyze. Protein folding into the correct shape is a spontaneous process. **Does this make sense?** The flexible extended chain has many more motions than the folded enzyme; the number of ways to distribute energy in the system decreases as the enzyme folds. On the basis of this information alone, we might predict that protein folding is nonspontaneous. A closer look reveals our need for a deeper understanding of entropy and spontaneity.

As a rule, we say that a spontaneous process is accompanied by an increase in the entropy of the universe. This is the **second law of thermodynamics**. Mathematically, the change in the entropy of the universe is greater than zero for a spontaneous process:

$$\Delta S_{\text{universe}} > 0$$

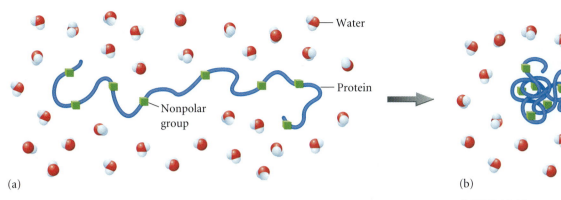

(a)

(b)

FIGURE 14.12

The folding of proteins is driven by entropy. The unfolded protein (a) disrupts the interactions of the water molecules. The nonpolar groups are tucked inside the protein, removing them from interaction with the solvent (b). The folding also increases the intermolecular forces of attraction between different regions of the protein and increases the number of interactions between solvent molecules.

However, recall the first law of thermodynamics (from Chapter 5), which says that the *energy* of the universe is constant ($\Delta E_{universe} = 0$). This contrasts with the second law, which implies that the *entropy* of the universe constantly increases. In other words, the number of possibilities for the distribution of the energy and matter of the universe constantly increases. This increase is related to the entropy changes in the system and surroundings (see Chapter 5 if you wish to review our definitions of the terms *universe, system,* and *surroundings*). Because the total entropy of the universe is the sum of the change in entropy for a particular system (ΔS_{system}) and the change in entropy of the surroundings ($\Delta S_{surroundings}$), we can describe the change in entropy of the universe as follows:

$$\Delta S_{universe} = \Delta S_{system} + \Delta S_{surroundings}$$

Visualization: Spontaneous Reactions

- If $\Delta S_{universe} > 0$, the process is spontaneous.
- If $\Delta S_{universe} < 0$, the process is nonspontaneous and is the reverse of the spontaneous process.
- If $\Delta S_{universe}$ is zero, we say that the process is neither spontaneous nor nonspontaneous but is at equilibrium, a condition of energetic stability that we will discuss shortly.

Because we take the sum of the change in entropy of the system and the change in entropy of the surroundings to obtain the change in entropy of the universe, ΔS_{system} could be a *negative* number and the overall process could still remain *spontaneous.* For example, if $\Delta S_{system} = -50$ J/K·mol and $\Delta S_{surroundings} = +80$ J/mol·K, then

$$\Delta S_{system} \quad + \quad \Delta S_{surroundings} \quad = \quad \Delta S_{universe}$$

$$-50 \text{ J/mol·K} + (+80 \text{ J/mol·K}) = +30 \text{ J/mol·K}$$

In this case, $\Delta S_{surroundings}$ increases more than ΔS_{system} decreases, so $\Delta S_{universe}$ increases and the process is spontaneous. Table 14.2 outlines the effects of $\Delta S_{universe}$ as a function of the change in entropy of the system and surroundings. Pick some

TABLE 14.2	$\Delta S_{universe}$ Is the Sum of ΔS_{system} and $\Delta S_{surroundings}$		
$\Delta S_{surroundings}$	ΔS_{system}	$\Delta S_{universe}$	**Spontaneity**
+	+	+	Spontaneous
+	−	?	Spontaneous if $\Delta S_{system} < \Delta S_{surroundings}$
−	+	?	Spontaneous if $\Delta S_{system} > \Delta S_{surroundings}$
−	−	−	Nonspontaneous

FIGURE 14.13

The unfolded protein reduces the interactions of the solvent molecules. By folding, it allows those favorable interactions to take place.

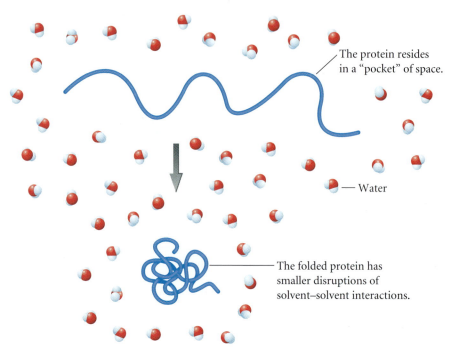

The protein resides in a "pocket" of space.

Water

The folded protein has smaller disruptions of solvent–solvent interactions.

sample values for ΔS_{system} and $\Delta S_{surroundings}$, as we did just above, to help clarify the outcomes in the table.

In the case of our folding protein, we must take into account the entropy of the system *and* that of the surroundings if we are to properly assess the change in entropy of the universe and determine whether the process is spontaneous. As an extended chain, the enzyme's nonpolar groups must interact with the aqueous cellular environment—the surroundings—as shown in Figure 14.13. The unfolded enzyme disrupts many of the interactions that occur among the solvent molecules (water). To reduce those disruptions within the surroundings, the folding enzyme tucks its *nonpolar* groups into the center of the globular structure, effectively removing their interaction with the water inside the cell. The folding allows the *polar* groups in the enzyme to interact with the *polar* water molecules in the surroundings. The water molecules are also able to interact with each other with minimal disruption. The number of microstates for the protein decreases during the process of folding, but the number of microstates for the water and the water–protein interaction increases. The result is a $\Delta S_{surroundings}$ that is more positive than the negative ΔS_{system}. And the overall process, the folding of a protein, is spontaneous ($\Delta S > 0$).

EXERCISE 14.3   **Reaction Spontaneity**

Determine whether the values shown below will produce a spontaneous process.

$$\Delta S_{system} = 140 \text{ J/mol·K}; \; \Delta S_{surroundings} = -155 \text{ J/mol·K}$$

Solution

In a spontaneous process the entropy of the universe increases, so we must add the values for the system and surroundings to answer the question.

$$\Delta S_{universe} = \Delta S_{system} + \Delta S_{surroundings}$$

$$= 140 \text{ J/mol·K} + (-155 \text{ J/mol·K})$$

$$= -15 \text{ J/mol·K}$$

Because our calculated value for the entropy of the universe is negative, the process is nonspontaneous.

PRACTICE 14.3

Determine whether the values for entropy in each of these cases will produce a spontaneous process.

a. $\Delta S_{system} = -23$ J/mol·K; $\Delta S_{surroundings} = -55$ J/mol·K

b. $\Delta S_{system} = 38$ J/mol·K; $\Delta S_{surroundings} = 59$ J/mol·K

c. $\Delta S_{system} = -84$ J/mol·K; $\Delta S_{surroundings} = 132$ J/mol·K

See Problems 17, 18, 29, and 30.

14.3   Temperature and Spontaneous Processes

Living organisms have been found all over the Earth, from the mouths of near-boiling geysers to the depths of the Arctic Ocean. These living organisms, just like the species that live near the tropics, require energy to survive. Many of them use glucose as a source of energy. In fact, some of the psychrotrophs (bacteria that can survive exposure to low temperatures) and psychrophiles (bacteria that thrive in low temperatures) employ a concentrated solution of glucose as an "antifreeze" because such a solution lowers their freezing point. Research by microbiologists and biochemists into the life processes of the psychrophiles indicates that these organisms, like a lot of other living things on the planet, break down glucose to produce energy through glycolysis. Does the temperature at which these organisms live affect the spontaneity of the reactions involved in glycolysis? To answer this question, we need to consider the signs on the change in entropy of the system (ΔS_{system}) and of the surroundings ($\Delta S_{surroundings}$). A positive change in the entropy of the universe is required for the process to be spontaneous.

In general, as compounds undergo changes in their physical states from solid to liquid to gas, the entropy of the *system* increases ($\Delta S_{system} = +$), as shown in Figure 14.16. Water molecules in an ice cube have a well-defined order in an ice cube. As the ice begins to melt, this well-defined order disappears and the water molecules increase their range of motions and, therefore, the number of microstates. The number of microstates that are possible in the liquid water suggests that the melting of ice corresponds to an *increase* in the entropy of the system. The same holds true when water is converted into steam. Conversely, the reverse of these processes (gas to liquid to solid) is typically accompanied by a *loss* of entropy ($\Delta S_{system} = -$).

What is the effect of energy transfer on ΔS? When steam condenses to liquid water, energy as heat flows out of the system and into the surroundings, and the kinetic energy of the particles in the surroundings increases. The motions of the atoms in the surroundings increase, and the sign of $\Delta S_{surroundings}$ is positive. On the other hand, if energy as heat flows from the surroundings to the system (liquid to vapor), we'd expect the kinetic

Water molecules gain increased freedom of motion in changing phase from solid to liquid.

Application

CHEMICAL ENCOUNTERS: Psychrotrophs and Psychrophiles

Video Lesson: Entropy and Temperature

NanoWorld / MacroWorld

Big effects of the very small: Industrial uses for the extremophiles

Imagine living at the bottom of the Arctic Ocean, thinking life was grand inside glacial ice, enjoying the weather near a thermal vent, or relaxing under the crush of 1 mile of bedrock. These conditions sound fairly extreme to us, but not to a class of bacteria known as the extremophiles. Some of these microorganisms thrive near the hot bubbling mud-pots of Yellowstone National Park, others in the sulfur-laden waters near a geothermal vent at the bottom of the Atlantic Ocean. Extremophiles, some examples of which are listed in Table 14.3, survive in conditions that we humans would find extreme.

Biochemists, microbiologists, and geologists from around the world study these creatures because of the extreme conditions in which they live and to learn more about the enzymes that continue to work under the equally extreme conditions inside them. For instance, millions of Americans are lactose-intolerant and have difficulty digesting the lactose in almost every dairy product, including milk and ice cream. Imagine if you could isolate beta-galactosidase (an enzyme that breaks down sugars like lactose into more easily digested compounds) from an extremophile that lived in icy environments. The extremophile's beta-galactosidase should be capable of working quite well in cold environments. By adding the isolated beta-galactosidase to milk and related products like ice cream, you could make lactose-free dairy products without having to heat them. Researchers at Pennsylvania State University have been able to show that this is possible by isolating a strain of bacteria, known as *Arthrobacter psychrolactophilus* that has a modified beta-galactosidase that works best when the temperature is 15°C and continues to work well when the temperature is as low as 0°C.

Thomas D. Brock isolated the first example of a true extremophile from hot springs, such as that shown in

TABLE 14.3	Classes of Extremophiles	
Class	**Extreme Environment**	**Locations Where They Live**
Acidophiles	Low pH	Sulfurous springs and acid mine drainage
Alkaliphiles	High pH	Alkaline lakes and basic soils
Anaerobes	Non-oxygen-containing environments	Fermenting juices
Barophiles	High pressure	Deep sea vents and deep within the Earth
Copiotrophs	High nutrient levels	Sugar solutions
Halophiles	High ion concentration	Saline lakes and salt deposits
Hyperthermophiles	Temperatures above 70°C	Hydrothermal vents, hot springs
Methanogens	Methane-rich	Deep-sea vents, oil deposits
Oligotrophs	Low nutrient levels	Desert, rocks
Psychrophiles	Low temperatures, typically below 10°C	Glaciers, Arctic Ocean, cold soils
Thermophiles	Temperatures above 50°C	Hydrothermal vents, hot springs

FIGURE 14.16

Entropy increases as a compound changes state from solid to liquid to gas. Note that the increased molecular motions allow more microstates to exist in the compound.

S_{solid} < S_{liquid} <<< S_{gas}

FIGURE 14.14

The Morning Glory Pool at Yellowstone National Park is named for the brightly colored thermophiles that flourish in this high-temperature environment.

FIGURE 14.15

Hydrothermal vent in the North Pacific west of Vancouver Island on the Juan de Fuca Ridge. Extremophiles such as *Pyrolobus fumarii* and *Methanopyrus* live on the sides of these chimneys. The "smoke" flowing from the vents is actually made up of minerals from the lava under the ocean floor.

Figure 14.14, in Yellowstone National Park in Wyoming. This bacterium, called *Thermus aquaticus,* grows most rapidly at temperatures near 70°C. Other thermophiles, or heat-loving bacteria, include *Sulfolobus acidocaldarius,* which lives in sulfur-laden hot springs at temperatures as high as 85°C, and *Pyrolobus fumarii,* which is isolated from deep-sea hydrothermal vents (Figure 14.15), grows only at temperatures above 90°C, and reproduces best at 105°C. These bacteria are of industrial interest because of their ability to grow at such high temperatures. For example, the enzyme Taq polymerase (isolated from *T. aquaticus*) is used in DNA fingerprinting because it can survive the severe temperature variations in the polymerase chain reaction used to make multiple copies of purified DNA (see Chapter 22).

The acid-tolerant extremophiles (acidophiles) are of interest because their enzymes are capable of operating in highly acidic environments. A potential application for the acidophile's enzymes is their addition to cattle feed, because they would work well in the acidic gut of an animal as an aid in digesting food. Their use would improve the usefulness of cheap food as a source of energy for the animals. The alkaliphiles (base-tolerant bacteria) thrive in basic soils such as those in the western United States and in Egypt. Proteases (enzymes that break down proteins) and lipases (enzymes that break down oils) isolated from these bacteria could find potential use in the detergent industry. Their addition to laundry detergents (which typically are basic) would improve the ability of the detergent to clean stains from clothing.

Investigators are currently searching, and finding, bacteria that live in environments we originally thought were sterile. After determining the types and properties of the enzymes that these bacteria possess, scientists are exploiting their industrial utility. The uses of these enzymes as catalysts to aid human life are endless. Will we find extremophiles that proliferate on Mars? . . . on Io, the volcanic innermost major moon of Jupiter? . . . on our own Moon? And, if so, what uses might we find for the enzymes they produce?

energy of the particles in the surroundings to decrease and the motions of the atoms (and, therefore, the entropy) in the surroundings to decrease also. In short, a flow of energy as heat *out of the system* and *into the surroundings* (an exothermic process) corresponds to a *positive sign for* $\Delta S_{\text{surroundings}}$. An endothermic process has an opposite effect on the surroundings; endothermic processes correspond to a negative sign for $\Delta S_{\text{surroundings}}$.

What is this exchange of energy to which we refer? If our process occurs under reversible conditions at a constant pressure, we can relate the energy of the process (q_{rev}) to the change in enthalpy of the process (ΔH). **Reversible conditions** occur when *the process is allowed to proceed in infinitesimally small steps*. At any point during the reaction, we could change the direction of the reaction with

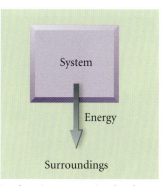

Exothermic processes involve the transfer of energy from the system to the surroundings.

merely slight modifications. *Often, reversible conditions exist during phase changes.* Quantitatively, we can summarize our statements by saying that the change in the entropy is equal to the change in enthalpy of the phase change (a reversible process) divided by the temperature:

$$\Delta S_{system} = \frac{q_{rev}}{T} = \frac{\Delta H_{system}}{T}$$

where the temperature (T) is reported in kelvins (K) and the enthalpy (ΔH_{system}) is reported in joules per mole. Because of some assumptions we've made to arrive at this equation, its use is limited to describing heat transfers when the temperature remains constant. For example, the equation works well for describing phase transitions but poorly for describing a reaction.

Because $q_{surroundings} = -q_{system}$, and $q_{system} = \Delta H_{system}$, the value of $\Delta S_{surroundings}$ can be obtained using a similar equation.

$$\Delta S_{surroundings} = \frac{-\Delta H_{system}}{T}$$

EXERCISE 14.4 Entropy Change at a Phase Change

Instead of carrying water in their backpacks, hikers can melt ice to make drinking water. They also boil water for drinking and food preparation. What is the entropy change of the system for melting 1 mol of ice at 0.00°C and 1 atm? What is the entropy change for melting 125 g of ice? The enthalpy of fusion at this phase change, $\Delta_{fus}H$ is 6.01 kJ/mol.

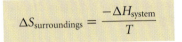

$$H_2O(s) \rightarrow H_2O(l) \qquad \Delta_{fus}H = 6.01 \text{ kJ/mol at } 0.0°C$$

First Thoughts

The entropy change for the system or surroundings can be calculated if we know the temperature and the enthalpy of the system. Because the units of entropy are usually reported in J/mol·K, we should convert the units for the enthalpy and the temperature to match, so our calculation is simplified. The second part of the question asks us to examine the specific entropy change for a quantity of water that is not equal to 1 mol. We may do this part of the calculation by dimensional analysis.

Solution

The change in the entropy of this reversible process can be calculated using the formula we just discussed. This process involves an increase in the entropy of the system.

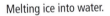

Melting ice into water.

$$\Delta S_{system} = \frac{\Delta H_{system}}{T} = \frac{6010 \text{ J/mol}}{273 \text{ K}} = 22.0 \text{ J/mol·K}$$

We've calculated the change in entropy ($\Delta S = 22.0$ J/mol·K) for the phase change, so we can use this value to determine the change in entropy for melting 125 g of water.

$$\Delta S_{system} = 125 \text{ g} \times \frac{1 \text{ mol}}{18.02 \text{ g}} \times \frac{22.0 \text{ J}}{\text{mol·K}} = \frac{153 \text{ J}}{\text{K}}$$

Further Insights

Reactions can occur with either an increase or a decrease of entropy. It is interesting to note that this change is based on the number of moles of the compounds in the reaction. If this relationship appears to be similar to what we observed in our discussion of enthalpy in Chapter 5, it should. Also, as we'll see later, entropy can be manipulated in ways that are very similar to those we used for enthalpy.

Moreover, because the temperature remains constant, we can calculate the value of $\Delta S_{surroundings}$. We find that the value is −22.0 J/mol·K. Therefore, the entropy change for the universe ($\Delta S_{universe}$) should be 0.0 J/mol·K ($\Delta S_{universe} = \Delta S_{system} + \Delta S_{surroundings}$). This process is neither spontaneous nor nonspontaneous. It is reversible.

PRACTICE 14.4

Calculate ΔS_{system} for each of these processes at 25°C. Assume that each is a reversible process.

$I_2(g) \rightarrow I_2(s)$ $\Delta_{sub}H = +62.4$ kJ

$H_2O(l) \rightarrow H_2O(g)$ $\Delta_{vap}H = +40.7$ kJ

See Problems 35 and 36.

HERE'S WHAT WE KNOW SO FAR

- Spontaneous processes are associated with an increase in the entropy of the universe.

- The rate of a spontaneous reaction is not related to its spontaneity.

- The reverse of a spontaneous process is a nonspontaneous process.

- The change in entropy of the universe can be calculated as the sum of the change in entropy of the system and the change in entropy of the surroundings.

- The entropy of the system can be calculated for reversible processes by dividing the enthalpy for the process by the temperature of the process.

14.4 Calculating Entropy Changes in Chemical Reactions

Metabolism (the biochemical reactions of an organism) releases energy. In this process, the potential energy stored in food is used by an organism, and some of the food is converted via chemical reactions into molecules that are needed for the organism to survive. All of these reactions are spontaneous in the body and vital to the living processes that occur at the cellular level. For instance, the body breaks down sucrose, perhaps contained in a sugar-laden gumdrop, into glucose and fructose. The glucose then enters a series of reactions, the glycolytic pathway being the first (Figure 14.9), as the body converts it into carbon dioxide and water. Along the way it produces ATP (Figure 14.10), a molecule used to store potential energy. Plants, on the other hand, synthesize glucose by combining carbon dioxide and water. This reaction is energetically uphill for the plant, so it uses the high-energy ATP molecule to drive the reaction to completion. Why is the breakdown of glucose energetically downhill, whereas the formation of glucose is uphill? In other words, why does it take energy to make glucose, and why is energy released when glucose is broken down? We can answer this question by examining the entropy changes in the combustion of glucose. *Qualitatively,* is the sign of

Application

CHEMICAL
ENCOUNTERS:
More Glycolysis

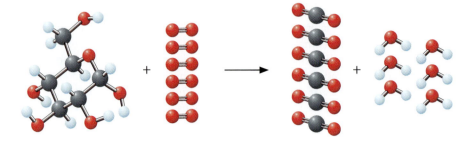

The reaction of glucose with oxygen generates carbon dioxide and water. Compare the number of molecules of gaseous products made to the number of molecules of gaseous reactants.

ΔS_{system} positive or negative? The reactants include 1 mol of glucose molecules and 6 mol of gaseous oxygen molecules.

$$C_6H_{12}O_6(s) + 6O_2(g) \rightarrow 6O_2(g) + 6H_2O(g)$$

The combustion proceeds as 7 mol of reactants are converted into 12 mol of products (6 mol of gaseous carbon dioxide and 6 mol of gaseous water). Prior to the reaction, there were a large number of possible microstates because of the large number (7 mol) of reactants. Because gases occupy a larger volume in a flask than do equivalent quantities of solids, the 6 mol of oxygen in our reaction occupy the majority of the locations within the flask as they rapidly travel within it. After the reaction, there are 12 mol of gas inside the flask. The number of microstates increased because of the larger number of gaseous products. The increase in the number of microstates is an increase in the entropy of the system (ΔS_{system} is positive), as shown in Figure 14.17. If we examine the reverse of this reaction (from the viewpoint of the glucose-producing plant), the entropy of the system is lowered because we are combining 12 mol of gaseous carbon dioxide and water to make 6 mol of gaseous oxygen and 1 mol of solid sugar.

What we've discussed is a method by which one can usually predict the sign of the entropy change accurately. As a rule, *the change in entropy of a reaction is positive if the number of gaseous molecules increases.* Although the number of moles of solid and liquid molecules contributes to the overall number of microstates, the large volume occupied by gaseous molecules contributes much more. By simply examining a reaction, we can predict the entropy change. The change is much harder to assess for reactions that do not involve gaseous molecules.

FIGURE 14.17

An increase in the number of gaseous molecules increases the entropy of the system.

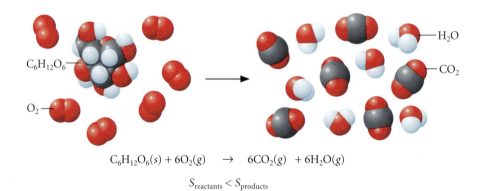

$$C_6H_{12}O_6(s) + 6O_2(g) \rightarrow 6CO_2(g) + 6H_2O(g)$$

$$S_{reactants} < S_{products}$$

EXERCISE 14.5 | **Predict the Sign**

1. Some camp stoves use butane as fuel. The combustion of butane is exothermic, providing heat to cook food and boil water. Predict the sign of ΔS_{system} for the combustion of butane.

$$2CH_3CH_2CH_2CH_3(g) + 13O_2(g) \rightarrow 8CO_2(g) + 10H_2O(g)$$

2. The Haber process, the combination of hydrogen and nitrogen gases to form ammonia gas (NH_3) is one of the most widely used manufacturing processes because of worldwide demand for ammonia-based fertilizer. Predict the sign of ΔS_{system} for the production of ammonia.

$$3H_2(g) + N_2(g) \rightarrow 2NH_3(g)$$

3. Barium hydroxide ($Ba(OH)_2 \cdot 8H_2O$) reacts with ammonium chloride (NH_4Cl) to form several products, including ammonia, water, and barium chloride. Predict the sign of ΔS_{system} for this reaction.

$$Ba(OH)_2 \cdot 8H_2O(s) + 2NH_4Cl(s) \rightarrow BaCl_2(aq) + 2NH_3(g) + 10H_2O(l)$$

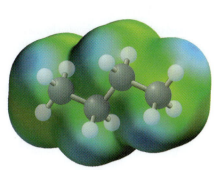

Butane

Solution

1. There are 18 mol of gaseous products and only 15 mol of gaseous reactants. Consequently, the number of microstates should increase for the reaction. This leads us to predict that the change in entropy should be a positive number for this reaction. Based on calculations that we'll discover later, $\Delta S_{system} = +789$ J/mol·K. This agrees with our prediction.

2. The Haber process takes 4 mol of gaseous reactants and forms 2 mol of gaseous products. The entropy change of the system is likely to be negative for this reaction. The actual ΔS_{system} value is −199 J/mol·K.

3. The reaction results in the formation of 2 mol of ammonia gas for each mole of barium hydroxide octahydrate that reacts with solid ammonium chloride. Consequently, we would predict that the change in system entropy would be positive. One factor to keep in mind is that the waters of hydration (that is, the eight H_2O molecules that are part of the barium hydroxide crystal) would be released as part of the process. These molecules would tend to raise the system entropy further as they join the liquid state. The actual ΔS_{system} value is 468 J/mol·K.

PRACTICE 14.5

Predict the sign of the change in entropy for each of these reactions.

$$H_2O(l) \rightarrow H_2O(g)$$

$$CH_3OH(l) + HCl(g) \rightarrow CH_3Cl(l) + H_2O(l)$$

See Problems 39–42, 45, and 47.

Qualitatively, our prediction of the sign of ΔS_{system} can be based on the number of gaseous molecules in the reaction. Quantitatively, the value for the change in the entropy (ΔS_{system}) is nearly as easy to determine. However, a problem arises. In order to calculate a change in a state function, we subtract the final value from the initial value. How do we determine the initial value of entropy for a compound?

FIGURE 14.18

Based on the distance traveled by the mountain climbers, Mt. McKinley appears to be taller than Mt. Everest. However, with our reference point in place, it is clear that Mt. Everest is much taller.

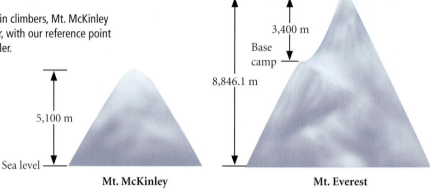

We can find the answer by returning to our Mt. Everest climbers from Section 14.1. Typically, such adventurers climb about 3,400 m (11,400 ft) from the base camp at the foot of the mountain to reach the summit. Using this information, can we accurately say that the mountain is the tallest in the world? Mt. McKinley in Alaska, with an ascent of 5,100 m (17,000 ft) appears to be taller. In order to accurately say that Mt. Everest is the tallest, we must consider that the base camp itself is 5,400 m above sea level. That is, we must establish a *reference point* from which to compare the heights, as we've done in Figure 14.18. In the same way, values of entropy are based on some reference point that is common to all compounds. The **third law of thermodynamics** establishes this point for entropy. The law states that the entropy of a pure perfect crystal at 0 K is zero. A "perfect" crystal, one in which all of the molecules are rigidly and uniformly aligned, has negligible kinetic energy at 0 K. In other words, if the excess kinetic energy of the crystal is zero, the crystal has zero entropy. We qualify the phrase *kinetic energy* with the word *excess* because there is still some atomic (electron) motion in a perfect crystal at 0 K.

We convert this perfect crystal into a collection of molecules at some higher temperature by adding kinetic energy. This increases the number of microstates, resulting in an increase in entropy, as illustrated in Figure 14.19. Under the standard conditions of 298 K and 1 atm, the compound has an associated standard molar entropy that we designate as $S°$. This value is the same as that for ΔS_{system} at this temperature and pressure. By measuring the change in entropy from 0 K to 298 K, researchers have determined the standard molar entropies of a wide variety of compounds (see Table 14.4).

Note that the magnitude of the standard molar entropies is larger for those molecules that are more complex: $S°_{H_2(g)} = 131$ J/mol·K and $S°_{H_2O(g)} = 189$ J/mol·K.

FIGURE 14.19

Adding kinetic energy to a perfect crystal at 0 K increases the number of microstates. The result: The entropy of a compound is always a positive number.
(a) Crystal at 0 K. (b) Crystal at some higher temperature.

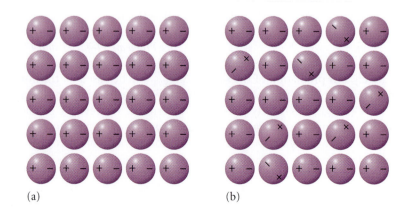

TABLE 14.4 | **Standard Molar Entropies and Free Energies for Selected Elements and Compounds at 25.0°C**

Substance	$\Delta_f H°$ (kJ/mol)	$\Delta_f G°$ (kJ/mol)	$S°$ (J/mol·K)	Substance	$\Delta_f H°$ (kJ/mol)	$\Delta_f G°$ (kJ/mol)	$S°$ (J/mol·K)
$Br_2(l)$	0	0	152	$C_2H_5OH(l)$	−278	−175	161
$Br_2(g)$	+31	+3	245	$CH_3Cl(g)$	−84	−60	234
$HBr(g)$	−36	−53	199	$HCl(g)$	−92	−95	187
$CaF_2(s)$	−1220	−1167	69	$HCl(aq)$	−168	−131	55
$CaCl_2(s)$	−796	−748	105	$H_2(g)$	0	0	131
$CaCO_3(s)$	−1207	−1129	93	$Fe_2O_3(s)$	−826	−740	90
$C(graphite)$	0	0	6	$N_2(g)$	0	0	192
$C(diamond)$	+2	+3	2	$NO(g)$	90	87	211
$CO_2(g)$	−394	−394	214	$NO_2(g)$	33	51	240
$CO_2(aq)$	−414	−386	118	$NH_3(g)$	−46	−16	192
$CH_4(g)$	−75	−50	186	$H_2O(g)$	−242	−229	189
$C_2H_2(g)$	+227	+209	201	$H_2O(l)$	−286	−237	70
$C_2H_4(g)$	+52	+68	220	$NaOH(s)$	−426	−379	64
$C_2H_6(g)$	−85	−33	230	$H_2SO_4(aq)$	−909	−745	20
$CH_3OH(l)$	−239	−166	127	$ZnO(s)$	−348	−318	44

As more and more atoms are incorporated within a compound, the entropy of the compound increases. Similarly, note that the standard molar entropies are different for compounds in different states. For example, liquid water ($S° = 70$ J/mol·K) has a lower entropy than gaseous water ($S° = 189$ J/mol·K). This is understandable, because the molecules in the gaseous state have more freedom of movement and hence have a greater possible distribution of energies.

Computationally, we use these standard molar entropy values in much the same way that we used enthalpy values to calculate $\Delta H°$ for a reaction. The basic principles of Hess's law (see Chapter 5), which we used to calculate the change in enthalpy for a reaction, also work for determining the change in the entropy of a reaction. To summarize, the net standard state entropy gain (or loss) for a reaction is calculated by subtracting the total standard state entropy change for the reactants from the total standard state entropy change for the products:

$$\Delta S° = \sum nS°_{products} - \sum nS°_{reactants}$$

where n is the stoichiometric coefficient of each of the compounds in the reaction, and Σ indicates that the standard state entropy ($S°$) is summed. Just as in the determination of the change in enthalpy of a reaction, $\Delta S°$ for a reaction depends on the way the reaction is written.

EXERCISE 14.6 | **Calculating $\Delta S°$**

In the oxidation of glucose at 25°C, the products are different from those we considered early in Section 14.4. In that discussion, glucose reacted at body temperature. Here, at 25°C, both oxygen and carbon dioxide are gases and water is a liquid. Use the table of thermodynamic values in the appendix to assess the change in entropy for this reaction under standard conditions.

$$C_6H_{12}O_6(s) + 6O_2(g) \rightarrow 6CO_2(g) + 6H_2O(l)$$

First Thoughts

If the water were written as a gas, we would predict a very large increase in the number of moles of gaseous products, leading to the prediction of a positive entropy change for the reaction. However, in this reaction as written, the number of moles of gas remains constant, and we are unable to assert that the entropy change would be positive from that standpoint alone. However, we might predict that the entropy of the system would increase slightly, because we are forming six molecules of a liquid where we had one molecule of solid reactants. We can use the summation equation to calculate the change in entropy for the reaction.

$$\Delta S° = \sum n S°_{products} - \sum n S°_{reactants}$$

Solution

From Table 14.4 we obtain $S°$ values for each of the compounds in the reaction. Using the equation shown above, we can calculate the value of $\Delta S°$:

$$6 \text{ mol } CO_2 \times 214 \text{ J/mol·K} = \quad 1284 \text{ J/K}$$
$$6 \text{ mol } H_2O \times 70 \text{ J/mol·K} = \quad \underline{420 \text{ J/K}}$$
$$\text{Total } S°_{products} = \quad 1704 \text{ J/K}$$

$$1 \text{ mol } C_6H_{12}O_6 \times 212 \text{ J/mol·K} = \quad 212 \text{ J/K}$$
$$6 \text{ mol } O_2 \times 205 \text{ J/mol·K} = \quad \underline{1230 \text{ J/K}}$$
$$\text{Total } S°_{reactants} = \quad 1442 \text{ J/K}$$

$$\text{Total } S°_{products} = \quad 1704 \text{ J/K}$$
$$-\text{Total } S°_{reactants} = \quad \underline{-1442 \text{ J/K}}$$
$$\Delta S° = \quad 262 \text{ J/K}$$

The reaction has an increase in entropy.

Further Insight

Does this answer make sense? Although the number of molecules of gas is the same for the reactants and the products, we can reasonably assert that the gain in entropy is due to the formation of six molecules of liquid from every molecule of solid. The effect is not as profound as for the formation of a gas ($\Delta S° = 262$ J/K versus $\Delta S° = 974$ J/K), but it does contribute, as in this case. Our answer makes sense.

PRACTICE 14.6

Predict the sign of $\Delta S°$ for each of these reactions. Then calculate the values of $\Delta S°$. You will need to use the table of thermodynamic values in the appendix. Do your calculations agree with your predictions?

$$CH_3OH(l) + HCl(g) \rightarrow CH_3Cl(g) + H_2O(g)$$
$$3O_2(g) \rightarrow 2O_3(g)$$

See Problems 46 and 48–52.

14.5 Free Energy

Application

CHEMICAL ENCOUNTERS: Pyruvate and Lactate

In exercising muscles, the level of oxygen often is a limiting reagent. As the muscles consume glucose to provide the energy needed to contract and relax, the limited supply of oxygen forces a buildup of pyruvate (see Figure 14.20). This molecule can continue on the typical glycolysis pathway toward complete oxidation, but with limited oxygen this does not happen. Instead, the exercising muscle obtains energy by converting pyruvate into lactate (a reduction!) by the pathway shown in Figure 14.21. The reduced form of a biological molecule known as NAD$^+$ (nicotinamide adenine dinucleotide) is oxidized in the process. (Remember that we must have an oxidation and a reduction for a redox reaction to occur.)

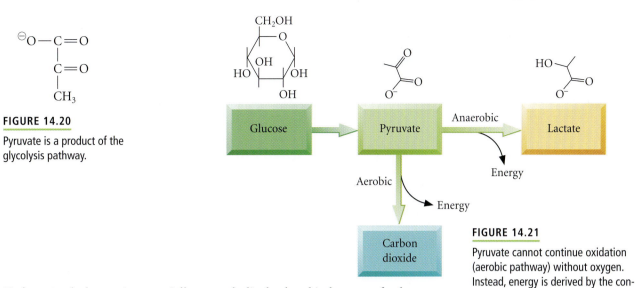

FIGURE 14.20

Pyruvate is a product of the glycolysis pathway.

FIGURE 14.21

Pyruvate cannot continue oxidation (aerobic pathway) without oxygen. Instead, energy is derived by the conversion of pyruvate to lactate (anaerobic pathway).

Unfortunately, lactate is potentially a metabolic dead end in humans; further catabolism of lactate may not occur. Eventually, the lactate can be changed back to pyruvate by the reverse reaction and then catabolized in the presence of oxygen. While early research indicated that the body signals the presence of excess lactate (and a decrease in oxygen in the muscles) with a burning sensation, recent results seem to contradict this claim.

The reaction of pyruvate to make lactate is energetically favorable. Some energy is harvested from the pyruvate. However, *qualitatively*, it is difficult to predict this by just looking at the reaction. Determining the spontaneity of the reaction by simply looking at the states of the reaction is also difficult.

$$H^{\oplus} + NADH + \underset{\substack{| \\ CH_3}}{\overset{COO^{\ominus}}{\underset{|}{C}=O}} \xrightarrow{\text{Lactate dehydrogenase}} \underset{\substack{| \\ CH_3}}{HO-\overset{COO^{\ominus}}{\underset{|}{C}}-H} + NAD^{\oplus}$$

After some calculations, we can determine the value of $\Delta S°$ (that is, ΔS_{system}), but we also have to determine the entropy of the surroundings ($\Delta S_{surroundings}$) in order to calculate the change in entropy of the *universe* (system + surroundings) and, therefore, the spontaneity of the reaction. A simpler and more useful way to determine reaction spontaneity was identified by Josiah Willard Gibbs (1839–1903), professor of mathematical physics at Yale from 1871 to 1903.

Gibbs showed that calculating a property he called the **free energy (G)** makes possible the straightforward determination of reaction spontaneity. The equation is indicative of two ideas that we have discussed.

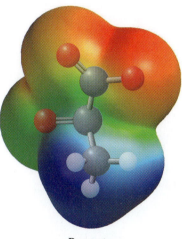

Pyruvate

■ *Idea 1:* An increase in the number of microstates available to the system favors an increase in the entropy of the universe.

■ *Idea 2:* An exothermic reaction increases the kinetic energy of the surroundings, therefore increasing the range of motions of its particles, leading to an increase in the entropy of the surroundings and favoring an increase in the entropy of the universe.

The **Gibbs equation** is $G = H - TS$

where H is the enthalpy of the system, T is the temperature in kelvins, and S is the entropy of the system. We are interested in the *change* in the free energy of the system, so the Gibbs equation becomes

$$\Delta G = \Delta H_{system} - T\Delta S_{system}$$

at constant temperature and pressure.

Video Lesson: Gibbs Free Energy

TABLE 14.5	Effect of Enthalpy, Entropy, and Temperature on the Spontaneity of a Process		
ΔH	ΔS	Low Temperature	High Temperature
+	+	$\Delta G = +$; nonspontaneous	$\Delta G = -$; spontaneous
+	−	$\Delta G = +$; nonspontaneous	$\Delta G = +$; nonspontaneous
−	+	$\Delta G = -$; spontaneous	$\Delta G = -$; spontaneous
−	−	$\Delta G = -$; spontaneous	$\Delta G = +$; nonspontaneous

This equation doesn't seem very useful at first. However, the value of the change in the free energy (ΔG) of a process is related to $\Delta S_{universe}$ by the equation shown below. In other words, the free energy change of a process can be used to determine the spontaneity of that process. Because of the negative sign in the relationship, negative free energy changes ($\Delta G = -$) imply spontaneous processes, and positive free energy changes ($\Delta G = +$) indicate nonspontaneous processes Table 14.5 lists the outcomes of the change in free energy based on the relationship between ΔH and ΔS.

$$\Delta S_{universe} = \frac{-\Delta G}{T}$$

Therefore, the Gibbs equation can be used to calculate the change in free energy for a process, because we know how to calculate ΔH and ΔS using tabulated values (see Table 14.4). All we need, then, is the temperature of the process and the application of some assumptions. First, as a requirement of the Gibbs equation, we must assume that the pressure remains constant. We also have to assume that both ΔH and ΔS are temperature independent, which is not always a good assumption, especially for ΔS.

Let's reexamine the conversion of pyruvate to lactate under anaerobic conditions. Qualitatively, we cannot use the change in the number of moles of gas from reactants to products to help us assess whether the reaction will be spontaneous, because there are no gases in the reaction.

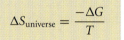

$$H^{\oplus} + NADH + \begin{matrix} COO^{\ominus} \\ | \\ C=O \\ | \\ CH_3 \end{matrix} \xrightarrow{\text{Lactate dehydrogenase}} \begin{matrix} COO^{\ominus} \\ | \\ HO-C-H \\ | \\ CH_3 \end{matrix} + NAD^{\oplus}$$

However, because the values of ΔH and ΔS have been measured for this reaction, we can determine the change in free energy. In this system, $\Delta H = -78.7$ kJ/mol and $\Delta S = -88.75$ J/mol·K. On the basis of these data alone, we identify the reaction as exothermic with a decrease in system entropy. What is the value of ΔG for the reaction at 298 K?

$$\Delta G = \Delta H - T\Delta S$$
$$\Delta G = -78,700 \text{ J/mol} - (298 \text{ K} \times -88.75 \text{ J/mol·K})$$
$$\Delta G = -52,253 \text{ J/mol} = -52.3 \text{ kJ/mol}$$

Our calculations show that the free energy of this reaction decreases. One thing to keep in mind during such calculations is that in order to use the Gibbs equation, we must modify the values of ΔH and ΔS so that they have the same units. In our calculation, we converted them into units of J/mol and J/mol·K.

We previously mentioned that the change in free energy is negatively related to the change in the entropy of the universe. That is, a negative value for the

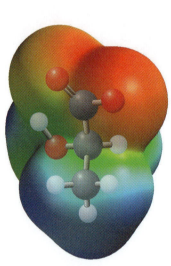

Lactate

change in free energy implies a spontaneous process. For a reaction, such as the conversion of pyruvate to lactate, in which the change in free energy is a negative value ($\Delta G = -$), the entropy of the universe has a positive value ($\Delta S_{universe} = +$). The reaction is spontaneous under standard conditions. We might not have predicted this solely on the basis of our examination of $\Delta S°$ for the reaction. Similarly, a process can be considered nonspontaneous if there is an increase in the free energy of the system ($\Delta G = +$). This means that if the reaction, as written, is nonspontaneous, the reverse reaction is spontaneous. For instance, the formation of lactate from pyruvate is spontaneous (a lowering of free energy), but the formation of pyruvate from lactate is nonspontaneous.

EXERCISE 14.7   Spontaneity and Biochemical Reactions

Yeast cells operate under anaerobic conditions to convert glucose to pyruvate, and pyruvate to ethanol and CO_2. Humans are unable to convert pyruvate to ethanol and must make lactate instead. What is the value of the change in the free energy (ΔG) for the conversion of glucose to ethanol at 25°C? (Assume that both the enthalpy and the entropy were also determined at 25°C.)

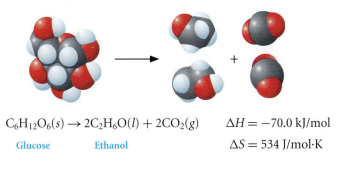

$$C_6H_{12}O_6(s) \rightarrow 2C_2H_6O(l) + 2CO_2(g) \qquad \Delta H = -70.0 \text{ kJ/mol}$$

Glucose   Ethanol $\Delta S = 534 \text{ J/mol·K}$

Solution

$$\Delta G = \Delta H - T\Delta S$$

$$\Delta G = -70{,}000 \text{ J/mol} - (298 \text{ K}) \times (534 \text{ J/mol·K})$$

$$\Delta G = -229{,}132 \text{ J/mol}$$

$$\Delta G = -229 \text{ kJ/mol}$$

Because the entropy is given in J/K·mol, the units of the enthalpy have been changed (from kilojoules to joules) so that the two values can be added easily. The reaction is spontaneous, with a rather large negative free energy change.

PRACTICE 14.7

Calculate ΔG at -10°C and at 45°C for the ice-to-water phase transition. Which do you predict to be spontaneous? Do your calculations agree with your predictions? (Assume that the enthalpy and entropy are temperature independent.)

$$H_2O(s) \rightarrow H_2O(l) \qquad \Delta H = 6.01 \text{ kJ/mol}$$

$$\Delta S = 22.0 \text{ J/mol·K}$$

See Problem 56.

A Second Way to Calculate ΔG

Just like enthalpy and entropy, free energy is a state function. Remember from Chapter 5 that when we are determining the value of a state function, it doesn't matter how we calculate it because the end result is independent of the path. To illustrate this, we return to the trek up Mt. Everest. Climbers can take either the more popular south route up the Khumbu icefall or the north route along the ridge to the summit. No matter which way the climbers ascend, no matter how many steps there are in each of the two routes, and no matter how long their ascent takes the climbers, if they make it to the top they have climbed to a point 29,035 ft above sea level. As we saw before, altitude change is a state function. It doesn't matter how two climbers arrive at the summit of Mt. Everest; their change in altitude from the base camp to the top is the same.

Because free energy is a state function, we can obtain ΔG by using the same method we used to calculate ΔH. Adding a series of reactions, each with a known free energy, is often a good way to arrive at a particular calculation of the free energy, because many reactions cannot be performed directly in the lab. For example, the combustion of carbon, as occurs in our charcoal grills, can never give just carbon monoxide but, rather, always results in a mixture of oxides. For another example, say we were interested in the calculation of ΔG for the formation of ice from steam at $-18°C$ (255 K) and 1 atm, another reaction that is difficult to do in the lab. We could sum the equations (and the free energy terms) that describe the stepwise conversion from gas to liquid and from liquid to ice to arrive at ΔG for the process. Note that the free energy terms are specific to this temperature:

$$H_2O(g) \rightarrow H_2O(l) \qquad \Delta G = -13.6 \text{ kJ/mol}$$
$$\underline{H_2O(l) \rightarrow H_2O(s) \qquad \Delta G = -0.39 \text{ kJ/mol}}$$
$$H_2O(g) \rightarrow H_2O(s) \qquad \Delta G = -14.0 \text{ kJ/mol}$$

Application

We see that the process of converting water vapor to ice has a negative change in free energy at $-18°C$. What does this value of ΔG tell us, and does it agree with what we expect for this system? Under these conditions (1 atm, $-18°C$), the formation of ice from steam is spontaneous, whereas the converse process, sublimation of solid to the vapor, is not.

The summation procedure is not limited to phase changes. For example, glycogen (a polymer of glucose used as storage for quick energy release) is degraded to glucose-1-phosphate (Glu-1-P) when the body signals that energy is needed. In order for this derivative of glucose to be used in glycolysis, it must first be converted to glucose-6-phosphate (Glu-6-P) by a two-step process. The hydrolysis reaction of each of these phosphates has been extensively studied and the free energy change noted. What is the net free energy change for the direct conversion of glucose-1-phosphate to glucose-6-phosphate? On the basis of our previous discussion, we can calculate ΔG:

$$\text{Glu-1-P} + \cancel{H_2O} \rightarrow \cancel{\text{glucose}} + \cancel{\text{phosphate}} \qquad \Delta G = -21 \text{ kJ/mol}$$
$$\cancel{\text{glucose}} + \cancel{\text{phosphate}} \rightarrow \text{Glu-6-P} + \cancel{H_2O} \qquad \Delta G = +14 \text{ kJ/mol}$$
$$\text{Glu-1-P} \qquad \rightarrow \qquad \text{Glu-6-P} \qquad \Delta G = -7 \text{ kJ/mol}$$

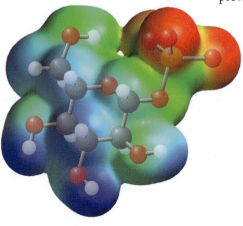

Glu-1-P

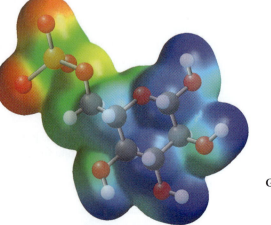

Glu-6-P

The net reaction is spontaneous. The reverse of this reaction, the conversion of glucose-6-phosphate into glucose-1-phosphate, is nonspontaneous with $\Delta G = +7$ kJ/mol. Does the result make sense? If the body needs energy, production of glucose-6-phosphate should be spontaneous.

As you can see, we can use the summation method for determining the spontaneity of a reaction in cases where the free energy of each in a series of reactions is known. In some cases, we can even measure the spontaneity of supposed reactions or processes that we do not wish to perform in the laboratory. The summation of a series of reactions makes possible the quick and easy calculation of the spontaneity of a reaction.

Free Energy of Formation

Video Lesson: Standard Free Energy Changes of Formation

We can also calculate the spontaneity of a reaction using standard free energies of formation ($\Delta_f G°$). The **standard molar free energy of formation ($\Delta_f G°$)** is the change in the free energy of 1 mol of a substance in its standard state as it is made from its constituent elements in their standard states. The equation relating the standard free energy of formation for any compound is analogous to the equation written for the standard enthalpy of formation. For example,

$$H_2(g) + \tfrac{1}{2}O_2(g) \rightarrow H_2O(l)$$

The standard molar free energy of formation, $\Delta_f G°$, for all elements in their standard states is zero, just as it is for the standard molar enthalpy of formation, $\Delta_f H°$. Recall from our previous discussion on the third law of thermodynamics that the standard molar entropy of an element, $S°$, is *not* equal to zero.

The tabulation of $\Delta_f G°$ values is given in Table 14.4 and in the appendix. Access to this and similar tables enables us to determine quickly the free energy change for an unknown reaction. In much the same manner that we use $\Delta S°$ and $\Delta H°$, we can find the change in the standard free energy of a reaction. The sum of the free energies of formation for the reactants is subtracted from the sum of the free energies of formation for the products:

$$\Delta G° = \sum n\Delta_f G°_{products} - \sum n\Delta_f G°_{reactants}$$

where n is the stoichiometric coefficient of each of the compounds in the reaction.

EXERCISE 14.8   **I'm Hungry. What's for Dinner? Free Energy and Spontaneity**

In a fasting organism, there is no more glucose left to make energy. Glycogen supplies become depleted. The organism must find a way to make molecules that can be catabolized for energy. One such source of energy is found in the amino acid alanine. This molecule is converted to pyruvate for use in the production of energy (remember that pyruvate is an intermediate during the complete oxidation of glucose). Given the equation for the production of pyruvic acid, the acidic form of pyruvate, from alanine, as well as the values of the free energy of formation for both, determine the spontaneity of this reaction under standard conditions.

First Thoughts

The existence of $\Delta_f G°$ values for both the starting materials and the products means that we can use the summation equation to calculate the standard free energy change for the reaction.

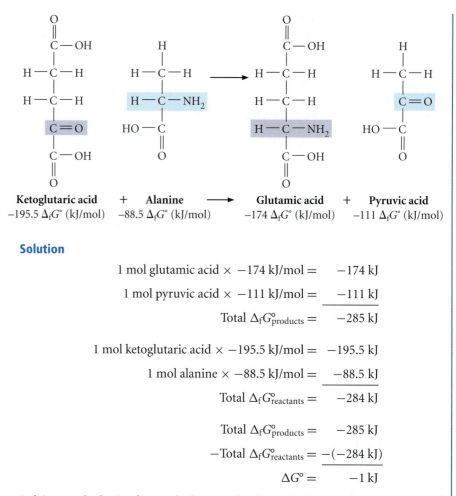

Ketoglutaric acid	+	Alanine	→	Glutamic acid	+	Pyruvic acid
$-195.5 \, \Delta_f G°$ (kJ/mol)		$-88.5 \, \Delta_f G°$ (kJ/mol)		$-174 \, \Delta_f G°$ (kJ/mol)		$-111 \, \Delta_f G°$ (kJ/mol)

Solution

$$1 \text{ mol glutamic acid} \times -174 \text{ kJ/mol} = \quad -174 \text{ kJ}$$

$$1 \text{ mol pyruvic acid} \times -111 \text{ kJ/mol} = \quad -111 \text{ kJ}$$

$$\text{Total } \Delta_f G°_{\text{products}} = \quad -285 \text{ kJ}$$

$$1 \text{ mol ketoglutaric acid} \times -195.5 \text{ kJ/mol} = -195.5 \text{ kJ}$$

$$1 \text{ mol alanine} \times -88.5 \text{ kJ/mol} = \quad -88.5 \text{ kJ}$$

$$\text{Total } \Delta_f G°_{\text{reactants}} = \quad -284 \text{ kJ}$$

$$\text{Total } \Delta_f G°_{\text{products}} = \quad -285 \text{ kJ}$$

$$-\text{Total } \Delta_f G°_{\text{reactants}} = -(-284 \text{ kJ})$$

$$\Delta G° = \quad -1 \text{ kJ}$$

Judging on the basis of our calculations, the change in free energy is negative and the reaction is spontaneous.

Further Insight

Pyruvate is so important to the production of energy that its formation is vital to the survival of an organism. Because of this, reactions that make pyruvate are typically spontaneous. One way to obtain pyruvate is from a source of alanine, as shown here. Where does the organism get alanine? Because alanine is an amino acid, we might predict that it comes from stores of excess alanine. However, there are few stores of this amino acid in the body, so it must come from another source. In particular, alanine can be obtained from the degradation of existing proteins (which contain alanine and other amino acids). Although there are other ways to make pyruvate, the only way to obtain pyruvate when an organism is starved is to consume proteins. *And the consumption of proteins means that muscle tissue is destroyed.* For this and other reasons, starvation diets are not the best way to lose weight.

PRACTICE 14.8

Use the table of thermodynamic values in the appendix to determine whether each of these oxidations is spontaneous. You may need to balance the equations.

$$C_6H_{12}O_6(s) + 6O_2(g) \rightarrow 6CO_2(g) + 6H_2O(g)$$

$$CH_3CH_2CH_3(g) + O_2(g) \rightarrow CO_2(g) + H_2O(g)$$

$$Mg(s) + N_2(g) \rightarrow Mg_3N_2(s)$$

See Problems 59–62.

HERE'S WHAT WE KNOW SO FAR

- Entropy and free energy are state functions.
- The change in entropy and free energy for a reaction can be determined by subtracting the sum of these quantities for the reactants from the sum for the products.
- Negative ΔG values are indicative of spontaneous processes.
- The Gibbs equation ($\Delta G = \Delta H - T\Delta S$) can be used to determine the spontaneity of a process.
- Like enthalpy changes, ΔS and ΔG for a multi-step process are simply the sum of the entropy or free energy changes for each individual process.

14.6 When $\Delta G = 0$; A Taste of Equilibrium

We've already mentioned that the complete oxidation of glucose in the body produces ATP. This molecule is a source of potential energy to force nonspontaneous reactions in the body to become spontaneous. For example, the preparation of glucose-6-phosphate from glucose and phosphate is nonspontaneous. Coupling the hydrolysis of ATP with the production of glucose-6-phosphate helps to make the overall process spontaneous. In this way, the ATP molecule is used to force a reaction to become spontaneous. Although the majority of the ATP in the body comes from the oxidation of glucose, ATP can be made in many other ways. One such way is through the reaction of two molecules of ADP (adenosine diphosphate), which can occur in vigorously active muscles. When the supply of glucose runs low and the level of ATP in the muscle drops, muscle cells gather energy by converting ADP to ATP and AMP. This enables the cells to continue their activity. Let's examine more closely the preparation of one molecule each of ATP and AMP (adenosine monophosphate) from two molecules of ADP. What is the first thing that you note about the reaction as written?

Application

CHEMICAL
ENCOUNTERS:
ATP Formation

$$2\text{ADP} \rightarrow \text{ATP} + \text{AMP} \qquad \Delta G \approx 0 \text{ kJ/mol}$$

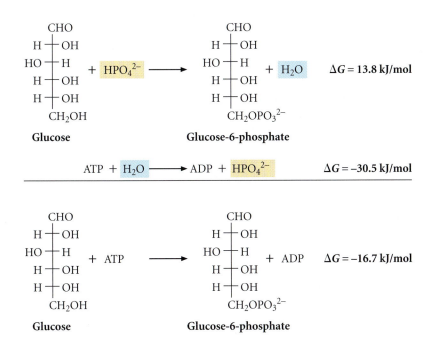

$$\text{ATP} + \text{H}_2\text{O} \longrightarrow \text{ADP} + \text{HPO}_4{}^{2-} \qquad \Delta G = -30.5 \text{ kJ/mol}$$

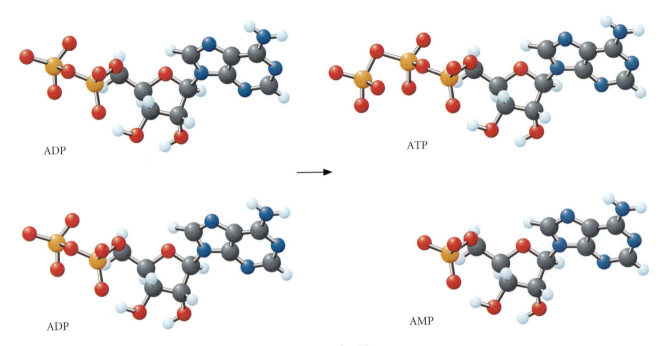

ADP → ATP

ADP   AMP

A phosphate group can be transferred from an ADP molecule to make ATP.

The change in the free energy of the reaction is zero. Is this reaction spontaneous? Is it nonspontaneous? We can look at the phase transition of water from solid to liquid to help sort all this out.

$$H_2O(s) \rightarrow H_2O(l)$$

At temperatures below the melting point of water, the reaction is spontaneous in the direction of $H_2O(s)$; that is, water freezes. At temperatures above the melting point of water, the reaction is spontaneous in the opposite direction; ice melts. At 0°C and 1 atm, the process is not spontaneous in either direction. At this temperature, the change in free energy for the process is zero ($\Delta G = 0$) and the reaction proceeds neither toward the products nor toward the reactants. We say that the reaction has reached **equilibrium**. We will focus on this important concept in Chapter 16, but for now, we need to know that a reaction at equilibrium hasn't stopped; it is just at a point where moving toward either the reactants or toward the products is not thermodynamically favored. The reaction continues to make both products and reactants at equal rates.

Remember that ΔG is *temperature-dependent* ($\Delta G = \Delta H - T\Delta S$), so the value of ΔG can be changed by adjusting the temperature. That is, we can melt ice or freeze water, depending on the temperature we pick. Active muscles do not change their temperature in order to drive the production of ATP from ADP, but this information can be used to calculate the temperature at which a process becomes spontaneous. Back to Mt. Everest. . . . The base camp experiences temperatures ranging from a high of about −3°C in July to a low of about −16°C in January. High up on the slopes of the mountain, the temperature drops even lower, to around −36°C. For trekkers huddled in a tent, it may take a lot of effort to get a propane stove to light (especially if the temperature drops below the boiling point of propane; see Figure 14.22). How cold must it get before the propane gas liquefies? To answer this question, we need to consider the phase transition shown below.

$$CH_3CH_2CH_3(l) \rightarrow CH_3CH_2CH_3(g) \qquad \Delta H = 15.1 \text{ kJ/mol}$$

$$\Delta S = 65.4 \text{ J/mol·K}$$

FIGURE 14.22

Trying to boil water at a camp on the Lhotse Face of Mt. Everest (23,500 ft).

 Application

When this reaction is at equilibrium, neither the forward nor the backward reaction will be spontaneous ($\Delta G = 0$). The temperature to which this corresponds is the boiling point. Because we know the equation that relates ΔG to the temperature, we can calculate the boiling point of this reaction.

$$\Delta G = \Delta H - T(\Delta S)$$

$$0 = 15,100 \ \text{J/mol} - T(65.4 \ \text{J/mol·K})$$

$$T(65.4 \ \text{J/mol·K}) = 15,100 \ \text{J/mol}$$

$$T = 231 \ \text{K}$$

$$T = -42°\text{C}$$

Our calculations suggest that if the temperature falls to $-42°$C ($-44°$F), the change in the free energy of the process will be zero. The reaction will be at equilibrium, and propane will be at its boiling point. If the temperature gets any lower than $-42°$C, the propane will be liquid, and the portable stove will become very difficult to light.

EXERCISE 14.9   **Temperature Dependence of ΔG**

In one of the final reactions to produce copper metal from its ore (chalcocite), copper(I) oxide reacts with carbon to produce gaseous carbon monoxide and copper metal. Above what temperature does the reaction become spontaneous?

$$Cu_2O(s) + C(s) \rightarrow 2Cu(s) + CO(g) \qquad \Delta H = 59 \ \text{kJ/mol}$$
$$\Delta S = 165 \ \text{J/mol·K}$$

Solution

$$\Delta G = \Delta H - T(\Delta S)$$

$$0 = 59,000 \ \text{J/mol} - T(165 \ \text{J/mol·K})$$

$$T(171 \ \text{J/mol·K}) = 59,000 \ \text{J/mol}$$

$$T = 358 \ \text{K}$$

$$T = 85°\text{C}$$

Above 85°C, this reaction becomes spontaneous ($\Delta G < 0$). So, at room temperature, the reaction doesn't proceed without outside assistance. In practice, this reaction is often run at very high temperatures to produce copper metal. The copper can be further purified by electrolytic deposition (see Chapter 19).

PRACTICE 14.9

What is the boiling point of methanol? . . . of water?

$$CH_3OH(l) \rightarrow CH_3OH(g) \qquad\qquad H_2O(l) \rightarrow H_2O(g)$$
$$\Delta_{vap}H = 35.27 \ \text{kJ/mol} \qquad\qquad \Delta_{vap}H = 40.66 \ \text{kJ/mol}$$
$$\Delta_{vap}S = 104.6 \ \text{J/mol·K} \qquad\qquad \Delta_{vap}S = 109.0 \ \text{J/mol·K}$$

See Problems 67 and 68.

Changes in Pressure Affect Spontaneity

As the temperature changes, the value of ΔG for a reaction changes. In fact, a nonspontaneous reaction can often be made spontaneous if the temperature is adjusted enough, such that $\Delta H - T\Delta S$ becomes negative (see Table 14.5). For a muscle cell in need of a quick energy fix, changing the temperature in order to make the conversion of ADP into ATP and AMP (adenosine monophosphate) spontaneous is not an option. Thankfully, temperature isn't the only variable that can change the spontaneity of a reaction. Pressure and concentration also play a major role in determining reaction spontaneity.

The entropy change of a reaction depends on its pressure. This occurs because there are fewer microstates in a compressed gas than there are if the pressure of the same sample is reduced by expanding the volume of the container. Mathematically, it has been shown that the pressure causes a change in the standard free energy of a reaction in accordance with the following equation:

$$\Delta G = \Delta G° + RT \ln Q_p$$

where ΔG is the change in the nonstandard state free energy

$\Delta G°$ is change in the standard state free energy

R is the universal gas constant, 8.3145 J/mol·K

T is the temperature in kelvins

Q_p is a term called the **pressure reaction quotient**

We'll discuss the pressure reaction quotient in much greater detail in Chapter 16. However, for purposes of our current discussion, you may consider the pressure reaction quotient, Q_p, to be *the ratio of the pressures of all of the gaseous products to the pressures of the gaseous reactants, raised to their respective stoichiometric coefficients.* For example, here is the equation for the Haber process for the formation of ammonia (part 2 of Exercise 14.5), along with the expression for its Q_p:

$$3H_2(g) + N_2(g) \rightarrow 2NH_3(g)$$

$$Q_p = \frac{P_{NH_3}^2}{P_{N_2} P_{H_2}^3}$$

For a reaction such as the fermentation of glucose, Q_p is equal to the pressure of carbon dioxide squared, because neither glucose nor ethanol is represented as a gas in the equation.

$$C_6H_{12}O_6(s) \rightarrow 2C_2H_6O(l) + 2CO_2(g)$$

Glucose   Ethanol

$$Q_p = \frac{P_{CO_2}^2}{1}$$

 Application

The brewmaster at a local brewpub knows about the effect of pressure on the fermentation of glucose. If the vat is left open to the atmosphere (1 atm), the yeast carries out the fermentation at 25°C. The reaction, as we saw earlier in this chapter, is spontaneous (−229 kJ/mol). If the brewmeister shuts the hatch on the vat and seals it, the formation of carbon dioxide begins to build up pressure as the glucose is catabolized. At some point, the pressure could reach 52.0 atm (assuming the vat is strong enough to hold that pressure). Is the reaction spontaneous at that pressure?

$$\Delta G = \Delta G° + RT \ln Q_p$$

$$\Delta G = -229{,}000 \text{ J/mol} + (8.3145 \text{ J/mol·K})(298 \text{ K}) \ln\left(\frac{(52.0)^2}{1}\right)$$

$$\Delta G = -229{,}000 \text{ J/mol} + 19{,}580 \text{ J/mol}$$

$$\Delta G = -209{,}420 \text{ J/mol}$$

$$\Delta G = -209 \text{ kJ/mol}$$

The reaction is still spontaneous, as you might expect for a reaction that produces energy for a living organism. However, the reaction has a lower free energy than it had at lower pressure.

Changes in Concentrations Affect Spontaneity

In a manner similar to that for gaseous reactions, the free energy of a reaction changes as the concentrations of the reactants and products change. **Does this make sense?** The equation relating the concentrations of reactants and products to the observed free energy change (ΔG) is

$$\Delta G = \Delta G° + RT\ln Q$$

Note that this equation is mathematically similar to the equation for determining the effect of changing pressure.

The greatest difference between this equation and the one presented in the previous section is the value of the **reaction quotient (Q)**. For reactions in *solution*, this reaction quotient is calculated by multiplying the initial molar concentrations of all of the products, raised to the power of their respective stoichiometric coefficients, divided by the initial molar concentrations of all of the reactants, raised to the power of their stoichiometric coefficients. Mathematically, for the reaction

$$rA + sB \rightarrow tC + uD$$

the reaction quotient is

$$Q = \frac{[C]_0^t[D]_0^u}{[A]_0^r[B]_0^s}$$

where $[\ ]_0$ = the initial molar concentration of the substance.

What does this mean? Reactions with large concentrations of reactants have a Q value less than 1. This means that $RT\ln Q$ is negative and that the observed free energy (ΔG) is less than the standard free energy ($\Delta G°$). In other words, large concentrations of reactants and small concentrations of products increase the spontaneity of the forward reaction (this situation decreases the value of the observed ΔG). When the concentration of reactants is small and the concentration of products is large ($Q > 1.0$), the value of $RT\ln Q$ is greater than zero and the observed ΔG is more positive than $\Delta G°$. In this case, the reaction is less spontaneous. Again, we will examine a similar relationship in much greater detail in Chapter 16 and will explore more deeply what happens when $\Delta G = \Delta G°$, the equilibrium condition.

Coupled Reactions

Many reactions, such as the formation of glucose-6-phosphate from glucose, are nonspontaneous. Metabolic sequences have developed to account for this. As we've seen in this chapter, biological reactions are often coupled with other reactions to produce a spontaneous process. Organisms typically couple a reaction that has a very large negative ΔG with reactions that have positive free energy changes, as shown in Figure 14.23. The result is a spontaneous reaction.

In glycolysis, there are four coupled reactions. Two of these reactions use the negative free energy change of ATP → ADP ($\Delta G = -30.5$ kJ/mol) to place phosphate groups on a glucose molecule before it is broken down. Then, as the resulting carbohydrate is catabolized, the organism uses the stored energy of the carbohydrate to drive the reverse of this reaction (ADP → ATP; $\Delta G = +30.5$ kJ/mol). In doing so, the glycolytic pathway consumes two molecules of ATP but produces four molecules of ATP using the stored energy of the carbohydrate. The net result is the accumulation of two molecules of ATP for each glucose molecule.

In this chapter we have examined why chemical reactions proceed and have shown that it is directly related to an increase in the entropy of the universe. We

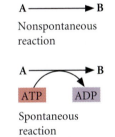

Nonspontaneous reaction

Spontaneous reaction

FIGURE 14.23

Biochemical reactions typically involve the coupling of a spontaneous reaction with a nonspontaneous reaction. Coupling the reactions makes the nonspontaneous reaction spontaneous.

 Application

Elevators are raised by a motor. To facilitate the motion of the elevator, a counterweight helps pull the elevator up. The motor and the counterweight couple their action to produce a working elevator.

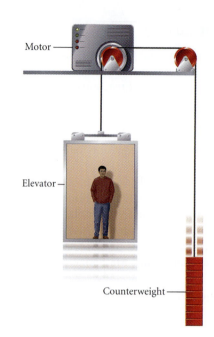

Motor

Elevator

Counterweight

routinely consider the spontaneity of a reaction as a function of the values of ΔH, ΔS, and reaction conditions such as temperature, pressure, and concentration. We've noted that the free energy change (ΔG) depends on all of these things.

Although we now know *whether* a reaction will proceed, we haven't discussed *how fast* reactions proceed. We know that the production of energy from glucose is a spontaneous process, as is the rusting of a nail. But the rates of these two reactions are quite different, with important effects on us and our surroundings. In the next chapter, we'll discover that the degree of spontaneity implies nothing about the speed of a reaction.

The Bottom Line

- The multiplicity of a system can be used to determine the behavior of the system. The most probable macrostate is the one that has the most contributing microstates. (Section 14.1)

- Spontaneous processes occur without outside assistance. The reverse of the spontaneous process is nonspontaneous. (Section 14.2)

- Entropy is a measure of how energy and matter can be distributed in a chemical system. Entropy is *not* disorder. (Section 14.2)

- The second law of thermodynamics says that the entropy of the universe (as we defined the term *universe* in Chapter 5) continues to increase. Any process that is spontaneous must correspond to an increase in the entropy of the universe ($\Delta S_{universe} > 0$). (Section 14.2)

- The third law of thermodynamics says that the entropy of a pure perfect crystalline material is zero. This law enables us to calculate the entropy of any compound in any state. (Section 14.4)

- The free energy, ΔG, can be calculated via the Gibbs equation ($\Delta G = \Delta H - T\Delta S$) to determine the spontaneity of a process. (Section 14.5)

- When $\Delta G = 0$, the system is at equilibrium. The forward and reverse reactions still proceed, but their rates are equal. (Section 14.6)

- Coupled reactions are used by the body to make nonspontaneous processes become spontaneous. The use of ATP in this manner assists in the overall production of energy from glucose. (Section 14.6)

Key Words

bioenergetics The study of the energy changes that occur within a living cell. (*p. 580*)

catabolism The biological degradation of molecules to provide an organism with smaller molecules and energy. (*p. 587*)

entropy A measure of how the energy and matter of a system are distributed throughout the system. (*p. 587*)

equilibrium The state of a reaction when $\Delta G = 0$. The reaction hasn't stopped; rather, the rates of the forward and reverse reactions are equal. (*p. 608*)

free energy (G) The state function that is equal to the enthalpy minus the temperature multiplied by the entropy. Used to determine the spontaneity of a process. (*p. 601*)

Gibbs equation $\Delta G = \Delta H - T\Delta S$ (*p. 601*)

glycolysis A series of biochemical reactions that convert glucose into two molecules of pyruvate. The process results in the formation of two molecules of ATP. (*p. 585*)

macrostate The macroscopic state of a system that indicates the properties of the entire system. (*p. 580*)

metabolism The biochemical reactions of an organism. (*p. 595*)

microstate The state of the individual components within a system. (*p. 581*)

multiplicity The number of microstates within a system. (*p. 587*)

nonspontaneous process A process that occurs only with continuous outside intervention. (*p. 585*)

pressure reaction quotient (Q_p) The ratio of the pressures of all of the gaseous products raised to their respective stoichiometric coefficients, divided by the pressures of all of the gaseous reactants raised to their stoichiometric coefficients. (*p. 610*)

reaction quotient (Q) The ratio of the initial molar concentrations of all of the products raised to the power of their respective stoichiometric coefficients, divided by the initial molar concentrations of all of the reactants raised to the power of their stoichiometric coefficients. (*p. 611*)

reversible conditions The conditions that occur when a process proceeds in a series of infinitesimally small steps. (*p. 593*)

second law of thermodynamics A spontaneous process is accompanied by an increase in the entropy of the universe; $\Delta S_{universe} > 0$. (*p. 588*)

spontaneous process A process that occurs without continuous outside intervention. (*p. 585*)

standard molar free energy of formation ($\Delta_f G°$) The free energy of a compound formed from its elements in their standard states. (*p. 605*)

thermodynamics The study of the changes in energy in a reaction. (*p. 586*)

third law of thermodynamics The entropy of a pure perfect crystal at 0 K is zero. (*p. 598*)

Focus Your Learning

The answers to the odd-numbered problems and some selected problems appear at the back of the book, as represented by the blue numbering.

Section 14.1
Probability as a Predictor of Chemical Behavior

Skill Review

1. Suppose that four pennies fall from your pocket to the floor. How many microstates for the four pennies would exist? (Consider only heads up versus tails up in your explanation.)

2. What is the probability that all four coins from Problem 1 would land heads up?

3. What is the probability that the four coins from Problem 1 would land two heads up and two tails up?

4. What is the most probable outcome for the four coins in Problem 1?

5. Contrast the meanings of the terms *macrostate* and *microstate*.

6. Define the term *multiplicity*.

7. One hundred chemistry students are about to take an exam. There are two equal 200-seat classrooms for the students. Knowing that students must have vacant seats next to them during the exam, describe at least two macrostate arrangements you could predict for the two rooms.

8. In the situation defined in Problem 7, how could the positions of the students allow you to use microstates to explain your answer?

Chemical Applications and Practice

9. Even a sample of only 1,000 atoms is too small to weigh accurately, and yet we can calculate a probable distribution of their numbers between two equal containers. How many microstates are possible if 1,000 atoms of He become distributed between two containers of equal size? Explain why an equal distribution becomes more probable as the number of He atoms increases.

10. A certain lecture hall is 25 m long, 12 m wide, and 15 m tall. The oxygen molecules necessary to sustain life in the room are, of course, free to circulate throughout the room. Explain, using probability and microstates, why a student sitting in the back should not be concerned that all the freely moving oxygen molecules might concentrate at the front row of the classroom.

11. Some powdered creamer is added to a cup of hot coffee. Compare the initial macrostate to the final macrostate as the creamer dissolves. Has the number of microstates for the creamer and coffee increased from the initial state to the final state? Explain the evidence you would use to explain your answer.

12. A glass of tea is cooled by placing ice cubes in it. However, after a few minutes, the ice has melted. Compare the initial macrostate to the final macrostate. Has the number of microstates for the ice and tea increased from the initial state to the final state? Explain the evidence you would use to explain your answer.

13. Which of these would you classify as increasing the number of microstates?
    a. Water in a glass evaporates.
    b. A precipitate of AgCl forms from the addition of a drop of a 0.10 M solution of AgNO$_3$ into a glass of chlorinated tap water.
    c. An icicle forms from a downspout.

14. Which of these would you classify as increasing the number of microstates?
    a. A chunk of ice melts when placed in a glass of tap water at room temperature.
    b. A glass of tap water freezes when placed in a refrigerator freezer compartment.
    c. The combustion of gasoline produces carbon dioxide and water vapor.

Section 14.2 Why Do Chemical Reactions Happen? Entropy and the Second Law of Thermodynamics

Skill Review

15. Which of these, if any, need outside intervention to take place? Describe the necessary intervention.
    a. Combustion (combination with oxygen) of a piece of paper in the air
    b. Making a hard-boiled egg
    c. Increasing muscle mass
    d. Formation of calcium carbonate (bathtub ring) from evaporating hard water

16. Which of these, if any, need outside intervention to take place? Describe the necessary intervention.
    a. The melting of butter on a warm pancake
    b. Opening this book to this page
    c. Paddling a canoe upstream
    d. Cooking eggs for breakfast

17. What is the relationship between entropy and a nonspontaneous process?

18. What is the relationship between entropy and rate of change?

19. What is the relationship between entropy and number of microstates?

20. What is the relationship between a nonspontaneous process and the number of microstates?

21. Suppose you were studying in a residence hall. Down the hall someone is preparing popcorn. The irresistible aroma eventually reaches your room. To take your mind off this, explain how the second law of thermodynamics enables you to detect this tasty temptation from so far away.

22. A scientist on the planet Zoltan expresses the third law of thermodynamics by saying that the entropy of any compound under standard state conditions is zero. Would this change the ΔS values for the reaction of pyruvate to lactate mentioned in the chapter? Explain.

Chemical Application and Practice

23. While pumping gasoline into your car, you might spill some gasoline on your hands. However, the gasoline quickly evaporates with a cooling sensation on your skin. How can this process be spontaneous if it requires the input of the energy from your skin?

24. In a previous problem you were asked about the freezing of water in a refrigerator. Water poured in a tray will spontaneously freeze in a freezer. Explain why this process arranges the water molecules in a more orderly crystal form and yet is still spontaneous.

25. Solid carbon dioxide is called dry ice. At room temperature, this material spontaneously changes from an orderly solid to a gas, bypassing the liquid phase entirely. When undergoing this change, the molecules absorb heat from the nearby surroundings. This, in turn, slows down the molecules of the surroundings. What can be said about the change in entropy for the carbon dioxide molecules relative to nearby air molecules? What can be said about the change in the entropy of the universe for this process?

Dry ice.

26. The entropy of molecules can also depend on their relative atomic motions. The atoms in molecules can twist, rotate, and vibrate around a chemical bond. For which molecule would you be able to predict more microstates: a molecule of nitrogen or a molecule of ammonia (NH$_3$)? Explain the basis of your choice.

27. The following equation illustrates a gaseous reaction that is possible in the atmosphere. Which direction in the reaction shows the greater entropy for the system? Explain your choice.

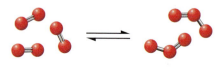

28. Another reaction common in the lower atmosphere is the formation of ozone from NO_2. Which direction in the reaction shows the greater entropy for the system? Explain your choice.

$$O_2(g) + NO_2(g) \rightleftharpoons O_3(g) + NO(g)$$

Section 14.3 Temperature and Spontaneous Processes

Skill Review

29. Classify each of these as representing either a positive change in entropy for the universe or a negative change in entropy for the universe, or indicate that you would need more information to determine this.
    a. $\Delta S_{sys} = +$; $\Delta S_{surr} = +$
    b. $\Delta S_{sys} = +$; $\Delta S_{surr} = -$
    c. $\Delta S_{sys} = -$; $\Delta S_{surr} = +$

30. Classify each of these as representing either a positive change in entropy for the universe or a negative change in entropy for the universe, or indicate that you would need more information to determine this.
    a. $\Delta S_{sys} = -$; $\Delta S_{surr} = -$
    b. $\Delta S_{sys} = +$; $\Delta S_{surr} = 0$
    c. $\Delta S_{sys} = -$; $\Delta S_{surr} = 0$

31. How does the multiplicity of a system change as the molecular motions increase in the system?

32. How does the molecular motion of the surroundings change as energy from the surroundings flows into a system?

33. How does the entropy of the surroundings change as energy flows out of a system?

34. Can we predict the change in entropy of the universe as energy flows into the system? Explain.

Chemical Applications and Practice

35. Mercury remains a liquid over a fairly wide temperature range, which is one reason why it has been used in thermometers. Classify each of these changes as a positive or a negative change in the entropy of a sample of mercury.
    a. Liquid mercury changes to a solid.
    b. Liquid mercury expands as it is heated.
    c. A small amount of mercury evaporates from the surface of a column of mercury in a sealed thermometer.

36. Chlorine gas (Cl_2) is a very toxic compound that can form as a result of the improper use of bleach and ammonia cleaning solutions. Classify each of these changes as a positive or a negative change in the entropy of a sample of this green gas.
    a. Chlorine gas condenses to a liquid.
    b. Chlorine gas is heated in a container.
    c. Chlorine gas deposits on a cold surface.

37. If you are enjoying an iced carbonated soft drink while studying, you may have noticed that the carbon dioxide used to carbonate the beverage bubbles slowly out of the beverage as it warms. Is this a positive or a negative entropy change for the CO_2? Explain.

38. In the bubbling of carbon dioxide mentioned in Problem 37, did the surroundings gain or lose heat? Is the sign for $\Delta S_{universe}$, at this temperature, positive or negative?

Section 14.4
Calculating Entropy Changes in Chemical Reactions

Skill Review

39. Propane stoves make use of the following combustion reaction:

    ___ $C_3H_8(g) +$ ___ $O_2(g) \rightarrow$ ___ $CO_2(g) +$ ___ $H_2O(g)$

    Balance the equation, report the number of moles of gas molecules on each side of the equation, and predict the change in entropy of the system.

40. Methane can be combusted by the following reaction:

    ___ $CH_4(g) +$ ___ $O_2(g) \rightarrow$ ___ $CO_2(g) +$ ___ $H_2O(g)$

    Balance the equation, report the number of moles of gas molecules on each side of the equation, and predict the change in entropy of the system.

41. Repeat the prediction for Problem 39, but assume that the reaction is performed under temperatures that cause the propane and water to be liquid, yet the oxygen and carbon dioxide remain as gases.

42. Repeat the prediction for Problem 40, but assume that the reaction is performed under temperatures that cause the methane and water to be liquid, yet the oxygen and carbon dioxide remain as gases.

43. Explain why the change in the number of moles of gases, from reactant to product, is typically more influential in determining the sign on entropy changes than the change in moles of liquids and solids.

44. The third law of thermodynamics specifies a particular condition for a perfect crystal. Why must the temperature be 0 K instead of 0°C?

45. Predict the sign on the entropy change for each of these reactions.
    a. $NH_3(g) + HCl(g) \rightarrow NH_4Cl(s)$
    b. $2HgO(s) \rightarrow 2Hg(l) + O_2(g)$
    c. $Cd(s) + \frac{1}{2}O_2(g) \rightarrow CdO(s)$

46. Using the values for ΔS^o found in the appendix, determine the actual values for the standard change in entropy for the reactions in Problem 45.

47. Predict the sign on the entropy change for each of these reactions.
    a. $2SO_2(g) + O_2(g) \rightarrow 2SO_3(g)$
    b. $2NH_3(g) \rightarrow N_2(g) + 3H_2(g)$
    c. $CO(g) + 2H_2(g) \rightarrow CH_3OH(l)$

48. Using the values for ΔS^o found in the appendix, determine the actual values for the standard change in entropy for the reactions in Problem 47.

Chemical Applications and Practices

49. Several states have implemented regulations for fuel alternatives in automobiles. One such gasoline alternative is ethanol (C_2H_5OH). Balance the following combustion reaction and determine the change in entropy for the reaction.

    ___ $C_2H_5OH(l) +$ ___ $O_2(g) \rightarrow$ ___ $CO_2(g) +$ ___ $H_2O(g)$

50. One important source of salt (NaCl) is the evaporation of ocean water. Calculate the change in the standard molar entropy as the dissolved ions precipitate the solid.

$$Na^+(aq) + Cl^-(aq) \rightarrow NaCl(s)$$

51. Graphite and diamond are both made only of carbon. However, pencils (incorporating the graphite form) sell for a few cents, whereas engagement rings (displaying diamonds) often sell for thousands of dollars. Using the appendix, determine the change in entropy for the conversion of diamond to graphite.

52. Using the appendix to note the standard $\Delta S°$ value for a metal, explain why the value for the metal alone is lower than the value for any of the listed compounds of that metal.

Section 14.5 Free Energy

Skill Review

53. Predict the relative magnitude of the temperature that would result in a spontaneous process in each of these combinations.
    a. $\Delta H = +; \Delta S = +$
    b. $\Delta H = +; \Delta S = -$

54. Predict the relative magnitude of the temperature that would result in a spontaneous process in each of these combinations.
    a. $\Delta H = -; \Delta S = +$
    b. $\Delta H = -; \Delta S = -$

55. Determine the value of ΔH in each of these.
    a. $\Delta G = -24.5$ kJ/mol; $\Delta S = 287$ J/mol·K; $T = 298$ K
    b. $\Delta G = 1.38$ kJ/mol; $\Delta S = 24$ J/mol·K; $T = 298$ K
    c. $\Delta G = 500.0$ kJ/mol; $\Delta S = -6439$ J/mol·K; $T = 325$ K
    d. $\Delta G = -24.5$ kJ/mol; $\Delta S = 187$ J/mol·K; $T = 39.9°C$

56. Determine the value of ΔG in each of these cases.
    a. $\Delta H = -20.5$ kJ/mol; $\Delta S = 259$ J/mol·K; $T = 298$ K
    b. $\Delta H = 350$ kJ/mol; $\Delta S = 73$ J/mol·K; $T = 157$ K
    c. $\Delta H = 299$ kJ/mol; $\Delta S = -639$ J/mol·K; $T = 325$ K
    d. $\Delta H = -4505$ kJ/mol; $\Delta S = 107$ J/mol·K; $T = 19.3°C$

57. Determine the value of ΔS in each of these cases.
    a. $\Delta G = -20.5$ kJ/mol; $\Delta H = 259$ kJ/mol; $T = 298$ K
    b. $\Delta G = 150$ kJ/mol; $\Delta H = -73$ kJ/mol; $T = 157$ K
    c. $\Delta G = 209$ kJ/mol; $\Delta H = -639$ kJ/mol; $T = 35.0$ K
    d. $\Delta G = -4505$ kJ/mol; $\Delta H = 107$ kJ/mol; $T = 135.8°C$

58. Determine the value of T in each of these cases.
    a. $\Delta G = -20.5$ kJ/mol; $\Delta H = 259$ kJ/mol;
       $\Delta S = 260$ J/mol·K
    b. $\Delta G = 150$ kJ/mol; $\Delta H = -73$ kJ/mol;
       $\Delta S = -290$ J/mol·K
    c. $\Delta G = 209$ kJ/mol; $\Delta H = 639$ kJ/mol; $\Delta S = 560$ J/mol·K
    d. $\Delta G = -4505$ kJ/mol; $\Delta H = 107$ kJ/mol;
       $\Delta S = 1160$ J/mol·K

59. Determine the spontaneity of each of these reactions at 298 K.
    a. $NH_3(g) + HCl(g) \rightarrow NH_4Cl(s)$
    b. $2HgO(s) \rightarrow 2Hg(l) + O_2(g)$
    c. $Cd(s) + \frac{1}{2}O_2(g) \rightarrow CdO(s)$

60. Determine the spontaneity of each of these reactions at 298 K.
    a. $2SO_2(g) + O_2(g) \rightarrow 2SO_3(g)$
    b. $2NH_3(g) \rightarrow N_2(g) + 3H_2(g)$
    c. $CO(g) + 2H_2(g) \rightarrow CH_3OH(l)$

Chemical Application and Practices

61. One of the authors of this text formerly drove an automobile that could best be described as blue and rust. Consider that the rust was formed, at least partially, by the reaction illustrated by this equation:

$$3O_2(g) + 4Fe(s) \rightarrow 2Fe_2O_3(s)$$

Assuming standard conditions, what would you calculate as the $\Delta G°$ for this reaction?

62. The impressive white cliffs of Dover in England have stood without significant change from their basic structure of calcium carbonate for eons. One possible reaction for the decomposition of calcium carbonate is

$$CaCO_3(s) \rightarrow CaO(s) + CO_2(g)$$

Use the $\Delta_f G°$ values of these compounds to determine the value of $\Delta G°$ for the reaction. Comment on how thermodynamics helps explain the relative stability of this beautiful landmark.

63. Sulfur dioxide and sulfur trioxide are important anhydrides for making sulfurous and sulfuric acid, respectively. When sulfur-containing coal is burned, sulfur gases pose environmental threats to air quality. Given the following two combustion reactions, determine the $\Delta G°$ value for the change from SO_2 to SO_3.

$$S(s) + \tfrac{3}{2}O_2(g) \rightarrow SO_3(g) \qquad \Delta G° = -371 \text{ kJ/mol}$$
$$S(s) + O_2(g) \rightarrow SO_2(g) \qquad \Delta G° = -300 \text{ kJ/mol}$$
$$SO_2(g) + \tfrac{1}{2}O_2(g) \rightarrow SO_3(g) \quad \Delta G° = ?$$

64. As we note earlier, graphite and diamond are both composed entirely of carbon atoms. Using the following reactions at 298 K, determine the value of $\Delta G°$ for the conversion of graphite into diamond.

$$C_{diamond}(s) + O_2(g) \rightarrow CO_2(g) \quad \Delta G° = -397 \text{ kJ/mol}$$
$$C_{graphite}(s) + O_2(g) \rightarrow CO_2(g) \quad \Delta G° = -394 \text{ kJ/mol}$$
$$C_{graphite}(s) \rightarrow C_{diamond}(s) \qquad \Delta G° = ??$$

65. As we saw in earlier in this chapter, the production of glucose requires the input of energy. This is typically shown as green plants using sunlight to help combine carbon dioxide with water to form glucose through photosynthesis. However, in some deep areas of the ocean, sunlight does not reach bacteria that may grow around thermal vents. These organisms have been shown to produce energy through oxidation of a sulfur compound, H_2S. Determine $\Delta H°$, $\Delta S°$, and $\Delta G°$ for this biochemically important reaction. Assume sulfur is in the standard form—rhombic allotrope.

$$H_2S(aq) + O_2(g) \rightarrow 2H_2O(l) + 2S(s)$$

66. Smog produced over a city is formed from trace amounts of automobile exhaust in a reaction with atmospheric oxygen. One of the steps in the formation of ozone is the oxidation of oxygen gas by nitrogen dioxide (a by-product of the combustion of gasoline). Determine $\Delta H°$, $\Delta S°$, and $\Delta G°$ for this reaction.

$$NO_2(g) + O_2(g) \rightarrow NO(g) + O_3(g)$$

Section 14.6
When ΔG = 0; A Taste of Equilibrium

Skill Review

67. Determine the temperature at which the indicated system reaches equilibrium.
a. $\Delta H° = 46.4$ kJ/mol; $\Delta S° = 27.6$ J/mol·K
b. $\Delta H° = 10.6$ kJ/mol; $\Delta S° = 77$ J/mol·K
c. $\Delta H° = 124$ kJ/mol; $\Delta S° = 295.5$ J/mol·K

68. Determine the temperature at which the indicated system reaches equilibrium.
a. $\Delta H° = 57.6$ kJ/mol; $\Delta S° = 17.4$ J/mol·K
b. $\Delta H° = 94$ kJ/mol; $\Delta S° = 306$ J/mol·K
c. $\Delta H° = 32.1$ kJ/mol; $\Delta S° = 552$ J/mol·K

69. At equilibrium, what is the pressure for the following process at 298 K?

$$CaCO_3(s) \rightleftharpoons CaO(s) + CO_2(g) \quad \Delta G° = 131 \text{ kJ/mol}$$

70. What is the free energy change for the following reaction at 298 K, given that the pressures of NH_3 and N_2 are 1.0 atm each and the pressure of H_2 is 2.0 atm? (*Hint*: Use the appendix to determine the free energy change under standard conditions.)

$$2NH_3(g) \rightleftharpoons N_2(g) + 3H_2(g)$$

Chemical Applications and Practices

71. At 1 atm of pressure the boiling point of pure water is 100.0°C. If 1000.0 g of water were brought to this point and then completely boiled away, what would you calculate as the entropy change for the water? (*Note*: The heat of vaporization for water is 40.7 kJ/mol.)

72. There is a historical approximation known as Trouton's rule. This approximation states that at the normal boiling point for nonpolar compounds, the standard molar entropy of vaporization is 87 J/mol·K. Using Trouton's rule, determine the approximate heat of vaporization of water. The actual molar entropy of vaporization for water is close to 110 J/mol·K. Explain why the Trouton value is so much different for polar substances such as water but is often found to be much closer to the actual value for nonpolar substances.

73. Under proper conditions, phase changes can be classified as reversible processes. The energy involved with melting or freezing of a sample is called the enthalpy of fusion. The enthalpy of fusion ($\Delta_{fus}H$) for the rare radioactive element actinium, $Z = 89$, is 10.50 kJ/mol. If the entropy of fusion is 9.6 J/mol·K, what would you calculate as the melting point for this silvery white metal?

74. One of the reasons why you are able to read this question is that the cellular functions necessary are supplied with energy from the molecule ATP (adenosine triphosphate), which couples with many biochemical reactions. The formation of ATP can be accomplished as follows:

$$ADP(aq) + H_2PO_4^-(aq) \rightleftharpoons ATP(aq) + H_2O(l)$$

$\Delta G°$ for the reaction is 31 kJ/mol. Using that information, what would you estimate for the reaction quotient, Q, at 25°C, for the reaction at equilibrium?

75. The gas commonly used in chemical laboratory burners is methane (CH_4). Shown here is the combustion reaction that takes place when you light your lab burner.

$$CH_4(g) + 2O_2(g) \rightleftharpoons CO_2(g) + 2H_2O(g)$$

a. Use $\Delta_f G°$ data to calculate the $\Delta G°$ value for the reaction.
b. What would the value of ΔG be if the pressure of O_2 were reduced to 0.20 atm from 1.00 atm, at room conditions, with the remaining gases at standard conditions?
c. Explain why, if ΔG has a negative value indicating a spontaneous change, we still have to bring a flame or spark to the system to get it to react in the lab.

76. Nitrogen gas and oxygen gas make up essentially 100% of our air. The following reaction has $\Delta G° = 174$ kJ. Fortunately, the two do not easily combine through this reaction under the Earth's atmospheric conditions.

$$N_2(g) + O_2(g) \rightleftharpoons 2NO(g)$$

a. What is the value of Q_p at 298 K for the reaction at equilibrium?
b. What temperature would be needed for the reaction to be spontaneous if the pressures were returned to 1.0 atm? (Assume that enthalpy and entropy values do not change significantly over this range.)
c. What would be the value of ΔG at 25°C if the pressure of both N_2 and O_2 were increased to 5.0 atm each, while the pressure of NO_2 was decreased to 0.50 atm?

Comprehensive Problems

77. Butane, burned in small portable lighters, combusts according to the following unbalanced equation:

$$\_\_\_ C_4H_{10}(g) + \_\_\_ O_2(g) \rightarrow \_\_\_ CO_2(g) + \_\_\_ H_2O(g)$$

First balance the equation, and then explain your answer to this question: "Does the combustion show an increase or a decrease in entropy for the system?"

78. Suppose a particularly noxious compound was entering a water supply at only one point. If it would stay at the entry point, it might easily be removed. However, this does not happen. Explain why environmental scientists must be aware of the second law of thermodynamics.

79. Because it was already known that a positive change in the entropy of the universe indicated a spontaneous reaction, why was it important to develop the Gibbs equation, which is also used to predict the spontaneity of a reaction? What advantage does the Gibbs equation offer?

80. In the ongoing research to find a cure or treatment for Alzheimer's disease, some investigators have focused on the appearance of plaque-like formations in brain cells. The solid masses form from protein fibers called beta-amyloids. Use Internet resources and journal sources to find out if the process is based on a positive or a negative value for the $\Delta S°$ of the plaque-forming reaction.

81. In calculations involving Gibbs energy and equilibrium, you must always be aware of the differences between ΔG and $\Delta G°$. When $\Delta G = 0$, at 25°C, what is true about Q?

82. A certain reaction is nonspontaneous at one temperature. If raising the temperature caused the reaction to become spontaneous, what could you conclude about the entropy change of the reaction? Explain your answer.

83. In the chapter, we note that metabolism of glucose results initially in the formation of two molecules of pyruvate for every molecule of glucose.
    a. According to Figure 14.8, what is the net production of ATP molecules for every molecule of glucose consumed?
    b. How many grams of pyruvate would result from the consumption of 0.057 g of glucose? (Assuming all of the glucose is converted into pyruvate.)
    c. This process, known as glycolysis, is spontaneous. How can these reactions be spontaneous if they also result in the generation of molecules with high-potential energy?

84. We know from Chapter 2 that about 76% of all chlorine atoms are chlorine-35 and about 24% are chlorine-37. Let's say that you were able to separate the two isotopes and form two samples of Cl_2, one that is pure ^{35}Cl–^{35}Cl and the other, ^{37}Cl–^{37}Cl. You then took each sample and reacted it with hydrogen gas to form HCl.

$$Cl_2 + H_2 \rightarrow 2HCl$$

Based on your understanding of thermodynamics, discuss which reaction—the one with the Cl-35 isotope or the one with the Cl-37 isotope—would have the greater enthalpy of reaction, entropy, and free energy.

85. "Is there life on other worlds?" has been asked for centuries, though we have made the greatest strides in obtaining data via NASA space problems within the past 40 years. How would the biochemical reactions we've studied in this chapter be affected on Venus with its surface temperature of over 400°C? What about the effect on Enceladus, a moon of Saturn which has a surface temperature of –330°C?

Thinking Beyond the Calculation

86. The oxidation of glucose was discussed in detail in this chapter. A similar compound, fructose, also undergoes oxidation in biological organisms to give (eventually) CO_2 and water:

$$\_\_\_ \; C_6H_{12}O_6(s) + \_\_\_ \; O_2(g) \rightarrow \_\_\_ \; CO_2(g) + \_\_\_ \; H_2O(l)$$

    a. Balance the equation and predict the numerical value of the change in entropy for this process.
    b. Use the appendix to calculate the values of $\Delta H°$, $\Delta S°$, and $\Delta G°$ for this reaction. Compare your calculated value for the entropy change to the predicted value. Assume that the thermodynamic values for fructose are equivalent to those for glucose.
    c. If the combustion reaction is done in the laboratory, the water is isolated as vapor. Recalculate the values of $\Delta H°$, $\Delta S°$, and $\Delta G°$ for the reaction where both products are gases.
    d. If 5.0 g of fructose is consumed, how many liters of $CO_2(g)$ would be produced? Where does this CO_2 go in a living organism? (Assume 25°C and 1 atm.)

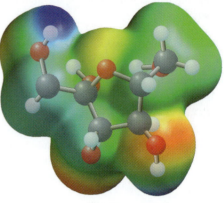

Fructose

    e. How much heat is liberated from the combustion of 5.0 g of fructose in the laboratory?
    f. In the first step of the catabolism of fructose, a phosphate is attached to the sugar unit:

$$C_6H_{12}O_6(aq) + ATP(aq) \rightarrow ADP(aq) + C_6H_{11}O_9P^{2-}(aq)$$
$$\Delta G = -16.7 \text{ kJ/mol}$$

We also know that ATP can be made from ADP and inorganic phosphate in the body:

$$ATP(aq) + H_2O(l) \rightarrow ADP(aq) + HPO_4^{2-}(aq)$$
$$\Delta G = -30.5 \text{ kJ/mol}$$

What is the free energy change for the reaction of fructose with phosphate?

$$C_6H_{12}O_6(aq) + HPO_4^{2-}(aq) \rightarrow C_6H_{11}O_9P^{2-}(aq) + H_2O(l)$$
$$\Delta G = ?$$

    g. Why would an organism have a need to utilize fructose in the same manner as glucose? What is a typical source of fructose?

Chemical Kinetics

Contents and Selected Applications

Chemical Encounters: Atrazine and the Environment

15.1 Reaction Rates

15.2 An Introduction to Rate Laws

15.3 Changes in Time—The Integrated Rate Law

Chemical Encounters: Decomposition of DDT

Chemical Encounters: Persistent Pesticides

15.4 Methods of Determining Rate Laws

15.5 Looking Back at Rate Laws

15.6 Reaction Mechanisms

Chemical Encounters: Metabolism of Methoxychlor

15.7 Applications of Catalysts

Chemical Encounters: Destruction of Ozone

Aerial spraying. No, this plane isn't parked in that field. It's flying inches above it, spraying to control pests and weeds. Pesticides and herbicides enhance crop production, result in greener lawns, and eliminate nasty pests from our homes. But once they're added to the environment, how long do they stick around?

Go to **college.hmco.com/pic/kelterMEE** for online learning resources.

In 2004, American farmers continued a decades-long increase in food production, harvesting over 700 million metric tons of wheat, corn, rice, and other grains. According to the U.S. Department of Agriculture, corn production in Iowa alone more than quadrupled, from 40 bushels per acre to 180 bushels per acre, in the 75 years between 1930 and 2004. Although the introduction of the tractor and other automated farm machinery has played a large role in this increase in production, the use of insecticides and herbicides (compounds used to kill unwanted plants) has had a substantial impact. No longer do insects and weeds run rampant through cornfields and destroy crop yields. Atrazine, a herbicide, is one of the agents most commonly sprayed onto the soil from which corn crops grow in order to control weeds; approximately 15 million pounds are used annually in Nebraska alone.

Application

CHEMICAL
ENCOUNTERS:
Atrazine and the
Environment

When it is introduced into a farmer's field, atrazine works well to control broadleaf weeds, such as pigweed, cocklebur, velvetleaf, and certain grass weeds, without harming the corn plants. What happens to the atrazine that doesn't land on weeds? Some of it travels into the soil, where microbes and water can degrade it into by-products. This degradation process has been studied extensively by scientists. One particular reaction, the hydrolysis of atrazine, in which the chlorine atom on atrazine is replaced with a hydroxy (—OH) group, is of particular interest to researchers. The product, hydroxyatrazine, is rapidly metabolized by microbes living in the soil and groundwater and is viewed by the U.S. Environmental Protection Agency as not harmful to humans. The length of time it takes atrazine to metabolize is largely determined by this initial hydrolysis reaction.

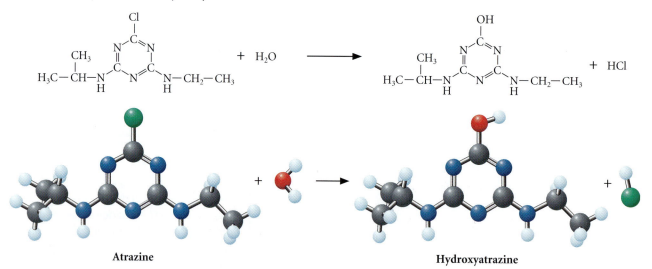

Atrazine

Hydroxyatrazine

Environmental chemists study the interaction of compounds and the environment, including the chemistry of the soil, water, and air. Their work proves that pesticides (a pesticide is any compound used to kill unwanted organisms) do not always rapidly disappear from the environment. For example, their analyses of lakes, rivers, and streams in states that use atrazine show that it persists in the environment for quite a long time. Depending on certain environmental and biological factors (including soil depth, temperature, and the presence of microorganisms, especially fungi), the

620

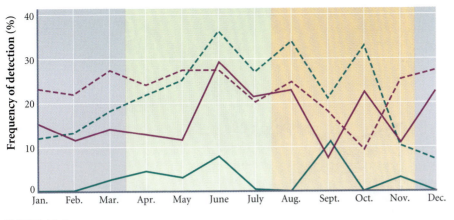

Little River—Agricultural basin
- - - - Herbicides
———— Pesticides

Lafayette Creek—Urban basin
- - - - Herbicides
———— Pesticides

The frequency of pesticide detections in a stream draining an agricultural basin was related to the agricultural cycle.

FIGURE 15.1

Atrazine and other herbicides persist in the environment long after they are first applied.

concentration of atrazine in *soil* can decline by half in less than a month, though about 60 days is typical. However, in natural *water* samples, atrazine typically degrades to one-half of its original concentration more slowly, in about 400 days. Because its degradation is relatively slow, atrazine is present in many water samples throughout the year, as shown in Figure 15.1. Laboratory studies, however, indicate that the hydrolysis of atrazine is spontaneous ($\Delta G < 0$ kJ/mol) and exothermic ($\Delta H = -35$ kJ/mol). **Does it make sense for a process to be so slow, compared to (for example) the combustion of propane in a propane torch, even if it is spontaneous?** As we noted in Chapter 14, the answer is a resounding yes. Values for free energy, enthalpy, and entropy are useful only in determining the thermodynamic properties of a reaction (that is, whether the reaction is spontaneous). The spontaneity of a reaction does not indicate anything about its rate. In order for us to determine how quickly the reaction occurs, we have to examine **chemical kinetics**—the study of the rates and mechanisms of chemical reactions, including the factors that influence these properties. To begin our study of chemical kinetics, we will briefly leave the farm fields and travel to the Olympic Games.

15.1   Reaction Rates

The Olympic Games rely heavily on the use of accurate timekeeping. In the bobsled and luge events, such as that shown in Figure 15.2, the accuracy of the timekeeping determines whether an athlete or team wins a gold or no medal at all. For example, at the 1998 Winter Olympic Games in Nagano, Japan, Silke Kraushaar of Germany placed first in the women's luge with a time of 50.617 s for her best run. The silver medal went to Barbara Niedernhuber, also from Germany, with a time of 50.625 s. The difference between these times is very small.

Let's examine the luge event more closely to help us introduce some new terms related to kinetics, the main topic of this chapter. The women's luge course at Nagano in 1998 was 1194 m long. Judging on the basis of her winning time, how fast did Silke Kraushaar travel? Speed is calculated by dividing the distance traveled by the change in time. Kraushaar traveled 1194 m in 50.617 s, so her **rate** of travel (speed) was 84.92 km/h (52.77 mi/h).

$$\text{Speed} = \frac{\text{distance}}{\text{time}} = \frac{1194 \text{ m}}{50.617 \text{ s}} \times \frac{1 \text{ km}}{1000 \text{ m}} \times \frac{3600 \text{ s}}{\text{h}} = \frac{84.92 \text{ km}}{\text{h}}$$

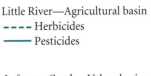 Application

 Tutorial: Reaction Rate and Concentration

FIGURE 15.2

In many Olympic events, fractions of a second determine the winners. Here, Armin Zoeggeler of Italy races at top speed on the luge track in the 2002 Winter Olympics in Salt Lake City, Utah.

Atrazine

But *top* speeds for luge and bobsled events routinely hit 135 km/h (84 mi/h)! What we've calculated is the **average rate** of her travel.

Chemical reactions also have rates. Whereas the speed of a bobsledder can be listed in kilometers per hour (or miles per hour), the speed of a reaction is often described in units of concentration, often molarity, per second. For instance, under certain laboratory conditions, a 0.50 *M* solution of the herbicide atrazine can be completely hydrolyzed in 24 h. We can calculate the rate of the reaction by dividing the concentration that is consumed in the reaction by the time it took to complete the hydrolysis.

$$\text{Average rate} = \frac{0.50\ M}{24\ \text{h}} \times \frac{1\ \text{h}}{3600\ \text{s}} = 5.8 \times 10^{-6}\ M/\text{s}$$

We can say that the rate of this reaction for the 24 h period is $5.8 \times 10^{-6}\ M/\text{s}$.

Determining the average rate of a reaction is one of the tasks accomplished in chemical kinetics. However, when we study kinetics, we have to be careful to note the difference between the *extent* of a reaction and the *rate* of the reaction. The extent of a reaction (which we'll cover in Chapter 16) is a measure of the completeness of a reaction. The rate of a reaction describes how quickly it gets to that point. For example, the combustion of methane to make the flame on your cooking stove is an essentially complete reaction that is also relatively fast. The oxidation of iron on a suspension bridge is also complete, but it is quite slow. *Kinetics deals only with "how fast or slow and by what route." Kinetics tells us nothing about the extent of a reaction.*

Instantaneous Rate, Initial Rate, and Average Rate

At the 2001 Grand Nationals in Chicago, Whit Bazemore in his Matco Tools Pontiac Firebird "funny car" finished the quarter-mile drag strip in 4.750 s (Figure 15.3). Using this information (0.2500 mi in 4.750 s), we can calculate the average speed of the funny car as 0.05263 mi/s, or 189.5 mi/h. However, the speed of the funny car at the start was much less than this (0 mi/h), and the speed at the finish line was a lot faster (323.3 mi/h) than the average speed. The rate of a chemical reaction changes throughout, much like the rate of travel of a funny car at a drag race. However, chemical reactions differ because they typically start rapidly and then slow down with time. In other words, at the start of a reaction, the rate of chemical change is typically fast. As the reaction nears its end, the rate is typically slow. Over a given time period (Δt), though, it has an average rate that describes how long it took to reach that certain point in the reaction.

$$\text{Average rate} = \frac{\Delta \text{concentration}}{\Delta t}$$

In short, just as the speed of a dragster changes during a race, *the rate of the reaction changes as the reaction proceeds.*

What if we consider a change in time that is negligible ($\Delta t \approx 0$)? When this happens, we are examining the **instantaneous rate** of a reaction. This is what we determine at the finish line in the funny car race (323.3 mi/h). It is also what we measure as the lights turn green at the start of the race (0 mi/h). At these points, and at any other point we pick, we are measuring the instantaneous rate. In a chemical reaction, the rate at the start of a reaction, when the reactant concentrations are greatest, is important. The instantaneous rate of the reaction measured at the start is referred to as the **initial rate** of reaction. Instantaneous rates can be measured if we have a plot of the reaction such as that shown in Figure 15.4. If, on the plot of our reaction, we draw a line that is tangent to the curve (that is, a line that just touches the curve and is going in the same direction *at that point*), as is done in Figure 15.5, and we measure the slope of that line, we get the instantaneous rate of the reaction at that specific time. This is a useful way to measure instantaneous rates.

Environmental chemists often measure the rate of a reaction. For example, the environmental hydrolysis of alachlor (also known as Lasso™, a common herbicide) occurs more rapidly than the hydrolysis of atrazine. However, the rate is still slow enough that new ways to decrease the time that the herbicide spends in the environment before chemically degrading are being sought. One such way includes the reduction of alachlor to an acetanilide via electrochemical techniques such as those that we will discuss in Chapter 19.

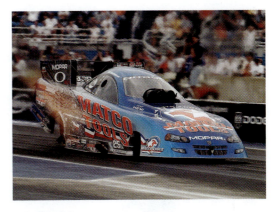

FIGURE 15.3

"Funny cars," with their characteristically powerful engines and oversized rear wheels, get ready to race the quarter-mile. Bazemore powered his Matco Tools Pontiac Firebird to a national record time of 4.750 s at a track record speed of 323.27 mph to lead the 16-car field.

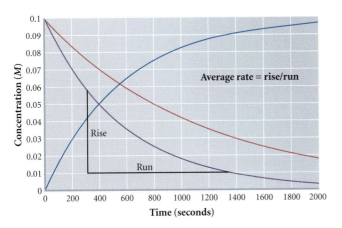

FIGURE 15.4

Calculating the average rate of a reaction from a sample plot of the reactant concentration versus time. Average rates and instantaneous rates differ in the length of time used to calculate them. The average rate of a reaction is calculated by dividing the rise by the run.

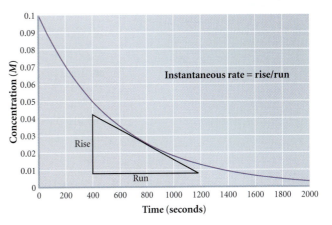

FIGURE 15.5

Calculating the instantaneous rate of a reaction from the same sample plot used to calculate the average rate in Figure 15.5. The slope of a tangent line can be used to find an instantaneous rate.

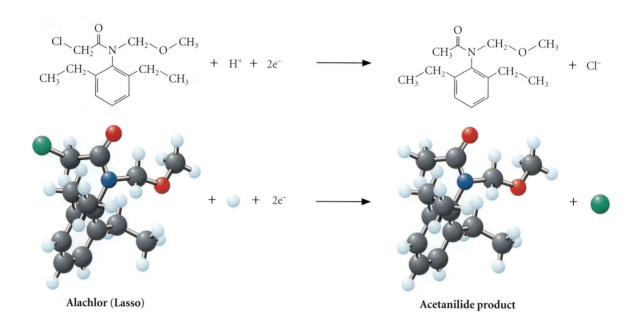

Alachlor (Lasso)　　　　　　　　　　　**Acetanilide product**

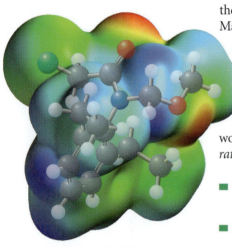

Alachlor

The rate of this reaction can be determined in the laboratory by measuring the change in the concentration of one of the compounds over a period of time. Mathematically, the rate of the reaction is equal to the rate of the disappearance of alachlor, the rate of appearance of the acetanilide, or the rate of the appearance of Cl^-, because they are all in a 1-to-1 mole ratio. The sign placed in the equation is used to indicate whether the compound being measured is disappearing (minus sign) or appearing (plus sign). Remember that the Δ symbol represents a change, measured by taking the final value minus the initial value. Even though the sign of the rate for the disappearance would appear to generate a negative value for the rate, it does not. In fact, *reaction rates are always positive.*

- Rate = rate of disappearance of alachlor = $\dfrac{-\Delta[\text{alachlor}]}{\Delta t}$

- Rate = rate of disappearance of H^+ = $\dfrac{-\Delta[H^+]}{\Delta t}$

- Rate = rate of appearance of product = $\dfrac{\Delta[\text{acetanilide}]}{\Delta t}$

- Rate = rate of appearance of Cl^- = $\dfrac{\Delta[Cl^-]}{\Delta t}$

The rate of the decomposition of hydrogen peroxide into water and oxygen can also be written using these rules. The rate of the reaction could be set equal to the rate of appearance of oxygen. In this case, the rate is simply the change in its

$$2H_2O_2(aq) \rightarrow 2H_2O(l) + O_2(g)$$

concentration divided by the change in time. *However, if we consider the rate of appearance of water as a measure of the rate of the reaction,* we must somehow note that two molecules of water are appearing for every molecule of $O_2(g)$ that is produced. This is done by dividing the rate by the mole ratio from the balanced equation. For instance, the rate of appearance of H_2O times $\dfrac{1 \text{ mol } O_2}{2 \text{ mol } H_2O}$ gives the rate of the reaction.

Our rate descriptions are

- Rate based on disappearance of $H_2O_2(g) = \dfrac{-\Delta[H_2O_2]}{\Delta t} \times \dfrac{1}{2}$

- Rate based on appearance of $O_2(g) = \dfrac{\Delta[O_2]}{\Delta t}$

- Rate based on appearance of $H_2O(l) = \dfrac{\Delta[H_2O]}{\Delta t} \times \dfrac{1}{2}$

EXERCISE 15.1 **Rate of the Haber Process**

Farmers apply ammonia to their cornfields during spring planting by injecting it directly into the soil. The ammonia they use is made by the Haber process; the chemical reaction for this process is indicated by the equation shown below. Describe the rate of the reaction in terms of the rate of appearance or disappearance of the components of the reaction.

$$N_2(g) + 3H_2(g) \rightarrow 2NH_3(g)$$

Solution

The rate of the reaction can be measured by measuring the changes in concentration of each of the species. Mathematically, the rate of the reaction of N_2 is equal to one-third the rate of disappearance of H_2, because 1 mol of N_2 disappears for every 3 mol of H_2 that react. The rate of the reaction of N_2 is also equal to one-half the rate of appearance of ammonia (NH_3), because 1 mol of N_2 disappears for every 2 mol of NH_3 that are produced. Note that the disappearance of H_2 is indicated with a minus sign and the appearance of NH_3 a plus sign.

$$\text{Rate} = -\frac{\Delta[N_2]}{\Delta t} = -\frac{1}{3}\frac{\Delta[H_2]}{\Delta t} = \frac{1}{2}\frac{\Delta[NH_3]}{\Delta t}$$

PRACTICE 15.1

What is the rate of the following reaction in terms of the rates of disappearance and appearance of the components of the reaction?

$$2HI(aq) \rightarrow H_2(g) + I_2(aq)$$

See Problems 5, 6, 9, 10, and 13.

We can examine this further by actually calculating the average rate of the reaction. Assume that an environmental chemist begins an experiment with a 0.400 M alachlor solution. After 10 days, the concentration of alachlor is determined to be 0.350 M, and the concentration of acetanilide and that of chloride are both 0.050 M. What is the average rate of the reaction? We know the final value (0.350 M alachlor at 10 days) and the initial value (0.400 M alachlor at 0 days). Although time can be used with almost any unit, we often report rates as M/s. So, we convert the time into seconds (10 days = 864,000 s) and fill in our equations, keeping in mind that the rate measures the change in concentration per unit change in time. In this example, however, we don't know the concentration of H^+ at either time, so we can't use it to determine the rate of the reaction. Note: It would be perfectly fine to report rates with units of $M/$day, $M/$hr, etc.

- $\text{Rate} = \text{disappearance of alachlor} = \dfrac{-(0.350\ M - 0.400\ M)}{(864{,}000\ s - 0\ s)}$

$$= 5.79 \times 10^{-8}\ M/s$$

The slope of a line that is tangent to the plot of concentration versus time for a reaction gives the instantaneous rate of the reaction. Instantaneous rates can be measured at any specific time.

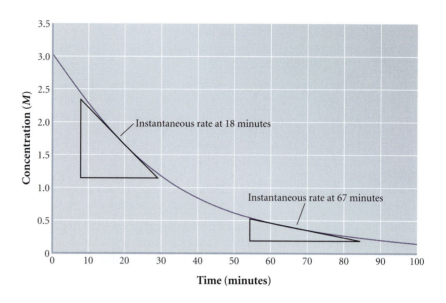

- Rate = appearance of acetanilide = $\dfrac{(0.050\ M - 0.000\ M)}{(864{,}000\ s - 0\ s)} = 5.79 \times 10^{-8}\ M/s$

- Rate = appearance of Cl^- = $\dfrac{(0.050\ M - 0.000\ M)}{(864{,}000\ s - 0\ s)} = 5.79 \times 10^{-8}\ M/s$

Note that no matter which calculation we complete, we should always obtain the same rate of reaction because of the 1-to-1-to-1 mole ratio of reactants and products.

Remember that average rates, instantaneous rates, and initial rates are calculated in the same manner. *The only difference is that the length of the time period is different.* Long changes in time determine the average rate over those long changes. Instantaneous rates are measured when the change in time (Δt) is very small (that is, $\Delta t \rightarrow 0$). Experimentally, average rates, instantaneous rates, and initial rates can be found by measuring the color, temperature, electrochemical voltage (Chapter 19), or some other physical property of the reactants or products that changes over time. Then by determining the slope of a line that is tangent to the curve drawn when the concentrations of reactants or products are plotted against time, the rate can be obtained. Initial rates are measured as instantaneous rates at the start of the reaction.

EXERCISE 15.2 The Rate of a Reaction

Methoxychlor is an important insecticide used to control parasites on livestock and a variety of pests on vegetables and fruits. Its breakdown by soil microbes may proceed through the following reaction. Calculate the average rate of this reaction in M/s of methoxychlor, assuming that the concentration of CH_3OH starts at 0.000 M and after 60.0 h climbs to 0.100 M.

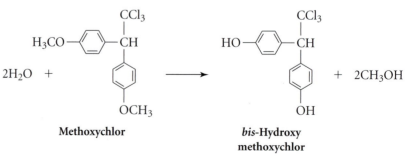

First Thoughts

The equation that describes this reaction includes two starting materials and two types of product molecules. We could use any of these compounds to determine the average rate of the reaction, but the question asks us to work with the formation of methanol to determine the rate. By examining the balanced equation, we see that the rate of methanol formation is two times the rate of the reaction of methoxychlor. That is, 2 mol of methanol are formed for every mole of methoxychlor that is consumed.

Solution

The rate of the reaction can be determined on the basis of the rate of appearance of methanol (CH_3OH). Because methanol is forming at twice the rate of the reaction of methoxychlor, the rate of the reaction of methoxychlor is one-half the rate of appearance of methanol.

$$\text{Rate} = \frac{1}{2}\left(\frac{\Delta[CH_3OH]}{\Delta t}\right) = \frac{1}{2}\left(\frac{(0.100\ M - 0.00\ M)}{(216,000\ s - 0.00\ s)}\right)$$

$$= \frac{1}{2}\left(\frac{0.100\ M}{216,000\ s}\right) = 2.31 \times 10^{-7}\ M/s$$

Further Insights

The rate we have calculated in this problem is an average rate of reaction, which can be determined by examining the length of time that was used to measure the rate over this time period. To measure the initial rate or the instantaneous rate of the reaction, we would need to measure the concentrations of methanol (or another of the compounds in the equation) during a much shorter reaction time. Currently, rates based on time changes on the order of *femtoseconds (10^{-15})* can be determined with laser-based measurement techniques.

PRACTICE 15.2

Calculate the average rate of the Haber process (Exercise 15.1), assuming that at 12.5 s the concentration of $H_2(g)$ was 0.355 M and at 83.3 s the concentration was 0.258 M.

See Problems 7 and 8.

15.2   An Introduction to Rate Laws

Video Lesson: An Introduction to Reaction Rates

We pointed out in the introduction to this chapter that under certain conditions, atrazine can degrade to one-half of its original concentration in 60 days. This means that in a sample containing 5.0×10^{-7} M atrazine, the concentration reduces to 2.5×10^{-7} M after 60 days. We can calculate the rate of this reaction using the concepts we learned in Section 15.1. **Can we predict the concentration of atrazine after 120 days?** This question highlights a complicating factor in kinetics: *The rate of a reaction typically decreases as the reaction progresses, because fewer reactant molecules exist after the reaction begins.* The lower concentration of reactant reduces the likelihood that molecules will interact to make products. As a result, determining concentrations of reactants and products at a certain time during a reaction requires greater understanding of how the reaction rate changes with time. In the end, we cannot say that the concentration of atrazine should be 0 M after 120 days (see Figure 15.6). Similarly, we cannot say that the concentration is 3.75×10^{-7} M after only 30 days.

Visualization: Reaction Rate and Concentration

   The decomposition of hydrogen peroxide can be used to illustrate how the rate of a reaction can be related to the reaction conditions. Hydrogen peroxide, a common staple in the home medicine cabinet, is used to clean cuts and scrapes

Application

FIGURE 15.6

The environmental decomposition of atrazine in groundwater. The half-life is the time it takes for a given concentration to be halved.

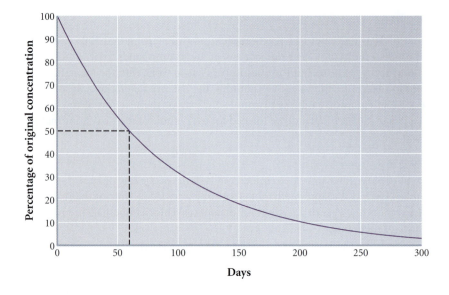

 Video Lesson: Rate Laws: How the Reaction Rate Depends on Concentration

because of its ability to oxidize microbes. We observe by experiment that the rate of the reaction depends on the concentration of hydrogen peroxide. As the concentration decreases, the rate of the reaction decreases.

$$\text{Rate} \propto [H_2O_2]$$

When we examine the relationship more closely, the following mathematical equation, called a **rate law**, emerges. This rate law states that the rate of the reaction of hydrogen peroxide is equal to the product of a constant (which we call the **rate constant**) times the concentration of H_2O_2.

$$2H_2O_2(aq) \rightarrow 2H_2O(l) + O_2(g)$$

$$\text{Rate} = k[H_2O_2]$$

From the rate law, we determine that the reaction is **first order** in hydrogen peroxide. That is, *the rate depends on the concentration of H_2O_2 raised to the first power*. The reaction is first order *overall* because the rate equation for the reaction is dependent only on the concentration of H_2O_2 to the first power—that is, it is linearly related. It is important to remember that the value of the rate constant, the compounds included in the rate law, and the orders of the compounds in the equation *can be found only experimentally*.

The rate law, as we will see, can be used to determine the rate of a reaction at any reactant concentration. It will also tell you which species are the most important contributors to the rate of a reaction. The typical rate law has the following form:

Video Lesson: Determining the Form of a Rate Law

$$\text{Rate} = k[A]^n[B]^m$$

where k is the rate constant, [A] and [B] are the concentrations of substances involved in the reaction, and n and m are the orders of the corresponding compounds. The values of n and m are measures of how dependent the rate is on the concentration of a particular reactant, and they *must be* experimentally determined. The reaction also can be described in terms of the overall order, which is calculated by adding n and m (and the exponents of any other reactants in the rate law). The order of a compound or the overall order of a reaction can certainly be negative or even a noninteger number. Again, it is important to remember that *the order of a compound in a reaction cannot be determined just by looking at the reaction*.

The reaction of nitrogen monoxide gas with hydrogen gas is

$$2NO(g) + 2H_2(g) \rightarrow 2H_2O(l) + N_2(g)$$

The experimentally determined rate law is

$$\text{Rate} = k[\text{NO}]^2[\text{H}_2]$$

The reaction is said to be second order in nitrogen monoxide, because the rate depends on [NO] to the second power. The rate is first order with respect to hydrogen gas, because the exponent is 1, which means there is a linear relationship between the rate and [H$_2$]. The reaction is third order (the sum of the individual orders, $2 + 1$) overall. A reaction can have fractional orders, though we will not deal with that here.

EXERCISE 15.3 **The Rate Law**

Environmental chemists are concerned with the damaging effects of compounds on the ozone layer. One such class of compounds is the nitrogen oxides, such as NO(g) and NO$_2$(g). Under certain conditions, the reaction of NO(g), an air pollutant released in automobile exhaust, with oxygen can produce N$_2$O$_4$(g). The rate law for this reaction was determined experimentally and is shown below. What is the reaction order of each of the compounds in the rate law? What is the overall order of the reaction?

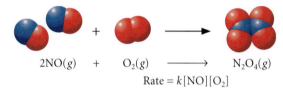

$$2\text{NO}(g) \quad + \quad \text{O}_2(g) \quad \longrightarrow \quad \text{N}_2\text{O}_4(g)$$
$$\text{Rate} = k[\text{NO}][\text{O}_2]$$

Solution

The experimentally determined rate law says that the reaction is first order in NO(g) and first order in O$_2$(g). Overall, then, the reaction is second order.

PRACTICE 15.3

The following reaction proceeds at 300°C. What is the reaction order of the compound in the following reaction with the given rate law? What is the order of the reaction?

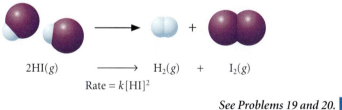

$$2\text{HI}(g) \quad \longrightarrow \quad \text{H}_2(g) \quad + \quad \text{I}_2(g)$$
$$\text{Rate} = k[\text{HI}]^2$$

See Problems 19 and 20.

Collision Theory

In Exercise 15.3 and Practice 15.3, the rate law is expressed as the concentration of the reactants raised to a power. This is not uncommon, but as we've mentioned before, it isn't true for reactions in general. Why do the orders of the rate law and the stoichiometric factors of a reaction often differ? Reactions are typically more complex than they appear when written on paper. To understand how this can cause the rate law to differ from what we may expect, we need to consult **collision theory**. This theory *describes how the rate of a reaction is related to the number of properly oriented collisions of the molecules involved.* Collision theory is heavily based on the kinetic molecular theory we discussed in Chapter 11.

Kinetic molecular theory says that the thermal motion of particles (the kinetic energy) can be used to explain how a gas behaves. For instance, the pressure

Visualization: Gas Phase Reaction of NO and Cl$_2$

Video Lesson: The Collision Model

Increasing the temperature of a reaction increases the kinetic energy of the components of the reaction. This means that the particles move faster and, as a result, have many more collisions than they do in the colder reaction. More collisions mean a faster reaction.

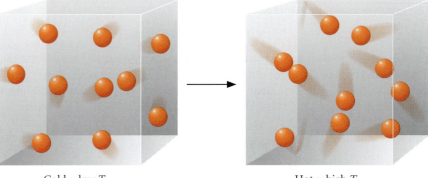

Cold = low T
Low kinetic energy

Hot = high T
High kinetic energy

of a gaseous system is related to the number of collisions of the molecules with the sides of their container in a given time. If we increase the number of molecules per unit volume, the number of collisions per second also increases, assuming the temperature is constant. The pressure of the system can be increased if we raise the temperature of the gas and leave the volume constant. This is because the molecules are moving faster (more kinetic energy) and engage in more collisions per second. In short, higher kinetic energy equals more collisions.

Collision theory, which is summarized in Table 15.1, requires that molecules collide in order to react. One of the more important statements in collision theory, as shown in Figure 15.7(a), says that the collisions must be energetic enough to make a product. The minimum energy required, the **activation energy (E_a)**, is specific to a particular reaction. However, an energetic collision alone is not enough to cause a reaction to occur. The collision must also occur between properly oriented molecules; see Figure 15.7(b). Because the equation we write to describe a reaction doesn't address all of these issues, rate laws are difficult to derive by simply examining the overall equation. We'll revisit this statement in Section 15.6.

TABLE 15.1	Collision Theory

A reaction occurs when the following conditions have been met:
- Molecules collide.
- Molecules have enough kinetic energy.
- Molecules are oriented properly.

Implications:
- Larger concentrations have faster reaction rates.
- Reactions with higher temperatures have faster rates.
- Rates depend on the number of properly oriented collisions.
- Predicting the rate of a reaction is difficult.

FIGURE 15.7

Collision theory. (a) Collisions must be energetic enough to be considered successful. (b) Successful collisions only occur between properly oriented molecules.

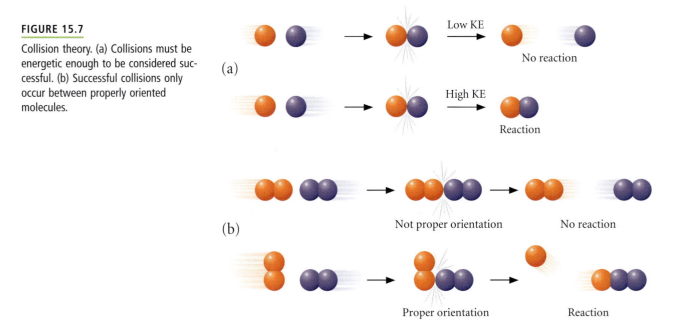

(a)

Low KE
No reaction

High KE
Reaction

(b)

Not proper orientation
No reaction

Proper orientation
Reaction

HERE'S WHAT WE KNOW SO FAR

- The rate of a reaction is the change in concentration (M) per unit time. Rates are always positive numbers, often reported in M/s.

- Average rates and instantaneous rates differ only in the time measurement. Instantaneous rates have $\Delta t \approx 0$.

- The rate law (Rate = $k[A]^n[B]^m$) indicates the relationship between the rate and the concentrations of reactants.

- The order of a reactant can be used to indicate quickly the relationship between the reactant concentration and the rate.

- Collision theory explains why the rate of a reaction typically decreases as time passes.

15.3   Changes in Time—The Integrated Rate Law

After the discovery in 1939 that DDT (1,1,1-trichloro-2,2-bis-(4′-chlorophenyl) ethane, or **d**ichloro**d**iphenyl**t**richloroethane) can be used to control mosquito-borne malaria, its use soared. Especially important was its use to protect soldiers who were fighting in the Pacific Rim countries in World War II. Since its discovery, DDT has been sprayed to eliminate insects from cotton crops, spiders from residences, and mosquitoes from towns all across the globe (Figure 15.8). Initial testing showed that the compound wasn't very toxic to mammals. However, because the metabolism of DDT is very slow, small amounts of DDT in the environment tended to accumulate in animals (including humans) until toxic levels were present. Evidence of this caused Sweden in 1970 and the United States in 1972 to ban the use of DDT as a pesticide, although it is still used in some other countries, such as Ethiopia and South Africa. Despite the 30-year ban on DDT use in the United States, the insecticide can still be found in the environment, mostly in waterways like that shown in Figure 15.9. The hazard of DDT accumulation in the food chain versus the benefit of saving human lives by preventing malaria in the populations of, for example, East African countries is still a topic of intense debate.

    In organisms that are resistant to DDT, an enzyme known as dehydrochlorinase converts DDT into dichlorodiphenyldiethene (DDE). Unfortunately, DDE can accumulate within birds and weaken their eggshells by interfering with the

Application

CHEMICAL ENCOUNTERS: Decomposition of DDT

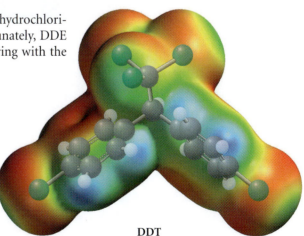

DDT

FIGURE 15.8

DDT and the mosquito. DDT is still one of the most cost-effective methods of controlling mosquito-borne malaria. Although many countries have banned its use because of DDT's persistence in the environment, many homes in central Africa are still sprayed inside and out.

FIGURE 15.9

Even though the use of DDT has been banned in the United States since the 1970s, DDT is still present in the environment. This plot of New York harbor shows that the sediments in the East River harbor still contain large quantities of DDT.

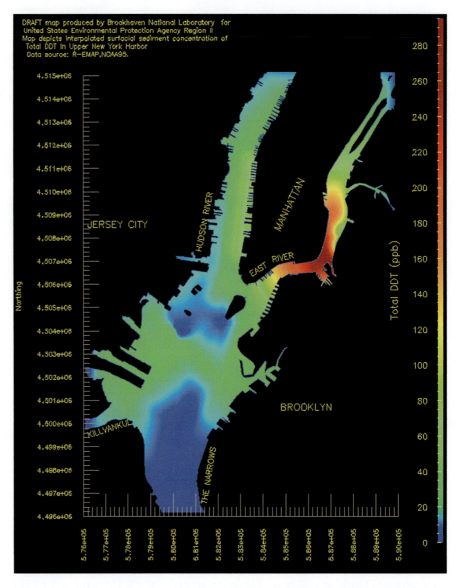

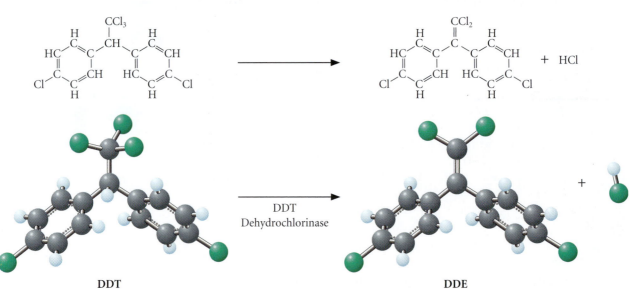

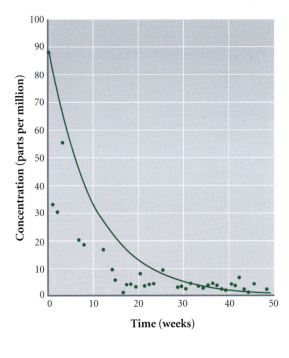

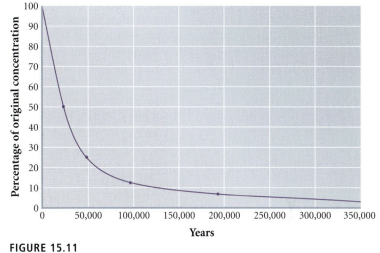

FIGURE 15.10

An experiment to illustrate the bioremediation (biological recycling) of DDT-contaminated ground by soil microbes and nutrients. The decomposition of DDT occurs rapidly at first and then slows. Note that the concentration of DDT is reported in parts per million (1 ppm of DDT = 1 μg of DDT per gram of soil). Although the curve appears to be placed incorrectly, it is not. Computer analysis of this set of data, and additional data not shown, produced this curve.

FIGURE 15.11

Plot of the radioactive decay of plutonium-239.

complex process of constructing the shells. This leads to shells that break under the normal pressures naturally associated with nesting. For this reason, several species of birds, such as the peregrine falcon, were nearly wiped out in the United States.

The rate of decomposition of DDT by soil microbes is plotted against time in Figure 15.10. Initially the rate is relatively high; then, as time passes, the rate begins to decline. **Does this make sense according to what we know about collision theory?** Yes. Reducing the number of reactant molecules reduces the number of collisions that would result in the formation of the product.

What if we're interested in determining what concentration of a chemical will exist after a certain amount of time has passed? For example, the Hanford Nuclear Reservation on the banks of the Columbia River just north of Richland, Washington, has produced 54 million gallons of radioactive plutonium waste over the past half-century. The waste is currently being stored in 177 underground tanks as a watery sludge. How long will it take for the radioactive waste stored at the Hanford Nuclear Reservation to decay to 1% of its current concentration? We could examine the plot shown in Figure 15.11 to figure this out, but we would need either to have a plot of the radioactive decay reaction or to know the rate law, the rate, and rate constant of the reaction. Can we determine the concentration at a specific time without knowing the rate or even having a plot of the reaction? Calculus comes to our rescue. With some manipulation of the rate law, we can make a more useful description of the rate of a reaction that will enable us to perform these calculations. These equations are referred to as the **integrated rate laws**.

Application

Integrated First-Order Rate Law

We discussed the decomposition of hydrogen peroxide to make oxygen and water near the start of Section 15.2. Experimentally, it has been determined that the rate law for this process describes a first-order reaction. Unfortunately, this doesn't tell us the concentration of H_2O_2 at any point during the reaction, unless we know the rate of the reaction at that time and the rate constant (k). In order to calculate the concentration of H_2O_2 at any point during the reaction, we can mathematically convert the rate law into an **integrated first-order rate law**.

$$2H_2O_2(aq) \rightarrow 2H_2O(l) + O_2(g)$$

$$\text{Rate} = k[H_2O_2]$$

The integrated rate law gets its name from the mathematical process, known as integration, that we follow to *generate the relationship between concentration and time*. When we integrate the rate law from $t = 0$ to some future time, the integrated rate law for the decomposition of hydrogen peroxide becomes

$$\ln\left(\frac{[H_2O_2]_t}{[H_2O_2]_0}\right) = -kt$$

where $[H_2O_2]_0$ is the concentration of hydrogen peroxide initially, $[H_2O_2]_t$ is the concentration of hydrogen peroxide at a particular time t, and k is the rate constant. In practice, we need know only three of the four values in the equation (time, rate constant, initial concentration, and concentration at time t) in order to find the fourth.

This equation is general for all first-order reactions, which we will designate using the general form

$$A \rightarrow \text{Products}$$

where A is any single reactant. Note that the stoichiometric coefficient for A is 1. The natural logarithm (ln) of the quotient of the final reactant concentration, $[A]_t$, divided by the initial concentration $[A]_0$ is equal to the negative of the rate constant, k, times the time, t, shown in the left-hand side of the box below. We can use this equation to calculate the time required to reach a given concentration. Alternatively, we can rearrange the equation to that shown on the right-hand side of the box if we wish to calculate the concentration at a particular time:

Video Lesson: First-Order Reactions

$$\ln\left(\frac{[A]_t}{[A]_0}\right) = -kt \qquad [A]_t = [A]_0 e^{-kt}$$

We can transpose this into an equation in the form $y = mx + b$ by recognizing that

$$\ln\left(\frac{[A]_t}{[A]_0}\right) = \ln[A]_t - \ln[A]_0$$

Substituting for $\ln\left(\dfrac{[A]_t}{[A]_0}\right)$ yields

$$\ln[A]_t - \ln[A]_0 = -kt$$

which enables us to come up with the final $y = mx + b$ form:

$$\ln[A]_t = -kt + \ln[A]_0$$
$$ y \; = \; mx + \quad b$$

where the y axis $= \ln[A]_t$

the x axis $= t$

the slope $m = -k$

the intercept $b = \ln[A]_0$

How do we use the first-order equation to find the concentration of hydrogen peroxide after 5.00 min if we have a solution in which the initial concentration,

$[H_2O_2]_0$, is equal to 0.100 M? The rate constant for this reaction was determined by experiment to be 3.10×10^{-3} s^{-1}. Using this equation, we can calculate the amount of peroxide still remaining after 5.00 min. We first *convert our time to match the units for the rate constant* (5.00 min = 3.00×10^2 s) and then insert our known values into the integrated first-order rate law.

$$\ln[H_2O_2]_t - \ln[H_2O_2]_0 = -kt$$
$$\ln[H_2O_2]_t - \ln(0.100\ M) = -(3.10 \times 10^{-3}\ s^{-1})(3.00 \times 10^2\ s)$$
$$\ln[H_2O_2]_t - \ln(0.100\ M) = -0.930$$
$$\ln[H_2O_2]_t + 2.3026 = -0.930$$
$$\ln[H_2O_2]_t = -3.2326$$
$$[H_2O_2]_t = e^{-3.2326} = 0.0395\ M$$

The concentration of hydrogen peroxide remaining after 5.00 min is 0.0395 M.

EXERCISE 15.4	Working with the Integrated First-Order Rate Law

Archaeologists near the Dead Sea in 1998 reported the discovery of a substance that they believe ancient peoples used as glue. A sample of *newly prepared* collagen exhibits 15.2 disintegrations per minute per gram (15.2 dis/min/g) of carbon. (A disintegration is the decomposition of a radioactive nucleus such as ^{14}C; see Chapter 21.) The ^{14}C decay rate for a sample of the *ancient* glue (made from collagen) was found to be 5.60 disintegrations per minute per gram (5.60 dis/min/g) of carbon. What is the age of the glue? The decomposition of ^{14}C is a first-order process with a rate constant of 1.209×10^{-4} $year^{-1}$.

First Thoughts

This problem shows that the integrated first-order rate law can be used with radioactive compounds to determine decay rates, concentrations, or times. In these cases, we consider the concentration of a reactant to be directly proportional to the number of disintegrations per minute per gram. In the laboratory, this method can be used to accurately determine the date of objects that are between 200 and 50,000 years old. Also note that we do not need to convert dis/min/g to dis/yr/g because the units cancel in the equation.

Solution

If we assume that fresh collagen has the same activity of ^{14}C that the ancient glue did when it was first made, we can calculate the length of time it would take a fresh sample of collagen to have the same activity of ^{14}C as the glue.

$$\ln\left(\frac{5.60\ \text{dis/min/g}}{15.2\ \text{dis/min/g}}\right) = -(1.209 \times 10^{-4}\ year^{-1})t$$
$$\ln(0.3684) = -(1.209 \times 10^{-4}\ year^{-1})t$$
$$-0.9985 = -(1.209 \times 10^{-4}\ year^{-1})t$$
$$t = 8259\ year = 8260\ year$$

The calculation reveals that it would take 8260 years for the activity in the fresh sample to decay to the level observed in the ancient piece of wood. This implies that the glue is 8260 years old.

Further Insight

Radiocarbon dating has been used extensively in determining the age of archaeological artifacts. The process relies on the assumption that the ratio of carbon-14 to carbon-12 in nature has always been constant. However, when an organism dies, the ratio begins to change as the radioactive carbon-14 decays. Controversy over the validity of this dating method has been addressed by compensating for small

Video Lesson: Radiochemical Dating

fluctuations in the original ratio. These fluctuations have been determined by dating objects with an age known by other methods. More information about radiocarbon dating can be found in Chapter 21.

Hanford's nuclear waste contains large quantities of plutonium (mostly ^{239}Pu). Researchers have determined that approximately 875 kg of solid waste plutonium are buried there. How long will it take for the mass of plutonium-239 to drop to 10% of its original value? Assume the rate constant that describes the decay of ^{239}Pu is 2.874×10^{-5} year^{-1}. How long will it take for the mass of plutonium-239 to reach half of its original concentration?

See Problems 33 and 34.

Visualization: Half-Life of Reactions

Video Lesson: A Kinetics Problem

Tutorial: Half-Life of Reactions

Application

CHEMICAL ENCOUNTERS: Persistent Pesticides

Half-life

The persistence of pesticides and herbicides in the environment is often reported as the amount of time that it takes for half of the original concentration to decompose. In general, any reaction can be reported in this manner by calculating the amount of time it takes for the reaction to proceed to *50% completion*. This value is known as the **half-life ($t_{1/2}$)** of the reaction (Figure 15.12). Perhaps you've heard this term used to express the rate of decay for a radioactive element, such as the radioactive waste stored at the Hanford Nuclear Reservation, or in accounts of the dating of ancient artifacts. The half-lives of pesticides and herbicides indicated in Table 15.2 are used to judge the safety of the compounds and to establish guidelines for the frequency of their application. How does the half-life fit in with our description of the integrated first-order rate law?

Consider the first-order hydrolysis of atrazine in groundwater. The rate constant for this reaction in water has been found to be 0.001733 day^{-1}. Note that we're using a rate constant with units of day^{-1} instead of s^{-1}. This will be important in the answer we generate from the integrated rate law equation.

$$\ln\left(\frac{[\text{atrazine}]_t}{[\text{atrazine}]_0}\right) = -kt$$

FIGURE 15.12

The half-life of a reaction is the amount of time required for the reaction to reach 50% of the original concentrations. In terms of radioactive decay or the decomposition of pesticides, the passage of one half-life reduces the concentration in half. Two half-lives reduce the concentration to one-quarter of the original.

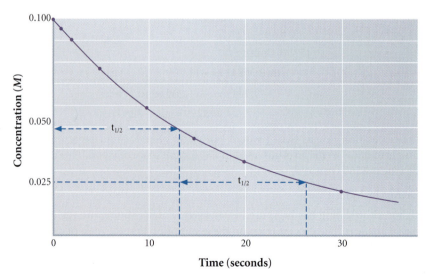

TABLE 15.2	Solubility and Half-life in Soil for Selected Pesticides

The sorption index is the ratio of pesticide concentration bound to soil particles divided by the concentration in the aqueous phase. Pesticides with low sorption indices are more likely to be leached into groundwater supplies, because a low proportion of each binds to the soil. The half-life is reported for the pesticide in sterile soil. 2,4-D, for example, is of greater environmental concern because of its sorption index value than is DDT.

Trade Name/ Brand Name	Water Solubility (ppm)	Sorption Index (higher = greater binding to the soil)	Soil Half-life (days)
2,4-D	890	20	10
Alachlor/Lasso	240	170	15
Atrazine/Aatrex	33	100	60
Dicamba/Banvel	400,000	2	14
Carbaryl/Sevin	110	300	10
Chlorsulfuron/Glean	7000	40	160
DDT	0.0055	24,000	3000
Diazinon	60	1000	40
Malathion/Cythion	145	1800	1
Metolachlor/Dual	530	200	90
Methoxychlor/Marlate	0.10	80,000	120
Pendimethalin/Prowl	<1	5000	90
Pronamide/Kerb	15	200	60
Terbacil/Sinbar	710	55	120
Terbufos/Counter	4.5	500	5
Trifluralin/Treflan	<1	8000	60

Source: Institute of Agriculture and Natural Resources, Univ. Nebraska Lincoln, *Factors That Affect Soil-Applied Herbicides*, 2005.

When the hydrolysis has consumed half of the original concentration of atrazine, $[\text{atrazine}]_t = \frac{1}{2}[\text{atrazine}]_0 = 0.5[\text{atrazine}]_0$. Substituting this into the equation above, we find that

$$\ln\left(\frac{0.5[\text{atrazine}]_0}{[\text{atrazine}]_0}\right) = -kt_{1/2}$$

The concentration of atrazine ($[\text{atrazine}]_0$) cancels. Simplifying the equation further yields

$$\ln 0.5 = -kt_{1/2}$$
$$-0.693 = -kt_{1/2}$$
$$t_{1/2} = \frac{0.693}{k}$$

Our derivation reveals that *the half-life ($t_{1/2}$) of a first-order reaction is, throughout the reaction, a constant that depends only on the rate constant and not on the concentration of the reactant.* This means that if we know the half-life for a particular first-order reaction, we can calculate the rate constant, and vice versa.

Substituting the rate constant for atrazine into this equation, we find that the half-life for the hydrolysis in water agrees with the experimental observation we noted at the start of this chapter. Half of the atrazine added to water in the environment disappears after 400 days.

$$t_{1/2} = \frac{0.693}{0.001733 \text{ day}^{-1}} = 4.00 \times 10^2 \text{ days}$$

Diazinon crystals are sprinkled around a home's foundation to kill and repel ants from the home. How long will it take for the diazinon to decompose, assuming first-order kinetics, to 25% of its original concentration in the soil? (From Table 15.2, the half-life of diazinon is 4.0×10^1 days.)

First Thoughts

The problem doesn't state the original concentration of the diazinon in the soil. However, we don't need to know the concentration to determine the answer, because we have the ratio of initial concentration to final concentration (that is, $[\text{diazinon}]_t = 0.25[\text{diazinon}]_0$).

Solution

Using the half-life, we first calculate the rate constant for the reaction:

$$t_{1/2} = \frac{0.693}{k}$$
$$4.0 \times 10^1 \text{days} = \frac{0.693}{k}$$
$$k = 0.0173 \text{ day}^{-1}$$

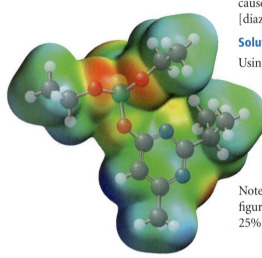

Diazinon

Note that the unit of the rate constant is reciprocal days and that we retain an extra figure in the value because we will use the data in the next part of the calculation. If 25% of the diazinon remains, then

$$\ln\left(\frac{0.25[\text{diazinon}]_0}{[\text{diazinon}]_0}\right) = -(0.0173 \text{ day}^{-1})t$$
$$\ln 0.25 = -0.0173 \text{ day}^{-1}t$$
$$-1.386 = -0.0173 \text{ day}^{-1}t$$
$$t = 8.0 \times 10^1 \text{ days}$$

Further Insight

There is an alternative method by which we can calculate the answer. One half-life reduces the original concentration by 50% (50% = 0.5 = ½). Two half-lives reduce the concentration to 25% (½ × ½ = ¼). Three reduce the concentration to 12.5% (½ × ½ × ½ = ⅛). Because we want to know the time required for the reaction to reduce the concentration of starting material to 25% of the initial concentration, we need two half-lives (4.0×10^1 days + 4.0×10^1 days = 8.0×10^1 days). Therefore, for a first-order process, the fraction remaining, $[A]_t/[A]_0$ equals 2^{-n}, where n = number of half-lives. For example, for $n = 3$ half-lives, $[A]_t/[A]_0 = 2^{-3}$ or 0.125.

PRACTICE 15.5

What is the length of time you would have to wait for alachlor to decompose to 25% of its original value? . . . to 6.25%? . . . to 0.001%?

See Problems 27 and 28.

Other Rate Laws

Not all reactions follow first-order kinetics. Other orders do exist, even noninteger orders. Reactions that take place on metal surfaces typically follow zero-order kinetics. Hydrogenation of vegetable oils, the decomposition of ammonia on a tungsten wire, and the reaction of N_2O with oxygen in your car's catalytic converter are reactions on metal surfaces. They follow zero-order kinetics.

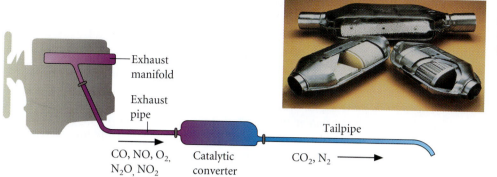

Catalytic converters catalyze the reaction of N_2O into N_2. The reaction is zero order.

Exhaust manifold

Exhaust pipe

Tailpipe

CO, NO, O_2, N_2O, NO_2

Catalytic converter

CO_2, N_2

In a **zero-order reaction**, the rate law does *not* depend on the concentration of any of the compounds in the reaction. The rate of the zero-order reaction is constant. No matter what the original concentration happens to be, the reaction always proceeds at the same rate.

$$\text{Rate of disappearance of A} = -\frac{\Delta[A]}{\Delta t}$$

$$\text{Rate} = k[A]^0 = k$$

The **integrated zero-order rate law**, determined by integrating the rate law over time, for a zero-order reaction is different from the integrated first-order rate law:

$$[A]_t = -kt + [A]_0$$

Substituting $0.5[A]_0$ for the concentration of A at time t, we can rearrange the equation to get the formula for the half-life of the zero-order reaction:

$$t_{1/2} = \frac{[A]_0}{2k}$$

The half-life of a zero-order reaction directly depends on the initial concentration of the reactant. Larger concentrations of the reactant will mean a larger half-life for the reaction. Note that the rate of a zero-order reaction is constant but that the half-life depends on the initial concentration of reactant. For example, consider a zero-order reaction where the rate constant $k = 2.5 \times 10^{-4}$ M/s. When $[A]_0 = 1.00$ M, the half-life of the reaction is 2000 s. If $[A]_0 = 0.25$ M, the half-life of the reaction is 500 s. The rate of the reaction (rate = k) is a constant, but the half-life changes as the concentration changes.

The **second-order rate law** can be more complicated, because there are two cases that fit the definition of a second-order reaction. In one of those cases, the reaction is second order in only one reactant:

$$\text{Rate} = k[A]^2$$

In the other case involving a second-order rate law, the reaction could be first order in two different species:

$$\text{Rate} = k[A][B]$$

In second-order reactions with only one reactant, the integrated rate law is determined using the method we have explored for the zero-order and first-order reactions. After integration and rearrangement, the **integrated second-order rate law** takes the form

$$\frac{1}{[A]_t} = kt + \frac{1}{[A]_0}$$

Video Lesson: Second-Order Reactions

The half-life of this type of second-order reaction can be determined by assuming that the concentration of A at some time, t, is equal to one-half the initial concentration ($[A]_t = 0.5[A]_0$). Simplifying the equation gives

$$t_{1/2} = \frac{1}{k[A]_0}$$

The half-life for a second-order reaction (involving only one compound in the rate law) is inversely dependent on the initial concentration of the reactant. Smaller initial concentrations of the reactant will mean a longer half-life for the reaction.

If the second-order reaction contains two species in the rate law, the integrated rate law becomes much more complicated. In fact, the math gets so complicated that chemists typically *manipulate the experimental conditions to reduce the amount of calculation.* Consider the reaction of one of the components of smog with ozone:

$$NO(g) + O_3(g) \rightarrow NO_2(g) + O_2(g)$$

$$\text{Rate} = k[NO][O_3]$$

The reaction is first order in NO, first order in O_3, and second order overall.

To calculate the concentration of a reactant, determine the rate constant, or find the time required to reach a certain concentration with this type of rate law, we conduct the reaction with a relatively small concentration of one of the reactants and a very large concentration of the other. **What effect does this have on the rate law?** Because one of the concentrations is very large, any change in its concentration is negligible, and *we can assume that its concentration remains constant through the course of the reaction.* An analogy is having someone who is rich spend $1 in a day (decreasing to 50 cents after one half-life!) from a fortune of $1 billion. The change is hardly noticeable. The fortune is essentially constant. Someone who had only $10, however, would notice the effect of spending $1 immediately. Having the concentration of one component much larger than that of the other reduces the rate law to a much simpler form. For example, if we perform the reaction with 0.100 M NO and 0.001 M O_3, the concentrations of the species after the reaction is complete can be determined:

$$[O_3]_f = 0.001\ M - 0.001\ M = 0.000\ M$$
$$[NO]_f = 0.100\ M - 0.001\ M = 0.099\ M \approx 0.100\ M$$

The O_3 is consumed in the reaction, but the concentration of NO is only slightly affected. We can make the assumption that the concentration doesn't change. Because the final concentration of NO has essentially remained constant, it can be combined with the rate constant. And the rate equation for the reaction can be reduced to

$$\text{Rate} = k\,[NO][O_3] = k\,[NO]_0\,[O_3]$$

and, because $k' = k\,[NO]$,

$$\text{Rate} = k'\,[O_3]$$

where k' is the new rate constant. What is the order of our new rate law? Because of our choice of the initial concentrations, the kinetics for this reaction has taken on the form of a first-order rate law. Because of our modification, we say that this is a **pseudo-first-order rate** equation.

We can do this manipulation with any reaction. By increasing the concentration of a particular reactant to a very large value, any order of a rate law can be simplified to a pseudo-first order rate equation. This enables us to study reactions no matter what order they appear to be. We must remember, however, that any modification of the rate equation means that the new rate constant is not the same rate constant as in the original reaction.

HERE'S WHAT WE KNOW SO FAR

- The integrated rate laws enable researchers to calculate one of four things about a reaction (rate, rate constant, initial concentration, or final concentration) if three of these are known.

- The half-life of a reaction indicates the time required for the reaction to reach 50% completion.

- Modifying a second-order reaction by using a large concentration of one of the reactants reduces the integrated rate law to a pseudo-first-order rate law.

15.4   Methods of Determining Rate Laws

Glyphosate herbicides are absorbed through the foliage of weeds, inhibiting a plant's ability to make new amino acids. Because these herbicides react relatively quickly with water ($t_{1/2} < 7$ days), the rate of their biological reaction in plants is important. Weed scientists study the rates of amino acid inhibition and environmental degradation in order to suggest modifications to improve glyphosate effectiveness (Figure 15.13). The greatest benefit of the glyphosate herbicides is attributable to their rapid degradation in the environment. The rapid degradation means this class of herbicide causes less harm to the environment than other herbicides. The rates of the degradation of these compounds have been measured by experiment, and the overall rate laws have been determined from those experiments.

We, too, can use the information from a chemical reaction to obtain the final rate law for a reaction. The first of two commonly used procedures that we will consider is the **method of initial rates**. In this appraoch, we measure and compare initial rates rather than comparing instantaneous rates later in the reaction. We use initial rates to determine the rate law because we can precisely measure and control the starting concentrations of the reactants and precisely identify the time for the reaction to reach a given point. In addition, limited side reactions (reactions that make products other than what we're interested in), rapid determination of the rate, and a well-defined time period make the comparison of multiple reactions experimentally straightforward.

We'll use as our example the reaction of nitrogen monoxide, $NO(g)$, with oxygen to make nitrogen dioxide, $NO_2(g)$, a component of smog. We can measure the initial rate of this reaction as we change the concentrations of both $NO(g)$ and $O_2(g)$. We conduct three separate reactions with different concentrations of reactants, measure the initial rate, and complete the table on the next page. Note that we have carefully chosen our starting concentrations to keep one reactant the same in two experiments while doubling the other reactant.

$$2NO(g) + O_2(g) \rightarrow 2NO_2(g)$$

We can determine the rate law for this reaction by assuming that the rate is based solely on the two reactants in the equation. We write

$$\text{Rate} = k[NO]^n[O_2]^m$$

where n and m are the orders of the two reactants. If we compare the rate of the third reaction to the rate of the second reaction, in which [NO] is constant and [O$_2$] changes, the equation simplifies dramatically. To do this, let's divide the rate law for the third experiment into the rate law for the second experiment.

$$\frac{\text{rate}_2}{\text{rate}_3} = \frac{k[NO]^n[O_2]^m}{k[NO]^n[O_2]^m}$$

Application

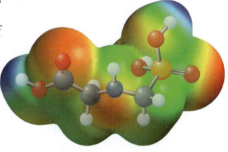

FIGURE 15.13

Roundup herbicide contains glyphosate. The general herbicide is sold for use on farms and in removing unwanted weeds around the yard.

Glyphosate

Automobile exhaust produces nitrogen compounds that react with oxygen.

Experiment	[NO] (M)	[O₂] (M)	Initial Rate (M/s)
1	0.0126	0.0125	1.41×10^{-2}
2	0.0252	0.0250	1.13×10^{-1}
3	0.0252	0.0125	5.64×10^{-2}

Next, we add the values from the table to the equation. Then we simplify the equation. Although the value of the rate constant, k, is not known, it should be the same in both reactions:

$$\frac{1.13 \times 10^{-1} M \cdot s^{-1}}{5.64 \times 10^{-2} M \cdot s^{-1}} = \frac{k(0.0252)^n (0.0250)^m}{k(0.0252)^n (0.0125)^m}$$

$$2 = \frac{(0.0250)^m}{(0.0125)^m}$$

$$2 = 2^m$$

$$m = 1$$

Why did we divide the second equation by the third? When one of the reactant concentrations is held constant ([NO], in this case) and the other concentration ([O₂]) is doubled, the effect on the rate is *due only to the change in [O₂]*. This effect is related to a power of 2 and reveals the order of the second reactant ($m = 1$). To continue, we divide the third rate law by the first because [O₂] is constant while [NO] changes, so *the effect on the rate is due only to the change in [NO]*. We substitute concentrations and solve:

$$\frac{5.64 \times 10^{-2} M \cdot s^{-1}}{1.41 \times 10^{-2} M \cdot s^{-1}} = \frac{k(0.0252)^n (0.0125)^1}{k(0.0126)^n (0.0125)^1}$$

$$4 = \frac{(0.0252)^n}{(0.0126)^n}$$

$$4 = 2^n$$

$$n = 2$$

We are nearly ready to write the rate law for the reaction because we know the order of the reactants. We are missing only the rate constant. We can solve for k by choosing one of the experiments we've completed and substituting the concentrations into the rate law. Solving for the rate constant, we obtain

$$\text{Rate} = k[NO]^2[O_2]$$

$$1.41 \times 10^{-2} = k(0.0126)^2 (0.0125)$$

$$k = 7.11 \times 10^3 \ M^{-2} \cdot s^{-1}$$

The method of initial rates requires that we can obtain and compare the initial rates of reactions. We need at least three reactions to solve for two unknown orders. And in every case, we've used *experimentally determined data* to calculate the rate law for the reaction.

Note that the units for the rate constant are different depending on the order of the rate law. Specifically, the units are $M^{-(\text{order}-1)} \cdot s^{-1}$.

EXERCISE 15.6 **Initial Rates**

The reaction of Cl_2 with NO occurs at a very rapid pace. Use the data in the table below to determine the rate law for the reaction. Then calculate the rate constant.

$$2NO(g) + Cl_2(g) \rightarrow 2NOCl(g)$$

Experiment	[NO] (M)	[Cl_2] (M)	Initial Rate (M/min)
1	0.10 M	0.10 M	0.18
2	0.10 M	0.20 M	0.36
3	0.20 M	0.20 M	1.45

Solution

The rate law for the reaction can be determined by using the method of initial rates. The overall rate law, based on the reaction, is

$$\text{Rate} = k[NO]^m[Cl_2]^n$$

The order of the reaction with respect to NO is

$$\frac{\text{Rate 3}}{\text{Rate 2}} = \frac{1.45 \ M \cdot \text{min}^{-1}}{0.36 \ M \cdot \text{min}^{-1}} = \frac{k(0.20)^m(0.20)^n}{k(0.10)^m(0.20)^n}$$

$$\frac{1.45}{0.36} = \frac{(0.20)^m}{(0.10)^m}$$

$$4 = 2^m$$

$$m = 2$$

The order of the reaction with respect to Cl_2 is

$$\frac{\text{Rate 2}}{\text{Rate 1}} = \frac{0.36 \ M \cdot \text{min}^{-1}}{0.18 \ M \cdot \text{min}^{-1}} = \frac{k(0.10)^m(0.20)^n}{k(0.10)^m(0.10)^n}$$

$$\frac{0.36}{0.18} = \frac{(0.20)^n}{(0.10)^n}$$

$$2 = 2^n$$

$$n = 1$$

The overall rate law and the value for the rate constant can then be calculated. Note that the rate is still expressed in concentration per unit time. Also note that while the orders in this rate law equal the coefficients in the equation for this example, this is not always true. We will deal with this in Section 15.6.

$$\text{Rate} = k[NO]^2[Cl_2]^1$$

$$k = \frac{0.18}{(0.10)^2(0.10)} = 1.8 \times 10^2 \ M^{-2} \cdot \text{min}^{-1}$$

PRACTICE 15.6

Use the initial rates for the reaction of carbon monoxide (CO) with hemoglobin (Hb) to determine the rate law and the rate constant. What is the overall order of the reaction?

Experiment	[Hb] (M)	[CO] (M)	Initial Rate (M/s)
1	$2.21 \times 10^{-6} \ M$	$1.00 \times 10^{-6} \ M$	0.619×10^{-6}
2	$4.42 \times 10^{-6} \ M$	$1.00 \times 10^{-6} \ M$	1.24×10^{-6}
3	$4.42 \times 10^{-6} \ M$	$3.00 \times 10^{-6} \ M$	3.71×10^{-6}

See Problems 57–60, 66, and 67.

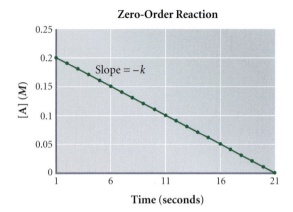

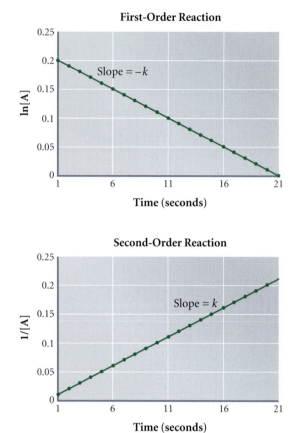

FIGURE 15.14

A linear relationship exists for the data if time is plotted versus [A] for zero-order reactions, versus ln[A] for first-order reactions, and versus 1/[A] for second-order reactions. Note that the slope of the line is equal to the negative of the rate constant for zero-order and first-order plots. The slope of the second-order plot is equal to the rate constant.

Visualization: Rate Laws

The second method for determining the rate law of a reaction has us examine only one reaction instead of a series of reactions. In this **method of graphical analysis** (also known as the method of integrated rate laws), we plot how the concentration of a reactant changes with time. Although this method does require us to measure the rate of the reaction as it proceeds over a long time period, little mathematical manipulation of the data is needed to determine the rate law. Various versions of the plot can be quickly constructed in order to establish a linear relationship between time and a measure of concentration that confirms one of our common rate laws. Figure 15.14 displays plots of zero-order, first-order, and second-order reactions. Each is a linear relationship that is derived from the integrated rate law. If our data are linear when plotted in one of these ways, it suggests that the data fit that model. Computer-based data acquisition can also help us to do statistical analyses to determine the model that best fits the data.

Here are the possible outcomes that we will consider:

■ The zero-order reaction produces a linear plot when the concentration of reactant is plotted against time. The slope of the line is equal to $-k$.

■ The first-order reaction produces a linear plot when the natural logarithm (ln) of the concentration of reactant is plotted against time. The slope in this case is also equal to the negative of the rate constant, $-k$.

■ The second-order reaction produces a linear relationship when the reciprocal concentration is plotted against time. The slope, which is equal to the rate constant, is positive in this case.

If the concentration of a reactant is followed as a function of time, this is the best method to use in determining the rate law. By graphing the data in three different ways, we can determine the overall order of the reaction (zero order, first order, or second order).

EXERCISE 15.7   **Determining the Reaction Order**

The decomposition of atrazine in the presence of titanium dioxide has been studied and the rate of the reaction measured. Plot the data shown in the table, and determine whether the reaction follows zero-order, first-order, or second-order kinetics.

Time (h)	[Atrazine] (M)
0	4.65×10^{-5}
3	2.98×10^{-5}
8	1.49×10^{-5}
15	6.98×10^{-6}
22	3.67×10^{-6}

First Thoughts

We must plot the data for each of the three models we have discussed: zero order, first order, and second order. If might be useful to add two columns to your table for ln[atrazine] and 1/[atrazine], respectively.

Solution

Using a graphical analysis software package (or three sheets of graph paper), we plot the data on three different graphs. In the first, the concentration is plotted versus time, the second relates the ln[atrazine] versus time, and the third illustrates 1/[atrazine] versus time. Examination of the results leads us to the conclusion that the middle plot—the one that relates first-order kinetics—is the appropriate graph. We choose this one because the data points seem to lie closest to the line of best fit, without being systematically curved. The reaction must be first order overall and first order in atrazine (the reactant).

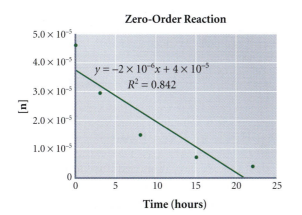

Zero-Order Reaction

$y = -2 \times 10^{-6}x + 4 \times 10^{-5}$
$R^2 = 0.842$

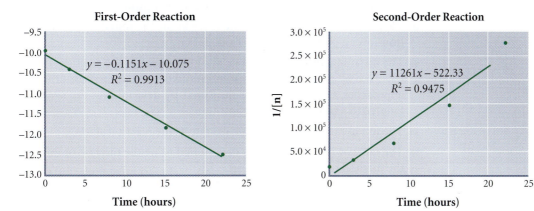

First-Order Reaction

$y = -0.1151x - 10.075$
$R^2 = 0.9913$

Second-Order Reaction

$y = 11261x - 522.33$
$R^2 = 0.9475$

Further Insights

There are a number of mathematical tools we can use to compute the best fit of the data to one of our reaction orders. One of these tools is the correlation coefficient, r, which tells how well any two measures are mathematically related to each other. A perfect directly linear relationship would yield a correlation coefficient, r, of 1. As infants grow, their height and weight give correlation coefficients tending toward 1. A perfect inverse relationship gives r values of -1. As we increase in age past about 45 years old, we inevitably run more slowly. Age and running speed in older runners are inversely related and give an r value that is closer to -1. If there is no relationship between variables, such as between human hair color and the number of strawberries people eat in July, we get an r value of 0. If a first-order model were to fit our data, it would have the highest correlation coefficient of the three possible models that we have introduced.

PRACTICE 15.7

Plot the data for the following reaction as was done in Exercise 15.7. Determine the overall order of this reaction. You may wish to use a graphing program to determine which graph gives the straightest line.

$$2NO_2(g) \rightarrow 2NO(g) + O_2(g)$$

Time (s)	[NO$_2$] (M)
0	0.0100
50	0.0079
100	0.0065
150	0.0055
200	0.0048
300	0.0038
400	0.0031

See Problems 53, 54, 61, 62, 65, and 68.

15.5 Looking Back at Rate Laws

A summary of the data that we have discussed thus far is appropriate. In short, here's what we know so far:

- Reaction rates determine the speed at which a reaction progresses but do not reveal anything about the extent to which they produce a product.

- We can measure the average rate if we know the initial and final concentrations over a particular time period.

- The instantaneous rate is the rate at a given point in the reaction. It can be determined by measuring the concentrations at points when the time difference approaches zero, or it can be measured by determining the slope of a line tangent to a plot of the rate versus time.

- The initial rate is the instantaneous rate as the reaction starts.

- The rate law is experimentally determined using the method of initial rates or the method of graphical analysis.

TABLE 15.3   Rate Laws

The value of the rate is reported in the units of concentration per unit time. The value of the rate constant is different for each of the three orders (zero order = M/s; first order = 1/s; and second order = 1/M·s).

Order	Rate Law	Integrated Rate Law	Half-life	Linear Plot
Zero	Rate $= k$	$[A]_t = -kt + [A]_0$	$t_{1/2} = \dfrac{[A]_0}{2k}$	$[A]$ versus t slope $= -k$
First	Rate $= k[A]$	$\ln\left(\dfrac{[A]_t}{[A]_0}\right) = -kt$	$t_{1/2} = \dfrac{0.693}{k}$	$\ln[A]$ versus t slope $= -k$
Second	Rate $= k[A]^2$	$\dfrac{1}{[A]_t} = kt + \dfrac{1}{[A]_0}$	$t_{1/2} = \dfrac{1}{k[A]_0}$	$\dfrac{1}{[A]}$ versus t slope $= +k$

- The half-life of a reaction is the time required for the concentration of a reaction to reach 50% of the initial value.

- Complex reaction orders can often be reduced to pseudo-first-order reactions by keeping the concentration of one of the reactants relatively large.

Specific information for the three rate orders that we have discussed is included in Table 15.3. There are many more overall orders for reactions than those listed here.

15.6   Reaction Mechanisms

At this point, we need to do some detective work. In Exercise 15.2 we determined the rate of metabolism of methoxychlor, an organochlorine insecticide, using actual values for the concentrations. However, upon deeper investigation, chemists have found that the reaction proceeds in two sequential steps. In the first step, the methoxychlor undergoes reaction with water and cytochrome P450 (an enzyme in the liver) to make mono-hydroxymethoxychlor. In a second step, the mono-hydroxymethoxychlor reacts with another molecule of water and cytochrome P450 and is converted into the product bis-hydroxymethoxychlor.

Application

CHEMICAL
ENCOUNTERS:
Metabolism of
Methoxychlor

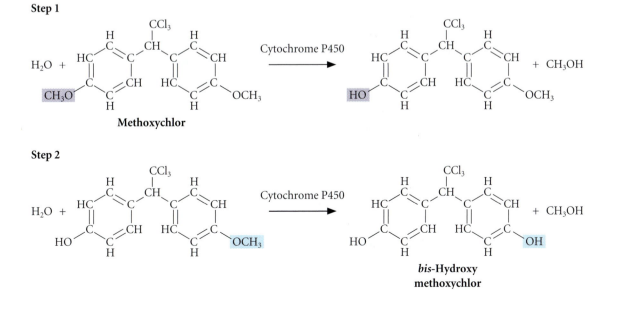

Step 1

Methoxychlor

Step 2

bis-Hydroxy
methoxychlor

Why is it useful to know the individual steps that make up the overall reaction?

Chemists often are interested in more detailed descriptions of chemical reactions, including how the reactions occur at the molecular level. We know from our previous discussions that chemical equations contain a wealth of information about a reaction. We can determine the spontaneity of the process, the enthalpy of the process, and the stoichiometry of the reactants by examining the chemical equation. However, the overall chemical equation doesn't show how the reactants collide to become products. To address this concern, investigators study a single reaction to determine exactly how each molecule moves during the course of the process. *The knowledge of exactly how a reaction proceeds is useful in predicting new reactions, determining the rates of those reactions, discovering new applications of chemistry, and learning how substances interact with humans and our environment.* Their study is part of the field of mechanistic chemistry.

A **mechanism** for a reaction is the set of steps that compounds take as they proceed from reactant to product. The mechanism of a reaction accounts for the *experimentally determined rate of the reaction* and is consistent with *the overall stoichiometry of the reaction.* In some cases, a mechanism is only one step. In others, there are a multitude of steps. However, we can't tell this by looking at the overall chemical equation.

Each of the single steps in a chemical reaction is called an **elementary step**. For example, as shown in Figure 15.15, the overall process of turning off the lights in a room is made up of two elementary steps: (1) walking to the light switch and (2) flipping the light switch. One of these steps, walking to the light switch, is slower. The other, flipping the switch, is so fast that its contribution to the time required to complete the overall process is negligible. Any calculations of the time required to turn off the lights, then, can be reduced to the time it takes to walk to the light switch. In other words, the time it takes to turn off the lights in the room is nearly equal to the length of time it takes to walk to the light switch. This slow step is known as the **rate-determining step**.

Within each elementary step, reactants come together and undergo successful collisions to make products. The number of molecules that collide in this process defines the **molecularity** of the step (Figure 15.16). A **unimolecular reaction** involves one molecule as the only reactant. When two molecules collide, the reaction is said to be a **bimolecular reaction**. Reactions that involve the collision of three molecules simultaneously (which are called **termolecular reactions**, or **trimolecular reactions**) are also known, but they are rare because they require that three molecules collide at the same time and in the proper orientation. In such cases, the decrease in the entropy associated with three molecules coming together at one time is often prohibitive (see Chapter 14). In general, an elementary step has only a single bond breakage or formation.

FIGURE 15.15

The elementary steps completed in turning off the lights.

Step 1: Walk to light
Slow

Step 2: Turn off light
Fast

Visualization: Decomposition of N_2O_5

FIGURE 15.16

Molecularity of reactions. The number of molecules that must collide at one time to produce the reaction determines the molecularity.

Video Lesson: Defining the Molecularity of a Reaction

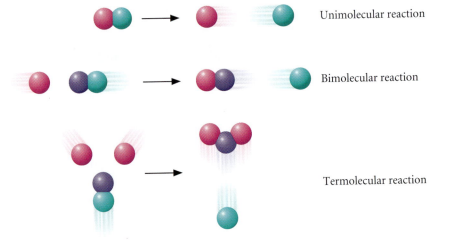

Unimolecular reaction

Bimolecular reaction

Termolecular reaction

If we know the elementary steps in a reaction, we can obtain a wealth of information about the rate of the reaction. Most important, *rate laws can be written directly from elementary steps*. Specifically, the rate orders for the reactants in an elementary step are given by the stoichiometric coefficients of the reactants in that step. Moreover, the rates of the elementary steps in a mechanism need not be the same—indeed, in most cases they are not. A mechanism with two elementary steps typically has a fast step and a slow step. The slow step in a mechanism is the one that will determine the rate of the overall reaction.

The reaction of NO_2 and F_2 gas is a useful case study:

$$2NO_2(g) + F_2(g) \rightarrow 2NO_2F(g)$$

Video Lesson: Determining the Rate Laws of Elementary Reactions

Video Lesson: Calculating the Rate Laws of Multistep Reactions

After careful experimentation, a mechanism involving two elementary steps has been suggested for the reaction. This first step was determined to be slow; the second was determined to be relatively fast:

$$NO_2(g) + F_2(g) \rightarrow NO_2F(g) + F(g) \quad \text{(slow)}$$
$$\underline{F(g) + NO_2(g) \rightarrow NO_2F(g) \qquad\qquad \text{(fast)}}$$

$$2NO_2(g) + F_2(g) \rightarrow 2NO_2F(g)$$

Note that the sum of the elementary steps gives the balanced equation. From these steps, we can determine the rate law for the overall reaction in a way that is similar to our discussion of the length of time required to turn off the lights. The slow elementary step will determine the rate of the reaction, so we can use this rate-determining step to write the rate law. Judging on the basis of the stoichiometric coefficients of the reactants in the slow step, the overall reaction is first order in NO_2, first order in F_2, and second order overall.

$$\text{Rate} = k[NO_2][F_2]$$

Not all mechanisms are so easy to work with. For instance, the reaction of NO with O_2 to make NO_2 is much more complicated than our first glance suggests. From our previous discussion in Section 15.4, we know the overall equation for the reaction:

$$2NO(g) + O_2(g) \rightarrow 2NO_2(g)$$

We saw that the experimentally determined rate equation for the reaction is

$$\text{Rate} = k[NO]^2[O_2]$$

This implies that the reaction is a termolecular process, which is unlikely because it would require three molecules to collide simultaneously. The mechanism of this reaction is known, and it has the elementary steps shown below.

$$2NO(g) \underset{k_{-1}}{\overset{k_1}{\rightleftharpoons}} N_2O_2(g) \qquad \text{(fast)}$$
$$N_2O_2(g) + O_2(g) \overset{k_2}{\longrightarrow} 2NO_2(g) \qquad \text{(slow)}$$

What is the meaning of the double arrows in the fast equation? These arrows indicate that the reaction can proceed in both the forward direction and the reverse direction at the same time. Moreover, each of these directions has a rate constant associated with it (k_1 and k_{-1}). The slow step in our mechanism is the reaction of $N_2O_2(g)$ with oxygen, and we could write the rate equation on the basis of this information. Note that we've specified the rate constants for both reactions:

$$\text{Rate} = k_1[NO]^2 \qquad\quad \text{for the fast step } (k_1 \gg k_2)$$
$$\text{Rate} = k_2[N_2O_2][O_2] \quad\ \text{for the slow step } (k_2 \ll k_1)$$

It follows that the rate law for the overall reaction should be

$$\text{Rate} = k_2[N_2O_2][O_2] \quad\ \text{for the overall reaction}$$

However, this doesn't agree with the experimentally determined rate law that we noted above:

$$\text{Rate} = k[NO]^2[O_2]$$

This difference arises because $N_2O_2(g)$ is an intermediate in the reaction. An **intermediate** is a compound that is formed and consumed during the course of a reaction. If we examine the overall reaction, we don't see the intermediate N_2O_2 as one of the reactants. Measuring the concentration of this species, then, could be difficult. We need to rewrite the rate equation so that the rate reflects only the compounds in the overall reaction. To deal with this, *we will assume that the fast reaction reaches equilibrium,* an assumption that greatly simplifies our determination of the overall rate law for the reaction. What does this assumption mean? Within a reaction that is at equilibrium, the rate of the forward reaction is equal to the rate of the reverse reaction.

$$2NO(g) \underset{k_{-1}}{\overset{k_1}{\rightleftharpoons}} N_2O_2(g) \qquad \text{(fast)}$$

$$k_1[NO]^2 = k_{-1}[N_2O_2]$$

Then we can rearrange our equation to solve for $[N_2O_2]$:

$$[N_2O_2] = \frac{k_1[NO]^2}{k_{-1}} = k'[NO]^2$$

where the new rate constant k' is equal to $\dfrac{k_1}{k_{-1}}$.*

We can now substitute for the concentration of the intermediate, $[N_2O_2]$, in our original rate equation and simplify:

$$\text{Rate} = k_2[N_2O_2][O_2]$$

$$[N_2O_2] = k'[NO]^2$$

$$\text{Rate} = k_2k'[NO]^2[O_2] = k''[NO]^2[O_2]$$

where the new rate constant k'' is equal to k_2k'.

This agrees with our experimentally determined rate equation. The mathematics used to convert our rate law containing the intermediate into the experimentally observed rate law are based on the assumption that the fast reaction has reached equilibrium. This assumption requires that the rate constant for the reverse of the fast reaction be much larger than the rate constant for the slow step.

EXERCISE 15.8 **Rate Laws**

Write the overall reaction, identify any intermediates, and write the rate law for the following proposed mechanism for the decomposition of $IBr(g)$ to $I_2(g)$ and $Br_2(g)$.

$$IBr(g) \rightarrow I(g) + Br(g) \qquad \text{(slow)}$$

$$IBr(g) + Br(g) \rightarrow I(g) + Br_2(g) \qquad \text{(fast)}$$

$$I(g) + I(g) \rightarrow I_2(g) \qquad \text{(fast)}$$

** **Note:** Alternatively, we could rearrange this equation to place the rate constants on one side and the concentrations of compounds on the other side:*

$$\frac{k_1}{k_{-1}} = \frac{[N_2O_2]}{[NO]^2}$$

This gives rise to something that looks remarkably similar to Q, the reaction quotient from Chapter 14. We'll explore this in much greater detail in Chapter 16.

Solution

The overall reaction is the sum of the elementary steps in the mechanism.

$$2IBr(g) \rightarrow I_2(g) + Br_2(g)$$

The intermediates are produced and consumed in the reaction. They are I(g) and Br(g). Because the first step in the mechanism is the slow step, it is rate determining. Therefore, the rate law for the reaction can be written directly from this step:

$$\text{rate} = k\,[IBr]$$

It is first order in IBr(g) and first order overall.

PRACTICE 15.8

Write the overall reaction, identify any intermediates, and write the rate law for the following proposed mechanism for the production of nitrogen dioxide (NO_2).

$$NO(g) + O_2(g) \;\rightleftharpoons\; NO_3(g) \quad \text{(fast)}$$

$$NO_3(g) + NO(g) \rightarrow 2NO_2(g) \quad \text{(slow)}$$

See Problems 85 and 86.

Transition State Theory

The destruction of ozone by atomic oxygen is one of the ways in which stratospheric ozone can be depleted:

$$O_3(g) + O(g) \rightarrow 2O_2(g)$$

By examining this reaction in the laboratory, we can determine the change in enthalpy ($\Delta H = -392$ kJ/mol) and measure the rate of the reaction. However, it is often helpful to examine a reaction by consulting a plot of the energies for the reactants, products, and any intermediates. If we make a graph of the energies as the reactants proceed along a **reaction coordinate** (the pathway describing the changes in each molecule in the reaction, even though they are happening at different times) to become product, we obtain a **reaction profile** like that shown in Figure 15.17. On the basis of our earlier discussion of collision theory, we know that the reactants must have enough energy to overcome a barrier known as the activation energy (E_a)—assuming that the reactants collide in the proper orientation. For the reaction of atomic oxygen and ozone, the barrier is rather small ($E_{a(\text{forward})} = 19$ kJ/mol) compared to the reverse reaction. The reaction profile illustrates this.

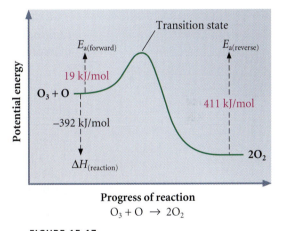

FIGURE 15.17

Reaction profile.

Visualization: Transition States and Activation Energy

The reaction profile is used to explore **transition state theory** and how it applies to a reaction. This theory describes how the bonds in the reacting molecules reorganize to represent the bonds in the products. At some point on the reaction coordinate the collision occurs, and the atoms in the reactants occupy a **transition state**. The collection of atoms at the transition state, called the **activated complex**, is very energetic at this point in the reaction—more so than the reactants, products, or intermediates. The activated complex is not a separate isolable compound. Rather, *it is a snapshot of the reaction at the point in time when the molecules have collided.* From the activated complex, the reaction could proceed to products, or the complex could dissociate back into the reactants.

The reaction profile shows the activation energy for the forward reaction, the activation energy for the backward reaction, and the overall change in the energy of the reaction (ΔE) in a graphical way. This change in energy (ΔE) is equal to ΔH

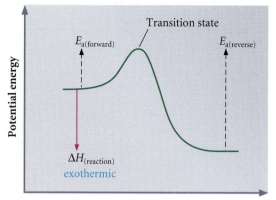

FIGURE 15.18

Exothermic reaction profile. The reactants are more energetic than the products in the endothermic reaction. $E_{a(reverse)}$ is larger than $E_{a(forward)}$.

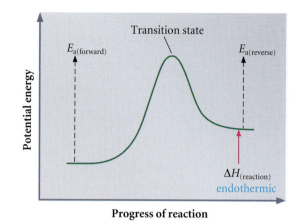

FIGURE 15.19

Endothermic reaction profile. The products are more energetic than the reactants in the endothermic reaction. $E_{a(forward)}$ is larger than $E_{a(reverse)}$.

for the reaction at constant pressure and volume. Even for cases where ΔE and ΔH are not equal (unequal volumes or pressures), the difference is usually very small. By looking at the reaction profile, we can tell whether a reaction is exothermic (Figure 15.18) or endothermic (Figure 15.19).

$$\Delta H_{reaction} \cong \Delta E = E_{a(forward)} - E_{a(reverse)}$$

To what is the activation energy related? In 1888, Svante Arrhenius (1859–1927), a Swedish chemist, studied how temperature affected the rate of a reaction. What came out of this work is a relationship between the activation energy and the rate of a reaction. His mathematical equation is

$$k = Ae^{-E_a/RT}$$

Video Lesson: The Arrhenius Equation

where k is the rate constant, A is the **frequency factor** that relates how many successfully oriented collisions occur in a particular reaction, E_a is the activation energy, T is the temperature in kelvins, and R is the universal gas constant (8.314 J/mol·K). This equation says that the rate constant of a reaction is related to the size of E_a. As E_a gets larger, the rate constant gets smaller and the rate of the reaction decreases.

This equation can be used to determine the rate of a reaction on the basis of the temperature of the reaction and the amount of energy the reactants require to make the activated complex. To use it, however, we must know the activation energy (E_a) *and* the frequency factor (A) for the reaction. Fortunately, by performing the reaction at two different temperatures, we can utilize this equation to calculate the activation energy without knowing the frequency factor. Alternatively, if we know the activation energy and the rate of reaction at a particular temperature, we can determine the rate of reaction at any other temperature. The equation that relates this calculation can be derived by taking the natural logarithm of the Arrhenius equation above, at two temperatures:

Video Lesson: Using the Arrhenius Equation

$$\ln\left(\frac{k_2}{k_1}\right) = \frac{E_a}{R}\left(\frac{1}{T_1} - \frac{1}{T_2}\right)$$

where k_1 and k_2 are the rate constants for the reaction obtained at two different temperatures, T_1 and T_2.

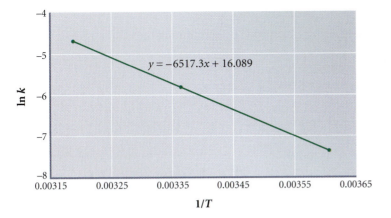

FIGURE 15.20

A plot of ln k versus 1/T gives a straight line whose slope is $-E_a/R$. Each data point on this plot corresponds to the measurement of the temperature and rate constant in separate reactions. The slope of the line of best fit (-6517.3) is equal to $-E_a/R$. Therefore, $E_a = 54.2$ kJ/mol for this reaction. If the axes went to zero, the intercept (16.089) corresponding to ln A would be found. The frequency factor is 9,710,000.

This equation provides a quick method to determine the energy of activation for a reaction, but the resulting answer can contain significant error, because its value is based on only two reactions. Alternatively, we can determine the value of E_a graphically by using a series of rate constants calculated at different temperatures. To see how this is done, let's examine the original equation produced by Arrhenius:

$$k = Ae^{-E_a/RT}$$

If we take the natural log of both sides of the equation, we get

$$\ln k = \ln A - \frac{E_a}{RT}$$

Rearranging this equation as shown enables us to determine the activation energy by plotting the rate constant of a reaction at several temperatures.

$$\ln k = -\frac{E_a}{R}\left(\frac{1}{T}\right) + \ln A$$
$$y\ \ =\ \ \ mx\ \ \ +\ \ b$$

If we construct an "Arrhenius plot" of ln k versus 1/T for a series of reactions, we obtain a straight line whose slope is equal to $-E_a/R$ and whose intercept is ln A, as shown in Figure 15.20. This method not only enables us to determine the energy of activation quite accurately but also provides a convenient way to determine the frequency factor A.

EXERCISE 15.9   **Energy Barrier**

The reaction of NO, a component in smog, with ozone has been extensively studied. Data for the temperature dependence are tabulated below. What is the activation energy for this reaction?

$$NO(g) + O_3(g) \rightarrow NO_2(g) + O_2(g)$$

Temperature (K)	k (1/M·s)
195	1.08×10^9
298	12.0×10^9

Solution

The activation energy can be calculated by substituting the values from the table into the equation:

$$\ln\left(\frac{k_2}{k_1}\right) = \frac{E_a}{R}\left(\frac{1}{T_1} - \frac{1}{T_2}\right)$$

$$\ln\left(\frac{12.0 \times 10^9}{1.08 \times 10^9}\right) = \frac{E_a}{8.314\ \text{J·mol}^{-1}\text{·K}^{-1}}\left(\frac{1}{195\ \text{K}} - \frac{1}{298\ \text{K}}\right)$$

$$2.408 = \frac{E_a}{8.314\ \text{J·mol}^{-1}\text{·K}^{-1}}\left(1.773 \times 10^{-3}\ \text{K}^{-1}\right)$$

$$20.02\ \text{J·mol}^{-1} = E_a\left(1.773 \times 10^{-3}\ \text{K}\right)$$

$$E_a = 11292\ \text{J·mol}^{-1} = 11.3\ \text{kJ·mol}^{-1}$$

PRACTICE 15.9

More data on the reaction of NO and O_3 are shown in the table below. Calculate E_a for each pair of reactions. Are the values the same? Explain.

Temperature (K)	k (1/M·s)
230	2.95×10^9
260	5.42×10^9
369	35.5×10^9

See Problems 83, 84, 87, 88, and 101.

Video Lesson: Catalysts and Types of Catalysts

15.7 Applications of Catalysts

Environmental chemists have comprehensively explored the rate of decomposition of atrazine in aqueous solutions. Because the rate in water is so slow ($t_{1/2} = 400$ days), they've spent time considering how the decomposition could be accelerated to clean up the environment more quickly. The exercise in the previous section showed that the rate of a reaction increases with an increase in temperature. That's one way to speed up a reaction. However, raising the temperature of groundwater, rivers, and lakes is not a feasible way to increase the rate of decomposition of herbicides and pesticides. Researchers at the University of Wisconsin have found a better way to enhance the rate of the decomposition. They have discovered that filtering atrazine-contaminated water through a container full of titanium(IV) oxide in the presence of ultraviolet light greatly increases the rate of decomposition (to $t_{1/2} = 15$ min). What does the titanium(IV) oxide do?

The titanium(IV) oxide is used as a **catalyst** in the decomposition of atrazine. A catalyst is a compound that, when added to a reaction mixture, changes the mechanism of the reaction to a new pathway with a lower activation energy. Rather than lowering the activation energy of the existing mechanism, it creates a new set of elementary steps whose rate-determining step has a lower energy of activation than the reaction would have without the catalyst. Because E_a *is lower, the new mechanism is faster, so we often say that a catalyst increases the rate of the reaction.* Another useful aspect of catalysts is that they can be recovered from the reaction; *catalysts are not consumed in a reaction.* Because of this, they don't appear in the net chemical equation. However, because they are involved in the reaction, they do appear in the mechanism. To show that a particular reaction involves a catalyst, we typically place it over or under the arrow in the overall equation. There are two types of catalysts, homogeneous catalysts and heterogeneous catalysts. The type of catalyst, as well as the reaction profile that results, depends on how the catalyst is mixed with the reaction.

A **homogeneous catalyst** is part of a reaction that is *catalyzed in a homogeneous mixture* (that is, the catalyst is intimately mixed with the reactants). For example, the destruction of ozone by chlorine radicals (a homogeneous catalyst) is illustrated by the net reaction of ozone with oxygen atoms:

$$O_3 + O \xrightarrow{\text{Cl}} 2O_2$$

The mechanism of this reaction consists of two elementary steps:

Step 1: The ozone molecule reacts with elemental chlorine (the catalyst) to make a molecule of ClO and a molecule of oxygen.

Step 2: The ClO reacts with an oxygen atom (the other reactant in the net equation) to make another molecule of oxygen and regenerate elemental chlorine.

The chlorine atom appears first as a reactant and later as a product; chlorine begins and escapes the reaction without being changed and therefore is not consumed. This is exactly what catalysts do. Is ClO a catalyst too? No. It is formed in the course of the reaction and then consumed before products are made. The ClO never escapes the reaction. As we saw earlier, this is exactly what happens to an intermediate.

The presence of chlorine atoms causes a tremendous increase in the rate of the decomposition of ozone. How does a catalyst cause a reaction's rate to increase? Consider the reaction profile, shown in Figure 15.21, for the first-order decomposition of hydrogen peroxide by a homogeneous catalyst. In the noncatalyzed reaction, the reaction follows a unimolecular mechanism with a clearly defined high-energy barrier (the activation energy) separating the reactants from the products. If a catalyst such as iodide is added to this reaction, the mechanism of the reaction changes. The iodide is the catalyst, and it is shown over the reaction arrow to identify it as such. **What is the role of OI⁻ in the reaction?** The reaction makes an intermediate, which is more stable than the activated complex. The solid line in the reaction profile in Figure 15.21 illustrates the new reaction.

What is the outcome of adding a catalyst? Remember that our catalyst has produced a new mechanism for the reaction. If we plot the reaction profile for this new mechanism, we can see that the activation energy is lower. This means that more molecules will have the

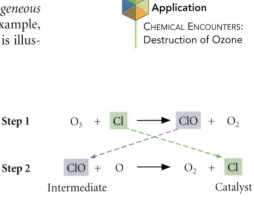

Application

CHEMICAL ENCOUNTERS:
Destruction of Ozone

Video Lesson: CIA
Demonstration: Elephant Snot

Visualization: Homogeneous
Catalysis

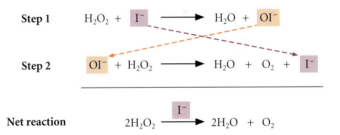

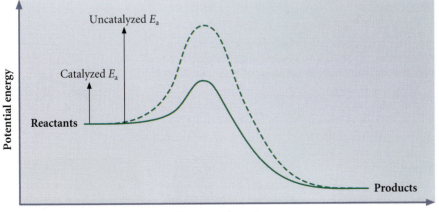

FIGURE 15.21

Decomposition of H_2O_2 with and without a homogeneous catalyst. The dotted line represents the reaction profile without the catalyst; the solid line indicates the effect of a homogeneous catalyst.

FIGURE 15.22

Heterogeneous catalysis of NO.

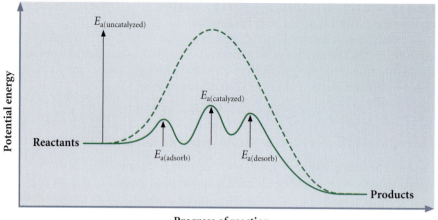

All of the different-colored heterogeneous catalysts in this collection work in a similar manner. The reactants must adsorb to the surface of the catalyst before the reaction can be catalyzed. After reaction, the products desorb from the catalyst. The reaction shown here illustrates the catalytic hydrogenation of ethylene to make ethane.

requisite energy to become activated complexes in the mechanism. Although the overall reaction enthalpy (ΔH) hasn't changed, the activation energy has been reduced dramatically. A reduction in the activation energy causes the rate of the reaction to increase.

In systems with a **heterogeneous catalyst**, the catalyst and the reactants are in *different* physical states. The catalyst is typically a metal in solid form, whereas the reactants are typically in gaseous, aqueous, or liquid form. The TiO_2 decomposition of atrazine is an example of this type of catalysis. The catalytic converter (usually made up of platinum, palladium, and/or rhodium metal) in your automobile is another. One of the reactions in the catalytic converter cuts down on smog-forming NO by reducing it to N_2. The general unbalanced reaction is

$$NO(g) \rightarrow N_2(g)$$

The heterogeneous catalyst works in a three-step process. In the first step shown in Figure 15.22, the reactant molecules **adsorb** to the catalyst—that is, they sticks to the catalyst's surface. Note that the word *adsorb* is different than the word **absorb**, which means being taken up or mixed into a substance. A small activation energy barrier must be crossed to achieve the surface binding of the reactant. Often, there is only a very small barrier for binding of a reactant to the surface of a catalyst. Then the reactants migrate around on the surface until they collide to make the product. A larger, yet still low, activation energy barrier exists for this step. In the final step, the products are **desorbed**, or released from the surface of the catalyst (the reverse of being adsorbed). The resulting reaction profile diagram has a characteristic shape.

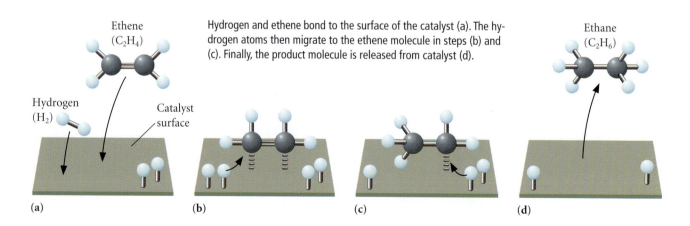

Hydrogen and ethene bond to the surface of the catalyst (a). The hydrogen atoms then migrate to the ethene molecule in steps (b) and (c). Finally, the product molecule is released from catalyst (d).

NanoWorld / MacroWorld

Big effects of the very small: Enzymes—nature's catalysts

The modern confectionery industry operates a booming business, helping to satisfy that sweet tooth in most of us. Fanciful desserts require sucrose as a sweetener. Fortunately, the American farmer can meet the large demand for sweeteners. For instance, sugar beet production in Colorado, Montana, Nebraska, and Wyoming yields 4.5 to 6 million tons of sucrose per year. Sugar cane production, primarily in Hawaii, Louisiana, and Florida, adds another 6 million tons to the total. But even more sugar is needed. One of the ways to meet the public's demand for sweeteners involves corn starch and a biological molecule known as D-glucose isomerase, shown in Figure 15.23. The product of these molecules is sweeter than sucrose alone. It is known as high-fructose corn syrup, and its use has surpassed that of sucrose in the confectionery industry.

Like other enzymes, D-glucose isomerase acts as a catalyst that speeds up a reaction. The enzymes work by binding selectively to a particular molecule, forcing it into just the right shape, and then assisting in the

FIGURE 15.23

High-fructose corn syrup is made from corn starch using an enzyme. D-Glucose isomerase catalyzes the reaction that converts glucose into fructose. The reaction proceeds much more rapidly with the enzyme than without it. The enzyme is shown here as a series of ribbons that represent the atoms that make up the strands of protein polymers. The strands loop and wind their way together to make a pocket that can catalyze the reaction of glucose to make fructose.

reaction that makes the product. They increase the value of A (the frequency factor from the Arrhenius equation) and decrease the energy of activation (E_a) at the same time. This activity arises because the backbone of the amino acid polymer weaves the enzyme into a structure similar to a catcher's mitt. Along the inside of the catcher's mitt (the active site of the enzyme) lie portions of the enzyme that are polar and portions that are nonpolar. The arrangements of the polar and nonpolar groups provide a template that exactly matches that of the molecule they bind (the reactant or substrate). When the substrate binds, the enzyme bends it into a conformation similar to that of the product of the reaction. Then, when the conformation is just right, the reaction takes place. After releasing the product, the enzyme returns to its original shape, ready to accept another substrate (Figure 15.24). The net result is an increase in the reaction rate without an increase in temperature, which is particularly useful in the food industry.

For example, lactase (an enzyme that converts lactose into glucose and galactose) is used in the dairy industry to make digestible milk products. Because many of these products spoil at temperatures warmer than those found in a refrigerator, the use of enzymes to speed the reaction without a temperature increase is quite helpful. After the lactase has been added to milk, lactose-intolerant people can drink all of the milk they want. And they owe their settled stomach to one of nature's catalysts.

FIGURE 15.24

Diagram of an enzyme-catalyzed reaction. The substrate binds to the active site on an enzyme, where the reaction is catalyzed.

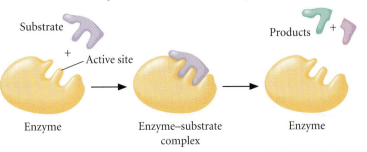

The principles of chemical kinetics are important not only to chemists but to anyone concerned with the rate of a process. The rate of decay at the landfill, the rate of ozone depletion, and the persistence of pesticides and herbicides in the environment can easily be determined by examining the rate laws associated with these processes. Our work in the lab can provide the rate laws for these reactions. And, after determining the thermodynamics of a reaction (Chapter 14), we can determine not only whether a reaction will take place but also how long it will take to complete.

The Bottom Line

- Reaction rates determine the speed at which a reaction progresses but do not reveal anything about the extent to which they produce a product. (Section 15.1)

- The average rate can be measured if you know the initial and final concentrations over a particular time period. (Section 15.1)

- The instantaneous rate is the rate at a given point in the reaction. It can be determined by measuring the concentrations at points when the time difference approaches zero, or it can be measured by determining the slope of a line tangent to a plot of the rate versus time. (Section 15.1)

- The initial rate of a reaction is the instantaneous rate as the reaction starts. (Section 15.1)

- The half-life of a reaction is the time required for the concentration of a reaction to reach 50% of the initial value. (Section 15.3)

- Complex reaction orders can often be reduced to pseudo-first-order reactions by keeping the concentration of one of the reactants large. (Section 15.3)

- The rate law is experimentally determined using the method of initial rates or the method of graphical analysis. (Section 15.4)

- Transition state theory, which is based on collision theory, describes the energy of a reaction during the course of a reaction. A plot of the reaction coordinate versus the energy can give meaningful information, such as the activation energy, the presence of any intermediates, the effect of a catalyst, and the enthalpy for the forward and reverse process. (Section 15.6)

- Catalysts speed up a reaction without being consumed in the reaction. (Section 15.7)

Key Words

absorb To be associated, through intermolecular forces of attraction, by being taken up or mixing into a substance. (*p. 656*)

activated complex The unstable collection of atoms that can break up either to form the products or to reform the reactants. (*p. 651*)

activation energy (E_a) The minimum energy of collision that reactants must have in order to successfully create the activated complex. (*p. 630*)

adsorb To be associated, through intermolecular forces of attraction, with a surface. (*p. 656*)

average rate The rate of a reaction measured over a long period of time. (*p. 622*)

bimolecular reaction A reaction involving the collision of two molecules in the rate-determining step. (*p. 648*)

catalyst A substance that participates in a reaction, is not consumed, and modifies the mechanism of the reaction to provide a lower activation energy. Catalysts increase the rate of reaction. (*p. 654*)

chemical kinetics The study of the rates and mechanisms of chemical reactions. (*p. 621*)

collision theory A theory that correlates the number of properly oriented collisions with the rate of the reaction. (*p. 629*)

desorbed Released from a surface. (*p. 656*)

elementary step A single step in a mechanism that indicates how reactants proceed toward products. (*p. 648*)

environmental chemist A scientist who studies the interactions of compounds in the environment. (*p. 620*)

first order A reaction is first order in a particular species when the rate of the reaction depends on the concentration of that species raised to the first power. (*p. 628*)

frequency factor A term in the Arrhenius equation that indicates the rate of collision and the probability that colliding reactants are oriented for a successful reaction. (*p. 652*)

half-life ($t_{1/2}$) The time required for a reaction to reach 50% completion. (*p. 636*)

herbicide Any compound used to kill unwanted plants. (*p. 620*)

heterogeneous catalyst A specific catalyst that exists in a different physical state than the reaction. (*p. 656*)

homogeneous catalyst A specific catalyst that exists in the same physical state as the compounds in the reaction. (*p. 655*)

initial rate The instantaneous rate of reaction when $t = 0$. (*p. 623*)

insecticide Any compound used to kill unwanted insects. (*p. 620*)

instantaneous rate The rate of reaction measured at a specific instant in time. The instantaneous rate is typically measured by calculating the slope of a line that is tangent to the curve drawn by plotting [reactant] versus time. (*p. 623*)

integrated first-order rate law An expression derived from a first-order rate law that illustrates how the concentrations of reactants vary as a function of time. (*p. 634*)

integrated rate law A form of the rate law that illustrates how the concentrations of reactants vary as a function of time. (*p. 633*)

integrated second-order rate law An expression derived from a second-order rate law that illustrates how the concentrations of reactants vary as a function of time. (*p. 639*)

integrated zero-order rate law An expression derived from a zero-order rate law that illustrates how the concentrations of reactants vary as a function of time. (*p. 639*)

intermediate A compound that is produced in a reaction and then consumed in the reaction. Intermediates are not indicated in the overall reaction equation. (*p. 650*)

mechanism The series of steps taken by the components of a reaction as they progress from reactants to products. (*p. 648*)

method of graphical analysis A method of determining the rate law for a reaction where the rate of a reaction is plotted versus time. Also known as the method of integrated rate laws. (*p. 644*)

method of initial rates A method of determining the rate law for a reaction where the initial rates of different trials of a reaction are compared. (*p. 641*)

molecularity A description of the number of molecules that must collide in the rate-determining step of a reaction. (*p. 648*)

pesticide Any compound used to kill unwanted organisms (whether they be animals, insects, or plants). (*p. 620*)

pseudo-first-order rate A modification of the second-order rate that enables one to use first-order kinetics. (*p. 640*)

rate The "speed" of a reaction, recorded in M/s. (*p. 621*)

rate constant A constant that is characteristic of a reaction at a given temperature, relating to the rate of disappearance of reactants. (*p. 628*)

rate-determining step The slowest elementary step in a reaction sequence. (*p. 648*)

rate law An equation that indicates the molecularity of a reaction as a function of the rate of the reaction. (*p. 628*)

reaction coordinate A measure that describes the progress of the reaction. (*p. 651*)

reaction profile A plot of the progress of a reaction versus the energy of the components. (*p. 651*)

second-order rate law The rate of a reaction is directly related either to the square of the concentration of a single reactant or to the product of the concentrations of two reactants. (*p. 639*)

termolecular (trimolecular) reaction A reaction in which three molecules collide in the rate-determining step of the reaction. (*p. 648*)

transition state The point along the reaction coordinate where the reactants have collided to form the activated complex. (*p. 651*)

transition state theory The theory that describes how the energy of activation is related to the rate of reaction. (*p. 651*)

unimolecular reaction A reaction in which one molecule alone is involved in the rate-determining step of the reaction. (*p. 648*)

zero-order reaction A reaction in which the rate is not related to the concentrations of any species in the reaction. (*p. 639*)

Focus Your Learning

The answers to the odd-numbered problems and some selected problems appear at the back of the book, as represented by the blue numbering.

Section 15.1 Reaction Rates

Skill Review

1. Use the analogy of a runner in a 10-km road race to define each of these pertinent reaction kinetics terms:
   a. Extent of reaction   c. Instantaneous rate
   b. Average rate   d. Initial rate

2. Use the analogy of firing a bullet from a gun to define each of these kinetics terms:
   a. Extent of reaction   c. Instantaneous rate
   b. Average rate   d. Initial rate

3. For the following reaction in a 4.00-L container, it was found that 1.00×10^{-4} mol of C_4H_8 reacted over a time period from 10:58 A.M. to 11:15 A.M.

$$C_4H_8(g) \rightarrow 2C_2H_4(g)$$

What is the average rate, in M/s, for this reaction? Is it possible to have an instantaneous rate faster than the average rate? Explain.

4. The reaction described in Problem 3 was performed a second time. Into a 1.5-L flask was placed 3.25 mol of $C_4H_8(g)$. After 56 min, 1.00 mol of $C_2H_4(g)$ was obtained. What is the average rate, in M/s, for this reaction?

5. Using the same reaction as in Problem 3 ($C_4H_8 \rightarrow 2C_2H_4$), compare the rate of appearance of C_2H_4 to the rate of disappearance of C_4H_8. Write out this relationship using the style presented in the chapter that depicts the rate, as it would be expressed based on either of the components.

6. The following reaction shows PH_3 decomposing into two products.

$$4PH_3(g) + 3O_2(g) \rightarrow 4P(g) + 6H_2O(g)$$

Write the rate expression that depicts the rate of disappearance of the reactant when compared to the appearance of each product.

7. Determine the rate of reaction, in M/s, for each of these systems over the time period indicated. Assume that the reaction is

$$A \rightarrow B$$

   a. $[A]_0 = 0.350\ M$; $[A]_t = 0.300\ M$; $t_0 = 0$ s; $t = 100$ s
   b. $[A]_0 = 1.522\ M$; $[A]_t = 0.350\ M$; $t_0 = 0$ min; $t = 15$ min
   c. $[A]_0 = 0.050\ M$; $[A]_t = 0.010\ M$; $t_0 = 0$ days; $t = 399$ days
   d. $[A]_0 = 0.280\ M$; $[A]_t = 0.140\ M$; $t_0 = 35$ min; $t = 2.5$ h

8. Determine the rate of reaction, in M/s, for each of these systems over the time period indicated. Assume that the reaction is

$$A \rightarrow 2B$$

   a. $[A]_0 = 0.350\ M$; $[A]_t = 0.280\ M$; $t_0 = 0$ s; $t = 100$ s
   b. $[A]_0 = 2.50\ M$; $[A]_t = 0.250\ M$; $t_0 = 0$ s; $t = 15$ min
   c. $[A]_0 = 1.750\ M$; $[A]_t = 0.010\ M$; $t_0 = 0$ days; $t = 250$ days
   d. $[A]_0 = 0.125\ M$; $[A]_t = 0.105\ M$; $t_0 = 15$ s; $t = 2.5$ min

9. In the general reaction shown below, the rate of disappearance of A is $4.5 \times 10^{-2}\ M/s$.

$$2A + 3B \rightarrow C + 2D$$

   a. What is the rate of disappearance of B?
   b. What is the rate of appearance of C?
   c. What is the rate of appearance of D?

10. In the general reaction shown below, the rate of disappearance of A is $1.85 \times 10^{-4}\ M/s$.

$$A + 2B \rightarrow 2C + 3D$$

   a. What is the rate of disappearance of B?
   b. What is the rate of appearance of C?
   c. What is the rate of appearance of D?

11. Either on a piece of graph paper or using a computer, plot the data given in the accompanying table, which are from a drag race. Then determine:
   a. The initial speed
   b. The instantaneous speed at $t = 8$ s
   c. The instantaneous speed at $t = 25$ s
   d. The average speed of the dragster
   e. In your own words, how does this compare to the typical plot of a reaction?

Distance (mi)	Time (s)
0	0
0.055	5
0.193	10
0.611	20
1.120	30

12. Plot the accompanying data on a piece of graph paper or using a computer. Then determine the following:
   a. The initial rate, in M/min, of the reaction
   b. The instantaneous rate, in M/min, at $t = 15$ min
   c. The instantaneous rate, in M/min, at $t = 45$ min
   d. The average rate, in M/min, of the reaction

[Pesticide]	Time (min)
$0.10000\ M$	0
$0.055294\ M$	15
$0.030575\ M$	30
$0.016906\ M$	45
$0.009348\ M$	60

Chemical Applications and Practice

13. The common disinfectant hydrogen peroxide (H_2O_2) can decompose according to the following balanced equation:

$$2H_2O_2(aq) \rightarrow 2H_2O(l) + O_2(g)$$

   a. How much faster is the appearance of H_2O than the appearance of O_2?
   b. Write the expression that defines the rate of the reaction based on the rate of disappearance of H_2O_2 and the rates of appearance of H_2O and O_2.

14. Reducing the level of nitrogen monoxide compounds emitted in automobile exhaust is a priority of environmentally concerned citizens. One response to this has been the development of catalytic converters. The NO in the exhaust passes over a rhodium-containing converter and is changed to the components already present in clean air.

$$2NO(g) \rightarrow N_2(g) + O_2(g)$$

If the rate of appearance of N_2 were 1.5×10^{-6} mol/s, what would you calculate as the rate of disappearance of NO?

15. The relationship between ozone and humans has a rather unique feature. Whether ozone benefits humans depends largely on proximity. Ozone is considered beneficial when it occurs in the upper atmosphere, but it can cause serious respiratory problems if it is in the atmosphere layer closest to us. Ozone can be formed by the following reaction:

$$O_2(g) + O(g) \rightarrow O_3(g)$$

This graphical representation shows the decomposition of oxygen over time. Redraw the graph and sketch a line that would depict the appearance of ozone over the same time period.

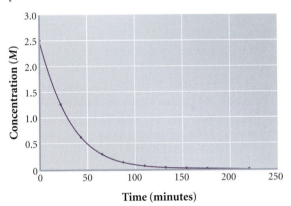

Time (minutes)

16. As we have seen, the reaction that is used to produce the agriculturally critical fertilizer ammonia combines nitrogen and hydrogen gas in the following manner:

$$3H_2(g) + N_2(g) \rightarrow 2NH_3(g)$$

a. Using the following graph of the disappearance of N_2 as a guide, sketch two additional lines that factor in the ratios of the different species for the disappearance of H_2 and appearance of NH_3, respectively.
b. If the disappearance of N_2 in a particular reaction were 1.2×10^{-3} mol/min, what would be the rate of appearance of NH_3?

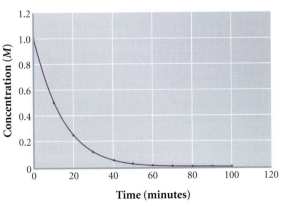

Time (minutes)

15.2 An Introduction to Rate Laws

Skill Review

17. Judging on the basis of collision theory, indicate whether each of these modifications would increase the rate of a reaction. Explain your answers.
a. increasing the temperature
b. decreasing the temperature
c. increasing the initial concentration of reactant
d. diluting the reaction with more solvent

18. Judging on the basis of collision theory, indicate whether each of these modifications would have an effect on the rate of a reaction. Explain your answers.
a. increasing the volume of the reaction
b. decreasing the number of reactant molecules
c. adding some product molecules to the reaction
d. removing the product molecules as they form

19. The following equation represents the rate law for the decomposition of an important pesticide (symbolized here as Pest).
a. What is the order of the reaction?
b. What is the meaning of k?
c. What effect would doubling the concentration of pesticide have on the rate of the reaction?
d. What effect would doubling the concentration of the pesticide have on the value of k?

$$Rate = k\,[Pest]^2[H^+]$$

20. Report the overall order of each of these reactions:
a. $Rate = k[NO][O_3]$ c. $Rate = k[H_2][Cl_2]^{1/2}$
b. $Rate = k[NO][H_2][H_2O]^{-1}$

21. Calculate the rate of the following reaction, if the rate constant is $3.95 \times 10^{-4}\ s^{-1}$ and $[A] = 0.509\ M$. The reaction is first order in A and zero order in B.

$$A + B \rightarrow 2C + D$$

22. Calculate the rate constant for a reaction if, when $[A] = 0.672\ M$, the rate of the reaction is $2.99 \times 10^{-3}\ M/s$. The reaction is second order in A.

$$A \rightarrow B$$

Chemical Applications and Practices

23. The hypochlorite ion (ClO^-) is present in commercial bleach. One aqueous reaction in which this ion participates is

$$3ClO^- \rightarrow ClO_3^- + 2Cl^-$$

Without any further information, give two reasons why we could not claim that the rate law is $Rate = k[ClO^-]$.

24. If the rate law for the hypochlorite reaction from the equation in Problem 23 ($3ClO^- \rightarrow ClO_3^- + 2Cl^-$), is $Rate = k[ClO^-]^2$, what would you report as the order of the reaction? If the concentration of hypochlorite were tripled, what effect would this have on the rate of the reaction (assuming no change in temperature)? The order of the reaction does not match the stoichiometry of the equation. What does this indicate?

Section 15.3
Changes in Time—The Integrated Rate Law

Skill Review

25. Determine the rate constant for each of these first-order reactions:
    a. decomposition of peroxyacetyl nitrate; $t_{1/2} = 1920$ s
    b. decomposition of sulfuryl chloride; $t_{1/2} = 525$ min
    c. radioactive decay of ^{40}K; $t_{1/2} = 1.25 \times 10^9$ years
    d. radioactive decay of ^{14}C; $t_{1/2} = 5730$ years

26. Determine the half-life for each of these first-order reactions:
    a. radioactive decay of ^{131}I; $k = 0.08619$ day^{-1}
    b. radioactive decay of ^{24}Na; $k = 0.0473$ h^{-1}
    c. decomposition of DDT; $k = 2.31 \times 10^{-4}$ day^{-1}
    d. metabolism of malathion; $k = 0.693$ day^{-1}

27. In the following first-order reaction, the half-life was determined to be 43 min.
    a. How long would it take for the concentration of A to drop to 50% of the original amount?
    b. To 25% of the original amount?
    c. To 10% of the original amount?

$$A \rightarrow B$$

28. The radioactive decay of ^{14}C is a first-order reaction.
    a. If a sample originally contains 1.59×10^{-5} M ^{14}C, how long will it take before the concentration is 0.795×10^{-5} M? The half-life of ^{14}C is 5730 years.
    b. How long will it take before the concentration is 1.00×10^{-6} M?

29. For each case below, calculate the concentration of A at the time indicated. The reaction is first order with a half-life of 3.95×10^2 s.
    a. $[A]_0 = 0.100$ M; $\Delta t = 50.0$ s
    b. $[A]_0 = 0.100$ M; $\Delta t = 100.0$ s
    c. $[A]_0 = 0.200$ M; $\Delta t = 50.0$ s
    d. $[A]_0 = 0.200$ M; $\Delta t = 20.0$ s

30. For each case below, calculate the time required to reach the concentration shown. The reaction is first order with a half-life of 1.25×10^{-3} s.
    a. $[A]_0 = 0.100$ M; $[A] = 0.010$ M
    b. $[A]_0 = 0.350$ M; $[A] = 0.095$ M
    c. $[A]_0 = 0.200$ M; $[A] = 0.010$ M
    d. $[A]_0 = 0.250$ M; $[A] = 0.100$ M

31. What was the initial concentration of A if $[A] = 0.0388$ M after 2.75 days? Assume that the half-life of the first-order reaction is 1.18 days.

$$A \rightarrow B$$

32. Determine the rate constant (in s^{-1}) for the first-order decomposition of the pesticide 2,4-D, if the initial concentration was 2.73×10^{-3} M and the concentration after 60.0 days was 8.44×10^{-4} M.

33. In the following cases of first-order reactions, something is wrong with the data. Determine which value is incorrect and indicate why.
    a. $[A]_0 = 0.100$ M; $[A]_t = 0.900$ M; $t_0 = 0$ s; $t = 10$ s
    b. $[A]_0 = 0.090$ M; $[A]_t = 0.010$ M; $t_0 = 10$ s; $t = 0$ s

34. In the following cases of first-order reactions, something is wrong with the data. Determine which value is incorrect and indicate why.
    a. $[A]_0 = 0.100$ M; $[A] = 0.050$ M; $t_0 = 0$ s; $t = 10$ s; $t_{1/2} = 24$ min
    b. $[A]_0 = 0.900$ M; $[A] = 0.450$ M; $t_0 = 0$ s; $t = 10$ s; $k = 0.085$ s^{-1}

35. What is the half-life of the decomposition of ammonia on a metal surface, a zero-order reaction, if $[NH_3] = 0.0333$ M initially and $[NH_3] = 0.0150$ M at $t = 450.0$ s?

36. Determine the concentration of ammonia from Problem 35 when $t = 800.0$ s.

37. What is the value of the rate constant for a zero-order reaction with $t_{1/2} = 3.55$ h? Assume the original concentration $[A]_0 = 0.100$ M.

38. What is the value of the rate constant for a second-order reaction with $t_{1/2} = 300.0$ days? Assume the original concentration $[A]_0 = 1.50$ M.

39. What is the half-life of a second-order reaction if the concentration of A drops to 10% of its original value of 2.00×10^{-3} M in 520 min?

$$2A \rightarrow B$$

40. What is the concentration of A after 3.5 h in the second-order reaction illustrated in Problem 39, if $[A]_0 = 0.100$ M and $k = 1.2 \times 10^{-4}$ M^{-1}·s^{-1}?

41. Determine the concentration of the reactant in each of these cases. Assume that the reaction is second order with $k = 0.0312$ M^{-1}·min^{-1}.
    a. $[A]_0 = 0.100$ M; $t = 10.0$ s
    b. $[A]_0 = 0.500$ M; $t = 10.0$ s
    c. $[A]_0 = 0.339$ M; $t = 200.0$ s
    d. $[A]_0 = 0.0050$ M; $t = 24$ days

42. Determine the concentration of the reactant in each of these cases. Assume that the reaction is second order with $k = 0.410$ M^{-1}·h^{-1}.
    a. $[A]_0 = 0.100$ M; $t = 10.0$ h
    b. $[A]_0 = 0.500$ M; $t = 10.0$ h
    c. $[A]_0 = 0.222$ M; $t = 1.75$ h
    d. $[A]_0 = 0.0010$ M; $t = 14$ days

43. A graphical plot of concentration of a reactant versus time during a reaction will reveal, for reactions above zero order, a changing rate. Use collision theory to explain why the rate slows over time.

44. When you examine the three integrated rate expressions mentioned in the chapter, you can easily note that all three contain the symbol k. What would be the units for k in each of the zero-order, first-order, and second-order rate expressions? (If necessary, use any time unit merely as "time.")

45. Suppose two students are discussing their recent chemistry lab experiment. The first student, Pablo, remarks that his first-order reaction has a half-life of 25 min. The other student, Peter, replies that coincidently, his second-order reaction also has a half-life of 25 min. Then both go back to repeat their experiments, but with different amounts of

starting materials. Neglecting experimental error, explain any differences, or lack of, that they might find this time.

46. a. What advantage does changing a reaction to pseudo-first-order kinetics give to an experimenter?
    b. Describe how to alter a kinetics experiment in order to study it in the pseudo-first-order kinetic model.
    c. Explain why the concentration of one component in the pseudo-first-order model can be made to be part of the specific rate constant of a reaction.

Chemical Applications and Practices

47. An agricultural chemist is attempting to detect the decomposition of a new herbicide. The chemist notes that after application, the compound decays from 100.0% potency to 75.0% potency over a time period of 1 week (168 h). Assume that the original concentration $[A]_0 = 1.00 \times 10^{-3} M$.
    a. What would be the specific rate constant if the reaction were first order with respect to the herbicide?
    b. What would be the specific rate constant if the reaction were second order with respect to the herbicide?
    c. What would be the specific rate constant if the reaction were zero order with respect to the herbicide?

48. Use the values obtained in Problem 47 to determine the half-life of the herbicide, assuming:
    a. first-order kinetics
    b. second-order kinetics
    c. zero-order kinetics

49. Assume that the fermentation of glucose by yeast, to produce ethanol, is a first-order process. Under certain conditions the value of the specific rate constant is 0.00205 h^{-1}. If the initial concentration of glucose were 0.980 M, what would be the concentration after 244 h of fermentation?

50. The precipitation of metal ions with sulfide is often used as an identifying technique for the metal ions. The production of hydrogen sulfide to be used in the precipitation, however, can be dangerous. Consequently, H_2S can be generated by placing thioacetamide (CH_3CSNH_2) in an aqueous acid solution. If the first-order decay constant for thioacetamide under those conditions were 0.46 min^{-1}, what would you calculate as the time required for 0.100 M thioacetamide to reach a concentration of 0.0100 M?

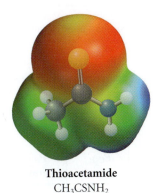

Thioacetamide
CH_3CSNH_2

51. All nuclear decay processes (alpha, beta, and gamma decay) follow first-order kinetics. The isotope americium-241 is used in many smoke detectors. It has a half-life of approximately 241 years. How long would it take the americium in

a smoke detector to decay (into neptunium) from its initial radiation level to 66.6% of its original value?

52. Explain why nuclear decay processes can be considered always to follow first-order kinetics?

15.4 Methods of Determining Rate Laws

Skill Review

53. Plot the following data and determine whether they follow zero-order, first-order, or second-order kinetics.

Time (s)	Concentration (M)
0	0.500
5	0.274
10	0.189
20	0.116
30	0.084

54. Plot the data in Problem 12 and determine whether they follow zero-order, first-order, or second-order kinetics.

55. Explain the change in the rate of the reaction in each case below.
$$Rate = k[A][B]$$
    a. We double [A].
    b. We double [B].
    c. We double [A] and [B].

56. Explain the change in the rate of the reaction in each case below.
$$Rate = k[A]^2[B]$$
    a. We double [A].
    b. We double [B].
    c. We triple [A].

57. Use the method of initial rates to determine the order of each component, along with the general rate law and the value of the rate constant, in the following hypothetical reaction. What is the overall order of the reaction?
$$A + B \rightarrow 2C$$

Experiment	[A] (M)	[B] (M)	Initial Rate (M/s)
1	0.10	0.10	0.222
2	0.10	0.20	0.444
3	0.20	0.20	0.444

58. Use the method of initial rates to determine the order of each component, along with the general rate law and the value of the rate constant, in the following hypothetical reaction. What is the overall order of the reaction?
$$A + B \rightarrow 2C$$

Experiment	[A] (M)	[B] (M)	Initial Rate (M/s)
1	0.10	0.10	0.286
2	0.10	0.20	0.143
3	0.20	0.20	0.286

59. Use the method of initial rates to determine the order of each component, along with the general rate law and the value of the rate constant, in the following hypothetical reaction. What is the overall order of the reaction?

$$A + 2B \rightarrow 2C$$

Experiment	[A] (M)	[B] (M)	Initial Rate (M/s)
1	0.10	0.10	0.105
2	0.10	0.20	0.420
3	0.20	0.20	0.840

60. Use the method of initial rates to determine the order of each component, along with the general rate law and the value of the rate constant in the following hypothetical reaction. What is the overall order of the reaction?

$$2A + 2B \rightarrow 2C$$

Experiment	[A] (M)	[B] (M)	Initial Rate (M/s)
1	0.050	0.10	0.074
2	0.10	0.20	0.888
3	0.050	0.20	0.222

61. Use the method of graphical analysis to determine the order for the hypothetical reaction $A \rightarrow C + D$. What is the rate constant (with appropriate units) for this reaction?

[A] (M)	Time (min)
0.432	0
0.385	1
0.291	3
0.197	5
0.103	7

62. Use the method of graphical analysis to determine the order for the hypothetical reaction $A \rightarrow B$. What is the rate constant (with appropriate units) for this reaction?

[A] (M)	Time (s)
0.100	0
0.050	62
0.025	124
0.013	186
0.0065	248

63. The following two diagrams represent two different containers. Within each container, two different compounds are placed.

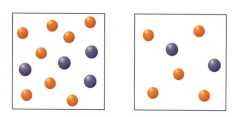

a. If the reaction about to take place is first order with respect to the large dark atoms and zero order with respect to the small light atoms, which graphical plot would yield a straight line for each component?

b. How would the overall reaction rate compare between the first and second containers?

c. What units would you assign to the rate constant for the reaction between the two components? (You may use seconds for the time unit.)

64. The following diagram represents a container with two gas components in their initial conditions. The reaction is zero order with respect to the large atoms and first order with respect to the small atoms. Prepare a similar diagram, but show what would be needed to produce a reaction that would be three times faster than the first (assuming no change in temperature or any other reaction conditions).

Chemical Applications and Practices

65. Municipal water supplies are often treated with chlorine compounds in order to take advantage of the oxidizing ability of chlorine to react with potential pollutants. In some industrial settings, other oxidizing substances are used. One example of such an application is use of the powerful oxidizing potential of iron(VI) compounds. Given the information below, what order would you report for the reaction of the iron(VI) compound? What would you report as the value for k?

[Fe^{6+}] (M)	Time (min)
0.100	0.00
0.0682	1.00
0.0517	2.00
0.0300	5.00
0.00920	10.00

66. The gas phase hydrogenation of ethene is shown in the following reaction:

$$C_2H_2(g) + 2H_2(g) \rightarrow C_2H_6(g)$$

a. If the following data were collected, using the method of initial rates, for four experiments at a fixed temperature, what would you determine as the rate law for the reaction?

Experiment	$P_{C_2H_2}$ (atm)	P_{H_2} (atm)	Initial Rate (atm/s)
1	0.10	0.10	11
2	0.20	0.10	22
3	0.10	0.20	22
4	0.20	0.05	11

b. What would you calculate as the rate constant for the reaction?

c. If the experiment were run once again, at the same temperature, with initial pressures of 0.020 atm for ethene and 0.020 atm for hydrogen, what would you determine as the rate?

67. Suppose the following represents the kinetic study of the oxidation of a new plastic stabilizer being proposed for use in automobile upholstery.

$$Stabilizer(aq) + H^+(aq) + O_2(g) \rightarrow \text{oxidation product}$$

Experiment	Stabilizer (aq) (M)	$H^+(aq)$ (M)	$O_2(g)$ (M)	Rate (M/s)
1	0.400	0.300	0.560	7.14×10^{-4}
2	0.100	0.500	0.200	4.55×10^{-5}
3	0.100	0.100	0.200	4.55×10^{-5}
4	0.400	0.300	0.750	1.28×10^{-3}
5	0.100	0.300	0.560	3.57×10^{-4}

a. What is the rate law for the reaction?

b. What is the rate constant for the reaction at this temperature?

c. What would you calculate as the rate of the reaction, at the same temperature, if the respective initial concentrations were 0.111, 0.200, and 1.00 for the stabilizer, hydrogen ion, and oxygen, respectively?

d. If the rate of the same reaction, at the same conditions, were determined to be 2.55×10^{-4} M/s when the initial concentration of stabilizer was 0.300 M with a hydrogen ion concentration of 0.100, what would the initial oxygen pressure have to be?

68. Brewed coffee left on a warming device will gradually change flavor as a consequence of several complex reactions that occur as the flavor oils decompose. Suppose a flavor chemist for Dr. Beans Coffee Company collected the following data on one such oil as it decomposed over time.

Concentration (M)	Time (min)
0.00128	0
0.00120	10.0
0.00113	20.0
0.00107	30.0
0.000781	100.0
0.000561	200.0

a. Using graphical techniques, determine the order of the decay process.

b. What is the rate constant for the decay?

c. Starting with the initial concentration, how long would it take for half the ingredient to decompose (half-life)?

69. The naturally occurring isotope of hydrogen known as tritium has two neutrons and one proton. The reaction between tritium and deuterium may provide the basis for controlled fusion reactions. The half-life of the radioactive tritium isotope is approximately 12 years. How much radioactive

tritium, starting with a sample producing a beta decay count rate of 545 cpm (counts per minute), would remain after 3 years?

70. Sodium hypochlorite (NaOCl) is the active ingredient in many aqueous commercial bleaching products. If the NaOCl added to a particular sewage treatment process had a half-life of 15 days, what percentage of the original concentration would remain after 1 year? (Assume first-order kinetics.)

15.5 Looking Back at Rate Laws

Skill Review

71. What mathematical effect would there be on the three graphs for first, second, and third order, respectively, if before plotting the terms on each y axis you first subtracted the initial concentration, ln of the initial concentration, and 1/(initial concentration), respectively, from each reading?

72. A researcher mistakenly plots the data from a first-order reaction using log [A] instead of ln [A]. What effect would this have on the resulting graph?

73. What sign do you expect for the slope of the line in a plot of [A] versus t for a zero-order reaction?

74. What sign do you expect for the slope of the line in a plot of ln[A] versus t for a first-order reaction?

75. Which indicates a faster reaction for a first-order process: $k = 1.66 \times 10^{-2}$ s^{-1} or $k = 8.95 \times 10^{-2}$ s^{-1}?

76. Which indicates a faster reaction for a first-order process: $t_{1/2} = 24$ days or $t_{1/2} = 15$ days?

Chemical Applications and Practices

77. One method of producing yellow lead chromate pigment is via a reaction of sodium chromate. Chromium(III) ions can be made to undergo a three-electron oxidation to the chromate ion (CrO_4^{2-}) with the addition of cerium(IV) ions. The rate law was found to be

$$Rate = k[Cr^{3+}]^2 [Ce^{4+}] [CrO_4^{2-}]^x$$

Suppose the concentration of each component in the rate law were doubled. If this caused the rate to quadruple, what would you calculate as the value of x?

78. An approximation that chemists may sometimes casually use is that increasing the temperature of a system by 10 degrees Celsius may double the rate. Suppose you wanted to increase a reaction rate by using only concentration changes. How much would a concentration have to change to cause a reaction to become ten times faster? (Assume the reaction is first order.)

15.6 Reaction Mechanisms

Skill Review

79. Explain the relationship among a reaction mechanism, elementary steps and the rate-determining step.

80. Explain how it is possible for the proposed mechanism of a reaction to be acceptable, yet perhaps not represent the actual occurrence of events in the mechanism of that reaction.

81. You are seated at your kitchen table. Describe each of the steps involved in answering the telephone in your home (be very detailed). Which of these steps is the rate-determining step for this process?

82. What is the rate-determining step in the process you use to go to your first-period chemistry class from your bedroom?

83. What is the activation energy for a hypothetical reaction (A → B) if $k = 1.74 \times 10^{-2}$ s^{-1} at 300 K and $k = 4.22 \times 10^{-2}$ s^{-1} at 400 K?

84. Calculate the rate constant of a first-order reaction at 35°C, given that $k = 8.5 \times 10^{-4}$ s^{-1} at 25°C and $E_a = 144$ kJ.

Chemical Applications and Practices

85. The aqueous reaction between hydrogen peroxide and iodide in an acid solution is represented here:

$$H_2O_2 + 3I^- + 2H^+ \rightarrow 2H_2O + I_3^-$$
$$\text{Rate} = k[H_2O_2][I^-][H^+]$$

One of the following mechanisms can be accepted for the above rate law, and one cannot. Select the acceptable mechanism and show proof for your selection as well as giving the basis for rejection of the other.

Mechanism A

$H^+ + I^- \rightleftharpoons HI$ (fast)

$H_2O_2 + HI \rightarrow H_2O + HOI$ (slow)

$HOI + H^+ + I^- \rightarrow H_2O + I_2$ (fast)

$I^- + I_2 \rightarrow I_3^-$ (fast)

Mechanism B

$H_2O_2 + I^- \rightarrow H_2O + OI^-$ (slow)

$H^+ + I^- \rightleftharpoons HI$ (fast)

$H^+ + OI^- + HI \rightarrow H_2O + I_2$ (fast)

$I_2 + I^- \rightarrow I_3^-$ (fast)

86. When one reactant produces two products, the reaction is said to be a disproportionation reaction. The following shows the disproportionation of the hypochlorite ion used to make commercial bleaching products:

$$3ClO^-(aq) \rightarrow 2Cl^-(aq) + ClO_3^-(aq)$$

On the basis of the following proposed mechanism, what would you write as the rate law for this interesting reaction?

$$ClO^- + ClO^- \rightarrow ClO_2^- + Cl^- \quad \text{(slow)}$$
$$ClO^- + ClO_2^- \rightarrow ClO_3^- + Cl^- \quad \text{(fast)}$$

87. The solvent acetone has many industrial uses and is also a common ingredient in nail polish remover. It does decompose to a weak acid, carbonic acid. At 10.0°C, the rate constant for that decomposition is approximately 6.4×10^{-5} L/mol·s. At 78°C, the rate constant for the reaction has a value of 2.03×10^{-1} L/mol·s.
   a. What is the activation energy for this reaction?
   b. What is the rate constant for this reaction at 37°C?

88. At 78°C, what would you calculate as the value of the Arrhenius frequency factor, A, for the decomposition of acetone (Problem 87)?

15.7 Applications of Catalysts

Skill Review

89. Provide clear and concise definitions for each term:
   a. Reaction intermediate
   b. Activated complex
   c. Homogeneous catalyst
   d. Heterogeneous catalyst

90. a. Explain how a catalyst increases the rate of a chemical reaction.
   b. Explain why a catalyst does not appear as part of the stoichiometry in a balanced equation.

91. Consider the following elementary steps for the decomposition of N_2O.

$$N_2O \rightarrow N_2 + O$$
$$N_2O + O \rightarrow N_2 + O_2$$

   a. What is the overall reaction?
   b. Indicate any intermediates or catalysts involved in the reaction.
   c. Experimental evidence reveals that the rate law for this reaction is Rate = $k[N_2O]$. Which of the steps is the rate-limiting step in the mechanism?

92. The following mechanism has been proposed for the reaction of ICl(g) and $H_2(g)$.

$$H_2 + ICl \rightarrow HI + HCl$$
$$HI + ICl \rightarrow I_2 + HCl$$

   a. What is the overall reaction?
   b. Indicate any intermediates or catalysts involved in the reaction.
   c. The first step in the mechanism is slow compared to the second. What is the rate law for the reaction?

93. A schematic mechanism for the reaction of lactose and lactase (an enzyme) to produce glucose and galactose is shown below.

$$\text{lactose} + \text{lactase} \rightarrow (\text{lactase–lactose})$$
$$(\text{lactase–lactose}) \rightarrow (\text{lactase–glucose–galactose})$$
$$(\text{lactase–glucose–galactose}) \rightarrow \text{lactase} + \text{glucose} + \text{galactose}$$

   a. What is the overall reaction?
   b. Indicate any intermediates or catalysts involved in the reaction.
   c. The first step in the mechanism is slow compared to the rest. What is the rate law for the reaction?

94. Each of the following represents an elementary step in a different mechanism. Classify each as unimolecular, bimolecular, or termolecular.
   a. $^{40}K \rightarrow {}^{40}Ar$
   b. $N_2 + Fe \rightarrow FeN_2$
   c. $2NO \rightarrow N_2O_2$
   d. $2NO + Cl_2 \rightarrow 2NOCl$

Chemical Applications and Practices

95. The industrial production of ammonia makes use of catalysts. One generalized mechanism may be presented as follows:

$$N_2(g) + Fe(s) \rightarrow FeN_2(s)$$
$$3H_2(g) + 3Fe(s) \rightarrow 3FeH_2(s)$$
$$FeN_2(s) + 3FeH_2(s) \rightarrow 4Fe(s) + 2NH_3(g)$$

   a. What is the overall stoichiometry of the reaction?
   b. What intermediates, if any, are present?
   c. What catalyst is present?
   d. Is the catalyst homogeneous or heterogeneous?

96. Diagram a reaction profile that illustrates the exothermic reaction whose mechanism is shown here. Next use a dotted line to show the same profile with any changes that would reflect the role of a heterogenous catalyst.

$$A(aq) + B(aq) \rightarrow C(aq) \quad \text{(slow)}$$
$$C(aq) \rightarrow D(g) \quad \text{(fast)}$$

Comprehensive Problems

97. In the reaction shown below, the initial concentration of H_2O_2 is 0.250 M, and 8 s later the concentration is 0.223 M. What is the initial rate of this reaction expressed in M/s and M/h?

$$2H_2O_2(aq) \rightarrow 2H_2O(l) + O_2(g)$$

98. What is the half-life for the first-order decomposition of dimethyl ether at 500°C if the rate constant for the reaction is 2.567×10^{-2} min^{-1}?

$$(CH_3)_2O \rightarrow CH_4 + H_2 + CO$$

99. A first-order reaction, $A \rightarrow B$, has a rate of 0.0875 M/s when $[A] = 0.250\ M$.
 a. What is the rate constant?
 b. What is the half-life for this reaction?

100. If the order of a reaction component were -1, what would be the effect of tripling the concentration of that component?

101. If the $\Delta H°$ value for a reaction were $+125$ kJ and the activation energy for the reverse of the reaction were $+75$ kJ, what would you calculate as the value for the activation energy for the forward reaction?

102. Consider the following reaction mechanism.

$$C_2H_6O(aq) + HCl(aq) \rightleftharpoons C_2H_7O^+(aq) + Cl^-(aq) \quad \text{(fast)}$$
$$C_2H_7O^+(aq) + Cl^-(aq) \rightarrow C_2H_5Cl(aq) + H_2O(l) \quad \text{(slow)}$$

 a. Write the overall reaction that is indicated by the mechanism.
 b. What would you predict as the rate law for the reaction?

103. Consider the following reaction mechanism.

$$C_2H_4(aq) + HCl(aq) \rightarrow C_2H_5^+(aq) + Cl^-(aq) \quad \text{(slow)}$$
$$C_2H_5^+(aq) + Cl^-(aq) \rightarrow C_2H_5Cl(aq) \quad \text{(fast)}$$

 a. Write the overall reaction that is indicated by the mechanism.
 b. What would you predict as the rate law for the reaction?

104. Here is the reaction of hydrogen with nitrous oxide:

$$2NO + 2H_2 \rightarrow 2H_2O + N_2$$

The rate law was determined by experiment to be

$$\text{rate} = k[NO]^2[H_2]$$

Is the following mechanism consistent with the rate law? Prove your answer.

$$NO + H_2 \rightleftharpoons N + H_2O \quad \text{(fast)}$$
$$N + NO \rightleftharpoons N_2O \quad \text{(fast)}$$
$$N_2O + H_2 \rightarrow N_2 + H_2O \quad \text{(slow)}$$

105. We pointed out in the text that there are several catalysts used in motor vehicle catalytic converters, including platinum, palladium, and rhodium. In diesel engines, new catalytic converters include the use of cerium(IV) oxide.
 a. Why do these particular metals work so well as heterogeneous catalysts?
 b. In your role as a chemical consultant for an automobile company, you are asked to choose one of the four listed catalysts as the most effective, based on your external research, which would you choose and why?

106. Under most conditions, the time it takes for a sample of methane to completely burn in oxygen is typically on the order of a few milliseconds. A half-life of 8.0 ms is typical for the complete combustion of one mole of methane to produce only $CO_2(g)$ and $H_2O(g)$. How much energy is given off after one half-life if all of the simplifying assumptions are valid?

107. Methane is a potent greenhouse gas that is stable in the upper atmosphere for about 9 to 15 years.
 a. Write the reaction that illustrates the complete combustion of methane.
 b. Why is the reaction of methane with oxygen so fast whereas the reaction in the upper atmosphere is so slow?

108. High-temperature reactions introduce a host of intermediate species even when the simplest substances react. Investigate and report on the intermediates generated at high temperature when hydrogen and oxygen react in the space shuttle's main engines.

Thinking Beyond the Calculation

109. Draw a hypothetical reaction profile for the Haber process involving 1 mol of nitrogen and 3 mol of hydrogen in a 1.0 L flask.

$$N_2(g) + 3H_2(g) \rightarrow 2NH_3(g)$$

 a. Use information you have learned in previous chapters to determine whether the reaction is exothermic or endothermic.
 b. Is this reaction spontaneous at room temperature? At what temperature is the reaction at equilibrium?
 c. On the reaction profile you drew, indicate what you'd expect to see if the reaction were homogeneously catalyzed.
 d. Draw what you'd expect to see if the reaction were heterogeneously catalyzed.
 e. If the catalyzed reaction were accomplished at 300°C using the quantities outlined in the start of this question, what would be the yield, in grams, of ammonia ($NH_3(g)$)? Assume the reaction produces only a 45% yield.
 f. If 10.0 kg of nitrogen and 30.0 kg of hydrogen were combined in the catalyzed reaction, what would be the theoretical yield, in kilograms, of ammonia?

16

Chemical Equilibrium

Just as this balance settles at a fixed position of lowest energy, chemical reactions "strive" toward a position of lowest free energy—that is, chemical equilibrium.

Contents and Selected Applications

16.1 The Concept of Chemical Equilibrium
Chemical Encounters: Myoglobin in the Muscles

16.2 Why Is Chemical Equilibrium a Useful Concept?
Chemical Encounters: Important Processes That Involve Equilibria

16.3 The Meaning of the Equilibrium Constant
Chemical Encounters: The Manufacture of Sulfuric Acid

16.4 Working with Equilibrium Constants

16.5 Solving Equilibrium Problems—A Different Way of Thinking

16.6 Le Châtelier's Principle
Chemical Encounters: Catalysts in Industry

16.7 Free Energy and the Equilibrium Constant

Go to **college.hmco.com/pic/kelterMEE** for online learning resources.

"I'm running late." We seem to hear that expression so often in our hyper-charged, have-to-do-it-now world. Running late might mean being stuck in traffic or on mass transit. There is not much you can do in a sea of cars and trucks. But sometimes you actually do run to get where you are going. As you struggle to make up time, you breathe more heavily than normal, desperately sucking air into your lungs. Your body's efforts to get more oxygen to the muscles are greatly enhanced by a protein in muscle cells called myoglobin.

When you are resting, the myoglobin becomes loaded up with oxygen molecules. These oxygen molecules can be quickly released when your muscle cells undergo activity. How does the myoglobin acquire and release oxygen when we need it?

Myoglobin (Mb), shown in Figure 16.1, is a protein with a molecular weight of about 16,900 g/mol. A key part of the protein is the **heme group**, shown in Figure 16.2. The heme contains an Fe^{2+} ion that can form a bond with one molecule of oxygen, creating a myoglobin–oxygen complex as follows:

$$Mb + O_2 \rightarrow MbO_2$$

When your cells need the oxygen, it is released from the complex by reversing the direction of the reaction:

$$MbO_2 \rightarrow Mb + O_2$$

The free myoglobin is then available to bind more oxygen, which it can later release as needed, and continue in this *reversible* cycle of bind–release–bind–release. Let's examine this reaction more closely to help us understand the concepts of reversible reactions and the essential details of reactions such as this, which are known as chemical equilibria.

Strenuous activities such as running require a sharp change in the amount of oxygen your muscles require. The interaction of myoglobin and oxygen in the muscles accommodates this need.

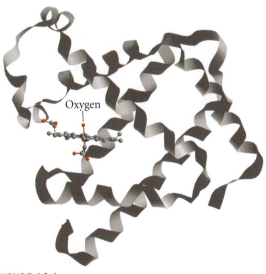

FIGURE 16.1

Myoglobin (Mb) is a protein with a molecular weight of about 16,900 g/mol. A key part of the protein is the heme group, highlighted on the left, which can bind to an oxygen molecule.

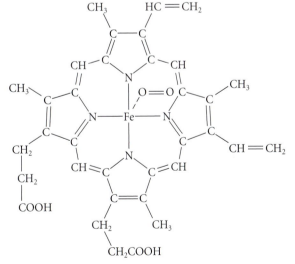

FIGURE 16.2

This drawing of a heme group highlights the Fe^{2+} ion, which can bind to one molecule of oxygen as shown.

Application

CHEMICAL ENCOUNTERS: Myoglobin in the Muscles

Video Lesson: The Concept of Equilibrium

Tutorial: The Equilibrium Condition and Equilibrium Constant

16.1 The Concept of Chemical Equilibrium

Why does myoglobin's reversible cycle of binding and releasing oxygen take place? We know from our understanding of thermodynamics (Chapter 14) that spontaneous reactions lower the free energy, G, of the chemicals involved. When we consider myoglobin and oxygen in living systems, we find that the two competing reactions exist. In one, myoglobin combines with oxygen. In the other, the MbO_2 complex releases oxygen. The end result is that the molecules in each reaction will never be completely converted into products, because these products are always being converted back into reactants. *In a closed system,* these reversible reactions will settle into a state in which the forward and reverse reactions occur at equal rates, where *no net change in the concentrations occurs,* even though both reactions continue. We describe this state as the position of chemical equilibrium, or just **equilibrium** (the plural is *equilibria*). What thermodynamic parameter determines when an equilibrium exists? *The concentrations will no longer change when the free energy of the system is at a minimum for the process.* At that point, the *change* in free energy of the system, ΔG, is 0 kJ/mol.

We symbolize a system that can settle into equilibrium by using double arrows, to represent the fact that both the forward and reverse reactions occur. For example, the myoglobin and oxygen reaction equilibrium is written

$$Mb + O_2 \rightleftharpoons MbO_2$$

As discussed in Chapter 15, we express the *molar* concentration of a substance by placing its chemical symbol in brackets, [ ], such as [Mb] or [MbO_2].

Let's examine what happens to the concentrations of the reactants and products as the minimum free energy is approached (see Figure 16.3.) To make things as straightforward as possible, we will consider the reaction occurring in a beaker, rather than in the human body where all kinds of complications (such as "running late") affect the chemistry.

Let's assume that we place oxygen in our beaker so that its concentration is $3.0 \times 10^{-4}\,M$. We would say $[O_2]_0 = 3.0 \times 10^{-4}\,M$, where the subscript 0 means the concentration at "time = 0," which is another way of saying the initial concentration. In fact, the amount of myoglobin normally present in muscle cells varies a great deal among, and within, biological species. In humans, a mid-range value is $[Mb]_0 = 2.0 \times 10^{-4}\,M$. We will assume for this discussion that we have a closed system (not the human body) and that there is no MbO_2 complex initially, so $[MbO_2]_0 = 0\,M$. As the reaction proceeds, oxygen binds to myoglobin. The forward reaction proceeds with a rate constant of k_1, known in this particular reaction as the **binding rate constant**.

$$Mb + O_2 \xrightarrow{k_1} MbO_2$$

FIGURE 16.3

Like a ball finding its lowest energy point in a valley, reactions find their lowest free energy position. This is the equilibrium position (*B*).

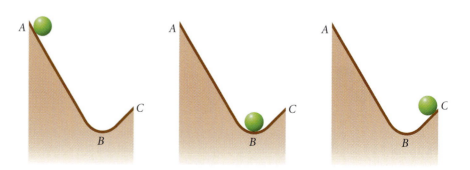

The concentration of oxygen *decreases*, as does that of free myoglobin, as shown in Figure 16.4. The concentration of the MbO_2 complex *increases* from zero as the reaction proceeds. As soon as the MbO_2 complex forms, however, it begins to dissociate back into Mb and O_2 with a rate constant of k_{-1}, known in this particular reaction as the **release rate constant**.

$$MbO_2 \xrightarrow{k_{-1}} Mb + O_2$$

Both reactions proceed until the free energy of the system is at a minimum, which occurs when *the rates of the forward and reverse reactions are equal*. That statement embodies the two key ways to define chemical equilibrium:

Definitions of Equilibrium

1. The free energy *change* of both forward and reverse reactions is zero ($\Delta G = 0$), and the free energy of the system is at its minimum.
2. The rates of the forward and reverse reactions are equal.

$$\text{rate}_f = \text{rate}_r$$
$$k_1[Mb][O_2] = k_{-1}[MbO_2]$$

This is shown in Figure 16.5. The forward and reverse reactions that bind and release oxygen are still occurring. In fact, they are occurring at the same rate. Therefore, *equilibrium is a dynamic process*. And there is no overall change in the concentration of Mb, O_2, or MbO_2 at equilibrium.

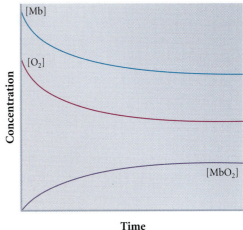

Time

FIGURE 16.4

When myoglobin and oxygen are mixed, the concentration of oxygen decreases, as does that of free myoglobin, as shown. The concentration of the MbO_2 complex increases from 0 *M*. As soon as the MbO_2 complex forms, however, it begins to dissociate back into Mb and O_2.

FIGURE 16.5

Equilibrium is a dynamic process. The forward and reverse reactions are occurring at equal rates. The result is no net change in concentration of Mb (the blue shapes), O_2 (the red dots), or MbO_2 at equilibrium.

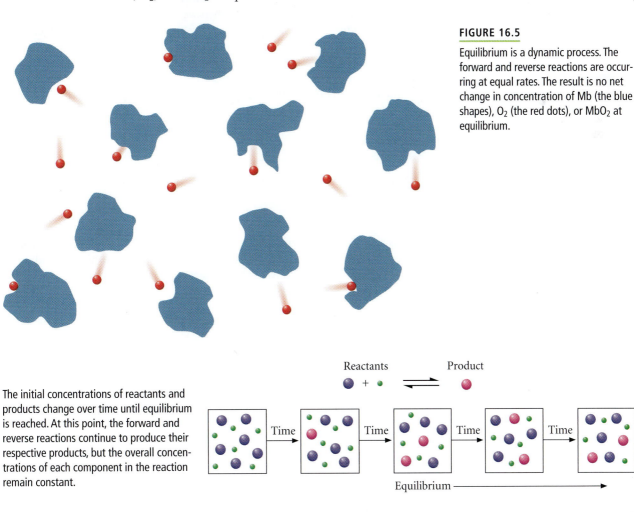

The initial concentrations of reactants and products change over time until equilibrium is reached. At this point, the forward and reverse reactions continue to produce their respective products, but the overall concentrations of each component in the reaction remain constant.

Reactants   Product

Equilibrium ⟶

EXERCISE 16.1 **Equilibrium Concentrations**

Hydrogen and iodine gases can be combined to form hydrogen iodide,

$$H_2(g) + I_2(g) \rightleftharpoons 2HI(g)$$

Given the following initial and final concentrations, calculate the equilibrium concentration of HI.

$$[I_2]_0 = 0.037 \ M \qquad [H_2]_0 = 0.121 \ M \qquad [HI]_0 = 0.099 \ M$$
$$[I_2] = 0.022 \ M \qquad [H_2] = 0.106 \ M \qquad [HI] = ??? \ M$$

First Thoughts

The mole ratios of reactants to product are 1-to-1-to-2, so 1 mol each of H_2 and I_2 react to produce 2 mol of HI in a constant volume. Because the volume remains constant, we can think of the reaction stoichiometry this way: 1 M H_2 and 1 M I_2 react to produce 2 M HI.

Solution

The concentrations of I_2 and H_2 both decrease by 0.015 M.

$$0.015 \ M \ H_2 \ (\text{or} \ I_2) \times \frac{2M \ HI}{1M \ H_2 \ (\text{or} \ I_2)} = 0.030 \ M \ HI \ \text{produced}$$

$$[HI] = [HI]_0 + 0.030 = 0.13 \ M$$

Further Insights

Much of the discussion in this chapter and in those that follow will consider how we can *predict* equilibrium concentrations when we know only about the system and the initial conditions. We will spend considerable time later in this chapter discussing how the ability to predict what will happen in a system is important in the chemical industry. For example, in the development of a manufacturing process, understanding how changes in conditions affect the equilibrium process enables the manufacturer to produce the largest possible yield of product, as well as save the company (and ultimately the consumer) money.

PRACTICE 16.1

Calculate the equilibrium concentration of HI if $[I_2]_0 = 0.050 \ M$, $[H_2]_0 = 0.044 \ M$, $[HI]_0 = 0.206 \ M$, and $[I_2] = 0.041 \ M$.

See Problem 21.

Video Lesson: The Law of Mass Action and Types of Equilibrium

It has been shown by experiment that the equilibrium concentrations of myoglobin, oxygen, and the MbO_2 complex are related by the ratio of the product to reactant concentrations known as the **mass-action expression** or, often, the **equilibrium expression**, of the reaction.

$$K = \frac{[MbO_2]}{[Mb][O_2]}$$

The value K is called the **equilibrium constant** for the reaction at a given temperature. We can get the equivalent result by separating the rate constants from the concentrations in the rate expressions as follows. At equilibrium, the rate of the forward and reverse reaction are equal.

$$k_1[Mb][O_2] = k_{-1}[MbO_2]$$

Rearranging the equation to put the rate constants on one side gives

$$\frac{k_1}{k_{-1}} = \frac{[MbO_2]}{[Mb][O_2]}$$

Therefore, the equilibrium constant, K, is the ratio of the forward rate constant to the reverse rate constant.

$$K = \frac{k_1}{k_{-1}}$$

The value for k_1 in the myoglobin–oxygen reaction at 20°C is $1.9 \times 10^7 \ M^{-1} \cdot s^{-1}$, and k_{-1} is $22 \ M^{-1} \cdot s^{-1}$, so we can calculate the value of K:

$$K = \frac{k_1}{k_{-1}} = \frac{1.9 \times 10^7}{22} = 8.6 \times 10^5$$

We typically would express a calculation such as this by including the units. However, the equilibrium constant does not have units. This is true because the actual thermodynamic definition of an equilibrium constant is based on the ratio of a substance's concentration to a standard state concentration. Even though the reasoning is subtle, the bottom line is that the equilibrium constant is typically given without units. In the reaction of myoglobin and oxygen, K is simply 8.6×10^5.

These last two results—that there exists a mass-action expression to find the equilibrium concentrations of Mb, O_2, and MbO_2, and that there exists an equilibrium constant for this expression at a particular temperature—can be extended to any reversible reaction. For example, in a general equilibrium (reactants A and B yielding products C and D) represented with stoichiometric coefficients m, n, p, and q,

$$mA + nB \rightleftharpoons pC + qD$$

The mass-action expression is

$$K = \frac{[C]^p [D]^q}{[A]^m [B]^n}$$

Exercise 16.2 explains why each equilibrium concentration is raised to the power of its coefficient in the mass-action expression.

EXERCISE 16.2   **The Mass-Action Expression**

The production of ammonia for use in fertilizers, via the Haber process, can be depicted as

$$H_2(g) + H_2(g) + H_2(g) + N_2(g) \rightleftharpoons NH_3(g) + NH_3(g)$$

Write the mass-action expression for the formation of ammonia.

First Thoughts

We did something unusual by breaking the equation down to list each individual molecule without the coefficients, a rather cumbersome way of displaying an equation. Our goal in doing this is to demonstrate how the product of the coefficients in the mass-action is displayed in exponential form.

Solution

If we base the mass-action expression on the reaction *exactly as it is written*, we get the following mass-action expression:

$$K = \frac{[NH_3][NH_3]}{[H_2][H_2][H_2][N_2]} = \frac{[NH_3]^2}{[H_2]^3 [N_2]}$$

Further Insights

The exponential form on the right-hand side is what we would have obtained had we written the Haber process in the usual way:

$$3H_2(g) + N_2(g) \rightleftharpoons 2NH_3(g)$$

This shows *why* the mass-action expression includes the equilibrium concentration of each substance raised to the power of its coefficient. If the coefficient is 1, as with N_2, it is not shown as an exponent, because raising anything to the power of 1 does not change its value.

PRACTICE 16.2

One of the ways to produce hydrogen gas industrially involves the reaction of methane with high-temperature steam. The reaction can be written as

$$CH_4(g) + H_2O(g) \rightleftharpoons CO(g) + H_2(g) + H_2(g) + H_2(g)$$

Write the mass-action expression for this reaction.

See Problems 5 and 6.

EXERCISE 16.3   Practice with Mass-Action Expressions

Write the mass-action expression for each of these reactions:

 a. $PCl_5(g) \rightleftharpoons PCl_3(g) + Cl_2(g)$
 b. $S_8(g) \rightleftharpoons 8S(g)$
 c. $Cl_2O_7(g) + 8H_2(g) \rightleftharpoons 2HCl(g) + 7H_2O(g)$

Solution

In each case, the mass-action expression for the general reaction

$$mA + nB \rightleftharpoons pC + qD$$

is of the form in which the equilibrium concentration of each reactant or product is raised to the power of its coefficient.

$$K = \frac{[C]^p [D]^q}{[A]^m [B]^n}$$

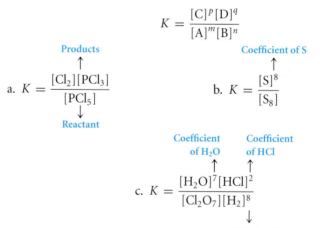

Note: Mathematically $[C]^p[D]^q = [D]^q[C]^p$, so the order in which we write these terms doesn't matter, as long as we keep each exponent with its term. For instance, the exponent p must remain with $[C]$.

PRACTICE 16.3

Write the equilibrium expression for each of these reactions:

 a. $NO(g) + O_3(g) \rightleftharpoons NO_2(g) + O_2(g)$
 b. $HCl(aq) + NH_3(aq) \rightleftharpoons NH_4Cl(aq)$
 c. $C_2H_2(g) + 2H_2(g) \rightleftharpoons C_2H_6(g)$

See Problems 9 and 10.

Concentration-based mass-action expressions are generally suitable representations for the equilibria that occur in chemical reactions. However, in some aqueous solutions, especially where the solute concentrations are high, chemists often make a correction using activities instead of molarities, to get the most meaningful results. For example, if we prepare a solution that contains 3.0 mol of hydrochloric acid (HCl) in a liter of water, we say that the HCl concentration is 3.0 molar. We expect this strong acid to dissociate into H^+ and Cl^- ions, and we therefore assert that in the solution, the concentration of each ion, H^+ and Cl^-, is 3.0 M. However, this is not completely true. Some of the hydrogen cations do interact with the chloride anions in solution, and the ions interact with the water solvent through ion–dipole interactions, as we discussed in Chapter 10. This means that the *effective* concentration is likely to be somewhat different from the intended concentration, especially in relatively concentrated solutions. This effective concentration of the solute is called its **activity**. We will generally not consider the impact of activity in our discussions, but it is important for you to know that such an idea exists and can be dealt with quantitatively.

HERE'S WHAT WE KNOW SO FAR

- Most reactions do not go to completion. Rather, they reach a point of minimum free energy. We call this the point the position of chemical equilibrium.

- At chemical equilibrium, the rates of the forward and reverse reactions are equal.

- Chemical equilibrium is a dynamic, not static, condition.

- A reaction at chemical equilibrium can be described by a mass-action expression for which there is an equilibrium constant, K, that depends on temperature.

16.2 Why Is Chemical Equilibrium a Useful Concept?

The reaction of myoglobin with oxygen is just one of countless examples of equilibrium processes in living systems. In fact, the chemistry of blood is filled with equilibria. The chemistry of the environment also involves equilibrium chemistry. As we will explore in Section 16.6, we can control the position of equilibrium—that is, we can make it possible for reactions to proceed almost all the way toward products or to reach a point at which mostly reactants exist. For example, we can select the conditions so that the greatest possible amount of ammonia is formed from hydrogen and nitrogen in the Haber process. We can also maximize the amount of sulfur trioxide that is formed from the reaction of sulfur dioxide with oxygen as part of the industrial-scale preparation of sulfuric acid. We can attempt to reduce the concentrations of acid in lakes and streams. We can learn about the impact of chlorofluorocarbons (CFCs) on stratospheric ozone levels. We can prepare pharmaceuticals to work in the most effective ways. We can analyze for the presence of an extraordinary variety of substances, from silver to steroids. We can do these things because we understand the fundamental ideas of equilibrium. Our ability to use equilibrium concepts to control the extent of chemical processes has the following vital implications:

1. Economic implications via the trillions of dollars of manufactured products prepared and sold by the chemical industry.

2. Environmental implications for the quality of air, water, and land in the closed system that is Earth.

3. Personal implications for our health.

Application

CHEMICAL ENCOUNTERS:
Important Processes
That Involve Equilibria

Achievement of chemical equilibrium occurs in living systems, chemical manufacturing, and the environment. Equilibrium figures in widely disparate processes related to the chemistry of blood, ammonia manufacture, steroid analysis, and lakes and waterways.

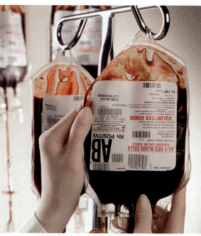

TABLE 16.1	**Chemical Equilibrium Processes**	
Category	**Process**	**Important Reaction**
Industrial	Contact process	$2SO_2(g) + O_2(g) \rightleftharpoons 2SO_3(g)$
Industrial	Haber process	$3H_2(g) + N_2(g) \rightleftharpoons 2NH_3(g)$
Biological	Myoglobin uptake of oxygen	$Mb(aq) + O_2(aq) \rightleftharpoons MbO_2(aq)$
Analytical	Chromatographic analysis	$A_{\text{mobile phase}} \rightleftharpoons A_{\text{stationary phase}}$
Analytical	Metal analysis with EDTA	$Ca^{2+} + EDTA^{4-} \rightleftharpoons CaEDTA^{2-}$
Environmental	Leaching of lead into water	$PbCO_3(s) + 2H^+(aq) \rightleftharpoons Pb^{2+}(aq) + CO_2(g) + H_2O(l)$

Table 16.1 lists some significant processes that involve chemical equilibria. Among them is chromatography, the subject of the accompanying boxed feature on equilibrium and chromatographic analysis.

16.3 The Meaning of the Equilibrium Constant

Application

CHEMICAL
ENCOUNTERS:
The Manufacture
of Sulfuric Acid

More sulfuric acid (H_2SO_4) is produced in the United States than any other chemical; 37.5 billion kg were produced in 2004. Sulfuric acid has a wide range of uses, the most important being the production of agricultural fertilizers. Its industrial and agricultural importance makes examining the manufacture of sulfuric acid an excellent way to illustrate the meaning of the equilibrium constant.

Most of the sulfuric acid produced in the United States is made via the **Contact process**, which began to be used on an industrial scale in 1880. The process is a set of steps performed on readily available materials.

Step 1: The process begins with the sulfur provided by minerals such as iron pyrite (FeS_2) or hydrogen sulfide gas (H_2S). The sulfur combines with oxygen from the air.

$$S(s) + O_2(g) \rightleftharpoons SO_2(g)$$

Step 2: In this step, sulfur trioxide (SO_3) is generated by the oxidation of SO_2 produced in step 1.

$$2SO_2(g) + O_2(g) \rightleftharpoons 2SO_3(g)$$

Sulfur dioxide gas is passed over a catalyst bed containing 6–10% vanadium(V) oxide (V_2O_5) at 600°C and then at 400°C. *Temperature has a major effect on the equilibrium of this reaction, as we will see.*

Step 3:   Sulfur trioxide (SO_3) is then combined with concentrated sulfuric acid and water to give *more* sulfuric acid in a net process that can be written as

$$SO_3(g) + H_2O(l) \rightleftharpoons H_2SO_4(aq)$$

Let's revisit step 2 of the contact process, the reaction of SO_2 and oxygen to make SO_3. The equilibrium constant, K, for this reaction at 27°C is 4.0×10^{24}. What does this value tell us about the relative amounts of reactants and products when the reaction reaches equilibrium? When the value for K is very large, the equilibrium lies essentially all the way to the products side. (When we say, "reactants and products side," we are always referring to these in terms of the forward reaction, written from left to right.) The reaction goes almost, but not fully, to completion. We can use a line chart, as shown to the right, to visualize the meaning of the equilibrium constant for this reaction by comparing the equilibrium position (E) to the starting position (S). We start with only reactants (R) and have no products (P). In practice, the contact process is not carried out at 27°C because it takes too long for this reaction to reach equilibrium at that temperature. In the chemical industry, time is money. To speed up the process, the temperature is raised, which lowers the equilibrium constant for this reaction. Unfortunately, for this and many other equilibria, *the extent of the reaction indicated by the equilibrium constant and the kinetics of the reaction (how fast it gets there), can force us into a compromise between reaction speed and reaction yield.* We will discuss the interplay of these variables and their effect on selecting reaction conditions in Section 16.6.

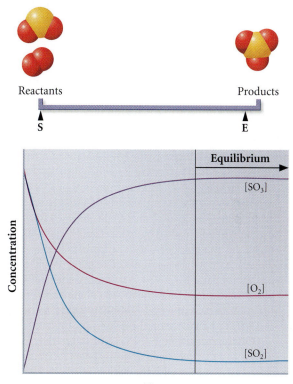

Every reaction has its own set of temperature-dependent values of K (which is also known as K_c when only concentrations are used to determine its value), the equilibrium constant. Our chapter-opening reaction of myoglobin with oxygen has $K = 8.6 \times 10^5$ at 20°C, which is fairly large. The opposite, a small equilibrium constant, is also possible. For example, an equation can be written that illustrates the formation of silver and chloride ions in a saturated solution of silver chloride (AgCl) from solid AgCl. Recall from Chapter 11 that a saturated solution is one in which no more solute will dissolve. The *small* equilibrium constant of 1.7×10^{-10} indicates that the *product* concentrations are quite small, $[Ag^+] = [Cl^-] = 1.3 \times 10^{-5} M$. This implies that very little silver chloride dissolves at 25°C before equilibrium (the saturated solution) is reached.

As the reaction proceeds, the concentration of product increases and the concentrations of reactants decrease. At some point, the concentrations of each component reach and maintain a constant value. This is the point when equilibrium is reached.

$$AgCl(s) \rightleftharpoons Ag^+(aq) + Cl^-(aq) \qquad K = 1.7 \times 10^{-10}$$

When the value for K is very small, the equilibrium lies nearly all the way to the reactants side. We can show this using our line chart. In this case, the equilibrium position is quite close to the silver chloride reactant, and we would say that silver chloride isn't very soluble in water.

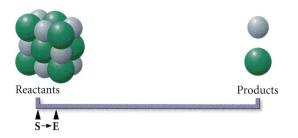

How do we know?

Equilibrium and chromatographic analysis

If you were to take a vote among chemists as to the single most important technique for finding out what you have, and how much of it there is in a sample, the winner might well be **chromatography**, which was the subject of the boxed feature in Chapter 12. This technique has been used in many chemical analyses. For instance, the analysis of steroids and other banned substances in the urine of baseball players, football players, and Olympic athletes is done by separating out the chemicals via chromatography. As another example, the compounds that make up gasoline can be separated out and identified chromatographically. The technique has even been used to identify the gases on the planet Venus.

In a chromatograph, the components of a sample can be separated on the basis of how they distribute themselves between two chemical or physical phases. Figure 16.6 shows the essential parts of a gas chromatograph. The sample to be analyzed is injected into the instrument and pushed by an inert gas (in **gas chromatography**) or by a liquid (in liquid chromatography) into a long tube known as a column. The gas or liquid that pushes the sample moves, so it is called the **mobile phase**. The sample and the mobile phase pass through the column packed with a **stationary phase**, so called because it stays in place on the column.

On the ride through the column, all of the components of the sample (called the analytes) interact physically or chemically with the stationary phase. Here is where our study of equilibria comes in. The interaction of each analyte, "A," with the mobile and stationary phases can be described by the reversible reaction

$$A_{\text{mobile phase}} \rightleftharpoons A_{\text{stationary phase}}$$

The equilibrium constant (called a **distribution constant**, K_D) has the mass-action expression

$$K_D = \frac{[A_{\text{stationary phase}}]}{[A_{\text{mobile phase}}]}$$

If an analyte interacts considerably with the stationary phase, the concentration of the analyte in the stationary phase will be greater than its concentration in the mobile phase. Therefore, the value of K_D will be a number greater than 1. In such cases, the analytes will slow down and take more time to travel through the column. If the analytes do not interact well with the stationary phase, the value of K_D will be small, and the analytes will move quickly through the column. For a given set of conditions in the chromatograph, each analyte will have

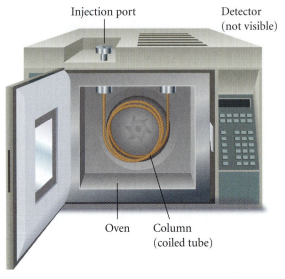

FIGURE 16.6

The essential parts of a gas chromatograph. There is an inlet connected to a column into which the sample is fed. The sample is then pushed through the column by a carrier gas such as helium (in gas chromatography) or by a liquid, often an aqueous solution (in liquid chromatography). This phase moves, so it is called the mobile phase. It passes through the column containing a stationary phase, so called because it stays in place on the column.

its own distribution constant (K_D) and will exit the column at a different time.

Figure 16.7 shows chromatograms of the compounds in refinery gas processed by a petroleum company, as well as the important compounds in a sample of caffeine and some "street drugs." The components in a sample from an athlete or a refinery can be identified by using chromatography. However, the method has even more uses. For instance, chromatography is often one of the first steps in many very sophisticated analyses. Once the analytes are separated in a chromatography instrument, they can immediately be fed into other instruments, such as a mass spectrometer or infrared spectrometer, to confirm their identity. In some cases, the information from the other instruments is used to determine the identity of an unknown component in a mixture. (Figure 16.8 shows two of the more important multistep analyses, which are commonly referred to as hyphenated techniques.) For this reason, chemical equilibrium, the basis of chromatography, is often the most important process in a multistep instrumental analysis.

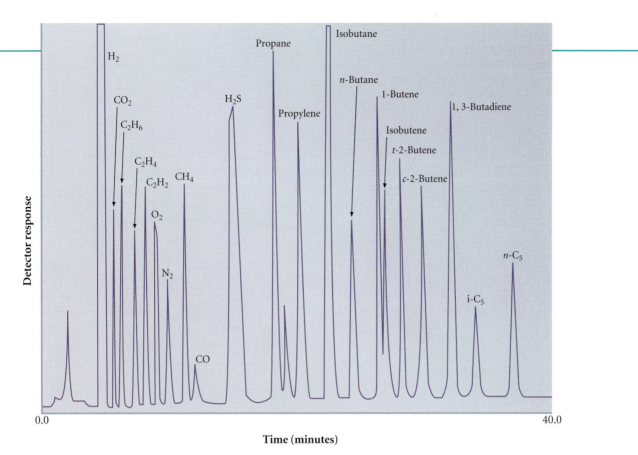

Time (minutes)

FIGURE 16.7

Chromatograms of compounds in refinery gas (top) and common drugs (left) show the importance of chromatography as a separation technique.

1 Methamphetamine	3 Amobarbital	5 Procaine	7 Nortriptyline
2 Nicotine	4 Caffeine	6 Cocaine	8 Heroin

FIGURE 16.8

Multistep analyses are commonly referred to as hyphenated techniques. They include techniques such as GC-MS (gas chromatography–mass spectrometry) and LC-NMR (liquid chromatography–nuclear magnetic resonance) (see photo). These techniques are used to separate, identify, and quantitate the solutes within a solution.

The value of the equilibrium constant for the ionization of sulfurous acid (H_2SO_3) indicates that a significant amount of un-ionized reactant remains at equilibrium. Note, as well, that when equilibrium is finally reached, the concentrations of each component in the reaction become constant.

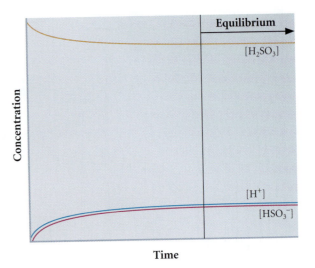

Reactants   Products

S   E

Let's examine another example of an equilibrium constant. The ionization of sulfurous acid (H_2SO_3) to form hydrogen ions and bisulfite ions (HSO_3^-) has a value for K of 0.0120 at 25°C.

$$H_2SO_3(aq) \rightleftharpoons H^+(aq) + HSO_3^-(aq) \qquad K = 0.0120$$

This is an intermediate value; that is, K is relatively close to 1. At equilibrium, there remain significant amounts of both reactants and products. When the value for K is close to 1, the equilibrium lies somewhere between the reactants and products sides. Again, our line chart helps us see the equilibrium position.

HERE'S WHAT WE KNOW SO FAR

- Processes that have very large values of K are those in which mostly products are present at equilibrium.
- Processes that have very small values of K have mostly reactants present at equilibrium.
- Processes with K values not too far from 1 have significant amounts of both reactants and products at equilibrium.

The meaning of the equilibrium constant is shown graphically in Figure 16.9. As we continue to build our understanding of equilibria, we will learn that the equilibrium position for a reaction is related to the equilibrium constant as well as to the temperature, the pressure, and changes in the concentration of the reactants and products. As we explore these relationships, our goal is not only to interpret the meaning of K but also to understand how we can manipulate our reaction conditions to produce what we want (products or reactants) under the best possible conditions. In industrial-scale manufacturing, such as the production of sulfuric acid, these "best possible conditions" can include chemical, environmental, *and* economic factors.

FIGURE 16.9

The relative equilibrium points of typical reactions based on their equilibrium constants. High values of K favor product formation. Small values of K favor the reactants side. Intermediate values favor a middle-of-the-road position.

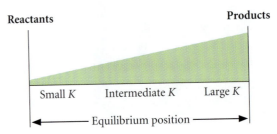

Application

EXERCISE 16.4 | Interpreting K

One of the most important chemicals in the toolbox of analytical chemists has the chemical name ethylenediaminetetraacetic acid, or, more simply, EDTA. A chemist would like to determine the concentration of several metal ions, including Co^{2+}, Zn^{2+}, Ni^{2+}, and Ca^{2+}, in a solution by titrating them (recall titrations from Section 4.3) with the ionic form of EDTA known as $EDTA^{4-}$ (shown below). The values of K for the reaction of the metal ion with $EDTA^{4-}$ are listed next to each cation. Judging solely on the basis of these values, and assuming the equilibrium is established very quickly, would EDTA be a reasonable choice to react almost completely with each of these metal ions?

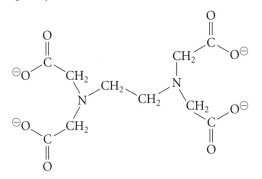

Ethylenediaminetetraacetate

(sample reaction) $Ca^{2+}(aq) + EDTA^{4-}(aq) \rightleftharpoons CaEDTA^{2-}(aq)$

Element	Cation	K
cobalt	Co^{2+}	2×10^{16}
zinc	Zn^{2+}	3×10^{16}
nickel	Ni^{2+}	4×10^{18}
calcium	Ca^{2+}	5×10^{10}

First Thoughts

What does a large value of the equilibrium constant mean? Recall our discussion in the text, where we noted that if K is very large, the equilibrium lies essentially all the way to the products side. This means that looking only at the value for K (and ignoring how quickly or slowly the reactions might occur) suggests that the reactions will be essentially complete.

Solution

Because each of the equilibrium constants are considerably larger than 1, $EDTA^{4-}$ will react essentially completely with these metals.

Further Insights

EDTA is a remarkable analytical reagent because of its ability to react with nearly every metal ion. Examples of its uses include determining water "hardness" by titrating to find the concentrations of calcium and magnesium in water, and determining the composition of metal alloys and metals in pharmaceuticals.

PRACTICE 16.4

Which of these reactions provide(s) mostly products? . . . mostly reactants?

a. $NH_4^+(aq) + H_2O(l) \rightleftharpoons NH_3(aq) + H_3O^+(aq)$ $\qquad K = 5.6 \times 10^{-10}$

b. $HF(aq) + H_2O(l) \rightleftharpoons H_3O^+(aq) + F^-(aq)$ $\qquad K = 7.2 \times 10^{-4}$

c. $HOCl(aq) + H_2O(l) \rightleftharpoons H_3O^+(aq) + OCl^-(aq)$ $\qquad K = 3.5 \times 10^{-8}$

d. $HClO_4(aq) + H_2O(l) \rightleftharpoons H_3O^+(aq) + ClO_4^-(aq)$ $\qquad K = 1.0 \times 10^7$

See Problems 25 and 26.
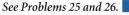

Homogeneous and Heterogeneous Equilibria

Let's look again at the reaction of the calcium and EDTA ions in the previous exercise. Compare that to the dissolution of silver chloride, as well as the Haber process for the combination of H_2 and N_2 to produce ammonia (NH_3), the focus of Exercise 16.2.

$$Ca^{2+}(aq) + EDTA^{4-}(aq) \rightleftharpoons CaEDTA^{2-}(aq)$$

$$3H_2(g) + N_2(g) \rightleftharpoons 2NH_3(g)$$

$$AgCl(s) \rightleftharpoons Ag^+(aq) + Cl^-(aq)$$

Note the phases of each substance. In the EDTA titration of calcium ion, the reactants and products are all in the aqueous phase. When all of the phases are the same in a reaction, we say that the reaction establishes a **homogeneous equilibrium**. The formation of ammonia is another example of homogeneous equilibrium because all of the substances are gases. On the other hand, the dissolution of silver chloride in water begins with a solid and forms ions in the aqueous phase. When there are different phases present in the reaction, we have a **heterogeneous equilibrium**.

The mass-action expressions for homogeneous and heterogeneous equilibria differ in one vital aspect, which we can examine by focusing on the formation of a saturated solution of silver chloride. The density of solid AgCl is 5.56 g/cm^3. We can think of density as a measure of the concentration, because it is the amount of substance per unit volume. Density is independent of the quantity of substance; therefore, the *concentration of pure AgCl is the same whether we have 10 g or 100 tons*. Constants, such as the concentration of any pure solid or liquid, are said to be part of the equilibrium constant and are not included in the mass-action expression. This is a general result that is valid for pure solids and pure liquids and is approximately correct for solvents such as water, but only when the solutes are very dilute. This means that for the dissolution of silver chloride, our original mass-action expression,

$$K = \frac{[Ag^+][Cl^-]}{[AgCl]}$$

can be simplified by recognizing that AgCl is a solid:

$$K = [Ag^+][Cl^-]$$

We can apply this same reasoning to derive the mass-action expression for one step in the large-scale cleanup of sulfur dioxide emitted from industrial smokestacks. In this process, SO_2 is converted to solid $CaSO_4$.

$$2CaO(s) + 2SO_2(g) + O_2(g) \rightleftharpoons 2CaSO_4(s)$$

$$K = \frac{1}{[SO_2]^2[O_2]}$$

16.4 Working with Equilibrium Constants

In this section, we begin to learn how we can work with reactions, mass-action expressions, and equilibrium constants. This is the next step in learning how to control experimental conditions *by using equilibria to achieve desired chemical outcomes*. As with our previous section, we begin by looking at the reaction of sulfur dioxide and oxygen.

Here are three ways of writing the balanced reaction of SO_2 with O_2:

$$2SO_2(g) + O_2(g) \rightleftharpoons 2SO_3(g) \quad \text{(reaction A)}$$
$$4SO_2(g) + 2O_2(g) \rightleftharpoons 4SO_3(g) \quad \text{(reaction B)}$$
$$SO_2(g) + \tfrac{1}{2}O_2(g) \rightleftharpoons SO_3(g) \quad \text{(reaction C)}$$

We multiplied each coefficient in reaction A by 2 to obtain reaction B. We then divided each coefficient by 4 to obtain reaction C. However, all the reactions are the same because the mole ratios within each reaction do not change. That is, the ratio of SO_2 to O_2 is 2 to 1 in each reaction. What about the mass-action expressions?

$$K \text{ for reaction A} = \frac{[SO_3]^2}{[SO_2]^2[O_2]}$$

$$K \text{ for reaction B} = \frac{[SO_3]^4}{[SO_2]^4[O_2]^2}$$

$$K \text{ for reaction C} = \frac{[SO_3]}{[SO_2][O_2]^{1/2}}$$

The mass-action expressions are all different! Does this change the value of K? Let's pick a set of typical contact process equilibrium concentration values and substitute them into each expression: $[O_2] = 1.5 \times 10^{-3}\ M$; $[SO_2] = 1.3 \times 10^{-14}\ M$; and $[SO_3] = 1.0 \times 10^{-3}\ M$. To calculate the equilibrium constant, we substitute the equilibrium concentrations of all substances into the mass-action expression and perform the math indicated by the expression.

$$K_{\text{reaction A}} = \frac{[SO_3]^2}{[SO_2]^2[O_2]} = \frac{[1.0 \times 10^{-3}]^2}{[1.3 \times 10^{-14}]^2[1.5 \times 10^{-3}]} = 3.94 \times 10^{24} \approx 4 \times 10^{24}$$

$$K_{\text{reaction B}} = \frac{[SO_3]^4}{[SO_2]^4[O_2]^2} = \frac{[1.0 \times 10^{-3}]^4}{[1.3 \times 10^{-14}]^4[1.5 \times 10^{-3}]^2} = 1.6 \times 10^{49}$$

$$K_{\text{reaction C}} = \frac{[SO_3]}{[SO_2][O_2]^{1/2}} = \frac{[1.0 \times 10^{-3}]}{[1.3 \times 10^{-14}][1.5 \times 10^{-3}]^{1/2}} = 2.0 \times 10^{12}$$

How do the values of K differ, and what is the relationship to the change in reaction coefficients? We doubled the coefficients going from reaction A to reaction B, and $K_{\text{reaction B}} = (K_{\text{reaction A}})^2$. In going from reaction B to reaction C we divided the coefficients by 4, and $K_{\text{reaction C}} = (K_{\text{reaction B}})^{1/4}$. In general, then, when you change the coefficients of a reaction by a factor of n,

$$K_{\text{new}} = (K_{\text{old}})^n$$

EXERCISE 16.5   **Reversing the Reaction**

We opened the chapter by discussing the importance of the interaction of human myoglobin with oxygen in the muscle cells.

$$Mb + O_2 \rightleftharpoons MbO_2 \qquad K = 8.6 \times 10^{-7}$$

Many biochemists and molecular biologists cite the "dissociation constant" for the reaction written in reverse:

$$MbO_2 \rightleftharpoons Mb + O_2$$

Given the relationship between the change in coefficients and the change in the value of K just discussed, can you determine the equilibrium constant for the reverse (dissociation) reaction?

Solution

Reversing the reaction requires us to take the inverse of the original mass-action expression. That is,

$$Mb + O_2 \rightleftharpoons MbO_2 \qquad \text{(original reaction)}$$

$$K = \frac{[MbO_2]}{[Mb][O_2]} \qquad \text{(original reaction)}$$

$$MbO_2 \rightleftharpoons Mb + O_2 \qquad \text{(reverse reaction)}$$

$$K = \frac{[Mb][O_2]}{[MbO_2]} \qquad \text{(inverse mass-action expression)}$$

Mathematically,

$$K_{\text{dissociation}} = (K_{\text{original}})^{-1}$$

$$K_{\text{dissociation}} = (8.6 \times 10^{-7})^{-1} = 1.2 \times 10^6$$

Video Lesson: Strategies for Solving Equilibrium Problems

Given the value of K in Exercise 16.5, what is the equilibrium constant for the following reaction?

$$\tfrac{1}{2} MbO_2 \rightleftharpoons \tfrac{1}{2} Mb + \tfrac{1}{2} O_2$$

See Problems 29, 30, 33, and 34.

Combining Reactions to Describe a Process

One of the most important questions that chemists answer is "How much do I have?" In the chemical industry, this issue is often related to quality control: "Is my vitamin C tablet actually filled with vitamin C?" "Does my aspirin tablet contain the right amount of aspirin?" "Does my potato chip have too much water in it?" A chemical titration is the chemist's tool often used to answer such questions. We mentioned titrations in Exercise 16.4, and much of Chapter 18 is devoted to the technique. However, two key features of a titration experiment are necessary for the titration to be effective. First, *the titration reaction has to be essentially complete*, meaning that the equilibrium constant, K, must be quite large. In addition, *a titration reaction should be fast, even when the concentrations of the reactants are low.*

Application

In the analytical laboratory, the concentration of acetic acid (CH_3COOH, known technically as ethanoic acid) can be measured by titration with a solution of sodium hydroxide of known concentration. Can we show that the equilibrium constant, which we will call $K_{\text{titration}}$, is very large at 25°C? The reaction is

$$CH_3COOH(aq) + OH^-(aq) \rightleftharpoons CH_3COO^-(aq) + H_2O(l)$$

The mass-action expression is

$$K_{\text{titration}} = \frac{[CH_3COO^-]}{[CH_3COOH][OH^-]}$$

We can consider this reaction as the sum of two separate equations for which we know the equilibrium constants at 25°C. Using the known equations, we can calculate the value for $K_{\text{titration}}$.

$$CH_3COOH(aq) + H_2O(l) \rightleftharpoons CH_3COO^-(aq) + H_3O^+(aq) \quad K_1 = 1.8 \times 10^{-5}$$

$$H_3O^+(aq) + OH^-(aq) \rightleftharpoons 2H_2O(l) \qquad\qquad\qquad K_2 = 1.0 \times 10^{14}$$

$$\overline{CH_3COOH(aq) + OH^-(aq) \rightleftharpoons CH_3COO^-(aq) + H_2O(l) \qquad K_{\text{titration}} = ?}$$

The mass-action expressions for the known reactions are

$$K_1 = \frac{[CH_3COO^-][H_3O^+]}{[CH_3COOH]} \qquad K_2 = \frac{1}{[H_3O^+][OH^-]}$$

If we multiply the two known mass-action expressions, we get

$$\frac{[CH_3COO^-][H_3O^+]}{[CH_3COOH]} \times \frac{1}{[H_3O^+][OH^-]} = \frac{[CH_3COO^-][H_3O^+]}{[CH_3COOH][H_3O^+][OH^-]} =$$

$$K_1K_2 = \frac{[CH_3COO^-]}{[CH_3COOH][OH^-]}$$

This is the mass-action expression for the titration reaction! For this process, then,

$$K_{titration} = K_1K_2 = (1.8 \times 10^{-5})(1.0 \times 10^{14}) = 1.8 \times 10^9$$

In short, when the overall reaction is the *sum* of other reactions, the overall equilibrium constant is the *product* of the equilibrium constants for these other reactions. In the case of the titration of acetic acid with sodium hydroxide, $K_{titration}$ is sufficiently large (and the reaction is quite fast). Therefore, the titration of acetic acid with sodium hydroxide should (and does!) work well.

Calculating Equilibrium Concentrations from *K* and Other Concentrations

We know from our previous discussion that it is possible to calculate the equilibrium constant by substituting equilibrium concentrations of reactants and products into the mass-action expression. We have one equation, the mass-action expression, and one unknown, K. In your day-to-day work as a chemist, biologist, medical technologist, or soil scientist, the more likely scenario is that the value of K is already known and you will need to calculate the equilibrium concentration of one or more of the reactants or products.

Let's use the Contact process to show how this is done for one unknown concentration when the other equilibrium concentrations and K are known. At 27°C, $K = 4.0 \times 10^{24}$ for the reaction

$$2SO_2(g) + O_2(g) \rightleftharpoons 2SO_3(g)$$

If the equilibrium concentrations of SO_2 and O_2 are 6.4×10^{-12} M and 1.2×10^{-3} M, respectively, what is the equilibrium concentration of SO_3? We have one equation with one unknown, so we can solve for $[SO_3]$.

$$K = \frac{[SO_3]^2}{[SO_2]^2[O_2]}$$

$$4.0 \times 10^{24} = \frac{[SO_3]^2}{[6.4 \times 10^{-12}]^2[1.2 \times 10^{-3}]}$$

$$[SO_3]^2 = 4.0 \times 10^{24} (6.4 \times 10^{-12})^2 (1.2 \times 10^{-3})$$

$$[SO_3] = \sqrt{0.197} = 0.443 \approx 0.44 \; M$$

Does our answer make sense? We ought to check our math, and we can do so by substituting the values back into the mass-action expression to verify the value of the equilibrium constant.

$$\frac{[SO_3]^2}{[SO_2]^2[O_2]} = \frac{[0.443]^2}{[6.4 \times 10^{-12}]^2[1.2 \times 10^{-3}]} = 4.0 \times 10^{24}$$

We can make such substitutions because the reaction is *at equilibrium*. When this is not the case, a different way of thinking, the subject of Section 16.5, will prove useful.

Using Partial Pressures

Video Lesson: Converting Between K_c and K_p

Most of the equilibrium constants that chemists employ are derived using concentrations measured in molarity because much of the chemistry we do is in solution. However, when working with gases, we can use partial pressures in units of atmospheres. **What is the relationship between the pressure of a gas and its concentration?** We can use the ideal gas equation, which holds well for gases in low concentration (see Chapter 10).

$$PV = nRT$$

$$\text{Concentration} = \frac{\text{moles}}{\text{volume}} = \frac{n}{V} = \frac{P}{RT}$$

The mass-action expression for the reaction of oxygen with sulfur dioxide can be written using molarity, as usual, or by substituting the equivalent variables from the ideal gas law $\left(\frac{P}{RT}\right)$ in place of the concentration.

$$2SO_2(g) + O_2(g) \rightleftharpoons 2SO_3(g)$$

$$K = \frac{[SO_3]^2}{[SO_2]^2[O_2]} = \frac{\frac{P_{SO_3}^2}{(RT)^2}}{\frac{P_{SO_2}^2}{(RT)^2} \times \frac{P_{O_2}}{RT}}$$

We can combine all of the RT terms and put them on the side with the pressure-based mass-action expression.

$$K = \frac{P_{SO_3}^2}{P_{SO_2}^2 \, P_{O_2}} (RT)^1$$

where K = the mass-action expression solved using only molar concentrations

P = the pressure in atmospheres

T = the temperature in kelvins

R = the universal gas constant

If we set K_p equal to the pressure-based mass-action expression and rearrange the equation, we get

$$K_p = \frac{P_{SO_3}^2}{P_{SO_2}^2 \, P_{O_2}}$$

$$K = K_p \, (RT)^1$$

We have introduced a new equilibrium constant, K_p, which is based on the mass-action expression for the partial pressures of substances in the reaction. K_p is often calculated with partial pressures, P, in units of atmospheres. However, remember that the actual definition of K involves the use of activities. Therefore, in the end, K_p has no units. As a matter of nomenclature, we use K_p when dealing with partial pressures and K (or, in some literature, K_c) when dealing with concentrations. *Note in this equation that the value of the exponent of RT indicates the difference in the number of moles of gas between reactants and products.* In general, then, we can write the relationship between K and K_p in two ways:

$$K_p = K(RT)^{\Delta n} \quad \text{or} \quad K = K_p(RT)^{-\Delta n}$$

where $\triangle n$ = the change in the number of moles of gas (products minus reactants)

$R = 0.08206$ L·atm·mol^{-1}·K^{-1}

T = temperature in kelvins

In the reaction of sulfur dioxide with oxygen, there is one fewer mole of gaseous products ($2SO_3$) than of reactants ($2SO_2$ and O_2), so the change in the number of moles of gas from reactant to product, Δn, is equal to -1. We can relate K to K_p in two ways, depending on which equilibrium constant we are given.

$$K_p = K(RT)^{-1} \quad \text{or} \quad K = K_p(RT)^1$$

At 673 K and 1 atm total system pressure, the value of K_p for the oxidation of SO_2 to SO_3 is 1.58×10^5, measured using the equilibrium partial pressure of each gas. We can calculate K using this information.

$$K = K_p(RT)^{-\Delta n}$$

$$K = (1.58 \times 10^5)(0.08206 \text{ L·atm·mol}^{-1}\text{·K}^{-1} \times 673 \text{ K})^{-(-1)} = 8.73 \times 10^6$$

EXERCISE 16.6   **Converting K to K_p**

Calculate K_p for the production of ammonia via the Haber process.

$$3H_2(g) + N_2(g) \rightleftharpoons 2NH_3(g) \qquad K = 0.060 \text{ (at 500°C)}$$

Solution

We lose 2 mol of gas going from reactants to products, so $\Delta n = -2$.

$$K_p = K(RT)^{\Delta n}$$

$$K_p = 0.060 \, (0.08206 \text{ L·atm·mol}^{-1}\text{·K}^{-1} \times 773 \text{ K})^{-2}$$

$$K_p = 0.060 \, (62.43)^{-2}$$

$$K_p = 1.5 \times 10^{-5}$$

PRACTICE 16.6

Calculate the value of K_p at 25°C for the reaction of chlorine and carbon monoxide.

$$CO(g) + Cl_2(g) \rightleftharpoons COCl_2(g) \qquad K = 3.7 \times 10^9 \text{ (at 25°C)}$$

See Problems 41 and 42.

16.5   Solving Equilibrium Problems— A Different Way of Thinking

The world is not at equilibrium. Its shifts are seen and felt in the massive upheavals of earthquakes, volcanoes, and hurricanes and in the smallest electronic interactions of atoms. Life itself is a process in which reversible reactions within us keep shifting their equilibrium positions, always chasing a moving target, in ways that make our survival possible. If one aspect of being human is our ability to understand our world, then one expression of this understanding is *the ability to manipulate our starting reaction conditions to control the position of equilibrium*. We cannot stop earthquakes, but we can reliably make ammonia, sulfuric acid, and a whole host of other chemicals. We can examine the complex interactions of carbon with the environment. We can understand the interaction of myoglobin and oxygen. We can pick the best conditions for chemical analyses. All these things are based on knowing how the conditions change from the starting point to equilibrium. How can we do this? We must first develop a different way of thinking.

This way of thinking is based on two questions: **"Given the chemical and physical conditions, how do I judge what is likely to happen?"** and **How can I use my** understanding to change these conditions to obtain the outcome I want?" Here are some judgments we must make when we combine substances:

1 What chemical reactions will occur?

2 Which among these are important? Which are unimportant?

3 Given the initial concentrations of substances, in which direction (toward the formation of more reactants or toward the formation of more products) is the reaction likely to proceed?

4 How can we solve for the amounts of substances of interest that are present at equilibrium?

5 How can we manipulate the conditions to maximize the concentration of the desired components and minimize that of the others?

Video Lesson: Predicting the Direction of a Reaction

A Case Study

We discussed the analysis of acetic acid (CH_3COOH) previously. Here, we will look at the equilibrium established when an acetic acid solution is prepared with an initial concentration of 0.500 *M*. Remember the overarching question: **"Given the chemical and physical conditions, how do I judge what is likely to happen?"** Specifically, what are the concentrations of all the substances in my flask at equilibrium? Let's approach this by answering the other questions, in order.

1 What chemical reactions will occur?

One reaction is the ionization of the acid as shown in reaction A, and this will supply CH_3COO^- and H^+. Another reaction that always occurs in aqueous solution is the ionization of water itself, also supplying H^+, as shown in reaction B.

A: $CH_3COOH(aq) \rightleftharpoons CH_3COO^-(aq) + H^+(aq)$ $\qquad K = 1.8 \times 10^{-5}$

B: $H_2O(l) \rightleftharpoons H^+(aq) + OH^-(aq)$ $\qquad K = 1.0 \times 10^{-14}$

2 Which among these are important? Which are unimportant?

The "important chemistry" is that which affects the outcome of the overall process. The unimportant chemistry (in this context) is that which has no significant bearing on the outcome. The extent of the reactions and their relative importance to us are indicated by their equilibrium constants. Even though neither reaction proceeds very far, the ionization of acetic acid is far more significant than the ionization of water (we will test this assertion later.) We can make this claim because the value of *K* for equation A is about 10^8 times larger than the value of *K* for equation B. The mass-action expression for the ionization of acetic acid (equation A) is

$$K = \frac{[CH_3COO^-][H^+]}{[CH_3COOH]} = 1.8 \times 10^{-5}$$

In many systems, it is entirely possible that more than one reaction is important to our calculations. However, in this chapter, we will typically deal with one important reaction in each process that we discuss.

3 Given the initial concentrations of substances, in which direction is the reaction likely to proceed?

We can predict the direction of change in a reaction by using the **reaction quotient** (*Q*), which we introduced in Section 14.6. This is the numerical outcome of the mass-action expression using *initial concentrations*, which are designated [ ]$_0$.

$$Q = \frac{[CH_3COO^-]_0[H^+]_0}{[CH_3COOH]_0}$$

Q is then compared to the equilibrium constant, K, in order to determine the direction in which the reaction will proceed. Keep in mind that Q deals with *initial conditions* and K deals with *equilibrium conditions*. Calculation of the reaction quotient indicates which way the reaction will proceed in order to establish equilibrium.

- If Q is equal to K, the system is *at equilibrium*.

- If Q is *greater than K*, there is too much product present. The system will *shift to the left* to reach equilibrium. Mathematically, the numerator is much larger, and the denominator much smaller, than the equilibrium values.

- If Q is *less than K*, there is too much reactant present. The system will *shift to the right* to reach equilibrium.

In our current example, the reaction must go toward formation of products ("to the right") because we initially have no products ($Q = 0$), and this will be the case in many equilibrium problems you will encounter.

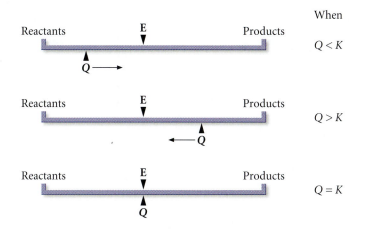

If we were to change the initial concentration of our substances, such that $[H^+]_0 = 0.0100\ M$, $[CH_3COO^-]_0 = 0.200\ M$ and $[CH_3COOH]_0 = 0.150\ M$, we could calculate Q for these initial concentrations.

$$Q = \frac{[CH_3COO^-]_0[H^+]_0}{[CH_3COOH]_0} = \frac{(0.200\ M)(0.0100\ M)}{(0.150\ M)} = 0.0133$$

What does this value tell us? Q is much greater than K (1.8×10^{-5}), so the reaction will shift toward the formation of reactants ("to the left"). Mathematically, this makes sense, because reducing the product concentration and increasing the reactant concentration will reduce the quotient of the mass-action expression. What does this indicate about the chemistry? If chemical change is an inevitable part of our world, it is especially useful to be able to control that chemical change in order to have a desirable impact on manufacturing, human biology, and the environment.

EXERCISE 16.7   **Predicting the Direction of Equilibrium**

Predict the direction in which the reaction of myoglobin and oxygen will proceed to reach equilibrium for each set of conditions.

$$Mb + O_2 \rightleftharpoons MbO_2 \qquad K = 8.6 \times 10^5$$

a. $Q = 1500$

b. $Q = 8.3 \times 10^6$

c. $[Mb]_0 = 2.04 \times 10^{-4}\ M$; $[O_2]_0 = 3.00 \times 10^{-6}\ M$; $[MbO_2]_0 = 2.50 \times 10^{-4}\ M$

d. $[Mb]_0 = 5.00 \times 10^{-5} M; \quad [O_2]_0 = 9.21 \times 10^{-7} M; \quad [MbO_2]_0 = 1.00 \times 10^{-4} M$

e. $[Mb]_0 = 1.0 \times 10^{-4} M; \quad [O_2]_0 = 2.25 \times 10^{-6} M; \quad [MbO_2]_0 = 1.94 \times 10^{-4} M$

First Thoughts

Keep in mind the significance of the reaction quotient, Q. It helps us see in which direction the reaction will go to reach equilibrium. When the value for K is small, it is true that the equilibrium position will favor the reactants. However, if you started off with *no* products at all, even such a reaction will move (slightly) to the right to form a little product. We can show this graphically:

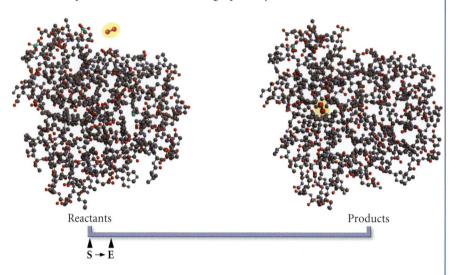

Reactants   Products

$S \rightarrow E$

Solution

a. $Q < K$, so the reaction will form more products and shift to the right.

b. $Q > K$, so the reaction will form more reactants and shift to the left.

c. $Q = \dfrac{[MbO_2]_0}{[Mb]_0[O_2]_0} = \dfrac{[2.50 \times 10^{-4}]}{[2.04 \times 10^{-4}][3.00 \times 10^{-6}]} = 4.08 \times 10^5$

$Q < K$, so the reaction will form more products and shift to the right.

d. $Q = \dfrac{[1.00 \times 10^{-4}]}{[5.00 \times 10^{-5}][9.21 \times 10^{-7}]} = 2.17 \times 10^6$

$Q > K$, so the reaction will form more reactants and shift to the left.

e. $Q = \dfrac{[1.94 \times 10^{-4}]}{[1.0 \times 10^{-4}][2.25 \times 10^{-6}]} = 8.6 \times 10^5$

$Q = K$, so the reaction is at equilibrium.

Further Insights

Food for thought: How might these shifts in the direction of the reaction toward equilibrium allow the muscles in the body to obtain needed oxygen during vigorous exercise?

PRACTICE 16.7

Under certain specific conditions, the formation of ozone from NO_2 has an equilibrium constant $K = 120$. Predict the direction of the following reaction, given the conditions in each case below.

$$NO_2(g) + O_2(g) \rightleftharpoons NO(g) + O_3(g) \qquad K = 120$$

a. $Q = 1.5 \times 10^5$

b. $[O_3] = 1.0\ M;\ [O_2] = 1.0\ M;\ [NO] = 0.50\ M;\ [NO_2] = 0.25\ M$

c. $[O_3] = 0.0010\ M;\ [O_2] = 2.55\ M;\ [NO] = 1.78 \times 10^{-6}\ M;\ [NO_2] = 5.4 \times 10^{-3}\ M$

d. $[O_3] = 0.033\ M;\ [O_2] = 0.019\ M;\ [NO] = 9.3\ M;\ [NO_2] = 0.044\ M$

See Problems 45–48.

4 **How can we solve for the amounts of substances of interest that are present at equilibrium?**

The present case study concerns a small value of K. After this, we will look at large and then intermediate values of K.

Small Value of K

We have a system that contains $0.500\ M\ CH_3COOH$, which ionizes with the relatively small K of 1.8×10^{-5}. What will be the equilibrium concentrations of acetate ions (CH_3COO^-) and hydrogen ions (H^+)?

$$CH_3COOH(aq) \rightleftharpoons CH_3COO^-(aq) + H^+(aq) \qquad K = 1.8 \times 10^{-5}$$

The lack of products initially present, along with the small equilibrium constant, suggests that the reaction will proceed to the right, but not very far. Our view of where we start and where we end looks something like this on our equilibrium line chart:

Reactants   Products

S → E

Tracking the changes that take place as we go from the start to equilibrium requires a bit of bookkeeping. Bookkeepers keep track of a company's finances using tables that show changes in financial data in an organized, clear way. We will do the same thing for our equilibrium data by setting up a table that contains the following rows:

Initial row: The "initial" row includes the initial concentration of each species.

$$CH_3COOH(aq) \rightleftharpoons CH_3COO^-(aq) + H^+(aq)$$

▶ initial	0.500 M	0 M	0 M

Change row: The "change" row describes the change in concentration of each species that occurs in order to reach equilibrium. This must take into account the stoichiometric ratios among reactants and products in the reaction (all 1-to-1 in this example) and the magnitude of the equilibrium constant, which in this case is small enough to indicate that the reaction will not go very far before equilibrium is reached. Because we don't yet know the amount of acetic acid that will ionize, we call it "x." The concentration of acetic acid will *decrease* by "x" M, and the acetate and hydrogen ion concentrations will *increase* by "x" M. Again, the stoichiometry of the equation is included in this line of the table.

$$CH_3COOH(aq) \rightleftharpoons CH_3COO^-(aq) + H^+(aq)$$

initial	0.500 M	0 M	0 M
▶ change	−x	+x	+x

Equilibrium row: When the free energy *change* of the reaction equals zero, such that the free energy itself is at a minimum, the reaction will be at equilibrium. The value for each substance in the "equilibrium" row equals the initial amount plus the change (so that $0.500 + "-x" = 0.500 - x$). This is sufficient for us to proceed with the problem, especially via programmable calculators, which can solve the necessary equations with a few keystrokes.

$$CH_3COOH(aq) \rightleftharpoons CH_3COO^-(aq) + H^+(aq)$$

initial	0.500 M	0 M	0 M
change	$-x$	$+x$	$+x$
▶ equilibrium	$0.500 - x$	$+x$	$+x$

Assumptions row: Whether or not we have a programmable calculator, we can get a deeper understanding of the equilibria in solution by making key assumptions. We can simplify the problem solving by assuming that because the value for K is small, the value of "x" will also be small—that is, there is negligible ionization compared to our initial acetic acid concentration. We'll assume that it will be so small as to be unimportant *compared to the initial concentration of 0.500 M*. Making that assumption enables us to say that $0.500 - x \approx 0.500$. However, because "x" is *not* unimportant compared to 0 M (any quantity is infinitely large compared to zero!), we cannot neglect "x" in the other columns of the table. The "assumptions" row shows the final equilibrium amounts with any assumptions we make.

$$CH_3COOH(aq) \rightleftharpoons CH_3COO^-(aq) + H^+(aq)$$

initial	0.500 M	0 M	0 M
change	$-x$	$+x$	$+x$
equilibrium	$0.500 - x$	$+x$	$+x$
▶ assumptions	0.500	x	x

The table that we have worked with is an expanded version of what is often called an ICE table, which stands for **I**nitial, **C**hange, **E**quilibrium table. Our expanded version includes the **A**ssumptions row, so we will call this an ICEA table. We will often use ICEA tables in our equilibrium problem solving because they give us an organized way to assess the changes in concentration that occur in our reactions.

We can now substitute the equilibrium concentrations generated in the assumptions row of our table into the mass-action expression and solve for "x."

$$K = \frac{[CH_3COO^-][H^+]}{[CH_3COOH]}$$

$$1.8 \times 10^{-5} = \frac{(x)(x)}{0.500}$$

$$1.8 \times 10^{-5} = \frac{x^2}{0.500}$$

$$9.0 \times 10^{-6} = x^2$$

$$3.0 \times 10^{-3} = x$$

Now, if we return the value of "x" to our ICEA table and solve for the assumptions row, we obtain the equilibrium concentrations of all species in the reaction, assuming our assumption is justified.

$$x = [CH_3COO^-] = [H^+] = 3.0 \times 10^{-3} \, M$$

The actual equilibrium concentration of acetic acid is $0.500 - 3.0 \times 10^{-3} = 0.497 \, M$, or $0.50 \, M$ to two significant figures. **Do our results make sense? That is, are they reasonable?** We can determine that our results are mathematically correct by substituting the concentration values back into the equilibrium expression to show that this results in K.

$$\frac{(3.0 \times 10^{-3})^2}{0.50} = 1.8 \times 10^{-5}$$

Was our assumption justified? In other words, was "x" negligible compared to the original acetic acid concentration? We say that our assumption is valid if the change as a result of the assumption is less than 5% of the original concentration.

$$\frac{x}{\text{original concentration}} \times 100\% \leq 5\%$$

Although we will use this "5% rule" in our work, professional chemists sometimes require a smaller tolerance, and sometimes a larger, depending on the process with which they are working. With our data,

$$\frac{3.0 \times 10^{-3}}{0.500} \times 100\% = 0.6\%$$

and using the 5% rule, we find that our assumption of negligible ionization (0.6%, in this case) was valid.

Let's revisit our equilibrium line chart and see what 0.6% ionization means in terms of the extent of the reaction.

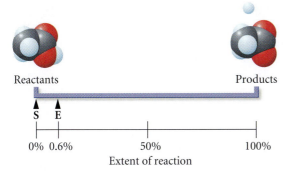

This confirms our initial thinking that with the very small value for K, the reaction would not proceed—acetic acid would not ionize—appreciably.

To be sure our calculations are valid, we should also test our other assumption, that the ionization of water is unimportant. This is a more complex issue that we will consider in Chapter 17.

$$H_2O \rightleftharpoons H^+ + OH^- \qquad K = 1.0 \times 10^{-14}$$

EXERCISE 16.8 **Concentration of Lead Ion in Saturated Lead Bromide**

Determine the concentration of lead ion in a saturated solution that would result from the dissolution of solid lead(II) bromide, $PbBr_2(s)$.

$$PbBr_2(s) \rightleftharpoons Pb^{2+}(aq) + 2Br^-(aq) \qquad K = 4.6 \times 10^{-6} \text{ at } 25°C$$

First Thoughts

What is the important chemistry that occurs in solution? The only important chemistry of interest is the dissolution of the lead bromide, which is a largely insoluble salt, judging by the small value of K. In which direction will the reaction proceed? Even though the equilibrium constant is quite small, the reaction will proceed (minimally) to the products side because there were no products present initially.

Reactants   Products

$$S \rightarrow E$$

Our equilibrium line chart is similar to that for the dissociation of acetic acid. What are the equilibrium concentrations of substances? We can use the same style of table as we did with the acetic acid dissociation. Our only modification arises because lead bromide is a pure solid, so it is not part of the equilibrium expression. We show this in the ICEA table below.

$$PbBr_2(s) \rightleftharpoons Pb^{2+}(aq) + 2Br^-(aq)$$

	PbBr$_2$(s)	Pb^{2+}	Br$^-$
initial	—	0 M	0 M
change	—	$+x$	$+2x$
equilibrium	—	x	$2x$

The concentration of the solid reactant is constant. In this table, "x" represents the amount of PbBr$_2$(s) that dissolves. This value is therefore equal to the solubility of the salt. Because the initial amount of each product was 0 M, the amount gained, "x" and "$2x$," is not negligible compared to 0 M; therefore, we made no assumptions. Note that we use "$2x$" to represent the change in [Br$^-$] because bromide has a stoichiometric factor of 2 in the equation. This stoichiometric factor will also come in to play when we write the mass-action expression.

Solution

How can we solve for the amounts of substances of interest that are present at equilibrium? We can substitute our concentrations into the mass-action expression for the equation:

$$K = [Pb^{2+}][Br^-]^2$$

$$4.6 \times 10^{-6} = (x)(2x)^2$$

$$4.6 \times 10^{-6} = 4x^3$$

$$[Pb^{2+}] = x = 1.048 \times 10^{-2}\,M \approx 1.0 \times 10^{-2}\,M$$

$$[Br^-] = 2x = 2.096 \times 10^{-2}\,M \approx 2.1 \times 10^{-2}\,M$$

We check our answer by solving for K.

$$K = (1.048 \times 10^{-2}) \times (2.096 \times 10^{-2})^2 = 4.6 \times 10^{-6}$$

This agrees well with the actual value of K.

Further Insights

Do our results make sense? The concentrations for the ions that we calculated are quite low, and that makes sense, given the very low value for K.

What is the concentration of lead ion in a saturated solution of $PbCl_2$?

$$PbCl_2(s) \rightleftharpoons Pb^{2+}(aq) + 2Cl^-(aq) \qquad K = 1.6 \times 10^{-5} \text{ at } 25°C$$

See Problems 49, 50, 55, 56, and 60.

In Exercise 16.8, only one important equilibrium needed to be considered. As our understanding of equilibrium deepens, we will find that more than one reaction may well be important. For example, although it wasn't illustrated, the chemical interaction of Pb^{2+} and Br^- with water were considered. In both of these cases, the interaction is minimal, so the dissolution of $PbBr_2$ was the only equilibrium of importance. However, if instead of $PbBr_2$ we had attempted to calculate the concentration of lead ion from the dissolution of PbS, the reaction of S^{2-} with water would have been very important, as illustrated by the equilibria below. In this case, both equations would need to be considered.

$$PbS(s) \rightleftharpoons Pb^{2+}(aq) + S^{2-}(aq) \qquad K = 7 \times 10^{-29}$$

$$S^{2-}(aq) + H_2O(l) \rightleftharpoons HS^- + OH^- \qquad K = 0.083$$

Including the formation of HS^- in our calculations would modify the solubility of lead(II) sulfide. We will learn how to deal quantitatively with multiple equilibria in Chapter 18.

Large Value of K

The reaction of sulfur dioxide and oxygen to form sulfur trioxide at 400°C serves as an excellent model with which to examine systems that have large equilibrium constants, in which the reaction is essentially complete.

$$2SO_2(g) + O_2(g) \rightleftharpoons 2SO_3(g) \qquad K = 8.7 \times 10^6$$

If the initial concentrations of SO_2 and O_2 are 1.0×10^{-3} M and 2.0×10^{-3} M, respectively, what is the equilibrium concentration of SO_3? Given the large equilibrium constant, we can assume that the reaction goes to completion, except for a small amount, "$2x$," that remains unreacted. Remember that the "$2x$" indicates the stoichiometry of the reaction. An equilibrium line chart can help us visualize the extent of reaction.

This is, in effect, a limiting reactant problem in which SO_2 is the limiting reactant. On the basis of the reaction stoichiometry and the large value of K, we can develop an ICEA table. We complete the first row of the table by writing the initial concentration of each species in the reaction.

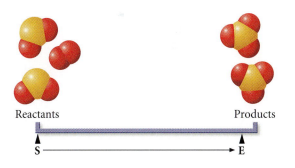

Reactants   Products

	$2SO_2(g)$	$+$	$O_2(g)$	$\rightleftharpoons$	$2SO_3(g)$
initial	1.0×10^{-3} M		2.0×10^{-3} M		0 M

The "change" row in our ICEA table will be a little different than before. We have said that the equilibrium constant is large, and this means the reaction will go essentially to completion. However, *we must hold a tiny amount of each reactant back because the reaction will not go all the way to products.* Let's indicate the small amount that remains behind as "$2x$" for SO_2 and "x" for O_2. That is the "$+2x$" and "$+x$" that we place in the respective concentrations in the change row. *Other than those small amounts that remain, all of the SO_2 that can react will do so.* That means, based on the reaction stoichiometry, that 1.0×10^{-3} M SO_2 and 0.5×10^{-3} M O_2 will react (less the "$2x$" and "x" amounts of reactants that remain) to form $1.0 \times 10^{-3}\, M\, SO_3$ (less the "$2x$" that is unreacted). The total $[SO_2]$ that will react, then, is *all of it* except for the $2x$ that remains. This is where we get the change of $-1.0 \times 10^{-3} + 2x$. In other words, all of it forms products except for the small amount, "$2x$," that does not react and remains. Because of the 2-to-1 mole ratio of SO_2 to O_2, $0.5 \times 10^{-3}\, M\, O_2$ will react with $1.0 \times 10^{-3}\, M\, SO_2$. And if $2x$ mol of SO_2 remains unreacted, x mol of O_2 will be unreacted. This is why the change in concentration of O_2 is $-0.5 \times 10^{-3} + x$.

	$2SO_2(g)$	$+$	$O_2(g)$	$\rightleftharpoons$	$2SO_3(g)$
initial	$1.0 \times 10^{-3}\, M$		$2.0 \times 10^{-3}\, M$		$0\, M$
change	$-1.0 \times 10^{-3} + 2x$		$-0.5 \times 10^{-3} + x$		$1.0 \times 10^{-3} - 2x$

The equilibrium row is, again, the result of adding together the initial and change entries in each column.

	$2SO_2(g)$	$+$	$O_2(g)$	$\rightleftharpoons$	$2SO_3(g)$
initial	$1.0 \times 10^{-3}\, M$		$2.0 \times 10^{-3}\, M$		$0\, M$
change	$-1.0 \times 10^{-3} + 2x$		$-0.5 \times 10^{-3} + x$		$1.0 \times 10^{-3} - 2x$
equilibrium	$2x$		$1.5 \times 10^{-3} + x$		$1.0 \times 10^{-3} - 2x$

In the last row of the table, the "assumptions" row, we make the assumption that "x" is small. If this is true, then $1.5 \times 10^{-3} + x \approx 1.5 \times 10^{-3}$. And, if *that* is true, then $1.0 \times 10^{-3} - 2x \approx 1.0 \times 10^{-3}$ must be true as well.

	$2SO_2(g)$	$+$	$O_2(g)$	$\rightleftharpoons$	$2SO_3(g)$
initial	$1.0 \times 10^{-3}\, M$		$2.0 \times 10^{-3}\, M$		$0\, M$
change	$-1.0 \times 10^{-3} + 2x$		$-0.5 \times 10^{-3} + x$		$1.0 \times 10^{-3} - 2x$
equilibrium	$2x$		$1.5 \times 10^{-3} + x$		$1.0 \times 10^{-3} - 2x$
assumptions	$2x$		1.5×10^{-3}		1.0×10^{-3}

We then write the mass-action expression, plug our assumptions row values into the equation, and solve for "x":

$$K = \frac{[SO_3]^2}{[SO_2]^2[O_2]} = \frac{(1.0 \times 10^{-3})^2}{(2x)^2(1.5 \times 10^{-3})} = 8.7 \times 10^6$$

$$4x^2 = 7.66 \times 10^{-11}$$

$$x = 4.38 \times 10^{-6}\, M$$

$$2x = [SO_2] = 8.75 \times 10^{-6} \approx 8.8 \times 10^{-6}\, M$$

Does this answer make sense? Substituting back into the mass-action expression shows that our equation is mathematically valid.

$$\frac{(1.0 \times 10^{-3})^2}{(8.75 \times 10^{-6})^2(1.5 \times 10^{-3})} = 8.7 \times 10^6\, M$$

Is our assumption valid? That is, is the value of "x" negligible compared to $1.5 \times 10^{-3}\, M$?

$$\frac{x}{\text{original concentration}} \times 100\% = \frac{4.38 \times 10^{-6}\, M}{1.5 \times 10^{-3}\, M} \times 100\% = 0.29\%$$

This easily passes the 5% test, so our assumption that "x" is negligible compared to the initial concentrations of O_2 and SO_3 is valid.

EXERCISE 16.9   **The Myoglobin–Oxygen System**

Solve for the equilibrium concentrations of myoglobin, oxygen, and the MbO_2 complex, given the following initial concentrations. Assume that the myoglobin-oxygen reaction is the most important reaction in this system. In other words, you may neglect the ionization of water because its equilibrium constant is relatively very small.

$$[Mb]_0 = 2.0 \times 10^{-4}\,M; \quad [O_2]_0 = 1.9 \times 10^{-5}\,M; \quad [MbO_2]_0 = 0\,M$$

$$Mb + O_2 \rightleftharpoons MbO_2 \qquad K = 8.6 \times 10^5$$

First Thoughts

The value of the equilibrium constant, K, is large, so the reaction will be nearly complete at equilibrium. We have a 1-to-1 mole ratio of myoglobin to oxygen, but we don't have enough oxygen ($[O_2]_0 = 1.9 \times 10^{-5}\,M$) to react with all of the myoglobin ($[Mb]_0 = 2.0 \times 10^{-4}\,M$). Therefore, *this is a limiting reactant problem, with oxygen as the limiting reactant—there is excess myoglobin.*

Solution

We set up the ICEA table, write the mass-action expression, and solve for "x." Our assumption, that "x" is small compared to 1.9×10^{-5}, will need to be checked.

	Mb	+	O_2	⇌	MbO_2
initial	$2.0 \times 10^{-4}\,M$		$1.9 \times 10^{-5}\,M$		$0\,M$
change	$-1.9 \times 10^{-5} + x$		$-1.9 \times 10^{-5} + x$		$1.9 \times 10^{-5} - x$
equilibrium	$1.81 \times 10^{-4} + x$		x		$1.9 \times 10^{-5} - x$
assumptions	1.81×10^{-4}		x		1.9×10^{-5}

$$K = \frac{[MbO_2]}{[Mb][O_2]} = \frac{(1.9 \times 10^{-5})}{(1.81 \times 10^{-4})(x)} = 8.6 \times 10^5$$

$$x = [O_2] = 1.22 \times 10^{-7}\,M \approx 1.2 \times 10^{-7}\,M$$

$$[Mb] = 1.81 \times 10^{-4} + x \approx 1.8 \times 10^{-4}\,M$$

$$[MbO_2] = 1.9 \times 10^{-5} - x \approx 1.9 \times 10^{-5}\,M$$

The value of "x" is well under 5% of our initial concentrations, so our assumption that "x" is negligible is valid. Checking our math yields

$$\frac{(1.9 \times 10^{-5})}{(1.81 \times 10^{-4})(1.22 \times 10^{-7})} = 8.6 \times 10^5 = K$$

Further Insights

A famous advertising campaign used to talk about pork as "the other white meat," the conventional white meat being chicken. Beef is redder than chicken or pork because of its myoglobin content in the muscles, which averages (give or take quite a bit) about 8 mg of myoglobin per gram of meat. Pork averages only about 2 mg of myoglobin per gram of meat, and chicken has about 1 to 3 mg of myoglobin per gram of meat. Dogs average about 7 mg/g, rats about 2 mg/g, and whales, who require relatively immense myoglobin concentrations to hold needed oxygen for their extended stays underwater, have levels between about 20 and 70 mg/g, depending on the species.

PRACTICE 16.9

Determine the equilibrium concentration of each compound in the following reaction at 25°C, given the data indicated.

$$[C_2H_4O_2]_0 = 0\ M; \quad [C_2H_3O_2^-]_0 = 0.100\ M; \quad [H^+]_0 = 0.0500\ M$$

$$C_2H_3O_2^-(aq) + H^+(aq) \rightleftharpoons C_2H_4O_2(aq) \qquad K = 5.6 \times 10^4 \text{ (at 25°C)}$$

See Problems 54, 59, 63, 65, 66, and 67.

Intermediate Value of K

Sulfurous acid dissociates in water to produce ions. The equilibrium constant is not very small, nor is it very large:

$$H_2SO_3(aq) \rightleftharpoons H^+(aq) + HSO_3^-(aq) \qquad K = 0.0120 \text{ (at 25°C)}$$

Other chemistry also occurs in an aqueous solution of sulfurous acid, including the ionization of water and the ionization of the HSO_3^- ion:

$$H_2O(l) \rightleftharpoons H^+(aq) + OH^-(aq) \qquad K = 1.0 \times 10^{-14} \text{ (at 25°C)}$$

$$HSO_3^-(aq) \rightleftharpoons H^+(aq) + SO_3^{2-}(aq) \qquad K = 1.0 \times 10^{-7} \text{ (at 25°C)}$$

The equilibrium constants of these reactions are quite small compared to the initial ionization of sulfurous acid, so we will focus only on the first equation. Let's calculate the equilibrium concentrations of all species. That is, if $[H_2SO_3]_0 = 1.50\ M$ and $[H^+]_0 = [HSO_3^-]_0 = 0\ M$, what is the equilibrium concentration of each species in solution?

Although the value of K is intermediate (close to 1), we will *assume* that we can *neglect dissociation of H_2SO_3* (that is, we assume that the dissociation reaction does occur but that the change in concentration, "x," is negligible compared to the concentrations of our initial substances).

	$H_2SO_3(aq)$	$\rightleftharpoons$	$H^+(aq)$	$+$	$HSO_3^-(aq)$
initial	1.50 M		0 M		0 M
change	$-x$		$+x$		$+x$
equilibrium	$1.50 - x$		x		x
assumptions	1.50		x		x

$$K = \frac{[H^+][HSO_3^-]}{[H_2SO_3]}$$

$$0.0120 = \frac{x^2}{1.50}$$

$$x = 0.134\ M = [H^+] = [HSO_3^-]$$

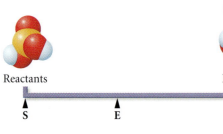

Reactants   Products

Before going on, we must test our assumption that "x" is negligible compared to 1.50 M, because our assumptions row in the ICEA table indicates that $1.50 - x = 1.50$. We do this by dividing "x" by 1.50 and multiplying by 100%. Unfortunately, "x" *(0.134) is 9% of 1.50! Our assumption is not valid.* This suggests that our equilibrium line chart might look something like that on page 698 (recall this same chart from Section 16.3), which shows more than a little forward reaction.

To arrive at the correct answer for this problem, we must solve the problem explicitly—that is, without making the assumption that the dissociation of H_2SO_3 is unimportant compared to its original concentration of 1.50 M. In order to solve the problem, we must substitute the *equilibrium* row of our ICEA table into the mass-action expression:

$$0.0120 = \frac{x^2}{1.50 - x}$$

Unfortunately, the presence of both an "x" and an "x^2" in the same equation complicates the math somewhat. One useful way to solve for "x" in these situations is to arrange the values so that we have a **quadratic equation** of the general form $ax^2 + bx + c$. For our example, we multiply both sides of the equation by $(1.50 - x)$:

Video Lesson: An Equilibrium Problem Using the Quadratic Equation

Video Lesson: Common Mathematical Functions

$$0.0120 \times (1.50 - x) = x^2$$

$$0.0180 - 0.0120x = x^2$$

Then we collect all of our terms onto one side of the equals sign by adding $0.0120x$ and subtracting 0.0180 from both sides of the equation. The result has set our equation equal to zero:

$$x^2 + 0.0120x - 0.0180 = 0$$

$$ax^2 + \quad bx \quad + \quad c \quad = 0$$

Comparing this to the general form for a quadratic equation, we obtain the values for the constants a, b, and c:

$$a = 1; \quad b = +0.0120; \quad c = -0.0180$$

Some programmable calculators will solve for the values of "x" in the equation by entering the constants a, b, and c. The other, perfectly satisfactory option is to employ the **quadratic formula**, the equation used to solve for "x":

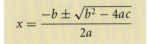

$$x = \frac{-b \pm \sqrt{b^2 - 4ac}}{2a}$$

Inserting our values for a, b, and c into this equation and solving, we obtain two values for x:

$$x = \frac{-b \pm \sqrt{b^2 - 4ac}}{2a}$$

$$x = \frac{-0.0120 \pm \sqrt{(0.0120)^2 - 4(1)(-0.0180)}}{2(1)}$$

$$x = \frac{-0.0120 \pm \sqrt{0.000144 + 0.072}}{2(1)}$$

$$x = \frac{-0.0120 \pm 0.2686}{2(1)}$$

$$x = 0.128, -0.140$$

The existence of two values for "x" would appear to be a problem. However, it is not. One of these values gives equilibrium concentrations that are physically impossible to obtain. Let's determine the concentrations in both cases and see why only one works:

If "x" = −0.140, then

$$[H_2SO_3] = 1.50 + 0.140 = 1.64\ M; \quad [H^+] = [HSO_3^-] = -0.140\ M$$

If "x" = 0.128, then

$$[H_2SO_3] = 1.50 - 0.128 = 1.37\ M; \quad [H^+] = [HSO_3^-] = 0.128\ M$$

When "x" = −0.140, we obtain negative values for both $[H^+]$ and $[HSO_3^-]$. This is impossible, so "x" must be 0.128. The correct answer to this problem must be $[H_2SO_3] = 1.37\ M$ and $[H^+] = [HSO_3^-] = 0.128\ M$. Using these values, we need to check our answer:

$$K = \frac{(0.128)^2}{1.37} = 0.0120$$

The invalid answer from the quadratic equation will generate a negative value for a concentration. We must never assume, however, that a negative value of "x" will always produce the invalid answer. In the end, the key to doing equilibrium problems is to *make assumptions when you can* to simplify the math. However, you must *test the assumptions that you make*. When the assumptions fail, we use the equilibrium row in the ICEA table to solve for "x" explicitly.

In Summary

We have seen that our ability to interpret the meaning of equilibrium constants and to make assumptions that simplify our problem solving are the ideas at the core of this different way of thinking about chemistry. We raised five questions at the beginning of this section:

1 What chemical reactions will occur?

2 Which among these are important? Which are unimportant?

3 Given the initial concentration of substances, in which direction (toward the formation of more reactants or toward the formation of more products) is the reaction likely to proceed?

4 How can we solve for the amounts of substances of interest that are present at equilibrium?

5 How can we create the conditions to maximize the concentration of the desired components and minimize that of the others?

We have now answered the first four of these questions. We are now ready for question 5, which concerns the human control of chemical processes—one big payoff of our study of equilibrium.

Tutorial: Le Châtelier's Principle

Visualization: Le Châtelier's Principle

Video Lesson: Le Châtelier's Principle

Video Lesson: CIA Demonstration: Shifting the Equilibrium of $FeSc_n^{2+}$

FIGURE 16.10

Henry Louis Le Châtelier (1850–1936), a French chemist, was a mining engineer before working as a professor. In addition to inspiring his work on thermodynamics, his interest in high temperatures, dating from his studies of mineralogy, led to the development of the oxyacetylene torch for cutting and welding steel.

16.6 Le Châtelier's Principle

We have discussed understanding chemical equilibrium as "a different way of thinking." Le Châtelier's principle, described by Henry Louis Le Châtelier (1850–1936; Figure 16.10) in 1884, extends this to *changes* in a system at equilibrium. Its implications are profound. This useful principle can be summarized as follows: *If a stress is applied to a system at equilibrium, it will change in such a way as to partially undo the applied stress and restore the equilibrium.* Although the system moves back *toward* its original set of equilibrium conditions, it never quite makes it, so there is a net change (often substantial) in equilibrium concentrations as a result of changes in the system. For people, the equivalent saying might be "Push me and I'll push back." This is shown graphically in Figure 16.11.

This principle has several implications for a system at equilibrium:

- If the *concentration* of a component is changed, the system will respond in such a way that the concentration returns toward (but doesn't make it to) its original equilibrium value.
- If the *pressure* of a system is changed, it will respond in such a way as to return the pressure toward (but not to) the original equilibrium value.
- If the *temperature* of a system is changed, it will respond by exchanging heat such that the temperature returns toward the original equilibrium value.

Let's look at the effect on equilibrium of each of these changes—concentration, pressure, and temperature—in more detail.

Changes in Concentration

If the concentration of a component is changed, the system will respond in such a way that the concentration returns toward its original equilibrium value. We began this chapter by discussing the interaction of oxygen and myoglobin to form a complex that stores and transports oxygen within muscle cells. How is the control of oxygen levels in the muscle cells indicative of this statement of Le Châtelier's principle? Let's look at it mathematically, using two separate sets of conditions. We saw one set of equilibrium conditions in Exercise 16.9. Recall the data for the reaction:

$$Mb + O_2 \rightleftharpoons MbO_2 \qquad K = 8.6 \times 10^5$$

$$[Mb] = 1.8 \times 10^{-4}\,M; \quad [O_2] = 1.2 \times 10^{-7}\,M; \quad [MbO_2] = 1.9 \times 10^{-5}\,M$$

Let's change the conditions so that the oxygen level in the blood increases to $1.0 \times 10^{-5}\,M$. How will the system respond to this change? We can use our reaction quotient, Q, to determine which way the reaction will go to reestablish equilibrium.

$$Q = \frac{[MbO_2]_0}{[Mb]_0[O_2]_0} = \frac{(1.9 \times 10^{-5})}{(1.8 \times 10^{-4})(1.0 \times 10^{-5})} = 1.1 \times 10^4$$

Because $Q < K$, the reaction will shift to the right (see Section 16.5) to produce more product and reduce the concentration of the reactants. Let's solve for the concentrations of myoglobin, oxygen, and the MbO_2 complex at this new equilibrium position. The ICEA table can be set up as follows:

	Mb	+	O_2	⇌	MbO_2
initial	$1.8 \times 10^{-4}\,M$		$1.0 \times 10^{-5}\,M$		$1.9 \times 10^{-5}\,M$
change	$-x$		$-x$		$+x$
equilibrium	$1.8 \times 10^{-4} - x$		$1.0 \times 10^{-5} - x$		$1.9 \times 10^{-5} + x$

When we solve, we find that "x" = $9.8 \times 10^{-6}\,M$, and

$$[Mb]_{new} = 1.7 \times 10^{-4}\,M; \quad [O_2]_{new} = 2.0 \times 10^{-7}\,M; \quad [MbO_2]_{new} = 2.9 \times 10^{-5}\,M$$

compared to the starting position,

$$[Mb]_0 = 1.8 \times 10^{-4}\,M; \quad [O_2]_0 = 1.0 \times 10^{-5}\,M; \quad [MbO_2]_0 = 1.9 \times 10^{-5}\,M$$

The reactant concentrations (myoglobin and oxygen) have decreased, and the concentration of the product (the myoglobin–oxygen complex) has increased. We added more reactant and the system shifted to the right to make more

FIGURE 16.11
Le Châtelier's principle is somewhat analogous to this situation. The wrestler in red pushes the defender out of position (applying a stress). The defender pushes back, trying to restore his original position, but not fully succeeding.

Video Lesson: The Effect of Changing Amounts on Equilibrium

Video Lesson: CIA Demonstration: Silver Chloride and Ammonia

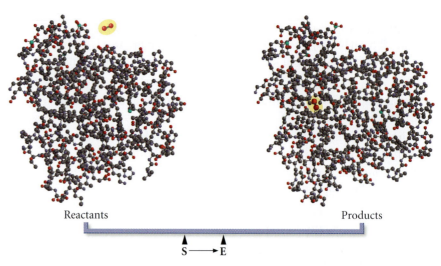

Reactants   Products

$$S \longrightarrow E$$

product. In the end, the system moved to restore the equilibrium. In our human analogy, the system was pushed and it pushed back. A snapshot of the changes on our equilibrium line chart is shown above.

During exercise, the body uses large amounts of oxygen. Some of that oxygen is supplied by the equilibrium position shifting to the left, in response to the fall in free oxygen levels that occurs because so much oxygen is being consumed. This releases more oxygen and free myoglobin, enabling athletes to run freely, for example. When the need for oxygen is less, the free myoglobin combines with oxygen (the reaction shifts to the right) and forms more of the MbO_2 complex. The complex can act as a storage location for oxygen, making it ready for release when it is needed.

Concentration changes make possible the industrial-scale manufacturing of chemicals. As the chemicals are produced in a reaction, they are removed, forcing the reaction to produce products continually in an "attempt" to restore the original equilibrium position. This is an important part of the contact process, in which the SO_3 that is produced from the reaction of SO_2 and O_2 is continually removed, forcing continued production of SO_3.

Changes in Pressure

Video Lesson: The Effect of Pressure and Volume on Equilibrium

Visualization: Equilibrium Decomposition of N_2O_4

Video Lesson: CIA Demonstration: NO_2/N_2O_4

If the pressure of a system is changed (at constant temperature and volume), it will respond in such a way as to return the pressure toward the original equilibrium value.

To illustrate the importance of this statement of Le Châtelier's principle, let's look at a reaction in which changes in pressure are meaningful—one that involves gases. Recall the reaction for the manufacture of ammonia from hydrogen and nitrogen via the Haber process:

$$3H_2(g) + N_2(g) \rightleftharpoons 2NH_3(g)$$

If the pressure of the system is substantially increased from 1 atm to 300 atm, how will it respond? The system will shift in a direction that will lower the pressure toward the original equilibrium value. Recall from our discussion of gases (Chapter 10) that the pressure of a gas is approximately proportional to the number of moles of a real gas. Fewer moles means lower pressure, so the reaction used in the Haber process will shift to the right when the pressure is increased. Typically, we observe changes in volume as the reaction seeks to accommodate pressure changes. As the volume is decreased, the pressure increases, and as the volume is increased, the pressure decreases.

An increase in pressure favors the side with fewer moles of gas.

A decrease in pressure favors the side with more moles of gas.

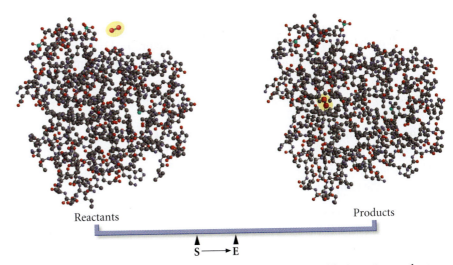

Reactants   Products

S ——→ E

product. In the end, the system moved to restore the equilibrium. In our human analogy, the system was pushed and it pushed back. A snapshot of the changes on our equilibrium line chart is shown above.

During exercise, the body uses large amounts of oxygen. Some of that oxygen is supplied by the equilibrium position shifting to the left, in response to the fall in free oxygen levels that occurs because so much oxygen is being consumed. This releases more oxygen and free myoglobin, enabling athletes to run freely, for example. When the need for oxygen is less, the free myoglobin combines with oxygen (the reaction shifts to the right) and forms more of the MbO_2 complex. The complex can act as a storage location for oxygen, making it ready for release when it is needed.

Concentration changes make possible the industrial-scale manufacturing of chemicals. As the chemicals are produced in a reaction, they are removed, forcing the reaction to produce products continually in an "attempt" to restore the original equilibrium position. This is an important part of the contact process, in which the SO_3 that is produced from the reaction of SO_2 and O_2 is continually removed, forcing continued production of SO_3.

Changes in Pressure

Video Lesson: The Effect of Pressure and Volume on Equilibrium

Visualization: Equilibrium Decomposition of N_2O_4

Video Lesson: CIA Demonstration: NO_2/N_2O_4

If the pressure of a system is changed (at constant temperature and volume), it will respond in such a way as to return the pressure toward the original equilibrium value.

To illustrate the importance of this statement of Le Châtelier's principle, let's look at a reaction in which changes in pressure are meaningful—one that involves gases. Recall the reaction for the manufacture of ammonia from hydrogen and nitrogen via the Haber process:

$$3H_2(g) + N_2(g) \rightleftharpoons 2NH_3(g)$$

If the pressure of the system is substantially increased from 1 atm to 300 atm, how will it respond? The system will shift in a direction that will lower the pressure toward the original equilibrium value. Recall from our discussion of gases (Chapter 10) that the pressure of a gas is approximately proportional to the number of moles of a real gas. Fewer moles means lower pressure, so the reaction used in the Haber process will shift to the right when the pressure is increased. Typically, we observe changes in volume as the reaction seeks to accommodate pressure changes. As the volume is decreased, the pressure increases, and as the volume is increased, the pressure decreases.

An increase in pressure favors the side with fewer moles of gas.

A decrease in pressure favors the side with more moles of gas.

result of changes in the system. For people, the equivalent saying might be "Push me and I'll push back." This is shown graphically in Figure 16.11.

This principle has several implications for a system at equilibrium:

- If the *concentration* of a component is changed, the system will respond in such a way that the concentration returns toward (but doesn't make it to) its original equilibrium value.

- If the *pressure* of a system is changed, it will respond in such a way as to return the pressure toward (but not to) the original equilibrium value.

- If the *temperature* of a system is changed, it will respond by exchanging heat such that the temperature returns toward the original equilibrium value.

Let's look at the effect on equilibrium of each of these changes—concentration, pressure, and temperature—in more detail.

Changes in Concentration

If the concentration of a component is changed, the system will respond in such a way that the concentration returns toward its original equilibrium value. We began this chapter by discussing the interaction of oxygen and myoglobin to form a complex that stores and transports oxygen within muscle cells. How is the control of oxygen levels in the muscle cells indicative of this statement of Le Châtelier's principle? Let's look at it mathematically, using two separate sets of conditions. We saw one set of equilibrium conditions in Exercise 16.9. Recall the data for the reaction:

$$Mb + O_2 \rightleftharpoons MbO_2 \qquad K = 8.6 \times 10^5$$

$$[Mb] = 1.8 \times 10^{-4}\,M; \quad [O_2] = 1.2 \times 10^{-7}\,M; \quad [MbO_2] = 1.9 \times 10^{-5}\,M$$

Let's change the conditions so that the oxygen level in the blood increases to $1.0 \times 10^{-5}\,M$. How will the system respond to this change? We can use our reaction quotient, Q, to determine which way the reaction will go to reestablish equilibrium.

$$Q = \frac{[MbO_2]_0}{[Mb]_0[O_2]_0} = \frac{(1.9 \times 10^{-5})}{(1.8 \times 10^{-4})(1.0 \times 10^{-5})} = 1.1 \times 10^4$$

Because $Q < K$, the reaction will shift to the right (see Section 16.5) to produce more product and reduce the concentration of the reactants. Let's solve for the concentrations of myoglobin, oxygen, and the MbO_2 complex at this new equilibrium position. The ICEA table can be set up as follows:

	Mb	+	O_2	⇌	MbO_2
initial	$1.8 \times 10^{-4}\,M$		$1.0 \times 10^{-5}\,M$		$1.9 \times 10^{-5}\,M$
change	$-x$		$-x$		$+x$
equilibrium	$1.8 \times 10^{-4} - x$		$1.0 \times 10^{-5} - x$		$1.9 \times 10^{-5} + x$

When we solve, we find that "x" $= 9.8 \times 10^{-6}\,M$, and

$$[Mb]_{new} = 1.7 \times 10^{-4}\,M; \quad [O_2]_{new} = 2.0 \times 10^{-7}\,M; \quad [MbO_2]_{new} = 2.9 \times 10^{-5}\,M$$

compared to the starting position,

$$[Mb]_0 = 1.8 \times 10^{-4}\,M; \quad [O_2]_0 = 1.0 \times 10^{-5}\,M; \quad [MbO_2]_0 = 1.9 \times 10^{-5}\,M$$

The reactant concentrations (myoglobin and oxygen) have decreased, and the concentration of the product (the myoglobin–oxygen complex) has increased. We added more reactant and the system shifted to the right to make more

FIGURE 16.11

Le Châtelier's principle is somewhat analogous to this situation. The wrestler in red pushes the defender out of position (applying a stress). The defender pushes back, trying to restore his original position, but not fully succeeding.

Video Lesson: The Effect of Changing Amounts on Equilibrium

Video Lesson: CIA Demonstration: Silver Chloride and Ammonia

TABLE 16.2	Mole Percent of Ammonia Present at Various Pressures and Temperatures					
	Pressure (atm)					
	1	10	50	100	300	1000
Temperature (°C)						
200	15.3	50.7	74.4	81.5	89.9	98.3
400	0.4	3.9	15.3	25.1	47.0	79.8
600	—	0.5	2.3	4.5	13.8	31.4

Source: Hocking, M. B., *Handbook of Chemical Technology and Pollution Control;* Academic Press: New York, 1998, p. 313.

How does this help in the manufacture of ammonia? Table 16.2 shows the mole percent of ammonia present at equilibrium at various pressures for the Haber process.

At every temperature, higher pressure means a shift toward ammonia. In practice, the industrial manufacture of ammonia is done between 100 and 300 atm.

Changes in Temperature

The reaction for the manufacture of ammonia is exothermic:

$$3H_2(g) + N_2(g) \rightleftharpoons 2NH_3(g) \qquad \Delta H° = -92.0 \text{ kJ}$$

How will the system respond if the temperature is raised from 200°C to 500°C at constant pressure? The change in enthalpy for the formation of ammonia is negative. Therefore, the reaction gives off heat. **What shift is necessary to attempt to restore the original temperature of the system?** The reaction will proceed in the direction that *absorbs* heat ("attempting" to lower the temperature) rather than proceeding in the direction that releases more heat, which would raise the system temperature. In the formation of ammonia, equilibrium would be restored with a shift to the left resulting in the formation of more reactants.

Exothermic reactions shift toward the left (the reactants side) when heated. (Note that this can be seen in Tables 16.2 and 16.3.)

Endothermic reactions shift toward the right (the products side) when heated.

There is another implication to the shift that occurs when a system at equilibrium is heated or cooled. The temperature itself is changing the concentrations of the components in the equation. The system is responding and we aren't adding or removing a component. This can happen only if *the equilibrium constant itself changes with temperature.* As we noted in Exercise 16.10, equilibrium constants are temperature dependent. Table 16.3 shows the change in K_p at various temperatures for the exothermic contact process addition of oxygen to SO_2 to form SO_3. Although the data in the table show that the value of K decreases as the temperature increases, there are many cases where K increases with increasing temperature. The direction of the change is dependent on the enthalpy of the process.

TABLE 16.3	Dependence of the Equilibrium Constant (K_p) on Temperature
$2SO_2(g) + O_2(g) \rightleftharpoons 2SO_3(g) \quad \Delta H = -198 \text{ kJ}$	
Temperature (°C)	K_p
400	1.6×10^5
500	2.3×10^3
600	91
700	6.9
800	0.84
900	0.15
1000	0.034
1100	0.0096

Source: Based on data from Hocking, M. B., *Handbook of Chemical Technology and Pollution Control;* Academic Press: New York, 1998, p. 260.

The position of equilibrium and, hence, the value of the equilibrium constant change when the temperature of the reaction is changed. The value of the equilibrium constant of an exothermic reaction decreases with an increase in temperature; the value of the equilibrium constant of an endothermic reaction increases with an increase in temperature.

For an exothermic reaction

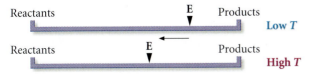

Low *T*

High *T*

For an endothermic reaction

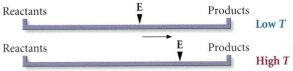

Low *T*

High *T*

Application

CHEMICAL ENCOUNTERS: Catalysts in Industry

Video Lesson: The Effects of Temperature and Catalysts on Equilibrium

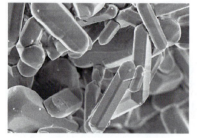

A look at metal catalysts through a scanning electron microscope reveals a porous crystal surface.

The Importance of Catalysts

Economics plays an important part in the chemical industry. A production process that reaches equilibrium more rapidly will allow more product to be manufactured. More product means the opportunity for increased sales. Catalysts—substances that substantially increase the rate of a reaction without being consumed—are therefore vital to the chemical process industries.

A catalyst does not affect the equilibrium position, but it allows the process to achieve equilibrium more rapidly than it would without a catalyst.

Fritz Haber used magnetite (Fe_3O_4) for the preparation of ammonia. Iron-related catalysts are still used for ammonia production (via the Haber process), although other metals can also be used, including osmium and mixtures of ruthenium and barium.

The nitrogen gas used in the synthesis of ammonia is obtained by distilling liquid air. The hydrogen gas for the Haber process is derived from the iron(II) oxide– and chromium(III) oxide–catalyzed reaction of carbon monoxide (derived from coal tar) with water. The equation illustrating the formation of hydrogen by this method is

$$CO(g) + H_2O(g) \xrightleftharpoons{FeO + Cr_2O_3} CO_2(g) + H_2(g) \qquad (500°C)$$

The nickel-catalyzed reaction of methane (from natural gas) and water vapor also produces hydrogen:

$$CH_4(g) + H_2O(g) \xrightleftharpoons{Ni,\ 400\ psi} CO(g) + 3H_2 \qquad (800°C)$$

This means that when the price of coal or natural gas rises, the cost of fertilizer can rise as well. This is another example of the relationship between industrial chemical processes and international economics.

The Contact process uses vanadium(V) oxide on pumice as the catalyst for the conversion of SO_2 to SO_3. Although the catalyst can last for up to 20 years, replacing it each year helps ensure efficiency in the reaction. Platinum is a better catalyst but costs much more. Platinum catalysts can also become less effective in the presence of impurities such as arsenic.

EXERCISE 16.10   **Le Châtelier's Principle**

The reaction for the combination of hydrogen gas and oxygen gas to give water vapor is

$$2H_2(g) + O_2(g) \rightleftharpoons 2H_2O(g) \qquad \Delta H° = -484\ kJ$$

Predict the effect of each of these changes to the system on the direction of equilibrium.

a. H_2O is removed as it is being generated.

b. H_2 is added.

c. The pressure on the system is decreased.

d. The system is cooled.

Solution

a. Moves to the right (removal of product forces Q to be less than K, and the system compensates by making more product).

b. Moves to the right (same as above, except Q is less than K because $[H_2]_0$ is larger than at the previous equilibrium).

c. Moves to the left because decreasing the pressure favors the side with more moles of gas.

d. Moves to the right because an exothermic reaction gives off heat. Cooling the system increases the temperature gradient between the system and the surroundings, therefore allowing a continued flow of heat to the surroundings.

PRACTICE 16.10

Predict the effect of each of these changes to the following system on the direction at equilibrium.

$$2C_2H_6O(aq) + 2CO_2(g) \rightleftharpoons C_6H_{12}O_6(aq) \qquad \Delta H = 70.0 \text{ kJ/mol}$$
Ethanol   Glucose

a. Heat is added.
b. Some ethanol (C_2H_6O) is removed from the reaction.
c. A catalyst is added.
d. More glucose is added to the flask.

See Problems 69, 70, and 74.

Le Châtelier's principle is one of the most important ideas in all of chemistry. Through its application, we understand and control reactions that have the most far-reaching importance, from the essential reactions in the human body to the production of billions of kilograms of material goods used in everything from fertilizers to plastics. Table 16.4 summarizes the impact of changing the concentration, pressure, or temperature, as well as the impact of the addition of a catalyst, on the direction of chemical equilibria.

TABLE 16.4   Impact of External Changes on the Equilibrium Direction

Change	System Response	Change in K
Concentration	Shifts toward restoring initial concentration	None
Pressure (of the system)		
Increase	Shifts toward side with fewer moles of gas	None
Decrease	Shifts toward side with more moles of gas	None
Temperature		
Exothermic reaction	Raising temperature shifts toward reactant	Decreases
	Lowering temperature shifts toward product	Increases
Endothermic reaction	Raising temperature shifts toward product	Increases
	Lowering temperature shifts toward reactant	Decreases
Catalyst	No change in direction; reaches equilibrium more rapidly	None

16.7 Free Energy and the Equilibrium Constant

We introduced the relationship between free energy and the reaction quotient, Q, in Section 14.6 of our thermodynamics chapter. We listed the quantitative relationship as

$$\Delta G = \Delta G° + RT \ln Q$$

where ΔG = the free energy of the reaction

$\Delta G°$ = the free energy at standard conditions

R = the universal gas constant, $8.3145 \text{ J·mol}^{-1}\text{·K}^{-1}$

T = the temperature in kelvins

Q = the reaction quotient illustrating the current concentrations in the reaction

Our thermodynamic definition says that at equilibrium, $\Delta G = 0$. Therefore, at equilibrium, the reaction quotient, Q, becomes the thermodynamic equilibrium constant, K_{eq}.

$$0 = \Delta G° + RT \ln(K_{eq})$$

Solving for $\Delta G°$, we obtain

$$\Delta G° = -RT \ln(K_{eq})$$

This equation is very useful. It enables us to calculate the equilibrium constant for a process if we know the standard free energy change for that process.

One word of caution is needed. The equilibrium constant in this equation requires that we use its thermodynamic definition. This means that all concentrations in the mass-action expression must be expressed in activities and that all gases must be expressed in fugacities (f). A fugacity is similar to an activity in that it measures the effective pressure of a gas. To simplify this, we'll assume that the activity values are close to the values of concentration expressed in molarity and that the fugacity values are close to the values of pressure expressed in atmospheres. Let's write the equilibrium constant expression for the environmental leaching of lead into acidic waters in order to illustrate this definition.

$$PbCO_3(s) + 2H^+(aq) \rightleftharpoons Pb^{2+}(aq) + CO_2(g) + H_2O(l)$$

$$K_{eq} = \frac{a_{Pb^{2+}} f_{CO_2} a_{H_2O}}{a_{PbCO_3} a_{H^+}^2} \approx \frac{[Pb^{2+}] P_{CO_2}}{[H^+]^2}$$

where: $a_{Pb^{2+}} \approx [Pb^{2+}]$ and $a_{H^+}^2 \approx [H^+]^2$

$f_{CO_2} \approx P_{CO_2}$ (the partial pressure of CO_2) in atmospheres

$a_{PbCO_3} \approx 1.0$ and $a_{H_2O} \approx 1.0$

In this equation, $PbCO_3$ is a solid and H_2O is a liquid. As with K and K_p, we consider their activity to be 1.0 and their concentration to be part of the equilibrium constant. As before, we simply remove them from the equation.

EXERCISE 16.11 **Free Energy and the Equilibrium Constant**

Using the standard free energy values given below, calculate the equilibrium constant at STP for the formation of $SO_3(g)$.

$$2SO_2(g) + O_2(g) \rightleftharpoons 2SO_3(g)$$

$\Delta G°$ (kJ/mol) -300 0 -371

Solution

$$\Delta_{rxn}G° = \Delta G°_{products} - \Delta G°_{reactants} = 2(-371 \text{ kJ/mol}) - 2(-300 \text{ kJ/mol})$$

$$\Delta_{reaction}G° = -142 \text{ kJ/mol}$$

In order to cancel units properly, we must convert $\Delta_{reaction}G°$ to J/mol.

$$\Delta_{reaction}G° = -142{,}000 \text{ J/mol}$$

$$\Delta G° = -RT \ln(K_{eq})$$

$$-142{,}000 \text{ J/mol} = -8.3145 \text{ J·mol}^{-1}\text{·K}^{-1} (298 \text{ K})\ln(K_{eq})$$

$$\ln(K_{eq}) = \frac{-142{,}000 \text{ J·mol}^{-1}}{-8.3145 \text{ J·mol}^{-1}\text{·K}^{-1}(298 \text{ K})} = 57.31$$

We can take the inverse natural log of both sides.

$$K_{eq} = e^{57.31} = 7.76 \times 10^{24}$$

When we compare this value to the value we mentioned for the reaction at 400°C ($K = 8.7 \times 10^6$), we notice that it is different. Why is it different? The equilibrium constant is highly temperature dependent. We've calculated it at 25°C, and so it should be different from K at 400°C.

PRACTICE 16.11

Use the thermodynamic data in the appendix to calculate the equilibrium constant for the ionization of sulfurous acid.

$$H_2SO_3(aq) \rightleftharpoons H^+(aq) + HSO_3^-(aq) \qquad K = 0.0120 \ (25°C)$$

How does the calculated value for K_{eq} compare to the value of K reported here? Explain any differences.

See Problems 77–82.

We will apply the concepts we learned in this chapter (including the nature of equilibrium, our new way of thinking about problem solving, and Le Châtelier's principle) to look at aqueous equilibria. Our immediate focus will be on a most important set of reactions—acid–base equilibria.

The Bottom Line

- Reactions can proceed reversibly toward the products or back toward the reactants. (Section 16.1)

- The point in a reaction at which there is no net change in the concentration of reactants or products is known as chemical equilibrium—or, often, simply as equilibrium. (Section 16.1)

- The free energy, G, of a reaction is at a minimum at equilibrium. (Section 16.1)

- The free energy change, ΔG, is equal to 0 at equilibrium. (Section 16.1)

- The rates of the forward and reverse reactions are equal at equilibrium. (Section 16.1)

- The mass-action expression relates the equilibrium concentrations of reactants and products in a reaction. (Section 16.1)

- The equilibrium constant is temperature dependent. (Sections 16.1, 16.6)

- The size of the equilibrium constant gives us information about the extent of a reaction. (Section 16.3)

- Modifying the coefficients of a reaction modifies the value of its equilibrium constant. (Section 16.4)

- The equilibrium constant can be converted for use with partial pressures or molarities. (Section 16.4)

- The equilibrium constant for the sum of chemical reactions is the mathematical product of the individual K values. (Section 16.4)

- We can use the equilibrium constant and mass-action expression to calculate the equilibrium concentration of substances in a reaction. (Section 16.5)

- Solving problems relating to reaction equilibria involves asking and answering a series of systematic questions. (Section 16.5)

- We can use the reaction quotient, Q, to assess which way a reaction will proceed to reach equilibrium. (Section 16.5)

- Le Châtelier's principle concerns the impact of changing the pressure, temperature, and concentration conditions of a reaction at equilibrium. (Section 16.6)

- A catalyst does not affect the equilibrium position. It changes the reaction mechanism in such a way as to speed up the reaction. (Section 16.6)

- The free energy change of a reaction can be determined from the equilibrium constant for that reaction and vice versa. (Section 16.7)

Key Words

activity The effective concentration of a solute in solution. (*p. 675*)

binding rate constant The rate constant that indicates the association of two molecules. (*p. 670*)

chromatography A chemical technique involving the partition of a solute between a stationary phase and a mobile phase. The technique can be used to separate or purify mixtures of solutes. (*p. 678*)

contact process An industrial method used to produce sulfuric acid from elemental sulfur. (*p. 676*)

distribution constant The equilibrium constant that describes the partitioning of a solute between two immiscible phases. (*p. 678*)

equilibrium A system of reversible reactions in which the forward and reverse reactions occur at equal rates, such that *no net change in the concentrations occurs,* even though both reactions continue. (*p. 670*)

equilibrium constant (*K*) The value of the equilibrium expression when it is solved using the equilibrium concentrations of reactants and products. (*p. 672*)

equilibrium expression The ratio of product to reactant concentrations raised to the power of their stoichiometric coefficients. This expression relates the equilibrium concentrations to the equilibrium constant. It is also known as the mass-action expression. (*p. 672*)

gas chromatography A specific chromatography technique in which the mobile phase is a gas and the stationary phase is a solid. (*p. 678*)

heme group A compound that, when bound to hemoglobin and iron cations, is responsible for binding oxygen. (*p. 669*)

heterogeneous equilibrium An equilibrium that results from reactants and products in different phases or physical states. (*p. 682*)

homogeneous equilibrium An equilibrium that results from reactants and products in the same phase, or physical state. (*p. 682*)

Le Châtelier's principle If a system at equilibrium is changed, it responds by returning toward its original equilibrium position. (*p. 700*)

mass-action expression The ratio of product to reactant concentrations raised to the power of their stoichiometric coefficients. This expression relates the equilibrium concentrations to the equilibrium constant. It is also known as the equilibrium expression. (*p. 672*)

mobile phase In chromatography, the phase that moves. (*p. 678*)

myoglobin (Mb) A biochemical compound responsible for storing and releasing oxygen in a living organism. (*p. 669*)

quadratic equation A mathematical equation written in the form $ax^2 + bx + c = 0$. (*p. 699*)

quadratic formula The method used to solve for *x* from the quadratic equation,

$$x = \frac{-b \pm \sqrt{b^2 - 4ac}}{2a} \quad (p. 699)$$

reaction quotient (*Q*) The ratio of product concentrations to reactant concentrations raised to the power of their stoichiometric coefficients for a reaction that is not at equilibrium. (*p. 688*)

release rate constant The rate constant of the reaction in which oxygen is released from the myoglobin-oxygen reaction. (*p. 671*)

stationary phase In chromatography, the phase that does not move. (*p. 678*)

Focus Your Learning

The answers to the odd-numbered problems and some selected problems appear at the back of the book, as represented by the blue numbering.

Section 16.1 The Concept of Chemical Equilibrium

Skill Review

1. The word *dynamic* refers to changes. Explain how the descriptive term *dynamic equilibrium* can be applied to a chemical system where the concentrations of reactants and products do not change.

2. Match each of these conditions of a chemical system to the appropriate description of the Gibbs free energy: +, −, or 0.
   a. System reacting toward products
   b. System at equilibrium
   c. System reacting toward reactants

3. When describing a reacting system, a scientist may say that the reaction "does not go to completion." Use free energy and equilibrium to explain the meaning of that phrase.

4. Assume the expression "rate₁" represents the rate of a reaction, and "rate₂" represents the expression for the rate of the reverse of that reaction. Which of these statements is *not* true of a reaction at equilibrium, and why?
   a. $\text{rate}_1/\text{rate}_2 = 1$
   b. $\text{rate}_2/\text{rate}_1 = -1$
   c. $\text{rate}_1/\text{rate}_2 = K^2$
   d. $\text{rate}_2 = \text{rate}_1 \times K$

5. Write the equilibrium expression for each of these reactions:
   a. $HCl(g) + C_2H_4(g) \rightleftharpoons C_2H_5Cl(g)$
   b. $CH_4(g) + 2O_2(g) \rightleftharpoons CO_2(g) + 2H_2O(g)$
   c. $2H_2(g) + O_2(g) \rightleftharpoons 2H_2O(l)$

6. Write the equilibrium expression for each of these reactions:
a. $CaCO_3(s) \rightleftharpoons CaO(s) + CO_2(g)$
b. $SO_3(aq) + H_2O(l) \rightleftharpoons H_2SO_4(aq)$
c. $4NH_3(g) + 7O_2(g) \rightleftharpoons 4NO_2(g) + 6H_2O(g)$

Chemical Applications and Practices

7. Phosphoric acid (H_3PO_4) is used in soft drinks and in producing fertilizers. As shown in the following reaction, phosphoric acid can be produced by the action of sulfuric acid on rocks that contain calcium phosphate.

$$Ca_3(PO_4)_2(s) + 3H_2SO_4(aq) \rightleftharpoons 3CaSO_4(aq) + 2H_3PO_4(aq)$$

Describe the system at equilibrium, using each of these three concepts:
a. Reaction rates
b. Concentration conditions
c. Gibbs free energy

8. Batteries in cars, watches, and the like all depend on a drive from reactants to products that produces electricity. When the production of electricity stops, we typically say that the battery is "dead." A chemical way to express this is to say that the battery has attained equilibrium. Explain why this chemical statement also describes why the battery no longer produces electricity.

9. Hydrogen chloride gas (used in the production of hydrochloric acid) can be produced directly by the combination of hydrogen and chlorine gas as follows:

$$H_2(g) + Cl_2(g) \rightleftharpoons 2HCl(g)$$

At equilibrium, $k_1[H_2][Cl_2] = k_{-1}[HCl]^2$. Write the mass-action expression (that is, the equilibrium expression) for the reaction.

10. Chlorofluorocarbons, including Freon-12, have been used in air conditioning units. However, their use is being phased out in most countries as a consequence of their breakdown into chlorine atoms that attack our planet's protective ozone layer. Atmospheric scientists studied the following reaction to understand the breakdown process:

$$CCl_2F_2(g) \text{ (Freon-12)} \rightleftharpoons CClF_2(g) + Cl(g)$$

At equilibrium, $k_1[CCl_2F_2] = k_2[CClF_2][Cl]$. Write the mass-action expression, i.e. equilibrium expression for the reaction.

Section 16.2
Why Is Chemical Equilibrium a Useful Concept?

Skill Review

11. In your own words, explain the economic outcome to the petroleum industry if equilibria could not be controlled.

12. What types of interactions might exist in a chromatographic system to make a solute interact strongly with the mobile phase instead of the stationary phase?

Chemical Applications and Practices

13. The following chromatogram was developed when ink from a black felt-tip pen was drawn on a plate containing a stationary phase. The plate was then dipped in solvent, and the solvent moved up the plate by capillary action. After development of the plate, it was obvious that the original black ink

had separated into the different dyes used to make the ink black.
a. Which color dye in the ink has the largest value of K_D?
b. Which color dye interacts least with the stationary filter paper?

14. The following is a chromatogram from a gas chromatography analysis of the vapor above a sample of gasoline. Each peak corresponds to a component in the gasoline vapor mixture. It is important to blend gasoline components for better performance at certain altitudes. From the chromatogram, which component has the lower value for K_D? Which has the most interaction with the stationary phase?

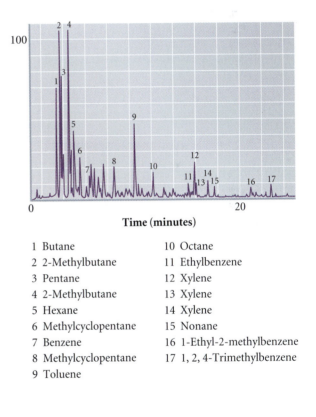

Time (minutes)

1 Butane	10 Octane
2 2-Methylbutane	11 Ethylbenzene
3 Pentane	12 Xylene
4 2-Methylbutane	13 Xylene
5 Hexane	14 Xylene
6 Methylcyclopentane	15 Nonane
7 Benzene	16 1-Ethyl-2-methylbenzene
8 Methylcyclopentane	17 1, 2, 4-Trimethylbenzene
9 Toluene	

Section 16.3 The Meaning of the Equilibrium Constant

Skill Review

15. For each of the following systems, an equilibrium expression is given. If any errors are present in the expressions, correct them and rewrite the equilibrium expression.
a. $2PbO(s) + O_2(g) \rightleftharpoons 2PbO_2(s)$ $K = [PbO_2]/[PbO]^2[O_2]$
b. $H_2O(l) + SO_3(g) \rightleftharpoons H_2SO_4(aq)$ $K = 1/[SO_3]$

16. For each of the following systems, an equilibrium expression is given. If any errors are present in the expressions, correct them and rewrite the equilibrium expression.
   a. $H_2CO_3(aq) \rightleftharpoons H_2O(l) + CO_2(g)$ $K = 1/[CO_2]$
   b. $H_2O(l) + NH_3(g) \rightleftharpoons NH_4^+(aq) + OH^-(aq)$
   $K = [NH_4^+][OH^-]/[NH_3][H_2O]$

17. Using the equilibrium line chart for the reaction of carbon monoxide and water:

$$CO(g) + H_2O(g) \rightleftharpoons CO_2(g) + H_2(g) \quad K = ????$$

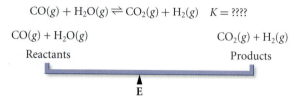

$CO(g) + H_2O(g)$ — Reactants

$CO_2(g) + H_2(g)$ — Products

E

   a. Estimate the value of K.
   b. Based on this information, what can you say about the relative concentrations of carbon monoxide and carbon dioxide when the reaction has reached equilibrium?

18. Based on the equilibrium line charts for the following hypothetic reactions, which reaction would produce fewer moles of product? Explain why this is possible. (Assume that each reaction has the same initial conditions.)

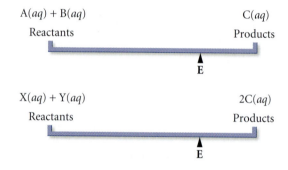

$A(aq) + B(aq)$ — Reactants

$C(aq)$ — Products

E

$X(aq) + Y(aq)$ — Reactants

$2C(aq)$ — Products

E

Chemical Applications and Practices

19. Some camp stoves and portable burners operate via the combustion of propane (C_3H_8). Balance the following combustion equation, and write the equilibrium expression for the reaction.

$$C_3H_8(g) + O_2(g) \rightleftharpoons CO_2(g) + H_2O(g)$$

20. Ethanol (C_2H_5OH) is widely used as a fuel additive in gasoline.
   a. Balance the following combustion equation, and write the equilibrium expression for this important reaction.

   $$C_2H_5OH(g) + O_2(g) \rightleftharpoons CO_2(g) + H_2O(g)$$

   b. Would you expect the value for K for this reaction, at combustion temperatures, to be large or small? Explain.

21. Hydrogen peroxide, in a dilute solution, is a very commonly used disinfectant. The following balanced equation shows the decomposition reaction for gas-phase hydrogen peroxide.

$$2H_2O_2(g) \rightleftharpoons 2H_2O(g) + O_2(g)$$

Given the following initial and final concentrations, determine the equilibrium concentration of O_2.

$[H_2O_2]_0 = 0.100;$ $[H_2O]_0 = 0.050;$ $[O_2]_0 = 0.025$

$[H_2O_2] = 0.050;$ $[H_2O] = 0.100;$ $[O_2] = ?$

22. The Haber process, at 450°C, has an equilibrium constant $K = 0.30$. What is the equilibrium concentration of hydrogen in a system containing the following equilibrium concentrations of nitrogen and ammonia?

$[N_2] = 0.100;$ $[NH_3] = 0.100;$ $[H_2] = ?$

23. Molybdenum, element 42, has many uses, perhaps most notably in the production of a strong type of steel. The reaction shown here is involved in one of the steps in recovering molybdenum from its natural ore.

$$2MoS_2(s) + 5O_2(g) \rightleftharpoons 2MoO_3(s) + 4SO(g)$$

Write the proper mass-action expression for the equilibrium constant for this reaction.

24. The following reaction shows the dissociation of magnesium hydroxide into ions. This compound can be found in several commercial antacids.

$$Mg(OH)_2(s) \rightleftharpoons Mg^{2+}(aq) + 2OH^-(aq)$$

Write the proper mass-action expression for the equilibrium constant for this reaction.

25. At 450°C, the Haber process has a K value of approximately 0.30. At 25°C, the reaction between NO from airplane exhaust and ozone has a K value of approximately 3.0×10^{34}. The ionization reaction of acetic acid in vinegar, at 25°C, is approximately 1.8×10^{-5}. Complete the following table and compare the three systems.

System	K	Reactant Favored	Product Favored	Mixture
Ammonia synthesis	0.30			
Ozone depletion	3.0×10^{34}			
Acid ionization	1.8×10^{-5}			

26. Aqueous solutions of hydrofluoric acid (HF) have the unique property of being able to dissolve glass. The ionization of HF in water has an equilibrium value, at 25°C, of approximately 3.5×10^{-4}. The decomposition of gaseous dinitrogen tetroxide ($N_2O_4(g)$), rocket fuel used on the lunar landers of the Apollo missions, has a K value of 0.133. The reaction of chlorine gas (Cl_2) and carbon monoxide (CO) to form phosgene ($COCl_2$), a deadly nerve gas, has a K value of 3.7×10^9 at 25°C. Complete the following table and compare the three systems.

System	K	Reactant Favored	Product Favored	Mixture
Ionization of HF	3.5×10^{-4}			
Decomposition	0.133			
Phosgene reaction	3.7×10^9			

27. Water hardness is a property of water that depends on the concentration of calcium and magnesium ions. One method used to determine these concentrations is titration with a known solution of EDTA. The goal in the titration is to produce a reaction between the dissolved metal ions and EDTA that converts essentially all the dissolved ions into compounds involving EDTA.

a. Explain why this is a desirable outcome in the analysis of water hardness.

b. Explain why K values on the order 10^{10} for the reactions between calcium and EDTA give us more confidence in the analysis than we would have if the K values were on the order of 10.

28. In aqueous solutions, acids dissociate to produce hydronium ions and a negative ion. If the reverse reaction is favored, the acid is considered weak. Judging on the basis of the following two acid dissociation reactions, which acid is the weaker acid?

Acetic acid (found in vinegar) $K = 1.8 \times 10^{-5}$

Benzoic acid (found in berries) $K = 6.5 \times 10^{-5}$

Section 16.4 Working with Equilibrium Constants

Skill Review

29. Water is a product in many chemical reactions. It can be directly made from its elements:

$$2H_2(g) + O_2(g) \rightleftharpoons 2H_2O(g)$$

a. Write the equilibrium expression for the reaction.

b. If the equilibrium concentrations for the compounds, at a specific temperature, were as follows, what would be the numerical value for K?

$$[H_2] = 0.134; \quad [O_2] = 0.673; \quad [H_2O] = 1.00$$

c. What is the value for K for the reverse reaction?

d. What is the value of K for the reaction of only 1 mol of hydrogen and 0.5 mol of oxygen to make 1 mol of water (that is, if we halve the coefficients in this reaction)?

30. If a solid bar of zinc is placed in a solution containing 1 M silver ions (Ag^+), a spontaneous reaction takes place producing zinc ions and silver metal with an equilibrium constant value of approximately 8×10^{52}.

a. What would you calculate as the equilibrium constant for the reverse reaction?

b. What would you calculate as the equilibrium constant if the stoichiometric coefficients were doubled?

Chemical Applications and Practices

31. When scientists seek information about air pollution in large cities, one of the reactions they study is

$$2NO(g) + O_2(g) \rightleftharpoons 2NO_2(g)$$

If K for the reaction, at 25°C, is approximately 1.7×10^{12} and the specific rate constant for the reverse reaction is approximately 6.6×10^{-12}, what would you calculate as the value of the rate constant for the forward reaction? (Units have been omitted for this problem.)

32. Ethyl acetate, a common laboratory solvent, can be prepared by the following reaction of acetic acid and ethanol:

$$C_2H_4O_2 + C_2H_6O \rightleftharpoons C_4H_8O_2 + H_2O$$

Aetic acid   Ethanol   Ethyl acetate

If K for the reaction, at 25°C, is 2.2 and the specific rate constant for the forward reaction is 4.22×10^{10}, what is the value

of the rate constant for the reverse reaction? (Units have been omitted for this problem.)

33. Producing industrially useful amounts of acetylene (C_2H_2) is important for more than acetylene torches. Acetylene is also used as a starting compound for many important polymers, such as vinyl chloride (used in PVC materials) and acrylonitrile.

a. Using the equation $CH_4(g) \rightleftharpoons \frac{1}{2}C_2H_2(g) + \frac{3}{2}H_2(g)$, write the appropriate mass-action equilibrium expression.

b. Rewrite the equation showing the stoichiometric ratios for the production of 1 mol of $C_2H_2(g)$. Write the appropriate mass-action equilibrium expression for this equation.

c. If the mathematical value for the equilibrium constant for the reaction in part a were known, how would it be mathematically converted to the equilibrium constant for the reaction in part b?

34. One of the reactions used to produce formaldehyde (CH_2O) is

$$CH_3OH(g) + \tfrac{1}{2}O_2(g) \rightleftharpoons CH_2O(g) + H_2O(g)$$

a. Write the mass-action equilibrium constant for the reaction.

b. Rewrite the reaction, keeping the same stoichiometric ratio, but showing the reaction utilizing 1 mol of oxygen. Write the mass-action equilibrium expression for this reaction.

c. How are the two equilibrium expressions mathematically related?

35. The reaction that allows many biochemical reactions to take place involves the breakdown of ATP (adenosine triphosphate) into ADP (adenosine diphosphate), as discussed in Chapter 14. The equilibrium constant of this reaction, at 37°C, is approximately 1.4×10^5. One of the early steps in the breakdown of glucose from food is the attachment of a phosphate group. The equilibrium constant for this process is approximately 4.7×10^{-3}. In living cells these two reactions are combined. What would be the equilibrium constant for the resulting combined reaction?

36. The production of tin is important because it has many practical uses, from plating iron objects to helping deliver fluoride in toothpaste as SnF_2. Combine the first two of the following reactions, and use the appropriate equilibrium constants, to obtain the equilibrium constant for the third reaction.

$$SnO_2(s) + 2CO(g) \rightleftharpoons Sn(s) + 2CO_2(g) \qquad K_1 = 14$$
$$CO(g) + H_2O(g) \rightleftharpoons CO_2(g) + H_2(g) \qquad K_2 = 1.3$$
$$SnO_2(s) + 2H_2(g) \rightleftharpoons Sn(s) + 2H_2O(g) \qquad K_3 = ?$$

37. Many compounds have more than one important use. Such is the case with the weak acid phenol (C_6H_5OH). It can be used, in dilute form, as an antiseptic and as a component in making some plastics. In water, phenol dissociates slightly as shown here:

$$C_6H_5OH(aq) \rightleftharpoons H^+(aq) + C_6H_5O^-(aq)$$

The equilibrium constant for this reaction is 1.3×10^{-10}. To analyze a solution containing phenol, you may carefully add measured amounts of aqueous NaOH. Recall that in water, the following reaction also takes place:

$$H^+ + OH^- \rightleftharpoons H_2O \qquad K = 1.0 \times 10^{14}$$

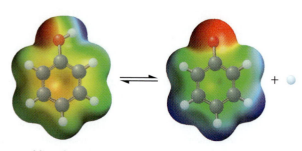

Adding these two equations produces a third that represents the reaction between sodium hydroxide and phenol. Using equilibrium constants, evaluate the feasibility of the analysis—that is, whether the reaction will proceed toward products enough for the analysis to be performed.

38. Procaine ($C_{13}H_{20}N_2O_2$) can be used to produce the anesthetic novocaine ($C_{13}H_{21}N_2O_2Cl$). Procaine is a weak base that undergoes the following reaction in aqueous solutions:

$$C_{13}H_{20}N_2O_2 + H_2O \rightleftharpoons OH^- + C_{13}H_{20}N_2O_2H^+$$

$$K = 7.1 \times 10^{-6}$$

If a solution of procaine were to be analyzed by adding carefully measured amounts of hydrochloric acid (HCl), explain how you could use the following reaction to assist in the evaluation of the viability of the analysis.

$$H^+ + OH^- \rightleftharpoons H_2O \qquad K = 1.0 \times 10^{14}$$

39. Using information from Problems 37 and 38, determine the equilibrium constant value for the reaction between phenol and procaine. Would it be a reasonable analytical strategy to employ phenol as a reactant to determine the concentration of procaine in solution?

40. Using the information from Problems 37 and 38, determine the equilibrium constant value for the reaction between HCl and NaOH. Would it be a reasonable analytical strategy to employ HCl as a reactant to determine the concentration of a NaOH solution?

41. At temperatures near 400°C, the K_p value for the synthesis of ammonia is 2.5×10^{-4}. At 400°C, what would be the approximate value of K for the following reaction?

$$3H_2(g) + N_2(g) \rightleftharpoons 2NH_3(g)$$

42. The reaction of nitrogen monoxide and chlorine gas has a large value for the equilibrium constant, K, at 298 K. However, it is often easier to measure the partial pressures of each component of a gaseous system than it is to measure their concentration. What is the value of K_p at this temperature?

$$2NO(g) + Cl_2(g) \rightleftharpoons 2NOCl(g) \quad K = 6.3 \times 10^4 \text{ (at 298 K)}$$

Section 16.5 Solving Equilibrium Problems— A Different Way of Thinking

Skill Review

43. Consider the acetic acid system described earlier in the text:

$$CH_3COOH(aq) \rightleftharpoons CH_3COO^-(aq) + H^+(aq) \quad K = 1.8 \times 10^{-5}$$

What is the equilibrium hydrogen ion concentration, $[H^+]$, given each of these initial concentrations?
a. $[CH_3COOH]_0 = 0.500\ M$; $[CH_3COO^-]_0 = 0.0\ M$
b. $[CH_3COOH]_0 = 0.100\ M$; $[CH_3COO^-]_0 = 0.100\ M$
c. $[CH_3COOH]_0 = 0.010\ M$; $[CH_3COO^-]_0 = 0.0\ M$

44. Consider the chemical system:

$$2NOCl(g) \rightleftharpoons 2NO(g) + Cl_2(g) \qquad K = 1.6 \times 10^{-5}$$

What is the equilibrium concentration of nitrogen monoxide, $[NO]$, given each of these initial concentrations?
a. $[NOCl]_0 = 0.500\ M$; $[Cl_2]_0 = 0.0\ M$
b. $[NOCl]_0 = 1.500\ M$; $[Cl_2]_0 = 2.00\ M$
c. $[NOCl]_0 = 2.25\ M$; $[Cl_2]_0 = 1.20\ M$

45. At some temperature, the reaction $H_2 + I_2 \rightleftharpoons 2HI$ has $K = 617$. Predict in which direction, forward or reverse, the reaction would proceed when:
a. $[H_2]_0 = 0.240\ M$; $[I_2]_0 = 0.080\ M$; $[HI]_0 = 0.20\ M$
b. $[H_2]_0 = 0.030\ M$; $[I_2]_0 = 0.100\ M$; $[HI]_0 = 1.50\ M$

46. At some temperature, the reaction $H_2 + I_2 \rightleftharpoons 2HI$ has $K = 617$. Predict in which direction, forward or reverse, the reaction would proceed when:
a. $[H_2]_0 = 0.990\ M$; $[I_2]_0 = 0.280\ M$; $[HI]_0 = 0.500\ M$
b. $[H_2]_0 = 0.250\ M$; $[I_2]_0 = 1.000\ M$; $[HI]_0 = 0.500\ M$

47. Using the value of $K = 8.6 \times 10^{-5}$ for the myoglobin and oxygen reaction described in the text, $Mb + O_2 \rightleftharpoons MbO_2$, indicate which of the systems described in the following table are at equilibrium. If an example is not at equilibrium, predict the direction of change (forward or reverse) that would take place to attain the equilibrium condition.

	$[Mb]\ (M)$	$[O_2]\ (M)$	$[MbO_2]\ (M)$
a.	3.5×10^{-4}	2.5×10^{-4}	0
b.	1.0×10^{-4}	1.0×10^{-4}	8.6×10^{-13}
c.	2.0×10^{-4}	1.5×10^{-4}	2.6×10^{-11}
d.	0	2.5×10^{-4}	1.0×10^{-10}

48. The decomposition of 2 moles of carbon dioxide into carbon monoxide and oxygen gas can occur under certain conditions. If, at a particular temperature, $K = 4.5 \times 10^{-5}$, predict the direction of change (forward or reverse) that would take place to attain the equilibrium condition. (*Hint:* Start by writing the balanced equation.)

	$[CO_2]\ (M)$	$[CO]\ (M)$	$[O_2]\ (M)$
a.	0.44	1.0×10^{-5}	1.0×10^{-5}
b.	1.0×10^{-4}	0.22	1.5×10^{-8}
c.	6.3×10^{-8}	3.9×10^{-2}	4.7×10^{-12}

49. Barium sulfate ($BaSO_4$) is a slightly soluble compound that has some medical applications when used to diagnose gastrointestinal problems. Because barium can be toxic, it is important to keep the concentration of barium ions at a minimum. Given the following reaction and equilibrium value, determine the maximum equilibrium concentration of $Ba^{2+}(aq)$ when in the presence of solid $BaSO_4$.

$$BaSO_4(s) \rightleftharpoons Ba^{2+}(aq) + SO_4{}^{2-}(aq) \quad K = 1.1 \times 10^{-10}$$

50. Many municipal water supplies are treated with fluoride ions (F^-) in an attempt to strengthen dental enamel. If the water contains significant amounts of calcium ions, a precipitation reaction can take place. When solid calcium fluoride is present in water, the following equilibrium is established:

$$CaF_2(s) \rightleftharpoons Ca^{2+}(aq) + 2F^-(aq) \qquad K = 4.0 \times 10^{-11}$$

What would be the equilibrium concentration of fluoride ion present in this equilibrium?

51. One of the least soluble compounds known is antimony sulfide (Sb_2S_3). If the concentration of antimony ion in a solution in which Sb_2S_3 was present were 2.2×10^{-19}, and no other solute was present, what would you calculate as the equilibrium constant for the following reaction?

$$Sb_2S_3(s) \rightleftharpoons 2Sb^{3+}(aq) + 3S^{2-}(aq)$$

52. The equilibrium constant for a reaction is very temperature dependent. Assume 0.060 is the equilibrium constant for the Haber process at a given temperature.

$$N_2(g) + 3H_2(g) \rightleftharpoons 2NH_3(g)$$

What is the equilibrium concentration of H_2 if the equilibrium concentrations of N_2 and NH_3 are found both to be 0.0010 M?

53. Using the same value of K for the ammonia synthesis reaction in Problem 52, solve for:
   a. The equilibrium concentration of NH_3 when $[N_2] = 0.0010\ M$ and $[H_2] = 0.010\ M$
   b. The equilibrium concentration of H_2 when $[NH_3] = 0.020\ M$ and $[N_2] = 0.015\ M$

54. Calculate the $[NO_2]$ given the equilibrium concentrations based on the reaction

$$2NO(g) + O_2(g) \rightleftharpoons 2NO_2(g) \qquad K = 1.71 \times 10^{12}$$

   a. $[NO] = 0.00020\ M$; $[O_2] = 0.000050\ M$; $[NO_2] = ?$
   b. $[NO] = 0.00010\ M$; $[O_2] = 0.000010\ M$; $[NO_2] = ?$

Chemical Applications and Practices

55. Codeine ($C_{18}H_{21}NO_3$), an analgesic drug obtained by prescription, produces the following reaction when added to water.

$$C_{18}H_{21}NO_3(aq) + H_2O(l) \rightleftharpoons OH^-(aq) + C_{18}H_{21}NO_3H^+(aq)$$
$$K = 1.6 \times 10^{-6}$$

   a. What reactions are taking place in the solution?
   b. Which among these are important? Which are unimportant?
   c. If a solution had the following concentrations, in which direction would it proceed?

   $[C_{18}H_{21}NO_3]_0 = 0.10\ M$; $[OH^-]_0 = 0\ M$;
   $[C_{18}H_{21}NO_3H^+]_0 = 0\ M$

   d. Once equilibrium was achieved, what would be the concentrations of the species mentioned in part c?

56. Acetylsalicylic acid ($C_9H_8O_4$, also known as aspirin) dissociates in water with an equilibrium constant, at 25°C, of 3.0×10^{-4}.

$$C_9H_8O_4(aq) \rightleftharpoons H^+(aq) + C_9H_7O_4^-(aq)$$

   a. If the initial concentration of $C_9H_8O_4$ were 0.10 M, what would you calculate as the amount of $C_9H_8O_4$ remaining at equilibrium?
   b. What percentage of $C_9H_8O_4$ reacted?
   c. Draw an equilibrium line chart that illustrates this system.

57. One method to obtain silver metal from impure lead samples is named after the work of Samuel Parkes (1761–1825). A silver-containing lead sample is melted, and zinc is added to the molten sample. The molten zinc makes a coating on the surface. The molten silver is approximately 300 times more soluble in the molten zinc than in the impure molten lead. (The zinc–silver mixture is later removed, and pure silver is obtained by distilling away the zinc.) An equilibrium constant can be written for the concentration of silver in the lead mixture versus the amount in the zinc: $K_D = [Ag_{(Zn)}]/[Ag_{(Pb)}]$. If the $[Ag_{(Zn)}]$ was 0.0010 M and the $[Ag_{(Pb)}]$ was 0.000011 M, would it be wise to wait to see whether more Ag could be extracted? Or has the extraction for this system reached a maximum? Explain your answer.

58. The solubility of a solute in one solvent compared to another can be used to our advantage. The ratio of the dissolved amounts of solute to solvent, called a partition coefficient, may be used in a similar way as the distribution constant described earlier. The K_D value for a compound between a water layer and an ether layer is 0.024. (*Note:* This represents the amount dissolved in ether divided by the amount dissolved in water.) Suppose that the compound was ether-extracted from a plant and is now 0.015 M in ether. What will be the molarity of the compound that will form, in water, when water is placed in contact with the ether?

59. At relatively high temperatures, the following reaction can be used to produce methyl alcohol:

$$CO(g) + 2H_2(g) \rightleftharpoons CH_3OH(l) \qquad K = 13.5$$

   a. If the concentration of CO, at equilibrium, were found to be 0.010 M, what would be the equilibrium concentration of hydrogen gas?
   b. Draw an equilibrium line chart that illustrates this system.

60. Pyruvic acid is produced as an intermediate during the metabolism of carbohydrates in cells; see Chapter 14. In water it undergoes the following reaction:

$$CH_3COCOOH(aq) \rightleftharpoons H^+(aq) + CH_3COCOO^-(aq)$$
$$K = 6.6 \times 10^{-3}$$

   a. If the equilibrium concentration of $CH_3COCOOH$ were found to be 0.0010 M, what would be the equilibrium concentration of CH_3COCOO^-?
   b. Draw an equilibrium line chart that illustrates this system.

61. At high temperatures, methane (CH_4) can be reacted with steam to produce carbon monoxide and hydrogen gas, as shown in the following reaction:

$$CH_4(g) + H_2O(g) \rightleftharpoons CO(g) + 3H_2(g)$$

If the equilibrium constant is 0.25, at a specific temperature and the equilibrium concentrations of $[CH_4] = 0.11\ M$, $[H_2O] = 0.28\ M$, and $[CO] = 0.75\ M$, what would you calculate as the $[H_2]$?

62. The aroma of rotten eggs can be partially attributable to the foul-smelling sulfur compound H_2S. Hydrogen sulfide decomposes according to the following reaction:

$$2H_2S(g) \rightleftharpoons 2H_2(g) + 2S(s)$$

The equilibrium constant for the process at a specific temperature is 0.020. If the initial concentration of H_2S were 0.0010 M, what would you determine to be the equilibrium concentrations of H_2S and H_2?

63. We discussed hydrogenation, the addition of hydrogen atoms to a carbon–carbon double or triple bond, in Chapter 12. One source of ethylene (C_2H_4) is the hydrogenation of acetylene (C_2H_2). The equilibrium constant for the reaction varies greatly with temperature. If the equilibrium constant for the following reaction is 4.2×10^{15}, what would be the equilibrium concentration of hydrogen gas in a system when the equilibrium concentrations of C_2H_2 and C_2H_4 were, respectively, 1.2×10^{-5} M and 0.025 M?

$$C_2H_2(g) + H_2(g) \rightleftharpoons C_2H_4(g)$$

64. Butanoic acid ($CH_3CH_2CH_2COOH$) has the aroma of spoiled butter. A better-smelling compound, an ester, can be made when butanoic acid is treated with methanol and an acid catalyst.

$$CH_3CH_2CH_2COOH + CH_3OH \rightleftharpoons$$
$$CH_3CH_2CH_2COOCH_3 + H_2O$$

Under certain conditions, the equilibrium constant for the reaction is 1.5×10^{-2}. What would be the equilibrium concentration of the product ester if the initial concentrations of each reactant were 0.500 M? Note that water is not the solvent and must be included in K.

65. The element vanadium is unusually resistant to corrosion. Alloyed with iron (approximately 5% vanadium) it produces a useful type of steel. One reaction used to obtain vanadium has a K value of 14 at 298 K.

$$VO^+(aq) + 2H^+(aq) \rightleftharpoons V^{3+}(aq) + H_2O(l)$$

Starting with $[VO^+]_0 = 0.15$ M and $[H^+] = 0.100$ M, what would you calculate as the equilibrium concentrations of $VO^+(aq)$, $H^+(aq)$, and $V^{3+}(aq)$?

66. Using the reaction in Problem 65, decide in which direction the reaction would proceed when the following concentrations were known. (Prove your answers.)
a. $[VO^{2+}] = 0.10$ M; $[H^+] = 0.25$ M; $[V^{3+}] = 0.85$ M
b. $[VO^{2+}] = .0025$ M; $[H^+] = 0.10$ M; $[V^{3+}] = 0.025$ M

67. Using the K value of 8.6×10^{-5} presented in the text for the myoglobin and oxygen reaction $Mb + O_2 \rightleftharpoons MbO_2$, what would you calculate as the $[Mb]$, $[O_2]$, and $[MbO_2]$ when the initial concentrations were as follows: $[Mb] = 0.00100$ M; $[O_2] = 0$ M, and $[MbO_2] = 0.000020$ M?

68. As was mentioned in Problem 32, ethyl acetate can be prepared by the reaction of acetic acid and ethanol.

$$C_2H_4O_2 + C_2H_6O \rightleftharpoons C_4H_8O_2 + H_2O \qquad K = 2.2 \text{ (at 25°C)}$$

What would you calculate as the equilibrium concentration of each component of the mixture if the initial concentrations of each component in the mixture were 0.100 M? (Assume that water is not the solvent.)

Section 16.6 Le Châtelier's Principle

Skill Review

69. Using Le Châtelier's principle, decide whether each of these changes would cause the equilibrium of the system presented to shift to the left, would cause it to shift to the right, or would have no effect on the equilibrium.

$$CH_4(g) + 2O_2(g) \rightleftharpoons CO_2(g) + 2H_2O(g)$$

(This is the exothermic reaction of burning methane in a lab burner.)

a. Removing CO_2 from the system
b. Adding heat to the system
c. Decreasing the volume of the container
d. Adding $H_2O(g)$ to the system
e. Adding inert He to the system

70. Calcium oxide (CaO) is also known as lime. It is one of the leading chemicals produced worldwide, thanks to its many uses in plant and animal foods, insecticides, paper making, and plaster products. It is produced, at high temperatures, from calcium carbonate.

$$CaCO_3(s) \rightleftharpoons CaO(s) + CO_2(g) \text{ (endothermic reaction)}$$

If the system were in a closed container, what effect (shift to the left, shift to the right, or no effect) would each of these changes have on the favored direction of the reaction?
a. Removing CO_2
b. Adding CaO
c. Raising the temperature
d. Enlarging the size of the container
e. Adding a suitable catalyst

Chemical Applications and Practices

71. The following equilibrium constant values are found for dissolving two solids in water to produce aqueous solutions of the ions shown. Note that the stoichiometry for the process is the same in both systems.

$$AgCl(s) \rightleftharpoons Ag^+(aq) + Cl^-(aq) \qquad K = 1.7 \times 10^{-10}$$
$$CuCl(s) \rightleftharpoons Cu^+(aq) + Cl^-(aq) \qquad K = 1.9 \times 10^{-7}$$

a. In which system would you find the greater number of moles of dissolved ions?
b. Would adding more solid to the other system increase the number of dissolved ions so that it might equal the total found in the choice for part a? Explain your reasoning.

72. The production of ethylene is important in the manufacture of polyethylene products. The following reaction shows how ethylene can be made from ethane by removing hydrogen from ethane.

$$CH_3CH_3(g) \rightleftharpoons CH_2CH_2(g) + H_2(g)$$

a. At 298 K, the equilibrium constant is 0.96. If a 1-liter container initially contained 0.10 M CH_3CH_3, what would be the equilibrium concentration of all three species?
b. After equilibrium is reached, an additional 0.010 mol of H_2 is injected into the container without changing its volume. What would be the concentration of all three species when equilibrium was once again restored?

73. A dilute solution of $Na_2Co(H_2O)_6Cl_4$ has a faint pink color and can be used as "invisible ink." When some of the loosely held water is driven off, with gentle heating, Na_2CoCl_4 forms, with a visible change to blue. If you stored your "invisible ink" solution in a refrigerator, would it appear pink or blue? Explain the basis of your answer.

74. The recognizable aromas of many fruits are due to a group of compounds known as esters, as we learned in Chapter 12. For example, the following reaction shows the production of pentyl acetate. (Assume that water is not the solvent.)

$$CH_3CH_2CH_2CH_2CH_2OH + CH_3COOH \rightleftharpoons$$

Pentanol   **Ethanoic acid**
(acetic acid)

$$CH_3CH_2CH_2CH_2CH_2OOCCH_3 + H_2O$$

Pentyl ethanoate
(pentyl acetate)

Which of these methods would increase the amount of product formed, and why?

a. Add water so that the pentyl ethanoate would dissolve better.

b. Add magnesium sulfate so it would react with any water formed.

75. One way to produce rubbing alcohol ($CH_3CHOHCH_3$) is from propene:

$$CH_3\text{—}CH\text{=}CH_2 + H_2O \underset{\text{Catalyst}}{\rightleftharpoons} CH_3\text{—}\overset{\overset{\displaystyle OH}{|}}{CH}\text{—}CH_3$$

Explain what the role of the catalyst is, in relation to the equilibrium, in this reaction.

76. Tungsten's unusually high melting point (over 3000°C) and its efficiency at producing light from electrical energy led to its use in light bulb filaments. It can be obtained via the following reaction:

$$WO_3(s) + 3H_2(g) \rightleftharpoons W(s) + 3H_2O(g)$$

Like most reactions in which a metal is obtained from another compound, this reaction is endothermic. Explain why pressure changes are not considered significant factors when tungsten is obtained in this manner.

Section 16.7 Free Energy and the Equilibrium Constant

Skill Review

77. Determine the equilibrium constant, K_{eq}, associated with reactions that have these free energy changes at 25°C.

a. $\Delta G° = -1.05$ J/mol
b. $\Delta G° = 0.230$ J/mol
c. $\Delta G° = 2.55$ kJ/mol
d. $\Delta G° = -9.80$ kJ/mol

78. Determine the free energy change, $\Delta G°$, for each of these equilibrium constants at 25°C.

a. $K_{eq} = 1.8 \times 10^{-5}$
b. $K_{eq} = 6.67 \times 10^{-1}$
c. $K_{eq} = 2.30 \times 10^{-2}$
d. $K_{eq} = 125$

79. What is the free energy change, $\Delta G°$, in kilojoules per mole, for the formation of methanol from carbon monoxide and hydrogen gas at 25°C?

$$CO(g) + 2H_2(g) \rightleftharpoons CH_3OH(l) \qquad K_{eq} = 13.5$$

80. Use the tables of $\Delta G°$ in the Appendix to determine the value of the equilibrium constant, K_{eq}, for the following reaction at 25°C.

$$CaCO_3(s) \rightleftharpoons CaO(s) + CO_2(g)$$

Chemical Applications and Practices

81. Hydrogen sulfide (H_2S), which is responsible for the odor of rotten eggs, decomposes by the reaction illustrated in Problem 62. Given the value of the equilibrium constant in Problem 62, calculate the free energy change, in kilojoules per mole, for the decomposition reaction at 25°C.

82. The ionization of HF in water (see Problem 26) has an equilibrium constant $K_{eq} = 3.5 \times 10^{-4}$ at 25°C. Calculate the free energy change, in kilojoules per mole, for the decomposition reaction.

Comprehensive Problems

83. Acids react by donating hydrogen ions. The strength of the acid reaction is determined by the ability the acid has to donate the H^+ to water. Symbolically, the reaction in water can be represented as follows:

$$HA \rightleftharpoons H^+ + A^-$$

The following is a list of some common weak acids followed by their water reaction equilibrium values. All are compared at the same temperature. Which acid in the list is the weakest? Which is the strongest?

Acetic acid, $HC_2H_3O_2$ (found in vinegar) $K = 1.8 \times 10^{-5}$

Formic acid, $HCHO_2$ (found in ants) $K = 1.8 \times 10^{-4}$

Benzoic acid, $HC_7H_5O_2$ (found in some berries)
$$K = 6.3 \times 10^{-5}$$

84. Scientists have investigated hydrazine (N_2H_4) and nitrogen monoxide (NO) in their study of rocket fuels. Use the following reaction to write the chemical equilibrium expression for this hydrazine reaction.

$$N_2H_4(g) + 2NO(g) \rightleftharpoons 2N_2(g) + 2H_2O(g)$$

85. Suppose there are two synthetic routes by which a pharmaceutical company can make the same prescription drug. Method A uses expensive starting materials but has a large value for K. Method B uses inexpensive starting materials but has a small value for K. You have been asked to discuss briefly, in a planning meeting, the various considerations involved in deciding between these two methods. What would you say about the arguments for making a profit with either method?

86. Is it possible for K_p to equal K? Explain your answer.

87. Using only whole-number coefficients, write a balanced chemical equation that represents the conversion of oxygen molecules into ozone. If the equilibrium constant for that reaction were 7×10^{-58}, what would you calculate as the equilibrium constant for the reaction that produces 1 mol of ozone?

88. The electrochemical reaction powering a nickel–cadmium rechargeable battery has an equilibrium constant value, at 298 K, of approximately 1.5×10^{11}, as written below. What is the value for the reverse reaction of this process? The reaction takes place in an alkaline solution.

$$Cd(s) + NiO_2(s) \rightleftharpoons Ni(OH)_2(s) + Cd(OH)_2(s)$$

89. An owner of a coffee shop, who happens to be a former chemistry student, notices that there are some similarities between customers (that is, people in the shop and those not entering the shop) and the reactants and products at equilibrium in a chemical system. What type of customer movement would represent equilibrium for the shop? What type of equilibrium would the profit-minded owner prefer, one with a large value of K or a small value of K? Explain. If Q were less than K, would it be a good business day or a poor business day? Explain.

$$K = \frac{[\text{customers}]}{[\text{people on the street}]}$$

90. As we have noted previously, the presence of fluoride in some municipal water systems can bring about the precipitation of CaF_2 if the concentration of Ca^{2+} is very high.

$$CaF_2(s) \rightleftharpoons Ca^{2+}(aq) + 2F^-(aq) \qquad K = 4.0 \times 10^{-11}$$

a. How many moles per liter of CaF_2 would dissolve in pure water?

b. If the concentration of Ca^{2+} in a water sample were 0.0050 M, as might be present in hard water samples, how many moles per liter of CaF_2 would dissolve?

91. The following reaction has historical importance as it was once used to produce a useful fuel called "water gas." The process involved using steam to convert coal into carbon monoxide and hydrogen gas.

$$C(s) + H_2O(g) \rightleftharpoons CO(g) + H_2(g) \quad K_p = 21 \text{ (at 1000°C)}$$

If the initial partial pressure of $H_2O(g)$ were 52 atm, what would you calculate as the equilibrium partial pressures of $H_2O(g)$; $CO(g)$ and $H_2(g)$?

92. Carbonic acid, present in carbonated beverages, can participate in two reactions that both involve removing a H^+ from the molecule.

$$H_2CO_3(aq) \rightleftharpoons H^+(aq) + HCO_3^-(aq) \quad K = 4.3 \times 10^{-7}$$
$$HCO_3^-(aq) \rightleftharpoons H^+(aq) + CO_3^{2-}(aq) \qquad K = 5.6 \times 10^{-11}$$

Starting with an initial concentration of H_2CO_3 of 0.10 M, what would you calculate as the equilibrium concentrations of $[H_2CO_3]$, $[HCO_3^-]$, $[H^+]$, and $[CO_3^{2-}]$? Be sure to justify any assumptions you made to solve the problem.

93. Examine the hypothetical reaction illustrated below. A snapshot of each reaction was taken at specific times during the course of the reaction. Which frame represents the first frame in which equilibrium has been reached? Explain your answer.

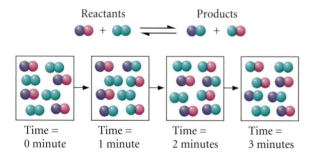

Reactants Products

Time = 0 minute | Time = 1 minute | Time = 2 minutes | Time = 3 minutes

94. We began this chapter discussing the interaction of myoglobin with oxygen. Hemoglobin is a much more familiar molecule, but we intentionally chose myoglobin rather than hemoglobin for our introduction. Look at the structure of hemoglobin in Figure 22.22 just before the start of Section 22.4. Why is myoglobin a more reasonable molecule than hemoglobin to choose for your introduction to equilibrium?

95. Look up the equilibrium constants for the solubility of lead(II) iodide and strontium oxalate in Table A5.4 in Appendix 5 at the back of the book.

a. Based on these values, draw equilibrium line charts that include the starting and equilibrium positions of the reaction.

b. Next, calculate the solubility of each.

c. Based on your answers, do your equilibrium line charts require any revision? Explain your answer.

96. Essay question: At the end of Section 16.5, we introduce the quadratic equation as a useful way to solve for concentrations when you have an intermediate value of an equilibrium constant, and your simple way of solving for the concentration does not pass the 5% rule. Given programmable calculators, many of which have quadratic equation solvers, is it ever worth introducing simplifying assumptions, or should you just solve all such equilibrium problems using the quadratic equation?

97. We note in Exercise 16.4 that EDTA-metal ion complexes have very high equilibrium constants. Can you surmise why this is so based on the structure of the $EDTA^{4-}$ ion shown in the exercise?

Thinking Beyond the Calculation

98. The industrial preparation of methanol, a potential gasoline substitute, is accomplished by the hydrogenation of carbon monoxide. The reaction is

$$CO(g) + 2H_2(g) \rightleftharpoons CH_3OH(g)$$

a. Calculate the thermodynamic parameters for this reaction using the Appendix. Is this reaction spontaneous at 25°C?

b. Using your calculated value of $\Delta G°$, determine the equilibrium constant for the reaction. Does this reaction favor products or reactants?

c. What is the value of K for the reverse reaction?

d. If the rate of the forward reaction was determined to be 3.56×10^{-12} M/s, what is the rate of the reverse reaction at equilibrium?

e. If a researcher began the synthesis of methanol by adding 1.5 mol of CO and 3.5 mol of H_2 to a 5.0-L flask, what would be the equilibrium concentration of methanol?

f. What effect, if any, would the addition of more CO have on the equilibrium concentration of methanol?

g. With added catalyst and removal of the methanol as it is formed in the reaction, the reaction can quickly produce 100% yield. If 1.0 L of $H_2(g)$ at 25°C cost $0.10 and 1.0 L of $CO(g)$ at 25°C cost $0.30, what would it cost to produce 1.0 L of $CH_3OH(l)$ at 25°C? Assume the expense associated with the experimental procedure is negligible. (*Hint:* Note that the problem asks for the cost to produce liquid methanol from gaseous reactants.)

17

Acids and Bases

Acids and bases are an integral part of our everyday life. For instance, the acidity of the local swimming pool is monitored daily. Adjustments to the pH can be made by adding chemicals.

Contents and Selected Applications

Chemical Encounters: Common Uses of Acids and Bases

17.1 What Are Acids and Bases?

17.2 Acid Strength

Chemical Encounters: Acids in Foods

17.3 The pH Scale

17.4 Determining the pH of Acidic Solutions

17.5 Determining the pH of Basic Solutions

17.6 Polyprotic Acids

Chemical Encounters: Production and Uses of Phosphoric and Sulfuric Acids

17.7 Assessing the Acid–Base Behavior of Salts in Aqueous Solution

Chemical Encounters: Acid–Base Properties of Amino Acids

17.8 Anhydrides in Aqueous Solution

Chemical Encounters: Acid Deposition and Acid-Neutralizing Capacity

Go to **college.hmco.com/pic/kelterMEE** for online learning resources.

"Home is where the heart is." Home is also where the phosphoric acid is and where other important acids are, including nitric, sulfuric, acetic, and even acetylsalicylic and ascorbic acids. Most homes also have products containing bases, among which we count ammonia and sodium hydroxide as the two most important.

Given the continuing increase in population, the resulting need for more food, the frantic pace of new home construction (see Figure 17.1), and the increase in sales of consumer products, we can understand why many acids and bases are among the top 20 chemicals produced in the United States, as shown in Table 17.1.

Why are acids and bases worth knowing about? As we will see in this chapter, they are a part of every household, the people in it, and the house itself. Acid–base reactions in our blood keep us alive. Amino acids combine to form proteins that are enzymes, hormones, and structural materials in our bodies. DNA and RNA are formed from nucleotides that are made by the chemical combination of phosphoric acid, a sugar, and a base. Our food is mostly acidic. Many of us take daily supplements of vitamin C (ascorbic acid) for its health benefits. We clean our homes with products that include both acids, such as acetic acid (CH_3COOH) and sulfuric acid (H_2SO_4) and bases, such as ammonia (NH_3) and

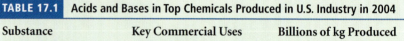

TABLE 17.1	Acids and Bases in Top Chemicals Produced in U.S. Industry in 2004	
Substance	**Key Commercial Uses**	**Billions of kg Produced**
Sulfuric acid	Fertilizer	37.5
Calcium oxide	Cement, paper	18.4*
Phosphoric acid	Fertilizers, animal feed	11.5
Ammonia	Fertilizers, plastics	10.8
Sodium carbonate	Glass, cleansers	10.3*
Sodium hydroxide	Industrial synthesis	9.5
Nitric acid	Fertilizers, explosives	6.7

*2002 data.
Source: *Chemical & Engineering News,* July 11, 2005, pp. 69–75.

FIGURE 17.1

Housing starts in the United States for the 16-year period ending in 2005. Acids and bases are a part of home construction, household furnishings, and appliances. This trend in new home construction indicates a robust demand for the industrial production of acids and bases.

TABLE 17.2	Common Acids and Bases and Examples of Their Uses
Acid or Base	**Example of Use**
Acetylsalicylic acid	Aspirin
Ammonia	Household cleaners, fertilizer, manufacture of fertilizers
Lauric acid, $CH_3(CH_2)_{10}COOH$	Key ingredient in toothpaste manufacture
Nitric acid	Key ingredient in acid precipitation
Nicotine (a base)	Tobacco products
Sodium hydroxide	Manufacture of soap, many industrial compounds

Sodium hydroxide
NaOH

Magnesium hydroxide
$Mg(OH)_2$

Acetic acid
CH_3OOH

Acetylsalicylic acid
$C_9H_8O_4$

Ascorbic acid
$C_6H_8O_6$

Acids and bases are worth knowing about because they are ubiquitous in the world both around us and within our bodies. Those household products containing bases are listed in blue; those containing acids are listed in red.

sodium hypochlorite (NaOCl). The walls of our house or apartment are likely to be made of gypsum, another name for calcium sulfate ($CaSO_4$), formed from the reaction of phosphate rock and sulfuric acid. Municipal water supplies are monitored, and (when required) chemically treated, to maintain safe acid concentrations.

Acids, such as aluminum sulfate ($Al_2(SO_4)_3$) and bases, such as sodium hydroxide (NaOH) and sodium carbonate (Na_2CO_3), are used in the manufacture of the paper on which you are now reading these ideas. Other acids and bases are used in the manufacture of plastics and polymer-based clothing, such as polyester and rayon. Computer chips are etched with hydrofluoric acid. Fertilizer for the garden is often made with ammonia-based compounds. Gold is isolated from its ores with a base known as sodium cyanide (NaCN). Toothpaste contains a base known as tetra-sodium pyrophosphate ($Na_4P_2O_7$). The list goes on and on and on. Whether the context we discuss is biological, medical, agricultural, environmental, or industrial, acids and bases are a part of it (see Table 17.2). How much of a part? As we will learn in Chapter 18, acids and bases are used to answer that as well.

17.1 What Are Acids and Bases?

Video Lesson: Arrhenius/ Brønsted–Lowry Definitions of Acids and Bases

There are several models we can use to define an acid. One of these is the Arrhenius model, named after the Swedish chemist Svante Arrhenius (1859–1927), in which an **acid** is any species that produces *hydrogen* ions in solution and a **base** is any species that produces *hydroxide* ions in solution. For example, nitric acid can be classified as an Arrhenius acid because it produces hydrogen ions in aqueous solution.

$$HNO_3(aq) \rightleftharpoons H^+(aq) + NO_3^-(aq)$$

Similarly, sodium hydroxide is an Arrhenius base because it produces hydroxide ions in aqueous solution.

$$NaOH(aq) \rightarrow Na^+(aq) + OH^-(aq)$$

In truth, hydrogen ions (H^+) do not exist free in solution. Rather, they *always* interact with the solvent, often water. Just as metal cations such as Na^+ are surrounded by water molecules through ion–dipole interactions, hydrogen cations (which are just protons because they have no neutrons or electrons) similarly

FIGURE 17.2

Data show that hydrogen ions interact with several water molecules. One possible structure involves a shell of 21 water molecules surrounding a H_3O^+ ion. (*Source:* Mark Johnson, Yale University.)

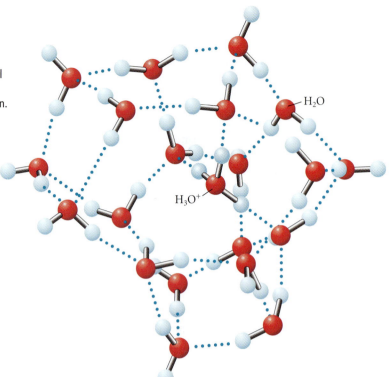

FIGURE 17.3

Johannes Nicolaus Brønsted (1879–1947) was a Danish physical chemist. In addition to studying quantum mechanics and developing an acid–base theory, he became interested in politics during the German occupation of Denmark in World War II. He was elected to the Danish parliament in 1947 but became ill and died before he could take office.

Visualization: Brønsted–Lowry Reaction

interact with water molecules that surround them. In fact, protons interact so strongly that they become directly incorporated into the fabric of one or more water molecules, and are easily passed among them. Experimental data show that each proton is immediately surrounded by at least four water molecules and is most likely surrounded by many more. Recent research by several groups in the U.S. suggests that, when H^+ is surrounded by twenty-one (21) water molecules, a most wonderful structure, shown in Figure 17.2, is formed! However, in room temperature aqueous solution, no single well-defined structure exists, because hydrogen bonds are being continually broken and formed. For simplicity, chemists typically refer to the hydrogen ion as either H^+ or H_3O^+ (the hydronium ion), understanding all the while that H^+ actually exists surrounded by several water molecules.

A more useful way of describing acids and bases is the Brønsted–Lowry model (named after the Danish scientist Johannes Nicolaus Brønsted (1879–1947, Figure 17.3) and the English chemist Thomas Lowry (1874–1936), who proposed the definition in 1923). In his model, a **Brønsted–Lowry acid** is defined as any species that *donates* a hydrogen ion (proton) to another species. This definition of an acid is applicable to *all solvents* in which protons can be exchanged, and, being more far-reaching than the Arrhenius model, it is the definition we will use. A **Brønsted–Lowry base** is any species that can *accept* a hydrogen ion (proton) from an acid. As an example of a Brønsted–Lowry acid and base, consider what happens when hydrogen chloride gas is dissolved in water.

$$HCl(g) \ + \ H_2O(l) \rightleftharpoons Cl^-(aq) \ + \ H_3O^+(aq)$$
Hydrochloric acid   Hydronium ion

Figure 17.4 shows the Lewis structures of the reactants and products. In the Brønsted–Lowry model, HCl is the acid and water is the base. The chloride ion (Cl^-) is called the **conjugate base** of HCl because it is the base that results after the HCl donates a proton to water. Similarly, H_3O^+ is the **conjugate acid** of water because it is the acid that results after the water accepts the proton from HCl. The word *conjugate* comes from the Latin word *conjugare*, which means "join together."

HCl and Cl^- are a conjugate acid–base pair.

H_3O^+ and H_2O are a conjugate acid–base pair.

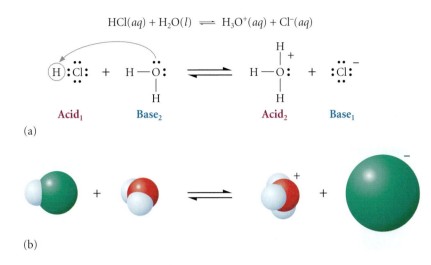

(a)

(b)

FIGURE 17.4

When hydrochloric acid gas is added to water, the HCl molecule ionizes and interacts with the solvent as shown here via Lewis structures (a) and space-filling models (b).

Note that a conjugate base is paired with an acid (Cl^- and HCl) and a conjugate acid is paired with a base (H_3O^+ and H_2O).

$$\underset{[Acid]}{HCl(g)} + \underset{[Base]}{H_2O(l)} \rightleftharpoons \underset{[Conjugate\ base]}{Cl^-(aq)} + \underset{[Conjugate\ acid]}{H_3O^+(aq)}$$

EXERCISE 17.1   **Acids, Bases, and Their Conjugates**

Complete the equation for the reaction of formic acid (HCOOH) and water. Identify the conjugate acid–base pairs.

$$HCOOH(aq) + H_2O(l) \rightleftharpoons$$

First Thoughts

Formic acid can donate a hydrogen ion to water. Is water an acid or a base in this process? The products will be the formate and hydronium ions. When deciding on the conjugate pairs, think about what each reactant forms when it either donates or accepts a hydrogen ion. We will denote the conjugate pairs as "1" and "2."

Solution

$$\underset{[Acid_1]}{HCOOH(aq)} + \underset{[Base_2]}{H_2O(l)} \rightleftharpoons \underset{[Conjugate\ base_1]}{HCOO^-(aq)} + \underset{[Conjugate\ acid_2]}{H_3O^+(aq)}$$

Further Insights

Conjugate acid–base pairs can exist in nonaqueous solvents (that is, the solvent is something other than water). For example, phenol (C_6H_6OH) acts as an acid when it reacts with the basic solvent ethylenediamine ($H_2NCH_2CH_2NH_2$) in the following way:

$$\underset{[Acid_1]}{C_6H_6OH} + \underset{[Base_2]}{H_2NCH_2CH_2NH_2} \rightleftharpoons \underset{[Conjugate\ base_1]}{C_6H_6O^-} + \underset{[Conjugate\ acid_2]}{H_2NCH_2CH_2NH_3^+}$$

PRACTICE 17.1

Complete the equation for the reaction of ethylenediamine with water. Identify the conjugate acid–base pairs.

$$H_2NCH_2CH_2NH_2(aq) + H_2O(l) \rightleftharpoons$$

See Problem 7.

FIGURE 17.5

Ammonia is an Arrhenius base because it generates OH^- in water. It is also considered a Brønsted–Lowry base because it accepts a proton from water.

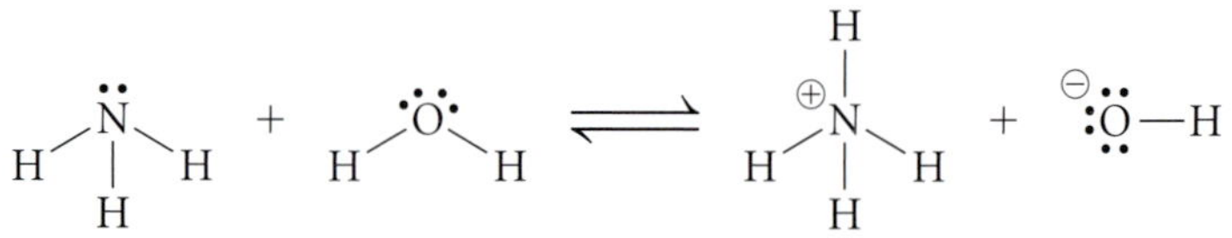

We previously mentioned that ammonia (NH_3) is an important chemical, and the 2004 worldwide production figure of 10.8 billion kg would seem to confirm this. Ammonia is a Brønsted–Lowry base because it accepts a hydrogen ion from water. The unshared pair of electrons on the nitrogen can form a covalent bond with the hydrogen, forming the ammonium ion, as shown in Figure 17.5.

Visualization: Ammonia Fountain

Video Lesson: CIA Demonstration: The Ammonia Fountain

EXERCISE 17.2 Ammonia as a Base

Identify the conjugate acid–base pairs in the reaction of ammonia with water. How does water behave differently in this reaction than when it reacts with HCl?

First Thoughts

This contrasts with Exercise 17.1 because ammonia (NH_3) is a base. Water takes on a very different role than in the previous exercise, that of an acid. Yet the concept of the conjugates is still the same—both ammonia and water will have conjugate pairs.

Solution

$$\underset{[\text{Base}_1]}{NH_3(aq)} + \underset{[\textbf{Acid}_2]}{H_2O(l)} \rightleftharpoons \underset{[\text{Conjugate acid}_1]}{NH_4^+(aq)} + \underset{[\textbf{Conjugate base}_2]}{OH^-(aq)}$$

Further Insights

We devote many words in this text to the important properties of water. Here we see another one. As the "universal solvent," water can be an acid or, as in Exercise 17.1, a base, depending on the solute. Any substance that can be an acid or a base is called amphiprotic. We will explore this property later in the chapter.

PRACTICE 17.2

Identify the conjugate pairs in the reaction of lauric acid ($CH_3(CH_2)_{10}COOH$) with water.

See Problems 3–6, 11, and 12.

Video Lesson: Lewis Acids and Bases

G. N. Lewis (remember him from Chapter 8 for his work representing substances via Lewis structures) proposed a *third* model of acids and bases in which a **Lewis base** donates a previously nonbonded pair of electrons (a lone pair) to form a coordinate covalent bond. Electrons are not transferred entirely but, rather, are shared to make a bond. Reduction and oxidation are not taking place when a Lewis base donates a pair of electrons. All Brønsted–Lowry acids are also Lewis acids, and all Brønsted–Lowry bases are also Lewis bases. However, many metal ions, such as Al^{3+}, are **Lewis acids**, because they can accept electrons, although they do not donate protons. For example, when Al^{3+} is in aqueous solution, it combines with water:

$$Al^{3+}(aq) + 6H_2O(l) \rightleftharpoons Al(H_2O)_6^{3+}(aq)$$

It then can act as a Brønsted–Lowry acid:

$$Al(H_2O)_6^{3+}(aq) + H_2O(l) \rightleftharpoons Al(H_2O)_5OH^{2+}(aq) + H_3O^+(aq)$$

In another example, we could identify borane (BH_3) in the reaction below as a Lewis acid and ammonia as a Lewis base. This reaction and the specific bonding patterns were discussed in detail in Chapter 8.

$$BH_3(g) + NH_3(g) \rightleftharpoons H_3B\text{—}NH_3(s)$$

Lewis acid   Lewis base

Tutorial: Acid Strength

We mentioned in the opening of the chapter that aluminum sulfate is added to wood pulp as it is formed into paper. The compound is used as a "sizing agent," a substance that prevents ink from spreading. Unfortunately, this acid slowly breaks down the cellulose in paper, causing it to turn light brown with age. That is why many newspapers and paperback books look old even after only weeks or months.

We will have more to say about Lewis acids in the next chapter. In our present discussion, because we are dealing primarily with water as a solvent, and because Lewis acids become Brønsted–Lowry acids in water, the Brønsted–Lowry model will be the most useful to us.

How do we know an acid when we see one? Even though nearly all common foods (including milk!) are acidic, we often associate acids specifically with citrus juices. Orange juice and grapefruit juice can taste sour. This is a characteristic property of acids. Bases, on the other hand, are slippery and often taste bitter.

We learned in Chapter 12 that hydrochloric acid reacts with zinc to produce hydrogen gas:

$$Zn(s) + 2HCl(aq) \rightarrow Zn^{2+}(aq) + 2Cl^-(aq) + H_2(g)$$

This is typical of another characteristic of acids: that they react with many metals to form solutions and often release hydrogen gas. In this sense, acids corrode metals. Bases often react with metal ions to produce insoluble hydroxides, such as $Fe(OH)_3$. A vital property that acids and bases share is that they modify the structure of some types of organic molecules, often found in plants, to cause color changes. In the next chapter, we will discuss how we use these color changes to help us analyze the amount of substances present in a sample. All of these things—sour or bitter, slimy, reactions with metals, color changes, and more—are properties that signal that a material is acidic or is basic (see Table 17.3).

Aluminum is a Lewis acid because it combines with water. Although aluminum sulfate is useful as a sizing agent in books and newspapers to prevent ink from spreading, the aluminum ion acting as an acid breaks down cellulose in paper, causing the pages to become yellow and brittle after only months.

TABLE 17.3	Properties of Acids and Bases
Acids	**Bases**
Taste sour	Taste bitter
Donate protons during an acid–base reaction	Accept protons during an acid–base reaction
React with some metals to produce hydrogen gas and metal ions	Form insoluble hydroxides with many metal ions

17.2   Acid Strength

We now know what acids and bases are, and we have seen some typical acid–base reactions. We have shown that acids and bases are present throughout our world and within ourselves. However, just as people come in all shapes and sizes, acids and bases come in different *strengths*. These differences have a profound impact on their chemical behavior and their uses.

We can begin to understand acid (and, by extension, base) strength by recalling that nearly all of our foods are acidic. Let's turn that statement around and see whether it still makes sense. If nearly all foods are acidic, are nearly all acids suitable as foods? That is, we know that citrus fruits contain citric acid ($C_6H_8O_7$) and that ingesting it in reasonable amounts is safe. Oranges and grapefruits also contain ascorbic acid ($H_2C_6H_6O_6$), which is also called vitamin C. Aspirin ($C_8H_8O_4$) is known chemically as acetylsalicylic acid. We eat these things. Some are necessary for good health (vitamin C); some can relieve headaches and may

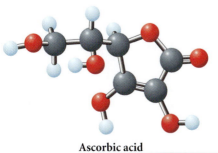
Ascorbic acid

Acid and base *strength* is an inherent property of a substance and is unaffected by dilution. Sulfuric acid, found in car batteries, is inherently strong, whereas citric acid, found in oranges, is inherently weak.

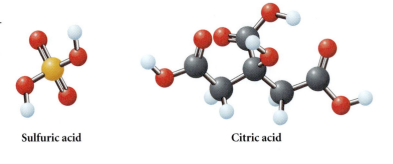

Sulfuric acid   **Citric acid**

even help prevent heart attacks (aspirin). Yet ingesting sulfuric acid (H_2SO_4) at the concentration found in a car battery would be lethal. This is also true of the hydrochloric acid that is used as a 30% solution in water to clean stains on concrete.

There are two issues here. The first is that sulfuric acid and hydrochloric acid are *strong* acids (a term defined below). Many acids found in foods are *weak* acids. The other point is that in batteries and concrete cleaners, the acids are relatively *concentrated*, whereas in an orange or other food, the acids are fairly *dilute*. Diluting the hydrochloric acid by a factor of 1000 with water would render it ineffective for cleaning concrete, even though *it would remain a strong acid*. Some acids are inherently strong and some are weak. Solute concentration, whether 10 *M* or 1×10^{-6} *M*—that is, whether relatively concentrated or dilute—doesn't change the inherent strength of the acids or bases.

Application

CHEMICAL ENCOUNTERS: Acids in Foods

Video Lesson: Weak Bases

Video Lesson: Strong Acids and Bases

Video Lesson: Weak Acids

Strong and Weak Acids

In aqueous solution, a **strong acid** is one that at 1 molar concentration *dissociates (or, in some cases, ionizes) essentially completely.* "Dissociates" means separates or, in this case, loses a hydrogen cation—a proton. Molecules in which the hydrogen atom is covalently bonded actually do ionize when they go into solution. *Whether the process is called dissociation or ionization, the result is the same: the production of H^+ in the solution.* A **weak acid** only partially ionizes (or dissociates) at 1 molar concentration in aqueous solution. We can reinforce these ideas by comparing a strong acid, nitric acid (HNO_3), to a weak acid, acetic acid (CH_3COOH). The ionization equations for 1 molar aqueous solutions of these two acids can be written as follows:

$$HNO_3(aq) + H_2O(l) \rightleftharpoons NO_3^-(aq) + H_3O^+(aq)$$

$$CH_3COOH(aq) + H_2O(l) \rightleftharpoons CH_3COO^-(aq) + H_3O^+(aq)$$

The ionization of nitric acid proceeds essentially all the way to products, so it is a strong acid. The reverse reaction will not occur to any extent because nitric acid donates a proton to water (forward direction) much more effectively than the hydronium ion (H_3O^+) donates a proton to nitrate ion (reverse direction). Acetic acid *only slightly dissociates* (typically less than 1%, depending on its initial concentration), so it is a weak acid. This occurs because H_3O^+ can donate a proton to the acetate ion (CH_3COO^-), the conjugate base of acetic acid, regenerating the initial reactants (see Figure 17.6).

The equilibrium constant for the dissociation of an acid is called the **acid dissociation constant (K_a)** and has the same relationship to the extent of an acid–base reaction as any other equilibrium constant to that of its own reaction. This is a key point:

The principles of equilibrium are consistent no matter what system we are working with.

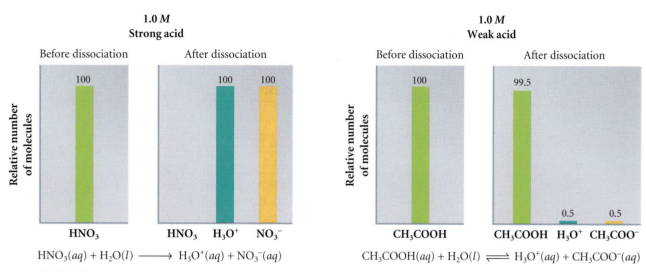

FIGURE 17.6

There is a competition between acetic acid (CH_3COOH) and hydronium ion (H_3O^+) to get rid of a hydrogen ion. The hydronium ion is a stronger acid, so the acetic acid does not dissociate to a great extent. The single-headed arrow in the ionization of nitric acid indicates that the reaction goes essentially to completion.

The equilibrium constant may be called K_a for acids, K_b for bases, or K_{sp} for solids; and there are others. *But their fundamental meaning as equilibrium constants, and how we use them in our problem solving, are the same.*

Table 17.4 lists several weak acids and their K_a values.

The mass-action expressions for the reactions shown for nitric and acetic acids with water can be written in the following "shorthand" form (neglecting the presence of water because it is the solvent, and neglecting activity effects).

Visualization: Acid Ionization Equilibrium

$$K_a = \frac{[NO_3^-][H^+]}{[HNO_3]}$$

$$K_a = \frac{[CH_3COO^-][H^+]}{[CH_3COOH]}$$

Nitric and other strong acids have K_a values much greater than 1, often regarded as infinity in aqueous solution. Weak acids, such as acetic acid ($K_a = 1.8 \times 10^{-5}$), have K_a values less than 1. We can use the equilibrium line chart that we introduced in Chapter 16 to see the extent of the reaction for HNO_3, the strong acid, compared to CH_3COOH, the weak acid.

There are differences in acid strength among strong acids. For example, $HClO_4$ is inherently stronger than HCl. However, this difference not observable in water, in which HCl, or any other strong acid, appears to be just as

TABLE 17.4	K_a of Selected Weak Acids	
Formula	**Name**	**K_a**
HSO_4^-	Hydrogen sulfate ion	1.2×10^{-2}
$HClO_2$	Chlorous acid	1.2×10^{-2}
HF	Hydrofluoric acid	7.2×10^{-4}
HNO_2	Nitrous acid	7.0×10^{-4}
$HC_3H_5O_3$	Lactic acid	1.3×10^{-4}
$HC_2H_3O_2$ (CH_3COOH)	Acetic acid	1.8×10^{-5}
$[Al(H_2O)_6]^{3+}$	Hydrated aluminum ion	1.4×10^{-5}
HOCl	Hypochlorous acid	3.5×10^{-8}
HCN	Hydrocyanic acid	6.2×10^{-10}
NH_4^+	Ammonium ion	5.6×10^{-10}
HOC_6H_5	Phenol	1.6×10^{-10}

Strong acid: nitric acid dissociation.

Weak acid: acetic acid dissociation.

strong as $HClO_4$, because their strength makes their reaction with water essentially complete. It is possible to use nonaqueous solvents to show differences in strength even among strong acids.

EXERCISE 17.3 **Relative Acid Strength**

Arrange the following acids in order from weakest to strongest. How is your placement of them related to their K_a values?

Acid	K_a
$HClO_2$	1.2×10^{-2}
CH_3COOH	1.8×10^{-5}
H_2SO_4	>1
HCN	6.2×10^{-10}

First Thoughts

Recall that a strong acid is one that dissociates (or ionizes) completely at 1 molar concentration, whereas a weak one does not. This is measured by the value of K_a.

Solution

The larger the value of K_a, the stronger the acid. Based on our values of K_a, the order of acids from weakest to strongest is $HCN < CH_3COOH < HClO_2 < H_2SO_4$.

Further Insights

The significance of the *magnitude* of the equilibrium constant is the same for acids (K_a) and bases (called "K_b") in a given solvent. For example, an acid with $K_a = 1 \times 10^{-5}$ has the same meaning for *the extent of reaction in water* as does a base with an equilibrium constant, K_b, of 1×10^{-5}, even though the specific reactions are different. This shows that understanding the concept of equilibrium goes well beyond a particular situation—a consistent theme in this discussion.

PRACTICE 17.3

The Appendix of this textbook has a list of many acids. On the basis of the K_a values, pick three acids that are weaker than hydrofluoric acid (HF) and three that are stronger than hypochlorous acid (HOCl).

See Problems 15 and 16.

Given the relative acid strengths of nitric acid and acetic acid, what might we conclude about the strengths of their conjugate bases, nitrate ions and acetate ions? Nitric acid is strong, so its ionization equilibrium to produce H^+ will lie all the way to the right. In other words, the nitrate ion is such a weak base that it stays as is and essentially doesn't go back to form nitric acid at all. Another way of expressing this is to say that in the tug of war for protons between water (which acts as a base on the reactant side) and nitrate (which acts as a base on the product side), the water wins because water is a much stronger base than the nitrate ion. This battle, and its outcome, are shown in Figure 17.7.

Let's consider acetic acid, which we call a weak acid because its acid ionization reaction doesn't go very far to the right in water. This means that most of the acetic acid stays as acetic acid when it is added to water. When water (on the reactant side) and acetate ion (on the product side) enter into the tug of war for protons, the acetate ion wins because it is a much stronger base than water. These reactions are shown in Figure 17.8. The bottom line is that *strong acids* have *very weak conjugate bases* and *weak acids* have *somewhat stronger conjugate bases*.

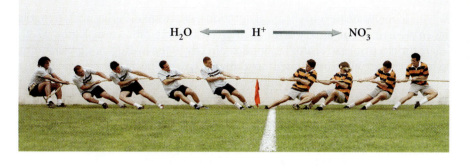

FIGURE 17.7

There is a "tug of war" for hydrogen ions between water and the nitrate ion. The water wins because water is a much stronger base than the nitrate ion.

$CH_3COOH + H_2O$ $CH_3COO^- + H_3O^+$ $CH_3COOH + H_2O$

FIGURE 17.8

In the reaction of acetate (CH_3COO^-) and water, the acetate wins because it is a stronger base than water.

EXERCISE 17.4 Strength of the Conjugates

For each of the following acids in water—hydrofluoric acid (HF, $K_a = 7.2 \times 10^{-4}$), acetic acid (CH_3COOH, $K_a = 1.8 \times 10^{-5}$), and hypochlorous acid (HOCl, $K_a = 3.5 \times 10^{-8}$):

a. Write the mass-action expression for dissociation of each acid.

b. Place the conjugate bases of these acids in order from strongest to weakest.

c. State which of the conjugate bases are stronger than water and which are weaker than water.

Solution

a. $K_a = \dfrac{[F^-][H^+]}{[HF]}$

$K_a = \dfrac{[CH_3COO^-][H^+]}{[CH_3COOH]}$

$K_a = \dfrac{[OCl^-][H^+]}{[HOCl]}$

b. Based on the K_a values, the acid strengths are ordered as follows:

$$HF > CH_3COOH > HOCl$$

The relative strengths of the conjugate bases are therefore in reverse order:

$$OCl^- > CH_3COO^- > F^-$$

c. All of these bases are stronger than water.

PRACTICE 17.4

Using Table 17.4, pick two acids with conjugate bases that are weaker than the acetate ion (CH_3COO^-).

See Problem 18.

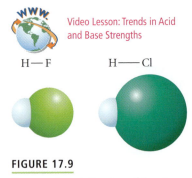

Video Lesson: Trends in Acid and Base Strengths

H—F   H——Cl

FIGURE 17.9

Compare the relative sizes of the atoms in HCl and HF.

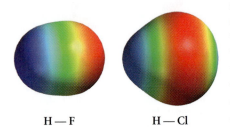

H — F   H — Cl

Electrostatic potential maps for HF and HCl indicate the location of electron density in the molecules. Note the color of the map near the hydrogen end of each molecule. The more intense blue color indicates less electron density around the hydrogen. How does this compare to the relative acidities for HF and HCl?

Why Do Acids Have Different Strengths?

As with so many answers to chemical questions, the key to differing acid strengths lies in structure. For binary acids such as HCl or HF (shown in Figure 17.9), where the electronegative atom is bonded directly to the hydrogen, smaller atoms have the valence electrons present in a smaller space. This *higher electron density* results in stronger bonds between the electronegative atom and hydrogen, which makes these acids weaker (less likely to donate the proton). That is why HF is weaker than HCl. However, if the sizes of the atoms bonded to hydrogen are about the same, *the acidity increases with increasing electronegativity of the atom bonded to hydrogen*, because the polarity of the bond also increases. This is why HF is a stronger acid than H_2O, which, though it is not binary, has two H—O bonds.

Consider a "generic" oxygen-containing compound with a central atom, A, as shown in Figure 17.10. If A has a high electronegativity, then it will have a tendency to form a covalent bond with oxygen, which is also highly electronegative, while weakening the bond between the oxygen and hydrogen. The hydrogen can then be easily removed, which means the compound is acidic. *The more electronegative the central atom (A), the more acidic the compound* (see Figure 17.11). Chlorine is more electronegative than sulfur, which, in turn, is more electronegative than phosphorus. This means that perchloric acid ($HClO_4$) is inherently stronger than sulfuric acid (H_2SO_4), which is stronger than phosphoric acid (H_3PO_4). We do not see the difference between perchloric and sulfuric acids in aqueous solution, but phosphoric acid is noticeably weaker in water than either of these other compounds.

For the same central atom (sulfur, for example), the higher the oxidation state, the higher the attraction for electrons and the stronger the covalent bond between the sulfur and oxygen atom. This tends to weaken the O—H bond in these compounds, as described above. This is why H_2SO_4 (with sulfur in the +6 oxidation state) is a stronger acid than H_2SO_3 (where sulfur is in the +4 oxidation state). For the same reason, HNO_3 is stronger than HNO_2, and the strength of so-called chlorine "oxoacids" is $HClO_4 > HClO_3 > HClO_2 > HClO$.

This model explains why a compound such as NaOH is basic. Let's look again at Figure 17.10, where "A" is Na. Sodium has a relatively low electronegativity and therefore will not form a strong covalent bond with the oxygen atom. The bond

FIGURE 17.10

In this "generic" oxygen-containing compound the central atom, A, is bonded to an oxygen atom, which is itself bonded to a hydrogen atom. If A is highly electronegative, it will weaken the bond between oxygen and hydrogen.

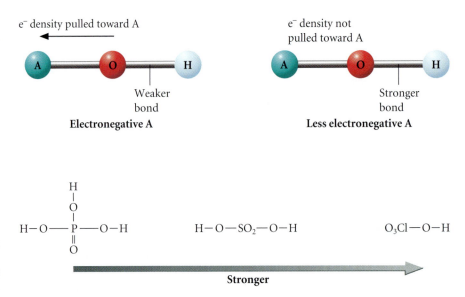

e^- density pulled toward A

Weaker bond

Electronegative A

e^- density not pulled toward A

Stronger bond

Less electronegative A

FIGURE 17.11

Chlorine is more electronegative than sulfur, which, in turn, is more electronegative than phosphorus. The result is that $HClO_4$ is inherently stronger than sulfuric acid (H_2SO_4), which is stronger than H_3PO_4. The listed structures are examples of resonance structures of each molecule.

$$H-O-\underset{\underset{O}{\|}}{\overset{\overset{H}{|}}{\underset{}{\overset{O}{|}}}}P-O-H \qquad H-O-SO_2-O-H \qquad O_3Cl-O-H$$

Stronger

between the oxygen and hydrogen is stronger and will remain intact. The hydroxide ion is therefore readily released to the solution, making NaOH a strong base in water. The same model holds for all Group IA bases.

EXERCISE 17.5 | Which Acid Is Stronger?

Given the two acids HIO and HIO_3 and two values for K_a, 0.17 and 2×10^{-11}, indicate which K_a value goes with which acid.

First Thoughts

How do we judge which of the two acids will be stronger? HIO_3 has more oxygen atoms than HIO. These additional oxygen atoms will draw electrons away from the O—H bond, weakening it. The acidic hydrogen will come off more easily from HIO_3 than the one in HIO.

Solution

HIO_3 is therefore the stronger acid. Its K_a value is 0.17. The K_a value for HIO is 2×10^{-11}.

Further Insights

It is possible to make acids that are amazingly strong by the addition of very electronegative groups to the central atom. For example, HSO_3F, shown at right, is one of several **superacids** that can be used to add hydrogen ions to organic molecules that are otherwise impervious to such reaction. Other superacids include HSO_3CF_3 and the recently synthesized $HCB_{11}H_6X_6$, where X = Cl or Br.

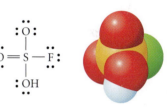

PRACTICE 17.5

Explain your answer to the following question: "Which acid is stronger, H_2S or HBr?"

See Problems 17, 19, and 20.

Will the hydrogen ion concentration always be greater in a solution of a strong acid than in a solution of a weak acid? We might envision that a solution of a strong acid generates a greater hydrogen ion concentration than a similar solution of a weak acid. However, recall our previous discussion of the terms *strong, weak, concentrated,* and *dilute*. The concentration of the acid, be it weak or strong, must be considered if we are to determine the concentration of hydrogen ions in solution. In fact, a concentrated solution of a weak acid can produce more hydrogen ions in solution than a dilute solution of a strong acid.

EXERCISE 17.6 | Concentrated Versus Dilute Strong Acids

Fish and other aquatic organisms are very sensitive to the acid concentration in their surroundings. Lake trout, for example, can flourish when the hydrogen ion concentration in a lake is 1×10^{-5} M, but they cannot survive if the hydrogen ion concentration becomes greater than 1×10^{-4} M. Indicate whether a 1000.0 L tank, filled to capacity, would have too great a hydrogen ion concentration for lake trout to survive if the tank contained:

a. 0.0365 g of HCl

b. 6180 g of boric acid (H_3BO_3). This much boric acid in 1000.0 L of water ionizes about 0.008% = 0.008 mol of H^+ produced per 100 mol of H_3BO_3 added. *Note:* We are strictly concerned with the acid concentration here, not with the possible impact of the boric acid itself on the fish.

First Thoughts

In order for the Lake trout to survive in the tank, the $[H^+]$ must be less than or equal to 1×10^{-4} M. Because there is 1000.0 L of water in the tank, the number of moles of hydrogen ion supplied by the acid in each case must not be more than

$$\frac{1 \times 10^{-4} \text{ mol } H^+}{1 \text{ L solution}} \times 1000.0 \text{ L solution} = 0.1 \text{ mol } H^+$$

Which of our acids supply more than this quantity of hydrogen ions to the solution (that is, which ones will kill the fish)? Keep in mind that HCl essentially completely ionizes in solution.

Solution

a. $0.0365 \text{ g HCl} \times \dfrac{1 \text{ mol HCl}}{36.5 \text{ g HCl}} \times \dfrac{1 \text{ mol } H^+}{1 \text{ mol HCl}} = 0.00100 \text{ mol } H^+$

   As this result shows, the strong acid solution is so dilute that the fish can survive.

b. $6180 \text{ g } H_3BO_3 \times \dfrac{1 \text{ mol } H_3BO_3}{61.8 \text{ g } H_3BO_3} \times \dfrac{0.008 \text{ mol } H^+}{100 \text{ mol } H_3BO_3} = 8 \times 10^{-3} \text{ mol } H^+$

   In this case, the fish would not die from the acid level caused by the addition of this very weak acid.

Further Insights

Aquatic life is very sensitive to acid concentration. Reproduction in salmon is affected when the hydrogen ion concentration is greater than 1×10^{-6} M, and snails cannot live in waterways that exceed that hydrogen ion concentration.

PRACTICE 17.6

What is the hydrogen ion concentration of a solution in a 500.0 L tank filled to capacity with an aqueous solution that contains 2.38 g of the strong acid HNO_3?

See Problems 21–24.

As we saw in the previous exercise, just because an acid is strong does not necessarily mean that the resulting hydrogen ion concentration due to this acid will be substantial. Instead, the equilibrium hydrogen ion concentration will depend on both the strength *and* the initial concentration of the acid. The acidity of solutions, whether food, shampoos, or samples of acid precipitation, is properly discussed in measures of *equilibrium hydrogen ion concentration*.

Visualization: pH Scale

Video Lesson: Hydronium, Hydroxide, and the pH Scale

17.3 The pH Scale

Although acid concentrations can be expressed in terms of their molarities, we can describe them more conveniently with a mathematical operator that gets rid of the exponent. This is done via a term, "p," that is the first part of the common expression **pH**. You may have heard pH used to describe the acidity of a waterway, as in "The pH of the water in the lake is about 4.5," or in advertisements for shampoo: "It will protect your hair; it is pH-balanced." **What is pH?** We can use this standard definition:

$$pH = -\log [H^+]$$

The term "p" is a mathematical operator that in practice means "take the negative base-10 logarithm of." The combined term "pH" is interpreted as "the negative base-10 logarithm of the hydrogen ion concentration." For example, if $[H^+] = 6.4 \times 10^{-2} M$, then $pH = -\log (6.4 \times 10^{-2} M)$. To solve this on the calculator, you would do the following:

1. Enter 6.4×10^{-2} into your calculator.

2. Press the "log" button. (The calculator display will now read "$-1.1938 \ldots$.")

3. Press the "$+/-$" key, to give the final answer of 1.19, rounded to two significant figures past the decimal point.

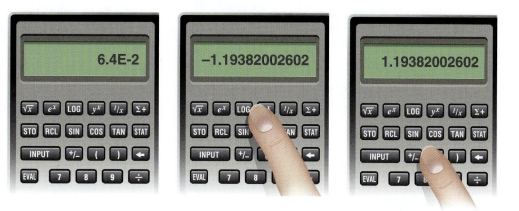

Video Lesson:
Common
Mathematical
Functions

This means that for a solution in which $[H^+] = 6.4 \times 10^{-2} M$, the pH is 1.19. In subsequent chapters we will deal with related terms such as pCl, which is roughly equal to $-\log [Cl^-]$, and pCa, which approximately represents $-\log [Ca^{2+}]$. The next exercise helps illustrate why "p" is a convenient operator with which to express concentrations.

The equation $(pH = -\log [H^+])$ can also be used in reverse. For instance, if we know that the pH of a solution is 1.19, we can calculate the $[H^+]$. Mathematically, the conversion is

$$pH = -\log[H^+]$$

$$1.19 = -\log[H^+]$$

$$-1.19 = \log[H^+]$$

$$10^{-1.19} = [H^+]$$

$$[H^+] = 6.4 \times 10^{-2} M$$

In general, we can rearrange our equation for calculating pH into a form that will enable us to calculate the $[H^+]$.

$$[H^+] = 10^{-pH}$$

We need to make a note about significant figures and logarithms. The exponent on a number written in scientific notation is not significant, as we have seen. The numbers to the left of the decimal point in a logarithm carry the same meaning as the exponent in scientific notation. That is, they are not considered significant figures. For instance, from our discussion above, each of

$$6.4 \times 10^{-2} M \qquad \text{and} \qquad pH = 1.19$$

possesses only two significant figures.

> **EXERCISE 17.7** **Using the Operator "p" to Determine Useful Quantities**
>
> Calculate the desired term.
>
> a. pH for $[H^+] = 3.2 \times 10^{-11}\ M$
> b. pK_a for $K_a = 1.8 \times 10^{-5}$
> c. pOH for $[OH^-] = 8.8 \times 10^{-13}\ M$
>
> d. pH for $[H^+] = 3.2 \times 10^{-10}\ M$
> e. $[H^+]$ for pH $= 12.73$
> f. $[OH^-]$ for pOH $= 5.08$
>
> **Solution**
>
> In parts a–d, we use the operator "p" and take "−log of" the indicated value. Note that the log of a value has no units; it is dimensionless. In parts e and f, we use 10^{-p} to determine the answer.
>
> a. pH $= -\log\,(3.2 \times 10^{-11}\ M) = 10.49$
> b. p$K_a = -\log\,(1.75 \times 10^{-5}\ M) = 4.74$
> c. pOH $= -\log\,(8.8 \times 10^{-13}\ M) = 12.06$
> d. pH $= -\log\,(3.2 \times 10^{-10}\ M) = 9.49$
> e. $[H^+] = 10^{-12.73} = 1.9 \times 10^{-13}\ M$
> f. $[OH^-] = 10^{-5.08} = 8.3 \times 10^{-6}\ M$

> **PRACTICE 17.7**
>
> Determine each of the desired terms.
>
> a. pH for $[H^+] = 4.61 \times 10^{-3}\ M$
> b. pK_a for $K_a = 2.77 \times 10^{-9}$
> c. pOH for $[OH^-] = 3.22 \times 10^{-6}\ M$
>
> d. $[H^+]$ for pH $= 3.92$
> e. $[H^+]$ for pH $= 1.49$
> f. $[OH^-]$ for pOH $= 9.93$
>
> *See Problems 25, 26, 33, and 34.*

Converting from $[H^+]$ to pH enables us to express acid concentration in a way that does not require exponents. The use of the log scale has another important implication. If we look again at parts a and d of Exercise 17.7, we see that the value of $[H^+]$ changes by a factor of 10 and the pH changes by one unit. This is the impact of taking the log of a number. *The pH will change by one unit for each power-of-10 change in $[H^+]$ concentration.* A *decrease* in $[H^+]$ from 1×10^{-5} to $1 \times 10^{-6}\ M$ is indicated by an *increase* in pH from 5 to 6. This power-of-10 relationship is shown in Figure 17.12, in which we begin with $[H^+] = 1 \times 10^{-8}\ M$ in water and increase by powers of 10 until we get to $[H^+] = 1\ M$.

FIGURE 17.12

Relationship between pH and [H⁺]. The pH unit is a logarithm term. Therefore, a 1-unit increase represents a 10-fold increase in [H⁺].

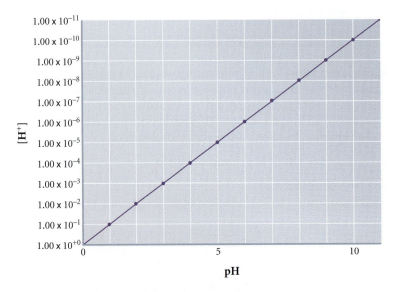

EXERCISE 17.8   **[H⁺] and pH: One Implication**

In Exercise 17.6, we discussed the sensitivity of aquatic species to the hydrogen ion concentration. Let's extend this discussion. Mussels can survive in waterways with a pH of 6.8. They cannot survive at pH 5.2. What is the ratio of the hydrogen ion concentrations in the two solutions? That is, how many times greater is one than the other?

First Thoughts

The pH scale is logarithmic. That is, each pH unit represents a difference in $[H^+]$ of a factor of 10. The pH values of 6.8 and 5.2 therefore have very different hydrogen ion concentrations. *In order to find the ratio of the concentrations, we must convert the pH of each solution into its hydrogen ion concentration. We cannot compare the pH values, themselves, as a measure of the hydrogen ion concentration ratio.* Which one has the higher hydrogen ion concentration, $[H^+]$? *The lower the pH, the higher the $[H^+]$,* so the pH 5.2 solution has a greater $[H^+]$ than the pH 6.8 solution.

Solution

$[H^+] = 10^{-pH}$, so for the waterway that has a pH = 6.80,

$$[H^+] = 10^{-6.80} = 1.6 \times 10^{-7}\ M$$

For the waterway that has a pH = 5.20,

$$[H^+] = 10^{-5.20} = 6.3 \times 10^{-6}\ M$$

The ratio of the hydrogen ion concentrations is the quotient of the two values:

$$\frac{6.3 \times 10^{-6}\ M}{1.6 \times 10^{-7}\ M} = 39$$

Further Insights

The waterway at pH 5.20 has an acid concentration that is about 40 times greater than that of the pH 6.80 waterway. This shows that seemingly small changes in pH can translate into large changes in hydrogen ion concentration. This is of particular importance in discussions involving the chemistry of life, where small changes in the pH of blood, for example, can be hazardous or fatal.

PRACTICE 17.8

If the hydrogen ion concentration of a body of water is 500 times that of another waterway, and the pH of the less acidic water is 8.84, what is the pH of the more acidic water?

See Problems 29 and 30.

Water and the pH Scale

Pure water can undergo **autoprotolysis**, proton transfer within only the solvent itself. Such a proton transfer would indicate that one of the water molecules is acting as a base and the other is acting as an acid.

$$H_2O(l) + H_2O(l) \rightleftharpoons H_3O^+(aq) + OH^-(aq) \qquad K_w = 1.00 \times 10^{-14} \text{ at } 24°C$$

The ability of a compound to act both as an acid and as a base (not necessarily in the same reaction) isn't unique to water. Those compounds capable of such a feat are called **amphiprotic**. They include, among many others, water, ammonia (NH_3), and, as we shall discover in Section 17.7, the amino acids that make up the proteins in the human body. In the autoprotolysis of water, the reaction is often simplified as

$$H_2O(l) \rightleftharpoons H^+(aq) + OH^-(aq) \qquad K_w = 1.00 \times 10^{-14} \text{ at } 24°C$$

 Visualization: Self-Ionization of Water

It is understood that the H^+ and OH^- really aren't naked ions; they are solvated by the aqueous solution. However, we can write the equation this way to simplify our equation. In any case, this equation has the mass-action expression

$$K_w = [H^+][OH^-] = 1.00 \times 10^{-14}$$

Taking the log of both sides gives

$$pK_w = pH + pOH = 14.00$$

In pure water, the only source of $[H^+]$ and $[OH^-]$ is water itself, so the concentrations of H^+ and OH^- (formed in a 1:1 ratio) will be equal. **What are the concentrations of H^+ and OH^- in pure water?** If we call each "x," then

$$[H^+][OH^-] = x^2 = 1.00 \times 10^{-14}$$

$$x = [H^+] = [OH^-] = 1.00 \times 10^{-7}\ M$$

This means that in pure water at 24°C, $[H^+] = [OH^-] = 1.00 \times 10^{-7}\ M$, and the solution will have pH = 7.0. We define this as **neutral pH** when the solvent is water and the system is at 24°C. This pH value also equals pOH because both $[H^+]$ and $[OH^-]$ are equal to $1.0 \times 10^{-7}\ M$ under these conditions. The K_w value is temperature dependent, as shown in Table 17.5.

Unless otherwise stated, we will assume a temperature of 24°C for the remainder of our discussion, so that $K_w = 1.00 \times 10^{-14}$. In aqueous solution, we can always determine pH, $[H^+]$, pOH, or $[OH^-]$ in a solution if any one of these factors is known.

TABLE 17.5 K_w **at Several Temperatures**

Temperature	K_w	pK_w
0°C	1.14×10^{-15}	14.94
10°C	2.92×10^{-15}	14.54
20°C	7.81×10^{-15}	14.11
24°C	1.00×10^{-14}	14.00
25°C	1.01×10^{-14}	14.00
30°C	1.47×10^{-14}	13.83
50°C	5.47×10^{-14}	13.26
60°C	9.71×10^{-14}	13.01

EXERCISE 17.9 **Water in a Pristine Lake**

Pure water has a pH = 7.0. But "pristine" rainwater (unaffected by pollutants from sources such as nitrogen oxides from automobile tailpipe emissions or sulfur oxides from industrial smokestack emissions) generally has a pH of between 5.5 and 6.0. This results from the dissolving and equilibration of carbon dioxide from the atmosphere and the subsequent release of a hydrogen ion to the water.

$$CO_2(g) + H_2O(l) \rightleftharpoons H_2CO_3(aq) \rightleftharpoons H^+(aq) + HCO_3^-(aq)$$

Many waterways in the United States have pH values significantly lower than this range as a consequence of acid precipitation, in which the stronger acids HNO_3 and H_2SO_4 combine with the water. We discussed the implications of this in Exercises 17.6 and 17.8. If the pH of the water in a pristine lake is 5.90, determine the value of $[H^+]$, $[OH^-]$, and pOH in this water.

Solution

There are several ways to do this problem. We'll show just one.

$$pOH = 14.00 - pH = 14.00 - 5.90 = 8.10$$

$$[H^+] = 10^{-pH} = 10^{-5.9} = 1.3 \times 10^{-6}\ M$$

$$[OH^-] = 10^{-pOH} = 10^{-8.10} = 7.9 \times 10^{-9}\ M$$

We can check the result by noting that $[H^+][OH^-] = K_w$ and

$$(1.3 \times 10^{-6})(7.9 \times 10^{-9}) = 1.0 \times 10^{-14}.$$

PRACTICE 17.9

What are the pH, $[H^+]$, and $[OH^-]$ values for a solution with pOH = 12.35?

See Problems 27 and 28.

The sum of the pH and pOH values is 14.00. This is the basis of the common pH scale, shown in Figure 17.13. Note that the hydrogen ion and hydroxide ion concentrations are inversely related. Aqueous solutions with a pH less than 7.0 are said to be **acidic**, and those with a pH greater than 7.0 are **basic**. Also, as the solutions become more acidic or basic, their pH moves farther away from 7.0, as shown in the figure. The pH values of some common substances are also listed in Table 17.6 and illustrated graphically in Figure 17.14.

FIGURE 17.13

The sum of the pH and pOH values is 14.00 in water at 24°C. This is the basis of the common pH scale, shown here. Note the inverse relationship between the concentration of hydrogen and that of the hydroxide ion.

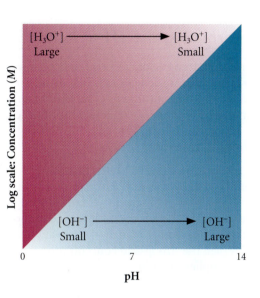

TABLE 17.6	pH of Some Common Substances	
Substance	**Contains This Acid or Base**	**pH**
Battery acid	Sulfuric acid	1.3
Stomach acid	Hydrochloric acid	1.5–3.0
Vinegar	Acetic acid	2.5
Wines	Tartaric acid	2.8–3.8
Apples	Malic acid	2.9–3.3
Food preservative	Benzoic acid	3.1
Cheese	Lactic acid	4.8–6.4
Blood	Carbonate ion and others	7.3–7.5
Baking soda	Bicarbonate ion	8.3
Detergents	Carbonate, phosphate ions	10–11
Milk of magnesia	Magnesium hydroxide	10.5
Drain cleaner	Sodium hydroxide	13+

Very basic

14.0 — Household lye
13.0 — Bleach
12.0
Ammonia — 11.0
10.0
9.0
Egg whites — 8.0 — Seawater
Distilled water — 7.0 — Swimming pool water
6.0
Pure rain — 5.0 — Egg yolks
Beer — 4.0 — Orange juice
Pickle processing — 3.0
Lemon juice — 2.0 — Vinegar
1.0 — Battery acid
0.0

Very acidic

FIGURE 17.14

pH of some common substances.

HERE'S WHAT WE KNOW SO FAR

- We now understand what is meant when we read or hear about acids and bases.
- We know the difference between strong and weak acids and the difference between concentrated and dilute acids.
- We have a frame of reference—the pH scale—with which we can categorize solutions of acids and bases.
- We now know where many common substances fit within this frame of reference.

At its heart, chemistry is about the *interactions* of substances. The applications of chemistry concern the *changes* that result from these interactions. To understand the nature of change in acid–base chemistry, we first need to consider how to determine the changes that occur when an aqueous acidic or basic solution is prepared. This will be the focus of the remainder of this chapter. In the next chapter, we will consider how changing a prepared acidic or basic solution affects its chemical behavior.

17.4 Determining the pH of Acidic Solutions

We have learned that strong acids essentially completely dissociate in aqueous solution ($K_a \gg 1$) and that weak acids as a rule dissociate only partially ($K_a < 1$). We also noted that the concentration of [H$^+$] (and therefore the pH) in solution will be determined by both the strength and the initial concentration of the acid.

Tutorial: Calculating pH of Strong Acid and Base Solutions

pH of Strong Acid Solutions

A monoprotic acid is an acid (such as HCl) that contains only one acidic hydrogen ion, or proton. H_2SO_4, by contrast, contains two acidic protons and is called a diprotic acid. In a strong monoprotic acid, the hydrogen ion concentration in aqueous solution roughly equals the initial concentration of the acid (neglecting activity effects). This is approximately true at any concentration greater than about 10^{-6} M. When the strong acid concentration is less than this, the autoprotolysis of water supplies relatively few hydrogen ions that exist in solution, and the pH becomes more difficult to calculate, as we will discuss later in this section. Even with an HCl solution of concentration 10^{-8} M or less, the pH is still held just below 7.0 because of the supply of H$^+$ from the autoprotolysis of water.

EXERCISE 17.10   Calculating the pH of a Strong Acid Solution

Your stomach contains hydrochloric acid. If we were going to prepare a solution that had a hydrogen ion concentration within the range of that in the stomach, we might work with a 3×10^{-2} M aqueous solution of HCl. What is the pH of this solution?

First Thoughts

This strong acid would essentially completely dissociate to give [H$^+$] $= 3 \times 10^{-2}\,M$ and [Cl$^-$] $= 3 \times 10^{-2}\,M$.

Solution

We can calculate the pH of this solution as follows:

$$\text{pH} = -\log[\text{H}^+] = -\log(3 \times 10^{-2}\,M) = 1.5$$

Further Insights

Does the answer make sense? We have a moderately concentrated strong acid, so we would expect the pH to be in the range of, perhaps, 1 to 3. We can narrow our expected answer down further by noting that [H$^+$] is between 10^{-1} and $10^{-2}\,M$, which means that the pH will be between 1 and 2. Our answer to this problem does make sense.

PRACTICE 17.10

What is the pH of a 0.0010 M HNO$_3$ solution? . . . of a 0.0000250 M HClO$_4$ solution?

See Problem 37.

Le Châtelier's Principle and the Supply of Hydroxide Ion in Acidic Solutions

Consider a strong acid at pH 3.0. On the basis of our previous discussion, we know that the pOH is 11.0, and $[OH^-] = 1 \times 10^{-11}$ M. Where does that small amount of hydroxide ion come from? The only source is the autoprotolysis of water:

$$H_2O(l) \rightleftharpoons H^+(aq) + OH^-(aq) \qquad K_w = 1.00 \times 10^{-14}$$

If the liquid were *pure* water, $[OH^-]$ would equal 1×10^{-7} M. However, the addition of H^+ from the strong acid imposed a stress on the aqueous system, to which it responded by lessening the extent of dissociation of water. To put it another way, when the acid was added, *the reaction shifted to the left to compensate.* This is an example of Le Châtelier's principle, which we discussed in Section 16.6, and is known as the **common-ion effect**. We added an ion *common* to one of the products (H^+), and the result was that the reaction did not proceed to the right as much as it would have without the acid. We will look at many of the practical outcomes of Le Châtelier's principle and the common-ion effect in the next chapter. The key in this introduction to acids and bases is that the addition to an aqueous solution of any substance that produces H^+ will reduce the supply of H^+ and OH^- from the autoprotolysis of water. Therefore, in all but the dilute acid solutions, the autoprotolysis of water as a source of H^+ is generally *negligible.*

For example, let's consider the pH of a 1.0×10^{-8} M solution of HCl, a strong acid. We might predict that the HCl produces 1.0×10^{-8} M H^+ in solution. We would be correct in our prediction.

$$HCl(aq) \rightarrow H^+(aq) \quad + \quad Cl^-(aq)$$
$$1.0 \times 10^{-8} M \rightarrow 1.0 \times 10^{-8} M \quad 1.0 \times 10^{-8} M$$

Determining the pH of this solution gives

$$pH = -\log[H^+]$$
$$pH = -\log(1.0 \times 10^{-8})$$
$$pH = 8.00$$

Does our answer make sense? *No it doesn't.* Adding HCl to water, even a very small amount, shouldn't cause the pH to increase! What have we forgotten to consider? The concentration of hydrogen ions in the solution is made up of all sources of $[H^+]$.

$$[H^+]_{total} = [H^+]_{HCl} + [H^+]_{water}$$

This means we should add all of the sources of hydrogen ion together and then determine the pH of the solution. What is the $[H^+]$ due to water? Recall our autoprotolysis equation:

$$H_2O(l) \rightleftharpoons H^+(aq) + OH^-(aq) \qquad K_w = 1.00 \times 10^{-14}$$

On the basis of this, we might be inclined to say that $[H^+]_{water} = [OH^-]_{water} = 1.0 \times 10^{-7}$ M. However, we must keep in mind the effect of Le Châtelier's principle, in which having hydrogen ions supplied by the HCl suppresses the ionization of H_2O. Therefore, $[H^+]_{water}$ will be less than 1.0×10^{-7} M. Although we won't go into the details here, it is possible to determine that in this solution, $[H^+]_{water} = 9.5 \times 10^{-8}$ M.

$$[H^+]_{total} = [H^+]_{HCl} + [H^+]_{water}$$
$$[H^+]_{total} = 1.0 \times 10^{-8} M + 9.5 \times 10^{-8} M$$
$$[H^+]_{total} = 1.05 \times 10^{-7} M$$
$$pH = 6.98$$

This is more reasonable than our initial answer of pH $= 8$ (after all, an acid solution should be acidic, not basic!), and it also shows the importance of Le Châtelier's principle as well as the autoprotolysis of water.

pH of Weak Acid Solutions

 Application

As we noted in Section 14.5, lactic acid ($HC_3H_5O_3$) is produced in our muscle cells when we work too strenuously to maintain aerobic respiration (respiration in the presence of sufficient oxygen). Recent evidence shows that contrary to long-time assumptions, lactic acid is a normal product of metabolism and not the barrier to athletic performance that was previously assumed. Lactic acid can be prepared commercially and is used in products ranging from biodegradable polymers to a spray to help extend the shelf life of beef strips. Its wide range of biological applications makes lactic acid a useful prototype for our discussion about the pH of weak acids.

 Tutorial: Calculating pH of Weak Acid and Base Solutions

Let's determine the pH of a 0.10 M solution of lactic acid. We considered the steps in accomplishing this task in Chapter 16. This is an equilibrium problem, and we may solve it using the same principles that we use to solve any other equilibrium problem.

Steps in Solving for the pH of a Weak Acid in Aqueous Solution

Step 1: Determine the equilibria, and the resulting species, that are in the solution.

There are two important equilibria occurring simultaneously in this solution: the ionization of lactic acid ($HC_3H_5O_3$, written here for simplicity as HL) and the autoprotolysis of water.

$$HL(aq) \rightleftharpoons H^+(aq) + L^-(aq) \qquad K_a = 1.38 \times 10^{-4}$$

$$H_2O(l) \rightleftharpoons H^+(aq) + OH^-(aq) \qquad K_w = 1.00 \times 10^{-14}$$

Step 2: Determine the equilibria that are the most important contributors to $[H^+]$ in the solution.

As we discussed previously, an acid that is much stronger than water will significantly depress the ionization of water in accordance with Le Châtelier's principle. Let's assume that this holds true for a relatively concentrated solution of the weak acid so that *we can safely neglect the autoprotolysis of water as a contributor of H^+*. We will test this assumption later.

Step 3: Write the equilibrium expression for the important contributors to $[H^+]$ (which usually means just the weak acid).

$$K_a = \frac{[L^-][H^+]}{[HL]}$$

Step 4: Set up a table of the initial and equilibrium concentrations of each pertinent species.

Weak acids are called weak because the extent of their ionization is normally quite small. As a first approximation, we may *assume* that "x" is far less than 0.100 M, or, equivalently, that $[HL] \approx [HL]_0$. We do this because it greatly simplifies our problem solving, but we will also test this assumption later. *The assumption will generally work if K_a is less than about $10^{-2} \times [HL]_0$*. In the present problem, $10^{-2} \times 0.10 = 1 \times 10^{-3}$. Our K_a of 1.38×10^{-4} is less than that, so our assumption is probably valid.

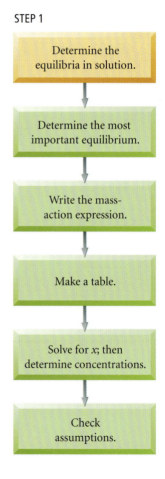

STEP 1

Determine the equilibria in solution.

Determine the most important equilibrium.

Write the mass-action expression.

Make a table.

Solve for x; then determine concentrations.

Check assumptions.

	HL	$\rightleftharpoons$	H^+	+	L^-
initial	0.10 M		0 M		0 M
change	$-x$		$+x$		$+x$
equilibrium	$0.10 - x$		$+x$		$+x$
assumptions	0.10		$+x$		$+x$

Step 5: Solve for the estimated concentration of each species.

$$1.38 \times 10^{-4} = \frac{x^2}{0.10}$$

$$x = [H^+] = [C_3H_5O_3^-] = 3.7 \times 10^{-3}\ M$$

$$[HC_3H_5O_3] = 0.10 - 3.7 \times 10^{-3} = 0.096\ M$$

Step 6: Check our assumption that the extent of ionization of lactic acid is negligibly small.

This is done using the "5% rule," which we introduced in Section 16.5. Also as discussed in that section, you can ignore the 5% simplification if you want to program your calculator to solve the quadratic equation in all cases. We present the 5% rule there, and here, because it shows that it is possible to use our understanding of chemistry to arrive at a solution strategy that is straightforward whether or not a programmable calculator is available.

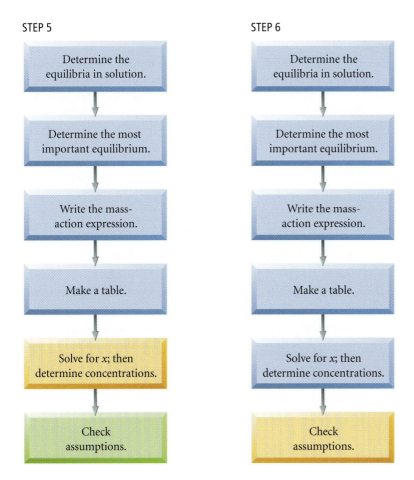

STEP 7

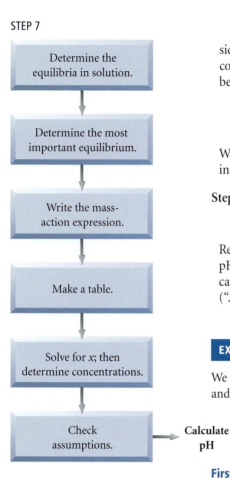

Determine the equilibria in solution.

Determine the most important equilibrium.

Write the mass-action expression.

Make a table.

Solve for x; then determine concentrations.

Check assumptions.

Calculate pH

If less than 5% of the original amount of the weak acid ionizes, this is considered "negligible." If more than 5% ionizes, then we must take this into account in our problem solving during Step 4. We will show how this is done below. What percentage of the lactic acid has ionized?

$$\% \text{ dissociated} = \frac{[C_3H_5O_3^-]}{HC_3H_5O_3} \times 100\% = \frac{3.7 \times 10^{-3}}{0.10} \times 100\% = 3.7\%$$

We can consider the ionization to be negligible because less than 5% of the original acid reacted to form products.

Step 7: Solve for the pH of the solution.

$$pH = -\log[H^+] = -\log(3.7 \times 10^{-3}\ M) = 2.43$$

Remember that the number of significant figures after the decimal point in the pH is equal to the number of significant figures in the pre-exponential term. Because the pre-exponential term is 3.7, we use two figures after the decimal point (".43"). The pH = 2.43.

EXERCISE 17.11   **pH of a Weak Acid**

We have mentioned aqueous hydrofluoric acid (HF) several times in this chapter, and we have found it to be a weak acid, $K_a = 7.2 \times 10^{-4}$. It is employed in the chemical industry to produce alkylated (see Section 12.3) compounds in gasoline and fluorocarbons used, for example, in refrigerants. Calculate the pH in a $1.0 \times 10^{-3}\ M$ HF solution.

First Thoughts

Our understanding of equilibrium tells us that even though the specific exercise is different from the one we just went over together, the questions that we ask about the chemical system remain the same, and therefore our approach is consistent. We understand our chemical system (and, in doing so, get a reasonable answer for the pH) by asking the right questions.

Solution

Step 1: Equilibria in solution.

The two important equilibria in the solution are the ionization of HF and the autoprotolysis of water.

$$HF(aq) \rightleftharpoons H^+(aq) + F^-(aq) \qquad K_a = 7.2 \times 10^{-4}$$
$$H_2O(l) \rightleftharpoons H^+(aq) + OH^-(aq) \qquad K_w = 1.00 \times 10^{-14}$$

Step 2: Determine the equilibria that are the most important contributors to $[H^+]$ in the solution.

We will assume that the HF ionization is the only important contributor to $[H^+]$ in solution because its K_a value is far greater than the K_w value of water.

Step 3: Write the equilibrium expression for the important contributors to $[H^+]$.

$$K_a = \frac{[F^-][H^+]}{[HF]}$$

Step 4: Table of concentrations.

	HF	$\rightleftharpoons$	H$^+$	+	F$^-$
initial	$1.0 \times 10^{-3}\ M$		$0\ M$		$0\ M$
change	$-x$		$+x$		$+x$
equilibrium	$1.0 \times 10^{-3} - x$		$+x$		$+x$
assumptions	1.0×10^{-3}		$+x$		$+x$

Although we assume that the ionization of HF is negligibly small (that is, "x" is essentially irrelevant compared to $1.0 \times 10^{-3}\ M$), the assumption is hazardous because K_a, 7.2×10^{-4}, is not less than about $10^{-2} \times$ [HF]$_0$ ($= 1.0 \times 10^{-5}\ M$). Rather, *it is nearly equal to [HF]$_0$!* Let's determine the implications of making the assumption.

Step 5: Solve.

$$7.2 \times 10^{-4} = \frac{x^2}{1.0 \times 10^{-3}}$$

$$x = [\text{H}^+] = [\text{F}^-] = 8.5 \times 10^{-4}\ M$$

Step 6: Check our assumption of negligible ionization.

$$\% \text{ dissociated} = \frac{[\text{F}^-]}{[\text{HF}]} \times 100\% = \frac{8.5 \times 10^{-4}}{1.0 \times 10^{-3}} \times 100\% = 85\%$$

Wow! This is certainly not negligible compared to 5%, so we may *not* claim that the ionization of HF is negligibly small. We must solve the following equation, which uses the "equilibrium" row of the ICEA table, rather than the "assumptions" row.

$$7.2 \times 10^{-4} = \frac{x^2}{1.0 \times 10^{-3} - x}$$

We can set up the quadratic equation by multiplying K_a, 7.2×10^{-4}, by [HF], $1.0 \times 10^{-3} - x$, to give

$$x^2 = 7.2 \times 10^{-7} - 7.2 \times 10^{-4}x$$

Setting the equation $= 0$ so we may solve for x yields

$$x^2 + (7.2 \times 10^{-4}x) - 7.2 \times 10^{-7} = 0$$

As we discussed in Section 16.5, we can solve the quadratic equation by substituting values for a, b, and c in the quadratic formula:

$$x = \frac{-b \pm \sqrt{b^2 - 4ac}}{2a}$$

In the present exercise,

$$a = 1 \qquad b = 7.2 \times 10^{-4} \qquad c = -7.2 \times 10^{-7}$$

Substituting into the quadratic formula yields

$$x = \frac{-7.2 \times 10^{-4} \pm \sqrt{(7.2 \times 10^{-4})^2 - 4(1)(-7.2 \times 10^{-7})}}{2(1)}$$

$$x = \frac{-7.2 \times 10^{-4} \pm 1.843 \times 10^{-3}}{2}$$

$$x = [\text{H}^+] = [\text{F}^-] = 5.62 \times 10^{-4}\ M \approx 5.6 \times 10^{-4}\ M$$

$$[\text{HF}] = 1.0 \times 10^{-3} - x = 1.0 \times 10^{-3} - (5.62 \times 10^{-4})$$

$$= 4.38 \times 10^{-4} \approx 4.4 \times 10^{-4}\ M$$

Step 7: Solve for the pH of the solution.

$$pH = -\log[H^+] = -\log(5.62 \times 10^{-4}) = 3.25$$

We can check our math by substituting back into the mass-action expression.

$$K_a = \frac{(5.62 \times 10^{-4})^2}{4.38 \times 10^{-4}} = 7.2 \times 10^{-4}$$

Our answer is confirmed.

Further Insights

We often assume, on the basis of the low value of K_w, that the autoprotolysis of water is unimportant, and we assumed that in Step 2 above. *Did this make sense?* What is the hydrogen ion concentration due to the autoprotolysis of water? It is equal to the hydroxide ion concentration due to water. That is,

$$[H^+]_{water} = [OH^-]_{water}$$

Also, water is the *only* source of hydroxide ion in this weak acid solution. Therefore,

$$[OH^-]_{total} = \frac{K_w}{[H^+]_{total}} = \frac{1.00 \times 10^{-14}}{5.62 \times 10^{-4}} = 1.8 \times 10^{-11}\,M = [OH^-]_{water} = [H^+]_{water}$$

This is so much smaller than $5.62 \times 10^{-4}\,M$ that it can be safely neglected. Our decision to neglect the autoprotolysis of water as a source of H^+ ions made sense.

PRACTICE 17.11

What is the pH of a 0.250 M acetic acid solution ($K_a = 1.8 \times 10^{-5}$)?

See Problems 39–50.

pH of a Mixture of Monoprotic Acids

If you ingest some vinegar, 3% acetic acid, does it significantly change the pH of your HCl-containing stomach? This question falls within the general topic of the pH of *a mixture of acids that differ greatly in strength*. Consider a solution that is 0.50 M HCl and 0.90 M acetic acid (CH_3COOH), $K_a = 1.8 \times 10^{-5}$. To solve for the pH of the mixture, we employ our usual approach. The relevant equilibria in the solution are

$$HCl(aq) \rightleftharpoons H^+(aq) + Cl^-(aq) \qquad\qquad K_a > 1$$

$$CH_3COOH(aq) \rightleftharpoons H^+(aq) + CH_3COO^-(aq) \qquad K_a = 1.8 \times 10^{-5}$$

$$H_2O(l) \rightleftharpoons H^+(aq) + OH^-(aq) \qquad\qquad K_w = 1.00 \times 10^{-14}$$

This says that there are three sources of hydrogen ion: HCl, CH_3COOH, and H_2O. We can write this in equation form:

$$[H^+]_{total} = [H^+]_{HCl} + [H^+]_{CH_3COOH} + [H^+]_{H_2O}$$

The hydrochloric acid is a substantially stronger acid than either the acetic acid or the water and is present in a concentration that is large enough so that we may make the *assumption* that neither the acetic acid nor the water ionizes significantly. In fact, the presence of hydrogen ion (the common ion!) from HCl further suppresses the acetic acid and water ionization in accordance with Le Châtelier's principle.

$$[H^+]_{total} = [H^+]_{HCl} \approx 0.50\,M$$

We will test this assumption later.

What is the hydrogen ion contribution due to the acetic acid? We know that $[H^+]_{total} = 0.50\,M$. We also know that the $[H^+]$ contributed by the acetic acid will be essentially equal to the $[CH_3COO^-]$. The concentration of acetic acid at equilibrium will be essentially equal to its initial concentration.

$$[CH_3COOH] = 0.90\ M$$

$$[H^+]_{acetic\ acid} = [CH_3COO^-] = \text{“}x\text{”}$$

We can use the mass-action expression for acetic acid ionization to solve for $[H^+]_{acetic\ acid}$.

$$K_a = \frac{[CH_3COO^-][H^+]}{[CH_3COOH]}$$

$$1.8 \times 10^{-5} = \frac{x(0.50)}{0.90}$$

$$x = [CH_3COO^-] = [H^+]_{acetic\ acid} = 3.2 \times 10^{-5}\ M$$

This shows that acetic acid is not a significant contributor to the hydrogen ion concentration of the solution, and although we did not quantify it here, the water is even less important. Further, if we had had an aqueous solution of only $0.90\ M$ CH_3COOH (without the HCl), the equilibrium concentration of the acetate ion, $[CH_3COO^-]$, would have been

$$K_a = \frac{[CH_3COO^-][H^+]}{[CH_3COOH]}$$

$$1.8 \times 10^{-5} = \frac{x^2}{0.90}$$

$$x = [H^+] = [CH_3COO^-] = 4.0 \times 10^{-3}\ M$$

This is nearly 100 times *higher* than the equilibrium acetate ion concentration of $3.2 \times 10^{-5}\ M$ in the HCl solution. This confirms what we said previously, that "the presence of hydrogen ion (the common ion!) from HCl further suppresses the acetic acid and water ionization in accordance with Le Châtelier's principle."

The main message to keep in mind from this discussion is that when you calculate the pH of a mixture of acids, you must look for the dominant equilibrium. This will determine the pH of the solution.

Application

17.5 Determining the pH of Basic Solutions

FIGURE 17.15

Ammonia and nicotine are both bases.

We have noted in this chapter the impact of acids and bases on ourselves and our world. A useful, though perhaps unfortunate, example concerns tobacco, which, according to the World Health Organization, prematurely kills well over 4 million people worldwide each year. Tobacco companies have used scientists' understanding of acid–base chemistry to increase by a factor of 100 the amount of nicotine that leaves solid burned tobacco particles and enters the gas phase to be absorbed by the lungs. The two important compounds involved in this chemistry are the nicotine itself ($C_{10}H_{14}N_2$) and ammonia, both of which are bases (see Figure 17.15). We will look at how ammonia can be used to increase the amount of gaseous nicotine that enters the lungs of smokers. As a first step in our analysis, we need to see how and why ammonia acts as a base.

Ammonia (NH_3) has many applications, including those shown in Table 17.1. It is the fourth-ranked industrial chemical, with 10.8 billion kg produced in 2004. Ammonia is quite soluble in water, acting as a weak base. Using the Arrhenius model of base behavior, it supplies hydroxide ions to the solution.

Ammonia

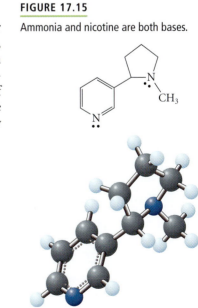

Nicotine

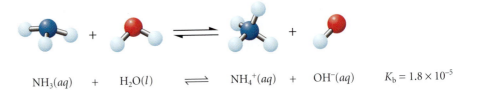

$$NH_3(aq) \quad + \quad H_2O(l) \quad \rightleftharpoons \quad NH_4^+(aq) \quad + \quad OH^-(aq) \qquad K_b = 1.8 \times 10^{-5}$$

Ammonia is typical of many bases in that it has an unshared pair of electrons on the nitrogen atom that can accept a proton from, in this case, the water molecule, as shown in the illustration.

Ammonia is considered a weak base because the K_b value is much less than 1. On the other hand, Group IA hydroxides, such as NaOH and KOH, *completely* dissociate in water ($K_b \gg 1$) to supply hydroxide ion and are considered strong bases.

$$NaOH(s) \rightarrow Na^+(aq) + OH^-(aq)$$

The solubility of Group IIA hydroxides increases from the essentially insoluble $Be(OH)_2$ to the more soluble (0.01 M) $Ca(OH)_2$ to the still more soluble (up to 0.1 M) $Ba(OH)_2$. Soluble Group IIA hydroxides, such as $Ca(OH)_2$, are strong bases like their Group IA counterparts, but they supply *two* moles of hydroxide for every mole of base that dissolves.

$$Ca(OH)_2(s) \rightleftharpoons Ca^{2+}(aq) + 2OH^-(aq)$$

Drawing on your understanding of equilibrium (that is, what are the right questions to ask in order to understand what happens in the solution?) and your sense of the pH scale, try the following two exercises dealing with the pH of a strong base solution and of the weakly basic ammonia solution. Even though we are now looking at bases rather than acids, the *equilibrium processes and the thinking involved are largely the same.*

EXERCISE 17.12 pH of a Strong Base

Calculate the pH of a 0.60 M aqueous solution of KOH.

First Thoughts

As a strong base, KOH completely dissociates. This is consistent with the behavior of Group IA and soluble Group IIA hydroxides.

$$KOH(s) \rightarrow K^+(aq) + OH^-(aq) \qquad K \gg 1$$

The other acid–base reaction that occurs in solution is the autoprotolysis of water, although it is unimportant as a source of OH^- because the K value is so small compared to that of the strong base.

$$H_2O(l) \rightleftharpoons H^+(aq) + OH^-(aq) \qquad K_w = 1.0 \times 10^{-14}$$

Solution

Potassium hydroxide is a strong base and will completely dissociate in aqueous solution.

$$KOH(s) \rightarrow K^+(aq) + OH^-(aq)$$

$$[OH^-] = 0.60 \ M$$

$$pOH = 0.22; pH = 13.78$$

Further Insights

Does the answer make sense? We expect the pH to be quite high, analogous to the very low pH of a strong acid, so our answer makes sense. In addition, the pH is considerably higher than with the same concentration of ammonia, a weak base. This is shown in the next exercise.

PRACTICE 17.12

Calculate the pH of a 1.06 M solution of NaOH. (*Hint:* It is possible to have aqueous solutions with a pH below 0 or above 14, although the actual pH is somewhat difficult to measure.)

See Problems 38 and 54.

EXERCISE 17.13   pH of a Weak Base

Determine the pH of a 0.600 M solution of ammonia.

First Thoughts

We may proceed as with any equilibrium problem, by listing the equilibria that occur in solution.

$$NH_3(aq) + H_2O(l) \rightleftharpoons NH_4^+ + OH^-(aq) \qquad K_b = 1.8 \times 10^{-5}$$
$$H_2O(l) \rightleftharpoons H^+(aq) + OH^-(aq) \qquad K_w = 1.0 \times 10^{-14}$$

The autoprotolysis of water is insignificant compared to the hydrolysis of ammonia. In fact, the OH^- produced by the ammonia reaction will further depress the autoprotolysis of water (via Le Châtelier's principle), so the OH^- and H^+ due to water will be less than in neutral solution, as discussed previously. The only relevant equilibrium expression is

$$K_b = \frac{[NH_4^+][OH^-]}{[NH_3]}$$

Solution

We set up the table of concentration changes in the usual fashion.

	NH$_3$	$\rightleftharpoons$	NH$_4^+$	+	OH$^-$
initial	0.600 M		0 M		0 M
change	$-x$		$+x$		$+x$
equilibrium	$0.600 - x$		$+x$		$+x$
assumptions	0.600		$+x$		$+x$

$$1.8 \times 10^{-5} = \frac{x^2}{0.600}$$
$$x = [NH_4^+] = [OH^-] = 3.3 \times 10^{-3} \, M$$
$$pOH = 2.48; \ pH = 11.52$$

Further Insights

Were we justified in neglecting "x"? The percent dissociation was

$$\frac{3.3 \times 10^{-6}}{0.600} \times 100\% = 0.5\%$$

Our assumption was valid. Were we justified in neglecting the autoprotolysis of water as a source of hydroxide ions? The only source of hydrogen ions in this solution is the autoprotolysis of water. If pH $= 11.52$, then $[H^+]_{water} = 3.0 \times 10^{-12} \, M$. This also equals $[OH^-]_{water}$, and this is insignificant compared to $3.3 \times 10^{-3} \, M$, the total $[OH^-]$. Does our answer make sense? To the extent that our pH reflects a moderately basic solution, our answer does make sense.

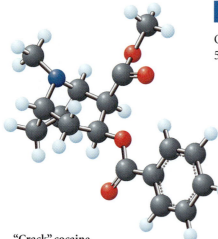

"Crack" cocaine

Calculate the pH of a 0.500 M solution of dimethylamine, $(CH_3)_2NH$, $K_b = 5.9 \times 10^{-4}$.

See Problems 51–53, 55, and 56.

One class of alkaloids, *ephedrines,* are used as dietary supplements to help in weight loss, to enhance body building, and to increase the user's energy level. In doses of more than 8 mg per serving, the compounds are considered dangerous for routine use. Another of the alkaloids, cocaine, has a severe psychological impact. It can be made basic to create the free-base form called "crack."

Issues and Controversies

Nicotine and pH Control in Cigarettes

Nicotine can exist in three forms, depending on the pH of its environment (see Figure 17.16). Below pH 3, it is in the diprotonated form. Between about pH 3 and pH 8, most of it is in the monoprotonated form.

If the pH is above 8, most of the nicotine exists in the volatile "free-base" form, which readily evaporates at the temperature of burning tobacco. This free-base is effectively absorbed by the lung tissue. Ammonia is added to tobacco leaves to raise the pH, making available more free-base nicotine for inhalation. The Food and Drug Administration has also determined that pH levels have been manipulated in smokeless tobacco products. The products for new users are at a relatively low pH, so there is not that much free-base nicotine available. Smokeless tobacco for "experienced" users has a higher pH, which leads to a higher level of nicotine absorption.

Nicotine is an **alkaloid**, one of many nitrogen-containing bases found in vegetables and other plants. Table 17.7 lists the structures and uses of some alkaloids. You might recognize the names of some of the compounds in the table. They often have a substantial impact on the central nervous system and brain.

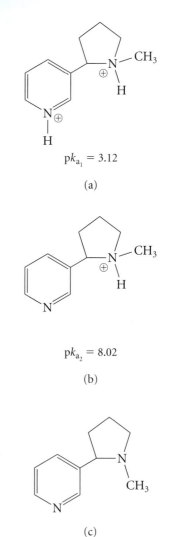

$pk_{a_1} = 3.12$

(a)

$pk_{a_2} = 8.02$

(b)

(c)

FIGURE 17.16

The three forms of nicotine. (a) The form that predominates below pH 3.12. (b) The form that predominates between pH 3.12 and 8.02. (c) The form that predominates above pH 8.02. This "free-base" form is readily vaporized and absorbed into the lungs when the cigarette burns.

17.6 Polyprotic Acids

Our theme is *impact*—the impact of acids and bases on the chemical industry, on our environment, and on ourselves. Phosphoric acid (H_3PO_4) fits well with that theme. It is different from other acids we have mentioned thus far, because it is a **polyprotic acid**. Like all polyprotic acids, phosphoric acid contains *more than one acidic hydrogen*. The structure of this **triprotic acid** (three acidic hydrogen atoms) acid is shown in Figure 17.17, with its three acidic hydrogen atoms highlighted. The acid is produced from "phosphate rock," which is largely composed of fluoroapatite ($Ca_5(PO_4)_3F$) and other compounds containing iron, calcium, silicon, aluminum, and fluorine.

Application

CHEMICAL ENCOUNTERS: Production and Uses of Phosphoric and Sulfuric Acids

Video Lesson: Examining Polyprotic Acids

FIGURE 17.17

Phosphoric acid (H_3PO_4) is triprotic acid because it contains three acidic hydrogen atoms.

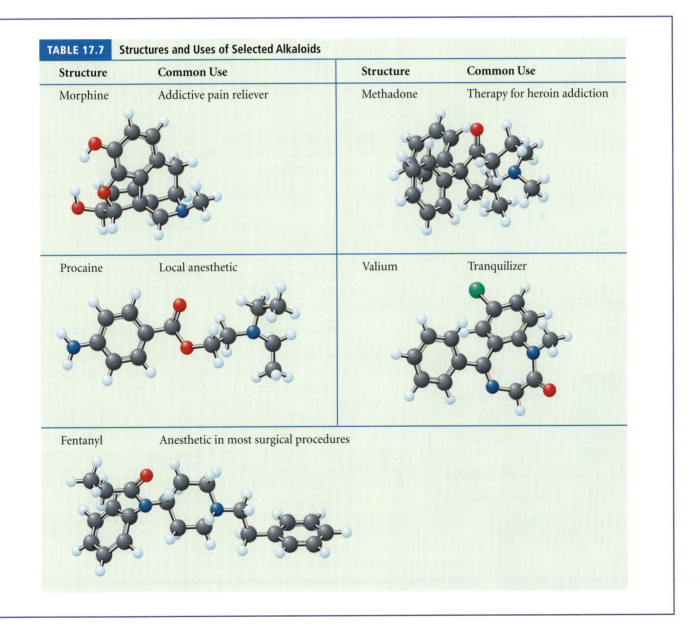

TABLE 17.7	**Structures and Uses of Selected Alkaloids**		
Structure	**Common Use**	**Structure**	**Common Use**
Morphine	Addictive pain reliever	Methadone	Therapy for heroin addiction
Procaine	Local anesthetic	Valium	Tranquilizer
Fentanyl	Anesthetic in most surgical procedures		

TABLE 17.8	Uses of Phosphoric Acid in Manufacturing

Fertilizer
Dentifrices
Soaps
Detergents
Fire control agents
Soft drinks
Incandescent light filaments
Corrosion inhibitors in metals
Organic chemicals such as ethylene and propylene

In 2003, over 33.3 million tons of marketable phosphate rock was mined in the United States, mostly from Florida and North Carolina, according to the U.S. Geological Survey. In fact, the U.S. production is about 24% of the world's phosphate rock production. Significant amounts of phosphates are also produced in China and in the Morocco and Western Sahara regions of Africa. This ore is converted into phosphoric acid for use in our society. Table 17.8 indicates the wide variety of uses of phosphoric acid. Phosphoric acid can be used to convert fluoroapatite into a soluble fertilizer, as shown in the following reaction:

$$2Ca_5(PO_4)_3F + 14H_3PO_4 \rightleftharpoons 10Ca(H_2PO_4)_2 + 2HF$$

In the human body, phosphate ion is important in maintaining the pH of blood in a fairly narrow range of 7.3 to 7.5. We will explore how it does so in the next chapter.

Phosphoric acid is one of two inorganic polyprotic acids that have a worldwide impact on our ability to convert what nature has given us into products that sustain and improve our quality of life. Sulfuric acid is the other.

Sulfuric acid is a diprotic acid that is prepared by the Contact process, as we discussed in Section 16.3. Sulfur is burned in oxygen to form sulfur dioxide, which is then converted into sulfur trioxide via a catalyst such as vanadium:

$$S(s) + O_2(g) \rightleftharpoons SO_2(g)$$

$$2SO_2(g) + O_2(g) \rightleftharpoons 2SO_3(g)$$

The sulfur trioxide is then combined with water to give sulfuric acid:

$$SO_3(g) + H_2O(l) \rightleftharpoons H_2SO_4(l)$$

Among the many uses for sulfuric acid is a century-old process for the conversion of phosphate rock to the fertilizer monocalcium phosphate, $Ca(H_2PO_4)_2 \cdot H_2O$. The reaction can be summarized as follows:

$$2Ca_5(PO_4)_3F + 7H_2SO_4 + 3H_2O \rightleftharpoons 3Ca(H_2PO_4)_2 \cdot H_2O + 7CaSO_4 + 2HF$$

Note that the product of this reaction is similar to the one produced by the treatment of phosphate rock with phosphoric acid.

Among the other uses of sulfuric acid is in the manufacture of phosphoric acid itself from fluoroapatite, with the resulting production of calcium sulfate dihydrate ($CaSO_4 \cdot 2H_2O$, also known as **gypsum**) under reaction conditions that are different from those in the previous reaction.

$$2Ca_5(PO_4)_3F + 10H_2SO_4 + 20H_2O \rightleftharpoons 10CaSO_4 \cdot 2H_2O + 6H_3PO_4 + 2HF$$

FIGURE 17.18

Calcium supplements often contain the tribasic phosphate ion. This supplement contains a mixture of calcium-containing minerals including $Ca_3(PO_4)_2$ and $CaCO_3$.

The gypsum produced in this reaction is quite valuable for use as Sheetrock and wallboard in the construction of homes. Furthermore, HF is a useful product in the glass industry.

In the reactions we have just shown, the phosphates and sulfates are present in a variety of forms: as polyprotic acids (H_3PO_4 and H_2SO_4); as a **monobasic salt** (it can accept one acidic hydrogen atom—$Ca(H_2PO_4)_2 \cdot H_2O$); and as a **dibasic salt** (it can accept two acidic hydrogen atoms—$CaSO_4$). Another common compound is calcium phosphate, $Ca_3(PO_4)_2$, a **tribasic salt** (it can accept three acidic hydrogen atoms). Calcium phosphate is one of several calcium salts, including calcium carbonate, $CaCO_3$, and calcium citrate, $Ca_3(C_6H_5O_7)_2 \cdot 4H_2O$, that many people take daily as a calcium supplement (Figure 17.18). Each of these species affects the acid concentration of solutions in predictable ways. Understanding these effects helps us to see how these acids are employed in manufacturing the products that we use.

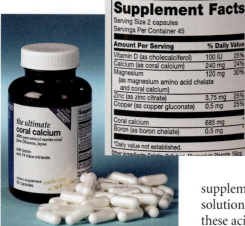

TABLE 17.9	K_a Values for Selected Polyprotic Acids			
Formula	Name	K_{a_1}	K_{a_2}	K_{a_3}
H_3PO_4	Phosphoric acid	7.4×10^{-3}	6.2×10^{-8}	4.8×10^{-13}
H_3AsO_4	Arsenic acid	5.0×10^{-3}	8.0×10^{-8}	6.0×10^{-10}
H_2CO_3	Carbonic acid	4.3×10^{-7}	5.6×10^{-11}	
H_2SO_4	Sulfuric acid	>1	1.2×10^{-2}	
H_2SO_3	Sulfurous acid	1.5×10^{-2}	1.0×10^{-7}	
H_2S	Hydrosulfuric acid	1.0×10^{-7}	1.0×10^{-15}	
$H_2C_2O_4$	Oxalic acid	6.5×10^{-2}	6.1×10^{-5}	
$H_2C_6H_6O_6$	Ascorbic acid	7.9×10^{-5}	1.6×10^{-12}	

The pH of Polyprotic Acids

Phosphoric acid is so common that it represents an important and useful model for our introduction to the pH of polyprotic acids. We begin by calculating the pH of 4.0 M aqueous phosphoric acid (neglecting activity effects).

As with any equilibrium problem, the first step is to establish the equilibria of the species in solution that contribute hydrogen ions. Phosphoric acid can donate three hydrogen ions per molecule, the second being more difficult to donate than the first because of the negative charge on the resultant dihydrogen phosphate anion ($H_2PO_4^-$). The third will be tougher still as a consequence of the greater charge on the monohydrogen phosphate anion (HPO_4^{2-}). This is reflected in the values for K_a of each step in the process, in which K_{a_1} is the equilibrium constant for the first acidic ionization, K_{a_2} refers to the second ionization, and K_{a_3} describes the third. Table 17.9 lists K_a values for several polyprotic acids. K_{a_1} values are not only larger than those of K_{a_2}, but they are often *considerably* so, and this is frequently true for acids. Water is also a source of hydrogen ions.

$$H_3PO_4 \rightleftharpoons H_2PO_4^- + H^+ \qquad K_{a_1} = 7.5 \times 10^{-3}$$

$$H_2PO_4^- \rightleftharpoons HPO_4^{2-} + H^+ \qquad K_{a_2} = 6.2 \times 10^{-8}$$

$$HPO_4^{2-} \rightleftharpoons PO_4^{3-} + H^+ \qquad K_{a_3} = 4.8 \times 10^{-13}$$

$$H_2O \rightleftharpoons OH^- + H^+ \qquad K_w = 1.0 \times 10^{-14}$$

The total concentration of hydrogen ion will equal the sum of the contributions from all the reactions.

$$[H^+]_{total} = [H^+]_{H_3PO_4} + [H^+]_{H_2PO_4^-} + [H^+]_{HPO_4^{2-}} + [H^+]_{H_2O}$$

Which reactions are *important* contributors of hydrogen ions to the solution? The K_{a_1} value is considerably larger than the others, so we may assume that it is the only important equilibrium. That is,

$$[H^+]_{total} \cong [H^+]_{H_3PO_4}$$

We will eventually want to prove that the other reactions are *not* important contributors if we make this assumption.

We may now solve the problem, determining the pH of a 4.0 M solution of H_3PO_4, as we have other equilibrium problems. We'll remember to test any assumptions we make along the way.

$$K_{a_1} = \frac{[H_2PO_4^-][H^+]}{[H_3PO_4]}$$

	H_3PO_4	$\rightleftharpoons$	H^+	+	$H_2PO_4^-$
initial	4.0 M		0 M		0 M
change	$-x$		$+x$		$+x$
equilibrium	$4.0 - x$		$+x$		$+x$
assumptions	4.0		$+x$		$+x$

$$7.4 \times 10^{-3} = \frac{x^2}{4.0}$$

$$x = [H_2PO_4^-] = [H^+] = 0.17\ M$$

$$pH = 0.76$$

$$[H_3PO_4] = [H_3PO_4]_0 - [H_2PO_4^-] = 4.0 - 0.17 = 3.83\ M \approx 3.8\ M$$

We have made the assumption that "x" is negligible relative to 4.0 M. In this case,

$$\frac{0.17}{4.0} \times 100\% = 4.3\%$$

Our assumption is valid.

We must now test our assumptions that the other equilibria were not important contributors of hydrogen ion to the solution. We may begin with water, knowing that

$$[H^+]_{water} = [OH^-]_{total} = \frac{K_w}{[H^+_{total}]} = \frac{1.0 \times 10^{-14}}{0.17} = 5.9 \times 10^{-14}\ M$$

Our assumption that water was a negligible source of hydrogen ion was fine (with thanks to Le Châtelier!). What about $[H^+]_{HPO_4^{2-}}$, which results from the loss of a hydrogen ion by $H_2PO_4^-$?

$$H_2PO_4^- \rightleftharpoons HPO_4^{2-} + H^+ \qquad K_{a_2} = 6.2 \times 10^{-8}$$

We showed above that $[H_2PO_4^-]$ is equal to 0.17 M. What is $[H^+]$ in this equation? It is roughly equal to the hydrogen ion concentration resulting from the first dissociation, that of H_3PO_4, $[H^+]_{total} \approx [H^+]_{H_3PO_4}$, and that is also equal to 0.17 M. We can substitute these values into the equilibrium expression to determine $[H^+]_{H_2PO_4^-}$ (which equals $[HPO_4^{2-}]$):

$$K_{a_2} = 6.2 \times 10^{-8} = \frac{[HPO_4^{2-}][H^+]}{[H_2PO_4^-]} = \frac{[HPO_4^{2-}](0.17)}{(0.17)}$$

Therefore,

$$[HPO_4^{2-}] = K_{a_2} = 6.2 \times 10^{-8}\ M$$

Also, $[H^+]_{HPO_4^{2-}}$ will equal $[HPO_4^{2-}]$, because they are both produced in the same reaction.

$$[H^+]_{HPO_4^{2-}} = 6.2 \times 10^{-8}\ M$$

This confirms the notion that only the first equilibrium, the dissociation of H_3PO_4, is an important contributor to $[H^+]_{total}$.

We can take this one step further and calculate the hydrogen ion concentration due to the dissociation of the dibasic anion HPO_4^{2-}:

$$HPO_4^{3-} \rightleftharpoons PO_4^{3-} + H^+ \qquad K_{a_3} = 4.8 \times 10^{-13}$$

In this case, $[H^+]_{total}$ is still 0.17 M and $[HPO_4^{2-}] = 6.2 \times 10^{-8}$ M. Substituting into the equilibrium expression for K_{a_3} yields

$$K_{a_3} = 4.8 \times 10^{-13} = \frac{[PO_4^{3-}][H^+]}{[HPO_4^{-2}]} = \frac{[PO_4^{3-}](0.17)}{6.2 \times 10^{-8}}$$

$$x = [PO_4^{3-}] = [H^+]_{PO_4^{3-}} = 1.8 \times 10^{-19} \ M$$

In summary, we have accounted for all of the phosphate species and all of the hydrogen ion.

[phosphate species] = $[H_3PO_4]$ + $[H_2PO_4^-]$ + $[HPO_4^-]$ + $[PO_4^{3-}]$

4.0 M = 3.83 M + 0.17 M + 6.2×10^{-8} M + 1.8×10^{-19} M

$[H^+]_{total}$ = $[H^+]_{H_3PO_4}$ + $[H^+]_{H_2PO_4^-}$ + $[H^+]_{HPO_4^{2-}}$ + $[H^+]_{H_2O}$

0.17 M ≈ 0.17 M + 6.2×10^{-8} M + 1.8×10^{-19} M + 5.9×10^{-14} M

We note that when the concentration of phosphoric acid is much greater than the value of K_{a_1}, only the dissociation of phosphoric acid itself contributes significantly to the hydrogen ion concentration. The concentration of each phosphate species changes in solution as the pH changes, with less of the acidic forms (H_3PO_4 and $H_2PO_4^-$) and more of the basic forms (HPO_4^{2-} and PO_4^{3-}) present at higher pH levels.

EXERCISE 17.14 | **Concentration of Species in a Polyprotic Acid Solution**

Oxalic acid ($H_2C_2O_4$), which is found in beet leaves, rhubarb, and spinach, is used in the bookbinding industry, as well as in dye and ink manufacturing. In the analytical laboratory it can be used as a primary standard against which to determine the molarity of sodium hydroxide. Using the information in Table 17.9, determine the pH and $[H_2C_2O_4]$ of a 1.40 M solution of oxalic acid.

$$H_2C_2O_4(aq) \rightleftharpoons HC_2O_4^-(aq) + H^+(aq) \qquad K_{a_1} = 6.5 \times 10^{-2}$$

$$HC_2O_4^-(aq) \rightleftharpoons H^+(aq) + C_2O_4^{2-}(aq) \qquad K_{a_2} = 6.5 \times 10^{-5}$$

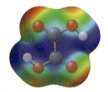

Oxalic acid
$C_2H_2O_4$

First Thoughts

The major species in solution are $H_2C_2O_4$ and H_2O. There are a number of equilibria that will occur in the solution. Judging on the basis of the values of K_{a_1}, K_{a_2}, and K_w, by far the most significant equilibrium will be the dissociation of $H_2C_2O_4$.

$$H_2C_2O_4(aq) \rightleftharpoons HC_2O_4^-(aq) + H^+(aq)$$

initial	1.40	0	0
change	−x	x	x
final	1.40 − x	x	x
assumptions	1.40	x	x

As always, we will test our "negligible ionization" assumption (1.40 ∼ x ≈ 1.40). However, because K_a is *not* less than $10^{-2} \times [H_2C_2O_4]_0$, our assumption may well not be valid.

Solution

$$K_{a_1} = \frac{[HC_2O_4^-][H^+]}{[H_2C_2O_4]} = 6.5 \times 10^{-2} = \frac{x^2}{1.40}$$

$$x = [H^+] = [HC_2O_4] = 0.302 \ M$$

Testing the 5% rule, we find that

$$\frac{0.302}{1.40} \times 100\% = 21.5\%!$$

This confirms our prediction based on K_a and $[H_2C_2O_4]_0$. Therefore, $[H_2C_2O_4] \neq [H_2C_2O_4]_0$ but rather equals "$1.40 - x$."

$$6.5 \times 10^{-2} = \frac{x^2}{1.40 - x}$$

We can solve using the quadratic formula, as we did in Section 16.5. Clearing the fraction and setting equal to zero, we get

$$x^2 + 0.065(x) - 0.091 = 0$$

where $a = 1$, $b = 0.065$, and $c = -0.091$. Solving yields

$$x = \frac{-0.065 \pm \sqrt{(0.065)^2 - 4(1)(-0.091)}}{2(1)}$$

$$x = 0.271 \ M$$

$$[H^+] = [HC_2O_4^-] = 0.271 \ M$$

$$[H_2C_2O_4] = 1.40 - 0.271 \approx 1.13 \ M$$

$$pH = -\log(0.271) = 0.57$$

Checking our math, we find that

$$K_{a_1} = \frac{(0.271)^2}{1.13} = 0.0650$$

Further Insights

Does our answer make sense? Although we classify oxalic acid as "weak" because its K_{a_1} value is less than 1 (and its K_{a_2} value is even smaller,) it is still stronger than many common weak acids such as acetic and citric acids. It is therefore reasonable that this relatively concentrated solution should have a low pH.

PRACTICE 17.14

Determine the pH of an aqueous 0.200 M H$_3$PO$_4$ solution.

See Problems 59, 60, 63, and 64.

We have seen that we can get a sense of what the pH of a solution will be by *assessing the competing equilibria that occur in solution.* Using this understanding, the pH of seemingly complex systems can be solved in a structured and meaningful fashion. We can extend this understanding to salts that contain an anion and cation that have acid–base behavior.

Video Lesson: Acid–Base
Properties of Salt Solutions

17.7 Assessing the Acid–Base Behavior of Salts in Aqueous Solution

Salts, such as NaCl, NH$_4$NO$_3$, and NaNO$_2$, are ionic compounds. When they dissociate in water they may exhibit acid–base behavior. The key questions you need to ask when assessing whether a salt will be acidic, basic, or neutral in aqueous solution are **What are the acid–base properties of the cation and anion parts of the salt?** and **Which is more influential, the acid strength of the cation or the base**

strength of the anion? Whichever is stronger will determine whether the salt solution is acidic or basic.

Sodium nitrite ($NaNO_2$) is an important example because it is a food additive that helps retard spoilage in meat and also is used in many industrial applications, including the production of nitrogen-containing dyes as well as anticorrosion agents. Its use in the food industry has been restricted by the Food and Drug Administration to 200 parts per million in meat and poultry that is ready for sale, because sodium nitrite was implicated in the 1970s as a possible precursor for some cancer-causing compounds. The salt essentially completely dissociates in aqueous solution to give Na^+ and NO_2^- ions. These are the main species in solution in addition to H_2O. **What are the acid–base properties of each of these species?** Remember the conjugate acid–base relationships:

- Strong acids and bases have weak conjugates. *Na^+ and other alkali and alkaline earth metal ions exhibit no important acid–base properties.*
- The nitrite ion (NO_2^-) is the conjugate base of the weak acid HNO_2 ($K_a = 4.6 \times 10^{-4}$). Remember that *weak* is a relative term. HNO_2 is weak compared to HNO_3, which has $K_a \gg 1$, but is far stronger than HCN ($K_a = 6.2 \times 10^{-10}$). *The NO_2^- ion will act as a weak base.*
- Water has relatively little acid–base effect.

Therefore, an aqueous solution of $NaNO_2$ should be slightly basic. We can now determine how basic by applying our understanding of equilibrium.

The Relationship of K_a to K_b

Consider an aqueous 0.500 *M* $NaNO_2$ solution. The process in which the nitrite ion (or any base) reacts with water to produce the conjugate acid and hydroxide ion is called **base hydrolysis**. The important equilibrium is

$$NO_2^-(aq) + H_2O(l) \rightleftharpoons HNO_2(aq) + OH^-(aq) \qquad K_b = ?$$

Note that this equation describes the equilibration of a base with water. What is the value for K_b? When we look in Appendix 5, we do not find an entry for NO_2^-. However, we do find a value for the K_a of nitrous acid:

$$HNO_2(aq) \rightleftharpoons H^+(aq) + NO_2^-(aq) \qquad K_a = 7.0 \times 10^{-4}$$

Is it possible to relate the two equilibria to get a value for K_b of the nitrite ion hydrolysis? The short answer is yes. If we take the expression for K_b and multiply by $\dfrac{[H^+]}{[H^+]}$, we get

$$K_b = \frac{[HNO_2][OH^-]}{[NO_2^-]} \times \frac{[H^+]}{[H^+]} = \frac{[HNO_2][OH^-][H^+]}{[NO_2^-][H^+]}$$

Because $K_w = [H^+][OH^-]$, we can substitute K_w into the equation, which gives

$$K_a = \frac{[HNO_2]K_w}{[NO_2^-][H^+]}$$

If you write the mass-action expression for K_a for the ionization of nitrous acid, you might note that $\dfrac{[HNO_2]}{[NO_2^-][H^+]}$ is equal to $1/K_a$. This means that the K_b expression may be rewritten as

$$K_b = \frac{K_w}{K_a}$$

or, as more often cited,

$$K_w = K_a \times K_b$$

This means that we can determine the K_a or K_b value for the conjugate of any weak acid or base, given its equilibrium constant. For the nitrite ion,

$$K_b = \frac{K_w}{K_a} = \frac{1.0 \times 10^{-14}}{7.0 \times 10^{-4}} = 1.4 \times 10^{-11}$$

We may now determine the pH of this weak base as we would any other weak base in solution. We will do this as Exercise 17.15.

EXERCISE 17.15 | pH of a Salt with a Cation with No Acidic Properties

Determine the pH of a 0.500 M NaNO$_2$ solution.

First Thoughts

As we discussed previously, the Na$^+$ cation has no acid–base properties, and the hydrolysis of the NO$_2^-$ ion will produce OH$^-$ ions, leading to a basic solution.

$$NO_2^-(aq) + H_2O(l) \rightleftharpoons HNO_2(aq) + OH^-(aq)$$

$$K_b = \frac{[HNO_2][OH^-]}{[NO_2^-]} = 1.4 \times 10^{-11}$$

We may now proceed as with any other weak-base problem.

$$NO_2^-(aq) + H_2O(l) \rightleftharpoons HNO_2(aq) + OH^-(aq)$$

initial	0.500	−0	0
change	−x	−x	x
final	0.500 − x	−x	x
with assumption	0.500	−x	x

In this exercise, "x" = [HNO$_2$] = [OH$^-$].

Solution

$$1.4 \times 10^{-11} = \frac{x^2}{0.500}$$

$$x = [OH^-] = [HNO_2] = 2.6 \times 10^{-6} \, M$$

This passes the 5% test (K_b is so small!).

$$pOH = -\log(2.6 \times 10^{-6}) = 5.59$$

$$pH = 14 - 5.59 = 8.41$$

Further Insights

Does the answer make sense? We have a weak base, and the pH is indicative of this. Therefore, our calculation is reasonable.

PRACTICE 17.15

Determine the pH of a 0.250 M sodium acetate (CH$_3$COONa) solution.

See Problems 75–82.

What happens when *both* the cation and the anion have acid–base properties? That is, what if the cation can react to supply hydrogen ion to the solution and the anion can supply hydroxide ion? In a broad sense, we can determine whether the solution will be acidic or basic from the *relative strengths of the acidic and basic parts of the salt*. For example, we can view ammonium cyanide (NH_4CN) as undergoing two important equilibrium reactions:

$$NH_4^+(aq) \rightleftharpoons NH_3(aq) + H^+(aq) \qquad K_a = \frac{K_w}{K_{b(NH_3)}} = 5.6 \times 10^{-10}$$

$$CN^-(aq) + H_2O(l) \rightleftharpoons HCN(aq) + OH^- \qquad K_b = \frac{K_w}{K_{a(HCN)}} = 1.6 \times 10^{-5}$$

Based on the equilibrium constants for the reactions, the CN^- ion is a much stronger base than the NH_4^+ ion is an acid, and we would therefore expect the solution to be basic. However, unlike the previous example using $NaNO_2$, in which only the nitrite ion had any acid–base behavior, here *both* the cation and the anion could contribute to the pH of the solution, and we must take both equilibria into account. In the sense that we have one substance that acts as an acid and one that acts as a base, this is similar to an amphiprotic substance. The derivation for the pH of a substance that has both acid and base properties, such as this type of salt or an amphiprotic substance such as Na_2HPO_4, is fairly complex. The resulting formula, however, is simple and useful:

$$[H^+] = (K_{a(NH_4^+)} \times K_{a(HCN)})^{1/2}$$

$$pH = \frac{1}{2}(pK_{a(NH_4^+)} + pK_{a(HCN)})$$

If the concentration of the substance is greater than 0.100 M (true with amphiprotic substances as well), the pH of the aqueous salt solution is approximately concentration independent. Let's use these equations to calculate the pH of a 0.800 M NH_4CN solution.

$$[H^+] = (K_{a(NH_4^+)} \times K_{a(HCN)})^{1/2} = (5.6 \times 10^{-10} \times 6.2 \times 10^{-10})^{1/2}$$
$$= 5.9 \times 10^{-10}\ M \qquad pH = 9.23$$

Amino acids are biologically vital compounds that have the ability to act as an acid and as a base within the same molecule. That is, one part of the molecule is acidic and a different part of the molecule is basic. Amino acids can polymerize into large units to form proteins such as hemoglobin (which transports oxygen), pepsinogen (which digests other proteins), and human growth hormone (which promotes normal growth). Amino acids are water soluble because they carry both positive and negative charge in aqueous solution. An example of how this can happen, involving the amino acid alanine, is shown in Figure 17.19. When dissolved in water, the carboxylic acid group loses a hydrogen ion, and the amine group gains a hydrogen ion. Why does this happen? The $—NH_3$ is a stronger base than the $—COO^-$, so the hydrogen ion will move from the COOH to the amine group. Ions that are doubly ionized in this way are called **zwitterions**. They can act in the same way as any other amphiprotic substance, donating a hydrogen ion to water or accepting one from water.

Application

CHEMICAL ENCOUNTERS:
Acid–Base Properties
of Amino Acids

Some amino acids.

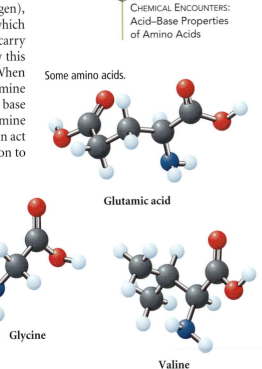

Glutamic acid

Glycine

Valine

FIGURE 17.19

Amino acids are soluble in water because they often exist as zwitterions—doubly charged amino acids. As shown here with alanine, this can occur because the $—NH_2$ is a stronger base than the $—COO^-$, so the hydrogen ion will move from the COOH to the amine group.

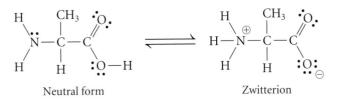

Neutral form   Zwitterion

EXERCISE 17.16 | **pH of Zwitterions**

Given the following reactions and equilibrium constants, calculate the pH of a solution of 0.200 M alanine.

$$CH_3(NH_3^+)COO^-(aq) \rightleftharpoons CH_3(NH_2)COO^-(aq) + H^+(aq) \qquad K_{a_2} = 1.4 \times 10^{-10}$$

$$CH_3(NH_3^+)COO^-(aq) + H_2O(l) \rightleftharpoons CH_3(NH_3^+)COOH(aq) + OH^-(aq)$$
$$K_{b_2} = 2.2 \times 10^{-12}$$

First Thoughts

The donation of hydrogen ion by alanine is favored over its acceptance of a hydrogen ion. We therefore would expect the solution to be somewhat, though not strongly, acidic.

Solution

We must find K_{a_1} for the second equation, just as in the ammonium cyanide case described previously.

$$K_{a_1} = \frac{K_w}{K_{b_2}} = \frac{1.0 \times 10^{-14}}{2.2 \times 10^{-12}} = 4.5 \times 10^{-3}$$

$$[H^+] = (K_{a_2} \times K_{a_1})^{1/2} = (1.4 \times 10^{-10} \times 4.5 \times 10^{-3})^{1/2} = 7.9 \times 10^{-7} \, M$$

$$pH = 6.10$$

Further Insights

At this pH, alanine exists primarily as the zwitterion and is therefore electrically neutral. This is called the **isoelectric pH**. Each amino acid has a different isoelectric point, depending on its own acid–base properties. Chemists and biologists make use of the different isoelectric points (Figure 17.20) to separate amino acids in an electric field by changing the pH, because each amino acid will respond to the electric field differently depending on its isoelectric point.

Does our answer make sense? We asserted that the solution should be somewhat acidic because the acid-producing equilibrium is favored over the base-producing equilibrium. Food for thought: What would you expect to be the pH if the value of K_a for the hydrogen ion–producing reaction in a particular salt *is the same as* the value of K_b for the base-producing equilibrium of the salt? Can you think of any examples of this?

FIGURE 17.20

Isoelectric focusing can be used to purify and identify a mixture of compounds. For example, a sample of proteins from a particular cultivar of barley can be separated into the individual "bands" shown here. The proteins have been stained purple with a dye.

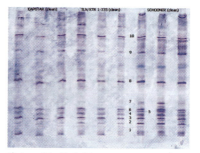

Photograph by Maria Sulman

PRACTICE 17.16

What is the pH of a 0.150 M alanine solution?

See Problems 83 and 84.

The principles that we have introduced for the evaluation of the pH of a salt can be extended from the two cases we have dealt with to a third case, in which the anion has no basic properties but the cation does have acidic properties. The thinking—that is, the questions that we raise—is the same. These questions about the nature of the equilibria in solution represent the unifying problem-solving theme in this chapter as well as the previous one dealing with chemical equilibrium.

Table 17.10 qualitatively summarizes the effect of the cation and of the anion on the pH of a salt.

TABLE 17.10	Effect of Cation and Anion of the Acidity of a Salt		
Cation	**Anion**	**Aqueous Solution**	**Example**
Acidic	Neutral	Acidic	NH_4NO_3
Neutral	Basic	Basic	Na_2CO_3
Neutral	Neutral	Neutral	NaCl
Acidic	Basic	Depends on the relative strength of each	

17.8 Anhydrides in Aqueous Solution

We started this chapter saying, "Home is where the heart is." The "home" that we talked about was not only our individual residence but ourselves and our world. In a sense, our final application, the set of reactions that cause acid deposition from the atmosphere, is a proper place to conclude the discussion, because the Earth is our communal home. To understand acid deposition, of which acid rain is one form, we need to understand the reactions of acidic and basic anhydrides (also known as acid and basic oxides) with water.

Basic anhydrides are binary compounds formed between metals with very low electronegativity and oxygen (see Section 17.2 for review of why these reactions would form bases). Strong bases are formed when Group IA and Group IIA anhydrides (*an* = "without" + *hydro* = "water") react with water. Metal hydroxides are often prepared this way instead of by reaction of the metal with water, which can sometimes be violent, as with the reaction of cesium with water to produce cesium hydroxide and hydrogen gas. Examples of anhydride and water reactions are

$$Li_2O(s) + H_2O(l) \rightleftharpoons 2LiOH(aq)$$

$$CaO(s) + H_2O(l) \rightleftharpoons Ca(OH)_2(aq)$$

Acid anhydrides are binary compounds formed between nonmetals and oxygen. Examples are SO_2, SO_3, NO_2, P_4O_{10}, and CO_2. These compounds react with water to form acids, their acid strength being related to the electronegativity of the nonmetal combined with oxygen. One example is the reaction of sulfur dioxide generated in industrial smokestacks with water:

$$SO_2(g) + H_2O(l) \rightleftharpoons H_2SO_3(aq)$$

The sulfurous acid generated in this reaction is not strong. However, dust in the air can catalyze the reaction between SO_2 and oxygen:

$$2SO_2(g) + O_2(g) \rightleftharpoons 2SO_3(g)$$

In a reaction analogous to the Contact process, the sulfur trioxide reacts with water vapor in the air to form sulfuric acid:

$$SO_3(g) + H_2O(l) \rightleftharpoons H_2SO_4(aq)$$

The acid that is formed can fall to Earth on a variety of surfaces, including snow, rain, and fog, and deposit on trees, lakes, and the like, and that is why we call the process **acid deposition**.

Nitrogen and oxygen released from the tailpipes of motorized vehicles during operation can react to form nitric oxide, which then slowly reacts with atmospheric oxygen to form nitrogen dioxide:

$$N_2(g) + O_2(g) \rightleftharpoons 2NO(g)$$

$$2NO(g) + O_2(g) \rightleftharpoons 2NO_2(g)$$

Application

CHEMICAL ENCOUNTERS:
Acid Deposition and
Acid-Neutralizing
Capacity

Acid deposition can severely damage the environment.

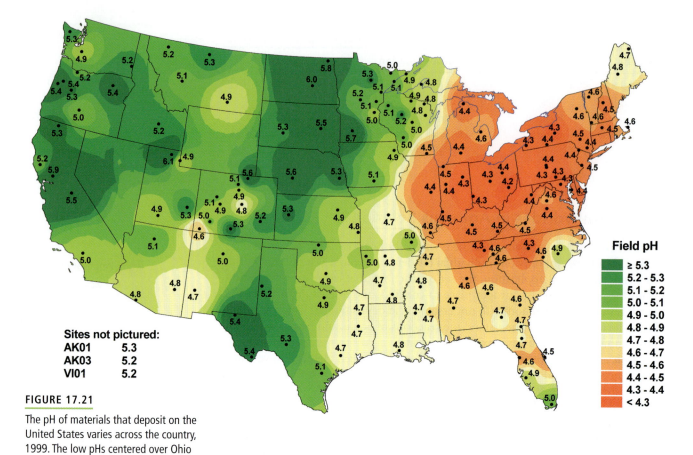

FIGURE 17.21

The pH of materials that deposit on the United States varies across the country, 1999. The low pHs centered over Ohio are probably due to industrial processes.

Sites not pictured:
AK01   5.3
AK03   5.2
VI01   5.2

Field pH

≥ 5.3
5.2 - 5.3
5.1 - 5.2
5.0 - 5.1
4.9 - 5.0
4.8 - 4.9
4.7 - 4.8
4.6 - 4.7
4.5 - 4.6
4.4 - 4.5
4.3 - 4.4
< 4.3

The nitrogen dioxide that is produced then reacts with water vapor to produce nitrous and nitric acids:

$$2NO_2(g) + H_2O(l) \rightleftharpoons HNO_2(aq) + HNO_3(aq)$$

The nitric acid adds to the problem of acid deposition. As can be seen in Figure 17.21, a lot of deposition that occurs throughout the United States has an unusually low pH. However, some places (for example, in the western United States) seem to be less affected by acid deposition than other places. Why?

The atmosphere acts as a large mixing chamber, and we've considered only the acidic inputs so far. Ammonia gas from agriculture and animal-feeding operations can react with water vapor to form aqueous ammonia, increasing the pH of the precipitation:

$$NH_3(g) + H_2O(l) \rightleftharpoons NH_4^+ + OH^-(aq)$$

There are more natural sources of ammonia gas in the western United States than sources of sulfur dioxide and nitrogen dioxide, so the pH of precipitation in this region is higher. The ability of some lakes to mitigate the effects of acid deposition has to do with **acid-neutralizing capacity**. This represents the theme of our next chapter: how and why acid–base and other types of reactions in aqueous solution are used in chemical analysis. This discussion will include acid–base neutralization, buffers, and titrations. Home is where the heart is. But we have shown that the home, and the people within it, are, chemically speaking, where the acids and bases also reside. Our global home, *Spaceship Earth,* is ever changing. It and we who occupy it owe much of this change to the acids and bases we have discussed here. In the next chapter, we will learn how.

The Bottom Line

- Acids and bases can be defined using three different models. (Section 17.1)

- Acids and bases come in different strengths. (Section 17.2)

- Both the strength and the initial concentration of the acid affect the acidity in solution at equilibrium. (Section 17.2)

- Acids and bases have conjugates pairs whose behavior is related to that of the acid or base from which they are derived. (Section 17.1)

- We use pH as our common measure of acidity. (Section 17.3)

- We can interconvert among H^+, OH^-, pH, and pOH for a given acidic or basic solution. (Section 17.3)

- We can calculate the pH of strong and weak acids and bases in aqueous solutions. (Sections 17.4 and 17.5)

- We can solve for the pH of a polyprotic acid or base, including salts. (Sections 17.6 and 17.7)

- K_a and K_b are related via K_w. (Section 17.7)

- The reaction of an acid anhydride with water results in an acid, and the reaction of a basic anhydride with water results in a base. (Section 17.8)

Key Words

acid According to the Arrhenius model, a species that produces hydrogen ions in solution. Compare the definitions of *Brønsted–Lowry acid* and *Lewis acid*. (*p. 719*)

acid deposition The precipitation of acidic compounds from the atmosphere. This includes wet deposition as rain and dry deposition of particles. (*p. 757*)

acid dissociation constant (K_a) The equilibrium constant for the dissociation of an acid. (*p. 724*)

acid-neutralizing capacity The capacity of a solution, such as lake water, to neutralize acidity. (*p. 758*)

acidic Having a pH less than 7.0 in aqueous solution at 24°C. (*p. 735*)

acid anhydrides Binary compounds formed between nonmetals and oxygen that react with water to form acids. (*p. 757*)

alkaloid Nitrogen-containing bases found in vegetables and other plants. (*p. 746*)

amphiprotic Having the ability to act as either an acid or a base in different circumstances. (*p. 733*)

autoprotolysis Proton transfer between molecules of the same chemical species, as in the autoprotolysis of water. (*p. 733*)

base According to the Arrhenius model, a species that produces hydroxide ions in solution. Compare the definitions of *Brønsted-Lowry base* and *Lewis base*. (*p. 719*)

base hydrolysis The process in which a base reacts with water to produce its conjugate acid and hydroxide ion. (*p. 753*)

basic Having a pH greater than 7.0 in aqueous solution at 24°C. (*p. 735*)

basic anhydrides Binary compounds that are formed between metals with very low electronegativity and oxygen and that react vigorously with water. (*p. 757*)

Brønsted–Lowry acid Any species that donates a hydrogen ion (proton) to another species. (*p. 720*)

Brønsted–Lowry base Any species that accepts a hydrogen ion (proton) from another species. (*p. 720*)

common-ion effect The addition of an ion common to one of the species in the solution, causing the equilibrium to shift away from production of that species. (*p. 737*)

conjugate acid The acid that results after accepting a proton. (*p. 720*)

conjugate base The base that results after donating a proton. (*p. 720*)

dibasic salt A salt that can accept two hydrogen ions. (*p. 748*)

diprotic acid An acid that contains two acidic protons. (*p. 736*)

gypsum Calcium sulfate dihydrate, $CaSO_4 \cdot 2H_2O$. (*p. 748*)

isoelectric pH pH value at which an amino acid is in the zwitterion form and so is electrically neutral overall. (*p. 756*)

Lewis acid Accepts a previously nonbonded pair of electrons (a lone pair) to form a coordinate covalent bond. (*p. 722*)

Lewis base Donates a previously nonbonded pair of electrons (a lone pair) to form a coordinate covalent bond. (*p. 722*)

monobasic salt A salt that can accept one hydrogen ion. (*p. 748*)

monoprotic acid An acid that contains one acidic proton. (*p. 736*)

neutral pH A pH of 7.0 in aqueous solution at 24°C. (*p. 734*)

pH A numerical value related to hydrogen ion concentration by the relationship $pH = -\log[H^+]$ or $pH = -\log[H_3O^+]$. (*p. 730*)

polyprotic acid An acid that can release more than 1 mol of hydrogen ions per mole of acid. (*p. 747*)

strong acid An acid that fully dissociates in water, releasing all of its acidic protons. (*p. 724*)

superacids Amazingly strong acids that can be used to add hydrogen ions to organic molecules that are otherwise impervious to such reaction. (*p. 729*)

tribasic salt A salt that can accept three acidic hydrogen atoms. (*p. 748*)

triprotic acid An acid that contains three acidic protons. (*p. 747*)

weak acid An acid that only partially dissociates in water. (*p. 724*)

zwitterion A molecular ion carrying both an acid group, which generates a negative ion, and a basic group, which generates a positive ion. (*p. 755*)

Focus Your Learning

The answers to the odd-numbered problems and some selected problems appear at the back of the book, as represented by the blue numbering.

Section 17.1 What Are Acids and Bases?

Skill Review

1. Explain how ammonia (NH_3) qualifies as a base in both the Brønsted–Lowry and Arrhenius acid–base models.

2. Explain how HCl and NaOH are classified in both the Brønsted–Lowry and Arrhenius acid–base models.

3. List the conjugate base of each of these acids.
   a. HNO_3 b. HBr c. H_2O d. $HClO_4$

4. List the conjugate base of each of these acids.
   a. HCl b. NH_4^+ c. CH_3OH d. H_2SO_4

5. List the conjugate acid of each of these bases.
   a. $NaOH$ b. NH_3 c. H_2O d. NaF

6. List the conjugate acid of each of these bases.
   a. KBr b. CH_3OH c. KNO_2 d. KSH

7. Acids react with active metals to produce hydrogen gas. Complete, balance, and name each of the products for these reactions.
   a. $Al(s) + HCl(aq) \rightarrow$ _____ + _____
   b. $Ca(s) + HNO_3(aq) \rightarrow$ _____ + _____
   c. $Na(s) + HCl(aq) \rightarrow$ _____ + _____
   d. $K(s) + HNO_3(aq) \rightarrow$ _____ + _____

8. Many bases react with metals to form insoluble hydroxides. Write the formula for each of these hydroxides.
   a. aluminum hydroxide   b. copper(II) hydroxide
   c. barium hydroxide   d. strontium hydroxide
   e. iron(III) hydroxide   f. calcium hydroxide

9. Being strong usually refers to the ability to carry out a task or produce an effect. (For example, a spice may have a strong flavor, and a weightlifter may display a lot of strength.) How is this term used when applied to acids? When an acid is said to be strong, what effect is being described?

10. What is the definition of a Lewis acid? Why are many metal ions capable of behaving as Lewis acids? Write out the reaction that depicts an aluminum cation behaving as a Lewis acid in an aqueous solution.

11. Balance and identify each of the species in the following reactions as either acid, base, conjugate acid, or conjugate base.
   a. $HCl + H_2O \rightarrow Cl^- + H_3O^+$
   b. $NaOH + CH_3COOH \rightarrow CH_3COONa + H_2O$
   c. $H_2SO_4 + Mg(OH)_2 \rightarrow MgSO_4 + H_2O$

12. Balance and identify each of the species in the following reactions as either acid, base, conjugate acid, or conjugate base.
   a. $HCOOH + NH_3 \rightarrow NH_4^+ + HCOO^-$
   b. $KOH + CH_3OH \rightarrow CH_3OK + H_2O$
   c. $H_3PO_4 + Ca(OH)_2 \rightarrow Ca_3(PO_4)_2 + H_2O$

Chemical Applications and Practices

13. One common antacid used in relief of upset stomachs is $Mg(OH)_2$. It helps to neutralize excess hydrochloric acid. Balance the following representative equation, and identify the conjugate acid–base pairs.

$$Mg(OH)_2 + HCl \rightarrow MgCl_2 + H_2O$$

14. Prizes in cereal boxes used to include little submarines that, when filled with baking soda ($NaHCO_3$) and placed in a cup of water containing a little vinegar (CH_3COOH), would rise and fall as though by magic. Balance the reaction of baking soda and vinegar, and identify the conjugate base and acid pairs.

$$CH_3COOH + NaHCO_3 \rightarrow CH_3COONa + H_2CO_3$$

Section 17.2 Acid Strength

Skill Review

15. Describe the characteristics of a strong acid with regard to each of the following:
   a. The numerical value of its K_a.
   b. The ability of its conjugate base to regain a H^+.
   c. The approximate percent dissociation of a 0.1 M solution.

16. Describe the characteristics of a weak acid with regard to each of the following:
   a. The numerical value of its K_a.
   b. The ability of its conjugate base to regain a H^+.
   c. The approximate percent dissociation of a 0.1 M solution.

17. Judging on the basis of electron density and electronegativity, which acid, HBr or HI, would you expect to be stronger? Explain your choice.

18. Which conjugate base, $Br^-(aq)$ or $I^-(aq)$, would you expect to be stronger? Explain your choice.

19. Using molecular structure and electronegativity, explain which acid, $HClO_4$ or $HBrO_4$, would be the weaker.

20. Which acid, phosphoric (H_3PO_4) or phosphorous (H_3PO_3), would you expect to be stronger? Explain your choice.

Chemical Applications and Practices

21. Formic acid, found in ants, and acetic acid, found in vinegar, have the K_a values 1.8×10^{-4} and 1.8×10^{-5}, respectively.
    a. Which acid has the stronger conjugate base?
    b. Which acid, if both were in 0.10 M solutions, would have the higher percent dissociation?

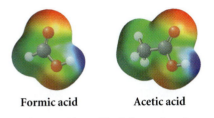

Formic acid   **Acetic acid**

22. The K_a value for an acid provides information about one type of reaction of the acid, its ability to provide H^+. However, a weak acid may have other very important reactions. For example, hydrofluoric acid, $K_a = 7.2 \times 10^{-4}$, has the ability to etch glass. Phenol, $K_a = 1.6 \times 10^{-10}$, can be used as a disinfectant.
    a. Which of these two is the weaker acid?
    b. Which has the stronger conjugate base?
    c. If both were 0.10 M, which would produce the greater concentration of H^+?

23. Propanoic acid is a weak acid that can be used to prepare a type of mold retardant. What is the value for K_a of the acid if, in a solution, the following equilibrium concentrations were found: [acid] = 0.10 M, [conjugate base] and $[H^+]$ = 0.0011 M?

24. At 25°C the K_b for ammonia (NH_3) is 1.8×10^{-5}. What is the hydroxide concentration when $[NH_3]$ = 0.103 M and $[NH_4^+]$ = 0.00205 M?

Section 17.3 The pH Scale

Skill Review

25. Calculate the pH of each of these solutions.
    a. $[H^+] = 4.55 \times 10^{-3} M$
    b. $[H^+] = 3.27 \times 10^{-6} M$
    c. $[H^+] = 8.11 \times 10^{-9} M$

26. Calculate the $[H^+]$ for each of these solutions.
    a. pH = 1.50   b. pH = 10.25
    c. pH = 5.38   d. pH = 7.00

27. Calculate the values missing from the following table.

	$[H^+]$ (M)	pH	$[OH^-]$ (M)	pOH
a.		4.42		
b.	0.0056			
c.			0.000078	
d.				10.10

28. Calculate the values missing from the following table.

	$[H^+]$ (M)	pH	$[OH^-]$ (M)	pOH
a.		12.50		
b.	0.000035			
c.			0.00388	
d.				3.75

29. If the pH value in an aqueous sample were doubled, what effect would be detected in the hydronium ion concentration?

30. What would be the effect on the hydroxide ion concentration if the pH were doubled?

Chemical Applications and Practices

31. Paper is produced from the processed fibers of trees. Part of the process of sulfate pulping uses sodium hydroxide. What would be the pH of a solution in a wood-pulping mill if it contained 10.0 g of OH^- ions for every 10.0 L of solution?

32. One method to increase oil production in areas drilled through limestone deposits is to use hydrochloric acid to increase drainage channels through the stone. If such a solution had a pH of 2.59, what would you calculate as the grams of HCl dissolved per liter of solution?

33. Two samples of rainwater are being analyzed for an environmental impact study. What is the hydronium concentration in each sample? What is the pH of each sample?
    a. 500.0 mL containing 1.55×10^{-5} mol of H^+
    b. 250.0 mL containing 7.25×10^{-6} mol of H^+

34. The pH of human blood must be maintained within a very narrow range to ensure proper health. The following blood samples were analyzed to determine their pH values. What would you calculate as the hydronium concentration in each?
    a. pH = 7.42   b. pH = 7.38   c. pH = 7.51

35. Assuming a negligible change in volume, how many moles of either OH^- or H^+ would have to be added to change the pH of 1.00 L of a solution from 4.35 to 5.85?

36. Formic acid has a pK_a value of 3.74. Benzoic acid has a pK_a value of 4.20. Compare the electrostatic potential maps of formic acid and benzoic acid. Which is the stronger acid? Which of the two acids would have the weaker conjugate base?

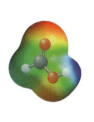 

Formic acid   **Benzoic acid**

Section 17.4 Determining the pH of Acidic Solutions

Skill Review

37. Determine the pH of each of these solutions of strong acids.
    a. 0.45 M HCl
    b. 0.045 M HCl
    c. 0.000487 M HNO$_3$
    d. 0.00026 M HBr

38. Determine the pH of each of these solutions of strong bases.
    a. 0.550 M NaOH
    b. 0.00089 M KOH
    c. 0.00388 M KOH
    d. 0.015 M KOH

39. Determine the pH of each of these solutions. (Use the table in the text to find the values for the appropriate K_a.)
    a. 0.45 M HOCl
    b. 0.0250 M CH$_3$COOH
    c. 0.18 M HF
    d. 0.0010 M HCOOH

40. Determine the pH of each of these solutions. (Use the table in the text to find the values for the appropriate K_a.)
    a. 0.299 M HOCl
    b. 0.18 M CH$_3$COOH
    c. 0.45 M lactic acid (HC$_3$H$_5$O$_3$)
    d. 0.050 M HCN

41. Determine the value of pK_a for:
    a. $K_a = 3.75 \times 10^{-5}$
    b. $K_a = 1.84 \times 10^{-2}$
    c. $K_a = 4.59 \times 10^{-8}$

42. Determine the value of K_a for:
    a. p$K_a = 3.50$
    b. p$K_a = 4.74$
    c. p$K_a = 6.17$

43. If the pH of a 0.015 M solution of codeine, a drug used in some pain relievers, is 10.19, what is the value of K_b for codeine (C$_{18}$H$_{21}$NO$_3$)?

44. What would be the resulting pH when a solution was made that was 0.0100 M in HCl and 0.100 M in HCN (hydrocyanic acid, a deadly poison, which has a K_a value of 6.2×10^{-10})?

Chemical Applications and Practices

45. Among the growth requirements for bacteria is the proper range of aqueous hydrogen ion concentration. Suppose a microbiologist was preparing growth media to study a specific bacterium. Examine both of the following situations and determine in which case the H$^+$(aq) would be greater than 1.0×10^{-6} M?
    a. A 0.500-L solution containing 1.00 g of benzoic acid ($K_a = 6.5 \times 10^{-5}$; molar mass = 122 g)
    b. 100.0 mL of 0.0001 M sulfuric acid

46. Benzoic acid is often used to prepare a preservative known as sodium benzoate. If the K_a value of benzoic acid is 6.5×10^{-5}, what is the hydrogen ion concentration when the acid concentration is 0.0040 M and the conjugate base concentration is 0.0024 M?

47. The following reaction depicts an industrial process to manufacture gaseous hydrogen fluoride.
    $$\text{CaF}_2(s) + \text{H}_2\text{SO}_4(aq) \rightarrow \text{CaSO}_4(aq) + 2\text{HF}(g)$$
    a. How many grams of HF can be made from 1.00 kg of fluorospar (CaF$_2$)?
    b. HF can then be used to prepare fluorocarbon compounds. What would be the H$^+$, F$^-$, and OH$^-$ concentrations in a solution that was 0.25 M in HF? (K_a of HF = 7.2×10^{-4})

48. An aspirin tablet may contain 325 mg of acetylsalicylic acid (p$K_a = 3.522$, 180.16 g/mol). What would be the approximate pH when two tablets were dissolved in 275 mL of water?

49. A vitamin C tablet may contain 500.0 mg of ascorbic acid (C$_6$H$_8$O$_6$). What would be the pH of a solution made from dissolving one such tablet in 355 mL of solution? (K_{a_1} of ascorbic acid = 8.0×10^{-5}) Does the 5% rule assumption apply in this example? Show your proof.

50. Benzoic acid ($K_a = 6.5 \times 10^{-5}$) and propionic acid ($K_a = 1.3 \times 10^{-5}$) can both be used to produce food preservatives. A 0.10 M solution of one of the acids has [H$^+$] = 0.00255 M. Which acid was used?

Section 17.5 Determining the pH of Basic Solutions

Skill Review

51. Determine the pH of each of these solutions.
    a. 0.100 M aniline ($K_b = 3.8 \times 10^{-8}$)
    b. 0.0100 M NaOH
    c. 0.250 M ammonia

52. Determine the pH of each of these solutions.
    a. 0.0333 M methylamine ($K_b = 5.9 \times 10^{-4}$)
    b. 0.0150 M Ca(OH)$_2$
    c. 0.016 M ammonia

Chemical Applications and Practices

53. Pyridine is a weak base that can be used to make a product used in some mouthwash preparations. The K_b value of pyridine is approximately 1.4×10^{-9}. What would be the hydroxide ion concentration, the pOH, and the pH of a solution that was 0.0010 M pyridine?

54. The base Ca(OH)$_2$ is known as slaked lime. It is widely used in the paper industry and in steel making. When properly heated, it gives off a bright light. In the 1800s, this light was used to illuminate some theaters. Actors began appearing in the "limelight."
    a. Write out the equation representing the dissociation of lime in water.
    b. Ca(OH)$_2$ is not very soluble in water. If 0.025 mol could dissolve per liter, what would you calculate as the pH of the solution?

55. The typical "fish aroma" is due to the production of amine compounds. What would be the K_b value of ethylamine (CH$_3$CH$_2$NH$_2$) if a 500.0 mL solution that contained 1.90 g of ethylamine had a pH of 11.87?

56. Metacaine is used to anesthetize groups of fish when scientific studies on them are conducted. The active ingredient of metacaine is a base known as ethyl 3-aminobenzoate. What would be the K_b value of ethyl 3-aminobenzoate (C$_9$H$_{11}$NO$_2$) if a 750.0 mL solution containing 1.00 g had a pH of 9.30?

Section 17.6 Polyprotic Acids

Skill Review

57. Write out the three hydrogen dissociation steps for phosphorous acid (H$_3$PO$_3$). What is the oxidation number of phosphorus in H$_3$PO$_3$?

58. Write out the two hydrogen dissociation steps for carbonic acid (H$_2$CO$_3$). What is the oxidation number of carbon in H$_2$CO$_3$?

59. The respective K_a values for the three dissociation steps of phosphoric acid are $7.4 \times 10^{-3}, 6.2 \times 10^{-8}$, and 4.8×10^{-13}. What would be the pH and $HPO_4{}^{2-}$ concentration of a $1.0\ M$ solution of H_3PO_4?

60. What are the pH and the $HSO_4{}^-$ concentration of a $0.750\ M$ solution of H_2SO_4?

Chemical Applications and Practices

61. Sulfurous acid (H_2SO_3) is a by-product of burning sulfur-containing coal. SO_2 is produced during the process and, when combined with water in the air, can produce H_2SO_3. Write the balanced equations that show the two-stage ionization of sulfurous acid. Identify the conjugate base produced in each stage.

62. Carbonic acid can be found in carbonated drinks. It forms when CO_2 reacts with water.
   a. Write the balanced equation that shows the formation of carbonic acid from dissolved carbon dioxide.
   b. Write out the equilibrium expressions for both ionization steps for this diprotic acid.
   c. Use Le Châtelier's principle to explain why increasing the pressure of CO_2 produces a lower pH in the solution.

63. Nicotine is dibasic because of the presence of two nitrogen atoms, each of which may accept hydrogen ions. The respective K_b values, at 25°C, are approximately 7.0×10^{-7} and 1.1×10^{-10}. What would be the pH of a solution that was $0.045\ M$ in nicotine.

64. Tartaric acid ($H_2C_4H_4O_6$) is a diprotic acid used in some baking preparations. For the successive hydrogen ionizations, $K_{a_1} = 9.2 \times 10^{-4}$ and $K_{a_2} = 4.3 \times 10^{-5}$. What would be the pH of a solution that was $0.10\ M$ in tartaric acid?

Section 17.7 Assessing the Acid–Base Behavior of Salts in Aqueous Solution

Skill Review

65. Arrange the following $0.10\ M$ solutions in order of decreasing pH: NH_4Cl, $NaCl$, $NaC_2H_3O_2$.

66. Which of these ions could produce a basic aqueous solution? S^{2-}, Cl^-, $NO_3{}^-$, $NO_2{}^-$, $CO_3{}^{2-}$, OCl^-.

67. Two acids, HX and HY, have pK_a values of 4.55 and 5.44, respectively. Which salt, NaX or NaY, will produce the more basic aqueous solution when prepared as $0.10\ M$?

68. Two sodium salts, symbolized as NaW and NaY, are completely dissolved to produce $0.20\ M$ solutions. The respective pH values of the two solutions are 8.55 and 9.55. Which acid, HW or HY, is stronger? Prove your choice.

69. What is isoelectric pH? Why is it useful information to know about amino acids?

70. If the isoelectric pH for an amino acid were above 7, what would that indicate about the relative values of its K_a and K_b?

71. The K_a value for the dissolved cation $Zn(H_2O)_6{}^{2+}$ is approximately 2.4×10^{-10}. What is the pH of a solution that is $0.10\ M$ $ZnCl_2$? (*Hint*: What happens when $ZnCl_2$ dissolves?)

72. Will each of these salts be acidic, basic, or neutral?
   a. KI   b. NH_4F c. $(NH_4)_3PO_4$

73. Determine the value of K_a if:
   a. $K_b = 4.26 \times 10^{-5}$
   b. $K_b = 8.36 \times 10^{-9}$
   c. $pK_a = 2.85$

74. Determine the value of K_b if:
   a. $K_a = 6.90 \times 10^{-3}$
   b. $K_a = 1.77 \times 10^{-12}$
   c. $pK_a = 4.74$

75. Determine the pH of a solution that is $0.050\ M$ in HCOONa.

76. Determine the pH of a solution that is $0.136\ M$ in KNO_2.

Chemical Applications and Practices

77. The salt ammonium chloride is used in some chemical "cold packs" to absorb heat and cool muscle wounds. Ammonium chloride can be produced from an acid–base reaction.
   a. Write out the reaction using an acid and a base that would produce ammonium chloride.
   b. Would the resulting solution of ammonium chloride be acidic, basic, or neutral? Explain.
   c. Determine the pH of a solution of ammonium chloride that is $0.136\ M$.

78. The salt sodium hypochlorite (NaOCl) can be used in some bleaching actions needed for disinfecting aqueous systems.
   a. Write out the balanced equation that represents the complete dissociation of the salt in water.
   b. Would this be likely to produce an acidic, basic, or neutral solution?
   c. Determine the pH of a solution that is $0.250\ M$ NaOCl.

79. Sodium carbonate, sometimes called soda ash, is used industrially in the manufacture of glass, paper, and soaps. What would be the pH of a $0.15\ M$ solution of sodium carbonate?

80. Sodium bicarbonate, also known as baking soda or sodium hydrogen carbonate, is produced industrially by the addition of carbon dioxide to soda ash. Sodium bicarbonate has widespread uses in antacids, paper manufacturing, and some fire extinguishers, as well as to remove some harmful gases during coal combustion. What would be the pH of a $0.15\ M$ solution of sodium bicarbonate?

81. Sodium benzoate, a salt of benzoic acid sometimes used as a food preservative, dissolves in water to produce the benzoate ion, $C_6H_5CO_2{}^-$. The K_a value for benzoic acid is approximately 6.5×10^{-5}.
   a. Write out the reaction of the benzoate ion in water and calculate the K_b value for the reaction.
   b. What would be the pH of a $0.010\ M$ solution of sodium benzoate?

82. Novocain is often used as a local anesthetic. The compound is actually a salt of the base procaine. Procaine has a K_b value of 7.13×10^{-6}.
   a. What would be the K_a of Novocain?
   b. What would be the pH of a $0.010\ M$ solution of Novocain?

83. Lysine is considered an essential amino acid. Essential amino acids are those that are not synthesized by humans and must, therefore, be part of a healthful diet. Lysine can be found in beans. The formula of lysine is given below. Rewrite the formula showing lysine as a zwitterion.

$$H_2N-CH_2-CH_2-CH_2-CH_2-\underset{\underset{\displaystyle COOH}{|}}{\overset{\overset{\displaystyle H}{|}}{C}}-NH_2$$

84. Glycine has the simplest structure of the amino acids.

$$H-\underset{\underset{\displaystyle H}{|}}{\overset{\overset{\displaystyle NH_2}{|}}{C}}-COOH$$

   a. Write out the reactions that show glycine acting as an acid and acting as a base.
   b. What would be the pH of a 0.10 M solution of glycine? (The approximate K_{a_2} and K_{b_2} values, at 25°C, needed are 2.0×10^{-10} and 2.2×10^{-12}.)

17.8 Anhydrides in Aqueous Solution

Skill Review

85. Give the structure of the anhydride of sulfuric acid.

86. Indicate the structure of the substance that would be the anhydride of $Ba(OH)_2$.

Comprehensive Problems

87. The hydrogen ion donated by Brønsted–Lowry acids is typically represented in aqueous solutions as $H_3O^+(aq)$.
   a. Explain the origin of the positive charge on this ion.
   b. What other forms could the hydrogen ion take in water?
   c. Would the shape of H_3O^+ be more likely to be flat or pyramidal? Explain.

88. Acetic acid, found in vinegar, has a small equilibrium constant. Hydrochloric acid is known as a strong acid. Which of these two representations depicts acetic acid? (*Note:* In the boxes, HX represents a general acid structure where X^- represents the conjugate base.)

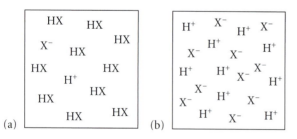

89. Is it possible for a concentrated solution of a weak acid to have the same level of hydronium ion concentration as a dilute solution of a strong acid such as HCl?

90. Water, which has hydrogen bonded to oxygen, can form both H^+ ions and OH^- ions. KOH, which also has hydrogen bonded to oxygen, produces only OH^- ions in solution. Using structure and electronegativity, explain why this situation occurs.

91. When comparing the strengths of strong acids, one must use a solvent other than water. One such solvent is acetic acid.

Explain why acetic acid would make a better solvent than water for comparing acid strength among strong acids.

92. At 25°C the K_w value for water is 1.0×10^{-14}. Using that value and the autoprotolysis of water, determine what percent of water molecules actually dissociate under these conditions.

93. Explain why the second ionization constant of a diprotic acid is typically much smaller than the first.

94. Only very pure phosphoric acid is used in food products. An older term for soft drinks is *phosphates*. This name was applied because pure phosphoric acid was used to produce a tart taste in some beverages and to help dissolve the other ingredients.
   a. What is the pH of a solution that is known to be 5.44 M in H_3PO_4?
   b. What would be the approximate concentration of $PO_4^{3-}(aq)$ in that solution?

95. a. Using the Internet reference http://www.fda.gov/medwatch/safety/1997/ephedr.htm explain why caffeine should not be mixed with stimulants that contain ephedrine alkaloids.
   b. Use other searches to obtain the formula of an ephedrine alkaloid.
   c. Is caffeine an alkaloid? What is meant by the term *alkaloid*?

96. Lactic acid is produced in muscle tissue during vigorous exercise. It is also found in spoiled milk. The K_a value of lactic acid ($CH_3CHOHCOOH$), at 25°C, is 1.3×10^{-4}. What is the $H^+(aq)$ concentration in a sample of cellular fluid that is 0.0000033 M in lactic acid and 0.0027 M in lactate ion?

97. You can answer the following question by looking at this link from the National Atmospheric Deposition Program (NADP), which contains a variety of data from its nationwide network of monitoring sites: http://nadp.sws.uiuc.edu/isopleths/annualmaps.asp. Look up the maps of pH from 1994, and then from 2005. How are the maps similar and different? Can you explain the reasons for any trends you see? (1994 data: http://nadp.sws.uiuc.edu/isopleths/maps1994/phlab.gif.) (2005 data: http://nadp.sws.uiuc.edu/isopleths/maps2005/phlab.gif.)

Thinking Beyond the Calculation

98. A salt of sorbic acid ($HC_6H_7O_2$) has been used as mold inhibitor. The salt most commonly used in this practice is potassium sorbate ($KC_6H_7O_2$).
   a. Will a solution of potassium sorbate produce a neutral, acidic, or basic solution?
   b. Write a balanced chemical reaction that describes the processes that take place when potassium sorbate is dissolved in water.
   c. The K_b value of the sorbate ion is 5.88×10^{-10}. What is the pH of a 0.0100 M solution of potassium sorbate?
   d. If a researcher wanted to make 500.0 mL of a solution of potassium sorbate with a pH of 9.44, how many grams of this compound would need to be added to the water?
   e. A solution is prepared that contains both 0.01000 M NH_3 and 0.01000 M potassium sorbate. What is the pH of the solution? What effect, if any, does the ammonia have on the pH of the solution (compare the answers to parts c and e).

Applications of Aqueous Equilibria

Contents and Selected Applications

Chemical Encounters: Industrial and Environmental Applications of Titrations

18.1 Buffers and the Common-Ion Effect

Chemical Encounters: Scrubbing Sulfur Dioxide Emissions

18.2 Acid–Base Titrations

Chemical Encounters: Anthocyanins and Universal Indicators

18.3 Solubility Equilibria

18.4 Complex-Ion Equilibria

Chemical Encounters: Commercial Uses of Aminopolycarboxylic Acid Chelating Agents

A chemical technician often performs titrations to analyze solutions for specific analytes.

Go to **college.hmco.com/pic/kelterMEE** for online learning resources.

Application

CHEMICAL
ENCOUNTERS:
Industrial and
Environmental
Applications of
Titrations

Water, so essential to life, makes up 70% of the Earth's surface and a nearly equal proportion of our own body mass. Water in living organisms is useful as a medium in which to dissolve the compounds necessary for life, and it also acts as a barrier to keep some compounds out of our bodies. As fundamental as it is to life, there are some everyday processes in which the presence of water can be harmful. Small amounts of water in your gas tank can reduce the efficiency of your car's engine; water in your motor oil increases the rate of decomposition of the lubricating properties (which can lead to the breakdown of the interior of your engine); and water in the coolant used in the manufacture of metal parts can damage the tooling machines, resulting in imperfections in the parts.

One job of people working as chemical technicians is to perform tests many times each day to determine the quality of the coolant, oil, or gasoline that their company uses or sells. Indeed, one of the many measures of quality oil is that it contains only a negligible amount of water. Technicians use the analytical method of titration, which we first discussed in Section 4.3, to measure the concentration of water in oil and to measure the quantity of a whole host of compounds and ions in water samples. A titration, shown in the photograph at the beginning of this chapter, is the controlled addition of just enough solution of *known concentration*, called a titrant, to react with essentially all of an analyte (the substance of interest) so that we can determine its concentration. Titrating oil to determine the amount of water present is only one of the multitudes of applications of titrations. These applications fall into several different categories, including those that

- cause a reduction or oxidation to occur in an analyte
- result in the formation of a precipitate
- form a complex ion
- involve an acid and a base reacting

Water, essential for life, is not as desirable in some consumer products such as motor oil, where it adversely affects the oil's lubricating properties.

Food and pharmaceutical manufacturers use titrations for quality control—making sure that the product contains what it is supposed to, and in the proper amounts. Environmental chemists use titrations for analyzing trace (very small) amounts of hazardous metals and other potentially harmful substances. Table 18.1 lists some analyses that are commonly accomplished using titrations. At the core of most titrations are many of the principles of aqueous equilibrium that we discussed in Chapters 16 and 17, and we will put those concepts to good use here. A good starting point is buffer solutions because they are commonplace both in industrial titration analyses and, more broadly, in biochemical systems (including us!).

TABLE 18.1	Selected Titration-Based Analyses
What the Titration Determines	**Primary Reagent**
Acidity	Sodium hydroxide
Alkalinity	Hydrochloric acid
Vitamin C	Iodine
Chloride ion	Silver nitrate
Water hardness	EDTA
Dissolved oxygen	Sodium thiosulfate
Salinity	Mercuric nitrate
Water	Iodine, sulfur dioxide, primary amines

18.1   Buffers and the Common-Ion Effect

The degree of "hardness" of water, a measure of the concentration of calcium (sometimes including magnesium and iron) in household water supplies, is determined by titration. When heated, the calcium ions in hard water form rock-hard carbonates and sulfates. The resulting solids, known as **boiler scale**, build up on the inside of pipes. In addition, ice cubes made with "hard" water melt to give an ugly-looking precipitate, as shown in Figure 18.1. Furthermore, very hard water has a bitter taste that many homeowners find unappealing.

The concentration of calcium ions in water can be determined by titration with ethylenediaminetetraacetic acid (EDTA), which we first discussed in Chapter 16 and will consider in greater detail later in this chapter as well as in Chapter 20. The equation describing the titration is

$$Ca^{2+}(aq) + EDTA^{4-}(aq) \rightleftharpoons CaEDTA^{2-}(aq) \qquad K = 5.0 \times 10^{10}$$

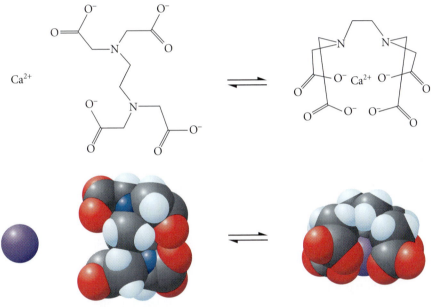

$Ca^{2+}(aq)$ $EDTA^{4-}(aq)$ $CaEDTA^{2-}(aq)$

Application

FIGURE 18.1

Hard water can deposit a white precipitate in your icy glass of water.

Visualization: Buffers

Video Lesson: An Introduction to Buffers

Tutorial: Buffered Solutions

In order to ensure that the EDTA exists as the tetraanion, the entire system must remain basic during the titration. An added buffer maintains the alkaline solution. As you will remember from our previous discussions, a **buffer** is a chemical system that is able to resist changes in pH. A buffer is the combination of a *weak acid and its conjugate base or of a weak base and its conjugate acid*. Buffers do *not* exist if a strong acid or strong base is paired with its conjugate. However, *buffers can accommodate the addition of strong acid and base, and can also withstand dilution, without large changes in the solution pH.*

One standard hard-water analysis protocol requires the addition of a buffer made from ammonia (a weak base) and ammonium chloride (a source of ammonia's conjugate acid). The required pH of the buffer for this analysis is 10. According to the protocol, this can be achieved when the initial concentrations are $[NH_3]_0 = 8.44\ M$ and $[NH_4^+]_0 = 1.27\ M$. How do these initial concentrations result in a solution with a pH close to 10?

To answer this question, we proceed as we do with any equilibrium process, by first examining the possible reactions that can take place in the aqueous solution. In doing so, we recognize that ammonium chloride (NH_4Cl) will dissociate

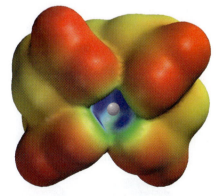

EDTA^{4-} complexes with a sodium ion. The complex involves each of the carboxylate oxygens (COO$^-$) and the two nitrogen atoms in the EDTA molecule. The resulting structure creates a pocket surrounding the sodium ion.

in water to give the ammonium ion (NH_4^+) with chloride (Cl^-) as a spectator ion. *The key reactants in the buffer are NH_4^+, NH_3, and H_2O.* Their specific reactions are given below.

Reaction 1: Ammonia, a weak base, reacts with water.

$$NH_3(aq) + H_2O(l) \rightleftharpoons NH_4^+(aq) + OH^-(aq) \qquad K_b = 1.8 \times 10^{-5}$$

Reaction 2: Ammonium ion, the conjugate acid of ammonia, reacts with water.

$$NH_4^+(aq) + H_2O(l) \rightleftharpoons NH_3(aq) + H_3O^+(aq) \qquad K_a = 5.6 \times 10^{-10}$$

(Recall from Section 17.7 that $K_a = K_w/K_b$ for NH_3.)

Reaction 3: Water undergoes autoprotolysis.

$$2H_2O(l) \rightleftharpoons H_3O^+(aq) + OH^-(aq) \qquad K_w = 1.0 \times 10^{-14}$$

The last reaction has a relatively small equilibrium constant, so its contribution to the pH of the solution is unimportant. Judging by their respective equilibrium constants, the base, ammonia, is stronger than its conjugate acid, ammonium ion, so we expect reaction 1 to be dominant. The reaction produces hydroxide ion, so we expect the solution to be basic.

The Impact of Le Châtelier's Principle on the Equilibria in the Buffer

Video Lesson: The Common Ion Effect

Reaction 1 is producing NH_4^+. However, in the buffer used to determine the hardness of water, we already have a relatively high initial concentration of ammonium ion ($[NH_4^+]_0 = 1.27\ M$). How will this affect the extent of reaction 1? Le Châtelier's principle (Sections 16.6 and 17.4) suggests that the presence of the ammonium ion (a "common ion," in this case) will shift the equilibrium of reaction 1 to the reactants side, as shown in Figure 18.2. Reaction 1 will be *drastically suppressed* because of the common-ion effect, an outcome of Le Châtelier's principle.

What about reaction 2? It is producing ammonia (NH_3). However, the initial concentration is quite high, $[NH_3]_0 = 8.44\ M$. As with reaction 1, the presence of a large concentration of ammonia will shift the reaction to the left, the reactants side (Figure 18.3). Reaction 2 will also be *drastically suppressed* because of the common-ion effect, an outcome of Le Châtelier's principle.

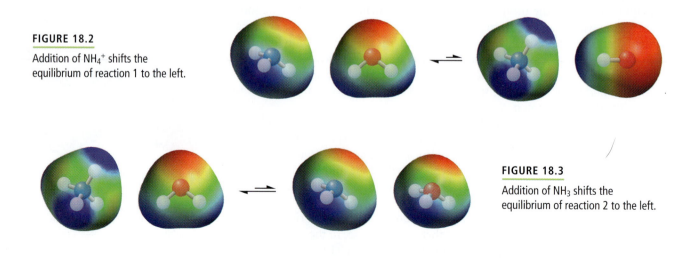

FIGURE 18.2

Addition of NH_4^+ shifts the equilibrium of reaction 1 to the left.

FIGURE 18.3

Addition of NH_3 shifts the equilibrium of reaction 2 to the left.

The impact of reactions 1 and 2 being suppressed (claims that we will prove after we solve for the pH of the buffer solution) is that we may assume that *the equilibrium concentrations of NH_3 and NH_4^+ are approximately equal to their initial concentrations.*

$$[NH_3] \approx [NH_3]_0 = 8.44\ M \quad \text{and} \quad [NH_4^+] \approx [NH_4^+]_0 = 1.27\ M$$

Because of this assumption, we can solve for the pH of this buffer solution using the mass-action expression derived from reaction 1, the hydrolysis of ammonia.

$$K_b = \frac{[NH_4^+][OH^-]}{[NH_3]} \qquad 1.8 \times 10^{-5} = \frac{(1.27)[OH^-]}{(8.44)}$$

$$[OH^-] = 1.2 \times 10^{-4}\ M$$

$$[H^+] = \frac{K_w}{[OH^-]} = \frac{1.0 \times 10^{-14}}{1.2 \times 10^{-4}} = 8.3 \times 10^{-11}\ M$$

$$pH = 10.08$$

Alternatively, we could have used the mass-action expression derived from reaction 2 to solve for the pH of the solution, as we will discover in Exercise 18.1.

What about our claim that reaction 1 was suppressed by the presence of the ammonium common ion? To show the impact of the common ion, let's calculate $[OH^-]$ in an 8.44 M NH_3 solution that has *no* NH_4^+—in other words, not a buffer, just a weak base—using the principles we learned in Chapters 16 and 17.

$$NH_3(aq) + H_2O(l) \rightleftharpoons NH_4^+(aq) + OH^-(aq) \qquad K_b = 1.8 \times 10^{-5}$$

$$K_b = \frac{[NH_4^+][OH^-]}{[NH_3]} \qquad 1.8 \times 10^{-5} = \frac{[NH_4^+][OH^-]}{(8.44)}$$

$$1.8 \times 10^{-5} = \frac{x^2}{8.44}$$

$$x = [NH_4^+] = [OH^-] = 0.012\ M$$

Our calculations reveal that the total hydroxide and ammonium ion concentrations, which are equal in this solution of weak base, are $[OH^-] = [NH_4^+] = 0.012\ M$. Moreover, when we compare the hydroxide ion concentration of the buffer, in which $[OH^-] = 1.2 \times 10^{-4}\ M$, to the value we just calculated for the weak base alone, we note that it is 100-fold less. *The presence of the ammonium ion in the buffer has suppressed reaction 1, the hydrolysis of ammonia, by 99%!* If we were to do a similar calculation with the ammonium ion, we would find that the buffer suppresses the acid dissociation of NH_4^+ by over 99.99%. Le Châtelier's principle has again proved its worth as a formidable part of the chemist's toolbox.

Our goal was to show how a mixture of these concentrations of ammonia and ammonium ion results in a buffer solution with a pH of about 10. Exercise 18.1 shows the implications of using a slightly different approach to achieve the same goal.

Video Lesson: CIA
Demonstration: Buffers in Action

EXERCISE 18.1   Alternative Route to the pH of the Buffer

Instead of calculating the pH of the system using reaction 1, calculate the pH of the buffer using reaction 2, the acid dissociation of the ammonium ion. What does your result tell you about solving for the pH of a buffer?

First Thoughts

We still assert that the extent of reaction in a buffer system is negligible; the starting position is the same as the equilibrium position, as shown by the equilibrium line chart.

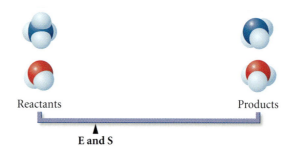

Reactants

Products

E and S

The implication is that the starting and equilibrium concentrations are essentially equal, so, as before,

$$[NH_3] \approx [NH_3]_0 = 8.44\ M$$

and

$$[NH_4^+] \approx [NH_4^+]_0 = 1.27\ M$$

Additionally, we are directed to solve the problem using reaction 2, so we will use its mass-action expression to solve for $[H^+]$ and pH.

Solution

$$NH_4^+(aq) + H_2O(l) \rightleftharpoons NH_3(aq) + H_3O^+(aq) \qquad K_a = 5.6 \times 10^{-10}$$

$$K_a = \frac{[NH_3][H_3O^+]}{[NH_4^+]} \qquad 5.6 \times 10^{-10} = \frac{(8.44)[H_3O^+]}{(1.27)}$$

$$[H^+] = 8.4 \times 10^{-11}\ M$$

$$pH = 10.08$$

Further Insights

Our answer is the same whether we use the mass-action expression for the hydrolysis of the base (ammonia) or the dissociation of its conjugate (ammonium ion). This is a useful outcome that is a result of the common-ion effect of suppressing both the acid reaction and the base reaction in the buffer. When working with buffers, we may use either the acid dissociation or the conjugate base hydrolysis mass-action expression, but we often work with the expression describing the reaction of *the stronger conjugate* (the ammonia hydrolysis, in this case).

PRACTICE 18.1

Determine the pH of a buffer made from 1.50 M NH$_3$ and 3.50 M NH$_4^+$.

See Problems 5 and 6.

EXERCISE 18.2 Practice Calculating the Initial pH of a Buffer

We have shown that an ammonia–ammonium ion buffer can have a pH of about 10. Are all buffers basic? To help answer the question, calculate the pH of a buffer that contains 0.200 M each of acetic acid (CH$_3$COOH) and its conjugate base, the acetate ion (CH$_3$COO$^-$). It is interesting, but nonetheless coincidental, that the K_a value for acetic acid is about the same as the K_b value for ammonia.

First Thoughts

We said in the last exercise that we normally use the reaction and mass-action expression for the stronger conjugate when calculating the pH of the buffer. The relevant conjugate reactions (using the shorthand form for the acid dissociation) and equilibrium constants are

$$CH_3COOH(aq) \rightleftharpoons H^+(aq) + CH_3COO^-(aq) \qquad K_a = 1.8 \times 10^{-5}$$

$$CH_3COO^-(aq) + H_2O(l) \rightleftharpoons CH_3COOH(aq) + OH^-(aq) \qquad K_b = 5.6 \times 10^{-10}$$

We will use the mass-action expression for the stronger conjugate, acetic acid. As with other buffer systems, we recognize that both reactions are suppressed as we have explained using Le Châtelier's principle, so *the equilibrium concentrations are about equal to the initial concentrations of the respective components.*

Solution

The mass-action expression for the reaction of acetic acid is

$$K_a = \frac{[CH_3COO^-][H^+]}{[CH_3COOH]}$$

Solving for $[H^+]$ and then pH, we find that

$$[H^+] = \frac{K_a[CH_3COOH]}{[CH_3COO^-]} = \frac{(1.8 \times 10^{-5})(0.200)}{(0.200)} = 1.8 \times 10^{-5}\,M$$

$$pH = 4.74$$

Further Insights

The ammonia–ammonium buffer has a basic pH, because the base, NH_3, is the stronger of the conjugate pairs. The acetic acid–acetate buffer system is acidic, because acetic acid is the stronger conjugate. This means that you can predict whether a buffer is likely to be acidic or basic judging by the strength of each of the conjugates. The acetic acid–acetate buffer cannot have a pH of 10.

PRACTICE 18.2

Would you predict that a buffer prepared from the mixture of 0.300 M formic acid (HCOOH) and 0.400 M sodium formate (HCOONa) would be acidic or basic? Why? Prove your assertion by calculating the pH of this buffer. K_a of formic acid = 1.8×10^{-4}.

See Problems 21 and 22.

We have seen from this discussion that it is possible to determine the approximate pH of a buffer. Extending that idea a step further, it is also possible to pick a buffer that will be in the pH range we want by looking at whether the acid or the base conjugate is the stronger. There is often much more to the selection of a buffer than just pH, because we must consider factors such as whether the buffer will interact with the substances we are studying and whether the buffer's presence in the reaction system has any unintended health consequences. Once these factors are taken into account, we need to know how to prepare the buffers in order to use them.

HERE'S WHAT WE KNOW SO FAR

- A buffer is a chemical system that is able to resist changes in pH.
- A buffer consists of the combination of a weak acid and its conjugate base, or of a weak base and its conjugate acid.
- A buffer contains a high enough concentration of each conjugate that acid–base equilibria are effectively suppressed, an outcome of Le Châtelier's principle.
- A buffer will be acidic or basic, depending on which of the conjugates is the stronger.
- It is possible to calculate the pH of the buffer system by applying the principles of equilibrium and acid–base chemistry that we learned in the last two chapters.

FIGURE 18.4

Phosphoric acid is often found in soft drinks.

Buffer Preparation

The chemical technician working in the food industry often uses a pH meter to determine the acidity of the food that is being prepared. Sodas, for instance, often have phosphoric acid added to them to provide a tart taste, and pH control is vital (Figure 18.4).

One of the most common uses of a buffer solution is to calibrate pH meters. This ensures that the reading on the meter accurately represents the pH of a solution with which we are working. Recipes exist for the preparation of standard calibration buffer solutions. One source for these recipes is the *The Pharmacopeia of the United States of America/The National Formulary*, a 2400-page compendium of "standards and specifications for materials and substances that are used in the practice of the healing arts." Organized in 1884 and produced and updated by medical and pharmaceutical experts ever since, the resulting *Pharmacopeia* ("USP" for short) contains standard analysis procedures for a great many substances.

The USP recipe to prepare calibration buffers in the range of 2.2 to 4.0 recommends using a solution of potassium hydrogen phthalate, or "KHP" ($KHC_6H_4(COO)_2$), which dissociates when added to water to form K^+ and HP^- ($HC_6H_4(COO)_2$). The conjugate acid of this weak base, phthalic acid ("H_2P") is generated by addition of hydrochloric acid (a source of H^+) to the HP^-.

$$HP^-(aq) + H^+(aq) \rightleftharpoons H_2P(aq) \qquad K \approx 900$$

The equilibrium position (minimum free energy) of the reaction of the weak base with the strong acid is so far toward the formation of products that we can say the reaction is essentially complete. That is, although we will normally write the reaction to show that it settles at some equilibrium point,

$$HP^-(aq) + H^+(aq) \rightleftharpoons H_2P(aq)$$

we may also write it to show that the reaction is essentially complete,

$$HP^-(aq) + H^+(aq) \rightarrow H_2P(aq)$$

Suppose we wish to prepare a buffer with pH = 3.0, and assume that we want the total concentration of phthalate species to be 0.0500 *M*. That is,

$$[H_2P] + [HP^-] = 0.0500 \, M$$

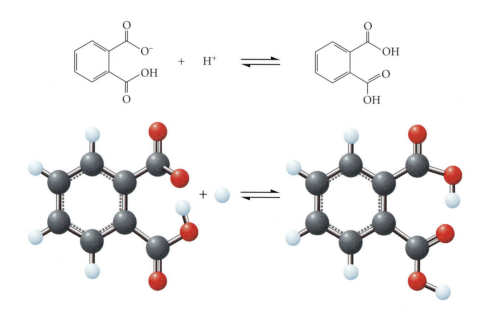

Once the proper ratio of conjugate acid to base is present, the buffer will be at pH = 3.00. **How do we find the proper ratio? How do we then find the final concentrations of H_2P and HP^-?**

As has been true throughout these three chapters on equilibrium (Chapters 16–18), the answer lies in writing down the relevant equilibrium reaction and its mass-action expression. For the phthalate buffer system, the important reaction is the dissociation of phthalic acid (H_2P) in water:

$$H_2P(aq) + H_2O(l) \rightleftharpoons HP^-(aq) + H_3O^+(aq) \qquad K_{a_1} = 1.12 \times 10^{-3}$$

We can write it using our shorthand form:

$$H_2P(aq) \rightleftharpoons HP^-(aq) + H^+(aq) \qquad K_{a_1} = 1.12 \times 10^{-3}$$

The mass-action expression for the reaction is

$$K_{a_1} = \frac{[HP^-][H^+]}{[H_2P]}$$

Dividing both sides by $[H^+]$ gives us an expression for the ratio of base to acid:

$$\frac{K_{a_1}}{[H^+]} = \frac{[HP^-]}{[H_2P]}$$

Solving, with the use of the pH = 3.00 ($[H^+] = 1.0 \times 10^{-3}\ M$),

$$\frac{(1.12 \times 10^{-3})}{(1.00 \times 10^{-3})} = \frac{(1.12)}{(1.00)} = \frac{[HP^-]}{[H_2P]}$$

we find that the ratio of the base, HP^-, to its conjugate acid, H_2P, is 1.12 to 1. We'll call this Equation 1.

$$[HP^-] = 1.12\ [H_2P] \qquad \text{(Equation 1)}$$

We said initially that we had set the sum of concentrations of the two species to be a total of 0.0500 M. We'll call this Equation 2.

$$[H_2P] + [HP^-] = 0.0500\ M \qquad \text{(Equation 2)}$$

This gives us two equations (Equations 1 and 2) and two unknowns ($[H_2P]$ and $[HP^-]$). We can solve Equation 2 by substituting $1.12[H_2P]$ in place of $[HP^-]$, as allowed by Equation 1:

$$[H_2P] + 1.12[H_2P] = 0.0500\ M$$
$$2.12[H_2P] = 0.0500\ M$$
$$\mathbf{[H_2P] = 0.0236\ M}$$

$$[HP^-] = 0.0500\ M - 0.0236\ M$$
$$\mathbf{[HP^-] = 0.0264\ M}$$

Therefore, in order to obtain a buffer at pH = 3.00, we'll have to make up a solution that is 0.0236 M in H_2P and 0.0264 M in KHP. We can check that these quantities are reasonable by substituting back into the mass-action expression and solving for $[H^+]$.

$$[H^+] = \frac{K_{a_1}[H_2P]}{[HP^-]} = \frac{(1.12 \times 10^{-3})(0.0236)}{(0.0264)}$$

$$[H^+] = 1.00 \times 10^{-3}\ M$$

This concentration of H^+ would produce a solution with a pH = 3.00.

| EXERCISE 18.3 | Practice with Buffer Preparation |

How many grams of sodium formate (HCOONa), molar mass = 68.01 g/mol, must be dissolved in a 0.300 M solution of formic acid (HCOOH), to make 400.0 mL of a buffer solution with a pH = 4.60? Assume that the volume of the solution remains constant when you add the sodium acetate.

First Thoughts

We are asked to find the mass of sodium formate needed to combine with formic acid to prepare the buffer. Let's develop a stepwise approach to the problem.

Step 1: As in prior examples involving buffers, we need to establish the ratio of the concentrations of the acid and the conjugate base, $\dfrac{[\text{HCOO}^-]}{[\text{HCOOH}]}$, needed to produce a buffer with a pH = 4.60. To solve for the ratio of concentrations of acid and conjugate base needed, we must first write the important processes that occur in the solution. Two equilibria are important in this buffer system:

$$\text{HCOOH}(aq) + \text{H}_2\text{O}(l) \rightleftharpoons \text{H}_3\text{O}^+(aq) + \text{HCOO}^-(aq) \qquad K_a = 1.8 \times 10^{-4}$$
$$\text{HCOO}^-(aq) + \text{H}_2\text{O}(l) \rightleftharpoons \text{HCOOH}(aq) + \text{OH}^-(aq) \qquad K_b = 5.6 \times 10^{-11}$$

As we discussed in Exercise 18.1, we may use either reaction to solve for the acid-to-conjugate-base ratio.

Step 2: Knowing that the concentration of formic acid is 0.300 M, we can then find the concentration of the conjugate base. Remember that the sodium salt will completely dissociate in solution, leaving the formate anion (HCOO^-) as the conjugate base and Na^+ as the spectator ion.

Step 3: We may then convert from the concentration of formate (HCOO^-) to the number of moles of HCOO^- in 400.0 mL of solution.

Step 4: Finally, we may convert from moles of formate (as the sodium formate salt) to grams of sodium formate.

A flowchart of the process looks like this:

$$\frac{[\text{HCOO}^-]}{[\text{HCOOH}]} \xrightarrow[\text{Step 1}]{\substack{\text{mass-action} \\ \text{expression}}} [\text{HCOO}^-] \xrightarrow[\text{Step 2}]{\substack{\text{volume} \\ \text{solution}}} \text{mol HCOO}^- \xrightarrow[\text{Step 3}]{\substack{\text{molar mass} \\ \text{HCOONa}}} \text{g HCOONa} \quad \text{Step 4}$$

Solution

Step 1:
$$K_a = \frac{[\text{HCOO}^-][\text{H}^+]}{[\text{HCOOH}]}$$

The pH of the solution is 4.60, so $[\text{H}^+] = 2.51 \times 10^{-5}\, M$. We are retaining an extra figure because this is an intermediate calculation.

$$1.8 \times 10^{-4} = \frac{[\text{HCOO}^-](2.51 \times 10^{-5})}{[\text{HCOOH}]}$$

$$\frac{[\text{HCOO}^-]}{[\text{HCOOH}]} = \frac{1.8 \times 10^{-4}}{2.51 \times 10^{-5}} = \frac{7.17}{1.00}$$

The concentration of formate ion is about 7.2 times as great as the concentration of formic acid.

Step 2:
$$[\text{HCOO}^-] = 7.17 \times [\text{HCOOH}]$$
$$[\text{HCOO}^-] = 7.17 \times 0.300\, M$$
$$[\text{HCOO}^-] = 2.15\, M$$

Alternatively, we could have combined steps 1 and 2 to find $[HCOO^-]$ directly by substituting $0.300\ M$ into the mass-action expression, as follows:

$$1.8 \times 10^{-4} = \frac{[HCOO^-](2.51 \times 10^{-5})}{0.300}$$

$$[HCOO^-] = 2.15\ M$$

Step 3: $\dfrac{2.15\ \text{mol HCOO}^-}{\text{L solution}} \times 0.4000\ \text{L solution} =$

$$0.860\ \text{mol HCOO}^- = 0.860\ \text{mol HCOONa}$$

Step 4: $0.860\ \text{mol HCOONa} \times \dfrac{68.01\ \text{g HCOONa}}{1\ \text{mol HCOONa}} = 58\ \text{g HCOONa}$

Further Insights

Does it make sense that the buffer system should be acidic? Formic acid (HCOOH) is stronger than its conjugate, so the acid dissociation reaction will dominate, resulting in an acidic buffer solution. This is a reasonable conjugate pair to use to prepare an acidic buffer solution. Our answer makes sense.

PRACTICE 18.3

How many grams of sodium formate (HCOONa) must be dissolved in a $0.150\ M$ solution of formic acid (HCOOH) to make $500.0\ mL$ of a buffer solution with $pH = 3.95$? Assume that the volume of the solution remains constant when you add the sodium formate.

See Problems 17, 18, 25, and 26.

EXERCISE 18.4   More Practice with Buffer Preparation

How many milliliters of $0.200\ M$ HCl must be added to $50.0\ mL$ of a $0.200\ M$ solution of NH_3 in order to prepare a buffer that has a pH of 8.60?

$$NH_3 + H_2O(l) \rightleftharpoons NH_4^+ + OH^- \qquad K_b = 1.8 \times 10^{-5}$$

First Thoughts

We have a solution of NH_3. We need to add enough strong acid, in the form of HCl, to convert NH_3 to NH_4^+ until the ratio of the weak base to its conjugate acid, $\dfrac{[NH_3]}{[NH_4^+]}$, will be the same as that in a pH 8.60 buffer.

As we think through a problem-solving strategy, let's write down what we know about the system.

- The K_b value for the hydrolysis of NH_3 is 1.8×10^{-5}.
- We know the mass-action expression for the formation of NH_4^+ from NH_3.
- We also know that there is $0.0100\ mol$ of NH_3 initially present in the solution:

$$0.0500\ \text{L NH}_3 \times \frac{0.200\ \text{mol NH}_3}{\text{L NH}_3\ \text{solution}} = 0.0100\ \text{mol NH}_3$$

Step 1: What is the ratio of the weak base to the conjugate acid, $\dfrac{[NH_3]}{[NH_4^+]}$, in the

pH $= 8.60$ buffer ($[H^+] = 2.5 \times 10^{-9}\ M$)? This ratio will be the first equation we need to solve in order to find the amounts of NH_3 and NH_4^+. The base hydrolysis equation contains $[OH^-]$ rather than $[H^+]$. We can solve for $[OH^-]$ from $[H^+]$ and K_w. We are keeping an extra figure in for the intermediate calculation.

$$[OH^-] = \frac{K_w}{[H^+]} = \frac{1.00 \times 10^{-14}}{2.51 \times 10^{-9}} = 3.98 \times 10^{-6}$$

We can then set up our mass-action expression to solve for $\dfrac{[NH_3]}{[NH_4^+]}$.

$$NH_3(aq) + H_2O(l) \rightleftharpoons NH_4^+(aq) + OH^-(aq) \qquad K_b = 1.8 \times 10^{-5}$$

$$K_b = \frac{[NH_4^+][OH^-]}{[NH_3]}$$

$$\frac{K_b}{[OH^-]} = \frac{[NH_4^+]}{[NH_3]}$$

Inverting both sides of the equation, we get

$$\frac{[OH^-]}{K_b} = \frac{[NH_3]}{[NH_4^+]}$$

Step 2: The sum of the moles of NH_3 and NH_4^+ in the final solution must equal the initial 0.100 mol of NH_3 used to prepare the original solution because we are *changing* the NH_3 to NH_4^+, *not getting rid of any* from the reaction flask. This is the second equation we need to solve to find the amounts of NH_3 and NH_4^+.

$$\text{mol } NH_4^+ + \text{mol } NH_3 = 0.0100 \text{ mol}$$

Step 3: Finally, we can find how much 0.200 M HCl will be needed to react with our original 0.100 mol of NH_3 to generate the proper number of moles of NH_4^+.

Solution

Step 1: Finding the ratio of $\dfrac{[NH_3]}{[NH_4^+]}$ in the solution.

$$\frac{[OH^-]}{K_b} = \frac{[NH_3]}{[NH_4^+]}$$

Substituting for $[OH^-]$ and K_b yields

$$\frac{3.98 \times 10^{-6}}{1.8 \times 10^{-5}} = \frac{0.221}{1.00} = \frac{[NH_3]}{[NH_4^+]}$$

Because the volume of the buffer solution is the same for NH_3 and NH_4^+, we can work with moles rather than molarity (the liters cancel out on top and bottom).

$$\text{mol } NH_3 = 0.221 \times (\text{mol } NH_4^+)$$

Step 2: Finding moles of NH_4^+ and NH_3 so that the sum equals 0.0100 mol.

$$\text{mol } NH_4^+ + \text{mol } NH_3 = 0.0100 \text{ mol}$$

$$\text{mol } NH_4^+ + [0.221 \times (\text{mol } NH_4^+)] = 0.0100 \text{ mol}$$

$$1.221 \times (\text{mol } NH_4^+) = 0.0100 \text{ mol}$$

$$\text{mol } NH_4^+ = 0.00819 \text{ mol}$$

$$\text{mol } NH_3 = 0.00181 \text{ mol}$$

As a check of your work, you will find that substituting these values back into the mass-action expression will give you K_b.

Step 3: Finding how many milliliters of 0.200 M HCl we need to add to produce 0.00819 mol of NH_4^+ from NH_3.

$$0.00819 \text{ mol HCl} \times \frac{1 \text{ L HCl solution}}{0.200 \text{ mol HCl}} \times \frac{1000 \text{ mL HCl solution}}{1 \text{ L HCl solution}} = 41 \text{ mL HCl solution}$$

Further Insights

Although we used the base hydrolysis of ammonia to get the initial conjugate acid–base ratio, we could have used the acid ionization of ammonium just as well, because we are dealing with a buffer solution. If we had done this, the solution in step 1 would have looked like this:

$$NH_4^+(aq) + H_2O(l) \rightleftharpoons NH_3(aq) + H_3O^+(aq)$$

$$K_a = \frac{[K_w]}{[K_b]} = \frac{(1.0 \times 10^{-14})}{(1.8 \times 10^{-5})} = 5.56 \times 10^{-10}$$

(We retain an extra figure in K_a, which we would drop at the end of the calculation.)

$$K_a = \frac{[NH_3][H_3O^+]}{[NH_4^+]}$$

$$\frac{[K_a]}{[H_3O^+]} = \frac{[NH_3]}{[NH_4^+]}$$

$$\frac{[NH_3]}{[NH_4^+]} = \frac{(5.56 \times 10^{-10})}{(2.51 \times 10^{-9})} = \frac{0.221}{1.00}$$

$$mol\ NH_3 = 0.221\ (mol\ NH_4^+)$$

This agrees with the answer we got in step 1.

PRACTICE 18.4

How many milliliters of 0.10 M HCl must be added to 25.0 mL of a 0.2000 M solution of NH_3 to prepare a buffer that has a pH of 8.80?

See Problems 11, 12, 27, and 28. ▌

In practice, even fairly dilute buffers with fairly low conjugate concentrations (not far above their equilibrium constants) will have pH values similar to those with equal ratios of higher conjugate concentrations. If the buffer is too dilute, other factors cause the pH to move toward neutrality.

The Henderson–Hasselbalch Equation: We Proceed, But with Caution

Video Lesson: The Henderson–Hasselbalch Equation

We have seen how the mass-action expression is used to find the approximate ratio of base to acid in a buffer. In 1902, seven years before Peter Sørenson coined the term *pH*, a Massachusetts physician named Lawrence Joseph Henderson, who was studying buffers in blood, published work relating $[H^+]$ to the acid and base concentrations in a buffer. For an acid, HA, and its conjugate base, A^-, Henderson noted that $[H^+] = \dfrac{K_a[HA]}{[A^-]}$, as we showed in Exercise 18.2 for the acetic acid–acetate ion buffer system. Written a slightly different way for visual clarity, this is

$$[H^+] = K_a \frac{[HA]}{[A^-]} = K_a \frac{[weak\ acid]}{[conjugate\ base]}$$

Note the relationship between the acid and the base. The acid is a weak acid, and the base is the conjugate base of that weak acid. In 1916 K.A. Hasselbalch (pronounced "hassle-back"), a physiologist at the University of Copenhagen, used

Sørenson's "new" pH term, taking the negative log of both sides of Henderson's equation to give

$$pH = -\log(K_a) - \log\left\{\frac{[\text{weak acid}]}{[\text{conjugate base}]}\right\}$$

This can be slightly modified by using pK_a (the same as $-\log(K_a)$) and changing the sign before the log term, which inverts the concentrations within the log term. This gives the final form of what is commonly known as the **Henderson–Hasselbalch equation**:

$$pH = pK_a + \log\left\{\frac{[\text{conjugate base}]}{[\text{weak acid}]}\right\}$$

The equation is not necessary; we can solve any buffer problem without it, including the ammonia–ammonium ion system and the KHP buffer system we discussed earlier in this section. However, scientists in the medical and biological professions use the equation because it can be a timesaver for calculating the pH.

We added the phrase "we proceed, but with caution" to the title of this subsection because, in spite of how common the use of the Henderson–Hasselbalch equation is, there are two reasons why you must be cautious with its use. First, a system *must* contain both an acid and its conjugate base in order for the equation to be valid. If you don't have a buffer system, using this equation will lead you to incorrect results. Second, and more troublesome for many students, *is the tendency to invert the ratio of acid to base within the log term*. Proceed with caution!

EXERCISE 18.5 **The Henderson–Hasselbalch Equation**

Use the Henderson–Hasselbalch equation to find the ratio of conjugate base to weak acid in an acetic acid–acetate buffer solution with a pH of 5.0. $K_a = 1.8 \times 10^{-5}$.

Solution

$$pH = pK_a + \log\left\{\frac{[\text{base}]}{[\text{acid}]}\right\}$$

$$5.0 = 4.74 + \log\left\{\frac{[CH_3COO^-]}{[CH_3COOH]}\right\}$$

$$0.26 = \log\frac{[CH_3COO^-]}{[CH_3COOH]}$$

$$10^{0.26} = \frac{[CH_3COO^-]}{[CH_3COOH]} = 1.8$$

This means there is 1.8 times as much acetate ion as acetic acid in a solution with a pH just above the pK_a.

PRACTICE 18.5

Determine the pH of an acetic acid–acetate buffer containing 0.500 M acetic acid and 0.250 M sodium acetate. $K_a = 1.8 \times 10^{-5}$.

See Problems 19–22.

The conclusion to Exercise 18.5 is important, because it says that *in order for the pH of the buffer to be higher than the pK_a, there must be more conjugate base than weak acid.* The opposite statement is also true (as illustrated in Practice 18.5); that is, pH values lower than the pK_a value require more conjugate acid than base. Take a quick look back at Exercises 18.1 through 18.3 to see that this is so. In Exercise 18.2, the concentrations of the weak acid and conjugate base are equal. When this is so, the pH is equal to the pK_a, and the buffer that results has the greatest possible capacity to neutralize strong acids and bases.

Buffer Capacity

The job of the chemical technician may include monitoring the emissions from a smoke stack. Recent laws enacted to protect the environment have made this a priority. To reduce the emissions, businesses often make use of a scrubber attached to the smoke stack. **Scrubbing** is used to remove sulfur dioxide from the smoke emitted by combustion of sulfur-rich coal. Chemically, the process can be accomplished by **flue-gas desulfurization**, in which a limestone slurry is combined with sulfur dioxide in a multistep process:

$$CaCO_3(s) + SO_2(g) \rightleftharpoons CaSO_3(s) + CO_2(g)$$

The calcium sulfite that is formed is further oxidized and hydrated to give gypsum ($CaSO_4 \cdot 2H_2O$), which is used to make wallboards for home construction, as discussed in Chapter 17.

$$CaSO_3(s) + \tfrac{1}{2}O_2(g) + 2H_2O(l) \rightleftharpoons CaSO_4 \cdot 2H_2O$$

To make this conversion of sulfur dioxide to calcium sulfate dihydrate more efficient, a mixture of formic acid (HCOOH) and sodium formate (HCOONa) is added to the scrubber. The low pH of this buffer (Exercise 18.3) enhances the reaction efficiency by converting some of the sulfite in solution to bisulfite, (HSO_3^-). Doing so allows the more soluble calcium bisulfite ($CaHSO_3$) to form, as well as minimizing calcium sulfite buildup on the processing machinery. **How can the formic acid–formate ion buffer system keep the system pH within a small range?** The **buffer capacity** of a system is a measure of the number of moles of strong acid or strong base that can be added while keeping the pH relatively constant. IUPAC defines "relatively constant" as between +/− 1 pH unit. Others have different parameters. Because this discussion serves as an introduction to acid–base titrations, we will define "relatively constant" as that point at which all of the conjugate acid (or base) is reacted.

Suppose we wish to determine the buffer capacity of 100.0 mL of a solution containing 0.200 *M* HCOOH and 0.400 *M* HCOONa. We first address this problem by noting all of the relevant equilibria in aqueous solution.

$$2H_2O(l) \rightleftharpoons H_3O^+(aq) + OH^-(aq) \qquad K = 1.0 \times 10^{-14}$$

$$HCOOH(aq) \rightleftharpoons HCOO^-(aq) + H^+(aq) \qquad K_a = 1.8 \times 10^{-4}$$

Again, the autoprotolysis of water is insignificant, based on the relative size of the equilibrium constant. Considering only the formic acid–formate equilibrium, the pH of this system is 4.05, shown by our calculation:

$$[H^+] = \frac{K_a[HCOOH]}{[HCOO^-]} = \frac{1.8 \times 10^{-4}(0.200)}{(0.400)}$$

$$[H^+] = 9.0 \times 10^{-5}\,M \qquad pH = 4.05$$

Let's look at the impact on the pH of adding strong acid and then strong base.

Application

CHEMICAL
ENCOUNTERS:
Scrubbing Sulfur
Dioxide Emissions

Video Lesson: Acidic Buffers
Video Lesson: Basic Buffers

Change 1: Addition of a strong acid*

What happens to the pH of our buffer if we add 10.0 mL of a 1.00 M HCl solution to 100.0 mL of buffer (total solution volume = 110.0 mL)? We can set the stage by calculating the moles of each component initially in solution.

$$\text{mol HCOOH}_{\text{initial}} = 0.1000 \text{ L HCOOH} \times \frac{0.200 \text{ mol HCOOH}}{\text{L HCOOH solution}}$$

$$= 0.0200 \text{ mol HCOOH}$$

$$\text{mol HCOO}^-_{\text{initial}} = 0.1000 \text{ L HCOO}^- \times \frac{0.400 \text{ mol HCOO}^-}{\text{L HCOO}^- \text{ solution}}$$

$$= 0.0400 \text{ mol HCOO}^-$$

We continue by finding out how many moles of HCl were added.

$$\text{mol HCl}_{\text{added}} = 0.0100 \text{ L} \times \frac{1.00 \text{ mol}}{\text{L}} = 0.0100 \text{ mol HCl}$$

The addition of the strong acid supplies H^+ to the solution, and this H^+ essentially completely reacts with $HCOO^-$ to form HCOOH. ($K = 5.6 \times 10^3$).

Visualization: Adding an Acid to a Buffer

How much HCOOH will be formed? We can organize our thinking using a table. At the top of our table we will write the equation that indicates the reaction of the added HCl (a source of H^+) with the conjugate base of formic acid ($HCOO^-$). We note that this is a *limiting-reactant* calculation in which the HCl is limiting and the $HCOO^-$ is in excess.

$$HCOO^-(aq) + H^+(aq) \rightleftharpoons HCOOH(aq) \qquad K = 5.6 \times 10^3$$

moles initial	0.0400		0.0200
moles added		0.0100	
change	−0.0100	−0.0100	+0.0100
moles at equilibrium	0.0300	≈0	0.0300

After addition of the HCl, which threw the system out of equilibrium, it returns again to its new equilibrium position, shown by

$$HCOOH(aq) \rightleftharpoons HCOO^-(aq) + H^+(aq)$$

We still have lots of conjugate acid and base—a buffer system, for which we can find the pH:

$$[H^+] = \frac{K_a[\text{HCOOH}]}{[\text{HCOO}^-]} = \frac{1.8 \times 10^{-4}(0.300)}{(0.300)} = 1.8 \times 10^{-4} \, M \qquad \text{pH} = pK_a = 3.74$$

The buffer has responded to the addition of HCl by having only a slightly lowered pH, from 4.05 to 3.74.

---

*The value for the equilibrium constant for the reaction of the strong acid with the formate anion ($HCOO^-$) to form HCOOH can be calculated by combining the following two equations, as we did in Section 16.4.

$$HCOO^-(aq) + H_2O(aq) \rightleftharpoons HCOOH(aq) + OH^-(aq) \qquad K_b = K_w/K_a = 5.6 \times 10^{-11}$$
$$\underline{H^+(aq) + OH^-(aq) \rightleftharpoons H_2O(aq) \qquad\qquad\qquad\qquad K = 1/K_w = 1.0 \times 10^{-14}}$$
$$HCOO^-(aq) + H^+(aq) \rightleftharpoons HCOOH(aq) \qquad K = (K_b \times 1/K_w) = 1/K_a = 5.6 \times 10^3$$

We use this method several times in this chapter to determine the equilibrium constant for the addition of strong acids or bases to weak bases or acids in a titration.

Change 2: Addition of a strong base

What happens to the pH of our *original* solution if we add 15.0 mL of a 1.00 M NaOH solution (total volume = 115.0 mL)? We already know how much formic acid (0.0200 mol) and formate ion (0.0400 mol) we started with. How many moles of NaOH were added?

$$\text{mol NaOH}_{added} = 0.0150 \text{ L NaOH solution} \times \frac{1.00 \text{ mol NaOH}}{\text{L NaOH solution}}$$

$$= 0.0150 \text{ mol NaOH}$$

The addition of the strong base supplies OH^- to the solution, and this OH^- essentially completely reacts with HCOOH to form $HCOO^-$ ($K = 1.8 \times 10^{10}$). Therefore, we will set up a table based on the reaction of OH^- with HCOOH.

$$\text{HCOOH}(aq) + \text{OH}^-(aq) \rightleftharpoons \text{HCOO}^-(aq) \quad K = 1.8 \times 10^{10}$$

moles initial	0.0200		0.0400
moles added		0.0150	
change	−0.0150	−0.0150	+0.0150
moles at equilibrium	0.0050	≈0	0.0550

We still have both acid and conjugate base in the buffer, although the acid concentration is a little low. We can solve for the pH.

$$[\text{H}^+] = \frac{K_a[\text{HCOOH}]}{[\text{HCOO}^-]} = \frac{1.8 \times 10^{-4}(0.0050)}{(0.0550)} = 1.64 \times 10^{-5} \, M \quad \text{pH} = 4.79$$

The pH of the buffer has gone up, but the solution still has the capacity to keep the pH close to the original value of 4.05.

Change 3: Exceeding the buffer capacity

At what point will the buffer no longer have the capacity to keep the pH reasonably close to the original value? In other words, when will the pH rise or fall sharply? How much will the pH change before stabilizing? The addition of 50.0 mL 1.00 M HCl solution to the *original* buffer solution will stress the system. Let's see how much, again using our table.

$$\text{HCOO}^-(aq) + \text{H}^+(aq) \rightleftharpoons \text{HCOOH}(aq) \quad K = 5.6 \times 10^3$$

moles initial	0.0400		0.0200
moles added		0.0500	
change	−0.0400	−0.0400	+0.0400
moles at equilibrium	≈0	0.0100	0.0600

How did we calculate the moles at equilibrium in this case? *This is still a limiting-reactant problem,* with the $HCOO^-$ instead of the H^+ limiting the amount of product formed. We have 0.0500 mol of H^+, but only 0.0400 mol of $HCOO^-$ with which to react! This means that 0.0400 mol will react to form the HCOOH product, and there will be 0.0100 mol of H^+ in excess.

What is the final pH of the resulting solution? At equilibrium, we have a solution that contains 0.0600 mol of HCOOH and 0.0100 mol of H^+. *The formic acid is so much weaker than the hydrochloric acid (judging on the basis of their K_a values) that its presence will not affect the final hydrogen ion concentration.* Only the H^+

from the HCl is important. Therefore, the pH is based solely on the $[H^+]$ due to the ionization of the strong acid. The total volume of 150.0 mL (0.1500 L) is the sum of the initial 100.0 mL of solution and the 50.0 mL HCl solution added to it.

$$[H^+] = \frac{(0.0100 \text{ mol})}{(0.1500 \text{ L})} = 0.0667 \ M \qquad pH = 1.18$$

This reveals that the buffer's ability to resist a change in pH has been exceeded; the pH has changed dramatically. In the original solution, the formate ion ($HCOO^-$) could react with up to 0.0400 mol of the strong acid without having any excess H^+ to greatly lower the pH. *This is its buffer capacity toward strong acid.* Similarly, the formic acid could react with up to 0.0200 mol of strong base without having any excess strong base to greatly raise the pH. *This is its buffer capacity toward strong base.* As we saw with the addition of too much HCl, when the buffer capacity is exceeded, the pH changes quite sharply. For a monoprotic acid or base, the buffer capacity is greatest when $[HA] = [A^-]$. When this occurs, the buffer can neutralize equal amounts of both strong base and strong acid.

EXERCISE 18.6   **Buffer Capacity**

Let's look at the buffer capacity of an ammonia–ammonium buffer system like the one we used for our calcium–EDTA analysis at the beginning of this chapter. We will start with 200.0 mL of a solution containing 0.500 M NH_3 and 0.200 M NH_4^+. What will be the initial pH of the buffer? Will we exceed the buffer capacity by adding 75.0 mL of 2.00 M NaOH? What will be the pH after that addition?

Solution

We have a total of 0.100 mol of NH_3 and 0.0400 mol NH_4^+. The solution has more conjugate base than acid, so the pH should be higher than the pK_a for ammonium, 9.26. We use the expression for K_a of NH_4^+ so that we can solve directly for $[H^+]$:

$$NH_4^+(aq) \rightleftharpoons NH_3(aq) + H^+(aq) \qquad K = 5.6 \times 10^{-10}$$

$$[H^+] = \frac{K_a[NH_4^+]}{[NH_3]} = \frac{5.6 \times 10^{-10}(0.200)}{(0.500)}$$

$$[H^+] = 2.2 \times 10^{-10} \ M \qquad pH = 9.65$$

We now take the system out of equilibrium by adding 75.0 mL of 2.00 M NaOH (0.150 mol of OH^- added). The OH^- will react with the weakly acidic ammonium ion to give more ammonia. Given the amount of OH^- added, what is the limiting reactant?

$$NH_4^+(aq) + OH^-(aq) \rightleftharpoons NH_3(aq) + H_2O(l) \qquad K = 5.6 \times 10^4$$

moles initial	0.0400		0.100
moles added		0.150	
change	−0.0400	−0.0400	+0.0400
moles at equilibrium	≈0	0.110	0.140

NH_4^+ is the limiting reactant, and we will have 0.110 mol of OH^- and 0.140 mol of NH_3 in excess. The $[OH^-]$ will determine the final pH because it is a much stronger base than NH_3. The total solution volume consists of the 200.0 mL with which we started and the 75.0 mL of strong base that we added, for a total of 275.0 mL.

$$[OH^-] = \frac{(0.0110 \text{ mol } OH^-)}{(0.2750 \text{ L } OH^- \text{ solution})} = 0.0400 \ M$$

$$pOH = 1.40 \qquad pH = 12.60$$

The excess of OH^-, and the resulting sharp increase in the pH, indicate that we have exceeded the buffer capacity.

PRACTICE 18.6

What is the pH of the ammonia–ammonium ion buffer in this exercise after the addition of 30.0 mL of 0.100 M HCl? ... after the addition of 50.0 mL of 1.50 M HCl?

See Problems 29 and 30.

EXERCISE 18.7   **Keeping the pH Within Specified Limits**

Our goal in buffer preparation is often to keep the pH within specific limits upon the addition of a strong acid or base. How many milliliters of 0.100 M HCl may be added to 100.0 mL of a buffer containing 0.150 M each of formic acid (HCOOH) and formate ion ($HCOO^-$) so that the pH will not change by more than 0.20?

First Thoughts

Our goal is to find out how many moles of HCl we can add to the system. We can then convert to milliliters of HCl via molarity. We can add only as many moles of HCl as will change the pH by only 0.20 unit. We can find this by determining the initial pH and the final pH and then calculating how the amounts of formic acid and formate ion change.

Solution

The initial amounts of formic acid and formate ion are equal:

$$\text{mol HCOOH} = \text{mol HCOO}^- = 0.1000 \text{ L} \times \frac{(0.150 \text{ mol})}{(1.00 \text{ L})} = 0.0150 \text{ mol of each}$$

$$[H^+] = \frac{K_a[\text{HCOOH}]}{[\text{HCOO}^-]} = \frac{1.8 \times 10^{-4}(0.0150)}{(0.0150)}$$

$$[H^+] = K_a = 1.8 \times 10^{-4} M$$

$$pH = pK_a = 3.74$$

When we add HCl solution, the pH is not to drop below 3.54 (0.20 from the original pH). The concentration of H^+ at this pH is $[H^+] = 10^{-3.54} = 2.88 \times 10^{-4} M$. We can solve for the ratio of acid to base:

$$\frac{[H^+]}{K_a} = \frac{2.88 \times 10^{-4}}{1.8 \times 10^{-4}} = \frac{[\text{HCOOH}]}{[\text{HCOO}^-]} = \frac{1.60}{1}$$

Therefore, retaining an extra figure in this immediate calculation,

$$[\text{HCOOH}] = 1.60[\text{HCOO}^-]$$

Because the volumes are equal,

$$\text{mol HCOOH} = 1.60(\text{mol HCOO}^-)$$

The total amount of formic acid and formate ion equals 0.0300 mol (it was originally 0.01500 mol each). We can substitute for HCOOH and solve.

$$1.60(\text{HCOO}^-) + \text{HCOO}^- = 0.0300 \text{ mol}$$

$$2.60(\text{HCOO}^-) = 0.0300 \text{ mol}$$

$$\text{mol HCOO}^- = \frac{0.0300 \text{ mol}}{2.60} = 0.0115 \text{ mol}$$

$$\text{mol HCOOH} = 1.60(0.0115 \text{ mol HCOO}^-) = 0.0184 \text{ mol}$$

Because our original amounts of HCOOH and HCOO⁻ were 0.01500 mol each, we can have an addition of 0.0035 mol of HCl and still have the pH stay within 0.20 of the original pH. We can now determine the maximum HCl solution volume.

$$\text{mL HCl} = 0.0035 \text{ mol HCl} \times \frac{1.000 \text{ L}}{0.1000 \text{ mol}} = 0.035 \text{ L} = 35 \text{ mL}$$

Further Insights

We see the importance of working in moles with buffer calculations. Keep in mind that this is possible only because the volumes of the conjugates cancel. This will be the case only when we are dealing with buffers.

PRACTICE 18.7

Solve the same problem using the Henderson–Hasselbalch equation instead of the mass-action expression for the buffer.

See Problems 29 and 30. ■

In Summary: What are the criteria for a suitable buffer?

A buffer should not react with the system it is buffering.

■ The pK_a of a buffer should be as close as possible to the pH you want to maintain.

■ The buffering capacity of a buffer must be sufficient to accommodate the addition of a strong acid or a strong base.

Food for Thought:
Are Strong Acids and Bases Buffers?

A buffer, as we noted before, is a mixture of a weak acid and its conjugate base or of a weak base and its conjugate acid. Is it possible that a solution containing only a strong acid (such as HCl) or a strong base (such as NaOH) could be a buffer? For example, 1 L of 0.10 M HCl has a pH of 1.0. The solution pH remains close to 1.0 even when some NaOH is added. Similarly, the pH of 0.10 M NaOH remains close to 13.0 even when some strong acid is added. In that sense, they meet the criterion of keeping the pH of the solution fairly constant upon addition of strong acid or base. However, buffers should also keep their pH constant when diluted, and this is where strong acids and bases fail as buffers. Dilution of a strong acid or a strong base solution causes the pH to change by roughly 1 unit with every 10-fold dilution. On the other hand, buffers made from conjugate acid–base pairs show very little change in pH with dilution. This is critical in biochemical processes, which require a fairly constant pH whatever the solution concentration.

The importance of buffers in medicine is illustrated by a class of drugs called antacids. These drugs are compounds that neutralize gastric acid and exert a buffering effect in the stomach for temporary relief of acid indigestion. Similarly, magnesium carbonate is added to certain brands of aspirin specifically to alter stomach pH. The magnesium carbonate, by neutralizing the "gastric juice," increases stomach pH. The net result is that the aspirin exists predominantly in its ionized form, as shown in Figure 18.5, and reduces the risk of aspirin-induced stomach bleeding or ulcers. Although this is desirable in some cases, it can reduce the amount of drug absorption. Unfortunately, antacids alone may also raise the pH of the stomach above the pK_a of other drugs, resulting in their ionization and in reduced absorption through the stomach. This reduced absorption means that the drug treatment is not effective. Consequently, many drugs bear the warning

NanoWorld / MacroWorld

Big effects of the very small: Buffers in biochemical studies and medicine

Enzymes, which catalyze nearly all the chemical reactions that occur in living organisms, are often highly sensitive to the hydrogen ion concentration of the environment in which they are found. Slight increases or decreases in pH can have a dramatic effect on an enzyme's ability to carry out its unique function. Consequently, living organisms typically have buffering systems in place to maintain a relatively constant pH. Technological advances in biochemistry and molecular biology have enabled chemists to prepare, purify, and study just about any enzyme they desire outside of its normal cellular environment, as long as they can maintain the pH at around 7. For example, members of the class of enzymes called cytochrome P450 are chemically active only between pH 6.5 and pH 8.5, with optimum activities usually occurring between pH 6.8 and pH 7.5. These enzymes are present in the human liver, where they help the body rid itself of foreign chemicals, including most pharmaceuticals.

In 1966, Norman Good and coworkers reported on the design of a dozen new buffers for use in biological research. These so-called **Good (or Good's) buffers** are now widely used, because they are fairly chemically stable in the presence of enzymes or visible light and do not interact with biological compounds. Moreover, they are easy to prepare. Several of the Good buffers (those shown with asterisks in Table 18.2) and boric acid are among the buffers most commonly used in the study of enzyme behavior. Their abbreviated names, structures, and pK_a values are listed in Table 18.2. An additional set of highly effective biological buffers were synthesized in the late 1990s and are now commercially available. Not too surprisingly, they are called "better buffers."

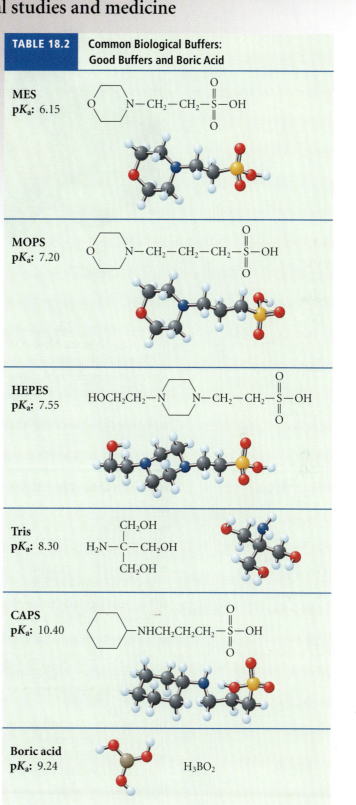

TABLE 18.2   **Common Biological Buffers: Good Buffers and Boric Acid**

MES
pK_a: 6.15

MOPS
pK_a: 7.20

HEPES
pK_a: 7.55

Tris
pK_a: 8.30

CAPS
pK_a: 10.40

Boric acid
pK_a: 9.24 H_3BO_2

FIGURE 18.5

The neutral and ionic forms of aspirin.

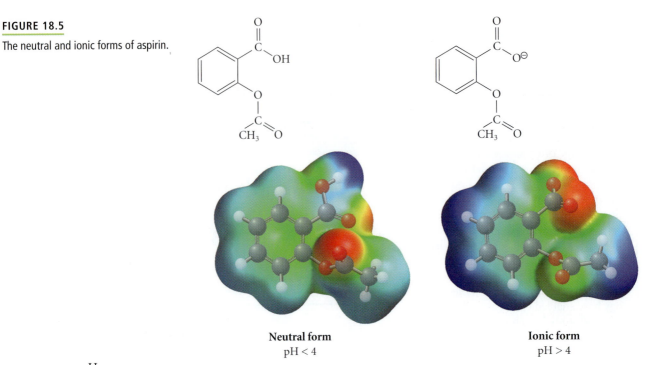

Neutral form
pH < 4

Ionic form
pH > 4

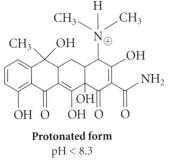

Protonated form
pH < 8.3

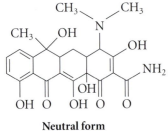

Neutral form
pH > 8.3

FIGURE 18.6

The ionic and neutral forms of tetracycline.

"Do not take antacids containing (hydroxides of) aluminum, calcium, or magnesium (e.g., Mylanta, Maalox) while taking this medication."

The buffering effects of antacids and other buffered medicines can give rise to a number of undesirable drug interactions with prescription medications. Tetracycline, for example, is an antibiotic drug that is absorbed primarily through the stomach lining in its protonated form, shown in Figure 18.6, thanks to the normally low pH of the stomach.

HERE'S WHAT WE KNOW SO FAR

- Many reactions are pH-sensitive and require buffers to control pH.
- A buffer resists change in pH upon addition of a strong acid or a strong base.
- The pH of a buffer is not changed when the solution is diluted.
- Buffers are typically composed of weak conjugate acid–base pairs.
- We can solve for the pH of buffers in a straightforward way by recognizing the importance of Le Châtelier's principle and the common-ion effect.
- We can calculate the approximate ratio of conjugate acid to base in order to prepare a buffer of a known pH.
- Solving for the pH of a buffer upon addition of strong acid or base is really solving a limiting-reactant problem.
- It is possible to exceed the buffer capacity, in which case the pH will move sharply higher (with excess base) or sharply lower (with excess acid).

We have seen that buffers are used to maintain the pH of all kinds of systems, ranging from municipal scrubbers to your body. Our overarching application of aqueous equilibria in this chapter is analysis of calcium in hard water via titration with EDTA. We are one step closer to completing our task. We next focus on titrations.

18.2   Acid–Base Titrations

We discussed the wide range of titrations, of which acid–base titration is one important type, in the opening section of this chapter. A small sample of the commercial, environmental and biological uses of acid–base titrations includes analysis of the acidity of food and drink, determination of the pH of water supplies, measurement of the solubility of pharmaceuticals, determination of amino acids in blood, and determination of the acidity or basicity (called the "total acid" or "total base" number) of motor oils.

In the lab, we typically set up an acid–base titration by monitoring the pH as shown in Figure 18.7. This normally includes a buret to accurately measure the volume of titrant delivered, a beaker or flask, and a calibrated pH meter. Industrial laboratories often use automated titrators to increase efficiency. The typical acid–base titrations fall into one of these main categories:

- strong-acid–strong-base titrations
- strong-acid–weak-base titrations
- weak-acid–strong-base titrations

A fourth type, weak-acid–weak-base titrations, is typically not used because the equilibrium constant for the overall reaction is not nearly as large as with the other systems, and the indication of the end of the titration is too gradual to tell us when the titration is complete.

FIGURE 18.7

A typical set-up for an acid–base titration monitored by a pH meter includes a buret to accurately measure the volume of titrant delivered, a beaker, and the calibrated pH meter. An automatic titrator is used when there are many titrations to be done.

Strong-Acid–Strong-Base Titrations

The determination of HCl molarity based on titration with NaOH is a common process throughout all levels of chemistry and from the academic to the industrial laboratory setting. Let's examine *the changes* that take place during a strong-acid–strong-base titration by assuming that we wish to titrate 50.00 mL of 0.1000 *M* HCl by adding known amounts of 0.2000 *M* NaOH. The results of the titration are graphically shown in Figures 18.8a–f, which illustrates the relationship between pH and volume of OH^- added.

Visualization: Acid–Base Titration

Video Lesson: Strong-Acid–Strong-Base Titration

Video Lesson: CIA Demonstration: Barium Hydroxide-Sulfuric Acid Titration

Tutorial: Titration Curves: Strong Acid with Strong Base

Part 1: Initial pH

We can calculate the pH of the initial 0.1000 *M* solution of HCl as we would that of any other strong acid:

$$pH = -\log[H^+] = -\log(0.1000) = 1.0$$

We enter this on the graph to the right (Figure 18.8a).

Part 2: Addition of 5.00 mL of NaOH solution

What will be the reaction of the strong acid with the strong base? We can write the reaction in molecular form:

$$HCl(aq) + NaOH(aq) \rightleftharpoons H_2O(l) + NaCl(aq)$$

However, the net ionic form gives a better sense of what is going on in the solution:

$$H^+(aq) + OH^-(aq) \rightleftharpoons H_2O(l) \qquad K = 1.0 \times 10^{14}$$

The equilibrium constant is very high and the reaction is fast, both good features to have when doing a titration. What, and how much, will be left over after addition of the NaOH titrant to the HCl solution? *This is a limiting-reactant problem*, and we can use the same type of table that we used when discussing

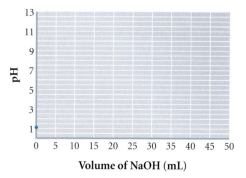

Volume of NaOH (mL)

FIGURE 18.8a

Each plot follows the pH changes as we add 0.2000 *M* NaOH solution to 50.00 mL of 0.1000 *M* HCl solution. The initial pH is shown here, followed in turn by the pH after addition of the listed volumes.

buffers to help us sort it all out. For example, the initial concentration of our strong acid is

$$[H^+] = 0.1000\ M = \frac{(0.1000\ \text{mol})}{(1.000\ \text{L})} = \frac{(0.1000\ \text{mmol})}{(1.000\ \text{mL})}$$

How much H^+ is in the solution? Using the concentration in moles per liter and the volume in liters, we find that

$$\frac{(0.1000\ \text{mol})}{(1.000\ \text{L})} \times 0.05000\ \text{L} = 0.005000\ \text{mol}\ H^+$$

We can calculate the amount of strong base, OH^-, added to the solution in the same way.

$$\frac{(0.2000\ \text{mol})}{(1.000\ \text{L})} \times 0.00500\ \text{L} = 0.001000\ \text{mol}\ OH^-$$

Putting our values in tabular format shows that we still have plenty of strong acid in the solution. Note that because water is the solvent, it will not enter into the mass-action expression, and we can ignore it in our calculations.

$$H^+(aq) + OH^-(aq) \rightleftharpoons H_2O(l) \qquad K = 1.0 \times 10^{14}$$

moles initial	0.005000	
moles added		0.001000
change	−0.001000	−0.001000
moles at equilibrium	0.004000	≈ 0

FIGURE 18.8b

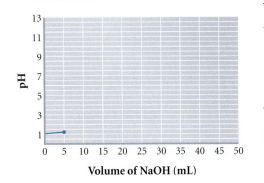

Volume of NaOH (mL)

We can calculate the pH, keeping in mind the total solution volume of 55.00 mL (50.00 mL of HCl solution + 5.00 mL of added NaOH solution).

$$[H^+] = \frac{(0.004000\ \text{mol})}{(0.05500\ \text{L})} = 0.0727\ M$$

$$pH = 1.14$$

The addition of the strong acid has raised the pH, but not much, because we still have plenty of excess acid. We have entered the data onto our graph in Figure 18.8b.

Part 3: Addition of a total of 12.50 mL of NaOH solution

Half of the H^+ is neutralized at this point.

$$H^+(aq) + OH^-(aq) \rightleftharpoons H_2O(l) \qquad K = 1.0 \times 10^{14}$$

moles initial	0.005000	
moles added		0.002500
change	−0.002500	−0.002500
moles at equilibrium	0.002500	≈ 0

The total volume is 62.50 mL (50.00 mL of HCl solution + 12.50 mL of NaOH solution), which results in a pH of 1.4, as shown here and in Figure 18.8c.

$$[H^+] = \frac{(0.0025000\ \text{mol})}{(0.06250\ \text{L})} = 0.0400\ M$$

$$pH = 1.40$$

FIGURE 18.8c

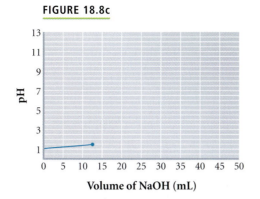

Part 4: Addition of a total of 24.00 mL of NaOH solution

By the time we add 24.00 mL of 0.2000 M NaOH to the acid, we have neutralized nearly all of the hydrogen ion, as shown in the following data table.

$$H^+(aq) + OH^-(aq) \rightleftharpoons H_2O(l) \quad K = 1.0 \times 10^{14}$$

moles initial	0.005000	
moles added		0.004800
change	−0.004800	−0.004800
moles at equilibrium	0.000200	≈ 0

The total volume is 74.00 mL (50.00 mL of HCl solution + 24.00 mL of NaOH solution), which results in a pH of 2.57, as shown.

$$[H^+] = \frac{(0.000200\ \text{mol})}{(0.07400\ \text{L})} = 2.70 \times 10^{-3}\ M$$

$$pH = 2.57$$

Adding a bit more, so that the volume of base added is 24.95 mL, results in a pH of 3.87, shown in Figure 18.8d. As we get very close to neutralizing the acid, the pH starts to rise sharply.

FIGURE 18.8d

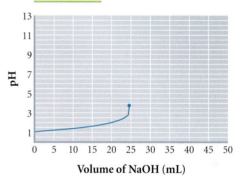

Part 5: Addition of a total of 25.00 mL of NaOH solution

At this point, all of the strong acid has been neutralized by the strong base. This is called the **equivalence point** of the titration, the exact point at which the reactant has been neutralized by the titrant.

$$H^+(aq) + OH^-(aq) \rightleftharpoons H_2O(l) \quad K = 1.0 \times 10^{14}$$

moles initial	0.005000	
moles added		0.005000
change	−0.005000	−0.005000
moles at equilibrium	≈ 0	≈ 0

We say that there are "≈ 0" hydrogen and hydroxide ions in the solution. *This really says that the amount is insignificant when compared to the original amounts of acid and base that were mixed.* How much is "≈ 0" in this neutral solution?

We have 75.00 mL of (ideally) pure water at equilibrium. The only reaction that is important in our solution at this point is the autoprotolysis of water. Finally, the equilibrium constant for this reaction has become very important.

FIGURE 18.8e

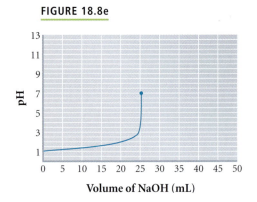

Volume of NaOH (mL)

Using the equation for this reaction, and the corresponding mass-action expression, we can solve for the concentration of H^+ in the solution.

$$H_2O(l) \rightleftharpoons H^+(aq) + OH^-(aq) \qquad K_w = 1.0 \times 10^{-14}$$

$$[H^+] = [OH^-] = 1.0 \times 10^{-7} M$$

$$pH = 7.00$$

The answer to our question here is $[H^+] = 1.0 \times 10^{-7} M$, which is insignificant compared to the initial concentration of strong acid and base. However, *in the absence of other sources of* H^+, *this is very significant*. The solution now has a neutral pH, as shown in Figure 18.8e. Note the sharp rise to the equivalence point, which makes it easy to identify.

Part 6: Addition of a total of 40.00 mL of NaOH solution

At this point, the additional strong base is just being added to water (neglecting the spectator ions Na^+ and Cl^-), so we would expect the pH to rise sharply. We are adding 15.00 mL of 0.2000 M NaOH, or 0.00300 mol, past the equivalence point, and our pH is calculated as shown for the total solution volume of 90.00 mL (50.00 mL of HCl solution + 40.00 mL of NaOH added).

$$[OH^-] = \frac{(0.003000 \text{ mol})}{(0.09000 \text{ L})} = 0.03333 \, M$$

$$pOH = 1.48$$

$$pH = 12.52$$

FIGURE 18.8f

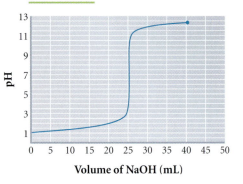

Volume of NaOH (mL)

The solution has become quite basic. The complete strong-acid–strong-base **titration curve** is shown in Figure 18.8f.

Summarizing, we have shown that a strong base will only slightly increase the pH of a strong acid until very close to the equivalence point, where it will rise sharply to pH = 7. After the equivalence point, the pH will continue its sharp increase to a point that depends on the concentration of the strong-base titrant.

A word of caution: *The pH at the equivalence point will equal 7 only in monoprotic strong-acid–strong-base titrations.* We will show why this is so immediately after Exercise 18.8.

EXERCISE 18.8	**Titrating Sodium Hydroxide with Hydrochloric Acid**

Calculate and draw a graph showing the relationship between the pH and the volume of HCl for the titration of 25.00 mL of 0.2500 M NaOH with the following total volumes of 0.1250 M HCl: a. 0 mL; b. 20.00 mL; c. 49.80 mL; d. 50.00 mL; e. 60.00 mL.

Solution

The net ionic reaction is the same as in the addition of NaOH to HCl, except that the reactant (now OH^-) and the titrant (now H^+) have switched roles.

$$OH^-(aq) + H^+(aq) \rightleftharpoons H_2O(l) \qquad K = 1.0 \times 10^{14}$$

a. 0 mL of acid added: The pH of this strong base can be found from the hydroxide ion concentration, which is equal to the initial concentration of the NaOH solution.

$$[OH^-] = 0.2500 \, M \qquad pOH = 0.60 \qquad pH = 13.40$$

b. 20.00 mL of acid added: We can use the same table setup that we used previously to calculate the moles of each after reaction. There are initially

$$\frac{(0.2500 \text{ mol})}{(1 \text{ L})} \times 0.02500 \text{ L} = 0.006250 \text{ mol OH}^-$$

in the solution. We add

$$\frac{(0.1250 \text{ mol})}{(1.000 \text{ L})} \times 0.02000 \text{ L} = 0.002500 \text{ mol of H}^+$$

in a total solution volume of 45.00 mL (25.00 mL of base + 20.00 mL of acid) = 0.04500 L.

$$\text{OH}^-(aq) + \text{H}^+(aq) \rightleftharpoons \text{H}_2\text{O}(l) \qquad K = 1.0 \times 10^{14}$$

moles initial	0.006250	
moles added		0.002500
change	−0.002500	−0.002500
moles at equilibrium	0.003750	≈ 0

$$[\text{OH}^-] = \frac{(0.003750 \text{ mol})}{(0.04500 \text{ L})} = 0.08333 \text{ } M$$

$$\text{pOH} = 1.08$$

$$\text{pH} = 12.92$$

c. 49.8 mL of acid added: Using the same type of calculations, we find that

$$\text{OH}^-(aq) + \text{H}^+(aq) \rightleftharpoons \text{H}_2\text{O}(l) \qquad K = 1.0 \times 10^{14}$$

moles initial	0.006250	
moles added		0.006225
change	−0.006225	−0.006225
moles at equilibrium	0.000025	≈ 0

$$[\text{OH}^-] = \frac{(0.0000250 \text{ mol})}{(0.07480 \text{ L})} = 3.34 \times 10^{-4} \text{ } M$$

$$\text{pOH} = 3.48$$

$$\text{pH} = 10.52$$

d. 50.00 mL of acid added:

$$\text{OH}^-(aq) + \text{H}^+(aq) \rightleftharpoons \text{H}_2\text{O}(l) \qquad K = 1.0 \times 10^{14}$$

moles initial	0.006250	
moles added		0.006250
change	−0.006250	−0.006250
moles at equilibrium	≈ 0	≈ 0

The solution is at the equivalence point, and pH = 7.0 for a strong-base–strong-acid titration.

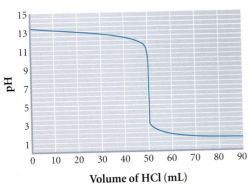

FIGURE 18.9

The titration curve for the addition of 0.125 M HCl to 25.00 mL of 0.250 M NaOH.

e. 60.00 mL of acid added: We have added 10.00 mL of excess acid, which is a total of 0.00125 mol. The total solution volume is 85.00 mL (25.00 mL of base + 60.00 mL of added acid) = 0.08500 L. We can calculate the pH from this information.

$$[H^+] = \frac{(0.001250 \text{ mol})}{(0.08500 \text{ L})} = 0.01471 \ M$$

$$pH = 1.83$$

The titration curve shown in Figure 18.9 has a shape similar to what we expected. It is fairly flat until close to the equivalence point, where it drops quite sharply and over a large pH range, so it is easily detectable.

PRACTICE 18.8

Calculate and draw a graph of the relationship between the pH and the volume of NaOH solution for the titration of 25.00 mL of 0.2500 M HCl with the following total volumes of 0.2500 M NaOH:

a. 0 mL   d. 25.00 mL
b. 10.00 mL   e. 30.00 mL
c. 20.00 mL   f. 40.00 mL

See Problems 33, 34, and 37.

Acid–Base Titrations in Which One Component Is Weak and One Is Strong

FIGURE 18.10a

Each plot follows the pH changes as we add 0.2000 M NaOH solution to 50.00 mL of 0.1000 M acetic acid solution. The pH values at each volume are superimposed on those from Figure 18.9 to show the difference in the nature of the titration curve between the strong acid and the weak acid.

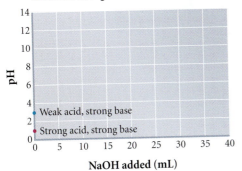

Although the general problem-solving strategy is the same when we titrate a weak acid with a strong base (or a weak base with a strong acid) as when we perform a strong-acid–strong-base titration, there is a critical difference, the K_a (or K_b) of the analyte, that affects both the pH and the pH change at the equivalence point. *Our ability to do a titration analysis depends on having a large, sharp break in the equivalence point.* We can best see this by comparing the data from our HCl–NaOH titration with those from the titration of 50.00 mL of a 0.1000 M acetic acid solution (CH$_3$COOH, $K_a = 1.8 \times 10^{-5}$) with a 0.2000 M sodium hydroxide solution. The amounts and concentrations are the same. Only the acid has been changed from strong to weak.

Part 1: Initial pH

We can determine pH in this weak acid as we would in any other weak acid, using principles we learned in Chapter 17.

$$CH_3COOH(aq) \rightleftharpoons CH_3COO^-(aq) + H^+(aq) \qquad K_a = 1.8 \times 10^{-5}$$

$$K_a = \frac{[H^+][CH_3COO^-]}{[CH_3COOH]} \qquad 1.8 \times 10^{-5} = \frac{x^2}{(0.1000)}$$

$$x = [CH_3COO^-] = [H^+] = 1.3 \times 10^{-3} \ M$$

$$pH = 2.87$$

We already see a difference in the titration curve (Figure 18.10a) because this weak acid pH is nearly 2 units higher than the initial pH of the strong acid, HCl, of the same concentration.

Part 2: Addition of 5.00 mL of NaOH solution

We will use the same thinking—and ask the same questions—to assess the chemistry here as in the strong-acid–strong-base titration. We begin with

$$0.005000 \text{ mol of acetic acid} \left(50.00 \text{ mL} \times \frac{0.1000 \text{ mol}}{L} \right),$$

to which we add 0.001000 mol of OH^- $\left(5.000 \text{ mL} \times \frac{0.2000 \text{ mol}}{L} \right).$

Question 1: *What will be the reaction of the acetic acid with the NaOH solution?* The net ionic form of the reaction is useful because spectator ions are not part of the acid–base chemistry.

$$CH_3COOH(aq) + OH^-(aq) \rightleftharpoons CH_3COO^-(aq) + H_2O(l) \qquad K = 1.8 \times 10^9$$

The equilibrium constant is the same whether we describe a reaction by giving its molecular or net ionic form. For this reaction, K is quite high and the reaction is fast.

Question 2: *What, and how much, will be left over after addition of the titrant?* We can use a table to clarify the amounts involved when the system reaches its new equilibrium position.

$$CH_3COOH(aq) + OH^-(aq) \rightleftharpoons CH_3COO^-(aq) + H_2O(l)$$

moles initial	0.005000		0
moles added		0.001000	
change	−0.001000	−0.001000	+0.001000
moles at equilibrium	0.004000	≈ 0	0.001000

Because we have both a weak acid and its conjugate base, *we have produced a buffer,* and we can solve for the pH as with any buffer system. Note that we have substituted moles into the equation instead of molarities because the total volume of the solution will cancel out.

$$[H^+] = \frac{K_a[CH_3COOH]}{[CH_3COO^-]} = \frac{(1.8 \times 10^{-5})(0.004000)}{(0.001000)} = 7.2 \times 10^{-5} M$$

$$pH = 4.14$$

We have entered the data point onto our graph in Figure 18.10b. How is the curve shaping up compared to that of the strong-acid–strong-base titration? Because the solution is a buffer at this point, this part of the titration is called the **buffer region**. We expect the pH to be relatively constant within the buffer region.

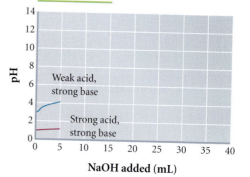

FIGURE 18.10b

Part 3: Addition of 12.50 mL of NaOH solution

Proceeding as we did in part b, we can generate the following table and solve for pH. Again, in our calculation of the hydrogen ion concentration, we needn't worry about the volumes of acid and base because they will cancel out during our calculation.

$$CH_3COOH(aq) + OH^-(aq) \rightleftharpoons CH_3COO^-(aq) + H_2O(l)$$

moles initial	0.005000		0
moles added		0.002500	
change	−0.002500	−0.002500	+0.002500
moles at equilibrium	0.002500	≈ 0	0.002500

FIGURE 18.10c

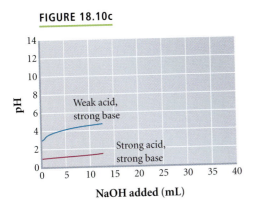

NaOH added (mL)

$$[H^+] = \frac{K_a[CH_3COOH]}{[CH_3COO^-]} = \frac{(1.8 \times 10^{-5})(0.002500)}{(0.002500)} = 1.8 \times 10^{-5}\ M$$

$$pH = 4.74$$

We are still in the buffer region, having neutralized precisely one-half of the acetic acid and forming an equal amount of acetate ion, the conjugate base. This is called the **titration midpoint**, at which the pH equals the pK_a of acetic acid. It is also possible to work backward, finding the pK_a of an acid from pH at the titration midpoint. As before, we have entered the data point onto our graph (see Figure 18.10c). How is the curve shaping up compared to that of the strong-acid–strong-base titration?

Part 4: Addition of a total of 24.00 mL of NaOH solution

Continuing, we generate the following table, which shows that we are still—though barely—in the buffer region. The pH is beginning to rise on the way to the equivalence point.

$$CH_3COOH(aq) + OH^-(aq) \rightleftharpoons CH_3COO^-(aq) + H_2O(l)$$

moles initial	0.005000		≈ 0
moles added		0.004800	
change	−0.004800	−0.004800	+0.004800
moles at equilibrium	0.000200	≈ 0	0.004800

$$[H^+] = \frac{K_a[CH_3COOH]}{[CH_3COO^-]} = \frac{(1.8 \times 10^{-5})(0.000200)}{(0.004800)} = 7.5 \times 10^{-7}\ M$$

$$pH = 6.12$$

Continue to keep an eye on the data comparison (Figure 18.10d) between this titration and the strong-acid–strong-base titration.

FIGURE 18.10d

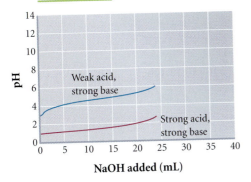

NaOH added (mL)

Part 5: Addition of a total of 25.00 mL of NaOH solution

The completed table shows that we are at the equivalence point. We converted all of the acetic acid to acetate ion.

$$CH_3COOH(aq) + OH^-(aq) \rightleftharpoons CH_3COO^-(aq) + H_2O(l)$$

moles initial	0.005000		≈ 0
moles added		0.005000	
change	−0.005000	−0.005000	+0.005000
moles at equilibrium	≈ 0	≈ 0	0.005000

How do we calculate the pH of this solution? To answer that, we need to go back to our most important questions: "What is in solution?" and "What equation describes their behavior?"

The answer to the first question is "0.005000 mol of acetate ion in a total of 75.00 mL of solution." Using the chemist's shorthand,

$$[CH_3COO^-] = \frac{(0.005000 \text{ mol})}{(0.07500 \text{ L})} = 0.06667 \text{ } M$$

We have neutralized all of the weak acid, and we now have a solution of its conjugate base. This acetate ion will hydrolyze, if only barely, to acetic acid.

$$CH_3COO^-(aq) + H_2O(l) \rightleftharpoons CH_3COOH(aq) + OH^-(aq)$$

We can calculate the equilibrium constant for the hydrolysis of this conjugate base of acetic acid, K_b, from K_w and the acetic acid K_a.

$$K_b = \frac{K_w}{K_a} = \frac{1.0 \times 10^{-14}}{1.8 \times 10^{-5}} = 5.6 \times 10^{-10}$$

We can now solve for the pH of the weak base.

$$K_b = \frac{[OH^-][CH_3COOH]}{[CH_3COO^-]}$$

$$5.6 \times 10^{-10} = \frac{[x^2]}{[0.06667]} \qquad x = [CH_3COOH] = [OH^-] = 6.1 \times 10^{-6} \text{ } M$$

$$pOH = 5.21$$

$$pH = 8.79$$

The pH of the weak base is, in fact, basic, as we would predict. Compare this to the pH = 7 equivalence point of the strong-acid–strong-base titration. There, only water was present at the equivalence point. In this titration, however, we have an equivalence point solution of a weak base. The pH for this titration did rise sharply near the equivalence point, making it easy to determine, as shown in Figure 18.10e.

FIGURE 18.10e

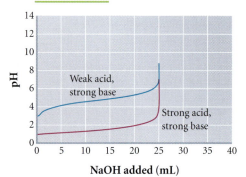

Part 6: Addition of a total of 40.00 mL of NaOH solution

We now add a 0.2000 M solution of a strong base to a very weak base. The weak base will not be an important contributor to the total hydroxide ion concentration because it is so weak. The hydroxide ion concentration will be strictly determined by the excess moles of OH^- and the total solution volume in which it is contained.

$$\text{excess mol } OH^- = 0.01500 \text{ L} \times \frac{(0.2000 \text{ mol})}{(1.000 \text{ L})} = 0.003000 \text{ mol}$$

$$[OH^-] = \frac{(0.003000 \text{ mol})}{(0.09000 \text{ L})} = 0.03333 \text{ } M$$

$$pOH = 1.48$$

$$pH = 12.52$$

FIGURE 18.10f

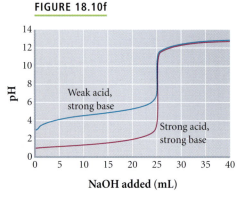

The pH has increased sharply. Compare this data point with that from the equivalent volume of strong base added to the strong acid in the previous titration, shown in Figure 18.10f. They are the same! In both cases, after the equivalence point, we added the same volume of the same concentration of strong base.

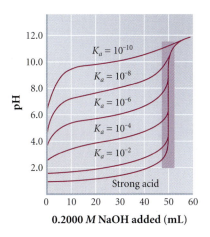

FIGURE 18.11

This plot shows the relationship between K_a value for 50.00 mL of a 0.2000 M acid and the change in pH at the equivalence point when titrated with a 0.2000 M NaOH solution. Notice two items in the plot. First, the concentration of the acid, not its relative strength, determines the equivalence point. Second, the larger the K_a value, the sharper the pH break at the equivalence point.

Summarizing the Key Ideas of the Weak-Acid–Strong-Base Titration Discussion

There are four key areas that define the titration curve and the information it yields.

Initial pH: This will be closer to neutral for weaker acids, further away for stronger acids.

Buffer region: This is where we have generated enough conjugate base, while still having the weak acid available to produce a buffer. This part of the titration curve will have a gently sloping pH until quite near the equivalence point. An important point in the buffer region is the titration midpoint (pH = pK_a). There is an inflection point at the titration midpoint that we do not see with strong-acid–strong-base titrations (compare Figures 18.9 and 18.11 in this regard).

Equivalence point: The starting weak acid has been exactly neutralized, and the solution is a weak base. The stronger the acid (the larger the K_a value), the sharper will be the break at the equivalence point. Figure 18.11 shows that when the K_a value of the weak acid is too small, the pH break at the equivalence point is too small to be of practical use.

Post-equivalence point: Excess strong base sharply raises the pH of the system.

EXERCISE 18.9 **Titrating a Weak Base with a Strong Acid**

To further compare systems, calculate the pH values and draw the curve for the titration of 50.00 mL of 0.1000 M ammonia with 0.2000 M of hydrochloric acid using the same volumes of titrant as in our previous titrations:

a. 0 mL d. 24.00 mL
b. 5.00 mL e. 25.00 mL
c. 12.50 mL f. 40.00 mL

Solution

a. Initial pH. The important reaction is the hydrolysis of ammonia.

$$NH_3(aq) + H_2O(l) \rightleftharpoons NH_4^+(aq) + OH^-(aq) \qquad K_b = 1.8 \times 10^{-5}$$

$$K_b = \frac{[NH_4^+][OH^-]}{[NH_3]}$$

$$1.8 \times 10^{-5} = \frac{x^2}{0.1000}$$

$$x = [OH^-] = 1.34 \times 10^{-3} \, M$$

$$pOH = 2.87$$

$$pH = 11.13$$

b. *Addition of 5.00 mL of HCl solution.* The system is thrown out of equilibrium by the addition of the HCl solution, much as the acetic acid system had to compensate for the addition of NaOH. We will generate a buffer, shown in the table.

$$NH_3(aq) + H^+(aq) \rightleftharpoons NH_4^+(aq) \qquad K_b = 1.8 \times 10^9$$

moles initial	0.005000		≈ 0
moles added		0.001000	
change	−0.001000	−0.001000	+0.001000
moles at equilibrium	0.004000	≈ 0	0.001000

When the system returns to equilibrium,

$$NH_3(aq) \rightleftharpoons NH_4^+(aq) + OH^-(aq) \qquad K_b = 1.8 \times 10^{-5}$$

$$[OH^-] = \frac{K_b[NH_3]}{[NH_4^+]} = \frac{(1.8 \times 10^{-5})(0.004000)}{(0.001000)} = 7.2 \times 10^{-5}\ M$$

$$pOH = 4.14$$

$$pH = 9.86$$

c. *Addition of 12.50 mL of HCl solution.* We can generate the following table and solve for pH.

$$NH_3(aq) + H^+(aq) \rightleftharpoons NH_4^+(aq) \qquad K_b = 1.8 \times 10^9$$

moles initial	0.005000		≈ 0
moles added		0.002500	
change	−0.002500	−0.002500	+0.002500
moles at equilibrium	0.002500	≈ 0	0.002500

When the system returns to equilibrium,

$$NH_3(aq) \rightleftharpoons NH_4^+(aq) + OH^-(aq) \qquad K_b = 1.8 \times 10^{-5}$$

$$[OH^-] = \frac{K_b[NH_3]}{[NH_4^+]} = \frac{(1.8 \times 10^{-5})(0.002500)}{(0.002500)} = 1.8 \times 10^{-5}\ M$$

$$pOH = pK_b = 4.74$$

$$pH = 9.26$$

d. *Addition of a total of 24.00 mL of HCl solution.* We are nearing the equivalence point. How do the various titration curves, shown in Figure 18.12 on page 798, compare?

$$NH_3(aq) + H^+(aq) \rightleftharpoons NH_4^+(aq) \qquad K = 1.8 \times 10^9$$

moles initial	0.005000		≈ 0
moles added		0.004800	
change	−0.004800	−0.004800	+0.004800
moles at equilibrium	0.000200	≈ 0	0.004800

When the system returns to equilibrium,

$$NH_3(aq) \rightleftharpoons NH_4^+(aq) + OH^-(aq) \qquad K = 1.8 \times 10^{-5}$$

$$[OH^-] = \frac{K_b[NH_3]}{[NH_4^+]} = \frac{(1.8 \times 10^{-5})(0.000200)}{(0.004800)} = 7.5 \times 10^{-7}\ M$$

$$pOH = 6.12$$

$$pH = 7.87$$

e. *Addition of a total of 25.00 mL of HCl solution.* We are at the equivalence point, at which we have added exactly the same number of moles of HCl as there were moles of ammonia at the start of the titration. We now have a solution that is weakly acidic as a consequence of the conversion of ammonia to the ammonium ion.

$$NH_3(aq) + H^+(aq) \rightleftharpoons NH_4^+(aq) \qquad K = 1.8 \times 10^9$$

moles initial	0.005000		≈ 0
moles added		0.005000	
change	−0.005000	−0.005000	+0.005000
moles at equilibrium	≈ 0	≈ 0	0.005000

$$[NH_4^+] = \frac{(0.005000 \text{ mol})}{(0.07500 \text{ L})} = 0.06667\ M$$

$$K_a = \frac{K_w}{K_b} = \frac{1.0 \times 10^{-14}}{1.8 \times 10^{-5}} = 5.6 \times 10^{-10}$$

We can now solve for the pH of the weak acid.

$$K_a = \frac{[H^+][NH_3]}{[NH_4^+]}$$

$$5.6 \times 10^{-10} = \frac{x^2}{0.06667}$$

$$x = [NH_3] = [H^+] = 6.1 \times 10^{-6}\ M$$

$$pH = 5.21$$

f. *Addition of a total of 40.00 mL of HCl solution.*

$$\text{mol excess } H^+ = 0.01500 \text{ L} \times \frac{(0.2000 \text{ mol})}{(1.000 \text{ L})} = 0.003000 \text{ mol}$$

$$[H^+] = \frac{(0.003000 \text{ mol})}{(0.09000 \text{ L})} = 0.03333\ M$$

$$pH = 1.48$$

PRACTICE 18.9

Calculate the pH values and draw the curve for the titration of 25.00 mL of 0.2500 M acetic acid with 0.2500 M sodium hydroxide using the same volumes of titrant as in our previous titrations: a. 0 mL; b. 5.00 mL; c. 12.50 mL; d. 24.00 mL; e. 25.00 mL; f. 40.00 mL. K_a (acetic acid) $= 1.8 \times 10^{-5}$

See Problems 35, 36, and 38.

FIGURE 18.12

This plot shows a comparison of the pH for strong-acid–strong-base, weak-acid–strong-base, and weak-base–strong-acid titrations using equal concentrations of acid and base.

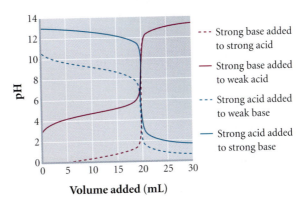

- - - Strong base added to strong acid

— Strong base added to weak acid

- - - Strong acid added to weak base

— Strong acid added to strong base

We have seen that a successful monoprotic acid–base titration has an analyte (what you are titrating) that, when titrated with a strong acid or strong base, has a large K. The reaction is typically fast and reproducible. Polyprotic acid and base titrations are common, and the analysis of their titration curves presents interesting challenges, although we will not deal with these here.

Indicators

The pH meter, discussed in Chapter 19, can provide very accurate readings of the pH. Making successful measurements using a pH meter, however, relies on its having been properly maintained, calibrated, and operated. For speed and simplicity in a titration, the chemical technician often relies on a chemical pH reporter. This class of compounds, known as **acid–base indicators** or **pH indicators**, visually indicates the change in pH as we approach, reach, and pass the equivalence point. Indicators themselves are conjugate acid–base pairs of organic molecules that change color as they change between their acid and base forms. Perhaps the best-known example is phenolphthalein, which changes from colorless to rose-pink as the pH changes from 8.5 to 9.5. Figure 18.13 shows that phenolphthalein actually has several structural changes (and therefore color changes) from very low to very high pH.

Why is phenolphthalein such a popular acid–base indicator? Its color change occurs in the pH region where many acids titrated with strong bases reach their

Video Lesson: Acid–Base Indicators

Video Lesson: CIA Demonstration: Natural Acid–Base Indicators

FIGURE 18.13

The structure of phenolphthalein changes as we increase the pH from very low to very high.

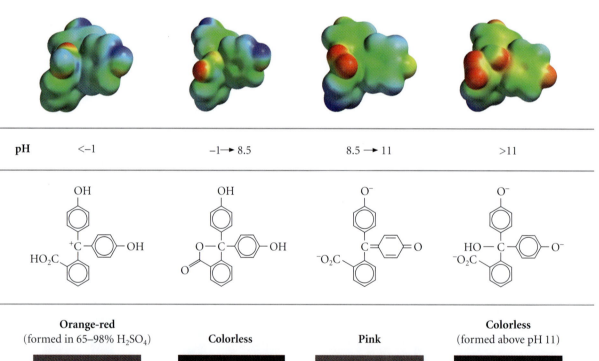

| pH | <−1 | −1 → 8.5 | 8.5 → 11 | >11 |

| **Orange-red** (formed in 65–98% H_2SO_4) | **Colorless** | **Pink** | **Colorless** (formed above pH 11) |

equivalence point. For example, acetic acid has an equivalence point, at which it is essentially all changed to acetate ion, at about pH 9, and the pH goes from about 6 to 11 within a small volume of titrant on either side of the equivalence point. The titration of a strong acid with base (or vice versa) changes very rapidly between the region of pH 6 to 11. An indicator that itself changes color anywhere in this area would, generally speaking, be a suitable reporter of the equivalence point.

In a given titration, how do we know which acid–base indicator to choose? We want to choose an indicator with a pK_a as close as possible to the pH at the equivalence point. A distinct color change is also useful. For example, the pK_a of phenolphthalein is about 9.5. This is well within the region of the pH at the equivalence point of titration of acetic acid by sodium hydroxide (pH ≈ 9, depending on acetic acid concentration). As a rule of thumb, *the color changes of pH indicators are visible to +/−1 pH unit on either side of the indicator pK_a.* This means that phenolphthalein will be completely colorless below a pH of 8.5 and will then be rose-pink from pH 8.5 to about pH 11. It will turn colorless again above pH 11. A selection of common pH indicators, their color changes, and their pH ranges is given in Figure 18.14.

During the chemical technician's analysis, only a few drops of an indicator solution are added to the analyte solution. Why so little? Because the indicator also undergoes an acid–base reaction. This means that in addition to adding the required volume to neutralize the analyte, we add just a bit more to cause the color change in the indicator. For example, if we titrate a solution of acetic acid

FIGURE 18.14

A selection of common pH indicators, their color changes, and their pH ranges. Our key criterion in selecting the proper indicator is that its pK_a should be as close as possible to the equivalence point of the titration in which it is used. Note that most indicators exhibit a color change over less than 2 pH units.

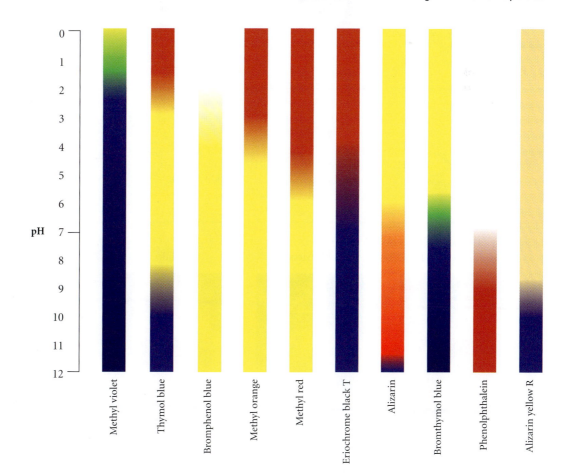

containing phenolphthalein indicator, we might need 35.27 mL of a sodium hydroxide solution to react with the acetic acid itself. Changing the indicator's structure (and therefore its color) might require *an additional 0.02 mL* of the strong base. The equivalence point is at 35.27 mL, but *the point at which you see the change in indicator color that tells you the titration is finished*, which is called the titration endpoint, is at 35.29 mL. Therefore, we would use 35.29 mL as our number for calculating the concentration of analyte. This could lead to very large errors if we had a lot of extra indicator in the solution. These errors are reduced if the endpoint is as close as possible to the equivalence point.

EXERCISE 18.10   **Picking an Indicator**

What would be a reasonable indicator for the titration of 0.10 *M* ammonia with 0.10 *M* HCl?

Solution

The key question is "What is the pH of the solution at the equivalence point?" When essentially all of the ammonia is converted to ammonium ion, the pH will be around 5.2, as we saw in Exercise 18.9, part e. As shown in Figure 18.14, several indicators change color around pH = 5.2, and methyl red appears to be a good choice.

PRACTICE 18.10

What would be a reasonable indicator for the titration of 0.10 *M* NaOH with 0.10 *M* HCl?

See Problems 39–42.

Some natural and commercially prepared indicators are made up of several colorful organic molecules, and they change color throughout the pH range. Notable among the natural indicators are the anthocyanins, which are responsible for most of the different colors found in vegetables and flowers. There are over 150 naturally occurring anthocyanins in foods such as the red cabbage that we cook and serve with dinner. The juice from the red cabbage can therefore be used as an indicator. Figure 18.15 shows the wide range of colors that can be obtained by adjusting the pH of red cabbage juice. Commercially prepared solutions of mixtures of indicators can mimic these color changes, but far more intensely, so far less is required. For instance, the commercially prepared "universal indicator" is a mixture of thymol blue, methyl red, bromthymol blue, and phenolphthalein. Because each indicator gains or loses protons at a different pH, the universal indicator has color changes over a wide pH range, as shown in Figure 18.16.

Application

CHEMICAL ENCOUNTERS:
Anthocyanins and
Universal Indicators

FIGURE 18.15

The spectrum of colors in each sample of red cabbage juice is caused by the changes in structure of the anthocyanins as the pH moves from 1 to 13.

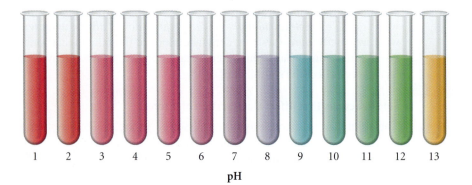

pH

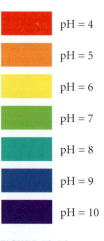

pH = 4

pH = 5

pH = 6

pH = 7

pH = 8

pH = 9

pH = 10

FIGURE 18.16

A typical universal indicator is a mixture of several common indicators.

HERE'S WHAT WE KNOW SO FAR

- Strong-acid–strong-base titrations show a relatively level pH until near the equivalence point, where the pH rises dramatically.

- Titration curves in which one component is weak and the other is strong contain four regions, including the initial pH, the buffer region, the equivalence point region and the post–equivalence point region.

- The buffer region contains a point at which one-half of the analyte has been converted to its conjugate. This is called the titration midpoint, and the pH at this point is equal to the pK of the analyte.

- The larger the pK of the analyte, the sharper will be the change in pH at the equivalence point.

- We can use an indicator to "see" the equivalence point of a titration.

- We add only a few drops of an indicator to the titration solution so that the equivalence point and titration endpoint can be as close together as possible.

Video Lesson: The Effects of pH on Solubility

18.3 Solubility Equilibria

The Pacific Ocean is an incredibly complex heterogeneous system. The bottom layers of this and other massive waterways are covered with a variety of soils and sediments, including **calcareous oozes**, calcium-containing detritus from dead single-celled, calcium-based sea life. One of the important compounds within the oozes is calcium carbonate ($CaCO_3$), some of which is in contact with ocean water, dissociating to form calcium and carbonate ions. The equation relating this dissociation is written so that the solid is a reactant and the dissolved ions are products:

Application

$$CaCO_3(s) \rightleftharpoons Ca^{2+}(aq) + CO_3^{2-}(aq)$$

The mass-action expression that can be used to determine the solubility of $CaCO_3$ is called the **solubility product** when the equation is written as shown above. The solubility product is equal to the product of the concentrations of the ions (remember that the $CaCO_3$ is a solid and is not written as part of the mass-action expression).

Video Lesson: The Solubility Product Constant

$$K_{sp} = [Ca^{2+}][CO_3^{2-}]$$

This equilibrium constant is called the **solubility product constant (K_{sp})** and has the same conceptual meaning as any other equilibrium constant along with its mass-action expression. Because the values of K_{sp} tend to be very small, the concentrations of ions are quite low and that activities are not important here. Table 18.3 lists representative K_{sp} values for some of the sparingly soluble salts.

FIGURE 18.17

Many processes, including the formation of bicarbonate ion and the reaction of hydrogen and hydroxide ions, affect the solubility of calcium carbonate in the ocean. These stromatolites are formations of calcium carbonate.

The difficulty with describing solubility using a single mass-action expression is that there are so many other processes that enter into the chemistry that our typically simple mass-action expression often just won't do. *The simple calculations we can perform do not always agree with what we observe in real systems.* Let's take a look at the calcium carbonate system in a somewhat nonmathematical approach as we discover the factors that affect the solubility of solids.

Side Reactions That Affect Our Reaction of Interest

The solubility of many ions, when dissolved in an aqueous system, is affected by side reactions. The solubility of calcium carbonate in a large ocean-based system is no exception (Figure 18.17).

TABLE 18.3 Selected K_{sp} Values at 25°C

Ionic Solid	K_{sp} (at 25°C)	Ionic Solid	K_{sp} (at 25°C)	Ionic Solid	K_{sp} (at 25°C)
Fluorides		$Hg_2CrO_4^*$	2×10^{-9}	$Co(OH)_2$	2.5×10^{-16}
BaF_2	2.4×10^{-5}	$BaCrO_4$	8.5×10^{-11}	$Ni(OH)_2$	1.6×10^{-16}
MgF_2	6.4×10^{-9}	Ag_2CrO_4	9.0×10^{-12}	$Zn(OH)_2$	4.5×10^{-17}
PbF_2	4×10^{-8}	$PbCrO_4$	2×10^{-16}	$Cu(OH)_2$	1.6×10^{-19}
SrF_2	7.9×10^{-10}			$Hg(OH)_2$	3×10^{-26}
CaF_2	4.0×10^{-11}	**Carbonates**		$Sn(OH)_2$	3×10^{-27}
		$NiCO_3$	1.4×10^{-7}	$Cr(OH)_3$	6.7×10^{-31}
Chlorides		$CaCO_3$	8.7×10^{-9}	$Al(OH)_3$	2×10^{-32}
$PbCl_2$	1.6×10^{-5}	$BaCO_3$	1.6×10^{-9}	$Fe(OH)_3$	4×10^{-38}
$AgCl$	1.6×10^{-10}	$SrCO_3$	7×10^{-10}	$Co(OH)_3$	2.5×10^{-43}
$Hg_2Cl_2^*$	1.1×10^{-18}	$CuCO_3$	2.5×10^{-10}		
		$ZnCO_3$	2×10^{-10}	**Sulfides**	
Bromides		$MnCO_3$	8.8×10^{-11}	MnS	2.3×10^{-13}
$PbBr_2$	4.6×10^{-6}	$FeCO_3$	2.1×10^{-11}	FeS	3.7×10^{-19}
$AgBr$	5.0×10^{-13}	Ag_2CO_3	8.1×10^{-12}	NiS	3×10^{-21}
$Hg_2Br_2^*$	1.3×10^{-22}	$CdCO_3$	5.2×10^{-12}	CoS	5×10^{-22}
		$PbCO_3$	1.5×10^{-15}	ZnS	2.5×10^{-22}
Iodides		$MgCO_3$	1×10^{-15}	SnS	1×10^{-26}
PbI_2	1.4×10^{-8}	$Hg_2CO_3^*$	9.0×10^{-15}	CdS	1.0×10^{-28}
AgI	1.5×10^{-16}			PbS	7×10^{-29}
$Hg_2I_2^*$	4.5×10^{-29}	**Hydroxides**		CuS	8.5×10^{-45}
		$Ba(OH)_2$	5.0×10^{-3}	Ag_2S	1.6×10^{-49}
Sulfates		$Sr(OH)_2$	3.2×10^{-4}	HgS	1.6×10^{-54}
$CaSO_4$	6.1×10^{-5}	$Ca(OH)_2$	1.3×10^{-6}		
Ag_2SO_4	1.2×10^{-5}	$AgOH$	2.0×10^{-8}	**Phosphates**	
$SrSO_4$	3.2×10^{-7}	$Mg(OH)_2$	8.9×10^{-12}	Ag_3PO_4	1.8×10^{-18}
$PbSO_4$	1.3×10^{-8}	$Mn(OH)_2$	2×10^{-13}	$Sr_3(PO_4)_2$	1×10^{-31}
$BaSO_4$	1.5×10^{-9}	$Cd(OH)_2$	2.5×10^{-14}	$Ca_3(PO_4)_2$	1.3×10^{-32}
		$Pb(OH)_2$	1.2×10^{-15}	$Ba_3(PO_4)_2$	6×10^{-39}
Chromates		$Fe(OH)_2$	1.8×10^{-15}	$Pb_3(PO_4)_2$	1×10^{-54}
$SrCrO_4$	3.6×10^{-5}				

*Contains Hg_2^{2+} ions. $K = [Hg_2^{2+}][X^-]^2$ for Hg_2X_2 salts, for example.

1. *Hydrolysis of carbonate ion.* The carbonate ion formed from the dissolution of calcium carbonate is a base that reacts with water to form bicarbonate ion and hydroxide ion.

$$CO_3^{2-}(aq) + H_2O(l) \overset{K_{b_1}}{\rightleftharpoons} HCO_3^-(aq) + OH^-(aq)$$

Because the carbonate ion is involved in this reaction, some of it is removed from the calcium carbonate solubility equilibrium. The net result, in accordance with Le Châtelier's principle, is that more of the calcium carbonate dissolves than we would predict.

2. *The interplay between the atmosphere and the ocean water.* Dissolved CO_2 from the air mixes with ocean water to form carbonic acid and, ultimately, hydrogen ion and bicarbonate ion.

$$CO_2(aq) + H_2O(l) \overset{K}{\rightleftharpoons} H_2CO_3(aq) \overset{K_{a_1}}{\rightleftharpoons} H^+(aq) + HCO_3^-(aq)$$

FIGURE 18.18

Even a system that seems as simple as the solubility of lead(II) iodide isn't. Most of the Pb^{2+} ions precipitate upon the addition of a small amount of iodide. However, a significant concentrate of Pb^{2+} remains.

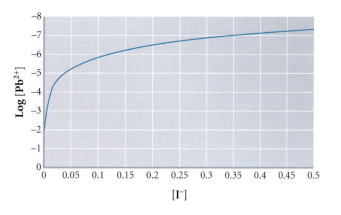

This interaction generates the bicarbonate ion, which influences the equilibrium shown at the bottom of page 803. The net result is to reduce the effect of the interaction of carbonate ions with water. The exact change depends on the amount of carbon dioxide dissolved in seawater.

3. *The formation of water from the reaction of hydrogen and hydroxide ions.* These are produced from the processes described in reactions 1 and 2.

$$H^+(aq) + OH^-(aq) \xrightleftharpoons{1/K_w} H_2O(l)$$

The effect of this reaction is an increase in the concentration of the bicarbonate ion generated from the carbon dioxide equilibrium.

As you can see, these three equilibria interact with others in the ocean as part of a remarkably complex system in which temperatures and concentrations change, making the calculation of calcium carbonate solubility at a given temperature most challenging.

Shifting our focus from the oceans to a more controlled setting, we find that side reactions can still confound apparently simple systems. Consider the solubility of lead(II) iodide (PbI_2) in distilled water. The amount of precipitated lead(II) iodide is related to the initial concentration of iodide as shown in Figure 18.18. *Not accounting for these changes can lead to massive errors in calculating the solubility of lead iodide.*

Molecular-Level Processes

We can take a simple view of the solubility of a salt such as calcium sulfate by considering only the dissociation reaction:

$$CaSO_4(s) \xrightleftharpoons{K_{sp}} Ca^{2+}(aq) + SO_4^{2-}(aq)$$

Our mass-action expression is

$$K_{sp} = [Ca^{2+}][SO_4^{2-}]$$

However, Meites, Pode, and Thomas wrote as early as 1966 that even in the absence of side reactions, the concentrations we would calculate would be wrong by about 60%, largely because of the tendency of the calcium and sulfate ions to stay together as individual **ion pairs**. When the small amount of calcium sulfate dissolves, most of it does not form individual ions. Rather, the ions associate intimately with each other. This is especially important in salts of highly charged (± 2 or ± 3) cations and anions. *The bottom line is that even in the absence of side reactions, such as the addition of H^+ to SO_4^{2-} to make HSO_4^- in acidic solution, there are several factors that affect solubility at the molecular level, making many such calculations challenging.* These factors include

1. *Formation of ion pairs,* as mentioned in the previous paragraph
2. *Ion activities,* a measure of the effective concentration of ions in solution
3. *Thermodynamic measures,* including enthalpy and entropy changes in the solution process

Some simple univalent (both cation and anion singly charged) systems do give reasonable answers when we do solubility calculations. In these cases, and others as well, the total number of moles of solute that dissolve per liter of solution is often called the **molar solubility**. For example, we can calculate the molar solubility of silver bromide (AgBr) by using its solubility product constant and mass-action expression.

$$AgBr(s) \rightleftharpoons Ag^+(aq) + Br^-(aq) \qquad K_{sp} = 5.0 \times 10^{-13}$$
$$K_{sp} = [Ag^+][Br^-]$$

We can set up our table, as we've done before. Although the solid AgBr will not enter into the mass-action expression, we'll include it in the K_{sp} ICE tables for the reasons mentioned below. In any case, the number of moles of silver bromide that will dissolve into solution and the number of moles of silver and bromide ions produced will all be equal, because there is a 1-to-1-to-1 mole ratio in the reaction. We can designate the molar solubility of AgBr as s, in which case the equilibrium concentrations, $[Ag^+]$ and $[Br^-]$, will also be s.

$$AgBr(s) \rightleftharpoons Ag^+(aq) + Br^-(aq)$$

initial	—		
change	$-s$	$+s$	$+s$
equilibrium	—	s	s

$$K_{sp} = [Ag^+][Br^-]$$
$$5.0 \times 10^{-13} = s^2$$
$$s = 7.1 \times 10^{-7} \, M = [Ag^+] = [Br^-]$$
$$s = \text{molar solubility of AgBr} = 7.1 \times 10^{-7} \, M$$

Qualitatively, what would we expect to happen to the solubility if we added a little sodium bromide, in which the added bromide is a common ion? According to Le Châtelier's principle, addition of an ion common to the product would push the dissociation reaction back to the left, further decreasing the solubility. Caution: If we add too much bromide ion, the solubility of silver bromide could actually *increase*, as a consequence of the formation of soluble species such as $AgBr_2^-$.

EXERCISE 18.11 **How Much Dissolves?**

One method of analyzing groundwater for nitrate requires that the chloride ions in the water sample be removed first. This is typically done by adding a solution of silver ions (Ag^+) in order to precipitate the sparingly soluble silver chloride salt (AgCl). What silver ion concentration is present when silver chloride is added to water?

$$AgCl(s) \rightleftharpoons Ag^+(aq) + Cl^-(aq) \qquad K_{sp} = 1.6 \times 10^{-10}$$

Solution

Setting up the table, we get

$$AgCl(s) \rightleftharpoons Ag^+(aq) + Cl^-(aq)$$

initial	—	0	0
change	$-s$	$+s$	$+s$
equilibrium	—	s	s

Then, as usual, we can find our equilibrium concentrations by solving the mass-action expression:

$$K_{sp} = [Ag^+][Cl^-]$$

$$1.6 \times 10^{-10} = (s)(s) = s^2$$

$$s = 1.3 \times 10^{-5} \, M$$

PRACTICE 18.11

Calculate the molar solubility of barium fluoride (BaF_2), $K_{sp} = 2.4 \times 10^{-5}$.

See Problems 49 and 50.

EXERCISE 18.12 | Calculating K_{sp}

The concentration of calcium ions in a saturated solution of calcium fluoride was found to be $2.15 \times 10^{-4} \, M$. What is the apparent value for the solubility product constant, K_{sp}?

First Thoughts

This problem is asking the same question as the previous exercise, but in the reverse direction. The problem gives us the molar solubility of calcium ions, so we'll examine the equilibrium expression to determine how to calculate K_{sp}.

Solution

The equilibrium under consideration is

$$CaF_2(s) \rightleftharpoons Ca^{2+}(aq) + 2F^-(aq)$$

The mole ratio of Ca^{2+} to F^- is 1-to-2, so if the equilibrium concentration of Ca^{2+} is 2.15×10^{-4}, that of F^- must be twice as large, or $[F^-] = 4.30 \times 10^{-4}$. We can substitute these values into the mass-action expression.

$$K_{sp} = [Ca^{2+}][F^-]^2$$

$$K_{sp} = (2.15 \times 10^{-4})(4.30 \times 10^{-4})^2 = 3.98 \times 10^{-11}$$

Further Insight

The question asked us to calculate the apparent K_{sp} value. This was done because there may be some side reactions, activity considerations, or other factors that affect the molar solubility of the calcium fluoride salt. In particular, we have neglected the fact the fluoride ion (F^-) is a Brønsted base, which will affect the solubility of the calcium fluoride. (Can you propose how?) In this case, we've calculated a value for K_{sp} that is similar to the actual value, $K_{sp} = 4.00 \times 10^{-11}$.

Video Lesson: Solubility and the Common Ion Effect

PRACTICE 18.12

Calculate the apparent value of K_{sp} for lead bromide ($PbBr_2$) if the concentration of bromide in a saturated solution is $2.1 \times 10^{-2} \, M$.

See Problems 53 and 54.

Video Lesson: Gravimetric Analysis

Application

Solubility, Precipitation, and Gravimetric Analysis

Chemical technicians can determine the concentration of substances in solution by causing them to form insoluble salt precipitates and weighing these precipitates or their related solids in a technique called **gravimetric analysis**. The quantitation of a sample on the basis of its mass is among the most powerful tools at the disposal of chemical technicians because good balances are both highly accurate and

precise, and weighing a sample is fast and inexpensive. For example, the amount of chloride in a sample is routinely determined by combining the chloride ion with silver ion, forming the solid silver chloride as described in the Exercise 18.11.

$$Cl^-(aq) + Ag^+(aq) \rightleftharpoons AgCl(s)$$

Other substances can be routinely determined by gravimetric analysis. Aluminum ion concentrations can be found via the formation of aluminum hydroxide ($Al(OH)_3$). Igniting (driving off water at high temperature) the aluminum hydroxide forms aluminum oxide (Al_2O_3), which can then be weighed. Aluminum can also be determined by reaction with 8-hydroxyquinoline (C_9H_7ON) to form $Al(C_9H_6ON)_3$ without subsequent ignition.

$$Al^{3+}(aq) + 3C_9H_7ON(aq) \rightleftharpoons Al(C_9H_6ON)_3(s) + 3H^+(aq)$$

The concentration of aluminum in solution, like many metals, can be determined by gravimetric analysis. Here, aluminum is reacted to form the 8-hydroxyquinoline salt, which is made pure by recrystallization and weighed.

The solid forms good crystals that can be weighed after drying.

Sulfur can be determined by reaction of the sulfate (SO_4^{2-}) with barium ion to form barium sulfate:

$$SO_4^{2-}(aq) + Ba^{2+}(aq) \rightleftharpoons BaSO_4(s)$$

Calcium concentrations can be measured by reaction to form calcium oxalate (CaC_2O_4).

$$Ca^{2+}(aq) + C_2O_4^{2-}(aq) \rightleftharpoons CaC_2O_4(s)$$

In each of these analyses there are complicating factors, such as the presence of other elements that can react with the precipitating agents, as well as the complex nature of the precipitation process, so the procedures are a bit more involved than the simple reactions suggest. In fact, acid–base and other equilibria are nearly always a vital part of chemistry. Despite all of these concerns, a host of elements can be determined via precipitation. Understanding solubility equilibria and how we can affect them makes the analyses all the more meaningful.

Precipitation is also used in **metal recovery**, in which dissolved metals are reclaimed from processing wastes. Many metals in industrial effluents (runoff from manufacturing) are worth recovering because they are environmental hazards or they waste finite metal resources. Recycling these metals saves money and the environment. Such metals, which include copper, mercury, lead, and zinc, are typically recovered as their sulfide salts, although some metals can be precipitated as their corresponding carbonate salt.

 Application

To Precipitate or Not to Precipitate

Often, the formation of a precipitate is not as obvious as simply mixing two solutions containing ions that form a sparingly soluble salt. Let's say our chemical technician is interested in mixing two solutions, one containing silver ions and one containing chloride ions, so that the final concentrations are $[Ag^+] = 1.0 \times 10^{-4}\ M$ and $[Cl^-] = 1.0 \times 10^{-4}\ M$. Will mixing these solutions cause the formation of the insoluble AgCl? To answer this question, we can perform a calculation using our mass-action expression. However, because the concentrations do not reflect equilibrium conditions, we will be calculating the reaction quotient (recall this from Section 16.5) related to the solubility product. We'll call this Q_{sp}.

$$AgCl(s) \rightleftharpoons Cl^-(aq) + Ag^+(aq) \qquad K_{sp} = 1.6 \times 10^{-10}$$

$$Q_{sp} = [Ag^+]_0\,[Cl^-]_0$$

$$Q_{sp} = (1.0 \times 10^{-4}\ M)\,(1.0 \times 10^{-4}\ M) = 1.0 \times 10^{-8}$$

The K_{sp} value for AgCl is 1.6×10^{-10}. This value is based on the equilibrium conditions of the sparingly soluble salt. The reaction quotient we calculated is 60 times greater than the equilibrium constant for the precipitation, so the AgCl precipitate forms. If the reaction quotient were smaller than the equilibrium value, no precipitate would form. To reiterate:

- If $Q_{sp} > K_{sp}$, a precipitate forms and continues to form until $Q_{sp} = K_{sp}$.
- If $Q_{sp} < K_{sp}$, no precipitate forms.

EXERCISE 18.13   Will It Make a Solid?

A chemical technician wishes to precipitate the lead ions in 100 mL of a water sample. If the water sample contains 3.10×10^{-10} M Pb^{2+}, and 100 mL of a solution containing 7.0×10^{-4} M Cl^- is added, will a precipitate form? K_{sp} for $PbCl_2 = 1.6 \times 10^{-5}$.

Solution

The solubility equilibrium is

$$PbCl_2(s) \rightleftharpoons Pb^{2+}(aq) + 2Cl^-(aq)$$

The mass-action expression is

$$K_{sp} = [Pb^{2+}][Cl^-]^2$$

When the two solutions are mixed, the total volume is doubled, to 200 mL. Therefore the final concentrations of the respective ions after the two solutions are mixed become

$$[Pb^{2+}] = 1.55 \times 10^{-10} \text{ } M$$

$$[Cl^-] = 3.5 \times 10^{-4} \text{ } M$$

and Q_{sp} is

$$Q_{sp} = (1.55 \times 10^{-10})(3.5 \times 10^{-4})^2 = 1.90 \times 10^{-17}$$

No, the chloride solution will not precipitate the lead in the sample.

PRACTICE 18.13

Will a precipitate form when equal volumes of the lead solution in the exercise above and 3.5×10^{-4} M sulfide ion are mixed? K_{sp} for PbS $= 7.0 \times 10^{-29}$.

See Problems 55 and 56. ∎

Acids, Bases, and Solubility

 Application

The chemical technician who works at the water treatment center is often responsible for treating waste water before it is returned to the environment. One such treatment is **sedimentation**, in which aluminum sulfate, $Al_2(SO_4)_3$ and calcium hydroxide, $Ca(OH)_2$, are added to help clarify and purify the wastewater (Figure 18.19). The aluminum and hydroxide ions in the solution form a gelatinous precipitate.

$$Al^{3+}(aq) + 3OH^-(aq) \rightleftharpoons Al(OH)_3(s) \qquad K_{sp} = 2 \times 10^{-32}$$

The solid settles, carrying with it some dissolved organic material, microorganisms and other undesirable substances in a process called **coagulation**. Iron(III) hydroxide can also be used in this way.

On the other hand, increasing the acidity of waterways can increase the concentration of undesirable metals. For example, lead can be found naturally as the insoluble sulfide, PbS. When acidic waters contact the natural lead sulfide, hydrogen ions compete via Le Châtelier's principle to form hydrogen sulfide, H_2S. This

FIGURE 18.19
The process of sedimentation can be used to clarify water for drinking.

allows the lead ion to enter the waterway as Pb^{2+}. The equilibrium constant for the process is not especially high ($K \approx 10^{-7}$), but the leaching of metals, including lead, mercury, cadmium and aluminum into waterways, even at low concentrations, is of concern.

18.4 Complex-Ion Equilibria

Video Lesson: The Formation of Complex Ions

We noted before that adding a common ion to a solution of a sparingly soluble salt affects its solubility. We said that if too much of the common ion were added, the sparingly soluble salt could dissolve instead of precipitate. This interesting phenomenon arises because of the formation of a chemical complex, which typically consists of one or more metal cations bonded to one or more Lewis bases known as **ligands** (recall that such bases donate electrons). Examples of ligands include Cl^-, F^-, OH^-, CN^-, NH_3, and H_2O. The complexes can be anions, as with $AgCl_4{}^{-3}$, or cations, such as $Co(NH_3)_6{}^{3+}$. Each of these ionic complexes has one or more ions of opposite charge to balance the total charge in the solution.

Introducing the Formation Constant

Our first example of a chemical complex begins with the zinc cation, which exists in aqueous solution bonded to four water molecules, written as $Zn(H_2O)_4{}^{2+}$. In an ammonia–ammonium ion buffer, an ammonia molecule replaces a water molecule.

$$Zn(H_2O)_4{}^{2+}(aq) + NH_3(aq) \rightleftharpoons ZnNH_3(H_2O)_3{}^{2+}(aq) + H_2O(l)$$

We will next simplify the expression by assuming the presence of water, as we often do in acid–base reactions, and simplify the expression.

$$Zn^{2+}(aq) + NH_3(aq) \rightleftharpoons Zn(NH_3)^{2+}(aq) \qquad K_{f_1} = 190$$

The equilibrium constant for the formation of the zinc–ammonia complex is called its **formation constant (K_f)** or **stability constant**, and conceptually, *it means the same thing as any other equilibrium constant*. Le Châtelier's principle teaches us that as we add more ammonia, more NH_3 ligands will form coordinate covalent bonds with the central atom, each step having its own formation constant.

$$Zn(NH_3)^{2+}(aq) + NH_3(aq) \rightleftharpoons Zn(NH_3)_2{}^{2+}(aq) \qquad K_{f_2} = 220$$

$$Zn(NH_3)_2{}^{2+}(aq) + NH_3(aq) \rightleftharpoons Zn(NH_3)_3{}^{2+}(aq) \qquad K_{f_3} = 250$$

$$Zn(NH_3)_3{}^{2+}(aq) + NH_3(aq) \rightleftharpoons Zn(NH_3)_4{}^{2+}(aq) \qquad K_{f_4} = 110$$

FIGURE 18.20

Here are the changes that occur in the equilibrium concentrations of the various zinc–ammonia complex ions as we increase the concentration of ammonia. The x-axis displays log [NH$_3$], so that each factor of 10 by which we change [NH$_3$] occupies equal space in the plot.

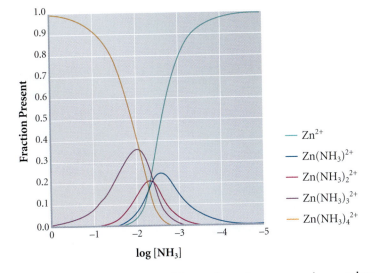

Application

CHEMICAL ENCOUNTERS: Commercial Uses of Aminopolycarboxylic Acid Chelating Agents

FIGURE 18.21

EDTA is a most important chelating agent. Solutions of EDTA are typically prepared as the disodium salt (Na$_2$EDTA), with EDTA^{2-} shown here as a Lewis dot structure (top) and in a free-energy-minimized configuration (bottom), including the two sodium ions.

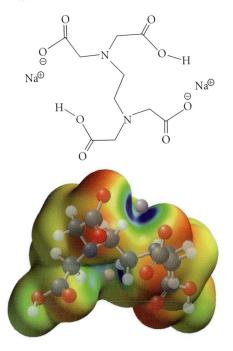

Figure 18.20 shows the distribution of the various zinc–ammonia complexes as the ammonia concentration is increased. Because the formation constants of the steps are so similar, several different zinc–ammonia species are typically present in solution, as shown in the figure. Going from Zn^{2+} to Zn(NH$_3$)$_4$$^{2+}$ does not give a clear, sharp endpoint, and in general, metal-ion concentrations cannot be analyzed by titration that involves a process with multiple formation constants. **Is it possible to have a titrant that will completely combine with a single reaction in a 1-to-1 mole ratio with the metal ion, in order to determine its concentration?** This is where EDTA, introduced as a titrant in Section 18.1, comes in.

Extending the Discussion to EDTA

EDTA is a very useful compound for titrations. It is typically used in its most basic form, shown in Figure 18.21. It has *six* pairs of electrons, one pair on each of the two nitrogen atoms and one pair on each of four oxygen atoms, which form between four and six coordinate covalent bonds to a single metal ion such as Ca^{2+}. Substances that form multiple bonds in this way are called **chelates** (the Greek word *chele* means "claw") or chelating agents because they grab the metal ion like a set of claws. These substances are further characterized by the number of coordinate covalent bonds they make to the metal ion. Ammonia (NH$_3$) is a **monodentate ligand** (one-toothed ligand). EDTA can be a **tetradentate ligand** (four-toothed ligand) or a **hexadentate ligand** (six-toothed ligand). In Chapter 16, we saw the very high equilibrium constants (formation constants) for the reactions of EDTA with metals, shown in Table 18.4. Note in the table how the higher charge on the Fe^{3+} results in a dramatically higher formation constant with EDTA compared to that of Fe^{2+}. We now see that the great stability of such metal–chelate complexes is a result of the **polydentate** nature of the ligand. Several industrially important polydentate chelating agents are listed in Table 18.5. These compounds are so useful that well over 150 million kg are used annually in a host of different products and applications, some of which are listed in Table 18.6.

Complex formation with EDTA and its chemical relatives can be used to determine the concentration of many metals, such as zinc, aluminum, nickel, cobalt, iron, and, in the study of the hardness of water, calcium, and magnesium. We can write the reaction of the calcium ion with EDTA in ionic equation form:

$$Ca^{2+}(aq) + EDTA^{4-}(aq) \rightleftharpoons CaEDTA^{2-}(aq)$$

or we can give it more visual clarity by giving Lewis structures as well as space-filling models, as shown in Section 18.1. Because the equilibrium

TABLE 18.4	Formation Constants of Some Metal–EDTA Complexes	
Element	**Cation**	K_f
silver	Ag^+	2.1×10^7
calcium	Ca^{2+}	5.0×10^{10}
cobalt	Co^{2+}	2.0×10^{16}
zinc	Zn^{2+}	3.0×10^{16}
iron(II)	Fe^{2+}	2.1×10^{14}
nickel	Ni^{2+}	3.6×10^{18}
bismuth	Bi^{3+}	8.0×10^{27}
iron(III)	Fe^{3+}	1.7×10^{24}
vanadium	V^{3+}	8.0×10^{25}

TABLE 18.5   Industrially Important Aminopolycarboxylic Acid Chelating Agents

Name and Abbreviation

Name and Abbreviation

Ethylenediaminetetraacetic acid, EDTA

Diethylenetriaminepentaacetic acid, DTPA

N-(hydroxyethyl)-ethylenediaminetriacetic acid, HEDTA

Nitrilotriacetic acid, NTA

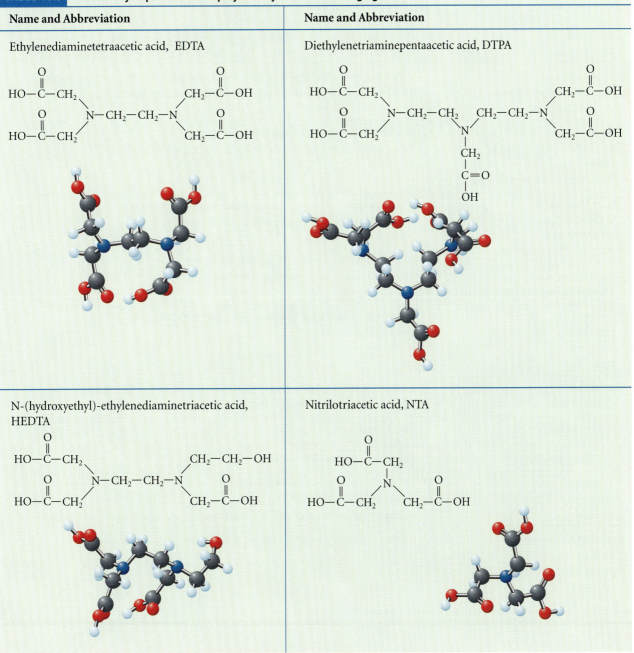

TABLE 18.6	Products and Applications of Chelating Agents
Application	**Benefits of Using Chelating Agents**

Foods and Beverages
  Canned seafood products
  Dressings, sauces, spreads
  Canned beans
  Beverages
	Protects the natural flavor, color, texture, and nutritive value of your food products
	Improves shelf life and consumer appeal

Cleaning Products
  Heavy-duty laundry detergents
  Hard-surface cleaners
	Better foaming, detergency, and rinsing in hard water
	Helps remove metal oxides and salts from fabrics
	Enhances shelf life by inhibiting rancidity, clouding, and discoloration
	Improved consumer appeal and product value
	Improved germicidal action

Personal Care Products
  Creams, lotions
  Bar and liquid soaps
  Shampoos
  Hair preparations
	Better lathering in shampoos and soaps, particularly in the presence of hard water
	Improves shelf life and consumer appeal
	Prevents softening, brown spotting, and cracking in bar soaps
	Improves stability of fragrances, fats, oils, and other water-soluble ingredients

Pharmaceuticals
  Treatment for lead poisoning
  Drug stabilization
	EDTA is approved by the FDA for use in treatment of heavy-metal poisoning
	Deactivates metal ions that interfere with drug performance

Pulp and Paper
  Mechanical pulp bleaching
  Chemical pulping
  Reduction of paper yellowing
	Higher brightness and/or lower bleaching costs
	Less need to overbleach to ensure specified brightness level

Water Treatment
  Boilers
  Heat exchangers
	Dissolves common types of scale during normal operation
	Improves process efficiency and reduces downtime
	Works over a wide range of temperatures, pH levels, and pressures

Metalworking
  Surface preparation
  Metal finishing and plating
	Improved product performance in hard water
	Improved high-temperature performance

Textiles
  Preparation
  Scouring
  Bleaching
	Less need to overbleach to ensure specified brightness level
	Dye shade stability

Agriculture Chelated micronutrients	Excellent water solubility makes metal chelants more readily utilized by plants than the inorganic forms of metals

Polymerization
  Styrene–butadiene polymerization
  PVC polymerization
	Stable polymerization rates
	Reduced polymer buildup in reactors
	Better polymer stability and shelf life

Photography
  Developers
  Bleaches
	Higher-quality prints and negatives
	Enhanced silver recovery
	Increased longevity of prints and negatives

Oilfield Applications
  Drilling
  Production
  Recovery
	Prevents plugging, sealing, precipitation by deactivating metal ions

Source: Dow Chemical, http://www.dow.com (accessed September 2005).

constant is so high, the reaction is essentially complete. This is important when designing a titration. It is also vital in another of EDTA's important uses: combining with metals in food products so that they are unavailable to participate in spoilage processes. To put it in a more formal way, the metal ions are **sequestered** (tied up) by EDTA.

The hard-water analysis at pH 10 allows EDTA to react with both calcium and magnesium ions in the water, which is a measure of "total hardness." At pH values above 12, magnesium ions precipitate as the hydroxide, and the analysis allows determination of the calcium ion concentration alone. Here again, as is our theme in this discussion, pH control is vital. Let's look more closely at the relationship of pH to the reaction. **Why should the titration of calcium with EDTA be more complete at basic than at acidic pH values?**

The Importance of the Conditional Formation Constant

Equilibrium represents a competition—a wrestling match between substances to acquire and release ions and molecules in their quest for energetic stability, as measured by the minimum system free energy. In an aqueous solution of calcium and EDTA, the primary competitor is hydrogen ion. The acidic H_4EDTA can lose four protons in stepwise fashion to form $EDTA^{4-}$.

$$H_4EDTA \rightleftharpoons H_3EDTA^- + H^+ \qquad K_{a_1} = 1.0 \times 10^{-2}$$

$$H_3EDTA^- \rightleftharpoons H_2EDTA^{2-} + H^+ \qquad K_{a_2} = 2.2 \times 10^{-3}$$

$$H_2EDTA^{2-} \rightleftharpoons HEDTA^{3-} + H^+ \qquad K_{a_3} = 6.9 \times 10^{-7}$$

$$HEDTA^{3-} \rightleftharpoons EDTA^{4-} + H^+ \qquad K_{a_4} = 5.5 \times 10^{-11}$$

The higher the pH, the greater the fraction of $EDTA^{-4}$ in the aqueous solution as hydrogen ions are sequentially removed from the molecule. Figure 18.22 shows that as the pH goes down (the solution is made more acidic), the fraction of EDTA present as $EDTA^{4-}$ (highlighted in the figure) diminishes drastically. The tendency to form protonated EDTA reduces the stability of the calcium–EDTA complex with reactions such as this:

$$CaEDTA^{2-} + H^+ \rightleftharpoons Ca^{2+} + HEDTA^{3-}$$

We show this reduction in stability via the **conditional formation constant (K')**, which takes into account the fractions of the free (uncomplexed) metal ion and the $EDTA^{4-}$.

$$K' = K_f \alpha_{Ca^{2+}} \, \alpha_{EDTA^{4-}}$$

where $\alpha_{Ca^{2+}}$ = fraction of free Ca^{2+} (1, in this case)
$\alpha_{EDTA^{4-}}$ = fraction of EDTA present as $EDTA^{4-}$

For example, with our buffer system at pH = 10.0, the formation constant, K_f, for the titration reaction is 5×10^{10} and $\alpha_{Ca^{2+}} = 1$. It is possible to calculate $\alpha_{EDTA^{4-}}$, but we will simply estimate it from Figure 18.22 as being roughly 0.10.

$$K' = K_f \alpha_{Ca^{2+}} \alpha_{EDTA^{4-}} = 5 \times 10^{10} \, (1)(0.10) = 5 \times 10^9$$

The conditional formation constant is still sufficiently high for the titration to be essentially complete. If, however, we calculate the conditional formation constant at pH = 3.0, at which point very little of the EDTA is present as $EDTA^{4-}$ and $\alpha_{EDTA^{4-}} = 2 \times 10^{-11}$, we get

$$K' = K_f \alpha_{Ca^{2+}} \, \alpha_{EDTA^{4-}} = 5 \times 10^{10} \, (1)(2 \times 10^{-11}) = 1$$

The conditional formation constant is far too low for the titration to be feasible.

FIGURE 18.22

As the pH is lowered, the fraction of $EDTA^{4-}$ sharply decreases, reducing the conditional stability constant of any metal–EDTA titration. This is one of several factors that is important to consider when selecting the best pH for this type of analysis.

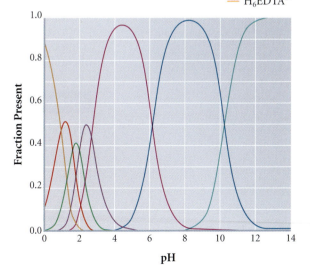

— $EDTA^{4-}$
— $HEDTA^{3-}$
— H_2EDTA^{2-}
— H_3EDTA^-
— H_4EDTA
— H_5EDTA^+
— H_6EDTA^{2+}

EXERCISE 18.14 **Conditional Formation Constant**

With some metals, the competition between EDTA and ammonia to bond to the metal ion can become important when we are judging the feasibility of a titration. If we were titrating zinc instead of calcium in our pH $= 10$ ammonia buffer, we would find that the zinc would form complex ions ranging from $Zn(NH_3)^{2+}$ to $Zn(NH_3)_4^{2+}$. This would make the fraction of free Zn^{2+}, $\alpha_{Zn^{2+}}$, less than 1, lowering the conditional formation constant.

$$Zn^{2+}(aq) + EDTA^{4-}(aq) \rightleftharpoons ZnEDTA^{2-}(aq) \qquad K_f = 3 \times 10^{16}$$

Calculate the conditional formation constant for the titration of zinc with EDTA in the pH $= 10$ buffer in which $\alpha_{Zn^{2+}} = 8 \times 10^{-6}$, and $\alpha_{EDTA^{4-}} = 0.10$. Is this titration still feasible in spite of the low fraction of uncomplexed zinc?

First Thoughts

We can solve for the conditional formation constant, K_f, just as we did with the calcium–EDTA system. Here, however, the fraction of free metal ion is very low.

Solution

$$K' = K_f \alpha_{Zn^{2+}} \alpha_{EDTA^{4-}} = 3 \times 10^{16} \, (8 \times 10^{-6})(0.10) = 2.4 \times 10^{10}$$

This value for K' is still quite large, and the titration works well.

Further Insights

Keep in mind that in spite of the various α and K terms we are working with, a key question with which we are concerned is "How much of each reactant is available to react?" In the case of the zinc ion, the answer is "Not much at all." With the EDTA, there is a much higher fraction available. To reinforce the point we have brought up before, it is the beauty of EDTA that even when the fraction of available reactant is low, the titration is still practical.

PRACTICE 18.14

Calculate the conditional formation constant, K_f, for the system in Exercise 18.14 at pH $= 3.0$.

See Problems 63 and 64.

In the titration of metals with EDTA, we have seen equilibrium principles all come together in one of the most important types of aqueous analyses at the disposal of the chemical technician. Recall Table 18.1 at the beginning of the chapter, in which we listed the calcium–EDTA titration-based hard-water analysis among several ones that are commonly done. Equilibrium principles make these processes ideal for use in many everyday venues. That EDTA titrations are so very common in industrial and academic settings is testimony to the universal utility of equilibrium theory and practice.

EXERCISE 18.15 **Data for a Calcium–EDTA Analysis**

Here are some data obtained from the titration of calcium ion in water with EDTA. How many milligrams of calcium are there per liter of water?

The EDTA solution was prepared by combining the disodium salt Na_2H_2EDTA with water. Its molarity is 0.01944 M. About 5 mL of an ammonia–ammonium ion buffer are combined with a 50.00 mL aliquot of the water. The resulting solution is titrated with EDTA, and it requires 31.88 mL to reach the Eriochrome Black T indicator endpoint, shown by a color change from wine red to blue.

Solution

We begin by finding the mass of calcium in the 50.00 mL aliquot.

Grams of Ca in the 50.00 mL aliquot

$$= 0.03188 \text{ L EDTA solution} \times \frac{0.01944 \text{ mol EDTA}}{1 \text{ L EDTA sol'n}} \times \frac{1 \text{ mol Ca}}{1 \text{ mol EDTA}} \times \frac{40.08 \text{ g Ca}}{1 \text{ mol Ca}}$$

$$= 0.02484 \text{ g Ca in the 50.00 mL aliquot}$$

To convert to the 1 L sample, we use dimensional analysis to scale up to the 1 L volume:

$$0.02484 \text{ g} \times \frac{1000 \text{ mL}}{50 \text{ mL}} = 0.497 \approx 0.500 \text{ g/L calcium}$$

This is equal to 500 mg/L, which is the same as 500 parts per million, or 500 ppm (recall the discussion about ppm in Section 4.2), of calcium in the solution. This corresponds to fairly hard water.

PRACTICE 18.15

How many milliliters of the EDTA solution described in this exercise would be needed to titrate a 50.00 mL sample of water containing 123.8 ppm Ca?

See Problems 65 and 66.

The Bottom Line

- A titration is a technique used to find out how much of a substance is in a solution. (Section 18.2)

- There are several types of titrations, including reduction–oxidation, precipitation, complex-formation, and acid–base titration. (Section 18.2)

- Many reactions are pH-sensitive and require buffers to control pH. (Section 18.1)

- A buffer resists change in pH upon addition of a strong acid or strong base or upon dilution. (Section 18.1)

- Buffers are typically composed of weak conjugate acid–base pairs. (Section 18.1)

- We can solve for the pH of buffers in a straight-forward way by recognizing the importance of Le Châtelier's principle and the common-ion effect. (Section 18.1)

- We can calculate the approximate ratio of conjugate acid to base in order to prepare a buffer of a known pH. Activity effects are important, and our acid-to-base ratio will probably need to be slightly adjusted to be at the desired pH. (Section 18.1)

- Solving for the pH of a buffer upon addition of strong acid or base is really solving a limiting-reactant problem. (Section 18.1)

- It is possible to exceed the buffer capacity, in which case the pH will go sharply higher (with excess base) or lower (with excess acid). (Section 18.1)

- Strong-acid–strong-base titrations show a relatively level pH until near the equivalence point, where the pH dramatically changes. (Section 18.2)

- Titration curves in which one component is weak and the other is strong contain four regions: the initial pH, the buffer region, the equivalence-point region and the post–equivalence point region. (Section 18.2)

- The buffer region contains a point at which one-half of the analyte has been converted to its conjugate. This is called the titration midpoint, and the pH is equal to the pK_a of the analyte. (Section 18.2)

- The larger the pK of the analyte, the sharper will be the change in pH at the equivalence point. (Section 18.2)

- The higher the concentration of the weak acid (or base) and the strong base (or acid), the sharper the endpoint. (Section 18.2)

- An indicator is used to visually detect the equivalence point of a titration. (Section 18.2)

- Only a few drops of an indicator are added to the titration solution so that the equivalence point and endpoint can be as close together as possible. (Section 18.2)

■ Solubility equilibria can often be complex, involving several side reactions and molecular-level processes that make calculations challenging. (Section 18.3)

■ The effects of ion-pairing, activity, and other thermodynamic considerations add to the challenge of properly calculating the concentration of dissolved salts in aqueous solution. (Section 18.3)

■ Gravimetric analysis is based on weighing the precipitate that includes the substance of interest. (Section 18.3)

■ The pH of an aqueous solution can significantly affect the solubility of the substances in that solution. (Section 18.3)

■ A chemical complex typically consists of one or more metal cations bonded to one or more Lewis bases. (Section 18.4)

■ The formation constant is a measure of the extent of reaction between a Lewis base and metal ion in aqueous solution. (Section 18.4)

■ EDTA is the primary example of a highly effective chelating agent. (Section 18.4)

■ The reaction of chelating agents and metal ions has a very high formation constant. (Section 18.4)

■ The analysis of calcium in hard water by EDTA titration is an important application of complex-ion equilibrium. (Section 18.4)

Key Words

acid–base indicator A compound that changes color on the basis of the pH of the solution in which it is dissolved. The color change is often a result of structural changes due to protonation or deprotonation of acidic groups within the compound. (*p. 799*)

analyte A solute whose concentration is to be measured by a laboratory test. (*p. 766*)

anthocyanins A naturally occurring class of compounds responsible for many of the colors of plants. These compounds often act as acid–base indicators. (*p. 801*)

boiler scale A buildup of calcium and magnesium salts within pipes and water heaters. Typically composed of calcium carbonate and magnesium carbonate. (*p. 767*)

buffer A solution containing a weak acid and its conjugate base or a weak base and its conjugate acid. Buffers resist changes in pH upon the addition of acid or base or by dilution. (*p. 767*)

buffer capacity The degree to which a buffer can "absorb" added acid or base without changing pH. (*p. 779*)

buffer region The region of a titration indicated by the presence of a weak acid or base and its conjugate. The pH changes little within this region. (*p. 793*)

calcareous oozes The calcium-containing detritus from dead single-celled, calcium-based sea life. (*p. 802*)

chelates Substances capable of associating through coordinate covalent bonds to a metal ion. Also known as chelating agents. (*p. 810*)

coagulation The precipitation of a solid along with some dissolved organic material, microorganisms and other undesirable substances (*p. 808*)

conditional formation constant (K') The formation constant that accounts for the free metal ion and its associated ligand. (*p. 813*)

equivalence point The exact point at which the reactant in a titration has been neutralized by the titrant. (*p. 789*)

flue-gas desulfurization A process used to remove sulfur dioxide (and other sulfur oxides) from combustion smoke. (*p. 779*)

formation constant (K_f) The equilibrium constant describing the formation of a stable complex. Typically, K_f values are large. Also known as the stability constant. (*p. 809*)

Good (Good's) buffers Buffers typically used in biochemical research because they are chemically stable in the presence of enzymes or visible light and are easy to prepare. (*p. 785*)

gravimetric analysis A laboratory technique in which the concentration of substances in solution is determined by forming insoluble salt precipitates and weighing them or their related solids. (*p. 806*)

Henderson–Hasselbalch equation A shorthand equation used to determine the pH of a buffer solution. $pH = pK_a + \log(\text{base}/\text{acid})$. (*p. 778*)

hexadentate ligand A ligand that makes six coordinate covalent bonds to a metal ion. (*p. 810*)

ion pair Ions in solution that associate as a unit. (*p. 804*)

ligand A compound that associates with a metal ion through coordinate covalent bonds. (*p. 809*)

metal recovery Recycling of metals from waste streams by complexation with chelating agents. (*p. 807*)

molar solubility The total number of moles of solute that dissolve per liter of solution. (*p. 805*)

monodentate ligand A ligand that makes one coordinate covalent bond to a metal ion. (*p. 810*)

pH indicator A compound that changes color on the basis of the acidity or basicity of the solution in which it is dissolved. Also known as an acid–base indicator. (*p. 799*)

polydentate Capable of forming several coordinate covalent bonds from a single ligand to a metal. (*p. 810*)

quality control The practice in industry of ensuring that the product contains what it is supposed to, and in the proper amounts. (*p. 766*)

scrubbing The process of removing harmful impurities from smokestack gases. (*p. 779*)

sedimentation The process of removing dissolved organic matter, heavy metals, and other impurities from water. (*p. 808*)

sequester To tie up an ion or compound by chelation and make it unavailable for use. (*p. 813*)

solubility product The mass-action expression for the solubility reaction of a sparingly soluble salt. (*p. 802*)

solubility product constant (K_{sp}) The constant that is part of the solubility product mass-action expression. Typically, K_{sp} values are much less than 1. (*p. 802*)

stability constant *See* formation constant. (*p. 809*)

tetradentate ligand A ligand that makes four coordinate covalent bonds to a metal ion. (*p. 810*)

titrant The solution being added to a solution of an analyte during a titration. (*p. 766*)

titration The process used to determine the exact concentration of an analyte. (*p. 766*)

titration curve A plot of the pH of the solution versus the volume (or concentration) of titrant. (*p. 790*)

titration endpoint The volume and pH at which the indicator has changed color during a titration. The endpoint is not always the same as the equivalence point. (*p. 801*)

titration midpoint The pH of the titration where the concentration of weak acid or base is equal to the concentration of its conjugate. At this point, the pH is equal to the pK_a of the analyte. (*p. 794*)

Focus Your Learning

The answers to the odd-numbered problems and some selected problems appear at the back of the book, as represented by the blue numbering.

Section 18.1 Buffers and the Common-Ion Effect

Skill Review

1. Write the equations that would describe the equilibria present in each of these solutions:
   a. 0.10 *M* NH$_3$
   b. 0.250 *M* Fe(OH)$_3$
   c. 0.125 *M* HCOONa

2. Write the equations that would describe the equilibria present in each of these solutions:
   a. 0.30 *M* PbS
   b. 0.150 *M* NH$_4$Cl
   c. 0.050 *M* CH$_3$COOH

3. Decide which, if any, of these pairs could be used to prepare a buffer solution.
   a. HF and NaF
   b. CH$_3$COOH and NH$_3$
   c. NH$_4$Cl and HF
   d. H$_2$SO$_4$ and NaHSO$_4$
   e. NH$_4$NO$_3$ and NH$_3$

4. Decide which, if any, of these pairs could be used to prepare a buffer solution.
   a. KBr and HBr
   b. NaOH and CH$_3$COOH
   c. HCOOH and HNO$_3$
   d. NaNO$_3$ and HNO$_3$
   e. CH$_3$COONa and HCl

5. Without using the Henderson–Hasselbalch equation, calculate the pH of a buffer made from each of these pairs. Assume that the concentrations given are those in the final mixture.
   a. 1.00 *M* NH$_3$ and 1.00 *M* NH$_4$Cl
   b. 4.50 *M* NH$_4$Cl and 0.50 *M* NH$_3$
   c. 2.50 *M* CH$_3$COOH and 0.75 *M* CH$_3$COONa

6. Without using the Henderson–Hasselbalch equation, calculate the pH of a buffer made from each of these pairs. Assume the concentrations given are those in the final mixture.
   a. 2.33 *M* NH$_3$ and 1.00 *M* NH$_4$Cl
   b. 2.50 *M* HCOOH and 1.50 *M* HCOONa
   c. 0.100 *M* CH$_3$COOH and 0.75 *M* CH$_3$COONa

7. Suppose an ammonia–ammonium buffer has pH of 10.1. Indicate the effect, if any, of each of these changes:
   a. Adding NH$_3$
   b. Adding NH$_4^+$
   c. Adding Cl$^-$

8. Suppose an acetic acid–acetate buffer has pH of 4.74. Indicate the effect, if any, of each of these changes:
   a. Adding NH$_3$
   b. Adding Na$^+$
   c. Adding HCl

9. Suppose you were to prepare a buffer solution using acetic acid and sodium acetate. What would be the molar ratio of acid to its conjugate when the pH was adjusted to each of these values?
   a. pH = 3.74
   b. pH = 4.74
   c. pH = 5.74

10. Suppose you were to prepare a buffer solution using ammonia and ammonium chloride. What would be the molar ratio of acid to its conjugate when the pH was adjusted to each of these values?
    a. pH = 10.10
    b. pH = 9.26
    c. pH = 8.40

11. How many milliliters of 0.20 *M* HCl would we have to add to 100.0 mL of 0.2500 *M* ammonia in order to prepare a buffer that has each of these pH values?
    a. pH = 9.26
    b. pH = 10.5
    c. pH = 8.5

12. How many milliliters of 0.150 *M* NaOH would we have to add to 100.0 mL of 0.100 *M* acetic acid in order to prepare a buffer that has each of these pH values?
    a. pH = 4.26
    b. pH = 3.75
    c. pH = 5.25

13. Indicate the approximate pH of a buffer made from equal concentrations of each of these pairs. You may need to use the appendix to determine K_a values.
    a. NH_3 / NH_4^+ b. CH_3COOH / CH_3COO^-
    c. $HCOOH$ / $HCOO^-$

14. Indicate the approximate pH of a buffer made from equal concentrations of each of these pairs. You may need to use the appendix to determine K_a values.
    a. $C_6H_5COOH/C_6H_5COO^-$ b. $CH_3NH_2/CH_3NH_3^+$
    c. H_3BO_3 / $H_2BO_3^-$

15. List and explain the factors that determine the pH of a buffered system.

16. List and explain the factors that determine the buffer capacity of a buffered system.

17. A buffer is prepared using chloroacetic acid ($ClCH_2COOH$, $K_a = 1.4 \times 10^{-3}$) and potassium chloroacetate ($ClCH_2COOK$).
    a. Write out the key equilibrium expressions.
    b. Calculate the pH of a solution made by diluting 1.5 g of potassium chloroacetate with 100.0 mL of 0.10 M chloroacetic acid.

18. A buffer is prepared using pyridine (C_5H_5N, $K_b = 1.7 \times 10^{-9}$) and pyridinium chloride (C_5H_5NHCl)
    a. Write out the key equilibrium expressions.
    b. Calculate the pH of a solution made by dissolving 2.50 g of pyridine and 1.25 g of pyridinium chloride into a solution with a final volume of 100.0 mL.

19. The K_b value of methylamine (CH_3NH_2) is 4.3×10^{-4}. The conjugate acid of this weak organic base is the methylammonium ion ($CH_3NH_3^+$, $K_a = 2.3 \times 10^{-11}$). Calculate the pH of a solution that is made from a solution containing 0.10 mol of methylamine and 0.20 mol of methylammonium ion, first using the K_a approach and then using the K_b approach.

20. The K_a value of benzoic acid (C_6H_5COOH) is 6.46×10^{-5}. The conjugate base of this weak organic acid is the benzoate anion ($C_6H_5COO^-$). Calculate the pH of a solution containing 0.025 mol of benzoic acid and 0.250 mol of benzoate anion, first using the K_a approach and then using the K_b approach.

21. Determine the pH of an ammonia–ammonium buffer ($K_b = 1.8 \times 10^{-5}$) with each of these concentrations:
    a. $[NH_3] = 0.10\ M$; $[NH_4^+] = 0.10\ M$
    b. $[NH_3] = 0.20\ M$; $[NH_4^+] = 0.050\ M$
    c. $[NH_3] = 1.50\ M$; $[NH_4^+] = 0.10\ M$
    d. $[NH_3] = 0.050\ M$; $[NH_4^+] = 0.750\ M$

22. Determine the pH of a phenol (C_6H_5OH) / phenoxide ($C_6H_5O^-$) buffer ($K_a = 1.28 \times 10^{-10}$) with these concentrations:
    a. $[C_6H_5OH] = 0.20\ M$; $[C_6H_5O^-] = 0.050\ M$
    b. $[C_6H_5OH] = 1.00\ M$; $[C_6H_5O^-] = 1.00\ M$
    c. $[C_6H_5OH] = 0.050\ M$; $[C_6H_5O^-] = 0.10\ M$
    d. $[C_6H_5OH] = 0.70\ M$; $[C_6H_5O^-] = 0.45\ M$

Chemical Applications and Practices

23. Physiologically important buffers help maintain proper pH levels within our cells. Although the actual buffer system is a complex mixture, we can focus on one particular system that involves phosphate ions. The pH of human blood must be maintained at approximately 7.40. What would you calculate as the ratio of dihydrogen phosphate ($H_2PO_4^-$) to monohydrogen phosphate (HPO_4^{2-}) at that pH? You will need to consult the acid dissociation table for the appropriate equilibrium value.

24. Another important buffer system for humans is formed between carbonic acid and the bicarbonate ion. Calculate the molar ratio of bicarbonate ion to carbonic acid present at pH 7.40. Obtain the necessary equilibrium constant from the table of acid dissociation constants.

25. When studying bacterial growth, microbiologists must determine the optimum pH range for maximum growth. Then, during subsequent culturing, this range can be maintained through proper application of buffer chemistry. The K_a value of formic acid ($HCOOH$) is 1.8×10^{-4}.
    a. If this acid and its salt, sodium formate ($HCOONa$), were selected as the main buffer in a bacteria growth medium, what would be the resulting pH of the media when 500.0 mL of 0.20 M formic acid was mixed with 0.45 g of sodium formate?
    b. Would this buffer be better equipped to resist changes in an acidic or basic direction? Explain.

26. Referring to the same situation as presented for the microbiologist in Problem 25, calculate the volume of 2.5 M NaOH that would be needed to neutralize the 0.20 M formic acid solution to prepare a formic acid–sodium formate buffer with a pH of 3.85.

27. Dairy products such as yogurt, buttermilk, and sour cream are made with the aid of bacteria that convert lactose (milk sugar) to lactic acid. During production, a yogurt sample may attain a pH of 4.00 as a consequence of the presence of lactic acid. If a lactic acid–potassium lactate buffer were produced with the following amounts, what would be the resulting pH? Lactic acid ($CH_3CH(OH)COOH$) = 0.020 mol; potassium lactate ($CH_3CH(OH)COOK$) = 0.015 mol; in 0.500 L. The K_a value of lactic acid is 1.4×10^{-4}.

28. Assume that the bacteria mentioned in Problem 27 produced an additional 0.010 g of lactic acid in a 0.500-L sample of yogurt that already contained [lactic acid] = 0.050 M and [lactate] = 0.050 M. What would be the resulting pH?

29. Propanoic acid (CH_3CH_2COOH) is naturally produced by *Propionibacter shermanii*, a bacterium responsible for the holes in Swiss cheese. The K_a value of propanoic acid is 1.3×10^{-5}. If propanoic acid and its sodium salt were chosen to prepare a buffer system, what would be the pH at which the buffer would have equal ability to resist acidic and basic changes?

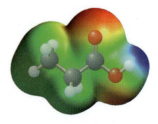

30. If 100.0 mL of a propanoic acid–propanoate buffer solution contained 0.50 mol of acid and 0.50 mol of propanoate, how many milliliters of 0.10 M NaOH would be required to exhaust the buffer capacity?

Section 18.2   Acid–Base Titrations

Skill Review

31. In each of these strong-acid–strong-base titrations, determine the volume of titrant that would effect a neutralization.
    a. 0.045 L of 0.23 M HCl titrated with 0.15 M NaOH
    b. 50.0 mL of 0.50 M NaOH titrated with 0.23 M HCl
    c. 20.0 mL of 0.20 M H_2SO_4 titrated with 0.15 M KOH
    d. 0.050 L of 0.10 M NaOH titrated with 0.23 M H_2SO_4

32. In each of these weak-acid–strong-base titrations, determine the volume of titrant that would effect a neutralization.
    a. 0.055L of 0.13 M CH_3COOH titrated with 0.15 M NaOH
    b. 50.0 mL of 0.50 M HCOOH titrated with 0.23 M NaOH
    c. 25.0 mL of 0.10 M $ClCH_2COOH$ titrated with 0.45 M KOH
    d. 0.045 L of 0.83 M C_6H_5COOH titrated with 0.70 M KOH

33. Determine the pH of the following titration at each of the points indicated. A 75.0 mL solution of 0.137 M NaOH is titrated with 0.2055 M HCl.
    a. initial pH
    b. after addition of 10.0 mL of HCl
    c. after addition of 25.0 mL of HCl
    d. after addition of 50.0 mL of HCl
    e. after addition of 100.0 mL of HCl

34. Determine the pH of the following titration at each of the points indicated. A 175-mL solution of 0.060 M HCl is titrated with 0.10 M NaOH.
    a. initial pH
    b. after addition of 10.0 mL of NaOH
    c. after addition of 50.0 mL of NaOH
    d. after addition of 105.0 mL of NaOH
    e. after addition of 150.0 mL of NaOH

35. Determine the pH of the following titration at each of the points indicated. A 50.0-mL solution of 0.100 M NH_3 is titrated with 0.125 M HCl.
    a. initial pH
    b. after addition of 10.0 mL of HCl
    c. after addition of 20.0 mL of HCl
    d. after addition of 40.0 mL of HCl
    e. after addition of 50.0 mL of HCl

36. Determine the pH of the following titration at each of the points indicated. A 100.0-mL solution of 0.017 M CH_3COOH ($K_a = 1.8 \times 10^{-5}$) is titrated with 0.025 M NaOH.
    a. initial pH
    b. after addition of 10.0 mL of NaOH
    c. after addition of 34.0 mL of NaOH
    d. after addition of 68.0 mL of NaOH
    e. after addition of 100.0 mL of NaOH

37. Perform the necessary calculations and sketch a titration curve diagram for the following strong-acid–strong-base titration: 25.0 mL of 0.250 M KOH using 0.150 M HNO_3 as the titrant.
    a. initial pH
    b. after adding 2.00 mL of HNO_3

    c. after adding 20.0 mL of HNO_3
    d. after adding 40.0 mL of HNO_3
    e. after adding 41.7 mL of HNO_3
    f. after adding 43.0 mL of HNO_3
    g. after adding 50.0 mL of HNO_3

38. Using 0.25 M NaOH as the titrant, calculate the pH of the resulting solution, and sketch the "pH versus volume of titrant" titration curve, for the neutralization of 50.0 mL of 0.10 M formic acid (HCOOH, $K_a = 1.8 \times 10^{-4}$) at each of these points:
    a. initial pH
    b. after adding 2.00 mL of NaOH
    c. after adding 10.0 mL of NaOH
    d. after adding 19.0 mL of NaOH
    e. after adding 20.0 mL of NaOH
    f. after adding 21.0 mL of NaOH
    g. after adding 30.0 mL of NaOH

39. From the list of indicators provided in the chapter, select the best choice for an indicator to use in each of these titrations:
    a. HCl analyte with NH_3 as the titrant
    b. Propanoic acid analyte with KOH as the titrant
    c. Nitric acid analyte with NaOH as the titrant

40. From the list of indicators provided in the chapter, select the best choice for an indicator to use in each of these titrations:
    a. Acetic acid analyte with NaOH as the titrant
    b. Ammonia analyte with HCl as the titrant
    c. Phenol analyte with NaOH as the titrant, K_a (phenol) $= 1.28 \times 10^{-10}$

41. Determine the color of each of the following indicators in their respective solutions.
    a. phenolphthalein; pH $= 2.5$
    b. bromthymol blue; distilled water
    c. methyl orange; 0.0056 M HCl
    d. methyl violet; 0.049 M NH_3

42. Determine the color of each of the following indicators in their respective solutions.
    a. alizarin; 0.025 M NaOH
    b. bromthymol blue; 0.15 M NH_3 and 0.15 M HCl
    c. thymol blue; 0.15 M HCOOH
    d. methyl red; 0.15 M acetic acid and 0.15 M acetate

Chemical Applications and Practices

43. Vinegar is a dilute solution of acetic acid in water. A 50.00-mL vinegar sample was found to require 20.0 mL of 0.15 M NaOH in order to change the phenolphthalein indicator to pink.
    a. What is the pH of the sample after the reaction?
    b. What is the molarity of the vinegar sample?
    c. What percent of the original vinegar solution is acetic acid (CH_3COOH, $K_a = 1.8 \times 10^{-5}$)? (Assume the density of the solution is 1.00 g/mL.)

44. A yogurt dessert was found to contain lactic acid by chemical analysis. Say 100.0 mL of the dessert required 22.43 mL of 0.0156 M NaOH in order to react completely with the acid.
    a. What is the pH of the sample after the reaction?
    b. What is the best indicator to use for this titration?
    c. What mass/volume percent of the dessert is lactic acid ($CH_3CH(OH)COOH$, $K_a = 1.4 \times 10^{-4}$)?

45. A chemist has isolated a potential acid–base indicator from a specific type of tealeaf.

a. Using the following data, determine the approximate pK_a of the indicator. The extracted compound shows a bright red color when in a solution that has a pH of 7.85. At a pH of 9.85, the color has shifted totally to green.

b. Using your estimated pK_a value, determine the ratio of the red form to the green form at a pH of 9.50.

c. If you used this new indicator in a titration of HCl with NaOH, would the resulting endpoint be accurately indicated? Explain why or why not.

46. A vinegar solution, which contains acetic acid as the active ingredient (CH_3COOH, $K_a = 1.8 \times 10^{-5}$), was mixed with some sodium acetate. The pH was found to be 4.15, and the concentration of acetic acid was determined to be 0.0125 M at equilibrium.

a. What is the equilibrium concentration of sodium acetate?

b. What color is bromthymol blue at this pH?

c. What is the pH if another 10.0 g of sodium acetate (CH_3COONa) is added to 500 mL of the solution? (Assume the volume of the solution does not change.)

Section 18.3 Solubility Equilibria

Skill Review

47. Write out the reaction that describes the dissolution of each of these sparingly soluble salts. Then write the corresponding mass-action expression for the equilibria.

a. AgI   b. Ag_2CrO_4 c. Al_2S_3 d. $Ca_3(PO_4)_2$

48. Write out the reaction that describes the dissolution of each of these sparingly soluble salts. Then write the corresponding mass-action expression for the equilibria.

a. $PbCl_2$ b. $NiCO_3$ c. MnS   d. $Zn(OH)_2$

49. Use the following data to calculate the molar solubility for each of these solids.

a. CuS, $K_{sp} = 8.5 \times 10^{-45}$

b. Ag_3PO_4, $K_{sp} = 1.8 \times 10^{-18}$

c. $FeCO_3$, $K_{sp} = 2.1 \times 10^{-11}$

50. Use the following data to calculate the molar solubility for each of these solids.

a. Ag_2S, $K_{sp} = 1.6 \times 10^{-49}$

b. $Fe(OH)_2$, $K_{sp} = 1.8 \times 10^{-15}$

c. MgF_2, $K_{sp} = 6.4 \times 10^{-9}$

51. Use the following data to calculate the solubility product constant (K_{sp}) for each of these solids.

a. NiS, $s = 5.5 \times 10^{-11}\ M$

b. $PbCrO_4$, $s = 1.41 \times 10^{-8}\ M$

c. Ag_2CO_3, $s = 2.0 \times 10^{-4}\ M$

52. Use the following data to calculate the solubility product constant (K_{sp}) for each of these solids.

a. CoS, $s = 2.2 \times 10^{-11}\ M$

b. $Zn(OH)_2$, $s = 2.24 \times 10^{-6}\ M$

c. $CaSO_4$, $s = 7.81 \times 10^{-3}\ M$

53. Use the table of K_{sp} values in the text to determine which of these salts has the greatest molar solubility.

a. CaF_2 b. BaF_2 c. MgF_2

54. Use the table of K_{sp} values in the text to determine which of these salts has the greatest molar solubility.

a. PbI_2 b. AgI   c. Ag_3PO_4

55. Indicate whether a solid will form in each of these mixtures. Start by finding the concentrations of each ion in the resulting solution.

a. 125 mL of 0.100 M $BaCl_2$ and 10.0 mL of 0.050 M Na_2SO_4

b. 35 mL of 0.0045 M $Ag(NO_3)$ and 100.0 mL of 0.0038 M NaCl

56. Indicate whether a solid will form in each of these mixtures. Start by finding the concentrations of each ion in the resulting solution.

a. 100.0 mL of 0.015 M K_2CO_3 and 100.0 mL of 0.0075 M $BaBr_2$

b. 25 mL of 0.57 M $Ni(NO_3)_2$ and 100.0 mL of 0.150 M NaOH

57. Iron(III) hydroxide has a very low K_{sp} value: 1.6×10^{-39}. Explain the effect of changing the pH of an iron(III) hydroxide solution on the solubility of this salt.

58. Copper(II) carbonate has a K_{sp} value of 2.5×10^{-10}. Explain the effect of changing the pH of a copper(II) carbonate solution on the solubility of this salt.

Chemical Applications and Practices

59. When sources of the fluoride ion were being considered for use in fluoridated toothpastes, compounds such as calcium fluoride (CaF_2) could have been among them. This salt has a K_{sp} value of approximately 4.0×10^{-11}. Write out the mass-action K_{sp} expression, and calculate the concentration of fluoride ion in a saturated solution.

60. To maintain good health, we need to have certain amounts of several dissolved metal ions. Zinc ions are critical for the role they play with several hundred enzymes that function in digestion, immune systems, and fertility. However, some foods bind the zinc ions and prevent them from being absorbed from food.

a. Write out the mass-action K_{sp} expression for zinc sulfate, the form of zinc in many vitamin supplements.

b. If phytic acid, which is found in some foods, reacted with zinc ions, how would this affect the solubility of zinc sulfate?

61. Calcium ions can be mineralized to form bones and teeth. Neglecting any other considerations, which of the following solids would provide the greatest number of calcium ions in a saturated solution? Use the mass-action expression for each to calculate the value for each. Explain the impact of alternative equilibria on your answers.

Calcium carbonate, $K_{sp} = 3.3 \times 10^{-9}$
Calcium iodate, $K_{sp} = 7.1 \times 10^{-7}$
Calcium phosphate, $K_{sp} = 1.2 \times 10^{-29}$

62. The cadmium $+2$ ion, which is considered a toxic heavy metal ion, has been a by-product of several mining operations. One method that could remove it from aqueous systems would be to tie the ions up as a cadmium hydroxide precipitate. The K_{sp} value of cadmium hydroxide is 2.5×10^{-14}.

a. What is the molar solubility of cadmium hydroxide?

b. Explain whether you believe that the presence of dissolved Cd^{2+} might be a greater problem at acidic or basic soil pH values.

Section 18.4 Complex-Ion Equilibria

Skill Review

63. Write equilibria equations that describe the stepwise formation of each of these complex ions.
    a. $Ag(NH_3)_2^+$ b. $Ni(NH_3)_6^{2+}$

64. Write equilibria equations that describe the stepwise formation of each of these complex ions.
    a. $CuBr_3^{2-}$ b. $Ag(S_2O_3)_2^{3-}$

65. How many milliliters of a 0.0156 M EDTA^{4-} solution would be needed to completely complex each of these metals in solution?
    a. 100.0 mL of 0.150 M Zn^{2+}
    b. 50.0 mL of 0.740 M Ca^{2+}
    c. 20.0 mL of 0.050 M Mg^{2+}

66. What would be the molar concentration of an EDTA^{4-} solution in each case if it took exactly 25.00 mL to react completely with each of these solutions?
    a. 30.0 mL of 0.250 M Zn^{2+}
    b. 40.0 mL of 0.700 M Ca^{2+}
    c. 20.0 mL of 0.150 M Mg^{2+}

Chemical Applications and Practices

67. As we noted in the chapter, the fraction of species for the EDTA^{4-} ion is critical when we consider EDTA–metal ion titrations. The available metal ion, uncomplexed with other ligands such as ammonia, can also be an important consideration. The formation constant for the Zn^{2+}–EDTA complex was given as 3×10^{16}. What would be the conditional formation constant if the fraction of free zinc ion were 1×10^{-4} and the fraction of EDTA^{4-} species were 0.05?

68. A 25.0-mL sample of water is treated with ammonia buffer and Eriochrome Black T indicator. The sample requires 15.0 mL of 0.0185 M EDTA solution to reach the endpoint. What is the concentration of Ca^{2+} ion in moles per liter and ppm?

69. A dilute solution of hydrated Cu^{2+} ions will appear blue and without a precipitate. However, the addition of some ammonium hydroxide will cause precipitation of some light blue copper(II) hydroxide, often within a deep blue solution. Further addition of ammonium hydroxide will dissolve the precipitate and form a dark blue solution of $Cu(H_2O)_2(NH_3)_4^{2+}$.
    a. What do these observations suggest about the relative value of the formation constant for $Cu(NH_3)_4^{2+}$?
    b. Write out the stepwise formation of the copper–ammonia complex.

70. Although very toxic, cyanide compounds such as KCN and NaCN can be used to extract gold from ores because the gold will form soluble complexes with cyanide ions. The cumulative formation constant for $Au(CN)_2^-$ is approximately 1.6×10^{38}. In this extraction process, the cyanide extracting solution must be kept very alkaline to prevent the formation of HCN. Very small amounts of gold can be extracted in this manner thanks to the very high value of the formation constant.
    a. Write out the mass-action expression for the formation constant of $Au(CN)_2^-$.
    b. If the concentration of CN^- were maintained at 0.010 M in an extract and the gold complex ion concentration were found to be 8×10^{-5} M, what would you estimate as the concentration of Au^+?

71. Under proper medical supervision, one treatment for lead poisoning may involve reaction with a soluble EDTA salt. This is called chelation therapy. If the following reactions were part of a successful treatment, which complex, the calcium–EDTA ion or the lead–EDTA ion, would you predict to have the higher formation constant?

$$CaEDTA^{2-}(aq) + Pb^{2+}(aq) \rightleftharpoons PbEDTA^{2-}(aq) + Ca^{2+}(aq)$$

72. How many milliliters of a solution of 0.0010 M EDTA^{4-} would be used to titrate the lead in 1000 mL of a 0.0020 M solution of $Pb(NO_3)_2$? (Assume a 1:1 reaction between Pb^{2+} and EDTA.)

Comprehensive Problems

73. Cite three processes in which the use of a buffer would be necessary to maintain the pH within a fixed range.

74. A buffer commonly found in biochemical labs and known as TRIS is $(CH_2OH)_3CNH_2$, $pK_b = 5.70$. If a 1.0-L solution were made with 0.15 mol of TRIS, how many grams of TRISH$^+$ chloride salt, $(CH_2OH)_3CNH_3Cl$, would have to be added to the solution to make a buffer with pH = 8.1?

75. a. A compound of highly oxidized iron may be used to break down certain environmental wastes. However, the reactions of the iron compound (K_2FeO_4) are highly pH dependent. In order to do feasibility studies of this compound for possible wastewater treatment, phosphate buffers could be used to maintain a fairly constant pH value. Obtain the K_a values of H_3PO_4, $H_2PO_4^-$, and HPO_4^{2-}. Which two would be the best combination to use if the study were to be done at pH = 8.20?
    b. What would be the ratio of the components at that pH?

76. The presence of lead ions in the environment can pose a hazard. One gravimetric method to test for the presence of lead in a sample is to precipitate the lead as lead sulfate. The K_{sp} value for lead sulfate is 1.3×10^{-8}.
    a. Write out the mass-action K_{sp} expression for this compound.
    b. If the sulfate concentration is made sufficiently high, lead ions will be almost completely precipitated from the solution. If a solution had a lead ion concentration of 0.0010 M initially, what concentration of lead would remain after precipitation if the sulfate concentration were maintained at 0.010 M?

77. a. Calcium, zinc, and cobalt all form +2 ions. However, from Table 18.4 we can see that the formation constant for calcium–EDTA is considerably less than that for the zinc-EDTA and cobalt–EDTA complexes. Explain why this contrast is logical.
    b. Explain why the formation constant for Fe^{2+}–EDTA is less than that for Fe^{3+}–EDTA.

78. By adjusting solution pH, a biologist may manipulate the charges on side chains of enzymes. A biologist studying the function of an enzyme responsible for a step in the conversion of atmospheric nitrogen to useable forms of nitrogen in a plant finds that the enzyme is neutral when the pH is 6.87.
    a. What is the significance of this pH?
    b. If the enzyme had an acid group with a $pK_a = 7.20$, what should the pH of the enzyme solution be to produce a molar ratio of protonated acid group to unprotonated acid group equal to 3:1?

79. Formic acid (HCOOH, $K_a = 1.77 \times 10^{-4}$) is a weak acid extracted from ants. Suppose that such an extraction resulted in 1.00 mL of volume. This 1.00 mL is then diluted to 50.0 mL with distilled water.
    a. If the resulting solution required 25.0 mL of 0.0010 M NaOH to neutralize, what would you calculate as the molarity of the formic acid solution?
    b. How many moles of formic acid were in the original 1.00-mL extract?

80. A biologist seeks to analyze a sample of lactic acid (CH₃CH(OH)COOH, $K_a = 1.4 \times 10^{-4}$) isolated from a tissue sample. The 20.0-mL sample required 12.5 mL of 0.086 M NaOH to neutralize. What was the molarity of the lactic acid sample? What was the initial pH, the pH midway to the equivalence point, and the pH at the equivalence point?

81. An aqueous solution starts as 0.20 M dissolved Fe(NO₃)₂. EDTA is added to produce a concentration of 0.10 M. Assume that the formation constant for the Fe^{2+}–EDTA⁴⁻ complex is 2×10^{14}.

a. Use the mass-action formation constant expression to determine the approximate concentration of Fe^{2+}.
b. If the solution is now made 0.20 M in hydroxide, is enough Fe^{2+} still present to cause precipitation of $Fe(OH)_2$?

82. Based on your understanding of equilibrium and acid–base titrations, can you estimate the equilibrium constant for the titration of acetic acid with ammonia? Would this combination be suitable for titration? Justify your answer.

83. In the chapter, we claim that strong acids and bases are not buffers in the sense that they resist changes in pH upon addition of a strong acid or base but are unable to resist changes in pH with the addition of water. If you were compelled to use 400.0 mL of a 2.00 M solution of HCl to keep the pH of a solution relatively constant, how much water would you be comfortable adding before the pH is no longer "constant" enough for your own use in the procedure? In other words, how constant is constant?

84. In Section 18.1, we discussed buffer preparation, using the example of the calibration buffer as described in the *Pharmacopeia*. If you had available only KHP (potassium hydrogen phthalate), HCl, and NaOH, how would you prepare 1.00 L of a pH 6.00 buffer? ($K_{a_2} = 3.9 \times 10^{-6}$.)

85. In an older method for the quantization of nitrate ion in water, the water sample is first treated with an aqueous solution of silver ions.
    a. What ion(s) is(are) typically in a sample of tap water that would react with the silver? Write the balanced net ionic reaction that would "remove" that ion from solution.
    b. Why is silver nitrate not used as the source of the silver ions? Suggest a possible compound that could be used to create a source of aqueous silver ions for this analysis.

Thinking Beyond the Calculation

86. A researcher has produced an extract from a tropical plant that contains a monoprotic acid.
    a. The researcher titrates 100.0 mL of the plant extract with 0.100 M NaOH. The titration midpoint is reached when 25.6 mL of NaOH has been added. The pH at this point was 3.58. What is the K_a value for the acid?
    b. Graph a titration curve for this titration by determining the pH at each of the following points along the curve: 0 mL, 10.0 mL, 25.6 mL, 50.0 mL, 75.0 mL, and 100.0 mL.
    c. Which indicator would work best in this titration?
    d. By evaporating the extract, the researcher determines that there is only 0.123 g of the acid in every 100.0 mL of plant extract. What is the molecular mass of the acid?

Electrochemistry

Contents and Selected Applications

19.1 What Is Electrochemistry?

19.2 Oxidation States—Electronic Bookkeeping

19.3 Redox Equations

19.4 Electrochemical Cells

Chemical Encounters: The Electric Eel

19.5 Chemical Reactivity Series

19.6 Not-So-Standard Conditions: The Nernst Equation

19.7 Electrolytic Reactions

Chemical Encounters: Electroplating

A patient undergoes open-heart surgery. The heart is a muscle that operates by developing a potential across the cell membrane.

Go to **college.hmco.com/pic/kelterMEE** for online learning resources.

"I admit the deed!—tear up the planks!— here, here!—it is the beating of his hideous heart!" (Edgar Allan Poe, "The Telltale Heart"). The heart is an incredible organ. Without it, there is no life. It collects blood from the extremities and pushes it to the lungs, where oxygen and carbon dioxide are exchanged. Then it pulls the blood back and pumps it out to the various regions of the body, where the dissolved oxygen is delivered to the cells. In a normal lifetime, the four chambers of the heart rhythmically contract and relax nearly 3 billion times. The heart is always beating. What causes this fabulous muscle to contract and relax?

Muscle cells contain different concentrations of sodium and potassium cations inside (18 mM Na$^+$, 166 mM K$^+$) and outside (135 mM Na$^+$, 5 mM K$^+$) the cell. This concentration difference arises as the cells work continuously to pump Na$^+$ ions out and K$^+$ ions in. The concentration gradient (a gradually increasing difference) across the cell membrane sets up an electrical potential—*a driving force to perform a reaction that results from a difference in electrical charge between two points.* The potential—in this case, a force to restore the ion concentrations to equality—is measured in volts (V), just as we measure the potential (voltage) of a battery. The muscle cell has a very small potential (~100 mV), but it is enough to prime the cell for contraction.

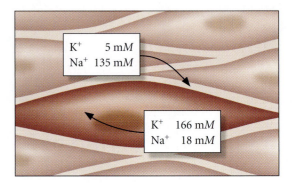

Just prior to a contraction, the muscle cell has changed the concentrations of potassium and sodium to produce a large gradient across its membrane.

Understanding how the concentrations of sodium and potassium ions contribute to the potential inside a heart cell is of interest to the cardiologist, a heart specialist. More broadly, the electron exchanges that occur in chemical reactions are of interest to electrochemists, who create or analyze systems that allow exchanges between chemical and electrical energy. This exchange occurs via *the gain of electrons (reduction) and the loss of electrons (oxidation) in reactions called redox reactions.* Knowing how redox reactions work, how they develop a potential, and how the electrons involved can be harnessed helps scientists understand biological systems (such as muscles and nerves). It also enables them to continue to develop and improve one of the more important inventions that enhances our lives on a daily basis: the battery. The principles behind the exchange between chemical and electrical energy rely on one fact: Chemical reactions can do work. In this chapter, we look at the interplay of electrical and chemical energy, along with some of its interesting and important uses. One of these uses is DNA profiling, the subject of the accompanying How Do We Know? discussion.

19.1 What Is Electrochemistry?

In 2003, over 593 million passengers boarded planes at U.S. airports. On those extremely rare occasions when a jetliner crashes, few may live to tell about it. However, each flight leaves a record of clues, better known as the "black box," for airline officials to decode. Known as a flight data recorder (FDR), the black box, a remarkable combination of engineering, data technology, and chemistry (Figure 19.1), is made rugged enough to survive a crash. So that it too can be found

How do we know?

DNA Profiling

Ionic compounds move when placed in an electric field. The Swedish chemist Arne Tiselius used this information in the 1930s to develop a separation technique. **Gel electrophoresis**, the use of electric fields to separate ions, is based on the application of an electric field across a gel-filled space. Ions, such as DNA fragments or proteins, are placed at one end of the gel, and a voltage is applied. The molecules with a net negative charge tend to migrate toward the positive pole as the positively charged molecules move toward the negative pole. The average rate of their movement is proportional to the average charge and to the average voltage applied. However, the rate of movement is inversely proportional to the size of the molecules. After a certain amount of time, the apparatus is disassembled and the gel stained to reveal the locations of specific compounds. The molar masses of DNA fragments or proteins can be estimated by comparing their locations with those of reference samples whose molar masses are known.

Gel electrophoresis is widely used in molecular biology for separating DNA, RNA, and proteins by size. After human DNA is broken into small pieces with enzymes, it can be analyzed using gel electrophoresis to provide evidence in criminal cases, to diagnose genetic disorders,

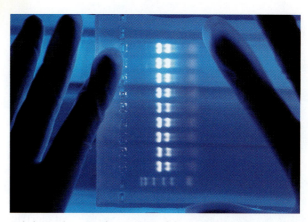

Gel electrophoresis. After staining of the completed gel, each vertical lane contains different-sized fragments of DNA.

and to solve paternity cases. The sizes of the DNA pieces differ among people, so a person's "DNA fingerprint" or "DNA profile" can be constructed. DNA profiling is also used by conservation biologists to determine genetic similarity among populations or individuals. Evolutionary biologists use DNA profile information to construct hypothetical family trees indicating the relationships among species.

---

after a crash, engineers must invest the same care in the design of the battery that powers the locator beacon. While constructing a battery to the exacting specifications of an FDR is challenging, the principles behind battery construction are nothing new. Like the FDR, for example, the battery in a cardiac pacemaker must also meet tough standards; it is expected to function reliably for up to ten years, to provide a steady supply of power, and not to leak chemicals into the person wearing it. The pacemaker battery is based on the same principles of electrochemistry as the FDR.

What is electrochemistry? Broadly speaking, **electrochemistry** is the study of the reduction and oxidation processes that occur at the interface between different phases of a system. Furthermore, electrochemistry typically involves reactions that take place at the surface of a solid. One important field of study in electrochemistry is called **electrodics**, the study of the interactions that occur between a solution of electrolytes and an electrical conductor, often a metal. Another important field within electrochemistry is **ionics**, the study of the behavior of ions dissolved in liquids.

Electrochemical processes typically take place in an **electrochemical cell**, a device that allows the exchange between chemical and electrical energy. Two types of electrochemical cells are possible. The **voltaic cell** (also called a **galvanic cell**) is named after Alessandro Volta (Figure 19.2). It is a type of electrochemical cell

FIGURE 19.1

The flight data recorder. This "black box" records data about each airline flight. The information it collects can be used to help solve the mysteries of a plane crash.

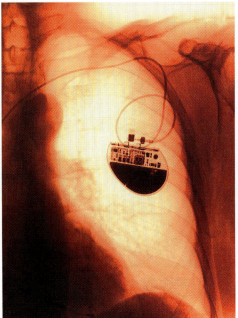

A view of a pacemaker in a patient. Note the battery at the lower portion of the image.

Video Lesson: Reviewing Oxidation–Reduction Reactions

Visualization: Voltaic Cell: Anode Reaction

Visualization: Voltaic Cell: Cathode Reaction

FIGURE 19.2

Count Alessandro Volta (1745–1827) invented the electrophorus, the first well-documented example of a voltaic cell, in 1775. He is also credited with the isolation of methane in 1778.

that *produces* electricity from a chemical reaction. Voltaic cells are commonly known as batteries, although technically speaking, a **battery** is two or more voltaic cells joined together in series. The second type of electrochemical cell is the **electrolytic cell**. This cell *requires* the addition of electrical energy to drive the chemical reaction under study. The industrial production of aluminum (see Section 19.7) is an electrolytic process.

All electrochemical cells require electron exchange that can be characterized in two parts, each known as a **half-reaction**, and the sum of the half-reactions equals the complete reaction observed in the cell. One half-reaction (an **oxidation reaction**) supplies the electrons, and a second (a **reduction reaction**) utilizes these electrons. For this reason, the reactions that take place in electrochemical cells are also known as redox reactions. The oxidation reaction occurs at an **electrode** (typically a metal surface that acts as a collector or distributor for the electrons) known as the **anode**. The reduction reaction takes place at the **cathode**.

How does a redox reaction differ from any other kind of reaction? If we were to place the components for each of the two half-reactions into the same beaker, we wouldn't note any difference. However, the half-reactions do not need intimate contact in order to produce products. As we will see later in this chapter, as long as we make sure that the half-reactions can exchange materials, the overall cell reaction will work. The driving force to complete the reaction—the cell potential—will ensure that the reaction proceeds and that we will be able to harness the power of the electrochemical cell. Our first task in understanding redox reactions and how to make use of the electron exchange is to know a redox reaction when we see it.

19.2 Oxidation States—Electronic Bookkeeping

The space shuttle (Chapter 5) requires electricity for lights, heaters, cameras, communications, and almost every other operation during its orbit of the Earth. In addition, the astronauts need water and oxygen to survive the trip. Electrochemists have discovered a way to provide both water for the crew and electricity for the ship. Their answer is the **fuel cell**, an electrochemical cell that utilizes the reaction of hydrogen with oxygen to produce electricity (Figure 19.3).

$$2H_2(g) + O_2(g) \rightarrow 2H_2O(g)$$

Although we see that this reaction produces water, our first glance at the reaction does not reveal a process that can produce electricity. But it can. Some chemical reactions (the redox reactions), such as this one, involve the shifting or transfer of electrons with one species *losing electrons* (**oxidation**) and another

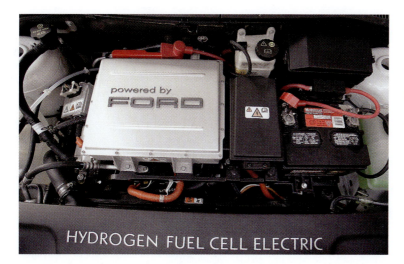

FIGURE 19.3

Under the hood of a Ford Focus hybrid that utilizes the hydrogen fuel cell to generate electrical power. Hydrogen could be the next big advance in alternative fuels. Many of the world's automobile manufacturers are working toward improving hybrid engines and educating consumers on the benefits of alternative fuels.

species *gaining these electrons* (**reduction**). Examining the **oxidation state** (Section 4.7) of the atoms helps us identify the redox reaction. All we must do is keep track of the electrons. A word of caution: There are times when the oxidation state of an atom has little physical meaning, and it can be misleading if the oxidation state is literally interpreted as an indication of where the electrons are found. For instance, when we determine the oxidation state of the oxygen atoms in formic acid (HCOOH), as we do in Exercise 19.1, we will find that both oxygen atoms have the same oxidation state. This should not be interpreted to mean that the same density of electrons will be found equally on both oxygen atoms. Yet even though that interpretation is wrong, oxidation states can be helpful in identifying the general distribution of the electrons in a substance.

As we discussed in Chapter 4, *the oxidation state of an atom in an element is zero*. In monatomic anions and cations, the oxidation state is written as a superscript to the right of the atom symbol, as is done in Ca^{2+} and Br^-. For compounds, oxidation states of atoms are small integers such as $+2, +7$, and -1 (and occasionally fractions) that indicate how an atom's electrons have shifted relative to the elemental state. In the fuel cell, both O atoms in O_2 and both H atoms in H_2 have an oxidation state of zero because these atoms are found in (neutral) elements. The oxidation state of oxygen in the product is -2, and that of each hydrogen atom is $+1$.

How did we know the oxidation state assignments for each atom in water? Electrochemists are familiar with chemical behavior. For example, they know that the hydrogen atom, which has a relatively low electronegativity, is assigned the oxidation state of $+1$ in all its compounds except metal hydrides such as NaH, in which it is -1 because the Group IA and Group IIA metals have even lower electronegativity values than hydrogen. With rare exceptions, all metals in compounds have positive oxidation states, as can be seen by the values in Figure 19.4. Nonmetals, however, can have either positive or negative oxidation states, depending on the compound in which they are found. Figure 19.5 presents a set of decision rules to assist you in assigning oxidation states.

What does an oxidation state mean? We can think of it as a measure of the electron density distribution in a particular compound. A positive oxidation state indicates that electrons have shifted *away from* that atom. A negative oxidation state means that electrons have shifted *toward* that atom. Because electrons are shared (to varying extents) between adjacent atoms, we generally expect nonzero oxidation states for each of the atoms if they are different elements. For example, in the water molecule, H is assigned as $+1$ and O is assigned as -2. The electrons

FIGURE 19.4

Oxidation states of the metals.

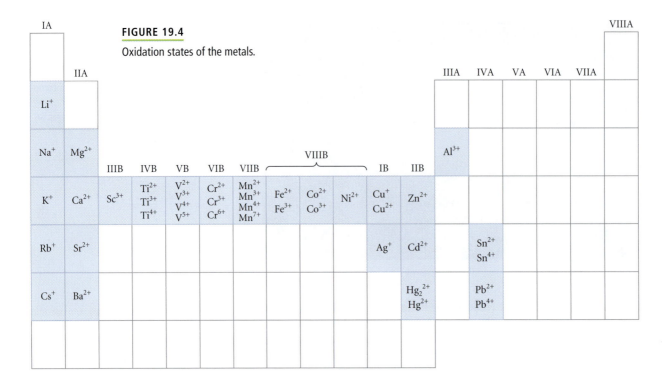

FIGURE 19.5

Decision rules for assigning oxidation states.

Video Lesson: Balancing Redox Reactions by the Oxidation Number Method

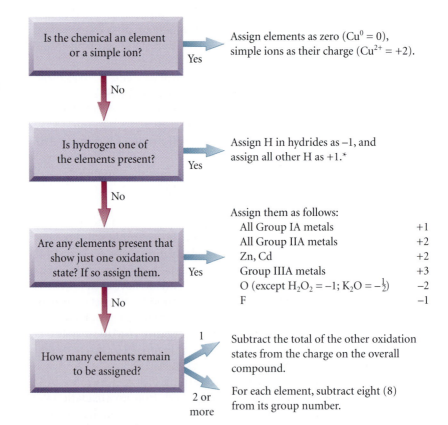

Is the chemical an element or a simple ion? → **Yes** → Assign elements as zero ($Cu^0 = 0$), simple ions as their charge ($Cu^{2+} = +2$).

No

Is hydrogen one of the elements present? → **Yes** → Assign H in hydrides as -1, and assign all other H as $+1$.*

No

Are any elements present that show just one oxidation state? If so assign them. → **Yes** → Assign them as follows:

All Group IA metals	$+1$
All Group IIA metals	$+2$
Zn, Cd	$+2$
Group IIIA metals	$+3$
O (except $H_2O_2 = -1$; $K_2O = -\frac{1}{2}$)	-2
F	-1

No

How many elements remain to be assigned? → **1** → Subtract the total of the other oxidation states from the charge on the overall compound.

→ **2 or more** → For each element, subtract eight (8) from its group number.

* How to recognize hydrides:

Hydrides have a metal first in their chemical formula.

For example: CaH_2, MgH_2, $LiAlH_4$, NaH

Hydrides contain no nonmetals.

For example, these are not hydrides: CH_4, $NaHCO_3$, NH_3, HCl

have shifted away from hydrogen toward the oxygen, as shown in the electrostatic density map in Figure 19.6. This is consistent with our understanding of electronegativity (Chapter 7). The oxidation states do not mean that hydrogen has a full +1 charge and oxygen a full −2 charge. Remember from Chapter 7 that water is a covalent molecule. We can talk more comfortably about oxidation states as representations of the charge of atoms in ionic compounds. Whether in covalent or ionic compounds, we will use oxidation state as a bookkeeping tool to see where electrons shift in chemical reactions.

FIGURE 19.6
The electrostatic potential map for water.

EXERCISE 19.1   Oxidation States

Assign an oxidation state to each of the elements in formic acid (HCOOH). Given the Lewis structure shown below, which of the electrostatic density maps that follow represents formic acid?

$$\begin{array}{c} O \\ \parallel \\ H-C-O-H \end{array}$$

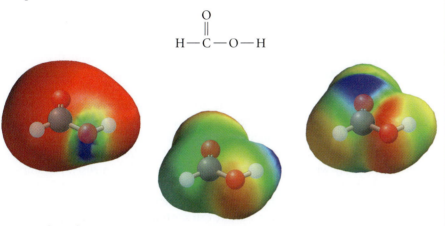

First Thoughts

We have two questions here: a technical one about the oxidation states and a more conceptual one concerning what the oxidation states imply about the electron distribution. In both cases, though, the assignment of the oxidation state and the electrostatic density map should be consistent with our understanding of electronegativity.

Solution

According to our rules for assigning oxidation states in Figure 19.5, each oxygen is assigned an oxidation state of −2, and each hydrogen is +1. For the neutral formic acid molecule, in which the sum of all of the oxidation states must be 0, we have

$$2H = +1 \times 2 = +2$$

$$2O = -2 \times 2 = -4$$

The total is $+2 + (-4) = -2$. The carbon atom must therefore have an oxidation state of +2. This means that the electron density in this molecule is focused more on the oxygen atoms than on the carbon or hydrogen atoms. Given the assignments of the oxidation states, the second electron density map is the most reasonable representation of the molecule. In the other two maps the charge density isn't in the correct location. In the first map, the electron density appears to be opposite of what we'd expect, given that the red color on the map indicates regions of high electron density. The third map again shows electron density that doesn't appear to match where we have predicted the electrons to reside in the molecule.

Further Insights

Carbon can have several oxidation states, depending on the atoms with which it bonds. When we talk about "complete combustion" of carbon, we mean the conversion to its highest possible oxidation state, +4, as is true in CO_2. For example, when glucose ($C_6H_{12}O_6$) is burned in air, the carbon atoms can undergo complete combustion to CO_2. However, they can also form partial combustion products, such as carbon monoxide (CO), in which the oxidation state of the carbon atom differs.

PRACTICE 19.1

Determine the oxidation states of carbon in glucose and carbon monoxide, the compounds discussed in the "Further Insights" section just above.

See Problems 5–8 and 44.

We began this section discussing the reaction of hydrogen and oxygen gases in a fuel cell. By comparing the reactants and products, we see that both H and O show a change in their oxidation state. The change indicated that a transfer of electrons from one species to another has occurred. *Any* chemical reaction in which atoms change their oxidation states is classified as an oxidation–reduction reaction, or redox reaction for short. When O_2 and H_2 react to form water, O_2 is reduced (it gains electrons) and its oxidation state decreases from 0 to −2. Similarly, H_2 is oxidized (it loses electrons), and its oxidation state increases from 0 to +1:

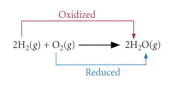

We can look at the compounds in this reaction in a different way. Oxygen (O_2) *causes the oxidation* of the hydrogen (H_2), so it is an **oxidizing agent**. In fact, oxygen stands as the premier oxidizing agent on planet Earth, both because of its abundance and because of its strong ability to accept electrons. We observe oxygen's effect every day when we metabolize glucose or explore a rusty, old shipwreck (rust results from the oxidation of Fe to Fe^{3+}). The hydrogen (H_2) in the fuel cell *causes the reduction* of the oxygen and is therefore a **reducing agent** (see Table 19.1). Looking at this another way, we can say that the oxidizing agent itself is reduced and the reducing agent itself is oxidized.

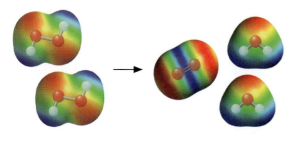

Hydrogen peroxide is sometimes used to bleach hair or disinfect wounds. Is its decomposition into oxygen and water a redox reaction? If so, which species is the oxidizing agent and which is the reducing agent?

$$2H_2O_2(l) \rightarrow O_2(g) + 2H_2O(l)$$

Occasionally, a chemical reaction such as this one employs a single reactant as both the oxidizing and the reducing agent. Such a reaction is known as a **disproportionation**. Using our decision rules in Figure 19.5, we can assign the following oxidation states to each atom.

$$\overset{+1\ -1}{2H_2O_2(l)} \rightarrow \overset{0}{O_2(g)} + \overset{+1\ -2}{2H_2O(l)}$$

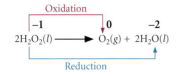

TABLE 19.1	Oxidizing and Reducing Agents	
Reactant	**What Happens**	**Examples**
Oxidizing agent	Gains electrons Is reduced	Element: O_2, O_3, and halogens Compound: H_2O_2 Ionic species (typically with a large positive oxidation state): MnO_4^-
Reducing agent	Loses electrons Is oxidized	Element: H_2 and metals Compound: BH_3 Ionic species: NaH, $LiAlH_4$

In this reaction, the oxygen atoms in hydrogen peroxide (like any peroxide) have an oxidation state of -1. The two products of the reaction contain oxygen atoms with different oxidation states, O_2 with a higher oxidation state (zero) and H_2O with oxygen in a lower oxidation state (-2). We say that the hydrogen peroxide is both the oxidizing agent and the reducing agent.

Not all reactions require the transfer of electrons. Many of the reactions we've discussed in this text thus far are not redox reactions. For example, the neutralization of sodium hydroxide by hydrochloric acid is not a redox reaction. We can tell because the oxidation states for each of the atoms remain the same on both sides of the equation:

$$\overset{+1\,-2\,+1}{NaOH} + \overset{+1\,-1}{HCl} \rightarrow \overset{+1\,-1}{NaCl} + \overset{+1\,-2}{H_2O}$$

Many elements show a variety of oxidation states, depending on the chemical species in which they are found. An example of this can be found in the nonmetal nitrogen, which has a range of possible oxidation states, as shown in Table 19.2 on page 832. The gases found in our atmosphere, both naturally and as pollutants, have different *positive* oxidation states of nitrogen, since the oxygen atom in each compound is more electronegative than nitrogen and is assigned an oxidation state of -2. For the same reason, nitrates (NO_3^-) and nitrites (NO_2^-) have nitrogen with positive oxidation states. However, when combined with a less electronegative element such as hydrogen, nitrogen is assigned a *negative* oxidation state. Accordingly, look for a negative oxidation state of N in ammonia and in compounds containing the ammonium ion, NH_4^+.

The **nitrogen cycle** (Figure 19.7) illustrates how nitrogen moves through its different oxidation states on our planet. You can find similar cycles where the oxidation states of sulfur and carbon change as these elements form different

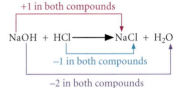

+1 in both compounds

$$NaOH + HCl \longrightarrow NaCl + H_2O$$

−1 in both compounds

−2 in both compounds

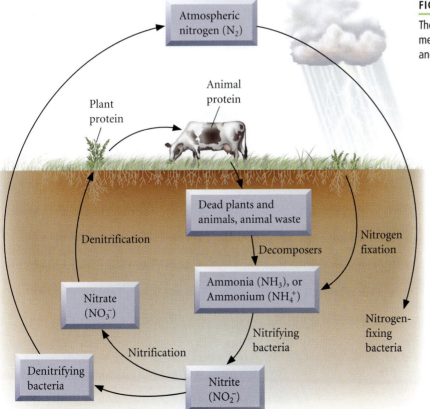

FIGURE 19.7

The nitrogen cycle. The nitrogen in the environment is found in many different compounds and with many different oxidation states.

TABLE 19.2   **The Oxidation State of Nitrogen**

Oxidation State	Formula	Name	Produced in Nature
+5	HNO_3 NO_3^-	Nitric acid Nitrate ion	From NO_2^- by nitrifying bacteria
+4	NO_2	Nitrogen dioxide	In air by oxidation of NO
+3	HNO_2 NO_2^-	Nitrous acid Nitrite ion	From NH_3 by nitrifying bacteria
+2	NO	Nitrogen monoxide (nitric oxide)	From N_2 by lightning or volcanoes
+1	N_2O	Nitrous oxide	From NO_2^- by denitrifying bacteria
0	N_2	Nitrogen	From N_2O by denitrifying bacteria
$-\frac{1}{3}$	N_3^-	Azide ion	(not found in nature)
−1	NH_2OH	Hydroxylamine	(not found in nature)
−2	N_2H_4	Hydrazine	(not found in nature)
−3	NH_3 NH_4^+	Ammonia Ammonium ion	From biological decay of proteins
	NH_4OH	Ammonium hydroxide	—

compounds. What is the point? Whereas oxidation states are fixed for a particular compound at a particular moment in time, chemicals in our world constantly undergo change. Oxidation states help us keep track of the changes that involve the important processes of oxidation and reduction.

EXERCISE 19.2   **The Electrochemistry of Smog**

Although oxygen and nitrogen do not react at low temperatures, they combine in a hot automobile engine to form the pollutant NO, with subsequent oxidation by atmospheric oxygen to form the poisonous brown gas NO_2, which is largely responsible for the smog found in urban areas on hot, sunny days:

$$N_2(g) + O_2(g) \rightarrow 2NO(g)$$
$$2NO(g) + O_2(g) \rightarrow 2NO_2(g)$$

Assign oxidation states to all the atoms involved in these two reactions. Which species are oxidized? Which are reduced? Do your answers make sense in terms of the relative electronegativity values of nitrogen and oxygen?

First Thoughts

Which of the two atoms, nitrogen or oxygen, is more likely to be reduced (gain electrons)? Oxygen is the more electronegative, so in this system, nitrogen will be oxidized. That is how we evaluate whether our answers make sense.

Solution

According to Figure 19.5, we may assign all atoms in elements (such as N_2 and O_2) an oxidation state of zero. In the compounds, we may assign oxygen as −2, because the compounds are not peroxides or superoxides. We assign nitrogen so that the sum of the oxidation states is zero, because NO and NO_2, as compounds, carry no charge.

Compound	Oxidation Number of N	Oxidation Number of O
N_2	0	−
O_2	−	0
NO	+2	−2
NO_2	+4	−2

The nitrogen has been successively oxidized going from N_2 to NO to NO_2, and the oxygen has been reduced as it changes from O_2 to its products.

Further Insights

We noted in Figure 19.5 that hydrogen, which is normally assigned an oxidation state of +1 in compounds, is assigned an oxidation state of −1 when combining with Group 1A metals. This can result in a very reactive reducing agent that is quite useful in organic chemical reactions. One example is NaH, mentioned earlier, which is used in the manufacture of many pharmaceuticals, perfumes, and other organic chemicals.

PRACTICE 19.2

Rank these chemicals in order of increasing oxidation state of sulfur: SO_2, H_2SO_4, S_8, and Li_2S.

See Problems 9–12.

19.3 Redox Equations

Ira Remsen (1846–1927), one of the co-discoverers of the artificial sweetener saccharine, was particularly interested in chemistry as a boy. As an adult, he told of a childhood visit to the doctor's office where he, when left alone in the examination room, set out to discover what was meant by something he had read in a chemistry book: "Nitric acid acts upon copper." To discover this for himself, he placed a pure copper penny on the exam table and poured nitric acid from the doctor's bottle onto the penny. Remsen continues,

> But what was this wonderful thing which I beheld? The cent was already changed, and it was no small change either. A greenish blue liquid foamed and fumed over the cent and over the table. The air in the neighborhood of the performance became dark red. A great colored cloud arose. This was disagreeable and suffocating—how should I stop this? I tried to get rid of the objectionable mess by picking it up and throwing it out the window, which I had meanwhile opened. I learned another fact—nitric acid not only acts upon copper but it acts upon fingers. The pain led to another unpremeditated experiment. I drew my fingers across my trousers and another fact was discovered. Nitric acid also acts upon trousers. Taking everything into consideration, that was the most impressive experiment, and, relatively, probably the most costly experiment I have ever performed.

Saccharine

The reaction that Remsen describes, which is illustrated in Figure 19.8 on page 834, is a redox reaction. Shown on the right is the unbalanced equation. How do we know that it is a redox reaction? Examine the oxidation state of copper. Copper metal (oxidation state = 0) is being oxidized to the copper(II) ion. Simultaneously, nitrogen in nitric acid is being reduced from +5 to +2 in forming nitrogen monoxide. The NO gas released by the reaction rapidly reacts with O_2 in the air to make $NO_2(g)$, the brown fumes that so alarmed the budding chemist.

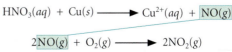

$$HNO_3(aq) + Cu(s) \longrightarrow Cu^{2+}(aq) + NO(g)$$

$$2NO(g) + O_2(g) \longrightarrow 2NO_2(g)$$

FIGURE 19.8

Nitric acid acts on copper. The spontaneous reaction is evident from the generation of a blue solution and a cloud of noxious brown gas. The gas results from the reaction of NO with oxygen in the air.

Video Lesson: Balancing Redox Reactions Using the Half-Reaction Method

Half-Reactions

Ira Remsen noted that this reaction proceeded spontaneously to generate a cloud of noxious fumes. Can we predict this spontaneity by examining the reaction equation, which tells us whether the driving force (the potential) is favorable for this reaction? To assist us in answering this question, we need to extract the oxidation and reduction reactions from the overall equation. These half-reactions, like so many half-reactions, are so well known that the potential for each has been measured and the results collected into a Table of Standard Reduction Potentials, such as Table 19.3. A more comprehensive table can be found in the Appendix.

What do you notice about these tables? One of the things is that all of the reactions are written as reduction reactions. That is, the reactions show the consumption of electrons to make products with less positive oxidation states. Standard Reduction Potentials tables can be used to determine the potential of a reaction, be it for the silver oxide battery found in a pacemaker or for the action of nitric acid on a copper penny. Moreover, knowing the potential of the half-reactions helps us determine the spontaneity of a redox reaction, as we will see later.

Each half-reaction in the table is balanced both atomically and electrically. Half-reactions are simply what they appear to be: half of an oxidation–reduction reaction that is occurring in aqueous solution. The half-reaction listed in the table for the reduction of copper shows the reactants (copper ions and electrons), the product (copper metal), and the **standard potential ($E°$)** of the half-reaction.

Video Lesson: Standard Reduction Potentials

Video Lesson: Electromotive Force

$$Cu^{2+}(aq) + 2e^- \rightarrow Cu(s) \qquad E° = +0.34\,V$$

The value of $E°$ is a measure of how strongly the reduced species on the right-hand side of the reduction half-reaction pulls electrons toward itself. The standard potential is measured in **volts**, the SI unit of electrical potential. It is sometimes referred to as the **electromotive force (emf)** of the half-cell or, more commonly, as the **voltage**.

The values listed for $E°$ are measured under a specific set of conditions:

- Any aqueous ion is present at a concentration (technically, activity) of 1.0 M. All gases are at a pressure of 1 bar (approximately 1 atm).

- The temperature is 25°C (298 K).

These conditions are "standard" for half-reactions and are indicated by the "°" in $E°$. If the conditions are not standard, the voltage will be different from that listed in the table (see Section 19.7), and the potential will be equal to E. Keep in mind that there are several "standards"! For example, standard conditions of temperature and pressure of gases (STP) refer to 0°C (273 K) as the standard temperature.

TABLE 19.3 Selected Standard Reduction Potentials

The selected potentials shown here were obtained under standard conditions (in aqueous solution, 25°C, all solutions 1.0 M, all gases 1.0 atm).

Shorthand Notation	Half-Cell Reaction	Standard Potential, $E°$ (V)
$Li^+(aq) \mid Li(s)$	$Li^+(aq) + e^- \rightarrow Li(s)$	−3.04
$Na^+(aq) \mid Na(s)$	$Na^+(aq) + e^- \rightarrow Na(s)$	−2.71
$Mg^{+2}(aq) \mid Mg(s)$	$Mg^{+2}(aq) + 2e^- \rightarrow Mg(s)$	−2.38
$Al^{3+}(aq) \mid Al(s)$	$Al^{3+}(aq) + 3e^- \rightarrow Al(s)$	−1.66
$H_2O(l) \mid H_2(g)$	$2H_2O(l) + 2e^- \rightarrow H_2(g) + 2OH^-(aq)$	−0.83
$Cd(OH)_2(s) \mid Cd(s)$	$Cd(OH)_2(s) + 2e^- \rightarrow Cd(s) + 2OH^-(aq)$	−0.81
$Fe^{2+}(aq) \mid Fe(s)$	$Fe^{2+}(aq) + 2e^- \rightarrow Fe(s)$	−0.44
$H^+(aq) \mid H_2(g)$	$2H^+(aq) + 2e^- \rightarrow H_2(g)$	0.00
$Fe^{3+}(aq) \mid Fe(s)$	$Fe^{3+}(aq) + 3e^- \rightarrow Fe(s)$	+0.04
$Cu^{2+}(aq) \mid Cu(s)$	$Cu^{2+}(aq) + 2e^- \rightarrow Cu(s)$	+0.34
$O_2(g) \mid OH^-(aq)$	$O_2(g) + 2H_2O(l) + 4e^- \rightarrow 4OH^-(aq)$	+0.40
$NiO_2(s) \mid Ni(OH)_2(s)$	$NiO_2(s) + 2H_2O(l) + 2e^- \rightarrow Ni(OH)_2(s) + 2OH^-(aq)$	+0.49
$Ag^+(aq) \mid Ag(s)$	$Ag^+(aq) + e^- \rightarrow Ag(s)$	+0.80
$HNO_3(aq) \mid NO(g)$	$3H^+(aq) + HNO_3(aq) + 3e^- \rightarrow NO(g) + 2H_2O(l)$	+0.96
$Br_2(l) \mid Br^-(aq)$	$Br_2(g) + 2e^- \rightarrow 2Br^-(aq)$	+1.07
$O_2(g) \mid H_2O(l)$	$O_2(g) + 4H^+(aq) + 4e^- \rightarrow 4H_2O(l)$	+1.23
$Cl_2(g) \mid Cl^-(aq)$	$Cl_2(g) + 2e^- \rightarrow 2Cl^-(aq)$	+1.36
$Au^{3+}(aq) \mid Au(s)$	$Au^{3+}(aq) + 3e^- \rightarrow Au(s)$	+1.50
$F_2(g) \mid F^-(aq)$	$F_2(g) + 2e^- \rightarrow 2F^-(aq)$	+2.87

We speak of standard reduction potential in terms of how strongly the species pulls electrons toward itself. Yet just as in a tug-of-war, *we must consider what we are pulling against*. We need a commonly used reference half-reaction with which to compare our reduction. Our reference is called the **standard hydrogen electrode reaction (SHE)**. This half-reaction, which is assigned the potential of zero volts, is also shown in the table as the reduction of H^+ to H_2. To say that the reduction of Cu^{2+} to Cu^0 has a voltage of +0.34 V, as in Table 19.3, really is to say that it has this voltage *compared to the reduction of H^+ described by the SHE reaction*. All potentials that we use in our discussions will be compared to the SHE reaction.

The potential of a half-reaction can be used to assess the spontaneity of the half-reaction. For instance, fluorine has a strong attraction for electrons. Recalling our discussion of ionization energy and electron affinity from Chapter 7, we might predict that adding electrons to F_2 should be more thermodynamically favorable than adding electrons to a Group IA metal cation such as Li^+. The half-reaction potentials for each reduction bear this out. Fluorine has a large positive reduction potential (+2.87 V), and lithium has a large negative reduction potential (−3.04 V). Michael Faraday (1791–1867), an English electrochemist, worked hard to illustrate how a favorable reaction could be related to the potential. Gibbs later was able to show this relationship mathematically as

$$\Delta G° = -nFE°$$

where n is the number of moles of electrons transferred in the reaction, and F is called Faraday's constant, which we'll discuss in a moment. A key feature of this equation is that the change in free energy, $\Delta G°$, and the cell potential, $E°$, have

Video Lesson: Using Standard Reduction Potentials

opposite signs (*n* and *F* are always positive). Because a negative value of free energy indicates a spontaneous process, *a positive value of cell potential must also indicate spontaneity.*

Faraday's constant is a unit of electric charge equal to the magnitude of charge on a mole of electrons:

$$1 \text{ faraday} = 1 \text{ F} = 96{,}485 \frac{\text{coulombs}}{\text{mol}} = 9.6485 \times 10^4 \frac{\text{C}}{\text{mol}}$$

On the basis of the relationship shown by Faraday, we can determine that 1 joule equals 1 coulomb·volt, and $1 \text{ volt} = 1 \dfrac{\text{joule}}{\text{coulomb}}$:

$$1 \text{ J} = 1 \text{ C} \cdot \text{V}$$

$$1 \text{ V} = \frac{\text{J}}{\text{C}}$$

EXERCISE 19.3   **Spontaneity and Potential**

Copper ions undergo reduction according to the following half-reaction:

$$\text{Cu}^{2+}(aq) + 2e^- \rightarrow \text{Cu}(s) \qquad E° = +0.34 \text{ V}$$

What is the free energy change associated with this process? Is this a spontaneous half-reaction?

Solution

The free energy change is negative; the half-reaction is spontaneous. However, this is only half of a redox reaction.

$$\Delta G° = -nFE°$$

$$= -(2 \text{ mol e}^-) \left(\frac{96485 \text{ C}}{\text{mol e}^-} \right) \left(\frac{+0.34 \text{ J}}{\text{C}} \right)$$

$$= -65609.8 \text{ J} = -66 \text{ kJ}$$

PRACTICE 19.3

The silver cell battery used in pacemakers utilizes the following reaction with a measured potential of 1.86 V. What is $\Delta G°$ for this reaction? Is this reaction spontaneous?

$$\text{Ag}_2\text{O}(s) + \text{Zn}(s) \rightarrow 2\text{Ag}(s) + \text{ZnO}(s)$$

See Problems 25 and 26.

Balancing Redox Reactions

To balance a redox reaction such as the one describing the action of nitric acid on copper, we first determine the identity of the half-reactions. It can be hard (if not seemingly impossible!) to balance a redox equation correctly using a trial-and-error approach (Chapter 3), so we often use a series of steps to accomplish the job (see Figure 19.9). To be fair, this method is just a device to make the balancing go more quickly, rather than a representation of what actually happens at the molecular level. In the nanoworld, electron transfer processes not only occur simultaneously rather than sequentially but also occur in a fairly complex way, with charges building up at the phase changes (the so-called interfaces) in solution. Here we will focus just on the technical aspects of balancing equations.

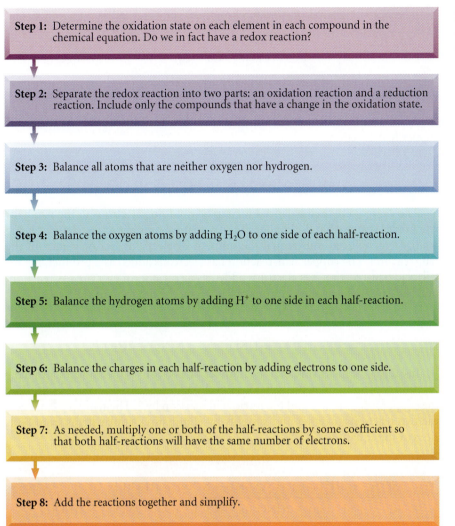

FIGURE 19.9
Algorithm for balancing redox reactions in acidic solution.

Step 1: Determine the oxidation state on each element in each compound in the chemical equation. Do we in fact have a redox reaction?

Step 2: Separate the redox reaction into two parts: an oxidation reaction and a reduction reaction. Include only the compounds that have a change in the oxidation state.

Step 3: Balance all atoms that are neither oxygen nor hydrogen.

Step 4: Balance the oxygen atoms by adding H_2O to one side of each half-reaction.

Step 5: Balance the hydrogen atoms by adding H^+ to one side in each half-reaction.

Step 6: Balance the charges in each half-reaction by adding electrons to one side.

Step 7: As needed, multiply one or both of the half-reactions by some coefficient so that both half-reactions will have the same number of electrons.

Step 8: Add the reactions together and simplify.

 Visualization: Copper Metal in Nitric Acid

Let's follow the algorithm in Figure 19.9 as we balance the copper–nitric acid redox reaction:

$$HNO_3(aq) + Cu(s) \rightarrow Cu^{2+}(aq) + NO(g)$$

Step 1 in Figure 19.9 indicates that we should determine whether we have a redox reaction. That is, does our oxidation state bookkeeping indicate that one species is undergoing oxidation and another is undergoing reduction? We determine that the equation describes a redox process by noting that the oxidation state of copper increases (from 0 to +2) while that of nitrogen decreases (from +5 to +2).

Step 2 indicates that we should separate the redox reaction into the two half-reactions. These half-reactions show the oxidation (copper to copper ion) and the reduction (nitric acid to nitrogen monoxide).

$$Cu(s) \rightarrow Cu^{2+}(aq)$$

$$HNO_3(aq) \rightarrow NO(g)$$

The numbers of atoms (not including H's and O's) are then balanced on both sides of each half-reaction in **Step 3**. No modification of our reactions is needed for this step.

In **Step 4** we balance the number of oxygen atoms by adding water molecules to the product side of the equation.

$$Cu(s) \rightarrow Cu^{2+}(aq)$$

$$HNO_3(aq) \rightarrow NO(g) + 2H_2O(l)$$

In **Step 5** the number of hydrogen atoms is then balanced by adding H^+ to the reactant side. Note that in both steps 4 and 5, we did not need to modify the copper half-reaction.

$$Cu(s) \rightarrow Cu^{2+}(aq)$$

$$3H^+(aq) + HNO_3(aq) \rightarrow NO(g) + 2H_2O(l)$$

The charges are then balanced in **Step 6** by adding electrons to the two half-reactions. We note that the change in the oxidation number of the element being oxidized or reduced must equal the number of electrons lost or gained in the half-reaction. The copper half-reaction has electrons added as a product; the nitric acid half-reaction has electrons added as a reactant. In every redox reaction, one half-reaction gets the electrons on the right, the other on the left.

$$Cu(s) \rightarrow Cu^{2+}(aq) + 2e^-$$

$$3e^- + 3H^+(aq) + HNO_3(aq) \rightarrow NO(g) + 2H_2O(l)$$

In **Step 7** we make sure that the number of electrons is the same in both half-reactions by multiplying the half-reactions by an integer. In this case, we multiply the copper reaction by 3 and the nitric acid reaction by 2. Then, in **Step 8,** we add the two reactions together and simplify by eliminating similar items from both sides of the equation.

$$3\{Cu(s) \rightarrow Cu^{2+}(aq) + 2e^-\}$$

$$2\,\{3e^- + 3H^+(aq) + HNO_3(aq) \rightarrow NO(g) + 2H_2O(l)\}$$

$$\overline{3Cu(s) + 6H^+(aq) + 2HNO_3(aq) \rightarrow 3Cu^{2+}(aq) + 2NO(g) + 4H_2O(l)}$$

The result is the balanced redox reaction. One observation and one question can reasonably arise. The observation is that even though the equation is electrically *balanced,* each side is not electrically *neutral.* That is, we have the same $+6$ charge on each side! Now for the question: Where are the negative charges that would make each side electrically neutral? The answer comes when we remember that we have written *net ionic* equations, rather than complete ionic or molecular equations. In this case, the molecular form comes by adding 6 mol of nitrate ion (NO_3^-) to both sides to give

$$3Cu(s) + 8HNO_3(aq) \rightarrow 3Cu(NO_3)_2(aq) + 2NO(g) + 4H_2O(l)$$

$$3Cu(s) + 6H^+(aq) + 2HNO_3(aq) \rightarrow 3Cu^{2+}(aq) + 2NO(g) + 4H_2O(l)$$
$$+ 6NO_3^- \qquad\qquad\qquad + 6NO_3^-$$
$$\overline{3Cu(s) + 8HNO_3(aq) \rightarrow 3Cu(NO_3)_2(aq) + 2NO(g) + 4H_2O(l)}$$

In other reactions, spectator ions such as Na^+ or Cl^- (if they are actually *in* the solution!) that are not listed in the net ionic equation make the system electrically neutral. The key point is that when we write equations in *net ionic* form, *we expect them to be electrically balanced, not electrically neutral.*

EXERCISE 19.4 **Balancing Redox Equations in Acidic Solutions**

Balance the following equation in acidic solution.

$$Cr_2O_7{}^{2-}(aq) + NO(g) \rightarrow Cr^{3+}(aq) + NO_3{}^-(aq)$$

Solution

Step 1: Determine the oxidation state on each element in each compound in the chemical equation. Do we in fact have a redox reaction?

$$Cr_2O_7{}^{2-}(aq) + NO(g) \rightarrow Cr^{3+}(aq) + NO_3{}^-(aq)$$

Oxidation states $\quad$ +6 −2 $\qquad$ +2 −2 $\qquad$ +3 $\qquad\quad$ +5 −2

Nitrogen is being oxidized from +2 to +5. Chromium is being reduced from +6 to +3. This is a redox reaction.

Step 2: Separate the redox reaction into two parts: an oxidation reaction and a reduction reaction. Include just the compounds that have a change in the oxidation state.

$$Cr_2O_7{}^{2-}(aq) \rightarrow Cr^{3+}(aq) \qquad \text{(reduction)}$$

$$NO(g) \rightarrow NO_3{}^-(aq) \qquad \text{(oxidation)}$$

Step 3: Balance all atoms except oxygen and hydrogen.

$$Cr_2O_7{}^{2-}(aq) \rightarrow 2Cr^{3+}(aq)$$

$$NO(g) \rightarrow NO_3{}^-(aq)$$

Step 4: Balance the oxygen atoms by adding H_2O to one side in each half-reaction.

$$Cr_2O_7{}^{2-}(aq) \rightarrow 2Cr^{3+}(aq) + 7H_2O(l)$$

$$2H_2O(l) + NO(g) \rightarrow NO_3{}^-(aq)$$

Step 5: Balance the hydrogen atoms by adding H^+ to one side in each half-reaction.

$$14H^+(aq) + Cr_2O_7{}^{2-}(aq) \rightarrow 2Cr^{3+}(aq) + 7H_2O(l)$$

$$2H_2O(l) + NO(g) \rightarrow NO_3{}^-(aq) + 4H^+(aq)$$

Step 6: Balance the charges in each half-reaction by adding electrons to one side.

$$6e^- + 14H^+(aq) + Cr_2O_7{}^{2-}(aq) \rightarrow 2Cr^{3+}(aq) + 7H_2O(l)$$

$$2H_2O(l) + NO(g) \rightarrow NO_3{}^-(aq) + 4H^+(aq) + 3e^-$$

Step 7: As needed, multiply one or both of the half-reactions by some coefficient so that the same number of electrons will appear in both half-reactions.

$$6e^- + 14H^+(aq) + Cr_2O_7{}^{2-}(aq) \rightarrow 2Cr^{3+}(aq) + 7H_2O(l)$$

$$2\{2H_2O(l) + NO(g) \rightarrow NO_3{}^-(aq) + 4H^+(aq) + 3e^-\}$$

Step 8: Add the reactions together and simplify. Note that the electrons on each side mathematically cancel, indicating the same number of electrons gained in the reduction as lost in the oxidation.

$$6e^- + 14H^+(aq) + Cr_2O_7{}^{2-}(aq) \rightarrow 2Cr^{3+}(aq) + 7H_2O(l)$$

$$\underline{4H_2O(l) + 2NO(g) \rightarrow 2NO_3{}^-(aq) + 8H^+(aq) + 6e^-}$$

$$6e^- + 14H^+(aq) + Cr_2O_7{}^{2-}(aq) + 4H_2O(l) + 2NO(g) \rightarrow$$
$$2Cr^{3+}(aq) + 7H_2O(l) + 2NO_3{}^-(aq) + 8H^+(aq) + 6e^-$$

$$= 6H^+(aq) + Cr_2O_7{}^{2-}(aq) + 2NO(g) \rightarrow 2Cr^{3+}(aq) + 2NO_3{}^-(aq) + 3H_2O(l)$$

The reaction of purple permanganate ion with methanol.

PRACTICE 19.4

Balance the following reaction in acidic solution.

$$ClO^-(aq) + H^+(aq) + Cu(s) \rightarrow Cl^-(aq) + H_2O(l) + Cu^{2+}(aq)$$

See Problems 35–38, 47, 48, and 51.

The process we have used to balance the reaction focuses on the use of acidic solutions. Chemical reactions can also occur in basic conditions. The method we present to balance basic reactions requires a little chemical sleight-of-hand but does work effectively. Essentially, we balance the reaction as though it were in an acidic solution (by adding H^+ as necessary) and then add a quantity of hydroxide ion (OH^-) as necessary, to both sides of the equation. Mathematically, we still have our equality. Chemically, we neutralize the H^+, getting water on one side and excess base on the other. *Although this does not depict what goes on in the solution* (we are starting with a base, not doing a titration!) we correctly end up with an excess of OH^- on one side or the other.

We will show the method by looking at the first step in a procedure to analyze methanol (CH_3OH) by reaction with the permanganate ion (MnO_4^-) in base. Before balancing, we have

$$CH_3OH(aq) + MnO_4^-(aq) \rightarrow CO_3^{2-}(aq) + MnO_4^{2-}(aq)$$

To balance the reaction in basic solution, *we first balance it as though it were in acidic solution.* Using our half-reaction technique, we get the following oxidation and reduction half-reactions:

$$2H_2O(l) + CH_3OH(aq) \rightarrow CO_3^{2-}(aq) + 8H^+(aq) + 6e^- \quad \text{(oxidation)}$$

$$e^- + MnO_4^-(aq) \rightarrow MnO_4^{2-}(aq) \quad \text{(reduction)}$$

We can multiply the reduction reaction by 6 to balance electrons and add the half-reactions to get the final equation, balanced in acidic solution:

$$2H_2O(l) + CH_3OH(aq) + 6MnO_4^-(aq) \rightarrow CO_3^{2-}(aq) + 8H^+(aq) + 6MnO_4^{2-}(aq)$$

To balance in basic solution, we add an amount of OH^- to each side equal to the amount of H^+.

$$2H_2O(l) + CH_3OH(aq) + 6MnO_4^-(aq) \rightarrow CO_3^{2-}(aq) + 8H^+(aq) + 6MnO_4^{2-}(aq)$$
$$+ 8OH^-(aq) \qquad\qquad\qquad + 8OH^-(aq)$$

$$8OH^-(aq) + 2H_2O(l) + CH_3OH(aq) + 6MnO_4^-(aq) \rightarrow CO_3^{2-}(aq) + 8H_2O(l) + 6MnO_4^{2-}(aq)$$

We have 8 water molecules on the right and 2 on the left. This leaves an excess of 6 water molecules on the right, to give us our final balanced equation in basic solution:

$$8OH^-(aq) + CH_3OH(aq) + 6MnO_4^-(aq) \rightarrow CO_3^{2-}(aq) + 6H_2O(l) + 6MnO_4^{2-}(aq)$$

EXERCISE 19.5 | **Balancing Redox Equations in Basic Solutions**

Balance the following equation in basic solution.

$$I_3^-(aq) + S_2O_3^{2-}(aq) \rightarrow I^-(aq) + SO_4^{2-}(aq)$$

Solution

We may first balance the equation in acidic solution, using our multistep procedure presented in Figure 19.9.

$$I_3^-(aq) + S_2O_3^{2-}(aq) \rightarrow I^-(aq) + SO_4^{2-}(aq)$$

The iodine changes from an oxidation state of $-\frac{1}{3}$ in I_3^- to -1 in I^-. (reduction)

The sulfur changes from an oxidation state of $+2$ in $S_2O_3^{2-}$ to $+6$ in SO_4^{2-}. (oxidation)

The balanced half-reactions in acidic solution are

$$2e^- + I_3^-(aq) \rightarrow 3I^-(aq)$$

$$5H_2O(l) + S_2O_3^{2-}(aq) \rightarrow 2SO_4^{2-}(aq) + 10H^+(aq) + 8e^-$$

We multiply the reduction equation by 4 to equalize the electrons gained and lost in the reduction and oxidation half-reactions

$$4\{2e^- + I_3^-(aq) \rightarrow 3I^-(aq)\} = 8e^- + 4I_3^-(aq) \rightarrow 12I^-(aq)$$

$$5H_2O(l) + S_2O_3^{2-}(aq) \rightarrow 2SO_4^{2-}(aq) + 10H^+(aq) + 8e^-$$

We add the two half-reactions to get the final, balanced equation in acidic solution.

$$5H_2O(l) + S_2O_3^{2-}(aq) + 4I_3^-(aq) \rightarrow 2SO_4^{2-}(aq) + 12I^-(aq) + 10H^+(aq)$$

To balance in base, we add $10OH^-$ to both sides.

$$5H_2O(l) + S_2O_3^{2-}(aq) + 4I_3^-(aq) \rightarrow 2SO_4^{2-}(aq) + 12I^-(aq) + 10H^+(aq)$$
$$+ 10OH^-(aq) \qquad\qquad\qquad\qquad\qquad + 10OH^-(aq)$$

We cancel our the extra waters to give the equation that is now balanced in base.

$$10OH^-(aq) + S_2O_3^{2-}(aq) + 4I_3^-(aq) \rightarrow 2SO_4^{2-}(aq) + 12I^-(aq) + 5H_2O(l)$$

Is the equation electrically and atomically balanced? Check to make sure that the charge is the same on both sides of the equation and that there is the number of atoms on each side. If one or both of these checks do not work, then we've made a mistake balancing the equation.

PRACTICE 19.5

Balance the following equation in basic solution.

$$MnO_4^-(aq) + Mn^{2+}(aq) \rightarrow MnO_2(s)$$

(*Hint:* MnO_2 can be written as the product in both an oxidation and a reduction half-reaction.)

See Problems 39–42 and 52.

Manipulating Half-Cell Reactions

We have learned to balance redox reactions by recognizing that they include electron loss and electron gain. Although calculating the spontaneity of an individual half-reaction may lead to the conclusion that the half-reaction is spontaneous, this is only half of the picture. Because a redox reaction is simply the sum of an oxidation half-reaction and a reduction half-reaction, we determine the spontaneity of the resulting reaction only by including both. Remember that we are using the SHE as our reference point in these calculations.

We can then say that for two half-reactions, the sum of their change in free energy should be equal to the free energy change for the complete redox reaction:

$$\Delta_{rxn}G° = \Delta G_1° + \Delta G_2°$$

If we substitute the equivalent expression $(-nFE°)$ for $\Delta G°$, the equation becomes

$$-n_{rxn}FE°_{rxn} = -n_1FE_1° + -n_2FE_2°$$

Because the value of F is a constant, and because the number of electrons was the same when we balanced the reaction ($n_1 = n_2 = n_{rxn}$), all terms but the cell potentials cancel:

$$E^\circ_{rxn} = E_1^\circ + E_2^\circ$$

This means that if the redox reactions are written so that one is a reduction and one is an oxidation, and the cell potential for the oxidation reaction relative to the SHE has the sign opposite that of its half-reaction potential when written as a reduction, then *the two half-reaction potentials are additive for a balanced redox equation.*

$$E^\circ_{rxn} = E^\circ_{red} + E^\circ_{ox}$$

From that conclusion, we can say, for example, that if copper is oxidized in a reaction, we write it as an oxidation, and just as we reverse the standard free energy value, ΔG°, when we reverse a reaction, we reverse the E°, as follows:

$$Cu^{2+}(aq) + 2e^- \rightarrow Cu(s) \qquad E^\circ = +0.34 \text{ V (reduction)}$$
$$Cu(aq) \rightarrow Cu^{2+}(aq) + 2e^- \qquad E^\circ = -0.34 \text{ V (oxidation)}$$

This notion of reversing the cell potential relative to the SHE when we reverse the reaction *gives a consistent picture of the thermodynamic relationship between ΔG° and E°.* In addition, it is a reminder that until 1953, when IUPAC changed its protocol to list half-reaction potentials as reductions, they were formerly listed as oxidations, with opposite signs relative to the SHE.

We know that the nitric acid—copper reaction that we introduced at the beginning of this section is spontaneous, as evidenced by its effect on Ira Remsen's lungs, hands, and pants. Let's see how we can manipulate the half-reactions to confirm this. In Table 19.3, we notice that one of the reactions is the reverse of the desired reaction.

From the balancing process steps 8 and 9:

$$3\{Cu(s) \rightarrow Cu^{2+}(aq) + 2e^-\}$$

$$\underline{2\{3e^- + 3H^+(aq) + HNO_3(aq) \rightarrow NO(g) + 2H_2O(l)\}}$$

$$3Cu(s) + 6H^+(aq) + 2HNO_3(aq) \rightarrow 3Cu^{2+}(aq) + 2NO(g) + 4H_2O(l)$$

From Table 19.3:

$$Cu^{2+}(aq) + 2e^- \rightarrow Cu(s) \qquad\qquad E^\circ = +0.34 \text{ V}$$
$$3H^+(aq) + HNO_3(aq) + 3e^- \rightarrow NO(g) + 2H_2O(l) \qquad E^\circ = +0.96 \text{ V}$$

To make the half-reactions from the table look like the half-reactions in the redox reaction, we need to reverse the copper reduction and write it as an oxidation. As we pointed out above, reversing the reaction also changes the sign on the potential of the reaction, or, in this case, the half-reaction.

$$Cu^{2+}(aq) + 2e^- \rightarrow Cu(s) \qquad E^\circ = +0.34 \text{ V (reduction)}$$
$$Cu(aq) \rightarrow Cu^{2+}(aq) + 2e^- \qquad E^\circ = -0.34 \text{ V (oxidation)}$$

However, *no modification of the potential is necessary when we double or triple a half-reaction because the number of electrons involved in the process also doubles or triples.* The free energy, however, does change when we multiply the equation, because $\Delta G^\circ = -nFE^\circ$, and if, as in this case, we triple n, ΔG° will triple as well. This also is consistent with our discussion of thermodynamics in Chapters 5 and 14.

$$Cu(aq) \rightarrow Cu^{2+}(aq) + 2e^- \qquad E^\circ = -0.34 \text{ V (oxidation)}$$
$$3Cu(s) \rightarrow 3Cu^{2+}(aq) + 6e^- \qquad E^\circ = -0.34 \text{ V (oxidation)}$$

Table 19.4 lists the key aspects of manipulating half-reactions.

TABLE 19.4	**Important Points for Potentials of Redox Reactions**

- Just as free energies can be combined for chemical reactions, an oxidation and a re-duction half-cell potential can be combined to produce a chemical equation.
- Just as you reverse the sign of $\Delta G°$ when you reverse a chemical reaction, you re-verse the sign of $E°$ when you reverse a half-cell reaction.
- Although $\Delta G°$ values depend on the coefficients in the chemical equation (that is, when you double the coefficients, you double $\Delta G°$), $E°$ values do not. In a half-reaction, if the coefficients change, the number of electrons, n, will change as well, in essence canceling the effect of the change to the coefficients.
- Because of the negative sign in $\Delta G° = -nFE°$, electrochemical cell reactions with a *positive* voltage are spontaneous. (Recall from Chapter 15 that the rate is not related to the spontaneity of the reaction.)

For the copper–nitric acid reaction, the reaction potential is determined by adding the nitric acid reduction ($+0.96$ V) to the copper oxidation (-0.34 V). The resulting potential is positive ($+0.62$ V), identifying the reaction as a spontaneous process, as is consistent with Ira Remsen's experience.

$$3\{Cu(s) \rightarrow Cu^{2+}(aq) + 2e^-\} \qquad\qquad E° = -0.34 \text{ V}$$

$$\underline{2\{3e^- + 3H^+(aq) + HNO_3(aq) \rightarrow NO(g) + 2H_2O(l)\} \qquad E° = +0.96 \text{ V}}$$

$$3Cu(s) + 6H^+(aq) + 2HNO_3(aq) \rightarrow 3Cu^{2+}(aq) + 2NO(g) + 4H_2O(l) \quad E°_{cell} = +0.62 \text{ V}$$

EXERCISE 19.6	**Dissolving Gold**

Nitric acid can be used to dissolve copper. Can nitric acid dissolve gold at standard conditions? The unbalanced reaction is shown here.

$$HNO_3(aq) + Au(s) \rightarrow Au^{3+}(aq) + NO(g)$$

First Thoughts

What we're really asking in this problem is "Is the reaction shown spontaneous?" To determine the spontaneity we can use Faraday's equation, which requires us to know the potential of the reaction. We can obtain this by combining properly balanced half-reactions, which have reduction potentials given in Table 19.3.

Solution

The two unbalanced half-reactions of interest are

$$Au(s) \rightarrow Au^{3+}(aq)$$

$$HNO_3(aq) \rightarrow NO(g)$$

We can balance them in acidic solution (because of the presence of nitric acid) and combine them to give

$$Au(s) \rightarrow Au^{3+}(aq) + 3e^- \qquad \text{(oxidation)}$$

$$\underline{3e^- + 3H^+(aq) + HNO_3(aq) \rightarrow NO(g) + 2H_2O(l) \qquad \text{(reduction)}}$$

$$3H^+(aq) + HNO_3(aq) + Au(s) \rightarrow NO(g) + Au^{3+}(aq) + 2H_2O(l)$$

Table 19.3 lists the potential relative to the SHE for each half-reaction written as a reduction. We can reverse the gold reduction reaction because the metal is being oxidized. We can use the potentials of the half-reactions to calculate the cell voltage.

$$Au(s) \rightarrow Au^{3+}(aq) + 3e^- \qquad E° = -1.50 \text{ V}$$

$$\underline{3e^- + 3H^+(aq) + HNO_3(aq) \rightarrow NO(g) + 2H_2O(l) \qquad E° = +0.96 \text{ V}}$$

$$3H^+(aq) + HNO_3(aq) + Au(s) \rightarrow NO(g) + Au^{3+}(aq) + 2H_2O(l) \quad E°_{cell} = -0.54 \text{ V}$$

The negative value for the cell voltage indicates that the reaction of nitric acid and gold is not spontaneous. Nitric acid doesn't dissolve gold.

Further Insights

You've probably noticed that you don't have to balance a redox reaction in order to determine the standard voltage of the reaction. This is because the voltage of a reaction is not dependent on the stoichiometry of the equation. Simply adding the half-reaction potentials (as long as one is an oxidation and the other a reduction) yields the voltage of the resulting reaction. However, it is good practice to balance the equation so that we can tell how many electrons are exchanged. We will need this information if our system is at nonstandard conditions, which is most of the time.

A process for dissolving gold was known during the early days of the alchemists. Because metals like gold and platinum wouldn't react with strong acids, such as nitric acid, they were known as the royal metals. However, there existed a solution that would dissolve gold and platinum. These metals were soluble in a mixture of one part nitric acid and three parts hydrochloric acid known as *aqua regia*, or royal water. Aqua regia is used commercially as a first step in separating platinum from gold that is combined with other metals in their ore mixtures. Both metals are initially dissolved and then selectively precipitated from solution.

PRACTICE 19.6

Determine the cell voltage at standard conditions for the following redox reactions:

$$CH_4(g) + H_2O(g) \rightarrow CO(g) + H_2(g)$$

$$Ag(s) + H_2O_2(aq) \rightarrow H_2O(l) + Ag^+(aq)$$

See Problems 27 and 28.

Video Lesson: Electrochemical Cells

Application

CHEMICAL ENCOUNTERS: The Electric Eel

Blood supply
Electrically conductive tissue
Nucleus
Papillae
Nerve
Connective tissue

19.4 Electrochemical Cells

It may be "shocking" to learn, but it's true: The electric eel is a formidable opponent. When startled, stepped on, or hunting for food, the eel can deliver up to 1 ampere (1 coulomb of charge per second) at 600 V, enough electrical energy to stun or even kill large animals. (To be accurate, the electric eel isn't an eel. It lives in fresh water and is really a fish.) How does the electric eel generate the electricity used in hunting?

The powerful shock that the eel can deliver is produced by 5000 to 10,000 specialized cells, called electroplates, in its tail (see Figure 19.10). Using biochemical processes, the eel charges up the electroplates in much the same manner as muscle and nerve cells are charged. Then, when a nerve impulse is sent to the tail, the electroplates discharge their stored potential. Can you think of an electrical storage device similar to the electric eel's that we use every day? The electrochemical cell seems to fit the definition. We'll explore the electrochemical cell in this section

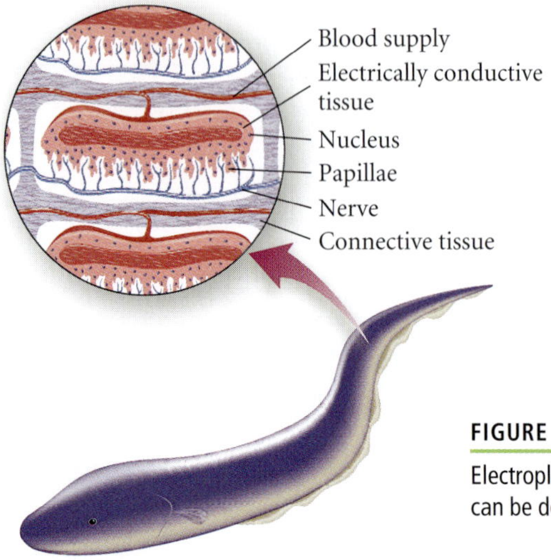

FIGURE 19.10

Electroplates in the tail of the electric eel develop a potential charge that can be delivered to shock its prey. (Drawing by Rick Simonson.)

and learn how we can make use of the electrical energy of chemical reactions.

Electrochemical Cells in the Laboratory

Redox reactions, such as the dissolving penny, can take place inside a beaker just like other reactions. However, we can harness the energy from the electrons that are exchanged in the redox reaction if we modify the experimental setup. Take a look at what happens when we separate the copper penny from the nitric acid shown in Figure 19.11. The brown fumes produced by the subsequent oxidation of NO to NO_2 by O_2 still waft from the beaker, and the copper ions are still generated. The redox reaction is still working. Examine the experimental setup closely to see why it works. We have

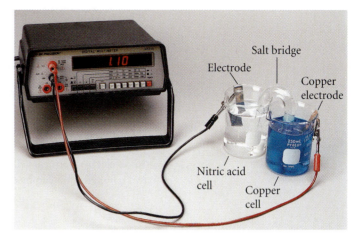

$$3Cu(s) + 6H^+(aq) + 2HNO_3(aq) \rightarrow 3Cu^{2+}(aq) + 2NO(g) + 4H_2O(l) \quad E°_{cell} = +0.62 \text{ V}$$

followed by the oxidation of NO to the poisonous NO_2 gas:

$$2NO(g) + O_2(g) \rightarrow 2NO_2(g) \qquad \Delta H° = -112 \text{ kJ}$$

The oxidation half-reaction is in the beaker on the right, and the reduction half-reaction is on the left. The two beakers are connected with a wire that transfers the electrons from beaker to beaker. At both ends of the wire is an electrode. The electrode in the oxidation reaction is called the anode, and the electrode in the reduction reaction is called the cathode. Remember, electrons are one of the products of the copper oxidation. They need someplace to go. Providing a wire for them to travel into the nitric acid reduction reaction keeps the reaction running. If we open the circuit on the wire, the reaction stops. The wire is an essential part of this electrochemical cell.

The tube labeled **salt bridge** in the diagram contains an electrolyte such as potassium chloride or sodium nitrate and *allows ions to pass from beaker to beaker*. The salt bridge is needed because as electrons move from the beaker on the right (the oxidation reaction) to the beaker on the left (the reduction reaction), a strong positive charge will develop at the anode as more of the copper penny becomes Cu^{2+}. A similar thing happens in the other beaker; the H^+ is being removed to make NO and water, and the cathode becomes negatively charged. The developing charges attract the electrons as they move away from the

FIGURE 19.11

The electrochemical cell. Ira Remsen's observations are still valid in this setup. The nitric acid still acts on the copper. Note the presence of the salt bridge and the electrodes.

Visualization: Electrochemical Half-Reactions in a Galvanic Cell

Visualization: Galvanic (Voltaic) Cells

FIGURE 19.12

Using the electrochemical cell to light a bulb.

The salt bridge.

beaker. Without the salt bridge, the electrons would stop this type of movement. We must have a salt bridge in our electrochemical cell design so that the entire system can remain electrically neutral. The finished electrochemical cell produces the same reaction that Ira Remsen observed (nitric acid acts upon copper), but the movement of the electrons through the wire can be used to light a bulb, as shown in Figure 19.12. We have harnessed the electrons as electrical energy.

EXERCISE 19.7   Alkaline Batteries

Some electrochemical cells are constructed using an alkaline electrolyte. Here are two half-cells that can be employed:

$$Cd(OH)_2(s) \rightarrow Cd(s)$$
$$NiO_2(s) \rightarrow Ni(OH)_2(s)$$

a. Based on the information in the standard reduction potentials table (Table 19.3), write the two oxidation−reduction reactions possible from these half-reactions, and calculate their cell potentials.

b. Which could be used as the basis of an electrolytic cell? . . . of a galvanic cell? What conditions are required to obtain these specific cell potentials? (Recall our discussion in Section 19.1.)

c. When used as a voltaic cell, this set of reactions makes a useful battery that can power portable appliances and tools. Judging on the basis of the elements that make up the galvanic cell, can you identify the name of this type of battery? What significance does the reverse reaction have?

Solution

a. The two half-cell reactions are

$$Cd(OH)_2(s) + 2e^- \rightarrow Cd(s) + 2OH^-(aq) \qquad E° = -0.81\ V$$
$$NiO_2(s) + 2H_2O(l) + 2e^- \rightarrow Ni(OH)_2(s) + 2OH^-(aq) \qquad E° = +0.49\ V$$

If the first reaction is reversed, $E°_{cell} = (+0.81\ V) + (+0.49\ V) = +1.30\ V$.

$$Cd(s) + NiO_2(s) + 2H_2O(l) \rightarrow Cd(OH)_2(s) + Ni(OH)_2(s) \qquad E° = +1.30\ V$$

If the second reaction is reversed, $E°_{cell} = (-0.81\ V) + (-0.49\ V) = -1.30\ V$.

$$Cd(OH)_2(s) + Ni(OH)_2(s) \rightarrow Cd(s) + NiO_2(s) + 2H_2O(l) \qquad E° = -1.30\ V$$

In both cases, the $OH^-(aq)$ and the electrons cancel when the half-cell reactions are added.

b. To obtain these voltages, the reactions must be run at 25°C at 1 atm pressure (standard conditions). The concentration of $OH^-(aq)$ must be 1 M (solids have a constant concentration). The first oxidation–reduction reaction could form the basis of a galvanic cell, because $E°_{cell}$ is positive. The second would be electrolytic, because $E°_{cell}$ is negative.

c. This voltaic cell is better known as a NiCad battery. The reverse electrolytic reaction proceeds well, because once formed, the reaction products remain in contact with the electrodes. Therefore NiCad batteries can be recharged by running the cell in reverse by applying a reverse voltage greater than the cell potential.

PRACTICE 19.7

Draw the electrochemical cell for each of the redox reactions given in Practice 19.6. What is the potential of the galvanic cell? . . . of the electrolytic cell?

See Problems 57–59. ■

Cell Notation

Because the balanced chemical equations require a lot of time to write, shorthand notation (called **cell notation**) is often used when describing electrochemical cells. It may look hard to do, but cell notation is as easy as knowing your ABCs—that is, anode, bridge, cathode. Consider the reaction that takes place in the silver oxide cell found in pacemaker batteries:

$$Ag_2O(s) + Zn(s) \rightarrow 2Ag(s) + ZnO(s) \qquad E° = +1.86 \text{ V}$$

A closer look at the half-reactions tells us that silver is being reduced and zinc is being oxidized. The silver metal must be the cathode, and the zinc metal must be the anode. We write each half-reaction, and then, without even balancing them, we construct the overall shorthand notation.

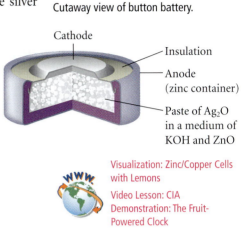

Cutaway view of button battery.

$$Zn(s) \rightarrow ZnO(s) \qquad \text{(oxidation, anode)} \qquad Zn(s) \,|\, ZnO(s)$$
$$Ag_2O(s) \rightarrow Ag(s) \qquad \text{(reduction, cathode)} \qquad Ag_2O(s) \,|\, Ag(s)$$

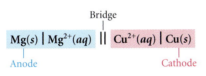

Bridge

$$\textbf{Zn}(s) \,|\, \textbf{ZnO}(s) \;\|\; \textbf{Ag}_2\textbf{O}(s) \,|\, \textbf{Ag}(s)$$

Anode   Cathode

EXERCISE 19.8 | Shorthand Notation

Write the shorthand cell notation for the following voltaic cell.

$$Cu^{2+}(aq) + Mg(s) \rightarrow Cu(s) + Mg^{2+}(aq)$$

Solution

Our goal is to write our ABCs—anode, bridge, and cathode. In order to do this, we need to know which half-reaction is the oxidation and which is the reduction. The copper ion gains two electrons, so it is reduced. The copper is the cathode. The magnesium loses two electrons, so it is oxidized. It is the anode. We now have enough information to write the cell notation:

Bridge

$$\textbf{Mg}(s) \,|\, \textbf{Mg}^{2+}(aq) \;\|\; \textbf{Cu}^{2+}(aq) \,|\, \textbf{Cu}(s)$$

Anode   Cathode

PRACTICE 19.8

Create a galvanic cell using these half-reactions, and write the cell notation. Check to make sure your reaction is written as a spontaneous redox reaction.

$$Fe^{2+}(aq) + 2e^- \rightarrow Fe(s)$$

$$Al^{3+}(aq) + 3e^- \rightarrow Al(s)$$

See Problems 55 and 56.

Batteries

Commercial batteries come in many shapes and sizes and are based on a wide variety of chemical processes (Table 19.5). The best battery for any application is usually chosen by considering power output, cost, convenience, size, and whether the battery will be rechargeable. The chemistry of commercial batteries is often rather complex compared to the simple electrochemical cells used in the teaching lab to convey the basic principles.

TABLE 19.5   Selected Batteries

Zinc–carbon battery Also known as a **standard carbon** battery. Zinc–carbon chemistry is used in all inexpensive AA, C, and D dry-cell batteries. The electrodes are zinc and carbon, with an acidic paste between them that serves as the electrolyte.

Alkaline battery Used in common Duracell and Energizer batteries. The electrodes are zinc and manganese oxide, with an alkaline electrolyte.

Lithium photo battery Lithium, lithium iodide, and lead-iodide are used in cameras because of their ability to supply power surges.

Lead–acid battery (rechargeable) Used in automobiles. The electrodes are made of lead and lead oxide with a strong acidic electrolyte.

Nickel–cadmium battery (rechargeable) The electrodes are nickel hydroxide and cadmium, with potassium hydroxide as the electrolyte.

Nickel–metal hydride battery (rechargeable) This battery is rapidly replacing nickel–cadmium because it does not suffer from the "voltage depression" that nickel-cadmiums do, in which repeated charging after only partial discharges prevents it from fully discharging.

Lithium–ion battery (rechargeable) With a very good power-to-weight ratio, this is often found in high-end laptop computers and cell phones.

Zinc–air battery This battery is lightweight and rechargeable.

Zinc–mercury oxide battery This is often used in hearing aids.

Silver–zinc battery This is used in aeronautical applications because the power-to-weight ratio is good.

Metal–chloride battery Used in electric vehicles.

Hydrogen fuel cell Used in electric vehicles and to power the space shuttle.

The Chemistry of Some Common Batteries

Nickel metal hydride (NiMH) rechargeable batteries are used in many cellular phones (Figure 19.13). During the charging phase, an external source of electricity causes water in the electrolyte (often aqueous potassium hydroxide) to react with a rare earth– or zirconium metal–based alloy at what will be the negative electrode of the battery when it is in operation. This generates hydrogen atoms that are absorbed into the alloy, and releases hydroxide ions:

$$\text{Alloy} + H_2O(l) + e^- \rightarrow \text{Alloy}-H(s) + OH^-(aq) \quad \text{(reduction)}$$

At the other electrode, which will be the positive electrode when the battery is powering the phone, nickel hydroxide reacts with hydroxide ions to form nickel oxyhydroxide, which has nickel in what for it is an unusual +3 oxidation state:

$$\text{Ni(OH)}_2(s) + OH^-(aq) \rightarrow \text{NiOOH} + H_2O + e^- \quad \text{(oxidation)}$$

When the battery is in use, the hydrogen atoms that were absorbed into the alloy at the negative electrode are released, combining with hydroxide ions to form water and supply the electrons that flow through a circuit to power the phone.

$$\text{Alloy}-H(s) + OH^-(aq) \rightarrow \text{Alloy} + H_2O(l) + e^- \quad \text{(oxidation)}$$

At the positive electrode, nickel oxyhydroxide is reduced back to nickel hydroxide by the electrons that arrive through the circuit, having done their work for us:

$$\text{NiOOH}(s) + H_2O(l) + e^- \rightarrow \text{Ni(OH)}_2(s) + OH^-(aq) \quad \text{(reduction)}$$

The cycle of charge and discharge can be repeated many times, to power all the talking and text messaging on the move that is such a pervasive part of modern life.

A typical nonrechargeable "alkaline" battery for a flashlight (Figure 19.14) uses the oxidation of zinc metal into zinc ions to generate the electrons for the electric current:

$$\text{Zn}(s) \rightarrow \text{Zn}^{2+}(aq) + 2e^- \quad \text{(oxidation)}$$

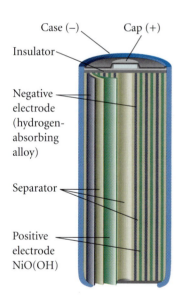

Case (−)   Cap (+)

Insulator

Negative electrode (hydrogen-absorbing alloy)

Separator

Positive electrode NiO(OH)

FIGURE 19.13

Nickel metal hydride cell.

FIGURE 19.14

A common dry cell battery.

FIGURE 19.15

A lithium ion battery.

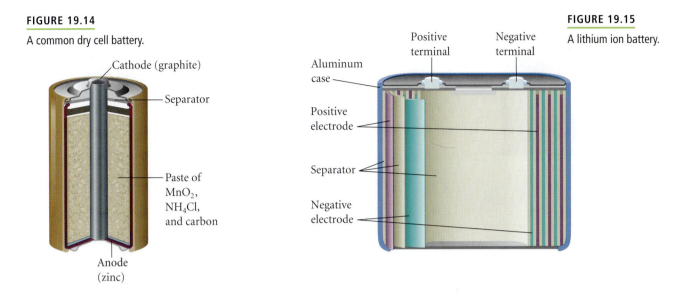

When electrons flow back into the battery at the other electrode, they combine with manganese dioxide:

$$2MnO_2(s) + 2H_2O(l) + 2e^- \rightarrow 2MnO(OH)(s) + 2OH^-(aq) \qquad \text{(reduction)}$$

Therefore, the indirect reaction of zinc with manganese dioxide is the source of the energy that lights the bulb.

Your laptop computer may be powered by a "lithium ion battery" (Figure 19.15). These batteries use lithium oxide mixed with other metal oxides as the positive electrode, and crystalline graphite with lithium ions intercalated within it as the negative electrode. Unlike most conventional batteries, however, the lithium ion battery is not powered by a redox reaction. Instead, lithium ions move back and forth within the battery during the cycle of charging and recharging, accompanied by electrons moving in the external circuit. During charging, the lithium ions are driven into the graphite cathode by application of the external electric current. When the battery is used as a source of power, the ions drift back to the lithium oxide anode. As the ions leave the cathode, electrons must travel around the external circuit, ensuring that there is no overall transfer of electric charge from cathode to anode as the battery is discharged.

19.5   Chemical Reactivity Series

Plumbers often use metal pipes to deliver water from the main water line to your sink faucet. In fact, plumbers used to use lead pipes, but because the lead in the pipe can leach into the water and cause heavy metal poisoning, that practice isn't followed anymore. Even though plastic pipes made from polyvinyl chloride (PVC) are much cheaper than metal pipes, there is still a demand for copper pipes. Why do plumbers use copper for pipes? The demand is mostly due to the durability of copper compared to plastic, but why don't they use iron, lithium, calcium, or magnesium pipes?

The Table of Standard Reduction Potentials (Table 19.3) provides not only cell potentials but also a *ranking* of reducing agents and oxidizing agents. This ranking is called a **reactivity series**. Table 19.6 shows a relative listing of some of the more common metals and includes hydrogen as a reference point. The strongest reducing agents (those elements that are most easily oxidized) are the most reactive metals and can be found at the top of the table: Li, Na, and Mg.

Copper pipes in a home transport water but don't corrode easily.

Sodium reacting with water.

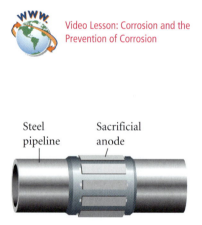
Reaction of copper with water.

Video Lesson: Corrosion and the Prevention of Corrosion

TABLE 19.6	Reactivity Series of the Metals		
	Ion	**Atom**	
Ions difficult to displace	Li^+	Li	**Metals that react with water**
	K^+	K	
	Ca^{2+}	Ca	
	Na^+	Na	
	Mg^{2+}	Mg	
	Al^{3+}	Al	
	Zn^{2+}	Zn	**Metals that react with acid**
	Fe^{2+}	Fe	
	Ni^{2+}	Ni	
	Pb^{2+}	Pb	
Ions easy to displace	H^+	H_2	
	Cu^{2+}	Cu	
	Ag^+	Ag	**Metals that are highly unreactive**
	Au^{3+}	Au	

Steel pipeline   Sacrificial anode

FIGURE 19.16

A sacrificial anode on a steel pipeline.

Because these metals are such good reducing agents, you would not want an earring made out of them; a better choice would be a relatively unreactive metal (and a weak reducing agent) such as gold, copper, or silver, found at the bottom of the table. "Fourteen-carat" gold (which means that $^{14}\!/_{24}$ of the sample is gold) is an alloy of gold, copper, and silver. "Twenty-four-carat" gold is pure gold. The reactivity series has social importance beyond its significance to earrings. Underwater steel pipelines, which contain substantial amounts of iron, electrochemically corrode (the metal deteriorates via oxidation) as a consequence of the interaction of the iron with water, salt, and oxygen dissolved in the water. This **corrosion** can be minimized by putting, for example, magnesium strips in direct contact with the pipeline, shown in Figure 19.16. The magnesium, which is more chemically active than iron, will preferentially oxidize, ideally leaving the iron in its elemental form. The magnesium, in effect, is sacrificed for the good of the pipeline and therefore is known as a **sacrificial anode**. Other sacrificial anodes include aluminum wrapped around steel in hot water heaters and zinc coating the propellers and rudders of ships.

EXERCISE 19.9	Which Is More Reactive?

Iron, especially in the form of stainless steel, can be used in jewelry, such as in the post of an earring. Where does iron fall on the activity series? Based on the potential for the oxidation of iron to iron(III) ion, why is iron suitable (or not suitable) for jewelry?

Solution

From Table 19.3 we see that the oxidation of metallic iron, Fe(*s*), has a half-reaction potential of +0.04 V. Metallic copper (−0.34 V), silver (−0.80 V), and gold (−1.50 V) have negative potentials. From this information we can conclude that iron is more reactive than copper, silver, or gold. In this sense *pure* iron would not be suitable, and you may know that iron spontaneously reacts to make rust when in contact with moisture and air. *Stainless steel,* however, is an alloy formulated to inhibit corrosion. Some formulas have as much as 18% chromium and 8% nickel added to the iron. The half-reaction potential for stainless steel is different, and it is difficult to predict reactivity from the $E°$ values for its constituent elements.

PRACTICE 19.9

Arrange the following in decreasing order of reactivity:

    Na   Al   Ca   Cu

See Problems 63–66.

19.6   Not-So-Standard Conditions: The Nernst Equation

Native copper and other copper objects can be cleaned with relatively dilute solutions of nitric acid. Concentrated nitric acid is too strong for the job, as Ira Remsen noted, so ancient coins should not be cleaned with concentrated nitric acid. A very dilute solution in the hands of a professional, however, can transform a 1600-year-old coin into a masterpiece that looks as new as the day it was minted. **How does lowering the concentration of nitric acid change the reactivity?** It does reduce the rate of the reaction (see Chapter 15), but is there an effect on the potential of the reaction as well? What about the temperature of the reaction? We know that in general, the rate of a reaction increases as the temperature is raised (also from Chapter 15). Does cold, dilute nitric acid still react with copper? We can understand these relationships, which are at *nonstandard* conditions, by considering a simpler system in which the copper ion reacts with zinc metal.

If we put a zinc strip into a solution of copper(II) chloride, we note that a dark film immediately begins to form on the zinc strip (Figure 19.17a on page 852). Within an hour, the entire zinc strip has oxidized into the solution, and we are left with a brown mass of copper metal (Figure 19.17b). We also note the disappearance of the blue color that is characteristic of Cu^{2+} in water. This is consistent with the following half-reactions and overall cell reaction:

$$Cu^{2+}(aq) + 2e^- \rightarrow Cu(s) \qquad\qquad E° = +0.34\ V$$

$$\underline{\qquad Zn(s) \rightarrow Zn^{2+}(aq) + 2e^- \qquad E° = +0.76\ V \qquad}$$

$$Cu^{2+}(aq) + Zn(s) \rightarrow Zn^{2+}(aq) + Cu(s) \qquad E°_{cell} = +1.10\ V$$

We know from Section 19.3 that we can relate the free energy of a system to the cell potential:

$$\Delta G = -nFE$$

We also discussed in Section 14.6 the relationship between free energy change and the standard free energy change:

$$\Delta G = \Delta G° + RT\ln Q$$

in which Q = the reaction quotient, the ratio of the concentrations (more properly, the activities) of the products over the reactants; R is the universal gas constant; and T is the temperature in kelvins. For our current example,

$$\Delta G = \Delta G° + RT\ln\frac{[Zn^{2+}]_0}{[Cu^{2+}]_0}$$

We can substitute for ΔG the expression for the cell potential, $\Delta G = -nFE$:

$$-nFE = -nFE° + RT\ln\frac{[Zn^{2+}]_0}{[Cu^{2+}]_0}$$

Dividing each term by $-nF$ enables us to solve for the cell potential at nonstandard conditions:

$$E = E° - \frac{RT}{nF}\ln\frac{[Zn^{2+}]_0}{[Cu^{2+}]_0}$$

FIGURE 19.17
Zinc metal reacting with copper(II) chloride solution.

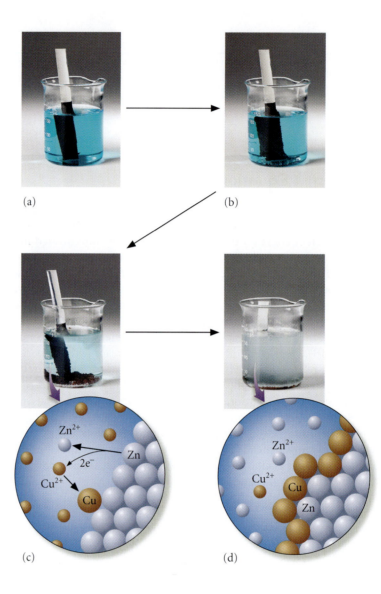

(a)　　(b)

(c)　　(d)

This is one form of the **Nernst equation**, named after Walther Hermann Nernst (1864–1941), a German chemist who studied the effect of concentration on the potential of an electrochemical cell. We can write a more general form of the equation, taking into account the reaction quotient for any process:

Video Lesson: The Nernst Equation

$$E = E° - \frac{RT}{nF} \ln Q$$

Because most reactions are conducted at standard temperature (25 °C, 298 K), we can calculate the term $\frac{RT}{F}$.

$$\frac{RT}{F} = \frac{8.3145 \text{ J·mol}^{-1}\text{·K}^{-1}(298 \text{ K})}{96485 \text{ C·mol}^{-1}} = 0.0257$$

which we can substitute into the Nernst equation in this form:

$$E = E° - \frac{0.0257}{n} \ln Q$$

Although our calculators can easily handle the natural logarithm ("ln") in the equation, we typically convert the equation to the more familiar base 10 "log" by multiplying by 2.3026 (that is, $\log(10) = 1$, and $\ln 10 = 2.3026$, so $\log = \ln/2.3026$).

To account for that, we multiply the coefficient, $\dfrac{0.0257}{n}$, by 2.3026, which gives us the *final, common form of the Nernst equation*:

$$E = E° - \frac{0.0592}{n} \log Q$$

Using this equation, we can determine the effect of lowering the concentration of nitric acid on the potential of the copper oxidation.

The Nernst equation can also be used to measure the concentration of a solution if the standard cell potential and the actual potential are known. Electrochemists take advantage of this use of the Nernst equation via ion-selective electrodes, in which the concentration of an ion such as chloride, ammonium, cadmium, nitrate, or hydrogen (in a pH electrode) is determined. And there's something else that's interesting about the Nernst equation. For example, let's consider the hypothetical redox reaction between copper metal and copper ions in solution. The potential of the reduction half-reaction and that of the oxidation half-reaction are the same at *standard conditions,* and, as we'd expect, no net potential should be noticed for an electrochemical cell containing these reactions.

$$Cu^{2+}(aq) + 2e^- \rightarrow Cu(s) \qquad\qquad E°_{red} = +0.34 \text{ V}$$
$$\underline{\qquad Cu \rightarrow Cu^{2+}(aq) + 2e^- \qquad\qquad E°_{ox} = -0.34 \text{ V}\qquad}$$
$$Cu(s) + Cu^{2+}(aq) \rightarrow Cu^{2+}(aq) + Cu(s) \qquad E°_{cell} = 0.00 \text{ V}$$

However, what would happen if we increased the concentration of the reactant copper ions to 2.0 *M* instead of the standard 1.0 *M*? Doing so changes the distribution of the species in the reaction. Using the Nernst equation, we can calculate the result of our modification.

$$E = E° - \frac{0.0592}{n} \log \left(\frac{1.0}{2.0} \right)$$
$$E = 0.00 - \frac{0.0592}{2} \log(0.5)$$
$$E = 0.00 - (-0.0089)$$
$$E = +0.0089 \text{ V}$$

FIGURE 19.18

The concentration cell. The voltage observed in the copper concentration cell is due to the differences in concentration of Cu^{2+} at the anode and cathode.

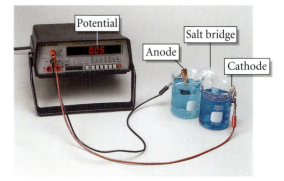

The electrochemical potential of the reaction is a nonzero value. By adjusting the concentrations of the products and reactants, we have created an electrochemical cell. This type of cell is called a **concentration cell** because the concentrations are driving the potential of the cell (Figure 19.18). This is the same type of potential that develops across the membrane of a muscle cell or nerve cell.

How does our cell notation change to reflect the fact that the conditions are not-so-standard? By indicating the concentrations in parentheses immediately after the species, we can immediately show how the reaction should be written. For example,

$$Cu(s) \mid Cu^{2+}(aq) \ (1.0 \ M) \parallel Cu^{2+}(aq) \ (2.0 \ M) \mid Cu(s)$$

EXERCISE 19.10   Heart Cell Potential

At the beginning of this chapter, we mentioned the electrical signals in the heart muscle. Given the differing concentrations of potassium ions inside and outside the heart cells, what is the electrochemical potential that corresponds to this concentration gradient? Assume that the electrochemical cell in the body can be represented by the following cell notation. (In truth, no elemental potassium exists in the human body. We use this concentration cell as a model merely to estimate the potential that is obtained in a heart muscle cell.)

$$K(s) \,|\, K^+(aq)\,(0.005\ M) \,\|\, K^+(aq)\,(0.166\ M) \,|\, K(s)$$

Solution

This concentration cell is based on the potassium half-reaction. The complete reaction is

$$K(s) + K^+(aq) \rightarrow K(s) + K^+(aq) \qquad E^\circ_{cell} = 0.00\ V$$

There is one electron involved in the reaction. Using the information from the Nernst equation, we get a potential of +0.090 V.

$$E_{cell} = E^\circ_{cell} - \frac{0.0592}{n} \log\left(\frac{[K^+_{product}]_0}{[K^+_{reactant}]_0}\right)$$

$$E_{cell} = 0.00 - \frac{0.0592}{1} \log\left(\frac{0.005}{0.166}\right)$$

$$E_{cell} = 0.00 - 0.0592 \log(0.03012)$$

$$E_{cell} = 0.00 - 0.0592(-1.521)$$

$$E_{cell} = 0.00 - (-0.09005)$$

$$E_{cell} = +0.090\ V$$

In the heart cell, the sodium gradient provides a potential of −0.052 V. The net potential across the membrane of the heart cell is +0.038 V.

PRACTICE 19.10

What is the potential of the following electrochemical cell written in shorthand cell notation? What is the half-reaction listed on the right-hand side? Is this a voltaic cell? (Assume that the pressure of hydrogen gas is 1.0 atm in each half-cell.)

$$H_2(g) \,|\, H^+(aq)\,(1.0\ M) \,\|\, H^+(aq)\,(0.10\ M) \,|\, H_2(g)$$

See Problems 69–72. ▮

When we listen to a battery-powered radio, the sound tends to get softer as the radio is used. What is happening inside the battery that causes this power loss? During the progress of the reaction inside the battery, the reactants are being used up as the products are being formed. According to the Nernst equation, as the value of Q gets larger, the modification to the standard cell potential (E°) gets more and more negative and closer and closer to zero. What is the result? As the reaction proceeds, the potential of the battery decreases as the system inches ever closer to equilibrium, and the music gets softer. At some point, the battery doesn't have enough voltage to run the radio. It has not yet reached equilibrium, but it is below the threshold that will allow the radio to operate. We say that the batteries are dead. As we have pointed out already, some batteries can be recharged, because their discharge reactions can be run in reverse. Rechargeable batteries can be charged only a finite number of times, typically in the range of 500 to 1000 times for household batteries, because the surfaces of the recharged electrodes do not form as cleanly as the original surface, and they eventually become too worn to be useful.

The Nernst Equation and the Equilibrium Constant

Video Lesson: Electrochemical Determinants of Equilibria

Measuring the standard cell potential is a very powerful way to solve for the equilibrium constant of a reaction. Here's how. When a redox reaction proceeds without any intervention, it will inevitably reach equilibrium. At that point the value of E_{cell} must become zero, and the free energy *change* for the process also becomes zero—our thermodynamic definition of equilibrium, introduced in Chapter 16. When this occurs, the Q value in the Nernst equation is equal to the equilibrium constant K. We can show this reasoning by using the following equations:

$$\Delta G = \Delta G° + RT \ln Q$$

At equilibrium, $\quad \Delta G = 0$

so $\quad\quad\quad 0 = \Delta G° + RT \ln K$ (note that "Q" has become "K")

$$\Delta G° = -RT \ln K$$

We know that $\quad \Delta G° = -nFE°$

Substituting $-nFE°$ into the previous equation yields

$$-RT \ln K = -nFE°$$

Solving for K, we get $\quad \ln K = \dfrac{nFE°}{RT}$

Calculating $\dfrac{F}{RT}$ at standard conditions (as we did for the Nernst equation), we find that

$$\ln K = 38.92 \, nE°$$

Converting from natural ln to base 10 log (by dividing by 2.3026, as we did in the Nernst equation, yields

$$\log K = 16.9 \, nE°$$

To clearly show the connection to the Nernst equation, we will invert 16.9 and put the result in the denominator.

$$\log K = \frac{nE°}{0.0592}$$

We can use the equation in this way. Alternatively, we can take the antilog of both sides and use it in this way:

$$K = 10^{\frac{nE°}{0.0592}}$$

Which form of the equation you use depends mostly on your comfort level. We can solve for the equilibrium constant either way. The key point is that *our understanding of the thermodynamic meaning of equilibrium enables us to relate cell potential to the equilibrium constant.* The rest is just manipulating equations to get where we want to go. Let's see how we use this relationship to determine the equilibrium constant for the copper–nitric acid reaction performed by Ira Remsen. Recall that the reaction is

$$3Cu(s) + 6H^+(aq) + 2HNO_3(aq) \rightarrow 3Cu^{2+}(aq) + 2NO(g) + 4H_2O(l) \quad E°_{cell} = +0.62 \text{ V}$$

$$K = 10^{\frac{nE°}{0.0592}}$$

$$K = 10^{\frac{6(+0.62)}{0.0592}} = 10^{62.8} \approx 10^{63}$$

Alternatively,

$$\log K = \frac{nE°}{0.0592} = \frac{6(+0.62)}{0.0592} = 62.8$$

Raising both sides to the power of 10 yields

$$K \approx 10^{63}$$

This is yet another confirmation that the reaction does, in fact, proceed toward products. As we have this discussion, however, please keep in mind the difference between thermodynamics and kinetics. Thermodynamics answers the question "Can a process occur spontaneously?" It says absolutely nothing about speed. Kinetics addresses the issues of rates and mechanisms. All that our calculations tell us is that nitric acid *can* react spontaneously with the copper penny. They don't say how fast. That's a question of kinetics.

EXERCISE 19.11   Equilibrium Constants and Cell Potential

Vanadium(V) ion can be reduced stepwise (that is, to V^{4+}, V^{3+} and, finally, V^{2+}) by reaction with a "Jones Reductor," a zinc–mercury amalgam. The reaction for the reduction of V^{5+} to V^{4+} ion includes the following two half-reactions:

$$VO_2^+(aq) + H^+(aq) \rightarrow VO^{2+}(aq) + H_2O(l) \qquad E° = +1.00 \text{ V}$$

$$Zn^{2+}(aq) \rightarrow Zn(s) \qquad E° = -0.76 \text{ V}$$

Calculate the equilibrium constant for this reaction.

First Thoughts

There are a number of steps to solving this problem. One way to figure out what we need to do in working forward is to start by working backward (where do we want to be, and how do we get there?). We want the value for K. In order to get that, we need the value for $E°$. In order to get *that*, we need to have a balanced redox equation for the reduction of VO_2^+ to VO^{2+}, in which V^{5+} is reduced to V^{4+}, and Zn^0 is oxidized to Zn^{2+}. Our order of operations, then, is

1. Balance the redox reaction.
2. Calculate the $E°$ value for the reaction.
3. Calculate the equilibrium constant, knowing the number of electrons exchanged in the reaction, along with the value for $E°$.

How might we assess whether the answer we calculate makes sense? At this point, because we know that the reduction of vanadium ion does occur, our equilibrium constant should be greater than 1. How much greater will depend on the cell voltage and on the number of electrons transferred in the process.

Solution

The balanced half-reactions and overall cell reaction are

$$2VO_2^+(aq) + 4H^+(aq) + 2e^- \rightarrow 2VO^{2+}(aq) + 2H_2O(l) \qquad E° = +1.00 \text{ V}$$

$$Zn(s) \rightarrow Zn^{2+}(aq) + 2e^- \qquad E° = +0.76 \text{ V}$$

$$2VO_2^+(aq) + 4H^+(aq) + Zn(s) \rightarrow 2VO^{2+}(aq) + Zn^{2+}(aq) + 2H_2O(l) \quad E° = +1.76 \text{ V}$$

$$K = 10^{\frac{nE°}{0.0592}}$$

$$K = 10^{\frac{2(+1.76)}{0.0592}} = 10^{59.5} \approx 10^{60}$$

Alternatively,

$$\log K = \frac{nE°}{0.0592} = \frac{2(+1.76)}{0.0592} = 59.5$$

Raising both sides to the power of 10 yields

$$K \approx 10^{60}$$

Further Insights

Does our answer make sense? The cell voltage is very high and positive in this two-electron transfer, so our equilibrium constant is large, indicating that the reduction of vanadium proceeds essentially to completion.

Here are pictures of the vanadium solutions, showing each ion from V^{5+} (leftmost picture) to V^{2+} (rightmost picture). We will discuss the reasons why transition metal ions in solution have color, and often change color when reduced or oxidized, in the next chapter.

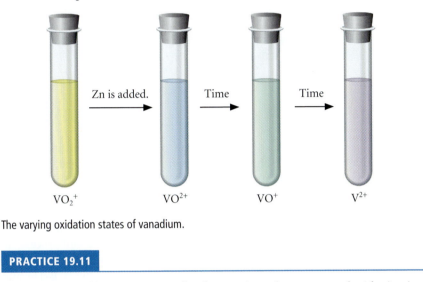

The varying oxidation states of vanadium.

<table>
<tr><td>PRACTICE 19.11</td></tr>
</table>

What is the equilibrium constant for the reaction of copper metal with zinc ion, discussed at the beginning of this section?

See Problem 68.

19.7 Electrolytic Reactions

Video Lesson: Electrolytic Cells
Tutorial: Electrolytic Cells

Some metals, including copper, gold, and silver, are found in their pure elemental state in the environment. On the other hand, aluminum metal, based on its reactivity (Section 19.5) is found only chemically combined in ores such as bauxite (hydrated aluminum oxide; $Al_2O_3 \cdot H_2O$ or $Al_2O_3 \cdot 3H_2O$). In fact, aluminum, largely in the form of the aluminum oxides and silicates, makes up 8.1% of the Earth's crust. However, we know that pure aluminum can be produced, because it is a major component in so many common products: the can in which we store our soda, the wrap in which we put our fish for freezing, and the lightweight bicycle we ride down the street. How is aluminum metal made from bauxite?

The basic process, called electrolysis, entails passing a current through a solution of metal ions in an electrochemical cell *in the direction opposite to the spontaneous reaction.* Doing so forces the nonspontaneous reaction to occur. This process, also known as **electrowinning**, is responsible for the manufacture and purification not only of aluminum but also of many other metals.

$$Al^{3+} + 3e^- \rightarrow Al$$

$$Cu^{2+} + 2e^- \rightarrow Cu$$

$$Ag^+ + e^- \rightarrow Ag \qquad etc....$$

Electrowinning is the most inexpensive method for making aluminum and magnesium metals. Electrowinning of metals such as aluminum and magnesium

FIGURE 19.19

The Hall–Heroult process. This process is still used today to satisfy the world's demand for aluminum products. The product, molten aluminum, is relatively dense and settles to the bottom of the electrochemical cell, where it is drawn off and poured into castings.

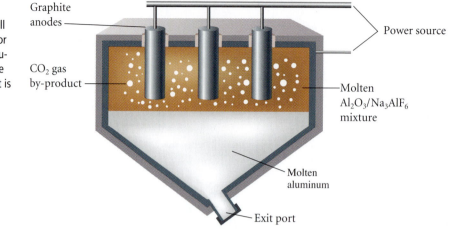

is the isolation of pure metals from a solution of metal ions, and it has been known for quite some time. Humphrey Davey, an English chemist, used this process to isolate metallic sodium from NaCl in 1807. However, the electrowinning of aluminum wasn't invented until 1886. Working independently, Charles M. Hall (an American chemist) and Paul Heroult (a French chemist) discovered that if bauxite ore was purified to alumina (Al_2O_3) and dissolved in molten cryolite (Na_3AlF_6), metallic aluminum could be made. In the **Hall–Heroult process**, an electrical current is passed into a molten mixture of alumina and cryolite to make molten aluminum (Figure 19.19). The overall reaction used in the Hall–Heroult process, shown below, is much simpler than the reactions that take place at the electrodes. Why can't an electric current be applied directly to an aqueous solution of Al^{3+} in order to produce aluminum metal? Because the potential is so large, instead of obtaining aluminum metal from the aqueous solution, the water undergoes electrolysis to form hydrogen and oxygen gases.

$$2Al_2O_3(l) + 3C(s) \rightarrow 4Al(l) + 3CO_2(g) \qquad E° \approx -2.1 \text{ V}$$

The negative potential tells us that this redox reaction is not spontaneous. In fact, the reverse reaction—that is, $4Al(l) + 3CO_2(g) \rightarrow 2Al_2O_3(l) + 3C(s)$— is quite spontaneous ($E° \approx +2.1$ V). Modifying the concentrations and temperatures helps a little to make the potential of the overall reaction less negative, via the Nernst equation, but not enough to make the reaction spontaneous. If we apply a positive potential to the reaction that is larger than the negative potential expressed by the electrochemical cell, we can force the reaction to go forward.

Examination of a simpler redox reaction can be helpful here. For example, in concept, if we wish to make the copper–iron redox reaction spontaneous, we must supply a potential of just over +0.78 V to the reaction to make up for the fact that $E°$ is −0.78 V.

$$Cu(s) + Fe^{2+}(aq) \rightarrow Fe(s) + Cu^{2+}(aq) \qquad E° = -0.78 \text{ V}$$

When we do this, however, we note experimentally that the reaction still doesn't proceed toward products. If we supply a still greater positive potential, the reaction does proceed. The extra voltage required is called the **overpotential** of the reaction. Overpotentials can be fairly high, especially when the products of the reaction are gases. Unfortunately, we can't easily predict what the overpotential for a particular reaction will be. Instead we measure the overpotential experimentally.

The Applications of Electrolysis

The average American uses 142 tin cans each year. From what are tin cans made? That might sound like a silly question, but "tin" cans are actually made of steel coated with a very thin layer of tin (Figure 19.20). Approximately 0.25% of the mass of a tin can is actually tin, and chromium is becoming more common as a coating on the steel. Without the coating, the steel would rust and the contents would spoil. The tin coating on a steel can isn't applied like paint on a house. How is the tin applied to the steel?

The most common use for electrolysis is **electroplating**, or depositing metals onto a conducting surface. The result is a coating that is very tightly integrated into the surface of the metal. Because of this tight integration into the metal surface, the coating resists flaking and peeling. This coating makes the item more attractive and imparts some corrosion resistance or chemical resistance to the surface. Electroplating is used in the manufacture of inexpensive jewelry and chrome bumpers for your car, but most common electroplating today occurs in the manufacture of tin cans.

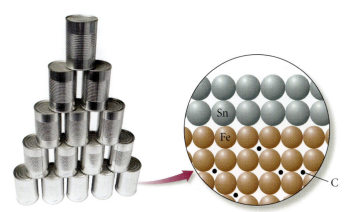

FIGURE 19.20

Tin cans are actually steel cans electroplated with a thin coat of tin. The tin resists corrosion and helps keep the contents fresh.

Application
CHEMICAL
ENCOUNTERS:
Electroplating

Calculations Involving Electrolysis

The process used to coat a steel can with tin is an example of electrolysis. How is it done? The steel can is hooked up to a power supply and dipped into a solution of tin ions (Sn^{2+}). A block of tin metal is also placed in the solution and connected to the power supply. Then a current is applied to the can (Figure 19.21). In the terminology of the electrolytic cell, the block of tin metal becomes the anode and the can becomes the site of reduction (the cathode) for tin ions.

Michael Faraday (remember him from our discussion on potentials and spontaneity) noted that the amount of current applied to a cell is directly proportional to the amount of metal that can be deposited in an electrolytic reaction. We can represent this mathematically as

$$g = \frac{A \cdot s \cdot (M)}{F} \left(\frac{\text{mol metal}}{\text{mol } e^-} \right)$$

where *A* is the number of amperes applied to the can (1 ampere = 1 coulomb of charge per second)
   s is the number of seconds that the current is applied
   <u>*M*</u> is the molar mass of the metal
   F is Faraday's constant (96,485 coulombs/mol electrons)
   the ratio (mol metal/mol e^-) is the mole ratio of the reduction half-reaction

It is helpful to look at this calculation as an extended unit conversion problem. Exercise 19.12 shows how this is done. Note that the unit conversion problem is the same as the equation shown above.

Video Lesson: The Stoichiometry of Electrolysis

FIGURE 19.21

Electroplating a tin can. The tin can acts as the cathode in the electrolysis experiment.

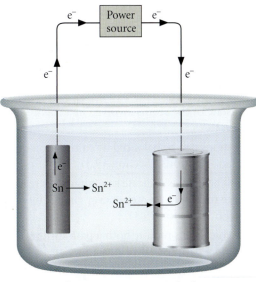

Anode   Cathode

To save money and still have a beautiful set of dinnerware, a chemist decides to electroplate the metal with gold. How many grams of gold will be electroplated on a fork if 2.5 A is applied to the fork for 20 s?

Solution

The number of grams can be calculated by starting with the current and performing a unit conversion. Note that the unit amperes (amps, A) can be written as coulombs/second (C/s). We also need to examine the reduction half-reaction from Table 19.3 to determine the number of electrons involved in the process.

$$Au^{3+}(aq) + 3e^- \rightarrow Au(s)$$

$$\frac{2.5\ C}{s} \times \frac{20\ s}{} \times \frac{1\ mol\ e^-}{96{,}485\ C} \times \frac{1\ mol\ Au}{3\ mol\ e^-} \times \frac{196.97\ g\ Au}{1\ mol\ Au} = 0.034\ g\ Au$$

$$\text{Amperes} \times \text{time} \times \frac{1}{\text{faraday}} \times \text{mole ratio} \times \underline{M} = \text{grams}$$

This means that our chemist would need to make sure to buy at least 0.034 g of gold per fork.

PRACTICE 19.12

How many grams of tin will be deposited from a solution of tin(II) nitrate on a steel can if 0.45 A are applied to the can for 1.5 h?

See Problems 73–78.

    The calculations can also be done in reverse. If we want to coat our steel can with a certain number of grams of tin, we can use either the equation or the unit conversion method to calculate the number of amps that need to be applied.

    As we've seen in this chapter, electrochemistry is a very useful topic, especially in today's society. Learning how muscles in our body begin a contraction can help us understand how the heart works and why it is so important to maintain sufficient levels of electrolytes during physical exercise. And as we become increasingly mobile, the demand for longer-lasting batteries will only increase. Batteries power our cell phones, our portable CD players, and even our cars. Electroplating the surfaces of many everyday items makes them appear expensive and confers resistance to corrosion. In fact, everywhere we look, there's electrochemistry!

The Bottom Line

- Redox reactions involve both a reduction and an oxidation half-reaction. (Sections 19.1 and 19.3)

- Redox reactions can be identified by determining the oxidation state of the atoms involved in a reaction. (Section 19.2)

- Redox reactions can be balanced by summation of balanced half-reactions. (Section 19.3)

- Positive cell potentials indicate a spontaneous reaction and are related to the free energy change by $\Delta G° = -nFE°$. (Section 19.3)

- Electrochemical cells require both a path for the electrons and a path for other ions. (Section 19.4)

- The oxidation reaction takes place at the anode. The reduction reaction takes place at the cathode. (Section 19.4)

- The Nernst equation relates the actual potential of a redox reaction to conditions other than the standard conditions. (Section 19.6)

- Half-reaction potentials can be used to determine the relative reactivity of metals. Organization of the metals in this fashion is known as a chemical reactivity series. (Section 19.5)

- Cell potentials enable us to calculate equilibrium constants. (Section 19.6)

- Electrowinning and electroplating are examples of electrolysis reactions. In electrolysis, a positive potential that includes the overpotential is applied to force the reaction to run in reverse. (Section 19.7)

Key Words

anode The electrode at which oxidation takes place. (*p. 826*)

battery Two or more voltaic cells joined in series. (*p. 826*)

cardiologist A heart specialist. (*p. 824*)

cathode The electrode at which reduction takes place. (*p. 826*)

cell notation The chemist's shorthand used to describe electrochemical cells. (*p. 847*)

concentration cell A cell in which different concentrations of identical ions on both sides of the cell provide the driving force for the reaction. (*p. 853*)

corrosion The deterioration of a metal as a consequence of oxidation. (*p. 850*)

disproportionation A reaction in which a single reactant is both the oxidizing and the reducing agent. (*p. 830*)

electrochemical cell A device that allows the exchange between chemical and electrical energy. (*p. 825*)

electrochemistry The study of the reduction and oxidation processes that occur at the meeting point of different phases of a system. (*p. 825*)

electrochemists Scientists who create or analyze systems that allow exchanges between chemical and electrical energy. (*p. 824*)

electrode A metal surface that acts as a collector or distributor for electrons. (*p. 826*)

electrodics The study of the interactions that occur between a solution of electrolytes and an electrical conductor, often a metal. (*p. 825*)

electrolytic cell A cell that requires the addition of electrical energy to drive the chemical reaction under study. (*p. 826*)

electromotive force (emf) A measure of how strongly a species pulls electrons toward itself in a redox process. Also known as voltage. (*p. 834*)

electroplating The process of depositing metals onto a conducting surface. (*p. 859*)

electrowinning The isolation of pure metals from a solution of metal ions. (*p. 857*)

Faraday's constant A unit of electric charge equal to the magnitude of charge on a mole of electrons. (*p. 836*)

fuel cell An electrochemical cell that utilizes continually replaced oxidizing and reducing reagents to produce electricity. (*p. 826*)

galvanic cell A cell that produces electricity from a chemical reaction. Also known as a voltaic cell. (*p. 825*)

gel electrophoresis The use of electric fields to separate ions. (*p. 825*)

half-reaction An equation that describes the reduction or oxidation part of a redox reaction. (*p. 826*)

Hall–Heroult process The most widely used process for the preparation of aluminum from bauxite. (*p. 858*)

ionics The study of the behavior of ions dissolved in liquids. (*p. 825*)

Nernst equation The equation used to determine the cell potential at nonstandard conditions. (*p. 852*)

nitrogen cycle The path that nitrogen follows through its different oxidation states on Earth. (*p. 831*)

overpotential The extra potential needed, above that which is calculated, in order to make an electrochemical process proceed. (*p. 858*)

oxidation The loss of electrons. (*p. 826*)

oxidation reaction In a redox reaction, the half-reaction that supplies electrons. (*p. 826*)

oxidation state A bookkeeping tool that gives us insight into the distribution of electrons in a compound. Also known as oxidation number. (*p. 827*)

oxidizing agent A substance that causes the oxidation of another substance. (*p. 830*)

potential The driving force (to perform a reaction) that results from a difference in electrical charge between two points. (*p. 824*)

reactivity series A ranking of the electrochemical reactivity of some elements. (*p. 849*)

redox reactions Chemical reactions in which reduction and oxidation occur. (*p. 824*)

reducing agent A substance that causes the reduction of another substance. (*p. 830*)

reduction The gain of electrons. (*p. 827*)

reduction reaction In a redox reaction, the half-reaction that acquires electrons. (*p. 826*)

sacrificial anode A material that will oxidize more easily than the one we seek to protect from oxidation. (*p. 850*)

salt bridge A device containing a strong electrolyte that allows ions to pass from beaker to beaker. (*p. 845*)

standard hydrogen electrode reaction (SHE) A reference half-reaction of the reduction of hydrogen ion to hydrogen gas, against which to compare our reduction. (*p. 835*)

standard potential ($E°$) The measure of the potential of a reaction at standard conditions. (*p. 834*)

volt The SI unit of potential. (*p. 834*)

voltage A measure of how strongly a species pulls electrons toward itself. Also known as electromotive force (emf). (*p. 834*)

voltaic cell A cell that produces electricity from a chemical reaction. Also known as a galvanic cell. (*p. 825*)

The answers to the odd-numbered problems and some selected problems appear at the back of the book, as represented by the blue numbering.

19.1 What Is Electrochemistry?

Skill Review

1. Explain why the descriptive term *battery* is often used, but is technically not correct, to describe these most commonly purchased sources of electricity.

2. Provide two other names for an electrochemical cell.

Chemical Applications and Practices

3. Every electrochemical cell is developed around two types of chemical reactions. Name and describe both of these connected reactions.

4. List three properties of electrochemical cells that are considered when designing an appropriate power source.

Section 19.2 Oxidation States—Electronic Bookkeeping

Skill Review

5. Determine the oxidation number of each atom in the structure of dimethylsulfoxide (DMSO).

6. Determine the oxidation number of each atom in the structure of periodic acid.

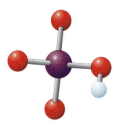

7. In which of the following compounds would the chlorine atom have the most positive oxidation number? In which would chlorine have the most negative oxidation number? Cl_2; ClO_2; $NaClO_4$; HCl

8. Which species in the following list shows the nitrogen atom in its most reduced form? Which depicts nitrogen in its most oxidized form? N_2; HNO_3; NH_3

9. Put the following compounds in order from the lowest to the highest oxidation number for nitrogen.

NO N_2O NO_2 N_2H_4 NH_3

10. Put the following compounds in order from the lowest to the highest oxidation number for carbon.

$C_6H_{12}O_6$ CO_2 CH_3OH CH_4 C_6H_6

11. Determine the oxidation state for each atom in the following compounds:
    a. $KMnO_4$ b. $LiMnO_2$ c. NH_4ClO_4

12. Determine the oxidation state for each atom in the following compounds:
    a. K_2MnCl_4 b. Cr_2O_3 c. $C_{12}H_{22}O_{11}$

13. Use the following four terms or expressions to identify each of the chemical situations indicated: (*is oxidized; is reduced; is an oxidizing agent; is a reducing agent*). You may use as many terms as apply.
    a. An atom has gained an electron.
    b. An atom increases its oxidation number.
    c. The oxidation number of an atom changes from −2 to −3.

14. Use the following four terms or expressions to identify each of the chemical situations indicated: (*is oxidized; is reduced; is an oxidizing agent; is a reducing agent*). You may use as many terms as apply.
    a. An atom decreases its oxidation number.
    b. An atom loses two electrons.
    c. The oxidation number of an atom changes from +3 to +5.

15. Supply the oxidation number of each atom, on both sides of the reaction arrow, in the equation

$$3Mg(s) + 2H_3PO_4(aq) \rightarrow Mg_3(PO_4)_2(aq) + 3H_2(g)$$

16. Supply the oxidation number of each atom, on both sides of the reaction arrow, in the equation

$$2AgNO_3(aq) + Cu(s) \rightarrow Cu(NO_3)_2(aq) + 2Ag(s)$$

Chemical Applications and Practices

17. The main active ingredient in commercial household bleach is the hypochlorite ion, ClO^-.
    a. What is the oxidation number of the Cl atom in the ion?
    b. Hypochlorite is known as a good oxidizing agent. Would that property indicate that the Cl tends to gain or lose electrons as it reacts? Explain the logic of your response.

18. The combustion of propane in a portable burner is a common redox reaction. Examine each component of the equation. Assign an oxidation number to each atom, and determine which component is acting as the reducing agent.

$$C_3H_8(g) + O_2(g) \rightarrow CO_2(g) + H_2O(g)$$

19. The brilliant red color of many fireworks is due to the presence of strontium. However, strontium metal is not typically found in its pure state. The following redox reaction depicts the isolation of strontium from molten strontium chloride.

$$SrCl_2(l) \rightarrow Sr(s) + Cl_2(g)$$

    a. Assign an oxidation number to each of the atoms in the reaction.
    b. What has been reduced in the reaction?

20. Sodium thiosulfate is familiar to photographers as "hypo." It helps dissolve some of the silver salts used in developing photographs. Aqueous solutions of sodium thiosulfate can undergo disproportionation reactions, as shown here:

$$S_2O_3{}^{2-}(aq) + H^+(aq) \rightarrow H_2O(l) + S(s) + SO_2(g)$$

    a. What is the change in the oxidation number of S in $S_2O_3{}^{2-}$ in the oxidation portion of the reaction?
    b. What is the change in the oxidation number of S in $S_2O_3{}^{2-}$ in the reduction portion of the reaction?

21. Four equations can be written to describe the rusting of iron:

$$O_2(aq) + 2H_2O(l) + 4e^- \rightarrow 4OH^-(aq)$$
$$Fe(s) \rightarrow Fe^{2+}(aq) + 2e^-$$
$$Fe^{2+}(aq) + 2OH^-(aq) \rightarrow Fe(OH)_2(s)$$
$$Fe(OH)_2(s) + OH^-(aq) \rightarrow FeO(OH)(s) + H_2O(l) + e^-$$

    a. Which of these are reduction or oxidation half-reactions?
    b. What other kind of equation is present here?
    c. Combine the equations to give the overall equation that describes the formation of rust, $FeO(OH)(s)$, from iron, $Fe(s)$.

22. In the chapter, we discussed the reaction of hydrogen and oxygen to give water. How is it possible for this reaction to proceed either as an explosion (as in the space shuttle main engines) or as a gentle, readily managed source of electricity?

Section 19.3 Redox Equations

Skill Review

23. The superscript on the symbol $E°$ refers to standard conditions for electrochemical reactions. What specifically does that indicate for the concentrations and pressure of the reacting system?

24. The voltage values (emf) given on a standard reduction table are referenced to a standard called SHE.
    a. To what do the letters refer?
    b. What is the voltage of the reference?

25. Complete each of these statements using the word *positive*, *negative*, *spontaneous*, or *nonspontaneous*.
    a. When $E°$ is negative, the electrochemical reaction is
    _____.
    b. When $E°$ is positive, the value for $\Delta G°$ is _____.
    c. When a reaction is spontaneous, the values for $\Delta G°$ will be _____, and the values for $E°$ will be _____.
    d. Nonspontaneous redox reactions have a _____ value for $E°$.

26. The typical battery used in a standard flashlight produces approximately 1.5 V. If the value for $\Delta G°$ of the reaction were $-289{,}500$ J, what would you calculate as the moles of electrons exchanged in the balanced redox reaction?

27. Combine these half-reactions in such a way that a galvanic cell results, and calculate the cell potential.

$$Fe^{3+} + e^- \rightarrow Fe^{2+} \qquad E° = 0.77 \text{ V}$$
$$Fe^{2+} + 2e^- \rightarrow Fe \qquad E° = -0.44 \text{ V}$$

28. Combine these half-reactions in such a way that a galvanic cell results, and calculate the cell potential.

$$Sn^{2+} + 2e^- \rightarrow Sn \qquad E° = -0.14 \text{ V}$$
$$Sn^{4+} + 2e^- \rightarrow Sn^{2+} \qquad E° = +0.15 \text{ V}$$

29. Calculate the free energy change for the cell in Problem 27.

30. Calculate the free energy change for the cell in Problem 28.

31. What is the free energy change for the following reaction at standard conditions?

$$PbO_2 + 4H^+ + 2Hg + 2Cl^- \rightarrow Pb^{2+} + 2H_2O + Hg_2Cl_2$$
$$E°_{cell} = 1.12 \text{ V}$$

32. What is the free energy change for the following reaction at standard conditions?

$$O_2 + 4H^+ + 2Ni \rightarrow 2H_2O + 2Ni^{2+} \qquad E°_{cell} = 1.46 \text{ V}$$

33. What are three considerations or aspects that must be "balanced" in a balanced redox reaction?

34. In this balanced redox reaction, chlorine is shown to replace bromide ions from a solution.

$$Cl_2(aq) + 2Br^-(aq) \rightarrow 2Cl^-(aq) + Br_2(aq)$$

    What is the value of n in the overall reaction?

35. Balance these half-reactions in acidic solution. Is each a reduction or an oxidation? Which substance is oxidized and which is reduced?
    a. $CO_2 \rightarrow H_2C_2O_4$ b. $Np^{4+} \rightarrow NpO_2{}^+$

36. Balance these half-reactions in acidic solution. Is each a reduction or an oxidation? Which substance is oxidized and which is reduced?
    a. $I_2 \rightarrow IO_3{}^-$ b. $NO_3{}^- \rightarrow NO$

37. Balance these redox equations in acidic solution:
    a. $Sn^{2+} + Cu^{2+} \rightarrow Sn^{4+} + Cu^+$
    b. $S_2O_3{}^{2-} + I_3{}^- \rightarrow S_4O_6{}^{2-} + 3I^-$
    c. $SO_3{}^- + Fe^{3+} \rightarrow SO_4{}^{2-} + Fe^{2+}$

38. Balance these redox equations in acidic solution:
    a. $Al + Cu^{2+} \rightarrow Al^{3+} + Cu$
    b. $UO_2{}^{2+} + Ag + Cl^- \rightarrow U^{4+} + AgCl$
    c. $H_2SO_4 + HBr \rightarrow SO_2 + Br_2$

39. Balance the redox equation that illustrates the reaction of solid copper and dichromate, first in acidic and then in basic solution.

$$Cu(s) + Cr_2O_7{}^{2-}(aq) \rightarrow Cu^{2+}(aq) + Cr^{3+}(aq)$$

40. Balance the redox equation that illustrates the reaction of permanganate and methanol, first in acidic and then in basic solution.

$$MnO_4{}^- + CH_3OH \rightarrow CO_3{}^{2-} + MnO_4{}^{2-}$$

41. Balance this redox reaction, first in acidic and then in basic solution.

$$ClO_4{}^- + I^- \rightarrow ClO^- + IO_3{}^-$$

42. Balance this redox reaction, first in acidic and then in basic solution.

$$Zn + NO_3{}^- \rightarrow Zn^{2+} + NH_3$$

Chemical Applications and Practices

43. Using the standard reduction potentials found in the appendix, locate the half-cell reaction for zinc. Zinc is often used in the production of dry cell batteries. It is also used to protect other metals from oxidation.
    a. What is the $E°$ value for this reaction?
    b. What would be the value of $\Delta G°$ for the half-reaction?
    c. What would be the $E°$ value if the reaction stoichiometry were doubled?

44. Splitting water into hydrogen gas and oxygen gas is one technique being investigated as a way to produce hydrogen gas for use in fuel cells. The reaction is shown here as

    $$2H_2O \rightarrow 2H_2 + O_2$$

    If the $E°$ value for this nonspontaneous reaction were approximately -2.00 V, what would you calculate as the value for $\Delta G°$?

45. Use the Standard Table of Reduction Potentials to answer the following questions:
    a. Which of the three alkali metal ions Na, K, and Li has the least potential to attract electrons?
    b. Of the three halogens which would be the strongest oxidizing agent, F_2, Cl_2, or Br_2?
    c. If you prepared a battery by connecting two of the following three half-cells, which combination could produce the highest potential? Zn, Cu, Ni

46. Suppose you attempted to build a battery using lead (and Pb^{2+}) with chromium (and Cr^{3+}).
    a. Write out the half-cell reduction reactions and their $E°$ values for each.
    b. Which of the two would provide the reduction reaction, and which would provide the oxidation reaction?
    c. What would be the total $E°$ of the overall redox reaction?
    d. How many moles of electrons would be exchanged in the overall reaction?

47. The oxidation of copper metal gives rise to a beautiful green patina. An equation that illustrates the oxidation of copper is shown here. Balance this redox reaction.

    $$Cu(s) + CO_2(g) \rightarrow CuO(s) + C_2O_4{}^{2-}(aq)$$

48. One technique used for the detection of ethanol involves the following redox reaction with dichromate ions. Balance this redox reaction.

    $$C_2H_5OH(aq) + Cr_2O_7{}^{2-}(aq) \rightarrow CH_3CO_2H(aq) + Cr^{3+}(aq)$$

49. As with acid–base reactions, careful addition of a solution containing an oxidizing agent to a solution containing a reducing agent can be used for titration analysis. Standardizing a solution of potassium permanganate to be used later in a redox titration often involves reaction with a known amount of sodium oxalate. Balance the redox reaction used in the standardization process.

    $$C_2O_4{}^{2-}(aq) + MnO_4{}^-(aq) \rightarrow CO_2(g) + Mn^{2+}(aq) + H_2O(l)$$

50. Once a standard solution of potassium permanganate is prepared, it can be used to determine the concentration of iron in an unknown sample. For example, the iron content of a small steel sample could be obtained through a titration reaction with a standard permanganate solution. The redox reaction that would take place in the analysis is shown here (unbalanced, and after the iron has been prepared as $+2$ ion). Balance the redox titration and identify the reducing agent.

    $$MnO_4{}^-(aq) + Fe^{2+}(aq) \rightarrow Mn^{2+}(aq) + Fe^{3+}(aq) + H_2O(l)$$

51. One step in a common method for the analysis of vitamin C in juice drinks is to oxidize the vitamin C (ascorbic acid) with aqueous I_2. Balance this redox reaction.

    $$C_6H_8O_6(aq) + I_2(aq) \rightarrow I^-(aq) + C_6H_6O_6(aq)$$

52. There are many half-cell reaction combinations that could be used to make batteries. One such reaction would be to use system consisting of silver(II) oxide and zinc. The reaction shown below would have to occur in a *basic* environment. Balance this redox reaction.

    $$Zn(s) + AgO(s) \rightarrow Zn(OH)_2(s) + Ag(s)$$

Section 19.4 Electrochemical Cells

Skill Review

53. Define the following terms in your own words: cell, half-reaction, galvanic cell, voltaic cell, electromotive force.

54. Compare and contrast an electrolytic cell with a galvanic cell on the basis of sign on the $E°$ value, ability to do work, chemical process taking place at the anode, and spontaneity.

55. The following notation refers to a specific voltaic cell. Identify the species that would serve as the anode and the species that is the oxidizing agent.

    $$Zn(s) \,|\, Zn^{2+}(aq) \,||\, Cu^{2+}(aq) \,|\, Cu(s)$$

56. The following notation refers to a specific voltaic cell. Identify the species that would serve as the anode and the species that is the oxidizing agent.

    $$Mg(s) \,|\, Mg^{2+}(aq) \,||\, Co^{2+}(aq) \,|\, Co(s)$$

57. The following schematic diagram represents two metals and their cations in separate beakers joined by a $NaNO_3$ salt bridge. If the metals were lead and silver (with their respective cations Pb^{2+} and Ag^+), which beaker would require the silver and which the lead if the nitrate ions were moving from left to right through the salt bridge? Justify your answer.

58. The following schematic diagram represents two metals and their cations in separate beakers joined by a $NaNO_3$ salt bridge. If the metals were copper and iron (with their respective cations Cu^{2+} and Fe^{3+}), which beaker would require the copper and which the iron if the nitrate ions were moving from left to right through the salt bridge? Justify your answer.

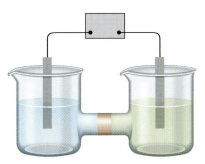

Chemical Applications and Practices

59. Diagram a battery consisting of two beakers and a salt bridge that makes use of the two half-reactions silver and gold (sort of expensive, but you're worth it).
    a. Predict the $E°$ value for the battery.
    b. What is the value of n for the balanced equation?
    c. Label the cathode.
    d. Exactly what species is acting as the oxidizing agent?
    e. Represent the battery in the ABC notation.

60. Search Internet references using the keywords "hydrogen fuel cell" and report on the basic operation of such a cell. Give particular emphasis to the electrolyte used in such a cell. Be sure to report the necessary URL for your references.

61. Small button-sized batteries can be made using mercury and zinc. If the two half-reactions are as follows, what would you report as the ABC notation for the battery? (Mercury in the batteries is considered a toxic substance and should be carefully recycled.)

$$Zn(s) \rightarrow ZnO(s)$$
$$HgO(s) \rightarrow Hg(l)$$

Note that the reactions are not balanced. The reaction takes place in a basic medium.

62. Write out the standard reduction half-cell reactions and $E°$ values for Zn, Ni, Pb, and $H_2(g)$.
    a. Select the two half-reactions that, when combined, would produce the battery with the greatest theoretical overall $E°$ value, and report that value.
    b. Which, if any, of those represented would have electrons move along an external wire, when connected with the hydrogen half-cell, *toward* the hydrogen electrode?

Section 19.5 Chemical Reactivity Series

Skill Review

63. Use the Standard Table of Reduction Potentials to answer the following questions.
    a. Which is the better reducing agent, Ba or Ca?
    b. Which is the better oxidizing agent, Pb^{2+} or Ni^{2+}?

c. In which direction, to the left or to the right, would the following reaction take place spontaneously? (Assume standard conditions.)

$$Ag + Fe^{3+} \rightarrow Fe^{2+} + Ag^+$$

64. Under standard conditions one of the following two reactions will *not occur* spontaneously. Use the Standard Table of Reduction Potentials to explain how you could correctly predict which one will not take place.

$$Fe + Sn^{2+} \rightarrow Fe^{3+} + Sn$$
$$Cu^{2+} + Fe \rightarrow Fe^{3+} + Cu$$

Chemical Applications and Practices

65. Suppose you have a shiny piece of metal that is unlabeled. It is known to be either aluminum or tin. You also have a solution of 1.0 M nickel(II) nitrate. Suggest a chemical test that you could use to determine the identity of the metal using the solution. Explain your expected results and the basis of your conclusion.

66. One technique to protect against oxidation of structures such as ship hulls and underground iron pipes is to place the structure in contact with a metal that will oxidize more easily than the iron in the structure. The more active metal is then referred to as a sacrificial anode. Neglecting such issues as cost and availability, would copper or zinc make the better sacrificial anode in an attempt to protect an iron-based structure? Explain the basis of your choice.

Section 19.6 Not-So-Standard Conditions: The Nernst Equation

Skill Review

67. The constant used in the Nernst equation, 0.0257, is derived from the combination of three other constants. Obtain the three values, assuming standard temperature, and derive the constant used in the equation. Some texts refer to the Nernst equation using the base 10 logarithmic scale. What would be the value of the constant in that scale?

68. The simplified equation presented here represents the redox reaction taking place inside the typical, nonalkaline flashlight battery. Use the principles described in the Nernst equation to answer the questions that follow.

$$Zn(s) + 2MnO_2(s) + 2NH_4^+(aq) \rightarrow$$
$$Zn^{2+}(aq) + Mn_2O_3(s) + 2NH_3(g) + H_2O(l)$$

    a. What would be the effect on the spontaneity of the reaction if the concentration of $NH_4^+(aq)$ were decreased?
    b. What would be the effect on the voltage of the cell if the concentration of $NH_4^+(aq)$ were decreased?
    c. What is the value of n in the equation?
    d. At equilibrium what is the value of E?

69. Calculate the value of E_{cell} for the reaction of iron and copper(II), given the specific concentrations listed.
    a. $Fe(s) \mid Fe^{2+}(0.10\ M) \parallel Cu^{2+}(0.10\ M) \mid Cu(s)$
    b. $Fe(s) \mid Fe^{2+}(1.5\ M) \parallel Cu^{2+}(0.10\ M) \mid Cu(s)$
    c. $Fe(s) \mid Fe^{2+}(0.10\ M) \parallel Cu^{2+}(1.5\ M) \mid Cu(s)$

70. Calculate the value of E_{cell} for the reaction of cobalt(II) chloride and zinc metal, given the specific concentrations listed.
    a. $Zn(s) \mid Zn^{2+}(0.10\ M) \parallel Co^{2+}(0.10\ M) \mid Co(s)$
    b. $Zn(s) \mid Zn^{2+}(2.5\ M) \parallel Co^{2+}(0.050\ M) \mid Co(s)$
    c. $Zn(s) \mid Zn^{2+}(0.010\ M) \parallel Co^{2+}(0.10\ M) \mid Co(s)$

Chemical Applications and Practices

71. Suppose that at night on a campout your flashlight dims and finally stops working. Your friend remarks, "I guess the battery is dead." In what three other chemical ways, using Gibbs free energy, equilibrium, and potential, could you state the same conclusion as your friend?

72. A quick source of hydrogen gas in the lab is to carefully place some solid magnesium in hydrochloric acid.

$$Mg(s) + 2HCl(aq) \rightarrow MgCl_2(aq) + H_2(g)$$

If the Mg^{2+} concentration were 1.0 M and the HCl concentration were 0.10 M (assume completely dissociated HCl), what would you calculate as the E value for the reaction at 25°C?

$$Mg(s) \,|\, Mg^{2+}(aq) \,||\, H^+(0.1\ M) \,|\, H_2(1\ atm)$$

Section 19.7 Electrolytic Reactions

Skill Review

73. Calculate the number of grams of gold electroplated onto a surface given these conditions. (Assume that the solution of Au^{3+} ions is concentrated enough to complete the electrolysis.)
    a. 1.25 A for 60 s   b. 2.11 A for 2.33 h   c. 0.75 A for 1 d

74. Calculate the time required to electroplate 1.0 g of tin metal onto a surface given these conditions. (Assume that the solution of Sn^{2+} ions is concentrated enough to complete the electrolysis.)
    a. 2.25 A   b. 0.11 A   c. 1.38 A

75. Alessandro Volta is given historical precedence in the discovery of the first battery in the sense that we think of batteries today. The "Voltaic pile" consisted of dissimilar metals joined by salt-moistened paper strips. Davy and his assistant Michael Faraday later refined this. At that time, early electroplating businesses began developing in England. If such a business silver-plated a teaspoon, would the teaspoon be the cathode or the anode in such a process? If the spoon received 0.33 g of silver after being plated for 1.0 h, what amperage was used?

76. In addition to isolating sodium, Sir Humphrey Davy isolated potassium, calcium, and magnesium from impure natural ore samples.
    a. If he used a current of 1.00 A for 1.00 h, how many grams of magnesium metal could he obtain from molten $MgCl_2$?
    b. Using the same electrical set up, how many grams of potassium metal could Sir Humphrey isolate from molten KCl?

Chemical Applications and Practices

77. In the typical lead storage battery found in most automobiles, lead is oxidized to Pb^{2+} (in the form of $PbSO_4$). In recharging, the reaction is reversed. If a battery were recharged for 30.0 min at a current of 8.00 A, how many grams of lead would be reduced, from $PbSO_4$ during the process?

78. Germanium has become a valuable metal in semiconductor fields. If a 1.00 A current were used for 1.00 h to plate out 0.677 g of germanium, what would you calculate as the oxidation number of the germanium ions in the plating solution?

Comprehensive Problems

79. Use the illustration below as the starting point to draw the electrolytic cell that would be used to plate copper metal onto a steel saucepan. Be sure to indicate the location of the cathode and the anode, the location of the copper electrode, and the steel saucepan.

80. The Nernst equation can be used just as well with half-cell reactions as with balanced redox reactions. An application of this is the use of potential differences in the acid level of solutions compared to a standard when producing the probes for pH meters. Using the half-reaction $2H^+(aq) + 2e^- \rightarrow H_2(g)$ at 1 atm, what would you calculate, using the Nernst equation, as the value for E in these situations?
    a. Pure water, pH = 7.00
    b. An acid solution with pH = 2.00
    c. A 0.10 M solution of nitric acid
    d. A 0.10 M solution of acetic acid, $K_a = 1.8 \times 10^{-5}$

81. Potassium ferrate (K_2FeO_4) is a powerful oxidizing agent ($E° = 2.20$ V relative to SHE in acidic solution) in which the iron(VI) ion is reduced to the iron(III) ion. It oxidizes water to oxygen gas.
    a. Determine the cell potential and equilibrium constant for the reaction at standard conditions for the oxidation of water by ferrate ion in acidic solution.
    b. Balance the reaction in basic solution.
    c. Calculate the cell potential, given the following initial concentrations:

    $$[FeO_4^{2-}] = 1.5 \times 10^{-3}\ M;\ [Fe^{3+}] = 1.1 \times 10^{-3}\ M;$$
    $$P_{O_2} = 8.3 \times 10^{-5}\ atm;\ pH = 2.8$$

    d. The reaction results in the production of a yellow solid that is especially apparent at high pH. Can you suggest what this might be? Further, can you suggest how this solid might help if the ferrate ion were used to treat wastewater contamination?

82. Potassium permanganate ($KMnO_4$) is a useful analytical reagent for determining the percentage of iron in an iron ore. The procedure includes dissolving the iron with HCl and then converting all of the iron in the ore to Fe^{2+} using several reagents. The titration of the resulting Fe^{2+} with MnO_4^- is

$$Fe^{2+} + MnO_4^- \rightarrow Fe^{3+} + Mn^{2+}$$

A sample of the original iron ore weighing 3.852 g was processed for titration in a total volume of 300.0 mL of solution. A 100.0 mL aliquot (accurately measured portion) was removed and titrated with a 0.1025 M $KMnO_4$ solution. A total of 23.14 mL was required to the light pink endpoint. What is the percent of iron in the ore sample?

83. We noted in Chapter 14, Thermodynamics, that we can have "standard conditions" at a given pressure, but the temperature can vary and must be stated as part of specifying your standard values. If we define our standard values at 1 bar pressure and 3000°C, how would that affect the values for our "new" standard reduction potentials compared to the ones we commonly use at 25°C?

84. We discuss rechargeable batteries in Section 19.4. You know from your own experience that these batteries can be recharged hundreds of times but eventually wear down and must be replaced. Why can't they be recharged an infinite number of times? In particular, investigate the way in which metals plate back onto the cathode when the recharging takes place.

85. Chrome plating is the process by which Cr^{6+} is electroplated on automobile surfaces to give them a nice luster. The chromium layer can be as little as 10.0 μm thick. How many grams of CrO_3 would be necessary to plate the back surface of a car mirror that has a coverable area of 150 cm^2? ($d = 7.2$ g/cm^3.)

86. We discuss fuel cells at several places in the chapter, but never actually calculate the electrochemical data for it. Calculate the $E°$, $\Delta G°$ and K at 298K for the hydrogen/oxygen fuel cell described in Section 19.2. Once you have done so, can you propose a fuel cell that might be more effective for use in automobiles? What are the criteria that you use to make that judgment?

Thinking Beyond the Calculation

87. A battery designer wishes to prepare a solution-based battery for use in a new automobile. The designer chooses iron and zinc as the two metals for study.

a. Draw the setup (using beakers and a salt bridge) that indicates the location of the iron and zinc electrodes, the iron(II) and zinc(II) solutions, and the flow of electrons.

b. Which half-reaction is the oxidation reaction?

c. What is the cell potential for the reaction if the initial concentrations of iron(II) and zinc(II) are 0.25 M at 25°C?

d. What would be the cell potential if the concentrations of both solutions in part c were increased to 1.0 M? . . . to 2.0 M?

e. If the battery is cooled to 10°C, is there a change in the cell potential? If so, what is the new potential?

f. Describe at least one advantage to using a battery made from these metals.

g. Describe at least one disadvantage to the use of iron in a battery.

20

Coordination Complexes

In 1434, Flemish master Jan van Eyck painted the Arnolfini Wedding. The pigments that he mixed and applied to the painting bring the scene of a newly married couple to life. The colors we see in this work are a result of the interaction of light with electrons in the d orbitals of transition metals.

Contents and Selected Applications

Chemical Encounters: Iron and Hemoglobin

20.1 Bonding in Coordination Complexes

20.2 Ligands

20.3 Coordination Number

20.4 Structure

20.5 Isomers

20.6 Formulas and Names

20.7 Color and Coordination Compounds

Chemical Encounters: Transition Metals and Color

20.8 Chemical Reactions

Chemical Encounters: Cytochromes

Go to **college.hmco.com/pic/kelterMEE** for online learning resources.

Many of the nutritional supplements we consume on a daily basis contain iron. Why? Iron is an essential nutrient with the U.S. recommended daily allowance of 10 mg for men and 15 mg for women. Even with supplements and recommendations from nutritional scientists, less than 30% of women aged 12–50 meet their daily recommended allowance of iron. Iron isn't a vitamin; rather, it's called a mineral—the term used for any inorganic element important for health. It turns out that few items claiming to be rich in iron actually contain the neutral metal. Instead, they typically contain iron ions in a salt, often iron sulfate. Why is iron so important to good health? Most of the iron in the body is contained in a protein known as hemoglobin within the red blood cells (Chapter 22). Specifically, iron ions are chemically bound in hemoglobin in a form called a coordination complex.

Within hemoglobin, the iron ions are bonded to other atoms in an elegant and complex structure known as a heme (see Figure 20.1). The arrangement of groups attached to the iron enables it to collect an oxygen molecule and later release it at an appropriate time. Close examination of the heme indicates that the conformational changes illustrated in Figure 20.2 occur during oxygen binding. We visited a simpler form of this complex in Chapter 16 when we discussed the

Application

CHEMICAL
ENCOUNTERS:
Iron and
Hemoglobin

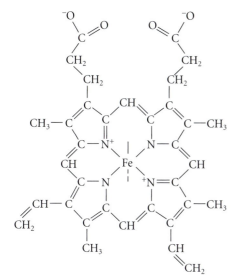

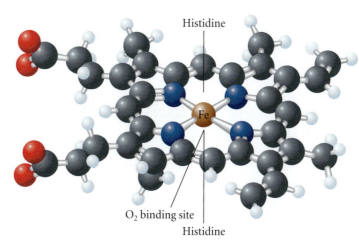

FIGURE 20.1

The iron is bonded to other atoms in an elegant and complex structure known as a heme. The hemoglobin protein uses two amino acid residues to hold the iron within the protein's structure. One of the amino acid residues can move out of the way to allow an oxygen molecule to bind to the iron.

equilibrium involving oxygen and myoglobin. Without this specific complex structure, whether in hemoglobin or myoglobin, iron would be incapable of playing the role of oxygen carrier.

In a typical **coordination complex** (or simply complex), a central metal atom or ion is *chemically bonded* to several other components. For example, natural water often contains iron in the form of a chemical combination of an Fe^{3+} ion and six water molecules, $[Fe(H_2O)_6]^{3+}$, as shown in Figure 20.3. The bonding, the structure, and the many biological, medical, and industrial applications of coordination complexes are the subject of this chapter.

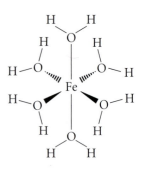

FIGURE 20.3

Natural water often contains iron in the form of a chemical combination of an Fe^{3+} ion and six water molecules, $[Fe(H_2O)_6]^{3+}$.

FIGURE 20.2

Conformational changes take place during the binding of oxygen to the iron ion in hemoglobin. These changes act like a switch in the hemoglobin protein, activating it and increasing its ability to bind to three other molecules of oxygen. In the structure shown here, the oxygen molecule (red) is bound in oxyhemoglobin. When oxygen is not bound to hemoglobin, the protein chains pull the iron so that the heme sits slightly above the center of the iron ion.

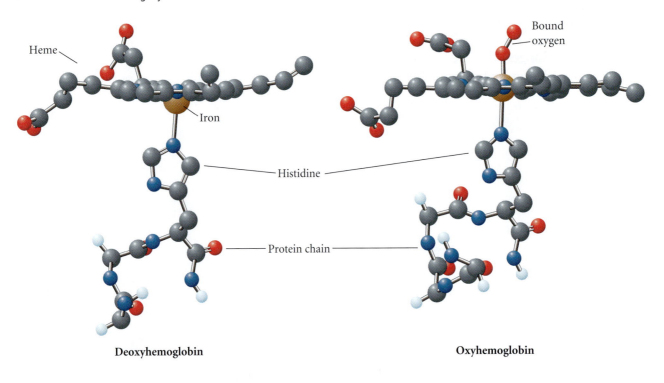

Deoxyhemoglobin

Oxyhemoglobin

20.1 Bonding in Coordination Complexes

A common feature of bonding within metal complexes is the **coordinate covalent bond**, which we discussed in Chapter 7. Recall that the coordinate covalent bond forms when *both bonding electrons originate from one atom*. This means that a molecule with a lone electron pair (a Lewis base) could potentially be a component of a coordination complex. Lewis bases that form a coordinate covalent bond with a metal or metal ion are known as **ligands**. In the aqueous iron complex found in natural water that contains iron, each water molecule donates a pair of electrons to the iron(III) ion and is therefore the Lewis base. The iron(III) ion behaves as a Lewis acid, an electron pair acceptor. The resulting bond between the

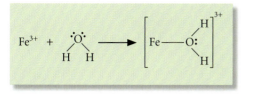

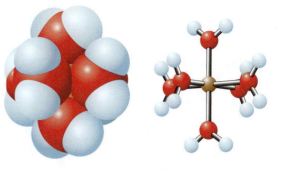

$[Fe(H_2O)_6]^{3+}$

oxygen of water and the iron(III) ion is a coordinate covalent bond. With *six* equivalent water molecules donating electron pairs to create *six* bonds to the central Fe^{3+} ion, the complete complex ion is $[Fe(H_2O)_6]^{3+}$.

In general, we can use the following equation to illustrate the formation of a coordination complex via donation of a lone pair of electrons from a ligand, L, to a metal center, M,

$$M + :L \rightarrow M\text{–}L$$

The chemistry of hemoglobin is an example of the formation of such a coordination complex. The iron ion in the hemoglobin molecule forms a bond with an oxygen molecule and ultimately transports the oxygen molecule from the lungs to cells throughout the body. Using our shorthand, we can represent the bonding with oxygen by the following equation:

where Hb represents the complex structure of the hemoglobin molecule. Where does the oxygen come from? In the lungs, the oxygen **coordinates** to the hemoglobin in a red blood cell—that is, *forms a coordinate covalent bond to it*. Under slightly different conditions in the cell, the oxygen is released in the reverse process and becomes available for respiration.

Carbon monoxide can also coordinate to hemoglobin. In fact, it does so more strongly than oxygen. A close look at the structure that results indicates that the lone pair of electrons on the carbon forms the coordinate covalent bond with the iron in hemoglobin. Equilibrium principles tell us that because the equilibrium constant for the reaction with CO is much greater (about 200 times greater) than that for the reaction with oxygen, the presence of small amounts of CO in the body limits the amount of O_2 that can be carried by hemoglobin. Excessive amounts of carbon monoxide in the body can result in suffocation. The National Institute of Occupational Safety and Health (NIOSH) cites an immediate danger to life with an exposure of 1200 parts per million of CO in air and specifies an 8-hour exposure limit of 35 ppm in air.

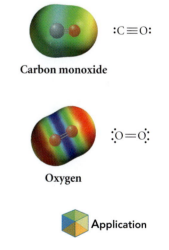

$:C\equiv O:$

Carbon monoxide

$:O = O:$

Oxygen

Application

HbFe $:C\equiv O:$ ⟶ HbFe—$C\equiv O:$

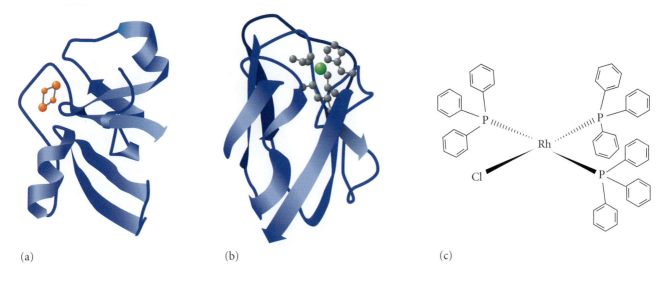

(a) (b) (c)

FIGURE 20.4

Three important coordination complexes: ferrodoxin, plastocyanin, and Wilkinson's catalyst. (a) An illustration of the protein ferrodoxin indicates the location of an Fe_2S_2 cluster—shown in orange. (b) A drawing of the protein plastocyanin contains a copper ion—green—held in place by the amino acids in the structure of the protein. (c) Wilkinson's catalyst contains a rhodium ion surrounded by triphenylphosphine ligands.

The coordinate covalent bond is very common in compounds involving metals and nonmetals. These types of compounds, as shown in Figure 20.4, are often found at critical reaction centers in biological systems and in many industrial processes. For example, cytochromes and ferredoxins are iron-containing coordination complexes in nonanimal biological systems, where they assist in reduction and oxidation reactions (see Chapter 19) during photosynthesis. In some plants, nitrogen is converted into a usable form by coordination of an N_2 molecule to a molybdenum ion in the enzyme nitrogenase. In industrial processes, metal ions can be used to form coordinate covalent bonds with small molecules in order to make the small molecule react in certain ways. Wilkinson's catalyst, shown in Figure 20.5, is a coordination complex of rhodium used to promote reaction of H_2 with alkenes (in polyunsaturated fats) to make margarine. In another industrial process used to make the class of molecules known as aldehydes (which are used in essential oils—volatile compounds that have odors characteristic of plants, polymer formation, and food additives), carbon monoxide forms a coordination

FIGURE 20.5

The reaction of hydrogen with an alkene in the presence of Wilkinson's catalyst. This reaction can be used to convert low-melting-point vegetable oils into semi-solid fats for use in making margarine.

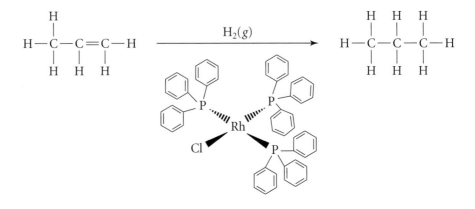

complex with a cobalt atom as it is transformed into the C=O fragment of an aldehyde.

The coordination complexes of transition metals have some particularly interesting properties. The color of the complex, such as that found in rubies, depends on the nature and number of ligands surrounding the metal complex. The magnetism of coordination complexes, such as the magnetism of the iron oxide used in videotape, depends on the nature and number of ligands surrounding the metal. The biochemical reactions that are possible with coordination complexes found in the body are also influenced by the ligands.

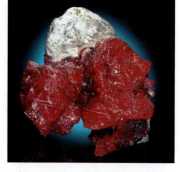

Rubies are minerals made of Al_2O_3 containing less than 1% Cr_2O_3 as an impurity.

20.2 Ligands

A coordination complex such as $[Fe(H_2O)_6]^{3+}$ consists of a central Fe^{3+} metal ion and coordinated water molecules. The species, such as the water molecules, that coordinate to the metal are called ligands. Table 20.1 lists some common ligands. Remember that ligands *donate* a lone pair of electrons to the metal or metal ion to form the coordinate covalent bond. A metal ion will tend to form coordinate covalent bonds with a specific number of ligands, often either four or six.

Video Lesson: Complexes and Ligands

EXERCISE 20.1 | Identifying Possible Ligands

Which of the following could act as ligands in forming coordination complexes?

a. CH_4 b. N_3^- c. BH_3 d. CS_2 e. Li^+

Solution

Because each has a nonbonded pair of electrons on one or more atoms, (b) and (d) could be ligands.

PRACTICE 20.1

Indicate the formula of a coordination complex that might result from the combination of these metals and ligands.

a. Fe^{2+} and $6Cl^-$ b. Ni^{2+} and $4NH_3$ c. Zn^{2+} and $6H_2O$

See Problems 3, 4, 9, and 10.

TABLE 20.1 | Selected Common Ligands and Their Names

The atoms shown in red donate a lone pair of electrons to form a coordinate covalent bond.

Monodentate ligands

Cl^-	chloro	Br^-	bromo	I^-	iodo
CN^-	cyano	NO_2^-	nitro	NO_3^-	nitrato*
SCN^-	thiocyanato	NO_2^-	nitrito	OSO_3^{2-}	sulfato
SSO_3^{2-}	thiosulfato	O^{2-}	oxido	F^-	fluoro
NH_3	ammine	H_2O	aqua	NO	nitrosyl
CO	carbonyl	OH^-	hydroxo		

Polydentate ligands

$H_2NCH_2CH_2NH_2$	ethylenediamine (en)
$(^-OOCCH_2)_2NCH_2CH_2N(CH_2COO^-)_2$	ethylenediaminetetracetato (EDTA)
$^-OOCCOO^-$	oxalato (ox)

*The nitrato ligand can be monodentate or bidentate.

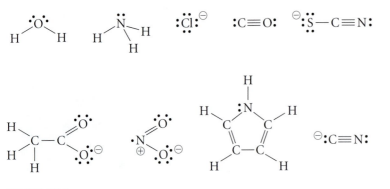

FIGURE 20.6

Examples of ligands.

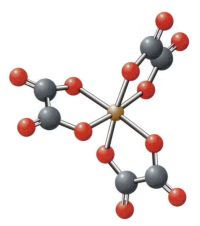

FIGURE 20.7

The oxalate ion is a bidentate ion. Each oxalate is capable of donating lone pairs of electrons from two different atoms to form two coordinate covalent bonds. Three oxalates can bind to one iron(III) ion.

Some of the molecules and ions that are commonly observed to act as ligands are shown in Figure 20.6. Oxygen, nitrogen, sulfur, and phosphorus atoms in molecules are common donor atoms in ligands because they possess a lone pair of electrons that can be donated to the metal center. Some ligands have nonbonded pairs of electrons in different locations throughout their structure. Often, any of these pairs of electrons can be used to create a coordinate covalent bond. In some cases, only one specific pair of electrons is used to create the bond. For instance, cyanide (CN^-) usually binds via the pair of electrons on the carbon atom. It does so in the "Prussian blue" dye used in dyes and paints, where the complex $[Fe(CN)_6]^{4-}$ is formed.

Ligands can also use more than one nonbonded pair of electrons to bond to a metal center. If more than one atom has a nonbonded pair of electrons, it may use each of those pairs to form independent coordinate covalent bonds to the metal. In such cases, if the ligand forms two bonds to the metal, we say it is **bidentate** ("two-toothed"); we call it **tridentate** ("three-toothed") if it forms three bonds, and so on. For example, the bidentate oxalate ion $C_2O_4^{2-}$ can bind to the iron(III) ion in this fashion, as shown in Figure 20.7. Bathroom stain removers often contain oxalate salts, because the oxalate ion will coordinate to iron ions in rust and help wash the stain away. Another common bidentate ligand that donates more than one pair of electrons is ethylenediamine $H_2NCH_2CH_2NH_2$ (often abbreviated en), in which each nitrogen has a lone pair of electrons that can be donated to the same metal ion.

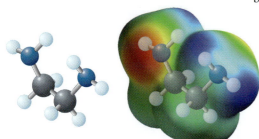

Ethylenediamine

$H_2N — CH_2 — CH_2 — NH_2$

FIGURE 20.8

The basic porphyrin structure, showing a metal at the center coordinated in six different positions. Compare this structure to that found in the hemoglobin protein from Figure 20.1.

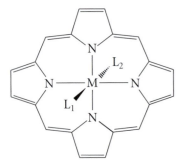

The **polydentate** ligands (those that form more than one coordinate covalent bond to the metal center) are known as **chelates** (from the Greek *chele* for "claw") because of the way they clamp onto the metal center and bond tightly. For example, the porphyrin ligand is a planar, **tetradentate** ligand found in hemoglobin, vitamin B_{12}, and chlorophyll, as shown in Figure 20.8. The porphyrin chelates to iron in hemoglobin, cobalt in vitamin B_{12}, and magnesium in chlorophyll. The chelates are so good at holding the metal firmly within their grasp that it is difficult to remove the iron from hemoglobin without destroying the hemoglobin molecule itself. Note that the porphyrin leaves two sites on the metal open so that it may bind to other ligands. It is the open locations on the metal within the porphryin chelate that make chemical reactions possible.

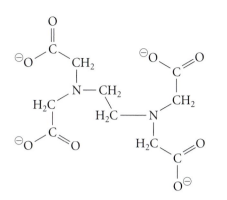

FIGURE 20.9

EDTA^{4-}

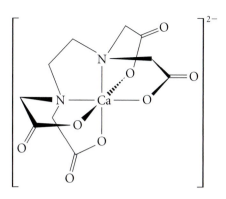

FIGURE 20.10

EDTA^{4-} coordinated to calcium ion.

The most important industrial ligand, discussed in Chapter 17 and 18, is ethylenediaminetetraacetate, which is abbreviated EDTA^{4-}. The structure of this tetraanion, shown in Figure 20.9, illustrates that both nitrogen atoms, and one oxygen atom from each end of the compound, have lone pairs of electrons. It acts as a chelate by forming up to six coordinate covalent bonds to a metal center, depending on the size of the metal ion. For example, EDTA^{4-} forms six bonds to calcium ions, as shown in Figure 20.10. The exceptionally large formation constant for the reaction of EDTA^{4-} with many metal ions illustrates that the formation of these bonds is quite favorable.

20.3   Coordination Number

The number of donor atoms to which a given metal ion bonds is known as its **coordination number**. Coordination numbers of 4 and 6 are most common, but coordination numbers as low as 2 and as large as 8 are not unusual. For example, in the film development process, excess silver ion is removed from photographic film by using the thiosulfate ion ($S_2O_3^{2-}$), which forms a coordination complex with a coordination number of 2 (see Figure 20.11). In another example, cis-platin, a prominent anticancer drug, has at its center a platinum(II) ion with a coordination number of 4, as shown in Figure 20.12. As we discussed earlier, when an Fe^{3+} ion is present in water, it is surrounded by 6 water molecules in a complex, $[Fe(H_2O)_6]^{3+}$. And iron's coordination number is 6. If iron forms a complex with the oxalate ligand, $[Fe(C_2O_4)_3]^{3-}$, its coordination number is still 6, because each oxalate ligand donates two electron pairs.

What makes a metal ion have a certain coordination number? It is primarily a function of the nature of metal ions. Large metal ions have space around them within which more ligands can fit. Not as many ligands can fit around a small metal ion. Other factors, such as the charges on the ligands and metal and the electron configuration of the metal ion, determine the most likely coordination number for a metal ion.

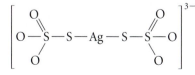

FIGURE 20.11

A two-coordinate complex of silver.

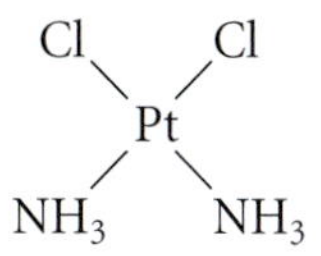

FIGURE 20.12

Four-coordinate cisplatin.

How do we know?

What is the nature of the structure, bonding, and reactivity in cisplatin?

Cancer is an often devastating disease. Research has made significant strides in understanding the disease in its many forms, including advances in detection, treatment, and care that have improved the quality of life for its victims. One prominent drug used in cancer treatment, especially testicular and ovarian cancers, is cisplatin, a coordination complex of platinum(II) that is shown in Figure 20.12. The anticancer activity of cisplatin was discovered as a result careful observation and basic interpretation of an experiment that had nothing to do with fighting cancer. But good science nonetheless led to the discovery.

Dr. Barnett Rosenberg at Michigan State University investigated the effect of an electric current on a culture of *Escherichia coli* bacteria. This bacterium, commonly found in the gastrointestinal tracts of many living creatures, is often used in initial biochemical studies because of its ready availability. Rosenberg observed that when an electric current was applied to solutions of the bacteria, cell division in the vicinity of a platinum electrode was inhibited. Studying this interesting result further, he recognized that it was not due to the electric current, which was the focus of the investigation. Noting that a compound known as *cis*-diamminedichloroplatinum (cisplatin) was being produced in the vicinity of the platinum electrode used for his experiment, Rosenberg reasoned that the cisplatin must be responsible for inhibiting the cell division. It was later determined that this compound, when given to cancer patients, can significantly reduce the size of their tumors and can even cause the disease to go into remission. Since 1970, the survival rate for testicular cancer patients has increased from 10% to over 90%.

How does cisplatin exhibit this remarkable biological activity? Cisplatin is a four-coordinate, square planar complex, as is commonly observed for many metal complexes with eight *d* orbital electrons. The platinum(II) metal center is fairly unreactive, which allows the neutral cisplatin complex to remain largely intact through injection, circulation, and penetration into the nucleus of a cancerous cell. The chloro ligands are eventually replaced by water molecules. This provides an opportunity for the platinum center to coordinate to DNA molecules. Ultimately, the platinum center binds to a nitrogen atom from each of two guanine units in a single DNA strand (see Chapter 22). The geometry of the ligands around the platinum center is vital to this biological activity. In fact, the chloro ligands must be *cis* to each other—on the same side of the complex. Once coordinated to the guanine nitrogens, the inert platinum center remains bound, tying two points on the DNA strand together. Such binding inhibits reproduction of DNA during cell division and restricts use of the DNA for normal cellular functions. Ultimately these effects hinder the growth of the cancer cells.

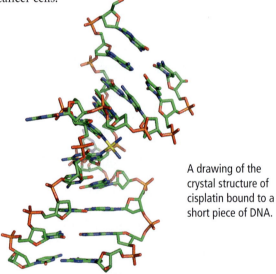

A drawing of the crystal structure of cisplatin bound to a short piece of DNA.

20.4 Structure

Mercury in the environment poses a particular threat to life, both to aquatic species and to those species that consume them. The most insidious form of mercury is dimethyl mercury, $[Hg(CH_3)_2]$. Transformed from mercury metal in the environment, dimethyl mercury is a very toxic product. The mercury(II) ion in this complex coordinates with the methyl groups, $:CH_3^-$, to form the complex $[H_3C—Hg—CH_3]$. Given the VSEPR rules, the complex is *linear*, with a C—Hg—C bond angle of 180°. This structure is typical of metals with a coordination number of 2.

Dimethylmercury
$CH_3 — Hg — CH_3$

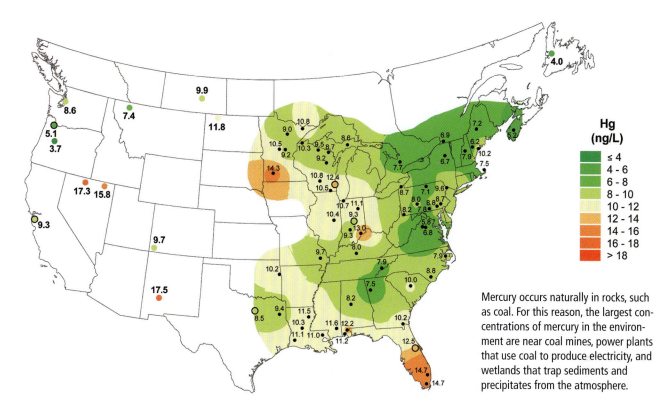

Hg (ng/L)

■	≤ 4
■	4 - 6
■	6 - 8
■	8 - 10
■	10 - 12
■	12 - 14
■	14 - 16
■	16 - 18
■	> 18

Mercury occurs naturally in rocks, such as coal. For this reason, the largest concentrations of mercury in the environment are near coal mines, power plants that use coal to produce electricity, and wetlands that trap sediments and precipitates from the atmosphere.

We observe that two geometries are common for metals that have a coordination number of 4. Those geometries, tetrahedral and square planar, are shown in Figure 20.13. The **tetrahedral** geometry is the one predicted by VSEPR (see Section 8.4) when there are four electron sets around the central atom. In this geometry, the bond angles are close to 109°. The **square planar** geometry arises as a consequence of the electron configuration of the metal ion. Square planar geometry, with bond angles of 90°, is often observed for metal ions such as Ni^{2+} and Pt^{2+} that have an outer nd^8 electron configuration. That is, they have eight electrons in their outermost d orbitals. An example of the square planar geometry is found in the anticancer drug cisplatin, which we introduced in the previous section. The square planar geometry is crucial in the interaction of this drug with DNA.

The final geometry common to the coordination compounds is the **octahedral** complex, an example of which is shown in Figure 20.14. In this geometry, the six ligands occupy equivalent positions around the metal center. For example, hexaaquairon(III), $[Fe(H_2O)_6]^{3+}$, includes six water ligands symmetrically surrounding the central Fe^{3+} ion. We can think of the ligands as sitting at positions on each end of the coordinate x, y, and z axes with bond angles of 90°. The iron in hemoglobin, shown in Figure 20.1, occupies a nearly octahedral coordination environment with four sites occupied by nitrogen atoms from a porphyrin ligand. One site is occupied by a histidine (an amino acid that makes up one part of the hemoglobin molecule), and the final site is occupied by a molecule of oxygen.

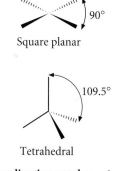

Square planar

109.5°

Tetrahedral

Coordination number = 4

FIGURE 20.13

Common geometries with a coordination number of 4.

Octahedral

Coordination number = 6

FIGURE 20.14

The octahedral geometry; all six positions in this geometry are identical.

EXERCISE 20.2   **Coordination Number and Geometry**

Give the coordination number of the central metal and predict the geometry of the following compounds.

a. $[Al(H_2O)_6]^{3+}$ b. $[Ag(CN)_2]^-$ c. $Na_2[PdCl_4]$ d. $[Co(ox)_3]^{3-}$

First Thoughts

Coordination Number: To address this question, we must first identify the ligands and the central metal to which they are attached. In part c, the formula is written as a neutral compound, not only as the coordination complex. The species in brackets represents the coordination complex. It is only the coordination complex that we should consider when answering a question about the coordination number. Remember that oxalate (ox) is a bidentate ligand.

Geometry: Once we know the coordination number of the complex, we can determine the geometry of the complex. For coordination number 2 or 6, the geometry is linear or octahedral, respectively. For four-coordinate complexes, either tetrahedral or square planar geometry is expected. Remember that metals with eight *d* orbital electrons typically form square planar complexes.

Solution

Coordination Number   a. 6   b. 2   c. 4   d. 6
Geometry   a. Octahedral   b. Linear   c. Square planar   d. Octahedral

Further Insights

Remember that the "coordination" of a ligand to a metal is a covalent bond, and the same concepts that you learned earlier for covalent bonds still apply. In short, the metal and ligand mutually attract an electron pair.

PRACTICE 20.2

Determine the coordination number and geometry for each of the following compounds.

a. $[Cr(CO)_6]$ b. $[Fe(H_2O)_6]^{2+}$ c. $K_2[TiCl_4]$ d. $[Cu(NH_3)_4]^+$

See Problems 17, 18, and 23–26.

20.5 Isomers

Video Lesson: Structures of Coordination Compounds and Isomers

Organic molecules (Chapter 12) are not unique in their ability to form isomeric structures. Coordination compounds can also form isomers. These types of isomers are divided into two main categories on the basis of whether there is a change in the structure or in the geometry of the complex. Structural differences are observed in the linkage isomers, ionization isomers, and coordination sphere isomers. Differences in geometric isomers arise from the nonspecific nature of the formation of a coordinate covalent bond between a ligand and a metal center.

Some ligands coordinate to a metal through one of several donor atoms. For example, the thiocyanate ion (SCN^-) can coordinate to a chromium(III) ion either through the nonbonded pair of electrons on the nitrogen or through the nonbonded pair of the electrons on the sulfur atom:

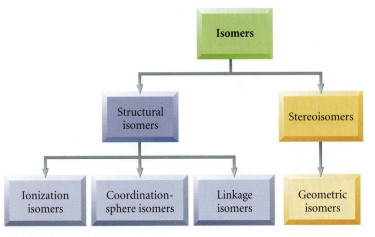

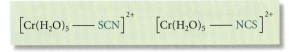

$$\left[Cr(H_2O)_5 \!-\! SCN\right]^{2+} \qquad \left[Cr(H_2O)_5 \!-\! NCS\right]^{2+}$$

Isomers in coordination compounds.

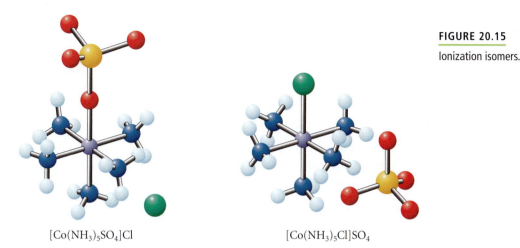

FIGURE 20.15
Ionization isomers.

[Co(NH₃)₅SO₄]Cl   [Co(NH₃)₅Cl]SO₄

These are called linkage isomers. They differ only in the atom that participates in the coordinate covalent bond with the metal.

Another isomer that is common among the coordination compounds is the ionization isomer. These are species in which a ligand and an ion (a counter ion) exchange roles, as with the cobalt complexes shown in Figure 20.15. Note that the counter ion and a ligand have changed positions in the two compounds. It is that change that makes these two compounds ionization isomers.

Coordination sphere isomers contain different ligands in the coordination spheres of cations and anions. In the coordination sphere isomers, both the cation and anion are coordination complexes. Examples are shown in Figure 20.16. Note that each of these complexes includes a cation that is a coordination complex and

FIGURE 20.16

Examples of coordination sphere isomers.

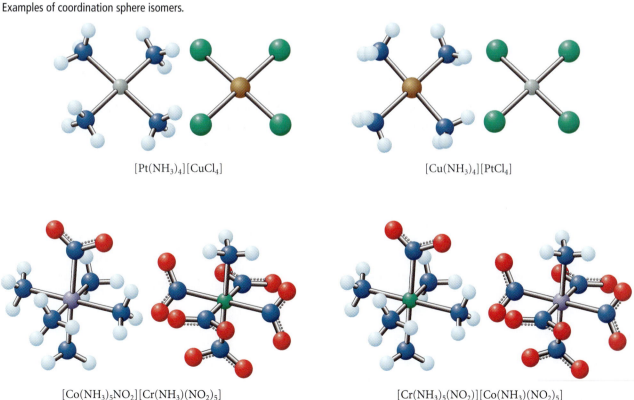

[Pt(NH₃)₄][CuCl₄]   [Cu(NH₃)₄][PtCl₄]

[Co(NH₃)₅NO₂][Cr(NH₃)(NO₂)₅]   [Cr(NH₃)₅(NO₂)][Co(NH₃)(NO₂)₅]

an anion that is a coordination complex. The metal ions in each case have switched location from the cation to the anion.

Geometric isomers, which we first discussed in Section 12.3, are substances in which all of the atoms are attached with the same connectivity, or bonds, but the geometric orientation differs. Square planar four-coordinate complexes of the general composition MA_2B_2 and octahedral complexes of the composition MA_2B_4, where M is the metal and A and B are different ligands, can exist as geometric isomers labeled either *cis* or *trans*. As shown in Figure 20.17 for the platinum complex $[Pt(NH_3)_2Cl_2]$, the *cis* **isomer** has two identical ligands next to each other. The *trans* **isomer** has two identical ligands on opposite sides of the metal. Octahedral complexes can exist as *cis* or *trans* isomers in complexes such as $[Co(NH_3)_4Cl_2]^+$ (Figure 20.18). In fact, in the days when Alfred Werner (1866–1919) first defined the nature of coordination complexes (for which he won the 1913 Nobel Prize in chemistry), he noted the two isomers of $[Co(NH_3)_4Cl_2]Cl$. He called one the Praseo complex and the other the Violeo complex. These two isomers were identified easily, and named to reflect their identification, because the difference in their color was quite evident. To be sure, the color of a coordination complex is markedly influenced by the geometric arrangement of the ligands around the metal center. Table 20.2 gives a summary of the various isomers common among coordination compounds.

TABLE 20.2	Isomers of Coordination Complexes	
Isomer	**Example**	**Explanation**
Linkage	$\left[Co(NH_3)_5 - ONO\right]^+$ $\left[Co(NH_3)_5 - NO_2\right]^+$	Ligand binds through different donor atoms
Ionization	$\left[Pt(NH_3)_3Cl\right]Br$ $\left[Pt(NH_3)_3Br\right]Cl$	Anion and ligand interchanged
Coordination sphere	$\left[Co(en)_3\right]\left[Cr(ox)_3\right]$ $\left[Cr(en)_3\right]\left[Co(ox)_3\right]$	Distribution of coordinating ligands differs
Geometric	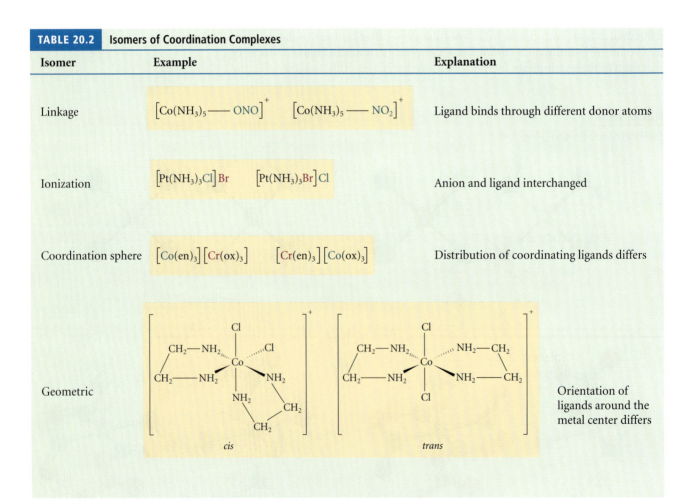	Orientation of ligands around the metal center differs

FIGURE 20.17

Cis and *trans* isomers are geometric isomers. In the *cis* isomer shown here, the two chloro ligands are on the same side of the complex. The chloro ligands are on opposite sides in the *trans* complex.

cis-**Diamminedichloroplatinum(II)**   *trans*-**Diamminedichloroplatinum(II)**

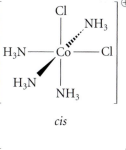

cis

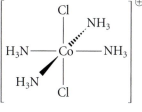

trans

FIGURE 20.18

Cis and *trans* isomers are also possible in octahedral complexes.

HERE'S WHAT WE KNOW SO FAR

- Coordination complexes are present in simple metal ions in solution and as the reaction center in many biological molecules.
- A Lewis base that donates a lone pair of electrons to a metal to form a coordinate covalent bond forms a coordination complex.
- The coordination numbers of various metal centers are commonly observed to be 2, 4, or 6.
- Common coordination geometries are linear, tetrahedral, square planar, and octahedral.
- A variety of isomer classes are observed for metal complexes.

20.6 Formulas and Names

Coordination compounds are quite varied in their structure, bonding, ability to form isomers, and other features. For this reason, assigning specific names to these compounds must be done with care. Just as in naming binary compounds (Chapter 2) and organic compounds (Chapter 12), IUPAC rules will guide us in constructing the proper name.

Formulas

Coordination complexes (the metal center and the ligands) can be neutral, anionic, or cationic in nature. To ensure continuity from one structure to another, we often write the formula of a coordination compound in accordance with a set of rules. Those rules are outlined in Table 20.3.

Nomenclature

Systematic names for coordination compounds are constructed in accordance with a set of rules. Those rules are outlined in Table 20.4. Let's use these rules to name a few examples. Consider that we are interested in naming the neutral complex [PtCl$_2$(NH$_3$)$_2$]. Step 1 from Table 20.4 indicates that we should name the compound using the rules in step 3. The ligands are named alphabetically before

TABLE 20.3	Rules for Writing Formulas for Coordination Compounds

1. The metal is written first, followed by the ligands.

2. Anionic ligands are written before the neutral ligands, each in alphabetical order.

3. The complex is enclosed in brackets.

4. Polyatomic ligands, such as NH$_3$ or NO$_2^-$, are enclosed in parentheses.

5. Ionic compounds containing coordination complexes are written in the traditional way: cation on the left, anion on the right.

TABLE 20.4   **Rules for Naming Coordination Compounds**

1. Name the cation, and then the anion. If the complex is neutral, name it using step 3.

2. Is the cation a complex ion?

3. *Yes* – name it using these rules, and then skip to step 5. *No* – skip to step 4.

   a. Name the ligands first using Table 20.1.

   b. Name the ligands in alphabetical order. Anionic ligands often end in "o," as in bromo, hydroxo, and sulfato.

   c. If more than one ligand of the same type is present, a prefix indicates the number of units. For simple ligands, the prefixes di-, tri-, tetra-, penta-, and hexa- are used. For complex ligands, the prefixes bis-, tris-, tetrakis-, pentakis-, and hexakis- are used.

   d. Name the metal last, with its oxidation number in parentheses in Roman numerals.

      (1) If the complex is negatively charged, the name of the metal ends in *-ate*.

      (2) If the complex is positively charged or neutral, no suffix is added to the metal name.

4. Name the cation using the conventions described in Chapter 3.

5. Is the anion a complex ion?

6. YES – Name it using the rules in step 3, and then stop. NO – Go to step 7.

7. Name the anion using the conventions described in Chapter 3, and then stop.

we name the metal. Thus the ammine ligand is named before the chloro ligand. Moreover, there are two of each of these ligands, so we should write down

diamminedichloro

without any spaces. Then, in step 3d, we write the name of the metal and its charge in parentheses immediately afterward. Because the charge on the entire complex is zero, and the complex is neutral, we do not add a suffix to the name:

diamminedichloroplatinum(II)

The name we have created for the formula indicates the number and type of each ligand, and the metal and its oxidation state. If we knew the three-dimensional arrangement of the atoms in this complex, we could also include that information in the name. For example, if we knew that the structure indicated a *cis* arrangement of the atoms, we could designate that by writing *cis*-diamminedichloroplatinum(II). Without that knowledge, though, we can provide only the name of the formula.

If we wish to name $[Co(NH_3)_6]Cl_3$, we can do so by following the rules. In step 1, we note that the compound is made up of a cationic complex and some anionic counter ions. Step 2 in Table 20.4 directs us to write the name of the complex (the cation). The ligands are named first, like this:

hexaammine

Then in step 3d we name the metal and its charge. Again, we do not add a suffix, because the complex is a cation:

hexaamminecobalt(III)

Finally, steps 6 and 7 tell us to add the counter ions by naming them as we did in Chapter 3. Note that we don't add a prefix to their name.

hexaamminecobalt(III) chloride

An anionic complex is not much different. Suppose we wished to name $K_2[NiCl_4]$. In step 1, we note that the anion is a complex and the cation is a counter ion. Accordingly, we name the cation first.

potassium

The complex anion is named using step 3. The ligands go first (note that there are four of them):

potassium tetrachloro

And the metal is part of an anion, so it gets the suffix -*ate*:

potassium tetrachloronickelate(II)

$$K_2\left[NiCl_4\right] \qquad \left[Co(NH_3)_6\right]Cl_3 \qquad \left[Co(NO_2)_2(NH_3)_4\right]_2SO_4$$

Examples of coordination compounds. See if you can name them all.

EXERCISE 20.3   Naming Compounds

Give the name for each of these compounds.

a. $[Cr(NH_3)_2(en)_2]SO_4$ b. $(NH_4)_3[Fe(CN)_6]$

First Thoughts

Naming coordination compounds requires us to identify the oxidation states of the metal center in addition to understanding and knowing the charges on the individual ligands and ions. To answer this question accurately, separate the compounds into their two halves (cation and anion). For the half containing the metal ion, determine its oxidation state, and then use the rules in Table 20.4 to name it.

Solution

a. diamminebis(ethylenediamine)chromium(II) sulfate

b. ammonium hexacyanoferrate(III)

Further Insight

The names for these compounds have a different sound to them, compared to the simpler compounds we worked with in Chapter 3. Just remember to follow the rules in Table 20.4, and be sure that you know the names of the ligands from Table 20.1.

PRACTICE 20.3

Name each of these compounds.

a. $[CrCl_2(NH_3)_4]_2SO_4$ b. $Na_2[Ni(CN)_4]$

Draw the formula for each of the following coordination compounds.

c. calcium hexafluoroferrate(II)

d. tetraamminedicarbonylmanganese(II) sulfate

See Problems 37–40.

20.7   Color and Coordination Compounds

Video Lesson: Color and Transition Metals

Many painters create images that both catch our eye and induce a feeling in our mind. The goal of the painter is often to portray more than just a staged scene. Van Gogh's painting *Starry Night* (see Figure 20.19) evokes a certain sense of wonder. The effect is generated by the brush strokes and the colors applied to the canvas. Are chemical compounds responsible for the color of paint? The answer is a resounding yes. Colors used in paintings are often constructed from ancient

FIGURE 20.19

Van Gogh's *Starry Night* uses the colors of transition metals to evoke the feeling of magic in the night sky.

formulations that include compounds of transition metals. One of the general characteristics of transition metals is the color of their compounds. Compare table salt ($NaCl$), baking soda ($NaHCO_3$), chalk ($CaCO_3$), and Epsom salts ($MgSO_4$) to ruby (with Cr_2O_3), emerald (with Cr_2O_3), and rust ($Fe_2O_3 \cdot nH_2O$, or hematite). Chromium was so named because of the variety and brilliance of the colors of its compounds. **What causes compounds that contain transition metals to have such striking colors?**

Transition Metals and Color

Because color is a common feature of transition metal compounds, it seems reasonable that there must be some common characteristics that give rise to their colors. The presence of color in most transition metal compounds can be attributed to the presence of partially filled *d* orbitals in the compounds and to the influence of the coordination environment on the energies of the *d* orbitals. Let's see how the color of transition metal compounds is related to fundamental atomic structure principles.

We perceive color when our eye detects light rays that differ from the ordinary distribution of those present in white light. For example, a sweater appears blue if white light strikes it, colors complementary to blue are absorbed, and the remaining light is reflected. A glass of fruit punch may appear red if white light strikes it and the colors complementary to red are absorbed. The remaining light, which appears red to our eye, is transmitted through the liquid and gives the liquid its red color. Color, then, is the array of light rays that our eyes observe being reflected from or transmitted through an object.

There are many objects that can absorb or transmit certain wavelengths of light but not be "colored." This occurs when the light being absorbed or transmitted is outside of the visible region (400–700 nm) of the electromagnetic spectrum. Recall from Chapter 6 that ultraviolet (UV) light has shorter wavelengths, typically defined as light in the wavelength range of 200–400 nm. A photon of UV light carries a lot of energy and may be damaging to substances it strikes. Infrared light has longer wavelengths than visible light.

Given our understanding of light (Chapter 6), we can begin to ask questions about how a compound can appear to be colored. **For example, what characteristic change occurs in a substance when light is absorbed?** Answering this question will enable us to understand why lime (CaO) is colorless but rust (Fe_2O_3) is colored. Several fundamental principles are involved.

Application

CHEMICAL ENCOUNTERS: Transition Metals and Color

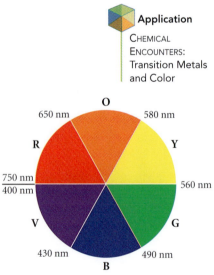

The color wheel indicates the range of wavelengths that corresponds to each color.

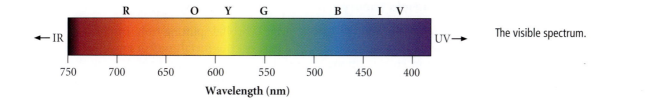

The visible spectrum.

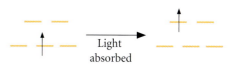

Absorption of a photon of visible light by a compound results in the excitation of an electron from a low energy orbital to a higher energy orbital within the substance. In order for the photon to be absorbed, the difference in orbital energy levels must match the energy of the photon absorbed. Finally, the excited electron must be able to be excited to the higher energy orbital—that is, the higher energy orbital must not be full. For example, when visible light strikes rust, an electron in a lower energy orbital absorbs a photon of blue light. The electron is excited to a higher energy orbital. (Usually the high-energy electron dissipates its energy as heat and relaxes back to its original energy level, ready to absorb again!) Orange light is reflected, and that is what our eyes detect. Rust is orange-colored.

In many transition metal compounds, of which rust is one, the same process occurs. What electronic transitions take place to allow the absorption of visible light? The d orbitals of the transition metal are often arranged such that they have a slight difference in energy. One example is shown in Figure 20.20. This difference in energy is roughly that of a photon in the visible region of the electromagnetic spectrum. If the d orbitals are partially filled, as they are in the transition metal ions, a photon can cause an electron in the lower energy orbitals to jump to an empty (or partially filled) higher energy d orbital.

Let's use this information to compare the colors of rust (Fe_2O_3) and chalk ($CaCO_3$). The iron(III) ion in rust has the electron configuration $[Ar]3d^5$. It contains partially filled d orbitals as its valence orbitals. These d orbitals on the metal, surrounded by a field of oxide ions, are split into different energy levels in ways that we will discuss in the next section. When visible light strikes rust, some wavelengths of light are absorbed, and an electron is excited to a higher empty energy level. The calcium ion in lime is isoelectronic with argon. It does not have partially filled d orbitals, cannot absorb visible light energy, and reflects all of the visible light to our eyes. It appears to be white, as shown in Figure 20.21.

FIGURE 20.20

A slight difference in the energy of the five d orbitals on a transition metal in a complex allows visible light to be absorbed. This causes an electron to be excited to the higher-energy d orbitals.

FIGURE 20.21

The color of lime and that of rust are a result of the absorption of specific wavelengths of light in the visible region of the electromagnetic spectrum. Calcium ions in lime do not possess electrons in d orbitals and cannot absorb light in the visible region. Iron ions in rust have d orbital electrons and do absorb visible light.

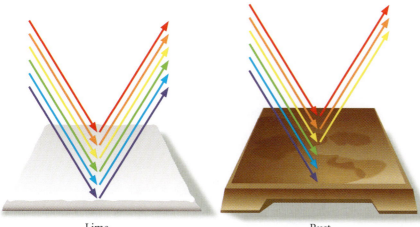

Lime   Rust

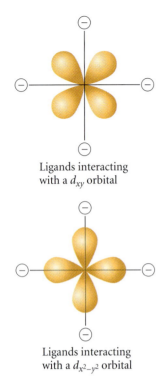

Ligands interacting with a d_{xy} orbital

Ligands interacting with a $d_{x^2-y^2}$ orbital

FIGURE 20.22

As ligands approach the d orbitals, they affect the relative energy levels of the $5d$ orbitals. Some, like the $d_{x^2-y^2}$ orbital, have an increased energy due to the repulsion of similar charges.

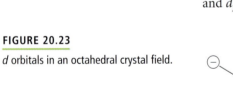

Video Lesson: Crystal Field Theory

Crystal Field Theory

We can understand the nature of the d orbital splitting from careful analysis of geometry and d orbital shape. In a free, gaseous metal atom or ion, there is no difference in energy among the orbitals of the same sublevel. We say that these orbitals, such as the $3d$ orbitals on gaseous iron, are degenerate. But something happens when we bring other atoms (such as the negatively charged anions) up close to the metal. They cause distortions in the orbitals of the metal. In a six-coordinate octahedral complex, for example, ions approach the central metal from each end of the three coordinate axes. Their orientation causes the orbitals on the metal to lose their degenerate nature.

Let's look just at the xy plane in an octahedral complex. The octahedral ligands are oriented along the x axis and the y axis. The d orbitals on a metal at the center of our plane are oriented in specific ways. Which electron would have lower energy, an electron in a $d_{x^2-y^2}$ orbital pointed directly at a negatively charged ligand or an electron in a d_{xy} orbital pointed between the negatively charged ligands? Using Figure 20.22 as a graphical guide, we can answer this question. Because like charges repel, an electron in a $d_{x^2-y^2}$ orbital would be higher in energy. We would represent this on an energy-level diagram as follows:

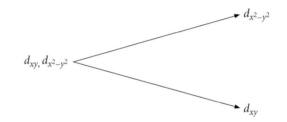

Octahedral Crystal Field Splitting

For a metal center in an octahedral crystal field, an electron in a $d_{x^2-y^2}$ orbital would be at higher energy than an electron in a d_{xy} orbital. We can extend this analysis to three dimensions and include all d orbitals as shown in Figure 20.23. Note that the $d_{x^2-y^2}$ and d_{z^2} orbitals are pointed right at the negatively charged ligands, but the d_{xy}, d_{xz}, and d_{yz} orbitals are pointed between the axes. The d_{xy}, d_{xz}, and d_{yz} orbitals will be at lower energy than the $d_{x^2-y^2}$ and d_{z^2} orbitals. Based on

FIGURE 20.23

d orbitals in an octahedral crystal field.

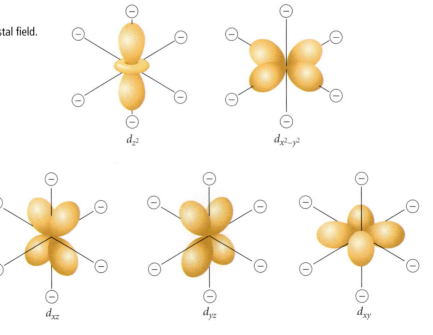

d_{z^2} $d_{x^2-y^2}$

d_{xz} d_{yz} d_{xy}

the symmetry of the system, the d_{xz}, d_{yz}, and d_{xy} orbitals are at the same energy level. Although it is less obvious, the $d_{x^2-y^2}$ and d_{z^2} orbitals are at the same energy level, which is higher than that of the d_{xy}, d_{xz}, and d_{yz} orbitals.

In an octahedral array of anions or ligands, the d orbital energies of the central metal are split into a lower energy set (d_{xz}, d_{xy}, d_{yz}) that we label the t_{2g} set and an upper energy set ($d_{x^2-y^2}$, d_{z^2}) that we label the e_g set, as shown in Figure 20.24. The energy difference between the t_{2g} orbitals and the e_g orbitals in the octahedral field, the **crystal field splitting energy**, is given the symbol Δ_o, where the subscript "o" indicates the splitting energy for an octahedral complex. The magnitude of Δ_o, the difference in energy between the t_{2g} and e_g orbitals, depends on the nature of the central metal and the nature of the ligands.

Tetrahedral Crystal Field Splitting

A tetrahedral ligand field can be viewed as one with the metal ion at the center of a cube and ions at alternate corners of a cube, with the coordinate axes going through the centers of the faces of the cube as shown in Figure 20.25. In this situation, none of the d orbitals of a central atom are pointed at the ions. The d_{xy}, d_{xz}, and d_{yz} lobes are closer to the ions than the $d_{x^2-y^2}$ and d_{z^2} lobes. Thus the d_{xy}, d_{xz}, and d_{yz} orbitals are at a higher energy than the $d_{x^2-y^2}$ and d_{z^2} orbitals. As shown in Figure 20.25, the tetrahedral crystal field splitting energy, Δ_t, where the subscript "t" indicates a tetrahedral complex, is qualitatively the inverse of the octahedral field splitting diagram. We label the lower orbital set "e" and the upper orbital set "t". As the comparison in Figure 20.26 shows, the magnitude of the splitting, Δ_t, is estimated to be 4/9 the size of Δ_o, so the magnitude of tetrahedral crystal field splitting is smaller than that of octahedral field splitting.

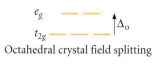

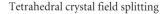

Octahedral crystal field splitting

Tetrahedral crystal field splitting

FIGURE 20.24

Crystal field splitting of metal d orbitals for octahedral and tetrahedral geometries.

FIGURE 20.26

The value of Δ_t is 4/9 that of Δ_o, the distance in energy from the $d_{x^2-y^2}$ to the d_{xy} orbital.

FIGURE 20.25

Tetrahedral crystal field splitting (arrows show the distance from orbital lobe to ions).

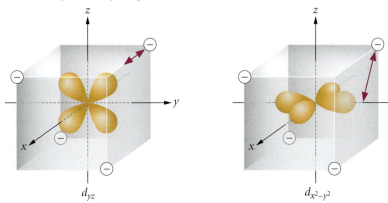

d_{yz} — $d_{x^2-y^2}$

Square Planar Crystal Field Splitting

We can define the square planar geometry as one with the four ligands in the xy plane as in an octahedral complex, but with the two opposing ligands on the z axis completely removed. The crystal field effect in the xy plane remains the same, but the effect in the z direction is removed. This results in stabilization of the d orbitals with z axis components. The resulting crystal field splitting diagram is shown in Figure 20.27. The difference in energy between d_{xy} and $d_{x^2-y^2}$ is still Δ_o. The other energy differences are considerably smaller.

FIGURE 20.27

Square planar coordination complex.

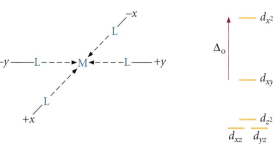

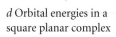

Ligand arrangement of a square planar complex

d Orbital energies in a square planar complex

Orbital Occupancy

The crystal field energy-level diagrams are the basis for understanding both physical and chemical properties of transition metal coordination complexes. However, as with all other chemicals, these properties are related to the electron configuration of the species. This configuration still follows the Aufbau rule, Pauli's exclusion principle, and Hund's rule of maximum multiplicity (Chapter 6). However, the splitting between the energy levels of the d orbitals becomes similar to the electrostatic repulsion that occurs when two electrons occupy the same orbital. This repulsion is called the **pairing energy (P)**. Depending on the values of Δ and P, it may be favorable to place an electron in a higher energy orbital rather than pair it with another electron in one orbital where the negatively charged electrons will repel each other.

Octahedral Complexes

An octahedral coordination complex containing a central metal atom with one d electron, such as Ti^{3+} or V^{4+}, will have one electron occupying one of the t_{2g} orbitals. For a central metal with a d^2 configuration, such as V^{3+}, one electron will occupy each of two t_{2g} orbitals, as shown in Figure 20.28. This configuration can be symbolized t_{2g}^2, where the superscript represents the number of electrons in the orbital set labeled t_{2g}. An octahedral complex with three d electrons would give rise to a t_{2g}^3 orbital occupancy. However, if another electron is added, it may either pair up with one of the electrons already in a t_{2g} orbital at an energy cost of $+P$, or occupy one of the e_g orbitals at an energy cost of $+\Delta_o$. The electron will tend to attain the lowest energy situation and hence will occupy the e_g orbital if Δ_o is less than the pairing energy, P. Therefore, depending on the values of Δ_o and P, the configuration could either be t_{2g}^4 with two unpaired electrons or $t_{2g}^3 e_g^1$ with four unpaired electrons, as shown in Figure 20.29.

These two configurations of electrons in the d orbitals are termed **low-spin**, (containing the minimum number of unpaired electrons, in this case two) or **high-spin** (with the maximum number of unpaired electrons, in this case four), respectively. For example, chromium complexes containing the chromium(II) ion (four d electrons) may be either high-spin with four unpaired electrons, as in $[Cr(H_2O)_6]^{2+}$, or low-spin with two unpaired electrons, as in $[Cr(CN)_6]^{4-}$.

Continuing with our orbital occupancy evaluation, a d^5 configuration could be either t_{2g}^5 (low-spin) or $t_{2g}^3 e_g^2$ (high-spin). At the d^6 configuration, electron pairing must occur, but either low-spin (t_{2g}^6) or high-spin ($t_{2g}^4 e_g^2$) may exist. A d^7 configuration could also be either low-spin ($t_{2g}^6 e_g^1$) or high-spin ($t_{2g}^5 e_g^2$). Like the electron configurations with one, two, or three electrons, only one option is available for occupancy of the t_{2g} and e_g orbital sets for the d^8 ($t_{2g}^6 e_g^2$), d^9 ($t_{2g}^6 e_g^3$), and d^{10} ($t_{2g}^6 e_g^4$) configurations.

$P > \Delta$

d^4 High-spin ($t_{2g}^3 e_g^1$)

$P < \Delta$

d^4 Low-spin (t_{2g}^4)

FIGURE 20.29

The relative magnitude of the pairing energy and the crystal field energy determine the electron configuration.

FIGURE 20.28

Electron distribution in d^2, d^4, d^6, d^8 octahedral coordination complexes.

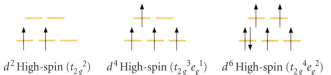

d^2 High-spin (t_{2g}^2) d^4 High-spin ($t_{2g}^3 e_g^1$) d^6 High-spin ($t_{2g}^4 e_g^2$) d^8 High-spin ($t_{2g}^6 e_g^2$)

d^4 Low-spin (t_{2g}^4) d^6 Low-spin (t_{2g}^6)

Tetrahedral Complexes

Although it would seem that similar high-spin and low-spin electron configurations would be possible for the $d^3 - d^6$ metals in tetrahedral complexes, such situations are not observed. The magnitude of Δ_t is usually less than the pairing energy P. Recall that Δ_t for a tetrahedral complex is only about ⅘ that of Δ_o for an octahedral complex. Because P is always greater than Δ_t, high-spin complexes are nearly always the only possibility for tetrahedral complexes.

Square Planar Complexes

Similar analyses could be done with square planar complexes. However, it turns out that most square planar complexes occur for metal centers and ligands that generate a low-spin configuration. For example, cisplatin (cis-$[PtCl_2(NH_3)_2]$) has the orbital occupancy shown in Figure 20.30.

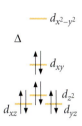

FIGURE 20.30

Electron distribution in square planar cis-$[PtCl_2(NH_3)_2]$.

EXERCISE 20.4   Drawing Crystal Field Diagrams

Draw crystal field splitting diagrams with electron occupancy for $[Mn(H_2O)_6]^{2+}$ (high-spin).

Solution

Manganese(II) has a d^5 configuration. With a coordination number of 6, an octahedral crystal field is expected. A high-spin d^5 ($t_{2g}{}^3 e_g{}^2$) configuration will result.

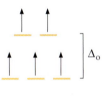

PRACTICE 20.4

How many unpaired electrons would you predict for the low-spin complex $[Fe(CN)_6]^{4-}$? Show the basis of your prediction using the crystal field splitting diagram for the iron metal ion.

See Problems 47–50.

The Result of d Orbital Splitting

A really dramatic example of color differences occurs in the two complexes of Co^{2+} shown in Figure 20.31. When water is coordinated to Co^{2+} in the $[Co(H_2O)_6]^{2+}$ ion, the color is red. When chloride ligands coordinate to Co^{2+} in $[CoCl_4]^{2-}$, the color is blue. **How does this fit into our understanding of color?** The valence electron configurations of all of these species are shown in Figure 20.32 on page 890. For $[Co(H_2O)_6]^{2+}$ with an octahedral ligand field, the red color arises when an electron is excited from a t_{2g} orbital to an e_g orbital by absorption of a photon of energy Δ_o. For $[CoCl_4]^{2-}$ with a tetrahedral ligand field, the blue color arises when an electron is excited from an e orbital to a t orbital by absorption of a photon of energy Δ_t. Because, in general, Δ_t is smaller than Δ_o, the photon absorbed by $[CoCl_4]^{2-}$ must be lower in energy than the photon absorbed by $[Co(H_2O)_6]^{2+}$. In this case, $[Co(H_2O)_6]^{2+}$ absorbs higher-energy blue light and appears red, and $[CoCl_4]^{2-}$ absorbs lower-energy red light and appears blue.

The absorption spectra in Figure 20.33, which show how much light of various wavelengths is absorbed, exhibit a maximum for $[CoCl_4]^{2-}$ at 660 nm (equivalent to 181 kJ/mol) and a maximum for $[Co(H_2O)_6]^{2+}$ at 510 nm (equivalent to 221 kJ/mol). Recall that longer wavelength corresponds to lower energy,

$$E = \frac{hc}{\lambda},$$ so the $[CoCl_4]^{2-}$ complex absorbs lower-energy photons. The key point

FIGURE 20.31

Absorption of light (arrow) by cobalt complexes. The red solution is a solution of $[Co(H_2O)_6{}^{2+}]$; the blue solution is the coordination of chloride with cobalt(II) in $[CoCl_4]^{2-}$. Note that different wavelengths of light are absorbed by these complexes.

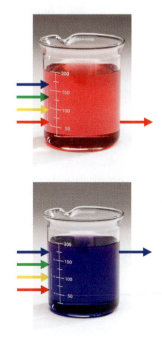

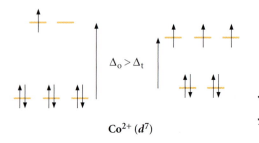

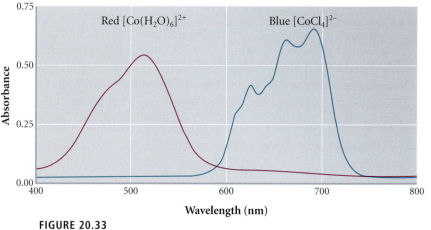

FIGURE 20.32

The electron configurations of two cobalt complexes, $[Co(H_2O)_6]^{2+}$ and $[CoCl_4]^{2-}$. The difference in the magnitude of the crystal field energy determines what wavelengths of light they absorb.

FIGURE 20.33

Absorption spectra of $[Co(H_2O)_6]^{2+}$ and $[CoCl_4]^{2-}$.

is that based on the model and with experimental evidence to support it, *the geometry of the ligand field around a metal center influences the size of* Δ.

The nature of the ligands around the metal center also influences the magnitude of Δ. Figure 20.34 shows solutions of three cobalt(III) complexes with different ligands around the cobalt center. The wavelength of light that is absorbed decreases, as shown in Figure 20.35, as the unique ligand varies from Cl^- to H_2O to NH_3. As the ligand varies, the value of Δ_o increases and the wavelength of maximum absorbance decreases. Using this type of information from a wide variety of complexes, a **spectrochemical series** was developed that illustrates the effect of a particular ligand on the value of Δ. The series is

$$Cl^- < F^- < OH^- < H_2O < NH_3 < NO_2^- < CN^- < CO$$

Conveniently, this series is generally the same for all metals. Knowing this information, we can predict that the wavelength of light absorbed by $[FeF_6]^{3-}$ would be longer (lower in energy) than the wavelength of light absorbed by $[Fe(H_2O)_6]^{3+}$ because a water ligand produces a larger value of Δ than the fluoride ligand.

 Application

Let's use this information to answer a practical question: Why is the blood in your veins dark red, whereas the blood in arteries is bright red? When blood in a vein is exposed to air, its color changes to bright red. Although the exact origin of this color change is fairly complex, it turns out that the *crystal field splitting increases when oxygen binds to the iron center*. When oxygen binds to the Fe^{2+} center, higher-energy light is absorbed, longer-wavelength light is transmitted by the hemoglobin, and the blood changes color. Oxyhemoglobin (with a larger crystal field splitting) absorbs higher-energy blue light and is red in the arteries;

FIGURE 20.34

The influence of coordinated ligands on color of coordination complexes. Each complex of cobalt(III) contains five NH_3 ligands and one other ligand. From left to right: $[Co(NH_3)_6]^{3+}$, $[Co(NH_3)_5(H_2O)]^{3+}$, and $[Co(NH_3)_5Cl]^{2+}$.

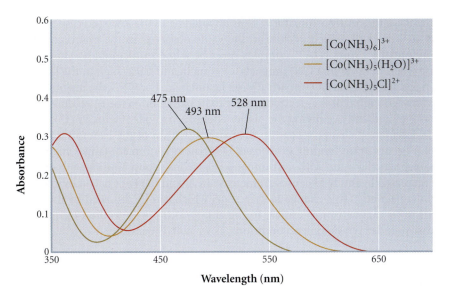

FIGURE 20.35

Absorption spectra of $[Co(NH_3)_6]^{3+}$, $[Co(NH_3)_5(H_2O)]^{3+}$, and $[Co(NH_3)_5Cl]^{2+}$, showing higher-energy, shorter-wavelength absorption in the order $[Co(NH_3)_6]^{3+} < [Co(NH_3)_5(H_2O)]^{3+} < [Co(NH_3)_5Cl]^{2+}$. (Spectra provided by Jerry Walsh.)

deoxyhemoglobin (with a smaller crystal field splitting) absorbs lower energies and appears bluish-red in the veins.

Magnetism

Video Lesson: Magnetic Properties and Spin

When magnetism is mentioned, it is common to think of bar magnets and their attraction for metal objects. This common form of magnetism shown by iron, nickel, and cobalt is known as **ferromagnetism**. However, there is a more common but also more subtle form of magnetism called **paramagnetism** (Chapter 9). This property exists in any substance that contains unpaired electrons. A substance that is paramagnetic is attracted to a magnetic field, though less strongly than in a ferromagnetic substance. A material that has all of its electrons paired exhibits **diamagnetism**, and these materials are very weakly repelled by a magnetic field.

The origin of paramagnetism is in the spin of the unpaired electrons in a substance. In the absence of a magnetic field, the spins of the unpaired electrons are randomly oriented. When these unpaired electrons are placed in a magnetic field, their spins align with the field and result in a net attraction. The experimental characterization of transition metal coordination complexes was greatly aided by measurement of the paramagnetism of various species. Measurement of the strength of the paramagnetism provides an experimental quantity called the **magnetic moment (μ)**. In many cases, μ is related to the number of unpaired electrons, n, and is nearly equal to or slightly greater than the theoretical value given by the formula

$$\mu = [n(n + 2)]^{1/2}$$

where $n =$ number of unpaired electrons.

The complex $[Mn(H_2O)_6]^{2+}$ has a magnetic moment of 5.9, whereas the magnetic moment of $[Mn(CN)_6]^{4-}$ is 2.2. Both have five d electrons. Why should they have such different magnetic moments? Using the relationship between magnetic moment and the number of unpaired electrons, $\mu = [n(n + 2)]^{1/2}$, we find that if $n = 5$, then μ should be 5.92 for the aqua complex. Thus the $[Mn(H_2O)_6]^{2+}$ complex is a high-spin d^5 complex, as shown in Figure 20.36.

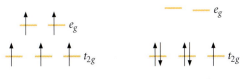

$Mn(OH_2)_6^{2+}\ d^5$ High-spin $\qquad$ $Mn(CN)_6^{4-}\ d^5$ Low-spin

FIGURE 20.36

High-spin and low-spin manganese complexes. In the high-spin case, the crystal field energy is less than the pairing energy. In the low-spin case, the magnitude of Δ is greater than the pairing energy.

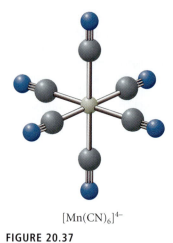

$[Mn(CN)_6]^{4-}$

FIGURE 20.37

Using the same relationship, if $n = 1$ (as it would be in the low-spin case), we expect μ to be 1.73, which is close to the value observed in the cyano complex. Therefore, the $[Mn(CN)_6]^{4-}$ complex must be a low-spin complex. This makes sense when we note that CN^- imparts a strong ligand field (high in the spectrochemical series) and H_2O imparts a weak ligand field. For CN^-, the magnitude of Δ is large enough that the electrons prefer to be paired up in the t_{2g} orbitals rather than unpaired and occupying the higher energy e_g orbitals (Figure 20.37).

A fascinating change related to the paramagnetism of the iron ion in hemoglobin occurs upon binding of oxygen. Deoxyhemoglobin contains a high-spin, paramagnetic Fe^{2+} ion with four unpaired electrons. Upon binding with oxygen, the increased ligand field causes an increase in Δ, and the Fe^{2+} becomes low-spin and diamagnetic. This change in spin state, which can be followed by measurement of its magnetism, is critical to the ability of the iron to bind oxygen and release it under physiological conditions, as we will discuss later.

20.8 Chemical Reactions

Aside from possessing interesting color and magnetism properties, transition metal coordination complexes also undergo useful chemical reactions. Of particular biological importance is a class of reactions known as ligand exchange reactions. This class of reactions includes the coordination of oxygen and its release from an iron ion in hemoglobin. Another class of reactions, known as electron transfer reactions, are also important in biological processes. This class of reactions is quite common in photosynthesis and respiration.

Ligand Exchange Reactions

Visualization: Nickel(II) Complexes

Why do coordination complexes form? All chemical reactions proceed when the total free energy of the system, ΔG, decreases. Because the free energy change depends on two factors, entropy and enthalpy, these must be considered in a reaction involving the association of a ligand with a metal. Consider the formation of hexaamminenickel(II) from the aqua complex:

$$[Ni(H_2O)_6]^{2+} + 6NH_3 \rightarrow [Ni(NH_3)_6]^{2+} + 6H_2O \qquad K = 4 \times 10^8$$

Because the two different ligands, NH_3 and H_2O, are similar in size and the same number of each are involved in the reaction, the change in entropy for the reaction is small. *The driving force for this reaction must then rely on the stability of the resulting coordinate covalent bonds.* In this example, the Ni^{2+}—N bond is stronger than the Ni^{2+}—O bond, so ΔG is relatively large and negative because of the favorable enthalpy change in nickel—nitrogen bond formation. This translates into a large equilibrium constant for the formation of $[Ni(NH_3)_6]^{2+}$ via the equation $\Delta G = -RT\ln K$.

Consider the following reaction of our nickel complex. Each mole of this complex can react with three moles of ethylenediamine to form $[Ni(en)_3]^{3+}$. Both complexes contain Ni^{2+}—N bonds. Does it make sense that this reaction should also have a very large equilibrium constant?

$$[Ni(NH_3)_6]^{2+} + 3en \rightarrow [Ni(en)_3]^{2+} + 6NH_3 \qquad K = 5 \times 10^9$$

There doesn't appear to be much change in the enthalpy of this reaction (the change in enthalpy is expected to be small because the bonds created in the product are similar to the bonds broken in the reactant), so we must focus on changes in entropy for this reaction. A very favorable entropy change is observed; the reaction produces three more moles of product than moles of reactant. This

FIGURE 20.38

Rates of ligand exchange. $[CoCl_4]^{2-}$ exchanges chloro ligands immediately on mixing.

exceptionally high favorability for forming complexes with polydentate ligands is known as the **chelate effect**.

Although we can predict the direction of a reaction by using thermodynamics, only an examination of the kinetics of the reaction will determine the rate of the reaction. Some coordination complexes exchange their ligands very rapidly and are referred to as **labile**; others do so more slowly and are referred to as **inert**. For example, the blue $[CoCl_4]^-$ complex rapidly exchanges chloro ligands with water to produce the red $[Co(H_2O)_6]^{2+}$ complex, both of which are shown in Figure 20.38. In contrast, the chromium complexes shown in Figure 20.39, green $[CrCl_2(H_2O)_4]^+$ and purple $[Cr(H_2O)_6]^{3+}$, take at least a day to exchange ligands. In the interaction between cisplatin and DNA molecules, it is important for the platinum complex to bond to the DNA and remain bonded long enough for the complex to be toxic to the system. The platinum–DNA complex is inert—and *must* be inert to exhibit the kind of anticancer activity that it does.

The kinetic and thermodynamic factors exhibited in the reactivity of coordination complexes are intriguingly complementary in hemoglobin. The Fe^{2+} metal center is kinetically labile, which is important for the rapid exchange of oxygen ligands:

$$HbFe^{2+} + O_2 \rightleftharpoons HbFeO_2^{2+} \qquad \text{rapid}$$

However, in spite of the fact that Fe^{2+} can exchange its ligands rapidly, the chelate effect of the tetradentate porphyrin ligand maintains stability of the iron–porphyrin portion of the complex.

$$HbFe^{2+} + 6H_2O \rightleftharpoons Hb + [Fe(H_2O)_6]^{2+} \qquad \text{equilibrium lies to the left}$$

The balance of chemical characteristics for molecules in living systems is exquisite. This is why the scientific challenge of generating a chemical substitute for hemoglobin to transport oxygen in the bloodstream has been formidable.

FIGURE 20.39

$[CrCl_2(H_2O)_4]^+$ exchanges ligands slowly, taking a day or more to fully exchange its ligands.

Electron Transfer Reactions

Application

CHEMICAL
ENCOUNTERS:
Cytochromes

Transition metals typically exhibit several stable oxidation states, and the +2 and +3 states are fairly common. Because many transition metals exhibit stability in two or more oxidation states, transition metal complexes can play important roles in electron transfer processes. For example, cytochromes are electron transfer agents in biological systems. Within a cytochrome protein, an iron ion coordinates to a porphyrin ring. The other sites on the octahedral complex are occupied by ligands that are part of the protein structure, as shown in Figure 20.40. During respiration and photosynthesis, the iron changes oxidation state from Fe^{2+} to Fe^{3+} to Fe^{2+} as the cytochrome shuttles electrons between two biological reaction sites (Figure 20.41).

FIGURE 20.40

Iron coordination in cytochrome electron transfer protein.

FIGURE 20.41

The role of iron (cytochrome) and copper (plastocyanin) oxidation states in electron transfer reactions in photosynthesis.

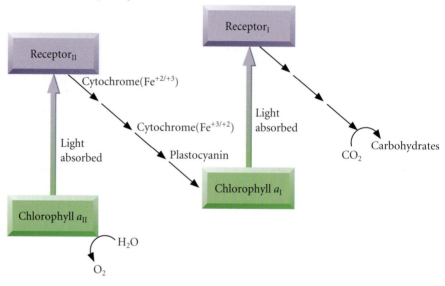

The Bottom Line

- Coordination complexes are present in simple metal ions in solution and as the reaction center in many biological molecules. (Section 20.1)

- A Lewis base that donates a lone pair of electrons to a metal to form a coordinate covalent bond acts as a ligand in producing a coordination complex. (Section 20.2)

- The coordination numbers of various metal centers are commonly observed to be 2, 4, or 6. (Section 20.3)

- Common coordination geometries are linear, tetrahedral, square planar, and octahedral. (Section 20.4)

- A variety of isomer classes are observed for metal complexes. (Section 20.5)

- The proper names and formulas for coordination complexes are specified by IUPAC rules. (Section 20.6)

- In transition metal complexes, the d orbitals are no longer degenerate but split into two or more energy levels, depending on coordination geometry. (Section 20.7)

- The electron configuration for octahedral complexes gives rise to high-spin and low-spin complexes for d^4 to d^7 metal centers. (Section 20.7)

- The color of many transition metal compounds arises when a photon of visible light is absorbed and an electron is excited to a higher energy d orbital. (Section 20.7)

- The order of ligands, in terms of their influence on the magnitude of the *d* orbital splitting and the energy of the photon of light absorbed, is defined as the spectrochemical series. (Section 20.7)

- The number of unpaired electrons determines the magnetic moment of the complex. (Section 20.7)

- Metal centers that exchange ligands rapidly are called labile; those that exchange ligands slowly are called inert. (Section 20.8)

- Chelate complexes have high formation constants because of an entropy effect. (Section 20.8)

- Because transition metal complexes often exhibit several oxidation states, transition metal complexes are good electron transfer (redox) agents. (Section 20.8)

Key Words

bidentate Capable of forming two coordinate covalent bonds to the same metal center. (*p. 874*)

chelate A polydentate ligand that forms strong metal–ligand bonds. (*p. 874*)

chelate effect An unusually large formation constant due to a favorable entropy change for the formation of a complex between a metal center and a polydentate ligand. (*p. 893*)

cis isomer An isomer containing two similar groups on the same side of the compound. (*p. 880*)

coordinate covalent bond A covalent bond that results when one atom donates both electrons to the bond. (*p. 870*)

coordinates Forms a coordinate covalent bond. (*p. 871*)

coordination complex A metal bonded to two or more ligands via coordinate covalent bonds. (*p. 870*)

coordination number The number of coordinate covalent bonds that form in a complex. (*p. 875*)

coordination sphere isomers Substances that contain different ligands in the coordination spheres of complex cations and complex anions. (*p. 879*)

crystal field splitting energy The difference in energy between the *d* orbital sets that arises because of the presence of ligands around a metal; symbolized by Δ. (*p. 887*)

diamagnetism The ability of a substance to be repelled from a magnetic field. This property arises because all of the electrons in the molecule are paired. (*p. 891*)

ferromagnetism Occurs when paramagnetic atoms are close enough to each other (such as in iron) that they reinforce their attraction to the magnetic field, so the whole is, in effect, greater than the sum of its parts. (*p. 891*)

geometric isomers Substances in which all of the atoms are attached with the same connectivity, or bonds, but the geometric orientation differs. (*p. 880*)

high-spin A coordination complex with the maximum number of unpaired electrons. (*p. 888*)

inert The opposite of labile; said of a compound that is very slow to exchange ligands. (*p. 893*)

ionization isomers Isomers that differ in the placement of counter ions and ligands. (*p. 879*)

labile The opposite of inert; said of a compound that exchanges ligands rapidly. (*p. 893*)

ligand A Lewis base that donates a lone pair of electrons to a metal center to form a coordinate covalent bond. (*p. 870*)

linkage isomers Isomers that differ in the point of attachment of a ligand to a metal. (*p. 879*)

low-spin A coordination complex with the minimum number of unpaired electrons. (*p. 888*)

magnetic moment (μ) The strength of the paramagnetism of a compound. Can be used to determine the number of unpaired electrons. (*p. 891*)

octahedral The geometry indicated by six electron groups (lone pairs and/or bonds) positioned symmetrically around a central atom such that the bond angles are 90°. (*p. 877*)

pairing energy (*P*) The energy required to spin-pair two electrons within a given orbital. (*p. 888*)

paramagnetism The ability of a substance to be attracted into a magnetic field. This attraction arises because of the presence of unpaired electrons within the molecule. (*p. 891*)

polydentate A ligand that contains two or more lone pairs on different atoms, each forming a coordinate covalent bond to a metal center. (*p. 874*)

spectrochemical series The series of ligands organized with respect to their effect on the value of the crystal field splitting energy. (*p. 890*)

square planar The geometry indicated by four electron groups (lone pairs and/or bonds) positioned symmetrically in a plane around a central atom such that the bond angles are 90°. (*p. 877*)

tetradentate Capable of forming four coordinate covalent bonds to the same metal center. (*p. 874*)

tetrahedral The geometry indicated by four electron groups (lone pairs and/or bonds) positioned symmetrically around a central atom such that the bond angles are 109°. (*p. 877*)

trans isomer An isomer containing two similar groups on opposite sides of the compound. (*p. 880*)

tridentate Capable of forming three coordinate covalent bonds to the same metal center. (*p. 874*)

Focus Your Learning

The answers to the odd-numbered problems and some selected problems appear at the back of the book, as represented by the blue numbering.

20.1 Bonding in Coordination Complexes

Skill Review

1. Define each of these terms in your own words:
   a. ligand   b. coordinate covalent bond   c. Lewis acid

2. Define each of these terms in your own words:
   a. Lewis base   b. complex   c. metal center

3. For each of these coordination complexes, state the number of ligands, the oxidation number of the coordinated metal, and the number of coordinate covalent bonds that are formed within the complex.
   a. $[Fe(H_2O)_6]^{2+}$ b. $[Co(NH_3)_6]^{2+}$
   c. $[Zn(NH_3)_4]^{2+}$ d. $[Pt(CO)_4]$

4. For each of these coordination complexes, state the number of ligands, the oxidation number of the coordinated metal, and the number of coordinate covalent bonds that are formed within the complex.
   a. $[Fe(CN)_6]^{3-}$ b. $[CuCl(NH_3)_3]^+$
   c. $[Ni(H_2O)_6]^{2+}$ d. $[Au(NH_3)_2]^{2+}$

5. Diagram a structure that depicts a metal ion (M) surrounded by four symmetrically arranged ligands (L).

6. Diagram a structure that depicts a metal ion (M) surrounded by six symmetrically arranged ligands (L). Why are coordination complexes rarely found with more than six ligands?

20.2 Ligands

Skill Review

7. Diagram the Lewis dot structure for ammonia (NH_3). Explain why this molecule is classified as a Lewis base.

8. Diagram the Lewis dot structure for methylamine (CH_3NH_2). Is this molecule a Lewis base?

9. Would you predict the following molecule to be a ligand? If so, indicate whether it would be monodentate, bidentate, tridentate, or tetradentate.

10. Would you predict the following molecule to be a ligand? If so, indicate whether it would be monodentate, bidentate, tridentate, or tetradentate.

11. Which of these are *not* likely act as ligands in forming coordination complexes?
    a. H_2O b. CN^- c. Ca^{2+} d. O_2 e. C_2H_6

12. Which of these can act as a ligand in forming coordination complexes?
    a. CH_4 b. Br_2 c. O^{2-} d. H_2O_2 e. Li^+

Chemical Applications and Practices

13. Explain why the structure of ethylenediamine (en) enables it to act as a bidentate ligand, whereas nitrogen gas, which also contains two nitrogen atoms, cannot act as a bidentate ligand.

14. $EDTA^{4-}$ can coordinate with metal ions at six different sites. $EDTA^{4-}$ is often used as a food preservative in such food products as mayonnaise. Explain how $EDTA^{4-}$ functions in such an effective manner.

15. The formula $Co(A_2B_2)$ represents a coordination complex with two ligands, A and B. The Co ion is coordinated through six sites. If two A ligands occupy two of those sites, what must be true about the bonding for ligand B?

16. Amino acids can act as bidentate ligands. This is due to the presence of both oxygen and nitrogen. For example, alanine (Ala) could form the metal complex $[Fe(ala)_3]^{3+}$. Use the alanine structure to draw a structure of the metal complex.

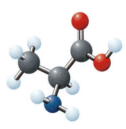

20.3 Coordination Number

Skill Review

17. Determine the oxidation state and the coordination number of the central metal in each of these coordination complexes:
    a. $[Cr(NH_3)_5Br]^{3+}$
    b. $[Mn(NH_2CH_2CH_2NH_2)_3]^{2+}$
    c. $[Cd(NH_2CH_3)_4]^{2+}$

18. Determine the oxidation state and the coordination number of the central metal in each of these coordination complexes:
    a. $[Co(NH_3)_6]^{3+}$ b. $[Pd(en)Cl_2]$ c. $[Mo(ox)_3]^{3-}$

19. Explain how it is possible that the central metals in these two complexes can have the same coordination number even though the number of ligands differs in the complexes.

$$[Fe(ox)_3]^{3-} \qquad [Co(SCN)_2(H_2O)_4]^+$$

20. Explain how it is possible that the central metal in these two complexes can have the same oxidation state even though the coordination number differs in the complexes.

$$[Fe(en)_3]^{3+} \qquad [ScI_4]^-$$

Chemical Applications and Practices

21. Many heavy-metal-containing salts are not very water soluble. One way to increase the solubility is to form a complex. Examine the structure of the bidentate oxinate ion shown here. If two of these complexed with a lead ion, what would be the coordination number of the lead?

Oxinate ligand

22. One method used to determine the "hardness" of water samples is to titrate the sample with EDTA. This reaction forms a complex with the calcium ions in the water. Using the structure of EDTA depicted in Figure 20.9, determine what the coordination number of the calcium ion would be if it combined with one disodium EDTA ion. Diagram the structure that justifies your answer.

20.4 Structure

Skill Review

23. Diagram the structures of the complexes listed in Problem 3, and identify the geometry (bond angles and overall shape) of each structure.

24. Diagram the structures of the complexes listed in Problem 4, and identify the geometry (bond angles and overall shape) of each structure.

25. Indicate the oxidation number, coordination number, and geometry for the cobalt ion in the following compound.

$$[CoCl(NO_2)(en)_2]Cl$$

26. Indicate the oxidation number, coordination number, and geometry for the silver ion in the following compound.

$$[Ag(NH_3)_4]Br_2$$

Chemical Applications and Practices

27. When nickel ore is refined, it must be removed from other metals. This can be done by forming a coordination complex between nickel and carbon monoxide. The highly poisonous nickel–carbon monoxide complex $Ni(CO)_4$ evaporates easily and can therefore be used to separate nickel from its impurities. What is the coordination number for nickel, and what two possible geometries could you predict for this structure?

28. The readily available electron pairs found in the oxygen, nitrogen, and some sulfur atoms of amino acids (Chapter 22) provide bonding sites for metals. The metal–amino acid combinations serve as the basis for many important biochemical processes. For example, a certain copper-containing enzyme utilizes an octahedral structure around a Cu^{2+} ion to assist in the transport of electrons within cells. Draw the structure of a copper(II) complex that would form if three glycine amino acids (shown below) formed coordinate covalent bonds with the copper. (Assume that each glycine is bidentate.)

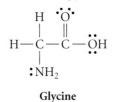

Glycine

20.5 Isomers

Skill Review

29. Define the type of isomer present, if isomerization exists, in each of these pairs of complexes.
    a. *trans*-$[Pt(NH_3)_2Cl_2]$ and *cis*-$[Pt(NH_3)_2Cl_2]$
    b. $[Pt(CN)_2(NH_3)_4]Cl_2$ and $[PtCl_2(NH_3)_4](CN)_2$
    c. $[Fe(H_2O)_6][CuBr_4]$ and $[Cu(H_2O)_6][FeBr_4]$

30. Define the type of isomer present, if isomerization exists, in each of these pairs of complexes.
    a. $[Cu(NH_3)_3(ONO)]$ and $[Cu(NH_3)_3(NO_2)]$
    b. $[Mn(H_2O)_4Cl_2]Br_2$ and $[Mn(H_2O)_4Br_2]Cl_2$
    c. *cis*-$[PdCl_2(NH_3)_2]$ and *trans*-$[PdCl_2(NH_3)_2]$

31. Diagram two square planar geometric isomers with the formula $[PtI_2(NH_3)_2]$. Label the *cis* isomer. It is not possible to diagram two tetrahedral geometric isomers of $[Pt(NH_3)_2I_2]$. Explain why.

32. Diagram all of the possible isomers of $[Co(NH_3)_2(SCN)_2]$.

33. How many different isomers of $[NiCl_3F_3]^{4-}$ can be drawn? Show the structure of each.

34. Show the structures of the coordination sphere isomers for $[Co(NH_3)_6][Cr(NO_2)_6]$.

Chemical Applications and Practices

35. Ethylenediamine (en) is a bidentate ligand, so it forms two attachments to metal ions in coordination complexes. However, the square planar complex $[Pt I_2(en)]$ exhibits only one type of geometric isomer. Draw possible structures for $[Pt I_2(en)]$ and for a complex between iron(III) and en.

36. Diagram the Lewis dot structure for the cyanate ion (OCN^-). Show how this ion would make it possible to have two different forms of the following complex: $[Co(OCN)(NH_3)_5]^{2+}$. What type of isomerism does this example illustrate?

20.6 Formulas and Names

Skill Review

37. Provide names for these complex ions:
    a. $[Co(CN)(en)_2(NH_3)]^{2+}$
    b. $[Cr(C_2O_4)_2(NH_3)_2]^-$
    c. $[Fe(NO_2)_6]^{3-}$
    d. $[CoCl_3(H_2O)_3]$

38. Provide names for these complex ions:
    a. $[Mn(en)_3]^{2+}$
    b. $[Ni(H_2O)_4(NH_3)_2]^{2+}$
    c. $[Cr(NO_2)_6]^{3-}$
    d. $[V(SCN)_2(H_2O)_4]$

39. Write the chemical formula for each of these compounds and complex ions:
    a. tetraammineaquachlorocobalt(III)
    b. *trans*-diaquabis(ethylenediamine)copper(II) chloride
    c. sodium tetrachlorocobaltate(II)
    d. pentacarbonylchloromanganese(I)

40. Write the chemical formula for each of these compounds and complex ions:
    a. tetraaquadichlorocopper(II)
    b. potassium *cis*-dibromooxalatoplatinate(II)
    c. tetraamminenickel(II) sulfate
    d. tetraaquathiosulfatoiron(III) nitrite

41. Which of these species would produce the greater number of ions per mole when dissolved in water?

$$K_2[Cr(C_2O_4)_2(H_2O)_2] \quad \text{or}$$
tetraamminediaquachromium(III) nitrate

42. Which of these species possesses the larger positive charge on the complex ion?

tetraaquacopper(II) nitrate   or
dichlorobis(ethylenediamine)iron(III) bromide

20.7 Color and Coordination Compounds

Skill Review

43. What is the electron configuration for each of these transition metal ions?
    a. Fe^{2+} b. Cr^{2+} c. Zn^{2+}

44. What is the electron configuration for each of these transition metal ions?
    a. Pd^{4+} b. Ag^+ c. Mn^{2+}

45. Consider the following two transition metal ions as free gaseous ions. Which would have the greater number of unpaired d electrons, Fe^{3+} or Cu^{2+}?

46. Which free ion has the greater number of unpaired d electrons, Ti^{2+} or Co^{2+}?

47. Draw the orbital diagram for the d orbitals in an octahedral complex containing each of these metal centers. (Assume that $P < \Delta_o$.)
    a. Fe^{3+} b. Co^{2+} c. Ni^{2+}

48. Draw the orbital diagram for the d orbitals in an octahedral complex containing each of these metal centers. (Assume that $P > \Delta_o$.)
    a. Mn^{2+} b. Fe^{2+} c. Cr^{+2}

49. Repeat Problem 47, but assume that the metal centers are involved in tetrahedral complexes. Although tetrahedral complexes typically have $\Delta_t > P$, what would you draw if the tetrahedral complex existed with $P > \Delta_t$?

50. Repeat Problem 48, but assume that the metal centers are involved in square planar complexes. Assume that $P > \Delta$ in this problem.

51. Draw the orbital diagram for the metal center in each of these complexes. Use the information in the spectrochemical series to assist you in placing the orbitals.
    a. $[FeCl_4]^-$ b. $[Co(CN)_6]^{3-}$ c. $[Mn(CO)_6]^+$

52. Draw the orbital diagram for the metal center in each of these complexes. Use the information in the spectrochemical series to assist you in placing the orbitals.
    a. $[CuF_6]^{4-}$ b. $[Ni(OH)_6]^{4-}$ c. $[Cr(NO_2)_6]^{4-}$

53. Which of the complexes in Problems 47 and 48 is(are) paramagnetic?

54. Calculate the magnetic moment for each of the complexes in Problems 47 and 48.

Chemical Applications and Practices

55. Which one of these complexes would you predict to absorb blue light: $[M(CN)_6^{2-}]$, $[M(H_2O)_6^{4+}]$, $[MCl_6^{2-}]$, or $[M(NH_3)_6^{4+}]$?

56. Which of these complexes would you predict to absorb the longest wavelength of visible light: $[M(CN)_6^{2-}]$, $[M(H_2O)_6^{4+}]$, $[MCl_6^{2-}]$, or $[M(NH_3)_6^{4+}]$?

57. The colors of common gemstones are due to the presence of transition metal ions. The color is produced when the metal ion absorbs visible light. Would you predict the common gemstones to have different "colors" under infrared light?

58. Coordination compounds with Zn^{2+} ions typically are white or colorless. Explain why this particular metal does not form brightly colored compounds the way many other transition metals do.

59. The crystal field theory provides an explanation of color in various coordination complexes. For example, $[Cr(H_2O)_6]^{3+}$ can be detected as a violet color when dissolved.
    a. What colors would the complex be absorbing?
    b. How many unpaired electrons does the central chromium ion have?
    c. The compound $[Cr(NH_3)_6]^{3+}$, when dissolved, appears yellow. Would you expect it to absorb light at a higher or lower frequency than $Cr(H_2O)_6]^{3+}$? Explain.
    d. Which ligand, NH_3 or H_2O, is causing the greater value of Δ_o?

60. Compare the two iron complexes $[Fe(H_2O)_6]^{3+}$ and $[Fe(CN)_6]^{3-}$.
    a. Which is more likely to be paramagnetic?
    b. Which is more likely to absorb light of greater energy?
    c. Which is more likely to be "high-spin"?

20.8 Chemical Reactions

Skill Review

61. Explain why the chelate effect typically provides for a very favorable entropy change when a ligand exchange reaction involves a complex going from a nonchelated complex to a chelate complex.

62. If a ligand exchange reaction produced a large positive ΔG value, would you expect the reaction to have a large or a small equilibrium constant? Justify your choice.

63. A common chemical demonstration is to change a light blue solution containing $[Cu(H_2O)_4]^{2+}$ quickly to a deep purple solution by changing $[Cu(H_2O)_4]^{2+}$ into $[Cu(NH_3)_4]^{2+}$ with the addition of ammonia to the solution. Would you consider the first compound labile or inert? Is the exchange of oxygen in hemoglobin considered to be representative of a labile or an inert complex?

64. Except through loss of blood, the level of iron in humans is fairly constant. One way that iron is moved throughout the body, particularly from the liver, is within a molecule known as ferritin. The iron(III) is held in a six-coordinate system through bonds to oxygen and nitrogen that are part of several amino acids. Would it be more logical for this molecule's iron site to be labile or inert with respect to other metals? Explain. (Remember, the terms *inert* and *labile* refer to kinetic considerations, not to equilibrium predictions.)

Comprehensive Problems

65. In addition to the coordination complexes of rhodium, cobalt, and molybdenum described at the start of the chapter, use other resources (the Internet or journals) to select another transition metal that has a catalytic role in a chemical reaction.

66. Visit the pharmacy section of a grocery store and list the metals found in a mineral supplement. Use a reference (the Internet or journals) to determine the primary biochemical function of two of the metals from your list.

67. If a complex were assembled from a Co^{3+} ion, four NH_3, and two Cl^-, would you expect to see a neutral compound, an anion, or a cation? Show a formula that justifies your answer.

68. The following complex contains an iron ion in the $+3$ state. However, the resulting charge has been omitted from the complex. Assign the charge for the complex, and indicate how many counter ions (either Na^+ or Cl^-) would have to be ionically bonded to the complex to form a neutral compound.
$$[FeBr_4(H_2O)_2]$$

69. The appearance of a visible color from a compound is associated with three fundamental phenomena. Describe a color-producing compound's properties associated with:
    a. electron excitation
    b. the energy of the photon of light being absorbed
    c. the relative occupancy of lower and higher level orbitals

70. If 35.7 g of the complex $Ca_3[Fe(C_2O_4)_3]_2$ formed as a result of using an oxalate-containing rust remover, how many grams of iron would be removed?

71. Polydentate ligands have been very effective in applications of soil chemistry. EDTA has been added to soil near citrus trees to concentrate iron. Some polydentates have been used to extract heavy metals from soil samples for further analysis. Some internal digestive functions use polydentate ligands to extract metals from foods we eat. Does this indicate that the relative equilibrium constant for these reactions is larger or smaller than one?

Thinking Beyond the Calculation

72. A new ligand is developed for use in a study to mimic the electron transport reactions of copper metal.

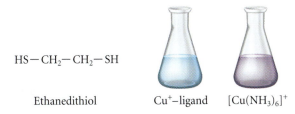

HS—CH₂—CH₂—SH

Ethanedithiol   Cu⁺–ligand   [Cu(NH₃)₆]⁺

a. Draw the octahedral complex that would result from the use of this ligand and copper(II) ions.
b. Draw the *d* orbital diagram for an octahedral complex of this ligand.
c. Compare the color of this octahedral complex to that of $[Cu(NH_3)_6]^{2+}$. Where does this ligand most likely fall in the spectrochemical series?
d. Under certain conditions, 1.00 g of copper(II) produces 3.96 g of the copper–ligand complex. What is the coordination number of the complex under these conditions?
e. The equilibrium constant for the ligand exchange of this new ligand and chloride ions is 1.45×10^{-4}. What does this indicate about the new ligand?

21

Nuclear Chemistry

Enrico Fermi used this very large cyclotron at the University of Chicago in the 1950s. The cyclotron can be used to prepare radioactive isotopes for medical imaging. The cyclotron generates fast-moving subatomic particles that can be directed at nonradioactive nuclei. The resulting collision produces radioactive nuclei such as fluorine-18, which can be used to help doctors observe a particular biological function within a patient's body.

Contents and Selected Applications

21.1 Isotopes and More Isotopes

21.2 Types of Radioactive Decay

21.3 Interaction of Radiation with Matter
Chemical Encounters: Radiation and Cancer

21.4 The Kinetics of Radioactive Decay

21.5 Mass and Binding Energy

21.6 Nuclear Stability and Human-made Radioactive Nuclides

21.7 Splitting the Atom: Nuclear Fission
Chemical Encounters: Nuclear Weapons
Chemical Encounters: Nuclear Reactors as a Vital Source of Electricity

21.8 Medical Uses of Radioisotopes
Chemical Encounters: Tracer Isotopes for Diagnosis

Go to **college.hmco.com/pic/kelterMEE** for online learning resources.

The odds are that you know somebody who has fought cancer. If you have talked with a friend or relative who is a cancer patient, you may have heard that radioactive substances are used in the process of producing images of internal organs. You may know that radiation can shrink a tumor or kill cancer cells. You may even be aware of the dramatic procedure using full-body radiation that is given before a bone marrow transplant. These applications illustrate how a health care team can use nuclear radiation for the benefit of a cancer patient.

Although radiation can save lives, it can also damage and kill. Back in the early 1900s, early radiologists held film plates in an X-ray beam to get pictures of their patients. These radiologists often developed sores on their hands that would not heal, and some eventually lost parts of their fingers. In August 1945, people in the Japanese cities of Hiroshima and Nagasaki were exposed to huge bursts of radiation from the only use of atomic bombs in warfare. Some died immediately, and others succumbed a few weeks later from a then-unknown disease that we now call radiation sickness. In the United States, those who worked in the nuclear weapons industry are now disproportionately contracting lung, stomach, lymphatic, and other cancers.

How can nuclear radiation both cure and cause cancer? The answer to this seeming paradox rests on our understanding of radioactivity (the release of particles and energy accompanying a nuclear change) and how it is produced from nuclear processes. In this chapter we will examine types of radioactive decay, touching on the mathematics of half-lives, the relationship between mass and energy, and the interactions of ionizing radiation with matter. We also will examine nuclear fission, a process that helps produce a whole host of substances useful in nuclear medicine. The story of radioactivity begins, however, at the tiny center of the atom—its nucleus.

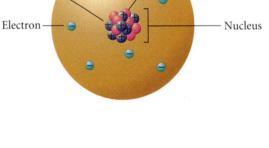

21.1 Isotopes and More Isotopes

Nuclei that occur naturally on Earth can be as small as the nucleus of a hydrogen atom (a single proton) or as large as uranium (92 protons plus 146 neutrons). All of these naturally occurring atoms, and all those that are human-made, are represented by writing their element symbol in the manner described in Chapter 2. The symbol is placed next to a superscripted mass number and a subscripted atomic number. Some of these nuclei are stable—that is, they do not spontaneously decompose—and others are radioactive, decomposing to other nuclei. As we will soon see, the size and makeup of the nucleus determine whether it is stable or radioactive.

Recall from Chapter 2 that we can use nuclide notation, in which we list the symbol for the element, accompanied by its atomic number and the mass number, to indicate the isotope of the element that we wish to describe. For example, consider the element potassium, whose atomic number (Z) is 19. As we know from the periodic table, this element has 19 protons. If the nucleus of a potassium atom has 21 neutrons, we represent this in nuclide notation as $^{40}_{19}K$. The number 40, the mass number, is the sum of 19 protons and 21 neutrons. Because "19 protons" is always potassium, we can more simply represent $^{40}_{19}K$ as ^{40}K, K-40 or

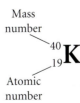

Mass number

Atomic number

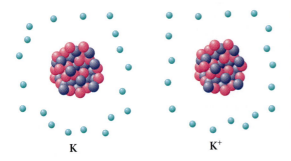

potassium-40. In nuclear chemistry, we are often not concerned with the number of electrons around a particular nucleus. We are interested only in the nucleus of the atom, so we often omit charges (even though they may exist). This means that in our chemist's shorthand for the discussion of changes in the nucleus, we'll consider $^{40}K^+$ as ^{40}K. The nucleus is our only focus here.

Potassium atoms come in several varieties, shown in Figure 21.1, some with fewer than 21 neutrons and some with more. These varieties are known as isotopes, and each isotope is known as a nuclide of that element, as we learned in Chapter 2.

FIGURE 21.1

Potassium has many isotopes, most of which are listed here. Those colored orange are radioactive. Natural abundances are provided for those isotopes that occur in nature. Note that 99.988% of the potassium atoms on Earth are *not* radioactive.

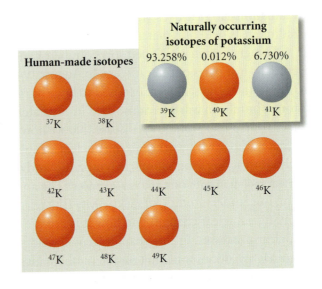

EXERCISE 21.1 **Decoding Isotopes**

Describe the differences and similarities in nuclear structure among the members of each of these sets:

   a. $^{12}_{6}C$, $^{13}_{6}C$, and $^{14}_{6}C$

   b. ^{40}Ar, ^{40}K, and ^{40}Ca

   c. $^{40}Ca^{2+}$ and ^{40}Ca

First Thoughts

The numbers at the bottom of each representation are simply a restatement of the element symbol. The mass number is the sum of the number of neutrons and protons. To find the number of neutrons in a particular nuclide, just subtract the atomic number from the mass number. Remember from Chapter 2 that numbers written on the upper right-hand side of the nuclide notation indicate a deviation in the number of electrons.

Solution

   a. $^{12}_{6}C$, $^{13}_{6}C$, and $^{14}_{6}C$ are isotopes of carbon, each containing six protons. But each has a different mass number and hence a different number of neutrons: 6, 7, and 8, respectively.

   b. ^{40}Ar, ^{40}K, and ^{40}Ca have the same mass number but different numbers of protons. Ar has 18p and 22n, K has 19p and 21n, and Ca has 20n and 20p.

   c. $^{40}Ca^{2+}$ and ^{40}Ca differ only in the number of electrons, so there is no difference in *nuclear* structure. The ^{40}Ca atom loses two electrons to form the $^{40}Ca^{2+}$ ion.

Further Insights

This exercise, while conceptually important, involves just notation and bookkeeping. The nuclide notations, by themselves, say little about the stability or radioactivity of an isotope. Far more useful and interesting are the topics of natural abundance, radioactivity, and half-life, which we discuss below.

PRACTICE 21.1

Indicate the number of protons and neutrons in each of the following nuclides.

a. ^{32}S

b. $^{23}_{11}$Na

c. radon-222

d. Tc-98

See Problems 7 and 8.

Potassium is considered an essential nutrient. Deficiencies in potassium can result in muscle pain, angina, and even heart problems. Eating bananas is one way in which we can ensure a healthy supply of potassium in our bodies. From Figure 21.1 we can see that three isotopes of potassium occur naturally: $^{39}_{19}$K, $^{40}_{19}$K, and $^{41}_{19}$K. However, whereas potassium-39 and potassium-41 possess stable nuclei, $^{40}_{19}$K is radioactive. This means that when we consume a banana, we get a measurable amount of radioactive potassium-40. How much? The natural abundance of potassium-40 is only 0.012%, or approximately 1 atom in 10,000. A typical banana has approximately 300 mg of potassium. Therefore, with each banana we eat, we ingest approximately 0.036 mg of radioactive potassium-40.

Although potassium has 18 known isotopes, most do not occur in nature and must be produced in a laboratory. Each of these isotopes behaves essentially the same in *chemical* reactions, which involve the interaction of electrons with other atoms. For example, on exposure to oxygen or moisture, potassium metal ionizes to form K^+. *All* isotopes of potassium behave in this manner. Because our planet is wet and blanketed in oxygen, *all* naturally occurring potassium isotopes are found in nature as K^+. Remember, however, that we often omit the charge when writing nuclides, and you are likely to see ^{40}K rather than $^{40}K^+$. The key idea here is that *we differentiate between chemical reactions (electron interactions between atoms) and nuclear processes (changes within the nucleus of an individual atom).* Potassium-39 is a stable nuclide, but potassium metal certainly is not a stable chemical when in the presence of water. On the other hand, argon is chemically inert, but it has isotopes that are radioactive.

Natural isotopic abundances for potassium or for any element can be found in many of the chemistry handbooks in the library. Table 21.1 offers a brief look at what you'll find there. As we just saw, the nuclei of elements present in nature are not necessarily stable. In fact, some elements, such as uranium and radon, exist naturally *only* in radioactive forms. Other elements, such as potassium and carbon, have both stable and radioactive isotopes. Still others, such as aluminum, have only one stable naturally occurring form.

Application

Sources of potassium.

TABLE 21.1	Natural Isotopic Abundances for Selected Elements		
Carbon		**Potassium**	
^{12}C	98.90%	^{39}K	93.2581%
^{13}C	1.10	^{40}K	0.0117
Oxygen		^{41}K	6.7302
^{16}O	99.762%	**Iron**	
^{17}O	0.038	^{54}Fe	5.8%
^{18}O	0.200	^{56}Fe	91.72
		^{57}Fe	2.2
Magnesium		^{58}Fe	0.28
^{24}Mg	78.99%		
^{25}Mg	10.00	**Silver**	
^{26}Mg	11.01	^{107}Ag	51.84%
		^{109}Ag	48.16
Aluminum			
^{27}Al	100%		

Video Lesson: The Nature of Radioactivity

21.2 Types of Radioactive Decay

There is no way for you hold a potassium atom in your hand and peer into its nucleus. But if you could keep an eye on a few ^{40}K atoms for a period of time (perhaps over a billion years), you would observe that some of the potassium atoms had been replaced by atoms of calcium. **How do we account for this nuclear sleight-of-hand?**

Beta-Particle Emission

The answer in this case is **beta-particle emission**, a type of radioactive decay. In beta emission, the *nucleus* of $^{40}_{19}$K ejects a **beta particle**, $^{0}_{-1}\beta$, that travels at 90% the speed of light. A new nucleus, $^{40}_{20}$Ca, is formed that has one more proton. Because calcium-40 has an energetically stable nucleus (you can verify this in a table of radioisotopes), no further *nuclear* reactions take place. For now, let's write the **nuclear equation** for the beta-minus emission, or "beta emission" for short, as

$$^{40}_{19}\text{K} \rightarrow {}^{40}_{20}\text{Ca} + {}^{0}_{-1}\beta$$

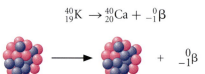

We note that the sums of the atomic masses, as well as the sums of the atomic numbers, are the same on both sides of our equation. The nuclide that results from the beta emission, $^{40}_{20}$Ca, is comparable in size to potassium. In addition, a tiny product is ejected with mass number zero and a negative charge—an electron. In the context of nuclear decay, this electron is called a beta emission particle. The subscript −1 in $^{0}_{-1}\beta$ may look strange as an atomic number. Interpret it as a "negative one charge" rather than a "minus one proton."

Beta emission, like many other of the nuclear reactions we will study, has the following features:

■ A nuclide of one element, through the process of **radioactive decay**, is transformed into a nuclide of another. The resulting **daughter nuclide** may be stable or radioactive.

■ The sum of the atomic numbers on one side of the nuclear equation is equal to the sum on the other. For the beta emission of ^{40}K, the sum of the mass numbers on each side of the equation is 40, and the sum of the atomic numbers on each side of the nuclear equation is 19.

■ Energy is released in radioactive decay reactions. **Gamma rays** of varying energy nearly always accompany the nuclear reaction.

How can an electron be emitted from a nucleus when there are no electrons in the nucleus? To see how this could happen, let's venture down from the macroworld into the nanoworld to look at the beta emission of a single neutron. Neutrons aren't something you can keep in a bottle in your chemistry lab. But if you had the proper specialized equipment, you could demonstrate that a neutron decays to form a proton, an electron, and an **antineutrino**, $^{0}_{0}\bar{\nu}$.

$$^{1}_{0}\text{n} \rightarrow {}^{1}_{1}\text{p} + {}^{0}_{-1}\beta + {}^{0}_{0}\bar{\nu}$$

$^{1}_{0}$n ●	$^{1}_{1}$p ●	+	$^{0}_{-1}\beta$ ●	+	$^{0}_{0}\bar{\nu}$
In nucleus	In nucleus		β expelled		Expelled

The net effect of beta decay is the *transformation of a neutron into a proton* with the release of an electron, an antineutrino, and energy. This is consistent with the beta decay we wrote for $^{40}_{19}$K, where the product nuclide has one more proton.

The antineutrino, $_0^0\bar{\nu}$, is an example of **antimatter**. Each antimatter particle has a mate in our world of "real" matter. For the antineutrino, this mate is the **neutrino**, or "little neutron"—a particle similarly hidden within the neutron. With no charge and probably very little mass, the neutrino required a sophisticated piece of scientific detective work to prove its existence. However, because this ghostly particle and its antimatter mate interact little if at all with matter, neutrinos typically are omitted from nuclear equations.

EXERCISE 21.2	**Beta Emissions In and Around Us**

A look at the world around us reveals a lot of naturally occurring radiation. Elements responsible for this radiation include carbon-14, potassium-40, and hydrogen-3 (tritium).

a. Write the nuclear equation for a beta emission by carbon-14.

b. If $_2^3\text{He}$ is formed via a beta emission, what radioisotope produced it?

First Thoughts

Beta emissions follow a set pattern: an increase of $+1$ in Z and no change in A.

Solution

a. $_6^{14}\text{C} \rightarrow _{-1}^{0}\beta + _7^{14}\text{N}$

b. $_1^3\text{H} \rightarrow _{-1}^{0}\beta + _2^3\text{He}$

Further Insights

We discussed the radioactivity of carbon-14 in Chapter 2. Knowing the rate of this nuclear reaction is part of a process that enables us to predict the age of an archaeological artifact.

PRACTICE 21.2

Write the nuclear equation for the beta emission of ^{60}Co.

See Problems 14, 15b, 15c, and 20.

Alpha-Particle Emission

There are two other common forms of radioactivity: alpha-particle emission and gamma-ray emission. To appreciate how these differ, let's revisit potassium-40. Imagine again that you were watching several individual atoms of the nuclide. No matter how long you waited, you would not observe an **alpha decay**, the emission of a helium nucleus ($_2^4\text{He}$) from a larger nucleus. Why not? Potassium-40, like many radioisotopes, does not exhibit this form of radioactivity. Why not? The simple answer is that the nucleus is made more energetically stable by beta emission than by alpha decay. Why? The answer to this is a little more complex and will be the focus of Sections 21.5 and 21.6.

Which elements decay by alpha emission? Radon is one example. The nuclear reaction for the alpha decay of radon-222 is

$$_{86}^{222}\text{Rn} \rightarrow _{84}^{218}\text{Po} + _2^4\text{He}$$

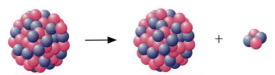

Energy is given off, along with an **alpha particle** ($_2^4\text{He}$), which is energetic but travels more slowly than a beta particle, at only about 5–10% of the speed of light.

You also may see $_2^4\text{He}^{2+}, _2^4\text{He}$ or simply α used to denote the alpha particle. Is alpha emission possible for a hydrogen or helium atom? No, because these atoms are too small to emit an alpha particle and still have any protons left for the remaining nucleus. In fact, alpha emission typically does not occur for small nuclei. It is far more common for elements above bismuth ($Z = 83$).

Gamma-Ray Emission

For both alpha and beta emissions, the nuclear equations we have written so far are not complete, because a gamma ray is usually produced as well. When we include this gamma ray ($_0^0\gamma$) in the alpha decay of radon, the nuclear equation becomes

$$_{86}^{222}\text{Rn} \rightarrow \ _{84}^{218}\text{Po} + \ _2^4\text{He} + \ _0^0\gamma$$

A gamma ray, as indicated in Figure 21.2, is a high-energy photon—a form of electromagnetic radiation with a short wavelength (typically less than a picometer) traveling at the speed of light. Loss of energy from the system, in the form of a gamma ray, is favorable because the products of the reaction would then have less energy than the starting material. In essence, the free energy of the system is lowered.

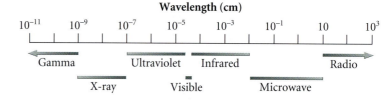

Wavelength (cm)

Gamma rays carry off excess energy in an amount depending on the particular nuclide. In some cases, lower-energy gamma rays are identical to X-rays; however, we mentally distinguish between the two by noting that the former originated *inside* the nucleus and the latter *outside* of it. For medical purposes, as we will see in the final section of this chapter, both X-rays and gamma rays can accomplish the same diagnostic and therapeutic tasks.

Few nuclides are pure or almost pure gamma emitters. A notable example is technetium-99m, where the m indicates a **metastable state**. It represents a misarrangement of protons and neutrons in a nucleus after a neutron has become a proton. In beta emitters, this arrangement occurs so rapidly that it appears to be simultaneous with the original beta emission. In some cases, however, it is slow enough to be obvious:

$$^{99m}\text{Tc} \rightarrow \ ^{99}\text{Tc} + \ _0^0\gamma$$

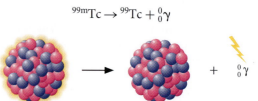

Technetium-99m is generated on demand in nuclear medicine imaging.

FIGURE 21.2

Gamma rays are at the high-energy end of the electromagnetic spectrum. They have *very* short wavelengths on the order of ten-trillionths of a meter or less. The energy of cosmic rays is comparable to that of gamma rays, but cosmic rays are particles and are not part of the electromagnetic spectrum.

EXERCISE 21.3	**Alpha Emissions In and Around Us**

Radon, discussed earlier, is present in the atmosphere at a concentration of about 1 part in 10^{21}. In some caves and basements, however, test kits like those shown in Figure 21.3 can detect it at much higher concentrations. Write nuclear equations for the alpha decay of radon-220, radon-222, and radon-219. A gamma ray accompanies each of these processes.

Solution

$$^{222}_{86}\text{Rn} \rightarrow {}^{218}_{84}\text{Po} + {}^{4}_{2}\text{He} + {}^{0}_{0}\gamma$$

$$^{220}_{86}\text{Rn} \rightarrow {}^{216}_{84}\text{Po} + {}^{4}_{2}\text{He} + {}^{0}_{0}\gamma$$

$$^{219}_{86}\text{Rn} \rightarrow {}^{215}_{84}\text{Po} + {}^{4}_{2}\text{He} + {}^{0}_{0}\gamma$$

PRACTICE 21.3

Write nuclear equations for the alpha decay of ^{218}Po and ^{230}Th.

See Problems 15a, 16a, 16c, 17a, and 19. ∎

FIGURE 21.3

Radon test kits can be purchased and used to measure the concentration of radon in a basement.

Other Types of Radioactive Decay

The three types of radioactive decay that we have discussed are the most common, but two other modes of nuclear decay are also important to a comprehensive picture of radioactive processes. **Electron capture (EC)** is the combination of an inner-orbital electron and a proton from the nucleus to form a neutron. The mass of the nuclide doesn't change during the process because a proton and a neutron are similar in mass, but the atomic number decreases by one as the proton is changed to a neutron. Typically, this process is accompanied by the emission of X-rays from the nuclide. The radioactive decay of iodine-125, which is used to diagnose problems with the pancreas and intestines, occurs by the process of electron capture:

$$^{125}_{53}\text{I} + {}^{0}_{-1}\beta \rightarrow {}^{125}_{52}\text{Te}$$

In **positron emission** $({}^{0}_{+1}\beta)$, a proton decays into a neutron and a positron. A neutrino accompanies this emission, and usually one or more gamma rays as well. A **positron**, or positive electron, is a particle that has the same mass as an electron but carries a charge of $+1$. Interestingly, the positron typically doesn't have a very long life, because when it comes into contact with an electron, the two particles combine to form two gamma rays. This type of radioactive decay is important in the lighter elements such as aluminum-26.

$$^{26}_{13}\text{Al} \rightarrow {}^{0}_{1}\beta + {}^{26}_{12}\text{Mg}$$

Table 21.2 lists the five types of radioactive decay that are important in understanding nuclear decay processes.

TABLE 21.2	Radioactive Decay Processes		
Type of Decay	**Emission**	**Change in Atomic Number**	**Change in Mass Number**
Alpha-particle emission	${}^{4}_{2}\text{He}$	−2	−4
Beta-particle emission	${}^{0}_{-1}\beta$	+1	0
Gamma-ray emission	${}^{0}_{0}\gamma$	0	0
Positron emission	${}^{0}_{+1}\beta$	−1	0
Electron capture	X-ray	−1	0

Decay Series

Radon, discussed in Exercise 21.3, is one example in which the products of the nuclear reaction are still radioactive. All isotopes of all elements past bismuth ($Z = 83$) are radioactive. So far we have described nuclear decay as though it were a one-step process. However, a nuclide may decay to form a second *radioactive* nuclide, which in turn may decay more than a dozen times in a stepwise progression toward stability that is called a **decay series**.

Consider the element uranium. Two natural isotopes exist for this element, ^{238}U and ^{235}U, with natural abundances of 99.28% and 0.72%, respectively. These isotopes exhibit a fairly extensive decay series. Both series consist of alpha and beta emissions, as shown in Figure 21.4. The accompanying gamma rays are not shown in these series. Why do the lines in the decay series zig-zag? Alpha decays

Uranium ore (known as yellow cake) can be refined into pellets of uranium metal.

FIGURE 21.4

The four natural decay series. Each decay series begins at the nuclide listed at the top right-hand side of the series and proceeds to the lower left-hand side by nuclear decay. The half-life of many of these transitions are indicated in red numbers. Gamma emissions accompany many of these decays.

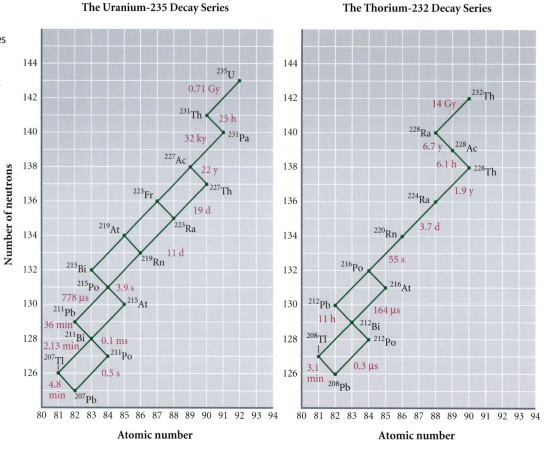

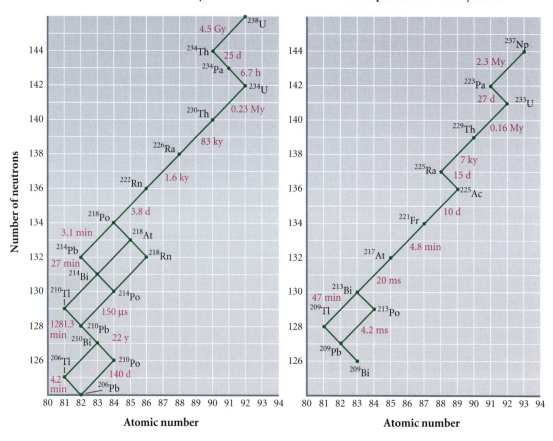

decrease Z by 2, thus moving the line to the left. But a beta-minus emission *increases Z* by 1 unit. If beta emission follows an alpha emission, there is a zag back to the right. The series that begins with ^{235}U ends with the stable lead-207 isotope, and the ^{238}U series forms ^{206}Pb. There are two other naturally occurring decay series, one that begins with ^{237}Np to form ^{209}Bi, and one in which ^{232}Th decays to ultimately form ^{208}Pb. Along the way, these decay chains provide dozens of radioisotopes that we typically find in our biosphere. This is one of the sources of naturally occurring radioactive isotopes. Of particular concern is radon, a radioactive gas that can collect in basements dug into soil that is rich in uranium ores. Is radon harmful to humans? Understanding the relationship between radioactivity and human health will help us answer this question.

 Application

HERE'S WHAT WE KNOW SO FAR

- Radioactivity results from the decay of an unstable nucleus.
- There are three main types of radioactive decay and three common forms of radioactivity.
- The alpha particle is a fast-moving helium nucleus.
- The beta particle is an electron ejected from the nucleus of an unstable atom.
- The gamma ray is a burst of high-energy electromagnetic radiation.
- A decay series is a stepwise progression of a radioactive nuclide toward stability.

21.3 Interaction of Radiation with Matter

Taking a walk on a crisp, sunny day is one of the pleasures of autumn. Any cloud that blocks the Sun is easily noticed. Not only does the shade reduce the amount of light hitting your eyes, but your skin also registers the change. The energy exchanges, from the infrared through the ultraviolet regions of the electromagnetic spectrum, are apparent and profound. However, when it comes to detecting the small amounts of alpha particles, beta particles, and gamma rays that bombard you on a daily basis, your senses don't help. For instance, you cannot detect the alpha particles that radioactive radon, an odorless, colorless, and tasteless gas, emits.

Though seemingly invisible, the different types of radiation form quite a nuclear arsenal, as summarized in Figure 21.5. We can envision alpha particles as the cannon balls of the group. With their greater size and +2 charge, they do not travel very far before they smash into other atoms. The typical collision results in the capture of two electrons to form a neutral helium atom. In contrast, beta particles, being smaller and traveling more rapidly, are the equivalent of high-velocity bullets. They are able to travel significantly longer distances before a collision, but they lack the punch of an alpha particle. Gamma rays, with no mass and no charge, are akin to laser weapons. They shoot great distances through matter, now and then searing something in their path. What about the

Collision of an alpha particle and a molecule results in the formation of an atom of helium.

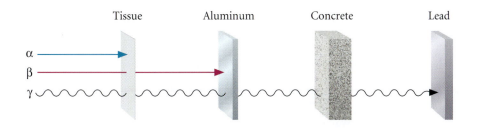

FIGURE 21.5

Alpha, beta, and gamma radiation differ in their penetration. Gamma rays are the most highly penetrating, and alpha particles the least.

TABLE 21.3	Differences in Penetration and Shielding for Different Types of Radiation				
Type	Examples	Penetration in Dry Air	Penetration in Skin or Tissue†	Shielded by	Q*
Alpha	uranium, plutonium, americium	2–4 cm	0.05 mm	Paper, air, clothing	20
Beta	potassium-40, cesium-137	200–300 cm	4–5 mm	Heavy clothing	~1
Gamma	technetium, cobalt-60	500 m	50 cm	Lead, concrete	~1
Fast neutrons	Accelerators	Several hundred feet	High	Water, plastic concrete	20

*Q is the relative biological effectiveness, a factor that indicates relative amounts of damage to living tissue.
†Alpha, beta, and gamma radiation all exhibit a range of energies. The degree of penetration depends on the actual energy.

antineutrinos that accompany beta emissions? Having neither a mass nor a charge, they are not absorbed and are considered harmless.

The type and the energy of the radiation dictate what must be used to shield us to the greatest extent possible (see Table 21.3). Alpha particles penetrate matter the least, being stopped by just a few centimeters of air, by the outer layer of your skin (which is mostly dead cells), or by a piece of paper. Beta particles penetrate more deeply and can pass through several pieces of paper, through a thin sheet of aluminum, or about a centimeter into your skin. In contrast, gamma rays can pass right through you. Shielding your body from them requires several inches of aluminum or lead, and even these may not do the job.

 Application

Use of radiation shield on a patient at the dentist's office.

 Application

CHEMICAL ENCOUNTERS: Radiation and Cancer

A visit to the dentist reveals an interesting feature of stopping damage from nuclear radiation. Before an X-ray of a patient's teeth is obtained, the patient is typically draped with a sheet of lead. Why is lead one of the materials employed to shield us from high-energy electromagentic radiation, such as X-rays or nuclear radiation? The high density of lead, 11.3g/cm^3, means that only two inches of lead will easily shield you from alpha and beta radiation, as well as from most gamma radiation. However, lead is not unique in this ability. Gold ($d = 19.3 \text{ g/cm}^3$) is more dense than lead and would work even better. Bricks or blocks of a moderately dense material such as concrete also would do the trick. However, no dentist is likely to place an apron of gold or a foot of concrete over your abdomen before taking dental X-rays. Given its density and price, lead is often the shielding medium of choice.

Although alpha, beta, and gamma radiation differ in penetration, they are alike in the effects they produce on the molecular level. All three are known as **ionizing radiation**; that is, they are capable of forming ions by knocking electrons out of atoms. The damage they cause is the reason for their detection with, for example, a Geiger counter or film badge.

The consequences of ionizing radiation can be negligible or severe, depending on how many molecules are damaged inside the body. Although small amounts of radiation typically lead to only negligible damage that can be repaired by the body, large doses of radiation can be life-threatening. **How does ionizing radiation cause damage?** Because between about 50% and 70% of your body is water, a scenario of particular interest occurs when ionizing radiation strikes a water molecule. The blow can knock off an electron to form a highly reactive species with an unpaired electron:

Electron density maps of water and the radical cation of water. Note the decrease in electron density around the oxygen end of the molecule.

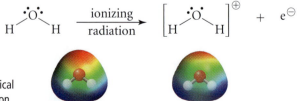

Water (neutral)   Water (cation radical)

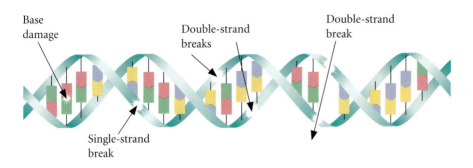

Base damage

Double-strand breaks

Double-strand break

Single-strand break

FIGURE 21.6

Examples of chromosomal damage from radiation.

The resulting radical cation (see Chapter 7) undergoes autoprotolysis to form hydroxyl free radicals:

$$H_2O^{.\oplus} + H_2O \rightleftharpoons H_3O^{\oplus} + {.}OH$$

The fate of the hydroxyl free radicals is particularly damaging to the cells within your body. If they encounter a molecule of DNA in a dividing cell, they may damage a section of the genetic code (Figure 21.6). The damage could result in death of the cell, triggering the formation of mutant proteins (proteins with a non-natural primary structure; see Chapter 22) or triggering an abnormal function of the cell leading to cancer. In short, nuclear radiation can be carcinogenic. Although we do not know exactly what cellular events ensue after a dose of ionizing radiation, two things seem clear: (1) The more radiation a person is exposed to, the greater the likelihood of that person's developing cancer. (2) These cancers may not show up until decades after the time of exposure.

Is there a threshold below which radiation is safe? Recent studies have convincingly demonstrated that the damage inflicted by low-level radiation upon workers in the nuclear industry and upon World War II nuclear bomb survivors have been greatly underestimated. Similarly, the link between fetal X-rays (an abandoned medical practice) and childhood cancer has been established. Fortunately, because cancer cells grow and divide, they also are susceptible to radiation. Carefully measured doses of radiation directed at cancerous cells can result in their death.

Overall, the biological effects of nuclear radiation depend on the quantity of energy transferred to the cells and tissues. In the United States, the **rem** or "roentgen equivalent in man" is the unit for estimating the damage. Other units that measure radiation include the **becquerel**, **curie**, **roentgen**, and **rad**. Of these, only the becquerel (Bq) is an SI unit. Scientists in most parts of the world employ two other SI units, the **sievert (Sv)** and the **gray (Gy)**, which are related to the rem and the rad, respectively, as shown in Table 21.4.

TABLE 21.4	**Units for Measuring Radiation**		
Measure of	**Name**	**Abbreviation**	**Definition**
Activity	becquerel*	Bq	1 disintegration per second
Activity	curie	Ci	1 curie = 3.7×10^{10} becquerel
Exposure	roentgen	R	1 roentgen = 2.58×10^{-4} couloumbs of charge per kg of air
Absorbed dose	radiation absorbed dose	rad	1 rad = 1×10^{-2} J of energy deposited per kg of tissue
Absorbed dose	gray*	Gy	1 gray = 100 rad
Dose equivalent	roentgen equivalent in man	rem	$Q \times$ absorbed dose
Dose equivalent	sievert*	Sv	1 sievert = 100 rem

*SI units

TABLE 21.5	Health Effects of Acute Radiation Exposure	
Exposure (rem)	Health Effect	Time to Onset
10	Burns, changes in blood chemistry	
50	Nausea	Hours
75	Vomiting, hair loss	2–3 weeks
100	Hemorrhage	
400	Death	Within 2 months
1000	Internal bleeding, death	Within 1–2 weeks
2000	Death	Within hours

The quantity of energy absorbed by tissues is directly related to the time of exposure to ionizing radiation. *In fact, time is an important part of the decision-making process in medical diagnosis and treatment.* For example, prostate cancer in elderly men is often treated by the implantation of metal "seeds" coated with ^{125}I or, more recently, ^{103}Pd. Why use these nuclides? They are sufficiently radioactive for only the time needed to control the cancer without creating new cancers. **How do we know how long they will be radioactive?** The most important measure that we use to judge the length of time a substance is radioactive is its half-life.

The U.S. Environmental Protection Agency suggests that the average person receives an annual dose of 0.3 rem of radiation from natural sources. Over the course of a lifetime, this is predicted to result in 5 or 6 deaths due to cancer per 10,000 people. This sounds shocking until we consider that the rate of deaths due to cancer from nonradioactive sources is predicted to be about 2000 people per 10,000. Larger doses received in one exposure have a much more deleterious effect on human health, as shown in Table 21.5. Acute exposures, such as those that result from accidents in nuclear power plants and those that resulted from the U.S. bombing of Japan in World War II, cause severe damage to the human body, often resulting in lifelong health problems and even death.

21.4 The Kinetics of Radioactive Decay

The half-life, $t_{1/2}$, of a radioactive isotope is the period of time it takes for exactly half of the original nuclei in a radioactive sample to decay. We discussed half-life in detail in Chapter 15. Table 21.6 shows that half-lives can vary widely among the radioactive isotopes. They can be as short as a few microseconds and as long as a few billion years.

How can the half-life tell us how much radioactivity remains after a given time? Remember from Chapter 15 that after one half-life, one-half of the sample has reacted and only one-half of the sample remains. After a second half-life, $\frac{1}{2} \times \frac{1}{2} = \frac{1}{4}$ of the sample is left. After 3 half-lives, $\frac{1}{2}$ of $\frac{1}{2}$ of $\frac{1}{2}$, or $(\frac{1}{2})^3 = \frac{1}{8}$, of the sample remains. More generally, then, for n half-lives, the fraction of the original sample remaining is $(\frac{1}{2})^n$. This trend is shown in Figure 21.7 and is valid for all radioactive decay processes.

Palladium-103, used to coat the seeds implanted in the prostate, decays by electron capture, in which an inner-orbital electron is captured by a proton in the nucleus to form a neutron. The half-life of Pd-103 is 16.97 days. How long will it take for the radiation to be diminished to 1.00% of its original value so it is considered safe for radiation workers, the prostate cancer patient, and his family? A look at Figure 21.7 indicates that this will take between 6 and 7 half-lives, or between 102 and

TABLE 21.6	Half-lives of Some Radioactive Elements	
Element	Nuclide	Half-life
nobelium	^{250}No	250 μs
technetium	^{99m}Tc	6.0 h
thallium	^{201}Tl	21.5 h
radon	^{222}Rn	3.8 d
iodine	^{131}I	8.040 d
palladium	^{103}Pd	16.97 d
cobalt	^{60}Co	5.271 y
hydrogen	^{3}H	12.3 y
carbon	^{14}C	5730 y
radium	^{226}Ra	1.6×10^3 y
uranium	^{238}U	4.5×10^9 y

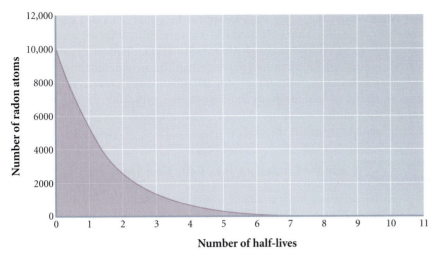

FIGURE 21.7

The kinetics of radioactive decay. The half-life of a radioisotope, such as radon, is the time required for half of the nuclei in the radioactive sample to decompose.

Visualization: Half-Life of Nuclear Decay

Video Lesson: Rates of Disintegration Reactions

119 days. Such approximations are often sufficient. When necessary, we can also solve explicitly for n.

$$(\tfrac{1}{2})^n = 1/100 = 0.0100$$

Taking the natural logarithm (ln) of both sides yields

$$n[\ln(\tfrac{1}{2})] = \ln(0.0100)$$

$$n(-0.693) = -4.605$$

$$n = 6.645 \text{ half-lives}$$

Solving for the time needed, we find that

$$t = 6.645 \text{ half-lives} \times 16.97 \text{ days/half-life}$$

$$= 113 \text{ days}$$

There is another approach based on the understanding that radioactive isotopes decay via first-order kinetics. Recall from Chapter 15 that the relationship among concentration, time, and half-life for any first-order process is shown by

$$\ln \frac{[A_t]}{[A_i]} = -kt$$

where A_t = the amount of substance remaining
 A_i = the initial amount of substance
 k = the first-order rate constant for the reaction (in this case, the decay)
 t = time

We also know that the rate constant for a first-order reaction can be determined by

$$k = \frac{0.693}{t_{1/2}}$$

This means that we can solve explicitly for the rate constant if we are given the half-life, $t_{1/2}$. In our example,

$$k = \frac{0.693}{t_{1/2}} = \frac{0.693}{16.97 \text{ day}} = 0.04084 \text{ day}^{-1}$$

Assuming that we start with $A_i = 1.000$, we can determine that if we only have 1% remaining, $A_t = 0.0100$. Then

$$\ln \frac{[0.0100]}{[1.00]} = -0.04084t$$

$$-4.605 = -0.04084t$$

$$113 = t$$

so $t = 113$ days, and we obtain the same answer (113 days) by either method.

Three palladium-103 "seeds," which are used for the treatment of prostate cancer, easily fit on the top of a penny.

But when would the *entire* Pd-103 sample be gone? This is a question we cannot *precisely* answer, although we can come very close. Why are we unable to tell when all of the Pd-103 is gone? After each half-life, half of the number of radioactive atoms remaining from a previous half-life are still remaining. Sooner or later, after a very large number (roughly 80) of half-lives have passed, only two radioactive atoms remain for every mole we started with. Half-life is a statistical measure. Probabilities, which don't apply with a sample size of two, will not accurately tell the rate of the reaction.

EXERCISE 21.4   Half-life Calculations: Here Today, Gone Tomorrow?

After exercising on a treadmill, a patient was given thallium-201 for a diagnostic scan of his heart (Figure 21.8). How long will it take for 95.0% of the thallium to have decayed? The half-life of thallium-201 is 21.5 hours.

Solution

As discussed previously, we can solve the problem in two ways. Using the first method, we find the number of half-lives that pass until 5.0%, or a fraction of 0.050, of the Tl-201 remains.

$$(\tfrac{1}{2})^n = 0.050$$

$$n[\ln(\tfrac{1}{2})] = \ln(0.050)$$

$$n(-0.693) = -3.00$$

$$n = 4.32 \text{ half-lives}$$

$$t = 4.32 \text{ half-lives} \times 21.5 \text{ hours/half-life} = 93 \text{ hours}$$

Using the second method, we proceed as follows:

$$k = \frac{0.693}{t_{1/2}} = \frac{0.693}{21.5 \text{ h}} = 0.0322 \text{ h}^{-1}$$

$$\ln\frac{[0.050]}{[1.00]} = -0.0322t$$

$$-3.00 = -0.0322t$$

$$93 \text{ hours} = t$$

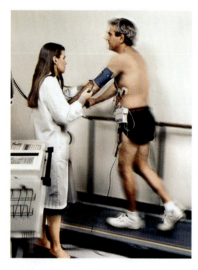

FIGURE 21.8

PRACTICE 21.4

The half-life of ^{198}Au is 2.69 days. How long would it take for 99% of a gold-198 sample to decay?

See Problems 39–42 and 47–48.

The previous problem involved percentages. But you also can work half-life problems given a starting mass or a starting number of atoms. Both are measures of the radioactivity present. Similarly, you can work half-life problems given a unit of activity such as the number of disintegrations per second (becquerels) or curies, because these also measure the amount of radioactivity and are proportional both to the mass and to the number of atoms. Such variations in units reflect the differing needs of real-world situations where you are likely to encounter radioisotopes.

The half-life of a radioactive isotope can also help you to determine how long the radioisotope will be useful or, perhaps, to weigh the hazards associated with that particular isotope. For example, would you rather be around a radioisotope that decayed to 1% of its activity in 10 seconds or one that did so in 10 centuries? For the latter, the rate of decay is much lower, and you would be bombarded with far fewer alpha particles, beta particles, or gamma rays in a given time period. This principle is important in the use of technetium in medical imaging. We have noted that technetium-99m is a gamma emitter. It has a half-life of 6.01 hours, which is long enough for a medical procedure, but short enough that the substance doesn't persist very long.

$$^{99m}_{43}\text{Tc} \rightarrow {}^{99}_{43}\text{Tc} + {}^{0}_{0}\gamma \qquad t_{1/2} = 6.01 \text{ h}$$

The product nuclide, $^{99}_{43}$Tc, is still radioactive. However, with its half-life of 213,000 years, the activity of Tc-99 as a beta emitter is low. In the two and a half weeks it takes for the majority of Tc-99 to be completely eliminated from the body, it does little damage.

21.5 Mass and Binding Energy

By now you might be a bit suspicious. Many nuclei are unstable and spontaneously decay. Significant amounts of energy are released during alpha, beta, and gamma emission—enough to ionize molecules or to kill cancer cells—but we haven't talked about the source of the energy. **Where does it come from, and how does this help explain the decay processes that nuclei undergo?**

The place to start to find an answer is with Einstein's famous equation,

$$E = mc^2$$

in which the constant, c, is equal to the speed of light. This relationship illustrates that a particular mass, when completely converted, is equivalent to a surprisingly large quantity of energy. Any chemical reaction that is accompanied by a loss in energy, such as in an exothermic reaction, actually also has a corresponding loss in mass. However, the mass losses are so minuscule that we cannot detect them using conventional instruments. In contrast, the changes in mass due to a nuclear reaction, though tiny, are quite measurable.

Let's explore this connection by examining the formation of nitrogen-14 from its individual nuclear particles:

$$7\text{p} + 7\text{n} \rightarrow {}^{14}\text{N nucleus}$$

A mole of nitrogen-14 nuclei (without any electrons) weighs 13.99540 g. What is the mass of 7 separate moles of protons (1.00727 g/mol) and 7 mol of neutrons (1.008665 g/mol)?

$$7 \text{ mol of protons} \times 1.00727 \text{ g/mol} = 7.05089 \text{ g}$$

$$7 \text{ mol of neutrons} \times 1.008665 \text{ g/mol} = 7.060655 \text{ g}$$

$$\text{Total mass} = 7.05089 \text{ g} + 7.060655 \text{ g} = 14.11154 \text{ g}$$

This *exceeds* the mass of a mole of nitrogen-14 nuclei by $(14.11154 \text{ g} - 13.99540 \text{ g}) = 0.11614$ g. This **mass defect**, the mass difference between the individual protons and neutrons and the composite nucleus, was used in binding the protons and neutrons together within the nucleus. The **binding energy**, expressed as a positive number, is the energy required to dismantle the nucleus into its individual protons and neutrons. To ensure that our calculations give us positive numbers for binding energy, we typically use Einstein's equation in a different form:

$$\Delta E = |\Delta m|c^2$$

where ΔE is the binding energy and $|\Delta m|$ is the absolute value of the change in mass *in kilograms*. We use kilograms because our SI unit of energy, the joule, is defined as kg·m/s². For nitrogen-14, using our mass defect of 0.11614 g ($= 1.1614 \times 10^{-4}$ kg), the binding energy can be calculated as

$$\Delta E = |\Delta m|c^2 = |-1.1614 \times 10^{-4} \text{ kg}| \times (2.9979 \times 10^8 \text{ m/s})^2 = 1.04 \times 10^{13} \text{ J}$$

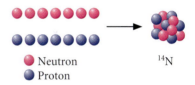

Video Lesson: Binding Energy

Seven protons and seven neutrons form the nucleus of nitrogen-14.

EXERCISE 21.5 **The Energy Advantage of Nuclear Reactions**

When bombarded with neutrons, uranium-235 can split to form bromine-87, lanthanum-146, and three neutrons. The masses reported here are those of the bare nuclei.

$$^{235}_{92}\text{U} + ^{1}_{0}\text{n} \rightarrow ^{87}_{35}\text{Br} + ^{146}_{57}\text{La} + 3^{1}_{0}\text{n}$$

grams/mole 234.9936 1.008665 86.90156 145.8944 3.0260

Calculate the mass defect and the energy released in kJ/mol.

Solution

The mass defect is the difference in mass between the ending and starting materials.

$$\Delta m = (86.90156 + 145.8944 + 3.0260) - (234.9936 + 1.008665)$$
$$= -0.1803 \text{ g/mol} = -1.803 \times 10^{-4} \text{ kg/mol}$$

The energy equivalent to this mass is $\Delta E = |\Delta m|c^2$.

$$\Delta E = |\Delta m|c^2 = |-1.803 \times 10^{-4} \text{ kg/mol}| \times (2.9979 \times 10^8 \text{ m/s})^2$$
$$= 1.621 \times 10^{13} \text{ J/mol} = 1.621 \times 10^{10} \text{ kJ/mol}$$

This is many orders of magnitude greater than the energy released by an equivalent mass of materials in a combustion reaction.

Even though this problem uses the masses of the bare nuclei, the problem could have been solved using the atomic masses, because the number of electrons on each side of the equation remains constant. In radioactive decay processes that involve a change in the number of protons, the mass of the electrons must also be considered in the calculation of the mass defect, when atomic masses are used.

PRACTICE 21.5

Calculate the energy released in the following process.

$$^{224}_{88}\text{Ra} \rightarrow ^{220}_{86}\text{Rn} + ^{4}_{2}\text{He}$$

See Problems 53 and 54.

Some atoms are more thermodynamically stable than others. A table of binding energies shows only that the values generally increase as the atoms get heavier. However, if you recalculate the binding energy for each atom and report the values *per nucleon* (proton or neutron), the stabilities pop right out at you. Nuclei with greater binding energies per nucleon are more stable. Note in Figure 21.9 that the lightest elements, those with mass numbers of 20 or less, have the lowest binding energies per nucleon. In comparison with iron, helium simply does not have enough nucleons to be as strongly glued together. The process that liberates energy on the sun, **fusion** (or nuclear fusion), is energetically favorable because lighter elements such as helium are joined to form heavier ones that have a more favorable binding energy per nucleon.

Nuclear fusion.

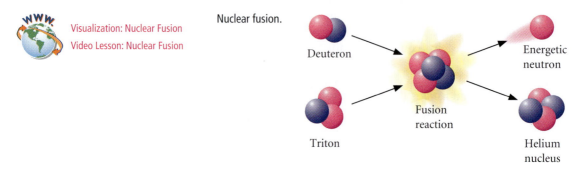

Deuteron

Triton

Fusion reaction

Energetic neutron

Helium nucleus

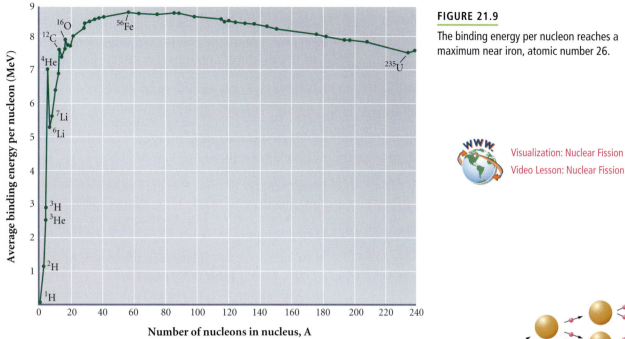

FIGURE 21.9

The binding energy per nucleon reaches a maximum near iron, atomic number 26.

Visualization: Nuclear Fission

Video Lesson: Nuclear Fission

The heaviest elements also have somewhat lower stability. Uranium, with so many protons and neutrons packed into its nucleus, has a lower binding energy per nucleon than iron or cobalt. As we will see in Section 21.7, it is energetically favorable to split heavier nuclei into smaller ones via **fission** (or nuclear fission) to form nuclei with a more favorable binding energy per nucleon.

Finally, look at the maximum of the curve and you will find elements with the most stable nuclei—those with mass numbers around 60, such as iron and nickel. Other elements also have surprisingly high binding energies given their mass number, such as ^{4}He (an alpha particle), ^{12}C, and ^{16}O. Alternatively, one might argue that the values for ^{6}Li and ^{14}N are surprisingly low.

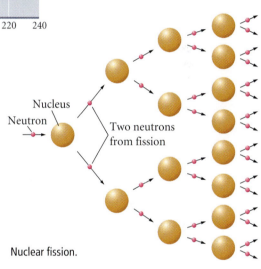

Nuclear fission.

21.6 Nuclear Stability and Human-made Radioactive Nuclides

"Here today, gone tomorrow" doesn't apply to most of the atoms that make up our world. That is fortunate for us. In fact, the majority of the atoms on our planet that are here today can be expected to be here tomorrow. Although it may be hard to locate a specific atom from one day to the next, you can be reasonably sure that it is still here. Why? Most atoms on Earth are *not* radioactive.

Which factors seem to affect nuclear stability? Measurements indicate that nature favors *even* numbers of protons. Elements such as helium, oxygen, iron, and lead that have even atomic numbers tend to be more abundant than their odd neighbors. For example, of the eight elements that make up over 99% of Earth's total mass, only one (aluminum) has an odd atomic number. Still more favored are nuclei that have even numbers of *both* protons and neutrons. Perhaps the most dramatic case is $^4_2$He, the alpha particle. Given this stability, it is not surprising that helium is the second most abundant element in the universe.

Experimental data confirm that certain *numbers* of either protons or neutrons (called "magic numbers") are favored: 2, 8, 20, 28, 50, 82, and 114. The elements helium ($Z = 2$), oxygen ($Z = 8$), calcium ($Z = 20$), and nickel ($Z = 28$) have

Oxygen (shown here in liquid form), calcium, and nickel have magic numbers of nucleons.

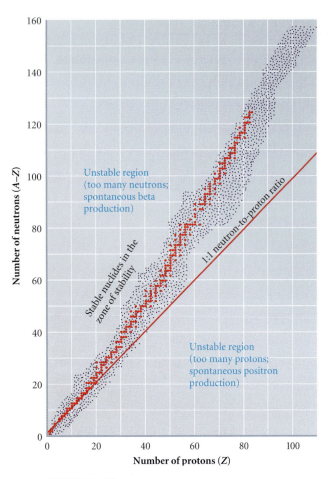

FIGURE 21.10

The band of known isotopes for each element with at least one stable isotope. n/p is the neutron-to-proton ratio. Note that the n/p ratio is greater than 1 for most stable isotopes.

Video Lesson: The Stability of Atomic Nuclei

more stable nuclei than their neighbors. Check the graph of binding energy per **nucleon** (proton or neutron in the nucleus) versus the mass number shown in Figure 21.9 to see that these elements sit at local maxima. The nuclides $^{16}_{8}\text{O}$ and $^{40}_{20}\text{Ca}$ are "doubly magic"; they have magic numbers for both protons and neutrons. In February 2000, a French research team created another doubly magic nucleus, Ni-48, which contains 28 protons and 20 neutrons.

When you examine a larger set of nuclides, other trends appear. There are 279 stable isotopes, a few naturally occurring radioactive elements, and hundreds of synthetic isotopes. Many of these find application in nuclear medicine. Figure 21.10 sketches the band of stable isotopes for each element with at least one stable isotope. Note the following points:

- At higher atomic numbers, stable nuclei have increasingly more neutrons than protons.

- Some radioactive elements have *too many neutrons* relative to the stable isotopes. These tend to decay by beta emission, where a neutron changes into a proton and an electron.

- Some radioactive elements have *too few neutrons* relative to the stable isotopes. These tend to decay by positron emission, where a proton changes into a neutron and a positron.

The plot ends at bismuth, $Z = 83$, for no elements beyond this have stable isotopes. Alpha emission is typical for heavier elements that are unstable, simply because too many nucleons are present, be they protons *or* neutrons. Table 21.7 lists some guidelines that are helpful in determining the type of decay for a particular nucleus.

The radioactive decay can occur by the process of positron emission, or **beta-plus emission**, $^{0}_{+1}\beta$. This process does not occur naturally on Earth and was not discovered until scientists began creating new isotopes in the laboratory. Physically this is done by slamming high-energy particles, ranging from protons and neutrons to atomic nuclei, into a target nucleus. This can result in changing the nucleus's mass number and giving it additional energy. In general, the laboratory process is

$$\text{Nucleus} + \text{small particle} \rightarrow \text{bigger nucleus}$$

The product nucleus that is formed may undergo radioactive decay.

For example, in 1930 phosphorus-30 was synthesized in the laboratory by the bombardment of an aluminum target with alpha particles, $^{4}_{2}\text{He}$:

$$^{27}_{13}\text{Al} + {}^{4}_{2}\text{He} \rightarrow {}^{30}_{15}\text{P} + {}^{1}_{0}\text{n}$$

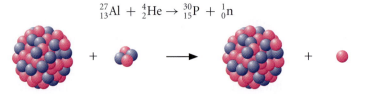

More modern work includes the formation of tiny amounts of superheavy elements, which we will define as those beyond atomic number 106. Recent (1999 and 2000) reactions include

$$^{208}_{82}\text{Pb} + {}^{86}_{36}\text{Kr} \rightarrow {}^{293}_{118}\text{Uuo} + {}^{1}_{0}\text{n} \qquad (t_{1/2} < 1 \text{ msec})$$

$$^{249}_{97}\text{Bk} + {}^{22}_{10}\text{Ne} \rightarrow {}^{267}_{107}\text{Bh} + 4{}^{1}_{0}\text{n} \qquad (t_{1/2} = 17 \text{ seconds})$$

TABLE 21.7	Predicting Nuclear Decay	
Type of Decay	**Reason for Instability**	**Change in n/p Ratio**
Alpha emission	Nucleus too heavy	Increase (small for heavy nuclides)
Beta emission	n/p too high (*)	Decrease
Positron emission	n/p too low (*)	Increase
Gamma	Too much energy Nucleus energetically excited	None
Electron capture	n/p too low (*)	Increase

*n/p represents the neutron-to-proton ratio.

Some nuclides are expected to be quite stable, others not so. The nature of the nucleus becomes clearer when we study the half-lives of these nuclides.

EXERCISE 21.6   Ten Tin Isotopes

Tin has more stable isotopes than any other element. Explain why you might expect this to be the case. Then examine the radioisotopes of tin and discuss their decay modes.

Stable isotopes	^{112}Sn, ^{114}Sn, ^{115}Sn, ^{116}Sn, ^{117}Sn, ^{118}Sn, ^{119}Sn, ^{122}Sn, ^{124}Sn (^{120}Sn, ^{118}Sn, and ^{116}Sn are the most abundant.)
Radioisotopes that decay by β^-	^{121}Sn, ^{123}Sn, ^{125}Sn, ^{126}Sn, ^{127}Sn (plus others with higher mass and short half-life)
Radioisotopes that decay by β^+ or by electron capture	^{110}Sn, ^{111}Sn, ^{113}Sn (plus others with lower mass and short half-life)

First Thoughts

Data like these exist for every element; in general, they are most conveniently accessed on the Web or in a chemistry handbook. You cannot reason out which isotopes actually exist. Although tin's location provides some guidance, you must still look them up.

Solution

For several reasons, tin would be expected to have a number of stable isotopes. First, tin is in the middle of the periodic table where nuclei are more stable and where a few extra neutrons do little to upset the balance of nuclear forces. Second, it has an atomic number of 50, one of the "magic numbers." Finally, 8 of the 10 isotopes have even numbers of protons and neutrons, and the even isotopes are the most abundant—another indication of their stability.

Further Insights

Isotopic stabilities hold some surprises. For example, a radioisotope may fall between a pair of stable isotopes. We noted this above for tin. It also happens for chlorine, where Cl-35 and Cl-37 are stable, but Cl-36, which has an odd number of both protons (17) and neutrons (19), is radioactive. Again, nuclear stability (or instability) arises from a combination of several factors and is therefore difficult to predict.

PRACTICE 21.6

Predict whether each of the following isotopes might be stable or radioactive.

a. ^{79}Br   b. ^{101}Ru   c. ^{136}Ba   d. ^{180}Ta

See Problems 59–62.

21.7 Splitting the Atom: Nuclear Fission

With the discovery of fission came the birth of new radioisotopes and of a new consciousness that the nuclear age was upon us. Nuclear fission was discovered in the 1930s through the work of scientists such as Enrico Fermi, Fritz Strassman, Otto Hahn, and Lise Meitner. Work on fission continued in the early 1940s in both Germany and the United States, culminating in the deployment by the United States of the first atomic bomb used in war, fueled by uranium-235, on the town of Hiroshima, Japan, on August 6, 1945, and of a second combat-based atomic bomb, fueled by plutonium-239, on Nagasaki, Japan, on August 9, 1945. The war ended shortly after the second bomb was dropped, but the nuclear age had just begun.

Why do ^{235}U and ^{239}Pu split and release energy? The answer lies in part in the thermodynamics of nuclear stability. Refer to Figure 21.9. Uranium nuclei are not so thermodynamically stable as iron, bromine, and other elements in the region of greatest stability near the top of the curve. The answer also lies in considering the precarious balance that exists in large nuclei such as uranium and plutonium. These nuclei are very heavy and are held together by the strong force between nucleons. However, opposing the strong force are the many proton–proton repulsions in an atom of this size. For some atoms, the injection of extra mass into the nucleus can tip the balance in favor of the proton repulsions and send the nucleus flying apart into two or more pieces. This is what happens with ^{235}U and ^{239}Pu. A neutron, with no charge, is an ideal particle to shoot into a nucleus. Once it slips into the nucleus, a heavier nuclide is formed that fragments in a matter of nanoseconds. Figure 21.11 illustrates the process for the fission of uranium-235.

$$^{239}_{94}\text{Pu} + {}^{1}_{0}\text{n} \rightarrow [^{240}\text{Pu}] \rightarrow {}^{70}_{30}\text{Zn} + {}^{167}_{64}\text{Gd} + 3{}^{1}_{0}\text{n} + \text{energy}$$

$$^{235}_{92}\text{U} + {}^{1}_{0}\text{n} \rightarrow [^{236}\text{U}] \rightarrow {}^{139}_{56}\text{Ba} + {}^{94}_{36}\text{Kr} + 3{}^{1}_{0}\text{n} + \text{energy}$$

Nuclear reactions such as these were initially of interest because the tremendous energy they released could be unleashed in a weapon. Since the violent birth of fission in 1945, these reactions have found a variety of other, more humanitarian

Lise Meitner (1878–1968) and Otto Hahn (1879–1968). As a woman in the male-dominated world of the early 1900s, Meitner worked as a physicist with Otto Hahn. She was responsible for the discovery of protactinium and was the first to explain nuclear fission correctly. Despite her contributions, Otto Hahn did not acknowledge her work when he received the 1944 Nobel Prize. In belated recognition of her work on radioisotopes, element 109 is named meitnerium.

FIGURE 21.11

A nucleus of ^{235}U undergoing nuclear fission.

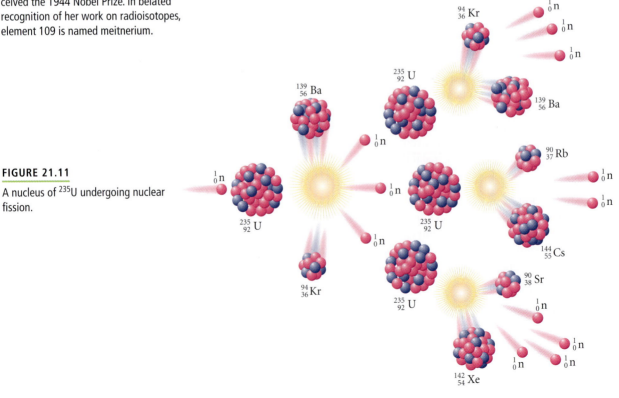

uses. Today, using non-weapons-grade fissionable fuel, they provide the energy in nuclear power plants worldwide, they power spacecraft and submarines, and they are the source of many isotopes used in nuclear medicine.

Nuclear Reactors as a Vital Source of Electricity

The 442 nuclear reactors currently in operation worldwide are alternatives to the pollution and greenhouse gas emissions caused by coal-burning and oil-burning power generation. The world's first nuclear power plant went on line in 1954 in the Russian city of Obninsk. This was immediately followed by the construction of a nuclear facility in Sellafield, England. It wasn't until 1957 that the first full-scale power plant in the United States began operation in Shippingport, Pennsylvania. Municipal power generation by nuclear fission is not new. However, the use of nuclear fission has been controversial because of safety concerns, the two most important being the possible accidental release of radiation, and the disposal and long-term (thousands, and perhaps millions, of years!) safeguarding of radioactive wastes. Accidental releases of radiation occurred on a relatively small scale at the Three Mile Island nuclear facility in Pennsylvania in 1979 and on a much larger scale just outside the Ukrainian town of Chernobyl, in 1986. Still, much of the world uses nuclear power to meet its energy needs, as shown in Figure 21.12.

The goal of any large power plant is to generate electricity by turning a turbine, which converts the mechanical energy into electricity. Steam, resulting from

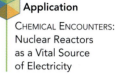

Application

CHEMICAL ENCOUNTERS:
Nuclear Reactors
as a Vital Source
of Electricity

FIGURE 21.12

Nuclear power plants are located in many countries of the world. The United States, France, Japan, and the Russian Federation possess the majority of the plants, with the capability to produce over 230,000 MW of energy annually.

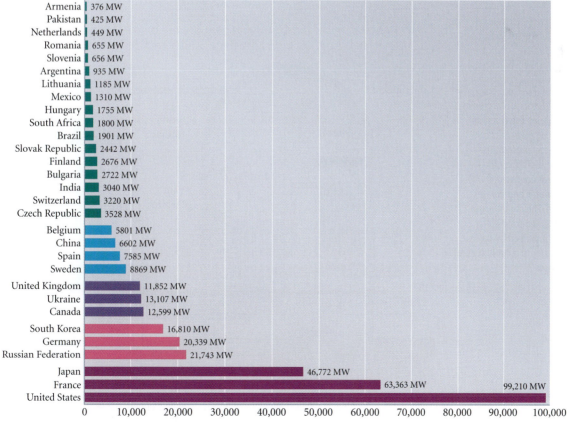

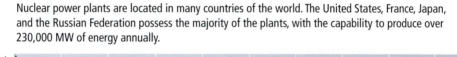

the heating of water, supplies the energy to turn the turbine. *The essential difference among the different types of power plants is the fuel source that creates the steam from water.* In conventional nuclear reactors, this energy is supplied by the controlled fission of ^{235}U (we say that the fission "goes critical").

Conventional nuclear reactors have three essential parts, as shown in Figure 21.13. The nuclear reactor part comprises 100 to 200 fuel rods that contain the fissionable uranium. A series of movable control rods, typically made of boron or cadmium, absorb neutrons as a way of controlling the rate of fission. The rods are located at the bottom of a pool of water, which acts as a moderator to slow these neutrons so that they can be captured by the uranium in the fuel rods. The other two parts of the system, the steam generator and the turbine and condenser, are common to many types of conventional power plants. Other types of nuclear reactors also exist. Breeder reactors, for example, are so named because they produce more fissionable fuel than they consume. In these reactors, relatively abundant ^{238}U is bombarded with neutrons to "breed" ^{239}Pu, along with the emission of β-particles.

$$^{238}_{92}\text{U} + {}^{1}_{0}\text{n} \rightarrow {}^{239}_{92}\text{U} \xrightarrow{\beta} {}^{239}_{93}\text{Np} \xrightarrow{\beta} {}^{239}_{94}\text{Pu}$$

Prototype large breeder reactors either have been built or are being built in China, France, Scotland, the United States, India, Japan, and the former Soviet Union. Breeder reactors are not currently used in the commercial production of energy because of the exceptionally long (24,400 years) half-life of ^{239}Pu. In these reactors, sodium metal is used as a coolant because it can transfer heat away from the reactor core much better than water and has a much higher boiling point, allowing it to remain liquid without being pressurized. Because of the high operating temperature of the liquid sodium, along with sodium's capacity to absorb neutrons (becoming radioactive after it travels by the core), sodium is viewed as particularly hazardous. The future of breeder reactors is, at the very least, uncertain.

FIGURE 21.13

The essential parts of a nuclear power plant. The nuclear reactor, submersed in a pool of water, contains the uranium-based fuel rods and a series of control rods to slow the fission process. The high-pressure water, heated by the fission, travels to a steam generator where the heat vaporizes water and creates high-pressure steam. The steam is used to turn a turbine and generate electricity. The steam is then condensed by passing large amounts of water in a cooling pond through the condenser.

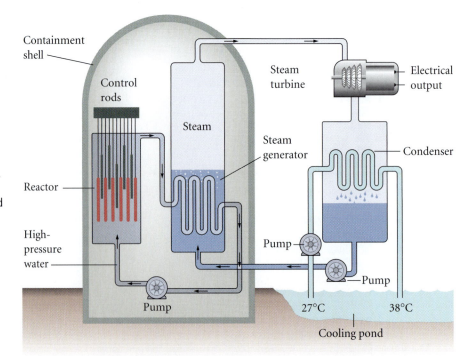

Here are the details of fission as we understand it today:

- Fission reactions release energy. The masses of the product nuclei are less than those of the starting nuclei, and the source of the energy is this mass difference. The energy released is orders of magnitude higher than that of "ordinary" chemical reactions.

- With rare exception, fission is not a naturally occurring process. We initiate it in some nuclei by brute force—that is, by smashing them with a high-energy neutron. The impact simply drives the nucleus apart. A few nuclei can be induced to undergo fission after they capture lower-energy neutrons. Furthermore, the only *naturally occurring* fissionable nucleus is ^{235}U, which is present in only 0.72% of all uranium atoms. The low availability of fissionable fuels slowed the development of fission. It also spurred the breeding of human-made fissionable fuels such as ^{233}U and ^{239}Pu.

- Once induced, fission releases more neutrons, typically 2 to 3 neutrons, per event. These daughter neutrons usually are traveling fast and may escape without further interaction with a fissionable nucleus. In this instance, we have a condition known as a **subcritical mass**. But if enough fissionable nuclei are nearby (a **critical mass**) or if the neutrons are slowed, these neutrons can continue the fission process in the absence of a neutron source. A self-sustaining **chain reaction** is possible. If too much fissionable material (a **supercritical mass**) is present, the reaction goes out of control.

- Nuclei can split in more than one way, forming a whole host of fission products. The split is usually into two or sometimes three fragments. This means that fission is "messy" because of the many products.

- Fission is also messy because the products are usually radioactive. They tend to be neutron-rich and beta emitters. In the case of nuclear reactors, the radioactive products end up in high-level and low-level nuclear waste—in other words, storage. In the case of weapons testing above ground, they result in nuclear fallout.

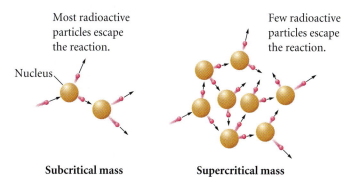

Most radioactive particles escape the reaction.

Nucleus

Subcritical mass

Few radioactive particles escape the reaction.

Supercritical mass

The conditions needed to sustain a chain reaction.

EXERCISE 21.7   **Fission: A Chain Reaction**

Draw a sketch to show why the fission of ^{235}U can be called a chain reaction. What factors do you think will influence whether the fission chain reaction will merely sustain itself or will be explosive?

First Thoughts

Fission of ^{235}U is initiated by neutrons and produces neutrons. This can result in a chain reaction if enough of the neutrons released are able to interact with more ^{235}U.

Solution

As shown here, the fission events quickly multiply, and the reaction becomes explosive. But if each fission event simply produced one more fission (instead of three), the chain reaction would simply sustain itself.

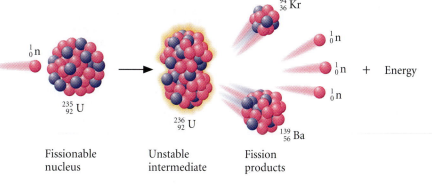

Fission of ^{235}U.

$^{94}_{36}Kr$

$^1_0 n$

$^1_0 n$ +   Energy

$^1_0 n$

$^{139}_{56}Ba$

$^1_0 n$

$^{235}_{92}U$

$^{236}_{92}U$

Fissionable nucleus

Unstable intermediate

Fission products

Further Insights

For a chain reaction to be explosive, two conditions must be met: (1) The fission reaction must produce more than one neutron, and (2) these neutrons must efficiently produce more fission events. ^{235}U produces 2 to 3 neutrons per event, so the first condition is met. The second is a bit tricky. First, you need a critical mass of ^{235}U if you want each fission event to induce one additional event. For ^{235}U, this is a fairly large (on the order of kilograms) amount. To be explosive, you need to have a mass of ^{235}U that is even greater. Second, you need to minimize the presence of substances such as ^{238}U that absorb neutrons and stop the reaction. Finally, you need to keep the ^{235}U in one place, which is difficult because the energy released tends to blow it apart. What are the implications? Although the principles of building a nuclear weapon are fairly simple, engineering one can be very difficult.

PRACTICE 21.7

Draw a sketch that would illustrate the fission reaction of ^{235}U that is just sustainable. (*Hint:* A sustainable reaction is not a supercritical chain reaction.)

See Problems 67 and 68.

Video Lesson: Applications
of Nuclear Chemistry

21.8 | Medical Uses of Radioisotopes

Now used almost routinely for diagnosis, radioisotopes enable us to image internal organs, bone, and tissue structures. We can watch biological processes such as oxygen uptake and brain activity. Radioisotopes also make possible the treatment of tumors without anesthesia or invasive surgery. There are more than 40,000 such procedures are carried out in the United States each day.

Application

CHEMICAL
ENCOUNTERS:
Tracer Isotopes
for Diagnosis

Tracer Isotopes for Diagnosis

We have already mentioned that you can image certain internal structures such as bones (or bullets or swallowed coins) using X-rays. The bones, by absorbing the incoming radiation, show up as shadows on the X-ray films or detectors. However, the heart, liver, and thyroid gland do not show up very well on X-ray films because they absorb so little of the X-rays.

Imaging these organs is the job of radioactive tracers, more technically called diagnostic **radiopharmaceuticals**. Radiopharmaceuticals are used in small amounts, so they release only low amounts of radiation into the body. They are considered safe to use for diagnostic purposes. Through introduction into the body by mouth, inhalation, or injection, the radiopharmaceutical can be used to outline the organs of interest. This imaging occurs by measuring the emitted radiation outside the patient. Either a "hot spot" (a tumor that preferentially absorbs the radioisotope) or a "cold spot" (a place where the surrounding tissues preferentially absorb the radioisotope) will produce the desired result. Using an alpha or beta emitter as a radioactive source would *not* work, because *these particles would be absorbed by the tissues inside the body before they could be detected outside the body.*

To understand imaging, we no longer can ignore the chemical form of the nuclide, as we have done throughout most of this chapter. The art of getting a good image lies in understanding the chemical behavior of the nuclide in the body. For example, although you can make elemental iodine, ^{131}I$_2$, from radioactive ^{131}I, you cannot feed this chemical to a patient to image the thyroid gland, because iodine is chemically reactive and would damage the mouth and stomach. Similarly, an organic fat-soluble compound containing iodine would tend to concentrate in the lymph system rather than travel via the bloodstream to the thyroid. The chemical form of choice for thyroid studies is the iodide ion, $^{131}I^-$, in the water-soluble form of sodium iodide, $Na^{131}I$, which can be swallowed in a salty "cocktail." Each radioactive substance must be prepared in a carefully tailored chemical form, because each organ in the body has a different chemical profile.

EXERCISE 21.8 | ^{123}I and ^{131}I : Cousins But Not Twins

Radioactive ^{131}I is used for both treatment and diagnostic scans, whereas radioactive ^{123}I is used only for diagnosis. Account for this difference after looking up the decay mode for each nuclide.

Solution

Both of these isotopes of iodine behave the same *chemically* in the body (they both are taken up by a normal thyroid gland), so the difference must lie in their *nuclear* properties. A table of radioactive decay shows that both ^{131}I and ^{123}I release gamma rays. Because gamma rays easily penetrate matter, they can be detected outside the body. Therefore, both can be used for diagnosis. For destruction of a tumor, you need alpha or beta particles that damage tissue at short distances. ^{131}I is a beta emitter; ^{123}I is not.

PRACTICE 21.8

Understanding the biological action of salicylic acid can be enhanced by imaging a patient fed with a radiopharmaceutical. Which of the following salicylic acid derivatives would *not* be a suitable choice for such a study?

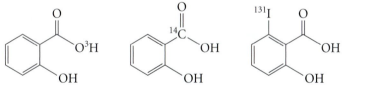

See Problems 73–76.

In many cases, technetium-99m is the nuclide of choice. The existence of technetium was predicted by Mendeleev as "eka-manganese," and it is the only non-rare-Earth element he foresaw that was not discovered in his lifetime. Although it was first produced in 1939 by Glen Seaborg and Emilio Segrè, it was not used in nuclear medicine until the 1960s. Technetium has a versatile chemistry and is dispensed in dozens of chemical compounds for imaging different parts of the body. Furthermore, technetium-99m emits only gamma rays, which are little absorbed by the body and therefore expose the patient to only a small radiation dose. Because it has such a short half-life, Tc-99m cannot be stored in a flask on a shelf. Rather, it usually is generated in the hospital from a source of molybdenum called a "molybdenum (or sometimes technetium) cow":

$$^{99}Mo \rightarrow ^{99m}Tc + _{-1}^{0}\beta + _{0}^{0}\gamma$$

$$^{99m}Tc \rightarrow ^{99}Tc + _{0}^{0}\gamma$$

The molybdenum-99 isotope used in the cow is not naturally occurring on Earth. It is generated as a fission product of uranium in nuclear reactors and shipped to hospitals worldwide. The ^{99m}Tc nuclide is separated from the molybdenum as it is needed in a process called "milking the cow." Figure 21.14 shows the earliest chromatographic column used to do a hands-on separation of ^{99m}Tc from ^{99}Mo.

FIGURE 21.14

This photograph shows the early column chromatography apparatus used to separate Tc-99m from the parent isotope (Mo-99). The molybdenum remains on the column, and the technetium passes through. The early work on technetium-99m was done at Brookhaven National Lab.

How do we know?

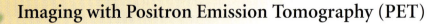

Imaging with Positron Emission Tomography (PET)

Why does nuclear medicine employ such exotic nuclei as technetium and palladium yet seemingly ignore the elements that make up most of our body, including hydrogen, oxygen, carbon, nitrogen, sulfur, and potassium? All of these elements play biochemical roles in the development of cells and tissues, yet none has been mentioned as a diagnostic or therapeutic tool. Why the omission?

Think back on what is needed to do imaging. First is a source of radiation that penetrates well enough to be detected outside the body. Gamma emitters usually are the nuclides of choice. Second, you need availability of the nuclide in sufficient quantity to do a study. Third, you need a half-life that is reasonably short, and if it is very short, the nuclide must be generated on site. Finally, you need to create a chemical form of the nuclide that will give either a "hot spot" or a "cold spot" in the area of medical interest.

Carbon-14, though biologically active, has too long a half-life and is a pure beta-minus emitter (see Figure 21.15). However, ^{11}C, with a n/p that is lower than those of the stable isotopes, is a positron emitter. Other positron emitters include ^{15}O, ^{13}N, and ^{30}S. Positrons themselves do not penetrate very far. But when a positron encounters an electron, which happens almost immediately, an annihilation occurs whereby the particle (positron) and antiparticle (electron) are converted into energy:

$$_{+1}^{0}\beta + e^- \rightarrow 2\,_0^0\gamma$$

A burst of energy is released as a positron and an electron collide.

The photons from the two gamma rays are emitted in exactly opposite directions. When a gamma detector is positioned both above and below the patient, if each one simultaneously records an event, then a positron was annihilated. By feeding the data from the detectors into a computer, it is possible to reconstruct an image of where the positron emission took place.

Positron emission imaging is better known as a PET scan, short for **positron emission tomography**. It is a

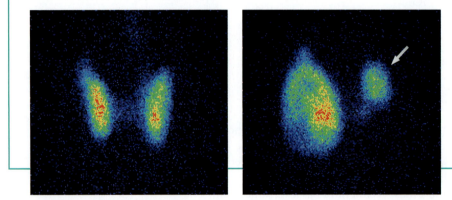

These thyroid scans were taken using radioactive iodine (I-123). The normal scan on the left shows uniform iodine uptake; the two thyroid lobes are similar in size. The lobe marked with an arrow in the photo on the right is not functioning properly, as is typical in thyroid cancer. A biopsy would follow to confirm the presence of cancer.

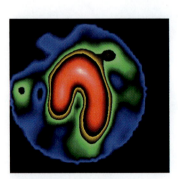

Tc image of heart muscle.

Today such processes are automated. The half-life of ^{99}Mo is a brief 67 hours, so it is shipped to medical suppliers for immediate distribution to hospitals.

One of the more widely used technetium compounds is sodium pertechnetate, $NaTcO_4$. The pertechnetate ion, TcO_4^-, has properties similar to those of the chloride ion, Cl^-, and concentrates in brain tumors, in the thyroid and salivary glands, and in areas of the body where blood is pooling (as happens in internal bleeding). Similarly, technetium pyrophosphate, TcP_2O_7, can be used to image the heart to see the extent of damage to heart muscle after a heart attack.

Although it is well developed, nuclear medicine is still a relatively young field; radioisotope tracers were developed in the 1930s and put into widespread

much trickier procedure than other types of nuclear imaging, largely because positron emitters tend to have half-lives on the order of minutes. To have enough radioactive material present, they must be produced nearby or on site at a hospital. In either case, technicians are needed to maintain the production equipment. PET scans also require the injection of radioactive material, which is not the case for MRI or CT scans.

The payoff with PET, though, is impressive. It produces "functional imaging"—that is, images of chemical processes in action. For example, it can record the brain in action during a seizure or when the patient is hearing music, thus decoding the neural pathways. Using glucose labeled with ^{11}C or with ^{18}F (see Figure 21.16), PET can reveal brain metabolism. Similarly, ^{18}F-labeled estrogen can be used within a patient to show how tumors grow. The color-enhanced real-time images are far more dramatic than the black-and-white slices produced by other means.

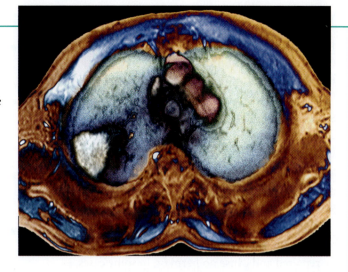

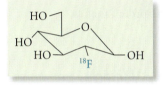

FIGURE 21.16

^{18}F-labeled 2-deoxyglucose, or FDG, is used to study glucose metabolism in the body. It can be used to differentiate benign tumors from cancerous ones, because the latter use glucose at a higher rate. After a patient consumes this radiopharmaceutical, tumors within her or his body show up as white spots where glucose metabolism is higher.

FIGURE 21.15

The isotopes of carbon.

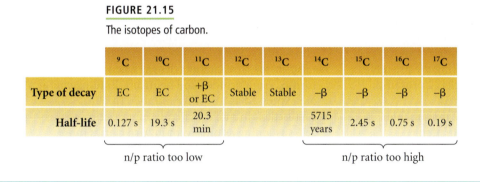

	^9C	^{10}C	^{11}C	^{12}C	^{13}C	^{14}C	^{15}C	^{16}C	^{17}C
Type of decay	EC	EC	+β or EC	Stable	Stable	−β	−β	−β	−β
Half-life	0.127 s	19.3 s	20.3 min			5715 years	2.45 s	0.75 s	0.19 s

n/p ratio too low   n/p ratio too high

use in the 1950s and 1960s. In the 1980s, the ready availability of computers to help process images led to explosive growth in the field that continues today. There are announcements of new techniques, new types of images, or methods that require lower amounts of radiation almost daily. (There are also cautions that costly scans are being used too routinely to warrant either the risks or the costs involved.) The odds are that you know somebody who has waged battle with cancer. Nuclear medicine is undoubtedly a part of that person's health history.

The Bottom Line

- Each element is composed of atoms containing the same number of protons. These may contain isotopes with differing numbers of neutrons. (Section 21.1)

- Some nuclear configurations are unstable. They decay in a stepwise progression toward stable nuclei. (Section 21.2)

- There are three main types of radioactive decay: alpha-particle emission, beta-particle emission, and gamma-ray emission. (Section 21.2)

- Ionizing radiation can interact with living tissue and cause damage to the DNA of a cell. This damage may be repaired and cause no harm or, in some cases, may lead to cancer. (Section 21.3)

- Radioactive decay occurs via first-order kinetics. (Section 21.4)

- Energy is released in radioactive decay processes as a consequence of the mass defect in nuclei. (Section 21.5)

- Nuclei with a "magic" number of protons and/or neutrons (2, 8, 20, 28, 50, or 82) are stable. Nuclei with even numbers of protons and/or neutrons are also more likely to be stable. (Section 21.6)

- Nuclear fission is the splitting of heavier nuclei into lighter ones. Nuclear fusion results when smaller nuclei combine into heavier nuclei. (Section 21.7)

- Radioisotopes can be used in medicine for imaging the body and for treating and eliminating cancerous tissues. (Section 21.8)

Key Words

alpha decay A type of radioactive decay wherein an alpha particle is emitted from the nucleus of a radioactive nuclide. Common for elements whose nuclei are larger than bismuth, alpha decay is often accompanied by the release of a gamma ray. (*p. 905*)

alpha particles (α particles) Particles emitted from the nucleus of a radioactive element during the process of alpha decay. They are helium nuclei (2 protons, 2 neutrons), with a +2 charge. (*p. 905*)

antimatter Particles that have the same mass as, but charges opposite to, corresponding matter. Antimatter particles such as the positron and anti-neutrino are similar to the electron and neutrino, respectively, but have opposite characteristics. (*p. 905*)

antineutrino A subatomic particle produced in beta-minus decay that has no charge, has essentially no mass, and interacts only rarely with matter. (*p. 904*)

becquerel (Bq) An SI unit of activity equivalent to one nuclear disintegration per second. (*p. 911*)

beta particles (β particles) Particles emitted from the nucleus of a radioactive atom during the process of beta decay. These particles are high-speed electrons. (*p. 904*)

beta-particle emission A naturally occurring type of radioactive decay wherein an electron is ejected at high speed from the nucleus, typical of nuclei that have too many neutrons to be energetically stable. An antineutrino accompanies this emission, and sometimes one or more gamma rays as well. Also known as beta emission. (*p. 904*)

beta-plus emission A type of radioactive decay wherein a positron is ejected at high speed from the nucleus, typical of nuclides that have too few neutrons. A neutrino accompanies this emission, and usually one or more gamma rays as well. Also known as positron emission. (*p. 918*)

binding energy The energy released when a nucleus is formed from protons and neutrons. Binding energies are expressed as a positive number. (*p. 915*)

chain reaction In nuclear chemistry, a reaction that is self-sustaining as one event in turn causes more events. (*p. 923*)

critical mass The amount of fissionable fuel needed to sustain a chain reaction. (*p. 923*)

curie (Ci) A larger unit of activity than the becquerel, equivalent to 3.7×10^{10} Bq. (*p. 911*)

daughter nuclide An isotope that is the product of a nuclear reaction. (*p. 904*)

decay series A series of nuclear reactions that a large nuclide undergoes as it changes from an unstable and radioactive nucleus to a stable nucleus. (*p. 907*)

electron capture (EC) A type of radioactive decay that occurs when an inner-core electron is captured by a proton from the nucleus to form a neutron. The process is usually accompanied by the emission of X-rays. (*p. 907*)

fission (or nuclear fission) A type of nuclear reaction wherein a large nucleus splits into two or three smaller nuclei with the release of energy. (*p. 917*)

fusion (or nuclear fusion) A type of nuclear reaction wherein small nuclei are joined to form a larger nucleus with the release of energy. Nuclear fusion powers the stars. (*p. 916*)

gamma rays (γ rays) A high-energy form of electromagnetic radiation that is emitted from the nucleus. Gamma rays sometimes accompany alpha and beta decays. (*p. 904*)

gray (Gy) A measure of absorbed radiation equal to 100 rad. (*p. 911*)

ionizing radiation Radiation such as alpha and beta particles, gamma rays, or X-rays that is capable of removing an electron from an atom or a bond when it interacts with matter. (*p. 910*)

mass defect (Section 21.5) The loss in mass that occurs when a nucleus is formed from its protons and neutrons. (*p. 915*)

metastable state An energetically unstable arrangement of protons and neutrons in a nucleus after a neutron has become a proton. (*p. 906*)

neutrino A subatomic particle that is produced in beta-plus decay and that has no charge, has essentially no mass, and interacts only rarely with matter. (*p. 905*)

nuclear equation An equation showing a nuclear transformation, where the atomic and mass numbers are provided. (*p. 904*)

nuclear radiation The particles and/or energy emitted during radioactive decay. (*p. 901*)

nucleon The name given to a particle (proton or neutron) that is part of a nucleus. For example, ^{13}C contains 13 nucleons, 6 protons, and 7 neutrons. (*p. 918*)

positron The antimatter equivalent of an electron. Positrons have a positive charge and the same mass as an electron. (*p. 907*)

positron emission See *beta-plus emission*. (*p. 907*)

positron emission tomography A medical imaging technique that images metabolic processes within the body. Also known as a PET scan. (*p. 926*)

rad A unit of energy absorbed by irradiated material equal to 0.01 J/kg of exposed material. (*p. 911*)

radioactive decay The process by which an unstable nucleus becomes more stable via the emission or absorption of particles and energy. (*p. 904*)

radioactivity The emission of radioactive particles and/or energy. (*p. 901*)

radiopharmaceuticals Compounds containing radioactive nuclides that are used for imaging studies in nuclear medicine. (*p. 924*)

rem A unit that measures the "equivalent dose" of radiation; that is, it takes into account the interaction of radiation with human tissue. The word stands for "roentgen equivalent in man." The rem is not an SI unit but is related to the SI unit, the sievert. (*p. 911*)

roentgen A unit used to measure exposure to radiation. One roentgen is equal to 2.58×10^{-4} C/kg of dry air at STP. (*p. 911*)

sievert (Sv) A unit that measures the "equivalent dose" of radiation; that is, it takes into account the interaction of radiation with human tissue. One sievert equals 100 rem. (*p. 911*)

subcritical mass A mass of a radioactive isotope that is too small to sustain a chain reaction. (*p. 923*)

supercritical mass A mass of a radioactive isotope that not only supports a chain reaction but causes the majority of the nuclei to undergo unfettered radioactive decay within a very short period of time, releasing huge amounts of energy. (*p. 923*)

▬ Focus Your Learning

The answers to the odd-numbered problems and some selected problems appear at the back of the book, as represented by the blue numbering.

Section 21.1 Isotopes and More Isotopes

Skill Review

1. What is the atom with smallest atomic number? The smallest mass number?

2. Can different elements both have the same number of protons? The same number of neutrons? Explain.

3. Can an atom have no neutrons? Explain.

4. Can an atom have no protons? No electrons? Explain.

5. Can a helium nuclide have a smaller mass number than a hydrogen nuclide? Explain.

6. Can a carbon nuclide have a smaller mass number than a nitrogen nuclide? Explain.

7. Identify the number of protons, neutrons, and electrons in each of the following isotopes.
   a. ^{12}N b. ^{124}Sb c. ^{152}Eu d. ^{9}Be

8. Identify the number of protons, neutrons, and electrons in each of the following isotopes.
   a. ^{7}Li b. ^{122}Cs c. ^{17}O d. ^{18}F

Chemical Applications and Practices

9. A person who weighs 60 kg (132 lb) has about 120 g of potassium in his or her body. How many grams of K-40 does this include? K-40 has a natural abundance of 0.0118%.

10. No isotopes of potassium are chemically *stable* in the presence of water and/or oxygen. Potassium-39 and potassium-41 are *stable* isotopes. Explain these different meanings of the word *stable*.

11. Fallout from a nuclear weapon includes the radioactive nuclide Sr-90.
    a. In the biosphere, which chemical form for Sr-90 is more likely, Sr^{2+} or Sr?
    b. Once Sr-90 lands downwind, it is extremely difficult to separate from the plants and soils. Suggest reasons why.

12. Strontium-90 from fallout gets into the food chain and eventually can end up in cows' milk. Can you remove the radioactivity from cows' milk by boiling it? Why or why not?

Section 21.2 Types of Radioactive Decay

Skill Review

13. Both gamma rays and infrared radiation are forms of electromagnetic radiation. How do they differ?

14. Beta decay involves the emission of a high-speed electron from an atom, yet the overall charge on the atom does not become less negative. Explain why.

15. Write nuclear equations for the following processes:
    a. An alpha particle (along with a gamma ray) is emitted from plutonium-239.
    b. Carbon-14 undergoes beta decay.
    c. Cesium-137 emits a beta particle with an accompanying gamma ray.

16. Write nuclear equations for the following processes:
    a. Plutonium-238 emits an alpha particle with an accompanying gamma ray.
    b. Radon-222 is produced from the decay of a radium isotope with the emission of a gamma ray.
    c. Radium-225 emits a gamma ray followed by an alpha particle.

17. Write nuclear equations for the following processes:
    a. Polonium-215 (with a gamma ray) is produced by an alpha emission.
    b. Strontium-90 decays by beta emission. Little or no gamma radiation is released.
    c. Tc-99 decays by beta-minus emission.

18. Write nuclear equations for the following processes:
    a. Nitrogen-14 is formed from a radioisotope of carbon.
    b. Cadmium-110 is formed from a radioisotope of silver.
    c. Technetium-99 is formed from Technetium-99m.

Chemical Applications and Practices

19. Samarium-146 is the lightest element found naturally on our planet to undergo alpha emission. Write the equation for this alpha decay. There is no accompanying gamma ray.

20. Iodine-131, used in medical imaging, undergoes beta decay. Write the nuclear equation for this reaction.

21. Darlene Hoffman, an award-winning nuclear chemist, postulated the existence of Pu-244 before it was discovered. Into which of the four natural decay series does it fit?

22. A chemistry source states that radon-219 is produced in the actinium-227 radioactive decay series. Into which of the four decay series mentioned in Section 21.2 does actinium-227 fit?

Section 21.3 Interaction of Radiation with Matter

Skill Review

23. Classify the following as ionizing or nonionizing radiation: cosmic rays, infrared radiation, gamma rays, visible light, microwaves, X-rays.

24. You can cook food using microwaves, and you can sterilize food using gamma rays. Why do these two types of radiation produce such different results?

25. Suppose that you had administered a gamma emitter to a patient in order to diagnose how well his or her heart was functioning. Name three things you could do to minimize your exposure to the radiation.

26. Which cells in your body are most susceptible to radiation? Why?

27. Explain the similarities and differences between:
    a. a curie and a becquerel
    b. a rad and a rem

28. Explain the similarities and differences between:
    a. a rem and a sievert
    b. a curie and a rem

29. Smoke detectors use only a small quantity of americium, less than 35 kBq. How many disintegrations per second is this?

30. Using the information in Problem 29, show that the result is comparable to 1 microcurie.

Chemical Applications and Practices

31. In the mid-1990s, a watch was advertised that glowed in the dark. The source of the glow was the radioisotope tritium interacting with a luminous paint. The annual dose for a person wearing the watch as estimated at 4.0 microsieverts.
    a. How many rem is this?
    b. Do you think this amount of radiation warrants concern?

Note the radiation symbol between the hour and minute hands and the H3 (^{3}H, or tritium) on the face of this watch.

32. In the previous problem, we noted that tritium, a radioisotope of hydrogen, was used in some watches.
    a. What mode of decay would you predict for tritium?
    b. Given that radiation escapes from the watch case, does this evidence support your prediction?

33. Three metals—aluminum, iron, and cadmium—are proposed as materials that could be used to shield nuclear radiation.
    a. What property of these materials would you need to look up to determine which one would have to be used in the greatest thickness?
    b. What else might you need to know about a substance before you use it in shielding?

34. If radon-222 gas decays in your lungs to produce solid polonium, which is then trapped there, how many decays does the polonium progress through before reaching the stable isotope of lead-206? How many alpha and beta particles are emitted in the process?

Section 21.4 The Kinetics of Radioactive Decay

Skill Review

35. Here are the decay plots for two different hypothetical radioactive nuclei. The plot denoted by the red line is A; that denoted by the blue line is B. From these graphs, determine:
    a. Which nuclide has the longer half-life?
    b. What is the half-life of the nuclide indicated by the red line?
    c. Which one has the higher activity?
    d. Which one would be more dangerous if swallowed?

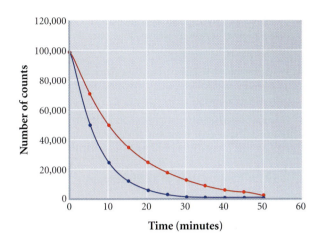

Time (minutes)

36. Using the plot in Problem 35, estimate the half-life of each nucleus. What does a shorter half-life imply?

37. If 75% of a radioactive sample is gone after 30 days, what is the half-life of the nuclide?

38. If 50% of a radioactive sample remains after 30 days, what is the half-life of the nuclide?

39. Tritium has a half-life of 12.3 years. What fraction of a sample of tritium will remain after 12.3 years? After 24.6 years?

40. What percent of the original sample of strontium-90 (half-life = 28.9 years) will remain
a. after 14 years?　　b. after 49.6 years?　　c. after 1000 years?

41. The half-life of strontium-90 is 28.9 years. How long would it take for the activity of a Sr-90 sample to diminish by 87.5%?

42. The half-life of strontium-85 (about 64 days) is considerably shorter than that of strontium-90. How long would it take for the activity of a sample of Sr-85 to diminish by 93.75%? Why do you think that Sr-85 is used diagnostically for bone scans, but Sr-90 is not?

Chemical Applications and Practices

43. A 10-mCi sample of a tracer isotope is used to diagnose blood flow from the heart. If it is desirable that nearly all of the radioactivity (99%) be gone after 3 days, approximately what half-life should the isotope have?

44. It is estimated that the nuclear accident at Chernobyl released 1.85×10^{18} becquerels, a large amount of radiation. How many curies is this? Why is it not easy to translate either of these values into the number of radioactive atoms present?

Chernobyl nuclear power plant.

45. A 0.1-microcurie sample of polonium-210 is used to demonstrate alpha decay in the lecture hall, because there is little accompanying gamma radiation. The half-life of this isotope is 138 days. About how often does an instructor need to buy a new source? State any assumptions you made in arriving at your answer.

46. The world uranium reserves are currently estimated at 3.4 million tonnes, where a tonne is a metric ton, or 1000 kg. Current knowledge places Earth at 4.5 billion years old. How much uranium was present at the time our world formed?

47. The half-life of plutonium-239 is about 24,000 years. After approximately how much time will over 99% of a sample of plutonium-239 have decayed? How can this element exist on our planet if its half-life is so short?

48. Plutonium oxide, used to power the Cassini space-probe to Saturn, was stored in corrosion-resistant materials designed to contain the fuel for 10 half-lives, or 870 years.
a. What is the half-life of Pu-238?
b. Why do you think the figure 10 half-lives was selected?
c. The Cassini batteries contained 72 lb (33 kg) of Pu-238 in the form of plutonium dioxide. How much Pu-238 (in kilograms and in grams) will remain after 870 years?

Section 21.5 Mass and Binding Energy

Skill Review

49. a. Why don't the masses of the neutrons and protons that make up an oxygen-16 atom add up to the mass of the oxygen nucleus?
b. Is this true for all isotopes of oxygen?

50. a. Is an atom of carbon likely ever to fall apart into its component protons, electrons, and neutrons?
b. Explain why or why not.

51. Explain how the mass defect and the binding energy are related.

52. Is mass conserved in nuclear reactions, such as alpha decay, that proceed spontaneously? Why or why not?

53. Calculate the mass defect and the resulting energy for the following nuclear reaction:
($^1_0$n = 1.008665 g/mol; ^{4}He = 4.002603 g/mol)
^{170}Ir (169.974970 g/mol) decays to ^{166}Re (165.965740 g/mol)

54. Calculate the mass defect and the resulting energy for the following nuclear reaction:
$^{40}_{19}$K (39.963999 g/mol) $\rightarrow$ $^{40}_{20}$Ca (39.962591 g/mol) $+ ^{\ 0}_{-1}\beta$

Chemical Applications and Practices

55. The masses of the neutral atoms involved for an alpha emission from U-238 are shown below. These values include the masses of the electrons.
a. Would you expect the U-238 or its decay products to have more mass?
b. What is the source/result of any difference in these masses?

$^4_2$He	4.002603 u
$^{238}_{92}$U	238.050784 u
$^{234}_{90}$Th	234.040945 u
e^-	0.005485799 u

56. Use the data in Problem 55 to calculate the mass defect for the process. What is the source/result of any difference in these masses?

57. Describe the similarities and differences between beta-minus and beta-plus emission.

58. The stable isotopes of carbon are ^{12}C and ^{13}C. What decay mode would you predict for ^{14}C?

Section 21.6 Nuclear Stability and Human-made Radioactive Nuclides

Skill Review

59. a. What does "doubly magic" mean, in reference to nuclides?
b. Give two examples of nuclei that are doubly magic and two examples of nuclides that have no magic numbers at all. How would you expect these nuclei to differ?

60. Suggest a reason why the heavier elements have proportionately more neutrons than the lighter ones.

Chemical Applications and Practices

61. Element 114 was recently discovered. Why was this element sought, but not the neighboring elements 115 and 113?

62. Some claims to the discovery of element 118 have been made. Give reasons why this element may be considered to have both nuclear and chemical stability.

Section 21.7 Splitting the Atom: Nuclear Fission

Skill Review

63. Why is alpha emission a better prediction for the mode of radioactive decay for uranium or plutonium than it is for iron, carbon, or hydrogen?

64. Answer the following questions for the fission of plutonium:
$$^{239}_{94}\text{Pu} + ^{1}_{0}\text{n} \rightarrow [^{240}_{94}\text{Pu}] \rightarrow ^{136}_{51}\text{Sb} + ^{100}_{43}\text{Tc} + 4\,^{1}_{0}\text{n} + \text{energy}$$
a. What is the significance of the fact that neutrons are produced?
b. What is the source of the energy in this equation?

65. Iron and cobalt are not expected to undergo nuclear fission. Why?

66. Would you expect helium to undergo nuclear fission? Will hydrogen-1 undergo fission?

67. Write nuclear reactions for the fission of ^{235}U to form:
a. ^{94}Kr and ^{139}Ba and neutrons
b. ^{80}Sr and ^{153}Xe and neutrons

68. An isotope of the element technetium can be produced by bombarding molybdenum-97 with deuterium nuclei. Two neutrons are also formed. Write the nuclear equation.

Chemical Applications and Practices

69. On our planet, both U-235 and U-238 occur naturally.
a. How do these isotopes differ?
b. What is the natural abundance of each?
c. Propose a reason why it is very difficult to separate these two isotopes.

70. Using "conventional" explosives such as TNT, you can make tiny explosive devices as well as huge ones. Is it possible to make a similarly tiny nuclear bomb? Why or why not?

71. When fission of U-235 or Pu-239 occurs, elements such as americium, californium, and berkelium are not found in the fallout. Explain why.

72. One particularly nasty component of nuclear fallout is strontium-90. Explore the reactivity of this particular isotope and explain why it may be harmful to living creatures.

Section 21.8 Medical Uses of Radioisotopes

Skill Review

73. What questions should you ask about a radionuclide to be injected for diagnostic purposes?

74. Why aren't alpha emitters useful for diagnostic scans, such as a scan of the heart or of the thyroid gland?

75. Technetium-99m samples should not be stored overnight but, rather, should be freshly prepared each day for diagnostic scans. Why?

76. Why is molybdenum-99 not used directly as a component of a radiopharmaceutical?

Chemical Applications and Practices

77. A patient was injected with 10 mg of fluorine-18–labeled glucose for a PET scan. Fluorine-18 has a half-life of 110 minutes and disintegrates by positron emission. Write the nuclear equation for the decay.

78. Using the information in Problem 77, determine the amount of time needed to reduce the radioactivity of fluorine-18 to 1/16 of its original activity.

79. In the chapter, it was mentioned that it takes about two and a half weeks for Tc-99 to be eliminated from the body. The Department of Energy reports that it takes approximately 60 hours for the body to eliminate half of the technetium. Are these two figures consistent with each other?

80. Gallium-67 citrate is used as a radiopharmaceutical for diagnosing tumors and infections.
a. What type of radioactive decay would you predict for this nuclide?
b. A typical activity of a radionuclide used in the treatment of an adult lymphoma is on the order of 10 mCi. After how many days would radiation levels drop to less than a millicurie?

Comprehensive Problems

81. Use the Internet to research the connection between smoking and exposure to the nuclides polonium-210 and lead-210.

82. For the same dose of radiation, which has a higher dose equivalent, strontium-90 or radon-222?

83. How do you know whether or not a gamma ray accompanies an alpha or a beta emission?

84. Write nuclear equations for the following processes:
a. A positron is emitted by oxygen-15.
b. Boron-11 is formed by positron emission.
c. A positron is emitted by chlorine-35.
d. Oxygen-18 is formed by positron emission.

85. One of the radioactive decay series is shown below. Identify the mode of radioactive decay at each step.

$$^{232}_{90}Th \rightarrow \, ^{228}_{88}Ra \rightarrow \, ^{228}_{89}Ac \rightarrow \, ^{228}_{90}Th \rightarrow \, ^{224}_{88}Ra \rightarrow \, ^{220}_{86}Rn \rightarrow$$

$$^{216}_{84}Po \rightarrow \, ^{212}_{82}Pb \rightarrow \, ^{212}_{83}Bi \rightarrow \, ^{212}_{84}Po \rightarrow \, ^{208}_{82}Pb$$

86. In the radioactive decay series in the previous problem, lead-212 was formed. Why didn't the decay series stop at lead?

87. In the radioactive decay series given in Problem 85, which elements are represented by the symbols Rn and Ra? Which one is a gas? Which one is a metal? Which one is chemically inert?

88. Radon is produced in three of the naturally occurring decay series. Which three? Which isotopes of radon are formed? How would you expect these isotopes to differ? How would you expect them to be the same?

89. Of the three radon isotopes mentioned in Problem 88, only radon-222 is a health hazard. Propose a reason why, and then research your answer to see whether you are correct.

90. Many tropical islands are volcanic in origin and contain uranium in the rocks and minerals beneath the soils. However, radon is less likely to be a problem in homes built in the tropics. Propose two reasons why (and more if you can).

91. How does the nucleus of a carbon atom compare in density with that of elemental lead ($d = 11.3$ g/cm^3)? To answer this question, calculate the volume of a ^{12}C nucleus, assuming that the nucleus is spherical and that the radius is 1.2×10^{-13} cm. The mass of the nucleus in ^{12}C is 11.96709 u, or 1.98718×10^{-23} g.

92. The transformation of elements into other elements also takes place in stars. Write the nuclear equation for the formation of oxygen-16 when carbon-12 is hit with an alpha particle. A gamma ray is also released in this reaction.

93. Does food irradiation make the food radioactive? Find an answer to this question using the resources of the World Wide Web. Cite your sources.

94. Look up on the Internet the current maximum allowed exposure of workers in the nuclear industry. The Department of Energy (DOE) sets this standard. How does this standard vary for some individuals?

95. Irene Curie and her mother Marie are not the only scientists to have won the Nobel Prize for their pioneering work in nuclear chemistry and physics. Others include Ernest Rutherford, Ernest O. Lawrence, and Emilio G. Segrè. Use the Internet to find out why these and/or other prizes for nuclear work were awarded.

96. We pointed out in the text that palladium-103 (half-life = 16.97 days) is used in the treatment of prostate cancer, replacing iodine-125 (half-life = 59.4 days), though both are widely used. Recently, Cs-131 (half-life = 9.7 days) has been used. Why are these different products being used? That is, what are the advantages and disadvantages of each?

97. Palladium-103 has a half-life of 16.97 days.
a. What is the half-life in years?
b. How many protons, neutrons, and electrons are in an atom of palladium-103 in PdCl$_2$?
c. A researcher develops cisplatin (Pd(NH$_3$)$_2$Cl$_2$) using the palladium-103 isotope. Assume that the process involves the direct conversion of palladium(II) chloride to cisplatin in one step, but the process itself requires 24 hours to complete. How many grams of radioactive cisplatin would remain if the researcher started with 100.0 g of pure palladium-130(II) chloride?

Thinking Beyond the Calculation

98. Americium oxide is typically used in household smoke detectors. A document reports that a gram of americium oxide provides enough active material for "more than 5000 household smoke detectors." The particular isotope used in this application is americium-241.
a. How much americium is present in a typical smoke detector?
b. Give two reasons why Am-240 and Am-242 would not be appropriate isotopes to use.
c. Americium-241 decays by alpha emission. Write the nuclear reaction for this process.
d. The half-life of americium-241 is 432.2 years. How long will it take the reactivity of a sample of this nuclide to drop to 1% of its original activity?
e. Beta-particle emission by americium-242 is found in 83% of the sample. The rest of an americium-242 sample decays by electron capture. Write nuclear reactions for these processes.

Smoke detectors contain about 150 millionths of a gram of americium-241, which is extracted from spent fuel rods.

22

The Chemistry of Life

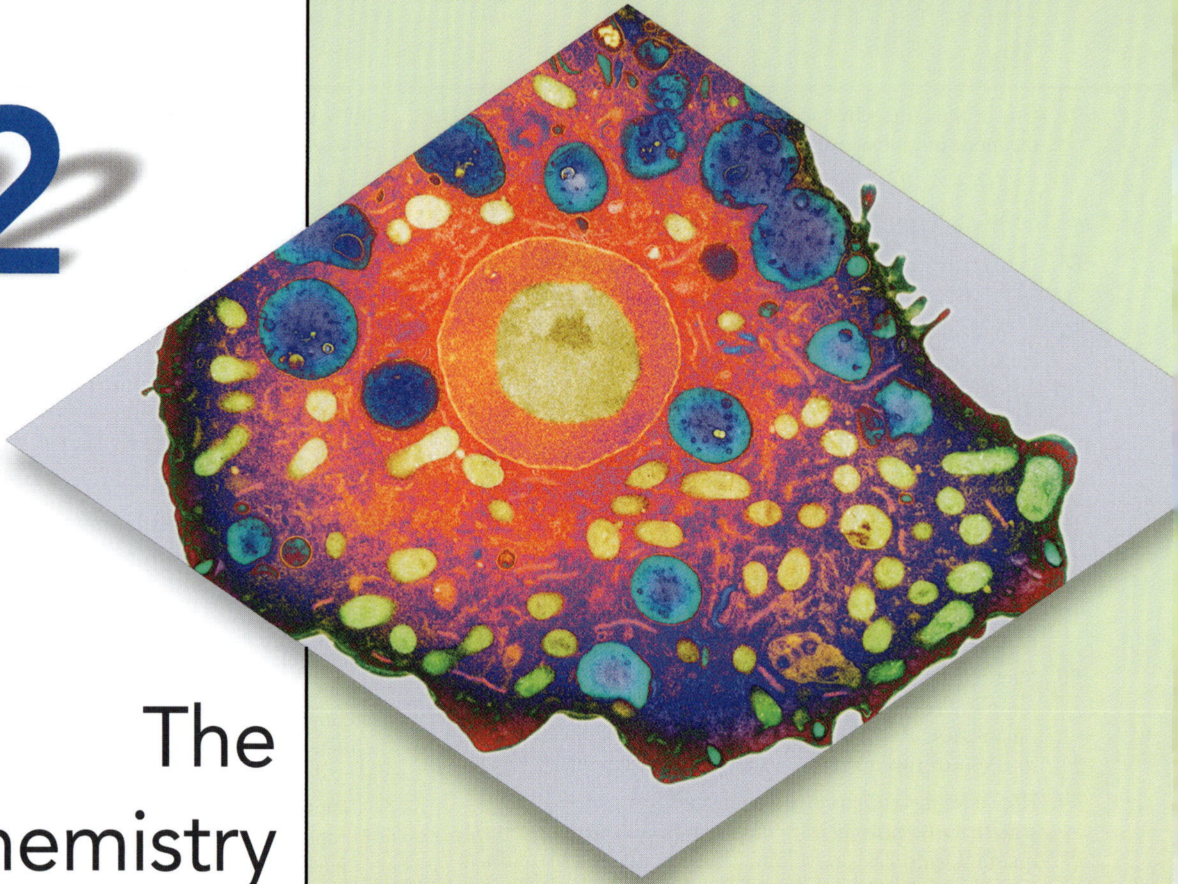

Contents and Selected Applications

22.1 DNA—The Basic Structure
Chemical Encounters: The Human Genome Project

22.2 Proteins

22.3 How Genes Code for Proteins

22.4 Enzymes

22.5 The Diversity of Protein Functions

22.6 Carbohydrates

22.7 Lipids

22.8 The Maelstrom of Metabolism

22.9 Biochemistry and Chirality

22.10 A Look to the Future

The cell is an intricate ballet of huge polymers, proteins, hormones, and lipids. Their dance keeps the cell alive. Shown here is a colored transmission electron micrograph of the protozoa responsible for causing meningoencephalitis (inflammation of the brain and its membranes). The nucleus (red) contains a large body known as a nucleolus (yellow), where RNA is made.

Go to **college.hmco.com/pic/kelterMEE** for online learning resources.

Life is an adventure of changes. From the moment of conception, through birth, growth, adolescence, maturity, aging, and death, we are sustained by an endless process of change. The chemistry of life is the chemistry of these changes, and as you might expect, it is exceptionally complex. The human body is one of the most intricate chemical systems we know, and yet there is a wonderful simplicity at the root of it all.

Our introduction to chemistry has given us the basic tools with which to examine the chemistry of life in some detail. At the root of this detail is a chemical we call DNA. This substance serves as the set of instructions that operate the chemical processes in living things. To understand the chemical reactions that take place in the body, we must understand the nature of DNA because *the chemistry of life is based on DNA and the chemical species it produces.*

Video Lesson:
Nucleic Acids

22.1 DNA—The Basic Structure

In the summer of the year 2000, scientists participating in the International Human Genome Project announced that they had reached the first major milestone in working out the structure of all the genetic material that is needed to make a human. They had deciphered a rough "first draft" of a complete set of human genes. Work will continue for many years to complete the draft and then to look at the different versions of genes that underlie many of the differences among us. The information gained about the structure of our genes will be used to develop new medicines and biotechnologies and to understand much more about how life works. What exactly are genes? **Why are they the focus of such intense scrutiny?**

To best uncover the meaning and significance of the gene, we will begin with DNA. **Deoxyribonucleic acid (DNA)** is the name for a series of polymers composed of chemicals called **nucleotides**. In DNA, each nucleotide is itself composed of a phosphate group bonded to a **deoxyribose** sugar group, bonded to one of four **nitrogenous organic bases** (schematically shown in Figure 22.1). The four nitrogenous organic bases, shown in Figure 22.2, are *adenine, guanine, thymine,* and *cytosine.*

DNA is a polymer of the four nucleotides bonded together in the manner shown in Figure 22.3. The phosphate groups on the nucleotides participate in condensation reactions to form *phosphodiester bonds* that bind the individual nucleotides into the DNA chain. Recall that condensation reactions result in the coupling of two molecules with the elimination of a molecule of water (Chapter 12).

Application

CHEMICAL ENCOUNTERS:
The Human Genome Project

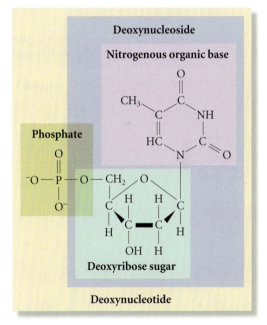

FIGURE 22.1

A nucleotide contains a nitrogenous organic base, a deoxyribose sugar, and a phosphate.

FIGURE 22.2

The four nitrogenous organic bases found in DNA.

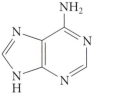

Adenine   Guanine   Cytosine   Thymine

935

FIGURE 22.3

The structure of DNA. Note how a strand of DNA naturally twists when the nitrogenous bases are on the same side of the molecule.

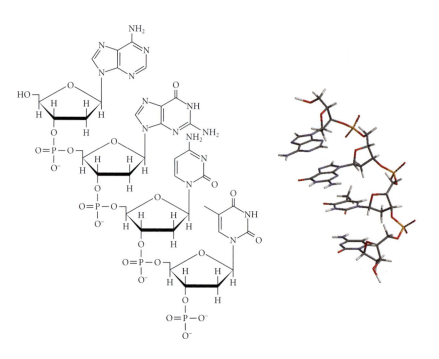

In DNA, the phosphate's —OH group is lost as the sugar's oxygen adds to the phosphorus atom (Figure 22.4).

The resulting condensation product is known as *single-stranded* DNA. However, the DNA that makes up genes is more complex in a very beautiful and significant way. This occurs when a second strand of DNA (upside down in relation to the first strand) interacts with the first strand. **Why does DNA occur as pairs of strands?** Double-stranded DNA is much more stable than a single strand, because the nitrogenous bases are able to participate in hydrogen bonds with each other. The hydrogen bonding results in a **base pair**. As a consequence of their relative size, the amount of space available between the two DNA strands, and the types of hydrogen bonds in the bases, adenine always base-pairs with thymine, and cytosine always base-pairs with guanine (Figure 22.5). We say that these pairs of bases are *complementary;* that is, they pair in a reciprocal fashion with one another.

To improve the hydrogen bonding, the two strands of DNA wind around each other to form a base-paired double helix, as shown in Figure 22.6. This structure had eluded scientists for quite some time until James Watson and Francis Crick

FIGURE 22.4

The condensation of two nucleotides and the elimination of water form the backbone of the DNA molecule.

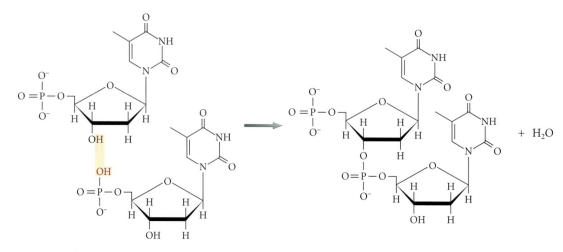

FIGURE 22.5

Base pairs in DNA, looking down the axis of the DNA molecule. Note the proximity of the oxygen atom to the amine group on the opposite base, and that of the nitrogen to the opposite amine.

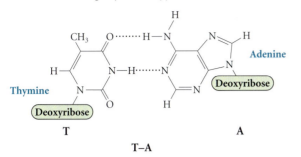

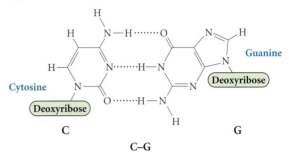

T–A   C–G

FIGURE 22.6

The double helix of DNA.

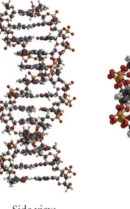

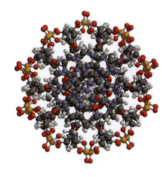

Side view   View down the axis

FIGURE 22.7

Rosalind Franklin (1920–1958) and the diffraction pattern she made that solved the double-helical structure of DNA. Her contributions to this discovery were never acknowledged during her life. She died of cancer at a very early age.

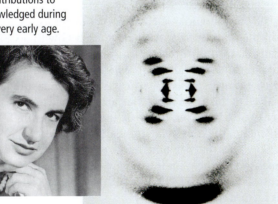

obtained the 1953 X-ray diffraction photograph shown in Figure 22.7. The image of this diffraction pattern was made by the chemist Rosalind Franklin and given to Watson and Crick without her knowledge. It turned out to be the key piece of data used to determine the structure of the double helix. In fact, the structure of the DNA double helix has become an icon of science. It was even the centerpiece of the 2004 Olympic Summer Games (Figure 22.8). Why is this structure so important? Within the double helix, we find the secret of how life is able to reproduce—and also the secret of how mere chemicals can contain the coded information that controls the structure and activities of all living things.

FIGURE 22.8

The opening ceremony of the 2004 Summer Olympic Games in Athens, Greece.

| EXERCISE 22.1 | What Exactly Is DNA? |

You will very often read and hear the term *molecules of DNA*. Scientists use the phrase routinely, and "DNA molecules" have become such a hot news story that they are referred to regularly in general magazines and newspapers and on TV. But when we closely examine the structure of DNA as we will do in this chapter, we will see that the phrase *a molecule of DNA* is not strictly accurate. Why not?

Solution

Several factors introduce complications here. First of all, the negative charges on the phosphate groups mean that DNA, in the form in which it exists in living things, is a multi-charged ion, not a molecule at all. If hydrogen ions combined with all these negative charges, it would become a molecule, but this is not the structure we observe in living things. Second, when people speak of "DNA molecules" they are often referring to double-helical DNA, rather than to the single-stranded form. The DNA double helix is held together by largely noncovalent forces from hydrogen bonds, so it is really a complex formed from two intertwined molecules (or multi-charged ions). It cannot properly be described as a molecule at all.

| PRACTICE 22.1 | |

Suppose that a new base, shown in the margin, was discovered by scientists. To which of the four nitrogenous organic bases would this new base be most likely to form a stable base pair?

See Problems 3 and 4.

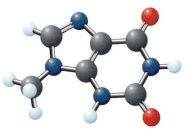

A new base for DNA.

DNA Replication—The Secret of Reproduction

One of the most important characteristics of living creatures (animal or plant) is that they make more of themselves. In other words, they "reproduce." Reproduction in this sense depends on the ability of double-helical DNA to copy itself or "replicate."

Because the bases form complementary pairs in a double strand of DNA, the sequence of nucleotide bases in one strand can be readily determined by examining the other strand of DNA. In other words, the A on one strand is paired with a T on the other, each T with an A, each G with a C, and each C with a G. This means that within a reproducing cell, the double helix can unravel and separate into individual strands. Then, *each strand can serve as the template on which a new complementary strand assembles,* as shown in Figure 22.9. The chemical reactions involved in this DNA **replication** are catalyzed by molecules within the cell known as enzymes. However, the structure of the existing DNA strands determines the sequence of the new DNA strands. Nucleotides carrying the appropriate bases are the only ones that can bind to the existing strand in a manner that allows the enzymes to link them up into a new DNA strand. The result of this process is two identical strands of DNA, one for the original cell and one for the new cell when it is formed.

| EXERCISE 22.2 | Using the Template |

If the DNA strand shown below serves as the template strand during DNA replication, what will be the sequence of the new double-helical DNA that results?

AATTGCGGGTCCGACC

Solution

Remember the rules of base-pairing. Only two types of base pairs are allowed: AT and GC (either way round), so the double-helical DNA will have the following sequence:

AATTGCGGGTCCGACC

TTAACGCCCAGGCTGG

PRACTICE 22.2

A small piece of one of the strands of DNA is added to a single strand of DNA. Indicate the correct alignment of the small piece with the long strand.

... AAAATGCTGGCATAGCGTTCCAGATACGGACTGACTGC...

CTATGC

See Problems 5 and 6.

DNA is located inside cells within a structure known as the nucleus. It is a beautiful chemical structure with the ability to act as a template on which new copies of itself are formed. **But what makes DNA so important to life?** The answer is that it carries coded information within its base sequence—information that can be decoded to specify which molecules are made in a cell. These molecules, known as **proteins,** then perform most of the chemical activities that actually sustain life.

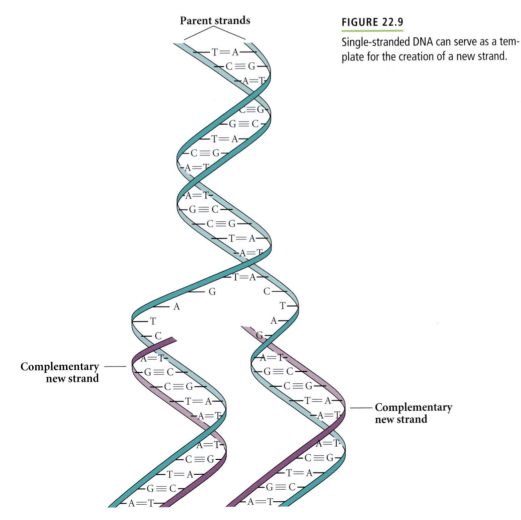

Parent strands

FIGURE 22.9

Single-stranded DNA can serve as a template for the creation of a new strand.

Complementary new strand

Complementary new strand

Video Lesson: Proteins

The amide bond

22.2 Proteins

Proteins are polymers that are formed when molecules called amino acids undergo condensation reactions. The amine on one amino acid forms an **amide bond** with the carboxylic acid on another amino acid. This process is repeated many times as the protein is formed. Some of the important amino acids found in living cells are shown in Figure 22.10. Note that all these amino acids contain very similar arrangements of the carboxylic acid and amine functional groups. The difference between them can be found in the group of atoms attached to the same carbon as the amine. The amino acids are often referred to as **residues** because they are what remains after the protein is broken up into its individual amino acids. Twenty different residues are common within your body, and each has been given a specific name.

FIGURE 22.10

Twenty common amino acids. The names of the amino acids are also shown with their three-letter abbreviation and one-letter code.

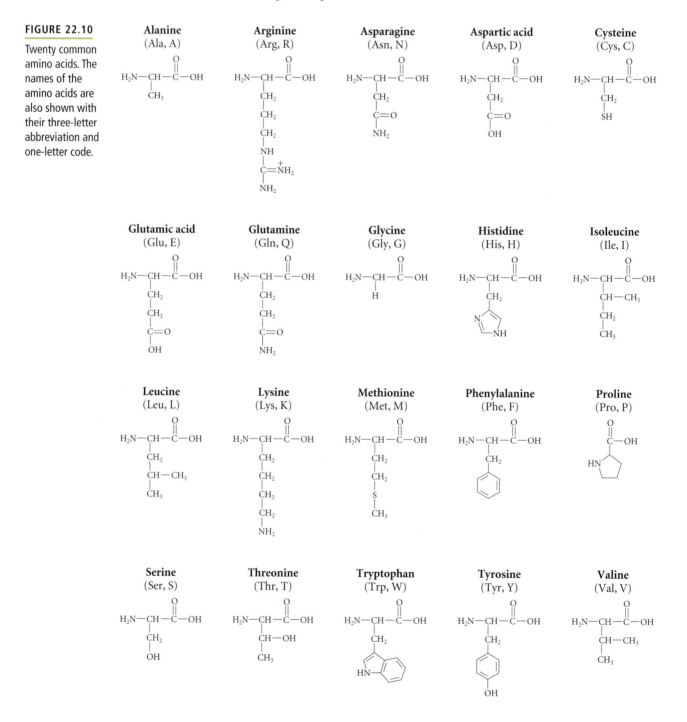

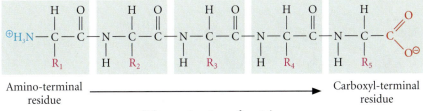

FIGURE 22.11

Primary structure of a protein. The primary structure is the order of the amino acid monomers in the strand of the protein.

Amino-terminal residue ⟶ Carboxyl-terminal residue

Primary structure of protein

Proteins are made in living cells by the sequential addition of amino acids to a lengthening **polypeptide** chain. This sequence of the protein constitutes the first level of structure of the protein. As indicated in Figure 22.11, it is often called the **primary structure** of the protein. Just like every individual person, each type of protein within a living cell is unique thanks to its amino acid sequence. The sequence of the amino acids determines the resulting shape of the protein *and* the resulting function of the protein.

We can now appreciate that both DNA and proteins are chemicals whose structure is determined by the *sequence* in which monomer units are bonded into polymers. Each DNA polymer has a *unique* base sequence (also know as its nucleotide sequence). Each protein has a unique amino acid sequence. The key to understanding how DNA can specify which proteins a living thing contains is to understand how *the base sequence of DNA can specify the amino acid sequence of a protein*. This is achieved by the operation of a simple chemical code.

22.3 How Genes Code for Proteins

A **gene** is a section of DNA that encodes a specific protein molecule. This means that the base sequence of the gene determines the amino acid sequence of the resulting protein. The whole process of converting the genetic code into a particular protein is known as *gene expression,* which occurs in two distinct phases: **transcription**, in which an RNA copy of the gene is made, and **translation**, in which the RNA copy directs production of the protein.

What is RNA? **Ribonucleic acid (RNA)** is a substance that bears a close resemblance to DNA. In fact, there are only two main differences between DNA and RNA (see Figure 22.12). The sugar in RNA is *ribose,* which carries one more oxygen atom than the deoxyribose found in

The basic components of the monomers that make up RNA.

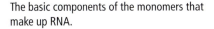

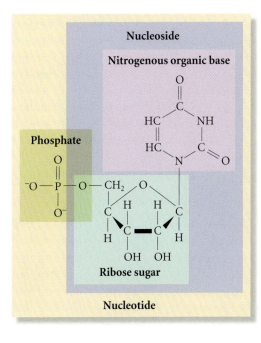

FIGURE 22.12

The difference between DNA and RNA lies in the ribose sugar.

DNA

RNA

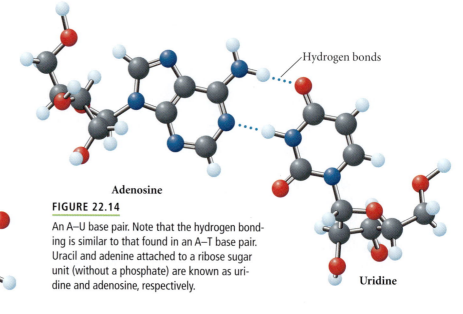

Adenosine

FIGURE 22.14

An A–U base pair. Note that the hydrogen bonding is similar to that found in an A–T base pair. Uracil and adenine attached to a ribose sugar unit (without a phosphate) are known as uridine and adenosine, respectively.

Hydrogen bonds

Uridine

FIGURE 22.13

The different monomer used in RNA, shown here without the attached phosphate group, is similar in structure to thymidine.

DNA. Also, in RNA the nitrogenous base *uracil*, shown in Figure 22.13, is found in place of the thymine of DNA. Uracil can participate in an A–U base pair as shown in Figure 22.14, in the same way that thymine can participate in an A–T base pair. The result is that RNA can form base pairs with a complementary strand of DNA.

Transcription—Making the Message

The process of gene transcription occurs in a very similar way to DNA replication. Inside the nucleus of the cell, the DNA unwinds into the individual strands of DNA, and a new, complementary strand of RNA is made. The complementary strand is manufactured using enzymes via condensation reactions in much the same manner as in the manufacture of DNA. The single-stranded RNA copy of a gene that is produced in transcription is known as **messenger RNA (mRNA)**. This copy carries the genetic "message" from the nucleus to the ribosomes (see below) in the cytoplasm, where the message is used (decoded) to make protein. mRNA gets its name from the fact that it takes the message held within the genes to the site of protein synthesis, as illustrated in Figure 22.15.

FIGURE 22.15

mRNA is made in the nucleus and then transferred to the cytoplasm for protein synthesis.

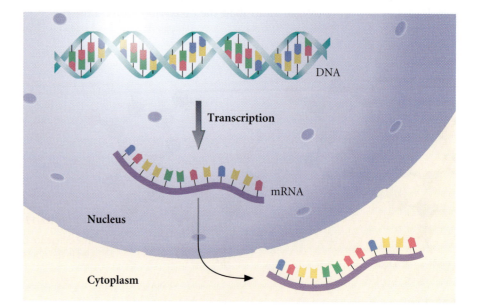

Translation—Making Proteins

Once at the site of protein synthesis, the mRNA is read by another class of RNA molecules known as **transfer RNA (tRNA)**. Each tRNA molecule includes a set of three nitrogenous bases that can form base pairs with three sequential bases of mRNA. Each set of three sequential bases within the mRNA is known as a **codon**. The codon binds to the complemetary **anticodon** on the tRNA. *What is the purpose of the tRNA?* Each tRNA also carries a specific amino acid at its amino acyl binding site. When the tRNA anticodon binds to the codon on the mRNA, the amino acid on that tRNA becomes incorporated into the growing protein chain. If the anticodon and codon are not complementary, the tRNA leaves without donating its amino acid to the protein. The code, indicating the relationship between each specific codon and amino acid is called the *genetic code* and is summarized in Table 22.1. The entire process of translation is outlined in Figure 22.16.

The mRNA contains a special "start" sequence that identifies the beginning of a protein. In fact, every protein that is made begins with the amino acid methionine. From that point onward, the amino acids are identified by each codon until the final "stop" codon is reached. Sometimes all of the mRNA is translated into a protein. Some genes make an mRNA that contains a lot of extraneous nucleotides.

"Stop" signal

Mature mRNA

Not all of the mRNA is used to create a protein.

TABLE 22.1	**The Genetic Code**

The three-letter codes listed are mRNA codons. For example, AAA is the mRNA codon for the amino acid lysine.

First letter		Second letter: U		Second letter: C		Second letter: A		Second letter: G		Third letter
U		UUU UUC	Phenyl-alanine	UCU UCC UCA UCG	Serine	UAU UAC	Tyrosine	UGU UGC	Cysteine	U C
		UUA UUG	Leucine			UAA UAG	Stop codon Stop codon	UGA UGG	Stop codon Tryptophan	A G
C		CUU CUC CUA CUG	Leucine	CCU CCC CCA CCG	Proline	CAU CAC	Histidine	CGU CGC CGA CGG	Arginine	U C A G
						CAA CAG	Glutamine			
A		AUU AUC AUA	Isoleucine	ACU ACC ACA ACG	Threonine	AAU AAC	Asparagine	AGU AGC	Serine	U C A G
		AUG	Methionine, initiation codon			AAA AAG	Lysine	AGA AGG	Arginine	
G		GUU GUC GUA GUG	Valine	GCU GCC GCA GCG	Alanine	GAU GAC	Aspartic acid	GGU GGC GGA GGG	Glycine	U C A G
						GAA GAG	Glutamic acid			

FIGURE 22.16

The overall process of translation.

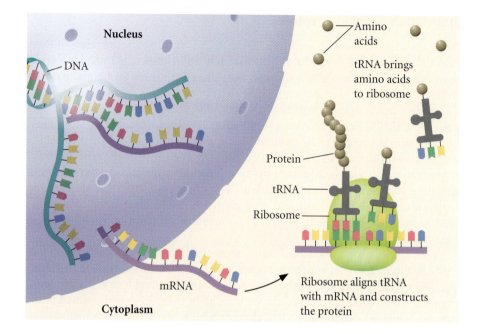

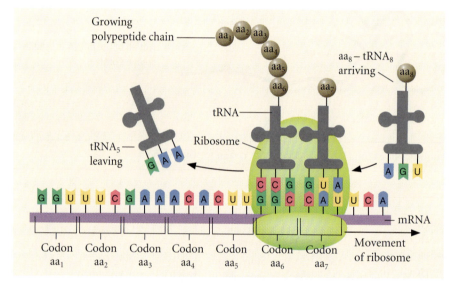

FIGURE 22.17

The ribosome slides along the mRNA. tRNAs then bring specific amino acids to the ribosome and allow the protein chain to be constructed. In this figure, the amino acids (aa) and the tRNAs are numbered to indicate their position.

To hold the molecules close together and to assist in catalyzing the process of adding amino acids into the protein, the process of translation takes place on giant assemblies of protein and RNA known as **ribosomes** (see Figure 22.17). As the ribosome moves along an mRNA molecule, the appropriate tRNAs are able to bind to special sites on the ribosome, allowing enzymes to link the amino acids they carry into a new protein chain. The end result is the translation of the base sequence of the gene (via the mRNA) into the amino acid sequence of the protein.

EXERCISE 22.3 **Examining the Genetic Code**

Examine the genetic code shown in Table 22.1. Do all codons encode amino acids? Does each codon specify a unique amino acid?

First Thoughts

Because there are four different bases and three positions, we can figure out the possible number of arrangements of these bases into the three positions.

Solution

There are 64 possible codons—different ways to arrange the four bases into sets of three. There are only about 20 amino acids found in proteins, so there must be more codons available than amino acids. Examining the genetic code reveals that each amino acid can be encoded by several alternative codons. It also reveals that three codons act as "stop" signals, indicating the point at which the synthesis of a protein chain should end.

Further Insights

Damage to DNA can result in changes to the genetic code. For example, a specific base within a codon could be changed as a result of the damage. Some of these changes have no effect on the protein encoded by the DNA, because they change a codon into one of the other codons specifying the same amino acid. Other such changes cause a different amino acid to appear in the encoded protein. This can cause problems with the resulting protein that is made using the mutant codon.

PRACTICE 22.3

What is the primary structure of a protein made using the following mRNA sequence?

AUGUGGCCAAAAUUGGACAUGUUCGACUAG

See Problems 15 and 16.

The human genome contains between about 20,000 and 30,000 genes. These genes are able to encode an even greater number of proteins, because the RNAs that are originally made from the genes can be edited to make many different proteins in a kind of enzymic "cut-and-paste" process. The way in which genes encode proteins is an astonishing demonstration of the power of chemistry to sustain the complex processes that underpin all life. To understand more fully just how powerfully genes influence the chemistry of life, we need to look at the proteins encoded by our genes and investigate the things these proteins can do.

Protein Folding

As soon as a protein chain begins to be formed, it starts to fold into a specific three-dimensional conformation. The folding process is governed principally by noncovalent interactions among the amino acids themselves, and also among the amino acids and the water molecules surrounding the protein. Localized regions within a polypeptide chain that fold in a particular way are examples of a protein's secondary structure. The most significant secondary structures are the *alpha helix* (α *helix*) and the *beta pleated sheet* (β *sheet*), shown in Figure 22.18. The α helix forms as a consequence of hydrogen bonds between the N—H and C=O groups of the polypeptide chain, holding the chain in the form of a helix. The β sheet is also held together by hydrogen bonds between N—H and C=O groups, but with the hydrogen bonding occurring between neighboring portions of a polypeptide chain.

Regions of specific secondary structure are linked by turns in the polypeptide chain, and by less ordered structures, to form the overall three-dimensional

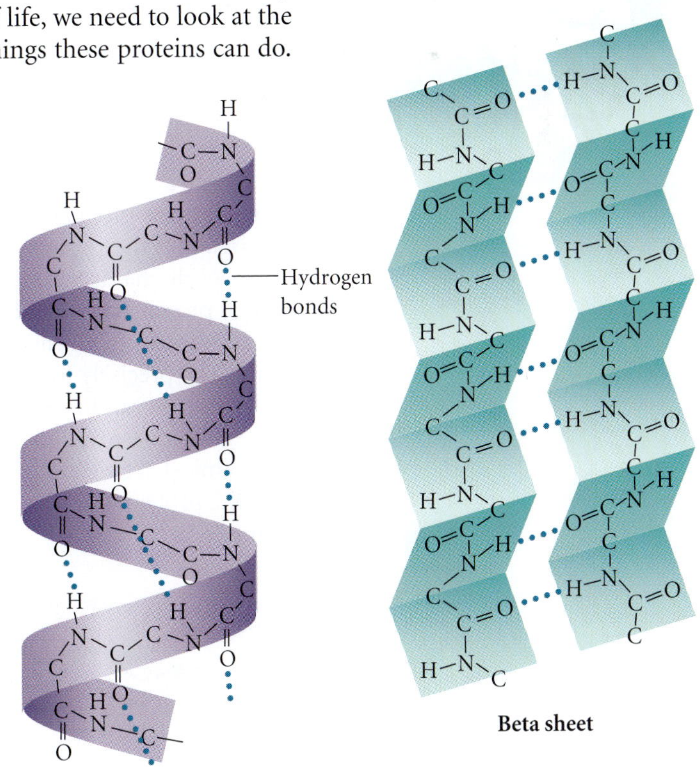

FIGURE 22.18

Two common types of secondary structures: the alpha helix and the beta sheet. These structures are held together by hydrogen bonding (shown as dotted lines).

Alpha helix

Beta sheet

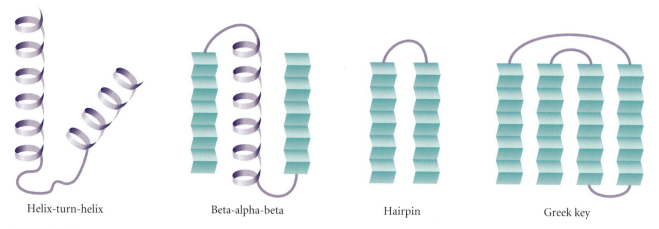

Helix-turn-helix   Beta-alpha-beta   Hairpin   Greek key

FIGURE 22.19

Some common tertiary structures. These structures are held in place by intermolecular forces of attraction such as London forces, dipole–dipole interactions, and hydrogen bonding.

FIGURE 22.20

Lactate dehydrogenase is a globular enzyme involved in the biochemical process of harvesting energy from sugars.

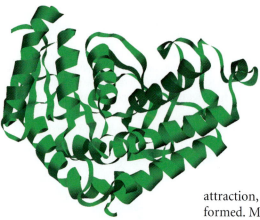

conformation of a protein. The folding of the secondary structures into a three-dimensional conformation is known as the protein's **tertiary structure**. Just as is true of α helices and β sheets, there are many common tertiary structures. Some of these are shown in Figure 22.19. However, many proteins have what can loosely be described as a "globular" tertiary structure; see the example shown in Figure 22.20. This is the general three-dimensional shape found for most of the proteins that act as enzymes. Other proteins have linear tertiary structures, largely composed of one or more extended helices or sheets. Many such proteins form strong fibers that contain several protein molecules intertwined. For example, the protein **collagen**, shown in Figure 22.21, gives strength to connective tissues such as tendons, ligaments, and bones. The structural strength of collagen is a result of three helical proteins wound around one another to form a triple helix.

When several polypeptide chains are held together by forces of attraction, as in collagen, we say that a protein with **quaternary structure** has been formed. Many globular proteins also possess quaternary structure. For example, hemoglobin, the protein that carries oxygen to your muscles, is made of four polypeptide chains that have folded around each other as shown in Figure 22.22. Not all proteins have quaternary structure. For example, myoglobin, the oxygen storage protein that resides in your muscles, consists of only a single polypeptide chain. The precisely folded structure of proteins can be disrupted, or **denatured**, by heat or by variations in the chemical surroundings, including changes in pH and in the concentration of salt ions. Denaturation reduces or destroys a protein's biochemical activity, depending on how severe it is. This explains, for example, why the protein albumin in an egg comes out of solution and turns white upon heating.

FIGURE 22.21

Collagen is a fibrous protein.

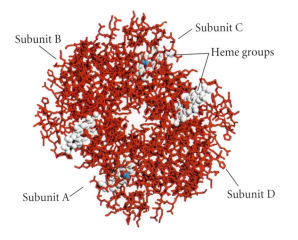

Myoglobin. The green ribbons represent the backbone of the chain of amino acids. Note the curled structures in the ribbons. They illustrate the alpha helix secondary structure in this protein.

FIGURE 22.22

Hemoglobin. There are four subunits, each with its own iron-containing heme, that come together to form hemoglobin. The subunits are held together to form the quaternary structure of the protein.

HERE'S WHAT WE KNOW SO FAR

- DNA is a polymer of nucleotides made from two complementary single strands wrapped around each other to form a double helix.
- Proteins are polymers of amino acids linked through an amide bond. They contain primary, secondary, tertiary, and sometimes quaternary structure.
- Proteins are made by translating the genetic code from mRNA. tRNA supplies the specific amino acids to the growing polypeptide chain. The construction occurs at the ribosomes within the cytoplasm of a cell.
- Primary structure is the sequence of amino acids that make up the protein.
- Secondary structures are the specific regions of a polypeptide chain that fold into an α helix or a β sheet.
- Tertiary structures result from folding of the secondary structures within a polypeptide chain into a three-dimensional shape. They can be globular or linear in arrangement.
- Quaternary structure results when two or more polypeptide chains are folded into a specific shape. Not all proteins have quaternary structure.

22.4 Enzymes

The covalent bond that links glucose and fructose together in a molecule of sucrose can be hydrolyzed by water, although the reaction is quite slow. In the laboratory, we can speed the reaction by adding a little acid to an aqueous solution of sucrose. The reaction can be monitored and the rate of the catalyzed reaction measured. **Why is this reaction important?** In the body, much of the "fuel" that we ingest is sucrose. This sugar is broken down into its components, glucose and fructose. As we noted in Chapter 14, both of the sugars are metabolized to provide energy to operate other biological processes and sustain life. Because these molecules are vital to our existence, any reaction that makes them, or metabolizes them, needs to be very rapid. Therefore, almost every chemical reaction within

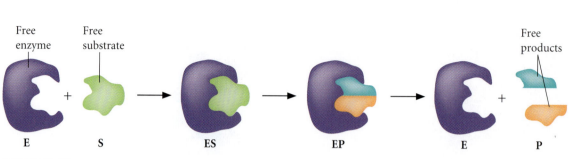

FIGURE 22.23

Substrates bind to the active site of an enzyme. The enzyme then catalyzes the reaction and releases the products. The enzyme is then primed for the next reaction.

our cells is quite fast; some are so fast that they must pause and wait for molecules to arrive at the reaction site. What type of compound makes these reactions fast? Specific types of proteins in the body, known as enzymes, catalyze these reactions.

A few key principles govern enzyme catalysis. This process is summarized in Figure 22.23.

- Enzymes contain specific sites to which certain chemicals can bind.

- The structure of the sites enhances the binding of specific molecules and ions through the same type of intermolecular forces of attraction that we discussed in Chapter 10.

- One of the binding sites on an enzyme includes the **active site** where the **substrate** binds.

- The enzyme next catalyzes the reaction of the substrate and produces the product, which is then released from the enzyme.

Cofactors, Coenzymes, and Protein Modifications

Regulation and control of the activity of an enzyme is often necessary. Imagine, for instance, what would happen if all of the sucrose you just consumed at lunch were immediately converted into energy. That energy would be used up quite quickly in the biological processes that keep you alive. After a very short time, your body would be out of energy, and the basic reactions that keep you functioning would stop. This suggests that the reactions within a cell must be regulated—turned on and off—so that they are performed only when needed. This regulation often involves chemical *modification* of the enzyme. The major modifications are as follows:

- Incorporation of specific metal ions at appropriate sites, to form *metalloproteins,* in which metal ions are tightly bound within the protein structure, or *metal-activated proteins,* which are activated by loosely bound metal ions picked up from the surrounding solution

- Incorporation of loosely bound small chemical groups called *coenzymes* at specific sites within the protein

- Incorporation of tightly bound *prosthetic groups,* often held to the protein by covalent bonds

- Covalent linking of phosphates to specific sites to form *phosphoproteins*

- Covalent linking of carbohydrates at specific sites to form *glycoproteins*

- Covalent bonding of lipids at specific sites to form *lipoproteins*

The need for many enzymes and other proteins to be modified in these ways explains many of our nutritional needs for specific minerals and vitamins. For instance, much of the iron that we must consume in our diet is needed to provide

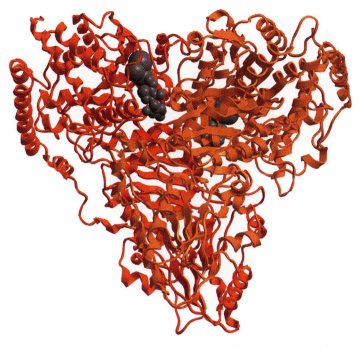

Vitamin B1 (thiamine) binds within the structure of yeast transketolase (a transferase enzyme). The thiamine is shown as a spacefilling model.

iron ions for incorporation into metalloproteins such as hemoglobin. Many of the vitamins we must consume in tiny amounts are needed to form the coenzymes that combine with specific enzymes and allow them to function.

The General Classes of Enzymatic Reactions

Enzymes catalyze many thousands of different chemical reactions. Fortunately, these reactions can be classified into just six categories on the basis of the general type of reaction that is being catalyzed.

- *Oxidoreductases* are enzymes that catalyze redox reactions.
- *Transferases* are enzymes that catalyze the transfer of groups from one molecule to another.
- *Hydrolases* are enzymes that catalyze hydrolysis reactions.
- *Lyases* are enzymes that catalyze elimination reactions that remove hydrogen atoms or functional groups and form alkenes ($C=C$) in the substrate.
- *Isomerases* are enzymes that catalyze the interconversion between isomers.
- *Ligases* are enzymes that catalyze the formation of new bonds linking substrates together.

EXERCISE 22.4 | **What Is It Doing?**

Examine the following reaction and classify the enzyme's activity into one of the six types of reactions common among enzymes.

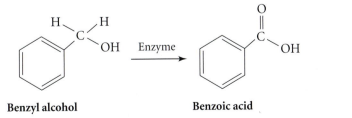

Benzyl alcohol   Benzoic acid

Solution continues on page 952.

NanoWorld / MacroWorld

Big effects of the very small: Vitamins and disease

Vitamins are small molecules that we need for proper health; most either are not normally synthesized in our bodies or are made in insufficient amounts. We humans obtain most of our needed vitamins from food. The exception to this is that our bodies can make vitamin D and vitamin K. In all, we need 13 vitamins to sustain our life: vitamins A, C, D, E, and K and the B vitamins (thiamine, riboflavin, niacin, pantothenic acid and biotin, vitamin B6, vitamin B12, and folate). Each of these vitamins plays an interesting role in the biochemistry of life. Details of these vitamins are shown in Table 22.2.

Vitamin D

Vitamin D (calciferol) promotes retention and absorption of calcium and phosphorus, primarily in the bones. Too much vitamin D in the body may have the opposite effect of taking calcium from the bones and depositing it in the heart or lungs, making them function less efficiently. Because vitamin D is essential for the body's utilization of calcium, a deficiency may result in severe loss of calcium and, consequently, a softening and weakening of bones (osteomalacia). Extreme vitamin deficiency gives rise to a disease known as rickets.

Like most vitamins, vitamin D may be obtained in the recommended amount with a well-balanced diet, including some enriched or fortified foods such as milk. In addition, the liver and kidneys manufacture vitamin D when the skin is exposed to sunshine. Deficiencies in this vitamin arise primarily from insufficient exposure to sunlight. That is why it is recommended that we get at least 10 to 15 minutes of sunshine three times a week.

Vitamin A

Vitamin A (retinol) is supplied by many foods of both animal and plant origin. Vegetables sources, such as carrots, pumpkin, and brocolli, actually contain a precursor of retinol called beta-carotene, which is

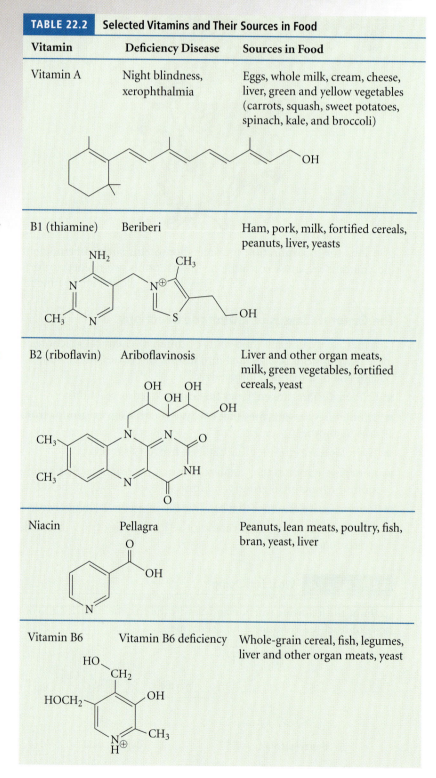

TABLE 22.2	Selected Vitamins and Their Sources in Food	
Vitamin	**Deficiency Disease**	**Sources in Food**
Vitamin A	Night blindness, xerophthalmia	Eggs, whole milk, cream, cheese, liver, green and yellow vegetables (carrots, squash, sweet potatoes, spinach, kale, and broccoli)
B1 (thiamine)	Beriberi	Ham, pork, milk, fortified cereals, peanuts, liver, yeasts
B2 (riboflavin)	Ariboflavinosis	Liver and other organ meats, milk, green vegetables, fortified cereals, yeast
Niacin	Pellagra	Peanuts, lean meats, poultry, fish, bran, yeast, liver
Vitamin B6	Vitamin B6 deficiency	Whole-grain cereal, fish, legumes, liver and other organ meats, yeast

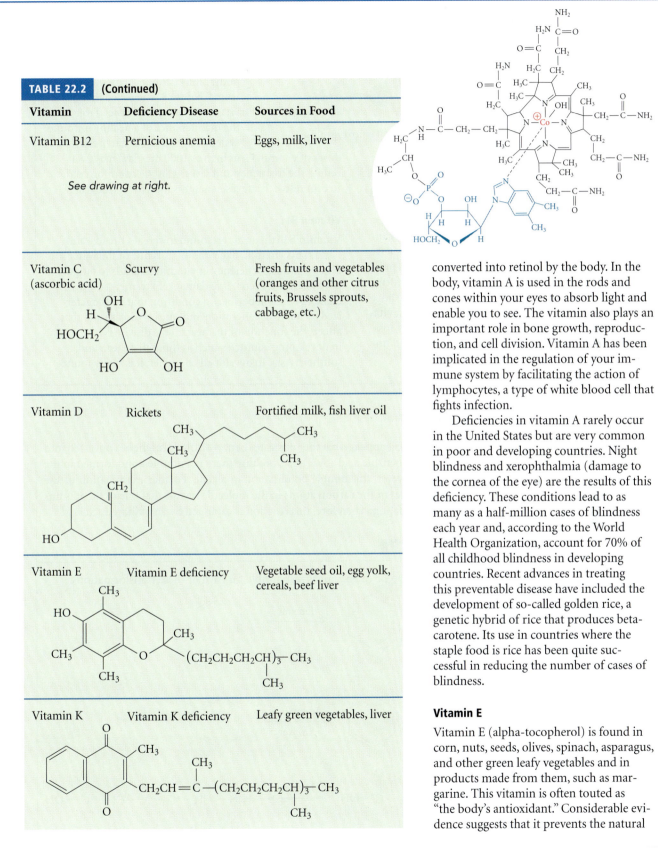

TABLE 22.2	(Continued)	
Vitamin	**Deficiency Disease**	**Sources in Food**
Vitamin B12	Pernicious anemia	Eggs, milk, liver
See drawing at right.		
Vitamin C (ascorbic acid)	Scurvy	Fresh fruits and vegetables (oranges and other citrus fruits, Brussels sprouts, cabbage, etc.)
Vitamin D	Rickets	Fortified milk, fish liver oil
Vitamin E	Vitamin E deficiency	Vegetable seed oil, egg yolk, cereals, beef liver
Vitamin K	Vitamin K deficiency	Leafy green vegetables, liver

converted into retinol by the body. In the body, vitamin A is used in the rods and cones within your eyes to absorb light and enable you to see. The vitamin also plays an important role in bone growth, reproduction, and cell division. Vitamin A has been implicated in the regulation of your immune system by facilitating the action of lymphocytes, a type of white blood cell that fights infection.

Deficiencies in vitamin A rarely occur in the United States but are very common in poor and developing countries. Night blindness and xerophthalmia (damage to the cornea of the eye) are the results of this deficiency. These conditions lead to as many as a half-million cases of blindness each year and, according to the World Health Organization, account for 70% of all childhood blindness in developing countries. Recent advances in treating this preventable disease have included the development of so-called golden rice, a genetic hybrid of rice that produces beta-carotene. Its use in countries where the staple food is rice has been quite successful in reducing the number of cases of blindness.

Vitamin E

Vitamin E (alpha-tocopherol) is found in corn, nuts, seeds, olives, spinach, asparagus, and other green leafy vegetables and in products made from them, such as margarine. This vitamin is often touted as "the body's antioxidant." Considerable evidence suggests that it prevents the natural

continued

NanoWorld/MacroWorld (*continued*)

oxidation of lipoproteins, which play an important role in the development of atherosclerosis, the disease process that leads to heart attacks and strokes.

4-Hydroxyproline

Vitamin C

Vitamin C (ascorbic acid) has many functions in the body, including assisting in the absorption of iron. This vitamin is found in citrus fruits, such as lemons and limes, and in most other vegetables. It is a component of enzymes involved in the synthesis of collagen. Without vitamin C, the enzymatic preparation of collagen still proceeds, but the resulting collagen protein doesn't include the modified amino acid known as hydroxyproline. Collagen with hydroxyproline forms strong intermolecular forces of attraction between strands of the collagen protein. Without hydroxyproline, the collagen cannot form these strong attractive forces. The result is nothing short of the breakdown of the human body. Deficiency in vitamin C, known as scurvy, is evidenced by sore joints, loss of teeth, and aching muscles. Severe deficiencies and untreated deficiencies result in respiratory failure.

Abnormally large doses of vitamin C appear to be harmless at the least, and they may be helpful. Because vitamin C is water soluble, excess vitamin C can be easily excreted from the body. Linus Pauling (remember him from Chapter 8) suggested that vitamin C may play a crucial role in the maintenance of a healthy immune system. He proposed that massive doses of vitamin C (as high as 1000 mg/day) may even reduce the frequency and severity of the common cold. Recent research suggests that this may, in fact, be true.

Vitamin B1

Vitamin B1 (thiamine) is found in cereals, breads, and pasta. This vitamin, like vitamin C, is one of the water soluble vitamins. Deficiency in vitamin B1 leads to fatigue, psychosis, and even nerve damage. Extremely severe deficiencies result in beriberi, a condition that is characterized by nerve degeneration and muscle disease and that affects the function of the heart. This disease is prevalent in developing countries such as those in eastern and southern Asia. In developed countries, it is rarely found.

Solution

At first glance, it appears that the substrate (benzyl alcohol) might be converted into the product (benzoic acid) by a lyase, because there is a double bond formed in the product. However, this double bond is not an alkene. Further examination of the oxidation states of the carbon atoms in the molecule reveals that one of them is undergoing oxidation. Therefore, this reaction is catalyzed by an oxidoreductase.

PRACTICE 22.4

Indicate what type of enzyme is responsible for each of these transformations.

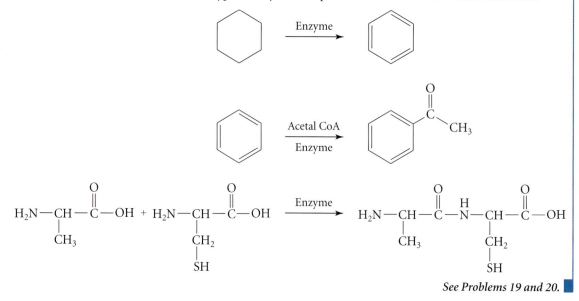

See Problems 19 and 20.

22.5 The Diversity of Protein Functions

One way of summarizing the relationship between genes and proteins—the two central categories of chemicals that sustain life—is to say that "genes hold the instructions while proteins do the work." In a human, our 20,000 to 30,000 genes encode the structure of proteins, and each protein has its own specialized task, its own little bit of "chemical work" to do. There is often simplicity at the heart of complexity, however, and proteins are no exception. Almost everything they achieve can be explained in terms of three fundamental activities. Selective binding, catalysis, and conformational change are the three keys to understanding the activities of proteins. Here is a brief list of the main tasks these three operational principles enable proteins to achieve.

Proteins as Enzymes

Some proteins contain active sites that bind substrates and catalyze reactions to make products, the main subject of Section 22.4.

Proteins as Transporters

Many proteins bind to specific chemicals at particular sites and then release their cargo at other sites. The protein hemoglobin, for example, binds to oxygen molecules in the lungs, transports the oxygen through the blood, and then releases it in the tissues of the body that need oxygen to survive.

Proteins as "Movers and Shakers"

What in your body makes you move? You move because you have muscles. Muscles are composed of filamentous *contractile proteins*. These are proteins that respond to stimuli from nerves to undergo conformational changes that result in their being ratcheted past one another and causing the whole structure of a muscle to contract. Reversal of the process enables the muscles to relax.

Proteins as Scaffolding and Structure

Much of the shape and structure of cells, tissues, organs, and the body as a whole is maintained by strong fibrous *structural proteins* that form fibers and pillars and sheets of material. Hair, skin, connective tissue, and an intricate skeleton of tubules within all cells are composed of structural proteins.

Proteins as Messengers

Proteins that are made at one place in the body can be transported through the blood to other parts of the body, where they bind to other chemicals to initiate specific biochemical responses. Some such proteins are known as hormones. The protein insulin, for example, is a hormone that is released from the pancreas and then binds to cells and assists in the uptake of glucose into these cells. A deficiency in this hormone or in its production causes diabetes.

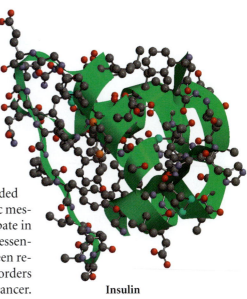

Insulin

Proteins as Receptors

The outermost portion of the cell, known as the cell membrane, is studded with proteins that act as receptor molecules. These receptors bind to specific messenger molecules, some of which are proteins like insulin, and then participate in acts of catalysis or conformational change that initiate the effect that the messenger elicits. Protein receptors also exist inside of cells. The interactions between receptors and messengers are vital in controlling cell growth. This makes disorders in these systems very significant in the onset of serious diseases such as cancer.

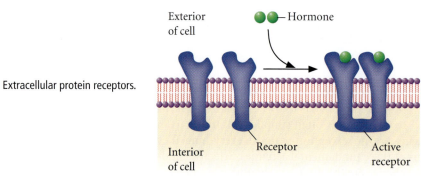

Extracellular protein receptors.

Proteins as Gates and Pumps

Cell membranes act as very selective barriers, through which only specific chemicals can pass, and only at appropriate rates and at appropriate times. Much of this selective movement of chemicals into and out of cells is maintained by proteins spanning the cell membrane. Some of these proteins can simply open or close a hole in the membrane, allowing selected chemicals to pass. Others act more like pumps; that is, they transport a chemical across a membrane in such a way that a significant difference can build up between the concentration of the chemical on one side of the membrane and its concentration on the other side.

Proteins as Controllers

A whole battery of proteins known as *regulatory proteins* serve the function of controlling other proteins, and controlling genes, by binding to them and switching them on or off. In most cells of the body, for example, only a small proportion of the genes in the cell nucleus must be active at any one time. Switching the right genes on and off at the correct times is largely achieved by the activities of regulatory proteins.

Proteins as Defenders

The immune system, which defends us against disease, is a network of interacting cells and chemicals, and many of its most significant chemicals are proteins. Most famous of all are the **antibodies**—proteins that selectively bind to foreign chemicals and assist in their elimination from the body. Allergies such as hay fever result when this system goes wrong by eliciting too strong a reaction against relatively harmless foreign chemicals. When it goes wrong by attacking the body's own tissues, serious autoimmune diseases such as rheumatoid arthritis can result.

Glorious Complexity

The actions of genes and proteins are part of a very complex picture of the biochemistry of life. All of that complexity is, at heart, simply chemistry, driven by the interactions of energy, charge, and force. However, genes and proteins aren't the only players. There are many other compounds involved in the smooth operation of a living being. Carbohydrates and lipids are two important classes of the compounds of life.

Video Lesson: Carbohydrates

22.6 Carbohydrates

Most people love carbohydrates. Runners eat copious amounts before a very long race. Snackers munch on carbohydrates in front of the TV. Toddlers often will drop everything else to get a small taste of a **carbohydrate**. **What are carbohydrates and why do our bodies require them?** This class of compounds shares the general

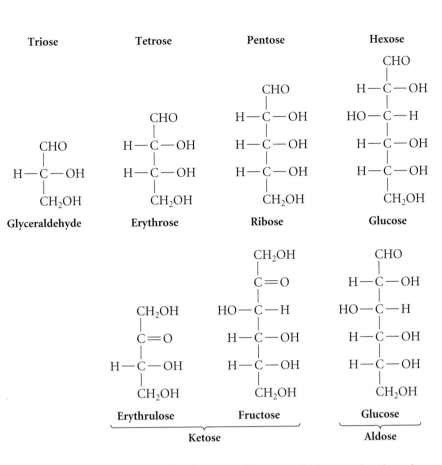

Classification of saccharides by number of carbon atoms.

Classification of saccharides by type of functional group.

formula $C_x(H_2O)_y$—historically, they were known as hydrates of carbon, hence their name. The simplest carbohydrates are **sugars**, which are known as *triose* sugars if they contain three carbon atoms per molecule, *tetrose* sugars with four carbons, *pentose* sugars with five carbon atoms per molecule, *hexose* sugars with six carbon atoms per molecule, and so on. In addition, the carbohydrates can be classified on the basis of the type of functional group (see Chapter 12). For example, glucose is an *aldose* because it contains an aldehyde functional group. Fructose is a *ketose* because it contains a ketone functional group. The glucose used in medical drip-feeding is an aldohexose (from aldose and hexose) and is the main form in which carbohydrate circulates in our blood.

The familiar table sugar we may add to coffee is sucrose and is formed when two simpler hexose sugars, glucose and fructose, become linked together in a condensation reaction, as shown in Figure 22.24. You'll immediately notice a difference between the sugars in Figure 22.24 and those from our previous discussion.

FIGURE 22.24

The condensation of glucose with fructose gives sucrose. Monosaccharides include glucose and fructose. Sucrose is a disaccharide.

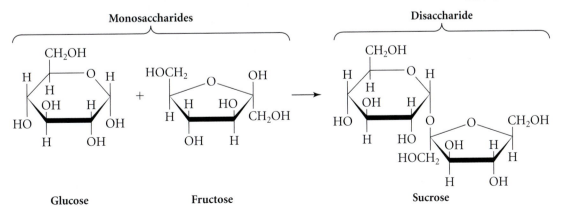

Why? Most carbohydrates undergo a condensation reaction within their own structures when they are placed in aqueous solution. The result is a group of cyclic structures that are more stable than the straight-chain carbohydrates.

Sugars such as glucose and fructose are also known as **monosaccharides** (for "single sugars"), whereas those like sucrose, made when two monosaccharides combine, are **disaccharides**. Note that two cyclic sugars, one glucose and one fructose, have come together to form the disaccharide, as shown in Figure 22.24.

One major role for carbohydrates in living things is to serve as energy storage compounds. For long-term storage, monosaccharides are linked up into giant polymers known as **polysaccharides**. The major storage polysaccharide in plants is *starch*, composed of many glucose monomers linked in the manner shown in Figure 22.25. Foods such as rice or potatoes supply us with energy largely thanks to the deposits of starch within them, whose role in the plant is to provide the energy needed to sustain its growth. Animals, including humans, contain a slightly different polymer of glucose called *glycogen*, in which the glucose monomers are linked together in a slightly different fashion, as shown in Figure 22.26. Another important polysaccharide is *cellulose*, which is a polymer of glucose that forms much of the framework of plant cell walls. The glucose monomers are linked together in a completely different manner, as shown in Figure 22.27. Humans lack the enzyme capable of hydrolyzing cellulose into glucose, so cellulose contains no nutritional value to us.

FIGURE 22.25

Starch is a polymer of glucose units linked together as shown.

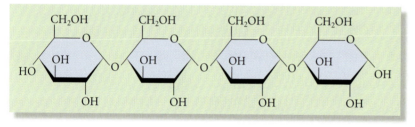

FIGURE 22.26

Glycogen is a polysaccharide.

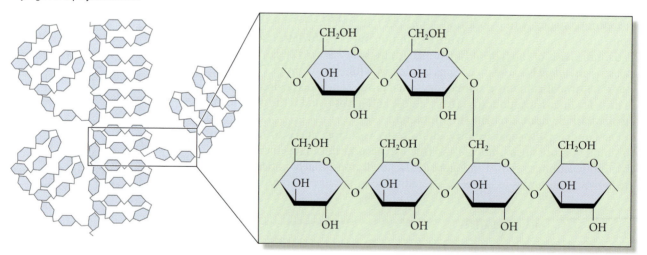

FIGURE 22.27

Cellulose is a polymer of glucose units linked together as shown. What is the difference between cellulose and starch (Figure 22.25)? Note the position of the OH on the glucose monomer at the right. It is pointed in a direction different from the direction it would point in starch.

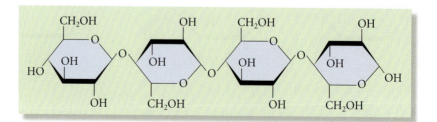

22.7 Lipids

Video Lesson: Lipids

Lipids are a very broad class of compounds that differ from most other classes of compounds. For example, the molecules known as carbohydrates exhibit similarity in structure and function. Lipids do not have similarity in structure or function. A compound is characterized as a **lipid** *solely on the basis of its solubility in nonpolar solvents.* Lipids include fats, oils, some hormones, and some vitamins. Lipids function as important energy storage molecules, act as intercellular signaling molecules, and also play many structural roles in living things. For example, the thin membranes around every living cell are composed of lipids.

Some of the simplest lipids are the **fatty acids**, which are long-chain carboxylic acids that may contain one or more C=C double bonds. Table 22.3 lists some of the common fatty acids. When three fatty acids combine with the tri-alcohol glycerol in an esterification reaction, a **triglyceride (triacylglycerol)** is formed. Figure 22.28 shows an example of this reaction. Triglycerides that are solid at room temperature are called **fats**, whereas those that are liquid at room temperature are **oils**. We explored this distinction in Chapter 10 when examining the effects of C=C double bonds on the melting points of fats and oils.

Triglycerides can be modified in various ways to form compound lipids, such as the **phospholipids** that are important components of cell membranes (see Figure 22.29). Living things also contain a great many derived lipids, which are fatty substances that can be made from simpler lipids but have their own unique structures. One of the major classes of derived lipids consists of the **steroids**, which share a characteristic four-ring structure and include cholesterol and various

Application

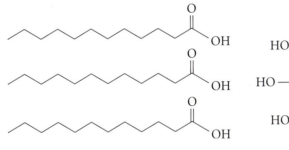

FIGURE 22.28

Triacylglycerides result from the condensation of three fatty acids and glycerol.

FIGURE 22.29

Sphingomyelin is one of the components of cellular membranes.

TABLE 22.3	Some Common Fatty Acids	
Common Name	**Formula**	**Melting Point**
Lauric acid	$CH_3(CH_2)_{10}COOH$	46°C
Myristic acid	$CH_3(CH_2)_{12}COOH$	55°C
Palmitic acid	$CH_3(CH_2)_{14}COOH$	64°C
Stearic acid	$CH_3(CH_2)_{16}COOH$	69°C
Oleic acid	$CH_3(CH_2)_7CH=CH(CH_2)_7COOH$	13°C
Linoleic acid	$CH_3(CH_2)_4CH=CHCH_2CH=CH(CH_2)_7COOH$	−5°C

Cholesterol

steroid hormones such as estrogen. Cholesterol has a bad reputation because of the link between excessive blood cholesterol levels and cardiovascular disease, but it is an essential component of cell membranes. Like almost every chemical associated with life, cholesterol is "good" in the right amounts and right places but "bad" in the wrong amounts and wrong places.

22.8 The Maelstrom of Metabolism

Chemists have a rather simple chart to look at on their walls; we call it the periodic table. Biochemists have a much more complex chart to admire. The biochemical pathways chart shown in Figure 22.30 is so complex that, even devoting a full book page to it, we struggle to make out the details. The most commonly used version is *more than 1 square meter in size*, and even at that scale it still looks dense and complex. The biochemical pathways chart summarizes the major sequences of chemical reactions of life, each particular sequence being called a **biochemical pathway**. The individual pathways are interconnected at various points, offering yet another example of glorious complexity. The chemical reactions enable the atoms needed for life to flow through all the different molecules and ions found in living things. The entire network of chemical reactions involved in life is called **metabolism**, and the individual chemicals are called **metabolites**.

Each arrow on the biochemical pathways chart represents a chemical reaction catalyzed by a specific enzyme. Looking at the chart, we can really begin to appreciate the true complexity of life, the reasons why we need around 20,000 to 30,000 genes (always remembering that many genes code for proteins other than enzymes), and the breathtaking achievement of the chemistry within each living thing. We can also appreciate the other breathtaking achievement: that in a little over 50 years, scientists have been able to sort this all out!

Breaking Things Down and Building Them Up Again

The food we eat is a mixture of water, carbohydrate, protein, lipids, vitamins, and minerals. It provides us with chemical raw materials needed to build the chemicals of our bodies, and it also acts a source of energy. To release the energy, and also to use much of our food as raw materials, our bodies must break the chemicals in the food down into simpler forms. This aspect of metabolism is called **catabolism**. Building the raw materials back up into our own carbohydrates, proteins, lipids, and other essential chemicals is the aspect of metabolism called **anabolism**.

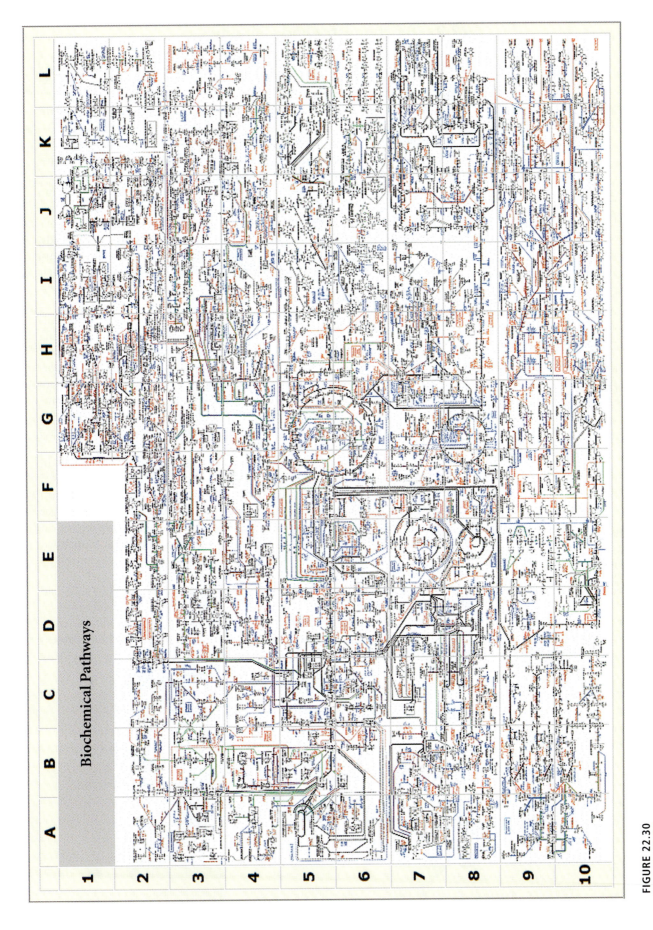

FIGURE 22.30
The biochemical pathways chart.

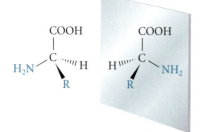

FIGURE 22.31

Mirror images of an amino acid.

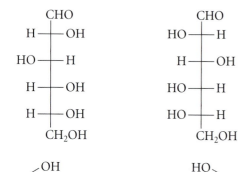

D-**Glucose**   L-**Glucose**

The two isomers of glucose (D-glucose and L-glucose). D-Glucose is the enantiomer that our bodies use for fuel.

Application

22.9 Biochemistry and Chirality

There is a crucial aspect of the chemistry of life that we have not even mentioned yet, and it is also an important aspect of the structure of many chemicals not necessarily involved in living systems. Many of the molecules in our bodies are chiral, a term we introduced in Chapter 12 (Section 12.14). For instance, the mirror image forms of the general structure of an amino acid are shown in Figure 22.31. These forms will be nonsuperimposable, provided that the R group is not a hydrogen atom. All amino acids except glycine are chiral, and the nonsuperimposable forms of molecules such as this are known as enantiomers.

Chirality is found in the body in a wide range of molecules, including amino acids, proteins, carbohydrates, and even DNA and RNA. The amino acids in proteins have, almost exclusively, the same handedness. That is, our bodies have evolved to utilize only one of the two enantiomers possible in the amino acids. The same is true for most of the carbohydrates in living things. The carbohydrates of life are similarly almost exclusive in their same handedness. Why is this so? This is another of the wonderful mysteries of life.

Why is chirality in chemicals important? Because it controls the way in which molecules interact. For example, the binding sites of enzymes are constructed from the chiral amino acids within protein molecules. *Therefore, enzymes themselves are chiral.* They will interact differently with the different enantiomers of the substrates to which they bind. In many cases, one enantiomer of the substrate will be able to bind to the protein, whereas the other will not fit into the active site. These distinctions can be highly significant.

From 1958 to 1962, large numbers of pregnant women in many countries began to take a drug called thalidomide, shown in Figure 22.32. This drug was prescribed in an effort to control the "morning sickness" that often accompanies the early stages of pregnancy. The large batches of thalidomide prepared on an industrial scale for distribution around the world contained both enantiomers. One of these was very effective in controlling morning sickness. Unfortunately, the other enantiomer caused a range of profound abnormalities in the developing fetus. Babies born to mothers who had taken thalidomide were badly deformed. These dangers became evident before thalidomide received approval for use in the United States, which was therefore fortunate enough never to experience the tragic wave of "thalidomide babies" that occurred in other countries.

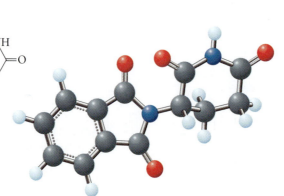

FIGURE 22.32

Thalidomide is a chiral molecule.

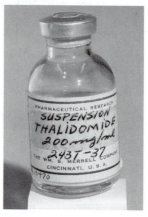

Thalidomide bottle from 1961.

Thalidomide has since been found to be useful in treating some aspects of various diseases, including leprosy, rheumatoid arthritis, AIDS, and some cancers, especially those related to blood marrow. It remains in use in carefully controlled circumstances, but never for the treatment of women who are, or ever may become, pregnant. Recently, a compound with a structure that is very similar to that of thalidomide has been identified as a potential treatment for myelodysplastic syndrome, a malignant disorder that affects blood cell production. The company manufacturing the drug is seeking approval from the U.S. Food and Drug Administration to market the compound under the name Revlimid. Initial studies have indicated that Revlimid has fewer of the harmful side effects caused by thalidomide, and its use could be beneficial to patients afflicted with the malignant blood disorder.

Many scientists were perplexed at why the safe enantiomer of thalidomide was not prepared and used in its pure form. Preparations of the safe enantiomer can indeed be made, but over time, the pure enantiomers spontaneously convert into a mixture of the two enantiomers, until an interconverting 50/50 mixture of both forms, known as a **racemic mixture**, is obtained. Unfortunately, the process of **racemization** occurs with many chiral molecules.

22.10   A Look to the Future

Talking about thalidomide reminds us that one major reason behind our interest in the chemistry of life is the desire to prevent and treat the diseases and discomforts that can afflict us. Why do we get ill and what can chemistry do about it? The major causes of illness and disease are

- *Infection*—occurs when our bodies become home to harmful microorganisms
- *Abnormal growth of tissues*—such as cancerous tumors and less damaging "benign" tumors
- *Abnormal production of important biochemicals*—such as hormones and the neurotransmitters that make it possible for signals to pass between nerves
- *Genetic diseases*—associated with one or more abnormal genes or larger abnormalities in genetic material
- *Aging*—the degeneration of maintenance and repair functions within the body

Application

Every doctor's shelf carries a large volume called a pharmacopeia (see Section 18.1), a book that lists the drugs available to fight all manner of diseases and includes instructions on dosage, uses, and possible adverse reactions. However, many of these drugs have limited use, exhibit lessening effectiveness, or are simply ineffective at treating or curing diseases and maladies that affect humans. The future of useful therapeutic agents relies on the preparation and implementation of alternatives to these exisiting drugs. Some of the important current advances in this area are noted here.

Antibiotics

Most of us are treated with **antibiotics** several times in our lives. These are chemicals that can inhibit the growth of microorganisms or even destroy them. The first antibiotic that was isolated and used on patients was penicillin (Figure 22.33). This compound inhibits the synthesis of bacterial cell walls by selectively binding to enzymes needed for such synthesis. Nowadays we have a vast armory of antibiotics that work in different ways, but very often by binding to and inhibiting specific enzymes. For example, the tetracycline antibiotics are all

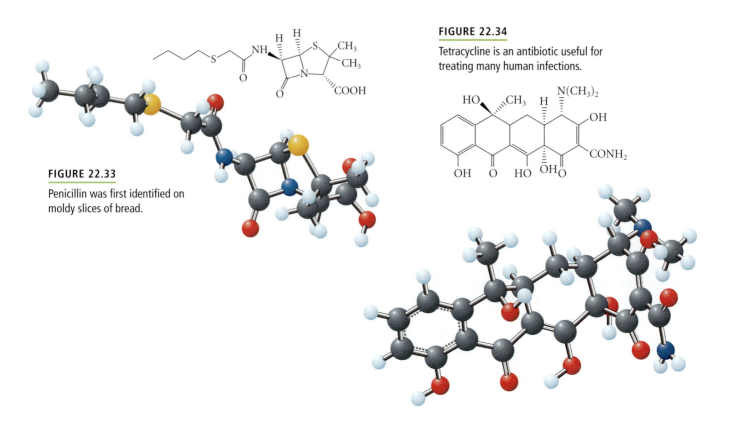

FIGURE 22.33

Penicillin was first identified on moldy slices of bread.

FIGURE 22.34

Tetracycline is an antibiotic useful for treating many human infections.

derived from the basic structure of tetracycline (Figure 22.34). They work by binding to and inhibiting enzymes that catalyze protein synthesis in bacteria.

Anticancer Agents

Application

One major strategy for treating cancer is to prevent the replication of DNA within cancerous cells. If DNA replication is prevented, the cells cannot multiply by cell division. Various drugs have been developed that achieve this effect by forming crosslinks between the two strands of DNA or by adding bulky groups to one strand, thus preventing the double helix from unwinding in the manner required for replication. One of the oldest anticancer drugs is mechlorethamine, shown in Figure 22.35, which forms crosslinks between the DNA strands. A more recent anticancer agent called m-amsacrine inserts into the DNA helix and causes the cell's natural maintenance enzymes to break the DNA strand into little pieces. m-Amsacrine is very effective in treating some forms of childhood leukemia.

FIGURE 22.35

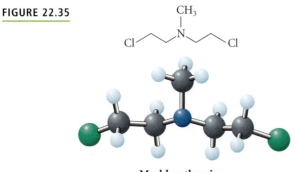

Mechlorethamine
(also known as nitrogen mustard)

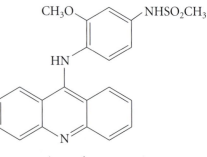

m-Amsacrine
(m-AMSA)

Hormones and Their Mimics

Insulin, injected under the skin, can be used to treat diabetic patients. Many other nonprotein hormones are also used to treat disease. One major category of such hormones consists of the steroid hormones, such as estrogen and testosterone. Natural and synthetic steroids are used to treat such conditions as cancer, rheumatoid arthritis, allergies, and a wide range of hormone-deficiency diseases. Synthetic steroids are also the active ingredients in contraceptive pills and in the hormone replacement therapy sometimes used to combat the effects of menopause in women.

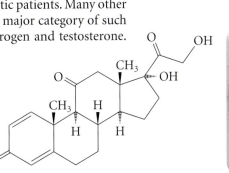

Prednisone is a synthetic hormone similar to cortisone that is used to treat Crohn's disease and a wide variety of other ailments. The basic structure of this compound reveals that it is a steroid.

Neurotransmitters

The nervous system is composed of billions of nerve cells that communicate with one another via the release of small organic compounds known as neurotransmitters. These chemicals bind to specific receptors on the surface of neighboring nerve cells and either activate or inhibit the transmission of nerve impulses. Many degenerative and psychiatric conditions are associated with abnormalities in neurotransmitter function. The pronounced tremors of Parkinson's disease, for example, are due to the degeneration of nerve cells that use a molecule known as dopamine as their neurotransmitter. The condition can be dramatically alleviated by administering a precursor of dopamine called L-Dopa, which is converted to dopamine by enzymes in the body.

One of the most significant neurotransmitters is serotonin (5-hydroxytryptamine). A large new range of drugs that appeared in the 1990s are useful because of their ability to modulate the effects of this neurotransmitter in various cells and tissues.

Another set of significant neurotransmitters are the endorphins, a class of short polypeptides. Long-distance runners know the calming effects of endorphins. They give rise to the (nonaddictive) "runner's high" that relieves stress for hours after a long run. Many of the endorphins have also been implicated in other significant biological effects. For instance, substance P shows remarkable activity in ocular wound repair.

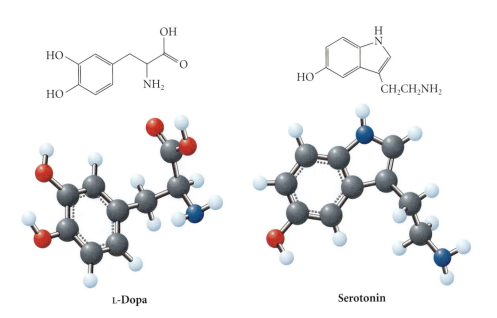

L-Dopa   Serotonin

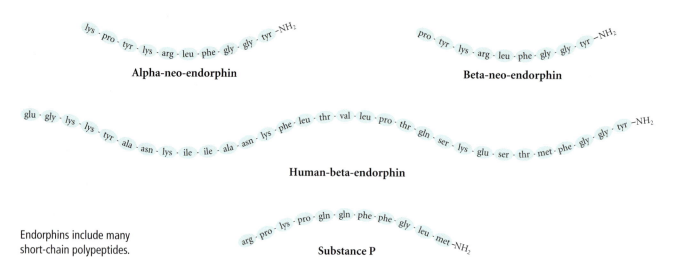

Alpha-neo-endorphin

Beta-neo-endorphin

Human-beta-endorphin

Endorphins include many short-chain polypeptides.

Substance P

Genetic Disease

Application

Genetic diseases are those that are clearly associated with one or more abnormal genes or larger abnormalities in genetic material. Genetic abnormalities give rise to sickle-cell anemia, retinitis pigmentosa, obesity, and a host of other diseases. **How does damage in the gene cause disease?** For each defective gene, there will be a correspondingly defective protein where at least one amino acid has been replaced with an incorrect amino acid. Unfortunately, it is also possible to have abnormalities in the **chromosomes** in which all of our DNA is packaged. Humans contain 23 pairs of chromosomes, one copy of each pair inherited from each parent. Having damaged, extra, or missing chromosomes can cause serious diseases.

One of the main driving forces behind the Human Genome Project that we mentioned at the beginning of this chapter is the desire to identify all the genes that are involved in genetic diseases. In doing so, scientists can learn exactly what has gone wrong with them and, in many cases, discover how to compensate for the defective proteins encoded by the genes. Eventually, we may even be able to develop ways to repair the genes and restore their normal state. The idea of actually repairing genetic damage is known as **gene therapy**. This advancement in science is currently in the early stages of development and trials. Someday, in the not too distant future, gene therapy may be performed on unborn babies to treat their diseases before we welcome them into this oh, so interesting world.

The Bottom Line

- DNA is a polymer of nucleotides made from two complementary single strands wrapped around each other to form a double helix. (Section 22.1)

- Proteins are polymers of amino acids linked through an amide bond (Section 22.2). They contain primary, secondary, tertiary, and sometimes quaternary structure. (Sections 22.2, 22.3)

- Proteins are made by translating the genetic code from mRNA. tRNA supplies the specific amino acids to the growing polypeptide chain. The construction occurs at the ribosomes within the cytoplasm of a cell. (Section 22.3)

- Enzymes are proteins that contain an active site and catalyze biochemical reactions. (Section 22.4)

- Carbohydrates function as structural features such as cell walls in plants and as energy storage molecules in all living organisms. (Section 22.6)

- Lipids are a diverse class of molecules with a variety of functions. (Section 22.7)

- Biological systems exploit the use of enantiomers. Most amino acids and carbohydrates in living organisms are only one of two possible enantiomers. (Section 22.9)

- The promising future activities in the chemistry of medicines include development of new antibiotics, new drugs to treat neurological disorders and cancer, and gene therapy. (Section 22.10)

Key Words

active site The location on an enzyme where the reaction of the substrate is catalyzed. (*p. 948*)

amide bond A bond constructed from the condensation of an amine and a carboxylic acid. Represented by the following general structure. (*p. 940*)

anabolism The metabolic process of constructing molecules from smaller molecules using high-energy molecules that have been made by catabolism. (*p. 958*)

antibiotics Substances that kill, harm, or retard the growth of microorganisms. (*p. 961*)

antibodies Large biological molecules that are produced during an immune response, travel to the site of infection, bind to foreign material, and signal the destruction of the foreign material. (*p. 954*)

anticodon The three-nucleic-acid-sequence on a tRNA polymer that recognizes and binds to the codon on an mRNA polymer. (*p. 943*)

base pair A pair of nucleic acids that are intermolecularly bound to each other through noncovalent forces of attraction—for example, the combination of A and T or of G and C in a DNA duplex. (*p. 936*)

biochemical pathway A sequence of biochemical reactions that lead to the consumption or production of a particular metabolite. (*p. 958*)

carbohydrates Compounds that have the general formula $C_x(H_2O)_y$. These compounds serve living organisms as structural molecules (cellulose), as cellular recognition sites, and for energy storage (starch). (*p. 954*)

catabolism The metabolic process of breaking down molecules to form smaller molecules and produce high-energy molecules. (*p. 958*)

chromosomes The packaged functional units of DNA. (*p. 964*)

codon A three-nucleic-acid sequence on an mRNA that is translated into a particular amino acid during protein synthesis. (*p. 943*)

collagen A structural protein that exhibits a linear tertiary structure formed from intertwined alpha helices. It is found in connective tissues in the body. (*p. 946*)

denature To destroy the biochemical activity of a protein. (*p. 946*)

deoxyribonucleic acid (DNA) A huge nucleotide polymer that has a double-helical structure. Each strand of the DNA polymer complements the other by forming base pairs. DNA provides instructions for the synthesis of proteins and enzymes that carry out the biological activities of the cell. (*p. 935*)

deoxyribose A dehydrated carbohydrate used in the construction of DNA. (*p. 935*)

disaccharide A carbohydrate made from the condensation of two simple sugars. (*p. 956*)

enantiomers A pair of molecules that are nonsuperimposable mirror images. (*p. 960*)

fat A triglyceride that is made from carboxylic acids with long carbon chains and has a relatively high melting point. Typcially derived from animal sources. (*p. 957*)

fatty acids Carboxylic acids that contain long carbon chains. Commonly serve as major building blocks for the production of cell membranes. (*p. 957*)

gene A specific region of the DNA polymer that codes (carries instructions) for a single protein. (*p. 941*)

gene therapy A relatively new area of biochemistry and medicine wherein attempts are made to cure or treat genetic diseases by modification of a patient's damaged genes. (*p. 964*)

hormones Small molecules, such as steroids, used by living organisms for intercellular communication. (*p. 953*)

lipids Biological compounds characterized by their ability to dissolve in nonpolar solvents. They include fatty acids, some vitamins, and the steroid hormones. (*p. 957*)

messenger RNA (mRNA) A short-lived single strand of ribonucleic acid that carries the information from a gene to the ribosomes where a protein is constructed. (*p. 942*)

metabolism The biochemical reactions inside living cells. (*p. 958*)

metabolite A product of each step in a biochemical pathway. (*p. 958*)

monosaccharide A small carbohydrate monomer. Typically contains between three and nine carbon atoms. (*p. 956*)

nitrogenous organic base The portion of a nucleotide where hydrogen bonding holds the strands of DNA together in a duplex. (*p. 935*)

nucleotides Monomers of the nucleic acids. A nucleotide contains a five-carbon sugar, a phosphate, and a nitrogenous organic base. (*p. 935*)

oil A triglyceride made from carboxylic acids with long carbon chains and a relatively low melting point. Typically derived from plant sources. (*p. 957*)

phospholipid A phosphate-modified lipid. (*p. 957*)

polypeptide A polymer of amino acids linked by amide bonds. (*p. 941*)

polysaccharide A polymer of carbohydrate monomers. (*p. 956*)

primary structure The sequence of amino acids within a protein. (*p. 941*)

protein A large polymer of amino acids. May contain one or more polypeptides and often involves complex and intricate folds of the chain. (*p. 939*)

quaternary structure The way in which two or more polypeptide chains have folded to make a protein. (*p. 946*)

racemic mixture A 1:1 mixture of two enantiomers. (*p. 961*)

racemization The reaction that describes the preparation of a racemic mixture from a single enantiomer. (*p. 961*)

regulation The control of the function of a gene or enzyme. (*p. 948*)

replication The process by which DNA is duplicated prior to cell division. (*p. 938*)

residue The portion of an amino acid that differs from other amino acids. (*p. 940*)

ribonucleic acid (RNA) A nucleotide polymer that contains the information from a single gene and either transfers that information to the ribosomes (mRNA) or recognizes and constructs the protein product (tRNA). (*p. 941*)

ribosomes Large molecular structures within the cytoplasm of a cell that aid the construction of a protein from mRNA, tRNA, and the amino acid building blocks. (*p. 944*)

secondary structure The specific folds within a polypeptide chain. Common secondary structures include the alpha helix and the beta pleated sheet. (*p. 945*)

steroids Molecules composed of three 6-member carbon rings and one 5-member carbon ring fused together. These compounds are part of the lipid fraction of a cell and function as structural materials and intercellular communicators. (*p. 957*)

substrate A reactant molecule that binds to an enzyme. (*p. 948*)

sugars The simplest carbohydrates. (*p. 955*)

tertiary structure The way in which secondary structures fold together within a polypeptide chain. (*p. 946*)

transcription The process by which mRNA is synthesized from a gene. (*p. 941*)

transfer RNA (tRNA) A nucleic acid polymer that recognizes the codon on mRNA and adds the corresponding amino acid to the growing protein. (*p. 943*)

translation The process by which a protein is synthesized from mRNA. (*p. 941*)

triglyceride (triacylglycerol) A molecule made from three fatty acids and a molecule of glycerol. Functions as structural material in cell membranes and as an energy storage molecule. (*p. 957*)

vitamins Small molecules that we need for proper health. Normally, vitamins either are not synthesized in our bodies or are made in insufficient amounts. (*p. 950*)

Focus Your Learning

The answers to the odd-numbered problems and some selected problems appear at the back of the book, as represented by the blue numbering.

Section 22.1 DNA—The Basic Structure

Skill Review

1. Define the term *DNA*.

2. Define the term *nucleic acid*.

3. Indicate the complement for the following DNA sequence:
   ATTAAAAAGGGACTA

4. Indicate the complement for the following DNA sequence:
   GGCGAATTAGCCCA

Chemical Applications and Practices

5. A researcher has determined that a particular sequence in double-stranded DNA can be cleaved by a laboratory method. The method is specific for breaking DNA in the middle of an ATTA sequence. How many breaks would occur if the following single strand of DNA were paired with its complement?

   ATTAAAGCCTAATTACCATAAT

6. A short segment of DNA is prepared as a way to locate a particular sequence in double-stranded DNA. To how many locations would the following segment of DNA bind in the double-stranded DNA shown?

   Segment: ATGCA

   DNA: TCGATTACGTATGCATTACGT

Section 22.2 Proteins

Skill Review

7. Cite two amino acids that differ by one carbon in their side groups.

8. Cite two amino acids that have acidic side groups.

9. Show the two products that could result from the condensation of valine and phenylalanine. Identify the amide bonds in each peptide.

10. Draw the three peptides that could be made from the condensation of two glycines and one proline. Identify the amide bonds in each peptide.

Section 22.3 How Genes Code for Proteins

Skill Review

11. Define the term *codon* in your own words.

12. Describe the process of transforming the information in a gene into a protein.

13. Explain how a mistake in the preparation of an mRNA could result in a protein that contains the wrong amino acid.

14. Is it possible to have a mistake in a gene without a mistake in the protein made by that gene? Explain.

Chemical Applications and Practices

15. A researcher isolates a fragment of a particular mRNA. Is this fragment at the beginning, end, or somewhere in the middle of the part that is translated into a protein?

UUUCGAAGUAUGGGUGGAGAUUCUCCCGCG

16. A particular genetic disorder causes an extra nucleotide to be inserted into the mRNA of a gene. Assuming that the entire mRNA is shown below and that the site of insertion is indicated, what is the primary structure of a protein that would result from the insertion of each of the four nucleotides?

AUG–AAU–UAU–CGG–AUU–UAA

17. The skeleton inside a cell is made up of microfilaments—polymers of the protein known as actin. Judging on the basis of the structure shown here, indicate whether actin is a globular or a fibrous protein?

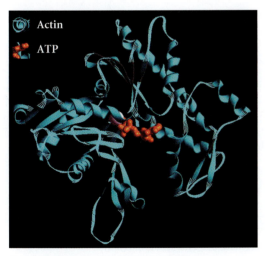

Actin
ATP

An actin protein holds a molecule of ATP.

18. Keratin is a protein used in defining and holding the structure of skin, hair, and claws. Judging on the basis of the structure shown here, indicate whether keratin is a globular or a fibrous protein?

Keratin

Section 22.4 Enzymes

Skill Review

19. Identify the class of enzyme based on the following reaction:

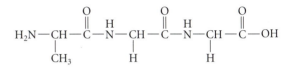

20. Identify the class of enzyme based on the following reaction:

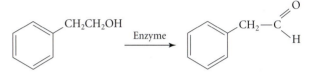

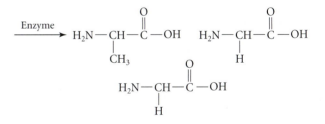

21. In some enzymatic reactions, the product is able to bind to the active enzyme and inhibit its activity. Why would this be a useful feature for an enzyme? Explain.

22. In some metabolic pathways (biochemical reactions involving multiple reactions), large concentrations of the final product can turn off the activity of the first enzyme in the pathway, as shown below. What advantage would this type of regulation have over that described in Problem 21? Explain.

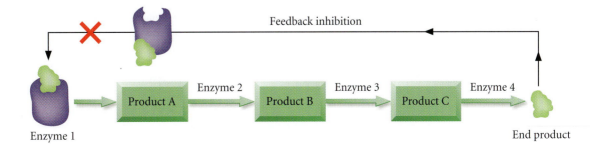

Feedback inhibition

Product A   Enzyme 2   Product B   Enzyme 3   Product C   Enzyme 4

Enzyme 1

End product

Section 22.6 Carbohydrates

Skill Review

23. Identify of the following carbohydrate as either a triose, a tetrose, a pentose, or a hexose.

$$HOCH_2 - \overset{\overset{\displaystyle H}{|}}{\underset{\underset{\displaystyle OH}{|}}{C}} - \overset{\overset{\displaystyle OH}{|}}{\underset{\underset{\displaystyle H}{|}}{C}} - CHO$$

24. Identify the following carbohydrate as either a triose, a tetrose, a pentose, or a hexose.

$$HOCH_2 - \overset{\overset{\displaystyle H}{|}}{\underset{\underset{\displaystyle OH}{|}}{C}} - \overset{\overset{\displaystyle O}{\|}}{C} - \overset{\overset{\displaystyle H}{|}}{\underset{\underset{\displaystyle OH}{|}}{C}} - \overset{\overset{\displaystyle OH}{|}}{\underset{\underset{\displaystyle H}{|}}{C}} - CH_2OH$$

25. Identify the carbohydrate in Problem 23 as either an aldose or a ketose.

26. Identify the carbohydrate in Problem 24 as either an aldose or a ketose.

27. Which simple carbohydrate do starch, glycogen, and cellulose have in common?

28. A very serious malady exists in humans who lack the enzyme responsible for hydrolyzing glycogen. Explain why this deficit would be harmful to health.

Chemical Applications and Practices

29. High-fructose corn syrup is a common ingredient in some fruit drinks. What is the most likely source of this fructose? Identify fructose as either a triose, a tetrose, a pentose, or a hexose. Identify fructose as an aldose or a ketose.

30. Maltose, shown below, is a disaccharide. Which two simple carbohydrates condense to make maltose?

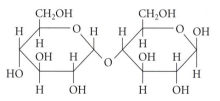

Section 22.7 Lipids

Skill Review

31. Explain the distinction between fats and oils.

32. A company wishes to manufacture margarine using vegetable oil. Explain how this is done.

33. Would you expect the triacylglycerol shown below to be a component of a fat or of an oil?

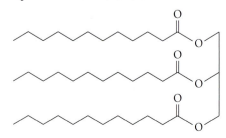

34. Would you expect the triacylglycerol shown below to be a component of a fat or of an oil?

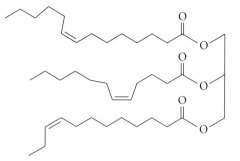

Chemical Applications and Practices

35. Triacylglycerols and phospholipids make up the cell membranes in living organisms. Which end of these molecules would you predict to associate closely with the outside of the cell (in contact with water)?

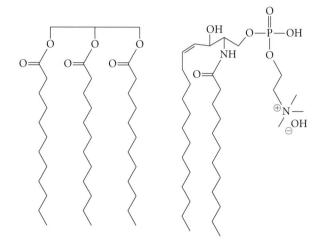

36. The interior of a cell is known as the cytoplasm. It contains many different types of molecules, but the bulk of the molecules are water. Explain how a cell membrane could be made from phospholipids if the exterior and interior of a cell are both made up of water molecules.

Section 22.9 Biochemistry and Chirality

Skill Review

37. Draw both enantiomers of serine and of phenylalanine.

38. Draw both enantiomers of glucose as a straight-chained molecule and as a cyclic monosaccharide.

39. What are the advantages of an enzyme's possessing a chiral active site?

40. If every enzyme possesses a chiral active site, why do both enantiomers of thalidomide exhibit biological activity?

Chemical Applications and Practices

41. Serotonin is a neurotransmitter involved in sleep, sensory perception, and the regulation of body temperature. From which amino acid is it probably likely manufactured in the body? Does serotonin have an enantiomer?

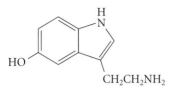

42. Nicotine is a common stimulant. In doses greater than 50 mg, it can be lethal to humans. Nicotine is excreted via urine by protonation, at which point the resulting cation is very soluble in water. Eating a meal causes the pH of urine to drop.
    a. Identify the chrial center in nicotine.
    b. Why would you predict that a smoker would crave a cigarette immediately after a meal?

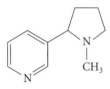

Comprehensive Problems

43. Indicate the mRNA and protein that would result from the transcription and translation of this DNA sequence.

    TACAGCGCTTAAATTCCGACGAATAA

44. What sequence of the gene would produce this polypeptide?

    gly-lys-glu-arg-lys-trp-trp

45. Explain why the melting points (shown in Table 22.3) increase from lauric acid, to myristic acid, to palmitic acid, to stearic acid.

46. Explain why the melting point of oleic acid is less than the melting point of stearic acid.

47. Would you predict the enantiomer of the compound we call table sugar to be sweet? Explain

48. Termites consume wood as food. What polysaccharide is present in wood? Knowing this, and knowing that termites lack the enzyme that can digest that particular polysaccharide, suggest what must be present within the termite's stomach in order for it to derive energy from its food.

Thinking Beyond the Calculation

49. Chocolate-covered cherries are made by coating a cherry with a paste of sucrose and the enzyme invertase. The cherries are stored after they are treated until the enzyme has had a chance to partially hydrolyze the sucrose. The resulting syrup is called invert sugar.
    a. Draw the reaction of sucrose and invertase.
    b. On the basis of your understanding of enzymes, would you predict that maltose would undergo a similar reaction with invertase.
    c. Genetic modification of invertase places an extra lysine in the protein. What codon and anticodon could produce this modification?
    d. A researcher wishes to manufacture 10.0 kg of invert sugar from sucrose. How many grams of sucrose would be required to accomplish this feat?
    e. If the same researcher were able to separate all of the glucose from the invert sugar, how many grams of fructose could be made from 15.5 kg of sucrose?
    f. The same researcher is interested in finding a hydrolase that will accept cellulose as the substrate. What benefit would this have to humans?

Appendixes

Appendix 1: Mathematical Operations

Appendix 2: Calculating Uncertainties in Measurements

Appendix 3: Thermodynamic Data for Selected Compounds at 298 K

Appendix 4: Colligative Property Constants for Selected Compounds

Appendix 5: Selected Equilibrium Constants at 298 K

Appendix 6: Water Vapor Pressure Table

Appendix 7: Standard Reduction Potentials at 298 K

Appendix 8: Common Radioactive Nuclei

Appendix 1

Mathematical Operations

In this discussion, we work with four mathematical tools that are essential parts of the chemist's toolbox: exponents, logarithms, the quadratic equation, and graphing linear functions. This is not intended to be a complete treatment. Rather, it is an introduction upon which you can build as you study chemistry.

A1.1 Working with Exponents

Numbers tell us how much we have. In chemistry, the numbers can be remarkably small, as in the mass of a single sodium atom (0.000 000 000 000 000 000 000 04 gram). They can also be very large, as in the total number of atoms of the universe. For instance, the number of atoms in the universe is estimated by the National Solar Observatory at Sacramento Peak to be, perhaps, 10 000 000 000 000 000 000 000 000 000 000 000 000 000 000 000 000 000 000 000 000 000 000 atoms.

These numbers are unwieldy to work with, so we convert them into *exponential notation,* a way of expressing numbers that uses the powers of 10. The key to exponential notation is to know what happens when we multiply 10 by itself any number of times. For example, when we multiply 10 by itself 3 times, what do we get?

$$10 \times 10 \times 10 = 1000$$

When we use exponential notation, we put the number of times we multiply 10 by itself as a superscript immediately after the 10. That is,

$$10 \times 10 \times 10 = 10^3 = 1000$$

We can expand this list indefinitely:

10	$10^1 = 10$
10×10	$10^2 = 100$
$10 \times 10 \times 10$	$10^3 = 1000$
$10 \times 10 \times 10 \times 10$	$10^4 = 10000$
$10 \times 10 \times 10 \times 10 \times 10 \times 10 \times 10 \times 10$	$10^8 = 100000000$
and so on	

Note that the number of zeros after the 1 is equal to the number of the superscript, which is called the *power* of 10. The number 10,000 is equal to 10^4, or, in words, "ten to the fourth power."

What would our number of atoms in the universe look like in scientific notation?

10 000 000 000 000 000 000 000 000 000 000 000 000 000 000 000 000 000 000 000 000 000 000 000 000 000 000 000 000 000 atoms $= 10^{79}$ atoms, or "ten to the 79th power" atoms

What if we have a number such as 60,000? We can split this into two numbers in such a way that one of the numbers is a power of 10. Thus 60,000 becomes $6 \times 10,000$. This is the same as 6×10^4. The number 6 is known as a *pre-exponential*. Using the pre-exponential and the exponential terms, we can express the number 823,000,000 in exponential notation as 8.23×10^8. It may also be expressed as

$$82.3 \times 10^7$$

$$823 \times 10^6$$

or even 8230×10^5 ($8230 \times 100,000$), although that is fairly awkward. In each case, however, we have reduced the exponent by 1 (a power of 10) as we increased the pre-exponential term by a factor of 10. The number remains the same—823,000,000—in each case.

We can also go from the large to the small:

1	$10^0 = 1$
1/10	$10^{-1} = 0.1$
$1/(10 \times 10)$	$10^{-2} = 0.01$
$1/(10 \times 10 \times 10)$	$10^{-3} = 0.001$
$1/(10 \times 10 \times 10 \times 10)$	$10^{-4} = 0.0001$
$1/(10 \times 10 \times 10 \times 10 \times 10 \times 10 \times 10 \times 10)$	$10^{-8} = 0.00000001$
and so on	

We note that the number of zeros after the decimal point is equal to the negative subscript minus 1. There are seven zeros after the decimal point in 10^{-8}.

Addition and Subtraction

If we wish to add or subtract numbers that are in exponential notation without using an electronic calculator, the numbers must be raised to the same power of 10. We can achieve this prior to the addition or calculation by moving the decimal place as appropriate. For example,

$$(3.11 \times 10^4) + (2.07 \times 10^5) = (3.11 \times 10^4) + (20.7 \times 10^4) = 23.8 \times 10^4 = 2.38 \times 10^5$$

$$(3.11 \times 10^4) - (9.50 \times 10^3) = (31.1 \times 10^3) - (9.50 \times 10^3) = 21.6 \times 10^3 = 2.16 \times 10^4$$

If we are using a calculator, we can just enter the numbers without any adjustment. We do this with the key labeled EXP or EE on most calculators. For example, to enter 3.11×10^4, we key in the following sequence:

$$\boxed{3}\,\boxed{\cdot}\,\boxed{1}\,\boxed{1}\,\boxed{\text{EXP}}\,\boxed{4}$$

To enter a number with a negative exponent, such as 3.11×10^{-4}, we follow the sequence

$$\boxed{3}\,\boxed{\cdot}\,\boxed{1}\,\boxed{1}\,\boxed{\text{EXP}}\,\boxed{-}\,\boxed{4}$$

Multiplication and Division

When numbers expressed in exponential notation are to be multiplied, the exponent parts of the numbers must be added together:

$(7.20 \times 10^4) \times (2.10 \times 10^3) = (7.20 \times 2.10) \times 10^{3+4} = 15.1 \times 10^7 = 1.51 \times 10^8$

When numbers expressed in exponential notation are to be divided, the exponent of the denominator is subtracted from the exponent of the numerator:

$$3.70 \times 10^2 / 2.20 \times 10^3 = (3.70/2.20) \times 10^{\,2-3} = 1.68 \times 10^{-1}$$

PRACTICE

a. Convert 345 000 000 into exponential form. $\qquad$ (Ans: 3.45×10^8)

b. Convert 1.43×10^{-3} into nonexponential form. $\qquad$ (Ans: 0.00143)

c. Calculate: $(2.56 \times 10^2) + (3.41 \times 10^3)$ $\qquad$ (Ans: 3.67×10^3)

d. Calculate: $(7.11 \times 10^6) - (2.50 \times 10^5)$ $\qquad$ (Ans: 6.86×10^6)

e. Calculate: $(8.26 \times 10^2) \times (1.70 \times 10^4)$ $\qquad$ (Ans: 1.40×10^7)

f. Calculate: $(6.32 \times 10^7)/(7.70 \times 10^4)$ $\qquad$ (Ans: 8.21×10^2)

A1.2 Working with Logarithms

Logarithms to the Base 10

As we saw in the previous section, we can use exponential notation to indicate the conversion of 10 into 1000 by raising 10 to the third power:

$$10^3 = 1000$$

Here 3 is referred to as the *logarithm* of 1000 to the base 10, because it is the exponent of the base number 10 that is needed to raise 10 to equal 1000. We can indicate this by writing $\log_{10} 1000 = 3$, which is read, "the logarithm to the base 10 of the number 1000 is 3."

Logarithms to the base 10 are known as *common* logarithms, and use of the base 10 is so common that it is not always explicitly stated. When you see a reference to the logarithm, or *log*, of a number without the base being specified, you should assume the base is 10.

Logarithms can be negative as well as positive. Here are some other examples:

logarithm of 10 000 to base 10 = 4	$\log_{10} 10\ 000 = 4$
logarithm of 100 to base 10 = 2	$\log_{10} 100 = 2$
logarithm of 1 to base 10 = 0 (because $10^0 = 1$)	$\log_{10} 1 = 0$
logarithm of 0.01 to base 10 = −2	$\log_{10} 0.01 = -2$
logarithm of 0.0001 to base 10 = −4	$\log_{10} 0.0001 = -4$

Most numbers are not the simple multiples of 10 used in our examples above. For more awkward numbers, such as 73.5, we use the LOG key on a calculator to find the value of the logarithm to the base 10. Get a scientific calculator, press the LOG key, and then enter 73.5 followed by the equals key. You will find that

$$\log_{10} 73.5 = 1.866$$

which tells us that $10^{1.866} = 73.5$.

This answer makes sense, because 73.5 is between 10 (which has a logarithm of 1, because $10^1 = 10$) and 100 (which has a logarithm of 2, because $10^2 = 100$). Accordingly, we expect the logarithm of 73.5 to be between 1 and 2 and to be closer to 2 than to 1. Because the integers to the left of the decimal in a logarithm only set the location of the decimal point, the number of digits after the decimal point in the logarithm should equal the number of significant figures in the original number. Hence the logarithm 1.866 only has three significant figures.

Now let's consider working in the opposite direction. Finding the number that corresponds to a given logarithm is known as obtaining an *antilogarithm* (or antilog). The calculator will do this for us when we either use the 10^x key or press the inverse function key (labeled INV or SHIFT) followed by the LOG key. Determine which method your calculator uses, and then confirm that

$$\text{antilog } 2 = 100$$
$$\text{antilog } 1.866 = 73.5$$

Natural Logarithms

The mathematical equations that describe the behavior of many natural phenomena frequently involve the constant *e*, which has a value of 2.71828. . . . Logarithms that have the number *e* as the base are referred to as *natural logarithms*. The natural logarithm (or natural log) of a number is the power to which *e* must be raised to equal that number. The natural logarithm is often symbolized ln to distinguish it from the log or $\log_{10}$ used for logarithms to the base 10.

You can use the LN key of your calculator to find the natural logarithm of any number. For example,

$$\ln 10.0 = 2.303 \text{ because } e^{2.303} = 10$$
$$\ln 73.5 = 4.297 \text{ because } e^{4.297} = 73.5$$

Natural antilogarithms can be obtained by either using the e^x key or pressing the inverse function key followed by the LN key. Use whichever system your calculator employs to confirm that

$$\text{natural antilogarithm of } 2.303 = 10$$
$$\text{natural antilogarithm of } 4.297 = 73.5$$

Mathematical Operations with Logarithms

The logarithm of a product equals the *sum* of the logarithms of the individual numbers being multiplied. This follows from the fact that logarithms are exponents, which we looked at in the previous section.

$$\text{logarithm}(a \times b) = \text{logarithm } a + \text{logarithm } b$$

By the same logic, the logarithm of the result of dividing a number by another is obtained by subtracting the logarithm of the denominator from that of the numerator:

$$\log(a/b) = \log a - \log b$$

And if we wish to raise a logarithm to a certain power, we use the following rules:

$$\log(a^n) = n \log a$$
$$\log(a^{1/n}) = (1/n) \log a$$

These rules apply no matter what base is used.

PRACTICE

Use your calculator to find the numerical answer to the following operations.

a. $\log_{10} 55.2$ (Ans: 1.742)

b. $\ln 27.6$ (Ans: 3.318)

c. antilog 1.522 (Ans: 33.3)

d. natural antilog 3.233 (Ans: 25.4)

A1.3   Solving the Quadratic Equation

A quadratic equation has the variable raised to the second power, and no higher. It has the general form

$$ax^2 + bx + c = 0$$

In this textbook, we use quadratic equations to solve some problems related to chemical equilibrium, including acid–base chemistry, solubility chemistry, and complex-ion chemistry (Chapters 16–18).

Although there are several ways to solve quadratic equations, we will use the *quadratic formula* in this text. The formula will always yield two answers (because second-order polynomial equations describe a parabola that has two y-axis values for every x-axis value), yet only one of these answers will be chemically reasonable. Whenever we use the quadratic equation, we will explain how you will know which of the two answers is reasonable. The quadratic formula is used to solve for the variable x and has the form

$$x = \frac{-b \pm \sqrt{b^2 - 4ac}}{2a}$$

where the constants a, b, and c are the numbers in the quadratic equation when it is written in general form.

For example, if we are solving the following equation, we need to determine the values of a, b, and c in order to use the quadratic formula.

$$x^2 + 0.0500x - 0.0300 = 0$$
$$ax^2 + \quad bx + c \quad = 0$$

Comparing this to the general form for a quadratic equation reveals that the values for the constants a, b, and c are

$$a = 1; \quad b = +0.0500; \quad c = -0.0300$$

Next, we insert our values for a, b, and c into the equadratic equation and solve. We obtain two values for x:

$$x = \frac{-b \pm \sqrt{b^2 - 4ac}}{2a}$$

$$x = \frac{-0.0500 \pm \sqrt{(0.0500)^2 - 4(1)(-0.0300)}}{2(1)}$$

$$x = \frac{-0.0500 \pm \sqrt{0.00250 + 0.120}}{2(1)}$$

$$x = \frac{-0.0500 \pm 0.350}{2(1)}$$

$$x = 0.150, -0.200$$

Just remember: Although we get two answers when we solve the quadratic equation, we will find in our chemistry work that only one of them will be chemically valid.

PRACTICE

What is the value of x in this quadratic equation?

$$x^2 + 0.0505x - 0.0435 = 0 \qquad \text{(Ans: +0.184, -0.235)}$$

A1.4 Graphing

In your study of chemistry and other aspects of science, you will frequently need to interpret or draw graphs that show the relationship between variables. For any value of one variable, the graph shows the corresponding value of the other variable. The variable whose value we monitor rather than control during an experiment is called the *dependent variable* and is usually displayed along the vertical axis, which is called the *y*-axis. The other variable, which we may be purposely varying during the course of an experiment, is called the *independent variable* and is displayed along the horizontal axis, which is called the *x*-axis. If we are monitoring the production of some chemical product with time, for example, the increasing concentration of the product would be plotted on the vertical *y*-axis, and time would be plotted on the horizontal *x*-axis.

Graphs can be linear (consisting of straight lines), can be made up of curves, or can consist of mixtures of linear and curved portions. Any straight-line graph is described by the general equation

$$y = mx + b$$

where *m* is the slope of the line and *b* is the value at the intercept of the line with the *y*-axis.

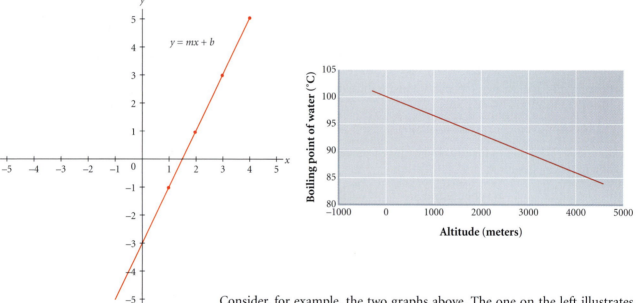

Consider, for example, the two graphs above. The one on the left illustrates the general principles, and on the right we show a real-world example. In the graph on the left, we can calculate the slope *m* as $(y_2 - y_1)/(x_2 - x_1) = 4/2 = 2$. And we can see that the intercept occurs at −3 on the *y*-axis. Therefore, the equation of the straight line is $y = 2x - 3$.

PRACTICE

The real world is rarely as neat and tidy as general principles. Calculate the equation of the straight line in the real-world example on the right, above. Note that in this case the slope has a negative value, and the *x*-axis does not cut across the *y*-axis at the zero level. These are both features you may encounter in graphs, but the equation can be readily calculated by using the method illustrated in the example above. (Ans: $y = -0.0033x + 100$)

Appendix 2

Calculating Uncertainties in Measurements

Error in measurements is an obstinate inevitability that all scientists (and every-one else, for that matter) must contend with. When we measure the distance to the store, or the mass of a raisin, we encounter a certain amount of variability. As we learned in Chapter 1, this variability is due not only to random error but also to systematic error. Even during the perfect measurement, where the systematic error is negligible, our measurement still contains some random error. Therefore, when scientists report a measurement, they make sure to include a statement about the level of confidence that they place in that number.

The statement of how *precise* a set of measurements are can be reported as the standard deviation(s) of the measurements. The mathematical definition of *standard deviation* is

$$s = \sqrt{\frac{\sum_{i=1}^{n}(x_i - \bar{x})^2}{n - 1}}$$

where x_i = each individual measurement
  $\bar{x}$ = the mean (average) of the measurements
  n = number of measurements taken

The equation indicates that we take the difference between each measurement and the mean, square it, and then add all these results together. The total is then divided by one less than the total number of measurements, and the square root of the result is taken.

The standard deviation can be used to determine how well the measurements agree with the literature value. This is typically established by calculating a *confidence limit* or a *confidence interval*. The confidence limit (C.L.) consists of a range of numbers that describes the precision of our set of measurements. The confidence interval (C.I.) describes the mean value and the confidence limit employed to generate a range of numbers surrounding the mean value of our measurements. This number is called a confidence interval because we have a great deal of confidence that the actual number lies within this range. As we increase our confidence in the measurement, the confidence limit increases in size. Conversely, as the confidence limit decreases in size with a set confidence level, our precision increases.

$$\text{C.L.} = \frac{ts}{\sqrt{n}}$$

$$\text{C.I.} = \bar{x} \pm \frac{ts}{\sqrt{n}}$$

where t is a measure of the confidence we place in the measurement.

The levels of confidence that are most commonly used in determining a confidence interval are the 90% and 95% levels. A 90% level of confidence means that in 10% of cases, the accepted value will lie outside our confidence interval. A 95% level of confidence indicates that the accepted value will lie outside our interval in only 5% of cases. For example, say an experiment produces three measurements such that the mean is 2.31 and the standard deviation is 0.48. The value of

TABLE A2.1	Values of t for Various Levels of Probability				
Number of Measurements	50%	90%	95%	99%	99.9%
2	1.000	6.314	12.706	63.657	636.619
3	0.816	2.920	4.303	9.925	31.98
4	0.765	2.353	3.182	5.841	12.924
5	0.741	2.132	2.776	4.604	8.610
6	0.727	2.015	2.571	4.032	6.869
7	0.718	1.943	2.447	3.707	5.959
8	0.711	1.895	2.365	3.500	5.408
9	0.706	1.860	2.306	3.355	5.041
10	0.703	1.833	2.262	3.250	4.781

t at the 90% confidence level for three measurements is 2.920 (see Table A2.1). Therefore, the confidence interval can be determined to be

$$\text{C.I.} = \bar{x} \pm \frac{ts}{\sqrt{n}}$$

$$= 2.31 \pm \frac{2.920 \times 0.48}{\sqrt{3}}$$

$$= 2.31 \pm 0.82$$

The researcher would report the result of the experiment as 2.31 ± 0.82. This means that in 90% of the cases, the accepted value should lie somewhere between 3.13 and 1.49. In 10% of the cases, the accepted value will lie outside these limits.

Occasionally, a set of measurements are collected for an experiment and one of the values appears to be erroneous. For example, measuring the distance from one town to another could give these data:

21.5 mi, 21.2 mi, 21.6 mi, 21.5 mi, 22.8 mi

Examination of the data leads us to believe that the last measurement is probably not a good measurement. But can we ignore it? No, we can't just throw it away without mathematically showing that it is an outlier.

This is accomplished by using a method known as the *Q-test*. In this method, a value of Q is calculated and compared to a Q value from a table based on our confidence level (see Table A2.2). Typically, the 90% confidence level is chosen

TABLE A2.2	Critical Values for the Rejection Quotient Q		
Number of Measurements	90% Confidence	95% Confidence	99% Confidence
3	0.941	0.970	0.994
4	0.765	0.829	0.926
5	0.642	0.710	0.821
6	0.560	0.625	0.740
7	0.507	0.568	0.680
8	0.468	0.526	0.634
9	0.437	0.493	0.598
10	0.412	0.466	0.568

(giving us only a 10% chance of being wrong). The rejection quotient, Q, is determined by solving

$$Q = \frac{d}{r} = \frac{gap}{range}$$

where the gap is the difference between the outlier and the next closest measurement, and the range is the difference between the largest and smallest measurements.

For our hypothetical set of distances, we could determine the value of Q and compare it to the tabulated value of Q for five measurements (0.642). If our calculation of Q is larger than this, we can safely discard the point.

$$Q = \frac{d}{r} = \frac{(22.8 - 21.6)}{(22.8 - 21.2)} = \frac{1.2}{1.6} = 0.75$$

Our calculation is larger than the value of Q from the table (Q at 90% is 0.642), and we have demonstrated the point to be an outlier. However, if the calculated Q is only slightly larger than the table value, it may be statistically safer *not* to reject the point.

Appendix 3

Thermodynamic Data for Selected Compounds at 298 K

The signs of the values for $\Delta_f H°$ and $\Delta_f G°$ are explicitly shown. The sign on all $S°$ values is positive unless otherwise noted. All values are rounded to the nearest whole number.

	$\Delta_f H°$ (kJ·mol⁻¹)	$\Delta_f G°$ (kJ·mol⁻¹)	$S°$ (J·K⁻¹·mol⁻¹)		$\Delta_f H°$ (kJ·mol⁻¹)	$\Delta_f G°$ (kJ·mol⁻¹)	$S°$ (J·K⁻¹·mol⁻¹)
Aluminum				Ba(OH)$_2$(s)	−946		
Al(s)	0	0	28	BaSO$_4$(s)	−1465	−1353	132
Al(l)	+11	+7	40	**Beryllium**			
Al(g)	+326	+286	164	Be(s)	0	0	10
Al$_2$O$_3$(s)	−1676	−1582	51	Be(g)	+324	+287	136
Al(OH)$_3$(s)	−1277			BeO(s)	−599	−569	14
AlCl$_3$(s)	−704	−629	111	Be(OH)$_2$(s)	−904	−815	47
Antimony				**Bismuth**			
Sb(s)	0	0	46	Bi(s)	0	0	57
SbH$_3$(g)	+145	+148	233	Bi(g)	+207	+168	187
Argon				**Bromine**			
Ar(g)	0	0	155	Br$_2$(l)	0	0	152
Arsenic				Br$_2$(g)	+31	+3	245
As(s)	0	0	35	Br$_2$(aq)	−3	+4	130
As(g)	+302	−261	174	Br(g)	+112	+82	175
As$_4$(g)	+144	+92	314	Br⁻(aq)	−121	−104	82
AsH$_3$(g)	+66	+69	223	HBr(g)	−36	−53	199
Barium				**Cadmium**			
Ba(s)	0	0	67	Cd(s)	0	0	52
Ba(g)	+180	+146	170	Cd(g)	+112	+77	168
BaO(s)	−582	−552	70	CdO(s)	−258	−228	55
BaCl$_2$(s)	−859	−810	124	CdCO$_3$(s)	−751	−669	92
BaCO$_3$(s)	−1219	−1139	112	Cd(OH)$_2$(s)	−561	−474	96

(Continued)

	$\Delta_f H°$ (kJ·mol^{-1})	$\Delta_f G°$ (kJ·mol^{-1})	$S°$ (J·K^{-1}·mol^{-1})
CdS(s)	−162	−156	65
CdSO$_4$(s)	−953	−823	123
Calcium			
Ca(s)	0	0	41
Ca(g)	+178	+144	155
CaC$_2$(s)	−63	−68	70
CaO(s)	−635	−604	40
CaCO$_3$(s)	−1207	−1129	93
CaF$_2$(s)	−1220	−1167	69
CaCl$_2$(s)	−796	−748	105
CaBr$_2$(s)	−683	−664	130
Ca(OH)$_2$(s)	−987	−899	83
Ca$_3$(PO$_4$)$_2$(s)	−4126	−3890	241
CaSO$_4$(s)	−1433	−1320	107
CaSiO$_3$(s)	−1630	−1550	84
Carbon			
C(s) (graphite)	0	0	6
C(s) (diamond)	+2	+3	2
C(g)	+717	+671	158
C$_2$(g)	+832	+776	119
CO(g)	−111	−137	198
CO$_2$(g)	−394	−394	214
CO$_2$(aq)	−414	−386	118
H$_2$CO$_3$(aq)	−700	−623	187
CCl$_4$(l)	−135	−65	216
CH$_3$Cl(g)	−84	−60	234
CS$_2$(l)	+90	+65	151
HCN(g)	+135	+125	202
HCN(l)	+109	+125	113
CH$_4$(g)	−75	−50	186
C$_2$H$_2$(g)	+227	+209	201
C$_2$H$_4$(g)	+52	+68	220
C$_2$H$_6$(g)	−85	−33	230
C$_3$H$_6$(g)	+21	+63	267
C$_3$H$_8$(g)	−104	−24	270
C$_4$H$_8$(g) (1-butene)	−0.1	+71	306
C$_4$H$_{10}$(g)	−126	−17	310
C$_5$H$_{12}$(g)	−146	−8	348
C$_5$H$_{12}$(l)	−173	−251	263
C$_6$H$_6$(l)	+49	+124	173
C$_6$H$_6$(g)	+83	+130	269
C$_6$H$_{12}$(l)	−156	+27	204
C$_6$H$_{14}$(l)	−199		204
C$_7$H$_{16}$(l) (heptane)	−224	+1	329
C$_8$H$_{18}$(l) (octane)	−250	+6	361
C$_8$H$_{18}$(l) (iso-octane)	−255		328
CH$_3$OH(l)	−239	−166	127
CH$_3$OH(g)	−201	−162	240
C$_2$H$_5$OH(l)	−278	−175	161

	$\Delta_f H°$ (kJ·mol^{-1})	$\Delta_f G°$ (kJ·mol^{-1})	$S°$ (J·K^{-1}·mol^{-1})
C$_2$H$_5$OH(g)	−235	−168	283
C$_6$H$_5$OH(s) (phenol)	−165	−51	146
HCOOH(l) (formic acid)	−425	−361	129
CH$_3$COOH(l) (acetic acid)	−484	−389	160
CH$_3$COOH(aq) (acetic acid)	−486	−396	179
H$_2$C$_2$O$_4$(s) (oxalic acid)	−827		115
C$_6$H$_5$COOH(s) (benzoic acid)	−385	−245	168
HCHO(g) (formaldehyde)	−109	−103	219
CH$_3$CHO(l) (acetaldehyde)	−192	−128	160
CH$_3$CHO(g) (acetaldehyde)	−166	−129	250
CH$_3$COCH$_3$(l) (acetone)	−248	−155	200
C$_6$H$_{12}$O$_6$(s) (glucose)	−1268	−911	212
C$_{12}$H$_{22}$O$_{11}$(s) (sucrose)	−2222	−1543	360
Cesium			
Cs(s)	0	0	85
Cs(g)	+76	+49	176
Chlorine			
Cl$_2$(g)	0	0	223
Cl$_2$(aq)	−23	7	121
Cl$^-$(aq)	−167	−131	57
HCl(g)	−92	−95	187
Chromium			
Cr(s)	0	0	24
Cr$_2$O$_3$(s)	−1128	−1047	81
CrO$_3$(s)	−579	−502	72
Copper			
Cu(s)	0	0	33
CuCO$_3$(s)	−595	−518	88
Cu$_2$O(s)	−170	−148	93
CuO(s)	−156	−128	43
Cu(OH)$_2$(s)	−450	−372	108
CuS(s)	−49	−49	67
Fluorine			
F$_2$(g)	0	0	203
F$^-$(aq)	−333	−279	−14
HF(g)	−271	−273	174
Hydrogen			
H$_2$(g)	0	0	131
H(g)	+217	+203	115
H$^+$(aq)	0	0	0
OH$^-$(aq)	−230	−157	−11
H$_2$O(l)	−286	−237	70
H$_2$O(g)	−242	−229	189
Iodine			
I$_2$(s)	0	0	116
I$_2$(g)	+62	+19	261
I$_2$(aq)	+23	+16	137
I$^-$(aq)	−55	−52	106

	$\Delta_f H°$ (kJ·mol^{-1})	$\Delta_f G°$ (kJ·mol^{-1})	$S°$ (J·K^{-1}·mol^{-1})
Iron			
Fe(s)	0	0	27
Fe$_3$C(s)	+21	+15	108
Fe$_{0.95}$O(s) (wustite)	−264	−240	59
FeO	−272	−255	61
Fe$_3$O$_4$(s) (magnetite)	−1117	−1013	146
Fe$_2$O$_3$(s) (hematite)	−826	−740	90
FeS(s)	−95	−97	67
FeS$_2$(s)	−178	−166	53
FeSO$_4$(s)	−929	−825	121
Lead			
Pb(s)	0	0	65
Pb(g)	+195	+162	175
PbO(s, yellow)	−217	−188	69
PbO(s, red)	−219	−189	66
PbO$_2$(s)	−277	−217	69
PbS(s)	−100	−99	91
PbSO$_4$(s)	−920	−813	149
Lithium			
Li(s)	0	0	29
Li(g)	+159	+127	139
Magnesium			
Mg(s)	0	0	33
Mg(g)	+148	+113	149
MgO(s)	−602	−570	27
MgCO$_3$(s)	−1096	−1012	65
MgCl$_2$(s)	−641	−592	90
Mg(OH)$_2$(s)	−925	−834	64
Mg$_3$N$_2$(s)	−461	−401	89
Manganese			
Mn(s)	0	0	32
MnO(s)	−385	−363	60
Mn$_3$O$_4$(s)	−1387	−1280	149
Mn$_2$O$_3$(s)	−971	−893	110
MnO$_2$(s)	−521	−466	53
MnO$_4^-$(aq)	−543	−449	190
Mercury			
Hg(l)	0	0	76
Hg(g)	+61	+32	175
Hg$_2$Cl$_2$(s)	−265	−211	193
HgCl$_2$(s)	−230	−184	144
HgO(s)	−91	−59	70
HgS(s, black)	−54	−48	88
Neon			
Ne(g)	0	0	146
Nickel			
Ni(s)	0	0	30

	$\Delta_f H°$ (kJ·mol^{-1})	$\Delta_f G°$ (kJ·mol^{-1})	$S°$ (J·K^{-1}·mol^{-1})
NiCl$_2$(s)	−316	−272	107
NiO(s)	−241	−213	38
Ni(OH)$_2$(s)	−538	−453	79
NiS(s)	−93	−90	53
Nitrogen			
N$_2$(g)	0	0	192
N(g)	+473	+456	153
NH$_3$(g)	−46	−16	192
NH$_3$(aq)	−80	−27	111
NH$_4^+$(aq)	−132	−79	113
HN$_3$(l)	264	327	141
HN$_3$(g)	294	328	239
NH$_4$NO$_3$(s)	−366	−184	151
NH$_4$Cl(s)	−314	−203	95
NH$_4$ClO$_4$(s)	−295	−89	186
NH$_4$Cl(s)	−314	−203	96
HNO$_3$(l)	−174	−81	156
HNO$_3$(aq)	−207	−111	146
NH$_2$OH(s)	−114		
NO(g)	+90	+87	211
NO$_2$(g)	+33	+51	240
N$_2$O(g)	+82	+104	220
N$_2$O$_4$(g)	+9	+98	304
N$_2$O$_5$(s)	−43	+114	178
N$_2$O$_5$(g)	+11	+115	356
N$_2$H$_4$(l)	+51	+149	121
N$_2$H$_3$CH$_3$(l)	+54	+180	166
Oxygen			
O$_2$(g)	0	0	205
O(g)	+249	+232	161
O$_3$(g)	+143	+163	239
Phosphorus			
P(s) (white)	0	0	41
P(s) (red)	−18	−12	23
P(s) (black)	−39	−33	23
P(g)	+315	+278	163
P$_2$(g)	+144	+104	218
P$_4$(g)	+59	+24	280
PH$_3$(g)	+5	+13	210
PF$_5$(g)	−1578	−1509	296
PCl$_3$(g)	−287	−268	312
PCl$_3$(l)	−320	−272	217
PCl$_5$(g)	−375	−305	365
PCl$_5$(s)	−444		
H$_3$PO$_3$(s)	−964		
H$_3$PO$_3$(aq)	−965		
H$_3$PO$_4$(s)	−1279	−1119	111
H$_3$PO$_4$(l)	−1267		

(Continued)

	$\Delta_fH°$ (kJ·mol^{-1})	$\Delta_fG°$ (kJ·mol^{-1})	$S°$ (J·K^{-1}·mol^{-1})
$H_3PO_4(aq)$	−1277	−1019	−222
$P_4O_{10}(s)$	−2984	−2697	229
Potassium			
$K(s)$	0	0	64
$K(g)$	+89	+61	160
$KF(s)$	−576	−538	67
$KCl(s)$	−437	−409	83
$KBr(s)$	−394	−381	96
$KI(s)$	−328	−325	106
$KClO_3(s)$	−391	−290	143
$KClO_4(s)$	−433	−304	151
$KOH(s)$	−425	−379	79
$KOH(aq)$	−481	−440	9
$KO_2(s)$	−283	−238	117
$K_2O(s)$	−361	−322	98
$K_2O_2(s)$	−496	−430	113
Silicon			
$Si(s)$	0	0	19
$Si(g)$	+456	+411	168
$SiO_2(s)$	+911	−857	42
$SiCl_4(l)$	−687	−620	240
Silver			
$Ag(s)$	0	0	43
$Ag(g)$	+285	+246	173
$Ag^+(aq)$	+105	+77	73
$AgBr(s)$	−100	−97	107
$AgCN(s)$	+146	+164	84
$AgCl(s)$	−127	−110	96
$Ag_2CrO_4(s)$	−712	−622	217
$AgI(s)$	−62	−66	115
$Ag_2O(s)$	−31	−11	121
$AgNO_3(s)$	−129	−33	141
$Ag_2S(s)$	−32	−40	146
Sodium			
$Na(s)$	0	0	51
$Na(g)$	+107	+77	153
$Na^+(aq)$	−240	−262	59
$NaH(s)$	−56	−33	40
$NaHCO_3(s)$	−948	−852	102
$NaCl(s)$	−411	−384	72
$NaBr(s)$	−361	−349	87
$NaI(s)$	−288	−286	99
$NaNO_2(s)$	−359		
$NaNO_3(aq)$	−467	−366	116
$NaOH(s)$	−426	−379	64
$NaOH(aq)$	−470	−419	50
$Na_2CO_3(s)$	−1131	−1048	136

	$\Delta_fH°$ (kJ·mol^{-1})	$\Delta_fG°$ (kJ·mol^{-1})	$S°$ (J·K^{-1}·mol^{-1})
$Na_2O(s)$	−416	−377	73
$Na_2O_2(s)$	−515	−451	95
Sulfur			
$S(s)$	0	0	32
$S(g)$	+279	+238	168
$S^{2-}(aq)$	+42	+86	+22
$S_2(g)$	+128	+79	228
$S_8(s)$	+102	+50	431
$H_2S(g)$	−21	−34	206
$H_2S(aq)$	−40	−28	121
$SF_6(g)$	−1209	−1105	292
$SO_2(g)$	−297	−300	248
$SO_3(g)$	−396	−371	257
$SO_4^{2-}(aq)$	−909	−745	20
$H_2SO_4(l)$	−814	−690	157
$H_2SO_4(aq)$	−909	−745	20
Tin			
$Sn(s)$ (white)	0	0	52
$Sn(s)$ (grey)	−2	+0.1	44
$Sn(g)$	+302	+267	168
$SnO(s)$	−286	−257	57
$SnO_2(s)$	−581	−520	52
$Sn(OH)_2(s)$	−561	−492	155
Titanium			
$TiCl_4(g)$	−763	−727	355
$TiO_2(s)$	−945	−890	50
Uranium			
$U(s)$	0	0	50
$UF_6(s)$	−2137	−2008	228
$UF_6(g)$	−2113	−2029	380
$UO_2(s)$	−1084	−1029	78
$UO_3(s)$	−1230	−1150	99
$U_3O_8(s)$	−3575	−3393	282
Xenon			
$Xe(g)$	0	0	170
$XeF_2(g)$	−108	−48	254
$XeF_4(s)$	−251	−121	146
$XeF_6(g)$	−294		
$XeO_3(s)$	402		
Zinc			
$Zn(s)$	0	0	42
$Zn(g)$	+131	+95	161
$ZnO(s)$	−348	−318	44
$Zn(OH)_2(s)$	−642		
$ZnS(s)$ (wurtzite)	−193		
$ZnS(s)$ (zinc blende)	−206	−201	58
$ZnSO_4(s)$	−983	−874	120

Appendix 4

Colligative Property Constants for Selected Compounds

	K_f (°C/m)	T_f (°C)	K_b (°C/m)	T_b (°C)
Acetic acid	3.9	16.2	3.1	117.9
Acetone	2.4	−94.8	1.7	56.5
Benzene	5.1	5.5	2.6	80.1
Camphor	40	175	5.6	204
Carbon disulfide	3.8	−112	2.4	46
Carbon tetrachloride	30	−23	5.0	76.7
Chloroform	4.7	−63.5	3.6	61.2
Cyclohexane	20.0	6.5	2.8	80.7
Cyclohexanol	40.8	20	3.5	161
Diethyl ether	1.8	−116.3	2.0	34.6
1,4-Dioxane	4.6	11.8	3.3	101.5
Ethanol	2.0	−117.3	1.2	78.5
Ethylene glycol	3.1	−13	2.3	197.3
Formic acid	2.8	8.4	2.4	100
Methanol		−98	0.80	64.7
Naphthalene	6.9	80	5.8	217.7
Phenol	7.4	43	3.0	182
Toluene	3.6	−94.5	3.3	110.7
Water	1.86	0.0	0.51	100.0

Appendix 5

Selected Equilibrium Constants at 298 K

TABLE A5.1   **Acid Ionization Constants (K_a) at 298 K**

All values except those of the strong acids (HI, HBr, HCl, H_2SO_4, $HClO_4$, HNO_3, and H_3O^+) are reported to two significant figures.

Name	Formula	K_a	Name	Formula	K_a
Acetic acid	CH_3COOH	1.8×10^{-5}	Hydronium ion	H_3O^+	1
Benzoic acid	C_6H_5COOH	6.5×10^{-5}	Hypobromous acid	$HBrO$	2.8×10^{-9}
Boric acid (K_{a_1})	H_3BO_3	5.8×10^{-10}	Hypochlorous acid	$HClO$	3.5×10^{-8}
Butanoic acid	$CH_3CH_2CH_2COOH$	1.5×10^{-5}	Hypoiodous acid	HIO	2.0×10^{-11}
Chlorous acid	$HClO_2$	1.2×10^{-2}	Iodic acid	HIO_3	1.7×10^{-1}
Formic acid	$HCOOH$	1.8×10^{-4}	Lactic acid	$CH_3CH(OH)COOH$	1.3×10^{-4}
Hydrazoic acid	HN_3	2.2×10^{-5}	Nitric acid	HNO_3	1×10^1
Hydrobromic acid	HBr	1×10^9	Nitrous acid	HNO_2	7.0×10^{-4}
Hydrochloric acid	HCl	1×10^7	Perchloric acid	$HClO_4$	4×10^1
Hydrofluoric acid	HF	7.2×10^{-4}	Phenol	C_6H_5OH	1.6×10^{-10}
Hydrogen cyanide	HCN	6.2×10^{-10}	Propanic acid	CH_3CH_2COOH	1.3×10^{-5}
Hydrogen peroxide	H_2O_2	2.4×10^{-12}	Sulfuric acid (K_{a_1})	H_2SO_4	1×10^2
Hydroiodic acid	HI	1×10^{11}	Thiocyanic acid	$HSCN$	1.3×10^{-1}

TABLE A5.2	Base Ionization Constants (K_b) at 298 K

All values are reported to two significant figures.

Name	Formula	K_b
Ammonia	NH_3	1.8×10^{-5}
Aniline	$C_6H_5NH_2$	3.8×10^{-8}
Diethylamine	$(CH_3CH_2)_2NH$	7.1×10^{-4}
Dimethylamine	$(CH_3)_2NH$	5.9×10^{-4}
Ethylamine	$CH_3CH_2NH_2$	6.4×10^{-4}
Hydrazine	H_2NNH_2	1.3×10^{-6}
Hydroxylamine	$HONH_2$	1.1×10^{-8}
Methylamine	CH_3NH_2	5.9×10^{-4}
Pyridine	C_5H_5N	1.7×10^{-9}
Triethylamine	$(CH_3CH_2)_3N$	5.6×10^{-4}
Trimethylamine	$(CH_3)_3N$	6.4×10^{-5}

TABLE A5.3	Polyprotic Acid Ionization Constants (K_{a1}, K_{a2}, and K_{a3}) at 298 K

The formula of the acid (HA) and that of its conjugate base (A^-) are indicated.

Acid	HA	A^-	K_a
Arsenic acid	H_3AsO_4	$H_2AsO_4^-$	5.0×10^{-3}
dihydrogen arsenate	$H_2AsO_4^-$	$HAsO_4^{2-}$	8.0×10^{-8}
hydrogen arsenate	$HAsO_4^{2-}$	AsO_4^{3-}	6.0×10^{-10}
Ascorbic acid (vitamin C)	$H_2C_6H_6O_6$	$HC_6H_6O_6^-$	7.9×10^{-5}
ascorbate	$HC_6H_6O_6^-$	$C_6H_6O_6^{2-}$	1.6×10^{-12}
Carbonic acid	H_2CO_3	HCO_3^-	4.3×10^{-7}
hydrogen carbonate	HCO_3^-	CO_3^{2-}	5.6×10^{-11}
Citric acid	$H_3C_6H_5O_7$	$H_2C_6H_5O_7^-$	8.4×10^{-4}
dihydrogen citrate	$H_2C_6H_5O_7^-$	$HC_6H_5O_7^{2-}$	1.8×10^{-5}
hydrogen citrate	$HC_6H_5O_7^{2-}$	$C_6H_5O_7^{3-}$	4.0×10^{-6}
Hydrogen sulfide	H_2S	HS^-	1.0×10^{-7}
hydrogen sulfide ion	HS^-	S^{2-}	1.0×10^{-15}
Oxalic acid	$H_2C_2O_4$	$HC_2O_4^-$	6.5×10^{-2}
hydrogen oxalate	$HC_2O_4^-$	$C_2O_4^{2-}$	6.1×10^{-5}
Phosphoric acid	H_3PO_4	$H_2PO_4^-$	7.4×10^{-3}
dihydrogen phosphate	$H_2PO_4^-$	HPO_4^{2-}	6.2×10^{-8}
hydrogen phosphate	HPO_4^{2-}	PO_4^{3-}	4.8×10^{-13}
Sulfuric acid	H_2SO_4	HSO_4^-	1.0×10^2
hydrogen sulfate	HSO_4^-	SO_4^{2-}	1.0×10^{-2}
Sulfurous acid	H_2SO_3	HSO_3^-	1.5×10^{-2}
hydrogen sulfite	HSO_3^-	SO_3^{2-}	1.0×10^{-7}

TABLE A5.4	Solubility Product Constants (K_{sp}) at 298 K					
Name	**Formula**	**K_{sp}**	**Name**	**Formula**	**K_{sp}**
Aluminium hydroxide	$Al(OH)_3$	2.0×10^{-32}	Iron(II) carbonate	$FeCO_3$	2.1×10^{-11}
Aluminium phosphate	$AlPO_4$	9.8×10^{-21}	Iron(II) fluoride	FeF_2	2.4×10^{-6}
Barium carbonate	$BaCO_3$	1.6×10^{-9}	Iron(II) hydroxide	$Fe(OH)_2$	1.8×10^{-15}
Barium chromate	$BaCrO_4$	8.5×10^{-10}	Iron(II) sulfide	FeS	3.7×10^{-19}
Barium fluoride	BaF_2	2.4×10^{-5}	Iron(III) hydroxide	$Fe(OH)_3$	1.6×10^{-39}
Barium hydroxide	$Ba(OH)_2$	5.0×10^{-4}	Iron(III) phosphate	$FePO_4$	9.9×10^{-16}
Barium iodate	$Ba(IO_3)_2$	4.0×10^{-9}	Lead(II) bromide	$PbBr_2$	4.6×10^{-6}
Barium molybdate	$BaMoO_4$	3.5×10^{-8}	Lead(II) carbonate	$PbCO_3$	1.5×10^{-15}
Barium phosphate	$Ba_3(PO_4)_2$	6.0×10^{-39}	Lead(II) chloride	$PbCl_2$	1.6×10^{-5}
Barium selenate	$BaSeO_4$	3.4×10^{-8}	Lead(II) chromate	$PbCrO_4$	2.0×10^{-16}
Barium sulfate	$BaSO_4$	1.5×10^{-9}	Lead(II) fluoride	PbF_2	4.0×10^{-8}
Barium sulfite	$BaSO_3$	5.0×10^{-10}	Lead(II) hydroxide	$Pb(OH)_2$	1.2×10^{-15}
Beryllium hydroxide	$Be(OH)_2$	6.9×10^{-22}	Lead(II) iodide	PbI_2	1.4×10^{-8}
Bismuth arsenate	$BiAsO_4$	4.4×10^{-10}	Lead(II) phosphate	$Pb_3(PO_4)_2$	1.0×10^{-54}
Bismuth iodide	BiI	7.7×10^{-19}	Lead(II) sulfate	$PbSO_4$	1.3×10^{-8}
Cadmium carbonate	$CdCO_3$	5.2×10^{-12}	Lead(II) sulfide	PbS	7.0×10^{-29}
Cadmium fluoride	CdF_2	6.4×10^{-3}	Lithium carbonate	Li_2CO_3	8.2×10^{-4}
Cadmium hydroxide	$Cd(OH)_2$	2.5×10^{-14}	Magnesium carbonate	$MgCO_3$	1.0×10^{-5}
Cadmium oxalate	CdC_2O_4	1.4×10^{-8}	Magnesium hydroxide	$Mg(OH)_2$	8.9×10^{-12}
Cadmium phosphate	$Cd_3(PO_4)_2$	2.5×10^{-33}	Magnesium oxalate	MgC_2O_4	4.8×10^{-6}
Cadmium sulfide	CdS	1.0×10^{-28}	Magnesium phosphate	$Mg_3(PO_4)_2$	1.0×10^{-24}
Calcium carbonate	$CaCO_3$	8.7×10^{-9}	Magnesium fluoride	MgF_2	6.4×10^{-9}
Calcium fluoride	CaF_2	4.0×10^{-11}	Manganese(II) carbonate	$MnCO_3$	8.8×10^{-11}
Calcium hydroxide	$Ca(OH)_2$	1.3×10^{-6}	Manganese(II) hydroxide	$Mn(OH)_2$	2.0×10^{-13}
Calcium iodate	$Ca(IO_3)_2$	7.1×10^{-7}	Manganese(II) oxalate	MnC_2O_4	1.7×10^{-7}
Calcium oxalate	CaC_2O_4	2.3×10^{-9}	Manganese(II) sulfide	MnS	2.3×10^{-13}
Calcium phosphate	$Ca_3(PO_4)_2$	1.3×10^{-32}	Mercury(I) bromide	Hg_2Br_2	1.3×10^{-22}
Calcium sulfate	$CaSO_4$	6.1×10^{-5}	Mercury(I) carbonate	Hg_2CO_3	9.0×10^{-15}
Cesium perchlorate	$CsClO_4$	4.0×10^{-3}	Mercury(I) chloride	Hg_2Cl_2	1.1×10^{-18}
Cesium periodate	$CsIO_4$	5.2×10^{-6}	Mercury(I) chromate	Hg_2CrO_4	2.0×10^{-9}
Chromium(III) hydroxide	$Cr(OH)_3$	6.7×10^{-31}	Mercury(I) fluoride	Hg_2F_2	3.1×10^{-6}
Cobalt(II) arsenate	$Co_3(AsO_4)_2$	6.8×10^{-29}	Mercury(I) iodide	Hg_2I_2	4.5×10^{-29}
Cobalt(II) carbonate	$CoCO_3$	1.0×10^{-10}	Mercury(I) oxalate	$Hg_2C_2O_4$	1.8×10^{-13}
Cobalt(II) hydroxide	$Co(OH)_2$	2.5×10^{-16}	Mercury(I) sulfate	Hg_2SO_4	6.5×10^{-7}
Cobalt(II) phosphate	$Co_3(PO_4)_2$	2.1×10^{-35}	Mercury(I) thiocyanate	$Hg_2(SCN)_2$	3.2×10^{-20}
Cobalt(II) sulfide	CoS	5.0×10^{-22}	Mercury(II) bromide	$HgBr_2$	6.2×10^{-20}
Cobalt(III) hydroxide	$Co(OH)_3$	2.5×10^{-43}	Mercury(II) iodide	HgI_2	2.9×10^{-29}
Copper(I) bromide	$CuBr$	6.3×10^{-9}	Mercury(II) hydroxide	$Hg(OH)_2$	3.0×10^{-26}
Copper(I) chloride	$CuCl$	1.7×10^{-7}	Mercury(II) oxide	HgO	3.6×10^{-26}
Copper(I) cyanide	$CuCN$	3.5×10^{-20}	Mercury(II) sulfide	HgS	1.6×10^{-54}
Copper(I) oxide	Cu_2O	2.0×10^{-15}	Neodymium carbonate	$Nd_2(CO_3)_3$	1.1×10^{-33}
Copper(I) iodide	CuI	1.3×10^{-12}	Nickel(II) carbonate	$NiCO_3$	1.4×10^{-7}
Copper(I) thiocyanate	$CuSCN$	1.8×10^{-13}	Nickel(II) hydroxide	$Ni(OH)_2$	1.6×10^{-16}
Copper(II) carbonate	$CuCO_3$	2.5×10^{-10}	Nickel(II) phosphate	$Ni_3(PO_4)_2$	4.7×10^{-32}
Copper(II) hydroxide	$Cu(OH)_2$	1.6×10^{-19}	Nickel(II) sulfide	NiS	3.0×10^{-21}
Copper(II) oxalate	CuC_2O_4	4.4×10^{-10}	Palladium(II) thiocyanate	$Pd(SCN)_2$	4.4×10^{-23}
Copper(II) phosphate	$Cu_3(PO_4)_2$	1.4×10^{-37}	Praseodymium hydroxide	$Pr(OH)_3$	3.4×10^{-24}
Copper(II) sulfide	CuS	8.5×10^{-45}	Radium sulfate	$RaSO_4$	3.7×10^{-11}
Europium(III) hydroxide	$Eu(OH)_3$	9.4×10^{-27}	Scandium fluoride	ScF_3	5.8×10^{-24}
Gallium(III) hydroxide	$Ga(OH)_3$	7.3×10^{-36}	Scandium hydroxide	$Sc(OH)_3$	2.2×10^{-31}

(Continued)

TABLE A5.4   (*Continued*)

Name	Formula	K_{sp}	Name	Formula	K_{sp}
Silver(I) acetate	$AgCH_3COO$	1.9×10^{-3}	Strontium phosphate	$Sr_3(PO_4)_2$	1.0×10^{-31}
Silver(I) arsenate	Ag_3AsO_4	1.0×10^{-22}	Strontium sulfate	$SrSO_4$	3.2×10^{-7}
Silver(I) bromide	$AgBr$	5.0×10^{-13}	Thallium(I) bromide	$TlBr$	3.7×10^{-6}
Silver(I) carbonate	Ag_2CO_3	8.1×10^{-12}	Thallium(I) chloride	$TlCl$	1.9×10^{-4}
Silver(I) chloride	$AgCl$	1.6×10^{-10}	Thallium(I) chromate	Tl_2CrO_4	8.7×10^{-13}
Silver(I) chromate	Ag_2CrO_4	9.0×10^{-12}	Thallium(I) hydroxide	$Tl(OH)_3$	1.7×10^{-44}
Silver(I) cyanide	$AgCN$	6.0×10^{-17}	Thallium(I) iodide	TlI	5.5×10^{-8}
Silver(I) hydroxide	$AgOH$	2.0×10^{-8}	Thallium(I) sulfide	Tl_2S	6.0×10^{-22}
Silver(I) iodide	AgI	1.5×10^{-16}	Tin(II) hydroxide	$Sn(OH)_2$	3.0×10^{-27}
Silver(I) oxalate	$Ag_2C_2O_4$	5.4×10^{-12}	Tin(II) sulfide	SnS	1.0×10^{-26}
Silver(I) phosphate	Ag_3PO_4	1.8×10^{-18}	Yttrium carbonate	$Y_2(CO_3)_3$	1.0×10^{-31}
Silver(I) sulfate	Ag_2SO_4	1.2×10^{-5}	Yttrium hydroxide	$Y(OH)_3$	1.0×10^{-22}
Silver(I) sulfite	Ag_2SO_3	1.5×10^{-14}	Zinc carbonate	$ZnCO_3$	2.0×10^{-10}
Silver(I) sulfide	Ag_2S	1.6×10^{-49}	Zinc fluoride	ZnF	3.0×10^{-2}
Strontium carbonate	$SrCO_3$	7×10^{-10}	Zinc hydroxide	$Zn(OH)_2$	4.5×10^{-17}
Strontium chromate	$SrCrO_4$	3.6×10^{-5}	Zinc oxalate	ZnC_2O_4	1.4×10^{-9}
Strontium fluoride	SrF_2	7.9×10^{-10}	Zinc selenide	$ZnSe$	3.6×10^{-26}
Strontium hydroxide	$Sr(OH)_2$	3.2×10^{-4}	Zinc sulfide	ZnS	2.5×10^{-22}
Strontium oxalate	SrC_2O_4	5.0×10^{-8}			

TABLE A5.5   Complex Ion Formation Constants (K_f) at 298 K

Complex Ion	K_f	Complex Ion	K_f	Complex Ion	K_f
$[Ag(CN)_2]^-$	5.6×10^{18}	$[Co(ox)_3]^{3-}$	1.0×10^{20}	$[HgI_4]^{2-}$	6.8×10^{29}
$[Ag(EDTA)]^{3-}$	2.1×10^7	$[Cr(EDTA)]^-$	1.0×10^{23}	$[Hg(ox)_2]^{2-}$	9.5×10^6
$[Ag(en)_2]^+$	5.0×10^7	$[Cr(OH)_4]^-$	8.0×10^{29}	$[Ni(CN)_4]^{2-}$	2.0×10^{31}
$[Ag(NH_3)_2]^+$	1.6×10^7	$[CuCl_3]^{2-}$	5.0×10^5	$[Ni(EDTA)]^{2-}$	3.6×10^{18}
$[Ag(SCN)_4]^{3-}$	1.2×10^{10}	$[Cu(CN)_2]^-$	1.0×10^{16}	$[Ni(en)_3]^{2+}$	2.1×10^{18}
$[Ag(S_2O_3)_2]^{3-}$	1.7×10^{13}	$[Cu(CN)_4]^{3-}$	2.0×10^{30}	$[Ni(NH_3)_6]^{2+}$	5.5×10^8
$[Al(EDTA)]^-$	1.3×10^{16}	$[Cu(EDTA)]^{2-}$	5.0×10^{18}	$[Ni(ox)_3]^{4-}$	3.0×10^8
$[Al(OH)_4]^-$	1.1×10^{33}	$[Cu(en)_2]^{2+}$	1.0×10^{20}	$[PbCl_3]^-$	2.4×10^1
$[Al(ox)_3]^{3-}$	2.0×10^{16}	$[Cu(CN)_4]^{2-}$	1.0×10^{25}	$[Pb(EDTA)]^{2-}$	2.0×10^{18}
$[Au(CN)_2]$	1.6×10^{38}	$[Cu(NH_3)_4]^{2+}$	1.1×10^{13}	$[PbI_4]^{2-}$	3.0×10^4
$[Bi(EDTA)]^{2-}$	8.0×10^{27}	$[Cu(ox)_2]^{2-}$	3.0×10^8	$[Pb(OH)_3]^-$	3.8×10^{14}
$[Ca(EDTA)]^{2-}$	5.0×10^{10}	$[Fe(CN)_6]^{4-}$	1.0×10^{37}	$[Pb(ox)_2]^{2-}$	3.5×10^6
$[Cd(CN)_4]^{2-}$	6.0×10^{18}	$[Fe(EDTA)]^{2-}$	2.1×10^{14}	$[Pb(S_2O_3)_3]^{4-}$	2.2×10^6
$[Cd(en)_3]^{2+}$	1.2×10^{12}	$[Fe(en)_3]^{2+}$	5.0×10^9	$[PtCl_4]^{2-}$	1.0×10^{16}
$[Cd(NH_3)_4]^{2+}$	1.3×10^7	$[Fe(ox)_3]^{4-}$	1.7×10^5	$[Pt(NH_3)_6]^{2+}$	2.0×10^{35}
$[Co(EDTA)]^{2-}$	2.0×10^{16}	$[Fe(CN)_6]^{3-}$	1.0×10^{42}	$[V(EDTA)]^-$	8.0×10^{25}
$[Co(en)_3]^{2+}$	8.7×10^{13}	$[Fe(EDTA)]^-$	1.7×10^{24}	$[Zn(CN)_4]^{2-}$	1.0×10^{18}
$[Co(NH_3)_6]^{2+}$	1.3×10^5	$[Fe(ox)_3]^{3-}$	2.0×10^{20}	$[Zn(EDTA)]^{2-}$	3.0×10^{16}
$[Co(ox)_3]^{4-}$	5.0×10^9	$[Fe(SCN)]^{2+}$	8.9×10^2	$[Zn(en)_3]^{2+}$	1.3×10^{14}
$[Co(SCN)_4]^{2-}$	1.0×10^3	$[HgCl_4]^{2-}$	1.2×10^{15}	$[Zn(NH_3)_4]^{2+}$	4.1×10^8
$[Co(EDTA)]^-$	1.0×10^{36}	$[Hg(CN)_4]^{2-}$	3.0×10^{41}	$[Zn(OH)_4]^{2-}$	4.6×10^{17}
$[Co(en)_3]^{3+}$	4.9×10^{48}	$[Hg(EDTA)]^{2-}$	6.3×10^{21}	$[Zn(ox)_3]^{4-}$	1.4×10^8
$[Co(NH_3)_6]^{3+}$	4.5×10^{33}	$[Hg(en)_2]^{2+}$	2.0×10^{23}		

Appendix 6

Water Vapor Pressure Table

T (°C)	Vapor Pressure (mm Hg)	T (°C)	Vapor Pressure (mm Hg)	T (°C)	Vapor Pressure (mm Hg)
0	4.58	22	19.80	50	92.57
5	6.54	23	21.05	60	149.4
10	9.21	24	22.36	70	233.7
11	9.84	25	23.74	80	355.1
12	10.52	26	25.20	90	525.7
13	11.23	28	28.34	95	634.0
14	11.99	30	31.82	96	657.7
15	12.79	33	37.74	97	682.2
20	17.54	37	47.09	98	707.4
21	18.62	40	55.35	99	733.4
				100	760.0

Appendix 7

Standard Reduction Potentials at 298 K

All half-reaction potentials are reported to two decimal places.

$E°$ (V)	Reduction Half-reaction
+2.87	$F_2(g) + 2e^- \rightarrow 2F^-(aq)$
+2.07	$O_3(g) + 2H^+(aq) + 2e^- \rightarrow O_2(g) + H_2O(l)$
+2.05	$S_2O_8^{2-}(aq) + 2e^- \rightarrow SO_4^{2-}(aq)$
+1.99	$Ag^{2+}(aq) + e^- \rightarrow Ag^+(aq)$
+1.82	$Co^{3+}(aq) + e^- \rightarrow Co^{2+}(aq)$
+1.78	$H_2O_2(aq) + 2H^+(aq) + 2e^- \rightarrow 2H_2O(l)$
+1.70	$Ce^{4+}(aq) + e^- \rightarrow Ce^{3+}(aq)$
+1.69	$Au^+(aq) + e^- \rightarrow Au(s)$
+1.69	$PbO_2(s) + 4H^+(aq) + SO_4^{2-}(aq) + 2e^- \rightarrow PbSO_4(s) + 2H_2O(l)$
+1.68	$MnO_4^-(aq) + 4H^+(aq) + 3e^- \rightarrow MnO_2(s) + 2H_2O(l)$
+1.67	$Pb^{4+}(aq) + 2e^- \rightarrow Pb^{2+}(aq)$
+1.63	$2HClO(aq) + 2H^+(aq) + 2e^- \rightarrow Cl_2(g) + 2H_2O(l)$
+1.60	$2HBrO(aq) + 2H^+(aq) + 2e^- \rightarrow Br_2(l) + 2H_2O(l)$
+1.60	$IO_4^-(aq) + 2H^+(aq) + 2e^- \rightarrow IO_3^-(aq) + H_2O(l)$
+1.51	$MnO_4^-(aq) + 8H^+(aq) + 5e^- \rightarrow Mn^{2+}(aq) + 4H_2O(l)$
+1.51	$Mn^{3+}(aq) + e^- \rightarrow Mn^{2+}(aq)$
+1.50	$Au^{3+}(aq) + 3e^- \rightarrow Au(s)$
+1.46	$PbO_2(s) + 4H^+(aq) + 2e^- \rightarrow Pb^{2+}(aq) + 2H_2O(l)$
+1.36	$Cl_2(g) + 2e^- \rightarrow 2Cl^-(aq)$
+1.33	$CrO_7^{2-}(aq) + 14H^+(aq) + 6e^- \rightarrow 2Cr^{3+}(aq) + 7H_2O(l)$
+1.24	$O_3(g) + H_2O(l) + 2e^- \rightarrow O_2(g) + 2OH^-(aq)$
+1.23	$O_2(g) + 4H^+(aq) + 4e^- \rightarrow 2H_2O(l)$
+1.21	$MnO_2(s) + 4H^+(aq) + 2e^- \rightarrow Mn^{2+}(aq) + 2H_2O(l)$
+1.20	$IO_3^-(aq) + 6H^+(aq) + 5e^- \rightarrow \frac{1}{2}I_2(s) + 3H_2O(l)$
+1.09	$Br_2(l) + 2e^- \rightarrow 2Br^-(aq)$
+1.00	$VO_2^+(aq) + 2H^+(aq) + e^- \rightarrow VO^{2+}(aq) + H_2O(l)$

(Continued)

$E°$ (V)	Reduction Half-reaction
+0.99	$AuCl_4^-(aq) + 3e^- \rightarrow Au(s) + 4Cl^-(aq)$
+0.97	$Pu^{4+}(aq) + e^- \rightarrow Pu^{3+}(aq)$
+0.96	$NO_3^-(aq) + 4H^+(aq) + e^- \rightarrow NO(g) + 2H_2O(l)$
+0.91	$2Hg^{2+}(aq) + 2e^- \rightarrow Hg_2^{2+}(aq)$
+0.89	$ClO^-(aq) + H_2O(l) + e^- \rightarrow Cl^-(aq) + 2OH^-(aq)$
+0.86	$Hg^{2+}(aq) + 2e^- \rightarrow Hg(l)$
+0.80	$NO_3^-(aq) + 2H^+(aq) + e^- \rightarrow NO_2(g) + H_2O(l)$
+0.80	$Ag^+(aq) + e^- \rightarrow Ag(s)$
+0.80	$Hg_2^{2+}(aq) + 2e^- \rightarrow 2Hg(l)$
+0.77	$Fe^{3+}(aq) + e^- \rightarrow Fe^{2+}(aq)$
+0.76	$BrO^-(aq) + H_2O(l) + 2e^- \rightarrow Br^-(aq) + 2OH^-(aq)$
+0.68	$O_2(g) + 2H^+(aq) + 2e^- \rightarrow H_2O_2(aq)$
+0.62	$Hg_2SO_4(s) + 2e^- \rightarrow 2Hg(l) + SO_4^{2-}(aq)$
+0.60	$MnO_4^{2-}(aq) + 2H_2O(l) + 2e^- \rightarrow MnO_2(s) + 4OH^-(aq)$
+0.56	$MnO_4^-(aq) + e^- \rightarrow MnO_4^{2-}(aq)$
+0.54	$I_2(s) + 2e^- \rightarrow 2I^-(aq)$
+0.52	$Cu^+(aq) + e^- \rightarrow Cu(s)$
+0.53	$I_3^-(aq) + 2e^- \rightarrow 3I^-(aq)$
+0.49	$NiOOH(aq) + H_2O(l) + e^- \rightarrow Ni(OH)_2(aq) + OH^-(aq)$
+0.45	$Ag_2CrO_4(s) + 2e^- \rightarrow 2Ag(s) + CrO_4^{2-}(aq)$
+0.40	$O_2(g) + 2H_2O(l) + 4e^- \rightarrow 4OH^-(aq)$
+0.36	$ClO_4^-(aq) + H_2O(l) + 2e^- \rightarrow ClO_3^-(aq) + 2OH^-(aq)$
+0.36	$[Fe(CN)_6]^{3-}(aq) + e^- \rightarrow [Fe(CN)_6]^{4-}(aq)$
+0.34	$Cu^{2+}(aq) + 2e^- \rightarrow Cu(s)$
+0.34	$Hg_2Cl_2(s) + 2e^- \rightarrow 2Hg(l) + 2Cl^-(aq)$
+0.25	$CO(g) + 6H^+(aq) + 6e^- \rightarrow H_2O(g) + CH_4(g)$
+0.22	$AgCl(s) + e^- \rightarrow Ag(s) + Cl^-(aq)$
+0.20	$SO_4^{2-}(aq) + 4H^+(aq) + 2e^- \rightarrow H_2SO_3(aq) + H_2O(l)$
+0.20	$Bi^{3+}(aq) + 3e^- \rightarrow Bi(s)$
+0.16	$Cu^{2+}(aq) + e^- \rightarrow Cu^+(aq)$
+0.15	$Sn^{4+}(aq) + 2e^- \rightarrow Sn^{2+}(aq)$
+0.07	$AgBr(s) + e^- \rightarrow Ag(s) + Br^-(aq)$
0.00	$Ti^{4+}(aq) + e^- \rightarrow Ti^{3+}(aq)$
0.00	$2H^+(aq) + 2e^- \rightarrow H_2(g)$
−0.04	$Fe^{3+}(aq) + 3e^- \rightarrow Fe(s)$
−0.08	$O_2(g) + H_2O(l) + 2e^- \rightarrow HO_2^-(aq) + OH^-(aq)$
−0.13	$Pb^{2+}(aq) + 2e^- \rightarrow Pb(s)$
−0.14	$In^+(aq) + e^- \rightarrow In(s)$
−0.14	$Sn^{2+}(aq) + 2e^- \rightarrow Sn(s)$
−0.15	$AgI(s) + e^- \rightarrow Ag(s) + I^-(aq)$
−0.23	$Ni^{2+}(aq) + 2e^- \rightarrow Ni(s)$
−0.28	$Co^{2+}(aq) + 2e^- \rightarrow Co(s)$
−0.34	$In^{3+}(aq) + 3e^- \rightarrow In(s)$
−0.34	$Tl^+(aq) + e^- \rightarrow Tl(s)$
−0.35	$PbSO_4(s) + 2e^- \rightarrow Pb(s) + SO_4^{2-}(aq)$
−0.37	$Tl^{3+}(aq) + e^- \rightarrow Tl^{2+}(aq)$
−0.40	$Cd^{2+}(aq) + 2e^- \rightarrow Cd(s)$
−0.40	$In^{2+}(aq) + e^- \rightarrow In^+(aq)$
−0.44	$Fe^{2+}(aq) + 2e^- \rightarrow Fe(s)$
−0.44	$In^{3+}(aq) + 2e^- \rightarrow In^+(aq)$
−0.48	$S(s) + 2e^- \rightarrow S^{2-}(aq)$
−0.49	$In^{3+}(aq) + e^- \rightarrow In^{2+}(aq)$
−0.50	$Cr^{3+}(aq) + e^- \rightarrow Cr^{2+}(aq)$

$E°$ (V)	Reduction Half-reaction
−0.61	$U^{4+}(aq) + e^- \rightarrow U^{3+}(aq)$
−0.73	$Cr^{3+}(aq) + 3e^- \rightarrow Cr(s)$
−0.76	$Zn^{2+}(aq) + 2e^- \rightarrow Zn(s)$
−0.81	$Cd(OH)_2(aq) + 2e^- \rightarrow Cd(s) + 2OH^-(aq)$
−0.83	$2H_2O(l) + 2e^- \rightarrow H_2(g) + 2OH^-(aq)$
−0.91	$Cr^{2+}(aq) + 2e^- \rightarrow Cr(s)$
−1.18	$Mn^{2+}(aq) + 2e^- \rightarrow Mn(s)$
−1.19	$V^{2+}(aq) + 2e^- \rightarrow V(s)$
−1.63	$Ti^{2+}(aq) + 2e^- \rightarrow Ti(s)$
−1.66	$Al^{3+}(aq) + 3e^- \rightarrow Al(s)$
−1.79	$U^{3+}(aq) + 3e^- \rightarrow U(s)$
−2.09	$Sc^{3+}(aq) + 3e^- \rightarrow Sc(s)$
−2.23	$H_2(g) + 2e^- \rightarrow 2H^-(aq)$
−2.37	$Mg^{2+}(aq) + 2e^- \rightarrow Mg(s)$
−2.37	$La^{3+}(aq) + 3e^- \rightarrow La(s)$
−2.48	$Ce^{3+}(aq) + 3e^- \rightarrow Ce(s)$
−2.71	$Na^+(aq) + e^- \rightarrow Na(s)$
−2.76	$Ca^{2+}(aq) + 2e^- \rightarrow Ca(s)$
−2.89	$Sr^{2+}(aq) + 2e^- \rightarrow Sr(s)$
−2.90	$Ba^{2+}(aq) + 2e^- \rightarrow Ba(s)$
−2.92	$Ra^{2+}(aq) + 2e^- \rightarrow Ra(s)$
−2.92	$Cs^+(aq) + e^- \rightarrow Cs(s)$
−2.92	$K^+(aq) + e^- \rightarrow K(s)$
−2.93	$Rb^+(aq) + e^- \rightarrow Rb(s)$
−3.05	$Li^+(aq) + e^- \rightarrow Li(s)$

Appendix 8

Common Radioactive Nuclei

More information can be found at the following websites:
http://www.epa.gov/radiation/radionuclides/index.html (accessed November 2005)
http://www.ndc.tokai.jaeri.go.jp/CN04/ (accessed November 2005)

Element	Nuclide	Half-life	Element	Nuclide	Half-life
Americium	^{240}Am	51 h	Plutonium	^{238}Pu	87.7 y
Americium	^{241}Am	432.2 y	Plutonium	^{239}Pu	2.44×10^5 y
Americium	^{242}Am	16 h	Plutonium	^{240}Pu	6.56×10^3 y
Carbon	^{14}C	5730 y	Polonium	^{210}Po	138 d
Cesium	^{137}Cs	30 y	Radium	^{226}Ra	1.6×10^3 y
Cobalt	^{58}Co	71 d	Radon	^{220}Rn	54.5 s
Cobalt	^{60}Co	5.271 y	Radon	^{222}Rn	3.8 d
Fluorine	^{18}F	110 min	Strontium	^{85}Sr	64 d
Gallium	^{67}Ga	78.25 h	Strontium	^{90}Sr	28.9 y
Gold	^{198}Au	2.69 d	Technetium	^{99}Tc	2.13×10^5 y
Hydrogen	^{3}H	12.3 y	Technetium	^{99m}Tc	6.0 h
Iodine	^{129}I	1.57×10^7 y	Thallium	^{201}Tl	21.5 h
Iodine	^{131}I	8.040 d	Thorium	^{232}Th	1.4×10^{10} y
Lead	^{210}Pb	22.3 y	Uranium	^{234}U	2.46×10^5 y
Molybdenum	^{99}Mo	67 h	Uranium	^{235}U	7.0×10^8 y
Nobelium	^{250}No	250 μs	Uranium	^{238}U	4.5×10^9 y
Palladium	^{103}Pd	16.97 d			

Answers to Practice Exercises and Selected Exercises

Chapter 1

P1.1 a. chemical; **b.** chemical; **c.** physical; **d.** physical; **e.** physical; **f.** chemical   **P1.2 a.** homogeneous; **b.** heterogeneous; **c.** heterogeneous; **d.** homogeneous; **e.** homogeneous; **f.** heterogeneous
P1.3 Answers to this problem may vary. The following solution is only one possibility.

Step 1: *Formulating a question:* The question is provided: "There is a dark liquid in my cup. What is the liquid and how did it get there?"
Step 2: *Finding out what is already known about your question.* You could ask people nearby if they knew what was in your cup, or if they had themselves poured the liquid in the cup. You may have your answer to the question after completing this step. Similar findings occur in scientific investigations in that a specific problem may have already been solved. Step 3: *Making observations.* Your first observations may be the odor you smell from the cup, the actual color and deepness of color of the liquid, and the thickness of the liquid in the cup. As a final step, you may wish to taste the liquid, which should provide the true identity of the liquid. Step 4: *Creating a hypothesis:* Here you would begin to posit an explanation of how the liquid got into your cup: "My friend Anne poured the coffee into my cup." Or "My nephew emptied his juice cup into my cup." To pose the proper question you need to both identify what is in the cup and how it got there. Step 5: *Designing and performing experiments.* You could create several experiments that could range from asking everyone around if they had seen what had happened to checking for fingerprints. If this is a repeated occurrence, you could hold a "stakeout."

Depending on what you find, you may need to change your hypothesis. For example, Anne might not have been to school or work that day and couldn't have done it. If it is a repeated occurrence and the same thing happened each time, you might have a reasonable theory that stated: "My nephew empties his grape juice into my cup each time he is here." Otherwise, if a reasonable explanation cannot be found, you might be limited to stating a law such as "Each Tuesday, coffee appears in my cup."   **P1.4** 104°C, 377 K   **P1.5** intensive
P1.6 3.0×10^1 cm^3 **P1.7 a.** 24600 g; **b.** 15.5 gal; **c.** $19.51CN
P1.8 1003 m^2, 1,555,200 in^2 **P1.9 a.** five; **b.** three or four; **c.** two; **d.** four; problem: two

1. "Chemistry is concerned with the systematic study of the matter of our universe. This study involves the composition, structure, and properties of matter." Because chemistry is a systematic study (following the scientific method) of our surroundings, it is a science.
3. Some possible answers include carpet, plastic soda bottles, computers, foods, inks, building materials, ceramic tile, sunglasses, *anything!*   **5.** The recipe will work only if we add 1 cup of water (no more, no less) to the mix. Too much water and the batter will be thin and watery; too little water and there might not be enough liquid to fully mix and bind the ingredients. Likewise, precise amounts are required for chemical reactions and processes to ensure that the correct products or processes are achieved.   **7.** An element is composed of only a single type of atom and cannot be further simplified by chemical or physical processes, whereas a compound is composed of more than one type of atom (or element) and can be separated into its component elements. Oxygen, carbon, and sodium are three examples of elements. Water, salt, and rust (iron oxide) are three examples of compounds.   **9.** Elemental oxygen exists as a diatomic molecule (it contains two atoms of oxygen). It is an element because there is only one type of atom involved. It is a molecule because more than one atom constitutes this natural form. It is not a compound because there are not two or more different types of atoms involved.   **11.** Both can be separated into simpler substances: the mixture by mechanical means into its component parts and the

compound by chemical means into its component elements.
13. a. element; **b.** compound: sodium and chlorine; **c.** compound: carbon, hydrogen, oxygen; **d.** element; **e.** compound: sulfur, copper, oxygen; **f.** element   **15. a.** both possible; **b.** heterogeneous; **c.** heterogeneous; **d.** homogeneous; **e.** heterogeneous; **f.** heterogeneous
17. distillation, reverse osmosis, electrodialysis, boiling/freezing
19. a. physical; **b.** chemical; **c.** physical; **d.** physical   **21.** It is a mixture of several gases: oxygen, nitrogen, argon, and others.   **23.** By stating that we should "Eat natural food, not chemicals," the writer implies that natural food is not made up of chemicals. All the food we eat is exactly that—chemicals. All everyday matter is composed of chemicals! What the writer was trying to convey is that we should eat natural food, not synthetically or artificially produced chemicals (or additives) or those foods produced with pesticides, artificial fertilizers, etc.   **25.** You would make *observations* to gather data about what is not working or why it is not working. Based on those observations you could develop a working *hypothesis* that was consistent with your observations that might explain what had happened. You would then *develop experiments* that would test your hypothesis. Depending on the outcome of the tests, you will have either solved the problem or ruled out your hypothesis, which would make a new hypothesis and experiments necessary.   **27.** Anything that involves observation or analysis of experimental data can be open to interpretation and can influence what hypotheses or further experiments might be developed. To some extent, finding out what is already known and what well-established theories exist should be the least ambiguous because they have been the most extensively tested and refined.   **29.** A hypothesis is a possible explanation for some observations, usually with little or no testing. With much testing, a hypothesis may eventually become a theory, which carries much more weight scientifically than a hypothesis.   **31.** Conflicting results can often be attributed to the fact that some variable is not the same in the two studies.   **33.** Some of the important questions: Are the fishes dying only in town? Are there places in the river where they are not dying? Are there contaminants in the water known to be lethal to fish? How do the levels of the contaminants vary with proximity to the industrial area, to the farmland, and to the town? Chemical analysis of the water at various locations, surveys of fish populations, and analysis of contaminants in the fish themselves are all tests that should be conducted.   **35.** 1 terameter > 1 kilometer > 1 millimeter > 1 nanometer   **37. a.** 1×10^5 g; **b.** 2.59×10^{-2} km; **c.** 77°F; **d.** 3.20×10^{-3} g; **e.** 9.11×10^3 pm; **f.** 37.0°C   **39. a.** 8.7×10^6 mg; **b.** 2.59 m; **c.** 374°F; **d.** 3.20×10^{-4} kL; **e.** 9.11×10^9 ns; **f.** 177°C   **41.** 20 containers   **43. a.** 3.27×10^5 m; **b.** 3.27×10^8 mm; **c.** 3.27×10^{11} μm; **d.** 3.27×10^{14} nm   **45.** 13.6 g/mL   **47.** 1.36 g/mL
49. a. m s^{-1}; **b.** m s^{-2}; **c.** m^3; **d.** J kg^{-1} K^{-1} **51.** ruler with four divisions between each number   **53.** 172°F, 162°F   **55.** 6.5×10^7 atoms
57. a. 7×10^1 ms; **b.** 2.8×10^5 μs;   **59.** The red tomato is denser. Two tomatoes with the same mass can have different densities because they have different volumes; the red tomato must have a smaller volume than the green tomato, even though they have the same mass.   **61.** The density of the water would decrease. We can write the formula Density = mass/volume. As the volume increases with no change in the mass, the density goes down.

63. 4 "just a bits"

65.

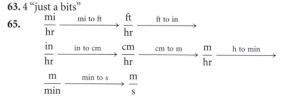

67. 22.6 g/cm³; 11 kg   **69. a.** 56 kph; **b.** 2.24 × 10⁴ cm³/s; **c.** 312 km/L; **d.** 5.125 × 10⁻³ lb/°F   **71. a.** 1.12 × 10⁷ mi/min; **b.** 1.61 × 10¹⁰ mi/day; **c.** 5.87 × 10¹² mi/gr   **73. a.** 0.12 lb; **b.** 3.8 L; **c.** 72.6 kg; **d.** 2.4 m; **e.** 4.0 × 10² cm; **f.** 51 km   **75.** 95 μ yr; 0.95 μcentury; 9.5 μdecade   **77.** 3.27 ×10⁻²² g/atom; 327 yg/atom   **79.** $1.27/s; $13,700/game   **81.** 51.4 kg   **83.** Neither accurate nor precise.   **85. a.** 18.170g; 15.412g; 13.871g; **b.** student 3; **c.** student 2; **d.** human error   **87. a.** 8; **b.** exact; **c.** 3; **d.** 2; **e.** 1; **f.** exact **89. a.** 0.7_0_0 cm; **b.** 0.1_0_1 kg; **c.** _1_00.0 cm; **d.** 100 m (ambiguous); **e.** 0.010_1_0 g   **91. a.** three; **b.** four; **c.** four; **d.** one; **e.** seven; **f.** six; **g.** one; **h.** five; **i.** six; **j.** four   **93. a.** 9.6; **b.** 8.57; **c.** 5.81; **d.** 63; **e.** 8 × 10³; **f.** 19; **g.** 5.8 × 10²; **h.** 71; **i.** 2.81   **95.** 105 cm   **97.** 10.2 g/cm³ **99. a.** The prefix *nano-* refers to the metric multiplier 10⁻⁹, implying technology on very small scales. **b.** 1 × 10⁻⁶ cm   **101.** One advantage is that genetic engineering of corn has greatly increased the yield per acre, and the disadvantages of some engineering include the decrease in genetic diversity of the corn and the risk of the development of pesticide-resistant insects or herbicide-resistant weeds as a consequence of the use of the corn. The genetic selection of crops is as old as agriculture.   **103.** Possible answers could include improved materials, clean water, and life-saving drugs.   **105.** One of the main energy problems that the United States faces is continuing growth of energy demands while oil, gas, and coal reserves are nearing depletion. Because these reserves influence the global energy market, this problem affects developing countries as well. The United States is in a much better position, with its resources, to tackle the problem than poorer countries are.   **107.** All aspects of daily life are likely to be affected.   **109.** Chemical changes: the incorporation of carbon dioxide and oxygen to form glucose in photosynthesis and the intake of nitrogen (from nitrates or other nitrogen-containing molecules) for protein and DNA formation. Physical changes: evaporation and condensation of water in the water cycle that provides moisture to the plants are two important physical changes.   **111.** Observationally, the sugar appears to break up and disappear into the water as it dissolves, while chemically, the molecules in the sugar crystal disperse among the molecules of the water.   **113. a.** Any measurement has some uncertainty, whereas an exact number is infinitely precise; **b.** 36 cans; **c.** 7,200 mL—not exact, 36 cans—exact.   **115.** Two extensive properties of seawater are its volume and mass. Two intensive properties of seawater are its density and temperature.   **117.** 0.0358 km/s   **119.** 1.46 ×10³ gal **121.** The original sources are Miller, S., 1953. "A production of amino acids under possible primitive earth conditions." *Science 117*: 528–529, and Miller, S., and H. Urey, 1959. "Organic compound synthesis on the primitive earth." *Science 130*: 245–251. You should be very careful when conducting Internet searches for this material. Much of what is posted is biased and often patently wrong in that the scientific method has not been properly used. This topic is very "controversial" and both sides are very passionate in the arguments. Although minor parts of the evolutionary theory are still being actively researched and scientists are still working to explain all the details, this is not an indication that the theory itself is wrong. The vast scientific consensus is that biological evolution is real and deserves its status as a theory in the strictest scientific sense.

Chapter 2

P2.1 196 g oxygen; 220 g water   **P2.2** nitrogen-14: 7p, 7n, 7e; neon-20: 10p, 10n, 10e; titanium-48: 22p, 26n, 22e; carbon-11: 6p, 5n, 6e; lithium-11: 3p, 4n, 3e; phosphorus-31: 15p, 16n, 15e   **P2.3** 65.40 amu

P2.4

Symbol	Protons	Neutrons	Electrons	Charge
$^{52}_{24}$Cr⁶⁺	24	28	18	+6
$^{39}_{19}$K¹⁺	19	20	18	+1
$^{79}_{35}$Br¹⁻	35	44	36	−1

P2.5 CsCl   **P2.6** SF₄; CCl₄; P₂O₅; phosphorus trichloride; dinitrogen monoxide; oxygen difluoride   **P2.7** MgCl₂; LiF; sodium bromide; lithium oxide   **P2.8** copper(II) chloride; chromium(VI) oxide; NiO; PdS₂   **P2.9** potassium permanganate; (NH₄)₂Cr₂O₇

1. The slow formation of the crystals on the branches from apparently clear air must indicate the presence of very small particles that are slowing adding to the crystal until their numbers are great enough to be seen.   **3.** Taking the word *dormitory* apart yields the following letters: d, i, m, o, o, r, r, t, y. These letters could be recombined to form the words *dim, dorm, door, toy, dot, rot, trim, try, moor, or, it,* and others. Each of these new words has a different meaning and function than the original word *dormitory*.   **5.** The law of conservation of mass means that atoms can't spontaneously appear or disappear in a chemical reaction; they have to go somewhere. **7.** Just like Democritus, the researcher trying to discover the age of the earth has to ask what processes are occurring and whether these processes have always been the same. The researcher must look for similarities in current and past processes to discover a common theme that can explain a way to determine the age of the earth. **9.** 5.7 g water; law of definite composition and law of conservation of mass   **11. a.** The law of conservation of mass; **b.** 247.8 g   **13.** 4.28 g O; 45.7 g Cu   **15.** If the ratio of 65 g Cu to 16 g O (4.1:1.0) represents a 1:1 ratio of atoms, you would expect a mass ratio of 8.1:1.0 for a 2 Cu : 1 O compound or a 2.0:1.0 mass ratio for a 1 Cu: 2 O compound.   **17.** CaCO₃ (CaCO₃ is 40% Ca; CaCl is 36% Ca)   **19.** The ratio of the amount of oxygen that combines with 2 g of hydrogen in water to the amount that combines with 2 g of hydrogen in hydrogen peroxide is 16:32, which is 1:2. This ratio is a small whole-number ratio, which agrees with the law of multiple proportions. **21.** Proton = +2; neutron = 0 **23.**

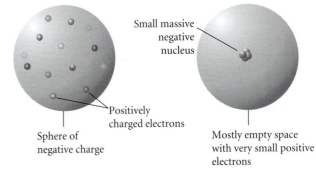

J. J. Thomson's Model   Rutherford's Model

Small massive negative nucleus

Positively charged electrons

Sphere of negative charge

Mostly empty space with very small positive electrons

25.

Isotope	Protons	Neutrons	Electrons	Charge
carbon-12	6	6	6	0
aluminum-27	13	14	10	+3
chlorine-35	17	18	18	−1

27. a. proton; **b.** neutron; **c.** neutrons or electrons   **29.** 1.6735 × 10⁻²⁴ g; 3.3484 × 10⁻²⁴ g   **31.** 0; 0   **33.** 0.54%   **35. a.** Fe; **b.** Te; **c.** Co; **d.** Ne   **37.** ₁H, ₅B, ₉F, ₁₃Al, ₂₀Ca, ₂₆Fe   **39.** ₁H, ₅B, ₐF, ₁₃Al,

$_{20}$Ca, $_{26}$Fe   **41.** H   **43.** +2, cation
45.

Symbol	Protons	Electrons	Neutrons
$^{24}_{11}$Na	11	11	13
$^{131}_{53}$I	53	53	78
$^{60}_{27}$Co	27	27	33
$^{51}_{24}$Cr	24	24	27
$^{32}_{15}$P	15	15	17

47. 8.000000 amu; 0.5000000 amu   **49.** 1.9926×10^{-23} g; 3.3247×10^{-23} g; 3.1550×10^{-23} g   **51.** Indium-113 = 4.2%; indium-115 = 95.8%   **53.** 32.06 amu

Mass Spectrum of Sulfur

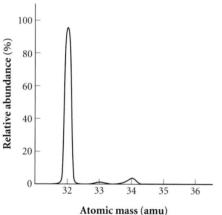

55. 140.1 amu

Mass Spectrum of Cerium

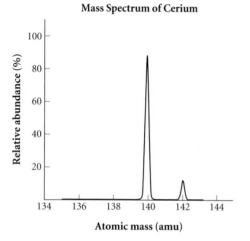

57. a. Be; **b.** Mn; **c.** Kr   **59.** 5;(F, Cl, Br, I, At)   **61.** sulfur – chalcogens (Group VIA); iodine – halogens (Group VIIA); helium – noble gases (Group VIIIA); beryllium – alkaline earth metals (Group IIA); francium – alkali metals (Group IA)   **63.** Metals are the elements that are usually shiny; gold, silver, and lead are metals. Additionally, silicon, as a metalloid, may be shiny.   **65.** Ca; Al; Al; Sr; Be   **67.** LiCl; BeCl$_2$; NaCl; CaCl$_2$; AlCl$_3$ **69.** MgBr$_2$, Mg^{2+} and Br$^-$; FeCl$_3$, Fe^{3+} and Cl$^-$; KI, K$^+$ and I$^-$; Na$_2$S, Na$^+$ and S^{2-} **71. a.** ionic compounds; **b.** molecules; **c.** ionic compounds   **73.** K$_2$O; WO$_2$ **75. a.** HO; **b.** C$_2$H$_5$; **c.** CF$_2$; **d.** CH$_2$O   **77.** SO$_2$ – sulfur dioxide;

N$_2$O$_5$ – dinitrogen pentoxide; Cl$_2$O – dichlorine monoxide; PCl$_3$ – phosphorus trichloride; CCl$_4$ – carbon tetrachloride   **79.** K$_2$O – potassium oxide; CaBr$_2$ – calcium bromide; Li$_3$N – lithium nitride; AlCl$_3$ – aluminum chloride; BaS – barium sílfide   **81.** CaBr$_2$ – calcium bromide; Fe(NO$_3$)$_3$ – iron(III) nitrate; CaSO$_4$ – calcium sulfate; NH$_4$Cl – ammonium chloride; NaCl – sodium chloride   **83.** copper(II) hydroxide – Cu(OH)$_2$; magnesium sulfate – MgSO$_4$; chromium(III) oxide – Cr$_2$O$_3$; sodium sulfite – Na$_2$SO$_3$; sulfur hexachloride – SCl$_6$; ammonium hydroxide – NH$_4$OH; carbon tetraiodide – CI$_4$; boron tribromide – BBr$_3$; aluminum hydroxide – Al(OH)$_3$; sodium acetate – NaCH$_3$COO   **85.** Mn$_2$(SO$_4$)$_5$; MnCl$_5$; Mn(NO$_2$)$_5$; Mn$_2$(CO$_3$)$_5$; Mn(HSO$_3$)$_5$ **87.** MnC$_2$O$_4$; Cu$_2$C$_2$O$_4$; Fe$_2$(C$_2$O$_4$)$_3$; Mn$_2$(C$_2$O$_4$)$_5$; Ti(C$_2$O$_4$)$_2$ **89.** (NH$_4$)$_2$CO$_3$ – ammonium carbonate; NaHCO$_3$ – sodium bicarbonate; Cu(HSO$_3$)$_2$ – copper (II) bisulfite; Ca(OH)$_2$ – calcium hydroxide; KMnO$_4$ – potassium permanganate; Na$_3$PO$_4$ – sodium phosphate; Mg(CN)$_2$ – magnesium cyanide; LiClO$_3$ – lithium chlorate   **91.** V = +5; Ti = +2; W = +6; Ag = +1; Ru = +3   **93.** The formula Fe(NO$_3$) does not give enough information to identify the salt, because it could be either Fe(NO$_3$)$_2$ or Fe(NO$_3$)$_3$. By using the compound in a chemical reaction, you should be able to determine its identity. The two forms of iron nitrate also have different physical properties that could be used to identify the salt.   **95.** The major objections to Dalton's original atomic theory have been that atoms do appear to be divisible (they can emit radiation) and there do appear to be atoms of the same element that are not identical (isotopes). Dalton's atomic theory would serve most of chemistry if we rewrote it as follows: Every substance is made of atoms.

- Atoms are divisible and consist of electrons, protons, and neutrons. The protons and neutrons occupy the center of the atom (the nucleus), and the electrons occupy the space around the nucleus.

- All atoms of any one element have the same number of protons and have the same chemical properties.

- The average atomic masses of different elements are different, and the masses of individual isotopes are related to the total number of protons, neutrons, and electrons in the atom.

- A chemical reaction rearranges the attachments between atoms in a compound.

97. The law of combining volumes would apply in both cases, in that both predict small whole-number ratios. The law of definite composition applies to both cases for the same reason. The important fact that Dalton did not account for was that elemental hydrogen and chlorine exist as diatomic molecules. Because hydrogen chloride contains one atom of each, two molecules of hydrogen chloride are formed from one molecule each of hydrogen and chlorine.   **99.** The number of balls per box and the total price of the box   **101.** 9.81×10^{-28} kg. This mass was lost as energy.   **103.** Yes, 721 mi   **105. a.** 4p$^+$, 4e$^-$; **b.** 13; **c.** 2.1768×10^{-26} kg   **107.** One condition that might have changed that would affect radiocarbon dating is the relative amount of carbon-13 available to be incorporated into the plants. If the amount in the past was greater than what is measured now, the amount of carbon-13 still present would be higher than expected, making the artifact appear not to be as old as it is. On the other hand, if the amount of carbon-13 in the past was lower than the current amount, the artifact would appear to be older than it is.

Chapter 3

P3.1 a. 299.4 amu; **b.** 120.38 amu; **c.** 227.14 amu
P3.2 3.06×10^{23} atoms   **P3.3** 4.78×10^{-4} mol; 9.0×10^{23} molecules   **P3.4** NaCl: 1.5×10^3 g, 0.15 g; H$_2$O: 4.7×10^2 g, 0.045 g; aspartame: 7.7×10^3 g; 0.74 g   **P3.5** C$_3$H$_6$O$_3$: 0.059 mol, 2.4×10^{-5} mol; H$_2$SO$_4$: 0.054 mol, 2.2×10^{-5} mol   **P3.6** 5.01×10^{22} molecules,

1.59×10^7 g, swimming pool   **P3.7** 52.45% K   **P3.8** CH, C_6H_6
P3.9 $C_2H_4O_2$ **P3.10** $Ca_3(PO_4)_2 + 3H_2SO_4 \rightarrow 3CaSO_4 + 2H_3PO_4$
P3.11 72.04 g   **P3.12** 678 kg   **P3.13** 71 g   **P3.14** 18.47 g
P3.15. 88.6 g

1. a. 28.01 amu; **b.** 60.09 amu; **c.** 17.03 amu; **d.** 158.12 amu;
e. 891.5 amu   **3.** $H_2O < CO < C_2H_4OH < C_6H_6 < CaCl_2$
5. a. saccharin = 183.19 amu; aspartame = 294.3 amu;
b. 1.607: 1.000; **c.** 26.1 g   **7.** $\$3 \times 10^{-21}$/atom   **9. a.** 10.8 g;
b. 3.271×10^{-10} g; **c.** 7.181×10^{-22} g; **d.** 7.321×10^{-23} g
11. a. 4.014×10^{23}; **b.** 4.896×10^{23}; **c.** 4×10^{24}; **d.** 7.7×10^{18}
13. a. 60.06 amu; **b.** 60.06 g; **c.** 6.006 g   **15.** 1.99×10^{21} molecules
17. 4.2×10^{23} units   **19. a.** 1.48 mol; **b.** 0.025 mol; **c.** 0.57 mol;
d. 7.5 mol   **21.** 245 g C   **23. a.** 8.14×10^{22} units; **b.** 9.03×10^{24}
units; **c.** 3.8×10^{21} units   **25. a.** 3.0×10^4 g; **b.** 287 g; **c.** 12 g
27. 3.1×10^{22} atoms   **29. a.** 9.274×10^{-23} g; **b.** 1.6×10^{20} atoms;
c. 2.7×10^{-4} mol   **31. a.** 2.5×10^{-5} mol; **b.** 1.5×10^{19} atoms
33. a. 1.30×10^3 g; **b.** 1.77 mol   **35.** 9.22×10^{-4} mol; 5.55×10^{20}
molecules; 4.44×10^{21} atoms   **37. a.** 85.63%; **b.** 37.48%; **c.** 92.26%;
d. 42.10%   **39.** $H_2S > SO_2 > H_2SO_3 > Na_2S_2O_4$ **41.** 85.40%,
23.14%, 61.81%   **43. a.** Na_2SO_4; **b.** $KMnO_4$; **c.** HNO_3 **45.** 18.75%
C, 31.29% Ca, 49.96% C   **47.** The mass percent of carbon in the saturated hydrocarbon would decrease relative to the unsaturated hydrocarbon. The percentage of carbon decreases because the total
mass of the compound increases while the mass of carbon is unchanged; when we divide by the larger total mass, the percentage
decreases   **49.** 8 violinists, 6 brasses, 2 cellos, 1 percussionist
51. CH_2 **53. a.** C_2H_2O; **b.** 84   **55.** $KAgC_2N_2$ **57.** $C_{18}H_{32}O_2$
59. a. 2,1,3,4; **b.** 1,2,1,2; **c.** 1,4,1,5; **d.** 1,6,3,2   **61.** $3CaCl_2 + 2Na_3PO_4$
$\rightarrow Ca_3(PO_4)_2 + 6NaCl$ **63.** $TiCl_4 + 2H_2O \rightarrow TiO_2 + 4HCl$
65. a. 2,1,1,1; **b.** 70.9 g   **67.** 0.639 g   **69. a.** 4,5,4,6; **b.** 59.9 g;
c. 54.0 g; **d.** 21.0 g   **71. a.** 3,2,1,6; **b.** 0.406 g; **c.** 0.358 g
73. a. 50 sandwiches, pickles   **75. a.** 129 g; **b.** 24%   **77.** 6.65 g
79. 11.2 g   **81.** 25.6%   **83.** $4Fe(s) + 3O_2(g) \rightarrow 2Fe_2O_3(s)$, 136 g O_2,
318 g Fe   **85.** $C_6H_{12}O_6 \rightarrow 2C_2H_5OH + 2CO_2$, 0.278 mol
87. The effect of a chemical on the body is often purely an effect of
the amount with one dose having a beneficial effect while another
dose having a detrimental effect. In the case of vitamin C, there is
strong evidence that small amounts are crucial for a healthy life;
however, large doses, while promoted by some, may possibly be
harmful   **89.** The atomic masses given on the periodic table can be
thought of in two ways: First, it is the total mass of 1 mole (6.022 ×
10^{23} atoms) of a substance (with all isotopes present), or second, it is
the weighted average mass of all that element's isotopes. In either
case, a single atom's mass is not the same as the average mass (unless
there is only one possible isotope of that element)   **91.** 19.8 g
93. 5.99×10^4 g   **95.** 83.2% C, 16.8% H, C_5H_{12} **97.** Since different
compounds will contain differing numbers of atoms of each ele-
ment, there is no reason that the coefficients should add up to equal
numbers on each side of a reaction. It is important that there be the
same number of atoms of each element on each side of the equation
99. 5.0 lb   **101.** 1.6 g   **103. a.** C_4H_5; **b.** C_8H_{10}; **c.** $2C_8H_{10} + 13O_2 \rightarrow$
$16CO + 10H_2O$; **d.** 24.5 g; **e.** O_2, 33.4 g CO; **f.** 4.19%

Chapter 4

P4.1 a. 0.17 M; **b.** 0.0202 M; **c.** 33.17 g   **P4.2** 3.8 mol, 12 mol
P4.3 a. 1.40 L; **b.** 2.4 L, 1.2 L   **P4.4** 7.69×10^{-6} M **P4.5** 19 mL
P4.6 0.0445 M **P4.7** $K_2CO_4(aq) + 2HNO_3(aq) \rightarrow H_2C_2O_4(aq) +$
$2KNO_3(aq)$; $2K^+(aq) + C_2O_4^{2-}(aq) + 2H^+(aq) + 2NO_3^-(aq) \rightarrow$
$H_2C_2O_4(aq) + 2K^+(aq) + 2NO_3^-(aq)$; $C_2O_4^{2-}(aq) + 2H^+(aq) \rightarrow$
$H_2C_2O_4(aq)$ **P4.8** $AgNO_3 + NaCl$ (forms $AgCl(s)$); $AgNO_3 + Na_2S$
(forms $Ag_2S(s)$); $AgNO_3 + ZnSO_4$ (forms $Ag_2SO_4(aq)$); $Na_2S +$
$ZnSO_4$ (forms $ZnS(s)$)   **P4.9** 17 g   **P4.10** 7.58 mL   **P4.11** K(+1),
Cl(−1); Fe(+3), O(−2); P(0); C(0), H(+1), Cl(−1); Al(0); P(+3),
Br(−1); H(+1), C(+2), N(−3)   **P4.12** No

1. Because the water molecule contains both partial positive charges
(on the hydrogens) and partial negative charges (on the oxygen), it
can interact favorably with both cations and anions.   **3.** The hydration sphere is the cage of water molecules that surrounds a charged
particle as it dissolves in water.   **5.** Water tends to dissolve those
compounds that have some type of charge on the molecule or ion;
however, oil molecules have very little, if any, charges on the molecule that water can attract. Oil, then, doesn't dissolve because it cannot interact favorably with water.   **7.** When some compounds dissolve, they form anions and cations in the water. Even though the
particles formed a neutral compound before dissolving, these ions
exist separately in the water and are free to move. Since the current
requires freely moving charges, the new ions in the water can carry
the current.   **9.** The apparatus could identify a strong electrolyte
with a brightly lit bulb, but could not distinguish between weak electrolyte and nonelectrolyte solutions that will not light the bulb.
11. c.   **13.** 0.300 mol   **15. a.** 5.68×10^{-4} M; **b.** 2.84×10^{-3} M;
c. 9.9×10^{-3} M **17. a.** 0.11 g; **b.** 0.136 g; **c.** 8.06 g   **19. a.** 0.221 M;
b. 0.442 M; **c.** 55.5 M **21.** 1.2×10^3 L   **23. a.** 0.80 L; **b.** 0.80 L;
c. 9.32 L   **25. a.** 2.5 ppm; **b.** 10.5 ppm; **c.** 41.7 ppm   **27. a.** 7.01 ppm;
b. 5.33×10^3 ppb; **c.** 0.0170%   **29.** Both have the same molarity.
31. 9.93 kg   **33.** 2.34 M **35.** 3.0×10^{-6} g; 1.4×10^{-8} M
37. 0.010 mol   **39. a.** 0.217 M; **b.** 0.0140 M; **c.** 0.22 M
41. 0.10 M $CuCl_2$
43. In words: Potassium hydroxide + hydrochloric acid $\rightarrow$
potassium chloride + water

Molecular equation: $KOH(aq) + HCl(aq) \rightarrow KCl(aq) + H_2O(l)$

Ionic equation: $K^+(aq) + OH^-(aq) + H^+(aq) + Cl^-(aq) \rightarrow$
$K^+(aq) + Cl^-(aq) + H_2O(l)$

Net ionic equation: $OH^-(aq) + H^+(aq) \rightarrow H_2O(l)$

45. Molecular equation: $2NaCl(aq) + Ca(NO_3)_2(aq) \rightarrow$
$2NaNO_3(aq) + CaCl_2(aq)$

Ionic equation: $2Na^+(aq) + 2Cl^-(aq) + Ca^{2+}(aq) + 2NO_3^-(aq) \rightarrow$
$2Na^+(aq) + 2NO_3^-(aq) + Ca^{2+}(aq) + 2Cl^-(aq)$

Net ionic equation: none

47. Brand A has higher conc.; Brand B has more vitamin C.
49. 0.05899 M **51.** $H^+ + OH^- \rightarrow H_2O$, 4.499 M **53. a.** $CaCO_3 +$
$2HCl \rightarrow CaCl_2 + H_2O + CO_2$; **b.** 0.125 g
55. a. $2C_4H_{10}(g) + 13O_2(g) \rightarrow 8CO_2(g) + 10H_2O(l)$, Redox
(Combustion); **b.** $Ca(OH)_2(aq) + 2HNO_3(aq) \rightarrow$
$Ca(NO_3)_2(aq) + 2H_2O(l)$, acid–base; Net ionic equation:
$H^+(aq) + OH^-(aq) \rightarrow H_2O(l)$; **c.** $Pb(NO_3)_2(aq) + 2NaCl(aq) \rightarrow$
$PbCl_2(s) + 2NaNO_3(aq)$, precipitation; Net ionic equation:
$Pb^{2+}(aq) + 2Cl^-(aq) \rightarrow PbCl_2(s)$ **57.** c., d., and e.
59. a. $BaCl_2(aq) + 2NaNO_3(aq) \rightarrow Ba(NO_3)_2(aq) + 2NaCl(aq)$;
Net ionic equation: none; **b.** $2Fe(NO_3)_3(aq) + 3(NH_4)_2SO_4(aq) \rightarrow$
$Fe_2(SO_4)_3(aq) + 6NH_4NO_3$; Net ionic equation: none;
c. $CaCl_2(aq) + K_2SO_4(aq) \rightarrow CaSO_4(s) + 2KCl(aq)$;
Net ionic equation: $Ca^{2+}(aq) + SO_4^{2-}(aq) \rightarrow CaSO_4(s)$
61. a. $Cu(NO_3)_2(aq) + 2KOH(aq) \rightarrow Cu(OH)_2(s) + 2KNO_3(aq)$;
Net ionic equation: $Cu^{2+}(aq) + 2OH^-(aq) \rightarrow Cu(OH)_2(s)$;
b. $3Na_2CO_3(aq) + 2AlCl_3(aq) \rightarrow 6NaCl(aq) + Al_2(CO_3)_3(s)$;
Net ionic equation: $3CO_3^{2-}(aq) + 2Al^{3+}(aq) \rightarrow Al_2(CO_3)_3(s)$;
c. $2(NH_4)_3PO_4(aq) + 3ZnCl_2(aq) \rightarrow 6NH_4Cl(aq) + Zn_3(PO_4)_2(s)$;
Net ionic equation: $2PO_4^{3-}(aq) + 3Zn^{2+}(aq) \rightarrow Zn_3(PO_4)_2(s)$
63. Answers may vary. **a.** $Ba(NO_3)_2 + Na_2S$; **b.** $Cu(NO_3)_2 + NaOH$;
c. $Pb(NO_3)_2 + Na_2SO_4$ **65.** First NaCl, next Na_2SO_4, then Na_2S
67. a. $AgNO_3(aq) + NaCl(aq) \rightarrow NaNO_3(aq) + AgCl(s)$; $Ag^+(aq)$
$+ Cl^-(aq) \rightarrow AgCl(s)$; **b.** 0.867 g AgCl; 0.653 g Ag^+ **69.** $H^+(aq) +$
$OH^-(aq) \rightarrow H_2O(l)$ **71. a.** 0.770 M; **b.** 17.2 mL   **73.** 0.220 M
75. a. $H_2CO_3 + 2NaOH \rightarrow 2H_2O + Na_2CO_3$; $H^+ + OH^- \rightarrow H_2O$;
b. 0.05633 M **77. a.** 0.0130 g; **b.** 0.0146 g; **c.** 0.0250 g   **79. a.** C;
b. F; **c.** O; **d.** P; **e.** O   **81. a.** O(−2), N(+5); **b.** O(−2), P(+5);

c. Cu(+2), C(+4), O(–2); **d.** N(0); **e.** H(+1), S(+4), O(–2)
83. Yes, Fe (+2 to +3) and Cr (+6 to +3)   **85.** Water is called the universal solvent because it dissolves so many molecules of varying sizes from small ionic compounds to very large proteins and DNA.
87. 43.4 mL   **89.** CuOH, $BaCO_3$, and Cu_2CO_3 **91.** $Ba(NO_3)_2(aq)$ + $Na_2SO_4(aq) \rightarrow BaSO_4(s) + 2NaNO_3(aq)$; $Ba^{2+}(aq) + 2\ NO_3^-(aq)$ + $2Na^+(aq) + SO_4^{2-}(aq) \rightarrow BaSO_4(s) + 2Na^+(aq) + 2NO_3^-(aq)$; 13.8 g   **93. a.** 0.016 M; **b.** 2.8×10^3 ppm; **c.** 2.61×10^{22} C atoms; **d.** 1.2×10^{25} molecules   **94. a.** soluble; **b.** $H_2C_2O_4(s) \rightarrow$ $H_2C_2O_4(aq) \rightarrow 2H^+(aq) + C_2O_4^{2-}(aq)$; **c.** CO_2; **d.** MnO_4^-; **e.** 10; **f.** MnO_4^- reduced, $C_2O_4^{2-}$ oxidized; **g.** 0.0736 g

Chapter 5

P5.1 Plants use energy in the environment (from the Sun, in the chemical bonds of water, carbon dioxide, minerals) to create sugars and starches. These sugars and starches, along with everything else that makes up a plant, are storage systems for chemical energy, which is the total of the kinetic and potential energies due to the motion and position of the atoms of the chemicals. Thus plants serve as storage depots of chemical energy for animals that eat them—and even for future fossil fuels as the plant material is converted into coal or oil.   **P5.2** −270 L·atm, $−2.70 \times 10^4$ J   **P5.3** +44.8 J   **P5.4** 2.9   **P5.5** −3220 kJ/mol   **P5.6** $−6.03 \times 10^2$ kJ   **P5.7 a.** 0.494 mol; **b.** 1.52 mol   **P5.8** +1973 kJ   **P5.9** −1532 kJ

1. At the top of one side, the skateboarder has only potential energy. As the skater begins down one side of the half-pipe, the potential energy begins to decrease as it is converted to the kinetic energy of the skater. As the skater hits the bottom of the half pipe, the kinetic energy is at a maximum and potential energy is at a minimum. As the skater begins to climb the opposite wall of the half-pipe, the kinetic energy decreases and some of it is converted into potential energy. At the top of the wall, the kinetic energy reaches zero and potential energy is at a maximum.   **3. a.** 5.4×10^2 J; **b.** 1.3×10^{-2} J; **c.** 1.1×10^{-20} J   **5. a.** potential; **b.** potential and kinetic; **c.** kinetic
7. Some of the kinetic energy in the particles is what is transferred between the system and the surroundings. The transfer is completed as the more energetic (higher-temperature) particles collide and transfer energy to the less energetic (lower-temperature) particles.
9. system: chemicals in combustion; surroundings: everything else; system is losing; $w = $ "−"   **11.** 36 J, lose   **13. a.** 90 J, lose; **b.** 930 J, lose; **c.** 9 kJ, lose; **d.** 44 kJ, lose   **15.** CO_2 is fully combusted (digested) and can release no further heat in these processes. **17.** Less force is required because of lower gravity and less atmospheric drag.   **19.** $q = $ "+"; $w = $ "−"   **21.** 19 m/s
23. 4.1×10^{-21} J   **25.** 2.6 Cal/g; 2.6 kcal/g; 11 kJ/g   **27.** 2.93 kJ
29. 1.06 kJ   **31. a.** 71.0°C; **b.** 71.0°C   **33.** 1.23×10^3 kJ   **35.** The size of a temperature *change* is the same in kelvins as in degrees Celsius.   **37.** part b   **39.** 0.239 g; 5.8 J, 0.0243 $J \cdot g^{-1} \cdot °C^{-1}$
41. 0.444 $J \cdot g^{-1} \cdot °C^{-1}$ **43.** 26.72 kJ/°C   **45. a.** 49.3 kJ/g; **b.** molar mass of sugar   **47.** 6.5 g   **49.** CH_4 **51.** 30.3°C   **53.** Any gases that are generated as a result of the chemical process will expand (or contract) until the pressure of the gas matches the atmospheric pressure. Because the entire process begins and ends at the same pressure, we can consider the process to be a constant-pressure process.   **55.** Enthalpy is equivalent to the amount of heat energy transferred in a constant-pressure process.   **57.** surroundings
59. 1 atm pressure; 1 M concentrations (25°C is common, but not standard)   **61. b.** forms 2 moles (not 1) of NH_3; **c.** describes phase change not formation; **d.** nonstandard state for H_2; **e.** and **f.** more than one product and nonelemental reactants   **63.** ΔH and ΔU differ by the term $P\Delta V$. Because there are 29 more moles of gas as products, ΔV and $P\Delta V$ are large, making ΔH and ΔU different.   **65. a.** $C(s) + \frac{1}{2}O_2(g) \rightarrow CO(g)$; **b.** exothermic; **c.** +110.5 kJ/mol   **67.** $C_4H_{10}(g) + \frac{13}{2} O_2(g) \rightarrow 4CO_2(g) + 5H_2O(l)$
69. 10 C(s, graphite) + $\frac{15}{2}$ $H_2(g)$ + $\frac{5}{2}N_2(g)$ + $\frac{13}{2}$ $O_2(g)$ + 3P(s, α white) → $C_{10}H_{15}N_5O_{13}P_3(s)$ **71. a.** +1012 kJ; **b.** −506 kJ; **c.** −2024 kJ   **73.** Arrange the reactions such that ΔH(total) =

ΔH_f(propane) $= −\Delta H_c$(propane) + 4 ΔH_c(hydrogen) + 3 ΔH_c(carbon)   **75.** −1273 kJ   **77. a.** −1367 kJ/mol; **b.** −2967 kJ   **79.** −5472 kJ/mol   **81.** −413 kJ   **83.** There are many possible answers for each type: solar cells, solar thermal systems, biomass conversion, hydroelectric systems, wind power, and geothermal systems. In addition to being renewable: several of these (solar, solar thermal, geothermal, wind, and biomass) can generate power on-site; biomass can use unwanted by-products of agriculture or even boost the prices of the commodities that are used; and all would probably reduce the overall production of pollution due to greenhouse gases or combustion by-products.   **85.** 5.62×10^{-21} J; less because the mass is lower.   **87.** It gets the reaction "over the hill" (that is, it gets the reaction started). Yes. Most reactions require more energy to get started. Then some energy is released, making the activation energy larger than the energy change.   **89. a.** exothermic, the heat that you feel is heat that has been released; **b.** surroundings; **c.** out; **d.** raised; **e.** no work except minor expansion of materials
91. 4.36 Cal/g; 1.82×10^4 J/g; 18.2 kJ/g; 231 m/s   **93. a.** 78.1 kJ; **b.** heat capacity of the mug or total heat absorbed by water and mug
95. 4.35 kJ/°C   **97.** 38 kJ/g   **99. a.** $2C_2H_6(g) + 7O_2(g) \rightarrow 4CO_2(g)$ + $6H_2O(g)$; **b.** −3070 kJ; **c.** 731.8 g; **d.** large increase in volume of gas can be explosive   **101. a.** −16.5 kJ; **b.** 0.0375 M   **103. a.** 0.52 M; **b.** $−3.6 \times 10^2$ kJ   **104. a.** +453 kJ; **b.** −11.6 kJ; **c.** exothermic; **d.** 2.07 g; **e.** −2.39 kJ; **f.** 28.8°C

Chapter 6

P6.1 403 nm; The wavelength should be shorter because of higher frequency.   **P6.2** 2.48×10^{-23} J   **P6.3** 1875.6 nm, 2625.8 nm, 7459.7 nm   **P6.4** 91.9 nm   **P6.5** 3.6×10^{-38} m   **P6.6** 2.00 × 10^{-3} kg·m·s^{-1}·mol^{-1}; 5.70×10^{-4} kg·m·s^{-1}·mol^{-1}; 3.99×10^{-4} kg·m·s^{-1}·mol^{-1}; Because the momentum is inversely proportional to the wavelength, there is a marked difference.
P6.7 2p   **P6.8** Ga: $1s^22s^22p^63s^23p^64s^23d^{10}4p^1$, or $[Ar]4s^23d^{10}4p^1$; Sr: $1s^22s^22p^63s^23p^64s^23d^{10}4p^65s^2$, or $[Kr]5s^2$

3. red tomato, yellow squash, green squash, and finally purple eggplant   **5. a.** 2.0×10^{19}/s; **b.** 1.3×10^{-14} J; **c.** 8.0×10^9 J/mol
7. infrared, 1.2×10^{13}/s, 8.0×10^{-21} J   **9.** based on 5 ft 6 in: 1.68 m, 1.68×10^9 nm, 1.77×10^{-16} light-years, meters are most convenient.
11. 0.364 m to 0.336 m, radio   **13.** 3.05 m   **15.** 2.94×10^{-19} J, 4.52×10^{-19} J   **17.** 177 kJ/mol, 272 kJ/mol   **19. a.** 3.52×10^{-19} J; **b.** 2.81×10^{-18} J   **21.** 6.85×10^{-19} J per photon, 1.03×10^{15}/s   **23.** 1.2×10^{-17} m, 2.5×10^{25}/s   **25.** 1.95×10^{18}/s, X-ray
27. 5.00×10^{-7} m, 5.00 × 10^3 Å, 5.00×10^{-5} cm, 1.97×10^{-5} in
29. While the whole number difference between the level is only one in each case, the emitted wavelengths are proportional to the difference in the inverse squares of the numbers. When using the inverse squares, the values come out much different.   **31.** for $n_i = 5$: 4341.6 Å, 6.91×10^{14}/s, 4.58×10^{-19} J; for $n_i = 6$: 4102.8 Å, 7.31×10^{14}/s, 4.84×10^{-19} J; for $n_i = 7$: 3971.1 Å, 7.55×10^{14}/s, 5.01×10^{-19} J
33. a. 1.40×10^{15}/s; **b.** UV **c.** Zinc   **35. a.** $−2.1786 \times 10^{-18}$ J; **b.** $−2.4207 \times 10^{-19}$ J; **c.** $−8.7144 \times 10^{-20}$ J; **d.** $−4.4461 \times 10^{-20}$ J
37. a. 2.0424×10^{-18} J; **b.** 1.9365×10^{-18} J; **c.** 4.9018×10^{-20} J; **d.** 5.0019×10^{-19} J   **39. a.** 121.57 nm; **b.** 1282.1 nm; **c.** 486.26 nm; **d.** 93.779 nm; **e.** 94.973 nm; **f.** 1875.6 nm   **41.** shortest (with $n_i = \infty$) is 1458 nm; longest (with $n_i = 5$) is 4052.2 nm   **43.** yes ($n = 6$ to $n = 2$)   **45.** 2.1786×10^{-18} J   **47.** Classically, matter has mass in discrete particles and can therefore have momentum; further, any possible value of energy (and momentum) or position is allowed. Both position and momentum can be measured (at the same time) to infinite precision. Waves can have specific but infinitely variable energies and amplitudes, but they do not have a specific location because they are spread out over space and time.   **49.** 1.9 × 10^{-22} kg·m·s^{-1} **51.** 8.4×10^6 kg·m·s^{-1} **53.** 1.23×10^{-27} kg·m·s^{-1}
55. Electrons have mass and can have discrete position.
57. 0.6648 nm   **59.** 3.8×10^{-33} m; A softball moving at the same speed, with its higher mass, would have a smaller wavelength.

61. Here p represents the momentum of the particle, whereas x represents its position. Because the uncertainties are multiplied, the more certain the measurement of one, the more *uncertain* the value for the other. **63.** $\Delta x \geq 5.5 \times 10^{-10}$ m; $\Delta x \geq 2.8 \times 10^{-10}$ m; Both are larger than the first Bohr radius. **65.** 1.0×10^{-24} kg·m·s^{-1} **67.** 0.60 kg·m·s^{-1}·mol^{-1} **69.** 1.1×10^{2} kg·m·s^{-1} **71.** The electron orbital describes the energy state and relative position in space of the electron within the atom. Since the position is described based on a probability, we cannot specify exactly where the electron is, but we can in some cases say where it is not and where it is most likely (but not required) to be. **73.** It is easier to start thinking about this problem in terms of the Bohr model. The Bohr orbits are stable because the wavelengths perfectly overlap with themselves after completing the orbit. In other words, they constructively interfere and become stable standing waves. At any other radius the electron wavelengths will not overlap correctly after completing the orbit and will destructively interfere. No stable orbit is possible. Since there are no stable orbits possible between two adjacent states, the transition from one state to another is abrupt and not gradual. The same thinking applies to the more complicated Schrödinger equation. **75.** Only sequence (e) is valid. Sequences (a), (b), and (d) are not possible because l can have only *positive* integer values that *must be less than* n. Sequence (c) is also not allowed because m_l must have a magnitude less than or equal to l. **77.** 5 sublevels, 25 orbitals **79.** 7 **81.** 50 **83.** Usually, the "shape" of an orbital refers to the surface within which 90% of the total probability of finding an electron is found. Very little of the electron probability occurs exactly on the surface, and the electron probability is spread throughout much of that space. By specifying these "shapes," we can better visualize where the most probable places for finding the electron are. **85.** A radial node exists at all angles at a given distance from the nuclear center. For example, the 2s orbital has a probability of 0 at a given distance from the nucleus. A planar node exists for a particular angle at all radial distances and usually passes through the nucleus. For example, the 2p orbitals (x, y, and z) have a node passing through the nucleus, separating the two parts of the orbital. **87. a.** Nothing actually touches in the conventional sense. When the atomic-sized needle tip comes close enough to the sample material, the wavefunction of the atom at the very tip of the needle overlaps the wavefunction of the nearest atoms in the material. With sufficient overlap, the electrons from the tip can move to the sample. This movement of electrons produces a current that is detected by the STM. **b.** Tunneling is the movement of a particle between two allowed spaces (or orbitals) through a space where it could not be classically. Classically, this current could not flow until the two physically touched. The closer the tip to the surface or an atom, the greater is the overlap of the orbitals and the greater the current that arises as the electrons move through space from one allowed orbital to the other. **89.** $m_s = +\frac{1}{2}$; electron spin quantum number **91.** C, O **93.** In the extreme example, we could place all electrons in the 1s orbital, choosing half with spin up and half with spin down. (However, we could also choose to have every electron spin up and to have all unpaired if the Pauli exclusion principle did not apply.) Other possibilities also exist. **95.** In multielectron atoms, the shapes of the orbitals are still very similar, the quantum numbers for the orbitals and the rules for using them remain the same, and the behavior of the nodes within the orbitals remain the same. The energies of the orbitals are changed! **97.** He is easier. Because the attraction will be higher in removing the second electron ($+2$ ion and -1 electron), ionizing He$^+$ is more difficult. **99. a.** $1s^2 2s^2 2p_x^1 2p_y^1 2p_z^1$ or $1s^2 2s^2 2p^3$; **b.** $1s^2 2s^2 2p^6$ **101. a.** Si; **b.** Cl; **c.** K; **d.** Sr **103.** $1s^2 2s^2 2p^6 3s^2 3p^6 4s^2 3d^1$. Because of the effects of shielding on the orbital energies, the 4s orbitals lie at lower energy than the 3d orbitals and fill first. **105.** K, Fe **107.** $1s^2 2s^2 2p^6 3s^2 3p^6 4s^1 3d^{10}$ **109. a.** 8.70×10^5; **b.** 6.71×10^8 mi/h **111.** 121.56 nm **113.** $\Delta r_{1,2} = 0.15875$ nm; $\Delta r_{3,4} = 0.37042$ nm; $\Delta r_{5,6} = 0.5821$ nm; The distances between successive levels continues to increase; the shells are becoming increasingly farther apart. **115.** In order to be stable, the electron must create a standing wave by perfectly overlapping itself after a single circumference. Perfect overlap occurs only if an integer number of waves lie on the circumference; therefore, n must be a positive integer. **117.** n, l, m_l: **a.** 1, 0, 0; **b.** 2, 1, (± 1 or 0); **c.** 3, 2, (± 2, ± 1 or 0) **119.** 311 kJ/mol **121. a.** $\nu = 4.32 \times 10^{14}$ Hz, $E = 2.86 \times 10^{-19}$ J; **b.** 2.25×10^{-34} oz **122. a.** 9.18×10^{-20} J; 5.53×10^4 J/mol; **b.** 1.39×10^{14}/s; **c.** IR; **d.** $n = 7$; **e.** 371.11 nm; uv; **f.** $2s^1$ or $2p^1$

Chapter 7

P7.1 a. Sn; **b.** Ta; **c.** H **P7.2** Metals: Li, Ni, Ce, Al, Po, Rb, and Cu; Heavy metals: Ce, Po, Rb, and Cu **P7.3 a.** Ba; **b.** Na; **c.** O; **d.** C **P7.4** 119, 168 **P7.5** Ra (lowest ionization energy) **P7.6** You would expect the energies to be high to start and to get higher with each electron. When atoms have either a filled orbital or a half-filled orbital, it takes more energy to ionize the electron than would otherwise be expected. When electrons are being ionized from a noble gas configuration (10 electrons and 2 electrons), there is a significant jump in the ionization energy, especially when electrons must be removed from the 1s orbital.

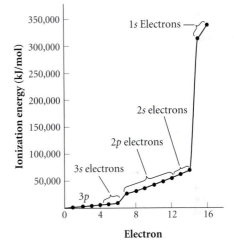

Electron Ionization Energies for Sulfur

P7.7 The trends are opposite one another. Those elements the most willing to part with an electron (lower ionization energy) will be the least willing to gain an electron (smaller electron affinity). **P7.8** Electronegativity generally increases as an element is closer to fluorine (upper left), so P is more electronegative than Na; Cl is more electronegative than Ne (noble gases generally don't form bonds and therefore have very low electronegativities); N is more electronegative than C.

1. $_{28}$Ni and $_{27}$Co; $_{52}$Te and $_{53}$I (126.9 g/mol); $_{18}$Ar and $_{19}$K; $_{90}$Th and $_{91}$Pa; $_{92}$U and $_{93}$Np. **3. a.** s; **b.** d; **c.** p; **d.** p; **e.** p; **f.** f; **g.** p **5. a.** p; **b.** s; **c.** d; **d.** d; **e.** s; **f.** p; **g.** f **7.** ~3090°C **9. a.** Ca; **b.** P; **c.** Br; **d.** Cs **11. a.** metal; **b.** metal; **c.** metalloid; **d.** metal; **e.** metalloid; **f.** nonmetal; **g.** nonmetal; **h.** metalloid **13.** Br, Hg **15. a.** 8.76g Ni, 16.7 g Cr; **b.** 8.99×10^{22} atoms Ni, 1.93×10^{23} atoms Cr **17.** Metalloid, $Sb_2S_3 + 3Fe \rightarrow 3FeS + 2Sb$, 4.47×10^8 mol Sb **19.** Li_2O; BeO; B_2O_3; CO_2; N_2O_5; O_2; and OF_2 **21.** Except for Po, the Group VIA elements are nonmetals forming anions and will form compounds with metals. **23.** 1, 0 **25.** New order: H, O, C after converting masses to moles. **27.** Many transition elements play a critical role as the "active site" within enzymes. **29.** It has a *completely* filled shell (a duet). **31.** Removing an electron from the filled s subshell in Ca is more difficult than removing an electron from the unfilled shell in K. **33.** When placed in the proper groups, both Te and I have the same number of valence electrons as their groups. **35.** From left to right: Na, S, Mg. **37.** 77 pm **39.** 94 pm **41.** K; Because K is lower in the same group, it has an additional filled shell, which makes it larger. **43.** 2.01 g **45. a.** From lowest to highest: Al, Si, P. For the second ionization energy, you would expect Al to be higher than Si or P because it would have to remove an electron from a filled 3s orbital, whereas Si and P continue to remove electrons from the unfilled 3p orbitals. (Rank: Si, P, Al); **b.** From lowest to highest: Kr, Ar, Ne. No change in order for the ranking by second ionization energy. **47.** A: metal, Group IIA; B: nonmetal, Group VIA or VIIA **49.** Group VIIIA, lower, first ionization energies decrease down the group. **51. a.** +1, +1; **b.** K$^+$; **c.** K **53.** Magnesium's first ionization removes an electron from a filled s orbital, which requires more energy than removing sodium's electron from an unfilled s orbital. The second ionization for sodium requires removing an electron from a completely filled shell, which takes even more energy, whereas magnesium's second electron is being removed from the now unfilled s orbital. **55. a.** Magnesium never forms a +3 ion, because removing a third electron would require breaking up a filled shell of electrons; **b.** Fluorine, seeking a filled shell, needs to gain an electron to achieve an octet configuration. A +1 fluorine ion is a move in the opposite direction; **c.** Hydrogen has only one electron to lose; it can never get to a +2 state; **d.** Aluminum, in losing three electrons, achieves an octet (filled shell) configuration; the octet configuration is more stable than other configurations. **57.** S, because Xe is a noble gas and doesn't spontaneously accept electrons. **59.** Cl **61.** Mg, Al, Na. No change. **63.** Fluorine is the smallest ion-forming element in the second period, which means that its electrons are the most tightly packed. Accepting an additional electron is not as favorable as would otherwise be expected because of the greater electron–electron repulsions in the small volume. **65. a.** +146 kJ/mol, endothermic; **b.** −502 kJ/mol, very exothermic and reactive **67.** Na, Li, As, S, F **69. a.** O; **b.** S; **c.** Br; **d.** O **71.** Change is 0.8 from Ga to Se; change is 0.3 from Zn to Sc. If the electronegativity difference were based on the number of protons, you might have expected a difference three times as large (nine protons rather than three protons). Electronegativity cannot be solely dependent on the atomic number. **73.** Mg, Na, Rb **75.** Metal: Lower ionization energy, more reactive; Nonmetal: generally opposite of metals **77. a.** all in Group IB; **b.** bottom; **c.** Yes; given that the most reactive elements are found toward the lower left and upper right of the periodic table (excluding the noble gases) and working to the center from the outside in toward Group IB, the metals in the activity series are found in the order that the periodic table would predict. **79.** O, with its higher electronegativity, will pull shared electrons more toward itself and therefore is the more negative of the two elements. **81.** A possible explanation for the lack of iron in the mantle relative to the crust is that during the early formation of the Earth, the very dense elements, such as iron, were drawn to the core and are therefore missing from the mantle. **83. a.** Ar; **b.** 5.62 × 10^{21} atoms **85.** Cu, Au **87. a.** Johan Dobereiner first identified groups of three elements that shared similar properties. These first "triads" form the basis for the current alkali metals group (Group IA), alkaline earth metals group (Group IIA), and halogens group (Group VIIA); **b.** John Newlands first placed elements in order of increasing mass and found that elements eight places apart share similar properties. He arranged the elements in octaves (periods of eight), a scheme that very much mirrors the s- and p-blocks of current periodic tables; **c.** Dmitri Mendeleev added more elements to his table, generally in mass order, that created vertical groups of elements with similar properties. Mendeleev left holes in his table and predicted elements would be discovered to fill those holes. Mendeleev's use of the table as a predictive tool laid the basis for the modern periodic table. **89. a.** carbon-steel; **b.** Carbon-steel contains iron and carbon only; other steels contain other elements. **91.** 115 pm. **93.** With the exception of He (Group VIIIA, two valence

electrons), there is very good correspondence between number of valence electrons and group number. The number of valence electrons is a periodic property;

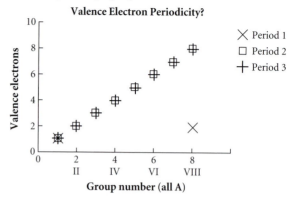

Valence Electron Periodicity?

× Period 1
□ Period 2
+ Period 3

b. The oxidation number for the first 18 elements also appears to be periodic. Even though there is a large break in the trend, the values repeat with successive periods, making the property periodic and predictable.

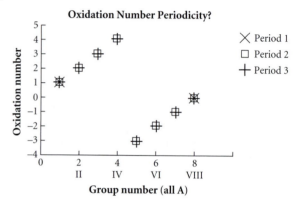

Oxidation Number Periodicity?

× Period 1
□ Period 2
+ Period 3

95. There are two competing trends in how the atomic number affects the size of the atom: (1) In a group, size increases with increasing atomic number. (2) In a period, size decreases with increasing atomic number. It is important to account for the relative positions (period and group) of the two atoms to be compared. **97.** The attraction of the increasingly more positive ion is larger, and the energy needed to remove electrons from the lower shells also increases. **99.** A configuration that half-fills the set of orbitals has additional stability. Arsenic has a valence configuration that is $4s^2 4p^3$, half-filling the p orbital set, whereas selenium has $4s^2 4p^4$ and doesn't have the additional stability. **101. a.** exothermic; **b.** Cl + e → Cl$^-$ + 350 kJ/mol **103. a.** $1s^2 2s^2 2p^6 3s^2 3p^6 4s^1 3d^5$; **b.** Cr; **c.** 4.50 × 10^{22} atoms **105.** silicon **107. a.** p$^+$ = e$^-$ = 87, n = 123; **b.** 2.66 g; **c.** francium-210 **108. a.** NaHCO$_3$ + HCl → H$_2$CO$_3$ + NaCl; **b.** H: Period 1, Group 1A; Na, Period 3, Group 1A; C, Period 2, Group IVA; O: Period 2, Group VIA; Cl: Period 3, Group VIIA; **c.** Na; **d.** H$_2$CO$_3$ → H$_2$O + CO$_2$; **e.** 25.3 ppm; **f.** 9.95 ppm; **g.** This reaction generates CO$_2$, which leavens the bread or pastry.

Chapter 8

P8.1 S^{2-}: $1s^2 2s^2 2p^6 3s^2 3p^6$, $\left[:\overset{\cdot\cdot}{\underset{\cdot\cdot}{S}}:\right]^{2-}$, Ar; F$^-$: $1s^2 2s^2 2p^6$, $\left[:\overset{\cdot\cdot}{\underset{\cdot\cdot}{F}}:\right]^-$, Ne; Mg^{2+}: $1s^2 2s^2 2p^6$, [Mg]$^{2+}$, Ne; Br$^-$: $1s^2 2s^2 2p^6 3s^2 3p^6 4s^2 3d^{10} 4p^6$, $\left[:\overset{\cdot\cdot}{\underset{\cdot\cdot}{Br}}:\right]^-$, Kr

P8.2 $[Na\cdot]\ [:\overset{\cdot\cdot}{O}:]\ [\cdot Na]\ \longrightarrow\ [Na]^+\ [:\overset{\cdot\cdot}{\underset{\cdot\cdot}{O}}:]^{2-}\ [Na]^+$

P8.3 $Mg^{2+} < Na^+ < Ne < F^- < O^{2-}$. **P8.4** $FeCl_3$ **P8.5** polar covalent; nonpolar covalent; ionic **P8.6** OCl: O (F.C. = −1), Cl (F.C. = 0); CH_3NH_2: all F.C. = 0

P8.7 $\begin{bmatrix} H & H \\ C=C \\ H & H \end{bmatrix}, \begin{bmatrix} H \\ C=N-H \\ H \end{bmatrix}$ **P8.8** $\Delta H = -22$ kJ/mol

P8.9 HNO_3: trigonal planar, 120°; CCl_4, tetrahedral, 109.5°; NH_3, trigonal pyramidal, 107° **P8.10** SO_2 **P8.11** N_2: nonpolar bond, no dipole; NH_3, polar bonds, net dipole:

$$\begin{bmatrix} N \\ H \quad | \quad H \\ H \end{bmatrix}$$

1. a. N; **b.** N; **c.** N **3.** C^{2+}, N^{3-}, and H^+ **5. a.** Group VIA; **b.** Group IIA; **c.** Group VA **7. a.** two extra, −2; **b.** three extra, −3; **c.** one extra, −1 **9. a.** $[:Se:]^{2-}$; **b.** $[:I:]^-$; **c.** $[Sr]^{2+}$; **d.** $[Sc]^{3+}$; **e.** $[Si:]^{2+}$ **11.** Na^+, F^-, Al^{3+} **13.** Ca^{2+}, Ar, and Cl^- **15.** Group IVA or Group IVB **17. a.** [Li·], [Na·], [K·]; **b.** Li_2O, Na_2O, and K_2O **19. a.** true; **b.** false; **c.** true; **d.** true **21.** To reduce its valence to an octet configuration, aluminum goes from the atom to the ion by losing three electrons: $[:Al] \rightarrow [Al]^{3+} + 3$ electrons. Oxygen forms oxide (and creates an octet) by gaining two electrons: $[:O:] + 2$ electrons $\rightarrow [:O:]^{2-}$. Chlorine forms chloride (and creates an octet) by gaining one electron: $[:Cl:] + 1$ electrons $\rightarrow [:Cl:]^-$. When forming a compound with chlorine, aluminum loses three electrons, which allows for the formation of three chlorides. All electrons have been accounted for, and the charges on the ions balance one another—$AlCl_3$ is the stable compound that forms, creating the 1:3 ratio. When forming a compound with oxygen, two aluminum atoms lose three electrons each (six in all), which allows for the formation of three oxides. All electrons have been accounted for, and the charges on the ions balance one another—Al_2O_3 is the stable compound that forms, creating the 2:3 ratio.

23. a. $I^{5+} < I < I^-$; **b.** $S^{6+} < S^{4+} < S^{2-}$; **c.** $C^+ < C < C^-$; **d.** $Fe^{3+} < Fe^{2+} < Fe$ **25.** LiCl > NaCl > KCl > RbCl > CsCl

27. $[K·] \quad [:Cl:] \longrightarrow [K]^+ \quad [:Cl:]^-$

29. Na_2O: $[Na]^+ [:O:]^{2-} [Na]^+$, NaOH: $[Na]^+ [:O:H]^-$

31. Mn^{5+} **33.** KBr, the distance between K^+ and Br^-, is smaller than Cs^+ and Br^-, so the attraction and energy are greater. **35.** Electron affinity gauges how much a particular atom (or molecule) "wants" to wholly gain an electron, with larger electron affinities indicating greater likelihood of gaining an electron. Electronegativity values gauge how much a particular atom will attract electrons from a shared covalent bond to itself. However, both electron affinity and electronegativity tend to increase toward the upper right of the periodic table (excluding the noble gases) and can predict the favorability of forming ionic compounds. Electronegativity is more useful, because its predictive capability extends to polar covalent bonding and dipoles. **37.** Within your table you should have a vertical arrow pointing up, indicating an increase in electronegativity as you go up a group, and a horizontal arrow pointing to the right, indicating an increase in electronegativity as you go to the right in a period.

39. H_2O: $[H-O-H]$, NO (radical): $[:N=O:]$, CO: [:C≡O:],

NO_2 (radical): $[:O-N=O:]$, HCl: $[H-Cl:]$,

PCl_2(radical): $[:Cl-P-Cl:]$, NBr_3: $\begin{bmatrix} :Br: \\ :Br-N: \\ :Br: \end{bmatrix}$

41. OH^-: $[:O-H]^-$, NO_2^-: $[:O-N=O:]^-$, Br^-: $[:Br:]^-$,

PO_4^{3-}: $\begin{bmatrix} :O: \\ :O=P-O: \\ :O: \end{bmatrix}^{3-}$, SO_3^{2-}: $\begin{bmatrix} :O: \\ :O=S: \\ :O: \end{bmatrix}^{2-}$,

CO_3^{2-}: $\begin{bmatrix} :O: \\ :O=C \\ :O: \end{bmatrix}^{2-}$, BrO_4^-: $\begin{bmatrix} :O: \\ :O=Br=O \\ :O: \end{bmatrix}^-$

43. C_2H_6: $\begin{bmatrix} H & H \\ H-C-C-H \\ H & H \end{bmatrix}$, C_3H_6: $\begin{bmatrix} H & H \\ H-C-C=C-H \\ H & H \end{bmatrix}$,

C_2H_4: $\begin{bmatrix} H & H \\ C=C \\ H & H \end{bmatrix}$, C_3H_8: $\begin{bmatrix} H & H & H \\ H-C-C-C-H \\ H & H & H \end{bmatrix}$,

C_4H_{10}: $\begin{bmatrix} H & H & H & H \\ H-C-C-C-C-H \\ H & H & H & H \end{bmatrix}$

45. $\begin{bmatrix} :O: \\ :O-N⊕ \\ :O: \end{bmatrix}^- \leftrightarrow \begin{bmatrix} :O: \\ :O=N⊕ \\ :O: \end{bmatrix}^- \leftrightarrows \begin{bmatrix} :O: \\ :O-N⊕ \\ :O: \end{bmatrix}^-$

hybrid structure: $\begin{bmatrix} :O: \\ :O—N \\ :O: \end{bmatrix}^-$

47. C_4H_4: $\begin{bmatrix} H & H \\ C=C \\ C=C \\ H & H \end{bmatrix} \leftrightarrow \begin{bmatrix} H & H \\ C-C \\ C-C \\ H & H \end{bmatrix}$ C_4H_6: $\begin{bmatrix} H & H \\ C-C-H \\ C-C-H \\ H & H \end{bmatrix}$

C_4H_8: $\begin{bmatrix} H & H \\ H-C-C-H \\ H-C-C-H \\ H & H \end{bmatrix}$

49. N_2O: $[:N≡N-O:] \leftrightarrow [:N=N=O:]$,

N_2: $[:N≡N:]$, N_2H_4: $\begin{bmatrix} H-N-N-H \\ H & H \end{bmatrix}$; $N_2 = 946$ kJ/mol;

N_2O = 418 kJ/mol; N_2H_6 = 160 kJ/mol

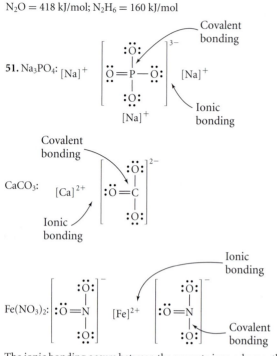

51. Na_3PO_4: Covalent bonding; Ionic bonding

$CaCO_3$: Covalent bonding; Ionic bonding

$Fe(NO_3)_2$: Ionic bonding; Covalent bonding

The ionic bonding occurs between the separate ions, whereas the covalent bonding is happening between the atoms of the polyatomic anions.   **53.** C—C < H—C < N—O < Ca—H < Ca—N

55. a. 0; **b.** no change; **c.** 0; **d.** −1   **57.**

59. The structure of NO_2 is [⊖:O:N::O:⊕]. In this diagram, the (*) are lone-pair electrons, (:) are bonding electrons, and (°) is a radical electron. The nonzero formal charges are labeled within the circles.   **61.** [:C≡N:]⁻ Because the formal charge of −1 is on the carbon, H^+ with its positive charge should attach to carbon and its negative charge.

63. The formal charge on sulfur is 0

(F.C. = 6 − 4 − 2 = 0), and the oxidation number on sulfur is +4. In the calculation of the oxidation number, oxygen is taken at −2 and hydrogen at +1, giving the sulfur a +4 oxidation number, but in the calculation of formal charges, both oxygen and hydrogen have a value of zero.

65.

67.

69. C_2H_6: , C_2H_4: ,

C_2H_2: [H—C≡C—H] The number of bonds between carbon increases from one to two to three, as shown above. The greater number of bonds between two atoms, the stronger the bond and the shorter the bond length. C_2H_2 will have the strongest (and shortest) bond, and C_2H_6 will have the longest (and weakest) bond.   **71.** ΔH = −2674 kJ/mol. This value is different from the direct determination because the values for the bond energies are taken from the average of the bond in several molecules and may not be exactly the same as the bond energies in the molecules of this reaction.
73. $\Delta H(H_2O)$ = −254 kJ/mol, $\Delta H(H_2O_2)$ = −153 kJ/mol. Because water releases more energy than hydrogen peroxide in its formation, it will be the more stable of the two molecules.   **75.** H_2O: bent; NO_2: bent; PCl_2: bent; NBR_3: trigonal pyramidal   **77.** NO_2^-, bent; PO_4^{3-}: tetrahedral; SO_3^{2-}: trigonal pyramidal; CO_3^{2-}: trigonal planar; BrO_4^-: tetrahedral   **79.** C_2H_6: tetrahedral at each; C_3H_6: tetrahedral on left, trigonal planar at the other two; C_2H_4: trigonal planar at each, C_3H_8: tetrahedral at each; C_4H_{10}: tetrahedral at each

81. The Lewis structure of ClO_2 is [:O—Cl=O:] (note the radical electron). At Cl, it has four electron groups (tetrahedral) but two lone pairs/radicals, making the molecular geometry bent.
83. $BeCl_2$ > $AlCl_3$ > CCl_4 > NCl_3 > $XeCl_4$

85. , see-saw

87. In N_2H_2, each nitrogen has three electron groups (trigonal planar) but one lone pair, so the molecular geometry is bent. The H—N—N bond angle will be slightly less than the normal 120° because of the additional lone-pair repulsion. In $N_2H_2^{2+}$, each nitrogen has only two electron groups and no lone pairs, making the geometry linear. The bond angle is then expanded from 120° in N_2H_2 to 180° in $N_2H_2^{2+}$.
89. C←H < C→N = C←B = C→Cl < C→O < C←Mg (arrow denote dipoles)

91. The structures are SF_6 and

SF_5 . SF_6 is octahedral (six electron groups around S) and nonpolar because all the S—F dipoles cancel. SF_5 has six electron groups but one lone pair, making the geometry square pyramidal. Removing one S—F dipole that helped make SF_6 nonpolar makes SF_5 polar, because not all the S—F dipoles cancel out.
93. A: ionic, B: nonpolar, C: polar   **95.** The structure of H_2O_2 is

[H—O—O—H] and is bent (see Problem 76). The O—H bonds are polar with an electronegativity difference of 1.4. Because the molecule is bent, even though the dipoles point in nearly opposite directions, the dipoles do not cancel and the molecule is polar. If the molecule were linear, the dipoles would cancel out, because they would be aligned and in opposite directions. The linear molecule would be nonpolar.   **97.** From left to right: H^+, H^-, and H
99. Because N_2 has a triple bond, its bond energy is very large

(941 kJ/mol). Breaking this bond in order to cause a reaction is very difficult and therefore unlikely. N_2 will emerge just as it entered. **101.** There is only a single choice for the anion (-2), so our choice will be based on which cation will have a weaker interaction. The weakest lattice attractions will come from ion pairs that are widely separated and have smaller charges. This set of ions (the small $+1$ and the larger $+3$) have these trends in opposition. The ion with the $+3$ charge has roughly twice the diameter, which is not a large enough increase in size to offset the larger charge. The $+3$ ion will have a larger ionic attraction. Therefore, the weakest interaction will be between the -2 anion and $+1$ cation. **103.** $+3$ cation with -2 anion **105. a.** The attraction that two charges particles have for one another is larger as the charges on each become larger, and the attraction is larger as the particles get closer to one another. Large attractive forces lead to high melting points. Neutral molecular species at most have partial charges from the dipoles, limiting the strength of attraction to other molecules. The lower attraction leads to lower melting points; **b.** For a high melting point, we should choose species that are small and have high charges. For a low melting point, we should choose low charges and large radii. By far the highest charge and lowest size belong to B^{3+}. For the anion we could choose N^{3-} or O^{2-}. Nitride has the higher charge but is bigger (171 pm versus 140 pm), but it is less than 1.25 times larger whereas its charge is 1.5 times larger than that of oxide. Highest melting point: BN. The largest ions are I^- and Cs^+; CsI should have the lowest melting point. **107. a.** Very few molecules can be found in only a diatomic form allowing for direct measurement of the bond energy. Inside larger molecules, the strength of the bond can vary depending on the other atoms and on how well the electrons are distributed within the molecule. In these cases, several measurements are made, and the average value of the bond energy is placed in the table; **b.** The presence of the three fluorine atoms with their high electronegativity will tend to shift the electrons from the other bonds toward the fluorine atoms. This has the potential to weaken the carbon–carbon bond. **109. a.** $2Fe + 3O_2 \rightarrow 2Fe_2O_3$; **b.** 5 unpaired; **c.** 69.94% **111. a.** No; **b.** 12.00%; **c.** $\ddot{O}=C=\ddot{O}$ **113. a.** 346.3 g/mol;

b.   **c.** The N in the ring

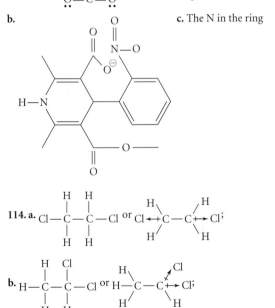

114. a.

b. (structure) $H-\overset{\overset{H}{|}}{\underset{\underset{H}{|}}{C}}-\overset{\overset{Cl}{|}}{\underset{\underset{H}{|}}{C}}-Cl$ or (structure with arrows)

c. Molecule in part a; **d.** $\begin{bmatrix} H & H \\ | & | \\ C=C \\ | & | \\ H & H \end{bmatrix}$; **e.** Yes, all bonds are nonpolar.;

f. 0.57 g

Chapter 9

P9.1 The configuration of the valence electrons in fluorine is $2s^2 2p^5$. Overlap of one of the $2p$ orbitals from each fluorine allows the electrons to be shared between the two atoms. The valence bond model of F—F shows a $2p$–$2p$ orbital overlap that constitutes the covalent bond.

P9.2 The bond formed in F_2 is from the overlap of two $2p$ orbitals, in HCl from $1s$ and $3p$ orbitals, and in Cl_2 from two $3p$ orbitals. Because both $2p$ orbitals are smaller than the $3p$ orbitals, the bond in F_2 is shorter than Cl_2 and because the $1s$ orbital is much smaller than the $3p$ orbital, the HCl bond is shorter than the Cl_2 bond. Cl_2 has the longest bond.

P9.3 a. Lewis dot structure model for OF_2, $\left[:\ddot{F}-\ddot{O}-\ddot{F}: \right]$, indicates that the central oxygen has two bonds and two lone pairs. Mixing the four orbitals (one $2s$ and three $2p$) on the oxygen atom allows us to have two lone pairs of equal energy and two bonds to the adjacent fluorine atoms. The oxygen atom possesses sp^3 hybridized orbitals. **b.** Lewis dot structure model for H_2S, $\left[H-\ddot{S}-H \right]$, indicates that the central sulfur has two bonds and two lone pairs. Mixing the four orbitals (one $3s$ and three $3p$) on the sulfur atom allows us to have two lone pairs of equal energy and two bonds to the adjacent hydrogen atoms. The sulfur atom possesses sp^3 hybridized orbitals. **c.** Lewis dot structure model for NH_4^+, $\begin{bmatrix} H \\ | \\ H-N-H \\ | \\ H \end{bmatrix}^{\oplus}$, indicates that the central nitrogen has four bonds. Mixing the four orbitals (one $2s$ and three $2p$) on the nitrogen atom allows us to have four bonds to the adjacent hydrogen atoms. The nitrogen atom possesses sp^3 hybridized orbitals.

P9.4 a. OF_2 geometry: sp^3 hybridized atoms adopt a tetrahedral geometry. Because two of the sp^3 orbitals contain lone pairs, the VSEPR model indicates that the molecule has an overall bent geometry. The bond angles should be less than 109.5° because the lone pairs repel each other more than the bonding pairs. **b.** H_2S geometry: sp^3 hybridized atoms adopt a tetrahedral geometry. Because two of the sp^3 orbitals contain lone pairs, the VSEPR model indicates that the molecule has an overall bent geometry. The bond angles should be less than 109.5° because the lone pairs repel each other more than the bonding pairs. **c.** NH_4^+ geometry: sp^3 hybridized atoms adopt a tetrahedral geometry. The bond angles should be 109.5°.

P9.5 A quick look at the Lewis dot structure for CH_2O, $\begin{bmatrix} :\ddot{O}: \\ || \\ H-C-H \end{bmatrix}$, shows that C will need three sigma bonds (one to O and two to H) and will use sp^2 hybridization to create the bonds. Oxygen also will be sp^2 hybridized to create the sigma bond to carbon and for the two lone pairs. Both carbon and oxygen have a half-filled p orbital that can be used to create a pi bond to complete the double bond (one sigma bond, one pi bond) between C and O. Formaldehyde has a total of one pi bond and three sigma bonds.

P9.6 To answer the question, a quick look at the Lewis dot structures will help:

CH_2NOH: $H-\overset{\overset{H}{|}}{C}=\ddot{N}-\ddot{\underset{..}{O}}-H$

CH_3NHOH: $H-\overset{\overset{H}{|}}{\underset{\underset{H}{|}}{C}}-\overset{\overset{H}{|}}{\ddot{N}}-\ddot{\underset{..}{O}}-H$

The shorter bond will be found in CH₂NOH as the NO bond is formed from the combination of sp^2 (on N) and sp^3 (on O) hybridizations. In CH₃NHOH, both N and O have sp^3 hybridization. We can further explain by noting that we can write a resonance structure for CH₃NOH in which a double bond exists between N and O, making the bond shorter:

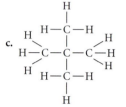

P9.7 We can get an idea of the shapes by first looking at the Lewis dot structures:

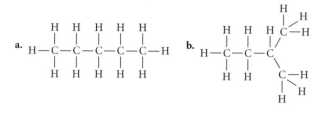

c.

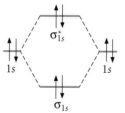

Pentane (structure a) has a straight chain structure and will be able to pack more closely together. It should have the most intermolecular attraction and the highest boiling temperature. The other two molecules are progressively more branched and can pack less tightly. They will boil at lower temperatures.

P9.8 The molecular orbital diagram for He₂ is shown here, including only the valence electrons ($1s^2$) from each He atom. Note that all the molecular orbitals are filled; equal numbers of bonding and antibonding orbitals are filled. The bond order will be zero, and He₂ is not stable.

P9.9 The molecular configuration for H₂ shows two valence electrons in the bonding σ_{1s} and none in the antibonding σ^*_{1s}. The bond order for H₂ is only 1. Through promotion of a single bonding electron to an antibonding orbital, the bond order would be zero and the molecular bond would be cleaved. The molecular configuration for O₂ (using only the valence $2s$ and $2p$ electrons, like the F₂ diagram) shows eight bonding electrons and four antibonding electrons, giving a bond order of 2. There are several possibilities for single photon absorptions. A transition from the π^* to σ^* MO causes no change in the bond order. A transition from the π to π^* MO decreases the bond order to 1, because this transition moves a bonding electron into an antibonding orbital. In either case, the bond order in oxygen remains higher.

1. a. The dashed line doesn't tell anything about the bond length, strength, or energy; it merely indicates that the connected atoms are bound in some manner. (It also doesn't tell anything about specific bond angles.); **b.** Using better models for bonding allows chemists to understand more fully the interaction between chemicals and how molecular shapes, bond strength, and polarity play a role in those interactions. **3.** $1s^22s^22p^63s^23p^2$; two bonds, not consistent

5. Cl: $1s^22s^22p^63s^23p^5$, one; Se: $1s^22s^22p^63s^23p^64s^23d^{10}4p^4$, two; B: $1s^22s^22p^1$, one **7.** $1s^22s^22p^2$; Carbon has only two unpaired electrons ($2p_x^12p_y^1$) to use in the formation of two covalent bonds. With only two electrons to form bonds, carbon cannot form the four bonds in CCl₄. **9. a.** H: $1s^1$; Br:$1s^22s^22p^63s^23p^64s^23d^{10}4p^5$; **b.** H: $1s$; Br: $4p$; **c.** This bond should be weaker than the H—F bond because the overlap of the much larger Br orbital with the small H orbital creates a weaker interaction and a weaker bond. **11.** Because each chlorine atom has a configuration of $1s^22s^22p^63s^23p^5$, one $3p$ orbital from each Cl is involved in creating the covalent bond. These orbitals are exactly the same size and energy, so the amount of overlap in the bond is substantial. **13.** Cl₂; Br₂; HBr **15.** The $1s$ orbital on H ($1s^1$), the $3p$ orbital on Cl ($1s^22s^22p^63s^23p^5$), and the $2p$ orbitals on O ($1s^22s^22p^4$) are likely candidates for bonding in HOCl. (Using the hybridization model, we realize that the sp^3 orbitals on oxygen are used in the molecule.) Because the same bonds are used by O in the bonds with H and Cl, the size of the orbitals on H (smaller) and Cl (larger) predict a larger O—Cl bond. **17.** KH; KH should have the shorter and stronger bond because the $4s$ orbital on K is smaller than the $6s$ orbital on Cs. **19.** The milk represents the s orbital and is involved in all hybridization schemes. The eggs represent the p orbitals and can be mixed with the milk such that one part milk combines with one, two, or three eggs.The character of the omelet representing the hybrid orbitals is a little different in each case, but all are similar in that each is still an omelet. Also, the size of the omelet increases (as does the size of the hybrid orbital) as the number of eggs (p orbitals) increases. **21. a.** six; **b.** We will still finish with generally the same shape for the hybrid orbital as for sp^3; however, the presence of more orbitals causes the angles between them to decrease to 90°, with an overall octahedral shape. Additionally, because the d orbitals are larger than the s or p orbitals, the sp^3d^2 hybrid orbitals will tend to be longer than any of the sp hybrid types. **23. a.** [:Si⁝] , sp^3;

b. tetrahedral **25.** H₂O = ~104.5°; NH₃ = ~107°; CH₄ = 109.5° **27.** The $3p$ orbitals used in the bonding by Cl are longer and larger than the $1s$ orbitals in H. The lengthening of the three C—Cl bonds relative to the C—H bonds changes the tetrahedral shape. Additionally, the Cl—C—Cl bond angles may increase slightly to accommodate the larger Cl atoms relative to H. **29.** two **31.** two, three, four **33. a.** sp^3; **b.** sp^3; **c.** sp^2; **d.** sp **35. a.** linear; **b.** trigonal planar; **c.** tetrahedron; **d.** trigonal bypyramid; **e.** octahedron

37.

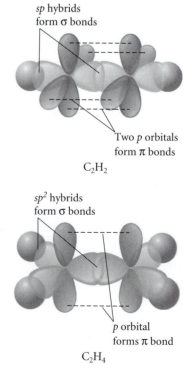

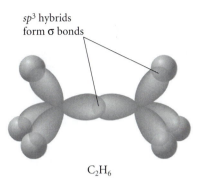

sp³ hybrids
form σ bonds

C₂H₆

The least amount of energy is required to break the bond in C_2H_6 because it has only a single bond between the carbon atoms.
39. a. sp^2, trigonal planar; **b.** sp^3, trigonal pyramidal; **c.** sp^3d^2, octahedral; **d.** sp^3, ring/crown   **41. a.** sp^3, tetrahedral; **b.** sp^3d^2, square planar; **c.** sp^3d, see-saw; **d.** sp^2, bent   **43.** For atoms bound to more than one other atom: O, sp^3; left C, sp^2; right C, sp^3; N, sp^3 **45.** Both Li_2 and K_2 will have the same bonding (because both have the same valence configuration of one *s* electron); however, the orbitals on K are larger. The overlap of the atomic orbitals from K in forming the covalent bond will be less than that on Li. K_2 would have the weaker bond.   **47.** sp^3, <109.5°   **49.** six, sp^2, 120°   **51.** left N: one lone pair, 120°, sp^2; right N: one lone pair, 109.5°, sp^3 **53.** sp^3, 109.5°
55. a.

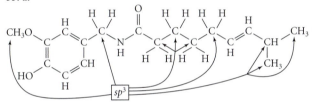

b. two; **c.** five; **d.** two
57.

Model	Scientist	General Summary
VSEPR	G. Lewis	Distribute bonding and nonbonding electron pairs.
VB	L. Pauling	Overlap and hybridization create new orbitals.
MO	E. Schrödinger	Subtract and add overlap to create π and σ bonds.

59. Because the *p* orbitals on sulfur are larger, the bond distance in S_2 is larger. Because the atoms are further apart than oxygen, the *p* orbitals are not able to create a good overlap to create the pi bond.

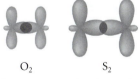

O₂   S₂

61. The greatest difference between bonding and antibonding orbitals is that bonding orbitals do not have a node perpendicular to the bond axis, whereas antibonding orbitals do. The bonding orbitals also have lower energies than antibonding orbitals made from the same atomic orbitals.   **63. a.** LCAO stands for linear combination of atomic orbitals. It is the method by which the molecular orbital theory creates its orbitals. The electron densities of the atomic orbitals

can be added or subtracted to create the new molecular orbitals; **b.** HOMO stands for highest-energy occupied molecular orbital. It is the orbital that has the highest energy among those orbitals that contain electrons; **c.** LUMO stands for lowest-energy unoccupied molecular orbital. It is the orbital that has the lowest energy among those orbitals that do not contain electrons.   **65. a.** The bond orders for the molecules shown below are as follows: N_2^-, 2.5; N_2, 3; and N_2^+, 2.5; **b.** N_2^- and N_2^+; **c.** $N_2 < N_2^- \approx N_2^+$

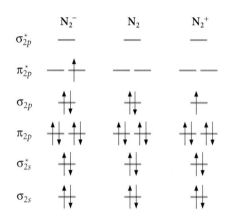

67. a. Ne; **b.** paramagnetic; **c.** 1.5   **69.** seven sigma bonds, two pi bonds   **71. a.** 1; **b.** 1.5; **c.** 1   **73. a.** 3; **b.** none   **75. a.** Orbital overlap occurs when the orbitals on two adjacent atoms share some space in common. When this occurs, the orbitals are free to combine and make bonds (or antibonds);
b.

s–s overlap   s–p overlap   p–p overlap

77. a. The Lewis dot structures (or skeleton structures) cannot show the single average structure that truly exists for molecules that have resonance structures. Some molecules will have bonds of order 1.5 or 1.33 that are difficult to draw without the resonance structures; **b.** This resonance structure has alternating positions for the double bonds inside the ring.

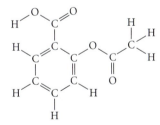

79. sp   **80.** 1-2-3 and 2-3-4 are both 180°.   **81.** 3 pi bonds, thirteen sigma bonds   **82.** carbon atoms 1 and 6   **87. a.** The carbons in the ring and the carbon attached to the ring have delocalized electrons; **b.** sp^2; **c.** *p* orbitals   **89.** All carbon atoms are sp^2.

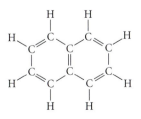

91. In the minimum-energy states: **a.** two; **b.** none; **c.** two; **d.** four
93. a. sp^3 hybridization, tetrahedral shape for orbitals, molecule has trigonal pyramidal geometry; **b.** sp^3d^2 hybridization, octahedral shape; **c.** sp^3 hybridization, tetrahedral shape for orbitals, molecule has bent geometry; **d.** sp^3d hybridization, trigonal bipyramid shape
95. NO_2^-: Nitrogen in both has sp^2 hybridization. Because nitrite has a lone pair in one of the sp^2 orbitals, the O—N—O bond angle in nitrite is reduced from the normal 120° that is found in the nitrate ion.
97. Each carbon in octane will be sp^3 hybridized with tetrahedral geometry and form a straight chain of carbons. Each C—C bond will be free to rotate, and the separate molecules can arrange themselves to closely associate with each other, making the formation of a solid more likely. 2,2,3,3-Tetramethylbutane has a branched structure. Each carbon is still sp^3 hybridized, but the center two carbons have three additional carbons attached to each. The molecule will be like a large ball, with little possible change in its geometry. This geometry makes close association between different molecules more difficult and the process of boiling easier. Octane will not boil as easily as 2,2,3,3-tetramethylbutane.

99. The structure of NH_2CHO is

$$\begin{bmatrix} & & & \overset{\displaystyle H}{\overset{|}{} } & \\ H-\overset{\displaystyle ..}{\underset{|}{N}}-C{=}O \\ & \overset{|}{H} & \end{bmatrix}$$. The carbon

is sp^2 hybridized (three sigma bonds, no lone pairs) and will be trigonal planar. Nitrogen is sp^3 hybridized (three sigma bonds, one lone pair) and will be tetrahedral. Because the carbon is sp^2 hybridized, the atoms connected to carbon lie in a single plane, but nitrogen has tetrahedral geometry, so its two hydrogens are not in the same plane as the rest of the molecule. The overall molecule is not planar and will be puckered.   **101. a.** C—C; **b.** 3.98×10^{-19} J; **c.** 239 kJ/mol; **d.** blue-green.   **103. a.** $C_{18}H_{34}O_2 + H_2 \rightarrow C_{18}H_{36}O_2$; **b.** 3.2 g; **c.** The

polar end ($\overset{\displaystyle O}{\underset{\displaystyle OH}{\overset{\displaystyle \|}{C}}}$) will interact with the polar water and point down.

104. a.

$$\begin{bmatrix} \overset{\displaystyle H}{\overset{|}{}} & \overset{\displaystyle H}{\overset{|}{}} \\ :N-N: \\ \underset{\displaystyle H}{\overset{|}{}} & \underset{\displaystyle H}{\overset{|}{}} \end{bmatrix}$$ **b.** trigonal pyramidal

around each N; **c.** sp^3; **d.** This compound is unlikely to have a color because it does not have a low-lying orbital to which an electron may make a transition. The sp^3 hybridization scheme has all four orbitals used in bond making or holding an electron pair. No empty pi or pi* are present; **e.** $N_2H_4 + O_2 \rightarrow N_2 + 2H_2O$ $\Delta H = -590$ kJ/mol.

Chapter 10

P10.1 6.7 m/s^2 **P10.2** 0.979 atm; 744 torr; 0.992 bar; 9.92×10^4 Pa
P10.3 1.2×10^2 torr   **P10.4** 0.75 mol   **P10.5** 1.3 L   **P10.6** 522 K
P10.7 It is polar and experiences attractive intermolecular forces.
P10.8 0.34 L   **P10.9** 0.036 mol   **P10.10** 0.734 atm   **P10.11** O_2: ideal, 24.4 atm; van der Waals, 23.9 atm; NH_3: ideal, 24.4 atm, van der Waals, 21.2 atm   **P10.12** 46.1 g/mol   **P10.13** 53.7 g
P10.14 18.5 kg   **P10.15** If the masses of two particles are different, but the particles have the same kinetic energy, their velocities must be different to offset the difference in mass. As the mass increases, the velocity must decrease in order for the kinetic energy to remain the same.   **P10.16** F_2 **P10.17** H_2 effuses 1.4 times as rapidly as He.

1. These forces are collectively known as van der Waals forces and are very small because of the large distances between the molecules and the rapid velocities.   **3.** High temperature and low pressure

5. 8.7 N   **7.** 8.4×10^2 N   **9.** 1.06×10^5 Pa; 106 kPa; 1.05 atm; 797 torr; 1.06 bar; 31.4 in Hg   **11.** 4.9 lb   **13.** 3.25×10^3 lb
15. 32.2 psi, 2.19 atm   **17.** 5.5×10^2 mm Hg   **19.** N_2: 0.656 atm; O_2: 0.574 atm   **21.** 758 mm Hg   **23.** Ne: 0.306 mm Hg; He: 3.12 mm Hg   **25.** 9.0 mol, 1.8 mol   **27.** 5.32 L, 10.1 L   **29.** 0.847 L, 1.34 L
31. 10.5 L   **33.** 594 K   **35.** 0.12 L   **37.** 0.144 mL
39. a. 0.08205 L·atm·mol^{-1}·K^{-1}, 8314 L·Pa·mol^{-1}·K^{-1}; **b.** 62.36 L·mm·Hg·mol^{-1} K^{-1}; **c.** 1.206 L·psi·mol^{-1}·K^{-1} **41.** 3.8×10^7 L
43. 22.4 L, 51.2 L   **45.** 0.193 mol   **47.** 17.1 g/mol   **49.** 0.755 g/L
51. 2.06×10^{21} molecules   **53.** 114 g/mol   **55.** 1.41×10^3 g/mol, No
57. 271×10^4 K   **59.** 8.47 L   **61.** 12.6 g/L   **63.** 0.306 atm
65. H_2 is the limiting reagent, 1.33 L   **67.** 33.9 g/mol   **69.** F_2
71. 529 L   **73.** According to the kinetic-molecular theory, the average kinetic energy of a particle is directly proportional to the temperature. The pressure of a gas inside the cylinder depends on both the number of collisions with the walls of the container and the speed of the molecules. As the temperature increases, the speed increases and both the number and the velocity of the collisions increase; therefore, the pressure is directly proportional to the temperature.   **75. a.** same; **b.** $Cl_2 < ClO_2$; **c.** ClO_2; **d.** ClO_2
77. a. 4.4×10^{-20} J; **b.** 1.7×10^{-20} J; **c.** 2.0×10^{-20} J   **79.** Ar $< C_2H_6$ $< H_2$ **81.** 411 m/s   **83.** CO_2: 376 m/s; H_2O: 588 m/s   **85.** Point 1: The marbles do not have negligible size relative either to the container or to the distance between marbles. Point 2: Unless the marbles are dirty and sticky, this should be true because the marbles will not be strongly attracted to each other. Point 3: If the marbles are continuously shaken during the demonstration, this should be true. Point 4: This should also be true during the demonstration if the marbles are continuously shaken. As the piston is moved, the number of particle collisions per surface area will change. A smaller volume will correspond to larger pressures. Point 5: This will be approximately correct. There will be some small energy loss (due to friction and collisional heating) as the marbles move and collide with each other and the container. Point 6: This will not be true in the analogy, but more violent shaking could be applied to show higher temperatures.   **87.** Xe $< SO_3 < CO_2 < N_2$ **89.** 72.7 g/mol   **91.** Ar   **93.** Because the masses are not exactly the same ($CO_2 = 44.01$ g/mol and $C_3H_8 = 44.09$ g/mol), there will be a slight difference in the effusion rates. It would be possible, though impractical, to try to separate these molecules via effusion.   **95.** 16 cm from the HCl end
97. 0.57% faster   **99. a.** 1.57×10^{13} L; **b.** 1.97×10^{13} L; **c.** 1.43×10^6 metric tons; **d.** 1.09×10^{12} mol   **101.** 4.75×10^{24} molecules
103. ideal: 21.8 atm; van der Waals: 22.1 atm   **105.** $\Delta T = 894$ K; $V_{final} = 0.916 V_{initial}$ **107.** 91.4 g; 338 balloons   **109.** One of the considerations is which property is easier to manage: a temperature of only 20 K or a high pressure of 400 atm. Each requires a different infrastructure to handle. High pressures require thick walls with excellent welding, whereas low temperatures require vacuum dewers (like a thermos), and any variation in temperature causes the hydrogen to boil and be lost.   **110. a.** dispersion only; **b.** hydrocarbons behave ideally except at high pressures or low temperatures; **c.** 0.0123 mol; **d.** C_3H_8; **e.** 0.453 L

Chapter 11

P11.1 The size of the London dispersion force between the molecules increases as the size of the molecule increases. Therefore, the forces between the molecules of dodecane are much larger than the forces between hexane molecules, which in turn are much larger than the forces between propane molecules. The larger the force, the closer the molecules and the stronger they will be held together. Dodecane is so strongly held that it forms a solid, and hexane is held strongly enough to cause it to condense to a liquid at room temperature.   **P11.2** Propylene glycol has a higher boiling point because the intermolecular forces will be higher.   **P11.3** Any molecule with greater forces will have a higher boiling point and a lower vapor

pressure. Examples include ethylene glycol and propylene glycol.
P11.4 Hexane should have lower overall forces than acetone and is likely to have the higher vapor pressure. **P11.5** 440 kJ
P11.6 80°C; Yes, lower external pressures make for lower boiling temperatures. **P11.7** CCl_4 and I_2 **P11.8** 47.1 g KOH, $\chi_{water} = 0.975$ **P11.9** 1.33×10^{-7} M **P11.10** No, 6.4 mL
P11.11 122.5 torr **P11.12 a.** 101.0°C; **b.** 100.7°C; **c.** 67.1°C
P11.13 a. 2.45 atm; **b.** 27.5 atm; **c.** 91.7 atm

1. Intermolecular forces are between molecules, while intramolecular forces are between atoms in a molecule and are stronger.
3. Intermolecular forces are due to the attraction of positive and negative charges, partial to full in magnitude, between molecules.
5. S—H = C—H < N—H < O—H **7. a.** intermolecular; **b.** intramolecular; **c.** intermolecular; **d.** intramolecular, intermolecular **9.** 0.101 nm, stronger and intramolecular; 0.175, intermolecular **11.** Larger forces lead to a higher boiling point. **a.** pentane, less compact and larger dispersion forces; **b.** cyclohexane, larger dispersion forces on rings than on chains; **c.** hexane, larger mass and dispersion forces; **d.** water, stronger forces (hydrogen bonding); **e.** pentane, larger mass and dispersion forces **13. a.** B → O; **b.** P → Cl; **c.** H → O **15. a.** dispersion, dipole–dipole, hydrogen bonding; **b.** dispersion; **c.** dispersion; **d.** dispersion, dipole–dipole, hydrogen bonding **17. a.** CF_4; **b.** H_2O **19. a.** yes; **b.** yes; **c.** no, lacks O, F, or N; **d.** no, lacks H bound to O, F, or N **21. a.** C—O bonds; **b.** dispersion, dipole–dipole **23.** hexane, because lower intermolecular forces (only dispersion versus hydrogen bonding in methanol) give a higher vapor pressure **25.** Kr, because it has the largest mass (and number of electrons) and is most polarizable, giving larger dispersion forces **27. a.** As the central atom becomes larger and more polarizable, the intermolecular forces increase. Higher forces lead to increasing boiling point, which is the trend seen down the group; **b.** It is the most polar and can hydrogen-bond.
29. The forces in water, including hydrogen bonding, are much larger than in methane, leading to higher energy requirements.
31. 2.5×10^6 J **33. a.** dispersion; **b.** gas; **c.** It is pressurized. **35.** The H—Te bond is not polar enough to allow for hydrogen bonding.
37. 48.0 kJ **39. a.** 231 K; **b.** 0.75 atm, 224 K; **c.** gas **41.** gas to solid
43. liquid to solid **45.** As the pressure increases, the gas will become a liquid at 1.4 atm. **47.** See the accompanying diagram; note that the pressure is plotted on a logarithmic scale.

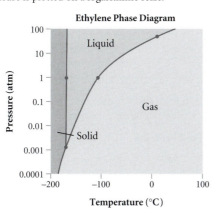

Ethylene Phase Diagram

49. Surface tension is a measure of how strongly the molecules of a substance interact and "pull" the surface molecules toward the center. Viscosity is a measure of how strongly the molecules interact and prevent the flowing of one molecule past another. Capillary action is a balance of two interactions: that of the substance with its container and that of the substance with itself. Molecules capable of strong intermolecular interactions often have strong interactions with other surfaces. **51.** The figure should show a convex surface (curved upward). **53.** From least to most viscous: gasoline, water, honey.

The order is the same as the increase in intermolecular forces—larger forces make for higher viscosity. **55.** The forces (and viscosity) are high because the molecules are large, with correspondingly large dispersion forces. **57. a.** positive; **b.** 49 g per 100 g water
59. Water is able to dissolve many different sizes and shapes of molecules, from very small salts to very large protein molecules.
61. a. yes; **b.** There is additional stability as a consequence of mixing.
63. A-water; B-heptane **65. a.** Solubility has limits, whereas miscibility is possible in any combination; **b.** They share type and size of intermolecular forces. **67. a.** 2.10 M; **b.** 0.222 M; **c.** 4.16 M
69. a. 2.01 m; **b.** 0.310 m; **c.** 0.217 m **71. a.** 0.266; **b.** 0.0555; **c.** 0.00318 **73.** NH_4Cl: 0.0424, 2.46 M; KNO_3: 1.96 g, 0.00278; $C_6H_{12}O_6$: 325 g, 7.21 M **75.** 0.037%, 2.1×10^{-3} M, 1.9×10^{-3} m **77.** 0.873 M **79. a.** 0.0130%; **b.** 1.30×10^5 ppb
81. a. left; **b.** right **83.** 5.1×10^{-4} M **85.** 0.00300 M
87. pentane > water > glycerol **89. a.** 100.76°C; **b.** 100.02°C; **c.** 100.39°C; **d.** 89.1°C **91. a.** −0.93°C; **b.** −0.20°C; **c.** 42.1°C; **d.** −2.23°C **93. a.** 3.66 torr; **b.** 3.9 torr; **c.** 4.01 torr; **d.** 4.39 torr
95. The vapor pressure is the equilibrium position at a specific temperature and does not depend on the surface area of the liquid. The area of the liquid affects how quickly that equilibrium can be reached, but the final value of the vapor pressure will be the same whether the liquid is in a cup or forms an ocean, as long as the container is closed. **97.** All these solutions have the same number of particles dissolved, whereas electrolytes may produce more or fewer ions than others. **99.** 5340 g

101. a.

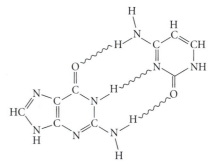

b. If the solution and body fluid do not have the same osmotic pressure, cell damage from shrinkage or rupture will result. **103.** A hydrogen bond is formed between a hydrogen on O or N and a second O or N without a hydrogen.

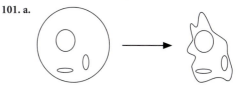

105. 0.868 g **107.** $CHBr_3$ because of its larger dispersion forces and dipole–dipole interactions **109.** Boiling occurs when the vapor pressure of the liquid matches the external pressure. Turning up the gas will not change the external pressure but only cause the boiling to be more vigorous.

111. 5.73 M **113. a.** 6; **b.** Yes, C_nH_{2n+2}; **c.** dispersion only
115. a. vaporization of water; **b.** 3.9×10^2 L; **c.** 688 kJ

116. a.

$$H-\underset{\underset{H}{|}}{\overset{\overset{H}{|}}{C}}-\underset{\underset{H}{|}}{\overset{\overset{H}{|}}{C}}-\underset{\underset{H}{|}}{\overset{\overset{H}{|}}{C}}-\underset{\underset{H}{|}}{\overset{\overset{H}{|}}{C}}-H$$

b. dispersion forces; **c.** yes, because the dispersion forces are small; **d.** no; **e.** 0.301 mol; **f.** The gas is under pressure, raising its boiling point; **g.** Pentane boils higher than butane, and propane boils lower than butane.

Chapter 12

P12.1

$CH_3 - CH_2 - CH_2 - CH_2 - CH_2 - CH_2 - CH_3$

$CH_3 - \underset{\underset{\displaystyle CH_3}{|}}{CH} - \underset{\underset{\displaystyle CH_3}{|}}{CH} - CH_2 - CH_3$

$CH_3 - \underset{\underset{\displaystyle CH_3}{|}}{\overset{\overset{\displaystyle CH_3}{|}}{C}} - \underset{\overset{\displaystyle CH_3}{|}}{CH} - CH_3$

$CH_3 - \underset{\overset{\displaystyle CH_3}{|}}{CH} - CH_2 - CH_2 - CH_2 - CH_3$

$CH_3 - \underset{\overset{\displaystyle CH_3}{|}}{CH} - CH_2 - \underset{\overset{\displaystyle CH_3}{|}}{CH} - CH_3$

$CH_3 - CH_2 - \underset{\underset{\displaystyle CH_3}{|}}{\overset{\overset{\displaystyle CH_3}{|}}{C}} - CH_2 - CH_3$

$CH_3 - CH_2 - \underset{\overset{\displaystyle CH_3}{|}}{CH} - CH_2 - CH_2 - CH_3$

$CH_3 - \underset{\underset{\displaystyle CH_3}{|}}{\overset{\overset{\displaystyle CH_3}{|}}{C}} - CH_2 - CH_2 - CH_3$

$CH_3 - CH_2 - \underset{\overset{\displaystyle |}{CH_2 - CH_3}}{CH} - CH_2 - CH_3$

P12.2

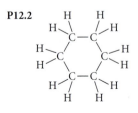

P12.3

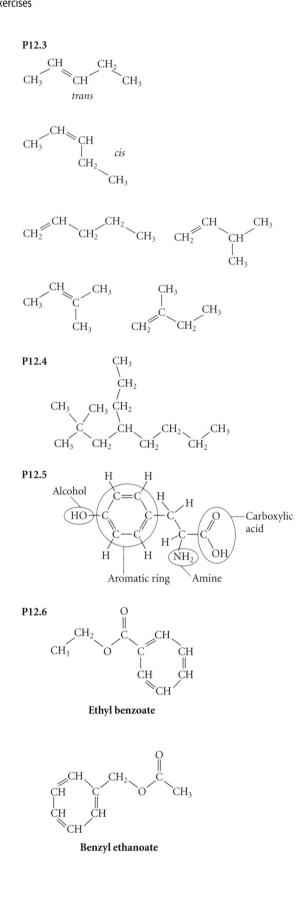

P12.7 Left molecule: second carbon from left; right molecule: carbon at the lower right   **P12.8** Yes, m-amsacrine will attack the DNA in healthy cells also.

P12.9

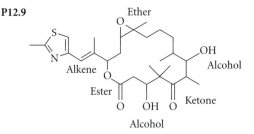

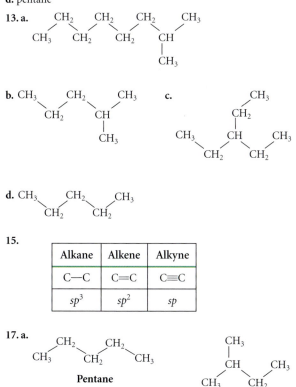

1. diamond (sp^3), graphite (sp^2), fullerenes (sp^2)   **3.** In diamond, all of the carbons are sp^3 hybridized (tetrahedral) and form an extensive three-dimensional network of carbon atoms bound together. The carbon atoms in graphite are all sp^2 (recall that sp^2 atoms form trigonal planes) hybridized and form extended planar surfaces that are only weakly bound via van der Waals forces between the planes. **5.** 2.01×10^{22} atoms   **7.** Aliphatic compounds are those with no double or triple bonds, while the aromatics are characterized by structures with several multiple bonds.   **9.** Because many biological compounds (proteins and others) contain sulfur and the source of the crude oil is those biological materials, many compounds resulting from the breakdown of this material also contain sulfur.
11. a. 2-methyloctane; **b.** 2-methylpentane; **c.** 3-ethylpentane; **d.** pentane

13. a.

CH₂   CH₂   CH₂   CH₃
CH₃   CH₂   CH₂   CH
                        |
                       CH₃

b. CH₃   CH₂   CH₃
         CH₂   CH
                  |
                 CH₃

c.
                        CH₃
                        CH₂
CH₃   CH   CH₃
         CH₂   CH₂

d. CH₃   CH₂   CH₃
         CH₂   CH₂

15.

Alkane	Alkene	Alkyne
C—C	C=C	C≡C
sp^3	sp^2	sp

17. a.

CH₂   CH₂
CH₃   CH₂   CH₃
Pentane

CH₃
CH   CH₃
CH₃   CH₂
2-Methylbutane

CH₃   CH₃
     C
CH₃   CH₃
2,2-Dimethylpropane

Each molecule is C_5H_{12}; **b.** pentane (36°C), 2-methylbutane (28°C), 2,2-dimethylpropane (9.5°C)   **19.** $C_{28}H_{58}$

21.

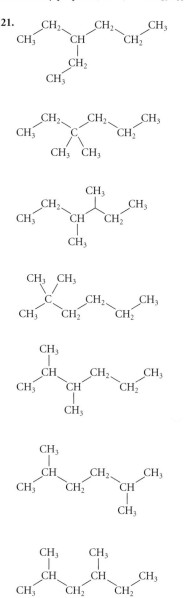

23. C_6H_{12} and C_7H_{14} **25.** left: *trans*; right: *cis*   **27.** $CaC_2 + 2H_2O \rightarrow Ca(OH)_2 + C_2H_2$ **29.** Cyclohexane: **a.** 12 H; **b.** sp^3; **c.** 109.5°; **d.** no double bonds; benzene: **a.** 6 H; **b.** sp^2; **c.** 120°; **d.** three double bonds;

e.

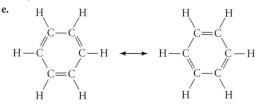

31. During a "run" of the gas chromatograph, a sample is injected into the instruments and heated to vaporize the sample. A carrier gas takes the sample into and through a column. Inside the column, the different molecules interact to varying extents with the materials

inside the column. The greater the interaction, the more slowly the molecules move through the column. The molecules are separated by being selectively slowed on the basis of their properties and are later detected upon exiting the column.   **33.** Boiling temperature; Larger and straighter molecules have higher boiling points.
35. C_8H_{18}; $2C_8H_{18} + 25O_2 \rightarrow 16CO_2 + 18H_2O$ **37.** C_7H_{16}
39. 87% isooctane and 13% heptane   **41.** $CH_4 + Br_2 \rightarrow CH_2Br_2 + H_2$

43.

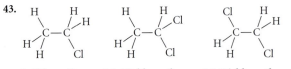

1-Chloroethane   1,1-Dichloroethane   1,2-Dichloroethane

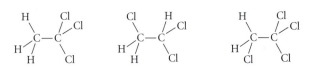

1,1,1-Trichloroethane 1,1,2-Trichloroethane 1,1,1,2-Tetrachloroethane

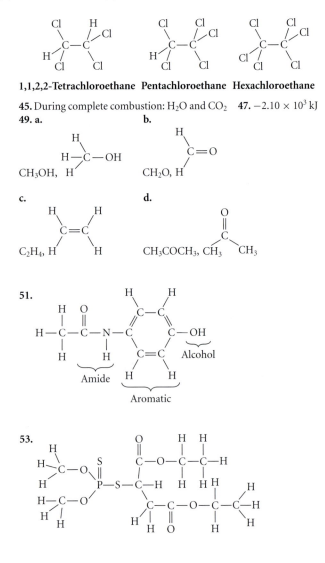

1,1,2,2-Tetrachloroethane Pentachloroethane Hexachloroethane

45. During complete combustion: H_2O and CO_2 **47.** -2.10×10^3 kJ
49. a.   **b.**

CH_3OH, CH_2O,

c.   **d.**

C_2H_4, CH_3COCH_3,

51.

Amide   Aromatic   Alcohol

53.

55. a. none; **b.** three; **c.** seven

57. a.

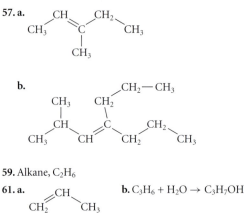

b.

59. Alkane, C_2H_6

61. a.   **b.** $C_3H_6 + H_2O \rightarrow C_3H_7OH$

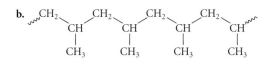

63. a.

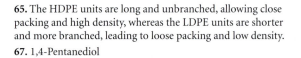

b.

65. The HDPE units are long and unbranched, allowing close packing and high density, whereas the LDPE units are shorter and more branched, leading to loose packing and low density.
67. 1,4-Pentanediol

69.   OH Secondary alcohol

Primary alcohol

71.

73.

75.   , Propanal:

77.   , 2-Butanol:

79. $CH_3CH_2COOH + H_2O \rightarrow CH_3CH_2COO^- + H_3O^+$

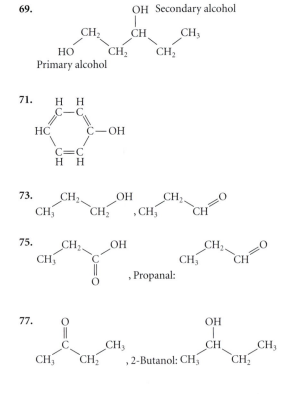

81.

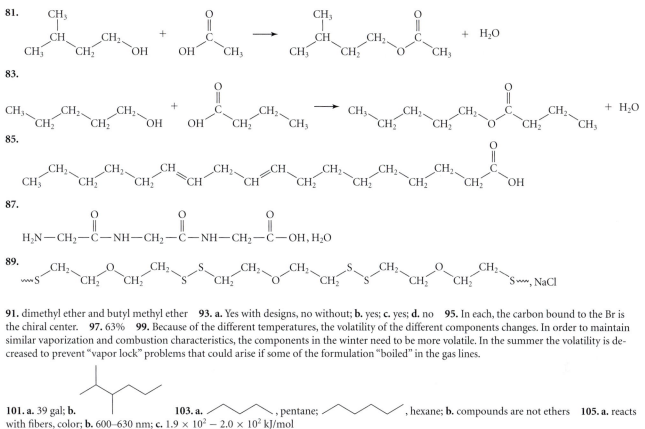

83.

85.

87.

89.

91. dimethyl ether and butyl methyl ether   **93. a.** Yes with designs, no without; **b.** yes; **c.** yes; **d.** no   **95.** In each, the carbon bound to the Br is the chiral center.   **97.** 63%   **99.** Because of the different temperatures, the volatility of the different components changes. In order to maintain similar vaporization and combustion characteristics, the components in the winter need to be more volatile. In the summer the volatility is decreased to prevent "vapor lock" problems that could arise if some of the formulation "boiled" in the gas lines.

101. a. 39 gal; **b.** 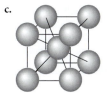   **103. a.** 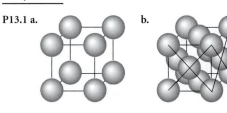, pentane; , hexane; **b.** compounds are not ethers   **105. a.** reacts with fibers, color; **b.** 600–630 nm; **c.** $1.9 \times 10^2 - 2.0 \times 10^2$ kJ/mol

106. a.

b. Polar molecule, hydrogen bonding, solid, acidic; **c.** 61.6%;

d.

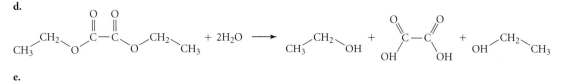

e.

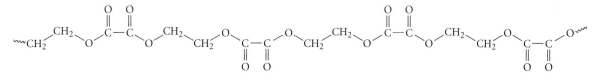

Chapter 13

P13.1 a.

b.

c.

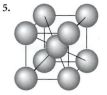

P13.2 47.6%   **P13.3** 11.35 g/cm^3 **P13.4** 1306 nm
P13.5 metal, glass/ceramic, ceramic, metal (or alloy)
P13.6 composite (magnetic media on plastic), plastic, composite (glass and plastic)   **P13.7** 4.11×10^{-7} mol; 4.15×10^4 layers for 1 g, 8.30×10^5 layers for 20 g; no

1. Crystalline solids have a regular arrangement of particles, whereas amorphous solids exhibit a random arrangement of particles.   **3.** A simple unit cell consists of atoms arranged within a parallelepiped (often at the corners, centered inside or on the faces and edges). The crystal lattice consists of rows and columns of the parallelepiped (unit cells) connected at the corners.

5.

7. four atoms **9.** A, metal; B, molecular; C, ionic **11. a.** 4.42 × 10^{-29} m³; **b.** 2.10 × 10^{-28} m³; **c.** 6.23 × 10^{-28} m³ **13.** 0.559 g **15.** 125 pm **17.** 142 pm, 32.0% **19.** 6.1 × 10^{23} atoms per mole **21.** Rh **23.** Na, Al, Ca, Cr, Bi **25.** one, three **27.** metal (2.5 kJ/mol), semiconductor (85 kJ/mol), insulator (450 kJ/mol) **29.** Since the valence electrons from the metals are free to travel in the delocalized orbitals spanning the entire metal (the electron gas), the remaining positive atomic cores are all attracted to the sea of electrons that surrounds them. Because all cores are attracted, it doesn't matter what the specific identity of the metal is; all metals can be accommodated in the metal alloy crystal lattice. **31.** If one considers a display of apples that is originally all one color (green, for example), alloys can be created in two ways: **a.** Remove some of the green apples and replace each, in that same position, with a red apple. This represents a substitutional alloy; **b.** Place some smaller red crabapples in the open spaces left between the green apples, without rearranging the apples or removing any apples. This represents an interstitial alloy. **33. a.** one; **b.** Sodium donates one orbital to the metal "molecular" orbitals. The number of molecular orbitals that are created must equal the number of atomic orbitals that were used, with half being bonding and the other half antibonding. Because each orbital can hold two electrons and sodium donates only one per atom, only half of the orbitals are filled. In other words, the lower half, all bonding, are filled. Any transition from the highest occupied molecular orbital to the lowest unoccupied molecular orbital will be a transition from a bonding orbital to an antibonding orbital. **35.** 2.42 × 10^{-19} J **37.** Indium, from Group IIIA, has one fewer valence electron than silicon. For each indium in the semiconductor, there is one fewer electron (one more *positive* hole), making the semiconductor p-type. **39.** Adding boron to carbon results in fewer electrons than pure carbon would have, creating openings in the valence band and a larger gap to the conduction band than in pure carbon. There are now lower-energy transitions available within the valence band. Because of these transitions, the electrons absorb some visible light, leaving behind the blue color we see. **41.** Ceramics are characterized by ionic bonding, which has a very large gap between the valence and conduction bands. A large amount of energy is required to bridge the gap, and very few (if any) electrons are promoted into the conduction band. Without electrons in the conduction band, no (or very little) current can be conducted through the ceramic. **43.** Because the bonding in glass is very disordered and random, different regions within the glass experience different localized bonding strengths. Different amounts of energy will be required to loosen up and melt these interactions, resulting in a range of melting temperatures. **45.** 1.2 m, 1.66 × 10^{-25} J

47.

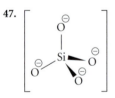

49. 7.51 g **51. a.** plastic or metal; **b.** plastic; **c.** metal or plastic **53.** All plastics are formed by the linking of several smaller monomer units that are then thermally molded and hardened in a particular shape. DNA is a polymer, but because it is not thermally molded and hardened, it is not a plastic. **55. a.** plastic or metal; **b.** glass and metal; **c.** plastic or metal; **d.** plastic, glass, ceramic, or metal **57.** A composite material used for automobile exteriors should be both strong and lightweight in order to provide protection in crashes and also to enhance the car's performance or economy of operation by reducing its weight. Composites tend to be expensive to produce. **59.** Most of the bonds in most polymers are sigma bonds and are strongly localized. The electrons are not free to conduct electricity.

61. 8.4 × 10^{20} atoms **63.** 7.67 × 10^{15} atoms **65.** 1.37 × 10^{-6} g, 4.12 × 10^{-6} g **67.** physical deposition or chemical-vapor deposition, depending on the exact process **69.** the size of holes within the zeolite **71.** Televisions are made from many plastics, metals, and glass and sometimes a large portion of lead. The plastics, metals, and lead are often recycled. **73.** The nucleic acids (DNA and RNA) make up one class of biopolymers; a second class consists of proteins (mostly enzymes); the third class consists of some carbohydrates, including cellulose, chitin, and starches. **75.** heterogeneous; It limits the formation of waste by-products. **77.** It is lighter, more insulating, and more flexible. **79.** Molecular solids do not conduct; ionic solids, when molten or dissolved, do conduct. **81.** Because so many metal atoms are in the bulk form of a metal, and because each donates an orbital to the overall molecular orbital, there are an extremely large number of bonding and antibonding orbitals produced. Because the molecular orbitals are so plentiful and close together, electrons can absorb and release almost any wavelength of light from the infrared to the ultraviolet. The result is the luster that we associate with metals. **83.** See http://www.dnr.state.oh.us/recycling/awareness/facts/tires/goodyear.htm **85.** contamination of groundwater and drinking water
86. a.

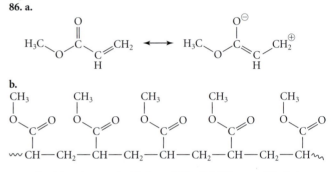

b.

c. reacts with water, may be biodegradable; **d.** 0.558 g; **e.** acidic, but does not react with water in ester cleavage

Chapter 14

P14.1 112 splits/256 combinations; 43.8% **P14.2** All but photosynthesis are spontaneous. **P14.3 a.** not spontaneous; **b.** spontaneous; **c.** spontaneous **P14.4** −209 J/K; 136 J/K **P14.5** The entropy change is positive for the first reaction and negative for the second. **P14.6** 109 J/K; −137 J/K **P14.7** 0.224 kJ/mol; −0.986 kJ/mol **P14.8** spontaneous (−2827 kJ); spontaneous (−2074 kJ); spontaneous (−401 kJ) **P14.9** 337.2 K; 373.0 K

1. 16 **3.** 37.5% **5.** A macrostate represents the total properties of the system. A microstate is a representation of a single way to arrange the parts of a system. **7.** One possibility is all students in one room with one seat between every student and the next. Another possibility is half the students in each room with vacant seats between students. **9.** $2^{1000} = 1.07 \times 10^{301}$. Because the number of microstates becomes very large, the number of individual microstates that correspond with either exact or very close to exact even distributions is much larger than the number of unequally distributed microstates. **11.** It has increased. We know that there have to be many more microstates available after mixing, because the process is spontaneous, indicating a final state with more available microstates. **13.** Water evaporation only **15. a.** At room temperature, burning a piece of paper requires the intervention of a heater or ignition source. **b.** Hard-boiling an egg requires the addition of heat. **c.** Muscle mass is not added without a lot of work, such as lifting weights. **17.** The entropy of the universe decreases in a nonspontaneous process. **19.** The greater the number of microstates, the greater the entropy in a system. **21.** The number of microstates

corresponding to an even dispersal of the aroma-producing molecules is much larger than the number of microstates in which the aroma remains contained in one part of the room. The spontaneous process is then the dispersal of the aroma corresponding to the greater number of microstates.   **23.** If the total entropy of the universe increases, the process is spontaneous. The conversion of the gasoline from a liquid to a gas involves a large increase in entropy, which offsets the small decrease in entropy due to the removal of heat (and therefore available microstates) from your hand.   **25.** The change in entropy of the carbon dioxide is very large and positive as the molecules go from solid to gas. Because the heat to sublime the CO_2 is taken from the environment, the nearby gases are cooled, lowering their entropy, which is a negative entropy change. Because the process is spontaneous, the entropy of the universe increases.
27. As the reaction proceeds from left to right, the number of moles of gas is reduced. Removal of a mole of gas corresponds to a large decrease in the entropy; therefore, a reaction from right to left will cause an increase in the entropy of the system.   **29. a.** positive; **b.** need more information; **c.** need more information   **31.** The more ways a molecule can move, the greater the multiplicity in a system.
33. The entropy of the surroundings increases.   **35. a.** negative; **b.** positive; **c.** positive   **37.** positive; For the CO_2 the transition from a dissolved species, confined to the liquid, to a gas with a much greater available volume, represents a large increase in entropy.
39. $\underline{1}C_3H_8(g) + \underline{5}O_2(g) \rightarrow \underline{3}CO_2(g) + \underline{4}H_2O(g)$; 6 mol of reactants gas; 7 mol of products gas; entropy increases.   **41.** $\underline{1}C_3H_8(l) + \underline{5}O_2(g) \rightarrow \underline{3}CO_2(g) + \underline{4}H_2O(l)$; 5 mol of reactant gas; 3 mol of product gas; entropy decreases.   **43.** Because the number of microstates that become available in a gas rather than a liquid or solid is so large, any conversion to a gas greatly influences the entropy. The change in the number of microstates of liquids and solids is much lower and influences the reaction much less.   **45. a.** negative; **b.** positive; **c.** negative   **46. a.** −284.5 J/K; **b.** 216.5 J/K; **c.** −99.5 J/K
49. $\underline{1}C_2H_5OH(l) + \underline{3}O_2(g) \rightarrow \underline{2}CO_2(g) + \underline{3}H_2O(g)$; 219 J/K
51. 4 J/K   **53. a.** high temperature; **b.** none   **55. a.** 61.0 kJ/mol; **c.** −1.59 × 10³ kJ/mol   **56. a.** −97.7 kJ/mol; **c.** 507 kJ/mol
57. a. 938 J/K·mol; **c.** −2.42 × 10⁴ J/K·mol   **58. a.** 1075 K; **c.** 769 K
59. a. spontaneous; **b.** nonspontaneous; **c.** spontaneous
61. −1.48 × 10³ kJ   **63.** −71 kJ/mol   **65.** −532 kJ/mol; −122 J/K; −446 kJ/mol   **67. a.** 1681 K; **b.** 138 K; **c.** 420 K   **69.** 1.09 × 10⁻²³ atm   **71.** 6.05 × 10³ J/K   **73.** 1094 K   **75. a.** −802 kJ/mol; **b.** −794 kJ; **c.** The negative value shows that the reaction is spontaneous at room temperature, but it says nothing about the rate at which the reaction proceeds. The additional heat from the spark or flame is necessary to get the reaction to happen at an appreciable rate.   **77.** $\underline{2}C_4H_{10}(g) + \underline{13}O_2(g) \rightarrow \underline{8}CO_2(g) + \underline{10}H_2O(g)$; As the reaction proceeds, 15 mol of gas is converted to 18 mol of gas. The increase of 3 mol of gas will cause an increase in entropy.   **79.** To predict spontaneity, we must calculate if the entropy of the universe increases. To complete this calculation, we must consider both the system and the surroundings. In other words, we need to account for the entire universe. Using Gibb's equation, we need only consider variables related to the system.   **81.** Under these conditions, Q becomes an equilibrium constant.   **83. a.** 4 formed, 2 used = 2 net ATP; **b.** 0.055 g; **c.** The reactants had higher potential energy.
85. At 400°C, every reaction will speed up including those reactions that cause the decomposition of biochemicals. Many proteins and other cellular structures are unstable and will decompose (cook!) at 400°C. At 330°C, reactions will slow nearly to a stop because there will not be enough energy available for reactions to occur even with the assistance of enzymatic catalysts.   **86. a.** $\underline{1}C_6H_{12}O_6(s) + \underline{6}O_2(g) \rightarrow \underline{6}CO_2(g) + \underline{6}H_2O(l)$; There should be an increase of entropy on the order of a few hundred J/K·mol; **b.** −2812 kJ; 262 J/K; −2875 kJ; **c.** −2548 kJ, 976 J/K, −2827 kJ; **d.** 4.07 L, the gas is removed via respiration; **e.** −71 kJ; **f.** 13.8 kJ/mol; **g.** If there were no glucose directly available to the organism, it would have to rely on other sources of energy, such as fructose. Fructose is naturally available in many fruits and vegetables; it also is a product in the hydrolysis of table sugar (sucrose).

Chapter 15

P15.1 $\text{Rate} = -\dfrac{1}{2}\dfrac{\Delta[\text{HI}]}{\Delta t} = \dfrac{\Delta[\text{H}_2]}{\Delta t} = \dfrac{\Delta[\text{I}_2]}{\Delta t}$
P15.2 4.6 × 10⁻⁴ M/s   **P15.3** Second order with respect to [HI]; second order overall   **P15.4** 8.01 × 10⁴ years; 2.41 × 10⁴ years
P15.5 30 days; 60 days; 2.5 × 10² days   **P15.6** Rate = (2.80 × 10⁵ $M^{-1} \cdot s^{-1}$) [CO]¹ [Hb]¹; second order   **P15.7** The second-order plot (inverse concentration versus time) shows a straight line, whereas the others do not. The reaction is second order.   **P15.8** The overall reaction is $2NO(g) + O_2(g) \rightarrow 2NO_2(g)$. The intermediate is $NO_3(g)$, and the rate law for the reaction is rate = $k[NO]^2[O_2]$.   **P15.9** First and second reactions: 10.1 kJ/mol; first and third reactions: 12.6 kJ/mol; second and third reactions: 13.8 kJ/mol. The values vary slightly as a consequence of small variations in experiments and mathematical rounding. When more than two data points are available, the preferred method is graphical (plot ln k versus 1/T), which gives a value of 12.8 kJ/mol.

1. a. Total distance run = extent of reaction; **b.** average speed (total distance over time) = average rate of reaction; **c.** exact speed the runner is moving at a particular moment (perhaps using a radar gun) = instantaneous rate; **d.** exact speed as the runner started the race (again, with the radar gun) = initial rate   **3.** 2.5 × 10⁻⁸ M/s; Yes, for many reactions, the initial rate of the reaction is fastest and the rate of the reaction slows as the reaction proceeds. In this case, the initial (and instantaneous) rate and others near the start of the reaction would be larger than the average rate.   **5.** C_2H_4 is produced at twice the rate;
$\text{Rate} = -\dfrac{\Delta[\text{C}_4\text{H}_8]}{\Delta t} = \dfrac{1}{2}\dfrac{\Delta[\text{C}_2\text{H}_4]}{\Delta t}$ **7. a.** 5 × 10⁻⁴ M/s;
b. 1.3 × 10⁻³ M/s; **c.** 1.2 × 10⁻⁹ M/s; **d.** 2.0 × 10⁻⁵ M/s
9. a. −6.8 × 10⁻² M/s; **b.** 2.2 × 10⁻² M/s; **c.** 4.5 × 10⁻² M/s
11. a. 0 mi/h; **b.** 112 mi/h; **c.** 187 mi/h; **d.** 134 mi/h; **e.** This plot is the opposite of most reactions, in which the amount of products is high at first and then levels off as the reaction slows down. Here the distance covered is lowest initially and is larger as the racer speeds up.
13. a. The rate of appearance of water is twice the rate of appearance of oxygen.

b. Rate = $-\dfrac{1}{2}\dfrac{\Delta[\text{H}_2\text{O}_2]}{\Delta t} = \dfrac{1}{1}\dfrac{\Delta[\text{O}_2]}{\Delta t} = \dfrac{1}{2}\dfrac{\Delta[\text{H}_2\text{O}]}{\Delta t}$

15.

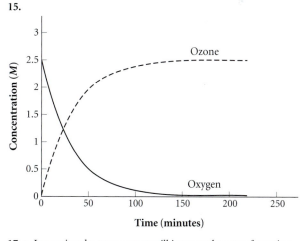

17. a. Increasing the temperature will increase the rate of reaction because that temperature increase (1) increases the number of collisions and (2) increases the speed of the molecules, which increases the energy in those collisions, making reaction more likely. **b.** Will not increase the rate: see part a above. **c.** Increasing the initial concentration of the reactants, will make collisions between those reactants more likely, thereby increasing the rate of the reaction. **d.** Will not increase the rate; lowering the concentration of the reactants will make collisions between those reactants less likely, thereby decreasing the rate of the reaction.   **19. a.** Third order; **b.** k is the rate

constant of the reaction. **c.** The reaction rate quadruples. **d.** no effect
21. 2.01×10^{-4} M/s **23.** Rate laws are experimentally determined, so there is no way, without additional information, to know what order the reaction is. Further, we cannot merely look at a reaction to determine the order of the reaction. **25. a.** 3.61×10^{-4} s^{-1}; **b.** 1.32×10^{-3} /min; **c.** 5.55×10^{-10}/year; **d.** 1.21×10^{-4}/year **27. a.** 43 min; **b.** 86 min; **c.** 143 min **29. a.** 0.0916 M; **b.** 0.0839 M; **c.** 0.183 M; **d.** 0.193 M **31.** 0.195 M **33. a.** $[A]_0 = 0.100$ M cannot be correct because the initial concentration cannot be lower than the later concentration. **b.** The time values here are reversed—that is, the initial time should be lower than the later time. **35.** 409 s for $[A]_0 = 0.0333$ M. **37.** 0.0141 M/h **39.** 8.65 $M^{-1}\cdot$min^{-1} **41. a.** 0.0999 M; **b.** 0.499 M; **c.** 0.327 M; **d.** 7.82×10^{-4} M **43.** Because collisions between the reactants must occur for a reaction to take place, and the number of reactants is decreasing with time, the number of collisions between reactants becomes rarer with time. A slower rate of collision results in a slower reaction rate. **45.** The half-life is constant for a first-order reaction and does not depend on the initial concentration. Pablo's experimental measurement should be consistent. The half-life of a second-order reaction is inversely proportional to the initial concentration. Peter's experiment will be different because the initial concentration, and therefore the half-life, will be different. **47. a.** 1.71×10^{-3} h^{-1}; **b.** 1.98 $M^{-1}\cdot$h^{-1}; **c.** 1.49×10^{-6} M/h **49.** 0.594 M **51.** 141 years **53.** The reaction is second order. **55. a.** rate doubles; **b.** rate doubles; **c.** rate quadruples **57.** Rate $= (2.2$ s$^{-1})$ [B]^1, first order **59.** Rate $= (1.0 \times 10^2$ $M^{-2}\cdot$s$^{-1})$ [A]^1 [B]^2, third order **61.** zero order, 0.047 M/min **63. a.** first-order plot [ln(large dark atoms) versus time]; **b.** The rate at left is double the rate at right. **c.** s^{-1} **65.** first order; 0.229 min^{-1} **67. a.** Rate $= k$ [Stabilizer]$^{0.5}$ [O$_2$]^2; **b.** 3.60×10^{-3} $M^{-1.5}\cdot$s^{-1}; **c.** 1.20×10^{-3} M/s; **d.** 0.360 M **69.** 84% (458 cpm) **71.** Because each of those values is the y intercept of the graph, each line would be moved vertically until the intercept was at $y = 0$. **73.** negative **75.** 8.95×10^{-2} s^{-1} **77.** $-1\cdot$ **79.** A *reaction mechanism* is a collection of single simple steps called *elementary steps* that represent what is thought to occur in a reaction. The slowest step in the mechanism usually determines the rate of the reaction and is called the *rate-determining step*. **81.** The answers will vary. One example: (1) Hear phone ring. (2) Decide to answer. (3) Push chair out from table. (4) Stand up. (5) Walk to phone. (6) Pick up phone. (7) Say "Hello," etc. Depending on how agile you are and how far the phone is from the table, you might choose step 3, 4, or 5 as the slow step. **83.** 8.84 kJ/mol **85.** Mechanism B, with its first-step slow step has a rate law of Rate $= k[H_2O_2][I^-]$, which does not have the proper hydrogen ion concentration dependence. The rate law for Mechanism A does match and is derived by setting the rates of the forward and reverse reactions in the first step equal, solving for the intermediate concentration, and placing that concentration into the rate law from the second step:

Rate $= \dfrac{k_1}{k_{-1}} k_2 [H_2O_2][H^+][I^-] = k'[H_2O_2][I^-][H^+]$

87. a. 97.9 kJ/mol; **b.** 2.4×10^{-3} $M^{-1}\cdot$s^{-1} **89. a.** A reaction intermediate is a compound that is formed and then consumed during the course of a reaction. **b.** An activated complex is the species or collection of atoms that exists at the transition state midway between reactants and products in a single elementary step. **c.** A homogeneous catalyst is a species in the same phase as the rest of the reactants and aids in the reaction by lowering the overall activation energy in the process of converting reactants to products. A homogeneous catalyst is destroyed, and then re-created, in the course of the reaction. **d.** A heterogeneous catalyst is a species or material in a different phase from the rest of the reactants; it aids in the reaction by lowering the overall activation energy in the process of converting reactants to products. **91. a.** $2N_2O \rightarrow 2N_2 + O_2$; **b.** O is an intermediate. There are no catalysts; **c.** first step **93. a.** Lactose $\rightarrow$ glucose + galactose; **b.** (Lactase–lactose) and (lactase–glucose–galactose) are intermediates. Lactase is the catalyst. **c.** Rate $= k$[Lactase][Lactose]

95. a. $N_2(g) + 3H_2(g) \rightarrow 2NH_3(g)$; **b.** FeH$_2$ and FeN$_2$; **c.** Fe(s); **d.** heterogeneous **97.** 1.7×10^{-3} M/s; 0.10 M/h **99. a.** 0.350 s^{-1}; **b.** 1.98 s **101.** 200 kJ **103. a.** $C_2H_4(aq) + HCl(aq) \rightleftharpoons C_2H_5Cl(aq)$; **b.** Rate $= k_1[C_2H_4][HCl]$ **105.** answers vary **107. a.** $CH_4(g) + 2O_2(g) \rightarrow CO_2(g) + 2H_2O(g)$; **b.** low densities, few collisions, low O$_2$ concentration **109.** See diagram; the uncatalyzed profile is represented by the solid line. **a.** exothermic ($\Delta H = -92$ kJ); **b.** spontaneous ($\Delta G = -32$ kJ); $T = 458$ K; **c.** See short-dashed line in diagram.

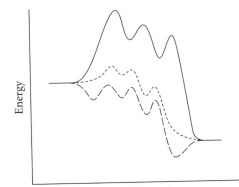

d. See long-dashed line in diagram. **e.** 15 g; **f.** 12.2 kg

Chapter 16

P16.1 0.224 M **P16.2** $K = \dfrac{[CO][H_2]^3}{[CH_4][H_2O]}$

P16.3 a. $K = \dfrac{[NO_2][O_2]}{[NO][O_3]}$; **b.** $K = \dfrac{[NH_4Cl]}{[HCl][NH_3]}$;

c. $K = \dfrac{[C_2H_6]}{[C_2H_2][H_2]^2}$

P16.4 a. reactants; **b.** reactants; **c.** reactants; **d.** products
P16.5 1.1×10^3 **P16.6** 1.5×10^8 **P16.7 a.** left; **b.** right; **c.** right; **d.** left **P16.8** 0.016 M **P16.9** $[H^+] = 1.8 \times 10^{-5}$ M, $[C_2H_3O_2^-] = 0.050$ M, $[C_2H_4O_2] = 0.050$ M **P16.10 a.** shift right; **b.** shift left; **c.** no change; **d.** shift left **P16.11** $K = 0.017$; The difference between the two values may lie in the difference between the activities and concentrations.

1. Even though the concentrations are not changing, the reactants and products are continuously reacting to form each other, making the equilibrium dynamic. **3.** A reaction that does not go to completion stops reacting before all of the reactants have been converted to products. In other words, the reaction reached equilibrium and a zero Gibbs energy change before all of the reactants were converted.

5. a. $K = \dfrac{[C_2H_5Cl]}{[HCl][C_2H_4]}$; **b.** $K = \dfrac{[CO_2][H_2O]^2}{[CH_4][O_2]^2}$; **c.** $K = \dfrac{1}{[H_2]^2[O_2]}$

7. a. At equilibrium, the forward and reverse reaction rates are equal. **b.** The concentrations are related via the equilibrium expression,

$K = \dfrac{[CaSO_4]^3 [H_3PO_4]^2}{[H_2SO_4]^3}$. **c.** At equilibrium, $\Delta G = 0$.

9. $\dfrac{k_1}{k_{-1}} = \dfrac{[HCl]^2}{[H_2][Cl_2]} = K +$

11. If equilibria could not be controlled, the amounts of products could not be maximized for profit. **13. a.** blue; **b.** yellow

15. a. $K = \dfrac{1}{[O_2]}$; **b.** $K = \dfrac{[H_2SO_4]}{[SO_3]} +$

17. a. 0.10 (any value slightly smaller than 1); **b.** [CO] > [CO$_2$]

19. $C_3H_8(g) + 5O_2(g) \rightleftharpoons 3CO_2(g) + 4H_2O(g)$,

$K = \dfrac{[CO_2]^3 [H_2O]^4}{[C_3H_8][O_2]^5}$

21. $0.050\ M$ **23.** $K = \dfrac{[SO]^4}{[O_2]^5}$

25. Values of $K \gg 1$ are product favored; values of $K \ll 1$ are reactant favored; and values of K near 1 produce mixtures. The ammonia synthesis produces a mixture, the ozone depletion is product favored, and the acid ionization is reactant favored.   **27. a.** In order to get a quantitative 1:1 reaction between EDTA and the metal, the reaction needs to go to completion. If the reaction does not go to completion, we will not have a firm idea of how much metal is in the solution. **b.** A $K = 10^{10}$ ensures complete reaction, whereas a $K = 10$ produces a mixture.

29. a. $K = \dfrac{[H_2O]^2}{[H_2]^2\,[O_2]}$; **b.** 82.8; **c.** 0.0121; **d.** 9.10   **31.** 11

33. a. $K = \dfrac{[C_2H_2]^{1/2}\,[H_2]^{3/2}}{[CH_4]}$; **b.** $2CH_4(g) \rightleftharpoons C_2H_2(g) + 3H_2(g)$;

$K = \dfrac{[C_2H_2]\,[H_2]^3}{[CH_4]^2}$; **c.** $K_b = (K_a)^2$ **35.** 6.6×10^2

37. Adding the two reactions results in a third reaction with $K = 1.3 \times 10^4$, which is a sufficiently large value to ensure the reaction proceeds enough toward the products for the investigators to do the analysis.   **39.** $K = 0.092$, the strategy is not reasonable.
41. 0.76   **43. a.** $3.0 \times 10^{-3}\ M$; **b.** $1.8 \times 10^{-5}\ M$; **c.** $4.2 \times 10^{-4}\ M$
45. a. forward; **b.** reverse   **47. a.** forward; **b.** no change; **c.** reverse;
d. reverse   **49.** $1.05 \times 10^{-5}\ M$ **51.** 1.7×10^{-93} **53. a.** $7.8 \times 10^{-6}M$;
b. $0.76\ M$ **55. a.** $H_2O \rightarrow H^+ + OH^-$ plus the forward and reverse codeine reaction; **b.** Only the forward codeine reaction is important;
c. forward; **d.** $[OH^-] = [C_{18}H_{21}NO_3H^+] =$
$4.0 \times 10^{-4}\ M$; $[C_{18}H_{21}NO_3] = 0.10\ M$ **57.** $Q < K$ for the conditions given, so more silver can be extracted.

59. a. $2.72\ M$;

b. Reactants ———————————————— Products

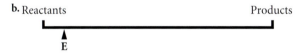

E

61. $0.22\ M$ **63.** $5.0 \times 10^{-13}\ M$
65. $[V^{3+}] = 0.0115\ M$, $[VO^+] = 0.138\ M$, $[H^+] = 0.0770\ M$
67. $[MbO_2] = 1.8 \times 10^{-12}\ M$, $[Mb] = 0.00102\ M$ $[O_2] = 0.000020\ M$
69. a. right; **b.** left; **c.** no effect; **d.** left; **e.** no effect   **71. a.** CuCl system; **b.** no effect because solids do not appear in K **73.** Because the reaction shifts toward the blue upon heating, we could assume that the reaction would shift in the opposite direction upon cooling—toward the pink.   **75.** A catalyst affects the speed of a reaction but does not affect the equilibrium position.   **77. a.** 1.00; **b.** 1.00;
c. 0.357; **d.** 52.2   **79.** –6.45 kJ/mol   **81.** 9.69 kJ/mol   **83.** The strongest is formic acid; the weakest is acetic acid.   **85.** The most important calculation would be the total cost per unit of product formed. Although Method A uses more expensive reactants, little would be wasted. With a much smaller equilibrium constant, Method B will produce less product per given amount of reactant. Much more reactant will be needed to produce the same amount of product as Method A, a disadvantage that offsets the lower cost of the reactants. The exact balance of reactant price versus amount of product formed will be the deciding factor.   **87.** 2.6×10^{-29}
89. An equilibrium condition would exist if the total number of people inside the shop remained the same. There would be the same number of people entering the shop as finishing their shopping and leaving. A larger value of K would indicate a greater number of people in the shop, which should be more profitable. If Q were less than K, there would be fewer people in the shop, indicative of a poor business day.   **91.** $[CO] = [H_2] = 24.2$ atm; $[H_2O] = 27.8$ atm
93. 1 min; All frames beginning at 1 min. have the same composition.
95. a. Reactants ———————————————— Products

E

b. $PbI_2\ s = 0.0017\ M$, $SrC_2O_4\ s = 0.00022\ M$; **c.** Because K_{sp} is a larger number for SrC_2O_4, you may have placed the E bar more toward the products. However, PbI_2 is actually more soluble.
97. With six sites, EDTA binds strongly, resulting in large K.
98. a. $\Delta G = -24.8$ kJ/mol; spontaneous; **b.** $K = 2.22 \times 10^4$, product favored; **c.** 4.5×10^{-5}; **d.** $3.56 \times 10^{-12}\ M/s$; **e.** 0.30 M;
f. The methanol concentration will increase. **g.** \$301

Chapter 17

P17.1 $H_2NCH_2CH_2NH_2\ (aq)$ + $H_2O\ (l)$ $\rightleftharpoons$ $H_2NCH_2CH_2NH_3^{\oplus}\ (aq)$ + $OH^{\ominus}\ (aq)$
      Base   Acid   Conjugate Acid   Conjugate Base

P17.2 $CH_3(CH_2)_{10}COOH\ (aq)$ + $H_2O\ (l)$ $\rightleftharpoons$ $CH_3(CH_2)_{10}COO^{\ominus}\ (aq)$ + $H_3O^{\oplus}\ (aq)$
        Acid   Base   Conjugate Base   Conjugate Acid

P17.3 Three possible acids weaker than HF ($K_a = 7.2 \times 10^{-4}$) are acetic acid (CH_3COOH, $K_a = 1.8 \times 10^{-5}$), phenol (C_6H_5OH, $K_a = 1.6 \times 10^{-10}$), and hypobromous acid ($HBrO$, $K_a = 2.8 \times 10^{-9}$).
Three acids that are stronger than HOCl ($K_a = 2.9 \times 10^{-8}$) are acetic acid (CH_3COOH, $K_a = 1.8 \times 10^{-5}$), hydrofluoric acid (HF, $K_a = 7.2 \times 10^{-4}$), and nitrous acid (HNO_2, $K_a = 4.0 \times 10^{-4}$).
P17.4 Any of these are correct: hydrogen sulfate ion, chlorous acid, hydrofluoric acid, nitrous acid, and lactic acid.   **P17.5** Because bromine is more electronegative than sulfur (2.8 vs. 2.5), more electrons will be pulled away from the hydrogen end of the bond in HBr than in H_2S. HBr should be a stronger acid than H_2S.   **P17.6** $7.55 \times 10^{-5}M$
P17.7 a. 2.336; **b.** 8.558; **c.** 5.492; **d.** $1.2 \times 10^{-4}\ M$; **e.** $3.2 \times 10^{-4}\ M$;
f. $1.2 \times 10^{-10}\ M$ **P17.8** 6.14   **P17.9** 1.65, 0.022 M, $4.5 \times 10^{-13}\ M$
P17.10 3.00, 4.602   **P17.11** 2.92   **P17.12** 14.025   **P17.13** 12.23
P17.14 1.46   **P17.15** 9.07   **P17.16** 6.10

1. In the Arrhenius model, a base produces OH^- in solution, which ammonia does in the reaction $NH_3 + H_2O \rightleftharpoons NH_4^+ + OH^-$. In the Brønsted–Lowry model, a base accepts a proton from another species (the acid), which NH_3 does from water in the formation of NH_4^+ in the reaction.   **3. a.** NO_3^-; **b.** Br^-; **c.** OH^-; **d.** ClO_4^-
5. a. H_2O; **b.** NH_4^+; **c.** H_3O^+; **d.** HF   **7. a.** $2Al(s) + 6HCl(aq) \rightarrow 2AlCl_3(aq) + 3H_2(g)$, aluminum chloride; **b.** $Ca(s) + 2HNO_3(aq) \rightarrow Ca(NO_3)_2(aq) + H_2(g)$, calcium nitrate; **c.** $2Na(s) + 2HCl(aq) \rightarrow 2NaCl(aq) + H_2(g)$, sodium chloride; **d.** $2K(s) + 2HNO_3(aq) \rightarrow 2KNO_3(aq) + H_2(g)$, potassium nitrate   **9.** A strong acid is one that dissociates completely in solution to form H^+.

11. a.
$HCl + H_2O \rightarrow Cl^-$ + H_3O^+
Acid   Base   Conj. base   Conj. acid

b.
$NaOH + CH_3COOH \rightarrow CH_3COONa + H_2O$
Base   Acid   Conj. base   Conj. acid

c.
$H_2SO_4 + Mg(OH)_2 \rightarrow MgSO_4$ + $2H_2O$
Acid   Base   Conj. base   Conj. acid

13.
$Mg(OH)_2$ + $2HCl$ $\rightarrow$ $MgCl_2$ + $2H_2O$
Base   Acid   Conj. base   Conj. acid

15. a. $K_a > 1$; **b.** The conjugate base will not regain H^+. **c.** 100%
17. HI; Even though bromine is slightly more electronegative than iodine, the iodide ion is much larger and has a much more diffuse electron density than the bromide ion. The attraction of H^+ for the larger I^- with its lower electron density is less, so the HI bond breaks more easily and HI is the stronger acid.   **19.** Because both acids have an equal number of oxygen atoms that contribute to pulling the

electrons from the bond to hydrogen, we must consider the electronegativity of the central halogen atom. Chlorine has a higher electronegativity than bromine and weakens the bond to hydrogen more than in $HBrO_4$; therefore, $HClO_4$ must be the stronger acid. **21. a.** acetic acid; **b.** formic acid **23.** 1.2×10^{-5} **25. a.** 2.342; **b.** 5.485; **c.** 8.091 **27. a.** $3.8 \times 10^{-5} M$, $2.6 \times 10^{-10} M$, 9.58; **b.** 2.25, $1.8 \times 10^{-12} M$, 11.75; **c.** $1.3 \times 10^{-10} M$, 9.89, 4.11; **d.** $1.3 \times 10^{-4} M$, 3.9, $7.9 \times 10^{-11} M$ **29.** The concentration is lowered. **31.** 12.769 **33. a.** 4.509; **b.** 4.538 **35.** 4.4×10^{-5} mol of OH^- **37. a.** 0.35; **b.** 1.35; **c.** 3.312; **d.** 3.59 **39. a.** 3.90; **b.** 3.17; **c.** 2.00; **d.** 3.47 **41. a.** 4.426; **b.** 1.735; **c.** 7.338 **43.** 1.6×10^{-6} **45.** both $(1.03 \times 10^{-3} M$ and $0.0010 M)$ **47. a.** 513 g; **b.** $[F^-] = [H^+] = 1.3 \times 10^{-2} M$, $[OH^-] = 7.7 \times 10^{-13} M$ **49.** pH $= 3.12$; The approximation cannot be used. **51. a.** 9.79; **b.** 12.00; **c.** 11.32 **53.** $[OH^-] = 1.18 \times 10^{-6} M$, pOH $= 5.92$, pH $= 8.07$ **55.** 7.1×10^{-4} **57.** $H_3PO_3 + H_2O \rightleftharpoons H_2PO_3^- + H_3O^+$; $H_2PO_3^- + H_2O \rightleftharpoons HPO_3^{2-} + H_3O^+$; $HPO_3^{2-} + H_2O \rightleftharpoons PO_3^{3-} + H_3O^+$; Phosphorus has an oxidation number of +3 in each of the species. **59.** $[HPO_4^{2-}] = 6.2 \times 10^{-8} M$; pH $= 1.08$

61. $H_2SO_3 + H_2O \rightleftharpoons H_3O^+ + \boxed{HSO_3^-}$; $HSO_3^- + H_2O \rightleftharpoons H_3O^+ + \boxed{SO_3^{2-}}$ **63.** 10.25 **65.** In order of decreasing pH: $NaC_2H_3O_2$, NaCl, NH_4Cl. **67.** NaY **69.** The isoelectric pH is the point where a chemical (usually an amino acid) is in its zwitterionic form (dual positive and negative charges) and is electrically neutral. Each amino acid will have its own isoelectric pH, which enables chemists to separate the amino acids electrically in a medium that has a pH gradient. **71.** 5.31 **73. a.** 2.34×10^{-10}; **b.** 1.20×10^{-6}; **c.** 1.4×10^{-3} **75.** 8.22 **77. a.** $NH_3 + HCl \rightarrow NH_4Cl$; **b.** acidic because of NH_4^+; **c.** 5.06 **79.** 11.72 **81. a.** $C_6H_5COO^- + H_2O \rightarrow C_6H_5COOH + OH^-$, $K_b = 1.5 \times 10^{-10}$; **b.** 8.09

83.

$H_3\overset{\oplus}{N}-CH_2-CH_2-CH_2-CH_2-\underset{\underset{COO^\ominus}{|}}{\overset{\overset{H}{|}}{C}}-NH_2$

85.

$\underset{O}{\overset{O}{\underset{\|}{\overset{\|}{S}}}}{=}O$

87. a. Because water is a neutral molecule, the addition of a positive proton to the structure results in a +1 charge. **b.** The hydrogen ion can associate with more than one unit of water to form $H_5O_2^+$, $H_7O_3^+$, etc. **c.** The Lewis structure of H_3O^+ has three bonds and one lone pair on the central oxygen. This molecule would have a tetrahedral electron geometry. Because the lone pair takes one corner of the tetrahedron, the molecular geometry is a trigonal pyramid.
89. Yes. Even though a strong acid is 100% dissociated, a low enough concentration of the acid could be made to match the acidity of a concentrated weak acid solution. **91.** Acetic acid is a stronger acid than water and therefore makes a weaker base than water. The strong acids will have a more difficult time dissociating in an acetic acid solution so that the differences in the strong acids can be detected by their varying dissociation in acetic acid. **93.** The larger positive charge left behind, along with the reduced pull of the electronegative atoms due to the extra electrons left from the first ionization, limits the extent of the second ionization. **95. a.** Ephedrine alkaloids act as a stimulant, as does caffeine. The combination of two stimulants greatly increases the chance of undesired side effects, including death.

b. The ephedrine structure:

c. Yes. Alkaloids are natural nitrogen-containing molecules. The nitrogen in the structure is an amine that has basic (alkaline)

properties; hence the name *alkaloid*. **97.** The range of low pH precipitation has decreased; a reduction in SO_2 and NO_x pollutants would decrease acid rain and increase pH levels. **98. a.** Basic; **b.** $KC_6H_7O_2(s) \rightarrow K^a(aq) + C_6H_7O_2^-(aq)$ followed by $C_6H_7O_2^- + H_2O \rightleftharpoons HC_6H_7O_2 + OH^-$; **c.** 8.38; **d.** 96.9 g; **e.** 10.63; The pH of the solution is much higher because of the presence of the ammonia.

Chapter 18

P18.1 8.88 **P18.2** acidic, 3.87 **P18.3** 8.2 g **P18.4** 37 mL **P18.5** 4.44 **P18.6** 9.61, 8.59 **P18.7** 34 mL **P18.8 a.** (0 mL, 0.6021); **b.** (10.00 mL, 0.9700); **c.** (20.00 mL, 1.5563); **d.** (25.00 mL, 7.000); **e.** (30.00 mL, 12.3565); **f.** (40.00 mL, 12.7611)

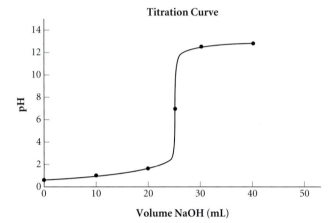

Titration Curve

P18.9 a. (0 mL, 2.67); **b.** (5.00 mL, 4.14); **c.** (12.50 mL, 4.74); **d.** (24.00 mL, 6.12); **e.** (25.00 mL, 8.92); **f.** (40.00 mL, 12.7611)

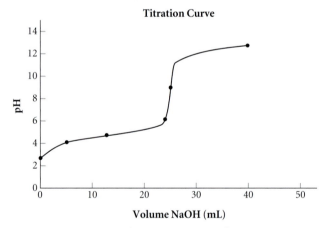

Titration Curve

P18.10 bromothymol blue **P18.11** 1.82×10^{-2} mol/L **P18.12** 4.6×10^{-6} **P18.13** Yes, $Q_{sp} = 2.71 \times 10^{-14} > K_{sp}$ **P18.14** 6×10^{-5} **P18.15** 7.945 mL

1. a. $NH_3 + H_2O \rightleftharpoons NH_4^+ + OH^-$; $2H_2O \rightleftharpoons H_3O^+ + OH^-$; **b.** $Fe(OH)_3 \rightleftharpoons Fe^{3+} + 3OH^-$; $Fe^{3+} + 6H_2O \rightleftharpoons Fe(H_2O)_6^{3+}$; $Fe(H_2O)_6^{3+} + H_2O \rightleftharpoons Fe(H_2O)_5(OH)^{2+} + H_3O^+$; $2H_2O \rightleftharpoons H_3O^+ + OH^-$; **c.** $HCOO^- + H_2O \rightleftharpoons HCOOH + OH^-$; $2H_2O \rightleftharpoons H_3O^+ + OH^-$ **3. a.** yes; **b.** no; **c.** no; **d.** no; **e.** yes **5. a.** 9.25; **b.** 8.30; **c.** 4.22 **7. a.** raises pH; **b.** lower pH; **c.** no effect **9. a.** 10; **b.** 1.0; **c.** 0.10 **11. a.** 62 mL; **b.** 6.7 mL; **c.** 106 mL **13. a.** 9.25; **b.** 4.74; **c.** 3.74 **15.** The final pH of a buffer is determined by the pK_a of the acid that sets the center of the possible range of pH for the buffer, whereas the exact ratio of the concentration of a conjugate base to its acid determines how far away the final pH is from the pK_a of the acid.

17. a. $K_a = \dfrac{[H^+][ClCH_2COO^-]}{[ClCH_2COOH]}$ and $K_b = \dfrac{[OH^-][ClCH_2COOH]}{[ClCH_2COO^-]}$;

b. 2.91   **19.** The two methods yield the same value: 10.34.   **21. a.** 9.26; **b.** 9.86; **c.** 10.43; **d.** 8.08   **23.** 0.64   **25. a.** 2.56; **b.** The buffer would be more resistant in the basic direction because there is more acid to react with added base.   **27.** 3.73   **29.** 4.88   **31. a.** 0.069 L; **b.** 0.11 L; **c.** 0.053 L; **d.** 0.011 L   **33. a.** 13.137; **b.** 12.99; **c.** 12.71; **d.** 7.00; **e.** 1.231   **35. a.** 11.13; **b.** 9.74; **c.** 9.26; **d.** 5.25; **e.** 1.903   **37. a.** 13.398; **b.** 13.342; **c.** 12.859; **d.** 11.59; **e.** 7.00; **f.** 2.53; **g.** 1.777

Titration Curve (KOH by HNO$_3$)

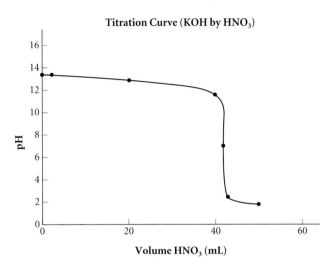

Volume HNO$_3$ (mL)

39. a. methyl orange or methyl red; **b.** phenolphthalein; **c.** bromthymol blue   **41. a.** colorless; **b.** green; **c.** red; **d.** blue   **43. a.** 8.69; **b.** 0.060 M; **c.** 0.36%   **45. a.** 8.85; **b.** 0.22; **c.** No, the endpoint will be at pH = 7.00, but the midpoint of color change will not occur until pH = 8.85.   **47. a.** $AgI(s) \rightleftharpoons Ag^+(aq) + I^-(aq)$; $K_{sp} = [Ag^+][I^-]$; **b.** $Ag_2CrO_4(s) \rightleftharpoons 2Ag^+(aq) + CrO_4^{2-}(aq)$; $K_{sp} = [Ag^+]^2[CrO_4^{2-}]$; **c.** $Al_2S_3(s) \rightleftharpoons 2Al^{3+}(aq) + 3S^{2-}(aq)$; $K_{sp} = [Al^{3+}]^2[S^{2-}]^3$; **d.** $Ca_3(PO_4)_2(s) \rightleftharpoons 3Ca^{2+}(aq) + 2PO_4^{3-}(aq)$; $K_{sp} = [Ca^{2+}]^3[PO_4^{3-}]^2$ **49. a.** 9.2×10^{-23} mol/L; **b.** 1.6×10^{-5} mol/L; **c.** 4.6×10^{-6} mol/L   **51. a.** 3.0×10^{-21}; **b.** 2.0×10^{-16}; **c.** 3.2×10^{-11} **53.** BaF_2 **55. a.** A solid forms; **b.** A solid forms.   **57.** Acid reacts with iron (III) hydroxide, increasing its solubility. Additional base lowers the solubility because of to the excess hydroxide ion in solution.   **59.** $K_{sp} = [Ca^{2+}][F^-]^2$, 4.3×10^{-4} mol/L   **61.** $Ca(IO_3)_2$; Acidic solutions enhance the solubility by reacting with the anions, which are weak bases.   **63. a.** $Ag^+(aq) + NH_3(aq) \rightleftharpoons Ag(NH_3)^+$; $Ag(NH_3)^+ + NH_3(aq) \rightleftharpoons Ag(NH_3)_2^+$; **b.** $Ni^{2+}(aq) + NH_3(aq) \rightleftharpoons Ni(NH_3)^{2+}(aq)$; $Ni(NH_3)^{2+}(aq) + NH_3(aq) \rightleftharpoons Ni(NH_3)_2^{2+}(aq)$; $Ni(NH_3)_2^{2+}(aq) + NH_3(aq) \rightleftharpoons Ni(NH_3)_3^{2+}(aq)$; $Ni(NH_3)_3^{2+}(aq) + NH_3(aq) \rightleftharpoons Ni(NH_3)_4^{2+}(aq)$; $Ni(NH_3)_4^{2+}(aq) + NH_3(aq) \rightleftharpoons Ni(NH_3)_5^{2+}(aq)$; $Ni(NH_3)_5^{2+}(aq) + NH_3(aq) \rightleftharpoons Ni(NH_3)_6^{2+}(aq)$ **65. a.** 962 mL; **b.** 2370 mL; **c.** 64 mL   **67.** 1.5×10^{-11} **69. a.** Because the K_{sp} value for $Cu(OH)_2$ is low (1.6×10^{-19}), the formation constant must be quite large to overcome the low solubility of the salt; **b.** $Cu^{2+}(aq) + NH_3(aq) \rightleftharpoons Cu(NH_3)^{2+}(aq)$; $Cu(NH_3)^{2+}(aq) + NH_3(aq) \rightleftharpoons Cu(NH_3)_2^{2+}(aq)$; $Cu(NH_3)_2^{2+}(aq) + NH_3(aq) \rightleftharpoons Cu(NH_3)_3^{2+}(aq)$; $Cu(NH_3)_3^{2+}(aq) + NH_3(aq) \rightleftharpoons Cu(NH_3)_4^{2+}(aq)$ **71.** lead–EDTA   **73.** blood chemistry, EDTA titrations of certain metals, growth of certain bacteria in a lab culture, and so on.   **75. a.** $H_2PO_4^-$ and HPO_4^{2-}; **b.** $[HPO_4^{2-}]/[H_2PO_4^-] = 9.8$ **77. a.** The zinc and cobalt ions are smaller than calcium, preventing strong coordination with all six EDTA binding sites. **b.** The higher charge of Fe^{3+} interacts more strongly with the –4 charge on EDTA.   **79. a.** 0.00050 M; **b.** 2.5×10^{-5} mol   **81. a.** 9.5×10^{-15} M; **b.** no   **83.** For a 0.1 pH change, 106 mL H_2O could be added.   **85. a.** Cl^-, $Ag^+ + Cl^- \rightarrow AgCl(s)$; **b.** $AgNO_3$ would change NO_3^- concentration. Silver acetate ($K_{sp} = 1.9 \times 10^{-3}$) may work.   **86. a.** 2.6×10^{-4}; **b.** (0 mL, 2.46); (10.0 mL, 2.97); (25.6 mL, 3.58); (50.0 mL, 5.20); (75.0 mL, 12.134); (100.0 mL, 12.387); **c.** thymol blue; **d.** 24.0 g/mol

Titration Curve (HA by NaOH)

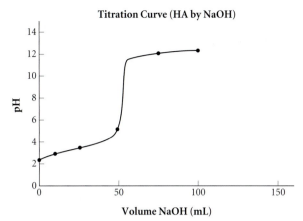

Volume NaOH (mL)

Answer 82b.

Chapter 19

P19.1 In $C_6H_{12}O_6$: H(+1), O(−2), C(0); In CO: O(−2), C(+2)
P19.2 $Li_2S < S_8 < SO_2 < H_2SO_4$ **P19.3** −359 kJ, spontaneous
P19.4 $2H^+(aq) + ClO^-(aq) + Cu(s) \rightarrow Cl^-(aq) + H_2O(l) + Cu^{2+}(aq)$ **P19.5** $2MnO_4^-(aq) + 3Mn^{2+}(aq) + 4OH^-(aq) \rightarrow 5MnO_2(s) + 2H_2O(l)$ **P19.6** −0.25 V, +0.98 V

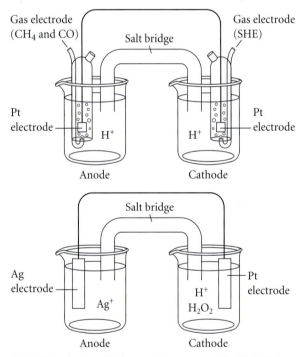

P19.7 The galvanic cell (above top) has a potential of 0.98 V, the electrolytic cell (above bottom) has a potential of −0.25 V.
P19.8 $Al(s) \mid Al^{3+}(aq) \parallel Fe^{2+}(aq) \mid Fe(s)$ **P19.9** Ca > Na > Al > Cu
P19.10 −0.0592 V, cathode, no   **P19.11** 1.45×10^{37} **P19.12** 1.5 g

1. A *battery* is a series of cells, most "batteries" are single cells.
3. The two reactions are a reduction reaction that involves the gain of electrons and an oxidation reaction that involves the loss of electrons.   **5.** O (−2), H (+1), S (−2), C (−1)   **7.** $NaClO_4$, HCl
9. $NH_3 < N_2H_4 < N_2O < NO < NO_2$ **11. a.** K (+1), O (−2), Mn (+7); **b.** Li (+1), O (−2), Mn (+3); **c.** H (+1), O (−2), N (−3), Cl (+7)   **13. a.** is reduced, is an oxidizing agent; **b.** is oxidized, is a reducing agent; **c.** is reduced, is an oxidizing agent   **15.** Reactants: Mg (0), H (+1), P (+5), O (−2); Products: Mg (+2), P (+5), O (−2), H (0)   **17. a.** +1; **b.** If ClO^- is a good oxidizing agent, it must be

itself reduced which is a gain of electrons. **19. a.** Reactants: Sr (+2), Cl (−1), Products: Sr (0), Cl (0); **b.** Sr^{2+} **21. a.** reduction: 1st reaction; oxidation: 2nd and 4th reactions; **b.** precipitation; **c.** $4Fe(s) + 3O_2(aq) + 2H_2O(l) \rightarrow 4FeO(OH)(s)$ **23.** concentrations = 1M, pressures = 1 atm **25. a.** non-spontaneous; **b.** negative; **c.** negative, positive; **d.** negative **27.** $2Fe^{3+} + Fe \rightarrow 3Fe^{2+}$, $E°_{cell} = +1.2\,V$ **29.** −228 kJ **31.** −216 kJ **33.** The redox atoms must be balanced, the non-redox atoms must be balanced, and the electrons (used to balance the charges) involved in the oxidation and reduction half-reactions must cancel. **35. a.** $2CO_2 + 2H^+ + 2e^- \rightarrow H_2C_2O_4$, reduction, CO_2 is reduced; **b.** $Np^{4+} + 2H_2O \rightarrow NpO^{2+} + 4H^+ + e^-$, oxidation, Np^{4+} is oxidized. **37. a.** $Sn^{2+} + 2Cu^{2+} \rightarrow Sn^{4+} + 2Cu^+$; **b.** $I_3^- + 2S_2O_3^{2-} \rightarrow S_4O_6^{2-} + 3I^-$; **c.** $Fe^{3+} + SO_3^- + H_2O \rightarrow SO_4^{2-} + Fe^{2+} + 2H^+$ **39.** acidic: $Cr_2O_7^{2-} + 14H^+ + 3Cu \rightarrow 3Cu^{2+} + 2Cr^{3+} + 7H_2O$; basic: $Cr_2O_7^{2-} + 7H_2O + 3Cu \rightarrow 3Cu^{2+} + 2Cr^{3+} + 14OH^-$ **41.** $ClO_4^- + I^- \rightarrow IO_3^- + ClO^-$ **43. a.** −0.76 V; **b.** 1.5×10^2 kJ; **c.** −0.76 V **45. a.** Li^+; **b.** F_2; **c.** $Zn + Cu^{2+} \rightarrow Cu + Zn^{2+}$, $E°_{cell} = 1.10\,V$ **47.** $2CO_2 + Cu + H_2O \rightarrow CuO + 2H^+ + C_2O_4^{2-}$ **49.** $2MnO_4^- + 16H^+ + 5C_2O_4^{2-} \rightarrow 10CO_2 + 2Mn^{2+} + 8H_2O$ **51.** $I_2 + C_6H_8O_6 \rightarrow C_6H_6O_6 + 2H^+ + 2I^-$ **53.** A *cell* is an electrochemical device consisting of an anode and cathode allowing for the exchange between chemical and electrical energy. A *half reaction* is the equation that describes the reduction or oxidation part of a redox reaction. A *galvanic cell* is a cell that produces electricity from a chemical reaction (also known as a *voltaic cell*). The *electromotive force* is a measure of how strongly a species pulls electrons towards itself in an redox process. **55.** Zn, Cu^{2+} **57.** The reduction potentials for lead ion and silver ion are −0.13 V and 0.80 V respectively. For a spontaneous reaction, the silver ion reaction will remain as a reduction and lead will be oxidized. If the nitrate ions are moving left to right through the salt bridge, the electrons are moving from the right cell to the left cell, which is the opposite of the standard convention. This requires that the left beaker be the cathode (gaining electrons) and the right beaker be the anode (losing electrons). Therefore, the oxidation of lead will occur in the right beaker and reduction of silver in the left beaker.

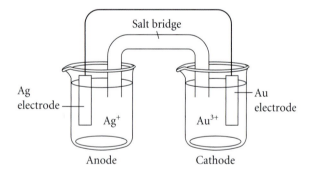
Ag electrode — Ag⁺ — Anode; Au⁺ / Au³⁺ — Au electrode — Cathode. Salt bridge.

59. a. 0.70 V; **b.** 3; **c.** shown above; **d.** Au^{3+}; **e.** $Ag \mid Ag^+ \parallel Au^{3+} \mid Au$ **61.** $Zn(s) \mid ZnO(s) \parallel HgO(s) \mid Hg(l)$ **63. a.** Ba; **b.** Pb^{2+}; **c.** left **65.** Placing the metal in the solution will reveal its identity. Only the oxidation potential of aluminum in nickel is positive enough to be spontaneous when coupled with the reduction of nickel. The tin will not react.

67. $\dfrac{RT}{F} = \dfrac{(8.3145\,\text{J/mol·K})(298.15\,\text{K})}{\left(96485\,\text{C/mol} \times \frac{1\,\text{J/V}}{1\,\text{C}}\right)} = 0.0257\,V$ Natural logarithms and base 10 logarithms are related by $\ln x = 2.303 \log x$, so we must multiply the value of 0.0257 V by 2.303 to give 0.0592 V. **69. a.** 0.78 V; **b.** 0.75 V; **c.** 0.81 V **71.** $\Delta G = 0$, the battery has reached equilibrium, and $E_{cell} = 0$ V **73. a.** 0.051 g; **b.** 12.0 g;

c. 44 g **75.** cathode, 0.33 A **77.** 15.5 g **79.** The steel pan is the cathode, copper metal can serve as the anode. **81. a.** $E° = 0.97$ V, $K = 3.98 \times 10^{196}$; **b.** $4FeO_4^{2-} + 10H_2O \rightarrow 20OH^- + 4Fe^{3+} + 3O_2$; **c.** 0.76 V; **d.** At high pH, $Fe(OH)_3$ forms. This solid removes Fe^{3+} from the water and forces the reaction to completion. **83.** ΔG changes as T changes, thus every E value would be adjusted accordingly. **85.** 0.21 g CrO_3 **87. a.**

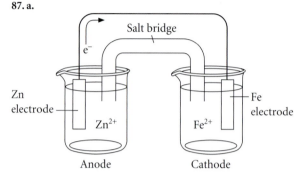
Salt bridge. Zn electrode — Anode — Zn^{2+}; Fe electrode — Cathode — Fe^{2+}.

b. $Zn \rightarrow Zn^{2+} + 2e^-$; **c.** 0.32 V; **d.** 0.32 V, 0.32 V; **e.** No. Because the concentrations are equal, the ln term goes to zero and the temperature has no effect. **f.** Avoids use of toxic metals like lead. **g.** Both Zn and Fe corrode in water.

Chapter 20

P20.1 a. $[FeCl_6]^{4-}$; **b.** $[Ni(NH_3)_4]^{2+}$; **c.** $[Zn(H_2O)_6]^{2+}$ **P20.2 a.** 6, octahedral; **b.** 6, octahedral; **c.** 4, tetrahedral; **d.** 4, tetrahedral **P20.3 a.** tetraamminedichlorochromium(IV) sulfate; **b.** sodium tetracyanonickelate(II); **c.** $Ca_2[FeF_6]$; **d.** $[Mn(NH_3)_4(CO)_2]SO_4$

P20.4 none unpaired

1. a. A ligand is a Lewis base that donates a lone pair of electrons to a metal center to form a coordinate covalent bond; **b.** A coordinate covalent bond is a covalent bond that results when one atom donates both electrons needed to form the bond; **c.** A Lewis acid accepts a pair of electrons from another atom in the formation of a coordinate covalent bond. **3. a.** 6, +2, 6; **b.** 6, +2, 6; **c.** 4, +2, 4; **d.** 4, 0, 4

5.

$$\begin{matrix} & L & \\ & | & \\ L \!-\! M \!\cdots\! L \\ & \searrow & \\ & L & \end{matrix} \quad \text{or} \quad \begin{matrix} L \cdots & M & \cdots L \\ L & & L \end{matrix}$$

7. $H-\overset{\cdot\cdot}{\underset{H}{N}}-H$; NH_3 is a Lewis base because it is capable of donating the electron pair to form a bond.

9. Yes; bidentate **11.** Ca^{2+} and C_2H_6 have no lone pairs and would not act as ligands. **13.** The structure of ethylenediamine is $NH_2CH_2CH_2NH_2$. Both NH_2 groups will have a lone pair and be able to act as a Lewis base. The carbon backbone is long enough to allow both nitrogen atoms to form a bond with the same metal atom. While N_2 does have two lone pairs, they are found at opposite ends of a linear molecule. It is not possible for both nitrogen atoms to form separate bonds with the same metal atom. **15.** B is bidentate **17. a.** +4, 6; **b.** +2, 6; **c.** +2, 4 **19.** Some of the ligands are bidentate; there can be more bonds than ligands. **21.** 4

23. a.

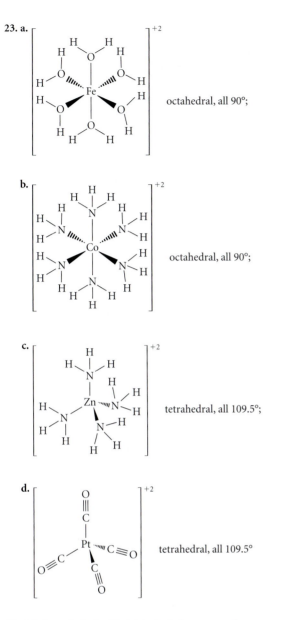

octahedral, all 90°;

b. octahedral, all 90°;

c. tetrahedral, all 109.5°;

d. tetrahedral, all 109.5°

25. +3, 6, octahedral   **27.** 4, tetrahedral or square planar
29. a. geometric; **b.** ionization; **c.** coordination sphere

31.

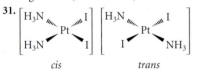

cis   *trans*

If the platinum is placed at the center of a hypothetical tetrahedron, each ligand would reside at one of the vertices. Because each vertex is directly connected to the other three (containing one ligand like itself and two that are different), it is not possible to create more than one unique arrangement. There can be no isomers.

33. Only two:

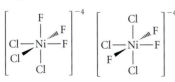

35.

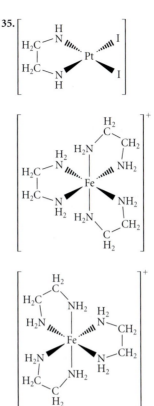

37. a. amminecyanobis(ethylenediamine) cobalt (III);
b. diamminedioxalatochromate(III); **c.** hexanitroferrate(III);
d. triaquatrichlorocobalt(III)
39. a. [CoCl(NH₃)₄(H₂O)]²⁺; **b.** *trans*-[Cu(H₂O)₂(en)₂]Cl₂;
c. Na₂[CoCl₄]; **d.** [MnCl(CO)₅]   **41.** Tetraamminediaquachromium(III) nitrate, [Cr(NH₃)₄(H₂O)₂](NO₃)₃   **43. a.** [Ar]3d⁶;
b. [Ar]3d⁴; **c.** [Ar]3d¹⁰   **45.** Fe³⁺

47. a. Fe³⁺ is d⁵: **b.** Co²⁺ is d⁷:

c. Ni²⁺ is d⁸:

49. a. Fe³⁺ is d⁵: **b.** Co²⁺ is d⁷:

c. Ni²⁺ is d⁸:

51. a. tetrahedral high spin Fe³⁺ (d⁵),

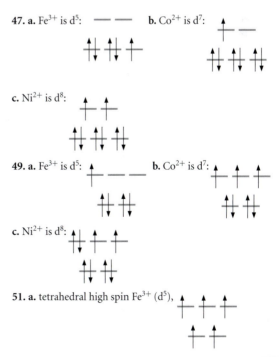
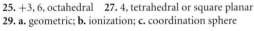

b. octahedral low spin Co^{3+} (d^6)

c. octahedral low spin Mn^+ (d^6)

53. all **55.** $M(CN)_6^{2-}$ **57.** No, The chemical species that form the majority of the common gemstones is the same for most stones and will have the same spectrum. **59. a.** yellow and red; **b.** 3; **c.** higher; According to the electrochemical series, ammine ligands cause larger splits in the d orbitals than does water; therefore, the ammine complex should absorb higher energy visible light; **d.** NH_3 **61.** Replacing two ligands with a single chelating ligand frees more species into solution, raising the entropy. **63.** Labile; labile **65.** Nearly all transition metals have important catalytic properties. **67.** Cation; $[Co(NH_3)_4Cl_2]^+$ is +1 because the total charges are $(+3) + 4 \times (0) + 2 \times (-1) = +1$. **69. a.** If there are some electrons in the d-orbitals of the metal center in a complex, but not enough electrons to fully fill the orbitals, electronic transitions are possible. Any configuration from d^1 to d^9 will be able to have its electrons promoted by light absorption; **b.** The energy of the photon being absorbed is determined by the splitting between the d-orbital levels, which is controlled by the ligands of the complex. The splitting increases in order with the spectrochemical series: $Cl^- < F^- < OH^- < H_2O < NH_3 < NO_2^- < CN^- < CO$; **c.** The intensity of the transition will increase with greater population in the lower states and lower population in the higher states. **71.** Larger K

72. a.

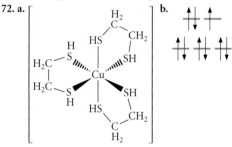

b.

c. Since the ammine complex has a violet color and the ethanedithiol complex is blue, the absorbed colors are green and yellow respectively. Green light lies at a higher energy than yellow; therefore the splitting caused by the ethanedithiol is less than that caused by ammine. The new ligand will fall before NH_3 in the spectrochemical series; **d.** 4; **e.** The ligand is inert.

Chapter 21

P21.1 a. 16, 16; **b.** 11, 12; **c.** 86, 136; **d.** 43, 55 **P21.2** $^{60}_{27}Co \rightarrow$ $^{0}_{-1}\beta + ^{60}_{28}Ni$ **P21.3** $^{218}_{84}Po \rightarrow ^{4}_{2}He + ^{214}_{82}Pb + ^{0}_{0}\gamma$, $^{230}_{90}Th \rightarrow ^{4}_{2}He + ^{226}_{88}Ra + ^{0}_{0}\gamma$ **P21.4** 17.8 days **P21.5** 5.588×10^8 kJ/mol **P21.6 a.** Stable; **b.** Stable; **c.** Stable; **d.** Stable or unstable: The numbers of protons and neutrons are odd, which argues for instability, but the mass is very close to the average periodic table mass, which argues for stability. It is in fact stable.

P21.7

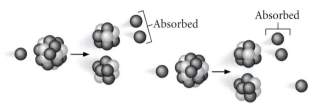

P21.8 The first and third structures are not suitable. This first has a radioactive tritium that is easily removed as an acidic hydrogen ion, and the third is not structurally the same as salicylic acid.

1. ^1_1H for both **3.** Yes, an atom can have no neutrons. Hydrogen, ^1_1H, has no neutrons—only a proton is required. **5.** Yes, ^2_2He has a smaller mass than ^3_1H. **7. a.** 7, 5, 7; **b.** 51, 73, 51; **c.** 63, 89, 63; **d.** 4, 5, 4 **9.** 0.014 g **11. a.** Sr^{2+}; **b.** Its chemical behavior is similar to that of calcium. **13.** Gamma rays have more energy per photon. **15. a.** $^{239}_{94}Pu \rightarrow ^{4}_{2}He + ^{235}_{92}U + ^{0}_{0}\gamma$; **b.** $^{14}_{6}C \rightarrow ^{0}_{-1}\beta + ^{14}_{7}N$; **c.** $^{137}_{55}Cs \rightarrow ^{0}_{-1}\beta + ^{137}_{56}Ba + ^{0}_{0}\gamma$ **17. a.** $^{219}_{86}Rn \rightarrow ^{4}_{2}He + ^{215}_{84}Po + ^{0}_{0}\gamma$; **b.** $^{90}_{38}Sr \rightarrow ^{0}_{-1}\beta + ^{90}_{39}Y$; **c.** $^{99}_{43}Tc \rightarrow ^{0}_{-1}\beta + ^{99}_{44}Ru + ^{0}_{0}\gamma$ **19.** $^{146}_{62}Sm \rightarrow ^{4}_{2}He + ^{142}_{60}Nd$ **21.** Th-232 **23.** Ionizing: cosmic rays, gamma rays, X-rays; Nonionizing: visible light, microwaves, IR radiation **25.** Use minimum gamma doses, wear protective clothing, apply the dose while in a different room. **27. a.** They differ only in the amount of disintegrations, (1 curie = 3.7×10^{10} Bq); **b.** A rem is the biologically adjusted version of rad. **29.** 3.5×10^4 disintegrations/s **31. a.** 0.40 mrem; **b.** This radiation is not unduly hazardous. **33. a.** Density; **b.** chemical reactivity, radiation hardness (does it weaken with exposure?), secondary radiation reactivity **35. a.** A; **b.** 10 min; **c.** A; **d.** A **37.** 15 days **39.** 50%, 25% **41.** 86.7 yr **43.** 0.45 days **45.** The sample would need to be replaced every other year (assuming that 5% of the original activity is still useful.) **47.** 1.6×10^5 yr; It is continuously generated by the decay of other long-lived radioactive nuclei. **49. a.** Some of the mass is lost as energy in the process of creating the nucleus; **b.** Yes. **51.** The binding energy, ΔE, is related to the mass defect, Δm, via $\Delta E = |\Delta m|c^2$. **53.** -0.006627 g/mol; 5.956×10^8 kJ/mol **55. a.** reactants **b.** The result is the release of energy and creation of binding energy of the products. **57.** Both processes involve the conversion between a neutron and proton. However, the beta-minus decay converts a neutron into proton, whereas the beta-plus decay converts a proton into a neutron. **59. a.** "Doubly magic" means that a nuclide has both a number of protons and a number of neutrons that are among the most stable numbers (2, 8, 20, 28, 50, 82, and 114); **b.** Oxygen-16 has 8 each of protons and neutrons, and 8 is one of the magic numbers. Helium-4 has 2 each of protons and neutrons, matching the magic number of 2. Carbon-12 has 6 each, and nitrogen-14 has 7 each. These two nuclides do not have a magic number of either protons or neutrons. In general, those nuclei without a magic number of protons or neutrons are more likely to be radioactive. **61.** Element 114 would have a magic number of protons and would probably be more stable and have a longer half-life before decaying. **63.** In order to reach the nearest stable nuclide, the heavy radioactive nuclides such as uranium and plutonium need to shed at lot of mass and can do so by emitting alpha particles. For nuclides of lighter elements, there are stable nuclides that can be created by converting protons and neutrons by beta decay and need not lose mass to achieve stability. **65.** Iron and cobalt have the most stable nuclei of all atoms, with the largest binding energy per nucleon. **67. a.** $^{235}_{92}U + ^{1}_{0}n \rightarrow ^{139}_{56}Ba + ^{94}_{36}Kr + 3^{1}_{0}n + $ energy; **b.** $^{235}_{92}U + ^{1}_{0}n \rightarrow ^{80}_{38}Sr + ^{153}_{54}Xe + 3^{1}_{0}n + $ energy **69. a.** The number of neutrons differs; **b.** U-235 is 0.72% and U-238 is 99.28% of the total. **c.** They are alike chemically and differ by only 1.3% in their total mass, which makes them difficult to separate using physical methods.　**71.** Fission creates smaller nuclei. Americium, californium, and berkelium are all heavier than uranium.　**73.** Some of the most important questions should be about the total radioactive dose expected, as well as the form of the radiation emitted. What is the half-life of the nuclide, and what is its expected persistence in the body?　**75.** Tc-99m has a half-life of 6 hours. After the night (12 hours) is over, only 1/4 of the original will remain. **77.** $^{18}_{9}F \rightarrow\ ^{0}_{+1}\beta +\ ^{18}_{8}O$　**79.** Yes, with a half-life of 60 hours, only 0.8% remains after 2.5 weeks.　**81.** The medical reference to see is Winters. T. H., & Franza, J. R., Radioactivity in Cigarette Smoke, *New England Journal of Medicine*, 1982; 306(6): 364–365. The radioactive isotopes lead-210 and polonium-210 are found in the tobacco leaves and cigarettes.　**83.** The gamma ray represents the energy that is given off in the reaction, which is possible only if a mass defect exists.　**85.** alpha, beta, beta, alpha, alpha, alpha, alpha, beta, beta, alpha　**87.** Rn is radon, an inert gas; and Ra is Radium, a reactive metal solid.　**89.** The main reason why Rn-222 is the most dangerous is that it is the only species with a long half-life. The other isotopes decay much more rapidly into solid species. Rn-222 has a half-life of 3.8 days, which is enough time for the gas to seep into houses and be inhaled by people inside.　**91.** The density of the nucleus is much greater: 2.8×10^{15} g/cm^3 compared to 11.3 g/cm^3 for lead. **93.** Websites may vary. The answer is NO!　**95.** A nice site with clickable names and more information can be found at http://www.slac.stanford.edu/library/nobel/.　**97. a.** 0.0465 yr; **b.** $p^+ = 46$, $n = 57$, $e^- = 44$; **c.** 114.4 g　**98. a.** <0.2 mg; **b.** Am-240 undergoes electron capture with a half-life of 51 hours. Not only does it not emit a particle, but nearly all would be gone very quickly—not a useful property for a smoke detector, which should last several years. Am-242 does emit beta particles 83% of the time, but it has a half-life of only 16 hours. Again, there would be little left in a very short time, making it also impractical for use in a smoke detector; **c.** $^{241}_{95}Am \rightarrow\ ^{4}_{2}He +\ ^{237}_{93}Np +\ ^{0}_{0}\gamma$; **d.** 2900 yr; **e.** $^{242}_{95}Am \rightarrow\ ^{0}_{-1}\beta +\ ^{242}_{96}Cm +\ ^{0}_{0}\gamma$, $^{242}_{95}Am +\ ^{0}_{-1}\beta \rightarrow\ ^{242}_{94}Pu +\ ^{0}_{0}\gamma$

Chapter 22

P22.1 Adenine　**P22.2** The sequence that pairs with our strand is GATACG, *or* we need to reverse the smaller strand (CGTATC), which would then bind with a sequence of GCATAG in the longer strand:

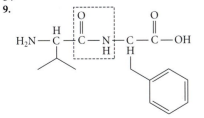

P22.3 Methionine–tryptophan–proline–lysine–leucine–aspartic acid–methionine–phenylalanine–aspartic acid　**P22.4** Lyase, transferase, ligase

1. DNA is a huge nucleotide polymer that has a double-helical structure. Each strand of the DNA polymer complements the other by forming base pairs. DNA provides instructions for the synthesis of proteins and enzymes that carry out the biological activities of the cell.　**3.** TAATTTTTCCCTGAT　**5.** Four　**7.** Valine and leucine or glycine and alanine
9.

11. A codon is a three-nucleic-acid sequence on an mRNA that is translated into a particular amino acid during protein synthesis. **13.** If the wrong nucleotide is used in the production of the mRNA or an extra nucleotide is added to or removed from the mRNA, a different three-nucleotide codon may result that could specify a different amino acid.　**15.** Middle　**17.** Globular　**19.** Oxidoreductase **21.** By binding to the enzyme and inhibiting further production, the product itself can ensure that its concentration does not rise too high or ensure that it is created only as needed.　**23.** Tetrose　**25.** Aldose **27.** Glucose　**29.** Fructose is one of the two sugars that results from the hydrolysis of sucrose. Fructose is classified as a hexose, and ketose. **31.** At room temperature, fats are solids and oils are liquids.　**33.** Fat **35.** The upper polar region

37.

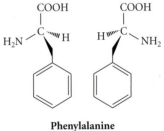

Serine

Phenylalanine

39. Because the body uses principally a single type of enantiomer, having chiral enzymes ensures that only the proper enantiomer is used in a protein or other product. The cell will not waste resources on a molecule it cannot use.　**41.** Tryptophan, no　**43.** RNA: AUG|UCG|CGA|AUU|UAA|GGC|UGC|UUA|UU; Protein: methionine-serine–arginine–isoleucine　**45.** Each of the molecules is saturated. As the chain length increases, the strength of the dispersion forces increases, requiring more energy and higher temperatures to melt. The longer chains therefore have higher melting points.　**47.** If the receptors that are responsible for detecting the sweetness of a food are sensitive only to a particular enantiomeric form, the other enantiomer is not sensed as sweet. However, if the receptors are *not* specifically sensitive to only the normal enantiomeric form, the other form could also taste sweet. The advantage to being able to taste the opposite enantiomer is that it may not be able to be metabolized—it would have no calories!

49. a.

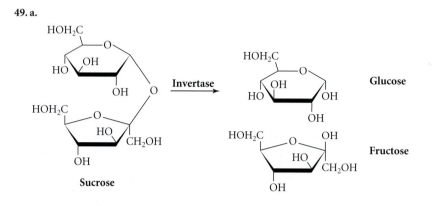

b. Because both reactions involve the hydrolysis, the enzyme may be able to catalyze the reaction; however, many enzymes have a very specific active site and therefore would not be able to catalyze the reaction; **c.** Lysine can be coded by the AAA or AAG codon on mRNA and by the UUU or UUC anticodon on tRNA; **d.** 9.50 kg; **e.** 8.16 kg; **f.** Nonedible fibers could be converted into edible food.

Credits

Text and Illustrations

Figures 1.1 and 1.2, p. 2: Reprinted with permission. Copyright 2002 American Chemical Society. **Figure 2.23, p. 62:** Reproduced with permission. Courtesy of Agilent Technologies, Inc. **Figure 2.26, p. 63:** "The Fingerprint of a Flavor" from "A Fragrant Feast" by Mark S. Lesney. *Today's Chemist at Work,* June 2002. Copyright © 2002 American Chemical Society. **Figure 5.8, p. 173:** Reprinted by permission of Dr. David Webb. **Figure 5.22, p. 193:** U.S. Department of Energy National Energy Technology Laboratory. National Methane Hydrate Program. **Figure 5.25, p. 197:** http://www.whitehouse.gov/news/release/2001/06/energyinit.html. **Figure 10.26, p. 430:** Reprinted courtesy of Forsyth County Environmental Affairs Department. **Figure 10.27, p. 432:** http://www.gcrio.org/ipcc/aa/05.html. United Nations Environment Programme World Meteorological Organization. **Figures 11.1 and 11.2, p. 441:** Reprinted by permission of UNESCO. **Figure 15.1, p. 621:** http://pubs.usgs.gov/circ/circ1151/nawqa91.6.html. **Figure 16.7, p. 679:** Reprinted courtesy of Robert Graves, Chromatographic Instruments Company. **Figure 17.1, p. 718:** http://www.census.gov/const/www/newresconstindex.html. **Figure 17.2, p. 720:** Science Magazine, May 21, 2004. Reprinted courtesy of Kenneth D. Jordan and Richard A. Christie. **Table 18.6, p. 812:** Reprinted by permission of The Dow Chemical Company. **Figure 21.12, p. 921:** http://www.iaea.org/programmes/a2/index.html.

Photos

Chapter 1

1: Jerry Walsh. **3:** (**top**) Paul Kelter. (**left center**) Teflon® is a registered trademark of DuPont. (**right center**) © Houghton Mifflin Company. All rights reserved. (**bottom**) Robert Bruschini, Bristol-Myers Squibb. **4:** (**top left**) Stockybyte Platinum/Getty Images. (**top center**) Photo Disc/Getty Images. (**top right**) © Houghton Mifflin Company. All rights reserved. (**bottom**) © Argus Fotoarchiv/Peter Arnold, Inc. **6:** © Houghton Mifflin Company. All rights reserved. **7:** (**top**) © Houghton Mifflin Company. All rights reserved. (**bottom**) NASA. **9:** (**top left**) Don Farrall/Photodic Green/Getty Images. (**bottom left and center**) © Houghton Mifflin Company. All rights reserved. (**top right**) Photodisc Green/Getty Images. (**bottom right**) Comstock Images/Alamy. **10:** © Bojan Brecelj/CORBIS. **12:** (**left and center**) © Houghton Mifflin Company. All rights reserved. (**right**) © Jim Olive/Peter Arnold, Inc. **13:** NASA. **14:** Hulton Archive/Getty Images. **15:** Courtesy of Bayer Healthcare LLC. **17:** (**both**) © Houghton Mifflin Company. All rights reserved. **18:** (**top**) © Houghton Mifflin Company. All rights reserved. (**bottom**) © Houghton Mifflin Company. All rights reserved. **19:** © Bettmann/CORBIS. **20:** (**both**) © Houghton Mifflin Company. All rights reserved. **21:** Burke/Triolo Productions/Brand X Pictures/Getty Images. **23:** (**all**) © Houghton Mifflin Company. All rights reserved. **24:** © Houghton Mifflin Company. All rights reserved. **25:** © Houghton Mifflin Company. All rights reserved. **29:** © Houghton Mifflin Company. All rights reserved. **31:** © Houghton Mifflin Company. All rights reserved. **35:** (**top**) © Richard Feldman/Phototake. (**bottom**) Jorge Uzon/Getty Images. **36:** (**top**) NASA. (**bottom**) © Institute for Molecular Manufacturing www.imm.org.

Chapter 2

45: © James L. Amos/CORBIS. **46:** NASA. **47:** (**top**) Hulton Archive/Getty Images. (**center**) Edgar Fahs Smith Collection, University of Pennsylvania Library. (**bottom**) Edgar Fahs Smith Collection, University of Pennsylvania Library. **48:** © Houghton Mifflin Company. All rights reserved. **49:** Courtesy of Manchester Literary and Philosophical Society. **50:** (**left**) Edgar Fahs Smith Collection, University of Pennsylvania Library. (**right**) © Houghton Mifflin Company. All rights reserved. **52:** (**top**) Canterbury Archaeological Trust. (**bottom**) © Paul A. Souders/CORBIS. **53:** (**top**) Keystone/Getty Images. (**bottom**) © Bettmann/CORBIS. **62:** (**top**) Courtesy of PerkinElmer Inc., Shelton, Connecticut. (**bottom**) Courtesy of PerkinElmer Inc., Shelton, Connecticut. **63:** Reprinted with permission from Culp, R. A., J. M. Legato, and E. Otero, 1998. Carbon isotope composition of selected flavoring compounds for the determination of natural origin by GC/IRMS. In C. J. Mussinan and M. J. Morello, eds., Flavor Analysis—Developments in Isolation and Characterization. American Chemical Society Symposium Series 705. 260–287. Copyright 1998 American Chemical Society. **65:** (**all**) © Houghton Mifflin Company. All rights reserved. **68:** © Houghton Mifflin Company. All rights reserved. **70:** (**top**) Charles Winters/Photo Researchers, Inc. **74:** © Houghton Mifflin Company. All rights reserved. **81:** © 1993 Richard Megna, Fundamental Photographs, NYC. **85:** Reprinted from R. C. Wiens, "Radiometric Dating: A Christian Perspective," http://www.asa3.org/ASA/resources/wiens.html. Original data from Stuiver et al., *Radiocarbon 40,* 1041–1083, 1998 and Beck et al., *Science 292,* 2453–2458. 2001.

Chapter 3

86: Will & Deni McIntyre/Photo Researchers, Inc. **89:** (**both**) © Houghton Mifflin Company. All rights reserved. **90:** (**all**) © Houghton Mifflin Company. All rights reserved. **91:** Edgar Fahs Smith Collection, University of Pennsylvania Library. **92:** © Houghton Mifflin Company. All rights reserved. **93:** © Houghton Mifflin Company. All rights reserved. **97:** © Houghton Mifflin Company. All rights reserved. **99:** (**left**) © Bob Krist/CORBIS. (**top center**) Isabelle Rozenbaum & Frederic Cirou/Photo Alto/Getty Images. (**bottom center**) Berit Myrekrok/Digial Vision/Getty Images. (**right**) Image Source/Getty Images. **101:** PDB ID: 1A3N; J. Tame, B. Vallone, 1998. **113:** © Houghton Mifflin Company. All rights reserved. **116:** (**left**) © Houghton Mifflin Company. All rights reserved. (**right**) © Royalty-Free/Corbis.

Chapter 4

124: Creatas Royalty Free/ Fotosearch Stock Photography. **125:** (**top**) © Royalty-Free/Corbis. (**bottom left**) AP/Wide World Photos. (**bottom right**) Eye of Science/Photo Researchers, Inc. **126:** (**both**) © Houghton Mifflin Company. All rights reserved. **129:** © Houghton Mifflin Company. All rights reserved. **130:** (**all**) Photo by Ken O'Donoghue. © Houghton Mifflin Company. All rights reserved. **135:** David Woodfall/Stone/Getty Images. **138:** (**all**) © Houghton Mifflin Company. All rights reserved. **140:** © Houghton Mifflin Company. All rights reserved. **141:** (**all**) © Houghton Mifflin Company. All rights reserved. **142:** (**top**) Michael Mosher. (**center**) Gordon M. Grant/Alamy. (**bottom**) Denver Instruments. **145:** Simon Fraser/Science Photo Library/Photo Researchers, Inc. **146:** (**top**) Nathan Miller/Center for Lithospheric Studies, University of Texas at Dallas. (**bottom**) © Houghton Mifflin Company. All rights reserved. **149:** © Houghton Mifflin Company. All rights reserved. **153:** (**left**) Jeff LePore/Photo Researchers. (**right**) Ross Barnett/Lonely Planet Images/Getty Images. **158:** (**top left**) USGS photo.

(**top right**) Photograph by Michele A. Justice. (**center**) Brookhaven National Laboratory. (**bottom**) blinkwinkel/Alamy.

Chapter 5

167: NASA. **168:** NASA. **172:** (**top**) Achim Sass/Westend61/Alamy. (**bottom**) NASA. **174:** AP/Wide World Photos. **175:** (**all**) © Houghton Mifflin Company. All rights reserved. **177:** NASA. **178:** NASA. **179:** (**top**) Douglas Pulsipher/Alamy. (**bottom**) © Houghton Mifflin Company. All rights reserved. **180:** (**both**) © Houghton Mifflin Company. All rights reserved. **181:** © Houghton Mifflin Company. All rights reserved. **187:** (**top**) Mark Baker/Reuters/Landov. (**bottom**) © Houghton Mifflin Company. All rights reserved. **188:** Kim Heacox/Peter Arnold Photography. **193:** AP Photo/Dean J. Koepfler. **196:** Archive of the Goldschmidt Gmbh. **197:** Goh Cha Hin/AFP/ Getty Images. **198:** (**top**) Peter Frischmuth/ Peter Arnold Photography. (**bottom left**) FPL Energy. (**bottom right**) Corbis Royalty-Free. **199:** (**top**) © Reuters CORBIS (**bottom**) John A. Turner/National Renewable Energy Laboratory.

Chapter 6

209: Courtesy: IBM Research, Almaden Research Center. Unauthorized use prohibited. **210:** Photo 24/Brand X Pictures/Getty Images. **212:** Courtesy of Casio. **213:** Science Photo Library/Photo Researchers, Inc. **214:** (**bottom**) National Telecommunications and Information Administration Office of Spectrum Management. (**top**) Junko Kimura/Getty Images. **215:** © Alfred Pasieka/Peter Arnold, Inc. **217:** (**left**) © Houghton Mifflin Company. All rights reserved. (**right**) © Richard Megna/Fundamental Photographs, NYC. **221:** Post Danmark. **224:** (**top**) Index Stock/Alamy. (**bottom**) © B. Runk/S. Schoenberger/Grant Heilman Photography. **225:** (**top**) © Peter Beck/CORBIS. (**bottom**) Science Photo Library/Photo Researchers Inc. **226:** © Royalty-Free/Corbis. **228:** © Bettmann/ CORBIS. **229:** NIST. **230:** © George D. Lepp/CORBIS. **232:** (**top**) © Bettmann/CORBIS. (**bottom**) © Hulton-Deutsch Collection/ CORBIS. **233:** © Royalty-Free/Corbis. **237:** Courtesy: IBM Research, Almaden Research Center. Unauthorized use prohibited. **242:** (**top**) © Bettmann/CORBIS. (**bottom**) © Gabe Palmer/CORBIS. **243:** Science Photo Library/Photo Researchers, Inc. **255:** © Royalty-Free/Corbis.

Chapter 7

259: Getty Images. **260:** (**top**) © Royalty-Free/Corbis. (**bottom**) Photodisc Green/Getty Images. **261:** © PEMCO-Webster & Stevens Collection; Museum of History and Industry, Seattle/CORBIS. **262:** (**all**) © Houghton Mifflin Company. All rights reserved. **264:** (**top**) © Houghton Mifflin Company. All rights reserved. (**bottom left**) University of Chicago Library. (**bottom right**) Edgar Fahs Smith Collection, University of Pennsylvania Library. **265:** © Houghton Mifflin Company. All rights reserved. **266:** © Royalty-Free/Corbis. **267:** (**left**) The Art Archive/British Library. (**right**) Mark Joseph/ Digital Vision/Getty Images. **268:** (**left right**) © Houghton Mifflin Company. All rights reserved. **270:** (**left and right**) © Houghton Mifflin Company. All rights reserved. **271:** (**all photos of metals**) © Houghton Mifflin Company. All rights reserved. (**bottom left**) AP/Wide World Photos. **272:** (**all**) © Houghton Mifflin Company. All rights reserved. **273:** (**top left and right**) © Houghton Mifflin Company. All rights reserved. (**bottom**) © Roger Ressmeyer/CORBIS. **274:** (**top photos, bottom left**) © Houghton Mifflin Company. All rights reserved. (**bottom right**) © 1993 Richard Megna, Fundamental Photographs, NYC. **275:** (**top**) Photodisc Blue/Getty Images. (**center**) David Buffington/Photodisc Green/Getty Images. (**bottom**) © Houghton Mifflin Company. All rights reserved. **276:** (**top**) Modified from the protein data bank structure, 1AG6. Primary reference: Xue, Y., Okvist, M., Hansson, O., Young, S. "Crystal structure of spinach plastocyanin at 1.7 A resolution," *Protein Sci.,*

1998, *7*, 2099. (**bottom left and right**) © Houghton Mifflin Company. All rights reserved. **277:** Elementree © Fernando Dufour. Photographer Joseph Donahue, Montreal, Quebec, Canada. **278:** Scott Camazine/Photo Researchers, Inc. **283:** (**all**) © Houghton Mifflin Company. All rights reserved. **289:** Wonderfile. **291:** Modified from the protein data bank structure, 1A3N and 1GZX. Primary reference: Paoli, M., Liddington, R., Tame, J., Wilkinson, A., Dodson, G. "Crystal Structure of T State Haemoglobin with Oxygen Bound at All Four Haems," *J. Mol. Biol.,* 1996, *256*, 775. **292:** (**left**) Chuck Pefley/Alamy. (**right**) David Toase/Photodisc Green/Getty Images. **293:** © Houghton Mifflin Company. All rights reserved. **299:** © Royalty-Free/Corbis. **301:** © Houghton Mifflin Company. All rights reserved.

Chapter 8

302: © Houghton Mifflin Company. All rights reserved. **304:** Lewis, Gilbert Newton. *Valence and the Structure of Atoms and Molecules.* New York: The Chemical Catalog Company, Inc., 1923. Othmer Library of Chemical History, Chemical Heritage Foundation, Philadelphia, PA. **307:** (**top**) Wonderfile. (**chalk**) Stockbyte Gold/Alamy. (**salt**) © Houghton Mifflin Company. All rights reserved. (**pigment**) Photograph by James Scherer. © Houghton Mifflin Company. All rights reserved. (**Milk of Magnesia**) Photograph by James Scherer. © Houghton Mifflin Company. All rights reserved. (**glass**) Ragnar Schmuck/fStop/Getty Images. (**baking soda**) © Houghton Mifflin Company. All rights reserved. (**salt**) © Royalty-Free/Corbis. (**toothpaste**) Ingram Publishing/Alamy. **310:** Science Photo Library/Photo Researchers, Inc. **316:** (**top**) Stockfood. (**bottom**) © Houghton Mifflin Company. All rights reserved. **317:** (**left**) © Houghton Mifflin Company. All rights reserved. (**top center**) Don Farrall/Photodisc Green/Getty Images. (**bottom center**) Robert Sullivan/Getty Images. (**right**) © Houghton Mifflin Company. All rights reserved. **318:** Bryan Mullennix/Getty Images. **323:** Collection of The Corning Museum of Glass, Corning, NY, gift of Mr. and Mrs. Richard Barons. **329:** © Houghton Mifflin Company. All rights reserved. **338:** © Houghton Mifflin Company. All rights reserved.

Chapter 9

355: Omikron/Photo Researchers, Inc. **363:** (**left**) Andrew Lawson Photography/Alamy. (**center**) Andrew Cowin/Alamy. (**right**) © Eric Crichton/CORBIS. **364:** © Houghton Mifflin Company. All rights reserved. **380:** © Richard Megna/Fundamental Photographs, NYC. **382:** (**left**) Library of Congress. (**right**) Courtesy of Eastman Kodak Company.

Chapter 10

394: Kevin Kelley/Getty Images. **396:** © Air Products and Chemicals, Inc. **400:** © Houghton Mifflin Company. All rights reserved. **401:** (**left**) © Houghton Mifflin Company. All rights reserved. (**right**) Courtesy of Cryovac, Inc. **402:** © Houghton Mifflin Company. All rights reserved. **403:** © Shepard Sherbell/CORBIS SABA. **404:** (**all**) © Houghton Mifflin Company. All rights reserved. **406:** (**left**) Courtesy of the Chemical Heritage Foundation Image Archives. *The Shannon Portrait of Hon. Robert Boyle F.R.S. (1627–1691),* 1689, Johann Kerseboom, oil on canvas. Chemical Heritage Foundation Collections, Philadelphia, Pennsylvania, U.S.A. Photo by Will Brown. (**center**) Courtesy of Carlos Castro-Acuna. (**left**) Michael Mosher. **407:** Michael Mosher. **408:** © Houghton Mifflin Company. All rights reserved. **409:** (**top**) Musee Lambinet, Versailles, France/Bridgeman Art Library. (**bottom**) Edgar Fahs Smith Collection, University of Pennsylvania Library. **419:** Patti McConville/Getty Images. **421:** (**left**) Bert Hardy/Getty Images. (**right**) © SSPL/The Image Works. **427:** © Houghton Mifflin Company. All rights reserved. **429:** NASA. **430:** BananaStock/Alamy. **431:** (**left**) © Royalty-Free/Corbis. (**right**) The source of this material is the Cooperative

DNA containing a cisplatin interstrand cross-link at 1.63 A resolution: hydration at the platinated site," from *Nucleic Acids Research*, Vol. 27 (8) 1999. By permission of Oxford University Press and Dr. Franck Coste. **877:** National Atmospheric Deposition Program. **884:** Digital Image © The Museum of Modern Art/Licensed by SCALA/Art Resource, NY. **889 (both)** © Houghton Mifflin Company. All rights reserved. **890:** © Houghton Mifflin Company. All rights reserved. **893: (all)** © Houghton Mifflin Company. All rights reserved.

Chapter 21

900: Fermilab. **903:** © Houghton Mifflin Company. All rights reserved. **907: (both)** © Houghton Mifflin Company. All rights reserved. **910:** © Tom Steward/CORBIS. **913:** Photograph courtesy of AstraZeneca Oncology. **914:** © Tom & Deen Ann McCarthy/CORBIS. **917:** © Houghton Mifflin Company. All rights reserved. **920:** AIP Emilio Segre Visual Archives, Brattle Books Collection. **925:** Courtesy of Brookhaven National Laboratory. **926: (top left and right)** © Dr. A. Leger/ISM/Phototake. **(bottom)** © ISM/Phototake. **927:** © Sovereign/ISM/Phototake. **930:** Courtesy of John DeArmond. **931:** © H. Windeck/UNEP/Peter Arnold, Inc. **933:** © Houghton Mifflin Company. All rights reserved.

Chapter 22

934: LSHTM/Photo Researchers, Inc. **937: (left)** © Jewish Chronicle Ltd/HIP/The Image Works. **(center)** Omikron/Photo Researchers, Inc. **(right)** Michael Steele/Getty Images. **946: (top)** PDB ID: 1LDG; C. Dunn, M. Banfield, J. Barker, C. Higham, K. Moreton, D. Turgut-Balik, L. Brady, J. J. Holbrook, *Nat. Struct. Biol.*, 1996, *3*, 11. **(bottom)** Based on PDB ID: 1BKV; R. Z. Kramer, J. Bella, P. Mayville, B. Brodsky, H. M. Berman, *Nat. Struct. Biol.*, 1999, 6, 454. **947: (left)** Kenneth Eward/BioGrafx/Photo Researchers, Inc. **(right)** PDB ID: 1MBN; H.C. Watson, *Prog. Stereochem.*, 1969, *4*, 299. **949:** PDB ID: 1TKA; S. Koenig, A. Schellenberger, H. Neef, G. Schneider, *J. Biol. Chem.*, 1994, *269*, 10879. **953:** PDB ID: 1BPH; O. Gursky, J. Badger, Y. Li, D.L.D. Caspar, *BIOPHYS. J.*, 1992, *63*, 1210. **959:** From Michal/Biochemical Pathways/0471331309. Reprinted with permission of John Wiley & Sons. **960:** Smithsonian Institution Photo. **963:** Courtesy of West-Ward Pharmaceuticals. **967: (top)** PDB ID: 1C1G; F. G. Whitby, G. N. Phillips Jr., *Proteins: Struct., Funct.*, 2000, *38*, 49. **(bottom)** © CORBIS.

Index/Glossary

Page numbers followed by *f* refer to figures. Page numbers followed by *t* refer to tables.

Absolute zero *The temperature obtained by extrapolation of a plot of gas volume versus temperature to the "zero volume" point. The lowest temperature possible, which is 0 K, or −273.15°C,* 409

Absorb *To be associated, through intermolecular forces of attraction, by being taken up or mixing into a substance,* 656

Absorption spectroscopy *The measurement of how atoms and molecules absorb electromagnetic radiation as a function of the wavelength or frequency of the radiation,* 216

Accuracy *The closeness of a measurement to the actual value,* 29–31

Acetaldehyde. *See* Ethanal

Acetanilide, 623–624, 625–626

Acetic acid (ethanoic acid)
  dehydration of, 104–105, 108–109
  formation of, 103–104, 519–520
  ionization of, 688–689, 691–693, 724–727
  in solution with hydrochloric acid, 742–743
  titration with sodium hydroxide, 684–685, 792–795
  uses of, 520
  as weak acid, 150, 724, 725
  as weak electrolyte, 129, 130*f*

Acetic acid–acetate buffer system, 770–771, 778

Acetone (propanone)
  boiling-point elevation of, 476*t*
  formation of, 519
  vapor pressure of, 454

Acetonitrile, 420, 513*t*

Acetylcellulose, 340

Acetylene (ethyne), 512*t*
  combustion of, 358
  formation of, 420, 421
  pi bonds in, 370
  uses of, 420, 501

Acetylsalicylic acid. *See* Aspirin

Achiral *A term used to describe molecules that have a superimposable mirror image,* 525

Acid anhydrides *Binary compounds formed between nonmetals and oxygen that react with water to form acids,* 757

Acid–base indicator *A compound that changes color on the basis of the pH of the solution in which it is dissolved. The color change is often a result of structural changes due to protonation or deprotonation of acidic groups within the compound,* 799–801

Acid–base reactions *The reaction between an acid and a base. The products are water and an ionic compound,* 144, 148–152
  conjugate acid–base pairs, 720–722, 753
  stoichiometry of, 151–152
  strong acids and bases, 148–149
  weak acids and bases, 150

Acid–base titrations, 779, 787–801
  applications for, 787
  buffer regions in, 793–794
  equivalence point of, 789–790
  pH indicators for, 799–801
  strong-acid–strong-base, 787–792
  titration curves, 790
  titration endpoint, 801
  titration midpoint, 794
  weak-acid–strong-base, 792–796
  weak-base–strong-acid, 796–798

Acid deposition *The precipitation of acidic compounds from the atmosphere. This includes wet deposition as rain and dry deposition of particles,* 757–758

Acid dissociation constant (K_a) *The equilibrium constant for the dissociation of an acid,* 724–726, A13

Acidic *Having a pH less than 7.0 in aqueous solution at 24°C,* 735

Acidic solutions
  balancing redox reactions in, 836–840
  common-ion effect in, 737–738
  equilibrium constant for, 724–726
  equilibrium hydrogen ion concentrations, 729–730
  with mixture of monoprotic acids, 742–743
  pH of strong, 736–738
  pH of weak, 738–742
  with polyprotic acids, 749–752
  salts, 752–757
  strong and weak acids, 724–730

Acid-neutralizing capacity *The capacity of a solution, such as lake water, to neutralize acidity,* 758

Acidophiles, 592*t*, 593

Acids *A compound that produces hydrogen ions (H^+) when dissolved in water. According to the Arrhenius model, a species that produces hydrogen ions in solution. Compare the definitions of Brønsted–Lowry acid and Lewis acid,* 128, 148
  Arrhenius, 719–720
  Brønsted–Lowry, 720–722
  characteristics of, 723
  conjugate, 720–722, 753
  diprotic, 736

  importance of, 718–719
  Lewis, 722–723
  list of common, 719*t*
  molecular structure of, 728–729
  monoprotic, 736, 742–743
  polyprotic, 747–752
  strength of, 130, 149–150, 723–730
  superacids, 729
  triprotic, 747
  See also Acid–base *entries;* Aqueous equilibria

Acrylonitrile, 516*t*, 567

Actinides *The second set of inner transition elements. The actinides include elements 90–103,* 275–276

Activated complex *The unstable collection of atoms that can break up either to form the products or to reform the reactants,* 651

Activation energy *The energy that is required to initiate a chemical reaction. The minimum energy of collision that reactants must have in order to successfully create the activated complex,* 174, 630, 651–654

Active site *The location on an enzyme where the reaction of the substrate is catalyzed,* 335, 948

Activity *The effective concentration of a solute in solution,* 675, 706

Activity series *A series of elements ranked in order of reactivity. Also known as a reactivity series,* 291–292, 849–850

Actual yield *The experimental quantity, in grams, of product obtained in a reaction,* 115

Addition polymer *A polymer formed by the process of addition polymerization,* 515–516

Addition polymerization *A type of polymerization involving addition reactions among the monomers involved,* 515–516, 522

Addition reaction *A type of reaction in which atoms are added to a double or triple bond,* 514–515

Adenine, 935, 936, 937*f*, 942

Adenosine diphosphate (ADP), 607–608

Adenosine monophosphate (AMP), 607–608

Adenosine triphosphate (ATP), 545, 585, 595, 607–608, 611

Adhesion *The interaction of a compound with the surface of another compound, such as the interaction of water with the surface of a drinking glass,* 463–464

Adrenaline (epinephrine), 87–88, 89

Adsorb *To be associated, through intermolecular forces of attraction, with a surface,* 656

Aerogel *A silica glass whose structure is mostly air; also known as ceramic foam,* 572–573

Agriculture
future applications, 35
pesticides, 620–621
use of chelates in, 812t

Alachlor (Lasso), 623–624, 625–626

Alanine, 605–606, 755–756, 940f

Alcohols *A chemical containing the hydroxyl (—OH) group,* 512t, 514
ethanol, 514, 516–517
list of selected, 518t
types of, 517

Aldehydes *Compounds that carry a carbonyl functional group (C=O) with at least one H atom bonded to the carbonyl carbon atom,* 512t, 517–519, 872–873

Aliphatic compounds *Compounds in which all of the carbon atoms are sp^3 hybridized and contain only single bonds,* 495t, 496

Alkali, 148

Alkali metals *The highly reactive metals found in Group IA of the periodic table, which form alkalis upon reaction with water,* 66, 270
commercial uses of, 270t
crystal lattice in, 545
electron configurations of, 250
ionization energies for, 282–283
properties of, 270, 279

Alkaline battery, 846, 848–849

Alkaline earth metals *The metals found in Group IIA of the periodic table,* 66t, 270
commercial uses of, 271t
electron configurations of, 250
in ionic compounds, 68, 74
properties of, 270–271

Alkaliphiles, 592t, 593

Alkaloids *Nitrogen-containing bases found in vegetables and other plants,* 746–747

Alkanes *Saturated hydrocarbons,* 496–500
branched-chain, 498–499
cyclic, 500, 502
naming of, 499–500
normal, 496–497
typical reactions of, 510–511

Alkenes *Hydrocarbons containing at least one C=C double bond,* 500–501, 502, 512t
formation of, 510–511
geometric isomers, 502–504
naming of, 502

Alkoxy anion *An anion that contains an alkyl group attached to an oxygen anion, such as CH$_3$O$^-$ or CH$_3$CH$_2$O$^-$,* 520

Alkyl group *A hydrocarbon group, such as —CH$_3$ or —C$_2$H$_5$, within the structure of an organic compound,* 504–506

Alkyl halide *A functional group in which a halogen atom is bonded to an alkyl group,* 511

Alkynes *Hydrocarbons containing a C≡C bond,* 501, 512t
formulas for, 101
naming of, 502

Allotropes *Forms of an element that have very different chemical and physical properties, such as the allotropes O$_2$ and O$_3$,* 71, 493

Alloys *A homogeneous mixture of metals. A solution of two or more metals,* 266, 267, 556, 559t

Allred–Rochow electronegativity scale, 320

Alpha decay *A type of radioactive decay wherein an alpha particle is emitted from the nucleus of a radioactive nuclide. Common for elements whose nuclei are larger than bismuth, alpha decay is often accompanied by the release of a gamma ray,* 905–906

Alpha helix (α helix), 945

Alpha particles (α particles) *Particles emitted from the nucleus of a radioactive element during the process of alpha decay. They are helium nuclei (2 protons, 2 neutrons), with a +2 charge,* 54–55, 905–906, 919t
penetration capability of, 909–910

Aluminosilicate, 560t

Aluminum
density of, 23
gravimetric analysis of, 807
as Lewis acid, 722, 723f
metallic bonding of, 304
oxidation of, 168, 586t
production of, 826, 857–858
reactivity of, 291–292, 857
sources and uses of, 265t, 267, 271
specific heat of, 181t
in thermite reaction, 196–197
thermodynamic data for, A9

Aluminum chloride, lattice enthalpy of, 316

Aluminum hydride
hybrid orbitals in, 367
molecular geometry of, 368

Aluminum hydroxide, 147

Aluminum perchlorate, 168

Aluminum sulfate
in paper formation, 719, 723
in water treatment, 808

Amalgam *A solution made from a metal dissolved in mercury,* 557, 856

American Academy of Orthopedic Surgeons, 557

American Chemical Society, 2

Amide bond *A bond formed by a condensation reaction between an amino group (—NH$_2$) and a carboxylic acid group (—COOH),* 523, 940

Amides, 513t

Amines *A compound containing the —NH$_2$ functional group,* 512t, 523, 940

Amino acids, 588
chirality of, 960
found in living cells, 940f
protein formation and, 940–945
water solubility of, 755–756

Amino functional group *A functional group containing an —NH$_2$ attached to an sp^3 hybridized carbon atom,* 523

Aminopolycarboxylic acid chelating agents, 811t

Ammonia
in aqueous solution, 722
as base, 150, 722, 723, 743–744
complex with zinc, 809–810
hybrid orbitals in, 367
hydrogen bonds in, 450f
mass-action expression for, 673–674
molecular geometry of, 340, 341, 368
nicotine and, 743, 746
oxidation states for, 831
pH of precipitation and, 758
pH of solution of, 745
production of, 156, 597, 625, 702–703, 704, 722
titration with hydrochloric acid, 796–798
uses of, 340, 395t, 718t, 719t, 743

Ammonia–ammonium ion buffer system, 767–770, 775–777, 782–783

Ammonium chloride, reaction with barium hydroxide, 597

Ammonium cyanide, 755

Ammonium nitrate, 97–98, 175

Ammonium phosphate, 47, 75

Ammonium sulfate, 97–98

Amorphous solid *A solid whose atoms, ions, or molecules occupy fairly rigid and fixed locations but that lacks a high degree of order over the long term,* 541

Amount of substance, as SI base unit, 17t, 20

Ampere (A) *The SI base unit of electric current,* 17t, 20

Amphiprotic *Having the ability to act as either an acid or a base in different circumstances,* 733, 755

Amplitude *The distance between the highest point and the midpoint (zero) of a wave,* 212

m-Amsacrine, 962

Anabolism *The metabolic process of constructing molecules from smaller molecules using high-energy molecules that have been made by catabolism,* 958

Analgesic drugs, 14–15, 335

Analyte *A solute whose concentration is to be measured by a laboratory test,* 766

Anastas, Paul, 571

Angstrom, Anders, 217

Angstrom *A unit of distance equal to 10^{-10} m. It is symbolized by Å,* 211

Angular *As a function of the angles that describe the orientation of the radius in spherical polar coordinates,* 234

Angular momentum quantum number *The quantum number, l, that describes the shape of the orbital; l can be any whole-number value from zero to n − 1,* 235, 243t

Anhydrides, in aqueous solution, 757–758

Anions *Ions with a negative charge,* 56, 286, 289
acidity of salt and, 756–757
alkoxy, 520
carboxylate, 520
electron configurations of, 305
Lewis dot symbols for, 305
size of, 311

Anode *The electrode at which oxidation takes place,* 826, 845
sacrificial, 850

Antacids, 784–786

Anthocyanins *A naturally occurring class of compounds responsible for many of the colors of plants. These compounds often act as acid–base indicators,* 801

Anthracene, 505f

Antibiotics *Substances that kill, harm, or retard the growth of microorganisms,* 961–962

Antibodies *Large biological molecules that are produced during an immune response, travel to the site of infection, bind to foreign material, and signal the destruction of the foreign material,* 954

Antibonding orbitals *An orbital that indicates a lack of electron density between adjacent nuclei (no bond exists when the orbital is occupied). The antibonding orbital arises from the subtraction of two overlapping atomic orbitals,* 375, 376, 378
in metals, 551–552

Anticodon *The three-nucleic-acid sequence on a tRNA polymer that recognizes and binds to the codon on an mRNA polymer,* 943

Antifreeze, 462, 477, 478

Antilogarithm, A4

Antimatter *Particles that have the same mass as, but charges opposite to, corresponding matter. Antimatter particles such as the positron and antineutrino are similar to the electron and neutrino, respectively, but have opposite characteristics,* 905

Antimony, 272t, A9

Antineutrino *A subatomic particle produced in beta-minus decay that has no charge, has essentially no mass, and interacts only rarely with matter,* 904–905, 909–910

Antlerite, 586

Aqua regia, 844

Aquatic systems
acid concentrations and, 729–730, 733
calcium carbonate dissociation, 802–804
impact of mercury on, 876–877

Aqueous *Water-based; also implies that a dissolved substance has a sphere of hydration,* 126

Aqueous equilibria
acid–base titrations, 787–801
acid strength and, 724–730
buffers and, 767–786
common-ion effect in, 737–738, 768–771
complex-ion equilibria, 809–815
Henderson–Hasselbalch equation and, 777–779
pH of acidic solutions and, 738–743
pH of basic solutions and, 744–745
pH of polyprotic acids and, 749–752
in salt solutions, 752–756
solubility equilibria, 802–809
See also Equilibrium; Solubility equilibria

Aqueous solutions *A homogeneous mixture of a solute dissolved in the solvent water,* 107, 465
anhydrides in, 757–758
dissolution process, 465–466
hydration, 466, 479
pH of, 735–746
solubility of substances in, 467–468
See also Solution stoichiometry; Solutions

Aramid, 567

Arginine, 940f

Argon, 903
lasers, 224t
thermodynamic data for, A9
uses of, 274t, 275, 312, 401

ArK excimer lasers, 224t

Arnolfini Wedding (van Eyck), 868

Aromatic hydrocarbons *Hydrocarbons that contain one or more aromatic rings,* 495t, 504

Arrhenius, Svante, 52, 652, 719

Arrhenius equation, 652

Arsenic
in drinking water, 135, 471
thermodynamic data for, A9
uses of, 272t

Arthrobacter psychrolactophilus, 592

Ascorbic acid. *See* Vitamin C

Asparagine, 940f

Aspartate, 940f

Aspirin (acetylsalicylic acid), 87, 719t
buffers in, 784, 786f
chemical formula for, 89, 90f, 303
as covalent bond, 304
formula mass of, 88, 90–91
Lewis dot structure for, 328
molar mass of, 92
properties of, 303, 723–724
synthesis of, 15, 103–104, 322

Astatine, 274t

Atmosphere, 292f, 394
changes in ozone concentration, 427–432
composition of, 71, 293–294, 394, 395
partial pressures in, 402–403

Atmosphere (atm) *A unit used to measure pressure. 1 atm = 760 mm Hg,* 399, 400t

Atomic bomb, 920

Atomic force microscopy (AFM), 570

Atomic mass units (amu) *The arbitrary unit of mass for elements based on the isotope carbon-12. One atom of carbon-12 is defined to have a mass of exactly 12.0000 amu,* 60–61, 87

Atomic number (Z) *The number of protons in the nucleus of an atom,* 56, 58
in nuclear equations, 904
nuclear stability and, 917–918
periodicity of, 264

Atomic orbitals. *See* Orbitals

Atomic radius *Half the distance between the nuclei in a molecule consisting of identical atoms. Also known as the covalent radius,* 280–281, 311

Atoms *The smallest identifying unit of an element,* 5
atomic number of, 56
Bohr's model of, 220–224
Dalton's atomic theory, 49–51
determining mass of, 59–63, 87–88
isotopes, 57–58
mass number of, 57
origins of word, 47
periodic table, 64–66
radioactivity and, 53–55
size of, 21, 280–281, 311
structure of, 51–57
total number in universe, A1, A2

ATP. *See* Adenosine triphosphate

Atrazine, 620–621, 636–637, 645, 654

Aufbau principle *As protons are added to the nucleus increasing the atomic number of the atom, electrons are added successively to the next highest energy orbital. The method of filling hydrogen atomic orbitals to get the ground-state electron configuration of a multielectron atom by adding the electrons one at a time until all are accommodated; from the German phrase for "building up,"* 245, 250, 261, 279

Automobiles
air bags, 419–420
catalytic converters, 277, 317, 430, 638–639, 656
gasoline, 317–318, 508–509
and increase of surface ozone, 430
solar-powered, 555

Autoprotolysis *Proton transfer between molecules of the same chemical species, as in the autoprotolysis of water,* 733, 737, 738

Average rate *The rate of a reaction measured over a long period of time,* 622–627

Average speed *The sum of the molecular speeds of each molecule divided by the number of molecules,* 424

Avogadro, Amadeo, 91, 404

Avogadro's law *Equal amounts of gases occupy the same volume at constant temperature and pressure,* 404–405, 413t, 423

Avogadro's number (N_A) *The number of particles (6.022×10^{23}) of a substance in 1 mol of that substance. By definition, Avogadro's number is equal to the number of carbon-12 atoms in exactly 12.0000 g of carbon-12*, 91, 92

Axial position *The position of a group when it is aligned along the z axis of a molecule*, 340

Baekeland, Leo, 563
Bakelite, 563
Balanced *Appropriate coefficients have been added such that the number of atoms of each element are the same in both reactants and products*, 103
Balances, 47, 668, 806
Balmer, Johann J., 217, 218
Balmer series, 218–219, 223
Band gap *A small energy gap that exists between the valence band and the conduction band*, 553–554
Band theory *A metal is a lattice of metal cations spaced throughout a sea of delocalized electrons*, 551–553
Bar *A unit used to measure pressure. 1.01325 bar = 1 atmosphere*, 399
Barium, 271t, A9
Barium hydroxide
　reaction with ammonium chloride, 597
　reaction with magnesium sulfate, 147
Barometer, 399
Bartlett, Neil, 329
Base hydrolysis *The process in which a base reacts with water to produce its conjugate acid and hydroxide ion*, 753
Base pair *A pair of nucleic acids that are intermolecularly bound to each other through noncovalent forces of attraction, for example, the combination of A and T or of G and C in a DNA duplex*, 936, 937f, 942
Bases *According to the Arrhenius model, a species that produces hydroxide ions in solution. Compare the definitions of Brønsted–Lowry base and Lewis base*, 129, 148
　Arrhenius, 719–720
　Brønsted–Lowry, 720–722
　characteristics of, 723
　conjugate, 720–722, 753
　equilibrium constants for, 725, A14
　importance of, 718–719
　Lewis, 722–723, 870
　list of common, 719t
　strength of, 130, 149–150
　See also Acid–base *entries;* Aqueous equilibria
Base units *The set of seven fundamental units of the International System*, 17–20
Basic anhydrides *Binary compounds that are formed between metals with very low*

electronegativity and oxygen and that react vigorously with water, 757
Basic *Having a pH greater than 7.0 in aqueous solution at 24°C*, 735
Basic solutions
　balancing redox reactions in, 840–841
　determining pH of, 743–746
　salts, 752–757
Batteries *Two or more voltaic cells joined in series*, 826, 847–849
　chemistry of common, 848–849
　decreasing potentials in, 854
　list of selected, 848t
　rechargeable, 848, 849, 854
　See also Electrochemical cells
Bauxite, 271, 420, 857, 858
Bazemore, Whit, 622, 623f
Becquerel (Bq) *An SI unit of activity equivalent to one nuclear disintegration per second*, 911
Bent geometry, 336–337, 340
Benzene, 71, 505f
　boiling-point elevation of, 476t
　freezing-point depression for, 478t
　miscibility of, 466
　molecular orbital for, 384–385
　properties of, 385, 467
　shorthand notations for, 385f
Beryllium
　electron configuration of, 246
　electronegativity of, 289
　Lewis dot symbol for, 305
　thermodynamic data for, A9
　uses of, 271t
Beryllium chloride, molecular geometry of, 337
Beryllium hydride, hybrid orbitals in, 366
Beta-galactosidase, 592
Beta-particle emission *A naturally occurring type of radioactive decay wherein an electron is ejected at high speed from the nucleus, typical of nuclei that have too many neutrons to be energetically stable. An antineutrino accompanies this emission, and sometimes one or more gamma rays as well. Also known as beta emission*, 904–905, 919t
Beta particles (β particles) *Particles emitted from the nucleus of a radioactive atom during the process of beta decay. These particles are high-speed electrons*, 54–55, 904
　penetration capability of, 909–910
Beta pleated sheet (β sheet), 945
Beta-plus emission *A type of radioactive decay wherein a positron is ejected at high speed from the nucleus, typical of nuclides that have too few neutrons. A neutrino accompanies this emission, and usually one or gamma rays as well. Also known as positron emission*, 918
Bidentate *Capable of forming two coordinate covalent bonds to the same metal center*, 874

Bimolecular reaction *A reaction involving the collision of two molecules in the rate-determining step*, 648
Binary covalent compounds *Compounds composed of only two nonmetal elements*, 72–73
Binary ionic compounds *Compounds typically composed of a metal and nonmetal element, which have ions that interact via electrostatic attractions*, 73–75
Binding energy *The energy released when a nucleus is formed from protons and neutrons. Binding energies are expressed as a positive number*, 915–917
Binding rate constant *The rate constant the indicates the association of two molecules*, 670
Biochemical pathway *A sequence of biochemical reactions that lead to the consumption or production of a particular metabolite*, 958–959
Bioenergetics *The study of the energy changes that occur within a living cell*, 580
Biological processes. *See* Chemistry of life
Biomass conversion *The release of energy from the chemical conversion—often simply the burning—of plants and trees*, 198
Biopolymer *A polymer of biological materials, such as DNA or cellulose*, 572
Bismuth, 272t, A9
Blood
　chemicals in, 126, 260
　color of, 890–891
　hemoglobin, 260, 291, 869–870
　as isotonic solution, 480
Body-centered cubic (bcc) unit cell *A unit cell built via the addition of an atom, ion, or molecule at the center of the simple cubic unit cell*, 543, 546, 549
Bohr, Niels, 220, 221f, 222
Bohr frequency condition, 222
Bohr's model of atomic structure, 220–224, 227–228, 234–235
Boiler scale *A buildup of calcium and magnesium salts within pipes and water heaters. Typically composed of calcium carbonate and magnesium carbonate*, 309–310, 767
Boiling *The process that occurs when the vapor pressure of a liquid is equal to the external vapor pressure*, 455
Boiling point *The temperature at which the pressure of a liquid's vapor is equal to the surrounding pressure*, 455
　hydrogen bonds and, 449–451
　impact of altitude on, 455–456
　molar mass and, 446–448, 506
　of selected compounds, 446t
Boiling-point elevation *The increase in the boiling point due to the presence of a dissolved solute*, 475–477, A13
Boltzmann, Ludwig, 582–583

Bomb calorimeter *Apparatus in which a chemical reaction occurs in a closed container, allowing the energy released or absorbed to be measured,* 184

Bond angles, 340, 357
   of coordination complexes, 876–877
   factors effecting, 341–342
   of hybrid orbitals, 367, 372

Bond dipole *The polarization of electrons in a bond that results in a separation of partial charges in the bond,* 343–344

Bond dissociation energy *The energy required to break 1 mol of bonds in a gaseous species. Also known as the enthalpy of bond dissociation,* 332–333

Bond energies
   covalent bonds, 331–334
   electronegativity and, 287
   ionic bonds, 313–316
   in valence bond model, 360–362

Bonding electron pairs *Pairs of electrons involved in a covalent bond. Also known as bonding pairs,* 340

Bonding orbitals *An orbital that indicates the presence of electron density between adjacent nuclei (a bond exists when the orbital is occupied). The bonding orbital arises from the addition of two overlapping atomic orbitals,* 375, 376, 378
   in metals, 551–552

Bonding pairs *Pairs of electrons involved in a covalent bond. Also known as bonding electron pairs,* 322
   multiple bonds, 325–327
   in VSEPR model, 337, 340, 341–342

Bond length *The average distance between the nuclei of bonded atoms,* 319, 332, 341
   of hybrid orbitals, 367t, 371–372
   in valence bond model, 360–362

Bond order *The number of electrons in bonding orbitals minus the number of electrons in antibonding orbitals, divided by 2. The bond order indicates the degree of constructive overlap between two atoms,* 377, 378

Borane, 723
Boric acid, 785
Borkenstein, Robert, 329
Born, Max, 313

Born–Haber cycle *A diagrammatic representation of the formation of an ionic crystalline solid using Hess's law. The cycle reveals the lattice enthalpy, which is difficult to obtain by direct measurement,* 313–314

Boron, 50f
   electron configuration of, 246, 365
   Lewis dot symbol for, 305
   uses of, 271t

Boron trifluoride, 330
   hybrid orbitals in, 365–366
   molecular geometry of, 338–339

Borosilicate, 560t
Bovine F1 ATP synthase, 545

Boyer, Paul D., 545
Boyle, Robert, 16, 406f

Boyle's law *The volume of a fixed amount of gas is inversely proportional to its pressure at constant temperature,* 15–16, 405–408, 413t, 422

Bragg, William Henry, 546
Bragg, William Lawrence, 546
Branched-chain alkanes, 498–499
Branched molecules, 447
Brass, 586t
Breathalyzer, 329
Breeder reactors, 922
Brochantite, 586
Brock, Thomas D., 592

Bromine
   boiling point of, 446
   thermodynamic data for, A9
   uses of, 274t

Bromine pentafluoride, molecular geometry of, 342
Bromochlorofluoromethane, 525
Brønsted, Johannes Nicolaus, 720

Brønsted–Lowry acids *Any species that donates a hydrogen ion (proton) to another species,* 720–722, 723

Brønsted–Lowry bases *Any species that accepts a hydrogen ion (proton) from another species,* 720–722

Bronze, 586t
Buckminsterfullerene, 493–494

Buffer capacity *The degree to which a buffer can "absorb" added acid or base without changing pH,* 779–784

Buffer region *The region of a titration indicated by the presence of a weak acid or base and its conjugate. The pH changes little within this region,* 793–794, 796

Buffers *A solution containing a weak acid and its conjugate base or a weak base and its conjugate acid. Buffers resist changes in pH upon the addition of acid or base or by dilution,* 767–786
   adding strong acid or base to, 780–782
   calibration, 772–773
   capacity of, 779–784
   common-ion effect in, 768–771
   criteria for suitable, 784
   defined, 767
   Henderson–Hasselbalch equation and, 777–779
   in medicine and biology, 784–786
   pH of, 767–771
   preparation of, 772–777

Buret *A piece of laboratory glassware used to measure volume and to add discrete amounts of solution in a repetitive fashion,* 140, 463–464

Butadiene, 357
1,3-Butadiene, 372, 381
Butane, 497t
   bond length in, 372
   combustion of, 106–107, 597

Butene, 502–503
1-Butene, 372, 501f, 502, 503
2-Butene, 503
Butylamine, 512t
Butyne, 101f
1-Butyne, 501f

Cadaverine, 523
Cadmium sulfide, 555

Calcareous oozes *The calcium-containing detritus from dead single-celled, calcium-based sea life,* 802

Calcium
   atomic mass of, 91
   gravimetric analysis of, 807
   in milk, 309
   reactivity of, 283f
   sources and uses of, 265t, 271t
   thermodynamic data for, A10
   titration with EDTA, 681, 767, 810, 813, 814–815

Calcium carbide, 420, 421f

Calcium carbonate, 748
   color of, 885
   as ionic compound, 307t
   in mortar, 558
   naming of, 75
   solubility of, 802–804

Calcium channel blockers, 331

Calcium chloride
   formation of, 309
   formula for, 69
   as ionic compound, 307t
   lattice enthalpy of, 315–316

Calcium dihydrogen phosphate, 98

Calcium fluoride
   formula for, 68
   lattice enthalpy of, 315–316

Calcium hydroxide, 558, 808
Calcium ion, 67, 261
Calcium oxide, 68, 271, 718t
Calcium phosphate, 748
Calcium sulfate, 719, 748, 804
Calcium sulfate hemihydrate, 557, 558
Calcium supplements, 748

Calorie (Cal) *(Note capital C) Unit of energy equal to 1000 calories (1 kilocalorie) and to 4184 joules,* 179–181

calorie (cal) *(Note lowercase c) Unit of energy equal to 4.184 joules,* 180

Calorimeter *An apparatus in which quantities of heat can be measured,* 183–185

Calorimetry *The study of the transfer of heat in a process,* 182–185

Calotype, 382
Cancer treatment, 528–530, 876, 901, 912, 962

Candela (cd) *The SI base unit of luminous intensity,* 17t, 20

Capacitors, thin film, 568

Capillary action *The effect seen when a liquid rises within a narrow tube as a consequence of adhesive forces being stronger than cohesive forces*, 463–464

Carbohydrates *Compounds that have the general formula $C_x(H_2O)_y$. These compounds serve living organisms as structural molecules (cellulose), as cellular recognition sites, and for energy storage (starch)*, 954–956, 960

Carbon
  allotropes of, 493–494
  band gap of, 553t
  combustion of, 190
  electron configuration of, 246
  isotopes of, 57–58, 902, 903t, 927f
  Lewis dot symbol for, 305
  molecular orbitals of, 552
  oxidation numbers for, 154, 829–830
  properties of, 272
  sources and uses of, 268t, 272t
  in steel, 267
  thermodynamic data for, A10
  See also Carbon-based compounds
Carbon-12, 59, 915f
  abundance of, 57, 61
  atomic mass of, 60
  biosignatures of, 46
Carbon-13, 57
Carbon-14, 57, 905, 926
Carbonate ion, Lewis dot structure for, 327–328
Carbon-based compounds, 492–530
  alcohols, 516–517
  aldehydes, 517–519
  carboxylic acids, 519–520
  chirality of, 525–527
  condensation polymers, 522–523
  crude oil, 494–496, 506–509
  esters, 520–522
  ethene, 514–516
  functional groups, 511–514
  hydrocarbons, 496–506
  ketones, 519
  modern medicine and, 527–530
  polyethers, 524
  See also Hydrocarbons
Carbon cycle *The natural process that describes how carbon atoms are moved among the land, sea, and atmosphere*, 431
Carbon dioxide
  in atmosphere, 35, 197, 431–432
  bond dipoles in, 345
  double bonds in, 326
  enthalpy of formation of, 188, 190
  lasers, 224t
  Lewis dot structure for, 325–326
  molecular geometry of, 337
  phase diagram for, 460–461
  process for removing, 113
  properties of, 70
  uses of, 395t, 401, 472–473
Carbon fibers, 566t, 567
Carbonic acid, 327

Carbon monoxide
  in hemoglobin, 871
  oxidation numbers for, 154
Carbon tetrachloride, 511f
  boiling-point elevation of, 476t
  solubility of, 468
  uses of, 510
Carbonyl functional group *A functional group containing the group $C{=}O$ attached to either hydrogen atoms (as in the aldehydes) or carbon atoms (as in the ketones)*, 517
Carboxylate anion *An anion resulting from the removal of a hydrogen ion from the oxygen in a carboxylic acid functional group*, 520
Carboxyl functional group *The —COOH functional group*, 519–520
Carboxylic acids *Organic acids that carry the carboxyl functional group (—COOH)*, 512t, 519–520, 940, 957
Cardiac defibrillator, 568
Cardiac pacemaker, 825, 826f, 836, 847
Cardiologist *A heart specialist*, 824
Carnot, Nicolas Léonard Sadi, 587
Carvone, 525, 526f
Casein, 309
Catabolism *The metabolic process of breaking down molecules to form smaller molecules and produce high-energy molecules*, 587, 958
Catalysts *A substance that participates in a reaction, is not consumed, and modifies the mechanism of the reaction to provide a lower activation energy. Catalysts increase the rate of reaction*, 272t, 654–658
  in food industry, 657
  heterogeneous, 656
  homogeneous, 655–656
  impact on equilibrium, 704, 705t
  transition elements as, 277
  zeolite, 572
Catalytic converters, 277, 317, 430, 638–639, 656
Catalytic cracking *A type of cracking wherein heat and a catalyst are used*, 508
Cathode *The electrode at which reduction takes place*, 826, 845
Cations *Ions with a positive charge*, 56, 286, 289
  acidity of salt and, 756–757
  electron configurations of, 305
  Lewis dot symbols for, 305
  size of, 311
Cell membranes
  components of, 953–954, 957
  difficulty of crossing, 343, 346
  structure of, 346
Cell notation *The chemist's shorthand used to describe electrochemical cells*, 847
Cellular respiration, 175
Cellulose
  in Green Chemistry, 572
  properties of, 6, 7, 956

Celsius, Anders, 19
Celsius scale, 19–20, 410
Cement, 558
Center for Applied Isotope Studies, 62
Ceramics *A nonmetallic material made from inorganic compounds*, 557–563
  aerogels, 572–573
  glass, 559–560
  properties of, 557–559
  superconductors, 561–563
  types of, 557, 558
Cesium, 216
  reaction with water, 757
  thermodynamic data for, A10
  uses of, 270t
Cesium fluoride, 70
Chadwick, James, 54
Chain reaction *In nuclear chemistry, a reaction that is self-sustaining as one event in turn causes more events*, 923–924
Chalcanthite, 586
Chalcogens *The elements found in Group VIA of the periodic table of elements*, 66t, 273
Chalk. *See* Calcium carbonate
Challenger space shuttle, 7
Chapman mechanism, 429
Charles, Jacques Alexander, 408, 409f
Charles's law *The volume of a fixed amount of gas is directly proportional to its temperature in kelvins at a constant pressure*, 408–411, 413t, 423
Chelate *A polydentate ligand that forms strong metal–ligand bonds*, 874–875
Chelates *Substances capable of associating through coordinate covalent bonds to a metal ion. Also known as chelating agents*, 280, 810–813
Chelate effect *An unusually large formation constant due to favorable entropy change for the formation of a complex between a metal center and a polydentate ligand*, 893
Chemical bonds *A sharing of electrons between two adjacent atoms. This sharing can be complete, partial, or ionic in nature*, 303–347
  chemical warfare and, 326–327
  compared with intermolecular forces, 444–445
  in coordination complexes, 870–873
  covalent bonds, 316–335
  hybridization, 363–374
  hydrogen bonds, 448–451
  ionic bonds, 306–316
  kinds of, 303–304
  Lewis dot symbols, 304–305
  molecular models of, 303–306
  molecular orbital theory, 374–381
  octet rule and, 306
  polarization of electrons, 343–345
  valence bond theory, 358–363
  VSEPR model, 335–343

Chemical changes *Changes that involve chemical reactions in which existing chemicals, the reactants, are changed into different chemicals, the products,* 6

Chemical energy *Energy that is stored in a substance as a result of the motions and positions of its atomic nuclei and their electrons,* 172. See also Energy

Chemical equation *A precise quantitative description of a reaction,* 103–108
 balancing, 104–107
 indicating phases in, 107
 limiting reagent in, 112–115
 molar mass in, 107–108
 molecular and ionic, 144–145
 percentage yields, 115–116
 stoichiometric ratios, 108–111
 using coefficients, 105

Chemical equilibrium. *See* Equilibrium

Chemical formula *A representation, in symbols, that conveys the relative proportions of atoms of the different elements in a substance,* 68–69
 for coordination complexes, 881
 in equations, 103

Chemical kinetics *The study of the rates and mechanisms of chemical reactions,* 620–658
 activation energy, 630, 651–654
 catalysts, 654–658
 collision theory, 629–630
 of coordination complexes, 893
 determining rate laws, 641–646
 half-life, 636–638, 912–915
 integrated rate laws, 631–640
 introducing rate laws, 627–630
 reaction mechanisms, 647–654
 reaction rates, 621–627
 summarizing rate laws, 646–647
 transition state theory, 651–653
 See also Rate laws

Chemical nomenclature *A system of rules used to assign a name to a particular substance,* 71–76
 alkyl groups, 504–506
 for binary covalent compounds, 72–73
 for binary ionic compounds, 73–75
 common names, 500
 for coordination complexes, 881–883
 IUPAC nomenclature, 500
 naming of alkanes, 499–500
 naming of alkenes and alkynes, 502
 polyatomic ions, 75–76

Chemical property *A characteristic of a substance that can be determined as it undergoes a change in its chemical composition,* 6–8

Chemical reactions *The combination or reorganization of chemicals to produce different chemicals,* 6, 103
 acid–base, 148–152
 addition, 514–515
 balancing equations, 103–116
 condensation, 520

exothermic vs. endothermic, 174–175
ion formation, 67
law of conservation of mass, 47
molecularity of, 648
vs. nuclear processes, 903
oxidation–reduction, 152–157
precipitation, 145–148
substitution, 510
titrations, 115

Chemicals. *See* Matter

Chemical-vapor deposition *A thin film prepared when a compound placed on a surface undergoes a reaction,* 569–570

Chemical warfare, 326–327

Chemistry *The systematic study of the composition, structure, and properties of the matter of our universe,* 4
 basic terminology, 4–11
 future challenges for, 34–36
 Green, 571–572
 scientific method in, 12–16
 units and measurements in, 16–34

Chemistry of life, 935–964
 adrenaline, 87–88
 bioenergetics, 580
 carbohydrates, 954–956
 carbon and, 492
 cell membranes, 343, 346
 cellular respiration, 175
 chirality and, 960–961
 cholesterol, 95–96
 dating origins of life, 46, 51–52
 diversity of protein functions, 953–956
 DNA, 935–939
 electrolytes, 129
 elements of life, 47, 65, 277–278
 enzymes, 947–952
 exercising muscles, 600–601, 669, 701–702, 738
 future challenges for, 36, 961–964
 genetic code for proteins, 941–947
 glycolysis pathway, 585
 heart cells, 824, 853–854
 human eye, 356–357
 lipids, 957–958
 metabolism, 595, 958–959
 photosynthesis, 111, 173, 872
 plant fertilization, 97–98
 proteins, 940–941
 vitamins, 950–952
 See also Medicine

Chemists
 career opportunities for, 36
 educational degrees, 2
 employment data for, 2
 work done by, 3–4

Chiral *A term used to denote molecules that have a nonsuperimposable mirror image,* 525–527, 572

Chiral center *A carbon attached to four different groups,* 525

Chirality *The quality of being chiral—that is, of having a nonsuperimposable mirror image,* 525–527, 960–961

Chloride ion, 261

Chlorine
 atomic mass of, 61
 in drinking water, 157–158
 gravimetric analysis of, 807
 interaction with light, 358
 ionization energy for, 282
 Lewis dot symbol for, 308
 mass spectrum of, 60*f*
 ozone depletion and, 655
 sources and uses of, 268*t*, 274*t*
 thermodynamic data for, A10

Chlorine gas
 production of, 405
 uses of, 395*t*

Chlorine trifluoride, molecular geometry of, 342

Chlorofluorocarbons (CFCs) *Compounds containing only carbon, chlorine, and fluorine. CFCs are typically used as solvents and refrigerants and are known to be harmful to the ozone layer,* 429
 as greenhouse gas, 431
 interaction with light, 380
 list of common, 429*t*
 ozone depletion and, 35, 341, 429–431

Chlorofluorohydrocarbons, 323

Chloroform, 345, 510, 511*f*

Chloromethane, 510

Chlorophyll, 173

Chocolate, 99–101, 462

Cholera, 157–158, 159*f*

Cholesterol, 95–96, 957–958

Cholesteryl benzoate, 550

Chromatography *A chemical technique involving the partition of a solute between a stationary phase and a mobile phase. The technique can be used to separate or purify mixtures of solutes,* 62–63, 507, 678–679

Chromium
 complexes, 884, 888
 electron configuration of, 249
 as heavy metal, 266
 in steel, 266, 267
 thermodynamic data for, A10

Chromosomes *The packaged functional units of DNA,* 964

cis *A molecule containing groups that are fixed on the same side of a bond that cannot rotate,* 503–504

cis **isomer** *An isomer containing two similar groups on the same side of the compound,* 880–881

Cisplatin, 529, 875, 876, 889

Citric acid, 723, 724*f*

Clapeyron, Benoit Paul Emile, 587

Classical mechanics *The physical description of macroscopic behavior based on Newton's equations of motion,* 225

Clausius, Rudolf Julius Emmanuel, 587

Clinoptilolite, 310

Closest packed structure *A structure wherein the atoms, molecules, or ions are arranged in the most efficient manner possible,* 544

Coagulation *The precipitation of a solid along with some dissolved organic material, microorganisms and other undesirable substances,* 808

Coal, 197, 198, 492

Cobalt
complexes, 889–890
in steel, 267

Cobalt chloride, boiling-point elevation of, 476

Cocaine, 746

Codeine, 88, 89f, 91

Codon *A three-nucleic-acid sequence on an mRNA that is translated into a particular amino acid during protein synthesis,* 943

Coefficients *The numbers placed in front of the substances in a chemical equation that reflect the specific numbers of units of those substances required to balance the equation,* 105

Coenzymes, 948

Cohesive forces *The collection of intermolecular forces that hold a compound within its particular phase,* 463–464

Coinage metals, 66t, 125, 547

Cold packs, 175

Collagen *A structural protein that exhibits a linear tertiary structure formed from intertwined alpha helices. It is found in connective tissues in the body,* 946

Colligative properties *Physical properties of a solution that depend only on the amount of the solute and not on its identity, including boiling-point elevation, freezing-point depression, vapor pressure lowering, and osmotic pressure,* 473–481
boiling-point elevation, 475–477
constants for selected compounds, A13
freezing-point depression, 478
osmosis, 479–480
vapor pressure lowering, 474–475

Collision theory *A theory that correlates the number of properly oriented collisions with the rate of the reaction,* 629–630

Color
changes, and acid–base chemistry, 723, 799–801
coordination complexes and, 873, 880, 883–892
human perception of, 356
as intensive property, 23t
molecular structure and, 233
photography, 382–383
transition metals and, 884–885

Columbia space shuttle, 7

Combined gas equation *The combination of Avogadro's law, Boyle's law, and Charles's law,* 411–412, 413t

Combustion
enthalpy of, 189–191, 334, 358
as redox reaction, 152, 153f

Common-ion effect *The addition of an ion common to one of the species in the solution, causing the equilibrium to shift away from production of that species,* 737–738, 768–771

Complete ionic equation *A chemical equation that indicates all of the ions present in a reaction as individual entities,* 144–145, 146

Complex-ion equilibria, 809–815
chelating agents and, 810–813
conditional formation constant, 813–814
formation constants, 809–810, 811t, A16

Composite material *A material made from two or more different substances,* 565, 568

Composition *The relative proportions of the elements in a compound,* 5
law of definite, 47–49

Compounds *A substance containing different elements chemically bonded together,* 5
covalent, 72–73
ionic, 67–70, 73–75
naming, 71–76

Concentration *An intensive property of a solution that describes the amount of solute dissolved per volume of solution or solvent. The typical concentration units include molarity, ppm, and w/w,* 131–140, 468–471
changes in, and equilibrium, 701–702, 705t
dilution of, 137–139
effective, in solution, 675
equilibrium, 672–675, 685–686
impact on electrochemical cells, 851–854
as intensive property, 131
measures based on mass, 470–471
measures based on moles, 468–470
molality, 468
molarity, 131–135, 468
mole fractions, 468
parts per million, per billion, per trillion, 135–137, 470
reaction rates and, 622, 627–628, 644
spontaneity and, 611
weight percent, 470

Concentration cell *A cell in which different concentrations of identical ions on both sides of the cell provide the driving force for the reaction,* 853–854

Concrete, 558

Condensation *The process of a gas undergoing a phase transition to a liquid; the opposite of evaporation,* 452–453

Condensation polymerization *Formation of a polymer by condensation reactions,* 522–523

Condensation reaction *A reaction in which two molecules become joined in a process accompanied by the elimination of a small molecule such as water or HCl,* 520, 935–936

Conditional formation constant (K′) *The formation constant that accounts for the free metal ion and its associated ligand,* 813–814

Conduction band *The collection of empty molecular orbitals on a metal,* 552–554

Conductor *A substance that contains a very small band gap,* 554
metals, 551, 553–555, 561–562
superconductors, 561–563

Cones, 356

Confidence interval, A7–A8

Confidence limit, A7–A8

Conjugate acid–base pairs, 720–722, 753
in pH indicators, 799
strength of, 726–727

Conjugate acid *The acid that results after accepting a proton,* 720–722

Conjugate base *The base that results after donating a proton,* 720–722

Conjugated π bonds *An extended series of alternating single and double bonds,* 374, 381, 383
delocalized, 384–385

Conjugation *The presence of conjugated π bonds—that is, a series of at least two double bonds alternating with single bonds,* 374

Constant-volume calorimetry *A form of calorimetry in which the reacting system is sealed within a chamber of fixed volume, and the only way the system can release or gain energy is by the exchange of heat with the surroundings,* 184

Contact process *An industrial method used to produce sulfuric acid from elemental sulfur,* 676–677, 704, 748, 757

Contractile proteins, 953

Conversion factor *A mathematical expression of the ratio of one unit to another, used to convert quantities from one system of units to another,* 25–28

Cooling tower *A large industrial apparatus used to condense gaseous vapor into liquid or to cool water that has been heated before it is discharged,* 471, 472f

Coordinate covalent bonds *A covalent bond that results from the donation of two electrons from one of the two atoms involved in the bond. The resulting bond is indistinguishable from other covalent bonds,* 330
chelates and, 810
in coordination complexes, 870–873

Coordinates *Forms a coordinate covalent bond,* 871

Coordination complexes *A metal bonded to two or more ligands via coordinate covalent bonds,* 869–895
bonding in, 870–873
color and, 880, 883–892
coordination numbers, 875

crystal field theory and, 886–891
electron transfer reactions of, 894
formation constant and, 809–810
formulas for, 881
isomers, 878–881
ligand exchange reactions of, 892–893
ligands, 809, 810, 873–875
magnetism and, 891–892
nomenclature for, 881–883
structure of, 876–878
Coordination number *The number of coordinate covalent bonds that form in a complex*, 875, 877–878
Coordination sphere isomers *Substances that contain different ligands in the coordination spheres of complex cations and complex anions*, 878, 879–880
Copper
atomic mass of, 60
emission spectrum for, 220f
as heavy metal, 266
oxidation products of, 586
pipes, 849
properties of, 5, 64
reaction with nitric acid, 833, 834f, 837–838, 842–843, 845–846, 855
reduction potential of, 834, 835, 850
specific heat of, 181t
structure of, 544
thermodynamic data for, A10
uses of, 265, 267
Copper chloride, reaction with zinc, 851–852
Copper ferrocyanide, 480
Core, Earth's, 292–293
Core electrons *Electrons in the configuration of an atom that are not in the highest principal energy shell*, 247
Correlation coefficient, 646
Corrosion *The deterioration of a metal as a consequence of oxidation*, 152, 153f, 291–292, 850
Cosmic rays, 906f
Coulomb, Charles Augustin, 314
Coulomb *The unit of electrical charge*, 55
Coulomb's law *The force between two particles is proportional to the product of the charges (Q) on each particle divided by the square of their distance of separation (d)*, 314–315
Coupled reactions, 611–612
Covalent bonds *A sharing of electrons between two adjacent atoms. This sharing can be complete of partial*, 70, 304, 316–335
bond energies for selected, 332t
coordinate, 330
description of, 317–319
electronegativity and, 319–320
energy of, 331–334
examples of, 317
formal charges and, 323–325
Lewis dot structures for, 322–323
multiple bonds, 325–327
nonpolar, 320–321

octet rule exceptions in, 329–331
polar, 319–321
resonance structures, 327–328
types of, 320–321
valence bond theory for, 359–362
Covalent radius *Half the distance between the nuclei in a molecule consisting of identical atoms. Also known as the atomic radius*, 280
Cracking *The breaking up of hydrocarbons into smaller hydrocarbons, using heat and in some cases steam and/or catalysts*, 508
Crick, Francis, 936–937
Critical mass *The amount of fissionable fuel needed to sustain a chain reaction*, 923
Critical point *The temperature and pressure at which the physical properties of the liquid phase and vapor phase of a substance become indistinguishable*, 460
Critical pressure *The highest pressure at which a liquid and gas coexist in equilibrium*, 460
Critical temperature *The highest temperature at which a liquid and a gas coexist in equilibrium*, 460
Crosslinking *Reactions that covalently bond two or more individual strands of a polymer together*, 563
18-crown-6, 280f
Crude oil, 491, 492
composition of, 494–496
fractional distillation of, 506–508
gas chromatography analysis of, 507
processing, 508–509
sources of, 494, 496t
uses of, 492, 496, 508t
Crust, Earth's, 292f, 293, 294t, 857
Crystal field splitting energy *The difference in energy between the d orbital sets that arises because of the presence of ligands around a metal; symbolized by Δ*, 887
Crystal field theory, 886–891
octahedral crystal field splitting, 886–887
result of d orbital splitting, 889–891
square planar crystal field splitting, 887
tetrahedral crystal field splitting, 887
Crystalline (or crystal) lattice *A highly ordered, three-dimensional arrangement of atoms, ions, or molecules into a solid*, 308–309, 542–543
Crystalline solids *A solid made from atoms, ions, or molecules in a highly ordered long-range repeating pattern*, 540–550
density of, 548–549
lattice structures, 542–543
liquid crystals, 550
metallic crystals, 543–545
X-ray crystallography of, 545–549
Cubic closest packed (ccp) structure *A closest packed structure made by staggering a second and third row of*

particles so that none of the rows line up, 544
Cubic decimeter (dm³) *The derived unit commonly used to measure volumes in the laboratory, see liter*, 22
Cubic lattices, 542–543
Cubic meter (m³) *The derived SI unit used for volume*, 22
Culp, Randy, 62
Cuprite, 586
Curie (Ci) *A larger unit of activity than the becquerel, equivalent to 3.7×10^{10} Bq*, 911
Cyanide leaching method, 126, 158
Cyanidin-based compounds, color of, 233
Cyclexanone, 513
Cyclic alkanes *Alkanes that possess a group of carbon atoms joined in a ring*, 500, 502
Cyclic molecules, 447
Cyclobutane, 500f, 502, 503
Cyclohexane, 500f
freezing-point depression for, 478t
solubility of, 467
Cyclopentane, 500f
Cyclopropane, 500f, 502
Cyclotron, 900
Cysteine, 940f
Cytochromes, 785, 872, 894
Cytosine, 935, 936, 937f

d orbital *An orbital with the quantum number l = 2*, 240
color and, 884, 885, 889–891
in coordination complexes, 884–891
Daguerreotype, 382
Dalton, John, 49–51, 401
Dalton's atomic theory *The theory, developed by John Dalton, that all substances are composed of indivisible atoms*, 49–51
Dalton's law of partial pressures *The total pressure of a mixture of gases is the simple sum of the individual pressures of all the gaseous components*, 401–402
Daughter nuclide *An isotope that is the product of a nuclear reaction*, 904, 923
Davey, Humphrey, 858
Davy, Edmund, 420
d-block elements, 261, 276–277
DDE (dichlorodiphenyldiethene), 631–633
DDT (dichlorodiphenyltrichloroethane), 631–633
10-Deacetylbaccatin III, 528–529
de Broglie, Louis, 228
de Broglie equation, 228–229, 230–231
Decane, 497t
Decay series *A series of nuclear reactions that a large nuclide undergoes as it changes from an unstable and radioactive nucleus to a stable nucleus*, 907–909
Degenerate orbitals *Orbitals that have equal energies*, 236, 239, 364

Degree Celsius (°C) *The unit of temperature on the Celsius scale*, 19

Degree Fahrenheit (°F) *The unit of temperature on the Fahrenheit scale*, 18–19

Dehydrogenation *The removal of hydrogen*, 510–511

Delocalized *Term used to describe a π system wherein the electron density in a molecule can be distributed among more than two atoms*, 233, 384–385

Democritus, 46–47

Denature *To destroy the biochemical activity of a protein*, 946

Density *The mass of a substance that is present in a given volume of the substance. The SI unit for measuring density is kilograms per cubic meter* (kg/m^3), 8, 22, 682

of crystalline solids, 548–549

as derived unit, 22–23

as intensive property, 23–24

of metals, 25

and molar mass of gases, 419

Dental amalgams, 557

Deoxyhemoglobin, 870f, 892

Deoxyribonucleic acid (DNA) *A huge nucleotide polymer that has a double-helical structure. Each strand of the DNA polymer complements the other by forming base pairs. DNA provides instructions for the synthesis of proteins and enzymes that carry out the biological activities of the cell*, 935–939

basic structure of, 70, 935–939

compared with RNA, 941–942

composition of, 47, 65

DNA profiling, 825

effects of radiation on, 911

impact of damage to, 945

location of, 939

as multi-charged ion, 938

nucleotide sequence of, 941

replication process, 938–939

Deoxyribose *A dehydrated carbohydrate used in the construction of DNA*, 935, 937f, 941

Dependent variable, A6

Deposition *The process of a gas undergoing a phase transition to a solid; the opposite of sublimation*, 452

Derived units *Units formed by the combination of SI base units*, 21–23

Desalination *The process that removes dissolved salts from seawater to make potable water*, 10–11, 480–481

Desorbed *Released from a surface*, 656

Deuterium, 57, 58t, 244

Diamagnetism *The ability of a substance to be repelled from a magnetic field. This property arises because all of the electrons in the molecule are paired*, 380, 891

Diamminedichloroplatinum(II), 881–882

Diamond
band gap of, 554
structure of, 493

Diatomic elements *Elements whose normal state is in the form of molecules composed of two atoms attached together (most notably H_2, N_2, O_2, F_2, Cl_2, Br_2, and I_2)*, 51f, 70, 71

MO diagrams for, 378–380

Diazinon, 638

Dibasic salt *A salt that can accept two hydrogen ions*, 748

Dichloroethane, 514

Dichloromethane, 385t, 510

Diet
organic foods, 62–63
power of peanuts, 180–181
quantitative chemistry and, 116
units of Calories, 179–181
vitamins, 116, 950–952

Diethylenetriaminepentaacetic acid (DTPA), 811t

Diethyl ether, 512t, 524

Diffraction pattern *A pattern of constructive and destructive interference after EMR passes through a solid material*, 546

Diffusion *The process by which two or more substances mix*, 424–427
entropy and, 588

Dilithium, 362, 378, 551

Dilution
ideal solutions and, 474
of solutions, 137–139

Dimensional analysis *An extremely useful method for performing calculations by using appropriate conversion factors and allowing units (dimensions) to cancel out, leaving only the desired answer in the desired units*, 26–28

2,2-Dimethylbutane, 446–447

2,3-Dimethylbutane, 446–447

Dimethyl mercury, 876

2,4-Dimethylpentane, 498f

2,2-Dimethylpropane, 498

2,5-Dimethylpyrazine, formulas for, 99–101

Dinitrogen tetroxide, reaction with methylhydrazine, 195–196

Diode lasers, 224t

Dipole–dipole forces, 448

Dipole moment *The polarization of electrons in a molecule that results in a net unequal distribution of charges throughout the molecule*, 344–345, 444

Diprotic *Can produce 2 mol of H^+ when it dissolves*, 151

Diprotic acid *An acid that contains two acidic protons*, 736

Disaccharide *A carbohydrate made from the condensation of two simple sugars*, 955f, 956

Disorder, vs. entropy, 588

Disproportionation *A reaction in which a single reactant is both the oxidizing and reducing agent*, 830–831

Dissociation constant, 683

Distribution constant *The equilibrium constant that describes the partitioning of a solute between two immiscible phases*, 678

DNA. *See* Deoxyribonucleic acid

Dobereiner, Johan, 263

Dodecane, boiling point of, 447–448

Dopamine, 963

Dope *To add an impurity to a pure semiconductor to alter its conductive properties*, 554

Double bond *A covalent bond consisting of two individual bonding pairs of electrons*, 326, 327

Double helix, 936–937

"Dover Boat," 51–52

Drug testing
chromatography for, 678, 679f
mass spectrometer for, 61, 62f

Ductile *Able to be pulled or drawn into a wire*, 552–553, 556

Duet rule *The exception to the octet rule involving the atoms H and He. A full valence shell for the atoms H and He*, 306, 323, 359

Dynamic equilibrium *A term sometimes used in chemistry as synonymous with equilibrium to emphasize that molecular-level equilibrium is not static*, 453

Earth
gravitational force of, 398
structure of, 292–294

EDTA. *See* Ethylenediaminetetraacetic acid

Effusion *The process by which a gas escapes through a small hole*, 424–427

Einstein, Albert, 47, 215, 222

Ekasilicon, 264

Electric current, 17t, 20

Electric eel, 844

Electrochemical cells *A device that allows the exchange between chemical and electrical energy*, 825–826, 844–849
cell notation, 847
commercial batteries, 846, 847–849
concentration cell, 853–854
electrolytic cells, 826
fuel cells, 826, 827f, 830
heart cells, 853–854
in laboratory, 845–846
voltaic cells, 825–826

Electrochemistry *The study of the reduction and oxidation processes that occur at the meeting point of different phases of a system*, 824–860
balancing redox reactions, 836–841
chemical reactivity series and, 849–850
defined, 824–826
determining standard voltage, 844
electrochemical cells, 825–826, 844–849
electrolytic reactions, 857–860

half-reaction potentials, 834–836
manipulating half-cell reactions, 841–844
Nernst equation and, 851–857
overpotential in, 858
oxidation states, 826–833
redox equations, 833–844
of smog, 832–833
standard hydrogen electrode reaction, 835
thermodynamics and, 834–836, 841–844, 855–856

Electrochemists *Scientists who create or analyze systems that allow exchanges between chemical and electrical energy,* 824

Electrode *A metal surface that acts as a collector or distributor for electrons,* 826, 845

Electrodialysis *The process of removing unwanted salts from a solution through a series of semipermeable membranes by applying an electrical charge,* 10–11

Electrodics *The study of the interactions that occur between a solution of electrolytes and an electrical conductor, often a metal,* 825

Electrolysis *A process in which an electric current passing through a solution causes a chemical reaction,* 397, 857–860
applications of, 859
calculations involving, 859–860

Electrolytes *A compound that produces ions when dissolved in water,* 129–131, 144, 542
in electrochemical cells, 845–846
van't Hoff factors for, 476, 477t

Electrolytic cell *A cell that requires the addition of electrical energy to drive the chemical reaction under study,* 826

Electromagnetic radiation *Energy that propagates through space. Examples include visible light, X-rays, and radiowaves,* 170, 211–216
commercial uses of, 215t
emission spectrum and, 216–220
energy of, 215–216
gamma rays, 906
photosynthesis and, 173
radio frequencies, 214
spectral range of, 213
wave–particle duality of, 230–231

Electromagnetic spectrum *The entire range of radiation as a function of wavelength, frequency, or energy,* 211

Electromotive force (emf) *A measure of how strongly a species pulls electrons toward itself in a redox process. Also known as voltage,* 834

Electron affinity *The energy change associated with the addition of an electron to an atom in the gaseous state,* 286–287

Electron capture (EC) *A type of radioactive decay that occurs when an inner-core electron is captured by a proton from the nucleus to form a neutron. The process is usually accompanied by the emission of X-rays,* 907, 919t

Electron configurations *The complete list of filled orbitals in an atom,* 245–250
for coordination complexes, 886–889
gas-phase, 248, 249
ground-state, 245–250
of ions, 305–306
noble gas, 275, 306, 317
octet rule and, 306
periodic table and, 246–250, 261–262

Electronegativity *The relative ability of an atom participating in a chemical bond to attract electrons to itself,* 287, 319
acid strength and, 728
bond dipoles and, 343–345
covalent bonds and, 317, 319–321
Pauling scale, 287–289, 320
periodicity of, 287–289

Electron-group geometry *The positions of the groups of electrons (lone pairs and bonding pairs) around a central atom in three dimensions,* 337–341

Electronic transition *A change in atomic or molecular energy level made by an electron bound in an atom or molecule,* 221–222

Electron shielding *The ability of electrons in lower energy orbitals to decrease the nuclear charge felt by electrons in higher energy orbitals,* 244–245, 310–311

Electrons *One of the subatomic particles of which atoms are composed. Electrons carry a charge of −1,*
discovery of, 52–53
energy levels of, 220–221
in ion formation, 67
properties of, 55, 56
wave nature of, 227–229
wave–particle duality of, 227

Electron spin, 242

Electron spin quantum number *The quantum number, m_s, that describes the spin of an electron. Also known as the spin angular momentum quantum number,* 236, 242, 243t

Electron subshell *The energy level occupied by electrons that share the same values for both n and l,* 235, 246

Electrophorus, 826f

Electroplates, 844

Electroplating *The process of depositing metals onto a conducting surface,* 125–126, 859–860

Electroweak force *The fundamental force responsible for the attraction between objects carrying opposite electric charges, for the repulsion between objects carrying the same electric charge, and for some transformations within subatomic particles. It is also responsible for the phenomena of magnetism and light,* 169t, 172

Electrowinning *The isolation of pure metals from a solution of metal ions,* 857–858

Elementary step *A single step in a mechanism that indicates how reactants proceed toward products,* 648–649

Elements *A substance that contains only one kind of atom. All elements are listed in the periodic table of the elements,* 5
in the environment, 292–294
of human life, 277–278
periodic table of, 64–66
reference form of, 187, 189
superheavy, 918

Emerald, 884

Emission spectroscopy *The measurement of how atoms and molecules give off electromagnetic radiation as a function of the wavelength or frequency of the radiation,* 216

Emission spectrum *A plot of the intensity of radiation as a function of wavelength or frequency in an emission experiment,* 216–217
for different metals, 220f
for hydrogen, 217–219, 222–223
solar, 216, 217f

Empirical formula *A chemical formula indicating the ratio in which elements are found to be present in a particular compound, regardless of the molecular structure of the compound,* 71, 99–102

Empirically derived *Derived from experiments and observations rather than from theory,* 219

Enantiomers *A pair of molecules that are nonsuperimposable mirror images,* 960

Endorphins, 963, 964f

Endoscope, 559–560

Endothermic reaction *A reaction that absorbs energy from the surroundings,* 175, 652, 703

End point *In a titration, the volume of the added reactant that causes a visual change in the color of the indicator,* 140, 801

Energy, 168–200
activation, 174
binding, 915–917
calorimetry for measuring, 182–185
chemical, 172
concept of, 169–178, 444
as derived unit, 22t
electromagnetic radiation, 170, 173, 215–216
enthalpy and, 186–191
as extensive property, 23t
heat and reaction profile diagrams, 174–175
heat capacity, 182–185
Hess's law, 191–197
human consumption of, 197–199
internal, 177–178
law of conservation of, 171, 178
vs. matter, 4–5

Energy (continued)
  potential vs. kinetic, 170–174
  renewable sources of, 198–199
  specific heat capacity, 181–185
  standard enthalpies of reaction, 187–190
  as state function, 191
  units of, 176, 179–181, 210
  work and, 169, 176–177
Energy levels *The allowed orbits that
    electrons may occupy in an atom,* 221
  in Bohr's model, 220–221, 234–235
  in multielectron atoms, 244–245
  periodicity of, 262
  wave functions and, 232–233
English system, of measurement, 16
Enthalpy *A thermodynamic quantity
    symbolized by H and defined as
    H = U + PV,* 186–188
  bond dissociation energies and,
    332–333
  of combustion, 334, 358
  Hess's law and, 191–197
  lattice, 313–315
  manipulating, 190–191
  reaction enthalpies from enthalpies of
    formation, 195–197
  standard enthalpies of reaction, 187–190
  standard enthalpy of combustion,
    189–191
  standard enthalpy of formation, 188–189
  as state function, 191–192
Enthalpy of bond dissociation ($\Delta_{diss}H$)
    *The enthalpy change related to breaking
    1 mol of bonds in a gaseous species. Also
    known as the bond dissociation energy,*
    332–333
Entropy *A measure of how the energy and
    matter of a system are distributed
    throughout the system,* 587–588
  calculating changes in, 595–600
  phase changes and, 591–595
  second law of thermodynamics and,
    588–590
  standard molar, 598–599
  third law of thermodynamics and, 598
Environmental chemist *A scientist who
    studies the interactions of compounds in
    the environment,* 620–621, 654
Environmental concerns
  acid deposition, 757–758
  automobiles, 430
  clean water, 135–137, 157–159, 470, 481
  future challenges, 35
  Green Chemistry and, 571–572
  greenhouse effect, 197, 431–432
  heavy metals, 145–146, 266
  mercury, 876–877
  nitrogen oxides, 757–758, 832–833
  ozone depletion, 35, 427–431, 629, 640,
    651, 653, 655
  sulfur oxides, 273
  thermal pollution, 471
  use of zeolites for, 310, 572
  water consumption, 441–443

Environmental Protection Agency (EPA),
    135, 157, 470, 620
Environment *The world around us,
    comprising the Earth and its
    atmosphere. We ourselves are part of the
    environment,* 292–294
Enzymes, 277, 278
  buffers and, 785
  in cells, 588, 589f, 590
  in chemistry of life, 947–952
  chirality of, 960
  classes of reactions for, 949
  in DNA replication, 938
  of extremophiles, 592–593
  proteins as, 953
  regulation of, 948–949
  structure of, 545, 946
Ephedrines, 746
Epothilones, 530
Epoxide *A compound containing a
    triangular ring of two carbon atoms
    linked by an oxygen atom, known as the
    epoxide group,* 524
Epoxy resins *Polymer resins containing the
    epoxide group, in which two carbon
    atoms and an oxygen atom form a cyclic
    ether ring,* 524
Equations. *See* Chemical equations
Equatorial position *The position of a
    portion of a molecule when it is arranged
    along the x-axis or y-axis of the
    molecule,* 340, 342
Equilibrium *The state of a reaction when
    $\Delta G = 0$. The reaction hasn't stopped;
    rather, the rates of the forward and
    reverse reactions are equal. A system of
    reversible reactions in which the forward
    and reverse reactions occur at equal
    rates, such that no net change in the
    concentrations occurs, even though
    both reactions continue,* 670
  calculating equilibrium concentrations,
    672, 685–686
  chromatographic analysis and, 678–679
  concentration changes, 701–702
  concept of, 608–609, 670–675
  determining important reactions, 688
  dynamic, 453
  effect of catalysts on, 704
  free energy and, 607–612, 670, 671, 692,
    705–707
  heterogeneous equilibrium, 682
  homogeneous equilibrium, 682
  ICEA tables, 691–692, 695–696, 701
  importance of, 675–676
  Le Châtelier's principle, 700–705
  mass-action expression, 672–675
  meaning of equilibrium constant,
    676–682
  predicting direction of reaction, 688–690
  pressure changes, 702–703
  significant processes involving, 676
  solving equilibrium problems, 687–700
  temperature changes, 703

  using partial pressures, 686–687
  vapor pressure, 453
  working with equilibrium constants,
    682–687
  See also Aqueous equilibria
Equilibrium concentrations, 672–675,
    685–686
Equilibrium constants (K) *The value of the
    equilibrium expression when it is solved
    using the equilibrium concentrations of
    reactants and products,* 672–673
  acid dissociation constant, 724–726
  calculating equilibrium concentrations
    from, 685–686
  in combined reactions, 684–685
  distribution constant, 678
  formation constant, 809–810
  free energy and, 705–707
  intermediate values of, 698–700
  large values of, 695–698
  list of selected, A13–A16
  meaning of, 676–682
  Nernst equation and, 855–857
  reaction quotient and, 688–689
  small values of, 691–695
  solubility product constant, 802
  using partial pressures, 686–687
  working with, 682–687
Equilibrium expression *The ratio of
    product to reactant concentrations raised
    to the power of their stoichiometric
    coefficients. This expression relates the
    equilibrium concentrations to the
    equilibrium constant. It is also known as
    the mass-action expression,* 672–675
Equilibrium vapor pressure *The pressure
    exerted by the vapor of a liquid or solid
    under equilibrium conditions,* 453
Equivalence point *The exact point at which
    the reactant in a titration has been
    neutralized (completely consumed) by
    the titrant,* 140, 789–790, 796, 800–801
Escherichia coli bacteria, 876
Esterification *The process by which an ester
    is formed when the —OH group of an
    alcohol participates in a condensation
    reaction with the —COOH group of a
    carboxylic acid,* 522
Esters *Compounds formed when an alcohol
    and a carboxylic acid become bonded as
    a result of a condensation reaction,* 512t,
    520–522
Estrogen, 958, 963
Ethanal (Acetaldehyde), 519–520
Ethane
  boiling point of, 446, 450–451
  bonding pairs in, 357
  combustion of, 156–157, 358
  formation of, 501, 656f
  reaction with steam, 106, 110
  structure of, 497
1,2-Ethanediol, 517, 518t
Ethanoate, 520f
Ethanoic acid. *See* Acetic acid

Ethanoic anhydride
    formation of, 104–105, 108–109
    in synthesis of aspirin, 103–104
Ethanol, 512*t*
    boiling-point elevation of, 476*t*
    boiling point of, 450–451
    composition of, 5
    in cough syrups, 316*f*, 317
    dissolution in water, 128
    formation of, 514, 516–517, 603, 610
    in gasoline, 509
    oxidation of, 519
    properties of, 6
    solubility of, 467
    specific heat of, 181*t*
Ethene (ethylene), 420*f*, 511, 512*t*
    bonding patterns in, 370–371
    combustion of, 358*t*
    as feedstock molecule, 514, 516
    hydrogenation of, 501, 514, 656*f*
    polymerization of, 514–515
Ethers *Any compound containing the ether functional group (—C—O—C—), 512t, 524*
Ethics in science, 336
Ethidium bromide, 309
Ethoxide, 520*f*
Ethylene. *See* Ethene
Ethylenediamine, 721, 874
Ethylenediaminetetraacetic acid (EDTA)
    complex formation with, 810–813, 875
    structure of, 810*f*, 811*t*
    titrations with, 681, 767, 810–815
Ethylene glycol
    boiling point of, 450–451, 477
    solubility of, 467
    structure of, 518*t*
    viscosity of, 462
Ethyl methyl ether, 524
Ethyne. *See* Acetylene
Evaporation *The process of a liquid undergoing a phase transition to a gas; the opposite of condensation, 452–453*
Exact number *A number that can be known with absolute certainty. Exact numbers possess an infinite number of significant figures, 28*
Excited state *Any higher energy state than the ground state characterized by the existence of an electron in an orbital that violates Hund's Rule and/or the Aufbau principle, 221–222*
Exothermic reaction *A reaction that releases energy into the surroundings, 174–175, 652, 703*
Expanded octet *The existence of more than eight electrons in the valence shell of an atom. Atoms of atomic number greater than 12 are capable of this feat. In the third row of the periodic table, it typically occurs in sulfur and phosphorus, 330, 368–369*
Experiments, 13*f*, 14
Exponential notations, A1–A3

Extensive property *A characteristic of a substance that is dependent on the quantity of that substance, 23–24*
Extremophiles, 592–593
Exxon Valdez, 507

f **orbital** *An orbital with quantum number l = 3, 241*
Face-centered cubic (fcc) unit cell *A unit cell built via the addition of an atom, ion, or molecule at the center of each side of the simple cubic unit cell, 543, 546, 547–549*
Fahrenheit, Gabriel Daniel, 19
Fahrenheit scale, 18–20, 21
Fairclough-Baity, Judith, 3, 12, 16, 22, 28, 31, 32
Faraday, Michael, 835, 859
Faraday's constant *A unit of electric charge equal to the magnitude of charge on a mole of electrons, 835–836, 859*
Fats *A triglyceride that is made from carboxylic acids with long carbon chains and has a relatively high melting point. Typically derived from animal sources, 373, 957*
Fatty acids *Carboxylic acids that contain long carbon chains. Commonly serve as major building blocks for the production of cell membranes, 373, 957*
    hydrogenated, 109, 110, 504
    list of common, 957*t*
f-block elements, 261, 276
Federal Communications Commission (FCC), 214
Fentanyl, 747*t*
Fermentation *The biological process that converts glucose into ethanol and carbon dioxide, 516–517*
Fermi, Enrico, 900, 920
Ferredoxins, 872
Ferromagnetism *A property of a compound that occurs when paramagnetic atoms are close enough to each other (such as in iron) that they reinforce their attraction to the magnetic field, such that the whole is, in effect, greater than the sum of its parts, 379–380, 891*
Fiber optic wire, 559–560
Fibers *A polymer whose chains are aligned in one direction, 563–567*
First ionization energy (I_1) *The energy required to remove one electron from each atom of an element in the gaseous state, 282*
First law of thermodynamics *The total change in the closed system's energy in a chemical process is equal to the heat flow (q) into the system and the work done (w) on the system: $\Delta U = q + w$. Energy is neither created nor destroyed; it is only transferred from place to place and converted from one form into another:*

$\Delta U_{system} + \Delta U_{surroundings} = 0$, 177–178, 589
First order *A reaction is first order in a particular species when the rate of the reaction depends on the concentration of that species raised to the first power, 628, 636–637, 644, 647t*
Fission (nuclear fission) *A type of nuclear reaction wherein a large nucleus splits into two or three smaller nuclei with the release of energy, 917, 920–924*
Flavor analysis, 62–63
Flight Data Recorder (FDR), 824–825
Flue-gas desulfurization *A process used to remove sulfur dioxide (and other sulfur oxides) from combustion smoke, 779*
Fluorapatite, 313, 315
Fluoride ion
    in drinking water, 137–139, 313
    electron configuration of, 305
    Lewis dot symbol for, 305
Fluorine
    boiling point of, 446
    electron configuration of, 246
    interaction with light, 358, 380
    oxidation numbers for, 155*t*
    reduction potential of, 835
    thermodynamic data for, A10
    uses of, 274*t*
Fluorine-18, 900
Fluoroapatite, 747, 748
Fluorosulfuric acid, 729
Food industry
    catalysts used in, 657
    flavor analysis, 62-63
    food packaging, 401, 407, 412, 426, 427*f*
    pH meters, 772
    use of chelates in, 812*t*
Force *The mass of an object multiplied by its acceleration, 397*
    calculating, 398
    as derived unit, 22*t*
    fundamental, 169
    pressure and, 398–399
    units of, 397–398
Formal charge *The difference between the number of valence electrons on the free atom and the number of electrons assigned to that atom in the molecule, 323–325, 326, 329*
Formaldehyde, 512*t*
    formation of, 517
    molecular geometry of, 341
Formalin, 517
Formamide, 513*t*
Formation constant (K_f) *The equilibrium constant describing the formation of a stable complex. Typically K_f values are large. Also known as the stability constant, 809–810, 811t, A16*
    conditional, 813–814
Formic acid, 512*t*
    in aqueous solution, 721
    electron density map of, 829*f*

Formic acid (continued)
freezing-point depression for, 478*t*
oxidation states for, 827, 829
Formic acid–formate ion buffer system,
774–775, 779–782, 783–784
Formula mass *The total mass of all the*
atoms present in the formula of an
ionic compound, in atomic mass units
(amu) or the mass of one mole of
formula units in grams per mole,
87–89, 92
Fossil fuel energy, 174, 197–198, 492
greenhouse effect and, 431–432
Fossils, 45, 46
Fractional distillation, 506–508
Francium, 270*t*
diameter of, 21
properties of, 284
Franklin, Rosalind, 937
Free energy (*G*) *The state function that is*
equal to the enthalpy minus the
temperature multiplied by the entropy.
Used to determine the spontaneity of a
process, 600–606
concentration and, 611
equilibrium and, 607–612, 670, 671, 692,
705–707
half-reaction potentials and, 835–836,
841–843, 851
pressure and, 609–610
temperature and, 608–609
Free energy of formation, 605–606
Free radicals *Chemical species that carry an*
unpaired electron and so are highly
reactive, 510
Freezing-point depression *The lowering*
of the melting point of a compound due
to the presence of a dissolved solute,
478, A13
Frequency *A descriptor of electromagnetic*
radiation. Defined as the number of
waves that pass a given point per second,
in units of 1/s (s^{-1}), 211–215
Frequency factor *A term in the Arrhenius*
equation that indicates the rate of
collision and the probability that
colliding reactants are oriented for a
successful reaction, 652
Fructose, 469, 585, 955–956
Fuel cells *An electrochemical cell that utilizes*
continually replaced oxidizing and
reducing reagents to produce electricity,
826, 827*f*, 830, 848*t*
Fuel sources
alternative, 334
as carbon based, 492
chemical challenges for, 35
fossil fuels, 197–198
human consumption of, 197–199, 492
hydrocarbons, 497
nuclear reactors, 921–922
photovoltaic devices, 555
renewable, 198–199
Fugacities (*f*), 706

Fullerenes *Forms of carbon based on the*
structure of buckminsterfullerene, C_{60},
493–494
Functional groups *Groups of atoms, or*
arrangements of bonds, that bestow a
specific set of chemical and physical
properties on any compound that
contains them, 511–514, 530
2-Furylmethanethiol, 88
Fusion (nuclear fusion) *A type of nuclear*
reaction wherein small nuclei are joined
together to form a larger nucleus with
the release of energy. Nuclear fusion
powers the stars, 916
Fusion, heat of, 457

Galactose, 469
Galena, 261, 262*f*
Gallium, 264, 271*t*
Galvanic cell *A cell that produces electricity*
from a chemical reaction. Also known as
a voltaic cell, 825–826
Gamma rays (γ rays) *A high-energy form of*
electromagnetic radiation that is emitted
from the nucleus. Gamma rays
sometimes accompany alpha and beta
decays, 54–55, 213*f*, 215, 904, 906, 919*t*
penetration capability of, 909–910
Gas chromatography (GC) *Analytical*
technique in which components of a
mixture are separated by their different
flow rates through a chromatography
column, driven by a carrier gas. A specific
chromatography technique in which the
mobile phase is a gas and the stationary
phase is a solid, 62–63, 507, 678
Gases *The phase of a substance characterized*
by widely spaced components exhibiting
low density, ease of flow, and the ability
to occupy an enclosed space in its
entirety, 10, 107, 443
amount–volume relationship, 404–405,
423
behavior and applications of, 394–432
behaviors of real, 416–417
characteristics of, 396–397, 422, 443
compressibility of, 396–397
effects of industrial use of, 427–432
effusion and diffusion of, 424–427
equilibrium constants for, 686
Gay-Lussac's observations of, 50–51,
408–409
ideal, 396, 402
industrial, 395
kinetic-molecular theory of, 422–424
measures of pressure, 397–401
molar mass and density of, 418–419
molecular speeds of, 423–425
partial pressures, 401–402
pressure–volume relationship, 15–16,
405–408, 422
solubility of, 471, 472–473
temperature–volume relationship,
408–411, 423

Gas laws, 13, 403–413
Avogadro's law, 404–405, 423
Boyle's law, 15–16, 405–408, 422
Charles's law, 408–411, 423
combined gas equation, 411–412
Dalton's law of partial pressures, 401–402
Graham's law of effusion, 426
ideal gas equation, 413–422
kinetic-molecular theory and, 422–424
van der Waals equation, 416–417
See also Ideal gas equation
Gas-liquid chromatography (GLC) *A form*
of gas chromatography in which the solid
phase is covered with a thin coating of
liquid, 507
Gasoline, 508*t*
chromatography analysis of, 678, 679*f*
composition of, 509
reforming of, 508–509
Gauge pressure, 399, 400
Gay-Lussac, Joseph Louis, 50–51, 408–409
Gel electrophoresis *The use of electric fields*
to separate ions, 825
Gene *A specific region of the DNA polymer*
that codes (carries instructions) for a
single protein, 941
Gene therapy *A relatively new area of*
biochemistry and medicine wherein
attempts are made to cure or treat
genetic diseases by modification of a
patient's damaged genes, 964
Genetic code
examining, 944–945
protein formation and, 941–945
structure of DNA, 935–939
summary of, 943*t*
transcription process, 942
translation process, 943–944
Genetic disease, 964
Geometric isomers *Molecules that differ in*
their geometry but have identical
chemical formula and points of
attachment. Substances in which all of
the atoms are attached with the same
connectivity, or bonds, but the geometric
orientation differs. See also cis; trans,
502–504, 880–881
Geothermal systems *Power-generating*
systems that involve drilling deep into
the Earth and exploiting the flow of heat
from the interior of the Earth out toward
the surface, 197*f*, 199
Germanium
band gap of, 553*t*, 554
properties of, 264, 268–269
uses of, 272
Gibbs, Josiah Willard, 601
Gibbs equation $\Delta G = \Delta H - T\Delta S$,
601–602
Gillespie, Ronald J., 337, 341
Glass *An amorphous solid ceramic material,*
559–560
Glass etching, 323, 359
Global positioning systems (GPS), 211–212

Global warming
 chemical challenges for, 35
 energy choice and, 197
 See also Environmental concerns
Globular proteins, 946
Glucose, 469
 combustion of, 184–185, 587, 595–596, 599–600
 dissolution in water, 128
 as energy source, 580, 591
 fermentation of, 603, 610
 in glycolysis pathway, 585, 587, 595, 607
 as molecular solid, 542
 as nonelectrolyte, 129
 as product of photosynthesis, 111
 properties of, 955–956
Glucose-1-phosphate, 604–605
Glucose-6-phosphate, 604–605, 607
D-glucose isomerase, 657
Glutamate, 940*f*
Glutamic acid, 755*f*
Glutamine, 940*f*
Glycerol
 structure of, 518*t*
 vapor pressure of, 474
Glycine, 755*f*, 940*f*
Glycogen, 956
Glycolysis *A series of biochemical reactions that convert glucose into two molecules of pyruvate. The process results in the formation of two molecules of ATP,* 585, 591, 595, 600–601, 604, 611
Glycoproteins, 948
Glyphosate herbicides, 641
Gold
 density of, 548–549
 electroplating with, 860
 extraction of, 124, 125–126, 557, 719
 process of dissolving, 844
 properties of, 5, 23, 64, 125, 260
 reactivity of, 283*f*, 289, 850
 structure of, 543*f*, 544
 wastewater from mining, 148, 149*f*, 158
Good (Good's) buffers *Buffers typically used in biochemical research because they are chemically stable in the presence of enzymes or visible light and are easy to prepare,* 785
Good, Norman, 785
Graham, Thomas, 425
Graham's law of effusion *The rates of the effusion of gases are inversely proportional to the square roots of their molar masses,* 426
Graphing, A6
Graphite, 493
Gravimetric analysis *A laboratory technique in which the concentration of substances in solution is determined by forming insoluble salt precipitates and weighing them or their related solids,* 806–807
Gravitational force *The fundamental force that causes all objects with mass to be*

attracted to one another, 169*t*, 172, 397–398
Gray (Gy) *A measure of absorbed radiation equal to 100 rad,* 911
Green Chemistry *The practice of chemistry using environmentally benign methods,* 571–572
Greenhouse effect *The re-radiation of energy as heat from gases in the atmosphere back toward Earth. An increase in the greenhouse effect leads to an increase in global temperatures,* 35, 431–432
Greenockite, 555
Ground state *The lowest energy state of an atom,* 221–222, 238, 245
Group *One of the vertical columns in the periodic table of the elements,* 64, 262, 269–278
Group IA, 65–66, 270. *See also* Alkali metals
Group IB, 66*t*
Group IIA, 66*t*, 270–271. *See also* Alkaline earth metals
Group IIIA, 68, 74, 271
Group IVA, 272
Group VA, 68, 272
Group VIA, 66*t*, 68, 273
Group VIIA, 66*t*, 274. *See also* Halogens
Group VIIIA, 65, 66*t*, 274–275. *See also* Noble gases
Guanine, 935, 936, 937*f*
Gypsum *Calcium sulfate dihydrate, $CaSO_4 \cdot 2H_2O$,* 719, 748, 779

Haber, Fritz, 313, 704
Haber process, 156, 597, 610, 625, 673, 702–703, 704
Hahn, Otto, 920
Half-life ($t_{1/2}$) *The time required for a reaction to reach 50% completion,* 636–638, 639, 640, 912–915
Half-reactions *An incomplete equation that describes the oxidation or reduction portion of a redox reaction,* 153, 826
 standard potential of, 834–836, A17–A19
Hall, Charles M., 858
Hall–Heroult process *The most widely used process for the preparation of aluminum from bauxite,* 493, 858
Halogens *The elements of Group VIIA of the periodic table,* 66*t*
 commercial uses of, 274*t*
 electron affinity values for, 286
 interaction with light, 358, 380
 properties of, 274
 in substitution reactions, 510
Halons *Compounds containing carbon and halogens. Halons are typically used in fire extinguishing applications but are considered harmful to the ozone layer,* 429
Hamilton, William, 232
Hamiltonian operator *A mathematical function that is used in the Schrödinger equation,* 232, 234

Hanford Nuclear Reservation, 633, 636
Hardness, of materials, 559*t*
Hard water, 309–310, 767, 814
Hasselbalch, K. A., 777–778
Heart disease, 331
Heat (q) *The energy that is exchanged between a system and its surroundings because of a difference in temperature between the two,* 174–175, 178
Heat capacity *The amount of heat needed to raise the temperature of any particular amount of a substance by 1°C,* 181–185
Heat desalination, 10, 11*f*
Heating curve *A plot of the temperature of a compound versus time or energy as heat is added at constant pressure. The plot indicates the specific temperature ranges for solid, liquid, and gas phases,* 456–458
Heat of fusion ($\Delta_{fus}H$) *The enthalpy change associated with the phase transition from liquid to solid,* 457
Heat of reaction *The energy as heat released or absorbed during the course of a reaction,* 186–187
Heat of solution ($\Delta_{sol}H$) *The enthalpy change associated with the dissolution of a solute into a solvent,* 466
Heat of vaporization ($\Delta_{vap}H$) *The enthalpy change associated with the phase transition from liquid to a gas,* 457–458
Heavy metals, 145–146, 266, 280
Heisenberg, Werner, 230
Heisenberg uncertainty principle *There is an ultimate uncertainty in the position and the momentum of a particle. Reducing the uncertainty in one increases the uncertainty in the other,* 229–230, 233
Helium
 atomic orbitals of, 244
 binding energy of, 916
 discovery of, 224
 electron configuration of, 246
 emission spectrum for, 220*f*
 molecular speed for, 424, 425
 properties of, 5, 65, 262, 275
 uses of, 274*t*, 275, 403
Helium–cadmium lasers, 224*t*
Heme group *A compound that, when bound to hemoglobin and iron cations, is responsible for binding oxygen,* 669, 869–870
Hemoglobin, 755
 chemistry of, 871
 formulas for, 101–102
 iron in, 260, 291, 869–870, 877, 892
 structure of, 946, 947*f*
Henderson, Lawrence Joseph, 777
Henderson–Hasselbalch equation *A shorthand equation used to determine the pH of a buffer solution. $pH = pK_a + \log(base/acid)$,* 777–779
Henry's law *The solubility of a gas is directly proportional to the pressure that the gas exerts above the solution,* 472–473

Heptane, 497t, 509

Herbicide *Any compound used to kill unwanted plants,* 619, 620, 641

Heroin, 15

Heroult, Paul, 858

Hertz (Hz), 211

Hess's law *Thermodynamic law stating that the enthalpy change of a chemical reaction is independent of the chemical path or mechanism involved in the reaction. This enables us to determine the enthalpy changes of reactions that might be very difficult to actually perform,* 191–197, 313

Heterogeneous catalyst *A specific catalyst that exists in a different physical state than the reaction,* 656

Heterogeneous equilibrium *An equilibrium that results from reactants and products in different phases or physical states,* 682

Heterogeneous mixture *A mixture that is not uniformly mixed, so there are different proportions of the components in different parts of the mixture,* 8–9

Hexaamminecobalt(III) chloride, 882

Hexaamminenickel(II), 892

Hexachlorobenzene, 136

Hexadentate ligand *A ligand that makes six coordinate covalent bonds to a metal ion,* 810

Hexagonal closest packed (hcp) structure *A closest packed structure made by staggering a second and third row of particles in such a way that the first and third rows line up,* 544

Hexane, 497t
  boiling point of, 446–447
  isomers of, 446–447, 498–499
  meniscus of, 464
  properties of, 385t

1,3,5-Hexatriene, 381

Hexose sugars, 955

High-density polyethylene (HDPE) *A form of polyethylene in which the molecules are linear, with no branch points,* 515

Highest-energy occupied molecular orbital (HOMO) *The most energetic molecular orbital that contains at least one electron,* 378

High-fructose corn syrup, 657

High-spin *A coordination complex with the maximum number of unpaired electrons,* 888

Histidine, 940f

Hoffman, Felix, 15, 322

Homogeneous catalyst *A specific catalyst that exists in the same physical state as the compounds in the reaction,* 655–656

Homogeneous equilibrium *An equilibrium that results from reactants and products in the same phase, or physical state,* 682

Homogeneous mixture *A mixture that is uniformly mixed, so that it has the same*
composition throughout. Also known as a solution, 8–9, 126. *See also* Solutions

Homologous series *A series of compounds sharing the same general formula,* 497, 501

Hormones *Small molecules, such as steroids, used by living organisms for intercellular communication,* 953, 963

Housing starts, 718f

Human eye, 355, 356–357

Human Genome Project, 935, 964

Human growth hormone, 755

Hund, Friedrich, 247

Hund's rule *When orbitals of equal energy are available, the lowest energy configuration for an atom has the maximum number of unpaired electrons with parallel spins,* 247, 370, 378

Hybridization *The mathematical combination of two or more orbitals to provide new orbitals of equal energy,* 363–374
  advantages and disadvantages of, 372–373
  algorithm for, 364t
  application of, 371–372
  definition of, 363–364
  predicting state of molecule from, 373
  shapes of hybrids, 367–368
  sigma and pi bonds, 369–371
  sp, sp^2, sp^3 orbitals, 364–367
  sp^3d orbitals, 368–369

Hydration *The interaction of water (as the solvent) with dissolved ions,* 466, 479

Hydration sphere *The shell of water molecules surrounding a dissolved molecule, ion, or other compound. This shell arises because of the force of attraction between the water molecules and the solute,* 127–128

Hydrazine, combustion of, 188–189

Hydrides, bonding in, 360, 361t

Hydrocarbons *Compounds composed of only hydrogen atoms and carbon atoms,* 496–506
  alkanes, 496–500, 510–511
  alkenes, 500–501
  alkyl groups, 504–506
  alkynes, 501
  aromatic, 504
  in crude oil, 495t, 507
  fractional distillation of, 506–508
  functional groups and, 511–514
  in gasoline, 509
  processing, 508–509
  saturated, 496
  unsaturated, 501

Hydrochloric acid
  applications for, 148–149
  dilution of, 139
  dissolution in water, 330, 720–721
  pH in solution, 736
  properties of, 149, 151
  reaction with sodium carbonate, 145
reaction with sodium hydroxide, 143, 144, 150–151, 787–792, 831
  reaction with zinc, 402
  in solution with acetic acid, 742–743
  strength of, 724, 725–726, 728
  titration with ammonia, 796–798

Hydrochlorofluorocarbons (HCFCs), 431

Hydrochlorothiazide (HCT), 94

Hydrocracking *A type of cracking that uses heat plus a catalyst in the presence of hydrogen gas,* 508

Hydroelectric systems *Power-generating systems in which the power of falling water is used to generate electricity,* 197f, 199

Hydrofluoric acid, 719
  dissociation of, 727
  pH of solution of, 740–742
  strength of, 728

Hydrofluorosilicic acid (HFSA), 137–139

Hydrogen
  as alternative fuel source, 334
  atomic orbitals of, 235, 236, 238–242, 244
  Bohr's model of, 220–224, 228
  combustion of, 358t
  diameter of, 21
  electron configuration of, 245–246, 359
  emission spectrum for, 217–219, 220f, 222–223
  in environment, 294
  industrial synthesis of, 110
  isotopes of, 57–58
  oxidation number of, 154, 155t, 827
  properties of, 270
  reaction with nitrogen, 156
  reaction with oxygen, 6–7, 10, 50–51
  as reducing agent, 830
  thermodynamic data for, A10
  uses of, 270t

Hydrogenation *The addition of a molecule of hydrogen to a compound,* 109–110, 501, 504

Hydrogen bond *A particularly strong intermolecular force of attraction between F, O, and/or N and a hydrogen atom,* 448–451, 463
  in DNA, 936, 938

Hydrogen chloride, 129

Hydrogen cyanide, bond dipoles in, 344–345

Hydrogen fluoride
  bond dipole in, 344
  electron density of, 318f, 344
  hydrogen bonds in, 450f
  Lewis dot structure for, 323, 324
  polar covalent bonds in, 318–319, 320
  uses of, 323
  valence bond model of, 359

Hydrogen fuel cell, 848t

Hydrogen gas
  combustion of, 168
  covalent bonding in, 318–319
  growing demand for, 110
  production of, 106, 360, 397, 402, 674

reaction with oleic acid, 109
uses of, 395*t*
Hydrogen iodide, 672
Hydrogen ion
    in acidic solutions, 128, 148, 719–720
    in Brønsted–Lowry acids and bases,
        720–722
    equilibrium concentration of,
        729–730
    in formation of water, 150–151
    pH and, 730–735
Hydrogen molecule
    Lewis dot structure for, 322–323
    MO diagram for, 378
    valence bond model of, 359
Hydrogen peroxide
    decomposition of, 624–625, 627–628,
        633–635, 655, 830–831
    formulas for, 71*f*
Hydrolases, 949
Hydronium ion. *See* Hydrogen ion
Hydrosphere, 292*f*, 293
Hydroxide ion
    in acidic solutions, 737
    in basic solutions, 129, 148, 719
    in formation of water, 150–151
    Lewis dot structure for, 325
Hydroxides, solubility of, 744
Hydroxyapatite, 313, 315, 557
Hydroxyatrazine, 620–621
N-(hydroxyethyl)-ethylenediaminetriacetic
    acid (HEDTA), 811*t*
Hydroxyl free radicals, 911
Hydroxyl group *The —OH functional
    group,* 514, 517
bis-Hydroxymethoxychlor, 647
Hypertonic *Containing a concentration of
    ions greater than that to which it is
    judged against,* 480
Hypochlorous acid, 727
Hypothesis *A tentative explanation for an
    observation—that is, a statement about
    what we think might be an explanation
    for an observation,* 12, 13*t*
Hypotonic *Containing a concentration of
    ions lower than that to which it is judged
    against,* 480

Ice
    fishing, 473
    structure of, 459, 461
ICEA tables, 691–692, 695–696, 701
Ideal gas constant *R, the constant used in the
    ideal gas equation. R = 0.08206 L·atm/
    mol·K,* 414, 479
Ideal gas equation *The equation relating the
    pressure, volume, moles, and temperature
    of an ideal gas,* 413–422, 686
    chemical reactions using, 419–422
    compressed gas and, 414–415
    physical change applications, 418–419
    pressure and volume deviations,
        416–417

Ideal gases *Gases that behave as though each
    particle has no interactions with any
    other particle and occupies no molar
    volume,* 396, 402
Ideal solution *A solution in which the
    properties of the solute and the solvent
    are not changed by dilution. This means
    that other than being diluted, combining
    solute and solvent in an ideal solution
    does not release or absorb heat, and the
    total volume in the solution is the sum
    of the volumes of the solute and
    solvent,* 474
Immune system, 954
Independent variable, A6
Indium, 271*t*
Induced dipole *A dipole produced in a
    compound as a consequence of its
    interaction with an adjacent dipole,* 446
Industrial gases *Gases used in industrial or
    commercial applications,* 395
    impact on atmosphere, 427–432
Inert *The opposite of labile; said of a
    compound that is very slow to exchange
    ligands,* 893
Inert gases *The Group VIIIA elements. Also
    known as noble gases,* 65, 274, 279–280.
    See also Noble gases
Infrared spectrophotometer, 345
Infrared radiation, 213*f*, 215
in Hg *A unit used to measure pressure.
    29.921 in Hg equals 1 atmosphere,* 399
Initial rate *The instantaneous rate of
    reaction when t = 0,* 623, 626, 627
    determining rate law from, 641–643
Inner transition elements *The elements in
    the f-block of the periodic table,
    consisting of the lanthanides and
    actinides,* 66, 250, 275–276
Insecticide *Any compound used to kill
    unwanted insects,* 620–621
Insoluble *Not capable of dissolving in a
    solvent to an appreciable extent,* 145
Instantaneous rate *The rate of reaction
    measured at a specific instant in time.
    The instantaneous rate is typically
    measured by calculating the slope of a line
    that is tangent to the curve drawn by
    plotting [reactant] versus time,* 623,
        626, 627
Insulators *A substance that contains a very
    large band gap,* 554
    aerogels, 573
    ceramic, 557–559
Insulin, 953, 963
Integrated first-order rate law *An expression
    derived from a first-order rate law that
    illustrates how the concentrations of
    reactants vary as a function of time,*
        633–636
Integrated rate laws *A form of the rate law
    that illustrates how the concentration of
    reactants vary as a function of time,*
        631–640

Integrated second-order rate law *An
    expression derived from a second-order
    rate law that illustrates how the
    concentrations of reactants vary as a
    function of time,* 639–640
Integrated zero-order rate law *An
    expression derived from zero-order rate
    law that illustrates how the
    concentrations of reactants vary as
    function of time,* 639
Intensive property *A characteristic of a
    substance that is independent of the
    quantity of that substance,* 23–24
Intermediate *A compound that is produced
    in a reaction and then consumed in the
    reaction. Intermediates are not indicated
    in the overall reaction equation,* 650
Intermolecular forces *A force of attraction
    between two molecules,* 444
    capillary action and, 463–464
    compared with chemical bonds, 444–445
    of gases, 396, 422
    hydrogen bonds, 448–451
    London forces, 445–448
    permanent dipole–dipole forces, 448
    phase changes and, 451–452
    in solids, 541–542
    surface tension and, 462–463
    van der Waals forces, 445
    vapor pressure, 452–455
    viscosity and, 462
    in water, 444–445, 451–458, 462–464
Internal energy (U) *The energy of a system
    defined as the sum of the kinetic and
    potential energies. Absolute internal
    energy is difficult to measure,*
        177–178
International System (SI) *The International
    System (Système International) of
    agreed-upon base units, derived units,
    and prefixes used to measure physical
    quantities,* 17–21. *See also* Measurement
International Telecommunications Union
    (ITU), 214
International Union of Pure and Applied
    Chemists (IUPAC), 71–72, 270, 500
Interstitial alloys *An alloy made by the
    addition of solute metals placed in the
    existing spaces within a solvent
    metal,* 556
Intravenous solutions, 480, 540
Iodine
    electron configuration of, 249
    radiopharmaceutical use of, 924–925, 926*f*
    thermodynamic data for, A10
    in titration, 140–143
    uses of, 274*t*
Iodine-125, 907
Ion–dipole interaction *An attraction
    between an ion and a compound with a
    dipole,* 466
Ionic bonds *A strong electrostatic attraction
    between ions of opposite charges,* 304,
        306–316

Ionic bonds (continued)
description of, 308
electronegativity differences in, 321
energy of, 313–316
examples of, 308–309
sizes of ions, 309–312

Ionic compounds *A compound composed of ions,* 67–70
charge of ions in, 68–69
dissolution in water, 127–128, 129, 130, 131*t*, 316–317
lattice enthalpies of, 313–316
list of important, 307*t*
naming binary, 73–75
properties of, 70, 343–346
salts as, 752
solubility rules for, 146, 147

Ionics *The study of the behavior of ions dissolved in liquids,* 825

Ionic solids *Solids made up of ionic compounds held together by strong electrostatic forces of attraction between adjacent cations and anions,* 541*t*, 542
solubility of, 472, 802–804

"ionic sponge," 310

Ionization energy *The energy needed to ionize an atom (or molecule or ion) by removing an electron from it,* 282–286
reactivity and, 282–283
for selected elements, 282*f*

Ionization isomers *Isomers that differ in the placement of counter ions and ligands,* 878, 879, 880*t*

Ionizing radiation *Radiation such as alpha and beta particles, gamma rays, or X-rays that is capable of removing an electron from an atom or a bond when it interacts with matter,* 910–912

Ion pair *Ions of opposite charges that exist in solution as an ionically bonded pair. Ions in solution that associate as a unit,* 308, 804–805

Ions *An electrically charged entity that results when an atom has gained or lost electrons,* 52, 56
cations and anions, 56
complex, 809–815
electron configurations of, 305–306
formation of, 67
Lewis dot symbols for, 305
list of common, 69
oxidation numbers for, 155*t*
polyatomic, 75–76
sizes of, 309–312
spectator, 144

Iron
binding energy of, 916, 917
coordination number for, 875
crystal structure of, 543
density of, 549
in environment, 293
in hemoglobin, 260, 291, 869–870, 871, 877, 892
isotopes of, 903*t*
in natural water, 870–871

properties of, 64, 260, 261
reactivity of, 289–290, 291–292, 850
rusting of, 152, 586*t*, 622, 850
specific heat of, 181*t*
in steel, 265–266, 267
thermodynamic data for, A11

Iron chloride, 74

Iron oxide
formation of, 152, 153, 154
as ionic compound, 307*t*
in thermite reaction, 196–197

Iron pyrite, 261, 262*f*

Iron(II) sulfide, 48

Irvine, Andrew, 582

Isoelectric pH *pH value at which an amino acid is in the zwitterion form and so is electrically neutral overall,* 756

Isoelectronic *Having the same electron configuration,* 305, 306

Isoleucine, 940*f*

Isomerases, 949

Isomers
coordination complex, 878–881
coordination sphere, 879–880
geometric, 502–504, 880–881
ionization, 879
linkage, 879
of normal alkanes, 498–499
stereoisomers, 525
structural, 498–499

Isooctane, 70–71, 498, 509

Isopentyl ethanoate, 521

Isosmotic *Possessing the same osmotic pressure,* 480

Isotonic *Containing the same concentration of ions,* 480

Isotopes *Forms of the same element that differ in the number of neutrons within the nucleus,* 57–58, 901–903
mass of, 59–63
nuclide notations and, 901–903

Janssen, Pierre, 224

Jones Reductor, 856

Joule (J) *The SI unit of energy. In terms of base units, 1 Joule = 1 kg·m²·s⁻²,* 22*t*, 176, 179

Joule, James Prescott, 176

Jupiter
gravitational force on, 398
moons of, 459, 461

Karl Fischer titration, 142

Kelvin (K) *The SI base unit of temperature,* 17*t*, 18–19

Kelvin, Lord, 19, 409

Kelvin scale, 19, 21, 409

Kerosene, 508*t*

Ketones *Compounds that have two carbon atoms bonded to a carbonyl carbon atom,* 512*t*, 519

Kevlar, 566*t*, 567

KHP buffer system, 772–773

Kilocalorie (kcal) *Unit of energy equal to 1000 calories and to 1 Calorie (note capital C),* 180

Kilogram (kg) *The SI base unit of mass,* 17–18

Kilograms per cubic meter (kg/m³) *The derived SI unit for density,* 22

Kilopascals, 398

Kinetic energy *The energy things possess as a result of their motion,* 170–174
calculating, 179
conductivity and, 553
of electron orbit, 220–221
entropy and, 598
temperature and, 408, 409, 422, 423

Kinetic-molecular theory *A theory explaining the relationship between kinetic energy and gas behavior. This theory makes possible derivation of the ideal gas laws,* 422–424
collision theory and, 629–630
effusion and diffusion in, 424–427
gas laws and, 422–423
molecular speeds in, 423–425

Knocking *The rapid explosion of a gas–air mixture as it is compressed in an automobile cylinder* before *the spark plug creates the spark that is intended to cause the mixture to explode,* 509

Kodak Research Laboratory, 382

Kolbe, Adolph Wilhelm Hermann, 336

Kraushaar, Silke, 621

Krypton, 58
lasers, 224*t*
uses of, 274*t*, 275

Labile *The opposite of inert; said of a compound that exchanges ligands rapidly,* 893

Lactase, 657

Lactate, 600–601, 602–603

Lactate dehydrogenase, 946*f*

Lactic acid, 95, 738–740

Lactose-intolerant, 592, 657

Lanthanides *The first set of inner transition elements. The lanthanides include elements 58–71,* 275–276

Laser *An acronym for "light amplification by stimulated emission of radiation,"* 224

Lattice energy *The amount of energy released as a solid ionic crystal is formed,* 314–316

Lattice enthalpy *The amount of energy required to separate 1 mol of a solid ionic crystalline compound into its gaseous ions,* 313–315

Lauric acid, 719*t*

Lavoisier, Antoine, 5, 47, 580

Law of combining volumes *Scientific law stating that when gases combine, they do so in small whole-number ratios, provided that all the gases are at the same temperature and pressure,* 50–51

Law of conservation of energy *Law stating that energy is neither created nor destroyed but is only transferred from place to place or converted from one form into another,* 171, 178

Law of conservation of mass *Scientific law stating that the mass of chemicals present at the start of a chemical reaction must equal the mass of chemicals present at the end of the reaction. Although not strictly true, this law is correct at the level of accuracy of all laboratory balances,* 47, 49

Law of definite composition *Scientific law stating that any particular chemical is always composed of its components in a fixed ratio, by mass,* 47–49

Law of multiple proportions *Scientific law stating that when the same elements can produce more than one compound, the ratio of the masses of the element that combine with a fixed mass of another element corresponds to a small whole number,* 49–50, 68

Lead
  in automobile gasoline, 317–318
  density of, 23
  as radiation shield, 910
  specific heat of, 181*t*
  thermodynamic data for, A11
  uses of, 272*t*
Lead–acid battery, 848*t*
Lead bromide, dissolution of, 693–694, 695
Lead crystal, 560*t*
Lead(II) iodide, 804
Lead ions, 69
Lead(II) sulfide, 146
Le Bel, Achille, 336
Le Châtelier, Henry Louis, 700
Le Châtelier's principle *If a system at equilibrium is changed, it responds by returning toward its original equilibrium position,* 700–705, 737–738, 768–771, 809
Length
  conversion factors for, 26
  as extensive property, 23
  as SI base unit, 17*t*, 18
Leucine, 47*f*, 940*f*
Lewis, G. N., 304, 359, 722
Lewis acid *Accepts a previously nonbonded pair of electrons (a lone pair) to form a coordinate covalent bond,* 722–723
Lewis base *Donates a previously nonbonded pair of electrons (a lone pair) to form a coordinate covalent bond,* 722–723, 870
Lewis dot structures *A drawing convention used to determine the arrangement of atoms in a molecule or ion,* 305
  bond dissociation energy and, 333
  for covalent bonds, 322–323
  formal charges and, 324
  limitations of, 336, 358
  resonance structures and, 327–328
  rules for drawing, 322*t*

Lewis dot symbols *A drawing convention used to determine the number of valence electrons on an atom or monatomic ion,* 304–305, 308–309
Ligand *A Lewis base that donates a lone pair of electrons to a metal center to form a coordinate covalent bond,* 809, 810, 870, 873–875. *See also* Coordination complexes
Ligand exchange reactions, 892–893
Ligases, 949
Light
  color and, 884–885, 890–891
  as electromagnetic radiation, 211
  in photographic reactions, 382–383
  polarized, and chiral molecules, 525–526
  quantization of, 210
  solar spectrum, 216, 217*f*, 224
  speed of, 211–212
  visible spectrum of, 213, 885*f*
Limiting reagent *The reactant that is consumed first, causing the reaction to cease despite the fact that the other reactants remain "in excess." Also known as the limiting reactant,* 112–115
Linear combination of atomic orbitals–molecular orbitals (LCAO–MO) theory *An approximation of molecular orbital theory wherein atomic orbitals are added together (both constructively and destructively) to make molecular orbitals,* 375, 386
Linear geometry, 338*f*, 339*t*
Linear molecules, 447
Linkage isomers *Isomers that differ in the point of attachment of a ligand to a metal,* 878, 879, 880*t*
Lipases, 593
Lipids *Biological compounds characterized by their ability to dissolve in nonpolar solvents. They include fatty acids, some vitamins, and the steroid hormones,* 957–958
Lipoproteins, 948
Liquefied petroleum gas (LPG), 184
Liquid chromatography, 678
Liquid crystals *A transitional phase that exists between the solid and liquid phases for some molecules. Typically, these molecules have long, rigid structures with strong dipole moments,* 550
Liquids *A phase of a substance characterized by closely held components. The properties of a liquid include a medium density, the ability to flow, and the ability to take the shape of the container that holds it by filling from the bottom up,* 10, 107, 443
  boiling points of, 455–456
  capillary action of, 463–464
  intermolecular forces in, 444–451
  properties of, 396, 397*f*, 443
  supercritical fluids, 460

  surface tension of, 462–463
  vapor pressure of, 452–455
  viscosity of, 462
Liter (L) *A commonly used unit for volume; equal to 1 dm³,* 22
Lithium
  band gap of, 553*t*, 554
  emission spectrum for, 220*f*
  ionization energy for, 282–283
  Lewis dot symbol for, 305
  molecular orbitals of, 551–552
  reactivity of, 279
  reduction potential of, 835
  thermodynamic data for, A11
  uses of, 270*t*
Lithium hydride, bond energies in, 360, 361*t*
Lithium hydroxide, reaction with carbon dioxide, 113
Lithium–ion battery, 848*t*, 849
Lithium photo battery, 848*t*
Lockyer, Norman, 224
Logarithms, A3–A4
London, Fritz, 446
London forces *The weakest of the intermolecular forces of attraction, characterized by the interaction of induced dipoles,* 446–448, 454
Lone pairs *Pairs of electrons that are not involved in bonding,* 322, 328
  in coordination complexes, 870–871
  in VSEPR model, 337, 340, 341–342
Low-density polyethylene (LDPE) *A form of polyethylene with branched-chain molecules that are shorter than the unbranched chains found in HDPE,* 515
Lowest-energy unoccupied molecular orbital (LUMO) *The least energetic molecular orbital that contains no electrons,* 378
Lowry, Thomas, 720
Low-spin *A coordination complex with the minimum number of unpaired electrons,* 888
Luminous intensity *Brightness, expressed in the SI unit candela (cd),* 17*t*, 20
Lyases, 949
Lye, 325
Lyman, Theodore, 218
Lyman series, 218*t*, 223
Lysine, 940*f*

Macrostate *The macroscopic state of a system that indicates the properties of the entire system,* 580–585
Macroworld *A term used to describe the "big world" of everyday experience, as opposed to the "nanoworld" of atoms, molecules, and ions, whose activities ultimately determine what happens in the macroworld,* 20
"magic numbers," 917–918
Magnesium
  ionization energy for, 285
  isotopes of, 903*t*

Magnesium (continued)
  production of, 857–858
  as sacrificial anode, 850
  structure of, 544
  thermodynamic data for, A11
  uses of, 271t
Magnesium carbonate, 784
Magnesium hydroxide
  as ionic compound, 307t
  precipitation of, 147
Magnesium oxide, formula for, 69
Magnesium sulfate, reaction with barium
    hydroxide, 147
Magnetic attraction, 169
Magnetic moment (μ) *The strength of the
    paramagnetism of a compound. Can be
    used to determine the number of
    unpaired electrons*, 891
Magnetic quantum number *The quantum
    number, m_l, that describes the
    orientation of the orbital; m_l can be any
    whole-number value from -1 to $+1$.
    Also known as the orbital angular
    momentum quantum number*, 235, 243t
Magnetic resonance imaging (MRI),
    242–243, 540, 561–563
Magnetism, 379–380, 873, 891–892
Magnetite, 704
Main groups *Groups IA through VIIIA of
    the periodic table. Groups IB through
    VIIIB are not main groups*, 66, 269
Malleable *Capable of being shaped*, 552, 556
Mallory, George Leigh, 582
Manganese
  complexes, 891–892
  in steel, 266, 267
  thermodynamic data for, A11
Mantle, Earth's, 292f, 293
Margarine, 872
Mars
  gravitational force on, 398
  meteorite from, 46
Marsden, Ernest, 53
Mass *A measure of the amount of matter in a
    body, determined by measuring its
    inertia (resistance to changes in its state
    of motion) and expressed in the SI base
    unit kilogram*, 17–18
  atomic, 59–63
  balancing in equations, 107–108
  binding energy and, 915–917
  conversion factors for, 26
  as extensive property, 23
  kinetic energy and, 170
  law of conservation of, 47, 49
  molar, 92
  molecular, 87–88
  rest, 231
  as SI base unit, 17–18
Mass-action expression *The ratio of product
    to reactant concentrations raised to the
    power of their stoichiometric coefficients.
    This expression relates the equilibrium
    concentrations to the equilibrium
    constant. It is also known as the*

equilibrium expression, 672–675,
    678, 706
  in acid–base reactions, 725
  for heterogeneous equilibrium, 682
  pressure-based, 686
  for titrations, 684–685
Mass defect *The loss in mass that occurs
    when a nucleus is formed from its
    protons and neutrons*, 915–916
Mass-energy equivalence, 47
Mass number (A) *The total number of
    protons and neutrons in the nucleus of
    an atom*, 57, 58
  in nuclear equations, 904
Mass percent *The percent of a component
    by mass. Mass percent =
    $\dfrac{total\ mass\ component}{total\ mass\ whole\ substance} \times 100\%$*,
    97–99, 135
Mass spectrometer *An instrument used to
    measure the masses and abundances of
    isotopes, molecules, and fragments of
    molecules*, 59–63, 100
Mass spectrum *The output of a mass
    spectrometer, in the form of a chart
    listing the abundance and masses of the
    isotopes, molecules, and fragments of
    molecules present*, 60
Materials science, 540–573
  ceramics, 557–563
  composite material, 565, 568
  future applications, 35, 571–573
  metals, 551–557
  plastics, 563–568
  structure of crystals and, 540–550
  thin films, 568–571
Mathematical operators, A1–A6
  exponential notations, A1–A3
  graphing, A6
  logarithms, A3–A4
  quadratic equation, A5
Matter *Anything that has a mass and
    occupies space*, 4
  classifications of, 8–10
  common states of, 10
  early explanations of, 46–49
  examples of, 4–5
Maximum contaminant level (MCL) *The
    highest acceptable level of a contaminant
    in a particular solution, according to
    the Environmental Protection
    Agency*, 470
Measurement, 16–34
  accuracy and precision in, 29–31
  base units, 17–20
  conversions and dimensional analysis,
    25–28
  derived units, 21–23
  of gas pressure, 397–401
  International System (SI), 17–21
  mathematical operators, A1–A6
  rounding off, 33
  significant figures, 31–33
  two major systems, 16
  uncertainty in, 28–29, A7–A9

Mechanisms *The series of steps taken by the
    components of a reaction as they
    progress from reactants to products*,
    647–654
  activation energy and, 651–654
  catalysts, 654–658
  elementary steps, 648–649
  intermediates, 650
Mechlorethamine, 962
Medicine, 303
  antibiotics, 961–962
  buffers in, 784–786
  calcium channel blockers, 331
  cancer treatment, 528–530, 876, 901,
    912, 962
  cardiac defibrillator, 568
  ceramic bone replacement, 557
  future applications, 34
  genetic disease, 964
  heart–lung machines, 3, 28
  hormones, 963
  intravenous solutions, 480, 540
  major causes of disease/illness, 961
  materials science and, 540
  modern drug discovery, 527–530
  neurotransmitters, 963
  nuclear, 901, 924–927
  pain control, 14–15, 335
  See also Chemistry of life
Meitner, Lise, 920
Melting point *The temperature of the phase
    transition as a compound changes from
    a solid to liquid*, 6, 8, 456
Mendeleev, Dmitri, 264–265, 410–411
Meningoencephalitis, 934
Meniscus *The concave or convex shape
    assumed by the surface of a liquid as it
    interacts with its container*, 463–464
Mercury, 464
  amalgams, 557
  in environment, 876–877
  thermodynamic data for, A11
Messenger RNA (mRNA) *A short-lived
    single strand of ribonucleic acid that
    carries the information from a gene to
    the ribosomes where a protein is
    constructed*, 942–943
Metabolism *The biochemical reactions
    inside living cells*, 595, 958–959
Metabolite *A product of each step in a
    biochemical pathway*, 958
Metal–chelate complexes, 810–811
Metal–chloride battery, 848t
Metal complexes. *See* Coordination
    complexes
Metal hydroxides, 757
Metallic bond *A bond in which metal
    cations are spaced throughout a sea of
    mobile electrons*, 304
Metallic radius *Half the distance between
    the nuclei of the atoms within the solid
    structure of a metal*, 280
Metalloids *A small number of elements,
    found at the boundary between metals
    and nonmetals in the periodic table, that*

have properties intermediate between those of metals and nonmetals. Also known as semimetals, 65, 268–269

Metalloproteins, 948, 949

Metal recovery *Recycling of metals from waste streams by complexation with chelating agents*, 807

Metals *The largest collection of elements in the periodic table. Metals exhibit such characteristic properties as ability to conduct electricity, shiny appearance, and malleability*, 64, 265, 551–557
  alloys, 556
  amalgams, 557
  band theory of, 551–553
  compared with ceramics, 559*t*
  conductivity of, 553–555, 561–562
  corrosion of, 152, 291–292, 850
  crystal lattice in, 543–545
  density of, 25
  diversity of steel, 267
  emission spectra for, 219, 220*f*
  heavy metals, 145–146, 266, 280
  in ionic compounds, 68–69
  oxidation numbers for, 155*t*, 827, 828*f*
  properties and uses of, 64, 65*f*, 265–267, 269*f*, 541*t*, 552–555
  reactions with acids, 723
  reactivity series of, 291–292, 849–850
  spontaneous oxidation of, 586

Metastable state *An energetically unstable arrangement of protons and neutrons in a nucleus after a neutron has become a proton*, 906

Meter (m) *The SI base unit of length*, 17*t*, 18

Meters per second, 22*t*

Methadone, 747*t*

Methane
  boiling point of, 445
  bond angles in, 357
  combustion of, 334, 358*t*, 622
  enthalpy of formation of, 193–194
  as greenhouse gas, 431
  hybrid orbitals in, 364–365
  intermolecular forces in, 444–445
  molecular geometry of, 339
  molecular speed for, 424, 425
  oxidation numbers for, 154
  pyrolysis of, 420
  sigma bonds in, 369–370
  sources of, 193
  structure of, 444, 496–497
  in substitution reactions, 510
  valence bond model of, 362, 363*f*

Methanol, 142, 517
  boiling-point elevation of, 476*t*
  Lewis dot structure for, 323–324
  oxidation of, 517
  properties of, 323
  reaction with permanganate, 840
  structure of, 518*t*
  vapor pressure of, 453

Methionine, 940*f*, 943

Method of graphical analysis *A method of determining the rate law for a reaction*

where the rate of a reaction is plotted versus time. Also known as the method of integrated rate laws, 644–645

Method of initial rates *A method of determining the rate law for a reaction where the initial rates of different trials of a reaction are compared*, 641–643

Methoxychlor, 626–627, 647

Methyl acetate, 512*t*

2-Methylbutane, 498*f*

Methylcyclopropane, 503

Methylene group *The —CH$_2$— group*, 497

Methyl ethyl ketone, 512*t*

3-Methylheptane, 499–500

Methylhydrazine, reaction with dinitrogen tetroxide, 195–196

Methyl methanoate, 520

2-Methylpentane, 498*f*

2-Methyl-2-propanol, 517, 518*t*, 519

2-Methyl-1-propene, 503

Methyl salicylate, 521

Methyl t-butyl ether (MTBE), 524

Metric system, 16. *See also* Measurement

Meyer, Julius Lothar, 264

Microstate *The state of the individual components within a system*, 581–585

Microwaves, 213

Milliliter (mL) *One thousandth of a liter*, 22

Miscible *Mixable. Two solvents that are miscible dissolve in each other completely*, 466

Mixture *A sample containing two or more substances*, 8–9

mm Hg *A unit used to measure pressure. 760 mm Hg represents the height of a column of mercury that can be supported by 1 atm pressure*, 399

Mobile phase *In chromatography, the phase the moves*, 678

Modified atmosphere packaging (MAP) *The replacement of the air within food packaging with gases in order to prolong the shelf life of the food*, 401, 412

Molality (m) *A concentration measure defined as moles of solute per kilogram of solvent*, 468, 469

Molar (M) *The "shorthand" method of describing molarity; as in "that is a 3 molar HCl solution,"* 132

Molar heat capacity *The heat capacity of one mole of a substance*, 182

Molarity (M) *A specific concentration term that reflects the moles of solute dissolved per liter of total solution*, 131–132, 468
  calculations involving, 132–135, 137, 469
  osmotic pressure and, 479

Molar mass *The total mass of all the atoms present in the formula of a molecule, in atomic mass units (amu) or in grams per mole*, 92
  boiling point and, 446–448
  of gases, 418–419, 424–426

Molar solubility *The total number of moles of solute that dissolve per liter of solution*, 805

Mole (mol) *The SI base unit of amount of substance. One mole of entities, such as 1 mole of atoms, is approximately equal to* 6.02214 × 10^{23} *entities*, 20, 92
  calculating, in solution, 133–134
  converting molecules and grams, 92–97
  as SI base unit, 17*t*, 20
  whole-number ratios and, 100

Molecular compounds
  naming binary, 72–73
  properties of, 343–346

Molecular equation *A chemical equation that shows complete molecules and compounds*, 144–145

Molecular formula *A chemical formula indicating the actual number of each type of atom present in one molecule of a compound*, 70–71, 99–102

Molecular geometry *The geometry of the atoms in a molecule or ion. The molecular geometry comes from the electron-group geometry but considers the lone pairs "invisible,"* 337
  of coordination complexes, 876–878
  examples of, 337–341
  of hybridized orbitals, 368

Molecular ions *Ions composed of two or more atoms attached together. Also known as polyatomic ions*, 75–76

Molecularity *A description of the number of molecules that must collide in the rate-determining step of a reaction*, 648

Molecular mass *The mass of one molecule, expressed in atomic mass units (amu) or the mass of one mole of molecules in grams per mole*, 87–88

Molecular models *Models used by chemists to explain the three-dimensional nature of molecules. Although many can be made from plastic or wood, these models can also be constructed using computers*, 303–304
  in chemist's toolbox, 386
  combining models, 384
  flat models, 336
  hybridization, 363–374
  information yielded from, 374
  Lewis dot structures, 305, 322–323
  molecular orbital theory, 374–381
  valence bond theory, 358–363
  VSEPR model, 337–343

Molecular orbitals *Electron orbitals that are appropriate for describing bonding between atoms in a molecule*, 244
  in crystal field theory, 886–889
  hybrid orbitals, 363–374
  key features in formation of, 375–377
  magnetism and, 379–380
  of metals, 551–553
  as replacing atomic orbitals, 376, 377
  shapes of, 377
  valence bond model of, 359–362

Molecular orbital (MO) diagram *A diagram used to illustrate the different molecular orbitals available on a molecule*, 377–381
Molecular orbital (MO) theory *A theory that mathematically describes the orbitals on a molecule by treating each electron as a wave instead of as a particle*, 374–381, 386
  bonding and antibonding orbitals, 375, 376
  bond order in, 377
  conjugated π systems, 374, 381, 383
  definition of, 375–377
  delocalized systems, 384–385
  and power of photons, 380–381
Molecular solids *Solids made up of molecules that are held together by intermolecular forces*, 541t, 542
Molecular speeds, 423–425
Molecules *Neutral compounds containing two or more atoms attached together*, 5, 70–71
  chiral, 525–527
  polar vs. nonpolar, 346
  shape of, 377
Mole fraction *A concentration measure defined as moles of solute per total moles of solution*, 468, 469
Molina, Mario, 429, 431
Molybdenum, 58
  atomic mass of, 60
  in steel, 266, 267
Molybdenum cow, 925–926
Monobasic salt *A salt that can accept one hydrogen ion*, 748
Monodentate ligand *A ligand that makes one coordinate covalent bond to a metal ion*, 810, 873t
Monomer *One of the small chemical units that combine with other monomers to form polymers*, 515–516
Monoprotic *Can produce one mol of H^+ when it dissociates*, 151
Monoprotic acid *An acid that contains one acidic proton*, 736, 742–743
Monosaccharide *A small carbohydrate monomer. Typically contains between three and nine carbon atoms*, 955f, 956
Morphine, 15, 335, 342–343, 747t
Mortar, 558
Most probable speed *The most likely speed of a molecule from among a group of molecules*, 424
Mulliken, Robert S., 375
Mulliken electronegativity scale, 320
Multielectron orbitals, 244–245
Multiple bonds, 325–327, 369–370
Multiplicity *The number of microstates within a system*, 581–585, 587–588
Myoglobin (Mb) *A biochemical compound responsible for storing and releasing oxygen in a living organism*, 669–673
  reaction with oxygen, 669–673, 689–690, 697–698, 701–702
  structure of, 669f, 946, 947f

NAD^+ (nicotinamide adenine dinucleotide), 600–601
Naming compounds. *See* Chemical nomenclature
Nanometers, 211
Nanotechnology *Technology that depends on manipulating materials and their chemistry at the very small, "nanoscale" level of individual particles or assemblies of small numbers of particles*, 36
Nanotube, 494f
Nanoworld *A term used to describe the "world of the very small" at the level of individual atoms, molecules, and ions, as opposed to the "macroworld" of everyday experience*, 20–21
Naphthalene, 478t, 505f
Naphthenes, 495t
National Cancer Institute, 528
National Telecommunications and Information Administration (NTIA), 214
Natrolite, 310f
Natural gas *A complex mixture of gases extracted from the Earth. The major component in many cases is methane*, 197, 198, 443–444, 497, 508t
Natural logarithms, A4
Nd:YAG lasers, 224t
Neon
  electron configuration of, 246, 305
  Lewis dot symbol for, 305
  light emissions from, 224
  properties of, 65
  thermodynamic data for, A11
  uses of, 274t, 275
Neoprene, 420
Nernst, Walther Hermann, 852
Nernst equation *The equation used to determine the cell potential at nonstandard conditions*, 851–857
Net ionic equation *A complete ionic equation written without the spectator ions*, 144–145, 146
Network covalent solids, 541t
Neurotransmitters, 963
Neutralization reactions, 148. *See also* Acid-base *entries*
Neutral pH *A pH of 7.0 in aqueous solution at 24°C*, 734
Neutrino *A subatomic particle that is produced in beta-plus decay and that has no charge, has essentially no mass, and interacts only rarely with matter*, 905
Neutrons *One of the subatomic particles of which atoms are composed, carrying no electrical charge and found in the nucleus*, 54–57
  beta decay of, 904–905
  fast, penetration capability of, 910f
Newlands, John, 263–264
Newton (N) *The basic SI unit of force. $1 N = 1 kg \cdot m \cdot s^{-2}$*, 22t, 397–398
Niacin, 950t
NiChrome, 561–562

Nickel
  complexes, 892
  sources and uses of, 265t
  in steel, 266, 267
  thermodynamic data for, A11
Nickel–cadmium battery, 846, 848t
Nickel(II) hydroxide, 147
Nickel–metal hydride battery, 848
Nicotine, 719t, 743, 746
Niépce, Joseph Nicéphore, 382
Nifedipine, 331, 335
Niobium, 267
Nitric acid
  acid deposition and, 758
  ionization of, 128, 719, 724–727
  reaction with copper, 833, 834f, 837–838, 842–843, 845–846, 855
  as strong acid, 724, 725
  uses of, 718t, 719t
Nitric anhydride, 50
Nitric oxide, 49–50, 757
Nitriles, 513t
Nitrilotriacetic acid (NTA), 811t
Nitrogen
  electron configuration of, 246, 247
  electronegativity of, 289
  Lewis dot symbol for, 305
  in ligands, 874
  oxidation state of, 831–833
  properties of, 5, 272
  reactions with oxygen, 49–50
  reaction with hydrogen, 156
  sources and uses of, 97–98, 268t, 272t
  thermodynamic data for, A11
Nitrogen-14, 915
Nitrogen cycle *The path that nitrogen follows through its different oxidation states on Earth*, 831
Nitrogen dioxide, 757–758, 832–833
Nitrogen gas
  in air bags, 419–420
  formation of, 704
  uses of, 395t, 401
Nitrogen molecule
  boiling point of, 445
  diamagnetism of, 380
  MO diagram for, 378–379
  triple bonds in, 326
Nitrogenous organic bases *The portion of a nucleotide where hydrogen bonding holds the strands of DNA together in a duplex*, 935
Nitrogen oxide
  catalysis of, 656
  oxidation states for, 832–833
  ozone depletion and, 629, 640, 653
  reaction rates for, 628–629, 641–643, 649–650
Nitrogen trichloride, 72
Noble gases *The unreactive elements in the rightmost group of the periodic table (Group VIIIA). Also known as inert gases*, 65, 66t, 274–275
  commercial uses of, 274t
  compounds of, 329–330

discovery of, 224
periodic properties of, 279–280, 287
reactivity and, 278
stable electron configurations of, 275, 306

Nodal planes *Flat, imaginary planes passing through bonded atoms where an orbital does not exist,* 370

Node *A point in the wave function where the amplitude is always zero,* 233, 234, 238–240

Nonane, 497*t*

Nonbonding molecular orbital *A molecular orbital that is not involved in bonding or antibonding,* 551

Nonelectrolyte *A compound that doesn't dissociate into ions when it dissolves,* 129–131

Nonmetals *A collection of elements on the right-hand side of the periodic table that shares certain characteristics that are in contrast to those of the metals, including being gases or dull, brittle solids at room temperature,* 65, 268, 269*f*
in ionic compounds, 68–69
oxidation states for, 827

Nonpolar covalent bond *A covalent bond with a very small difference in electronegativity between the bonded atoms. The result is minimal or no charge separation in the bond,* 320–321

Nonpolar molecules *A molecule in which there is no net charge separation and no net dipole moment,* 345, 346
boiling points of, 446–448

Nonspontaneous process *A process that occurs only with continuous outside intervention,* 585–586

Normal alkanes *Alkanes that have no branched chains or rings of carbon atoms,* 496–497

Normal boiling point *The temperature at which a liquid boils when the pressure is 1 atm,* 455

n-type semiconductor *A semiconductor that contains electrons (negative charges) in the conduction band,* 554

Nuclear charge
ion size and, 310–312
shielding and, 244–245
Nuclear chemistry, 901–928
chain reactions, 923–924
vs. chemical reactions, 903
common radioactive nuclei, A19
fission, 917, 920–924
fusion, 916
health effects of radiation exposure, 901, 911–912
interaction of radiation with matter, 909–912
ionizing radiation, 910–912
isotopes, 901–903
kinetics of radioactive decay, 912–915
mass and binding energy in, 915–917
medical use of radioisotopes, 924–927

nuclear reactors, 921–922
nuclear stability, 917–919
nuclide notations, 901–903
positron emission tomography, 926–927
structure of atom and, 53–55
types of radioactive decay, 54–55, 904–909
units for measuring radiation, 911
See also Radioactive decay

Nuclear equation *An equation showing a nuclear transformation, where the atomic and mass numbers are provided,* 904–906

Nuclear fission. *See* Fission
Nuclear fusion. *See* Fusion
Nuclear magnetic resonance spectrophotometer (NMR), 562–563
Nuclear medicine, 924–927
Nuclear power plants
cooling towers, 471, 472*f*
as electricity source, 197*f*, 921–922

Nuclear radiation *The particles and/or energy emitted during radioactive decay,* 901

Nuclear spin, 242–243
Nuclear stability, 917–919
Nuclear Structure Laboratory, 284

Nucleon *The name given to a particle (proton or neutron) that is part of a nucleus. For example,* ^{13}C *contains 13 nucleons, 6 protons, and 7 neutrons,* 918

Nucleotides *Monomers of the nucleic acids. A nucleotide contains a five-carbon sugar, a phosphate, and a nitrogenous organic base,* 935

Nucleus, 53–55

Nuclide notation *Shorthand notation used to represent atoms, listing the symbol for the element, accompanied by the atomic number and the mass number of the atom,* 57, 67, 901–903

Nyholm, Sir Ronald, 337
Nylon, 523, 524*f*, 565, 566*t*

Observations, 12, 13*t*
Obsidian, 559

Octahedral *The geometry indicated by six electron groups (lone pairs and/or bonds) positioned symmetrically around a central atom such that the bond angles are 90°,* 338*f*, 339*t*, 877, 880, 888

Octahedral crystal field splitting, 886–887
Octane, 497*t*
boiling point of, 445–446
vapor pressure of, 454
viscosity of, 462

Octane rating *A number assigned to motor fuel (gas) on the basis of its "anti-knock" properties,* 509

1,3,5,7-Octatetraene, 381

Octet *Eight electrons in the valence shell of an atom.*
expanded, 330, 368–369
stable, 275, 279

Octet rule *The existence of eight electrons in the valence shell of an atom, creating a stable atom or ion,* 306, 323, 326
exceptions to, 329–331
valence bond theory and, 359

Oil
as energy source, 197, 198
measuring quality of, 766
See also Crude oil

Oils *A triglyceride made from carboxylic acids with long carbon chains and a relatively low melting point. Typically derived from plant sources,* 373, 957

Oleic acid, 109, 373
Olympic Games, 621–622, 937
Opioid receptors, 335, 343
Opsin, 356
Orbital angular momentum quantum number, 235

Orbitals *The volume of space to which the electron is restricted when it is in a bound atomic or molecular energy level,* 233
degenerate, 236, 364
d orbitals, 240
electron shielding, 244–245
f orbitals, 241
hybridization of atomic, 363–374
introduction to, 233–234
in multielectron atoms, 244–245
overlapping, 359–362
p orbitals, 239
quantum numbers and *s* orbitals, 234–237
s orbitals, 238–239
See also electron configurations; molecular orbitals

Organic chemistry *The chemistry of carbon-containing compounds, especially those derived from living things or fossil fuels,* 372–373, 492. *See also* Carbon-based compounds; Hydrocarbons

Organic compounds *Carbon-based compounds,* 492

Organic foods, 62–63

Osmosis *The flow of solvent into a solution through a semipermeable membrane,* 10–11, 479–480

Osmotic pressure *The pressure required to prevent the flow of a solvent into a solution through a semipermeable membrane,* 479–480, 540

Overpotential *The extra potential needed, above that which is calculated, in order to make an electrochemical process proceed,* 858

Oxalate ion, 874
Oxalic acid, pH of solution of, 751–752

Oxidation *The process of losing electrons. Such a substance is said to be oxidized,* 152, 826

Oxidation number *A bookkeeping tool that gives us insight into the distribution of electrons in a compound. Also known as oxidation state,* 153–156, 324. *See also* Oxidation state

Oxidation reaction *In a redox reaction, the half-reaction that supplies electrons,* 826

Oxidation–reduction reactions *Reactions that involve the transfer of electrons from one species to another. Also known as redox reactions,* 144, 152–157
  mnemonics for, 152
  oxidation numbers, 153–156
  See also Redox reactions

Oxidation state *A bookkeeping tool that gives us insight into the distribution of electrons in a compound. Also known as oxidation number,* 153–156
  acid strength and, 728
  defined, 153–156, 827–829
  in electrochemistry, 826–833
  rules for assigning, 155–156, 828f

Oxidized *The species that has lost electrons in a redox reaction,* 152

Oxidizing agent *A substance that causes the oxidation of another substance,* 830–831

Oxidoreductases, 949, 952

Oxy-acetylene welding, 501

Oxygen
  allotropes of, 71
  as diatomic element, 71
  electron configuration of, 246
  in hemoglobin, 869–870, 871
  isotopes of, 57–58, 903t
  Lewis dot symbol for, 305
  in ligands, 874
  oxidation number of, 154, 155t, 827
  as oxidizing agent, 830
  paramagnetism of, 379–380
  properties of, 5, 260, 273
  reactions with nitrogen, 49–50
  reaction with hydrogen, 6–7, 10, 50–51
  reaction with myoglobin, 669–673, 689–690, 697–698, 701–702
  reaction with ozone, 651
  reaction with sulfur dioxide, 682–683, 686–687, 695–697
  reference form of, 187
  solubility of, 468, 471
  thermodynamic data for, A11
  uses of, 268t, 273t, 395t

Oxygen molecule, MO diagram for, 379

Oxyhemoglobin, 291f, 870f

Ozone
  depletion of atmospheric, 35, 427–431, 629, 640, 651, 653, 655
  formation of, 428–429
  increase in surface, 430
  methods for monitoring, 403
  molecular geometry of, 340–341, 342
  properties of, 71, 273

***p* orbital** *An orbital with quantum number l = 1,* 239

Pain control, 14–15

Pairing energy (*P*) *The energy required to spin-pair two electrons within a given orbital,* 888

Palladium-103, half-life of, 912–914

Parallelepiped, 542, 547

Paramagnetism *The ability of a substance to be attracted into a magnetic field. This attraction arises because of the presence of unpaired electrons within the molecule,* 379–380, 891

Parkinson's disease, 963

Partial pressure *The pressure exerted by a single component of a gaseous mixture,* 401–402
  equilibrium constant using, 686–687

Parts per billion (ppb) *One gram of solute per billion grams of solution. A concentration defined as the mass of a solute per billion times that mass of solution,* 135–137, 470

Parts per million (ppm) *One gram of solute per million grams of solution. A concentration defined as the mass of a solute per million times that mass of solution,* 135–137, 470

Parts per trillion (ppt) *One gram of solute per trillion grams of solution. A concentration defined as the mass of a solute per trillion times that mass of solution,* 135–137, 470

Pascal (Pa) *The basic SI unit of pressure. 1 Pa = 1 N·m^{-2},* 22t, 398

Pascal, Blaise, 398

Patinas, 586

Pauli, Wolfgang, 242

Pauli exclusion principle *In a given atom, no two electrons can have the same set of the four quantum numbers –n, l, m$_l$, and m$_s$,* 242, 378

Pauling, Linus, 19, 116, 287, 359, 363

Pauling electronegativity scale, 287–289, 320

p-block elements, 261

Penicillin, 961, 962f

Pentadecane, boiling point of, 447–448

2,2,4,6,6-Pentamethylheptane, boiling point of, 447–448

Pentane, 497t
  isomers of, 498
  viscosity of, 462

Pentose sugars, 955

Pepsinogen, 755

Percentage yield *The actual yield of a reaction divided by the theoretical yield and then multiplied by 100%,* 115–116

Percent by mass. *See* Mass percent

Perchloric acid, strength of, 725–726, 728

Period *One of the horizontal rows in the periodic table of the elements,* 64, 262

Periodicity *The recurrence of characteristic properties as we move through a series, such as the elements of the periodic table,* 279–280
  atomic size, 280–281
  electron affinity, 286–287
  electronegativity, 287–289
  ionization energies, 282–286
  reactivity, 289–292

Periodic table of the elements *A table presenting all the elements, in the form of horizontal periods and vertical groups; shown on the inside front cover of this book,* 5, 56, 64f
  atomic number in, 58
  common ions in, 69f
  electron configurations and, 246–250, 261–262
  elements of Earth, 292–294
  elements of life, 277–278
  groups in, 269–278
  historical development of, 263–265
  metals, 265–267
  nonmetals, 268
  recent modifications to, 270
  semimetals, 268–269
  structure of, 64–66, 260–265
  transition elements, 275–276

Pesticide *Any compound used to kill unwanted organisms (whether they be animals, insects, or plants),* 619, 620–621, 623, 626, 631, 637t

Petroleum ether, 508t

Pfund series, 218t

pH *A numerical value related to hydrogen ion concentration by the relationship pH = −log[H$^+$] or pH = −log[H$_3$O$^+$],* 730–731
  of acid deposition in U.S., 758
  of basic solutions, 743–746
  of buffers, 767–771
  of common substances, 735t
  defined, 730–731
  meters, calibrating, 772
  of mixture of monoprotic acids, 742–743
  neutral pH, 734
  pH scale, 730–735
  of polyprotic acids, 749–752
  of salt solutions, 752–757
  of strong acidic solutions, 736–738
  of water, 733–735
  of weak acidic solutions, 738–742

pH indicator *A compound that changes color on the basis of the acidity or basicity of the solution in which it is dissolved. Also known as an acid–base indicator,* 799–801

pH meter, 772, 799

Pharmacognocist *A person who studies natural drugs, including their biological and chemical components, botanical sources, and other characteristics,* 303

Pharmacopeia of the United States of America/The National Formulary, The, 772, 961

Phase *A part of matter that is chemically and physically homogenous,* 107

Phase boundary *On a phase diagram, a set of specific temperatures and pressures at which a compound undergoes a phase transition,* 460

Phase changes
  entropy change and, 591–595

heating curve for water, 456–458
vapor pressure and, 452–455
in water, 451–452
Phase diagram *A plot of the phases of a compound as a function of temperature and pressure*, 459–461
Phenanthrene, 505*f*
Phenol
freezing-point depression for, 478*t*
manufacture of, 572
reaction with ethylenediamine, 721
structure of, 518*t*
Phenolphthalein, 799–800
Phenylalanine, 940*f*
Phillips electronegativity scale, 320
Phonons *The induced vibrations produced in a metal lattice*, 561
Phosgene, 333, 326–327
Phosphate group, in DNA, 935–936
Phosphate rock, 747–748
Phosphodiester bonds, 935
Phospholipids *A phosphate-modified lipid*, 957
Phosphoproteins, 948
Phosphoric acid
pH of aqueous, 749–751
reaction with sodium hydroxide, 114–115, 151
strength of, 728
structure of, 747
uses of, 718*t*, 748
Phosphorus
in ligands, 874
thermodynamic data for, A11–A12
uses of, 98, 268*t*, 272
Phosphorus-30, 918
Phosphorus pentachloride
hybrid orbitals in, 368–369
molecular geometry of, 340
Photodissociation *A chemical reaction where light, of sufficient energy, causes the cleavage of a bond in a molecule*, 428
Photography, 382–383
Photons *A single unit of electromagnetic radiation*, 210
energy of single, 215
reactions with molecules, 356, 358, 380–381
wave–particle duality of, 230–231
Photoreceptor cells, 355, 356–357
Photosynthesis, 111, 173, 872
Photovoltaic devices *A device in which light energy can be used to generate an electric current*, 555
Photovoltaic systems (PVs) *Fabricated systems that convert the energy of sunlight into electricity. Also known as solar cells*, 198
Phthalate buffer system, 772–773
Physical changes *Changes in the physical state of a substance, such as changes between the solid, liquid, and gaseous states, that do not involve the formation of different chemicals*, 6

Physical deposition *The process of sublimation and deposition used to place a thin film on a substrate*, 568–569
Physical property *A characteristic of a substance that can be determined without changing its chemical composition*, 6–8
Pi bond (π bond) *A covalent bond resulting from side-to-side overlap of orbitals. The pi bond possesses a single nodal plane along the axis of the bond*, 370–371, 372
conjugated, 374, 381, 383
delocalized, 384–385
Pitchblende, 426
Planck, Max, 215
Planck's constant *A fundamental constant equal to 6.62608×10^{-34} J·s*, 215, 228
Plants
fertilization of, 97–98
photosynthesis in, 111, 173, 872
synthesis of glucose, 595
Plaster, 557, 558
Plastics *A thermally moldable polymer*, 563–568
Plastocyanin, 872*f*, 894
Platinum
catalysts, 704
coordination complexes, 876, 880, 881–882
Plum pudding model, 53
Plutonium-239
in atomic bomb, 920
in nuclear reactors, 922
radioactive decay of, 633, 636
Polar covalent bond *A covalent bond in which the atoms differ in electronegativity, resulting in an unequal sharing of electrons between two adjacent atoms*, 320–321, 341
Polarity *The location of a compound on a scale from polar to nonpolar. The dipole moment is a measure of the polarity of a compound*, 345, 346
Polarizability *The extent to which electrons can be shifted in their location by an electric field*, 446–447
Polarization *A molecule or bond containing electrons that are unequally distributed*, 319
bond dipoles, 343–344
of covalent bonds, 319–321
dipole moments, 344–345
Polarized *A molecule or bond that contains an unequal distribution of electrons*, 319
Polar molecule *A molecule in which there exists a charge separation resulting in a net dipole moment*, 345, 346
Pollution. *See* Environmental concerns
Polonium, 273*t*, 542, 543*f*
Polyacrylonitrile (PAN), 516*t*, 567
Polyamides *Polymers whose monomers are bonded together by amide bonds*, 523

Polyatomic ions *Ions composed of two or more atoms covalently attached together. Also known as molecular ions*, 75–76
Polydentate *A ligand that contains two or more lone pairs on different atoms, each forming a coordinate covalent bond to a metal center*, 810–811, 873*t*, 874–875, 893
Polyesters *Polymers in which the monomers are held together by repeated ester linkages*, 522, 565, 566*t*
Polyethenes *Organic polymers made when huge numbers of ethene molecules participate in addition reactions with themselves to generate long chain-like molecules. Also known as polyethylenes*, 514–515
Polyethers, 524
Polyethylenes *Organic polymers made when huge numbers of ethene molecules participate in addition reactions with themselves, to generate long chain-like molecules*, 370, 514–515, 564*t*
Polyethylene terephthalate (PETE), 522, 564*t*
Polymers *A chemical composed of large molecules made by the repeated bonding together of smaller monomer molecules*, 515
addition, 515
biopolymers, 572, 935, 940, 941
condensation, 522–523
fibers, 563–567
pi bonds and, 370–371
plastics, 563–568
polyethers, 524
use of chelates in, 812*t*
Polypeptide *A polymer of amino acids linked by amide bonds*, 941, 945–946
Polypropylene, 516*t*, 564*t*
Polyprotic acids *An acid that can release more than 1 mol of hydrogen ions per mole of acid*, 747–752
ionization constants for, A14
pH of, 749–752
titrations, 799
Polysaccharide *A polymer of carbohydrate monomers*, 956
Polystyrene, 516*t*
Polyurethane, 565*t*
Polyvinyl chloride (PVC), 514, 516*t*, 849
Porphyrin ligand, 874
Positron *The antimatter equivalent of an electron. Positrons have a positive charge and the same mass as an electron*, 907
Positron emission *A type of radioactive decay wherein a positron is ejected at high speed from the nucleus, typical of nuclides that have too few neutrons. A neutrino accompanies this emission, and usually one or gamma rays as well. Also known as beta-plus emission*, 907, 918, 919*t*

Positron emission tomography *A medical imaging technique that images metabolic processes within the body. Also known as a PET scan,* 926–927

Potassium
electron configuration of, 247
as essential nutrient, 903
in heart cells, 824, 853–854
ionization energy for, 282–283
isotopes of, 902, 903
nuclide notation for, 901–902
reactivity of, 279, 289
thermodynamic data for, A12
uses of, 270*t*

Potassium-40, 903, 904, 905
Potassium chloride, 129, 321
Potassium dichromate, 329
Potassium hydride, bond energies in, 360, 361*t*
Potassium hydrogen phthalate (KHP), 772–773
Potassium hydroxide
heat of solution for, 466
pH of solution of, 744–745
Potassium tetrachloronickelate(II), 882–883

Potential *The driving force (to perform a reaction) that results from a difference in electrical charge between two points,* 824

Potential energy *The energy things possess as a result of their positions, such as their position in a gravitational or electromagnetic field,* 170–174

Pounds, 398

Pounds per square inch (psi) *A unit of pressure. 14.7 psi = 1 atm,* 399

Pounds per square inch gauge (psig) *A unit of pressure expressing the pressure of a gas in excess of the standard atmospheric pressure. Also known as gauge pressure. 14.7 psig = 2 atm,* 399

Power of 10, A2
Praseo complex, 880

Precipitate *Any solid material that forms within a solution; the action describing the formation of a solid,* 145

Precipitation reactions *A reaction involving the formation of a solid that isn't soluble in the reaction solvent,* 144, 145–148
for extracting gold, 125–126
gravimetric analysis, 806–807
identifying, 147
metal recovery, 807
predicting, 146
stoichiometry of, 147–148

Precision *The reproducibility of a measurement,* 29–31

Prednisone, 963*f*
Prefix, 17

Pressure *The force per unit area exerted by an object on another. The force exerted per unit area by a gas within a closed system,* 398, 452
boiling point and, 455–456
conversion of units of, 400

as derived unit, 22*t*
deviations of real gases, 416–417
equilibrium and, 702–703, 705*t*
impact on solubility, 472–473
law of partial, 401–402
measures of gas, 397–401
osmotic, 479–480
in phase diagrams, 459–460
relationship with volume, 15–16, 405–408, 422
spontaneous processes and, 609–610
standard, 399–400
vapor, 452–455
–volume work, 176–177, 186

Pressure reaction quotient (Q_p) *The ratio of the pressures of all of the gaseous products raised to their respective stoichiometric coefficients, divided by the pressures of all of the gaseous reactants raised to their stoichiometric coefficients,* 610

Pressure swing adsorption, 106

Primary alcohols *Alcohols that have no more than one carbon atom bonded to the carbon atom that carries the hydroxyl group,* 517

Primary structure *The sequence of amino acids within a protein,* 941

Principal quantum number *The quantum number, n, that describes the energy level of the orbital; n can be any whole-number value from 1 to infinity,* 234–235, 236, 243*t*

Principal shell *The energy level that is occupied by electrons with the same value for the principal quantum number (n),* 235, 246

Probability
chemical behavior and, 580–585
entropy and, 587–588
Probability function, 233–234
Procaine, 747*t*

Products *The materials that are produced in a chemical reaction. The substances located on the right-hand side of a chemical equation,* 47, 103

Proline, 940*f*
Propane, 497*t*
combustion of, 184–185, 189
reaction with steam, 110
Propanethiol, 513*t*
1,2,3-Propanetriol, 517, 518*t*
2-Propanol, 517, 518*t*, 519
Propanone. *See* Acetone
Propene, 511, 516*t*
1-Propene, 501*f*
Propylene glycol, 451
Propyne, 501*f*
Prosthetic groups, 948
Proteases, 593

Proteins *A large polymer of amino acids. May contain one or more polypeptides and often involves complex and intricate folds of the chain,* 939

amino acid sequence of, 941
denaturation of, 946
diverse functions of, 953–956
enzymes, 947–952
folding of, 588, 589*f*, 590, 945–947
formation of, 940–945
primary structure of, 941
quaternary structure of, 946
secondary structure of, 945–946
tertiary structure of, 946

Protons *One of the subatomic particles of which atoms are composed, carrying an electrical charge of +1 and found in the nucleus,* 53–54, 55
atomic number and, 56–58
Proust, Joseph Louis, 47–48

Pseudo-first-order rate *A modification of the second-order rate that enables one to use first-order kinetics,* 640

Psychrophiles, 591
Psychrotrophs, 591

p-type semiconductor *A semiconductor that is electrically conductive if we apply a small electric field. The semiconductor contains positive holes in the valence band,* 554

Putrescine, 523
Pyrene, 505*f*
Pyridine, 142
Pyrolysis, 420
Pyruvate, 585, 600–601, 602–603, 605–606

Q-test, A8–A9

Quadratic equation *A mathematical equation written in the form $ax^2 + bx + c = 0$,* 699, A5

Quadratic formula *The method used to solve for x from the quadratic equation,* $x = \dfrac{-b \pm \sqrt{b^2 - 4ac}}{2a}$, 699, A5

Quality control *The practice in industry of ensuring that the product contains what it is supposed to, and in the proper amounts,* 766

Quantitative chemistry, 86–116
Avogadro's number, 91
chemical equations, 103–108
counting by weighing, 89–92
everyday controversies of, 116
finding formulas, 99–102
formula mass, 87–89
limiting reagent, 112–115
mass percent, 97–99
moles, 92–97
percentage yields, 115–116
stoichiometric ratios, 108–111

Quantum *A single unit of matter or energy,* 210

Quantum chemistry *The study of chemistry on the atomic and molecular scale,* 210–251
atomic orbitals, 234–241
Balmer's analysis, 218–220

Bohr's model of atom, 220–224, 227–228
de Broglie equation, 228–229
electromagnetic radiation, 211–216
electron and nuclear spin, 242–243
electron configurations, 245–250
emission spectra, 216–220
Heisenberg uncertainty principle, 229–230
introduction to, 210
mathematical language of, 232–234
of multielectron atoms, 244–245
periodic table and, 250
scanning tunneling microscopes and, 229, 237–238
wave functions, 232–234
wave–particle duality, 225–231
"Quantum corral," 209
Quantum mechanics *The physical laws governing energy and matter at the atomic scale,* 210
Quantum number *A number used to arrive at the solution of an acceptable wave function that describes the properties of a specific orbital,* 218, 221, 222
angular momentum, 235
electron spin, 236
magnetic, 235
Pauli exclusion principle, 242
principal, 234–235
significance of, 236–237, 243
s orbitals and, 234–237
Quantum tunneling, 237–238
Quaternary structure *The way in which two or more polypeptide chains have folded to make a protein,* 946
Questions, in scientific method, 12, 13*t*

Racemic mixture *A 1:1 mixture of two enantiomers,* 961
Racemization *The reaction that describes the preparation of a racemic mixture from a single enantiomer,* 961
Rad *A unit of energy absorbed by irradiated material equal to 0.01 J/kg of exposed material,* 911*t*
Radial *Along the radius in spherical polar coordinates,* 234
Radiation *The particles and/or energy emitted during radioactive decay,* 54–55
Radical *A molecule that contains an unpaired nonbonding electron,* 331, 358
Radioactive decay *The process by which an unstable nucleus becomes more stable via the emission or absorption of particles and energy,* 54–55
alpha-particle emission, 905–906
beta-particle emission, 904–905
decay series, 907–909
electron capture, 907
gamma-ray emission, 906
half-life, 636–638, 912–915
integrated rate laws and, 635
of plutonium, 633
positron emission, 907, 918

predicting type of, 918, 919*t*
See also Nuclear chemistry
Radioactive nuclei, list of common, A19
Radioactivity *The emission of radioactive particles and/or energy,* 53–55, 901
Radiocarbon dating, 51–52, 57, 635–636
Radio frequencies, 214
Radiopharmaceuticals *Compounds containing radioactive nuclides that are used for imaging studies in nuclear medicine,* 924–927
Radio waves, 213, 214
Radium, 271*t*
Radon
alpha decay of, 905, 906–907
half-life of, 913*f*
properties of, 909
uses of, 274*t*
Ramsey, William, 224
Raoult, Francois-Marie, 474
Raoult's law *Describes the change in vapor pressure of a solvent as solute is added to a solution,* 474, 479
Rate *The "speed" of a reaction, recorded in M/s,* 621–622
Rate constant *A constant that is characteristic of a reaction at a given temperature, relating to the rate of disappearance of reactants,* 628
activation energy and, 652
binding, 670
half-life and, 636–637
release, 671
Rate-determining step *The slowest elementary step in a reaction sequence,* 648
Rate laws *An equation that indicates the molecularity of a reaction as a function of the rate of the reaction,* 628
collision theory and, 629–630
elementary steps and, 649
half-life and, 636–638
integrated first-order, 634–636
integrated rate laws, 631–640
integrated second-order, 639–640
integrated zero-order, 639
introduction to, 627–630
methods of determining, 641–646
pseudo-first-order, 640
summary of, 646–647
Reactants *The materials that combine in a chemical reaction. The substances located on the left-hand side of a chemical equation,* 47, 103
limiting reagent, 112–115
Reaction coordinate *A measure that describes the progress of the reaction,* 651
Reaction mechanisms. *See* Mechanisms
Reaction profile *A plot of the progress of a reaction versus the energy of the components,* 174–175, 651–652
Reaction quotient (Q) *The ratio of the initial molar concentrations of all of the products raised to the power of their*

respective stoichiometric coefficients, divided by the initial molar concentrations of all of the reactants raised to the power of their stoichiometric coefficients.
equilibrium constant and, 688–689, 701, 705–706
free energy and, 611, 705
pressure, 610
standard potential and, 851–852, 854
Reaction rates, 621–627. *See also* Chemical kinetics; Rate laws
Reactivity *The ability of an element or compound to participate in chemical reactions. Reactivity is an imprecise but useful term, particularly when we are discussing relative reactivities under defined conditions,* 289
of alkali metals, 270
ionization energies and, 282–283
periodicity of, 279, 289–292
of radicals, 358
Reactivity series *A series of elements ranked in order of reactivity. Also known as an activity series,* 291–292, 849–850
Redox reactions *Chemical reactions in which reduction and oxidation occur,* 153, 824, 833–844
balancing, 836–841
in electrochemical cells, 826, 845–849
half-reaction potentials, 834–836, A17–A19
half-reactions, 153, 826
identifying, 156–157
manipulating half-cell reactions, 841–844
oxidation states, 826–833
Reduced *The species that has gained electrons in a redox reaction,* 152
Reducing agent *A substance that causes the reduction of another substance,* 830–831, 849–850
Reduction *The process of gaining electrons. Such a substance is said to be reduced,* 152, 827
Reduction reaction *In a redox reaction, the half-reaction that acquires electrons,* 826
Reference form *The most stable form of the element at standard conditions,* 187, 188, 189
Reforming *A process that changes a mixture of predominantly straight-chain hydrocarbons into a mixture containing more aromatic and branched-chain hydrocarbons,* 508–509
Regulation *The control of the function of a gene or enzyme,* 948–949
Regulatory proteins, 954
Release rate constant *The rate constant of the reaction in which oxygen is released from the myoglobin-oxygen reaction,* 671

Rem *A unit that measures the "equivalent dose" of radiation; that is, it takes into account the interaction of radiation with human tissue. The word stands for "roentgen equivalent in man." The rem is not an SI unit but is related to the SI unit, the sievert,* 911

Remsen, Ira, 833

Renewable sources of energy *Energy sources that can be rapidly replaced by natural processes,* 198–199

Replication *The process by which DNA is duplicated prior to cell division,* 938–939

Residue *The portion of an amino acid that differs from other amino acids,* 940

Resonance hybrid *An equal or unequal (based on experimental evidence) combination of all of the resonance structures for a molecule,* 328

Resonance structure *A model of a molecule in which the positions of the electrons have changed, but the positions of the atoms have remained fixed,* 327–328, 341, 372

Rest mass *The mass a particle has when it is stationary,* 231

11-*cis*-Retinal, 355, 356–357, 374, 381

trans-Retinal, 356–357

Reverse osmosis *The process of purifying water by passing it through a semipermeable membrane. Dissolved solutes are unable to pass through the membrane. A purification process wherein pure solvent is forced (with pressure) to flow through a semipermeable membrane away from a solution,* 10–11, 480–481

Reversible conditions *The conditions that occur when a process proceeds in a series of infinitesimally small steps,* 593–594

Revlimid, 961

Rhodium, 872

Rhodopsin, 355, 356–357

Ribonucleic acid (RNA) *A nucleotide polymer that contains the information from a single gene and either transfers that information to the ribosomes (mRNA) or recognizes and constructs the protein product (tRNA),* 941–945

compared with DNA, 941–942

messenger (mRNA), 942–943

transfer (tRNA), 943–944

Ribose, 941

Ribosomes *Large molecular structures within the cytoplasm of a cell that aid the construction of a protein from mRNA, tRNA, and the amino acid building blocks,* 944

Ritalin (methylphenidate hydrochloride), 95

Ritz-Paschen series, 218t

RNA. *See* Ribonucleic acid

Rods, 356

Roentgen *A unit used to measure exposure to radiation. One roentgen is equal to 2.58×10^{-4} C/kg of dry air at STP,* 911t

Root-mean-square (rms) speed *The square root of the sum of the squares of the individual speeds of particles divided by the number of those particles,* 424–425

Rosenberg, Barnett, 876

Rosuvastatin, 545

Rowland, Sherwood, 429, 431

Rubidium, 216, 247, 270t

Ruby, 873, 884

lasers, 224t

Rust, color of, 884, 885

Rutherford, Ernest, 53–54

Rydberg, Johannes, 218

Rydberg formula, 218, 223

s **orbital** *An orbital with quantum number $l = 0$,* 234–237, 238–239

Saccharide, 955f

Saccharine, 833

Sacrificial anode *A material that will oxidize more easily than the one we seek to protect from oxidation,* 850

Salicylic acid, 14–15, 103–104

Salt bridge *A device containing a strong electrolyte that allows ions to pass from beaker to beaker,* 845–846

Salt River-Pima Maricopa Indian Community, 193

Salts

in aqueous solution, 752–757

sparingly soluble, 802, 803t, 807–808

Saturated *When considering a pure substance—the surrounding atmosphere contains the maximum possible amount of vapor, expressed as the vapor pressure. When considering solution—the solution contains the maximum concentration of dissolved solute,* 452–453, 467

Saturated fats, 496

Saturated hydrocarbons *Hydrocarbons that do not contain double or triple bonds,* 496

s-block elements, 261

Scandium, 264

Scanning electron microscopes (SEM), 229

Scanning tunneling microscopes (STM), 209, 229, 237–238, 377, 570

Schrödinger, Erwin, 232f, 375

Schrödinger equation *The mathematical expression that relates the wave function to the energy in a quantum system,* 232–233, 250, 375, 377

Scientific laws *Concise descriptions of the behavior of the natural world,* 13

Scientific method *A reliable way to find out things about nature by making use of appropriate combinations of these key activities: making observations, gathering data, proposing hypotheses, performing experiments, interpreting the results of those observations and experiments, checking to ensure that the results are repeatable, publishing the results, establishing scientific laws, and formulating theories,* 12–16

Scientific notation, 31

Scrubbing *The process of removing harmful impurities from smokestack gases,* 779

Seaborg, Glen, 925

Seawater

composition of, 465, 468, 470

desalination of, 10–11, 480–481

Second (s) *The SI base unit of time,* 17t, 18

Secondary alcohols *Alcohols that have two carbon atoms bonded to the carbon atom that carries the hydroxyl group,* 517, 519

Secondary structure *The specific folds within a polypeptide chain. Common secondary structures include the alpha helix and the beta pleated sheet,* 945–946

Second ionization energy (I_2) *The energy required to remove one electron from each singly charged +1 ion of an element in the gaseous state,* 282

Second law of thermodynamics *A spontaneous process is accompanied by an increase in the entropy of the universe; $\Delta S_{universe} > 0$,* 588–590

Second-order rate law *The rate of a reaction is directly related either to the square of the concentration of a single reactant or to the product of the concentrations of two reactants,* 639–640, 644, 647t

Sedimentation *The process of removing dissolved organic matter, heavy metals, and other impurities from water,* 808, 809f

See-saw structure, 342

Segrè, Emilio, 925

Selenium

band gap of, 553t

in steel, 266

uses of, 273t

Semiconductors *A metal that contains a small band gap,* 65, 269, 553–555

Semimetals *A small number of elements found at the boundary between metals and nonmetals in the periodic table that have characteristics midway between those of the metals and those of the nonmetals. Also known as metalloids,* 65, 268–269

Semipermeable membrane *A thin membrane (typically made from a polymer) with pores big enough to allow solvent to pass through, but small enough to restrict the flow of dissolved solute particles,* 479

Sequester *To tie up an ion or compound by chelation and make it unavailable for use,* 813

Serine, 940f

Serotonin, 963

SI *The International System (Système International) of agreed-upon base units, derived units, and prefixes used to measure physical quantities,* 17–21. *See also* Measurement

Sievert (Sv) *A unit that measures the "equivalent dose" of radiation; that is, it takes into account the interaction of radiation with human tissue. One sievert equals 100 rem,* 911

Sigma bond (σ bond) *A covalent bond resulting from end-on overlap of orbitals. The sigma bond does not possess a nodal plane along the axis of the bond,* 369–371

Significant figures *Those specific numbers in a measurement whose values we can trust,* 31–33

Silica glass, 559, 560*f*, 572–573

Silicon, 58
　band gap of, 553–554
　doping of, 554, 555
　properties of, 65, 268–269
　thermodynamic data for, A12
　uses of, 266, 272

Silicon dioxide, 293, 463

Silver
　coordination number for, 875
　isotopes of, 903*t*
　oxidation of, 586*t*
　properties of, 64
　reactivity of, 850
　specific heat of, 181*t*
　structure of, 544
　thermodynamic data for, A12

Silver bromide
　molar solubility of, 805
　in photography, 382–383

Silver cell battery, 836, 847

Silver chloride
　precipitation of, 807–808
　saturated solution of, 677, 682
　solubility of, 145

Silver–zinc battery, 848*t*

Simple cubic unit cell *An arrangement of particles within a crystalline solid comprised of six particles occupying the corners of a cubic box,* 542, 546

Skou, Jens C., 545

Soda-lime glass, 560*t*

Sodalite, 310*f*

Sodium
　electron configuration of, 247
　emission spectrum for, 220*f*
　in heart cells, 824
　ionization energy for, 282–283, 285
　Lewis dot symbol for, 308
　in nuclear reactors, 922
　production of, 858
　properties of, 270
　reactivity of, 279, 289
　sources and uses of, 265*t*, 270*t*
　thermodynamic data for, A12

Sodium azide, 419–420

Sodium benzoate, 93

Sodium bicarbonate, 303, 307*t*

Sodium carbonate
　as ionic compound, 307*t*
　reaction with hydrochloric acid, 145
　uses of, 718*t*, 719

Sodium chloride, 274
　crystal structure of, 308, 541
　dissolution in water, 127–128, 316, 465–466, 472
　as electrolyte, 129
　formation of, 308–309
　formula for, 68
　formula mass of, 88
　heat of solution for, 466
　as ionic compound, 70, 73, 304, 306–307, 321, 542
　in IV solutions, 540
　lattice enthalpy of, 313–314, 316
　oxidation numbers for, 154
　sources and uses of, 306–307

Sodium cyanide, 719

Sodium fluoride, 307*t*

Sodium hydride, bond energies in, 360, 361*t*

Sodium hydroxide
　applications for, 149
　calculating molarity of, 132–133
　dissolution in water, 129, 719
　properties of, 132
　reaction with hydrochloric acid, 143, 144, 150–151, 787–792, 831
　reaction with phosphoric acid, 114–115
　as strong base, 149
　titration with acetic acid, 684–685, 792–795
　uses of, 718*t*, 719

Sodium ion, 56, 261
　electron configuration of, 305

Sodium nitrate, in aqueous solution, 753–754

Sodium pertechnetate, 926

Soft drinks, 472–473, 772

Solar cells *Fabricated systems that convert the energy of sunlight into electricity. Also known as photovoltaic systems or PVs,* 198, 199, 555

Solar spectrum, 216, 217*f*, 224

Solar thermal systems *Fabricated systems that convert the energy of sunlight into heat,* 198

Solids *A phase of a substance characterized by high density, rigid shape, and the inability to flow,* 107, 443
　properties of, 396, 397*f*, 443
　types of, 541–542

Solid state lasers, 224*t*

Solubility *A property of a solute; the mass of a solute that can dissolve in a given volume (typically 100 mL) of solvent,* 467
　in aqueous solution, 467–468
　of ionic compounds, 146, 147
　polarity and, 346
　pressure effects on, 472–473
　temperature effects on, 471–472

Solubility equilibria, 802–809
　forming precipitates, 806–808
　gravimetric analysis, 806–807
　impact of side reactions, 802–804
　molar solubility, 805
　molecular processes impacting, 804–805

　wastewater treatment, 808–809
　See also Aqueous equilibria

Solubility product *The mass-action expression for the solubility reaction of a sparingly soluble salt,* 802

Solubility product constant (K_{sp}) *The constant that is part of the solubility product mass-action expression. Typically, K_{sp} values are much less than 1,* 802, 803*t*, A15–A16

Soluble *The ability of a substance to dissolve within a solution,* 145

Solute *The minor component in a solution,* 465

Solutes *Molecules, ions, or atoms that are dissolved in a solvent to form a solution,* 126–127

Solutions *A homogeneous mixture of solute and solvent,* 8–9, 126–127, 465
　aqueous, 465–468
　boiling-point elevation of, 475–477
　colligative properties, 473–481
　effective concentration of solutes in, 675
　factors effecting solubility, 471–473
　freezing-point depression for, 478
　ideal, 474
　isotonic, 480
　osmosis of, 479–480
　saturated, 467–468
　solubility of substances in, 467–468
　vapor pressure of, 474–475

Solution stoichiometry, 124–159
　acid-base reactions, 128–129, 148–152
　applications for, 140
　concentration, 131–140, 468–471
　dilution, 137–139
　electrolytes, 129–131
　measuring water content, 142
　molarity, 131–135
　molecular and ionic equations, 144–145
　oxidation–reduction reactions, 152–157
　parts per million, per billion, per trillion, 135–137
　precipitation reactions, 145–148
　protecting fresh water, 157–159
　stoichiometric analysis, 140–143
　titration, 140–143
　water as versatile solvent, 126–131

Solvation *The interaction of a solvent with its dissolved solute,* 466

Solvent *A compound that typically makes up the majority of a homogeneous mixture of molecules, ions, or atoms; dissolves the solute,* 126
　water as universal, 126–131, 465–468

Sørenson, Peter, 777

sp^3d hybrid orbitals, 368–369

Space shuttle
　chemistry of, 6–7, 9–10, 167, 168, 172, 177, 188–189, 195–197
　cleaning air in spacecraft, 113
　fuel cell power, 826, 827*f*
　insulation material, 557–559, 573
　orbiting speed of, 221

Specific heat *The amount of heat needed to raise the temperature of one gram of a substance by one degree Celsius (or one kelvin) when the pressure is constant. Also known as the substance's specific heat capacity,* 181, 456

Specific heat capacity (*c*) *The amount of heat needed to raise the temperature of one gram of a substance by one degree Celsius (or one kelvin) when the pressure is constant. (Also known as the substance's specific heat.),* 181–185

Spectator ions *Ions that do not participate in a reaction,* 144

Spectrochemical series *The series of ligands organized with respect to their effect on the value of the crystal field splitting energy,* 890–891

Spectroscopy *The measurement of how atoms and molecules interact with electromagnetic radiation as a function of the wavelength or frequency of the radiation,* 216

Sphingomyelin, 346, 957*f*

sp hybrid orbitals, 364–367

Spinach plastocyanin, 276*f*

Spin angular momentum quantum number, 236

Spontaneous process *A process that occurs without continuous outside intervention,* 585–587
concentration and, 611
entropy and, 588–590
free energy and, 601–606, 670
half-reaction potentials and, 834–836
pressure and, 609–610
temperature and, 591–595, 608–609

Sputtering *The process of producing a thin film on a substrate by using high voltage to move a material from a negative terminal (the cathode) to a positive terminal (the anode),* 569

Square planar *The geometry indicated by four electron groups (lone pairs and/or bonds) positioned symmetrically in a plane around a central atom such that the bond angles are 90°,* 877, 880, 889

Square planar crystal field splitting, 887

Stability constant. *See* Formation constant

Stainless steel, 850

Standard ambient temperature and pressure (SATP) *1 bar and 298 K,* 400

Standard atmosphere *Exactly 1 atm, the pressure that supports 760 mm of mercury,* 399, 400*t*

Standard carbon battery, 848*t*

Standard deviation, A7

Standard enthalpy of combustion ($\Delta_c H°$) *The enthalpy change when one mole of a substance in its standard state is completely burned in oxygen gas. Also known as the substance's standard heat of combustion,* 189–191, 334

Standard enthalpy of formation ($\Delta_f H°$) *The enthalpy change for the formation of one mole of a substance in its standard state from its elements in their reference form. Also known as standard heat of formation,* 188–189, 195–197

Standard enthalpy of reaction ($\Delta_{rxn} H°$) *Enthalpy change of a reaction in which all of the reactants and products are in their standard states,* 187–190, 195–197

Standard heat of combustion, 189–190

Standard heat of formation, 188–189

Standard hydrogen electrode reaction (SHE) *A reference half-reaction of the reduction of hydrogen ion to hydrogen gas, against which to compare our reduction,* 835, 841

Standard molar entropy, 598–599

Standard molar free energy of formation ($\Delta_f G°$) *The free energy of a compound formed from its elements in their standard states,* 605

Standard potential ($E°$) *The measure of the potential of a reaction at standard conditions,* 834–836, A17–A19
equilibrium constant and, 855–857
Nernst equation and, 851–854

Standard pressure *Defined as 1 bar, although chemists typically use 1 atm,* 399–400

Standard solution *A solution with a well-defined and known concentration of solute,* 140

Standard state *The state of a chemical under a set of standard conditions, usually at 1 atmosphere of pressure and a concentration of exactly 1 molar for any substances in solution. Standard states are often reported at 25°C,* 187

Standard temperature *For gases, defined to be 0°C, or 273 K,* 400

Standard temperature and pressure (STP) *1 atm and 273 K,* 400, 834

Standing wave *The constructive interference of two or more waves that results in the presence of nodes in fixed locations,* 226, 233

Starch, 6, 7, 956

Starry Night (Van Gogh), 883, 884*f*

State *The physical appearance of a chemical, typically as a solid, liquid, or gas,* 10

State function *A property of a system that depends only on its present state, not on the path by which it reached that state,* 191–192

Stationary phase *In chromatography, the phase that does not move,* 678

Steam cracking *A form of cracking in which heat, steam, and a catalyst are used,* 508

Steam power, 197

Stearic acid, 109, 373

Steel
composition and uses of, 266, 267, 556
pipelines, 850
stainless, 850
in tin cans, 859

Stereoisomers *Structural isomers that differ in their three-dimensional arrangements of atoms, rather than in the order in which the atoms are bonded,* 525, 527

Steroids *Molecules composed of three 6-member carbon rings and one 5-member carbon ring fused together. These compounds are part of the lipid fraction of a cell and function as structural materials and intercellular communicators,* 957–958, 963

Stibnite, 68

St. John–Bloch electronegativity scale, 320

Stoichiometric ratios *The mole ratios relating how compounds react and form products,* 108–111

Stoichiometry *The study and use of quantitative relationships in chemical processes,* 108. *See also* Solution stoichiometry

Stoney, G. Johnstone, 52

Straight-chain alkanes *Molecules that contain only hydrogen atoms and sp^3 hybridized carbon atoms arranged in linear fashion,* 497

Strassman, Fritz, 920

Strong acids *An acid that fully dissociates in water, releasing all of its acidic protons,* 149, 724–730
concentrated vs. dilute, 729–730
pH of, 736–738

Strong bases *A base that completely dissociates in solution,* 149

Strong electrolyte *Any compound that completely dissociates in solution,* 129–131, 144

Strong nuclear force *The force that holds protons and neutrons together in atomic nuclei,* 169*t*

Strontium, 271*t*

Strontium fluoride, 70

Structural isomers *Molecules with the same formula but different structures,* 498–499, 525, 585

Structural proteins, 953

Styrene, 516*t*

Styrofoam, 564*t*

Subcritical mass *A mass of a radioactive isotope that is too small to sustain a chain reaction,* 923

Sublimation *The process of a solid undergoing a phase transition to a gas; the opposite of deposition,* 452

Substance *A type of matter that has a fixed composition,* 6

Substance P, 963, 964*f*

Substitutional alloys *An alloy made by directly replacing solvent atoms with solute atoms,* 556

Substitution reactions *Reactions in which one or more atoms (often hydrogen atoms) are replaced by other types of atoms (often halogens or a hydroxyl group),* 510

Substrate *A reactant molecule that binds to an enzyme*, 948

Sucrose, 585, 657
  composition of, 5
  as covalent bond, 316
  properties of, 6, 93–94, 97, 955–956
  solubility of, 145
  in solution, 126–127, 316

Sugars *The simplest carbohydrates*, 955–956

Sulfate-reducing bacteria, 146

Sulfur
  electronegativity of, 289
  gravimetric analysis of, 807
  in ligands, 874
  properties of, 261
  sources and uses of, 268*t*, 273
  specific heat of, 181*t*
  thermodynamic data for, A12

Sulfur dioxide, 273
  cleanup of, 682, 779
  reaction with oxygen, 676, 677, 682–683, 686–687, 695–697
  reaction with water, 757
  uses of, 395*t*, 401

Sulfuric acid, 95, 273
  as acid deposition, 757
  in Breathalyzer, 329
  Lewis dot structure for, 329
  production of, 676–677, 748
  reaction with sodium hydroxide, 151
  strength of, 724, 728
  uses of, 718*t*, 748

Sulfur ion, 261, 305

Sulfurous acid
  formation of, 757
  ionization of, 680, 698–700

Sulfur tetrafluoride, molecular geometry of, 342

Sulfur trioxide, formation of, 677, 695–697

Sun, luminous intensity of, 20

Superacids *Amazingly strong acids that can be used to add hydrogen ions to organic molecules that are otherwise impervious to such reaction*, 729

Superconductors *A ceramic whose resistance to the flow of electrons drops to zero as the temperature is reduced*, 561–563

Supercritical fluid *This phase is seen only when the pressure and temperature of the substance are greater than the critical pressure and critical temperature. It is characterized with a density midway between that of a liquid and a gas, a viscosity similar to a gas, and the ability to completely fill the volume of a closed container*, 460

Supercritical mass *A mass of a radioactive isotope that not only supports a chain reaction but causes the majority of the nuclei to undergo unfettered radioactive decay within a very short period of time, releasing huge amounts of energy*, 923

Superheavy elements, 918

Superoxide, 330–331

Surface tension *A measure of the energy per unit area on the surface of a liquid*, 462–463

Surroundings *Everything in contact with a chemical system. Energy may flow between the surroundings and system. Strictly speaking, the surroundings of a chemical system comprise the entire universe except the system*, 168, 169–170
  entropy changes in, 589–594

Symmetry *The property associated with orbitals that have similar size and shape*, 375–377, 378

Synthesis gas, 106

System *Any set of chemicals whose energy change we are interested in*, 168, 169–170
  entropy changes in, 589–594

Table salt. *See* Sodium chloride

Table sugar. *See* Sucrose

Taq polymerase, 593

Taxol, 3, 71, 528–530

Technetium-99m, 906, 925–926

Technetium pyrophosphate, 926

Teflon, 3

Tellurium
  electron configuration of, 249
  in steel, 266, 267
  uses of, 273*t*

Temperature
  absolute zero, 409
  conversions, 19–20, 21
  equilibrium and, 677, 703, 705*t*
  impact on solubility, 471–472
  as intensive property, 23*t*, 24
  kinetic energy and, 408, 409, 422, 423
  osmotic pressure and, 479
  in phase diagrams, 459–460
  reaction rates and, 630, 652–653
  relationship with volume, 408–411, 423
  as SI base unit, 17*t*, 18–20
  spontaneous processes and, 591–595, 608–609
  standard, 400
  vapor pressure and, 453
  viscosity and, 462

Tenorite, 586

Tensile strength *Resistance to breaking when a substance is stretched*, 563

Teratogens, 527

Termolecular (trimolecular) reaction *A reaction in which three molecules collide in the rate-determining step of the reaction*, 648

Tertiary alcohols *Alcohols that have three carbon atoms bonded to the carbon atom that carries the hydroxyl group*, 517, 519

Tertiary structure *The way in which secondary structures fold together within a polypeptide chain*, 946

Testosterone, 963

Tetrachloroethylene, 461

Tetrachloromethane, 510

Tetracosane, 446

Tetracycline, 786, 961–962

Tetradentate *Capable of forming four coordinate covalent bonds to the same metal center*, 874

Tetradentate ligand *A ligand that makes four coordinate covalent bonds to a metal ion*, 810

Tetraethyl lead, 317–318, 321

Tetrahedral *The geometry indicated by four electron groups (lone pairs and/or bonds) positioned symmetrically around a central atom such that the bond angles are 109°*, 338*f*, 339*t*, 877, 889

Tetrahedral crystal field splitting, 887

Tetrahydrofuran, 385*t*

Tetrasodium pyrophosphate, 719

Tetrose sugars, 955

Thalidomide, 526–527, 960–961

Thallium, 271*t*

Thallium-201, 914

Theobroma cacao, 99

Theoretical yield *The maximum amount of any chemical, in grams, that could be produced in a chemical reaction. This value can be calculated from the equation for the reaction*, 115

Theory *A trusted explanation of an observation, based on a hypothesis that has been tested in experiments*, 13

Thermal cracking *A form of cracking that uses heat alone*, 508

Thermal pollution *An artificial increase in the temperature of water*, 471

Thermite reaction, 196–197

Thermochemistry *The study of energy exchange in chemical processes*, 168

Thermodynamic equilibrium constant, 706

Thermodynamics *The study of the changes in energy in a reaction*, 168, 586
  calculating entropy changes, 595–600
  of coordination complexes, 892–893
  coupled reactions, 611–612
  data for selected compounds, A9–A12
  electrochemistry and, 834–836, 841–844, 855–856
  entropy, 587–588
  equilibrium, 607–612
  first law of, 177–178, 589
  free energy, 600–606
  second law of, 588–590
  spontaneous processes, 585–587, 591–595, 670
  third law of, 598
  See also Enthalpy; Entropy

Thermophiles, 592*t*, 593

Thin films, 568–571

Thiol, 513*t*

Third law of thermodynamics *The entropy of a pure perfect crystal at 0 K is zero*, 598

Thomson, Joseph John (J. J.), 52–53

Threonine, 940*f*

Thymine, 935, 936, 937*f*

Time
  reaction rates and, 631–640
  as SI base unit, 17t, 18
Tin
  band gap of, 553t
  isotopes of, 919
  sources and uses of, 265t, 272t
  thermodynamic data for, A12
Tin cans, 859
Tiselius, Arne, 825
Titanium
  density of, 25
  sources and uses of, 265t, 267
  thermodynamic data for, A12
Titanium (IV) oxide, 654
Titrant *The solution being added to a solution of an analyte during titration*, 766
Titration *The process of adding one reactant to an unknown amount of another until the reaction is complete; used to determine the concentration of an unknown solute*, 115, 140–143, 766
  common analyses using, 766
  equilibrium constant for, 684–685
  features of effective, 684
  Karl Fischer titration, 142
  using buffers, 767–768
  using EDTA, 681, 767, 810–813
  of vitamin C in solution, 140–141
  See also Acid–base titrations
Titration curve *A plot of the pH of the solution versus the volume (or concentration) of titrant*, 790
Titration endpoint *The volume and pH at which the indicator has changed color during a titration. The endpoint is not always the same as the equivalence point*, 140, 801
Titration midpoint *The pH of the titration where the concentration of weak acid or base is equal to the concentration of its conjugate. At this point, the pH is equal to the pK_a of the analyte*, 794
Toluene, 334, 385t
  in gasoline, 509
  miscibility of, 466
  properties of, 467
Torr *A unit used to measure the pressure of gases. 1 torr = 1 mm Hg*, 399
Torricelli, Evangelista, 399
Total Ozone Mapping Spectrometer (TOMS), 429f
Total synthesis *The preparation of a molecule from readily available starting materials. Typically the total synthesis involves many steps*, 528
trans *A molecule containing groups that are fixed on opposite sides of a bond that cannot rotate*, 503–504
***trans* isomer** *An isomer containing two similar groups on opposite sides of the compound*, 880–881

Transcription *The process by which mRNA is synthesized from a gene*, 941, 942
Transferases, 949
Transfer RNA (tRNA) *A nucleic acid polymer that recognizes the codon on mRNA and adds the corresponding amino acid to the growing protein*, 943–944
Transition elements *Metal elements from Groups IB through VIIIB found in the middle of the periodic table*, 66, 275–276
  electron configurations of, 250
  valence electrons in, 269
Transition metal complexes, 873, 884–885, 888. *See also* Coordination complexes
Transition state *The point along the reaction coordinate where the reactants have collided to form the activated complex*, 651
Transition state theory *The theory that describes how the energy of activation is related to the rate of reaction*, 651–653
Transition temperature *The temperature of a superconductor at which the resistance to the flow of electrons becomes negligible*, 562
Translation *The process by which a protein is synthesized from mRNA*, 941, 943–944
Tribasic salt *A salt that can accept three acidic hydrogen atoms*, 748
Trichloromethane, 510
Tridentate *Capable of forming three coordinate covalent bonds to the same metal center*, 874
Trifluoromethyl sulfur pentafluoride, 431
Triglyceride (triacylglycerol) *A molecule made from three fatty acids and a molecule of glycerol. Functions as structural material in cell membranes and as an energy storage molecule*, 373, 957
Trigonal bipyramidal geometry, 338f, 339t, 342
Trigonal planar geometry, 338f, 339t
2,2,4-Trimethylpentane (isooctane), 70–71, 498f, 509
2,3,3-Trimethylpentane, 454
Trinitrotoluene (TNT), 88, 89f
Trios sugars, 955
Triple bond *A covalent bond in which three electron pairs (six electrons) are shared between adjacent atoms*, 326
Triple point *The temperature and pressure at which the gas, liquid, and solid phases of a substance are in equilibrium*, 460
Triple superphosphate, 98
Triprotic *Can produce 3 mol of H^+ when it dissolves*, 151
Triprotic acid *An acid that contains three acidic protons*, 747
Tritium, 57, 58t, 244, 905
Tryptophan, 940f
Tungsten, 265t, 267
Tyndall, John, 559
Tyrosine, 940f

Ultraviolet radiation, 213f, 215, 884
Uncertainty *A measure of the lack of confidence in a measured number*, 28–29, A7–A9
Unimolecular reaction *A reaction in which one molecule alone is involved in the rate-determining step of the reaction*, 648
Unit cells *The repeating pattern that makes up a crystalline solid*, 542–543
  X-ray crystallography of, 545–549
United States Customary System (USCS), 16
United States Geological Survey (USGS), 306–307
Universal indicator, 801
Universal solvent *Water is often called the universal solvent because of its ability to dissolve so many chemicals to form homogeneous mixtures*, 126, 465
Universe *The space comprising both the system and the surroundings*, 168
  entropy of, 589–590
Unsaturated hydrocarbons *Hydrocarbons that contain C=C or C≡C bonds*, 501
Uracil, 942
Uranium
  atomic mass of, 60
  binding energy of, 917
  decay series for, 907–909
  enrichment of, 426
  isotopes of, 426, 907
  structure of, 543f
  thermodynamic data for, A12
Uranium-235
  in atomic bomb, 920
  decay series for, 908f
  fission of, 923–924
  in nuclear reactors, 922
U.S. Geological Survey (USGS), 443f
U.S. National Renewable Energy Laboratory, 198

Valence band *The collection of filled molecular orbitals on a metal*, 552–554
Valence bond theory *The theory that all covalent bonds in a molecule arise from overlap of individual valence atomic orbitals. A modification of this theory allows valence atomic orbitals to include hybridized orbitals*, 358–363, 386
  application of, 360–362
  definition of, 359
  hybrid orbitals and, 363–374
  problems with, 362, 363f, 372
Valence electrons *The electrons in an atom, ion, or molecule that are in the highest principal energy shell. The electrons that occupy the outer-most energy level of an atom, which interact with those of other atoms*, 247, 269
  in covalent bonding, 317
  in Lewis dot symbols, 304–305
  in octet rule, 306

Valence shell electron-pair repulsion model *A model that proposes the three-dimensional shapes of molecules on the basis of the number of electron groups attached to a central atom. Also known as VSEPR,* 335–343

bond angles and electron pairs in, 341–343

description of, 337

examples of, 337–341

limitations of, 357

Valine, 755*f*, 940*f*

Valium, 747*t*

Vanadium

electron configuration of, 249

reduction of, 856–857

in steel, 267

Vanadium oxide, 704

van der Waals, Johannes Diderik, 416–417, 445

van der Waals equation *The equation that corrects the gas laws for gases that deviate from ideal behavior,* 416–417

van der Waals forces *The intermolecular forces of attraction that result in associations of adjacent substances,* 445

Vanilla, GC/MS analysis of, 62–63

van't Hoff, Jacobus Henricus, 336, 476, 479

van't Hoff factor *A factor that modifies the colligative properties on the basis of their ability to dissociate in solution,* 476, 477*t*

Vaporization, heat of, 457–458

Vapor pressure *The pressure of vapor above a liquid in a closed system,* 452–455, A17

Vapor pressure lowering, 474–475

Velocity

as derived unit, 22*t*

kinetic energy and, 170

Venable, Francis, 420

Vinyl chloride, 420, 516*t*

Violeo complex, 880

Viscosity *A measure of resistance to flow,* 462

Visible spectrum, 213, 885*f*

Vitamins *Small molecules that we need for proper health. Normally, vitamins either are not synthesized in our bodies or are made in insufficient amounts,* 950–952

Vitamin A (retinol), 287–288, 950–951

Vitamin B1 (thiamine), 949*f*, 950*t*, 952

Vitamin B2 (riboflavin), 950*t*

Vitamin B6, 950*t*

Vitamin B12, 951*t*

Vitamin C (ascorbic acid), 89, 723, 951*t*

in human body, 116, 952

molecular formula of, 140*f*

titration of, in solution, 140–141

water solubility of, 287–288

Vitamin D (calciferol), 950, 951*t*

Vitamin E (alpha-tocopherol), 951–952

Vitamin K, 951*t*

Volt *The SI unit of potential,* 834

Volta, Alessandro, 825, 826*f*

Voltage *A measure of how strongly a species pulls electrons toward itself. Also known as electromotive force (emf),* 834

Voltaic cell *A cell that produces electricity from a chemical reaction. Also known as a galvanic cell,* 825–826. *See also* Electrochemical cells

Volume

conversion factors for, 26

as derived unit, 22

deviations of real gases, 416–417

as extensive property, 23*t*

law of combining, 50–51

in pressure–volume work, 176–177, 186

relationship with amount, 404–405, 423

relationship with pressure, 15–16, 405–408, 422

relationship with temperature, 408–411, 423

of unit cells, 547–548

using molarity to calculate, 134–135

Volumetric glassware, 138

VSEPR model *A model that proposes the three-dimensional shapes of molecules on the basis of the number of electron groups attached to a central atom. See* Valence shell electron-pair repulsion model

Walker, John E., 545

Warner, John, 571

Water

autoprotolysis of, 733, 737, 738

boiling-point elevation of, 476*t*

boiling point of, 19, 445

different names for, 71

dipole–dipole forces in, 448

dipole moment in, 345, 444

dissolution of salt in, 465–466

electrolysis of, 397

electron density map of, 127*f*, 829*f*

enthalpy of formation of, 188

formation of, 6, 7*f*, 10, 50–51, 150–151

formula for, 70

freezing-point depression for, 478*t*

heating curve for, 456–458

hybrid orbitals in, 367

hydrogen bonds in, 448–450

impact of radiation on, 910–911

intermolecular forces in, 444–445

iron in, 870–871

meniscus of, 463–464

molecular shape of, 336–337, 340, 357, 368, 444

oxidation numbers for, 154, 827–829

phase changes of, 451–452

phase diagram for, 459–460, 461

pH scale and, 733–735

polar covalent bonds in, 321

properties of, 465, 766

protecting fresh water, 157–159

reaction with phosgene, 333

simultaneous detection of elements in, 219–220

specific heat of, 181*t*, 456, 457

stability as liquid, 444, 445, 449

standard state of, 187

structure of ice, 459, 461

surface tension of, 462–463

testing for small amounts of, 142

as universal solvent, 126–131, 465–468

vapor pressure of, 452–453, 474, A17

Water consumption, 441–443

Water treatment and standards

chlorine usage, 157–158

desalination process, 10–11, 480–481

with fluoride ion, 137–139, 313

for hard water, 309–310, 767, 814

industrial wastewater, 145–146, 148, 158, 807

measuring contaminants, 135–136, 157–159

sedimentation, 808–809

use of chelates, 812*t*

Watson, James, 936–937

Wave energy, 172–174

Wave function *The mathematical equation describing a system, such as an electron, atom, or molecule, that contains all physical information that can be obtained for the system by quantum mechanics,* 232–234

Wavelength *A descriptor of electromagnetic radiation. Defined as the distance from the top of one crest to the top of the next crest of the electromagnetic wave,* 211–215

Wave–particle duality *The quantum mechanical theorem that states that all things have both wave and particle natures simultaneously and that both of these natures can be observed on the atomic scale,* 225–227

of electrons, 227–229

of protons, 230–231

Waves

classical mechanics view of, 226–227

quantization and, 227–229

standing, 226

Weak acids *An acid that only partially dissociates in water,* 150, 724–730

equilibrium constants for, 725*t*

pH of, 738–742

Weak bases *A base that partially dissociates in solution,* 150

Weak electrolyte *Any substance that only partially dissociates in solution,* 129–131

Weight *A measure of the gravitational force exerted on a body,* 17–18

counting by, 89–92

Weight percent *A concentration unit based on the mass of the solute per total mass of solution, reported in percent,* 470

Werner, Alfred, 880

Whole-number ratios, 100

Wilkinson's catalyst, 872

Willow bark, 14

Willson, Thomas, 420

Wind power *The use of the energy of the wind to generate electricity*, 172, 197, 199

Work (w) *Force acting on an object over a distance*, 169, 178

pressure–volume, 176–177, 186

Xenon

compounds containing, 369

thermodynamic data for, A12

uses of, 274t, 275

Xenon tetrafluoride

hybrid orbitals in, 369

Lewis dot structure for, 329–330

molecular geometry of, 342

Xenophanes of Colophon, 46

X-ray crystallography, 377

of unit cells, 545–549

X-rays, 213f, 215–216, 561, 906, 910, 924

Xylene, 509

Zeolites *A porous solid with a well-defined structure, typically made of aluminum and silicon. Zeolites are capable of binding to ions of specific radius based on the relative size of its pores*, 310, 572

Zero-order reactions *A reaction in which the rate is not related to the concentrations of any species in the reaction*, 638–639, 644, 647t

Zinc

atomic mass of, 63

reaction with copper chloride, 851–852

reaction with hydrochloric acid, 402

structure of, 544

thermodynamic data for, A12

titration with EDTA, 814

Zinc–air battery, 848t

Zinc–ammonia complex, 809–810

Zinc–carbon battery, 848t

Zinc–mercury amalgam, 856

Zinc–mercury oxide battery, 848t

Zinc sulfide, 553t

Zirconium, 267

Zwitterions *A molecular ion carrying both an acid group, which generates a negative ion, and a basic group, which generates a positive ion*, 755–756

Applications and Chemical Encounters

Chapter	Page	Title	Chapter	Page	Title
12	494	CHEMICAL ENCOUNTERS: Crude Oil—the Basic Resource	16	704	CHEMICAL ENCOUNTERS: Catalysts in Industry
13	540	Intravenous saline solution	17	718	CHEMICAL ENCOUNTERS: Common Uses of Acids and Bases
13	545	The structure of adenosine triphosphate (ATP) and the development of drugs	17	724	CHEMICAL ENCOUNTERS: Acids in Foods
13	546	CHEMICAL ENCOUNTERS: X-ray Crystallography	17	738	The uses of lactic acid
13	550	Liquid crystals	17	743	Regulating nicotine with acid–base chemistry
13	551	"Silver" fillings for cavities	17	747	CHEMICAL ENCOUNTERS: Production and Uses of Phosphoric and Sulfuric Acids
13	555	CHEMICAL ENCOUNTERS: Photovoltaic Devices			
13	557	CHEMICAL ENCOUNTERS: Dental Amalgams	17	753	Sodium nitrite and the preservation of food
13	557	Ceramics and bone replacement	17	755	CHEMICAL ENCOUNTERS: Acid–Base Properties of Amino Acids
13	559	Ceramics and the insulation of the space shuttle	17	757	CHEMICAL ENCOUNTERS: Acid Deposition and Acid-Neutralizing Capacity
13	559	Glass and the fiber optic wire			
13	561	CHEMICAL ENCOUNTERS: Magnetic Resonance Imaging	18	766	CHEMICAL ENCOUNTERS: Industrial and Environmental Applications of Titrations
13	567	Aramid polymers and the bulletproof vest	18	767	Using EDTA to measure water hardness
13	568	CHEMICAL ENCOUNTERS: Heart Defibrillators	18	779	CHEMICAL ENCOUNTERS: Scrubbing Sulfur Dioxide Emissions
13	570	CHEMICAL ENCOUNTERS: "Green Chemistry"	18	801	CHEMICAL ENCOUNTERS: Anthocyanins and Universal Indicators
14	580	CHEMICAL ENCOUNTERS: Bioenergetics	18	802	Calcium carbonate and the ocean
14	585	CHEMICAL ENCOUNTERS: Glycolysis	18	806	Gravimetric analysis
14	586	The patina on copper and bronze statues	18	807	Recovering metals from industrial effluents
14	588	The spontaneity of protein folding	18	808	Treating waste water through sedimentation
14	591	CHEMICAL ENCOUNTERS: Psychrotrophs and Psychrophiles	18	810	CHEMICAL ENCOUNTERS: Commercial Uses of Aminopolycarboxylic Acid Cheating Agents
14	595	CHEMICAL ENCOUNTERS: More Glycolysis			
14	600	CHEMICAL ENCOUNTERS: Pyruvate and Lactate	19	844	CHEMICAL ENCOUNTERS: The Electric Eel
14	604	Converting water vapor to ice	19	859	CHEMICAL ENCOUNTERS: Electroplating
14	607	CHEMICAL ENCOUNTERS: ATP formation	20	869	CHEMICAL ENCOUNTERS: Iron and Hemoglobin
14	608	Temperature and lighting a portable propane stove	20	871	Carbon monoxide and hemoglobin
14	610	Pressure and the fermentation of glucose in brewing	20	884	CHEMICAL ENCOUNTERS: Transition Metals and Color
14	611	The reactions of glycolysis	20	890	The colors of blood
15	620	CHEMICAL ENCOUNTERS: Atrazine and the Environment	20	894	CHEMICAL ENCOUNTERS: Cytochromes
15	627	Hydrogen peroxide and disinfecting	21	903	The sources of potassium
15	631	CHEMICAL ENCOUNTERS: Decomposition of DDT	21	909	Radioactivity and human health
15	633	The decay of radioactive waste	21	910	Using lead to protect against x-ray radiation
15	636	CHEMICAL ENCOUNTERS: Persistent Pesticides	21	910	CHEMICAL ENCOUNTERS: Radiation and Cancer
15	641	Glyphosate herbicides	21	920	CHEMICAL ENCOUNTERS: Nuclear Weapons
15	647	CHEMICAL ENCOUNTERS: Metabolism of Methoxychlor	21	921	CHEMICAL ENCOUNTERS: Nuclear Reactors as a Vital Source of Electricity
15	655	CHEMICAL ENCOUNTERS: Destruction of Ozone	21	924	CHEMICAL ENCOUNTERS: Tracer Isotopes for Diagnosis
16	670	CHEMICAL ENCOUNTERS: Myoglobin in the Muscles	22	935	CHEMICAL ENCOUNTERS: The Human Genome Project
16	675	CHEMICAL ENCOUNTERS: Important Processes That Involve Equilibria	22	957	Triglycerides as fats and oils
			22	960	Thalidomide and its dangers
16	676	CHEMICAL ENCOUNTERS: The Manufacture of Sulfuric Acid	22	961	Chemistry and the treatment of illness
16	681	EDTA and the concentration of metal ions	22	962	Treating cancer by preventing DNA replication
16	684	Determining the concentration of acetic acid	22	963	Treating Parkinson's disease with L-Dopa
			22	964	Diseases caused by abnormalities in genetic material

SI Units

Physical Quantity	Name of Unit	Symbol
Amount of substance	mole	mol
Electric current	ampere	A
Length	meter	m
Luminous intensity	candela	cd
Mass	kilogram	kg
Temperature	kelvin	K
Time	second	s

SI Unit Prefixes

Multiple	Prefix	Name
10^{24}	Y	yotta
10^{21}	Z	zetta
10^{18}	E	exa
10^{15}	P	peta
10^{12}	T	tera
10^{9}	G	giga
10^{6}	M	mega
10^{3}	k	kilo
10^{-1}	d	deci
10^{-2}	c	centi
10^{-3}	m	milli
10^{-6}	μ	micro
10^{-9}	n	nano
10^{-12}	p	pico
10^{-15}	f	femto
10^{-18}	a	atto
10^{-21}	z	zepto
10^{-24}	y	yocto